## SELECTED TOPICS
## IN
# INORGANIC CHEMISTRY

### For B.Sc. (Honours) and M.Sc. Students

**Dr. WAHID U. MALIK**
*Ph.D., D.Sc.*
*Formerly HOD, Chemistry*
*IIT Roorkee*

**Dr. G.D. TULI**
*M.Sc., Ph.D.*
*Formerly Principal, Shyam Lal College*
*Delhi University, Delhi*

**Dr. R.D. MADAN**
*M.Sc., Ph.D.*
*Formerly Reader, Department of Chemistry*
*N.R.E.C College, Khurja*

S. CHAND
PUBLISHING

**S Chand And Company Limited**
**(ISO 9001 Certified Company)**

# S Chand And Company Limited

**(ISO 9001 Certified Company)**

***Head Office:*** Block B-1, House No. D-1, Ground Floor, Mohan Co-operative Industrial Estate, New Delhi – 110 044 | Phone: 011-66672000

S. CHAND PUBLISHING

***Registered Office:*** A-27, 2nd Floor, Mohan Co-operative Industrial Estate, New Delhi – 110 044 Phone: 011-49731800

www.**schandpublishing.com**; e-mail: **info@schandpublishing.com**

**Branches**

Chennai : Ph: 23632120; chennai@schandpublishing.com
Guwahati : Ph: 2738811, 2735640; guwahati@schandpublishing.com
Hyderabad : Ph: 40186018; hyderabad@schandpublishing.com
Jalandhar : Ph: 4645630; jalandhar@schandpublishing.com
Kolkata : Ph: 23357458, 23353914; kolkata@schandpublishing.com
Lucknow : Ph: 4003633; lucknow@schandpublishing.com
Mumbai : Ph: 25000297; mumbai@schandpublishing.com
Patna : Ph: 2260011; patna@schandpublishing.com





*First Edition 1976*
*Subsequent Editions and Reprints 1976, 83, 84, 86, 87, 88, 89, 90, 91, 93, 94, 95, 96, 98, 99, 2000, 2001 (Twice), 2002, 2003, 2004, 2005, 2006, 2007, 2008 (Twice), 2010, 2011, 2012, 2013 (Twice)*
*Eighth Revised Edition 2014*
*Reprints 2016, 2017, 2018 (Thrice), 2019, 2020, 2021*

***Reprint 2022***

**ISBN:** 978-81-219-0600-5 **Product Code:** H6ICH68CHEM10ENAH14O

PRINTED IN INDIA

By Vikas Publishing House Private Limited, Plot 20/4, Site-IV, Industrial Area Sahibabad, Ghaziabad – 201 010 and Published by S Chand And Company Limited, A-27, 2nd Floor, Mohan Co-operative Industrial Estate, New Delhi – 110 044.

# PREFACE TO THE EIGHTH EDITION

The text of the present edition of the book has been throughly revised and updated in every respect.

**Main Features**

The following chapters are new additions in the book

- Oxidation and Reduction
- Hard and Soft Acids and Bases (HSAB)
- Elements and their Compounds of *1st* (*3d*), *2nd* (*4d*) and *3rd* (*5d*) Transition Series
- Preparation and Applications of Complexes.
- Splitting of Metal Ion d-Orbitals in Other Crystal Fields
- John-Teller Effect and Distortion in Octahedral and Tetrahedral Complexes
- Electronic Absorption Spectra of Transition Metal Complexes
- Magnetic Properties of Transition Metal Complexes
- Organometallic Compounds
- Bioinorganic Chemistry
- Silicones and Phosphazenes

We sincerely hope that the present edition will be warmly received by our learned fellow-teachers and students. Suggestions for further improvement of the book will be highly appreciated.

**Acknowledgements**

We express our sincere thanks to our students and fellow-teachers who have been sending us their valuable suggestions for the improvement of the text of the book. We are extremely grateful to the Management Team and Editorial Department of S.Chand and Company Pvt. Ltd. for all help and support that we have received from them during the publication of the book.

One of the authors, Dr. R.D. Madan, is sincerely grateful to **Anshu Lochab,** Assistant Professor, Rukmini Devi Institute of Advanced Studies, Rohini, Delhi-85 for rendering valuable help in reshaping the chapter: *Bioinorganic Chemistry*.

**AUTHORS**

**To our Fellow-teachers and Students**

We humbly request our students and teachers to send their constructive criticism and suggestions to Dr. R.D. Madan in order to enable us to incorporate them in the next edition to make this book error-free and more useful from users point of view.

Dr. R.D. Madan
37, Bhagirathi Appartments
Sector-9, Rohini, Delhi-110085
Phone : 011-27860313

# PREFACE TO THE FIRST EDITION

During the last few years, the Boards of Studies in Chemistry of the various universities in India have been trying to overhaul and modernise the syllabi at the undergraduate and postgraduate level. There has accordingly been a distinct change in approach and content with the result that a shortage of suitable texts written on these lines was being keenly felt. The authors seek justification in presenting the present volume in the sincere attempts that they have made to fill up the gap.

Considerable thought has been given to the selection of topics for discussion. As the book has been written primarily for the use of B.Sc. (Hons.) and M.Sc. students, most of the topics deal with the theoretical aspects of Inorganic Chemistry. Great care has been taken to elucidate the fundamentals and the approach to discussion is modern throughout. Particular mention may be made in this connection of the Chapters on Structure of the Atom, Chemical Bonding, Inner Transistion Elements and Coordination Chemistry.

A special feature of the book is that the text has been illustrated with a large number of line diagrams and the data presented in the form of numerous tables for reference and comparison.

In the preparation of the text, standard works and reviews by renowed authors have been freely consulted and the references given chapterwise at the end of the book will be found useful by those who wish to make a more detailed study of the topics discussed.

The authors wish to thank all those who helped in the preparation of the text and illustrations. Since drawbacks and better methods of presentation are always there, suggestions for improvement and criticism will be gratefully received and acknowledged.

November 20, 1975

**AUTHORS**

## CONTENTS

# Some Important Points

**[I] Aufbau Build-up Principle.**

Sequence in which orbitals are filled with electrons is 1s, 2*s*, 2*p*, 3*s*, 3*p*, 4*s*, 3*d*, 4*p*, 5*s*, 4*d*, 5*p*, 6*s*, 4*f*, 5*d*, 6*p*, 7*s*, 5*f*, 6*d*, 7*p*.....

**[II] Electronic Configurations of Noble Gases**

Helium, He (Z = 2) = $1s^2$
(1st period) = $[He]_2$

Neon, Ne(10) = $1s^2, 2s^2, 2p^6$
(2nd period) = $[He]_2\, 2s^2\, 2p^6$
= $[Ne]_{10}$

Argon, Ar (18) = $1s^2, 2s^2, 2p^6, 3s^2, 3p^6$
(3rd period) = $1s^2, 2s^2 2p^6, 3s^2\, 3p^6$
= $[Ne]_{10}\, 3s^2\, 3p^6$
= $[Ar]_{18}$

Krypton, Kr (36) = $1s^2, 2s^2, 2p^6, 3s^2, 3p^6, 4s^2, 3d^{10}, 4p^6$
(4th period) = $1s^2, 2s^2 2p^6, 3s^2 3p^6\, 3d^{10}, 4s^2\, 4p^6$
= $[Ar]_{18}\, 3d^{10}\, 4s^2\, 4p^6$
= $[Kr]_{36}$

Xenon, Xe(54) = $1s^2, 2s^2, 2p^6, 3s^2, 3p^6, 4s^2, 3d^{10}, 4p^6, 5s^2, 4d^{10}, 5p^6$
(5th period) = $1s^2, 2s^2 p^6, 3s^2 3p^6 3d^{10}, 4s^2 4p^6 4d^{10}, 5s^2 5p^6$
= $[Kr]_{36}\, 4d^{10}\, 5s^2\, 5p^6$
= $[Xe]_{54}$

Randon, Rn (86) = $1s^2, 2s^2, 2p^6, 3s^2, 3p^6, 4s^2, 3d^{10}, 4p^6, 5s^2, 4d^{10}, 5p^6, 6s^2, 4f^{14}, 5d^{10}, 6p^6$
(6th period) = $1s^2, 2s^2\, 2p^6, 3s^2\, 3p^6\, 3d^{10}, 4s^2\, 4p^6\, 4d^{10}\, 4f^{14}, 5s^2\, 5p^6\, 5d^{10}, 6s^2\, 6p^6$
= $[Xe]_{54}\, 4f^{14}\, 5d^{10}\, 6s^2\, 6p^6$
= $[Rn]_{86}$

Ununoctium, Uno (118)= $[Rn]_{86}\, 5f^{14}\, 6d^{10}\, 7s^2\, 7p^6$
(7th period) = $[Uuo]_{118}$

**[III] Electronic Configurations of the Elements.**

**[A] Elements of 1st period ($n$ = 1) ($H_1$ and $He_2$) (1*s*)**

**1*s*-block elements**

Hydrogen, H ( Z = 1) = $1s^1$
Helium, He (2) = $1s^2$
= $[He]_2$

**[B] Elements of 2nd period ($n$ = 2) ($Li_3$ to $Ne_{10}$) (2$s$, 2$p$)**

**2$s$-block elements**

Lithium, Li (Z = 3) = $[He]_2\ 2s^1$
Beryllium, Be (4) = $[He]_2\ 2s^2$

**2$p$-block elements**

Boron, B (5) = $[He]_2\ 2s^2\ 2p^1$
Carbon, C (6) = $[He]_2\ 2s^2\ 2p^2$
Nitrogen, N (7) = $[He]_2\ 2s^2\ 2p^3$
Oxygen, O (8) = $[He]_2\ 2s^2\ 2p^4$
Fluorine, F (9) = $[He]_2\ 2s^2\ 2p^5$
Neon, Ne (10) = $[He]_2\ 2s^2\ 2p^6$
= $[Ne]_{10}$

**[C] Elements of 3rd period ($n$ = 3) ($Na_{11}$ to $Ar_{18}$) (3$s$, 3$p$)**

**3$s$-block elements**

Sodium, Na (Z = 11) = $[Ne]_{10}\ 3s^1$
Magnesium, Mg (12) = $[Ne]_{10}\ 3s^2$

**3$p$-block elements**

Aluminium, Al (13) = $[Ne]_{10}\ 3s^2\ 3p^1$
Silicon, Si (14) = $[Ne]_{10}\ 3s^2\ 3p^2$
Phosphorus, P (15) = $[Ne]_{10}\ 3s^2\ 3p^3$
Sulphur, S (16) = $[Ne]_{10}\ 3s^2\ 3p^4$
Chlorine, Cl (17) = $[Ne]_{10}\ 3s^2\ 3p^5$
Argon, Ar (18) = $[Ne]_{10}\ 3s^2 3p^6$
= $[Ar]_{18}$

**[D] Elements of 4th period ($n$ = 4) ($K_{19}$ to $Kr_{36}$) (4$s$, 3$d$, 4$p$)**

**4$s$-block elements**

Potassium, K (Z = 19) = $[Ar]_{18}\ 4s^1$
Calcium, Ca (20) = $[Ar]_{18}\ 4s^2$

**3$d$-series elements (TEs)**

Scadium, Sc (21) = $[Ar]_{18}\ 4s^2\ 3d^1$
= $[Ar]_{18}\ 3d^1\ 4s^2$
Titanium, Ti (22) = $[Ar]_{18}\ 4s^2\ 3d^2$
= $[Ar]_{18}\ 3d^2\ 4s^2$
Vanadium, V (23) = $[Ar]_{18}\ 4s^2\ 3d^3$
= $[Ar]_{18}\ 3d^3\ 4s^2$
Chromium, Cr* (24) = $[Ar]_{18}\ 4s^1\ 3d^5$
= $[Ar]_{18}\ 3d^5\ 4s^1$

Manganese, Mn (25) = $[Ar]_{18}\ 4s^2\ 3d^5$
= $[Ar]_{18}\ 3d^5\ 4s^2$
Iron, Fe (26) = $[Ar]_{18}\ 4s^2\ 3d^6$
= $[Ar]_{18}\ 3d^6\ 4s^2$
Cobalt, Co (27) = $[Ar]_{18}\ 4s^2\ 3d^7$
= $[Ar]_{18}\ 3d^7\ 4s^2$
Nickel, Ni (28) = $[Ar]_{18}\ 4s^2\ 3d^8$
= $[Ar]_{18}\ 3d^8\ 4s^2$
Copper, Cu* (29) = $[Ar]_{18}\ 4s^1\ 3d^{10}$
= $[Ar]_{18}\ 3d^{10}\ 4s^1$
Zinc, Zn (30) = $[Ar]_{18}\ 4s^2\ 3d^{10}$
= $[Ar]_{18}\ 3d^{10}\ 4s^2$

**4*p*-block elements**

Gallium, Ga (31) = $[Ar]_{18}\ 4s^2\ 3d^{10}\ 4p^1$
= $[Ar]_{18}\ 3d^{10}\ 4s^2\ 4p^1$
Germanium, Ge (32) = $[Ar]_{18}\ 4s^2\ 3d^{10}\ 4p^2$
= $[Ar]_{18}\ 3d^{10}\ 4s^2\ 4p^2$
Arsenic, As (33) = $[Ar]_{18}\ 4s^2\ 3d^{10}\ 4p^3$
= $[Ar]_{18}\ 3d^{10}\ 4s^2\ 4p^3$
Selenium, Se (34) = $[Ar]_{18}\ 4s^2\ 3d^{10}\ 4p^4$
= $[Ar]_{18}\ 3d^{10}\ 4s^2\ 4p^4$
Bromine, Br (35) = $[Ar]_{18}\ 4s^2\ 3d^{10}\ 4p^5$
= $[Ar]_{18}\ 3d^{10}\ 4s^2\ 4p^5$
Krypton, Kr (36) = $[Ar]_{18}\ 4s^2\ 3d^{10}\ 4p^6$
= $[Ar]_{18}\ 3d^{10}\ 4s^2\ 4p^6$
= $[Kr]_{36}$

**[E] Elements of 5th period ($n = 5$) ($Rb_{37}$ to $Xe_{54}$) (5*s*, 4*d*, 5*p*)**

**5*s*-block elements**

Rubidium, Rb (Z = 37) = $[Kr]_{36}\ 5s^1$
Stronsium, Sr (38) = $[Kr]_{36}\ 5s^2$

**4*d*-series elements (TEs)**

Ytterium, Y (39) = $[Kr]_{36}\ 5s^2\ 4d^1$
= $[Kr]_{36}\ 4d^1\ 5s^2$
Zirconium, Zr (40) = $[Kr]_{36}\ 5s^2\ 4d^2$
= $[Kr]_{36}\ 4d^2\ 5s^2$
Niobium, Nb* (41) = $[Kr]_{36}\ 5s^1\ 4d^4$
= $[Kr]_{36}\ 4d^4\ 5s^1$
Molybdenum, Mo* (42) = $[Kr]_{36}\ 5s^1\ 4d^5$
= $[Kr]_{36}\ 4d^5\ 5s^1$

Technetium, Tc (43) $= [Kr]_{36}\ 5s^2\ 4d^5$
$= [Kr]_{36}\ 4d^5\ 5s^2$

Ruthenium, Ru* (44) $= [Kr]_{36}\ 5s^1\ 4d^7$
$= [Kr]_{36}\ 4d^7\ 5s^1$

Rhodium, Rh* (45) $= [Kr]_{36}\ 5s^1\ 4d^8$
$= [Kr]_{36}\ 4d^8\ 5s^1$

Palladium, Pd* (46) $= [Kr]_{36}\ 5s^0\ 4d^{10}$
$= [Kr]_{36}\ 4d^{10}\ 5s^0$

Silver, Ag* (47) $= [Kr]_{36}\ 5s^1\ 4d^{10}$
$= [Kr]_{36}\ 4d^{10}\ 5s^1$

Cadmium, Cd (48) $= [Kr]_{36}\ 5s^2\ 4d^{10}$
$= [Kr]_{36}\ 4d^{10}\ 5s^2$

**5*p*-block elements**

Indium, In (49) $= [Kr]_{36}\ 5s^2\ 4d^{10}\ 5p^1$
$= [Kr]_{36}\ 4d^{10}\ 5s^2\ 5p^1$

Tin, Sn (50) $= [Kr]_{36}\ 5s^2\ 4d^{10}\ 5p^2$
$= [Kr]_{36}\ 4d^{10}\ 5s^2\ 5p^2$

Antimony, Sb (51) $= [Kr]_{36}\ 5s^2\ 4d^{10}\ 5p^3$
$= [Kr]_{36}\ 4d^{10}\ 5s^2\ 5p^3$

Tellurium, Te (52) $= [Kr]_{36}\ 5s^2\ 4d^{10}\ 5p^4$
$= [Kr]_{36}\ 4d^{10}\ 5s^2\ 5p^4$

Iodine, I (53) $= [Kr]_{36}\ 5s^2\ 4d^{10}\ 5p^5$
$= [Kr]_{36}\ 4d^{10}\ 5s^2\ 5p^5$

Xenon, Xe (54) $= [Kr]_{36}\ 5s^2\ 4d^{10}\ 5p^6$
$= [Kr]_{36}\ 4d^{10}\ 5s^2\ 5p^6$
$= [Xe]_{54}$

**[F] Elements of 6th period (*n* = 6) ($Ce_{55}$ to $Rn_{86}$) (6*s*, 4*f*, 5*d*, 6*p*)**

**6*s*-block elements**

Cerium, Cs (Z = 55) $= [Xe]_{54}\ 6s^1$

Borium, Ba (56) $= [Xe]_{54}\ 6s^2$

**5*d*-series element (TE)**

Lanthanum, La* (57) $= [Xe]_{54}\ 6s^2\ 4f^0\ 5d^1$
$= [Xe]_{54}\ 4f^0\ 5d^1\ 6s^2$

**4*f*- block elements (Lathanides)**

Cerium, Ce (58) $= [Xe]_{54}\ 6s^2\ 4f^1\ 5d^1$
$= [Xe]_{54}\ 4f^1 5d^1\ 6s^2$

Praseodymium, Pr (59) $= [Xe]_{54}\ 6s^2\ 4f^3\ 5d^0$
$= [Xe]_{54}\ 4f^3\ 5d^0\ 6s^2$

Neodymium, Nd (60) $= [Xe]_{54}\, 6s^2\, 4f^4\, 5d^0$
$= [Xe]_{54}\, 4f^4\, 5d^0\, 6s^2$

Promethium, Pm (61) $= [Xe]_{54}\, 6s^2\, 4f^5\, 5d^0$
$= [Xe]_{54}\, 4f^5\, 5d^0\, 6s^2$

Samarium, Sm (62) $= [Xe]_{54}\, 6s^2\, 4f^6\, 5d^0$
$= [Xe]_{54}\, 4f^6\, 5d^0\, 6s^2$

Europium, Eu (63) $= [Xe]_{54}\, 6s^2\, 4f^7\, 5d^0$
$= [Xe]_{54}\, 4f^7\, 5d^0\, 6s^2$

Gadolinium, Gd* (64) $= [Xe]_{54}\, 6s^2\, 4f^7\, 5d^1$
$= [Xe]_{54}\, 4f^7\, 5d^1\, 6s^2$

Terbium, Tb (65) $= [Xe]_{54}\, 6s^2\, 4f^9\, 5d^0$
$= [Xe]_{54}\, 4f^9\, 5d^0\, 6s^2$

Dysprosium, Dy (66) $= [Xe]_{54}\, 6s^2\, 4f^{10}\, 5d^0$
$= [Xe]_{54}\, 4f^{10}\, 5d^0\, 6s^2$

Holmium, Ho (67) $= [Xe]_{54}\, 6s^2\, 4f^{11}\, 5d^0$
$= [Xe]_{54}\, 4f^{11}\, 5d^0\, 6s^2$

Erbium, Er (68) $= [Xe]_{54}\, 6s^2\, 4f^{12}\, 5d^0$
$= [Xe]_{54}\, 4f^{12}\, 5d^0\, 6s^2$

Thullium, Tm (69) $= [Xe]_{54}\, 6s^2\, 4f^{13}\, 5d^0$
$= [Xe]_{54}\, 4f^{13}\, 5d^0\, 6s^2$

Ytterbium, Yb (70) $= [Xe]_{54}\, 6s^2\, 4f^{14}\, 5d^0$
$= [Xe]_{54}\, 4f^{14}\, 5d^0\, 6s^2$

Lutetium, Lu (71) $= [Xe]_{54}\, 6s^2\, 4f^{14}\, 5d^1$
$= [Xe]_{54}\, 4f^{14}\, 5d^1\, 6s^2$

**5*d*-series elements (TEs)**

Hafnium, Hf (72) $= [Xe]_{54}\, 6s^2\, 4f^{14}\, 5d^2$
$= [Xe]_{54}\, 4f^{14}\, 5d^2\, 6s^2$

Tantalum, Ta (73) $= [Xe]_{54}\, 6s^2\, 4f^{14}\, 5d^3$
$= [Xe]_{54}\, 4f^{14}\, 5d^3\, 6s^2$

Tungsten, W (74) $= [Xe]_{54}\, 6s^2\, 4f^{14}\, 5d^4$
$= [Xe]_{54}\, 4f^{14}\, 5d^4\, 6s^2$

Rhenium, Re (75) $= [Xe]_{54}\, 6s^2\, 4f^{14}\, 5d^5$
$= [Xe]_{54}\, 4f^{14}\, 5d^5\, 6s^2$

Osmium, Os (76) $= [Xe]_{54}\, 6s^2\, 4f^{14}\, 5d^6$
$= [Xe]_{54}\, 4f^{14}\, 5d^6\, 6s^2$

Iridium, Ir (77) $= [Xe]_{54}\, 6s^2\, 4f^{14}\, 5d^7$
$= [Xe]_{54}\, 4f^{14}\, 5d^7\, 6s^2$

Platinum, Pt* (78) $= [Xe]_{54}\, 6s^1\, 4f^{14}\, 5d^9$
$= [Xe]_{54}\, 4f^{14}\, 5d^9\, 6s^1$

Gold, Au* (79) = $[Xe]_{54}\ 6s^1\ 4f^{14}\ 5d^{10}$
= $[Xe]_{54}\ 4f^{14}\ 5d^{10}\ 6s^1$
Mercury, Hg (80) = $[Xe]_{54}\ 6s^2\ 4f^{14}\ 5d^{10}$
= $[Xe]_{54}\ 4f^{14}\ 5d^{10}\ 6s^2$

**6*p*-block elements**

Thallium, Tl (81) = $[Xe]_{54}\ 6s^2\ 4f^{14}\ 5d^{10}\ 6p^1$
= $[Xe]_{54}\ 4f^{14}\ 5d^{10}\ 6s^2\ 6p^1$
Lead, Pb (82) = $[Xe]_{54}\ 6s^2\ 4f^{14}\ 5d^{10}\ 6p^2$
= $[Xe]_{54}\ 4f^{14}\ 5d^{10}\ 6s^2\ 6p^2$
Bismuth, Bi (83) = $[Xe]_{54}\ 6s^2\ 4f^{14}\ 5d^{10}\ 6p^3$
= $[Xe]_{54}\ 4f^{14}\ 5d^{10}\ 6s^2\ 6p^3$
Polonium, Po (84) = $[Xe]_{54}\ 6s^2\ 4f^{14}\ 5d^{10}\ 6p^4$
= $[Xe]_{54}\ 4f^{14}\ 5d^{10}\ 6s^2\ 6p^4$
Astatine, At (85) = $[Xe]_{54}\ 6s^2\ 4f^{14}\ 5d^{10}\ 6p^5$
= $[Xe]_{54}\ 4f^{14}\ 5d^{10}\ 6s^2\ 6p^5$
Radon, Rn (86) = $[Xe]_{54}\ 6s^2\ 4f^{14}\ 5d^{10}\ 6p^6$
= $[Xe]_{54}\ 4f^{14}\ 5d^{10}\ 6s^2\ 6p^6$
= $[Rn]_{86}$

**[G] Elements of 7th period (*n* = 7) ($Fr_{87}$ to $Uuo_{118}$) (7*s*, 5*f*, 6*d*, 7*p*)**

**7*s*-block elements**

Francium, Fr (Z = 87) = $[Rn]_{86}\ 7s^1$
Radium, Ra (88) = $[Rn]_{86}\ 7s^2$

**6*d*-series element (TE)**

Actinium, Ac (89) = $[Rn]_{86}\ 7s^2\ 5f^0\ 6d^1$
= $[Rn]_{86}\ 5f^0\ 6d^1\ 7s^2$

**5*f*-block elements (Actinides)**

Thorium, Th (90) = $[Rn]_{86}\ 7s^2\ 5f^0\ 6d^2$
= $[Rn]_{86}\ 5f^0\ 6d^2\ 7s^2$
Protoactinium, Pa (91) = $[Rn]_{86}\ 7s^2\ 5f^2\ 6d^1$
= $[Rn]_{86}\ 5f^2\ 6d^1\ 7s^2$
Uranium, U (92) = $[Rn]_{86}\ 7s^2\ 5f^3\ 6d^1$
= $[Rn]_{86}\ 5f^3\ 6d^1\ 7s^2$
Neptunium, Np (93) = $[Rn]_{86}\ 7s^2\ 5f^4\ 6d^1$
= $[Rn]_{86}\ 5f^4\ 6d^1\ 7s^2$
Plutonium, Pu (94) = $[Rn]_{86}\ 7s^2\ 5f^6\ 6d^0$
= $[Rn]_{86}\ 5f^6\ 6d^0\ 7s^2$
Americium, Am (95) = $[Rn]_{86}\ 7s^2\ 5f^7\ 6d^0$
= $[Rn]_{86}\ 5f^7\ 6d^0\ 7s^2$

Curium, Cm (96) = $[Rn]_{86}\ 7s^2\ 5f^7\ 6d^1$
= $[Rn]_{86}\ 5f^7\ 6d^1\ 7s^2$

Berkelium, Bk (97) = $[Rn]_{86}\ 7s^2\ 5f^9\ 6d^0$
= $[Rn]_{86}\ 5f^9\ 6d^0\ 7s^2$
or = $[Rn]_{86}\ 5f^8\ 6d^1\ 7s^2$

Califormium, Cf (98) = $[Rn]_{86}\ 7s^2\ 5f^{10}\ 6d^0$
= $[Rn]_{86}\ 5f^{10}\ 6d^0\ 7s^2$

Einsteinium, Es (99) = $[Rn]_{86}\ 7s^2\ 5f^{11}\ 6d^0$
= $[Rn]_{86}\ 5f^{11}\ 6d^0\ 7s^2$

Fermium, Fm (100) = $[Rn]_{86}\ 7s^2\ 5f^{12}\ 6d^0$
= $[Rn]_{86}\ 5f^{12}\ 6d^0\ 7s^2$

Mendelevium, Md* (101) = $[Rn]_{86}\ 7s^2\ 5f^{13}\ 6d^0$
= $[Rn]_{86}\ 5f^{13}\ 6d^0\ 7s^2$

Nobelium, No (102) = $[Rn]_{86}\ 7s^2\ 5f^{14}\ 6d^0$
= $[Rn]_{86}\ 5f^{14}\ 5d^0\ 7s^2$

Lawrencium, Lr (103) = $[Rn]_{86}\ 7s^2\ 5f^{14}\ 6d^1$
= $[Rn]_{86}\ 5f^{14}\ 6d^1\ 7s^2$

### 6*d*-series elements (TEs)

Rutherfordium or Kurchatovium (Unnilquadium), Rf or Ku or Uuq (104)
= $[Rn]_{86}\ 7s^2\ 5f^{14}\ 6d^2$
= $[Rn]_{86}\ 5f^{14}\ 6d^2\ 7s^2$

Hahnium or Dubnium (Unnilpentium), Ha or Db or Unp (105)
= $[Rn]_{86}\ 7s^2\ 5f^{14}\ 6d^3$
= $[Rn]_{86}\ 5f^{14}\ 6d^3\ 7s^2$

Seabargium (Unnilhexium), Sg or Unh (106)
= $[Rn]_{86}\ 7s^2\ 5f^{14}\ 6d^4$
= $[Rn]_{86}\ 5f^{14}\ 6d^4\ 7s^2$

Neilsbohrium or Bahrium (Unnilseplium), Ns or Bh or Uns (107)
= $[Rn]_{86}\ 7s^2\ 5f^{14}\ 6d^5$
= $[Rn]_{86}\ 5f^{14}\ 6d^5\ 7s^2$

Hassnium (Unnilotium), Hs or Uno (108)
= $[Rn]_{86}\ 7s^2\ 5f^{14}\ 6d^6$
= $[Rn]_{86}\ 5f^{14}\ 6d^6\ 7s^2$

Meitnerium (Unnilennium), Mt or Une (109)
= $[Rn]_{86}\ 7s^2\ 5f^{14}\ 6d^7$
= $[Rn]_{86}\ 5f^{14}\ 6d^7\ 7s^2$

Darmstadtium (Ununnilium) Ds or Uun (110)
= $[Rn]_{86}\ 7s^2\ 5f^{14}\ 6d^8$
= $[Rn]_{86}\ 5f^{14}\ 6d^8\ 7s^2$

Roentgenium (Unununiun), Rg or Uuu (111)
$= [Rn]_{86}\, 7s^1\, 5f^{14}\, 6d^{10}$
$= [Rn]_{86}\, 5f^{14}\, 6d^{10}\, 7s^1$

Copermisium (Ununbium), Cn or Uub (112)
$= [Rn]_{86}\, 7s^2\, 5f^{14}\, 6d^{10}$
$= [Rn]_{86}\, 5f^{14}\, 6d^{10}\, 7s^2$

### 7*p*-block elements

Ununtrium, Uut (113) $= [Rn]_{86}\, 7s^2\, 5f^{14}\, 6d^{10}\, 7p^1$
$= [Rn]_{86}\, 5f^{14}\, 6d^{10}\, 7s^2\, 7p^1$

Ununquadium, Uuq (114) $= [Rn]_{86}\, 7s^2\, 5f^{14}\, 6d^{10}\, 7p^2$
$= [Rn]_{86}\, 5f^{14}\, 6d^{10}\, 7s^2\, 7p^2$

Ununpentium, Uup (115) $= [Rn]_{86}\, 7s^2\, 5f^{14}\, 6d^{10}\, 7p^3$
$= [Rn]_{86}\, 5f^{14}\, 6d^{10}\, 7s^2\, 7p^3$

Ununhexium, Uuh (116) $= [Rn]_{86}\, 7s^2\, 5f^{14}\, 6d^{10}\, 7p^4$
$= [Rn]_{86}\, 5f^{14}\, 6d^{10}\, 7s^2\, 7p^4$

Ununseptium, Uus (117) $= [Rn]_{86}\, 7s^2\, 5f^{14}\, 6d^{10}\, 7p^5$
$= [Rn]_{86}\, 5f^{14}\, 6d^{10}\, 7s^2\, 7p^5$

Ununoctium, Uuo (118) $= [Rn]_{86}\, 7s^2\, 5f^{14}\, 6d^{10}\, 7p^6$
$= [Rn]_{86}\, 5f^{14}\, 6d^{10}\, 7s^2\, 7p^6$

**Note:** Elements with asterisk (*) have exceptional electronic configurations.

**[IV]** **1.** Pairing energy for *m* electron pairs present in $t_{2g}^p\, e_g^q$ configuration

= Total number electron pairs present in $t_{2g}^p\, e_g^q$ configuration (*m*)

× Pairing energy for one electron (*P*)

$= m \times P\ cm^{-1}$

**2.** $\Delta_o = 10\ Dq$

**3.** Value of crystal field stabilisation energy (CFSE) for $t_{2g}^p\, e_g^q$ configuration of $d^x$ ion ($x = p + q$) in octahedral complex ($\Delta_o$) is given by:

CFSE = $-\,0.4\,\Delta_o$ × No. of electrons in $t_{2g}$ set of orbitals + $0.6\,\Delta_o$ × No. of electrons in $e_g$ set of orbitals

$= -\,0.4\,\Delta_o \times p + 0.6\,\Delta_o \times q$
$= (-\,0.4\,p + 0.6\,q)\,\Delta_o$
$= (-\,0.4\,p + 0.6\,q) \times 10\,Dq$ $(\because \Delta_o = 10\,Dq)$
$= (-\,4p + 6q)\,Dq$

**4.** Between the two complexes, the complex having less positive or more negative value of CFSE is more stable

**5.** (*a*) When $P > \Delta_o$, complex is high spin (HS) or spin free.

(*b*) When $P < \Delta_o$, complex is low spin (LS) or spin paired.

(*c*) When $P > \Delta_o$, complex in the HS state is more stable.

(*d*) When $P < \Delta_o$, complex in LS state is more stable

**6.** (*i*) **Weak ligands** are those :

(*a*) Which are not able to force the electrons to pair (VBT)

(*b*) Which form outer orbital octahedral complexes (VBT)

(*c*) Which give smaller value of energy gap ($\Delta_o$) between $t_{2g}$ and $e_g$ sets of orbitals (CFT)

(*d*) Which form HS octahedral complexes, *i.e.* complexes formed by weak ligands have more number of unpaired electrons.

(*e*) In case of weak ligands the energy of $t_{2g}$ and $e_g$ sets of orbitals is supposed to be almost the same. Hence the filling of these orbitals with electrons takes place in accordance with Aufbau principle.

(*ii*) **Strong ligands** are those:

(*a*) Which can force the electrons to pair (VBT)

(*b*) Which form inner orbital octahedral complexes (VBT)

(*c*) Which give larger value of energy gap ($\Delta_o$) between $t_{2g}$ and $e_g$ sets of orbitals (CFT)

(*d*) Which form LS octahedral complexes, *i.e.* complexes formed by strong ligands have less number of unpaired electrons

(*e*) In case of strong ligands the energy of $t_{2g}$ and $e_g$ sets of orbitals is different. Hence the filling of these orbitals with electrons does not take place in accordance with Aufbau principle.

**7.** For square planar complexes:

(*a*) of $Co^{2+}$ ($3d^7$), $Ni^{2+}$ ($3d^8$) and $Cu^{2+}$ ($3d^9$) ions

$\Delta_{sp} = 1.746\Delta_o = 17.46\ Dq$ $(\because \Delta_o = 10\ Dq)$

(*b*) of $Pd^{2+}$ ($3d^8$) and $Pt^{2+}$ ($5d^8$) ions

$\Delta_{sp} = 1.3\ \Delta_o$

(*c*) $\Delta_{sp} > \Delta_o$ $(\because \Delta_{sp} = 1.3\ \Delta_o)$

**8.** For tetrahedral complexes:

(*a*) $\Delta_t = 0.45\ \Delta_o$

(*b*) $\Delta_t < \Delta_o$ $(\because \Delta_t = 0.45\ \Delta_o)$

**9.** 1 kJ $mol^{-1}$ = 83.6 $cm^{-1}$

**10.** 1 K cal $mol^{-1}$ = 350 $cm^{-1}$

**11.** Frequency is expressed as wave number ($\bar{\nu}$)

**12.** Wave number ($\bar{\nu}$) is expressed in ($m^{-1}$) and wave length ($\lambda$) is expressed in $A°$ or nm.

**13.** Wave length ($\lambda$) = Inverse of wave number ($\bar{\nu}$)

$$= \frac{1}{\text{Wave number } (\bar{\nu})} \text{ cm}$$

## [V] Greek Alphabets

| | | | | | |
|---|---|---|---|---|---|
| Α | α | Alpha | Ν | ν | Nu |
| Β | β | Beta | Θ | ξ | Xi |
| Γ | γ | Gamma | Ο | ο | Omicron |
| Δ | δ | Delta | Π | π | Pi |
| Ε | ε | Epsilon | Ρ | ρ | Rho |
| Ζ | ξ | Zeta | Σ | σ | Sigma |
| Η | η | Eta | Τ | τ | Tau |
| Θ | θ | Theta | Υ | υ | Upsilon |
| Ι | ι | Iota | Φ | ϕ, φ | Phi |
| Κ | κ | Kappa | Χ | χ | Chi |
| Λ | λ | Lambda | Ψ | ψ | Psi |
| Μ | μ | Mu | Γ | ω | Omega |

## [VI] Some Commonly Used Prefixes with the Base Units

**The prefixes with the base units that are commonly used are given in Table 2.**

**Table 2. Some commonly used prefixes with the base units.**

| PREFIX | SYMBOL FACTOR | MULTIPLYING | EXAMPLE |
|---|---|---|---|
| deci | d | $10^{-1}$ | 1 decimetre (dm) = $10^{-1}$ m = 0.1 m |
| centi | c | $10^{-2}$ | 1 centimetre (cm) = $10^{-2}$ m = 0.01 m |
| milli | m | $10^{-3}$ | 1 millimetre (mm) = $10^{-3}$ m |
| micro | μ | $10^{-6}$ | 1 micrometre (μm) = $10^{-6}$ m |
| nano | n | $10^{-9}$ | 1 nanometre (nm) = $10^{-9}$ m |
| pico | p | $10^{-12}$ | 1 picometre (pm) = $10^{-12}$ m |
| femto | f | $10^{-15}$ | 1 femtometre (fm) = $10^{-15}$ m |
| atto | a | $10^{-18}$ | 1 atto metre (am) = $10^{-18}$ m |
| zepto | z | $10^{-21}$ | 1 zepto metre (zm) = $10^{-21}$ m |
| yocto | y | $10^{-24}$ | 1 yocto metre (ym) = $10^{-24}$ m |
| deka | da | $10^{1}$ | 1 dekametre (dam) = $10^{1}$ m = 10 m |
| hecto | — | $10^{2}$ | 1 hectometre = $10^{2}$ m |
| kilo | k | $10^{3}$ | 1 kilometer (km) = $10^{3}$ m = 1000 m |
| myria | — | $10^{4}$ | — |
| mega | M | $10^{6}$ | 1 megametre (Mm) = $10^{6}$ m |
| giga | G | $10^{9}$ | 1 gigametre (Gm) = $10^{9}$ m |
| tera | T | $10^{12}$ | 1 terametre (Tm) = $10^{12}$ m |
| peta | P | $10^{15}$ | 1 petametre (Pm) = $10^{15}$ m |
| exa | E | $10^{18}$ | 1 exametre (Em) = $10^{18}$ m |
| zeta | Z | $10^{21}$ | 1 zeta metre (Zm) = $10^{21}$ m |
| yotta | Y | $10^{24}$ | 1 yotta metre (Ym) = $10^{24}$ m |

## [VII] Some Important Points to Remember About S.I. Units

(*i*) The unit named after a scientist is started with a small letter and not with a capital letter, *e.g.*, unit of force is written as newton and not as Newton. Likewise unit of heat or work is written as joule and not as Joule.

(*ii*) Symbols of the units do not have a plural ending like '*s*'. For example we have 10 cm and not 10 cms.

(*iii*) Words and symbols should not be mixed, *e.g.*, we should write either joules per mole or J $mol^{-1}$ and not joules $mol^{-1}$.

(*iv*) Prefixes are used with the base units, *e.g.*, kilometer means 1000 m because meter is the base unit.

***Exception.*** Though kilogram is the base unit of mass, yet prefixes are used with gram because in kilogram, kilo is already a prefix.

(*v*) A unit written with a prefix and a power is a power for the complete unit, *e.g.*, $cm^3$ means $(centimeter)^3$ and not centi $(meter)^3$ *i.e.*, powers also operate on the prefixes. Thus, 1 $cm^3$ = $(10^{-2}\ m)^3 = 10^{-6}\ m^3$ and not $10^{-2}\ m^3$.

(*vi*) Unit combinations are represented by leaving a small space between them, *e.g.*, joule second as J s and not as Js or metre per second as $m\ s^{-1}$ and not as $ms^{-1}$.

(*vii*) Temperatures in Kelvin are expressed as 25 K, 30 K, etc. and not as 25°K, 30°K etc.

(*viii*) In explaining the text, the word 'mole' is used but when used as a unit, it is written as 'mol'.

## [VIII] Length Conversion Factors

(*i*)

| | *Å (angstrom)* | *cm (centimeter)* | *m (meter)* | *nm (nanometre)* | *pm (picometre)* |
|---|---|---|---|---|---|
| 1Å = | 1 | $10^{-8}$ | $10^{-10}$ | $10^{-1}$ | $10^{2}$ |
| 1 cm = | $10^{8}$ | 1 | $10^{-2}$ | $10^{7}$ | $10^{10}$ |
| 1 m = | $10^{10}$ | $10^{2}$ | 1 | $10^{9}$ | $10^{12}$ |
| 1 nm = | $10^{1}$ | $10^{-7}$ | $10^{-9}$ | 1 | $10^{3}$ |
| 1 pm = | $10^{-2}$ | $10^{-10}$ | $10^{-12}$ | $10^{-3}$ | 1 |

(*ii*) 1 nm = 1 millimicron (1 mμ) = $10^{-7}$ cm = 10 Å

(*iii*) 1 micron (1μ) = $10^{-4}$ cm

(*iv*) 1 $cm^{-1} = 10^2\ m^{-1}$

(*v*) 1 $m^{-1} = 10^{-2}\ cm^{-1}$

(*vi*) 1 km = 1000 m = $10^3$ m

(*vii*) 1 Å = 0.1 mμ = 4.11 $cm^{-1}$

(*viii*) 1 kK (Kilo Kaiser) = 1000 $cm^{-1}$

## [IX] Energy Conversion Factors

(*i*) Avogadro's number = $6.023 \times 10^{23}$

(*ii*) 1 cal = 4.184 J

(*iii*) 1 cal = $10^{-3}$ k cal

(*iv*) 1 J = $10^{-3}$ kJ = $10^{7}$ ergs

(*v*) 1 eV $atom^{-1} = 3.83 \times 10^{-20}$ cal $atom^{-1}$ = 23.06 k cal $mol^{-1}$

(*vi*) 1 eV atom$^{-1}$ = $1.602 \times 10^{-19}$ J atom$^{-1}$ = 96.49 k J mol$^{-1}$
(*vii*) 1 amu = 931. 48 MeV
(*viii*) 1 erg = $10^{-7}$ J
(*ix*) (1 cal > 1J > 1 erg > 1eV)

## [X] Relation between Different Units of Volume

$$1\ L = 10^{-3}\ m^3 = 1\ dm^3 = 10^3\ cm = 1000\ mL$$

## [XI] Relation between Different Units of Mass

$$1\ kg = 1000\ g = 10^3\ g$$
$$1\ g = 1000\ mg = 10^3\ mg$$

## [XII] Some Number Measures

(*i*) **Couple** means 2
(*ii*) **Dozon** means 12
(*iii*) **Gross** means 12 dozones or 12 × 12 = 144 pieces
(*iv*) **Great gross** = 12 grosses
(*v*) **Score** = 20 = 12 × 12 dozones = 12 × 12 × 12 pieces = 1728 pieces

## [XIII] Roman Numerals

| | | | |
|---|---|---|---|
| I | One | XVII | Eighteen |
| II | Two | XIX | Nineteen |
| III | Three | XX | Twenty |
| IV | Four | XXX | Thirty |
| V | Five | XL | Forty |
| VI | Six | L | Fifty |
| VII | Seven | LX | Sixty |
| VIII | Eight | LXX | Seventy |
| IX | Nine | LXXX | Eighty |
| X | Ten | XC | Ninety |
| XI | Eleven | C | Hundred |
| XII | Twelve | | |
| XIII | Thirteen | | |
| XIV | Fourteen | | |
| XV | Fifteen | | |
| XVI | Sixteen | | |
| XVII | Seventeen | | |

**1**

CHAPTER

# Oxidation and Reduction

## Covalency, Oxidation Number and Oxidation State

**Covalency** *of an element represents (i) the number of H-atoms with which an atom of that element can combine, (ii) the number of single bonds which an atom of that element can form, (iii) the number of electrons its atom is able to share with other element. Thus in every case covalency of an element is a pure number and has no plus or minus sign associated with it.* For example, covalency of nitrogen in $NH_3$ is 3.

***Oxidation number*** *of an element is defined as the formal charge which an atom of that element appears to have when electrons are counted. Oxidation number of an atom may be positive or negative.* For example, (*i*) oxidation number of K and Br in KBr is +1 and –1 respectively. (*ii*) Oxidation number of N in $NH_3$ is –3.

There may be a difference between the magnitude of covalency and oxidation number of the same element in various compounds. For example, in each of the following compounds, since C-atom shares four pairs of electrons with other atoms, the covalency of carbon is 4 while the oxidation number of carbon in these compounds is – 4, – 2, 0, + 2 and + 4 respectively.

| *Compound:* | $CH_4$ *Methane* | $CH_3Cl$ *Methyl chloride* | $CH_2Cl_2$ *Methylene chloride* | $CHCl_3$ *Chloroform* | $CCl_4$ *Carbon tetra-chloride* |
|---|---|---|---|---|---|
| Covalency of carbon : | 4 | 4 | 4 | 4 | 4 |
| Oxidation number of carbon : | – 4 | –2 | 0 | +2 | +4 |

*In ionic compounds the oxidation state of an element is the same as the charge on the ion formed from the atom of the element.* For example, in KBr, K is said to be in +1 oxidation state while Br is said to be in –1 oxidation state.

***Oxidation state*** *of an element is its oxidation number per atom.* For example, the oxidation state of Al in $Al_2O_3$ = + 6/2 = +3

## Differences Between Oxidation Number (O.N.) and Valency

The differences between oxidation number and valency has been shown in a tabular form as follows:

| *Oxidation Number (O.N.)* | *Valency* |
|---|---|
| (1) O.N. of an element is the number of excess charges that an atom of the element has or appears to have in its combined state, assuming that each electron belongs either to one or another atom. | (1) Valency of an element is its combining capacity and is expressed as the number of H-atoms or Cl atoms or double the number of oxygen atoms that combine with one atom of the element. |
| (2) With O.N. plus (+) or minus (–) signs is attached as it is the charge left on the atom of the element. For example, O.N. of C in CO = +2, while O.N. of C in $CH_3Cl$ = –2. | (2) Since valency is the combining capacity of the element, no plus or minus sign is attached to it. For example, valency of C is 4. |
| (3) O.N. of an element varies from compound to compound. *i.e.*, O.N. of an element depends on the nature of the compound in which the element is present. For example, O.N. of C in $CH_4$, $C_2H_6$, $C_2H_4$, $C_2H_2$ and $CH_2Cl_2$ is –4, –3, –2, –1 and 0 respectively. | (3) Valency of an element is usually fixed. For example, the valency of carbon in its compound is 4, *i.e.* carbon is tetravalent in its compounds. |
| (4) O.N. of an element may be whole number or it may be fractional. For example, O.N. of Mn in $Mn_3O_4$ is $+\frac{8}{3}$. | (4) Valency of an element is always whole number and is never fractional. For example, valency of Mn is 2. |
| (5) O.N. of an element can be zero. For example O.N. of carbon in $CH_2Cl_2$ is 0. | (5) Valency of an element is never zero (exception is noble gases), since it is the combining capacity of the element. |

## Rules for Calculating Oxidation Number

In order to calculate the oxidation number (O.N.) of an element in a given molecule/ion, the following rules are followed. These are arbitrary rules

(*i*) The oxidation number of an element in the free or uncombined state is zero.

(*ii*) Fluorine, the most electronegative element, has an oxidation number of –1 in all its compound.

(*iii*) The oxidation number of hydrogen is +1 in all its compounds except in the ionic metal hydrides where its O.N. is –1. For example, the O.N. of H in $H_2O$ is +1 while its O.N. in NaH is equal to –1.

(*iv*) The O.N. of oxygen is generally equal to –2 except in $F_2O$ in which the O.N. of oxygen is equal to +2. In $H_2O_2$ molecule whose Lewis structure is :Ö··Ö: (with H under each O) or :Ö—Ö: (with H–bonded under each O), the electron pair shared between H and O atoms is counted with O-atom, since it is more electronegative than H-atom. Therefore the number of electrons counted with each O-atom is equal to seven instead of its outermost shell six electrons. Therefore, O-atom in $H_2O_2$, molecule appears to have –1 charge on it, *i.e.* O.N. of O-atom in $H_2O_2$ molecule (or its derivatives like $BaO_2$ *etc.*) is equal to –1.

(*v*) The net charge on a given ion is equal to the sum of the oxidation numbers of all the atoms present in the ion.

(*vi*) The oxidation number of a neutral molecule is always zero and is equal to the sum of oxidation numbers of the individual atoms, each multiplied by the number of atoms of the element in the molecule.

## Solved Examples

**Example 1. Determine oxidation state (O.S.) of: (*i*) C-atom in carbon suboxide ($C_3O_2$), (*ii*) Br-atom in tribromo octaoxide ($Br_3O_8$) (*iii*) S-atom in tetrathionate ion ($S_4O_6^{2-}$), (*iv*) Fe-atom in magnetic oxide ($Fe_3O_4$), (*v*) Mn-atom in $Mn_3O_4$, (*vi*)Pb atom in red lead ($Pb_3O_4$) (*vii*) O-atom in $KO_3$, (*viii*) Fe-atom in ferriferrocynide molecule, $Fe_4[Fe(CN)_6]_3$, (*ix*) Fe-atom in $Fe_{0.4}O$.**

**Ans.** (*i*) **O.S. of C-atom in carbon suboxide, $C_3O_2$.** The structure of $C_3O_2$ (carbon suboxide) is:

$$\overset{-2}{O} = \overset{+2}{C} = \overset{0}{C} = \overset{+2}{C} = \overset{-2}{O}$$

Since a double bond is formed by mutual sharing of two electrons and oxygen is more electronegative than carbon, each of the two terminal oxygen atoms has an oxidation state of –2 and the carbons to which they are attached have an oxidation state of +2. Further, since whenever a covalent bond is formed between similar atoms, each of the atoms is given an oxidation state of zero. Therefore, O.S. of the central carbon atom in $C_3O_2$ is zero.

Thus, the two carbon atoms are present in +2 oxidation state each, whereas the third one is present in zero oxidation state. Therefore, the average oxidation state of carbon in $C_3O_2 = (2 \times 2 + 1 \times 0) / 3 =$ **4/3 (Ans.)**

However, *in reality, the terminal carbon atoms have an oxidation state of* +2 *while the middle carbon has an oxidation state of zero.*

(*ii*) **O.S of Br-atom in tribromoocta oxide, $Br_3O_8$.** The structure of tribromooctaoxide ($Br_3O_8$) is:

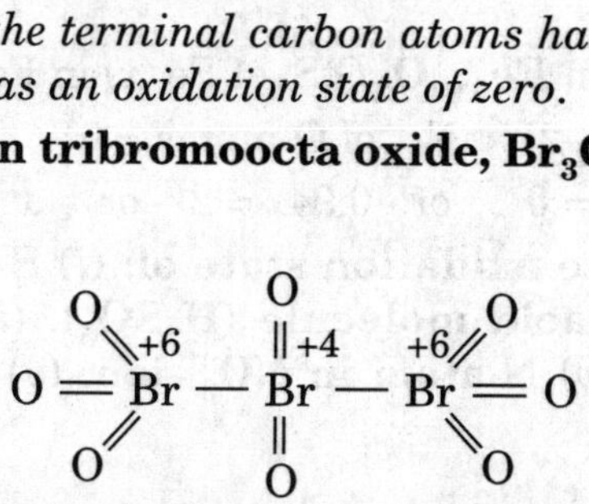

Following the argument discussed under $C_3O_2$, we can easily find out that terminal bromine atoms are present in +6 oxidation state while the middle one has an oxidation state of +4. Thus the average oxidation number of Br in $Br_3O_8$ turns out to be [2 (+ 6) + (+ 4)] / 3 = **16/3. (Ans.)**

(*iii*) **O.S. of S-atom in tetrathionate ion, $S_4O_6^{-2}$.** The structure of $S_4O_6^{-2}$ ion is:

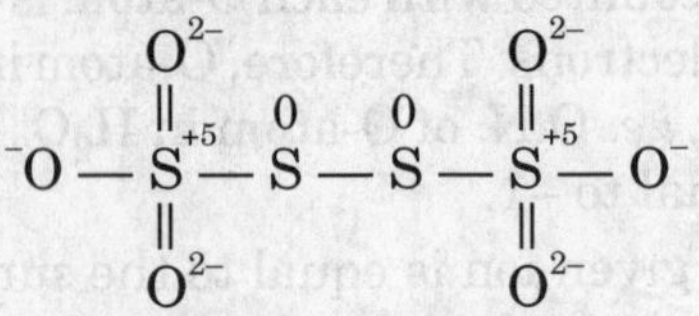

Since each of the two terminal sulphur atoms is connected to two oxygen atoms by a double bond and to one oxygen atom by a single bond, the oxidation state of each of these terminal sulphur atoms is +5. Since two central sulphur atoms are linked to each other by a single bond and each sulphur is further attached to similar species on either side, the electron pair forming the S – S bond remains in the centre and hence each of the two central sulphur atoms has an oxidation state of zero. Thus, the average oxidation state of the four sulphur atoms is [2 × 5 + 2 × 0]/4 = 2.5 **(Ans).**

(*iv*) **O.S. of Fe in magnetic oxide, $Fe_3O_4$.** $Fe_3O_4$ is known to be a mixture of FeO and $Fe_2O_3$ having the composition FeO. $Fe_2O_3$. Therefore, oxidation number of Fe in FeO is +2 while in $Fe_2O_3$ it is +3. The average value, however, is [1 (+ 2) + 2 (+ 3)]/3 = **8/3 (Ans.)**

(*v*) **O.S. of Mn in $Mn_3O_4$.** $Mn_3O_4$ is regarded as a mixture of MnO and $MnO_2$ having the composition, 2MnO.$MnO_2$. Therefore, the oxidation state of Mn in MnO is +2 while that in $MnO_2$ it is + 4. The average is [2 (+ 2) + (+ 4)]/3 = **8/3 (Ans.)**

(*vi*) **O.S. of Pb in red lead, $Pb_3O_4$.** $Pb_3O_4$ is considered to be a mixture with the composition 2PbO. $PbO_2$. Therefore, the oxidation state of Pb in PbO is + 2 while in $PbO_2$, it is +4. Here, again the average value is [2 (+ 2) + 1 (+ 4)]/3 = **8/3. (Ans.)**

(*vii*) **O.S. of O-atom in $KO_3$.** Let the oxidation state of O-atom in $KO_3$ be $x$. Hence :

$$(+1) \times 1 + 3 \times x = 0 \quad \text{or} \quad \boldsymbol{x = -1/3} \textbf{ (Ans.)}$$

(*viii*) **O.S. of Fe atom in $Fe_4[Fe(CN)_6]_3$.** In $Fe_4[Fe(CN)_6]_3$ four Fe atoms are in +3 O.S. and three Fe atoms are in +2 O.S. Thus the average O.S. of Fe atom

$$= \frac{4\times(+3)+3\times(+2)}{7} = \frac{\mathbf{18}}{\mathbf{7}} \textbf{ (Ans.)}$$

(*ix*) **O.S. of Fe-atom in $Fe_{0.94}O$.** O.S. of Fe ($x$) in $Fe_{0.94}O$ is given by:

$x$ × No. of Fe atoms + (–2) × No. of O-atoms = 0

or $x \times 0.94 + (-2) \times 1 = 0$ or $0.94x = 2$ or $x = 2/0.94 =$ **2.12 (Ans.)**

**Example 2. Calculate oxidation state of: (*i*) S-atom in Caro's acid or peroxomono sulphuric acid molecule ($H_2SO_5$). (*ii*) N-atom in $NH_4NO_3$. (*iii*) Cr-atom in $CrO_5$. (*iv*) N-atom in $NO_3^-$ ion. (*v*) Cl-atoms in bleaching**

**powder molecule ($CaOCl_2$). (*vi*) Fe-atom in $K_4$ [$Fe(CN)_6$]. (*vii*) C-atom in HCN. (*viii*) C-atom in HNC. (*ix*) S-atom in oleum molecule ($H_2S_2O_7$) (*x*) C-atom in HCOOH molecule. (*xi*) S-atom in sodium thiosulphate molecule ($Na_2S_2O_3$). (*xii*) S-atom in sodium tetrathionate ($Na_2S_4O_6$). (*xiii*) Na atom in NaHg amalgam. (*xiv*) S-atom in Marshall's acid or peroxdisulphuric acid molecule ($H_2S_2O_8$). (*xv*) C-atom in sucrose molecule ($C_{12}H_{22}O_{11}$).**

**Ans.** (*i*) By conventional method, the O.N. of S (*x*) in $H_2SO_5$ is given by:

$$2(+1) + x + 5(-2) = 0 \text{ or } \boldsymbol{x = +8}$$

This is impossible because the maximum O.N. of S cannot be more than six since it has only six electrons in the valence shell. This fallacy is overcome if we calculate the O.N. of S by chemical bonding method.

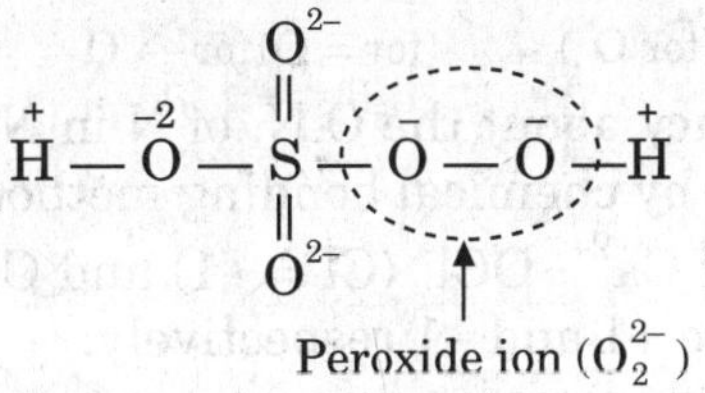


$$2 \times (+1) \quad + x + \quad 2(-1) \quad + \quad 3 \times (-2) = 0 \quad \text{or} \quad \boldsymbol{x = +6}$$

(for H) (for S) for (O – O) (for other O atoms)

(*ii*) Oxidation numbers of two N atoms in $NH_4NO_3$ are different. $NH_4NO_3$ is made of $NH_4^+$ and $NO_3^-$

In $NH_4^+$ ion, $x + 4(+1) = +1$ $\therefore$ O.N. of N is –3

In $NO_3^-$ ion, $x + 3(-2) = -1$ $\therefore$ O.N. of N is +5

$$\text{Average O.S. of N-atom} = \frac{+5+(-3)}{2} = +1 \textbf{ (Ans.)}$$

(*iii*) According to conventional method, O.N. of Cr atom (*x*) in $CrO_5$ molecule is given by:

$$x + 5 \times (-2) = 0, \quad \therefore \quad \boldsymbol{x = +10}$$

This is impossible because the maximum O.N. of Cr cannot be more than six since it has $3d^5\,4s^1$ outer orbital configuration.

The correct value of O.N. of Cr atom is calculated with the help of the structure of $CrO_5$ given below:

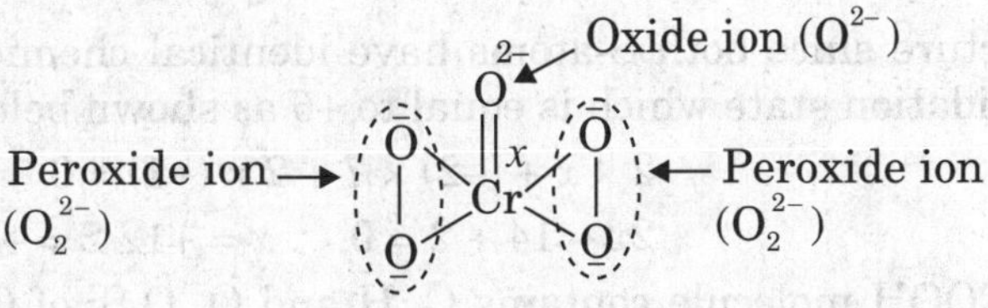

This structure shows that the value of $x$ is given by:

$$x + 2 \times (-2) \quad + \quad 1 \times (-2) = 0$$

Due to two peroxide ions ($O_2^{2-}$)    Due to one oxide ion ($O^{2-}$)

$$x - 4 - 2 = 0 \quad \text{or} \quad \boldsymbol{x = +6} \textbf{ (Ans.)}$$

(*iv*) According to conventional method, O.N. of N is $NO_3^- = x + 3(-2) = -1$ or $x = +5$

According to chemical bonding method, $^-O{-}\overset{x}{N}(=O^{2-})(\rightarrow O^{2-})$

$$x + 1(-1) \quad + \quad 1(-2) \quad + \quad 1(-2) = 0 \qquad \text{or } x + 5$$

(for $O^-$)    (for = O) for → O

Thus, there is no fallacy about the O.N. of N in $NO_3^-$ whether one calculates by conventional method or by chemical bonding method.

(*v*) $CaOCl_2$ is made of $Ca^{2+}$, $OCl^-$ (Cl = +1) and $Cl^-$ (Cl = –1). Hence, O.N. of two Cl atoms in $CaOCl_2$ are +1 and –1 respectively.

(*vi*) In $K_4[Fe(CN)_6]$ the individual O.N. of C and N are not known but for calculating O.N. of Fe, the sum of O.N. of C and N in $(CN)^-$ unit is taken to be –1. Thus, O.N. of Fe is calculated as $4(+1) + x + 6(-1) = 0$ or $\boldsymbol{x = +2}$. **(Ans.)**

(*vii*) The structure of HCN molecule is H–C ≡ N. Since N-atom in more electronegative than C-atom, each C – N covalent bond gives in O.S. of –1 to N-atom, Now since there are three C – N covalent bonds in HCN molecule, O.S. of N-atom = –3. Hence HCN can be represented as $\overset{+1}{H}{-}\overset{x}{C}\equiv\overset{-3}{N}$. Hence $(+1) \times 1 + x \times 1 + (-3) \times 1 = 0$ or $x =$ **+2 (Ans.)**

(*viii*) The structure of HNC molecule is $H{-}N\overset{\rightarrow}{=}C$. Since N-atom (donor atom) is more electronegative than C-atom (accepter atom), O.S. of N-atom = –3. Thus, HNC can be represented as $\overset{+1}{H}{-}\overset{-3}{N}\overset{\rightarrow}{=}\overset{x}{C}$. Hence $(+1) \times 1 + (-3) \times 1 + x \times 1 = 0$ or $x$ = **+ 2 (Ans.)**

(*ix*) The structure of $H_2S_2O_7$ molecule is given below:

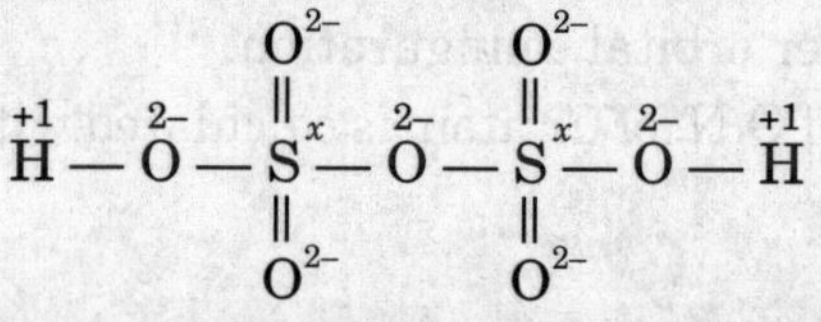

In this structure since both S-atoms have identical chemical bonding, they have the same oxidation state which is equal to +6 as shown below:

$$2 \times x + (-2) \times 7 + 2 \times (+1) = 0$$

or $$2x - 14 + 2 = 0, \therefore x = +12/2 = \textbf{+6 (Ans.)}$$

(*x*) Since HCOOH molecule contains C, H and O, O.S. of C-atom ($x$) in this molecule can be calculated by using the formula:

No. of H-atoms × O.S. of H-atom + No. of O-atoms × O.S. of O-atom + No. of C-atoms × O.S. of C-atom = 0

Hence in case of HCOOH molecule the above relation can be written as:

$$2 \times (+1) + 2 \times (-2) + 1 \times x = 0$$

$$\therefore \quad \mathbf{x = +2 \ (Ans.)}$$

(*xi*) Conventional method gives O.S. to S-atom in $Na_2S_2O_3$ molecule = + 2 as shown below :

$$2 \times (+1) + 2x + 3 \times (-2) = 0$$

(Due to Na) (Due to O)

$$\therefore \quad x = +2$$

This oxidation state (= +2) is wrong since the two S-atom in $Na_2S_2O_3$ molecule are in different oxidation states because of the fact that when $Na_2S_2O_3$ is heated with dil. $H_2SO_4$, one S-atom gets precipitated and the other S-atom is converted into $SO_2$. The correct O.S. of S-atom can be calculated from the structure of $Na_2S_2O_3$ which is given below.

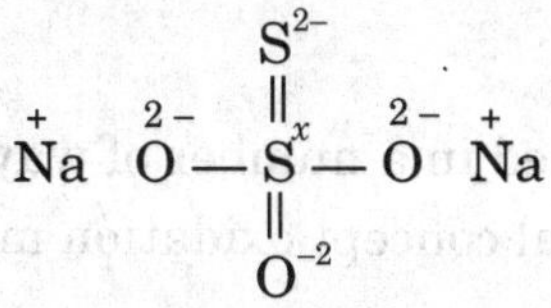

In the structure given above each of the two Na atoms has an O.S. = +1, each of three O-atoms has O.S. = –2. Terminal S-atom has O.S. = –2. Let the O.S. of the central S-atom be $x$. Thus:

$(+1) \times 2 \quad + \quad (-2) \times 3 \quad + \ (-2) \times 1 \quad + \quad x = 0$

(For Na) (For O-atom) (For terminal S-atom)

$+ 2 - 6 - 2 + x = 0, \ \therefore \ \mathbf{x = +6}$

Thus:

$$\left.\begin{aligned} \text{O.S. of terminal S-atom} &= -2 \\ \text{O.S. of central S-atom} &= +6 \end{aligned}\right\} \textbf{(Ans.)}$$

$$\text{Average O.S. of S-atom} = \frac{1 \times (-2) + 1 \times (+6)}{2} = \frac{-2+6}{2} = \mathbf{+2 \ (Ans.)}$$

(*xii*) The structure of $Na_2S_4O_6$ molecule is given below:

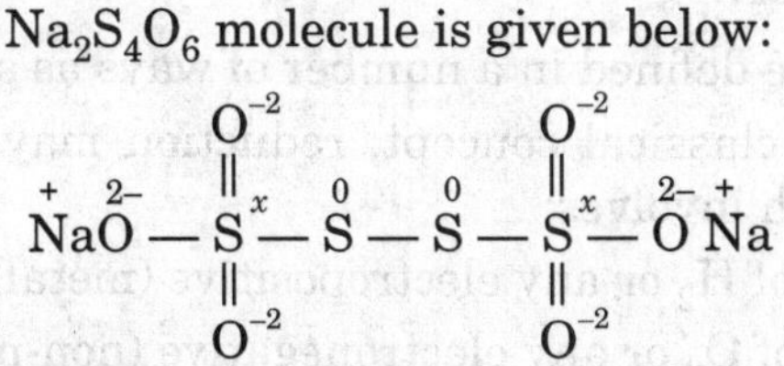

In this structure each of the two central S-atoms which are linked together has O.S. = 0. Let the O.S. of each of the two remaining S-atoms be $x$. Hence:

$2 \times (+1) \quad + 6 \times (-2) \quad + \ 2 \times 0 \quad + \ 2 \times x = 0$

(For Na atoms) (For O-atoms) (For central S-atoms)

$$+2 - 12 + 0 + 2x = 0$$

$$\therefore \quad x = +10/2 = +5$$

Thus O.S. of each central S-atom = 0

In the structure each of the two H-atoms has O.S. = –1, each of the six terminal O-atoms has O.S. = –2. This structure has one peroxide ion ($O_2^{-2}$) whose O.S. = –2. In this ion each O-atom has O.S = –1. Thus:

$$\underset{\text{(Due to H-atoms)}}{2 \times (+1)} \quad + \quad \underset{\text{(Due to six O-atoms)}}{6 \times (-2)} \quad \underset{\text{(Due to peroxide ion, } O_2^{-2})}{-2} \quad +2x = 0$$

∴ $x = + 12/2 =$ **+6 (Ans.)**

(*xv*) Since the given molecule ($C_{12}H_{22}O_{11}$) contains H, C and O, O.S. of C-atom ($x$) is given by:

$$x \times n_C + (-2) \times n_O + (+1) \times n_H = 0$$

or $x \times 12 + (-2) \times 11 + (+1) \times 22 = 0$

or $12x - 22 + 22 = 0$

∴ $\boldsymbol{x = 0}$ **(Ans.)**

## What is Oxidation?

Oxidation can be defined in a number of ways as shown below.

(*i*) According to classical concept oxidation may be defined as a chemical reaction which involves:

(*a*) addition of $O_2$ or any electronegative (non-metallic) element.

(*b*) removal of $H_2$ or any electropositive (metallic) element.

(*ii*) According to electronic concept (modern concept), oxidation is a chemical reaction in which:

(*a*) Oxidation number (O.N.) of the element increases.

(*b*) An atom or an ion loses one or more electrons and thus the valency of the atom or ion is increased.

(*c*) An atom or an ion loses one or more electrons and is changed into more electropositive or less electronegative state.

Since electrons are lost in oxidation, oxidation is also called **de-electronation**.

## What is Reduction?

Reduction can be defined in a number of ways as shown below.

(*i*) According to classical concept, reduction may be defined as a chemical reaction which involves:

(*a*) addition of $H_2$ or any electropositive (metallic) element.

(*b*) removal of $O_2$ or any electronegative (non-metallic) element.

(*ii*) According to electronic concept (modern concept), reduction is a chemical reaction in which:

(*a*) Oxidation number (O.N.) of the element decreases.

(*b*) An atom or an ion gains one or more electrons and thus the valency of the atom or ion is decreased.

(*c*) An atom or an ion loses one or more electrons and is changed into less electropositive or more electronegative state.

Since electrons are gained in reduction, reduction is also called **electronation**.

## Oxidising Agent (O.A.) or Oxidant

(*i*) According to classical concept, an O.A. is a substance which gives $O_2$ or any other electronegative element to another substance in a reaction

(*ii*) According to electronic concept:

(*a*) An O.A. is a substance (atom, ion or molecule) which contains an element that gains electrons in a reaction:

(*b*) An O.A. is a substance which contains an element whose O.N. decreases in a chemical reaction.

(*c*) An O.A. is a substance which oxidises some other substance and is itself reduced to lower valency state by gaining electrons.

(*d*) An O.A. is an electron acceptor.

## Reducing Agent (R.A.) or Reductant

(*i*) According to classical concept, a R.A. is a substance which releases $H_2$ or any other electropositive element in a chemical reaction.

(*ii*) According to electronic concept:

(*a*) A R.A. is a substance which contains an element that loses electrons in a reaction.

(*b*) A R.A. is a substance which contains an element whose O.N. increases in a chemical reaction.

(*c*) A R.A. is a substance which reduces some other substance and is itself oxidised to higher valency state by losing electrons.

(*d*) A R.A. is an electron donor.

## Balancing Redox Equation by Oxidation Number Method

This method is based on the principle that any increase in oxidation number must be compensated by a decrease. This method consists of the following steps:

(*i*) Note the elements which undergo change in oxidation numbers.

(*ii*) Select the suitable coefficients for the oxidising and reducing agents so that the total decrease in oxidation number of the oxidising agent becomes equal to the total increase in the oxidation number of the reducing agent.

## Solved Examples

The following examples illustrate the above method.

**Example 1. Balance the following equation by oxidation number method:**

$$CuO + NH_3 \longrightarrow Cu + N_2 + H_2O$$

**Solution.** The given equation shows that the oxidation number (O.N.) of Cu decreases from +2 (in CuO) to 0 (in Cu) while that of N increases from – 3 (in $NH_3$) to 0 (in $N_2$) and hence :

Increase in O.N = 0 – (–3) = 3

(Oxidation)

$$\underset{(Cu = +2)}{CuO} + \underset{(N = -3)}{NH_3} \rightarrow \underset{(Cu = 0)}{Cu} + \underset{(N = 0)}{N_2} + H_2O$$

Decrease in O. N. = 2 – 0 = 2

(Reduction)

In order to equalise the total increase in O.N. (= 3) to the total decrease in O.N. (= 2), we should have three atoms of Cu for every two atoms of N and hence the equation should be written as:

$$3CuO + 2NH_3 \longrightarrow 3Cu + N_2 + H_2O$$

Now in order to balance O-atoms we should add $3H_2O$ molecules to the right hand side. Thus:

$$3CuO + 2NH_3 \longrightarrow 3Cu + N_2 + 3H_2O$$

This is the balanced equation.

**Example 2. Balance the following equation by oxidation number method.**

$$\mathbf{K_2Cr_2O_7 + H_2SO_4 + FeSO_4 \longrightarrow K_2SO_4 + Cr_2(SO_4)_3 + Fe_2(SO_4)_3}$$

***(Lucknow B.Sc.1981)***

**Solution.** The given equation shows that O.N. of Cr decreases from +6 (in $K_2Cr_2O_7$) to +3 [in $Cr_2(SO_4)_3$] while that of Fe increases from +2 (in $FeSO_4$) to +3 [in $Fe_2(SO_4)_3$] and hence:

Increase in O.N = 3 – 2 = 1

(Oxidation)

$$\underset{(Cr + +6)}{K_2Cr_2O_7} + H_2SO_4 + \underset{(Fe = +2)}{FeSO_4} \longrightarrow K_2SO_4 + \underset{(Cr = +3)}{Cr_2(SO_4)_3} + \underset{(Fe + 3)}{Fe_2(SO_4)_3}$$

Decrease in O.N. = 6 – 3 = 3

(Reduction)

In order to equalise the total increase in O.N. (= 1) to the total decrease in O.N. (= 3) we should have one atom of Cr for every three atoms of Fe or two atoms of Cr for every six atoms of Fe and hence above equation should be written as:

$$K_2Cr_2O_7 + H_2SO_4 + 6FeSO_4 \rightarrow K_2SO_4 + Cr_2(SO_4)_3 + Fe_2(SO_4)_3$$

In order to balance Fe atoms we should write the above equation as:

$$K_2Cr_2O_7 + H_2SO_4 + 6FeSO_4 \rightarrow K_2SO_4 + Cr_2(SO_4)_3 + 3Fe_2(SO_4)_3$$

Now in order to balance O-atoms we add $7H_2O$ molecules to the right hand side and to balance H-atoms we write $7H_2SO_4$ in place of $H_2SO_4$ on the left hand side. Thus the equation in its balanced form becomes:

$$K_2Cr_2O_7 + 7H_2SO_4 + 6FeSO_4 \rightarrow K_2SO_4 + Cr_2(SO_4)_3 + 3Fe_2(SO_4)_3 + 7H_2O$$

## Balancing Redox Equations by Ion-electron Method—By the Use of Half-reactions.

This method was developed in 1927 by Jatte and Lamer and various steps involved in the method are as follows:

(*i*) Break up the complete equation in two half reactions, one for the change undergone by the reducing agent and the other for the change undergone by the oxidising agent.

(*ii*) Balance each half reaction as to the number of atoms of each element. For this purpose :

(*a*) Balance the atoms other than H and O for each half-reaction by using simple multiples.

(*b*) In neutral and acid solutions, $H_2O$ and $H^+$ are added for balancing oxygen and hydrogen atoms. First balance the oxygen atoms. For each excess oxygen atom on one side of the equation, add one $H_2O$ to the other side. Now use $H^+$ to balance hydrogen atoms.

(*c*) In alkaline solutions, $OH^-$ may be used. For each excess on one side, balance is secured by adding one $H_2O$ to the same side and $2H^-$ to the other side. If hydrogen is still unbalanced, balance is secured by adding one $OH^-$ for each excess hydrogen on the same side as the excess and one $H_2O$ to the other side.

(*iii*) Equalise the charges on both sides by adding electrons to the side deficient in negative charges.

(*iv*) Multiply one or both half-reactions by a suitable number so that on adding the two equations, the electrons are balanced.

(*v*) Add the two balanced half reactions and cancel any terms common to both sides. Also see that all electrons cancel.

## Solved Examples

Various steps of the ion-electron method can be illustrated by considering the following examples:

**Example 1. Balance the following redox reaction by ion-electron method:**
**$Fe^{2+} + MnO_4^- + H^+ \rightarrow Mn^{2+} + Fe^{3+} + H_2O$ (*Udaipur B.Sc. 1978; Kanpur 1980*)**

**Solution.** Obviously the given redox reaction takes place in acidic medium and can be broken into the following two half-reactions:

$$\underset{(Mn = +7)}{MnO_4^-} \rightarrow \underset{(Mn +2)}{Mn^{2+}} \quad \text{......\textit{Reduction half-reaction}}$$

and

$$\underset{(Fe = +2)}{Fe^{2+}} \rightarrow \underset{(Fe = +3)}{Fe^{3+}} \quad \text{......\textit{Oxidation half-reaction}}$$

*For reduction half-reaction.* (*i*) For balancing O-atoms add $4H_2O$ to right hand side to get:

$$MnO_4^- \rightarrow Mn^{2+} + 4H_2O$$

(*ii*) For balancing H-atoms add $8H^+$ to the left hand side to get:

$$MnO_4^- + 8H^+ \rightarrow Mn^{2+} + 4H_2O$$

(*iii*) For balancing the charges add $5e^-$ to the left hand side to get:

$$MnO_4^- + 8H^+ + 5e^- \rightarrow Mn^{2+} + 4H_2O \quad ...(i)$$

*For oxidation half-reaction.* Balance the charges on both sides by adding 1$e^-$ to the left hand side to get:

$$Fe^{2+} \rightarrow Fe^{3+} + e^-$$

or $$5Fe^{2+} \rightarrow 5Fe^{3+} + 5e^- \quad ...(ii)$$

On adding equations (*i*) and (*ii*), we get

$$MnO_4^- + 8H^+ + 5Fe^{2+} \rightarrow Mn^{2+} + 4H_2O + 5Fe^{3+}$$

which is the balanced equation.

**Example 2. Balance the following equation by using ion-electron method:**

$$\mathbf{Cr_2O_7^{2-} + Fe^{2+} + H^+ \rightarrow Cr^{3+} + Fe^{3+} + H_2O}$$

***(Mysore B.Sc. 1980; Kanpur 1980; Raj. 1975; Udaipur 1981)***

**Solution.** Obviously the given redox reaction takes place in acidic medium and can be broken into the following two half-reactions:

$$Cr_2O_7^{2-} \rightarrow Cr^{3+} \quad ......\textit{Reduction half-reaction}$$
$$(Cr = +6) \quad (Cr = +3)$$
$$Fe^{2+} \rightarrow Fe^{3+} \quad ......\textit{Oxidation half-reaction}$$
$$(Fe = +2) \quad (Fe = +3)$$

*For reduction half-reaction.* (*i*) Balance the atoms other than oxygen and hydrogen:

$$Cr_2O_7^{2-} \rightarrow 2Cr^{3+}$$

(*ii*) For balancing O-atoms add 7$H_2O$ to the right hand side to get:

$$Cr_2O_7^{2-} \rightarrow 2Cr^{3+} + 7H_2O$$

(*iii*) For balancing H-atoms add 14$H^+$ to the left hand side to get:

$$Cr_2O_7^{2-} + 14H^+ \rightarrow 2Cr^{3+} + 7H_2O$$

(*iv*) For balancing the charges add 6$e^-$ to the left hand side to get:

$$Cr_2O_7^{2-} + 14H^+ + 6e^- \rightarrow 2Cr^{3+} + 7H_2O \quad ...(i)$$

*For oxidation half-reaction.* Balance the charges on both the sides by adding 1$e^-$ to right hand side to get:

$$Fe^{2+} \rightarrow Fe^{3+} + e^-$$

or $$6Fe^{2+} \rightarrow 6Fe^{3+} + 6e^- \quad ...(ii)$$

On adding equations (*i*) and (*ii*), we get:

$$Cr_2O_7^{2-} + 6Fe^{2+} + 14H^+ \rightarrow 2Cr^{3+} + 6Fe^{3+} + 7H_2O$$

This is the equation in its balanced form.

## What is Electrode Potential?

When a metal rod is partially immersed in the solution of one of its salts, there are two possibilities, (i) the metal is oxidised and the metal ions formed leave the rod and pass into the solution leaving the electrons behind and hence make the metal rod negatively charged, (ii) the metal ions from the solution deposit on the metal rod by gaining electrons rendering the metal rod positively charged. Soon, an equilibrium is established between the rod and the metal ions in the solution and some potential difference develops between them, which depends on the concentration of salt solution and temperature. This potential difference is known as **electrode potential**.

## Measurement of Single Electrode Potential

Single electrode potential due to an electrode reaction cannot be measured experimentally. Only difference between potentials of two electrodes can be measured. In order to determine the single electrode potential directly by experiments, this electrode must be coupled with some reference electrode having zero electrode potential and the difference be determined. However, **no electrode with true zero potential is known.** Due to this reason, a standard hydrogen electrode is taken as the reference electrode whose potential is assumed to be zero at all temperatures.

## Galvanic Cell and emf of the Cell

The system obtained by connecting two electrodes having different electrode potentials is known as a cell and the difference of potentials which causes a current to flow from the electrode of higher potential to the one of lower is called the electromotive force (emf) of the cell and is expressed in volts.

A very common cell is obtained by connecting zinc rod immersed in zinc sulphate solution and a copper rod immersed in copper sulphate solution. At the zinc electrode, zinc ions pass into solution leaving two electrons per atom on the zinc rod. On the other hand, copper ions from the solution tend to deposit on copper electrode rendering it positively charged. When the two electrodes are connected, the current (electrons) move from zinc to copper through the external circuit. The chemical reaction taking place in the cell may be represented as follows:

At zinc electrode (Anode) $\underset{\text{Reductant}}{Zn} \xrightleftharpoons{\text{Oxidation}} \underset{\text{Oxidant}}{Zn^{2+}} + 2e^- \quad E^\circ_{ox} = +0.76\ V$

At copper electrode (Cathode) $\underset{\text{Oxidant}}{Cu^{2+}} + 2e^- \xrightleftharpoons{\text{Reduction}} \underset{\text{Reductant}}{Cu} \quad E^\circ_{red} = +0.34\ V$

We know that the reaction in which electrons are lost is known as oxidation and the reaction that involves taking of the electrons with consequent decrease in positive valence must be reduction. Obviously, the oxidation state increases on oxidation, whereas positive oxidation state decreases on reduction. Thus, oxidation takes place at zinc electrode (anode) and reduction takes place at the copper electrode (cathode). A reaction representing a single process is known as half-reaction. The separate half-reactions taking place at copper and zinc electrodes have been shown above. Here the electrode potential corresponding to an oxidation half-reaction is called oxidation potential, $E^\circ_{ox}$ and that corresponding a reductions half-reaction as reduction potential, $E^\circ_{red}$. Here it is important to note that separation of a cell reaction into oxidation and reduction half-reactions is only conceptual and any half-reaction does not take place unless it is coupled with some other half-reaction forming a cell. Thus, both the electron donating and electron accepting reactions, *i.e.*, oxidation and reduction reactions take place simultaneously and the total electrochemical reaction is the sum of the two half-reactions. Thus, the electrochemical reaction for the above zinc-copper cell is obtained by adding the two half-reactions,

$$Zn + Cu^{2+} \longrightarrow Zn^{2+} + Cu$$

The total electromotive force of the cell is the algebraic sum of both the electrode potentials. Thus, the *emf* ($E^{\circ}_{cell}$) of the above cell:

$$E^{\circ}_{cell} = 0.76 + 0.34 = 1.10 \text{ V}$$

It has been observed that the sign of the potential of an electrode is the same as the sign of the reduction potential of the half-reaction, whereas the sign of oxidation potential is opposite. Hence International Union of Pure and Applied Chemistry (IUPAC) in 1953 recommended that only reduction potentials be called electrode potentials. Since reduction is reverse of oxidation, oxidation potential for a reaction will be the same as the reduction potential for the reverse reaction with sign changed.

Thus, it is desirable to write all half-reactions as reduction reactions as given below in which the oxidised form is always written on the left hand side.

$$Ox + ne \longrightarrow Red$$

Here Ox and Red represent the oxidised and reduced forms. The overall reaction is then the difference of the two reduction half-reactions. It must be emphasised that no electron is evolved in the final reaction. So, the number of electrons in both the half-reactions must match. Thus, the first reaction given on the last page is written as below with the oxidant on the left hand side.

$$Zn^{2+} + 2e^{-} \longrightarrow Zn \qquad E^{\circ}_{red} = -0.76 \text{ V}$$

The standard electrode potential for the reaction is its standard reduction potential. Reduction potential of copper may be represented by the following equation,

$$Cu^{2+} + 2e \longrightarrow Cu \qquad E^{\circ}_{red} = +0.34 \text{ V}$$

Standard reduction potential $E^{o}$ for the overall reaction is the difference of the standard reduction potentials for both the reduction half-reactions into which it can be divided. The reaction occurring in the Zn-Cu cell.

$$Zn + Cu^{2+} \longrightarrow Zn^{2+} + Cu$$

may be represented as:

$$Zn \mid Zn^{2+} \mid\mid Cu^{2+} \mid Cu$$

where | and | | represent interface and salt bridge respectively. The standard potential of the above cell ($E^{\circ}_{cell}$) is given by:

$$E^{\circ}_{cell} = 0.34 - (-0.76) = 1.10 \text{ V}$$

## Electrochemical Series or e.m.f. Series

### What is Electrochemical Series?

The list of elements or ions arranged in the decreasing order (more positive → less positive → zero → less negative → more negative) of their standard reduction potential values, as in Table 1.1, is called **electrochemical** or **e.m.f. series.**

**Table 1.1. Standard reduction potentials of some electrodes (half-cells) – Electro chemical series**

| *Half-cell (Electrode or couple)* | *Half-cell reaction (Reduction reaction)* | *Standard reduction potential (volts) (E° values)* |
|---|---|---|
| | Strongest oxidising agent — Weakest reducnig agent | |
| $F_2/2F^-$ | $F_2 + 2e^- \longrightarrow 2\bar{F}$ | +2.85 |
| $H_2O_2$ / $H_2O$ | $H_2O_2 + 2H^+ 2e^- \longrightarrow 2H_2O$ | +1.77 |
| $MnO_4^-/MnO_2$ | $MnO^-_4 + 4H^+ + 3e^- \longrightarrow MnO_2 + 2H_2O$ | +1.69 |
| $Au^+/Au$ | $Au^+ + e^- \longrightarrow Au$ | +1.68 |
| $HClO_3/HClO$ | $HClO_3 + 4H^+ + 4e^- \longrightarrow HClO + 2H_2O$ | +1.64 |
| $HClO$ / $Cl_2$ | $2HClO + 2H^+ + 2e^- \longrightarrow Cl_2 + 2H_2O$ | +1.64 |
| $Ce^{4+}$ / $Ce^{3+}$ | $Ce^{4+} + e^- \longrightarrow Ce^{3+}$ | +1.61 |
| $2BrO_3^-$/ $Br_2$ | $2BrO_3^- + 12H^+ + 10e^- \longrightarrow Br_2 + 6H_2O$ | +1.52 |
| $MnO_4^-$/ $Mn^{2+}$ | $MnO_4^- + 8H^+ + 5e^- \longrightarrow Mn^{2+} + 4H_2O$ | +1.51 |
| $Au^{3+}$ / $Au$ | $Au^{3+} + 3e^- \longrightarrow Au$ | +1.42 |
| $Cl_2/2Cl^-$ | $Cl_2 + 2e^- \longrightarrow 2Cl^-$ | +1.36 |
| $Cr_2O_7^{2-}$ / $2Cr^{3+}$ | $Cr_2O_7^{2-} + 14H^+ + 6e^- \longrightarrow 2Cr^{3+} + 7H_2O$ | +1.33 |
| $2IO_3^-/I_2$ | $2IO_3^- + 12H^+ + 10e^- \longrightarrow I_2 + 6H_2O$ | +1.20 |
| $Pt^{2+}$ / $Pt$ | $Pt^{2+} + 2e^- \longrightarrow Pt$ | +1.20 |
| $IO_3^-$ / $I^-$ | $IO_3^- + 6H^+ + 6e^- \longrightarrow I^- + 3H_2O$ | +1.09 |
| $Br_2$ / $2Br^-$ | $Br_2 + 2e^- \longrightarrow 2Br^-$ | +1.06 |
| $Hg^{2+}$ / $Hg$ | $Hg^{2+} + 2e^- \longrightarrow Hg$ | +0.85 |
| $Ag^+$ / $Ag$ | $Ag^+ + e^- \longrightarrow Ag$ | +0.80 |
| $Fe^{3+}$ / $Fe^{2+}$ | $Fe^{3+} + e^- \longrightarrow Fe^{2+}$ | +0.77 |
| $O_2/H_2O_2$ | $O_2 + 2H^+ + 2e^- \longrightarrow H_2O_2$ | +0.68 |
| $H_3AsO_4$ / $H_3AsO_3$ | $H_3AsO_4 + 2H^+ + 2e^- \longrightarrow H_3AsO_3 + H_2O$ | +0.56 |
| $I_2/2I^-$ | $I_2 + 2e^- \longrightarrow 2I^-$ | +0.53 |
| $Cu^+$ / $Cu$ | $Cu^+ + 2e^- \longrightarrow Cu$ | +0.52 |
| $Cu^{2+}$ / $Cu$ | $Cu^{2+} + 2e^- \longrightarrow Cu$ | +0.34 |
| $Bi^{3+}$ / $Bi$ | $Bi^{3+} + 3e^- \longrightarrow Bi$ | +0.32 |
| $Cu^{2+}$ / $Cu^+$ | $Cu^{2+} + e^- \longrightarrow Cu^+$ | +0.15 |
| $Sn^{4+}$ / $Sn^{2+}$ | $Sn^{4+} + 2e^- \longrightarrow Sn^{2+}$ | +0.15 |
| $S_4O_6^{2-}$ / $S_2O_3^{2-}$ | $S_4O_6^{2-} + 2e^- \longrightarrow 2S_2O_3^{2-}$ | +0.09 |
| $2H^+$ / $H_2$ | $2H^+ + 2e^- \longrightarrow H_2$ | +0.00 (By definition) |

Oxidising power decreasing ↓

Reducing power increasing ↓

E° values decreasing [– ← 0 ← +] ↓

| *Half-cell (Electrode or couple)* | *Half-cell reaction (Reduction reaction)* | *Standard reduction potential (volts)* |
|---|---|---|
| $Pb^{2+} / Pb$ | $Pb^{2+} + 2e^- \longrightarrow Pb$ | –0.13 |
| $Sn^{2+}/Sn$ | $Sn^{2+} + 2e^- \longrightarrow Sn$ | –0.14 |
| $Ni^{2+} / Ni$ | $Ni^{2+} + 2e^- \longrightarrow Ni$ | –0.24 |
| $Co^{2+} / Co$ | $Co^{2+} + 2e^- \longrightarrow Co$ | –0.28 |
| $Cd^{2+} / Cd$ | $Cd^{2+} + 2e^- \longrightarrow Cd$ | –0.40 |
| $Cr^{3+} / Cr^{2+}$ | $Cr^{3+} + e^- \longrightarrow Cr^{2+}$ | –0.41 |
| $Fe^{2+} / Fe$ | $Fe^{2+} + 2e^- \longrightarrow Fe$ | –0.44 |
| $S/S^{2-}$ | $S + 2e^- \longrightarrow S^{2-}$ | –0.51 |
| $Cr^{3+} / Cr$ | $Cr^{3+} + 3e^- \longrightarrow Cr$ | –0.71 |
| $Zn^{2+} / Zn$ | $Zn^{2+} + 2e^- \longrightarrow Zn$ | –0.76 |
| $Mn^{2+} / Mn$ | $Mn^{2+} + 2e^- \longrightarrow Mn$ | –1.05 |
| $Al^{3+} / Al$ | $Al^{3+} + 3e^- \longrightarrow Al$ | –1.67 |
| $Be^{2+} / Be$ | $Be^{2+} + 2e^- \longrightarrow Be$ | –1.70 |
| $La^{3+} / La$ | $La^{3+} + 3e^- \longrightarrow La$ | –2.37 |
| $Mg^{2+} / Mg$ | $Mg^{2+} + 2e^- \longrightarrow Mg$ | –2.38 |
| $Na^+ / Na$ | $Na^+ + e^- \longrightarrow Na$ | –2.71 |
| $Ca^{2+} / Ca$ | $Ca^{2+} + 2e^- \longrightarrow Ca$ | –2.76 |
| $Sr^{2+} / Sr$ | $Sr^{2+} + 2e^- \longrightarrow Sr$ | –2.89 |
| $Ba^{2+} / Ba$ | $Ba^{2+} + 2e^- \longrightarrow Ba$ | –2.90 |
| $Cs^+ / Cs$ | $Cs^+ + e^- \longrightarrow Cs$ | –2.923 |
| $K^+ / K$ | $K^+ + e^- \longrightarrow K$ | –2.925 |
| $Li^+ / Li$ | $Li^+ + e^- \longrightarrow Li$ | –3.045 |
| $\frac{3}{2} N_2 / HN_3$ | $\frac{3}{2} N_2 + H^+ + e^- \longrightarrow HN_3$ | –3.09 |
| | Weakest oxidising agent; Strongest reducing agent | |
| | Oxidising power decreasing ↓; Reducing power increasing ↓ | E° values decreasing [– ← 0 ← +] ↓ |

## Positive and Negative E° Value of Metals

Let us consider the following half-reaction:

$$\underset{OA}{M^{n+}} + ne^- \underset{\text{Oxidation}}{\overset{\text{Reduction}}{\rightleftharpoons}} \underset{RA}{M^{\circ}}$$

If value of $E^o$ of $M^{n+}$ / M couple is positive, metal ion ($M^{n+}$) has a tendency to be reduced to elemental metal ($M^0$), *i.e.* if the value of E° for $M^{n+}/M^{\circ}$ couple is positive, the metal (M°) has an ability to act as an oxidising agent (OA). If the value of $E^o$ for $M^{n+}$ / $M^o$ couple is negative, the metal ($M^0$) has a tendency to be oxidised to elemental metal (M°), *i.e.* if the value of E° for $M^{n+}$ / $M^o$ couple is negative, the metal ($M^0$) has an ability to act as a reducing agent (RA).

## Uses of Redox Potentials

**I. Calculation of Standard e.m.f. of a Golvanic Cell ($E^o_{cell}$)**

The value of $E^o_{cell}$ is given by any of the following relations:

$$\mathbf{E^o_{cell} = (E^o_{red})_{cathode} - (E^o_{red})_{anode}} \quad ...(i)$$

Now since $E^o_{cell}$ is positive, $(E^o_{red})_{cathode}$ should be greater than $(E^o_{red})_{anode}$. Hence

$$E^o_{cell} = \text{Higher } E^o_{red} \text{ value} - \text{Lower } E^o_{red} \text{ value}$$

or
$$\mathbf{E^o_{cell} = (E^o_{red})_{Higher} - (E^o_{red})_{Lower}} \quad ...(ii)$$

## Solved Examples

**Example 1. A galvanic cell is composed of Zn/$Zn^{2+}$ and Cu/$Cu^{2+}$ electrodes having $Zn^{2+}$ / Zn = –0.763 V and $Cu^{2+}$ / Cu = +0.337 V.**

**(*i*) Calculate the value of $E°_{cell}$**

**(*ii*) State whether the cell will work or not.**

**Ans.** (*i*) $E^o_{cell} = (E^o_{red})_{Higher} - (E^o_{red})_{Lower} = E^o_{Cu^{2+}/Cu} - E^o_{Zn^{2+}/Zn}$

$= 0.337 - (-0.763) =$ **1.100 V (Ans.)**

(*ii*) Since $E^o_{cell}$ is positive, the cell obtained by coupling the two electrodes will work.

**Example 2. E° values of some half-cells are given below. Calculate the value of $E°_{cell}$ of the cell obtained by coupling the two half-cells.**

***$Zn^{2+}$ | Zn (–0.763 V), $Mg^{2+}$ | Mg ( –2.37 V), $Ni^{2+}$ | Ni (–0.25 V), $Ag^+$ | Ag ( 0.799 V)***

**Ans.** Following combinations of two half-cells are possible.

(*i*) If $Zn^{2+}$ | Zn and $Mg^{2+}$ | Mg half-cells are combined, then

$E^o_{cell}$ = Larger $E^o$ value – Smaller $E^o$ value

$= -0.763 - (-2.37) = -0.763 + 2.370 =$ **1.607 V (Ans.)**

(*ii*) If $Zn^{2+}$ | Zn and $Ni^{2+}$ | Ni half-cells are combined together, then

$E^o_{cell} = -0.25 - (-0.763) = -0.250 + 0.763 =$ **0.513 V (Ans.)**

(*iii*) If $Mg^{2+}$ | Mg and $Ag^+$ | Ag half-cells are combined, then

$E^o_{cell} = 0.799 - (-2.370) = 0.799 + 2.370 =$ **3.169 V (Ans.)**

(*iv*) If $Ni^{2+}$ | Ni and $Ag^+$ | Ag are coupled, then

$E^o_{cell} = 0.799 - (-0.250) = 0.799 + 0.250 =$ **1.049 V (Ans.)**

(*v*) If $Zn^{2+}$ | Zn and $Ag^+$ | Ag are coupled, then

$E°_{cell} = 0.799 - (-0.763) = 0.799 + 0.763 =$ **1.562 V (Ans.)**

(*vi*) $Mg^{2+}$ | Mg and $Ni^{2+}$ | Ni half-cells are coupled, then

$$E^\circ_{cell} = -0.25 - (-2.37) = -0.25 + 2.37 = \mathbf{2.12\ V\ (Ans.)}$$

**Example 3. Given: $Sn^{4+} + 2e^- \rightarrow Sn^{2+}$, E° = 0.154 V**

**$Fe^{3+} + e^- \rightarrow Fe^{2+}$, E° = 0.771 V, Determine cell voltage.**

**Ans.** $E^\circ_{cell}$ = Higher E° value – Lower E° Value = 0.771 – 0.154 = **0.617 V (Ans.)**

**Example 4. Give the cell reaction and calculate cell potential formed by combining the following half-reactions:**

$$\mathbf{Cr_2O_7^{2-} + 14H^+ + 6e^- \longrightarrow 2Cr^{3+} + 7H_2O, E^\circ = +1.33\ V} \quad \ldots (i)$$

$$\mathbf{S_2O_8^{2-} + 2e^- \longrightarrow 2SO_4^{2-}, E^\circ = +2.01\ V} \quad \ldots (ii)$$

**Ans.** First of all multiply equation (*ii*) by 3 to equalise the number of electrons in both the equations:

$$Cr_2O_7^{2-} + 14H^+ + 6e^- \rightarrow 2Cr^{3+} + 7H_2O, E^\circ = +1.33\ V \text{ (Lower value)} \quad \ldots(i)$$

$$3S_2O_8^{2-} + 6e^- \rightarrow 6SO_4^{2-}, E^\circ = +2.01\ V \text{ (Larger value)} \quad \ldots (iii)$$

Cell reaction is obtained by subtracting equation having lower E° value from the equation having larger E° value. Thus:

$$3S_2O_8^{2-} + 6e^- - Cr_2O_7^{2-} - 14H^+ - 6e^- \rightarrow 6SO_4^{2-} - 2Cr^{3+} - 7H_2O$$

or $$3S_2O_8^{2-} + 2Cr^{3+} + 7H_2O \rightarrow Cr_2O_7^{2-} + 14H^+ + 6SO_4^{2-}$$

This is the required cell reaction.

Cell potential = Larger E° value – Lower E° value

= +2.01 – (+1.33) = **0.68 V (Ans.)**

## II. To Predict the Feasibility of a Given Redox Reaction

If the value of $E^\circ_{reaction}$ is positive, the given reaction would be feasible, *i.e.* the reaction would proceed in the forward direction (from left to right). If the value of $E^\circ_{reaction}$ is negative, the given reaction would proceed from right to left (*i.e.* in the backward direction).

### Solved Examples

**Example 1. Calculate $E^\circ_{reaction}$ of the following reaction and predict whether this reaction is feasible or not. Given that : $E^\circ_{Cu^{2+}/Cu}$ = +0.337 V and $E^\circ_{Zn^{2+}/Zn}$ = –0.763 V.**

**Ans.** The answer of this questions can be given in the following two ways:

(*i*) Oxidation

$$Zn\ (s) + Cu^{2+} \rightarrow Zn^{2+} + Cu\ (s)$$

Reduction

The above reaction can be broken into the following two half-reactions.

$$Zn\ (s) \xrightarrow{\text{Oxidation}} Zn^{2+} + 2e^-, E^\circ_{0x} = +0.763\ V$$

$$Cu^{2+} + 2e^- \xrightarrow{\text{Reduction}} Cu\,(s),\ E^o_{red} = +0.337\ V$$

Redox reaction: $Zn\,(s) + Cu^{2+} \longrightarrow Zn^{2+} + Cu\,(s)$,

$$E^o_{redox} = E^o_{ox} + E^o_{red} = +0.763 + 0.337 = \mathbf{1.1\ V}$$

Now since $E^o_{redox}$ is positive, the given redox reaction is spontaneous.

(*ii*) $E^o_{redox}$ = Higher E° value – Lower E° value = 0.337 – (–0.763)

= 0.337 + 0.763 = **+1.1 V**

Since $E^o_{redox}$ is positive, the given redox reaction is feasible.

**Example 2. Given $Ce^{4+} + e^- \rightarrow Ce^{3+}$, $E^o = +1.61$ V** ...(*a*)

$$\mathbf{MnO_4^- + 8H^+ + 5e^- \longrightarrow Mn^{2+} + 4H_2O,\ E^o = +1.51\ V} \quad ...(b)$$

**State whether the following redox reaction is feasible or not.**

$$\mathbf{5Ce^{4+} + Mn^{2+} + 4H_2O \longrightarrow 5Ce^{3+} + MnO_4^- + 8H^+}$$

**Ans.** The answer of this question can be given in the following two different ways.

(*i*) Since E° value for reduction half-reaction (*a*) is greater than that for reduction reaction (*b*), $Ce^{4+}$ is stronger OA than $MnO^-_4$. Hence $Ce^{4+}$ can oxidise $Mn^{2+}$ to $MnO^-_4$ and is itself is reduced to $Ce^{3+}$ as is evident from the given redox reaction. Thus the given redox reaction is feasible.

(*ii*) The given redox reaction can be written as:

Oxidation

$$5Ce^{4+} + Mn^{2+} + 4H_2O \rightarrow 5Ce^{3+} + MnO_4^- + 8M^+$$

Reduction

$$Mn^{2+} + 4H_2O \xrightarrow{\text{Oxidation}} MnO_4^- + 8H^+ + 5e^-,\ E^o_{ox} = -1.51\ V$$

$$5Ce^{4+} + 5e^- \xrightarrow{\text{Reduction}} 5Ce^{3+},\ E^o_{red} = +1.61\ V$$

*On adding*: $Mn^{2+} + 4H_2O + 5Ce^{4+} \longrightarrow MnO_4^- + 8H^+ + 5Ce^{3+}$,

$$E^o_{redox} = E^o_{ox} + E^o_{red} = -1.51 + 1.61 = \mathbf{+0.10\ V}$$

Since $E^o_{redox}$ is positive, the given reaction is feasible.

The value of $E°_{redox}$ can also be calculated as follows.

$E^o_{redox}$ = Higher E° value – Lower E° value

= $E^o$ value for reaction (*a*) – E° value for reaction (*b*)

= [+ 1.61 – (+ 1.51)] V

= **+0.10 V**

## III. To Predict the Feasibility of a Metal Displacement Reaction

Let us see whether Zn metal can displace Cu from $CuSO_4$ solution or not, i.e., whether the following metal-metal displacement reaction would occur spontaneously or not.

$$Zn\,(s) + CuSO_4\,(aq) \longrightarrow ZnSO_4\,(aq) + Cu\,(s)$$

$$Zn\,(s) + Cu^{2+}\,(aq) \longrightarrow Zn^{2+}(aq) + Cu\,(s)$$

In the above reaction since Zn (s) displaces Cu from $CuSO_4$ (*aq*), Zn (s) is called **displacing metal** and Cu is called **displaced metal**. In the above reaction, since Zn (s) is oxidised to $Zn^{2+}$ (*aq*), Zn (*s*) acts as a RA. Similarly, since $Cu^{2+}$ (aq) is reduced to Cu (s), $Cu^{2+}$ (aq) is called OA. Thus the above reaction is a redox reaction which can also be written as:

$$\underset{\substack{\text{Displacing}\\ \text{metal}\\ \text{(RA)}}}{Zn\,(s)} + \underset{\substack{\text{Displaced}\\ \text{metal}\\ \text{(OA)}}}{Cu^{2+}\,(aq)} \longrightarrow Zn^{2+}\,(aq) + Cu\,(s)$$

The above reaction will take place spontaneously, if one of the following conditions is fulfilled.

(*i*) $E^o_{red}$ value of the metal acting as RA should be lower than $E^o$ value of the metal acting as OA. *i.e.*,

$$E^o_{RA} < E^o_{OA}$$

(*ii*) $E^o$ value of the displacing metal should be lower than $E^o$ value of the displaced metal. Thus any metal having lower value of $E^o$ can displace a metal with higher $E^o$ value from its aqueous solution.

In the electrochemical series, since $E^o$ values decrease down the series, a metal can displace any other metal placed above it in the series from the aqueous salt solution of the latter.

(*iii*) $(E^o_{OA} - E^o_{RA})$ or ($E^o$ value of the displaced metal $-E^o$ value of the displacing metal) should be a positive quantity.

(*iv*) The value of $E^o_{cell}$ corresponding to the above redox reaction which is equal to $(E^o_{cathode} - E^o_{anode})$ should be a positive quantity, i.e.,

$$E^o_{cell} = E^o_{cathode} - E^o_{anode} = \text{A positive quantity.}$$

**Solved Example**

**Arrange the following metals in the order in which they can displace each other. Fe, Cu, Zn, Ag, Mg and Al. Given that $Fe^{2+}$ / Fe = –0.44 V, $Cu^{2+}$ / Cu = + 0.34 V, $Zn^{2+}$ / Zn = –0.76 V, $Ag^{+}$/Ag = +0.80 V, $Mg^{2+}$ / Mg = –2.37 V and $Al^{3+}$ / Al = –1.66 V.**

**Solution.** The procedure consists of the following steps:

(*a*) Arrange E° values in the decreasing order as shown below:

| | | | | | | |
|---|---|---|---|---|---|---|
| *Metals*: | Ag | Cu | Fe | Zn | Al | Mg |
| *E° values*: | +0.80 | +0.34 | –0.44 | –0.76 | –1.66 | –2.37 |

——————— E° values decrease ———————→

(*b*) We know that in given metal-metal dispacement reaction, the displacing metal can displace all those metals from their salt solutions which have higher E° values than the E° value of the displacing metal itself. Thus the displacing metal should have lower value of E°. As a result:

(*i*) Since E° value of Mg is the least (most negative), this metal can displace all other metals (*viz*., Al, Zn, Fe, Cu and Ag) from their salt solutions which have higher E° values.

(*ii*) Al can displace only Zn, Fe, Cu and Ag from their salt solutions. Here it should be noted that Al cannot displace Mg since A1 has higher E° value (less negative) than that of Mg (Al = –1.66 V, Mg = –2.37 V).

(*iii*) Zn can displace Fe, Cu and Ag from their salt solutions. Zn cannot displace Al and Mg since this metal has higher E° value (less negative value).

(*iv*) Fe can displace only Cu and Ag from their salt solutions.

(*v*) Cu can displace only Ag but cannot displace any of other metals.

(*vi*) Ag cannot displace any of the given metals, since this metal has the highest E° value.

## IV. Oxidising Power of Oxidants and Reducing Power of Reductant.

We know that in a reduction half-reaction, the species written on LHS (*i.e.* the species which accepts electrons) is the oxidant and the species written on RHS (*i.e.* the species which loses or releases electrons) is the reductant. For example:

$$\underset{\text{(OA)}}{\underset{\text{Oxidant}}{Ag^+}} + e^- \longrightarrow \underset{\text{(RA)}}{\underset{\text{Reductant}}{Ag}}$$

If we are given many reduction half-reactions and they are arranged in the decreasing order of their E° values (more positive → less positive → zero → less negative → more negative), then the following conclusions may be drawn.

(*i*) With the decrease in E° values, the tendency of oxidants to accept electrons decreases and hence oxidising power of oxidants decreases.

(*ii*) With the decrease in E° values, the tendency of reductants to lose electrons increases and hence reducing power of reductants increases.

### Solved Examples

**Example 1. Reduction potentials of some reduction half-reactions are given below.**

$$Mg^{2+} + 2e^- \longrightarrow Mg,\ E° = -2.37\ V$$
$$Ag^+ + e^- \longrightarrow Ag,\ E° = +0.799\ V$$
$$Cd^{2+} + 2e^- \longrightarrow Cd,\ E° = -0.403\ V$$
$$Mn^{2+} + 2e^- \longrightarrow Mn,\ E° = -1.18\ V$$

**State (*i*) which metals can reduce $Cd^{2+}$ to Cd metal (*ii*) which metal cannot reduce $Cd^{2+}$ to Cd (iii) which metal can reduce $Ag^+$ to Ag.**

**Ans.** Arrange the given reduction half-reactions in the decreasing order of their E°values.

| Oxidising power of ions decreases ↓ | Reaction | Reducing power of metals increases ↓ |
|---|---|---|
| | $Ag^+ + e^- \longrightarrow Ag,\ E° = +0.799V$ | |
| | $Cd^{2+} + 2e^- \longrightarrow Cd,\ E° = -0.403V$ | |
| | $Mn^{2+} + 2e^- \longrightarrow Mn,\ E° = -1.18V$ | |
| | $Mg^{2+} + 2e^- \longrightarrow Mg,\ E° = -2.371$ | |

The decreasing order of E° values shows that oxidising power of ions decreases as $Ag^+ > Cd^{2+} > Mn^{2+} > Mg^{2+}$ and hence reducing power of the metals increases as Ag < Cd < Mn < Mg. This order shows that:

(*i*) Mn and Mg both can reduce $Cd^{2+}$ to Cd

(*ii*) Ag cannot reduce $Cd^{2+}$ to Cd.

(*iii*) Cd can reduce $Ag^+$ to Ag.

**Example 2. Reduction half reactions involving halogens are given below:**

$$F_2 + 2e^- \longrightarrow 2F^-, E° = +2.85\ V$$
$$Cl_2 + 2e^- \longrightarrow 2Cl^-, E° = +1.36\ V$$
$$Br_2 + 2e^- \longrightarrow 2Br^-, E° = +1.06\ V$$
$$I_2 + 2e^- \longrightarrow 2I^-, E° = +0.53V$$

**(*i*) Arrange the hologens in the increasing order of their oxidising power (*ii*) which halide ions can be oxidised by $Cl_2$ (*iii*) which halogen can oxidise $I^-$ ion to $I_2$.**

**Ans.** (*i*) Since $E°$ values decrease from $E^o_{F_2/2F^-}$ couple to $E^o_{I_2/2I^-}$ couple, oxidising power of halogens (oxidants) deceases as $F_2 > Cl_2 > Br_2 > I_2$

(*ii*) Since $Cl_2$ is stronger oxidant than $Br_2$ and $I_2$ both, this oxidant (*i.e.* $Cl_2$) can oxidise $Br^-$ and $I^-$ both to $Br_2$ and $I_2$ respectively.

(*iii*) Since $Br_2$ is stronger oxidant than $I_2$, this oxidant can oxidise $I^-$ to $I_2$.

**Example 3. Is it possible for acidified $Cr_2O_7^{2-}$ to oxidise $Cl^-$ and $CO^{2+}$.**

**Given $E^o_{Cl_2/2Cl^-} = 1.3595\ V$, $E^o_{Co^{3+}/Co^{2+}} = 1.81\ V$ and $E^o_{Cr_2O_7^{2-}/2Cr^{3+}} = 1.33V$.**

**Ans.** Arrange the given E° values in the decreasing order.

$$Co^{3+} \longrightarrow Co^{2+}, E° = 1.8100\ V$$
$$Cl_2 \longrightarrow 2Cl^-, E° = 1.3595\ V$$
$$Cr_2O_7^{2-} \longrightarrow 2Cr^{3+}, E° = 1.300\ V$$

Now since $Cr_2O_7^{2-} \rightarrow 2Cr^{3+}$ has the lowest E° value, $Cr_2O_7^{2-}$ is the weakest OA. Being the weakest OA, $Cr_2O_7^{2-}$ ion is not able to oxidise $Co^{2+}$ to $Co^{3+}$ and $Cl^-$ to $Cl_2$.

**Example 4. Will permanganate ion ($MnO_4^-$) oxidise $H_2O$ to $O_2$ in acidic solution. Given that:**

$$MnO_4^- + e^- \longrightarrow MnO_4^{2-}, E° = 0.564\ V \quad ...(i)$$
$$\frac{1}{2}O_2 + 2H^+ + 2e^- \longrightarrow H_2O\ (l), E° = 1.23\ V \quad ...(ii)$$

**Ans.** Since E° value for reaction (*i*) is smaller than that for reaction (*ii*), $MnO^-_4$ ion is a weaker OA than $O_2$. Hence $MnO^-_4$ ion is not able to oxidise $H_2O$ (*l*) to $O_2$

**Example 5. Two electrochemical cells and their reduction potentials are given below**

**(*a*) $Mg | Mg^{2+} || Ag^+ | Ag$** (Mg: −2.37V; Ag: + 0.799V)  **(*b*) $Cd | Cd^{2+} || Mn^{2+} | Mn$** (Cd: −0.403V; Mn: −1.18 V)

**Explain (*i*) which electrode in each cell acts as cathode and which electrode acts as anode (ii) which way the electrons flow in the outer circuit. (iii) which metal is the most reducing.**

**Ans.** (*i*) In electrochemical cell (*a*), since $Ag^+$ | Ag electrode has positive value of E°, this electrode acts as cathode and the other electrode acts as anode. Thus:

$$\underset{\substack{\text{Anode} \\ E^\circ = -2.37V}}{Mg \mid Mg^{2+}} \mid\mid \underset{\substack{\text{Cathode} \\ E^\circ = +0.799V}}{Ag^+ \mid Ag}$$

In electrochemical cell (*b*), since Cd | $Cd^{2+}$ electrode has less negative value of E°, this electrode acts as cathode. Thus:

$$\underset{\substack{\text{Anode} \\ E^\circ = -1.18\ V}}{Mn \mid Mn^{2+}} \mid\mid \underset{\substack{\text{Cathode} \\ E^\circ = -0.403\ V}}{Cd^{2+} \mid Cd}$$

(*ii*) Flow of electrons takes places from anode to cathode in the outer circuit in both the cells.

(*iii*) In the given electrochemical cells, $Mg^{2+}$, $Ag^+$, $Mn^{2+}$ and $Cd^{2+}$ ions act as OAs and Mg, Ag, Mn and Cd act as RAs. Since $Mg^{2+}$ / Mg electrode has the most negative E° value, $Mg^{2+}$ ion is the weakest (poorest) OA and hence Mg is the strongest RA.

**Example 6. (i) Which one of $Ni^{2+}$ or $I_2$ is a stronger oxidising agent ?**

$$(E^o_{Ni^{2+}/Ni} = -0.25\ V,\ E^o_{I_2/I^-} = +0.54\ V)$$

***(ii)* Which one of Ni or $H_2S$ is a better reducing agent?**

$$(E^o_{Ni^{2+}/Ni} = -0.25\ V,\ E^o_{S/H_2S} = +0.14\ V)$$

***(iii)* Will $O_2$ react with HI or with HCl?**

$$(E^o_{O_2/OH^-} = +0.82\ V,\ E^o_{I_2/I^-} = +0.54\ V,\ E^o_{Cl_2/Cl^-} = +1.36\ V)$$

**Ans.** (*i*) In the given couples $Ni^{2+}$ and $I_2$ act as oxidising agents. Since the value of $E^o_{I_2/I^-}$ couple ($E^o$ = +0.54 V) is greater than the value of $E^o_{Ni^{2+}/Ni}$ couple (E° = –0.25V), $I_2$ is stronger oxidising agent than $Ni^{2+}$.

(*ii*) In the given couples Ni and $H_2S$ act as reducing agents. Now since the value of $E^o_{Ni^{2+}/Ni}$ couple (= –0.25 V) is smaller than the value of $E^o_{S/H_2S}$ Couple (= + 0.14V), Ni is a better reducing agent than $H_2S$.

(*iii*) Write E° value for all the three couples in the decreasing order:

$$Cl_2 \longrightarrow Cl^-,\quad E^\circ = +1.36\ V$$
$$O_2 \longrightarrow OH^-,\ E^\circ = +0.82\ V$$
$$I_2 \longrightarrow I^-,\quad E^\circ = +0.54\ V$$

Since $E^o_{O_2/OH^-}$ is greater than $E^o_{I_2/I^-}$, $O_2$ is stronger agent than $I_2$. Hence $O_2$ will oxidise $I^-$ (HI) to $I_2$. Since $E^o_{O_2/OH^-}$ is smaller than $E^o_{Cl_2/Cl^-}$, $O_2$ will not oxidise $Cl^-$ (HCl) to $Cl_2$.

## V. Calculation of Equilibrium Constant (K) of a Redox Reaction

The standard potentials help us to calculate the value of equilibrium constant (K) of a given redox reaction and there from to decide if the given redox reaction will proceed to completion. Let us consider the following examples to illustrate this point.

## Solved Examples

**Example 1. (*i*) Calculate the value of equilibrium constant (K) of the following redox reaction.**

$$\mathbf{Ni^{2+} + Co \rightleftharpoons Ni + Co^{2+} \ (E° = 0.027\ V,\ n = 2)}$$

**(*ii*) Ascertain whether the given reaction will proceed to completion.**

**Ans.** (*i*) Standard free energy change (ΔG°) for the given reaction is given by:

$$\Delta G° = -nE°F \quad ...(i)$$

Here $n$ = No. of electrons involved in the given redox reaction = 2

E° = Standard potential for the given reaction = 0.027 V

F = Faraday constant.

We also know that:

$$\Delta G° = -RT \ln K \quad ...(ii)$$

On comparing equations (*i*) and (*ii*), we get

$$-nE°F = -RT \ln K$$

or $$nE° = \frac{RT}{F} \ln K$$

or $$nE° = 0.0591 \log K$$

or $$\boxed{\log K = \frac{nE°}{0.0591}}$$

In the given case, since $n$ = 2 and E° = 0.027 V,

$$\log K = \frac{2 \times 0.027}{0.0591} = 0.915$$

$$\therefore \quad \mathbf{K = 8.22\ (Ans.)}$$

(*ii*) Since the value of K is not much low, the given redox reaction is expected to proceed to completion.

**Example 2. (*i*) Calculate the value of equilibirium constant (K) of the following redox reaction.**

$$\mathbf{Zn\ (s) + Cu^{2+} \longrightarrow Zn^{2+} + Cu\ (s)}$$

**Given that $Cu^{2+}$ | Cu = +0.337 V and $Zn^{2+}$ / Zn = –0.763 V**

**(*ii*) Ascertain whether the above reaction will go to completion.**

**Ans.** (*i*) We know that E° for the given reaction

$$= \text{Larger E° value} - \text{Smaller E° Value}$$

$$= E^o_{Cu^{2+}/Cu} - E^o_{Zn^{2+}/Zn} = 0.337 - (-0.763)$$

$$= 0.337 + 0.763 = \mathbf{1.1\ V}$$

For the given reaction $n$ = 2

$$\therefore \quad \log K = \frac{nE^o}{0.0591} = \frac{2 \times 1.1}{0.0591} = 37$$

$$\therefore \quad \mathbf{K = 10^{37}}$$

(*ii*) Since $K$ is very large (= $10^{37}$), the given redox reaction goes to completion.

**Example 3. Predict whether the following reaction will proceed to completion or not.**

**(i) Reduction of $Fe^{3+}$ ions by $Sn^{2+}$ ion**

**(ii) Reaction between Sn metal and $Pb^{2+}$ ions**

**E° values for different couples are as:**

**$Sn^{4+}/Sn^{2+}$ = 0.15 V, $Fe^{3+}$ / $Fe^{2+}$ = 0.77 V, $Sn^{2+}/Sn$ = –0.136 V and $Pb^{2+}/Pb$ = –0.126 V**

**Ans.** (*i*) The reduction of $Fe^{3+}$ ions by $Sn^{2+}$ ion can be represented as:

$$2Fe^{3+} + Sn^{2+} \rightleftharpoons 2Fe^{2+} + Sn^{4+} \; (n = 2)$$

$$E^o \text{ value for this reaction} = E^o_{Higher} - E^o_{Lower}$$

$$= E^o_{Fe^{3+}/Fe^{2+}} - E^o_{Sn^{4+}/Sn^{2+}}$$

$$= 0.77 - 0.15 = 0.62 \text{ V}$$

$$\therefore \quad \log K = \frac{nE^o}{0.0591} = \frac{2 \times 0.62}{0.0591} = 21$$

$$\therefore \quad \mathbf{K = 10^{21}}$$

Since the value of equilibrium constant (K) is very high, the given reaction will proceed to completion.

(*ii*) Reaction between Sn and $Pb^{2+}$ can be represented as:

$$Sn + Pb^{2+} \rightleftharpoons Sn^{2+} + Pb \; (n = 2)$$

$$E^o \text{ value for this reaction} = E^o_{Higher} - E^o_{Lower}$$

$$= E^o_{Pb^{2+}/Pb} - E^o_{Sn^{2+}/Sn} = -0.126 - (-0.136)$$

$$= 0.01 \text{ V}$$

$$\log K = \frac{nE^o}{0.0591} = \frac{2 \times 0.01}{0.0591} = 0.3$$

$$\therefore \quad \mathbf{K = 2}$$

Since the value of K is very small, the reaction between *S*n and $Pb^{2+}$ will not proceed to completion. Instead both products and reactants will be found in significant amount in equilibrium.

**Example 4. State whether the following redox reaction proceeds to completion.**

$$MnO_4^- + 8H^+ + 5Fe^{2+}a \longrightarrow Mn^{2+} + 5Fe^{3+} + 4H_2O$$

***Given: $MnO_4^-$ / $Mn^{2+}$ = 1.510 V and $Fe^{3+}$ / $Fe^{2+}$ = 0.771 V.***

**Ans.** The given redox reaction can be broken into the following two half-reactions:

$$MnO_4^- + 8H^+ + 5e^- \xrightarrow{\text{Reduction}} Mn^{2+} + 4H_2O, \; E^o_{red} = +1.150 \text{ V}$$

$$5 \times [Fe^{2+} \xrightarrow{\text{Oxidation}} Fe^{3+} + e^-], \; E^o_{ox} = -0.771 \text{ V}$$

$$MnO_4^- + 8H^+ + 5Fe^{2+} \longrightarrow Mn^{2+} + 4H_2O + 5Fe^{3+} \; (n = 5).$$

$$E^o_{redox} = 1.510 - 0.771 = 0.739 \text{ V}$$

$$\therefore \quad \log K = \frac{n \times E^{o}_{cell}}{0.059} = \frac{5 \times 0.739}{0.059} = 62.6$$

$$\therefore \quad \mathbf{K = 10^{62.6}}$$

Since the value of K is very large, $Fe^{2+}$ ions are oxidised to a large extent in the presence of acidified $MnO_4^-$ ions.

## VI. Standard Electrode Potentials and Metal Extraction Procedures

Standard electrode potentials, as we have already seen, are measures of the reducing ability of the metals or conversely the ease or difficulty of reducing the metal ion to elementary states. The couple with greatest negative potential evidently has a metal ions which is the most difficultly reducible to elementary state. On the other hand the couple with the highest positive potential has a metal ion which is the most easily reducible. For the most difficultly reducible ion we need the most powerful reducing metal, and for the most easily reducible ion the least powerful one will do. From these considerations metal ions can be divided into four categories:

(*a*) couples with potential more negative than –1.5 volt.
(*b*) couples with negative potential in the range –1.5 to –0.75 volt.
(*c*) couples with negative potentials in the range –0.75 to 0.0 volt.
(*d*) couples with potentials more positive than 0.0 volt.

For metal ions of category (*a*) electrolysis is used. For elements of category (*b*) an element with a pronounced affinity for oxygen is used to reduce the metal oxide. For example, reduction of $Cr^{3+}$ ion is done by Al in thermite process. Carbon reduction is employed for couples of category (*c*) and simple heating is enough for elements of category (*d*).

## Analysis of Redox Cycle

Redox cycles are the cycles involving enthalpy change ($\Delta H$) and energy absorbed or evolved in other steps.

### Oxidation Cycle of Alkali Metals

Let us consider oxidation cycle of alkali metals. The tendency of an alkali metal, M(*s*) to lose an electron in solution to form hydrated cation, $M^+(aq)$ is a measure of the standard oxidation potential ($E_{ox}$) of the metal. Thus, $E_{ox}$ involves the following process.

$$M(s) + aq \xrightarrow[\text{aqueous solution}]{\text{Oxidation in}} M^+(aq) + e^-, E = E^o_{ox} \quad \ldots (i)$$

$E^o_{ox}$ values of alkali metals are Li / $Li^+$ = 3.04 V, Na / $Na^+$ = 2.71 V, K/$K^+$ = 2.92V, Rb/$Rb^+$ = 2.99 V and Cs /$Cs^+$ = 3.02 V. It may be seen that Li has the highest value of $E^o_{ox}$

The oxidation reaction (*i*) given above takes place through the following steps:

(*a*) **Sublimation of M(s) to M(g).** In this step M(*s*) is converted in M(*g*) (sublimation). In this process energy equal to sublimation energy ($\Delta H_{sub}$) of M(*s*) is required (+ sign)

$$M(s) \xrightarrow{\text{Sublimation}} M(g),\ \Delta H = \text{Energy required} = +\Delta H_{sub}$$

(*b*) **Ionisation of M (g) to form M⁺ (g)**. In this step, M(*g*) formed in step (*a*) loses an electron and is converted into $M^+(g)$ ion. In this process energy equal to ionisation energy (IE) of M(*g*) is required (+ sign).

$$M(g) \xrightarrow[(-e^-)]{\text{Ionisation}} M^+(g) + e^-,\ \Delta H = \text{Energy required} = +IE$$

(*c*) **Hydration of $M^+(g)$ to form $M^+(aq)$.** In this step, $M^+(g)$ formed in step (*b*) gets hydrated in aqueous solution to form hydrated ion, $M^+(aq)$. In this process energy equal to hydration energy ($\Delta H_{hyd}$) of $M^+(g)$ is released (– sign)

$$M^+(g) + aq \xrightarrow{\text{Hydration}} M^+(aq),\ \Delta H = \text{Energy released} = -\ \Delta H_{hyd}$$

On adding equations given at (*a*), (*b*) and (*c*), we get:

$$M(s) + aq \xrightarrow[\text{aqueous solution}]{\text{Oxidation in}} M^+(aq) + e^-,\ \Delta H = +\Delta H_{sub} + IE - \Delta H_{hyd} \quad \ldots (ii)$$

Since reaction (*ii*) is the same as reaction (*i*), $E^o_{ox}$ should be given by:

$$E^o_{ox} = \Delta H_{sub} + IE - \Delta H_{hyd}$$

Sublimation energies ($\Delta H_{sub}$ values) of alkali metals are quite low and are almost of the same magnitude. IE of Li is the highest among alkali metals. Since $Li^+$ (*g*) ion is the smallest in size of all other $M^+(g)$ ions, this ion has the maximum value of $\Delta H_{hyd}$. Although IE of Li is the highest among alkali metals, it is more than compensated by the large value of $\Delta H_{hyd}$ of $Li^+(g)$. This explains why among alkali metals Li has the highest positive value of $E^o_{ox}$ (= 3.04V). Highest positive value of $E^o_{ox}$ for Li shows that the oxidation reaction,

$$Li\,(s) + aq \xrightarrow{\text{Oxidation}} Li^+\,(aq) + e^-$$

takes place with the greatest ease. In other words we can say that among alkali metals, Li (*s*) is most easily (or most readily) oxidised to $Li^+$ (a*q*) in aqueous solution or Li (*s*) in aqueous solution acts as the strongest reducing agent.

It follows from the above discussion that the highest positive value of $E^o_{ox}$ for Li is due to highest value of $\Delta H_{hyd}$ for $Li^+(g)$ ion.

In order to work out the value of E for reaction (*i*) which is a measure of oxidation potential of alkali metals, oxidation reaction (*i*) and the three steps *viz* (*a*), (*b*) and (*c*) may be shown as a cycle (Born-Haber cycle) given in Fig 1.1. This cycle is called **oxidation cycle of alkali metals**.

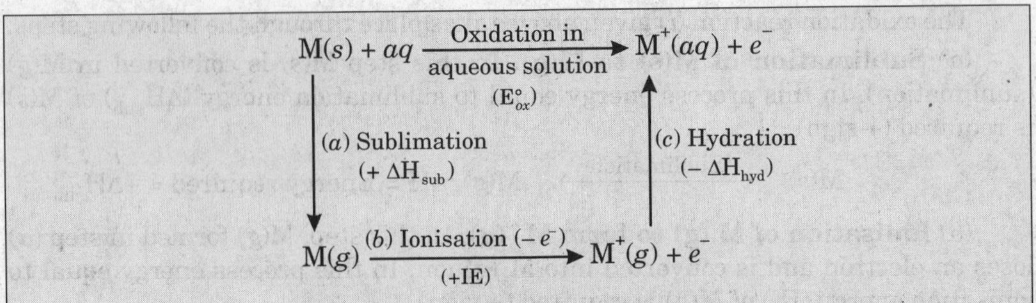


**Fig. 1.1** Oxidation cycle of alkali metals

## Reduction Cycle of Halogens

Halogens ($X_2$) have a tendency to accept an electron in aqueous solution to form hydrated halide ion, $X^-(aq)$. In this process energy is required (+ sign)

$$\frac{1}{2}X_2\,(s, l \text{ or } g) + e^- + aq \xrightarrow[\text{aqueous solution}]{\text{Reduction in}} X^-(aq),\ \Delta H = \text{Energy required} = +\,\Delta E$$

The magnitude of $\Delta E$ is a measure of oxidising potential (i.e. oxidising strength) of halogens in aqueous solution.

In the above reaction, if halogen is a solid, then the above reaction is written as:

$$\frac{1}{2}X_2\,(s) + e^- + aq \xrightarrow[\text{aqueous solution}]{\text{Reduction in}} X^-(aq),\ \Delta H = \Delta E \qquad \ldots (i)$$

Reaction (i) takes place through the following steps:

***(a)* Conversion of $\frac{1}{2}$ $X_2$ *(s)* to $\frac{1}{2}$ $X_2$ *(l)*.** In this process heat equal to heat of fusion $\left(\frac{1}{2}\Delta H_f\right)$ of $\frac{1}{2}X_2(s)$ is required (+ sign)

$$\frac{1}{2}X_2(s) \xrightarrow{\text{Fusion}} \frac{1}{2}X_2(l),\ \Delta H = \text{Heat required} = +\frac{1}{2}\Delta H_f$$

***(b)* Conversion of $\frac{1}{2}X_2(l)$ to $\frac{1}{2}X_2(g)$.** $\frac{1}{2}X_2(l)$ formed in step *(a)* is evaporated to form $\frac{1}{2}X_2(g)$. Heat required (+ sign) in this process is heat of vaporisation $\left(\frac{1}{2}\Delta H_v\right)$

$$\frac{1}{2}X_2(l) \xrightarrow{\text{Vaporisation}} \frac{1}{2}X_2(g),\ \Delta H = \text{Heat required} = +\frac{1}{2}\Delta H_v$$

***(c)* Conversion of gaseous molecule, $\frac{1}{2}X_2(g)$ to gaseous atom, *X (g)*.** In this process heat equal to heat of dissociation $\left(\frac{1}{2}\Delta H_d\right)$ is required (+ sign)

$$\frac{1}{2}X_2(g) \xrightarrow{\text{Dissociation}} X(g),\ \Delta H = \text{Heat required} = +\frac{1}{2}\Delta H_d$$

***(d) Conversion of gaseous atom, X (g) to gaseous ion, $X^-(g)$.*** In this process, energy equal to electron affinify (EA) of X (*g*) is liberated (– sign)

$$X(g) + e^- \longrightarrow X^-(g),\ \Delta H = \text{Energy liberated} = -\,\text{EA}$$

(*e*) **Conversion of gaseous ion, $X^-(g)$ into hydrated ion, $X^-(aq)$.** In this process heat energy equal to heat of hydration ($\Delta H_{hyd}$) is liberated (–sign)

$$X^-(g) + aq \xrightarrow{\text{Hydration}} X^-(aq),\ \Delta H = \text{Heat liberated} = -\Delta H_{hyd}$$

On adding reactions given at (*a*), (*b*), (*c*), (*d*) and (*e*), we get:

$$\frac{1}{2}X_2(s) + e^- + aq \longrightarrow X^-(aq),\ \Delta H = +\frac{1}{2}\Delta H_f + \frac{1}{2}\Delta H_v + \frac{1}{2}\Delta H_d - EA - \Delta H_{hyd} \quad ...(ii)$$

Since reaction (*ii*) is the same as reaction (*i*), $\Delta E$ should be given by:

$$\Delta E = +\frac{1}{2}\Delta H_f + \frac{1}{2}\Delta H_v + \frac{1}{2}\Delta H_d - EA - \Delta H_{hyd} \quad ...(iii)$$

Reduction reaction (*i*) and steps *viz* (*a*), (*b*), (*c*), (*d*) and (*e*) can be shown as a cycle which is given in Fig. 1.2. This cycle is called **reduction cycle of halogens.**

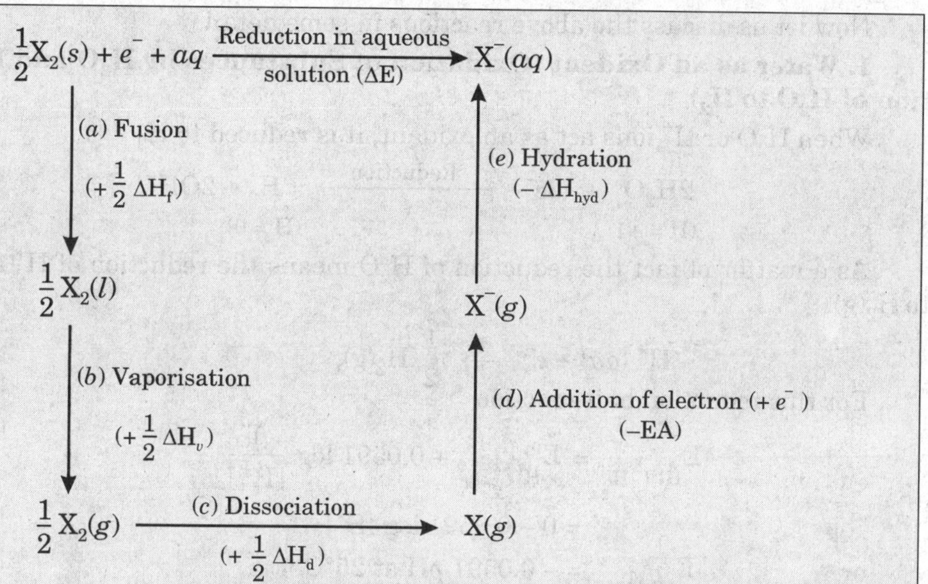


**Fig. 1.2** Reduction cycle of halogens

Since $F_2$ and $Cl_2$ are gases at room temperature, heat of fusion ($\Delta H_f$) and heat of vaporisation ($\Delta H_v$) are omitted. Similarly, since $Br_2$ is a liquid, heat of fusion ($\Delta H_f$) is omitted.

If we put the values of $\Delta H_f$ (only in case of $I_2$), $\Delta H_v$ (only in case of $Br_2$ and $I_2$), $\Delta H_d$, EA and $\Delta H_{hyd}$ in equation (*iii*) we get $\Delta E$ values (in KJ $mol^{-1}$) for different halogens as $F_2$ = –768, $Cl_2$ = –607, $Br_2$ = –575 and $I_2$ = –494. Since these values decrease from $F_2$ to $I_2$, $F_2$ has the maximum tendency to accept an electron and hence is the strongest OA. Although $Cl_2$ has maximum value of EA (EA values for halogens are: $F_2$ = –332 kJ $mol^{-1}$, Cl = –348, $Br_2$ = –324 and $I_2$ = –295), it is not the strongest OA. This is because of the fact that $Cl_2$ does not have the maximum value of $\Delta E$. This discussion shows that the oxidising power of halogens in aqueous solution does not depend on their EA values but it depends on their $\Delta E$ values.

## Redox Stability of Water

There are species which are stable in water since these species neither oxidise or reduce water nor undergo disproportionation in water.

On the other hand, there are compounds which are unstable in water, since these compounds undergo any of the following reactions in water.

(*i*) The compounds may evolve $H_2$ by reacting with $H_2O$. In these reactions the compounds are oxidised by $H_2O$ and $H_2O$ itself is reduced to $H_2$. Thus, in these reactions *$H_2O$ acts as on OA.*

(*ii*) The compounds may evolve $O_2$ by reacting with $H_2O$. In these reactions the compounds are reduced and $H_2O$ itself is oxidised to $O_2$. Thus, in these reactions *$H_2O$ acts as a RA.*

(*iii*) The compounds may *undergo disproportionation*. In these reactions compounds of higher and lower oxidation states are formed.

Now let us discuss the above reactions in some detail.

**1. Water as an Oxident (Oxidation of Substances by $H_2O$ and Reduction of $H_2O$ to $H_2$)**

When $H_2O$ or $H^+$ ions act as an oxidant, it is reduced to $H_2$

$$\underset{(H = +1)}{2H_2O} + 2e^- \xrightarrow{\text{Reduction}} \underset{(H = 0)}{H_2} + 2OH^-$$

As a matter of fact the reduction of $H_2O$ means the reduction of $H^+(aq)$ ions to $H_2(g)$.

$$H^+ (aq) + e^- \longrightarrow \frac{1}{2} H_2(g),$$

For this reduction half-reaction

$$E_{H^+/\frac{1}{2}H_2} = E^o_{H^+/\frac{1}{2}H_2} + 0.0591 \log \frac{1}{[H^+]}$$

$$= 0 - 0.0591 \log [H^+]$$

or $$E_{H^+/\frac{1}{2}H_2} = -\,0.0591 \text{ pH at } 25°C \qquad ...(i)$$

When $H_2O$ or $H^+$ ions (aqueous acid) react with alkali metals or alkaline earth metals (except Be), metals are oxidised to their corresponding hydroxides and $H_2O$ or $H^+(aq)$ is reduced to $H_2(g)$.

$$\underset{(M = 0)}{M} + \underset{(H = +1)}{H_2O} \longrightarrow \underset{(M= +1)}{MOH} + \underset{(H = 0)}{\frac{1}{2}H_2}$$

or $$M + H^+ (aq) \longrightarrow M^+ + \frac{1}{2}H_2(g)$$

The above type of reaction is also feasible when M is a first transition series metal like Sc, Ti, V, Cr, Mn, etc.

$$Sc + 3H^+ \longrightarrow Sc^{3+} + \frac{1}{2}H_2$$

**2. Water as a Reductant (Reduction of Substances by $H_2O$ and Oxidation of $H_2O$ to $O_2$)**

When $H_2O$ acs as a reductant, it is oxidised to $O_2$.

$$\underset{(O=-2)}{2H_2O} \xrightarrow{\text{Oxidation}} 4H^+ + \underset{(O=0)}{O_2} + 4e^-, E^o_{ox} = -1.23 \text{ V}$$

If this oxidation half-reaction is written as reduction half-reaction, then we get:

$$O_2 + 4H^+ + 4e^- \xrightarrow{\text{Reduction}} 2H_2O, \; E^o_{red} = +1.23 \text{ V}$$

For this reduction half-reaction

$$E = E^o_{red} - 0.0591 \text{ pH}$$

At pH = 0, $E = 1.23 - 0.0591 \times 0 = 1.23$ V

When $F_2$ reacts with $H_2O$, $F_2$ is reduced to $F^-$ ion and $H_2O$ is oxidised to $O_2$.

$$\underset{(F=0)}{2F_2} + \underset{(O=-2)}{2H_2O} \longrightarrow \underset{(F=-1)}{4F^-} + 4H^+ + \underset{(O=0)}{O_2}$$

$H_2O$ also reduces $Co^{3+}$ to $Co^{2+}$ and $Ce^{4+}$ to $Ce^{3+}$.

$$4Co^{3+} + 2H_2O \longrightarrow 4Co^{2+} + 4H^+ + O_2$$

$$4Ce^{4+} + 2H_2O \longrightarrow 4Ce^{3+} + 4H^+ + O_2$$

There are many oxidising agents whose E° values are greater than +1.23 V, but these oxidising agents are not capable of bringing about oxidation of $H_2O$ to $O_2$.

**3. Disproportionation Reactions**

Diproportionation reaction is a redox reaction in which the oxidation number (O.N.) of an element increases as well as decreases simultaneously, leading to the formation of two products one of which has the element in lower oxidation state and the other product has the element in higher oxidation state. In other words the species which undergoes disproportionation acts as an oxidant as well as a reductant.

**Examples** (*i*) Following reaction is a disproportionation reaction, since $Cu^+$ is oxidised to $Cu^{2+}$ and is also reduced to Cu°. Thus $Cu^+$ acts as an oxidant as well as a reductant.

$$\underset{(Cu=+1)}{2Cu^+} \longrightarrow \underset{(Cu=+2)}{Cu^{2+}} + \underset{(Cu=0)}{Cu^0}$$

The above reaction can be broken into the following two half-reactions.

$$Cu^+ \xrightarrow{\text{Oxidation}} Cu^{2+} + e^-, \; E^o_{ox} = E^o_{Cu^+/Cu^{2+}} = -0.16 \text{ V}$$

$$Cu^+ + e^- \xrightarrow{\text{Reduction}} Cu^0, \; E^o_{red} = E_{Cu^+/Cu^o} = +0.52 \text{ V}$$

---

*On adding:* $2Cu^+ \longrightarrow Cu^{2+} + Cu^o, \; E^o_{redox} = E^o_{ox} + E^o_{red} = -0.16 + 0.52$

$= 0.36$ V

Since $E^o_{redox}$ is positive, the disproportionation reaction is spontaneous.

(*ii*) Following reaction is a dispropotionation reaction shown by hypochlorous acid (HClO).

$$\underset{(Cl=+1)}{HClO} \longrightarrow \underset{(Cl=0)}{2Cl_2} + \underset{(Cl=+5)}{HClO_3} + 2H_2O$$

This reaction can be broken into the following two half-reactions

$$\underset{(Cl = +1)}{HClO} + 2H_2O \xrightarrow{\text{Oxidation}} \underset{(Cl = +5)}{HClO_3} + 4H^+ + 4e^-, E^\circ = E^\circ_{ox} = -1.43\ V$$

$$2\times[\underset{(Cl = +1)}{2HClO} + 2H^+ + 2e^- \xrightarrow{\text{Reduction}} \underset{(Cl = 0)}{Cl_2} + 2H_2O, E^\circ = E^\circ_{red} = +1.63\ V]$$

*On adding:* $5HClO \longrightarrow HClO_3 + 2Cl_2 + 2H_2O,$

$$E^o_{redox} = E^o_{ox} + E^o_{red} = -1.43 + 1.63 = \mathbf{+0.20\ V}$$

Since $E^o_{redox}$ is positive, the dispropotionation reaction shown above is a spontaneous reaction.

**Redox Stability Field of Natural Waters**

Waters obtained from lakes and rivers are called natural waters. pH = 4 and pH = 9 are the limits of pH values that commonly prevail in natural waters

A curve obtained by plotting reduction potentials of $H^+$ ions to give $H_2$ and standard oxidation potentials of $H_2O$ to give $O_2$ with the change in pH of solution is shown in Fig. 1.3. This Figure has been obtained without considering the hydrogen overvoltage or oxygen overvoltage.

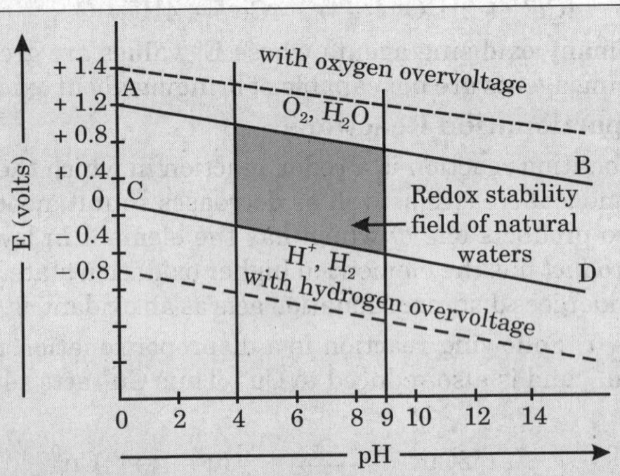


**Fig. 1.3** The redox stability field of natural waters.

This figure contains the shaded area. This area depicts the range of value of E° and pH in which water is thermodynamically stable towards oxidation and reduction, *i.e.* in the shaded area $H_2O$ is neither oxidised nor reduced. This is, therefore, called **redox stability field of natural waters.** *Thus the redox stability field of water is the region of pH and redox potential in which water is stable towards both oxidation and reduction.* It may be seen from the figure that for the stability field area pH limits are at pH = 4 and pH = 9 which have been shown by vertical lines. Redox potential limits are represented by upper (AB) and lower (CD) slopping lines. The oxidising agents whose redox potentials lie above the upper AB line can oxidise $H_2O$ to liberate $O_2$ and the reducing agents whose redox potentials lie below the lower CD line can reduce water (*i.e.* $H^+$ ions) to liberate $H_2$. Thus any species whose E value is more than [1.23 V – (0.591 V) pH] can oxidise $H_2O$ to $O_2$ and the species having E value more negative than [–(0.0591 V) pH] can reduce $H^+$ ions to $H_2$.

If hydrogen overvoltage and oxygen voltage are taken into consideration, the stability field of water gets extended on the upper as well as lower side of AB and CD lines respectively and redox potential limits are represented by the broken lines.

## Comproportionation Reactions

Comproportionation reactions are simply the reverse of disproportionation reactions. In these reactions an element in its two different oxidation states combine together to form a product which has the element in intermediate oxidation state. For example,

$$\underset{(Cu = +1)}{2Cu^+} \longrightarrow \underset{(Cu = 0)}{Cu^o} + \underset{(Cu = +2)}{Cu^{2+}} \quad (E^o = +0.37\,V)$$

is a disproportionation reaction and hence the reaction

$$\underset{(Cu = 0)}{Cu^o} + \underset{(Cu = +2)}{Cu^{2+}} \longrightarrow \underset{(Cu = +1)}{2Cu^+} \quad (E^o = -0.37\,V)$$

is a comproportionation reaction. It may be noted that since $E^o$ for this reaction (= –0.37 V) is negative, this comproportionation reaction is not a spontaneous reaction. However the following reaction is a spontaneous comproportionation reaction

$$Ag^{2+} + Ag^0 \longrightarrow 2Ag^+ + (E^o = +1.18\,V)$$

## Auto-oxidation

Auto-oxidation is a process in which a substance undergoes oxidation and reduction both by itself. It may be noted that in disproportionation reaction the same thing happens but in case of auto-oxidation this reaction occurs slowly without the action of heat, light or electricity.

**Examples** (*i*)

$$\underset{(N = +3)}{3\,HNO_2} \longrightarrow \underset{(N = +5)}{HNO_3} + \underset{(N = +2)}{2NO}$$

or

$$3NO_2^- + 2H^+ \longrightarrow NO_3^- + 2NO + H_2O$$

This reaction can be broken into the following two half-reactions

$$NO_2^- + H_2O \xrightarrow{\text{Oxidation}} NO_3^- + 2H^+ + 2e^-$$

$$NO_2^- + 2H^+ + e^- \xrightarrow{\text{Reduction}} NO + H_2O\,] \times 2$$

*On adding:* $3NO_2^- + 2H^+ \longrightarrow NO_3^- + 2NO + H_2O$

(*ii*)

$$\underset{\substack{\text{Manganic acid}\\(Mn = +6)}}{H_2MnO_4} \longrightarrow \underset{\substack{\text{Paramanganic}\\\text{acid}\\(Mn = +7)}}{2HMnO_4} + \underset{(Mn = +4)}{MnO_2} + 2H_2O$$

or

$$3MnO_4^{2-} + 4H^+ \longrightarrow 2MnO_4^- + MnO_2 + 2H_2O$$

This reaction can also be broken into the following two half-reactions.

$$MnO_4^{2-} \xrightarrow{\text{Oxidation}} MnO_4^- + e^-\,] \times 2$$

$$MnO_4^{2-} + 4H^+ + 2e^- \xrightarrow{\text{Reduction}} MnO_2 + 2H_2O$$

*On adding:* $3MnO_4^{2-} + 4H^+ \longrightarrow 2MnO_4^- + MnO_2 + 2H_2O$

*(iii)*

$$\underset{(Cl = +1)}{3HClO} \longrightarrow \underset{(Cl = +5)}{HClO_3} + \underset{(Cl = -1)}{2HCl}$$

or

$$3ClO^- \longrightarrow ClO_3^- + 2Cl^-$$

This reaction is broken into the following two half-reactions

$$ClO^- + 2H_2O \xrightarrow{\text{Oxidation}} ClO_3^- + 4H^+ + 4e^-$$

$$[ClO^- + 2H^+ + 2e^- \xrightarrow{\text{Reduction}} Cl^- + H_2O\,] \times 2$$

*On adding :* $3ClO^- \longrightarrow ClO_3^- + 2Cl^-$

## Oxidation of Metals by Atmospheric Oxygen

Let us consider the following oxidation reactions.

*(i)* **Oxidation of $Fe^{2+}$ *(aq)*.** Let us consider a solution containing $Fe^{2+}$ *(aq)* ions. Suppose this solution is out of contact of atmospheric oxygen. Since E° value for $Fe^{3+} \mid Fe^{2+}$ couple (= +0.77 V) lies within the stability field of water, $Fe^{2+}$ *(aq)* ions should be stable in water. In other words we can say that the oxidation of metallic iron (Fe) by $H_2O$ or $H^+$ ions should not proceed beyond $Fe^{2+}$ *(aq)* because further oxidation of $Fe^{2+}$ *(aq)* to $Fe^{3+}$ *(aq)* is undesirable by +0.77 V.

However, if the solution containing $Fe^{2+}$ *(aq)* ions is exposed to atmospheric $O_2$, then $Fe^{2+}$ *(aq)* is oxidised or not to $Fe^{3+}$ *(aq)* by $O_2$ can be predicted. If $Fe^{2+}$ *(aq)* is oxidised to $Fe^{3+}$*(aq)*, then the following radox reaction should be feasible.

$$\underset{(Fe = +2)}{«««««««^{2+}\; aq} + \underset{(O = 0)}{{}_2\, g} + {}^+\, aq \longrightarrow \underset{(Fe = +3)}{4Fe^{3+}(aq)} + \underset{(O = -2)}{2H_2O(l)}$$

This reaction would be feasible if $E^o_{redox}$ for this reaction is positive. This redox reaction can be broken into the following two half-reactions

$$4 \times [Fe^{2+}(aq) \xrightarrow{\text{Oxidation}} Fe^{3+}(aq) + e^-],\ E^o_{ox} = -0.77\ V$$

$$O_2(g) + 4H^+(aq) + 4e^- \xrightarrow{\text{Reduction}} 2H_2O(l),\ E^o_{red} = +1.23\ V - 0.0591pH$$

*On adding:* $4Fe^{2+}(aq) + O_2(g) + 4H^+(aq) \longrightarrow 4Fe^{3+}(aq) + 2H_2O(l),$

$$E^\circ_{redox} = E^o_{ox} + E^o_{red}$$

$$= -0.77 + 1.23 - 0.0591\ pH$$

If pH = 0, then $E^\circ_{redox} = +0.46$ V

Since the value of $E^o_{redox}$ is positive, the above redox reaction is spontaneous, *i.e.* the oxidation of $Fe^{2+}$*(aq)* to $Fe^{3+}$*(aq)* by atmospheric $O_2$ is spontaneous. Now since the value of $E^o_{redox}$ (= +0.46 V) is not large, the oxidation of $Fe^{2+}$*(aq)* is slow in the absence of catalyst.

*(ii)* **Oxidation of Cu(s).** Whether Cu(*s*) is oxidised or not to $Cu^{2+}$*(aq)* by atmospheric $O_2$ in damp environment can be decided by calculating the value of $E^o_{redox}$ for the following redox reaction.

$$2Cu(s) + O_2(g) + 4H^+(aq) \longrightarrow 2Cu^{2+}(aq) + 2H_2O(l) \quad ...(i)$$

This reaction can be broken into the following two half-reactions

$$2 \times [Cu(s) \xrightarrow{\text{Oxidation}} Cu^{2+} [(aq) + 2e^-],\ E^o_{ox} = -0.34\ V$$

$$O_2(g) + 4H^+(aq) + 4e^- \xrightarrow{\text{Reduction}} 2H_2O(l),\ E^o_{red} = 1.23\ V - 0.0591\ pH$$

*On adding* : $2Cu(s) + O_2(g) + 4H^+(aq) \longrightarrow 2Cu^{2+}(aq) + 2H_2O(l)$,

$$E^o_{redox} = -0.34 + 1.23 - 0.0591\ pH$$

If $\quad pH = 0$, then

$$E^o_{redox} = -0.34 + 1.23 - 0.0591 \times 0 = +0.89\ V$$

Since the value of $E^o_{redox}$ is positive, the oxidation of Cu (*s*) to $Cu^{2+}(aq)$ in solution having pH = 0 as represented by equation (*i*) is feasible

If the atmosphere contains $CO_2$ and $SO_2$, $Cu^{2+}$ ions produced by the atmospheric oxidation of Cu is converted in $CuCO_3$ and $CuSO_4$ respectively.

It is due to the formation of $CuSO_4$ and $CuCO_3$ that copper roofs get covered with a green layer which contains a mixture of $CuSO_4$ and $CuCO_3$ in deep environment.

(*iii*) **Oxidation of Mg and Al.** Mg and Al are oxidised to their respective oxides in presence of air and $H_2O$. These oxides are formed as impervious layers on the surface of these metals. These layers protect the metals from further attack of the reagents on them. Thus, due to the formation of these layers the metals become **passive** or are **passivated**.

## Latimer Diagrams (Reduction Potential Diagrams)

### What is Latimer Diagram?

Latimer diagram of an element presents the standard reduction potential values ($E^o$ values) for a series of compounds of that element in which the element is in different oxidation states. Different species given in the Latimar diagram have arrows in between them and $E^o$ values involving the two immediate (adjacent) species are written over the arrows. The oxidation state of the element concerned is written under the species. The species having the element in the highest O.S. is written on the extreme left and the species having the element in the lowest O.S. is written on the extreme right. Thus, on proceeding from left to right in a Latimer diagram, the O.S. of the element concerned goes on decreasing. Latimer diagram contains the notations for the reduction half-reactions. In a given adjacent pair of species, the left hand species is reduced to right hand species.

**Examples** (*i*) Latimer diagram for chlorine in **acidic solution** is written as:

$$\underset{Cl = +7}{ClO_4^-} \xrightarrow{+1.20\ V} \underset{+5}{ClO_3^-} \xrightarrow{+1.18\ V} \underset{+3}{HClO_2} \xrightarrow{+1.65V} \underset{+1}{HClO} \xrightarrow{+1.67\ V} \underset{0}{Cl_2} \xrightarrow{+1.36\ V} \underset{-1}{Cl^-}$$

In this diagram, the notation (change) $ClO^-_4 \xrightarrow{+1.20\ V} ClO^-_3$ represents the following reduction-half-reaction in acidic medium.

$$ClO_4^- + 2H^+ + 2e^- \longrightarrow ClO_3^- + H_2O,\ E^o = +1.20\ V$$

Reduction half-reactions for other notations (changes) can be written as:

$$ClO_3^- + 3H^+ + 2e^- \longrightarrow HClO_2 + H_2O, E° = +1.18 V$$
$$HClO_2 + 2H^+ + 2e^- \longrightarrow HClO + H_2O, E° = +1.65 V$$
$$2HClO + 2H^+ + 2e^- \longrightarrow Cl_2 + 2H_2O, E° = +1.67 V$$
$$Cl_2 + 2e^- \longrightarrow 2Cl^-, E° = +1.36 V$$

(*ii*) Latimer diagram for chlorine in **basic medium** is written as:

$$\underset{Cl=+7}{ClO_4^-} \xrightarrow{+0.37V} \underset{+5}{ClO_3^-} \xrightarrow{+0.30V} \underset{+3}{ClO_2^-} \xrightarrow{+0.68V} \underset{+1}{ClO^-} \xrightarrow{+0.42V} \underset{0}{Cl_2} \xrightarrow{+1.36V} \underset{-1}{Cl^-}$$

Reduction half-reductions for different changes shown in this diagram are given below:

$$ClO_4^- + 2H^+ + 2e^- \longrightarrow ClO^-_3 + H_2O, E° = +0.37 V$$
$$ClO_3^- + 2H^+ + e^- \longrightarrow ClO^-_2 + H_2O, E° = +0.30 V$$
$$ClO_2^- + 2H^+ + 2e^- \longrightarrow ClO^- + H_2O, E° = +0.68 V$$
$$2ClO^- + 4H^+ + 2e^- \longrightarrow Cl_2 + 2H_2O, E° = +0.42 V$$
$$Cl_2 + 2e^- \longrightarrow 2Cl^-, E° = +1.36 V$$

(*iii*) Fe exists in different oxidation states like +6, +3, +2 and 0. Latimer diagram showing these oxidation states can be written as:

$$\underset{Fe=+6}{FeO_4^{2-}} \xrightarrow{+2.20 V} \underset{+3}{Fe^{3+}} \xrightarrow{+0.77 V} \underset{+2}{Fe^{2+}} \xrightarrow{-0.47 V} \underset{0}{Fe^0}$$

Reduction half-reactions for the changes shown in the Latimar diagram are:

$$FeO_4^{2-} + 8H^+ + 3e^- \longrightarrow Fe^{3+} + 4H_2O, E^o = +2.20 V$$
$$Fe^{3+} + e^- \longrightarrow Fe^{2+}, E^o = +0.77 V$$
$$Fe^{2+} + 2e^- \longrightarrow Fe°, E^o = -0.47 V.$$

(*iv*) Latimer diagram for oxygen is given below:

$$\underset{O=0}{O_2} \xrightarrow{0.70 V} \underset{-1}{H_2O_2} \xrightarrow{+1.76 V} \underset{-2}{H_2O}$$

Reduction half-reactions for $O_2 \longrightarrow H_2O_2$ and $H_2O_2 \longrightarrow H_2O$ changes are:

$$O_2 + 2H^+ + 2e^- \longrightarrow H_2O_2, E^o = +0.70 V$$
$$H_2O_2 + 2H^+ + 2e^- \longrightarrow 2H_2O, E^o = +1.76 V$$

## Applications of Latimer Diagrams

## I. To Predict the Feasibility of a Disproportionation Reaction

With the help of Latimer diagram we can predict whether a given species will disproportionate or not into lower and higher oxidation state.

Let us consider the following Latimer diagram of a metal (M)

$$\underset{M=+2}{M^{2+}} \xrightarrow{E_1^0} \underset{+1}{M^+} \xrightarrow{E_2^0} \underset{0}{M^0}$$

With the help of this diagram we can predict whether or not $M^+$ will disproportionate into M° (lower O.S. of M) and $M^{2+}$ (higher O.S. of M). The reaction showing the disproportionation of $M^+$ into $M^0$ and $M^{2+}$ can be written as:

$$\underset{(M = +1)}{2M^+} \longrightarrow \underset{(M = 0)}{M^0} + \underset{(M = +2)}{M^{2+}} \quad ...(i)$$

The two reduction half-reactions corresponding to the given Latimer diagram are:

$$M^{2+} + e^- \longrightarrow M^+, E^o = E_1^0 \quad ...(ii)$$

$$M^+ + e^- \longrightarrow M^0, E = E_2^0 \quad ...(iii)$$

Disproportionation reaction (*i*) can be obtained by subtracting reaction (*ii*) from reaction (*iii*):

$$M^+ + e^- - M^{2+} - e^- \longrightarrow M^0 - M^+, E^o = E_2^0 - E_1^0$$

or

$$2M^+ \longrightarrow M^0 + M^{2+}, E^o = E_2^0 - E_1^0 \quad ...(iv)$$

Disproportionation reaction (*iv*) is feasible only when $(E_2^0 - E_1^0)$ value is positive. $(E_2^0 - E_1^0)$ will be positive only when

$$E_2^0 > E_1^0 \text{ or } E^0_{M^+/M^0} > E^0_{M^{2+}/M^+}$$

This discussion shows *that a species given in a Latimer diagram will diproportionate into its two adjacent neighbours, if* $E^o$ *value of the species lying on the right side* $(E_2^0)$ *is higher than* $E^o$ *value of the species lying on the left side* $(E_1^0)$.

## Solved Examples

**Example 1. With the help of the Latimer diagram of oxygen given as:**

$$\underset{O = 0}{\mathbf{O_2}} \xrightarrow{+0.70\ V} \underset{-1}{\mathbf{H_2O_2}} \xrightarrow{+1.76\ V} \underset{-2}{\mathbf{H_2O}}$$

**(a) Write the reduction half-reactions corresponding to this diagram.**

**(b) Predict whether $H_2O_2$ will disprotionate into $O_2$ and $H_2O$ or not**

**Ans.** (*a*) Reduction half-reaction for $O_2 \rightarrow H_2O_2$ change is:

$$O_2 + 2H^+ + 2e^- \longrightarrow H_2O_2, E^o = +0.70 \text{ V}$$

Reduction half-reaction for $H_2O_2 \longrightarrow H_2O$ change is:

$$H_2O_2 + 2H^+ + 2e^- \longrightarrow 2H_2O, E^o = 1.76 \text{ V}$$

(*b*) Since $E^o_{H_2O_2/H_2O}$ (= +1.76 V) is greater than $E^o_{O_2/H_2O_2}$ (+0.70 V), disproportionation of $H_2O_2$ into $O_2$ and $H_2O$ is feasible. Also since $E^o_{H_2O_2/H_2O} - E^o_{O_2/H_2O}$ = 1.76 – 0.70 = +1.06 V is a positive quantity, the disproportionation of $H_2O_2$ to $O_2$ and $H_2O$ is feasible.

**Example 2. Latimer diagram of iron is given as:**

$$\underset{Fe = +3}{\mathbf{Fe^{3+}}} \xrightarrow{+0.771\ V} \underset{+2}{\mathbf{Fe^{2+}}} \xrightarrow{+0.44\ V} \underset{0}{\mathbf{Fe^0}}$$

**Predict whether the disproportionation of $Fe^{2+}$ into $Fe^{3+}$ and $Fe^o$ is feasible or not.**

**Ans:** Since $E^o_{Fe^{2+}/Fe^0}$ (= +0.44V) is smaller than $E^o_{Fe^{3+}/Fe^{2+}}$ (= +0.771 V), the disproportionation of $Fe^{2+}$ to $Fe^{3+}$ and $Fe^o$ is not feasible. Also since $E^o_{Fe^{2+}/Fe^0} - E^o_{Fe^{3+}/Fe^{2+}}$

$= 0.440 - 0.771 = -0.331$ V is a negative quantity, the disproportionation of $Fe^{2+}$ to $Fe^{3+}$ and $Fe^0$ is not feasible.

## II. To Calculate E° Value for Any Non-adjascent Couple

The procedure for calculating E° value for the reduction of one species to the other (the two species are non-adjacent species) can be understood by studying the following examples.

**Example 1. Calculate the value of E° for the reduction of $Fe^{3+}$ to Fe° with the help of the following Latimer diagram.**

$$\mathbf{Fe^{3+} \xrightarrow{+0.771\ V} Fe^{2+} \xrightarrow{-0.44\ V} Fe^0}$$

**Solution.** The given diagram can be shown as:

$$Fe^{3+} \xrightarrow[\Delta G_1^0 = -n_1 F E_1^0]{E_1^0 = +0.771V} Fe^{2+} \xrightarrow[\Delta G_2^0 = -n_2 F E_2^0]{E_2 = -0.44V} Fe^0$$

$$Fe^{3+} \longrightarrow Fe^0 : \quad \Delta G^0 = -(n_1 + n_2) FE^0$$

Here $\Delta G_1^0$, $\Delta G_2^0$ and $\Delta G^0$ are the energy changes for different half-cell reactions. $n_1$ and $n_2$ are the number of electrons involved in the following half-cell reactions.

$$Fe^{3+} + e^- \longrightarrow Fe^{2+}\ (n_1 = 1)$$
$$Fe^{2+} + 2e^- \longrightarrow Fe^0\ (n_2 = 2)$$

Obviously the value of $\Delta G^0$ is equal to the sum of $\Delta G_1^0$ and $\Delta G_2^0$ *i.e.*,

$$\Delta G^0 = \Delta G_1^0 + \Delta G_2^0$$

or $$-(n_1 + n_2) FE^o = -n_1 FE_1^0 + (-n_2 FE_2^0)$$

or $$(n_1 + n_2)E^o = n_1 E_1^0 + n_2 E_2^0$$

or $$\boxed{E^0 = \frac{n_1 E_1^0 + n_2 E_2^0}{n_1 + n_2}}$$

On putting the values of $n_1$, $n_2$, $E°_1$ and $E°_2$ we get:

$$E^o = \frac{1 \times 0.771 + 2 \times (-0.44)}{1+2} = \frac{0.771 - 0.880}{3}$$

$$= \frac{-0.109}{3} = \mathbf{-0.036\ V}$$

**Example 2. Latimer diagram for Mn system is given as:**

$$\mathbf{MnO_4^- \xrightarrow{+0.56V} MnO_4^{2-} \xrightarrow{+2.26V} MnO_2 \xrightarrow{+0.95V} Mn^{3+} \xrightarrow{+0.51V} Mn^{2+}}$$

**Calculate E° value for (a) $MnO_4^{2-} - Mn^{2+}$ and (b) $MnO_4^- - Mn^{3+}$ couples.**

**Ans.** (*a*) The Latimer diagram for $MnO_4^{2-} \rightarrow Mn^{2+}$ change can be written as:

$$MnO_4^{2-} \xrightarrow[\Delta G_1^0 = -n_1FE_1^0]{E_1^0 = +2.26\ V} MnO_2 \xrightarrow[\Delta G_2^0 = -n_2FE_2^0]{E_2^0 = +0.95V} Mn^{3+} \xrightarrow[\Delta G_3^0 = -n_3FE_3^0]{E_3^0 = +0.51V} Mn^{2+}$$

$$\Delta G^0 = -(n_1 + n_2 + n_3)\ FE^0$$

$n_1$, $n_2$ and $n_3$ are the number of electrons involved in the following reduction half-reactions:

$$MnO_4^{2-} + 4H^+ + 2e^- \longrightarrow MnO_2 + 2H_2O\ (n_1 = 2,\ E_1^0 = +2.26\ V)$$

$$MnO_2 + 4H^+ + e^- \longrightarrow Mn^{3+} + 2H_2O\ (n_2 = 1,\ E_2^0 = +0.95\ V)$$

$$Mn^{3+} + e^- \longrightarrow Mn^{2+}\ (n_3 = 1,\ E_3^0 = +0.51\ V)$$

Obviously:

$$\Delta G^\circ = \Delta G_1^0 + \Delta G_2^0 + \Delta G_3^0$$

or $$-(n_1 + n_2 + n_3)FE^\circ = -n_1FE_1^0 - n_2FE_2^0 - n_3FE_3^0$$

or $$E^\circ = \frac{n_1E_1^0 + n_2E_2^0 + n_3E_3^0}{n_1 + n_2 + n_3}$$

$$= \frac{2\times2.26 + 1\times0.95 + 1\times0.51}{2+1+1} = \mathbf{1.49\ V\ (Ans.)}$$

(*b*) The Latimer diagram for $MnO_4^- \rightarrow Mn^{3+}$ change can be written as:

$$MnO_4^- \xrightarrow[n_1 = 1]{E_1^0 = +0.56V} MnO_4^{2-} \xrightarrow[n_2 = 2]{E_2^0 = +2.26V} MnO_2 \xrightarrow[n_3 = 3]{E_3^0 = 0.95V} Mn^{3+}$$

$$E^0 = ?,\ n = n_1 + n_2 + n_3 = 1 + 2 + 1 = 4$$

$$\therefore \quad E^\circ = \frac{n_1E_1^0 + n_2E_2^0 + n_3E^0}{n_1 + n_2 + n_3} = \frac{1\times0.56 + 2\times2.26 + 1\times0.95}{4}$$

$$= \frac{6.03}{4} = \mathbf{1.5\ V\ (Ans.)}$$

## Solved Problems Based on Latimer Diagram

**Example 1. Consider the following Latimer diagram and predict the stability or unstability of $Tl^+$ with respect to disproportionation.**

$$Tl^{3+} \xrightarrow{+1.26\ V} Tl^+ \xrightarrow{-0.34\ V} Tl^0$$

**Ans.** Since $E^\circ$ value for the right side $Tl^+/Tl^0$ couple ($E^\circ = -0.34$ V) is lower than $E^\circ$ value for the left side $Tl^{3+}/Tl^+$ couple ($E^\circ = +1.26$ V), *$Tl^+$ will not disproportionate into $Tl^0$ and $Tl^{3+}$. In other words $Tl^+$ is stable with respect to disproportionation.*

**Example 2. Latimer diagram for amercium is:**

$$AmO_2^{2+} \xrightarrow{+1.70V} AmO_2^{+} \xrightarrow{+0.86V} Am^{4+} \xrightarrow{+2.62V} Am^{3+} \xrightarrow{-0.207V} Am^{0}$$

($AmO_2^{2+} \rightarrow Am^{3+}$: + 1.726 V; $AmO_2^{+} \rightarrow Am^{3+}$: + 1.724 V)

**Ans.** $AmO_2^+$ will disproportionate into $Am^{3+}$ (right hand neighbour) and $AmO_2^{2+}$ (left hand neighbour), since E° value for $AmO_2^+ / Am^{3+}$ couple (E° = +1.724 V) is greater than E° value for $AmO_2^{2+} / AmO_2^+$ couple (E° = +1.70 V). Similarly, $Am^{4+}$ will disproportionate into $Am^{3+}$ (right hand species) and $AmO_2^+$ (left hand species), since E° value for $Am^{4+} / Am^{3+}$ couple (E° = +2.62 V) is greater than E° value for $AmO_2^+ / Am^{4+}$ couple (E° = +0.86 V).

**Example 3. Draw Latimer diagram for the following reduction half-reactions and calculate the value of E° for the reduction of $Cu^{2+}$ to Cu.**

$$Cu^{2+} + e^- \longrightarrow Cu^+,\ E^\circ = +0.15\ V$$
$$Cu^+ + e^- \longrightarrow Cu^0,\ E^\circ = +0.50\ V$$

**Ans.** Latimer diagram for the given half-reactions is:

$$Cu^{2+} \xrightarrow{+0.15\ V} Cu^+ \xrightarrow{+0.015\ V} Cu^0$$

This diagram can be shown as:

$$Cu^{2+} \xrightarrow[n_1 = 1]{E_1^0 = +0.15V} Cu^+ \xrightarrow[n_2 = 2]{E_2^0 = +0.50V} Cu^0$$

$$Cu^{2+} \longrightarrow Cu^0: \quad E^0 = ?,\ n = n_1 + n_2$$

$$\therefore \quad E^\circ = \frac{n_1 E_1^0 + n_2 E_2^0}{n_1 + n_2} = \frac{1 \times 0.15 + 1 \times 0.50}{1+1} = \frac{0.65}{2} = \mathbf{0.321\ V}$$

**Example 4. The Latimer diagram is:**

$$ClO^- \xrightarrow{+0.42\ V} Cl_2 \xrightarrow{+1.36\ V} Cl^-$$

**Calculate the value of E° for the reduction of $ClO^-$ to $Cl^-$**

**Ans.** First of all we should calculate the number of electrons ($n$) for $ClO^- \rightarrow Cl_2$ and $Cl_2 \rightarrow Cl^-$ changes.

$$ClO^- + 2H^+ + e^- \longrightarrow \frac{1}{2}Cl_2 + H_2O\ (n_1 = 1,\ E^\circ = +0.42V)$$

$$\frac{1}{2}Cl_2 + e^- \longrightarrow Cl^-\ (n_2 = 1,\ E^\circ = +1.36\ V)$$

Thus, the given diagram can be written as:

$$ClO^- \xrightarrow[n_1 = 1]{E^0 = +0.42V} Cl_2 \xrightarrow[n_2 = 1]{E^0 = +1.36V} Cl^-$$

$$ClO^- \longrightarrow Cl^-: \quad n = n_1 + n_2 = 1 + 1 = 2,\ E^0 = ?$$

**Obviously:**

$$n \times E^o_{ClO^-/Cl^-} = n_1 \times E^o_{ClO^-/Cl_2} + n_2 \times E^o_{Cl_2/Cl^-}$$

or
$$2 \times E^o_{ClO^-/Cl^-} = 1 \times 0.42 + 1 \times 1.36$$

$\therefore$
$$E^o_{ClO^-/Cl^-} = \frac{1.78}{2} = \mathbf{0.89\ V\ (Ans.)}$$

**Example 5. Calculate the $E^o$ value for the half-cell reaction, $MnO_4^- + 4H^+ + 3e^- \rightarrow MnO_2 + 2H_2O$. Given that:**

$$MnO_4^- + 8H^+ + 5e^- \longrightarrow Mn^{2+} + 4H_2O;\ E^o_{red} = +1.51\ V$$

$$Mn^{2+} + 2H_2O \longrightarrow MnO_2 + 4H^+ + 2e^-;\ E^o_{ox} = -1.23\ V$$

**Ans.** Latimer diagram corresponding to the given half-cell reactions can be written as:

$$\underset{Mn\,=\,+7}{MnO_4^-} \xrightarrow[n_1 = 3]{E_1^0 = ?} \underset{+4}{MnO_2} \xrightarrow[n_2 = 2]{E_2^0 = +1.23V} \underset{+2}{Mn^{2+}}$$

$$E^0 = 1.51V,\ n = n_1 + n_2 = 3 + 2 = 5$$

$\therefore$
$$E^o = \frac{n_1E_1^o + n_2E_2^o}{n_1 + n_2} = \frac{3 \times E_1^o + 2 \times 1.23}{5}$$

or
$$1.51 = \frac{3E_1^o + 2.46}{5}$$

or
$$7.55 = 3E_1^o + 2.46$$

$\therefore$
$$E_1^o = \frac{7.55 - 2.46}{3} = \mathbf{1.69\ V\ (Ans.)}$$

**Example 6. Calculate $E^o$ value for the half-cell reaction,**

$$\frac{1}{2}Cl_2 + 3H_2O \longrightarrow ClO_3^- + 6H^+ + 5e^-.\ \textbf{Given that:}$$

$$Cl^- + 3H_2O \longrightarrow ClO_3^- + 6H^+ + 6e^-;\quad E_{ox} = -1.45\ V$$

$$2Cl^- \longrightarrow Cl_2 + 2e^-;\quad E_{ox} = -1.36\ V$$

**Ans.** All the given half-reactions should be written as reduction half-reactions as shown below:

(*i*) $ClO_3^- + 6H^+ + 5e^- \rightarrow \frac{1}{2}Cl_2 + 3H_2O,\ E^o = ?,\ n = 5$

(*ii*) $ClO_3^- + 6H^+ + 6e^- \rightarrow Cl^- + 3H_2O,\ E^o = +1.45\ V,\ n = 6$

(*iii*) $Cl_2 + 2e^- \rightarrow 2Cl^-$, or $\frac{1}{2}Cl_2 + e^- \rightarrow Cl^-, E^o = +1.36, n = 1$

The Latimer diagram corresponding to the above reduction half-reactions is

$$\underset{+5}{ClO_3^-} \xrightarrow[n_1 = 5]{E_1^0 = ?} \underset{0}{\tfrac{1}{2}Cl_2} \xrightarrow[n_2 = 1]{E_2^0 = +1.36V} \underset{-1}{Cl^-}$$

$$E = +1.45V,\ n = n_1 + n_2 = 5 + 1 = 6$$

$$\therefore \qquad E = \frac{n_1 E_1^o + n_2 E_2^o}{n_1 + n_2}$$

$$\text{or} \qquad 1.45 = \frac{5 \times E_1^o + 1 \times 1.36}{6}$$

$$\text{or} \qquad E_1^0 = \frac{1.45 \times 6 - 1.36}{5} = \frac{8.70 - 1.36}{5} = \frac{7.34}{5} = 1.488 \approx \mathbf{1.47\ V\ (Ans.)}$$

**Example 7. Latimer digram for iron is given as:**

$$\mathbf{FeO_4^{2-} \xrightarrow{+2.20\ V} Fe^{3+} \xrightarrow{0.77\ V} Fe^{2+} \xrightarrow{-0.445\ V} Fe^0}$$

**Determine the value of $E^o_{FeO_4^{2-}/Fe^{2+}}$**

**Ans.** The Latimer diagram can be written as:

$$FeO_4^{2-} \xrightarrow[n_1 = 3]{E_1^0 = +2.20V} Fe^{3+} \xrightarrow[n_2 = 1]{E_2^0 = +0.77V} Fe^{2+} \xrightarrow{-0.445V} Fe^0$$

$$E^0 = ?,\ n = n_1 + n_2 = 3 + 1 = 4$$

The values of $n_1$ and $n_2$ are 3 and 1 respectively as shown below:

$$FeO_4^{2-} + 8H^+ + 3e^- \longrightarrow Fe^{3+} + 4H_2O \ (n_1 = 3,\ E_1^o = +2.20\ V)$$

$$Fe^{3+} + e^- \longrightarrow Fe^{2+} \ (n_2 = 1,\ E_2^o = +0.77\ V)$$

$$\therefore \qquad E^o = \frac{n_1 E_1^o + n_2 E_2^o}{n_1 + n_2} = \frac{3 \times 2.20 + 1 \times 0.77}{4}$$

$$= \frac{6.60 + 0.77}{4} = \frac{7.37}{4} = \mathbf{1.84\ V\ (Ans.)}$$

**Example 8. With the help of the following Latimer diagram calculate $E^o$ value for (i) NO/$N_2O$ and (ii) $N_2O$ / $N_2$ couples.**

$$\mathbf{\underset{N\ =\ +3}{HNO_2} \xrightarrow{0.983\ V} \underset{+2}{NO} \xrightarrow{?} \underset{+1}{N_2O} \xrightarrow{?} \underset{0}{N_2}}$$

**NO → $N_2$: 1.68 V**

**$HNO_2$ → $N_2O$: 1.29 V**

**Ans.** First of all we should calculate the number of electrons ($n$) involved in the reduction half-reactions for $HNO_2$ | NO, NO | $N_2O$ and $N_2O$ / $N_2$ couples.

(*a*) *$HNO_2$ / NO couple:*

$$\underset{(N=+3)}{HNO_2} + H^+ + e^- \longrightarrow \underset{(N=+2)}{NO} + H_2O\ (n_1 = 1,\ E^0 = 0.983\ V)$$

(*b*) *$NO/N_2O$ couple:*

$$\underset{(N=+2)}{NO} + H^+ + e^- \longrightarrow \underset{(N=+1)}{\frac{1}{2}N_2O} + \frac{1}{2}H_2O(n_2 = 1, E^\circ = ?)$$

(*c*) *$N_2O$ / $N_2$ couple:*

$$\underset{(N=+1)}{\frac{1}{2}N_2O} + H^+ + e^- \longrightarrow \underset{(N=0)}{\frac{1}{2}N_2} + \frac{1}{2}H_2O\ (n_3 = 1, E^\circ = ?)$$

**Thus, the given Latimer diagram can be written as:**

$$n_5 = n_2 + n_3 = 1 + 1 = 2,\ E^0 = 1.68V\ (NO \rightarrow N_2)$$

$$HNO_2 \xrightarrow[n_1 = 1]{E^0 = +0.983V} NO \xrightarrow[n_2 = 1]{E^0 = ?} N_2O \xrightarrow[n_3 = 1]{E^0 = ?} N_2$$

$$n_4 = n_1 + n_2 = 2,\ E^0 = 1.29V\ (HNO_2 \rightarrow N_2O)$$

(*i*) E° value for $NO/N_2O$ couple is given by:

$$n_4 \times E^\circ_{HNO_2/N_2O} = n_1 \times E^o_{HNO_2/NO} + n_2 \times E^o_{NO/N_2O}$$

or $$2 \times 1.29 = 1 \times 0.983 + 1 \times E^o_{NO/N_2O}$$

$\therefore$ $E^o_{NO/N_2O} = 2 \times 1.29 - 0.983 =$ **1.59 V (Ans.)**

(*ii*) $E^o$ value for $N_2O/N_2$ couple is given by:

$$n_5 \times E^o_{NO/N_2} = n_2 \times E^o_{NO/N_2O} + n_3 \times E^o_{N_2O/N_2}$$

or $$2 \times 1.68 = 1 \times 1.59 + 1 \times E^o_{N_2O/N_2}$$

$\therefore$ $$E^o_{N_2O/N_2} = 2 \times 1.68 - 1.59$$

$$= \mathbf{1.77\ V\ (Ans.)}$$

**Example 9. Consider the following Latimer diagram in acidic medium.**

$$\underset{+7}{MnO_4^-} \xrightarrow{+0.56\ V} \underset{+6}{MnO_4^{2-}} \xrightarrow{+2.26\ V} \underset{+4}{MnO_2} \xrightarrow{+0.95\ V} \underset{+3}{Mn^{3+}}$$

$$\xrightarrow{+1.51\ V} \underset{+2}{Mn^{2+}} \xrightarrow{-1.19\ V} \underset{0}{Mn^0}$$

**(*i*) Write the reduction half-reaction for $MnO_4^- - Mn^{2+}$ pair**

**(*ii*) Find out the $E^o$ value for the above pair.**

**Ans.** (*i*) The reduction half-reaction for $MnO_4^- - Mn^{2+}$ pair is

$$MnO_4^- + 8H^+ + 5e^- \longrightarrow Mn^{2+} + 4H_2O$$

(*ii*) First of all we should calculate the number of electrons (*n*) involved in the reduction half-reactions for the pairs upto $Mn^{2+}$.

$$MnO_4^- + e^- \rightarrow MnO_4^{2-}\ (n_1 = 1, E^\circ = +0.56\ V)$$

$$MnO_4^{2-} + 4H^+ + 2e^- \longrightarrow MnO_2 + 2H_2O\ (n_2 = 2,\ E^\circ = +2.26\ V)$$

$$MnO_2 + 4H^+ + e^- \longrightarrow Mn^{3+} + 2H_2O\ (n_3 = 1,\ E^\circ = +0.95\ V)$$

$$Mn^{3+} + e^- \longrightarrow Mn^{2+}\ (n_4 = 1,\ E^\circ = +1.51\ V)$$

Thus, the given Latimer diagram in its modified form can be written as:

$$MnO_4^- \xrightarrow[n_1=1]{+0.56\ V} MnO_4^{2-} \xrightarrow[n_2=2]{+2.26\ V} MnO_2 \xrightarrow[n_3=1]{+0.95\ V} Mn^{3+} \xrightarrow[n_4=1]{+1.51\ V} Mn^{2+} \xrightarrow{-1.14\ V} Mn^0$$

$$n_5 = n_1 + n_2 + n_3 + n_4 = 5,\ E^0 = ?\quad (MnO_4^- \to Mn^{2+})$$

$E^o$ value of $MnO_4^- - Mn^{2+}$ pair is given by:

$$n_5 \times E^o_{MnO_4^-/Mn^{2+}} = n_1 \times E^o_{MnO_4^-/MnO_4^{2-}} + n_2 \times E^o_{MnO_4^{2-}/MnO_2} + n_3 \times E^o_{MnO_2/Mn^{3+}} + n_4 \times E^o_{Mn^{3+}/Mn^{2+}}$$

or $$5 \times E^o_{MnO_4^-/Mn^{2+}} = 1 \times 0.56 + 2 \times 2.26 + 1 \times 0.95 + 1 \times 1.51$$

$\therefore$ $$E^o_{MnO_4^-/Mn^{2+}} = \frac{0.56+4.52+0.95+1.51}{5} = \frac{7.54}{5} = \mathbf{1.51\ V\ (Ans.)}$$

**Example 10. (*i*) Represent the following reduction half-reactions by Latimer diagram.**

**(*a*)** $NO_3^- + 2H^+ + e^- \longrightarrow \frac{1}{2}N_2O_4 + H_2O\ (n = 1,\ E^\circ = 0.803\ V)$

**(*b*)** $\frac{1}{2}N_2O_4 + H^+ + e^- \longrightarrow HNO_2\ (n = 1,\ E^\circ = +1.065\ V)$

**(*c*)** $HNO_2 + H^+ + e^- \longrightarrow NO + H_2O\ (n = 1,\ E^\circ = +0.983\ V)$

**(*d*)** $NO + 2H^+ + 2e^- \longrightarrow \frac{1}{2}N_2 + H_2O\ (n = 2,\ E^\circ = 1.080\ V)$

**(*ii*) Calculate E° value for $NO_3^-/HNO_2$ and $NO_3^-/NO$ couples.**

**Ans.** (*i*) All the given reduction reactions can be represented by the following Latimer diagram

$$n_5 = n_1 + n_2 + n_3 = 3,\ E^0 = ?\quad (NO_3^- \to NO)$$

$$\underset{+5}{NO_3^-} \xrightarrow[n_1=1]{0.803\ V} \underset{+4}{N_2O_4} \xrightarrow[n_2=1]{1.065\ V} \underset{+3}{HNO_2} \xrightarrow[n_3=1]{0.983\ V} \underset{+2}{NO} \xrightarrow{1.680\ V} \underset{0}{N_2}$$

N = +5, +4, +3, +2, 0

$$n_4 = n_1 + n_2 = 1 + 1 = 2,\ E^0 = ?\quad (NO_3^- \to HNO_2)$$

(*ii*) $E^o_{NO_3^-/HNO_2}$ is given by:

$$n_4 \times E^o_{NO_3^-/HNO_2} = n_1 \times E^o_{NO_3^-/N_2O_4} + n_2 \times E^o_{N_2O_4/HNO_2}$$

or $$2 \times E^o_{NO_3^-/HNO_2} = 1 \times 0.803 + 1 \times 1.065$$

$\therefore$ $$E^o_{NO_3^-/HNO_2} = \frac{0.803+1.065}{2} = \mathbf{0.934\ V\ (Ans.)}$$

$E^o_{NO_3^-/NO}$ is given by:

$$n_5 \times E^o_{NO_3^-/NO} = n_1 \times E^o_{NO_3^-/N_2O_4} + n_2 \times E^o_{N_2O_4/HNO_2} + n_3 \times E^o_{HNO_2/NO}$$

or $\quad 3 \times E^o_{NO_3^-/NO} = 1 \times 0.803 + 1 \times 1.065 + 1 \times 0.983$

$$\therefore \quad E^o_{NO_3^-/NO} = \frac{0.803 + 1.065 + 0.983}{3} = \mathbf{0.950\ V\ (Ans.)}$$

**Example 11. Write reduction half-reactions for all the pairs present in the following Latimer diagram**

$$\underset{Cl = +7}{ClO_4^-} \xrightarrow{+1.20\,V} \underset{+5}{ClO_3^-} \xrightarrow{+1.18\,V} \underset{+3}{HClO_2} \xrightarrow{+1.65V} \underset{+1}{HClO} \xrightarrow{+1.67\,V} \underset{0}{Cl_2} \xrightarrow{+1.36\,V} \underset{-1}{Cl^-}$$

**Ans.** (*i*) *$ClO_4^- - ClO_3^-$ pair*:

$$ClO_4^- + 2H^+ + 2e^- \rightarrow ClO_3^- + H_2O\ (n = 2, E° = +1.20\ V)$$

(*ii*) *$ClO_3^- - HClO_2$* pair:

$$ClO_3^- + 3H^+ + 2e^- \rightarrow HClO_2 + H_2O\ (n = 2, E^o = +1.18\ V)$$

(*iii*) *$HClO_2 - HClO$ pair:*

$$HClO_2 + 2H^+ + 2e^- \rightarrow HClO + H_2O\ (n = 2, E^o = +1.65\ V)$$

(*iv*) *$HClO - Cl_2$ pair*:

$$HClO + H^+ + e^- \rightarrow \frac{1}{2}Cl_2 + H_2O\ (n = 1, E^o = +1.67\ V)$$

(*v*) *$Cl_2 - Cl^-$ pair*:

$$\frac{1}{2}Cl_2 + e^- \rightarrow Cl^-\ (n = 1, E^o = +1.36\ V)$$

**Example 12. Latimer diagram for chlorine in basic medium is given as:**

$$ClO_4^- \xrightarrow{+0.36\,V} ClO_3^- \xrightarrow{+0.33V} ClO_2 \xrightarrow{+0.66V} ClO^- \xrightarrow{-0.42\,V} \frac{1}{2}Cl_2 \xrightarrow{+1.36\,V} Cl^-$$

**Predict whether the disproportionation of:**

**(*a*) $ClO_3^-$ to $ClO^-$ and $ClO_4^-$ is feesible or not.**

**(*b*) $ClO^-$ to $ClO_3^-$ and $Cl^-$ is feasible or not.**

**(*c*) $ClO^-$ to $Cl_2$ and $ClO_2$ is feasible or not.**

**Ans.** The given question shows that the given Latimer diagrem should be shown as:

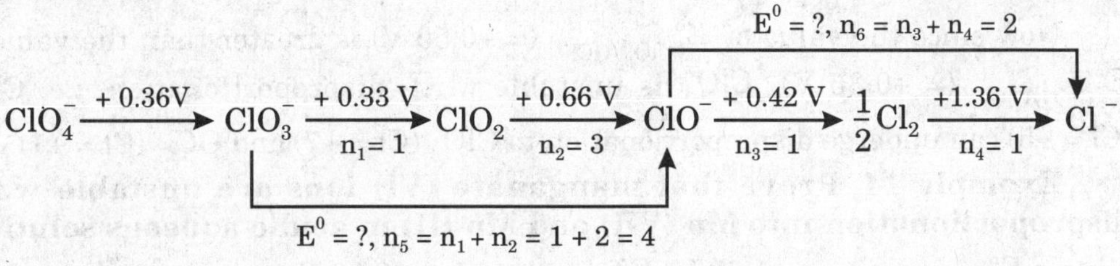


To answer this question we should calculate the values of $E°_{ClO_3^-/ClO^-}$ and $E^o_{ClO^-/Cl^-}$

$E^\circ_{ClO_3^-/ClO^-}$ is given by:

$$n_5 \times E^\circ_{ClO_3^-/ClO^-} = n_1 \times E^\circ_{ClO_3^-/ClO_2} + n_2 \times E^\circ_{ClO_2/ClO^-}$$

*or* $$4 \times E^\circ_{ClO_3^-/ClO^-} = 1 \times 0.33 + 3 \times 0.66$$

$\therefore$ $$E^\circ_{ClO_3^-/ClO^-} = \mathbf{0.57\ V}$$

$E^\circ_{ClO^-/Cl^-}$ is given by:

$$n_6 \times E^\circ_{ClO^-/Cl^-} = n_3 \times E^\circ_{ClO^-/\frac{1}{2}Cl_2} + n_4 \times E^\circ_{\frac{1}{2}Cl_2/Cl^-}$$

*or* $$2 \times E^\circ_{ClO^-/Cl^-} = 1 \times 0.42 + 1 \times 1.36$$

$\therefore$ $$E^\circ_{ClO^-/Cl^-} = \mathbf{0.89\ V}$$

In the modified form the Latimer diagram should be shown as:

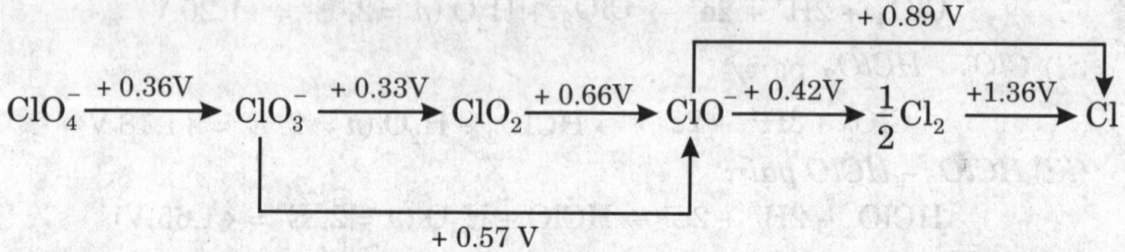


(*a*) The disproportionation of $ClO_3^-$ into $ClO^-$ and $ClO_4^-$ is feasible, since E° value on the RHS of $ClO_3^-$ (= 0.57 V) is greater than E° value on the LHS of $ClO_3^-$ (=0.36 V).

(*b*) Disproportionation of $ClO^-$ to $ClO_3^-$ and $Cl^-$ is feasible, since E° value on the RHS of $ClO^-$ (= 0.89 V) is greater than E° value on LHS of $ClO^-$ (= 0.57 V)

(*c*) The disproportionation of $ClO^-$ into $Cl_2$ and $ClO_2$ is not feasible because E° value on RHS of $ClO^-$ (= 0.42 V) is smaller than E° value on LHS of $ClO^-$ (= 0.66 V)

**Example 13. With the help of the following reduction half-reactions predict whether $ClO_3^-$ is unstable w.r.t. desproportionation.**

$$\mathbf{ClO_4^- + 2H^+ + 2e^- \longrightarrow ClO_3^- + H_2O,\ E^\circ = +0.36V}$$
$$\mathbf{ClO_3^- + 4H^+ + 4e^- \longrightarrow OCl^- + 2H_2O,\ E^\circ = +0.50V}$$

**Ans.** Latimer diagram corresponding to the given reduction half-reactions can be drawn as:

$$\underset{\text{Cl} = +7}{ClO_4^-} \xrightarrow{+0.36V} \underset{+5}{ClO_3^-} \xrightarrow{+0.50V} \underset{+1}{OCl^-}$$

Now since the value of $E^\circ_{ClO_3^-/OCl^-}$ (= +0.50 V) is greater than the value of $E^\circ_{ClO_4^-/ClO_3^-}$ (= +0.36 V), $ClO_3^-$ is unstable w.r.t. disproportionation, *i.e.* $ClO_3^-$ (Cl = +5) can undergo disproportionation to $ClO_4^-$ (Cl = +7) and $OCl^-$ (Cl = +1)

**Example 14. Prove that manganate (VI) ions are unstable w.r.t. disproportionation into Mn (VII) and Mn (II) in acidic aqueous solution. Given $E^\circ_{MnO_4^{2-}/Mn^{2+}}$ = +1.63 V, $E^\circ_{MnO_4^-/MnO_4^{2-}}$ = +0.90 V.**

**Ans.** This question can be answered in the following two ways:

(*i*) Managanate (VI) ion is $MnO_4^{2-}$ and Mn (VII) is present in $MnO_4^-$. The disproportionation of $MnO_4^{2-}$ ion into $MnO_4^-$ and $Mn^{2+}$ can be represented by the redox reaction.

$$\underset{(Mn = +6)}{5MnO_4^{2-}} + 8H^+ \longrightarrow \underset{(Mn = +7)}{4MnO_4^-} + \underset{(Mn = +2)}{Mn^{2+}} + 4H_2O$$

This redox reaction can be broken into the following two reduction half-reactions.

$$4 \times \left[MnO_4^{2-} \xrightarrow{\text{oxidation}} MnO_4^- + e^-, E°_{ox} = -0.90\ V\right]$$

$$MnO_4^{2-} + 8H^+ + 4e^- \xrightarrow{\text{reduction}} Mn^{2+} + 4H_2O, E°_{red} = +1.63\ V$$

*On adding:* $5MnO_4^{2-} + 8H^+ \rightarrow 4MnO_4^- + Mn^{2+} + 4H_2O$,

$$E° = E°_{ox} + E°_{red} = -0.90 + 1.63 = +0.73\ V$$

Since E° value is positive, the above disproportionation reaction is feasible.

(*ii*) Construct a Latimer diagram containing $MnO_4^{2-}$ (Mn = +6), $MnO_4^-$ (Mn = + 7) and $Mn^{2+}$ (Mn = + 2)

$$\underset{Mn = +7}{MnO_4^-} \xrightarrow{E°_1 = +0.90\ V} \underset{+6}{MnO_4^{2-}} \xrightarrow{E°_2 = +1.63\ V} \underset{+2}{Mn^{2+}}$$

Since $E_2^\circ$ (= +1.63 V) > $E_1^0$ (= +0.90 V), disproportionation of $MnO_4^{2-}$ into $MnO_4^-$ and $Mn^{2+}$ is feasible.

**Example 15. From the considerations of the following potentials decide if $Mn^{2+}$ is likely to disproportionate in solution.**

$$\mathbf{Mn^{2+} + 2e^- \longrightarrow Mn, E° = -1.18\ V}$$

$$\mathbf{Mn^{2+} \longrightarrow Mn^{3+} + e^-, E° = -1.51\ V}$$

**Ans.** The disporportionation of $Mn^{2+}$ to Mn° and $Mn^{3+}$ is not feasible. Hence $Mn^{2+}$ ion remains stable in solution.

**Explanation.** Latimer diagram having $Mn^{3+}$, $Mn^{2+}$ and Mn° is:

$$Mn^{3+} \xrightarrow{+1.51\ V} Mn^{2+} \xrightarrow{-1.18\ V} Mn°$$

Since E° value for $Mn^{2+}$ / Mn° couple (E° = –1.18 V) (right hand couple) is less than E° value of $Mn^{3+}$ / $Mn^{2+}$ couple (left hand couple) (E° = +1.51 V), disproportionation of $Mn^{2+}$ to Mn° and $Mn^{3+}$ is not possible.

Also since E° value of right hand side couple – E° value of left hand side couple = –1.18 – 1.51 = –2.69 V is a negative quatity (= –2.69 V), the disproportionation of $Mn^{2+}$ to Mn° and $Mn^{3+}$ is not feasible.

**Example 16. The potentials for $Fe^{2+}$ / Fe and $Fe^{3+}$ / $Fe^{2+}$ are – 0.41 V and +0.77 V respectively. Predict whether $Fe^{2+}$ will disproportionate to $Fe^{3+}$ and Fe.**

**Ans.** The Latimar diagram for $Fe^{3+}$, $Fe^{2+}$ and Fe° can be given as:

$$Fe^{3+} \xrightarrow{+0.77\ V} Fe^{2+} \xrightarrow{-0.41\ V} Fe°$$

Now E° for $Fe^{2+}$ / Fe° couple (E° = –0.41 V) is smaller than that for $Fe^{3+}$ / $Fe^{3+}$ couple (E° = +077 V), the disproptionation of $Fe^{2+}$ to $Fe^{3+}$ and Fe is not feasible.

**Example 17. From the reducion potentials of the following two reduction half-reactions determine the E° value of the half-reactions: $Fe^{3+} + e^- \rightarrow Fe^{2+}$**

$$Fe^{2+} + 2e^- \longrightarrow Fe^\circ,\ E^\circ = -0.44\ V$$
$$Fe^{3+} + 3e^- \longrightarrow Fe^\circ,\ E^\circ = -0.036\ V$$

**Ans.** Latimer diagam having Fe, $Fe^{2+}$ and $Fe^{3+}$ can be written as:

$$\underset{+3}{Fe^{3+}} \xrightarrow[n_1 = 1]{E_1^0 = ?} \underset{+2}{Fe^{2+}} \xrightarrow[n_2 = 2]{E_2^0 = -0.44V} \underset{0}{Fe^0}$$

$$E^0 = -0.036\ V,\ n = n_1 + n_2 = 1 + 2 = 3$$

$$\therefore \quad E^\circ = \frac{n_1 E^\circ_1 + n_2 E^\circ_2}{n_1 + n_2} = \frac{1 \times E^\circ_1 + 2 \times (-0.44)}{3} = \frac{E^\circ_1 - 0.88}{3}$$

or $\quad -0.036 = \dfrac{E^\circ_1 - 0.88}{3}$

or $\quad -0.036 \times 3 + 0.88 = E^\circ_1$

or $\quad E^\circ_1 =$ **+0.772 V (Ans.)**

**Example 18. The two half-reactions of $Au^+$ / Au and $Au^{3+}$ / $Au^+$ couples and their E° values are given below.**

$$Au^+ + e^- \longrightarrow Au,\ E^\circ = 1.7\ V$$
$$Au^{3+} + 2e^- \longrightarrow Au^+,\ E^\circ = 1.4\ V$$

**Calculate the value of E° for the reduction half-reaction:**

$$Au^{3+} + 3e^- \longrightarrow Au$$

**Ans.** In all the given reduction half-reactions Au is present as Au°, $Au^+$ and $Au^{3+}$. Thus Latimer diagram having these species can be written as:

$$\underset{+3}{Au^{3+}} \xrightarrow[n_1 = 2]{E_1^0 = 1.4V} \underset{+1}{Au^{+}} \xrightarrow[n_2 = 1]{E_2^0 = 1.7V} \underset{0}{Au^0}$$

$$E^0 = ?,\ n = n_1 + n_2 = 2 + 1 = 3$$

Since $Au^{3+} \rightarrow Au^+$ is a two-electron change, $n_1 = 2$. Similarly $Au^+ \rightarrow Au^\circ$ is a one-electron charge ($n_2 = 1$)

$$\therefore \quad E^\circ = \frac{n_1 E^\circ_1 + n_2 E^\circ_2}{n_1 + n_2} = \frac{2 \times 1.4 + 1 \times 1.7}{3} = \frac{2.8 + 1.7}{3} = \frac{4.5}{3} = \mathbf{1.5\ V\ (Ans.)}$$

**Example 19. Latimer diagram for vanadium element in acidic medium is shown below.**

$$V(OH)_4^+ \xrightarrow{+1.00V} VO^{2+} \xrightarrow{+0.359V} V^{3+} \xrightarrow{-0.256V} V^{2+} \xrightarrow{-1.186\ V} V^0$$

$$V(OH)_4^+ \xrightarrow{-0.254V} V^0$$

**Write the reduction half-reactions correspoding to the changes shown in the diagram.**

**Ans.** The reduction reactions corresponding for this diagram are given as:

$$V(OH)_4^+ + 2H^+ + e^- \longrightarrow VO^{2+} + 3H_2O;\ E° = 1.00\ V$$
$$VO^{2+} + 2H^+ + e^- \longrightarrow V^{3+} + H_2O;\quad E° = +0.359\ V$$
$$V^{3+} + e^- \longrightarrow V^{2+};\quad E° = -\ 0.256\ V$$
$$V^{2+} + 2e^- \longrightarrow V;\quad E° = -\ 1.186\ V$$
$$[V(OH)_4]^+ + 4H^+ + 5e^- \longrightarrow V + 4H_2O;\quad E° = -\ 0.254\ V$$

**Example 20. Calculate the electrode potential of $Cu^+$ / Cu half-cell. Given that the standard reduction protentials of $Cu^{2+}$ / Cu and $Cu^{2+}$ / $Cu^+$ are 0.337 V and 0.153 V respectively.**

**Ans.** In the given half-cells Cu is present as Cu°, $Cu^+$ and $Cu^{2+}$. Thus Latimer diagram containing Cu°, $Cu^+$ and $Cu^{2+}$ ions can be given as:

$$\underset{+2}{Cu^{2+}} \xrightarrow[n_1 = 1]{E_1^0 = 0.153V} \underset{+1}{Cu^+} \xrightarrow[n_2 = 1]{E_2^0 = ?} \underset{0}{Cu^0}$$
$$E^0 = 0.337\ V,\ n = n_1 + n_2 = 1 + 1 = 2$$

The values of $n_1$ and $n_2$ can be calculated as follows:

$$Cu^{2+} + e^- \longrightarrow Cu^+\ (n_1 = 1, E°_1 = 0.153\ V)$$
$$Cu^+ + e^- \longrightarrow Cu°\ (n_2 = 1, E°_2 = ?)$$

$$\therefore \qquad E° = \frac{n_1E°_1 + n_2E°_2}{n_1 + n_2} = \frac{1 \times 0.153 + 1 \times E°_2}{2}$$

or $$0.337 = \frac{0.153 + E_2^\circ}{2}$$

$$\therefore \qquad E°_2 = 0.337 \times 2 - 0.153 = \mathbf{0.521\ V\ (Ans.)}$$

**Example 21. The standard electrode potentials (E°) for $OCl^-$ / $Cl^-$ and $\frac{1}{2}Cl_2$ / $Cl^-$ respectively are 0.94 V and 1.36 V. Calculate E° value for $OCl^-$ / $\frac{1}{2}Cl_2$ electrode.**

**Ans.** Oxidation number of Cl-atom in different species is as: $OCl^- = +1$, $Cl^- = -1$ and $\frac{1}{2}Cl_2 = 0$. Latimer diagram containing these species can be written as:

$$\underset{+1}{OCl^-} \xrightarrow[n_1 = 1]{E_1^0 = ?} \underset{0}{Cl_2} \xrightarrow[n_2 = 1]{E_2^0 = +1.36V} \underset{-1}{Cl^-}$$
$$E^0 = 0.95\ V,\ n = n_1 + n_2 = 1 + 1 = 2$$

Values of $n_1$ and $n_2$ are 1 and 1 respectively as shown below.

$$OCl^- + 2H^+ + e^- \longrightarrow \frac{1}{2}Cl_2 + H_2O\ (n_1 = 1, E°_1 = ?)$$

$$\frac{1}{2}Cl_2 + e^- \longrightarrow Cl^- \; (n_2 = 1, E°_2 = +1.36 \text{ V})$$

$$\therefore \quad E° = \frac{n_1 E°_1 + n_2 E°_2}{n_1 + n_2}$$

$$\text{or} \quad 0.95 = \frac{1 \times E°_1 + 1 \times 1.36}{2}$$

$$\therefore \quad E°_1 = 2 \times 0.95 - 1.36 = \mathbf{0.52\ V\ (Ans.)}$$

**Example 22. E° value for the reduction half-reactions: $Sn^{2+} + 2e^- \rightarrow Sn$ and $Sn^{4+} + 2e^- \rightarrow Sn^{2+}$ are –0.136 V and 0.15 V respectively. Calculate E° value for the reaction, $Sn^{4+} + 4e^- \rightarrow Sn$ and comment on the ease of reducing $Sn^{2+}$ to Sn and $Sn^{4+}$ to Sn.**

**Ans.** In all the given reduction half-reactions, Sn is present as Sn°, $Sn^{2+}$ and $Sn^{4+}$. Thus, Latimer diagram having these species can be shown as:

$$\underset{+4}{Sn^{4+}} \xrightarrow[n_1 = 2]{E^0_1 = +0.15V} \underset{+2}{Sn^{2+}} \xrightarrow[n_2 = 2]{E^0_2 = -0.136V} \underset{0}{Sn^0}$$

$$Sn^{4+} \rightarrow Sn^0: \quad E^0 = ?, \; n = n_1 + n_2 = 2 + 2 = 4$$

$$\therefore \quad E° = \frac{n_1 E°_1 + n_2 E°_2}{n_1 + n_2} = \frac{2 \times 0.15 + 2 \times (-0.136)}{4}$$

$$= \frac{0.300 - 0.272}{4} = \frac{0.028}{4} = \mathbf{0.007\ V\ (Ans.)}$$

Now since E° value for the reduction half-reaction, $Sn^{2+} + 2e^- \rightarrow Sn$ is negative (E° = –0.136 V) and that for the reduction half-reaction, $Sn^{4+} + 4e^- \rightarrow Sn$ is positive (E° = +0.007 V), $Sn^{4+}$ is a stronger O.A. than $Sn^{2+}$. In other words, $Sn^{4+}$ can more easily be reduced to Sn than $Sn^{2+}$ can be reduced to Sn.

**Example 23. From the reduction potentials of the following sequence of reactions predict whether $MnO^-_4$ or $MnO_4^{2-}$ is the better oxidant, if the reduction product is $Mn^{2+}$.**

$$\mathbf{MnO_4^- \xrightarrow{0.564V} MnO_4^{2-} \xrightarrow{2.26V} MnO_2 \xrightarrow{0.95V} Mn^{3+} \xrightarrow{1.51V} Mn^{2+}}$$

**Ans.** In order to answer this question we will have to determine E° value for $MnO^-_4 \rightarrow Mn^{2+}$ and $MnO_4^{2-} \rightarrow Mn^{2+}$ changes.

$$MnO_4^- \xrightarrow[n=1]{+0.564V} MnO_4^{2-} \xrightarrow[n=2]{+2.26V} MnO_2 \xrightarrow[n=1]{+0.95V} Mn^{3+} \xrightarrow[n=1]{+1.51V} Mn^{2+}$$

$$MnO_4^{2-} \rightarrow Mn^{2+}: \quad E^0 = ?, \; n = 2 + 1 + 1 = 4$$

$$MnO_4^- \rightarrow Mn^{2+}: \quad E^0 = ?, \; n = 1 + 2 + 2 + 1 + 1 = 5$$

$$\therefore \quad E° \text{ for } MnO^-_4 / Mn^{2+} \text{ couple} = \frac{1 \times 0.564 + 2 \times 2.26 + 1 \times 0.95 + 1 \times 1.51}{1 + 2 + 1 + 1} = \mathbf{1.5\ V}$$

$$E° \text{ for } MnO_4^{2-} / Mn^{2+} \text{ couple} = \frac{2\times2.26+1\times0.95+1\times1.51}{2+1+1} = \mathbf{1.7\ V}$$

Since E° value for $MnO_4^{2-} / Mn^{2+}$ couple (= +1.7 V) is more positive than that for $MnO_4^- / Mn^{2+}$ couple (E° = 1.5 V), $MnO_4^{2-}$ is a better OA than $MnO^-_4$.

## Frost Diagrams

Suppose an element X forms a radox couple, X (N) | X (0) in which X (N) is the oxidised form of X and X (0) is the reduced form of X. In X (N), N is the oxidation number (O.N.) of X and in X (0), 0 (zero) is the oxidation number of X, *i.e.* in X (N) | X (0), X (0) means that X is present in elemental state. Frost diagram of an element X is the plot between the product of oxidation number (N) and reduction potential (E°) of X (N) | X (0) couple against oxidation number (N) of the element. Thus, Frost diagram of an element is a plot between NE° and N. Now ΔG° = – NE°F or NE° = – ΔG° / F. This relation shows that since F is Faraday constant, NE° is proportional to –ΔG°. Thus, the plot of NE° *versus* N is the same as the plot of –ΔG° *versus* N. To draw a Frost diagram, NE° (or – ΔG°/F) is taken on *y*-axis (abscissa) and N is taken on *x*-axis (ordinate). N is the number of electrons required to reduce the oxidant to zero oxidation state (elemental state) and E° is the corresponding standard reduction potential. When an element forms a number of species in different oxidation states, then E° value for each species involved in reduction half-reaction, $X^{N+} + Ne^- \longrightarrow X°$ is determined (*i.e.* E° value for the couple $X^{N+} / X°$ is determined) and NE° is ploted against oxidation number (N). Frost diagram is obtained by joining these points obtained for each species of that element.

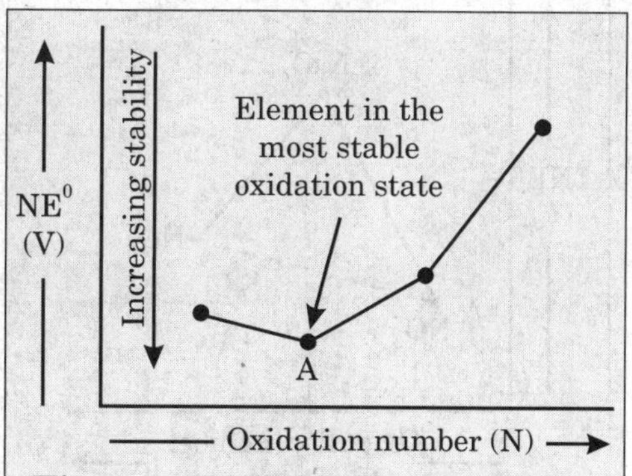

**Fig. 1.4.** Plot of *NE°* vs *N* showing the most stable oxidation state of an element

It may be noted that in the Frost diagram of an element the species occupying the lowest position in the diagram has the element in its most stable oxidation state. For example in Fig 1.4 the species shown as A has the element in the most stable oxidation state.

In Frost diagram shown in Fig 1.5, since the couple corresponding to points C and D have higher value of E° than the couple corresponding to points B and C, the species represented by the point D is likely to undergo reduction while the species represented by point B is liable to undergo oxidation.

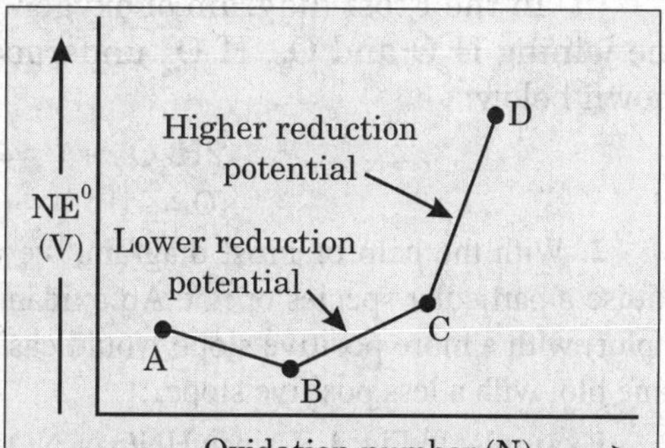

**Fig. 1.5.** Frost diagram depicting higher and lower reduction potential of a species.

## Applications of Frost Diagrams

**1.** With the help of Frost diagram we can predict whether a species given in Frost diagram is stable or unstable to disproportionation *i.e.* whether a species given in Frost diagram will undergo or will not undergo disproportionation. It has been observed that if the point of a species in a Frost diagram lies *above the line connecting the two adjacent points of two species, it would undergo disproportionation. If the point of a species in a Frost diagram lies below the line connecting the two adjacent points of two species, the species will not undergo disproportionation.*

**Examples** (*i*) In the Frost diagram of nitrogen given in Fig 1.6, since $N_2O_4$ molecule lies above the line joining $NO_3^-$ and $NO_2^-$, $N_2O_4$ is unstable w.r.t. disproportionation, *i.e.* $N_2O_4$ undergoes disproportionation into $NO^-_3$ and $NO^-_2$ ions as shown below.

$$\underset{(N = +4)}{N_2O_4} + H_2O \longrightarrow \underset{(N=+3)}{HNO_2} + \underset{(N = +5)}{HNO_3}$$

or

$$\underset{(N = +4)}{N_2O_4} + O^{2-} \longrightarrow \underset{(N=+3)}{NO_2^-} + \underset{(N = +5)}{NO_3^-}$$

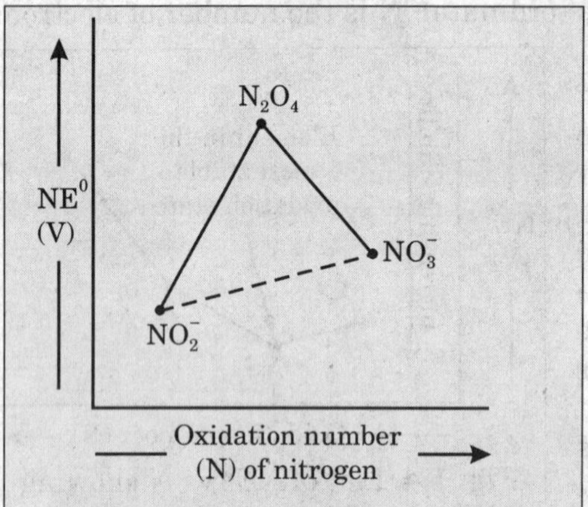


**Fig. 1.6** Frost diagram of nitrogen.

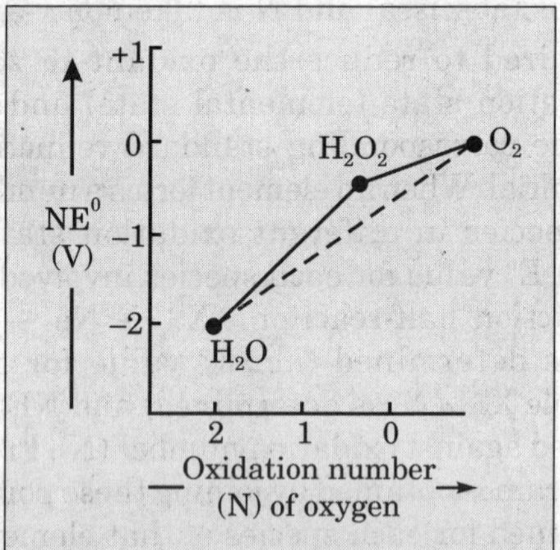


**Fig. 1.7** Frost diagram of oxygen

(*ii*) In the Frost diagram of oxygen given in Fig 1.7, since $H_2O_2$ lies above the line joining $H_2O$ and $O_2$, $H_2O_2$ undergoes disproportionation into $H_2O$ and $O_2$ as shown below:

$$\underset{(O = -1)}{2H_2O_2} + \longrightarrow \underset{(O = -2)}{2H_2O} + \underset{(O = 0)}{O_2}$$

**2.** With the help of Frost diagram, we can predict whether a particular oxidant would oxidise a particular species or not. An oxidant occupying a position in Frost diagram (NE° *vs* N plot) with a more positive slope would easily oxidise a species occupying a position in the same plot with a less positive slope.

**Example.** In Fig. 1.8, since $HNO_3$ – NO line has more positive slope than Cu – $Cu^{2+}$ line, $HNO_3$ would easily oxidise Cu to $Cu^{2+}$ and would itself be reduced to NO.

$$\underset{(N = +5)}{8HNO_3} + \underset{(Cu = 0)}{3Cu} \rightarrow \underset{(Cu = +2)}{3Cu(NO_3)_2} + \underset{(N = +2)}{2NO} + 4H_2O$$

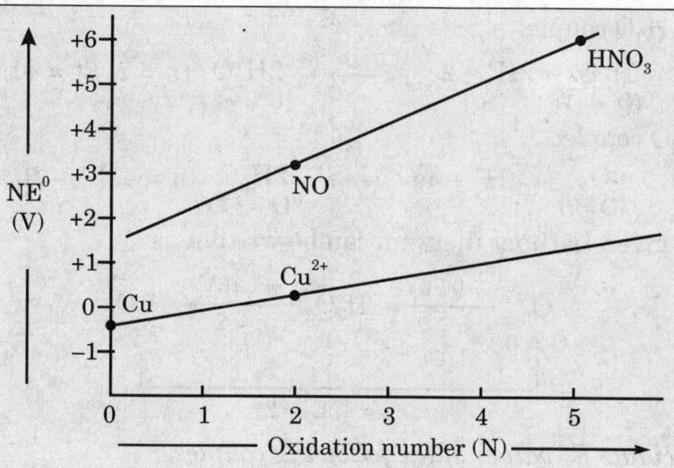


**Fig.1.8** Plot of *N*E° vs oxidation number (N) for $HNO_3$ – NO and Cu – $Cu^{2+}$ systems

**3.** E° value of a couple in a Latimer diagram can be calculated, if E° values of other couples are given.

**Example: Latimar diagram of oxygen is given as:**

$$\underset{O\,=\,0}{\mathbf{O_2}} \xrightarrow{\mathbf{0.70\ V}} \underset{-1}{\mathbf{H_2O_2}} \xrightarrow{\mathbf{1.78\ V}} \underset{-2}{\mathbf{H_2O}}$$

**Calculate E° value for $O_2$ / $H_2O$ couple.**

**Ans.** E° value for $O_2$ / $H_2O$ couple is given by:

NE° value for $O_2$ / $H_2O$ couple = NE° value $O_2$ / $H_2O_2$ couple + NE° value for $H_2O_2$ / $H_2O$ couple.

or $$(-2-0)\ E^\circ_{O_2/H_2O} = (-1-0) \times 0.70 + [-2-(-1)] \times 1.78$$

or $$-2E^\circ_{O_2/H_2O} = -0.70 - 1.78$$

or $$E^\circ_{O_2/H_2O} = \frac{-2.48}{-2} = \mathbf{1.24\ V\ (Ans.)}$$

## Solved Problems Based on Frost Diagrams

**Problem 1. Draw the Frost diagram from the following Latimer diagram and predict whether $H_2O_2$ is stable or unstable w.r.t. disproportionation.**

$$\underset{O\,=\,0}{\mathbf{O_2}} \xrightarrow{\mathbf{+0.70\ V}} \underset{-1}{\mathbf{H_2O_2}} \xrightarrow{\mathbf{+1.76\ V}} \underset{-2}{\mathbf{H_2O}}$$

**Ans.** The construction of Frost diagram with the help of the given Latimer diagram consists of the following steps:

*(i) To determine the number of electrons (n) involved in reduction half-reactions of the following couples.*

(*a*) $O_2$ / $H_2O_2$ couple:

$$\underset{(O\,=0)}{O_2} + 2H^+ + 2e^- \longrightarrow \underset{(O\,=\,-1)}{H_2O_2} \quad (n = 2,\ E^\circ = +0.70V)$$

(*b*) $H_2O_2$ / $H_2O$ couple:

$$\underset{(O = -1)}{H_2O_2} + 2H^+ + 2e^- \longrightarrow \underset{(O = -2)}{2H_2O} \quad (n = 2, E^\circ = +1.76 \text{ V})$$

(*c*) $O_2$ / $H_2O$ couple:

$$\underset{(O = 0)}{O_2} + 4H^+ + 4e^- \longrightarrow \underset{(O = -2)}{2H_2O} \quad (n = 4, E^\circ = ?)$$

Thus, the given Latimer diagram can be written as:

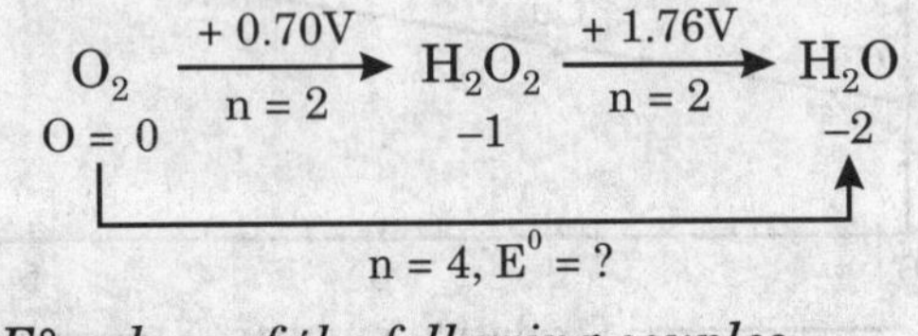


(*ii*) *To calculate E° values of the following couples.*

(*a*) $E^\circ_{O_2/H_2O_2} = +0.70$ V (Given)

(*b*) $E^\circ_{H_2O_2/H_2O} = +1.76$V (Given)

(*c*) $4 \times E^\circ_{O_2/H_2O} = 2\times E^\circ_{O_2/H_2O_2} + 2\times E^\circ_{H_2O_2/H_2O}$

$4\times E^\circ_{O_2/H_2O} = 2 \times 0.70 + 2 \times 1.76$

or $\quad E^\circ_{O_2/H_2O} = \dfrac{0.70+1.76}{2} = \mathbf{1.23\ V}$

(*iii*) *To calculate NE° values for the following couples.*

NE° value for $O_2/O_2$ couple = 0 × 0 = 0

NE° value for $O_2/H_2O_2$ couple = –1 × 0.70 = –0.70 V

NE° value for $O_2/H_2O$ couple = –2 × 1.23 = – 2.46 V

(*iv*) *To write the coordinates for $O_2$, $H_2O_2$ and $H_2O$*

coordinates for $O_2$ = 0, 0

coordinates for $H_2O_2$ = N, NE° = –1, –0.70 V

coordinates for $H_2O$ = N, NE° = –2, –2.46 V

(*v*) Taking NE° values along y-axis and oxidation number (N) of oxygen along *x*-axis, we plot N, NE° points for $O_2$, $H_2O_2$ and $H_2O$ molecules as (0, 0), (–1, –0.70 V) and (–2, –2.46 V) respectively. When these points are joined, Frost diagram of oxygen is obtained (See Fig 1.9).

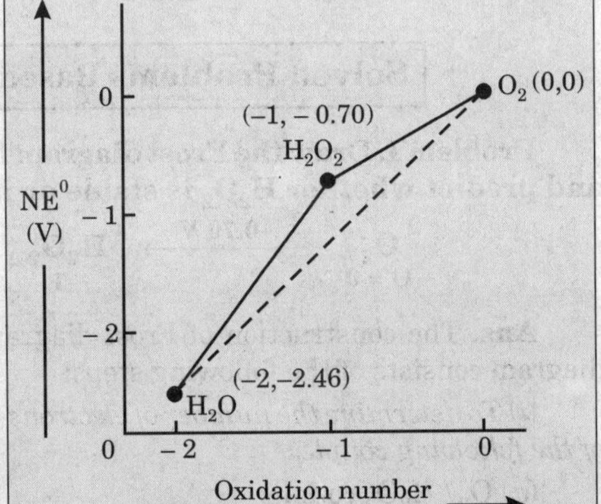


**Fig. 1.9** Frost diagram for oxygen in acidic medium

In the Frost diagram of oxygen, since $H_2O_2$ lies **above** the line joining the points of $H_2O$ and $O_2$, this molecule is unstable w.r.t. disportionation, *i.e.* $H_2O_2$ disproportionates into $H_2O$ and $O_2$.

$$\underset{(O=-1)}{2H_2O_2} \longrightarrow \underset{(O=-2)}{2H_2O} + \underset{(O=0)}{O_2}$$

**Problem 2. (*i*) Construct the Frost diagram from the following Latimer diagram**

$$\underset{Tl\,=\,+3}{\mathbf{Tl^{3+}}} \xrightarrow{+1.26\text{ V}} \underset{+1}{\mathbf{Tl^{+}}} \xrightarrow{-0.34\text{ V}} \underset{0}{\mathbf{Tl^{\circ}}}$$

**(*ii*) Predict the stability or unstability of $Tl^+$**

**(*iii*) Which of these species is strong oxidising agent.**

**Ans.** (*i*) The construction of Frost diagram with the help of the given Latimer diagram consists of the following steps:

(A) *To determine the number of electrons (n) in the reduction half-reactions of the following couples.*

(*a*) $Tl^{3+}$ / $Tl^+$ couple

$Tl^{3+} + 2e^- \longrightarrow Tl^+$ ($n = 2$, E° = +1.36 V)

(*b*) $Tl^+$ / Tl° couple

$Tl^+ + e^- \longrightarrow Tl^\circ$ ($n = 1$, E° = –0.34 V)

(*c*) $Tl^{3+}$ / Tl° couple

$Tl^{3+} + 3e^- \rightarrow Tl^\circ$ ($n = 3$, E° = ?)

Thus, the given Latimer diagram can be written as:

$$\underset{Tl\,=\,+3}{Tl^{3+}} \xrightarrow[n\,=\,2]{+\,1.26V} \underset{+1}{Tl^{+}} \xrightarrow[n\,=\,1]{-\,0.34V} \underset{0}{Tl^{0}}$$

$$n = 3,\ E^0 = ?\ \ (Tl^{3+} \rightarrow Tl^0)$$

(B) *To calculate E° values of the following couples*

(*a*) $E^\circ_{Tl^{3+}/Tl^\circ}$ = +1.26 V (Given)

(*b*) $E^\circ_{Tl^{+}/Tl^\circ}$ = –0.34 V (Given)

(*c*) $3\times E^\circ_{Tl^{3+}/Tl^\circ} = 2\times E^\circ_{Tl^{3+}/Tl^{+}} + 1\times E^\circ_{Tl^{+}/Tl^\circ}$

or $\quad 3\times E^\circ_{Tl^{3+}/Tl^\circ} = 2 \times 1.26 + 1 \times (-0.34)$

or $\quad E^\circ_{Tl^{3+}/Tl^\circ} = \dfrac{2.18}{3} = 0.726 \approx \mathbf{0.73\ V}$

(C) *To calculate NE° values for the following couples*

NE° value for Tl° / Tl° couple = 0 × 0 = 0

NE° value for $Tl^{3+}$ / Tl° couple = +3 × $E^\circ_{Tl^{3+}/Tl}$ = 3 × 0.73 = 2.19 V

NE° value for $Tl^+$ / Tl° couple = 1 × $E^\circ_{Tl^{+}/Tl^\circ}$ = 1 × (–0.34) = – 0.34V

(D) *To write the coordinates of Tl°, $Tl^+$ and $Tl^{3+}$*

Coordinates of Tl° = 0, 0

Coordinates of $Tl^+$ = N, NE° = +1, –0.34 V

Coordinates of $Tl^{3+}$ = N, NE° = +3, +2.19 V

On plotting the above coordinates of Tl°, $Tl^+$ and $Tl^{3+}$ we get the Frost diagram shown in Fig. 1.10.

(*ii*) In the Frost diagram, since the point for $Tl^+$ lies **below** the line connecting the points of Tl° and $Tl^{3+}$ (Tl° – $Tl^{3+}$ line), $Tl^+$ is more stable than Tl° and $Tl^{3+}$ both. Thus, $Tl^+$ is the most stable of all the three species.

(*iii*) $Tl^+$ is strong OA because the slope of the line connecting it with either oxidation state is larger and positive.

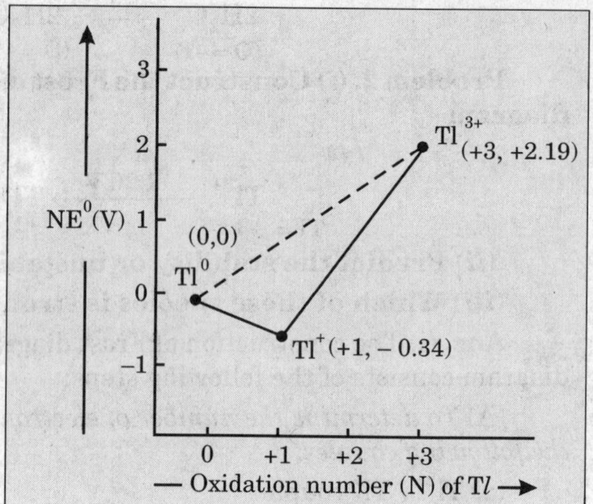


**Fig. 1.10** Frost diagram of Tl

**Problem 3. Consider the Frost diagram given in Fig. 1.11 for manganese in acidic medium. What conclusion can be drawn from this diagram ?**

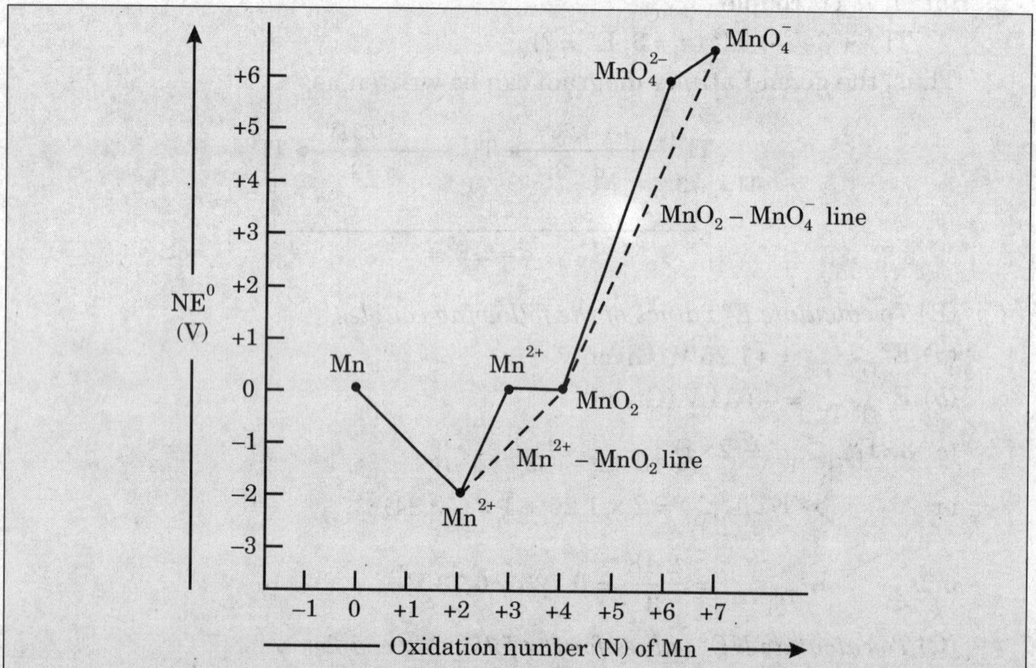


**Fig. 1.11** Frost diagram for Mn in acidic medium

**Ans.** Following conclusions can be drawn from the given diagram.

(*i*) The lower a species is present on the diagram, the more thermodynamically stable it is. Thus, Mn (II) is the most stable species.

(*ii*) $Mn^{3+}$ lies above the line joining $Mn^{2+}$ and $MnO_2$. Therefore, it will disproportionate as per the following equation:

$$2Mn^{3+} + 2H_2O \longrightarrow Mn^{2+} + MnO_2 + 4H^+$$

(*iii*) Manganate ion ($MnO_4^{2-}$) lies above the line joining $MnO_2$ and $MnO_4^-$. Therefore, it will disporportionate as per the following reaction:

$$3MnO_4^{2-} + 4H^+ \longrightarrow MnO_2 + 2MnO^-_4 + 2H_2O$$

(*iv*) The species on a concave curve shows no tendency towards disproportionation. Thus, $MnO_2$ does not proportionate.

(*v*) Any species located on the upper right side of the diagram will be a strong oxidising agent. Thus, $MnO^-_4$ is a strong oxidising agent.

(*vi*) A species located on the upper left side of the diagram will work as a reducing agent. Thus, Mn metal is a moderate reducing agent.

**Problem 4. Draw Frost diagram from the following Latimer diagram in acidic medium.**

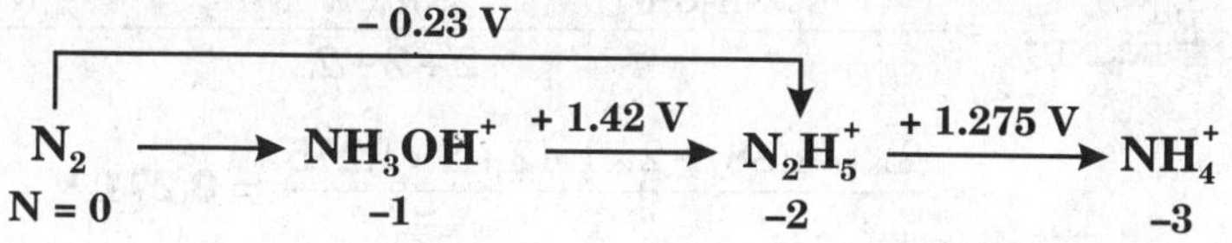


**Which of the four oxidation states (0, –1, –2, –3) of nitrogen is the most stable.**

**Ans.** The formation of Frost diagram from Latimer diagram consists of the following steps:

(*i*) *To calculate the number of electrons (n) involved in different reduction half-reactions*

(*a*) $N_2 / NH_3OH^+$ couple. Reduction half-reaction for this couple is:

$$\underset{(N = 0)}{N_2} + 2H_2O + 4H^+ + 2e^- \rightarrow \underset{(N = -1)}{2NH_3OH^+} \quad (E° = ?, n = 2)$$

(*b*) $NH_3OH^+ / N_2H_5^+$ couple. Reduction half-reaction for this couple is:

$$\underset{(N = -1)}{2NH_3OH^+} + H^+ + 2e^- \longrightarrow \underset{(N = -2)}{N_2H_5^+} + 2H_2O \ [E° = +1.42 \text{ V (given)}, n = 2]$$

(*c*) $N_2H_5^+ / NH^+_4$ couple. Reduction half-reaction for this couple is:

$$\underset{(N = -2)}{N_2H_5^+} + 3H^+ + 2e^- \longrightarrow \underset{(N = -3)}{2NH_4^+} \ [n = 2, E° = +1.275 \text{ V (given)}]$$

(*d*) $N_2 / N_2H_5^+$ couple. Reduction half-reaction for this couple is:

$$\underset{(N = 0)}{N_2} + 5H^+ + 4e^- \longrightarrow \underset{(N = -2)}{N_2H_5^+} \ [E° = -0.23 \text{ V (given)}, n = 4)$$

(*e*) $N_2 / NH^+_4$ couple. Reduction half-reaction for this couple is:

$$\underset{(N = 0)}{N_2} + 8H^+ + 6e^- \longrightarrow \underset{(N = -3)}{2NH_4^+} \ (E^0 = ?, n = 6)$$

Thus, the given Latimer diagram can be written as:

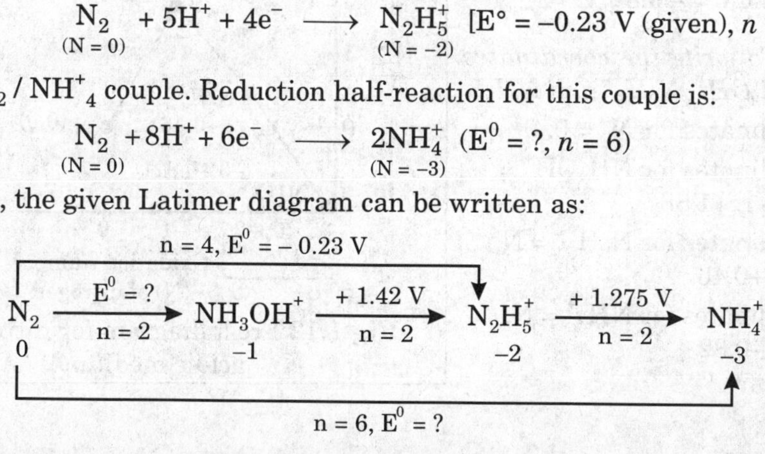

*(ii) To calculate E° values for different couples.*

(*a*) $N_2 / NH_3OH^+$ couple:

$$4\times E^\circ_{N_2/N_2H_5^+} = 2\times E^\circ_{N_2/NH_3OH^+} + 2\times E^\circ_{NH_3OH^+/N_2H_5^+}$$

$$4 \times (-0.23) = 2\times E^\circ_{N_2/NH_3OH^+} + 2 \times 1.42$$

$$\therefore \quad E^\circ_{N_2/NH_3OH^+} = -1.88 \text{ V}$$

(*b*) $N_2 / N_2H_5^+$ couple: E° value for this couple is given equal to –0.23 V

(*c*) $N_2 / NH_4^+$ couple:

$$E^\circ_{N_2/NH_4^+} = \frac{2\times E^\circ_{N_2/NH_3OH^+} + 2\times E^\circ_{NH_3OH^+/N_2H_5^+} + 2\times E^\circ_{N_2H_5^+/NH_4^+}}{2+2+2}$$

$$= \frac{2\times(-1.88)+2\times1.42+2\times1.275}{6} = \mathbf{0.271\ V}$$

Now, the Latimer diagram can shown as:

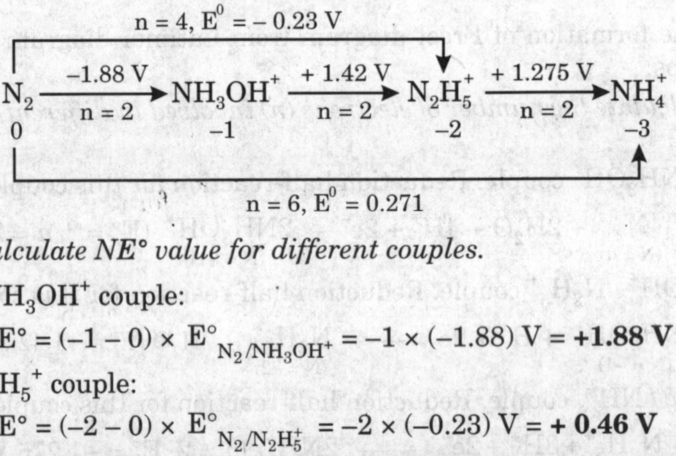


*(iii) To calculate NE° value for different couples.*

(*a*) $N_2 / NH_3OH^+$ couple:

$$NE^\circ = (-1 - 0) \times E^\circ_{N_2/NH_3OH^+} = -1 \times (-1.88) \text{ V} = \mathbf{+1.88\ V}$$

(*b*) $N_2/N_2H_5^+$ couple:

$$NE^\circ = (-2 - 0) \times E^\circ_{N_2/N_2H_5^+} = -2 \times (-0.23) \text{ V} = \mathbf{+\ 0.46\ V}$$

(*c*) $N_2 / NH^+_4$ couple:

$$NE^\circ = (-3 - 0) \times E_{N_2/NH_4^+}$$
$$= -3 \times 0.271 \text{ V} = \mathbf{-\ 0.82\ V}$$

*(iv) To write the coordinates for $N_2$, $NH_3OH^+$, $N_2H_5^+$ and $NH_4^+$*

Coordinates for $N_2$ = 0, 0

Coordinates for $NH_3OH^+$ = N, NE° = –1, +1.88

Coordinates for $N_2H_5^+$ = N, NE° = –2, +0.46

Coordinates for $NH_4^+$ = N, NE° = –3, –0.82

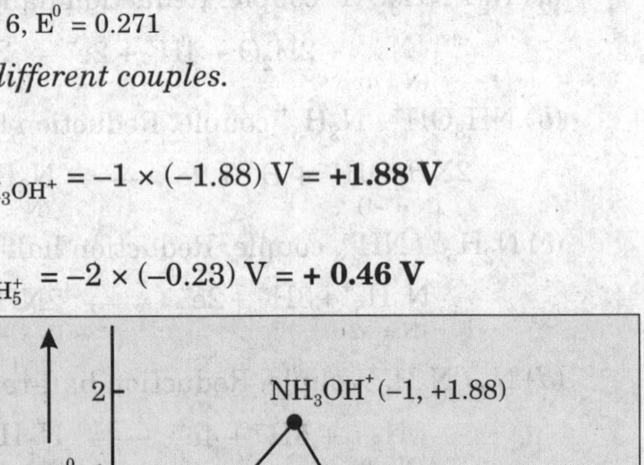


**Fig 1.12** Frost diagram for nitrogen in acidic medium

(*v*) Take NE° values along y-axis and oxidation number (N) of nitrogen along x-axis, we plot N, NE° points for $N_2$, $NH_3OH^+$, $N_2H_5^+$ and $NH_4^+$. When these points are joined, Frost diagram of nitrogen is obtained. (See Fig 1.12).

Since $NH_4^+$ ion lies at the minimum point in the diagram, O.S. of N in $NH_4^+$ ion which is equal to –3 is the most stable oxidation state of nitrogen.

**Problem 5. Give the interpretation of Frost diagram for $N_2$ under standard conditions (pH = 0) given in Fig 1.13**

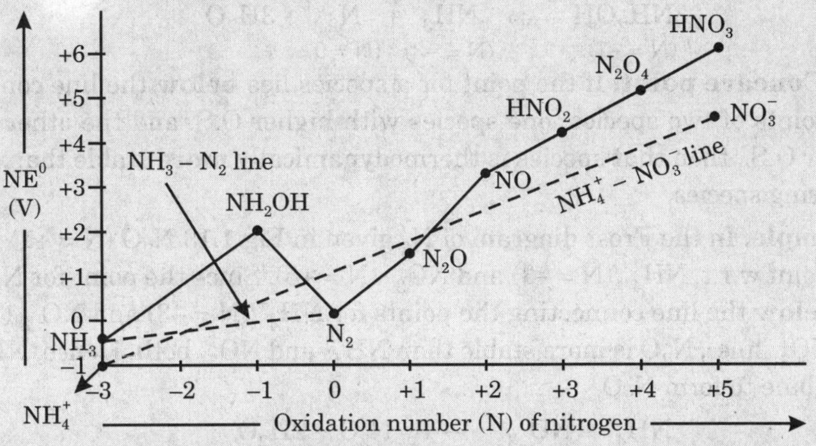


**Fig. 1.13** Frost diagram for nitrogen.

**Ans. Interpretation of Frost diagram.** This diagram provides a qualitative information about the relative stability of oxidation state of nitrogen w.r.t. other oxidation states of the same element. The position of an oxidation state on the curve can be used to explain its tendency to undergo disproportionation. In connection with the diagram the following points may be noted:

(*i*) **Minimum point.** Oxidation state of nitrogen at the minimum point in the curve is the most stable oxidation state of nitrogen. For example, – 3 is the most stable oxidation state of nitrogen among all other oxidation states of this element, since this oxidation state lies at the minimum point in the curve. Under appropriate conditions, all other oxidation states of nitrogen tend to attain this most stable oxidation state. Minimum point in the curve represents maximum decrease in free energy.

(*ii*) **Maximum point.** Oxidation state of nitrogen at the maximum point in the curve is the least stable oxidation state of nitrogen. Hence, the species present at the maximum point has a tendency to undergo reduction. In other words, the species present at the maximum point is strong OA and, depending on conditions, is reduced to the species of the low valent oxidation state. Maximum point in the curve represents high free energy change.

(*iii*) **Convex point.** If the point for a species lies **above** the line connecting the two neighbouring species, one species with higher O.S. and the other with lower O.S., then the species under consideration is thermodynamically **less stable** than the two neighbouring species. Hence, the species decomposes (or disproportionates) into a species with higher O.S. and a species with lower O.S.

**Example.** In the Frost diagram of $N_2$ given in Fig 1.13 $NH_2OH$ (N = –1) forms a convex point w.r.t. $NH_3$ (N = –3) and $N_2$ (N = 0). Since the point for $NH_2OH$ (N = –1) lies **above** the line connecting the points for $NH_3$ (N = –3) and $N_2$(N = 0) ($NH_3 - N_2$ line), $NH_2OH$ is thermodynamically **less stable** than $NH_3$ and $N_2$ both. Hence, $NH_2OH$ decomposses (disproportionates) readily into $NH_3$ and $N_2$.

$$\underset{(N=-1)}{3NH_2OH} \longrightarrow \underset{(N=-3)}{NH_3} + \underset{(N=0)}{N_2} + 3H_2O$$

(*iv*) **Concave point.** If the point for a species lies **below** the line connecting the two points of two species, one species with higher O.S. and the other species with lower O.S., then that species is thermodynamically **more** stable than the two neighbouring species.

**Example.** In the Frost diagram of $N_2$ given in Fig 1.13 $N_2O$ (N = +1) forms a concave point w.r.t. $NH_4^+$ (N = –3) and $NO_3^-$ (N = +5). Since the point for $N_2O$ (N = +1) lies **below** the line connecting the points for $NH_4^+$ (N = –3) and $NO^-_3$ (N = +5) ($NH_4^+ - NO_3^-$ line), $N_2O$ is more stable than $NH_4^+$ and $NO_3^-$ both. Hence, $NH_4^+$ and $NO^-_3$ combine to form $N_2O$

$$NH_4^+ + NO^-_3 \longrightarrow N_2O + 2H_2O$$

(*v*) **Linear points.** When three consecutive points of three species lie more or less on a straight line, then all the three species would be almost equally stable. In other words, no one species is completely converted into the other, instead all the three species exist in equilibrium. The reason for this is that all the species falling on a straight line have the same slope. Hence E° values for all the reactions would be the same. Consequently, there is no change in the free energy of these species and an equilibrium between various species is maintained.

**Example.** Since NO, $HNO_2$ and $N_2O_4$ lie on the same line, these species are equally stable.

## Pourbaix Diagrams

### What is Pourbaix Diagram?

A Pourbaix diagram is a map between potential and pH for different species of an element. This map depicts the conditions of potential and pH under which a given metal species is thermodynamically stable towards oxidation or reduction in water. Thus, Pourbaix diagram is also called **potential–pH diagram**. Pourbaix diagram is also called **predominance diagram**. This name is based on the fact that area bound by the lines of particular pH and E (volts) values is predominated mainly by the species mentioned within the particular area.

Pourbaix diagram has electrode potentials (E values) on y-axis (ordinate). Electrode potentials start from negative potentials at lower end of y-axis to positive potentials at upper end of y-axis. Since positive potentials indicate strong oxidants and negative potentials indicate strong reductants, Pourbaix diagram has strong oxidants at the top and strong reductants at the bottom. Strong acids are to the left (low pH) and strong bases are to the right (high pH) of x-axis.

## Pourbaix Diagram for Iron

For this element, the Pourbaix diagram depicts the stability ranges in terms of potential and pH for species such as $Fe^{3+}$, $Fe^{2+}$, $Fe(OH)_3$, $Fe(OH)_2$ etc. found in natural waters. The Pourbaix diagram showing predominance areas of different forms of Fe with varying pH and potential is given in Fig 1.14

Some reactions of iron occurring in natural waters are discussed below.

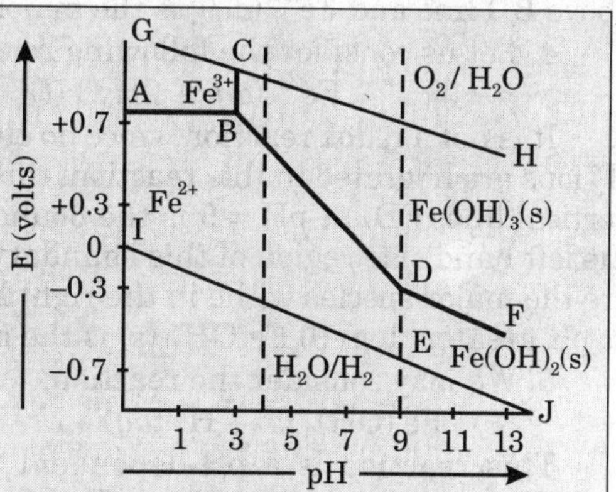


**Fig. 1.14** Pourbaix diagram for iron species in natural waters.

**1.** The following reduction half-reaction can be considered.

$$Fe^{3+}(aq) + e^- \longrightarrow Fe^{2+}(aq),\ E° = +0.77\ V$$

Since this reaction does not involve $H^+$ ions, the potential of this reaction is independent of pH, *i.e.* there is no change in the value of E with the increase in pH value. E *vs* pH plot gives a horizontal line AB at pH = 0.7 V. This line is the boundary line above which (*i.e.* in presence of an O.A. having E value more positive than 0.7 V) $Fe^{3+}$ ions are the major species and below which $Fe^{2+}$ ions are the major species.

**2.** The following reaction may be considered.

$$\underset{(Fe = +3)}{Fe^{3+}(aq)} + \underset{(H = +1)}{3H_2O(l)} \longrightarrow \underset{(Fe = +3)}{Fe(OH)_3(s)} + \underset{(H = +1)}{3H^+(aq)}$$

Since there is no electron involved and hence no change in O.S. of any of the elements in this reaction, this reaction is not a redox reaction. In this reaction, since $H^+$ ions are formed, this is a pH-dependent reaction. The value of E decreases continuously with the increase in pH value. At low pH value, $Fe^{3+}(aq)$ ions are the main species while at high pH value, $Fe^{3+}(aq)$ ions are precipitated as $Fe(OH)_3(s)$. The vertical line (BC) at pH ~ 2 is the boundary line for $Fe^{3+}(aq)$ ions and $Fe(OH)_3(s)$. To the left of this boundary line (*i.e.* below pH ~ 2) $Fe^{3+}(aq)$ ios are the major species and to the right of this boundary line (i.e. above pH ~ 2), $Fe(OH)_3(s)$ is the main species.

**3.** Reduction $Fe(OH)_3(s)$ to $Fe^{2+}(aq)$ can be considered.

$$Fe(OH)_3(s) + 3H^+(aq) + e^- \longrightarrow Fe^{2+}(aq) + 3H_2O(l)$$

Since $H^+$ ions are involved in this reaction, it is a pH-dependent redox reaction. The variation of E with change in pH is given by Nernst equation.

$$E = E° - 0.0591 \log \frac{[Fe^{2+}]}{[H^+]^3}$$

or $$E = E° - 0.0591 \log [Fe^{2+}] - (3 \times 0.0591) \frac{1}{[H^+]}$$

or $$E = E° - 0.0591 \log [Fe^{2+}] - 0.1773\ pH$$

As pH increases (i.e. the medium gets more and more basic), the potential continues to drop. The slopy line (BD) is the boundary line for Fe $(OH)_3$ (*s*) and $Fe^{2+}$ (*aq*). This line drops linearly above pH~ 2. Fe $(OH)_3$ (*s*) is he major species above BD line and $Fe^{2+}$ (*aq*) is the major species below BD line.

**4.** Let us consider the following reaction.

$$Fe^{2+} (aq) + 2H_2O (l) \longrightarrow Fe(OH)_2 (s) + 2H^+ (aq)$$

It is not a redox reaction, since no electrons are involved in this reaction. Since $H^+$ ions are liberated in this reaction, this reaction is a pH-dependent reaction. The vertical line (ED) at pH = 9 is the boundary line for $Fe^{2+}$ (*aq*) and Fe $(OH)_2$(*s*). In the left hand side region of this boundary line (*i.e.* at pH less than 9), $Fe^{2+}$(*aq*) ions are the major species while in the right hand side region of this boundary line (*i.e.* at pH greater than 9) $Fe(OH)_2$(*s*) is the major species.

**5.** We may consider the reaction.

$$Fe(OH)_3 (s) + H^+ (aq) + e^- \longrightarrow Fe(OH)_2 (s) + H_2O (l)$$

This reaction is a pH-dependent reaction in which both $Fe(OH)_3$ (*s*) and Fe $(OH)_2$ (*s*) are stable at pH > 9. The variation of E with the change in pH is given by Nernst equation.

$$E = E° - 0.0591 \text{ pH}$$

The stability of one species over the other depends on E value. Since $H^+$ ion in the above reaction appears on the LHS, the separation of the two species is given by sloping DF line which slopes downwards. This line separates the regions in which Fe $(OH)_2$ (*s*) and Fe $(OH)_3$(*s*) are stable.

In Fig 1.14, two sloping lines, GH (upper line) and IJ (lower line) indicate the boundaries for the stability field of $H_2O$. If E value of a couple is more positive than the upper line (GH), then $H_2O$ will be oxidised to $O_2$ ($2H_2O \rightarrow O_2 + 4H^+ + 4e^-$). On the other hand, if E is less than the lower line (IJ), then $H_2O$ or $2H^+$ will be reduced to $H_2$ [$2H_2O(l) + 2e^- \rightarrow H_2(g) + 2OH^- (aq)$ or $2H^+ (aq) + 2e^- \rightarrow H_2(g)$]. It is clear from the figure that all couples representing redox reactions in water fall within the stability field of water, *i.e.* no species can oxidise or reduce $H_2O$.

## Pourbaix Diagram for Manganese

Pourbaix diagram for Mn is given in Fig. 1.15 This diagram shows the stability region for the species of Mn such as $MnO_4^-$, $MnO_2$, $Mn_2O_3$, $Mn(OH)_2$, $Mn^{2+}$ and elemental Mn.

Let us consider the following reactions:

**1.** The following reaction can be considered

$$MnO_4^- + 2H_2O + 3e^- \longrightarrow MnO_2 + 4OH^-, E° = +0.588 \text{ V}$$

The slanting line (AB) shown at the top of the diagram is the boundary line for $MnO_4^-$ and $MnO_2$. Potential (E) drops sharply with the increasing pH showing the instability of $MnO_4^-$ ion.

**2.** Let us consider the reaction.

$$Mn^{2+} + 2e^- \longrightarrow Mn, E° = -1.18 \text{ V}$$

Since this reaction does not involve any $H^+$ ions, this reaction is a pH independent reaction. Horizontal line (CD) is the boundary line for $Mn^{2+}$ and Mn. This line begins at E $\simeq$ –1.2 V and remains horizontal to pH upto ~7.5. Thus the horizontal line, CD has constant value of E for $Mn^{2+}$ / Mn couple.

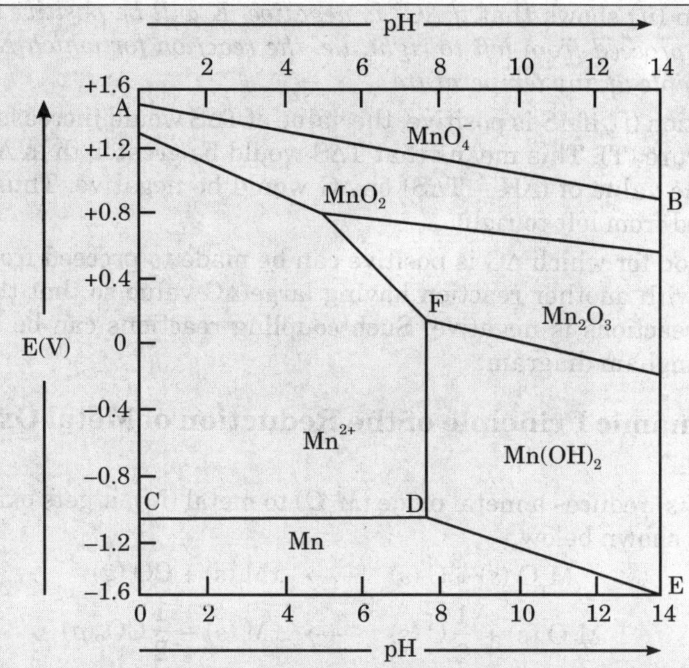


**Fig. 1.15** Pourbaix diagram showing predominance areas of different forms of manganese with varying pH and potential.

**3.** Let us consider the reaction

$$Mn\,(OH)_2 + 2e^- \longrightarrow Mn + 2OH^-, E° = -1.565\ V$$

The potential value of CD line drops sharply with increasing pH and CD line becomes sharply slanting line (DE). At higher pH, Mn $(OH)_2$ becomes the major species. Thus, DE line is the boundary line for Mn $(OH)_2$ and Mn.

**4.** The following reaction can be considered.

$$Mn\,(OH)_2\,(aq) \longrightarrow Mn^{2+}\,(aq) + 2OH^-\,(aq)$$

The diagram shows that $Mn^{2+}$ is stable over a wide range of pH and potential range. The vertical line (DF) at pH ~7.5 indicates that Mn $(OH)_2$ and $Mn^{2+}$ exist in equilibrium.

**5.** Let us consider the following reaction.

$$Mn_2O_3\,(s) + 3H_2O + 2e^- \longrightarrow 2Mn\,(OH)_2\,(s) + 2OH^-\,(aq), E° = -0.234\ V$$

It may be seen from the diagram that at pH greater than 7.5, $Mn_2O_3$ remains in equilibrium with Mn $(OH)_2$.

## Thermodynamic Principles of Metallurgy

Gibbs free energy change (ΔG), entropy change (ΔS) at a temperature (T) for a reaction are related as:

$$\Delta G = \Delta H - T\,\Delta S \qquad \ldots (i)$$

ΔG° and equilibrium constant (K) of a reaction are related as:

$$\Delta G° = -\,RT \log K$$

or $$-\Delta G° = RT \log K \qquad \ldots (ii)$$

Equation (*ii*) shows that *if* $\Delta G°$ *is negative, K will be positive and hence the reaction will proceed from left to right, i.e. the reaction for which* $\Delta G°$ *is negative would be feasible at any temperature.*

In equation (*i*), if $\Delta S$ is positive, the value of $T\Delta S$ would increase on increasing the temperature (T). This means that $T\Delta S$ would be greater than $\Delta H$ ($T\Delta S > \Delta H$) and hence the value of ($\Delta H - T\Delta S$) or $\Delta G$ would be negative. Thus, the reaction would proceed from left to right.

A reaction for which $\Delta G$ is positive can be made to proceed from left to right by coupling with another reaction having large $\Delta G$ value so that the value of $\Delta G$ for the two reactions is negative. Such coupling reactions can be understood by studying Ellingham diagram.

## Thermodynamic Principle of the Reduction of Metal Oxides by C(s) and CO(g)

When C(*s*) reduces a metal oxide ($M_xO$) to metal (M), it gets oxidised to CO(*g*) or $CO_2$ (*g*) as shown below.

$$M_xO\,(s) + C\,(s) \longrightarrow xM\,(s) + CO\,(g) \qquad \ldots (i)$$

$$M_xO\,(s) + \frac{1}{2}C\,(s) \longrightarrow xM\,(s) + \frac{1}{2}CO_2(g) \qquad \ldots (ii)$$

Both these reactions are redox reactions

Oxidation of C(*s*) to CO (*g*) and that of M(*s*) to $M_xO$ (*s*) can be represented by the following oxidation reactions.

$$C(s) + \frac{1}{2}O_2\,(g) \longrightarrow CO\,(g),\ \Delta G° = \Delta_f G°\,(C \rightarrow CO) \qquad \ldots (a)$$

$$xM\,(s) + \frac{1}{2}O_2\,(g) \longrightarrow M_xO\,(s),\ \Delta G° = \Delta_f G°\,(M \rightarrow M_xO) \qquad \ldots (b)$$

Subtract equation (*b*) from equation (*a*) to get equation (*i*)

$$C(s) + M_xO\,(s) \longrightarrow xM(s) + CO(g) \qquad \ldots (i)$$

$$\Delta G° = \Delta_r G° = \Delta_f G°\,(C \rightarrow CO) - \Delta_f G°\,(M \rightarrow M_xO)$$

Now we know that reaction (*i*) would be feasible only when the value of $\Delta_r G°$ or [$\Delta_f G°$ (C $\rightarrow$ CO) – $\Delta_f G°$ (M $\rightarrow$ $M_xO$)] is negative. Since $\Delta_f G°$ (C $\rightarrow$ CO) and $\Delta_f G°$ (M $\rightarrow$ $M_xO$) both have negative values, in order to make the value of $\Delta_r G°$ negative the value of $\Delta_f G°$ (C $\rightarrow$ CO) should be more negative than the value of $\Delta_f G°$ (M $\rightarrow$ $M_xO$).

Oxidation of C(*s*) to $CO_2$ (*g*) and that of M(*s*) to $M_xO$ (*s*) can be represented by the following oxidation reactions.

$$\frac{1}{2}C(s) + \frac{1}{2}O_2(g) \longrightarrow \frac{1}{2}CO_2\,(g),\ \Delta G° = \Delta_f G°\,(C \rightarrow CO_2) \qquad \ldots (c)$$

$$xM\,(s) + \frac{1}{2}O_2(g) \longrightarrow M_xO(s),\ \Delta G° = \Delta_f G°\,(M \rightarrow M_xO) \qquad \ldots (b)$$

Subtract equation (b) from equation (c) to get equation (*ii*)

$$\frac{1}{2}C(s) + M_xO\,(s) \longrightarrow xM(s) + \frac{1}{2}CO_2\,(g) \qquad \ldots (ii)$$

$$\Delta G° = \Delta_r G° = \Delta_f G°\,(C \rightarrow CO_2) - \Delta_f G°\,(M \rightarrow M_xO)$$

Obviously in order to make the value of $\Delta_r G°$ negative, the value of $\Delta_f G°$ ($C \rightarrow CO_2$) should be more negative than the value of $\Delta_f G°$ ($M \rightarrow M_xO$).

When CO (*g*) reduces a metal oxide ($M_xO$) to metal (M), it gets oxidised to $CO_2(g)$ as shown below.

$$M_xO(s) + CO(g) \longrightarrow xM(s) + CO_2(g) \qquad ...(iii)$$

It is also a redox reaction.

Oxidation of CO (*g*) to $CO_2(g)$ and that of M (*s*) to $M_xO$ (*s*) can be represented by the following oxidation reactions.

$$CO(g) + \frac{1}{2}O_2(g) \longrightarrow CO_2(g),\ \Delta G° = \Delta_f G°(CO \rightarrow CO_2) \qquad ...(d)$$

$$xM(s) + \frac{1}{2}O_2(g) \longrightarrow M_xO(s),\ \Delta G° = \Delta_f G°(M \rightarrow M_xO) \qquad ...(b)$$

Subtract equation (*b*) from equation (*d*) to get equation (*iii*)

$$CO(g) + M_xO(s) \longrightarrow xM(s) + CO_2(g) \qquad ...(iii)$$

$$\Delta G° = \Delta_r G° = \Delta_f G°(CO \rightarrow CO_2) - \Delta_f G°(M \rightarrow M_xO)$$

Obviously in order to make the value of $\Delta_r G°$ negative, the value of $\Delta_f G°$ ($CO \rightarrow CO_2$) should be more negative than the value of $\Delta_f G°$ ($M \rightarrow M_xO$).

Above discussion shows that the redox reactions (*i*), (*ii*) and (*iii*) given as:

$$M_xO(s) + C(s) \longrightarrow xM(s) + CO(g) \qquad ...(i)$$

$$M_xO(s) + \frac{1}{2}C(s) \longrightarrow xM(s) + \frac{1}{2}CO_2(g) \qquad ...(ii)$$

$$M_xO(s) + CO(g) \longrightarrow xM(s) + CO_2(g) \qquad ...(iii)$$

would be feasible only when the value of heat of formation of CO from C [represented as $\Delta_f G°(C \rightarrow CO)$], heat of formation of $CO_2$ from C [represented as $\Delta_f G°$ ($C \rightarrow CO_2$)] and heat of formation of $CO_2$ from CO [represented as $\Delta_f G°$ ($CO \rightarrow CO_2$)] is more negative than the value of heat of formation of $M_xO$ (oxide) from M (metal) [represented as $\Delta_f G°$ ($M \rightarrow M_xO$)].

## Thermodynamic Principle of the Reduction of a Metallic Oxide (BO) by another Metal (A)

Let us consider the following redox reaction

$$\underbrace{A + BO}_{\text{Reactants}} \longrightarrow \underbrace{AO + B}_{\text{Products}} \qquad ...(i)$$

In this reaction A and B are two different metals. BO and AO are the oxides of metals B and A respectively. Metal A reduces BO (oxide of metal B) to metal B and is itself oxidised to its oxide, AO.

The redox reaction (*i*) would be feasible, if free energy change ($\Delta_r G°$) for this reaction is negative. $\Delta_r G°$ is given by:

$\Delta_r G°$ for reaction (*i*) = Free energy of formation of products

– Free energy of formation of reactants.

= [Free energy of formation of oxide (AO)

– Free energy of formation of metal (B)]

– [Free energy of formation of oxide (BO)

– Free energy of formation of metal (A)]

Now since A and B metals are in the elemental state, free energy of formation of both these metals is zero.

$\therefore$ $\Delta_r G°$ for reaction (*i*) = Free energy of formation of AO – Free enegy of formation of BO

$$= (\Delta_f G°)_{AO} - (\Delta_f G°)_{BO}$$

Thus, we can say that if the value of $\Delta_r G°$ or $[(\Delta_f G°)_{AO} - (\Delta_f G°)_{BO}]$ is negative, reaction (*i*) would be feasible. We know that since $(\Delta_f G°)_{AO}$ and $(\Delta_f G°)_{BO}$ both are negative, $\Delta_r G°$ is given by:

$$\Delta_r G° = -(\Delta_f G°)_{AO} - [(-\Delta_f G°)_{BO}] = -(\Delta_f G°)_{AO} + (\Delta_f G°)_{BO}$$

This relation shows that in order to make $\Delta_r G°$ negative, the magnitude of $(\Delta_f G°)_{AO}$ should be greater than $(\Delta_f G°)_{BO}$ or the value of $(\Delta_f G°)_{AO}$ should be more negative than that of $(\Delta_f G°)_{BO}$.

Above discussion shows that the redox reaction,

$$A + BO \rightarrow AO + B$$

is feasible or A can reduce BO to B, if any of the following conditions is fulfilled.

(*i*) $\Delta_r G°$ for the above reaction should be negative.

(*ii*) $[(\Delta_f G°)_{AO} - (\Delta_f G°)_{BO}]$ is negative.

(*iii*) $(\Delta_f G°)_{AO}$ should have more negative value than $(\Delta_f G°)_{BO}$ value. Note that $(\Delta_f G°)_{AO}$ and $(\Delta_f G°)_{BO}$ both have negative values. If negative sign is not considered, then the magnitude of $(\Delta_f G°)_{AO}$ should be greater than the magnitude of $(\Delta_f G°)_{BO}$.

## Solved Examples

**Example 1. The value of $\Delta_f G°$ for the formation of $Cr_2O_3$ is –540 KJmol$^{-1}$ and that of $Al_2O_3$ is –827 KJmol$^{-1}$. Is the reduction of $Cr_2O_3$ possible with Al?**

**Ans.** The reduction of $Cr_2O_3$ with Al can be represented by the equation:

$$Cr_2O_3(s) + 2Al(s) \xrightarrow{\Delta} 2Cr(s) + Al_2O_3(s) \quad ...(i)$$

The above reaction is feasible, if the value of $\Delta_r G°$ for this reaction is negative. Now, let us calculate the value of $\Delta_r G°$ for this reaction.

Write reactions showing the formation of $Cr_2O_3(s)$ and $Al_2O_3$ (*s*) oxides as shown below. Note that in the formation of these oxides only one mole of $O_2$ has been used.

$$\frac{4}{3}Al\,(s) + O_2\,(g) \longrightarrow \frac{2}{3}Al_2O_3\,(s),\ \Delta_f G° = -827\text{ KJ} \quad ...(a)$$

$$\frac{4}{3}Cr\,(s) + O_2\,(g) \longrightarrow \frac{2}{3}Cr_2O_3\,(s),\ \Delta_f G° = -540\text{ KJ} \quad ...(b)$$

Now, subtract equation (*b*) from equation (*a*) so that we may get equation (*i*).

$$\frac{4}{3}Al\,(s) - \frac{4}{3}Cr\,(s) \longrightarrow \frac{2}{3}Al_2O_3\,(s) - \frac{2}{3}Cr_2O_3\,(s)$$

or

$$\frac{4}{3}Al\,(s) + \frac{2}{3}Cr_2O_3\,(s) \longrightarrow \frac{2}{3}Al_2O_3\,(s) + \frac{4}{3}Cr\,(s)$$

or $2Al(s) + Cr_2O_3(s) \longrightarrow Al_2O_3(s) + 2Cr(s), \Delta G_r^\circ = -827 - (-540) = -287$ kJ

Since, the value of $\Delta_r G^\circ$ for reaction (*i*) is negative, reaction (*i*) is feasible *i.e.* the reduction of $Cr_2O_3$ by Al is possible.

This question can also be answered in the following two ways:

(*i*) Since the magnitude of $\Delta_f G^\circ$ of $Al_2O_3$ (= 827 KJ) is greater than the magnitude of $\Delta_f G^\circ$ of $Cr_2O_3$, the given reaction is feasible.

(*ii*) We know that $(\Delta_f G^\circ)_{Al_2O_3} - (\Delta_f G^\circ)_{Cr_2O_3} = -827 - (-540) = -287$ kJ. Since this value is negative, the given reaction is feasible.

**Example 2. With the help of the oxidation reactions:**

$$2Zn(s) + O_2(g) \rightarrow 2ZnO(s), \Delta G^\circ = -360 \text{ KJ mol}^{-1}$$

$$2C(gr) + O_2(g) \rightarrow 2CO(g), \Delta G^\circ = -460 \text{ KJ mol}^{-1}$$

**Predict whether (i) Zn can be oxidised to ZnO by CO (ii) ZnO can be reduced to Zn by graphite.**

**Ans.** (*i*) Oxidation of Zn to ZnO by CO can be represented by the following redox reaction.

$$Zn(s) + CO(g) \longrightarrow ZnO(s) + C(s) \quad \ldots(i)$$

Now, write reactions showing the formation of ZnO (*s*) and CO (*g*) oxides as shown below. Note that in the formation of these oxides only one mole of $O_2$ has been used

$$2Zn(s) + O_2(g) \longrightarrow 2ZnO(s), \Delta_f G^\circ = -360 \text{ KJ mol}^{-1} \quad \ldots(a)$$

$$2C(gr) + O_2(g) \longrightarrow 2CO(g), \Delta_f G^\circ = -460 \text{ KJ mol}^{-1} \quad \ldots(b)$$

Now, subtract equation (*b*) from equation (*a*) so that we may get equation (*i*)

$$2Zn(s) - 2C(gr) \longrightarrow 2ZnO(s) - 2CO(g)$$

or $$2Zn(s) + 2CO(g) \longrightarrow 2ZnO(s) + 2C(gr)$$

or $Zn(s) + CO(g) \longrightarrow ZnO(s) + C(gr), \Delta G_r^\circ = -360 - (-460) = +100 \text{ KJ mol}^{-1}$

Since, $\Delta_r G^\circ$ is positive for the above reaction, the oxidation of Zn (*s*) to ZnO (*s*) by CO (*g*) is not feasible.

(*ii*) The reduction of ZnO to Zn by graphite can be represented by the following redox reaction.

$$ZnO(s) + C(gr) \longrightarrow Zn(s) + CO(g) \quad \ldots(ii)$$

Now, subtract equation (*a*) from equation (*b*) so that we may get equation (*ii*).

$$2C(gr) - 2Zn(s) \longrightarrow 2CO(g) - 2ZnO(s)$$

or $C(gr) + ZnO(s) \longrightarrow Zn(s) + CO(g), \Delta_r G^\circ = -460 - (-360) = -100 \text{ KJ mol}^{-1}$

Since the value of $\Delta G_r^\circ$ for reaction (*ii*) is negative, this reaction is feasible *i.e.* ZnO can be reduced to Zn by graphite.

## Ellingham Diagrams

When standard free energy of formation of metallic oxides (*i.e.* $\Delta_f G^\circ$) per mole of $O_2$ are ploted against absolute temperature (T), we get $\Delta_f G^\circ$ vs T plots which are called **Ellingham diagrams**. Similar diagrams can also be drawn for metallic sulphides and halides, using one mole of sulphur and halogen in each case.

Ellingham diagram ($\Delta_f G°$ vs T plots) showing the formation of some metallic oxides is given in Fig 1.16

Let us study $\Delta G_f°$ during the formation of a metal oxide (MO).

$$2M\,(s) + O_2\,(g) \longrightarrow 2MO\,(s) \quad \ldots(i)$$

Since $O_2$ is a gas and the metal (M) and metal oxide (MO) both are solids, $O_2$ has higher entropy than M and MO. Thus in the above oxidation reaction there would be a decrease in entropy. In other words, ΔS for the above reaction would be negative with increase in temperature (T). Obviously with the increase in temperature (T), the term TΔS in the equation.

$$\Delta G = \Delta H - T\Delta S \quad \ldots(ii)$$

would become more and more negative. Since the term TΔS is subtracted from ΔH in the above equation, *ΔG would become less and less negative or would increase with an increase in temperature (T)*. In other words, $\Delta_f G°$ vs T lines have positive slopes for most of the metal oxides as shown in Fig. 1.16. This means that most of the $\Delta_f G°$ vs T curves for metal oxides slope upwards from left to right.

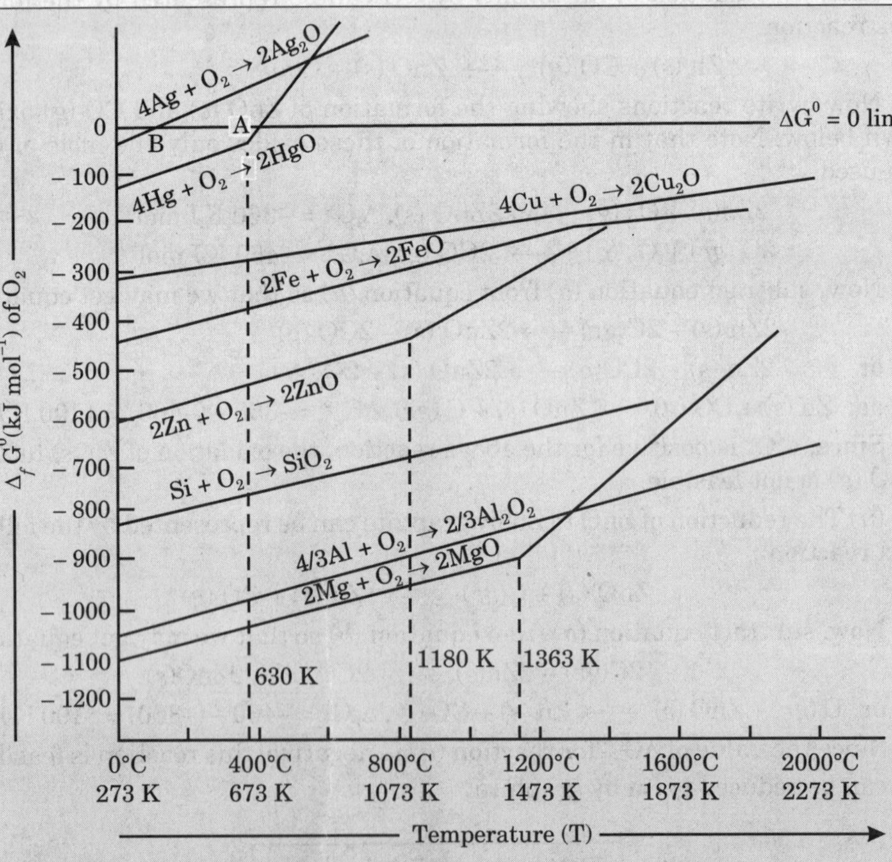


**Fig. 1.16** Ellingham diagram ($\Delta_f G°$ vs T plots) showing the formation of metal oxides of some metals. On *x*-axis (T-axis) temperature (T) becomes more and more positive on moving from left to right while on *y*-axis ($\Delta_f G°$ axis) values of $\Delta_f G°$ become less and less negative on moving from bottom to top.

We have said above that $\Delta_f G°$ vs T plots are straight lines. The plots remain straight lines only upto the temperature at which the metals remain in the solid state. When some change in the solid state of metals takes place (*e.g.* solid $\xrightarrow{\text{Melting}}$ liquid, liquid $\xrightarrow{\text{Vaporisation}}$ gas), abrupt change in the positive slope of the curve (line) takes place. For example, since the b.pts. of Hg, Zn and Mg are 630 K, 1180 K and 1363 K respectively, an abrupt increase in the positive slope of Hg $\rightarrow$ HgO, Zn $\rightarrow$ ZnO and Mg $\rightarrow$ MgO lines takes place at their b.pts. as shown in Fig. 1.16.

Now, if it is assumed that ΔH given in equation (*ii*) does not vary with the change of temperature (T), then the differentiation of equation (*ii*) w.r.t. T gives.

$$\frac{d(\Delta G)}{dt} = -\Delta S \qquad \ldots (iii)$$

This relation shows that if ΔS for an oxidation reaction is negative, then the slope of the line (plot) obtained by plotting $\Delta_f G°$ vs *T* will be positive.

From Ellingham diagram given in Fig. 1.16 the following points may be noted.

(*i*) Hg $\rightarrow$ HgO and Ag $\rightarrow$ $Ag_2O$ lines cross the $\Delta G° = 0$ line at A and B points. The temperatures corresponding to these points are easily attainable. Therefore, these oxides decompose easily to give Hg and Ag and $O_2$ respectively. Thus, Hg and Ag can be obtained simply by thermal decomposition of HgO and $Ag_2O$ respectively.

(*ii*) At any given temperature, metals for which free energy of formation ($\Delta_f G°$) of their oxides is more negative can reduce those metal oxides for which $\Delta_f G°$ is less negative. In other words, at any given temperature, any metal can reduce the oxide of other metals which lie above it in the Ellingham diagram.

Let us consider the following redox reaction

$$A + BO \longrightarrow AO + B \qquad \ldots (iv)$$

In this reaction metal (A) reduces BO (oxide of metal, B) to metal B and is itself oxidised to its oxide (AO). This reaction is feasible only when A $\rightarrow$ AO line lies below B $\rightarrow$ BO line in Ellingham diagram.

**Examples:** (*i*) Mg can reduce $Al_2O_3$, ZnO, FeO and $Cu_2O$ to their respective metal, since Mg $\rightarrow$ MgO curve lies below each of the curves *viz* Al $\rightarrow$ $Al_2O_3$ Zn $\rightarrow$ ZnO, Fe $\rightarrow$ FeO and Cu $\rightarrow$ $Cu_2O$ as is evident from Ellingham diagram given in Fig. 1.16.

(*ii*) Al can reduce ZnO, FeO and $Cu_2O$ to their respective metals, since Al $\rightarrow$ $Al_2O_3$ curve lies below each of the curves namely Zn $\rightarrow$ ZnO, Fe $\rightarrow$ FeO and Cu $\rightarrow$ $Cu_2O$ curves as is evident from Fig. 1.16.

(*iii*) Zn can reduce FeO and $Cu_2O$ to Fe and Cu respectively, since Zn $\rightarrow$ ZnO curve lies below Fe $\rightarrow$ FeO and Cu $\rightarrow$ $Cu_2O$ curves both as is evident from Fig. 1.16.

Examples given above show that the tendency of Mg, Al, Zn and Fe to reduce the oxides of other metals is in the order: Mg > Al > Zn > Fe > Cu.

## Limitation of Ellingham Diagrams

(*i*) Ellingham diagrams simply suggest whether the reduction of metal oxide by a given RA (e.g. other metal, C or CO) is possible or not. These diagrams do not tell anything about the rate at which the reduction reactions take place.

(*ii*) These diagrams are based on the assumption that the reactants and products are in equilibrium. But, this is not always true because the reactants/products may be solid.

## Reducing Behaviour of Carbon and CO : Extraction of Iron

When carbon reduces a metal oxide to metal, it gets oxidised to $CO_2$ or CO as shown below.

$$\underset{1\text{ mole}}{C(s) + O_2(g)} \longrightarrow \underset{1\text{ mole}}{CO_2(g)} \qquad \ldots(i)$$

$$2C(s) + \underset{1\text{ mole}}{O_2(g)} \longrightarrow \underset{2\text{ moles}}{2CO(g)} \qquad \ldots(ii)$$

CO formed in reaction (*ii*) may react with $O_2$ to form $CO_2$.

$$\underbrace{\underset{2\text{ moles}}{\text{«««««}g} + \underset{1\text{ mole}}{{}_2\,g}}_{3\text{ moles}} \longrightarrow \underset{2\text{ moles}}{{}_2\,g} \qquad \ldots(iii)$$

In reaction (*i*), since one mole of $CO_2$ (*g*) is produced from one mole of $O_2(g)$, the entropy change ($\Delta$S) for this reaction is practically **zero** and hence $\Delta G°$ for this reaction remains independent of temperature (T). As a result, $\Delta G°$ *vs* T line for $C \rightarrow CO_2$ oxidation reaction is a **horizontal line** as shown in Fig. 1.17. The horizontal line is parallel to temperature axis.

In reaction (*ii*), since two moles of CO (*g*) are produced from one mole of $O_2$ (*g*) this reaction is accompanied by an increase in entropy, i.e. entropy change ($\Delta$S) for this reaction is **positive.** Hence, $\Delta G (= \Delta H - T\Delta S)$ goes on becoming more and more negative as temperature increases. As a result, $\Delta G°$ *vs* T line for $C \rightarrow CO$ reaction slopes downwards as shown in Fig. 1.17. In other words, we can say that the slope of the line for $C \rightarrow CO$ reaction is **negative.**

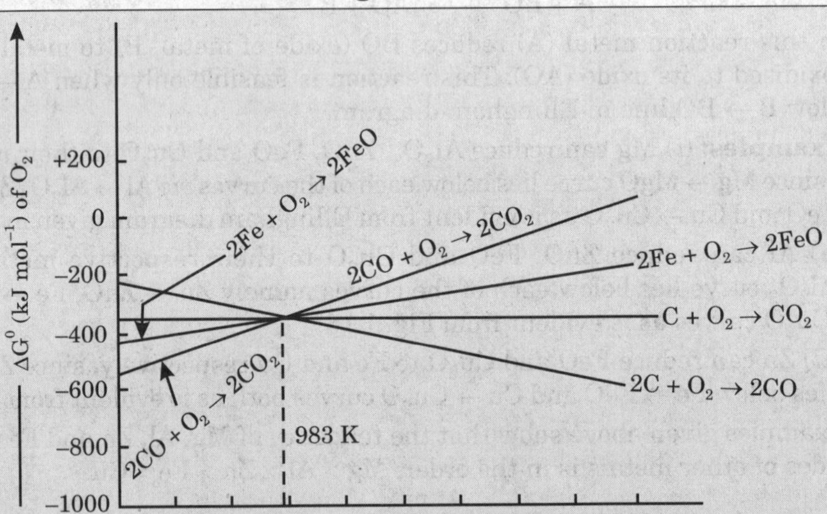


**Fig 1.17** Ellingham diagram showing the formation of CO ($2C + O_2 \rightarrow 2CO$), $CO_2$ ($C + O_2 \rightarrow CO_2$ and $2CO + O_2 \rightarrow 2CO_2$) and FeO ($2Fe + O_2 \rightarrow 2FeO$)

In reaction (*iii*), since two moles of $CO_2$ (*g*) are produced from three moles of reactants, $\Delta S$ for this reaction is **negative** and hence $\Delta G$ goes on becoming more and more positive as the temperature increases. As a result, $\Delta G°$ *vs* T line for $CO \rightarrow CO_2$ reaction slopes **upwards** as shown in Fig 1.17 In other words, we can say that the slope of the line for $CO \rightarrow CO_2$ reaction is **positive**.

In all the above three reactions, since C and CO are oxidised, these species can be used for reducing metal oxides to their corresponding metals. Out of C and CO which is a better RA depends on the temperature at which the reduction takes place It may be seen from Fig. 1.17 that $C \rightarrow CO_2$, $C \rightarrow CO$ and $CO \rightarrow CO_2$ curves cross each other at about 983 K. Below 983 K, since the oxidation of C to $CO_2$ ($C + O_2 \rightarrow CO_2$) and of CO to $CO_2$ ($2CO + O_2 \rightarrow 2CO_2$) is energetically more favourable than the oxidation of C to CO, below this temperature C and CO both can easily be oxidised to $CO_2$ and hence C and CO both can be used to reduce metal oxides. Now, since below 983 K,CO can more easily be oxidised to $CO_2$ than C can be oxidised to $CO_2$, below this temperature CO acts as a more effective reducing agent than C. Above 983 K, since the oxidation of C to CO ($2C + O_2 \rightarrow 2CO$) is more favourable than the oxidation of C to $CO_2$ and CO to $CO_2$, above this temperature C is a better RA than CO.

Fig. 1.17 also contains $Fe \rightarrow FeO$ curve. Below 983 K, since $CO \rightarrow CO_2$ curve lies below $Fe \rightarrow FeO$ curve, in the temperature range 500 – 900 K (this temperature range is below 983 K) in blast furnace, CO reduces haematite ($Fe_2O_3$) and magnetite ($Fe_3O_4$) (these are iron ores) to FeO and CO is oxidised to $CO_2$.

$$\underset{\text{Haematite}}{Fe_2O_3}\ (s) + 3CO\ (g) \xrightarrow{500\text{–}900K} 2FeO\ (s) + CO_2(g)$$

$$\underset{\text{Magnetite}}{Fe_3O_4}\ (s) + CO(g) \xrightarrow{500\text{–}800K} 3FeO\ (s) + CO_2\ (g))$$

FeO (*s*) obtained in the above reactions is reduced to Fe (*s*) by CO and CO is oxidised to $CO_2$.

$$FeO\ (s) + CO\ (g) \xrightarrow{1123K} Fe\ (s) + CO_2\ (g)$$

Since the downward sloping $C \rightarrow CO$ curve will eventually cross and will lie below all metal → metal oxide curves at high temperature (> 983 K), carbon can reduce all metal oxides to metal and carbon itself will be oxidised to CO. This implies that carbon can be used to reduce any metal oxide at high temperature. Sometimes it is difficult to attain high temperatures. Moreover, some extracting metals form carbides with carbon at high temperatures.

Fig. 1.17 shows that above 983 K, since $C \rightarrow CO$ curve lies below $Fe \rightarrow FeO$ line, carbon can reduce FeO to Fe and itself is oxidised to CO.

$$FeO\ (s) + C(s) \xrightarrow{>983K} Fe\ (s) + CO\ (g)$$

The reduction of FeO by C as represented by the above equation is feasible, since $\Delta_r G°$ or $[(\Delta_f G°)_{CO} - (\Delta_f G°)_{FeO}]$ for the above reaction is negative. Negative value of $\Delta_r G°$ is because of the fact that the magnitude of $(\Delta_f G°)_{CO}$ is greater than that of $(\Delta_f G°)_{FeO}$. Negative sign associated with these terms has not been considered.

**Extraction of Zn from ZnO.**

Ellingham diagram given in Fig. 1.18 shows that Zn → ZnO curve lies below all the three curves namely C → CO, C → $CO_2$ and CO → $CO_2$ until the boiling point (b.pt.) of Zn (= 1180 K) is reached. Above 1180 K, ΔG° for C → CO reaction decreases (becomes more negative) while that for Zn → ZnO reaction increases (becomes less negative) very rapidly. Zn → ZnO curve intersects C → $CO_2$ curve (C → $CO_2$ curve is a horizonal line parallel to temperature axis) at 1270 K, This means that above 1270 K, ΔG° for Zn → ZnO reaction is higher (less negative) than that for C → CO reaction, *i.e.* above 1270 K, C → CO curve lies below Zn → ZnO curve. Hence, above 1270 K, ZnO is essily reduced to Zn by coke. In this reaction carbon is oxidised to CO.

$$ZnO(s) + C(s) \xrightarrow{>1270\ K} Zn(s) + CO(g)$$

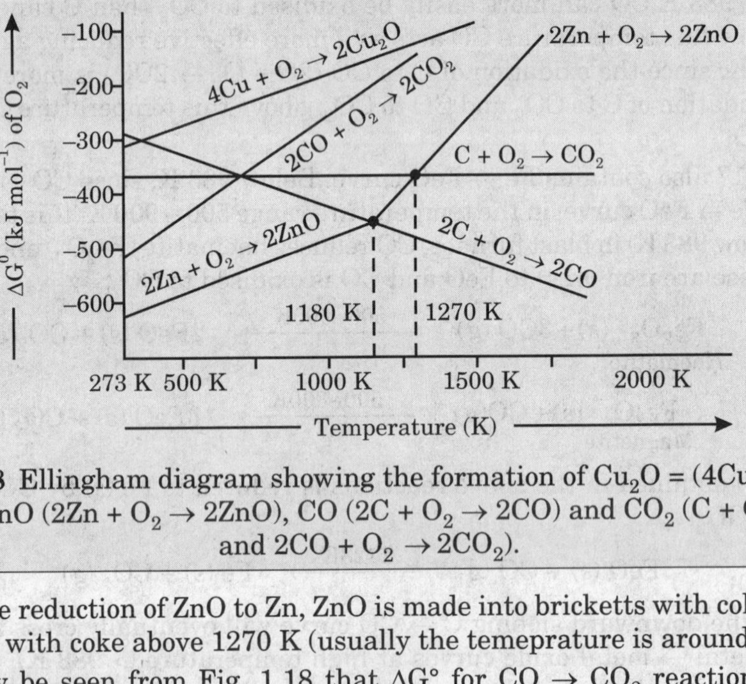


**Fig. 1.18** Ellingham diagram showing the formation of $Cu_2O$ = ($4Cu + O_2 \rightarrow 2Cu_2O$), ZnO ($2Zn + O_2 \rightarrow 2ZnO$), CO ($2C + O_2 \rightarrow 2CO$) and $CO_2$ ($C + O_2 \rightarrow CO_2$ and $2CO + O_2 \rightarrow 2CO_2$).

For the reduction of ZnO to Zn, ZnO is made into bricketts with coke and clay and heated with coke above 1270 K (usually the temepratures is around 1673 K)

It may be seen from Fig. 1.18 that ΔG° for CO → $CO_2$ reaction is higher (less negative) than that for Zn → ZnO reaction, *i.e.* CO → $CO_2$ curve lies above Zn → ZnO curve. This means that CO cannot be used for reducing ZnO to Zn.

Since boiling point of Zn is low (= 1180 K), Zn will be produced in the vapour form at a temperature around 1275 K.

The fact that ZnO can be reduced to Zn by carbon can be explained as follows on the basis of free energy change ($\Delta_r G°$) in the following redox reaction

$$ZnO(s) + C(s) \rightarrow Zn(s) + CO(g)$$

This reaction is freasible if $\Delta_r G°$ for this reaction is negative. We know that $\Delta_r G°$ for this reaction is given by:

$$\Delta_r G° = \Delta_f G° \text{ of products} - \Delta_f G° \text{ of reactants}$$

$$= [\Delta_f G° \text{ of } Zn(s) + \Delta_f G° \text{ for } CO(g)] - [\Delta_f G° \text{ of } ZnO(s) + \Delta_f G° \text{ of } C(s)]$$

Since the values of $\Delta G°$ for CO (*g*) and ZnO (*s*) are equal to – 460 KJ and –360 KJ resepectively,

$$\Delta_r G° = [0 + (-460)] - [-360 + 0]$$

$$= -460 + 360 = -\mathbf{100\ KJ}$$

Since the value of $\Delta_r G°$ is negative (= –100 KJ), the redox reaction given above is feasible.

## Extraction of Cu from $Cu_2O$

The Ellingnam diagram given in Fig. 1.18 shows that $Cu \rightarrow Cu_2O$ curve lies almost at the top while $CO \rightarrow CO_2$ and $C \rightarrow CO$ curves lie much below it particularly in the temperaure range 500 – 600 K. This means that it is easy to reduce $Cu_2O$ to metallic Cu by heating the oxide with coke. In this reaction C is oxidised to CO.

$$Cu_2O + C \longrightarrow 2Cu + CO$$

## Extraction of Non-Metals by Oxidation

Non-metals occur in nature in the combined state in the reduced form. These elements are, therefore, generally extracted by the oxidation of their compounds. For example:

(*i*) $Cl_2$ is isolated by the oxidation of brine (NaCl) in water.

$$\underset{(Cl = -1)}{2Cl^-(aq)} + \underset{(H = +1)}{2H_2O\ (l)} \xrightarrow{\text{Oxidation}} \underset{«N«««««}{2OH^-(aq)} + \underset{\substack{\text{(At anode)} \\ (H = 0)}}{H_2(g)} + \underset{\substack{\text{(At cathode)} \\ (Cl = 0)}}{Cl_2(g)} \quad ...(i)$$

The value of $\Delta G_r°$ for this reaction is equal to +422 KJ $mol^{-1}$. Since this value is positive, the oxidation of $Cl^-(aq)$ to $Cl_2$ (*g*) cannot be carried out by ordinary chemical method. The oxidation is accomplished by electrolysis. In the oxidation of $Cl^-(aq)$ to $Cl_2$ (*g*) two electrons are involved [$2Cl^-(aq) \rightarrow Cl_2\ (g) + 2e^-$]. Thus $n = 2$. The potential (E°) required for the oxidation of $Cl^-$ (*aq*) to $Cl_2$ (*g*) is given by:

$$\Delta G° = -nFE°$$

or $$422 \times 1000\ J = -2 \times 96500 \times E°$$

or $$E° = -2.86\ V$$

However, due to over voltage, actual electrolysis requires an external emf which is greater than 2.86 V. During electrolysis $Cl_2$ is liberated at cathode and $H_2$ is obtained at anode. NaOH is obtained in solution.

(*ii*) Electrolysis of $H_2O$ gives $O_2$ and $H_2$.

(*iii*) Electrolysis of aqueous fluoride solution gives $O_2$ rather than $F_2$. Hence, $F_2$ is obtained by the electrolysis of fused mixture of anhydrous KF and HF.

(*iv*) Heavier halogens ($Br_2$ and $I_2$) are obtained by the action of $Cl_2$ on aqueous bromide / iodide solution.

$$2KBr + Cl_2 \longrightarrow 2KCl + Br_2$$

$$2KI + Cl_2 \longrightarrow 2KCl + I_2$$

## Principles Involved in the Extraction of Elements from Their Ores.

The methods used for the extraction of elements from their ores depends mainly on the properties of the elements to be extracted and the chemical composition of the ore. The ores which are mostly used are oxides, carbonates, sulphides, sulphates, halides etc. Let us discuss the metallurgical principles for the extraction of metals from the above said ores.

**1. Oxide ores.** Metals like Fe, Zn, Sn, Bi, Cr, Mn etc occur as oxides in nature. These metals are obtained by reducing their oxide ores by suitable reducing agents. For example:

(*i*) $Fe_2O_3 + 3C$ (*charcoal*) $\xrightarrow{\Delta}$ $2Fe + 3CO$

$$ZnO + C \xrightarrow{\Delta} Zn + CO$$

$$SnO_2 + 2C \xrightarrow{\Delta} Sn + 2CO$$

$$Bi_2O_3 + 3C \xrightarrow{\Delta} 2Bi + 3CO$$

(*ii*) The metals which have a tendency to react with carbon at the reduction temperature are obtained by reducing their oxides with Al powder (Gold Schmidth - Thermite process). For example:

$$Cr_2O_3 + 2Al \longrightarrow 2Cr + Al_2O_3$$

$$3Mn_3O_4 + 8Al \longrightarrow 9Mn + 4Al_2O_3$$

(*iii*) Al can be obtained by the electrolytic reduction of alumina ($Al_2O_3$). Since the melting point of $Al_2O_3$ is very high (= 2210°C), $Al_2O_3$ is not electrolysed directly at such a high temperature, but a solution of $Al_2O_3$ in molten cryolite ($3NaF \cdot AlF_3$ or $N_3AlF_6$) (m.p.t. = 1000° C) is electrolysed. The reactions taking place during electrolysis are given below.

$$Al_2O_3\ (Solution) \xrightarrow{\text{Ionisation}} 2Al^{3+} + 3O^{2-}$$

$$2Al^{3+} + 6e^- \xrightarrow[\text{at cathode}]{\text{Reduction}} 2Al$$

$$3O^{2-} \xrightarrow[\text{at anode}]{\text{Oxidation}} \frac{3}{2}O_2 + 6e^-$$

**2. Carbonate ores.** Carbonate ores are first heated to convert them into oxides. Oxides thus formed are reduced to metals by suitable reducing agents as discussed above. For example:

$$ZnCO_3 \xrightarrow{\Delta} ZnO + CO_2$$

$$PbCO_3 \longrightarrow PbO + CO_2$$

$$FeCO_3 \longrightarrow FeO + CO_2$$

**3. Sulphide ores.** Sulphide ores generally have the impurities of S, As, Sb etc. Such ores are heated in air to convert them into oxide which is then reduced with C or Al to get the metal. For example:

(*i*) *Zinc blende (ZnS)* is heated in supply of air.

$$2ZnS + 3O_2 \longrightarrow 2ZnO + 2SO_2$$

$$ZnO + C \longrightarrow Zn + CO$$

(*ii*) *Copper pyrites ($CuFeS_2$) ore* is roasted (oxidation) in a reverberatory furnace when the ore is converted into a mixture of FeS and $Cu_2S$ (copper glance)

$$2CuFeS_2 + O_2 \xrightarrow{\text{Roasting}} \underbrace{2FeS + Cu_2S}_{\text{Mixture}} + SO_2 \quad \ldots (i)$$

FeS and $Cu_2S$ thus obtained are practially oxidised to FeO and $Cu_2O$

$$2FeS + 3O_2 \longrightarrow 2FeO + 2SO_2$$

$$2Cu_2S + 3O_2 \longrightarrow 2Cu_2O + 2SO_2$$

FeO is removed by forming ferrous silicate, $FeSiO_3$(slag) with silica ($SiO_2$)

$$FeO + SiO_2 \longrightarrow FeSiO_3$$

$Cu_2O$ reacts with unroasted (unreacted) $Cu_2S$ to give metallic Cu,

$$2Cu_2O + Cu_2S\ (\textit{unreacted}) \longrightarrow 6Cu + SO_2$$

$Cu_2O$ can also be reduced to Cu by carbon,

$$Cu_2O + C \longrightarrow 2Cu + CO$$

Here, it may be noted that $Cu_2S$ formed in reaction (*i*) cannot be directly reduced to Cu by C or $H_2$. This means that the following reactions are not feasible.

$$\left.\begin{array}{l} 2Cu_2S + C \rightarrow 4Cu + CS_2 \\ Cu_2S + H_2 \rightarrow 2Cu + H_2S \end{array}\right\} \text{These reactions are not feasible}$$

**Explanation.** Since the value of $\Delta_f G°$ for the formation of $CS_2$ and $H_2S$ (products) are not more negative than the value of $\Delta_f G°$ for formation of $Cu_2S$ (reactant), above reactions can not proceed from left to right.z(*iii*) *Galena (PbS)* is heated in air at 500 – 600°C so that some PbS is converted into PbO and some PbS is converted into $PbSO_4$.

$$2PbS + 3O_2 \longrightarrow 2PbO + 2SO_2$$

$$PbS + 2O_2 \longrightarrow PbSO_4$$

Now, excess of galena (PbS) is added to the charge. Temperature is increased so that PbO and $PbSO_4$ formed as above combine with the galena (PbS) to give Pb.

$$2PbO + PbS \longrightarrow 3Pb + SO_2$$

$$PbSO_4 + PbS \longrightarrow 2Pb + 2SO_2$$

(*iv*) *Cinnabar (HgS)* is heated in a current of air to get Hg.

$$HgS + O_2 \xrightarrow{\Delta} Hg + SO_2$$

**4. Halide ores.** (*i*) Water soluble ionic chlorides of alkali metals and alkaline earth metals are very stable and hence do not decompose easily to give metals. Such chlorides in their fused state are, therefore, electrolysed to get the metals at cathode.

$$NaCl \xrightarrow{\text{Ionisation}} Na^+ + Cl^-$$

$$2Cl^- \longrightarrow Cl_2\ (\textit{At anode}) + 2e^-$$

$$2Na^+ + 2e^- \longrightarrow 2Na\ (\textit{At cathode})$$

(*ii*) Insoluble chlorides like that of Ag, horn silver (AgCl) (ore of Ag) is brought into solution by treating it with a dil.solution of NaCN.

$$AgCl + 2NaCN \longrightarrow \underset{\substack{\text{Sod. dicyano-}\\ \text{orgentate (I)}\\ \text{(soluble complex)}}}{Na[Ag(CN)_2]} + NaCl$$

Na $[Ag(CN)_2]$ solution obtained as above is treated with Zn scrap to obtain metallic silver (Ag)

$$2Na[Ag(CN)_2] + Zn \longrightarrow Na_2[Zn(CN)_4] + 2Ag$$

## UNIVERSITY QUESTIONS

**1.** Write notes on: (*i*) Redox potential data and their analysis (*ii*) Latimer diagrams and their utility. (*R.U. 2008*)

**2.** What do you understand by Pourbaix diagram? Discuss various stable oxidation states of Mn (*R.U. 2006*)

**3.** What are Frost diagrams? Discuss the importance of these diagrams with the help of exampless (*MDSU 2006*)

**4.** (*a*) What is disproportionation reaction? How would you determine the disproportionation reaction? Explain with example.
(*b*) Explain Frost diagram. (*Pd. Ravishankar shukla 2010*)

**5.** Give an example of molecule in which the oxidation state of oxygen is +2. (*Himachal Pradesh 2010*)

**6.** Explain why $H_2O_2$ is a better oxidising agent than $H_2O$. (*Kanpur 2010*)

**7.** Distinguish between oxidation number and valency. (*Pune 2009*)

**8.** What is Latimer diagram? Explain with example. (*Barktullah 2010*)

**9.** (*a*) What do you understand by Frost diagram? Explain the utility of Latimer diagram in the construction of a Frost diagram. Interpret different types of points found in a Frost diagram.
(*b*) What is redox potential? How does the reactivity of elements depend upon their redox potentials? (*Rajasthan 2010*)
(*c*) Write short note on the following –
(*i*) Disproportionation and comproportionation (*ii*) Oxidation and reduction by water (*iii*) Latimer diagrams (*iv*) Stability field of water. (*Rajasthan 2010*)

**10.** Choose the correct answer: In oxidation – reduction reaction –
(*a*) Only oxidation performs (*b*) Only reduction performs
(*c*) Both reactions perform (*d*) None of these (*Barkatullah 2010*)

**11.** (*i*) Which of the following metals show more than one oxidation state:
(*a*) Na (*b*) Mg (*c*) Fe (*d*) Al
(*ii*) Valency of sulphur in sulphuric acid is:
(*a*) 2 (*b*) 4 (*c*) 6 (*d*) 8 (*H.N. Bahuguna Garhwal 2010*)

**12.** Write short notes on the following: (*a*) Frost diagram (*b*) Use of redox potential data in extraction of elements (*Ranchi 2010*)

**13.** (A) Explain the extraction of elements by reduction method.
(B) What is Latimer diagram? Explain Latimer diagram of chlorine in acidic solution.
(C) Explain the extraction of elements by oxidation method.
(D) Explain the Pourbaix diagram of manganese.
(E) Draw a Frost diagram for nitrogen under basic condition. (*Nagpur 2010, 2012*)

**14.** Differentiate between Latimer and Frost diagram. (*GND Amritsar 2010*)

**15.** Comment on the oxidation state of *Tl* in $TlI_3$ and $K[TlI_4]$ (*Vidya Sagar 2012*)

# 2 CHAPTER Non-aqueous Solvents

## Introduction

Most of the chemical reactions familiar to us take place in aqueous solution, since water, due to high value of its dielectric constant, has an increased ability to dissolve the ionic compounds and many other substances. However there are many chemical reactions which cannot be carried out in aqueous solution but can be studied in a non-aqueous medium. A large number of non-aqueous solvents have been discovered in the last few years. These solvents have sizeable value of their dielectric constant and hence can dissolve many substances. Examples of such solvents are anhydrous liq. $NH_3$, liq. $SO_2$, anhydrous HF, liq. $N_2O_4$ etc.

## Classification of Solvents.

Solvents have been classified in a number of ways, depending on the properties of the solvents. The most convenient classification of solvents is based on the electrolytic characteristics of the solvents. This classification has been shown below:-

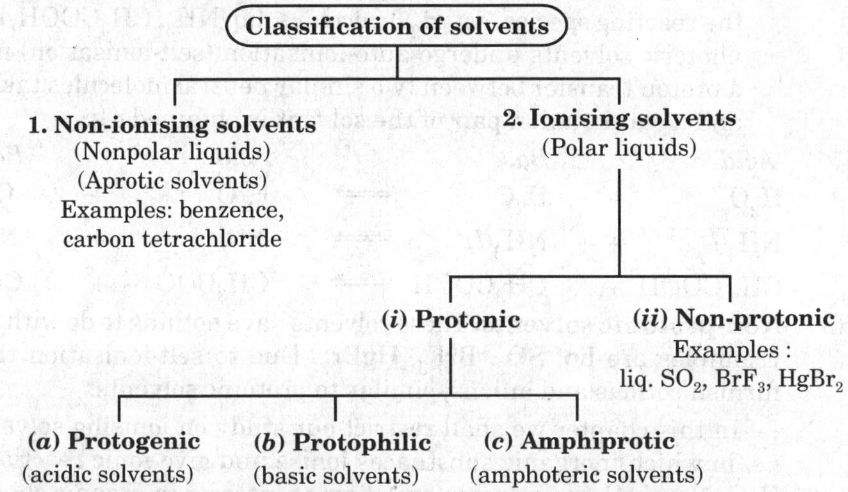


The classification shown above gives the following types of solvents:

1. **Non-ionising solvents (Non-polar solvents)** : These solvents are also called *aprotic solvents (or non-protonic or non-protic solvents)*. These solvents have no hydrogen in their structure. Examples are $C_6H_6$, $CCl_4$, etc.

These solvents have very little dielectric constant. They dissolve non-polar substances.

2. **Ionising solvents (Polar solvents)** Examples of such solvents are $H_2SO_4$, $H_2F_2$, $CH_3$ COOH, $H_2O$, liq. $NH_3$, liq. $SO_2$, pyridine, $BrF_3$, $HgBr_2$, $CH_3OH$, etc. Because of their polar nature, these solvents have associated structures. Non-polar substances do not dissolve in these solvents because of the **squeezing effect** produced by their associated structure. However ionic and covalent polar substances dissolve in them. The dissolution of ionic substances is due to their greater solvation energy than the lattice energy of the salts. Several covalent substances dissolve forming H-bonds.

**Classification of Ionising Solvents**

(*i*) **Protonic solvents:** These solvents have hydrogen in their structure. These may be of the following types:

(*a*) *Proto-genic solvents (Acidic solvents).* These solvents have a tendency to donoate protons. Examples are $H_2SO_4$, $H_2F_2$, $CH_3COOH$. These solvents are also called acidic solvents.

(*b*) *Proto-philic solvents (Basic solvents).* These solvents have a tendency to accept protons. Examples are liq. $NH_3$, pyridine, ethylenedimine, etc. These solvents are also known as basic solvents.

(*c*) *Amphi-protic or amphoteric solvents.* These show dual character *i.e.*, they can lose as well as accept protons, depending on the nature of the reacting species, *e.g.* $H_2O$, alcohols, liq. $NH_3$, $CH_3COOH$, etc. Amphoteric solvents undergo auto-ionisation (self-ionisation) in which a proton transfer between two similar neutral molecules takes place and a cation-anion pair of the solvent is obtained *e.g.*,

| *Acid* | | *Base* | | *Acid* | | *Base* |
|---|---|---|---|---|---|---|
| $H_2O$ | + | $H_2O$ | $\rightleftharpoons$ | $H_3O^+$ | + | $OH^-$ |
| $NH_3(l)$ | + | $NH_3(l)$ | $\rightleftharpoons$ | $NH_4^+$ | + | $NH_2^-$ |
| $CH_3$ COOH | + | $CH_3COOH$ | $\rightleftharpoons$ | $CH_3COO H_2^+$ | + | $CH_3COO^-$ |

(*ii*) **Non-protonic solvents.** These solvents have nothing to do with protons. Examples are liq. $SO_2$, $BrF_3$, $HgBr_2$. Due to self-ionisation they also furnish cotions and anions, similar to protonic solvents.

In this chapter we shall restrict our study on ionising solvents only *i.e.* in which inorganic substnaces ionise and give ionic reactions as in $H_2O$. Non-ionising solvents are of great interest in organic chemistry.

## Physical Properties of Solvents and their Role in Chemical Reactions

Every liquid cannot be used as a solvent in a chemical reaction. Here we, therefore, concentrate ourselves on the properties of a solvent which makes the solvent suitable for using it as a good solvent. Here we shall also compare the properties of some non-aqueous solvents with the properties of $H_2O$.

Some important properties of a solvent which make the solvent a useful solvent in chemical reactins are discussed below.

1. **Melting point and boiling point.** Most of the chemical reactions are carried out in liquid phase. Hence the melting point and boiling point of a solvent indicates the range of temperature within which the solvent can be used. Melting points and boiling points (°C) of some solvents are given below:

| |
|---|
| $H_2O$ = 0.0, 100; $NH_3$ = –77.7, – 33.4; $SO_2$ = –75.5, –10.2; |
| HF = – 89.3, 19.5 and $CH_3$ COOH = 16.16, 118.1. |

These values show that $H_2O$ has a very convenient liquid temperature range between 0 and 100°C. $CH_3COOH$ can act as a solvent at ordinary temperature range (16.16°C to 118.1°C). $NH_3$ and $SO_2$ exist as gases at ordinary temperature and pressure and hence act as solvents only at low temperatures.

2. **Heat of fusion( $\Delta H_{fus}$ ) and heat of vaporisation ( $\Delta H_{vap}$ ).** The heat absorbed by one mole of a substance to change from solid to liquid state is called its molar heat of fusion.

Solid (*One mole*) → Liquid, ΔH = Heat of fusion = + $\Delta H_{fus}$

Similarly, the heat absorbed by one mole of a substance to change from liquid to vapour state is called its molar heat of vaporisation.

Liquid (*One mole*) → Vapour, ΔH = Heat of vaporisation = + $\Delta H_{vap}$

The heats of fusion and vaporisation of a solvent indicate the nature and strength of forces with which the molecules of the solvent are held together in the solid or in the liquid state. A high value of $\Delta H_{vap}$ of a liquid indicates that the intermolecular forces in the liquid are strong. The ratio of $\Delta H_{vap}$(in joules) and boiling point , $T_b$ (in K) of a liquid is a constant which is known as **Trouton constant,** *i.e.*

$$\frac{\Delta H_{vap}\text{(Joules)}}{T_b\text{(K)}} = \text{Troutant constant}$$

For normal liquids the value of Troutant constant is about 90 $JK^{-1}$. Such liquids do not have any bonds between their molecules. Thus these liquids exist as independent molecules.

Value of Troutant constant greater than 90 $JK^{-1}$ for liquid (solvent) indicates that the molecules of the liquid are associated. Since the common solvents like $H_2O$, $NH_3$ and HF have Troutant constant value greater than 90 $JK^{-1}$, these solvents are associated solvents. These liquids are polar. The values of $\Delta H_{fus}$ and $\Delta H_{vap}$ (in kJ $mol^{-1}$) for $H_2O$, $NH_3$, $SO_2$ and HF are given below.

| |
|---|
| $H_2O$ =6.02, 40.65; $NH_3$ = 5.65, 23.34; |
| $SO_2$ = 7.40, 24.93 and HF = 4.58, 30.28. |

Above values show that the values of $\Delta H_{fus}$ for $H_2O$ and $NH_3$ are almost the same. This means that the forces which hold the molecules of $H_2O$ and $NH_3$ are of the same magnitude. Thus these solvents will serve as similar solvents. It may also be noted that since the value of $\Delta H_{fus}$ of $SO_2$ is comparatively high and that of HF is comparatively low, the force holding the molecules of $SO_2$ together is stronger and that holding the molecules of HF is weaker.

3. **Dielectric constant (ε).** Dielectric constant (ε) of a solvent determines the ability of the solvent to dissolve polar and non-polar substances in it. We know that the coulombic force (F) between a cation and an anion of an ionic compound is given by the expression:

$$F = \frac{q_1 q_2}{\varepsilon(r_1 + r_2)^2}$$

In this expression $q_1$ and $q_2$ are the charges on cation and anion respectively, $r_1$ and $r_2$ are the radii of the two ions and ε is the dielectric constant of the solvent. The value of ε depends on the nature of the solvent in which the ionic compound is dissolved. The value of dielectric constant of some solvents are given below.

$H_2O$ = 78.5 (25°C), $NH_3$ = 22.0 (– 33. 5°C),
$SO_2$ = 17.3(–16°C), $CH_3COOH$ = 9.7 (18°C),
HF = 83.6 (0°C), $N_2O_4$ = 2.42 (0°C)

It may be clear from the above expression that if ε of a solvent is large, F would be small *i.e.* if ε is large, small amount of energy would be required to separate the ions and hence it would be easy for a solvent having a high value of ε to dissolve an ionic compound in it. For example, since anhydrous HF and $H_2O$ have high value of ε ,these are the best solvents for ionic compounds. On the other hand, since liq. $NH_3$ and liq. $SO_2$ have low values of ε, these solvents show smaller ability to dissolve ionic compounds especially those containing multi-charged ions. Thus carbonates, sulphates and phosphates which contain multi-charged ions are insoluble in liq. $NH_3$ and liq.$SO_2$.

4. **Dipole moment.** Greater is the polarity of the bond in a solvent molecule, greater is the charge separation and higher will be the value for dipole moment. Substances having high dipole moment values are good solvents for polar solutes. This is because of the fact that greater is the polartiy of a solvent molecule, greater is the solvation energy released on dissolution of a solute. Dipole moment value of a solvent also gives an idea about the extent of association of the molecules of a liquid and hence its liquid range. Dipole moment values (in D) of $H_2O$, $NH_3$ and $SO_2$ are given below:

$H_2O$ = 1.85, $NH_3$ = 1. 47, $SO_2$ = 1.61

5. **Viscosity** : Viscosity gives a measure of the fluidity of the solvent. Solvents like water, carbon tetrachloride have low viscoscities and flow rapidly under ordinary temperature. In solvents of low viscosities, the operations such as precipitation, crystallization, filteration, etc. can be easily carried out without any difficulty. With increasing viscotiy of a liquid, the difficulty of such operations increases. Solvents like anhydrous sulphuric acid have higher viscosities and this reduces their usefulness as solvent, Viscosity of $H_2O$, $NH_3$ and $SO_2$ is 1.00,0. 241 and 0.009 respectivly.

6. **Proton affinity:** It is applicable for protonic solvents only. It greatly affects the behaviour of a solute in a given solvent. $NH_3$ has greater proton affinity than $H_2O$. Hence acetamide ($CH_3\ CO\ NH_2$) which behaves as a very weak base in aqueous solution shows acidic properties in liq. $NH_3$

$$CH_3\ CONH_2 + H_2O \rightleftharpoons CH_3\ CONH_3^+ + OH^-$$

$$CH_3\ CONH_2 + NH_3 \rightleftharpoons CH_3\ CONH^- + NH_4^+$$

Proton affinity (in kJ $mol^{-1}$) for $H_2O$ and $NH_3$ are 760 and 865 respectively. The properties of $H_2O$, $NH_3$ and $SO_2$ other than those mentioned above are given below:

| Property | Solvents | | |
|---|---|---|---|
| | $H_2O$ | $NH_3$ | $SO_2$ |
| Equivalence conductance ($ohm^{-1}$) | $6 \times 10^{-8}$ (25°C) | $5\times10^{-9}$ (133.4°C) | $1\times10^{-7}$ |
| Critical temperature (°C) | 374 | 132.4 | 157.5 |
| Critical pressure | 217.7 | 112.0 | 77.8 |
| Density (g/cc) | 0.96 | 0.68 | 1.46 |

## Types of Chemical Reactions Taking Place in Non-aqueous Solvents

### 1. Metathetical (Precipitation) Reactions

The reactions in which precipitate is formed by mixing two solutions of two compounds are called metathetical or precipitation reactions. Thus precipitation reactions are normally double decomposition. The formation of a precipitate in different solvents depends on the solubilites of the products in those solvents. For example the precipitate of AgCl is obtained by mixing $BaCl_2$ and $AgNO_3$ in aqueous medium.

$$BaCl_2 + 2\ AgNO_3 \xrightarrow{\text{Water}} AgCl \downarrow + Be\ (NO_3)_2 \quad \dots (i)$$

In liq. $NH_3$, the above reaction is reversed, *i.e.*, in liq. $NH_3$ ppt. of $BaCl_2$ is obtained by combining AgCl with Ba $(NO_3)_2$ in liq. $NH_3$.

$$2\ AgCl + Ba\ (NO_3)_2 \xrightarrow{\text{liq. } NH_3} BaCl_2 \downarrow + 2\ Ag\ NO_3$$

Thus we see that reaction (*i*) is reversed on changing the solvent.

### 2. Acid-Base Reactions

We have already said that the ionic solvents are polar compounds and undergo self-ionisation. Self-ionisation of some important solvents is given below:

| *Solvent* | | *Solvent cation (Acid ion)* | | *Solvent anion (Base ion)* |
|---|---|---|---|---|
| $2H_2O$ | $\rightleftharpoons$ | $H_3O^+$ | + | $OH^-$ |
| $2NH_3$ | $\rightleftharpoons$ | $NH_4^+$ | + | $NH_2^-$ |
| $2BrF_3$ | $\rightleftharpoons$ | $BrF_2^+$ | + | $BrF_4^-$ |
| $2HgBr_2$ | $\rightleftharpoons$ | $HgBr^+$ | + | $HgBr_3^-$ |
| $2SO_2$ | $\rightleftharpoons$ | $SO^{2+}$ | + | $SO_3^{2-}$ |
| $2H_2SO_4$ | $\rightleftharpoons$ | $H_3SO_4^+$ | + | $HSO_4^-$ |
| $2CH_3COOH$ | $\rightleftharpoons$ | $CH_3COOH_2^+$ | + | $CH_3COO^-$ |
| $2HClO_4$ | $\rightleftharpoons$ | $H_2ClO_4^+$ | + | $ClO_4^-$ |
| $3HF$ | $\rightleftharpoons$ | $H_2F^+$ | + | $HF_2^-$ |
| $N_2O_4$ | $\rightleftharpoons$ | $NO^+$ | + | $NO_3^-$ |

An acid- base reaction in a non-aqueous solvent can be explained on the basis of solvent system concept of acids and bases. According to this concept an acid is a substance which contains the solvent cation and a base is a substance which contains the solvent anion. For example since self-ionisation of liq. $NH_3$ gives $NH_4^+$ (solvent cation - acid ion) and $NH_2^-$ (solvent anion-base ion), all ammonium salts will behave as acids in liq. $NH_3$ and all soluble amides will act as bases in liq. $NH_3$. Hence the reaction between $NH_4Cl$ (acid) and $KNH_2$ (base) in liq. $NH_3$ shown below is an acid-base reaction.

$$NH_4Cl \text{ (Acid)} + KNH_2 \text{ (Base)} \rightleftharpoons KCl \text{ (Salt)} + 2NH_3 \text{ (Solvent)}$$

Some examples of acid-base reaction in different solvents are given below:

**In liq. Br $F_3$ :** *Self-ionisation :* $2BrF_3$ (Solvent) $\rightleftharpoons$ $BrF^+_2$ (Solvent cation – acid ion) + $BrF_4^-$ (Solvent anion – base ion)

$SnF_4$ reacts with $BrF_4^-$ ion (base) to produce the solvent cation, $BrF_2^+$. Hence $SiF_4$ acts as an acid in liq. $BrF_3$.

*Acid-base reaction:* $SnF_4$(Acid) + $BrF_4^-$(Base) $\rightarrow SnF_6^{2-} + BrF_2^+$(Solvent cation)

**In liq HF** : *Self-ionisation*: 3HF (Solvent) $\rightleftharpoons$ $H_2F^+$ (Solvent cation – acid ion) + $HF_2^-$ (Solvent anion – base ion)

Acetone [$(CH_3)_2$ CO] and $HNO_3$ both behave as bases in liq $H_2F_2$, since these compounds give $HF_2^-$ (base anion) in liq. HF

*Acid-base reactions :* $(CH_3)_2CO$ (Base) + 2HF $\rightleftharpoons$ $(CH_3)_2COH^+$ + $HF_2^-$ (Base ion)

$$HNO_3 \text{ (Base)} + 2\,HF \rightleftharpoons H_2NO_3^+ + HF_2^- \text{ (Base ion)}$$

**Acidic character of $CH_3COOH$ in $H_2O$ and liq. $NH_3$** : When $CH_3COOH$ reacts with $H_2O$ and liq. $NH_3$ (solvents), $H_2O$ and liq. $NH_3$ both accept a proton from $CH_3COOH$ to prdouce $H_3O^+$ and $NH_4^+$ ions. Due to the production of these ions, $CH_3COOH$ behaves as an acid in both the solvents.

$$\underset{\substack{\text{Acid}\\\text{(weak acid)}}}{CH_3COOH} + \underset{\text{Base}}{H_2O} \rightleftharpoons \underset{\text{Acid}}{H_3O^+} + \underset{\text{Base}}{CH_3COO^-}$$

$$\underset{\substack{\text{Acid} \\ \text{(strong acid)}}}{CH_3COOH} + \underset{\text{Base}}{\ddot{N}H_3} \rightleftharpoons \underset{\text{Acid}}{NH_4^+} + \underset{\text{Base}}{CH_3COO^-}$$

Now since the proton accepting ability of $H_2O$ is less than that of liq. $NH_3$, $CH_3COOH$ behaves as weak acid in aqueous solution and acts as a strong acid in liq. $NH_3$.

**Comparison of acidic strength of $HClO_4$, HCl, HBr, HI and $HNO_3$ in aqueous solution:** When the given compounds (represented as H X) react with $H_2O$, proton is donated by HX to $H_2O$ to produce $H_3O^+$ because HX molecule has stronger proton donating propety than $H_2O$. Due to the production of $H_3O^+$ ion, $HClO_4$, HCl, HBr, HI and $HNO_3$ behave as acids in aqueous medium.

$$\underset{\substack{\text{Stronger} \\ \text{proton donor}}}{HX} + H_2O \longrightarrow H_3O^+ + X^-$$

In aqueous medium all the given acids dissociate (or ionise) completey and hence behave as **equally strong acids.** Thus $H_2O$ acts as a levelling solvent for the given acids. Here it should be noted that HX and $H_2O$ both are *protonic substances*.

**3. Solvation Reactions: Formation of Solvates**

Solvation reaction is a general reaction in which a solute (a cation, an anion or a neutral molecule) reacts with one or more molecules of a solvent (*e.g.* $H_2O$, liq. $NH_3$, liq. $SO_2$, etc.) to form a product in which the solute and solvent species are attached to each other by a H-bond or by a coordinate bond. The product formed is called **solvate.** Solvate is an **addition compound** and hence is also called an **adduct.** The addition compound contains **solvent of crystallisation.**

In the formation of a solvate, the solvent acts as a Lewis base while the solute species behaves as a Lewis acid. When the solvent used is water, the solvation reaction is called **hydration** and the addition compound formed is called **hydrate.** Hydrate contains one or more molecules of water as water of crystallisation. Similarly when the solvent is liq. $NH_3$, the reaction is called **ammoniation** and the addition compound formed is called **ammoniate.** Ammoniate contains one or more molecules of ammonia as **ammonia of crystallisation.**

**4. Solvolytic Reactions: Solvolysis**

Solvolytic reactions are the reactions in which the solvent molecules react with the solute molecule (salt) or ion in the way which consists of the following steps:

(*a*) The solvent molecule undergoes auto-ionisation (self-ionisation) to give solvent cations and solvent anions.

(*b*) The solute (salt) splits into solute cations and solute anions. The solute cations or solute anions interact with the salvent cations or solvent anions. Due to this interaction the concentration of the solvent cations or solvent anions is increased.

When $H_2O$ and $NH_3$ are used as solvents in the solvolytic reactions, these reactions are called **hydrolysis** and **ammonolysis** (or **ammonolytic reactions**) respectively.

## Liquid Ammonia (Liq. $NH_3$)

### Introduction

Liq. $NH_3$ is a protonic solvent, since it gives a proton ($H^+$) on ionisation ($NH_3 \rightarrow H^+ + NH_2^-$). It is a highly useful solvent and a number of chemical reactions are carried out in this medium. The value of dielectric constant of liq. $NH_3$ (= 22.0 at –34°C) is much smaller than that of water (= 78.5 at 25°C) and hence liq. $NH_3$ has decreased ability to dissolve ionic compounds. The liquid range for liq. $NH_3$ is from –34.0°C to –77°C (b.pt= –34.0°C. m.pt = –77.7°C).

### Solubility of Various Substances in Liq. $NH_3$

(*a*) **Solubility of ionic compounds (inorganic salts).** We know that ammonia has a low value of its dielectric constant (= 22 at –34°C). This low value suggests that liq. $NH_3$ has a poor ability to dissolve ionic compounds. Ammonium salts (*e.g.* $NH_4NO_3$, $NH_4SCN$, $CH_3COONH_4$, etc.) and most of the nitrites, nitrates, cyanides, thiocyanates, perchlorates, are soluble in liq. $NH_3$. Salts containing highly charged ions (*e.g.*, oxides, sulphides, sulphates, phosphates and carbonates) are insoluble. Fluorides and chlorides (except $Be^{2+}$ and $Na^+$ chlorides) are practically insoluble, bromides are less soluble while iodides are freely soluble. Thus the solubility of the halides of a given metal increases in going from fluoride to iodide (MF < MCl < MBr < MI). Most of the metal amides (except those of alkali metals) are insoluble. The salts of some metals (*e.g.* $Ni^{2+}$, $Cu^{2+}$, $Zn^{2+}$, etc.) react with liq. $NH_3$ and form ammine complexes.

(*b*) **Solubility of non-ionic compounds (organic compounds).** Halogen compounds, alcohols, ketones, esters, simple ethers, amines, phenol and its derivatives etc. are solouble. Alkanes are insoluble and alkenes and alkynes are slightly soluble. In this sense, liq. $NH_3$ is a better solvent for non ionic and non-polar compounds (organic compounds).

(*c*) **Solibility of non-metals.** The non-metals like S, P, $I_2$, Se, etc. are soluble and they react with the solvent.

(*d*) **Solubility of alkali metals and alkaline earth metals.** All the alkali metals and alkaline earth metals (excepting Be) are soluble in liq. $NH_3$. 100g of liq. $NH_3$ dissolves 10.9g of Li, 24.8 g of Na, 46.8 g of K (all at –33°C) and 334g of Cs (at –50°C). Thus we see that the solubility of alkali metals in liq. $NH_3$ increases as we pass from Li to Cs (Li < Na < K < Cs). The solution obtained by dissolving alkali metals in liq. $NH_3$ has the following characteristics:

(*i*) *Colour of the solution.* The colour of the dilute solution is deep blue. As the concentration of the solution increases, the blue colour changes to bronze which disappears on the further increase of the concentration of the solution. The colour of the solution is independent of the metal dissolved in liq. $NH_3$. The blue colour of the dilute solution is due to the formation of ammoniated metal cation and ammoniated electron in it.

$$M \longrightarrow M^+ + e^-$$

$$M^+ + xNH_3 \longrightarrow \underset{\text{Ammoniated cation}}{[M(NH_3)_x]^+}$$

$$e^- + yNH_3 \longrightarrow \underset{\text{Ammoniated electron}}{[e\,(NH_3)_y]^-}$$

(*ii*) *Decomposition of very dilute solution.* Very dilute solution of the alkali metals in liq. $NH_3$ is meta stable, since in the presence of a catalyst like $Fe_2O_3$ its ammoniated electron undergoes decomposition.

$$[e(NH_3)_y]^- \xrightarrow{Fe_2O_3} NH_2^- + \frac{1}{2}H_2 + (y-1)NH_3$$

(*iii*) *Evaporation of the solution.* When alkali metal-ammonia solution is evaporated, alkali metal dissolved in the solution is recovered.

(*iv*) *Paramagnetic character.* Dilute solution of alkali metals in liq. $NH_3$ is paramagnetic which is due to the presence of unpaired ammoniated electrons in the solution. The susceptibility of the solution falls down as the concentration of the solution increases and ultimately the solution becomes diamagnetic. Paramagnetism reappears at higher concentration of the solution

(*v*) *Electrical conductivity.* The electrical conductivity of very dilue solutions is high and is in the range of fully ionised slats in water. The high conductivity is due to the presence of ammoniated electrons in the solution. It decreases with concentration upto about 0.05 M, but then rises sharply. The concentrated bronze solution has conductivities in the range of metals.

(*vi*) *Density*. Dilute solution has very low density. Low density is due to the fact that ammoniated electrons present in the solution are supposed to occupy the cavities or holes within ammonia molecules.

(*vii*) *Reducing property of dilute solution.* We have said above that the blue solution of alkali metals in liq. $NH_3$ contains ammoniated electrons. These ammoniated electrons reduce various substances. Thus dilute solutions of alkali metals are reducing agents. In these reactions the alkali metal is oxidised to some alkali metal salt. The reducing property of alkali metals in liq. $NH_3$ has been used to synthesise useful compounds which are otherwise difficult to prepare. For example many covalent hydries (e.g. $PH_3$, $AsH_3$, etc.) react with metal-ammonia solutions to produce metal salts that are stable in liq. $NH_3$.

$$PH_3 + Na \xrightarrow{\text{liq. } NH_3} Na\,PH_2 + \tfrac{1}{2}\,H_2$$

$$PH_3 + 2Na \xrightarrow{\text{liq. } NH_3} Na_2\,PH + H_2$$

$$As\,H_3 + Na \xrightarrow{\text{liq. } NH_3} Na\,As\,H_2 + \tfrac{1}{2}\,H_2$$

In organic synthesis metal-ammonia solutions react with conventional transition metal complexes to give complexes in which the central metal atom has O.N. = 0. For example.

$$\underset{(Cr\ =\ +3)}{K_3[Cr(CN)_6]} + 3K \xrightarrow{liq.NH_3} \underset{(Cr\ =\ 0)}{K_6[Cr(CN)_6]}$$

$$\underset{(Ni\ =\ +2)}{K_2[Ni(CN)_4]} + 2K \xrightarrow{liq.NH_3} \underset{(Ni\ =\ 0)}{K_4[Ni(CN)_4]}$$

$$\underset{(Pt\ =\ +2)}{K_2[Pt(CN)_4]} + 2K \xrightarrow{liq.NH_3} \underset{(Pt\ =\ 0)}{K_4[Pt(CN)_4]}$$

## Auto-ionisation of Liq. $NH_3$

Liq. $NH_3$ ionises in the following three ways:

(*a*)

$$NH_3 \rightleftharpoons H^+ + NH_2^-$$

$$H^+ + NH_3 \rightleftharpoons NH_4^+$$

$$NH_3 + NH_3 \rightleftharpoons NH_4^+ + NH_2^-$$

or $\underset{\text{Solvent}}{2NH_3} \rightleftharpoons \underset{\text{(Solvent cation : acid ions)}}{NH_4^+ \textit{(Ammonium ions)}} + \underset{\text{(Solvent anions: base ions)}}{NH_2^- \textit{(Amide ions)}}$

$$K_b = [NH_4^+][NH_2^-] = 1.9 \times 10^{-33} \text{ at } -50° C$$

(*b*) $\underset{\text{Solvent}}{3NH_3} \rightleftharpoons \underset{\text{(Solvent cation : acid ions)}}{2NH_4^+ \textit{(Ammonium ions)}} + \underset{\text{(Solvent anions: base ions)}}{NH^{2-} \textit{(Imide ions)}}$

(*c*) $\underset{\text{Solvent}}{4NH_3} \rightleftharpoons \underset{\text{(Solvent cation : acid ions)}}{3NH_4^+ \textit{(Ammonium ions)}} + \underset{\text{(Solvent anions: base ions)}}{N^{3-} \textit{(Nitride ions)}}$

## Ammono Acids and Ammono Bases

Different modes of auto-ionisation of liq. $NH_3$, given above, indicate that, according to solvent system concept, in liq. $NH_3$, any substance that gives $NH_4^+$ ions (which are solvent cations) will act as an acid in liq. $NH_3$, while that which produces $NH_2^-$ or $NH^{2-}$ or $N^{3-}$ ions (which are solvent anions) will behave as a base in liq. $NH_3$. The compounds which give $NH_4^+$ ions in liq. $NH_3$ are called **ammono acids** while those which give $NH_2^-$ or $NH^{2-}$ or $N^{3-}$ ions in this solvent are called **ammono bases**.

## Examples of Ammono Acids

Ammonium salts [*e.g.* $NH_4Cl$, $NH_4Br$, $NH_4I$, $NH_4NO_3$, $(NH_4)_2S$, etc.], organic amides (*e.g.*, urea, acetamide, sulphamide, etc.), acetic acid, sulphamic acid, etc., all act as ammono acids in liq. $NH_3$, since they produce $NH_4^+$ ions in this solvent as shown below:

$$\underset{(X = Cl, Br, I)}{NH_4X} \rightleftharpoons X^- + NH_4^+$$

$$\underset{\text{Urea}}{H_2N.CO.NH_2} + NH_3 \longrightarrow H_2N.CO.NH^- + NH_4^+$$

$$\underset{\text{Acetamide}}{CH_3CONH_2} + NH_3 \longrightarrow CH_3CONH^- + NH_4^+$$

$$\underset{\text{Sulphamide}}{H_2N.SO_2.NH_2} + NH_3 \longrightarrow H_2N.SO_2.NH^- + NH_4^+$$

$$\underset{\text{Acetic acid}}{CH_3COOH} + NH_3 \longrightarrow CH_3COO^- + NH_4^+$$

$$\underset{\substack{\text{Sulphamic acid}\\ \text{(Dibasic acid)}}}{H_2N.SO_2.OH} + 2NH_3 \longrightarrow H^-N.SO_2.O^- + 2NH_4^+,$$

It may be noted that since $H_2N.SO_2.OH$ produces two $NH_4^+$ ions, this acid behaves as a dibasic acid in liq. $NH_3$.

## Examples of Ammono Bases

$KNH_2$, PbNH and BiN are the examples of ammono bases, since these compounds give $NH_2^-$, $NH^{2-}$ and $N^{3-}$ ions respectively in liq. $NH_3$.

$$KNH_2 \xrightarrow{\text{liq.}NH_3} K^+ + NH_2^- \textit{(Amide ion)}$$

$$PbNH \xrightarrow{\text{liq.}NH_3} Pb^{2+} + NH^{2-} \textit{(Imide ion)}$$

$$BiN \xrightarrow{\text{liq.}NH_3} Bi^{3+} + N^{3-} \textit{(Nitride ion)}$$

## Reactions of Ammono Acids

1. **Neutralisation reactions:** These reactions have been discussed on the subsequent pages.
2. **Reaction with metals:** Acids give out $H_2$ with metals in aqueous solution. For example:

$$Co + H_2SO_4 \xrightarrow{H_2O} CoSO_4 + H_2$$

or $$Co + 2H^+ \text{(Acid cation)} \xrightarrow{H_2O} Co^{2+} + H_2$$

In a similar way, ammonium salts which act as acids in liq. $NH_3$ give $H_2$ when they react with metals in liq. $NH_3$. For example:

(a) $$Co + 2NH_4NO_3 \textit{(Ammono acid)} \xrightarrow{\text{liq.}NH_3} Co(NO_3)_2 + 2NH_3 + H_2$$

or $$Co + 2NH_4^+ \textit{(Acid cation)} \xrightarrow{\text{liq. }NH_3} Co^{2+} + 2NH_3 + H_2O$$

(b) $$2Na + 2NH_4^+ \textit{(Acid cation)} \xrightarrow{\text{liq.}NH_3} 2Na^+ + 2NH_3 + H_2$$

3. **Decolourisation of blue solution.** Ammonium salts decolourise blue solutions of alkali metals in liq. $NH_3$.
4. **Metathetical (or precipitation) reactions:** When an ammono acid like $NH_4Cl$, $NH_4Br$, $NH_4I$, $(NH_4)_2S$ etc. is mixed with the liq. $NH_3$ solution of a metalic salt like $Sr(NO_3)_2$, $Zn(NO_3)_2$, $AgNO_3$, $Cu(NO_3)_2$, $Cd(NO_3)_2$, KI etc.,

the salt of the metal is precipitated. For example:

$$\underset{\substack{(X = Cl,\, Br)\\ \text{Ammono acid}}}{2NH_4X} + Sr(NO_3)_2 \rightleftharpoons SrX_2 \downarrow + 2NH_4NO_3$$

$$2NH_4I + Zn(NO_3)_2 \xrightleftharpoons{\text{liq. }NH_3} ZnI_2 \downarrow + 2NH_4NO_3$$

$$(NH_4)_2S + 2AgNO_3 \xrightleftharpoons{\text{liq. }NH_3} Ag_2S \downarrow + 2NH_4NO_3$$

$$(NH_4)_2S + \underset{(M = Cu,\, Cd)}{M(NO_3)} \xrightleftharpoons{\text{liq. }NH_3} MS \downarrow + 2NH_4NO_3$$

$$NH_4Cl + KI \xrightleftharpoons{\text{liq. }NH_3} KCl \downarrow + NH_4I$$

5. **Reaction with metal oxides and metal hydroxides.** Ammono acids are able to dissolve metal oxides and metal hydroxides. For example:

$$\underset{\text{Ammono acid}}{2NH_4NO_3} + CaO \longrightarrow Ca(NO_3)_2 + 2NH_3 + H_2O$$

$$\underset{\text{Ammono acid}}{NH_4Cl} + NaOH \longrightarrow NaCl + NH_3 + H_2O$$

6. **Decomposition of salts:** Ammono acids can decompose salts of weaker acids. For example:

$$Mg_2Si + 4NH_4^+ \longrightarrow 2Mg^{2+} + 4NH_3 + SiH_4$$

## Reactions of Ammono Bases

1. Extremely strong bases such as $H^-$ and $O^{2-}$ are levelled to $NH_2^-$ ion in liq. $NH_3$.

$$H^- + NH_3 \longrightarrow NH_2^- + H_2$$

$$O^{2-} + NH_3 \longrightarrow NH_2^- + OH^-$$

2. **Precipitation reactions.** When an ammono base like $KNH_2$ is mixed with the liq. $NH_3$ solution of a metallic salt like $AgNO_3$, $PbI_2$, $BiI_3$, etc, the amide, imide or nitride of the metal is precipitated. For example

$$KNH_2(\textit{Ammono base}) + AgNO_3 \xrightarrow{\text{liq.}NH_3} KNO_2 + AgNH_2 \downarrow (\textit{Amide})$$

$$2KNH_2 + Zn(NO_3)_2 \xrightarrow{\text{liq.}NH_3} 2KNO_3 + Zn(NH_2)_2 \downarrow$$

$$3KNH_2 + AlCl_3 \xrightarrow{\text{liq.}NH_3} 3KCl + Al(NH_2)_3 \downarrow$$

$$KNH_2 + PbI_2 \xrightarrow{\text{liq.}NH_3} KI + HI + Pb(NH) \downarrow (\textit{Imide})$$

$$KNH_2 + BiI_3 \xrightarrow{\text{liq.}NH_3} KI + 2HI + BiN \downarrow (\textit{Nitride})$$

3. **Complex formation reactions:** When metal salts [e.g $Zn(NO_3)_2$, $AlCl_3$, etc.] and metal amides [ e.g. $Zn(NH_2)_2$, Ag $MH_2$, $Al(NH_2)_3$, etc.] react with excess of $KNH_2$ in liq. $NH_3$, soluble amido complexes are formed. For example:

$$Zn(NO_3)_2 + 4KNH_2(\text{Excess}) \xrightarrow{\text{liq.}NH_3} \underset{\text{Soluble amido complex}}{K_2[(Zn(NH_2)_4]} + 2KNO_3$$

$$Al\,Cl_3 + 4KNH_2(\text{Excess}) \xrightarrow{\text{liq.}NH_3} K[Al(NH_2)_4] + 3KCl$$

$$Zn\,(NH_2)_2 + 4KNH_2(\text{Excess}) \xrightarrow{\text{liq.}NH_3} K_2[Zn(NH_2)_4]$$

$$Ag\,NH_2 + KNH_2(\text{Excess}) \xrightarrow{\text{liq.}NH_3} K[Ag(NH_2)_3]$$

$$Al(NH_2)_3 + KNH_2(\text{Excess}) \xrightarrow{\text{liq.}NH_3} K[Al(NH_2)_4]$$

**4.** $NH_2^-$ ion can pick up protons from very weak acids and form salts. For example:

$$HC \equiv CH + KNH_2 \longrightarrow HC \equiv \bar{C}\,K^+ + NH_3$$

$$\text{or } HC \equiv CH + N\bar{H}_2 \longrightarrow HC \equiv \bar{C} + NH_3$$

$$NH_2C(NH)NH_2 + KNH_2 \longrightarrow NH_2C(NH)N\bar{H}K^+ + NH_3$$

$$\text{or } NH_2C(NH)NH_2 + NH^-{}_2 \longrightarrow NH_2C(NH)N\bar{H} + NH_3$$

## Chemical Reactions Taking Place in Liq. $NH_3$

Chemical reactions that can take in liq. $NH_3$ are of the following types:

**1. Metathetical (Pecipitation) Reactions**

Examples of some precipitation reactions taking place in liq. $NH_3$ are given below.

(*i*) **Chlorides** are precipitated by the following reactions:

$$2NH_4Cl(\textit{Ammono acid}) + Sr(NO_3)_2 \xrightarrow{\text{liq.}NH_3} SrCl_2 \downarrow + 2NH_4NO_3$$

$$NH_4Cl(\textit{Ammono acid}) + KI \xrightarrow{\text{liq.}NH_3} KCl \downarrow + NH_4I$$

$$NH_4Cl(\textit{Ammono acid}) + LiNO_3 \xrightarrow{\text{liq.}NH_3} LiCl \downarrow + NH_4\,NO_3$$

$$KNO_3 + AgCl \xrightarrow{\text{liq.}NH_3} KCl \downarrow + AgNO_3$$

$$Ba\,(NO_3)_2 + 2\,AgCl \xrightarrow{\text{liq }NH_3} BaCl_2 \downarrow + 2Ag\,NO_3$$

(*ii*) **Bromides and iodides** are precipitated by the following reactions.

$$Sr\,(NO_3)_2 + 2NH_4\,Br\,(\textit{Ammono acid}) \xrightarrow{\text{liq. }NH_3} Sr\,Br_2 \downarrow + 2NH_4\,NO_3$$

$$Zn\,(NO_3)_2 + 2\,NH_4\,I\,(\textit{Ammono acid}) \longrightarrow ZnI_2 \downarrow + 2NH_4\,NO_3$$

(*iii*) **Amides** ($NH_2^-$), **imides** ($NH^{2-}$) and **nitrides** ($N^{3-}$) can be precipitated by the following reactions.

$$KNH_2(\textit{Ammono base}) + AgNO_3 \xrightarrow{\text{liq.}NH_3} KNO_2 + AgNH_2 \downarrow (\textit{Amide})$$

$$KNH_2 + PbI_2 \xrightarrow{\text{liq.}NH_3} KI + HI + Pb(NH) \downarrow (\textit{Imide})$$

$$KNH_2 + BiI_3 \xrightarrow{\text{liq.}NH_3} KI + 2HI + BiN \downarrow (\textit{Nitride})$$

$$2KNH_2 + 3HgI_2 \xrightarrow{\text{liq.}NH_3} 2KI + 4HI + Hg_3N_2(\textit{Nitride})$$

(*iii*) **Sulphides** are presipitated by the following reactions.

$$(NH_4)_2S(\textit{Ammono acid}) + 2AgNO_3 \xrightarrow{liq.NH_3} Ag_2S \downarrow + 2NH_4NO_3$$

$$(NH_4)_2S(\textit{Ammono acid}) + M(NO_3)_2 \xrightarrow{liq.NH_3} MS \downarrow + 2NH_4NO_3$$

$(M = Ba, Cu, Cd)$

(*iv*) **Barium alcoholate,** $Ba(OC_2H_5)_2$ can be precipitated by the following reaction :

$$K(OC_2H_5) + Ba(NO_3)_2 \xrightarrow{liq.NH_3} Ba(OC_2H_5)_2 \downarrow + 2KNO_3$$

**2. Acid-Base Neutralisation (Salt Formation) Reactions**

In liq. $NH_3$ acid - base neutralisation reaction is that in which a compound containing or making available $NH_4^+$ ion (*ammono acid*) combines with a compound containing or making available $NH_2^-$ or $NH^{2-}$ or $N^{3-}$ ion (*ammono bases*) to form the salt and the solvent (*i.e.* $NH_3$), *e.g.*,

$$NH_4Cl\ (\textit{ammono acid}) + KNH_2\ (\textit{ammono base}) \xrightarrow{liq.NH_3} KCl\ (\textit{salt}) + 2NH_3\ (\textit{solvent})$$

Actually the above reaction takes place through the following steps:

$$NH_4Cl \xrightarrow{liq.NH_3} NH_4^+ + Cl^-$$

$$KNH_2 \xrightarrow{liq.NH_3} K^+ + NH_2^-$$

$$NH_4Cl + KNH_2 \xrightarrow{liq.NH_3} K^+ + Cl^- + NH_4^+ + NH_2^-$$

or $NH_4Cl\ (\textit{ammono acid}) + KNH_2\ (\textit{ammono base}) \xrightarrow{liq.NH_3} KCl\ (\textit{salt}) + 2NH_3\ (\textit{solvent})$

or $NH_4^+ + Cl^- + K^+ + NH_2^- \xrightarrow{liq.NH_3} K^+ + Cl^- + 2NH_3$

or $NH_4^+(\textit{solvent cation}) + NH_2^-\ (\textit{solvent anion}) \xrightarrow{liq.NH_3} 2NH_3(\textit{solvent})$ ...(*i*)

Equation (*i*) indicates that neutralisation reaction in liq. $NH_3$ can also be defined as the combination of solvent cation ($NH_4^+$) and solvent anion ($NH_2^-$) to form the un-ionised solvent (i.e. liq. $NH_3$).

Other examples of acid-base neturalisation reactions taking place in liq. $NH_3$ are:

| *Ammono acid* | | *Ammono base* | | *Ammono salt* | | *Solvent* |
|---|---|---|---|---|---|---|
| $NH_4X$ ($X = Cl, NO_3$) | + | $NaNH_2$ | $\xrightarrow{liq.NH_3}$ | $NaX$ | + | $2NH_3$ |
| $2NH_4X$ ($X = Cl, I$) | + | $PbNH$ | $\xrightarrow{liq.NH_3}$ | $PbX_2$ | + | $3NH_3$ |

$$\underset{(X = Cl, I)}{3NH_4X} + BiN \xrightarrow{liq.NH_3} BiX_3 + 4NH_3$$

$$2CH_3COONH_4 + Zn(NH_2)_2 \xrightarrow{liq.NH_3} Zn(CH_3COO)_2 + 4NH_3$$

$Zn^{2+}$ ions obtained by the dissociation of $Zn(CH_3COO)_2$ react with excess of liq. $NH_3$ to form $[Zn(NH_3)_4]^{2+}$

$$Zn^{2+} + 4NH_3 \longrightarrow [Zn(NH_3)_4]^{2+}$$

**Comparison of the process of neutralisation reaction in aqueous solution and liq. $NH_3$.** *In aqueous solution the process of neutralisation reaction of a strong acid by a strong base is the combination of $H_3O^+$ (Solvent cation – acid ion) and $OH^-$ (solvent anion – base ion) to form un-ionised $H_2O$ (solvent).* For example

$$HCl\ (Acid) + KOH\ (Base) \xrightarrow{H_2O} KCl\ (Salt) + H_2O\ (Solvent)$$

or $$H^+ + Cl^- + K^+ + OH^- \xrightarrow{H_2O} K^+ + Cl^- + H_2O$$

or $$H^+ + OH^- \longrightarrow H_2O$$

or $$H^+ + H_2O + OH^- \longrightarrow 2H_2O$$

or $$H_3O^+(Acid\ ion) + OH^-(Base\ ion) \longrightarrow 2H_2O(Solvent)$$

A similar thing happens in case of liq. $NH_3$. *In liq. $NH_3$ the process of neutralisation reaction is the combination of $NH_4^+$ (solvent cation – acid ion) and $NH^{2-}$ or $NH^{2-}$ or $N^{3-}$ ion (solvent anion – base ion) to form the un-ionised $NH_3$ (solvent).* For example the neutralisation reaction between $NH_4Cl$ and $KNH_2$ can be shown as:

$$NH_4Cl\ (Ammono\ acid) + KNH_2\ (Ammono\ base) \xrightarrow{liq.NH_3} KCl\ (Salt) + 2\ NH_3\ (Solvent)$$

or $$NH_4^+ + Cl^- + K^+ + NH_2^- \xrightarrow{liq.NH_3} K^+Cl^- + 2NH_3$$

or $$NH_4^+(Ammono\ acid) + NH_2^-(Ammono\ base) \longrightarrow 2NH_3(Solvent)$$

Thus we see that the role of $NH_4Cl$ in liq. $NH_3$ is the same as that of HCl in aqueous solution. Similarly the role of $KNH_2$ in liq. $NH_3$ is the same as that of KOH in aqueous solution. Thus $NH_4Cl$ may be regarded as a strong acid and $KNH_2$ as a strong base in liq. $NH_3$.

Above discussion shows that substances producing $H_3O^+$ ions in $H_2O$ are acids. Similarly, ammonium salts giving $NH_4^+$ ions in liq. $NH_3$ are acids. The substances producing $OH^-$ ions in $H_2O$ are bases and the substances containing $NH_2^-$ or $NH^{2-}$ or $N^{3-}$ ions are bases in liq. $NH_3$.

The course of neutralisation reactions occuring in liq. $NH_3$ may be followed by means of conductometric titrations, and in certain cases through indicators. For example phenolphthalein is colourless in liq. $NH_3$ but develops an intense red colour in presence of alkali amides ($MNH_2$)

### 3. Amphoteric Behaviour of $Zn(NH_2)_2$ in Liq.$NH_3$

Zn $(NH_2)_2$ dissolves in ammono acid as well as in ammono base in liq. $NH_3$ and hence is said to show amphoteric character in liq. $NH_3$

$$Zn(NH_2)_2 + 2NH_4Cl\ (\textit{Ammono acid}) \xrightarrow{\text{liq.}NH_3} ZnCl_2\ (\textit{Salt}) + 4NH_3\ (\text{Solvent})$$

$$Zn(NH_2)_2 + 2Na\,NH_2\ (\textit{Ammono base}) \xrightarrow{\text{liq.}NH_3} Na_2[Zn(NH)_2] + 2NH_3$$

Above reactions can be compared with the following reactions in aqueous solution. In these reactions $Zn(OH)_2$ shows amphoteric character since it dissolves in HCl (acid) as well as in NaOH (base)

$$Zn(OH)_2 + 2HCl \xrightarrow{H_2O} ZnCl_2 + 2H_2O$$

$$Zn(OH)_2 + 2NaOH \xrightarrow{H_2O} Na_2[ZnO_2] + 2H_2O$$

**4. Ammonation Reactions : Formation of Ammoniates (Solvation Reactions – Formation of Solvates)**

In the general discussion of solvation reactions we have already said that solvation reaction in which liq. $NH_3$ is used as solvent is called **ammonation reaction** and the solvate formed is called **ammoniate.**

In the formation of ammoniates liq. $NH_3$ (solvent) acts as a Lewis base and the solute behaves as a Lewis acid. The formation of some ammoniates in liq. $NH_3$ has been shown below. These ammoniates may be 1 : 2 or 1 : 1 adducts.

| *Solute (Lewis acid)* | | *Solvent (Lewis base)* | | *Ammoniates* |
|---|---|---|---|---|
| $SO_3$ | + | $2NH_3$ | $\longrightarrow$ | $SO_3.2NH_3$ (*1 : 2 adduct*) |
| $SiF_4$ | + | $2NH_3$ | $\longrightarrow$ | $SiF_4.\ 2NH_3$ (*1 : 2 adduct*) |
| $BF_3$ | + | $NH_3$ | $\longrightarrow$ | $BF_3.\ NH_3$ (*1 : 1 adduct*) |

The formation of $NH_4^+$ by the action of liq. $NH_3$ on $H_2O$ is also an example of ammoniation.

$$H_3N{:} + H{-}\ddot{O}{-}H \longrightarrow H_3N{:}{-}{-}{-}H{-}\ddot{O}{-}H \longrightarrow NH_4^+ + OH^- \quad \text{or } [NH_4]^+[OH]^-$$

More examples of ammoniates formed by liq. $NH_3$ with inorganic salts are $MgI_2.\ 6NH_3$, $NiI_2.\ 6NH_3$, $NaI.\ 4NH_3$ etc. In the formation of these ammoniates, $NH_3$ molecules (solvent molecules) are coordinated to the metal ion of the inorganic salt (solute).

**5. Ammonolysis or Ammonolytic Reactions (Solvolysis or Solvolytic Reactions)**

We have already said that the solvolytic reactions taking place in aqueous medium ($H_2O$) and liq. $NH_3$ are called **hydrolysis** and **ammonlysis** (or **ammonolytic reactions**) respectively. In hydrolysis $H_2O$ is used as a solvent while in ammonolysis liq. $NH_3$ is used as a solvent.

**Examples of hydrolysis** (*i*) *Hydrolysis of $CuSO_4$ and $SO_2Cl_2$.* In the hydrolysis of $CuSO_4$ and $SO_2Cl_2$ (solutes), the concentrationof $H_3O^+$ (solvent cations) is increased as shown below:

$$4H_2O\ (Solvent) \rightleftharpoons 2H_3O^+\ (Solvent\ cations) + 2OH^-\ (Solvent\ anions)$$

$$CuSO_4(Solute) \longrightarrow Cu^{2+}(Solute\ cation) + SO_4^{2-}$$

$$Cu^{2+}(Solute\ cation) + 2OH^-(Solvent\ anion) \longrightarrow Cu(OH)_2$$

---

$$On\ adding: CuSO_4(Salt) + 4H_2O\ (Solvent) \longrightarrow \underset{\text{Weak base}}{Cu(OH)_2} + SO_4^{2-} + 2H_3O^+$$

$$4H_2O \rightleftharpoons 2H_3O^+ + 2OH^-$$

$$SO_2Cl_2(Solute) \longrightarrow SO_2^+\ (Solute\ cation) + 2Cl^-$$

$$SO_2^{2+} + 2OH^- \longrightarrow SO_2(OH)_2$$

---

$$On\ adding: SO_2Cl_2 + 4H_2O \longrightarrow SO_2(OH)_2 + 2Cl^- + 2H_3O^+$$

(*ii*) *Hydrolysis of $AlCl_3$, $Cl_2$ and $OPCl_3$* can be represented by the following equations.

$$AlCl_3 + 2H_2O \longrightarrow AlCl_2(OH) + Cl^- + H_3O^+$$

$$Cl_2 + 2H_2O \longrightarrow Cl(OH) + Cl^- + H_3O^+$$

$$OPCl_3 + 6H_2O \longrightarrow OP(OH)_3 + 3Cl^- + 3H_3O^+$$

(*iii*) *Hydrolysis of alkali metal hydride* (*e.g.*, NaH) can be shown as:

$$NaH + H_2O \xrightarrow{Water} NaOH + H_2$$

or $$H^- + H_2O \xrightarrow{Water} OH^- + H_2O$$

**Examples of ammonolysis** (*i*) *Ammonolysis of $SnCl_4$, $SiCl_4$ and $SO_2Cl_2$.* In the ammonolysis of $SnCl_4$ and $SO_2Cl_2$ (*solute*) the concentration of $NH_4^+$ (solvent cations) is increased as shown below:

(*a*) $$8\ NH_3\ (Solvent) \rightleftharpoons 4NH_4^+(Solvent\ cations) + 4NH_2^-(Solvent\ anions)$$

$$SnCl_4(Solute) \longrightarrow Sn^{4+}(Solute\ cations) + 4Cl^-$$

$$Sn^{4+}\ (Solute\ cations) + 4NH_2^-(Solvent\ anions) \longrightarrow Sn(NH_2)_4$$

---

$$On\ adding: SnCl_4(Solute) + 8NH_3(Solvent) \longrightarrow Sn(NH_2)_4 + 4Cl^- + 4NH_4^+$$

(*b*) $SiCl_4 + 8\ NH_3 \longrightarrow Si\ (NH_2)_4 + 4Cl^- + 4\ NH_4^+$

(*c*) $$4NH_3 \rightleftharpoons 2NH_4^+ + 2NH_2^-$$

$$SO_2Cl_2 \longrightarrow SO_2^{2+} + 2Cl^-$$

$$SO_2^{2+} + 2NH_2^- \longrightarrow SO_2(NH_2)_2$$

---

$$On\ adding: SO_2Cl_2 + 4NH_3 \longrightarrow SO_2(NH_2)_2 + 2Cl^- + 2NH_4^+$$

(*ii*) *Ammonolysis of alkali metal hydrides and oxides.* The ammonolysis of alkali metal hydrides and oxides gives alkali metal amides ($MNH_2$). In these reactions the concentration of $NH_2^-$ ions (solvent anions) is increased.

$NaH\ (Solute) + NH_3\ (Solvent) \longrightarrow NaNH_2 + H_2$

or $H^- + NH_3 \longrightarrow NH_2^- + H_2$

(In water: $H^- + H_2O \longrightarrow OH^- + H_2$)

$Na_2O$ (*Solute*) + $2NH_3$ (*Solvent*) $\longrightarrow 2\,NaNH_2 + H_2O$

or $O^{2-} + 2NH_3 \longrightarrow 2NH_2^- + H_2O$

(*iii*) *Ammonolysis of* $TiCl_4$. Ammonolysis of $TiCl_4$ (salt) in presence of excess of liq. $NH_3$ produces Ti $(NH_2)_4$. This reaction takes place in steps shown below. In each step Cl-atom of the salt undergoing ammonolysis is replaced by $-NH_2$ group.

$$TiCl_4 + NH_3 \longrightarrow TiCl_3(NH_2) + HCl$$
$$TiCl_3(NH_2) + NH_3 \longrightarrow TiCl_2(NH_2)_2 + HCl$$
$$TiCl_2(NH_2)_2 + NH_3 \longrightarrow TiCl(NH_2)_3 + HCl$$
$$TiCl(NH_2)_3 + NH_3 \longrightarrow Ti(NH_2)_4 + HCl$$

*On adding* : $TiCl_4 + 4NH_3 \longrightarrow Ti(NH_2)_4 + 4HCl$

(*iv*) *Ammonolysis of alkyl and aryl halides:* The ammonolysis of alkyl halides, RX(R = alkyl group, X = Cl, Br, I) takes place slowly at the boiling point of liq. $NH_3$. In this reaction mixture of primary ($RNH_2$), secondary ($R_2NH$) and tertiary ($R_3N$) amines is obtained.

$$RX + 2NH_3 \longrightarrow R(NH_2) + NH_4^+ + X^-$$
$$2RX + 3NH_3 \longrightarrow R_2(NH) + 2NH_4^+ + 2X^-$$
$$3RX + 4NH_3 \longrightarrow R_3N + 3NH_4^+ + 3X^-$$
$$C_6H_5Cl + 2NH_3 \longrightarrow C_6H_5NH_2 + NH_4^+ + Cl^-$$

(*v*) *Ammonolysis reactions shown by* $Hg_2Cl_2$, $HgCl_2$, $BX_3$, $AlCl_3$, $PbI_2$, $SiH_3Cl$, $Cl_2$, $OPCl_3$ *and* $PCl_3$ can be represented by the following reactions. In all these reactions the concentration of $NH_4^+$ is increased.

$$Hg_2Cl_2 + 2NH_3 \longrightarrow Hg(NH_2)Cl + Hg + NH_4^+ + Cl^-$$
$$HgCl_2 + 2NH_3 \longrightarrow Hg(NH_2)Cl + NH_4^+ + Cl^-$$
$$BX_3 + 6NH_3 \longrightarrow B(NH_2)_3 + 3NH_4^+ + 3X^-$$
$$AlCl_3 + 2NH_3 \longrightarrow AlCl_2(NH_2) + NH_4^+ + Cl^-$$

(*In water:* $AlCl_3 + 2H_2O \longrightarrow AlCl_2(OH) + H_3O^+ + Cl^-$)

$$2PbI_2 + 6NH_3 \longrightarrow PbI(NH_2) + PbI(NH_2)_2 + 3NH_4^+ + 3I^-$$
$$3(SiH_3)Cl + 4NH_3 \longrightarrow (SiH_3)_3N + 3NH_4^+ + 3Cl^-$$
$$Cl_2 + 2NH_3 \longrightarrow Cl(NH_2) + NH_4^+ + 3Cl^-$$

(*In water:* $Cl_2 + 2H_2O \longrightarrow Cl(OH) + H_3O^+ + Cl^-$

$$OPCl_3 + 6NH_3 \longrightarrow OP(NH_2)_3 + 3NH_4^+ + 3Cl^-$$

(*In water:* $OPCl_3 + 6H_2O \longrightarrow OP(OH)_3 + 3H_3O^+ + 3Cl^-$)

$$PCl_3 + 6NH_3 \longrightarrow P(NH_2)_3 + 3NH_4^+ + 3Cl^-$$

$$HCl + NH_3 \longrightarrow NH_4^+ + Cl^-$$

$$HI + NH_3 \longrightarrow NH_4^+ + I^-$$

**6. Complex Formation Reactions**

(*i*) Many metal salts [*e.q.* $Zn(NO_3)_2$, $AlCl_3$, etc.] react wth excess of $KNH_2$ solution in liq $NH_3$ to form soluble amido complexes. For example:

$$(a)\ \underset{}{Zn(NO_3)_2} + \underset{\substack{\text{Ammono base}\\\text{(excess)}}}{4KNH_2} \xrightarrow{\text{liq.}NH_3} \underset{\substack{\text{Amido complex}\\\text{(soluble)}}}{K_2[Zn(NH_2)_4]} + 2KNO_3$$

$$\text{or } Zn^{2+} + \underset{\text{(excess)}}{4NH_2^-} \xrightarrow{\text{liq.}NH_3} [Zn(NH_2)_4]^{2-}$$

The above reaction actually takes place through the following steps:

$$Zn(NO_3)_2 + 2KNH_2 \xrightarrow{\text{liq.}NH_3} \underset{\text{Insoluble}}{Zn(NH_2)_2} + 2KNO_3$$

$$Zn(NH_2)_2 + 2KNH_2(\text{Excess}) \xrightarrow{\text{liq.}NH_3} K_2[Zn(NH_2)_4]$$

*On adding* : $Zn(NO_3)_2 + 4KNH_2 \xrightarrow{\text{liq.}NH_3} K_2[Zn(NH_2)_4] + 2KNO_3$

The above complex formation reaction that takes place in liq. $NH_3$ can be compared with the following complex formation reaction that occurs in aqueous medium.

$$Zn(NO_3)_2 + \underset{\text{Base(excess)}}{4NaOH} \xrightarrow{H_2O} \underset{\text{Hydroxo complex(soluble)}}{Na_2[Zn(OH)_4]} + 2NaNO_3$$

$$\text{or}\quad Zn^{2+} + 4OH^- \xrightarrow{H_2O} [Zn(OH)_4]^{2-} \text{ or } [ZnO_2^{2-} + 2H_2O]$$

The above reaction also takes place through the following two steps:

$$Zn(NO_3)_2 + NaOH \xrightarrow{H_2O} \underset{\text{Insoluble}}{Zn(OH)_2} + 2NaNO_3$$

$$Zn(OH)_2 + 2NaOH(\textit{Excess}) \xrightarrow{H_2O} Na_2[Zn(OH)_4]$$

*On adding* : $Zn\ (NO_3)_2 + 4\ NaOH \xrightarrow{H_2O} Na_2\ [Zn\ (OH)_4] + 2\ NaNO_3$

$$(b)\ AlCl_3 + 4KNH_2(\textit{Excess}) \xrightarrow{\text{liq.}NH_3} \underset{\text{Soluble amido complex}}{K[Al(NH_2)_4]} + 3KCl$$

$$\text{or}\quad Al^{3+} + 4NH_2^-(\textit{Excess}) \xrightarrow{\text{liq.}NH_3} [Al(NH_2)_4]^-$$

The above reaction takes place thorugh the following two steps.

$$AlCl_3 + 3KNH_2 \xrightarrow{\text{liq.}NH_3} \underset{\text{Insoluble}}{Al(NH_2)_3} + 3KCl$$

$$Al(NH_2)_3 + KNH_2\ (\textit{Excess}) \xrightarrow{\text{liq.}NH_3} K[Al\ (NH_2)_4]$$

*On adding* : $AlCl_3 + 4\ KNH_2 \longrightarrow K\ [Al(NH_2)_4] + 3KCl$

This reaction can be compared with the reaction of $AlCl_3$ with excess of NaOH taking place in aqueous solution.

$$AlCl_3 + 4NaOH\ (Excess) \xrightarrow{H_2O} Na\,[Al\,(OH)_4] + 3\,NaCl$$

or $$Al^{3+} + 4OH^- \xrightarrow{H_2O} \underset{\text{Soluble hydroxo complex}}{[Al(OH)_4]^- \text{ or } [AlO_2^- + 2H_2O]}$$

This reaction also takes place through the following two steps

$$\underset{\text{Base}}{AlCl_3 + 3NaOH} \xrightarrow{H_2O} \underset{\text{ppt.}}{Al(OH)_3} + 3NaCl$$

$$\underset{\text{ppt}}{Al(OH)_3} + \underset{\text{Base (excess)}}{NaOH} \xrightarrow{H_2O} \underset{\text{Soluble hydroxo complex}}{Na[Al(OH)_4]}$$

*On adding*: $$AlCl_3 + \underset{\substack{\text{Base}\\ \text{(excess)}}}{4NaOH} \xrightarrow{H_2O} 3NaCl + Na[Al(OH)_4]$$

(*ii*) Many metal amides [*e.g.* Zn $(NH_2)_2$, Ag $NH_2$, Al $(NH_2)_3$, etc]. imides and nitrides dissolve in the excess of $KNH_2$ in liq. $NH_3$ to form soluble amido complexes. For example.

(*a*) $$Zn(NH_2)_2 + 2KNH_2\ (Excess) \xrightarrow{liq.NH_3} \underset{\text{Soluble amido complex}}{K_2\,[Zn\,(NH_2)_4]}$$

The above reaction can be compared with the reactio of $Zn(OH)_2$ with excess of NaOH taking place in aqueous medium.

$$Zn(OH)_2 + 2NaOH\ (Excess) \xrightarrow{H_2O} \underset{\text{Soluble hydroxo complex}}{Na_2\,[Zn(OH)_4]}$$

(*b*) $$AgNH_2 + KNH_2\ (Excess) \xrightarrow{liq.NH_3} K\,[Ag\,(NH_2)_2]$$

(*c*) $$Al(NH_2)_3 + KNH_2\ (Excess) \xrightarrow{liq.NH_3} K\,[Al\,(NH_2)_4]$$

**7. Redox Reactions**

Redox reactions in which liq. $NH_3$ is involved can be studied under the following heads:

(*i*) **Redox reactions in which liq. $NH_3$ serves as a medium only.** Liquid $NH_3$ provides a medium in which various redox reactions can be carried. For example:

(*a*) $KMnO_4$ acts as a weak oxidising agent in liq. $NH_3$. It oxidises $KNH_2$ to $N_2$ in liq. $NH_3$ and is itself reduced to $K_2MnO_4$.

$$\underset{(Mn=+7)}{6KMnO_4} + \underset{(N=-3)}{6KNH_2} \xrightarrow{liq.NH_3} \underset{(Mn=+6)}{6K_2MnO_4} + 6NH_3 + \underset{(N=0)}{N_2}$$

(*b*) Iodine also acts as a weak oxidising agent in liq. $NH_3$. It oxidises potassium ammonostannite, $K_4[Sn(NH_2)_6]$ to potassium ammonostannate, $K_2[Sn(NH_2)_6]$ and is itself reduced to KI.

$$\underset{(I=0)}{I_2} + \underset{(Sn=+2)}{K_4[Sn(NH_2)_6]} \xrightarrow{liq.NH_3} \underset{(Sn=+4)}{K_2[Sn(NH_2)_6]} + \underset{(I=-I)}{2KI}$$

(*ii*) **Redox reactions in which liq. $NH_3$ (N= –3, H = +1) itself acts as a reduing agent.** Liq. $NH_3$ reduces various substances like $O_2$, CuO, etc. and is itself oxidised to NO (N = +2), $N_2$ (N = 0) etc. Examples are:

(*a*) $$\underset{(N=-3)}{4NH_3} + \underset{(O=0)}{5O_2} \xrightarrow{Pt} \underset{(N=+2)}{4NO} + \underset{(O=-2)}{6H_2O}$$

(*b*) $$\underset{(N=-3)}{2NH_3} + \underset{(Cu=+2)}{3CuO} \longrightarrow \underset{(N=0)}{N_2} + \underset{(Cu=0)}{3Cu} + 3H_2O$$

(*iii*) **Redox reactions in which alkali metals in liq. $NH_3$ act as reducing agents.** Liq.$NH_3$ can dissolve the alkali metals in it. The solution thus obtained contains ammoniated electrons. These electrons reduce various substances. Thus alkali metals in liq. $NH_3$ are able to reduce a variety of substances and are themselves oxidised to some alkali metal salt. Following examples illustrate the reducing property of alkali metals in liq. $NH_3$.

(*a*) Sodium in liq. $NH_3$ reduces ammonium salts (*e.g.*, $NH_4Br$) to $H_2$.

$$\underset{(Na=0)}{2Na} + \underset{(H=+1)}{2NH_4Br} \xrightarrow{liq.NH_3} \underset{(Na=+1)}{2NaBr} + 2NH_3 + \underset{(H=0)}{H_2}$$

(*b*) Potassium in liq. $NH_3$ reduces nitrous oxide ($N_2O$) to $N_2$.

$$\underset{(K=0)}{2K} + NH_3 + \underset{(N=+1)}{N_2O} \xrightarrow{liq.NH_3} \underset{(K=+1)}{KNH_2} + KOH + \underset{(N=0)}{N_2}$$

(*c*) Sodium in liq. $NH_3$ reduces elemental sulphur to sodium sulphide ($Na_2S$)

$$\underset{(S=0)}{S} + \underset{(Na=0)}{2Na} \xrightarrow{liq.NH_3} \underset{(Na=+1,S=-2)}{Na_2S}$$

(*d*) Li and Na in liq. $NH_3$ reduce elemental oxygen to their peroxide ($Li_2O_2$ or $Na_2O_2$) while other alkali metals give their peroxide ($M_2O_2$) as well as superoxide ($MO_2$).

$$\underset{(K=0)}{2K} + \underset{(O=0)}{O_2} \xrightarrow{liq.NH_3} \underset{(K=+1,O=+1)}{K_2O_2} \quad \text{(peroxide - white)}$$

$$\underset{(K=0)}{K} + \underset{(O=0)}{O_2} \xrightarrow{liq.NH_3} \underset{K=+1,O=-1/2}{KO_2}\text{(superoxide – yellow)}$$

(*e*) Na in liq. $NH_3$ reduces CuI to Cu

$$\underset{(Cu=+1)}{CuI} + \underset{(Na=0)}{Na} \xrightarrow{liq.NH_3} \underset{(Cu=0)}{Cu} + \underset{(Na=+1)}{NaI}$$

(*f*) K in liq. $NH_3$ reduces $KMnO_4$ to MnO and is itself oxidised to $KNH_2$.

$$\underset{(K=0)}{30K} + 20NH_3 + \underset{(Mn=+7)}{6KMnO_4} \longrightarrow \underset{(Mn=+2)}{6MnO} + \underset{(K=+1)}{18KNH_2} + 18KOH + 3H_2 + N_2$$

(*g*) Na in liq. $NH_3$ reduces $ZnI_2$ to $NaZn_4$ and is itself oxidised to NaI.

$$\underset{(Zn=+2)}{4ZnI_2} + \underset{(Na=0)}{9Na} \xrightarrow{liq.NH_3} \underset{(Zn=-1/4)}{NaZn_4} + \underset{(Na+1)}{8NaI}$$

(*h*) Excess of K in liq. $NH_3$ reduces $K_2[Ni(CN)_4]$ and $[Pt(NH_3)_4]Br_2$ to $K_4[Ni(CN)_4]$ and $[Pt(NH_3)_4]^0$ respectively.

$$\underset{(Ni = +2)}{K_2[Ni(CN)_4]} \xrightarrow[\text{liq.}NH_3(-33°C)]{\text{Excess of K in}} \underset{(Ni = 0)}{K_4[Ni(CN)_4]}$$

$$\underset{(Pt = +2)}{[Pt(NH_3)_4]Br_2} \xrightarrow[\text{liq.}NH_3(-33°C)]{\text{Excess of K in}} \underset{(Pt = 0)}{[Pt(NH_3)_4]^0} + 2KBr$$

**8. Dissociation of Weak Acids by Liq. $NH_3$**

Since liq. $NH_3$ has strong proton accepting ability, it accepts a proton from weak acids like $CH_3COOH$.

$$NH_3\ (l) + CH_3COOH \longrightarrow NH_4^+ + CH_3COO^-$$

$CH_3COOH$ is completely ionised in liq. $NH_3$ and hence acts as a strong acid in liq. $NH_3$.

**9. Reactions in Which $NH_3$ Accepts a Proton : Pyrolysis**

$NH_3$ is a powerful proton acceptor to form $NH_4^+$ ion. For example:

(*i*) $NH_3$ picks up a proton from $H_2O$ to form $NH_4^+$ and $OH^-$ ions

$$NH_3 + H\,OH \longrightarrow NH_4^+ + OH^-$$

The high solubility of $NH_3$ in $H_2O$ or of $H_2O$ in $NH_3$ is due to the above reaction.

(*ii*) $CH_3COOH$ is a weak acid in $H_2O$ but shows strong acidic character in liq. $NH_3$ because $NH_3$ picks up a proton from $CH_3COOH$ to form $NH_4^+$ ion.

$$NH_3 + CH_3COOH \longrightarrow NH_4^+ + CH_3COO^-$$

(*iii*) Urea ($NH_2CONH_2$) and acetamide ($CH_3CONH_2$) are weakly basic in aqueous solution but show acidic properties in liq. $NH_3$, since these compounds can donate a proton to $NH_3$ to produce $NH_4^+$ ion.

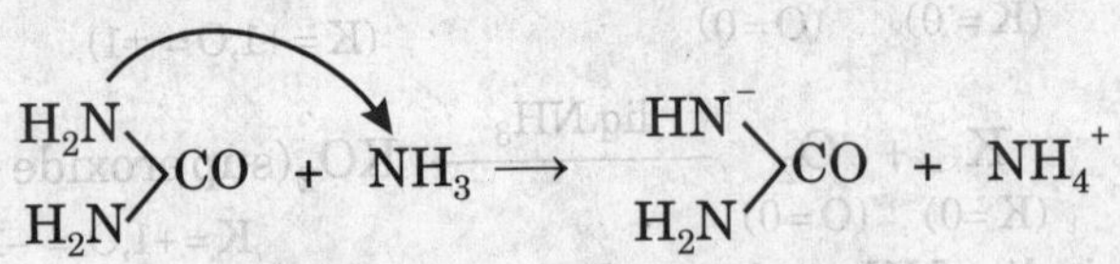

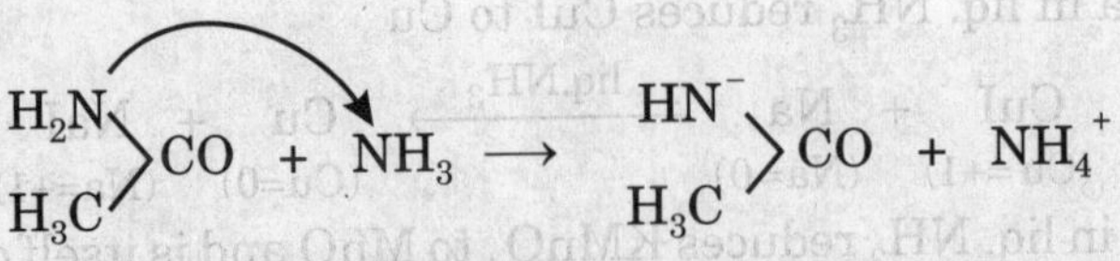

(*iv*) Sulphanic acid ($NH_2.\ SO_2.\ OH$) behaves as a dibasic acid in liq. $NH_3$

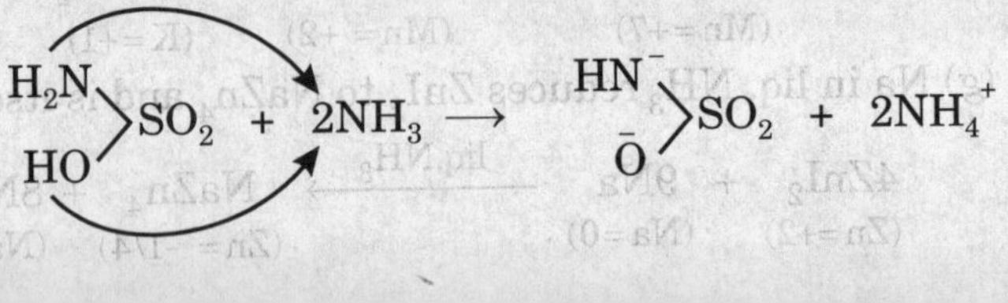

Here it may be noted that $NH_2 . SO_2 . OH$ acts as a monobasic acid in aqueous solution, since only OH group present in this compound donates a proton to $H_2O$

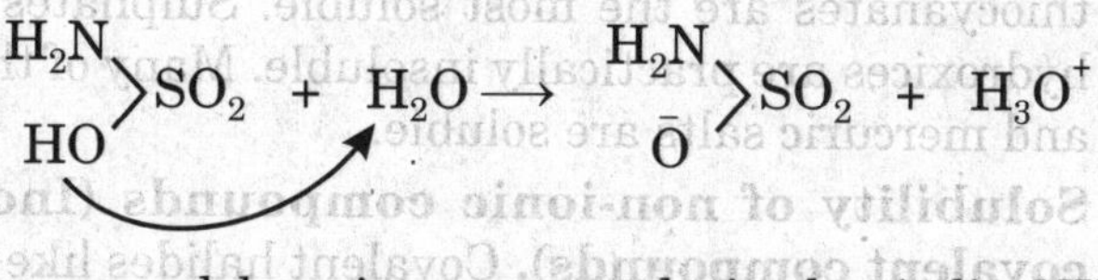

(*v*) $AgNH_2$ behaves as a weak base in aqueous solution but in liq. $NH_3$ it behaves as an acid.

$$AgNH_2 + 2NH_3 \longrightarrow [Ag\,(NH_2)_2\,]^- + NH_4^+$$

In all the above examples the concentration of $NH_4^+$ in liq. $NH_3$ is increased. Hence, according to Cady and Esley concept of acids and bases, all these substances behave as acids in liq. $NH_3$

## Advantages of Using Liquid Ammonia as a Solvent

(*i*) We have seen above that alkali metals, without reacting with liq. $NH_3$, are soluble in this solvent. The dissolved alkali metals can be recovered by evaporating the alkali metal-liq. $NH_3$ solution.

(*ii*) The alkali metal-liq. $NH_3$ solutions contain ammoniated electron and hence these solutions act as strong reducing agents.

(*iii*) We have seen under the study of precipitation reactions taking place in liq. $NH_3$ that these reactions can be used to precipitate metallic halides, sulphides, alcoholates, amides, imides and nitrides.

## Disadvantages of Using Liq. $NH_3$ as a Solvent

(*i*) Low temperature or high pressure is necessary while working with liq. $NH_3$. This is because of the fact that the liquid range for liq. $NH_3$ is from –34.0°C to –77.7°C.

(*ii*) Liq. $NH_3$ is hygroscopic in nature and hence all the reactions must be carried out in the sealed tube.

(*iii*) Liq. $NH_3$ has an offensive odour and hence the use of liq. $NH_3$ as a solvent and as a reaction medium requires special technique.

# Liquid Sulphur Dioxide (Liq. $SO_2$)

## Introduction

Liq. $SO_2$ is a non-protonic solvent, because it cannot give a proton on self ionisation. Under normal temperature and pressure, sulphur dioxide is a gas but can be readily liquefied. It has a wide liquid range from –10.0°C to –75.5°C (b.pt = –10.0° and m.pt. = –75.5°C) and hence can be used as a solvent. Its dielectric constant is low (= 17.4 at – 20°C) and hence it is a poor solvent for ionic compounds but acts as a good solvent for covalent compounds.

## Solubility of Various Substances in Liq. $SO_2$

(*a*) **Solubility of ionic compounds (Inorganic salts).** Iodides and thiocyanates are the most soluble. Sulphates, sulphides, oxides and hydroxices are practically insoluble. Many of the ammonium, thallium and mercuric salts are soluble.

(*b*) **Solubility of non-ionic compounds (Inorganic and organic covalent compounds).** Covalent halides like IBr, $BCl_3$, $AlCl_3$, $AsCl_3$, $PBr_3$, $CCl_4$, $SiCl_4$ and $SnCl_4$ are soluble in liq. $SO_2$. Organic compounds like ammines, ethers, alcohols, benzene, alkenes, pyridine, quinoline, halogen derivatives and acid chlorides are soluble, Alkanes are insoluble.

(*c*) **Solubility of metals.** Metals are insoluble in liq. $SO_2$.

(*d*) **Conductivity of salt solutions in liq. $SO_2$.** It has been found that the conductivity of electrolyte solutions of salts containing $Na^+$, $NH_4^+$, $K^+$, $(CH_3)_3\,S^+$ and $(CH_3)_4\,N^+$ cations increases with the increase in the size of these cations. Thus the increasing order of conductivity is as:

$$Na^+ < NH_4{}^+ < K^+ < (CH_3)_3\,S^+ < (CH_3)_4\,N^+.$$

The conductivity of some anions increases in the order

$$SCN^- < Cl\,O_4{}^- < Cl^- < I^-$$

## Auto-ionisation of liq. $SO_2$

Self-ionisation of liq. $SO_2$ takes place as follows:

$$SO_2 \rightleftharpoons SO^{2+} + O^{2-}$$

$$SO_2 + O^{2-} \rightleftharpoons SO_3^{2-}$$

$$\underbrace{SO_2 + SO_2}_{\text{Solvent}} \rightleftharpoons \underset{\text{(Solvent cations: acid ions)}}{SO^{2+}\,(\textit{Thionyl ions})} + \underset{\text{(Solvent anions : base ions)}}{SO_3^{2-}\,(\textit{Sulphite ions})}$$

When we compare that self-ionisation of liq. $SO_2$ with that of water and liq. $NH_3$, we find that $SO^{2+}$ ion is analogous to $H_3O^+$ and $NH_4^+$ ions while $SO_3^{2-}$ ion is analogous to $OH^-$ and $NH_2^-$ ion produced by the self-ionisation of $H_2O$ and liq. $NH_3$ respectively. Thus all those compounds which contain or make available $SO^{2+}$ ions(solvent cations–acid ions) will act as acids in liq. $SO_2$. Similarly all those compounds which contain or make available $SO_3^{2-}$ ions, will behave as bases in liq. $SO_2$.

The above discussion shows that since $SOCl_2$, $SOBr_2$, $SO(SCN)_2$, etc, give $SO^{2+}$ ions in liq. $SO_2$, they behave as acids in this solvent. On the other hand, the substances *viz.* $Cs_2SO_3$, $[N(CH_3)_4]_2\,SO_3$, $K_2SO_3$ which give $SO_3^{2-}$ ions in this solvent act as bases.

## Chemical Reactions Taking Place in Liq. $SO_2$

The chemical reactions that take place in liq. $SO_2$ can be of the following types:

**1. Metathetical (or Precipitation) Reactions**

Some of the procriptitation reactions occuring in liq. $SO_2$ are given below:

(*i*) Thionyl chloride ($SOCl_2$) in liq. $SO_2$ has helped to prepare many compounds. For example:

$$2KBr + SOCl_2 \xrightarrow{liq.SO_2} 2KCl\downarrow + SOBr_2$$

$$2KI + SOCl_2 \xrightarrow{liq.SO_2} 2KCl\downarrow + SOI_2$$

$$NH_4(SCN) + SOCl_2 \xrightarrow{liq.SO_2} 2NH_4Cl\downarrow + SO(SCN)_2$$

$$2Ag(CH_3COO) + SOCl_2 \xrightarrow{liq.SO_2} 2AgCl\downarrow + SO(CH_3COO)_2$$

(*ii*) Examples of other precipitation reactions that can be carried in liq. $SO_2$ are:

$$SbCl_3 + 3LiI \xrightarrow{liq.SO_2} SbI_3\downarrow + 3LiCl$$

$$PbF_2 + Li_2SO_4 \xrightarrow{liq.SO_2} PbSO_4\downarrow + 2LiF$$

$$AlCl_3 + 3NaI \xrightarrow{liq.SO_2} AlI_3\downarrow + 3NaCl$$

$$BaI_2 + Zn(CNS)_2 \xrightarrow{liq.SO_2} Ba(CNS)_2\downarrow + ZnI_2$$

$$KI + [(CH_3)_3NH]Cl \longrightarrow KCl\downarrow + [(CH_3)_3NH]I$$

**2. Acid-Base Neturalisation (Salt Formation) Reactions**

In liq. $SO_2$, acid-base reaction is that in which a compound containing or making available $SO_2^{2+}$ ion (*acid*) combines with a compound containing or making available $SO_3^{2-}$ ion (*base*) to form the salt and the solvent (*i.e.* $SO_2$), e.g.

$$SOCl_2\ (acid) + Cs_2SO_3\ (base) \xrightarrow{liq.SO_2} 2CsCl\ (salt) + 2SO_2\ (solvent)$$

Actually the above reaction takes place through the following steps:

$$SOCl_2 \xrightarrow{liq.SO_2} SO^{2+} + 2Cl^-$$

$$Cs_2SO_3 \xrightarrow{liq.SO_2} 2Cs^+ + SO_3^{2-}$$

$$SOCl_2 + Cs_2SO_3 \xrightarrow{liq.SO_2} 2Cs^+Cl^- + SO^{2+} + SO_3^{2-}$$

or $SOCl_2\ (acid) + Cs_2SO_3\ (base) \xrightarrow{liq.SO_2} 2CsCl\ (salt) + 2SO_2\ (solvent)$

or $SO^{2+} + 2Cl^- + 2Cs^+ + SO_3^{2-} \xrightarrow{liq.SO_2} 2Cs^+ + 2Cl^- + 2SO_2$

or $SO^{2+}\ (solvent\ cation) + SO_3^{2-}\ (solvent\ anion) \xrightarrow{liq.SO_2} 2SO_2(solvent)$ ... (*i*)

It may be seen from equation (*i*) that neutralisation reaction in liq. $SO_2$ can also be defined as the combination of the solvent cation ($SO^{2+}$) and solvent anion ($SO_3^{2-}$) to produce the solvent ($SO_2$).

The acid - base reaction between $SOCl_2$ (acid) and $Cs_2SO_3$ (base) given above has been explained on the basis of self-ionisation of liq.$SO_2$ shown below:

$$2SO_2(\textit{Solvent}) \longrightarrow SO^{2+}(\textit{Acid ion}) + SO_3^{2-}(\textit{Base ion})$$

However, there is no experimental support for the existence of $SO^{2+}$ ions. Thus in order to explain the acid – base reaction the self-ionisation of liq. $SO_2$ is not required. The acid-base reaction can be explained on the basis of coordination model according to which the reaction proceeds through the following steps:

$$SOCl_2\ (\textit{Acid}) + Cs_2SO_3\ (\textit{Base})$$
$$\downarrow$$
$$2Cs^+ + SO_3^{2-} + SOCl_2$$
$$\rightleftharpoons$$
$$2Cs^+ + (SO_3 \cdot SOCl_2)^{2-}$$
$$\rightleftharpoons$$
$$2Cs^+ + SO_3 \cdot SO + 2Cl^-$$
$$\downarrow$$
$$2SO_2\ (\textit{Solvent}) + 2CsCl\ (\textit{Salt})$$

Examples of some other acid-base neutralisation reactions taking place in liq. $SO_2$ are given below:

| *Acid* | | *Base* | | *Salt* | | *Solvent* |
|---|---|---|---|---|---|---|
| $SO(SCN)_2$ | + | $K_2SO_3$ | $\xrightarrow{\text{liq.}SO_2}$ | 2K (SCN) | + | $2SO_2$ |
| $SOBr_2$ | + | $[N(CH_3)_4]_2\,SO_3$ | $\xrightarrow{\text{liq.}SO_2}$ | $2[N(CH_3)_4]$ Br | + | $2SO_2$ |

**3. Amphoteric Behaviour of Salts in liq. $SO_2$.**

(*i*) Amphoteric behaviour is shown by Zn (II) and Al (III) salts in liq. $SO_2$ as is evident by the reaction of $K_2SO_3$ with $ZnCl_2$, $ZnSO_3$, $Al_2(SO_3)_2$ and $AlCl_3$.

*Acidic character of $ZnCl_2$*: $\underset{\text{Acid}}{ZnCl_2} + K_2SO_3 \xrightarrow{\text{liq.}SO_2} ZnSO_3 + 2kCl$

*Basic character of $ZnSO_3$*: $\underset{\text{Base}}{ZnSO_3} + K_2SO_3 \xrightarrow{\text{liq.}SO_2} K_2[Zn(SO_3)_2]$

*Acidic character of $AlCl_3$*: $\underset{\text{Acid}}{2AlCl_3} + 3K_2SO_3 \xrightarrow{\text{liq.}SO_2} Al_2(SO_3)_3 + 6KCl$

*Basic character of $Al_2(SO_3)_3$*: $\underset{\text{Base}}{Al_2(SO_3)_3} + 3K_2SO_3 \xrightarrow{\text{liq.}SO_2} 3K_3[Al(SO_3)_3]$

(*ii*) $AlCl_3$ reacts with NaOH which acts as a base in aqueous solution to form the ppt. of $Al(OH)_3$. $AlCl_3$ also reacts with $(N(CH_3)_4]_2\,SO_3$ which acts as a base in liq. $SO_2$ to form the ppt. of $Al_2(SO_3)_3$.

$$AlCl_3 + 3NaOH \xrightarrow{H_2O} \underset{ppt.}{Al(OH)_3 \downarrow} + 3NaCl$$

or $$AlCl_3 + 3OH^- \xrightarrow{H_2O} Al(OH)_3(ppt) + 3Cl^-$$

$$2\,AlCl_3 + 3\,[N(CH_3)_4]_2SO_3 \xrightarrow{liq.SO_2} \underset{ppt.}{Al_2(SO_3)_3 \downarrow} + 6\,[N(CH_3)_4]Cl$$

or $$2\,AlCl_3 + 3SO_3^{2-} \xrightarrow{liq.SO_2} Al_2(SO_3)_3(ppt) + 6Cl^-$$

The ppt. of $Al(OH)_3$ dissolves in the excess of NaOH (base in aqueous solution) in aqueous medium to form the soluble hydroxo complex, $Na[Al(OH)_4]$. Similarly the ppt. of $Al_2(SO_3)_3$ dissolves in the excess of $[N(CH_3)_4]_2\ SO_3$ (base in liq. $SO_2$) in liq. $SO_2$ to form the soluble complex, $[N\ (CH_3)_4]_3^{\ 3+}\ [Al\ (SO_3)_3]^{3-}$

$$Al(OH)_3(ppt) + NaOH(base) \xrightarrow{H_2O} \underset{\text{Soluble hydroxo complex}}{Na[Al(OH)_4]}$$

$$Al_2(SO_3)_3\ (ppt) + 3\ [Na(CH_3)_4]_2SO_3(base) \xrightarrow{liq.\,SO_2} \underset{\text{Soluble couplex}}{2\,[N(CH_3)_4]_3\ [Al(SO_3)_3]}$$

From the soluble complex, $Na[Al\ (OH)_4]$, $Al(OH)_3$ can be precipitated by adding HCl (acid in aqueous solution) in aqueous solution.

$$\underset{\text{Soluble hydroxo complex}}{Na\ [Al\ (OH)_4]} + HCl\ (Acid) \xrightarrow{H_2O} NaCl + H_2O\ (Solvent) + Al\ (OH)_3\ (ppt)\downarrow$$

Similarly, from the soluble complex $[N(CH_3)_4]\ [Al\ (SO_3)_3]$, $Al_2(SO_3)_3$ can be precipitated by adding $SO_2Cl_2$ (acid in liq. $SO_2$)

$$\underset{\text{Soluble complex}}{2\ [N(CH_3)_4]\ [Al(SO_3)_3]} + 3SOCl_2 + (Acid) \xrightarrow{liq.\ SO_2} 6[N(CH_3)_4]\ Cl$$
$$+ 6SO_2\ (Solvent) + Al_2\ (SO_3)_2\ (ppt)\downarrow$$

(*iii*) $GaCl_3$ also shows amphoteric behaviour in liq. $SO_2$. This can be explained in the same way.

**4. Solvation Reactions: Formation of Solvates**

Like water and liq. $NH_3$. liq. $SO_2$ also forms additon compounds with solutes. These addition compounds are called solvates and contain one of more molelcules of sulphur dioxide as sulphur dioxide of crystallisation. Examples of some solvates formed in liq. $SO_2$ are LiI. $2SO_2$, KBr. $4SO_2$, MI.$4SO_2$ (M = Na, K or Rb), $MI_2$. $4SO_2$ (M= Ca, Sr or Ba), $AlCl_3$, $2SO_2$, KCNS, $nSO_2$ ($n$ = 0.5, 1 or 2), $[(CH_3)_4N]\ Cl.SO_2$, $[(CH_3)_4N]_2\ 2SO_4.3SO_2$.

In the formation of solvates, $SO_2$ molecule acts as a Lewis acid [electron pair acceptor) and the solute acts as a Lewis base (electron pair donor). With the decrease in the Lewis base character, the solubilities of alkali metal halides in liq. $SO_2$ also decrease in the order $I^- > Br^- > Cl^- > F^-$.

**5. Solvolytic Reactions: Solvolysis in Liq. $SO_2$**

Solvolytic reactions of the salts in liq. $SO_2$ are much less common than in many protonic solvents (*e.g.* $H_2O$, $NH_3$). Some such reactions are given below. In these reactions concentration of $SO^{2+}$ (solvent cation) or $SO_3^{2-}$ (solvent anion) is increased. For example:

(*a*) $CH_3COONH_4$ undergoes solvolysis reaction in liq. $SO_2$. Here the concentration of $SO_3^{2-}$ ions is increased

$$2CH_3COONH_4 + 2SO_2(liq) \longrightarrow (NH_4)_2SO_3 + (CH_3COO)_2SO$$
$$(CH_3COO)_2SO \longrightarrow SO_2 + (CH_3CO)_2O$$
$$2CH_3COONH_4 + 2SO_2(liq) \longrightarrow \underset{\text{Base}}{(NH_4)_2SO_3} + (CH_3CO)_2O + SO_2$$

(*b*) Some covalent halides like $PCl_5$, $UCl_6$, $WCl_6$, $AsCl_5$, etc. undergo solvolysis reactions in liq. $SO_2$ and form oxyhalides. In all these reactions the concentration of $SO^{2+}$ ion is increased.

$$PCl_5 + SO_2(liq.) \longrightarrow POCl_3 + SOCl_2$$
$$UCl_6 + 2SO_2(liq.) \longrightarrow UO_2Cl_2 + 2SOCl_2$$
$$WCl_6 + SO_2(liq.) \longrightarrow WOCl_4 + SOCl_2$$
$$AsCl_5 + SO_2(liq.) \longrightarrow AsOCl_3 + SOCl_2$$
$$NbCl_5 + SO_2(liq.) \longrightarrow NbOCl_3 + SOCl_2$$
$$Zn(C_2H_5)_2 + 2SO_2(liq.) \longrightarrow ZnSO_3 + SO(C_2H_5)_2$$
$$(\textit{In water}: Zn(C_2H_5)_2 + 2H_2O \longrightarrow Zn(OH)_2 + 2C_2H_6)$$

**6. Complex Formation Reactions**

(*i*) When the solutions of the salts like $AlCl_3$, $ZnCl_2$ etc. in liq $SO_2$ react with the excess of a compound containing $SO_3^{2-}$ ions (*e.g.* $K_2SO_3$), soluble sulphito complexes are formed. For example:

(*a*) $$2AlCl_3 + 6K_2SO_3\ (\textit{Excess}) \xrightarrow{\text{liq.}SO_2} \underset{\text{Soluble sulphito complex}}{2K_3[Al(SO_3)_3]} + 6\,KCl$$

or $$Al^{3+} + 3SO_3^{2-}\ (\textit{Excess}) \xrightarrow{\text{liq.}SO_2} [Al(SO_3)_3]^{3-}$$

The above reaction takes place through the following two steps:

$$2AlCl_3 + \underset{\text{Base}}{3K_2SO_3} \xrightarrow{\text{liq.}SO_2} \underset{\text{ppt}}{Al_2(SO_3)_3} + 6KCl$$

$$\underset{\text{ppt}}{Al_2(SO_3)_3} + \underset{\text{Base}}{3K_2SO_3} \xrightarrow{\text{liq.}SO_2} \underset{\text{Soluble sulphito complex}}{2K_3[Al(SO_3)_3]}$$

*On adding*: $$2AlCl_3 + \underset{\text{Base(excess)}}{6K_2SO_3} \xrightarrow{\text{liq.}SO_2} 6KCl + 2K_3[Al(SO_3)_3]$$

or $$2Al^{3+} + 6SO_3^{2-} \xrightarrow{\text{liq.}SO_2} 2[Al(SO_3)_3]^{3-}$$

or $Al^{3+} + 3SO_3^{2-} \xrightarrow{\text{liq. } SO_2} [Al(SO_3)_3]^{3-}$

(*b*) $ZnCl_2 + 2K_2SO_3(\textit{Excess}) \xrightarrow{\text{liq } SO_2} K_2[Zn(SO_3)_2] + 2KCl$
Soluble sulphito complex

or $Zn^{2+} + 2SO_3^{2-}\ (\textit{Excess}) \longrightarrow [Zn(SO_3)_2]^{2-}$

This reaction takes place through the following steps.

$$\underset{\text{Base}}{ZnCl_2 + K_2SO_3} \xrightarrow{\text{liq.}SO_2} \underset{\text{ppt}}{ZnSO_3} + 2KCl$$

$$\underset{\text{ppt}}{ZnSO_3} + \underset{\text{Base(excess)}}{K_2SO_3} \xrightarrow{\text{liq.}SO_2} \underset{\text{Soluble sulphito complex}}{K_2[Zn(SO_3)_2]}$$

---

On *adding*: $ZnCl_2 + \underset{\text{Base (excess)}}{2K_2SO_3} \xrightarrow{\text{liq.}SO_2} 2KCl + K_2[Zn(SO_3)_2]$

or $Zn^{2+} + 2SO_3^{2-}(\textit{Excess}) \longrightarrow [Zn(SO_3)_2]^{2-}$

(*ii*) Solubility of $I_2$ in liq. $SO_2$ is greatly increased when KI or RbI is added to it. This is due to the formation of complex compounds viz. $KI_3$ or $RbI_3$.

$$\underset{(M=K,Rb)}{MI} + I_2 \xrightarrow{\text{liq.}SO_2} \underset{\text{Soluble complex}}{MI_3}$$

(*iii*) When $HgI_2$ or $CdI_2$ is dissolved in liq. $SO_2$ in presence of KI, the solubility of these iodides is increased due to the formation of $K_2[MI_4]$ which is a complex compound.

$$\underset{(M=Hg,Cd)}{MI_2} + 2KI \xrightarrow{\text{liq.}SO_2} \underset{\text{Soluble complex compound}}{K_2[MI_4]}$$

(*iv*) Other reactions showing the formation of hexahalo complexes in liq. $SO_2$ by the action of covalent halides (*e.g.* $SbCl_3$, $SbCl_5$, etc.) on alkali metal halides (*e.g.* KCl) or NOCl or $[N(CH_3)_4]Cl$ are given below:

$$SbCl_3 + 3KCl \xrightarrow{\text{liq.}SO_2} K_3[SbCl_6]$$

$$SbCl_5 + KCl \xrightarrow{\text{liq.}SO_2} K_3[SbCl_6]$$

$$SbCl_5 + NOCl \xrightarrow{\text{liq.}SO_2} NO[SbCl_6]$$

$$SbCl_5 + [N(CH_3)_4]Cl \xrightarrow{\text{liq.}SO_2} [N[(CH_3)_4][SbCl_6]$$

$$SbCl_5 + CH_3COCl \xrightarrow{\text{liq.}SO_2} [CH_3CO]^+[SbCl_6]^-$$

*Hexahalo complexes*

**7. Redox Reactions**

Liq. $SO_2$ itself does not act as a strong oxidising /reducing agent. It serves only as a medium to form any redox reactions. For example:

(*a*) A sulphite ($R_2SO_3$) in liq. $SO_2$ reduces iodine to iodide and itself oxidised to sulphate ($R_2SO_4$).

$$\underset{(S=+4)}{2R_2SO_3} + \underset{(I=0)}{I_2} \xrightarrow{\text{liq.}SO_2} \underset{(S=+6)}{R_2SO_4} + \underset{(I=-1)}{2RI} + SO_2$$

(*b*) $SbCl_5$ in liq.$SO_2$ oxidises KI to free $I_2$ and is itself reduced to $SbCl_3$.

$$\underset{(Sb=+5)}{5SbCl_5} + \underset{(I=-1)}{I_2} \xrightarrow{\text{liq.}SO_2} \underset{(I=0)}{3I_2} + \underset{(Sb=+3)}{SbCl_3} + \underset{(Sb=+3)}{2K_3[SbCl_6]}$$

(*c*) $FeCl_3$ in liq.$SO_2$ oxidises KI to free $I_2$ and is itself reduced to $FeCl_2$.

$$\underset{(Fe=+3)}{2FeCl_3} + \underset{(I=-1)}{2KI} \xrightarrow{\text{liq.}SO_2} \underset{(Fe=+2)}{2FeCl_2} + \underset{(I=0)}{I_2} + 2KCl$$

**8. Organic Reactions in Liquid Sulphur Dioxide.**

Organic reactions where relatively inert and non-inflammable solvents are required, liquid $SO_2$ serves as the best medium. Some of the important reactions are:

(*i*) **Sulphonation:** Good yields have been reported for the sulphonation of aromatic compounds in liquid $SO_2$.

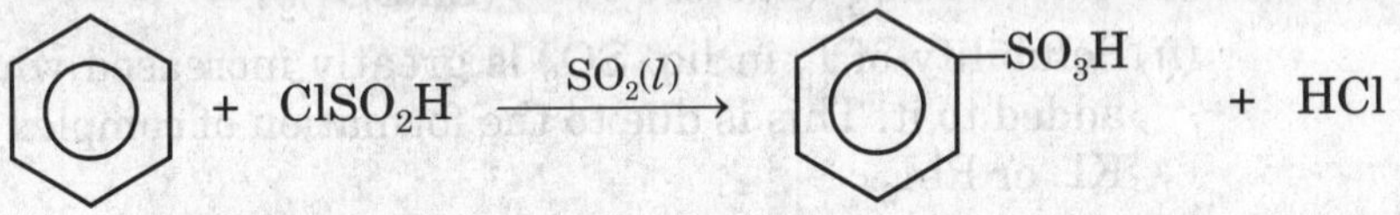


Chloro sulphonic acid Benzene sulphonic acid

(*b*) **Bromination:** Both for additon and substituion reactions with bromine,liquid $SO_2$ has been used.

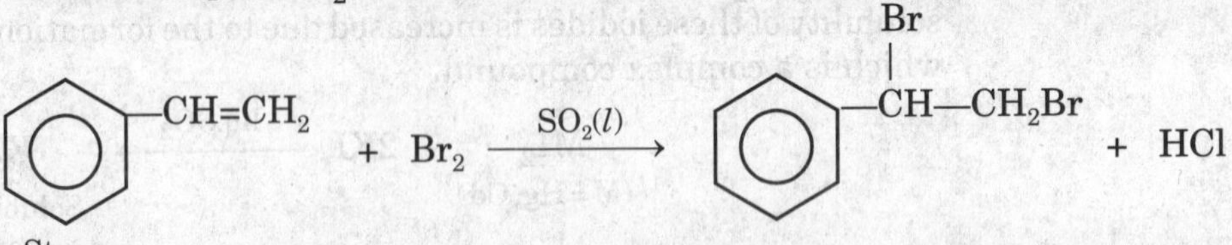


Styrene

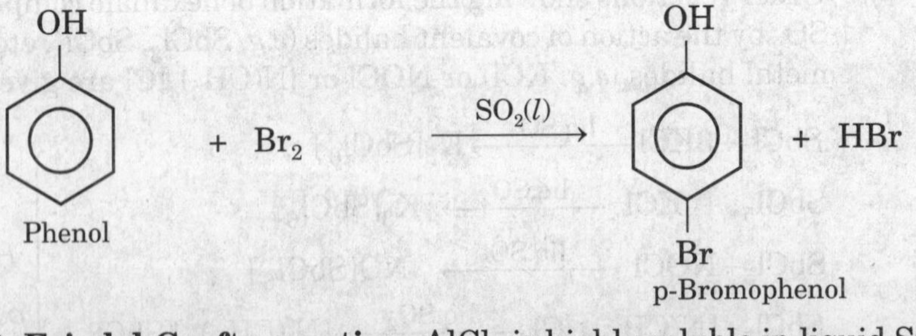


Phenol

p-Bromophenol

(*c*) **Friedel-Crafts reaction.** $AlCl_3$ is highly soluble in liquid $SO_2$ and is, therefore, a preferred solvent over ether which is highly inflammable.

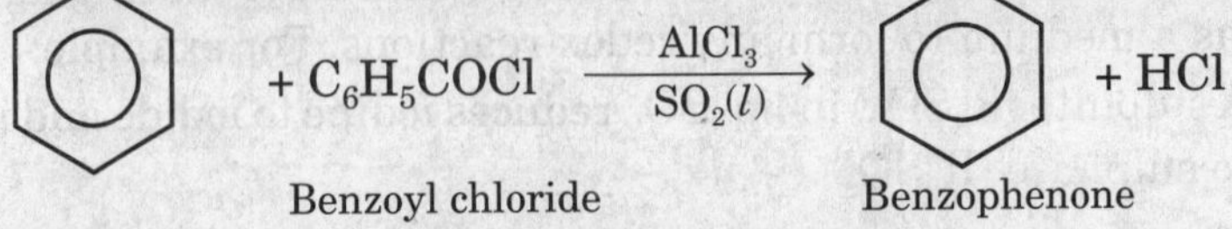


Benzoyl chloride Benzophenone

## UNIVERSITY QUESTIONS

**1.** (*a*) What is non-aqueous solvent? Describe three different types of reactions taking place in liq. $NH_3$.

(*b*) Write short note on: Liq. $SO_2$ as a solvent

(*Pd. Ravishankar shukla 2010*)

**2.** (*a*) Explain why $SO_2$ is a better solvent for organic compounds.

(*b*) What are the non-aqueous solvents ? Discuss their classification.

(*c*) In what respect, liq. $NH_3$ is a better solvent than $H_2O$ ?

(*Himachal Pradesh 2010*)

(*d*) Complete the reactions

(*i*) $HgI_2 + KI \xrightarrow{\text{Liq.}SO_2}$ .... (*ii*) $SOCl_2 + KI \xrightarrow{\text{Liq.}NH_3}$ ....

**3.** Discuss the chemistry of liquid $NH_3$ as a solvent. What are the advantages and disadvantages of liq. $NH_3$ as a solvent? (*Kanpur 2010*)

**4.** Write down the nature of products obtained when AgCl is treated with $Ba(NO_3)_2$ in liq. $NH_3$. (*Pune 2010*)

**5.** (*a*) Describe the six comparative chemical reactions in water and liquid ammonia.

(*b*) How does the viscosity change the character of a solvent?

(*c*) Give two examples to show that by changing the solvent the course of a chemical reaction can be changed. (*Rajasthan 2010*)

**6.** Choose the correct answer: Aprotic solvent is–

(*a*) HCl (*b*) $CH_3$ COOH

(*c*) $CCl_4$ (*d*) HCN (*Barkatulhah 2010*)

**7.** (*a*) With reference to liquid ammoinia, give its auto-ionisation reaction and two balanced equations for each of the following:

(*i*) Acid-base reaction (*ii*) Solvolysis reaction.

(*b*) What do you understand by aprotic solvent and protogenic solvents? Give examples.

(*c*) Explain the complex formation and precipitation reaction in liquid $SO_2$.

(*Barkatulhah 2010*)

**8.** Explain why liq. $NH_3$ is a better solvent for covalent compounds.

(*Mumbai 2010*)

**9.** (*i*) Explain acid-base reaction of liquid $NH_3$ or liquid $SO_2$.

(*ii*) Give one example of protic and aprotic solvent

(*Gujrat Technological University 2010*)

**10.** (*i*) What are non-aqueous solvlents ? Discuss the acid-base, redox and precipitation reactions occurring in liquid sulphur dioxide.

(*ii*) Describe the general characteristics of non-aqueous solvents.

(*iii*) Discuss the following reactions in liquid sulphur dioxide:

(*a*) Acid-Base reactions

(*b*) Complex formation reactions

(*c*) Precipitation reactions

(*d*) Reaction with organic compounds

(*e*) Redox reactions. (*Ranchi 2010*)

**11.** (*a*) Discuss the classification of non-aqueous solvents on the basis of "Like dossolves Like".

(*b*) Discuss acid-base reaction in liquid $NH_3$ and liquid $SO_2$ giving one example of each. (*Nagpur 2010*)

**12.** Why the electrical conductivity of liq.$NH_3$ increases on addition of $NH_4Cl$? (*GND Amritsar 2010*)

**13.** (*a*) What happens when:

(*i*) $BiI_3$ reacts with $KNH_2$ in liquid ammonia

(*ii*) NaH is dissolved in liquid ammonia.

(*b*) Give reasons

(*i*) Metal - ammonia solutions are extremely good conductors of electricity.

(*ii*) HCl is a strong acid both in $H_2O$ and liquid ammonia.

(*iii*) Acetic acid behaves as weak acid in aqueous solution but shows strong acidic properties in liq. $NH_3$.

(*c*) Write note on "Levelling and differentiating solvents".

(*d*) What are protic and sprotic solvents? Explain giving suitable examples. (*Utkal 2010*)

**14.** (*a*) Why AgCl dissolves in liq. $NH_3$ ? (*GND Amritsar 2010*)

(*b*) Why liq. $SO_2$ is a better solvent for organic compounds?

**15.** (*i*) Give an account of following types of reactions occuring in liq. $SO_2$:

(*a*) Acid – base reactions (*b*) Solvolysis reaction (*c*) Precipitation reactions (*d*) Alkali metal solution in liquid $NH_3$

(*ii*) $K_2SO_3$ is a base and $SOCl_2$ is an acid in liq. $NH_3$. Explain (*Lucknow 2010*)

**16.** (*a*) Describe the role of the following physical properties on the role of solvent in chemical reaction (*i*) M. pt and B. pt. (*ii*) Heat of fusion and heat of vaporisation.

(*b*) Mention some important advantages which liq.$SO_2$ possesses as *a* solvent.

(*c*) AgCl is soluble in liq. $NH_3$ but insoluble in water. Explain. (*Kurkshetra 2011*)

**17.** (*a*) Describe the following reactions in liq. $NH_3$.

(*i*) Solvolytic reactions

(*ii*) Acid-base neutralisation reactions

(*b*) Account for the following:

(*i*) $K_2SO_3$ is soluble in liq $SO_3$.

(*ii*) $SO_2$ acts both as Lewis acid and as Lewis base

(*iii*) Liq. $NH_3$ exhibits association whereas liq.$SO_2$ does not

(*Kurukshetra 2011*)

**18.** (*a*) Explain the folloiwng:

(*i*) Solution of Na in liq. $NH_3$ is blue and good reducing agent.

(*ii*) The electrical conductivity of liq. $NH_3$ is increased when $NH_4Cl$ is dissolved in it.

(*b*) Discuss the following reactions in liq. $SO_2$ with suitable examples:

(*i*) Acid-base reactions

(*ii*) Complex formation reactions. (*MD Rohtak 2011*)

**19.** Complete the following reactions:

(*i*) $SnI_4 + 4KNH_2 \xrightarrow{\text{Liq.}NH_3} ....$

(*ii*) $Cr_2SO_3 + SOCl_2 \xrightarrow{\text{Liq. }SO_2} ....$

(*iii*) $SbCl_5 + NaCl \xrightarrow{\text{Liq. }SO_2} ....$ (*MD Rohtak 2011*)

**20.** (*a*) Discuss the metathetical reactions in liq. $NH_3$ and liq. $SO_2$ with one example each

(*b*) Classify the solvents on the basis of donation and acceptance of $H^+$ ion.

(*Nagpur 2012*)

**21.** (*a*) What are non-aqueous solvents? Give their characteristics and classification based on protonic concept. Compare the solvent reactions of liq. $NH_3$ and $H_2O$, giving suitable examples

(*b*) Alkali metals give blue solution in liq. $NH_3$ Explain.

(*Ch. Charan Singh, Meerut 2012*)

**22.** (*i*) Give chemical reaction of $KNH_2$ with a Zn(II) salt in liq. $NH_3$

(*ii*) Give an account of the following reactions occuring in liq. $SO_2$.

(*a*) Acid-base reactions (*b*) Solvoysis reactions (*c*) Precipitation reactions

(*Lucknow 2013*)

# 3
CHAPTER
# Acids and Bases

## Introduction

Throughout the history of chemistry, various acid-base concepts have been proposed and used. In this section, we shall briefly discuss the different concepts put forth from time to time. Each of these can be applied with advantage in appropriate circumstances. In a given situation, the chemist uses the concept that suits his purpose best.

## I. Classical Concept (Operational Definitions)

The earliest classification of substances into acids and bases was based upon the characteristic properties possessed by each one of them. For example,

**Acid** was defined as a substance whose aqueous solution possessed the following characteristic properties:

(*i*) conducts electricity (*ii*) reacts with active metals like zinc, magnesium etc. to give hydrogen (*iii*) turns blue litmus red (*iv*) has a sour taste. (*v*) whose acidic properties disappear on reaction with a base.

**Base** was defined as a substance whose aqueous solution possessed the following characteristic properties:

(*i*) conducts electricity (*ii*) turns red litmus blue (*iii*) has a bitter taste (*iv*) has a soapy (slippery) touch (*v*) whose basic properties are destroyed on reaction with an acid.

The above definitions of acids and bases are called *operational definitions as they are based upon certain operations* (*i.e.*, tests) to be performed on the substances. The above definitions of acids and bases have been replaced by *conceptual definitions* (put forward by Arrhenius, Bronsted-Lowry and Lewis) which go into the causes of the observed behaviour based upon structure and composition of the substances.

## II. Arrhenius Concept: Water Ion Concept

To explain the observed properties of electrolytes, Arrhenius in 1884 put forward a theory, popularly known as '*Arrhenius theory of ionization*'. The main idea contained in this theory was that when an electrolyte is dissolved in water, it dissociates into positively and negatively charged ions.

Based upon this theory, Arrhenius gave the definitions of acids and bases as follows:

## Acids

An acid is defined as a substance which contains hydrogen and which when dissolved into water gives hydrogen ions ($H^+$)

## Arrhenius Classification of Acids

On the basis of Arrhenius definition of acids, acids have been classified as strong acids and weak acids.

**1. Strong acids.** Substances like HCl, $HNO_3$ and $H_2SO_4$ containing hydrogen, when dissolved in water dissociate completely into $H^+$ ions and the negative ions as follows:

$$HCl \xrightarrow{\text{Water}} H^+ + Cl^-$$

$$HNO_3 \xrightarrow{\text{Water}} H^+ + NO_3^-$$

$$H_2SO_4 \xrightarrow{\text{Water}} 2H^+ + SO_4^{2-}$$

Such acids are called **strong acids.**

**2. Weak acids.** Substances like acetic acid ($CH_3COOH$), carbonic acid ($H_2CO_3$), phosphoric acid ($H_3PO_4$), etc., when dissolved in water dissociate into ions to a small extent. Thus, an equilibrium is set up between the ions and the undissociated molecules as follows:

$$CH_3COOH \xrightleftharpoons{\text{Water}} CH_3COO^- + H^+$$

$$H_2CO_3 \xrightleftharpoons{\text{Water}} 2H^+ + CO_3^{2-}$$

$$H_3PO_4 \xrightleftharpoons{\text{Water}} 3H^+ + PO_4^{3-}$$

Such acids are called **weak acids**. In general, we can write

$$\underset{\text{(Acid)}}{HA} \xrightleftharpoons{\text{Water}} H^+ + A^-$$

In fact, all the properties of an acid are due to the $H^+$ ions which it furnishes in aqueous solution.

## Formation of hydronium ion ($H_3O^+$) in aqueous solution

$H^+$ ion is simply a proton which is very small in size (radius – $10^{-13}$ cm). It has a strong electric field. It takes up a lone pair of electrons from water molecule 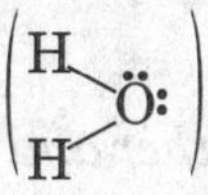 to form 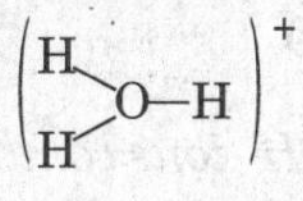 *i.e.*, $H_3O^+$, called **hydronium ion**.

Hence we simply represent $H^+$ ion as $H_3O^+$ (*aq*). Thus the dissociation of an acid in aqueous solution can be represented as:

$$HCl\ (\textit{Strong acid}) + H_2O \longrightarrow H_3O^+\ (aq) + Cl^-\ (aq)$$

$$CH_3COOH\ (\textit{Weak acid}) \rightleftharpoons H_3O^+\ (aq) + CH_3COO^-\ (aq)$$

## Bases

A base is defined as a substance which contains hydroxyl groups and which when dissolved in water gives hydroxide ions ($OH^-$)

## Classification of Arrhenius Bases

On the basis of Arrhenius definition of bases, bases have been classified as strong bases and weak bases.

**1. Strong bases.** Substances like NaOH and KOH containing the –OH groups, when dissolved into water, dissociate **completely** to give $OH^-$ ions as follows:

$$NaOH \xrightarrow{\text{Water}} Na^+ + OH^-$$

$$KOH \xrightarrow{\text{Water}} K^+ + OH^-$$

These are called **strong bases.**

(*ii*) **Weak bases.** Substances like $NH_4OH$, Ca $(OH)_2$, Mg $(OH)_2$, Al $(OH)_3$, etc. dissociate to a small extent as follows:

$$NH_4OH \xrightleftharpoons{\text{Water}} NH_4^+ + OH^-$$

$$Ca(OH)_2 \xrightleftharpoons{\text{Water}} Ca^{2+} + 2OH^-$$

$$Al(OH)_3 \xrightleftharpoons{\text{Water}} Al^{3+} + 3OH^-$$

These are called **weak bases**.

In aqueous solution $OH^-$ ion is generally represented as $OH^-$ (*aq*). Hence the dissociation of bases in aqueous solution may be represented as:

$NaOH$ (*strong base*) + $H_2O \rightarrow Na^+(aq) + OH^-$ (*aq*)

$NH_4OH$ (*weak base*) + $H_2O \rightleftharpoons NH^+{}_4\,(aq) + OH^-\,(aq)$

## Neutralisation Reaction According to Arrhenius Concept

According to Arrhenius concept of acids and bases, neutralisation reaction is a process in which an acid and a base react together to form salt and $H_2O$.

**Example** $\underset{\text{Acid}}{HCl} + \underset{\text{Base}}{NaOH} \longrightarrow \underset{\text{Salt}}{NaCl} + H_2O$

Actually this reaction takes place through the following steps:

$$\underbrace{H^+ + Cl^-}_{\text{Acid}} + \underbrace{Na^+ + OH^-}_{\text{Base}} \longrightarrow \underbrace{Na^+ + Cl^-}_{\text{Salt}} + H_2O$$

or $$H^+ + OH^- \longrightarrow H_2O$$

*Thus neutralisation is a process in which $H^+$ and $OH^-$ ions combine together and form unionised molecules of water ($H_2O$).* As a result of this process, the characteristic properties of acids and bases are destroyed.

## Basicity or Protocity of an Arrehenius Acid.

*The number of $H^+$ (or $H_3O^+$) ions given by one molecule of an Arrehenius acid is known as the basicity or protocity of the acid.* For example the basicity of HCl, $H_2SO_4$ and $H_3PO_4$ is one, two and three respectively as shown below:

$$HCl \rightleftharpoons H^+ + Cl^-, H_2SO_4 \rightleftharpoons 2H^+ + SO_4$$

$$H_3PO_4 \rightleftharpoons 3H^+ + PO_4^{3-}$$

## Monobasic (Monoprotonic) and Polybasic (Polyprotonic) Acids

The acids giving one, two and three $H^+$ ions in aqueous solution are called *monobasic (monoprotonic) acids* (*e.g.* HCl, HF, $HNO_3$ $CH_3COOH$ etc.), *dibasic (diprotonic) acids* [*e.g.* $H_2SO_4$, $H_2CO_3$, $H_2S$, $(COOH)_2$ etc.] and *tribasic (triprotonic) acids* (e.g. $H_3PO_4$ etc). Dibasic and tribasic acids are called *polybasic or polyprotonic acids.*

## Acidity or Hydroxity of an Arrhenius Base.

*The number of $OH^-$ ions given by one molecule of an Arrehenius base is known as the acidity or hydroxity of the base.* For example the acidity of NaOH, $Ca(OH)_2$ and $Al(OH)_3$ is one, two and three respectively as shown below:

$$NaOH \rightleftharpoons Na^+ + OH^-, Ca(OH)_2 \rightleftharpoons Ca^{2+} + 2\,OH^-$$

$$Al(OH)_3 \rightleftharpoons Al^{3+} + 3OH^-$$

## Monoacidic (Monohydroxic) and Polyacidic (Plyhydroxic) Bases

The bases giving one, two and three $OH^-$ ions in aqueous solution are called *monoacidic (monohydroxic) bases* (e.g. NaOH, $NH_4OH$ etc), *diacidic (dihydroxic) bases* [e.g. Ca $(OH)_2$, Mg $(OH)_2$ etc] and *triacidic (trihydroxic) acids* [*e.g.* Al $(OH)_3$, $Fe(OH)_3$ etc]. Diacidic and triacidic bases are called *polyacidic or polyhydroxic bases.*

## Relative Strength of Acids and Bases According to Arrhenius Concept

According to Arrhenius concept, greater is the number of $H^+$ ions produced by an acid in aqueous solution, stronger is the acid. Similarly greater is the number of $OH^-$ ions produced by a base in aqueous solution, stronger is the base.

We know that greater is the value of acid dissociation constant of an acid ($K_a$), greater is the amount of $H^+$ ions produced by the dissociation of the acid in aqueous solution and hence stronger is the acid. Thus greater is the value of $K_a$ of an acid, greater is the strength of weak acids. Similarly greater in the value of base dissociation constant of a base ($K_b$), greater is the amount of $OH^-$ produced by the dissociation of the base in aqueous solution, greater is the strength of the weak bases.

## Utility of Arrhenius Concept

(*i*) Since the reaction representing neutralisation process involves the combination $H^+$ and $OH^-$ ions, the approximately constant molar heat of neutralisation would be expected. Thus the constant heat of neutralisation of a strong acid by a strong base is readily understandable in terms of this concept.

(*ii*) This concept has offered a means of correlating catalytic behaviour with the concentration of the $H^+$ ion.

(*iii*) The Arrhenius concept of acids and bases is able to explain a number of phenomena like neutralization, salt hydrolysis, strength of acids and bases etc.

## Limitations of Arrhenius Concept

(*i*) *Nature of $H^+$ ions and $OH^-$ ions:* According to Arrhenius concept, acids and bases were defined as substances which gave $H^+$ ions and $OH^-$ ions respectively in aqueous solution. But as discussed above, these ions cannot exist as such in the aqueous solution but exist as hydrated ions, written as $H^+$ (*aq*) and $OH^-$ (*aq*).

(*ii*) *Inability to explain acidic and basic character of certain substances:* Arrehenius concept demands that an acid must contain hydrogen and a base must contain hydroxyl ($OH^-$) groups. However, a number of substances like $NH_3$, $Na_2CO_3$, CaO etc., are known to be basic but do not contain any hydroxyl groups. Similarly, a number of substances like $CO_2$, $SO_2$, $SO_3$ etc., are known to be acidic but do not contain any hydrogen.

This limitation is, however, overcome if water is supposed to play an important role. This is clear from the following reactions:

$$NH_3\,(g) + H_2O \rightleftharpoons NH^+{}_4(aq) + OH^-\,(aq)$$

$$Na_2CO_3\,(s) + H_2O \rightleftharpoons NaHCO_3(aq) + NaOH\,(aq)$$

$$Na_2CO_3\,(s) + H_2O \rightleftharpoons NaHCO_3(aq) + Na^+\,(aq) + OH^-\,(aq)$$

$$CaO + H_2O \rightleftharpoons Ca^{2+}\,(aq) + 2\,OH^-\,(aq)$$

$$CO_2 + H_2O \rightleftharpoons CO_3{}^{2-}\,(aq) + 2H^+\,(aq)$$

$$SO_2 + H_2O \rightleftharpoons SO_4{}^{-2}\,(aq) + 2H^+\,(aq)$$

(*iii*) *Inability to explain the reaction between an acid and base in absence of water:* Two important examples are:

$$\underset{Base}{NH_3(g)} + \underset{Acid}{HCl(g)} \longrightarrow \underset{Salt}{NH_4Cl(s)}$$

$$CaO\,(s) + SO_3(g) \longrightarrow CaSO_4(s)$$

(*iv*) According to this concept, HCl is regarded as an acid only when dissolved in $H_2O$ and not in some other solvent such as $C_6H_6$ or when it exists in the gaseous form.

(*v*) It cannot account for the acidic and basic character of the materials in non-aqueous solvents, e.g., $NH_4NO_3$ in liq. $NH_3$ acts as an acid, though it does not give $H^+$ ions. Similarly many organic materials and $NH_3$, which do not have $OH^-$ ions at all, are actually known to show basic character.

(*vi*) The neutralisation process is limited to those reactions which can occur in aqueous solutions only, although reactions involving salt formation do occur in many other solvents and even in the absence of solvents.

(*vii*) It cannot explain the acidic character of certain salts such as $AlCl_3$ in aqueous solution.

## III. Bronsted-Lowry Concept: Proton Donor-Acceptor Concept

**Definition** – To overcome the limitations of Arrhenius concept, Bronsted (a Danish chemist) and Lowry (an English chemist) independently and simultaneously, in the year 1923, proposed a broader concept of acids and bases.

According to this concept, an acid is defined as any hydrogen containing material (a molecule or a cation or an anion) that can release a proton ($H^+$) to any other substance, whereas a base is any substance (a molecule or a cation or an anion) that can accept a proton from any other substance. In short, *an acid is a proton-donor and a base is a proton-acceptor*. As an example consider the reaction between HCl and $H_2O$ as given below:

one proton ($H^+$) going to $H_2O$ — one proton ($H^+$) going to $Cl^-$

$$\underset{\substack{\text{Acid}\\(H^+\text{ donor})}}{HCl} + \underset{\substack{\text{Base}\\(H^+\text{ acceptor})}}{H_2O} \rightleftharpoons \underset{\substack{\text{Acid}\\(H^+\text{ donor})}}{H_3O^+} + \underset{\substack{\text{Base}\\(H^+\text{ acceptor})}}{Cl^-}$$

In this reaction HCl donates a proton to $H_2O$ and is, therefore, an acid. Water, on the other hand, accepts a proton from HCl, and is, therefore, a base. In the reverse reaction which at equilibrium proceeds at the same rate as the forward reaction, $H_3O^+$ ion donates a proton to $Cl^-$ ion and hence $H_3O^+$ ion is an acid. $Cl^-$ ion, because it accepts a proton from $H_3O^+$ ion, is a base.

### Points to Be Noted

(*i*) A substance acts as a Bronsted acid, *i.e.* gives a proton only when another substance to accept the proton (*i.e.* base), donated by the acid is also present. For example $CH_3COOH$ does not act as an acid in $C_6H_6$ solution because the latter does not accept the proton donated by $CH_3COOH$.

(*ii*) According Bronsted-Lowry concept the presence of hydroxyl groups ($OH^-$) is not essential for a substance to act as a base. The only requirement is that the substance should have tendency to accept a proton.

### Examples of Bronsted-Lowry Acids and Bases

As already stated, Bronsted-Lowry acids and bases may be molecular, cationic and anionic species. Examples of such species are given below:

| *Type of Species* | *Acids* | *Bases* |
|---|---|---|
| *Molecular* | HCl, HBr, $HClO_4$, $HNO_3$, $H_2SO_4$, $H_3PO_4$, $H_2O$ | $NH_3$, $N_2H_4$, amines, $H_2O$, $NH_2OH$ |
| *Cationic* | $NH_4^+$, $[Fe(H_2O)_6]^{3+}$, $[Al(H_2O)_6]^{3+}$, $H_3O^+$ | $[Fe(H_2O)_5(OH)]^{2+}$, $[Al(H_2O)_5(OH)]^{2+}$ |
| *Anionic* | $HCO_3^-$, $HSO_4^-$ | $Cl^-$, $Br^-$, $I^-$, $OH^-$, $CO_3^{2-}$, $SO_4^{2-}$, $H^-$, $NH_2^-$ $OH^-$, $HS^-$ |

## Conjugate Acid-Base Pairs

In order to understand the concept of conjugate acid-base pair let us consider the following equilibrium.

$$\underset{\text{Acid}}{HCl} + \underset{\text{Base}}{CH_3COOH} \rightleftharpoons \underset{\text{Acid}}{CH_3COOH_2^+} + \underset{\text{Base}}{Cl^-}$$

$(H^+)$ $(H^+)$

If we consider the forward reaction of this equilibrium, we find that since HCl donates a proton ($H^+$) to $CH_3COOH$, HCl acts as a Bronsted acid. Since $CH_3COOH$ accepts a proton from HCl, $CH_3COOH$ acts as a Bronsted base. If we consider the reverse reaction of this equilibrium, we find that since $CH_3COOH_2{}^+$ donates a proton to $Cl^-$ ion, $CH_3COOH^+{}_2$ acts as a Bronsted acid. Since $Cl^-$ ion accepts a proton donated by $CH_3COOH_2{}^+$, it acts as a Bronsted base In the given equilibrium the two acid-base pairs namely HCl (acid) – $Cl^-$(base) and $CH_3COOH_2^+$ (acid) – $CH_3COOH$ (base) are called conjugate acid-base pairs.

One conjugate acid (HCl)– base ($Cl^-$) pair

$$\underset{\substack{\text{Acid}\\(H^+\text{ ion donor})}}{HCl} + \underset{\substack{\text{Base}\\(H^+\text{ ion acceptor})}}{CH_3COOH} \rightleftharpoons \underset{\substack{\text{Acid}\\(H^+\text{ ion donor})}}{CH_3COOH_2^+} + \underset{\substack{\text{Base}\\(H^+\text{ ion acceptor})}}{Cl^-}$$

Other conjugate acid ($CH_3COOH_2^+$)– base ($CH_3COOH$) pair

From the above acid-base equilibrium the following points may be noted:

(*i*) There is one acid and one base on each side of the equilibrium. If HCl (acid) is designated as $Acid_1$ or $A_1$, then its conjugate base *viz* $Cl^-$ ion may be designated as $Base_1$ or $B_1$. Similarly $CH_3COOH$(base) is designated as $Base_2$ or $B_2$ and its conjugate acid viz $CH_3COOH_2{}^+$ is designated as $Acid_2$ or $A_2$. Thus the above equilibrium can also be written as:

$$\underset{Acid_1(A_1)}{HCl} + \underset{Base_2(B_2)}{CH_3COOH} \rightleftharpoons \underset{Acid_2(A_2)}{CH_3COOH_2^+} + \underset{Base_1\ (B_1)}{Cl^-}$$

(*ii*) Each acid-base reaction has two conjugate acid-base pairs.

(*iii*) Conjugate base of a given acid is obtained by eliminating (removing) a proton from the acid. Similarly, a conjugate acid of a given base is obtained by adding a proton to the base. Thus:

$$\text{Conjugate acid} \xrightarrow{-H^+} \text{Conjugate base} + H^+ \qquad \ldots(a)$$

$$\text{Conjugate base} + H^+ \rightarrow \text{Conjugate acid} \qquad \ldots(b)$$

Above two relations *viz* (*a*) and (*b*) can be combined to get:

$$\textbf{Conjugate Acid} \underset{+H^+}{\overset{-H^+}{\rightleftharpoons}} \textbf{Conjuagate Base}$$

## Example of Conjugate Acid-Base Pairs in Some Acid-Base Reactions

Examples of conjugate acid-base pairs in some acid-base reactions are given below:

*Conjugate acid-base pair* ($Acid_1$ – $Base_1$)

*Conjugate acid-base pair* ($Base_2$ – $Acid_2$)

| *$Acid_1$* ($H^+$ ion donor) | | *$Base_2$* ($H^+$ ion acceptor) | | *$Acid_2$* ($H^+$ ion donor) | | *$Base_1$* ($H^+$ ion acceptor) |
|---|---|---|---|---|---|---|
| $HClO_4$ | + | $H_2O$ | $\rightleftharpoons$ | $H_3O^+$ | + | $ClO_4^-$ |
| $H_2SO_4$ | + | $H_2O$ | $\rightleftharpoons$ | $H_3O^+$ | + | $HSO_4^-$ |
| $HCN$ | + | $H_2O$ | $\rightleftharpoons$ | $H_3O^+$ | + | $CN^-$ |
| $HCl$ | + | $H_2O$ | $\rightleftharpoons$ | $H_3O^+$ | + | $Cl^-$ |
| $HNO_3$ | + | $H_2O$ | $\rightleftharpoons$ | $H_3O^+$ | + | $NO_3^-$ |
| $[Al(H_2O)_6]^{3+}$ | + | $H_2O$ | $\rightleftharpoons$ | $H_3O^+$ | + | $[Al\ (H_2O)_5\ (OH)]^{2+}$ |
| $[Fe(H_2O)_6]^{3+}$ | + | $H_2O$ | $\rightleftharpoons$ | $H_3O^+$ | + | $[Fe(H_2O)_5\ (OH)]^{2+}$ |
| $NH_4^+$ | + | $H_2O$ | $\rightleftharpoons$ | $H_3O^+$ | + | $NH_3$ |
| $H_2O$ | + | $CN^-$ | $\rightleftharpoons$ | $HCN$ | + | $OH^-$ |
| $H_2O$ | + | $CO_3^{2-}$ | $\rightleftharpoons$ | $HCO_3$ | + | $OH^-$ |
| $H_2O$ | + | $NH_3$ | $\rightleftharpoons$ | $NH_4^+$ | + | $OH^-$ |
| $HCl$ | + | $NH_3$ | $\rightleftharpoons$ | $NH_4^+$ | + | $Cl^-$ |
| $CH_3COOH$ | + | $NH_3$ | $\rightleftharpoons$ | $NH_4^+$ | + | $CH_3COO^-$ |
| $HSO_4^-$ | + | $NH_3$ | $\rightleftharpoons$ | $NH_4^+$ | + | $SO_4^{2-}$ |

## Acid-Base Neutralisation Reaction According to Bronsted-Lowry Concept

According to this concept, an acid-base neutralisation reaction is that in which a Bronsted acid reacts with a Bronsted base and their conjugate base and conjugate acid respectively are produced. In the neutralisation reaction, the acid (HA) gives up its proton ($H^+$) and produces its conjugate base ($A^-$). The proton given by HA is sccepted by the base (B) which is thereby converted into its conjugate acid, $BH^+$. Thus according to Bronsted-Lowry concept, the neutralisation reaction between the acid, HA and the base, B can be represented as:

$$\underset{\text{Acid}}{HA} + \underset{\text{Base}}{B} \rightleftharpoons \underset{\text{Conjugate acid}}{BH^+} + \underset{\text{Conjugate base}}{A^-} \qquad (H^+ \text{ transferred from HA to B})$$

Thus we see that HA (acid) $-A^-$(base) and $BH^+$ (acid) – B (base) are two conjugate acid-base pairs in the above acid-base neutralisation reaction.

## Amphiprotic Substances

Molecules or ions that can lose as well accept proton(s) are called amphiprotic substances, *i.e.*, the molecules or ions that can act as Bronsted acids (loss of proton) as well as Bronsted bases (gain of proton) are called amphiprotic substances.

**Illustration.** As an example, let us consider the behaviour of $H_2O$ in the following acid-base reactions:

$$\underset{\text{Acid}_1}{\boxed{H_2O}} + \underset{\text{Base}_2}{NH_3} \rightleftharpoons \underset{\text{Acid}_2}{NH_4^+} + \underset{\text{Base}_1}{OH^-} \ldots \quad \text{(A)}$$

$(H^+)$

$(H^+)$

$$HCl + \boxed{H_2O} \rightleftharpoons H_3O^+ + Cl^- \cdots \quad \text{(B)}$$

In reaction (A) $H_2O$ molecule loses a proton ($H_2O \xrightarrow{-H^+} OH^-$) and hence acts as an acid while in reaction (B) this molecule gains a proton from HCl ($H_2O \xrightarrow{+H^+} H_3O^+$) and hence behaves as a base. Thus since $H_2O$ can lose as well as gain a proton, it behaves as an acid as well as a base. In other words, we say that $H_2O$ is an amphiprotic substance.

Examples of other species (molecules or ions) which act as amphiprotic substances are given in the following acid-base reactions. Amphiprotic substances have been shown in a box.

| | *Acid*$_1$ | | *Base*$_2$ | | *Acid*$_2$ | | *Base*$_1$ |
|---|---|---|---|---|---|---|---|
| (*i*) (*a*) | $\boxed{CH_3COOH}$ | + | $H_2O$ | $\rightleftharpoons$ | $H_3O^+$ | + | $CH_3COO^-$ |
| (*b*) | HF | + | $\boxed{CH_3COOH}$ | $\rightleftharpoons$ | $CH_3COOH_2^+$ | + | $F^-$ |
| (*ii*) (*a*) | $\boxed{H_2SO_4}$ | + | $2H_2O$ | $\rightleftharpoons$ | $2H_3O^+$ | + | $SO_4^{2-}$ |
| (*b*) | HF | + | $\boxed{H_2SO_4}$ | $\rightleftharpoons$ | $H_3SO_4^+$ | + | $F^-$ |
| (*iii*) (*a*) | $\boxed{H_2N.CO.NH_2}$ | + | $NH_3$ | $\rightleftharpoons$ | $NH_4^+$ | + | $H_2N.CO.NH^-$ |
| (*b*) | $H_2SO_4$ | + | $\boxed{H_2N.CO.NH_2}$ | $\rightleftharpoons$ | $H_2N.CO.NH_3^+$ | + | $HSO_4^-$ |
| (*iv*) (*a*) | $\boxed{HNO_3}$ | + | $H_2O$ | $\rightleftharpoons$ | $H_3O^+$ | + | $NO_3^-$ |
| (*b*) | HF | + | $\boxed{HNO_3}$ | $\rightleftharpoons$ | $H_2NO_3^+$ | + | $F^-$ |
| (*v*) (*a*) | $\boxed{HF}$ | + | $HNO_3$ | $\rightleftharpoons$ | $H_2NO_3^+$ | + | $F^-$ |
| (*b*) | $HClO_4$ | + | $\boxed{HF}$ | $\rightleftharpoons$ | $H_2F^+$ | + | $ClO_4^-$ |
| (*vi*) (*a*) | $\boxed{HSO_4^-}$ | + | $H_2O$ | $\rightleftharpoons$ | $H_3O^+$ | + | $SO_4^{2-}$ |
| (*b*) | $H_3O^+$ | + | $\boxed{HSO_4^-}$ | $\rightleftharpoons$ | $H_2SO_4$ | + | $H_2O$ |

| | | | | |
|---|---|---|---|---|
| (*vii*) (*a*) $\boxed{HS^-}$ | $+ NH_3$ | $\rightleftharpoons$ | $NH_4^+$ | $+ S^{2-}$ |
| (*b*) $H_2O$ | $+ \boxed{HS^-}$ | $\rightleftharpoons$ | $H_2S$ | $+ OH^-$ |

## Applications of the Dual Behaviour of Water ($H_2O$)

Acid-base reactions (A) and (B) given above show that since $H_2O$ acts both as Bronsted acid [*i.e.* proton donor] as shown in reaction (A)] and as Bronsted base [*i.e.* proton acceptor as shown in reaction (B)], $H_2O$ shows dual behaviour. The dual behaviour of $H_2O$ has been used to explain the following.

(*i*) **Aqueous solution of $Na_2CO_3$ is alkaline.** When $Na_2CO_3$ is dissolved in $H_2O$, $CO_3^{2-}$ ions are obtained. Since $CO_3^{2-}$ ion is a strong proton-acceptor, it accepts a proton from $H_2O$ to produce $HCO_3^-$ and $OH^-$ ions. Finally $NaHCO_3$ and NaOH are produced. Since NaOH is a strong base, the aqueous solution of $Na_2CO_3$ shows alkaline nature.

$$Na_2CO_3 \rightleftharpoons 2Na^+ + CO_3^{2-}$$

$(H^+)$

$$\underset{\text{Bronsted acid}}{H_2O} + \underset{\text{Bronsted base}}{CO_3^{2-}} \rightleftharpoons HCO_3^- + OH^-$$

$$Na_2CO_3 + H_2O \rightleftharpoons 2Na^+ + HCO^-_3 + OH^-$$

or $$Na_2CO_3 + H_2O \rightleftharpoons NaHCO_3 + \underset{\text{strong base}}{NaOH}$$

(*ii*) **Aqueous solutions of the salts of $Fe^{3+}$ $Al^{3+}$ etc., are acidic**. Aqueous solutions of the salts of $Fe^{3+}$, $Al^{3+}$ etc. get hydrated to form hydrated metal cations, e.g.

$$\left.\begin{aligned} Fe^{3+} + 6H_2O &\rightleftharpoons [Fe\,(H_2O)]^{3+} \\ Al^{3+} + 6H_2O &\rightleftharpoons [Al\,(H_2O)_6]^{3+} \end{aligned}\right\} \text{Hydrated metal cation}$$

The hydrated metal cations are strong proton donors and hence donate a proton to $H_2O$ to produce $H_3O^+$ ions which make the solution acidic

$(H^+)$

$$\underset{\text{Bronsted acid}}{[Fe(H_2O)_6]^{3+}} + \underset{\text{Bronsted base}}{H_2O} \rightleftharpoons [Fe(H_2O)_5(OH)]^{2+} + H_3O^+$$

$(H^+)$

$$\underset{\text{Bronsted acid}}{[Al(H_2O)_6]^{3+}} + \underset{\text{Bronsted base}}{H_2O} \rightleftharpoons [Al(H_2O)_5(OH)]^{2+} + H_3O^+$$

It may be noted from the above examples that in example (*i*) $H_2O$ acts as a Bronsted acid while in example (*ii*), $H_2O$ acts as a Bronsted base. Thus $H_2O$ shows dual behavior.

## In a Conjugate Acid-Base Pair Strong Acid has its Weak Conjugate Base and Vice Versa

We know that according to Bronsted-Lowry concept, an acid-base reaction has two conjugate acid-base pairs. Each of the two conjugate acid-base pairs is a *strong acid-weak base pair* or *a weak acid-strong base pair*. Thus in a conjugate acid-base pair a strong acid has is weak conjugate base and vice versa. This fact can be proved by considering the following two acid-base equilibiria.

(*i*) $\underset{\text{Acid}}{HCl} + \underset{\text{Base}}{H_2O} \rightleftharpoons \underset{\text{Acid}}{H_3O^+} + \underset{\text{Base}}{Cl^-}$ (each step transfers $(H^+)$)

In this reaction HCl and $H_3O^+$ are two acids and $H_2O$ and $Cl^-$ in are two bases. Suppose this reaction *proceeds from left to right*. Hence HCl (acid) has stronger tendency to donate a proton ($H^+$) to $H_2O$ (base) than $H_3O^+$ (acid) which donates a proton to $Cl^-$ ion (base). Thus HCl is a stronger acid than $H_3O^+$. In other words we can say that *HCl is a strong acid and $H_3O^+$ is a weak acid*. Further, since the reaction proceeds from left to right, $H_2O$ (base) has stronger tendency to accept a proton from HCl (and) than $Cl^-$ ion (base) which accepts a proton from $H_3O^+$ (acid). Thus $H_2O$ is a stronger base than $Cl^-$ ion. In other words we can say that *$H_2O$ is a strong base and $Cl^-$ ion $Cl^-$ ion is a weak base*. Above discussion shows that the acid-base reaction given above can be written as:

$\underset{\text{Strong acid}}{HCl} + \underset{\text{Strong base}}{H_2O} \rightleftharpoons \underset{\text{Weak acid}}{H_3O^+} + \underset{\text{Weak base}}{Cl^-}$ (each step transfers $(H^+)$)

Weak acid ($H_3O^+$) – strong base ($H_2O$) pair

Strong acid (HCl) – weak base ($Cl^-$) pair

In this reaction the longer half arrow (⟶) indicates of the direction of the reaction. It may be seen that if an acid-base reaction proceeds from left to right, the reacting acid and base both are strong and the acid and base produced both are weak.

(*ii*) $\underset{\text{Acid}}{CH_3COOH} + \underset{\text{Base}}{H_2O} \rightleftharpoons \underset{\text{Acid}}{H_3O^+} + \underset{\text{Base}}{CH_3COO^-}$ (each step transfers $(H^+)$)

In reaction (*ii*), $CH_3COOH$ and $H_3O^+$ are two acids, and $H_2O$ and $CH_3COO^-$ are two bases. Suppose this reaction proceeds from *right to left*. Hence $H_3O^+$ (acid) has stronger tendency to donate a proton to $CH_3COO^-$ (base) than $CH_3COOH$ (acid) which donates a proton to $H_2O$ (base). Thus $H_3O^+$ is a strong acid and $CH_3COOH$ is a weak acid. Further, since the reaction proceeds from right to left, $CH_3COO^-$ (base) has stronger tendency to accept a proton from $H_3O^+$ (acid) than $H_2O$ (base) which accepts a proton from $CH_3COOH$ (acid). Thus $CH_3COO^-$ is a strong base and $H_2O$ is a weak base. Above discussion shows that the acid-base reaction (*ii*) can be written as:

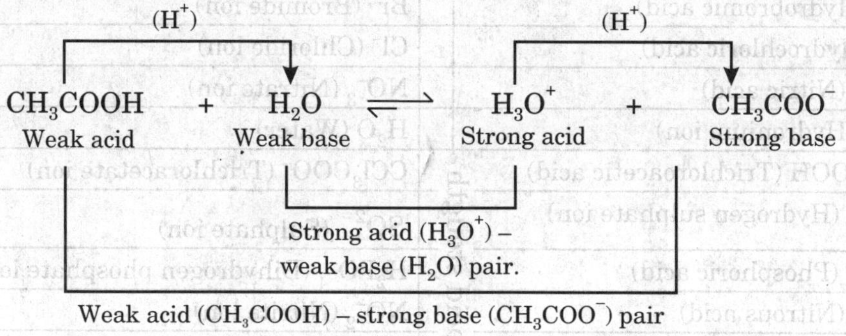


In the above reaction, the longer half arrow (↼——) indicates the direction of the acid-base reaction. It may be seen that if an acid-base reaction proceeds from right to left, the reacting acid and base both are weak and the acid and base produced both are strong.

Above discussion shows that *in a conjugate acid-base pair in an acid-base reaction, if the acid is strong, the conjugate base of this acid would be weak and vice versa*. In other wards we can say that *an acid-base reaction has one strong acid-weak base pair and one weak acid-strong base pair*.

## Relative Order of Basic Strength of Conjugate Bases of Given Conjugate Acids

We have seen that in a conjugate acid-base pair, if the acid is strong, its conjugate base would be weak. Similarly, if the acid is weak, its conjugate base would be strong.

**Illustrative example** Conjugate bases of HF, HCl, HBr and HI are $F^-$, $Cl^-$, $Br^-$ and $I^-$ respectively. We know that acidic strength of HF, HCl, HBr and HI in aqueous solution *increases* as HF < HCl < HBr < HI. This order shows that HF is the weakest acid and HI is the strongest acid. On the basis of the above rule, the conjugate base of the weakest acid, HF (i.e. $F^-$ ion) would be the strongest base and the conjugate base of the strongest acid, HI (*i.e.* $I^-$ ion) would be the weakest base. Thus the basic strength of the conjugate bases (*i.e.* $F^-$, $Cl^-$, $Br^-$ and $I^-$) would be in the decreasing order $F^- > Cl^- > Br^- > I^-$. It may be noted that the relative order of basic strength of bases ($F^-$, $Cl^-$, $Br^-$ and $I^-$) *is opposite to relative order of acidic strength of acids (HF, HCl, HBr and HI)*

In Table 3.1, since the acid strength of the acids given in 1st column **decreases** from $HClO_4$ to $CH_4$, the basic strength of the conjugate bases (given in *2nd* column) of the acids **increases** from $ClO_4^-$ (conjugate base of $HClO_4$) to $CH^-_3$ (conjugate base of $CH_4$)

**Table 3.1 Relative order of acid strength of acids and relative order of base strength of their conjugate bases.**

| *Acid* | *Conjugate Base* |
|---|---|
| $HClO_4$ (Per chloric acid) | $ClO^-_4$ (Perchlorate ion) |
| $H_2SO_4$ (Sulphuric acid) | $HSO^-_4$ (Hydrogen sulphate ion) |
| HI (Hydriodic acid) | $I^-$ (Iodide ion) |
| HBr (Hydrobromic acid) | $Br^-$ (Bromide ion) |
| HCl (Hydrochloric acid) | $Cl^-$ (Chloride ion) |
| $HNO_3$ (Nitric acid) | $NO^-_3$ (Nitrate ion) |
| $H_3O^+$ (Hydronium ion) | $H_2O$ (Water) |
| $CCl_3COOH$ (Trichloroacetic acid) | $CCl_3COO^-$ (Trichloracetate ion) |
| $HSO^-_4$ (Hydrogen sulphate ion) | $SO_4^{2-}$ (Sulphate ion) |
| $H_3PO_4$ (Phosphoric acid) | $H_2PO^-_4$ (Dihydrogen phosphate ion) |
| $HNO_2$ (Nitrous acid) | $NO^-_2$ (Nitrite ion) |
| HF (Hydrofluoric acid) | $F^-$ (Fluoride ion) |
| HCOOH (Formic acid) | $HCOO^-$ (Formate ion) |
| $CH_3COOH$ (Acetic acid) | $CH_3COO^-$ (Acetate ion) |
| $H_2CO_3$ (Carbonic acid) | $HCO^-_3$ (Hydrogen carbonate ion) |
| $H_2S$ (Hydrogen sulphide) | $HS^-$ (Hydrogen sulphide ion) |
| $NH_4^+$ (Ammonium ion) | $NH_3$ (Ammonia) |
| HCN (Hydrogen cyanide) | $CN^-$ (Cyanide ion) |
| $C_6H_5OH$ (Phenol) | $C_6H_5O^-$ (Phenoxide ion) |
| $HS^-$ (Hydrogen sulphide ion) | $S^{2-}$ (Sulphide ion) |
| $H_2O$ (Water) | $OH^-$ (Hydroxide ion) |
| $C_2H_5OH$ (Ethanol) | $C_2H_5O^-$ (Ethoxide ion) |
| $NH_3$ (Ammonia) | $NH^-_2$ (Amide ion) |
| $H_2$ (Hydrogen) | $H^-$ (Hydride ion) |
| $CH_4$ (Methane) | $CH^-_3$ (Methide ion) |

(Acid column: Decreasing acid stength ↓; Conjugate Base column: Increasing base strength ↓)

## Relative Strength of Acids and Bases According to Bronsted-Lowry Concept

According to Bronsted-Lowry concept,the relative strength of two acids (acid strength) is compared by comparing their tendency to donate proton ($H^+$). Similarly the relative strength of two bases (base strength) is compared by comparing their tendency to accept proton. This discussion shows that the acids which can donate a proton ($H^+$) to a base comparatively more easily or readily are said to be strong acids while the acids which donate a proton with great difficulty are said to be weak acids.

Similarly the bases that can accept a proton quite easily are said to be strong bases while the bases that can accept a proton with great difficulty are said to be weak bases.

## Some Concepts

### 1. Electron-Withdrawing Groups (EWGs) and Electron-Donating Groups (EDGs)

EWGs are those which decrease the electron density and EDGs are those which increase the electron density. Examples of EWGs are $-NO_2$, $-CN$, $-COOH$, $-X$ (halogen), $-OH$, $-C_6H_5$ etc. and examples of EDGs are $-R$ (alkyl), $-NH_2$, $-OR$, $-OCH_3$, $-COO^-$ etc. EWG is also called **electron attracting group** and EDG is also called **electron repelling group** or **electron releasing group**.

### 2. Inductive Effect

Inductive effect may be defined as the effect transferred by a charged atom at one point in an ion or a molecule to another point in the same ion or molecule.

Inductive effect caused by electron-withdrawing groups (EWG) (EWG is also called *electron attracting group*) decreases the electron density (*i.e.* disperses the negative charge) on the C-atom with which EWG is attached (See Fig. 3.1) and hence is called electron-withdrawing inductive effect. This effect is expressed as –I effect. – sign indicates that EWG decreases the electron density. *–I effect increases the acidic strength of acids and decreases the basic strength of bases.*

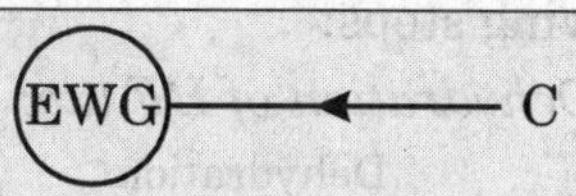


**Fig. 3.1** Electron-withdrawing group (EWG) decreases the electron density on C-atom.

Examples of species which have electron-withdrawing inductive effect (–I effect) are $-NO_2$, $-CN$, $-COOH$, $-X$ (halogen), $-OH$, $-C_6H_5$ etc. –I effect of these groups decreases in the order:

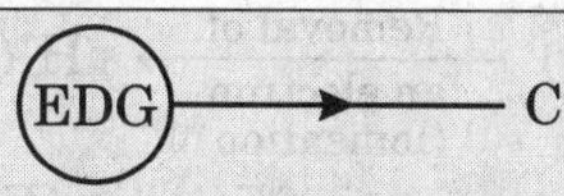


**Fig. 3.2** Electron-donating group (EDG) increases the electron density on C-atom

$$-NO_2 > -CN > -COOH > -F > -Cl > -Br > -I > -OH > C_6H_5$$

Inductive effect caused by electron-donating groups (EDG) (EDG is also called *electron repelling or electron releasing group*) increases the electron density (*i.e.* intensifies the negative charge) on C-atom with which EDG is attached (See Fig. 3.2) and hence is called electron-donating inductive effect. This effect is represented as +I effect. + sign indicates that EDG increases the electron density. *+I effect decreases the acid strength of acids and increases the basic strength of bases.*

Examples of species which have electron-donating inductive effect (+I effect) are $-NH_2$, $-OR$ (aloxy group), $-R$ (alkyl group), $-OCH_3$ (methoxy group), $-COO^-$ etc. +I effect of some alkyl groups increases in the order:

$$\underset{\text{Methyl}}{-CH_3} < \underset{\text{Ethyl}}{CH_3-CH_2-} < \underset{\text{Isopropyl}}{(CH_3)_2CH-} < \underset{\text{t-butyl}}{(CH_3)_3C-}$$

In general, +I effect of alkyl groups is in the order: 1° < 2° < 3°.

### 3. Heat of Ionisation ($\Delta H_I$) of MX Molecules in Aqueous Solution

Heat of ionisation ($\Delta H_I$) of HX molecule in aqueous solution is the amount of heat released (–sign) when HX molecule undergoes ionisation to give $H^+$ and $X^-$ ions in aqueous solution. Thus:

$$HX + aq \xrightarrow[(+aq)]{\text{Hydration}} HX\,(aq)$$

$$HX\,(aq) \xrightarrow{\text{Ionisation}} H^+(aq) + X^-(aq),$$

$\Delta H$ = Heat released = Heat of ionisation of HX (*aq*) = $-\Delta H_I$ of HX (*aq*)

**Calculation of $\Delta H_I$ for HX (*aq*) molecules** As an example let us calculate the value of $\Delta H_I$ for HF (*aq*) molecule. We know that HF, in aqueous solution, ionises to give $H^+(aq)$ and $F^-(aq)$ ions as shown below.

(A) $HF\,(aq) \xrightarrow{\text{Ionisation}} H^+\,(aq) + F^-(aq)$, $\Delta H$ = Heat *released*
$= -\Delta H_I$ of HF (*aq*)

As a matter of fact, the ionisation reaction shown above takes place through the following steps:

(*a*) *Dehydration of HF (aq) to HF (g)*

$HF\,(aq) \xrightarrow{\text{Dehydration}} HF(g)$, $\Delta H$ = Heat *absorbed*
= Heat of dehydration of HF (*aq*) = $+\Delta H_1$

(*b*) *Dissociation of HF (g) into H (g) and F (g)*

$HF\,(g) \xrightarrow{\text{Dissociation}} H(g) + F\,(g)$, $\Delta H$ = Heat *absorbed*
= Heat of dissociation of HF (*g*) = $+\Delta H_2$

(*c*) *Conversion of H(g) into $H^+$ (g)*

$H(g) \xrightarrow[\text{an electron (Ionisation)}]{\text{Removal of}} H^+(g) + e^-$, $\Delta H$ = Heat *absorbed* = IE of H (*g*) = $+\Delta H_3$

(*d*) *Conversion of F(g) into $F^-$ (g)*

$F(g) + e^- \xrightarrow[\text{electron}]{\text{Addition of an}} F^-(g)$, $\Delta H$ = Heat *released* = EA of F(g) = $-\Delta H_4$

(*e*) *Hydration of $H^+$(g) to $H^+$(aq)*

$H^+(g) + aq \xrightarrow{\text{Hydration}} H^+(aq)$, $\Delta H$ = Heat *released*
= Heat of hydration of $H^+$ (*g*) = $-\Delta H_5$

(*f*) *Hydration of $F^-$(g) to $F^-$ (aq)*

$F^-(g) + aq \xrightarrow{\text{Hydration}} F^-(aq)$, $\Delta H$ = Heat *released*
= Heat of hydration of $F^-$ (*g*) = $-\Delta H_6$

On adding the equations given at (*a*), (*b*), (*c*), (*d*), (*e*) and (*f*), we get:

$$HF\,(aq) + 2aq \longrightarrow H^+\,(aq) + F^-\,(aq),$$

or

$$HF(aq) \longrightarrow H^+\,(aq) + F^-\,(aq),\ \Delta H = ?$$

Since the above equation is the same as equation (A), $\Delta H$ for the above reaction is $\Delta H_I$, i.e. $\Delta H_I$ is equal to the sum of $\Delta H$ for reactions (*a*), (*b*), (*c*), (*d*), (*e*) and (*f*). Thus:

$$H_I \text{ for } HF(aq) = \Delta H_1 + \Delta H_2 + \Delta H_3 + (-\Delta H_4) + (-\Delta H_5) + (-\Delta H_6) = \Delta H_1 + \Delta H_2 + \Delta H_3 - \Delta H_4 - \Delta H_5 - \Delta H_6 \quad \dots (i)$$

If the values of $\Delta H_1$, $\Delta H_2$, $\Delta H_3$, $\Delta H_4$, $\Delta H_5$ and $\Delta H_6$ for HF molecule are put in equation (*i*), we get the value of $\Delta H_I$ for HF molecule which is equal to

–12 KJ $mol^{-1}$. Similarly putting the values of $\Delta H_1$ , $\Delta H_2$, $\Delta H_3$, $\Delta H_4$, $\Delta H_5$ and $\Delta H_6$ for HCl, HBr and HI molecules in equation (i), the values of $\Delta H_I$ for these molecules have been found equal to –59, –63 and –57 kJ $mol^{-1}$.

All the steps shown at (*a*), (*b*), (*c*), (*d*), (*e*), and (*f*) can be shown in a collective way in the form of a cycle known as Born-Haber cycle (See Fig. 3.3).

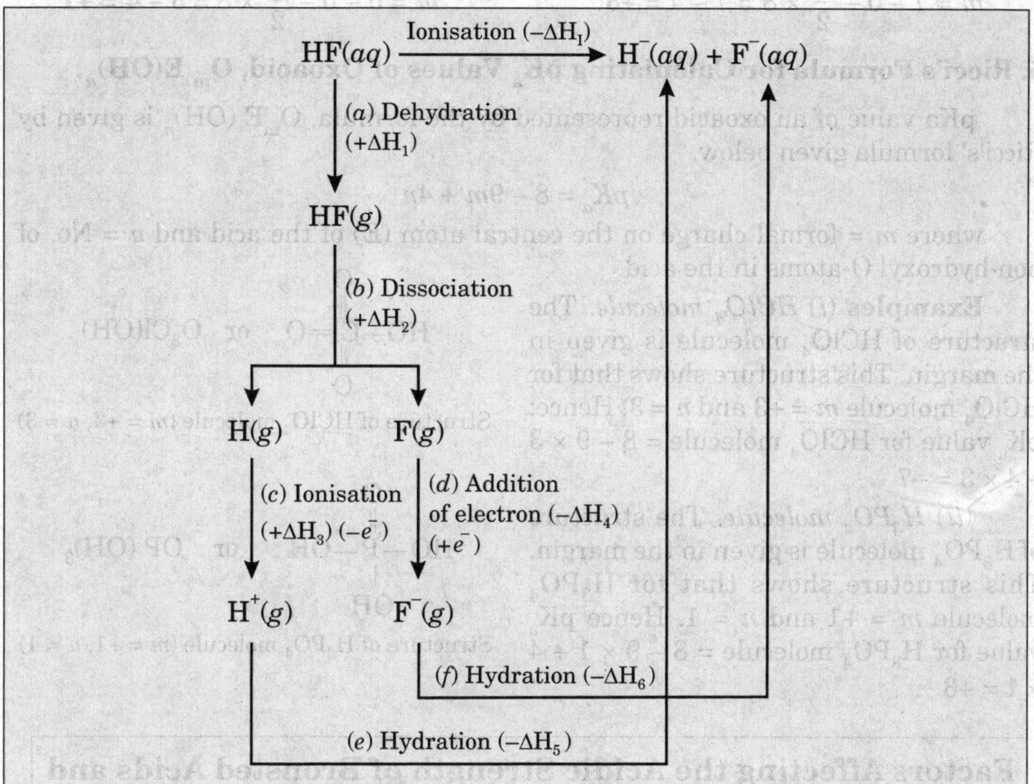


**Fig. 3.3** Different steps involved in the ionisation of HF molecule in aqueous solution.

## 4. To Calculate the Formal Charge (*m*) of the Central Atom in Some Oxoacids.

Formal charge (m) of the central atom of an oxoacid is given by:

m = No. of electrons present in the valence shell of the central atom – No. of unused electrons on the central atom – $\frac{1}{2}$ × No. of electrons used in forming σ-bonds between the central atom and O-atoms.

The value of *m* for some oxoacids has been calculated as follows;

(*i*) **HClO** :C̤̈l—OH

$$m = 7 - 6 - \frac{1}{2} \times 2 = \mathbf{0}$$

(*ii*) **$HClO_2$** O═C̤̈l—OH

$$m = 7 - 4 - \frac{1}{2} \times 4 = 7 - 6 = \mathbf{+1}$$

(*iii*) **$HClO_3$** O═C̤l(═O)—OH

$$m = 7 - 2 - \frac{1}{2} \times 6 = 7 - 2 - 3 = \mathbf{+2}$$

(*iv*) **$HClO_4$** 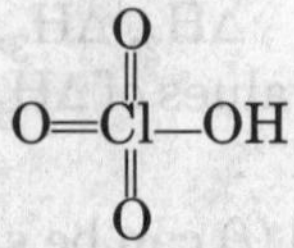

(*v*) **$H_3PO_4$** 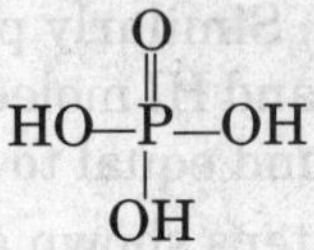

$m = 7 - 0 - \frac{1}{2} \times 8 = 7 - 4 = \mathbf{+3}$

$m = 5 - 0 - \frac{1}{2} \times 8 = 5 - 4 = \mathbf{+1}$

### 5. Ricci's Formula for Calculating $pK_a$ Values of Oxoacid, $O_m E(OH)_n$.

pKa value of an oxoacid represented by the formula, $O_m E (OH)_n$ is given by Ricci's' formula given below.

$$pK_a = 8 - 9m + 4n$$

where $m$ = formal charge on the central atom (E) of the acid and $n$ = No. of non-hydroxyl O-atoms in the acid.

**Examples** (*i*) *$HClO_4$ molecule.* The structure of $HClO_4$ molecule is given in the margin. This structure shows that for $HClO_4$ molecule $m$ = +3 and $n$ = 3. Hence: $pK_a$ value for $HClO_4$ molecule = 8 – 9 × 3 + 4 × 3 = –7

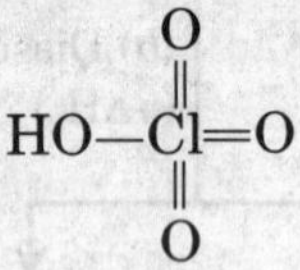 or $O_3Cl(OH)$

Structure of $HClO_4$ molecule ($m$ = +3, $n$ = 3)

(*ii*) *$H_3PO_4$ molecule.* The structure of $H_3PO_4$ molecule is given in the margin. This structure shows that for $H_3PO_4$ molecule $m$ = +1 and $n$ = 1. Hence $pK_a$ value for $H_3PO_4$ molecule = 8 – 9 × 1 + 4 × 1 = +3

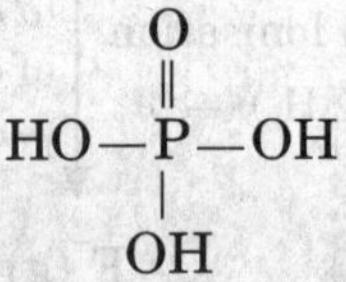 or $OP(OH)_3$

Structure of $H_3PO_4$ molecule ($m$ = +1, $n$ = 1)

## Factors Affecting the Acidic Strength of Bronsted Acids and Basic Strength of Bronsted Bases

Some of the factors which affect the acidic strength of Bronsted acids and basic strength of Bronsted bases are discussed below:

### 1. Effect of Solvent

The affect of solvent on the acidic and basic strength is discussed under the following heads:

**(*i*) Effect of solvent on the acid strength of Bronsted acids: Concept of levelling solvent and levelling effect.** The relative order of acid strength of various Bronsted acids depends on the extent to which the various acids ionise in a given solvent which acts as a base. The solvent acts as a base, since it accepts a proton donated by the Bronsted acid. The acid strength of a Bronsted acid is its capacity to donate a proton to a base (solvent). As an example let us compare the acid strength the strong acids *viz* $HClO_4$, HBr, $H_2SO_4$, HCl and $HNO_3$ in water solvent. All these acids ionise in $H_2O$ (water acts as a base) almost completely to produce $H_3O^+$ ion.

| *Acid* | | *Base* *(Solvent)* | | *Acid* *(Cation of the solvent)* | | *Base* |
|---|---|---|---|---|---|---|
| $HClO_4$ | + | $H_2O$ | $\longrightarrow$ | $H_3O^+$ | + | $ClO_4^-$ |
| HBr | + | $H_2O$ | $\longrightarrow$ | $H_3O^+$ | + | $Br^-$ |
| $H_2SO_4$ | + | $H_2O$ | $\longrightarrow$ | $H_3O^+$ | + | $HSO_4^-$ |
| HCl | + | $H_2O$ | $\longrightarrow$ | $H_3O^+$ | + | $Cl^-$ |
| $HNO_3$ | + | $H_2O$ | $\longrightarrow$ | $H_3O^+$ | + | $NO_3^-$ |

Since all the acids ionise in $H_2O$ (solvent) to the same extent (or since all the acids are *completely ionised* to produce $H_3O^+$), these strong acids in aqueous solution appear almost equally strong, *i.e.* in aqueous solution these acids behave as being equally strong. Obviously the relative acid strength of these acids cannot be measured in aqueous solution. This phenomenon due to which acids appear to be equally strong is known as **levelling effect** of the solvent (Here the solvent is $H_2O$) on the acids. The solvent (i.e. $H_2O$) in which the acids appear equally strong is known as **levelling solvent.** Obviously $H_2O$ acts as a levelling solvent for strong acids mentioned above.

The solvent in which given acids / bases ionise completely and appear to be equally strong is called a levelling solvent for the acid and bases. For example

(*a*) Liq. $NH_3$ acts as a levelling solvent for the acids HCl (strong acid) and $CH_3COOH$ (weak acid), since these acids ionise completely in liq $NH_3$ to give $NH_4^+$ (ammoniated proton) cation (cation of the solvent, $NH_3$)

| *Acid* | | *Base* *(Levelling solvent)* | | *Acid* *(Cation of the solvent)* | | *Base* |
|---|---|---|---|---|---|---|
| HCl (*strong*) | + | $NH_3$ | $\xrightarrow[\text{ionisation}]{\text{Complete}}$ | $NH_4^+$ | + | $Cl^-$ |
| $CH_3COOH$ (*weak*) | + | $NH_3$ | $\xrightarrow[\text{ionisation}]{\text{Complete}}$ | $NH_4^+$ | + | $CH_3COO^-$ |

Although HCl is a strong acid and $CH_3COOH$ is a weak acid, yet these acids due to their complete ionisation in lig. $NH_3$ appear to be equally strong in $NH_3$ solvent which acts as a levelling solvent for these acids. The acid strength of HCl and $CH_3COOH$ cannot be measured in the levelling solvent for these acids.

(*b*) $CH_3COOH$ acts as a levelling solvent for the bases $C_6H_5NH_2$ (weak base) and NaOH (strong base), since these bases ionise completely in $CH_3COOH$ to give $CH_3COO^-$ anion (anion of the solvent, $CH_3COOH$)

| *Acid* *(Levelling solvent)* | | *Base* | | *Acid* | | *Base* *(Anion of the solvent)* |
|---|---|---|---|---|---|---|
| $CH_3COOH$ | + | $C_6H_5NH_2$(*weak*) | $\xrightarrow[\text{ionisation}]{\text{Complete}}$ | $C_6H_5NH_3^+$ | + | $CH_3COO^-$ |
| $CH_3COOH$ | + | NaOH (*strong*) | $\xrightarrow[\text{ionisation}]{\text{Complete}}$ | $NaOH_2^+$ | + | $CH_3COO^-$ |

Although $C_6H_5NH_2$ is a weak base and NaOH is a strong base, yet these bases, due to their complete ionisation in $CH_3COOH$ appear to be equally strong in $CH_3COOH$ which acts as a levelling solvent for these bases. The base strength of $C_6H_5NH_2$ and NaOH cannot be measured in the levelling solvent for these bases.

***(ii)* Effect of solvent on the base strength of bronsted bases** The relative order of the base strength of various Bronsted bases depends on the extent to which the various bases ionise in a given solvent which acst as an acid. The solvent acts as an acid, since it donates a proton to the base. The base strength of a Bronsted base is its capacity to accept a proton from the acid (solvent). As an example let us compare the base strength of the strong bases *viz* $O^{2-}$ and $NH_2^-$ in $H_2O$ solvent. Both these bases ionise in solvent $H_2O$ (water acts as an acid) almost completely to produce $OH^-$ ions

| *Acid (Levelling solvent)* | | *Base* | | *Acid* | | *Base (Anion of the solvent)* |
|---|---|---|---|---|---|---|
| $H_2O$ | + | $O^{2-}$ (*strong*) | $\xrightarrow[\text{ionisation}]{\text{Complete}}$ | $OH^-$ | + | $OH^-$ |
| $H_2O$ | + | $NH_2^-$ (*strong*) | $\xrightarrow[\text{ionisation}]{\text{Complete}}$ | $NH_3$ | + | $OH^-$ |

Since both the bases ionise in $H_2O$ (solvent) to the same extent (or since both the bases are completely ionised to produce $OH^-$), these strong bases in aqueous solution appear almost equally strong, i.e. in aqueous solution these bases behave as being equally strong. Obviously the relative base strength of these bases cannot be measured in aqueous solution. It should be easy to understand that $H_2O$ acts as a levelling solvent for the strong bases.

**Explanation of levelling effect of a solvent on the basis of solvent system concept of acids and bases** According to solvent-system concept of acids and bases, the acids which are stronger than the characteristic cation of the solvent are completely ionised in the solvent to produce the characteristic cation of the solvent and hence the acids appear equally strong in that solvent.

If $H_2O$ is used as solvent, then the characteristic cation of this solvent is $H_3O^+$. Similarly according to solvent-system concept of acids and bases, the bases which are stronger than the characteristic anion of the solvent are completely ionised in the solvent to produce the characteristic anion of the solvent and hence the bases appear equally strong in that solvent. If $H_2O$ is used as solvent, then the characteristic anion of this solvent is $OH^-$.

**Examples** (*i*) We have shown above that the strong acids *viz* $HClO_4$, HBr, $H_2SO_4$, HCl and $HNO_3$ appear equally strong in solvent, $H_2O$. This behaviour of acids is due to the fact that, since these acids are stronger than the characteristic cation ($H_3O^+$) of the solvent ($H_2O$), these acids are completely ionised in $H_2O$ solvent to produce $H_3O^+$ which is the characteristic cation ($H_3O^+$) of the solvent ($H_2O$). Due to the complete ionisation of these acids, these acids appear equally strong in solvent ($H_2O$).

(*ii*) We have shown above that the strong bases namely $O^{2-}$ and $NH_2^-$ appear equally strong in solvent, $H_2O$. This behaviour of bases is due to the fact that since these bases are stronger than the characteristic anion ($OH^-$) of the solvent ($H_2O$), these bases are completely ionised in $H_2O$ to produce $OH^-$ which is the characteristic anion of the solvent ($H_2O$). Due to the complete ionisation of the bases, these bases appear equally strong in solvent ($H_2O$).

**Differentiating solvents.** The solvents in which the acids and bases ionise *to different extents* so that the relative strengths of the acids and bases can be measured are known as differentiating solvents for these acids and bases.

**Examples** (*i*) $CH_3COOH$ is a differenting solvent for strong acids like $HClO_4$, $HBr$, $H_2SO_4$, $HCl$ and $HNO_3$, since in $CH_3COOH$ solvent these acids ionise to different extents.

Thus these acids remain in equilibirium in $CH_3COOH$ solvent. In other words these acids do not ionise completely in $CH_3COOH$ to produce $CH_3COOH_2^+$ cations.

| *Acid* | | *Base* (*Differentiating solvent*) | | *Acid* (*Cation of the solvent*) | | *Base* |
|---|---|---|---|---|---|---|
| $HClO_4$ (Weak) | + | $CH_3COOH$ | $\rightleftharpoons$ | $CH_3COOH_2^+$ | + | $ClO_4^-$ |
| $HBr$ (Weak) | + | $CH_3COOH$ | $\rightleftharpoons$ | $CH_3COOH_2^+$ | + | $Br^-$ |

In the present case since $HClO_4$ and $HBr$ are weaker acids than the characteristic cation ($CH_3COOH_2^+$) of the solvent ($CH_3COOH$), these acids remain in equilibrium in $CH_3COOH$ solvent. Hence the acid strength of these acids can be measured in $CH_3COOH$ solvent. The given acids have been arranged in the following decreasing order of their acid strength.

$$HClO_4 > HBr > H_2SO_4 > HCl > HNO_3$$

(*ii*) Now let us consider the ionisation of the bases like $NH_3$ and $HCOO^-$ in $H_2O$ (solvent). These bases do not get ionised completely in $H_2O$ to produce $OH^-$ ions but exist in equilibrium in aqueous solution.

| *Acid* (*Differentiating solvent*) | | *Base* | | *Acid* | | *Base* (*Anion of the solvent*) |
|---|---|---|---|---|---|---|
| $H_2O$ | + | $NH_3$ (Weak) | $\rightleftharpoons$ | $NH_4^+$ | + | $OH^-$ |
| $H_2O$ | + | $HCOO^-$ (Weak) | $\rightleftharpoons$ | $HCOOH$ | + | $OH^-$ |

Thus we see that $NH_3$ and $HCOO^-$ ionise in solvent ($H_2O$) to different extent. Hence the basic strength of these bases can be measured in $H_2O$. Consequently $H_2O$ acts as a differentiating solvent for $NH_3$ and $HCOO^-$ bases. In the present case since $NH_3$ and $HCOO^-$ are weaker bases than the characteristic anion ($OH^-$) of the solvent ($H_2O$), these bases remain in equilibrium in aqueous solution.

**Behaviour of a given species as an acid or as a base in different solvents.** According to solvent-system of acid and bases, whether a given species will behave as an acid or as a base in the given solvent depends on the acidic or basic character of the solvent. For example:

(*i*) In $H_2O$ since $CH_3COOH$ gives $H_3O^+$ cations (cations of the solvent, $H_2O$) of *small concentration*, this acid behaves as a weak acid in $H_2O$.

$$\underset{Acid}{CH_3COOH} + \underset{\substack{Base\\(Solvent)}}{H_2O} \rightleftharpoons \underset{\substack{Acid\\(\text{Small concentration})}}{H_3O^+} + \underset{Base}{CH_3COO^-}$$

(*ii*) In liq. $NH_3$, since $CH_3COOH$ gives $NH^+_4$ cations (cations of the solvent, $NH_3$) of *considerable concentration*, $CH_3COOH$ behaves as a strong acid in liq. $NH_3$.

$$\underset{Acid}{CH_3COOH} + \underset{\substack{Base\\(Solvent)}}{NH_3} \rightleftharpoons \underset{\substack{Acid\\(\text{Considerable concentration})}}{NH^+_4} + \underset{Base}{CH_3COO^-}$$

(*iii*) In $H_2SO_4$, since $CH_3COOH$ gives $HSO^-_4$ ions (anions of the solvent, $H_2SO_4$) of *considerable concentration*, $CH_3COOH$ behaves as a base in $H_2SO_4$.

$$\underset{\substack{Acid\\(Solvent)}}{H_2SO_4} + \underset{Base}{CH_3COOH} \rightleftharpoons \underset{Acid}{CH_3COOH^+_2} + \underset{\substack{Base\\(\text{Considerable}\\\text{concentration})}}{HSO^-_4}$$

(*iv*) Urea ($H_2NCONH_2$) behaves as a weak acid in $NH_3$ but as a strong base in $H_2SO_4$

$$\underset{\text{Acid}}{H_2NCONH_2} + \underset{\text{Base (solvent)}}{NH_3} \rightleftharpoons \underset{\text{Acid}}{NH_4^+} (\text{Small concentration}) + \underset{\text{Base}}{H_2NCONH^-}$$

$$\underset{\text{Acid (solvent)}}{H_2SO_4} + \underset{\text{Base}}{H_2NCONH_2} \rightleftharpoons \underset{\text{Acid}}{H_2NCONH_3^+} + \underset{\text{Base}}{HSO_4^-} \ (\text{Considerable concentration})$$

## 2. Effect of the Dielectric Constant (ε) and Polarity of the Solvent

The acidic and basic strength of an ionic species (solute) dissolved in a solvent increase with an increase in the dielectric constant (ε) or polarity of the solvent in which the species has been dissolved.

**Explanation.** According to coulomb's law, the force of attraction (F) between two ions of a solute having opposite charges $q_1$ and $q_2$ is given by:

$$F = \frac{q_1 q_2}{\Sigma r^2},$$

where ε is dielectric constant of the solvent and $r$ is the distance between the ions. This relation shows that in solvents having lower value of ε, the force of attraction (F) between the ions of the solute would be relatively large and consequently the extent of dissociation of the solute into ions in the solvent would be small. On the other hand, in solvent having high value of ε, the force of attraction (F) between the ions of the solute would be relatively small and hence the extent of dissociation of the solute would be large. From this discussion, we can conclude *that the acidic strength of acids (solute) and basic strength of bases (solute) would increase with the increase in the value of ε of the solvent.*

Now since the dielectric constant (ε) and dipole moment are directly related with each other, *greater is the dielectric constant or polarity of the solvent, greater would be the acidic or basic strength of the species dissolved in that solvent.*

## 3. Effect of Electron Withdrawing and Electron-Donating Substituents or Groups

Electron-withdrawing and electron-donating nature of substituents present in acids and bases have marked effect on the relative strength of acids and bases.

An electron withdrawing substituent or group (EWG) present in a species has electron-withdrawing inductive effect (–I effect). The presence of EWG in a species increases the acidic strength of the species. An electron-donating substitutent or group (EDG) present in a species has electron-donating inductive effect (+I effect). The presence of EDG in a species decreases the acidic strength of the species.

Similarly the presence of EWG in a species decreases the basic strength of the species and the presence of an EDG increases the basic strength of the species.

How the presence of an EWG or EDG in an acid or a base affects the acidic strength of acids and basic strength of bases can be explained by considering the following examples:

***(i) Formic acid (HCOOH) is a stronger acid than acetic acid ($CH_3COOH$).*** **This can be explained as follows:**

The structures of these acids are given below:

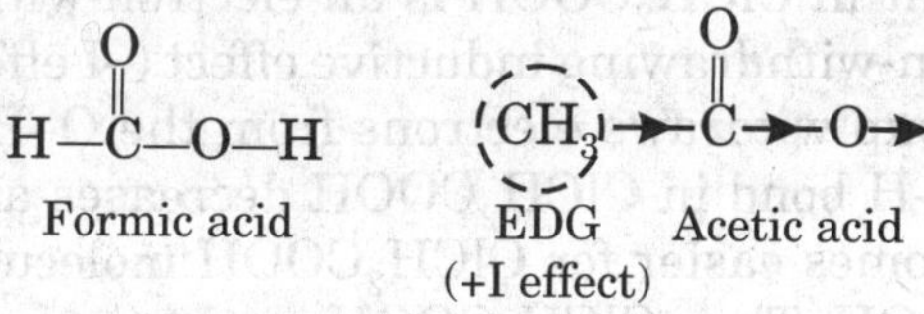


In $CH_3COOH$ molecule, methyl group ($-CH_3$) is an electron-donating group (EDG) and hence has electron-donating inductive effect (+I effect). Because of +I effect of $-CH_3$ group present in $CH_3COOH$, this group donates electrons to O–H bond in $CH_3COOH$ molecule. As a result, the electron density on O-atom in O–H bond in $CH_3COOH$ increases. Due to the increase in electron density on O-atom, it becomes more difficult for $CH_3COOH$ molecule to release $H^+$ ions as compared to HCOOH. Thus HCOOH can release $H^+$ ions more easily as compared to $CH_3COOH$ and is a stronger acid than $CH_3COOH$.

Further, +I effect of $CH_3$ group intensifies (increases) the magnitude of negative charge (electron density) on carboxylate anion ($CH_3COO^-$) present in $CH_3COOH$ and thus makes $CH_3COO^-$ anion of $CH_3COOH$ less stable as compared to $HCOO^-$ anion present in HCOOH. This decreases the acidic strength of $CH_3COOH$.

$$\underset{\substack{\text{More stable}\\\text{(weaker acid)}}}{CH_3COOH} \longrightarrow \underset{\text{Less stable}}{CH_3COO^-} + H^+$$

$$\underset{\substack{\text{Less stable}\\\text{(stronger acid)}}}{HCOOH} \longrightarrow \underset{\text{More stable}}{HCOO^-} + H^+$$

(*ii*) *Acidic strength of $CH_3$-COOH, $CH_3 - CH_2 -$ COOH, $(CH_3)_2$ CH – COOH and $(CH_3)_3$ C – COOH decreases in the order: $CH_3 -$ COOH > $CH_3 - CH_2 -$ COOH > $(CH_3)_2$ CH – COOH > $(CH_3)_3$C – COOH.* This order can be explained as follows:

We know that greater is the magnitude of +I effect of an alkyl group, lower is the acidic strength of the acid containing the alkyl group. Thus since + I effect of alkyl groups namely $CH_3$ –, $CH_3 - CH_2$ –, $(CH_3)_2$ CH –, $(CH_3)_3$C – *increases* in the order: $CH_3$– < $CH_3$– $CH_2$– < $(CH_3)_2$ CH – < $(CH_3)_3$C–, the relative acid strength of carboxylic acids containing these alkyl groups decreases in the same order as shown below.

$$CH_3COOH > CH_3-CH_2-COOH > (CH_3)_2CH-COOH > (CH_3)_3C-COOH$$

Acetic acid; Propionic acid; Isobutyric acid (2-methyl propionic acid) or dimethyl acetic acid; Trimethyl acetic acid (2, 2-dimethyl propionic acid)

(*iii*) *Chloroacetic acid (Cl – $CH_2$ – COOH) is a stronger acid than acetic acid ($CH_3$COOH).* This can be explained as follows:

The structures of these acids are given below.

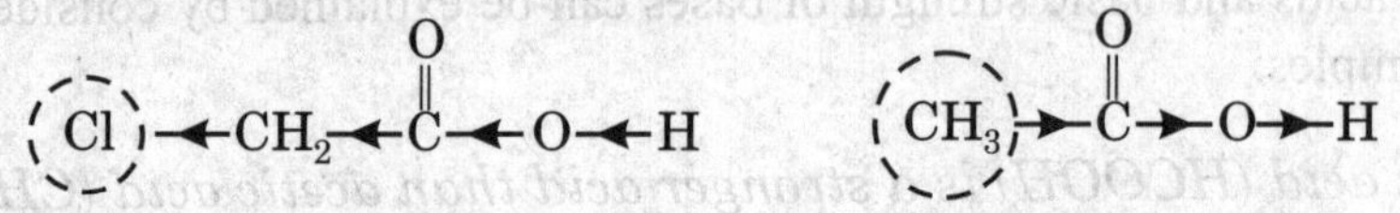

EWG (–I effect) Chloroacetic acid EDG (+I effect) Acetic acid

–Cl group present in Cl$CH_2$COOH is an electron-withdrawing group (EWG) and hence has electron-withdrawing inductive effect (–I effect). Because of –I effect of –Cl group, this group witdraws electrons from the O–H bond. As a result, the electron density in O-H bond in Cl$CH_2$COOH decreases and O–H bond weakens. In other words it becomes easier for Cl$CH_2$COOH molecule to release $H^+$ ions as compared the $CH_3$COOH. Thus Cl$CH_3$COOH is a stronger acid than $CH_3$COOH.

Further –I effect of –Cl group disperses (decreases) the magative charge (electron density) on Cl$CH_2COO^-$ ion as compared to Cl$CH_2$COOH and thus makes Cl$CH_2COO^-$ ion more stable as compared Cl$CH_2$COOH

$$\underset{\substack{\text{Less stable} \\ \text{(Stronger acid)}}}{Cl-CH_2-COOH} \longrightarrow \underset{\text{More stable}}{Cl-CH_2-COO^- + H^+}$$

Since –$CH_3$ group in $CH_3$COOH is an electron-donating group (EDG), it has electron-donating inductive effect (+I effect). Because of +I effect of –$CH_3$ group, this group intensities (increases) the negative charge (electron density) on $CH_3COO^-$ ion as compared to $CH_3$COOH and thus makes $CH_3COO^-$ ion less stable as compared to $CH_3$COOH.

$$\underset{\substack{\text{More stable} \\ \text{(Weaker acid)}}}{CH_3COOH} \longrightarrow \underset{\text{Less Stable}}{CH_3COO^- + H^+}$$

It is because of –I effect of –Cl group and +I effect of –$CH_3$ group that Cl $CH_2$ COOH is a stronger acid than $CH_3$COOH.

Above discussion can be summarised by stating that –Cl group in $ClCH_2COOH$ decreases (–I effect) the negative charge on $ClCH_2COO^-$ and hence increases the stability of $ClCH_2COO^-$ as compared to $ClCH_2COOH$. On the other hand, $-CH_3$ group increases (+I effect) the negative charge on $CH_3COO^-$ ion and hence decreases the stability of $CH_3COO^-$ in as compared to $CH_3COOH$.

*(iv) The acidic strength of $BH_3$, $BF_3$ and $BMe_3$ molecules increases in the order: $BMe_3 < BH_3 < BF_3$.* This can be explained as follows:

The central B-atom in each of the compounds has vacant 2p orbital and hence each of these compounds acts as Lewis acid. Since methyl group (Me) is an electron-donating group, the presence of this group in $Me_3B$ decreases the acidity of $Me_3B$ w.r.t. that of $BH_3$. Similarly, since $F^-$ ion is an electron-withdrawing group, it will increase the acidity of $BF_3$ as compared to that $BH_3$. Thus $Me_3B$ is the weakest acid, $BF_3$ is the strongest acid and $BH_3$ has intermediate acidic strength, *i.e.* the order of acidic strength is $Me_3B < BH_3 < BF_3$.

*(v) $C_6H_5COOH$ is a stronger acid than $CH_3COOH$.* This can be explained as follows:

$C_6H_5COOH$ is stronger acid than $CH_3COOH$ because phenyl group ($-C_6H_5$) present in $C_6H_5COOH$ is an EWG whereas $-CH_3$ group present in $CH_3COOH$ is an EDG.

*(vi) Basic strength of $NH_3$ and alphatic amines ($RNH_2$, $R_2NH$ and $R_3N$) in gas phase (non-aqueous solvent) increases in the order : $NH_3 < RNH_2 < R_2NH < R_3N$.* This can be explained as follows:

Due to the presence of electron-donating alkyl groups in amines, the electron density on N-atom increases (+I effect) as we proceed as $NH_3 \rightarrow RNH_2$ (1° amine) $\rightarrow R_2NH$ (2° amine) $\rightarrow R_3N$ (3° amine). Due to the increase in electron density on N-atom, it becomes more easy for amines to donate the lone pair of electrons than $NH_3$ as we move from $RNH_2$ to $R_3N$.

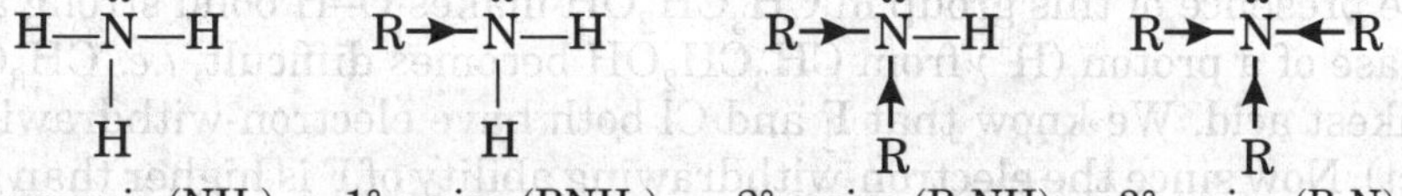

Ammonia ($NH_3$) 1° amine ($RNH_2$) 2° amine ($R_2NH$) 3° amine ($R_3N$)

Thus the basic strength of amines increases as :

$$NH_3 < RNH_2 < R_2NH < R_3N$$

or $$NH_3 < 1° \text{ amine} < 2° \text{ amine} < 3° \text{ amine}$$

Since the solvation effect, *i.e.* stabilisation of conjugate acid due to H-bonding is absent in amines in gas phase, the basic strength of amines in gas phase depends only on +I effect of alkyl groups.

*(vii) $C_6H_5NH_2$ is a weaker base than $NH_3$.* This can be explained as follows:

Since $-C_6H_5$ group (phenyl group) is an electron-withdrawing group, this group decreases the electron density on N-atom in $C_6H_5NH_2$ molecule. The decrease in electron density on N-atom makes $C_6H_5NH_2$ weaker base than $NH_3$.

*(viii) Diphenylamine [$(C_6H_5)_2NH$] is a weaker base than aniline ($C_6H_5NH_2$).* This can be explained as follows:

Diphenylamine, $(C_6H_5)_2NH$ and aniline, $C_6H_5NH_2$ both contain negative $-C_6H_5$ groups which are electron-withdrawing groups. $(C_6H_5)_2NH$ contains two $-C_6H_5$ groups and $C_6H_5NH_2$ has only one $-C_6H_5$ group. Thus in $(C_6H_5)_2$ NH *lp* of electrons on N-atom is distributed over two $-C_6H_5$ groups while in $C_6H_5$ $NH_2$ *lp* of electrons on N-atom is distributed only over one $-C_6H_5$ group. Hence in $(C_6H_5)_2$ $\ddot{N}H$ molecule the electron density of *lp* of electrons on N-atom is decreased. The decrease in electron density makes $(C_6H_5)_2$ NH weaker base than $C_6H_5NH_2$.

## 4. Effect of the Magnitude of –I Effect Exerted by Electron-Withdrawing Substituent

**Examples** (*i*) In case of α-halo acetic acids, greater is the magnitude of electron-withdrawing inductive effect (–I effect) of halogen group, greater is the acidic strength of halo acetic acids.

We know that α-haloacetic acids are $FCH_2COOH$ (fluoroacetic acid), $ClCH_2COOH$ (chloroacetic acid), $BrCH_2COOH$ (bromoacetic acid) and $ICH_2COOH$(iodoacetic acid). All the acids have halogen group which has – I effect. Now since electronegativity of halogen atoms decreases as F > Cl > Br > I, the magnitude of –I effect exerted by halogens also decreases in the same order (F > Cl > Br > I). Due to the decrease in the magnitude of –I effect of halogens, acidic strength of halo acetic acids decreases as:

$$\underset{K_a = 260 \times 10^{-5}}{FCH_2COOH} > \underset{K_a = 136 \times 10^{-5}}{ClCH_2COOH} > \underset{K_a = 125 \times 10^{-5}}{BrCH_2COOH} > \underset{K_a = 67 \times 10^{-5}}{ICH_2COOH}$$

(*ii*) The acidic strength of $CH_3CH_2OH$, $CF_3CH_2OH$ and $CCl_3$ $CH_2OH$ decreases in the order : $CF_3CH_2OH > CCl_3$ $CH_2OH > CH_3CH_2OH$. This decreasing order can be explained as follows.

Since $CH_3 - CH_2$ group (ethyl group) is an electron-releasing group (+I effect), the presence of this group in $CH_3CH_2OH$ makes O–H bond strong and hence the release of a proton ($H^+$) from $CH_3CH_2OH$ becomes difficult, *i.e.* $CH_3CH_2OH$ is the weakest acid. We know that F and Cl both have electron-withdrawing ability (–I effect). Now since the electron-withdrawing ability of F is higher than that of Cl, O – H bond in $CF_3CH_2OH$ weakens and the release of a proton ($H^+$) from this acid becomes comparatively more easy than from $CCl_3CH_2OH$. Consequently $CF_3CH_2OH$ is stronger acid than $CCl_3CH_2OH$. $CF_3CH_2OH$ is stronger acid than $CCl_3CH_2OH$. $CH_3CH_2COOH$ is the weakest acid. Thus the order of acidity of the given acids is:

$$CF_3CH_2OH > CCl_3CH_2OH > CH_3CH_2OH$$

## 5. Effect of the Number of Electron-Withdrawing Substituents

**Examples** (*i*) In case of chloro acetic acids, higher is the number of electron-withdrawing groups (–Cl groups) present in the chain attached with carboxyl group, greater is the acidic strength of chloroacetic acid.

It has been observed that the acidic strength of the three chloroacetic acids increases in the following order:

$CH_2ClCOOH$ < $CHCl_2COOH$ < $CCl_3COOH$

Monochloro acetic acid — Dichloro acetic acid — Trichloro acetic acid

We know that since –Cl group is an electron-withdrawing group (EWG), it has –I effect. Due to –I effect of –Cl group, the dipersal of negative charge on carboxylate ion increases with the increase in the number of –Cl groups. This increases the stability of carboxylate ion and hence the acidic strength of the corresponding chloro acetic acid increases.

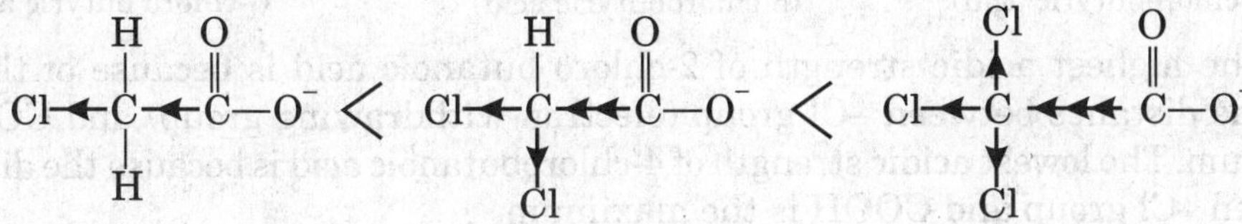

Monochloro acetate ion (One-Cl group) — Dichloroacetate ion (Two – Cl groups) — Trichloroacetate ion (Three-Cl groups)

——— Stability of carboxylate ion increases ———→

$CH_2ClCOOH$ < $CHCl_2COOH$ < $CCl_3COOH$

(Cl = 1) (Cl = 2) (Cl = 3)

——— Acidic strength increases ———→

Since $CH_3COOH$ has no –Cl group, it is the weakest acid of all the chloroacetic acids. Thus:

$CH_3COOH$ < $CH_2ClCOOH$ < $CHCl_2COOH$ < $CCl_3COOH$

(Cl = 0) (Cl = 1) (Cl = 2) (Cl = 3)

——— Acidic strength increases ———→

(*ii*) The basic strength (basicity) of ammonia ($NH_3$), phenylalamine ($C_6H_5NH_2$), diphenyl amine [$(C_6H_5)_2NH$] and triphenylamine [$(C_6H_5)_3N$] decreases in the order:

$$NH_3 > C_6H_5NH_2 > (C_6H_5)_2NH > (C_6H_5)_3N$$

This decreasing order can be explained as follows:

**Explanation.** $NH_3$ has no electron-withdrawing group and hence has the maximum basicity. On the other hand, each of the remaining substituted amines has one, two and three electron-withdrawing phenyl ($-C_6H_5$) groups. The presence of these groups decreases the basicity of the substituted amines. Greater is the number of electron-withdrawing groups, lesser is the basicity. Thus the basicity of the given compounds is in the order:

$$NH_3 > NH_2(C_6H_5) > NH(C_6H_5)_2 > N(C_6H_5)_3$$

## 6. Effect of the Distance of the Electron-Withdrawing Substituent from Carboxyl Group in an Aliphatic Carboxylic Acid

As the distance of the electron-withdrawing substituent from the carboxyl group present in an aliphatic carboxylic acid increases, –I effect exerted by the substituent decreases and hence the dispersal of negative charge of carboxylate ion becomes less pronounced. This decreases the stability of the carboxylate ion and, therefore, the acid strength of the acids decreases.

**Example.** The acid strength of chloro substituted butanoic acids decreases in the following order:

$$\underset{\gamma}{\overset{4}{CH_3}}-\underset{\beta}{\overset{3}{CH_2}}-\underset{\alpha}{\overset{2}{CH}}(Cl)-\overset{1}{COOH} > \underset{\gamma}{\overset{4}{CH_3}}-\underset{\beta}{\overset{3}{CH}}(Cl)-\underset{\alpha}{\overset{2}{CH_2}}-\overset{1}{COOH} > \underset{\gamma}{\overset{4}{CH_2}}(Cl)-\underset{\beta}{\overset{3}{CH_2}}-\underset{\alpha}{\overset{2}{CH_2}}-\overset{1}{COOH}$$

2-chloro butanoic acid (α-chlorobutyric acid) — 3-Chlorobutonoic acid (β-chlorobutyric acid) — 4-Chlorobutanoic acid (γ-chloro butyric acid)

The highest acidic strength of 2-chloro butanoic acid is because of the fact that the distance between –Cl group (electron-withdrawing group) and COOH is minimum. The lowest acidic strength of 4-chlorobutanoic acid is because the distance between –Cl group and COOH is the maximum.

## 7. Effect of Steric Overcrowding of the Substituents

**Examples** (*i*) The basic strength of tiralkyl amines decreases in the order:

$$\ddot{N}(CH_3)_3 > \ddot{N}(CH_3CH_2)_3 > \ddot{N}(CH(CH_3)_2)_3$$

or Trimethyl amine > Triethylamine > Triisoprophyl amine

**Explanation.** As the bulk of the substituent goes on increasing, the sferic overcrowding of the substituent around the electron pair on N-atom of amines also increases. As a result, the approach of an electron acceptor (acid) towards the electron pair on N-atom goes on becoming more and more difficult and hence basic strength of amines goes on decreasing.

(*ii*) $(CH_3)_3$ N is less basic than $(CH_3)_2$ N

**Explanation.** $(CH_3)_3$ N is less basic than $(CH_3)_2N$ because the sterric overcrowding of the three $CH_3$ groups (electron-releasing groups) round N-atom in $(CH_3)_3N$ molecule makes the bonding of N-atom of this molecule with proton relatively difficult.

## 8. Solvation Effects

We have already shown that the basic strength of 1°,2° and 3° amines in gas phase decreases in the order (Here we have considered methyl amines)

$$(CH_3)_3N > (CH_3)_2NH > CH_3NH_2$$

or 3°amine > 2° amine > 1° amine

but in aqueous solution the basic strength of the above amines decreases in the order:

$$(CH_3)_2NH > CH_3NH_2 > (CH_3)_3 N$$

or 2° amine > 1° amine > 3° amine.

or Secondary amine > Primary amine > Tertiary amine

**Explanation.** The above decreasing order of basic strength of amines in aqueous solution can be explained as follows:

The basic strength of an amine in aqueous solution depends on the stability of ammonium cation or conjugate acid formed by the amine by accepting a proton

($H^+$) from $H_2O$. The stability of the ammonium cation, in turn, depends on the following factors.

*(a) +I effect of alkyl groups.* As the number of alkyl groups (here alkyl groups are $CH_3$ groups) increases, due to +I effect of alkyl groups, the dispersal of positive charge on the ammonium cation increases. Due to the increase in the dispersal of positive charge, the basic strength of amines also increases in the order.

$$\underset{1^\circ \text{ amine}}{CH_3NH_2} < \underset{2^\circ \text{amino}}{(CH_3)_2N} < \underset{3^\circ \text{ amine}}{(CH_3)_3N}$$

*(b) Stability of ammonium cations derived from amines due to H-bonding (Solvation effect).* The formation of ammonium cations obtained by the reaction of three amines with $H_2O$ ($H_2O$ acts as an acid towards amines) can be shown as:

| *Amine (Base)* | *Solvent (Acid)* | *Ammonium cation having H–bond (Acid)* | *Base* |
|---|---|---|---|

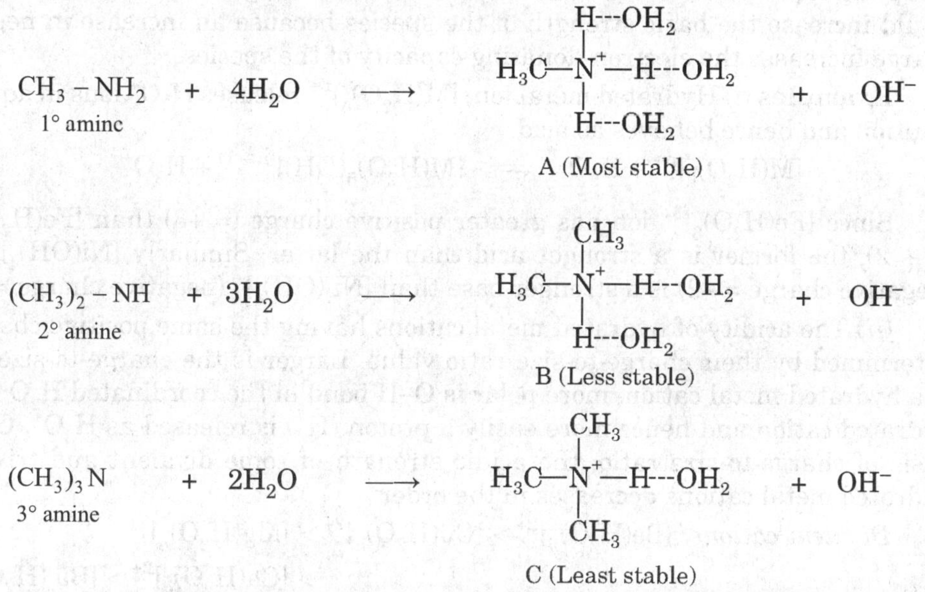

Ammonium cations derived from $CH_3NH_2$, $(CH_3)_2NH$ and $(CH_3)_3N$ amines are $CH_3\overset{+}{N}H_3(H_2O)_3$ (A), $(CH_3)_2\overset{+}{N}H_2(H_2O)_2$(B) and $(CH_3)_3\overset{+}{N}H(H_2O)$(C) respectively. These cations are obtained due to H-bonding between H-atom of amines and O-atom of $H_2O$ molecules. The number of H-bonded H-atoms in these cations decreases as 3(in A) > 2 (in B) > 1(in C). Due to the decrease in the number of H-bonded H-atoms, the stability of A, B and C cations decreases in the same order: A (*Most stable*) > B > C (*Least stable*)

If we consider the combined effect of +I effect and H-bonding effect due to solvation, the relative order of basic strength of methyl amines, in aqueous solution is found to decrease as:

$$(CH_3)_2NH > CH_3NH_2 > (CH_3)_3N$$

or

$$2^\circ\text{amine} > 1^\circ \text{ amine} > 3^\circ \text{ amine}$$

This order is true only when alkyl group is $CH_3$ group.

**Steric hinderance of alkyl groups to H-bonding.** If the alkyl group is bigger than $CH_3$ group (*e.g.* ethyl, propyl etc), there will be some steric hinderance to H-bonding. As a result, the relative order of basic strength of amines in aqueous solution is found to decrease as:

$$\underset{2^\circ\text{amine}}{(CH_3CH_2)_2NH} > \underset{3^\circ\text{ amine}}{(CH_3CH_2)_3N} > \underset{1^\circ\text{ amine}}{(CH_3CH_2)NH_2}$$

When alkyl group is a $CH_3$ group, there is no hinderance of this group to H-bonding.

## 9. Charge on the Species and Charge-to-Size Ratio: Acidic Strength of Hydrated Metal Ions

An increase in the positive charge on a species would increase the acidic strength of the species because the increase in positive charge increases the electron-accepting capacity of the species. An increase in the negative charge on a species would increase the basic strength of the species because an increase in negative charge increases the electron-donating capacity of the species.

**Examples** (*i*) Hydrated metal ion, $[M(H_2O)_6]^{n+}$ releases $H_3O^+$ ions in aqueous solution and hence behaves as acid.

$$[M(H_2O)_6]^{n+} + H_2O \rightleftharpoons [M(H_2O)_5(OH)]^{(n-1)} + H_3O^+$$

Since $[Fe(H_2O)_6]^{3+}$ ion has greater positive charge (= +3) than $[Fe(H_2O)_6]^{2+}$ (= + 2), the former is a stronger acid than the latter. Similarly $[Ni(OH)_4]^{2-}$ ion (negative charge = –2) is a stronger base than $[Ni\ (OH)_4]^-$ (negative charge = –1)

(*ii*) The acidity of hydrated metal cations having the same positive charge is determined by their charge-to-size ratio value. Larger is the charge-to-size ratio of a hydrated metal cation, more polar is O–H bond in the coordinated $H_2O$ in the hydrated cation and hence more easily a proton ($H^+$) is released as $H_3O^+$. On the basis of charge-to-size ratio, the acidic strength of some divalent and trivalent hydrated metal cations decreases in the order:

*Divalent cations:* $[Be(H_2O)_n]^{2+} > [Cu(H_2O)_n]^{2+} > [Mg(H_2O)_n]^{2+} > [Ca(H_2O)_n]^{2+} > [Ba\ (H_2O)n]^{2+}$

*Trivalent cations:* $[Fe(H_2O)_6]^{3+} > [V(H_2O)_6]^{3+} > [Cr\ [H_2O]_6]^{3+} > [Sc\ (H_2O)_6]^{3+} > [Al(H_2O)_6]^{3+} > [La\ (H_2O)_6]^{3+}$

## 10. Effect of Electronegativity (EN)

Following examples show how the electronegativity of elements present in given species affects the acidic strength of the species.

**Examples.** (*i*) The acidic strength of $CH_4$, $NH_3$, $H_2O$ and HF increases in the under:

$$CH_4\ (\textit{weakest acid}) < NH_3 < H_2O < HF(\textit{strongest acid})$$

This increasing order can be explained on the basis of electronegativity values of the non-metals present in the given hydrides. Higher is the electronegativity of the non-metal in the given hydride, easier would be the release of proton ($H^+$) from the hydride. In other words, the acid strength (proton releasing capacity) of hydrides would increase with the increase in electronegativity of non-metal. Now since electronegativity of C, N, O and F increases as C < N < O < F, the acidic strength of the hydrides also increases in the same order ($CH_4 < NH_3 < H_2O < HF$).

The increase in acidic strength from $CH_4$ to HF is supported by the successive increase in acid dissociation constant values of these hydrides as shown below.

$$CH_4 = 10^{-58},\ NH_3 = 10^{-35},\ H_2O = 10^{-14},\ HF = 10^{-4}$$

In general, in a period since the electronegativity of elements increases on moving from left to right, the acidic strength of hydrides also increases in the same direction as shown below.

——— Acidic strength increases ———→

| | | | |
|---|---|---|---|
| $CH_4$ | $NH_3$ | $H_2O$ | HF |
| | | $H_2S$ | HCl |
| | | $H_2Se$ | HBr |
| | | | HI |

(*ii*) Acidic strength of hypohalous acids (HClO, HBrO and HIO) decreases as the electronegativity of halogen atoms decreases from F to I. Similarly, the acidic strength of perhalic acids ($HClO_4$, $HBrO_4$ and $HIO_4$) also decreases with the decrease of EN from Cl to I. Thus:

*Hypohalous acids:* HClO > HBrO > HIO

— EN of halogen decreases →

— Acidic strength decreases →

*Perhalic acids:* $HClO_4 > HBrO_4 > HIO_4$

— EN of halogen decreases →

— Acidic strength decreases →

(*iii*) Acidic strength of $HNO_3$, $HPO_3$ and $HAsO_3$ decreases in the order: $HNO_3 > HPO_3 > HAsO_3$. The decrease is due to the decrease in EN values of the central atoms (N > P > As) as we move down the group from N to As.

(*iv*) $H_2SO_3$ is stronger acid than $H_2SeO_3$ ($H_2SO_3 > H_2SeO_3$) because of greater electronegativity of S-atom than Se-atom.

(*v*) Acidic strength of $H_3PO_4$, $H_3AsO_4$ and $H_3SbO_4$ decreases in the order:

$H_3PO_4 > H_3AsO_4 > H_3SbO_4$. The decrease is due to the decrease in EN value of the central atoms (P > As > Sb) as we move down the group from P to Sb.

## 11. Effect of Oxidation Number (O.N.) of the Central Atom.

(*i*) In oxoacids, higher is the O.N. of the central atom, greater is the acidic strength of the acid. Higher O.N. increases the electron-pulling capacity of the acid and makes the release of $H^+$ ions from the acid easier. This is the reason why acidic strength of various oxoacids increases with the increase of O.N. of the central atom. For example:

(*a*) $H\overset{+1}{Cl}O < H\overset{+3}{Cl}O_2 < H\overset{+5}{Cl}O_3 < H\overset{+7}{Cl}O_4$

(*b*) $H_2\overset{+4}{S}O_3 < H_2\overset{+6}{S}O_4$

(*c*) $H\overset{+1}{N}O < H\overset{+3}{N}O_2 < H\overset{+5}{N}O_3$

(*ii*) Acidic strength of oxo acids of Mn increases in the order shown below. The increasing order is due to the increase in O.N. of Mn atom as we move from $H_2MnO_2$ to $HMnO_4$.

$$\overset{+2}{H_2MnO_2} < \overset{+3}{H_3MnO_3} < \overset{+4}{H_4MnO_4} < \overset{+5}{H_3MnO_4} < \overset{+6}{H_2MnO_4} < \overset{+7}{HMnO_4}$$

| Base | Weak base | Amphoteric | Strong acid | Strong acid | Very Strong acid |
|---|---|---|---|---|---|

O.N. of Mn atom increases ⟶

Acidic strength increases ⟶

(*iii*) Acidic strength of $HMnO_4$, $H_2MnO_4$ and $H_3MnO_4$ decreases as $\overset{+7}{HMnO_4} > \overset{+6}{H_2MnO_4} > \overset{+5}{H_3MnO_4}$ due to the decrease in O.N. of Mn atom.

(*iv*) Acidic strength of $H_2SO_3$, $H_2SO_4$, $HClO_3$ and $HClO_4$ increases as $\overset{+4}{H_2SO_3} < \overset{+5}{HClO_3} < \overset{+6}{H_2SO_4} < \overset{+7}{HClO_4}$. This increasing order is due to the increase in O.N. of the central atom from S to Cl.

(*v*) The acidic strength of $H_2CO_3$, $H_3PO_4$, $H_2SO_4$ and $HClO_4$ also increases in the order $\overset{+4}{H_2CO_3} < \overset{+5}{H_3PO_4} < \overset{+6}{H_2SO_4} < \overset{+7}{HClO_4}$. This increasing order is also due to the increase in O.N. of the central atom from C to Cl.

(*vi*) The basic strength (basicity) of oxo anions *viz*. $ClO^-$, $ClO_2^-$, $ClO_3^-$ and $ClO_4^-$ decreases as:

$$[\overset{+1}{ClO}]^- > [\overset{+3}{ClO_2}]^- > [\overset{+5}{ClO_3}]^- > [\overset{+7}{ClO_4}]^-$$

Basic strength decreases ⟶

The decrease in basic strength is due to the increase in O.N. of Cl atom from +1 (in $ClO^-$) to +7 (in $ClO_4^-$).

## 12. Effect of the Number of –OH Groups Per Unhydroxyl O-atom

Greater is the number of –OH groups per unhydroxyl (unprotonated) O-atom in an oxoacid, lesser is the acidic strength of the acid. For example:

(*i*) $H_2S_2O_7$ is a stronger acid than $H_2SO_4$.

**Explanation.** The structures of $H_2S_2O_7$ and $H_2SO_4$ are given in Fig. 3.4.

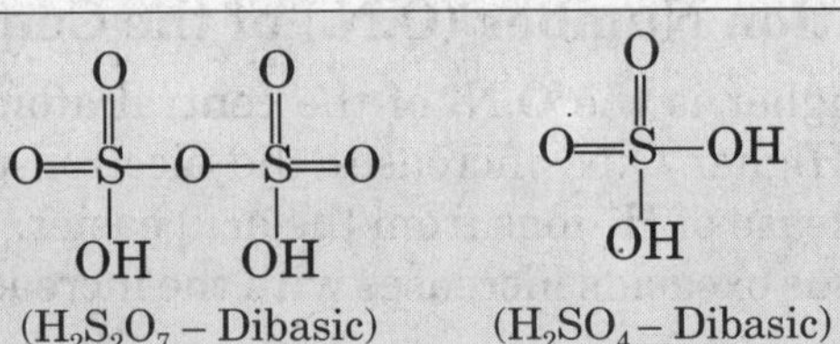

($H_2S_2O_7$ – Dibasic) ($H_2SO_4$ – Dibasic)

**Fig 3.4** Structure of $H_2S_2O_7$ and $H_2SO_4$ molecules.

These structures show that the number of –OH groups per unhydroxyl O-atom in $H_2S_2O_7$ and $H_2SO_4$ is equal to 2/4 = 0.5 and 2/2 = 1.0 respectively. Since this number is less in $H_2S_2O_7$ than in $H_2SO_4$, $H_2S_2O_7$ is a stronger acid than $H_2SO_4$. ($H_2S_2O_7 > H_2SO_4$)

(*ii*) The acid strength of $H_3PO_4$, $H_4P_2O_7$ and $H_5P_3O_{10}$ increases in the order:

$$H_3PO_4 < H_4P_2O_7 < H_5P_3O_{10}$$

**Explanation:** The structures of $H_3PO_4$, $H_4P_2O_7$ and $H_5P_3O_{10}$ are given in Fig 3.5.

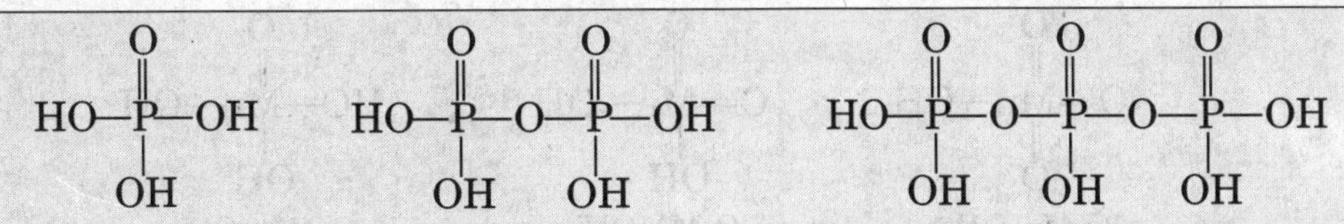

($H_3PO_4$–Tribasic) ($H_4P_2O_7$–Tetrabasic) ($H_5P_3O_{10}$–Pentabasic)

**Fig. 3.5** Structure of $H_3PO_4$, $H_4P_2O_7$ and $H_5P_3O_{10}$ molecules

These structures show that the number of –OH groups per unprotonated O-atoms in these acids is as $H_3PO_4 = 3$, $H_4P_2O_7 = 4/2 = 2$, $H_5P_3O_{10} = 5/3 = 1.66$. Now since this number *decreases* as $H_3PO_4 > H_4P_2O_7 > H_5P_3O_{10}$, the acidic strength of these acids *increases* as $H_3PO_4 < H_4P_2O_7 < H_5P_3O_{10}$.

(*iii*) The acidic strength of $H_3PO_2$, $H_3PO_3$ and $H_3PO_4$ decreases in the order $H_3PO_2 > H_3PO_3 > H_3PO_4$.

**Explanation** The structures of $H_3PO_2$, $H_3PO_3$ and $H_3PO_4$ are given in Fig. 3.6.

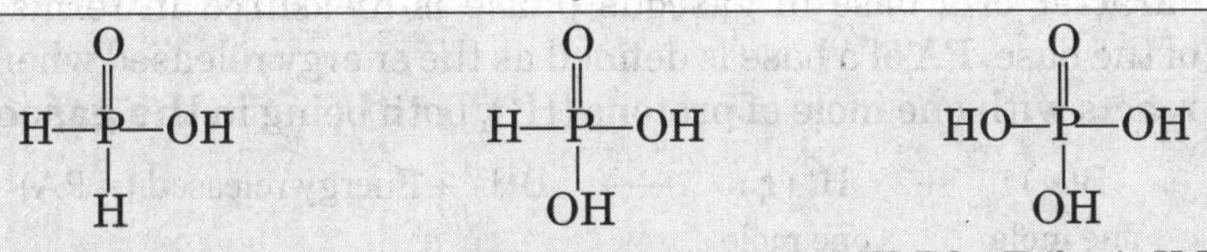

($H_3PO_2$–Monobasic) ($H_3PO_3$–Dibasic) ($H_3PO_4$–Tribasic $H_3PO_2$)

**Fig. 3.6** Structure of $H_3PO_2$, $H_3PO_3$ and $H_3PO_4$ molecules.

These structures show that the number of –OH groups per unhydroxyl O-atom in $H_3PO_2$, $H_3PO_3$ and $H_3PO_4$ is equal to 1, 2 and 3 respectively. Since the number of –OH groups increases as $H_3PO_2 < H_3PO_3 < H_3PO_4$, the acidic strength of these acids decreases as $H_3PO_2 > H_3PO_3 > H_3PO_4$.

(*iv*) The acidic strength of $ClO_3(OH)$, $ClO_2(OH)$, $SO(OH)_2$ and $SO_2(OH)_2$ decreases in the order $ClO_3(OH) > ClO_2\ (OH) > SO_2(OH)_2 > SO\ (OH)_2$.

**Explanation.** The structures of the given acids are given in Fig. 3.7.

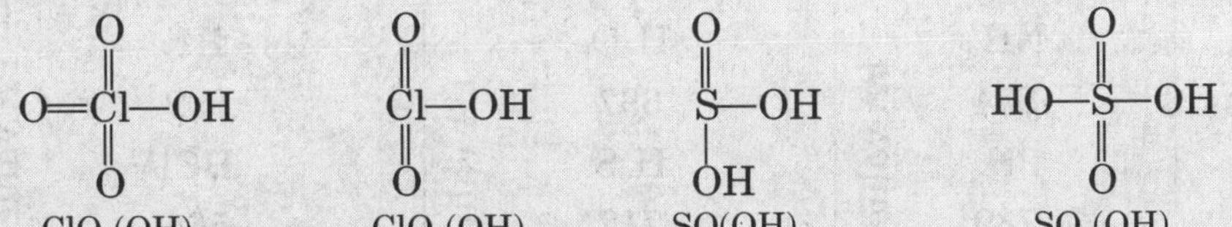

$ClO_3(OH)$ $ClO_2(OH)$ $SO(OH)_2$ $SO_2(OH)_2$

**Fig. 3.7** Structure of $ClO_3(OH)$, $ClO_2$ (OH), $SO(OH)_2$ and $SO_2(OH)_2$ molecules

The structures of the given acids shown in Fig. 3.7 show that the number (*n*) of –OH groups per unhydroxyl O-atom in these acids is $ClO_3(OH) = 1/3 = 0.33$, $ClO_2(OH) = 1/2 = 0.50$, $SO(OH)_2 = 2$ and $SO_2(OH)_2 = 2/2 = 1$. Since the value of *n* for these acids increases as $ClO_3(OH) < ClO_2\ (OH) < SO_2(OH)_2 < SO(OH)_2$, the acidic strength of these acids decreases as $ClO_3\ (OH) > ClO_2\ (OH) > SO_2\ (OH)_2 > SO(OH)_2$. This order shows that $ClO_3$ (OH) is the strongest acid and SO $(OH)_2$ is the weakest acid.

(*v*) The acidic strength of $O_3Mn(OH)$, $O_2Mn(OH)_2$ and $OMn\ (OH)_3$ decreases as $O_3Mn\ (OH) > O_2Mn\ (OH)_2 > OMn\ (OH)_3$

**Explanation.** The structures of the given acids are given in Fig. 3.8.

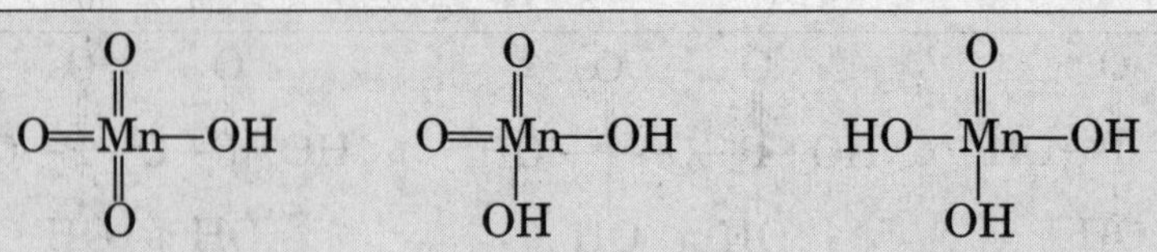

$O_3Mn(OH)$ $O_2Mn(OH)_2$ $OMn(OH)_3$

**Fig. 3.8** Structure of $O_3Mn\ (OH)$, $O_2Mn\ (OH)_2$ and $OMn\ (OH)_3$ molecules

These structures show that the number of –OH groups (*n*) per unhydroxyl O-atom in these acids is equal to $O_3Mn(OH) = 1/3$, $O_2Mn(OH)_2 = 2/2 = 1$ and $OMn\ (OH)_3 = 3$. Since the value of *n* increases as $O_3Mn(OH) < O_2Mn\ (OH)_2 < OMn\ (OH)_3$, the acidic strength of these acids decreases as $O_3Mn\ (OH) > O_2Mn\ (OH)_2 > OMn\ (OH)_3$.

## 13. Effect of Proton Affinity (PA) Values of Bases in Gaseous Phase

Basic character of a base in gaseous phase is measured in terms of proton affinity (PA) of the base. PA of a base is defined as the energy released when one mole of a base (B) reacts with one mole of protons ($H^+$), both being in **the gaseous state**.

$$\underset{\text{one mole}}{B(g)} + \underset{\text{one mole}}{H^2(g)} \longrightarrow BH^+ + \text{Energy released } (= \text{PA})$$

As is the practice with EA, PA is also expressed in KJ / mole with a positive sign. Higher is the value of PA of a species, higher is the amount of energy released and hence greater is the basic character of the species.

**Basic strength of binary hydrides.** PA values of the binary hydrides of the elements of groups 15, 16 and 17 are given in Table 3.2.

**Table 3.2 : Proton affinities (in KJ / mole) of some binary hydrides.**

| *Group VA (15)* | *Group VIA (16)* | *Group VIIA (17)* |
|---|---|---|
| ——— PA values decrease ———→ | | |
| $NH_3$ | $H_2O$ | HF |
| 854 | 697 | 399 |
| $PH_3$ | $H_2S$ | HCl |
| 789 | 712 | 564 |
| $AsH_3$ | $H_2Se$ | HBr |
| 750 | 717 | 589 |
| | | HI |
| | | 628 |
| —PA values decrease ↓ | —PA values increase ↓ | —PA values increase ↓ |

It may be seen that the PA values of binary hydrides decrease on moving from left to right in the hydrides of group 15 to 17. Thus $NH_3$ has greater affinity for a proton ($H^+$) to form $NH_4^+$ ion than $H_2O$ to from $H_3O^+$ or HF to form $HF_2^+$. As a result

Bronsted basicity of binary hydrides decreases along a period with the decrease of their PA values.

As we move down group 15, PA values of the trihydrides decrease. Thus since PA value of $NH_3$ (= 854 KJ/mole) is greater than that of $PH_3$ (= 789 KJ/mole), $NH_3$ would hold on a proton better than $PH_3$

As we move down the hydrides of the elements of groups 16 and 17, PA values increase. This means that the basic strength of the hydrides also increases down the groups. Thus the eagerness of HCl molecule to take up a proton to form $H_2Cl^+$ is greater than that of HF to form $H_2F^+$.

PA values (in KJ/mole) of $NH_3$, $CH_3NH_2$, $(CH_3)_2NH$ and $(CH_3)_3N$ are 854, 920, 950 and 970 respectively. Since these values increase from $NH_3$ to $(CH_3)_3N$, the basic strength of these amines increases in the order: $NH_3 < (CH_3)NH_2 < (CH_3)_2NH < (CH_3)_3N$.

Conjugate bases of HF, HCl, HBr and HI are $F^-$, $Cl^-$, $Br^-$ and $I^-$ respectively. PA values (in KJ/mole) of these conjugate bases are 1554, 1395, 1354 and 1315 respectively. Since these PA values decrease as $F^- > Cl^- > Br^- > I^-$, the basicity of these conjugate bases also decreases in the same order: $F^- > Cl^- > Br^- > I^-$. Due to the decrease in the basicity of conjugate bases, the acidity of conjugate acids of these bases (HF, HCl, HBr and HI are the conjugate acids of $F^-$, $Cl^-$, $Br^-$ and $I^-$) increase in the order: HF < HCl < HBr < HI.

Conjugate bases of $H_2O$, $H_2S$ and $H_2Se$ are $OH^-$, $SH^-$ and $SeH^-$ respectively. PA values (in KJ/mole) of these conjugate bases are 1635, 1478 and 1420 respectively. Since these PA values decrease as $OH^- > HS^- > HSe^-$, the basicity of the conjugate bases also decreases in the same order : $OH^- > SH^- > SeH^-$. Due to the decrease in the basicity of conjugate bases, the acidity of conjugate acids of these bases ($H_2O$, $H_2S$ and $H_2Se$ are the conjugate acids of $OH^-$, $SH^-$ and $SeH^-$ respectively) increase in the order: $H_2O < H_2S < H_2Se$.

## 14. $K_a$ Values of Acids and $K_b$ Values of Bases

In aqueous solution, values of dissociation (or ionisation) constant of acids ($K_a$) are used to determine the relative strength of acids. The value of dissociation constant of an acid in aqueous solution is represented as $K_a$. $K_a$ value of a weak acid (HA) in aqueous solution is given by equation (*i*) which is derived as follows. $K_a$ is called **acidity constant.**

$$\underset{\text{Acid}}{HA} + \underset{\text{Base}}{H_2O} \rightleftharpoons \underset{\text{Acid}}{H_3O^+} + \underset{\text{Base}}{A^-}$$

The equilibirium constant (K) for he above equilibirium is given by

$$K = \frac{[H_3O^+][A^-]}{[HA][H_2O]}$$

or $$\frac{[H_3O^+][A^-]}{[HA]} = K \times [H_2O]$$

Now since $[H_2O]$ = Constant, $$\frac{[H_3O^+][A^-]}{[HA]} = K \times \text{Constant} = K_a \quad \ldots (i)$$

The value of $K_a$ describes the relative acid strength of a weak acid. Equation (*i*) shows that larger is the value of acid dissociation constant ($K_a$) of an acid, higher is the concentration of $H_3O^+$ ions and consequently stronger is the acid For example since $K_{CH_3COOH}$ value (= $1.8 \times 10^{-5}$) for $CH_3COOH$ is higher than $K_{HCN}$ for HCN (= $7.1 \times 1.0^{-10}$), $CH_3COOH$ is stronger acid than HCN.

$pK_a$ value is the negative logarithm of acidity constant ($K_a$), *i.e. $pK_a$ value is the negative logarithm of acidity constant ($K_a$), i.e.*

$$pK_a = -\log K_a$$

This relation shows that:

(*a*) Smaller is the value of $pK_a$, larger is the value of $K_a$ and hence stronger is the acid. Thus stronger acids have larger $K_a$ values and smaller $pK_a$ values.

(*b*) Larger is the value of $pK_a$, smaller is the value of $K_a$ and hence weaker is the acid. Thus weaker acids have smaller value of $K_a$ and larger value of $pK_a$.

**Table 3.3 Dissociation constants of some acids and bass at 25°C and their relative strength**

| *Acids* | *Dissociation* | *$K_a$ or $K_b$* | *Relative strength* |
|---|---|---|---|
| **Acids** | | | |
| HCl | $HCl \rightleftharpoons H^+ + Cl^-$ | Very large | Very strong |
| $HNO_3$ | $HNO_3 \rightleftharpoons H^+ + NO_3^-$ | Very large | Very strong |
| $H_2SO_4$ | $H_2SO_4 \rightleftharpoons H^+ + HSO_4^-$ | Large | Very strong |
| $HSO_4^-$ | $HSO_4^- \rightleftharpoons H^+ + SO_4^{2-}$ | $1.3 \times 10^{-2}$ | Strong |
| HF | $HF \rightleftharpoons H^+ + F^-$ | $6.7 \times 10^{-4}$ | Weak |
| $CH_3COOH$ | $CH_3COOH \rightleftharpoons CH_3COO^- + H^+$ | $1.8 \times 10^{-5}$ | Weak |
| $H_2CO_3$ ($H_2O + CO_2$) | $H_2CO_3 \rightleftharpoons H^+ + HCO_3^-$ | $4.4 \times 10^{-7}$ | Weak |
| $H_2S$ | $H_2S \rightleftharpoons H^+ + HS^-$ | $1.0 \times 10^{-7}$ | Weak |
| $NH_4^+$ | $NH_4^+ \rightleftharpoons H^+ + NH_3$ | $5.7 \times 10^{-10}$ | Weak |
| $HCO_3^-$ | $HCO_3^- \rightleftharpoons H^+ + CO_3^{2-}$ | $4.7 \times 10^{-11}$ | Weak |
| $H_2O$ | $H_2O \rightleftharpoons H^+ + OH^-$ | $1.8 \times 10^{-16}$ | Very weak |
| **Bases** | | | |
| NaOH | $NaOH \rightleftharpoons Na^+ + OH-$ | Very high | Very strong |
| KOH | $KOH \rightleftharpoons K^+ + OH^-$ | Very high | Very strong |
| Ca $(OH)_2$ | $Ca(OH)_2 \rightleftharpoons Ca^{2+} + 2OH^-$ | High | Strong |
| $NH_4OH$ | $NH_4OH \rightleftharpoons NH_4^+ OH^-$ | $1.81 \times 10^{-5}$ | Weak |
| $CH_3NH_2$ | $CH_3NH_2 + H_2O \rightleftharpoons CH_3NH_3^+ + OH^-$ | $4.38 \times 10^{-4}$ | Weak |
| $C_6H_5NH_2$ | $C_6H_5NH_2 + H_2O \rightleftharpoons C_6H_5NH_3^+ + OH^-$ | $4.70 \times 10^{-10}$ | Very weak |

The value of dissociation (or ionisation) constant of a base in aqueous solution is represented as $K_b$. $K_b$ value of a base (B) is given by equation (*ii*) which is derived as follows:

$$\underset{\text{Acid}}{H_2O} + \underset{\text{Base}}{B} \rightleftharpoons \underset{\text{Acid}}{BH} + \underset{\text{Base}}{OH^-}$$

The equilibirium constant (K) for the above equilibirium is given by:

$$K = \frac{[BH][OH^-]}{[B][H_2O]}$$

or
$$\frac{[BH][OH^-]}{[B]} = K \times [H_2O] = K \times \text{Constant}$$

or
$$\frac{[BH]\,[OH^-]}{[B]} = K_b \qquad \ldots(ii)$$

This equation shows that larger is the value of base dissociation constant ($K_b$) of a base, higher is the concentration of $OH^-$ ions and consequently strong is the base. For example since $K_{C_2H_5OH}$ value ($= 5.60 \times 10^{-4}$) is higher than $K_{NH_4OH}$ value ($= 1.81 \times 10^{-5}$), $C_2H_5OH$ is stronger base than $NH_4OH$. Stronger bases have lower $pK_b$ values and weaker bases have larger $pK_b$ values.

Values of dissociation constants (at 25° C) and relative strengths of some acids and bases are given in Table 3.3.

## 15. Resonance Effects

In addition to –I and + I effects, electron-withdrawing and electron-donating groups also have resonance effects. Electron-withdrawing resonance effect is represented as –R effect. –R effect increases the acidic strength of acids. In all the three nitrobenzoic acids, $-NO_2$ group which is an electron-withdrawing group has strong electron-withdrawing resonance effect (–R effect) as well as electron-withdrawing inductive effect (–I effects). Electron-donating resonance effect is represented as +R effect. +R effect decreases the acidic strength of acids. In all the three chlorobenzoic acids, –Cl group which is an electron-donatng group, has electron-donating resonance effect (+R effect) as well as electron-withdrawing inductive effect (–I effect).

How resonance concept affects the acidic strength of acids can be explained by considering the following examples:

**(*i*) Comparison of acidic strength of $C_6H_5COOH$ and nitrobenzoic acids**

Nitrobenzoic acids have –COOH and $-NO_2$ groups in the benzene ring. $-NO_2$ group present in these acids is an electron withdrawing group. This group may be present at any of the three positions *viz ortho* (*o*), *meta* (*m*) and *para* (*p*) in the benzene ring. Thus we have *o*–, *m*– and *p*-nitrobenzoic acids. It has been observed that $-NO_2$ group which is an electron-withdrawing group, placed at any of the three positions (*o* –, *m*– and *p*– positions) usually increases the acidic strength of nitrobenzoic acid, but $-NO_2$ group placed at *o*- and *p*-positions increases the acidic strength of the acids to a greater extent than when this group (i.e. $-NO_2$ group) is placed at *m*– position.

Further, when $-NO_2$ group is placed at *o*– position, the acid strength of *o*–nitrobenzoic acid increases to a greater extent than when $-NO_2$ group is placed at *p*-position (**ortho effect**), i.e. acid strength of *o*-nitrobenzoic acid is greater than that of *p*-notrobenzoic acid. Thus the acid strength of nitrobenzoic acids *increases* as :

*m-nitrobenzoic acid < p-nitrobenzoic acid < o-nitrobenzoic acid.*

As a matter of fact, the acid strength of nitrobenzoic acids is much more than that explained above. The extra increase in acid strength is due to the **resonance effect** which has been explained below.

Various resonance structures of benzene containing $-NO_2$ group are given below in Fig. 3.9.

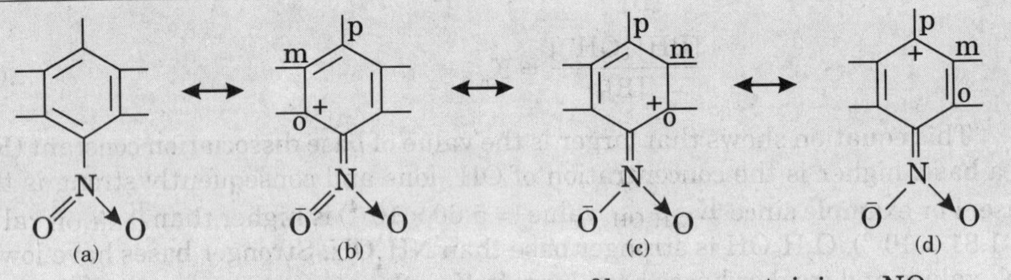


**Fig. 3.9** Various resonance structures of benzene containing $-NO_2$ group which is an electron-withdrawing group.

It may be seen from the resonance structures that *ortho* position (designated as *o*) (w.r.t. $-NO_2$ group) in structures (*b*) and (*c*) and *para* position (designated as *p*) in structure (*d*) are electron-deficient centres. Electron deficiency at *ortho* and *para* positions is caused by resonance effect. *Meta* position in all the four structures retains the original electron density. Therefore, the release of a proton ($H^+$) from COOH placed at *ortho* or *para* position w.r.t $-NO_2$ group would be easier. In other words, o-and *p*-nitrobenzoic acids would be stronger than $C_6H_5COOH$ itself and *m*-nitrobenzoic acid would be practically of the same strength as $C_6H_5COOH$.

It has been observed that *o*-nitrobenzoic acid is stronger than p-nitrobenzoic acid, although the electron deficiency caused by resonance effect at *o*-and *p*-positions is the same. Higher acidic strength of *o*-nitrobenzoic acid as compared to *p*-nitrobenzoic acid can be explained as follows:

In case of nitrobenzoic acids, electron-withdrawing inductive effect (–I effect) as well as electron-withdrawing resonance effect (–R effect) operate together. In case of *o*-and *p*-nitrobenzoic acids –R effect is the same but –I effect is different in these acids. This is because – I effect of $-NO_2$ group decreases with the increase in the number of bonds separating $-NO_2$ group from COOH group. Hence *o*-nitrobenzoic acid is stronger than p-nitrobenzoic acid.

Above discussion shows that the relative order of acid strength of the three nitrobenzoic acids follows the order:

*o - nitrobenzoic acid* > *p - nitrobenzoic acid* > *m-nitrobenzoic acid* > $C_6H_5COOH$
($pK_a = 2.17$) ($pK_a = 3.43$) ($pK_a = 3.45$) ($pK_a = 4.20$)

(*ii*) **Comparison of acidic strength of $C_6H_5COOH$ and its derivatives having electron-donating group at *o*–, *m*– and *p*–positions.** Benzoic acid having an electron-donating group at *o*-and *p*-positions has lower acidic strength than $C_6H_5COOH$ having an electron-donating group at m-position. This is due to resonance effect according to which electron density at *o*- and *p*-positions increases and hence the release of a proton from COOH group is more difficult.

(*iii*) **Phenol ($C_6H_5OH$) is acidic but ethanol ($C_2H_5OH$) is neutral.** Phenols are acidic in nature since they turn blue litmus red and react with aqueous alkalies to form *metal phenoxides or metal phenates* ($C_6H_5O^-$)

$$\underset{\text{Phenol}}{C_6H_5\,OH} + NaOH \rightarrow \underset{\text{Sod. phenoxide or sod. phenate}}{C_6H_5O^-Na^+} + H_2O$$

It may be noted that alcohols (ROH) neither turn blue litmus red nor react with NaOH. Thus alcohols are neutral.

$$ROH + NaOH \rightarrow \text{No reaction}$$

The fact that phenol is acidic and ethanol is neutral can be explained as follows on the basis of resonance stabilisation concept.

Phenol ($C_6H_5OH$) molecule is supposed to exist in the following resonance structures: I, II, III, IV and V.

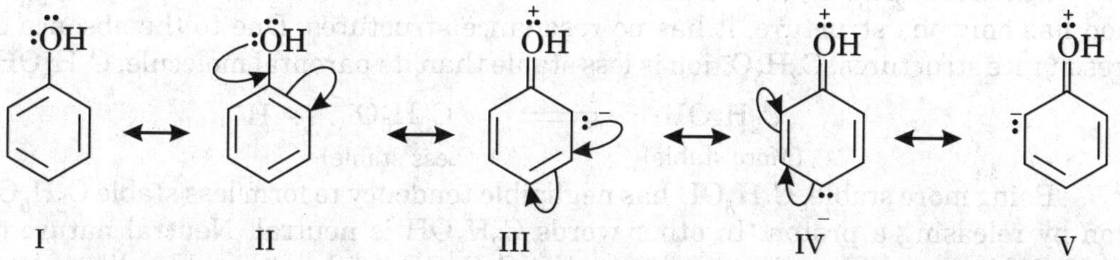

I II III IV V

It may be seen that O-atom of O–H bond in resonance structures III, IV and V has a partial positive charge. Due to the presence of positive charge, O-atom pulls the shared electron pair of O–H bond and thus the release of a proton is facilitated as shown below.

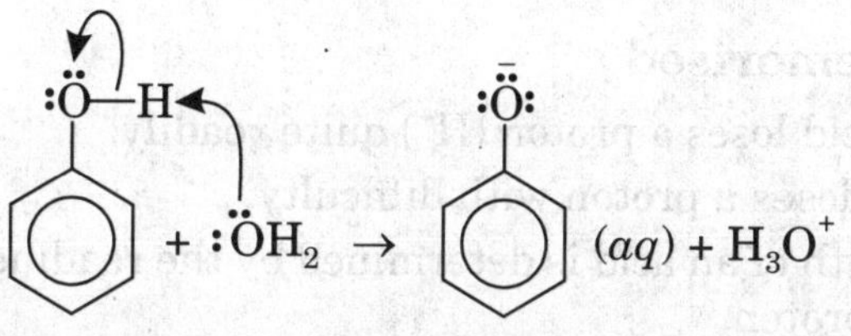

Phenol ($C_6H_5OH$) Phenoxide ion ($C_6H_5O^-$)

Phenoxide ion ($C_6H_5O^-$) also has the following resonance structures: A, B, C, D and E.

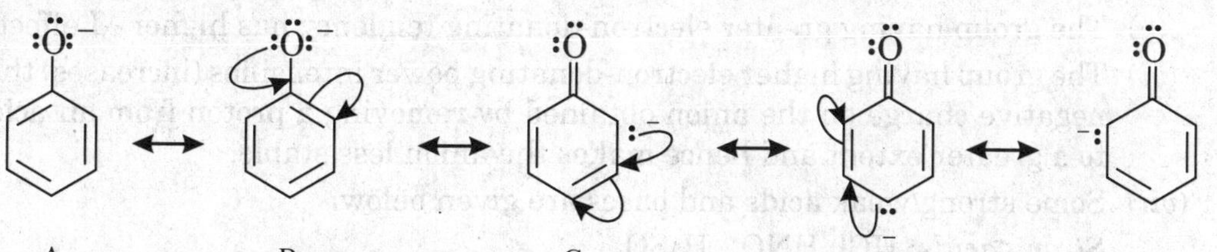

A B C D E

The occurrence of resonance in $C_6H_5OH$ molecule and $C_6H_5O^-$ ion indicates that both these species acquire stability due to resonance, but $C_6H_5O^-$ ion acquires greater stability than $C_6H_5OH$ molecule. This is because all the resonance structures of $C_6H_5O^-$ ion carry only negative charge which is localised. Thus resonance structures of $C_6H_5\bar{O}$ ion donot have a separation of positive and negative charges. On the other hand, the resonance structures III, IV and V of $C_6H_5OH$ molecule show that there is a separation of positive and negative charges in these structures. Thus resonance structures III, IV and V of $C_6H_5OH$ molecule do not contribute much to the stability of $C_6H_5OH$ molecule.

Due to the absence of charge separation in $C_6H_5O^-$ ion, this ion has greater stability than $C_6H_5OH$ molecules, *i.e.* $C_6H_5OH$ molecule is less stable than $C_6H_5O^-$ ion. Being less stable, $C_6H_5OH$ molecule has great tendency to form more stable $C_6H_5O^-$ ion by releasing a proton.

$$\underset{\text{(Less stable)}}{C_6H_5OH} \longrightarrow \underset{\text{(More stable)}}{C_6H_5O^-} + H^+$$

The tendency of $C_6H_5OH$ molecule to release a proton shows that $C_6H_5OH$ has acidic character.

Ethanol ($C_2H_5OH$) loses a proton to form ethoxide ion ($C_2H_5O^-$). Since $C_2H_5O^-$ ion has only one structure, it has no resonance structures. Due to the absence of resonance structures, $C_2H_5O^-$ ion is less stable than its parental molecule, $C_2H_5OH$.

$$\underset{\text{(More stable)}}{C_2H_5OH} \rightleftharpoons \underset{\text{(Less stable)}}{C_2H_5O^-} + H^+$$

Being more stable, $C_2H_5OH$ has negligible tendency to form less stable $C_2H_5O^-$ ion by releasing a proton. In other words $C_2H_5OH$ is neutral. Neutral nature of $C_2H_5OH$ is also evident from the facts that $C_2H_5OH$ neither turns blue litmus red not reacts with NaOH solution to form the salt.

Above discussion shows that $C_6H_5OH$ has acidic character and $C_2H_5OH$ is neutral because $C_6H_5O^-$ ion is stabilised by resonance but $C_2H_5O^-$ ion is not stabilised.

## Points to Be Memorised

(*i*) Stronger acid loses a proton ($H^+$) quite readily.

(*ii*) Weak acid loses a proton with difficulty.

(*iii*) Acid strength of an acid is determined by the readiness with which the acid donates a proton.

(*iv*) Acid strength of an acid will be increased by any factor which increases the stability of the anion of the acid or which promotes the loss of a proton from the acid.

(*v*) The group having greater electron-donating tendency has higher +I effect.

(*vi*) The group having higher electron-donating power intensifies (increases) the negative charge on the anion obtained by removing a proton from an acid to a greater extent and hence makes the anion less stable.

(*vii*) Some strong/weak acids and bases are given below.

*Strong acids:* HCl, $HNO_3$, $H_2SO_4$

*Weak acids:* $CH_3COOH$, $H_2CO_3$ $H_3PO_4$, HF, $NH_3$
*Strong bases:* NaOH, KOH, Mg $(OH)_2$
*Weak bases:* $NH_4OH$, $Ca(OH)_2$, Al $(OH)_3$, $Cu(OH)_2$

(*viii*) In keeping with Arrlenius concept and Lowry concept, the relative strength of acids can be measured in terms of proton releasing ability of acids.

## Relative Order of Acidic and Basic Strength of Some Species

Let us discuss the relative order of acidic /basic strength of the following molecules / ions.

### 1. Relative Order of Basic Strength (Basicity) of Conjugate Bases of $CH_4$, $NH_3$, $H_2O$ and HF.

Conjugate bases of the given species are obtained by removing a proton ($H^+$) from the given species. Thus $CH^-{}_3$, $NH^-{}_2$, $OH^-$ and $F^-$ are conjugate bases of the given species. The basic strength of $CH^-{}_3$, $MH^-{}_2$, $OH^-$ and $F^-$ has been found in the following *decreasing order.*

$$CH_3^- > NH_2^- > OH^- > F^-$$

⟵ Basic strength decreases ⟶

This decreasing order can be explained as follows:

(*a*) $CH_4$, $NH_3$, $H_2O$ and HF are conjugate acids of $CH_3^-$, $NH_2^-$, $OH^-$ and $F^-$ conjugate bases. We Know that in a given conjugate acid-base pair, if the acid is strong, its conjugate base would be weak, Similarly, if the base is strong, its conjugate acid would be weak. Now since the acidic strength of $CH_4$, $NH_3$, $H_2O$ and HF (conjugate acids) is in the increasing order.

$$CH_4 < NH_3 < H_2O < HF,$$

⟵ Acid strength increases ⟶

the basic strength of $CH^-{}_3$, $NH^-{}_2$, $OH^-$ and $F^-$ which are conjugate bases of $CH_4$, $NH_3$, $H_2O$ and HF would be in the decreasing order:

$$CH_3^- > NH_2^- > OH^- > F^-$$

⟵ Basic strength decreases ⟶

(*b* It may be seen from the structures of the given bases that:

(*i*) in $CH^-{}_3$ ion 3/4 of the volume of carbon atom carrying the negative charge is occupied by three H-atoms. Thus the negative charge on C-atom exists only on 1/4 volume of the carbon atom.

(*ii*) in $NH_2^-$ ion 2/3 of the volume of N-atom carrying negative charge is occupied by two H-atoms. Thus the negative charge on N-atom exists only on 1/3 volume of N-atom.

(*iii*) in $OH^-$ ion the negative charge on O-atom exists on more volume of O-atom

(*iv*) in $F^-$ ion the negative charge on $F^-$ atom exists on the whole volume of F-atom.

Thus the volume available to electron increases as we pass from $CH_3^-$ to $F^-$ ion. The increase in this volume increases the electron delocalisation from $CH_3^-$ to

$F^-$ ion. As the electron delocalisation increases, the electron density and hence the attraction for a proton decreases. The decrease in the attraction for a proton means that the basicity of these conjugate bases decreases from $CH_3^-$ to $F^-$ as shown below:

$CH_3^-$ (*Strongest base*) > $NH_2^-$ > $OH^-$ > $F^-$ (*Weakest base*).

## 2. Relative Order of Basic Strength of the Hydrides of Group VA (15) Elements ($NH_3$, $PH_3$, $AsH_3$, $SbH_3$, $BiH_3$)

The hydrides formed by the elements of group 15 can be represented by the general formula, $H_3M$(M = N, P, As, Sb, Bi).

We know that the central atom, M in $MH_3$ hydrides has a *lp* of electrons in one of the four $sp^3$ hybrid orbitals. These molecules therefore have a tendency to donate the *lp* of electrons to the Lewis acids like $H^+$, $BF_3$, $Ag^+$, $Cu^+$, $Cd^{2+}$ etc, i.e. $MH_3$ molecules show Lewis basic character due to the presence of *lp* of electrons on the central atom. As we move down the group from $NH_3$ to $BiH_3$, the basic character of $MH_3$ molecules decreases as:

$$NH_3 > PH_3 > AsH_3 > SbH_3 > BiH_3$$

This order shows that $NH_3$ is distinctly basic and $PH_3$ is weaker base than $NH_3$. $AsH_3$, $SbH_3$ and $BiH_3$ do not show basic character.

***Explanation:*** Since the central atoms in $MH_3$ molecules are different, the decrease in basic character from $NH_3$ to $BiH_3$ can be explained on the basis of the size of the central atom. As the size of the central atom increases from N to Bi, the *lp* of electrons on the cental atom occupies larger volume and hence the electron density due to *lp* of electrons round the central M-atom in $MH_3$ molecules to donate its *lp* of electrons to a Lewis acid also decreases. Thus the basic character decreases from $NH_3$ to $BiH_3$ ($NH_3 > PH_3 > AsH_3 > SbH_3 > BiH_3$). $NH_3$ is the strongest base due to the smallest size of N-atom. Since P-atom is larger in size than N-atom, $PH_3$ is weaker base than $NH_3$. Since the size of As, Sb and Bi is very large, the trihydrides of thes elements ($AsH_3$, $SbH_3$ and $BiH_3$) do not show basic character. These hydrides are neutral.

## 3. Relative Order of Acidic Strength of Hydrides of Group VIA (16) Elements ($H_2O$, $H_2S$, $H_2Se$, $H_2Te$)

The hydrides formed by the elements of group 16 can be represented by the formula, $H_2M$ (M = O, S, Se, Te). $M_2H$ molecule act as weak, diprotic or dibasic acids and thus dissociate in aquous solution in two stages to give $H^+$ ions.

$$H_2M + aq \rightleftharpoons H^+ + HM^-$$

$$HM^- + aq \rightleftharpoons H^+ + M^{2-}$$

The acidic strength of $H_2M$ molecules is in the following increasing order:

$$H_2O < H_2S < H_2Se < H_2Te.$$

This increasing order can be explained as follows:

**Explanation.** As the atomic size increases down the group, the distance between the central atom (M) and hydrogen increases and the cleavage of M – H bond (M = O, S, Se, Te etc). becomes earier. As a result the tendency to release hydrogen as proton increases down the group (least in case of oxygen and maximum in case of tellurium).

The increasing order of acidic strength is also supported by the following facts:

(*i*) As we move from $H_2O$ to $H_2Te$, the magnitude of single bond dissolution energy decreases down the group ($H_2O$ = 464 KJ $mol^{-1}$, $H_2S$ = 339, $H_2Se$ = 276, $H_2Te$ = 247). With the decrease in bond dissociation energy, the acidic strength of $H_2M$ molecules increases.

(*ii*) As the values of $K_a^I$ for $H_2M$ molecules increases down the group ($K_a{}^I$ values for $H_2M$ molecules are $H_2O = 1 \times 10^{-14}$, $H_2S = 1 \times 10^{-7}$, $H_zSe$ = $1.7 \times 10^{-4}$, $H_2Te = 2.3 \times 10^{-3}$), the acidic strength of $H_2M$ molecules also increases in the same direction.

## 4. Relative Order of Acid Strength of Halogen Acids (HX) in Aqueous Solution (HF, HCl, HBr, HI)

In the gaseous state, hydrogen halides (HX) are covalent and hence do not show any acidic character. However, in **aqueous solution**, HX molecules undergo ionisation to give $H_3O^+$ and $X^-$ ions and, therefore, behave as acids.

$$HX + aq \longrightarrow H^+(aq) + X^-(aq)$$

or

$$HX + H_2O \longrightarrow H_3O^+ + X^-$$

The acidic strength of hydrogen halides in aqueous solution follows the following increasing order:

HF (*Weakest acid*) < HCl < HBr < HI (*Strongest acid*) ... (*i*)

——— Acid strength (in aqueous solution) increases ———→

This order shows that HF is the weakest acid in aqueous solution and HI is the strongest acid in aqueous solution.

The above increasing order of acid strength of HX molecules in aqueous solution cannot be explained on the basis of electronegativity difference between X and H atoms as discussed below.

Since F atom has the highest value of electronegativity and I-atom has the minimum value, electronegativity difference between X and H atoms in HF, HCl, HBr and HI molecules decreases as:

H–F > H–Cl > H–Br > H–I

Due to the decreasing order of electronegativity difference the percent ionic character in HF, HCl, HBr and HI molecules also decreases in the same order (HF > HCl > HBr > HI). Since the percent ionic character decreases from HF to HI, acidic strength of HX molecules also decreases in the same order. Thus the acidic strength of HX molecules decreases as:

HF > HCl > HBr > HI ...(*ii*)

——— Acidic strength decreases ———→

It may be seen that the order of acid strength shown at (*i*) above is the reverse of the order shown at (*ii*) which is based on electronegativity values of X and H atoms.

The increasing order of acid strength [shown at (*i*)] in aqueous solution can be explained as follows:

(*i*) *Explanation based on the dissociation energy of H–X bond.* Lesser is the magnitude of dissociation energy of H–X bond, more easily HX molecule will dissociate in aqueous solution to give $H_3O^+$ and $X^-$ ions and hence greater is the acid strength of HX molecule in aqueous solution. Now since bond dissociation energy (in KJ/mol) of H – X bond decreases from H – F to H – I (H – F = 574, H – Cl = 432, H – Br = 363 and H – I = 295), acid strength of halogen acids **increases** in the order:

$$HF < HCl < HBr < HI$$

*(ii) Explanation based on heats of ionisation ($\Delta H_I$) of HX molecules in aqueous solution.* We have already stated that heat of ionisation of HX molecules in aqueous solution is the energy **released** in the ionisation of HX(*aq*) into $H^+$(*aq*) into $H^+$(*aq*) ions.

$$HX\,(aq) \xrightarrow[\text{Acid strength}]{\text{Ionisation}} H^+\,(aq) + X^-(aq)$$

$\Delta H$ = Heat released = Heat of ionisation of HX (*aq*) = – $\Delta H_I$ of HX (*aq*)

Obviously larger is the value of $\Delta H_I$ (donot consider negative sign) for a given HX (*aq*) molecule, greater is the feasibility of ionisation of HX (*aq*) molecule into $H^+$(*aq*) and $X^-$ (*aq*) ions and hence greater is the acid strength of HX (*aq*) molecule.

Now we know that the values of $\Delta H_I$ (in KJ $mol^{-1}$) for HF, HCl, HBr and HI molecules are:

| HF | HCl | HBr | HI |
|---|---|---|---|
| –12 | –59 | –63 | –57 |

—— $\Delta H_I$ values increase ——→ (HF to HBr)

(Do not consider negative sign)

Since the value of $\Delta H_I$ for HF is the lowest, HF is the weakest acid. Further, since $\Delta H_I$ values increase from HF to HBr, the acid strength of HF, HCl and HBr is in the order: HF < HCl < HBr. We know that the order of acid strength of HX molecules in aqueous solution is HF < HCl < HBr < HI. The highest acid strength of HI in aqueous solution cannot be explained on the basis of its $\Delta H_I$ value. This discrepancy is explained on the basis of entropy changes ($\Delta S$) accompanying the process of ionisation.

## 5. Relative Order of Basic Strength of Conjugate Bases of HF, HCl, HBr and HI ($F^-$, $Cl^-$, $Br^-$, $I^-$)

Conjugate bases of HF, HCl, HBr and HI are $F^-$, $Cl^-$, $Br^-$ and $I^-$ respectively, since these have been obtained by removing $H^+$ ion from HF, HCl, HBr and HI respectively. The basic strength of conjugate bases has been found in the following decreasing order:

$$F^- > Cl^- > Br^- > I^-$$

—— Basic strength decreases ——→

This decreasing order can be explained as follows:

(*a*) *Explanation based a the relative order of acid strength of HF, HCl, HBr and HI.* HF, HCl, HBr and HI are the conjugate acids of $F^-$, $Cl^-$, $Br^-$ and $I^-$. We know that since the acidic strength of HF, HCl, HBr and HI (conjugate acids) is in the *increasing order*

$$HF < HCl < HBr < HI,$$

—— Acid strength increases ——→

the basic strength of $F^-$, $Cl^-$, $Br^-$ and $I^-$ which are the conjugate bases of HF, HCl, HBr and HI should be in the *decreasing* (opposite) *order*

$$F^- > Cl^- > Br^- > I^-$$

——— Basic strength decrease ———→

(*b*) *Explanation based on the size of halide ions.* We know that the size of $F^-$, $Cl^-$, $Br^-$ and $I^-$ ions increases on moving from $F^-$ to $I^-$ ($F^- < Cl^- < Br^- < I^-$). Due to the increase in size of $X^-$ ions, the volume available to electron on $X^-$ ion increases from $F^-$ to $I^-$. The increase in the volume increases the delocalisation of the electron on $X^-$ ion and hence the electron density on $X^-$ ion decreases from $F^-$ to $I^-$. As a result, the proton accepting power (*i.e.* basicity) of $X^-$ ions decreases in the order:

$$F^- > Cl^- > Br^- > I^-$$

——— Basicity decreases ———→

## 6. Relative Order of Acid Strength of Oxoacids of Chlorine (HClO, $HClO_2$, $HClO_3$, $HClO_4$)

Oxoacids formed by Cl are HClO, $HClO_2$, $HClO_3$ and $HClO_4$. The acidic strength of these acids increases in the order:

$$HClO < HClO_2 < HClO_3 < HClO_4$$

——— Acidic strength increases ———→

This increasing order can be explained as follows:

In all the four oxoacids, since the electronegativity (EN) of the central atom (*i.e.* Cl atom) is the same, acidic strength of the acids does not depend on EN value of Cl-atom Now since O.N. of Cl-atom in the given acids is different, acid strength of the acids depends on O.N. of Cl-atom. Acid strength increases with the increase of O.N. of Cl-atom. As O.N. of Cl-atom increases from +1 (in HClO) to +7 (in $HClO_4$), acid strength increases from HClO to $HClO_4$.

$$\overset{+1}{HClO} < \overset{+3}{HClO_2} < \overset{+5}{HClO_3} < \overset{+7}{HClO_4}$$

——— O.N. of Cl increases ———→

——— Acidic strength increases ———→

## 7. Relative Order of Basic Strength of Conjugate Bases of HClO, $HClO_2$, $HClO_3$ and $HClO_4$ ($ClO^-$, $ClO_2^-$, $ClO_3^-$, $ClO_4^-$)

Conjugate bases of HClO, $HClO_2$ $HClO_3$ and $HClO_4$ are $ClO^-$, $ClO_2^-$, $ClO_3^-$ and $ClO^-_4$, since these have been obtained by removing a proton ($H^+$) from the given acids. The basic strength of these conjugate bases **decreases** in the order:

$$ClO^- > ClO^-_2 > ClO^-_3 > ClO^-_4$$

——— Basic strength decreases ———→

**Explanation.** This decreasing order can be explained as follows:

(*a*) *Explanation based on oxidation number (O.N.) of Cl-atom.* In the given oxoanions, since electronegativity (EN) of Cl-atom in all the four oxoanions remains the same, the basic strength of these anions decreases due to the increase in O.N of Cl-atom from +1 in $ClO^-$ to +7 in $ClO_4^-$. Thus:

$$\left[\overset{+1}{ClO}\right]^-, \left[\overset{+3}{ClO_2}\right]^-, \left[\overset{+5}{ClO_3}\right]^-, \left[\overset{+7}{ClO_4}\right]^-$$

——— EN of Cl-atom remains the same ⟶

——— O.N. of Cl-atom increases from +1 to +7 ⟶

——— Basicity of oxoanions decreases ⟶

(*b*) *Explanation based on the stability of oxoanions.* We know that the stability of the given oxoanions increases as:

$$ClO^- < ClO^-_2 < ClO^-_3 < ClO^-_4$$

——— Stability increases ⟶

Now since $ClO^-$ ion is the least stable, it has the maximum tendency to accept a proton ($H^+$) to form its conjugate acid, HClO. Thus $ClO^-$ in is the strongest base. On the other hand, since $ClO_4^-$ ion has the maximum stability (most stable), this ion has the minimum tendency to accept a proton to form its conjugate acid, $HClO_4$. Thus $ClO^-_4$ ion is the weakest base. Hence:

$$ClO^-, \quad ClO^-_2, \quad ClO^-_3, \quad ClO^-_4$$

——— Stability increases ⟶

——— Basic strength decreases ⟶

(*c*) *Explanation based on the electron density on $ClO^-$, $ClO_2^-$, $ClO_3^-$ and $ClO_4^-$ ions.* With the increase in the number of oxygen atoms in the conjugate bases, the delocalisation of the π bond becomes more and more extended. This results in decrease in electron density Consequently, proton attraction and basicity also decrease (Fig. 3.10) in the order:

$ClO^-$ (*Strongest base*) > $ClO_2^-$ > $ClO_3^-$ > $ClO_4^-$ (*Weakest base*)

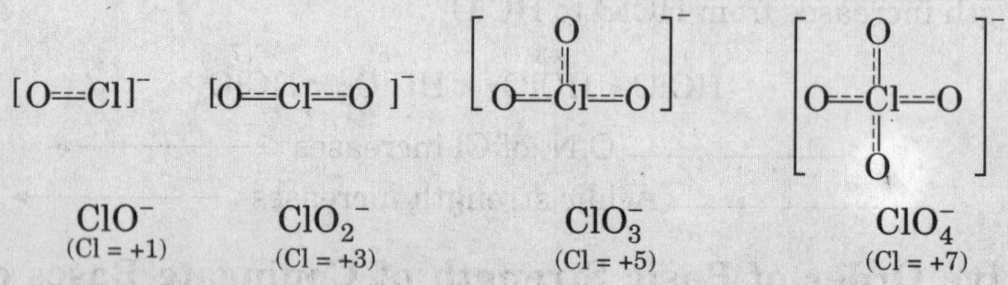


— Oxidation number of chlorine increasing ⟶

— Number of oxygen atoms increasing ⟶

— Delocalized π bond becomes more extended ⟶

— Electron delocalization increasing ⟶

— Electron density decreasing ⟶

— Attraction for proton decreasing ⟶

— Basicity decreasing ⟶

**Fig. 3.10** The basicity of $ClO^-$, $ClO_2^-$, $ClO_3^-$ and $ClO_4^-$ ions decreases as we pass from $ClO^-$ ion to $ClO_4^-$ ion through $ClO_2^-$ and $ClO_3^-$, ions.

(*d*) *Explanation based on the relative order of acidic strength of HClO, $HClO_2$, $HClO_3$ and $HClO_4$.* HClO, $HClO_2$, $HClO_3$ and $HClO_4$ are the conjugate acids of $ClO^-$,

$ClO^-_2$, $ClO^-_3$ and $ClO^-_4$ (conjugate bases). We know that since the acidic strength of HClO, $HClO_2$, $HClO_3$ and $HClO_4$ (conjugate acids) is in the increasing order:

$$HClO < HClO_2 < HClO_3 < HClO_4,$$

——————— Acidic strength increases ——→

the basic strength of $ClO^-$, $ClO^-_2$, $ClO^-_3$ and $ClO^-_4$ which are conjugate bases should be in the decreasing order (opposite order).

$$ClO^- > ClO^-_2 > ClO^-_3 > ClO^-_4$$

——————— Basic strength decreases ——→

**8. Relative order of Acid Strength of Hypohalous Acids (HClO, HBrO, HIO) and Perhalic Acids ($HClO_4$, $HBrO_4$, $HIO_4$).**

(*a*) Acid strength of hypohalous acids (HClO, HBrO and HIO) decreases in the order:

$$HClO > HBrO > HIO$$

——————— Acid strength decreases ——→

The acid strength of HClO, HBrO and HIO decreases with the decrease in electronegativity (EN) of halogen atom (central atom) from Cl to I as explained below.

The structure of HClO, HBrO and HIO molecules are H – O – Cl, H – O – Br and H – O – I respectively. Since O-atom is more electronegative than X-atom (X = Cl, Br, I), this atom (i.e. O-atom) attracts the shared pair of electrons of O – X bond towards itself as shown below.

$$H \leftarrow O - Cl, \quad H - O \leftarrow Br, \quad H - O \leftarrow I$$

Now since EN value of X-atom decreases from Cl to I (Cl > Br > I), the shared pair of electrons of O — X bond moves close and closer towards O-atom and hence electron density on O-atom increases from HClO to HIO. In other words electron density (ED) on O-atom of HClO molecule is the lowest and that on O-atom of HIO is the highest as shown below.

ED on this atom is lowest ↓          ED on this atom is highest ↓

H—O ÷ Cl      H—O ÷ Br,      H—O ÷ I

↑ This electron pair is farthest away from O-atom      ↑ This electron pair is closest to O-atom

Since ED on O-atom in H–O–Cl is the lowest, this atom (i.e. O-atom) attracts the shared pair of electrons of H–O bond toward itself *most strongly*. On the other hand, since ED on O-atom in H–O–I is the highest, this atom (i.e. O-atom) attracts the shared pair of electrons of H–O bond towards itself *least strongly*. Thus the tendency of O-atom to attract the shared pair of electrons in H–O bond towards itself decreases from H–O–Cl to H–O–I as shown below.

H⟶⟶⟶O—Cl      H⟶⟶O—Br      H⟶O—I.

As a result, O–H bond breaks most readily in H–O–Cl and least easily in H–O–I. In other words, we can say that as the electronegativity of halogen atom decreases from Cl to I, the possibility of the rupture of O–H bond to give $H^+$ ions also decreases and hence the acid strength of HClO, HBrO and HIO decreases as :

$HClO > HBrO > HIO$

Acid strength decreases ⟶

(*b*) Acid strength of perhalic acids ($HClO_4$, $HBrO_4$ and $HIO_4$) also decreases with the decrease in electronegativity from Cl to I.

$HClO_4 > HBrO_4 > HIO_4$

Acid strength decreases ⟶

The decrease in acid strength can be explained as follows:

As the electronegativity of halogen atom decreases from Cl to I, the tendency of $XO_3$ group in $HXO_4$ or $O_3X-O-H$ molecule to withdraw the electrons of O–H bond towards itself decreases and hence acid strength of perhalic acids decreases in the same order, *i.e.*

$HClO_4 > HBrO_4 > HIO_4$

Acid strength decreases ⟶

## Pauling's Empirical Rule: Acidic Strength of Oxoacids

Pauling has introduced an empirical rule for estimating the acidic strength of oxoacids having the general formula, $XO_n(OH)_m$ where $n$ is the number of unpropotionated O-atoms and $m$ is the number of –OH groups. $pK_1$ value of an oxoacid is approximately given by:

$$pK_1 = 7 - 5\,n$$

Thus if for an oxoacid $n = 0$, then $pK_1 = 7$ and hence the acid is weak. As the value of $n$ increases, $pK_1$ becomes negative and the acid becomes strong. As the number of unprotonated O-atoms ($n$) increases, the overall inductive effect on the dissociation of the proton of –OH group increases and hence acid strength increases.

**Examples.** The number of unprotonated O-atoms ($n$) in some oxoacids has been calculated in Fig. 3.11.

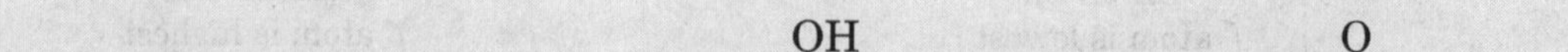


**Fig. 3.11** Calculation of the number ($n$) of unprotonated O-atoms in some oxoacids.

Since $n = 0$ for HClO and $H_3BO_3$, both these acids are very weak acids. Since the value of $n$ for $H_2SO_4$ ($n = 2$) is greater than that for $H_2SO_3$ ($n = 1$), the former is a stronger acid than the latter. Similarly, since $HNO_3$($n = 2$) has greater value of $n$ than $HNO_2$ ($n = 1$), the former is a stronger acid than the latter.

## Reaction Between a Stronger Acid and the Salt of a Weaker Acid

If we have two acids one of which is a stronger acid and the other is a weaker acid, then the reaction between the stronger acid and the salt of the weaker acid produces the weaker acid and salt of the stronger acid. For example:

(*i*) Since $CH_3OH$ is a stronger acid than $HC \equiv CH$, in the reaction between $CH_3OH$ (stronger acid) and $HC \equiv \bar{C}\overset{+}{Na}$ (salt of weaker acid, $HC \equiv CH$), stronger acid (i.e. $CH_3OH$) donates a proton ($H^+$) to the salt and weaker acid, $HC \equiv CH$ (*i.e.* parental acid of salt) is produced.

$$\underset{\substack{\text{Stronger acid}\\ \text{(Acid)}}}{CH_3OH} + \underset{\substack{\text{Salt of weaker}\\ \text{acid, } HC\equiv CH\\ \text{(Base)}}}{HC\equiv\bar{C}Na^+} \xrightarrow{H^+} \underset{\substack{\text{Weaker acid}\\ \text{(Acid)}}}{HC\equiv CH} + \underset{\substack{\text{Salt of stronger}\\ \text{acid } CH_3OH\\ \text{(Base)}}}{CH_3\bar{O}Na^+}$$

(*ii*) Since $H_2O$ is a stronger acid than $CH_3 - CH_2 - OH$, in the reaction between $H_2O$ (stronger acid) and $CH_3–CH_2– \bar{O}Na^+$ (salt of weaker acid, $CH_3 - CH_2 - OH$), the stronger acid (*i.e.* $H_2O$) donates a proton ($H^+$) to the salt and weaker acid, $CH_3 - CH_2 - OH$ (i.e. parental acid of the salt) is produced.

$$\underset{\substack{\text{Stronger}\\ \text{acid}\\ \text{(Acid)}}}{H_2O} + \underset{\substack{\text{Salt of weaker acid,}\\ CH_3-CH_2-OH\\ \text{(Base)}}}{CH_3-CH_2-\bar{O}Na^+} \xrightarrow{+H^+} \underset{\substack{\text{Weaker acid}\\ \text{(Acid)}}}{CH_3-CH_2-OH} + \underset{\substack{\text{Salt of stronger}\\ \text{acid, } H_2O\\ \text{(Base)}}}{NaOH}$$

In general :

| **Stronger acid + Salt of weaker acid → Weaker acid + Salt of stronger acid** |
|---|

## Reaction Between a Weaker Acid and the Salt of a Stronger Acid

The reaction between a weaker acid and the salt of a stronger acid can not take place. This can be explained by considering the following reaction.

$$\underset{\text{Weaker acid}}{H_2O} + \underset{\substack{\text{Salt of stronger}\\ \text{acid, } C_6H_5OH}}{C_6H_5O^-Na^+} \not\longrightarrow \underset{\text{Stronger acid}}{C_6H_5OH} + \underset{\substack{\text{Salt of weaker}\\ \text{acid, } H_2O}}{NaOH}$$

In the above reaction, $H_2O$ is a weaker acid than $C_6H_5OH$. This reaction will not take place, since $H_2O$, being a weaker acid, is not able to donate a proton ($H^+$) to $C_6H_5O^-Na^+$ to form $C_6H_5OH$ (strong acid)

In general, the following reaction cannot take place.

| **Weaker acid + Salt of stronger acid $\not\longrightarrow$ Stronger acid + Salt of weaker acid.** |
|---|

### Advantages of Bronsted – Lowry concept over Arrhenins Concept.

(*i*) According to Bronsted-Lowry concept, not only molecules but even the anions can also act as acids or bases. For example in the following reaction $CO_3^{2-}$ ion acts as a base

$$\underset{\text{Acid}}{H_2O} + \underset{\text{Base}}{CO_3^{2-}} \rightleftharpoons \underset{\text{Acid}}{HCO_3^-} + \underset{\text{Base}}{OH^-}$$

(*ii*) It can explain the basic character of the substances like $Na_2CO_3$, $NH_3$ etc. (which do not contain $OH^-$ group and hence were not bases according to Arrhenius definition) on the basis that they are proton acceptors.

(*iii*) It can explain the acid-base reactions in the non-aqueous medium or even in the absence of a solvent For example in the following acid-base reaction no aqueous solution is required.

$$\underset{\text{Acid}}{HCl} + \underset{\text{Base}}{NH_3} \rightleftharpoons \underset{\text{Acid}}{NH_4^+} + \underset{\text{Base}}{Cl^-}$$

**Limitation of Bronsted-Lowry Concept**

Although Bronsted-Lowry concept has a wider scope than the Arrhenius concept but it has some limitations, e.g.,

(*i*) It cannot explain the reactions between acidic oxides like $CO_2$, $SO_2$, $SO_3$ etc. and the basic oxides like CaO, BaO, MgO etc. which take place even in the absence of the solvent, e.g.,

$$CaO + SO_3 \longrightarrow CaSO_4$$

Obviously, there is no proton transfer involved in this reaction.

(*ii*) Substances like $BF_3$, $AlCl_3$ etc., do not have any hydrogen and hence cannot give a proton but are known to behave as acids.

(*iii*) A given substance may behave as an acid in one solvent and may behave as a base in some other solvent. For example $CH_3COOH$ behaves as an acid in liquid $NH_3$ and as a base in $H_2SO_4$ solvent as is evident from the following equilibria.

$$\underset{\text{Acid}}{CH_3COOH} + \underset{\text{Base}}{NH_3(\text{solvent})} \rightleftharpoons \underset{\text{Acid}}{NH_4^+} + \underset{\text{Base}}{CH_3COO^-}$$

$$\underset{\text{Acid}}{H_2SO_4(\text{solvent})} + \underset{\text{Base}}{CH_3COOH} \rightleftharpoons \underset{\text{Acid}}{CH_3COOH_2^+} + \underset{\text{Base}}{HSO_4^-}$$

The dual behavior of $CH_3COOH$ cannot be explained by Bronsted-Lowry Concept.

## IV. Lux–Flood Concept

**Definition**

This concept was proposed by Lux (1939) and extended by Flood (1947). This concept explains the formation of salts by the combination of acids and bases in the absence of a solvent. The formation of salt involves the transfer of an oxide ion ($O^{2-}$) from the base to the acid. The compound which loses (donates) an $O^{2-}$ ion acts as a base and the species which accepts the $O^{2-}$ ion acts as an acid. In other words *base is $O^{2-}$ ion donor and the acid is $O^{2-}$ ion acceptor.* Thus:

$$\text{Base} \rightleftharpoons \text{Acid} + O^{2-}$$

**Examples:** *Base* *Acid*

$$CaO \rightleftharpoons Ca^{2+} + O^{2-}$$

$$SO_4^{2-} \rightleftharpoons SO_3 + O^{2-}$$
$$2SO_4^{2-} \rightleftharpoons S_2O_7^{2-} + O^{2-}$$
$$SiO_3^{2-} \rightleftharpoons SiO_2 + O^{2-}$$
$$ZnO_2^{2-} \rightleftharpoons ZnO + O^{2-}$$
$$4PO_4^{3-} \rightleftharpoons P_4O_{10} + 6O^{2-}$$
$$TiO_2 \rightleftharpoons TiO^{2+} + O^{2-}$$
$$Nb_2O_5 \rightleftharpoons 2NbO_2^{+} + O^{2-}$$

## Acid-Base Reactions.

According to Lux concept, the following reactions are acid-base reactions.

| *Base* ($O^{2-}$ *ion donor)* | | *Acid* ($O^{2-}$ *ion acceptor)* | | *Salt* |
|---|---|---|---|---|
| (*i*) $CaO$ | + | $CO_2$ | $\longrightarrow$ | $CaCO_3$ |
| (*ii*) $CaO$ | + | $SiO_2$ | $\longrightarrow$ | $CaSiO_3$ (slag) |
| (*iii*) $Na_2O$ | + | $ZnO$ | $\longrightarrow$ | $Na_2ZnO_2$ |
| (*iv*) $PbO$ | + | $SO_3$ | $\longrightarrow$ | $PbSO_4$ |
| (*v*) $6Na_2O$ | + | $Pb_4O_{10}$ | $\longrightarrow$ | $4Na_3PO_4$ |
| (*vi*) $TiO_2$ | + | $Na_2S_2O_7$ | $\longrightarrow$ | $(TiO)SO_4 + Na_2SO_4$ |
| or $TiO_2$ | + | $S_2O_7^{2-}$ | $\longrightarrow$ | $(TiO)SO_4 + SO_4^{2-}$ |
| (*vii*) $Nb_2O_5$ | + | $Na_2S_2O_7$ | $\longrightarrow$ | $(NbO_2)_2SO_4 + Na_2SO_4$ |
| or $Nb_2O_5$ | + | $S_2O_7^{2-}$ | $\longrightarrow$ | $(NbO_2)_2 SO_4 + SO_4^{2-}$ |
| (*viii*) $FeO$ | + | $SiO_2$ | $\longrightarrow$ | $FeSiO_3$ |

In the above acid-base reactions the oxides *viz* $CaO$, $Na_2O$, $PbO$, $TiO_2$, $Nb_2O_5$ and $FeO$ act as bases, since these oxides donate $O^{2-}$ ion to $CO_2$, $SiO_2$, $ZnO$, $SO_3$, $Pb_4O_{10}$ and $Na_2S_2O_7$ (or $S_2O_7^{2-}$) which act as acids because these species accept $O^{2-}$ ion given by bases. The transfer of $O^{2-}$ ion from the base to the acid in each of the above reactions has been explained below:

(*i*) $\underset{\text{Base}}{CaO} \longrightarrow Ca^{2+} + O^{2-}$

$\underset{\text{Acid}}{CO_2} + O^{2-} \longrightarrow CO_3^{2-}$

*On adding*: $CaO + CO_2 \longrightarrow Ca^{2+} + CO_3^{2-}$

or $\underset{\text{Base}}{CaO} + \underset{\text{Acid}}{CO_2} \longrightarrow \underset{\text{Salt}}{CaCO_3}$

(*ii*) $\underset{\text{Base}}{CaO} \longrightarrow Ca^{2+} + O^{2-}$

$\underset{\text{Acid}}{SiO_2} + O^{2-} \longrightarrow SiO_3^{2-}$

$CaO + SiO_2 \longrightarrow Ca^{2+} + SiO_3^{2-}$

or $\underset{\text{Base}}{CaO} + \underset{\text{Acid}}{SiO_2} \longrightarrow \underset{\text{Salt}}{CaSiO_3}$

(*iii*) $\underset{\text{Base}}{Na_2O} \longrightarrow 2Na^+ + O^{2-}$

$\underset{\text{Acid}}{ZnO} + O^{2-} \longrightarrow ZnO_2^{2-}$

$Na_2O + ZnO \longrightarrow 2Na^+ + ZnO_2^{2-}$

or $\underset{\text{Base}}{N_2O} + \underset{\text{Acid}}{ZnO} \longrightarrow \underset{\text{Salt}}{Na_2ZnO_2}$

(*iv*) $\underset{\text{Base}}{PbO} \longrightarrow Pb^{2+} + O^{2-}$

$\underset{\text{Acid}}{SO_3} + O^{2-} \longrightarrow SO_4^{2-}$

$PbO + SO_3 \longrightarrow Pb^{2+} + SO_4^{2-}$

or $\underset{\text{Base}}{PbO} + \underset{\text{Acid}}{SO_3} \longrightarrow \underset{\text{Salt}}{PbSO_4}$

(*v*) $\underset{\text{Base}}{Na_2O} \longrightarrow 2Na^+ + O^{2-}\ ] \times 6$

$\underset{\text{Acid}}{P_4O_{10}} + 6O^{2-} \longrightarrow 4PO_4^{3-}$

$6Na_2O + Pb_4O_{10} \longrightarrow 12Na^+ + 4PO_4^{3-}$

or $\underset{\text{Base}}{6Na_2O} + \underset{\text{Acid}}{P_4O_{10}} \longrightarrow \underset{\text{Salt}}{4Na_3PO_4}$

(*vi*) $\underset{\text{Base}}{TiO_2} \rightarrow TiO^{2+} + O^{2-}$

$\underset{\text{Acid}}{Na_2S_2O_7} + O^{2-} \longrightarrow SO_4^{2-} + Na_2SO_4$

$TiO_2 + Na_2S_2O_7 \longrightarrow TiO^{2+} + SO_4^{2-} + Na_2SO_4$

or $\underset{\text{Base}}{TiO_2} + \underset{\text{Acid}}{Na_2S_2O_7} \longrightarrow \underset{\text{Salt}}{(TiO)SO_4} + Na_2SO_4$

or $\underset{\text{Base}}{TiO_2} + \underset{\text{Acid}}{S_2O_7^{2-}} \longrightarrow \underset{\text{Salt}}{(TiO)SO_4} + SO_4^{2-}$

(*vii*) $\underset{\text{Base}}{Nb_2O_5} \longrightarrow 2NbO_2^+ + O^{2-}$

$\underset{\text{Acid}}{Na_2S_2O_7 + O^{2-}} \longrightarrow SO_4^{2-} + Na_2SO_4$

$Nb_2O_5 + Na_2S_2O_7 \longrightarrow 2NbO_2^+ + SO_4^{2-} + Na_2SO_4$

or $\underset{\text{Base}}{Nb_2O_5} + \underset{\text{Acid}}{Na_2S_2O_7} \longrightarrow \underset{\text{Salt}}{(NbO_2)_2SO_4} + Na_2SO_4$

or $\underset{\text{Base}}{Nb_2O_5} + \underset{\text{Acid}}{S_2O_7^{2-}} \longrightarrow \underset{\text{Salt}}{(NbO_2)_2SO_4} + SO_4^{2-}$

(*viii*) $\underset{\text{Base}}{FeO} \longrightarrow Fe^{2+} + O^{2-}$

$\underset{\text{Acid}}{SiO_2 + O^{2-}} \longrightarrow SiO_3^{2-}$

$FeO + SiO_2 \longrightarrow Fe^{2+} + SiO_3^{2-}$

or $\underset{\text{Base}}{FeO} + \underset{\text{Acid}}{SiO_2} \longrightarrow \underset{\text{Salt}}{FeSiO_3}$

It is obvious from the above Lux acid-base reactions that these reactions are *non-protonic acid-base reactions.*

## Amphoteric Behavior of ZnO and $Al_2O_3$.

According to Lux concept ZnO and $Al_2O_3$ both show amphoteric behavior, since these oxides behave as acids by accepting an $O^{2-}$ ion and behave as bases by donating an $O^{2-}$ ion as shown below:

$\underset{\text{Acid}}{ZnO}$ (*$O^{2-}$ ion acceptor*) $+ O^{2-} \rightleftharpoons ZnO_2^{2-}$

$\underset{\text{Base}}{ZnO}$ (*$O^{2-}$ ion donor*) $\rightleftharpoons Zn^{2+} + O^{2-}$

$\underset{\text{Acid}}{Al_2O_3}$ (*$O^{2-}$ ion acceptor*) $+ O^{2-} \rightleftharpoons 2AlO_2^-$

$\underset{\text{Base}}{Al_2O_3}$ (*$O^{2-}$ ion donor*) $\rightleftharpoons 2Al^{3+} + 3O^{2-}$

Amphoteric behaviour of ZnO and $Al_2O_3$ is evident from the following acid-base reactions.

(*i*) *ZnO as an acid:* $\underset{\text{Base}}{Na_2O} + \underset{\text{Acid}}{ZnO} \longrightarrow \underset{\text{Salt}}{Na_2ZnO_2}$

In this reaction ZnO behaves as an acid because it accepts an $O^{2-}$ ion donated by $Na_2O$ which acts as a base.

$$\underset{\text{Base}}{Na_2O} \longrightarrow 2Na^+ + O^{2-}$$

$$\underset{\text{Acid}}{ZnO + O^{2-}} \longrightarrow ZnO_2^{2-}$$

$$Na_2O + ZnO \longrightarrow 2Na^+ + ZnO_2^{2-}$$

or
$$\underset{\text{Base}}{Na_2O} + \underset{\text{Acid}}{ZnO} \longrightarrow \underset{\text{Salt}}{Na_2ZnO_2}$$

*(ii) ZnO as a base:* $\underset{\text{Base}}{ZnO} + \underset{\text{Acid}}{S_2O_7^{2-}} \longrightarrow \underset{\text{Salt}}{ZnSO_4} + SO_4^{2-}$

In this reaction ZnO behaves as a base since it donates an $O^{2-}$ ion to $S_2O_7^{2-}$ ion which acts as an acid.

$$\underset{\text{Base}}{ZnO} \longrightarrow Zn^{2+} + O^{2-}$$

$$\underset{\text{Acid}}{S_2O_7^{2-} + O^{2-}} \longrightarrow SO_4^{2-} + SO_4^{2-}$$

$$ZnO + S_2O_7^{2-} \longrightarrow Zn^{2+} + SO_4^{2-} + SO_4^{2-}$$

or
$$\underset{\text{Base}}{ZnO} + \underset{\text{Acid}}{S_2O_7^{2-}} \longrightarrow \underset{\text{Salt}}{ZnSO_4} + SO_4^{2-}$$

*(iii) $Al_2O_3$ as an acid:* $\underset{\text{Base}}{Na_2O} + \underset{\text{Acid}}{Al_2O_3} \longrightarrow \underset{\text{Salt}}{2NaAlO_2}$

In this reaction $Al_2O_3$ behaves as an acid since it accepts an $O^{2-}$ ion donated by $Na_2O$ which acts as a base.

$$\underset{\text{Base}}{Na_2O} \longrightarrow 2Na^+ + O^{2-}$$

$$\underset{\text{Acid}}{Al_2O_3 + O^{2-}} \longrightarrow 2AlO_2^-$$

$$Na_2O + Al_2O_3 \longrightarrow 2Na^+ + 2AlO_2^-$$

or
$$\underset{\text{Base}}{Na_2O} + \underset{\text{Acid}}{Al_2O_3} \longrightarrow \underset{\text{Salt}}{2NaAlO_2}$$

*(iv) $Al_2O_3$ as a base:* $\underset{\text{Base}}{Al_2O_3} + \underset{\text{Acid}}{3SiO_2} \longrightarrow \underset{\text{Salt}}{Al_2(SiO_3)_3}$

In this reaction $Al_2O_3$ behaves as a base because it donates an $O^{2-}$ ion to $SiO_2$ which acts as an acid.

$$\underset{\text{Base}}{Al_2O_3} \longrightarrow 2Al^{3+} + 3O^{2-}$$

$$\underset{\text{Acid}}{SiO_2 + O^{2-}} \longrightarrow SiO_3^{2-}\ ] \times 3$$

$$Al_2O_3 + 3SiO_2 \longrightarrow 2Al^{3+} + 3SiO_3^{2-}$$

or $$\underset{\text{Base}}{Al_2O_3} + \underset{\text{Acid}}{3SiO_2} \longrightarrow \underset{\text{Salt}}{Al_2(SiO_3)_3}$$

## Lux Concept is Also Covered by Lewis Concept

According to Lewis concept, an acid is an electron pair acceptor and a base is an electron pair donor. According to Lux concept, an acid is $O^{2-}$ ion acceptor and a base is $O^{2-}$ ion donor. Thus, according to Lux concept, an electron pair is shifted from the base to the acid *via* $O^{-2}$ ion. In other words we say that Lux concept is also covered by Lewis concept.

## Extension of Lux-Flood Concept

Lux-Flood concept has been extended and according to this extension, the substance that gives up (loses) an anion like halide ($X^-$), sulphide ($S^{2-}$) etc is called a base and the substance that gains the anion is called an acid. For example.

$$\underset{\text{Base}}{3NaF} + \underset{\text{Acid}}{AlF_3} \xrightarrow{\text{High temperature}} \underset{\text{Salt}}{Na_3[AlF_6]}$$

The formation of $Na_3[AlF_6]$ (salt) can be explained as follows:

$$\underset{\text{Base}}{3NaF} \longrightarrow 3Na^+ + 3F^-$$

$$\underset{\text{Acid}}{AlF_3} + 3F^- \longrightarrow [AlF_6]^{3-}$$

$$3NaF + AlF_3 \longrightarrow 3Na^+ + [AlF_6]^{3-}$$

or $$\underset{\text{Base}}{3NaF} + \underset{\text{Acid}}{AlF_3} \longrightarrow \underset{\text{Salt}}{Na_3^+[AlF_6]^{3-} \text{ or } Na_3[AlF_6]}$$

## Applications of Lux Concept

Lux definition of acids and bases is applicable in ceramics and metallurgy. For example:

(*i*) In the metallurgy of iron, $SiO_2$ (acidic oxide) present in iron ore is removed by adding CaO (basic oxide) in blast furnace.

$$\underset{\text{Acid}}{SiO_2\ (\textit{Acidic oxide})} + \underset{\text{Base}}{CaO\ (\textit{Basic oxide})} \longrightarrow \underset{\text{Salt (slag)}}{CaSiO_3}$$

(*ii*) In the metallurgy of Cu from its sulphide ore, the impurity of FeS is removed by mixing the ore with $SiO_2$ and heating it in reverberatory furnace in the supply of air.

$$2FeS\ (\textit{Impurity}) + 3O_2\ (\textit{From air}) \rightarrow 2FeO + 2SO_2$$

$$\underset{\text{Acid}}{SiO_2} + \underset{\text{Base}}{FeO} \rightarrow \underset{\text{Salt (Slag)}}{FeSiO_3}$$

(*iii*) The basic natue of clay ($CaSO_4$) and phosphates [$Ca_3(PO_4)_2$] is revealed from their reactions with $SiO_2$ where more volatile oxides are removed and the oxide ion ($O^{2-}$), left behind, is transferrd to $SiO_2$.

$$CaSO_4 + SiO_2 \longrightarrow CaSiO_3 + SO_3\uparrow$$
$$Ca_3(PO_4)_2 + 3SiO_2 \longrightarrow 3CaSiO_3 + P_2O_5\uparrow$$

(*iv*) The oxide ion ($O^{2-}$) acceptor nature of metal ions has been used to explain the decomposition (instability) of metal carbonates and sulphates. For example the instability (decomposition) of $Fe_2(CO_3)_3$ as shown below

$$Fe_2(CO_3)_2 \ (\textit{unstable}) \rightleftharpoons 2Fe^{3+} + 3CO_3^{2-}$$

indicates that $Fe^{3+}$ ion is strongly acidic and hence rapidly accepts $O^{2-}$ ion from $CO_3^{2-}$ ion to form $Fe_2(CO_3)_3$. Thus $Fe_2(CO_3)_3$ is unstable.

## Acidity Scale

On the basis of Lux-Flood concept, an acidity scale has been prepared. In this scale, each oxide is given acidity parameter value ($\alpha$). The acidity parameter of an acid and a base is represented as $\alpha A$ and $\alpha B$ respectively. The difference, $(\alpha_A - \alpha_B)$ has been found to be equal to $\sqrt{\Delta H}$ where $\Delta H$ is the enthalpy change in the reaction (heat of reaction) between the acid and the base.

$$\alpha_A - \alpha_B = \sqrt{\Delta H}$$

**Example.** Since the heat of reaction ($\Delta H$) involving the combination of CaO (base) and $SiO_2$ (acid) is –86 KJ/mole, $\alpha_A - \alpha_B = \sqrt{85} = 9$ units (approximate value). The values of $\alpha$ for some oxides are given below:

*Basic oxides:* $Na_2O$ = – 12.5, $Rb_2O$ = –15.0, $Cs_2O$ = –15.2, SrO = –9.4.

*Amphoteric oxides:* $H_2O$ = 0.0, ZnO = –3.2, $Al_2O_3$ = –2.0

*Acidic oxides:* $SiO_2$ = 0.9, $N_2O_5$ = 9.3, $P_4O_{10}$ = 7.5, $SO_2$ = 7.1, $SO_3$ = 10.5, $MoO_3$ = 5.2, $Mn_2O_7$ = 9.6, $Cl_2O_7$ = 11.5

From the values given above the following points may be noted:

(*i*) The value of $\alpha$ of basic oxides is negative. More negative is the value of $\alpha$, more is the basic character of the oxide. Thus $Cs_2O$ is the most basic oxide.

(*ii*) The value of $\alpha$ of amphoteric oxides is close to zero. For example $\alpha$ for $H_2O$, ZnO and $Al_2O_3$ is as $H_2O$ = 0.0, ZnO = –3.2 and $Al_2O_3$ = –2.0.

(*iii*) The value of $\alpha$ for acidic oxides is positive. Higher is the positive value of $\alpha$, more is the acidic character of the oxide. Thus $Cl_2O_7$ is the most acidic oxide.

## V. Solvent-System (Auto-ionisation) Concept

This concept was put forth by Franklin. This concept is based on the dissociation of solvents.

### Auto-ionisation of Solvents.

Before learning the definition of acids and bases on the basis of solvent system concept we should have a knowledge of the nature of ions obtained by the auto-ionisation (self ionisation or dissociation) of some protonic and non-protonic solvents as shown below:

**Protonic solvents:**

| *Solvent* | | *Solvent cation (Acid ion)* | | *Solvent anion (Base ion)* |
|---|---|---|---|---|
| $2H_2O$ | $\rightleftharpoons$ | $H_3O^+$ | + | $OH^-$ |
| $2NH_3$ | $\rightleftharpoons$ | $NH_4^+$ | + | $NH_2^-$ |
| $2CH_3COOH$ | $\rightleftharpoons$ | $CH_3COOH^+{}_2$ | + | $CH_3COO^-$ |
| $2H_2SO_4$ | $\rightleftharpoons$ | $H_3SO_4^+$ | + | $HSO_4^-$ |

**Non-protonic solvents:**

| *Solvent* | | *Solvent cation (Acid ion)* | | *Solvent anion (Base ion)* |
|---|---|---|---|---|
| $2BrF_3$ | $\rightleftharpoons$ | $BrF_4^+$ | + | $BrF_2^-$ |
| $2N_2O_4$ | $\rightleftharpoons$ | $NO^+$ (*Nitrosyl ion*) | + | $NO_3^-$ |
| $COCl_2$ | $\rightleftharpoons$ | $COCl^+$ | + | $Cl^-$ |
| $POCl_3$ | $\rightleftharpoons$ | $POCl_2^+$ | + | $Cl^-$ |
| $2SO_2$ | $\rightleftharpoons$ | $SO^{2+}$ (*Thionyl ion*) | + | $SO_3^{2-}$ |

## Definition of Acids and Bases

According to solvent system concept, an acid is a substance which, by dissolution in the solvent, gives the solvent cation (acid ion) obtained by the auto-ionisation of the solvent itself. A base is a substance which, by dissolution in the solvent, gives the solvent anion (base ion) obtained by the ionisation of the solvent itself.

In general we can say that any species that gives solvent cations (acid ions) in a given solvent behaves as an acid in that solvent. Similarly, any species that gives solvent anions (base ions) in a given solvent behaves as a base in that solvent.

**Examples** (*i*) We know that liq. $NH_3$, on ionisation, gives $NH_4{}^+$ (acid ion) and $NH^-{}_2$ (base ion) ions.

$$2NH_3\ (Solvent) \rightleftharpoons NH^+{}_4\ (Acid\ ion) + NH_2^-\ (Base\ ion)$$

Hence, since $NH_4Cl$ gives $NH_4^+$ ion (acid ion) and $KNH_2$ gives $NH^-{}_2$ ion (base ion) in $NH_3$ solvent, these salts behave as acid and base respectively in this solvent.

$$NH_4Cl\ (Acid) \xrightleftharpoons{Liq.\ NH_3} NH^+{}_4\ (Acid\ ion) + Cl^-$$

$$KNH_2\ (Base) \xrightleftharpoons{Liq.\ NH_3} K^+ + NH^-{}_2\ (Base\ ion)$$

(*ii*) We know that $COCl_2$ solvent, on ionisation, gives $COCl^+$ (acid ion) and $Cl^-$ (base ion) ions

$$COCl_2\ (Solvent) \rightleftharpoons COCl^+\ (Acid\ ion) + Cl^-\ (Base\ ion)$$

Now since $COCl^+\ (AlCl_4)^-$ gives $COCl^+$ ion (acid ion) and $CaCl_2$ gives $Cl^-$ ion (base ion) in $COCl_2$ solvent, $COCl^+\ (AlCl_4)^-$ behaves as an acid and $CaCl_2$ behaves as a base in $COCl_2$ solvent.

$$COCl^+\ (AlCl_4)^-\ (Acid) \xrightleftharpoons{COCl_2\ solvent} COCl^+\ (Acid\ ion) + AlCl^-{}_4$$

$$CaCl_2\ (Base) \xrightleftharpoons{COCl_2\ solvent} Ca^{2+} + 2Cl^-\ (Base\ ions)$$

(*iii*) We know that $POCl_3$ solvent, on ionisation, gives $POCl^+{}_2$ (acid ion) and $Cl^-$ (base ion)

$POCl_3$ (*Solvent*) $\rightleftharpoons$ $POCl^+{}_2$ (*Acid ion*) + $Cl^-$ (*Base ion*)

Now since $FeCl_3$ gives $POCl_2{}^+$ ion (acid ion) in $POCl_3$ solvent, this salt behaves as an acid this solvent.

$FeCl_3$ (*Acid*) + $POCl_3$ (*Solvent*) $\longrightarrow$ $FeCl^-{}_4$ + $POCl^+{}_2$ (*Acid ion*)

Since $(CH_3)_4\, N^+Cl^-$ gives $Cl^-$ ion (base ion) in $POCl_3$ solvent, this salt behaves as a base in this solvent.

$$(CH_3)_4\overset{+}{N}Cl^-\ (Base) \xrightarrow{POCl_3\ \text{solvent}} (CH_3)_4\overset{+}{N} + Cl^-(Base\ ion)$$

## Acid-Base Neutralisation Reactions According to Solvent System Concept

According to solvent system concept of acids and bases, acid-base neutralisation reaction between an acid and a base is the combination of solvent cation (acid ion) and solvent anion (base ion) to form the solvent. This is clear from the following examples:

**Examples (*i*)** We know that, on self ionisation, liq. $NH_3$ (solvent) gives $NH_4{}^+$ (acid ion) and $NH^-{}_2$ (base ion) ions.

$2NH_3$ (*Solvent*) $\rightleftharpoons$ $NH^+{}_4$ (*Acid ion*) + $NH^-{}_2$ (*Base ion*)

Further in this solvent, since $NH_4Cl$ gives $NH^+{}_4$ ion (acid ion) and $KNH_2$ gives $NH^-{}_2$ ion (base in), the acid-base reaction between $NH_4Cl$ (*acid*) and $KNH_2$ (*base*) is the combination of $NH_4{}^+$ (solvent cation) and $NH^-{}_2$ (solvent anion) as shown below.

$$NH_4Cl\ (Acid) + KNH_2\ (Base) \xrightleftharpoons{\text{Liq. } NH_3} KCl\ (Salt) + 2NH_3\ (Solvent)$$

or $\quad NH_4{}^+ + \underbrace{Cl^- + K^+} + NH_2^- \rightleftharpoons KCl + 2NH_3$

or $\quad NH_4^+ + KCl + NH_2^- \rightleftharpoons KCl + 2NH_3$

or $\quad NH_4^+$ (*Solvent cation or acid ion*) + $NH_2^-$ (*Solvent anion or base ion*)
$\rightleftharpoons 2NH_3$ (*Solvent*)

(*ii*) We know that, on self ionisation $POCl_3$ (solvent) gives $POCl^+{}_2$ (acid ion) and $Cl^-$ (base ion) ions.

$POCl_3$ (*Solvent*) $\rightleftharpoons$ $POCl^+{}_2$ (*Acid ion*) + $Cl^-$ (*Base ion*)

Further in this solvent, since $FeCl_3$ gives $POCl^+{}_2$ ion (acid ion) and $(CH_3)_4N^+Cl^-$ gives $Cl^-$ ion (base ion), the acid-base reaction between $FeCl_3$ (acid) and $(CH_3)_4\, N^+Cl^-$ (base) is the combination of $POCl^+{}_2$ (solvent cation) and $Cl^-$ (solvent anion) to form $POCl_3$ (solvent)

$POCl^+{}_2$ (*Solvent cation*) + $Cl^-$ (*Solvent anion*) $\rightarrow$ $POCl_3$ (*Solvent*)

(*iii*) We know that, on self-ionisation, $COCl_2$ (solvent) gives $COCl^+$ (acid ion) and $Cl^-$ (base ion) ions

$COCl_2$ (*Solvent*) $\rightleftharpoons$ $COCl^+$ (*Acid ion*) + $Cl^-$ (*Base ion*)

Further, since in this solvent, $CO\overset{+}{C}l\,(AlCl_4)^-$ gives $COCl^+$ ion (acid ion) and $CaCl_2$ gives $Cl^-$ ion (base ion) the acid-base reaction between $COCl^+\,(AlCl_4)^-$ (acid)

and $CaCl_2$ (base) is the combination of $COCl^+$ (solvent cation) and $Cl^-$ (solvent anion) to form $COCl_2$ (solvent)

$COCl^+$ (*Solvent cation*) + $Cl^-$ (*Solvent anion*) → $COCl_2$(*Solvent*)

Thus we see that acid-base neutralisation reaction between an acid and a base proceeds in the direction opposite to that of the auto-ionisation of the solvent.

The acid-base neutralisation reaction between $NH_4Cl$ (acid) and $KNH_2$ (base) in liq. $NH_3$ is analogous to that between HCl (acid) and NaOH (base) in $H_2O$

| *Acid* | | *Base* | | *Salt* | | *Solvent* |
|---|---|---|---|---|---|---|
| $NH_4Cl$ | + | $KNH_2$ | $\xrightarrow{\text{Liq. } NH_3}$ | KCl | + | $2NH_3$ |
| HCl | + | NaOH | $\xrightarrow{\text{In } H_2O}$ | NaCl | + | $H_2O$ |

Newtralisation reactions in some protonic and non-protonic solvens are given below:

| | | *Acid* | *Base* | | *Salt* | *Solvent* |
|---|---|---|---|---|---|---|
| *In liq. $NH_3$* | : | $NH_4Cl$ | + $NaNH_2$ | ⇌ | NaCl | + $2NH_3$ |
| *In $N_2O_4$* | : | NOCl | + $NaNO_3$ | ⇌ | NaCl | + $N_2O_4$ |
| *In $CH_3COOH$* | : | HCl | + $CH_3COONa$ | ⇌ | NaCl | + $CH_3OOH$ |
| *In liq. $SO_2$* | : | $SOCl_2$ | + $[N(CH_3)_4]_2SO_3$ | ⇌ | $2[N(CH_3)_4]Cl$ | + $2SO_2$ |
| *In liq. $C_2H_5OH$* | : | $C_2H_5OH_2Br$ | + $KOC_2H_5$ | ⇌ | KBr | + $2C_2H_5OH$ |
| *In liq. $COCl_2$* | : | $COCl (AlCl_4)$ | + $CaCl_2$ | ⇌ | $Ca (AlCl_4)_2$ | + $2COCl_2$ |

## Amphoteric Behaviour of $Zn(NO_3)_2$ in liq. $NH_3$

$Zn(NO_3)_2$ shows amphoteric behavior in liq. $NH_3$, since in this solvent, this compound can react with ammono acid as well as with ammono base as shown below:

$Zn (NO_3)_2 + 2NH_4Cl$ (*Ammono acid*) $\xrightarrow{\text{Liq. } NH_3}$ $ZnCl_2$ (*Salt*) + $2NH_4NO_3$

$Zn (NO_3)_2 + 4KNH_2$ (*Ammono base*) $\xrightarrow{\text{Liq. } NH_3}$ $K_2[Zn (NH_2)_4]$ + $2KNO_3$

*Amido complex*

*(Soluble)*

## Utility of the Concept

(*i*) The solvent-system concept can be used to explain the acid-base reaction occurring in aqueous and non-acqueous solvents (protonic and non-protonic solvents).

(*ii*) The hydrolysis of non-metal halides (e.g. $POCl_3$) is analogous to the ammonolysis and alcoholysis reactions of these halides. In these reactions $Cl^-$ ion of the halide is replaced by the base ion abtained by the ionisation of the solvent. For example:

$$2H_2O \rightleftharpoons H_3O^+ + OH^-$$
$$2NH_3 \rightleftharpoons NH^+_4 + NH^-_2$$
$$ROH \rightleftharpoons H^+ + OR^-$$

*Hydrolysis:* $OPCl_3 + 3H_2O \rightleftharpoons OP(OH)_3 + 3HCl$

*Ammonolysis* : $OPCl_3 + 3NH_3 \rightleftharpoons OP(NH_2)_3 + 3HCl$

*Alcoholysis* : $OPCl_3 + 3ROH \rightleftharpoons OP(OR)_3 + 3HCl$

### Limitations of the Concept

(*i*) Acid-base reaction is dictated by the mode of ionisation of the solvent and the nature of the ions obtained by the auto-ionisation of the solvent.

(*ii*) Acid-base reactions taking place in the absence of solvent cannot be explained by solvent-system concept.

(*iii*) The self-ionisation of liq. $SO_2$ is supposed to take place as follows:

$$2SO_2\ (\textit{Solvent}) \rightleftharpoons SO^{2+}\ (\textit{Acid ion}) + SO_3^{2-}\ (\text{Base ion}) \qquad ...(i)$$

Thus since $SOCl_2$ gives $SO^{2+}$ ion (acid ion) in liq. $SO_2$, this salt behaves as an acid in liq. $SO_2$

$$SOCl_2\ (\textit{Acid}) \xrightleftharpoons{\text{Liq. } SO_2} SO^{2+}\ (\textit{Acid ion}) + 2Cl^-$$

Also since $Cs_2SO_3$ gives $SO_3^{2-}$ ion (*base ion*) in liq. $SO_2$, this salt behaves as a base in liq. $SO_2$

$$Cs_2SO_3\ (\textit{Base}) \xrightleftharpoons{\text{Liq. } SO_2} 2Cs^+ + SO_3^{2-}\ (\textit{Base ion})$$

Since $SOCl_2$ is an acid and $Cs_2SO_3$ is a base in liq. $SO_2$, the neutralisation reaction between $SO_2Cl_2$ (acid) and $Cs_2SO_3$ (base) is the combination between $SO^{2+}$ (solvent cation) and $SO_3^{2-}$ (solvent anion) to produce $SO_2$ solvent

$$SO^{2+}\ (\textit{Solvent cation}) + SO_3^{2-}\ (\textit{Solvent anion}) \rightarrow SO_2\ (\textit{Solvent}) \qquad ...(ii)$$

The above reaction does occur between $SOCl_2$ (acid) and $Cs_2SO_3$ (base). Since the self-ionisation of liq. $SO_2$ as represented by equation (*i*) has not been established at all, the occurrence of acid-base reaction (*ii*) cannot be explained on the basis of self ionisation of $SO_2$.

## VI. Lewis Concept : Electron Pair Acceptor-Donor Concept

In the early 1930s, G.N. Lewis defined an acid as a molecule or an ion that can accept an electron pair from some other substance and a base as a molecule or an ion which can donate an electron pair (lone pair of electrons) to some other substance. *In other words, a Lewis acid is an electron pair acceptor and a Lewis base is an electron pair donor.*

Since Lewis acid is an electron pair acceptor, it is an electron-deficient species. Similarly since Lewis base is an electron pair donor, it is an electron rich species.

Since a Lewis acid loves to accept the electron pair (s), it is also called an *electrophilic* (electro = electron, philic = lover) *reagent* or simply *electrophile.* Similarly, since a Lewis base loves to donate the electron pair (s) to a nucleus, in fact to a positively charged ion, it is a *nucleus lover* or *nucleus-seeker* and hence is also called *nucleophilic* (nucleo = nucleus, philic = lover) *reagent* or simply *nucleophile*.

Lewis acid should contain one or more vacant orbitals into which electron pair (s) donated by the Lewis base can be accommodated.

*An acid qualifying Lewis concept is called Lewis acid and a base qualifying Lewis concept is termed as Lewis base.*

## Neutralisation Reaction According to Lewis Concept

According to Lewis concept of acids and bases, the neutralisation reaction is that in which a Lewis acid reacts with a Lewis base and forms a compound which is called *an adduct or complex compound*. This compound contains (Lewis base $\rightarrow$ Lewis acid) coordinate bond. For example:

(*i*) $BF_3 + \,:NH_3 \longrightarrow [H_3N \rightarrow BF_3]$

Lewis acid    Lewis base    Adduct

(Lewis base $\rightarrow$ Lewis acid) coordinate bond results by the overlap of the filled orbital on the Lewis base with the vacant orbital on the Lewis acid.

(*ii*) $H^+$(*Proton*) $+ :NH_3 \longrightarrow H^+ \leftarrow NH_3$ or $NH_4^+$

Lewis acid    Lewis base    Adduct

Evidently in the above reaction proton ($H^+$) accepts one electron pair from $:NH_3$ molecule and is, therefore, an acid, whereas $:NH_3$ molecule which donates an electron pair, is a base. The adduct is $NH_4^+$ ion.

## Lewis Acids and Lewis Bases in Coordination Chemistry

In coordination chemistry, the central metal ion, $M^{n+}$ is regarded as a Lewis acid (*i.e.* electrophile), since it accepts an electron pair while the ligand, *L* is regarded as Lewis base (*i.e.* nucleophile), since it donates an electron pair. The compound formed by the combination of metal cation ($M^{n+}$) and ligand (L) is called a *complex* or an *adduct*.

Electron pair being donated by L to $M^{n+}$

$M^{n+} + \,:L \rightarrow [M^{n+} \leftarrow L]^{n+}$

Lewis acid (Electrophile or Electron pair acceptor)    Lewis base (Nucleophile or Electron pair donor)    Complex or adduct

$M^{n+}$ and L are linked together by a coordinate covalent bond from L to $M^{n+}$ ($M^{n+} \leftarrow$ L bond). Thus the reaction between a metallic cation, $M^{n+}$ and a ligand, L shown above can be considered as an acid-base reaction according to Lewis concept.

The metallic cation (Lewis acid) must have at least one vacant orbital in its valence-shell in which it can accommodate the electron pair donated by the ligand (Lewis base) and the ligand should have at least one lone pair of electrons so that it may be donated to the metallic cation.

## Classification of Lewis Acids

Any Lewis acid must contain at lease one empty orbital in the velence shell of one of its atoms to accept an electron pair from a Lewis base. Lewis ocids may be classified as:

1. **Molecules containing a central atom with incomplete octet**. Molecules having a central atom with incomplete octet are electron deficient and hence can accept a *lp* of electrons donated by Lewis base. Thus such molecules act as Lewis acids. Examples of such molecules are $BF_3$, $BCl_3$, $AlCl_3$, $Me_3B$, $SO_3$ etc.

(*i*) 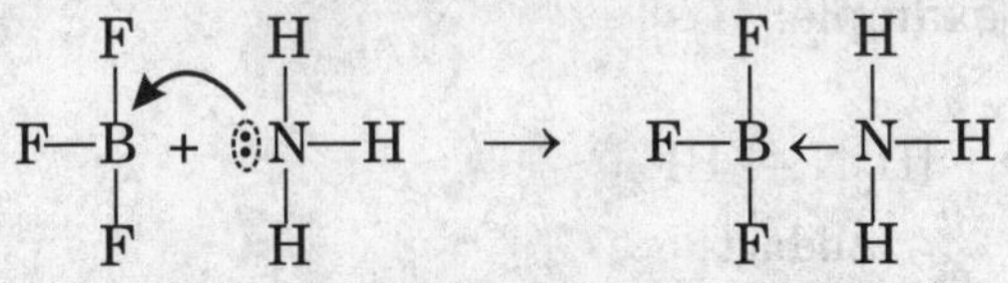

or $BF_3 + :NH_3 \rightarrow F_3B \leftarrow NH_3$
Lewis acid, Lewis base, Adduct

(*ii*) $:\ddot{Cl}^{-} + AlCl_3 \longrightarrow [:\ddot{Cl} \rightarrow AlCl_3]^{-}$ or $[AlCl_4]^{-}$
Lewis base, Lewis acid

(*iii*) $:\ddot{F}^{-} + BF_3 \longrightarrow [:\ddot{F} \rightarrow BF_3]^{-}$ or $[BF_4]^{-}$

(*iv*) 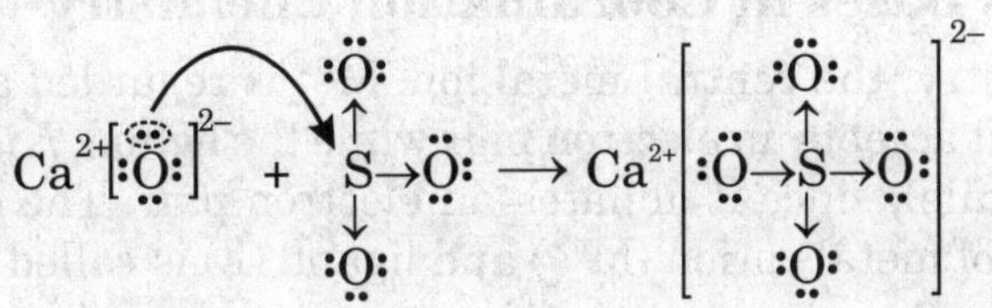

or $CaO + SO_3 \longrightarrow CaSO_4$
Lewis base, Lewis acid

(*v*) 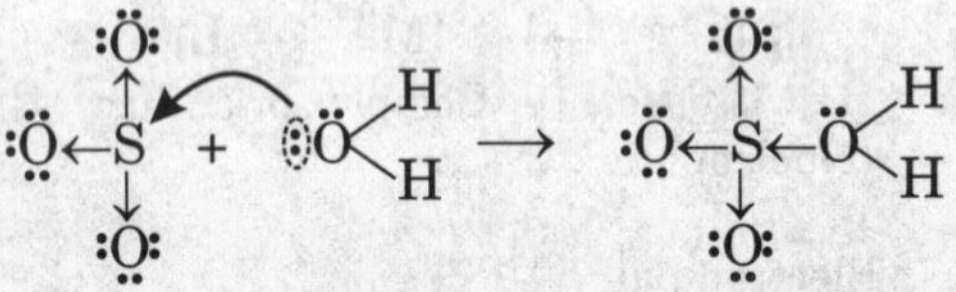

or $SO_3 + H_2O \rightarrow O_3S \leftarrow OH_2$ or $H_2SO_4$
Lewis acid, Lewis base

(*vi*) 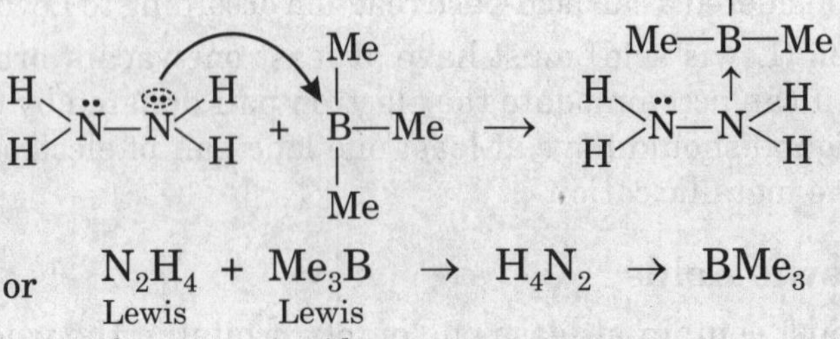

or $N_2H_4 + Me_3B \rightarrow H_4N_2 \rightarrow BMe_3$
Lewis base, Lewis acid

(vii) 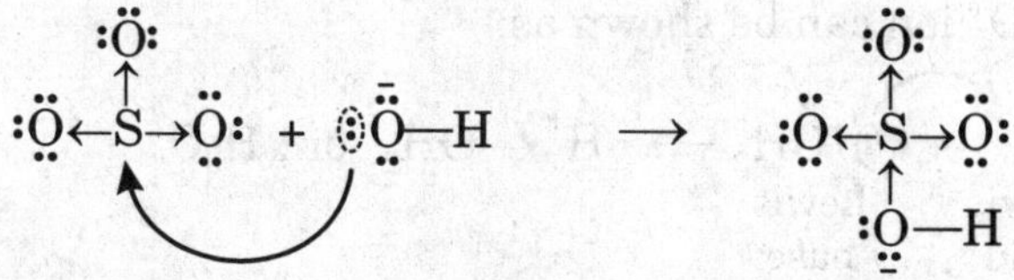

or $SO_3$ + $OH^-$ ⟶ $HSO_4^-$
Lewis acid, Lewis base, Bisulphate ion

**2. Molecules containing central atom with empty d-orbitals:** The empty d-orbitals of the central atom in molecules like $SiX_4$ (X = F, Cl etc.), $SbF_5$ etc. can accommodate electron pair (s) donated by Lewis base like halide ion ($F^-$, $Cl^-$ etc.), $SO_2$ molecule etc. Examples are given below:

(i) 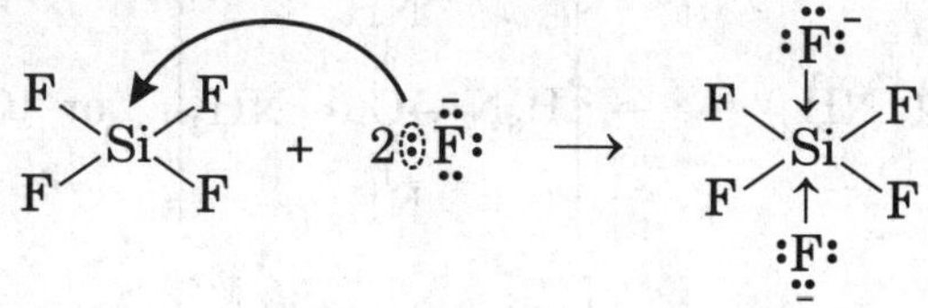

or $SiF_4$ + $2F^-$ ⟶ $[SiF_6]^{2-}$
Lewis acid, Lewis base

(ii) $SiCl_4$ + $2Cl^-$ ⟶ $[SiCl_6]^{2-}$
Lewis acid, Lewis base

(iii) 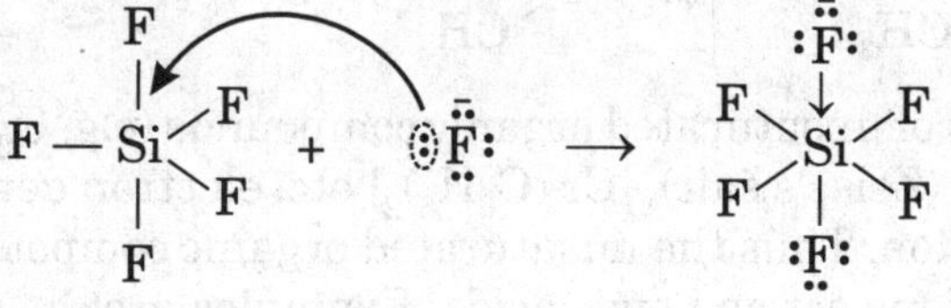

or $SbF_5$ + $F^-$ ⟶ $[SbF_6]^-$
Lewis acid, Lewis base

(iv) 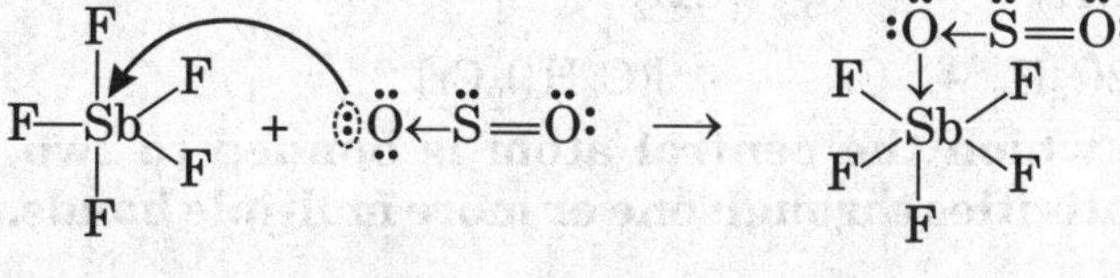

or $SbF_5$ + $SO_2$ ⟶ $F_5Sb \leftarrow SO_2$
Lewis acid, Lewis base

(v) $H^+$ ion has one vacant 1s orbital. Hence it can accept an electron pair donated by $NH_3$ molecule (Lewis base) to form $NH_4^+$ ion. Thus in the reaction between $H^+$ and $NH_3$, $H^+$ ion acts as a Lewis acid.

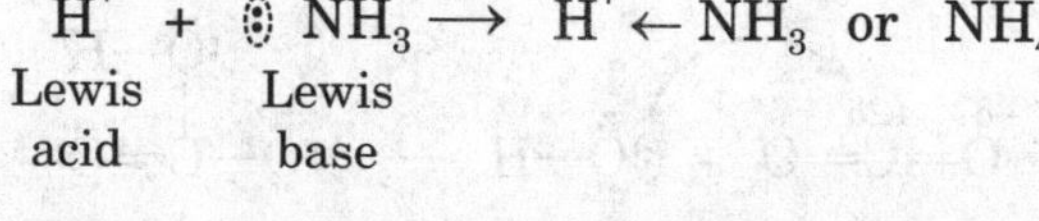

The formation of $H_3O^+$ ion can be shown as:

$$\underset{\text{Lewis acid}}{H^+} + \underset{\text{Lewis base}}{:\ddot{O}H_2} \longrightarrow H^+ \leftarrow OH_2 \text{ or } H_3O^+$$

**3. Simple cations.** Theoretically all simple cations are potential Lewis acids. Reactions of some cations as Lewis acids with Lewis bases are shown below. It will be seen that these reactions are identical with those which produce Werner complexes.

*Lewis acid* *Lewis base* *Adduct or addition compound*

*Ammonation*: $Ag^+ + 2(:NH_3) \longrightarrow [NH_3 \rightarrow Ag \leftarrow NH_3]^+$ or $[Ag(NH_3)_2]^+$

$$Cu^{2+} + 4(:NH_3) \longrightarrow \left[\begin{array}{c} NH_3 \\ \downarrow \\ H_3N \rightarrow Cu \leftarrow NH_3 \\ \uparrow \\ NH_3 \end{array}\right]^{2+} \text{ or } [Cu(NH_3)_4]^{2+}$$

*Hydration:* $Co^{3+} + 6(:OH_2) \longrightarrow \left[\begin{array}{c} H_2O \searrow \quad \swarrow OH_2 \\ H_2O \rightarrow Co \leftarrow OH_2 \\ H_2O \nearrow \quad \nwarrow OH_2 \end{array}\right]^{3+}$ or $[Co(H_2O)_6]^{3+}$

*Alcoholation:* $Li^+ + :O\begin{smallmatrix} \diagup H \\ \diagdown CH_3 \end{smallmatrix} \longrightarrow \left[Li^+ \leftarrow O\begin{smallmatrix} \diagup H \\ \diagdown CH_3 \end{smallmatrix}\right]^+$ or $[CH_3OH\ Li]^+$

In coordination complexes of unsaturated organic compounds (e.g. $C_2H_4$, $C_2H_2$, $C_6H_6$ etc) e.g. K [Pt $Cl_3$ ($C_2H_4$)] (Zeise's salt), [Cr $(C_6H_6)_2$] etc. electron density from $\pi$-bond is donated to the metal ion. Thus the unsaturated organic compounds serve as Lewis bases and the metal ions act as Lewis acids. Examples are:

$$\begin{array}{c} CH_2 \\ \| \\ CH_2 \end{array} + Ag^+ \longrightarrow \left[\begin{array}{c} CH_2 \\ \| \\ CH_2 \end{array} \rightarrow Ag\right]^+$$

$$2C_6H_6 + Cr \longrightarrow [(C_6H_6)_2Cr]$$

**4. Molecules in which the central atom is bonded to two atoms of different electronegativities through one or more multiple bonds.** Examples are given below:

(*i*) In $CO_2$ molecule (O = C = O), since O-atom is more electronegative than C-atom, this atom pulls the *pi* electrons towards itself leaving a partial positive charge (+δ) on C-atom. As a result C-atom becomes electron-deficient centre $(\overset{-\delta}{O} = \overset{+2\delta}{C} = \overset{-\delta}{O})$ and hence acts as an electron acceptor (Lewis acid) towards Lewis base like $OH^-$ ion which will donate its electron pair to C-atom to form bicarbonate ion, $HCO_3^-$.

$$\overset{-\delta}{O}=\overset{+2\delta}{C}=\overset{-\delta}{O} + :\ddot{\underset{..}{O}}-H \longrightarrow :\ddot{\underset{..}{O}}-\underset{}{\overset{\overset{:\ddot{O}-H}{|}}{C}}=\ddot{\underset{..}{O}}:$$

or $CO_2$ + $OH^-$ ⟶ $HCO_3^-$
Lewis acid Lewis base

(*ii*) $SO_2$ also acts as a Lewis acid towards $OH^-$ ion and $NR_3$ molecule

(*a*) 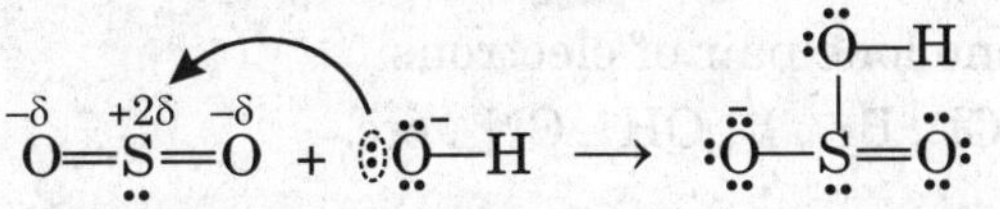

or $SO_2$ + $OH^-$ ⟶ $HSO_3^-$
Lewis acid Lewis base Biosulphate ion

(*b*) 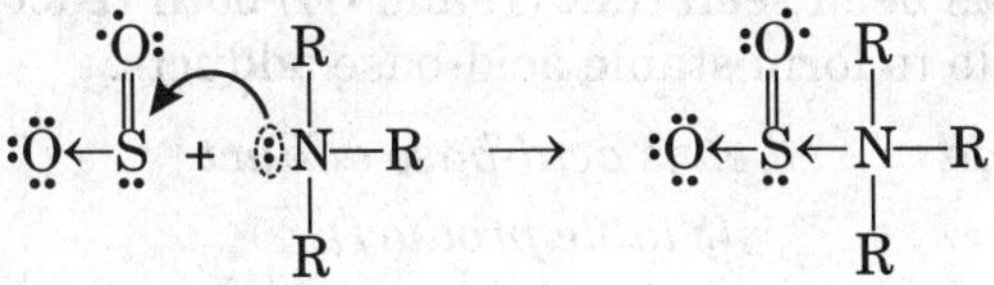

or $SO_2$ + $NR_3$ ⟶ $O_2S \leftarrow NR_3$
Lewis acid Lewis base

5. **Elements with an electron sextent.** Oxygen and sulphur atoms contain six electrons in their valence shell and can, therefore, be regarded as Lewis acids. The oxidation of $SO_3^{2-}$ to $SO_4^{2-}$ ion by oxygen and to $S_2O_3^{2-}$ ion by sulphur are the acid-base reactions.

*Lewis base* *Lewis acid* *Adduct*

(*i*) (*ii*) 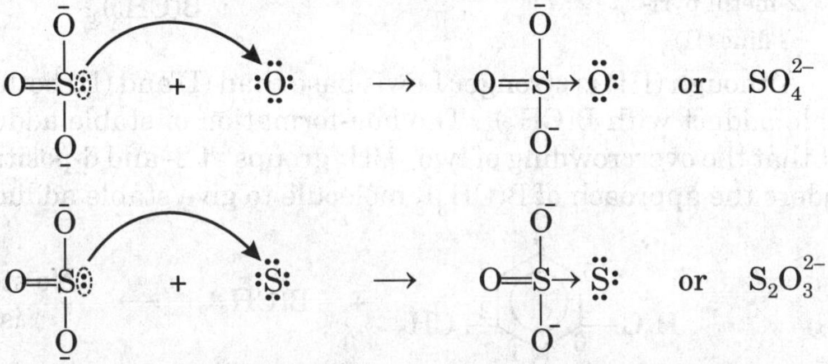

The formation of pyridine oxide ($C_5H_5N \rightarrow O$) is an example of Lewis acid-base reaction in which pyridine ($C_5H_5N$) acts as a Lewis base and O-atom acts as a Lewis acid.

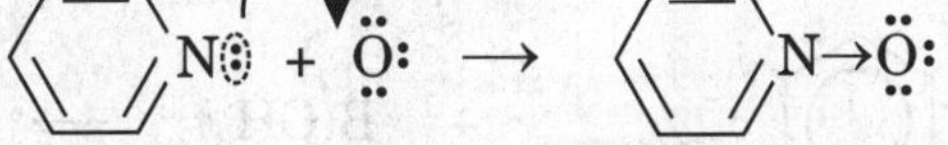

## Classification of Lewis Bases.

Lewis bases can be classified as:

(*i*) Neutral molecules like $NH_3$, $R - NH_2$, $R - \ddot{\underset{\cdot\cdot}{O}}H$, $H - \ddot{\underset{\cdot\cdot}{O}} - H$ etc., in which one of the atoms has got at least one lone pair of electrons.

(*ii*) All negative ions like $F^-$, $Cl^-$, $Br^-$, $I^-$, $OH^-$, $CN^-$, etc.

## Steric Effect of Substituents in Lewis Acid-Base Reactions

It has been observed that due to the presence of a *lp* of electrons on N-atom, pyridine (I), 2-methyl pyridine (II) and 2, 6-dimethyl pyridine (III) show basic properties. Due to the inductive effect of $-CH_3$ groups, their basic property increases in the order : (I) < (II) < (III). It has been seen that (I) and (II) both react with the electron-deficient $B(CH_3)_3$ molecule to form stable acid-base adducts.

*Lewis base* *Lewis acid* *Lewis acid-base adduct (Stable product)*

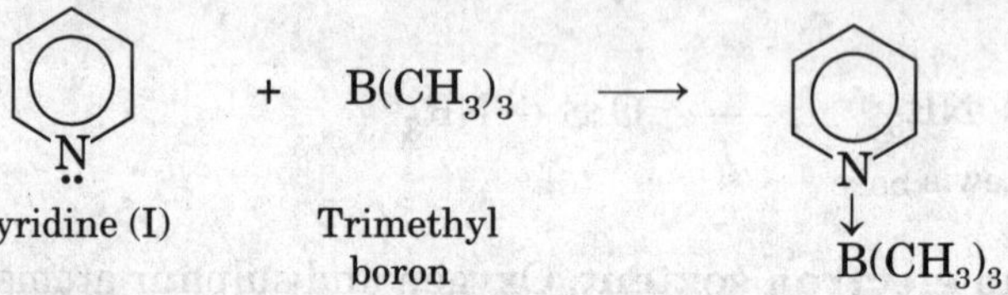

Pyridine (I) Trimethyl boron

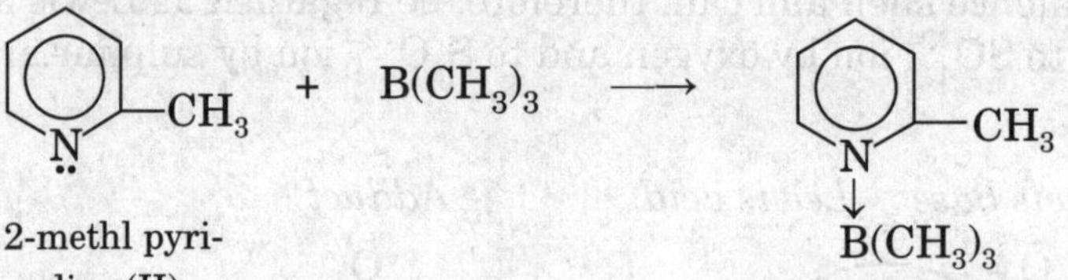

2-methl pyridine (II)

Although (III) is stronger Lewis base than (I) and (II) both, yet it does not form stable adduct with $B(CH_3)_3$. The non-formation of stable adduct is because of the fact that the overcrowding of two $-CH_3$ groups at 2- and 6-positions in (III) molecule hinders the approach of $B(CH_3)_3$ molecule to give stable adduct.

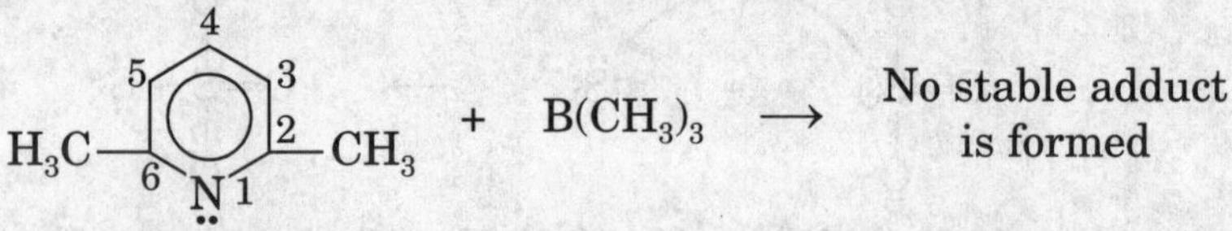

2, 6-dimetlyl pyridine (III)

2-t-butyl pyridine (IV) also does not form stable acid-base adduct with $B(CH_3)_3$ because of the steric effect of t-butyl group, $-C(CH_3)_3$ at the 2-position of the pyridine.

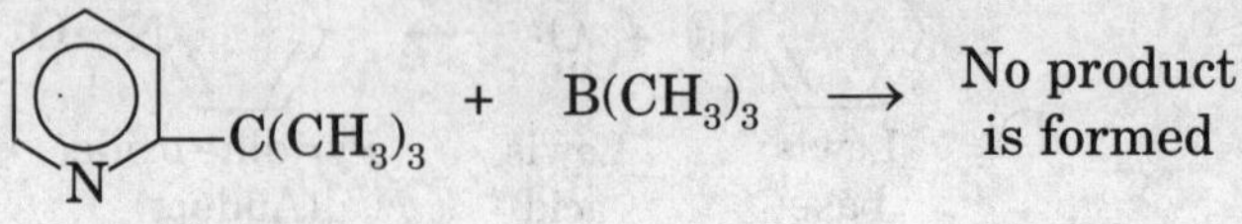

2-t-butyl pyridine (IV)

## Arrhenius Acids vs Lewis Acids

Arrhenius considered those substances as acids which could donate proton ($H^+$). Since proton can accept an electron pair from a base, *all Arrhenius acids are also Lewis acids*. For example:

$$\underset{\text{Arrhenius acid}}{HCl} \longrightarrow H^+ + Cl^-$$

$$\underset{\text{Lewis base}}{H_3N:} + \underset{\text{Lewis acid}}{H^+} \longrightarrow H_3N \rightarrow H+ \text{ or } NH_4^+$$

## Bronsted Bases vs Lewis Bases

According to Bronsted a base is a substance which accepts proton while according to Lewis, a base is any species which can donate an electron pair. Bronsted base can accept proton only when it has an unshared pair of electrons in it. *Thus all Bronsted bases are also Lewis bases*. This fact can be made evident from the following neutralisation reaction in which $NH_3$ behaves both as Bronsted base as well as Lewis base.

$$H^+ + \underset{\substack{\text{Bronsted base}\\ \text{or Lewis base}}}{:NH_3} \longrightarrow H^+ \leftarrow NH_3 \text{ or } NH_4^+$$

In this reaction, since $NH_3$ molecule accepts a proton ($H^+$), it behaves as a Bronsted base. At the same time, since $NH_3$ molecule also donates an electron pair to $H^+$ ion, this molecule (i.e. $NH_3$ molecule) behaves as a Lewis base.

## Bronsted Acids vs Lewis Acids

There are many Lewis acids (e.g. $SO_3$, halides of B, Al, Fe (III) and Zn) which cannot be treated as Bronsted acids. The central atom of these acids is electron-deficient and hence can accept a pair of electrons from a base to complete its octet.

Similarly HCl which is a Bronsted acid cannot be called a Lewis acid, since it has no vacant orbital to accept an electron pair from a base.

## Solvent System vs Lewis System

Solvent system considers $NH^+{}_4$ ion as an acid and $NH^-_2$ ion as a base in liq. $NH_3$ as shown below:

$$\underset{\textit{Acid}}{NH_4^+} + \underset{\textit{Base}}{NH_2^-} \rightleftharpoons \underset{\textit{Acid}}{NH_3} + \underset{\textit{Base}}{NH_3}$$

In the above reaction, $NH_2^-$ ion donates a lone pair of electrons to $NH_4^+$ ion to produce two molecules of $NH_3$. Since $NH_4^+$ ion is nothing but an ammoniated proton and the proton ($H^+$) can accept a lone pair of electrons from a base (*i.e.* $NH_2^-$ ion), $NH_4^+$ and $NH_2^-$ ions which are acid and base respectively in liquid $NH_3$ are also Lewis acid and Lewis base.

## Factors Affecting the Relative Strength of Lewis Acids

Relative strength of Lewis acids depends mainly on the following factors:

**1. Inductive effect of EWG and EDG** The inductive effect caused by EWG increases the acid strength of Lewis acid while inductive effect caused by EDG decreases the acid strength of Lewis acid. For example if H in $BH_3$ is replaced by F which is an EWG, acidic strength increases. Thus $BF_3$ is a stronger Lewis acid and than $BH_3$ ($BF_3 > BH_3$). On the other hand, if H in $BH_3$ is replaced by Me group which is an EDG, the acidic strength decreases. Thus $Me_3B$ is weaker Lewis acid than $BH_3$ ($Me_3B < BH_3$) Thus the overall acidic strength of $BH_3$, $BF_3$ and $BMe_3$ increases as:

$$Me_3B < BH_3 < BF_3.$$

**Note.** Experimentally it has been found that the acid strength of $BF_3$, $BCl_3$ and $BBr_3$ molecules which are Lewis acids increases as:

$$BF_3 < BCl_3 < BBr_3 \quad ...(i)$$

This increasing order cannot be explained on the basis of electron-withdrawing inductive effect (–I effect) of –F, –Cl and –Br groups present in the trihalide molecules as shown below.

Since –I effect of halide groups decreases as –F > – Cl > – Br, the electron-accepting power (Lewis acid strength) of B-atom in $BX_3$ molecules (X = F, Cl, Br) should also increase in the order:

$$BF_3 > BCl_3 > BBr_3 \quad ...(ii)$$

This decreasing order which is based on –I effect of halide groups is opposite to that shown at (*i*) which is experimentally observed order. Thus we see that the experimentally observed increasing order of acidic strength of $BF_3$, $BCl_3$ and $BBr_3$ shown at (*i*) cannot be explained on the basis of inductive effect.

**2. Electronegativity.** Tetrahalides of Sn ($SnCl_4$, $SnBr_4$ and $SnI_4$) are Lewis acids. The acid strength of these acids decreases as $SnCl_4 > SnBr_4 > SnI_4$. This decreasing order is due to the decrease in electronegativity of halogen atoms from Cl to I.

**Note.** It has been found experimentally that the Lewis acid strength of $BX_3$ molecules (X = F, Cl, Br) which are Lewis acids decreases as:

$$BF_3 < BCl_3 < BBr_3 \quad ...(i)$$

This order cannot be explained on the basis of electronegativity (EN) of halogen atoms. Due to the decrease in EN values of halogens from F to Br (F > Cl > Br), the Lewis acid strength of $BX_3$ molecules should decrease as:

$$BF_3 > BCl_3 > BBr_3 \quad ...(ii)$$

This order is opposite to the order shown at (i) which has been found experimentally. Thus we see that the experimentally observed increasing order of acidic strength of $BF_3$, $BCl_3$ and $BBr_3$ shown of (*i*) cannot be explained on the basis of electronegativity values of halogen atoms.

**3. Back donation in $BF_3$, $BCl_3$ and $BI_3$.** It has been found experimentally that the Lewis acid strength of $BF_3$, $BCl_3$ and $BBr_3$ molecules increases in the order:

$$BF_3 < BCl_3 < BBr_3$$

This order can neither be explained on the basis of inductive effect of halogen groups nor on the basis of EN values of halogen atoms. This order has been explained on the basis of the formation of dative bond or $p\pi - p\pi$ back bond (back donation) in $BX_3$ molecules. The structures of $BF_3$, $BCl_3$ and $BBr_3$ molecules are given in Fig. 3.12.

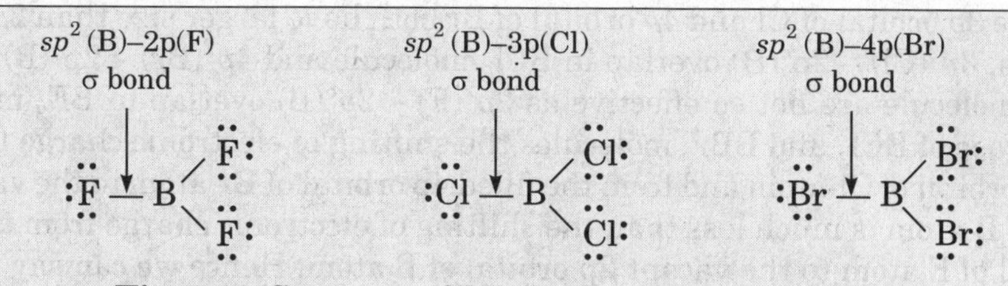


**Fig. 3.12** Structure of $BF_3$, $BCl_3$ and $BBr_3$ molecules

In the structure of each of trihalide molecules, *B*-atom is $sp^2$ lybridised [B* = $2s^1\,2p^1 2p^1 2p^0$] or $(sp)^1\,(sp)^1\,(sp)^1\,2p^0$]. Obviously *B*-atom has three singly-filled *sp* hybrid orbitals and one empty 2*p* orbital in its valence shell. In $BF_3$ molecule each of the three B–F σ bonds is formed by the overlap between singly-filled $sp^2$ hybrid orbital on B-atom and singly-filled 2*p* orbital on F-atom. (F = $2s^2\,2p^2\,2p^2\,2p^1$). Thus each of the three B–F bonds in $BF_3$ molecules is $sp^2$(B) – 2*p*(F) σ-bond. In case of $BCl_3$ and $BBr_3$ molecules B – Cl and B – Br σ-bonds are $sp^2$(B) – 3*p*(Cl) and $sp^2$(B) – 4*p*(*Br*) σ-bonds. (Cl = $3s^2\,3p^2\,3p^2\,3p^1$, Br = $4s^2\,4p^2\,4p^2\,4p^1$).

Now let us take the example of $BF_3$ molecule to explain tbc formation of $p\pi - p\pi$ back bond in this molecule. We have already stated F-atom has one completely-filled 2*s* orbital and two completely-filled 2*p* orbitals (F = $2s^2\,2p^2\,2p^2\,2p^1$). One of the filled 2*p* orbitals on F-atom overalps with the vacant 2*p* orbital (acceptor orbital) on B-atom and forms an additional F → Bπ bond which is a $2p^2$ (*F*) → $2p°$ (B)π bond. This bond is represented as $p\pi - p\pi$ bond and is called dative bond or $p\pi - p\pi$ back bond. Obviously $p\pi - p\pi$ bond is formed by the donation of an electron pair from the filled 2*p* orbital of F-atom to the vacant 2*p* orbital of B-atom (see Fig. 3.13). This is called **back donation**. Filled 2*p* orbital of F-atom is called **donor** 2*p* orbital and vacant 2*p* orbital on B-atom is called **acceptor** 2*p* orbital.

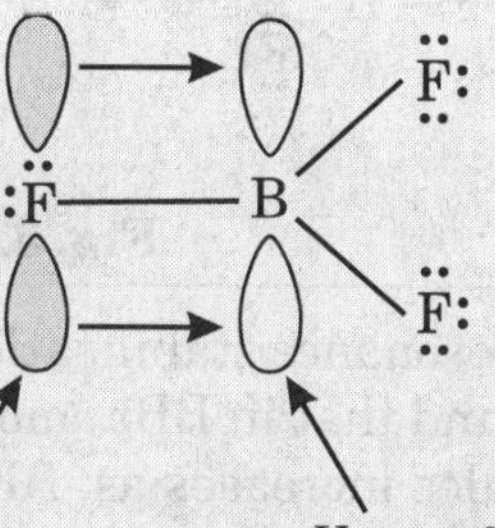


**Fig. 3.13** Formation of $(p\pi - p\pi)$ back bonding in $BF_3$ molecule

Due to the formation of $p\pi - p\pi$ back bond, electronic charge from the filled 2*p* orbital of F-atom is shifted to the vacant 2*p* orbital of B-atom. The shifting of electronic charge from F-atom to B-atom decreases the electron-accepting power (electron-accepting tendency) or Lewis acid strength of B-atom in $BF_3$ molecule.

The back bonding in $BF_3$ molecule is supported by the fact that the observed B – F bond length (=130pm) in $BF_3$ molecule is much less than the sum of the covalent radii of B and F atoms (B = 85 pm and F = 64 pm, sum = 85 + 64 = 149 pm). Here it should be noted that since 2*p* orbtial (filled) of F atom and 2*p* orbital (empty) of B atom have almost the same size, the overlap between these two 2p orbitals to form $2p^2$ (F) → $2p°$ (B)π bond is very effective. In case of $BCl_3$ and $BBr_3$ molecules $p\pi - p\pi$ back bonds are $3p^2$(Cl) → $2p°$(B) and $4p^2$ (*Br*) → $2p°$(B) π bonds respectively.

Now since $3p$ orbital of Cl and $4p$ orbital of Br both have larger size than $2p$ orbital of B-atom, $3p^2$(Cl) – $2p°$(B) overlap in $BCl_3$ molecule and $4p^2$(Br) – $2p°$(B) overlap in $BBr_3$ molecule are not so effective as $2p^2$(F) – $2p°$(B) overlap in $BF_3$ molecule. Thus in case of $BCl_3$ and $BBr_3$ molecules the shifting of electronic charge from the filled $3p$ orbital of Cl-atom and from the filled 4p orbital of Br-atom to the vacant 2p orbital of B-atom is much less than the shifting of electronic charge from the filled 2p orbital of F atom to the vacant 2p orbital of B atom. Hence we can say that the electron-accepting power of B-atom in $BCl_3$ and $BBr_3$ is more than that of B-atom in $BF_3$. In other words, the electron-accepting power or Lewis acid strength of the three boron trihalides increases in the order:

$$BF_3 < BCl_3 < BBr_3$$

The increasing order of Lewis acid strength of $BX_3$ molecules from $BF_3$ to $BBr_3$ can also be explained on the basis of resonance stabilisation of $BX_3$ molecules. Back bonding in $BF_3$ molecule can oscillate among the three B – F bonds in $BF_3$ molecule as shown in Fig. 3.14.

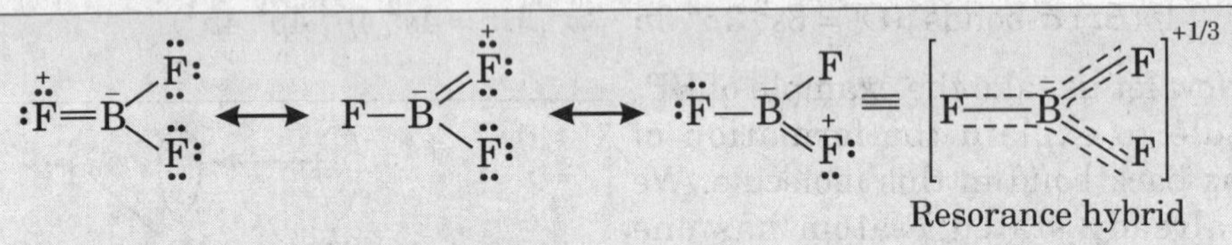


**Fig. 3.14** Resonance forms of $BF_3$ molecule

Resonance stabilisation of $BF_3$ molecule is large while that in $BCl_3$ molecule is small and that in $BBr_3$ molecule is still smaller. Hence Lewis acid strength of $BX_3$ molecules increases as: $BF_3 < BCl_3 < BBr_3$.

**4. Lewis acidic strength of simple cations.** The Lewis acid strength or coordinating ability of the simple cations which, according to Lewis, are Lewis acids, increases with (*a*) an increase in the positive charge carried by the cation (*b*) an increase in the nuclear charge for atoms in any period of the periodic table (*c*) a decrease in ionic radius (*d*) a decrease in the number of shielding electron shells.

Evidently the acid strength of simple cations increases for the elements on moving from left to right in a period and from bottom to top in a group of periodic table. Thus:

$Fe^{2+} < Fe^{3+}$ (positive charge increases from +2 to +3)

$K^+ < Na^+$ (on moving form bottom to top in a group)

$Li^+ < Be^{2+}$ (on moving from left to right in a period)

——— Strength of Lewis acids increasing ⟶

## Lewis Acid Strength of Metal Cations in Gas Phase

Acidity of metal cations in gas phase can be measured from the values of EA of metal cation. Lower is the value of EA of a given metal cation, lower is the acidity of the cation. For example since the value of EA of $K^+$ ion (= 420 KJ / mol) is lower than that of $Ca^{2+}$ ion (= 1150 KJ/mol), the former (*i.e* $K^+$ cation) is a weaker acid than the latter (*i.e.* $Ca^{2+}$ cation)

## Relative Strength of Lewis Bases

As a we move down a group, the basic strength of Lewis bases decreases. For example the basic strength of $(CH_3)_3N$, $(CH_3)_3P$ and $(CH_3)_3As$ which are Lewis bases decreases in the order:

$$(CH_3)_3\,N > (CH_3)_3P > (CH_3)_3\,As$$

On moving from left to right in a period, the basic strength of Lewis bases decreases. For example:

$$(CH_3)_3N > (CH_3)_2O > (CH_3)\,F$$

The decrease in basic strength is due to the increase in electronegativity of the central atoms from N to F.

## Behaviour of $NH_4^+$ and $NH_2^-$ Ions According to Solvent System and Lewis System

Solvent system considers $NH_4^+$ ion as an acid and $NH_2^-$ ion as a base in liquid $NH_3$ as shown below:

| *Acid* | | *Base* | | *Acid* | | *Base* |
|---|---|---|---|---|---|---|
| $NH_4^+$ | + | $NH_2^-$ | $\rightleftarrows$ | $NH_3$ | + | $NH_3$ |

Now since $NH_4^+$ ion is an ammoniated proton, the above equation can also be written as:

$$\underbrace{NH_3 + H^+}_{\substack{NH_4^+ \\ \text{Lewis acid}}} + \underset{\text{Lewis base}}{:\bar{N}H_2} \longrightarrow NH_3 + NH_3$$

In this reaction, since $NH_2^-$ ion donates an electron pair to $H^+$ ion, $NH_2^-$ ion acs as a Lewis base and $H^+$ ion or $NH_4^+$ ion which accepts an electron pair behaves as a Lewis acid. Thus we see that $NH_4^+$ and $NH_2^-$ ions which behave as acid and base respectively in liq. $NH_3$ also behave as Lewis acid and Lewis base.

## Utility of Lewis Concept

(*i*) This concept also includes those reactions in which no protons are involved.

(*ii*) Lewis concept is more general than the Bronsted-Lowry concept (*i.e.* protonic concept) in that acid-base behaviour is not dependent on the presence of one particular element or on the presence or absence of a solvent.

(*iii*) It explains the long accepted basic properties of metallic oxides and acidic properties of non-metallic oxides.

(*iv*) This theory also includes many reactions such as gas-phase, high temperature and non-solvent reaction as neutralisation processes.

(*v*) The Lewis approach is, however, of great value in cases where the protonic concept is inapplicable, for example, in reaction between acidic and basic oxide in fused state.

## Drawback of Lewis Concept

Lewis concept suffers from the following drawbacks.

(*i*) Lewis concept is so general that it considers every reaction forming a coordinate bond to be acid-base reaction. This however, may not be always true, since according to this concept, even some metals are acids, e.g., nickel is acid because it forms the coordination compound with CO, *i.e.*, Ni (: C ≡ O :)$_4$, called nickel tetracarbony.

(*ii*) The necessary requirement in Lewis concept is the formation of a coordinate bond between the acid and base. However, the well known acids like HCl and $H_2SO_4$ do not form any coordinate bond and, therefore, should not be acids according to this concept.

(*iii*) Acid-base reactions are usually fast but formation of coordination compound is slow. Hence, it does not fit in the acid-base concept.

(*iv*) The catalytic activity of an acid is due to $H^+$ (*aq*) ion. Since the presence of hydrogen is not an essential requirement for a Lewis acid, many Lewis acids will not have this property.

(*v*) The drawback of Lewis concept compared to Arreheneins concept or Bornsted concept is that there is no simple method of measuring the relative strength of bases as electron donors or of acids as electron acceptors.

## Summary: Definition of Acids and Bases According to Different Concepts

(*i*) **Arrhenins concept (Water ion system)** An acid is a species that gives $H^+$ or $H_3O^+$ ions in aqueous solution. A base is a species that gives $OH^-$ ions in aqueous solution.

(*ii*) **Bronsted-Lowry concept (Proton donor-acceptor system).** An acid is a species which can donate a proton ($H^+$ ion) to other substance, *i.e.* an acid is a proton donor. A base is species which can accept a proton donated by an acid, *i.e.* base is a proton acceptor.

(*iii*) **Lux-Flood concept:** The species that loses (donates) an $O^{2-}$ ion behaves as a base and the species that accepts an $O^{2-}$ ion is an acid, *i.e.* a base is $O^{2-}$ ion donor and an acid is $O^{2-}$ ion acceptor.

(*iv*) **Solvent-system concept**. An acid is a substance which, by dissolution in the solvent, gives the solvent cation (acid ion) obtained by the auto-ionisation of the solvent itself. A base is a substance which, by dissolution in the solvent, gives the solvent anion (base ion) obtained by the ionisation of the solvent itself.

(*v*) **Lewis concept (Electron donor-acceptor system).** An acid is a species that can accept an electron pair, *i.e.* an acid is an electron pair acceptor. A base is a species that can donate an electron pair, *i.e.* base is an electron pair donor.

## UNIVERSITY QUESTIONS

1. Write a short note on "Arrhenius concept of acids and bases." (*Ranchi 2010*)
2. Write conjugate acids and bases of $H_2O$ and $HCO^-_3$ (*Kota 2006*)
3. (*a*) Arrange the following conjugate bases in the increasing order of their basic strength : $ClO_4^-$, $ClO^-$, $ClO_3^-$ and $ClO^-_2$.

(*b*) Arrange the following acids in the increasing order of their acidic strength: $CH_4$, $NH_3$, $H_2O$ and HF.

(*c*) Explain why $OH^-$ iom is a strong base where as $Cl^-$ ion is a weak base. (*MDSU 2006*)

**4.** (*a*) Discuss the Bronsted-Lowry acid-base principle.

(*b*) Write short note on: (*i*) Conjugate acid-base pair (*ii*) The relation of dipole moment and dielectric constant of a solvent.

(*Pd. Ravishankar Shukla 2010, Kanpur 2010*)

**5.** (*a*) Explain the Bronsted concept of acids and bases. How is it superior to Arrhenius concept? What are its limitations?

(*b*) What you understand by acid strength? (*Rajasthan 2010*)

**6.** What is Bronsted-Lowry concept of acids and bases? Classify the following as Bronsted acids and Bronsted bases? $NH_4^+$, $NO_3^-$, $H_3O^+$ and $Br^-$.

(*Mumbai 2010*)

**7.** Explain amphoteric substances? (*Pune 2009*)

**8.** The conjugate base of $[Al(H_2O)_6]^{3+}$ is ... (*H.N. Bahuguna Garhwal 2010*)

**9.** Explain Bronsted-Lowery concept of acids and bases with examples.

(*Nagpur 2010*)

**10.** What are levelling and differentiating solvents? (*Rajasthan 2010*)

**11.** Explain Lux-Flood acids and bases with examples (*Pune 2010, Nagpur 2010*)

**12.** Identify Lewis acids and Lewis bases from the following.

(*a*) $BF_3$, $NH_3$, $Ag^+$, $F^-$, $NH_4^+$, $CO_3^{2-}$ $AlCl_3$, $CH_3 - O - CH_3$.

(*b*) $PCl_5$, $CO_2$, $BF_3$, $C_5H_5N$, $Na_2SO_3$ (*R.U. 2006, Kota Uni 2006*)

**13.** (*a*) Explain conjugate acid and conjugate base.

(*b*) What is Lewis concept of acids and bases: Classify the following as Lewis acids and bases: $NH_3$, $SO_2$, $BCl_3$, $Fe^{3+}$, $SiCl_4$. (*Lucknow 2010*)

**14.** How does Lewis acid-base principle differ from Bronsted-Lowry acid-base principle ? (*Pd. Ravishankar Shukla 2010, Kanpur 2010*)

**15.** Define Lewis acids. (*Himachal Pradesh 2010*)

**16.** Write a note on Lewis concept of acids and bases. (*Ranchi 2010*)

**17.** What do you understand by the solvent system concept of acids and bases? Cite one example. (*GND Amritsar 2010*)

**18.** Give explanatory note on Usanovich's concept of acidity and basicity

(*Vidya Sagar 2012*)

**19.** Discuss with examples (*i*) Lewis concept of acids and bases (*ii*) Bronsted and Lowry concept of acids and bases. (*Ch. Charan Singh, Meerut 2012*)

**20.** (*i*) HI is stronger acid then HF. Explain.

(*ii*) Explain the concept of Lewis acids and bases. Which among the following is a Lewis acid and Lewis base: $TlCl_4$, CO, $(C_2H_5)_2O$, $Fe^{3+}$, $Cl^-$

(*iii*) Explain Lux Flood theory of acids and bases (*Lucknow 2013*)

# 4 CHAPTER Hard and Soft Acids and Bases (HSAB)

## Hard and Soft Acids and Bases

According to Lewis concept of acids and bases, a Lewis and Lewis base reaction is the combination of a Lewis acid and a Lewis base to form complex ion (adduct). In this reaction the filled orbital on Lewis base and a vacant orbital on Lewis acid combine together to form a stable complex.

Coordination chemists studied the relative reactivities of a large number of ligands (Lewis bases) with different metal ions (Lewis acids) and found that certain ligands form complexes with heavier metal ions like $Ag^+$, $Hg^{2+}$, $Pd^{2+}$, $Pt^{2+}$, etc. while other ligands form complexes with metal ions like $Be^{2+}$, $Al^{3+}$, $Ti^{4+}$, etc.

Based on the preferential tendency of ligands (Lewis bases) and metal ions (Lewis acids) to form stable complexes, the metal ions as well as ligands have been classified by Pearson (1963) into two categories each. Metal ions (Lewis acids) are categorised as **hard acids** and **soft acids,** and the ligands (Lewis bases) are categorised as **hard bases** and **soft bases.**

There are Lewis acids and Lewis bases which are neither too hard nor too soft. such acids and bases are called **border line acids** and **border line bases** respectively.

**Table 4.1. Some common hard, soft and border line Lewis acids**

| *Hard Acids (a type metal ions)* | *Soft Acids (b type metal ions)* | *Border Line Acids* |
|---|---|---|
| $H^+$, $Li^+$, $Na^+$, $K^+$, $Mn^{2+}$, $Co^{2+}$ | $Cu^+$, $Ag^+$, $Au^+$, $Tl^+$, $Hg^+$, $CH_3Hg^+$ | $Fe^{2+}$, $Co^{2+}$, $Ni^{2+}$, $Cu^{2+}$ |
| $Be^{2+}$, $Mg^{2+}$, $Ca^{2+}$, $Sr^{2+}$, $Ni^{2+}$ | $Pt^{2+}$, $Cd^{2+}$, $Pd^{2+}$, $Hg^{2+}$, $Ir^{2+}$, $Pt^{4+}$, $CH_3^+$ | $Zn^{2+}$, $Sb^{3+}$, $Bi^{3+}$, $NO^+$ |
| $Al^{3+}$, $Sc^{3+}$, $Ga^{3+}$, $In^{3+}$, $La^{3+}$ | $Ti^{3+}$, $BH_3$, $GaCl_3$, $GaBr_3$ | $SO_2$, $B(CH_3)_3$, $GaH_3$ |
| $Cr^{3+}$, $Co^{3+}$, $Fe^{3+}$, $As^{3+}$, $Ce^{3+}$ | $GaI_3$, $InCl_3$, $Cs^+$, $Rh^{2+}$ | |
| $Si^{4+}$, $Ti^{4+}$, $Zr^{4+}$, $Th^{4+}$, $Pu^{4+}$ | $I^+$, $Br^+$, $HO^+$, $RO^+$, $CH_3Cd^+$, $C_2H_5Hg^+$ | |
| $UO_2^{2+}$, $VO_2^{2+}$, $MoO^{3+}$, $Cu^{2+}$ | | |
| $BF_3$, $BCl_3$, $AlCl_3$, $AlH_3$, $Al(CH_3)_3$, $Zn^{2+}$ | $I_2$, $Br_2$, ICN | |
| $Cl^{3+}$, $Cl^{7+}$, $I^{5+}$, $I^{7+}$ | O, CI, Br, I, N | |
| $RCO^+$, $CO_2$, $NC^+$, HX | M° (Metal atom) | |
| HX(Hydrogen bonding molecules) | $CH_2$, Carbenes | |

Some common metal ions which act as hard, soft or border line acids are given in Table 4.1. According to Ahrland and Chatt hard acids are ***a*** type metal ions or acceptors while soft acids are ***b*** type metal ions or acceptors.

List of some common hard, soft and border line bases is given in Table. 4.2.

**Table 4.2. Some common hard, soft and border line bases**

| *Hard bases* | *Soft bases* | *Border line bases* |
|---|---|---|
| $NH_3$, $RNH_2$, $N_2H_4$, $H^+$, $CF_3^+$ | $H^-$, $R^-$, $C_2H_4$, $C_6H_6$, $CH_3^+$ | $C_6H_5NH_2$, $C_5H_5N$ |
| $N_2O$, $OH^-$, $O^{2-}$, ROH, $RO^-$, $R_2O$ | $CN^-$, RNC, CO, $R_2S$ | $N_2$, $N_3^-$, $NO_2^-$ |
| $CH_3COO^-$, $CO_3^{2-}$, $NO_3^-$ | $S_2O_3^{2-}$, $SCN^-$, $R_3P$ | $Br^-$ |
| $PO_4^{3-}$, $SO_4^{2-}$, $ClO_4^-$, $F^-$, $Cl^-$ | $R_3As$, $I^-$, $SH^-$, $S^{2-}$, $SO_3^{2-}$ | |

## Characteristics of Hard Acids

(*i*) Ahrland and Chatt have termed hard acid metal ions as class ***a*** metals or acceptors.

(*ii*) In general, alkali metal ions, alkaline earth metal ions, lighter and more highly charged metal ions are hard acids.

(*iii*) Hard acids have generally high positive oxidation states or high positive charge.

(*iv*) Hard acids have smaller acceptor atoms.

(*v*) Hard acids do not contain large number of valence electrons and hence their valence electrons cannot be easily distorted, *i.e.* hard acids cannot be easily polarised.

(*vi*) Hard acids have high EN.

(*vii*) Hard acids bind strongly to proton.

(*viii*) Hard acids usually form stable complexes with bases that bind well with protons.

(*ix*) The tendency of different donor atom (ligands) to form complexes with hard acids follows the order:

$$N >> P > As > Sb$$

$$O >> S > Se > Te$$

$$F > Cl > Br > I$$

## Characteristics of Soft Acids

(*i*) Ahrland and Chatt have termed soft acid metal ions as class ***b*** metals or acceptors.

(*ii*) In general soft acids are heavier transition metal ions.

(*iii*) Soft acids have large acceptor atoms.

(*iv*) Soft acids have usually low (or even zero) positive oxidation state.

(*v*) Soft acids have large number of valence electrons and therefore, their valence electrons can easily be distorted, *i.e.* soft acids can be easily polarised.

(*vi*) Soft acids have low EN.

(*vii*) Soft acids form stable complexes with bases that are highly polarisable, are good reducing agents and not necessarily good bases towards the proton.

(*viii*) The tendency of different donor atoms (ligands) to form complexes with soft acids follows the order:

$$N << P < As < Sb$$
$$O << S < Se < Te$$
$$F < Cl < Br < I$$

## Important Points

(*i*) If donor atoms of most common bases (ligands) are arranged in the increasing order of their electronegativity, the following order is obtained.

| As | P | < | C | Se | S | I | < Br | < N | Cl | < O | < F |
|---|---|---|---|---|---|---|---|---|---|---|---|
| 2.0 | 2. 1 | < | 2.5 | 2.4 | 2.5 | 2.5 | < 2. 8 | < 3.0 | 3.0 | 3.5 | 4.0 |

Soft acids will form more stable complexes with left hand donor atoms of this series and hard acids will form more stable complexes with right hand donor atoms of this series.

(*ii*) It should be noted that all hard acids are not equally hard. Similarly, all soft acids are not equally soft. For example $Cu^+$ ion is a harder acid than $Mn^{2+}$ ion because of smaller size of $Cu^+$ ion. Similarly, $Li^+$ ion is a harder acid than $Cs^+$ ion because of smaller size of $Li^+$ ion

(*iii*) Metals and non-metals in lower oxidation states behave as soft acids while the same species in higher oxidation states behave as hard acids.

## Characteristics of Hard Bases

(*i*) Hard bases are usually anions or neutral molecules.

(*ii*) Donor atoms present in hard bases have high electronegativity (EN).

(*iii*) Donor atoms of hard bases have low polarisability, *i.e.* donor atoms of hard bases cannot be easily polarised.

(*iv*) Donor atoms are hard to oxidise, *i.e.* donor atoms cannot be easily oxidised.

(*v*) Hard bases hold on to their valence electrons strongly. Thus, hardness of hard bases is associated with a highly held electron shell with little tendency to polarise.

(*vi*) Hard bases have completely filled low energy orbitals.

(*vii*) Hard bases strongly link to the proton.

(*viii*) Hard bases prefer to coordinate with hard acids.

## Characteristics of Soft Bases

(*i*) Soft bases are usually anions or neutral molecules.

(*ii*) Donor atoms in soft bases have low EN.

(*iii*) Donor atoms have high polarisability, *i.e.* donor atoms of soft bases can be easily polarised.

(*iv*) Soft bases can easily be oxidised.

(*v*) Soft bases hold on to their valence electrons rather loosely. Thus softness of soft bases is associated with a loosely bound polarisable electron shell.

(*vi*) Soft bases contain empty low energy orbitals.

(*vii*) Soft bases prefer to combine with soft acids.

It should be noted that all the bases containing the same donor atom are not equally hard. For example the hardness of $C_6H_5$ $NH_2$, $(C_6H_5)_2$ NH and $NH_3$ bases is in the order: $NH_3 > C_6H_5$ $NH_2 > (C_6H_5)_2N$. In fact, $(C_6H_5)_2$ N is a soft base.

## Characteristics of Border Line Acids and Border Line Bases

The properties (*e.g.* size, oxidation states, polarisability, etc.) of border line acids and border line bases are neither too low nor too high. These properties are intermediate between the properties of hard/soft acids and hard /soft bases.

## Pearson's HSAB Principle

In 1963 Pearson proposed a rule for predicting the feasibility of a given reaction. This rule states:

A given reaction ***is* feasible** in which a hard acid reacts with a hard base and a soft acid reacts with a soft base, *i.e.* according to HSAB principle, only that reaction is feasible (*i.e.* only that reaction proceeds from left to right) in which the products are formed by *hard acids-hard base* and *soft acid-soft base combinations.* In other words, we can say that the reaction in which the products are formed by *hard acid-soft base* or *soft acid-hard base combinations is not feasible.*

OR

A given reaction **is *feasible*** in which the products formed in the reaction are either hard acid-hard base or soft acid-soft base combinations.

OR

A given reaction **is *not feasible*** in which the products formed in the reaction are either hard acid-soft base or soft acid-hard base combinations.

**Illustrative Examples :** (*i*) The following reaction is feasible, since a hard acid ($Li^+$) combines with a hard base ($F^-$) and a soft acid ($Cs^+$) combines with a soft base ($I^-$) to form Li F (hard – hard combination) and CsI (soft – soft combination) respectively.

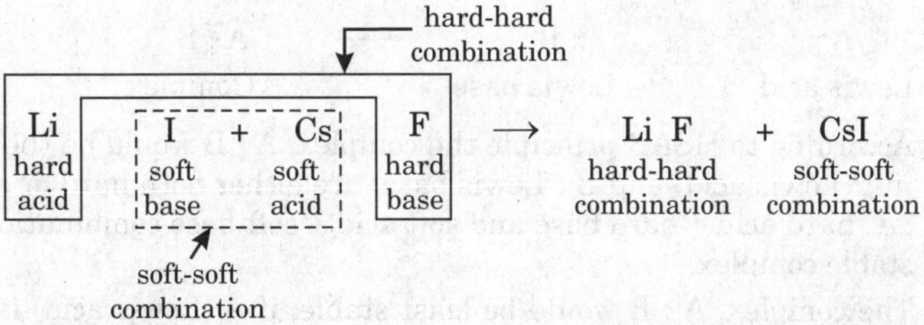


The reaction given above is feasible since the products (LiF and CsI) formed are hard acid ($Li^+$) – hard base ($F^-$) and soft acid ($Cs^+$) – soft base ($I^-$) combinations

(*ii*) The following reaction is not feasible, since the products ($HgF_2$ and $BeI_2$) formed are soft acid ($Hg^{2+}$) - hard base ($F^-$) and hard acid ($Be^{2+}$) – soft base ($I^-$) combinations

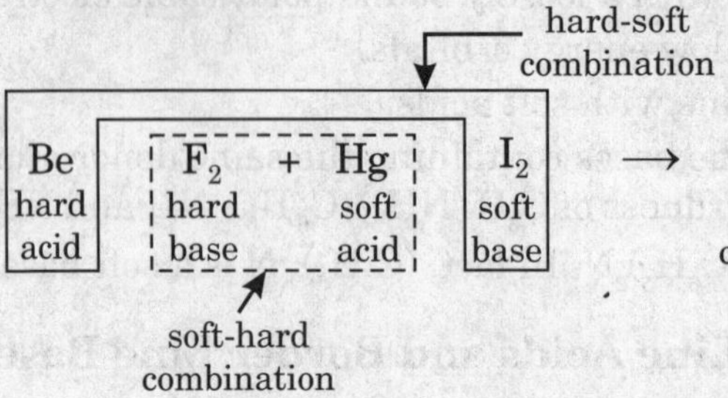


(*iii*) The following reaction is feasible, since soft acid($Ag^+$) combines with a soft base ( $I^-$) to form $AgI_2^-$ (product) which is a soft acid ($Ag^+$) – soft base ($I^-$) combination.

$$\underset{\text{Soft acid}}{Ag^+} + \underset{\text{Soft base}}{2I^-} \longrightarrow \underset{\text{Soft–soft}}{AgI_2^-} \text{ (stable)}$$

(*iv*) The following reaction is not feasible, since soft acid ($Ag^+$) combines with a hard base ($F^-$). According to HSAB principle, this type of combination does not give any product.

$$\underset{\text{Soft acid}}{Ag^+} + \underset{\text{Hard base}}{2F^-} \longrightarrow \underset{\text{Soft – hard}}{AgF_2^-} \text{ (unstable)}$$

Also, since the product ($AgF_2^-$) is a soft acid ($Ag^+$) – hard base ($F^-$) combination, the reaction is not feasible.

## Applications of HSAB Principle

HSAB principle has several applications some of which are discussed below:

**1. Relative Stability of Complexes**

Pearson (1963) has explained the relative stability of complexes on the basis of HSAB principle. Consider the formation of A : B complex by the combination of A (Lewis acid) and : B (Lewis base) as shown below:

$$\underset{\text{Lewis acid}}{A} + \underset{\text{Lewis base}}{:B} \longrightarrow \underset{\text{Complex}}{A:B}$$

According to HSAB principle the complex, A : B would be the most stable, if A (Lewis acid) and B ( Lewis base) are either both hard or are both soft, *i.e.*, hard acid + hard base and soft acid + soft base combinations give most stable complex.

The complex, A : B would be least stable, if A (Lewis acid) is hard and B (Lewis base) is soft or *vice versa*, *i.e.* hard acid + soft base and soft acid + hard base combinations give least stable complex.

**Examples** (*i*) $AgI_2^-$ is a stable complex ion but $AgF_2^-$ is an unstable complex ion. This can be explained as follows:

Since $Ag^+$ ion is soft acid and $I^-$ ion is a soft base, the combination of these ions gives stable complex ion, $AgI_2^-$

$$\underset{\text{Soft acid}}{Ag^+} + \underset{\text{Soft base}}{2I^-} \xrightarrow{\text{Soft–soft interaction}} \underset{\text{Stable complex ion}}{AgI_2^-}$$

Since $Ag^+$ ion is a soft acid and $F^-$ ion is a hard base, the combination of these ions gives unstable complex ion, $AgF_2^-$

$$\underset{\text{Soft acid}}{Ag^+} + \underset{\text{Hard base}}{2F^-} \xrightarrow{\text{soft–hard interaction}} \underset{\text{Unstable complex ion}}{AgF_2^-}$$

(*ii*) $[Co\,F_6]^{3-}$ is more stable than $[CoI_6]^{3-}$. This can be explained as follows:

$$\underset{\text{Hard acid}}{\overset{3+}{Co}} + \underset{\text{Hard base}}{6F^-} \xrightarrow{\text{Hard–hard interaction}} \underset{\text{More stable complex ion}}{[CoF_6]^{3-}}$$

$$\underset{\text{Hard acid}}{\overset{3+}{Co}} + \underset{\text{Soft base}}{6I^-} \xrightarrow{\text{Hard–soft interaction}} \underset{\text{Less stable complex ion}}{[CoF_6]^{3-}}$$

(*iii*) Complex ions of a given metal ion having a combination of soft ligands (soft bases) and hard ligands (hard bases ) are unstable while the complex ions of the same metal ion having a combination of soft ligands or hard ligands are stable.

**Example :** $[Co\,(CN)_5\,F]^{3-}$ and $[Co(NH_3)_5\,I]^{2+}$ are unstable complex ions while $[Co\,(CN)_5\,I]^{3-}$ and $[Co\,(NH_3)_5\,F]^{2+}$ are stable complex ions. This observation is due to the fact that $CN^-$ and $I^-$ both are soft ligands (soft bases) and $NH_3$ and $F^-$ both are hard ligands (hard bases).

**2. Formation of Complexes with Ambidentate Ligands**

HSAB Principle can explain the formation of complexes with ambidentate ligands. *An ambidentate ligand is a unidentate (monodentate) ligand which can coordinate to the metal ion through either of its two coordinating atoms:* For example, $SCN^-$ ion is an ambidentate ligand, since it can coordinate to the metal ion either through S-atom or through N-atom. When this ligand gets coordinated to the metal ion through S-atom, *thiocyanato complexes* are formed. When this ligand is coordinated to the metal ion through N-atom, *isothiocyanato complexes* are formed. In this ligand N-atom is a *hard base* and S-atom is a *soft base.* This ligand forms thiocyanato complexes of $[M(SCN)_4]^{2-}$ type with $Rh^{2+}$, $Ir^{2+}$, $Pd^{2+}$, $Pt^{2+}$, etc. cations (*soft acids*) by coordinating through S-atom (*soft base*) to form M – SCN bond. This ligand forms isothiocyanato complexes of $[M(NCS)_4]^{2-}$ type with $Co^{2+}$, $Ni^{2+}$, $Cu^{2+}$ and $Zn^{2+}$ cations (*hard acids*) by coordinating through N-atom (*hard base*) to form M – NCS bond.

**Example:** We have said above that $Co^{2+}$ ion forms isothiocyanato complex, $[Co(NSC)_4]^{2-}$ while $Pd^{2+}$ ion forms thiocyanate complex, $[Pd\ (SCN)_4]^{2-}$. The formation of these complexes can be explained on the basis of HSAB principle. According to this principle, since $Co^{2+}$ ion is a *hard acid,* it prefers to coordinate to *hard base N-atom* of the ligand to form $[Co\ (NSC)_4]^{2-}$ ion which contains Co – NCS bond (*hard acid-hard base interaction*). Since $Pd^{2+}$ ion is a *soft acid,* it prefers to coordinate to *soft base* S-atom of the ligand to form $[Pd\ (SCN)_4]^{2-}$ which contain Pd – SCN bond (*Soft acid-soft base interaction*).

**3. Predication of Hardness and Softness of an Acid and a Base.**

HSAB principle is useful in predicting whether a given base or an acid is hard or soft. Suppose we want to predict whether a given base B present in HB is a hard base or a soft base. For this let us consider the reaction between $CH_3Hg^+$ and HB. $CH_3Hg^+$ is a soft acid. In HB, $H^+$ ion is a hard acid and B may be a soft base or a hard base.

If B is a soft base, then the reaction between $CH_3Hg^+$ and HB can be written as:

$$\underset{\text{soft acid}}{CH_3Hg^+} + \underset{\text{hard acid}}{H^+}\underset{\text{soft base}}{B} \longrightarrow \underset{\text{soft-soft}}{CH_3Hg^+B} + \underset{\text{hard acid}}{H^+} \quad ...(i)$$

If B is a hard base, then the reaction between $CH_3\ Hg^+$ and HB can be written as:

$$\underset{\text{soft acid}}{CH_3Hg^+} + \underset{\text{hard acid}}{H^+}\underset{\text{hard base}}{B} \longrightarrow \underset{\text{soft-hard}}{CH_3Hg^+B} + \underset{\text{hard acid}}{H^+} \quad ...(ii)$$

It may be noted that the products formed in both the reactions are the same. Now let us decide which of the above two reactions is feasible. This can be decided as follows:

In reaction (*i*), since $CH_3\ Hg^+$ B (product) is formed by soft acid ($CH_3\ Hg^+$) – soft base (B) interaction, this reaction **is feasible** in accordance with HSAB principle. In other words we say that if reaction (*i*) is feasible (*i.e.* roceeds from left to right), the base , B is a **soft base.**

In reaction (*ii*), since $CH_3\ Hg^+$B (product) is formed by soft acid ($CH_3Hg^+$) – hard base (B) interaction, this reaction is **not feasible** as suggested by HSAB principle. In other words we say that if the reaction (*ii*) is not feasible (*i.e.* does not proceed from left to right), the base, B is a **hard base.**

Now let us consider reaction (*ii*) proceeding from right to left as written below:

$$\underset{\text{soft acid}}{CH_3Hg^+}\underset{\text{hard base}}{B} + \underset{\text{hard acid}}{H^+} \longrightarrow \underset{\text{soft acid}}{CH_3Hg^+} + \underset{\text{hard-hard}}{H^+B} \quad ...(iii)$$

In reaction (*iii*), since $H^+$B (product) is formed by hard acid ($H^+$) – hard base (B) interaction, this reaction **is feasible** (HSAB principle). In other words we say that if reaction (*ii*) proceeds from right to left, B is a **hard base.**

**Solved Examples.** (*i*) *Predict whether $R_2S$ is a soft base or a hard base in the following reaction which proceeds from left to right.*

$$CH_3Hg^+ + H^+R_2S \longrightarrow CH_3Hg^+SR_2 + H^+$$

**Ans:** We know that $CH_3Hg^+$ is a soft acid and $H^+$ is a hard acid. Since the reaction proceeds from left to right,the formation of $CH_3Hg^+ SR_2$ (product) should take place through soft acid ($CH_3Hg^+$) – soft base ($R_2S$) interaction (HSAB principle). For this interaction $R_2S$ must be a **soft base.** Thus the given reaction can be written as:

| | | | | | | |
|---|---|---|---|---|---|---|
| $CH_3Hg^+$ | + $H^+$ | $R_2S$ | $\longrightarrow$ | $CH_3Hg^+SR_2$ | + | $H^+$ |
| soft acid | hard acid | soft base | | soft soft | | hard acid |

(*ii*) *Predict whether $NH_3$ is a soft base or a hard base in the following reaction which proceeds from left to right.*

$$CH_3Hg^+ NH_3 + H^+ \longrightarrow NH_4^+ + CH_3Hg^+$$

**Ans.** We know that $CH_3Hg^+$ and $H^+$ are soft acid and hard acid respectively. Since this reaction proceeds from left to right, the formation of $NH_4^+$ (product) should take place through hard acid ($H^+$) – hard base ($NH_3$) interaction (HSAB principle). For this interaction $NH_3$ must be a **hard base.** Thus the given reaction can also be written as:

| | | | | | | |
|---|---|---|---|---|---|---|
| $CH_3Hg^+$ | $NH_3$ | + $H^+$ | $\longrightarrow$ | $H^+ NH_3$(or $NH_4^+$) | + | $CH_3Hg^+$ |
| soft acid | hard base | hard acid | | hard-hard | | soft acid |

**4. Predicting Feasibility of a Reaction.**

According to HSAB principle only those reactions are feasible (*i.e.* only those reactions proceed from left to right) in which products are formed by hard – hard and soft – soft combinations. The reactions in which products are formed by hard – soft and soft – hard combinations are not feasible.

**Examples:** (*i*) The following reaction proceeds from left to right.

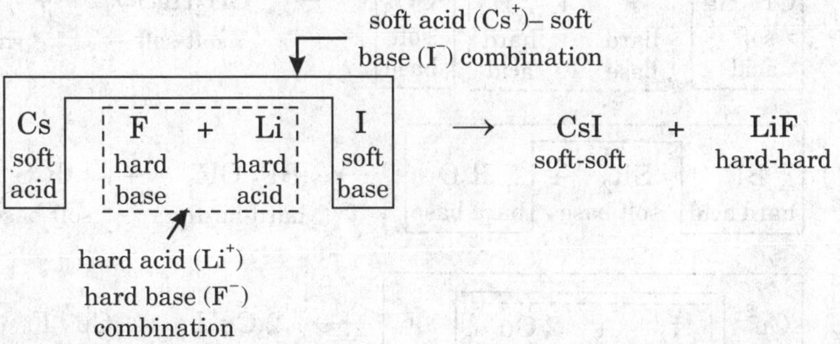


**Explanation.** In CsF, $Cs^+$ (soft acid) is combined with $F^-$ (hard base) while in LiI, $Li^+$ (hard acid) is combined with $I^-$ (soft base). CsI (soft + soft) and LiF (hard + hard) (products) are obtained by soft – soft and hard – hard combinations.

(*ii*) The following reaction proceeds in the forward direction

$$\underset{\text{soft acid}}{Hg}\underset{\text{hard base}}{F_2} + \underset{\text{hard acid}}{Be}\underset{\text{soft base}}{I_2} \longrightarrow \underset{\text{hard-hard}}{BeF_2} + \underset{\text{soft-soft}}{HgI_2}$$

**Explanation:** In $HgF_2$, $Hg^{2+}$ (soft acid) is combined with $F^-$ (hard base) while in $BeI_2$, $Be^{2+}$ (hard acid) is combined with $I^-$ (soft base). $BeF_2$ (hard-hard) and $HgI_2$ (soft-soft) (products) are obtained by hard-hard and soft-soft combinations.

(*iii*) Other examples of reactions which proceed from left to right are given below:

(*a*) $$\underset{\text{hard acid}}{BF_3}\underset{\text{soft base}}{H^-} + \underset{\text{soft acid}}{BH_3}\underset{\text{hard base}}{F^-} \longrightarrow \underset{\text{soft-soft}}{BH_4^-} + \underset{\text{hard-hard}}{BF_4^-}$$

(*b*) $$\underset{\text{soft acid}}{Cd}\underset{\text{hard base}}{CO_3} + \underset{\text{hard acid}}{Na_2}\underset{\text{soft base}}{S} \longrightarrow \underset{\text{hard-hard}}{Na_2CO_3} + \underset{\text{soft-soft}}{CdS(ppt)}$$

In this reaction the ppt. of CdS is obtained by the combination of $Cd^{2+}$ ion (soft acid) and $S^{2-}$ ion (soft base).

(*c*) $$\underset{\text{soft acid}}{C_2H_5Hg^+}\;\underset{\text{hard base}}{Cl^-} + \underset{\text{soft base}}{R_2S} \longrightarrow \underset{\substack{\text{soft-soft}\\\text{(complex ion)}}}{[C_2H_5Hg(R_2S)]^+} + Cl^-$$

In this reaction, $C_2H_5Hg^+$ (soft acid) combines with $R_2S$ (soft base) to form the complex ion, $[C_2H_5Hg(R_2S)]^+$.

(*d*) $$\underset{\text{soft acid}}{CH_3Hg}\underset{\text{hard base}}{F} + \underset{\text{hard acid}}{H}\underset{\text{soft base}}{SO_3^-} \longrightarrow \underset{\text{soft-soft}}{CH_3HgSO_3^-} + \underset{\text{hard-hard}}{HF}$$

(*e*) $$\underset{\text{hard acid}}{BF_3}\;\underset{\text{soft base}}{SR_2} + \underset{\text{hard base}}{R_2O} \longrightarrow \underset{\text{hard-hard}}{BF_3\,OR_2} + \underset{\text{soft base}}{R_2S}$$

(*f*) $$\underset{\text{hard acid}}{Cu^{2+}}\underset{\text{soft base}}{I_2} + \underset{\text{soft acid}}{2\,Cu^+}\underset{\text{hard base}}{F} \longrightarrow \underset{\text{soft-soft}}{2\,Cu^+I} + \underset{\text{hard-hard}}{Cu^{2+}F_2}$$

**5. Direction of Reaction**

Consider the following reactions

(*a*) $\underset{\text{hard acid}}{H^+} + \underset{\text{soft acid}}{CH_3Hg^+}\ \underset{\text{hard base}}{OH^-} \longrightarrow \underset{\text{hard -hard}}{H^+\ OH^-}\ (\text{or } H_2O) + \underset{\text{soft acid}}{CH_3Hg^+}$

(*b*) $\underset{\text{hard acid}}{H^+} + \underset{\text{soft acid}}{CH_3Hg^+}\ \underset{\text{soft base}}{SH^-} \longrightarrow \underset{\text{hard–soft}}{H^+SH^+}\ (\text{or } H_2S) + \underset{\text{soft acid}}{CH_3Hg^+}$

In reaction (*a*), since $H^+$ $OH^-$ or $H_2O$ (product) is obtained by hard acid ($H^+$) – hard base ($OH^-$) interaction, this reaction proceeds from left to right.

In reaction (*b*), since $H^+$ $SH^-$ or $H_2S$ (product) is obtained by hard acid ($H^+$) – soft base ($SH^-$) interaction which is against HSAB principle, this reaction proceeds from right to left.

**6. Reaction of HgO and HgS with HCl**

HgO reacts with HCl to form $HgCl_2$ and $H_2O$ but HgS does not react with HCl.

**Explanation.** The reaction of HgO and HgS with HCl can be shown as follows:

(*a*) $\underset{\text{soft acid}}{Hg^{2+}}\ \underset{\text{hard base}}{O^{2-}} + \underset{\text{hard acid}}{2H^+}\ \underset{\text{soft base}}{Cl^-} \longrightarrow \underset{\text{soft–soft}}{HgCl_2} + \underset{\text{hard–hard}}{2H^+\,O^{2-}\text{or } H_2O}$

(*b*) $\underset{\text{soft acid}}{Hg^{2+}}\ \underset{\text{soft base}}{S^{2-}} + \underset{\text{hard acid}}{2H^+}\ \underset{\text{soft base}}{Cl^-} \longrightarrow \underset{\text{soft-soft}}{HgCl_2} + \underset{\text{hard-soft}}{2H^+S^{2-}\text{ or } H_2S}$

In reaction (*a*), since $HgCl_2$ (product) is obtained by soft acid [$Hg^{2+}$] – soft base ($Cl^-$) interaction and $2H^+$ $O^{2-}$ or $H_2O$ (another product) is obtained by hard acid ($H^+$) – hard base ($O^{2-}$) interaction, both the interactions are in accordance with HSAB principle. Thus reaction (*a*) is feasible, *i.e.* HgO reacts with HCl and forms $HgCl_2$ and $H_2O$.

On the other hand, in reaction (*b*), since one product ($HgCl_2$) is obtained by soft acid ($Hg^{2+}$) – soft base ($Cl^-$) interaction and the other product ($H_2S$) is formed by hard acid ($H^+$) – soft base ($S^{2-}$) interaction, reaction (*b*) is not feasible, *i.e.* HgS does not react with HCl.

**7. Order of Stability of Hydrogen Halides, HX (X = F, Cl, Br, I).**

In HX molecule, $H^+$ ion is a *hard acid*. Halide ions ($X^-$) are *hard bases*. The hardness of halide bases *decreases in the order* $F^- > Cl^- > Br^- > I^-$. This order shows that $F^-$ ion is the *hardest ion* and $I^-$ ion is the *least hard* (*i.e. softest*) ion. Hence HF which is a hard acid ($H^+$) – hard base ($F^-$) combination is the *most stable*. On the other hand , since HI is a *hard acid* ($H^+$) – *soft base* ($I^-$) combination, HI is the *least stable*. Thus, the order of stability of HX molecules is as:

$$HF > HCl > HBr > HI$$

**8. Occurence of Metals in Nature**

We know that some metals occur as carbonates, oxides and halides in nature while some metals occur as sulphides in nature. Now:

(*i*) Since $Mg^{2+}$, $Ca^{2+}$, $Mn^{2+}$, $Al^{3+}$, $Fe^{3+}$, $Cr^{3+}$, etc. metal ions are *hard acids*, these metal ions prefer to combine with $CO_3^{2-}$ $O^{2-}$ and $X^-$ (halide) ions which are *hard bases* (*hard acid – hard base interaction*) and thus these metal ions occur as *carbonates, oxides* and *halides* in nature.

(*ii*) Since $Cu^+$, $Ag^+$, $Pb^{2+}$, $Hg^{2+}$, $Pd^{2+}$, etc. metal ions are *soft acids*, these metal ions prefer to combine with $S^{2-}$ ion which is a *soft base* (*soft acid - soft base interaction*) and thus occur as sulphides in nature.

(*iii*) Since $Ni^{2+}$, $Fe^{2+}$, $Zn^{2+}$, $Pb^{2+}$, $Co^{2+}$, etc. metal ions are *border line acids*, these metal ions occur in nature as *oxides* or *carbonates* (hard bases) as well as sulphides (*soft bases*)

It should be clear from the above discussion that since $Mg^{2+}$, $Ca^{2+}$ and $Al^{3+}$ cations are *hard acids* and $S^{2-}$ ion is a *soft base,* these cations are never found as sulphides in nature.

**9. Piosoning of Metal Catalysts.**

There are some metals like Ni, Pt, Pd, Cr, Mo, etc. (*soft acids*) which act as catalysts. These metal catalysts can strongly adsorb on their surface the substances like CO, unsaturated hydrocarbons (olefins, arenes, dienes, alky alkylenes), phosphine and arsine containing ligands which are *soft bases.* The adsorption of soft bases on the surface of the metal catalysts is called poisoning of the metal catalysts, since the adsorption of the ligands by the metals catalysts blocks the active sites of the metal catalysts. The adsorption of soft bases (*i.e.* poisoning of the metal catalysts) takes place through soft acid (metal catalysts) – soft base (ligands) interaction. Since metal catalysts are soft acids, these are not affected by hard bases or ligands containing N, O or F.

## Theoretical Basis of Hardness and Softness of Acids and Bases

Several theories have been put forth to explain the basis of hard acid – hard base and soft acid – soft base interactions. However, no single theory is completely satisfactory. These interactions may be of the following types:

**1. Hard acid - hard base electrostatic (ionic) interaction.** According to the most accepted view, the hard acid – hard base interaction is an electrost atic (ionic) interaction, *i.e.* the bonding between hard acid and hard base is predominantly ionic . For example, the interaction between hard acids like $Li^+$, $Na^+$, $K^+$, etc. and hard bases like $OH^-$, $F^-$, $O^{2-}$, etc. gives ionic compounds.

Now we know that if both the ions (*i.e.* positive and negative ions) of an ion pair are smaller in size, the inter-nuclear distance between such ions would also be smaller and hence the attraction between such two ions would be greater. Consequently, the compound resulted from hard acid – hard base interaction would be highly stable.

**2. Soft acid - soft base covalent interaction.** We know that most of the soft acids are transition metal ions which contain 6 to 10 electrons in their *d*-orbitals. These *d*-electrons of the metal ions (soft acids) have polarising power and soft bases like $S^{2-}$, $I^-$, etc. are easily polarisable.

As a result, bonding between soft acids and soft bases is largely covalent (Recall covalent bonding is established between $Li^+$ and $I^-$ ions due to polarisation of $I^-$ ion by $Li^+$ ion)

3. **π-bonding in soft acid – soft base interaction.** In soft acid – soft base interactions, π-bonding plays an important role. Soft acids are generally metals in low oxidation states and have a tendency to form pi bonds with soft bases. Soft bases are good pi bonding ligands.

## Miosons Relation

Miosons and coworkers (1967) have proposed a relation with the help of which we can predict whether a given species is hard or soft. This relationship is:

$$pK = -\log K = aX + bY + C$$

where K = equilibrium constant for the dissociation of metal-ligand (*i.e.* acid-base) complex. X and Y = parameters for the metal ions (acids). These are called *metal* or *acid parameters*. $a$ and $b$ = parameters for the ligands (*i.e.* bases). These are called *ligand parameters*. $C$ = a constant required to adjust the value of $pK$ in such a way that all of them lie in the same scale.

Values of acid parameter (Y) for some metal ions which may be hard acids as well as soft bases are given below:

*Hard acids:* $Li^+ = 0.36$, $Al^{3+} = 0.70$, $Mg^{2+} = 0.87$, $Na^+ = 0.93$, $Ca^{2+} = 1.62$, $Fe^{3+} = 2.37$, $Ca^{2+} = 2.56$, $Cs^+ = 2.73$, $Co^{2+} = 2.93$

*Soft acids:* $Sn^{2+} = 3.17$, $Tl^{3+} = 3.2$, $Cu^+ = 3.45$, $Pb^{2+} = 3.58$, $Tl^+ = 3.78$, $Hg^{2+} = 4.25$, $Au^+ = 5.95$

If the value of Y is less than 2.8, the acid is hard and if the value of Y is more than 3.2, the acid is soft.

The value of $b$ (ligand parameter) increases as we move from hard bases to soft bases, *i.e.* with the increase in the value of $b$, the softness of bases increases as shown below.

| $OH^-$ | $NH_3$ | $Cl^-$ | $Br^-$ | $I^-$ | $S_2O_3^{2-}$ |
|---|---|---|---|---|---|
| 0.40 | 1.04 | 2.49 | 5.58 | 7.17 | 12.4 |
| Hard bases | | | Soft bases | | |

Softness of bases increases ⟶

If the value of $b$ is less than 3, the base is hard and if the value of $b$ is more than 5,the base is soft.

## Acid-Base Strength and Hardness and Softness

Hardness or softness of an acid or a base has no relationship with acidic strength of an acid and basic strength of a base. For example: (*i*) $OH^-$ and $F^-$ both are *hard bases* but $OH^-$ ion is much stronger (about $10^{13}$ times) base than $F^-$ ion. (*ii*) $SO_3^{2-}$ and $Et_3P$ both are *soft bases* but $Et_3P$ is much stronger (about $10^7$ times) than $SO_3^{2-}$ ion towards $CH_3Hg^+$ which is a *soft acid*. (*iii*) $SO_3^{2-}$ ion is a stronger base than $F^-$ ion (*iv*) $OH^-$ ion is much stronger than $SO_3^{2-}$ ion.

There are many reactions which proceed from left to right, but these reactions cannot be explained on the basis of HSAB principle, since in these reactions the products *are not formed* by hard-hard or soft-soft interactions, rather the products are formed either by hard-soft or soft-hard interaction. Such reactions are explained on the basis of relative basic strength of bases involved in these reactions.

**Examples:** (*i*) The following reaction proceeds from lefts to right.

$$\underset{\substack{\text{soft base}\\\text{(stronger base)}}}{SO_3^{2-}} + \underset{\text{hard acid}}{H^+} \underset{\substack{\text{hard base}\\\text{(weaker base)}}}{F^-} \longrightarrow \underset{\text{hard–soft}}{HSO_3^-} + \underset{\text{hard base}}{F^-}$$

The formation of $HSO_3^-$ (product) cannot be explained on the basis of HSAB principle, since the product ( $HSO_3^-$ ) is formed by hard acid ($H^+$) – soft base ($SO_3^{2-}$) interaction. According to HSAB principle $HSO_3^-$ (product) should have been formed either by hard acid-hard base or by soft acid-soft base interaction.

**How to explain this reaction?** The feasibility of the above reaction is explained on the basis of relative basic strength of $SO_3^{2-}$ and $F^-$ bases. Since $SO_3^{2-}$ ion is stronger base than $F^-$ ion, $SO_3^{2-}$ ion displaces $F^-$ ion from HF and forms $HSO_3^-$ which is a hard acid ($H^+$) – soft base ( $SO_3^{2-}$ ) – product

(*ii*) The following reaction proceeds from left to right.

$$\underset{\substack{\text{hard base}\\\text{(stronger base)}}}{OH^-} + \underset{\text{soft acid}}{CH_3Hg^+} \underset{\substack{\text{soft base}\\\text{(weaker base)}}}{SO_3^{2-}} \longrightarrow \underset{\text{soft–hard}}{CH_3Hg^+OH^-} + \underset{\text{soft base}}{SO_3^{2-}}$$

The formation of $CH_3HgOH$ (product) cannot be explained on the basis of HSAB principle, since the product ($CH_3HgOH$) is formed by soft acid ($CH_3Hg^+$) – hard base ($OH^-$) interaction. According to HSAB principle. $CH_3HgOH$ (product) should have been formed either by hard acid – hard base or by soft acid – soft base interaction.

**How to explain this reaction?** The feasibility of the above reaction is explained on the basis of relative basic strength of $OH^-$ and $SO_3^{2-}$ bases.

Since $OH^-$ ion is stronger base than $SO_3^{2-}$ ion, this ion displaces $SO_3^{2-}$ ion from $CH_3HgSO_3^-$ and $CH_3HgOH$ which is a soft acid ($CH_3$ $Hg^+$) – hard base ($OH^-$) product.

(*iii*) The reaction

$$\underset{\text{soft acid}}{CH_3Cd} \quad \underset{\substack{\text{hard base}\\\text{(stronger base)}}}{OH} + \underset{\substack{\text{soft base}\\\text{(weaker base)}}}{SO_3^{2-}} \not\longleftrightarrow \underset{\text{soft–soft}}{CH_3Cd\,SO_3^-} + \underset{\text{hard base}}{OH^-}$$

*is not feasible* (*i.e.* this reaction does not occur in the forward direction), since $SO_3^{2-}$ base is weak base than OH ion and hence $SO_3^{2-}$ ion is not able to displace $OH^-$ ion from $CH_3Cd$ OH to form $CH_3Cd$ $SO_3^-$ (product).

We can conclude from the above discussion that Lewis acid – Lewis base reactions can be explained by two independent principles which depend on (*i*) softness and hardness of acids and bases, and (*ii*) acidic strength of acids and basic strength of bases.

## Variation of Hardness and Softness of Cations in a Group

We know that the size of the cations of the elements of a group having the same charge increases with the increase of atomic number. Due to the increase in the size, the hardness of cations decreases and softness increases. For example:

(*a*) $Li^+ > Na^+ > K^+ > Rb^+ > Cs^+$

Hardest (Least soft) … Least hard (Softest)

(*b*) $Cu^+ > Ag^+ > Au^+$

Hardest (Least soft) … Least hard (Softest)

## Variation of hardness and softness of cations in a period

When we move from left to right in a period, the charge on cations increases and hence hardness of cations increases. For example:

$Na^+ < Mg^{2+} < Al^{3+} < Si^{4+}$

least hard (softest) … hardest (least soft)

Thus in a period, alkali metal cations are the softest.

## Cations with High Charge are Hard Acids

The cations with high charge like $Mn^{7+}$, $Cr^{6+}$, $Os^{8+}$, $Pt^{6+}$, $I^{7+}$, $S^{6+}$ etc. are hard acids. This nature of these cations is confirmed by the fact that these cations combine with hard bases like $O^{2-}$ and $F^-$ and form stable compounds and complex ions. Examples of compounds and complex ions are $MnO_4^-$, $CrO_4^{2-}$, $OsO_4$, $PtF_6$, $IF_7$ and $SF_6$

## Transition Metal Ions in their Low Oxidation States are Soft Acids

Transition metal ions in their low oxidation state act as soft acids and hence combine with soft bases like CO, CNR, $PR_3$, etc. to form stable complexes. Examples of some stable complexes are Ni $(CO)_4$, $Cr(CO)_6$, $Ni(CNR)_4$, $Ni(PCl_3)_4$, Fe $(CO)_5$ etc.

## Many Soft Acids are Toxic

Many soft acids are toxic, For example:

(*i*) $Hg^{2+}$ and $CH_3Hg^+$ which are soft acids can bind with crystein S in enzymes and can thus destroy the enzymetic functions.

(*ii*) $Pb^{2+}$ and $Cd^{2+}$ ions which are soft border line acids are quite toxic, since these metal ions react with enzyme-SH groups and thus spoil the enzyme function.

## Limitations of HSAB Principle

(*i*) HSAB principle is unable to explain the reactions which take place through *hard acid- soft base* and *soft acid- hard base* interactions. For example, consider the following reactions which take place spontaneously in the forward direction.

(*a*) $SO_3^{2-} + H^+ F^- \longrightarrow HSO_3^- + F^-$

| $SO_3^{2-}$ | $H^+$ | $F^-$ | $HSO_3^-$ | $F^-$ |
|---|---|---|---|---|
| soft base (stronger base) | hard acid | hard base (weaker base) | hard-soft | hard base |

(*b*) $OH^- + CH_3Hg\ SO_3^- \longrightarrow CH_3HgOH + SO_3^{2-}$

| $OH^-$ | $CH_3Hg$ | $SO_3^-$ | $CH_3HgOH$ | $SO_3^{2-}$ |
|---|---|---|---|---|
| hard base (stronger base) | soft acid | soft base (weaker base) | soft-hard | soft base |

On the basis of HSAB principle, reaction (*a*) should not be expected to proceed from left to right, since the formation of $HSO_3^-$ (product) does not take place through soft acid-soft base or hard acid-hard base interaction. The formation of $HSO_3^-$ occurs through hard acid ($H^+$) and soft base $\left(SO_3^{2-}\right)$ combination which is against HSAB principle. The occurrence of this reaction can be explained as follows: $SO_3^{2-}$ ion is a *stronger soft base* and $F^-$ ion is a *weaker hard base* ($SO_3^{2-} > F^-$). Being stronger, $SO_3^{2-}$ ion (stronger base) displaces $F^-$ ion (weaker base) from HF to form $HSO_3^-$.

On the basis of HSAB principle, reaction (*b*) should not proceed from left to right, since the formation of $CH_3$ HgOH (product) does not take place through soft acid-soft base or hard acid-hard base interaction. The formation of $CH_3HgOH$ occurs through soft acid ($CH_3\ Hg^+$) – hard base ($OH^-$) combination which is against HSAB principle. The occurence of this reaction can be explained as follows:

$OH^-$ ion is a *stronger hard base* and $SO_3^{2-}$ ion is a *weaker soft base* ($OH^- > SO_3^{2-}$). Being stronger, $OH^-$ ion (stronger base) displace $SO_3^{2-}$ ion (weaker base) from $CH_3\ HgSO_3^-$ to form $CH_3HgOH$.

It should be clear from the above discussion that the relative strength of bases ($SO_3^{2-} > F^-$ and $OH^- > SO_3^{2-}$) plays an important role in explaining the occurrence of these reactions. HSAB principle cannot be used to explain these reactions.

(*ii*) These are many reaction which, according to HSAB principle, should proceed from left to right. Actually these reactions do not take place in the forward direction.

**Examples** (*a*) According to HSAB principle, the reaction:

$CH_3Cd\ OH + SO_3^{2-} \longrightarrow CH_3Cd\ SO_3^- + OH^-$

| $CH_3Cd$ | $OH$ | $SO_3^{2-}$ | $CH_3Cd\ SO_3^-$ | $OH^-$ |
|---|---|---|---|---|
| soft acid | hard base (stronger base) | soft base (weaker base) | soft-soft | hard base |

should proceed from left to right to produce $CH_3CdSO_3^-$ as shown above by soft acid ($CH_3Cd^+$) - soft base ($SO_3^{2-}$) combination. As a matter of fact this reaction does not occur to give $CH_3CdSO_3^-$. The non-occurence of this reaction is explained on the basis of relative strength of $OH^-$ and $SO_3^{2-}$ bases. Since $SO_3^{2-}$ ion is weaker base than $OH^-$ ion, this ion (*i.e.,* $SO_3^{2-}$ ion) is not able to displace stronger $OH^-$ base from $CH_3$ Cd OH to produce $CH_3CdSO_3^-$

(*b*) According to HSAB principle, the reaction:

| $CH_3^+$ | $+ H^+$ | $H^-$ | $\longrightarrow$ | $CH_3^+H^-$ or $CH_4$ | $+ H^+$ |
|---|---|---|---|---|---|
| soft acid (weaker acid) | hard base (stronger base) | soft base | | soft-soft | hard acid |

should proceed from left to right to produce $CH_4$ as shown above by soft acid ($CH_3^+$) -soft base ($H^-$) combination. As a matter of fact this reaction does not occur due to its unfavourable entropy change (= + 359 kJ /mole) to give $CH_4$, since $CH_3^+$ is a weaker acid than $H^+$ and hence is not able to displace stronger $H^+$ acid from $H^+ H^-$ to produce $CH_4$.

## Symbiosis

The term symbiosis was introduced by Jorgenson in 1868.

According to Jorgenson, soft ligands have a tendency to combine with a centre (metal ion) which is already linked with soft ligands. Similarly hard ligands have a tendency to combine with a centre which is already associated with hard ligands. Thus the term symbiosis indicates the tendency of ligands to combine with a centre (metal ion) which is already linked with ligands.

**Examples:** Let us consider the following reactions:

(*i*) $BF_3 + F^- \longrightarrow BF_4^-$

In $BF_3$ molecule, B-atom is associated with three hard ligands ($F^-$ ions). Due to symbiosis tendency of $F^-$ ion (hard ligand), this ligand (*i.e.* $F^-$ ion) prefers to combine with $F^-$ ions (hard ligands) present in $BF_3$ and thus forms $BF_4^-$ ion. Thus the formation of $BF_4^-$ ion can be shown as:

$$\underset{\substack{\uparrow \\ \text{hard} \\ \text{ligands}}}{BF_3} + \underset{\substack{\text{hard} \\ \text{ligand}}}{F^-} \longrightarrow BF_3F^- \text{ or } BF_4^-$$

(*ii*) $BH_3 + H^- \longrightarrow BH_4^-$

Since H-atoms present in $BH_3$ molecule are soft ligands and $H^-$ ion is also a soft ligand, $BF_4^-$ ion is formed

$$\underset{\substack{\uparrow \\ \text{soft} \\ \text{ligands}}}{BH_3} + \underset{\substack{\text{soft} \\ \text{ligand}}}{H^-} \longrightarrow BH_3H^- \text{ or } BH_4^-$$

(*iii*) $BF_3H^- + BH_3F^- \longrightarrow BF_4^- + BH_4^-$

In $BF_3$, B-atom is linked with three hard ligands ($F^-$ ions) and $H^-$ present in $BF_3H^-$ is a soft ligand. In $BH_3$, B-atom is linked with three soft ligands (H-atoms) and $F^-$ ion present in $BH_3F^-$ is a hard ligand. In this reaction, $F^-$ ion (hard ligand) present in $BH_3F^-$ prefers to combine with B-atom of $BF_3$ which is already linked with F-atoms (hard ligands) to form $BF_4^-$. Similary $H^-$ ion (soft ligand) present in $BF_3H^-$ prefers to combine with B-atom of $BH_3$ which is already assoicated with H-atom (soft ligand). Thus, the formation of $BF_4^-$ and $BF_4^-$ ions can be shown as:

B $F_3$ (↑ hard ligands) | $H^-$ (soft ligand) + B $H_3$ (↑ soft ligand) | $F^-$ (hard ligand) $\longrightarrow$ $BF_3F^-$ (hard-hard) + $BH_3H^-$ (soft)

or $BF_3H^- + BH_3F^- \longrightarrow BF_4^- + BH_4^-$

(*iv*) Fluroinated methanes ($CH_3F$ and $CHF_3$) react in the same way as $BF_3H^-$ and $BH_3F^-$ react together as shown above:

$CH_3$ (↑ soft ligand) F (↑ hard ligand) + $CF_3$ (↑ hard ligand) H (↑ soft ligands) $\longrightarrow$ $CH_3H$ (soft – soft) + $CF_3F$ (hard – hard)

or $CH_3F + CF_3H \longrightarrow CH_4 + CF_4$

(*v*) $BF_3 + NH_3 \longrightarrow F_3B.NH_3$ (adduct)

In $BF_3$ molecule, B-atom is associated with three hard ligands (F atoms) and $NH_3$ molecule is also a hard ligand. Hence, $NH_3$ (hard ligand) gets associated with B-atom of $BF_3$ molecule to form $F_3B.NH_3$ adduct.

$BF_3$ (↑ hard ligands) + $NH_3$ (↑ hard ligand) $\longrightarrow$ $F_3B.NH_3$ (Adduct) (hard hard)

## Relation of Electronegativity (EN) of Acids and Bases with Their Hardness and Softness

Hardness and softness of acids and bases are related with their EN values. In general, a species having high EN is hard and the species with low EN value is soft. For example *since $Li^+$, $Na^+$, etc. ions have high EN, these ions are hard acids*. On the other hand, since *transition metal ions in low oxidation states like $Cu^+$, $Ag^+$. $Au^+$, etc. have low EN, those ions are soft acids.* On similar basis the hardness and softness of bases can also be explained.

The relationship between hardness/softness of acids and bases and their EN values helps us to explain the fact that $CF_3$ group is harder than $CH_3$ group and $BF_3$ molecule is harder than $BH_3$ molecule. This is because the presence of highly electronegative groups makes the central atom more electronegative.

## UNIVERSITY QUESTIONS

**1.** Explain the application of HSAB principle in predicting the feasibility of a reaction and stability of complexes with suitable examples.
(*Osmania 2010*)

**2.** (*a*) What do you understand by hard-soft acid-base principle. On the basis of this principle explain stability of complex complexes.
(*b*) Write note on: Relation of hardness-softness with electronegativity
(*Pd. Ravishankar Shukla 2010)*

**3.** (*a*) What in symbiosis? Give the process of $BF_4^-$ complex formation from $BF_3$.
(*b*) What is HSAB principle? Write important uses with examples.
(*c*) Give five examples each of hard and soft acids.
(*d*) Classify hard and soft acids and bases and write their properties along with five examples of each. (*Kanpur 2010)*

**4.** (*i*) Explain the following on the basis of HSAB principle.
(*a*) Reaction between $CdSO_4$ and $Na_2S$
(*b*) $AgI_2^-$ ion is more stable than $AgF_2^-$ ion
(*c*) $CoF_6^{3-}$ ion is more stable than $CoI_6^{3-}$ ion.
(*ii*) State the direction in which the following reaction takes place
$BeI_2 + HgF_2 \rightarrow BeF_2 + HgI_2$
(*iii*) Explain HSAB principle. Discuss its uses.
(*iv*) Classify the following as hard acids or as hard bases. $H^+$, $F^-$, $Cl^-$, $NH_3$, $N_2H_4$, $Al^{3+}$, $Ca^{2+}$, $BF_3$ and $Be^{2+}$.
(*v*) Describe symbiosis.
(*vi*) Mention two examples of each of hard, soft and borderline acids.
(*Pune 2009*)

**5.** (*a*) What are hard and soft bases? Explain HSAB principle with suitable examples.
(*b*) Explain: (*i*) Trimethylamine is more basic than $NH_3$. (*ii*) $NF_3$ is much weaker base than $NH_3$. (*iii*) Cl-OH is an acid whereas Na-OH is base.
(*c*) What do you understand by labelling effect of solvent?
(*Himachal Pradesh 2010)*

**6.** Explain "Pearson's principle of hard and soft acids and bases. Give any three of its applications. (*Mumbai 2010*)

**7.** (*a*) Explain why
(*i*) $BF_3$ is a hard acid but $BH_3$ is a soft acid
(*ii*) Ca and Mg are present in nature in the form of carbonates and Cu and Ag are present in the form of sulphides
(*iii*) $[CoF_6]^{3-}$ is more stable than $[CoI_6]^{3-}$
(*iv*) $[Rh(NH_3)_5(SCN)]^{2+}$ is less stable than $[Rh(NH_3)_5(NCS)]^{2+}$
(*v*) HgO is soluble in HCl but HgS is insoluble.

(*b*) Explain HSAB concept involved in the separation of $Cu^{2+}$ and $Cd^{2+}$ in qualitative analysis.
(*c*) What are the charactersitics of hard and soft acids and bases?
(*d*) How is electronegativity related with hardness and softness of acids and bases?
(*e*) What are the limitations of HSAB theory? (*Rasjasthan 2010*)

**8.** (*a*) Explain Person's HSAB principle.
(*b*) Out of $I^{5+}$ and $I^{+}$, which one is soft acid?
(*H.N. Bahuguna Garhwal 2010*)

**9.** (*a*) Define and explain symbiosis with reference to fluoride and hydride.
(*b*) Discuss the contribution of $\pi$ bonding in soft and soft base interaction.
(*c*) How does electronegativity affect hardness and softness?

**10.** Explain Pearson's HSAB principle. (*Andhra 2010*)

**11.** (A) Explain with examples: (*i*) Hard and soft acids, and (*ii*) symbiosis.
(B) Ca exists in nature in the form of $CaCO_3$, but not as CaS. Explain.
(*Nagpur 2010*)

**12.** Explain Pearson's HSAB concept of acids and bases giving suitable examples.
(*Utkal 2010*)

**13.** (*a*) " Hard acids prefer to bind to hard bases and soft acids prefer to bind and soft bases" Explain it with reason and suitable examples.
(*b*) Classify the following as hard/soft acids and bases:
$Na^+$, $Ag^+$, $BF_3$, carbenes, $H_2O$, CO , $NO_3^-$,$S_2O_3^{2-}$, $H^-$, $I^-$, $Ca^{2+}$, $CN^-$,
(*c*) Discuss Pearson's HSAB concept with examples.
(*b*) Explain the theoretical basis of hardness and softness.
(*BJB college, Orissa, Nagpur 2012*)

**14.** (*a*) What do you understand by Pearson's HSAB concept?
(*b*) What are hard and soft acids and bases? Explain in detail.
(*c*) Explain symbiosis. (*G.N.D. Amritsar 2010, Nagpur 2012*)

**15.** Describe Pearson's concept for hard and soft acids and bases. Use it (*i*) to explain why do small and highly charged transition metal ions do not form metal carbonyls and (*ii*) to predict the direction of the following reaction:
$$BeI_2 + HgF_2 \longrightarrow HgI_2 + BeF_2$$
(*Lucknow 2010*)

**16.** (*a*) Explain HSAB principle of acids and bases. What type of bonds are formed in hard-hard and soft-soft combinations? Discuss the applications of HSAB principle with suitable examples.
(*b*) Arrange the following acids in the decreasing order of their hardness, giving proper reasons: $H^+$, $Be^{2+}$, $Cu^{2+}$ and $Ag^+$ (*Ch. Charan Singh, Meerut 2012*)

**17.** (*i*) On the basis of HSAB principle explain that $CH_3COF$ is more stable than $CH_3COCl$.
(*ii*) Describe the main characteristics of Pearson's ASAB principle. Explain the following on the basis of this concept: (*a*) Stability of complex compounds (*b*) Poisoning of catalysts (*c*) Occurence of metals in nature.
(*Lucknow 2013*)

# 5 CHAPTER

# General Characteristics of Transition Metals (*d*- Block Elements)

## Introduction

The elements form $Sc_{21}$ to $Zn_{30}$ (ten elements), $Y_{39}$, to $Cd_{48}$ (10 elements) $La_{57}$, $Hf_{72}$ to $Hg_{80}$ (ten elements), $Ac_{89}$, Rf $_{104}$ to $Cn_{112}$ (ten elements) are called transition metals. Since the last electron in the atoms of these elements enters *d*-sub-shell belonging to penultimate shell, [*i.e.* (*n*-1)th shell], these elements are also called *d*-block elements. Thus in these elements the last electron goes to $(n-1)\,d$ sub-shell.

## Position of *d*-block Elements in the Periodic Table

The position of *d*- block elements in the periodic table has been shown in shaded portion in shaded portion in Fig. 5.1.

*s*-Block Elements ←→ | ← *p*-Block Elements →

| *n* ↓ | IA 1 | | | | | | | | | | | | | | | | | Zero 18 |
|---|---|---|---|---|---|---|---|---|---|---|---|---|---|---|---|---|---|---|
| 1 | 1 H | IIA 2 | | | | | | | | | | | IIIA 13 | IVA 14 | VA 15 | VIA 16 | VIIA 17 | 2 He |
| 2 | 3 Li | 4 Be | ← *d*-Block Elements (TransitionElements)→ | | | | | | | | | | 5 B | 6 C | 7 N | 8 O | 9 F | 10 Ne |
| 3 | 11 Na | 12 Mg | IIIB 3 | IVB 4 | VB 5 | VIB 6 | VIIB 7 | ← VIII → 8 | 9 | 10 | IB 11 | IIB 12 | 13 Al | 14 Si | 15 P | 16 S | 17 Cl | 18 Ar |
| 4 | 19 K | 20 Ca | 21 Sc | 22 Ti | 23 V | 24 Cr | 25 Mn | 26 Fe | 27 Co | 28 Ni | 29 Cu | 30 Zn | 31 Ga | 32 Ge | 33 As | 34 Se | 35 Br | 36 Kr |
| 5 | 37 Rb | 38 Sr | 39 Y | 40 Zr | 41 Nb | 42 Mo | 43 Tc | 44 Ru | 45 Rh | 46 Pd | 47 Ag | 48 Cd | 49 In | 50 Sn | 51 Sb | 52 Te | 53 I | 54 Xe |
| 6 | 55 Cs | 56 Ba | 57-71 La* | 72 Hf | 73 Ta | 74 W | 75 Re | 76 Os | 77 Ir | 78 Pt | 79 Au | 80 Hg | 81 Ti | 82 Pb | 83 Bi | 84 Po | 85 At | 86 Rn |
| 7 | 87 Fr | 88 Ra | 89-103 Ac** | 104 Rf | 105 Db | 106 Sg | 107 Bh | 108 Hs | 109 Mt | 110 Ds | 111 Rg | 112 Cn | (113) | (114) | (115) | (116) | (117) | (118) |

← f-Block Elements [Group III B (3)] →

| | | | | | | | | | | | | | | |
|---|---|---|---|---|---|---|---|---|---|---|---|---|---|---|
| Lanthanides* | 58 Ce | 59 Pr | 60 Nd | 61 Pm | 62 Sm | 63 Eu | 64 Gd | 65 Tb | 66 Dy | 67 Ho | 68 Er | 69 Tm | 70 Yb | 71 Lu |
| Actinide** | 90 Th | 91 Pa | 92 U | 93 Np | 94 Pu | 95 Am | 96 Cm | 97 Bk | 98 Cf | 99 Es | 100 Fm | 101 Md | 102 No | 103 Lr |

**Fig 5.1** Position of d-block elements in the periodic table.

From this figure the following points may be noted:

(*i*) *d*-block elements lie in between *s*- and *p*- block elements *i.e.* these elements are located in the middle of the periodic table.

(*ii*) *d*-block elements are present in *4th* ($n$= 4), *5th* ($n$= 5), *6th* ($n$ = 6) and *7th* ($n$= 7) periods. Each period has ten elements. The elements present in these periods are given below:

*4th* period ($n$ = 4) : $S_{C_{21}}$ to $Zn_{30}$
*5th* period ($n$ = 5) : $Y_{39}$ to $Cd_{48}$
*6th* period ($n$ = 6) : $La_{57}$, $Hf_{72}$ to $Hg_{80}$
*7th* period ($n$ = 7) : $Ac_{89}$, $Rf_{104}$ to $Cn_{112}$

(*iii*) *d*-block elements are present in IIIB (3), IVB (4), VB (5), VIB (6), VIIB (7), VIII (8, 9, 10), IB (11) and IIB (12) groups.

## Electronic Configurations and Definition

The electronic configurations of the atoms of *d* -block elements are given in Table 5.1. In this table $[Ar]_{18}$, $[Kr]_{36}$, $[Xe]_{54}$ and $[Rn]_{86}$ indicate the electronic configurations of Ar, Kr, Xe and Rn respectively, which are given below:

$[Ar]_{18}$ = 2,8, 8 (Three - shells)
$[Kr]_{36}$ = 2,8, 18, 8 (Four shells)
$[Xe]_{54}$ = 2,8, 18, 18,8 (Five shells)
$[Rn]_{86}$ = 2, 8, 18, 32, 18, 8 (Six shells)

The configurations given in Table 5.1 show that:

(*i*) *d*-block elements can be defined as those elements in which the last electron (differentiating electron) enters ($n$ – 1) *d* orbitals (*i.e. d*- orbitals of the penultimate or last but one shell) or in which ($n$–1) *d* orbitals are progressively filled up with electrons.

(*ii*) *d*- block elements are also defined as those elements whose two outer-most shells are incomplete (*i.e.* partially-filled).

(*iii*) Although most the transition elements have electronic configurations as proposed by aufbau principle of order of energy of different orbitals, yet there are many transition elements whose observed V.S.E. Cs. are different from their expected configurations. Examples of such elements are given below:

| *Series* | *Elements* | *Expected V.S. E.C.* | *Observed V.S. E.C.* |
|---|---|---|---|
| 1st (3*d* series) | $Cr_{24}$ | $3d^4\ 4s^2$ | $3d^5\ 4s^1$ |
| | $Cu_{29}$ | $3d^9\ 4s^2$ | $3d^{10}\ 4s^1$ |
| 2nd (4d series) | $Nb_{41}$ | $4d^3\ 5s^2$ | $4d^4\ 5s^1$ |
| | $Mo_{42}$ | $4d^4\ 5s^2$ | $4d^5\ 5s^1$ |
| | $Ru_{44}$ | $4d^6\ 5s^2$ | $4d^7\ 5s^1$ |
| | $Rh_{45}$ | $4d^7\ 5s^2$ | $4d^8\ 5s^1$ |
| | $Pd_{46}$ | $4d^8\ 5s^2$ | $4d^{10}\ 5s^0$ |
| | $Ag_{47}$ | $4d^9\ 5s^2$ | $4d^{10}\ 5s^1$ |
| 3rd (5d series) | $Pt_{78}$ | $4f^{14}\ 5d^8\ 6s^2$ | $4f^{14}\ 5d^9\ 6s^1$ |
| | $Au_{79}$ | $4f^{14}\ 5d^9\ 6s^2$ | $4f^{14}\ 5d^{10}\ 6s^1$ |
| 4th (6d series) | $Rg_{111}$ | $5f^{14}\ 6d^9\ 7s^2$ | $5f^{14}\ 6d^{10}\ 7s^1$ |

**Table 5.1. Electronic configurations of d-block (transition) elements. The elements shown in shaded portion have anomalous configurations.**

| | III B 3 | IV B 4 | V B 5 | VI B 6 | VII B 7 | ← VIII → 8 | 9 | 10 | I B 11 | II B 12 |
|---|---|---|---|---|---|---|---|---|---|---|
| **3d (1st) series (4th period)** | | | | | | | | | | |
| $[Ar]_{18}$ + | 21 Sc $3d^1$ $4s^2$ | 22 Ti $3d^2$ $4s^2$ | 23 V $3d^3$ $4s^2$ | 24 Cr $3d^5$ $4s^1$ | 25 Mn $3d^5$ $4s^2$ | 26 Fe $3d^6$ $4s^2$ | 27 Co $3d^7$ $4s^2$ | 28 Ni $3d^8$ $4s^2$ | 29 Cu $3d^{10}$ $4s^1$ | 30 Zn $3d^{10}$ $4s^2$ |
| **4d (2nd) series (5th period)** | | | | | | | | | | |
| $[Kr]_{36}$ + | 39 Y $4d^1$ $5s^2$ | 40 Zr $4d^2$ $5s^2$ | 41 Nb $4d^4$ $5s^1$ | 42 Mo $4d^5$ $5s^1$ | 43 Tc $4d^5$ $5s^2$ | 44 Ru $4d^7$ $5s^1$ | 45 Rh $4d^8$ $5s^1$ | 46 Pd $4d^{10}$ $5s^0$ | 47 Ag $4d^{10}$ $5s^1$ | 48 Cd $4d^{10}$ $5s^2$ |
| **5d (3rd) series (6th period)** | | | | | | | | | | |
| $[Xe]_{54}$ + | 57 La $4f^0$ $5d^1$ $6s^2$ | 72 Hf $4f^{14}$ $5d^2$ $6s^2$ | 73 Ta $4f^{14}$ $5d^3$ $6s^2$ | 74 W $4f^{14}$ $5d^4$ $6s^2$ | 75 Re $4f^{14}$ $5d^5$ $6s^2$ | 76 Os $4f^{14}$ $5d^6$ $6s^2$ | 77 Is $4f^{14}$ $5d^7$ $6s^2$ | 78 Pt $4f^{14}$ $5d^9$ $6s^1$ | 79 Au $4f^{14}$ $5d^{10}$ $6s^1$ | 80 Hg $4f^{14}$ $5d^{10}$ $6s^2$ |
| **6d (4th) series (7th period)** | | | | | | | | | | |
| $[Rn]_{86}$ + | 89 Ac $5f^0$ $6d^1$ $7s^2$ | 104 Rf $5f^{14}$ $6d^2$ $7s^2$ | 105 Db $5f^{14}$ $6d^3$ $7s^2$ | 106 Sg $5f^{14}$ $6d^4$ $7s^2$ | 107 Bh $5f^{14}$ $6d^5$ $7s^2$ | 108 Hs $5f^{14}$ $6d^6$ $7s^2$ | 109 Mt $5f^{14}$ $6d^7$ $7s^2$ | 110 Ds $5f^{14}$ $6d^8$ $7s^2$ | 111 Rg $5f^{14}$ $6d^{10}$ $7s^1$ | 112 Cn $5f^{14}$ $6d^{10}$ $7s^2$ |

The observed V.S.E. Cs of the above elements show that $Cr_{24}$, $Cu_{29}$, $Nb_{41}$, $Mo_{42}$, $Ru_{44}$, $Rh_{45}$, $Ag_{47}$, $Pt_{78}$, $Au_{79}$, and $Rg_{111}$ (10 elements) have only one electron in *ns* orbital and $Pd_{46}$ has no electron in its *ns* orbital.

**Explanation:** The irregularities in electronic configurations can be explained on the basis of the fact that in these elements (*n*–1) *d* subshell and *ns* subshell have very small difference in energy. Hence the incoming electron may enter into *ns* or (*n* –1) *d* subshell. In some cases the irregular (abnormal) electronic configurations are due to nucleus -electron and electron-electron forces.

Since the V.S.E.Cs of $Cr_{24}$ and $Mo_{42}$ have half-filled 3d/4d orbitals, these configurations are more stable than their expected configurations. Similarly, since V.S.E.Cs of $Cu_{29}$, $Pd_{46}$, $Ag_{47}$, $Rg_{111}$ and $Au_{79}$ have completely filled 3*d*/ 4*d*/ 5*d*/ 6*d* orbitals, these configurations are more stable than their expected configurations. Half-filled and completely-filled *d*-orbitals (*i.e.* $d^5$ and $d^{10}$ configurations) are more stable than other *d*- orbitals because the distribution of 5 and 10 electrons in *d* orbitals is symmetrical in $d^5$ and $d^{10}$ configurations. *d*- orbitals having symmetrical distribution of 5 and 10 electrons have lower energy and hence are more stable.

(*iv*) In $La_{57}$ one electron goes to 5d orbitals before filling 4f orbitals. This is against aufbau order. $La_{57}$ has no electron in 4f orbitals but all the nine elements *viz.* $Hf_{72}$ to $Hg_{80}$ which are placed after $La_{57}$ in 5*d* series (6th period) have 14 electrons in 4*f* orbitals ($4f^{14}$ configuration). In $Ac_{89}$ one electron goes to 6*d* orbitals before 5*f* orbitals. This is also against anfbau order. $Ac_{89}$ has no electron in 5*f* orbitals but all the nine elements *viz.* $Rf_{104}$ to $Cn_{112}$ which are placed after $Ac_{89}$ in 6*d* series (7*th* period) have 14 electrons in 5*f* orbitals ($5f^{14}$ configuration).

(*v*) Though Cu, Ag and Au have completely filled *d*-orbitals, yet these elements are transition elements.

(*vi*) Zn, Cd and Hg are generally not considered as transition elements, since these elements in their common oxidation state of +2 have completely-filled *d*- orbitals ($Zn^{2+} = 3d^{10}$, $Cd^{2+} = 4d^{10}$ and $Hg^{2+} = 5d^{10}$). Also these elements do not show properties of transition elements to any appreciable extent except for their ability to form complexes. In order to maintain a rational classification of elements, they are generally studied with the *d*-block elements.

(*vii*) General electronic configurations of the atoms of the elements namely $Sc_{21}$ to $Zn_{30}$ (ten elements), $Y_{39}$ to $Cd_{48}$ (ten elements), $La_{57}$ and $Ac_{89}$ can be represented as [Noble gas] $(n–1)\,d^{1–10}\,ns^{1,2}$ ($n$ = 4,5, 6 or 7). General electronic configuration of the elements from $Hf_{72}$ to $Hg_{80}$ (nine elements) and from $Rf_{104}$ to $Cn_{112}$ (nine elements) can be represented as [Noble gas] $(n-2)\,f^{14}$ $(n-1)d^{\,2–10}\,ns^2$ ($n$ = 6 or 7). Here [Noble gas] represents the electronic configuration of the noble gas present in the previous period.

## Classifications of *d*-block Elements in 3*d*, 4*d*, 5*d* and 6*d* Series (Four Series)

The electronic configurations of *d*-block elements as given in Table 5.1 show that depending on whether the last electron enters 3*d*, 4*d*, 5*d* or 6*d* orbitals, these elements can be grouped into the following four series:

1. **3*d*-series (1st series: 4th period).** This series contains ten elements *viz.* $Sc_{21}$ to $Zn_{30}$. These elements are present in 4th period. In the atoms of these elements the last electron goes to 3d orbitals, *i.e.,* in this series 3*d* orbitals are progressively filled up with electrons as we move from $Sc_{21}$ to $Zn_{30}$. It may be noted that the configurations of $Cr_{24}$ and $Cu_{29}$ (two elements) are anomalous, since $Cr_{24}$ has 5 electrons (instead of 4) in 3*d* orbitals and Cu has 10 electrons (instead of 9) in these orbitals. Thus the correct electronic configurations of $Cr_{24}$ and $Cu_{29}$ are $[Ar]_{18}\,3d^5\,4s^1$ [instead of $[Ar]_{18}\,3d^4\,4s^2$] and $[Ar]_{18}\,3d^{10}\,4s^1$ [instead of $[Ar]_{18}\,3d^9\,4s^2$] respectively. Complete and valence-shell electronic configurations of the atoms of 3*d*-series elements can be written as follows: Here $[Ar]_{18}$ = 2,8, 8 (*three shells*).

*Complete configuration* = $[Ar]_{18}\ 3d^{1-10}\ 4s^{1-2}$
= 2, 8, (8 +1 to 10), 1 or 2 (four shells)
= 2, 8, ( 9 to 18), 1 or 2
= 2, 8, $3s^2p^6\ d^{1-10}$, $4s^{1,2}$
*Valence-shell configuration* = $3d^{1-10}\ 4s^{1,2}$

2. **4d-series (2nd series:5th period).** Like 3*d*-series, this series also has ten elements namely $Y_{39}$ to $Cd_{48}$. These elements are present in 5th period. In this series, the differentiating electron occupies 4*d* orbitals, *i.e.*, the elements of this series involve the progressive filling of 4*d* orbitals as we proceed from $Yb_{39}$ to $Cd_{49}$. In this series there are more elements which have anomalous configurations. The elements having anomalous configurations are $Nb_{41}$, $Mo_{42}$, $Ru_{44}$, $Rh_{45}$, $Pd_{46}$ and $Ag_{47}$ (*six elements*). These anomalous configurations are explained on the basis of nuclear-electron and electron-electron forces existing in these atoms. Complete and valence-shell electronic configurations of the atoms of 4*d*- series elements can be written as follows. Here $[Kr]_{36}$ = 2, 8, 18, 8 (*four shells*).

   *Complete configuration* = $[Kr]_{36}\ 4d^{1-10}\ 5s^{0-2}$
   = 2, 8, 18, (8 + 1 to 10), 0 to 2 (*five shells*)
   = 2, 8, 18, ( 9 to 18), 0 to 2
   = 2, 8, 18, $4s^2p^6\ d^{1-10}$, $5s^{0-2}$
   *Valence-shell configuration* = $4d^{1-10}$, $5s^{0-2}$

3. **5*d*-series (3rd series: 6th period).** Like 3*d* and 4*d* series, this series also consists of *ten* elements which are $La_{57}$ (one element) and $Hf_{72}$ to $Hg_{80}$ (nine elements). The elements of this series involve the gradual filling of 5*d* orbitals. In between $La_{57}$ and $Hf_{72}$, there are 14 elements *viz*. $Ce_{58}$ to $Lu_{71}$ which are called *lanthanides* or *lanthanones*. These 14 elements involve the progressive filling of 4*f* orbitals and hence do not belong to 5*d* series. Thus at $Lu_{71}$, 4*f* orbitals are completely-filled. Consequently at $La_{57}$, 4f orbitals are vacant ($4f^0$ configuration) while in the remaining nine elements ($Hf_{72}$ to $Hg_{80}$) 4*f* orbitals are completely-filled ($4f^{14}$ configuration). The elements namely $Pt_{78}$ and $Au_{79}$ (*two elements*) have anomalous configurations. Complete and valence-shell electronic configurations of the atoms of 5*d*- series elements can be written as follows. Here $[Xe]_{54}$ = 2, 8, 18, 18, 8 (*five shells*)

   *Complete configuration of* $La_{57}$ = $[Xe]_{54}\ 4f^0\ 5d^1\ 6s^2$
   = 2, 8, 18, (18 + 0), (8 +1), 2
   = 2, 8, 18, 18, 9, 2 ( *Six shells*)
   ∴ Valence shell configuration of $La_{57}$ = $4f^0\ 5d^1\ 6s^2$

   *Complete configuration of the elements from* $Hf_{72}$ *to* $Hg_{80}$ = $[Xe]_{54}\ 4f^{14}\ 5d^{2-10}\ 6s^{1,2}$
   ∴ *Valence shell configuration of the elements from* $Hf_{72}$ *to* $Hg_{80}$ = $4f^{14}5d^{2-10}\ 6s^{1,2}$

4. **6*d*- series (4th series: 7th period).** This series has ten elements which are $Ac_{89}$ (one element) and $Rf_{104}$ to $Cn_{112}$ (nine elements). These elements involve the gradual filling of 6*d* orbitals. In between $Ac_{89}$ and $Rf_{104}$ there are 14 elements *viz*. $Th_{90}$ to $Lr_{103}$ which are called *actinides* or *actinones*. These 14

elements involve the progressive filling of 5*f* orbitals and hence do not belong to 6*d* series. Thus at $Lr_{103}$ 5*f* orbitals are completely-filled. Consequently at $Ac_{89}$, 5*f* orbitals are vacant ($5f^0$ configuration) while in the remaining elements *viz*. from $Rf_{104}$ to $Cn_{112}$ (nine elements), 5*f* orbitals are completely-filled ($5f^{14}$ configuration). Complete and valence-shell configurations of the atoms of 6*d*- series elements can be written as follows. Here $[Rn]_{86}$ = 2,8, 18,32, 18, 8 (*six shells*)

*Complete configurations of* $Ac_{89} = [Rn]_{86} 5f^0 6d^1 7s^2$
= 2, 8, 18, 32, (18 + 0), (8 + 1), 2
= 2, 8, 18, 32, 18, 9, 2

∴ V.S. configuration of $Ac_{89} = 5f^0 6d^1 7s^2$

*Complete configuration of the elements from* $Rf_{104}$ *to* $Cn_{112} = [Rn]_{86} 5f^{14} 6d^{2-10} 7s^{1,2}$

∴ *V.S. configuration of the elements from* $Rf_{104}$ *to* $Cn_{112} = 5f^{14} 6d^{2-10} 7s^{1,2}$

## Electronic Configuration of *d*-block Elements in their Respective Groups

1. **Group 3 (III B) : $Sc_{21}$, $Y_{39}$, $La_{57}$ and $Ac_{89}$.** V.S.E.Cs of these elements can be represented as $(n-1)\,d^1 ns^2$ (*n* = 4, 5, 6 or 7). For $Sc_{21}$, $Y_{39}$, $La_{57}$ and $Ac_{89}$, *n* = 4, 5, 6 and 7 respectively.
2. **Group 4 (IV B) : $Ti_{22}$, $Zr_{40}$, $Hf_{72}$ and $Rf_{104}$.** V.S.E.Cs of $Ti_{22}$ and $Zr_{40}$ can be represented as $(n-1)\,d^2\,ns^2$ (*n* = 4 or 5) while V.S.E. Cs of $Hf_{72}$ and $Rf_{104}$ can be represented as $(n-2)\,f^{14}\,(n-1)\,d^2\,ns^2$ (*n* = 6 or 7). For $Ti_{22}$, $Zr_{40}$, $Hf_{72}$ and $Rf_{104}$, *n* = 4,5 6 ad 7 respectively.
3. **Group 5 (V B) : $V_{23}$, $Nb_{41}$, $Ta_{73}$ and $Db_{105}$.** V.S.E. Cs of $V_{23}$ and $Nb_{41}$ can be represented as $(n-1)d^{3,4}\,ns^{1,2}$ (*n* = 4 or 5) while that of $Ta_{73}$ and $Db_{105}$ can be represented as $(n-2)f^{14}\,(n-1)d^3\,ns^2$ (*n* = 6 or 7). For $V_{23}$, $Nb_{41}$, $Ta_{73}$ and $Db_{105}$, *n* = 4, 5, 6 and 7 respectively.
4. **Group 6 (VI B) : $Cr_{24}$, $Mo_{42}$, $W_{74}$ and $Sg_{106}$.** V.S.E. C of $Cr_{24}$ and $Mo_{42}$ can be represented as $(n-1)\,d^5\,ns^1$ (*n* = 4 or 5) while that of $W_{74}$ and $Sg_{106}$ can be represented as $(n-2)f^{14}\,(n-1)\,d^4\,ns^2$ ( *n*= 6 or 7). The value of *n* for these elements is as *n* = 4 for $Cr_{24}$, *n*= 5 for $Mo_{42}$, *n* = 6 for $W_{74}$ and *n* = 7 for $Sg_{106}$.
5. **Group 7 (VII B) : $Mn_{25}$, $Tc_{43}$, $Re_{75}$ and $Bh_{107}$.** V.S.E. C of the atoms of $Mn_{25}$ and $Tc_{43}$ can be represented as $(n-1)\,d^5\,ns^2$ (*n* = 4 for Mn and *n*= 5 for Tc) while that of $Re_{75}$ and $Bh_{107}$ can be represented as $(n-2)\,f^{14}\,(n-1)\,d^5\,ns^2$ (*n* = 6 for Re and *n* = 7 for Bh).
6. **Groups, 8, 9, and 10 (VIII).** Group VIII consists of the following three transition trials.

   *Group number:* 8 9 10
   *Ferrous metals*: $Fe_{26}$, $Co_{27}$ and $Ni_{28}$
   *Light platinum metals*: $Ru_{44}$, $Rh_{45}$ and $Pd_{46}$
   *Heavy platinum metals*: $Os_{76}$, $Ir_{77}$ and $Pt_{78}$

   The elements of the three triads have more horizontal similarities than vertical similarities. Hence the three elements of each triad are considered as the members of the same group. The configurations of these elements given in Table 5.1 show that $Ru_{44}$, $Rh_{45}$ and $Pt_{78}$ have one electron in their 5*s*, 5*s* and 6*s* orbital respectively while $Pd_{46}$ has no electron in its 5*s* orbital.

7. **Group 11 (IB): $Cu_{29}$, $Ag_{47}$, and $Au_{79}$.** V.S.E.C of the atoms of $Cu_{29}$ and $Ag_{47}$ can be represented as $(n\text{–}1)d^{10}\ ns^1$ ($n$= 4 for Cu and $n$= 5 for Ag) while that of the atom of $Au_{79}$ is represented as $4f^{14}\ 5d^{10}\ 6s^1$.
8. **Group 12 (IIB): $Zn_{30}$, $Cd_{48}$ and $Hg_{80}$.** V.S.E.C of the atoms of all these elements can be represented as $(n-1)\ d^{10}\ ns^2$ ( $n$ = 4 for *Zn*, $n$ = 5 for *Cd* and $n$ = 6 for Ag)

## General Characteristics

The transition elements show several properties. Some of these properties and their trends of variation are discussed below:

### I. Atomic Radii

Atomic radii of the first three series of *d*-block elements are given in Table 5.2

**Table 5.2. Atomic radii (pm) of d-block elements**

| | | | | | | | | | |
|---|---|---|---|---|---|---|---|---|---|
| **3d (1st) series (4th period)** | | | | | | | | | |
| IIIB 3 | IVB 4 | VB 5 | VIB 6 | VIIB 7 | ⟵ VIII ⟶ 8 | 9 | 10 | IB 11 | IIB 12 |
| $Sc_{21}$ 162 | $Ti_{22}$ 147 | $V_{23}$ 134 | $Cr_{24}$ 127 | $Mr_{25}$ 126 | $Fe_{26}$ 126 | $Co_{27}$ 125 | $Ni_{28}$ 124 | $Cu_{29}$ 128 | $Zn_{30}$ 138 |
| **4d (2nd) series (5th period)** | | | | | | | | | |
| $Y_{39}$ 180 | $Zr_{40}$ 160 | $Nb_{41}$ 146 | $Mo_{42}$ 139 | $Tc_{43}$ 136 | $Ru_{44}$ 134 | $Rh_{45}$ 134 | $Pd_{46}$ 137 | $Ag_{47}$ 144 | $Cd_{48}$ 154 |
| **5d (3rd) series (6th period)** | | | | | | | | | |
| $La_{57}$ $Ce_{58}$ – Lu 187 165 156 | $Hf_{72}$ 158 | $Ta_{73}$ 146 | $W_{74}$ 139 | $Re_{75}$ 137 | $Os_{76}$ 135 | $Ir_{77}$ 136 | $Pt_{78}$ 138 | $Au_{79}$ 144 | $Hg_{80}$ 157 |

(*i*) **Variation of atomic radii in a given series (period).** Following points may be noted:

(*a*) *The atomic radii of the transition metals lie in between those of s-and p-block elements.*

(*b*) *Generally the atomic radii of d-block elements in a series decrease with increase in atomic number but the decrease in atomic size is small after midway*. For example, for elements of 1st transition series the atomic radii decrease gradually from Sc to Mn but from Fe to Cu these values remain practically constant.

**Explanation:** For the elements from Sc to Mn the atomic radii decrease. This decrease is because of the gradual increase in nuclear charge with the increase in atomic number. The increased nuclear charge makes the atom to shrink in size and hence the size of the atom decreases. However, since the electrons added to $3d$ orbitals screen the $4s$ electron ($s$), the attraction between the nucleus and the $4s$ electron (s) decreases, *i.e.,* due to the screening effect caused by $3d$ electrons the magnitude of nuclear charge decreases and consequently the atomic radii for the elements from Fe to Cu remain almost uncharged.

(*c*) *At the end of the period, there is a slight increase in the atomic radii.* For example, atomic radius of Zn is higher than that of Cu.

**Explanation:** Towards the end of each series, there is an increase in electron-electron repulsion between the electrons being added to $3d$ orbitals. This increase in repulsion becomes greater than that of the attraction between the nucleus and the $4s$ electron($s$). Because of the greater magnitude of electron-electron repulsion, the electron cloud of Zn expands and hence its size becomes greater than that of Cu.

(*d*) *There is no major change in atomic radii on going from Fe to Cu.*

**Explanation:** Upto Mn ($3d^5$ $4s^2$) electron-electron repulsions do not come into picture because there is no pairing of electrons in $3d$ orbitals. However, in case of the elements from Fe to Zn, the pairing of electrons in $3d$ orbitals takes place and hence electron-electron repulsions are produced. Now the increase in electron-electron repulsion makes the atom to *expand* in size while the nuclear charge makes the atom to *shrink* in size. In the elements from Fe to Cu, these opposing effects almost counterbalance each other and hence there is no major change in size on going from Fe to Cu (similarly from Ru to Ag in 2nd series and from Os to Au in 3rd series). In the last elements of each period, *viz.*, Zn, Cd and Hg, the electron-electron repulsion becomes greater than the attraction between the nucleus and outermost $s$ electrons (*i.e.* nuclear charge) and, therefore, there is an increase in size.

(*ii*) **Variation of atomic radii in a given group.** The covalent radii of the atoms of the elements of group III B (Sc, Y and La) increase (as expected) from top to bottom as the atomic number increases. When we compare these values for the elements of other groups of transition elements, it is found that these values for the elements of 2nd transition series are higher (as expected) than those for the elements of 1st transition series, but these values for the elements of 3rd transition series are almost equal to those for the elements of 2nd transition series. The similarity in the covalent radii for the elements of 2nd and 3rd series is due to *lanthanide contraction* which takes place among lanthanides ($Ce_{58}$ to $Lu_{71}$=14 elements). These 14 lanthanides are present between $La_{57}$ (group IIIB) and $Hf_{72}$ (group IVB) in 3rd transition series.

Due to the similarity in the covalent radii of the elements of 2nd and 3rd transition series, these elements show close similarity in their properties while the elements of 1st and 2nd transition series do not exhibit so close similarity. For example, the solubility of the salts of the elements of 2nd and 3rd transition series are very close and hence the separation of these elements by precipitation method is difficult. Examples of the pairs of elements which show similar properties and hence are difficult to separate are Zr — Hf, Nb — Ta, Mo — W, Ru — Os, Rh — Ir, Pd — Pt and Ag—Au. The above discussion shows that due to lanthanide contraction the elements of 2nd and 3rd transition series resemble each other much more than the elements of 1st and 2nd transition series.

**Table 5.3. Ionic radii (pm) of the common cations derived from d-block elements**

| $M^{n+}$ ion | III B 3 | IV B 4 | V B 5 | VI B 6 | VII B 7 | ← VII B → 8 | 9 | 10 | I B 11 | II B 12 |
|---|---|---|---|---|---|---|---|---|---|---|
| **3d (1st) series (4th period)** | | | | | | | | | | |
| | $Sc_{21}$ | $Ti_{22}$ | $V_{23}$ | $Cr_{24}$ | $Mn_{25}$ | $Fe_{26}$ | $Co_{27}$ | $Ni_{28}$ | $Cu_{29}$ | $Zn_{30}$ |
| $M^+$ | – | – | – | – | – | – | – | – | 91 | – |
| $M^{2+}$ | 81 | 91 | 88 | 84 | 80 | 76 | 74 | 72 | 69 | 74 |
| $M^{3+}$ | 88.5 | 76 | 74 | 69 | 66 | 64 | 63 | 63 | – | – |
| $M^{4+}$ | – | 74.5 | 72 | 68 | 60 | – | – | – | – | – |
| $M^{5+}$ | – | – | 68 | 63 | – | – | – | – | – | – |
| $M^{6+}$ | – | – | – | 58 | – | – | – | – | – | – |
| **4d (2nd) series (5th period)** | | | | | | | | | | |
| | $Y_{39}$ | $Zr_{40}$ | $Nb_{41}$ | $Mo_{42}$ | $Tc_{43}$ | $Ru_{44}$ | $Rh_{45}$ | $Pd_{50}$ | $Ag_{47}$ | $Cd_{48}$ |
| $M^+$ | – | – | – | – | – | – | – | – | 129 | – |
| $M^{2+}$ | – | – | – | – | – | – | – | 50 | 108 | 97 |
| $M^{3+}$ | 90 | – | 86 | 83 | – | 82 | 69 | 90 | 89 | – |
| $M^{4+}$ | – | 80 | 82 | 79 | 78.5 | 76 | 74 | 75.5 | – | – |
| $M^{5+}$ | – | – | 70 | 75 | 74 | 70 | 69 | – | – | – |
| $M^{6+}$ | – | – | – | 73 | – | – | – | – | – | – |
| $M^{7+}$ | – | – | – | – | 70 | – | – | – | – | – |
| **5d (3rd) series (6th period)** | | | | | | | | | | |
| | $La_{57}$ | $Hf_{72}$ | $Ta_{73}$ | $W_{74}$ | $Re_{75}$ | $Os_{76}$ | $Ir_{77}$ | $Pt_{78}$ | $Au_{79}$ | $H_{80}$ |
| $M^+$ | – | – | – | – | – | – | – | – | 137 | 133 |
| $M^{2+}$ | – | – | – | – | – | – | – | 94 | – | 110 |
| $M^{3+}$ | 106 | – | – | – | – | – | 82 | – | 99 | – |
| $M^{4+}$ | – | 80 | 86 | 80 | 77 | 77 | 76.5 | 76.5 | – | – |
| $M^{5+}$ | – | – | 73 | 76 | 72 | 71 | 71 | 71 | 71 | – |
| $M^{6+}$ | – | – | – | 68 | 69 | – | – | – | – | – |
| $M^{7+}$ | – | – | – | – | 67 | – | – | – | – | – |

## II. Ionic Radii

Ionic radii of the common cations derived from transition elements have been listed in Table 5.3.

From Table 5.3 the following points may be noted:

(*i*) Since the transition metals show many oxidation states (variable oxidation states), the radii of the ions in different oxidation states are also different.

(*ii*) In general the ionic radii of different cations of the same element in different oxidation states decrease with the increase in oxidation state. For example, $Ti^{2+} > Ti^{3+} > Ti^{4+}$ ; $Cr^{2+} > Cr^{3+} > Cr^{4+} > Cr^{5+} > Cr^{6+}$ ; $Cu^{+} > Cu^{2+}$ etc.

(*iii*) Ionic radii of the cations of different elements in the same oxidation state generally decrease with the increase in nuclear charge or atomic number. Thus the ionic radii of the bivalent cations derived from the elements of 1st transition series decrease as we move from left to right in the series, *i.e.* $Ti^{2+} > V^{2+} > Cr^{2+} > Mn^{2+} > Fe^{2+} > Co^{2+} > Ni^{2+} > Cu^{2+}$.

(*iv*) These values for the cations of the elements of group III B (*i.e.* $Sc^{3+}$, $Y^{3+}$ and $La^{3+}$) increase as $Sc^{3+} < Y^{3+} < La^{3+}$. These values for the cations derived from most of the elements of 3rd transition series are almost the same as those of the corresponding cations derived from the elements of 2nd transition series. The similarity in their ionic radii is due to the presence of 14 lanthanides ($Ce_{58}$ to $Lu_{71}$) between $La_{57}$ (group III B) and $Hf_{72}$ (group IV B).

## III. Metallic Character and Related Properties

All the transition elements show metallic character, *i.e.*, all these elements are metals.

**Explanation.** Since the transition elements have low values for their ionisation energies, *ns* electrons can easily be lost by them to form $M^{+}$ or $M^{2+}$ cations and thus these elements show metallic character.

As usual, the metallic character of these elements increases on descending the group, *e.g.*, basic character of the pentaoxides of the elements of group V B ($V_2O_5$, $Nb_2O_5$ and $Ta_2O_5$) increases as $V_2O_5 < Nb_2O_5 < Ta_2O_5$. Thus $V_2O_5$ is amphoteric (but more basic than acidic) while $Nb_2O_5$ and $Ta_2O_5$ are more basic than acidic.

The metallic character of the transition elements is evident from the following properties of these elements.

(*i*) **Electrical and thermal conductivity.** These elements are good conductors of electricity and heat because of the existence of metallic bonding (*i.e.,* presence of free mobile electrons) among the atoms of these elements. Cu, Ag and Au show exceptionally high thermal and electrical conductivity.

(*ii*) **Hardness.** These elements are hard (*i.e.,* cannot be cut with a knife) and brittle. Their hardness is due to the formation of covalent bonding and metallic bonding between metal atoms. Thus covalent and metallic bonding both exist in transition metals. Overlapping of partially filled *d*- orbitals of metal atoms gives covalent bonding and valence *s* -electrons cause metallic bonding. Greater is the number of unpaired *d*-electrons, greater is the possibility of covalent bond formation and hence greater is the strength of covalent bonding or hardness of the elements. Since Cr, Mo and W have maximum number of unpaired electrons (= 6), these metals are very hard. In case of Zn, Cd and Hg, since $(n-1)\,d$ orbitals are completely filled (Zn = $3d^{10}\,4s^2$, Cd = $4d^{10}\,5s^2$, Hg = $5d^{10}\,6s^2$), the formation of covalent bonding in these metals it not expected. Hence these metals are not very hard. As a matter of fact, Hg is a liquid and soft like alkali metals.

(*iii*) **Crystal structure.** These elements have simple hexagonal close-packed (*hcp*), cubic close packed (*ccp*) or body-centred cubic (*bcc*) lattices which are characteristic of metals. Hence these metals have high coordination number of 12 corresponding to *fcc* and *hcp* structures and 8 corresponding to *bcc* structure.

(*iv*) **Melting and boiling points.** The melting and boiling points of transition elements are generally very high. High values are due to strong covalent bonding in transition metals. Covalent bonding is due to strong overlapping between $(n-1)\,d$ orbitals.

(*v*) These elements are lustrous malleable, ductile and have high tensile strength.

## IV. Enthalpies of Atomisation ($\Delta H^\circ_a$)

Enthalpies of atomisation of transition metals are given in Table 5.4.

**Table 5.4. Enthalpies of atomization (KJ mol$^{-1}$) of transition metals.**

| | | | | | | | | | |
|---|---|---|---|---|---|---|---|---|---|
| **3d - series elements** | | | | | | | | | |
| Sc | Ti | V | Cr | Mn | Fe | Co | Ni | Cu | Zn |
| 326 | 473 | 575 | 397 | 281 | 416 | 425 | 430 | 339 | 126 |
| **4d - series elements** | | | | | | | | | |
| Y | Zr | Nb | Mo | Tc | Ru | Rh | Pd | Ag | Cd |
| 410 | 611 | 724 | 659 | 649 | 650 | 577 | 381 | 286 | 111 |
| **5d - series elements** | | | | | | | | | |
| La | Hf | Ta | W | Re | Os | Ir | Pt | Au | Hg |
| 435 | 618 | 781 | 837 | 791 | 790 | 690 | 566 | 368 | 61 |

From Table 5.4 the following points may be noted.

(*a*) The metals of 3rd series metals have high enthalpies of atomisation.

**Explanation:** Enthalpy of atomisation of a metal is the amount of heat required to break the metal lattice to get free atoms. Since transition metals contain a large number of strong interatomic attractions (*i.e.*, metallic bonds) and hence have high enthalpies of atomisation.

(*b*) The metals of the second (4*d*) and third (5*d*) series have greater enthalpies of atomisation than the corresponding elements of the first (3*d*) series. This is an important factor for their having much more frequent metal-metal bonding in their compounds.

(*c*) In the metals of 3*d* series the enthalpy of atomisation of Zn is the lowest ( = 126 KJ mol $^{-1}$).

**Explanation:** Excepting Zn, all other metals of 3*d* series have one or more unpaired electrons. Since Zn has no unpaired electron (Zn = $3d^{10} 4s^2$, $n$ =0), the atomic inter-metallic bonding (metal - metal bonding) is the weakest in Zn. Hence enthalpy of atomisation of Zn is the lowest.

## V. Melting and Boiling Points

The melting and boiling points of transition metals are given in Table 5.5

**Table 5.5. Melting and boiling points (°C) of d-block elements**

| | | | | | | | | | | |
|---|---|---|---|---|---|---|---|---|---|---|
| **3d - series elements:** | | | | | | | | | | |
| | Sc | Ti | V | Cr | Mn | Fe | Co | Ni | Cu | Zn |
| M.P.(°C) | 1539 | 1668 | 1900 | 1975 | 1245 | 1536 | 1495 | 1453 | 1083 | 419.5 |
| B.P.(°C) | 2730 | 3260 | 3450 | 2665 | 2150 | 3000 | 2900 | 2730 | 2595 | 906.00 |
| **4d - seriers elements:** | | | | | | | | | | |
| | Y | Zr | Nb | Mo | Tc | Ru | Rh | Pd | Ag | Cd |
| M.P.(°C) | 1509 | 1852 | 2415 | 2610 | 2200 | 2500 | 1966 | 1552 | 960.8 | 320.9 |
| B.P.(°C) | 2927 | 1500 | 3300 | 5560 | — | 4900 | 4500 | 3980 | 2210 | 765 |
| **5d - seriers elements:** | | | | | | | | | | |
| | La | Hf | Ta | W | Re | Os | Ir | Pt | Au | Hg |
| M.P.(°C) | 920 | 2222 | 2996 | 3410 | 3180 | 2700 | 2454 | 1769 | 1063 | –38.4 |
| B.P.(°C) | 3470 | 5400 | 5425 | 5930 | 5900 | 5500 | 5300 | 4530 | 2920 | 357 |

Following points may be noted:

(*i*) *These metals have very high melting and boiling points as compared to those of representative elements.*

**Explanation:** The high melting and boiling points of transition metals are due to the close packed structure of the metals and covalent bonding in these metals. Covalent bonding is caused by the strong overlapping between ($n$-1)$d$ orbitals. It is due to the closed packed structure of the metals that the atoms of these metals are held together by strong metallic bonds. Therefore, it requires considerable amount of energy to break the metallic bonds in order to melt the metal. Consequently, these metals have very high melting and boiling points.

(*ii*) *In a series the melting points first increase, attain a maximum value and then steadily decease as the atomic number increases.*

**Explanation:** The strength of the metallic bonds depends upon the number of unpaired electrons. Greater the number of unpaired electrons, stronger is the metallic bonding. As we have seen earlier, in a particular transition series, the number of unpaired electrons in ($n$–1) $d$- subshell increases upto the middle *i.e.,* upto $d^5$ configuration. Therefore, the strength of the metallic bonds and hence the melting and boiling points increase upto $d^5$

configuration, *i.e.,* upto the middle of a series. Beyond $d^5$ configuration, the electrons start pairing up and the number of unpaired electrons decrease steadily on moving further in the given series. This progressively decreases the strength of the metallic bonds and hence the melting and boiling points decrease progressively after the middle of the series on moving further in the given series. In order to understand this more clearly, let us consider the elements of the first transition series. As the atomic number increases, the number of unpaired electrons in the 3*d*- subshell increases upto chromium (Sc has 1, Ti has 2, V has 3 and Cr has 5 unpaired electrons). Therefore, the strength of metallic bonds and hence the melting and boiling points increases from Sc to Cr. Chromium has the maximum melting point in the first transition series. After Cr, the number of unpaired electrons goes on decreasing (Fe has 4, Co has 3, Ni has 2 unpaired electrons). Therefore, the strength of metallic bonds and hence the melting and boiling points decrease in going from Cr to Cu.

(*iii*) Zn, Cd and Hg have no unpaired electrons. Hence the metallic bonds in these metals are weak. That is why these metals are soft and have low melting and boiling points. *Hg is a liquid at room temperature and has melting point of –38.4°C.*

(*iv*) *Chromium is a typical hard metal while mercury is a liquid.*

**Explanation:** The difference in the physical state of Cr and Hg can be explained on the basis of the strength of metallic bonding in the lattices of these metals. Cr has 5 unpaired electrons in *d*-orbitals (Cr = $3d^5\,4s^1$) while Hg has no unpaired electrons in *d*-orbitals (Hg = $5d^{10}6s^2$). Hence metallic bonds in Cr metal are very strong and metallic bonding in Hg is very weak. Due to strong metallic bonds, Cr is a typical hard metal and due to very weak metallic bonds, Hg is a liquid.

(*v*) Since Sc, Ti and V show an increasing trend in their m.pts., it may be concluded that all the 3*d* and 4*s* electrons present in these metals take part in metallic bonding.

(*vi*) Fe has higher m.pt. than Cu. This is because Fe has four unpaired electrons in its 3*d* orbitals while Cu has only one unpaired electron in 4*s* orbital. Hence metallic bonds in Fe are much stronger than those in Cu.

(*vii*) It may be noted that tungsten (W) has the highest melting point among the *d*-block elements.

(*viii*) Lower values in case of Mn and Tc can be explained on the basis of the fact that both these elements have exactly half-filled *d*-orbitals. As a result in both the elements electronic configuration is stable and hence electrons are held tightly by the nucleus so that the delocalisation is less and the metallic bond is much weaker that in preceeding element.

## VI. Atomic Volumes and Densities

The atomic volumes (in ml) and density (in g/cm³) of transition elements are given below in Table 5.6. The values given against *ml* are atomic volumes while those given against $g/cm^3$ represent densities.

**Table 5.6. Atomic volumes (ml) and densities (g/cm³) of d-block elements.**

| | IIIB | IVB | VB | VIB | VII B | ← | VIII | → | IB | IIB |
|---|---|---|---|---|---|---|---|---|---|---|
| **3d (1st) series (4th period)** | | | | | | | | | | |
| | $Sc_{21}$ | $Ti_{22}$ | V | Cr | Mn | Fe | Co | Ni | Cu | $Zn_{30}$ |
| ml | 15.02 | 10.60 | 8.35 | 7.23 | 7.39 | 7.10 | 6.70 | 6.60 | 7.10 | 9.20 |
| g/cm³ | 3.43 | 4.51 | 6.10 | 7.19 | 7.43 | 7.86 | 8.90 | 8.90 | 8.96 | 7.14 |
| **4d (2nd) series (5th period)** | | | | | | | | | | |
| | $Y_{39}$ | $Zr_{40}$ | Nb | Mo | Tc | Ru | Rh | Pd | Ag | $Cd_{48}$ |
| ml | 19.8 | 14.1 | 10.8 | 9.4 | 11.5 | 8.3 | 8.3 | 8.9 | 10.4 | 13.1 |
| g/cm³ | 4.47 | 6.49 | 8.40 | 10.2 | 10.50 | 12.20 | 12.40 | 12.00 | 10.51 | 8.65 |
| **5d (3rd) series (6th period)** | | | | | | | | | | |
| | $La_{71}$ | $Hf_{72}$ | Ta | W | Re | Os | Ir | Pt | Au | $Hg_{80}$ |
| ml | 20.50 | 13.60 | 10.91 | 9.53 | 8.85 | 8.43 | 8.54 | 9.10 | 10.22 | 14.80 |
| g/cm³ | 6.17 | 13.10 | 16.60 | 19.30 | 21.00 | 22.59 | 22.61 | 22.40 | 19.30 | 13.60 |

Atmoic volumes of *d* - block elements are much lower than those of the elements of *s*-and *p*- blocks.

We know that density and atomic volume are inversely proportional to each other. All the transition metals have low values for their atomic volumes and hence their densities are high. The atomic volumes are low because the electrons are added to $(n-1)$ *d* orbitals and not to *ns* orbital. The increased nuclear charge is partly screened by $(n-1)$ *d* electrons and *ns* electrons are strongly atracted by the nucleus. Consequently the densities of transition metals are high.

**Variation of atomic volumes in a given series.** With a few exceptions, atomic volumes of the elements in a series decrease as we move from left to right in a given series. The decrease in atomic volumes is due to the decrease in atomic radii of the elements. The decrease in atomic radii is due to the increase in nuclear charge.

**Variation of density in a given series.** With a few exceptions, density of elements increases along a series. The increase in density is due to the decrease in atomic radii and increase in atomic mass. In the first transition series density of Zn is lower than that of Cu (Zn = 7.14 g/cm³, Cu = 8.96 g/cm³). This is due to larger volume of Zn as compared to that of Cu (Cu = 7.10 ml, Zn = 9.20 ml).

Among *d*- block elements, Ir has the highest density and Sc has the lowest density (Ir= 22.61 g/cm³, Sc= 3.43 g/cm³). Os and Ir have almost the same density (Os = 22.59 g/cm³, Ir = 22.61 g/cm³).

**Variation of density in a given sub-group.** Density of elements belonging to the same subgroup increases on moving down the sub-group. The densities of the elements of 2nd transition series are only slightly higher than those of the corresponding elements of 1st transition series while these values for the transition elements from $Hf_{72}$ to $Hg_{80}$ (elements of 3rd transition series) are almost double these values for the elements from $Zr_{40}$ to $Cd_{48}$ respectively (elements of 2nd transition series) Note that the density of $La_{57}$ (= 6.17) is not double that of the value for $Y_{39}$

(= 4.47). This may be seen from the density values of the following four pairs of elements given below. In each pair one element belongs to *2nd* series and the other element belongs to *3rd* transition series.

| | | | | |
|---|---|---|---|---|
| **2nd series** | Nb<br>8.4 | Mo<br>10.2 | Tc<br>10.5 | Ag<br>10.51 |
| **3rd series** | Ta<br>16.61 | W<br>19.3 | Re<br>21.0 | Au<br>19.30 |

The variation of densities of transition series elements in a given sub-group, as discussed above, can be explained as follows:

(*a*) The atomic weights of the elements of 3rd transition series are almost double the atomic weights of the corresponding elements of 2nd transition series. This makes the densities of the element of 3rd transition series almost double those of the corresponding elements of 2nd transition series.

(*b*) Because of lanthanide contraction occurring in lanthanides, the atomic sizes of the elements of 3rd transition series coming after $Lu_{71}$ (*i.e.*, elements from $Hf_{72}$ to $Hg_{80}$) become very small. Consequently the packing of the atoms in their metallic crystals become so much compact that their densities become very high. Thus the densities of the element of 3rd transition series become almost double the densities of the corresponding elements of 2nd transition series.

## VII. Ionisation Energies

The values of first ionisation energy ($IE_1$) of *1st* (3*d* series), *2nd (4d* series) and *3rd* (5*d* series) are given Table 5.7 while the values of *1st* ($IE_1$), *2nd* ($IE_2$) and *3rd* ($IE_3$) ionisation energies of the elements belonging to 1st transition series (3*d* series) are listed in Table 5.8.

**Table 5.7. Ionisation energies (KJmol$^{-1}$) *d*-block elements**

| **3d (1st) series (4th period)** | | | | | | | | | |
|---|---|---|---|---|---|---|---|---|---|
| IIIB<br>3 | IVB<br>4 | VB<br>5 | VIB<br>6 | VIIB<br>7 | ←— VIII —→<br>8 | 9 | 10 | IB<br>11 | IIB<br>12 |
| $Sc_{21}$<br>631 | $Ti_{22}$<br>658 | $V_{23}$<br>650 | $Cr_{24}$<br>652 | $Mr_{25}$<br>717 | $Fe_{26}$<br>762 | $Co_{27}$<br>758 | $Ni_{28}$<br>735 | $Cu_{29}$<br>745 | $Zn_{30}$<br>906 |
| **4d (2nd) series (5th period)** | | | | | | | | | |
| $Y_{39}$<br>616 | $Zr_{40}$<br>660 | $Nb_{41}$<br>664 | $Mo_{42}$<br>685 | $Tc_{43}$<br>702 | $Ru_{44}$<br>711 | $Rh_{45}$<br>720 | $Pd_{46}$<br>805 | $Ag_{47}$<br>731 | $Cd_{48}$<br>867.7 |
| **5d (3rd) series (6th period)** | | | | | | | | | |
| $La_{57}$ $Ce_{58}$ – $Lu_{71}$<br>538.1 | $Hf_{72}$<br>654 | $Ta_{73}$<br>761 | $W_{74}$<br>770 | $Re_{75}$<br>760 | $Os_{76}$<br>840 | $Ir_{77}$<br>880 | $Pt_{78}$<br>870 | $Au_{79}$<br>890.1 | $Hg_{80}$<br>1007 |

**Table 5.8. Ionisation energies of the elements of first transition series**

| Ionisation energy | Sc | Ti | V | Cr | Mn | Fe | Co | Ni | Cu | Zn |
|---|---|---|---|---|---|---|---|---|---|---|
| $IE_1$ ($kJ\ mol^{-1}$) | 631 | 658 | 650 | 652 | 717 | 762 | 758 | 736 | 745 | 906 |
| $IE_2$ (kJ mol–¹) | 1245 | 1320 | 1376 | 1635 | 1513 | 1563 | 1647 | 1756 | 1961 | 1736 |
| $IE_3$ ($kJ\ mol^{-1}$) | 2450 | 2721 | 2873 | 2994 | 3258 | 2963 | 3237 | 3400 | 3560 | 3838 |

From Tables 5.7 and 5.8 the following conclusions have been drawn.

(*i*) *The ionisation energies of transition metals are quite high and lie between those of s-block and p-block elements. They are higher than those of s-block elements and lower than those of p-block elements.*

**Explanation:** The atomic radii and nuclear charge of transition metals lie between those of *s*-block and *p*-block elements; they possess smaller atomic radii and higher nuclear charge as compared to those of s-block elements. But reverse is the trend when compared with *p*-block elements. That is why the ionisation energies of transition elements lie between those of *s*-block and *p*-block elements.

The fact that IE values of transition elements lie between those of s-and p-block elements indicates that transition elements are less electropositive than s-block elements. The compounds of transition elements in which a smaller valency is exercised are ionic in nature and those in which higher valency is exercised are covalent in nature.

*(ii) The first ionisation energy increases with increase in atomic number across a given transition series, although the increase is not very regular.*

**Explanation:** The increase is ionisation energy with increase in atomic number across a given series is due to an increase in the nuclear charge with increase in the atomic number. The added electrons enter into ($n$ – 1) d-subshell and shield the valence electrons from the nucleus. Thus, the effect of the increased nuclear charge is opposed by the screening effect of ($n$ – 1) $d$-electrons. That is why increase in ionisation energy with increase in atomic number is rather slow and not very regular. For example IE values for the first four elements namely Sc, Ti, V and Cr differ only slightly from one another. Similarly IE values for Fe, Co, Ni and Cu are fairly close to one another.

(*iii*) *In each series the elements of group II B (Zn, Cd and Hg) have the highest values for their ionisation energy.* This is due to the extra stability associated with their completely-filled $(n-1)d$ and $ns$ orbitals ($Zn_{30} = 3d^{10} 4s^2$, $Cd_{48} = 4d^{10} 5s^2$ and $Hg_{80} = 5d^{10}6s^2$).

(*iv*) If we compare IE values of 5d elements with those of 3$d$ and 4$d$ elements, we find that *these values for 5d elements are higher than those for 3d and 4d elements.* This can be explained as follows.

**Explanation.** We know that 5d elements ($Hf_{72}$ to $Hg_{80}$) have 4f electrons. These 4f electrons have poor shielding effect on the nucleus of 5d elements. As a result, the outer valence electrons of 5$d$ elements experience greater effective nuclear charge than the outer valence electrons of 3$d$ and 4$d$ elements. Therefore, 5$d$ elements have higher IE values than 3$d$ and 4$d$ elements.

(*v*) *2nd and 3rd IE values of the elements, in general, increase along a period. However, there are some exceptions to this trend.* These exceptions are given below:

(*a*) $IE_2$ values for Cr and Cu are higher than those of their neighbours.

**Explanation** V.S.E.C. of Cr and Cu are Cr = $3d^5 4s^1$ and Cu = $3d^{10} 4s^1$. $IE_2$ values of these elements involve the removal of an electron from $3d^5$ configuration of $Cr^+$ ion and an electron from $3d^{10}$ configuration of $Cu^+$ ion.

$$\underset{3d^5 4s'}{Cr^\circ} \xrightarrow[IE_1]{-4s^1} \underset{3d^5}{Cr^+} \xrightarrow[IE_2]{-3d^1} \underset{3d^4}{Cr^{2+}}$$

$$\underset{3d^{10} 4s^1}{Cu} \xrightarrow[IE_1']{-4s^1} \underset{3d^{10}}{Cu^+} \xrightarrow[IE_2]{-3d^1} \underset{3d^9}{Cu^{2+}}$$

Now since $3d^5$ and $3d^{10}$ configurations are stable configurations, the removal of an electron from these configurations requires large amount of energy. As a result, $IE_2$ values for Cr and Cu are very high.

(*b*) Third ionization energy of Mn is very high because the 3rd electron has to be removed from the stable half-filled 3d orbital. (Mn = $3d^5 4s^2$). Third ionization energy for change from $Fe^{2+}$ to $Fe^{3+}$ (Fe = $3d^6 4s^2$) is very small because loss of 3rd electron gives the stable configuration of $3d^5$. Hence, $IE_3$ for Fe << $IE_3$ for Mn.

(*vi*) *The high values of 3rd ionization energies for Cu, Ni and Zn explain why they have a maximum oxidation state of +2.*

(*vii*) *$IE_1$ value of Cu is higher than that of alkali metals but $IE_2$ and $IE_3$ values for Cu are lower than those for alkali metals.*

**Explanation.** E.C. of Cu atom is [Ar]$3d^{10}4s^1$ while that of alkali metals it is [Noble gas] $ns^1$. Since the filled 3d sub-shell ($3d^{10}$) in Cu has much less shielding effect on $4s^1$ electron in Cu than [noble gas] configuration has on $ns^1$ electron in alkali metals. Hence $4s^1$ electron in Cu is more tightly bound than $ns^1$ electron in alkali metals. As a result, $IE_1$ for Cu is higher than that for alkali metals. For $IE_2$ and $IE_3$, electrons have to be removed from the diffused $3d^{10}$ configuration in Cu but in alkali metals the electrons have to be removed from a [noble gas] configuration. Hence $IE_2$ and $IE_3$ for alkali metals are higher.

## VIII. Enthalpy of Hydration ($\Delta H_{hyd}$)

Enthalpy of hydration ($\Delta H_{hyd}$) for $M^{2+}(g)$ ions is the energy released in the hydration reaction.

$M^{2+}(g) + aq \rightarrow M^{2+}(aq) + \Delta H_{hyd}$

$\Delta H_{hyd}$ values for $M^{2+}(g)$ ions (M = Ti to Zn) are given below:

| | |
|---|---|
| $Ti^{2+}(g)$ | = –1866kJ mol⁻¹, $V^{2+}(g)$ = –1895, $Cr^{2+}(g)$ = –1925, $Mn^{2+}(g)$ = –1862, |
| $Fe^{2+}(g)$ | = –1998, $Co^{2+}(g)$ = –2079, $Ni^{2+}(g)$ = –2121, $Cu^{2+}(g)$ = –2121, |
| $Zn^{2+}(g)$ | = –2059 |

It may be seen that $\Delta H_{hyd}$ values are high.

**Explanation:** High values of enthalpy of hydration of transition metal ions are due to the small size and large nuclear charge of the metal. This is because when we move along any transition series, the nuclear charge increases and size decreases.

## IX. Standard Oxidation Potential ($E^\circ_{ox}$) Values and Reducing Properties of Transition Elements in Aqueous Solution

$E^o_{ox}$ value of a solid transition metal, M(*s*) is a measure of its tendency to get oxidised to $M^{2+}(aq)$ or $M^{3+}(aq)$ in aqueous solution

$$M(s) + aq \xrightarrow{\text{Oxidation}} M^{2+}(aq) + 2e^-$$

$$M(s) + aq \xrightarrow{\text{Oxidation}} M^{3+}(aq) + 3e^-$$

The enthalpy change ($\Delta H$) taking place in the oxidation half-reaction,

$$M(s) + aq \longrightarrow M^{2+}(aq) + 2e^-, \Delta H = ? \quad \text{...(A)}$$

is a measure of oxidation potential of $M(s) / M^{2+}(aq)$ electrode. Oxidation reaction (A) actually takes place through the following steps:

(*a*) In this step individual atoms of the solid metal, M(*s*) *absorb* energy equal to sublimation energy of M(s) ($\Delta H_{sub}$) and are converted into gaseous state, M(*g*). In this conversion, since energy is absorbed (endothermic process), $\Delta H_{sub}$ has positive sign.

$$M(s) \rightarrow M(g), \Delta H = \text{Energy absorbed} = +\Delta H_{sub} \quad \text{...}(a)$$

Stronger is the metallic bonding in a particular metal, greater will be the energy required to separate the individual atoms of the metal.

(*b*) In this step the individual gaseous metal atoms, M (*g*) absorb energy equal to sum of $IE_1$ and $IE_2$ of M(*g*) and are oxidised to $M^{2+}(g)$ ions. In this process, since energy is required (endothermic process), ($IE_1 + IE_2$) has positive sign.

$$M(g) \rightarrow M^{2+}(g) + 2e^-, \; \Delta H = \text{Energy absorbed} = + (IE_1 + IE_2) \quad \text{...}(b)$$

(*c*) In this step, the gaseous metal ion, $M^{2+}(g)$ gets hydrated (because these reactions are generally carried out in aqueous medium) to give hydrated ion, $M^{2+}(aq)$. In this step, since energy equal to enthalpy of hydration of $M^{2+}(g)$ ($\Delta H_{hyd}$) is *released* (exothermic process), $\Delta H_{hyd}$ is given a negative sign.

$$M^{2+}(g) + aq \rightarrow M^{2+}(aq), \Delta H = \text{Enthalpy of hydration} = -\Delta H_{hyd} \quad \text{...}(c)$$

Smaller is the size of $M^{2+}(g)$ ion, greater will be the value of $\Delta H_{hyd}$.

On adding equations (*a*), (*b*) and (*c*) given above, we get:

$$M(s) + aq \rightarrow M^{2+}(aq) + e^-$$

This equation is the same as equation (A). Since the addition of equations (*a*), (*b*) and (*c*) gives equation (A), the total enthalpy change ($\Delta H$) for reaction (A) must be equal to the sum of $+\Delta H_{sub}$, $+(IE_1 + IE_2)$ and $-\Delta H_{hyd}$.

$$\Delta H = \Delta H_{sub} + (IE_1 + IE_2) - \Delta H_{hyd} \quad \text{...(B)}$$

Relation (B) shows that:

(*a*) The electrode potential ($E^\circ_{ox}$) for the oxidation half-reaction:

$$M(s) \rightarrow M^{2+}(aq) + 2e^-$$

depends on the net effect of $\Delta H_{sub,}$ ($IE_1 + IE_2$) and $\Delta H_{hyd}$.

(*b*) Smaller is the value of $\Delta H$ for a particular oxidation state of a metal ion in aqueous solution, greater is the stability of that oxidation state, *i.e.*,

$$\text{Stability of an oxidation state} \; \alpha \frac{1}{\Delta H}$$

All the three steps given at (*a*), (*b*) and (*c*) can be shown in the form of a cyclic process which is called Born-Haber cycle. This cycle is shown in Fig. 5.2.

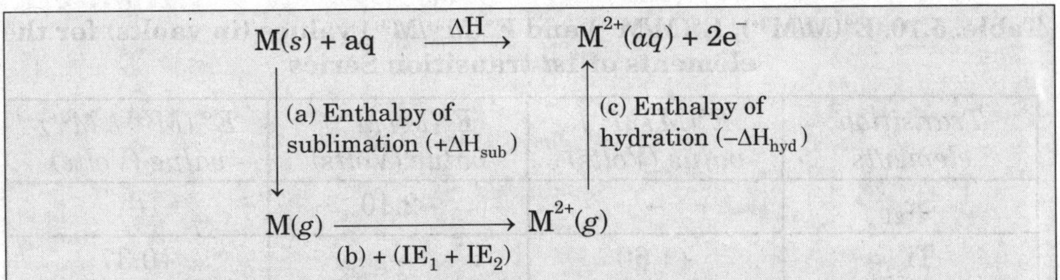


**Fig. 5.2** Born-Haber cycle for the conversion of M(*s*) to $M^{2+}$(*aq*) in aqueous solution.

The values of $\Delta H_{sub}$, $IE_1$, $IE_2$ and $\Delta H_{hyd}$ (thermochemical data) of the transition elements of 1st transition series are given in Table 5.9. These values can be used for calculating E° for M(*s*)/$M^{2+}$(*aq*) electrodes.

**Table. 5.9. Thermochemical data (KJ $mol^{-1}$) for the elements of 1st transition series**

| *Element, M(s)* | $\Delta_{sub}$ | $IE_1$ | $IE_2$ | $\Delta H_{hyd}$ *of* $M^{2+}$ *(g) ions* |
|---|---|---|---|---|
| Ti | 473 | 658 | 1320 | –1866 |
| V | 575 | 650 | 1376 | –1895 |
| Cr | 397 | 652 | 1635 | –1925 |
| Mn | 281 | 717 | 1513 | –1862 |
| Fe | 416 | 762 | 1563 | –1998 |
| Co | 425 | 758 | 1647 | –2079 |
| Ni | 430 | 736 | 1756 | –2121 |
| Cu | 339 | 745 | 1961 | –2121 |
| Zn | 126 | 906 | 1736 | –2059 |

E° (M/$M^{2+}$), E° (M/$M^{3+}$) and E°($M^{2+}$/$M^{3+}$) values for the elements of 1*st* transition series are given in Table 5.10.

From E° values given in Table 5.10 the following points may be noted.

(*i*) Since E° (M/$M^{2+}$) values for the metals, excepting Cu, are positive, these metals are capable of evolving $H_2$ gas from acid solution.

$$M(s) + 2H^+ (aq) \rightarrow M^{2+}(aq) + H_2(g)$$

Here it may be noted that Cr metal does not evolve $H_2$ gas from acid solution. This behaviour of Cr is because of the fact that Cr metal gets protected from the attack of the acid solution by a thin impervious layer of an inert (non-reactive) oxide, $Cr_2O_3$. Ti and V are passive to dilute non-oxidising acids at room temperature.

(*ii*) Since E°(M/$M^{2+}$) values are positive, transition metals, excepting Cu, have a strong tendency to get oxidised to $M^{2+}$(*aq*) ions in aqueous solution, *i.e.*, transition elements act as strong reducing agents in aqueous solution. Higher is the value of $E°_{ox}$, stronger is the reducing agent.

**Table. 5.10. E°($M/M^{2+}$), E°($M/M^{3+}$) and E°($M^{2+}/M^{3+}$) values (in vaolts) for the elements of 1*st* transition Series**

| *Transition elements* | *E°(M / $M^{2+}$) value (Volts)* | *E°(M / $M^{3+}$) value (Volts)* | *E° ($M^{2+}$ / $M^{3+}$) value (Volts)* |
|---|---|---|---|
| $Sc_{21}$ | – | –2.10 | – |
| $Ti_{22}$ | +1.60 | – | +0.37 |
| $V_{23}$ | +1.18 | +0.26 | +0.26 |
| $Cr_{24}$ | + 0.91 | +0.74 | +0.41 |
| $Mn_{25}$ | +1.18 | –1.57 | –1.57 |
| $Fe_{26}$ | +0.44 | –0.77 | –0.77 |
| $Co_{27}$ | +0.28 | –1.97 | –1.97 |
| $Ni_{28}$ | +0.25 | – | – |
| $Cu_{29}$ | –0.34 | – | – |
| $Zn_{30}$ | +0.76 | – | – |

(*iii*) With the exception of a few elements, E°($M/M^{2+}$) values keep on decreasing on moving from Sc to Zn. This decreasing trend of $E°_{ox}$ values indicates that the reducing ability of transition elements is decreasing from $Sc_{21}$ to $Zn_{30}$. This decreasing reducing ability is due to the decreasing size and increasing ionisation energies from $Sc_{21}$ to $Zn_{30}$. The decreasing reducing ability of the elements from $Sc_{21}$ to $Zn_{30}$ shows that these elements (M) have a decreasing tendency to form $M^{2+}$ ions.

(*iv*) There is no regularity in E°($M/M^{2+}$) values. The irregularity in these values is because of the fact that, as we have seen earlier, E° values depend on the sum of $\Delta H_{sub}$, ($IE_1 + IE_2$) and $\Delta H_{hyd}$. The irregularity is due to the irregular variation in the values of $\Delta H_{sub}$ and ($IE_1 + IE_2$).

(*v*) E° (Cu / $Cu^{2+}$) value for Cu is negative because of the reason that the sum of $\Delta H_{sub}$ and ($IE_1 + IE_2$) is not balanced by $\Delta H_{hyd}$. Negative value of E° for Cu shows that this metal is not able to liberate $H_2$ gas from acid solution. This metal can, however, reduce oxidising acids like $HNO_3$ and $H_2SO_4$.

(*vi*) Since E° (Cu/$Cu^{2+}$) has negative value, the tendency of Cu (*s*) to get oxidised to $Cu^{2+}$ (*aq*) in aqueous solution is extremely small and hence Cu is a poor reducing agent. Thus the following oxidation half-reaction proceeds to an extremely small extent.

$$\text{Cu (s) + aq} \xrightarrow{\text{Oxidation}} Cu^{2+}(aq) + 2e^-$$

(*vii*) Mn has high value of E° (Mn/$Mn^{2+}$) because Mn ($3d^5 4s^2$) has high tendency to get oxidised to $Mn^{2+}$ which has a stable half-filled configuration ($Mn^{2+} = 3d^5$). Similarly Zn ($3d^{10}$ $4s^2$) also has high value of E°(Zn/$Zn^{2+}$) because this metal has high tendency to get oxidised to $Zn^{2+}$ ion which has stable fully-filled configuration ($Zn^{2+} = 3d^{10}$).

$$\underset{3d^5 4s^2}{Mn} \longrightarrow \underset{\substack{3d^5 \\ \text{(stable)}}}{Mn^{2+}} + 2e^-, \quad \underset{3d^{10} 4s^2}{Zn} \longrightarrow \underset{\substack{3d^{10} \\ \text{(stable)}}}{Zn^{2+}} + 2e^-$$

(*viii*) It has been observed that $Mn^{3+}$ and $Co^{3+}$ both are strong oxidising agents in aqueous solution and $Co^{3+}$ ion is a stronger OA than $Mn^{3+}$ ion. This can be explained as follows:

When $Mn^{3+}$ and $Co^{3+}$ ions act as oxidising agents, these ions are reduced to Mn and Co respectively.

$$Mn^{3+} + 3e^- \longrightarrow Mn, \; E° (Mn^{3+} / Mn) = +1.57 \text{ V}$$

$$Co^{3+} + 3e^- \longrightarrow Co, \; E° (Co^{3+} / Co) = +1.97 \text{ V}$$

Since both E° values are highly positive, both the ions are easily reduced to the metals. Hence both the ions act as strong oxidising agents. Now since E° ($Co^{3+}$ / Co) value is higher than E° ($Mn^{3+}$ / Mn) value, $Co^{3+}$ ion is more easily reduced to Co than $Mn^{3+}$ gets reduced to Mn. Thus $Co^{3+}$ ion is a stronger OA than $Mn^{3+}$ ion.

(*ix*) Negative value of E° ($Mn^{2+}$ / $Mn^{3+}$) [$Mn^{2+} \longrightarrow Mn^{3+} + e^-$, E° ($Mn^{2+}$ / $Mn^{3+}$) = –1.57 V] shows that $Mn^{2+}$ ion is not oxidised to $Mn^{3+}$ ion. In other words we can say that $Mn^{2+}$ ion is stable. The stability of $Mn^{2+}$ ion is also due to the presence of stable $3d^5$ configuration in $Mn^{2+}$ ion.

(*x*) $Co^{3+}$ ion is a better OA than $Co^{2+}$ ion. This can be explained as follows.

When $Co^{3+}$ and $Co^{2+}$ ions act as oxidising agents, they are reduced to $Co^{2+}$ and Co respectively. $E°_{red}$ values for the reduction half-reactions are:

(*a*) $Co^{3+} + e^- \longrightarrow Co^{2+}, \; E°_{red} = +1.97V$

(*b*) $Co^{2+} + 2e^- \longrightarrow Co, \; E°_{red} = -0.28V$

Now since $E°_{red}$ value for reduction half-reaction (*a*) is higher than that for reduction half-reaction (*b*), $Co^{3+}$ ion is more easily reduced to $Co^{2+}$ ion than $Co^{2+}$ ion is reduced to Co. Thus $Co^{3+}$ ion is a better oxidising agent.

(*xi*) E°($V^{2+}$ / $V^{3+}$) value of V is comparatively low.

$$\underset{3d^3 (t_{2g}^3 e°_g)}{V^{2+}} \longrightarrow \underset{3d^2}{V^{3+}} + e^-, \; E°(V^{2+} / V^{3+}) = +\,0.26 \text{ V}$$

Since $V^{2+}$ ion has half-filled $t^3_{2g}$ configuration (stable configuration), $V^{2+}$ ion has little tendency to get oxidised to $V^{3+}$ ion. Hence E° value for the above oxidation half-reaction is very low.

(*xii*) Since E°($Fe^{3+}$ / $Fe^{2+}$) value is low [$Fe^{3+} + e^- \rightarrow Fe^{2+}$, E° ($Fe^{3+}$ / $Fe^{2+}$) = +0.77 V], $Fe^{3+}$ ion is not reduced to $Fe^{2+}$ ion, *i.e.*, $Fe^{3+}$ ion is stable. The stability of $Fe^{3+}$ ion is also because of the stable $3d^5$ configuration of $Fe^{3+}$ ion.

(*xiii*) Since E° ($Ti^{2+}$ / $Ti^{3+}$), E° ($V^{2+}$ / $V^{3+}$) and E° ($Cr^{2+}$ / $Cr^{3+}$) values are positive, $Ti^{2+}$, $V^{2+}$ and $Cr^{2+}$ are easily oxidised to $Ti^{3+}$, $V^{3+}$ and $Cr^{3+}$ ions respectively. Thus these ions are strong reducing agents and hence can liberate $H_2$ gas from an acid, *e.g.*

$$2Cr^{2+} (aq) + 2H^+ (aq) \rightarrow 2Cr^{3+} (aq) + H_2(g).$$

## X. Variable Oxidation States

Various oxidation states shown by transition series elements are given in Table 5.11.

**Table 5.11. Commonly shown oxidation states of transition metals (very rare oxidation states are given in parentheses and most common ones are enclosed in boxes for first transition Series)**

| *1st series* | *Sc* | *Ti* | *V* | *Cr* | *Mn* | *Fe* | *Co* | *Ni* | *Cu* | *Zn* |
|---|---|---|---|---|---|---|---|---|---|---|
| | | | | | | | | | +1 | |
| | | (+2) | +2 | +2 | **+2** | **+2** | **+2** | **+2** | **+2** | **+2** |
| | +3 | +3 | +3 | **+3** | +3 | **+3** | **+3** | (+3) | | |
| | | **+4** | +4 | +4 | +4 | +4 | +4 | (+4) | | |
| | | | **+5** | +5 | (+5) | | | | | |
| | | | | **+6** | +6 | (+6) | | | | |
| | | | | | **+7** | | | | | |

| *2nd series* | *Y* | *Zr* | *Nb* | *Mo* | *Tc* | *Ru* | *Rh* | *Pd* | *Ag* | *Cd* |
|---|---|---|---|---|---|---|---|---|---|---|
| | | | | | | | | | +1 | |
| | | | (+2) | +2 | +2 | +2 | +2 | +2 | (+2) | +2 |
| | +3 | (+3) | (+3) | +3 | | +3 | +3 | (+3) | (+3) | |
| | | +4 | (+4) | +4 | (+4) | +4 | +4 | +4 | | |
| | | | +5 | +5 | (+5) | (+5) | | | | |
| | | | | +6 | | (+6) | (+6) | | | |
| | | | | | +7 | (+7) | | | | |
| | | | | | | (+8) | | | | |

| *3rd series:* | *La* | *Hf* | *Ta* | *W* | *Re* | *Os* | *Ir* | *Pt* | *Au* | *Hg* |
|---|---|---|---|---|---|---|---|---|---|---|
| | | | | | (+1) | | | | +1 | +1 |
| | | | (+2) | +2 | (+2) | +2 | +2 | +2 | | +2 |
| | +3 | (+3) | (+3) | (+3) | +3 | +3 | +3 | (+3) | +3 | |
| | | +4 | (+4) | +4 | +4 | +4 | +4 | +4 | | |
| | | | +5 | +5 | +5 | | | (+5) | | |
| | | | | +6 | (+6) | +6 | (+6) | (+6) | | |
| | | | | | +7 | | | | | |
| | | | | | | +8 | | | | |

From Table 5.11 the following points may be noted.

(1) *With the exception of first and last transition element in each series, all other elements show more than one oxidation states (O.S.) (i.e. variable or several oxidation states)*

**Explanation:** As we have seen earlier, the electronic configuration of transition elements are of the type $(n-1)d^{1-10}\ ns^{1-2}$. The energies of $(n-1)$ *d*-orbtials and *ns*-orbitals are very close and there is not much difference in them. Therefore, in case of transition elements, both $(n-1)d$-and *ns*-orbitals are available for bond formation. +1 and +2 oxidation states involve the participation of *ns*-electrons but higher oxidation states such as +3, +4, +5, +6 etc., involve the use of $(n-1)$ *d*-electrons in addition to *ns*-electrons for bond formation. In the excited state, $(n-1)$ *d*-electrons become bonding *i.e.*, they are free to take part in bond formation and thus enable the atom to show a variety of oxidation states. For example, the outer electronic configuration of Sc is $3d^1\ 4s^2$. It shows the oxidation state +2 when it uses its two 4s-electrons. The oxidation state + 3 is shown when it uses its one 3*d*-electron in addition to two 4s-electrons. Similarly, Ti ($3d^2\ 4s^2$) exhibits the oxidation state +2 when it uses only its two 4s-electrons, +3 when it uses one 3*d*-electron in addition to the two 4*s*-electrons and +4 when it uses two 3*d*-electrons in addition to the two 4*s*-electron for bonding. In short, we say that since the energies of $(n-1)$ *d* and *ns* orbitals are close, electrons from both orbitals can participate in bonding and hence transition metals show variable oxidation states.

(2) *The most common oxidation state of the element belonging to the first transition series (except Sc) is +2 which arises from the loss of two 4s-electrons. This indicates that after scandium, d-orbitals become more stable than the s-orbital.*

(3) *Stabilisation of low (unstable) oxidation states*. Transition metals also show low oxidation states like –1, 0, +1, etc. These oxidation states are unstable. However these states get stabilised when the metals form compounds or complexes with π-acid ligands like CO, NO, $N_2$, $PR_3$, $AsR_3$, CNR (isocyanide), $CN^-$, $C_6H_6$, toluene, α, α′-dipyridyl, o-phenanthroline, ethylene, etc. For example in +1 O.S. Cu forms stable CuCl. Ni and Fe form carbonyls Ni $(CO)_4$ and Fe $(CO)_5$ respectively in which Ni and Fe have zero oxidation state.

(4) *Highest oxidation state*. (*a*) The highest (maximum) oxidation state for the first five elements (Sc to Mn) is equal to the sum of the electrons in 4*s* and 3*d* orbitals. On the other hand, the highest oxidation state for the remaining five elements (Fe to Zn) is not equal to the sum of the electrons in 4s and 3*d* orbitals. The stability of higher oxidation states of the elements lying after Mn decreases successively.

(*b*) The high oxidation states of many transition metals like V (V), Cr (VI), Mn (IV), Mn (VII), Fe (III), Fe (VI), Co (III), Ni (III), Cu (III), Ag(II), Ag (III) etc. are oxidising (*i.e.*, unstable). These unstable oxidation states which are high oxidation states get stabilised by forming complexes with small highly electronegative ligands like $O^{2-}$, $F^-$, $Cl^-$, per iodae ($IO_6^{5-}$), tellurate ($TeO_6^{6-}$) etc.

(*c*) *Within a group, the maximum oxidation state increases with atomic number*. For example, the common oxidation states of iron (group 8) are +2 and +3 but ruthenium and osmium belonging to the same group (*i.e.*, group 8) form compounds in the +4, +6 and +8 oxidation states also.

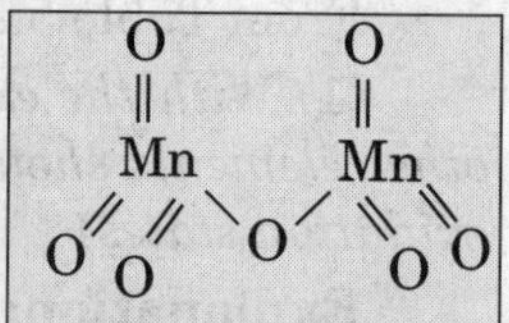

Structure of $Mn_2O_7$ molecule

(*d*) *The highest oxidation states of transition metals are found in their fluorides and oxides*. This is because fluorine and oxygen are the most electronegative elements and hence both these elements stabilise the highest oxidation state. Oxygen stabilises the highest O.S. even more than fluorine, e.g. highest fluoride of Mn is $MnF_4$ whereas the highest oxide of Mn is $Mn_2O_7$. The reason for greater stabilising power of oxygen is the ability of this element to form multiple bonds with the transition metal atoms. This fact is evident from the structure of $Mn_2O_7$ given in the margin.

(*e*) *The highest oxidation state shown by any transition element is eight*. This O.S. is shown by Ru and Os.

(5) *Ionic / covalent character of compounds of a given metal in various oxidation states*. Element in lower oxidation states like +2 and +3 form ionic compounds while the elements in higher oxidation states form mostly covalent compounds which are formed by the sharing of *d*-electrons. For example $\overset{2+}{V}Cl_2$ is ionic, $VCl_3$ is less ionic and $VCl_4$ is covalent. $MnO_4^-$ ion has Mn in +7 O.S. In this ion the bonds between Mn and oxygen are covalent.

**Explanation.** As the oxidation state of the transition metal increases, the charge density on the metal also increases. This results in the increase of the polarisation of the anion charge cloud by the metal and hence covalent character increases (*See Fajan's Rules*).

(6) *Acidic / basic character of the compounds of a given transition metal in various oxidation states*. We have said earlier that with the increase in oxidation state of a given transition metal, the covalent character of its compounds increases. The increase in the covalent character also increases the acidic character of the compound. Thus the compounds of a given transition metal become more and more acidic as the oxidation state of the transition metal increases. This fact becomes evident when we study the following:

(*a*) The variation of the basic, amphoteric and acidic character of the oxides of V, Cr and Mn is as shown below:

| | | | | | | |
|---|---|---|---|---|---|---|
| (*i*) *Oxides of vanadium:* | VO | $V_2O_3$ | $VO_2$ | $V_2O_5$ | | |
| *Oxidation state of vandium* | +2 | +3 | +4 | +5 | | |
| *Nature of the oxides:* | Basic | Basic | Amphoteric | Acidic | | |
| (*ii*) *Oxides of Cr:* | CrO | $Cr_2O_3$ | $CrO_2$ | $CrO_3$ | | |
| *Oxidiation state of Cr:* | +2 | +3 | +4 | +6 | | |
| Nature of the oxides: | Basic | Amphoteric | Amphoteric | Acidic | | |
| (*iii*) *Oxides of Mn*: | MnO | $Mn_3O_4$ | $M_2O_3$ | $MnO_2$ | $MnO_3$ (Unstable) | $Mn_2O_7$ |
| *Oxidation state of Mn*: | +2 | +8/3 | +3 | +4 | +6 | +7 |
| *Nature of the oxides*: | Basic | Amphoteric | Amphoteric | Amphoteric | Weakly Acidic | Strong Acidic |

The variation of acidic, basic and amphoteric character of the oxides as shown above indicates that the oxides of a given transition metal in lower oxidation states are basic. Those of the same metal in intermediate oxidation states are amphoteric while those of the metal in higher oxidation states are acidic.

(*b*) The acidic character of $TiCl_2$, $TiCl_3$ and $TiCl_4$ increases in the order:

$$\overset{2+}{Ti}Cl_2 < \overset{3+}{Ti}Cl_3 < \overset{4+}{Ti}Cl_4 .$$

Similarly the acidic character of $VCl_2$, $VCl_3$ and $VOCl_3$ is in the order:

$$\overset{2+}{V}Cl_2 < \overset{3+}{V}Cl_3 < \overset{5+}{V}OCl_3$$

(*c*) Acidic character of $Mn(OH)_2$, $Mn(OH)_3$, $Mn(OH)_4$, $H_2MnO_4$ and $HMnO_4$ is in the order:

$\overset{2+}{Mn}(OH)_2$ (base of medium strength) $< \overset{3+}{Mn}(OH)_3$ (weak base) $< \overset{4+}{Mn}(OH)_4$ (amphoteric hydroxide) $< H_2\overset{6+}{Mn}O_4$ (strong acid) $< H\overset{7+}{Mn}O_4$ (very strong acid).

(7) *No. of various oxidation states shown by elements*. When we proceed from Sc ($Sc = 3d^1 4s^2$, $n = 1$) to Mn ($Mn = 3d^5 4s^2$, $n = 5$) in the series, the number of unpaired electrons ($n$) increases from 1 (in Sc) to 5 (in Mn) (do not consider Cr which has 6 unpaired electrons) and hence the number of various oxidation states shown by the elements increases from Sc to Mn. On proceeding further from Mn to *Zn* ($Zn = 3d^{10} 4s^2$, $n = 0$) the number of unpaired electrons decreases from 5 in Mn to zero in Zn and hence the number of various oxidation states shown by these elements decreases from Mn to Zn. Thus we find that the number of oxidation states shown by Mn is maximum while that shown by the elements lying at the right and left of Mn goes on decreasing when we proceed in either direction of Mn. The lesser number of oxidation states on the extreme ends are either due to too few electrons to lose or share (e.g, Sc, Ti) or too many *d*-electrons so that fewer orbitals are available to share electrons with others (e.g., Cu, Zn).

(8) *Oxidation state of Zn*. 3*d* orbitals in $Zn^{2+}$ ion ($Zn^{2+} = 3d^{10} 4s^0$), being completely-filled, are stable and hence do not lose any more electrons. Thus Zn atom cannot show any oxidation state higher than +2.

(9) *Relative stability of various oxidation states of a given element*. (*a*) The stability of a given oxidation state of a transition metal depends on the nature of the element with which it combines.

(*b*) The relative stability of various oxidation states of a given element can be explained on the basis of the stability of $d^0$, $d^5$ and $d^{10}$ orbitals. For example: $Ti^{4+}$ ($3d^0$. $4s^0$) is more stable than $Ti^{+3}$ ($3d^1 . 4s^0$) because of the presence of $3d^0$ orbital in $Ti^{4+}$ ion. $Mn^{2+}$ ($3d^5 . 4s^0$) is more stable than $Mn^{3+}$ ($3d^1 . 4s^0$) since $Mn^{2+}$ ion has $3d^5$ orbital.

(*c*) The stability of +2 oxidation state increases as we proceed from Sc to Zn in the series, *i.e.*, the reducing power of $M^{2+}$ ions decreases from left to right. For example, the salts of $V^{2+}$ (vanadous salts) and $Cr^{2+}$ (chromous salts) are powerful reducing agents while the salts of other $M^{2+}$ ions are not reducing agents because of the presence of half-filled 3*d* orbitals ($Mn^{2+} = 3d^5$). $Ni^{2+}$ and $Cu^{2+}$ do not have reducing property.

(*d*) The stability of +3 oxidation state decreases on proceeding from Sc to Cu (Zn does not show +3 oxidation state) in the series, *i.e.*, *the* oxidising power of $M^{3+}$ ions increases along the series. +3 is a stable oxidation stable for Sc. $Ti^{3+}$ ion acts as a reducing agent, since it is readily oxidised to $Ti^{4+}$ ion. $V^{3+}$ and $Cr^{3+}$ ions do not act as oxidising agents (*i.e.*, these ions are stable), since they do not get reduced to cations with lower oxidation states. $Mn^{3+}$ ion acts as an oxidising agent (*i.e,* this ion is unstable), since it is reduced to $Mn^{2+}$ ion. $Fe^{3+}$ ion ($3d^5 4s^0$) is stable because of the presence of stable half-filled 3*d*-orbitals in it.

(*e*) The stability of a given high oxidation state of different transition metals of the same sub group increases as we move down the group, *e.g.* the compounds of V (+5) are oxidising agents (*i.e.*, unstable) while those of Nb (+5) and Ta (+5) are stable. Similarly paramagnets containing Mn (+7) are strong oxidising agents while perrhenates containing Re (+7) are stable.

(*f*) The stability of a given lower oxidation state of different transition metals of a given sub group decreases as we descend in the sub-group.

(10) The variability of oxidation states of transition elements,which is due to incompletely filled *d*-orbitals. arises in such a way that successive oxidation states differ by unity, e.g. $V^{II}$, $V^{III}$, $V^{IV}$, $V^{V}$. This is unlike non-transition elements where oxidation states normally differ by two units, e.g., $Pb^{II}$, $Pb^{IV}$, $Sn^{II}$, $Sn^{IV}$, etc.

(11) In a group of *d*-block elements, the higher oxidation states are more stable for heavier elements. For example, in group 6, Mo (VI) and W(VI) are found to be more stable than Cr (VI). For this reason, Cr (VI) in the form of dichromate is a stronger oxidizing agent in acidic medium whereas $MoO_3$ and $WO_3$ are not. This is again unlike p-block elements where lower oxidation states are more stable for heavier members due to inert pair effect.

**Factors on which the Stability of Different Oxidation States of Transition Metals Depend.**

Following are the factors on which the stability of different oxidation states of transition metals depend.

**1. E° values.** With the help of E° values we can explain the relative stability of $Cu^{2+}$ ion as compared to $Cu^+$ ion. E° values for the reduction of $Cu^{2+}$ ion to $Cu^+$ and of $Cu^+$ to Cu are given below

(a) $Cu^{2+} + e^- \longrightarrow Cu^+$, E° = +0.15 V

(b) $Cu^+ + e^- \longrightarrow Cu°$, E° = + 0.52 V

Since E° value for reduction reaction (a) is less than E° value for reduction reaction (b), $Cu^{2+}$ ion has less tendency to be reduced to $Cu^+$ than $Cu^+$ ion has tendency to be reduced to Cu°. Thus $Cu^{2+}$ ion is more stable than $Cu^+$ ion in aqueous solution. In other words we can say that Cu is more stable in +2 oxidation state than in +1 oxidation state.

**2. Nature of solvent.** Another factor which affects the stability of different oxidation states is the nature of solvent with which the ions having different oxidation states are in contact. For example, in water $Cr^{3+}$ ion is stable but $Cr^{2+}$ ion is unstable in the same solvent. Similarly, $Fe^{3+}$ ion is quite stable in water but $Fe^{2+}$ ion is unstable in aerated water as it undergoes oxidation in it.

**3. Ionisation energy.** Lesser is the value of ionisation energy (IE) of an ion, greater is the stability of the compound in that particular oxidation state. For example we can explain the following:

*(i) Compounds of Ni in +2 oxidation state are more stable than those of Pt in the same (i.e. +2) oxidation state.* This can be explained as follows.

When a metal (M) is converted into $M^{2+}$ ion (O.S. of M = +2), energy is required which is equal to ($IE_1 + IE_2$). For Ni and Pt ($IE_1 + IE_2$) values are given below.

$Ni \rightarrow Ni^{2+} + 2e^-$, Energy required = $IE_1 + IE_2 = 2.49 \times 10^{23}$ KJ mol$^{-1}$

$Pt \rightarrow Pt^{2+} + 2e^-$, Energy required = $IE_1 + IE_2 = 2.66 \times 10^3$ KJ mol$^{-1}$

Now since ($IE_1 + IE_2$) value for Ni is less than ($IE_1 + IE_2$) value for Pt, compounds of Ni in +2 oxidation state are more stable than those of Pt in the same (*i.e.* +2) oxidation state.

*(ii) $K_2[PtCl_6]$ having Pt in +4 oxidation state (O.S.) is stable while $K_2[Ni\ Cl_6]$ having Ni in the same (i.e. +4) O.S. is unknown.* This can be explained as follows.

When Pt and Ni is converted into $Pt^{4+}$ and $Ni^{4+}$ respectively, energy equal to ($IE_1 + IE_2 + IE_3 + IE_4$) is required as shown below.

$Pt \rightarrow Pt^{4+} + 4e^-$, Energy required = $IE_1 + IE_2 + IE_3 + IE_4 = 9.36 \times 10^3$ KJ mol$^{-1}$

$Ni \rightarrow Ni^{4+} + 4e^-$, Energy required = $IE_1 + IE_2 + IE_3 + IE_4 = 11.29 \times 10^3$ KJ mol$^{-1}$

Now since ($IE_1 + IE_2 + IE_3 + IE_4$) value for Pt is less than ($IE_1 + IE_2 + IE_3 + IE_4$) value for Ni, compounds of Pt in +4 O.S. are more stable than those of Ni in the same (*i.e.* +4) O.S. As a result $K_2[PtCl_6]$ which contains Pt in +4 O.S. is stable while $K_2[NiCl_6]$ which has Ni in +4 O.S. is unstable, *i.e.* is not known.

## XI. Catalytic Properties

Many transition metals and their compounds are used as catalysts.

**Explanation:** Following explanations have been given to account for the catalytic properties of the metals and their compounds.

**(*i*) Presence of unpaired electrons in *d*-orbitals:** The catalytic properties of the transition elements are probably due to the presence of *unpaired electrons in their incomplete d-orbitals* and hence possess the capacity to absorb and re-emit wide range of energies. This makes the required *energy of activation* available.

**(*ii*) Intermediate compound formation theory:** In some cases, the transition element (*i.e.* catalyst) with varying oxidation states may form intermediate compounds (*Intermediate compound formation theory of catalysis*) with one of the reactants. These intermediates provide a new path with lower activation energy. Further, they get decomposed on reaction with the other reactant regenerating the catalyst (*i.e.*, transition metal).

$$\underset{\text{One reactant}}{X} + \underset{\text{Transition metal}}{\text{Catalyst}} \longrightarrow \underset{\text{Intermediate unstable compound}}{X \cdot \text{Catalyst}}$$

$$X.\ \text{Catalyst} + \underset{\text{Other reactant}}{Y} \longrightarrow \underset{\text{Product}}{XY} + \text{Catalyst}$$

*Illustrative examples:* (*a*) How the change in oxidation state of the transition metal helps the reaction through the formation of intermediate may be seen from the reaction between $SO_2$ and $O_2$ to form $SO_3$ in presence of $V_2O_5$.

$$\underset{(V=+5)}{V_2O_5} + SO_2 \longrightarrow \underset{\text{Divanadium tetraoxide } (V = +4)}{V_2O_4} + SO_3$$

$$\underset{(V = +4)}{2V_2O_4} + O_2 \longrightarrow \underset{(V=+5)}{2V_2O_5}$$

In the above reaction, vanadium changes its oxidation state from +5 to +4 and again to +5.

(*b*) Another example may be quoted of the reaction between $I^-$ and $S_2O_8^{2-}$ ions in presence of Fe (III) as catalyst.

$$2I^- + S_2O_8^{2-} \longrightarrow I_2 + 2SO_4^{2-}$$

The reaction is believed to take place as follows:

$$2Fe^{3+} + 2I^- \longrightarrow 2Fe^{2+} + I_2$$

$$2Fe^{2+} + S_2O_8^{2-} \longrightarrow 2Fe^{3+} + 2SO_4^{2-}$$

Thus we can say that the transition metal and their compounds act as good catalysts in chemical reactions because of variable oxidation states exhibited by the transition metals.

(*iii*) **Adsorption theory:** In other cases, the transition metal provides a suitable large surface area with free valencies on which the reactants are absorbed. As a result, the concentration of the reactants on the surface of the catalyst increases. Hence, the rate of reaction increases (*Adsorption theory*).

**Examples of catalysts.** (*i*) The *finely-divided nickel* can adsorb fairly large quantity of hydrogen especially at high temperatures. Due to this property, finely-divided Ni is used as a catalyst in various reactions like those shown below:

(*a*) Reduction of unsaturated ethylenic compounds (*Sabatier process*).

$$CH_2 = CH_2 + H_2 \xrightarrow{\text{Ni(410 K)}} CH_3—CH_3$$

(*b*) *In the hydrogenation of oils*. In this process the unsaturated oils are converted into saturated fats (hardening of oils) in presence of Ni.

(*c*) The following hydrogenation reactions are catalysed by heated nickel.

$$CO + 3H_2 \longrightarrow CH_4 + H_2O$$

$$2NO + 5H_2 \longrightarrow 2NH_3 + 2H_2O$$

(*ii*) *Platinum black* is used as catalyst in the preparation of HCHO from $CH_3OH$ and in the decomposition of $H_2O_2$. Copper is also used as a catalyst for converting $CH_3OH$ to HCHO.

$$2CH_3OH + O_2\,(\textit{From air}) \xrightarrow{\text{Pt black}} 2HCHO + 2H_2O$$

$$2H_2O_2 \xrightarrow{\text{Pt black}} 2H_2O + O_2$$

$$2CH_3OH + O_2\,(\textit{From air}) \xrightarrow{\text{Cu (770 K)}} 2HCHO + 2H_2O$$

(*iii*) *Platinsed asbestos and $V_2O_5$* are used as catalyst in the oxidation of $SO_2$ to $SO_3$ required for the manufacture of $H_2SO_4$ by contact process.

(*iv*) *Platinised gauge* is used as a catalyst in the oxidation of $NH_3$ to NO required for the manufacture of $HNO_3$ by Ostwald's process.

$$4NH_3 + 5O_2 \xrightarrow[800°C]{\text{Pt gauge}} 4NO + 6H_2O$$

(*v*) Finely divided Fe mixed with Mo (as promotor) is used for catalysing the combination of $N_2$ and $H_2$ to form $NH_3$ by Harber's process.

$$N_2 + 3H_2 \xrightleftharpoons[770\,K]{\text{Finely divided Fe +Mo as prmotor}} 2NH_3$$

(*vi*) Fenton's reagant ($FeSO_4 + H_2O_2$) is used for oxidising alcohols to aldehydes.

(*vii*) Pd is used for hydrogenation of phenol to cyclohexanone.

(*viii*) Ziegler-Natta catalyst ($TiCl_4$) is used for the manufacture of polythene.

(*ix*) $V_2O_3$ is used as a catalyst for converting ethylene to ethyl alcohol.

(*x*) $VOCl_3$ is used as catalyst in olefine polymerisation including ethylene propylene rubber.

(*xi*) NiO catalyses the following reaction which is used for the manufacture of sulphur from coal gas.

$$Fe_2O_3 + 3H_2S \xrightarrow{\text{NiO}} 2FeS + 3H_2O + S$$

(*xii*) $MnO_2$ is used to catalyse the decomposition of $H_2O_2$ solution.

(*xiii*) Co (II) salts catalyse the decomposition of bleaching powder ($ClO^-$) as Co (II) salts can easily be oxidised to Co (III) salts.

$$Co^{2+} + OCl^- + H_2O \longrightarrow Co^{3+} + Cl^- + 2OH^-$$

$$2Co^{3+} + 2OH^- \longrightarrow 2Co^{2+} + H_2O + \frac{1}{2}O_2$$

(*xiv*) Vanadium as $V_2O_5$ is used in the manufacture of $H_2SO_4$ by contact process.

(*xv*) *Cobalt* as cobalt-thorium is used in the Fischer – Tropsch process in the synthesis of gasoline (petrol).

(*xvi*) *Nickel* in the finely divided state is used in the hydrogenation reactions, *e.g.*, hydrogenation of oils into fats.

(*xvii*) *Platinum* as spongy platinum is used in the manufacture of $H_2SO_4$ by contact process and in the oxidation of ammonia to nitric acid by Ostwald's process.

## XII. Colour of Transition Metal Complex Ions

Most of the transition metal complexes are coloured. We know that the sun light is composed of seven colours *viz.* violet, indigo, blue, green, yellow, orange and red (VIBGYOR) - These colours are present in the visible region of light. When sun light falls on a transition metal complex ion, the ion absorbs light of a specific colour from the visible region and transmits (emits) the remaining portion of the incident sun light. The transmitted light has colour which is different from that of the absorbed light. *The colour of the transmitted light is, in fact, the colour of the complex ion.* The colour of the absorbed light and that of the transmitted light are called *complementary colours to each other*. Table 5.12 contains a list of colours of the absorbed light and transmitted light.

**Table. 5.12. Wavelength (λ) and wave number (v) of absorbed light**

| *Wavelength (λ) of absorbed light (Å)* | *Wave number of absorbed light, $\bar{v}$ ($cm^{-1}$)* | *Colour of absorbed light* | *Colour of transmitted light* |
|---|---|---|---|
| 4000—4350 | 25,000—22,990 | Violet | Yellow-green |
| 4350—4800 | 22,990—20,830 | Blue | Yellow |
| 4800—4900 | 20,830—20,410 | Green-blue | Orange |
| 4900—5000 | 20,410—20,000 | Blue-green | Red |
| 5000—5600 | 20,000—17,800 | Green | Purple |
| 5600—5800 | 17,800—17,240 | Yellow-green | Violet |
| 5800—5950 | 17,240—16,810 | Yellow | Blue |
| 5950—6050 | 16,810—16,530 | Orange | Green-blue |
| 6050—7500 | 16,530—13,330 | Red | Blue-green |

In the first column of Table 5.12 the wavelength (λ) of the absorbed light has been shown in Å(1Å = $10^{-8}$ cm) and in the second column the frequency of the absorbed light has been shown as wave number ($\bar{v}$) in $cm^{-1}$.

In 3rd column of Table 5.12 we have given the colour of the absorbed light and in 4*th* column we have given the colours of their corresponding transmitted light (*i.e.* complementary colours of the absorbed light).

Thus if we know the colour of the light absorbed by a complex ion, the colour of the complex ion can be predicted with the help of 4th column. For example:

(*a*) Since the solution of $[Cu(H_2O)_6]^{2+}$ ion absorbs the radiation of *yellow colour* and transmits the radiation of *blue colour*, this complete ion has *blue colour*.

(*b*) Aqueous solution of $[Ti(H_2O)_6]^{3+}$ ion absorbs radiation of *green colour* and transmits *purple radiation*. Hence $[Ti(H_2O)_6]^{3+}$ ion has *purple colour*.

(*c*) $[Co(H_2O)_6]^{2+}$ ion absorbs radiation of blue-green and transmits the radiation of red colour. Hence this ion is red in colour. Here it should be noted that the complex ion absorbs only that portion of the incident light whose wavelength lies in the visible region of electromagnetic radiation. Visible region lies between 4000Å and 8000Å.

## Factors On Which Colour Depends

Following are the important factors on which the colour of complex ions depends:

1. **Nature of metal ion.** Since the complex ions *viz* $[Ti(H_2O)_6]^{3+}$ and $[Co(H_2O)_6]^{2+}$ have different metal ions, these ions have different colours. Note that these ions have the same ligands. Their different colours are due to the fact that they absorb radiations of different colours and hence transmit different colours. $[Ti(H_2O)_6]^{3+}$ ion absorbs green radiation and transmits violet (almost purple) radiation. Thus this ion looks purple to our eyes. On the other hand $[Co(H_2O)_6]^{2+}$ ion absorbs blue green radiation and transmits red radiation. Hence, this ion appears red.

2. **Nature of ligands.** Since the complex ions *viz* $[Ni(H_2O)_6]^{2+}$ and $[Ni(NH_3)_6]^{2+}$ ions have different ligands, these ions have different colours. Their different colours are because of the presence of different ligands in them. It is due to the presence of different ligands that they absorb different colours and hence transmit different colours. $[Ni(H_2O)_6]^{2+}$ ion absorbs red radiation and transmits blue radiation. Thus $[Ni(H_2O)_6]^{2+}$ ion is blue green. On the other hand $[Ni(NH_3)_6]^{2+}$ ion absorbs yellow radiation and transmits blue radiation. Thus $[Ni(NH_3)_6]^{2+}$ ion is blue in colour.

Now consider the complexes of cobalt with different ligands such as $[Co(H_2O)_6]^{3+}$, $[Co(NH_3)_6]^{3+}$ and $[Co(CN)_6]^{3-}$. The crystal field splitting of the ligands is in the order $H_2O < NH_3 < CN^-$ (from spectrochemical series) and the excitation energies will be in the order $[Co(H_2O)_6]^{3+} < [Co(NH_3)_6]^{3+} < [Co(CN)_6]^{3-}$. The wavelengths absorbed will be in the opposite order as $E = \frac{hc}{\lambda}$. The complementary colours will be transmitted as shown below in the table and are different for different ligands.

| Complex | $[Co(H_2O)_6]^{3+}$ | $[Co(NH_3)_6]^{3+}$ | $[Co(CN)_6]^{3-}$ |
|---|---|---|---|
| Colour absorbed | Orange | Blue | Violet |
| Colour transmitted | Blue | Yellow orange | Yellow |

Also the colour of the complex changes with the concentration of the ligand. To illustrate this, take $[Ni(H_2O)_6]^{2+}$ complex (obtained on dissolving nickel chloride in water) and add more and more of bidentate ligand ethane –1, 2-dammine (*en*) into it. The following changes in colour take place:

| *Ni²⁺: (en) molar ratio* | *Reaction and colour changes* |
|---|---|
| 1 : 1 | $[Ni(H_2O)_6]^{2+}(aq) + en\,(aq) \rightarrow [Ni(H_2O)_4(en)]^{2+}(aq) + 2H_2O$<br>Green → Pale blue |
| 1 : 2 | $[Ni(H_2O)_4(en)]^{2+}(aq) + en\,(aq) \rightarrow [Ni(H_2O)_2(en)_2]^{2+}(aq) + 2H_2O$<br>Pale blue → Blue / Purple |
| 1 : 3 | $[Ni(H_2O)_2(en)_2]^{2+}(aq) + en\,(aq) \rightarrow [Ni(en)_3]^{2+}(aq) + 2H_2O$<br>Blue/Purple → Violet |

It is important to note that in the absence of a ligand, crystal field splitting does not occur and hence the substance is colourless. For example, $[Ti(H_2O)_6]Cl_3$ is purple but anhydrous $TiCl_3$ is colourless. Similarly, $CuSO_4.5H_2O$ is blue but anhydrous $CuSO_4$ is colourless.

## Explanation of Colour of Complexes on the Basis of CFT

Transition metal complexes whose central atom has partially-filled *d*-orbitals are coloured. In other words we can say that the transition metal ions having one or more unpaired electrons (*n*) in their *d*-orbitals are coloured.

Colours of some hexahydrated complex ions, $[M^{n+} (H_2O)_6]^{n+}$ whose central metal ion ($M^{n+}$) has unpaired electrons in *d*-orbitals are given in Table 5.13.

**Table 5.13. Colours of some hexahydrated transition metal complex ions, $[M(H_2O)_6]^{n+}$. For the sake of convenience, $6H_2O$ molecules have not been shown (*n* = No. of unpaired electrons).**

| *Ions* | *Outer electronic configuration* | *Colour of the ion* |
|---|---|---|
| $Ti^{3+}$ | $3d^1(n = 1)$ | Violet (almost purple) |
| $V^{3+}$ | $3d^2(n = 2)$ | Green |
| $V^{2+}$, $Cr^{3+}$ | $3d^3(n = 3)$ | Violet |
| $Cr^{2+}$ | $3d^4(n = 4)$ | Blue |
| $Mn^{3+}$ | $3d^4(n = 4)$ | Violet |
| $Mn^{2+}$ | $3d^5(n = 5)$ | Pink |
| $Fe^{3+}$ | $3d^5(n = 5)$ | Yellow |
| $Fe^{2+}$ | $3d^6(n = 4)$ | Green |
| $Co^{3+}$ | $3d^6(n = 4)$ | Pink |
| $Co^{2+}$ | $3d^7(n = 3)$ | Pink (red) |
| $Ni^{2+}$ | $3d^8(n = 2)$ | Blue-green |
| $Cu^{2+}$ | $3d^9(n = 1)$ | Blue |

As an example the purple colour of the octahedral hexahydrate complex ion, $[Ti (H_2O)_6]^{3+}$ can be explained as follows on the basis of CFT.

In the given complex ion Ti is present as $Ti^{3+}$ ion whose valence-shell electronic configuration in $3d^1$. According to crystal field theory when six water molecules (which act as ligands) approach the central $Ti^{3+}$ ion to form the octahedral complex ion, $[Ti (H_2O)_6]^{3+}$, the five *d*-orbitals of $Ti^{3+}$ ion split into $t_{2g}$ ($d_{xy}$, $d_{yz}$ and $d_{zx}$) and $e_g$($d_{x^2-y^2}$ and $d_{z^2}$) sets of orbitals. This phenomenon is called **crystal field splitting.** $t_{2g}$ set of orbitals has lower energy than $e_g$ set of orbitals and hence $d^1$ electron of $Ti^{3+}$ ion resides in $t_{2g}$ set and $e_g$ set remains vacant. Thus the ground state electronic configuration of $Ti^{3+}$ ion can be written as $t_{2g}^1 e_g^0$.

$[Ti (H_2O)_6]^{3+}$ ion absorbs green radiation and transmits purple radiation. Hence this complex ion has purple colour. The absorption of green radiation takes place at wave number ($\overline{v}$) of about 20,000 $cm^{-1}$ as is evident from the visible absorption spectrum of $[Ti (H_2O)_6]^{3+}$ ion given in Fig. 5.3.

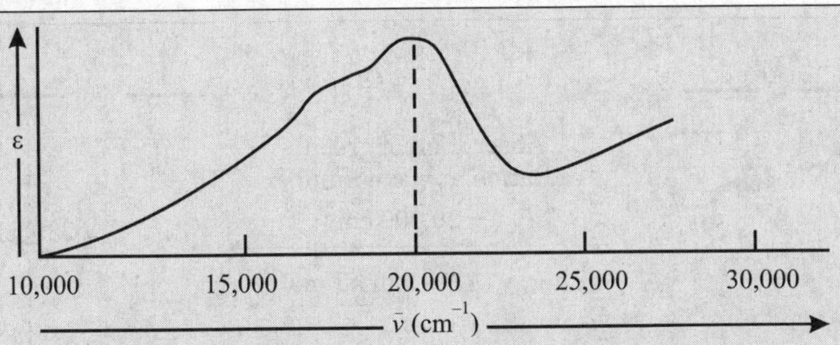


**Fig. 5.3** Crystal field absorption band of $[Ti(H_2O)_6]^{3+}$ in aqueous solution.

The energy ($\Delta_0$) associated with 20,000 cm$^{-1}$ is about 239 kJ mol$^{-1}$ as shown below.

We know that $\Delta_0$ is given by :

$$\Delta_0 = \frac{h(Js) \times c\,(ms^{-1}) \times N_o(\text{mol}^{-1})}{\lambda(m)} \quad ...(i)$$

Here $h$ = Planck's constant = $6.626 \times 10^{-34}$ Js, $c$ = Velocity of light = $3 \times 10^8$ ms$^{-1}$, No = Avogadro's number = $6.023 \times 10^{23}$ mol$^{-1}$ and $\lambda$ = wavelength ($m$).

Wavelength ($\lambda$) is given as 20,000 cm$^{-1}$. This value must be converted into m. Thus:

$$20{,}000 \text{ cm}^{-1} = \frac{1}{20{,}000}\text{cm} = \frac{1}{2\times10^4}\text{ cm} = 0.5 \times \frac{1}{10^4}\text{cm}$$

$$= 0.5\times\frac{1}{10^4}\times\frac{1}{10^2}\text{m} = 0.5 \times 10^{-6}\text{ m}$$

Now put the values of $h$, $c$, No and $\lambda$ in equation ($i$) to get the value of $\Delta_0$

$$\therefore \quad \Delta_0 = \frac{(6.626\times10^{-34}\text{ Js})\times(3\times10^8\text{ ms}^{-1})\times(6.023\times10^{23}\text{ mol}^{-1})}{0.5\times10^{-6}\text{ m}}$$

$$= \frac{6.626\times3\times6.023\times10^{-34}\times10^8\times10^{23}}{0.5\times10^{-6}}\text{J mol}^{-1}$$

$$= \frac{6.626\times3\times6.023\times10^3}{0.5}\text{ J mol}^{-1}$$

$$= \frac{6.626\times3\times6.023\times10^3}{0.5\times10^3}\text{KJ mol}^{-1} = \frac{119.72}{0.5}\text{ KJ mol}^{-}$$

$$= 239.44 \text{ KJ mol}^{-1} \approx 239 \text{ KJ mol}^{-1}$$

This energy (= 239 kJ mol$^{-1}$) is close to the *crystal field spitting energy*, $\Delta_0$ (also called *excitation energy*) (Excitation energy is the energy difference between $t_{2g}$ and $e_g$ orbitals) and therefore, the electron present in $t_{2g}$ orbitals absorbs 239 KJ mol$^{-1}$ and is excited to occupy the vacant $e_g$ orbitals ($t^1_{2g} \rightarrow e^0_g$ electron transition) so that the electronic configuration of $Ti^{3+}$ ion in the excited state becomes $t^0_{2g}\, e^1_g$ (See Fig 5.4). Thus we see that the purple colour of $[Ti(H_2O)_6]^{3+}$ ion is due to the excitation of an electron from the *d*-orbitals of lower energy ($t_{2g}$ orbitals) to the *d*-orbitals of higher energy ($e_g$ orbitals).

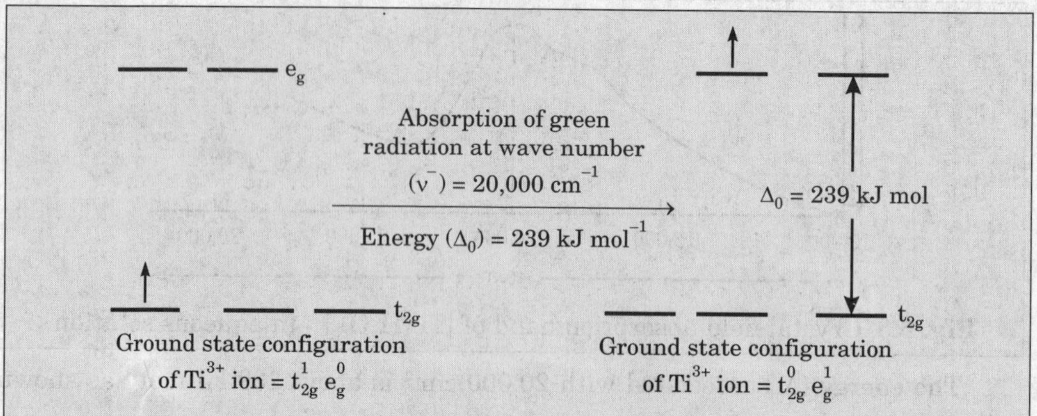


**Fig. 5.4** Promotion of an electron from $t_{2g}$ orbitals to $e_g$ orbitals when $[Ti\,(H_2O_6)]^{3+}$ absorbs green radiation at wave number ($\bar{v}$) of 20,000 cm$^{-1}$.

Above discussion which is based on CFT shows that $d-d$ electronic transition is possible only in case of those transition metal ions whose $d$-orbitals have unpaired electrons. This implies that only those ions are coloured which have unpaired electrons.

The transition metal complex ions whose central atom contains empty ($d^0$ configuration) or completely-filled orbitals ($d^{10}$ configuration) are colourless. This is because of the fact that since the $d$-orbitals of the central metal ion do not contain any electron (empty $d$-orbitals) or are completely-filled (*i.e.*, contain all the electrons in the paired state), $d-d$ electron transition is not possible and hence no light of any colour or wavelength is absorbed by such ions in the visible region. Thus the transition metal ions containing empty $d$-orbitals [e.g. $Sc^{3+}$ and $Ti^{4+}$ ($TiO_2$) ions] or completely-filled $d$-orbitals [e.g., $Cu^+$, $Ag^+$, $Zn^{2+}$ ($ZnSO_4$), $Cd^{2+}$, $Hg^{2+}$ etc.] are colourless. In other words we can say that the transition metal ions having $d^0$ or $d^{10}$ configuration are colourless, *i.e.* the transition metal ions having $n = 0$ are colourless.

## Relation Between Crystal Field Splitting Energy ($\Delta_0$) and Wavelength of the Absorbed Radiation

The magnitude of $\Delta_0$ ($\Delta_0$ is the energy difference between $t_{2g}$ and $e_g$ orbitals) is inversely proportional to that of the wavelength of the absorbed radiation. This means that a complex ion having higher value of $\Delta_0$ will absorb the radiation of lower wavelength and *vice versa*. This fact is evident from the following examples of octahedral complex ions of $Co^{3+}$ and $Ni^{2+}$ ions.

(*a*) Since the value of $\Delta_0$ for $[Co\,(CN)_3]^{3-}$ ion is higher than that of $[Co\,(NH_3)_6]^{3+}$, $[Co\,(CN)_6]^{3-}$ ion absorbs violet radiation which has lower wavelength. On the other hand, $[Co\,(NH_3)_6]^{3+}$ ion absorbs blue green radiation which has higher wavelength. Thus $[Co\,(CN)_6]^{3-}$ ion appears yellow-green in solution and $[Co\,(NH_3)_6]^{3+}$ ion appears orange or pink in solution.

(*b*) Since $[Ni\,(H_2O)_6]^{2+}$ ion has higher value of $\Delta_0$ than $[Ni\,(NH_3)_6]^{2+}$ ion, $[Ni(H_2O)_6]^{2+}$ ion absorbs yellow colour which has lower wavelength while $[Ni\,(NH_3)_6]^{2+}$ ion absorbs red colour which has higher wavelength. Thus,

Ni $(H_2O)_6]^{2+}$ ion absorbs red colour which has higher wavelength. Thus, [Ni $(H_2O)_6]^{2+}$ ion in solution has blue colour but [Ni $(NH)_3)_6]^{2+}$ ion in solution has blue green colour.

## XIII. Magnetic Properties

(Also see chapter "*Magnetic properties of transition metal complexes*")

On the basis of their behavior in magnetic field, the transition elements and their compounds have been classified into the following three main categories *viz. paramagnetic substances, dimagnetic substances and ferromagnetic substances*.

### 1. Paramagnetic Substances

These are the species (atom, ion or molecule) which are attracted by the magnetic field. When a paramagnetic substance is weighed in air on a Gouy magnetic balance and magnetic field is applied to it, the weight of the substance increases in magnetic field. The property of paramagnetism is due to the presence of unpaired electrons in the species.

**Magnetic Moment ($\mu$)**

We know that an electron in a species revolves in orbits round the nucleus and also spins around its own axis. The revolving of the electron round the nucleus is called *orbital motion* of the electron and the spinning of the electron round its own axis is called *spin motion* of the electron. Both these motions produce magnetic moments which are called *orbital magnetic moment* and *spin magnetic moment* respectively. The observed (*i.e.,* experimental) magnetic moment value of a given species is the sum of the above said two magnetic moments. In other words we can say that both the motions of the electron (spin motion and orbital motion) make a combined contribution to the value of magnetic moment of a species. The contribution made by orbital motion of the electron to the magnetic moment value is called *orbital contribution*.

It has been observed that in case of the complexes of the elements of *2nd* and *3rd* transition series (*i.e.,* *4d* and 5 *d* series elements) both types of motion of the electrons make a contribution to the magnetic moment of the complexes. Thus the magnetic moment of such complexes is given by equation (*i*) and hence is represented as $\mu_{S+L}$

$$\mu_{S+L} = \sqrt{4S(S+1)+L(L+1)} \text{ B.M.} \quad ...(i)$$

In equation (*i*)

S = Sum of the spin quantum numbers

and L = Sum of the orbital angular momentum quantum numbers

The magnetic moment of a species in which the orbital motion of the electron is quenched is called *spin only magnetic moment*, since this magnetic moment is only due to spin motion of the electrons. Spin only magnetic moment is denoted by $\mu_S$. Since orbital motion is quenched, the value of $\mu_S$ is given by equation (*ii*) which is obtained by putting L = 0 in equation (*i*).

$$\mu_S = \sqrt{4S(S+1)} \text{ B.M.} \quad ...(ii)$$

In equation (*ii*) *S* is the sum of spin quantum numbers (*s*) for unpaired electrons.

The value of $S$ used in equation (*ii*) is given by:

$$\boldsymbol{S = n \times \frac{1}{2}} \quad \ldots(iii)$$

where $n$ = No. of unpaired electrons.

With he help of equation (*iii*), equation (*ii*) becomes

$$\mu_S = \sqrt{4 \times \frac{n}{2}\left(\frac{n}{2}+1\right)}\ BM = \sqrt{2n\left(\frac{n+2}{2}\right)}\ BM$$

or

$$\boldsymbol{\mu_S = \sqrt{n(n+2)}\ \text{BM}} \quad \ldots(iv)$$

$\mu_S$ is the spin only magnetic moment value. Hence it is also represented as $\mu_{\text{spin only}}$. Thus:

$$\boldsymbol{\mu_{\text{spin only}} = \sqrt{n(n+2)}\ \text{BM}} \quad \ldots(v)$$

With the help of equation (*v*) the value of $\mu_{\text{spin only}}$ has been calculated for $n$ = 1, 2, 3, 4 and 5. These values in BM are 1.73 BM (for $n$ = 1), 2.84 (for $n$ = 2), 3.87 (for $n$ = 3), 4.90 (for $n$ = 4) and 5.92 BM (for $n$ = 5). Here it should be noted that equations (*ii*) and (*v*) are used for calculating magnetic moment value only of those transition metal complexes in which the orbital motion of the electrons are completely or partially quenched or suppressed.

Spin only magnetic moment values for the complexes formed by the elements of 1st transition series (3*d* series elements) have been calculated with the help of equation (*ii*) or (*iv*) and are given in Table 5.14.

**Table 5.14. Paramagnetic moments of some ions of first transition series**

| *Ions* | *Outer configuration* | *No. of unpaired electrons* | *Magnetic moment (BM)* | |
|---|---|---|---|---|
| | | | *Calculate* | *Observed* |
| $Sc^{3+}$, $Ti^{4+}$ | $3d^0$ | 0 | 0 | 0 |
| $Ti^{3+}$, $V^{4+}$ | $3d^1$ | 1 | 1.73 | 1.75 |
| $Ti^{2+}$, $V^{3+}$ | $3d^2$ | 2 | 2.84 | 2.76 |
| $V^{2+}$, $Cr^{3+}$ $Mn^{4+}$ | $3d^3$ | 3 | 3.87 | 3.86 |
| $Cr^{2+}$, $Mn^{3+}$ | $3d^4$ | 4 | 4.90 | 4.80 |
| $Mn^{2+}$, $Fe^{3+}$ | $3d^5$ | 5 | 5.92 | 5.96 |
| $Fe^{2+}$, $Co^{3+}$ | $3d^6$ | 4 | 4.90 | 5.0-5.5 |
| $Co^{2+}$ | $3d^7$ | 3 | 3.87 | 4.4-5.2 |
| $Ni^{2+}$ | $3d^8$ | 2 | 2.84 | 2.9-3.4 |
| $Cu^{2+}$ | $3d^9$ | 1 | 1.73 | 1.8-2.2 |
| $Cu^{+}$, $Zn^{2+}$ | $3d^{10}$ | 0 | 0 | 0 |

It may be seen from the Table that in case of ions having $d^\circ$ to $d^6$ configurations the observed (experimental) magnetic moment values agree fairly well with the calculated values. This agreement is due to the fact that in case of these ions the orbital motion of the electrons (orbital magnetic moment) is either completely or partially quenched. The quenching of orbital motion of electrons is due to electric field of the surrounding ligands (*i.e.*, crystal field) around the metal ion.

It may also be seen from the Table that in the later half of the series the observed magnetic moment values slightly exceed the calculated values (spin only values). These differences are due to some contribution from the orbital magnetic moments to the observed magnetic moment values.

## How to Determine the Paramagnetic and Diamagnetic Nature of a Given Substance?

The paramagnetic / diamagnetic nature of a given substance can be determined by weighing the substance in a *Gouy magnetic balance* in the absence and presence of magnetic field as shown in Fig. 5.5. The substance is taken in a small test tube and weighed in air (*i.e.*, in the absence of magnetic field) as shown at (*a*) of Fig. 5.5 and then weighed between the poles of a magnet. If the substance shows an increase in its weight, this implies that the substance is attracted by the magnetic field and hence is paramagnetic [See (*b*) of Fig. 5.5]. On the other hand, if the substance shows a decrease in its weight, it implies that the substance is repelled by the magnetic field and hence is diamagnetic. [See (*c*) of Fig. 5.5]

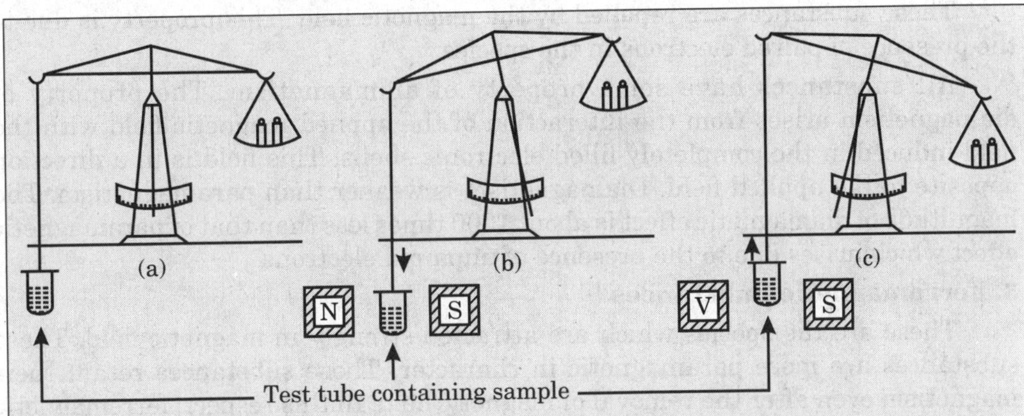


**Fig. 5.5** Gouy magnetic balance used to determine the pramagnetic and diamagnetic nature of a given substance. (*a*) weight of the substance in the absence of magnetic field (*b*) weight of the substance increases when placed between the poles of a magnet – paramagnetic substance (*c*) weight of the substance decreases when placed between the poles of a magnet – diamagnetic substance.

The change in weight (*i.e.*, increase or decrease in weight) is used for calculating the magnetic moment of the substance.

### Uses of Magnetic Moment Values

The experimental magnetic moment value of a given complex compound can be used to determine the number of unpaired electrons occupying the *d*-orbitals of the central metal ion of a given complex compound. The number of unpaired electrons determined in this way can be used to predict:

(*i*) Whether a given 4-coordinated complex compound is square planar ($dsp^2$ hybridisation) or tetrahedra ($sp^3$ hybridization).

(*ii*) Whether a given 6-coordinated complex compound (octahedral) results form $d^2sp^3$ hybridisation (inner orbital octahedral complex) or $sp^3d^2$ hybridisation (outer orbital octahedral complex).

This point would be clear when we consider the geometries of $[Ni(CN)_4]^{2-}$, $[NiCl_4]^{2-}$, $[Fe(CN)_6]^{3-}$ and $[FeF_6]^{3-}$ ions $[Ni(CN)_4]^{2-}$ ion has $n = 0$ (diamagnetic) and is square planar ($dsp^2$ hybridisation) while $[NiCl_4]^{2-}$ ion has $n = 2$ (paramagnetic) and is tetrahedral ($sp^3$ hybridisation). $[Fe(CN)_6]^{3-}$ ion has $n = 1$ (paramagnetic) and is inner-orbital octahedral in shape ($d^2sp^3$ hybridisation) but $[FeF_6]^{3-}$ ion $n = 5$ (paramagnetic) and is outer-orbital octahedral in shape ($sp^3d^2$ hybridisation).

Since $CN^-$ ligands are strong ligands, on the basis of CFT, the distribution of $3d^5$ electrons of $Fe^{3+}$ ion ($Fe^{3+} = 3d^5$) in $[Fe(CN)_6]^{3-}$ in $t_{2g}$ and $e_g$ orbitals is as $t_{2g}^5 e_g^\circ$ ($n = 1$). Thus $[Fe(CN)_6]^{3-}$ is a *low spin octahedral complex*. On the other hand, since $F^-$ ligands are weak ligands, the distribution of $3d^5$ electrons of $Fe^{3+}$ ion in $[FeF_6]^{3-}$ in $t_{2g}$ and $e_g$ orbitals is as $t_{2g}^3 e_g^2$ ($n = 5$). Thus $[FeF_6]^{3-}$ is a *high spin octahedral complex*.

**2. Diamagnetic Substances**

These substances are repelled by the magnetic field. This property is due to the presence of paired electrons in the species.

All substances have some property of diamagnetism. The property of diamagnetism arises from the interaction of the applied magnetic field with the field induced in the completely-filled electronic shells. This field is in a direction opposite to the applied field. Diamagnetism is weaker than paramagnetism. The magnitude of diamagnetic effect is about 1000 times less than that of paramagnetic effect which arises due to the presence of unpaired electrons.

**3. Ferromagnetic Substances**

These are the species which are attracted strongly in magnetic field. These substances are more paramagnetic in character. These substances retain their magnetism even after the removal of magnetic field. In this respect, ferromagnetic substances differ from paramagnetic substances, since paramagnetic substances lose their paramagnetism after the removal of magnetic field. Since Fe, Co and Ni have sufficient number of unpaired electrons (Fe = 4, Co = 3, Ni = 2), these elements are ferromagnetic.

## XIV. Electronegativity (EN)

Electronegativity (EN) of 3*d*, 4*d* and 5*d* series transition elements are given in Table 5.15.

**Table 5.15. Electronegativity values of 3*d*, 4*d* and 5*d* series transition elements**

| | | | | | | | | | | |
|---|---|---|---|---|---|---|---|---|---|---|
| *3d-series:* | Sc | Ti | V | Cr | Mn | Fe | Co | Ni | Cu | Zn |
| | 1.3 | 1.5 | 1.05 | 1.65 | 1.55 | 1.8 | 1.8 | 1.8 | 1.8 | 1.6 |
| *4d-series:* | Y | Zr | Nb | Mo | Tc | Ru | Rh | Pd | Ag | Cd |
| | 1.2 | 1.4 | 1.6 | 1.8 | 2.9 | 2.9 | 2.2 | 2.2 | 1.9 | 1.9 |
| *5d-series:* | La | Hf | Ta | W | Re | Os | Ir | Pt | Au | Hg |
| | 1.1 | 1.3 | 1.7 | 1.7 | 1.9 | 2.2 | 2.01 | 2.2 | 2.4 | 1.9 |

From this table the following points may be noted:

(*i*) As in case of main group elements, EN values of 3*d* series elements increase with the increase in atomic number along the period. However, the increase is relatively small and not regular. This is due to the involvement of inner *d*-orbitals.

(*ii*) EN value of Mn is lower than its neighbours namely Cr and Fe (Mn = 1.55, Cr = 1.65, Fe = 1.80). Lower value of EN for Mn is due to the presence of half-filled 3*d* orbitals in this element (Mn = $3d^5 4s^2$). Similarly EN value of Zn is lower than that of Cu (Zn = 1.6, Cu = 1.8). Lower EN value for Zn is due to the presence of completely filled 3*d* orbitals in Zn (Zn = $3d^{10}\ 4s^2$ ).

(*iii*) EN values for Co, Ni and Cu are more or less similar.

EN of an atom is not a fixed quantity but depends on the following factors.

(*i*) **Oxidation state.** Electronegativity of an atom increases with the increase in its positive oxidation state, since with the increase in positive oxidation state the tendency of the atom to attract electrons towards itself increases. As a result electronegativtiy also increases.

**Examples:** Fe (III) has higher electronegative than Fe (II). Similarly Cu (II) has a higher electronegativity than Cu (I). Fe (III) = 1.96, Fc (II) = 1.83, Cu (II) = 2.0, Cu (I) = 1.9

(*ii*) **Hybridisation.** Since *s*-electron is nearer the nucleus due to its greater penetration power, the electronegativity of an atom increases with the increase in s-character of hybrid orbital. Thus since *s*-character in *sp*, $sp^2$, $sp^3$ and $sp^3d^2$ hybrid orbitals decreases as sp > $sp^2$ > $sp^3$ > $sp^3d^2$, the electronegativity of an element having these hybrid orbitals decreases as sp > $sp^2$ > $sp^3$ > $sp^3d^2$. For example since $Cr^{2+}$ ion in $[CrO_4]^{2-}$ ion is $sp^3$ hybridised and $Cr^{3+}$ ion in $[Cr(NH_3)_6]^{3+}$ ion is $sp^3d^2$ hybridised, the former complex ion has higher electronegativity than the latter.

(*iii*) **Positive charge on the complex ion.** $[Cr(NH_3)_6]^{3+}$ and $[Cr(C_2O_4)_3]^{3-}$ both have Cr in +3 oxidation state. Since $[Cr(NH_3)_6]^{3+}$ ion has +3 charges and $[Cr(C_2O_4)_3]^{3-}$ ion has –3 charges, the former complex has higher electronegativity than the latter complex.

## XV. Formation of Complex Compounds

The transition metal atoms or the cations derived from them have an ability to form complex compounds with certain molecules (e.g., CO, NO, $NH_3$ etc.) or ions (e.g., $F^-$, $Cl^-$, $CN^-$, etc). These molecules or ions are called ligands. These ligands contain one or more lone pairs of electrons which they can easily donate to the transition metal atom or cation and thus complete the vacant orbitals of these atoms or cations through the formation of L → M coordinate bonds. The tendency of transition metal atom or cation to form the complex compounds is due to its following properties:

(*a*) Due to the small size of the transition metal atoms or their cations and high effective nuclear charge, they have a high positive charge density on them. This high positive change density makes the atoms or cations to attract (accept) the lone pair of electrons from the ligands.

(*b*) The transition metal cations or atoms have vacant *d*-orbitals [(*n* – 1) *d* orbitals] in which they can accommodate the lone pairs of electrons donated by the ligands and thus can form L → M coordinate bonds.

**Stability of Complex Compounds**

(*i*) Stability of complex compounds decreases with the increase in atomic number of the central transition metal.

(*ii*) Transition metals in high oxidation state form stable complexes with the small highly electronegative and basic ligands like $F^-$, $Cl^-$, $NH_3$, $IO_6^{5-}$, $TeO_6^{6-}$, etc. while these metals in low oxidation states (e.g., +1, 0, –1) give stable complexes with $\pi$-acid ligands like CO, NO, $CN^-$, $PCl_3$, $C_6H_6$, etc. The complex compounds formed by transition metal atoms in low oxidation states contain L $\rightleftharpoons$ M coordinate bond. Also see "*Variable Oxidation States*".

(*iii*) When a given transition metal forms complexes with the same ligand in different oxidation states, the complex compound having the metal in higher oxidation state is more stable, e.g. $[Co^{3+}(NH_3)_6]^{3+}$ is more stable than $[Co^{2+}(NH_3)_6]^{2+}$. Greater stability is due to the fact that the metal cation having higher charge has greater power to attract the lone pair of electrons on the ligand. In other words we can say that since the metal cation with high charge is smaller in size, it can attract the lone pair of electrons of the ligand more strongly and hence can give more stable complex ion.

## XVI. Formation of Interstitial Compounds

We know that in the close-packed layers, the metal atoms touch each other only at some of the points and a considerable part of the total space remains empty between the atoms of the metallic structure. These empty spaces are called interstitial spaces / sites / holes or simply interstices. These holes can accommodate into them smaller non-metallic atoms like H, B, C, N, Si, etc. which, being of suitable size, can fit into them. The substances obtained by putting the smaller non-metallic atoms into the holes existing between the atoms of the metals are called **interstial compounds**. Transition metals have many empty holes in their structure.

**Examples of intersterile compounds.** Examples of interstitial compounds are the hydrides, bromides, carbides, nitrides, oxides, etc. given by transition metals.

(*i*) The analysis of FeO has shown that the actual composition of this oxide varies from $Fe_{0.84}O$ to $Fe_{0.94}O$. In order to show the intestial nature of FeO, a bar is placed over its formula *i.e.* FeO is represented as $\overline{FeO}$ *or* as $Fe_{0.84}O \rightarrow Fe_{0.94}O$ *or* as $Fe_{0.84-0.94}O$.

(*ii*) Other examples of interstial compounds are: $VSe_{0.98} \rightarrow VSe_{1.2}$ or $VSe_{0.98-1.2}$, $VSe_{1.2-1.6}$, $VSe_{1.6-2.0}$, $TiC_{0.6-1.0}$, $TiO_{1.9-2.0}$, $ZrN_{0.56-1.00}$, $Fe_{0.88-1.00}S$ etc.

**Characteristics of interstitial compounds.** Some of the important characteristics of interstitial compounds are as follows.

(*i*) They are very hard and rigid, e.g., steel which is an interstitial compound of Fe and C is quite hard. Similarly, some borides are as hard as diamond.

(*ii*) They have high melting points which are higher than those of the pure metals.

(*iii*) They show conductivity like that of the pure metal.

(*iv*) They acquire chemical inertness.

(*v*) Since interstitial compounds do not have definite composition, these compounds are **non-stoichiometric compounds**.

## XVII. Alloy Formation

*Alloys are homogeneous solid solutions of two or more metals obtained by melting the components and then cooling the melts.* These are formed by metals whose atomic radii differ by not more than 15% so that the atoms of one metal can easily take up the positions in the crystal lattice of the other.

Now, as transition metals have similar atomic radii and other characteristics, hence they form alloys very readily. Alloys are generally harder, have higher melting points and more resistant to corrosion than the individual metals.

The most commonly used are the *ferrous alloys*. The metals chromium, vanadium, molybdenum, tungsten and manganese are used in the formation of aloy steels and stainless steels. Some alloys of transition metals with non-transition metals are also very popular, e.g., brass (Cu + Zn) and bronze (Cu + Sn).

## XVIII. Some Other Properties

General properties like atomic radii, ionic radii, metallic character, etc. of transition metals (3*d*, 4*d* and 5*d* series elements) have already been discussed in the previous pages of this chapter. Some other properties of these elements are given in Table 5.16.

**Table 5.16. Some important properties of the elements of 1*st*, 2*nd* and 3*rd* transition series**

| *Element* | *Properties* | | |
|---|---|---|---|
| | *Crystal structure* | *Heat of fusion (KJ mole$^{-1}$)* | *Heat of vaporisation (KJ mol$^{-1}$)* |
| **1st transition series elements (3*d* - series elements)** | | | |
| $Sc_{21}$ | fcc | 15.9 | 338.9 |
| $Ti_{22}$ | hcp | 15.5 | 445.6 |
| $V_{23}$ | bcc | 17.6 | 443.6 |
| $Cr_{24}$ | bcc | 13.8 | 305.4 |
| $Mn_{25}$ | bcc | 14.6 | 224.7 |
| $Fe_{26}$ | bcc, fcc | 15.3 | 353.9 |
| $Co_{27}$ | hcp, fcc | 15.2 | 389.1 |
| $Ni_{28}$ | fcc | 17.6 | 380.7 |
| $Cu_{29}$ | fcc | 13.0 | 338.9 |
| $Zn_{30}$ | hcp | 7.4 | 114.6 |

(*Continued on the next page*)

| | | | |
|---|---|---|---|
| **2nd transition series elements (4*d*-series elements)** | | | |
| $Y_{39}$ | hcp | 11.28 | 389 |
| $Zr_{40}$ | hcp, bcc | 16.7 | 502 |
| $Nb_{41}$ | bcc | 26.7 | – |
| $Mo_{42}$ | bcc | 27.6 | 535 |
| $Tc_{43}$ | hcp | 23.0 | 502 |
| $Ru_{44}$ | hcp | 25.5 | 603 |
| $Rh_{45}$ | fcc | 21.7 | 577 |
| $Pd_{46}$ | fcc | 16.7 | 381 |
| $Ag_{47}$ | fcc | 11.3 | 286 |
| $Cd_{48}$ | hcp | 6.1 | 180 |
| **3rd transition series elements (5*d*-series elements)** | | | |
| $La_{57}$ | hcp, fcc | 10 | 939 |
| $Hf_{72}$ | hcp | 21.7 | 648 |
| $Ta_{73}$ | bcc | 28.4 | 753 |
| $W_{74}$ | bcc | 33.7 | 774 |
| $Re_{75}$ | hcp | 33.0 | 636 |
| $Os_{76}$ | hcp | 26.8 | 690 |
| $Ir_{77}$ | fcc | 27.6 | 648 |
| $Pt_{78}$ | fcc | 21.7 | 565 |
| $Au_{79}$ | fcc | 12.7 | 363 |
| $Hg_{80}$ | rhombic | 2.37 | 58 |

**Isomorphism**

Isomorphism is most common among transition metal compounds, e.g. $FeSO_4 . 7H_2O$ and Zn $SO_4 . 7H_2O$. This is due to similarity in the size of their metal ions.

**Comparison Between the Elements of 3d Series and Those of 4d and 5d Series**

***(i)* Atomic/ionic size.** The atomic as well as ionic radii of the *3d* series elements are always smaller than those of *4d* and *5d* series elements but due to lanthanide contraction, the atomic as well as ionic radii of the *4d* and *5d* series elements almost have identical size.

***(ii)* Coordination number.** The elements of the first series (3*d*) rarely show coordination number greater than six. Coordination numbers 1, 8 or higher are quite common with elements of second (4*d*) and third (5*d*) series. This may be due to their bigger size, so that compounds with higher coordination number may be formed.

***(iii)* Reactivity of elements.** The reactivity of the transition elements generally decreases in going from first transition series to the third transition series. The elements of the first transition series are much more reactive than those of the second and third series. The less reactivity may be due to their higher ionization energies, lower hydration energies, higher sublimation energies, etc.

***(iv)* Lower oxidation state.** The elements of the first transition series have at least one stable lower oxidation state which furnishes stable hydrated cations and oxy salts. On the other hand the second and third series elements form compounds in their higher oxidation states. The hydrated cations and oxysalts for 4*d* and 5*d* series elements generally do not exist.

***(v)* Maximum oxidation state.** The maximum oxidation state in each series is exhibited by iron-group elements. Iron shows a maximum of oxidation state of +6 in $FeO_4^{2-}$ and Ru and Os exhibit maximum oxidation state of +8 in their tetra oxide, $MO_4$.

***(vi)* Nature of complexes.** The four coordinate complexes of the 3*d* series elements, with few exceptions, are tetrahedral; for example, $Ni^{2+}$, $Co^{2+}$, $Cu^{2+}$ form tetrahedral as well as square planar complexes. The four-cordinate complexes of 4d and 5d series elements are square planar.

The elements of 3*d* series form both low and high spin complexes but due to high crystal field splitting energies of the elements of 4*d* and 5*d* series form only low-spin complexes.

Thus we see that the elements of 1st transition series form low spin or high spin complexes depending on the strength of the ligand field. Heavier transition elements form low spin complexes irrespective of the strength of the ligand field.

***(vii)* Stability of complexes.** Elements of the 3*d* series form more stable complexes with N, O and F donors, whereas the elements of 4*d* and 5*d* series form more stable complexes with P, S and heavier halogen donors.

***(viii)* Application of spin-only formula.** The magnetic data for the 3*d* series elements can be interpreted by spin-only formula. On the other hand, the magnetic data for 4*d* and 5*d* series elements cannot be interpreted by this formula only.

***(ix)* Metal-metal bonding in compounds, melting points and enthalpy of atomisation**. In the same group of *d*-block elements, 4*d* and 5*d* transition elements have larger size than 3*d* elements. Hence the valence electrons of 4*d* and 5*d* transition elements are less tightly held. As a result, 4*d* and 5*d* transition elements can form metal-metal bond more frequently. That is why m. pts. and enthalpies of atomisation of 4*d* and 5*d* series elements are higher than those of 3*d* series elements.

## Comparison Between the Transition Metals Placed in the Same Vertical Columns.

***(i)* Electronic configuration.** The transition elements present in the same vertical column generally have the same electronic configuration. However, some elements have anomalous configurations. For example: Group 5 (VB) – $Nb_{41} = 4d^4 5s^1$; Group 6 (VI B) – $Cr_{24} = 3d^5 4s^1$, $Mo_{42} = 4d^5 5s^1$; Group 8 (VIII) – $Ru_{44} = 4d^7 5s^1$; Group 9 (VIII) – $Rh_{45} = 4d^8 5s^1$; Group 10 (VIII) – $Pd_{46} = 4d^{10} 5s^0$, $Pt_{78} = 5d^9 6s^1$ and Group 11 (IB) – $Cu_{29} = 3d^{10} 4s^1$, $Ag_{47} = 4d^{10} 5s^1$, $Au_{79} = 5d^{10} 6s^1$. Thus we see that in the same vertical column, a number of elements in the three series have the unexpected configurations.

(*ii*) **Oxidation states.** First row transition metals form a number of stable complexes in which the metals have +2 and +3 oxidation state. For example $[Cr^{3+}Cl_6]^{3-}$ is a stable complex in which Cr is in +3 oxidation state. Corresponding complexes of Mo (III) and W (III) are not known. Similarly $[Co^{3+}(NH_3)_6]^{3+}$ is a stable complex of $Co^{3+}$ ion but corresponding complexes of Rh (III) and Ir(III) are not known. Some compounds of second and third row elements exist in high oxidation state but such compounds of the first row elements do not exist. For example, Ru (*2nd* series element) forms $RuO_4$ (O.S. of Ru = +8), W, Re, Os and Pt (*3rd* series elements) form $WF_6$ (O.S. of W = +6), $ReF_7$ (O.S. of Re = +7) $OsO_4$ (O.S. of Os = + 8) and $PtF_6$ (O.S. of Pt = +6). Such compounds in which the transition metals are in high oxidation states are not formed by the transition metals of *1st* series For the elements of *1st* row (*3d* series), +2 and +3 oxidation states are more common but for the elements of *2nd* (*4d*) and *3rd* (*5d*) series (heavier elements) higher oxidation states are more common.

(*iii*) **Ionisation enthalpies (IE).** Some of the elements of *4d* series have higher values of IE while some of the elements of this series have lower values of IE than the elements of 3*d* series in the same vertical column. However, the elements of *5d* series have higher IE values than the elements of *3d* and *4d* series placed in the same vertical column. This is because of weak shielding of the nucleus by 4*f* electrons in lanthanides lying between $La_{57}$ and $Hf_{72}$.

(*iv*) **Atomic size.** The size of the atoms of *4d* series elements is larger than the size of the atoms of the corresponding elements of 3*d* series while the size of the atoms of the elements of *5d* series are nearly the same as the size of the atoms of the corresponding elements of *4d* series due to lanthanide contraction.

### Differences Between d-Block Elements and Alkali and Alkaline Earth Metals (s-Block Elements)

The differences between *d*-block elements and alkali and alkaline earth (s-block elements) are given in Table 5.17.

**Table 5.17. Differences between d-block elements and alkali and alkaline earth metals**

| *d-block metals* | *Alkali and alkaline earth metals* |
|---|---|
| (*i*) They are less metallic. | (*i*) They are more metallic. |
| (*ii*) The last electron in them enters *d*-orbitals. | (*ii*) The last electron in them enters s-orbital. |
| (*iii*) Most of them form complexes. | (*iii*) Very few of them form complexes. |
| (*iv*) Their salts are mostly coloured. | (*iv*) Their salts are white. |
| (*v*) They are usually used as catalysts. | (*v*) Very few of them are used as catalysts. |

**Comparison of Transition Elements with Non-transition Elements (Representative Elements)**

A comparison of important properties of transition elements (*d*-block elements) with those of non-transition elements (s- and *p*-block elements) is given in Table 5.18. There are , however, two generalizations that will apply equally to both the transition and non-transition elements: (*i*) High oxidation states are usually stabilized in alkaline media. (*ii*) The capacity of the halogens to bring out high oxidation states varies in the order : F > Cl > Br > I.

**Table 5.18. Comparison of the important properties of transition elements (*d*-block elements) with those of non-transition elements (*s*- and *p*-block elements).**

| *Transition elements (d-block elements)* | *Non-transition elements (s- and p-block elements)* |
|---|---|
| (*i*) All the transition elements are metals. | (*i*) Non-transition elements may be metals, metalloids or non-metals. The metallic character in a group increases with the increase in atomic weight. |
| (*ii*) The atomic radii show irregular decrease in a series of transition elements. The decrease is not pronounced. | (*ii*) The atomic radii show regular and pronounced decrease along a period of non-transition elements. |
| (*iii*) The ionization energies show irregular increase in a series of transition elements. The increase is not pronounced. | (*iii*) The ionization energies show regular and pronounced increase on moving across a period of non-transition elements |
| (*iv*) They form a large number of complexes. For example: $[Ag(NH_3)_2]^+$ $[Cu(NH_3)_4]^{2+}$ etc. | (*iv*) They do not form many complexes. |
| (*v*) Transition elements or their compounds show catalytic activity. For example, nickel catalyses hydrogenation reactions and $V_2O_5$ is used as a catalyst for the oxidation of $SO_2$ to $SO_3$. | (*v*) They lack catalytic activity. |

(*Continued on the next page*)

| | |
|---|---|
| (*vi*) They form interstitial compounds with the non-metals like H, C and N. These compounds may have non-stoichiometric ratios. For example : $TiH_{1.7}$, $VH_{0.56}$, etc. | (*vi*) They do not form non-stoichiometric compounds. |
| (*vii*) In transition elements the differentiating electron enters $(n-1)d$ orbitals. | (*vii*) In non-transition elements (*s*-and *p*-block elements) the differentiating electron occupies *ns* orbital (s-block elements) or *np* orbitals (p-block elements). |
| (*viii*) V.S.E.C. of transition elements is $(n-1)\ d^{1-10}\ ns^{0-2}$. Thus in transition electrons the valence electrons are present in the s-orbital of outermost (ultimate) shell and in *d*-orbitals of the penultimate shell. | (*viii*) V.S.E.C. of non-transition elements is $ns^{1-2}$ (*s*-block elements) or $ns^2np^{1-6}$ (*p*-block elements). Thus in non-transition elements the valence electrons are present in the outermost shell. |
| (*ix*) Transition elements have partially filled $(n-1)\,d$ orbitals. | (*ix*) Representative elements do not have any *d*-orbitals in their outermost shell. |
| (*x*) Transition elements show variable oxidation states because, in addition to *ns* electrons, the electrons of $(n-1)d$ orbitals can also participate in bonding. For example Mn has V.S.E.C. as $3d^5 4s^2$. Hence it shows oxidation states of +2, +3, +4, +6 and +7. | (*x*) *s*-block elements do not show variable oxidation states because they lose *ns* electrons and acquire stable configuration of noble gas. Some p-black elements do show variable oxidation states because *ns* and *np* electrons may be excited to give different numbers of half-filled orbitals, *e.g.* sulphur shows oxidation states of +2, +4 and +6. |
| (*xi*) Transition elements are hard metals, have high melting points and form coloured paramagnetic compounds. | (*xi*) *s*-block elements are soft metals, have low melting points and form colourless diamagnetic compounds. |

| | |
|---|---|
| (*xii*) Transition elements of the same series have similar properties though they are placed in different groups. | (*xii*) *s*- and *p*-block elements present in the same group have similar properties but these elements placed in different groups have different properties. |
| (*xiii*) Hydrides and oxides of transition elements are non-stoichiometric compounds. | (*xiii*) Hydrides and oxides are simple compounds. |

## Important Compounds of First Row Transition Metals

### A. Oxides

Oxides formed by the transition metals of *1st* transition series are given in Table 5.19.

**Table 5.19. Oxides of metals of first transition series in different oxidation sates (mixed oxides)**

| *Oxidation No.* | *Sc (3)* | *Ti (4)* | *V (5)* | *Cr (6)* | *Mn (7)* | *Fe (8)* | *Co (9)* | *Ni (10)* | *Cu (11)* | *Zn (12)* |
|---|---|---|---|---|---|---|---|---|---|---|
| +7 | — | — | — | — | $Mn_2O_7$ | — | — | — | — | — |
| +6 | — | — | — | $CrO_3$ | $MnO_3$ (unstable) | — | — | — | — | — |
| +5 | — | — | $V_2O_5$ | — | — | — | — | — | — | — |
| +4 | — | $TiO_2$ | $V_2O_4$ | $CrO_2$ | $MnO_2$ | — | — | — | — | — |
| +3 | $Sc_2O_3$ | $Ti_2O_3$ | $V_2O_3$ | $Cr_2O_3$ | $Mn_2O_3$ | $Fe_2O_3$ | — | — | — | — |
| | | | | | $Mn_3O_4$ | $Fe_3O_4$ | $Co_3O_4$ | — | — | — |
| +2 | — | TiO | VO | (CrO) | MnO | FeO | CoO | NiO | CuO | ZnO |
| +1 | — | — | — | — | — | — | — | — | $Cu_2O$ | — |

**Preparation.** Oxides are generally obtained by heating the metals with $O_2$.

**Properties.** (*i*) $Mn_3O_4$, $Fe_3O_4$ and $Co_3O_4$ are mixed oxides. These oxides can be represented by a general formula, $M_3O_4$ (M = Mn, Fe and Co). In these oxids one M atom is in +2 O.S. and each of the two remaining M atoms has +3 O.S. Thus $M_3O_4$ is a combination of MO and $M_2O_3$ *i.e.* $M_3O_4 \equiv \overset{2+}{M}O \cdot \overset{3+}{M}_2O_3$.

(*ii*) *Variation in acidic, basic and amphoteric character of oxides of a given metal in various oxidation states.* The acidic, basic and amphoteric character of the oxides has been shown below. (*a*), (*b*) and (*c*) written after the formulae of the oxides indicate their acidic, basic and amphoteric character respectively.

Sc : $Sc_2O_3$ (*b*)

Ti : TiO (*b*), $Ti_2O_3$(*b*), $TiO_2$(*c*)

V : VO(*b*), $V_2O_3$(*b*), $VO_2$(*c*), $V_2O_5$(*a*)

Cr : CrO(*b*), $Cr_2O_3$(*c*), $CrO_2$(*c*), $CrO_3$(*a*)

Mn : MnO (*b*), $Mn_3O_4$(*c*), $Mn_2O_3$ (*c*), $MnO_2$ (*c*), $MnO_3$ (unstable) (*c*), $Mn_2O_7$(*a*)

Fe : FeO (*b*), $Fe_2O_3$(*b*), $Fe_3O_4$(*b*),
Co : CoO (*b*)
Ni : NiO (*b*)
Cu : $Cu_2O$ (*b*), CuO (*c*)
Zn : ZnO (*c*)

In general, the oxides in the lower oxidation states of the metals are basic and in their higher oxidation states they are acidic whereas in the intermediate oxidation state, the oxides are amphoteric. For example, the behaviour of the oxides of manganese may be represented as follows:

| $\overset{+2}{MnO}$ | $\overset{+8/3}{Mn_3O_4}$ | $\overset{+3}{Mn_2O_3}$ | $\overset{+4}{MnO_2}$ | $\overset{+7}{Mn_2O_7}$ |
|---|---|---|---|---|
| Basic | Amphoteric | Amphoteric | Amphoteric | Acidic |

——— Acidic character increases ⟶

Similarly, CrO is basic while $Cr_2O_3$ is amphoteric.

(*iii*) *Solubility of oxides*. The basic and amphoteric oxides dissolve in non-oxidi sing acids and form hexaaquo ions, $[M(H_2O)_6]^{n+}$. Some of these oxides dissolve in acids and bases, forming oxometallic salts such as potassium dichromate, permanganate and manganate. The acidic oxides like $CrO_3$ and $Mn_2O_7$ dissolve in water to form the oxy acids. These oxides also reacts with bases (*e.g.*, KOH) to form their oxo salts.

$$CrO_3 + H_2O \longrightarrow H_2CrO_4 \text{ (Chromic acid)}$$
$$2CrO_3 + H_2O \longrightarrow H_2Cr_2O_7 \text{ (Dichromic acid)}$$
$$Mn_2O_7 + H_2O \longrightarrow 2HMnO_4 \text{ (Permanganic acid)}$$
$$CrO_3 + 2KOH \longrightarrow K_2CrO_4 + H_2O$$
$$2\,CrO_3 + 2KOH \longrightarrow K_2CrO_4 + H_2O$$
$$Mn_2O_7 + 2KOH \longrightarrow 2KMnO_4 + H_2O$$

(*iv*) *Reducing properties of oxides*. As we move along a given transition series, +2 oxidation state of the metals becomes more and more stable and hence the tendency of these metals to lose electrons becomes more. Thus we can say that the reducing property of the oxides increases as we move along a given transition series. VO and CrO as strong reducing agents. As we move down a group of transition metals, the higher oxidation state becomes more stable. For example, $OsO_4$ (Os = +8) and $RuO_4$ (Ru = +8) both have been isolated ($RuO_4$ is unstable and decomposes on standing) while $FeO_4$ (Fe = +8) has not been isolated.

(*v*) *Variation of ionic character of oxides of a given metal in various oxidation states*. As the oxidation number of the metal increases, ionic character of oxides decreases, e.g., $Mn_2O_7$ is a covalent green oil. Even $CrO_3$ and $V_2O_5$ have low melting points.

$$\overset{+2}{MnO} \quad \overset{+8/3}{Mn_3O_4} \quad \overset{+3}{Mn_2O_3} \quad \overset{+4}{Mn_2} \quad \overset{+7}{Mn_2O_7}$$

——— Ionic character decreases ⟶

(*vi*) *Trends in the stability of higher oxidation states of transition metals in their oxides*. The following points may be noted:

(*a*) The highest oxidation sate of a transition metal upto Mn in its oxide is equal to its group number. For example, $Sc_2O_3$ (Grp = 3, O.S. of Sc = +3), $TiO_2$ (Grp = 4, O.S. of Ti = +4), $V_2O_5$ (Grp = 5, O.S. of V = +5), $CrO_3$(Grp = 6, O.S. of Cr = +6) and $Mn_2O_7$ (Grp = 7, O.S. of Mn = +7).

(*b*) Fe shows its highest O.S. of +3 in $Fe_2O_3$. However, Fe also shows +6 O.S. in ferrate ion, $FeO_4^{2-}$ which is formed in alkaline medium. This ion is unstable because it decomposes into $Fe_2O_3$ and $O_2$.

(*c*) Some transition metals in their higher oxidation states are stabilised by forming oxocations and oxo anions. Examples of such ions are:

*Oxocations:* $VO_2^+$ (V = +5), $VO^{2+}$ (V = +4), $TiO^{2+}$(Ti = +4)

*Oxoanions:* $MnO^-_4$ (Mn = +7), $CrO_4^{2-}$ (Cr = +6), $Cr_2O_7^{2-}$ (Cr = +6), $VO^{3-}_4$ (V = +5). $VO_4^{3-}$ ion is called *vandate ion*.

## B. Halides

Halides formed by the transition metals of *1st* transition series are given in Table 5.20.

**Table 5.20. Stable halides of 3*d* series elements in different oxidation states. (X = F, Cl, Br, I. Exceptions are given in brackets)**

| *Oxidation No.* | *Ti* | *V* | *Cr* | *Mn* | *Fe* | *Co* | *Ni* | *Cu* | *Zn* |
|---|---|---|---|---|---|---|---|---|---|
| +6 | — | — | $CrF_6$ | — | — | — | — | — | |
| +5 | — | $VF_5$ | $CrF_5$ | — | — | — | — | — | — |
| +4 | $TiX_4$ | $VX_4$ ($F^-$, $Cl^-$, $Br^-$) | $CrX_4$ | $MnF_4$ | — | — | — | — | — |
| +3 | $TiX_3$ | $VX_3$ | $CrX_3$ | $MnF_3$ | $FeX_3$ (F, $Cl^-$, $Br^-$) | $CoF_3$ | — | — | |
| +2 | $TiX_2$ ($Cl^-$, $Br^-$, $I^-$) | $VX_2$ ($Cl^-$, $Br^-$, $I^-$) | $CrX_2$ | $MnF_2$ | $FeX_2$ | $CoX_2$ | $NiX_2$ | $CuX_2$ ($F^-$, $Cl^-$) | |
| +1 | — | — | — | — | — | — | — | CuX ($Cl^-$, $Br^-$, $I^-$) | |

**Preparation.** Transition metals react with halogens at high temperatures and form halides. The reactions of the metals with halogens have high activation energies and hence require high temperatures. The order of the tendency of the halogens to react with a given metal decreases from F to I (F > Cl > Br > l). Transition metals in higher oxidation states give fluorides.

**Properties.** (*i*) Transition metal halides have low volatility and high susceptibility to hydrolysis.

(*ii*) Although V shows an O.S. of +5 in $VF_5$, other halides of V undergo hydrolysis to form oxyhalides, $VOX_3$ in which O.S.of V is also equal to +5.

(*iii*) In lower oxidation states, fluorides are unstable. $CuF_2$, $CuCl_2$ are known but Cu $I_2$ is not known. Cu (II) halides oxidise iodide ($I^-$) to iodine ($I_2$).

$$\underset{(Cu=+2)}{2Cu^{2+}} + \underset{(I=1)}{4I^-} \longrightarrow \underset{\substack{\text{Cuprous iodide} \\ (Cu=+1)}}{Cu_2I_2} + \underset{(I=0)}{I_2}$$

(*iv*) In +2 oxidation state all the metals (except Sc) form dihahides, $MX_2$.

(*v*) Many copper (I) halides are unstable in aquous solution and undergo *disproportionation* as follows:

$$2Cu^+ \longrightarrow Cu^{2+} + Cu$$

The greater stability of $Cu^{2+}$ (*aq*) than $Cu^+$ (aq) is due to muh more negative enthalpy of hydration for $Cu^{2+}$(*aq*) than $Cu^+$ (aq). This more than compensates for the high second ionization enhalpy of Cu.

(*vi*) *Ionic/covalent character*. Transition metal fluorides are ionic in nature whereas chlorides and bromides are usually covalent. For example, $CuF_2$ is ionic but $CuCl_2$ and $CuBr_2$ both are covalent and have infinite chains.

**Explanation:** As the electronegativity of halogen decreases from F to Br, the ionic character of M – X bond in MX molecules decreases as M – F > M – Cl > M – I. Hence fluorides are ionic and chlorides and bromides are covalent.

(*vii*) The order of reactivity of the halogen in the formation of the halides of transition metals is:

$$F > Cl > Br > 1$$

(*viii*) As fluorine is the most electronegative element, the metals are oxidised to their highest oxidation states with fluorine.

(*ix*) *Structure*. Structure of Cu (II) chloride is given below in Fig. 5.6.

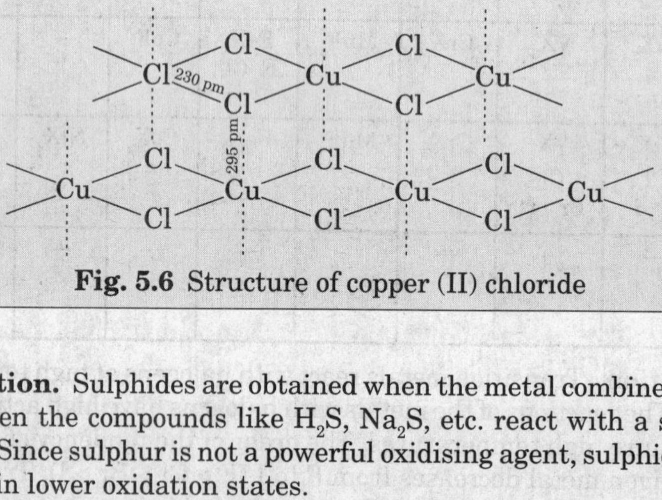


**Fig. 5.6** Structure of copper (II) chloride

## C. Sulphides

**Preparation.** Sulphides are obtained when the metal combines directly with sulphur or when the compounds like $H_2S$, $Na_2S$, etc. react with a solution of the metallic salts. Since sulphur is not a powerful oxidising agent, sulphides are formed by the metals in lower oxidation states.

**Properties.** (*i*) Transition metal sulphides are highly coloured or black.

(*ii*) They are insoluble in water.

(*iii*) In some cases, when a sulphide is treated with an oxidising agent, sulphur gets precipitated and sulphide is oxdised to sulphate.

**D. Salts Containing Oxoanions of Transition Metals**

Important examples of salts containing oxoaninons of the metals are potassium chromate ($K_2CrO_4$), pot. dichromate ($K_2Cr_2O_7$) and pot. permanganate ($KMnO_4$). These salts contain oxoanions *viz* $CrO_4^{2-}$, $Cr_2O_7^{2-}$ and $MnO_4^-$ respectively. $CrO_4^{2-}$ and $Cr_2O_7^{2-}$ ions contain Cr in +6 oxidation state while $MnO^-_4$ ion has Mn in +7 oxidation state. Manganate ion ($MnO_4^{2-}$) has Mn in +6 oxidation state.

## UNIVERSITY QUESTIONS

1. Explain the following properties of d-block elements:
   (*i*) Magnetic behaviour (*ii*) Catatytic properties (*iii*) Complex formation (*iv*) Interstrial compound formation. (*Ravishankar Shukla 2010*)
2. Explain the following: (*i*) Paramagnetism is exhibited by most transition elements. (*ii*) Cr and Cu have irregular electronic configuration. (*iii*) Zn, Cd and Hg are not typical transition elements. (*iv*) First ionisation energy of Hg is higher than that of Cd. (*v*) $Cu^{2+}$ is more stable than $Cu^+$. (*vi*) The densities of *3rd* transition series metals are double those of the second transition series metals. (*vii*) Zr and Hf have similar properties. (*Ravishankar Shukla 2010*)
3. Describe the paramagnetism and diamagnetism in transition metal complexes. (*Pd. Ravishankar Shukla, 2010*)
4. Complexes of $Cu^{2+}$ are more stable than those of $Cu^+$. (*Kanpur 2010*)
5. Explain the following: (*i*) Zn, Cd and Hg have low m.pts. unlike other *d*-block elements. (*ii*) $Fe^{3+}$ ion is more stable than $Fe^{2+}$. (*iii*) Hg is a liquid. (*iv*) Compounds of *s*- and *p*-block are generally colourless but those of transition metals are coloured. (*v*) Transition metals are poor reducing agents, though they possess very high electrode potential. Explain. (*Himachal Pradesh 2010*)
6. Explain why $[FeF_6]^{3-}$ is colourless whereas $[CoF_6]^{3-}$ is coloured. (*Himachal Pradesh 2010*)
7. Explain *d-d* transition. (*H.N. Bahuguna Garhwal 2010*)
8. Explain why transition metals generally work as catalyst. (*Kanpur 2010*)
9. (*a*) 4*d* and 5*d* transition metals resemble much with each other. Explain.
   (*b*) Write down four characteristics of transition elements including electronic configuration.
   (*c*) Name the elements that comprise 2*nd* transition series. (*Pune 2009*)
10. (*a*) What are the transition elements? Explain why Zn, Cd and Hg are not considered as transition elements.
    (*b*) On the basis of electronic configuration of first row transition series elements, discuss the variation of ionic and atomic radii and ionisation potential within the series.
    (*c*) Write electronic configuration of the elements having the following atomic numbers : 40, 41, 46, 72, 76, 78

(*d*) Discuss the stereochemistry of second and third transition series elements and compare them with their lighter congeners.

(*Rajasthan 2010*)

**11.** Write note on *d-d* transition. (*Barkatullah 2010*)

**12.** Choose the correct answer: (*i*) *d-d* transition is: (*a*) Ligand-ligand transition (*b*) Metal-metal transition (*c*) Ligand-metal transition (*d*) None of the above (*ii*) The colour of aqueous solution of $[Ti(H_2O)_6]^{+3}$ is: (*a*) White (*b*) Green (*c*) Yellow (*d*) Purple (*Barkatullah 2010*)

**13.** (*i*) Write down the electronic configuration of $Co^{3+}$ ion.

(*ii*) An example of non-stoichometric compound is: (*a*) $Al_2O_3$ (*b*) $Fe_2O_4$ (*c*) $NiO_2$ (*d*) PbO

(*iv*) Correct ground state electronic configuration of chromium is:

(*a*) [Ar] $3d^5 4s^1$ (*b*) [Ar] $3d^4 4s^2$ (*c*) [Ar]$3d^6 4s^0$ (*d*) [Ar]$4d^5 4s^1$

(*H.N. Bahuguna Garhwal 2010*)

**14.** (*i*) Disuss the general characteristic properties of transition elements.

(*ii*) Explain why Cu, Ag, Au, Zn, Cd and Hg are included amongst transition elements.

(*iii*) Various oxidation states exhibited by the third transition series.

(*Ranchhi 2010*)

**15.** What are the transition metals? Why are they so called? Discuss magnetic properties of Cr, Fe, Co and Ni. (Atomic numbers of Cr, Fe, Co and Ni are 24, 26, 27 and 28 respectively.) (*Magadh 2010*)

**16.** Discuss the following properties of *d*-block elements. (*a*) Catalytic properties (*b*) Oxidation states and (*c*) Ability to form complex compounds.

(*Andhra 2010*)

**17.** (*a*) Discuss the element of 3*d*-transition series with reference to their: (*i*) Magnetic properties and (*ii*) Complex formation tendency.

(*b*) Discuss the variable oxidation states of 3*d*-transition series elements.

(*c*) Explain the colour properties of first transition series elements.

(*d*) Write electronic configuration of 3*d*-block elements.

(*e*) Discuss 3*d*-block elements with respect to catalytic activity.

(*f*) Write electronic configurations of 5*d*-block elements. (*Nagpur 2010*)

**18.** Why Zr-Hf and Nb-Ta pair elements have similar ionic radii? Explain.

(*Nagpur 2010*)

**19.** Explain the stability of various oxidation states of *d*-block elements.

(*A.P. University 2010*)

**20.** (*a*) Transition metals have a strong tendency to form complexes compared to main group elements. Justify your answer.

(*b*) Why transition metals are less reactive than alkali metals.

(*GND Amritsar 2010*)

**21.** (*a*) Write and explain the electronic configuration of Cu.

(*b*) Discuss the following properties of 3*d* series elements. (*i*) Coloured ions (*ii*) Magnetic properties. (*Lucknow 2010*)

**22.** Explain why Zr/Hf and Mo/W occur in nature in the form of pair. (*Lucknow 2010*)

**23.** Compare the size of first, second and third row transition elements. (*Madras April 2011*)

**24.** (*a*) Calculate magnetic moment of $Cr^{3+}$ ion by spin-only formula.

(*b*) Write a short note on "*Paramagnetism*" (*M.D. Rohtak 2011*)

**25.** (*a*) How do transition metals give coloured and paramagnetic ions?

(*b*) 4*s* orbital is filled prior to 3d orbital but on ionisation, 4*s* electrons are removed first. Why ? (*MD. Rohtak 2011*)

**26.** Give reasons for the following:

(*a*) IE values of 3d elements do not vary much with increasing atomic number.

(*b*) Transition elements exhibit the property of alloy formation.

(*c*) IE of Hg is higher than that of Cd. (*Kurushetra 2011*)

**27.** Calculate the number of unpaired electrons in $Ni^{2+}$ and $Fe^{3+}$ ions. (*Kurushetra 2011*)

**28.** Ni, Pd and Pt belong to the same group in the periodic table but their electronic configurations are different. Explain (*Vidya Sagar 2012*)

**29.** (*a*) Discuss the elements of 1st transition series with respect to (*i*) Colour (*ii*) Atomic and ionic radii. (*iii*) Complex formation (*iv*) Catalytic properties

(*b*) Write electronic configuration of $Mn^{2t}$, Cr, $Cu^{+}$ and Fe

(*c*) Explain why $Fe^{3+}$ is more stable than $Fc^{2+}$ and Zn shows only +2 oxidation state (*Nagpur 2012*)

**30.** (*a*) Compare the stereochemistry of Ni, Pd and Pt complexes

(*b*) Compare Cr, Mo and W with respect to their magnetic property. (*Nagpur 2012*)

**31.** (*i*) Explain the following : (*a*) First ionisation energies of 5*d* transition elements are higher than those of 3d *and* 4d transition elements. (*b*) *d*-block elements generally form coloured compounds. (*c*) Many transition elements exhibit + 2 oxidation state in their compounds.

(*ii*) Write the names of 1*st* transition series elements with their electronic configuration. (*Lucknow 2013*)

# 6 CHAPTER

# Chemistry of Elements of First Transition Series (3*d* - Series Elements)

## 3d-Series Elements

We have seen in the previous chapter that 1st transition series (3d-series) contains ten elements namely $Sc_{21}$, $Ti_{22}$, $V_{23}$, $Cr_{24}$, $Mn_{25}$, $Fe_{26}$, $Co_{27}$, $Ni_{28}$, $Cu_{29}$ and $Zn_{30}$ which are present in **4th** period of the periodic table as shown in Fig. 6.1.

| Group | 1 | 2 | 3 | 4 | 5 | 6 | 7 | 8 | 9 | 10 | 11 | 12 | 13 | 14 | 15 | 16 | 17 | 18 |
|---|---|---|---|---|---|---|---|---|---|---|---|---|---|---|---|---|---|---|
| 1 | 1 H | | | | | | | | | | | | | | | | | 2 He |
| 2 | 3 Li | 4 Be | | | | | | | | | | | 5 B | 6 C | 7 N | 8 O | 9 F | 10 Ne |
| 3 | 11 Na | 12 Mg | | | | | | | | | | | 13 Al | 14 Si | 15 P | 16 S | 17 Cl | 18 Ar |
| 4 | 19 K | 20 Ca | 21 Sc | 22 Ti | 23 V | 24 Cr | 25 Mn | 26 Fe | 27 Co | 28 Ni | 29 Cu | 30 Zn | 31 Ga | 32 Ge | 33 As | 34 Se | 35 Br | 36 Kr |
| 5 | 37 Rb | 38 Sr | 39 Y | 40 Zr | 41 Nb | 42 Mo | 43 Tc | 44 Ru | 45 Rh | 46 Pd | 47 Ag | 48 Cd | 49 In | 50 Sn | 51 Sb | 52 Te | 53 I | 54 Xe |
| 6 | 55 Cs | 56 Ba | 57-71 La* | 72 Hf | 73 Ta | 74 W | 75 Re | 76 Os | 77 Ir | 78 Pt | 79 Au | 80 Hg | 81 Ti | 82 Pb | 83 Bi | 84 Po | 85 At | 86 Rn |
| 7 | 87 Fr | 88 Ra | 89-103 Ac** | 104 Rf | 105 Db | 106 Sg | 107 Bh | 108 Hs | 109 Mt | 110 | 111 | 112 | (113) | (114) | (115) | (116) | (117) | (118) |

← f-block elements →

| | | | | | | | | | | | | | | |
|---|---|---|---|---|---|---|---|---|---|---|---|---|---|---|
| Lanthanides* | 58 Ce | 59 Pr | 60 Nd | 61 Pm | 62 Sm | 63 Eu | 64 Gd | 65 Tb | 66 Dy | 67 Ho | 68 Er | 69 Tn | 70 Yb | 71 Lu |
| Actinides** | 90 Th | 91 Pa | 92 U | 93 Np | 94 Pu | 95 Am | 96 Cm | 97 Bk | 98 Cf | 99 Es | 100 Fm | 101 Md | 102 No | 103 Lr |

**Fig. 6.1** Position of 1st transition series elements (3d-series elements) in the periodic table.

These elements are also called 3d- series elements because the differentiating electron enters in *3d* orbitals of the atoms of these elements. General characteristics of these elements along with the elements of 2nd and 3rd transition series have already been discussed in detail in the previous chapter. Important physical properties of these elements have been given in Table 6.1 for the sake of reference.

In this chapter we shall describe the chemistry of all the ten elements of 1st transition series and their compounds.

**Table 6.1: Important physical properties of the elements of 1st transition series (3*d*- series elements)**

| *Property* | *Sc* | *Ti* | *V* | *Cr* | *Mn* | *Fe* | *Co* | *Ni* | *Cu* | *Zn* |
|---|---|---|---|---|---|---|---|---|---|---|
| (*i*) Atomic number | 21 | 22 | 23 | 24 | 25 | 26 | 27 | 28 | 29 | 30 |
| (*ii*) Outer electronic configuration | $3d^1 4s^2$ | $3d^2 4s^2$ | $3d^3 4s^2$ | $3d^5 4s^1$ | $3d^5 4s^2$ | $3d^6 4s^2$ | $3d^7 4s^2$ | $3d^8 4s^2$ | $3d^{10} 4s^1$ | $3d^{10} 4s^2$ |
| (*iii*) Atomic radii (pm) | 162 | 147 | 134 | 127 | 126 | 125 | 126 | 124 | 128 | 138 |
| (*iv*) Ionic radii (pm) | | | | | | | | | | |
| $M^{2+}$ ion | 81 | 91 | 88 | 84 | 80 | 76 | 74 | 72 | 69 | 74 |
| $M^{3+}$ ion | 88.5 | 76 | 74 | 69 | 66 | 64 | 63 | 63 | — | — |
| (*v*) Crystal structure | hcp | hcp | bcc | bcc | bcc | bcc, fcc | hcp, fcc | fcc | fcc | hcp |
| (*vi*) Atomic volume (ml) | 15.02 | 10.60 | 8.35 | 7.23 | 7.39 | 7.10 | 6.70 | 6.60 | 7.10 | 9.20 |
| (*vii*) Density (g/cm$^3$) | 3.43 | 4.51 | 6.10 | 7.19 | 7.43 | 7.86 | 8.90 | 8.90 | 8.96 | 7.14 |
| (*vii*) Atomic mass (a.m.u) | 44.96 | 47.90 | 50.95 | 52.01 | 54.94 | 55.85 | 58.94 | 58.71 | 63.54 | 65.38 |
| (*ix*) Enthalpy of atomisation (KJ mol$^{-1}$) | 326 | 473 | 575 | 397 | 281 | 416 | 425 | 430 | 339 | 126 |
| (*x*) Melting points (°C) | 1539 | 1668 | 1900 | 1975 | 1245 | 1536 | 1495 | 1453 | 1083 | 419.5 |
| (*xi*) Boiling points(°C) | 2730 | 3260 | 3460 | 2665 | 2150 | 3000 | 2900 | 2730 | 2595 | 906.0 |
| (*xii*) Ionisation energies, I.E. (KJ mol$^{-1}$) | | | | | | | | | | |
| $IE_1$ | 631 | 658 | 650 | 652 | 717 | 762 | 758 | 736 | 745 | 906 |
| $IE_2$ | 1245 | 1320 | 1376 | 1635 | 1513 | 1563 | 1647 | 1756 | 1961 | 1736 |
| $IE_3$ | 2450 | 2721 | 2873 | 2994 | 3258 | 2963 | 3237 | 3400 | 3560 | 3838 |
| (*xiii*) Standard oxidation Potential (volts) | | | | | | | | | | |
| $M/M^{2+}$ | – | +1.60 | + 1.18 | +0.91 | +1.18 | + 0.44 | + 0.28 | + 0.25 | –0.34 | +0.76 |
| $M/M^{3+}$ | – 2.10 | – | + 0.26 | +0.74 | –1.57 | – 0.77 | –1.97 | – | – | – |
| (*xiv*) Stable oxidation states | + 3 | + 4 | + 3, + 4, + 5 | + 2,+ 3 + 6 | + 2, +3, + 4, +7 | +2, + 3, | +2, +3 | +2 | +1, +2 | +2 |
| (*xv*) Electronegativity | 1.3 | 1.5 | 1.05 | 1.65 | 1.55 | 1.8 | 1.8 | 1.8 | 1.8 | 1.6 |
| (*xvi*) Heat of fusion (KJ mol$^{-1}$) | 15.9 | 15.5 | 17.6 | 13.8 | 14.6 | 15.3 | 15.2 | 17.6 | 13.0 | 7.4 |
| (*xvii*) Heat of vaporisation (KJ mol$^{-1}$) | 338.9 | 445.6 | 443.6 | 305.4 | 224.7 | 353.9 | 389.1 | 380.7 | 338.9 | 114.6 |

## Scandium, $Sc_{21}$

### Properties

(i) Scandium is a silvery white metal with a metallic lustre. However, it gets tarnished in air due to the formation of $Sc_2O_3$ on the surface.

(ii) Scandium is mononuclidic; its only natural (stable) isotope is $^{45}Sc$. Twelve radioactive isotopes are known but all have short half lives. The longest half-life of 85 days is shown by $^{46}Sc$; the others have in minutes or seconds.

(iii) It crystallizes on the hexagonal close packed (hcp) system having lattice parameters $a = 330.9$ pm, $c = 527.3$ pm and *axial ratio* $(c/a) = 1.593$

(iv) Its m.p. is 1812 K and b.p. is 3015 K

(v) Scandium ($3d^1 4s^2$) loses 3 electrons to form $Sc^{3+}$ ions (elec. config. $3d^0$). $3d$ subshell being empty scandium compounds are colourless and diamagnetic.

The sum of the first three ionization energies for scandium is a little less than the sum for aluminium. Properties of scandium are, therefore, similar to those of aluminium in a number of respects. Scandium resembles lanthanides (*f*- block elements) also in several respects though there are some points of difference as well.

## Comparison of the Properties of Scandium and Aluminium

(i) **Formation of $M^{3+}$ ions.** Outer electronic configuration of scandium is $3d^1 4s^2$ and that of aluminium is $3s^2 3p^1$. Both of them lose three electrons to give trivalent ions, $Sc^{3+}$ and $Al^{3+}$. Both these ions have a noble gas configuration as shown below:

Sc : [Ar] $3d^1 4s^2$ $\quad$ $Sc^{3+}$ : [Ar]

Al : [Ne] $3s^2 3p^1$ $\quad$ $Al^{3+}$ : [Ne]

(ii) Both form colourless and diamagnetic compounds.

(iii) **Reaction with water.** Both react with water vigorously liberating hydrogen.

$$2Sc + 3H_2O \longrightarrow Sc_2O_3 + 3H_2$$
$$2Al + 3H_2O \longrightarrow Al_2O_3 + 3H_2$$

(iv) **Nature of oxides.** Both $Sc_2O_3$ and $Al_2O_3$ are insoluble in water. However, $Al_2O_3$ is amphoteric whereas $Sc_2O_3$ is basic.

(v) **Formation of double salts.** Scandium and aluminium both form double sulphates: $K_2SO_4\ Sc_2\ (SO_4)_s.\ xH_2O$ and $K_2SO_4\ Al._2\ (SO_4)_3\ 24H_2O$. The two differ in size of their $M^{3+}$ ions. $Sc^{3+}$ ion is much larger in size than $Al^{3+}$ ion.

### Comparison of Scandium with Lanthanides

Scandium resembles lanthanides in the formation of insoluble salts, oxalate, phosphate, carbonate and fluoride.

However, scandium differs from lanthanides in the size of their ions. Ionic radius of $Sc^{3+}$ is 81 pm whereas ionic radii of lanthanides are in the range of 85–106 pm.

### Important Points of Difference:

(i) **Nature of oxides.** Scandium oxide is less basic than oxides of lanthanides.

(ii) **Nature of chloride, nitrate and sulphate.** Scandium chloride, $ScCl_3$ is volatile and sublimes at a much lower temperature than lanthanide chlorides, scandium nitrate readily decomposes and sulphate is readily soluble in cold as well as hot water. These differ from corresponding salts of lanthanides.

(*iii*) **Double salt.** Scandium sulphate forms a double salt like $K_2SO_4 . Sc_2 (SO_4)_3$ $24H_2O$ and no such double sulphate is given by lanthanides.

(*iv*) **Formation of thiocyanate.** Scandium salts on treatment with KSCN give a yellow precipitate of $Sc(SCN)_3$. Salts of the lanthanides do not give any precipitate of sulphocyanide.

(*v*) **Hydrolysis.** $Sc^{3+}$ ions being smaller in size than the corresponding lanthanide ions, have a greater tendency to hydrolyse.

(*vi*) **Formation of complexes.** $Sc^{3+}$ ion on account of its smaller size forms complex ions more readily than the lanthanide ions. The fluroride is readily soluble in excess HF and $NH_4F$ to form definite fluoro complexes, *e.g.* $ScF_6^{3-}$.

(*vii*) **Polarizing power.** On account of its smaller size, $Sc^{3+}$ has greater polarizing power and greater tendency to form covalent linkages in compounds.

## Compounds of Scandium

In various compounds (oxides, halides and oxyhalides) of scandium its oxidation state is +3.

### 1. Scandium Hydroxide

Scandium hydroxide does not appear to have a definite formula. It is considered to be a *hydrous oxide*, $Sc_2O_3 . nH_2O$ not $Sc(OH)_3$. The hydroxide readily dissolves in conc. NaOH and $Na_3Sc (OH)_6 . 2H_2O$ can be crystallized.

It is basic in nature but less basic than $Ca (OH)_2$ as well as hydroxides of other metals of this group. This is because the basic nature of elements increases with increase in size from Sc and La. The basicity of oxides and hydroxides of the elements of this group, therefore, increases as we move down the column from Sc to La and lanthanum hydroxide is the strongest base of this group.

### 2. Scandium Carbonate

Normal scandium carbonate does not exist. Addition of alkali carbonate to a solution of scandium salt incompletely precipitates the double carbonate. Degree of hydration of precipitates such as R5Sc(CO3)4 (where R = alkyl groups) varies with conditions of production, and these are soluble in cold water. Such solutions undergo hydrolysis and subsequently precipitate a basic salt or hydroxide.

### 3. Scandium Nitrate

The tetrahydrate $Sc(NO_3)_3 . 4H_2O$ prepared by dissolution of the oxide in nitric acid crystallizes from neutral solution in deliquescent prisms which are soluble in organic solvents. No double nitrate forms are known and the simple nitrate loses water at 370 K and decomposes to the oxide at 520 K. Ultrasonic spectroscopy has shown that the nitrate is hydrolyzed in solution.

**4. Scandium Sulphate**

$Sc_2(SO_4)_3$. $5H_2O$ is the normal form of the solid scandium sulphate at room temperature. Two to three moles of water are lost at 370 K and the remainder at 520 K. The anhydrous salt decomposes progressively to the oxide at about 1400 K. The anhydrous sulphate may also be prepared by addition of concentrated sulphuric acid to a solution of the nitrate. It is white powder, readily soluble in water, and unlike the lanthanum group sulphates, its solubility does not decrease with increase in temperature.

**5. Scandium Halides**

Preparation of scandium halides ($ScX_3$ where X = F, Cl, Br or l) is quite easy and these are of great value in metal production. The compounds are all hygroscopic solids with varying degree of hydration and with the exception of the fluoride, soluble in water. Fluoride dissolves in excess fluoride ion to form the soluble fluoride complex $ScF_6^{3-}$.

The hydrated salts are usually prepared by dissolving the oxide or hydroxide in the corresponding acid and crystallizing. Scandium iodide prepared in this way, however, decomposes readily and is best prepared as the anhydrous compound and then recrystalized from water. The anhydrous halides and preferably *not* prepared by direct thermal dehydration of the hydrates as basic salts are inevitably formed. Dehydration with ammonium halides has been practised with some success, but direct interaction of the elements or reaction in non-aqueous media is preferred. Fluoride production can be easily accomplished by the reaction of oxide with hydrogen fluoride at 1070 K.

$$Sc_2O_3 + 6HF \xrightarrow{1070K} 2ScF_2 + 3H_2O$$

**6. Scandium Oxide**

The white powder (m.p. 3370 K) of scandium oxide ($Sc_2O_3$) is obtained by oxidative ignition of the metal or its compounds including nitrate, hydroxide and sulphate, or by hydrolysis of the halides. At 1170. 1270 K, the oxide is not hygroscopic and finds use as weighing form in gravimetric determination of the metal. A strongly ignited oxide is only slightly attacked by acids but concentrated hydrochloric acid or nitric acid dissolves it.

Only the sesquioxide, $Sc_2O_3$, is known to be stable though spectroscopic evidences are there for monoxide, suboxide and peroxide. The sequioxide has a cubic structure. The cube contains 16 moles and the metal atoms are 6-coordinate with 2 oxygen atoms missing at the cube corners. At high pressure, the sesquioxide transforms to the β-structure.

**7. Scandium Hydride**

Scandium forms a hydride (an interstitial compound) which is not quite stoichiometric, the number of hydrogen atoms per atom of scandium being a little less than 3. Scandium hydride is, however, generally represented by the formula $ScH_3$. The exact composition depends upon the temperature and pressure of hydrogen. It contains hydrogen in the form of hydride ion, $H^-$. It, therefore, reacts with water liberating hydrogen.

**8. Scandium Carbide, $ScC_2$.**

It is prepared by heating the scandium oxide with carbon at a high temperature of about 1300 K

$$Sc_2O_3 + 7C \rightarrow 2ScC_2 + 3CO$$

It reacts with water liberating acetylene. This suggests that it is an acetylide. At one time the carbide was thought to be $Sc^{3+}(C\equiv C)^{3-}$ but it has been shown by conductivity and magnetic measurements that scandium carbide contains $Sc^{3+}$ ions, $C_2^{3-}$ ions and free electrons.

**9. Complexes**

Size of an ion determines its tendency to form complexes. The smaller the size of an ion, the greater is its tendency to form complexes. Ionic radius of $Sc^{3+}$ ion is smaller than that of other ions of this group and has, therefore, the greatest tendency to form complexes. It forms complex species like $[Sc(OH)_6]^{3-}$, $[ScF_6]^{3-}$, $[Sc(DMSO)_6(ClO_4)]_3$ agents such as oxalic acid, citric acid, acetylacetone and edta.

## Titanium, $Ti_{22}$

### Physical Properties

Metallic titanium has a silvery lustre and when polished resembles steel appearance. Amorphous form is a dark grey powder resembling iron (reduced) in appearance. It is brittle at ordinary temperature but at low red heat it is readily malleable. The pure metal is very ductile. Its m.pt. and b.pt are 1668°C and 3260°C. The fused metal is very hard and crystalline and scratches quartz.

### Chemical Properties

1. **Action of air.** On exposure to air at ordinary temperature, titanium metal reacts with $O_2$ and $N_2$ of the air and forms a thin film (coating) of $TiO_2$ and TiN on the surface of the metal. This coating prevents the metal from corrosion. This coating remains stable in air up to 400°C.
2. **Action of halogens.** The metal readily combines with $Cl_2$ (at 350°C), $Br_2$ and $I_2$ to form the tetrahalides, $TiX_4$.
3. **Combination with non-metals.** On being heated, the metal combines with $H_2$, B, C and $N_2$ to produce $TiH_2$, TiB, TiC and TiN. These are interstitial compounds and hard. These have refractory properties.
4. **Action of acids and alkalies.** The metal remains passive towards the action of acids and alkalies. This behaviour is due to the formation of a protective film of $TiO_2$ on the surface of the metal. The metal dissolves in HF to give $[TiF_6]^{2-}$ ion.

$$Ti + 4HF \longrightarrow TiF_4 + 2H_2$$
$$TiF_4 + 2F^- \longrightarrow [TiF_6]^{2-}$$

The metal also dissolves in HCl and $H_2SO_4$ to give soluble salts.

$$2Ti + 6HCl \longrightarrow 2TiCl_3 + 3H_2$$
$$Ti + 4H_2SO_4(conc.) \longrightarrow Ti(SO_4)_2 + 2SO_2 + 4H_2O$$

With $HNO_3$, metatitanic acid ($H_2TiO_3$) is obtained

$$Ti + 4HNO_3 \longrightarrow H_2TiO_3 + 4NO_2 + H_2O$$

**5. Action on steam.** Steam is decomposed by the metal

$$Ti + 2H_2O\ (steam) \longrightarrow TiO_2 + 2\ H_2O$$

**6. Action on $TiCl_4$.** $TiCl_4$ is reduced to $TiCl_2$ when heated with Ti metal

$$TiCl_4 + Ti \xrightarrow{\Delta} 2\ TiCl_2$$

**Uses of Titanium Metal, its Compounds and its Alloys**

(*i*) Due to the strength, lightness, corrosion, resistance and low thermal expansion of Ti metal, it is used in high speed aircraft, naval and military projects and also in chemical industry.

(*ii*) It is used to remove the last traces of gases like $O_2$, $H_2$ and $N_2$ from evacuated vessels.

(*iii*) A titanium cathode is used in special photoelectric cell.

(*iv*) Pure Ti metal is used in making incandescent filament.

(*v*) A number of titanium alloys have been found to be very useful in industry. For example: (*a*) Ferrotitanium alloy is used in steel industry as a scavanger to remove adsorbed $O_2$, $H_2$ and $N_2$, and as a cleansing and deoxidising agent in the metallurgical operations. Ferrotitanium is obtained in the electric furnace by reducing ilmenite ore, $FeTiO_3$ with carbon or Al metal.

$$\underset{\text{Ilmenite}}{FeTiO_3} + 3C\ (or\ 2Al) \longrightarrow \underbrace{Fe + Ti +}_{\text{Ferrotitanium alloy}} 3CO\ (or\ Al_2O_3)$$

(*b*) The alloy of Ti with Cr, Fe and Mo are used for making rails, gears, propellers, car wheels etc. (*c*) The alloys of Ti with Co and Ni are used as filaments in vacuum tubes in place of Pt — Ir alloys which are very costly.

(*vi*) Among its compounds, its carbide is used as refractory material, its nitride in high speed tools and its oxide ($TiO_2$) is used as a white pigment in paints under the name of *titan white*.

## Electronic Configuration and Oxidation States of Ti

V.S.E.C. of Ti is $3d^2\ 4s^2$. Ti shows mainly +2, +3 and +4 oxidiation states. +4 oxidation state is the most stable oxidation state. In some of its complexes, Ti shows 0 and +1 oxidiation state. In $[Ti\ (dipy)_3]^\circ$, Ti shows zero oxidation state.

## Compounds of Titanium (II)

TiO and $TiCl_2$ are the important compounds of Ti (II)

1. **Titanium Oxide, TiO**

It is formed by heating $TiO_2$ with Ti metal or carbon.

$$TiO_2 + Ti \longrightarrow 2\ TiO$$

$$TiO_2 + C \longrightarrow 2\,TiO + CO$$

It is a golden-yellow crystalline substance. It is a basic oxide and dissolves in acids, evolving $H_2$. TiO reduces $H_2O$ to $H_2$.

$$TiO + H_2O \longrightarrow TiO_2 + H_2$$

It has ionic character.

**2. Titanium Dichloride, $TiCl_2$**

It is prepared:

(*i*) by reducing $TiCl_4$ with Ti metal

$$TiCl_4 + Ti \longrightarrow 2TiCl_2$$

(*ii*) by the disproportionation of $TiCl_3$

$$\underset{(Ti = +3)}{2\,TiCl_3} \longrightarrow \underset{(Ti = +2)}{TiCl_2} + \underset{(Ti = +4)}{TiCl_4}$$

$TiCl_4$ is volatile and hence can easily be removed by distillation.

**Properties.** (*i*) It is a black powder and is volatile.

(*ii*) When heated in air, it catches fire.

(*iii*) With dry $NH_3$ at red heat, $Ti_3N_4$ is formed.

$$3TiCl_2 + 4NH_3 \longrightarrow Ti_3N_4 + 6HCl + 3H_2O$$

(*iv*) $TiCl_2$ undergoes disproportionation to form $TiCl_4$ and Ti

$$2\,TiCl_2 \longrightarrow TiCl_4 + Ti$$

## Compounds of Ti (III)

Ti (III) compounds are called **titanous compounds.** These compounds are prepared by reducing the corresponding Ti (IV) compounds with Zn and dil. HCl. The reduction can also be done electrolytically.

Ti (III) compounds are strong reducing agents as they are oxidised to Ti (IV) compounds, *e.g.*

$$\underset{(Ti = +3)}{Ti_2O_3} + O \longrightarrow \underset{(Ti = +4)}{2\,TiO_2}$$

Thus Ti (III) compounds can reduce $MnO_4^-$ ion to $Mn^{2+}$ ion

$$\underset{(Ti = +3)}{5Ti^{3+}} + \underset{(Mn = +7)}{MnO_4^-} + H_2O \longrightarrow \underset{(Ti = +4)}{5TiO^{2+}} + \underset{(Mn = +2)}{Mn^{2+}} + 2H^+$$

**1. Titanium Trichloride or Titanous Chloride, $TiCl_3$**

**Preparation.** It is the most important compound of Ti(+3). It is prepared by reducing $TiCl_4$ with $H_2$ in an electrically heated spiral in a water cooled tube or with Ag at 200°C.

$$2TiCl_4 + H_2 \longrightarrow 2TiCl_3 + 2HCl$$

$$TiCl_4 + Ag \longrightarrow TiCl_3 + AgCl$$

$TiCl_3$ can also be obtained by dissolving Ti metal in HCl.

$$2Ti + 6HCl \longrightarrow 2TiCl_3 + 3H_2$$

**Properties and uses.** Anhydrous $TiCl_3$ is a violet powder. It disproportionates into $TiCl_4$ and $TiCl_2$.

$$2TiCl_3 \longrightarrow TiCl_4 + TiCl_2$$

It is a powerful reducing agent since it reduces chlorates and perchlorates to chlorides, $Fe^{3+}$ to $Fe^{2+}$, $Cu^{2+}$ to Cu, and persulphate to $SO_4^{2-}$. Due to this property $TiCl_3$ is used in volumetric analysis of the above mentioned compounds . An aqueous solution of $TiCl_3$, when added to a dilute solution of a gold salt, produces an intense violet colour, due to the reduction of the gold salt followed by the absorption of colloidal gold on hydrous $TiO_2$. This reaction has been used as a very delicate test for gold. $TiCl_3$ is also used for removing certain dyes from the goods so that they may be redyed. It is also used for removing strains.

2. **Titanium Sesquioxide, $Ti_2O_3$**

It is obtained by heating $TiO_2$ with $H_2$ at 1000°C.

$$2TiO_2 + H_2 \longrightarrow Ti_2O_3 + H_2O$$

It dissolves in $H_2SO_4$ forming titanous sulphate. It has *corundum structure*.

3. **Complex Compounds**

Ti (III) forms a number of complexes both *cationic* as well as *anionic*. Examples are $[TiF_6)^{3-}$ [called hexafluorotitanate (III)] $[TiCl_5(H_2O)]^{2-}$ [pentachloroaquotitanate (III)] etc. Hydrated titanium trichloride exists in two forms having the formulae $[Ti\ [H_2O]_6]\ Cl_3$ and $[Ti\ (H_2O)_5Cl]\ Cl_2$. In aqueous solution, Ti (III) exists as blue – violet $[Ti\ (H_2O)_6]^{3+}$ complex ion.

**Uses of Ti (III) compounds.** (*i*) $Ti^{3+}$ salts are used for the volumetric estimation of $Fe^{3+}$ ions. In this estimation $NH_4SCN$ is used as an indicator. It gives red colour as long as $Fe^{3+}$ is present in solution.

$$FeCl_3 + TiCl_3 + H_2O \longrightarrow FeCl_2 + TiOCl_2 + 2HCl$$

(*ii*) $Ti^{3+}$ salts are also used for the estimation of organic nitrites. In this estimation methylene blue is used as indicator. When even a slight excess of $Ti^{3+}$ is generated, methylene blue is reduced and becomes colourless.

$$RNO_2 + 6Ti^{3+} + 4H_2O \longrightarrow RNH_2 + 6TiO_2^{2+} + 6H^+$$

## Compounds of Ti (IV)

Compounds of Ti (IV) are called **titanic compounds.**

1. **Titanium Dioxide or Titanic Dioxide or Tiania, $TiO_2$.**

This oxide occurs in nature as rutile which is an important ore of tianium.

**Preparation.** **(*i*) *From Ti (OH)$_4$*.** $TiO_2$ can be obtained by calcining the ppt. of Ti $(OH)_4$ formed by adding $NH_4OH$ to a solution of $TiCl_4$.

$$TiCl_4 + 4NH_4OH \longrightarrow \underset{ppt}{Ti(OH)_4} + 4NH_4Cl$$

$$Ti(OH)_4 \xrightarrow{\Delta} TiO_2 + 2H_2O$$

***(ii) From ilmenite ore, $FeTiO_3$.*** $TiO_2$ can be obtained from $FeTiO_3$ ore by the following processes:

(*a*) When ilmenite ore is heated with HCl in a current of $Cl_2$, $TiO_2$ is obtained.

$$\underset{\text{Ilmenite}}{2FeTiO_3} + 4HCl + Cl_2 \xrightarrow{\Delta} FeCl_3 + 2TiO_2 + 2H_2O$$

$$\underset{\text{Ilmenite}}{FeTiO_3} + Na_2S \xrightarrow{\text{Fuse}} \underbrace{FeS + TiO_2}_{\text{Insoluble}} + \underset{\text{Soluble}}{2Na_2O}$$

(*b*) The powdered ilmenite ore is fused with $Na_2S$. The product obtained consists of FeS, $TiO_2$ and $Na_2O$. Now when the product consisting of FeS, $TiO_2$ and $Na_2O$ is extracted with $H_2O$, FeS and $TiO_2$ remain insoluble and $Na_2O$ goes into solution. The insoluble portion containing FeS and $TiO_2$ is treated with $H_2SO_3$ in presence of atmospheric $O_2$ when FeS is converted into soluble $FeS_2O_3$ but $TiO_2$ remains unaffected.

$$FeS + H_2SO_3 + \frac{1}{2}O_2 \ (\textit{From air}) \longrightarrow \underset{\text{Soluble}}{FeS_2O_3 + H_2O}$$

$$TiO_2 + H_2SO_3 + \frac{1}{2}O_2 \ (\textit{From air}) \longrightarrow \text{No reaction}$$

(*c*) *Chlorine process:* When ilmenite ore is mixed with carbon and heated in a current of $Cl_2$ at 360°, a mass is obtained which consists of $FeCl_3$ and $TiCl_4$. Now when this mass is heated at 550°C. volatile $TiCl_4$ distils over. $TiCl_4$ thus obtained is hydrolysed by $H_2O$ to get $TiO_2$ or is heated with $O_2$ to get pure $TiO_2$ and $Cl_2$.

$$\underset{\text{Ilmenite}}{2FeTiO_3} + 2C + 7Cl_2 \xrightarrow{350°C} \underbrace{2FeCl_3 + 2TiCl4}_{} + 3CO_2$$

$$\downarrow 550C$$

$$TiCl_4 \text{ distils over}$$

$$TiCl_4 + 2H_2O \xrightarrow{\text{Hydrolysis}} TiO_2 + 4HCl$$

$$TiCl_4 + 2O_2 \xrightarrow{\Delta} TiO_2 + 2Cl_2$$

(*d*) *Sulphate process:* The powdered ilmenite is digested with excess of conc. $H_2SO_4$ when $FeSO_4$, $Fe_2(SO_4)_3$ $TiOSO_4$ (titanyl sulphate) are formed. The resulting product is extracted with water and the solution is then treated with scrap iron to reduce ferric sulphate to ferrous sulphate. The solution is filtered and transferred to vacuum evaporator. During this process crystals of $FeSO_4 . 7H_2O$ are formed which are separated and the solution containing $TiOSO_4$ is boiled with excess water to get hydrated titanium dioxide. The precipitate is filtered, washed and calcined in a rotary kiln to get white pigment grade titanium dioxide. The overall reaction is given below.

$$TiOSO_4 + H_2O \longrightarrow TiO_2 + H_2SO_4$$

***(iii) From $TiCl_4$.*** The spraying of liq. $TiCl_4$ into moist air produces white smokes of $TiO_2$.

$$\begin{array}{lcl} TiCl_4 + 4H_2O & \longrightarrow & Ti(OH)_4 + 4HCl \\ Ti(OH)_4 & \xrightarrow{\Delta} & TiO_2 + 2H_2O \\ \hline TiCl_4 + 2H_2O & \longrightarrow & TiO_2 + 4HCl \end{array}$$

**Properties.** (*i*) $TiO_2$ occurs in nature in three crystalline forms which are *rutile*, *anatase* and *brookite*. Rutile is the commonest form and forms tetragonal transparent to opaque, red or yellowish crystals. This form is stable above 1040°C. The anatase modification also crystallises in the tetragonal system but with the different axial ratio. This form is stable up to 860°C. Brookite form crystallises in rhombic form and is stable between 860–1040°C. The rutile form of $TiO_2$ is considered to be the most stable. The rutile or brookite forms are metastable below 860°C and thus are converted into anatase by suitable heat treatment.

$$\text{Anatase} \xrightleftharpoons{860°C} \text{Brookite} \xrightleftharpoons{1040°C} \text{Rutile}$$

(*ii*) Natural $TiO_2$ is impure due to the presence of iron oxide as an impurity contaminated with it. Pure $TiO_2$ is colourless (amorphous white powder) when cold, and yellowish when hot.

(*iii*) $TiO_2$ is insoluble in water.

(*iv*) *Reduction:* (*a*) When $TiO_2$ is heated with Ti-metal or in a carbon crucible, $TiO_2$ is reduced to TiO.

$$TiO_2 + Ti \longrightarrow 2TiO$$

$$TiO_2 + C \longrightarrow 2TiO + CO$$

(*b*) When reduced by $H_2$ at 1000°C, $Ti_2O_3$ is obtained.

$$2TiO_2 + H_2 \longrightarrow Ti_2O_3 + H_2O$$

(*c*) $TiO_2$ melts at about 1800°C and loses $O_2$ to form $Ti_2O_3$.

$$4TiO_2 \xrightarrow{1800°C} 2Ti_2O_3 + O_2$$

(*d*) When heated with carbon, Ca or Al, $TiO_2$ gets reduced to Ti-metal.

$$TiO_2 + C \xrightarrow{2500°\,C} Ti + CO_2$$

$$3\,TiO_2 + 4Al \longrightarrow 3Ti + 2Al_2O_3$$

$$TiO_2 + 2Ca \longrightarrow Ti + 2CaO$$

(*v*) *Action of $Cl_2$ on a mixture of $TiO_2$ and carbon.* When $TiO_2$ is heated with carbon in a stream of $Cl_2$, $TiCl_4$ is obtained.

$$TiO_2 + C + 2Cl_2 \xrightarrow{\Delta} TiCl_4 + CO_2$$

(*vi*) *Action of $Cl_2$.* On heating $TiO_2$ in a current of dry $Cl_2$, it forms $TiCl_4$ with the evalution of $O_2$.

$$TiO_2 + 2Cl_2\,(dry) \xrightarrow{\Delta} TiCl_4 + O_2$$

(*vii*) *Amphoteric nature.* $TiO_2$ is an amphoteric oxide and hence reacts with acids to form *titanyl salts* ($TiO^{2+}$) and with alkalies to form *titanates* ($TiO_3^{2-}$).

$$TiO_2 + H_2SO_4 \longrightarrow TiOSO_4 + H_2O$$
$$TiO_2 + 2KOH \longrightarrow K_2TiO_3 + H_2O$$

(*viii*) *Action of vapours of* $CCl_4$. On passing the vapours of $CCl_4$ over $TiO_2$ at high temperature, $TiCl_4$ is formed.

$$TiO_2 + CCl_4\ (vapours) \longrightarrow TiCl_4 + CO_2$$

(*ix*) *Action of HF*. $TiO_2$ dissolves in HF to form hydro fluotitanic acid, $H_2[TiF_6]$.

$$TiO_2 + 6HF \longrightarrow H_2[TiF_6] + 2H_2O$$

**Structure:** $TiO_2$ has rutile structure. This structure has been shown in Fig. 6.2. It has coordination number 6:3 in which each titanium is surrounded octahedrally by six oxygen atoms while each oxygen atom is surrounded by three titanium atoms disposed towards the corners of an equilateral triangle as shown in Fig. 6.2.

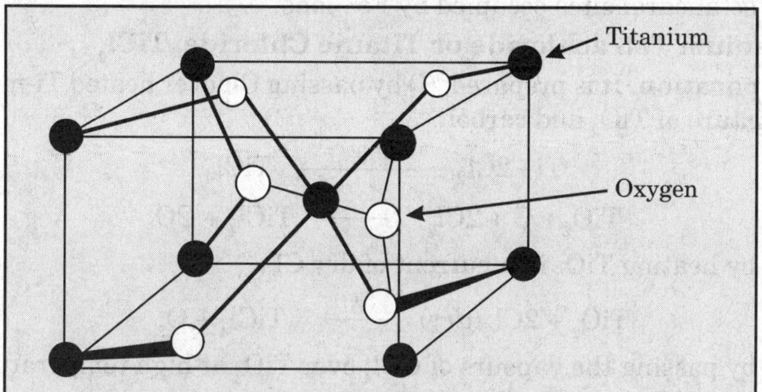


**Fig. 6.2** Rutile stucture of $TiO_2$

**Uses**. (*i*) Because of its high opacity, inertness towards $H_2S$, high covering power, light weight, non-poisonous nature, ease of wetting and dispersion in oil as well as resistance to high temperature, it is used as a white pigment under the name *titan white.* Mixed with $BaSO_4$ it is sold under the trade name *titanox.* (*ii*) It can also be used for interior protection purpose. When applied externally the film gets chalked out. (*iii*) It is also used (*a*) in making non-greasable paper (*b*) in the manufacture of arc lamp electrodes (*c*) in gas mantles (*d*) for the preparation of glazes (*e*) in making components for use in radio and television sets (*f*) as an additive in the manufacture of fibre glass (*g*) in the manufacture of artificial gems (*h*) in enamelling of porcelain goods (*i*) in the surface of white leather (*j*) in rubber industry to soften rubber (*k*) as a whitener in cosmetics (*e.g.* in face powder, cold creams etc.) (*l*) in preparing printing and decorating inks.

**2. Titanates ($TiO_3^{2-}$)**

Ti (IV) compounds having titanate ion ($TiO_3^{2-}$) are called **titanates.**

(*a*) **Sodium titanate, $Na_2TiO_3$.** It is prepared by fusing $Na_2O$ with $TiO_2$. When $Na_2TiO_3$ is heated in an atmosphere of $H_2$, non-stoichiometric compounds having the approximate fermulae $Na_{0.20-0.25}\ TiO_2$ are obtained. These compounds are called **titanium bronzes.** These compounds have high electrical conductivity and a metallic bluish lustre.

(*b*) **Magnesium titanate, Mg $Ti_2O_4$.** $MgTi_2O_4$ has high refractive index. Hence its crystals show great brilliance, even greater than diamond. Since it is soft, it is not used in jewellery. It is used as a semi-precious artificial gem.

This titanate has spiral structure (MgO $Ti_2O_3$) in which $O^{2-}$ ions form ccp arrangement, $Mg^{2+}$ ions occupy the octahedral sites and $T^{4+}$ ions occupy the tetrahedral sites.

(*c*) **Calcium titanate, $CaTiO_3$.** This titanate occurs naturally as *perovskite.* The structure of perovskite has *ccp* arrangement of $O^{2-}$ ions and $Ca^{2+}$ ions in which *1/4th* of octahedral sites formed by $O^{2-}$ ions only.

(*d*) **Iron titanate, $FeTiO_3$.** This titanate occurs naturally as *ilmenite*. The crystal structure of ilmenite comprises *hcp* arrangement of $O^{2-}$ ions with *1/3rd* of octahedral sites occupied by $Tc^{4+}$ ions and another *1/3rd* of octahedral sites occupied by $Fe^{3+}$ ions.

### 3. Titanium Tetrachloride or Titanic Chloride, $TiCl_4$

**Preparation.** It is prepared: (*i*) by passing $Cl_2$ over heated Ti-metal or over a mixture of $TiO_2$ and carbon.

$$Ti + 2Cl_2 \xrightarrow{350°C} TiCl_4$$

$$TiO_2 + C + 2Cl_2 \longrightarrow TiCl_4 + CO_2$$

(*ii*) by heating $TiO_2$ in a current of dry $Cl_2$.

$$TiO_2 + 2Cl_2\,(dry) \xrightarrow{\Delta} TiCl_4 + O_2$$

(*iii*) by passing the vapours of $CCl_4$ over $TiO_2$ at high temperature.

$$TiO_2 + CCl_4\,(vapours) \longrightarrow TiCl_4 + CO_2$$

(*iv*) from ilmenite ore, $FeTiO_3$. See "*Extraction of Ti-metal*".

**Properties.** (*i*) It is colourless strongly fuming liquid and has m.pt. = –30°C and b.pt. =136.4°C.

(*ii*) *Hydrolysis.* (*a*) The spraying of liq. $TiCl_4$ into moist air results in its hydrolysis, forming dense white smokes of $TiO_2$. These smokes have irritating nature.

$$TiCl_4 + 2H_2O \longrightarrow TiO_2 + 2HCl$$

It is due to this property that $TiCl_4$ is used for producing white smokescreens and for making writings in the sky in army operations.

(*b*) In presence of a little amount of water, $TiCl_4$ is hydrolysed to $TiOCl_2$.

$$TiCl_4 + H_2O \longrightarrow TiOCl_2 + 2HCl$$

(*iii*) *Reduction.* (*a*) It can be reduced to $TiCl_2$ by Ti metal.

$$TiCl_4 + Ti \longrightarrow 2TiCl_2$$

(*b*) When reduced by $H_2$ or Ag, $TiCl_3$ is obtained.

$$2TiCl_4 + H_2 \longrightarrow 2TiCl_3 + 2HCl$$

$$TiCl_4 + Ag \longrightarrow TiCl_3 + AgCl$$

(*c*) When reduced by Mg (*Kroll process*), Na (*I.C.I. process*) or NaH, Ti metal is obtained.

$$TiCl_4 + 2Mg \longrightarrow Ti + 2MgCl_2$$

$$TiCl_4 + 4Na \longrightarrow Ti + 4NaCl$$

$$TiCl_4 + 4NaH \longrightarrow Ti + 4\,NaCl + 2H_2$$

(*iv*) *Action of $NH_4Cl$.* Yellow crystals of *ammonium chloro titanate,* $(NH_4)_2\,[TiCl_6].2H_2O$ is obtained.

$$TiCl_4 + 2NH_4Cl + 2H_2O \longrightarrow (NH_4)_2\,[TiCl_6].2H_2O$$

(*v*) *Action of acids.* (*a*) Conc. HCl dissolves $TiCl_4$, forming chloro complexes like $H[TiCl_5\,(H_2O)]$, $H_2\,[TiCl_6]$ etc.

$$TiCl_4 + HCl + H_2O \longrightarrow H\,[TiCl_5\,(H_2O)]$$

$$TiCl_4 + 2HCl \longrightarrow H_2\,[TiCl_6]$$

(*b*) Anhydrous HF gives, $TiF_4$.

$$TiCl_4 + 4HF \longrightarrow TiF_4 + 4HCl$$

(*c*) The action of HI on $TiCl_4$ produces $TiI_4$.

(*d*) When $TiCl_4$ is treated with $H_2SO_4$ and the solution is evaporated in vacuum, $TiSO_4$ is obtained.

(*vi*) *Lewis acid character*. It is a strong Lewis acid (*i.e.*, electron pair acceptor) and hence can accept electron pairs from donor molecules like $NH_3$ and $PCl_5$, forming the addition compounds like $TiCl_4.\,6NH_3$, $TiCl_4.\,8NH_3$ and $TiCl_4.\,PCl_5$ etc. $TiCl_4$ also forms addition compounds with alcohols and ethers of the type $(R_2O)\,TiCl_4.\,(R_2O)_2\,TiCl_4$ etc.

**Uses.** It is used : (*i*) for the production of dense white irritating smokescreens and for making writings in the sky in army operations (*ii*) as a catalyst in the manufacture of polyethylene (*iii*) in the preparation of $TiCl_2$ and $TiCl_3$ (*iv*) in the production of Ti metal by reducing it with Mg, Na or NaH (*v*) in the manufacture of rutile gems and organic titanium compounds.

### 4. Titanium Nitrate, $Ti\,(NO_3)_4$

**Preparation.** Titanium (IV) nitrate can be prepared by the action of $N_2O_5$ on titanium (IV) chloride.

$$TiCl_4 + 2N_2O_5 + 2H_2O \longrightarrow Ti(NO_3)_4 + 4HCl$$

**Properties.** (*i*) It is a white volatile compound.

(*ii*) It melts at 58°C.

(*iii*) It is highly reactive towards organic substances. While reacting with organic substances, it causes inflammation or explosion. The highly reactive nature of titanium nitrate is probably due to the presence of very reactive nitrate $(NO_3)$ radical in its molecule.

**Structure.** The structure of $Ti(NO_3)_4$ is shown in Fig. 6.3. In its structure two oxygen atoms from each nitrate group are bonded to titanium atom exhibiting its bidentate nature.

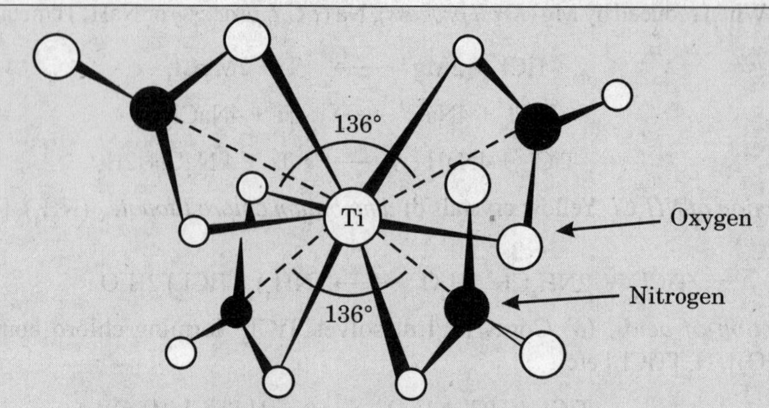


**Fig. 6.3** Structure of titanium (IV) nitrate.

The bidentate nitrate groups form a tetrahedron around titanium bonded through two oxygen atoms giving a dodecahedral structure. In this compound titanium attains a coordination number of 8.

**5. Titanyl Salts ($TiO^{2+}$)**

$Ti^{4+}$ ion exists as titanyl ion ($TiO^{2+}$) in aqueous solution and forms a number of salts which are called *titanyl salts*. Titanyl sulphate ($TiOSO_4$) is an important titanyl salt. This salt can be crastallised out from the aqueous solution containing $TiO^{2+}$ and $SO_4^{2-}$ ions.

**6. Complexes of Ti (IV).**

(*a*) ***Peroxo complex,* $[Ti(O_2)(OH)(H_2O)_n]^+$.** When $H_2O_2$ is added to the solution of $TiO_2^{2+}$ or $[Ti\ (OH)_2]^{2+}$, an orange colour is produced due to the formation of peroxo complex of $Ti^{3+}$ ion whose probable formula is $[Ti(O_2)(OH)(H_2O)_n]^+$. In this complex ion, peroxo group $\left(O_2^{2-}\right)$ is attached with $Ti^{4+}$ ion.

(*b*) **Penta-coordinated complexes.** $Et_4NCl$ reacts with $TiCl_4$ to give 5-coordinated complex $[Et_4N]^+\ [TiCl_5]^-$

(*c*) **Hexa-coordinated complexes.** $TiCl_4$ and $TiF_4$ react with HCl and HF to form $H_2\ [TiCl_6]$ and $H_2[TiF_6]$ respectively.

$$TiCl_4 + 2HCl \longrightarrow H_2\ [Ti\ Cl_6]$$

$$TiF_4 + 2HF \longrightarrow H_2\ [TiF_6]$$

$TiX_4$ also reacts with amines, phosphines and arsines to form $TiX_4 . 2PR_3$

## Organometallic Compounds of Titanium

When $TiCl_4$ is added to $Al(C_2\ H_5)_3$ in hexane, it yields a brown solid known as **Ziegler-Natta catalyst** which is employed in the conversion of ethene to straight chain polymer polyethene. The probable mechanism is as follows:

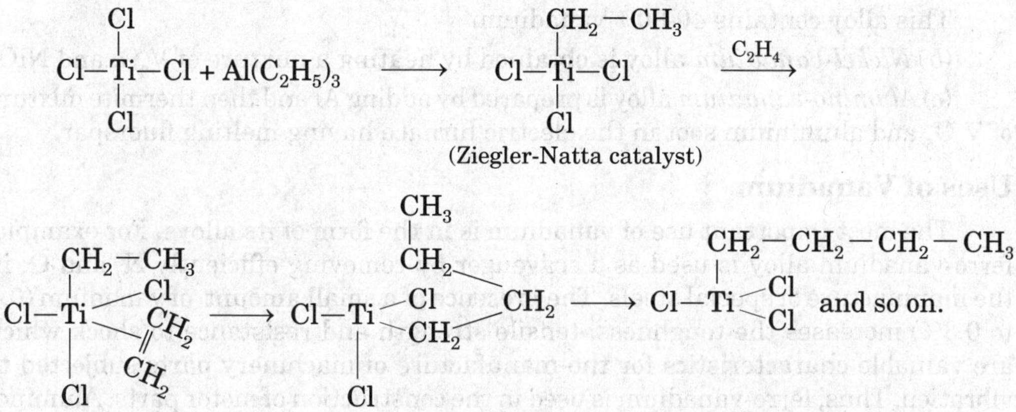


Several stable cyclopentadienyls, such as $[Ti(\eta^5-(C_5H_5)_2(CO)_2]$, $Ti(\eta^5-C_5H_5)_2]$ etc., are known. More interesting are the compounds in which Ti is attached to four $C_5H_5$ rings. The structure of Ti $(C_5H_5)_4$ is given in Fig. 6.4.

Some alkyls and aryls of Ti are also known but these are quite unstable.

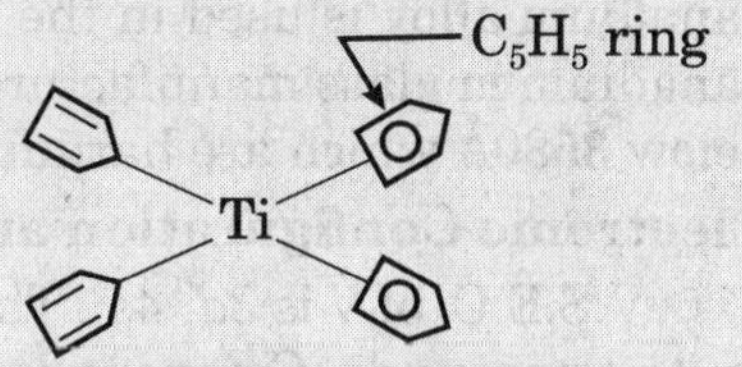


**Fig. 6.4** Structure of cyclopentadienyl of titanium, Ti $(C_5H_5)_4$

## Vanadium, $V_{23}$

### Properties of Vanadium Metal

It is a steel-grey hard metal which melts above 1700°C and boils at 3450°C. When pure, it is malleable as well as ductile. When heated in air, the metal is converted into $V_2O_5$. When heated above 1400°C, it forms $VO_2$. When treated with excess of dry $Cl_2$, $VCl_4$ is formed. Bromine water has no action on vanadium metal. HCl and $H_2SO_4$ both have no action on the metal. However, with hot and conc. $H_2SO_4$ it gives $SO_2$. With HF it gives $H_2$. It is easily attacked by $HNO_3$ and forms $V(NO_3)_3$. Aqua regia attacks the metal very vigorously. Fused alkalies give orthovanadates ($Na_3VO_4$), with the liberation of $H_2$.

$$2V + 6NaOH + 2H_2O \longrightarrow 2Na_3VO_4 + 5H_2$$

Oxidising agents like chloric acid, perchloric acid, bromic acid, $KClO_3$, bromate, iodate and perchlorate convert vanadium into $V_2O_5$. Vanadium metal reduces the solution of $AgNO_3$, $AuCl_3$, $PtCl_4$ and $IrCl_4$. It combines with non-metals like $N_2$, C, S and Si to form nitride, carbide, sulphide and silicide respectively. Vanadium combines with a number of metals like Fe, Cu, Co, Ni and Al forms alloys. For example:

(*a*) *Ferro-vanadium alloy* is obtained by heating $V_2O_5$ with iron ore ($Fe_2O_3$) and coke in presence of a flux (lime or fluorspar) in an electric furnace.

$$V_2O_5 + 5C \longrightarrow 2V + 5CO\uparrow$$

$$Fe_2O_3 + 3C \longrightarrow 2Fe + 3CO\uparrow$$

This alloy contains 40-80% vanadium.

(*b*) *Nickel-vanadium* alloy is obtained by heating a mixture of $V_2O_5$ and NiO.

(*c*) *Alumino-vanadium* alloy is prepared by adding Al and then thermite mixture of $V_2O_5$ and aluminium soot in the electric furnace having melting fluorspar.

## Uses of Vanadium

The most important use of vanadium is in the form of its alloys. For example, ferro-vanadium alloy is used as a scavenger by removing efficiently $N_2$ and $O_2$ in the manufacture of special steels. The presence of a small amount of vanadium (0.1 to 0.3%) increases the toughness, tensile strength and resistance to shock which are valuable characteristics for the manufacture of machinery parts subjected to vibration. Thus, ferro-vanadium is used in the construction of motor parts. Alumino-vanadium alloy is used in the manufacture of aircraft parts.The addition of 0.2% vanadium in glass manufacture is recommended for cutting of all the wavelengths below 3580Å which are harmful for the eyes and fade fabrics.

## Electronic Configuration and Oxidation States of V

V.S.E.C. of V is $3d^3\,4s^2$. This element shows +2, +3, +4 and +5 oxidation states in its compounds. Compounds in lower oxidation state are good reducing agents. They are ionic in character and are coloured. Compounds in +5 O.S. are generally *colourless.*

When a compound of *V* in +5 O.S. is reduced with Zn, it undergoes stepwise reduction to give compounds in +4, +3 and +2 oxidation states. The compounds in these oxidation states are *blue*, *green* and *violet* respectively.

## Compounds of Vanadium (II)

$V^{2+}$ ion is called **vanadous ion.**

1. **Vanadium Oxide, VO**

VO is prepared by reducing $V_2O_5$ with $H_2$ at about 2000°

$$V_2O_5 + 3H_2 \xrightarrow{-\ 2000°C} 2VO + 3H_2O$$

VO is a blackish solid with metallic lustre. It is soluble in $H_2O$. When an alkali is added to the aqueous solution of VO, the ppt. of $V(OH)_2$ is obtained. The addition of an acid to its aqueous solution gives a water soluble violet complex, $[V(H_2O)_6]^{2+}$.

VO has NaCl type structure with considerable degree of lattice defects. Because of these defects, VO conducts electricity.

2. **Vanadium (II) Chloride, $VCl_2$**

This compound is obtained by reducing $VCl_4$ vapours with hydrogen in a red hot tube.

$$VCl_4 + H_2 \longrightarrow VCl_2 + 2HCl$$

It forms bright green lustrous leaflet crystals. It is soluble in water. The vanadous ion $V^{2+}$ is a strong reducing agent. Thus, the violet solution of $VCl_2$ soon turns green because of its oxidation to trivalent state with evolution of hydrogen (reduction of water).

$$2VCl_2 + 2H_2O \longrightarrow 2VOCl + 2HCl + H_2$$

**3. Vanadium (II) Sulphate, $VSO_4 . 7H_2O$.**

It is prepared by the electrolytic reduction of vanadium pentoxide dissolved in sulphuric acid in the presence of sulphur trioxide till the solution turns violet.

It is soluble in water and forms red-violet monoclinic crystals which are isomorphous with $FeSO_4 . 7H_2O$. The double salt, $VSO_4 . (NH_4)_2 SO_4 . 6H_2O$ is isomorphous with Mohr's salt., $FeSO_4 . (NH_4)_2 SO_4 . 6H_2O$.

It is a *strong reducing agent.* As such the violet coloured aqueous solution containing $V^{2+}$ ion in the form of $[V(H_2O)_6]^{2+}$ complex ion becomes green due to oxidation of $[V(H_2O)_6]^{2+}$ ion to $[V(H_2O)_6]^{3+}$ ion with the liberation of hydrogen (reduction of water).

$$\underset{\text{Violet}}{[V(H_2O)_6]^{2+}} + H^+ \longrightarrow \underset{\text{Green}}{[V(H_2O)_6]^{3+}} + \frac{1}{2}H_2$$

**4. Complexes of Vanadium (II).**

Apart from the already mentioned $[V(H_2O)_6]^{2+}$ complex ion, some other V(II) complexes are also known. Examples are $K_4[V(CN)_6]$ and $[V(en)_3]Cl_2$. V(II) complexes, in general, are not very stable.

## Compounds of Vanadium (III)

Compounds of vanadium (III) are prepared by the reduction of compounds of vanadium with higher oxidation states.

**1. Vanadium Trichloride, $VCl_3$.**

It is obtained by the reduction of $VCl_4$.

$$2VCl_4 + H_2 \longrightarrow 2VCl_3 + 2HCl$$

Vandium trichloride forms beautiful peach-blossom coloured crystal leaflets. It is non-volatile. It is readily hydrolysed by water. Vanadium trichloride is easily oxidised by atmospheric air and is, therefore, *reducing in character.* $VCl_3$ dissolves in water giving $[V(H_2O)_6]^{3+}$ and $Cl^-$ ions. With amines, it forms complexes which have trigonal bipyramidal structure. Well known example is $[VCl_3(NR_3)_2]$.

$VCl_3$ is polymaric in nature. Each V atom is surrounded octahedrally by six Cl atoms. These Cl atoms form bridges between V atoms.

**2. Vanadium trioxide $VI_3$.**

It is also obtained by reducing $VI_4$ with hydrogen. It is thermally less stable than $VCl_3$ and readily disproportionates as follows:

$$2VI_3 \longrightarrow VI_2 + VI_4$$

**3. Vanadium (III) Oxide, $V_2O_3$.**

It is obtained by reduction of $V_2O_5$ with carbon or hydrogen at high temperatures.

$$V_2O_5 + 2H_2 \longrightarrow V_2O_3 + 2H_2O$$

It is readily oxidised and thus acts as a strong *reducing agent.*

$V_2O_3$ is *basic* in nature and dissolves in acids to give solutions of complex hexaaqua ion, $[V(H_2O)_6]^{3+}$. $V_2O_3$ is actually a non-stoichiometric compound with composition varying from $V_2O_{2\cdot75}$ to $V_2O_3$.

**4. Vanadium (III) Sulphate, $V_2(SO_4)_3$.**

It is obtained by the electrolytic reduction of a solution of $V_2O_5$ in $H_2SO_4$ till a green solution results.

The green solution, on evaporation, deposits green lustrous needles having the composition $V_2(SO_4)_3 \cdot H_2SO_4 \cdot 12H_2O$. This compound, on heating at 180°C, loses water and $SO_3$, leaving a yellow powder of the composition $V_2(SO_4)_3$, which is insoluble in water.

Aqueous solutions containing $V^{3+}$ ions are remarkably stable in the presence of sulphuric acid. They get oxidised slowly when exposed to atmosphere.

Vanadium (III) sulphate is a powerful reducing agent.

**5. Vanadium (III)Complexes.**

Vanadium (III) forms a large number of complexes, most of them being anionic. Examples are $[V(C_2O_4)_3]^{3-}$, $[V(CN)_6]^{3-}$ and $[V(NCS)_6]^{3-}$. A well known example of cationic complex is $[V(H_2O)_6]^{3+}$.

## Compounds of Vanadium (IV)

Important compounds of vanadition in which vanadium has +4 oxidation state are $VO_2$, $VX_4$ and vanadyl compounds. Vanadyl compounds are also called **oxovanadium (IV) compounds.** $VO_2$ and $VX_4$ are covalent compounds while oxo compounds are ionic in character and contain the blue **vanadyl ion $VO^{2+}$**.

**1. Vanadium (IV) Oxide, $VO_2$ or $V_2O_4$.**

This oxide is also called **hypovanadic oxide.** It is made by reducing $V_2O_5$ with $SO_2$ or by fusing $V_2O_5$ with oxalic acid , $(COOH)_2$

$$N_2O_5 + SO_2 \longrightarrow SO_3 + V_2O_4$$

$$V_2O_5 + (COOH)_2 \longrightarrow V_2O_4 + 2\,CO_2 + H_2O$$

$V_2O_4$ is an amphoteric oxide. With acids it gives *vanadyl salts* ($VO^{2+}$) and with alkalies it forms *hypovanadites* ($V_4O_9^{2-}$) which give blue solutions.

$$V_2O_4 + 4\,HCl \longrightarrow 2\,VOCl_2 + 2H_2O$$

$$2V_2O_4 + 2KOH \longrightarrow K_2V_4O_9 + H_2O$$

$V_2O_4$ reacts with alkaline earth oxides at high temperatures in vacuum to form vanadates of the types $MVO_3$, $M_2VO_4$ and $M_2VO_5$ (M = Ca, Sr or Ba).

**2. Vanadium Tetrachloride, $VCl_4$.**

It is obtained as a brown liquid by passing chlorine over heated vanadium or by passing vapours of $VOCl_3$ and $Cl_2$ over heated carbon.

$$V + 2Cl_2 \longrightarrow VCl_4$$

$$2VOCl_3 + 2C + Cl_2 \longrightarrow 2VCl_4 + 2CO$$

It is an unstable compound and decomposes even at room temperature to give the trichloride and chlorine.

$$2VCl_4 \longrightarrow 2VCl_3 + Cl_2$$

It is immediately hydrolysed by water.

$$VCl_4 + H_2O \longrightarrow VOCl_2 + 2HCl$$

Electron-diffraction study shows that $VCl_4$ vapour is tetrahedral. It acts like a Lewis acid and forms a large number of adducts and substitution products which are neutral or anionic. The products are usually six-coordinate compounds. However, lower coordinate compounds are also obtained when reduction of $VCl_4$ takes place. The ligand *o*-phenylenebisdimethylarsine (diars) forms eight-coordinate compound $VCl_4$(diars)$_2$ (structure shown in Fig. 6.5.). The product is isomorphous with the titanium analogue. The ionic compound $[PCl_4^+]$ $[VCl_5^-]$ containing five-coordinate $VCl_5^-$ is formed when $VCl_4$ interacts with $PCl_5$.

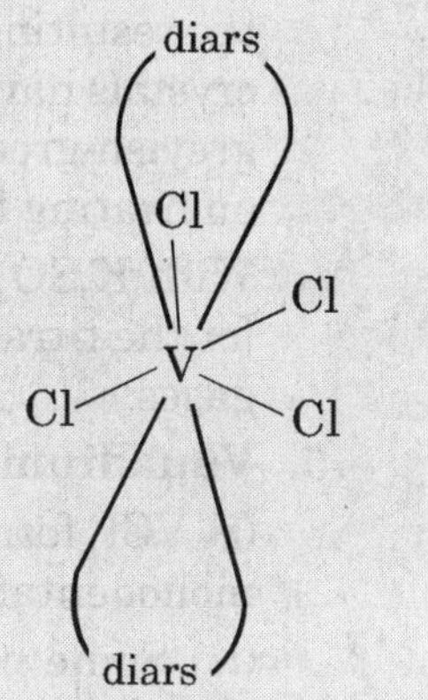


**Fig. 6.5** Structure of 8-coordinate $VCl_4$ (diars)

**3. Vanadyl Chloride, $VOCl_2$.**

**Preparation.** (*i*) It can be prepared by the reduction of $VOCl_3$ with hydrogen at 300°C.

$$2VOCl_3 + H_2 \xrightarrow{300°C} 2VOCl_2 + 2HCl$$

(*ii*) Pure compound is prepared by heating $VOCl_3$ with metallic zinc in a sealed tube.

$$2VOCl_3 + Zn \longrightarrow 2VOCl_2 + ZnCl_2$$

(*iii*) It can be prepared by the action of HCl on $V_2O_5$.

$$V_2O_5 + 6HCl \longrightarrow 2VOCl_2 + 3H_2O + Cl_2$$

(*iv*) It can be prepared by the hydrolysis of $VCl_4$.

$$VCl_4 + H_2O \longrightarrow VOCl_2 + 2HCl$$

**Properties** (*i*) The pure compound exists as grassy green crystals.

(*ii*) *Hydrolysis.* It is hydrolysed to yellow brown $VO(OH)_2$ with dilute sodium carbonate solution.

$$VOCl_2 + 2H_2O(aq) \xrightarrow{Na_2CO_3(aq)} VO(OH)_2(aq) + 2HCl$$

(*iii*) *Action with alkalis.* It reacts with strong alkalis to form vanadates.

$$VOCl_2 + 4NaOH \longrightarrow \underset{\text{Sodium vanadate}}{Na_2VO_3} + 2NaCl + 2H_2O$$

**Uses.** Vanadyl chloride is used commercially in dyeing under the name *Divanadyl tetrachloride*. It is a strong reducing agent.

**Structure.** $VOCl_2$ has been found to contain double (V = O) bond formed by the overlap of $\pi$ electrons of oxygen into the vacant *d*-orbitals of vanadium. It has trigonal planar structure (Fig. 6.6)

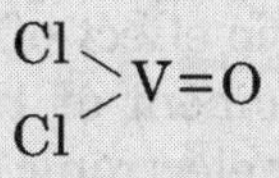


**Fig. 6.6** Trigonal planar structure of $VOCl_2$ molecule.

**4. Vanadyl Sulphate, $VOSO_4$.**

This salt exists in two modifications; one is soluble while the other is insoluble in water. When a solution of $V_2O_5$ in $H_2SO_4$ is reduced by means of $SO_2$ and the resulting solution is evaporated, bright blue crystals are obtained. These crystals have the composition, $2VOSO_4 . H_2SO_4 . xH_2O$. The pure substance is greyish-green crystalline powder which changes to blue soluble modification on heating to 130°C with a little water.

With $K_2SO_4$, it forms a blue double salt, $K_2SO_4 . VOSO_4 . xH_2O$. It is used for the titration of weak oxidising agents and for removing traces of $O_2$ from gases.

**5. Vanadium (IV) Complexes.**

(*i*) $VCl_4$ forms complexes of the types $VCl_4L$ and $VCl_4L_2$ in which L is a monodentate ligand. These complexes are readily hydrolysed by $H_2O$.

(*ii*) Some of the complexes of V (IV) contain vanadyl ion, $VO^{2+}$. These complexes are called $VO^{2+}$ complexes. Examples of such complexes are: $[VO(bipy)_2Cl]^+$, $[VO(acac)_2]^\circ$, $[VOCl_2(NMe_3)_2]^0$, $[VO(NCS)_4]^{2-}$, $[VOX_4]^{2-}$. *bipy* is *bipyridyl* and *acac* is *acetyl acetonate anion*

The strucutres of $[VO(acac)_2]^0$ and $[VOCl_2(NMe_3)_2]^0$ are given in Fig. 6.7 and Fig. 6.8 respectively.

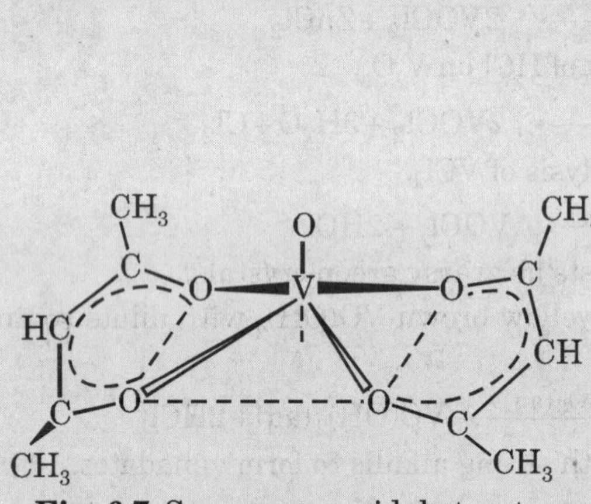


**Fig. 6.7** Square pyramidal structure of $[VO(acac)_2]^0$

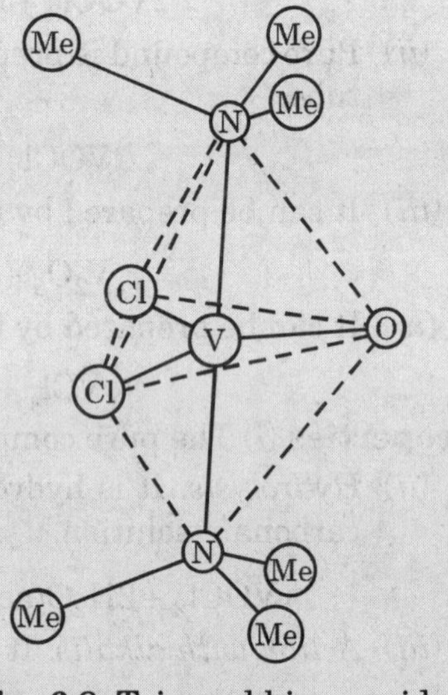


**Fig. 6.8** Trigonal bipyramidal structure of $[VOCl_2(NMe_3)_2]^\circ$

$[VO(acac)_2]^0$ has square pyramidal structure in which O-atom occupies the apical position. $[VOCl_2(NMe_3)_2]^0$ has trigonal bipyramidal structure due to a dominant steric effect of the bulky ligand molecule, $NMe_3$. $VO^{2+}$ complexes have a magnetic moment of 1.73 BM (spin only value) corresponding to one unpaired electron. Their electronic spectra are complicated due to a strong $\pi$-bonding between V and O.

## Compounds of Vanadium (V)

Vanadium forms a limited number of compounds in which it shows +5 oxidation state. Following are important compounds of vanadium in +5 oxidation state.

**1. Vanadium Pentoxide, $V_2O_5$.**

This oxide is also called *vanadic anhydride*, since it can be supposed to be anhydride of orthovanadic acid, $H_3VO_4$.

$$2H_3VO_4 \xrightarrow{-3H_2O} V_2O_5 + 3H_2O$$

**Preparation.** $V_2O_5$ is obtained: (*i*) by heating ammonium meta vanadate ($NH_4VO_3$) or by adding dil. $H_2SO_4$ to its aqueous solution.

$$2NH_4VO_3 \xrightarrow{\Delta} V_2O_5 + 2NH_3 + H_2O$$

$$2NH_4VO_3 + H_2SO_4 \longrightarrow V_2O_5 + (NH_4)_2SO_4 + H_2O$$

$NH_4VO_3$ used in the above reactions can be obtained from any of the ores of vanadium like carnotite, vanadinite and patronite.

(*ii*) by the hydrolysis of vanadium oxytrichloride ($VOCl_3$).

$$2VOCl_3 + 3H_2O \longrightarrow V_2O_5 + 6HCl$$

(*iii*) by heating vanadium in an atmosphere of $O_2$ or in air.

$$4V + 5O_2 \longrightarrow 2V_2O_5$$

Along with $V_2O_5$, little amount of lower oxides is also formed.

**Properties. (*i*) Physical state and dissociation.** $V_2O_5$ is a yellowish-red poisonous powder having m.pt = 670°C. It is slightly soluble in $H_2O$ and the resulting pale yellow solution is acidic in nature. It undergoes reversible dissociation in the temperature range of 700 –1125° C.

$$2V_2O_5 \rightleftharpoons 2V_2O_4 + O_2$$

**(*ii*) Action of $Cl_2$.** When $V_2O_5$ is heated in $Cl_2$ at elevated temperature or when $Cl_2$ is passed over heated mixture of $V_2O_5$ and carbon, $VOCl_3$ is obtained.

$$2V_2O_5 + 6Cl_2 \xrightarrow{\Delta} 4VOCl_3 + 3O_2$$

$$V_2O_5 + 3C + 3Cl_2 \longrightarrow 2VOCl_3 + 3CO$$

**(*iii*) Action of acids.** (*a*) $V_2O_5$ dissolves in HCl to give salts like vanadium oxydichloride ($VOCl_2$) and $VCl_5$.

$$V_2O_5 + 6HCl \longrightarrow 2VOCl_2 + 3H_2O + Cl_2$$

The liberation of $Cl_2$ indicates that $V_2O_5$ has oxidised HCl to $Cl_2$, *i.e.*, $V_2O_5$ can be used as an oxidising agent. $V_2O_5$ also reacts with HCl in the following manner:

$$V_2O_5 + 10HCl \longrightarrow 2VCl_5 + 2H_2O$$

(*b*) With $HNO_3$, $V_2O_5$ gives metavanadic acid, $HVO_3$.

$$V_2O_5 + 2HNO_3 \longrightarrow 2HVO_3 + N_2O_5$$

(*iv*) **Action of $H_2O_2$ in alkaline medium**. When $H_2O_2$ reacts with $V_2O_5$ in alkaline medium, peroxy vanadate $[V(O_2)_4]^{3-}$ (blue-violet) is obtained.

$$V_2O_5 + 8H_2O_2 + 6OH^- \longrightarrow 2[V(O_2)_4]^{3-} + 11H_2O$$

(*v*) **Action of carbon**. When heated with carbon at high temperature, vanadium carbide (VC) is obtained.

$$2V_2O_5 + 4C \xrightarrow{\Delta} 4VC + 5O_2$$

(*vi*) **Action of $Na_2CO_3$**. When $V_2O_5$ is melted with $Na_2CO_3$, sodium orthovanadate ($Na_3VO_4$) is obtained.

$$V_2O_5 + 3Na_2CO_3 \longrightarrow 2Na_3VO_4 + 3CO_2\uparrow$$

(*vii*) **Reduction**. (*a*) Potassium reduces $V_2O_5$ to hypovanadium oxide or vanadium monoxide ($V_2O_5$ or VO).

$$V_2O_5 + 6K \longrightarrow V_2O_2 + 3K_2O$$

(*b*) $H_2$, carbon and S reduce $V_2O_5$ to vanadium trioxide, $V_2O_3$.

$$V_2O_5 + 2H_2 \xrightarrow{\Delta} V_2O_3 + 2H_2O$$
$$V_2O_5 + S \longrightarrow V_2O_3 + SO_2$$
$$V_2O_5 + 2C \xrightarrow{\Delta} V_2O_3 + 2CO$$

(*c*) $SO_2$, $H_2S$ and $H_2C_2O_4$ reduce $V_2O_5$ to vanadium tetroxide, $V_2O_4$ or $VO_2$.

$$V_2O_5 + SO_2 \longrightarrow V_2O_4 + SO_3$$
$$V_2O_5 + H_2S \longrightarrow V_2O_4 + H_2O + S$$
$$V_2O_5 + H_2C_2O_4 \longrightarrow V_2O_4 + H_2O + 2CO_2$$

(*d*) Al-powder reduces $V_2O_5$ to V-metal.

$$3V_2O_5 + 10Al \longrightarrow 5Al_2O_3 + 6V$$

(*e*) When $V_2O_5$ is heated with excess of Ca in presence of $CaCl_2$ in a steel bomb at 900-950°C (*Marden and Rich, 1927*), pure vanadium metal is obtained.

$$V_2O_5 + 5Ca \xrightarrow[\text{of } CaCl_2]{\text{Heat in presence}} 2V + 5CaO$$

(*f*) When a solution of $V_2O_5$ in $H_2SO_4$ is allowed to undergo electrolytic reduction, a green solution is obtained. This solution, on evaporation, deposits green lustrous needles of $V_2(SO_4)_3.H_2SO_4.12H_2O$. These needles, on being heated to 180°C, lose $13H_2O$ and $SO_3$, leaving a yellow powder of $V_2(SO_4)_3$.

$$V_2O_5 + 4H_2SO_4 + 9H_2O \xrightarrow[\text{reduction}]{\text{Electrolytic}} \underbrace{V_2(SO_4)_3.H_2SO_4.12H_2O}_{} + O_2$$

$$\downarrow 180°C$$

$$\underset{\text{yellow powder}}{V_2(SO_4)_3} + 3H_2O + SO_3$$

(*g*) When a solution of $V_2O_5$ in $H_2SO_4$ is reduced by means of $SO_2$ and the resulting solution is evaporated, bright blue crystals are obtained. Probably these crystals have the composition, $2VOSO_4 . H_2SO_4 . xH_2O$.

**(*viii*) Formation of various types of vanadates at different pH values.** $V_2O_5$ dissolves in strong alkalies (pH = 14) to give colourless ortho vanadate ($VO_4^{3-}$).

$$V_2O_5 + 6NaOH \longrightarrow 2Na_3VO_3 + 3H_2O$$

or

$$V_2O_5 + 6OH^- \longrightarrow 2VO_4^{3-} + 3H_2O$$

When this solution containing $VO_4^{3-}$ ion is progressively acidified, *i.e.*, when the pH value of the solution is decreased from 14, $VO_4^{3-}$ ion gets polymerised to give more complicated vanadate ions and in acidic solution (pH = 6.8) pentavanadate, $[V_5O_{14}]^{3-}$ can be isolated. With further decrease in pH value (pH < 6.8) hydrated vanadium pentoxide gets precipitated. Two definite hydrates namely $V_2O_5 .H_2O$ or $H_2V_2O_6$ or $HVO_3$ (meta vanadic acid) and $2V_2O_5 . H_2O$ or $H_2V_4O_{11}$ (tetravanadic acid) have been identified. In strongly acidic solution (pH < 2), $VO_2^+$ is formed. A typical schem is shown below:

$$V_2O_5 \xrightarrow{+OH^-(pH=14)} \underset{\text{Ortho vanadate (colourless)}}{[VO_4]^{3-}} \xrightarrow{pH=12} [VO_3(OH)]^{2-} \xrightarrow{pH=10}$$

$$\underset{\text{Colourless}}{[V_2O_6(OH)]^{3-}} \xrightarrow{pH=9} \underset{\text{Colourless}}{[V_3O_9]^{3-}} \xrightarrow{pH=6.8} \underset{\text{Pentavanadate}}{[V_5O_{14}]^{3-}} \xrightarrow{pH<6.8}$$

$$\underset{\text{Hydrated pentaoxide}}{mV_2O_5.nH_2O} \xrightarrow{pH=2.2} [V_{10}O_{28}]^{6-} \xrightarrow{pH<2} [VO_2]^+$$

It may be seen that vanadium in all above mentioned ions is in +5 oxidation state.

**(*ix*) Formation of ferro-vanadium alloy.** When a mixture of $V_2O_5$ and haematite ore ($Fe_2O_3$) is reduced by carbon or Al-powder in an electric furnace (using carbon electrodes) in the presence of a flux *viz.* lime or flourspar, ferro-vanadium alloy is obtained in the molten state.

*Reduction by carbon:*
$$\begin{cases} V_2O_5 + 5C \rightarrow 2V + 5CO\uparrow \\ Fe_2O_3 + 3C \rightarrow 2Fe + 3CO\uparrow \end{cases}$$

*Reduction by Al-powder:*
$$\begin{cases} 3V_2O_5 + 10Al \rightarrow 6V + 5Al_2O_3 \\ Fe_2O_3 + 2Al \rightarrow 2Fe + Al_2O_3 \end{cases}$$

*Formatioin of ferro-vanadium alloy: V + Fe + C* → Ferro-vanadium alloy

The molten alloy and the slag are removed by separate pipes. The alloy formed by Al-powder may contain small amounts of C, Si and other metals as impurities.

Ferro-vanadium alloy is used as a scavenger (*i.e.,* clearing agent) in steel industry, since vanadium present in the alloy removes $O_2$ and $N_2$, if dissolved in the molten steel in the form of its oxide ($V_2O_5$) and nitride (VN). Also at the same time, vanadium dissolves in the molten steel and thus (*a*) it increases the resistance of the steel towards shock, strain, vibration etc. (*b*) it imparts a higher tensile strength to the steel; (*c*) it improves the machinery qualities of the steel; and (*d*) it improves the ductility.

**Uses of $V_2O_5$.** $V_2O_5$ is used: (*i*) As a catalyst in a number of reactions. Some such reactions are: (*a*) Oxidiation of $SO_2$ to $SO_3$ used in the manufacture of $H_2SO_4$ by contact process. (*b*) Oxidation of anthrancene to anthraquinone. (*c*) Oxidation of phenanthrene to phenanthraquinone. (*d*) Oxidation of toluen to benzaldehyde to benzoic acid. (*e*) Oxidation of ethyl alcohol to acetaldehyde. (*ii*) In the preparation of vanadium glass. This glass is prepared by adding 0.02% vanadium as $V_2O_5$ to ordinary glass. This glass cuts out completely the harmful actinic rays which injure the eye and brings about fadings of organic compounds. Such glass is used particularly in textile store windows. Such glasses protect the eyes against daylight glare. (*iii*) In the preparation of ferro-vanadium alloy which is used as a scavenger for removing $O_2$ dissolved in molten cast iron used for the manufacture of special steels.

**Structure of $V_2O_5$ molecule.** In $V_2O_5$, the unit cell is $VO_6$ octahedron which is strongly distorted, almost to a trigonal bipyramid by an additional O-atom.

### 2. Vanadium Pentafluoride, $VF_5$

It is the only pentahalide of vanadium. It is obtained by heating V with $F_2$ at 300°C

$$2V + 5F_2 \longrightarrow 2VF_5$$

It can also be obtained by the following reaction:

$$2VF_4 \xrightarrow[N_2]{600°C} VF_5 + VF_3$$

$VF_5$ gets hydrolysed to $VOF_3$ by moisture and finally to hydrated pentoxide, $V_2O_5 . 3H_2O$

$$VF_5 + H_2O \longrightarrow VOF_3 + 2HF$$

$$2VOF_3 + 6H_2O \longrightarrow V_2O_5 .3H_2O + 6HF$$

$VF_5$ forms addition compounds with suitable donor molecules, *e.g.* $VF_5(XeF_6)_2$. When heated in air, $VF_5$ changes to $VOF_3$.

$$2VF_5 \xrightarrow{O_2(Air)} 2VOF_3 + 2F_2$$

$VF_5$ is a powerful oxidising agent, better than $IF_5$. In the gaseous state, $VF_5$ is monomeric and probably has trigonal bipyramidal structure. In the solid and liquid state, $VF_5$ has polymeric structure which consists of $VF_5$ octahedral units joined together the rough V-F-V bridges as shown in Fig. 6.9.

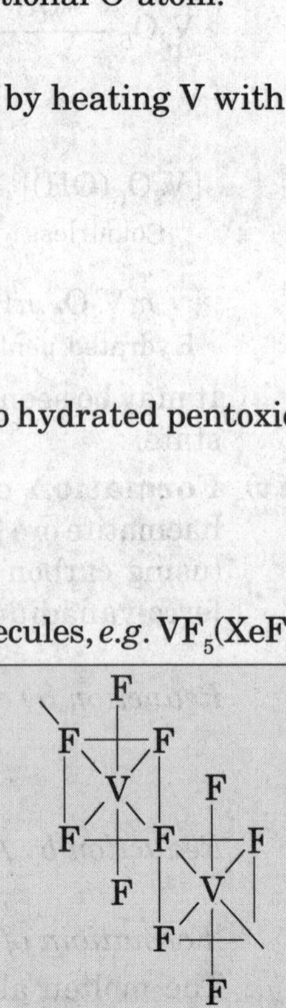


**Fig. 6.9** Polymeric structure of $VF_5$ in solid and liquid state

**3. Vanadium Oxotrihalides, $VOX_3$.**

Vanadium oxotrichloride ($VOCl_3$) and vanadium oxotribromide ($VOBr_3$) can be obtained by halogenating $V_2O_3$.

$$V_2O_3 + 3X_2\ (X = Cl, Br) \rightarrow 2VOX_3 + O_2$$

$VOCl_3$ can also be prepared by passing $Cl_2$ over a mixture of $V_2O_5$ and carbon at red heat.

$$V_2O_5 + 3C + 3Cl_2 \longrightarrow 2VOCl_3 + 3CO$$

$VOF_3$ can be prepared by treating $VOCl_3$ with HF in cold.

$$VOCl_3 + 3HF \longrightarrow VOF_3 + 3HCl.$$

Vanadium oxotrihalides are simple covalent compounds and get hydrolysed readily. For example:

$$VOCl_3 + 2H_2O \longrightarrow HVO_3 + 3HCl.$$

**4. Vanadic Acids and their Salts**

These acids may be regarded as having been formed by adding one, two or three $H_2O$ molecules to one molecule of $V_2O_5$ as shown below.

$$V_2O_5 + H_2O \longrightarrow 2HVO_3\ \textit{(Metavanadic acid)}$$

$$V_2O_5 + 2H_2O \longrightarrow H_4V_2O_7\ \textit{(Pyrovanadic acid)}$$

$$V_2O_5 + 3H_2O \longrightarrow 2H_3VO_4\ \textit{(Orthovanadic acid)}$$

The salts of these acids are important. Some important salts are discussed below:

(*i*) **Ammonium metavanadate, $NH_4VO_3$:** It is obtained (*i*) by adding an ammonium salt like $NH_4Cl$ to a solution of $NaVO_3$, (*ii*) by treating metavanadic acid, $HVO_3$ with $NH_3$.

$$NaVO_3 + NH_4Cl \longrightarrow NH_4VO_3 + NaCl$$

$$HVO_3 + NH_3 \longrightarrow NH_4VO_3$$

On large scale, $NH_4NO_3$ can be obtained from any of the ores of vanadium like carnotite [$2K(UO_2)VO_4 . 3H_2O$], vanadite [$Pb_5(VO_4)_3\,Cl$] and patronite ($V_2S_5$) (*See the extraction of vanadium*)

(*ii*) **Sodium pyrovanadate, $Na_4V_2O_7$.** In weaker alkaline solutions, sodium vanadate ($Na_3VO_4$) changes to sodium pyrovanadate ($Na_4V_2O_7$).

$$2Na_3VO_4 \xrightarrow[\text{solution}]{\text{Alkaline}} Na_4V_2O_7 + Na_2O$$

or $$2VO_4^{3-} \longrightarrow V_2O_7^{4-} + O^{2-}$$

On heating with NaOH, $Na_2V_2O_7$ is changed into $Na_3VO_4$.

$$Na_2V_2O_7 + 2NaOH \longrightarrow 2Na_3VO_4 + H_2O$$

or $$V_2O_7^{4-} + 2OH^- \longrightarrow 2VO_4^{3-} + H_2O$$

(*iii*) **Sodium orthovanadate, $Na_3VO_4 . 12H_2O$.** It is prepared: (*i*) by dissolving $V_2O_5$ in NaOH, followed by crystallisation or by treating sodium pyrovanadate, $Na_4V_2O_7$ with NaOH, (*ii*) by melting $V_2O_5$ with $Na_2CO_3$.

$$V_2O_5 + 6NaOH \longrightarrow 2Na_3VO_4 + 3H_2O$$

$$Na_4V_2O_7 + 2NaOH \longrightarrow 2Na_3VO_4 + H_2O$$
$$V_2O_5 + 3Na_2CO_3 \longrightarrow 2Na_3VO_4 + 3CO_2\uparrow$$

It is soluble in $H_2O$. It is isomorphous with sodium orthophosphate dodecahydrate, $Na_3PO_4 . 12H_2O$. This compound is used: (*i*) as a catalyst in the dyeing of leather to black colour, (*ii*) in making good quality, fire-flowing blue-black ink and marking inks, (*iii*) in volumetric analysis for certain oxidation-reduction reactions. The estimations by means of this reagent include metals such as Fe, Mo, Sn and Cu, and organic compounds such as citric acid, lactic acid, tartaric acid, aldehydes, ketones and sugars.

5. **Vandium (V) Complexes:**

Anionic complexes of the type $M[VOF_4]$, $M_2[VOF_5]$ and $M_2[VO_2F_4]$ where M= Na, K etc. are formed by the interaction of $V_2O_5$ with alkali metal fluorides in HF solution. $K_2[VF_7]$ and $Na_3[VOF_6]$ can be prepared as follows:

$$VF_5 + 2KF \longrightarrow K_2[VF_7]$$
$$VOF_3 + 3NaF \longrightarrow Na_3[VOF_6]$$

## Chromium, $Cr_{24}$

### Physical Properties

Chromium is a bluish-white lustrous metal and cryalline in nature. It is capable of taking good polish on its surface and this polish produces a brilliant surface and lasts long because of the formation of an invisible and then self-protective oxide coating. This coating is not tarnished even after long exposure to the atmosphere. The sp. gravity is 7.19, m.pt. =1875°C and b.pt. =2665°C. The metal is malleable, ductile and paramagnetic.

The pure metal is soft but becomes hard when it contains small amounts of carbon as impurity. It is resistant to corrosive agents.

### Chemical Properties

1. **Action of air.** Cr is extremely stable and is not affected by moist air, rain, sea-water, $H_2S$, $SO_2$ etc. The inertness is probably due to the formation of a protective thin oxide film on the surface of the metal. In oxy-hydrogen flame, the metal burns at about 2000°C brilliantly and forms $Cr_2O_3$.

$$2Cr + 3H_2O \xrightarrow{2000°C} Cr_2O_3 + 3H_2O$$

2. **Action of steam.** Cr has no action on water at ordinary temperature but at red heat steam is decomposed by the metal.

$$2Cr + 3H_2O\,(\textit{steam}) \xrightarrow{\text{Red heat}} Cr_2O_3 + 3H_2$$

3. **Action of halogens.** $F_2$ and dry $Cl_2$ combine with heated metal and give Cr (III) halides.

$$2Cr + \underset{(X=F,\ Cl)}{3X_2} \xrightarrow{\Delta} 2CrX_3$$

4. **Action of alkalies.** Alkalies have no action on the metal at ordinary temperature.

5. **Action of $NH_3$ or $N_2$.** When heated in $NH_3$ or $N_2$, CrN is formed.

$$2Cr + 2NH_3 \longrightarrow 2CrN + 3H_2$$

6. **Reducing properties.** In the finely-divided state, Cr acts as a good reducing agent, because in this state Cr readily reacts with $O_2$.

7. **Action of acids.** (*a*) In the absence of air, dil. HCl and dil. $H_2SO_4$ both dissolve the metal due to the formation of blue solutions containing $CrCl_2$ and $CrSO_4$ respectively. $H_2$ is evolved in both the cases.

$$Cr + 2HCl\,(dil.) \longrightarrow \underset{\text{Blue solution}}{CrCl_2} + H_2$$

$$Cr + H_2SO_4\,(dil.) \longrightarrow \underset{\text{Blue solution}}{CrSO_4} + H_2$$

In the presence of air or $O_2$, $CrCl_2$ (blue solution) is very easily oxidised to green $CrCl_3$.

$$\underset{\text{Blue}}{4CrCl_2} + 4HCl + O_2 \longrightarrow \underset{\text{Green}}{4CrCl_3} + 2H_2O$$

(*b*) Conc. $H_2SO_4$ reacts with metal on heating and forms $Cr_2(SO_4)_3$ and evolves $SO_2$.

$$2Cr + 6H_2SO_4\,(conc.) \longrightarrow Cr_2(SO_4)_3 + 3SO_2 + 6H_2O$$

(*c*) Dil. $HNO_3$ has no action on the metal while conc. $HNO_3$ renders the metal passive due to the formation of a protective oxide film on the surface of the metal.

The reaction of Cr metal with dil. HCl and dil, $H_2SO_4$ can be co-related with the following oxidation potential values:

$$Cr \xrightarrow{+0.91V} Cr^{2+} \xrightarrow{+0.41V} Cr^{3+}$$
$$Cr \xrightarrow{+0.74V} Cr^{3+}$$

The values of $E^o_{ox}$ given above show that

(*a*) Since $E^o_{ox}$ values for $Cr/Cr^{2+}$ and $Cr^{2+}/Cr^{3+}$ electrodes are positive, being equal to 0.91 V and 0.41V respectively, Cr has a tendency to get oxidised to $Cr^{2+}$ and $Cr^{2+}$ ion has a tendency to get oxidised to $Cr^{3+}$.

(*b*) Since $E^o_{ox}$ value for $Cr/Cr^{2+}$ electrode is higher than that for $Cr/Cr^{3+}$ electrode, Cr is a better reducing agent in equilibrium with $Cr^{2+}$ solutions than with $Cr^{3+}$ solutions and Cr metal readily dissolves in dil. HCl and $H_2SO_4$, forming Cr (II) salts and liberating $H_2$. Cr(II) salts so formed are easily oxidised to Cr (III) salts as expected from $E^o_{ox}$ value for $Cr^{2+}/Cr^{3+}$ electrode.

**Uses of Chromium**

1. **Preparation of chrome steels: Alloys of chromium.** Chromium is used in the preparation of various types of chrome steels and alloys. Examples of such chrome steels are *stainless steel, chrome steel* (Cr =2.5 – 4.0%, C = 1.0 – 1.5%), *chrome-vanadium steel, chrome-nickel steel* etc. The composition, properties and uses of some important chrome steels are given

in Table 6.2. The presence of chromium in its alloys imparts greater strength, resistance towards corrosion, hardness and brightness to the chrome steels. Ferrochrome is an important alloy of chromium which is used for making special steels. For this purpose ferrochrome is added to the steel. Special steel thus obtained has special properties. Alloys to Cr with Ni containing about 20% of Cr are used in making electrical resistances for heaters, electrical furnaces etc., since these alloys resist oxidation effectively even at red heat. Cr also forms alloys which are not acted up by acids.

**Table 6.2. Alloys of chromium**

| *Name of the alloy* | *Composition* | *Properties and uses* |
|---|---|---|
| (*i*) Ferrochrome | Cr =65%, Fe = 35%. It is prepared by heating the chromite ore, FeO $Cr_2O_3$ with carbon in an electric furnace. $FeO.\,Cr_2O_3 + 4C \xrightarrow{\Delta} \underbrace{Fe + 2Cr}_{\text{Ferrochrome}} + 4CO\uparrow$ | It is used in the manufacture of chrome-steel. Chrome-steel is drill proof and is, therefore, used in the manufacture of burglar-proof safes. |
| (*ii*) Stainless steel | Cr=18%, Fe= 73%, Ni= 8%, Carbon = remaining part | It is used for cutlery and house hold wares. |
| (*iii*) Chromium-vanadium steel | Cr = 1% V = 0.15% | It is very hard and strong and is used for gears, locomotive wheels, oxide and bearings etc. |
| (*iv*) Nichrome | Cr = 15%, Ni = 60%, Fe = 25% | It is resistant to atomspheric corrosion and can withstand high temperature. It is used in resistance wires for electrical heating. |
| (*v*) Stellite | Cr = 20 – 35%, Co = 40 –80%, Ni = 0.25%, C = 0.75 – 2.5%, W= remaining part. | It is extremely hard, bright and very resistant to corrosion and is used in cutlery, surgical instruments, cutting and high speed tools etc. |
| (*vi*) Tungsten steel | Cr = 3 – 8% W = 14 – 20% | It is used for high speed tool. |

2. **Chrome plating.** Cr is widely used for plating metals (*chrome plating*), since it is resistant to corrosion. Even very thin layer of Cr on an article imparts a bright bluish-white appearance to the article to be plated and makes the article resistant to corrosion. The layer of Cr on the article is very hard.

## Electronic Configuration and Oxidation States of Cr

V.S.E.C. of Cr atom is $3d^5\ 4s^1$. This element exhibits a large number of oxidation states ranging from 0 to 6. Thus Cr shows 0, +1, +2, +3, +4, +5 and +6 oxidation states. 0 and +1 are unusual oxidation states. +2, +3 and +6 are the stable oxidation states shown by Cr in its compounds. These oxidation states arise from the transference or sharing of one or more of the 3*d* electrons as well. Compounds of Cr in +2 and +3 oxidation state are called **chromous** and **chromic** compounds respectively.

## Compounds of Cr (II)

### Preparation.

Chromium (II) compounds are called chromous compounds. These can be obtained by reducing chromic salts with Zn in presence of $CO_2$ or some other inert atmosphere to prevent oxidation of $Cr^{2+}$ ion by atmospheric $O_2$.

Fluorides, chlorides and bromides of Cr (II) can be obtained by heating Cr with HF, HCl or HBr. CrS and $CrI_2$ may be obtained by heating Cr with S and $I_2$ respectively. Cr(II) acetate which exists as a dimeric molecule $[CrCH_3COO.\ H_2O]_2$ is prepared by the interaction of chromous chloride solution with a concentrated solution of sodium acetate in an atmosphere of hydrogen. The red coloured compound is soluble in acids and is used for preparation of other chromous salts.

In its dimeric structure (Fig.6.10) four acetate groups bridge between the two Cr atoms. The existence of a strong metal – metal bond is indicated by the short distance of 236 pm, and the diamagnetic nature of the compound. Copper (II) acetate dihydrate has a similar structure but the Cu—Cu distance is larger (264 pm) and the spin moment is not completely quenched at room temperature. This compound is diamagnetic because of the coupling of the two spins of two unpaired electrons on two Cr-atoms.

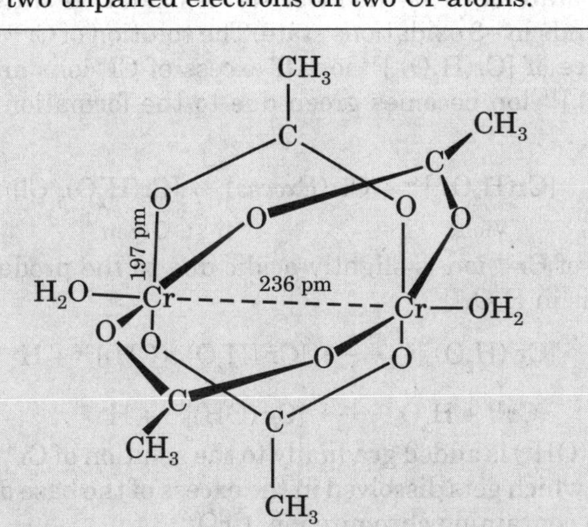


**Fig. 6.10** Structure of dimeric molecule, $[Cr(CH_3COO)_2.\ H_2O]_2$

**Properties.**

(*i*) Cr (II) compounds are not very stable. However, some of the hydrated salts like $CrCl_2 . 4H_2O$ and $CrSO_4 . 7H_2O$ are relatively more stable.

(*ii*) $Cr^{2+}$ ion has beautiful blue colour in aqueous solution.

(*iii*) Since $E^o_{ox}$ for the oxidation half-reaction $Cr^{2+} \rightarrow Cr^{3+} + e^-, E^0_{ox} = +0.4\,V$ is positive (= +0.4 V), chromous salts are easily oxidised by suitable oxidising agents. In other words, chromous salts are powerful reducing agents.

(*a*) $Cr^{2+}$ ion can be oxidised by $H^+$. Hence acidic solutions of Cr (II) salts evolve $H_2$.

$$2Cr^{2+} + 2H^+ \longrightarrow 2Cr^{3+} + H_2$$

(*b*) $O_2$ oxidises $Cr^{2+}$ ion to $Cr^{3+}$ ion in aqueous or acidic solution.

$$4Cr^{2+} + O_2 + 4H^+ \longrightarrow 4Cr^{3+} + 2H_2O, E^0 = +1.22V$$

(*c*) $Cr(OH)_2$ is oxidised to $Cr(OH)_3$ by $O_2$ of the air.

$$4Cr(OH)_2 + 2H_2O + O_2 \longrightarrow 4Cr(OH)_3$$

(*iv*) $CrCl_2$ and $CrSO_4$ are soluble in water. CrS and $CsCO_3$ are only slightly soluble while $Cr(CH_3COO)_2$ is moderately soluble. Soluble substances are ionised in aqueous solutions.

(*v*) Chromous salts resemble ferrous salts because $Cr^{2+}$ and $Fe^{2+}$ ions are similar in size. However, chromous salts are stronger reducing agents than ferrous salts.

## Compounds of Cr (III)

**+3** oxidation state of Cr is the most stable oxidation state. Cr forms a large number of compounds in +3 oxidations state. The solution of $Cr^{3+}$ ion is usually violet due to the presence of $[Cr(H_2O)_6]^{3+}$ ion. If excess of $Cl^-$ ions are added, the violet colour of $[Cr(H_2O)_6]^{3+}$ ion becomes green due to the formation of chloro complex. $[Cr(H_2O)_5Cl]^{2+}$

$$\underset{\text{Violet}}{[Cr(H_2O)_6]^{3+}} + Cl^- (\textit{Excess}) \rightarrow \underset{\text{Green}}{[Cr(H_2O)_5 Cl]^{2+}} + H_2O$$

The solution of $Cr^{3+}$ ion is slightly acidic due to the production of $H^+$ ions as result of hydrolysis in $H_2O$.

$$[Cr(H_2O)_6]^{3+} \longrightarrow [Cr(H_2O)_5(OH)]^{2+} + H^+$$

or

$$Cr^{3+} + H_2O \longrightarrow [Cr(OH)]^{2+} + H^+$$

When a base ($OH^-$) is added gradually to the solution of $Cr^{3+}$ ion, a green ppt. of $Cr(OH)_3$ is formed which gets dissolved in the excess of the base due to the formation of soluble solution containing chromite ion, $CrO_2^-$.

$$Cr^{3+} + 3OH^- \longrightarrow Cr(OH)_3 \; (\textit{Green ppt})$$

$$Cr(OH)_3 + OH^- \longrightarrow CrO_2^- + 2H_2O$$

An alkaline solution of $Cr^{3+}$ ion containing chromite ion, $CrO_2^-$ (Cr = +3) is easily oxidised to a solution containing chromate ion, $CrO_4^{2-}$(Cr = +6) by moderately oxidising agents like $H_2O_2$.

$$\underset{(Cr=+3)}{2CrO_2^-} + 2OH^- + \underset{(O=-1)}{3H_2O_2} \longrightarrow \underset{(Cr=+6)}{2CrO_4^{2-}} + \underset{(O=-2)}{4H_2O}$$

Chromic salts resemble Fe(III) and Al (III) salts in forming alums.

Here we shall discuss $Cr_2O_3$, $Cr_2O_3.\ 2H_2O$ and $CrCl_3$ which are the compounds of Cr (III).

## 1. Chromic Oxide or Chromium Sesquioxide, $Cr_2O_3$

**Preparation.** This oxide is prepared by the reactions given below:

(*a*) $2Cr(OH)_3 \xrightarrow{\Delta} Cr_2O_3 + 3H_2O$

(*b*) $(NH_4)_2Cr_2O_7 \xrightarrow{\Delta} Cr_2O_3 + 4H_2O + N_2\uparrow$

(*c*) $K_2Cr_2O_7 + 2NH_4Cl \xrightarrow{\Delta} Cr_2O_3 + 2KCl + 4H_2O + N_2\uparrow$

(*d*) $Na_2Cr_2O_7 + S \xrightarrow{\text{Heat in an iron pot}} Cr_2O_3 + Na_2SO_4$

(*e*) $4CrO_2Cl_2(vapours) \xrightarrow{\text{Pass vapour in red hot tube}} Cr_2O_3 + O_2 + 4Cl_2$

(*f*) $4Hg_2CrO_4 \xrightarrow{\Delta} 8Hg + 2Cr_2O_3 + 5O_2$

(*g*) $4CrO_3 \xrightarrow{250°-400°C} 2Cr_2O_3 + 3O_2$

(*h*) $4K_2Cr_2O_7 \xrightarrow{\text{Heat strongly}} 4K_2CrO_4 + 2Cr_2O_3 + 3O_2$

$Cr(OH)_3$ required in reaction (*a*) is obtained by adding an alkali to a chromic salt. In reactions (*c*) and (*d*) the residue of $Cr_2O_3$ is washed with water to remove KCl and $Na_2SO_4$ respectively and then dried.

**Properties.** $Cr_2O_3$ is a green powder with very high melting point and is very refractory. It dissolves in fused borax or glass and gives a green colour and hence is used in lining glass and painting porcelain. It is not decomposed by heat nor reduced by $H_2$. It is reduced to metal when heated with carbon (better with Al powder) at a high temperature. It is an amphoteric oxide because it forms salts with acids and chromites when fused with alkalies.

$$Cr_2O_3 + 3H_2SO_4 \longrightarrow Cr_2(SO_4)_3 + 3H_2O$$

$$Cr_2O_3 + 2NaOH \longrightarrow 2NaCrO_2 + H_2O$$

When it is fused with an alkali in presence of air or an oxidising agent, a chromate is formed.

$$2Cr_2O_3 + 8NaOH + 3O_2 \longrightarrow 4\,Na_2CrO_4 + 4H_2O$$

If ignited too strongly, it becomes passive towards both acid and base. When fused with an oxide of MO type, it gives a crystalline compound of the composition, $MO.Cr_2O_3$.

When $Cl_2$ is passed over a mixture of $Cr_2O_3$ and carbon and heated strongly, anhydrous $CrCl_3$ is obtained.

$$CrO_3 + 3C + 3Cl_2 \xrightarrow{\Delta} 2CrCl_3 + 3CO$$

When $S_2Cl_2$ vapour is passed over hot $Cr_2O_3$, anhydrous $CrCl_3$ is obtained.

$$2Cr_2O_3 + 6S_2Cl_2 \xrightarrow{\Delta} 4CrCl_3 + 3SO_2 + 9S$$

**Uses.** $Cr_2O_3$ is used as a green pigment for oil paints and in lining glass and painting porcelain.

**2. Hydrated Chromic Oxide, $Cr_2O_3$. $2H_2O$ : Guigret's Green**

It is a powder which has brilliant green colour and hence is used as a green pigment. It is prepared by fusing equimolecular amounts of $K_2Cr_2O_7$ and boric acid and treating the product with water. It also contains some amount of boric acid.

**3. Chromic Chloride, $CrCl_3$**

**Preparation.** Anhydrous $CrCl_3$ is obtained:

(*i*) by passing $Cl_2$ over heated Cr.

$$2Cr + 3Cl_2 \longrightarrow 2CrCl_3$$

(*ii*) by passing $Cl_2$ over a mixture of $Cr_2O_3$ and carbon heated to redness.

$$Cr_2O_3 + 3C + 3Cl_2 \xrightarrow{\Delta} 2CrCl_3 + 3CO$$

(*iii*) by passing $S_2Cl_2$ vapour over heated $Cr_2O_3$

$$2Cr_2O_3 + 6S_2Cl_2 \longrightarrow 4CrCl_3 + 3SO_2 + 9S$$

(*iv*) by heating $K_2Cr_2O_7$ with conc. HCl.

$$K_2Cr_2O_7 + 14HCl \xrightarrow{\Delta} 2KCl + 2CrCl_3 + 7H_2O + 3Cl_2$$

**Properties.** (*i*) It is a pinkish-violet solid which is almost insoluble in cold water but readily dissolves in water in presence of traces of $CrCl_2$ or a reducing agent like CuCl or $SnCl_2$ and gives a green solution.

(*ii*) It volatalises at 1065°C and dissociates when heated above 1300°C

$$2CrCl_3 \rightleftharpoons 2CrCl_2 + Cl_2$$

(*iii*) *Hydrate isomerism.* Hexahydrate, $CrCl_3$. $6H_2O$ exists in three crystalline forms. Each of them is soluble in water. They have different colours. All of them react with $AgNO_3$ solution to give the white ppt. of AgCl. Coordination number of $Cr^{3+}$ ion in each form is six and hence each has octahedral geometry. These three forms are discussed below:

(*a*) *Violet form. $[Cr(H_2O)_6]Cl_3$ (β-variety).* This form does not lose water when treated with conc. $H_2SO_4$ and all the three $Cl^-$ ions can be precipitated as AgCl by means of $AgNO_3$ solution. The above properties show that in this form all the six $H_2O$ molecules are present in the coordination sphere and all the three Cl-atoms are present outside the coordination sphere. Thus this form can be represented by the strucutre, $[Cr(H_2O)_6]Cl_3$. All the three Cl-atoms are ionisable.

$$[Cr(H_2O)_6]Cl_3 \rightleftharpoons [Cr(H_2O)_6]^{3+} + 3Cl^-$$

$$3Cl^- + 3Ag^+ \longrightarrow 3AgCl\downarrow$$

All the six $H_2O$ molecules act as ligands and this form has no water of crstallisation.

(*b*) *Blue-green form, [Cr(H$_2$O)$_5$Cl]Cl$_2$.H$_2$O (γ-variety)*. This form loses only one $H_2O$ molecule when treated with conc. $H_2SO_4$ and two $Cl^-$ ions can be precipitated as AgCl by means of $AgNO_3$ solution. These properties show that in this form five $H_2O$ molecules and Cl-atom are ionisable. Thus this form can be represented as $[Cr(H_2O)_5Cl]\ Cl_2.H_2O$

$$[Cr(H_2O)_5Cl]\ Cl_2{\cdot}H_2O \rightleftharpoons [Cr(H_2O_5)Cl]^{2+} + 2Cl^- + H_2O$$

$$2Ag^+ + 2Cl^- \longrightarrow 2AgCl\downarrow$$

Five $H_2O$ molecules and one $Cl^-$ ion act as ligands and one $H_2O$ molecule behaves as water of crystallisation.

(*c*) *Green form, [Cr(H$_2$O)$_4$Cl$_2$]Cl.2H$_2$O (α-variety)*. This form loses two $H_2O$ molecules on treatment with conc.$H_2SO_4$ and only one $Cl^-$ ion can precipitated as AgCl by $AgNO_3$ solution. These properties show that in this form four $H_2O$ molecules and two Cl-atoms are present in the bracket and only one Cl-atom is present outside the bracket. This one Cl-atom is ionisable. Thus this form has its structure as $[Cr(H_2O)_4Cl_2].2H_2O$.

$$[Cr(H_2O)_4Cl_2]Cl{\cdot}2H_2O \rightleftharpoons [Cr(H_2O)_4Cl_2]^+ + Cl^- + 2H_2O$$

$$Cl^- + Ag^+ \longrightarrow AgCl\downarrow$$

Four $H_2O$ molecules and two $Cl^-$ ions act as ligands and two $H_2O$ molecules behave as water of crystallisation.

The above discussion shows that since in $CrCl_3.6H_2O$ water molecules can be placed either inside the coordination sphere as ligands or outside the sphere as water of hydration, $CrCl_3.6H_2O$ shows *hydrate isomerism*. All the three forms are called *hydrate isomers*.

In solution, these isomers exist in equilibrium as shown below:

$$[Cr(H_2O)_4Cl_2]^+ \underset{-Cl^-}{\overset{+H_2O}{\rightleftharpoons}} [Cr(H_2O)_5Cl]^{2+} \underset{-Cl^-}{\overset{+H_2O}{\rightleftharpoons}} [Cr(H_2O)_6]^{3+}$$

$[Cr(H_2O)_4Cl_2]^+$ ion present in $[Cr(H_2O)_4\ Cl_2]\ Cl.H_2O$ (green – α variety) exists in two geometrical isomers *viz., cis* and *trans* isomers whose structures are given in Fig. 6.11.

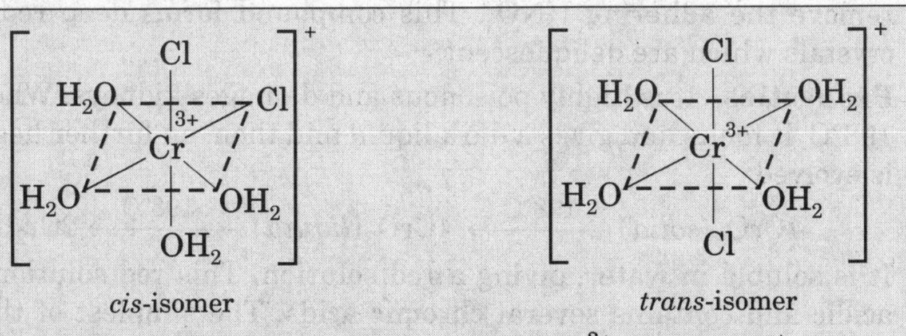


*cis*-isomer *trans*-isomer

**Fig. 6.11** Cis and trans isomers of $[Cr^{3+}(H_2O)_4Cl_2]^+$ ion

In *cis*-isomer the two $Cl^-$ ions occupy the adjacent (*i.e.* cis) positions of the octahedron while in *trans*-isomer these ions (*i.e.* $Cl^-$ ions) have opposite (trans) positions.

## Compounds of Cr (IV)

Chromium dioxide ($CrO_2$) and $CrF_4$ are the compounds of Cr in +4 oxidation state.

$CrO_2$ is obtained by reducing $CrO_3$ by $H_2$ and $CrF_4$ can be prepared by the combination of Cr and $F_2$ at elevated temperature.

$$CrO_3 + H_2 \longrightarrow CrO_2 + H_2O$$

$$Cr + 2F_2 \xrightarrow{\Delta} CrF_4$$

## Compounds of Cr (V)

Compounds of Cr in +5 oxidation state are rare. These compounds dissociate to give Cr (III) or Cr (IV) compounds which are more stable than Cr (V) compounds. $CrF_5$ and $[Cr(O_2)_4]^{3-}$ are the examples of Cr(V) compounds. $[Cr(O_2)_4]^{3-}$ contains peroxide ion, $O_2^{2-}$ $\left(H_2O_2 \longrightarrow 2H^+ + O_2^{2-}\right)$.

## Compounds of Cr (VI)

Important compounds formed by Cr in +6 oxidation state are *chromic trioxide* ($CrO_3$), *chromium oxychloride* ($CrO_2Cl_2$) and *chromic acids* and their salts like potassium chromate ($K_2CrO_4$) and potassium dichromate ($K_2Cr_2O_7$)

1. **Chromium Trioxide or Chromic Acid Anhydride, $CrO_3$**

This oxide is the anhydride of unknown chromic acid, $H_2CrO_4$ and hence is also called *chromic acid anhydride*.

**Preparation.** It is obtained as deep red needle-like crystals when a well-cooled solution of $K_2Cr_2O_7$ is treated with conc. $H_2SO_4$.

$$K_2Cr_2O_7 + H_2SO_4 \longrightarrow 2\,CrO_3 + K_2SO_4 + H_2O$$

or

$$Cr_2O_7^{2-} + H_2SO_4 \longrightarrow 2CrO_3 + SO_4^{2-} + H_2O$$

The crystals of $CrO_3$ are filtered by means of a Buchner funnel containing asbestos, washed with pure $HNO_3$ and dried in a current of warm air to remove the adhering $HNO_3$. This compound forms deep red needle-like crystals which are deliquescent.

**Properties.** It is highly poisonous and damages kidneys. When heated to 198°C, it melts and gives a dark liquid and then, on further heating, $O_2$ gas is evolved.

$$4CrO_3(solid) \xrightarrow{198°C} 4CrO_3(liquid) \xrightarrow{>198°C} 2Cr_2O_3 + 3O_2$$

It is soluble in water, giving a red solution. This red solution is strongly acidic and contains several chromic acids. The simplest of these acids is *chromic acid,* $H_2CrO_4$.

$$CrO_3 + H_2O \rightleftharpoons H_2CrO_4$$

It is a powerful oxidising agent and its oxidising action is based on the following equation:

$$2CrO_3 \longrightarrow Cr_2O_3 + O$$

When $CrO_3$ oxidises any substance, the oxide itself is reduced to Cr(+3) salt. Thus it oxidises $SO_2$, acidified $H_2S$, $FeSO_4$, $SnCl_2$ as shown by the following equations:

$$2CrO_3 + 3SO_2 \longrightarrow Cr_2(SO_4)_3$$

$$2CrO_3 + 3H_2S + 3H_2SO_4 \longrightarrow Cr_2(SO_4)_3 + 6H_2O + S$$

$$2CrO_3 + 6FeSO_4 + 6H_2SO_4 \longrightarrow Cr_2(SO_4)_3 + 3Fe_2(SO_4)_3 + 6H_2O$$

$$2CrO_3 + 3SnCl_2 + 12HCl \longrightarrow 2CrCl_3 + 3SnCl_4 + 6H_2O$$

When $CrO_3$ is dissolved in calculated amount of $NH_3$, ammonium dichromate, $(NH_4)_2Cr_2O_7$ is formed.

$$2CrO_3 + 2NH_3 + H_2O \longrightarrow (NH_4)_2Cr_2O_7$$

$CrO_3$ reacts with gaseous HCl to give chromyl chloride, $CrO_2Cl_2$.

$$CrO_3 + 2HCl\,(gas) \longrightarrow CrO_2Cl_2 + H_2O$$

It dissolves in alkalies to give chromates $(CrO_4^{2-})$.

$$CrO_3 + 2NaOH \longrightarrow Na_2CrO_4 + H_2O$$

**Uses.** The mixture of $K_2Cr_2O_7$ and conc. $H_2SO_4$ is called **chromic acid mixture**. This mixture is used as an oxidising agent in volumetric analysis and for cleaning glass apparatus. $CrO_3$ dissolved in glacial acetic acid is used as an oxidising agent and as a disinfectant to wash foul ulcers and sores.

2. **Chromium Oxychloride or Chromyl Chloride, $CrO_2Cl_2$**

**Preparation:** $CrO_2Cl_2$ is prepared by the following methods:

(*i*) It is obtained as a dark red liquid when a dry mixture of $K_2Cr_2O_7$ and NaCl is distilled with conc. $H_2SO_4$ and the resultant vapours of $CrO_2Cl_2$ are condensed.

$$K_2Cr_2O_7 + 4NaCl + 3H_2SO_4(conc.) \xrightarrow{\Delta} 2CrO_2Cl_2 + K_2SO_4 + 2Na_2SO_4 + 3H_2O$$

(*ii*) It is also prepared by adding small quantity of conc. $H_2SO_4$ to a cold solution of $CrO_3$ in conc. HCl.

$$CrO_3 + 2HCl(conc.) \longrightarrow CrO_2Cl_2 + H_2O$$

The solution mixture is separated into two layers. $CrO_2Cl_2$ is present in the lower layer from which $CrO_2Cl_2$ can be separated by a separating funnel.

(*iii*) $CrO_2Cl_2$ can also be prepared by treating chromic acid, $H_2CrO_4$ or $CrO_2(OH)_2$ with HCl gas.

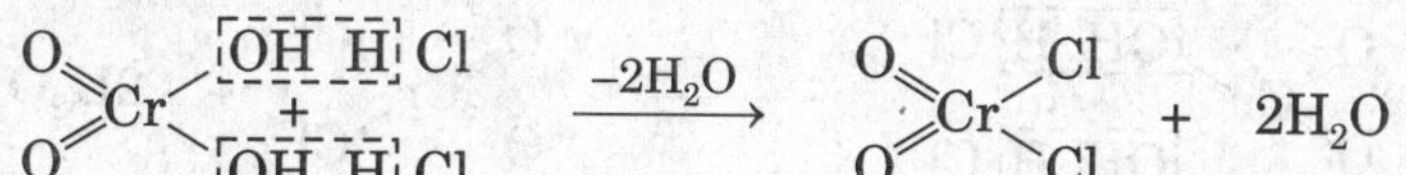

or $$CrO_2(OH)_2 + 2HCl \longrightarrow CrO_2Cl_2 + 2H_2O$$

**Physical properties** (*i*) It is a deep red liquid having b.pt. = 116.7°C and m.pt = –96.5°C.

(*ii*) It is miscible with organic solvents like $CCl_4$ and $CS_2$

**Chemical properties** (*i*) *Hydrolysis.* It is hydrolysed by water, forming chromic acid ($H_2CrO_4$) and HCl.

$$CrO_2Cl_2 + 2H_2O \longrightarrow H_2CrO_4 + 2HCl$$

(*ii*) *Stability.* It is photosensitive but otherwise it is stable. When heated, $Cr_2O_3$ is obtained.

$$4CrO_2Cl_2 \xrightarrow{\Delta} 2Cr_2O_3 + 4Cl_2 + O_2$$

(*iii*) *Action of $(CH_3COO)_2$ Pb.* The solution of $CrO_2Cl_2$ gives a yellow precipitate of $PbCrO_4$ with $(CH_3COO)_2Pb$.

$$CrO_2Cl_2 + 2H_2O + (CH_3COO)_2\,Pb \longrightarrow \underset{\text{Yellow ppt}}{PbCrO_4} + 2HCl + 2CH_3COOH$$

(*iv*) *Oxidising properties.* It is a mild oxidising agent. A solution of $CrO_2Cl_2$ in $CS_2$ at 25–40°C oxidises toluene ($C_6H_5CH_3$) to benzaldehyde ($C_6H_5CHO$) (*Etard's reaction*).

$$CrO_2Cl_2 \longrightarrow CrCl_2 + 2O$$

$$C_6H_5CH_3 + 2O \longrightarrow C_6H_5CHO + H_2O$$

(*v*) *Action of NaCl or KCl solution.* When the vapours of $CrO_2Cl_2$ are passed into the saturated solution of NaCl or KCl, red crystals of the *metal chloro chromate* ($MCrO_3Cl$) are obtained. $MCrO_3Cl$ is called **Petigot's salt**.

$$CrO_2Cl_2 + MCl + H_2O \longrightarrow \underset{\text{Petigot's salt}}{MCrO_3Cl} + 2HCl$$
$$(M = Na, K)$$

(*vi*) *Action of NaOH.* Yellow solution of $Na_2CrO_4$ is obtained

$$CrO_2Cl_2 + 4NaOH \longrightarrow \underset{\text{Yellow solution}}{Na_2CrO_4} + 2NaCl + 2H_2O$$

(*vii*) *Action of $F_2$ or HF.* In both cases chromyl fluoride ($CrO_2F_2$) is formed.

$$CrO_2Cl_2 + F_2 \longrightarrow CrO_2F_2 + Cl_2$$

$$CrO_2Cl_2 + 2HF \longrightarrow CrO_2F_2 + 2HCl$$

**Uses.** It is used: (*i*) as an intermediate product in the identification of $Cl^-$ ions, (*ii*) as an excellent oxidising agent in organic chemistry, *e.g.*, its solution in $CS_2$ is used to oxidise toluene to benzaldehyde.

**Structure.** $CrO_2Cl_2$ molecule can be regarded as having been obtained by removing two $H_2O$ molecules from the combination of one molecule of chromic acid, $H_2CrO_4$ or $CrO_2(OH)_2$ and two molecules of HCl. Thus:

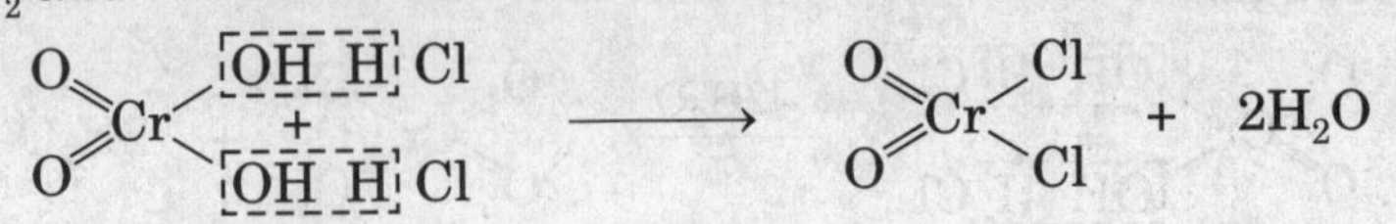

It may be concluded from the above discussion that $CrO_2Cl_2$ is a dichloro derivative of $H_2CrO_4$ or $CrO_2(OH)_2$.

Chromyl chloride, $CrO_2Cl_2$ has a distorted tetrahedral structure (Fig. 6.12).

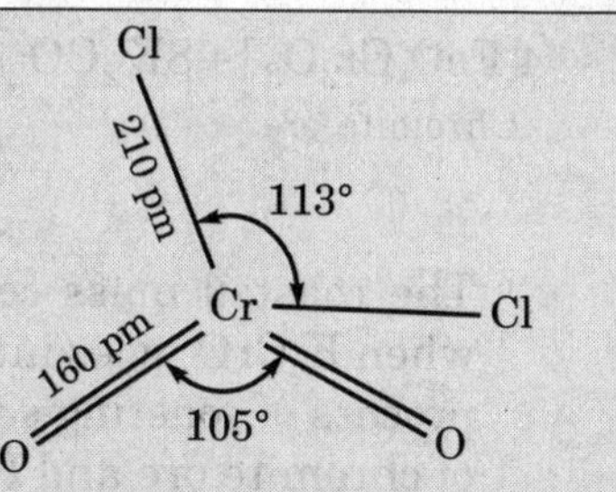


**Fig. 6.12** Distorted tetrahedral structure of $CrO_2Cl_2$

### 3. Chromic Acids

$CrO_3$ dissolves readily in water to give a red solution which is strongly acidic. The acidic nature of the aqueous solution is due to the fact that the solution has several chromic acids.

$$\underset{}{H_2O} \xrightarrow{+CrO_3} \underset{\text{Chromic acid}}{H_2CrO_4} \xrightarrow{+CrO_3} \underset{\text{Dichromic acid}}{H_2Cr_2O_7} \xrightarrow{CrO_3} \underset{\text{Trichromic acid}}{H_2Cr_3O_{10}}$$

$$\xrightarrow{+CrO_3} \underset{\text{Tetrachromic acid}}{H_2Cr_4O_{13}} \xrightarrow{+CrO_3} .......$$

Higher chromic acids can be supposed to have been formed by the condensation of 2, 3 or more $H_2CrO_4$ molecules with the elimination of water molecules. Thus:

$$2H_2CrO_4 \xrightarrow{-H_2O} H_2Cr_2O_7 + H_2O$$

$$3H_2CrO_4 \xrightarrow{-2H_2O} H_2Cr_3O_{10} + 2H_2O$$

$$3H_2CrO_4 \xrightarrow{-3H_2O} H_2Cr_4O_{13} + 3H_2O$$

### 4. Chromates $\left(CrO_4^{2-}\right)$

Chromic acid ($H_2CrO_4$) forms only normal salts of $M_2^+CrO_4$ and $M^{2+}CrO_4$ type, and no acid salts of $MHCrO_4$ type. These salts are called *chromates.* Chromates are soluble in water and give brilliant yellow solution. When acidified, the yellow colour of the solution changes to orange due to the conversion of $CrO_4^{2-}$ ion into $Cr_2O_7^{2-}$ ion.

$$\underset{\text{Yellow}}{2CrO_4^{2-} + 2H^+} \longrightarrow \underset{\text{Orange}}{Cr_2O_7^{2-}} + H_2O$$

Examples of important chromates are $K_2CrO_4$, $Na_2CrO_4$, $PbCrO_4$, basic lead chromate [$Pb(OH)_2$. $PbCrO_4$], basic zinc chromate [$Zn(OH)_2$. $ZnCrO_4$. $H_2O$] etc.

The detailed discussion of $K_2CrO_4$ is given below.

### 5. Potassium Chromate $K_2CrO_4$

**Commercial preparation.** (*i*) On commercial scale it is prepared from chromite ore, $FeO.Cr_2O_3$. The ore is first concentrated by Wilfley table method (gravity separation method). The concentrated ore is mixed with excess of $K_2CO_3$ and roasted in a reverberatory furnace at a high temperature in the presence of free supply of air. By this operation the chromite ore, $FeO.Cr_2O_3$ is converted into soluble potassium chromate, $K_2CrO_4$.

$$\underset{\text{Chromite ore}}{4[FeO.Cr_2O_3]} + 8K_2CO_2 + 7O_2(\textit{From air}) \xrightarrow{\Delta} \underbrace{\underset{\text{Soluble}}{8K_2CrO_4} + \underset{\text{Insoluble}}{2Fe_2O_3}}_{\text{Roasted mass}} + 8CO_2 \uparrow$$

The roasted mass containing $K_2CrO_4$ and $Fe_2O_3$ is extracted with water when $K_2CrO_4$ goes into solution and $Fe_2O_3$ is left behind as a residue. In the process of roasting some quantity of CaO is also added so that the mixture of chromite ore and $K_2CO_3$ may become porous.

(*ii*) $K_2CrO_4$ can also be prepared from $K_2Cr_2O_7$ . To the water solution of $K_2Cr_2O_7$ calculated amount of KOH or $K_2CO_3$ is added. On evaporating the resulting solution, lemon-yellow crystals of anhydrous potassium chromate are obtained.

$$K_2Cr_2O_7 + 2KOH \longrightarrow 2K_2CrO_4 + H_2O$$

$$K_2Cr_2O_7 + K_2CO_3 \longrightarrow 2K_2CrO_4 + CO_2 \uparrow$$

**Properties.** (*i*) $K_2CrO_4$ forms lemon yellow crystals having m.pt.= 968°C. $K_2CrO_4$ is isomorphous with $K_2SO_4$.

(*ii*) It is highly soluble in water and the resulting solution is yellow in colour.

(*iii*) *Action of $H_2SO_4$.* When $H_2SO_4$ is added to the yellow solution of $K_2CrO_4$, yellow colour changes to orange-red due to the formation of $K_2Cr_2O_7$.

$$2K_2CrO_4 + H_2SO_4 \longrightarrow K_2Cr_2O_7 + K_2SO_4 + H_2O$$

(*iv*) *Action of heat.* When heated above 670°C, it changes to a red modification which melts at 970°C.

(*v*) *Oxidising properties.* Acidified solution of $K_2CrO_4$ acts as an oxidising agent.

$$2K_2CrO_4 + 5H_2SO_4 \longrightarrow 2K_2SO_4 + Cr_2(SO_4)_3 + 5H_2O + 3O$$

Thus acidified solution of $K_2CrO_4$ liberates $I_2$ from KI solution and oxidises $H_2SO_4$ to S, $SO_2$ to $H_2SO_4$, HCl to $Cl_2$ etc.

$$2K_2CrO_4 + 5H_2SO_4 + 3H_2O \longrightarrow 2K_2SO_4 + Cr_2(SO_4)_3 + 8H_2O + 3S$$

(*vi*) *Precipitation reactions.* $K_2CrO_4$ precipitates chromates from different salts, *e.g.*, $Ag_2CrO_4$ (*brick-red*) from Ag salt solution, *yellow* $BaCrO_4$ from Ba salt and *yellow* $PbCrO_4$ from Pb salt solutions etc.

$$CrO_4^{2-} + 2Ag^+ \longrightarrow \underset{\text{Brick-red ppt}}{Ag_2CrO_4 \downarrow}$$

$$\underset{(M=Ba,Pb)}{CrO_4^{2-} + M^{2+}} \longrightarrow \underset{\text{Yellow ppt}}{MCrO_4 \downarrow}$$

(*vii*) *Action on neutral solution of zinc salt.* When $K_2CrO_4$ is added to the neutral solution of zinc salt, basic zinc chromate, $Zn(OH)_2.\ ZnCrO_4.\ H_2O$ is obtained.

$$K_2CrO_4 + 2Zn^{2+} + H_2O \longrightarrow Zn(OH)_2.ZnCrO_4.H_2O + 2K^+$$

(*viii*) *Action of $H_2O_2$.* When $H_2O_2$ is added to a solution of $K_2CrO_4$ containing dil. $H_2SO_4$ and a small amount of ether, an intense blue colour is produced which persists in etharal solution. This blue colour is due to the formation of chromium diperoxide, $Cr(O_2)_2O$ or $CrO_5$.

$$K_2CrO_4 + H_2SO_4 + 2H_2O_2 \xrightarrow{\text{Ether}} K_2SO_4 + CrO_5 + 3H_2O$$

or $$CrO_4^{2-} + 2H^+ + 2H_2O_2 \xrightarrow{\text{Ether}} CrO_5 + 3H_2O$$

**Uses.** $K_2CrO_4$ is used (*i*) as a reagent in the laboratory, (*ii*) *as* an indicator in Mohr's volumetric method of silver estimation, (*iii*) as an inhibitor for corrosion, (*iv*) for the preparation of basic zinc chromate, $Zn(OH)_2 . ZnCrO_4 . H_2O$ which is used as a yellow pigment.

**6. Dichromates ($Cr_2O_7^{2-}$)**

The salts of dichromic acid, $H_2Cr_2O_7$ are called dichromates. In presence of dil. $H_2SO_4$, $Cr_2O_7^{2-}$ ion acts as a strong oxidising agent as it gets reduced to $Cr^{3+}$ ion.

$$Cr_2O_7^{2-} + 14H^+ + 6e^- \longrightarrow 2Cr^{3+} + 7H_2O$$

$Cr_2O_7^{2-}$ ion reacts with conc. $H_2SO_4$ to give a solution of red $CrO_3$.

$$Cr_2O_7^{2-} + H_2SO_4(conc.) \longrightarrow 2CrO_3 + H_2O + SO_4^{2-}.$$

$(NH_4)_2Cr_2O_7$, $Na_2Cr_2O_7$ and $K_2Cr_2O_7$ are the examples of important dichromates. Here we shall discuss the preparation, properties and uses of $K_2Cr_2O_7$.

**7. Potassium Dichromate, $K_2Cr_2O_7$**

**Laboratory preparation.** $K_2Cr_2O_7$ is prepared in the laboratory by adding the calculated quantity of $H_2SO_4$ to a saturated solution of $K_2CrO_4$.

$$2K_2CrO_4 + H_2SO_4 \longrightarrow K_2Cr_2O_7 + K_2SO_4 + H_2O$$

**Manufacture:** $K_2Cr_2O_7$ is manufactured from chromite ore, FeO., $Cr_2O_3$. The method consists of the following steps:

(*a*) *Concentration of the chromite ore.* See "*Extraction of Cr*".

(*b*) *Conversion of the concentrated chromite ore (FeO. $Cr_2O_3$) into $Na_2CrO_4$: Roastieng of the concentrated ore.*

$$\underset{\text{Chromite ore}}{4(FeO.Cr_2O_3)} + 8Na_2CO_3 + 7O_2(\textit{From air}) \xrightarrow{900-1000°C} \underbrace{\underset{\text{Soluble}}{8Na_2CrO_4} + \underset{\text{Insoluble}}{2Fe_2O_3}}_{\text{Roasted mass}} + 8CO_2 \uparrow$$

For the details of this step see "*Extraction of Cr.*'

(*c*) *Conversion* of *$Na_2CrO_4$ into $Na_2Cr_2O_7$*.

$$2Na_2CrO_4 + H_2SO_4 \longrightarrow \underset{\text{more soluble}}{Na_2Cr_2O_7} + \underset{\text{less soluble}}{Na_2SO_4} + H_2O$$

For the details of this step see "*Extraction of Cr.*"

(*d*) *Conversion of $Na_2Cr_2O_7$ into $K_2Cr_2O_7$.* Hot solution of $Na_2Cr_2O_7$ obtained as above in step (*c*) is treated with the calculated quantity of KCl whereby less soluble $K_2Cr_2O_7$ separates out from the solution on crystallisation.

$$Na_2Cr_2O_7 + 2KCl \longrightarrow \underset{\text{More soluble}}{2NaCl} + \underset{\text{Less soluble}}{K_2Cr_2O_7}$$

**Physical properties.** (*i*) $K_2Cr_2O_7$ forms organge -red prismatic crystals. Its sp. gra. = 2.676 and m.pt. =396°C.

(*ii*) *Solubility*. It is moderately soluble in cold water but freely saluble in hot water. It is insoluble in alcohol.

**Chemical properties.** (*i*) *Action of heat.* When heated to a white heat, it decomposes to give potassium chromate, $K_2CrO_4$ and $O_2$ gas is evolved.

$$4K_2Cr_2O_7 \xrightarrow{\Delta} 4K_2CrO_4 + 2Cr_2O_3 + 3O_2\uparrow$$

(*ii*) *Action of alkalies and acids.* When an alkali is added to an orange-red solution of $K_2Cr_2O_7$, a yellow solution results. The yellow colour of the solution is due to the formation of $K_2CrO_4$.

$$\underset{\text{Orange-red}}{K_2Cr_2O_7} + 2KOH \longrightarrow \underset{\text{Yellow}}{2K_2CrO_4} + H_2O$$

$$\underset{\text{Orange-red}}{K_2Cr_2O_7} + 2NaOH \longrightarrow \underset{\text{Yellow}}{2K_2CrO_4} + H_2O + Na_2CrO_4$$

or $$Cr_2O_7^{2-} + 2OH^- \longrightarrow 2CrO_4^{2-} + H_2O$$

or $$2Cr_2O_7^{2-} + 4OH^- \longrightarrow 4CrO_4^{2-} + 2H_2O \quad \text{... }(a)$$

On acidifying, the yellow colour of $K_2CrO_4$ again changes to range-red due to the reformation of $K_2Cr_2O_7$.

$$\underset{\text{Yellow}}{2K_2CrO_4} + H_2SO_4 \longrightarrow \underset{\text{Orange-red}}{K_2Cr_2O_7} + K_2SO_4 + H_2O$$

or $$2CrO_4^{2-} + 2H^+ \longrightarrow Cr_2O_7^{2-} + H_2O$$

or $$4CrO_4^{2-} + 4H^+ \longrightarrow 2Cr_2O_7^{2-} + 2H_2O \quad \text{... }(b)$$

Equation (*a*) and (*b*) show that $Cr_2O_7^{2-}$ and $CrO_4^{2-}$ ions exist in equlibrium which are inter-convertible by changing the pH of the solution. Thus the reaction shown by (*a*) and (*b*) can be combined and can be written as

$$\underset{\text{Orange–red solution}}{2Cr_2O_7^{2-}} \underset{+4H^+(-2H_2O)}{\overset{+4OH^-(-2H_2O)}{\rightleftharpoons}} \underset{\text{Yellow solution}}{4CrO_4^{2-}}$$

Thus we see that the acidic solution contains $Cr_2O_7^{2-}$ ions while alkaline solution has $CrO_4^{2-}$ ions.

On adding an alkali, the conctration of $H^+$ ions decreases and hence the above reaction proceeds in the forward direction producing yellow solution containing $CrO_4^{2-}$ ion. On the other hand when an acid is added, the concentration of $H^+$ ions increases and hence the above reaction proceeds in the backward direction producing the orange-red solution containing $Cr_2O_7^{2-}$ ions.

(*iii*) *Precipitation reactions.* The above discussion shows that the orange-red solution of $K_2Cr_2O_7$ actually contains both $Cr_2O_7^{2-}$ and $CrO_4^{2-}$ ions. Thus when the orange-red solution of $K_2Cr_2O_7$ is treated with a metallic salt solution [*e.g.* $Pb(NO_3)_2$ solution], theoretically $PbCrO_4$ or $PbCr_2O_7$ may be precipitated.

$$2Pb(NO_3)_2 + K_2Cr_2O_7 + H_2O \longrightarrow \underset{\text{Less soluble}}{2PbCrO_4} + 2KNO_3 + 2HNO_3$$

$$Pb(NO_3)_2 + K_2Cr_2O_7 \longrightarrow \underset{\text{More soluble}}{PbCr_2O_7} + 2KNO_3$$

Since $PbCrO_4$ has less solublity product than $PbCr_2O_7$, it is only $PbCrO_4$ (chrome yellow) which actually gets precipitated when $K_2Cr_2O_7$ solution is added to the solution of $Pb(NO_3)_2$, *i.e.* $PbCr_2O_7$ is not precipitated.

(*iv*) *Action of conc. $H_2SO_4$ and HCl.* (*a*) When a well-cooled solution of $K_2Cr_2O_7$ is treated with conc. $H_2SO_4$, red crystals of $CrO_3$ are obtained.

$$K_2Cr_2O_7 + H_2SO_4 \longrightarrow 2CrO_3 + K_2SO_4 + H_2O$$

or $$Cr_2O_7^{2-} + H_2SO_4 \longrightarrow 2CrO_3 + SO_4^{2-} + H_2O$$

(*b*) On heating the mixture of $K_2Cr_2O_7$ and conc. $H_2SO_4$, $O_2$ is evolved.

$$2K_2Cr_2O_7 + 8H_2SO_4 \xrightarrow{\Delta} 2K_2SO_4 + 2Cr_2(SO_4)_3 + 8H_2O + 3O_2$$

(*c*) $K_2Cr_2O_7$ reacts with conc. HCl to evolve $Cl_2$.

$$K_2Cr_2O_7 + 14HCl \xrightarrow{\Delta} 2KCl + 2CrCl_3 + 7H_2O + 3Cl_2$$

(*d*) When a solution containing 3 parts of powdered $K_2Cr_2O_7$, 4 parts of warm conc.HCl and a little water is cooled, red crystals of pot. chlorochromate, $KCrO_3Cl$ are formed. This compound is called **Petigot's salt**.

$$K_2Cr_2O_7 + 2HCl \longrightarrow KCrO_3Cl + H_2O$$

(*v*) *Chromyl chloride test.* When solid $K_2Cr_2O_7$ is heated with conc. $H_2SO_4$ and a soluble chloride like NaCl, deep red vapours of chromyl chloride, $CrO_2Cl_2$ are produced.

$$K_2Cr_2O_7 + 4NaCl + 3H_2SO_4(\textit{concentrated}) \xrightarrow{\Delta} \underset{\text{Red vapours}}{2CrO_2Cl_2} + K_2SO_4 + 2Na_2SO_4 + 3H_2O$$

The above reaction is used for the identification of $Cl^-$ ion. It is called chromyl chloride test for $Cl^-$ ion. This reaction has also been used for the preparation of $CrO_2Cl_2$.

(*vi*) *Oxidising properties.* $K_2Cr_2O_7$ is a powerful oxidising agent. In presence of dil. $H_2SO_4$, one molecule of this compound gives three atoms of oxygen as shown below:

$$K_2Cr_2O_7 + 4H_2SO_4\ (\textit{dil}) \longrightarrow K_2SO_4 + Cr_2(SO_4)_3 + 4H_2O + 3O$$

It terms of electronic concept, $Cr_2O_7^{2-}$ ion takes electrons in the acidic medium and hence it reduce to $Cr^{3+}$ ion. Thus $Cr_2O_7^{2-}$ acts as an oxidising agent in acidic medium.

$$Cr_2O_7^{2-} + 14H^+ + 6e^- \longrightarrow 2Cr^{3+} + 7H_2O$$

**Note.** Both $Na_2Cr_2O_7$ and $K_2Cr_2O_7$ are oxidising agents, but $K_2Cr_2O_7$ is preferred because it is not hygroscopic and can be used as a primary standard.

Thus in presence of dil. $H_2SO_4$, $K_2Cr_2O_7$ liberates $I_2$ from KI solution, oxides Fe(+2) salts to Fe(+3) salts, oxidises $H_2S$ to S, oxidises $SO_2$ to $H_2SO_4$, sulphites to sulphates, halogen acids to halogens and arsenites to arsenates.

(*a*) $K_2Cr_2O_7 + 7H_2SO_4 + 6KI \longrightarrow 4K_2SO_4 + Cr_2(SO_4)_3 + 7H_2O + 3I_2$

$$Cr_2O_7^{2-} + 14H^+ + 6I^- \longrightarrow 2Cr^{3+} + 7H_2O + 3I_2$$

(*b*) $K_2Cr_2O_7 + 7H_2SO_4 + 6FeSO_4 \longrightarrow K_2SO_4 + Cr_2(SO_4)_3 + 7H_2O + 3Fe_2(SO_4)_3$

or $Cr_2O_7^{2-} + 14H^+ + 6Fe^{2+} \longrightarrow 2Cr^{3+} + 7H_2O + 6Fe^{3+}$

(*c*) $K_2Cr_2O_7 + 4H_2SO_4 + 3H_2S \longrightarrow K_2SO_4 + Cr_2(SO_4)_3 + 7H_2O + 3S$

or $Cr_2O_7^{2-} + 8H^+ + 3H_2S \longrightarrow 2Cr^{3+} + 7H_2O + S$

(*d*) $K_2Cr_2O_7 + 4H_2SO_4 \longrightarrow K_2SO_4 + Cr_2(SO_4)_3 + 4H_2O + 3O$

$$[SO_2 + O + H_2O \longrightarrow H_2SO_4] \times 3$$

---

$$3SO_2 + K_2Cr_2O_7 + H_2SO_4 \longrightarrow K_2SO_4 + Cr_2(SO_4)_3 + H_2O$$

or $3SO_2 + Cr_2O_7^{2-} + 2H^+ \longrightarrow 2Cr^{3+} + 3SO_4^{2-} + H_2O$

The above reaction has been used to prepare chrome alum, $K_2SO_4.\ Cr_2(SO_4)_3.\ 24H_2O$.

(*e*) $3Na_2SO_3 + K_2Cr_2O_7 + 4H_2SO_4 \longrightarrow 3Na_2SO_4 + K_2SO_4 + Cr_2(SO_4)_3 + 4H_2O$

or $3SO_3^{2-} + Cr_2O_7^{2-} + 8H^+ \longrightarrow 3SO_4^{2} + 2Cr^{3+} + 4H_2O$

(*f*) $K_2Cr_2O_7 + 14HCl \longrightarrow 2KCl + 2CrCl_3 + 7H_2O + 3Cl_2$

or $Cr_2O_7^{2-} + 14H^+ + 6Cl^- \longrightarrow 2Cr^{3+} + 7H_2O + 3Cl_2$

(*g*) Acidified solution of $K_2Cr_2O_7$ also oxidises primary alcohols ($CH_3OH$) to aldehydes and then to acids. ($CH_3COOH$)

$$K_2Cr_2O_7 + 4H_2SO_4 \longrightarrow K_2SO_4 + Cr_2(SO_4)_3 + 4H_2O + 3O$$

$$C_2H_5OH + O \longrightarrow CH_3CHO + H_2O$$

$$\xrightarrow{+O} CH_3COOH$$

(*h*) Acidified solution of $K_2Cr_2O_7$ oxidises chloroform ($CHCl_3$) to carbonyl chloride ($COCl_2$) and aniline to *p*-benzoquinone.

$$2CHCl_3 + 3O \longrightarrow 2COCl_2 + Cl_2 + H_2O$$

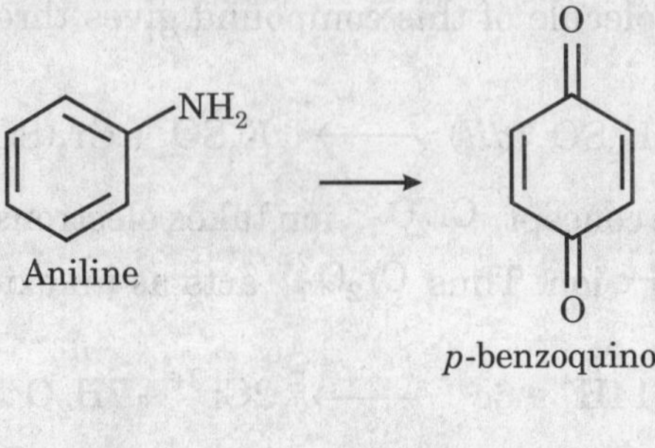

Aniline

*p*-benzoquinone

(*vii*) Action of $H_2O_2$ (*a*) When a solution of $K_2Cr_2O_7$ containing dil. $H_2SO_4$ and some ether, $(C_2H_5)_2O$ is treated with $H_2O_2$ and the solution mixture is shaken, the layer of ether becomes blue due to the formation of chromic diperoxide, $CrO_5$ or $CrO(O_2)_2$.

$$K_2Cr_2O_7 + H_2SO_4 + 4H_2O_2 \xrightarrow{\text{Ether}} K_2SO_4 + 2CrO_5 + 5H_2O$$

or $$Cr_2O_7^{2-} + 2H^+ + 4H_2O_2 \xrightarrow{\text{Ether}} 2CrO_5 + 5H_2O$$

(*b*) When neutral or slightly acidic solution of $K_2Cr_2O_7$ is treated with $H_2O_2$, *violet diperoy* chromate, $K^+[CrO(O_2)_2(OH)]^-$ is obtained.

$$K_2Cr_2O_7 + 6H_2O_2 \longrightarrow 2K[CrO(O_2)_2(OH)] + 5H_2O + O_2$$

or $$Cr_2O_7^{2-} + 6H_2O_2 \longrightarrow 2[CrO(O_2)_2(OH)]^- + 5H_2O + O_2$$

(*viii*) *Action of boric acid.* When a mixture having equimolecular amounts of $K_2Cr_2O_7$ and boric acid is fused and the product so formed is treated with water, hydrated chromic oxide, $Cr_2O_3 . 2H_2O$ (called Guigret's green) is obtained. This oxide has brilliant colour and is used as a green pigment.

(*ix*) *Action of $K_2CO_3$.* When $K_2Cr_2O_7$ solution is treated with the calculated amount of $K_2CO_3$ and the resulting solution is evaporated, lemon-yellow crystals of anhydrous $K_2CrO_4$ are obtained.

$$K_2Cr_2O_7 + K_2CO_2 \longrightarrow 2K_2CrO_4 + CO_2$$

This reaction has been used for the preparation of $K_2CrO_4$.

(*x*) *Action of $NH_4Cl$.* When $K_2Cr_2O_7$ is heated with $NH_4Cl$, chromic oxide ($Cr_2O_3$) is obtained.

$$K_2Cr_2O_7 + NH_4Cl \xrightarrow{\Delta} Cr_2O_3 + 2KCl + 4H_2O + N_2\uparrow$$

(*xi*) *Action of carbon.* When the solution containing $K_2Cr_2O_7$ and carbon is heated, $Cr_2O_3$ is obtained.

$$K_2Cr_2O_7 + 3C + H_2O \xrightarrow{\Delta} Cr_2O_3 + 2KOH + 3CO\uparrow$$

(*xii*) *Action of $K_4[Fe(CN)_6]$.* When $K_2Cr_2O_7$ is fused with dry $K_4[Fe(CN)_6]$, pot. cyanate, KCNO is obtained.

$$2K_2Cr_2O_7 + 2K_4[Fe^{2+}(CN)_6] \longrightarrow 2K_2O + 2Cr_2O_3 + 6KCNO + 2K_2Fe^+[Fe^{3+}(CN)_6]$$

(*xiii*) *Action of $H_2SO_4$.* (*a*) With cold $H_2SO_4$, red crystals of $CrO_3$ are obtained.

$$K_2Cr_2O_7 + H_2SO_4\ (\text{cold}) \longrightarrow 2CrO_3 + K_2SO_4 + H_2O$$

(*b*) When $K_2Cr_2O_7$ is heated with $H_2SO_4$, $O_2$ gas is evolued.

$$K_2Cr_2O_7 + 8H_2SO_4 \xrightarrow{\Delta} 2K_2SO_4 + 2Cr_2(SO_4)_3 + 8H_2O + 3O_2\uparrow$$

**Uses:** $K_2Cr_2O_7$ is used

(*i*) in dyeing, calico printing and chrome tanning

(*ii*) as an oxidising agent in laboratory.

(*iii*) as a volumetric reagent in the laboratory for the estimation of $Fe^{2+}$, $I^-$, etc., ions.

(*iv*) in photography for hardening gelatin film.

(*v*) a mixture of $K_2Cr_2O_7$ and conc. $H_2SO_4$ is used as a cleaning agent in the laboratory for glass vessels.

(*vi*) for the preparation of various compounds of chromium such as $K_2CrO_4$, $PbCrO_4$ (chrome yellow), $Pb(OH)_2$. $PbCrO_4$ (basic lead chromate, called chrome red), $Zn(OH)_2$. $ZnCrO_4$ (basic zinc chromate), chrome alum [$K_2SO_4$. $Cr_2(SO_4)_3.24H_2O$], chromyl chloride ($CrO_2Cl_2$), $Cr_2O_3.2H_2O$ (hydrated chromic oxide) etc. These compounds have industrial applications.

## Uses of $K_2Cr_2O_7$ in Analytical Chemistry

The uses of $K_2Cr_2O_7$ in analytical chemistry can be studied under the following heading.

**A. Uses in qualitative analysis.** $K_2Cr_2O_7$ is used in the detection of $SO_3^{2-}$ and $Cl^-$ radicals.

(*a*) *Detection of* $SO_3^{2-}$ *radical.* We have seen that acidified solution of $K_2Cr_2O_7$ oxides $SO_2$ to $K_2SO_4$. This property of $K_2Cr_2O_7$ is used to detect $SO_3^{2-}$ ion in a given inorganic salt.

The procedure is that the sulphite is heated with dil $H_2SO_4$ in a test tube when $SO_2$ gas which has suffocating smell comes out of the test tube.

$$Na_2SO_3 + H_2SO_4 \longrightarrow Na_2SO_4 + H_2O + SO_2.$$

Now when a piece of filter paper moistened with acidified $K_2Cr_2O_7$ is brought on the mouth of test tube, paper becomes green due to the formation of green $Cr_2(SO_4)_3$ which is produced by the reduction of $K_2Cr_2O_7$ to $Cr_2(SO_4)_3$ (green) by $SO_2$ and $SO_2$ is oxidised by acidified $K_2Cr_2O_7$ to $K_2SO_4$.

$$K_2Cr_2O_7 + H_2SO_4 + 3SO_2 \longrightarrow \underset{\text{Green}}{Cr_2(SO_4)_3} + K_2SO_4 + H_2O$$

$$\text{or} \quad Cr_2O_7^{2-} + 2H^+ + 3SO_2 \longrightarrow 2Cr^{3+} + 3SO_4^{2-} + H_2O$$

(*b*) *Detection of* $Cl^-$ *radical by chromyl chloride test.* We have seen that when solid $K_2Cr_2O_7$ mixed with a soluble chlordie (*e.g.*, NaCl) is heated with conc. $H_2SO_4$, deep red vapours of chromyl chloride ($CrO_2Cl_2$) are produced.

$$K_2Cr_2O_7 + 4NaCl + 3H_2SO_4 \xrightarrow{\Delta} \underset{\text{Red vapours}}{2CrO_2Cl_2} + K_2SO_4 + Na_2SO_4 + 3H_2O$$

On passing these vapours in NaOH solution, yellow solution, due to the formation of sodium chromate ($Na_2CrO_4$), is obtained. Now this yellow solution is acidified with $CH_3COOH$ and is added $(CH_3COO)_2$ Pb to it. This treatment gives yellow ppt. of lead chromate ($PbCrO_4$).

$$\underset{\text{Red vapours}}{CrO_2Cl_2} + 4NaOH \longrightarrow \underset{\text{Yellow solution}}{Na_2CrO_4} + 2NaCl + H_2O$$

$$Na_2CrO_4 + (CH_3COO)_2Pb \longrightarrow \underset{\text{Yellow ppt}}{PbCrO_4} + CH_3COOH$$

B. **Uses in quantitative analysis.** We have seen that the acidified solution of $K_2Cr_2O_7$ oxidises ferrous salts to ferric salts.

$$6FeSO_4 + K_2Cr_2O_7 + 7H_2SO_4 \longrightarrow 3F_2(SO_4)_3 + Cr_2(SO_4)_3 + 7H_2O + K_2SO_4$$

$$\text{or} \quad 6Fe^{2+} + Cr_2O_7^{2-} + 14H^+ \longrightarrow 6Fe^{3+} + 2Cr^{3+} + 7H_2O$$

This property of $K_2Cr_2O_7$ has been used for the determination of the following:

(*a*) Determination of the amount of ferrous ammonium sulphate (FAS) (also called Mohr's salt) present per litre of its given solution. The procedure is that a known volume (say 20 ml) of the given FAS solution containing some dil. $H_2SO_4$ is taken in a conical flask and this solution is titrated with a standard solution of $K_2Cr_2O_7$ taken in a burette (intermediate solution). Potassium ferricyanide, $K_3[Fe(CN)_6]$ placed on a tile is used as an external indicator.

Before the end point, the reaction mixture gives deep blue colour with $K_3[Fe(CN)_6]$ solution drop placed on the tile, since $Fe^{2+}$ ions (present in the reaction mixture) which have not been oxidised to $Fe^{3+}$ ions by the acidified $Cr_2O_7^{2-}$ ions react with $[Fe(CN)_6]^{3-}$ ion and produce the blue coloured ferro-ferricyanide, $Fe_3^{2+}[Fe^{3+}(CN)_6]_2$ (I) which is called Turnbull's blue.

$$3Fe^{2+} + 2\,[Fe(CN)_6]^{3-} \rightleftharpoons \underset{\text{(I)}}{Fe_3^{2+}\,[Fe^{3+}(CN)_6]}$$

(*b*) Determination of percentage of ferric ammonium sulphate, $Fe_2(SO_4)_3.(NH_4)_2SO_4.24H_2O$ (ferric alum)

(*c*) Determination of the percentage purity of a given sample of ferrous ammonium sulphate.

(*d*) Determination of the percentage of ferrous iron present in a given sample of an iron ore like haematite ($Fe_3O_4$) or spanthic iron ($FeCO_3$).

In all the above determinations $K_3[Fe(CN)_6]$ (external indicator) or diphenyl amine (internal indicator) is used as an indicator.

## Structures of Chromate $\left(CrO_4^{2-}\right)$ and Dichromate $\left(Cr_2O_7^{2-}\right)$ Ions

$CrO_4^{2-}$ ion has tetrahedral geometry due to $sp^3$ hybridisation of $Cr^{6+}$ ion. In this ion all the four Cr – O bonds are equivalent. *X*-ray study of $(NH_4)_2Cr_2O_7$ crystal has shown that $Cr_2O_7^{2-}$ ion has two tetrahedral $CrO_4^{2-}$ ions which are linked together through O-atom. Thus this O-atom acts as a bridge between two $CrO_4^{2-}$ tetrahedral ions and hence is called *bridging O-atom.* Since $CrO_4^{2-}$ ion is tetrahedral and results from $sp^3$ hybrdisation of $Cr^{6+}$ ion, each Cr-atom in $Cr_2O_7^{2-}$ is also $sp^3$ hybridised and hence is linked with four O-atoms which are placed at the four corners of the tetrahedron. Excepting the bridging O-atom, all the remaining six O-atoms are called terminal O-atoms. Thus this ion has two bridging Cr — O bonds and six terminal Cr—O bonds. The two bridging Cr — O bonds are longer (bond length =179 pm) than the other six equivalent terminal Cr— O bonds (bond length = 163 pm). Cr—O—Cr bond angle =126°. The structures of $CrO_4^{2-}$ and $Cr_2O_7^{2-}$ are given in Fig. 6.13.

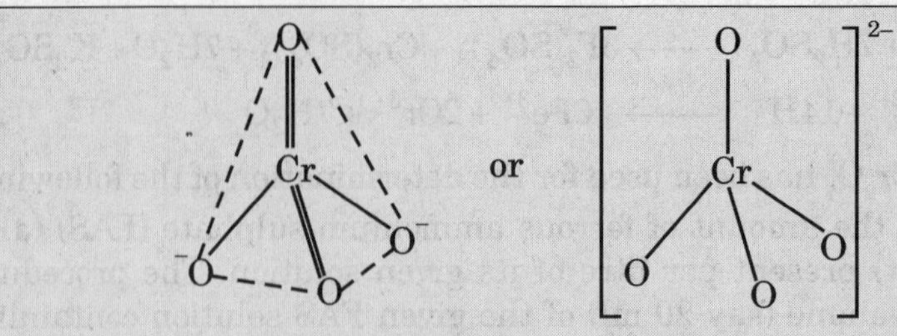


(*a*) Structure of chromate ion, $CrO_4^{2-}$

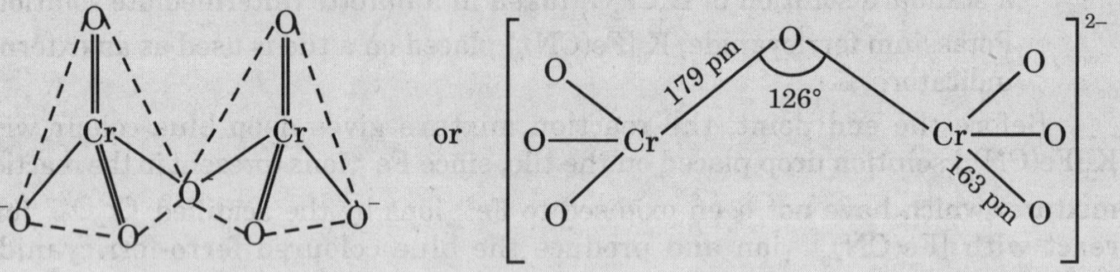


(*b*) Structure of dichromate ion, $Cr_2O_7^{2-}$

**Fig. 6.13** Structures of $CrO_4^{2-}$ and $Cr_2O_7^{2-}$ ions.

We have already said that both the ions have $Cr^{6+}$ ion which is a d° ion. Thus the yellow and orange colours of $CrO_4^{2-}$ and $Cr_2O_7^{2-}$ ions (respectively) are not due to *d-d* transition. Their colours are due to charge transfer.

## Complexes of Cr (II)

Examples of complexes of Cr(II) are $[Cr(NH_3)_6]^{2+}$, $[Cr(CN)_6]^{4-}$, $[(Cr(H_2O)_6]^{2+}$ etc. $[Cr(NH_3)_6]^{2+}$ is formed by the direct combination of $CrCl_2$ and $NH_3$. Most of these complexes have distorted octahedral geometry in which four equatorial ligands are at equal distances from the central $Cr^{2+}$ ion while the two axial ligands are at longer distances.

These complexes are usually *high spin* or *spin free* ($Cr^{2+} = 3d^4 = t_{2g}^3 eg^1$ with $n = 4$) but $[Cr(CN)_6]^{4-}$ ion is *low spin* or *spin paired* ($Cr^{2+} = 3d^4 = t_{2g}^4 e_g^0$ with $n = 2$) and have lower value of magnetic moment. Thus we see that high spin complexes result from $sp^3d^2$ hybridisation of $Cr^{2+}$ ion (outer orbital complexes) and the low spin complexes result from $d^2sp^3$ hybridisation of $Cr^{2+}$ ion (inner orbital complexes). Some complexes of Cr(II) such as $[Cr(NH_3)_6]^{2+}$ are relatively more stable than Cr(III) complexes.

## Complexes of Cr (III)

A large number of complexes of $Cr^{3+}$ *ion* (*chromic* ion) are known in which Cr element has C.N. = 6. Examples of such complexes are:

$[Cr(NH_3)_6]^{3+}$, $[Cr(NH_3)_5Cl]^{2+}$, $[Cr(NH_3)_4Cl_2]^+$ $[Cr(NH_3)_3Cl_3]^0$, $[Cr(ox)_3]^{3-}$, $[Cr_3O(CH_3COO)_6(H_2O)_3]^+$, $[Cr(H_2O)_6]^{3+}$, $[CrCl_6]^{3-}$, $[Cr(CN)_6]^{3-}$, $[Cr(en)_3]^{3+}$, $[Cr(en)_2Cl_2]^+$, $[CrF_6]^{3-}$, $[Cr(H_2O)_5Cl]^{2+}$, $NH_4^+[Cr(NH_3)_2(NCS)_4]^-$ etc.
Most of these complexes are octahedral due to $d^2sp^3$ hybridisation (inner orbital octahedral) of $Cr^{3+}$ ion. These complexes are paramagnetic due to the presence of three unpaired electrons present in $t_{2g}$ orbitals. Thus these complexes are high spin octahedral complexes.

Tetrathiocyanate diammine Cr (III) ion, $[Cr^{3+}(NH_3)_2(NCS)_4]^-$ present in Reineck's salt $NH_4^+[Cr(NH_3)_2(NCS)_4]^-$ has octahedral geometry shown in Fig. 6.14. This complex can be made by adding slowly $(NH_4)_2Cr_2O_7$ to molten $NH_4SCN$.

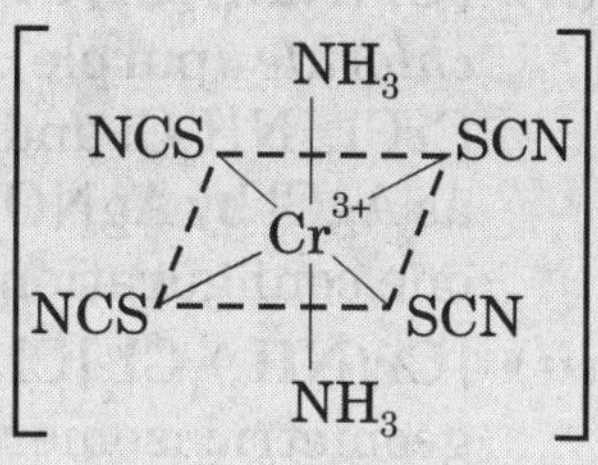


**Fig. 6.14** Octahedral geometry of $[Cr(NH_3)_2(SCN)_4]^-$ ion

$[Cr(en)_3]^{3+}$ ion contains three symmetrical bidentate ligands. Hence this ion is of $[M(AA)_3]$ type and exhibits optical isomerism. Its optical isomers are related to each other as object and mirror image. The structures of *d*-and *l*-forms (optical isomers) of $[Cr(en)_3]^{3+}$ ion have been given in chapter on "*Isomerism in Coordination Compounds*".

Complex ion, $[Cr(en)_2Cl_2]^+$ exists in *cis* and *trans* isomers whose structures have been given in the chapter on " *Isomerism in Coordination Compounds*".

The complex cation, $[Cr_3O(CH_3COO)_6(H_2O)_3]^+$ is called *chromium* (III) *basic acetate ion*. The structure of this ion is given in Fig. 6.15. It may be seen from the figure that this cation has three Cr-atoms which are placed in triangular plane. These Cr-atoms are linked with O-atom placed at the centre of the triangular plane. Two $CH_3COO^-$ ions make two bridges between two Cr-atoms across each edge of the triangular plane. The three $H_2O$ molecules occupy the remaining three octahedral sites on Cr-atoms.

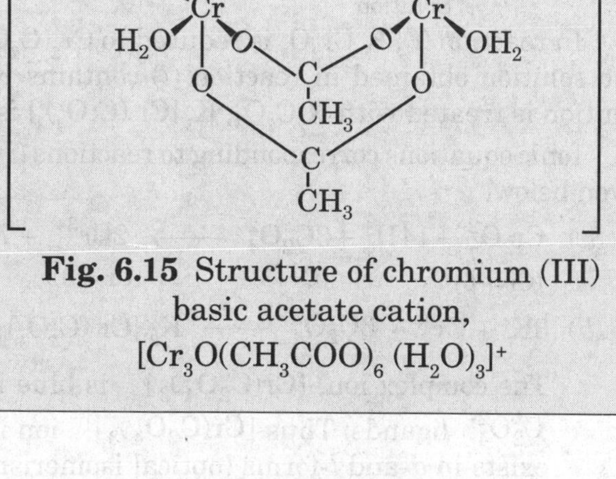


**Fig. 6.15** Structure of chromium (III) basic acetate cation, $[Cr_3O(CH_3COO)_6(H_2O)_3]^+$

**Ammine complexes of Cr (III).** Cr (III) forms the following ammine complexes:

(*i*) **$[Cr(NH_3)_6]Cl_3$.** This complex is also called *leuto chromic chloride* (yellow). It is made by treating cold concentrated solution of $CrCl_2$, $NH_4Cl$ and $NH_3$. All the three $Cl^-$ ions can be precipitated as AgCl by $AgNO_3$. $[Cr(NH_3)_6]^{3+}$ ion has octahedral geometry. This ion can have only one configuration.

(*ii*) **$[Cr(NH_3)_5Cl]Cl_2$.** This complex is also called *purpureo chromic chloride* (purple). It is made by bubbling air through a solution of $CrCl_2$, $NH_4Cl$ and $NH_3$ in water. Only two $Cl^-$ ions can be precipitated as AgCl by $AgNO_3$. $[Cr(NH_3)_5Cl]^+$ ion is octahedral and can have only one configuration.

(*iii*) **$[Cr(NH_3)_4Cl_2]Cl$.** It has violet colour. $[Cr(NH_3)_4Cl_2]^+$ ion has two geometric isomers *viz cis* and *trans*. The structures of these isomers are the same as those of $[Co(NH_3)_4Cl_2]^+$ ion shown in the chapter on "*Isomerism in Coordination Compounds*". Only one $Cl^-$ ion can be precipitated as AgCl with $AgNO_3$.

(*iv*) **$[Cr(NH_3)_3Cl_3]^\circ$.** It is a neutral complex, since it does not undergo ionisation. It has *cis* (or fac) and *trans* (or mer) isomers (geometric isomers). The structures of these isomers are the same as those of *cis* and *trans* isomers of $[Co(NH_3)_3Cl_3]^\circ$.

**Oxalate complexes of Cr (III).** Potassium trioxalate chromate (III), $K_3[Cr^{3+}(C_2O_4)_3].3H_2O$ and potassium dioxalato diaquochromate (III), $K[Cr^{3+}(C_2O_4)_2(H_2O)_2]$ are important oxalate complexes of Cr(III).

$K_3[Cr(C_2O_4)_3]$ is obtained by the following reactions:

(*i*) $$\underset{(Cr=+6)}{K_2Cr_2O_7} + \underset{(C=+3)}{7H_2C_2O_4} \longrightarrow \underbrace{K_2C_2O_4 + \underset{(Cr=+3)}{Cr_2(C_2O_4)_3}}_{\text{Solution}} + \underset{(C=+4)}{6CO_2} + 7H_2O$$

(*ii*) $$\underbrace{K_2C_2O_4 + Cr_2(C_2O_4)_3}_{\text{Solution}} + 2K_2C_2O_4 \longrightarrow 2K_3[Cr(C_2O_4)_3]$$

In reaction (*i*), $K_2Cr_2O_7$ is reduced to $Cr_2(C_2O_4)_3$ and $H_2C_2O_4$ is oxidised to $CO_2$. The solution obtained in reaction (*i*) contains $K_2C_2O_4$ and $Cr_2(C_2O_4)_3$. When this solution is treated with $K_2C_2O_4$, $K_3[Cr(C_2O_4)_3]$ is formed as shown in reaction (*ii*).

Ionic equations corresponding to reactions (*i*) and (*ii*) are (*a*) and (*b*) respectively given below.

(*a*) $$\underset{(Cr=+6)}{Cr_2O_7^{2-}} + 14H^+ + \underset{(C=+3)}{3C_2O_4^{2-}} \longrightarrow \underset{(Cr=+3)}{2Cr^{3+}} + 7H_2O + \underset{(C=+4)}{6CO_2}$$

(*b*) $$3K^+ + Cr^{3+} + 3C_2O_4^{2-} \longrightarrow K_3[Cr(C_2O_4)_3]$$

The complex ion, $[Cr(C_2O_4)_3]^{3-}$ is blue in colour and has three bidentate $C_2O_4^{2-}$ ligands. Thus $[Cr(C_2O_4)_3]^{3-}$ ion is of $[M(AA)_3]$ type. Hence this ion exists in *d*-and *l*-forms (optical isomerism) whose structures are shown in Fig. 6.16.

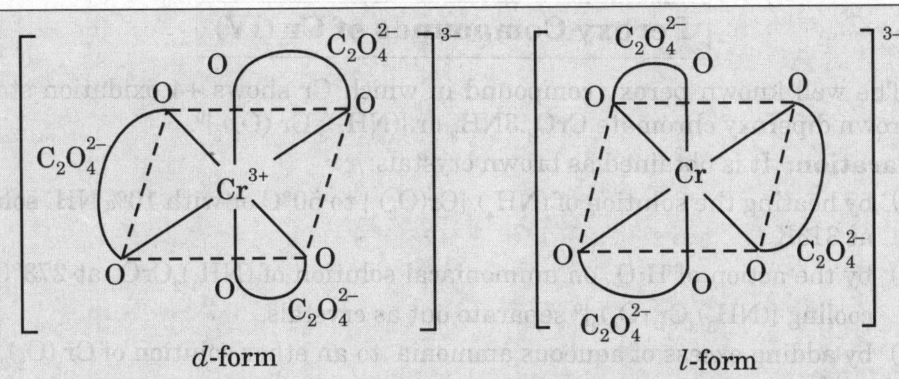


**Fig. 6.16** *d*-and *l*-forms (optical isomers) of $[Cr^{3+}(C_2O_4)_3]^{3-}$ ion.

The complex anion, $[Cr^{3+}(C_2O_4)_2(H_2O)_2]^-$ exists in *cis*- and *trans*-isomers (geometrical isomers) whose structures are shown in Fig. 6.17.

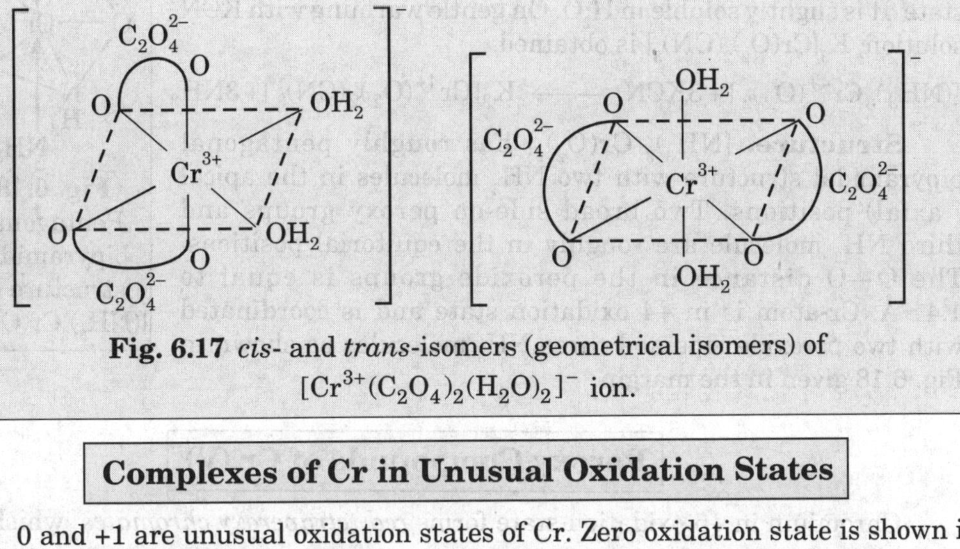


**Fig. 6.17** *cis*- and *trans*-isomers (geometrical isomers) of $[Cr^{3+}(C_2O_4)_2(H_2O)_2]^-$ ion.

## Complexes of Cr in Unusual Oxidation States

0 and +1 are unusual oxidation states of Cr. Zero oxidation state is shown in carbonyls like $Cr(CO)_6$. Similarly $Cr(C_6H_6)_2$ also has Cr in zero oxidation state. This compound is formed as follows:

$$3CrCl_3 + 6C_6H_6 + Al + AlCl_3 \longrightarrow 3[Cr(\eta^6 - C_6H_6)_2]^+[AlCl_4]^-$$

$$2[Cr(\eta^6 - C_6H_6)_2]^+[AlCl_4]^- + Na_2S_2O_4 + 4NaOH \longrightarrow$$

$$2[Cr(\eta^6 - C_6H_6)_2] + 2Na_2SO_4 + 2NaCl + 2AlCl_3 + 2H_2O$$

Cr atom in $Cr(C_6H_6)_2$ is *sandwiched* between the two $C_6H_6$ rings. Cr atom also has zero oxidation state in $\eta^5 - C_5H_5$ complex. Cr has +1 oxidation state in $[Cr(dipyridyl)_3]^+ ClO_4^-$. In $\left[Cr_2^{-1}(CO)_{10}\right]^{2-}$ and $\left[Cr^{-2}(CO)_5\right]^{2-}$ complex ions Cr shows –1 and –2 oxidation state respectively.

## Peroxy Compounds of Cr (IV)

The well known peroxy compound in which Cr shows +4 oxidation state is red-brown diperoxy chromate $CrO_4.3NH_3$ or $[(NH_3)_3 Cr (O_2)_2]^0$.

**Preparation.** It is obtained as brown crystals:

(*i*) by heating the solution of $(NH_4)_3[Cr(O_2)_4]$ to 50°C or with 10% $NH_3$ solution at 313K.

(*ii*) by the action of $H_2O_2$ on ammoniacal solution of $(NH_4)_2CrO_2$ at 273°C. On cooling $[(NH_3)_3Cr (O_2)_2]^0$ separate out as crystals.

(*iii*) by adding excess of aqueous ammonia to an ether solution of $Cr (O_2)_2O$.

$$Cr^{+6}(O_2)_2O + 3NH_4OH \longrightarrow [(NH_3)_3Cr^{4+}(O_2)_2]^0 + 3H_2O + \frac{1}{2}O_2$$

**Properties.** The compound is stable at room temperature but explodes on heating. Its paramagnetic value is equal to 2.8 B.M. which indicates that the compound has two unpaired electrons on Cr-atom and hence the metal is in +4 oxidation state. It is slightly soluble in $H_2O$. On gentle warming with KCN solution, $K_3[Cr(O_2)_2(CN)_3]$ is obtained.

$$[(NH_3)_3Cr^{4+}(O_2)_2] + 3KCN \longrightarrow K_3[Cr^{4+}(O_2)_2(CN)_3] + 3NH_3$$

**Structure.** $[NH_3)_3 Cr(O_2)_2]$ has roughly pentagonal bipyramidal structure with two $NH_3$ molecules in the apical ( axial) positions. Two broad side-on peroxy groups and third $NH_3$ molecule are roughly in the equitorial positions. The O—O distance in the peroxide groups is equal to 1.41 Å. Cr-atom is in +4 oxidation state and is coordinated with two peroxide ions and three $NH_3$ molecules as shown in Fig. 6.18 given in the margin.

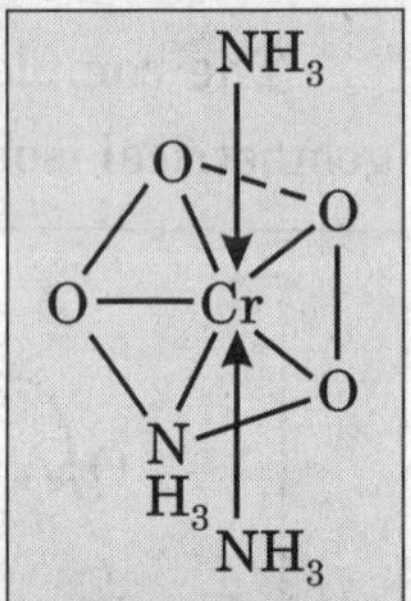


**Fig. 6.18** Pentagonal bipyramidal structure of $[(NH_3)_3Cr(O_2)_2]^0$

## Peroxy Compounds of Cr (V)

Chromium in +5 oxidation state forms *red tetraperoxy chromates* which can be represented by the general formula, $M_3^+[CrO_8]^{3-}$ or $M_3^+[Cr(O_2)_4]^{3-}$ $(M = K, NH_4)$.

**Preparation.** These compounds are obtained by treating alkaline solution of a chromate with 30% $H_2O_2$ solution below 0°C. For example:

$$2K_2CrO_4 + 2KOH + 7H_2O_2 \longrightarrow 2K_3[Cr(O_2)_4] + 8H_2O$$

or $$2CrO_4^{2-} + 2OH^- + 7H_2O_2 \longrightarrow 2[\ddot{C}r(O_2)_4]^{3-} + 8H_2O$$

**Properties.** (*i*) These compounds are paramagnetic due to the presence of one unpaired electron per Cr-atom. The presence of one unpaired electron per Cr-atom indicates that Cr is present in +5 oxidation state in these compounds ($Cr = 3d^5\,4s^1$, $Cr^{5+} = 3d^1 4s^0$)

(*ii*) These compounds are stable at room temperature and also for some days in aqueous solution.

(*iii*) Under neutral or alkaline conditions these compounds decompose to give chromate ($CrO_4^{2-}$) and $O_2$, but in acidic solution they produce Cr (III) salts and $O_2$.

$$4K_3[Cr(O_2)_4] + 2H_2O \longrightarrow 4K_2CrO_4 + 4KOH + 7O_2$$

or $$4[Cr(O_2)_4]^{3-} + 2H_2O \longrightarrow 4CrO_4^{2-} + 4OH^- + 7O_2$$

$$2[K_3Cr(O_2)_4] + 6H_2SO_4 \longrightarrow Cr_2(SO_4)_3 + 3K_2SO_4 + 6H_2O + 5O_2$$

or $$[Cr(O_2)_4]^{3-} + 12H^+ \longrightarrow 2Cr^{3+} + 6H_2O + 5O_2$$

(*iv*) When solution of $(NH_4)_3[Cr(O_2)_4]$ is heated to 50° or when it is treated with 10% $NH_3$ solution at 313 K, *red brown diperoxy chromate,* $[(NH_3)_3 Cr^{+4} (O_2)_2]^0$ is obtained.

$$(NH_4)_3[Cr^{+5}(O_2)_4] \xrightarrow{50°C} \underset{\text{Red-brown diperoxy chromate}}{[(NH_3)_3Cr^{4+}(O_2)_2]^0}$$

$$(NH_4)_3[Cr(O_2)_4] + NH_3(10\%) \longrightarrow [(NH_3)_3Cr(O_2)_2]^0$$

(*v*) $K_3[Cr(O_2)_4]$ is isomorphous with $K_3[V(O_2)_4]$, $K_3[Nb(O_2)_4]$ and $K_3[Ta(O_2)_4]$ and $K_3(Ta(O_2)_4)$.

**Structure of $CrO_8^{3-}$ ion.** Conductivity and freezing point measurements have shown that $CrO_8^{3-}$ ion in $K_3[CrO_8]$ has four broad side-on-peroxide groups, $(—O—O—)^{2-}$ which are arranged round $Cr^{5+}$ ion ($CrO_8^{3-} \equiv [Cr(O_2)_4]^{3-}$). The O—O distance in the peroxide group is 1.41Å. $[Cr(O_4)_2]^{3-}$ ion has dodecahedral structure which is shown in Fig. 6.19. $[Cr(O_2)_4]^{3-}$ ion is paramagnetic due to the presence of one unpaired electron present in $Cr^{5+}$ ion ($Cr^{5+} = 3d^1 4s°$)

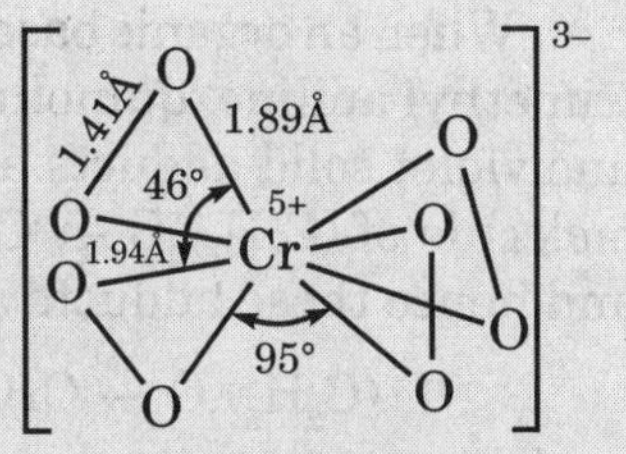


**Fig. 6.19** Dodecahedral structure of $[Cr(O_2)_2]^{3-}$ ion.

## Peroxy Compounds of Cr (VI)

Chromium in +6 oxidation state forms the following peroxy compounds.

**1. Chromic Peroxide or Chromic Diperoxide or Blue Peroxy Chromate, $CrO_5$ or $CrO(O_2)_2$**

**Preparation.** When a solution of $K_2CrO_4$ or $K_2Cr_2O_7$ containing dil. $H_2SO_4$ and some ether $(C_2H_5)_2O$ is treated with $H_2O_2$ and the solution mixture is shaken, the layer of ether becomes blue due to the formation of chromic diperoxide, $CrO_5$.

$$K_2CrO_4 + H_2SO_4 + 2H_2O_2 \longrightarrow K_2SO_4 + CrO_5 + 3H_2O$$

or $$CrO_4^{2-} + 2H^+ + 2H_2O_2 \longrightarrow \underset{\text{(blue)}}{CrO_5} + 3H_2O$$

$$K_2Cr_2O_7 + H_2SO_4 + 4H_2O_2 \longrightarrow K_2SO_4 + 2CrO_5 + 5H_2O$$

or $$Cr_2O_7^{2-} + 2H^+ + 4H_2O_2 \longrightarrow 2CrO_5 + 5H_2O$$

**Properties.** (*i*) *Decomposition of $CrO_5$.* Blue colour of ether solution of $CrO_5$ is unstable and hence fades rapidly. The fading of the colour is due to the fact that $CrO_5$, in presence of $H_2SO_4$, decomposes to give $Cr_2(SO_4)_3$ with evolution of $O_2$.

$$4CrO_5 + 6H_2SO_4 \longrightarrow 2Cr_2(SO_4)_3 + 6H_2O + 7O_2$$

$CrO_5$ has two peroxide groups ($O_2^{2-}$) and one oxide group ($O^{2-}$) and hence is formulated as $CrO(O_2)_2$.

(*ii*) *Action of $H_2O_2$.* When an alkaline solution of $CrO_5$ is treated with $H_2O_2$, *blue pot. peroxy chromate,* $K_2Cr_2O_{12}.\ 2H_2O$ and *red pot. peroxychormate,* $K_2Cr_2O_{16}$ are obtained.

$$2CrO_5 + H_2O_2 + 2KOH \longrightarrow K_2Cr_2O_{12}.\ 2H_2O\ (blue)$$

$$2CrO_5 + 3H_2O_2\ (excess) + 6KOH \longrightarrow K_2Cr_2O_{16}(red) + 6H_2O$$

(*iii*) *Action of organic bases.* When the blue coloured solution containing $CrO(O_2)_2$ is shaken with ether, $(C_2H_5)_2O$, an addition compound (adduct), $(C_2H_5)_2O \rightarrow CrO(O_2)_2$ is obtained.

$$(C_2H_5)_2O + CrO(O_2)_2 \longrightarrow \underset{\text{Ether adduct}}{[(C_2H_5)_2O \rightarrow CrO(O_2)_2]}$$

When an organic base like pyridine ($C_5H_5N$), 2, 2'-dipyridine, piperidine, aniline, trimethyl aniline, quinoline etc., is added to the ether solution of $CrO(O_2)_2$, crystalline violet solid adducts are obtained. In the formation of these adducts, $(C_2H_5)_2O$ molecule of $(C_2H_5)_2O \rightarrow CrO(O_2)_2$ adduct is replaced by the organic base molecule and hence these adducts are formulated as $L \rightarrow CrO(O_2)_2$ (L = organic base).

$$\underset{\text{Ether adduct}}{(C_2H_5)_2O \rightarrow CrO(O_2)_2} + L \longrightarrow L \rightarrow CrO(O_2)_2 + (C_2H_5)_2O$$

These adducts are non-electrolytes and insoluble in water but soluble in organic solvents like $C_6H_6$.

The pyridine adduct, $C_5H_5N \rightarrow CrO(O_2)_2$ is monomeric, soluble in benzene and is diamagnetic in nature.

*Structure of $C_5H_5$ N $\rightarrow CrO(O_2)_2$ adduct.* This adduct has pentagonal pyramidal structure in which oxide ion ($O^{2-}$) is placed at the apex while the two peroxy groups ($O_2^{2-}$) and N-atom of $C_5H_5N$ molecule are placed at the corners of the base of the pentagon. Cr-atom is located at the centre of the pentagonal base. (See Fig. 6.20). Thus we see that Cr-atom is coordinated with two peroxide groups ($O_2^{2-}$), one $NC_5H_5$ molecule and one oxide ion ($O^{2-}$). The O—O distance in the peroxide groups is equal to 1.41Å. Here Cr is in +6 oxidation state.

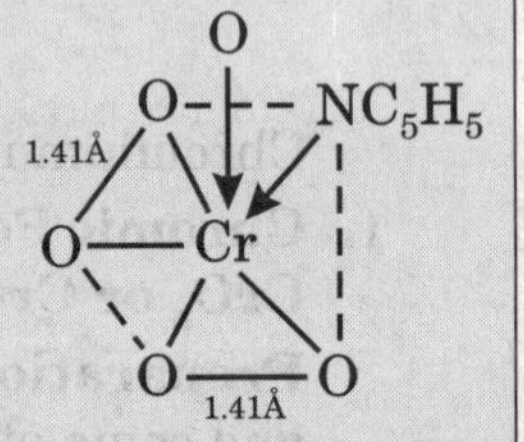


**Fig. 6.20**
Pentagonal pyramidal structure of $C_5H_5N \rightarrow CrO(O_2)_2$

(*iv*) *Action of aqueous ammonia.* When an excess of aqueous ammonia is added to the ether solution of $CrO_5$, *red-brown diperoxy chromate,* $[(NH_3)_3\ Cr^{+4}(O_2)_2]^0$ is obtained.

$$Cr^{+6}(O_2)_2O + 3NH_4OH \longrightarrow [(NH_3)_3Cr^{4+}(O_2)_2]^0 + 3H_2O + \frac{1}{2}O_2$$

**Structure of $CrO_5$ molecule.** Formerly the blue compound ($CrO_5$) was assumed to be perchromic acid ($HCrO_5$) and hence its structure was represented as

$$HO—O—\overset{\overset{O}{\|}}{\underset{\underset{O}{\|}}{Cr}}=O$$

which shows that the oxidation state of Cr-atom is + 7 which is not possible. Schwarz and Giese in 1932 showed that the blue-coloured compound is a peroxide of chromium (VI) which contains two peroxide groups, $O_2^{2-}$ and one oxide ion, $O^{2-}$ and hence is formulated as $CrO(O_2)_2$. The presence of two peroxide groups is evident from the following experimental observations:

(*a*) Dilute acids liberate 7 atoms of active oxygen per two moles of $CrO_5$ rather than 6 atoms of active oxygen as expected from $H_2CrO_5$ formulation.

$$2CrO_5 \longrightarrow Cr_2O_3 + 7O$$
$$2HCrO_5 \longrightarrow Cr_2O_3 + H_2O + 6O$$

(*b*) Titration of $CrO_5$ with $KMnO_4$ in acidic medium has shown that the compound consumes four equivalents of $KMnO_4$. This observation shows that $CrO_5$ has two peroxy groups $(—O—O—)^{2-}$ or $O_2^{2-}$.

$$CrO_5 + 2H_2O \longrightarrow CrO_3 + 2H_2O_2$$
$$2KMnO_4 + 3H_2SO_4 + 5H_2O_2 \longrightarrow K_2SO_4 + 2MnSO_4 + 8H_2O + 5O_2$$

(Note: 1 mole of $H_2O_2 \equiv 2$ equivalents of $KMnO_4$).

Thus the structure of $CrO_5$ can be written as CrO $(—O—O—)_2$ or $CrO(O_2)_2$ and should be called chromic diperoxide. It has a structure in which Cr-atom is hexavalent and shows hexacovalency. It may be seen from Fig. 6.21 that each peroxy group assumes two coordination positions around Cr-atom and thus a three-membered strained ring is formed at Cr-atom.

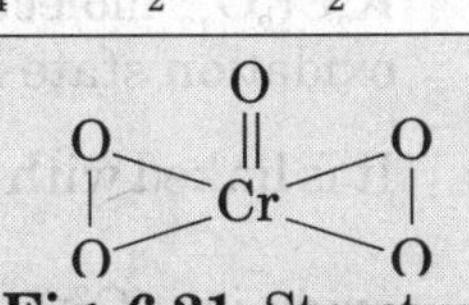

**Fig. 6.21** Structure of chromic diperoxide molecule, $CrO_5$ or $CrO(O_2)_2$.

**2. Violet Diperoxy Chromates, $M^+[CrO(O_2)_2(OH)]^-$ ($M^+ = K^+, NH_4^+$ or $Tl^+$)**

These compounds are obtained by treating neutral or slightly acidic solutions of $K^+$, $NH_4^+$ or $Tl^+$ dichromates with $H_2O_2$, *e.g.*,

$$K_2Cr_2O_7 + 6H_2O_2 \longrightarrow 2K[CrO(O_2)_2(OH)] + 5H_2O + O_2$$

or

$$Cr_2O_7^{2-} + 6H_2O_2 \longrightarrow 2[CrO(O_2)_2(OH)]^- + 5H_2O + O_2$$

These are violet compounds, are diamagnetic and violently explosive. The presence of two peroxy groups in these compounds has been established analytically and the magnetic properties show that Cr is in +6 oxidation state. Analysis shows the presence of one H-atom in the molecule. Since the molecular conductivity of the ammonium salt in aqueous solution is about 100 ohm$^{-1}$ cm$^2$, it is a uni-univalent electrolyte. Like $CrO(O_2)_2 \cdot NC_5H_5$ adduct, $[CrO(O_2)_2(OH)]^-$ ion also has pentagonal pyramidal structure with oxide oxygen at the apex.

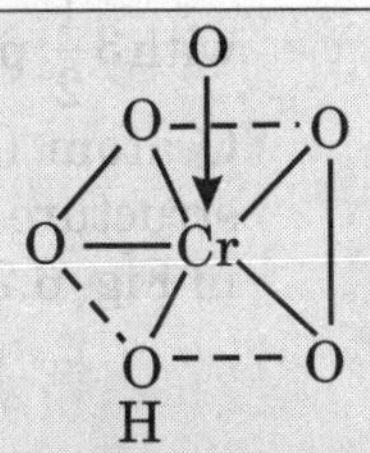

**Fig. 6.22** Structure of $[CrO(O_2)_2(OH)]^-$ ion.

Two peroxy groups and OH group are placed at the corners of the pentagonal pyramidal base while Cr-atom is located at the centre (See Fig. 6.22).

This structure shows that Cr-atom is coordinated with two peroxide groups ($O_2^{2-}$), one OH group and one oxide oxygen atom (Cr → O bond).

**3. Blue Potassium Peroxy Chromate, $K_2Cr_2O_{12}$. $2H_2O$**

**Preparation.** It is obtained by the action of $H_2O_2$ on alkaline solution of $CrO_5$.

$$2CrO_5 + H_2O_2 + 2KOH \longrightarrow K_2Cr_2O_{12}.2H_2O \text{ (Blue)}$$

**Properties.** This salt decomposes at room temperature to form the anhydrous $K_2Cr_2O_{12}$. It is diamagnetic in nature. When this salt is treated with an alkali or dil. acid, it decomposes to give $CrO_4^{2-}$ or $Cr^{3+}$

$$2K_2Cr_2O_{12} + 4KOH \longrightarrow 4K\,CrO_4 + 2H_2O + 5O_2$$

or
$$2Cr_2O_{12}^{2-} + 4OH^- \longrightarrow 4CrO_4^{2-} + 2H_2O + 5O_2$$

$$K_2Cr_2O_{12} + 4H_2SO_4 \longrightarrow Cr_2(SO_4)_3 + K_2SO_4 + 4H_2O + 4O_2$$

or
$$Cr_2O_{12}^{2-} + 8H^+ \longrightarrow 2Cr^{3+} + 4H_2O + 4O_2$$

**Structure of $K_2Cr_2O_{12}$ molecule.** Schwarz and Giese have shown that in $K_2Cr_2O_{12}$ molecule each Cr-atom is in +6 oxidation state and shows *hexacovalency*. It is linked with $2\frac{1}{2}$ peroxo groups and one oxide ion ($O^{2-}$). Potassium is attached with $O^{2-}$ ion. Thus the structure of $K_2Cr_2O_{12}$ molecule is as shown in Fig. 6.23. This structure clearly shows that the given molecule can be represented as $K^+O^-(O_2)_2\,Cr—O—O—Cr(O_2)_2O^-K^+$.

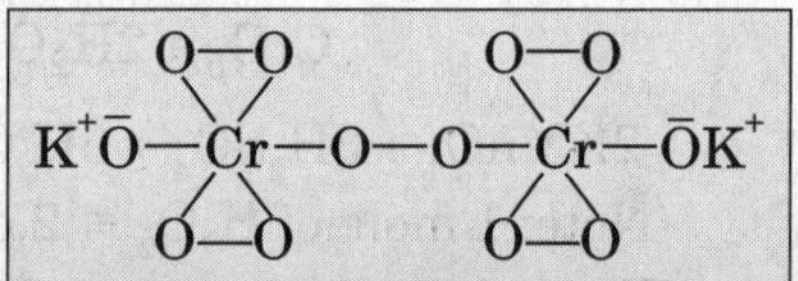

**Fig. 6.23** Structure of $K_2Cr_2O_{12}$ molecule.

**4. Red Potassium Peroxy Chromate, $K_6\,Cr_2O_{16}$**

**Preparation.** It is prepared by the action of $H_2O_2$ (excess) on an alkaline solution of $CrO_5$.

$$2CrO_5 + 3H_2O_2(\textit{excess}) + 6KOH \longrightarrow K_6Cr_2O_{16}\ (\textit{red}) + 6H_2O$$

**Structure.** In $K_2Cr_2O_{16}$ molecule each Cr-atom is in +6 oxidation state and exhibits *hexacovalency*. It is linked with $3\frac{1}{2}$ peroxo groups and one neutral O-atom (Cr = O bond). Thus the structure of this molecule is as shown in Fig. 6.24.

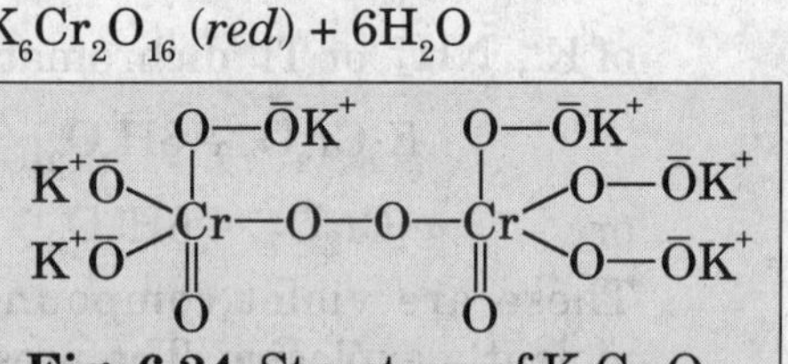

**Fig. 6.24** Structure of $K_6Cr_2O_{16}$ molecule.

## Manganese, $Mn_{25}$

### Properties of Manganese Metal

Pure manganese is silvery white but is usually very pink tinge metal in industrial samples. It exists in three forms which are: (*a*) α-form which is stable below 742°C and has complex cubic strucutre, (*b*) β-form which is stable between 742°C and 1190°C and like α-form, has complex cubic structure and (*c*) γ-form which is stable between 1190°C and 1244°C and forms face-centred cubic tetragonal crystals. Finely divided Mn, when heated in air, $Mn_3O_4$ is obtained.

$$3Mn + 2O_2 \longrightarrow Mn_3O_4$$

Dil.HCl and $H_2SO_4$ react with Mn to form Mn (+2) salts and $H_2$ gas is liberated.

$$Mn + 2H^+ \longrightarrow Mn^{2+} + H_2\uparrow$$

Conc. $H_2SO_4$ gives $MnSO_4$ and $SO_2$ is evolved

$$Mn + 2H_2SO_4(conc.) \longrightarrow MnSO_4 + 2H_2O + SO_2$$

With slightly more conc. $HNO_3$, NO is obtained

$$3Mn + 8HNO_3 \longrightarrow 3Mn(NO_3)_2 + 4H_2O + 2NO$$

With moderately conc. $HNO_3$, $NO_2$ is obtained

$$Mn + 4HNO_3 \longrightarrow Mn(NO_3)_2 + 2H_2O + 2NO_2$$

Hot water gets decomposed by the action of Mn

$$Mn + 2H_2O\ (hot) \longrightarrow Mn(OH)_2 + H_2$$

When heated with halogens, sulphur, nitrogen and carbon, halides, sulphide, nitrides ($Mn_5N_2$ and $Mn_3N_2$) and carbide are formed.

**Uses of Manganese**

(*i*) Manganese forms a number of alloys with other metals. These alloys are very useful, *e.g.*, **spiegeleisen**, an alloy of Mn (Mn=20–30%, C =5%, rest is Fe) is used in the manufacture of steel.

(*ii*) Manganese compounds are very useful, *e.g.*, (*a*) $MnO_2$ is used as an oxidising agent in the production of hydroquinone and also used in dry cells (*b*) Manganese lactate, $Mn(C_3H_5O_3).3H_2O$ and manganese pyro phosphate $Mn(H_2PO_2)_2.H_2O$ are used in medicines.

## Electronic Configuration and Oxidation States of Mn.

V.S.E.C. of Mn is $3d^5\,4s^2$. Mn shows a large number of oxidation states varying from –1 to +7. Thus Mn shows –1, 0, +1, +2, +3, +4, +5, +6 and +7 oxidation states in its compounds. –1, 0 and +1 are unusual oxidation states. Mn has –1 oxidation state in the complex anion, $[Mn(CO)_5]^-$ which is quite stable. Mn has zero oxidation state in complexs $[Mn_2(CO)_{10}]$ and $K_6[Mn(CN)_6].2NH_3$. +1 oxidation state is found in $K_5[Mn(CN)_6]$ which is unstable.

## Chemistry of Mn (II)

### 1. Manganous Compounds

Compounds of Mn (II) are called *manganeous compounds.* The value of $E^o_{ox}$ for $Mn/Mn^{2+}$ couple is 1.05 V.

$$Mn\ (aq) \longrightarrow Mn^{2+}(aq) + 2e^-,\ E^o_{ox} = 1.05\ V$$

Since $E^{o}_{ox}$ value is fairly high, Mn is readily oxidised to $Mn^{2+}$ ion in acid solution. $Mn^{2+}$ ion is stable, since this ion has half-filled 3*d* sub-shell ($Mn^{2+} = 3d^5\ 4s^o$) which adds to the stability of $Mn^{2+}$ ion.

The value of $E^{o}_{ox}$ for $Mn^{2+}/Mn^{3+}$ couple is –1.51 V

$$Mn^{2+}(aq) \longrightarrow Mn^{3+}(aq) + e^-,\ E^{o}_{ox} = -1.51\ V$$

Since $E^{o}_{ox}$ value is negative, $Mn^{2+}$ ion is not oxidised to $Mn^{3+}$ ion. This means that the neutral or acidic solution of $Mn^{2+}$ ions remains unaffected even by the prolonged exposure to $O_2$. In other words we say that $Mn^{2+}$ ion is not a reducing agent.

## 2. Complexes of Mn (II)

(*a*) **Octahedral complexes.** Some of the hydrated salts of Mn(II) have complex cation. For example $Mn(ClO_4)_2 . 6H_2O$ contains the complex cation, $[Mn(H_2O)_6]^{2+}$. The action of $NH_3$ on Mn(II) salts gives $[Mn(NH_3)_6]^{2+}$. These complexes are octahedral and have pink colour . They have 5 unpaired electrons in 3*d* orbitals. Therefore, the magnetic moment values for these complexes are in the range of 5.8 – 6.1 BM. These complexes are highly stable and are not affected by oxidising as well as by reducing agents.

The octahedral complex anions, $[Mn(CN)_6]^{4-}$ and $[Mn^{2+}(CN)_5(NO)]^{3-}$ are low spin complex ions and hence have only one unpaired electron. These complexes are not stable and hence get oxidised readily, *e.g.*

$$\left[Mn^{2+}(CN)_6\right]^{4-} \longrightarrow \left[Mn^{3+}(CN)_6\right]^{3-} + e^-$$

(*b*) **Tetrahedral complexes.** The tetrahalo complex anions, $[MnX_4]^{2-}$ have tetrahedral geometry. These anions are associated with large cations like $R_4N^+$, $R_4P^+$. These complexes have green-yellow colour. These complexes are not stable in aqueous solution. They are stable in solvents having low dipole moments such as chloroform, nitrobenzene etc. These complexes have 5 unpaired electrons in 3*d* orbitals. Hence these complexes are high spin complexes having magnetic moment values being in the range of 5.8 – 6.1 BM.

# Chemistry of Mn (III)

## 1. Manganic Compounds

Compounds of Mn in +3 oxidation state are called *manganic compounds.* E° value for $Mn^{3+}/Mn^{2+}$ couple is 1.51 V.

$$Mn^{3+}(aq) + e^- \longrightarrow Mn^{2+}(aq),\ E° = +1.51\ V$$

Since E° value is positive, $Mn^{3+}$ ion gets reduced easily in aqueous solution. Hence $Mn^{3+}$ ion is not stable in aqueous solution. Thus $Mn^{3+}$ ion acts as a strong oxidising agent. It is so strong that it can evolve $O_2$ even from $H_2O$. Above discussion shows that $Mn^{3+}$ ion is stable only in solid insoluble compounds and complex ions. For example, $MnPO_4$ and $Mn_2O_3$ are insoluble compounds.

In aqueous solution $Mn^{3+}$ ion undergoes disproportionation to form $Mn^{2+}$ and $MnO_2(s)$ (Mn = +4)

$$2Mn^{3+} + 2H_2O \longrightarrow Mn^{2+} + MnO_2(s) + 4H^+$$

$Mn^{3+}$ ion can be stabilised by preventing the above disproportionate reaction. This can be done by converting $Mn^{3+}$ ion into its complex ions $[Mn(CN)_6]^{3-}$ and $[Mn(C_2O_4)_3]^{3-}$

$$Mn^{3+} + 6\,CN^- \longrightarrow [Mn(CN)_6]^{3-}\ (\textit{LS complex})$$

$$Mn^{3+} + 3C_2O_4^{2-} \longrightarrow [Mn(C_2O_4)_3]^{3-}\quad (\textit{HS complex})$$

## 2. Complexes of Mn (III)

Complexes of $Mn^{3+}$ ion are quite stable. $[Mn(CN)_6]^{3-}$ and $[Mn(C_2O_4)_3]^{3-}$ are well known examples of complexes of $Mn^{3+}$ ion.

# Chemistry of Mn (IV)

Mn in +4 oxidation state exists primarily in water-insoluble manganese dioxide, $MnO_2$. Thus there is very little aqueous chemistry of Mn(IV). $MnO_2$ which is a familar compound of Mn(IV) is discussed below.

## 1. Manganese Dioxide, $MnO_2$.

**Preparation.** The pure $MnO_2$ is prepared by heating $Mn(NO_3)_2$ while the hydrated form, $MnO_2.H_2O$ (called manganous acid, $H_2MnO_3$) is prepeared by the alkaline oxidation of Mn(II) salt by sodium hypochlorite or bleaching powder.

$$Mn(NO_3)_2 \xrightarrow{\Delta} MnO_2 + 2NO_2$$

$$Mn^{2+} + 3OH^- + ClO^- \longrightarrow MnO_2.H_2O + Cl^-$$

**Properties.** It is a black powder, insoluble in water and a good conductor of electricity. When heated above 530°C in air, it loses $O_2$ to form $Mn_3O_4$. It acts as a basic oxide towards cold acids like HCl, forming salts containing Mn in +4 oxidation state. With conc. $H_2SO_4$, $MnSO_4$ is formed and $O_2$ is evolved. $HNO_3$ has no action on $MnO_2$.

$$3MnO_2 \xrightarrow{>530°C} Mn_3O_4 + O_2$$

$$MnO_2 + 4HCl\,(\textit{cold and concentrated}) \longrightarrow MnCl_4 + 2H_2O$$

$$2MnO_2 + 2H_2SO_4\,(\textit{concentrated}) \longrightarrow 2MnSO_4 + 2H_2O + O_2$$

In the presence of excess of $H^+$ ions, $MnO_2$ acts as a strong oxidising agent. In showing this behaviour $Mn^{4+}$ changes to $Mn^{2+}$ ion.

$$\underset{(Mn=+4)}{MnO_2} + 4H^+ + 4e^- \longrightarrow \underset{(Mn=+2)}{Mn^{2+}} + 2H_2O$$

Thus in presence of hot dil. $H_2SO_4$, $MnO_2$ oxidises oxalic acid to $CO_2$, Fe(II) salt to Fe(III) salt and chloride ($Cl^-$) to $Cl_2$

$$C_2O_4^{2-} + MnO_2 + 4H^+ \longrightarrow Mn^{2+} + 2H_2O + 2CO_2$$

$$2Fe^{2+} + MnO_2 + 4H^+ \longrightarrow 2Fe^{3+} + Mn^{2+} + 2H_2O$$

$$MnO_2 + 4H^+ + 2Cl^- \longrightarrow Mn^{2+} + 2H_2O + Cl_2$$

In presence of air or oxidising agents (*e.g.* $KNO_3$, $KClO_3$) $MnO_2$ reacts with fused KOH to give potassium manganate, $K_2MnO_4$.

$$2MnO_2 + 4KOH + O_2 \longrightarrow 2K_2MnO_4 + 2H_2O$$

$MnO_2$ is reduced to Mn by Al metal.

$$2MnO_2 + 4Al \longrightarrow 3\,Mn + 2Al_2O_3.$$

**Uses.** $MnO_2$ is used: (*i*) as a depolariser in Lechanehe cell and dry cells. In these cells $MnO_2$ oxidies $H_2$ liberated at the cathode and thus counteracts polarisation, (*ii*) as a decolouriser in glass, (*iii*) in forming with ferric oxide, a dark brown glaze which is used in pottery, (*iv*) as a catalyst in the preparation of $O_2$ from $KClO_3$, (*v*) as a drier in paints and varnishes, (*vi*) in the preparation of $Cl_2$ in the laboratory, (*vii*) in dyestuff industry and in varnish manufacture, (*viii*) in the manufacture of metallic manganese, maganates, permanganates and a number of its salts.

**Structure.** Since in $MnO_2$, Mn is in +4 state, the structure of $MnO_2$ is $O{=}Mn{=}O$ which has also been supported by magnetic susceptibility.

### 2. Complexes of Mn (IV)

$K_2[MnF_6]$, $K_2[MnCl_6]$, $K_2[Mn(CN)_6]$ etc are the examples of complexes of Mn (IV). These are octahedral complexes.

## Chemistry of Mn (V)

Mn in +5 oxidation state is very unstable. So far only $K_3MnO_4$ has been prepared by reacting the aqueous solution of $KMnO_4$ with $K_2SO_3$.

$$KMnO_4 + K_2SO_3 + H_2O \longrightarrow K_3MnO_4 + H_2SO_4$$

$K_3MnO_4$ is bright blue in colour and is quite unstable.

## Chemistry of Mn (VI)

Mn in +6 oxidation state exists in manganates which contain manganate ion, $MnO_4^{2-}$. Manganates are the salts of manganic acid, $H_2MnO_4$ which is unknown even in solution.

**Preparation of Manganates ($MnO_4^{2-}$)**

Manganates can be obtained by the following methods:

(*i*) **From pyrolusite mineral, $MnO_2$: Large scale preparation.** When finely powdered mineral ($MnO_2$) is fused with KOH or $K_2CO_3$ in presence of air or an oxidising agent like $KNO_3$, $KClO_3$ etc., green coloured mass containing $K_2MnO_4$ is obtained.

(*a*) $2MnO_2 + 4KOH + O_2 \longrightarrow 2K_2MnO_4 + 2H_2O$

or $2MnO_2 + 4OH^- + O_2 \longrightarrow 2MnO_4^{2-} + 2H_2O$

(*b*) $2MnO_2 + 2K_2CO_3 + O_2 \longrightarrow 2K_2MnO_4 + 2CO_2$

or $2MnO_2 + 2CO_3^{2-} + O_2 \longrightarrow 2MnO_4^- + 2CO_2$

(*c*) $MnO_2 + 2KOH + KNO_3 \longrightarrow K_2MnO_4 + KNO_2 + H_2O$

or $MnO_2 + 2OH^- + NO_3^- \longrightarrow MnO_4^{2-} + NO_2^- + H_2O$

(*d*) $2MnO_2 + 6KOH + KClO_3 \longrightarrow 3K_2MnO_4 + KCl + 3H_2O$

or $3MnO_2 + 6OH^- + ClO_3^- \longrightarrow 3MnO_4^{2-} + Cl^- + 3H_2O$

**The fused mass is dissolved in small quantity of water and green solution thus obtained is allowed to evaporate in vacuum when dark green crystals of $K_2MnO_4$ are obtained.**

(*ii*) **From permanganate ($MnO_4^-$).** When a permanganate ($MnO_4^-$) is boiled with an alkali hydroxide solution, a solution containing manganate ($MnO_4^{2-}$) is obtained.

$$4KMnO_4 + 4KOH \longrightarrow 4K_2MnO_4 + 2H_2O + O_2$$

or $4MnO_4^- + 4OH^- \longrightarrow 4MnO_4^{2-} + 2H_2O + O_2$

**General Properties of Manganates:**

(*i*) **Stability.** Manganates are stable only in highly alkaline solutions which have green colour. On adding large quantity of $H_2O$ or on acidifying the solution, the green $MnO_4^{2-}$ ion (Mn=+6) undergoes disproportionation to $MnO_2(s)$ (Mn = +4) (*brown ppt.)* and permanganate ion, $MnO_4^-$ (Mn = +7).

$$3K_2MnO_4 + 2H_2O \longrightarrow MnO_2(s) + 2KMnO_4 + 4KOH$$

or $3MnO_4^{2-} + 2H_2O \longrightarrow MnO_2(s) + 2MnO_4^- + 4OH^-$

$$3K_2MnO_4 + 2H_2SO_4 \longrightarrow MnO_2(s) + 2KMnO_4 + 2K_2SO_4 + 2H_2O$$

or $3MnO_4^{2-} + 4H^+ \longrightarrow MnO_2(s) + 2MnO_4^- + 2H_2O$

$MnO_4^{2-}$ is changed to $MnO_2$ and $MnO_4^-$ ion even on passing $CO_2$ in its solution.

$$3K_2MnO_4 + 2H_2O + 4CO_2 \longrightarrow MnO_2 + 2K MnO_4 + 4KHCO_3$$

or $3MnO_4^{2-} + 2H_2O + 4CO_2 \longrightarrow MnO_2 + MnO_4^- + 4HCO_3^-$

(*ii*) **Oxidation of $MnO_4^{2-}$ ion to $MnO_4^-$ ion.** (*a*) When $Cl_2$ or $O_3$ is passed through the solution of a manganate ($MnO_4^{2-}$), it is oxidised to permanganate ion ($MnO_4^-$).

$$2K_2MnO_4 + Cl_2 \longrightarrow 2KMnO_4 + 2KCl$$

or $2MnO_4^{2-} + Cl_2 \longrightarrow 2MnO_4^- + 2Cl -$

$$2K_2MnO_4 + O_3 + H_2O \longrightarrow 2KMnO_4 + 2KOH + O_2$$

or $2MnO_4^{2-} + O_3 + H_2O \longrightarrow 2MnO_4^- + 2OH^- + O_2$

(*b*) On passing an electric current through the solution of a manganate ($MnO_4^{2-}$), it gets oxidised to $MnO_4^-$ ion at anode (anodic oxidation) and $H_2$ is evolved at cathode.

$$MnO_4^{2-} \longrightarrow MnO_4^- + e^- \text{(At anode)}$$

$$\left.\begin{aligned} H^+ + e^- &\longrightarrow H \\ H + H &\longrightarrow H_2 \end{aligned}\right\} \text{(At cathode)}$$

(*iii*) Manganates are powerful oxidising agents.

(*iv*) Manganates are isomorphous with the corresponding sulphates, *e.g.* $K_2MnO_4$ is isomorphous with $K_2SO_4$ and similarly $Na_2MnO_4 . 10H_2O$ is isomorphous with $Na_2SO_4 . 10H_2O$.

(*v*) **Action of heat: Reduction.** At red heat $K_2MnO_4$ (Mn= +6) is decomposed into pot. permanganite, $K_2MnO_3$ (Mn= +4) and $O_2$ is evolved.

$$2K_2MnO_4 \xrightarrow{\Delta} 2K_2MnO_3 + O_2$$

or $$2MnO_4^{2-} \xrightarrow{\Delta} 2MnO_3^{2-} + O_2$$

**Uses of Manganates:**

$Na_2MnO_4 .10H_2O$ is used as a disinfectant.

## Chemistry of Mn (VII): Potassium Permanganate, $KMnO_4$

Mn in +7 oxidation state is found in permanganate ion, $MnO_4^-$. Potassium permanganate ($KMnO_4$) is a salt containing $MnO_4^-$ ion. $KMnO_4$ is a salt of permanganic acid, $HMnO_4$ which is an unstable acid and exists only in solution.

## Preparation of $KMnO_4$ From Pyrolusite Ore ($MnO_2$): Industrial Preparation.

Preparation of $KMnO_4$ from pyrolusite ore consists of the following steps:

**(1)** ***Oxidation of $MnO_2$ (Mn= +4) to $K_2MnO_4$ (Mn = + 6).*** The finely-powdered pyrolusite ore ($MnO_2$) is fused with KOH or $K_2CO_3$ in presence of air or with KOH in presence of an oxidising agent like $KNO_3$, $KClO_3$ etc. so that the fused mass containing green-coloured $K_2MnO_4$ is obtained.

$$2MnO_2 + 4KOH + O_2 \longrightarrow K_2MnO_4 + 2H_2O$$

or $$2MnO_2 + 4OH^- + O_2 \longrightarrow 2MnO_4^{2-} + 2H_2O$$

$$2MnO_2 + 2K_2CO_3 + O_2 \longrightarrow 2K_2MnO_4 + 2CO_2$$

or $$2MnO_2 + 2CO_3^{2-} + O_2 \longrightarrow 2MnO_4^{2-} + 2CO_2$$

$$MnO_2 + 2KOH + KNO_3 \longrightarrow K_2MnO_4 + KNO_2 + H_2O$$

or $$MnO_2 + 2OH^- + NO_3^- \longrightarrow MnO_4^{2-} + NO_2^- + H_2O$$

$$3MnO_2 + 6KOH + KClO_3 \longrightarrow 2K_2MnO_4 + KCl + 3H_2O$$

or $$3MnO_2 + 6OH^- + ClO_3^- \longrightarrow 3MnO_4^{2-} + Cl^- + 3H_2O$$

(2) ***Conversion of $K_2MnO_4$ to $KMnO_4$.*** Fused mass (green coloured) of $K_2MnO_4$ obtained as above is extracted with $H_2O$ and the solution, after filtration, is converted into $KMnO_4$.

(*a*) by the disproportionation of $K_2MnO_4$ (Mn = +6) to $MnO_2$ (Mn = +4) and $KMnO_4$ (Mn = +7) by acidifying the solution of $K_2MnO_4$ with $H_2SO_4$ or by passing $CO_2$ gas through the solution.

$$3K_2MnO_4 + 2H_2SO_4 \longrightarrow K_2SO_4 + 2KMnO_4 + 2H_2O + MnO_2\downarrow$$

$$\text{or} \quad 3MnO_4^{2-} + 4H^+ \longrightarrow 2MnO_4^- + 2H_2O + MnO_2\downarrow$$

$$3K_2MnO_4 + 4CO_2 + 2H_2O \longrightarrow 2KMnO_4 + MnO_2 + 4KHCO_3$$

$$\text{or} \quad 3MnO_4^{2-} + 4CO_2 + 2H_2O \longrightarrow 2MnO_4^- + MnO_2 + 4HCO_3^-$$

(*b*) by oxidising $K_2MnO_4$ (Mn = +6) to $KMnO_4$ (Mn= +7). This can be done:

(*i*) by passing $Cl_2$ or $O_3$ through the solution of $K_2MnO_4$.

$$2K_2MnO_4 + Cl_2 \longrightarrow 2KMnO_4 + 2KCl$$

$$\text{or} \quad 2MnO_4^{2-} + Cl_2 \longrightarrow 2MnO_4^- + 2Cl^-$$

$$2K_2MnO_4 + H_2O + O_3 \longrightarrow 2KMnO_4 + 2KOH + O_2$$

$$\text{or} \quad 2MnO_4^{2-} + H_2O + O_3 \longrightarrow 2MnO_4^- + 2OH^- + O_2$$

(*ii*) by electrolysing the solution of $K_2MnO_4$ in a cell (Ni: anode, Fe : cathode). On passing the electric current the following reactions take place in the cell.

*Ionisation*: $[K_2MnO_4 \rightleftharpoons K^+ + KMnO_4^-] \times 2$

*At anode*: (*Oxidation*) $[KMnO_4^- \longrightarrow KMnO_4 + e^-] \times 2$

*At cathode*:( *Reduction*) $[K^+ + e^- \longrightarrow K] \times 2$

$$2K + 2H_2O \longrightarrow 2KOH + H_2\uparrow$$

On adding all the above equations, we get the final equation as:

$$2K_2MnO_4 + 2H_2O \longrightarrow \underset{\text{Solution}}{2KMnO_4} + 2KOH + H_2\uparrow$$

The solution of $KMnO_4$ obtained as above is concentrated by evaporation so that crystals of $KMnO_4$ are obtained.

## Properties of $KMnO_4$

Potassium permanganate forms dark purple lustrous crystals giving deep pink colour in solution. It is isomorphous with potassium perchlorate. Its solubility in water at 20° is only about 7% whilst it dissolves more at higher temperatures (25% at 63°C). It gives off $O_2$ at 200°C.

$$2KMnO_4 \longrightarrow K_2MnO_4 + MnO_2 + O_2$$

At red heat $K_2MnO_4$ formed as above is decomposed into permanganite, $K_2MnO_3$ and $O_2$.

$$2K_2MnO_4 \longrightarrow 2K_2MnO_3 + O_2$$

When an alkaline soltution of $KMnO_4$ is heated, $O_2$ is evolved.

$$4KMnO_4 + 4KOH \xrightarrow{\Delta} 4K_2MnO_4 + 2H_2O + O_2$$

or $$4MnO_4^- + 4OH^- \longrightarrow 4MnO_4^{2-} + 2H_2O + O_2$$

With *well-cooled* conc. $H_2SO_4$, it gives $Mn_2O_7$ which decomposes on warming.

$$2KMnO_4 + 2H_2SO_4 \longrightarrow Mn_2O_7 + 2KHSO_4 + H_2O$$

$$2Mn_2O_7 \longrightarrow 4\,MnO_2 + 3O_2$$

With *warmed* conc. $H_2SO_4$ oxygen gas is evolved.

$$KMnO_4 + 6H_2SO_4 \longrightarrow 2K_2SO_4 + 4MnSO_4 + 6H_2O + 5O_2$$

When heated in a current of $H_2$, solid $KMnO_4$ gives KOH, MnO and steam.

$$2KMnO_4 + 5H_2 \longrightarrow 2KOH + 2MnO + 4H_2O$$

A mixture of $KMnO_4$ with sulphur and charcoal burns when heated, and explodes like a gun powder when rubbed. A well-powdered mixture of $KMnO_4$ and oxalic acid catches fire spontaneously in a few seconds. When a drop of glycerine is poured over powdered $KMnO_4$, the same thing happens.

$KMnO_4$ reacts with $H_2SiF_6$ and gives $HMnO_4$

$$2KMnO_4 + H_2SiF_6 \longrightarrow K_2SiF_6 + 2HMnO_4$$

## Oxidising Properties of $KMnO_4$

$KMnO_4$ is a powerful oxidising agent. The course of reaction depends on whether the oxidatin takes place in *alkaline, neutral* or *acid* solution.

(*a*) **In alkaline solution.** In alkaline solution, $MnO_4^-$ ion (Mn = +7) is first reduced to $MnO_4^{2-}$ ion (Mn = +6) and then to insoluble $MnO_2$ (Mn = +4). The solution first changes from purple (due to $MnO_4^-$ ) to green (due to $MnO_4^{2-}$ ) and finally it becomes colourless ($MnO_2$).

$$\underset{\substack{\text{Purple}\\(Mn=+7)}}{2KMnO_4 + 2KOH} \longrightarrow \underset{\substack{\text{Green}\\(Mn=+6)}}{2K_2MnO_4} + H_2O + O \quad \text{... (i)}$$

$$\underset{\substack{\text{Green}\\(Mn=+6)}}{2K_2MnO_4} + 2H_2O \longrightarrow \underset{\substack{\text{Colourless}\\(Mn=+4)}}{2MnO_2} + 4KOH + 2O \quad \text{... (ii)}$$

---

*On adding*: $$\underset{\substack{\text{Purple}\\(Mn=+7)}}{2KMnO_4} + H_2O \xrightarrow{\text{Alkali}} \underset{\substack{\text{Colourless}\\(Mn=+4)}}{2MnO_2} + 2KOH + 3O \quad \text{... (iii)}$$

This is the complete equation which represents the oxidising action of $KMnO_4$ in alkaline medium.

From equation (*iii*) it is evident that two molecules of $KMnO_4$ in alkaline solution give three oxygen atoms (*i.e., six equivalents*) and therefore, *the equivalent weight of $KMnO_4$ as an oxidising agent in alkaline solution is one-third of its molecular weight*.

Equations (*i*), (*ii*) and (*iii*) can also be written in the ionic forms as follows:

$$\underset{\substack{\text{Purple}\\(Mn=+7)}}{2MnO_4^-} + 2OH^- \longrightarrow \underset{\substack{\text{Green}\\(Mn=+6)}}{2MnO_4^{2-}} + H_2O$$

$$\underset{\substack{\text{Green}\\(\text{Mn}=+6)}}{2MnO_4^{2-}} + 2H_2O \longrightarrow \underset{\substack{\text{Colourless}\\(\text{Mn}=+4)}}{2MnO_2} + 4OH^- + 2O$$

---

*On adding*: $$\underset{\substack{\text{Purple,}\\(\text{Mn}=+7)}}{2MnO_4^-} + H_2O \xrightarrow{\text{Alkali}} \underset{\substack{\text{Colourless}\\(\text{Mn}=+4)}}{2MnO_2} + 2OH^- + 3O$$

or $$MnO_4^- + 2H_2O + 3e^- \longrightarrow MnO_2 + 4OH^-$$

Thus in alkaline medium, $KMnO_4$ oxidises:

(*i*) *Iodides ($I^-$) to iodates ( ($IO_3^-$)*

$$2KMnO_4 + H_2O \longrightarrow 2MnO_2 + 2KOH + 3O \text{ (Reduction reaction)}$$

$$KI + 3O \longrightarrow KIO_3 \text{(\textit{Oxidation reaction})}$$

---

$$2KMnO_4 + H_2O + KI \longrightarrow 2MnO_2 + 2KOH + KIO_3$$

OR

$$2\times[MnO_4^- + 2H_2O + 3e^- \longrightarrow MnO_2 + 4OH^-] \text{ (\textit{Reduction half-reaction})}$$

$$I^- + 6OH^- \longrightarrow IO_3^- + 3H_2O + 6e^- \text{ (\textit{Oxidation half-reaction})}$$

---

$$2MnO_4^- + H_2O + I^- \longrightarrow 2MnO_2 + 2OH^- + IO_3^-$$

(*ii*) $NH_3(N=-3)$ *to* $N_2(N=0)$

$$2KMnO_4 + H_2O \longrightarrow 2MnO_2 + 2KOH + 3O \text{ (Reduction reaction)}$$

$$2NH_3 + 3O \longrightarrow N_2 + 3H_2O \text{ (\textit{Oxidation reaction})}$$

---

$$2KMnO_4 + 2NH_3 \longrightarrow 2MnO_2 + 2KOH + N_2 + 2H_2O$$

OR

$$2\times[MnO_4^- + 2H_2O + 3e^- \longrightarrow MnO_2 + 4OH^-] \text{ (\textit{Reduction half - reaction})}$$

$$2NH_3 + 6OH^- \longrightarrow N_2 + 6H_2O + 6e^- \text{ (\textit{Oxidation half - reaction})}$$

---

$$2MnO_4^- + 2NH_3 \longrightarrow 2H_2O + 2OH^- + N_2 + 2MnO_2$$

(*iii*) *Nitrotoluene to nitrobenzoic acid*

$$2KMnO_4 + H_2O \longrightarrow 2MnO_2 + 2KOH + 3O \text{ (Reduction reaction)}$$

$$C_6H_4\begin{matrix}\diagup CH_3\\ \diagdown NO_2\end{matrix} + 3O \longrightarrow C_6H_4\begin{matrix}\diagup COOH\\ \diagdown NO_2\end{matrix} + H_2O \text{ (\textit{Oxidation reaction})}$$

---

$$2KMnO_4 + \underset{\text{Nitrotoluene}}{C_6H_4\begin{matrix}\diagup CH_3\\ \diagdown NO_2\end{matrix}} \longrightarrow 2MnO_2 + 2KOH + \underset{\text{Nitrobenzoic acid}}{C_6H_4\begin{matrix}\diagup COOH\\ \diagdown NO_2\end{matrix}}$$

OR

$$2\times[MnO_4^- + 2H_2O + 3e^- \longrightarrow MnO_2 + 4OH^-]\ \text{(Reduction reaction)}$$

$$C_6H_4(CH_3)(NO_2) + 6OH^- \longrightarrow C_6H_4(COOH)(NO_2) + 4H_2O + 6e^-\ \textit{(Oxidation reaction)}$$

---

$$2MnO_4^- + C_6H_4(CH_3)(NO_2) \longrightarrow 4MnO_2 + C_6H_4(COOH)(NO_2) + 2OH^-$$

(*iv*) *Ethylene* ($H_2C = CH_2$) *to ethylene glycol* ($HO - CH_2 - CH_2 - OH$)

$$2KMnO_4 + H_2O \longrightarrow 2MnO_2 + 2KOH + 3O\ \textit{(Reduction reaction)}$$

$$3 \times [H_2C = CH_2 + O + H_2O \longrightarrow HO - CH_2 - CH_2 - OH]\ \textit{(Oxidation reaction)}$$

---

$$2KMnO_4 + 4H_2O + 3H_2C = CH_2 \longrightarrow 2MnO_2 + 2KOH + 3HO{-}CH_2{-}CH_2{-}OH$$

OR

$$2\times[MnO_4^- + 2H_2O + 3e^- \longrightarrow MnO_2 + 4OH^-]\ \textit{(Reduction reaction)}$$

$$3\times [H_2C = CH_2 + 2OH^- \longrightarrow HO{-}CH_2{-}CH_2{-}OH + 2e^-]\ \textit{(Oxidation reaction)}$$

---

$$2MnO_4^- + 4H_2O + 3H_2C = CH_2 \longrightarrow 2MnO_2 + 2OH^- + 3HO - CH_2 - CH_2 - OH$$

(*b*) **In neutral solution.** In neutral solution, $KMnO_4$ is directly reduced to $MnO_2$.

$$2KMnO_4 + H_2O \longrightarrow 2KOH + 2MnO_2 + 3O$$

or $$2MnO_4^- + H_2O \longrightarrow 2OH^- + 2MnO_2 + 3O$$

or $$MnO_4^- + 2H_2O + 3e^- \longrightarrow MnO_2 + 4OH^-$$

This equation is the same as for alkaline medium. Evidently, in *neutral medium as well the equivalent weight of* $KMnO_4$ *is one-third of its molecular weight*. It is obvious from the equation representing the oxidising action of $KMnO_4$ in neutral medium that an alkali is generated in the reaction. This alkali makes the medium alkaline even if we start with the neutral solution. It is for this reason that the equations representing the oxidising property of $KMnO_4$ in alkaline and neutral media are the same. Thus in neutral medium, $KMnO_4$ oxidises:

(*i*) *Manganous salt (e.g.* $MnSO_4$*) to insoluble* $MnO_2$ ($Mn^{2+} \longrightarrow Mn^{4+}\ O_2$)

$$2KMnO_4 + H_2O \longrightarrow 2KOH + 2MnO_2 + 3O$$

$$[MnSO_4 + O + H_2O \longrightarrow MnO_2 + H_2SO_4] \times 3$$

$$2KOH + H_2SO_4 \longrightarrow K_2SO_4 + 2H_2O$$

---

$$2KMnO_4 + 4H_2O + 3MnSO_4 + H_2SO_4 \longrightarrow 5MnO_2 + 3H_2SO_4 + K_2SO_4 + 2H_2O$$

or $$2KMnO_4 + 2H_2O + 3MnSO_4 \longrightarrow 5MnO_2 + 2H_2SO_4 + K_2SO_4$$

OR

$$[MnO_4^- + 2H_2O + 3e^- \longrightarrow MnO_2 + 4OH^-] \times 2 \text{ (Reduction reaction)}$$

$$[Mn^{2+} + 2H_2O \longrightarrow MnO_2 + 4H^+ + 2e^-] \times 3 \text{ (Oxidation reaction)}$$

$$2MnO_4^- + 10H_2O + 3Mn^{2+} \longrightarrow 5MnO_2 + 8OH^- + 12H^+$$

$$\text{or } 2MnO_4^- + 10H_2O + 3Mn^{2+} \longrightarrow 5MnO_2 + 8H_2O + 4H^+$$

$$\text{or } 2MnO_4^- + 2H_2O + 3Mn^{2+} \longrightarrow MnO_2 + 4H^+$$

*(ii) Sodium thiosulphate, $Na_2S_2O_3$ (S = +2) to $Na_2SO_4$ (S = +6) [$S_2O_3^{2-} \rightarrow SO_4^{2-}$]*

$$[2KMnO_4 + H_2O \longrightarrow 2KOH + 2MnO_2 + 3O] \times 4 \text{ (Reduction reaction)}$$

$$[Na_2S_2O_3 + 2KOH + 4O \longrightarrow Na_2SO_4 + K_2SO_4 + H_2O] \times 3 \text{ (Oxidation reaction)}$$

$$8KMnO_4 + H_2O + 3Na_2S_2O_3 \longrightarrow 8MnO_2 + 2KOH + 3Na_2SO_4 + 3K_2SO_4$$

OR

$$[MnO_4^- + 2H_2O + 3e^- \longrightarrow MnO_2 + 4OH^-] \times 8$$

$$[5H_2O + S_2O_3^{2-} \longrightarrow 2SO_4^{2-} + 10H^+ + 8e^-] \times 3$$

$$8MnO_4^- + 16H_2O + 15H_2O + 3S_2O_3^{2-} \longrightarrow 8MnO_2 + 32OH^- + 6SO_4^{2-} + 30H^+$$

$$\text{or } 8MnO_4^- + 31H_2O + 3S_2O_3^{2-} \longrightarrow 8MnO_2 + 30H_2O + 2OH^- + 6SO_4^{2-}$$

$$8MnO_4^- + H_2O + 3S_2O_3^{2-} \longrightarrow 2MnO_2 + 2OH^- + 6SO_2^{2-}$$

*(iii) Nitrogen dioxide, $NO_2$ (N = +4) to $HNO_3$ (N = +5) ($NO_2 \rightarrow NO_3^-$]*

$$2KMnO_4 + H_2O \longrightarrow 2KOH + 2MnO_2 + 3O \text{ (Reduction reaction)}$$

$$[2NO_2 + H_2O + O \longrightarrow 2HNO_3] \times 3 \text{ (Oxidation reaction)}$$

$$2KMnO_4 + 4H_2O + 6NO_2 \longrightarrow 2KOH + 2MnO_2 + 6HNO_3$$

OR

$$MnO_4^- + 2H_2O + 3e^- \longrightarrow MnO_2 + 4OH^- \text{ (Reduction reaction)}$$

$$[NO_2 + H_2O \longrightarrow NO_3^- + 2H^+ + e^-] \times 3 \text{ (Oxidation reaction)}$$

$$MnO_4^- + 5H_2O + 3NO_2 \longrightarrow MnO_2 + 4OH^- + 3NO_3^- + 6H^+$$

$$\text{or } MnO_4^- + 5H_2O + 3NO_2 \longrightarrow MnO_2 + 4H_2O + 3NO_3^- + 2H^+$$

$$\text{or } MnO_4^- + H_2O + 3NO_2 \longrightarrow MnO_2 + 3NO_3^- + 2H^+$$

(*c*) **In acidic medium.** In acidic medium (dil $H_2SO_4$), $MnO_4^-$ ion is reduced to $Mn^{2+}$ ion and five O-atoms are rendered available from two $MnO_4^-$ ions.

$$2KMnO_4 + 3H_2SO_4 (\textit{Dilute}) \longrightarrow K_2SO_4 + 2MnSO_4 + 3H_2O + 5O$$

$$\text{or } \underset{\substack{\text{Purple} \\ (Mn=+7)}}{2MnO_4^- + 6H^+} \longrightarrow \underset{\substack{\text{Colourless} \\ (Mn=+2)}}{2Mn^{2+}} + 3H_2O + 5O$$

The reduction of $MnO_4^-$ ion into $Mn^{2+}$ ion as shown above can also be represented by the following ionic equation:

$$\underset{\substack{\text{Purple}\\(\text{Mn}=+7)}}{2MnO_4^-} + 8H^+ + 5e^- \longrightarrow \underset{\substack{\text{Colourless}\\(\text{Mn}=+2)}}{2Mn^{2+}} + 4H_2O$$

It is obvious that *in acid solution the equivalent weight of $KMnO_4$ is one-fifth of its molecular weight.*

In acidic medium $KMnO_4$ oxidises:

(*i*) *KI to $I_2$ ($I^- \rightarrow I_2$)*

$$2KMnO_4 + 3H_2SO_4 \longrightarrow K_2SO_4 + 2MnSO_4 + 3H_2O + 5O$$

$$[2KI + H_2SO_4 + O \longrightarrow K_2SO_4 + H_2O + I_2] \times 5$$

$$\underset{\text{O.A}}{2KMnO_4} + \underset{\text{R.A}}{10KI} + 8H_2SO_4 \longrightarrow \underset{\text{R.A}}{2MnSO_4} + \underset{\text{OA}}{5I_2} + 8H_2O + 6K_2SO_4$$

OR

$$2\times[MnO_4^- + 8H^+ + 5e^- \longrightarrow Mn^{2+} + 4H_2O]\ (\textit{Reduction half-reaction})$$

$$5\times[2I^- \longrightarrow I_2 + 2e^-]\ (\textit{Oxidation half-reaction})$$

$$\underset{\text{O.A.}}{2MnO_4^-} + 16H^+ + \underset{\text{R.A}}{10I^-} \longrightarrow \underset{\text{R.A}}{2Mn^{2+}} + 8H_2O + \underset{\text{O.A}}{5I_2}$$

(*ii*) *Ferrous salts to ferric salts ($Fe^{2+} \rightarrow Fe^{3+}$)*

$$2KMnO_4 + 3H_2SO_4 \longrightarrow K_2SO_4 + 2MnSO_4 + 2MnSO_4 + 3H_2O + 5O$$

$$[2FeSO_4 + H_2SO_4 + O \longrightarrow Fe_2(SO_4)_3 + H_2O] \times 5$$

$$\underset{\text{O.A.}}{2KMnO_4} + 8H_2SO_4 + \underset{\text{R.A.}}{10FeSO4} \longrightarrow K_2SO_4 + \underset{\text{R.A.}}{2MnSO_4} + \underset{\text{O.A.}}{5Fe_2(So_4)_3} + H_2O$$

OR

$$MnO_4^- + 8H^+ + 5e^- \longrightarrow Mn^{2+} + 4H_2O\ (\textit{Reduction half - reaction})$$

$$5\times[Fe^{2+} \longrightarrow Fe^{3+} + e^-]\ (\textit{Oxidation half-reaction})$$

$$\underset{\substack{\text{Purple}\\\text{O.A.}}}{MnO_4^-} + 8H^+ + \underset{\text{R.A}}{5Fe^{2+}} \longrightarrow \underset{\substack{\text{Colourless}\\\text{R.A}}}{Mn^{2+}} + \underset{\text{O.A}}{5Fe^{3+}} + 4H_2O$$

The above reaction has been utilised to determine the strength of a given ferrous salt solution (*e.g.*, ferrous ammonium sulphate solution) by titrating its known volume with pre-standardised $KMnO_4$ solution. In this titration $KMnO_4$ solution acts both as an intermediate solution and as a self indicator. When all $MnO_4^-$ ions (purple) have been reduced to $Mn^{2+}$ ions (colourless) by $Fe^{2+}$ ions, the mixture solution becomes colourless due to the production of colourless $Mn^{2+}$ ions. Now if more quanity of $KMnO_4$ solution is added to $Fe^{2+}$ ions solution, the colour of the reaction mixture becomes pink due to the pink colour of the added $KMnO_4$ solution. This state indicates end point, *i.e.*, end point is indicated by the production of pink colour.

(*iii*) *Oxalic acid ($H_2C_2O_4$) or oxalate (($C_2O_4^{2-}$) to $CO_2$*

$$2KMnO_4 + 3H_2SO_4 \longrightarrow K_2SO_4 + 2MnSO_4 + 3H_2O + 5O$$

$$[H_2C_2O_4 + O \longrightarrow 2CO_2 + H_2O] \times 5$$

$$\underset{\text{O.A. (Mn=+7)}}{2KMnO_4} + 3H_2SO_4 + \underset{\text{R.A.(C=+3)}}{5H_2C_2O_4} \longrightarrow K_2SO_4 + \underset{\text{R.A. (Mn=+2)}}{2MnSO_4} + 8H_2O + \underset{\text{O.A.(C=+4}}{10CO_2}$$

OR

$$2\times[MnO_4^- + 8H^+ + 5e^- \longrightarrow Mn^{2+} + 4H_2O] \text{ (Reduction half-reaction)}$$

$$5\times[C_2O_4^{2-} \longrightarrow 2CO_2 + 2e^-] \text{ (Oxidation half - reaction)}$$

$$\underset{\substack{\text{O.A.}\\ \text{purple}}}{2MnO_4^-} + 16H^+ + \underset{\text{R.A.}}{5C_2O_4^{2-}} \longrightarrow \underset{\substack{\text{R.A.}\\ \text{Colourless}}}{2Mn^{2+}} + 8H_2O + \underset{\text{O.A.}}{10CO_2}$$

The above reaction has been used to determine the strength of a given solution of oxalic acid or an oxalate by titrating its known volume with pre-standardised $KMnO_4$ solution. $KMnO_4$ acts both as an intermediate solution and as a self indicator.

The above reaction has also been used to detect $C_2O_4^{2-}$ ion when present alone or present along with $F^-$ ion in a given inorganic mixture. The procedure is that $Na_2CO_3$ extract is acidified with $CH_3COOH$ and then $CaCl_2$ solution is added so that a mixture of $CaC_2O_4$ and $CaF_2$ (both are white precipitates) is obtained.

$$Na_2C_2O_4 + CaCl_2 \longrightarrow \underset{\text{White ppt.}}{CaC_2O_4} + 2NaCl$$

$$NaF + CaCl_2 \longrightarrow \underset{\text{White ppt.}}{CaF_2} + 2NaCl$$

Both $CaC_2O_4$ and $CaF_2$ are sparingly soluble in $CH_3COOH$. This mixture is put into dil. $H_2SO_4$ when $CaC_2O_4$ is converted into oxalic acid ($H_2C_2O_4$), but $CaF_2$ remains as such. Now when a few frops of $KMnO_4$ solution (purple) are added to the solution containing $H_2C_2O_4$, its purple colour gets discharged due to the formation of colourless $Mn^{2+}$ ions. These ions are formed by the reduction of $MnO_4^-$ ions (purple) to $Mn^{2+}$ ions (colourless) by $C_2O_4^{2-}$ ions as shown above.

(*iv*) *$H_2S$ to S* ($S^{2-} \rightarrow S^0$)

$$2KMnO_4 + 3H_2SO_4 \longrightarrow K_2SO_4 + 2MnSO_4 + 3H_2O + 5O$$

$$[H_2S + O \longrightarrow H_2O + S]\times 5$$

$$\underset{\text{O.A.}}{2KMnO_4} + 3H_2SO_4 + \underset{\text{R.A.}}{5H_2S} \longrightarrow 2MnSO_4 + 8H_2O + 5S + K_2SO_4$$

OR

$$2+[MnO_4^- + 8H^+ + 5e^- \longrightarrow Mn^{2+} + 4H_2O] \text{ (Reduction half - reaction)}$$

$$5\times[S^2 \longrightarrow S^0 + 2e^-] \text{ (Oxidation half-reaction)}$$

$$\underset{\text{O.A.}}{2MnO_4^-} + 16H^+ + \underset{\text{R.A.}}{5S^{2-}} \longrightarrow 2Mn^{2+} + 8H_2O + 5S^0 \downarrow$$

(*v*) *Nitrite to nitrate* ($NO_2^- \rightarrow NO_3^-$)

$$2KMnO_4 + 3H_2SO_4 \longrightarrow K_2SO_4 + 2MnSO_4 + 3H_2O + 5O$$

$$[KNO_2 + O \longrightarrow KNO_3]\times 5$$

$$\underset{\text{O.A}}{2KMnO_4} + 3H_2SO_4 + \underset{\text{R.A.}}{5KNO_2} \longrightarrow K_2SO_4 + 2MnSO_4 + 3H_2O + 5KNO_3$$

OR

$$2\times[MnO_4^- + 8H^+ + 5e^- \longrightarrow Mn^{2+} + 4H_2O]\ (\textit{Reduction reaction})$$
$$5\times[NO_2^- + H_2O \longrightarrow NO_3^- + 2H^+ + 2e^-]\ (\textit{Oxidation reaction})$$

$$\underset{\text{O.A.}}{2MnO_4^-} + 6H^+ + \underset{\text{R.A.}}{5NO_2^-} \longrightarrow 2Mn^{2+} + 3H_2O + 5NO_3^-$$

*(vi) Arsenite, $AsO_3^{3-}$ (As = +3) to arsenate, $AsO_4^{3-}$ (As = + 5)*

$$2KMnO_4 + 3H_2SO_4 \longrightarrow K_2SO_4 + 2MnSO_4 + 3H_2O + 5O$$
$$[Na_3AsO_3 + O \longrightarrow Na_3AsO_4]\times 5$$

$$2KMnO_4 + 3H_2SO_4 + 5Na_3AsO_3 \longrightarrow K_2SO_4 + 2MnSO_4 + 3H_2O + 5Na_3AsO_4$$

OR

$$[MnO_4^- + 8H^+ + 5e^- \longrightarrow Mn^{2+} + 4H_2O]\times 2(\textit{Reduction reaction})$$
$$[AsO_3^{3-} + H_2O \longrightarrow AsO_4^{3-} + 2H^+ + 2e^-]\times 5(\textit{Oxidation reaction})$$

$$2MnO_4^- + 6H^+ + 5AsO_3^{3-} \longrightarrow 2Mn^{2+} + 3H_2O + 5AsO_4^{3-}$$

*(vii) Sulphite, $SO_3^{2-}$ (S = +4) to sulphate, $SO_4^{2-}$ (S = + 6)*

$$2KMnO_4 + 3H_2SO_4 \longrightarrow K_2SO_4 + 2MnSO_4 + 3H_2O + 5O$$
$$[Na_2SO_3 + O \longrightarrow Na_2SO_4]\times 5$$

$$\underset{\text{O.A.}}{2KMnO_4} + 3H_2SO_4 + \underset{\text{R.A.(S=+4)}}{5Na_2SO_3} \longrightarrow K_2SO_4 + 2MnSO_4 + 3H_2O + \underset{\text{(S = +6)}}{5Na_2SO_4}$$

OR

$$[MnO_4^- + 8H^+ + 5e^- \longrightarrow Mn^{2+} + 4H_2O]\times 2(\textit{Reduction reaction})$$
$$[SO_3^{2-} + H_2O \longrightarrow SO_4^{2-} + 2H^+ + 2e^-]\times 5\ (\textit{Oxidation reaction})$$

$$2MnO_4^- + 6H^+ + 5SO_3^{2-} \longrightarrow 2Mn^{2+} + 3H_2O + 5SO_4^{2-}$$

*(viii) Halogen acid, HX (X = –1) to the correspongding halogen, $X_2$ (X=0) ($X^- \rightarrow X_2$)*

$$KMnO_4 + 3H_2SO_4 \longrightarrow K_2SO_4 + 2MnSO_4 + 3H_2O + 5O$$
$$[2HX + O \longrightarrow H_2O + X_2] \times 5$$

$$\underset{\text{O.A.}}{2KMnO_4} + 3H_2SO_4 + \underset{\text{R.A. (X=–1)}}{10HX} \longrightarrow K_2SO_4 + 2MnSO_4 + 8H_2O + \underset{\text{(X = 0)}}{5X_2}$$

OR

$$2\times[MnO_4^- + 8H^+ + 5e^- \longrightarrow Mn^{2+} + 4H_2O]\ (\textit{Reduction half-reaction})$$
$$5\times[2X^- \longrightarrow X_2 + 2e^-]\ (\textit{Oxidation reaction})$$

$$2MnO_4^- + 16H^+ + 10X^- \longrightarrow 2Mn^{2+} + 8H_2O \times 5X_2$$

The oxidation of HCl by $KMnO_4$ has been utilised for the preparation of $Cl_2$ gas. $KMnO_4$ alone can also oxidise conc. HCl to $Cl_2$.

$$2KMnO_4 + 16HCl \longrightarrow 2MnCl_2 + 2KCl + 8H_2O + 5Cl_2$$

*(ix) $H_2O_2$ (O = –1) to $O_2$ (O = 0)*

$$2KMnO_4 + 3H_2SO_4 \longrightarrow K_2SO_4 + 2MnSO_4 + 3H_2O + 5O$$
$$[H_2O_2 + O \longrightarrow H_2O + O_2] \times 5$$

$$\underset{\text{O.A.}}{2KMnO_4} + 3H_2SO_4 + \underset{\text{R.A.}(O=-1)}{5H_2O_2} \longrightarrow K_2SO_4 + 2MnSO_4 + 8H_2O + \underset{(O=0)}{5O_2}$$

OR

$$2[MnO_4^- + 8H^+ + 5e^- \longrightarrow Mn^{2+} + 4H_2O] \times 2 \text{ (Reduction half-reaction)}$$
$$5 \times [H_2O_2 \longrightarrow O_2 + 2H^+ + 2e^-] \text{ (Oxidatio half-reaction)}$$

$$2[MnO_4^- + 6H^+ + 5H_2O_2 \longrightarrow 2Mn^{2+} + 8H_2O + 5O_2$$

(*x*) *$SO_2$ (S = +4) to $H_2SO_4$ (S = +6) ($SO_2 \rightarrow SO_4^{2-}$)*

$$KMnO_4 + 3H_2SO_4 \longrightarrow K_2SO_4 + 2MnSO_4 + 3H_2O + 5O$$
$$[SO_2 + H_2O + O \longrightarrow H_2SO_4] \times 5$$

$$2KMnO_4 + 5SO_2 + 2H_2O \longrightarrow K_2SO_4 + 2MnSO_4 + 2H_2SO_4$$

OR

$$2\times[MnO_4^- + 8H^+ + 5e^- \longrightarrow Mn^{2+} + 4H_2O] \text{ (Reduction reaction)}$$
$$5\times[SO_2 + 2H_2O \longrightarrow SO_4^{2-} + 4H^+ + 2e^-] \text{ (Oxidationr reaction)}$$

$$2MnO_4^- + 5SO_2 + 2H_2O \longrightarrow 2Mn^{2+} + 5SO_4^{2-} \times 4H^+$$

(*xi*) *Hydrazine, $N_2H_4$ (N = –2) to $N_2$ (N = 0)*

$$[2KMnO_4 + 3H_2SO_4 \longrightarrow K_2SO_4 + 2MnSO_4 + 3H_2 + 5O] \times 2$$
$$[N_2H_4 + 2O \longrightarrow N_2 + 2H_2O] \times 5$$

$$4KMnO_4 + 6H_2SO_4 + 5N_2H_4 \longrightarrow 2K_2SO_4 + 4MnSO_4 + 16H_2O + 5N_2$$

OR

$$4\times[MnO_4^- + 8H^+ + 5e^- \longrightarrow Mn^{2+} + 4H_2O] \text{ (Reduction reaction)}$$
$$5\times[N_2H_4 \longrightarrow N_2 + 4H^+ + 4e^-] \text{ (Oxidation reaction)}$$

$$4MnO_4^- + 12H^+ + 5N_2H_4 \longrightarrow 4Mn^{2+} + 16H_2O + 5N_2$$

(*xii*) *Hydrazoic acid $HN_3$ (N = –1/3) to $N_2$ (N = 0)*

$$2KMnO_4 + 3H_2SO_4 \longrightarrow K_2SO_4 + 2MnSO_4 + 3H_2O + 5O$$
$$2HN_3 + O \longrightarrow 3N_2 + H_2O] \times 5$$

$$2KMnO_4 + 3H_2SO_4 + 10\,HN_3 \longrightarrow K_2SO_4 + 2MnSO_4 + 8H_2O + 15N_2$$

OR

$$2\times[MnO_4^- + 8H^+ + 5e^- \longrightarrow Mn^{2+} + 4H_2O] \text{ (Reduction reaction)}$$
$$5\times[2HN_3 \longrightarrow 3N_2 + 2H^+ + 2e^-] \text{ (Oxidation reaction)}$$

$$2MnO_4^- + 6H^+ + 10HN_3 \longrightarrow 2Mn^{2+} + 8H_2O + 15N_2$$

(*xiii*) *Nitric oxide, NO (N = +2) to $HNO_3$ (N = +5)*

$$[2KMnO_4 + 3H_2SO_4 \longrightarrow K_2SO_4 + 2MnSO_4 + 3H_2O + 5O] \times 3$$
$$[2NO + 3O + H_2O \longrightarrow 2HNO_3] \times 5$$
$$6KMnO_4 + 9H_2SO_4 + 10NO \longrightarrow 4H_2O + 3K_2SO_4 + 6MnSO_4 + 10HNO_3$$

OR

$$3\times[MnO_4^- + 8H^+ + 5e^- \longrightarrow Mn^{2+} + 4H_2O] \text{(Reduction reaction)}$$
$$5\times[NO + 2H_2O \longrightarrow HNO_3 + 3H^+ + 3e^-] \text{(Oxidation reaction)}$$
$$3MnO_4^- + 9H^+ + 5NO \longrightarrow 2H_2O + 3Mn^{2+} + 5HNO_3$$

(*xiv*) *Potassium ferrocyanide, $K_4[Fe^{2+}(CN)_6]$ to potassium ferricyanide, $K_3[Fe^{3+}(CN)_6]$*

$$2KMnO_4 + 3H_2SO_4 \longrightarrow K_2SO_4 + 2MnSO_4 + 3H_2O + 5O$$
$$[2K_4[Fe^{2+}(CN)_6] + O + H_2O \longrightarrow 2K_3[Fe^{3+}(CN)_6] + 2KOH] \times 5$$
$$2KMnO_4 + 3H_2SO_4 + 10K_4[Fe^{2+}(CN)_6] \longrightarrow 2H_2O + K_2SO_4 + 2MnSO_4 + 10K_3[Fe^{3+}(CN)_6] + 10KOH$$

OR

$$[MnO_4^- + 8H^+ + 5e^- \longrightarrow Mn^{2+} + 4H_2O] \text{(Reduction reaction)}$$
$$5\times[(Fe^{2+}(CN)_6]^{4-} \longrightarrow [Fe^{3+}(CN)_6]^{3-} + e^- \text{(Oxidation reaction)}$$
$$MnO_4^- + 8H^+ + 5[Fe^{2+}(CN)_6]^{4-} \longrightarrow Mn^{2+} + 4H_2O + 5[Fe^{3+}(CN)_6]^{3-}$$

(*xv*) *Sod.thiosulphate, $Na_2S_2O_3$(S = + 2) to sod. dithionate, $Na_2S_2O_6$ (S = + 5)*

$$[2KMnO_4 + 3H_2SO_4 \longrightarrow K_2SO_4 + 2MnSO_4 + 3H_2O + 5O] \times 3$$
$$[Na_2S_2O_3 + 3O \longrightarrow Na_2S_2O_6] \times 5$$
$$6KMnO_4 + 6H_2SO_4 + 5Na_2S_2O_3 \longrightarrow 3K_2SO_4 + 6MnSO_4 + 9H_2O + 5Na_2S_2O_6$$

OR

$$6\times[MnO_4^- + 8H^+ + 5e^- \longrightarrow Mn^{2+} + 4H_2O] \text{(Reduction reaction)}$$
$$5\times[S_2O_3^{2-} + 3H_2O \longrightarrow S_2O_6^{2-} + 6H^+ + 6e^-] \text{(Oxidation reaction)}$$
$$6MnO_4^- + 18H^+ + 5S_2O_3^{2-} \longrightarrow 9H_2O + 6Mn^{2+} + 5S_2O_6^{2-}$$

**Uses of $KMnO_4$**

$KMnO_4$ is used: (*i*) as an oxidising agent in the laboratory as well as in industry; (*ii*) in volumetric estimations of ferrous salts, oxalic acid, $H_2O_2$ etc.; (*iii*) as a disinfectant and a germicide on account of its oxidising property; (*iv*) in qualitative analysis for detecting halides and oxalates; (*v*) in the manufacture of saccharin, benzoic acid, acetaldelyde etc.; (*vi*) for bleaching waxes; (*vii*) for the detection of double bond in organic compounds. For this purpose alkaline $KMnO_4$ solution under the name *Bayer's reagent* is used; (*viii*) as a source of reactive form of $MnO_2$ which is used in place of pyrolusite in the manufacture of dry cells.

**Structure of Permanganate Ion, $MnO_4^-$.**

In $MnO_4^-$ ion, Mn is present as $Mn^{7+}$ ion. $Mn^{7+}$ ion in $MnO_4^-$ ion undergoes $sp^3$ hybridisation and hence $MnO_4^-$ ion has tetrahedral geometry. (Fig. 6.25). In this ion four O-atoms are arranged tetrahedrally around Mn atom. Since $Mn^{7+}$ ion has $d^0$ configuration, the deep purple colour of $KMnO_4$ does not arise due to *d–d* transition. Purple colour arises from charge-transfer from O to Mn. This charge transfer reduces the O.S. of Mn from +7 to +6 momentarily. Since this charge transfer takes place from O to Mn, it is called ligand → metal charge transfer.

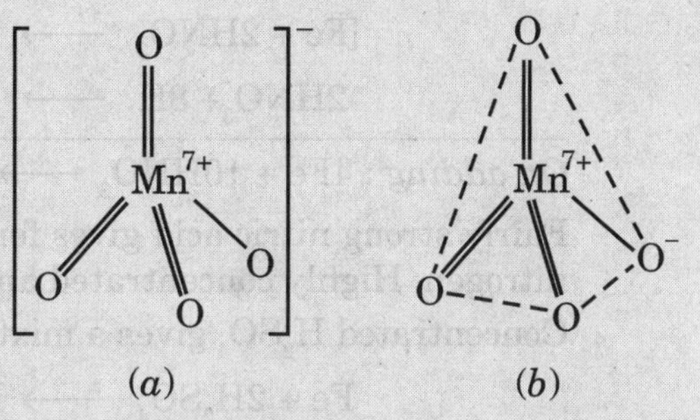

**Fig. 6.25** (*a*) Structure of permanganate ion, $MnO_4^-$ (*b*) Tetrahedral geometry of $MnO_4^-$ ion

## Iron, $Fe_{26}$

## Physical Properties of Iron

Pure iron possesses a grey colour and is malleable, ductile and a good conductor of heat and electricity. It is the most magnetic of all metals but loses this property above 766°C. It melts at 1536°C, boils at 3000°C and its density is 7.86.

### Chemical Properties

1. **Action of air or oxygen.** When iron is strongly heated in air or oxygen, ferrosoferric oxide (tri ferric tetroxide), $Fe_3O_4$ is formed.

$$3Fe + 2O_2 \longrightarrow Fe_3O_4$$

Iron wire burns when heated in oxygen or in oxy-coal gas flame.

2. **Action of moisture: Rusting.** Iron when exposed to moisture or water, or allowed to remain damp, is covered with a reddish yellow film or compound. This is known as *rust* and the phenomenon is known as *rusting*. Rust is a hydrated iron oxide with an approximate chemical composition $2Fe_2O_3 \cdot 3H_2O$. Pure iron does not rust in pure air or pure water.

3. **Action of water (steam).** Iron decomposes steam.

$$3Fe + 4H_2O \rightleftharpoons Fe_3O_4 + 4H_2$$

4. **Action of acids.** Since the value of $E^o_{ox}$ for the oxidation half-reaction,

$Fe \longrightarrow Fe^{2+} + 2e^-$ is positive (=0.44V), Fe is easily oxidised to $Fe^{2+}$ by dil. acids forming $H_2$ and Fe (II) salts.

$$Fe + 2H^+ \longrightarrow Fe^{2+} + H_2$$

Thus Fe reacts with dil. $H_2SO_4$ and HCl to form Fe(II) salt and $H_2$. For example:

$$Fe + H_2SO_4 \longrightarrow FeSO_4 + H_2$$

$$Fe + 2HCl \longrightarrow FeCl_2 + H_2$$

Dilute nitric acid gives a mixture of ferrous nitrate and ammonium nitrate.

$$[Fe + 2HNO_3 \longrightarrow Fe(NO_3)_2 + 2H] \times 4$$

$$2HNO_3 + 8H \longrightarrow NH_4NO_3 + 3H_2O$$

*On adding* : $4Fe + 10HNO_3 \longrightarrow 4Fe(NO_3)_2 + NH_4NO_3 + 3H_2O$

Fairly strong nitric acid gives ferric nitrate and a mixture of the oxides of nitrogen. Highly concentrated and pure nitric acid makes it passive.

Concentrated $H_2SO_4$ gives a mixture of ferrous and ferric sulphates.

$$Fe + 2H_2SO_4 \longrightarrow FeSO_4 + SO_2 + 2H_2O$$

$$2FeSO_4 + 2H_2SO_4 \longrightarrow Fe_2(SO_4)_3 + 2H_2O + SO_2$$

5. **Action of alkalies, halogens and sulphur.** Alkalies have no action of iron. Halogens and sulphur combine with the heated metal forming halides and sulphide respectively.

$$2Fe + 3Cl_2 \longrightarrow 2FeCl_3$$

$$Fe + S \longrightarrow FeS$$

6. **Displacement of less electropositive metals.** For example, it displaces copper from a solution of copper sulphate.

$$Fe + Cu^{2+} \longrightarrow Fe^{2+} + Cu$$

7. **Passivity.** Concentrated nitric acid renders iron *passive*. When dipped in strong nitric acid, chromic acid, chloric acid or hydrogen peroxide, iron becomes chemically inert and does not react further, *i.e.*, it does not liberate hydrogen with dilute acids. When scratched or hammered the activity is restored. The passivity of iron is due to the formation of very thin invisible coating of oxide on the metal surface.

## Electronic Configuration and Oxidation States of Fe

The V.S.E.C. of Fe is $3d^6\ 4s^2$. Fe shows –2, 0, +1, +2, +3, +4, +5 and +6 oxidation states in its compounds. +2 and +3 are the common oxidation states and –2, 0 and +1 are the unusual oxidation states. –2 oxidation state is shown by Fe in the complex ion, $[Fe^{-2}(CO)_4]^{2-}$. Zero oxidation state is exhibited by Fe in its carbonyls like $Fe(CO)_5$, $Fe_2(CO)_9$ and $Fe_3(CO)_{12}$. +1 oxidation state is shown by Fe in $[Fe^{+}(H_2O)_5(NO^{+})]^{2+}$ in which NO is present as $NO^{+}$. Fe is present in +4 oxidation state in potassium tetraoxoferrate (IV), $K_4[FeO_4]$.

## Chemistry of Fe (II)

Compounds of Fe in +2 oxidation state are called **ferrous compounds.** Compunds of Fe (+2) give $Fe^{2+}$ ions (ferrous ions) in aqueous solution. This ion is pale green. All Fe(+2) salts are, therefore, pale green in colour. In aqueous solution, $Fe^{2+}$ ion can readily be oxidised to $Fe^{3+}$ ion even by the atmospheric $O_2$.

$$4Fe^{2+} + O_2 + 4H^+ \longrightarrow 4Fe^{3+} + 2H_2O$$

It is due to this property that $Fe^{2+}$ ion is a strong reducing agent. Thus this ion reduces $Hg^{2+}$ to $Hg_2^{2+}$, $Sn^{4+}$ to $Sn^{2+}$, $Au^{3+}$ to Au, acidified solution of $MnO_4^-$ to $Mn^{2+}$, $Cr_2O_7^{2-}$ to $Cr^{3+}$, $HNO_3$ to NO etc.

$$2Hg^{2+} + 2Fe^{2+} \longrightarrow Hg_2^{2+} + 2Fe^{3+}$$

$$Sn^{4+} + 2Fe^{2+} \longrightarrow Sn^{2+} + 2Fe^{3+}$$

$$Au^{3+} + 3Fe^{2+} \longrightarrow Au + 3Fe^{3+}$$

$$MnO_4^- + 8H^+ + 5Fe^{2+} \longrightarrow Mn^{2+} + 7H_2O + 6Fe^{3+}$$

$$Cr_2O_7^{2-} + 14H^+ + 6Fe^{2+} \longrightarrow 2Cr^{3+} + 7H_2O + 3Fe^{3+}$$

$$NO_3^- + 4H^+ + 3Fe^{2+} \longrightarrow NO + 2H_2O + 3Fe^{3+}$$

Examples of simple compounds containing Fe in +2 oxidation state are FeO, $Fe(OH)_2$, $FeSO_4 . 7H_2O$ (Green vitriol), $FeCl_2$, FeS, $NH_4SO_4 . FeSO_4 . 6H_2O$ (Mohr's salt).

## Complexes of Fe (II)

Fe (II) which has $3d^6$ configuration forms a number of complexes. Most of these complexes are octahedral and high spin having $t_{2g}^4\ e_g^2$ configuration with $n$ = 4. Low spin octahedral complexes with $t_{2g}^6\ e_g^o$ configuration having $n$ = 0 are formed only with strong ligands like $CN^-$ and ortho-phenanthroline.

Aqua octahedral complex, $[Fe(H_2O)_6]^{2+}$ is *paramagnetic* with $n$= 2. $H_2O$ ligands present in $[Fe(H_2O)_6]^{2+}$ can be replaced by $CN^-$ ligands to form ferrocyanide ion, $[Fe(CN)_6]^{4-}$.

$Fe^{2+}$ ion forms a red coloured octahedral complex, $(Fe\ (o\text{-}phen)_3]^{2+}$ with bidentate ligand, ortho-phenanthroline (o-phen). This complex is used for calorimetric determination of Fe. Since ortho-phenanthroline is a strong ligand, the complex $[Fe(o\text{-}phen)_3]^{2+}$ is a LS octahedral complex.

The ligand, *o-phen* is also used as *ferroin* indicator in the redox titration of Fe (II) which gets converted into Fe(III) by an oxidising agent. The complex ion, $[Fe(o\text{-}phen)_3]^{2+}$ is highly stable and does not oxidise until an excess of oxidising agent is present. In other words, the red colour given by ferroin with Fe(II) does not fade unless the whole of Fe(II) is oxidised to Fe (III).

The bidentate ligand, 2, 2 bipyridyl (bipy) forms stable complex with Fe(II) whose composition is $[Fe(bipy)_3]^{2+}$. It has octahdedral geometry. With bidentate ligand, ethylene diamine (en), $Fe^{2+}$ ion forms a series of stable complexes *viz* $[Fe\ (en)_3]^{2+}$, $[Fe(en)_2\ (H_2O)_2]$ and $[Fe(en)\ (H_2O)_4]^{2+}$. These are octahedral complexes.

Some tetrahedral complexes of Fe(II) are also known. They are mostly halide complexes of the type, $[FeX_4]^{2-}$ with large cations. $[FeCl_4]^{2-}$ ion has been well characterised.

Some important complexes of Fe (II) are discussed below:

**1. Hexammine Ferrate (II) Chloride, $[Fe^{2+}(NH_3)_6]Cl_2$**

$FeCl_2$ combines with $NH_3$, giving $[Fe(NH_3)_6]Cl_2$.

$$FeCl_2 + 6NH_3 \longrightarrow [Fe(NH_3)_6]Cl_2$$

$[Fe(NH_3)_6]^{2-}$ ion has octahedral geometry.

**2. Sodium Pentacyano Nitrosyl Ferrate (II), $Na_2[Fe^{2+}(CN)_5(NO^+)].2H_2O$ (Also Called Sod. Nitroprusside)**

In this complex compound five $CN^-$ ions and one nitrosyl ion ($NO^+$) act as ligands. It is also an octahedral complex.

**Preparation.** (*i*) When a conc. solution of $K_4[Fe(CN)_6]$ is heated with 50% $HNO_3$, $K_3[Fe(CN)_6]$ along with $KNO_3$ and NO is obtained. On cooling, $KNO_3$ crystallises out and the liquor which contains $K_3[Fe(CN)_6]$ and NO is neturalised by $Na_2CO_3$ and then evaporated to crystallisation. This gives ruby-red crystals of sodium nitroprusside.

$$3K_4[Fe(CN)_6] + 4HNO_3 \longrightarrow 3K_3[Fe(CN)_6] + NO + 3KNO_3 + 2H_2O$$

or $$3[Fe(CN)_6]^{4-} + 4H^+ + NO_3^- \longrightarrow 3[Fe(CN)_6]^{3-} + NO + 2H_2O$$

$$2K_3[Fe(CN)_6] + 2NO + 3Na_2CO_3 \longrightarrow 2Na_2[Fe(CN)_5(NO)] + 2NaCN + 3K_2CO_3$$

(*ii*) It may also be prepared by the action of $NaNO_2$ on $Na_4[Fe(CN)_6]$.

$$Na_4[Fe(CN)_6] + NaNO_2 + H_2O \longrightarrow Na_2[Fe(CN)_5(NO)] + NaOH$$

**Properties.** (*i*) *Physical state*. It is a ruby-red crystalline solid which is readily soluble in water. Its aqueous solution decomposes on standing. It is, therefore, freshly prepared when required.

(*ii*) *Action of sulphides*. When a freshly prepared solution of $Na_2[Fe(CN)_5(NO)]$ is added to an alkaline solution of a sulphide (*e.g.* $Na_2S$ but not $H_2S$), a purple or violet colour is obtained due to the formation of a complex ion, $[Fe^{2+}(CN)_5(NO^+)(S^{-2})]^{4-}$ which does not last for long.

$$Na_4S + Na_2[Fe^{2+}(CN)_5(NO^+)] \longrightarrow \underset{\text{Purple or violet coloured solution}}{Na_4[Fe^{2+}(CN)_5(NO^+)(S^{-2})]}$$

or $$S^{-2} + [Fe^{2+}(CN)_5(NO^+)]^{2-} \longrightarrow [Fe^{2+}(CN)_5(NO^+)(S^{-2})]^{4-}$$

The production of purple or violet colour is used to confirm the presence of $S^{2-}$ ion in a given inorganic salt.

(*iii*) *Action of alkali sulphites and thiosulphates*. Alkali sulphites give a rose-red colour due to the formation of $Na_4[Fe(CN)_5(NO)(SO_3)]$

$$Na_2SO_3 + Na_2[Fe(CN)_5(NO)] \longrightarrow \underset{\text{Rose-red colour}}{Na_4[Fe(CN)_5(NO)(SO_3)]}$$

Alkali thiosulphates do not show this reaction. Thus the above reaction can be used to distinguish sulphites from thiosulphates.

(*iv*) *Action of $AgNO_3$*. A flesh coloured complex is formed.

$$2AgNO_3 + Na_2[Fe(CN)_5(NO)] \longrightarrow \underset{\substack{\text{Flesh-coloured complex} \\ \text{(brown ppt)}}}{Ag_2[Fe(CN)_5(NO)]} + 2NaNO_3$$

(*v*) *Action of aldehydes and ketones*. Aldehydes and ketones (*e.g.* acetone and methyl ethyl ketone) give red colour with alkaline solution of sodium nitroprusside.

(*vi*) *Action of an alkali*. $Na_2[Fe(CN)_5(NO)]$ is converted into sodium hexacyanate ferrate (II), $Na_4[Fe(CN)_6]$

$$6Na_2[Fe(CN)_5(NO)] + 14NaOH \longrightarrow 5Na_4[Fe(CN)_6] + Fe(OH)_2 + 6NaNO_3 + 6H_2O$$

**Uses.** Sodium nitroprusside is used (*i*) as a reagent in the laboratory for the detection of $S^{2-}$ ion, aldehydes and ketones, (*ii*) for distinguishing sulphites from thiosulphates.

**Structure of $[Fe^{2+}(CN)_5(NO^+)]^{2-}$ ion.** NO is present as $NO^+$ in the given complex ion and Fe is present as $Fe^{2+}$. Since C.N. of $Fe^{2+}$ ion is 6, the given complex ion has octahedral geometry. The complex is diamagnetic ($n$= 0). Thus all the six electrons of $Fe^{2+}$ ion ($Fe^{2+} = 3d^6$) are present in $t_{2g}$ orbitals, and $e_g$ orbitals remain vacant ($t_{2g}^6 e_g^0$). As a result, the octahedral geometry of the complex ion results from $d^2sp^3$ hybridisation.

### 3. Potassium Ferrocyanide or Potassium Hexacyano Ferrate (II), $K_4[Fe^{2+}(CN)_6]$

**Preparation.** (*i*) *Laboratory preparation*. In laboratory $K_4[Fe(CN)_6]$ is obtained by adding KCN to $FeSO_4$ (green vitriol) solution when a precipitate of $Fe(CN)_2$ is first formed and then gets dissolved in the excess of KCN solution and forms $K_4[Fe(CN)_6]$.

$$FeSO_4 + 2KCN \longrightarrow Fe(CN)_2\downarrow + K_2SO_4$$

$$Fe(CN)_2 + 4KCN \longrightarrow F_4[Fe(CN)_6]$$

The solution on concentration gives the lemon yellow crystals of $K_4[Fe(CN)_6]$.

(*ii*) When $K_3[Fe(CN)_6]$ is reduced by $H_2O_2$ in alkaline medium, $K_4[Fe(CN)_6]$ is obtained. In this reaction $H_2O_2$ is oxidised to $O_2$.

$$2K_3[Fe(CN)_6] + 2KOH + H_2O_2 \longrightarrow 2K_4[Fe(CN)_6] + 2H_2O + O_2$$

or $$2[Fe(CN)_6[^{3-} + 2OH^- + H_2O_2 \longrightarrow 2[Fe(CN)_6]^{4-} + 2H_2O + O_2$$

(*iii*) *Commercial method: From crude coal gas*. On large scale, $K_4[Fe(CN)_6]$ is obtained from crude gas. This gas contains 0.3% of HCN as impurity. The following two methods are used:

(*a*) Crude coal gas is passed over hydrated iron oxide [mixture of $Fe(OH)_2$ and $Fe(OH)_3$], when prussian blue, $Fe_4^{3+}[Fe^{2+}(CN)_6]_3$ is obtained.

$$Fe(OH)_2 + 2HCN \longrightarrow Fe(CN)_2 + 2H_2O$$

$$Fe(OH)_3 + 3HCN \longrightarrow Fe(CN)_3 + 3H_2O$$

$$3Fe(CN)_2 + 4Fe(CN)_3 \longrightarrow \underset{\text{Prussian blue}}{Fe_4^{3+}[Fe^{2+}(CN)_6]_3}$$

$Fe_4[Fe(CN)_6]_3$, when heated with $Ca(OH)_2$, gives calcium ferrocyanide, $Ca_2[Fe^{2+}(CN)_6]$

$$Fe_4^{3+}[Fe^{2+}(CN)_6]_3 + 6Ca(OH)_2 \longrightarrow 3Ca_2[Fe^{2+}(CN)_6] + \underset{\text{ppt}}{4Fe(OH)_3}$$

$Ca_2[Fe(CN)_6]$ is extracted with $H_2O$ and then treated with $K_2CO_3$ solution when $K_4[Fe(CN)_6]$ is obtained.

$$Ca_2[Fe(CN)_6] + 2K_2CO_3 \longrightarrow K_4[Fe(CN)_6] + \underset{ppt}{2CaCO_3}$$

$CaCO_3$ is removed by filtration and the filtrate is evaporated to get the crystals of $K_4[Fe(CN)_6]$.

(*b*) Coal gas, when passed through a washer containing KOH and $FeSO_4$, gives $K_4[Fe(CN)_6]$.

$$[HCN + KOH \longrightarrow KCN + H_2O] \times 6$$

$$6KCN + FeSO_4 \longrightarrow K_4[Fe(CN)_6] + K_2SO_4$$

$$6HCN + 6KOH + FeSO_4 \longrightarrow K_4[Fe(CN)_6] + K_2SO_4 + 6H_2O$$

**Properties.** 1. Potassium ferrocyanide forms lemon-yellow crystal which are soluble in water. This compound contains $3H_2O$ as water of crystallisation.

2. *Action of heat.* When heated to 42°C, all the three $H_2O$ molecules are lost. On further heating, it gives iron carbide ($FeC_2$), KCN, $N_2$. etc.

$$K_4Fe(CN)_6 \xrightarrow{\Delta} FeC_2 + 4KCN + N_2$$

$$K_4Fe(CN)_6 \xrightarrow{\Delta} Fe + 4KCN + (CN)_2$$

3. *Action of $H_2SO_4$.* (*a*) When $K_4Fe(CN)_6$ is heated with dil. $H_2SO_4$, hydrocyanic acid (HCN) is produced.

$$K_4Fe(CN)_6 + 3H_2SO_4(dil.) \longrightarrow 2K_2SO_4 + FeSO_4 + 6HCN$$

(*b*) On heating with conc. $H_2SO_4$, CO gas is evolved.

$$K_4Fe(CN)_6 + 6H_2SO_4(conc.) + 6H_2O$$
$$\longrightarrow 2K_2SO_4 + 3(NH_4)_2SO_4 + FeSO_4 + 6CO$$

4. *Action of Zn (II) salts.* With Zn (II) salts [*e.g.* $(CH_3COO)_2$ Zn, $ZnSO_4$ etc.] bluish-white ppt. of zinc ferro-cyanide, $Zn_2[Fe^{2+}(CN)_6]$ is obtained, *e.g.*,

$$2(CH_3COO)Zn + K_4[Fe(CN)_6] \longrightarrow Zn_2[Fe(CN)_6] + 4CH_3COOK$$

$$2ZnSO_4 + K_4[Fe(CN)_6] \longrightarrow Zn_2[Fe(CN)_6] + 2K_2SO_4$$

or $$2Zn^{2+} + [Fe(CN)_6]^{4-} \longrightarrow \underset{\text{Bluish white ppt}}{Zn_2[Fe^{2+}(CN)_6]}$$

The above reaction has been used in the identification of $Zn^{2+}$ ion present in a given inorganic salt.

**5.** *Action of Fe(III) salts: Formation of blue precipitate of prussian blue.* See Prussian blue discussed on the subsequent pages of this chapter.

**6.** *Action of Fe (II) salts.* When $K_4[Fe(CN)_6]$ solution is added to the solution of a Fe(II) salt, bluish-white ppt. of pot. ferrous ferro hexacyanide, $K_2Fe^{2+}[Fe^{2+}(CN)_6]$ is obtained.

$$K_4[Fe^{2+}(CN)_6] + Fe^{2+}Cl_2 \longrightarrow \underset{\text{Bluish–white ppt.}}{K_2Fe^{2+}[Fe^{2+}(CN)_6]} + 2KCl$$

or $$[Fe(CN)_6]^{4-} + Fe^{2+} \longrightarrow [Fe^{2+}\{Fe^{2+}(CN)_6\}]^{2-}$$

The bluish-white ppt. of $K_2Fe[Fe(CN)_6]$ slowly becomes blue due to its oxidation by air to Prussian blue, $KFe^{3+}[Fe^{2+}(CN)_6]$

$$K_2Fe^{2+}[Fe^{2+}(CN)_6] \xrightarrow{\text{Air oxidation}} \underset{\text{Prussian blue}}{KFe^{3+}[Fe^{2+}(CN)_6]} + K$$

**7.** *Action of Cu (II) salts.* Cu (II) salts (*e.g.*, $CuSO_4$) give a red-brown ppt. of cupric ferro hexacyanide, $Cu_2^{2+}[Fe^{2+}(CN)_6]$.

$$2CuSO_4 + K_4[Fe(CN)_6] \longrightarrow Cu_2[Fe^{2+}(CN)_6] + 2K_2SO_4$$

$$\text{or} \quad 2Cu^{2+} + [Fe(CN)_6]^{4-} \longrightarrow \underset{\text{Red-brown ppt}}{Cu_2^{2+}[Fe^{2+}(CN)_6]}$$

This reaction is used for the detection of $Cu^{2+}$ ions present in an inorganic salt.

**8.** *Action of Ag (I) salts.* White ppt. of silver ferro cyanide, $Ag_4[Fe(CN)_6]$ is obtained.

$$K_4[Fe(CN)_6] + 4AgNO_3 \longrightarrow Ag_4[Fe(CN)_6] + 4KNO_3$$

$$\text{or} \quad [Fe(CN)_6]^{4-} + 4Ag^+ \longrightarrow \underset{\text{White ppt}}{Ag_4[Fe(CN)_6]}$$

**9.** *Action of Ca(II) salt + $NH_4Cl$.* When a Ca (II) salt (*e.g.*, $CaCl_2$) solution containing some $NH_4Cl$ is treated with $K_4[Fe(CN)_6]$, a white ppt. of cal. ammonium pot. ferro-cyanide, $Ca(NH_4)K[Fe(CN)_6]$ is obtained.

$$CaCl_2 + NH_4Cl + K_4[Fe(CN)_6] \longrightarrow 3KCl + \underset{\text{White ppt}}{Ca(NH_4)K[Fe(CN)_6]}$$

$$\text{or} \quad Ca^{2+} + NH_4^+ + K^+ + [Fe(CN)_6]^{4-} \longrightarrow Ca(NH_4)K[Fe(CN)_6]$$

This reaction is used in the detection of $Ca^{2+}$ ions present in an inorganic salt.

**10.** *Action of fused $K_2Cr_2O_7$.* When fused $K_2Cr_2O_7$ reacts with dry $K_4Fe(CN)_6$, potassium cyanate (KCNO) is produced.

$$2K_2Cr_2O_7 + 2K_4[Fe(CN)_6] \longrightarrow 6KCNO + 2K_2O + 2Cr_2O_3 + K_2Fe^+[Fe^{3+}(CN)_6]$$

**11.** *Action $K_2CO_3$ and sulphur.* When $K_4Fe(CN)_6$ is heated with a mixture of ($K_2CO_3$ +S), potassium thiocyanate (KCNS) is obtained.

$$K_4Fe(CN)_6 + K_2CO_3 + 6S \xrightarrow{\Delta} 6KCNS + CO_2 + FeO$$

**12.** *Reducing properties: Oxidation of $K_4[Fe^{2+}(CN)_6]$ to $K_3[Fe^{3+}(CN)_6]$.* It is a reducing agent and hence (*a*) reduces many oxidising agents like $Cl_2$, $O_3$, acidified solution of $H_2O_2$ and $KMnO_4$ to chloride ($Cl^-$), $O_2$, $H_2O$ and $MnSO_4$ respectively. In all these reactions $K_4[Fe^{2+}(CN)_6]$ is oxidised to $K_3[Fe^{3+}(CN)_6]$.

$$2K_4[Fe^{2+}(CN)_6] + Cl_2 \longrightarrow 2KCl + 2K_3[Fe^{3+}(CN)_6].$$

$$\text{or} \quad 2[Fe^{2+}(CN)_6]^{4-} + Cl_2^0 \longrightarrow 2Cl^- + 2[Fe^{3+}(CN)_6]^{3-}$$

$$2K_4[Fe(CN)_6] + O_3 + H_2O \longrightarrow 2K_3[Fe(CN)_6] + 2KOH + O_2$$

$$\text{or} \quad 2[Fe(CN)_6]^{4-} + O_3 + H_2O \longrightarrow 2[Fe(CN)_6]^{3-} + 2OH^- + O_2$$

$$2K_4[Fe(CN)_6] + H_2SO_4 + H_2O_2 \longrightarrow 2K_3[Fe(CN)_6] + K_2SO_4 + 2H_2O$$

$$\text{or} \quad 2[Fe(CN)_6]^{4-} + 2H^+ + H_2O_2 \longrightarrow 2[Fe(CN)_6]^{3-} + 2H_2O$$

$$5K_4[Fe(CN)_6] + 4H_2SO_4 + KMnO_4 \longrightarrow 5K_3[Fe(CN)_6] + MnSO_4 + 4H_2O$$

or $5\,[Fe(CN)_6]^{4-} + 8H^+ + MnO_4^- \longrightarrow 5[Fe(CN)_6]^{3-} + MN^{2+} + 4H_2O$

(*b*) Electrolytic oxidation of slightly alkaline solution of $K_4Fe(CN)_6$ gives $K_3Fe(CN)_6$

$$2K_4\,[Fe(CN)_6] + 2H_2O \longrightarrow 2K_3\,[Fe(CN)_6] + 2KOH + H_2$$

or $$2[Fe(CN)_6]^{4-} + 2H_2O \longrightarrow 2[Fe(CN)_6]^{3-} + 2OH^- + H_2$$

(*c*) When a concentrated solution of $K_4Fe(CN)_6$ is heated with 50% $HNO_3$, $K_4Fe(CN)_6$ gets oxidised to $K_3Fe(CN)_6$ and $HNO_3$ is reduced to NO.

$$3K_4\,[Fe(CN)_6] + 4HNO_3 \longrightarrow 3K_3[Fe(CN)_6] + NO + 3KNO_3 + 2H_2O$$

or $$3[Fe(CN)_6]4^- + 4H^+ + NO_3^- \longrightarrow 3[Fe(CN)_6]^{3-} + NO + 2H_2O$$

(*d*) When a solution of $K_4[Fe\,(CN)_6]$ and $KNO_2$ is added to dil. $CH_3COOH$, it is oxidised to $K_3[Fe(CN)_6]$ and $KNO_2$ is reduced to NO. This reaction is used for the preparation of NO.

$$2K_4[Fe(CN)_6] + KNO_2 + 2CH_3COOH \longrightarrow 2K_3[Fe(CN)_6] + 2CH_3COOK + H_2O + NO$$

or $$[Fe(CN)_6]^{4-} + NO_2^- + 2H^+ \longrightarrow [Fe(CN)_6]^{3-} + NO + H_2O$$

**Uses.** $K_4Fe\,(CN)_6$ is used:

(*i*) for the preparation of CO gas

(*ii*) for the preparation of Prussian blue pigments.

(*iii*) for the manufacture of $K_3[Fe(CN)_6]$ as shown under its reducing properties.

(*iv*) as a reagent in laboratory for the detection of $Cu^{2+}$, $Fe^{3+}$, $Zn^{2+}$ and $Ca^{2+}$ ions present in an inorganic salt.

(*v*) to distinguish between ferrous and ferric salts.

(*vi*) for making blue points.

**Structure of $[Fe^{2+}(CN)_6]^{4-}$ ion.** This ion has octahedral geometry which arises from $d^2sp^3$ hybridisation of $Fe^{2+}$ ion ($Fe^{2+} = 3d^6$). Thus it is an inner orbital octahedral complex. According to CFT the configuration of $Fe^{2+}$ ion is $t_{2g}^6 e_g^0\,(n = 0)$. Thus this ion is diamagnetic. For detailed discussion of the structure see chapter on *Valence Bond Theory of Coordination Compounds.*

**4. Potassium Ferrous Ferro Hexacyanide, $K_2Fe^{2+}[Fe^{2+}\,(CN)_6]$**

It is obtained as a bluish-white ppt. by adding Fe(II) salt (*e.g.*, $FeCl_2$) to $K_4\,[Fe(CN)_6]$

$$FeCl_2 + K_4\,[Fe\,(CN)_6] \longrightarrow K_2Fe[Fe(CN)_6] + 2KCl$$

$K_2Fe[Fe(CN)_6]$ is oxidised by air to Prussian blue, $KFe^{3+}\,[Fe^{2+}\,(CN)_6]$

$$K_2Fe[Fe(CN)_6 \xrightarrow{\text{Air oxidation}} KFe[Fe(CN)_6] + K$$

## Chemistry of Fe (III)

Compounds of Fe in +3 oxidation state are called **ferric compounds.**

**Preparation.**

These compounds are obtained when the corresponding Fe(+2) compounds are oxidised.

**Properties.**

(*i*) **Hydrolysis.** Due to its small size and high charge density, $Fe^{3+}$ ion undergoes hydrolysis to give $H^+$ ions. The production of $H^+$ ions makes the aqueous solution of $Fe^{3+}$ ions acidic.

$$Fe^{3+} + H_2O \rightleftharpoons [Fe(OH)]^{2+} + H^+$$

The value of equilibrium constant of the above reaction at ordinary temperature has been found to be equal to $9 \times 10^{-4}$.

(*ii*) ***Ionisation.*** Fe(+3) compounds are much less ionised than Fe(+2) compounds. For example:

(*a*) $FeCl_3$ is so little ionised in aqueous solution that this solution does not give any test for $Fe^{3+}$ and $F^-$ ions.

(*b*) It is also due to little ionisation of $Fe(OH)_3$ that this compound is less basic than $Fe(OH)_2$.

(*iii*) ***Reduction.*** $Fe^{3+}$ ion can be reduced to $Fe^{2+}$ ion by reducing agents like H, $H_2S$, $Sn^{2+}$ etc.

$$Fe^{3+} + H \longrightarrow Fe^{2+} + H^+$$

$$2Fe^{3+} + H_2S \longrightarrow 2Fe^{2+} + 2H^+ + S$$

$$2Fe^{3+} + Sn^{2+} \longrightarrow 2Fe^{2+} + Sn^{4+}$$

Examples of salts containing $Fe^{3+}$ ion are $Fe_2O_3$, Fe $(OH)_3$, $FeF_3$, $FeCl_3$, $Fe_2S_3$, $Fe_2(SO_4)_3$, $(NH_4)_2$ $SO_4$. $Fe_2(SO_4)_3 \cdot 24H_2O$ (ammonium iron alum), $K_2SO_4$. $Fe_2(SO_4)_3$. $24H_2O$ (potassium iron alum) etc.

## Complexes of Fe (III)

$Fe^{3+}$ ion forms even a larger number of complexes than $Fe^{2+}$ ion. Most of the complexes have C.N. = 6 and hence are octahedral. $Fe^{3+}$ ion ($3d^5$ ion) is high spin in most of its octahedral complexes. High spin arrangement of five 3*d* electrons of $Fe^{3+}$ ion gives $t_{2g}^3 e_g^2$ configuration to $Fe^{3+}$ ion with $n = 5$. Any $d-d$ transition will therefore be spin-forbidden, since that involves both promoting an electron as well as reversing its spin. Consequently,the absorption spectra is expected to consist of weak spin forbidden bands in the visible region of the spectrum. Hexaaqua $[Fe(H_2O)_6]^{3+}$ ion is thus pale yellow in colour, even in the absence of any complexing ions. The magnetic moments are close to the spin-only value of 5.9 BM.

Strong ligands such as $CN^-$, $SCN^-$, bipy, etc., form low spin complexes. The magnetic moments are low, have considerable orbital contribution and are temperature dependent. These complexes show intense colours from $d-d$ transitions.

Iron (III) forms a variety of cationic, neutral and anionic complexes. An interesting feature of its coordination chemistry is marked preference for O-donor as opposed to N-donor ligands. Majority of its complexes are octahedral, but tetrahedral and square planar complexes are also known.

Iron (III) has very low affinity for ammine ligands. Ammines of Fe(III) are unstable and dissociate in water. Addition of aqueous ammonia to Fe(III) precipitates the hydroxide. Chelating ligands such as 2, 2′. bipyridine and 1, 10-phenanthroline which produce ligand fields strong enough to induce spin pairing form stable complexes.

Fe (III) exhibits greater affinity for O-donor ligands. With oxalate the tri-oxalate complex $[Fe(C_2O_4)_3]^{3-}$ is formed. It is green, very stable towards dissociation but is photosensitive due to oxidation of the oxalate ion by Fe(III) and hence decomposes to $Fe(C_2O_4)$ and $CO_3$.

The blood-red colour produced by adding $SCN^-$ to an aqueous solution of Fe(III) is mainly due to the ion $[Fe(SCN)(H_2O)_5]^{2+}$. This serves as a test for the detection of Fe(III).

Halide complexes are also known. The stability decreases from $F^-$ to $Br^-$. Fluoro complexes are the most stable and the predominant species in aqueous solution is $[FeF_5(H_2O)]^{2-}$. However in the solid state higher species $[FeF_6]^{3-}$ is also known. Other halide complexes are $FeCl_5^{2-}$, $FeCl_6^{3-}$, $Fe_2Cl_9^{3-}$, $[FeBr_4]^-$, etc. In chloro-complexes, tetrahedral geometry is favoured.

$[FeF_6]^{3-}$ ion has octahedral geometry. Since $F^-$ ions are weak ligands, the configuration of $Fe^{3+}$ (= 3 $d^5$) is not disturbed in $[FeF_6]^{3-}$ ion. Thus this complex ion has five unpaired electrons and is paramagnetic. Octahedral geometry results from $sp^3d^2$ hybridisation. According to CFT the configuration of $Fe^{3+}$ ion is $t_{2g}^3e_g^2$ $(n = 5)$

A vast majority of Fe (III) complexes are high spin. Ligands such as bipy, phen, $CN^-$ cause spin pairing. The low spin $[Fe(CN)_6]^{3-}$, best known as red crystalline potassium salt, is prepared by oxidation of $[Fe(CN)_6]^{4-}$ with $Cl_2$. However, it is interesting to note that the hexacyanoferrate (II), $[Fe(CN_6]^{4-}$, is more stable than hexacyanoferrate (III) $[Fe(CN)_6]^{3-}$. In the former case $CN^-$ ions are tightly held and the complex is non-toxic, whereas in the latter labile $CN^-$ ligands render it toxic.

A few complexes of $Fe^{3+}$ with C.N. = 4 are tetrahderal. $[FeCl_4]^-$ is the most important tetrahedral complex ion. The tetrahedral complex ions, $[FeX_4]^-$ are particulary stable, if the cation attached with $[FeX_4]^-$ anion is bigger in size. Examples of such tetrahedral complexes are $[(t-Bu)_4 N]^+ [Fe X_4]^-$ where $X^- = Cl^-$, $Br^-$ and $I^-$.

Here we shall discuss $H_3 [Fe (CN)_6]$ and $K_3 [Fe (CN)_6]$ in detail.

1. **Hydroferricyanic Acid, $H_3[Fe^{3+} (CN)_6]$**

   It is obtained by decomposing $Pb_3[Fe(CN)_6]_2$ with dil. $H_2SO_4$, filtering and evaporating.

$$Pb_3[Fe(CN)_6]_2 + 3H_2SO_4 \longrightarrow 2H_3 [Fe(CN)_6] + 3PbSO_4$$

   $H_3[Fe(CN)_6]$forms brown needles.

2. **Potassium Ferricyanide or Potassium Hexacyano Ferrate (III), $K_3 [Fe^{3+} (CN)_6]$**

**Preparation.** $K_3[Fe(CN)_6]$ may be prepared by oxidising $K_4[Fe(CN)_6]$ to $K_3[Fe(CN)_6]$. The oxidation can be done by the following methods:

(*i*) by the oxidation of $K_4[Fe(CN)_6]$ by $Cl_2$, $O_3$, acidified solution of $H_2O_2$ or $KMnO_4$.

$$2K_4 [Fe(CN)_6] + Cl_2 \longrightarrow 2KCl + 2K_3[Fe (CN)_6]$$

or $$2[Fe(CN)_6]^{4-} + Cl_2 \longrightarrow 2Cl^- + 2 [Fe(CN)_6]^{3-}$$

$$2K_4[Fe(CN)_6] + O_3 + H_2O \longrightarrow 2K_3[Fe(CN)_6] + 2KOH + O_2$$

or $[Fe(CN)_6]^{4-} + O_3 + H_2O \longrightarrow 2[Fe(CN)_6]^{3-} + 2OH^- + O_2$

$2K_4[Fe(CN)_6] + H_2SO_4 + H_2O_2 \longrightarrow 2K_3[Fe(CN)_6] + K_2SO_4 + 2H_2O$

or $[Fe(CN)_6]^{4-} + 2H^+ + H_2O_2 \longrightarrow 2[Fe(CN)_6]^{3-} + 2H_2O$

$5K_4[Fe(CN)_6] + 4H_2SO_4 + KMnO_4 \longrightarrow 5K_3[Fe(CN)_6] + MnSO_4 + 4H_2O$

or $5[Fe(CN)_6]^{4-} + 8H^+ + MnO_4^- \longrightarrow 5[Fe(CN)_6]^{3-} + Mn^{2+} + 4H_2O$

(*ii*) by the electrolytic oxidation of slightly alkaline solution of $K_4[Fe(CN)_6]$

$2K_4[Fe(CN)_6] + 2H_2O \longrightarrow 2K_3[Fe(CN)_6] + H_2 + 2KOH$

or $2[Fe(CN)_6]^{4-} + 2H_2O \longrightarrow 2[Fe(CN)_6]^{3-} + H_2 + 2OH^-$

(*iii*) by heating a concentrated solution of $K_4[Fe(CN)_6]$ with 50% $HNO_3$.

$3K_4[Fe(CN)_6] + 4HNO_3 \xrightarrow{\Delta} 3K_3[Fe(CN)_6] + NO + 3KNO_3 + 2H_2O$

or $3[Fe(CN)_6]^{4-} + 4H^+ + NO_3^- \longrightarrow 3[Fe(CN)_6]^{3-} + NO + 2H_2O$

**Properties.** 1. It is a dark red crystalline solid. It is readily soluble in water and gives a yellow solution. In solution, it ionisies as:

$$K_3[Fe(CN)_6] \rightleftharpoons 3K^+ + \underset{\text{Complex ion}}{[Fe^{3+}(CN)_6]^{3-}}$$

The complex ion is very stable. This ion does not give the usual tests of $Fe^{3+}$, since six $CN^-$ ions are strongly linked with $Fe^{3+}$ in an octahedral manner.

**2.** *Oxidising properties: Reduction of* $K_3[Fe^{3+}(CN)_6]$ *to* $K_4[Fe^{2+}(CN)_6]$. $K_3[Fe(CN)_6]$ oxidises a number of substances in alkaline as well as in acidic medium and is itself reduced to $K_4[Fe(CN)_6]$.

(*i*) *In alkaline medium.* In this medium, $K_3[Fe(CN)_6]$ is reduced to $K_4[Fe(CN)_6]$, liberating nascent oxygen which oxidises the substances.

$2K_3[Fe^{3+}(CN)_6] + 2KOH \longrightarrow 2K_4[Fe^{2+}(CN)_6] + H_2O + O$

or $2[Fe^{3+}(CN)_6]^{3-} + 2OH^- \longrightarrow 2[Fe^{2+}(CN)_6]^{4-} + H_2O + O$

or $[Fe^{3+}(CN)_6]^{3-} + e^- \longrightarrow [Fe^{2+}(CN)_6]^{4-}$

Thus, $K_3[Fe(CN)_6]$ in alkaline medium oxidises $H_2S$ to S, PbO (*lithrage*) to $PbO_2$, MnO to $MnO_2$ ($Mn^{2+}O + O \rightarrow Mn^{4+}O_2$), $Cr_2O_3$ to $K_2CrO_4$, ($Cr_2^{3+}O_3 + 4KOH + O \rightarrow 2K_2Cr^{+6}O_4 + 2H_2O$), $CrCl_3$ to $K_2CrO_4$, $SO_2$ to $H_2SO_4$ ($S^{4+}O_2 + H_2O \rightarrow H_2S^{6+}O_4$), $H_2O_2$ to $O_2$ ($H_2O_2^{1-} + O \rightarrow H_2O + O_2^0$), $N_2H_4$ (hydrazine) to $N_2$ ($N_2^{2-}H_4 + 2O \rightarrow N_2^0 + 2H_2O$). As an example, let us see how the oxidation of $H_2S$ to S can be represented by various chemical equations given below:

$2K_3[Fe^{3+}(CN)_6] + 2KOH \longrightarrow 2K_4[Fe^{2+}(CN)_6] + H_2O + O$ (Reduction reaction)

$H_2S^{-2} + O \longrightarrow H_2O + S^0$ (*Oxidation reaction*)

$2K_3[Fe^{3+}(CN)_6] + 2KOH + H_2S^{-2} \longrightarrow 2K_4[Fe^{2+}(CN)_6] + 2H_2O + S^0$ (*Complete equation*)

OR

$2[Fe^{3+}(CN)_6]^{3-} + 2OH^- \longrightarrow 2[Fe^{2+}(CN)_6]^{4-} + H_2O + O$

$H_2S^{-2} + O \longrightarrow H_2O + S^0$

$2[Fe^{3+}(CN)_6]^{3-} + 2OH^- + H_2S^{-2} \longrightarrow 2[Fe^{2+}(CN)_6]^{4-} + 2H_2O + S^0$

OR

$$[Fe^{3+}(CN)_6]^{3-} + e^- \longrightarrow [Fe^{2+}(CN)_6]^{4-}] \times 2$$

$$S^{-2} \longrightarrow S^0 + 2e^-$$

$$2[Fe^{3+}(CN)_6]^{3-} + S^{-2} \longrightarrow 2[Fe^{2+}(CN)_6]^{4-} + S^0$$

Similar equations can be written for other oxidation reactions.

(*a*) $2K_3[Fe(CN)_6] + 2KOH + M^{2+}O \longrightarrow 2K_4[Fe(CN)_6] + H_2O + M^{4+}O_2$
(M = Pb, Mn)

or $2[Fe^{2+}(CN)_6]^{3-} + 2OH^- + M^{2+}O \longrightarrow 2[Fe(CN)_6]^{4-} + H_2O + M^{4+}O_2$

(*b*) $6K_3[Fe(CN)_6] + 10KOH + Cr_2^{3+}O_3 \longrightarrow 6K_4[Fe(CN)_6] + 5H_2O + 2K_2Cr^{+6}O_4$

or $6[Fe^{3+}(CN)_6]^{3-} + 10OH^- + Cr_2^{3+}O_3 \longrightarrow 6[Fe^{2+}(CN)_6]^{4-} + 5H_2O + 2[Cr^{+6}O_4]^{2-}$

(*c*) $3K_3[Fe(CN)_6] + 8KOH + CrCl_3 \longrightarrow 3K_4[Fe(CN)_6] + 4H_2O + K_2CrO_4 + 3KCl$

or $3[Fe^{3+}(CN)_6]^{3-} + 8OH^- + Cr^{3+}Cl_3 \longrightarrow 3[Fe^{2+}(CN)_6]^{4-} + 4H_2O + [Cr^{+6}O_4]^{2-} + 3Cl^-$

(*d*) $2K_3[Fe(CN)_6] + 2KOH + SO_2 \longrightarrow 2K_4[Fe(CN)_6] + H_2SO_4$

or $2[Fe^{3+}(CN)_6]^{3-} + 2OH^- + S^{4+}O_2 \longrightarrow 2[Fe^{2+}(CN)_6]^{4-} + H_2S^{6+}O_4$

(*e*) $2K_3[Fe(CN)_6] + 2KOH + H_2O_2 \longrightarrow 2K_4[Fe(CN)_6] + 2H_2O + O_2$

or $2[Fe^{3+}(CN)_6]^{3-} + 2OH^- + H_2O_2^{-1} \longrightarrow 2[Fe^{2+}(CN)_6]^{4-} + 2H_2O + O_2^0$

(*f*) $4K_3[Fe(CN)_6] + 4KOH + N_2H_4 \longrightarrow 4K_4[Fe(CN)_6] + 4H_2O + N_2$

or $4[Fe^{3+}(CN)_6]^{3-} + 4OH^- + N_2^{2-}H_4 \longrightarrow 4[Fe(CN)_6]^{4-} + 4H_2O + N_2^0$

(*ii*) *In acidic medium.* In acidic medium, $K_3[Fe(CN)_6]$ oxidises HCl to $Cl_2$ and $H_2S$ to S.

(*a*) $2K_3[Fe(CN)_6] + 8HCl \longrightarrow 2H_4[Fe(CN)_6] + 6KCl + Cl_2$

or $2[Fe(CN)_6]^{3-} + 2Cl^- \longrightarrow 2[Fe(CN)_6]^{4-} + Cl_2$

(*b*) $6K_3[Fe^{3+}(CN)_6] + 3H_2S \longrightarrow 4K_4[Fe^{2+}(CN)_6] + K_2Fe^{2+}[Fe^{2+}(CN)_6] + 6HCN + 3S$

**3.** *Action of Fe (II) salts : Formation of blue precipitate of Turnbull's blue.* See " Turnbull's Blue" discussed on the subsequent pages of this chapter.

**4.** *Action of Fe (III) salts.* Redish-brown coloured ferric ferri cyanide, $Fe^{3+}[Fe^{3+}(CN)_6]$ is obtained.

$$K_3[Fe(CN)_6] + FeCl_3 \longrightarrow Fe[Fe(CN)_6] + 3KCl$$

or $[Fe(CN)_6]^{3-} + Fe^{3+} \longrightarrow Fe[Fe(CN)_6]$
Reddish-brown

**5.** *Action on $CoCl_3$.* $K_3[Fe(CN)_6]_3$ gives a reddish-brown ppt. of $Co_3[Fe(CN)_6]_3$ with $CoCl_3$ solution.

$$3CoCl_3 + 3K_3[Fe(CN)_6] \longrightarrow Co_3^{3+}[Fe^{3+}(CN)_6]_3 + 9KCl$$

Cobaltic ferric cyanide complex (reddish-brown ppt.)

**6.** *Action of $Na_2CO_3$ in presence of NO.* When $K_3[Fe(CN)_6]$ is treated with $Na_2CO_3$ and NO gas is passed through it and the solution is evaporated to crystallisation, ruby-red crystals of sodium nitroprusside, $Na_2[Fe^{2+}(CN)_5(NO^+)]$ are obtained.

$$2K_3[Fe(CN)_6] + 3Na_2CO_3 + 2NO \longrightarrow 2Na_2[Fe(CN)_5(NO)] + 2NaCN + 3K_2CO_3$$

**Uses.** $K_3[Fe(CN)_6]$ is used:

(*i*) to detect $Fe^{2+}$ ion present in an inorganic salt.

(*ii*) for distinguishing ferrous and ferric salts.

(*iii*) as an external indicator in the determination of the strength of ferrous salt (ferrous ammonium sulphate) solution by titrating it with standard solution of $K_2Cr_2O_7$.

(*iv*) in making blue prints.

(*v*) In making sodium nitroprusside, $Na_2[Fe(CN)_5(NO)]$ which is used to detect $S^{2-}$ ion.

**Structure of $[Fe(CN)_6]^{3-}$ ion.** This ion has octahedral geometry corresponding to $d^2sp^3$ hybridisation of $Fe^{3+}$ ion. Due to presence of one unpaired electron, the ion is paramagnetic. For the detailed discussion of its structure see chapter on *Valence Bond Theory of Coordination Compounds.*

## Complex Compounds Having Iron in +2 as well as in +3 Oxidation State

Prussian blue and Turnbull's blue are the examples of complex compounds which contain $Fe^{2+}$ as well as $Fe^{3+}$ ions. These are discussed below:

### Formulae of Prussian Blue and Turnbull's Blue

Formerly it was proposed that these two substances have different composition and have different formula, but now it has been established by analysis that both these substances are the names of the same substance and hence have the same chemical formula. Some chemists say that each of them is *potassium ferric hexacyano ferrate (II)* or *potassium ferric ferro hexacyanide*, $KFe^{3+}[Fe^{2+}(CN)_6]$ while others are of the opinion that each of them is *ferric hexacyano ferrate* (II) or *ferric ferro hexacyanide*, $Fe_4^{3+}[Fe^{2+}(CN)_6]_3$. Both have blue colour but the blue colour of Turnbull's blue is less intense than the blue colour of Prussian blue.

## Preparation of Prussian Blue, $K^+Fe^{3+}[Fe^{2+}(CN)_6]$

Prussian blue having the formula, $KFe^{3+}[Fe^{2+}(CN)_6]$, is obtained as a blue precipitate by adding Fe(III) salt solution (*e.g.*, $FeCl_3$) to the solution of $K_4[Fe^{2+}(CN)_6]$.

$$Fe^{3+}Cl_3 + K_4[Fe^{2+}(CN)_6] \longrightarrow \underset{\text{Blue ppt. of Prussian blue}}{KFe^{3+}[Fe^{2+}(CN)_6]} + 3KCl$$

or $$Fe^{3+} + [Fe^{2+}(CN)_6]^{4-} \longrightarrow [Fe^{3+}\{Fe^{2+}(CN)_6\}]^-$$

Prussian blue having the formula, $Fe_4^{3+}[Fe^{2+}(CN)_6]_3$ can be prepared by the following chemical equation:

$$4FeCl_3 + 3K_4[Fe(CN)_6] \longrightarrow \underset{\text{Blue ppt. of Prussian blue}}{Fe_4^{3+}[Fe^{2+}(CN)_6]_3} + 12KCl$$

or $$4Fe^{3+} + 3[Fe^{2+}(CN)_6]^{4-} \longrightarrow Fe_4^{3+}[Fe^{2+}(CN)_6]_3$$

If the amount of Fe(III) salt solution being added to $K_4[Fe (CN)_6]$ solution is less than the equivalent amount, then the Prussian blue is obtained in its colloidal solution.

The formation of Prussian blue by the reaction between Fe(III) salt solution and $K_4 [Fe^{2+} (CN)_6]$ has been utilised for the identification of $Fe^{3+}$ ion in inorganic analysis.

### Preparation of Turnbull's Blue, $K^+Fe^{3+} [Fe^{2+}(CN)_6]$

Turnbull's blue having the formula $KFe^{3+}[Fe^{2+}(CN)_6]$ can be obtained as a blue precipitate by adding Fe(II) salt solution (*e.g.* $FeCl_2$, $FeSO_4$ etc) to the solution of $K_3 [Fe^{3+} (CN)_6]$.

$$Fe^{2+} Cl_2 + K_3[Fe^{3+}(CN)_6] \longrightarrow KFe^{3+}[Fe^{2+}(CN)_6] + 2KCl$$

$$FeSO_4 + K_3[Fe(CN)_6] \longrightarrow KFe[Fe(CN)_6] + K_2SO_4$$

Blue ppt. of Turnbull's blue

Both the above equations can be written in ionic form as:

$$Fe^{2+} + [Fe^{3+}(CN)_6]^{3-} \longrightarrow [Fe^{3+}\{Fe^{2+}(CN)_6\}]^-$$

The formation of Turnbull's blue having the formula $[Fe_4^{3+}[Fe^{2+}(CN)_6]_3$ can be shown as:

$$4FeCl_2 + 3K_3 [Fe(CN)_6] \longrightarrow Fe_4[Fe(CN)_6]_3 + 8KCl + K$$

or $$4Fe^{2+} + 3[Fe(CN)_6]^{3-} + K^+ \longrightarrow Fe_4[Fe(CN)_6]_3 + K$$

Blue ppt. of Turnbull's blue

The formation of Turnbull's blue by the action of $K_3[Fe(CN)_6]$ on Fe(II) salt is utilised to detect the presence of $Fe^{2+}$ ion in a salt. The use of $K_3[Fe(CN)_6]$ as an external indicator in the determination of the strength of FAS solution (ferrous ammonium sulphate solution) by titrating it (*i.e.*, FAS solution) with standard solution of $K_2Cr_2O_7$ is based on the reaction between Fe(II) salt solution and $K_3[Fe(CN)_6]$.

### Properties of $K^+Fe^{3+} [Fe^{2+} (CN)_6]$

(*i*) It is a deep blue crystalline solid. It is insoluble in water and gives colloidal dispersion in pure water but is coagulated by the addition of salt. It is also insoluble in dil. acids. It is soluble in oxalic acid and gives a deep blue colour which is used in the manufacture of ink.

(*ii*) *Oxidation*. When $KFe[Fe(CN)_6]$ is oxidised by means of $Cl_2$ or $HNO_3$, it gives a green pigment which is called **Berlin green,** $Fe^{3+} [Fe^{3+}(CN)_6]$ (Ferri ferri cyanide).

$$2KFe^{3+}[Fe^{2+}(CN)_6] + Cl_2 \longrightarrow 2K + Fe^{3+}[Fe^{3+}(CN)_6]$$

Berlin green

(*iii*) *Decomposition*. When treated with caustic alkalies, it gets decomposed to form potassium ferro cyanide, $K_4 [Fe (CN)_6]$ and a brown precipitate of $Fe(OH)_3$.

$$KFe [Fe(CN)_6] + 3KOH \longrightarrow K_4[Fe(CN)_6] + Fe(OH)_3\downarrow$$

Brown ppt

**Uses of K $Fe^{3+}$ [$Fe^{2+}$(CN)$_6$]**

It is used as a dye for blue print paper, for ink and as a bluing agent in laundry industry.

**Properties of $Fe_4^{3+}[Fe^{2+}(CN)_6]_3$**

(*i*) It is blue powder and has metallic lustre.

(*ii*) *Action of conc. $H_2SO_4$*. When boiled with conc. $H_2SO_4$, it gets decomposed to form CO and HCN.

(*iii*) *Action of an alkali or $Ca(OH)_2$ solution*. When boiled with an alkali or $Ca(OH)_2$. solution, it decomposes to give the precipitate of $Fe(OH)_3$.

$$Fe_4[Fe(CN)_6]_3 + 12KOH \longrightarrow 3K_4[Fe(CN)_6] + 4Fe(OH)_3 \downarrow$$

$$Fe_4[Fe(CN)_6]_3 + 6Ca(OH)_2 \longrightarrow 3Ca_2[Fe(CN)_6] + 4Fe(OH)_3 \downarrow$$

**Uses of $Fe_4^{3+}[Fe^{2+}(CN)_6]_3$**

It is employed as a permanent blue pigment which has great covering power.

## Structure of $K^+Fe^{3+}[Fe^{2+}(CN)_6]$ Molecule

It has simple cubic structure in which iron atoms occupy each corner of the cube and $CN^-$ ions occupy the middle of each side of the cube (Fig.6.26). In this arrangement, half of the iron atoms have six carbon atoms as the nearest neighbours while the other half have six nitrogen atoms as the nearest neighbours. Potassium atoms occupy only half the cubic centres.

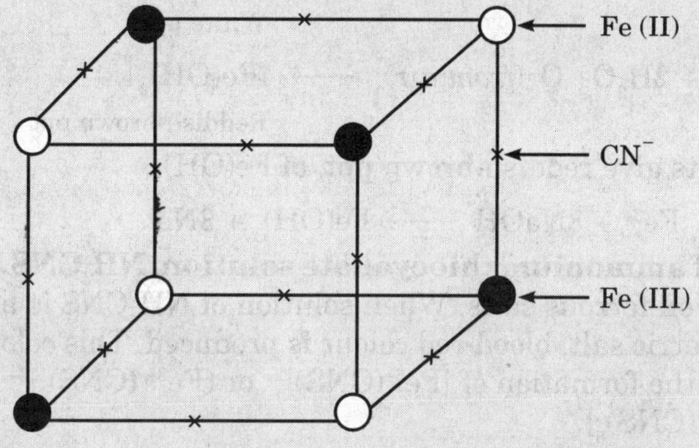


**Fig. 6.26** Structure of $KFe^{3+}[Fe^{2+}(CN)_6]$molecule (Prussian blue).

**Comparison Between Prussian Blue and Turnbull's Blue**

(*i*) Both are the same substances and hence have identical chemical formula.

(*ii*) Both have blue colour, but the blue colour of Turnbull's blue is less intense than the blue colour of Prussian blue. Less intense colour of Turnbull is due to the presence of white $KFe^{2+}[Fe^{3+}(CN)_6]$ in it. White colour of $KFe^{2+}[Fe^{3+}(CN)_6]$ makes the blue colour of Turnbull's less intense.

(*iii*) Both are obtained by different methods. Prussian blue is obtained by adding Fe(III) salt solution to that of $K_4[Fe^{2+}(CN)_6]$ while Turnbull's blue is prepared by adding Fe(II) salt solution to that of $K_3[Fe^{3+}(CN)_6]$.

**Differences Between Ferrous and Ferric Salts**

1. **Colours of the anhydrous and hydrated salts.** Anhydrous ferrous and ferric salts are white. Hydrated ferrous salts are light green while hydrated ferric salts are yellow or reddish-brown.
2. **Colour of the solution in HCl.** The solution of ferrous salts in HCl is light green while that of the solution of ferric salts in this acid is yellow.
3. **Nature of the hydroxides.** $Fe(OH)_2$ is a weak base, light green in colour and can easily be oxidised. $Fe(OH)_3$ is much weaker base and shows amphoteric character.
4. **Oxidising and reducing property.** Ferrous salts are strong reducing agents and hence can reduce acidified $KMnO_4$, $K_2Cr_2O_7$ etc. On the other hand, ferric salts are mild oxidising agents and hence can oxidise KI, $SO_2$ etc.
5. **Action of nitric oxide, NO.** When aqueous solution of a ferrous salt (*e.g.*, $FeSO_4$) reacts with NO, brown-black solution is obtained. This solution is due to the formation of nitroso ferrous cation, $[Fe^+(NO^+)]^{2+}$; *e.g.*,

$$FeSO_4 + NO \longrightarrow [Fe^+(NO^+]^{2+}\ SO_4^{2-}$$

or $$Fe^{2+} + NO \longrightarrow [Fe^+(NO^+)^{2+}$$

Ferric salt has no action on NO.

6. **Action of NaOH solution.** In the absence of air ferrous salts give a white ppt. of $Fe(OH)_2$. This ppt. is oxidised by air to reddish-brown ppt. of $Fe(OH)_3$.

$$Fe^{2+} + 2NaOH \longrightarrow \underset{\text{White ppt}}{Fe(OH)_2\downarrow} + 2Na^+$$

$$4Fe(OH)_2 + 2H_2O + O_2\,(\textit{from air}) \longrightarrow \underset{\text{Reddish-brown ppt}}{4Fe(OH)_3\downarrow}$$

Ferric salts give reddish-brown ppt. of $Fe(OH)_3$.

$$Fe^{3+} + 3NaOH \longrightarrow Fe(OH)_3 + 3Na^+$$

7. **Action of ammonium thiocyanate solution, $NH_4CNS$.** This reagent has no action on ferrous salts. When solution of $NH_4CNS$ is added to the solution of a ferric salt, blood-red colour is produced. This colour is supposed to be due to the formation of $[Fe^{3+}(CNS)]^{2+}$ or $(Fe^{3+}(CNS)_6]^{3-}$ or $Fe^{3+}(SCN)_3$ or $Fe^{3+}[Fe^{3+}(CNS)_6]$.

$$Fe^{3+} + CNS^- \longrightarrow [Fe^{3+}(CNS)]^{2+}\ (\textit{blood-red colour})$$

$$Fe^{3+} + 6CNS^- \longrightarrow [Fe^{3+}(CNS)_6]^{3-}\ (\textit{blood-red colour})$$

$$Fe^{3+} + 3SCN^- \longrightarrow Fe^{3+}(SCN)_3\ (\textit{blood-red colour})$$

$$2Fe^{3+} + 6SCN^- \longrightarrow [Fe^{3+}(Fe^{3+}(CNS)_6]\ (\textit{blood-red colour})$$

8. **Action of $K_3[Fe(CN)_6]$.** (*a*) When $K_3[Fe(CN)_6]$ solution is added to the solution of a Fe (II) salt, blue ppt. of Turnbull's blue is obtained. Turnbull's blue can be represented by the formula, $K\,Fe^{3+}[Fe^{2+}(CN)_6]$ or $Fe_4^{3+}[Fe^{2+}(CN)_6]_3$.

$$FeCl_2 + K_3[Fe(CN)_6] \longrightarrow KFe^{3+}[Fe^{2+}(CN)_6] + 2KCl$$

or $$Fe^{2+} + [Fe(CN)_6]^{3-} \longrightarrow [Fe^{3+}\{Fe^{2+}(CN)_6\}]^-$$

$$4FeCl_2 + 3K_3[Fe(CN)_6 \longrightarrow Fe_4[Fe(CN)_6]_3 + 8KCl + K$$

or $$4Fe^{2+} + K^+ + 3[Fe^{3+}(CN)_6]^{3-} \longrightarrow Fe_4^{3+}[Fe^{2+}(CN)_6]_3 + K$$

(*b*) When $K_3[Fe(CN)_6]$ solution is added to the solution of a Fe(III) salt, reddish-brown coloured ferric-ferri cyanide, $Fe^{3+}[Fe^{3+}(CN)_6]$ is obtained.

$$FeCl_3 + K_3[Fe(CN)_6] \longrightarrow Fe^{3+}[Fe^{3+}(CN)_6] + 3KCl$$

or $$Fe^{3+} + [Fe(CN)_6]^{3-} \longrightarrow \underset{\text{Reddish-brown}}{Fe^{3+}[Fe^{3+}(CN)_6]}$$

(*ix*) **Action of $K_4[Fe(CN)_6]$.** (*a*) When *$K_4[Fe(CN)_6]$* solution is added to that of a Fe(II) salt, bluish-white ppt. of ferrous ferro hexacyanide, $K_2Fe^{2+}[Fe^{2+}(CN)_6]$ is obtained.

$$K_4[Fe(CN)_6] + FeCl_2 \longrightarrow K_2Fe[Fe(CN)_6] + 2KCl$$

or $$[Fe^{2+}(CN)_6]^{4-} + Fe^{2+} \longrightarrow [Fe^{2+}\{Fe^{2+}(CN)_6\}]^{2-}$$

Bluish-white ppt.of $K_2Fe[Fe(CN)_6]$ slowly becomes blue due to its oxidation by air to Prussian blue, $KFe^{3+}[Fe^{2+}(CN)_6]$.

$$K_2[Fe^{2+}[Fe^{2+}(CN)_6] \xrightarrow{\text{Air oxidation}} KFe^{3+}[Fe^{2+}(CN)_6] + K$$

(*b*) When $K_4[Fe(CN)_6]$ solution is added to that of Fe(III) salt, blue ppt. of Prussian blue is obtained. Prussian blue may have the formula $KFe^{3+}[Fe^{2+}(CN)_6]$ or $Fe_4^{3+}[Fe^{2+}(CN)_6]_3$.

$$FeCl_3 + K_4[Fe(CN)_6] \longrightarrow KFe^{3+}[Fe^{2+}(CN)_6] + 3KCl$$

or $$Fe^{3+} + [Fe(CN)_6]^{4-} \longrightarrow Fe^{3+}[\{Fe^{2+}(CN)_6\}]^-$$

$$4FeCl_3 + 3K_4[Fe(CN)_6] \longrightarrow [Fe_4^{3+}[Fe^{2+}(CN)_6]_3 + 12KCl$$

or $$4Fe^{3+} + 3[Fe(CN)_6]^{4-} \longrightarrow Fe_4[Fe(CN)_6]_3$$

## Complex of Fe (I) : $[Fe^+(NO^+)(H_2O)_5]SO_4$

Nitrosonium pentaquo iron (I) sulphate, $[Fe^+(H_2O)_5(NO^+)]SO_4$ (also called nitroso ferrous sulphate) is an important example of the complex compound containing iron in +1 oxidation state. Five $H_2O$ molecules and one $NO^+$ ion act as ligands. It is an octahedral complex. This compound is formed in ring test which is done for nitrates and nitrites.

### Preparation

When dil. freshly prepared solution of $FeSO_4$ is added to the aqueous solution of a nitrate (*e.g.* $KNO_3$) and conc. $H_2SO_4$ is poured carefully along the sides of the test tube without disturbing the solution, brown ring of the brown complex, $[Fe(H_2O)_5(NO)]SO_4$ is formed at the junction of the two layers of nitrate and $H_2SO_4$.

$$2KNO_3 + 4H_2SO_4 + 8FeSO_4 + 6H_2O \longrightarrow K_2SO_4 + 3Fe(SO_4)_3 + \underset{\text{Brown complex}}{2[Fe(H_2O)_5(NO)]SO_4} \quad ...(i)$$

The formation of the brown complex, $[Fe(H_2O)_5(NO)]SO_4$ takes place through the following steps:

(*i*) $HNO_3$ is produced by the action $H_2SO_4$ on $KNO_3$. $HNO_3$ (N = + 5) thus produced oxidises $FeSO_4$ (Fe = +2) to $Fe_2(SO_4)_3$ (Fe = +3) in presence of $H_2SO_4$ and is itself reduced to NO (N = +2)

$$2KNO_3 + H_2SO_4 \rightarrow K_2SO_4 + 2HNO_3$$

$$6FeSO_4 + 3H_2SO_4 + 2HNO_3 \longrightarrow 4H_2O + 2NO + 3Fe_2(SO_4)_3$$

$$2KNO_3 + 6FeSO_4 + 4H_2SO_4 \longrightarrow K_2SO_4 + 2NO + 4H_2O + 3Fe_2(SO_4)_3 \quad ...(a)$$

(*ii*) NO formed as above combines with unreacted $FeSO_4$ and forms $[Fe(H_2O)_5(NO)]SO_4$

$$NO + FeSO_4 + 5H_2O \longrightarrow [Fe(H_2O)_5(NO)]SO_4 \quad ...(b)$$

The net reaction is represented by equation (*i*) can be obtained by adding equations (*a*) and (*b*) as shown below.

$$2KNO_3 + 6FeSO_4 + 4H_2SO_4 \longrightarrow K_2SO_4 + 2NO + 4H_2O + 3Fe_2(SO_4)_3$$

$$2 \times [NO + FeSO_4 + 5H_2O \longrightarrow [Fe(H_2O)_5(NO)]SO_4$$

$$2KNO_3 + 8FeSO_4 + 4H_2SO_4 + 6H_2O \longrightarrow K_2SO_4 + 3Fe_2(SO_4)_3 + 2[Fe(H_2O)_5(NO)SO_4$$

Brown complex

The formation of brown complex ion, $[Fe(H_2O)_5(NO)]^{2+}$ can be represented by the following ionic equation:

$$\underset{\text{From nitrate}}{NO_3^-} + 4H^+ + \underset{\text{From } FeSO_4}{4Fe^{2+}} + 3H_2O \longrightarrow 3Fe^{3+} + \underset{\text{Brown complex ion}}{[Fe(H_2O)_5(NO)]^{2+}}$$

The above ionic equation can be abtained by adding the following ionic equations which have been derived from molecule equations (*a*) and (*b*) respectively.

$$NO_3^- + 3Fe^{2+} + 4H^+ \longrightarrow NO + 2H_2O + 3Fe^{3+}$$

$$NO + Fe^{2+} + 5H_2O \longrightarrow [Fe(H_2O)_5(NO)]^{2+}$$

$$\underset{\text{From nitrate}}{NO_3^-} + 4H^+ + \underset{\text{From } FeSO_4}{4Fe^{2+}} + 3H_2O \longrightarrow 3Fe^{3+} + \underset{\text{Brown complex ion}}{[Fe(H_2O)_5(NO)]^{2+}}$$

The brown-black colour is due to the $Fe^+$—$NO^+$ charge transfer bands occurring in the visible region.

**Properties**

$[Fe(NO)(H_2O)_5]SO_4$ is stable only at low temperature and decomposes on heating with the evolution of NO.

$$[Fe(NO)(H_2O)_5]^{2+} \overset{\Delta}{\rightleftharpoons} Fe^{2+} + NO + 5H_2O$$

This reaction is employed for the purification of NO and its separation from other gases.

**Uses**

The formation of $[Fe(NO)(H_2O)_5]SO_4$ has been utilised in the identification of $NO_2^-$ and $NO_3^-$ ions by ring test.

**Structure**

The magnetic moment of $[Fe(NO)(H_2O)_5]^{2+}$ has been found to be equal to 3.9 B.M. This value shows that Fe has three unpaired electrons, *i.e.*, Fe is present as $Fe^+$ which has its valence-shell configuration as $3d^6 4s^1 = 3d^7 4s^0$. The presence of three unpaired electrons can be explained only when it is assumed that $Fe^+$ is $sp^3d^2$ hybridised in $[Fe(NO)(H_2O)_5]^{2+}$. Thus $[Fe(H_2O)_5(NO)]^{2+}$ ion is an outer orbital octahedral complex. NO is present as $NO^+$ ion. According to CFT the distribution of seven $3d$ electrons of $Fe^+$ ion in $t_{2g}$ and $e_g$ orbitals is $t_{2g}^5 e_g^2$ which gives $n = 3$.

## Cobalt, $Co_{27}$

**Physical Properties**

(*i*) Cobalt is a silver-white lustrous metal, slightly harder than iron. It is malleable and ductile.

(*ii*) It is more magnetic than all other metals except Fe.

(*iii*) It has a density equal to 8.9. It melts at 1495°C and boils at 2900°C. It can take up a good polish.

(*iv*) The finely-divided Co absorbs 59-153 volumes of $H_2$ and thus resembles Ni and Pd. The finely-dividied Co is pyrophoric.

**Chemical Properties**

1. **Action of air.** Dry air or moist air has no action on cobalt but the metal reacts with $O_2$ to form $Co_3O_4$ when red hot.

$$3Co + 2O_2 \xrightarrow{\text{Red hot}} Co_3O_4$$

2. **Action of steam.** Co metal decomposes steam at red heat into $H_2$.

$$Co + H_2O(\textit{steam}) \xrightarrow{\text{Red hot}} CoO + H_2$$

3. **Action of acids.** (*a*) Dil.HCl and $H_2SO_4$ form the salts and $H_2$ is evolved. *e.g.*,

$$Co + 2HCl \longrightarrow CoCl_2 + H_2$$

(*b*) Dil. $HNO_3$ dissolves the metal and renders it passive due to the formation of a protective thin layer of the metallic oxide on the surface of the metal.

4. **Action of alkalies.** Even conc. alkali has no action on the metal. The metal is dissolved in fused caustic potash at 550°C.

5. **Action of CO.** Finely-divided Co combines with CO at 150°C and under 30 atmospheric pressure to form orange-red $Co_2(CO)_8$.

6. **Action of $NH_3$.** $Co_4N_2$ is formed.

$$4Co + 2NH_3 \xrightarrow{\Delta} Co_4N_2 + 3H_2$$

7. **Action of other elements.** Cobalt readily combines with $H_2$, S, P and B but $N_2$ has no action on it.

## Uses of Cobalt and its Compounds

(*i*) Cobalt is used in electroplating.

(*ii*) Cobalt is used for the preparation of various types of alloys which have been put to many uses. For example:

(*a*) *Cobalt steel*. The alloy contains 35% of Co and is used for making permanent magnets.

(*b*) *Stellite*. It is an alloy of Co, Cr and W. It is very hard and is resistant towards corrosion. It is used for making surgical instruments.

(*c*) *Cobalt-chrome or festel metal*. It is an alloy of Co, Cr and Fe and is used for making resistance wire like ni-chrome.

(*iii*) Cobalt oxides and silicates are used for colouring glass, pottery and glazes, Thenard's blue ($CoO. Al_2O_3$), Rinmann's blue (CoO. ZnO) and smelt ($CoO. K_2O. 3SiO_2$) are used as pigments.

## Electronic Configuration and Oxidation States of Co

V.S.E.C. of Co-atom is $3d^7 4s^2$. Co shows –1, 0, +2 and +3 oxidation states in its compounds. –1 and zero are the unusual oxidation states shown by Co. Co shows –1 oxidation state in complexes formed by it with pi acid ligands. $K[Co(CO)_4]$, $[Co(CO)_3(NO)]^\circ$ and $[Co(H)(N_2)(PPH_3)_3]^\circ$ are the complexes having Co in –1 oxidation state. $[Co(H)N_2(PPH_3)_3]^\circ$ has trigonal bipyramidal geometry. Co is present in zero oxidation state in $[Co_2(CO)_8]$, $K_4[Co(CN)_4]$ and $[Co(PMe_3)_4]^\circ$. + 2 and +3 are important oxidation states of cobalt.

## Chemistry of Co (II)

Compounds of Co in +2 oxidation state are called **cobaltous compounds.** $Co^{2+}$ ion ($3d^7 4s^0$) is more stable than $Fe^{2+}$ ion ($3d^6 4s^0$), since this ion cannot easily be oxidised to $Co^{3+}$ ion. However, in basic solutions, Co (+2) compounds are not stable, since in this medium $Co^{2+}$ ion can easily be oxidised to $Co^{3+}$ ion. In aqueous solution, $Co^{2+}$ ion exists as $[Co(H_2O)_6]^{2+}$ ion and has pink colour. When heated, it turns blue due to dehydration. Thus $CoCl_2$ is blue when anhydrous and turns pink when hydrated ($CoCl_2. 6H_2O$). Most of Co(+2) compounds are soluble in water. When a base is added to a solution having $Co^{2+}$ ion, $Co(OH)_2$ is obtained as a blue ppt. Important examples of Co(+2) copounds are CoO, $Co(OH)_2$, $CoCl_2.6H_2O$, $Co(NO_3)_2.6H_2O$, $CoSO_4. 7H_2O$. $Co_3O_4$ is a combination of CoO and $Co_2O_3$. Here we shall discuss $CoO_2$, $Co_3O_4$, $Co(OH)_2$ and $CoCl_2$.

### 1. Cobaltous Oxide, CoO

**Preparation.** It is prepared:

(*i*) by heating $Co(OH)_2$

$$Co(OH)_2 \xrightarrow{\Delta} CoO + H_2O$$

(*iii*) by passing steam over red hot cobalt.

$$Co + H_2O(steam) \rightleftharpoons CoO + H_2O$$

**Properties.** (*i*) It is an olive-green powder.

(*ii*) *Action of heat*. When heated at 600–700°C in air, it gives $Co_3O_4$

$$3CoO + \frac{1}{2}O_2 \longrightarrow Co_3O_4$$

(*iii*) *Basic nature.* CoO is a basic oxide and hence gives cobaltous salts ($Co^{2+}$) with acids.

(*iv*) *Combination with oxides.* CoO combines with various oxides to form coloured compounds. For example, CoO combines with $Al_2O_3$ and forms *cobalt metal aluminate,* $CoO.Al_2O_3$ or $Co(AlO_2)_2$ (also called *Thenard's blue*). Similarly, CoO gives *cobalt zincate*, CoO.ZnO or $CoZnO_2$ (also called *Rinmann's green*) with ZnO. *Smelt*, CoO. $K_2O.3SiO_2$ can be obtained by mixing the three oxides. These coloured substances are used as pigments. The formation of these coloured double oxides makes the basis of cobalt nitrate test.

**Uses.** CoO is used (*i*) in making enamels and also for decorating china ware blue, (*ii*) in the preparation of pigments like Thenard's blue, Rinmann's green etc.

**2. Cobalto-cobaltic Oxide or Co (+2, +3) Oxide, $Co_3O_4$ or $CoO.Co_2O_3$**

**Preparation.** It is prepared:

(*i*) by strongly heating (below 1000°C) cobaltous nitrate

(*ii*) by heating CoO at 600 – 700°C in air

$$3CoO + \frac{1}{2}O_2 \longrightarrow Co_3O_4$$

(*iii*) by heating cobaltic oxide, $Co_2O_3$.

(*iv*) by heating $Co(OH)_2$ in presence of air

$$6Co(OH)_2 + O_2 \ (\textit{From air}) \longrightarrow 2Co_3O_4 + 6H_2O$$

**Properties.** It is black powder. Unlike $Fe_3O_4$, it is non-magnetic. It dissolves in acids to form cobaltous salts, *e.g.*

$$Co_3O_4 + 8HCl \longrightarrow 3CoCl_2 + 4H_2O + Cl_2$$

**Structure.** $Co_3O_4$ has normal spinal structure and is represented as $Co^{2+}O.Co_2^{3+}O_3$ .$Co^{2+}$ ions are present in tetrahedral and $Co^{3+}$ ions are present in octahedral sites created due to *ccp* arrangement of oxide ($O^{2-}$) ions.

**3. Cobaltous Hydroxide, $Co(OH)_2$.**

This compound is obtained as a blue ppt. by adding a base to the solution containing $Co^{2+}$ ions.

$$Co^{2+} + 2OH^- \longrightarrow Co(OH)_2 \downarrow$$

When $Co(OH)_2$ is heated in presence of air, cobalt (II) + cobalt (III) oxide, $Co_3O_4$ is obtained.

$6Co(OH)_2 + O_2 \rightarrow 2Co_3O_4 + 6H_2O$

**4. Cobaltous Chloride, $CoCl_2$.**

When a solution of CoO in HCl is crystallised, ruby-red monoclinic crystals of hydrated salt. $CaCl_2.6H_2O$ are obtained. When $CoCl_2.6H_2O$ is heated, it changes colour. This property of changing colour is used in synthetic inks. Anything written with $CoCl_2$ solution on white paper is invisible ordinarily but becomes visible on heating.

## Complexes of Co (II)

Co (II) forms a number of complexes which are not so stable as the corresponding Co(III) complexes. Co (II) forms stable 4- and 6-coordinate complexes. These complexes have tetrahedral and octahedral geometry respectively. Some square planar and 5-coordinate complexes are also known.

When a solution of $NH_4SCN$ is added to an aqueous solution containing $Hg^{2+}$ and $Co^{2+}$ ions, $Hg[Co(NCS)_4]$ is obtained. This complex has tetrahedral geometry and has blue colour. It is used for calibrating magnetic balance.

Most octahedral and tetrahedral cobalt (II) complexes are paramagnetic. They are more labile. Low-spin octahedral complexes such as $[Co(CN)_6]^{4-}$ are unstable to oxidation. A number of cobalt (II) complexes can be isolated without being oxidised provided the solutions are alkaline and the ligand is not a strong field ligand.

Acetylacetone forms the orange $[Co(acac)_2(H_2O)_2]$ which has an octahedral structure. This can be dehydrated to $[Co(acac)_2]$ which attains octahedral coordination by forming a tetrameric species $[Co(acac)_2]_4$ whose structure is shown in Fig. 6.27. [Co edta) $(H_2O)]^{2+}$ is analogous to the 7-coordinate Mn (II) and Fe(II) complexes of similar type but cobalt is only 6-coordinate. edta filling only five coordinate positions due to one of its oxygen atoms being far away and hence unstable to coordinate. The sixth position is filled by $H_2O$. $[Co(NH_3)_6]^{2+}$ is easily oxidised in solution to the low-spin $[Co(NH_3)_6]^{3+}$ by oxygen of air. Bidentate N-donor ligands such as en, bipy form octahedral cationic complexes such as $[Co(L–L)_3]^{2+}$ which are more stable to oxidation.

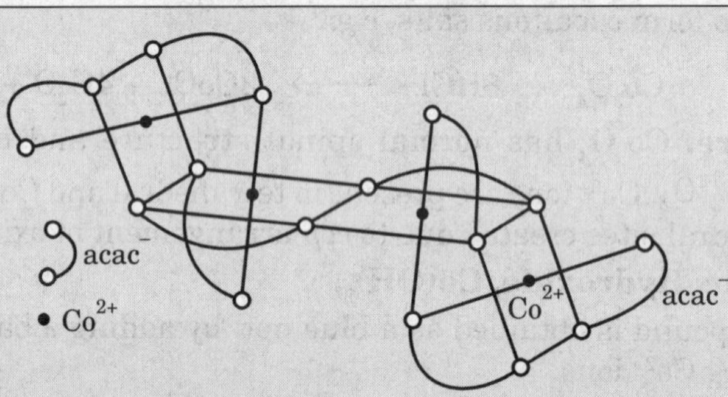


**Fig. 6.27** Tetrameric structure of $[Co(acac)_2]_4$

Cobalt (II) forms more tetrahedral complexes than any other transition metal ion, except zinc (II). This is due to the fact that a $d^7$ configuration is favourable for the adoption of a tetrahedral than an octahedral geometry. However, the difference in the stability between the two geometries of cobalt (II) is small so that in aqueous solutions of $[Co(H_2O)_4]^{2+}$ there is always some $[Co(H_2O)_6]^{2+}$ in equilibrium with it.

The anionic complexes of Co(II) have a tetrahedral structure, $[CoX_4]^{2-}$ ($X=Cl^-$, $Br^-$, $I^-$, $CNS^-$). These are blue or violet. Neutral complexes $[CoL_2X_2]$ are known. A very characteristic distinction between the octahedral and tetrahedral complexes is that of colour. Octahedral complexes are generally pink whereas the tetrahedral ones are blue. For example, $[CoCl_4]^{2-}$ ion absorbs in the red portion of the visible region and the observed colour is blue. $[Co(H_2O)_6]^{2+}$, on the other hand, is pink.

## Structure of $[Co^{2+}(NO_2)_6]^{4-}$ and $[Co^{2+}(CN)_6]^{4-}$ ions

Both these complex ions are the complexes of Co(II). These ions are paramagnetic corresponding to the presence of one unpaired electron and have octahedral geometry. The V.S.E.C of $Co^{2+}$ ion in its free state is $3d^7$. The distribution of seven $3d$ electrons in $t_{2g}$ and $e_g$ orbitals is $t_{2g}^5 e_g^2 (n = 3)$. When $Co^{2+}$ ion gets coordinated with six strong field ligands $(NO_2^- \text{ and } CN^-)$ to form the octahedral complex ions, one of the two electrons from $e_g$ orbitals gets shifted to $t_{2g}$ orbitals and the remaining one $e_g$ electron goes to occupy the *higher energy* 5s orbital. Thus the configuration of $Co^{2+}$ ion in the given complex ions becomes $t_{2g}^6\, e_g^o\, 5s^1$ ($n$ = 1). The vacant two $3d$ orbitals of $e_g$ set are used in $d^2\, sp^3$ hybridisation. Thus both the complex ions are low spin ($n$ = 1) inner orbital octahedral complex ions. Both the complex ions are labile and hence are easily oxidised to $[Co^{3+}(NO_2)_6]^{3-}$ and $[Co^{3+}(CN)_6]^{3-}$ respecitvely by air and $H_2O_2$.

$$[Co^{2+}L_6]^{4-} (L = NO_2^- \text{ or } CN^-) \xrightarrow{\text{Oxidation}} \left[Co^{3+}L_6\right]^{3-} + e^-$$

The oxidation of $[Co^{2+}L_6]^{4-}$ ions to $[Co^{3+}L_6]^{3-}$ ions takes place, since $5s$ orbital is of very high energy and hence 5s electron can easily be lost during oxidation. Since $[Co\,L_6]^{4-}$ ions can easily be oxidised by air, these complexes should be prepared in an inert atmosphere.

## Chemistry of Co (III)

Compounds of Co in +3 oxidation state are called cobaltic compounds. $Co^{3+}$ ion is a strong oxidising agent, since it gets reduced to $Co^{2+}$ ion easily because of the fact that the value of E° for the reduction half-reaction:

$$Co^{3+}(aq) + e^- \rightarrow Co^{2+}(aq)$$

is quite high, being equal to 1.84 V. Thus $Co^{3+}$ can oxidise $H_2O$ to $O_2$ rapidly.

$$4Co^{3+} + 2H_2O \longrightarrow 4Co^{2+} + O_2(g) + 4H^+$$

Important simple salts of $Co^{3+}$ are $CoF_3$, $Co_2(SO_4)_3 . 8H_2O$ etc.

## Complexes of Co (III)

We have already said that $Co^{3+}$ ion is unstable, since it gets reduced easily to $Co^{2+}$ ion. $Co^{3+}$ ion can be made stable by complexing it. It is then not easily reduced. Cobalt (III) forms numerous complexes. Almost all cobalt (III) complexes are octahedral, and diamagnetic. Only $[Co(F_6]^{3-}$ is high spin and paramagnetic. Compared to cobalt (II), complexes in this oxidation state are kinetically inert.

Cobalt (III) exhibits affinity for N-donor ligands and majority of its complexes are with ammonia, en, $NO_2^-$ group, nitrogen bonded NCS groups, etc. Many cation complexes are known for cobalt (III) in which $NH_3$ groups are partly replaced by other ligands, *e.g.*, $[Co(NH_3)_5(H_2O]^{3+}$ (red), *cis*- $[Co(NH_3)_4Cl_2]^+$ (blue), *trans isomer* is green.

Complexes of Co(III) are kinetically inert and therefore indirect methods of preparation are employed. Generally the ligand is added to an aqueous solution of cobalt (II) and the Co (II) complex thus formed is oxidised by an appropriate oxidising agent (mostly oxygen by drawing a current of air through the solution for several hours).

Interaction of Co(II) salt solution and KCN in the presence of an oxidant yields a pale yellow $K_3[Co(CN)_6]$ which is extremely stable and non-toxic. It has a $t_{2g}^6 e_g^\circ$ configuration and is diamagnetic.

Co (III) complexes with O-donor ligands are generally less stable than those with N-donors, However, $[Co(acac)_3]$, $[Co(oxinate)_3]$ are quite stable.

Co (III) has a tendency to form polynuclear complexes. A number of polynuclear ammine complexes are known which have $OH^-$ (hydroxo), $O_2^{2-}$ (peroxo), $—NH_2^-$ (amido) and $—NH^{2-}$ (imido) groups as bridging ligands.

Examples of important complexes formed by Co(III) are $Na_3[Co(NO_2)_6]$ $K_3[Co(NO_2)_6]$, $K_3[Co(CN)_6]$, $K_3[Co(en)_3]$, $[Co(NH_3)_6]Cl_3$ etc. These are octahedral complexes and are highly stable.

Here we shall discuss $Na_3[Co(NO_2)_6]$, $K_3[Co(NO_2)_6]$ and $[Co(NH_3)_6]Cl_3$.

**1. Sodium Hexanitrocobaltate (III) or Sodium Cobaltinitrite, $Na_3[Co(NO_2)_6]$**

**Preparation.** This compound is obtained as a yellow precipitate by the action of Co(II) salt (*e.g.*, $CoCl_2$) on $NaNO_2$ in presence of $CH_3COOH$.

$$CoCl_2 + 7NaNO_2 + 2CH_3COOH \longrightarrow 2CH_3COONa + 2NaCl + NO + H_2O + \underset{\text{Yellow ppt.}}{Na_3[Co(NO_2)_6]} \quad ...(A)$$

or $$Co^{2+} + 3Na^+ + 7NO_2^- + 2H^+ \longrightarrow Na_3[Co(NO_2)_6] + H_2O + NO$$

Actually the above reaction takes place through the following steps:

(*a*) $$CoCl_2 + 2NaNO_2 \longrightarrow Co(NO_2)_2 + 2NaCl$$

(*b*) $$CH_3COOH + NaNO_2 \longrightarrow CH_3COONa + HNO_2$$

$HNO_2$ produced in reaction (*b*) oxidises $Co(NO_2)_2$ formed in reaction (*a*) to $Co(NO_2)_3$ and is itself reduced to NO.

(*c*) $$\underset{(N=+3)}{2HNO_2} + \underset{(Co=+2)}{Co(NO_2)_2} \longrightarrow \underset{(Co=+3)}{Co(NO_2)_3} + \underset{(N=+2)}{NO} + H_2O$$

$Co(NO_2)_3$ formed in reaction (*c*) combines with $NaNO_2$ to form yellow ppt. of $Na_3[Co(NO_2)_6]$

(*d*) $$Co(NO_2)_3 + 3NaNO_2 \longrightarrow \underset{\text{Yellow ppt.}}{Na_3[Co(NO_2)_6]}$$

Now multiply equation (*b*) by 2 and then add all the equations to get equation (*A*).

On treating $Na_3[Co(NO_2)_6]$ with alcohol, yellow crystals of $Na_3[Co(NO_2)_6]$ are obtained.

**Properties.** (*i*) It forms yellow crystals which are soluble in water. Its aqueous solution ionises as:

$$Na_3[Co(NO_2)_6] \rightleftharpoons 3Na^+ + [Co(NO_2)_6]^{3-}$$

(*ii*) *Action of ammonium and potassium salts.* With aqueous solution of ammonium and potassium salts, this compound gives a yellow ppt. of ammonium cobaltinitrite, $(NH_4)_3[Co(NO_2)_6]$ and potassium cobalinitrite (called *fischer's salt*), $K_3[Co(NO_2)_6]$ respectively. These complexes are slightly soluble in $H_2O$.

$$\underset{}{Na_3[Co(NO_2)_6]} + \underset{(M=NH_4,K)}{3MCl} \longrightarrow \underset{\text{Yellow ppt}}{M_3[Co(NO_2)_6]} + 3NaCl$$

or $$[Co(NO_2)_6]^{3-} + 3M^+ \longrightarrow \underset{\text{Yellow ppt}}{M_3[Co(NO_2)_6]}$$

The above reaction is used for the detection and estimation of $K^+$ and $NH_4^+$ ions.

**Uses.** $Na_3[Co(NO_2)_6]$ is used (*i*) for the detection and estimation of $K^+$ and $NH_4^+$ ions, (*ii*) for the preparation of $K_3[Co(NO_2)_6]$ (*Fischer's salt*) which is an important compound.

**Structure of $[Co(NO_2)_6]^{3-}$ ion.** In $[Co(NO_2)_6]^{3-}$ ion, Co is present as $Co^{3+}$ ion whose V.S.E.C. is $3d^6$. The given complex ion has octahedral geometry. Magnetic studies of the complex ion have shown that this ion has no unpaired electrons and hence the ion is diamagnetic. It is due to the diamagnetic nature of $[Co(NO_2)_6]^{3-}$ ion that all the six electrons in $3d$ orbitals of $Co^{3+}$ ion are paired. As a reult, the octahedral geometry of the ion results from $d^2sp^3$ hybridisation. In other words, we can say that the complex ion is an inner orbital octahedral complex ion. According to CFT the distribution of six $3d$ electrons in $t_{2g}$ and $e_g$ orbitals is as $t_{2g}^6 e_g^0 (n = 0)$.

**2. Potassium Hexanitrocobaltate (III) or Potassium Cobaltinitrite, $K_3[Co(NO_2)_6]$**

$K_3[Co(NO_2)_6]$ is also called ***Fischer's salt.***

**Preparation.** It is prepared: (*i*) by the action of Co (II) salt solution [e.g., $Co(NO_3)_2$, $CoCl_2$ etc.] on $KNO_2$ in presence of $CH_3COOH$.

(*a*) $$2CH_3COOH + Co(NO_3)_2 + 7KNO_2 \longrightarrow 2CH_3COOK + 2KNO_3 + H_2O + NO + K_3[Co(NO_2]_6$$

or $$2H^+ + Co^{2+} + 3K^+ + 7NO_2^- \longrightarrow H_2O + NO + K_3[Co(NO_2)_6] \text{ (Yellow ppt)}$$

(*b*) $$2CH_3COOH + CoCl_2 + 7KNO_2 \longrightarrow 2CH_3COOK + 2KCl + H_2O + NO \; K_3[Co(NO_2)_6]$$

or $$2H^+ + Co^{2+} + 3K^+ + 7NO_2^- \longrightarrow H_2O + NO + K_3[Co(NO_2)_6] \text{ (Yellow ppt)}$$

(*ii*) by the action of aqueous solution of potassium salt on $Na_3[Co(NO_2)_6]$.

$$Na_3[Co(NO_2)_6 + 3KCl \longrightarrow K_3[Co(NO_2)_6] + 3NaCl$$

**Properties.** (*i*) $K_3[Co(NO_2)_6]$ is a yellow crystalline solid and is insoluble in water.

(*ii*) $[Co(NO_2)_6]^{3-}$ ion is decomposed by alkalies to form $Co(OH)_3$ while strong acids decompose it into oxides of nitrogen.

**Uses.** This compound is used (*i*) as a yellow pigment under the name of *Indian yellow* or *cobalt yellow,* (*ii*) in printing porcelain blue, (*iii*) Reactions given at (*a*) and (*b*) are utilised in the detection of $Co^{2+}$ ion in a given inorganic salt, (*iv*) as a test for $K^+$ ions. $K_3[Co(NO_2)_6]$ is, therefore, used as a laboratory reagent for detecting and estimating potassium salts.

### 3. Hexammine Cobalt (III) Chloride or Hexammine Cobaltic Chloride, $[Co(NH_3)_6]Cl_3$

When a vigorous stream of air is drawn for several hours through a solution of $CoCl_2$ containing ammonia, ammonium chloride and some activated charcoal, good yield of hexammine cobalt (III) chloride is obtained.

$$4CoCl_2 + 4NH_4Cl + 20NH_3 + O_2 \longrightarrow 4[Co(NH_3)_6]Cl_3 + 2H_2O$$

In the absence of charcoal, replacement usually occurs to give, for example, $[Co(NH_3)_5\ Cl]^{2+}$ and $[Co(NH_3)_4\ (CO_3)]^+$. This forms orange -yellow crystals. Conductance measurement indicates the presence of four ions in its solution, namely one $[Co(NH_3)_6]^{3+}$ ion and three $Cl^-$ ions as shown below:

$$[Co(NH_3)_6]Cl_3 \rightleftharpoons [Co(NH_3)_6]^{3+} + 3Cl^-$$

**Structure of $[Co(NH_3)_6]^{3+}$ ion.** V.S.E.C. of $Co^{3+}$ ion is $3d^6$. Since $NH_3$ ligands are strong field ligands, the six $3d$ electrons get paired in $[Co(NH_3)_6]^{3+}$ ion. Thus the complex ion has all the six $3d$ electrons in the paired state ($n$ =0). This ion is, therefore, diamagnetic. As a result, the complex ion results from $d^2sp^3$ hybridisation (inner orbital octahedral geometry).

According to CFT, the distribution of six $3d$ electrons of $Co^{3+}$ ion in $t_{2g}$ and $e_g$ orbitals is as $t_{2g}^6 e_g^0 (n = 0)$.

## Nickel, $Ni_{28}$

### Physical Properties

Nickel is a soft silvery white metal, quite malleable and ductile. It is slightly magnetic and possesses a high melting and boiling point (m.pt. =1453°C, b.pt =2730°C). Like iron and cobalt, it absorbs a large volume of hydrogen gas (one volume of the metal absorbing 17 volumes of the gas).

### Chemical Properties

The action of air on the metal is slow at the ordinary temperature, but it vigorously burns in oxygen. It is attacked by chlorine, when nickel chloride, $NiCl_2$ is formed. Nickel decomposes steam at a red heat, forming NiO and $H_2$. Dilute hydrochloric and sulphuric acid solutions have a very slow action on this metal. But it readily dissolves in nitric acid and aqua regia.

$$3Ni + 8HNO_3 \longrightarrow 3Ni(NO_3)_2 + 2NO + 4H_2O$$

When Ni is kept for some time under the surface of fuming $HNO_3$, the metal becomes passive.

Ni decomposes steam at red heat

$$Ni + H_2O(steam) \xrightleftharpoons{\text{Red heat}} NiO + H_2$$

The metal remains unaffected by caustic alkalies, even by fused alkalies, Nickel crucibles are, therefore, commonly used for alkali fusion in qualitative experiments. When CO gas is passed over nickel kept at about 50°C, a volatile colourless liquid, nickel tetracarbonyl, $Ni(CO)_4$ is formed. The formation of $Ni(CO)_4$ finds an industrial application in the purification (or extraction) of nickel by Mond's process.

**Uses of Nickel**

(*i*) Due to high melting point, great resistance to corrosion, acids and alkalies, nickel is largely used in making crucibles, dishes and other chemical wares and also in electrical industry.

(*ii*) It is used for nickel plating in which $NiSO_4.7H_2O$ or $NiSO_4.(NH_4)_2SO_4.6H_2O$ solution is used as an electrolyte.

(*iii*) Finely divided nickel is used as a catalyst in the hydrogenation reactions like,

$$R.CH = CH.R + H_2 \longrightarrow R.CH_2 - CH_2.R$$

$$CO + 3H_2 \longrightarrow CH_4 + H_2O$$

$$2NO + 5H_2 \longrightarrow 2NH_3 + 2H_2O$$

(*iv*) $Ni(CO)_4$ is used to get pure nickel (99.9% pure) by heating it at 180°C.

(*v*) Nickel metal combines with many other metals and forms alloys which find a number of applications. Some important alloys of Ni and their uses are given below:

(*a*) *Invar steel.* It contains 35% Ni and 0.3% carbon and has a negligible coefficient of expansion. It is used in surveying instruments, pendulums, chronometer components, etc.

(*b*) *Monel metal*. (70% Ni, 30% Cu). It is resistant to corrosion and is used in chemical plants.

(*c*) *Constantan*. (40% Ni, 60% Cu). It is used in electrical resistances.

(*d*) *Nichrome.* (80% Ni, 20% Cr). It is very resistant to oxidation and is therefore, used in making heating elements in electric heaters.

(*e*) *Nickel silver or German silver.* (10-30% Ni with Cu and Zn). It is used for making cheap ornaments, domestic utensils and tableware.

(*f*) *Coinage alloy* (Cu = 75%: Ni = 25%) It is used for making coins.

**Electronic Configuration and Oxidation States of Ni**

The V.S.E.C of Ni is $3d^8 4s^2$. Ni shows –1, 0, +1, +2, +3 and +4 oxidation states in its compounds. Ni has –1 oxidation state in the complex anion, $[Ni_2(CO)_6]^{2-}$. Ni has zero oxidation state in $[Ni(CO)_4]°$, $[Ni(PF_3)_4]°$, $[Ni(PF_3)_2(CO)_2]$ and $K_4[Ni(CN)_4]$. Compounds containing Ni in +1 and +4 oxidation states are very rare. $K_4[Ni_2(CN)_6]$ has Ni in +1 oxidation state while $K_2[NiF_6]$ has Ni in +4 oxidation state. $K_2[NiF_6]$ is prepared by reacting a mixture of $NiCl_2$ and KCl with $F_2$. This complex is strongly oxidising and liberates $O_2$ from $H_2O$. Ni (IV) also occurs in certain oxide systems.

## Aqueous Chemistry of Ni(II)

+2 oxidation state in Ni is the only stable oxidation state. In aqueous solution, $Ni^{2+}$ ion exists as $[Ni(H_2O)_6]^{2+}$ which has green colour. Thus the salts of $Ni^{2+}$ ion are green or blue.

$Ni^{2+}$ ion gets precipitated as $Ni(OH)_2$ which is apple-green when treated with NaOH or KOH.

$$Ni^{2+} + 2OH^- \longrightarrow Ni(OH)_2$$

This apple-green ppt. dissolves in aqueous ammonia or ammonium salts to give violet-blue solutions which contain $[Ni(NH_3)_6]^{2+}$.

$Ni(OH)_2$ is not oxidised by atmospheric $O_2$ or $H_2O_2$. When treated with a mild oxidising agent like $Br_2$ in alkaline medium, it is converted into black powder to $N_2O_3 . H_2O$. When treated with strong oxidising agent like $Cl_2$ or hypochlorite ($ClO^-$), a black ppt. of $NiO_2 . xH_2O$ is obtained.

$$Ni(OH)_2 + OCl^- \longrightarrow NiO_2 + H_2O + Cl^-$$

Compounds of Ni in +2 oxidation state are called **nickelous compounds.** NiO, $Ni(OH)_2$, $NiCl_2 . 6H_2O$, $NiSO_4 . 7H_2O$, $NiSO_4 . (NH_4)_2 SO_4 . 6H_2O$, $Ni(NO_3)_2 . 6H_2O$, $NiCO_3 . 6H_2O$ and NiS are important compounds of Ni in +2 oxidation state.

Ni $(CH)_2$ is a green solid. It is precipitated when a Ni(II) salt is treated with caustic alkali.

$$Ni^{2+} + 2OH^- \longrightarrow Ni\,(OH)_2\downarrow$$

The ppt. of $Ni(OH)_2$ dissolves in the excess of aqueous $NH_3$ to form the blue solution which contains the complex ion. $[Ni\,(NH_3)_6]^{2+}$ . When an alkaline solution of hypochlorite ($ClO^-$) is added to the suspension of $Ni(OH)_2$ in water, a black ppt. is obtained which is mainly hydrated $NiO_2$.

$$Ni(OH)_2 + ClO^- \longrightarrow NiO_2 + H_2O + Cl^-$$

On being heated, $Ni(OH)_2$ gives NiO.

$$Ni(OH)_2 \xrightarrow{\Delta} NiO + H_2O$$

## Complexes of Ni (II)

Nickel (II) forms stable complexes with coordination number four and six. Compounds of higher coordination number are rare. The principal geometries of its complexes are octahedral and square planar. Some trigonal bipyramidal, square pyramidal and tetrahedral complexes are also known.

Complexes with ammonia, ethylenediamine, $H_2O$ are all octahedral. In contrast to the aqua $[Ni(H_2O)_6]^{2+}$ complex which is green, the ammine complex $[Ni(NH_3)_6]^{2+}$ is intense blue so is the $[Ni(en)_3]^{2+}$ ion. All these are paramagnetic since the $d^8$ ion has two unpaired electrons.

Ni(II) forms a green complex with acetylacetone which attains octahedral coordination by polymerization. This is obtained as the monomeric octahedral *trans*-dihydrate $[Ni(acac)_2\,(H_2O)_2]$ from aqueous solutions. On dehydration it forms the trimer $[Ni(acac)_2]_3$ (Fig. 6.28).

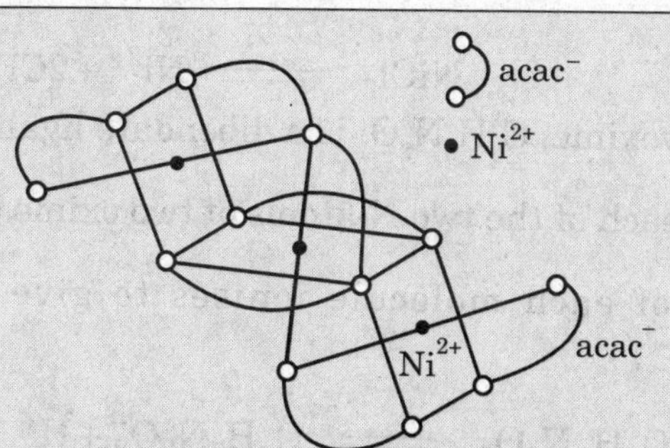


**Fig. 6.28** Structure of $[Ni(acac)_2]_3$.

$Ni^{2+}$ ion forms a 6-coordinated complex, $[Ni(diars)_2I_2]°$. $Ni^{2+}$ ion present in this complex is a $d^8$ ion. The distribution of $d^8$ electrons in $t_{2g}$ and $e_g$ orbitals is as $r_{2g}^6\ e_g^2$. In this configuration, $t_{2g}$ orbitals are symmetrically filled. Although $e_g$ orbitals are half-filled, these orbitals are said to be asymmetrically filled since both $e_g$ electrons are present in $d_z{}^2$ orbital. Thus $t_{2g}^6 e_g{}^2$ configuration is $t_{2g}^6 d_{z^2}^2\ d_{x^2-y^2}^0$ or $d_{xy}^2 d_{yz}^2 d_{xz}^2 d_{z^2}^2 d^0{}_{x^2-y^2}$ ($n$ =0). Hence the complex, $[Ni(diars)_2\ I_2]°$ has elongated distorted octahedral geometry with two longer axial bond lengths. The elongation of axial bond lengths in the complex is so large that the complex assumes square planar geometry.

The presence of both $e_g$ electrons in $d_{z^2}$ orbital is because of the fact that the order of energy of various $d$ -orbitals in elongated distorted octahedral complex, $[Ni(diars)_2\ I_2]°$ is as:

$$d_{yz} = d_{zx} \big| d_{xy} \big| d_{z^2} \big| d_{x^2-y^2}$$
$$\xrightarrow{\hspace{2em}\text{Energy increasing}\hspace{2em}}$$

Above discussion shows that the addition of two coordinating $I^-$ ligands to the square plane of diamagnetic square planar complex, $[Ni(diars)_2]^{2+}$ gives rise to tetragonal distortion to $[Ni(diars)_2I_2]°$.

Nickel (II) also forms square planar complexes which include the yellow $[Ni\ (CN)_4]^{2-}$, the most stable anion complex of $Ni^{2+}$ ($K_3$= 3 ×$10^{31}$). It is obtained by the action of KCN on Ni (II) salts. The insoluble $Ni(CN)_2$, first formed, dissolves in an excess KCN:

$$NiSO_4 + 2KCN \longrightarrow Ni(CN)_2 + K_2SO_4$$
$$Ni(CN)_2 + 2KCN \longrightarrow K_2[Ni(CN)_4]$$

$Ni^{2+}$ forms a complex with dimethyl glyoxime known as *bis(dimethyl glyoximato) nickel (II)*, $[Ni(C_4\ H_7\ N_2O_2)_2]°$ or $[Ni\ (dmg)_2]°$. The formation of $[Ni(dmg)_2]°$ can be explained as follows:

Dimethyl glyoxime has the molecular formula, $C_4H_8N_2O_2$ or $(C_4H_7N_2O_2)H$ or dmgH and its structural formula is as shown below:

When dimethyl glyoxime is added to a nickel (II) salt (*e.g.*, $NiCl_2$) containing $NH_4OH$, red ppt. of *bis* (dimethyl glyoximato) nickel (II), $[Ni(C_4H_7N_2O_2)_2]^0$ or $[Ni(dmg)_2]$ is obtained. The formation of this complex compound takes place through the following steps:

(*a*) $NiCl_2 \rightleftharpoons Ni^{2+} + 2Cl^-$

(*b*) Dimethyl glyoxime, $C_4H_8N_2O_2$ is a dibendate ligand and has one OH group attached to each of the two N-atoms of two oxime groups $\left[ \rangle N-OH \right]$. One OH group of each molecule ionises to give $H^+$ and glyoxime ion, $C_4H_7N_2O_2^-$.

$$C_4H_8N_2O_2 \rightleftharpoons C_4H_7N_2O_2^- + H^+$$

or

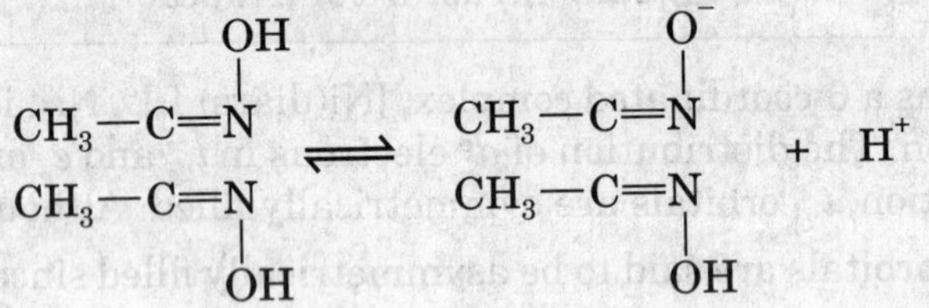

or $$dmgH \rightleftharpoons dmg^- + H^+$$

Now two *dmg*⁻ ions combine with one $Ni^{2+}$ ion to form electrically neutral *bis* (dimethyl glyoximato) nickel (II), $[Ni(C_4H_7N_2O_2)_2]^0$ or $[Ni(dmg)_2]^0$.

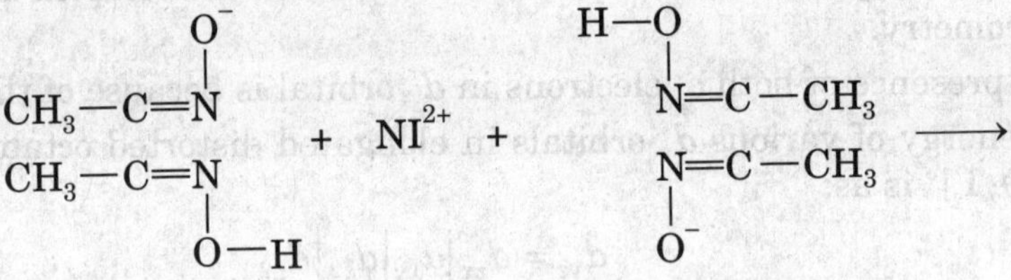

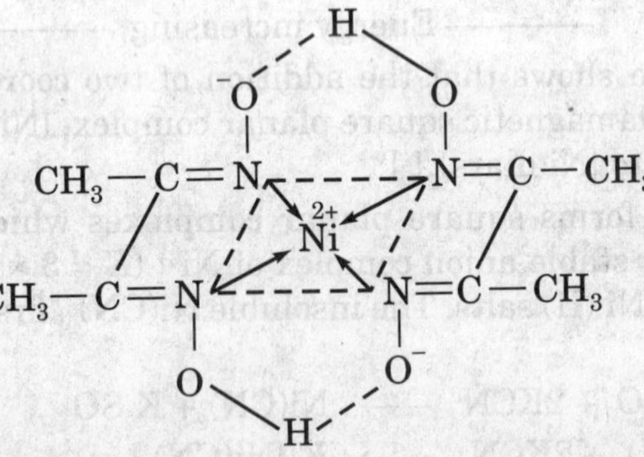

or $$C_4H_7N_2O_2^- + Ni^{2+} + C_4H_7N_2O_2^- \longrightarrow [Ni(C_4H_7N_2O_2)_2]^0$$

or $$dmg^- + Ni^{2+} + dmg^- \longrightarrow [Ni(dmg)_2]^0$$

Bis–(dimethyl glyoximato) nickel (II) (red ppt)

*Bis* (dimethyl glyoximato) nickel (II), $[Ni(C_4H_7N_2O_2)_2]^0$ or $[Ni(dmg)_2]^0$ is a chelate complex. It is electrically neutral and covalent in character. Its covalent character accounts for its low solubility in water. It may be seen from the structure that the coordination of *dmg*⁻ ion (ligand) to $Ni^{2+}$ ion occurs through both its N-atoms making two 5-membered rings. Further two 6-membered rings are also formed by H-bonding between NOH and ON. H-bonding stabilises the structure of the complex.

This complex compound has square planar geometry corresponding to $dsp^2$ hybridisation of $Ni^{2+}$ ion. In this complex compound $Ni^{2+}$ ion has its valence-shell configuration as $3d^8 = t_{2g}^6\, 3d_{x^2-y^2}^2\, 3d_{z^2}^0$. The compound has no unpaired electrons, *i.e.* all the eight electrons are present as paired electrons and hence the complex compound is diamagnetic.

The formation of this complex compound is used in detecting as well as in estimating nickel and also in distinguishing Ni from Co.

Several tetrahedral Ni (II) complexes are known. The simplest of these are the blue halide complexes, $[NiX_4]^{2-}$ (X = Cl, Br, I) obtained from ethanolic solutions by large cations such as $[NR_4^+]$.

The square planar complexes are magnetic, whereas the tetrahedral are paramagnetic due to two unpaired electrons.

A number of nickel (II) complexes exist, especially with ligands containing both O- and N-donors, which have magnetic moments greater than zero but below the values expected for two unpaired electrons. It is now recognized that such complexes originate from the following possible equilibria:

1. *Square planar–tetrahedral equilibria.* In the complex $NiL_2X_2$, when L is triphenylphosphine, tetrahedral structures result, whereas with trialkylphosphines square planar complexes are formed.
2. *Monomer-polymer equilibria.* In many cases polymerization of 4-coordinate complexes takes place to form species in which nickel ions become 5- or 6-coordinated. With acetylacetone the trimeric structure (Fig 6.28) is adopted in order to give a 6-coordinate nickel.
3. *Square planar – octahedral equilibria.* Square planar complexes of the type $NiL_4$ on being dissolved in coordinating solvents such as pyridine, form octahedral complexes by adding on two solvent molecules.

$$ML_4 + L' = ML_4\,L'$$

$$ML_4\,L' + L' = ML_4\,L'_2$$

Only a 5-coordinate species is formed when L = L'= CN, but in most cases the equilibria strongly form the 6-coordinate molecule having *trans* structures and two unpaired electrons. Examples of this behaviour are provided by complexes of the type $[Ni(L–L)\,X_2]$ where L–L is substituted ethylenediamine and X is a variety of anions.

## Chemistry of Ni (III)

$Ni_2O_3 \cdot 2H_2O$ contains Ni in +3 oxidation state. When $Ni(OH)_2$ is oxidised with $Br_2$ in alkaline medium, $Ni_2O_3 \cdot 2H_2O$ gets precipitated.

A number of complexes having Ni in +3 oxidation state are known. Well known examples of such complexes are $[Ni\,(en)_2\,Cl_2]\,Cl$ and $[Ni\,(PEt_3)_2\,Br_2]\,Br$. Former complex is octahedral while the latter is trigonal bipyramidal.

## Copper, $Cu_{29}$

### Physical Properties of Copper

(*i*) It is a reddish coloured lustrous metal.

(*ii*) It is highly malleable and ductile.

(*iii*) It has specific gravity 8.9.

(*iv*) It has high melting and boiling points (m.pt. 1083°C, b.pt. 2580°C).

(*v*) It is good conductor of heat and electricity (next to silver).

(*vi*) Its atomic radii is 1.17Å and ionisation potential is 7.72 eV.

(*vii*) Molten copper absorbs sulphur dioxide which it gives out on cooling.

### Chemical Properties of Copper

1. **Reactivity.** Copper is chemically less reactive as it has high heat of sublimation and high ionisation potential.

2. **Action of air and moisture.** Copper is not affected by dry air at ordinary temperatures but when exposed to air and moisture, it is covered by a green thin film which was formerly supposed to be the basic carbonate but consists of *bronchantite*, $CuSO_4 . 3Cu(OH)_2$ or *atacamite*, $CuCl_2 . 3Cu(OH)_2$ due to presence of $SO_2$ and HCl in the atmosphere.

$$8Cu + 6H_2O + 2SO_2 + 5O_2 \longrightarrow 2\,[CuSO_4 . 3\,Cu(OH)_2]$$

$$2Cu + H_2O + CO_2 + O_2 \longrightarrow \underset{\text{Basic carbonate}}{CuSO_3 . Cu(OH)_2}$$

The layer protects the rest of the metal from further action.

3. **Action of $O_2$.** When heated in oxygen, it first forms red cuprous oxide ($Cu_2O$) which changes into black cupric oxide (CuO). If the temperature is raised above 1100°C, cupric oxide changes back to cuprous oxide.

$$4Cu + O_2 \longrightarrow 2Cu_2O \text{ (red)}$$

$$2Cu_2O + O_2 \longrightarrow 4CuO \text{ (black)}$$

$$2CuO \xrightarrow{>1100°C} Cu_2O + \frac{1}{2}O_2$$

4. **Action of water.** Water at ordinary temperature has no action on copper. However, when steam is passed over red hot copper, hydrogen is evolved.

$$Cu + H_2O \xrightarrow{\text{Red heat}} CuO + H_2$$

5. **Action of acids.** (*a*) Non-oxidising acids (HCl and dil. $H_2SO_4$) do not have any effect on copper. However, they dissolve the metal in presence of air.

$$2Cu + 4HCl \text{ (dil. or conc.)} + O_2 \longrightarrow 2CuCl_2 + 2H_2O$$

$$2Cu + 2H_2SO_4 \text{ (dil.)} + O_2 \longrightarrow 2CuSO_4 + 2H_2O$$

(*b*) Copper is readily affected by oxidising acids (conc. $H_2SO_4$ and dilute and conc. $HNO_3$). $SO_2$ is evolved on heating with conc. $H_2SO_4$.

$$H_2SO_4 \longrightarrow H_2O + SO_2 + O$$

$$Cu + O + H_2SO_4 \longrightarrow CuSO_4 + H_2O$$

$$\overline{Cu + 2H_2SO_4 \longrightarrow CuSO_4 + SO_2 + 2H_2O}$$

NO is evolved on heating with dil . $HNO_3$

$$2HNO_3 \longrightarrow H_2O + 2NO + 3[O]$$

$$[Cu + 2HNO_3 + O \longrightarrow CU(NO_3)_2 + H_2O] \times 3$$

$$3Cu + 8HNO_3 \text{(Dilute)} \longrightarrow 3Cu(NO_3)_2 + 2NO + 4H_2O$$

$NO_2$ is evolved on heating with conc. $HNO_3$

$$2HNO_3 \longrightarrow H_2O + 2NO_2 + O$$

$$Cu + 2HNO_3 + O \longrightarrow Cu(NO_3)_2 + H_2O$$

$$Cu + 4HNO_3 \text{(Concentrated)} \longrightarrow Cu(NO_3)_2 + 2NO_2 + 2H_2O$$

(*c*) Copper metal reacts with hydrobromic acid (HBr) and hydroidic acid (HI) forming complexes

$$2Cu + 2HBr \longrightarrow 2CuBr + H_2$$

$$2CuBr + 2HBr \longrightarrow H_2[Cu_2Br_4]$$

(*d*) Copper plants are slowly attacked by acetic acid (vinegar) in presence of air to form basic copper acetate (verdigris).

$$2Cu + 2CH_3COOH + O_2 \longrightarrow \underset{\text{Basic copper accetate (verdigris)}}{Cu(CH_3COO)_2 . Ca(OH)_2}$$

6. **Action of ammonia.** (*a*) Copper is dissolved by aqueous ammonia in the presence of air oxygen with the formation of a complex.

$$2Cu + 8NH_3 + 2H_2O + O_2 \longrightarrow 2[Cu(NH_3)_4](OH)_2$$

(*b*) However, when ammonia gas is passed over red hot copper, the nitrogen is absorbed by the metal leaving hydrogen free.

7. **Action of halogens.** Copper combines with halogens directly.

$$Cu + Cl_2 \xrightarrow{\text{Heat}} CuCl_2$$

$$Cu + Br_2 \xrightarrow{\text{Heat}} CuBr_2$$

$$Cu + I_2 \xrightarrow{\text{Heat}} CuI_2$$

Moist chlorine rapidly attacks copper even at ordinary temperatures.

8. **Reducing properties.** Copper reduces oxides of nitrogen to elementary nitrogen and ferric ion to ferous ion.

$$2Cu + 2NO \longrightarrow 2CuO + N_2$$

$$2Fe^{3+} + Cu \longrightarrow 2Fe^{2+} + Cu^{2+}$$

The etching of designs on copper plates using ferric chloride is based on this property.

$$2FeCl_3 + Cu \longrightarrow CuCl_2 + 2FeCl_2$$

9. **Displacement reactions.** Copper displaces silver, gold and platinum metals from their salt solutions.

$$2AgNO_3 + Cu \longrightarrow 2Ag + Cu(NO_3)_2$$

$$2Cu + 2AuCl_3 \longrightarrow 3CuCl_2 + 2Au$$

**Uses of copper**

(*i*) Being a good conductor of electricity, copper is extensively used for making electric cable and other electric appliances.

(*ii*) Copper is used for making utensils, containers, kettles, evaporating pans, calorimeters, coins, etc. as copper is not attacked by water or air.

(*iii*) It is used in electroplating and electrotyping.

(*iv*) Copper is alloyed with gold and silver for making coins and jewellery.

(*v*) Copper is used for the preparation of a number of copper salts like $CuSO_4$, $CuCl_2$, etc.

(*vi*) Copper is used for making a number of useful alloys. Some important alloys of Cu are given in Table 6.3.

**Table 6.3. Some important alloys of copper**

| *Alloy* | *Percentage composition* | *Uses* |
|---|---|---|
| (*i*) Brass | Cu = 80, Zn = 20 | For making utensils, condenser tubes parts of machinery etc. |
| (*ii*) Bronze or copper bronze | Cu = 80, Zn = 10, Sn = 10 | For making cooking utensils, statues. |
| (*iii*) Aluminium bronze | Al = 95, Cu = 5 | Coins, picture frames, cheap jewellery |
| (*iv*) Gun metal | Cu = 90, Sn = 10 | For making gun barrels |
| (*v*) Bell metal | Cu = 80, Sn = 20 | For making bells, gongs etc. |
| (*vi*) Constantan | Cu = 60, Ni = 40 | For electrical apparatus |
| (*vii*) German silver | Cu = 60, Zn = 20, Ni = 20 | For making silver ware, resistance wire |
| (*viii*) Monel metal | Cu = 30, Ni =67, Fe, and Mn = 3 | For making acid pumps and acid containers |
| (*ix*) Phosphor bronze | Cu = 95, Sn = 4.8, P = 0.2 | For making springs, electrical equipment |
| (*x*) Gold-copper alloy | Au = 90, Cu = 10 | For making gold coins, jewellery, watch, spectacle rims etc. |

## Biological Improtance of Copper

Copper is an essential constituent of many metallo proteins and enzymes. The copper-containing enzymes play an important role in the pigmentation of skin, functioning of brain and in iron metabolism in the human body. The deficiency of copper in the human system results in inability to utilise the iron stored in the liver. The approximate quantity of iron in human body is around 100 mg.

The excess of copper in human body is, however, harmful. A disease, known as **Wilson disease**, upsets the copper metabolism of the human system in such a way that it gets deposited in excess in the liver, kidneys and brain. The patient suffering from such a disease is treated with controlled dose of EDTA which forms a water-soluble complex with $Cu^{2+}$ ion and gets excreted through urine.

Haemocyanin is a copper-containing protein which is present in the blood of snails, lobsters, etc., and acts as oxygen-carrier.

Plastocynanin, containing copper, is present in chloroplasts of green plants and plays an important role in the process of photosynthesis as an electron carrier.

A copper-containing protein, azurin, is found in certain bacteria.

## Electronic Configuration and Oxidation States of Cu

V.S.E.C of Cu is $3d^6\ 4s^1$ which shows that Cu shows +1 ($Cu+ = 3d^{10}$) and +2 ($Cu^{2+} = 3d^9$) oxidation states in its compounds. Compounds in +1 O.S. are called **cuprous compounds** while those in +2 O.S. are called **cupric compounds.** Cu (I) compounds are not stable, since these compounds undergo disproportionation in aqueous solution.

$$2Cu^+(aq) \rightleftharpoons Cu^{2+}(aq) + Cu^o\ (s)$$

Cu (I) compounds are stable only when they are insoluble and complex compounds. Compounds of Cu in +2 O.S. are more stable.

## Aqueous Chemistry of Cu (I)

(*i*) We know that V.S.E.C. of $Cu^+$ ion is $3d^{10}$. Since all the $3d$ electrons in $3d^{10}$ configuration are paired, compounds of Cu(I) are diamagnetic.

(*ii*) $Cu^+$ ion undergoes disproportionation in solution.

$$2\ Cu^+\ (s) \longrightarrow Cu^{2+}\ (aq) + Cu^0\ (s)$$

Above reaction can be broken into the following two oxidation half-reactions.

(*a*) $Cu(s) \longrightarrow Cu^+\ (aq) + e^-,\ E^o_{ox} = -0.52\,V$

(*b*) $Cu^+(aq) \longrightarrow Cu^{2+}(aq) + e^-, E^o_{ox} = -0.18V$

Since $E^o_{ox}$ value for reaction (*b*) is less negative than that for reaction (*a*), $Cu^+$ ions have greater tendency to change into $Cu^{2+}$ ions. Hence even if $Cu^+$ ions are produced by reaction (*a*), they are converted into $Cu^{2+}$ ions by reaction (*b*). When insoluble $Cu_2O$ in which Cu is present in +1 O.S. is dissolved in $H_2SO_4$, $CuSO_4$ (Cu = +2) and not $Cu_2SO_4$ (Cu = +1) is formed.

$$Cu_2O + H_2SO_4 \longrightarrow CuSO_4 + Cu(s) + H_2O$$

or

$$\underset{(Cu=+1)}{Cu_2O} + 2H^+ \longrightarrow \underset{(Cu=+2)}{Cu^{2+}(aq)} + \underset{(Cu=0)}{Cu(s)} + H_2O$$

(*iii*) We have said above that $Cu^+$ ion is unstable because of its disproportionation tendency. However, $Cu^+$ ion can be stabilised by the formation of insoluble CuCl when brought in contact with $Cl^-$ ions.

$$2Cu^+ + 2Cl^- \longrightarrow CuCl\ (s)$$

CuCl formed dissolves in the excess of $Cl^-$ ions to form the complex ion, $[CuCl_2]^-$.

$$CuCl + Cl^- \longrightarrow [CuCl_2]^-$$

(*iv*) When KI is added to a solution of $CuSO_4$, $I^-$ ion reduces $Cu^{2+}$ ion to $Cu^+$ which reacts with more $I^-$ ions to form the ppt. of cuprous iodide, $Cu_2I_2$.

$$2Cu^{2+} + 2I^- \longrightarrow 2\,Cu^+ + I_2$$

$$2Cu^+ + 2I^- \longrightarrow Cu_2\,I_2\,(ppt)$$

$$2Cu^{2+} + 4I^- \longrightarrow Cu_2\,I_2\,(ppt) + I_2$$

This reaction is used for estimating Cu volumetrically.

## Aqueous Chemistry of Cu (II)

(*i*) Although most of anhydrous Cu(II) salts are colourless, yet they are blue when hydrated. This is due to the presence of blue hydrated Cu (II) ion, $[Cu(H_2O)_4]^{2+}$.

(*ii*) Solutions of Cu(II) salts are acidic, since they undergo hydrolysis to give $H^+$ ions.

$$Cu^{2+} + H_2O \longrightarrow [Cu(OH)]^+ + H^+$$

(*iii*) Since $Cu^{2+}$ ions ($Cu^{2+} = 3d^9$) has unpaired electron (=1), Cu (II) compounds are paramagnetic.

## Compounds of Cu (I)

1. **Cuprous Oxide or Copper (I) Oxide, $Cu_2O$**

   **Preparation.** It is obtained by the reduction of CuO by Cu or by thermal decomposition of CuO.

   **Properties.** The oxide is soluble in $NH_3$ to give the complex ion, $[Cu(NH_3)_2]^{2+}$. With KCN, it gives the complex compounds $K_3\,[Cu(CN)_4]$, $K_2\,[Cu(CN)_3]$ and $K[Cu(CN)_2]$. $K_3[Cu(CN)_4]$ is the major complex that is formed.

2. **Cuprous Chloride, CuCl**

   **Preparation.** It is obtained by the following reactions:

$$CuCl_2 + Cu \xrightarrow[\text{of conc. HCl}]{\text{Heat in presence}} 2CuCl$$

$$2CuCl_2 + 2H \xrightarrow{\text{Reduction}} 2CuCl + 2HCl$$

$$2CuCl_2 + SO_2 + 2H_2O \xrightarrow{\text{Reduction}} 2CuCl + H_2SO_4 + 2HCl$$

$$2CuCl_2 \xrightarrow{450°C} 2CuCl + Cl_2$$

**Properties.** (*i*) Cuprous chloride is a colourless crystalline substance, insoluble in water but soluble in HCl and $NH_4OH$. When exposed to air, it is slowly oxidised to green basic cupric chloride, $3CuO.\ CuCl_2.\ 3H_2O$.

With HCl it gives the complex ion, $[CuCl_4]^{3-}$

$$CuCl + 3Cl^- \longrightarrow [CuCl_4]^{3-}$$

With $NH_4OH$, it gives $[Cu(NH_3)_4]Cl$

$$CuCl + 4NH_4Cl \longrightarrow [Cu(NH_3)_4]Cl + 4H_2O$$

(*ii*) *Action of oxidising agents.* With oxidising agents, it is oxidised into cupric chloride.

$$2CuCl + 2HCl + O \longrightarrow 2CuCl_2 + H_2O$$

(*iii*) An acidified solution of cuprous chloride absorbs oxygen and carbon monoxide while an ammoniacal solution absorbs acetylene.

$$4CuCl + 4HCl + O_2 \longrightarrow 4CuCl_2 + 2H_2O$$

This reaction is utilised in the manufacture of chloride by *deacon's process.*

$$CuCl + 2CO \longrightarrow \underset{\text{Addition compound}}{CuCl.\ 2CO}$$

This addition compound decomposes on heating, regenerating carbon monoxide. This reaction is, therefore, employed in gas analysis for separation of carbon monoxide.

$$2CuCl + \underset{\text{Acetylene}}{C_2H_2} \longrightarrow \underset{\text{Copper acetylide}}{C_2Cu_2} + 2HCl$$

Acetylene can be regenerated by treating the acetylide ($C_2Cu_2$) with strong HCl. The reaction is, therefore, used in the *purification and separation of acetylene.*

(*iv*) *Action of alkalies.* Cuprous chloride solution gives with alkalies a yellow ppt. of cuprous oxide which gradually changes to a red colour.

$$2CuCl + 2NaOH \longrightarrow \underset{\text{Yellow ppt changing to red}}{Cu_2O} + 2NaCl + H_2O$$

(*v*) *Action of $H_2S$.* With $H_2S$, a black precipitate of cuprous sulphide, $Cu_2S$ is produced.

$$CuCl + H_2S \longrightarrow \underset{\text{Black ppt}}{Cu_2S} + 2HCl$$

(*vi*) *Action of heat.* On heating, white CuCl becomes deep blue at 450 K and melts to a deep green liquid.

**Uses.** Ammoniacal solution of cuprous chloride is used in gas analysis for absorbing acetylene while HCl solution of it is used for absorbing CO and oxygen.

**Structure.** CuCl is a covalent compound and has tetrahedral zinc blende (ZnS) structure with the metal atom tetrahedrally surrounded by Cl atoms. $Cu^+$ ions occupy half of the suitable tetrahedral sites in a face centred cubic lattice formed by $Cl^-$ ions. The each $Cl^-$ ions is also surrounded tetrahedrally by four $Cu^+$ ions (Fig. 6.29).

In the vapour state, CuCl has cyclic trimer ring structure which has alternating Cl and Cu(I) ions. (See Fig. 6.30).

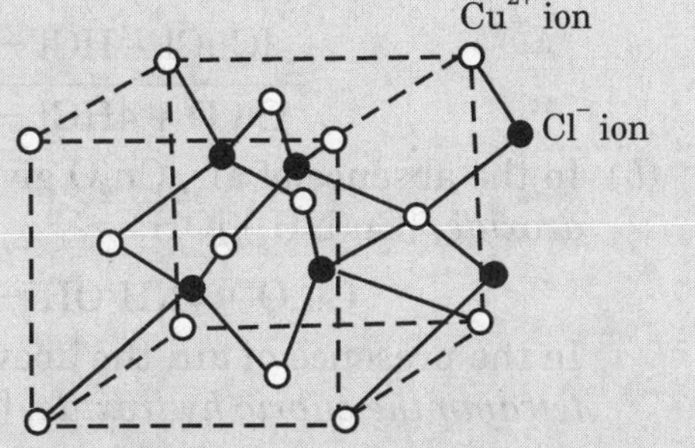


**Fig. 6.29** Tetrahedral zinc blende structure of CuCl

## Complexes of Cu (I)

Copper (I) complexes show varying coordination numbers. When the coordination number is 2, the structure is linear, with coordination number 4 the stereochemistry is tetrahedral and in a few cases a planar coordination of three monodentate ligand is known. Typical examples are $[Cu(NH_3)_2]^+$, $K_3[Cu(CN)_4]$, $K_2[Cu(CN)_3]$, $K[Cu(CN)_2]$ etc. In $K[Cu(CN)_2]$ the metal does not obey the apparent coordination number of two; instead the compound has a trigonal planar structure such that copper (I) has a coordination number three. Two CN groups are linked to copper via carbon and one via nitrogen. (See Fig. 6.31)

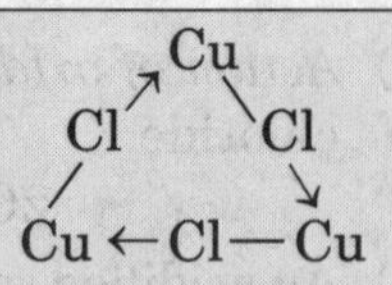

**Fig. 6.30** Cyclic trimeric ring structure of CuCl

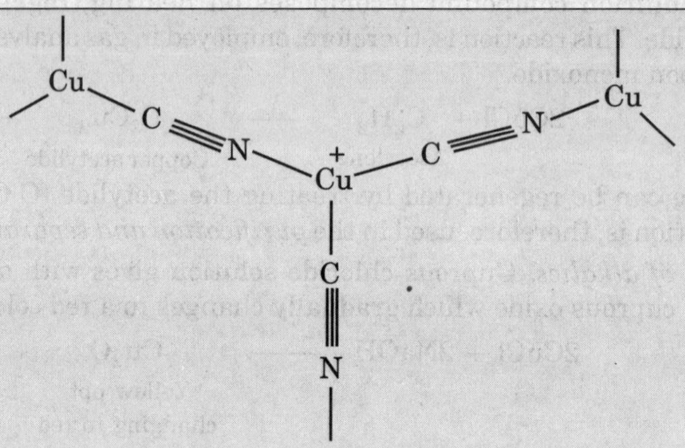

**Fig. 6.31** Trigonal planar structure of $[Cu(CN)_2]^-$ ion

Stable complexes of copper (I) with substituted phosphines, arsines and sulphides have been synthesised, *e.g*: $[Cu(R_3P)_2]ClO_4$, $[CuI.bpy.\ R_3P]$, $[Cu\{SC(NH_2)_2\}]X$ etc.

Some complexes formed by Cu(I) are discussed below:

(*a*) Conc. HCl dissolves $Cu_2O$ forming CuCl which dissolves in the excess of HCl to produce *hydrocuprochloric acid*, $H[CuCl_2]$.

$$Cu_2O + 2HCl \longrightarrow 2CuCl + H_2O$$
$$[CuCl + HCl \longrightarrow H\,[CuCl_2]] \times 2$$
$$Cu_2O + 4HCl \longrightarrow 2H\,[CuCl_2] + H_2O$$

(*b*) In the absence of air, $Cu_2O$ gives colourless soluble *diammine cuprous hydroxide*, $[Cu(NH_3)_2]OH$.

$$Cu_2O + 4NH_4OH \longrightarrow 2[Cu\,(NH_3)_2]OH + 3H_2O$$

In the presence of air, the above cuprous complex compound is oxidised to *tetrammine cupric hydroxide*, $[Cu(NH_3)_4](OH)_2$.

(*c*) $Cu_2O$ dissolves in alkali cyanide (*e.g.* KCN) solution, forming a colourless complex compound, $K_2[Cu(CN)_3]$.

$$Cu_2O + 6\,KCN + H_2O \longrightarrow 2K_2[Cu(CN)_3] + 3KOH$$

(*d*) CuCl reacts with KCl and forms $K[CuCl_2]$.

$$CuCl + KCl \longrightarrow K[CuCl_2]$$

(*e*) CuCl dissolves readily in $NH_4OH$ forming a colourless solution which contains $[Cu(NH_3)_2]\ Cl.H_2O$ in the absence of any trace of $O_2$.

$$CuCl + 2NH_4OH \longrightarrow \underset{\text{Solution}}{[Cu(NH_3)_2]Cl} + 2H_2O$$

The solution of $[Cu(NH_3)_2]Cl$ turns deep blue when exposed to air due to the formation of $[Cu(NH_3)_4]Cl_2$.

(*f*) $CuSO_4$ reacts with KCN to form $Cu(CN)_2$ which decomposes to give $Cu_2(CN)_2$ and $(CN)_2$. $Cu_2(CN)_2$ dissolves in the excess of KCN and forms *pot. cuprous cyanide,* $K_3[Cu(CN)_4]$.

$$[CuSO_4 + 2KCN \longrightarrow Cu(CN)_2 + K_2SO_4] \times 2$$
$$2Cu(CN)_2 \longrightarrow Cu_2(CN)_2 + (CN)_2$$
$$Cu_2(CN)_2 + 6\ KCN \longrightarrow 2K_3\ [Cu(CN)_4]$$
$$\overline{2CuSO_4 + 10KCN \longrightarrow 2K_3\ [Cu(CN)_4] + 2K_2SO_4 + (CN)_2}$$

(g) When $CuSO_4$ reacts with $Na_2S_2O_3$ (hypo), blue colour of $CuSO_4$ gets discharged and *sodium cuprous thiosulphate,* $Na[Cu(S_2O_3)]$ is formed.

$$CuSO_4 + Na_2S_2O_3 \longrightarrow CuS_2O_3 + Na_2SO_4$$
$$2CuS_2O_3 + 2Na_2S_2O_3 \longrightarrow 2Na\ [Cu\ (S_2O_3)] + Na_2S_4O_6.$$

$Na\ [Cu(S_2O_3)]$, on boiling or on acidification gives a ppt. of $Cu_2S$.

$$2Na\ [Cu\ (S_2O_3)] + H_2O \longrightarrow Cu_2S + Na_2S_2O_3 + H_2SO_4$$

## Compounds of Cu (II)

### 1. Cupric Chloride ($CuCl_2 . 2H_2O$)

**Preparation:** (*i*) The metal or cupric oxide or basic copper carbonate is dissolved in conc. HCl. The resulting solution on crystallisation gives green crystals of hydrated cupric chloride.

$$2Cu + 4HCl + O_2 \longrightarrow 2CuCl_2 + 2H_2O$$
$$CuO + 2HCl \longrightarrow CuCl_2 + H_2O$$
$$Cu(OH)_2 . CuCO_3 + 4HCl \longrightarrow 2CuCl_2 + 3H_2O + CO_2$$

(*ii*) Anhydrous cupric chloride is obtained as dark brown mass when copper metal is heated in excess of chlorine gas or by heating hydrated cupric chloride in HCl gas at 150° C.

$$Cu + Cl_2 \xrightarrow{\Delta} CuCl_2$$
$$CuCl_2 . 2H_2O \xrightarrow[\text{HCl gas}]{150°C} CuCl_2 + 2H_2O$$

**Properties:** (*i*) It is deliquescent compound and is readily soluble in water. The dilute solution is blue but concentrated solution is, however, green. It changes to yellow when conc. HCl is added. The blue colour is due to complex cation $[Cu(H_2O)_4]^{2+}$ and yellow colour due to complex anion, $[CuCl_4]^{2-}$ and green when both are present.

(*ii*) The aqueous solution is acidic due to its hydrolysis.

$$CuCl_2 + 2H_2O \rightleftharpoons Cu(OH)_2 + 2HCl$$

(*iii*) The anhydrous salt on heating forms $Cu_2Cl_2$ and $Cl_2$.

$$CuCl_2 \longrightarrow Cu_2Cl_2 + Cl_2$$

Hydrated salt on strong heating gives CuO, $Cu_2Cl_2$, HCl and $Cl_2$.

$$3CuCl_2.2H_2O \longrightarrow CuO + Cu_2Cl_2 + 2HCl + Cl_2 + 5H_2O$$

(*iv*) It is readily reduced to $Cu_2Cl_2$ by copper turnings, or $SO_2$ gas, or nascent hydrogen (obtained by the action of HCl on Zn) or $SnCl_2$.

$$CuCl_2 + Cu \longrightarrow Cu_2Cl_2$$

$$2CuCl_2 + SO_2 + 2H_2O \longrightarrow Cu_2Cl_2 + 2HCl + H_2SO_4$$

$$2CuCl_2 + 2H \longrightarrow Cu_2Cl_2 + 2HCl$$

$$2CuCl_2 + SnCl_2 \longrightarrow Cu_2Cl_2 + SnCl_4$$

(*v*) A pale blue precipitate of basic cupric chloride, $CuCl_2$. $3Cu(OH)_2$ is obtained when NaOH is added.

$$CuCl_2 + 2NaOH \longrightarrow Cu(OH)_2 + 2NaCl$$

$$CuCl_2 + 3Cu(OH)_2 \longrightarrow CuCl_2.\ 3Cu(OH)_2$$

(*vi*) It dissolves in ammonium hydroxide forming a deep blue solution. On evaporating this solution deep blue crystals of tetrammine cupric chloride are obtained.

$$CuCl_2 + 4NH_4OH \longrightarrow Cu(NH_3)_4Cl_2.H_2O + 3H_2O$$

**Structure.** $CuCl_2$ exists in extended chain form, in crystals known from X-ray analysis.

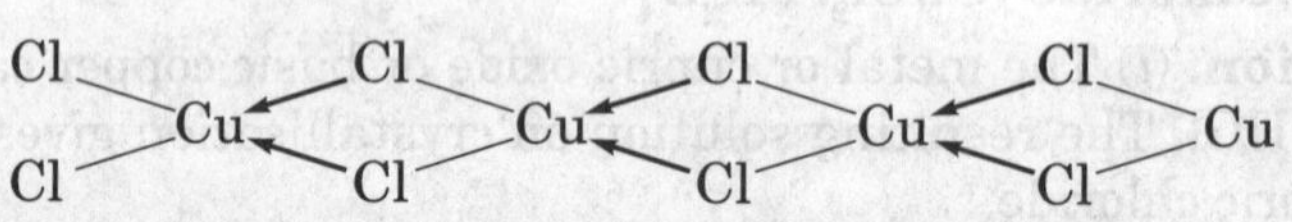

**Uses:** (*i*) It is used as an oxygen carrier in the preparation of organic dyestuffs.

(*ii*) It was used in Deacon's process as a catalyst for manufacturing chlorine gas.

(*iii*) It is used in medicine.

**2. Copper (II) Acetate Monohydrate, $Cu(CH_3(COO)_2.\ H_2O$**

**Preparation.** It is prepared by dissolving copper carbonate or copper oxide in acetic acid and evaporating the solution to crystallisation.

$$2CuCO_3 + 4CH_3COOH \longrightarrow [Cu(CH_3COO)_2.\ H_2O]_2 + 2CO_2$$

$$CuO + CH_3COOH \longrightarrow Cu(CH_3COO)_2.\ H_2O$$

**Properties.** (*i*) It is dark green crystalline solid soluble in alcohol and water.

(*ii*) It is antiferromagnetic having magnetic moment of 1.43 BM at 25°C.

(*iii*) It is dimeric in crytalline form and undergoes dissociation in solution.

**Structure.** Cu (II) acetate, $Cu(CH_3COO)_2$. $H_2O$ is a dimeric molecule in both solid and in solution. Hence it is formulated as $[Cu(CH_3COO)_2]_2$. $2H_2O$ or $Cu_2(CH_3COO)_4$. $2H_2O$. In the structure of dimer, each of the two Cu (II) ions is coordinated in a square plane to four O-atoms of four bridging acetate ($CH_3COO^-$) groups. Each Cu(II) ion is also attached with O-atom of $H_2O$ molecule. The sixth coordination position of each Cu(II) ion is occupied by the other Cu(II) ion (See Fig. 6.32 (*a*)]. Cu (II)–Cu (II) distance is very short. Cu (II) – Cu (II) bond results from the lateral overlap of $d_{x^2-y^2}$ orbitals of the two Cu (II) ions [See Fig. 6.32. (*b*)]. The bond thus obtained is called δ-bond which is different from σ and π bond. The formation of this bond leads to the quenching of the magnetic moment of this compound. Its magnetic moment value is 1.43 BM which is less than the expected value for one electron unpaired electron.

On the basis of CFT the distribution of $d^9$ electrons in Cu(II) ion in tetragonally distorted octahedral complex is :

$$d_{xz}^2 = d_{yz}^2 \left| d_{xy}^2 \right| d_{z^2}^2 \left| d_{x^2-y^2}^1 \right. (n = 1)$$

Energy increasing ⟶

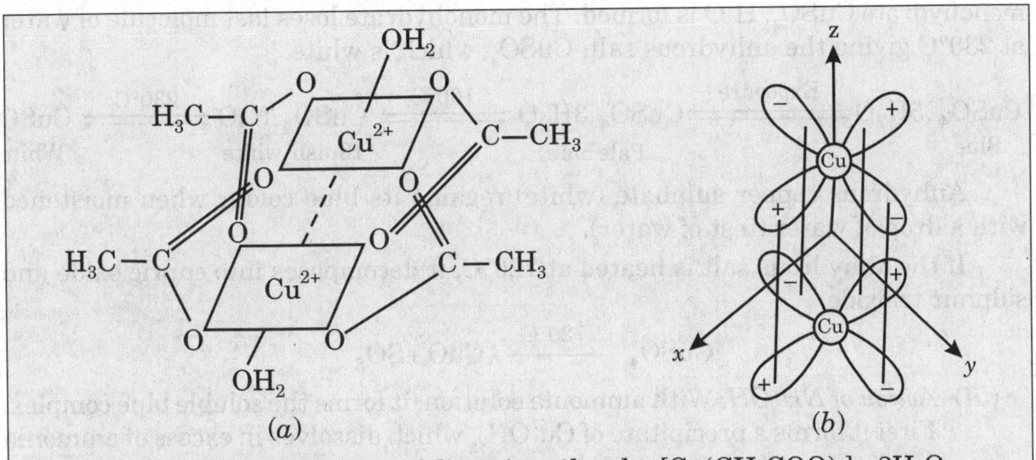


**Fig. 6.32** (*a*) Structure of dimeric molecule, $[Cu(CH_3COO)_2]_2$. $2H_2O$
(*b*) Formation of δ -bond by the lateral overlap between two $3d_{x2-y2}$ orbitals on two Cu (II) ions in $[Cu(CH_3COO)_2]_2$. $2H_2O$ molecule.

### 3. Basic Cupric Acetate or Verdigris, $Cu(CH_3COO)_2$. $Cu(OH)_2$

**Preparation.** It is prepared by exposing thin sheets of copper to the action of vinegar, $CH_3COOH$ and air.

$$2Cu + 2CH_3COOH + O_2 \longrightarrow Cu\,(CH_3COOH)_2.\ Cu(OH)_2$$

**Properties.** It is greenish blue powder. It is sparingly soluble in cold water but dissolves in dilute acids and ammonia.

**Uses.** (*i*) Because of its fairly good covering power, it is used as green pigment, (*ii*) used in dyeing and calico-printing, (*iii*) used in making insecticides and fungicides, (*iv*) when verdigris is mixed with $As_2O_3$, cupric aceto arsenite, (called ***paris green***) $Cu_3(AsO_3)_2$. $Cu(CH_3COO)_2$ which is poisonous is obtained.

**4. Copper (II) Sulphate or Blue Vitriol ($CuSO_4 . 5H_2O$).**

It exists as cupric sulphate pentahydrate, which is also known as **blue vitriol** or **Nila Thotha.**

**Preparation.** Chemical equations involved in the **laboratory preparation** of copper sulphate are given below:

$$CuO + H_2SO_4 \longrightarrow CuSO_4 + H_2O$$

$$Cu(OH)_2 + H_2SO_4 \longrightarrow CuSO_4 + 2H_2O$$

$$CuCO_3 + 2H_2SO_4 \longrightarrow CuSO_4 + H_2O + CO_2$$

On a **commercial scale,** it can be prepared by the action of hot dilute $H_2SO_4$ on copper in the presence of air.

$$2Cu + 2H_2SO_4 + O_2 \longrightarrow 2CuSO_4 + 2H_2O$$

**Properties.** (*i*) The anhydrous salt is *colourless* but the salt is *blue* in colour. It readily dissolves in water but is insoluble in ethyl alcohol.

(*ii*) *Action of heat:* $CuSO_4 . 5H_2O$ crystals effloresce on exposure and converted into a pale blue powder, $CuSO_4 . 3H_2O$. When heated to 100°C, bluish white monohydrate $CuSO_4 . H_2O$ is formed. The monohydrate loses last molecule of water at 230°C giving the anhydrous salt, $CuSO_4$, which is white.

$$\underset{\text{Blue}}{CuSO_4 . 5H_2O} \xrightleftharpoons{\text{Exposure}} \underset{\text{Pale blue}}{CuSO_4 . 3H_2O} \xrightleftharpoons{100°C} \underset{\text{Bluish white}}{CuSO_4 . H_2O} \xrightleftharpoons{230°C} \underset{\text{White}}{CuSO_4}$$

Anhydrous copper sulphate (white) regains its blue colour when moistened with a drop of water (test of water).

If the anhydrous salt is heated at 720°C, it decomposes into cupric oxide and sulphur trioxide.

$$CuSO_4 \xrightarrow{720°C} CuO + SO_3$$

(*iii*) *Action of $NH_4OH$:* With ammonia solution, it forms the soluble blue complex. First it forms a precipitate of $Cu(OH)_2$ which dissolves in excess of ammonia solution.

$$CuSO_4 + 2NH_4OH \longrightarrow Cu(OH)_2 + (NH_4)_2 SO_4$$

$$Cu(OH)_2 + 2NH_4OH + (NH_4)_2SO_4 \longrightarrow \underset{\text{Tetrammine cupric sulphate}}{[Cu(NH_3)_4] SO_4} + 4H_2O$$

The complex is known as ***Schwitzer's reagent*** which is used for dissolving cellulose in the manufacture of artificial silk.

(*iv*) *Action of alkalies:* Alkalies form a pale blue precipitate of copper hydroxide.

$$CuSO_4 + 2NaOH \longrightarrow Cu(OH)_2 + Na_2SO_4$$

(*v*) *Action of potassium iodide:* First cupric iodide ($CuI_2$) is formed which decomposes to give white cuprous iodide ($Cu_2I_2$) and iodine.

$$[CuSO_4 + 2KI \longrightarrow CuI_2 + K_2SO_4] \times 2$$

$$2CuI_2 \longrightarrow Cu_2I_2 + I_2$$

$$2CuSO_4 + 4KI \longrightarrow Cu_2I_2 + 2K_2SO_4 + I_2$$

Above equation can be written in its ionic form.

$$\underset{OA}{2Cu^{2+}} + \underset{RA}{2I^-} \longrightarrow 2Cu^+ + I_2$$

The liberation of iodine in this reaction is *quantitative.* Therefore, this reaction is used to estimate copper *volumetrically* by titrating the molecular iodine formed in the process against a standard solution of sodium thiosulphate.

*(vi) Action of potassium cyanide:* First cupric cyanide is formed which decomposes to give cuprous cyanide and cyanogen gas. Cuprous cyanide dissolves in excess of potassium cyanide to form a complex, potassium cupro-cyanide $K_3[Cu(CN)_4]$.

$$[CuSO_4 + 2KCN \longrightarrow Cu(CN)_2 + K_2SO_4] \times 2$$

$$2Cu(CN)_2 \longrightarrow Cu_2(CN)_2 + (CN)_2$$

$$Cu_2(CN)_2 + 6KCN \longrightarrow 2K_3[Cu(CN)_4]$$

$$2CuSO_4 + 10KCN \longrightarrow 2K_3[Cu(CN)_4] + 2K_2SO_4 + (CN)_2$$

*(vii) Action of potassium ferrocyanide:* Reddish brown precipitate of cupric ferrocyanide is formed (Test of $Cu^{2+}$ ion).

$$2CuSO_4 + K_4Fe(CN)_6 \longrightarrow Cu_2Fe(CN)_6 + 2K_2SO_4$$

*(viii) Addition of electropositive metals:* Electropositive elements like zinc and iron precipitate copper from a solution of copper sulphate.

$$CuSO_4 + Fe \longrightarrow Cu + FeSO_4$$

$$CuSO_4 + Zn \longrightarrow Cu + ZnSO_4$$

*(ix) Action of $H_2S$:* When $H_2S$ is passed through copper sulphate solution, a black precipitate of copper sulphide is formed.

$$CuSO_4 + H_2S \longrightarrow CuS + H_2SO_4$$

The black precipitate dissolves in conc. $HNO_3$.

$$3CuS + 8HNO_3 \longrightarrow 3Cu(NO_3)_2 + 2NO + 3S + 4H_2O$$

*(x) Action of potassium sulphocyanide:* Cupric sulphocyanide is formed.

$$CuSO_4 + 2KCNS \longrightarrow Cu(CNS)_2 + K_2SO_4$$

If $SO_2$ is passed through the solution, a white precipitate of cuprous sulphocyanide is formed.

$$2CuSO_4 + 2KCNS + SO_2 + 2H_2O \longrightarrow Cu(CNS)_2 + K_2SO_4 + 2H_2SO_4.$$

This is the general method for obtaining cuprous compounds.

*(xi) Formation of double sulphates:* Copper sulphate forms double salts with alkali sulphate (*e.g.* $K_2SO_4 \cdot CuSO_4 \cdot 6H_2O$) and with ammonium sulphate (*e.g.* $(NH_4)_2SO_4 \cdot CuSO_4 \cdot 6H_2O$).

*(xii) Action of sodium thiosulphate:* Sodium thiosulphate reacts with copper sulphate to form cupric thiosulphate which is reduced by sodium thiosulphate. The cuprous compound thus formed dissolves in excess of sodium thiosulphate to form a complex.

$$CuSO_4 + Na_2S_2O_3 \longrightarrow CuS_2O_3 + Na_2SO_4$$

$$2CuS_2O_3 + Na_2S_2O_3 \longrightarrow Cu_2S_2O_3 + Na_2S_4O_6$$

$$3Cu_2S_2O_3 + 2Na_2\,S_2O_3 \longrightarrow \underset{\text{Sodium cuprothiosulphate}}{Na_4\,[Cu_6(S_2O_3)_5]}$$

**Uses:** (*i*) Copper sulphate is used for the preparation of other copper compounds.

(*ii*) It finds use in electroplating, electrotyping, calico printing and dyeing.

(*iii*) It is used in agriculture as a fungicide and germicide. ***Bordeaux mixture*** consisting of copper sulphate and lime is used to kill moulds and fungi on vines, trees, potatoes, etc.

(*iv*) It is used as a laboratory reagent especially in the preparation of Fehling's solution.

(*v*) It finds use as an antiseptic in medicine.

(*vi*) It is extensively used in electric batteries.

**Structure of $CuSO_4$. $5H_2O$.** In the structure of lattice of $CuSO_4$. $5H_2O$, one of the five $H_2O$ molecules is different from the remaining four $H_2O$ molecules. This suggests that $CuSO_4$. $5H_2O$ should be formulated as $Cu(H_2O)_4\ SO_4$. $H_2O$. Four $H_2O$ molecules are coordinated to $Cu^{2+}$ ion in a square plane while the fifth $H_2O$ molecule is attached to two coordinated $H_2O$ molecules and $SO_4^{2-}$ ion by H-bonding. Thus fifth $H_2O$ molecule is placed deep inside the crystal lattice of $CuSO_4$. $5H_2O$ (See Fig. 6.33).

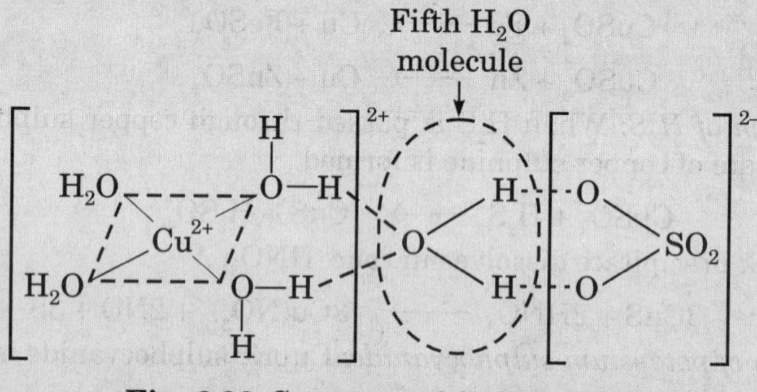


**Fig. 6.33** Structure of $CuSO_4$. $5H_2O$

## Complexes of Cu(II)

$Cu^{2+}$ ion forms various types of complexes some of which are discussed below:

(*i*) $Cu(CN)_2$ dissolves in the excess of KCN, forming *pot. cupricyanide*, $K_2[Cu(CN)_4]$.

$$Cu(CN)_2 + 2KCN \longrightarrow K_2[Cu(CN)_4]$$

(*ii*) $Cu(CN)_2$ reacts with pryridine (*py*) and produces the complex compound, $[Cu(py)_2](CN)_2$.

$$Cu(CN)_2 + 2py \longrightarrow [Cu(py)_2](CN)_2$$

(*iii*) With $NH_4OH$, $CuSO_4$ gives soluble tetrammine, cupric sulphate, $[Cu(NH_3)_4]SO_4$.

$$CuSO_4 + 4NH_4OH \longrightarrow [Cu(NH_3)_4]SO_4 + 4H_2O$$

$$\text{or } Cu^{2+} + 4NH_4OH\ (\textit{Excess}) \longrightarrow [Cu(NH_3)_4]^{2+} + 4H_2O$$

(*iv*) In aqueous medium all copper (II) salts are present as $[Cu(H_2O)_6]^{2+}$ ions. Addition of other ligands leads to successive replacement of these coordinated water molecules. The fifth and sixth stepwise stability constants are decidedly smaller than the first four for copper(II) complexes. With ammonia replacement in aqueous medium leads to $[Cu(H_2O)_2(NH_3)_4]^{2+}$ (commonly written as $[Cu(NH_3)_4]^{2+}$. In liquid ammonia the hexammime is formed.

(*v*) A variety of mixed ligand complexes have been reported *e.g.*: $[Cu(Hbig)(bpy)]X_2$ (X = halogen), $[Cu(Hbig)\ o\text{-}phen)]X_2$, [Cu(gly) (Hbig)]X, (Cu(SB)(bpy)], Cu(SB)(o-phen)] ($H_2SB$ = tridentate dibasic schiff base ONO donors). All these complexes are believed to assume distorted octahedral structures particularly in coordinating solvents like water, pyridine etc.

(*vi*) When a base ($OH^-$) is added to a solution containing $Cu^{2+}$ ions, light blue ppt. of $Cu(OH)_2$ is formed.

$$Cu^{2+} + 2OH^- \longrightarrow Cu(OH)_2 \downarrow$$

With aqueous $NH_3$, $Cu(OH)_2$ gives an intense blue solution due to the formation of complex ion, $[Cu(NH_3)_4]^{2+}$

$$Cu(OH)_2(s) + 4NH_3 \longrightarrow \underset{\text{Solution}}{[Cu(NH_3)_4]^{2+}} + 2OH^-$$

$[Cu(NH_3)_4]^{2+}$ gets decomposed by acids and also by heating.

$$[Cu(NH_3)_4]^{2+} + 4H^+ \longrightarrow Cu^{2+} + 4NH_4^+$$

$$[Cu(NH_3)_4]^{2+} \xrightarrow{\Delta} Cu^{2+} + 4NH_3$$

## Complexes of Cu (III)

+3 oxidation state of Cu is unstable, but this state of $Cu^{3+}$ ion ($Cu^{3+} = d^8$ ion) has been stabilised in some complexes as shown below.

(*i*) More recently a series of hexafluorocuperates (III) of the type $M_3[CuF_6]$ (M= Na, K, Rb, Cs) and $M_2M'[CuF_6]$ (M = K, Rb, Cs, M′= Li, Na, K, Rb, Cs) have been prepared. These complex compounds of Cu(III) have been obtained by the fluorination (under high pressure) of appropriate chlorocuperate (II) or alkali halide/$CuCl_2$ mixture. All these complexes are green micro crystalline solids and are paramagnetic ($\mu \simeq$ 2.9 to 3.0 BM). This value of $\mu$ indicates that Cu (III) complexes have octahedral stereochemistry.

(*ii*) Hexaoxotellurate (VI) ion, $\left[Te^{+6}O_6\right]^{6-}$ and hexaoxoiodate (VII) ion, $\left[I^{+7}O_6\right]^{5-}$ have been found to stabilise Cu(III) by forming the complexes $Na_7H_2[Cu^{3+}(TeO_6)_2].12H_2O$, $K_7[Cu^{3+}(IO_6)_2].7H_2O$ and $Na_7[Cu^{3+}(IO_6)_2].12H_2O$. $(Te^{+6}O_6]^{6-}$ and $[I^{+7}O_6]^{5-}$ ions have Te and I atoms in their highest oxidation state (Te = +6, I = +7). $Na_7H_2[Cu(TeO_6)_2].12H_2O$ is obtained (*a*) by the anodic oxidation of Cu in strongly alkaline solution of potassium hexaoxotellurate (VI), $K_6$ [$TeO_6$], (*b*) by the oxidation of Cu (II) in strongly alkaline hexaoxotellurate (VI) by $K_2S_2O_8$.

$K_7$ [Cu ($IO_6$)$_2$] . $7H_2O$ and $Na_7$ [Cu ($IO_6$)$_2$]. $12H_2O$ are obtained by anodic oxidation of an alkaline solution of Cu (II) in the presence of excess of hexaoxoiodate (VII)

$Na_7$ $H_2$ [Cu (Te $O_6$)$_2$]. $12H_2O$ has *maroon colour* and is *diamagnetic*. $K_7$ [Cu($IO_6$)$_2$].$7H_2O$ and $Na_7$[Cu($IO_6$)$_2$]. $12H_2O$ are *red coloured diamagnetic* crystals. [Cu($IO_6$)$_2$]$^{7-}$ ion present in both these crystals has the structure shown in Fig. 6.34.

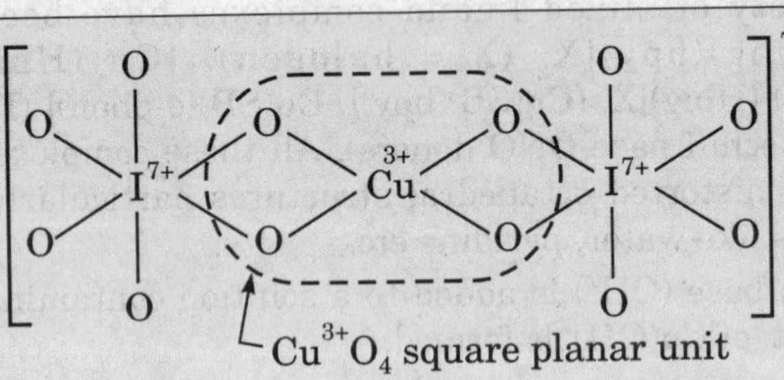


**Fig. 6.34** Structure of $\left[Cu^{3+}(IO_6)_2\right]^{7-}$ ion.

The diamagnetic nature of the above said complexes results from $dsp^2$ hybridisation of $Cu^{3+}$ ion ($3d^8$ ion) in square planar unit, $Cu^{3+}$ $O_4$ present in the structure of [$Cu^{3+}$ ($IO_6$)$_2$]$^{7-}$ ion. $dsp^2$ hybridisation has been shown below.

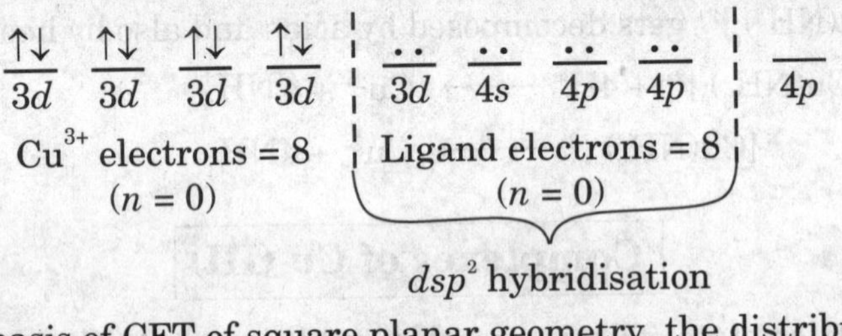


On the basis of CFT of square planar geometry, the distribution of $d^8$ electrons of $Cu^{3+}$ ion in five *d*-orbitals is as:

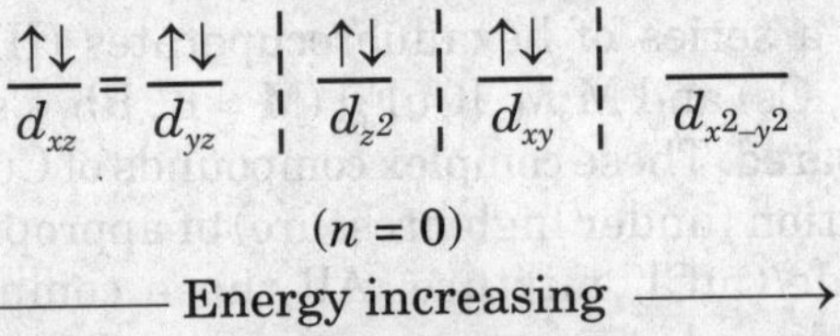


## Zinc, $Zn_{30}$

### Physical Properties of Zinc

Zinc is a bluish-white metal, brittle at the room temperature. It loses its brittleness at about 120°C but regains it at 200°C when it can be easily powdered. Its specific gravity is 7.14, melting point 419.5°C and boiling point 906°C. It is a good conductor of heat and electricity. At 100° –150°C it becomes ductile and malleable.

**Chemical Properties of Zinc**

1. **Action of air.** Zinc is stable in air. In moist air a protective covering of basic zinc carbonate is formed. When heated to 500°C, it catches fire with a bluish-white flame forming zinc oxide which is very light and is called **philosopher's wool**.

2. **Action of water.** Pure zinc is not acted upon by water. Impure zinc decomposes boiling water slowly and steam rapidly.

$$Zn + 2H_2O \longrightarrow Zn(OH)_2 + H_2$$

Zn-Cu couple decomposes water at ordinary temperature evolving hydrogen.

3. **Action of acids.** Very pure zinc reacts very slowly with acids whereas commercial zinc reacts readily with all acids. The reason for this has not been definitely established. Most probably a layer of condensed hydrogen forms on the surface of pure zinc and stops the action of the acid.

Impure zinc contains particles of other metals as impurity which are electronegative to zinc and set up a galvanic action and form a number of tiny voltaic cells similar to that of copper and zinc.

(*a*) *Hydrochloric acid:* Dilute or concentrated hydrochloric acid forms zinc chloride with the evolution of hydrogen.

$$HCl + Zn \longrightarrow ZnCl_2 + H_2$$

(*b*) *Sulphuric acid:* Dilute acid reacts with commercial zinc evolving hydrogen. Hot and concentrated acid attacks zinc giving off sulphur dioxide.

$$Zn + H_2SO_4\ (\textit{Dilute}) \longrightarrow ZnSO_4 + H_2$$

$$Zn + 2H_2SO_4\ (\textit{Hot and concentrated}) \longrightarrow ZnSO_4 + SO_2 + 2H_2O$$

Moderately concentrated acid (20 per cent) when heated with the metal evolves hydrogen sulphide and precipitates a little sulphur.

$$4Zn + 5H_2SO_4\ (\textit{Moderately concentrated}) \xrightarrow{\Delta} 4ZnSO_4 + H_2S + 4H_2O$$

$$3Zn + 4H_2SO_4\ (\textit{Moderately concentrated}) \xrightarrow{\Delta} 3ZnSO_4 + S + 4H_2O$$

(*c*) *Nitric acid.* Zn reacts with very dil. $HNO_3$(6%) and is oxidised to $Zn(NO_3)_2$. $HNO_3$ is reduced to $NH_4NO_3$.

$$4Zn + 10HNO_3\ (\textit{Very dilute}) \longrightarrow 4Zn(NO_3)_2 + NH_4NO_2 + 3H_2O$$

or $$4Zn + NO_3^- + 10H^+ \longrightarrow 4Zn^{2+} + \overset{+}{NO_4} + 3H_2O$$

Zn reacts with cold and dil. $HNO_3$ (20%) and is oxidised to $Zn(NO_3)_2$. $HNO_3$ is reduced to $N_2O$.

$$4Zn + 10HNO_3\ (\textit{Cold and dilute}) \longrightarrow 4Zn\ (NO_3)_2 + 2N_2O + 5H_2O$$

or $$Zn + 2NO_3^- + 10H^+ \longrightarrow 4Zn^{2+} + N_2O + 5H_2O$$

Zn reacts with cold and conc. $HNO_3$ (70%) and is oxidised to $Zn\ (NO_3)_2$. $HNO_3$ is reduced to $NO_2$.

$$Zn + 4HNO_3\ (\textit{Cold and conc.}) \longrightarrow Zn(NO_3)_2 + 2NO_2 + 2H_2O$$

or $$Zn + 2NO_3^- + 10H^+ \longrightarrow -2Zn^{2+} + 2NO_2 + 2H_2O$$

Zn also reacts with cold and moderately conc. $HNO_3$ (50%). In this reaction Zn is oxidised to $Zn(NO_3)_2$ and NO gas is evolved.

$$3Zn + 8HNO_3 \longrightarrow 3Zn(NO_3)_2 + 4H_2O + 2NO$$

or $$3Zn + 2NO_3^- + 8H^+ \longrightarrow 3Zn^{2+} + 2NO + 4H_2O$$

4. **Action of alkalies.** Zinc dissolves in hot and concentrated caustic alkalies forming alkali zincates with the evolution of hydrogen.

$$Zn + 2NaOH \longrightarrow Na_2ZnO_2 + H_2$$

5. **Displacement of metals.** Most metals being less electropositive than zinc are precipitated from their salt solutions. The reaction is utilized in the precipitation of silver and gold from the complex cyanide solution in the *cyanide process*.

$$Zn + CuSO_4 \longrightarrow ZnSO_4 + Cu$$

$$Zn + 2NaAu(CN)_2 \longrightarrow Na_2Zn(CN)_4 + 2Au\downarrow$$

6. **Action of non-metals.** Zinc reacts with halogens and sulphur directly on heating.

$$Zn + Cl_2 \longrightarrow ZnCl_2$$

$$Zn + S \longrightarrow ZnS$$

7. **Action of ammonia.** Zinc nitride is formed.

$$3Zn + 2NH_3 \longrightarrow Zn_3N_2 + 3H_2$$

8. **Reducing action.** It acts as a powerful reducing agent, *e.g.*, it reduces ferric salts to ferrous salts and nitrate to ammonia.

$$Fe_2(SO_4)_3 + Zn \longrightarrow 2FeSO_4 + ZnSO_4$$

$$NaNO_3 + 4Zn + 7NaOH \longrightarrow 4Na_2ZnO_2 + 2H_2O + NH_3$$

Zinc in presence of an acid is an excellent reducing agent, the general reaction being:

$$Zn + 2HCl + XO \longrightarrow ZnCl_2 + H_2O + X$$

It has reducing power much greater than that of ordinary hydrogen. The greater chemical activity is explained by the fact that the whole of the liberated energy in the reaction does not appear as heat, but a part is associated with the hydrogen molecules increasing its chemical energy. Zinc amalgam is an effective reducing agent.

9. **Action of alkyl halides.** Zn, like Mg, forms addition compounds with alkyl halides.

$$Zn + C_2H_5I \longrightarrow ZnC_2H_5I$$

### Uses of Zinc

(*i*) Zinc is extensively employed for galvanizing iron, *i.e.*, for the manufacture of iron sheets for roofing, iron gutters, utensils, buckets, pipes etc.

(*ii*) For sherardizing iron for the manufacture of conduit pipes, laying telephone cables, electric wire etc.

(*iii*) Zinc is used in the laboratory for the preparation of hydrogen gas and as a powerful reducing agent with dilute hydrochloric or sulphuric acid or concentrated solution of a caustic alkali and also as zinc dust or zinc-copper couple.

(*iv*) It is also used as cathode plate in dry cells and electric batteries and in producing smokescreens in war.

(*v*) For desilverizing lead in Parke's process and for the precipitation of silver and gold in cyanide process.

(*vi*) In the form of zinc dust it is used for the manufacture of dyestuffs, paints drugs and face powders, snow, creams, etc.

(*vii*) Many zinc compounds find extensive uses, *e.g.*, the oxide in the manufacture of paints, filling rubber and in medicine; the sulphate, in preserving hides, calico printing, in medicine as eye lotion and paints etc.

(*viii*) For the preparation of useful alloys like *brass, muntz metal, German silver and die casting metal.*

(*a*) *Brass:* It is yellow, malleable and ductile alloy containing 60 to 80 per cent copper and 20 to 40 per cent zinc.

(*b*) *Muntz metal*: It is an alloy containing 60 per cent copper and 40 per cent zinc.

(*c*) *German silver:* It is a shining white alloy containing 25-50 per cent copper, 25-35 per cent zinc and 10 to 35 per cent nickel. It is used for preparing cheap quality ornaments and utensils.

(*d*) *Die casting metal*: Zinc alloyed with 4 per cent aluminium, up to 3.5 per cent copper and 0.05 per cent manganese is used for the production of die castings.

## Aqueous Chemistry of Zn (II)

Zn shows +2 O.S. in all its compounds.

(*i*) $Zn^{2+}$ ion ($Zn^{2+} = 3d^{10}$) is colourless due to the presence of $3d^{10}$ configuration in $Zn^{2+}$ ion.

(*ii*) $Zn^{2+}$ ion undergoes hydrolysis in $H_2O$ to give acidic solution due to the formation of $H^+$ ions.

$$Zn^{2+} + H_2O \xrightarrow{\text{Hydrolysis}} [Zn(OH)]^+ + H^+$$

(*iii*) When a base ($OH^-$) is added to the solution of $Zn^{2+}$ ions, ppt of $Zn(OH)_2$ is obtained.

$$Zn^{2+} + 2OH^- \longrightarrow [Zn(OH)]_2 \downarrow$$

$Zn(OH)_2$ is amphoteric, since it dissolves in acids as well as in alkalies.

$$Zn(OH)_2 + 2HCl \longrightarrow ZnCl_2 + 2H_2O$$

$$Zn(OH)_2 + 2NaOH \longrightarrow \underset{\text{(Soluble)}}{Na_2[Zn(OH)_4]}$$

## Complexes of Zn(II)

(*a*) $ZnCl_2$ gives hydrated complex ion, $[Zn(H_2O)_4]^{2+}$ in $H_2O$

$$ZnCl_2 + 4H_2O \longrightarrow [Zn(H_2O)_4]Cl_2$$

or $$Zn^{2+} + 4H_2O \longrightarrow [Zn(H_2O)_4]^{2+}$$

(*b*) $Zn(OH)_2$ dissolves in aqueous $NH_3$ and KCN solution to form the complex ions, $[Zn(NH_3)_4]^{2+}$ and $[Zn(CN)_4]^{2-}$ respectively.

$$Zn(OH)_2 + 4NH_3 \longrightarrow (Zn(NH_3)_4]\ (OH)_2$$

or $$Zn^{2+} + 4NH_3 \longrightarrow (Zn(NH_3)_4]^{2+}$$

$$Zn(OH)_2 + 4KCN \longrightarrow K_2\ [(Zn\ (CN)_4] + 2KOH$$

or $$Zn^{2+} + 4CN^- \longrightarrow [Zn(CN)_4]^{2-}$$

$[Zn(CN)_4]^{2-}$ ion is more stable than $[Zn(NH_3)_4]^{2+}$ as is evident from the values of the equilibrium constants (*K*) of decomposition reaction of these complex ions.

$$[Zn(CN)_4]^{2-} \rightleftharpoons Zn^{2+} + 4CN^-, K_{[Zn(CN)_4]^{2-}} = 1.2 \times 10^{-18}$$

$$[Zn(NH_3)_4]^{2+} \rightleftharpoons Zn^{2+} + 4NH_3, K_{[Zn(NH_3)_4]^{2+}} = 3.4 \times 10^{-10}$$

Both the complex ions have *tetrahedral structure*. $Zn^{2+}$ ion also forms complex ion, $[Zn(NH_3)_6]^{2+}$ which is known in crystals only and has *octahedral structure*.

**Biological Importance of Zinc**

Zinc is very important for proper functioning of the enzyme system of animals. Human body contains about 2g of zinc which is next only to iron (about 4g). There are more than twenty zinc-containing enzymes in the body which are responsible for proper absorption of $CO_2$ by red blood cells in muscles and other tissues and for maintaining proper *p*H in the muscles and tissues.

Some of the zinc-containing enzymes play an important role in the digestion of proteins by animals. Besides this, zinc enzymes are also involved in the energy release processes, sugar metabolism and metabolism of alcohol in the human body.

## UNIVERSITY QUESTIONS

1. (*a*) How is Ti extracted from ilmenite?
   (*b*) Give the technical applications of Ti metal and its compounds. (*Kanpur 2010*)
2. Give the preparation, properties and uses of $TiO_2$. [*Lucknow 2010*]
3. Write the preparation, properties and uses of ammonium venadate. (*Utkal -2010*)
4. What are different ores of chromium? Discuss the preparation, properties and uses of chromyl chloride. (*Magadh 2010*)
5. How will you convert chromite ore into sodium chromate? (*Utkal 2010*)
6. Discuss different oxidation states of Cr. [*Lucknow* 2010)

**7.** (*a*) Discuss the main ores of manganese. How is it extracted from its ores?

(*b*) What is the product when $KMnO_4$ oxidises toluene in acidic medium? [*Himachal Pradesh 2010*)

**8.** (*a*) Outline the method for the extraction of Mn-metal from its ores. Write all the equations of reactions involved.

(*b*) What happens when $KMnO_4$ is added to $MnSO_4$ solution in neutral medium? (Write the reaction involved) (*Utkal 2010*)

**9.** (*a*) Why Mn (II) compounds show the highest paramagnetic behaviour among the bivalent ions of the first transition series?

(*b*) Draw the structure of $Mn_2O_7$. (*GND Amritsar 2010*)

**10.** Explain the most stable oxidation state of Mn. [*Lucknow 2010*)

**11.** Discuss the chemistry of Fe, Co and Ni with reference to their periodic position, oxidation states, oxides, complexes and uses. [*Ranchi 2010*]

**12.** (*a*) Discuss electrolytic method of refining the impure cobalt.

(*b*) Give the structure of sodium cobaltinitrite. (*Himachal Pradesh 2010*)

**13.** What are coinage metals? (*Andhra 2010*)

**14.** (*i*) How does Cr occur in nature? Starting from chromite ore, how will you obtain:

(*a*) Pure Cr (*b*) Ferro-chrome (*c*) $K_2Cr_2O_7$ (*d*) $CrO_2Cl_2$

(*ii*) Describe the preparation, properties and uses of the following compounds.

(*a*) $TiCl_4$ (*b*) $KMnO_4$ (*c*) $Na_3[Co(NO_2)_6]$ (*Ch. Charan Singh, Meerut 2012*)

# Chemistry of Elements of Second Transition Series (4*d*-Series Elements)

## 4*d*-Series Elements

We have already shown in Chapter 5 that 2*nd* transition series (4*d*-series) contains ten elements namely $Y_{39}$, $Zr_{40}$, $Nb_{41}$, $Mo_{42}$, $Tc_{43}$, $Ru_{44}$, $Rh_{45}$, $Pd_{46}$, $Ag_{47}$ and $Cd_{48}$ which are present in 5*th* period of the periodic table as shown in Fig 7.1

| Group | 1 | 2 | 3 | 4 | 5 | 6 | 7 | 8 | 9 | 10 | 11 | 12 | 13 | 14 | 15 | 16 | 17 | 18 |
|---|---|---|---|---|---|---|---|---|---|---|---|---|---|---|---|---|---|---|
| 1 | 1 H | | | | | | | | | | | | | | | | | 2 He |
| 2 | 3 Li | 4 Be | | | | | | | | | | | 5 B | 6 C | 7 N | 8 O | 9 F | 10 Ne |
| 3 | 11 Na | 12 Mg | | | | | | | | | | | 13 Al | 14 Si | 15 P | 16 S | 17 Cl | 18 Ar |
| 4 | 19 K | 20 Ca | 21 Sc | 22 Ti | 23 V | 24 Cr | 25 Mn | 26 Fe | 27 Co | 28 Ni | 29 Cu | 30 Zn | 31 Ga | 32 Ge | 33 As | 34 Se | 35 Br | 36 Kr |
| 5 | 37 Rb | 38 Sr | 39 Y | 40 Zr | 41 Nb | 42 Mo | 43 Tc | 44 Ru | 45 Rh | 46 Pd | 47 Ag | 48 Cd | 49 In | 50 Sn | 51 Sb | 52 Te | 53 I | 54 Xe |
| 6 | 55 Cs | 56 Ba | 57-71 La* | 72 Hf | 73 Ta | 74 W | 75 Re | 76 Os | 77 Ir | 78 Pt | 79 Au | 80 Hg | 81 Ti | 82 Pb | 83 Bi | 84 Po | 85 At | 86 Rn |
| 7 | 87 Fr | 88 Ra | 89-103 Ac** | 104 Rf | 105 Db | 106 Sg | 107 Bh | 108 Hs | 109 Mt | 110 Ds | 111 Rg | 112 Cn | (113) | (114) | (115) | (116) | (117) | (118) |

f-block elements

| | | | | | | | | | | | | | | |
|---|---|---|---|---|---|---|---|---|---|---|---|---|---|---|
| Lanthanides* | 58 Ce | 59 Pr | 60 Nd | 61 Pm | 62 Sm | 63 Eu | 64 Gd | 65 Tb | 66 Dy | 67 Ho | 68 Er | 69 Tn | 70 Yb | 71 Lu |
| Actinide** | 90 Th | 91 Pa | 92 U | 93 Np | 94 Pu | 95 Am | 96 Cm | 97 Bk | 98 Cf | 99 Es | 100 Fm | 101 Md | 102 No | 103 Lr |

**Fig. 7.1** Position of 2*nd* transition series elements (4*d*-series elements) in the periodic table.

These elements are also called 4*d-series elements* because the differenting electron enters in 4*d*-orbitals of the atoms of these elements. General characteristics of these elements along with those of 1*st* and 3*rd* transition series have already been discussed in chapter 5. Important physical properties of these elements have been given in Table 7.1 for the sake of reference. In this chapter we shall discuss the chemistry of all the ten elements of 2*nd* transition series and their compounds.

**Table 7.1. Important physical propertes of the elements of 2*nd* transition series (4*d*-series elements)**

| Property | Y | Zr | Nb | Mo | Tc | Ru | Rh | Pd | Ag | Cd |
|---|---|---|---|---|---|---|---|---|---|---|
| (*i*) Atomic number | 39 | 40 | 41 | 42 | 43 | 44 | 45 | 46 | 47 | 48 |
| (*ii*) Outer electronic configuration | $4d^15s^2$ | $4d^25s^2$ | $4d^45s^1$ | $4d^55s^1$ | $4d^55s^2$ | $4d^75s^1$ | $4d^85s^1$ | $4d^{10}5s^0$ | $4d^{10}5s^1$ | $4d^{10}5s^2$ |
| (*iii*) Atomic radii (pm) | 180 | 160 | 149 | 139 | 136 | 134 | 134 | 137 | 144 | 154 |
| (*iv*) Ionic radii (pm) | 90 ($Y^{3+}$) | 80 ($Zr^{4+}$) | 70 ($Nb^{5+}$) | 62 ($Mo^{6+}$) | – | 82 ($Ru^{3+}$) | 69 ($Rh^{3+}$) | 50 ($Pd^{2+}$) | 129 ($Ag^+$) | 97 ($Cd^{2+}$) |
| (*v*) Crystal structure | hcp | hcp, bcc | bcc | bcc | hcp | hcp | fcc | fcc | fcc | hcp |
| (*vi*) Atomic volume (ml) | 19.8 | 14.1 | 10.8 | 9.4 | 11.5 | 8.3 | 8.3 | 8.9 | 10.4 | 13.1 |
| (*vii*) Density (g/cm$^3$) | 4.47 | 6.49 | 8.40 | 10.2 | 10.50 | 12.20 | 12.40 | 12.00 | 10.51 | 8.65 |
| (*viii*) Atomic mass (amu) | 88.91 | 90.22 | 92.91 | 95.94 | 98 | 101.1 | 102.9 | 106.4 | 107.9 | 112.4 |
| (*ix*) Enthalpy of atomisation (KJ mol$^{-1}$) | 410 | 611 | 724 | 659 | 649 | 650 | 577 | 381 | 286 | 111 |
| (*x*) Melting points (°C) | 1509 | 1852 | 2415 | 2610 | 2200 | 2500 | 1966 | 1552 | 960.8 | 320.9 |
| (*xi*) Boiling points (°C) | 2927 | 1500 | 3300 | 5560 | – | 4900 | 4500 | 3980 | 2210 | 765 |
| (*xii*) Ionisation energy (KJ mol$^{-1}$) | 616 | 660 | 664 | 685 | 702 | 711 | 720 | 805 | 731 | 867.7 |
| (*xiii*) Stable oxidation states | +3 | +4 | +5 | +6 | +4 | +3 | +3 | +2 | +1 | +2 |
| (*xiv*) Electronegativity | 1.2 | 1.4 | 1.6 | 1.8 | 2.9 | 2.9 | 2.2 | 2.2 | 1.9 | 1.9 |
| (*xv*) Heat of fusion (KJ mol$^{-1}$) | 11.28 | 16.7 | 26.7 | 27.6 | 23.0 | 25.5 | 21.7 | 16.7 | 11.3 | 6.1 |
| (*xvi*) Heat of vaporisation (KJ mol$^{-1}$) | 389 | 502 | – | 535 | 502 | 603 | 577 | 381 | 286 | 180 |

## Yttrium, $Y_{39}$

## Properties of Yttrium

Yttrium is a silvery-white metal. Some of its physical properties are given in Table 7.1.

## Compounds of Yttrium

**1. Oxides and hydroxides** Yttrium forms the oxide $Y_2O_3$. $Y(OH)_3$ is obtained as gelatinous precipitate when alkali is added to an aqueous solution of $Y^{3+}$ salt. $Y(OH)_3$ is basic in nature.

**2. Halides.** With the exception of $YF_3$, all halides are soluble in water. These are prepared by heating the metal with the halogens.

**3. Hydrides.** Yttrium forms a non-stoichiometric hydride with the formula approximately equal to $YH_3$.

**4. Complexes.** Yttrium forms a number of complexes with EDTA, acac$^-$, NCS$^-$ with formulae, [Y(EDTA)], [Y(acac)$_3$ . $H_2O$], [Y(NCS)$_6$]$^{3-}$ respectively. Complexes of $Y^{3+}$ with coordination numbers 8 and 10 are also known.

**5. Organometallic compounds.** Yttrium forms $Y(C_5H_5)_3$ when $YCl_3$ reacts with $NaC_5H_5$. The bonding in $Y(C_5H_5)_3$ is predominately ionic.

## Zirconium, $Zr_{40}$

### Comparison between Zr and Ti

We know that Zr and Ti are placed in the same group *viz* group 4(IVB). These two elements can be compared in the following points:

(*i*) Important difference between Zr and Ti is that the lower oxidation states of Zr (+3 and +2) are not so stable as those of Ti.

(*ii*) Zr is much more soft than Ti.

(*iii*) Zr resembles Ti in its passivity and resistance to corrosion and to the action of acids.

(*iv*) Like Ti, Zr forms interstial borides, carbides, nitrides, etc.

(*v*) Due to increased atomic size, oxide of Zr is more basic than that of Ti.

(*vi*) As for Ti, the most stable oxidation state of Zr is +4.

### Compounds of Zirconium

**1. Oxides. $ZrO_2$ . $nH_2O$** is obtained as a gelatinous precipitate by adding an alkali to zirconium (IV) solutions. On strong heating, the hydrous oxides gives **$ZrO_2$**. The hydrous oxide is also obtained by vapour phase oxidation of $ZrCl_4$. This is done by passing the vapour of $ZrCl_4$ mixed with air through a special flame produced by burning a hydrocarbon in excess of oxygen.

**2. Halides. $ZrCl_4$** is made by passing chlorine over the dioxide heated with carbon. It gets easily hydrolysed yielding oxochloride.

$$ZrCl_4 + H_2O \longrightarrow ZrOCl_2 + 2HCl$$

The oxochloride crystallises from the solution as the octahydrate, $ZrOCl_2 . 8H_2O$. $ZrCl_4$ resembles $TiCl_4$ in most of its properties.

**$ZrBr_4$** and **$ZrI_4$** are also known. These are made by direct combination of the elements and are similar in properties to $ZrCl_4$ . $ZrBr_4$ is yellow while $ZrI_4$ is red-brown in colour.

**3. Zirconyl compounds.** Zirconyl compounds contain zirconyl ion, $ZrO^{2+}$. These compounds exist in solution.

$ZrO(NO_3)_2$, $ZrOSO_4$, etc. are important compounds. In the solid state, $ZrO(NO_3)_2$ contains an oxy-bridged chain as shown in Fig. 7.2.

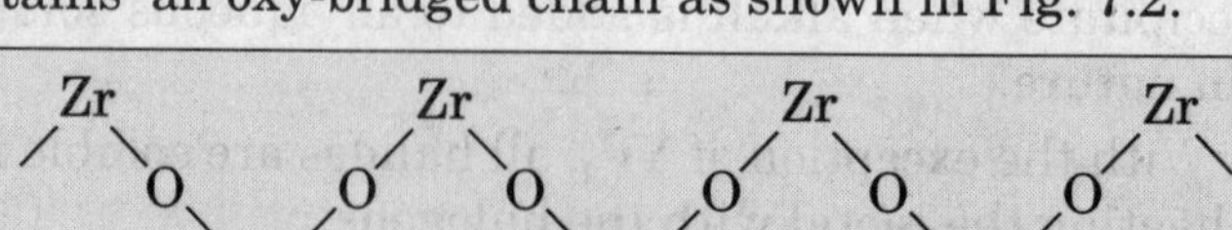


**Fig. 7.2** Oxygen-bridged chain structure of $ZrO(NO_3)_2$ in the solid state

Zirconyl nitrate solution is used for the removal of phosphate ions in qualitative analysis.

**4. Complexes of zirconium.** Apart from the complex ions, $[ZrX_6]^{2-}$ (X = halogen), some 7-coordinated complex ions of Zr are also known. Well known example is $[ZrF_7]^{3-}$. $Na_3[ZrF_7]$ has pentagonal bipyramidal structure while $(NH_4)_3$ $[ZrF_7]$ has capped trigonal prismatic structure. $Na_4[ZrF_8]$ has bi disphenoid structure. In this complex coordination number of Zr is 8.

## Niobium ($Nb_{41}$) and Tantalum ($Ta_{73}$)

Atomic radii of $Nb_{41}$ and $Ta_{73}$ are identical. Hence these elements occur together and have similar properties. Due to their similar properties, these elements are discussed together.

### Properties and Uses

Both these metals are moderately hard grey metals with high melting points (Nb, 2741 K; Ta, 3253 K). These are stable in air and tarnished only superficially when heated. When heated in current of chlorine, they burn vigorously at red heat forming chlorides. They unite with other non-metals (e.g., $O_2$, $N_2$, S, Se) directly when heated forming oxides, nitrides, sulphides, etc. Hydrofluoric acid attacks them slowly and all other acids including aqua-regia have no action upon them. The best solvent is a mixture of hydrofluoric and nitric acids which gives solution containing fluorocomplexes. The metals are slowly attacked by fused alkalis.

The complete insolubility of the metals in oxidizing acids is due to their becoming passive in solutions of oxidizing agents.

Due to their resistance to corrosion, these meals are used in surgical instruments, and certain manufacturing plants. Niobium is used in chromium – nickel stainless steel and its carbide in carbide tools. Because tantalum is unreactive and is not rejected by human body, it is used for making metal plates, screws and wires needed for repairing fractured bones in very bad shape.

## Compounds with +5 Oxidation State

### 1. Oxides

$Nb_2O_5$ and $Ta_2O_5$ are white powders obtained by dehydration of their hydrous oxides (so-called niobic and tantalic acids) or by roasting their compounds with other elements like S, C.

They are relatively inert substances and can be brought in solution either by dissolving in HF or by fusing them with NaOH, $Na_2CO_3$ or $NaHSO_4$.

$$Nb_2O_5 + 3Na_2CO_3 \longrightarrow \underset{\text{Sod. orthoniobate}}{2Na_3NbO_4} + 3CO_2$$

$Ta_2O_5$ is more sluggish as compared to $Nb_2O_5$.

## 2. Halides

The pentafluorides are prepared by heating the metals, pentoxide and pentachloride in fluorine. These are white volatile solids.

The oxofluorides are obtained by heating the pentafluorides with $SiO_2$.

$$2TaF_5 + SiO_2 \longrightarrow 2TaOF_3 + SiF_4$$

The pentafluorides ($MF_5$) consist of cyclic tetramers of octahedral $MF_6$ units with two *cis* fluorine atoms acting as bridging groups with other octahedra. (See fig. 7.3)

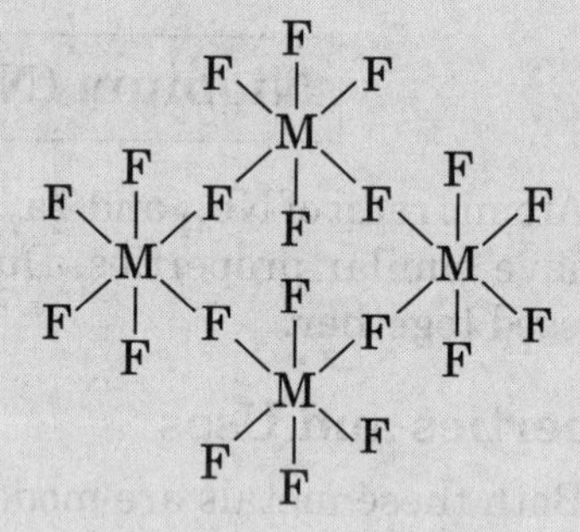


**Fig. 7.3** Cyclic tetrameric structure of $MF_5$ molecules (M = Nb or Ta)

$NbF_5$ and $TaF_5$ react with $F^-(aq)$ to give $[MF_6]^-$ and $[MF_7]^{2-}$ (M is Nb or Ta). With very high concentration of $F^-$ ions in aqueous solutions, $TaF_5$ forms $[TaF_8]^{3-}$ and $NbF_5$ forms $[NbOF_6]^{3-}$.

Other pentahalides can be prepared by direct combination of the elements at elevated temperatures or by heating pentoxides with carbon in presence of the halogen.

$$2Ta + 5I_2 \xrightarrow{570\text{–}670K} 2TaI_5$$

$$Nb_2O_5 + 5C + 5Cl_2 \xrightarrow{670K} 2NbCl_5 + 5CO$$

All the six pentahalides are yellow to brown or purple-red solids. These are soluble in organic solvents like ether, $CCl_4$, etc. In presence of water they are readily hydrolysed as follows:

$$2TaCl_5 + 5H_2O \longrightarrow Ta_2O_5 + 10HCl$$

The chlorides are effective Friedel-Crafts catalyts like anhydrous $AlCl_3$. They behave as Lewis acids and form 1 : 1 complexes with $Cl^-$ ions and other donor molecules like ethers and dialkyl sulphides.

The pentachlorides ($MCl_5$) form monomeric trigonal bipyramids in the vapour form but the solid $NbCl_5$ and $TaCl_5$ are dimeric with two octahedra joined by sharing two corners. (See Fig 7.4)

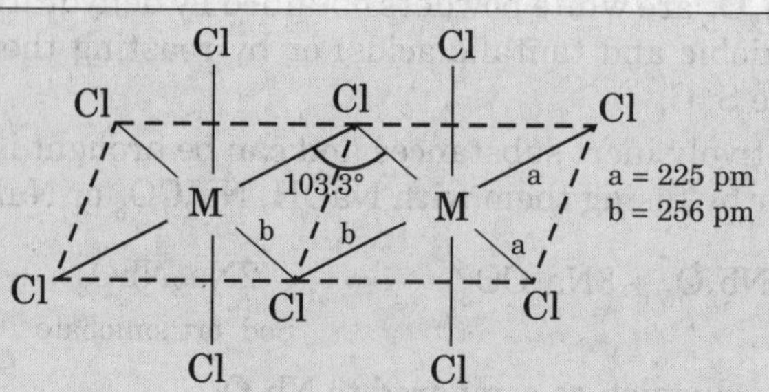


**Fig 7.4** Dimeric structure of $MCl_5$ molecules (M = Nb or Ta).

### 3. Oxohalides

Some known oxohalides of the two metals are $NbOCl_3$, $NbOBr_3$, $NbOI_2$, $NbO_3I$, $TaOCl_3$ and $TaOBr_3$. These are prepared by controlled oxidation of pentahalides.

$$2NbCl_5 + O_2 \longrightarrow 2NbOCl_3 + 2Cl_2$$
$$2TaCl_5 + O_2 \longrightarrow 2TaOCl_3 + 2Cl_2$$

These are volatile solids but not as volatile as the pentahalides. Oxochlorides are white in colour whereas oxobromides are yellow.

These are monomeric in the vapour state and undergo hydrolysis in water to give hydrous pentoxides (gelatinous white precipitates).

$TaOF_3$ can be prepared by heating $TaF_5$ with $SiO_2$.

$$2TaF_5 + SiO_2 \longrightarrow 2TaOF_3 + SiF_4$$

## Compounds with +4 Oxidation State

### 1. Oxides

$NbO_2$ is prepared by heating $Nb_2O_5$ in a current of hydrogen.

$$Nb_2O_5 + H_2 \longrightarrow 2NbO_2 + H_2O$$

$TaO_2$ is prepared by heating $Ta_2O_5$ with carbon.

$$Ta_2O_5 + C \longrightarrow 2TaO_2 + CO$$

Both of them are dark-grey powders which are insoluble in water and acids but soluble in hot aqueous alkalis.

### 2. Halides

The halides of Nb and Ta in +4 oxidation states are obtained by reduction of pentahalides. Reduction is brought about by heating the pentahalide with aluminium or the metal itself. The oxidation state of the product is determined by the extent of reduction which depends upon: (*i*) temperature and (*ii*) time. The known halides in +4 state are $NbF_4$, $NbCl_4$, $NbBr_4$, $NbI_4$, $TaCl_4$, $TaBr_4$ and $TaI_4$.

These are all dark solids. Except $NbF_4$, all of them are volatile above 570 K. $NbF_4$ is paramagnetic while others are diamagnetic. $NbF_4$ is made up of regular octahedra joined in a chain by their edges. The structure of $NbI_4$ also consists of chain of octahedra joined by their edges but the Nb atoms are displaced from the centre of the $NbI_6$ octahedron and occur in pairs. This allows weak Nb – Nb bonds of length 320 pm.

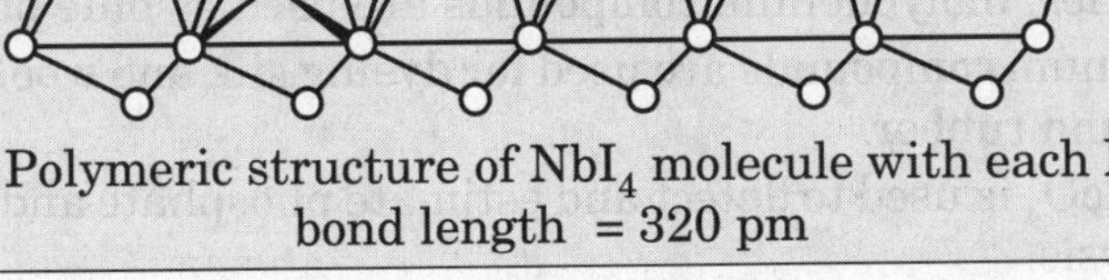


**Fig. 7.5** Polymeric structure of $NbI_4$ molecule with each Nb – Nb bond length = 320 pm

$NbCl_4$ has a similar structure with Nb – Nb bonds of length 306 pm.

## Compounds with +3 and +2 Oxidation State

Some known halides in +3 and +2 oxidation state obtained by reduction of pentahalides are:

**+3 State :** $NbF_3$, $NbCl_{2.67}$, $NbCl_{2.33}$, $NbI_3$;
$TaCl_{2.9-3.1}$, $TaBr_{2.9-3.1}$, $TaI_3$

**+2 State :** $NbF_{2.6}$, $NbCl_{2.67}$, $NbCl_{2.33}$, $NbBr_{2.67}$
$NbBr_2$, $NbI_{2.67}$, $NbI_2$;
$TaCl_{2.5}$, $TaBr_{2.5}$, $TaI_{2.33}$

As seen from their formulae given above, majority of them are non-stoichiometric compounds. Tantalum does not give any fluoride in these lower oxidation states.

## Molybdenum, $Mo_{42}$

### Properties of Molybdenum

Massive Mo has typical silver white appearance and metallic properties. Mo has density 10.21 and high m.pt. (= 2610°C). The metal does not combine at ordinary temperature with non-metals but at elevated temperature it readily combines with $O_2$, S, C, $N_2$ and halogens to form $MoO_3$, $MoS_2$, MoC, MoN, $MoF_6$, $MoCl_5$ etc.

$$Mo + 3F_2 \longrightarrow MoF_6$$
$$2Mo + 5Cl_2 \longrightarrow Mo_2Cl_{10}$$
$$2Mo + 3Br_2 \longrightarrow 2MoBr_3$$

$NH_3$ has no action on Mo metal even when it is hot. $H_2S$ attacks the metal at 1200°C to form the sulphide. It is oxidised to $MoO_3$ by $CO_2$.

$$Mo + 3CO_2 \longrightarrow MoO_3 + 3CO$$

It is only slowly attacked by acids. With hot and conc. $HNO_3$ it is oxidised to $MoO_3$.

$$Mo + 6HNO_3 \longrightarrow MoO_3 + 6NO_2 + 3H_2O$$

Fused caustic potash oxidises the metal to $MoO_3$. The metal dissolves in $AgNO_3$ with separation of Ag.

### Uses of Molybdenum and its Compounds

(*i*) Molybdenum finds most important use in the production of alloy steels and high speed tools.

(*ii*) Molybdenum is used in thermo-couples for high temperature measurement.

(*iii*) Both in radio valves and bulbs molybdenum is used as filament support.

(*iv*) Mo is used in the manufacture of standard weights.

(*v*) In potteries, molybdenum compounds are used as blue pigments.

(*vi*) Molybdenum compounds are used for dyeing silk and wool and for colouring leather and rubber.

(*vii*) $(NH_4)_2MoO_4$ is used to detect and estimate phosphate and arsenic in chemical analysis.

(*viii*) $MoO_3$ is used as a catalyst in many chemical reactions.

## Compounds of Mo

Mo forms compounds in –2, –1, 0, +1, +2, +3, +4, +5 and +6 oxidation states. Most of the compounds are covalent. The metal also forms interstial compounds.

## Compounds of Mo (VI)

Following are important compounds of Mo in +6 oxidation state.

### 1. Molybdenum Trioxide, $MoO_3$

**Preparation.** It is prepared: (*i*) by heating Mo metal or $MoS_2$ in air or $O_2$. (*ii*) by heating $MoO_2$ with $HNO_3$, (*iii*) by heating normal ammonium molybdate, $(NH_4)_2MoO_4$ alone or by heating ammonium paramolybdate, $(NH_4)_6 Mo_7O_{24} . 4H_2O$ alone or with $HNO_3$.

$$2Mo + 3O_2 \longrightarrow 2MoO_3$$
$$2MoS_2 + 7O_2 \longrightarrow 2MoO_3 + 4SO_2$$
$$MoO_2 + 2HNO_3 \longrightarrow MoO_3 + H_2O + 2NO_2$$
$$(NH_4)_2MoO_4 \xrightarrow{\Delta} 2NH_3 + H_2O + MoO_3$$
$$(NH_4)_6Mo_7O_{24}4H_2O \xrightarrow{\Delta} 7H_2O + 6NH_3 + 7MoO_3$$
$$(NH_4)_6Mo_7O_{24} . 4H_2O + 6HNO_3 \xrightarrow{\Delta} 7H_2O + 6NH_4NO_3 + 7MoO_3$$

**Properties.** At room temperature $MoO_3$ is colourless but on heating it becomes yellow. Although $MoO_3$ is regarded as the anhydride of molybdic acid, $H_2MoO_4$, it does not form this acid directly. It is more soluble in cold water than in hot water. $MoO_3$ dissolves in alkalies to form *normal molybdates* of $M_2$ $[MoO_4]$ type.

$$MoO_3 + 2MOH \longrightarrow M_2[MoO_4]$$

A solution of $MoO_3$ in $HNO_3$, when evaporated, gives a white crystalline powder of molybdic acid, $H_2MoO_4$. When heated with dry HCl at about 150-200°C, it gives a pale yellow water-soluble sublimate of the composition, $MoO_3.2HCl$ or $MoO(OH)_2Cl_2$. When $MoO_3$ is dissolved in an alkali sulphide (e.g., $Na_2S$) and then the solution so obtained is acidified, $MoS_3$ is formed.

$$MoO_3 \xrightarrow{+Na_2S} Na_2MoS_4 \xrightarrow{+H^+} MoS_3$$

On being reduced by $H_2$, it gives $MoO_2$ and then Mo metal.

$$MoO_3 + H_2 \xrightarrow{500-470°C} MoO_2 + H_2O; MoO_2 + 2H_2 \xrightarrow{600°C} Mo + 2H_2O$$

With hot conc. $NH_4OH$, normal ammonium molybdate, $(NH_4)_2MoO_4$ is formed.

$$MoO_3 + 2NH_4OH \longrightarrow (NH_4)_2MoO_4 + H_2O$$

Now when the solution containing $(NH_4)_2MoO_4$ is evaporated, crystals of *ammonium paramolybdate*, $(NH_4)_6$ $[Mo_7O_{24}]$ or $6NH_3 . 7MoO_3 . 3H_2O$ are obtained.

$$7MoO_3 + 6NH_4OH \longrightarrow (NH_4)_6 [Mo_7O_{24}] + 3H_2O$$

## 2. Molybdic Acid, $H_2MoO_4 \cdot H_2O$

It is obtained by treating ammonium molybdate with $HNO_3$ and allowing the solution to crystallise in *cold*. It is a yellow crystalline powder slightly soluble in water, giving an acidic solution. This acid is much weaker than $H_2CrO_4$. It dissolve in alkalies, forming molybdates.

## 3. Ammonium Molybdate, $(NH_4)_2MoO_4$

This compound is obtained by the treatment of $MoO_3$ with $NH_4OH$ and evaporating the solution to crystallisation. It is a white crystalline powder, soluble in water and insoluble in alcohol. $(NH_4)_2MoO_4$ easily loses $NH_3$ while $(NH_4)_6[Mo_7O_{24}]$ is quite stable in air. The solution of ammonium molybdate in dil. $HNO_3$ slowly deposits yellow α-molybdic acid, $H_2MoO_4.H_2O$ which, on warming at 70°C with $H_2O$ forms white β-molybdic acid, $H_2MoO_4$. The two acids have different X-ray spectra and vapour pressure. When a solution of ammonium molybdate and HCl is dialysed, colloidal molybdic acid is obtained. On heating the salt alone or with $HNO_3$, $MoO_3$, is obtained (See the preparation of $MoO_3$). When a phosphate or an arsenate is heated with conc. $HNO_3$ and $(NH_4)_2MoO_4$ solution, yellow precipitate of ammonium phosphomolybdate, $(NH_4)_3PO_4 \cdot 12MoO_3$ or ammonium arsenomolybdate, $(NH_4)_3AsO_4.12MoO_3$ is obtained.

$$Na_3PO_4 + 12(NH_4)_2MoO_4 + 24HNO_3 \longrightarrow (NH_4)_3PO_4 \cdot 12MoO_3 + 12H_2O + 3NaNO_3 + 21NH_4NO_3$$

or $$Na_3PO_4 + 12(NH_4)_2MoO_4 + 24HNO_3 \longrightarrow (NH_4)_3[PMo_{12}O_{40}] + 12H_2O + 3NaNO_3 + 21NH_4NO_3$$

or $$PO_4^{3-} + 12MoO_4^{2-} + 24NO_3^- \longrightarrow \underset{\text{Phospho-12 molybdate}}{[PMo_{12}O_{40}]^{3-}} + 12H_2O$$

$$AsO_4^{3-} + 12MoO_4^{2-} + 24NO_3^- \longrightarrow \underset{\text{Phospho-12 arsenate}}{[PAs_{12}O_{40}]^{3-}}$$

On the basis of the above reactions, $(NH_4)_2MoO_4$ is used in the laboratory for the detection and estimation of $PO_4^{3-}$ and $AsO_4^{3-}$ ions. When $As_2S_3$ or $As_2S_5$ is boiled with conc. $HNO_3$, arsenic acid, $H_3AsO_4$ is produced which gives a yellow precipitate of ammonium arsenomolybdate, $(NH_4)_3AsO_4 \cdot 12MoO_3$ with $(NH_4)_2MoO_4$.

$$As_2S_3 + 10HNO_3 \longrightarrow 2H_3AsO_4 + 10NO_2 + 2H_2O + 3S$$

$$As_2S_5 + 10HNO_3 \longrightarrow 2H_3AsO_4 + 10NO_2 + 2H_2O + 5S$$

$$H_3AsO_4 + 12(NH_4)_2MoO_4 + 21HNO_3 \longrightarrow \underset{\text{yellow ppt.}}{(NH_4)_3AsO_4 \cdot 12MoO_3} + 21NH_4NO_3 + 12H_2O$$

Ammonium molybdate has *not* been assigned the simple formula, $(NH_4)_2MoO_4$ but is represented as $(NH_4)_2Mo_7O_{24} \cdot 4H_2O$

$(NH_4)_2MoO_4$ is used: (*i*) as analytical reagent for the detection and estimation of $PO_4^{3-}$ and $AsO_4^{3-}$ ions (*ii*) as a catalyst for dehydrogenation in petroleum and coal technology.

### 4. Polymolybdates

The salts of molybdic acid are called molybdates. These salts give polymolybdates on acidification. The composition of polymolybdate formed depends upon the pH of the solution as shown below:

$$MoO_4^{2-} \xrightarrow{pH = 6} [Mo_7O_{24}]^{6-} \xrightarrow{pH=2} [Mo_8O_{26}]^{4-} \xrightarrow{pH = 1} MoO_3.2H_2O$$

The polymolybdates obtained by the condensation of $MoO_4^{2-}$ units alone are called **isopolymolybdates.** $[Mo_7O_{24}]^{6-}$ is an example.

The structure of the polymolybdates contains octahedral $MoO_6$ units. These octahedral $MoO_6$ units are linked together by sharing their corners and their edges. The way in which the octahedral $MoO_6$ units are linked together determines the type of polymolybdate formed.

### 5. Heteropolymolybdic Acids and their Salts

Heteropolymolybdates are formed when molybdate solutions containing other oxo anions, usually phosphate ($PO_4^{3-}$), silicate ($SiO_4^{4-}$) and arsenate ($AsO_4^{4-}$) are allowed to crystallise together. The ratio of $MoO_3$ to $PO_4^{3-}$ or $SiO_4^{4-}$, etc., condensed together is usually 12:1 or 6:1.

The structure of heteropolymolybdates consists of tetrahedral $PO_4^{3-}$, $SiO_4^{3-}$ or $BO_4^{3-}$ ions surrounded by $MoO_6$ octahedral units. A well known example of heteropolymolybdates is the product obtained by mixing solutions of $PO_4^{3-}$ and ammonium polymolybdate, *i.e.* $(NH_4)_3 [PO_4Mo_{12}O_{36}]$.

### 6. Mixed Oxides

When $MoO_3$ is fused with the oxides of alkali and alkaline earth metals, mixed oxides are produced. The structures of these mixed oxides are quite complicated.

### 7. Halides

(*a*) **$MoF_6$.** This is another important compound of Mo (VI). It is formed by direct combination of the elements. It forms colourless crystals melting at 17°C.

It reacts slowly with NaF to give a crystalline solid of composition $MoF_6.2NaF$. It is more stable than $CrF_6$.

(*b*) **$MoCl_6$ and $MoBr_6$.** The dark coloured $MoCl_6$ and $MoBr_6$ are well known. These solids hydrolyse easily to produce molybdic acids.

### 8. Oxyhalides

The oxyhalides of Mo (VI) are also known. Examples are $MoOF_4$, $MoOCl_4$, etc.

## Compounds of Mo (V)

Compounds of Mo (V) produce deep red colouration with $NH_4CNS$ due to the formation of $Mo(OH)_2(CNS)_3$.

### 1. Molybdenum Pentoxide, $Mo_2O_5$

It is obtained by heating molybdenum powder with molybdenum trioxide in nitrogen at 750°C.

$$Mo + 5MoO_3 \longrightarrow 3Mo_2O_5$$

$Mo_2O_5$ is a violet black powder and dissolves in acids on warming.

## 2. Halides

(*a*) **$MoF_5$.** It is prepared by the action of fluorine diluted with nitrogen on the metal at 400°C. It is also formed by the reduction of hexafluoride by the carbonyl of the metal.

$$5MoF_6 + Mo(CO)_6 \xrightarrow{25°C} 6MoF_5 + 6CO$$

$MoF_5$ is a crystalline solid. It has cyclic structure in which four $MoF_6$ units are joined together through Mo – F – Mo bridges to form a ring.

(*b*) **$MoCl_5$.** It is prepared by heating the metal in chlorine. It is highly deliquescent and is hydrolysed in water to give the oxochloride $MoOCl_3$.

$$MoCl_5 + H_2O \longrightarrow MoOCl_3 + 2HCl$$

In solution (e.g. in benzene) $MoCl_5$ exists as a monomer, but $MoCl_5$ exists as a dimer ($Mo_2Cl_{10}$) in the solid state. $Mo_2Cl_{10}$ does not contain any Mo – Mo bond.

## 3. $Mo_2S_5$

This compound is black powder.

# Compounds of Mo (IV)

## 1. $MoO_2$

It is obtained by heating the trioxide at 450°C in hydrogen.

$$2MoO_3 + 2H_2 \longrightarrow 2MoO_2 + 2H_2O$$

$MoO_2$ is a violet powder. It is oxidised by $HNO_3$ to $MoO_3$. $MoO_2$ has polymeric structure which contains Mo – Mo bonds.

## 2. Halides

(*a*) **$MoF_4$** is obtained by treating $MoF_5$ with molybdenum carbonyl at –75°C when a compound of composition $Mo_2F_9$ is obtained which on heating to 150°C gives $MoF_4$. $MoF_4$ is a non-volatile solid.

(*b*) **$MoCl_4$** is obtained by refluxing the solution of $MoCl_5$ in benzene, in a reducing atmosphere. It is also obtained on treating $MoO_2$ with carbon tetrachloride at 250°C.

$$MoO_2 + CCl_4 \longrightarrow MoCl_4 + CO_2$$

$MoCl_4$ is a dark red solid.

## 3. $MoS_2$

It is naturally occurring *molybdenite ore*.

## Compounds of Mo (III)

### 1. $Mo(OH)_3$

It is obtained as a black ppt. by adding alkali to a solution of molybdate reduced by Zn. It dissolves in acids and forms salts.

### 2. Halides

(a) **$MoF_3$** is obtained by treating $MoF_6$ with Mo at about 400°C. It is a yellowish-brown non-volatile solid.

**(b) $MoCl_3$** can be prepared by heating $MoCl_5$ in $H_2$ at 250°C. It is a dark red solid, insoluble in cold water. It forms stable complex ion, $[MoCl_6]^{3-}$ when it reacts with $Cl^-$ ion.

### 3. $M_2S_3$

Is obtained as a ppt on adding $(NH_4)_2S$ to a solution containing a Mo (III) salt.

## Compounds of Mo (II)

### 1. $MoCl_2$

It is obtained on heating the trichloride in an atmosphere of carbon dioxide.

$$2MoCl_3 \longrightarrow MoCl_2 + MoCl_4$$

The tetrachloride, being more volatile, distils off. It can also be prepared by passing chlorine over molybdenum heated at 120°C. It is an amorphous yellow solid, insoluble in water.

### 2. $MoBr_2$

This dihalide is similar to $MoCl_2$ in most of its properties. This bromide has the formula, $Mo_6Br_{12}$ which is actually $[Mo_6Br_8]^{4+}$ $Br_4^-$. $[Mo_6Br_8]^{4+}$ ion consists of $Mo_6$ octahedral unit which contains Mo – Mo bonds and Mo – Br – Mo bridges.

## Molybdenum Blues

These are blue oxides of Mo and are prepared by the reduction of acidified solutions of molybdates by $H_2S$ or $SO_2$. They contain oxides and hydroxides. Their compositions vary, although the colour in each case remains blue. Thus, they have the compositions $MoO_{2.33}$. $H_2O$, $Mo_{2.5}(OH)_{2.5}$, $MoO_2$ (OH), etc. Some molybdenum blues are crystalline while others are amorphous. Examples are:

(i) *Crystalline molybdenum blues*: $MoO_{2.0}$ (OH) (Mean O.S. of Mo = 5), $MoO_{2.6}$ . $nH_2O$ (O.S. = 5.20), $MoO_{2.5}$ $(OH)_{0.25}$ (O.S.= 5.50), $MoO_{2.38}$.$H_2O$ (O.S. = 5.76).

(ii) *Amorphous molbdenum*: $MoO_{2.75}$ . $nH_2O$(O.S. = 5.50), $MoO_{2.88}$.n $H_2O$ (O.S. 5.76)

On heating molybdenum blue above 350°C, only $H_2O$ molecules are lost while the OH groups remain attached. This may suggest that OH groups are involved in bridging of different groups. Formation of molybdenum blue is utilised in the detection and calorimetric determination of Mo. Phosphorus is indirectly estimated by this method. Because of the colloidal nature of molybdenum blue, it is easily absorbed on vegetable and animal fibres and hence is used for dyeing those fibres.

## Compounds of Mo in +1, 0, –1 and –2 Oxidation States

The organometallic compound $(C_6H_6)_2 Mo^+$ has Mo in +1 oxidation state whereas metal carbonyls $Mo(CO)_6$, $[Mo(CO)_{10}]^-$ and $[M(CO)_5]^{2-}$ contain the metal in zero, –1 and –2 oxidation states respectively. This is in agreement with the general statement that pi-acceptor ligands stabilise lower oxidation states of the metal.

## Interstitial Compounds of Mo

Mo combines with many non-metals like B, C, N and Si and form interstitial compounds. The composition of these compounds correspond to $Mo_2C$, MoN, MoC, $Mo_2N$, $Mo_2B$, $MoB_2$ and $Mo_2B_5$. These compounds are hard, refractory and chemically inert. It is due to these properties that these compounds are used in the construction of Kilns, jet engines, gas turbines and for the lubrication of cutting tools.

## Technetium, $Tc_{43}$

It is an unstable element which is a product of artificial nuclear transmutations. The longest-lived isotope of technetium is $^{99}Tc$ (half-life period = $2.12 \times 10^5$ years) and is a fission product of uranium. The other isotopes $^{97}Tc$ and $^{95}Tc$ have half-lives more than $10^5$ years. Small quantities (a few mg) of $^{99}Tc$ can be conveniently prepared by irradiation of molybdenum with neutrons.

$$^{98}_{42}Mo + ^{1}_{0}n \longrightarrow ^{99}_{42}Mo \longrightarrow ^{99}_{43}Tc + ^{0}_{-1}e$$

## Compounds of Technetium

### 1. Oxides

Different oxides of technetium are $TcO_2$ (black), $TcO_3$(purple) and $Tc_2O_7$ (yellow) in which oxidation states of the elements are +4, +6 and +7 respectively.

$TcO_2$ is the most stable and is prepared by heating ammonium pertechnate, $NH_4TcO_4$. It sublimes on heating at 1270K. $TcO_3$ is obtained by heating $TcO_2Br$. $Tc_2O_7$ (mp. 570 K) can be prepared by heating the metal or its other oxides to 470 K in air.

### 2. Pertechnic Acid and Pertechnates

Pertechnic acid is obtained by dissolving $Tc_2O_7$ in water.

$$Tc_2O_7 + H_2O \longrightarrow 2HTcO_4$$

Dark red crystals of anhydrous acid are obtained on evaporation of the solution obtained.

Dissolution of $Tc_2O_7$ in alkalis gives pertechnates (e.g., $NaTcO_4$) which are more stable than permanganates and do not possess any oxidizing properties.

## 3. Halides and Oxohalides

Known halides of the element are $TcCl_2$, $TcF_6$ and $TcCl_6$. The hexafluoride is a golden yellow crystalline solid (m.p. 306 K) and can be prepared by heating the metal in a current of fluorine.

Technetium oxofluoride, $TcO_3F$ is a yellow solid (m.p. 291.3 K) which can be obtained by the action of fluorine on $TcO_2$ at 320 K.

The oxidation states and stereochemistry of the element are summarized in Table 7.2.

**Table 7.2. Oxidation states and stereochemistry of technetium**

| *Oxidation state and configuration* | *Coordination no.* | *Geometry* | *Example* |
|---|---|---|---|
| $Tc^0$, $d^7$ | 6 | Octahedral | $Tc(CO)_{10}$ |
| $Tc^{II}$, $d^5$ | 6 | Octahedral | $Tc(diars)_2Cl_2$ |
| $Tc^{III}$, $d^4$ | 6 | Octahedral | $[Tc(diars)_2Cl_2]^+$ |
| $Tc^{IV}$, $d^3$ | 6 | Octahedral | $TcCl_4$ |
| $Tc^V$, $d^2$ | 6 | Octahedral | $[Tc(NCS)_6]^-$ |
| $Tc^{VII}$, $d^0$ | 4 | Tetrahedral | $TcO_4^-$ |

# Ruthenium, $Ru_{44}$

## Properties and Uses of Ruthenium

It is a fairly hard, greyish white brittle metal which is not attacked by acids and alkalies. However, it dissolves when fused with alkalies in presence of an oxidizing agent like $KNO_3$ or $KClO_3$ to form ruthenate, e.g., $K_2RuO_4$.

It combines with oxygen, sulphur on heating to form binary compounds $RuO_2$ and $RuS_2$. Similarly with fluorine it gives $RuF_3$ and with chlorine it forms $RuCl_3$.

Ruthenium forms hard inert alloys with platinum used in fountain-pen tips and gramophone needles.

### Compounds of Ruthenium

Ru forms compounds in –2, 0, +2, +3, +4, +6 and +8 oxidation states. –2 and 0 (zero) oxidation states are unusual oxidation states. Examples of compounds formed by Ru in the above oxidation states are given below.

| *Oxidation states* | *Compounds* |
|---|---|
| –2 | $[Ru(CO)_4]^{2-}$ |
| zero | $Ru(CO)_5$, $Ru_2(CO)_9$, $Ru_3(CO)_{12}$ |
| +2 | $RuCl_2$, $RuO$, $[Ru(NH_3)_6]^{2+}$ |

| | |
|---|---|
| | $[Ru(CN)_6]^{4-}$, $[Ru^{2+}(NO^+)Cl_5]^{2-}$, |
| | $[Ru^{2+}(NO^+)(CN)_5]^{2-}$ |
| +3 | $[Ru(NH_3)_6]^{3+}$, $[Ru(NH_3)_5(H_2O)]^{3+}$ |
| | $[Ru(NH_3)_5N_2]^{3+}$ |
| +4 | $RuO_2$ |
| +6 | $[RuO_4]^{2-}$ |
| +8 | $RuO_4$ |

The formation of $[Ru(CO)_4]^{2-}$, $Ru(CO)_5$, $Ru_2(CO)_9$ and $Ru_3(CO)_{12}$ show that π-acid ligand CO present in these compounds stabilises the lower oxidation states (–2 and zero) of Ru. Complexes of Ru in +2 oxidation state such as $[Ru(NH_3)_6]^{2+}$ and $[Ru(CN)_6]^{4-}$ are more stable than Ru (II) simple compounds.

π-acid ligand, NO forms a large number of mixed nitroso complexes with Ru (II). Examples are $[Ru(NO)Cl_5]^{2-}$ and $[Ru(NO)(CN)_5]^{2-}$. In these ions Ru – NO linkage is either straight (Ru — N = O) or bent (Ru—N=O, with N=O bent)

Ru (III) complexes are stable. $[Ru^{3+}(NH_3)_5N_2]^{3+}$ is prepared as follows.

$$[Ru(NH_3)_5(H_2O)]^{3+} \xrightarrow{NaN_3} [Ru(NH_3)_5N_2]^{3+}$$

$RuO_2$(Ru = +4) is a black solid. $RuO_4$ is a yellow solid and is a powerful oxidising agent. It dissociates in alkalies to form ruthernate $(RuO_4^{2-})$

$$\underset{(Ru\,=\,+8)}{2RuO_4} + 4OH^- \longrightarrow \underset{(Ru\,=\,+6)}{2RuO_4^{2-}} + 2H_2O + O_2$$

$RuO_4^{2-}$ is similar to $MnO_4^-$

Ruthenocene, $(C_5H_5)_2Ru$ which is a sandwich compound and is similar to ferrocene, $(C_5H_5)_2Fe$ has also been prepared.

# Rhodium, $Rh_{45}$

## Properties of Rhodium

Rhodium is a silvery white metal which is soft and ductile. It lacks reactivity, e.g., it combines with oxygen or chlorine only at red heat. It dissolves in hot concentrated sulphuric acid and in hot concentrated hydrochloric acid containing sodium chlorate at 400-425 K.

Being inert, it is used in plating scientific instruments. Its alloy with platinum is employed in making scientific apparatus, crucibles, thermocouples, etc, which are required to withstand high temperatures.

## Compounds of Rhodium

Rh forms compounds in –1, 0, +1, +2, +3, +4, +5 and +6 oxidation states.

(*i*) +6 is the highest O.S. of Rh. This O.S. is shown by Rh in $RhF_6$ which is an unstable and highly reactive compound.

(*ii*) Rh shows +5 O.S. in $RhF_5$ which is very rare and unstable. This compound is a cyclic tetramer, $(RhF_5)_4$ in which four $RhF_6$ octahedra form a ring through Rh — F — Rh bonds. The structure of the tetramer, $(RhF_5)_4$ is similar to the tetrameric structure of $NbF_5$ and $TaF_5$.

(*iii*) $RhF_4$ in which Rh has +4 O.S. is unstable. $RhO_2$ is formed by the oxidation of an aqueous solution of Rh (III) compounds by a strong oxidising agent such as $NaBiO_3$ in presence of an alkali.

$$K_2[\overset{+4}{Rh}Cl_6] \text{ and } K_2[\overset{+4}{Rh}F_6] \text{ are slightly more stable than } RhF_4 \text{ and } RhO_2.$$

(*iv*) The +3 oxidation state of Rh is more stable than any other oxidation state of the metal. All the trihalides are known and are found to be quite stable. These halides, however, cannot be represented by simple structures. Their structures contain Rh $X_6$ octahedra fused with one another through Rh – X – Rh bonds in such a manner that the aggregate structure is $RhX_3$. However $RhCl_3 . 3H_2O$ is monomeric and has octahedral structure.

The complexes of Rh (III) are also very stable. Most of these are diamagnetic in character. The complex cation, $[Rh(H_2O)_6]^{3+}$ is quite stable in aqueous solution, whereas $[Co\,(H_2O)_6]^{3+}$ is oxidising. This indicates that Rh (III) salt solutions in water are more stable than Co (III) salt solutions. The anion $[RhCl_6]^{3-}$ is also quite stable.

The pentahalide and heptahalide complex ions of Rh (III) such as $[RhBr_5]^{2-}$ and $[RhBr_7]^{4-}$ are known but these ions are not very stable.

The basic acetate ion, $[Rh_3O\,(CH_3COO)_6\,(H_2O)_3]^+$ has Rh in +3 oxidation state. The structure of this ion is the same as that of basic acetate ion of Cr (III), $[Cr_3O\,(CH_3COO)_6\,(H_2O)_3]^+$.

$[Rh\,(R_3P)_2(H)Cl_2]^+$ ion has Rh in +3 oxidation state. This ion is stable due to the presence of π-acid ligands, $(R_3P)_2$ in this ion.

(*v*) The +2 oxidation state of Rh in not very stable. However, the pi acceptor ligands stabilise this oxidation state. Rhodocene. Rh (II) $(n^5 - C_5H_5)_2$ is similar to ferrocene and cobaltocene in structure but it is much less stable.

(*vi*) The +1 oxidation state of Rh is quite common in complexes containing pi acid ligands such as $PR_3$ and CO.

The complex $[RhCl\,(PPh_3)_3]$ is known as **Wilkinson's catalyst.** It has a square planar structure and is red-violet in colour. It is obtained by refluxing $RhCl_3.3H_2O$ with $PPh_3$. It is used as a catalyst in the OXO process in which CO and $H_2$ are added to the end double bond of alkenes forming aldehydes which on oxidation give alcohols.

(*vii*) The carbonyls, $[Rh_6(CO)_{16}]$, $[Rh_4(CO)_{12}]$ etc. contain Rh in zero oxidation state.

(*viii*) The complex ion, $[Rh(CO)_4]^-$ contains Rh in –1 oxidation state.

## Palladium, $Pd_{46}$

### Physical Properties of Palladium

Palladium is a lustrous silvery-white metal which is malleable and ductile. Its density is 12.0 and is lighter than platinum and gold. It is as hard as platinum and takes a good polish. Its sp. heat is 0.05928. It melts at 1552°C and boils at 3980°C. This metal has the characteristic property of absorbing gases, more specially hydrogen. Freshly prepared palladium absorbs 370 volumes of hydrogen at room temperature and 650 volumes at 100°C. Spongy palladium absorbs the gas upto 980 volumes. When palladium metal is heated in oxy-hydrogen flame, a greenish vapour of palladium is evolved. Palladium black, dry or suspended in water or alcohol, absorbs 4000 volumes, and colloidal palladium about 5000 volumes of acetylene per volume of palladium.

## Chemical Properties of Palladium

Palladium is not tarnished by air and on heating a blue oxide layer is formed on the metallic surface. When heated to dull redness in oxygen, palladium sponge is oxidised to palladous oxide which decomposes on heating above 807°C. On heating with sodium peroxide, a soluble palladite is formed.

Palladium combines with fluorine at a dull red heat, forming $PdF_3$. Spongy palladium at dull redness forms palladous chloride. Palladium is also attacked by chlorine or bromine water. Hydrochloric acid attacks the metal only in presence of air or free chlorine. When palladium foil is evaporated to dryness with an alcoholic solution of iodine, palladous iodide is formed.

The metal combines with sulphur with incandascence. On heating the metal in a current of sulphur dioxide, palladium sulphide is formed. The metal also dissolves in the fused potassium hydrogen sulphate forming palladium sulphate.

Palladium also dissolves in nitric acid at ordinary temperature more easily when alloyed with copper or silver. It also dissolves in boiling concentrated sulphuric acid but easily soluble in aqua regia.

$$Pd + 4HNO_3\,(\text{conc.}) \longrightarrow Pd(NO_3)_2 + 2H_2O + 2NO_2$$

$$Pd + 2H_2SO_4 \longrightarrow PdSO_4 + SO_2 + 2H_2O$$

Palladium combines on heating with phosphorus to form its phosphite, with antimony it forms antimonides, PdSb, $Pd_3Sb$ and $PdSb_2$.

On fusion with potassium hydrogen sulphate, palladous sulphate is formed; with potassium pyrosulphate, $K_2S_2O_7$, palladous sulphate and $SO_2$ are formed. With ammonium sulphate or nitrate, the metal forms a diamine. Palladium is not oxidised on fusion with oxidising fusion mixture ($Na_2CO_3$ and $KNO_3$).

## Uses of Palladium

(*i*) Palladium and its compounds are used as catalyst in different types of reactions such as oxidation, reduction, chlorination, isomerisation, polymerisation, synthetic reactions, hydrogenations of oils etc. Usually, palladous oxide, palladium black an colloidal platinum are used for this purpose. The use of Pd-metal as a catalyst in hydrogenation reactions is due to the fact that in presence of Pd, molecular hydrogen gets dissociated into atomic hydrogen. Since at atomic hydrogen is more reactive, the reaction is brought about very speedily and very efficiently.

Palladium catalysts also find wide application in gas analysis, manufacture of synthetic petroleum, manufacture of HCN, acetylene and alkaline earth formates.

(*ii*) It is used for the purification of hydrogen, in dentistry and jewellery as a substitute for gold and platinum.

(*iii*) Palladium is also used in astronomical and other scientific instruments.

(*iv*) It is used as a coating for mirrors for search lights. Its salts are used in photography.

## Compounds of Palladium

Pd forms compounds in zero, +2, +3 and +4 oxidation states.

## Palladium (0) Compounds

Palladium shows its zero oxidation state in its complexes formed by $\sigma$ donor and $\pi$ acceptor ligands. One such complex is $[Pd(CO)(Ph_3)_3]$. $Pd(CO)_4$ in which Pd has zero oxidation state does not exist.

## Palladium (II) Compounds

Palladium forms a large number of compounds in which it has the oxidation state of +2. The compounds of Pd in +2 oxidation state are called ***palladous compounds***. Some of the important compounds of Pd (II) are described below.

### 1. Palladous Oxide, PdO

It is obtained by heating spongy palladium at 800°C in oxygen. It is a black powder. It is insoluble in acids and even in aqua regia. It is strong oxidising agent as on heating it decomposes into the metal and oxygen.

### 2. Palladous Hydroxide, $Pd(OH)_2$

It is obtained as a precipitate on heating a solution of $Pd(NO_3)_2$. Unlike PdO, $Pd(OH)_2$ dissolves readily in acids. It starts decomposing when heated above 550°C.

### 3. Palladous Chloride, $PdCl_2$

$PdCl_2$ is the most important chloride. It is obtained by the action of $Cl_2$ on Pd at red heat. It is a dark red solid which is highly soluble in water. Its aqueous solution, on evaporation, gives the crystals of $PdCl_2 \cdot 2H_2O$. It is reduced by $SO_2$ or CO to Pd metal. When heated at red heat, PdCl is obtained.

$$2PdCl_2 \xrightarrow{\Delta} 2PdCl + Cl_2$$

The action of $F_2$ on $PdCl_2$ gives $PdF_3$. Solution of $PdCl_2$ in HCl contains the complex ion, $[PdCl_4]^{2-}$.

$$PdCl_2 + 2HCl \longrightarrow H_2[PdCl_4] \rightleftharpoons 2H^+ + [PdCl_4]^{2-}$$

A solution of $PdCl_2$ mixed with that of KCl, on evaporation, gives the crystals of pot. tetrachloro palladate (II), $K_2[PdCl_4]$. The action of KBr on $PdCl_2$ gives $PdBr_2$.

$$PdCl_2 + 2KBr \longrightarrow PdBr_2 + 2KCl$$

When the solution of $PdCl_2$ is supersaturated with excess of $NH_3$ and concentrated by heating, crystals of $[Pd(NH_3)_4]Cl_2.H_2O$ are obtained.

$$PdCl_2 + 4NH_3 + H_2O \longrightarrow [Pd(NH_3)_4]Cl_2.H_2O$$

When the crystals of $[Pd(NH_3)_4]Cl_2 . H_2O$ are dissolved in water and a small amount of HCl is added, a yellow precipitate of $[Pd(NH_3)_2Cl_2]$ is obtained.

$$[Pd(NH_3)_4]Cl_2 . H_2O + 2HCl \longrightarrow [Pd(NH_3)_2Cl_2] + 2NH_4Cl + H_2O$$

When KCN is added to a solution of $PdCl_2$, $Pd(CN)_2$ is obtained.

$$PdCl_2 + 2KCN \longrightarrow Pd(CN)_2 + 2KCl$$

$PdCl_2$ exists as a linear polymer in the solid state.

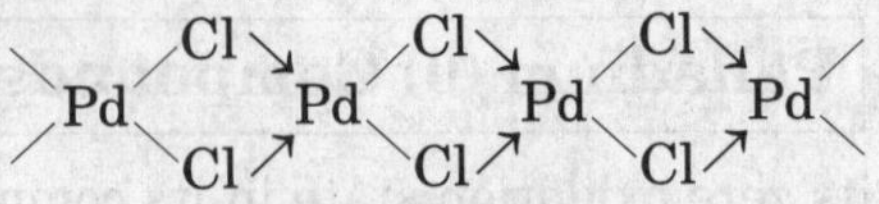

Another form of $PdCl_2$ is hexamer, $(PdCl_2)_6$ which is not soluble in water. This form of $PdCl_2$ is called $\beta$-$PdCl_2$. The structure of this hexamer comprises of a complicated three-dimensional pattern of Pd and Cl atoms in which each Pd atom is surrounded by four Cl-atoms in a square planar geometry. All Cl-atoms are bridging type.

## 4. Palladous Nitrate, $Pd(NO_3)_2$

It is obtained by dissolving palladium metal or palladous hydroxide in dilute nitric acid. The solution on evaporation yields brown deliquescent prism-like crystals.

## 5. Palladous Sulphate, $PdSO_4 . 2H_2O$

This is obtained by dissolving palladium hydroxide in dilute sulphuric acid. It is a brown deliquescent solid.

## 6. Palladous Cyanide, $Pd(CN)_2$

It is obtained as a precipitate on adding KCN to a solution of $PdCl_2$. The ppt. dissolves on adding excess of KCN to give the complex compound, $K_2[Pd(CN)_4] . 3H_2O$

## Palladium (II) Complexes

A number of complex compounds of divalent palladium are known. Palladium ion, $Pd^{2+}$, in aqueous solution is hydrated and exists as the complex, $[Pd(H_2O)_4]^{2+}$. It also forms a complex with ammonia, $[Pd(NH_3)_4]^{2+}$. A solution of palladous chloride in hydrochloric acid contains the divalent complex anion, $[PdCl_4]^{2-}$.

Tetraamminepalladium (II) chloride. $[Pd(NH_3)_4]Cl_2$, is a pink solid and dischlorodiamminepalladium (II), $[Pd(NH_3)_2Cl_2]$ is a yellow solid. The latter exists in two forms:

*cis*-form *trans*-form

The *cis* form is commonly known as **cis platin** which is used in cancer therapy.

We know that the addition of suitable ligands perpendicular to the square plane of certain diamagnetic square planar Ni (II) complexes may give rise to tetragonal distortion to the complex [See the case of [Ni (diars)$_2$I$_2$]$^0$]. Now since Pd has higher nuclear charge than Ni, the magnitude of tetragonal distortion in octahedral complexes of Pd (II) is even more severe. As a result, the spin state of square planar Pd (II) complexes remains unperturbed by incoming additional ligands.

## Palladium (III) Compounds

Palladium (III) compounds are rather uncommon. Amongst the few compounds known, $PdF_3$ is important. This is obtained by heating the metal in a current of fluorine at 500°C. It is also obtained by the action of $F_2$ on $PdCl_2$ at 200–250°C. It is a black crystalline powder. It is hygroscopic in nature. With cold or hot water, it forms $Pd(OH)_2$.

$$4PdF_3 + 10H_2O \longrightarrow 4Pd(OH)_2 + 12HF + O_2$$

## Palladium (III) Complexes

The complex compounds containing Pd in +3 oxidation state, $Na_3[PdF_6]$ and $Na[PdF_4]$ are known to exist, but these are unstable complex compounds.

## Palladium (IV) Compounds

Very few compounds of Pd(IV) are known. The complexes of Pd (IV) are more stable than the Pd(IV) compounds. Well known complexes of Pd (IV) are $Na_2[PdX_6]$ and $Pd[X_4(NH_3)_2]$ (X = F, Cl, Br).

## Organometallic Compounds of Palladium

$PdCl_2$ and $K_2PdCl_4$ react with olefine to produce complexes of the type $[PdCl_3(C_2H_4)]^-$, Pd $[Cl_2(C_2H_4)_2]$ etc. The Pd-olefin bonding in these complexes is unique in the sense that olefin acts both as σ donor through its filled pi bonding orbitals and π acceptor through its vacant pi antibonding orbital. The reactions involving $PdCl_2$ and olefins are of great commercial importance.

$$C_2H_4 + PdCl_2 + H_2O \longrightarrow \underset{\text{Ethanal}}{CH_3CHO} + Pd + 2HCl$$

$$C_2H_4 + CO + H_2 \xrightarrow{PdCl_2} \underset{\text{Propanoic acid}}{CH_3CH_2COOH}$$

## Silver, $Ag_{47}$

### Physical Properties of Silver

**1. *Silver is lustrous white metal.*** Very thin layer of the metal has a bluish tinge. Powdered silver is grey and looks earthy in appearance. It is the best conductor of heat and electricity. This metal is highly malleable and ductile. Its m.p. is 960.8°C, b.p. = 1955° and sp. gravity = 10.47. Its molecule is mnoatomic in the vapour state.

**2. *Spitting or sprouting of silver.*** The molten silver absorbs about 20 times its volume of $O_2$ and rejects most of it on cooling. The evolution of $O_2$ is often violent enough to spurt some of the metal away from the main mass and hence is termed as spitting or sprouting of silver. This can be controlled by covering the molten metal with a layer of charcoal.

## Chemical Properties of Silver

**1. *Action of S and $H_2S$.*** The action of S and $H_2S$ on Ag metal gives $Ag_2S$.

$$2Ag + S \longrightarrow Ag_2S; \quad 2Ag + H_2S \longrightarrow Ag_2S + H_2$$

**2. *Action of $SO_2$.*** (*a*) $SO_2$ reacts with Ag at high temperature, forming $Ag_2SO_4$ and $Ag_2S$.

$$4Ag + 2SO_2 \longrightarrow Ag_2SO_4 + Ag_2S$$

(*b*) In presence of $O_2$ and at a temperature below 1085°C, $Ag_2SO_4$ is formed.

$$2Ag + SO_2 + O_2 \longrightarrow Ag_2SO_4$$

**3. *Action of halogens.*** On heating, Ag combines with halogens forming silver halides. $Br_2$ and $I_2$ combine very slowly.

$$2Ag + X_2 \longrightarrow 2AgX$$

**4. *Action of alkali cyanides.*** In presence of $O_2$, Ag dissolves in alkali cyanide solution by forming complex compound, *e.g.*,

$$4Ag + 8KCN + 2H_2O + O_2 \longrightarrow \underset{\text{Soluble complex compound}}{4K[Ag(CN)_2]} + 4KOH$$

**5. *Displacement of Ag from its salt solution by other metals.*** The displacement of Ag from its salt solution by a metal, M can be represented as:

$$\underset{\text{OA}}{Ag^+} + \underset{\text{RA}}{M} \longrightarrow Ag + M^+$$

In this reaction $Ag^+$ ion acts as an OA and M metal behaves as a RA. This reaction is feasible, if E° value of OA (*i.e.* $E^\circ_{Ag^+/Ag}$ value) is higher than $E^\circ_{M^+/M}$ value. In other words the above displacement reaction will take place in the forward direction, if $E^\circ_{M^+/M}$ value is lower than $E^\circ_{Ag^+/Ag}$ value. Since $E^\circ_{Ag^+/Ag} = 0.80V$, the metals like Cu, Zn etc. whose E° values ($E^\circ_{Cu^{2+}/Cu} = +0.34V$, $E^\circ_{Zn^{2+}/Zn} = -0.76V$) are lower than 0.80V, these metals can displace Ag from its salt solution. On the other hand, the metals like Hg, Pt, Au etc. whose E° values ($E^\circ_{Hg^{2+}/Hg} = +0.85V$, $E^\circ_{Pt^{2+}/Pt} = +1.20V$, $E^\circ_{Au^{3+}/Au} = +1.50V$) are higher than 0.80 V, these metals cannot displace Ag from its salt solution.

**6. *Liberation of $H_2$ from dil. acids.*** The liberation of $H_2$ from dil. acids (*e.g.* dil. HCl) by Ag can be represented by the equation.

$$\underset{\text{RA}}{Ag} + \underset{\text{OA}}{2H^+} \longrightarrow 2Ag^+ + H_2$$

This reaction is feasible, if $E^\circ_{H^+/\frac{1}{2}H_2}$ value is higher than $E^\circ_{Ag^+/Ag}$ value. Now since $E^\circ_{H^+/\frac{1}{2}H_2} = 0$ and $E^\circ_{Ag^+/Ag} = +0.80V$, $E^\circ_{H^+/\frac{1}{2}H_2}$ is not greater than $E^\circ_{Ag^+/Ag}$. Hence the liberation of $H_2$ from dil. acids by Ag is not possible or Ag cannot liberate $H_2$ from dil. acids.

**7. *Action of air and water.*** Air has no action on the metal but the air contaminated with $H_2S$ makes the surface of the metal black. This is due to the formation of a thin film of $Ag_2S$ on the surface of the metal.

$$4Ag + 2H_2S + O_2 \longrightarrow 2Ag_2S + 2H_2O$$

**8. *Action of conc. acids.*** (*a*) Hot and conc. $H_2SO_4$ dissolves Ag and gives $SO_2$.

$$2Ag + 2H_2SO_4 \longrightarrow Ag_2SO_4 + 2H_2O + SO_2$$

(*b*) Conc. HCl in presence of an oxidising agent forms AgCl.

$$2Ag + 2HCl + O \longrightarrow 2AgCl + H_2O$$

(*c*) When hot and conc. $HNO_3$ reacts with Ag, the metal is oxidised to $AgNO_3$ and $HNO_3$ gets reduced to $NO_2$.

$$2HNO_3 \longrightarrow 2NO_2 + H_2O$$
$$2Ag + O \longrightarrow Ag_2O$$
$$Ag_2O + 2HNO_3 \longrightarrow 2AgNO_3 + H_2O$$

$$2Ag + 4HNO_3 \longrightarrow 2AgNO_3 + 2NO_2 + 2H_2O$$

or $$Ag + 2HNO_3 \longrightarrow AgNO_3 + NO_2 + H_2O$$

OR

$$NO_3^- + 2H^+ + e^- \longrightarrow NO_2 + H_2O \quad \textit{(Reduction reaction)}$$
$$Ag \longrightarrow Ag^+ + e^- \quad \textit{(Oxidation reaction)}$$

$$NO_3^- + 2H^+ + Ag \longrightarrow NO_2 + H_2O + Ag^+$$

(*d*) When dil. $HNO_3$ reacts with Ag, the metal is oxidised to $AgNO_3$ but the acid gets reduced to NO.

$$2HNO_3 \longrightarrow H_2O + 2NO + 3O$$
$$[2Ag + O \longrightarrow Ag_2O] \times 3$$
$$[Ag_2O + 2HNO_3 \longrightarrow 2AgNO_3 + H_2O] \times 3$$

$$6Ag + 8HNO_3 \longrightarrow 6AgNO_3 + 2NO + 4H_2O$$

or $$3Ag + 4HNO_3 \longrightarrow 3AgNO_3 + NO + 2H_2O$$

OR

$$NO_3^- + 4H^+ + 3e^- \longrightarrow NO + 2H_2O \quad \textit{(Reduction reaction)}$$
$$3 \times [Ag^\circ \longrightarrow Ag^+ + e^-] \quad \textit{(Oxidation reaction)}$$

$$NO_3^- + 4H^+ + 3Ag \longrightarrow NO + 2H_2O + 3Ag^+$$

**9. *Action of fused alkalies.*** The metal is not attacked by fused alkalies.

## Uses of Silver

(*i*) Silver is used for making jewellery, coins and silver wares. Being soft, this metal cannot be used as such for this purpose. Hence it is alloyed with Cu (Cu = 7.5%) so that it becomes hard and tough. (*ii*) It is used for the manufacture of chemical wares as it not attacked by fused alkalies. (*iii*) Thin leaves of the metal are used as a tonic in Ayurvedic and Unani systems of medicine. These leaves are also used for decorative purposes. (*iv*) It is used for coating the metallic articles, in silvering of mirrors and in silver voltmeters. (*v*) Colloidal silver which has been stabilised by the addition of albumin, has been used as non-irritating disinfectant in eye diseases. (*vi*) Silver salts are used in photography and marking inks. Ag and $AgNO_3$ are used as analytical reagents. (*vii*) Ag is used in filling dental cavities.

## Compounds of Silver (I)

+1 oxidation state of Ag is its normal oxidation state. Examples of compounds formed by Ag (I) are : $Ag_2O$, $Ag_2O_2$ (silver peroxide), AgX (X = F, Cl, Br or I), $Ag_2S$, $Ag_2SO_4$, $Ag_2CO_3$, $AgHCO_3$, $AgNO_2$, $AgNO_3$, $AgNH_2$, AgCN, Ag CNS (silver thiocyanate) etc.

All Ag(I) compounds are diamagnetic since $Ag^+$ ion in these compounds has $d^{10}$ configuration in which all the electrons are in the paired state.

$AgNO_3$ which is also called **Lunar caustic** is prepared by dissolving silver in dilute $HNO_3$ and evaporating to crystallisation:

$$3Ag + 4HNO_3 \longrightarrow 3AgNO_3 + 2H_2O + NO$$

This forms colourless crystals. Like all heavy metal nitrates it decomposes on heating:

$$2AgNO_3 \longrightarrow 2Ag + 2NO_2 + O_2$$

The compound is not only very soluble in water but also in many organic solvents such as pyridine, aniline, acetonitrile etc.

The chloride, bromide and the iodide are precipitated from aqueous silver salts by the addition of the appropriate halide ion. The fluoride is soluble and is obtained by neutralising $Ag_2O$ with HF and crystallising. The solubility in aqueous medium falls in the order Cl > Br > I. The chloride and the bromide dissolve in ammonia to give the soluble complex $[Ag(NH_3)_2]X$ (X = Cl, Br) but the iodide is only sparingly soluble. The solubility of silver halides in $NH_4OH$ decreases in the order: AgCl > AgBr > AgI. But all the three halides dissolve in excess KCN.

## Comparison of the Solubility of Ag (I) Halides

It has been observed that AgF is soluble in water while AgBr and AgI are less soluble than AgCl. This observation can be explained as follows:

**Explanation.** Solubility of Ag (I) halides depends on the magnitude of lattice energy of these halides. Greater is the value of lattice energy, lesser is the solubility of the halide.

The number of electrons present in the halide ions of AgF, AgCl, AgBr and AgI are as $F^- = 9 + 1 = 10$, $Cl^- = 17 + 1 = 18$, $Br^- = 35 + 1 = 36$ and $I^- = 53 + 1 = 54$.

Lattice energies of Ag (I) halides depends on the number of electrons present halide ions. As a general rule, larger is the number of electrons in halide ion, the stronger is its van der Waals attraction for the neighbouring ions and hence higher is the lattice energy of Ag(I) halide having the halide ion with larger number of electrons.

Since $Br^-$ and $I^-$ ions have larger number of electrons ($Br^-$ = 36, $I^-$ = 54) than $Cl^-$ ion (= 18), AgBr and AgI have higher values of lattice energy than AgCl. Consequently, AgBr and AgI are less soluble than AgCl. Since $F^-$ ion has the smallest number of electrons (= 10), AgF has the lowest value of lattice energy and hence is soluble in $H_2O$.

## Complexes of Ag (I)

Like copper (I), silver (I) is also a soft acid and therefore, prefers to bind better with soft donors. Its most usual coordination number is 2 (linear) with *sp* hybridisation. In a few cases coordination number 3 ($sp^2$) and coordination number 4 (tetrahedral $sp^3$) are known. The linear coordination is achieved with an L–Ag–L angle = 180°, so that a didentate chelating ligand cannot help to achieve this coordination. Thus AgCl reacts with ethylenediamine, $NH_2 - (CH_2)_2 - NH_2$ to give a dimeric species in which $NH_2 - (CH_2)_2 - NH_2$ acts as a bridging ligand instead of chelating ligand. Thus the structure of dimeric species is Cl — Ag ← $NH_2$—$(CH_2)_2$—$NH_2$→Ag—Cl. Tetrahedral complexes $[AgL_4]ClO_4$(L = $R_3P$; phosphites) and thiourea (L) complexes, $[AgL_2]X$, are also known.

When AgCl dissolves in aqueous $NH_3$, alkali cyanides and $Na_2S_2O_3$ solution, complex ions $[Ag(NH_3)_2]^+$, $[Ag(CN)_2]^-$ and $[Ag(S_2O_3)_2]^{3-}$ are obtained.

$$Ag^+ + 2NH_3 \longrightarrow [Ag(NH_3)_2]^+$$
$$Ag^+ + 2CN^- \longrightarrow [Ag(CN)_2]^-$$
$$Ag^+ + 2S_2O_3^{2-} \longrightarrow [Ag(S_2O_3)_2]^{3-}$$

The structures of dicyanoargentate (I) ion, $[Ag(CN)_2]^-$ and dithio-sulphatoargentate (I) ion, $[Ag(S_2O_3)_2]^{3-}$ are given below in Fig. 7.6.

$[N \equiv C–Ag–C \equiv N]^-$

Dicyanoargentate (I) ion, $[Ag(CN)_2]^-$

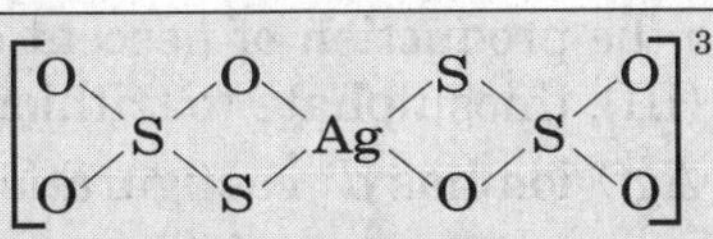

Dithiosulphato argentate (I) ion, $[Ag(S_2O_3)_2]^{3-}$

**Fig 7.6** Structures of $[Ag(CN)_2]^-$ and $[Ag(S_2O_3)_2]^{3-}$ ions

$[Ag(CN)_3]^{2-}$ is obtained when excess of $CN^-$ ions react with AgCl

$$Ag^+ + 3CN^- \rightarrow [Ag(CN)_3]^{2-}.$$

AgI dissolves in liq. $NH_3$, a solution of KCN and a strong solution of KI to form the complexes $[Ag(NH_3)_2I]$, K $[Ag(CN)_2]$ and $K[AgI_2]$ respectively.

In presence of air, $Ag_2S$ dissolves in NaCN to form the complex, $Na[Ag(CN)_2]$.

With excess of the solution of $Na_2S_2O_3$, $AgNO_3$ gives the soluble complex, $[Ag(S_2O_3)_2]^{3-}$

$$Ag^+ + 2S_2O_3^{2-} \longrightarrow [Ag(S_2O_3)_3]^{3-}$$

AgSCN reacts with the excess of KSCN to form $K[Ag(SCN)_2]$

$$AgSCN + KSCN \longrightarrow K[Ag(SCN)_2]$$

$Ag^+$ ion probably exists as $[Ag(H_2O)_2]^+$ in aqueous solution.

AgBr is soluble in solutions containing high concentration of $S_2O_3^{2-}$ ion due to the formation of the complex, $[Ag(S_2O_3)_2]^{3-}$

$$AgBr + 2S_2O_3^{2-} \longrightarrow [Ag\,(S_2O_3)_2]^{3-} + Br^-$$

or $$Ag^+ + 2S_2O_3^{2-} \longrightarrow [Ag(S_2O_3)_2]^{3-}$$

This reaction is used in fixing the image on photographic plate since $Na_2S_2O_3$ dissolves away the whole of unaffected AgBr from the exposed film.

## Compounds of Ag (II)

A few compounds of Ag(II) have been reported but these are highly unstable. $Ag^{2+}$ is reduced readily to $Ag^+$ ion in these compounds. Hence Ag(II) compounds act as strong oxidising agents. +2 oxidation state of Ag is not stable at all but this oxidation state for Cu is important. AgO and $AgF_2$ are important compounds of Ag (II).

### 1. Silver (II) Oxide, AgO

This compound is formed by passing $O_2$ over Ag. AgO is also prepared by the oxidation of $AgNO_3$ by $K_2S_2O_8$ in an alkaline medium.

$$4AgNO_3 + 2K_2S_2O_8 + 8NaOH \longrightarrow 4AgO + K_2SO_4 + 3Na_2SO_4 + 2NaNO_3 + 2KNO_3 + 4H_2O$$

AgO(Ag = +2) can also be obtained by oxidising $Ag_2O$ (Ag = +1) with NaClO in alkaline medium. AgO readily decomposes to $Ag_2O$ and gives nescent oxygen which acts as a strong oxidising agent.

$$2AgO \longrightarrow Ag_2O + O$$

Due to the production of nescent oxygen, AgO oxidises iodide to iodine, iron (II) to iron (III), thiosulphate to trithionate and sulphate.

Since $Ag^{2+}$ ion has $d^9$ configuration, AgO should be paramagnetic due to the presence of one unpaired electron in $d^9$ configuration. Structural study of AgO has shown that the crystal lattice of AgO contains two kinds of Ag which are $Ag^+$ and $Ag^{3+}$

Thus AgO is a mixed oxide and hence is represented as $Ag_2^+O$. $Ag_2^{3+}O_3$ or $Ag_2^+\,Ag_2{}^{3+}\,(O^{2-})_4$ or $Ag^+\,Ag^{3+}\,(O^{2-})_2$. $Ag^+$ ion is a $d^{10}$ ion and this ion in square planar coordination with $O^{2-}$ ions will be diamagnetic.

### 2. Ag(II) Fluoride, $AgF_2$

This compound is obtained by reacting silver or silver (I) fluoride with fluorine at ~ 400°C. It is a dark brown substance which reacts with water liberating oxygen and ozone. It is a powerful oxidant which oxidises ethanol to acetaldehyde, chromium (III) to chromate, manganeses (II) to permanganate etc. It is also a powerful fluorinating agent converting $SO_2$ and $CO_2$ to $SO_2F_2$ and $COF_2$ respectively. $CCl_4$ is converted to $CF_4$. It is paramagnetic with a room temperature moment of 2.0 B.M. It is antiferromagnetic with a Neel temperature of 163°K.

## Complexes of Ag (II)

$[Ag\,(py)_4]^{2+}$, $[Ag\,(o\text{-}phen)_2]^{2+}$, $[Ag\,(bipy)]^{2+}$, $M_2[AgF_4]$ (M = K, Rb, Cs) etc, are important complexes of Ag (II).

Blue $M_2AgF_4$ (M = K, Rb, Cs) are obtained by heating appropriate proportions of MF and $AgF_2$ in argon. The compounds have magnetic moments in the range 1.5—1.7 B.M.

Suitable ligands for the purpose of stabilisation of silver (II) are pyridine, bipyridine, o-phenanthroline, picolinic acid and other pyridine carboxylic acids. Addition of aqueous $K_2S_2O_8$ to an aqueous pyridine solution of silver (I) sulphide gives orange crystals of $[Ag(py)_4]S_2O_8$. Complexes $[Ag(bpy)_2]^{2+}$, $[Ag(o\text{-}phen)_2]^{2+}$ are also similarly obtained. With terpyridyl (terpy), complexes of the type [AgX (terpy)]X(X $=ClO_3^-$, $ClO_4^-$, $NO_3^-$) have been isolated. These compounds are all orange to red. These are good oxidants and are paramagnetic ($\mu$ = 1.8 B.M).

**Structure.** Complexes of Ag (II) have square planar geometry which results from $dsp^2$ hybridisation of $Ag^{2+}$ ion ($Ag^{2+} = 3d^9$). This hybridisation has one unpaired electron as shown in Fig. 7.7. On the basis of CFT of square planar geometry, the distribution of $d^9$ electrons of $Ag^{2+}$ ion in four sets of orbitals obtained by the splitting of five *d* orbitals can be shown as:

$$(d^2_{yz} = d^2_{xz}),\ (d^2_{z^2}),\ (d^2_{xy}),(d^1_{x^2-y^2})$$
$$\xrightarrow{\text{Energy increasing}}$$

This distribution also gives $n = 1$

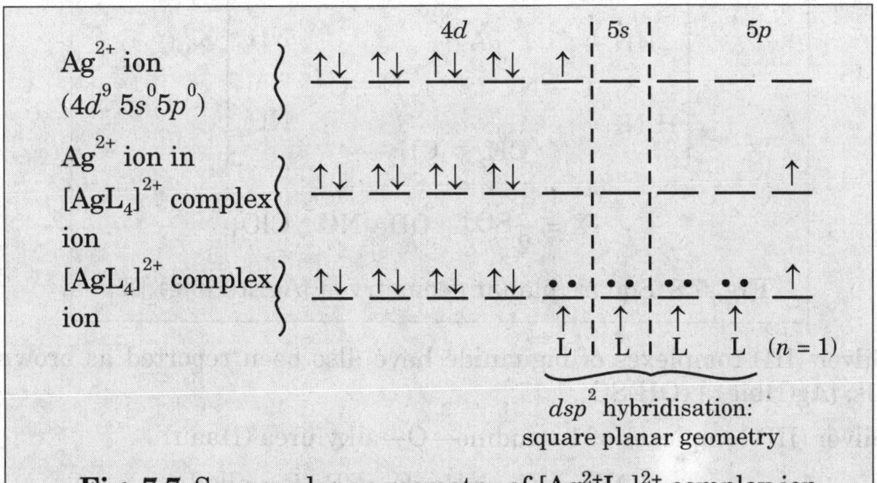


**Fig. 7.7** Square planar geometry of $[Ag^{2+}L_4]^{2+}$ complex ion

## Complexes of Ag (III)

Like copper (III), silver (III) is also a $d^8$ system and is an unusual oxidation state of the element. As with copper (III), fluorine reacts with an mixture of KF and AgF on heating to produce yellow $KAgF_4$. With CsF the complex $CsAgF_4$ is obtained. The fluoro complexes are diamagnetic. A square planar stereochemistry with $dsp^2$ hybridisation or a crystal field splitting of the type $(d_{xz}^2, d_{yz}^2), \left(d_{z^2}^2\right), (d_{xy}^2), (d^0_{x^2-y^2})$ is suggested. The complex is very sensitive to moisture. Stabilisation of silver (III) has also been achieved by hexaoxoiodate (VII) and hexaoxotellurate (VI) groups. Peroxodisulphate (μ-peroxohexaoxodisulphate) oxidation of strongly basic solutions containing $TeO_6^{6-}$ or $IO_6^{5-}$ and $Ag^+$ gives dark red diamagnetic complexes: $K_6H[Ag(IO_6)_2]$. $10H_2O$, $Na_6H_3[Ag(TeO_6)_2]$. $18H_2O$. Their structures and hybridisation schemes are comparable to those of copper (III).

A series of very stable cationic silver (III) complexes with the tetradentate ethylenedibiguanide ligand (Et $(Hbig)_2$) has been prepared. The complex sulphate is obtained as a dark orange substance by potassium μ-peroxo-hexaoxodisulphate oxidation of a solution containing silver sulphate and ethylenedibiguanide sulphate. The base is violet and produces on neutralisation with $HClO_4$ or nitric acid the complex perchlorate or nitrate salt. The nitrate salt is remarkably stable and can be recrystallised from hot dilute nitric acid solution. The complex $[AgEt(Hbig)_2]X_3$ is square planar since it is diamagnetic (Fig 7.8). An X-ray crystal structure study shows a primary square coordination around Ag(III) and a coordination of one or two counter anions at a significantly longer distance. The molar conductance of the salts in aqueous solution is comparable to those of other tri-univalent electrolytes. These are all strong oxidants and liberate two equivalents of iodine from acidified iodide solution.

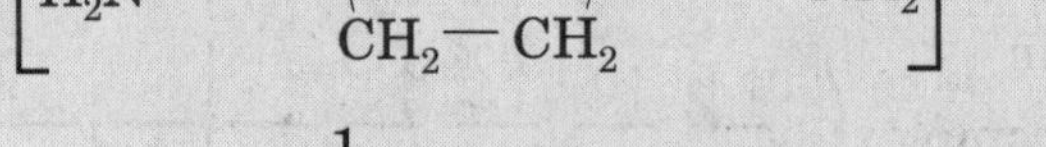


$$X = \frac{1}{2}SO_4^{2-}, OH^-, NO_3^-, ClO_4^-$$

**Fig. 7.8** Square planar geometry of $[AgEt(Hbig)_2]X_3$

Silver (III) complexes of biguanide have also been reported as brown silky crystals: $[Ag(Hbig)_2]$ $(OH)SO_4$.

Silver (III) complexes of 1-amidino—O—alkylurea (Haau)

$NH_2$—C(=NH)—NH—C(=NH)—OR ; R = alkyl

have also been isolated as red/orange crystals $[Ag(aau)_2]X$ ($X = OH^-$; $\frac{1}{2}SO_4^{2-}$). These complexes behave as two-electron oxidising agents towards iron (II) sulphate or acidified potassium iodide. These are diamagnetic. A square planar configuration is indicated. A silver (III) complex of biuret $[Ag(H_2bi)_2](NO_3)_3$ has been prepared by μ-peroxohexaoxodisulphate oxidation of a silver (I) complex followed by precipitation with sodium nitrate. Square planar $Ag^{III}N_4$ has been proposed.

Silver (III) complex of o-phenylenebisdimethylarsine $[Ag(diars)_2](ClO_4)_3$ has been obtained as yellow solid by conc. $HNO_3$ oxidation of the silver (I) complex $[Ag(diars)_2]ClO_4$ in $HClO_4$.

**Question. AgCl is much less soluble than KCl. Explain.**

**Ans.** AgCl and KCl both have the same anion *viz* $Cl^-$ ion but the cation in these chlorides is different. AgCl has $Ag^+$ cation and KCl has $K^+$ cation. $Ag^+$ ion has 47 – 1 = 46 electrons while $K^+$ ion has 19 – 1 = 18 electrons. Now since $Ag^+$ ion in AgCl has larger number of electrons than $K^+$ ion in KCl, AgCl has higher value of lattice energy than KCl (AgCl = 895.4 KJ $mol^{-1}$, KCl = 702.9 KJ $mol^{-1}$). Due to higher value of lattice energy, greater amount of energy will be needed to break the crystal of AgCl than that of KCl. Consequently AgCl is much less soluble than KCl.

## Cadmium, $Cd_{48}$

### Properties of Cadmium

Cadmium is a soft bluish white metal with a low melting point of 594 K and a boiling point of 1040 K. It is slightly more malleable and ductile and heavier than zinc.

$$Cd = Cd^{2+} + 2e^-;\ E° = -0403 \text{ volt } (\textit{Reduction potential})$$

Cadmium is electropositive though not as strongly electropositive as zinc (E° = 0.763 volt). Hence in many of its chemical properties, it resembles zinc but it is comparatively less reactive. It is very slowly affected by moist air and does not react with cold water. It decomposes steam at red heat. It combines with halogens to form $CdCl_2$, $CdBr_2$ and $CdI_2$, and with sulphur to form CdS. It liberates hydrogen from non-oxydizing acids with the formation of corresponding salts.

$$Cd + 2H^+ \longrightarrow Cd^{2+} + H_2$$

With dilute nitric acid it liberates nitric oxide.

$$3Cd + 2NO_3^- + 8H^+ \longrightarrow 3Cd^{2+} + 2NO + 4H_2O$$

It is, however, not amphoteric like zinc. Its sulphide can be precipitated from weakly acidic solutions, while zinc sulphide cannot be precipitated from acidic solutions.

$$Cd^{2+} + H_2S \longrightarrow CdS \text{ (yellow)} + 2H^+$$

### Uses of Cadmium

(*i*) It is used in the preparation of a number of important fusible alloys, e.g., Wood's alloy, containing four parts of Bi, two of Pb, and one each of Sn and Cd, which melts at 333 K. It is added to aluminium to increase its stability and to silver to decrease staining. It is also used in antifriction alloys and solders.

(*ii*) It is also used as a plating metal.

(*iii*) Cadmium sulphide and cadmium selenide are used in pigments.

(*iv*) Cadmium rods are used in atomic reactors to slow down the nuclear reactions since it is a good absorber of neutrons.

(*v*) It is also used in Weston standard E.M.F. cell, which gives an E.M.F. of 1.019 volts at 288 K. The change of temperature does not affect the voltage to any appreciable extent. The electrodes in this cell consist of metallic mercury and cadmium amalgam and the electrolyte is a standard solution of cadmium and mercurous sulphates.

## Compounds of Cadmium (II)

### 1. Cadmium Oxides

Cadimum oxide, CdO is obtained as a dark brown powder by heating the metal in air or by igniting its hydroxide, carbonate, or the nitrate.

Yellow cadmous oxide, $Cd_2O$ is obtained by fusing the metal with cadmium chloride and treating the product with water when cadmous hydroxide is precipitated. Cadmous hydroxide on gentle heating loses water and gives cadmous oxide.

$$Cd + CdCl_2 \longrightarrow 2CdCl$$

$$CdCl + H_2O \longrightarrow CdOH\downarrow + HCl$$

$$2CdOH \longrightarrow Cd_2O + H_2O$$

### 2. Cadmium Hydroxide, Cd $(OH)_2$

It is obtained as a white precipitate by adding an alkali to the solution of a cadmium salt.

$$Cd^{2+} + 2OH^- \longrightarrow Cd(OH)_2$$

It is insoluble in caustic alkalis but dissolves in ammonia due to the formation of the complex hydroxide, $[Cd(NH_3)_4](OH)_2$.

$$Cd\,(OH)_2 + 4NH_3 \longrightarrow [Cd(NH_3)_4]^{2+} + 2OH^-$$

### 3. Cadmium Sulphide, CdS

It is obtained as a bright yellow precipitate by passing hydrogen sulphide through a solution of cadmium salt, provided the solution is not strongly acidic.

$$Cd^{2+} + H_2S \longrightarrow CdS + 2H^+$$

It does not dissolve in yellow ammonium sulphide (distinction from arsenious sulphide).

Cadmium sulphide is used as a yellow pigment in making oil and water colours. Mixed with barium sulphate ($CdS + BaSO_4$) it is used as a pigment under the name **cadmipone**.

## 4. Cadmium Sulphate, $3CdSO_4 . 8H_2O$

It is prepared by dissolving cadmium oxide in sulphuric acid and crystallizing it from the solution. It forms colourless monoclinic prisms soluble in water. It forms double salts of the type $K_2SO_4 . CdSO_4 . 6H_2O$ which are isomorphous with the corresponding double sulphates of bivalent metals, e.g., zinc, magnesium, etc.

## 5. Cadmium Carbonate, $CdCO_3$

It is obtained by the addition of ammonium carbonate to cadmium chloride solution and dissolving the precipitate formed in the requisite amount of ammonia solution. On concentrating the solution on a water bath, tiny crystals of normal cadmium carbonate are deposited. Alkali carbonates give a precipitate of an indefinite composition when added to the solution of a cadmium salt. It is used for detection of sulphide in mixture analysis.

## 6. Cadmium Chloride, $2CdCl_2.5H_2O$

It can be prepared by dissolving the metal, its oxide or carbonate, in hydrochloric acid and crystallizing it from the solution. It is neither deliquescent nor is hydrolysed in aqueous solutions but is efflorescent.

It is poor electrolyte. This behaviour is ascribed to the formation of autocomplexes.

$$CdCl_2 \rightleftharpoons Cd^{2+} + 2Cl^-$$

$$Cd^{2+} + 2Cl^- + CdCl_2 \rightleftharpoons Cd[CdCl_4]$$

It is, however, decomposed by hot and concentrated suphuric acid.

## 7. Cadmium Iodide, $CdI_2$

It is soluble in alcohol and is even a poorer electrolyte than cadmium chloride. It probably forms a complex salt in which cadmium exists in the complex anion.

$$2CdI_2 \rightleftharpoons Cd[CdI_4] \rightleftharpoons Cd^{2+} + [CdI_4]^{2-}$$

On account of the formation of this soluble complex, insoluble salts of cadmium, e.g., cadmium sulphide dissolve in potassium iodide solution. On diluting the solution, the complex begins to dissociate. Cadmium iodide is used in photography.

## 8. Cadmium Cyanide, $Cd(CN)_2$

When potassium cyanide is added to the solution of a cadmium salt, a precipitate of cadmium cyanide, $Cd(CN)_2$, is first formed which dissolves in excess of the precipitant forming the complex salt, $K_2 [Cd(CN)_4]$. This is not very stable and gives a precipitate with $H_2S$.

## Chemistry of Cd (I)

Cd forms $Cd_2^{2+}$ ion which is called dicadmium (I) ion. In this ion Cd has +1 oxidation state. Dimercury (I) ion, $Hg_2^{2+}$ is quite stable but $Cd_2^{2+}$ ion is very unstable. This ion is formed only in the fused melts of the type $(Cd - Cd)^{2+} 2Cl^-$ or $Cd - Cd Cl_2$.

## Complexes of Cd (II)

$Cd^{2+}$ ion has $d^{10}$ configuration. Since *d*-orbitals are completely filled, $Cd^{2+}$ ion does not form π-acid complexes. +2 charge on $Cd^{2+}$ does not favour transfer of electron density from $Cd^{2+}$ ion to π-acid ligands. Thus no carbonyls, nitrosyls etc. are formed by Cd (II). However, usual complexes are formed by Cd (II). These complexes have L → M σ donation.

Commonest complexes formed by Cd (II) are 4-coordinated and have tetrahedral geometry corresponding to $sp^3$ hybridisation. Examples of such complexes are $[Cd(OH)_4]^{2-}$, $[Cd(NH_3)_4]^{2+}$, $[Cd(CN)_4]^{2-}$, $[CdCl_4]^{2-}$, $[CdI_4]^{2-}$, $[CdCl_2(R_3P)_2]^0$ etc.

Like $Zn(OH)_2$, $Cd(OH)_2$ also dissolves in $NH_3$ due to the formation of tetrammine cadmium (II) complex, $[Cd(NH_3)_4]^{2+}$

$$Cd(OH)_2 + 4NH_3 \longrightarrow [Cd\,(NH_3)_4]^{2+} + 2OH^-$$

Cd $(OH)_2$ dissolves in *very concentrated alkali* to form hydroxocadmiate, $[Cd(OH)_4]^{2-}$.

$$Cd\,(OH)_2 + 2OH^- \rightarrow [Cd(OH)_4]^{2-}.$$

$CdCl_2$ and $CdI_2$ both are poor electrolytes. They probably form soluble auto complexes, $Cd^{2+}\,[CdX_4]^{2-}$ (X = Cl, I). In these complexes Cd is present as complex anion, $[CdX_4]^{2-}$.

$$2CdX_2\ (X = Cl, I) \rightleftharpoons Cd^{2+}\,[CdX_4]^{2-}$$

When KCN is added to a solution of Cd (II) salt, white ppt. of Cd $(CN)_2$ is formed. This ppt. dissolves in the excess of KCN to form the soluble complex, $K_2[Cd(CN)_4]$.

$$2KCN + Cd^{2+} \longrightarrow \underset{\text{White ppt.}}{Cd(CN)_2} + 2K^+$$

$$Cd(CN)_2 + 2KCN\ (\textit{Excess}) \longrightarrow \underset{\text{Soluble complex}}{K_2[Cd(CN)_4]}$$

$K_2[Cd\,(CN)_4]$ is not very stable and gives a yellow ppt. of CdS with $H_2S$.

The complex formed by Cd (II) with thiocyanate ligand is formulated as $[Cd(SCN)_4]^{2-}$. In this complex $Cd^{2+}$ ion is coordinated to S-atom of the ligand. Note that the complex formed by Zn (II) is formulated as $[Zn\,(NCS)_4]^{2-}$ which indicates that Zn is linked to the ligand through its N-atom. The complex anion, $[Cd^{2+}Cl_5]^{3-}$ is a 5-coordinated ion and has trigonal bipyramidal geometry.

The complexes $[Cd\,(NH_3)_6]^{2+}$, $[Cd(en)_3]^{2+}$ etc. have octahedral geometry.

## Organometallic Compounds of Cadmium

Organo cadmium compounds are prepared from the appropriate Grignard reagents:

$$CdX_2 + 2RMgX \longrightarrow CdR_2 + 2MgX_2$$

They are thermally less stable than their zinc counterparts, but generally less reactive, and hence their most important use is to prepare ketones from acid chlorides.

## UNIVERSITY QUESTIONS

**1.** Write the electronic configuration of Yttrium. *(Andhra 85)*

**2.** (*a*) State how Zr is extracted from its ores. Describe a method for obtaining ultrapure Zr from its pure

(*b*) Compare the properties of Zr with Hf.

(*c*) Describe preparation, properties and uses of $ZrCl_2$.

*(Delhi B.Sc., (Hons,) (86, Raj. 85)*

**3.** How do Nb and Ta occur in nature? How are the metals extracted from their ores ? Give important properties and uses of the metals. *(Kerala 86)*

**4.** (*a*) Write a note on pentavalent molybdenum.

(*b*) Give an account of oxides of Mo.

(*c*) Name important ores of Mo. How is the metal extracted? What are the uses of the metal ?

(*d*) Discuss the structure of 12-hetero-polyanions of Mo.

(*e*) Give preparation and uses of ammonium molybdate.

*(Utkal 86, MDU 84, Bharthidasan 85, 86)*

**5.** How does Ru occur in nature ? How is the metal extracted ?

**6.** How does Rh occur in nature ? How is it extracted ? Give important properties of the metal.

**7.** (*a*) How does Pd occur in nature ? How is it extracted ? Give important properties of the metal.

(*b*) Give the preparation, properties and uses of $PdCl_2$. Write the structure of anhydrous Pd (II) chloride.

**8.** (*a*) Explain cyanide process for the extraction of Ag metal from its sulphide ore.

(*b*) Explain (*i*) AgCl is soluble in $NH_3$ (*ii*) Ag $NO_3$ reacts with $NH_3$. (*iii*) AgCl is soluble in saturated solution of KI but AgI is insoluble in saturated KCl.

(*c*) Name two important ores of Ag.

*(Punjab 86; Gauwahati 86; Delhi 85; Indore 85)*

**9.** (*a*) How is Cd extracted from its ore ? State its properties and uses.

(*b*) Write the formula of hydrated cadmium sulphate. Explain its use in standard cadmium cell. *(Guwahati 82; Delhi 83)*

**10.** Write a note on Wilkinson's Catalyst. *(Ch. Charan Singh, Meerut 2012)*

8

CHAPTER

# Chemistry of Elements of Third Transition Series (5*d*-Series Elements)

## 5*d*-Series Elements

We have already shown in chapter 5 that *3rd* transition series (*5d*-series) contains ten elements namely $La_{57}$, $Hf_{72}$, $Ta_{73}$, $W_{74}$, $Re_{75}$, $Os_{76}$, $Ir_{77}$, $Pt_{78}$, $Au_{79}$ and $Hg_{80}$ which are present in *6th* period of the periodic table as shown in Fig 8.1

| Group | 1 | 2 | 3 | 4 | 5 | 6 | 7 | 8 | 9 | 10 | 11 | 12 | 13 | 14 | 15 | 16 | 17 | 18 |
|---|---|---|---|---|---|---|---|---|---|---|---|---|---|---|---|---|---|---|
| 1 | 1 H | | | | | | | | | | | | | | | | | 2 He |
| 2 | 3 Li | 4 Be | | | | | | | | | | | 5 B | 6 C | 7 N | 8 O | 9 F | 10 Ne |
| 3 | 11 Na | 12 Mg | | | | | | | | | | | 13 Al | 14 Si | 15 P | 16 S | 17 Cl | 18 Ar |
| 4 | 19 K | 20 Ca | 21 Sc | 22 Ti | 23 V | 24 Cr | 25 Mn | 26 Fe | 27 Co | 28 Ni | 29 Cu | 30 Zn | 31 Ga | 32 Ge | 33 As | 34 Se | 35 Br | 36 Kr |
| 5 | 37 Rb | 38 Sr | 39 Y | 40 Zr | 41 Nb | 42 Mo | 43 Tc | 44 Ru | 45 Rh | 46 Pd | 47 Ag | 48 Cd | 49 In | 50 Sn | 51 Sb | 52 Te | 53 I | 54 Xe |
| 6 | 55 Cs | 56 Ba | 57-71 La* | 72 Hf | 73 Ta | 74 W | 75 Re | 76 Os | 77 Ir | 78 Pt | 79 Au | 80 Hg | 81 Ti | 82 Pb | 83 Bi | 84 Po | 85 At | 86 Rn |
| 7 | 87 Fr | 88 Ra | 89-103 Ac** | 104 Rf | 105 Db | 106 Sg | 107 Bh | 108 Hs | 109 Mt | 110 | 111 | 112 | (113) | (114) | (115) | (116) | (117) | (118) |

← f-block elements →

| | | | | | | | | | | | | | | |
|---|---|---|---|---|---|---|---|---|---|---|---|---|---|---|
| Lanthanides* | 58 Ce | 59 Pr | 60 Nd | 61 Pm | 62 Sm | 63 Eu | 64 Gd | 65 Tb | 66 Dy | 67 Ho | 68 Er | 69 Tm | 70 Yb | 71 Lu |
| Actinide** | 90 Th | 91 Pa | 92 U | 93 Np | 94 Pu | 95 Am | 96 Cm | 97 Bk | 98 Cf | 99 Es | 100 Fm | 101 Md | 102 No | 103 Lr |

**Fig. 8.1** Position of *3rd* transition series elements (5*d*-series elements) in the periodic table.

These elements are also called 5d-series elements because the differentiating electron enters in 5*d* orbitals of the atoms of these elements. General characteristics of these elements along with those of *1st* and *2nd* transitions series have already been discussed in chapter 5. Important physical properties of these elements have been given in Table 8.1 for the sake of reference.

**Table 8.1 Important physical properties of the elements of 3rd transition series (5d-series elements)**

| *Property* | La | Hf | Ta | W | Re | Os | Ir | Pt | Au | Hg |
|---|---|---|---|---|---|---|---|---|---|---|
| (*i*) Atomic number | 57 | 72 | 73 | 74 | 75 | 76 | 77 | 78 | 79 | 80 |
| (*ii*) Outer electronic configuration | $4f^{o}$ $5d^1$ $6s^2$ | $4f^{14}$ $5d^2$ $6s^2$ | $4f^{14}$ $5d^3$ $6s^2$ | $4f^{14}$ $5d^4$ $6s^2$ | $4f^{14}$ $5d^5$ $6s^2$ | $4f^{14}$ $5d^6$ $6s^2$ | $4f^{14}$ $5d^7$ $6s^2$ | $4f^{14}$ $5d^9$ $6s^1$ | $4f^{14}$ $5d^{10}$ $6s^1$ | $4f^{14}$ $5d^{10}$ $6s^2$ |
| (*iii*) Atomic radii (pm) | 187 | 158 | 146 | 139 | 137 | 135 | 136 | 138 | 144 | 157 |
| (*iv*) Ionic radius (pm) | 106 ($La^{3+}$) | 80 ($Hf^{4+}$) | 73 ($Ta^{5+}$) | 68 ($W^{6+}$) | 67 ($Re^{7+}$) | 77 ($Os^{4+}$) | — | 76.5 ($Pt^{4+}$) | 137 ($Au^{+}$) | 110 ($Hg^{2+}$) |
| (*v*) Crystal structure | hcp,fcc | hcp | bcc | bcc | hcp | hcp | fcc | fcc | fcc | rhombic |
| (*vi*) Atomic volume (ml) | 20.50 | 13.60 | 10.91 | 9.53 | 8.85 | 8.43 | 8.54 | 9.10 | 10.22 | 14.80 |
| (*vii*) Density (g/cm$^3$) | 6.17 | 13.10 | 16.60 | 19.30 | 21.00 | 22.59 | 22.61 | 22.40 | 19.30 | 13.60 |
| (*viii*) Atomic mass (amu) | 138.9 | 178.5 | 180.9 | 183.8 | 186.2 | 190.2 | 192.2 | 195.1 | 197.0 | 200.6 |
| (*ix*) Enthalpy of atomisation (KJ mol$^{-1}$) | 435 | 618 | 781 | 837 | 791 | 790 | 690 | 566 | 368 | 61 |
| (*x*) Melting points (°C) | 920 | 2222 | 2996 | 3410 | 3180 | 2700 | 2454 | 1769 | 1063 | – 38.4 |
| (*xi*) Boiling points (°C) | 3470 | 5400 | 5425 | 5930 | 5900 | 5500 | 5300 | 4530 | 2920 | 357 |
| (*xii*) Ionisation energy (KJ mol$^{-1}$) | 538.1 | 654 | 761 | 770 | 760 | 840 | 880 | 870 | 890.1 | 1007 |
| (*xiii*) Stable oxidation states | +3 | +4 | +5 | +6 | +2 to +7 | +4, +6 | +3, +4 | +2, +4 | +1, +3 | +2 |
| (*xiv*) Electronegativity | 1.1 | 1.3 | 1.7 | 1.7 | 1.9 | 2.2 | 2.01 | 2.2 | 2.4 | 1.9 |
| (*xv*) Heat of fusion (KJ mol$^{-1}$) | 10 | 21.7 | 28.4 | 33.7 | 33.0 | 26.8 | 27.6 | 21.7 | 12.7 | 2.37 |
| (*xvi*) Heat of vaporisation (KJ mol$^{-1}$) | 939 | 648 | 753 | 774 | 636 | 690 | 648 | 565 | 363 | 58 |

Since $La_{57}$ has many similarities with lanthanides, the chemistry of this element and its compounds will be discussed with the chemistry of lanthanides and their compounds. Thus in the present chapter we shall discuss the chemistry of only nine elements of 5*d*-series *viz* $H_{72}$ to $Hg_{80}$.

## Hafnium, $Hf_{72}$

The atomic and ionic radii of Zr and Hf are identical and, therefore, their chemistry is remarkably similar. Hf forms similar compounds as Zr. There is a very close similarity in properties of the compounds of Hf and Zr.

## Tantalum, $Ta_{73}$

We know that $Nb_{41}$ and $Ta_{73}$ are present in the same group (Group 5). These elements have identical atomic radii and hence have identical properties. As a result, these elements have already been discussed together in chapter 7.

## Tungsten, $W_{74}$

### Properties of Tungsten

Massive tungsten possesses typical white appearance and the metallic properties. Its density and m.pt. are 19.3 and 3410°C respectively. The metal is not acted upon by dil. and conc. acids and solutions of alkalies. It is, however, readily attacked by fused alkalies in presence of oxidising agent like $KNO_3$. The metal does not combine with elements at ordinary temperature but at high temperature it interacts with $O_2$, S, $N_2$, halogens, C and $H_2$.

$$2W + 3O_2 \longrightarrow 2WO_3; \quad W + 2S \longrightarrow WS_2; \quad W + C \longrightarrow WC$$

$$2W + C \longrightarrow W_2C; \quad 4W + N_2 \longrightarrow 2W_2N; \quad 2W + N_2 \longrightarrow 2WN$$

$$W + 3X_2 \longrightarrow WX_6\ (X = F, Cl, Br); \quad W + I_2 \longrightarrow WI_2$$

### Uses of Tungsten and its Compounds

The metal is used in making high speed tool steels, alloy steels, filaments of electric lamps, pen-points, strings of musical instruments, contact points, telegraphic keys, gramophone needles, voltage regulators etc. Surgical instruments are manufactured from its alloy namely *stellite* which contains W along with Co and Cr.

## Compounds of Tungsten

We know that Cr, Mo and W are present in the same group (Group 6). Hence W, like Cr, shows +2, +3, +4, +5 and +6 oxidation states in its compounds. Compounds of W in +6 oxidation state are important. Some compounds of W are discussed below.

### 1. Oxides

**(*a*) Tungsten trioxide or tungstic oxide, $WO_3$. Preparation.** (*i*) It is obtained by the following methods:

$$\underset{\text{Tungstic acid}}{H_2WO_4} \xrightarrow{\text{Dehydration}} WO_3 + H_2O$$

$$2W + 3O_2 \xrightarrow{\Delta\,(500°C)} 2WO_3$$

$$2WS_2 + 7O_2\ (\textit{From air}) \xrightarrow{\text{Roasting}} 2WO_3 + 4SO_2$$

$$\underset{\text{Amm. tungstate}}{(NH_4)_2WO_4} \xrightarrow{\Delta} WO_3 + 2NH_3 + H_2O$$

$$Na_2WO_4 + 2HNO_3 \longrightarrow WO_3 + 2NaNO_3 + H_2O$$

(*ii*) $WO_3$ can also be obtained from wolframite ore, (Fe, Mn) $WO_4$.

**Properties.** (*i*) It is a soft lemon-yellow coloured powder, but turns orange on heating. It melts at 1470° to a green liquid. It is insoluble in water and acids (excepting HCl).

$$WO_3 + 4HF \longrightarrow \underset{\text{Hydrofluotungstic acid}}{H_2WO_2F_4} + H_2O$$

(*ii*) It dissolves in alkalies to form tungstate.

$$WO_3 + 2NaOH \longrightarrow Na_2WO_4 + H_2O$$

(*iii*) When heated with $H_2$, it is reduced to W metal *via* $W_2O_5$ and $WO_2$.

$$\underset{\text{Yellow}}{WO_3} \xrightarrow{H_2} \underset{\text{Blue}}{W_2O_5} \xrightarrow{H_2} \underset{\text{Brown}}{WO_2} \xrightarrow{H_2} W$$

When heated with C, Mg etc., W metal is obtained directly. When heated with $Cl_2$ it gives volatile oxychloride, $WO_2Cl_2$ while with $H_2S$ it gives $WS_3$.

$$WO_3 + 3C \longrightarrow 3CO + W$$
$$WO_3 + 3Mg \longrightarrow 3MgO + W$$
$$2WO_3 + 2Cl \longrightarrow 2WO_2Cl_2 + O_2$$
$$WO_3 + 3H_2S \longrightarrow WS_3 + 3H_2O$$

(*iv*) When fused with alkali or alkaline earth oxides, $WO_3$ gives mixed oxides. The structures and properties of these mixed oxides are very similar to those of the corresponding mixed oxides of Mo.

**Uses.** It is used (*i*) as a mordant in dyeing and ceramic industry (*ii*) as a pigment in oil and water colours.

(*b*) **Tungsten dioxide, $WO_2$.** When $WO_3$ is heated mildly in a current of $H_2$, it changes to $WO_2$. $WO_2$ is a brown powder. It is soluble in water. It has complicated structure containing W–W bonds.

## 2. Tungstic Acid, $H_2WO_4$ and Tungstates

A precipitate of this acid may be obtained by treating a solution of any soluble tungstate with acid. The precipitate of the acid may be white ($H_2WO_4 \cdot H_2O$) or yellow ($H_2WO_4$) according as the precipitation has been made from cold solution or hot solution.

$$(NH_4)_2WO_4 + H_2SO_4 \longrightarrow \underset{\text{Yellow ppt}}{H_2WO_4} + (NH_4)_2SO_4$$

The hydrated acid ($H_2WO_4.H_2O$) is somewhat soluble in water but the anhydrous acid ($H_2WO_4$) is neither soluble in $H_2O$ nor in any acid, excepting HF. With HF hydrofluotungstic acid, $H_2WO_2F_4$ is formed. With alkalies, $H_2WO_4$ forms tungstates. On heating $WO_3$ is obtained.

$$H_2WO_4 + 4HF \longrightarrow H_2WO_2F_4 + 2H_2O$$
$$H_2WO_4 + 2NaOH \longrightarrow Na_2WO_4 + 2H_2O$$

$$H_2WO_4 \xrightarrow{\Delta} WO_3 + H_2$$

The composition of various tungstic acids is as : *meta tungstic acid* — $WO_3.H_2O$ or $H_2WO_4$; *para tungstic acid* — $WO_3.2H_2O$ or $H_2WO_4.H_2O$ or $H_2WO_5$; *ortho tungstic acid* — $WO_3.3H_2O$ or $H_2WO_4 . 2H_2O$ or $H_6WO_6$.

$CrO_4^{2-}$ ion is a strong oxidising agent but $WO_4^{2-}$ ion is a very mild oxidising agent.

On lowering the pH of tungstate solution, **isopolytungstates** are formed. These compounds have complicated structures containing the $WO_6$ octahedral units joined in a variety of ways by sharing the corners or edges of the octahedral units.

When $WO_4^{2-}$ concondenses with $PO_4^{3-}$, $SiO_4^{4-}$ and $BO_4^{3-}$, hetropolytungstates are formed. A well known example of a heteropolytungstic acid is phosphotungstic acid, $H_3PO_4.12WO_3$.

Tungstic acid is used in ceramic industry and in making pigments.

### 3. Halides

The known halides given by tungsten are: (*i*) *Fluoride:* $WF_6$ (*ii*) *Chlorides*: $WCl_6$, $WCl_6$, $WCl_4$, $WCl_2$. (*iii*) *Bromides:* $WBr_6$ (*very unstable*), $WBr_5$, $WBr_2$ (*iv*) *Iodides*: $WI_4$, $WI_2$. Of these $WF_6$ and $WCl_6$ are the most important.

$WF_6$ can be prepared by any of the following methods:

$$W + 3F_2 \longrightarrow WF_6$$
$$WCl_6 + 6HF\ (anyhydrous) \longrightarrow WF_6 + 6HCl$$
$$WCl_6 + 2AsF_3 \longrightarrow WF_6 + 2AsCl_3$$
$$WCl_6 + 3SbF_5 \longrightarrow WF_6 + 2SbF_3Cl_2$$

It is colourless or slightly yellow coloured gas. It is hydrolysed by $H_2O$, forming $WO_3$.

$$WF_6 + 3H_2O \longrightarrow WO_3 + 6HF$$

It is extremely volatile diamagnetic compound. It attacks all metals except Pt and Au. In presence of traces of moisture, it attacks glass because of the production of HF (due to hydrolysis). It behaves a Lewis acid and forms addition compounds with $NH_3$, pyridine (py) and methylamine ($CH_3 . NH_2$) (*e.g.*, $WF_6 . 4NH_3$, $WF_6.3py$, $WF_6 . 3CH_2NH_6$). With NaF it gives crystalline solid, $WF_6 . 2NaF$ in which W is 8-coordinated. With $SbCl_6$ it gives $WCl_6$.

$WCl_6$ can be prepared by the following reactions:

$$W + 3Cl_2 \xrightarrow{300°C} WCl_6$$

$$WS_2 + 4Cl_2 \xrightarrow{\Delta} WCl_6 + S_2Cl_2$$

$$2WO_3 + 6Cl_2 \xrightarrow{\text{In } CCl_4 \text{ in a sealed tube at } 280°} 2WCl_6 + 3O_2$$

Pure $WCl_6$ is steel blue coloured compound, slowly attacked by cold water, but rapidly by hot $H_2O$.

$$WCl_6 + 4H_2O \longrightarrow H_2WO_4 + 6HCl$$

Other reactions of $WCl_6$ are given below:

$$WCl_6 + 5HBr \longrightarrow WBr_5 + HCl + \frac{1}{2}Cl_2$$

$$WCl_6 + 2NH_3 \longrightarrow WN_2 + 6HCl$$

$$WCl_6 \xrightarrow{H_2} WCl_5 \longrightarrow WCl_4 \longrightarrow WCl_2$$

## 4. Tungsten Bronzes

On reducing alkali metal tungstates (e.g., $Na_2WO_4$ or $K_2WO_4$) with tungsten, *deeply coloured* susbstances are obtained. These substances are called tungsten bronzes. Tungsten bronzes are also obtained when (*a*) alkali metal tungstates are heated in a current of $H_2$ (*b*) tungstates are reduced electrolytically (*c*) $WO_3$ is reduced with alkali metal in vapour phase. Tungsten bronzes are unreactive but are good conductors of electricity. They have the general formula $M_xWO_3$ (where M = Li, Na or K). The value of $x < 1$. These are, evidently, nonstoichiometric substances in which $M^+$ ion is always present in interstitial sites created by three-dimensional network formed due to corner sharing of $WO_6$ octahedra.

Tungsten bronzes are highly coloured. Their colour depends on the amount ($x$) of alkali or alkaline earth metal ions. Different amounts of alkali metal or alkaline earth metal ions produce varying extents of lattice defects and hence impart different colours to the tungsten bronzes. Thus the colour of $Na_xWO_3$ varies from gold yellow ($x = 0.93$) to blue-violet ($x = 0.32$) as shown below.

| | | | | |
|---|---|---|---|---|
| $x =$ | 0.93 | 0.64 | 0.46 | 0.32 |
| *Colour =* | *Gold yellow* | *Orange* | *Red-violet* | *Blue-violet* |

These are resistant to the action of acids excepting HF. In the presence of an alkali, these are oxidised by $O_2$.

The electrons produced due to ionisation of alkali metal M to $M^+$ ions in the tungsten bronzes are completely delocalised over the whole of the lattice and are, therefore, free to move throughout the lattice. As a result these bronzes become conducting. The conducting ability of the bronzes decreases with increase in temperature.

Tungsten bronzes are largely used as metallic paints and pigments.

## 5. Tungsten Blues

Like molybdenum blues, tungsten also forms tungsten blues by similar methods Like molybdenum blues, they have varied compositions as $WO_{2.67}(OH)_{0.33}$ and $WO_{2.7}(OH)_{0.1}$.

# Rhenium, $Re_{75}$

Rhenium was first detected by its X-ray spectrum only in 1925 by Nodolack, Berg and Tache. They isolated it from molybdenite. It is now recovered from flue dust obtained during roasting of molybdenum sulphide ores.

## Compounds of Rhenium

### 1. Oxides

Some important oxides of rhenium are $ReO_2$ (brown), $ReO_3$ (red) and $Re_2O_7$(yellow) in which the oxidation states of the metal are +4, +6 and +7 respectively.

$ReO_2$ can be prepared by heating the heptoxide with rhenium at 870 K. At higher temperatures the dioxide reverts back to the heptoxide and metallic Re.

$ReO_3$ can be prepared by heating the heptoxide with rhenium at 570 K. On strong heating it disproportionates as follows:

$$3ReO_3 \longrightarrow Re_2O_7 + ReO_2$$

The structure of $ReO_3$ molecule is given in Fig. 8.2. In this structure each Re metal atom (Re) lies at the centre of an octahedron formed by six oxygen atoms (O).

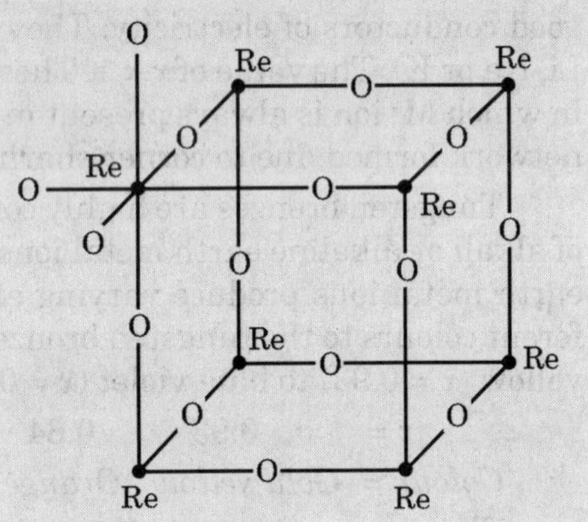

**Fig. 8.2** Structure of $ReO_3$ molecule

### 2. Perrhenic Acid and Perrhenates

Perrhenic acid is obtained by dissolving the heptoxide in water.

$$Re_2O_7 + H_2O \longrightarrow 2HReO_4$$

It can also be prepared by oxidation of the metal by heating with nitric acid or by passing a current of chlorine in a suspension of $ReO_2$.

$$2ReO_2 + 3Cl_2 + 4H_2O \longrightarrow 2HReO_4 + 6HCl$$

Anhydrous perrhenic acid has not so far been isolated. The solution on heating gives back the heptoxide. It is a strong monobasic acid which reacts with metals like Mg, Zn and Fe evolving hydrogen. It is not strong oxidizing agent but can oxidize HBr to $Br_2$ and HI to $I_2$. Sodium and ammonium rhenates are colourless stable salts very soluble in water. Potassium rhenate is, however, much less soluble. Being very stable, these are not used in oxidation reactions. Ammonium rhenate on heating in nitrogen gives $Re_3N$ and the metal.

### 3. Halides and Oxohalides

A number of halides of rhenium in which the element is in +2, +3, +4, +5, +6 and +7 oxidation states, are known. List of halides is given in Table 8.2.

**Table 8.3. Halides formed by rhenium**

| $Re^{(II)}$ | $Re^{(III)}$ | $Re^{(IV)}$ | $Re^{(V)}$ | $Re^{(VI)}$ | $Re^{(VII)}$ |
|---|---|---|---|---|---|
| | | $ReF_4$<br>blue<br>sub > 575 K | $ReF_5$<br>greenish-<br>yellow<br>m.p. 321 K | $ReF_6$<br>pale yellow<br>m.p. 292 K | $ReF_7$<br>pale yellow |
| | $Re_3Cl_9$<br>dark red<br>$Re_3Br_9$<br>red-brown | $ReCl_4$<br>black | $ReCl_5$<br>dark red-<br>brown<br>m.p. 534 K | $ReCl_6$<br>green-black<br>m.p. 302 K | |
| $ReI_2$<br>black | $ReI_3$<br>black | $ReBr_4$<br>dark brown<br>$ReI_4$<br>black | $ReBr_5$<br>dark brown | | |

Re also forms oxyhalides. Important examples are $ReOF_3$ (Re = +5), $ReOBr_4$ (Re = +6), $ReOF_5$ (Re = +7) and $ReO_2F_3$ (Re = +7).

**$ReBr_3$** is obtained by direct combination of the elements at 300°C. It is a dark green solid. It has a trimeric structure. The lines between the atoms in this structure do not show the exact bond order.

**$ReCl_5$** is also made by heating the metal in chlorine. When heated in an inert atmosphere, it decomposes into $ReCl_3$ which is a violet crystalline solid. It reacts with water to give perrhenic acid.

$$3ReCl_5 + 8H_2O \longrightarrow HReO_4 + 2ReO_2 \text{ (hydrated)} + 15HCl$$

**$ReBr_4$** and **$ReI_4$** are obtained by reducing perrhenic acid with HBr and HI, respectively.

**Structure of trihalides**. Trihalides of Re are trimeric and hence are represented as $(ReX_3)_3$ (X = Cl, Br or I). In their structure, three Re atoms form an equilateral triangle. Three X atoms act as bridging groups along the edges of the triangle, and the remaining six X atoms are coordinated to the Re atom (Fig. 8.3). The simplified form of the structure of $(ReX_3)_3$ or $Re_3X_9$ unit shown in Fig. 8.3(*a*) can be depicted as given in Fig. 8.3(*b*). It may be seen from the structure of $(ReX_3)_3$ or $Re_3X_9$ unit that each Re atom which is placed at the vertices of the triangle is linked with four X atoms. In this structure there is one unfilled coordination position on each Re atom which is filled in the solid by further X bridging with other $Re_3X_9$ units (Fig. 8.4) or by additional ligands L (Fig. 8.5) in Re (III) complexes.

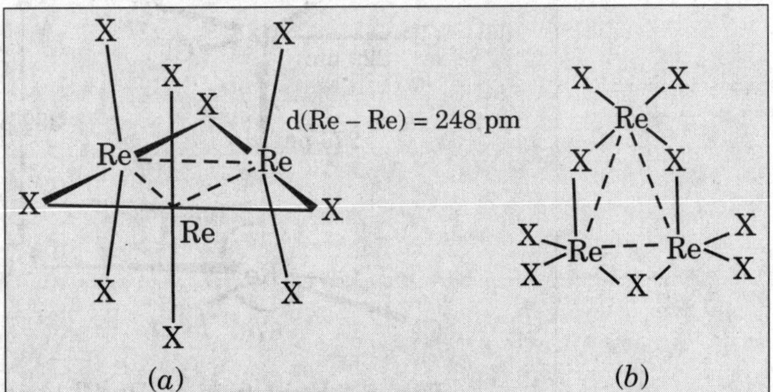


(*a*) (*b*)

**Fig. 8.3** The structure of $Re_3X_9$ unit.

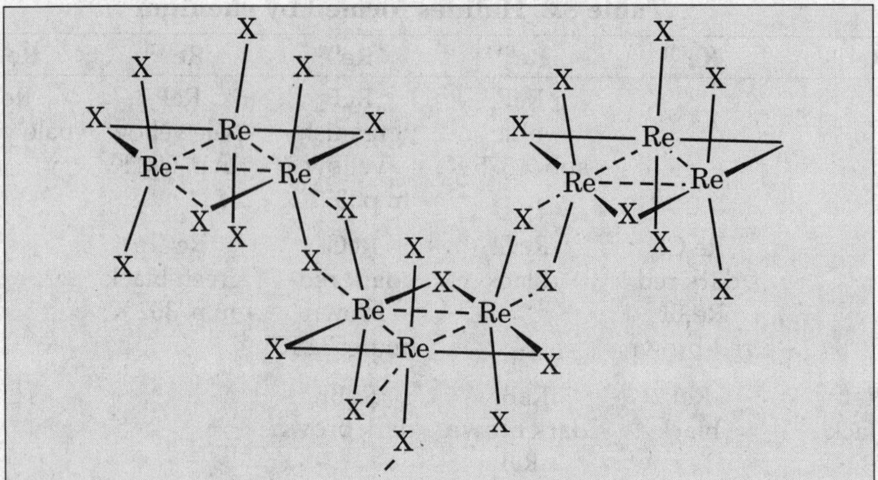


**Fig. 8.4** Further X bridging of $Re_3X_9$ solid with other $Re_3X_9$ units.

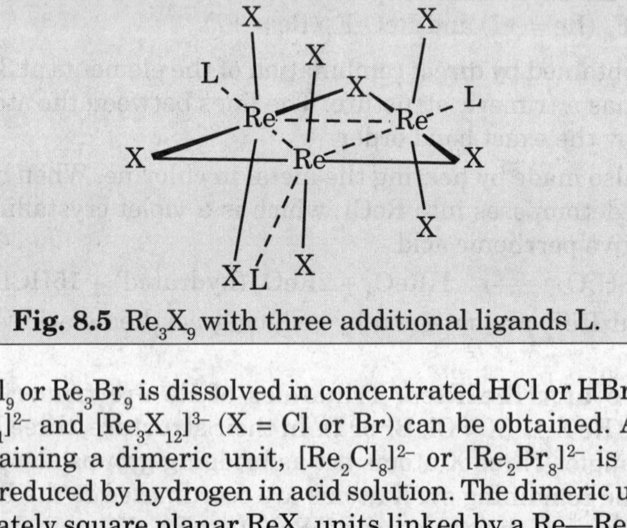


**Fig. 8.5** $Re_3X_9$ with three additional ligands L.

When $Re_3Cl_9$ or $Re_3Br_9$ is dissolved in concentrated HCl or HBr, the complexes $[Re_3X_{10}]^-$, $[Re_3X_{11}]^{2-}$ and $[Re_3X_{12}]^{3-}$ (X = Cl or Br) can be obtained. A different type of complex containing a dimeric unit, $[Re_2Cl_8]^{2-}$ or $[Re_2Br_8]^{2-}$ is obtained when perrhenates are reduced by hydrogen in acid solution. The dimeric unit is composed of two approximately square planar $ReX_4$ units linked by a Re—Re bond (Fig. 8.6).

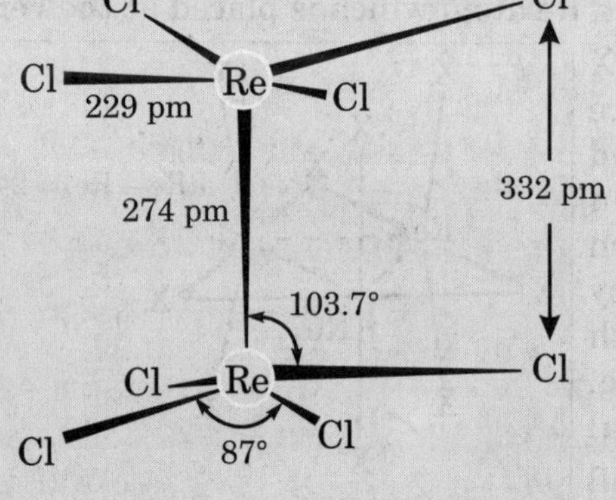


**Fig. 8.6** Structure of $[Re_2Cl_8]^{2-}$ ion.

### 4. Compounds in Unusual Oxidation States

Re has zero oxidation state in $[Re_2(CO)_{10}]$ and +1 oxidation state in $[ReCl(CO)_5]$.

## Osmium, $Os_{76}$

### Physical Properties of Osmium

The compact metal is bluish-grey, and in the finely divided state it is bluish-black. Osmium is a hard and brittle metal. It is the heaviest element known, its sp. gr. being 22.6. It has the highest m.pt. equal to 2700°C and very high boiling point, 5500°C. It is very difficult to prepare very pure metal.

### Chemical Properties of Osmium

Compact metal is not affected by air or oxygen even at 100°C, but is oxidised at higher temperatures. Finely divided osmium undergoes slow oxidation in air even at room temperature which is indicated by the smell of its oxidation product, the tetroxide $OsO_4$. The metal burns on strong heating in air forming $OsO_4$, in some cases the spongy osmium even glows.

The metal powder reacts with fluorine at 100°C or above with incandescence, forming tetra, hexa and octa fluorides. With chlorine it forms a dark green dichloride, $OsCl_2$ and a red mass of the tetrachloride, $OsCl_4$ at 650°C. Bromine and iodine do not react with the metal.

It is not attacked by hydrochloric acid in the absence of oxygen, but is oxidised by $HNO_3$ and $H_2SO_4$. The metal is dissolved readily in a concentrated solution of sodium or potassium hypo-chlorite.

Osmium burns brightly in sulphur vapour forming its sulphide. It combines with phosphorous vapour with incandescence, forming its phosphite. It forms alloys with metals such as copper, silver, gold, zinc, mercury, iron, cobalt, nickel etc.

When osmium is fused with potassium hydroxide in oxygen or air, potassium osmate, $K_2OsO_4$, is formed.

$$2Os + 4KOH + 3O_2 \longrightarrow 2K_2OsO_4 + 2H_2O$$

### Uses of Osmium

Osmium is used in making hard alloys for fountain pens, phonograph needles and for sparking points. It is also used in lamp mantles. A solution of $OsO_4$ known as osmic acid blackens organic materials due to reduction to metallic state and is, therefore, used in differentiating tissues under microscope.

## Compounds of Osmium

Osmium forms compounds in +2, +3, +4, +6 and +8 oxidation states. Most important compounds are those in which Os is in +4 and +6 oxidation state. Examples of some compounds in different oxidation states are given below.

+2 $\longrightarrow$ $OsO$, $OsCl_2$, $OsI_2$

+3 $\longrightarrow$ $Os_2O_3$, $OsCl_3$, $OsI_3$

+4 $\longrightarrow$ $OsO_2$, $OsF_4$, $OsCl_4$, $OsI_4$

$+6 \longrightarrow$ $OsO_3$, $OsF_6$, $[OsO_4]^{2-}$ (Osmate ion)

$+8 \longrightarrow$ $OsO_4$, $OsF_8$

–2 and zero are the unusual oxidation states of Os. Like ruthenium, osmium also forms a number of complexes with π acid ligands such as CO in which osmium is present in –2 or zero oxidation state. Examples are $[Os\ (CO)_4]^{2-}$, $[Os\ (CO)_5]$, $[Os_3\ (CO)_{12}]$, etc.

Important compounds of Os are discussed below.

## 1. Osmium Dioxide, $OsO_2$

It is obtained by the controlled reduction of $OsO_4$ or by fusing $K_2\ [OsCl_6]$ with sodium carbonate. It is also obtained when salts of tetravalent osmium are heated with sodium carbonate in an atmosphere of $CO_2$. The dioxide is a greyish-black powder, forming copper-red octahedral crystals. It decomposes when heated above 450°C:

$$2OsO_2 \rightleftharpoons 2Os + 2O_2$$

It is insoluble in water and acids. When heated in air, it forms $OsO_4$.

## 2. Osmium Tetroxide, $OsO_4$

It is the volatile oxide of osmium and has a characteristic smell and the vapours are poisonous and attack eyes and nose. It is formed by heating osmium in oxygen. The finely divided osmium inflames in air at 400°C forming $OsO_4$. It is pale-yellow solid with chlorine like irritating odour. The tetroxide is commonly known as **osmic acid,** whereas, actually it is the anhydride of perosmic acid. Its aqueous solution is weaklyacidic. It forms unstable addition compounds with strong alkalies. It is a strong oxidising agent and gets reduced easily even by dust or grease to $OsO_2$.

$OsO_4$ is used as a catalyst in Organic chemistry, particularly for the addition of OH groups at C – C double bond on treatment with $H_2O_2$. It is a good catalyst for several other reactions. Aqueous solutions of $OsO_4$ are used in histology for staining the tissues. It hardens the tissues without distorting them.

## 3. Osmium Hexafluoride, $OsF_6$

It is formed as one of the products when fluorine reacts with osmium. It is a yellow, volatile solid which melts at 32.2°C. It dissolves in water to form clear solution which undergoes hydrolysis to form $OsO_4$.

## 4. Osmium Octafluoride, $OsF_8$

It is formed as vapours when the metal is heated in fluorine at 250°C. The vapours are cooled by solid $CO_2$. It is a lemon yellow solid which melts at 34.5°C and boils at 47.5°C. It is hydrolysed by water and alkaline solutions and reacts readily with organic substances.

## 5. Osmium Tetrachloride, $OsCl_4$

It may be prepared by the action of chlorine on osmium at 700°C. It is insoluble in all common solvents but is slowly attacked by water and oxidising acids. With water it forms hydrated osmium dioxide, $OsO_2 \cdot nH_2O$. It is a black crusty solid and vaporises without melting.

## Iridium, $Ir_{77}$

### Properties of Iridium

It is a silvery-white metal which is brittle but fairly hard. It is resistant to the action of acids. In bulk state it does not dissolve even in aqua regia. It dissolves in a mixture of conc. HCl and sodium chlorate in a sealed tube (i.e., under pressure) at about 150°C.

Ir is used as an additive for hardening of Pt.

## Compounds of Iridium

### Ir (VI) Compounds

Ir combines with $F_2$ at 600°C to form $IrF_6$. $IrF_6$ is quite unstable

### Ir(V) Compounds

Ir forms $IrF_5$ in which metal shows +5 oxidation state. The structure of $IF_5$ is similar to that of $RhF_5$ and $MoF_5$. It is better represented as $(IrF_5)_4$.

### Ir (IV) Compounds

When Ir is heated above 1000°C in a current of $O_2$, $IrO_2$ is obtained. Ir also forms complexes in +4 oxidation state. Examples of such complexes are $[IrCl_6]^{2-}$, $[IrCl_3(H_2O)_3]^+$, $[IrCl_4(H_2O)_2]^\circ$, $[IrCl_5(H_2O)]^-$ and $[Ir(C_2O_4)_3]^{2-}$. The last complex is optically active.

### Ir (III) Compounds

Ir combines with $Cl_2$ at red heat to form $ICl_3$. All the trihalides of Ir are insoluble in water. These are polymeric containing Ir–halogen–Ir bonds in their structures. The hydrated trihalides are monomeric and soluble in $H_2O$.

When an alkali is added to an aqueous solution of $Ir^{3+}$, hydrated $I_2O_3$ is obtained as precipitate.

Iridium (III) basic acetate ion, $[Ir_3^{3+}O(CH_3COO)_6(H_2O_3]^+$ has a structure which

is similar to the basic acetate of Rh (III) and Cr (III).

### Ir (II) Compounds

$IrCl_2$ has polymeric structure with Ir–Cl–Ir bridges in its structure. Some phosphine complexes of Ir (II) also known.

### Ir (I) Compounds

**Vaska's compound** which is *trans* $[Ir^+Cl(CO)(PPh_3)_2]^\circ$ is yellow in colour and is capable of getting coordinated to aerial $O_2$. The yellow colour turns orange when it takes $O_2$. When flushed with $N_2$, it releases $O_2$ and again turns yellow.

### Ir (–I) and Ir (o) Compounds

Ir has –1 and zero oxidation state in $[Ir(PF_3)_4]^-$ and $[Ir_4(CO)_{12}]^0$ respectively.

## Platinum, $Pt_{78}$

### Physical Properties of Platinum

Platinum is a soft and silver-white metal. It has very high density equal to 22.4. Its m.pt. and b.pt. are 1769°C and 4530°C respectively. Excepting Ag and Au, it is the most malleable and ductile of all the metals. It can be rolled into foils and drawn into wires. It softens before melting. Due to this property, this metal can be welded. Its coefficient of expansion (= 0.000,009 per °C) is almost the same as that of glass. It is due to this property that Pt wires and foils can be sealed into glass without causing any cracks on cooling. The hardness of the metal increases by the presence of Ir metal. The metal is a good conductor of electricity. Molten Pt metal absorbs $O_2$ which is given out on cooling. This phenomenon is called *spitting*.

### Chemical Properties of Platinum

**1. Action of air or $O_2$.** Platinum is a noble metal and hence is not affected by air or $O_2$ even on heating the metal.

**2. Action of acids.** Pure metal does not react with any of the acids. However, pure metal reacts with aqua regia and forms $H_2[PtCl_6]$.

$$2HNO_3 + 8HCl + Pt \longrightarrow 2NOCl + 4H_2O + H_2[PtCl_6]$$

$$HNO_3 + 18HCl + 3Pt \longrightarrow 4NO + 8H_2O + 3H_2[PtCl_6]$$

HF has no action on platinum. The impure metal is attacked by $H_2SO_4$.

**3. Action of elements.** Platinum metal combines with many elements like $F_2$, $Cl_2$, P, S, As, etc. and forms the compounds like $PtF_4$, $PtCl_4$. $Pt_3P_5$, PtS, $Pt_2As_3$ etc. respectively.

With carbon, Pt forms platinum carbide which is brittle. Hence Pt-crucible should not be heated in a smoky flame. When $Cl_2$ is passed through a solution of spongy platinum in conc. HCl, $H_2[PtCl_6]$ is obtained.

$$\underbrace{Pt(\textit{spongy platinum}) + 2HCl(\textit{conc.}) + Cl_2}_{\text{Solution}} \longrightarrow H_2[PtCl_6]$$

**4. Action of metals.** Pt combines directly with many metals like Sn, Pb, Ag, Zn, etc. but not with Hg and forms alloys. Hence the above said metals should not be heated in platinum crucibles.

**5. Action of compounds.** Pt metal is attacked by various compounds like fused alkalies, $Na_2O_2$, alkali cyanides, alkali sulphides and nitrates. Such compounds should, therefore, not be heated in Pt crucible. With $CS_2$ vapour the metal forms $PtCS_2$. At 100°C, the metal forms $PtCl_4.2NOCl$ with NOCl. $Na_2CO_3$ has no action on the metal.

#### Uses of Platinum

***(i)* In scientific instruments and laboratory apparatus.** Due to its remarkable properties like high melting point and corrosion resistance, platinum is largely used to manufacture scientific instruments and laboratory apparatus such as crucibles, dishes, basins, etc. Pt is also used for contacts in electrical apparatus and for making electrodes.

***(ii)* As catalyst.** Different forms of platinum are used as catalyst in many manufactural and laboratory processes. Examples are given below:

(*a*) *Platinum gauge* is used as a catalyst in the oxidation of $NH_3$ to NO required for the manufacture of $HNO_3$ by Ostwald's process.

$$4NH_3 + 5O_2 \xrightarrow[800°C]{\text{Pt-gauge}} 4NO + 6H_2O$$

(*b*) When mixture of $H_2$ and $Br_2$ or $I_2$ is passed over red hot finely-divided platinum, HBr and HI is obtained.

$$H_2 + X_2 \text{ (X = Br or I)} \rightleftharpoons 2HX$$

Since *finely-divided platinum* can absorb several hundred times its own volume of hydrogen gas, it is used in many hydrogenation or dehydrogenation reactions. Large quantity of platinum metal is also used in the petroleum, synthetic fibre and glass industry.

***(iii)* In jewellery.** Pt is also used in jewellery as a setting for diamonds. For this purpose the metal is alloyed with metals like Ir, Ru or Cu. Pt—Ir alloy is also used in pen points, standard weights, coins, etc.

***(iv)* In dentistry.** The metal is used for making artificial teeth, pins, dental filling and support for dentures.

(*v*) Pt-salts are used in printing and toning in photography. Pt wire is also used in the qualitative analysis in chemistry. In the medical fields many instruments like X-ray tubes, radium applicators, lypodermic needles require platinum.

## Compounds of Platinum

+2 and +4 are the main oxidation states of Pt in which this metal forms simple salts and complex compounds. Compounds in +2 and +4 oxidation state are called ***platinous*** and ***platinic compounds***. Zero, +1 +5 and +6 are unusual oxidation states of Pt.

(*i*) Pt has zero oxidation state in complex compounds $[Pt(CO)_2(PPh_3)_2]^0$ and $[P(PF_3)_4]^0$. $[P(PF_3)_4]^0$ is obtained by the action of $PF_3$ on $PtCl_2$ in presence of Cu (as halogen acceptor) at 100°C.

(*ii*) PtCl has Pt in +1 oxidation state. This salt is not known in pure state. PtCl is obtained by heating $PtCl_2$ with Pt at 581-583°C

$$PtCl_2 + Pt \longrightarrow 2PtCl.$$

(*iii*) Tetrameric pentafluoride, $(PF_5)_4$ is the only compound of Pt(V). This compound is obtained as deep red product by the fluorination of $PtCl_2$ at 350°C. It has m.pt. = 80°C.

(*iv*) $PF_6$ is the only compound of Pt (VI). This is obtained as *dark red vapours* when Pt wire is heated in $F_2$. Its m.pt. = 61°C and b.pt. = 69°C. It is very unstable and is virulent oxidant. Barlett was successful in extracting electrons from $O_2$ molecule and Xe atom to produce the crystalline $O_2^+[PtF_6]^-$ and $Xe[PtF_6]$.

## Compounds of Pt (II)

$PtCl_2$ and $PtI_2$ are important compounds of Pt (II). $PtBr_2$ is brown.

### 1. Platinous Chloride or Platinum (II) Chloride, $PtCl_2$

It is obtain by heating $PtCl_4$ at 580°C.

$$PtCl_4 \xrightarrow{\Delta} PtCl_2 + Cl_2$$

It is a green-grey powder which is insoluble in water. With hydrochloric acid and potassium chloride, the following complex compounds are found.

$$PtCl_2 + 2HCl \longrightarrow H_2[PtCl_4]$$
$$PtCl_2 + 2KCl \longrightarrow K_2[PtCl_4]$$

**Structure:** $PtCl_2$ forms a metal cluster compound, $(PtCl_2)_6$ in which six Pt atoms occupy the six vertices of an octahedron and twelve edges of the octahedron are occupied by twelve bridging Cl-atoms.

### 2. Platinum (II) Iodide, $PtI_2$

It is obtained by heating $PtCl_2$ with KI

$$PtCl_2 + 2KI \longrightarrow PtI_2 + 2KCl$$

$PtI_2$ is a black insoluble powder.

## Complexes of Pt (II)

Most of the complexes of Pt (II) are low spin square planar complexes. Some examples of such complexes are $[PtCl_4]^{2-}$, $[PtCl_3\ (NH_3)]^-$, $[PtCl_2(NH_3)_2]^0$, $[PtCl\ (NH_3)_3]^+$, $[Pt\ (NH_3)_4]^{2+}$, $[PtI_4]^{2-}$, $[PtI_3\ (R_3P)]^-$, $[Pt\ I_2\ (R_3P)_2]^0$, $[PtCl_2\ (NH_3)(NO_2)]^-$, $[PtCl_3\ (NO_2)]^{2-}$, $[Pt(CN)_4]^{2-}$, $[Pt\ (acac)_2]^0$, $[Pt\ (dmg)_2]^0$ etc.

Complexes of $[PtL_4]^{2+}$ type can be oxidised by halogens ($X_2$) to six co-ordinated *trans* dihalo Pt (IV) complexes, *trans* $[Pt^{4+}\ L_4X_2]^{2+}$.

$$[Pt^{2+}X_4]^{2+} + X_2 \rightarrow trans\ [Pt^{4+}\ L_4X_2]^{2+}$$

Some Pt (II) complexes are discussed below.

### 1. Hydrogen Tetrachloro Platinate (II) or Tetra Chloroplatinous Acid, $H_2[PtCl_4]$

In this complex compound Pt is in +2 oxidation state. Here the coordination number of Pt is 4. It is obtained by dissolving $PCl_2$ in HCl or by the reduction of $H_2PtCl_6$ with $SO_2$.

$$PtCl_2 + 2HCl \longrightarrow H_2PtCl_4$$
$$H_2PtCl_6 + 2H_2O + SO_2 \longrightarrow H_2PtCl_4 + H_2SO_4 + 2HCl$$

The acid does not crystallise but forms a dark-brown amorphous solid on evaporation. Its salts are called **tetrachloroplatinate (II) complexes**. Potassium tetrachloroplatinate (II) is an important salt of $H_2PtCl_4$.

## 2. Potassium Tetrachloroplatinate (II), $K_2[PtCl_4]$

It is obtained by reducing chloroplatinate, $K_2PtCl_6$ by cuprous chloride.

$$K_2PtCl_6 + Cu_2Cl_2 \longrightarrow 2CuCl_2 + K_2PtCl_4$$

The complex forms dark red crystals and is used in platinum printing in photography. The corresponding ammonium, silver or lead salts are also known. $[PtCl_4]^{2-}$ ion is diamagnetic and has square planar geometry corresponding to $dsp^2$ hybridisation of $Pt^{2+}$ ion. $[PtCl_4]^{2-}$ ion is also found in **magnus green salt**, $[Pt(NH_3)_4]^{2+}[PtCl_4]^{2-}$.

## 3. Potassium Tetracyano Platinate (II) or Potassium Platinocyanide, $K_2[Pt(CN)_4].3H_2O$

It is also a complex compound of Pt in +2 oxidation state. It is obtained as yellow needles by mixing solutions of $PtCl_2$ and KCN in proper proportions and evaporating the solution to crystallisation.

$$PtCl_2 + 4KCN \longrightarrow K_2[Pt(CN)_4] + 2KCl$$

$K_2[Pt(CN)_4]$ is used in photography.

## 4. Barium Tetracyano Platinate (II) or Barium Platinocyanide, $Ba[Pt(CN)_4] . 4H_2O$

In this complex compound Pt is in +2 oxidation state. It is prepared by warming $H_2[PtCl_6]$ solution with baryta water, $Ba(OH)_2$ and HCN passing in $SO_2$ until the solution becomes colourless.

$$H_2[PtCl_6] + 5Ba(OH)_2 + 4HCN + SO_2 \longrightarrow Ba[Pt(CN)_4] + \underset{\text{ppt}}{BaSO_4} + 3BaCl_2 + 8H_2O$$

$BaSO_4$ is filtered off and the filtrate is crystallised to get $Ba[Pt(CN)_4]. 4H_2O$. It is a lemon-yellow powder which is soluble in water. It is used for fluorescent screens in X-ray work.

$[Pt(CN)_4]^{2-}$ ion is diamagnetic and, like $[PtCl_4]^{2-}$, this ion also has square planar geometry.

## 5. Cis-and Trans-diammine Dichloroplatinum (II), $[Pt(NH_3)_2Cl_2]^0$.

This planar complex exists in two geometrical isomers: *cis* planar and *trans* planar forms. These have been prepared by the following reactions.

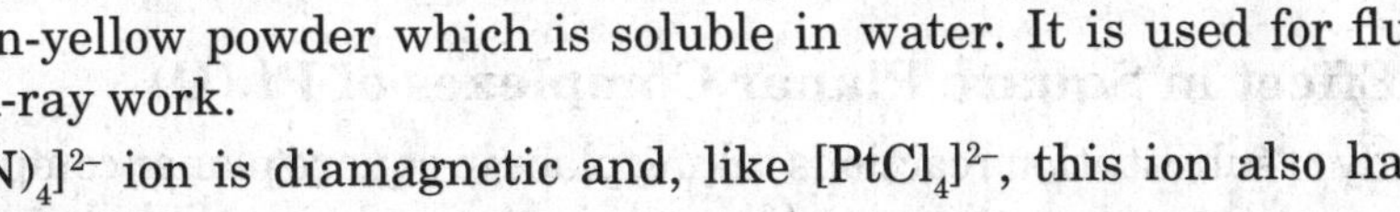

*cis*-planar (yellow)

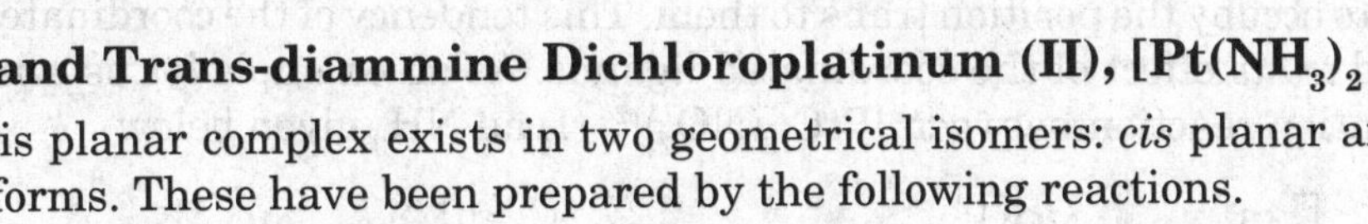

*trans*-planar (yellow)

*Cis*-$[Pt^{2+}Cl_2(NH_3)_2]^0$ is called ***cisplatin*** which is used in the treatment of some types of cancer. $[Pt^{2+}Cl_2\,(NH_3)(NO_2)]^-$ also exists in *cis* and *trans* forms whose structures are given below:

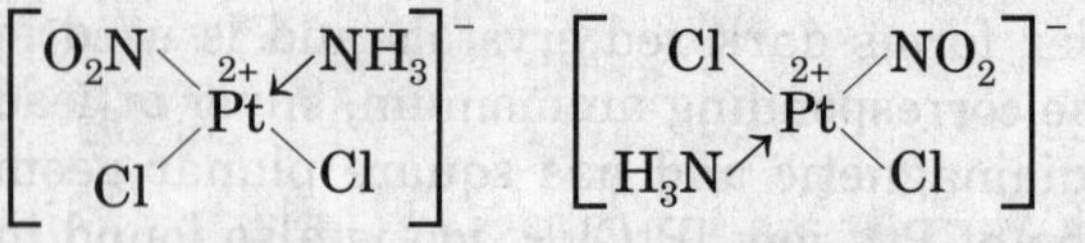

*cis-form* *trans-form*

**Structure.** We have said above that 4-co-ordinated complexes of Pt(II) have square planar geometry. This geometry results from *dsp*$^2$ hybridisation of $Pt^{2+}$ ion which is a $5d^8$ ion. *dsp*$^2$ hybridisation is shown is Fig. 8.7.

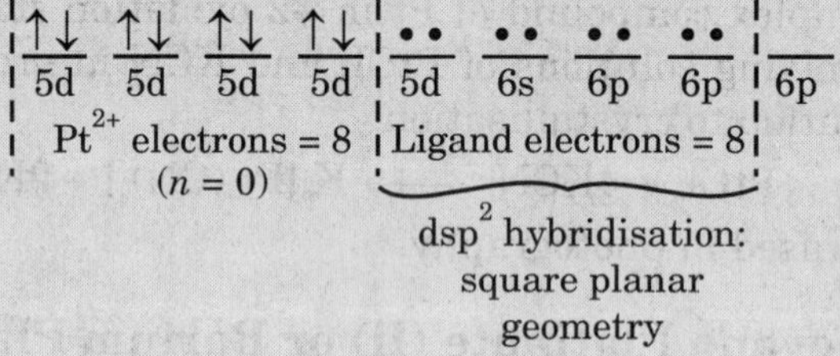

**Fig. 8.7** dsp$^2$ hybridisation of $Pt^{2+}$ ion its square planar complex.

On the basis of CFT of square planar complexes the distribution of $d^8$ electrons of $Pt^{2+}$ ion can be shown as:

$$d^2_{z^2} \;\vdots\; d^2_{xz} = d^2_{yz} \;\vdots\; d^2_{xy} \;\vdots\; d^0_{x^2-y^2}$$

$$\xrightarrow{\quad\text{Energy increasing}\quad}$$

It may be seen that square planar complexes of Pt (II) are diamagnetic since $n = 0$.

## Trans Effect in Square Planar Complexes of Pt (II)

Study of substitution reactions taking place in square planar complexes of Pt (II) has shown that certain coordinated ligands in these complexes induce the incoming groups to occupy the position trans to them. This tendency of the coordinated ligands is called *trans* effect of the coordinated ligands. As an example let us consider the substitution reaction between $[PtCl_3(NO_2)]^{2-}$ (I) ad $NH_3$ given below.

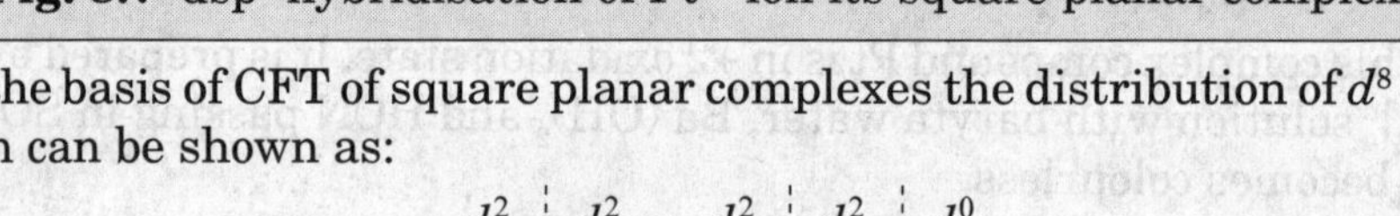

(I) *trans*-form

Theoretically in this reaction, the incoming group $NH_3$ can replace any $Cl^-$ ligand in (I) to form $[Pt^{2+}Cl_2\,(NH_3)\,(NO_2)]^-$, but experiments have shown that $NO_2^-$ ligand present in (I) has greater *trans* effect than $Cl^-$ ligand [$NO_2^-$ and three $Cl^-$ ligands are the coordinated ligands present in (I)]. As a result, the incoming $NH_3$ group will replace that group from (I) which has less *trans* effect than $NO_2^-$ ligand and also is placed *trans* to $NO_2^-$ ligand. Thus in this substitution reaction $Cl^2$ will be replaced by the incoming $NH_3$ group to form *trans* $[PtCl_2(NH_3)(NO_2)]^-$. Note that only $\overset{2}{Cl}$ is placed *trans* to $NO_2^-$. $\overset{1}{Cl}$ and $\overset{3}{Cl}$ are not placed *trans* to $NO_2^-$.

## Compounds of Pt (IV)

All the four halides of Pt (IV), Pt $X_4$ are known. Pt $F_4$ is obtained by the fluorination of $PtBr_4$. $PtCl_4$, $PtBr_4$ and $PtI_4$ are formed by the direct interaction of the halogens with Pt at elevated temperatures. The temprature for the preparation of $PtCl_4$, $PtBr_4$ and $PtI_4$ is ~250°C, ~150°C and ~150°C respectively. Tetrahalides have different colours as shown: $PtF_4$–yellow brown, $PtCl_4$–red brown, $PtBr_4$-brown black, $PtI_4$-brown black. $PtCl_4$ is an important tetrahalide of Pt. It is described below. $PtCl_4$ is also called **platinic chloride**.

**Preparation.** $PtCl_4$ is obtained:

(*i*) by heating $H_2[PtCl_6]$ at 370°C in a current of $Cl_2$ or at 165°C in $H_2$.

$$H_2[PtCl_6] \xrightarrow{\Delta} PtCl_4 + 2HCl$$

(*ii*) by the action of $Cl_2$ on Pt-metal at 200–300°C.

$$Pt + 2Cl_2 \longrightarrow PtCl_4$$

**Properties.** (*i*) It forms reddish-brown slightly hygroscopic crystalline mass.

(*ii*) *Hydrolysis.* It dissolves in water, giving yellowish-red solution which is strongly acidic. The acidic nature of the solution is due to the formation of a complex acid, $H_2[PtCl_4(OH)_2]$.

$$PtCl_4 + 2H.OH \longrightarrow H_2[PtCl_4(OH)_2]$$

On electrolysis, this acid gives the complex anion on the anode showing platinum is discharged at the anode as a part of $[PtCl_4(OH)_2]^{2-}$ ion.

$$H_2[PtCl_4(OH)_2] \rightarrow 2H^+ \text{ (At } cathode) + [PtCl_4(OH)_2]^{2-} \text{ } (At\ anode)$$

(*iii*) *Oxidising property: Reduction of $PtCl_4$.* (*a*) $PtCl_4$ oxidises KI to iodine and is itself reduced to $PtCl_2$. Thus $PtCl_4$ liberates $I_2$ from KI solution.

$$2KI + PtCl_4 \longrightarrow I_2 + 2KCl + PtCl_2$$

or $$2I^- + Pt^{+4} \longrightarrow I_2^0 + Pt^{2+}$$

(*b*) When $PtCl_4$ reacts with hydroxyl amine ($NH_2OH$) in acidic medium, it gets reduced to Pt-metal. Hydrazine ($N_2H_4$) also reduces $PtCl_4$ to Pt-metal. $NH_2OH$ (N = –1) gets oxidised to $N_2O$ (N = +1) while $N_2H_4$ (N = –2) gets oxidised to $N_2$(N = 0).

$$PtCl_4 + 2NH_2OH \longrightarrow Pt + N_2O + 4HCl + H_2O$$
$$PtCl_4 + N_2H_4 \longrightarrow Pt + N_2 + 4HCl$$

(*iv*) *Action of HCl and alkali metal halides.* Complexes of $[PtX_6]^{2-}$ type (X = Cl, Br, I) are obtained.

$$PtCl_4 + 2HCl \longrightarrow H_2[PtCl_6]$$
$$PtCl_4 + 2NaCl \longrightarrow Na_2[PtCl_6]$$
$$PtCl_4 + 2KCl \longrightarrow K_2[PtCl_6]$$
$$PtCl_4 + 2NH_4Cl \longrightarrow (NH_4)_2[PtCl_6]$$

(*v*) *Action of heat*. When heated at 853K, $PtCl_2$ is obtained.

$$PtCl_4 \xrightarrow{\Delta} PtCl_2 + Cl_2$$

(*vi*) *Action of KOH*. Pt $(OH)_2$ gets precipitated when $PtCl_4$ reacts with KOH.

$$PtCl_4 + 4KOH \longrightarrow Pt(OH)_4\downarrow + 4KCl$$

(*vii*) *Action of KCN*. Potassium tetra cyano platinate (II), $K_2[Pt(CN)_4]$ along with cynogen, $(CN)_2$ are produced.

$$PtCl_4 + 6KCN \longrightarrow K_2[Pt(CN)_4] + 4KCl + (CN)_2\downarrow$$

**Uses.** $PtCl_4$ is used for the preparation of $H_2[PtCl_6]$. Both these compounds are very useful compounds.

## Complexes of Pt (IV)

### 1. Hydrogen hexachloro platinate (IV) or hexachloro platinic acid, $H_2[PtCl_6]$

**Preparation.** It is prepared:

(*i*) by the action of HCl on $PtCl_4$ or Pt $(OH)_4$.

$$2HCl + PtCl_4 \longrightarrow H_2[PtCl_6]$$

$$6HCl + Pt(OH)_4 \longrightarrow H_2[PtCl_6] + 4H_2O$$

(*ii*) by the action of aqua regia $(HNO_3 + 3HCl)$ on pure platinum metal. NOCl or NO are produced.

(*a*)

$$[HNO_3 + 3HCl \longrightarrow NOCl + 2H_2O + 2Cl] \times 2$$

$$Pt + 4Cl \longrightarrow PtCl_4$$

$$PtCl_4 + 2HCl \longrightarrow H_2[PtCl_6]$$

*On adding*: $2HNO_3 + 8HCl + Pt \longrightarrow 2NOCl + 4H_2O + H_2[PtCl_6]$

(*b*)

$$[HNO_3 + 3HCl \longrightarrow NO + 2H_2O + 3Cl] \times 4$$

$$[Pt + 4Cl \longrightarrow PtCl_4] \times 3$$

$$[PtCl_4 + 2HCl \longrightarrow HPtCl_6] \times 3$$

*On adding*: $4HNO_3 + 18HCl + 3Pt \longrightarrow 4NO + 8H_2O + 3H_2[PtCl_6]$

The solution is evaporated with HCl till $HNO_3$ is completely removed from the reaction mixture. On cooling, the red-brown crystals of $H_2PtCl_6 . 6H_2O$ are left behind.

(*iii*) by passing $Cl_2$ through a solution of spongy platinum in conc. HCl.

Pt (*spongy platinum*) + $2HCl + Cl_2 \longrightarrow H_2[PtCl_6]$

**Properties.** (*i*) $H_2PtCl_6$ is a red-brown crystalline solid that deliquesces in air and is readily soluble in $H_2O$, alcohol and ether. It is associated with six $H_2O$ molecules, $H_2[PtCl_6] . 6H_2O$.

(*ii*) *Electrolysis*. When the aqueous solution of $H_2PtCl_6$ is electrolysed, hydrogen is liberated. This hydrogen reduces $H_2PtCl_6$ to liberate Pt-metal at cathode.

$$H_2PtCl_6 \rightleftharpoons 2H^+ + [PtCl_6]^{2-}$$

$$2H^+ + 2e^- \longrightarrow H_2$$

$$H_2[PtCl_6] + 2H_2 \xrightarrow{\text{Reduction}} \text{Pt (\textit{At cathode})} + 6HCl$$

(*iii*) *Action of heat*. (*a*) When heated at 165°C in $H_2$ or at 370°C in $Cl_2$, $PtCl_4$ is formed. This reaction has been used to prepare $PtCl_4$.

$$H_2[PtCl_6] \xrightarrow[Cl_2]{370°C} PtCl_4 + 2HCl$$

(*b*) When heated to 390°C in $Cl_2$, $PtCl_3$ is formed.

$$2H_2[PtCl_6] \xrightarrow[Cl_2]{390°C} 2PtCl_3 + 4HCl + Cl_2$$

(*c*) The heating of $H_2[PtCl_6]$ in $Cl_2$ at 580°C gives $PtCl_2$.

$$H_2[PtCl_6] \xrightarrow[Cl_2]{580°C} PtCl_2 + 2HCl + Cl_2$$

(*iv*) *Reduction: Oxidising properties*. $H_2[PtCl_6]$ can be reduced by various substances. The nature of the products obtained by reduction depends on the nature of the reducing agent used. For example:

(*a*) Nascent hydrogen obtained by the action of Zn on HCl or formic acid (HCOOH) reduces the acid to platinum black.

$$2Zn + 4HCl \longrightarrow 2ZnCl_2 + 4H$$

$$H_2[PtCl_6] + 4H \longrightarrow Pt\ (\textit{platinum black}) + 6HCl$$

$$H_2[PtCl_6] + 2HCOOH \longrightarrow Pt\ (\textit{platinum black}) + 2CO_2\uparrow$$

(*b*) Hydrazine ($NH_2$—$NH_2$) reduces the acid to colloidal platinum in presence of some protective colloid. $NH_2$—$NH_2$ (N = –2) is oxidised to $N_2$ (N = 0).

$$H_2[PtCl_6] + NH_2\text{—}NH_2 \longrightarrow Pt\ (\textit{colloidal platinum}) + N_2 + 6HCl$$

(*c*) When asbestos is soaked in 10% solution of $H_2[PtCl_6]$, dried and heated to 500°C, $H_2PtCl_6$ gets reduced to fnely-divided platinum which is distributed throughout the asbestos fibre. The asbestos fibre having finely-divided platinum in it is called *platinised asbestos*.

(*d*) KI solution reduces $H_2PtCl_6$ to $PtI_2$ and is itself oxidised to $I_2$.

$$H_2[PtCl_6] + 4KI \longrightarrow I_2 + PtI_2 + 4KCl + 2HCl$$

(*e*) When $H_2PtCl_6$ is treated with $Ba(OH)_2$ and HCN and $SO_2$ gas is passed through the reaction mixture till the solution becomes colourless, $Ba[Pt(CN)_4] \cdot 3H_2O$ is obtained in solution.

$$H_2[PtCl_6] + 5Ba(OH)_2 + 4HCN + SO_2$$

$$\longrightarrow Ba[Pt\ (CN)_4]\ (\textit{solution}) + BaSO_4\ (\textit{ppt}) + 3BaCl_2$$

$BaSO_4$ is filtered off and the filtrate containing $Ba[Pt(CN)_4]$ is crystallised to get the crystals of $Ba[Pt(CN)_4]$. In the above reaction $H_2[PtCl_6]$ (Pt = +4) is reduced to $Ba[Pt\ (CN)_4]$ (Pt = +2) and $SO_2$ (S = +4) is oxidised to $BaSO_4$ (S = +6). Thus here $SO_2$ acts as a reducing agent.

(*f*) The aqueous solution of $SO_2$ reduces $H_2[PtCl_6]$ to $H_2[PtCl_4]$ and is itself oxidised to $H_2SO_4$.

$$H_2[PtCl_6] + SO_2 + H_2O \longrightarrow H_2[PtCl_4] + H_2SO_4 + 2HCl$$

(*v*) *Formation of salts of the acid: Hexachloro platinates*. $H_2[PtCl_6]$ is a strong dibasic acid and ionises as:

$$H_2[PtCl_6] \rightleftharpoons 2H^+ + [PtCl_6]^{2-}$$

This mode of ionisation shown above indicates that $H_2[PtCl_6]$ forms salts of the type $M_2^+[PtCl_6]^{2-}$ ($M^+ = Li^+, Na^+, K^+, NH_4^+, Ag^+$ etc.). These salts are called hexachloro platinates which contain hexachloro platinate ion, $[PtCl_6]^{2-}$.

(*vi*) *Action of $H_2S$.* When $H_2S$ is passed into a hot solution of $H_2PtCl_6$, platinic sulphide ($PtS_2$) is obtained as a black precipitate.

$$H_2[PtCl_6] + 2H_2S \longrightarrow PtS_2 \text{ (black ppt.)} + 6HCl$$

(*vii*) ***Formation of addition compound with organic bases.*** The organic bases like $C_6H_5NH_2$ form a stable crystalline addition compound with $H_2[PtCl_6]$. The formation of such compounds is used in the gravimetric estimation of these bases.

$$2C_6H_5NH_2 + H_2[PtCl_6] \longrightarrow \underset{\text{Addition compound}}{2C_6H_5NH_2.H_2[PtCl_6]}$$

**Uses.** $H_2PtCl_6$ is used:

(*i*) for the preparation of a number of compounds of platinum (*e.g.* $PtCl_4$)
(*ii*) for the determination of the molecular weight of bases.
(*iii*) for the gravimetric estimation of organic bases.
(*iv*) for the preparation of hexachloro platinates.
(*v*) for preparing platinum black, colloidal platinum, platinised asbestos.
(*vi*) as a tanning agent in photography.

## 2. Hexachloroplatinates (IV)

Hexachloroplatinic acid, $H_2[PtCl_6]$ is a strong dibasic acid and ionises as:

$$H_2[PtCl_6] \rightleftharpoons 2H^+ + [PtCl_6]^{2-}$$

Thus this acid gives the stable salts of the type $M_2^+[PtCl_6]$ ($M^+ = Li^+, Na^+, K^+, NH_4^+, Ag^+$ etc.) which are called *hexachloro platinates* (IV) of $M^+$. Examples of such salts are $Li_2[PtCl_6]$ . $6H_2O$. $Na_2[PtCl_6]$ . $6H_2O$, $K_2[PtCl_6]$, $(NH_4)_2[PtCl_6]$, $Ag_2[PtCl_6]$ etc.

$Li_2[PtCl_6]$ and $Na_2[PtCl_6]$ are obtained by the action of LiOH or NaOH on $H_2[PtCl_4]$.

$$2LiOH + H_2[PtCl_6] \longrightarrow Li_2[PtCl_6] + 2H_2O$$

$$2NaOH + H_2[PtCl_6] \longrightarrow Na_2[PtCl_6] + 2H_2O$$

$Na_2[PtCl_6]$ can also be obtained by the action of $PtCl_4$ on NaCl.

$$PtCl_4 + 2NaCl \longrightarrow Na_2[PtCl_6]$$

$Ag_2[PtCl_6]$ is obtained as a yellow precipitate when $H_2[PtCl_6]$ is treated with $AgNO_3$ solution.

$$H_2[PtCl_6] + 2AgNO_3 \longrightarrow Ag_2[PtCl_6] + 2HNO_3$$

$Li_2[PtCl_6]$ . $6H_2O$ and $Na_2[PtCl_6]$ . $6H_2O$ are soluble in water.

Here we shall discuss in some detail $K_2[PtCl_6]$ and $(NH_4)_2[PtCl_6]$.

(***i***) **Potassium hexacholoro platinate (IV), $K_2[PtCl_6]$:** It is obtained as a yellow precipitate by the action of KCl on $PtCl_4$ or on $H_2[PtCl_6]$.

$$2KCl + PtCl_4 \longrightarrow K_2[PtCl_6]$$

$$2KCl + H_2[PtCl_6] \longrightarrow K_2[PtCl_6] + 2HCl$$

**Properties.** (*i*) This compound is poorly soluble in $H_2O$. It is yellow in colour.

(*ii*) *Action of an alkali*. When $K_2[PtCl_6]$ is treated with KOH, $K_2[Pt(OH)_6]$ is obtained.

$$K_2[PtCl_6] + 6KOH \longrightarrow K_2[P(OH)_6] + 6KCl$$

or $$[PtCl_6]^{2-} + 6OH^- \longrightarrow [Pt(OH)_6]^{2-} + 6Cl^-$$

(*iii*) *Action of heat*. When heated, it gives spongy platinum.

$$K_2[PtCl_6] \xrightarrow{\Delta} 2KCl + 2Cl_2 + Pt \text{ (spongy platinum)}$$

(*iv*) *Reduction*. It gets reduced to potassium tetrachloro platinate (II), $K_2[PtCl_4]$ by CuCl.

$$K_2[PtCl_6] + 2CuCl \longrightarrow 2CuCl_2 + K_2[PtCl_4]$$

**Uses.** $K_2[PtCl_6]$ is used in photography.

**(*ii*) Ammonium hexachloro platinate (IV), $(NH_4)_2$ $[PtCl_6]$ (also called ammonium chloroplatinate).** When $H_2[PtCl_6]$ is treated with neutral solution of $NH_4Cl$ or $NH_4OH$, yellow crystalline ppt. of $(NH_4)_2[PtCl_6]$ is obtained.

$$H_2[PtCl_6] + 2NH_4Cl \longrightarrow \underset{\text{Yellow ppt}}{(NH_4)_2[PtCl_6]} + 2HCl$$

$$H_2[PtCl_6] + 2NH_4OH \longrightarrow \underset{\text{Yellow ppt}}{(NH_4)_2[PtCl_6]} + 2H_2O$$

$(NH_4)_2[PtCl_6]$ can also be obtained from the platini-ferrous sands and gravels present in the river beds and also from the non-volatile residue obtained at the end of Mond's process.

**Properties.** (*i*) It is a yellow substance and is poorly soluble in $H_2O$.

(*ii*) *Action of heat.* When heated to redness, spongy platinum is obtained.

$$(NH_4)_2 [PtCl_6] \xrightarrow{\Delta} 2NH_4Cl + 2Cl_2 + Pt \text{ (spongy platinum)}$$

(*iii*) *Action of liquid $NH_3$*. Hexammine platinum (IV) chloride, $[Pt (NH_3)_6] Cl_4$ is obtained.

$$(NH_4)_2 [PtCl_6] + 6NH_3 \longrightarrow [Pt(NH_3)_6]Cl_4 + 2NH_4Cl$$

(*iv*) *Reduction*. Hydrazine ($N_2H_4$) reduces $(NH_4)_2[PtCl_6]$ to platinum metal and is itself oxidised to $N_2$. Some of the platinum produced gets deposited on the inside of the tube as a platinum mirror.

$$(NH_4)_2 [PtCl_6] + N_2H_4 \longrightarrow Pt + N_2 + 4HCl + 2NH_4Cl$$

**Uses.** $(NH_4)_2[PtCl_6]$ is used (*i*) in the extraction of platinum metal, (*ii*) for the detection and estimation of potassium and ammonium and for the equivalent weight determination of organic bases.

**Structure of $[PtCl_6]^{2-}$ ion.** This ion has octahedral geometry which results from $d^2sp^3$ hybridisation of $Pt^{4+}$ ion. Thus the complex ion is an inner orbital octahedral complex ion. V.S.E.C. of $Pt^{4+}$ ion in this complex ion is $t_{2g}^6 e_g^0$ which gives $n = 0$. Thus the complex ion is diamagnetic.

## 3. Ammine Complexes

Many ammine 6-coordinated cationic complexes of Pt (IV) are also known. Examples are $[Pt(NH_3)_6]^{4+}$, $[Pt(NH_3)_4Cl_2]^{2+}$, $[Pt(NH_3)_5Cl]^{3+}$ . *trans* form of $[Pt\,L_4\,X_2]^{2+}$ can be obtained by oxidising square planar Pt (II) complex, $[PtL_4]^{2+}$ with halogen $(X_2)$

$$[PtL_4]^{2+} + X_2 \rightarrow trans - [PtL_4X_2]^{2+}$$

# Organometallic Compounds of Platinum

Platinum forms a number of organometallic compounds. The first organometallic compound was reported in 1825. It is known as **Zeis's salt** and has the formula $K[Pt(C_2H_4)Cl_3]\,.\,H_2O$.

It may be pointed out that $C_2H_4$ in this compound acts as σ donor and pi acceptor to Pt (II) with the result that $C_2H_4$ ligates, perpendicularly to the plane containing $Pt^{2+}$ and three $Cl^-$ ions. The Zeis's salt is commercially employed as a catalyst in the synthesis of a number of organic compounds.

# Gold, $Au_{79}$

## Physical Properties of Gold

Gold is a lustrous yellow metal which melts at 1063°C to green liquid and boils at 2920°C. Its specific gravity is 19.3. It is the most malleable and ductile of all metals and is a good conductor of heat and electricity. With mercury, it forms an amalgam.

## Chemical Properties of Gold

**1. Action of air and water.** Neither dry or moist air nor water has any action on gold at any temperature. It is due to this property that gold is called *noble metal*.

**2. Action of acids.** (*a*) Selenic acid $(H_2SeO_4)$ reacts with Au at 225°C, forming auric selenate, $Au_2(SeO_4)_3$.

$$2Au + 6H_2SeO_4 \longrightarrow Au_2(SeO_4)_3 + 3SeO_2 + 6H_2O$$

(*b*) Metal dissolves in aqua regia forming $H\,[AuCl_4]$.

$$[HNO_3 + 3HCl \longrightarrow 2H_2O + NOCl + 2Cl] \times 3$$

$$[Au + HCl + 3Cl \longrightarrow H[AuCl_4]] \times 2$$

$$3HNO_3 + 11HCl + 2Au \longrightarrow 2H[AuCl_4] + 6H_2O + 3NOCl$$

(*c*) HF even mixed with $HNO_3$ has no action on the metal.

(*d*) Very conc. and hot $HNO_3$ dissolves only a few milligrams of the metal per litre.

**3. Action of halogens.** $Cl_2$, $Br_2$ and $I_2$ form $AuX_3$ when Au is heated with these halogens.

$$2Au + 3X_2 \xrightarrow{\Delta} 2AuX_3$$

**4. Action of alkali cyanide solutions.** In presence of $O_2$, complex compound, $M[Au(CN)_2]$ (M = alkali metal) is formed.

$$4Au + 8KCN + 2H_2O + O_2 \longrightarrow \underset{\text{Soluble}}{4K[Au(CN)_2]} + 4KOH$$

**5. Action of alkaline sulphides and thiosulphates.** These compounds dissolve the metal at high temperature.

**6. Action of sulphur.** Sulphur has no action on the metal.

**7. Action of oxidising agents.** Oxidising agents like $H_2O_2$, alkali metal nitrates, chlorates, perchlorates, permanganates and bleaching powder etc. along with HCl dissolve the metal.

### Uses of Gold

(*i*) In coinage, jewellery and ornamental vessels.

(*ii*) In gold plating.

(*iii*) Gold leaf is used for decorations, lettering and in medicine as tonic.

(*iv*) Purple of Cassium which is a combination of gold with colloidal stannic acid is used for making ruby red glass and high class pottery.

(*v*) Gold is used in making white gold which is a gold-nickel alloy. This alloy resembles platinum in appearance and is used in jewellery.

## Purity or Fineness of Gold: Concept of Carat

We have already said that gold is used in making ornaments, jewellery and watches. Since pure gold is very soft, to make it harder, it is mixed with small amount of Cu or Ag for making jewellery. The purity or fineness of gold in a given ornament is generally expressed as *carats*. Pure gold is 24 carats. These days ornaments are generally made of 22 carats.

A 22 carat gold means that the sample of gold contains 22 parts of gold and $24 - 22 = 2$ parts of Cu or Ag. Thus the percentage of gold in 22 carats = $\frac{22}{24} \times 100$ = 91.6 and % of Cu or Ag = 100 – 91.6 = 8.4.

## Compounds of Gold

Main oxidation states of Au are +1 and +3. Most of the compounds formed by Au have Au is +1 and +3 oxidation states. Compounds having Au in +1 oxidation state are called *aurous compounds* while those containing Au in +3 oxidation state are known as *auric compounds*. In general auric compounds are more stable and more common than aurous compounds. All gold compounds are easily decomposed. They are easily reduced to free gold metal either by weak reducing agents or simply by heating.

In addition to +1 and +3 oxidation states, Au also forms compounds in –1, +2 and +5 oxidation states.

## Compounds of Au (– I)

When Au and Cs are fused together in equal proportions, a mass is obtained. This mass is netiher in alloy nor conducts electricity. This mass is a salt having composition as $Cs^+ Au^-$. In this salt Au has –1 oxidation states.

## Compounds of Au (I) : Aurous Compounds

$Au^+$ ion exists in aqueous solution in very minute concentration, since this ion undergoes disproportionation in aqueous solution.

$$3Au^+ \rightleftharpoons Au^{3+} + 2Au$$

Thus the presence of $Au^+$ ions cannot be detected in aqueous solution. $Au^+$ ion becomes stable when it forms a complex compound. Thus $K[Au(CN)_2]$ is highly stable in solution.

Examples of compounds containing Au in +1 oxidation state are given below:

(*a*) *Simple compounds:* $Au_2O$, Au (OH), AuX (X = Cl, Br, I), $Au_2S$, AuCN etc.

(*b*) *Complex compounds:* $K[Au(CN)_2]$, $Na_3[Au(S_2O_3)_2] \cdot \frac{1}{2}H_2O$ etc. 2-coordinate complex compounds given by Au(+1) are *linear* involving *sp* hybridisation whereas 4-coordinate complexes are probably *tetrahedral* resulting from $sp^3$ hybridistion. Both the types of complexes are *diamagnetic*. Here we shall discuss the following compounds of Au (+1).

### 1. Aurous chloride, AuCl

**Preparation.** It is prepared by heating auric chloride ($AuCl_3$) to 175°C.

$$AuCl_3 \xrightarrow{175°C} AuCl + Cl_2$$

**Properties.** It is a light-yellow powder which is insoluble in water. When heated above 175°C, it gets decomposed into metallic gold and free $Cl_2$.

$$2AuCl \xrightarrow{>175°C} 2Au + Cl_2$$

When treated with $H_2O$ at ordinary temperature, it is decomposed into metallic Au and $AuCl_3$.

$$3AuCl \longrightarrow AuCl_3 + 2Au$$

AuCl combines with alkali metal chlorides (*e.g.*, KCl) to form chloroaurates (I), $M[AuCl_2]$. When treated with $Cl_2$, $AuCl_3$ is formed.

$$AuCl + KCl \longrightarrow K[AuCl_2] \text{ [\textit{Pot. dichloroaurate (I)}]}$$
$$AuCl + Cl_2 \longrightarrow AuCl_3$$

### 2. Potassium Dicyano Aurate (I) or Potassium Aurocyanide, $K[Au(CN)_2]$

**Preparation.** It is obtained by the following methods:

(*a*) Gold is dissolved in aqua regia and pecipitated with ammonia forming fulminating gold. It is washed and dissolved in potassium cyanide and crystallised.

(*b*) Fine gold is treated with potassium cyanide solution in contact with atmosphere oxygen:

$$4Au + 8KCN + O_2 + 2H_2O \longrightarrow 4KAu(CN)_2 + 4KOH$$

**Properties.** It is a colourless substance and soluble in water. When heated with HCl at 50°C, yellow crystals of AuCN are produced.

$$KAu(CN)_2 + HCl \longrightarrow KCl + HCN + AuCN$$

When Zn is added, gold is precipitated.

$$KAu(CN)_2 + Zn \longrightarrow K_2[Zn(CN)_2] + \underset{ppt}{2Au}$$

**Uses.** It is used in gold plating.

### 3. Sodium Dithiolsulphato Acurate (I) or Sodium Auro Thiosulphate, $Na_3[Au(S_2O_3)_2]$

Just as hypo forms a complex with silver chloride, a similar complex is obtained by adding auric chloride to hypo solution.

$$4Na_2S_2O_3 + AuCl_3 \longrightarrow Na_3[Au(S_2O_3)_2] + Na_2S_4O_6 + 3NaCl$$

Hypo is side by side oxidised to sodium tetrathionate as in iodine reaction.

It crystallises in colourless needles, fairly soluble in water. It is not reduced by $FeSO_4$, oxalic acid or $SnCl_2$. It is used in medicine as *sanochrysin* (Greek *chrusos*).

## Compounds of Au (II)

Although Au (II) has a tendency to undergo disproportionation into Au (I) and Au (III), yet Au (II) forms a limited number of complexes in +2 oxidation state. Tetra *n*-butylammonium *bis* (maleonitriledithiolate) aurate (II), $(n\text{-}Bu_4N)_2[Au^{2+}(mnt)_2]$ has been prepared by the tetrahydroborate reduction of the corresponding Au (III) complex. Its magnetic moment which is equal to 1.85 BM shows that Au present in the complex compound is a $d^9$ ion ($n = 1$). Some authors claim that the unpaired electron resides on the ligand (radical) rather than on gold.

The structure of the complex is given in Fig. 8.8.

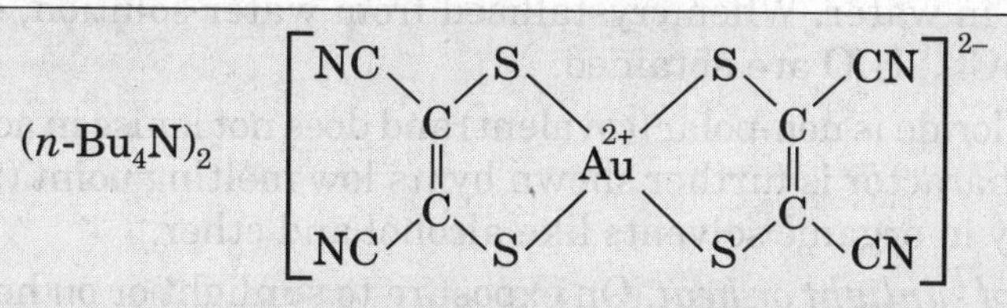


**Fig. 8.8** Structure of $(n - Bu_4N)_2[Au(mnt)_2]$ complex

## Compounds of Au (III)

These compounds are covalent in nature and do not have $Au^{3+}$ ions in pure aqueous solutions. Auric compounds are powerful oxidising agents. Examples of auric compounds are given below:

(*a*) *Simple compounds:* $Au_2O_3$, $Au(OH)_3$, $AuX_3$ (X = Cl, Br, I), $Au_2S_3$, $Au(CN)_3$, $Au(NH_2)(NH)$ etc.

(*b*) *Complex compounds:* $K[Au(CN)_4]$, $Na[AuO_2]$, $H[AuX_4]$ (X = Cl, Br, I), $[Au(NH_3)_4](NO_3)_3$, $[Au(NH_3)_4](C_2O_4)(NO_3)$. These are 4-coordinated complex compounds. These are diamagnetic and square planar resulting from *dsp*$^2$ hybridisation. 5-coordinated complexes are *trigonal bipyramidal* while 6-coordinated complexes are *octahedral*. Here we shall discuss the following compounds in some detail.

## 1. Auric Chloride, $AuCl_3$

**Preparation.** (*i*) $AuCl_3$ is prepared by treating AuCl with water or with $Cl_2$.

$$3AuCl \longrightarrow AuCl_3 + 2Au$$

$$AuCl + Cl_2 \longrightarrow AuCl_3$$

(*ii*) When metallic gold is dissolved in aqua regia, a bright yellow solution of $HAuCl_4$ is obtained.

$$2Au + \underbrace{3HNO_3 + 11HCl}_{\text{Aqua regia}} \longrightarrow \underbrace{2HAuCl_4}_{\text{Solution}} + 6H_2O + 3NOCl$$

On evaporating the solution, yellow deliquescent crystals of tetrahydrate chloroauric acid, $HAuCl_4 . 4H_2O$ deposite. This is called *gold chloride* of commerce. When these yellow crystals are heated at 200°C, deep red crystals of anhydrous auric chloride, $AuCl_3$ are obtained.

$$HAuCl_4 . 4H_2O \xrightarrow{200°C} AuCl_3 + HCl + 4H_2O$$

(*iii*) The preparation of *anhydrous auric chloride* can also be done as follows: Gold is first reacted with chlorine to give auric chloride which is dissolved in a little water. This is gently heated, in order to convert any contaminated aurous chloride into metallic gold which is thrown down as a precipitate and is filtered out, and then the solution evaporated and heated to 150°C. This gives anhydrous auric chloride. It is soluble in ether.

(*iv*) $AuCl_3$ can also be obtained by the action of moist $Cl_2$ on gold.

$$2Au + 3Cl_2\ (moist) \longrightarrow 2AuCl_3$$

**Properties.** (1) Anhydrous $AuCl_3$ is a deep red crystalline substance which is freely soluble in water. When crystallised from water solution, orange crystals of dihydrate, $AuCl_3.2H_2O$ are obtained.

(2) Auric chloride is non-polar (covalent) and does not ionise in aqueous solution. The non-polar character is further shown by its low melting point (290°), volatility and its solubility in organic solvents like alcohol and ether.

(3) *Action of sunlight or heat*. On exposure to sunlight or on heating at 450 K, $AuCl_3$ decomposes into AuCl (yellow powder) and $Cl_2$.

$$AuCl_3 \longrightarrow AuCl + Cl_2$$

(4) *Action of conc. $NH_3$*. When concentrated ammonia is added to $AuCl_3$, fulminating gold, $AuN_2H_3$ or $HN = Au - NH_2$ is obtained.

$$AuCl_3 + 2NH_3 \longrightarrow HN = Au - NH_2 + 3HCl$$

(5) *Decomposition of $AuCl_3$ solution*. Solution of $AuCl_3$ is not very stable, since it decomposes to form AuCl and $H_2O_2$.

$$AuCl_3 + 2H_2O \longrightarrow AuCl + 2HCl + H_2O_2$$

(6) *Reduction*. (*i*) $AuCl_3$ can be reduced to free gold by various reducing agents. For example:

(*a*) *Reduction by $SnCl_2$ and preparation of colloidal gold or purple of Cassius*. When $AuCl_3$ solution is treated with $SnCl_2$, $AuCl_3$ is reduced to colloidal particles of Au and $SnCl_2$ is oxidised to $SnCl_4$.

$$2AuCl_3 + 3SnCl_2 \longrightarrow 2Au + 3SnCl_4 \qquad ...(i)$$

$SnCl_4$ formed undergoes hydrolysis to form stannic acid, $H_4SnO_4$ or $Sn(OH)_4$ which absorbs colloidal particles of Au and purple of Cassius is obtained.

$$SnCl_4 + 4H_2O \longrightarrow H_4SnO_4 \text{ or } Sn(OH)_4 + 4HCl$$

or $$SnCl_4 + 4H_2O \longrightarrow SnO_2 . 2H_2O + 4HCl$$

or $$3SnCl_4 + 12H_2O \longrightarrow 3SnO_2 . 6H_2O + 12HCl \quad ...(ii)$$

On adding equation (*i*) and (*ii*) we get

$$2AuCl_3 + 3SnCl_2 + 6H_2O \longrightarrow \underbrace{2Au + 3SnO_2}_{\text{Purple of Cassius}} + 12HCl$$

This equation shows that purple of Cassius may be considered as the colloidal form of $SnO_2$ containing absorbed colloidal gold. Purple of Cassius is generally used for making ruby glass and high glass pottery.

(*b*) *Reduction by hydrazine* ($N_2H_4$) *and hydroxylamine* ($NH_2OH$). $N_2H_4$ and $NH_2OH$ both reduce $AuCl_3$ to free Au metal in acidic medium.

$$4AuCl_3 + 2N_2H_4 \longrightarrow 4Au + 3N_2 + 12HCl$$

$$4AuCl_3 + 6NH_2OH \longrightarrow 4Au + 3N_2O + 12HCl + 3H_2O$$

(*c*) *Reduction by* $FeSO_4$. When $AuCl_3$ solution is treated with $FeSO_4$, gold is precipitated and $FeSO_4$ is oxidised to $Fe_2(SO_4)_3$.

$$AuCl_3 + 3FeSO_4 \longrightarrow Au\downarrow + Fe_2(SO_4)_3 + FeCl_3$$

This reaction has been utilised in the extraction of gold from auriferrous quarts by chlorination process.

(*d*) *Reduction by Ag metal*. When $AuCl_3$ solution is treated with Ag, Au is replaced by Ag and Au is precipitated.

$$AuCl_3 + 3Ag \longrightarrow 3AgCl + Au\downarrow$$

The above reaction is used for toning process used in photography. In this process the photographic plate obtained after fixing is dropped in $AuCl_3$ solution. In this process silver grains of the image are replaced by those of gold (purple), as is evident from the above chemical equation.

(*e*) *Reduction by formaldehyde*. When $AuCl_3$ is reduced by formaldehyde, Faraday's gold is obtained. It is red, brown or purple in colour.

(*f*) *Reduction by KI solution*. $AuCl_3$ (Au = +3) is reduced to AuI (Au = +1) by KI solution (I = – 1). In this reaction KI is oxidised to $I_2$ (I = 0)

$$AuCl_3 + 3KI \longrightarrow AuI + 3KCl + I_2$$

(*g*) *Reduction by* $H_2S$ *gas*. $AuCl_3$ (Au = +3) is reduced to AuS (Au = +2) by $H_2S$ gas (S = –2). In this reaction $H_2S$ is oxidised to $H_2SO_4$ (S = +6).

$$8AuCl_3 + 9H_2S + 4H_2O \longrightarrow 8AuS + H_2SO_4 + 24HCl$$

(7) *Formation of complex compounds*. $AuCl_3$ reacts with various substances and forms complex compounds. For example:

(*a*) With excess of HCl, $H[AuCl_4]$ is obtained. With alkali chloride (e.g., NaCl), well-defined crystalline, $Na[AuCl_4]$ is obtained. $Na[AuCl_4]$ is a salt of $H[AuCl_4]$.

$$AuCl_3 + HCl\ (excess) \longrightarrow H[AuCl_4]$$

$$AuCl_3 + NaCl \longrightarrow Na[AuCl_4]$$

(*b*) When conc. solution of $AuCl_3$ is treated with KCN, solution which contains K [Au $(CN)_4$] is obtained. On evaporating and cooling the solution, colourless crystals of $K[Au(CN)_4] \cdot \frac{3}{2} H_2O$ are obtained.

$$AuCl_3 + 4KCN \longrightarrow K[Au\,(CN)_4] + 3KCl$$

(*c*) When a solution of $AuCl_3$ is treated with $Na_2S_2O_3$ (hypo), $Na_3\,[Au\,(S_2O_3)_2]$ is obtained.

$$AuCl_3 + 4Na_2S_2O_3 \longrightarrow Na_3\,[Au(S_2O_3)_2] + 2Na_2S_4O_6 + 3NaCl$$

(*d*) When KOH or NaOH is added to a solution of $AuCl_3$, $Au(OH)_3$ is precipitated which dissolves in the excess of the alkali forming potassium or sodium aurate (III), $M[AuO_2]$ (M = K or Na).

$$AuCl_3 + 3NaOH \longrightarrow \underset{\text{ppt}}{Au(OH)_3} + 3NaCl$$

$$Au(OH)_3 + NaOH \longrightarrow Na\,[AuO_2] + 2H_2O$$

**Structure.** The vapour density of auric chloride shows that it is a dimeric molecule, $Au_2Cl_6$.

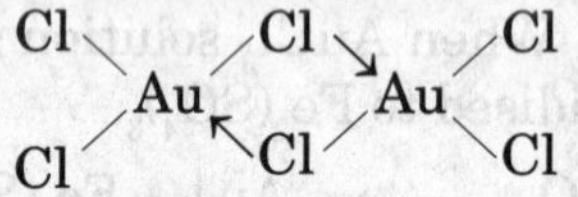

**Uses.** Hydrochloroauric acid, $H\,[AuCl_4] \cdot 4H_2O$ under the name of *gold chloride* of commerce is used in photography as a toning agent. $AuCl_3$, when dissolved in organic solvents, is called **liquid gold** which is used in paints.

## 2. Fulminating Gold or Gold Fulminate, Au $(NH_2)$ (NH)

**Preparation.** It is prepared:

(*i*) by treating auric chloride solution with ammonia:

$$AuCl_3 + 2NH_3 \longrightarrow HN = Au - NH_2 + HCl$$

(*ii*) by treating auric hydroxide with ammonia.

$$Au\,(OH)_3 + 2NH_3 \longrightarrow HN = Au - NH_2 + 3H_2O$$

**Properties.** (*i*) It is an olive green powder of variable composition.

(*ii*) The dry mass explodes violently on heating or percussion, forming Au, $N_2$, $NH_3$ and steam.

$$2Au\,(NH_2)\,(NH) \longrightarrow 2Au + N_2 + 2NH_3$$

(*iii*) It dissolves in HCl to form $H[AuCl_4]$ and $NH_4Cl$

$$Au\,(NH_2)\,(NH) + 6HCl \longrightarrow H[AuCl_4] + 2NH_4Cl$$

(*iv*) It readily dissolves in alkali cyanide solutions. It is converted to non-explosive compounds by the action of $H_2S$ and by careful heating with S. It is reduced to Au with explosion by $SnCl_2$.

**Structure.** Raschig (1986) formulated it as $HN = Au—NH_2 \cdot \frac{3}{2} H_2O$. According to Weiltz (1915) the fulminating gold produced by the action of $NH_3$ on $AuCl_3$ solution contains some chlorine and, therefore, he suggested the following structure to this compound.

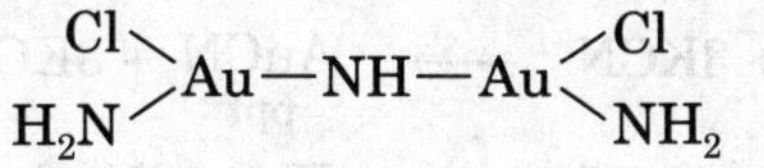

## 3. Tetrachloroauric Acid or Hydrogen Tetrachloro Aurate (III), $H[AuCl_4]$

**Preparation.** (*i*) When metallic gold is dissolved in aqua regia, a bright yellow solution of $HAuCl_4$ is obtained.

$$2Au + 3HNO_3 + 11HCl \longrightarrow \underset{\text{Solution}}{2HAuCl_4} + 6H_2O + 3NOCl$$

Actually the above reaction takes place through the following steps:

$$[HNO_3 + 3HCl \longrightarrow 2H_2O + NOCl + 2Cl] \times 3$$
$$[Au + 3Cl \longrightarrow AuCl_3] \times 2$$
$$[AuCl_3 + HCl \longrightarrow H[AuCl_4]] \times 2$$
$$\overline{2Au + 3HNO_3 + 11HCl \longrightarrow 2H[AuCl_4] + 6H_2O + 3NOCl}$$

On evaporating the solution, yellow deliquescent crystals of tetrahydrate chloroauric acid, $HAuCl_4 . 4H_2O$ deposit. This is called *gold chloride* of commerce.

(*ii*) When $AuCl_3$ reacts with excess of HCl, $H[AuCl_4]$ is obtained.

$$HCl\ (excess) + AuCl_3 \longrightarrow H[AuCl_4]$$

(*iii*) When fulminating gold, Au $(NH_2)$ (NH) reacts with HCl, $H[AuCl_4]$ is produced.

$$Au\ (NH_2)\ (NH) + 6HCl \longrightarrow H[AuCl_4] + 2NH_4Cl$$

**Properties.** (*i*) It is soluble in water and donor organic solvents.

(*ii*) *Action of heat*. When the crystals of $H[AuCl_4]$ are heated in a current of $Cl_2$, HCl and $AuCl_3$ (deep red crystals) are obtained.

$$H[AuCl_4] \xrightarrow[\text{of } Cl_2]{\text{Heat in a current}} AuCl_3 + HCl$$

(*iii*) *Reduction*. It is easily reduced to metallic gold by reducing agents like $FeSO_4$, $SnCl_2$ and even by $H_2$.

$$HAuCl_4 + 3FeSO_4 \longrightarrow Fe_2(SO_4)_2 + FeCl_3 + HCl + Au$$
$$2HAuCl_4 + 3SnCl_2 \longrightarrow 3SnCl_4 + 2HCl + 2Au$$
$$2HAuCl_4 + 3H_2 \longrightarrow 8HCl + 2Au$$

(*iv*) *Action of KI*. Reaction with KI precipitates AuI.

$$HAuCl_4 + 3KI \longrightarrow \underset{\text{ppt}}{AuI} + 3KCl + HCl + I_2$$

(*v*) *Reaction with alkalies*. Alkalies precipitate $Au(OH)_3$ which dissolves in excess of the alkali solution to form soluble complex, Na $[Au\ (OH)_4]$.

$$HAuCl_4 + 4NaOH \longrightarrow \underset{\text{ppt}}{Au(OH)_3} + 4NaCl + H_2O$$
$$Au(OH)_3 + NaOH \longrightarrow \underset{\text{Soluble complex}}{Na[Au(OH)_4]}$$

(*vi*) *Reaction with KCN*. First AuCN is obtained as precipitate which in excess of KCN dissolves due to the formation of soluble complex, $K[Au(CN)_2]$.

$$HAuCl_4 + 3KCN \longrightarrow \underset{\text{ppt}}{AuCN} + 3KCl + HCl + (CN)_2$$

$$AuCN + KCN \longrightarrow \underset{\text{Soluble complex}}{K[Au(CN)_2]}$$

*(vii) Reaction with $H_2S$.* With $H_2S$ the acid yields a precipitate which is probably a mixture of $Au_2S$ and $Au_2S_3$.

*(viii) Formation of tetrachloroaurates.* $H[AuCl_4]$ is a monobasic acid and gives salts of $M[AuCl_4]$ type (M = alkali metal). These sales are called *tetrachloroaurates* which contain tetra chloro aurate (III) ion, $[AuCl_4]^-$. Alkali tetrachloro aurates can be obtained by the action of an alkali chloride on $H[AuCl_4]$. Alkali tetrachloro aurates can be obtained by the action of an alkali chloride on $H[AuCl_4]$ or on $AuCl_3$.

$$\underset{(M = Na, K)}{MCl} + H[AuCl_4] \longrightarrow M[AuCl_4] + HCl$$

$$MCl + AuCl_3 \longrightarrow M[AuCl_4]$$

$[AuCl_4]^-$ ion is diamagnetic and has square planar geometry arising from $dsp^2$ hybridisation of Au atom.

**Uses.** The acid and its sodium and potassium salts are used in photography for toning and printing.

## 4. Tetrabromoauric Acid or Hydrogen Tetrabromo aurate (III), $H[AuBr_4]$

This compound is obtained by heating $AuBr_3$ with HBr. It is a red solid and soluble in water.

## 5. Potassium Tetracyano Aurate (III) or Potassium Auricyanide, $K[Au(CN)_4]$

**Preparation.** A hot concentrated solution of $AuCl_3$ when mixed with KCN gives $K[Au(CN)_4]$ is solution.

$$AuCl_3 + 4KCN \longrightarrow \underset{\text{Solution}}{K[Au(CN)_4]} + 3KCl$$

When the solution is evaporated and cooled, colourless crystals of $K[Au(CN)_4].\frac{3}{2}H_2O$ are obtained.

**Properties.** The crystals of $K[Au(CN)_4]$ are colourless and give a ppt. of $Ag[Au(CN)_4]$ when treated with a solution of $AgNO_3$.

$$K[Au(CN)_4] + AgNO_3 \longrightarrow Ag[Au(CN)_4] + KNO_3$$

**Uses.** $K[Au(CN)_4]$ is used in gold plating.

## Complexes of Au (III)

Substitution reactions of $AuCl_4^-$ leads to the formation of many types of complexes of Au (III). Examples of such complexes are $[Au\,py_2Cl_2]Cl$, $[Au(o\text{-}phen)Cl_2]Cl$, $[Au(diars)_2]Cl_3$ etc.

A few complexes with higher coordination numbers have also been prepared. Examples of such complexes are [Au (diars)$_2$I$_2$]I, [Au (diars)$_2$I]I$_2$ etc. All the complexes are coloured yellow to orange. The 4-coordinated Au (III) compounds are square planar (*dsp*$^2$ hybridisation) and are diamagnetic.

## Compounds of Au (V)

$AuF_5$ is a compound of Au (V). It is dark red in colour. It has diamagnetic character. It has polymeric structure.

## Organo metallic Compounds of Gold

A number of organometallic compounds of gold are known. [R$_2$Au (CN)]$_4$ and [R$_2$AuCl]$_2$ are the examples of such compounds. [R$_2$Au (CN)]$_4$ is a tetrameric compound and [R$_2$AuCl]$_2$ is a dimeric compound. In [R$_2$Au (CN)]$_4$ four CN$^-$ ions act as bridging ligands while in [R$_2$AuCl]$_2$ two Cl$^-$ ions act as bridging ligands. (See Fig. 8.9)

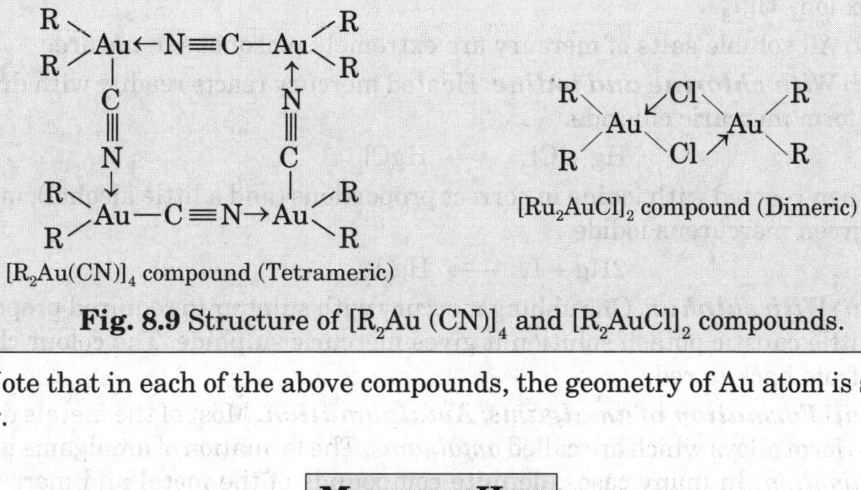

[R$_2$Au(CN)]$_4$ compound (Tetrameric)

[Ru$_2$AuCl]$_2$ compound (Dimeric)

**Fig. 8.9** Structure of [R$_2$Au (CN)]$_4$ and [R$_2$AuCl]$_2$ compounds.

Note that in each of the above compounds, the geometry of Au atom is square planar.

## Mercury, Hg$_{80}$

### Physical Properties of Mercury

(*i*) Mercury is a silver-white shining heavy liquid with a density of 13.6.

(*ii*) It freezes at 234.13 K and boils at 632. 6 K giving mono-atomic vapours which are poisonous in nature.

(*iii*) It is a fair conductor of heat and electricity and has a regular coefficient of expansion.

### Chemical Properties of Mercury

(*i*) ***Oxidation States.*** Mercury shows a variable oxidation number; +1 in mercurous salts and +2 in mercuric salts. It is, however, divalent even in mercurous salts, mercurous ion being (—Hg—Hg—)$^{2+}$.

(*ii*) ***Action of air.*** Mercury tarnishes slowly in air unless heated to about 570 K when it forms mercuric oxide which decomposes to give mercury again at a higher temperature.

$$2Hg + O_2 \xrightarrow{570\,K} 2HgO$$

(*iii*) ***Action of water and alkalis.*** Mercury is not attacked by water or alkalis.

(*iv*) ***Action of acids.*** Dilute acids have no action on mercury except nitric acid which gives mercurous nitrate and nitric oxide.

$$6Hg + 8HNO_3 \longrightarrow 3Hg_2(NO_3)_2 + 2NO + 4H_2O$$

Hot concentrated sulphuric acid gives sulphur dioxide with mercury whereas concentrated nitric acid gives mercuric nitrate and nitrogen dioxide.

$$2Hg\,(excess) + 2H_2SO_4 \longrightarrow Hg_2SO_4 + SO_2 + 2H_2O$$

$$Hg + 2H_2SO_4\,(excess) \longrightarrow HgSO_4 + SO_2 + 2H_2O$$

$$Hg + 2NO_3^- + 4H^+ \longrightarrow Hg^{2+} + 2NO_2 + 2H_2O$$

Mercury is also acted upon by hydriodic acid owing to the formation of the complex ion, $HgI_4^{2-}$.

(*v*) All soluble salts of mercury are extremely poisonous in nature.

(*vi*) ***With chlorine and iodine.*** Heated mercury reacts readily with dry chlorine to form mercuric chloride.

$$Hg + Cl_2 \longrightarrow HgCl_2$$

When reacted with iodine in correct proportions (and a little alcohol), mercury forms green mercurous iodide.

$$2Hg + I_2 \longrightarrow Hg_2I_2$$

(*vii*) ***With sulphur.*** On rubbing mercury with sulphur in required proportions and a little caustic potash solution it gives mercuric sulphide. The colour changes slowly from back to red.

(*viii*) ***Formation of amalgams: Amalgamation.*** Most of the metals dissolve in Hg to form alloys which are called *amalgams*. The formation of amalgams is called *amalgamation*. In many cases, definite compounds of the metal and mercury are formed which remain dissolved in excess of mercury, e.g., $NaHg_2$, $Au_2Hg$, MgHg, etc.

When a clean piece of sodium is pressed under the surface of mercury, it dissolves with a bright flash forming the amalgam (containing 1-2% of sodium). When sodium amalgam is placed in a concentrated solution of ammonium chloride, mercury swells up to the extent of nearly thirty times its original volume. This butter like mass of soluble mercury is **ammonium amalgam** Probably, sodium of sodium amalgam combines with the chlorine of ammonium chloride leaving free ammonium radical, $NH_4^+$ which forms amalgam with mercury. On heating the swelling of mercury subsides with evolution of hydrogen and ammonia. Ammonium amalgam is quite stable below 0°.

Amalgams of silver, gold, tin, lead, copper, bismuth, cadmium, magnesium, etc, can be prepared by rubbing the metal with mercury.

Copper displaces mercury to form a mercuric salt solution and amalgamates with the displaced mercury which causes a thin white film on its surface.

$$HgCl_2 + Cu \longrightarrow Hg + CuCl_2$$

$$Cu + Hg \longrightarrow Cu.Hg \text{ (Copper amalgam)}$$

Iron amalgam may be obtained indirectly by placing sodium or zinc amalgam in a solution of ferrous sulphate whereby iron is precipitated and forms amalgam with mercury.

$$FeSO_4 + 2Na \longrightarrow Na_2SO_4 + Fe$$

Silver-tin amalgam is prepared by rubbing an alloy of silver and tin in the ratio of 3:1 with mercury. This amalgam is used for dental filling.

If a little mercury is poured into a silver nitrate solution, a tree-like growth of silver amalgam is produced. This is called 'mercury tree'.

## Uses of Mercury

(*i*) It finds use in thermometer, barometers, manometers and high vacuum air pumps.

(*ii*) It is used in the extraction of gold and silver by the amalgamation process.

(*iii*) Mercury is used in the preparation of its alloys with other metals called amalgams. Amalgams of tin, silver and gold are used in dentistry and sodium amalgam as an important reducing agent.

(*iv*) In Castner-Kellner cell for the manufacture of caustic soda, mercury is used as an electrode.

(*v*) Mercury vapour lamps give a blue light, rich in ultraviolet rays and mercury are rectifiers are used in charging batteries.

(*vi*) In the manufacture of its salts and in medicine.

## +1 Oxidation State of Hg : Mercurous ion or Dimercury (I) Ion, $Hg_2^{2+}$

The ion having Hg in +1 oxidation is called **mercurous ion**. This ion consists of two $Hg^+$ ions ($Hg^+ = 5d^{10}6s^1$) linked together by a covalent bond which is formed by the overlap of two singly-filled 6*s* orbitals on two $Hg^+$ ions. Thus mercurous ion is formulated as $Hg^+ - Hg^+$ or $Hg_2^{2+}$ and not $Hg^+$ and is called **dimercury (I) ion**. Thus mercurous ion is a **diatomic divalent ion**, $Hg_2^{2+}$. Due to the presence of $Hg^+ - Hg^+$ bond, $Hg_2^{2+}$ ion is diamagnetic.

Evidence for the structure, $Hg_2^{2+}$ ion for mercurous ion comes from a number of sources discussed below.

(*i*) The conductivity of mercurous nitrate solution was found similar to that of $Pb(NO_3)_2$. Now since the ionisation of $Pb(NO_3)_2$ gives divalent $Pb^{2+}$ ion as shown below

$$Pb(NO_3)_2 \longrightarrow Pb^{2+} + 2NO_3^-,$$

the ionisation of mercurous nitrate should also give $Hg_2^{2+}$ ion. In other words the mercurous ion present in mercurous nitrate should be represented as $Hg_2^{2+}$.

$$Hg_2(NO_3)_2 \longrightarrow Hg_2^{2+} + 2NO_3^-$$

(*ii*) All metal nitrates have the same Raman spectrum due to covalent bonding in $NO_3^-$ ion but mercurous nitrate has an extra line in its spectrum. This extra line is due to Hg–Hg bonding in $Hg_2^{2+}$ ion.

(*iii*) X-ray diffraction study has shown that the unit cell of the crystal of mercurous chloride has linear molecules Cl–Hg–Hg–Cl. This structure shows that mercurous ion in mercurous chloride has Hg–Hg bond.

## Disproportionation of $Hg_2^{2+}$ Ion

$Hg_2^{2+}$ ion undergoes disproportionation in aqueous solution. The disproportionation of $Hg_2^{2+}$ ion (Hg = +1) in aqueous solution to Hg (Hg = 0) and $Hg^{2+}$ (Hg = +2) ion can be represented by the chemical equation.

$$Hg_2^{2+}(aq) \rightleftharpoons Hg(s) + Hg^{2+}(aq)$$

This reaction proceeds towards right to a very small extent only. If some reagent which may precipitate $Hg^{2+}$ is added to the aqueous solution of $Hg_2^{2+}$, then more amount of $Hg_2^{2+}$ will disproportionate into Hg and $Hg^{2+}$. If the reagent to be added is in excess, then almost the entire amount of $Hg_2^{2+}$ disproportionates into Hg and a compound containing $Hg^{2+}$. For example:

(*i*) When $H_2S$ is passed through aqueous solution of $Hg_2^{2+}$ ions, then HgS (Hg = +2) and Hg (Hg = 0) is obtained as precipitate.

$$\underset{(Hg = +1)}{Hg_2^{2+}} + H_2S\ (\textit{Precipitating reagent}) \longrightarrow \underset{(Hg = +2)}{HgS(ppt)} + 2H^+ + \underset{(Hg = 0)}{Hg}$$

(*ii*) When NaOH is added to the aqueous solution of $Hg_2^{2+}$ ions, then HgO and Hg are obtained.

$$\underset{(Hg = +1)}{Hg_2^{2+}} + 2OH^- \longrightarrow \underset{(Hg = +2)}{HgO} + H_2O + \underset{(Hg = 0)}{Hg}$$

Other disproportionation reactions shown by $Hg_2^{2+}$ ion are given below

(*i*) When $Hg_2O$ is warmed or is exposed to light, it readily decomposes to form HgO and Hg

$$\underset{(Hg = +1)}{Hg_2O} \xrightarrow{\Delta} \underset{(Hg = +2)}{HgO} + \underset{(Hg = 0)}{Hg}$$

(*ii*) When $Hg_2Cl_2$ is heated strongly, $HgCl_2$ and Hg are obtained.

$$\underset{(Hg = +1)}{Hg_2Cl_2} \xrightarrow{\Delta} \underset{(Hg = +2)}{HgCl_2} + \underset{(Hg = 0)}{Hg}$$

## Compounds of Mercury

Compounds formed by $Hg_2^{2+}$ ion are called **mercurous compounds**. In these compounds Hg is in +1 oxidation state. Compounds formed by $Hg^{2+}$ ion are called

**mercuric compounds**. In these compounds Hg is in +2 oxidation state. Let us discuss the following compounds of Hg.

## 1. Mercury (I) Oxide or Mercurous Oxide, $Hg_2O$

It is obtained as a brownish black precipitate when sodium hydroxide is added to a mercurous salt solution. On warming or on exposure to light, it readily decomposes into mercury and mercuric oxide. It is regarded a mixture of finely divided mercury and mercuric oxide (HgO. Hg) as indicated by its X-ray diagram.

## 2. Mercury (II) Oxide or Mercuric Oxide, HgO

**Preparation.** It is produced as a red powder by decomposing the nitrate by heat or by heating mercury for a long long time in air at 570 K.

$$2Hg(NO_3)_2 \longrightarrow 2HgO + 4NO_2 + O_2$$

$$2Hg + O_2 \rightleftharpoons 2HgO$$

It may also be obtained by mixing boiling solution of mercuric chloride and potassium carbonate.

$$HgCl_2 + K_2CO_3 \longrightarrow HgO + 2KCl + CO_2$$

A yellow variety of mercuric oxide is obtained by adding sodium hydroxide to mercuric nitrate solution.

$$Hg^{2+} + 2OH^- \longrightarrow \underset{\text{(Yellow)}}{HgO} + H_2O$$

**Properties** (*i*) On heating, the yellow variety changes to red variety but on strong heating it decomposes to give oxygen.

$$2HgO \longrightarrow 2Hg + O_2$$

The difference between the red and yellow varieties only is of the particle size, the particles of yellow variety are smaller.

(*ii*) On warning yellow variety of HgO with aqueous $NH_3$, yellow powder of *Millon's base* is obtained.

$$2HgO + NH_4OH \longrightarrow \underset{\substack{\text{Million's base} \\ \text{(Yellow powder)}}}{\left[O\begin{matrix}\diagup Hg \diagdown \\ \diagdown Hg \diagup\end{matrix}NH_2\right]^+} OH^-.\ H_2O$$

(*iii*) The oxide is weakly basic and dissolves to form Hg (II) salts.

**Structure.** The oxide has zig-zag chain structure (Fig. 8.10), –O–Hg–O–Hg–O–with HgOHg angle = 109° and OHgO angle = 179°. Evidently Hg (II) utilises *sp* hybridisation.

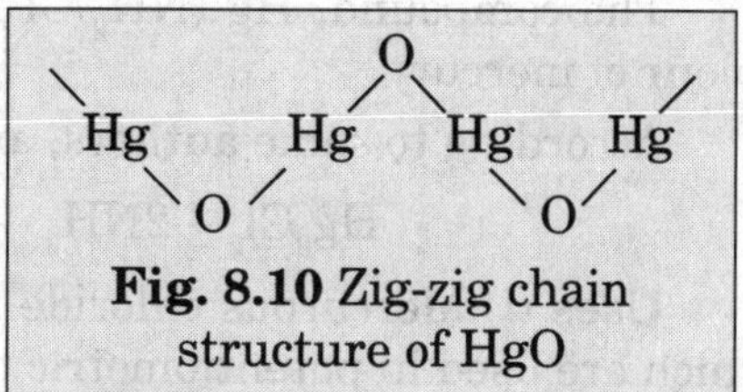

**Fig. 8.10** Zig-zig chain structure of HgO

**Uses.** The red variety of HgO finds applications as a pigment for painting out the surface of the ship bottoms. The poisonous nature of HgO prevents marine growth.

## 3. Mercury (I) Chloride or Mercurous Chloride or Calomel, $Hg_2Cl_2$.

**Preparation.** Calomel is prepared —

(*i*) By subliming a mixture of mercuric chloride and metallic mercury in appropriate proportions.

$$HgCl_2 + Hg \rightleftharpoons Hg_2Cl_2$$

The sublimate is treated with cold dilute nitric acid to remove any metallic mercury and then washed thoroughly to remove any mercuric chloride.

(*ii*) By adding an excess of hydrochloric acid or any other chloride to any soluble mecurous salt, e.g., mercurous nitrate.

$$Hg_2(NO_3)_2 + 2HCl \longrightarrow Hg_2Cl_2 + 2HNO_3$$

$$Hg_2(NO_3)_2 + 2NaCl \longrightarrow Hg_2Cl_2 + 2NaNO_3$$

**Properties** (*i*) It is a white powder insoluble in water. However, it dissolves in chlorine water forming mercuric chloride.

$$Hg_2Cl_2 + Cl_2 \longrightarrow 2HgCl_2$$

(*ii*) It decomposes on heating to give $HgCl_2$ and Hg.

$$Hg_2Cl_2 \xrightarrow{\text{Heat}} HgCl_2 + Hg$$

(*iii*) $Hg_2Cl_2$ (white ppt) reacts with gaseous $NH_3$ as well as with $NH_4OH$ to give a black residue which consists of black Hg and white ppt. of mercuric amido chloride, $[Hg^{2+}(NH_2)Cl]$ (complex compound).

$$Hg_2Cl_2 + 2NH_3\,(g) \longrightarrow \underbrace{\underset{\text{Black}}{Hg} + \underset{\text{White ppt}}{[Hg^{2+}(NH_2)Cl]}}_{\text{Black residue}} + NH_4Cl$$

This reaction takes place through the following steps:

$$Hg_2Cl_2 \xrightarrow{\text{Disporportionation}} HgCl_2 + Hg$$

$$HgCl_2 + 2NH_3 \longrightarrow Hg(NH_2)Cl + NH_4Cl$$

$$Hg_2Cl_2 + 2NH_3 \longrightarrow [Hg + [Hg(NH_2)Cl]\downarrow + NH_4Cl$$

$$Hg_2Cl_2 + 2NH_4OH \longrightarrow \underbrace{Hg + [Hg(NH_2)]\,Cl}_{\text{Black residue}} + NH_4Cl + 2H_2O$$

Thus we see that metallic black Hg is formed due to the disproportionation of $Hg_2Cl_2$ to $HgCl_2$ and Hg.

The compound, $Hg(NH_2)Cl$ is white but its colour is masked by the black colour of mercury.

According to some authors, an addition compound, $Hg_2Cl_2.\,2NH_3$ is formed.

$$Hg_2Cl_2 + 2NH_3 \longrightarrow Hg_2Cl_2\,.\,2NH_3$$

**Uses** (*i*) Mercurous chloride is used for making standard calomel electrodes which are used in potentiometric titrations.

(*ii*) $Hg_2Cl_2$ is also used as a purgative in medicine.

## 4. Mercury (II) chloride or Mercuric chloride or Corrosive Sublimate, $HgCl_2$

On account of its corrosive property mercuric chloride is also known as *corrosive sublimate*.

**Preparation.** (*i*) It is prepared by passing dry chlorine over heated mercury.

$$Hg + Cl_2 \xrightarrow{\text{Heat}} HgCl_2$$

(*ii*) It is also obtained by treating mercury (II) oxide with hydrochloric acid.

$$HgO + 2HCl \longrightarrow HgCl_2 + H_2O$$

(*iii*) It is commercially prepared by heating mercuric sulphate with an equal quantity of sodium chloride in the presence of a little manganese dioxide. Mercuric sulphate needed is obtained by evaporating a solution of mercury in hot concentrated sulphuric acid.

$$Hg + 2H_2SO_4 \longrightarrow HgSO_4 + SO_2 + 2H_2O$$

$$HgSO_4 + 2NaCl \longrightarrow HgCl_2 + Na_2SO_4$$

The mixture is heated in a long neck, round-bottom flask when mercuric chloride sublimes and collects in the upper cooler parts. Manganese dioxide prevents the formation of any mercurous chloride.

**Properties.** (*i*) It is a white crystalline solid sparingly soluble in cold water but soluble in hot water. The solubility can be increased by adding $Cl^-$ ions when a soluble tetrachloromercurate (II) complex ion is formed.

$$HgCl_2 + 2Cl^- \longrightarrow [HgCl_4]^{2-}$$

(*ii*) It also readily dissolves in organic solvents such as ethanol, ether, etc. thereby suggesting its covalent nature.

(*iii*) It is a dangerous poison, the antidote being white of an egg which is coagulated by the salt in the system and is eliminated by the system with salt absorbed in it.

(*iv*) *With stannous chloride*, it gives a white precipitate of mercurous chloride ($Hg_2Cl_2$) which is finally reduced to metallic mercury (grey).

$$2HgCl_2 + SnCl_2 \longrightarrow Hg_2Cl_2 + SnCl_4$$

$$Hg_2Cl_2 + SnCl_2 \longrightarrow 2Hg + SnCl_4$$

(*v*) When copper turnings are placed in contact with an aqueous solution of mercury (II) chloride, a shining grey film of mercury deposits over copper turnings.

$$HgCl_2 + Cu \longrightarrow Hg + CuCl_2$$

(*vi*) *With potassium iodide: Formation of Nesseler's solution*. $HgCl_2$ gives a scarlet precipitate soluble in excess of potassium iodide forming Nessler's solution.

$$HgCl_2 + 2KI \longrightarrow \underset{\text{Scarlet ppt.}}{HgI_2} + 2KCl$$

$$HgI_2 + 2KI\ (\textit{Excess}) \longrightarrow \underset{\substack{\text{Clear solution}\\\text{(Nessler's solution)}}}{K_2HgI_4}$$

(*vii*) It is reduced to mercurous chloride by sulphur dioxide and arsenites.

$$2HgCl_2 + SO_2 + 2H_2O \longrightarrow Hg_2Cl_2 + H_2SO_4 + 2HCl$$

(*viii*) Mercuric chloride dissolves with evolution of heat in concentrated hydrochloric acid and the solution on cooling deposits crystals of $HgCl_4^{2-}$.

(*ix*) With chlorides of alkali metals, it forms crystalline compounds of the type $KHgCl_3$, $Na_2HgCl_4$, etc.

(*x*) Addition of an alkali to mercuric chloride solution results in the partial precipitation of mercuric oxide according to the reversible reaction.

$$HgCl_2 + 2NaOH \rightleftharpoons HgO + 2NaCl + H_2O$$

(*xi*) Mercuric chloride combines with phosphorus pentachloride to form the volatile crystalline compound, $3HgCl_2 . 2PCl_5$.

(*xii*) On boiling a solution of mercuric chloride with mercuric oxide, a number of mercuric oxychlorides are formed. e.g., $2HgCl_2.HgO$ (red), $HgCl_2.2HgO$ (black); $HgCl_2 . 3HgO$ (yellow); $HgCl_2.4HgO$ (black).

(*xiii*) *Action of* $NH_3$ (*a*) In presence of concentrated solution of $NH_4Cl$, the addition of $NH_3$ to $HgCl_2$ solution forms *fusible white precipitate of diammine mercury (II) chloride*, $Hg(NH_3)_2Cl_2$

$$HgCl_2 + 2NH_3 \longrightarrow \underset{\text{(Fusible white ppt)}}{\underset{\text{Diammine mercury (II) chloride}}{Hg(NH_3)_2Cl_2 \downarrow}}$$

(*b*) In the absence of $NH_4Cl$, the addition of dil. solution of $NH_3$ to $HgCl_2$ solution gives *infusible ppt. of mercuric amido chloride*, $Hg(NH_2)Cl$.

$$HgCl_2 + 2NH_3 \text{ (Dil. solution)} \longrightarrow \underset{\text{(Infusible ppt)}}{\underset{\text{Mercuric amido chloride}}{Hg(NH_2)Cl}} + NH_4Cl$$

The fusible white precipitate, $Hg(NH_3)_2Cl_2$ has tetrahedral geometry ($sp^3$ hybridisation). The infusible white precipitate, $Hg(NH_2)Cl$ contains amide links in an infinite chain, $—Hg \leftarrow \underset{H_2}{N} - Hg \leftarrow \underset{H_2}{N}—$ (See Fig. 8.11). N-atoms in this chain are tetrahedral but N←Hg–N link is linear and chlorine lies in between the chains.

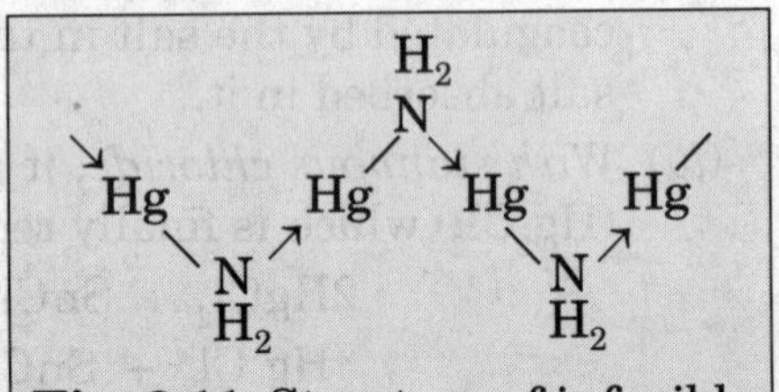


**Fig. 8.11** Structure of infusible white precipitate, $Hg(NH_2)Cl$.

(*xiv*) *Action of* $Hg(CN)_2$. When a mixture of $HgCl_2$ and $Hg(CN)_2$ is heated, cynogen $(CN)_2$ is set free.

$$HgCl_2 + Hg(CN)_2 \longrightarrow Hg_2Cl_2 + (CN)_2$$

**Uses.** (*i*) It is violent poison, the fatal dose being 0.2-0.5g. A 0.1% solution of mercuric chloride is used as antiseptic for sterilizing hands and instruments in surgery.

(*ii*) It is also used as a preservative for wood and hides.

(*iii*) Mercuric chloride also finds use in the preparation of dry cells.

## 5 Mercury (I) Bromide or Mercurous Bromide, $Hg_2Br_2$.

It is obtained as a white insoluble powder by adding a soluble bromide to mercurous nitrate solution.

## 6 Mercury (II) Bromide or Mercuric Bromide, $HgBr_2$.

It is obtained by treating mercury with bromine, or by dissolving mercuric oxide in hydrobromic acid. It is less soluble in water than mercuric chloride, but forms soluble complexes with sodium chloride and bromide, *e.g.*, $HgBr_2.2NaCl$ and $HgBr_2.2NaBr$.

## 7 Mercury (I) Iodine or Mercurous Iodide, $Hg_2I_2$.

It is formed as a green powder by rubbing excess of mercury and iodine. On heating it decomposes into mercuric iodide and mercury.

## 8 Mercury (II) Iodide or Mercuric Iodide, $HgI_2$.

It is obtained as a yellow precipitate soon turning scarlet, when potassium iodide solution is added to any mercuric salt solution.

$$HgCl_2 + 2KI \longrightarrow \underset{\text{Scarlet ppt.}}{HgI_2} + 2KCl$$

**Properties.** (*i*) Mercuric iodide exists in two forms, *i.e.* red and yellow. The yellow form is stable above 400 K while the red form is stable below this temperature.

$$\underset{\text{Red}}{HgI_2} \underset{\text{Below 400 K}}{\overset{\text{Above 400 K}}{\rightleftharpoons}} \underset{\text{Yellow}}{HgI_2}$$

(*ii*) It readily dissolves in excess of potassium iodide solution to form *potassium mercuric tetraiodide*, $K_2[HgI_4]$ (Complex compound).

$$HgI_2 + 2KI\ (\textit{Excess}) \longrightarrow K_2[HgI_4]$$

An alkaline solution of $K_2HgI_4$ is called *Nessler's reagent* and is used to detect the presence of $NH_3$ or $NH_4^+$ ion with which it gives a brown precipitate due to the formation of iodide of *Million's base*.

(*iii*) *Action of heat*. Violet vapours of $I_2$ are obtained.

$$HgI_2 \xrightarrow{\Delta} Hg + I_2$$

## 9. Million's Base

Million's base has been given the following structures.

$$\left[O{<}{}^{Hg}_{Hg}{>}NH_2\right]^+ OH^-.H_2O, \quad \left[{}^{Hg}_{Hg}{>}NH\right]^{2+} (OH)_2.H_2O, \quad [Hg{-}N{=}Hg]^+OH^-.2H_2O$$

(I) (II) (III)

Generally in our further discussion we shall be using structure (I) for Million's base.

Million's base can be obtained by the action of aqueous $NH_3$ (i.e. $NH_4OH$) on yellow HgO.

$$2HgO + NH_4OH \longrightarrow \left[O\begin{smallmatrix}Hg\\ \\ Hg\end{smallmatrix}NH_2\right]^+ OH^-.H_2O$$

Million's base
(Yellow powder)

Iodide of Millon's base is obtained by replacing OH group present in the strcuture of Millon's base. Structure of iodide of Millon's base is given in the margin.

$$\left[O\begin{smallmatrix}Hg\\ \\ Hg\end{smallmatrix}NH_2\right]^+ I^-$$

(Iodide of Million's base)

## 10. Nesseler's Reagent

$HgI_2$ readily dissolves in the excess of KI solution due to the formation of ***potassium mercuric tetraiodide complex***, $K_2[HgI_4]$. The alkaline solution of this complex is called Nesseler's reagent

$$Hg\,I_2 + 2KI\ (Excess) \longrightarrow K_2[HgI_4]\ (Soluble)$$

Nesseler's reagent

**Uses of Nesseler's reagent.** Nesseler's reagent, $K_2[HgI_4]$ containing KOH or NaOH is used to detect the presence of $NH_3$ or $NH_4^+$ ion with which it gives a brown precipitate of iodide of Million's base. In this reaction the active species is $[HgI_4]^{2-}$ ion.

$$(i)\ 2K_2[HgI_4] + NH_3 + 3KOH \longrightarrow \left[O\begin{smallmatrix}Hg\\ \\ Hg\end{smallmatrix}NH_2\right]^+ I^- + 7KI + 2H_2O$$

Iodide of Million's base (Brown ppt)

$$\text{or } 2[HgI_4]^{2-} + NH_3 + 3OH^- \longrightarrow \left[O\begin{smallmatrix}Hg\\ \\ Hg\end{smallmatrix}NH_2\right] I + 7I^- + 2H_2O$$

$$(ii)\ 2K_2[HgI_4] + 4NH_4Cl + 4KOH \longrightarrow \left[O\begin{smallmatrix}Hg\\ \\ Hg\end{smallmatrix}NH_2\right]^+ I^- + 3NH_4I + 4KI + 4KCl + 3H_2O$$

Reaction (*ii*) takes place through the following steps:

$$(a)\quad 4NH_4Cl + 4KOH \longrightarrow 4KCl + 4H_2O + 4NH_3$$

$$(b)\ 2K_2[HgI_4] + 4NH_3 + H_2O \longrightarrow \left[O\begin{smallmatrix}Hg\\ \\ Hg\end{smallmatrix}NH_2\right] I + 4KI + 3NH_4I$$

---

$$2K_2[HgI_4] + 4NH_4Cl + 4KOH \longrightarrow \left[O\begin{smallmatrix}Hg\\ \\ Hg\end{smallmatrix}NH_2\right] I + 3NH_4I + 4KI + 4KCl + 3H_2O$$

The ionic equations corresponding to the above reactions can be written as :

$$NH_4^+ + OH^- \rightarrow NH_3 + H_2O$$

$$2[HgI_4]^{2-} + NH_3 + H_2O \longrightarrow \left[O\langle{}^{Hg}_{Hg}\rangle NH_2\right] I + 7I^- + 3H^+$$

$$2[HgI_4]^{2-} + NH_4^+ + OH^- \longrightarrow \left[O\langle{}^{Hg}_{Hg}\rangle NH_2\right] I + 7I^- + 3H^+$$

## 11. Mercury (II) Sulphide or Mercuric Sulphide, HgS.

It occurs native as cinnabar. It is obtained as a black precipitate by passing hydrogen sulphide through a solution of mercuric chloride.

$$HgCl_2 + H_2S \longrightarrow HgS + 2HCl$$

This black precipitate turns red on sublimation.

It can also be prepared by subliming mercury with sulphur or by rubbing mercury and sulphur with a little sodium hydroxide or potassium hydroxide solution, when the black sulphide first formed gradually turns red and crystalline.

It is insoluble in water, boiling concentrated hydrochloric acid or dilute nitric acid but dissolves in aqua regia or concentrated hydrochloric acid in presence of potassium chlorate with $Cl_2$, $HgCl_2$ is obtained.

$$HgS + Cl_2 \longrightarrow HgCl_2 + S$$

It is also soluble in concentrated solution of sodium sulphide or potassium sulphide forming thiosalts, e.g., $K_2HgS_2 . 5H_2O$. It is decomposed by boiling concentrated sulphuric acid.

$$HgS + 2H_2SO_4 \longrightarrow HgSO_4 + SO_2 + S + 2H_2O$$

On heating, it decomposes into mercury and sulphur which burns in air to give sulphur dioxide; mercury may also be converted to mercuric oxide.

$$HgS + O_2 \longrightarrow Hg + SO_2$$

It is also reduced to mercury on heating with iron filings, sodium carbonate etc.

The red variety is used as a pigment and in Ayurvedic medicines (*Makaradhwaja* and *Rasa-sindur*). $Hg_2S_2$ does not exist, since $Hg_2S_2$ readily disproportionates to Hg and HgS.

## Complexes of Mecury

If we consider one positive charge on each of the two Hg atoms in $Hg_2^{2+}$ ion ($Hg_2^{2+} \equiv Hg^+ - Hg^+$), we would expect $Hg_2^{2+}$ ion to have much smaller tendency for the formation of complexes. Smaller complex formation tendency of $Hg_2^{2+}$ ion is also because of the fact that $Hg_2^{2+}$ ion compounds readily undergo disproportionation to Hg(II) and Hg (0). $Hg_2^{2+}$ ion forms complexes with ligands oxalate, suicinate etc. $Hg^{+2}$ ion has moderately strong tendency for complexation. The most stable complexes formed by $Hg^{2+}$ ion are those with halogen, N, P and S as ligand atoms.

The stability of halide complexes varies as $Cl^- < Br^- < I^-$. $Hg^{2+}$ ion has little tendency to form fluoride complexes.

Examples of some complexes formed by $Hg^{2+}$ ion are given below.

## 1. 2-Coordinated Complexes

Examples of such complexes are $HgX_2$ etc. These complexes have linear geometry which results from *sp* hybridisation of $Hg^{2+}$ ion.

## 2. 4-Coordinated Complexes

Important examples of such complexes are $[HgX_4]^{2-}$, $[HgX_2(NH_3)_2]$, $[Hg(NH_3)_4]^{2+}$, $[Hg(CN)_4]^{2-}$, $[HgCl_2\ (Ph_3AsO_2)]$, $[HgX_2\ .\ R_2S]_2$. These complexes have tetrahedral geometry corresponding to $sp^3$ hybridisation of $Hg^{2+}$ ion. The complex, $[HgX_2.\ R_2S]_2$ attains 4-coordinated tetrahedral stereochemistry with halogen bridges as shown in Fig. 8.12.

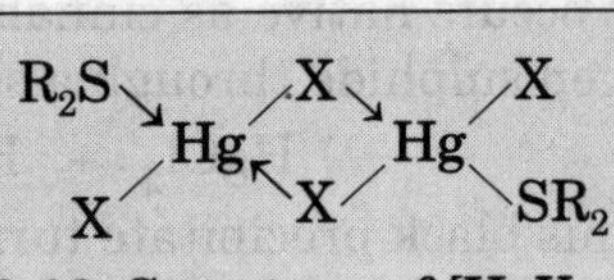


**Fig. 8.12** Structure of $[HgX_2\ .\ R_2S]_2$

## 3. 6-Coordinated Complexes

$[Hg\,(en)_3]^{2+}$ is an important example of 6-coordinated complex. It has octahedral geometry due to $sp^3d^2$ hybridisation.

## 4. Mercur-ammine Complexes

Hg forms a number of complexes with $NH_3$. Some of them are discussed below.

(*i*) The preparation of *fusible white precipitate of diammine mercury (II) chloride*, $Hg\,(NH_3)_2Cl_2$ and of *infusible white ppt. of mercuric amido chloride*, $Hg\,(NH_2)Cl$ has already been given under "*Action of $NH_3$ on $HgCl_2$*". Chemical equations for their preparation are given below.

$$HgCl_2 + 2NH_3 \xrightarrow[NH_4Cl]{\text{In the presence of conc. solution of}} \underset{\text{Diammine mercury (II) chloride (Fusible white ppt)}}{Hg(NH_3)_2Cl_2 \downarrow}$$

$$HgCl_2 + 2NH_3\ (\textit{Dil. solution}) \xrightarrow{\text{In the absence of } NH_4Cl} \underset{\text{Mercuric amide chloride (Infusible white ppt)}}{Hg(NH_2)Cl} + NH_4Cl$$

(*ii*) On warming yellow HgO with aqueous $NH_3$ (*i.e.* $NH_4OH$), yellow powder of Millon's base is obtained.

$$2HgO + NH_4OH \longrightarrow \underset{\text{Millon's base (yellow powder)}}{\left[O\begin{matrix}Hg\\Hg\end{matrix}NH_2\right]^+ OH^-.H_2O}$$

When Millon's base reacts with an acid, salt of the base is obtained.

(*iii*) When $Hg_2Cl_2$ is treated with gaseous or with aqueous $NH_3$, a black solid is obtained. This solid contains a mixture of infusble white ppt of *mercuric amido chloride,* $Hg(NH_2)Cl$ and black finely divided Hg. It is due to the presence of black Hg in the solid mixture that the solid has black colour.

$$Hg_2Cl_2 + 2NH_3\,(g) \longrightarrow \underbrace{\underset{\text{White}}{Hg(NH_2)Cl} + \underset{\text{Black}}{Hg}}_{\text{Black solid}} + NH_4Cl$$

$$Hg_2Cl_2 + 2NH_4OH \longrightarrow \underbrace{Hg(NH_2)Cl + Hg}_{\text{Black solid}} + NH_4Cl + 2H_2O$$

## Organo Metallic Compounds of Mercury

A large number of organomercury compounds are known, namely of the type RHgX and $HgR_2$. They may be obtained by the action of sodium amalgam on RX.

$$2Hg + 2RX \longrightarrow HgR_2 + HgX_2$$

RHgX are crsytalline solids, and $HgR_2$ are extremely toxic liquids or low melting solids. Compounds with $X = F^-, NO_3^-$ or $SO_4^{2-}$ are water soluble and ionic. Others are essentially covalent. These compounds are much more stable to air and water than the corresponding zinc compounds. They find use as fungicides, pesticides and drugs.

These organometallic compounds can be easily hydrolysed and, therefore, these compounds act as intermediates in the synthesis of aldehydes and alcohols from acetylenes and olefins.

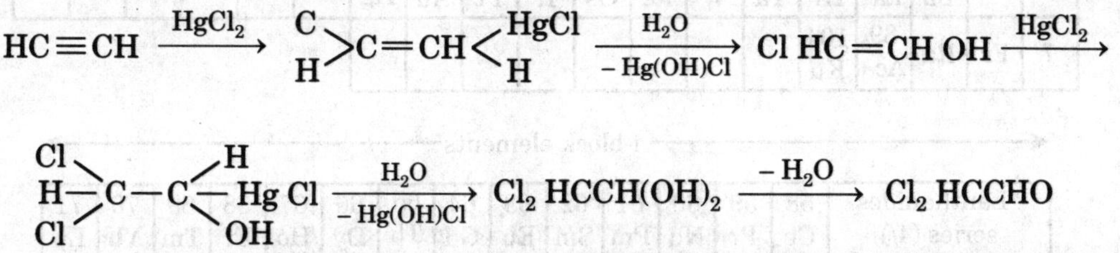


## UNIVERSITY QUESTIONS

1. Write a note on (*i*) Tungsten Bronzes (*ii*) Tungsten Blues.
2. Discuss the structure of rhenium trihalides.
3. Write the preparation and properties of $OsO_4$.
4. Write the formula of Vaskas compound.
5. Discuss the trans effect in square planar complexes of Pt (II).
6. Write a note on organometallic compounds of gold.
7. Mercurous ion is $Hg_2^{2+}$ and not $Hg^+$. Explain.
8. What is Nesseler's reagent? What are its uses ?

# 9 CHAPTER Chemistry of Lanthanides (4f-Block Elements)

## Inner-transition Elements or f-Block Elements

The elements from $Ce_{58}$ to $Lu_{71}$ (14 elements) and from $Th_{90}$ to $Lr_{103}$ (14 elements) are called **inner-transition elements**. These elements have been shown at a separate place at the bottom of the periodic table as depicted in Fig. 9.1.

Groups

Transition Elements

| Period | 1 | 2 | 3 | 4 | 5 | 6 | 7 | 8 | 9 | 10 | 11 | 12 | 13 | 14 | 15 | 16 | 17 | 18 |
|---|---|---|---|---|---|---|---|---|---|---|---|---|---|---|---|---|---|---|
| 1 | H | | | | | | | | | | | | | | | | | He |
| 2 | Li | Be | | | | | | | | | | | B | C | N | O | F | Ne |
| 3 | Na | Mg | | | | | | | | | | | Al | Si | P | S | Cl | Ar |
| 4 | K | Ca | 21 Sc | 22 Ti | 23 V | 24 Cr | 25 Mn | 26 Fe | 27 Co | 28 Ni | 29 Cu | 30 Zn | Ga | Ge | As | Se | Br | Kr |
| 5 | Rb | Sr | 39 Y | 40 Zr | 41 Nb | 42 Mo | 43 Tc | 44 Ru | 45 Rh | 46 Pd | 47 Ag | 48 Cd | In | Sn | Sb | Te | I | Xe |
| 6 | Cs | 56 Ba | 57 La* | 72 Hf | 73 Ta | 74 W | 75 Re | 76 Os | 77 Ir | 78 Pt | 79 Au | 80 Hg | Ti | Pb | Bi | Po | At | Rn |
| 7 | Fr | Ra | 89 Ac** | 104 Ku | | | | | | | | | | | | | | |

f-block elements

| | | | | | | | | | | | | | | |
|---|---|---|---|---|---|---|---|---|---|---|---|---|---|---|
| *Lanthanides series (4*f*) | 58 Ce | 59 Pr | 60 Nd | 61 Pm | 62 Sm | 63 Eu | 64 Gd | 65 Tb | 66 Dy | 67 Ho | 68 Er | 69 Tm | 70 Yb | 71 Lu |
| **Actinide series (5*f*) | 90 Th | 91 Pa | 92 U | 93 Np | 94 Pu | 95 Am | 96 Cm | 97 Bk | 98 Cf | 99 Es | 100 Fm | 101 Md | 102 No | 103 Lr |

**Fig. 9.1** Position of lanthanides in the periodic table.

The first series of 14 elements lies between $La_{57}$ and $Hf_{72}$ in 6th period while the other series of 14 elements lies between $Ac_{89}$ and $Ku_{104}$ in 7th period (incomplete period). Both the series are present in IIIB (3) group.

The elements of 1st series are called **lanthanides** or **lanthanones** (rare earth elements) while those of the 2nd series are called **actinides** or **actinones**. The name lanthanides and actinides has been given, since these elements follow $La_{57}$ and $Ac_{89}$ respectively with which these elements show close similarities.

Since the last electron (differentiating electron) in the atoms of these elements enters the *f*-sub-shell belonging to anti-penultimate shell [*i.e.*, $(n-2)$th shell], these elements are also called *f*-**block elements**. Thus the last electron enters $(n-2)f$ sub-shell in inner-transition elements.

Valence-shell electronic configuration of the atoms of *f*-block elements can be represented as $(n-2)f^{0-14}.(n-1)d^{0-2}.ns^2$ which shows that in these elements the outermost three shells are partially-filled while the remaining inner shells are completely-filled.

## Classification of f-Block Elements

Inner transition elements (f-block elements) have been classified into the following two series.

**(*i*) Lanthanide series ($Ce_{58}$ to $Lu_{71}$).** The 14 elements from $Ce_{58}$ to $Lu_{71}$ are the members of this series, *i.e.* 14 elements from $Ce_{58}$ to $Lu_{71}$ are called **lanthanides**. These elements are called lanthanides, since these elements are placed after lanthanum ($La_{57}$) in group 3 and period 6 of the periodic table (See Fig. 9.1) and show close similarities with $La_{57}$.

Since lanthanides are present in *6th* period ($n = 6$) of the periodic table, these elements have **six** shells in their electronic configuration. Thus the last electron in these elements enters (6–2) *f* or 4*f* sub-shell. Thus lanthanides are also called **4f block elements,** *i.e., lanthanides are a series of 14 elements ($Ce_{58}$ to $Lu_{71}$) in which the last electron goes to 4f sub-shell or to f sub-shell of 4th shell.*

It has been observed that all the 15 elements from $La_{57}$ to $Lu_{71}$ have similar properties. Hence the study of lanthanides consists of the study all the 15 elements from $La_{57}$ to $Lu_{71}$ together. $La_{57}$ is the prototype of lanthanides.

Lanthanides are also called **lanthanones or rare earths**. The name rare earth was given to them because they were originally extracted from oxides for which ancient name was *earth* and which were considered to be rare. The term *rare-earths* is avoided now because many of these elements are no longer rare but are abundant.

**(*ii*) Actinide series ($Th_{90}$ to $Lr_{103}$)** The 14 elements from $Th_{90}$ to $Lr_{103}$ are the members of this series, *i.e.* 14 elements from $Th_{90}$ to $Lr_{103}$ are called actinides. These elements are called actinides, since these elements are placed after actinium ($Ac_{89}$) in group 3 and period 7 of the periodic table (See Fig. 9.1) and show close similarities with $Ac_{89}$.

Since actinides are present in *7th* period ($n = 7$) of the periodic table, these elements have **seven** shells in their electronic configuration. Thus the last electron in these elements enters (7 – 2) *f* or 5*f* sub-shell. Thus actinides are also called **5f block elements**, *i.e., actinides are a series of 14 elements ($Th_{90}$ to $Lr_{103}$) in which the last electron goes to 5f sub-shell or to f sub-shell of 5th shell.*

Since *f*-orbitals can accomodate 14 electrons, each of the two series of inner transition elements consists of 14 elements.

In this chapter we shall make a detailed discussion of lanthanides.

## Position of Lanthanides in the Periodic Table

All the fifteen lanthanides have atomic weights between those of barium (atomic number, Z = 56) and hafnium (Z = 72) and, therefore, must be placed between these two elements as was also proved by Moseley.

Barium has exactly the same outer electronic configuration as calcium and strontium and resembles them very closely. In a similar way hafnium is similar to zirconium (Z = 40). Therefore Ba must be placed below Sr (Group II A) and Hf below Zr (Group IV B), thus leaving only one place between them, which lies exactly below Y (Z = 39, Group III B).

Since all the fifteen lanthanides resemble one another in many respects, these must be placed in the same group. These elements also resemble $Y_{39}$ because of the following additional similarities : (*a*) Owing to lanthanide contraction (to be discussed subsequently), the ionic radius of $Y^{3+}$ ion is almost similar to that of $Er^{3+}$ ($r_{Y^{3+}} = 0.93$Å, $r_{Er^{3+}} = 0.96$ Å). (*b*) Y generally occurs in nature associated with the ores of heavier lanthanides and resembles Tb (III) and Dy (III) in its compounds. It, therefore, became necessary to accommodate all the fifteen lanthanides together at one place. This has been done by placing the first element *viz.* $La_{57}$ below $Y_{39}$ and remaining fourteen elements (*viz.* $Ce_{58}$ to $Lu_{71}$) have been placed separately in the lower part of the periodic table.

## Occurrence of Lanthanides

The lanthanides are not so scare in the earth's crust as the name rare-earths implies. Potentially their sources are unlimited. Each lanthanide mineral contains all lanthanides, although some in traces only, excepting Pm which is an unstable radioactive element.

Major minerals of lanthanides are: (*i*) Monazite (*ii*) Bastnaesite and (*iii*) Cerite

**Classification of Rare-Earth Minerals.** The rare-earth minerals, like the earths themselves, fall into two distinct classes.

(1) *Cerium Group Minerals*. These are also called *cerite-earth minerals* or simply *cerium earths* and are rich in cerite-earths. This group of minerals contains largely elements of atomic numbers from 57 to 63 (i.e. $La_{57}$ — $Eu_{63}$), the light lanthanide elements. These are thus light rare earths.

(2) *Yttrium Group Minerals*. These are also called *yttrium-earth minerals* or simply *yttrium earths* and are rich in gadolinite earths. This group contains largely yttrium and lanthanides of atomic numbers from 64 to 71 ($Gd_{64}$—$Lu_{71}$), the heavier lanthanides. These are thus heavy rare-earths.

Besides rare-earth minerals proper, there are many others, particularly the phosphates, arsenates, molybdates and tungstates of Ca, Sr, Fe and Pb which contain small amounts of rare-earths. In some of these minerals, Sm, Gd and Eu are present in much larger quantities than in rare-earth minerals.

## Extraction of Lanthanides From Monazite

The extraction of lanthanides from monazite sand consists of two steps: (*i*) *To get a mixture of lanthanides from monazite sand* (*ii*) *Separation of lanthanides from the mixture obtained at* (*i*).

## I. To Get a Mixture of Lanthanides from Monazite Sand

Monazite sand contains a mixture of phosphates of lanthanides, Th, Zr, etc. along with the impurities. Monazite sand is powdered and digested with conc. $H_2SO_4$ at 200°C so that a paste mass is obtained. This pasty mass consists of sulphates of lanthanides etc. and $H_3PO_4$. The pasty mass is treated with cold water when silica ($SiO_2$) and zircon ($ZrSiO_4$) are obtained as residue and the solution contains sulphates of lanthanides and thorium plus $H_3PO_4$. Thorium sulphate present in solution is removed as insoluble thorium phosphate. On diluting and adjusting the pH of the solution in alkaline range, solution contains only the sulphates of lanthanides. This solution is treated with $Na_2SO_4$ so that double sulphates of light lanthanides (La to Sm) are obtained as residue and simple sulphates of heavy lanthanides (Gd to La) go into solution (See flow sheet).

Hot solution of NaOH is added to the ppt. of double sulphates of light lanthanides to get a mixture of oxides and hydroxides of lanthanides. This mixture of oxides and hydroxides is washed to remove $Na_2SO_4$ and then dried in air at 100°C whereby a residual solid having $La_2O_3$ (17%), $CeO_2$ (50%), $Pr_2O_3$(8%), $Nd_2O_3$(20%) and $Sm_2O_3$ (5%). This mixture of light lanthanide oxides is reacted with dil. $HNO_3$ when all the lanthanide oxides (except $CeO_2$) get dissolved and filtered out. The crude $CeO_2$ is dissolved in 85% $HNO_3$ and then treated with excess of dil. $H_2SO_4$ so that a ppt. of red basic nitrate, $Ce(OH)(NO_3)_2.2H_2O$ is obtained. After the removal of $Ce(OH)(NO_3)_2.2H_2O$, the lanthanides present in filtrate are separated from each other.

**Flow sheet.** *To get a mixture of sulphates of lanthanides from monazite sand.*

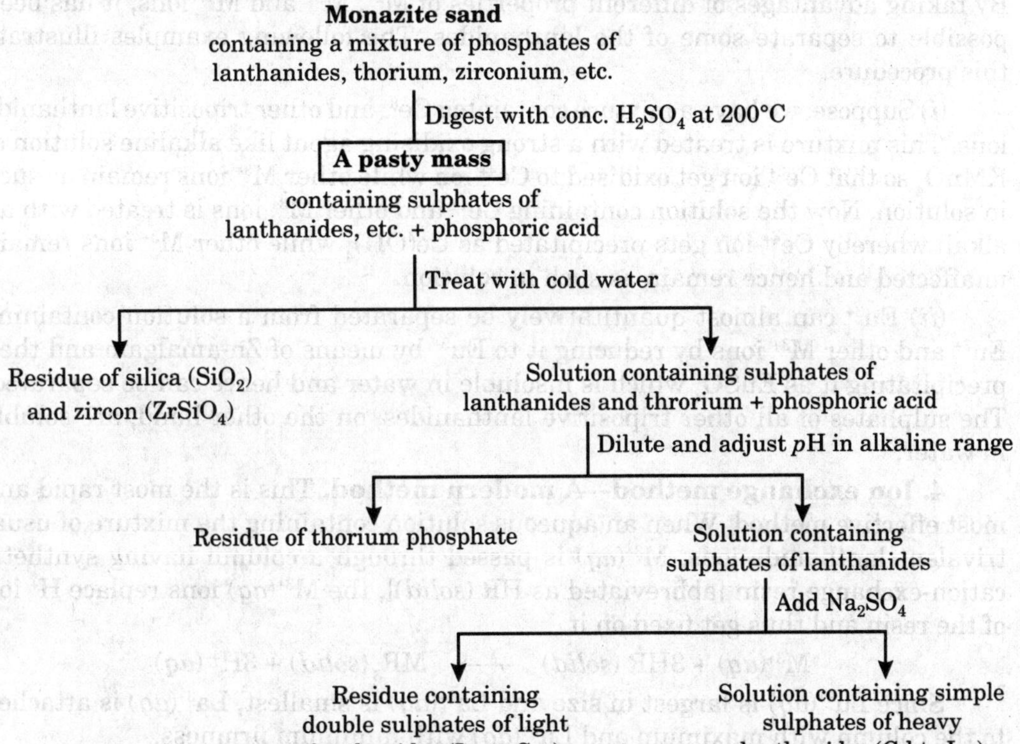

## II. Separation of Lanthanides from Lanthanides Mixture

All the lanthanide elements form tripositive ions ($M^{3+}$ ions) which are almost identical in size. The chemical properties of lanthanides which depend on the size and charge of $M^{3+}$ ions are almost the same. Due to the similarity in chemical properties, it is difficult to separate the lanthanides from each other. However, the following methods have been used to separate them.

**1. Repeated fractional crystallisation method.** Simple salts of lanthanides like nitrates, sulphates, oxalates, bromates, perchlorates and double salts such as $2M(NO_3)_3.3Mg(NO_3)_2.\ 24H_2O$ crystallise well and form well-defined crystals. Since the solubility of these simple and double salts decreases from La to Lu, these lanthanides can be separated from each other by repeating the fractional crystallisation a number of times. In the separation of $Nd(NO_3)_3$ from $Pr(NO_3)_3$ the use of non-aqueous solvent like ether has been made.

**2. Fractional precipitation method.** When NaOH is added to a solution of lanthanide nitrates, $Lu(OH)_3$ which is the *weakest base* and has the *lowest solubility product* is precipitated *first* while $La(OH)_3$ which is the *strongest base* and has the *highest solubility product* remains dissolved and precipitates out *last* of all. By dissolving and precipitating the hydroxides for a number of times it is possible to get the complete separation of lanthanides.

**3. Change in oxidation state method.** Some of the lanthanides show +2 and +4 oxidation states in addition to the usual +3 state which is the common state for all the lanthanides. The properties of $M^{2+}$ and $M^{4+}$ ions are different from those of $M^{3+}$ ions. A change in oxidation state results in a remarkable change in properties. By taking advantages of different properties of $M^{2+}$, $M^{3+}$ and $M^{4+}$ ions, it has been possible to separate some of the lanthanides. The following examples illustrate this procedure.

(*i*) Suppose, we have a mixture containing $Ce^{3+}$ and other tripositive lanthanide ions. This mixture is treated with a strong oxidising agent like alkaline solution of $KMnO_4$ so that $Ce^{3+}$ ion get oxidised to $Ce^{4+}$ ion while other $M^{3+}$ ions remain as such in solution. Now the solution containing $Ce^{4+}$ and other $M^{3+}$ ions is treated with an alkali whereby $Ce^{4+}$ ion gets precipitated as $Ce(OH)_4$ while other $M^{3+}$ ions remain unaffected and hence remain as such in solution.

(*ii*) $Eu^{3+}$ can almost quantitatively be separated from a solution containing $Eu^{3+}$ and other $M^{3+}$ ions by reducing it to $Eu^{2+}$ by means of Zn-amalgam and then precipitating it as $EuSO_4$ which is insoluble in water and hence can be separated. The sulphates of all other tripositive lanthanides, on the other hand, are soluble in water.

**4. Ion exchange method—A modern method.** This is the most rapid and most effective method. When an aqueous solution containing the mixture of usual trivalent lanthanide ions, $M^{3+}(aq)$ is passed through a column having synthetic cation-exchange resin [abbreviated as HR (*solid*)], the $M^{3+}(aq)$ ions replace $H^+$ ion of the resin and thus get fixed on it.

$$M^{3+}(aq) + 3HR\ (solid) \longrightarrow MR_3\ (solid) + 3H^+\ (aq)$$

Since $Lu^{3+}(aq)$ is largest in size and $La^{3+}(aq)$ is smallest, $La^{3+}(aq)$ is attached to the column with maximum and $Lu^{3+}(aq)$ with minimum firmness.

In order to recover the $M^{3+}$ ions fixed on the resin, the column is *eluted* (*i.e.*, leached) with a citric acid—ammonium citrate solution (called **eluant** or **eluante**). During elution process $NH_4^+$ ions replace $M^{3+}$ ions and M-citrate complexes are formed.

$$MR_3 + 3NH_4^+ \longrightarrow 3NH_4R + M^{3+}$$
$$M^{3+} + \text{citrate ions} \longrightarrow \text{M-citrate complex}$$

We have seen that since $La^{3+}$ (*aq*) is attached to the resin with maximum and $Lu^{3+}$(*aq*) with minimum firmness, *Lu-citrate complex comes out of the column first* and *La-citrate complex comes out last*. In actual practice the process has to be repeated several times by careful control of concentration of citric acid–ammonium citrate solution.

A typical elution curve of lanthanide ions using citric acid–ammonium citrate solution as the eluent is shown in Fig. 9.2.

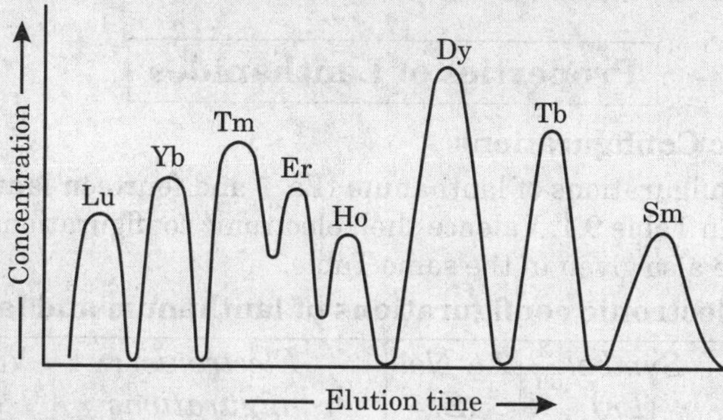


**Fig. 9.2** Elution curve of lanthanide ions using citric acid-ammonium citrate solution as the eluent. (Ion exchange method for the separation of lanthanides)

**5. Solvent extraction method.** This method makes the use of the difference in the value of partition coefficient of lanthanides between two solvents. $La(NO_3)_3$ and $Gd(NO_3)_3$ have been separated from each other by this method. The partition coefficient of Gd $(NO_3)_3$ between water and tributyl phosphate in kerosene is different from that of $La(NO_3)_3$ between the same solvents. This means that $Gd(NO_3)_3$ can be separated from $La(NO_3)_3$ by continuous extraction with water from a solution of these salts in tributyl phosphate in kerosene.

**6. Fractional decomposition of oxy salts.** On heating, simple oxy salts of lanthanides like oxalates, nitrates, acetates, etc. decompose to give oxides. The decomposition temperature of oxy salts depends on the basicity of lanthanides. Thus oxides of different lanthanides can be obtained by heating the oxy salts of the metals at different temperatures.

**7. Complex formation method.** The stability of complex ions of tripositive lanthanides ($M^{3+}$) increases with the increase of atomic number of lanthanides. Due to the difference in stability of complex ions of a given type, the rate and extent of particular reactions are also different. This fact has been used in the separation of lanthanides. For example, if oxalate or chromate ion is added to the solution of complexed lanthanide ions, the complex cation which is the least stable, is precipitated first.

## Production of Lanthanide Metals

The following methods may be used for this purpose:

**1. Electrolysis of fused chlorides.** This method is similar to that used in the metallurgy of Ca by the electrolysis of $CaCl_2$.

**2. Reduction of anhydrous chlorides with Na.** Lighter lanthanides such as La, Ce and Gd can easily be prepared by the reduction of their anhydrous chlorides with Na at 100°C.

$$LaCl_3 + 3Na \xrightarrow{100°C} La + 3NaCl$$

**3. Reduction of anhydrous fluorides and chlorides with Mg or Ca.** Heavier lanthanides such as Lu are prepared when their anhydrous fluorides and chlorides are reduced by Ca or Mg metals at a temperature above 1000°C, since the fluorides are less volatile than the chlorides and consequently the loss caused by evaporation in case of fluorides is small.

## Properties of Lanthanides

### 1. Electronic Configurations

Electronic configurations of lanthanum ($La_{57}$) and fourteen lanthanides ($Ce_{58}$ to $Lu_{71}$) are given in Table 9.1. Valence shell electronic configurations (V.S.E.C.) of these elements are also given in the same Table.

**Table 9.1. Electronic configurations of lanthanum and lanthanides.**

| *Name of the elements* | *Symbol (Ln)* | *At. No. (Z)* | *Electronic configurations* | *Valence shell electronic configurations* |
|---|---|---|---|---|
| Lanthanum | La | 57 | $[Xe]_{54}\,4f^{\circ}\,5d^16s^2$ | $4f^{\circ}5d^16s^2$ |
| Cerium | Ce | 58 | $[Xe]_{54}\,4f^25d^06s^2$ | $4f^25d^06s^2$ |
| Praseodymium | Pr | 59 | $[Xe]_{54}\,4f^35d^06s^2$ | $4f^35d^06s^2$ |
| Neodymium | Nd | 60 | $[Xe]_{54}\,4f^45d^06s^2$ | $4f^45d^06s^2$ |
| Promethium | Pm | 61 | $[Xe]_{54}\,4f^55d^06s^2$ | $4f^55d^06s^2$ |
| Samarium | Sm | 62 | $[Xe]_{54}\,4f^65d^06s^2$ | $4f^65d^06s^2$ |
| Europium | Eu | 63 | $[Xe]_{54}\,4f^75d^06s^2$ | $4f^75d^06s^2$ |
| Gadolinium | Gd | 64 | $[Xe]_{54}\,4f^75d^16s^2$ | $4f^75d^16s^2$ |
| Terbium | Tb | 65 | $[Xe]_{54}\,4f^95d^06s^2$ | $4f^95d^06s^2$ |
| Dysprosium | Dy | 66 | $[Xe]_{54}\,4f^{10}5d^06s^2$ | $4f^{10}5d^06s^2$ |
| Holmium | Ho | 67 | $[Xe]_{54}4f^{11}5d^06s^2$ | $4f^{11}5d^06s^2$ |
| Erbium | Er | 68 | $[Xe]_{54}4f^{12}5d^06s^2$ | $4f^{12}5d^06s^2$ |
| Thulium | Tm | 69 | $[Xe]_{54z}4f^{13}5d^06s^2$ | $4f^{13}5d^06s^2$ |
| Ytterbium | Yb | 70 | $[Xe]_{54}4f^{14}5d^06s^2$ | $4f^{14}5d^06s^2$ |
| Lutetium | Lu | 71 | $[Xe]_{54}4f^{14}5d^16s^2$ | $4f^{14}5d^16s^2$ |

After the completion of *6s* orbital at $Ba_{56}$ [$Ba_{56}$ = $[Xe]_{54}6s^2$] in *6th* period and group 2, the next electron (*i.e. 57th* electron) in $La_{57}$, according to Aufbau principle, must enter *4f* orbitals to give $[Xe]_{54}$ $4f^{1}5d^{0}6s^{2}$ configuration to $La_{57}$, but in $La_{57}$ the energy of 5*d* orbitals is lower than that of *4f* orbitals as shown in Fig. 9.3. Consequently *57th* electron in $La_{57}$ enters 5*d* orbitals instead of *4f* orbitals. Thus the electronic configuration of $La_{57}$ is $[Xe]_{54}$ $4f^{0}5d^{1}6s^{2}$ and not $[Xe]_{54}$ $4f^{1}5d^{0}6s^{2}$ as predicted by Aufbau principle.

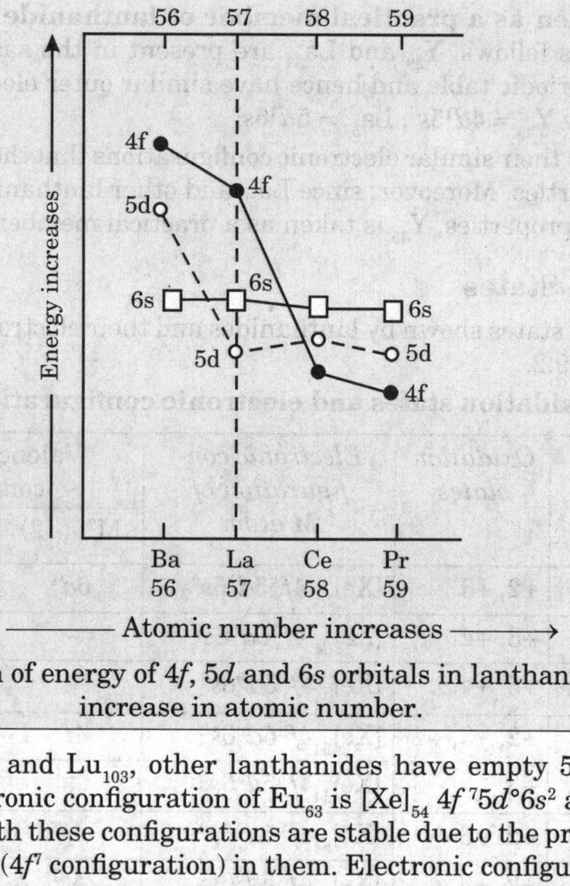


**Fig. 9.3** Variation of energy of *4f*, *5d* and *6s* orbitals in lanthanides with the increase in atomic number.

Excepting $Gd_{64}$ and $Lu_{103}$, other lanthanides have empty 5d orbitals ($5d^{0}$ configuration). Electronic configuration of $Eu_{63}$ is $[Xe]_{54}$ $4f^{7}5d^{0}6s^{2}$ and that of $Gd_{64}$ is $[Xe]_{54}$ $4f^{7}5d^{1}6s^{2}$. Both these configurations are stable due to the presence of stable half-filled 4f orbitals ($4f^{7}$ configuration) in them. Electronic configuration of $Yb_{70}$ is $[Xe]_{54}$ $4f^{14}$ $5d^{0}$ $6s^{2}$ and that of $Lu_{71}$ is $[Xe]_{54}4f^{14}$ $5d^{1}$ $6s^{2}$. Both these configurations are stable due to the presence of stable completely filled *4f* orbitals in them.

Since lanthanides are present in *6th* period of the periodic table, these elements have six shells in their electronic configurations. Thus the complete electronic configuration of these elements can be written as:

| *Shells:* | 1st | 2nd | 3rd | 4th (Ante penultimate) | 5th (Penultimate) | 6th (Ultimate or outermost) |
|---|---|---|---|---|---|---|
| *No. of electrons:* | 2 | 8 | 18 | 20 to 32 | 8 or 9 | 2 |
| or | 2 | 8 | 18 | $4s^{2}4p^{6}4d^{10}4f^{2-14}$ | $5s^{2}\,5p^{6}5d^{0-1}$ | $6s^{2}$ |

The above configuration can be written as

$$[Xe]_{54}\, 4f^{2-14}\, 5d^{0-1}\, 6s^2$$

Hence valence shell electronic configuration can be written as

$$4f^{2-14}\, 5d^{0-1}\, 6s^2$$

**$Y_{39}$ is taken as a practical member of lanthanide series.** This fact can be explained as follows. $Y_{39}$ and $La_{57}$ are present in the same group (group 3 or III B) of the periodic table and hence have similar outer electronic configurations as shown below $Y_{39} = 4d^1 5s^2$; $La_{57} = 5d^1 6s^2$

It is due to their similar electronic configurations that these two elements have identical properties. Moreover, since $La_{57}$ and other lanthanides ($Ce_{58}$ to $Lu_{71}$) also have identical properties, $Y_{39}$ is taken as a practical member of lanthanide series.

## 2. Oxidation States

Oxidation states shown by lanthanides and their electronic configurations are given in Table 9.2.

**Table 9.2. Oxidation states and electronic configurations of lanthanides**

| *Elements* | *Oxidation states* | *Electronic configuration of M atom* | Valence shell electronic configurations of: | | |
|---|---|---|---|---|---|
| | | | $M^{2+}$ (+2) | $M^{3+}$ (+3) | $M^{4+}$ (+4) |
| $La_{57}$ | +2, +3 | $[Xe]_{54}\, 4f^{\circ}5d^1 6s^2$ | $5d^1$ | $4f^0$ | – |
| $Ce_{58}$ | +3, +4 | $[Xe]_{54}\, 4f^2 5d^0 6s^2$ | – | $4f^1$ | $4f^0$ |
| $Pr_{59}$ | +3, +4 | $[Xe]_{54} 4f^3 5d^0 6s^2$ | – | $4f^2$ | $4f^1$ |
| $Nd_{60}$ | +2, +3, +4 | $[Xe]_{54} 4f^4 5d^0 6s^2$ | $4f^4$ | $4f^3$ | $4f^2$ |
| $Pm_{61}$ | +3 | $[Xe]_{54} 4f^5 5d^0 6s^2$ | – | $4f^4$ | – |
| $Sm_{62}$ | +2, +3 | $[Xe]_{54} 4f^6 5d^0 6s^2$ | $4f^6$ | $4f^5$ | – |
| $Eu_{63}$ | +2, +3 | $[Xe]_{54} 4f^7 5d^0 6s^2$ | $4f^7$ | $4f^6$ | – |
| $Gd_{64}$ | +3 | $[Xe]_{54} 4f^7 5d^1 6s^2$ | – | $4f^7$ | – |
| $Tb_{65}$ | +3, +4 | $[Xe]_{54} 4f^9 5d^0 6s^2$ | – | $4f^8$ | $4f^7$ |
| $Dy_{66}$ | +3, +4 | $[Xe]_{54} 4f^{10} 5d^0 6s^2$ | – | $4f^9$ | $4f^8$ |
| $Ho_{67}$ | +3 | $[Xe]_{54} 4f^{11} 5d^0 6s^2$ | – | $4f^{10}$ | – |
| $Er_{68}$ | +3 | $[Xe]_{54} 4f^{12} 5d^0 6s^2$ | – | $4f^{11}$ | – |
| $Tm_{69}$ | +2, +3 | $[Xe]_{54} 4f^{13} 5d^0 6s^2$ | $4f^{13}$ | $4f^{12}$ | – |
| $Yb_{70}$ | +2, +3 | $[Xe]_{54} 4f^{14} 5d^0 6s^2$ | $4f^{14}$ | $4f^{13}$ | – |
| $Lu_{71}$ | +3 | $[Xe]_{54} 4f^{14} 5d^1 6s^2$ | – | $4f^{14}$ | – |

From this table the following points may be noted:

(*i*) **+3 oxidation state.** +3 oxidation state is shown by all lanthanides. This is the most stable oxidation state. This oxidation state is obtained by the loss of two 6*s* electrons and one more electron from 5*d* orbital in case of La or from 4*f* orbitals in case of other elements. Thus the E.C. of $La^{3+}$ ion is $[Xe]_{54}\, 4f^{\circ}5d^{\circ}6s^{\circ}$ which is highly stable configuration. It is, therefore, difficult to remove another electron from this configuration. Hence La cannot show an oxidation state higher than +3.

$Gd_{64}$ and $Lu_{71}$ also show +3 oxidation state. The electronic configurations of $Gd^{3+}$ and $Lu^{3+}$ ions are given below.

$$Gd^{3+} = [Xe]_{54} 4f^7; \quad Lu^{3+} = [Xe]_{54}\, 4f^{14}$$

Both these configurations are very stable due to the presence of half-filled 4*f* orbitals ($4f^7$ configuration) in $Gd^{3+}$ ion and fully-filled 4*f* orbitals ($4f^{14}$ configuration) in $Lu^{3+}$ ion. $M^{3+}$ ions formed by the lanthanides other $La^{3+}(4f^{\circ})$, $Gd^{3+}(4f^7)$ and $Lu^{3+}(4f^{14})$ do not have $4f^{\circ}$ or $4f^7$ or $4f^{14}$ orbitals in their configuration. $Ce^{3+}$ and $Pr^{3+}$ ions are the examples of such $M^{3+}$ ions. The E.C. of these ions are given below.

$$Ce^{3+} = [Xe]_{54} 4f^1; \quad Pr^{3+} = [Xe]_{54} 4f^2$$

The stability of +3 oxidation state of $Ce_{58}$ and $Pr_{59}$ cannot be explained on the basis of stability associated with $4f^{\circ}$, $4f^7$ and $4f^{14}$ configurations.

***(ii)* +2 and +4 oxidation states.** Many lanthanides show +2 and +4 oxidation states which are less stable than +3 oxidation state which is common oxidation state of all lanthanides. Examples of lanthanides which show +2 or +4 oxidation state are given below.

(*a*) $Eu_{63}$ and $Yb_{70}$ acquire $f^7$ and $f^{14}$ configurations when they form $Eu^{2+}$ and $Yb^{2+}$ ions as shown below.

$$Eu^{2+} = [Xe]_{54} 4f^7; \quad Yb^{2+} = [Xe]_{54} 4f^{14}$$

(*b*) $Ce_{58}$ and $Tb_{65}$ acquire $f^{\circ}$ and $f^7$ configurations when they form $Ce^{4+}$ and $Tb^{4+}$ ions as shown below.

$$Ce^{4+} = [Xe]_{54} 4f^{\circ}; \quad Tb^{4+} = [Xe]_{54} 4f^7$$

(*c*) $Ce_{58}$, $Nd_{60}$, $Sm_{62}$ and $Tm_{69}$ show +2 oxidation state but $Ce^{2+}$, $Nd^{2+}$, $Sm^{2+}$ and $Tm^{2+}$ ions do not have $4f^{\circ}$, $4f^7$ or $4f^{14}$ configurations as shown below.

$$Ce^{2+} = [Xe]_{54} 4f^2; \quad Nd^{2+} = [Xe]_{54} 4f^4$$
$$Sm^{2+} = [Xe]_{54} 4f^6, \quad Tm^{2+} = [Xe]_{54} 4f^{13}$$

(*d*) $Pr_{59}$, $Nd_{60}$ and $Dy_{66}$ show + 4 oxidation state but $Pr^{4+}$, $Nd^{4+}$ and $Dy^{+4}$ ions do not have $4f^0$, $4f^7$ and $4f^{14}$ configurations as shown below.

$$Pr^{4+} = [Xe]_{54}\, 4f^1; \quad Nd^{4+} = [Xe]_{54} 4f^2; \quad Dy^{4+} = [Xe]_{54} 4f^8$$

It is evident from the above discussion that although +2, +3 and +4 oxidation states of many lanthanides can be explained on the basis of the concept of special stability associated with $4f^{\circ}$, $4f^7$ and $4f^{14}$ configurations, yet there are many lanthanids whose +2, +3 and +4 oxidation states cannot be explained on the basis of the above concept. In case of these lanthanides, their oxidation states are explained on the basis of highly complicated thermodynamic and kinetic factors.

We know that +3 oxidation state of lanthanides is the most stable oxidation state. Greater stability of $M^{3+}$ ions as compared to $M^{2+}$ and $M^{4+}$ ions is evident from the experimental fact that $Sm^{2+}$ ion has a tendency to get oxidised to $Sm^{3+}$ ion and $Ce^{4+}$ ion has a tendency to get reduced to $Ce^{3+}$ ion in aqueous solution.

$$\underset{(RA)}{2Sm^{2+}} + \underset{(H = +1)}{2H_2O} \longrightarrow 2Sm^{3+} + OH^- + \underset{(H = 0)}{H_2}$$

$$\underset{(OA)}{Ce^{4+}} + Fe^{2+} \longrightarrow Ce^{3+} + Fe^{3+}$$

Thus $sm^{2+}$ is a good RA and $Ce^{4+}$ is a good OA in aqueous solution. $Pr^{4+}$ and $Tb^{4+}$ are even more powerful oxidising agents.

## 3. Atomic and Ionic Radii: Lanthanide Contraction

Atomic radii of Ln atoms and ionic radii of $Ln^{3+}$ ions are given below in Table 9.3. It may be seen that as we move along lanthanide series, there is a steady decrease in the value of radii.

*The steady decrease in atomic ($Eu_{63}$ has abnormally high value) and ionic radii of lanthanide elements with increasing atomic number is called lanthanide contraction.*

**Table 9.3. Atomic and ionic radii (pm) of $La_{57}$ and lanthanides ($Ce_{58}$ to $Lu_{71}$)**

| *Lanthanides* | $La_{57}$ | $Ce_{58}$ | $Pr_{59}$ | $Nd_{60}$ | $Pm_{61}$ | $Sm_{62}$ | $Eu_{63}$ | $Gd_{64}$ | $Tb_{65}$ | $Dy_{66}$ | $Ho_{67}$ | $Er_{68}$ | $Tm_{69}$ | $Ys_{70}$ | $Lu_{71}$ |
|---|---|---|---|---|---|---|---|---|---|---|---|---|---|---|---|
| Atomic radii of Ln atoms | 187 | 183 | 182 | 181 | 181 | 180 | 199 | 180 | 178 | 177 | 176 | 175 | 174 | 173 | 156 |
| Ionic radii of $Ln^{3+}$ ions | 106 | 103 | 101 | 99 | 98 | 96 | 95 | 94 | 92 | 91 | 89 | 88 | 87 | 86 | 85 |

## Cause of Lanthanide Contraction

We know that as we proceed from one element to the next one in lanthanide series, the nuclear charge (*i.e.*, atomic number) increases by +1 at each next element. Thus as we move from Ce to Lu, the attraction between the nucleus and the outermost shell electron increases gradually at each step. It is also known that as we move from Ce to Lu, the addition of extra electron takes place to 4*f* orbitals. Since 4*f* orbitals have very diffused shape, the electrons in these orbitals are not able to shield (decrease) effectively the attraction of the nucleus for the electrons in the outer-most shell as the atomic number of lanthanides increases. Thus it is only due to the gradual increase in the nuclear charge (*i.e.*, increase in the attraction between the nucleus and the outer-most shell electrons) that the size of the lanthanide atoms and $M^{3+}$ ions decreases gradually with atomic number. The above discussion shows that it is due to the poor shielding effect of 4*f* electrons and gradual increase in the nuclear charge that the lanthanide contraction takes place among lanthanides.

## Consequences of Lanthanide Contraction

Lanthanide contraction plays an important role in determining the chemistry of lanthanides and heavier transition series elements. Some important consequences of lanthanide contraction are discussed below:

(*i*) **The atomic radii of the two elements in the pairs viz Zr—Hf, Nb—Ta, Mo—W etc. are almost the same.** This fact is a direct consequence of lanthanide contraction as explained below. The atomic (covalent) radii of the elements of *1st*,

*2nd* and *3rd* transition series are given in Table 9.4. This table shows that the atomic radius of La (element of *3rd* transition series) is greater than that of Y (element of *2nd* transition series) (La = 187 pm, Y = 180 pm). On the basis of this trend, the atomic radii of the two elements present in each of the pairs *viz* Zr-Hf, Nb-Ta, Mo-W etd. should be in the order Hf > Zr, Ta > Nb, W > Mo etc, but the atomic radii of the two elements present in each pair given above has been found to be almost the same as is evident from Table 9.4. *The similarity in atomic radii of the two elements is due to lanthanide contraction seen in lanthanides ($Ce_{58}$ to $Lu_{71}$) which lie between $La_{57}$ and $Hf_{72}$*. Thus we can say that, *due to lanthanide contraction, the atomic radius of any element of 3rd transition series following $La_{57}$ is nearly the same as that of the corresponding element lying in the same group of 2nd transition series.*

**Table 9.4. Atomic radii (pm) of transition series element**

| Groups ⟶ | 3 | 4 | 5 | 6 | 7 | 8 | 9 | 10 | 11 | 12 |
|---|---|---|---|---|---|---|---|---|---|---|
| Elements of *1st* transition series | $Sc_{21}$ 162 | $Ti_{22}$ 147 | V 134 | Cr 127 | Mn 126 | Fe 126 | Co 125 | Ni 124 | $Cu_{29}$ 124 | $Zn_{30}$ 138 |
| Elements of *2nd* transition series | $Y_{39}$ 180 | $Zr_{40}$ 160 | Nb 146 | Mo 139 | Tc 136 | Ru 134 | Rh 134 | Pd 137 | $Ag_{47}$ 144 | $Cd_{48}$ 154 |
| Elements of *3rd* transition series | $La_{57}$ 187 | *$Hf_{72}$ 158 | Ta 146 | W 139 | Re 137 | Os 135 | Ir 136 | Pt 138 | $Au_{79}$ 144 | $Hg_{80}$ 157 |

**Lanthanide contraction*

(*ii*) **Similarity in properties of the two elements in the pairs viz Zr-Hf, Nb-Ta, Mo-W etc**, We have said above that, due to lanthanide contraction, the atomic size of the two elements present in the pairs *viz* Zr-Hf, Nb-Ta, Mo-W etc. are almost the same. Due to similarity in their atomic size, the properties of the two elements in each of the above pairs are very similar.

(*iii*) **Difficulty in the separation of lanthanides.** We know that the properties of metal ions is determined by their size and charge. Now since, on proceeding from $Ce_{58}$ to $Lu_{71}$, the change in the size of $Ln^{3+}$ ions is very small and these ions have the same charge (= +3), chemical properties of lanthanides are almost identical. Due to identical properties, the separation of lanthanides from one another in the pure state is difficult.

(*iv*) **Comparison between the densities of the elements of 2nd and 3rd transition series.** The densities of d-block elements are given in Table 9.5. It may be seen from the table that the densities of elements belonging to the same sub-group increases on moving down the sub-group. The densities of the elements of 2nd transition series are only slightly higher than those of the corresponding elements of 1st transition series while these values for the transition elements from $Hf_{72}$ to $Hg_{80}$ (elements of 3rd transition series) are almost double these values for the elements from $Zr_{40}$ to $Cd_{48}$ respectively (elements of 2nd transition series.) Note that the density of $La_{57}$ (= 6.17) is not double that of the value for $Y_{39}$ (= 4.47).

The variation of densities of transition series element in the given sub-group, as discussed above, can be explained as follows on the basis of lanthanide contraction.

Because of lanthanide contraction occurring in lanthanides, the atomic sizes of the elements of 3rd transition series coming after $La_{57}$ (i.e., elements from $Hf_{72}$ to $Hg_{80}$) become very small. Consequently the packing of the atoms in their metallic crystals become so much compact that their densities become very high. Thus the densities of the elements of 3rd transition series become almost double the densities of the corresponding elements of 2nd transition series.

**Table 9.5. Densities (g/cm³) of d-block elements**

| Group ⟶ | III B | | IV B | V B | VI B | VII B | ⟵ | VIII | ⟶ | IB | II B |
|---|---|---|---|---|---|---|---|---|---|---|---|
| Elements of 1st transition series | $Sc_{21}$ 3.01 | | $Ti_{22}$ 4.51 | V 6.10 | Cr 7.19 | Mn 7.43 | Fe 7.86 | Co 8.90 | Ni 8.90 | Cu 8.96 | $Zn_{30}$ 7.14 |
| Elements of 2nd transition series | $Y_{39}$ 4.47 | | $Zr_{40}$ 6.49 | Nb 8.40 | Mo 10.21 | Tc 11.50 | Ru 12.20 | Rh 12.40 | Pd 12.00 | Ag 10.51 | $Cd_{48}$ 8.65 |
| Elements of 3rd transition series | $La_{57}$ 6.17 | $Ce_{58}$-$Lu_{71}$ Lanthanides | $Hf_{72}$ 13.10 | Ta 16.60 | W 19.30 | Re 21.00 | Os 22.60 | Ir 22.50 | Pt 22.40 | Au 19.30 | $Hg_{80}$ 13.60 |

(*v*) **Basic character of hydroxides, $M(OH)_3$ decreases from $La(OH)_3$ to $Lu(OH)_3$.** Due to lanthanide contraction, the size of +3 lanthanide ions ($M^{3+}$ ions) decreases regularly with increase in atomic number. As a result of this decrease in size, the covalent character between $M^{3+}$ ion and $OH^-$ ions increases from $La(OH)_3$ to $Lu(OH)_3$. (*Fajan's rules*). Therefore, the basic character of the hydroxides decreases with increase in atomic number. Consequently, $La(OH)_3$ is most basic while $Lu(OH)_3$ is the least basic.

Hydroxides of lanthanides are stronger bases than $Al(OH)_3$ but weaker than $Ca(OH)_2$.

## 4. Colour of Tripositive Lanthanide Ions ($M^{3+}$ Ions)

Most of the trivalent cations of lanthanide elements are coloured in the solid as well as in aqueous solution while only a few ions are colourless. (See Fig. 9.4)

It may be seen from Fig. 9.4 that the colour depends on the number of electrons present in 4*f* orbitals. The ion having *n* electrons in 4*f* orbitals has the same colour as the ion which has (14 – *n*) electrons in 4*f* orbitals. For example $La^{3+}$ ($4f^0$) which has no electron in its 4*f* orbitals ($n = 0$) is colourless and $Lu^{3+}$ ion ($4f^{14}$) which has (14 – 0) = 14 electrons in 4*f* orbitals is also colourless. Similarly $Pr^{3+}$ ion ($4f^2$) which has two electrons in its 4*f* orbitals ($n = 2$) and $Tm^{3+}$ ion ($4f^{12}$) which has (14 – 2) = 12 electrons in its 4*f* orbitals have the same colour (green). (See Fig. 9.3).

**Origin of colour.** Colour of lanthanide tripositive ions is due to *f*–*f* transition. The absorption bands in the visible region of electronic spectra of the ions in their compounds arise due to the absorption of light in the visible range. This results in the transition of the electrons of the ions from the *lower energy 4f*-orbitals to the *higher energy 4f*-orbitals. Thus lanthanide ions have colour which is complementary to the colour of the absorbed light. This type of electronic transition, which takes place due to absorption of light, is called **f–f transition**. Evidently *absorption bands seen in the electronic spectra arise due to electronic transition within 4f orbitals.*

*f*–*f* transitions of lanthanides are more forbidden than *d*–*d* transition of transition metal ions, because 4*f* electrons of lanthanide ions are much less affected by the ligand electrons than the electrons in *d*-orbitals of transition metal ions in their complexes. Hence the selection rules are more strictly followed for transitions in the compounds of lanthanides than in the compounds of transition metals.

The individual absorption bands of the spectra of tripositive lanthanide ions are **very sharp**. Many of these bands are **line like** and become even narrower as the temperature is lowered. This is due to the fact that the electrons present in 4*f*-orbitals are effectively shielded by the overlying electrons present in 5*s* and 5*p* orbitals of the tripositive lanthanide ions.

The colourless ions absorb either in the ultraviolet ($Ce^{2+}$, $Gd^{3+}$ ions) or in the infrared region ($Yb^{3+}$ ion). The $Ln^{2+}$ ions absorb strongly in the ultraviolet. The only $Ln^{4+}$ ion which is stable in aqueous solution, is $Ce^{4+}$ ion. This ion absorbs in the blue ultraviolet regions.

**Fig. 9.4** Colour of $M^{3+}$ ions having *n* electrons in 4*f* orbitals is almost the same as that of $M^{3+}$ ions having (14 – *n*) electrons in 4*f* orbitals.

*Laporte-permitted bands* due to the transitions of $4f^n \longrightarrow 5d^1$ type have been observed in $Ce^{3+}$, $Tb^{3+}$, $Sm^{2+}$, $Eu^{2+}$ and $Yb^{2+}$. These bands are *strong* and *broader*, since the transition is considerably influenced by the chemical environment. A charge transfer phenomena is also observed in certain lanthanide ions, *e.g.* the orange-red colour of $Ce^{4+}$ is due to the *electronic transition from the ligand orbital to the f-orbital of cerium*. Compounds of $Eu^{3+}$ with reducing anions are yellow *due to electron transfer from the metal*.

## 5. Magnetic Properties of Tripositive Lanthanide Ions ($M^{3+}$ ions)

We know that 4*f* orbital in $La^{3+}$ ion is empty ($4f^{0}$ configuration) and that in $Lu^{3+}$ ion is completely-filled ($4f^{14}$ configuration). Thus since all the electrons present in 4f orbitals in $Lu^{3+}$ ion are paired, $Lu^{3+}$ ion shows diamagnetic character. Due to the absence of any electron in 4*f* orbitals of $La^{3+}$ ion, this ion is also diamagnetic. The remaining $M^{3+}$ ions are paramagnetic, since 4*f* orbitals in the remaining $M^{3+}$ ions are partly filled.

**To Calculate the value of $\mu_{eff}$ for $M^{3+}$ ions.** In $M^{3+}$ ions, since 4f orbitals are well shielded from the surroundings by the overlying 5*s* and 5*p* orbitals, the electric field of ligands surrounding $M^{3+}$ ions does not restrict (destroy or quench) the orbital motion of electrons present in partly filled 4*f* orbitals of $M^{3+}$ ions. Thus, in $M^{3+}$ ions, in addition to spin motion, orbital motion of electrons also contributes to the value of $\mu_{eft}$ of $M^{3+}$ ions, *i.e.* both types of motion of electrons contribute to the value of $\mu_{eft}$ of $M^{3+}$ ions. 4*f*-electrons are free to undergo L and S coupling to give overall (total) *angular momentum quantum number*, which decides the value of $\mu_{eff}$ for $M^{3+}$ ions. $\mu_{eft}$ for $M^{3+}$ ions is thus represented as $\mu_J$ which is given by:

$$\mu_{eff} = \mu_J = g\sqrt{J(J+1)}\ \text{BM} \qquad \ldots(i)$$

In the above equation *g* is called **Lande splitting factor** which is given by:

$$g = 1 + \frac{S(S+1) - L(L+1) + J(J+1)}{2J(J+1)}$$

$$= 1 + \frac{S(S+1) - L(L+1)}{2J(J+1)} + \frac{J(J+1)}{2J(J+1)}$$

$$= 1 + \frac{S(S+1) - L(L+1)}{2J(J+1)} + \frac{1}{2}$$

$$= 1 + \frac{S(S+1) - L(L+1)}{2J(J+1)} + 0.5$$

or

$$g = 1.5 + \frac{S(S+1) - L(L+1)}{2J(J+1)} \qquad \ldots(ii)$$

$\mu_J$ values of $M^{3+}$ ions calculated from equation (*i*) are given in Table 9.6 Experimental values ($\mu_{exp}$) are also given for the sake of comparison.

**Table 9.6. Magnetic moment (calculated and experimental) values of tripositive lanthanide ions ($M^{3+}$ ions).**

| *$M^{3+}$ ions* | *Ground state term,* $^{2S+1}L_J$ | *J* | *g* | *Calculated value,* $\mu_J = g\sqrt{J(J+1)}$ B.M. | *Experimental value,* $\mu_{exp}$ *(B.M.)* |
|---|---|---|---|---|---|
| $La^{3+}$ ($4f^0$) ($n = 0$) | $^1S_0$ (S = 0, L = 0) | 0 | – | 0.0 | Diamagnetic |
| $Ce^{3+}$ ($4f^1$) ($n = 1$) | $^2F_{5/2}\left(S=\frac{1}{2}, L=3\right)$ | 5/2 | 6/7 | 2.54 | 2.3 – 2.5 |
| $Pr^{3+}$ ($4f^2$) ($n = 2$) | $^3H_4$(S = 1, L = 5) | 4 | 4/5 | 3.58 | 3.4 – 3.6 |
| $Nd^{3+}$ ($4f^3$) ($n = 3$) | $^4I_{9/2}\left(S=\frac{3}{2}, L=6\right)$ | 9/2 | 8/11 | 3.62 | 3.5 – 3.6 |
| $Pm^{3+}$ ($4f^4$) ($n = 4$) | $^5I_4$ (S = 2, L = 6) | 4 | 3/5 | 2.68 | – |
| $Sm^{3+}$ ($4f^5$) ($n = 5$) | $^6H_{5/2}\left(S=\frac{5}{2}, L=5\right)$ | 5/2 | 2/7 | 0.86 | 1.5 – 1.6 |
| $Eu^{3+}$ ($4f^6$) ($n = 6$) | $^7F_0$ (S = 3, L = 3) | 0 | – | 0.0 | 3.4 – 3.6 |
| $Gd^{3+}$ ($4f^7$) ($n = 7$) | $^8S_{7/2}\left(S=\frac{7}{2}, L=0\right)$ | 7/2 | 2 | 7.94 | 7.8 – 8.0 |
| $Tb^{3+}$ ($4f^8$) ($n = 6$) | $^7F_6$ (S = 3, L = 3) | 6 | 3/2 | 9.72 | 9.4 – 9.6 |
| $Dy^{3+}$ ($4f^9$) ($n = 5$) | $^6H_{15/2}\left(S=\frac{5}{2}, L=5\right)$ | 15/2 | 4/3 | 10.63 | 10.4 – 10.5 |
| $Ho^{3+}$ ($4f^{10}$) ($n = 4$) | $^5I_8$ (S = 2, L = 6) | 8 | 5/4 | 10.6 | 10.3 – 10.5 |
| $Er^{3+}$ ($4f^{11}$) ($n = 3$) | $^4I_{15/2}\left(S=\frac{3}{2}, L=6\right)$ | 15/2 | 6/5 | 9.57 | 9.4 – 9.6 |
| $Tm^{3+}$ ($4f^{12}$) ($n = 2$) | $^3H_6$ (S = 1, L = 5) | 6 | 7/6 | 7.63 | 7.1 – 7.4 |
| $Yb^{3+}$ ($4f^{13}$) ($n = 1$) | $^2F_{7/2}\left(S=\frac{1}{2}, L=3\right)$ | 7/2 | 8/7 | 4.50 | 4.4 – 4.9 |
| $Lu^{3+}$ ($4f^{14}$) ($n = 0$) | $^1S_0$ (S = 0, L = 0) | 0 | – | 0.0 | Diamagnetic |

It may be seen from Table 9.6 that the calculated values of $\mu_J$ of some $M^{3+}$ ions are in good agreement with the experimental values.

In Fig. 9.5 experimental values of magnetic moments (in BM) of $M^{3+}$ ions are plotted against the number of unpaired electrons ($n$) in $4f$ orbitals. From this figure it may be seen that $La^{3+}$ ion ($4f^{\circ}$) is diamagnetic ($\mu$ = 0), since it has no electron in $4f$ orbitals. $\mu_{exp}$ value increases upto $Nd^{3+}$ ion and then decreases to 1.47 BM for $Sm^{3+}$ ion. It starts rising again and becomes maximum at about 10.5 BM for $Dy^{3+}$. It again starts decreasing and becomes zero (diamagnetic) at $Lu^{3+}$ ion due to $4f^{14}$ configuration.

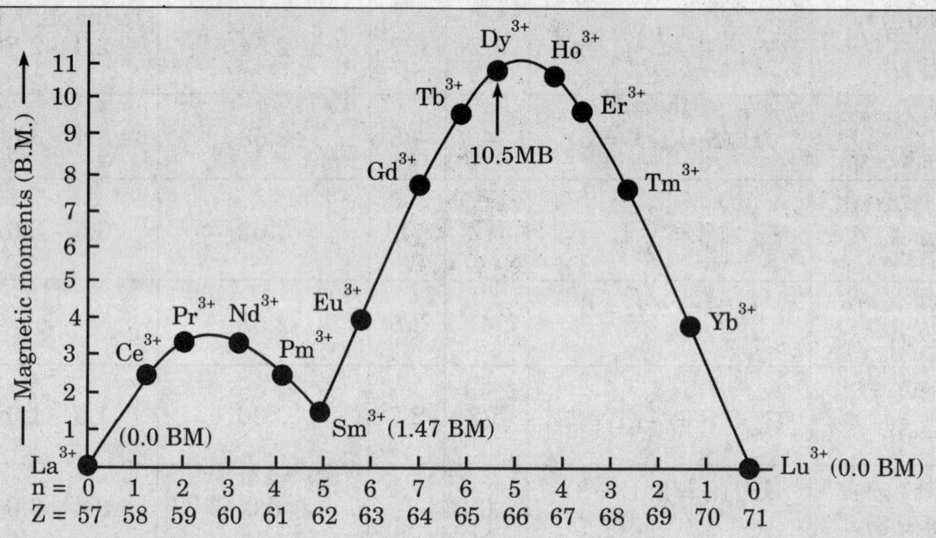


**Fig. 9.5** Plot between the experimental values of magnetic moments (in BM) of $M^{3+}$ ions vs number of unpaired electrons ($n$) and atomic number (Z).

The type of paramagnetism which is found in $M^{3+}$ ions is that in which the energy difference ($\Delta E$) between the lowest energy J level of the metal ion and the adjascent excited J state is **larger** than thermal energy ($k$T) at room temperature. Thus since the excited states are much above the ground state, it is only the lowest energy state which is occupied by metal ions. Other states are not occupied by metal ions. Note that in a magnetic field of strength H, each J level is split into 2J + 1 states in which each level is separated from its neighbours by g$\beta$H. In case of such metal ions the value of $\mu_{eff}$ is given by equation ($i$).

Equation ($i$) is not applicable for $Sm^{3+}$ and $Eu^{3+}$ ions, since the lowest J state (ground state) is close to the excited state and hence the energy separation or gap ($\Delta E$) $\approx$ $k$T. In such cases, $\mu_{eff}$ calculated by taking J of the ground state only will not give the correct value of $\mu_{eff}$ for $Sm^{3+}$ and $Eu^{3+}$ ions.

The value of $\mu_{eff}$ calculated from equation ($i$) for $Sm^{3+}$ and $Eu^{3+}$ ions are 0.86 BM and 0.0 BM respectively (See Table 9.6). In the calculation of these values we have used J = 5/2 for $Sm^{3+}$ and J = 0 for $Eu^{3+}$. Total values of J for these ions are as :

$$Sm^{3+} = \frac{15}{2}, \frac{13}{2}, \frac{11}{2}, \frac{9}{2}, \frac{7}{2}, \frac{5}{2}$$

$$Eu^{3+} = 6, 5, 4, 3, 2, 1, 0$$

Experimental values of $\mu_{eff}$ for $Sm^{3+}$ and $Eu^{3+}$ ions are 1.6 BM and 3.60 BM respectively. These values show that the value of J for $Sm^{3+}$ is not equal to 5/2. In case of $Eu^{3+}$ ion, if we put J = 2 (and not J = 0) in equation (*i*), we get $\mu_{eff}$ = 3.67 B.M.

## 6. Physical Properties

All the lanthanides are soft, malleable and ductile, and have low tensile strength. They are not good conductors of heat and electricity. In general, the atomic volumes and densities of these elements increase with the increase in atomic number.

Lanthanides have high melting and boiling points. However, they donot exhibit any regular trend with rise in atomic number. Lanthanides have low ionisation energies which compare well with those of the alkaline earth metals particularly calcium.

## 7. Properties Dependent on Standard Oxidation Potential Values

The standard oxidation potentials ($E_{ox}$ values) of lanthanides (M) for the oxidation half-reaction,

$$M(s) \longrightarrow M^{3+}(aq) + 3e^-$$

are given below.

$\longrightarrow$ $E^\circ_{ox}$ values in volts decrease $\longrightarrow$

La = 2.52, Ce = 2.48, Pr = 2.46, Nd = 2.43, Pm = 2.42, Sm = 2.41, Eu = 2.40, Gd = 2.39, Tb = 2.39, Dy = 2.35, Ho = 2.32, Er = 2.30, Tm = 2.28, Yb = 2.27 Lu = 2.25.

These $E^\circ_{ox}$ values explain the following properties of lanthanides:

***(i)* Reducing property.** Since lanthanides have positive $E^\circ_{ox}$ values, these elements (M) have a strong tendency to lose their three electrons to form $M^{3+}$ ions and hence act as strong reducing agents. Due to the decrease of $E^\circ_{ox}$ values from La to Lu, the reducing power of lanthanides also decreases from La to Lu.

(*ii*) **Electropositive character.** Since $E^\circ_{ox}$ values are high, lanthanides can readily lose their electrons and hence show strong electropositive (or metallic) character. With the decrease of $E^\circ_{ox}$ values from La to Lu, the electropositive character of these elements decreases from La to Lu.

***(iii)* Liberation of $H_2$ from $H_2O$.** Due to highly electropositive character of lanthanides, these elements slowly react with cold water and rapidly with hot water to liberate $H_2$ from it.

## 8. Solubility of Compounds of Lanthanides

The nitrates, chlorides, sulphates, perchlorates and salts of oxy acids of lanthanides are soluble but the oxalates, carbonates and fluorides are insoluble in water. Note that the sulphates of the elements of group 2 are insoluble in water.

## 9. Formation of Double Salts

Lanthanides form a number of double salts. Important double salts formed by lanthanides are given below (M is the lanthanide element) (*i*) *Carbonates, e.g.* $K_2CO_3 \cdot Ce_2(CO_3)_3 \cdot 3H_2O$ (*ii*) *Nitrates, e.g.* $3\,Mg(NO_3)_2 \cdot 2M(NO_3)_3 \cdot 24H_2O$ (*iii*) *Sulphates, e.g.* $(NH_4)_2SO_4 \cdot M_2(SO_4)_3 \cdot 8H_2O$.

## 10. Chemical Reactivity

Lanthanides differ from one another only in the number of 4f electrons. Since these electrons are very effectively shielded from interaction with other elements by the overlying 5s, 5p and 6s electrons, they *show very little difference in their chemical reactivity*. Some of the chemical properties of lanthanides are given below.

(*i*) Lanthanides are highly reactive, silvery-white metals. These metals tarnish readily on exposure to air. In the finely-divided state, these metals burn in air to form sesquioxide ($M_2O_3$). Ce forms $CeO_2$. Ytterbium (Yb) resists the action of air even at 1000° due to the formation of a protective coating of its oxide.

(*ii*) Lanthanides combine with $H_2$ on heating. The hydrides formed are of $MH_2$ and $MH_3$ type. These hydrides are stable.

(*iii*) Lanthanides react with non-metals like S, $X_2$, P, $N_2$, C and Si to form the corresponding compounds.

(*iv*) Lanthanides decompose $H_2O$ to evolve $H_2$. Evolution of $H_2$ takes place slowly in cold and rapidly on heating.

## 11. Formation of Complexes

Although tripositive lanthanide cations have a high charge equal to +3 on them, yet their size is so large that their charge-to-radius ratio becomes so small that these ions have very poor tendency to form complexes. Common ligands with which $M^{3+}$ ions form stable complexes are: (*i*) *chelating oxygen containing ligands* like EDTA, citric acid, oxalic acid, acetylacetone, (*ii*) *nitrogen containing ligands* like ethylenediamine, NCS, etc.

Bonding between $M^{3+}$ ions and the coordinating ligands mainly depends on the electronegativity of the bonding atom of the ligands. The following order of the bond formation of monodentate ligands has been observed : $F^- < OH^- < H_2O < NO_3^- < Cl^-$ etc.

Complex formation in aqueous solution is possible only with those ligands which bind to metal through O-atoms. For example carboxylate anions ($RCOO^-$) and β-diketones are such type of ligands. Due to the resonating structures, the chelate ring is stabilised to such extent that it cannot be replaced by $OH^-$ or $H_2O$. But the complexes in which the ligands are bonded to the metal through N and S dissociate in aqueous solution. Hence such complexes are prepared in non-aqueous solutions.

Generally coordination number (C.N.) of lanthanide ions in their complexes ranges from 6 to 9. Maximum C.N. of lanthanide ions in complexes formed by monodentate ligands like $F^-$, $H_2O$, $Cl^-$ etc. is 9. Bidentate ligands form complexes in which C.N. of lanthanide ions is generally 6, 7 and 8.

Complexes formed by lanthanides in +4, +3 and +2 oxidation states are discussed below.

(*i*) **Complexes of lanthanides in +4 oxidation state.** Ce is the only lanthanide which forms complexes in +4 oxidation state. Pr, Nd and Dy form some fluoro complexes like $Na_2[PrF_6]$, $Cs_3[NdF_7]$ and $Cs_3$ [Dy $F_7$]. Cerric ammonium nitrate, $(NH_4)_2[Ce(NH_3)_6]$ and cerric ammonium sulphate, $(NH_4)_4[Ce(SO_4)_4]$ are important complexes of $Ce^{+4}$. These compounds are soluble in water. Iodates, β-diketonates etc. also form stable complexes having Ce in +4 oxidation state. $Ce^{4+}$ also forms complex ions like $[CeF_8]^{4-}$, $[CeF_6]^{2-}$, $[CeCl_6]^{2-}$ etc.

***(ii)* Complexes of lanthanides in +3 oxidation state.** Since +3 is the most stable oxidation state of lanthanides, lanthanides form the maximum number of complexes having lanthanides in +3 oxidation state. All the lanthanide ions form complex cation, $[M(H_2O)_n]^{3+}$ where $n$ is generally 8 or 9.

$M^{3+}$ ions form complex ions with various organic and inorganic anions like oxalate, citrate, tartrate, nitrate and sulphate. In all these complexes C.N. of $M^{3+}$ ion is quite high. For example C.N. of $Ce^{3+}$ ion in $[Ce(NO_3)_5]^{2-}$ and $[Ce(NO_3)_6]^{3-}$ is 5 and 6 respectively.

β-diketonate anion $(RCOCHCOR)^-$ reacts with $M^{3+}$ ions to form [M (β-diketonate)$_4$]$^-$, [M (β-diketonato)$_3$ L] (L = $H_2O$, pyridine etc.) and M (β-diketonato)$_3$]. These complexes are more stable than all other types of $M^{3+}$-complexes.. $[EDTA]^{4-}$ anion also forms complex anions of $[M(EDTA).\ 3H_2O]^-$ type with all $M^{3+}$ ions.

Although nitrogen containing monodentate ligands do not form stable complexes with $M^{3+}$ ions, bidentate ligands having two donor N-atoms form stable chelates. For example, ethylene diamine (en) forms complex cation, $[M(en)_4]^{3+}$ with $M^{3+}$ ions in polar organic solvents.

*(iii)* **Complexes of lanthanides in +2 oxidation state**. Complexes having lanthanides in +2 oxidation state are rare.

## Compounds of Lanthanides in +2 Oxidation State

Compounds of $Sm^{2+}$, $Eu^{2+}$ and $Yb^{2+}$ have been characterised. Compounds of these ions are obtained by *(i)* the reduction of fused trihalides or oxides with the corresponding metal *(ii)* electrolytic reduction of $Eu^{3+}$ and $Yb^{3+}$ in aqueous solution *(iii)* by thermal decomposition of anhydrous trihalides ($2MX_3 \rightarrow 2MX_2 + X_2$). Of the divalent compounds of lanthanides those of $Eu^{2+}$ are the most stable. All compounds of $M^{2+}$ ions decompose in $H_2O$ with evolution of $H_2$.

$$2M^{2+} + 2H_2O \rightarrow 2M^{3+} + 2OH^- + H_2$$

Compounds of $Sm^{2+}$, $Eu^{2+}$ and $Yb^{2+}$ ions exist in solution. These ions are oxidised to $M^{3+}$ ions in aqueous solution

$$Sm^{2+}(aq) \longrightarrow Sm^{3+}(aq) + e^-,\ E^o_{ox} = 1.55\ V$$

$$Eu^{2+}(aq) \longrightarrow Eu^{3+}(aq) + e^-,\ E^o_{ox} = 0.43\ V$$

$$Yb^{2+}(aq) \longrightarrow Yb^{3+}(aq) + e^-,\ E^o_{ox} = 1.15\ V$$

$E^o_{ox}$ values given above show that reducing strength of $M^{2+}$ ions is in the order:

$$Sm^{2+} > Yb^{2+} > Eu^{2+}$$

$Sm^{2+}$ and $Yb^{2+}$ ions are rapidly oxidised to $M^{3+}$ ions by $H_3O^+$ (acidic solution) while $Eu^{2+}$ ion is fairly stable and is only slowly oxidised by $H_3O^+$.

$$\underset{(M^{2+} = Sm^{2+},\ Yb^{2+})}{2M^{2+}} + 2H_3O^+ \longrightarrow 2M^{3+} + 2H_2O + H_2$$

All these ions are rapidly oxidised in presence of $O_2$.

$$\underset{(M^{2+} = Sm^{2+},\ Eu^{2+},\ Yb^{2+})}{4M^{2+}} + 4H_3O^+ + O_2 \longrightarrow 4M^{3+} + 6H_2O$$

The compounds of $Sm^{2+}$, $Eu^{2+}$ and $Yb^{2+}$ which are insoluble in $H_2O$ are not oxidised by $H_2O$, while hydrated water-soluble compounds of $Sm^{2+}$ and $Yb^{2+}$ are oxidised by their water. Hydrated water-soluble compounds of $Eu^{2+}$ are more stable.

Compounds of Nd and Tm in +2 oxidation state are known only as solid halides. These compounds are immediately oxidised with air.

### 1. Dihalides ($MX_2$)

These compounds are obtained by the reaction between molten $MX_3$ and elemental lanthanides. Difluorides ($MF_2$) of all lanthanides are iso structural with each other and with $CaF_2$. $SmCl_2$ and $EuCl_2$ are nearly identical but differ from $YbCl_2$.

### 2. Divalent Chalco-genides, MZ (Z = O, S, Se, Te)

These compounds have been prepared for all lanthanides except for Pm, mostly by direct combinations. These compounds are almost black with the exception of SmZ, EuZ, YbZ, TmSe and Tm Te. They have high metallic conductivity. Crystals of these compounds have cubic NaCl type structure.

### 3. MO oxides

MO oxides are obtained by the reduction of $M_2O_3$ oxides with the metal at elevated temperature.

## Compounds of Lanthanides in +3 Oxidation State

Nearly all known anions form the compounds with $M^{3+}$ cation. These compounds are stable in solid as well as in solution state. Compounds of $M^{3+}$ cation with the anions such as $OH^-$, $CO_3^{2-}$, $SO_4^{2-}$, $C_2O_4^{2-}$, $NO_3^-$ etc. decompose on heating, give first basic salts and finally oxides. Hydrated salts that contain thermally stable anions such as $F^-$, $Cl^-$, $Br^-$, $PO_4^{3-}$ etc. also give similar products on heating because of hydrolysis.

Compounds of $M^{3+}$ cation with the anions $Cl^-$, $Br^-$, $I^-$, $NO_3^-$, $CH_3COO^-$, $BO_3^{3-}$ are generally soluble in water while those with $F^-$, $OH^-$, $O^{2-}$, $C_2O_4^{2-}$, $CO_3^{2-}$, $CrO_4^{2-}$, $PO_4^{3-}$ etc. are generally insoluble.

### 1. Trihalides, $MX_3$

Fluorides are precipitated by the addition of HF or a soluble fluoride to a $M^{3+}$ salts solution. The fluorides particularly of heavier lanthanides, are sparingly soluble in HF due to the formation of fluoro complexes.

The anhydrous chlorides can be prepared by the direct combination of the elements on heating. These are best prepared by heating the oxides ($M_2O_3$) with carbonyl chloride ($COCl_2$) or $NH_4Cl$.

$$M_2O_3 + 3COCl_2 \longrightarrow 2MCl_3 + 3CO_2$$

$$M_2O_3 + 6NH_4Cl \xrightarrow{300°} 2MCl_3 + 3H_2O + 6NH_3$$

The anhydrous chlorides cannot be obtained from the hydrated chlorides, since these lose HCl on heating to give the oxychlorides (MOCl) more easily than they lose $H_2O$. ($ScCl_3 . 3H_2O$ and $CeCl_4$ $2H_2O$, however, give $Sc_2O_3$ and $CeO_2$). The chlorides are non-volatile, deliquescent solids soluble in $H_2O$ and alcohol. The hydrated chlorides are formed by dissolving the oxides or carbonates in HCl and concentrating the solution to crystallising point. They crystallise from solution usually as hexahydrates, $MCl_3 . 6H_2O$.

Bromides and iodides are rather similar to the chlorides. Iodides of the first few lanthanides are orthorhombic while those of the remaining lanthanides are hexagonal.

## 2. Oxides, $M_2O_3$

Excepting $Ce_2O_3$, $Pr_2O_3$ and $Tb_2O_3$, other oxides are obtained by: (*i*) burning lanthanides in $O_2$ (*ii*) igniting carbonates, nitrates and salts containing oxoanions (e.g. $CO_3^{2-}$, $C_2O_4^{2-}$, $SO_4^{2-}$ etc.) in air.

$$2M + 3O_2 \longrightarrow M_2O_3$$
$$M_2(CO_3)_3 \longrightarrow M_2O_3 + 3CO_2$$
$$4M(NO_3)_3 \longrightarrow 2M_2O_3 + 2NO_2 + 3O_2$$

$M_2O_3$ oxides are strongly basic and their basic character decreases as atomic number increases. For example $La_2O_3$ is strongly basic while $Lu_2O_3$ is least basic. Oxides are soluble in $H_2O$ and form $M(OH)_3$. Oxides dissolve in aqueous acids to give solution which contains $[M(H_2O)_x]^{3+}$ ion. $M(OH)_3$ can be prepared by prolonged treatment of $M_2O_3$ with cone NaOH at high temperature and pressure.

## 3. Trivalent chalco-genides, $M_2Z_3$ (Z = S, Se, Te)

$M_2S_3$, $M_2Se_3$ and $M_2Te_3$ have been obtained for most of lanthanides. These compounds are obtained by (*i*) the direct combination of elements (ii) the reaction of $H_2S$ or $H_2Se$ on lanthanide metals (*iii*) by the reduction of oxo salts. In general these compounds are stable in dry air but are hydrolysed in presence of moisture. If heated in air, they (especially sulphides) are oxidised to basic salts of the corresponding oxo-anion and they are attacked by acids with the evolution of $H_2Z$.

## 4. Nitrides

Nitrides are prepared by direct combination of elements at about 1200°C or by the action of $N_2$ or $NH_3$ on hydrides of lanthanides.

## 5. Carbonates, $M_2(CO_3)_3$

The normal carbonates can be prepared by passing $CO_2$ into an aqueous solution of $M(OH)_3$. They can also be prepared by adding $Na_2CO_3$ solution to $M^{3+}$ salt solution. The carbonates are insoluble in $H_2O$, but dissolve in acids with liberation of $CO_2$ and forming $M^{3+}$ - salts.

## 6. Nitrates, $M(NO_3)_3$

The hydrated nitrates, $M(NO_3)_3 . 6H_2O$ are obtained by the evaporation of aqueous solutions. These compounds are soluble in $H_2O$, alcohols, ketones and esters.

## 7. Phosphates and Oxalate

These compounds are insoluble in water. All lanthanides are quantitatively precipitated as oxalates from $M^{3+}$ solution containing $C_2O_4^{2-}$ ion. The precipitate on drying and ignition gives $M_2O_3$.

## 8. Double Salts

Lanthanide salts form a large number of double salts. The most important double salts are:

(*i*) Double nitrates such as $2M(NO_3)_3 . 3M^{II}(NO_3)_2 . 24H_2O$ ($M^{II}$ = Mg, Zn, Ni, Mn) and $M(NO_3)_3 . 2NH_4NO_3 . 4H_2O$.

(*ii*) Double sulphates such as $M_2(SO_4)_3 . 3M_2^I SO_4 . 12H_2O$ ($M^I$ = alkali metal)

The double sulphates, $M_2(SO_4) . 3Na_2SO_4 . 12H_2O$ where M = La – Eu are only sparingly soluble in $Na_2SO_4$, while those where M = Gd – Lu are appreciably soluble. Thus a separation of lanthanides into two groups: **Cerium group** having $La_{57}$ – $Eu_{63}$ lanthanides and **Yttrium group** having $Gd_{64}$–$Lu_{71}$ lanthanides is possible. Since the double salts crystallise well, these are sued to separate the rare-earths from one another. In the above description M represents lanthanide atom.

## Compounds of Lanthanides in +4 Oxidation State

Chemistry of compounds in +4 oxidation state is mainly the chemistry of Ce (IV) compounds. Double salts like $Ce(NO_3)_4 . 2NH_4NO_3$ and $Ce(SO_4)_2 . 2(NH_4)_2SO_4 . 2H_2O$ have also been prepared.

The standard oxidation potentials at 25°C, in acid solution, of $Ce^{4+}$ and $Pr^{4+}$ ions are given as under:

$$Ce^{3+} \rightleftharpoons Ce^{4+} + e^-, E^\circ_{ox} = +1.74 \text{ V}$$

$$Pr^{3+} \rightleftharpoons Pr^{4+} + e^-, E^\circ_{ox} = +2.86 \text{ V}$$

$E^\circ_{ox}$ values show that Ce (IV) and Pr (IV) are strong oxidising agents, the latter being by far the stronger of the two. $Ce(SO_4)_2$ is generally used in volumetric analysis. $Ce^{4+}$ ion is readily reduced to $Ce^{3+}$ ion.

The tetravalent ions of Ce are stable in the solid state as well as in solution; $Pr^{IV}$, $Nd^{IV}$, $Tb^{IV}$ and $Dy^{IV}$ are stable only in solution.

## Uses of Lanthanides

(*i*) Lanthanides are used in *metallo-thermic reactions* due to their extraordinary reducing property (Ce is a stronger reducing agent than Al). Lanthanido-thermic processes can yield sufficiently pure Nb, Zr, Fe, Co, Ni, Mn, Y, W, U, B and Si. These metals are also used as de-oxidising agents particularly in the manufacture of Cu and its alloys.

(*ii*) **Uses of mish-metals.** Alloys of lanthanides are known as *mish-metals*. The major constituents of mish-metals are Ce (45—50%), La (25%), Nd (5%) and small quantities of other lanthanide metals and Fe and Ca impurities.

Mish-metals are used for the production of different brands of steel like heat resistant, stainless and instrumental steels. The addition of 0.75% of mish-metal to steel raises its yield point and its workability in heated state and improves its resistance to oxidation Misch metal is an excellent scavenger for adsorbed oxygen and sulphur in metallurgy.

Mg-alloys containing about 30% misch metal and 1% Zr are useful in making parts of jet engine. When alloyed with 30% iron, it is sufficiently pyrophoric to be useful in lighter flints.

## Uses of the Lanthanide Compounds

The uses of the compounds of lanthanides can broadly be classified as follows:

**1. Non-nuclear applications.** The following uses are important.

(*a*) **Ceramic applications.** $CeO_2$, $La_2O_3$, $Nd_2O_3$ and $Pr_2O_3$ are widely used for decolorising glass. Approximately 1% $CeO_2$ is used in the manufacture of protective transparent glass blocks to be used in nuclear technology because these blocks are not affected by pro-longed exposure to radiations. Because lanthanide oxides can absorb ultra-violet rays, these are used as additives in glasses for special purposes, e.g. for making (*i*) sun-glasses (by adding, $Nd_2O_3$) (*ii*) goggles for glass blowing and welding work ($Nd_2O_3 + Pr_2O_3$) (*iii*) glasses protecting eyes from neutron radiation ($Gd_2O_3 + Sm_2O_3$) etc.

The addition of more than 1% $CeO_2$ to glass gives it a brown colour. $Nd_2O_3$ and $Pr_2O_3$ give respectively red and green colours. ($Nd_2O_3 + Pr_2O_3$) gives a blue colour.

(*b*) **Refractories.** CeS (m.pt. = 2000°C) is used in the manufacture of a special type of crucibles which are used for melting metals in a reducing atmosphere at temperatures upto 1800°C. Borides, carbides and nitrides of lanthanides are also used as refractories.

(*c*) **Abrasives.** Lanthanide oxides are used as abrasives for polishing glasses, *e.g.* the mixture of oxides, $CeO_2$ (47%), $La_2O_3 + Nd_2O_3 + Pr_2O_3$ (51%) + $SiO_2$, CaO, $Fe_2O_3$ etc. (≃ 2%) which is called polirite has been used for polishing glasses.

(*d*) **Paints.** Lanthanide compounds are used in the manufacture of lakes, dyes and paints for porcelain, *e.g.* cerium molybdate gives light yellow colour, cerium tungstate gives greenish blue colour and salts of Nd give red colour.

(*e*) **In textiles and leather industries.** Ceric salts are used for dying in textile industries and as tenning agents in leather industries. $Ce(NO_3)_4$ is used as a mordant for alizarin dyes. Chlorides and acetates of lanthanides make the fabrics water-proof and acid resistant.

(*f*) **In medicine and agriculture.** Dimals which are salicylates of Pr and Nd are used as germicides. Cerium salts are used for the treatment of vomiting and sea-sickness. Salts of Er and Ce increase the red-blood corpuscles and haemoglobin content of blood.

In agriculture lanthanide compounds are used as insecto-fungicides and as trace elements in fertilisers.

(*g*) **In lamps.** Salts of La, Ce, Eu and Sm are used as activators of luminophores. They are used in the manufacture of gas mantles, in the coatings of luminescent lamps and for painting the screens of cathode-ray tubes.

(*h*) **In analytical chemistry.** $Ce(SO_4)_2$ is used as an oxidising agent in volumetric titrations.

Radio-isotopes of lanthanides (*e.g.* $La^{140}$, $Ce^{142}$, $En^{152}$, $Tb^{160}$ etc.) are used in the study of co-precipitation, chromatographic separations etc.

(*i*) **Catalytic applications.** Certain compounds of lanthanides are employed for the hydrogenation, dehydrogenation and oxidation of various organic compounds. Cerium phosphate finds use as a catalyst in petroleum cracking.

(*j*) **Electronic applications.** Ferrimagnetic garnets of the type $3M_2O_3.5Fe_2O_3$ are employed in microwave devices.

**2. Nuclear applications.** Lanthanide elements and their compounds find many important nuclear uses, *e.g.*, in nuclear fuel control, shielding and fluxing devices. $Pr^{147}$ is used in the production of atomic battery.

**Comparison Between *d*- and *f*-Block Elements** (*particularly between 3d and 4f-block elements*)

Some of the differences are given below:

**1. Entry of extra electron.** In case of *d*-block elements the extra electron enters $(n-1)$ *d*-orbital while in case of *f*-block elements the extra electron occupies $(n-2)$ *f*-orbitals.

**2. Energy of orbitals.** $(n-1)$ d-orbitals have higher enegy than $(n-2)$ *f*-orbitals

**3. Cause of paramagnetism.** The maximum value of magnetic moment (μ value) for *d*-block elements is due to the presence of 5 unpaired electrons while the maximum value of μ for *f*-block elements is due to the presence of 7 unpaired electrons. Further, in case of *d*-block elements, since *d*-orbitals project out, these are not well shielded from the ligands surrounding the transition metal atoms or ions, *i.e.*, the electrons residing in *d*-orbitals are able to interact with the surrounding ligands. Thus in case of transition metals, the orbital motion of the electron is quenched and hence the observed paramagnetism in these elements is only due to the spin motion of the electron. Due to the quenching of orbital motion, the value of μ is calculated from the spin-only formula, $\mu = \sqrt{n(n+2)}$.

Now consider the case of tripositive lanthanide ions. In case of these ions since 4*f*-orbitals are quite deeply seated, these are well shielded from the ligands surrounding these ions by the overlapping 5*s* and 5*p* orbitals, *i.e.*, the electrons residing in 4*f* orbitals are not able to interact with the surrounding ligands. Thus in case of lanthanides, the orbital motion of the electron is not quenched as in case of *d*-block elements and hence the observed paramagnetism in these elements is due to both the motions of the electron *viz.* orbital motion and spin motion. In other words, the value of μ for $M^{3+}$ ion cannot be calculated by using spin-only formula. Rather μ is calculated from : $\mu = g\sqrt{J(J+1)}$.

**4. Nature of absorption bands.** *d*-block elements give broad bands in the absorption spectra in the visible region while, excepting $La^{3+}$ and $Lu^{3+}$, all the remaining tripositive lanthanide ions give sharp almost line-like bands in the absorption spectra.

**5. Oxidation states.** *d*-block elements show much greater number of oxidation states while *f*-block elements, particularly lanthaides, show a limited number of oxidation states.

**6. Formation of complex compounds.** *d*-block elements form complexes more easily while *f*-block elements, particularly lanthanides, do so with great difficulty.

**7. Coordination geometry.** Because of their larger size, lanthanide ions generally have higher coordination numbers than transitional metal ions. Coordination numbers greater than 6 for transition metal ions are only formed with difficulty because of the strong ligand repulsion in the coordination spheres.

# Lanthanum, $La_{57}$

## Occurrence

Lanthanum occurs alongwith cerium and neodymium, etc., as oxide in a Swedish mineral called *cerite*. It also occurs as $La_2(CO_3)_3.8H_2O$ in the mineral *lanthanite*.

## Preparation

Lanthanum is obtained in relatively pure state by reduction of its trichloride with calcium at 1000°C.

**Properties.** Lanthanum is a silvery-white metal but gets tarnished on exposure to air. It is malleable and ductile and can be hammered into foils and drawn into wires. It burns easily to give the sesquioxide, $La_2O_3$. It reacts with water, slowly in cold but rapidly on heating, liberating hydrogen in each case. This is due to its high oxidation potential and hence strong electropositive character.

Lanthanum reacts with hydrogen yielding the hydrides of composition $LaH_2$ and $LaH_3$. The reaction is exothermic but initial heating to 300-400°C is often required. The hydrides are highly stable even upto 900°C.

Lanthanum also reacts with several other non-metals including oxygen, halogens, sulphur, nitrogen, phosphorus, carbon, silicon, etc. When heated in oxygen, it burns energetically yielding the oxide, $La_2O_3$. When heated in nitrogen or ammonia, it gives a black nitride, LaN, which is decomposed in water evolving ammonia. Thus, when heated in air, lanthanum gives a mixture of the oxide, $La_2O_3$ and the nitride, LaN.

# Compounds of Lanthanum

Lanthanum forms stable compounds in +3 oxidation state. Its important compounds are discussed below:

## 1. Lanthanum Hydrides, $LaH_2$ and $LaH_3$

Lanthanum reacts with $H_2$, forming dihydride ($LaH_2$) and trihydride ($LaH_3$). At the initial stage of the reaction between La and $H_2$, heating to 300–400°C is required, but later on heat is evolved, *i.e.*, the reaction is exothermic. Even upto 900°C both the hydrides are stable.

## 2. Lanthanum Oxide, $La_2O_3$

It is obtained by heating lanthanum carbonate or oxalate. It is snow-white powder. When put into water, it is converted into $La(OH)_3$ and a large amount of heat is evolved. This behaviour of $La_2O_3$ is similar to that shown by CaO. It is sparingly soluble in water. It shows basic character and dissolves in dil. acids. On heating a mixture of $La_2O_3$ and carbon in a current of $Cl_2$, anhydrous $LaCl_3$ is obtained. $La_2O_3$, on being treated with dil. $H_2SO_4$, gives $La_2(SO_4)_3$. $La_2O_3$ reacts with dil. $HNO_3$ to form $La(NO_3)_3$.

$La_2O_3$ is used (*i*) in the manufacture of crookes glass, since it can absorb ultraviolet rays to some extent (*ii*) in the preparation of $LaCl_3$, $La_2(SO_4)_3$, $La(NO_3)_3$ etc.

**3. Lanthanum Halides, $LaX_3$**

(*a*) $LaF_3$ is obtained as a white gelatinous precipitate when HF reacts with a lanthanum salt. If the solution containing HF and lanthanum salt is allowed to stand for sometime, crystals of $LaF_3 \cdot 3H_2O$ are obtained. It is insoluble in dil. but soluble in conc. mineral acids.

(*b*) When a mixture of $La_2O_3$ and carbon is heated in a current of $Cl_2$, anhydrous $LaCl_3$ is obtained. It is a white deliquescent powder and is soluble in water. Crystals of heptahydrate ($LaCl_3 \cdot 7H_2O$) can be obtained from the aqueous solution of $LaCl_3$.

(*c*) $LaBr_3$ and $LaI_3$ are also known and are prepared by similar methods.

**4. Lanthanum Sulphate, $La_2(SO_4)_3$**

Anhydrous sulphate is obtained by reacting $La_2O_3$ with dil. $H_2SO_4$. It is a white hygroscopic powder with a density of 3.6. It is soluble in ice-cold water. On adding $K_2SO_4$ to the solution of $La_2(SO_4)_3$, a double sulphate, $La_2(SO_4)_3 \cdot 3K_2SO_4$ is obtained. When a concentrated solution of $La_2(SO_4)_3$, is warmed, sparingly soluble *nona-hydrate* $La_2(SO_4)_3 \cdot 9H_2O$ separaes out.

## 5. Lanthanum Nitrate, La $(NO_3)_3 \cdot 6H_2O$

It is obtained by treating $La_2O_3$ with dil. $HNO_3$. It gets separated from water as hexahydrate, La $(NO_3)_3 \cdot 6H_2O$. When concentraed solution of La $(NO_3)_3$ and $NH_4NO_3$ are mixed, crystals of $La(NO_3)_3 \cdot 2NH_4NO_3 \cdot 4H_2O$ (double salt) are obtained.

## 6. Lanthanum Carbonate, $La_2(CO_3)_3$

This salt occurs in nature as *lanthanite*, $La_2(CO_3)_3 \cdot 8H_2O$. When $(NH_4)_2CO_3$ is added to a solution of $La(NO_3)_3$, $La_2(CO_3)_3$ is obtained as a white ppt. When the solution containing the two salts is allowed to stand, crystals of $La_2(CO_3)_3 \cdot (NH_4)_2CO_3$ (double salt) separate out.

## Compounds of Cerium

In aqueous solution, cerium exists in +3 and +4 oxidation states. Its salts in +3 and +4 salt are called *cerous* and *ceric* salts respectively. Ceric compounds are important. In these compounds Ce is in +4 oxidation state. Ce (IV) compounds may be prepared by the oxidation of $Ce^{3+}$ ion in the aqueous solution. Since oxidation of $Ce^{3+}$ ions is accompanied by the increase in positive charge and decrease in size, the polarising power of $Ce^{4+}$ ions becomes so high that it has a pronounced tendency to attract water molecules to form the hydrate ion $[Ce(H_2O)_4]^{4+}$. This ion is a fairly strong acid and, except at very low pH, dissociates to give ceric hydroxide and protons. Thus, the cerium (IV) salts are readily hydrolysed, if the medium is not strongly acidic.

The simple hydrated tetrapositive cerium ion is a stronger oxidising agent than $MnO_4^-$ ion. In solutions containing sulphate and nitrate ions cerium (IV) is more often complexed with the anion but is still a strog oxidising agent. The cerium (IV) compounds are, therefore, valuable reagents in analytical chemistry.

Following are the important compounds of Ce (IV).

## 1. Cerium (IV) Oxide, $CeO_2$

Ceric dioxide is obtained by burning the metal in oxygen.

$$Ce + O_2 \longrightarrow CeO_2$$

It can also be prepared by heating cerous oxide, hydroxide, carbonate or nitrate in presence of air.

$$2Ce_2O_3 + O_2 \longrightarrow 4CeO_2$$
$$4Ce(OH)_3 + O_2 \longrightarrow 4CeO_2 + 6H_2O$$
$$2Ce_2(CO_3)_3 + O_2 \longrightarrow 4CeO_2 + 6CO_2$$
$$2Ce(NO_3)_3 + O_2 \longrightarrow 2CeO_2 + 6NO_2 + 2O_2$$

Ceric oxide is a white powder with a yellow tinge. On heating, it turns lemon yellow. It is a very stable compound which melts above 1500°C. Ceric oxide prepared by the above methods is an inert substance with fairly high 7.2 density. It is insoluble in HCl and $HNO_3$. However, concentrated $H_2SO_4$ converts it to sulphate.

$$CeO_2 + 2H_2SO_4\ (\text{Conc.}) \longrightarrow Ce(SO_4)_2 + 2H_2O$$

If the concentration of $H_2SO_4$ is moderate, cerous sulphate is formed.

$$2CeO_2 + 3H_2SO_4 \longrightarrow Ce_2(SO_4)_3 + 3H_2O + 1/2\ O_2$$

When it is heated strongly, a mixed oxide of the composition $Ce_4O_7$ ($Ce_2O_3$.2$CeO_2$) is obtained. When heated again in presence of air, the mixed oxide changes to $CeO_2$. It has calcium fluorite structure.

## 2. Ceric Hydroxide, $Ce(OH)_4$ or $CeO_2.2H_2O$

It is prepared by adding $NH_4OH$ or NaOH to the solution of ceric sulphate or ceric chloride in HCl

$$Ce(SO_4)_2 + 4NH_4OH \longrightarrow Ce(OH)_4 + 2(NH_4)_2SO_4$$

It can also be prepared by treating an oxidising agent with cerous salt in neutral medium.

$$6Ce(NO_3)_3 + KBrO_3 + (n + 9)H_2O \longrightarrow 6CeO_2 . nH_2O + KBr + 18HNO_3$$

It is a yellow coloured gelatinous substance which dissolves easily in acids. On fusion with $Na_2O$ or NaOH, sodium cerate is obtained:

$$CeO_2 + Na_2O \longrightarrow Na_2CeO_3$$
$$CeO_2 + 2NaOH \longrightarrow Na_2CeO_3 + H_2O$$

## 3. Ceric Halides

The only stable halide of tetravalent cerium is the fluoride, $CeF_4$. It can be prepared by the direct reaction of cerium and fluorine:

$$Ce + 2F_2 \longrightarrow CeF_4$$

$CeF_4$ can also be prepared by the reaction of $Ce^{4+}$ ions with $F^-$ ions. It can also be synthesised by the reaction of fluorine with $CeF_3$ or $CeCl_3$:

$$2CeF_3 + F_2 \longrightarrow 2CeF_4$$
$$2CeCl_3 + 4F_2 \longrightarrow 2CeF_4 + 3Cl_2$$

Ceric tetrafluoride is a white coloured solid insoluble in water. It can be heated under vacum upto 400°C without any change. But when heated in hydrogen gas at 300°C, trifluoride is obtained:

$$2CeF_4 + H_2 \longrightarrow 2CeF_3 + 2HF$$

The chloride $CeCl_4$ has not been isolated because chloride ions are readily oxidised into chlorine. Thus, cerium (IV) oxide reacts with concentrated HCl to give cerous chloride and chlorine:

$$2CeO_2 + 8HCl \longrightarrow 2CeCl_3 + Cl_2 + 4H_2O$$

## 4. Ceric Sulphate, Ce $(SO_4)_2$

Ceric sulphate can be prepared by treating ceric oxide with concentratd sulphuric acid

$$CeO_2 + 2H_2SO_4 \longrightarrow Ce(SO_4)_2 + 2H_2O$$

It can also be prepared by fusing $CeO_2$ with $KHSO_4$:

$$4KHSO_4 + CeO_2 \longrightarrow 2K_2SO_4 + Ce(SO_4)_2 + 2H_2O$$

Ceric sulphate is a deep yellow crystalline powder which is insoluble in $H_2SO_4$. It readily forms yellow solution when dissolved in water.

When sulphuric acid is added to the freshly prepared aqueous solution of ceric sulphate, the colour changes to deep red. This is due to the formation of $H_4[Ce\,(SO_4)_4]$ complex acid.

When Ce $(SO_4)_2$ is crystallised from solution, hydrated crystals which contain varying amount of water are obtained. The amount of water of crystallisation depends upon the temperature of crystallisation. For example, when crystallised above 42.5°C the crystals of the composition $Ce(SO_4)_2 . H_2O$ are obtained, between 42.5 – 43°C $Ce(SO_4)_2 . 8H_2O$ and below 3°C $Ce(SO_4)_2 . 12H_2O$ is obtained.

When aqueous solution of ceric sulphate is diluted, it gets hydrolysed and sparingly soluble basic salts are precipitated.

Ceric sulphate is a good oxidising agent. It oxidises nitrite to nitrate, thiosulphate to tetrathionate:

$$2Ce(SO_4)_2 + KNO_2 + H_2O \longrightarrow KNO_3 + H_2SO_4 + Ce_2\,(SO_4)_3$$
$$2Ce(SO_4)_2 + 2Na_2S_2O_3 \longrightarrow Ce_2(SO_4)_3 + Na_2S_4O_6 + Na_2SO_4$$

A number of organic compounds like oxalic acid, hydrazine, methanol, formic acid, acetic acid, tartaric acid, hydroxylamine and anthracene are readily oxidised:

$$2Ce\,(SO_4)_2 + H_2C_2O_4 \longrightarrow Ce_2(SO_4)_3 + H_2SO_4 + 2CO_2$$
$$4Ce(SO_4)_2 + 4NH_2 . NH_2 \longrightarrow 2Ce_2\,(SO_4)_3 + 2(NH_4)_2\,SO_4 + 2N_2$$

### 5. Ceric Ammonium Sulphate, $Ce(SO_4)_2 \cdot 2(NH_4)_2SO_4 \cdot 2H_2O$ or $(NH_4)_4[Ce(SO_4)_4] \cdot 2H_2O$

This compound is also called *ammonium tetrasulphatocerate* (IV) and hence is represented as $(NH_4)_4[Ce(SO_4)_4] \cdot 2H_2O$. It is prepared by adding $(NH_4)_2SO_4$ to $Ce(SO_4)_2$ solution containing excess of dil. $H_2SO_4$ to prevent the hydrolysis, followed by crystallisation. This double salt is more soluble than $Ce(SO_4)_2$ and is, therefore, prepared for use as a volumetric reagent. Ionic equations showing the quantitative estimations of $C_2O_4^{2-}$, $NO_2^-$, $I^-$, $H_2O_2$, etc. made with this salt are given below:

$$C_2O_4^{2-} + 2Ce^{4+} \longrightarrow 2Ce^{3+} + 2CO_2$$

$$NO_2^- + 2Ce^{4+} + H_2O \longrightarrow 2Ce^{3+} + 2H^+ + NO_3^-$$

$$2I^- + 2Ce^{4+} \longrightarrow 2Ce^{3+} + I_2$$

$$H_2O_2 + 2Ce^{4+} \longrightarrow 2Ce^{3+} + 2H^+ + O_2$$

### 6. Basic Ceric Nitrate, $Ce(NO_3)_3(OH)$

Normal ceric nitrate is not known in solid state. Basic ceric nitrate is prepared by the reaction of ceric hydroxide with concentrated nitric acid. The solution on evaporation gives solid basic ceric nitrate:

$$Ce(OH)_4 + 3HNO_3 \longrightarrow Ce(NO_3)_3(OH) + 3H_2O$$

It is a yellow crystalline solid. When dissolved in water it gives acidic solution due to hydrolysis. When $HNO_3$ or alkali metal nitrates are added to this yellow solution, the colour deepens and a red crystalline salt of the composition $M^I_2[Ce(NO_3)_6]$ separates out. With nitrates of bivalent metals, salts of $M^{II}[Ce(NO_3)_6] \cdot 8H_2O$ type are obtained. These double salts contain $[Ce(NO_3)_6]^{2-}$ ion. Thus, its well known double salt ceric ammonium nitrate is better named as ammonium hexanitratocerate (IV), $(NH_4)_2[Ce(NO_3)_6]$.

## UNIVERSITY QUESTIONS

1. f-block elements are called inner transition elements. Why? *(Pune 2010)*
2. Lanthanides and actinides are known as 'inner transiton elements'. Explain. *(Mumbai 2010)*
3. Explain the magnetic properties of lanthanides. *(Ranchi 2010)*
4. (*i*) Explain the oxidation states of lanthanons on the basis of their electronic configuration.
   (*ii*) Explain 'lanthanon contraction' and any two of its consequences. *(Mumbai 2010)*
5. (*a*) What are lanthanides? Why they are called so?
   (*b*) What are consquences of lanthanides contraction? *(Ranchi 2010)*
6. Explain how is the observed oxidation state of lanthanides related to their electronic configuration? *(Pd. Ravishankar Shukla 2010)*
7. (*a*) What is lanthanide contraction? Discuss the cause of the contraction.
   (*b*) What is Ziegler-Natta catalyst? How the catalyst helps in the stereo-gular polymerization of alkene?
   (*c*) Why +3 oxidation state of lanthanides is more stable? *(Himachal Pradesh 2010)*
8. What is lanthanide contraction. Explain. *(Pune 2010)*

**9.** Write descriptive notes on: (*i*) Magnetic moments of tripositive ions of lanthanides (*ii*) Oxidation states of lanthanides (*iii*) Colour and absorption spectra of lanthanides. *(Rajasthan 2010)*

**10.** Explain why maximum paramagnetic moment is observed in $Dy^{3+}$ and $Ho^{3+}$ and not in $Gd^{3+}$ ion. *(Mumbai 2010)*

**11.** (*i*) What is lanthanide contraction? What are its consequences?
(*ii*) Explain any one method for separation of lanthanides. *(Andhra 2010)*

**12.** (A) (*i*) What are lanthanides? Discuss its position in periodic table.
(*ii*) Describe ion exchange method for the separation of lanthanides.
(B) Discuss the complex formation tendency of lanthanides.
(C) Discuss any two consequences of lanthanide contraction.
(D) Discuss oxidation states of lanthanides.
(E) Write an electronic configuration of lanthanides.
*(Nagpur 2010, Rajsthan 2010)*

**13.** Choose the correct answer: The most common oxidation state of lanthanides is (*a*) +2 (*b*) +4 (*c*) +3 (*d*) +1 *(Barktullah 2010)*

**14.** What is lanthanide contraction? Explain. *(Barktullah 2010, Lucknow 2010)*

**15.** How will you separate lanthanides by ion exchange method.
*(A.P. University 2010)*

**16.** Why lanthanides do not form oxocations? *(GND Amritsar 2010)*

**17.** Explain the following: (*i*) $Ln^{3+}$ ion is more stable than $Ln^{2+}$ (*ii*) 4*f* series of elements are called lanthanides, although there is no 4*f* electron in case of $La_{57}$. *(Kanpur 2010)*

**18.** (*i*) Explain why 4*f* electrons do not take part in bonding?
(*ii*) What are lanthanides? Explain the ion exchange method for the separation of lanthanides.
(*iii*) Explain the magnetic properties of lanthanides. *(Madras, April 2011)*

**19.** Promethium and its compounds do not occur in nature. Explain why.
*(Rajasthan 2010)*

**20.** (*a*) Discuss lanthanides with reference to : (*i*) Electronic configuration (*ii*) Oxidation states
(*b*) What is lanthanide contraction? How does it affect the basicity of lanthanides. *(Nagpur 2012)*

**21.** (*a*) Give two examples of lanthanide ions which have the same number of unpaired *f*-electrons, and the same colour
(*b*) Which is thermally more stable and why? $La(NO_3)_3$ and $Tm(NO_3)_3$.
*(Kurukhetra 2012)*

**22.** What are lanthanides? Discuss their following properties: (*i*) Oxidation states (*ii*) Magnetic properties (*iii*) Formation of coloured ions.
*(Ch. Charan Singh, Meerut 2012)*

**23.** (*i*) Explain the following:
(*a*) +3 oxidation state is the characteristic oxidation state of lanthanides, though their atoms contain only 2($6s^2$) outermost electrons
(*b*) La, Gd and Lu show +3 oxidation state whereas other lanthanides can show +2 and + 4 oxidation states. *(Kurukshetra 2012)*

**24.** (*i*) Explain why Zr/Hf and Mo/W occur in the form of pair.
(*ii*) What is lanthamide contraction and what are its consequences?
(*iii*) How are lanthanide ions separated from each other by ion exchange method. *(Lucknow 2013)*

# 10 CHAPTER Chemistry of Actinides (5f-Block Elements)

## What are Actinides?

The 14 elements from $Th_{90}$ to $Lr_{103}$ are called *actinides* or actinodes. These elements are called actinides, since these elements are placed after actinum ($Ac_{89}$) in group 3 and period 7 of the periodic table, and show close similarities with $Ac_{89}$.

Since actinides are present in 7th period ($n$ = 7) of the periodic table, these elements have seven shells in their electronic configuration.

Thus the last electron in these elements enters (7–2) *f* or 5*f* sub-shell. Thus actinides are also called **5f block elements,** *i.e.* actinides are a series of 14 elements ($Th_{90}$ to $Lr_{103}$) in which the last electron goes to 5*f* sub-shell or to *f* sub-shell of *5th shell*.

Since *f*-sub-shell can accommodate 14 electrons, actinide series of elements has 14 elements.

It has been observed that all the 15 elements from $Ac_{89}$ to $Lr_{103}$ have similar chemical and physical properties. Hence the study of actinides consists of the study of all the 15 elements from $Ac_{89}$ to $Lr_{103}$ together. Just as $La_{57}$ is the prototype of lanthanides, $Ac_{89}$ is the prototype of actinides.

## Position of Actinides in the Periodic Table

The study of the position of the actinides in the periodic table can be made under the following two heads:

(*i*) **Prior to the discovery of the trans-uranium elements** (*i.e.* before 1940). Prior to the discovery of trans-uranium elements (trans-uranium elements are the elements lying beyond uranium in the periodic table), the naturally occurring heaviest known elements namely $Th_{90}$, $Pa_{91}$, and $U_{92}$, were placed below $Hf_{72}$, $Ta_{73}$ and $W_{74}$ in IVB, VB and VIB groups of the periodic table, because these elements showed +4, +5 and +6 oxidation states and resembled Hf, Ta and W respectively in many of their properties. The then undiscovered trans-uranium elements with atomic numbers 93 to 100 were thus expected to occupy the positions in the periodic table below $Re_{75}$, $Os_{76}$, $Ir_{77}$, $Pt_{78}$, $Au_{79}$, $Hg_{80}$, $TI_{81}$ and $Pb_{82}$ respectively as shown below:

| *IVB(4)* | *VB(5)* | *VIB (6)* | *VIIB(7)* | *VIII(8)* | | | *IB(11)* | *IIB(12)* | *IIIA(13)* | *IVA(14)* |
|---|---|---|---|---|---|---|---|---|---|---|
| — | — | — | — | — | — | — | — | — | — | — |
| $Hf_{72}$ | $Ta_{73}$ | $W_{74}$ | $Re_{75}$ | $Os_{76}$ | $Ir_{77}$ | $Pt_{78}$ | $Au_{79}$ | $Hg_{80}$ | $Tl_{81}$ | $Pb_{82}$ |
| $Th_{90}$ | $Pa_{91}$ | $U_{92}$ | 93 | 94 | 95 | 96 | 97 | 98 | 99 | 100 |

(*ii*) **Following the discovery of the trans-uranium elements**. The discovery of the element neptunium ($Np_{93}$) came in 1940 and this discovery was followed shortly by the discovery of plutonium ($Pu_{94}$) in 1941. The tracer chemical experiments with $Np_{93}$ and $Pu_{94}$ showed that the chemical properties of these two elements very much resemble those of $U_{92}$ and not at all those of $Re_{75}$ and $Os_{76}$. On this basis in 1944, $U_{92}$, $Np_{93}$ and $Pu_{94}$ were placed below $W_{74}$ in group VI B. It was thought that the undiscovered elements with atomic numbers 95 and 96 should resemble $U_{92}$, $Np_{93}$ and $Pu_{94}$ in chemical properties. Hence all the five elements *viz.* $U_{92}$, $Np_{93}$, $Pu_{94}$ and the undiscovered two elements with atomic numbers 95 and 96 should be placed below $W_{74}$ in group VIB as shown below.

| *IVB(4)* | *VB(5)* | *VIB(6)* |
|---|---|---|
| — | — | — |
| $Hf_{72}$ | $Ta_{73}$ | $W_{74}$ |
| $Th_{90}$ | $Pa_{91}$ | $U_{92}$, $Np_{93}$, $Pu_{94}$, 95, 96 |

This assumption, however, proved to be wrong, since the experiments directed towards the discovery of elements with atomic numbers 95 and 96 on the pattern of discovery of $Np_{93}$ and $Pu_{94}$ failed. Later on, in the same year (1944) Seaborg suggested that all the elements having atomic number greater than 89 ($Ac_{89}$) constitute a second series of inner-transition elements. This series is similar to lanthanide series. The elements of this new series which have atomic number greater than 89 were called *actinides*. The elements of this series were placed below lanthanides out of the main body of the periodic table as shown below in Fig. 10.1.

| Group | 1 | 2 | 3 | 4 | 5 | 6 | 7 | 8 | 9 | 10 | 11 | 12 | 13 | 14 | 15 | 16 | 17 | 18 |
|---|---|---|---|---|---|---|---|---|---|---|---|---|---|---|---|---|---|---|
| 1 | 1 H | | | | | | | | | | | | | | | | | 2 He |
| 2 | 3 Li | 4 Be | | | | | | | | | | | 5 B | 6 C | 7 N | 8 O | 9 F | 10 Ne |
| 3 | 11 Na | 12 Mg | | | | | | | | | | | 13 Al | 14 Si | 15 P | 16 S | 17 Cl | 18 Ar |
| 4 | 19 K | 20 Ca | 21 Sc | 22 Ti | 23 V | 24 Cr | 25 Mn | 26 Fe | 27 Co | 28 Ni | 29 Cu | 30 Zn | 31 Ga | 32 Ge | 33 As | 34 Se | 35 Br | 36 Kr |
| 5 | 37 Rb | 38 Sr | 39 Y | 40 Zr | 41 Nb | 42 Mo | 43 Tc | 44 Ru | 45 Rh | 46 Pd | 47 Ag | 48 Cd | 49 In | 50 Sn | 51 Sb | 52 Te | 53 *l* | 54 Xe |
| 6 | 55 Cs | 56 Ba | 57-71 La* | 72 Hf | 73 Ta | 74 W | 75 Re | 76 Os | 77 Ir | 78 Pt | 79 Au | 80 Hg | 81 Tl | 82 Pb | 83 Bi | 84 Po | 85 At | 86 Rn |
| 7 | 87 Fr | 88 Ra | 89-103 Ac** | 104 Rf | 105 Db | 106 Sg | 107 Bh | 108 Hs | 109 Mt | 110 | 111 | 112 | (113) | (114) | (115) | (116) | (117) | (118) |

← *f*-block elements →

| | | | | | | | | | | | | | | |
|---|---|---|---|---|---|---|---|---|---|---|---|---|---|---|
| Lanthanides* | 58 Ce | 59 Pr | 60 Nd | 61 Pm | 62 Sm | 63 Eu | 64 Gd | 65 Tb | 66 Dy | 67 Ho | 68 Er | 69 Tm | 70 Yb | 71 Lu |
| Actinide** | 90 Th | 91 Pa | 92 U | 93 Np | 94 Pu | 95 Am | 96 Cm | 97 Bk | 98 Cf | 99 Es | 100 Fm | 101 Md | 102 No | 103 Lr |

**Fig. 10.1** Position of actinides in the periodic table

The position of actinides shown above was confirmed by the discovery of the elements from atomic numbers 95 to 103 in 1961 and by the electronic configuration of these elements.

## Sources of Actinides

Thorium is mostly obtained from monaxite sands while uranium is mostly obtained from pitchblende and carnotite. Actinium and protoactinium occur in all uranium ores as decay products of $^{235}U$. None of the elements heavier than uranium occur in nature.

## Properties of Actinides

### 1. Electronic Configurations

Electronic configurations of actinium ($Ac_{89}$) and fourteen actinides ($Th_{90}$ to $Lr_{103}$) are given in Table 10.1. Valence shell electronic configurations (VSEC) of these elements are also given in the same table.

**Table 10.1. Eelectronic configurations of actinium ($Ac_{89}$) and fourteen actinides ($Th_{90}$ to $Lr_{103}$)**

| *Name of the elements* | *Symbol (M)* | *At.No.(Z)* | *Electronic configurations* | *Valence-shell electronic configurations* |
|---|---|---|---|---|
| Actinium | Ac | 89 | $[Rn]_{86} 5f^{\circ} 6d^1 7s^2$ | $5f^{\circ} 6d^1 7s^2$ |
| Thorium | Th | 90 | $[Rn]_{86} 5f^{\circ} 6d^2 7s^2$ | $5f^{\circ} 6d^2 7s^2$ |
| Protactinium | Pa | 91 | $[Rn]_{86} 5f^2 6d^1 7s^2$ | $5f^2 6d^1 7s^2$ |
| Uranium | U | 92 | $[Rn]_{86} 5f^3 6d^1 7s^2$ | $5f^3 6d^1 7s^2$ |
| Neptunium | Np | 93 | $[Rn]_{86} 5f^4 6d^1 7s^2$ | $5f^4 6d^1 7s^2$ |
| Plutonium | Pu | 94 | $[Rn]_{86} 5f^6 6d^0 7s^2$ | $5f^6 6d^0 7s^2$ |
| Americium | Am | 95 | $[Rn]_{86} 5f^7 6d^0 7s^2$ | $5f^7 6d^0 7s^2$ |
| Curium | Cm | 96 | $[Rn]_{86} 5f^7 6d^1 7s^2$ | $5f^7 6d^1 7s^2$ |
| Berklium | Bk | 97 | $[Rn]_{86} 5f^9 6d^0 7s^2$ | $5f^9 6d^0 7s^2$ |
| Californium | Cf | 98 | $[Rn]_{86} 5f^{10} 6d^0 7s^2$ | $5f^{10} 6d^0 7s^2$ |
| Einsteinium | Es | 99 | $[Rn]_{86} 5f^{11} 6d^0 7s^2$ | $5f^{11} 6d^0 7s^2$ |
| Fermium | Fm | 100 | $[Rn]_{86} 5f^{12} 6d^0 7s^2$ | $5f^{12} 6d^0 7s^2$ |
| Mendelevium | Md | 101 | $[Rn]_{86} 5f^{13} 6d^0 7s^2$ | $5f^{13} 6d^0 7s^2$ |
| Nobelium | No | 102 | $[Rn]_{86} 5f^{14} 6d^0 7s^2$ | $5f^{14} 6d^0 7s^2$ |
| Lawrencium | Lr | 103 | $[Rn]_{86} 5f^{14} 6d^1 7s^2$ | $5f^{14} 6d^1 7s^2$ |

After the completion of 7s orbital in $Ra_{88}$ [$Ra_{88}$ = $[Rn]_{86}$ $7s^2$] in 7th period and group 2, the next electron (*i.e. 89th electron*) in $Ac_{89}$, according to Aufbau principle, must enter $5f$ orbitals to give $[Rn]_{86}$ $5f^1 6d^0 7s^2$ configuration to $Ac_{89}$, but in $Ac_{89}$ the energy of $6d$ orbital is lower than that of $5f$ orbitals as shown in Fig. 10.2. Consequently *89*th electron in $Ac_{89}$ enters $6d$ orbitals instead of $5f$ orbitals. Thus the electronic configuration of $Ac_{89}$ is $[Rn]_{86}$ $5f^0$ $6d^1$ $7s^2$ and not $[Rn]_{86}$ $5f^1 6d^0 7s^2$ as suggested by Aufbau principle.

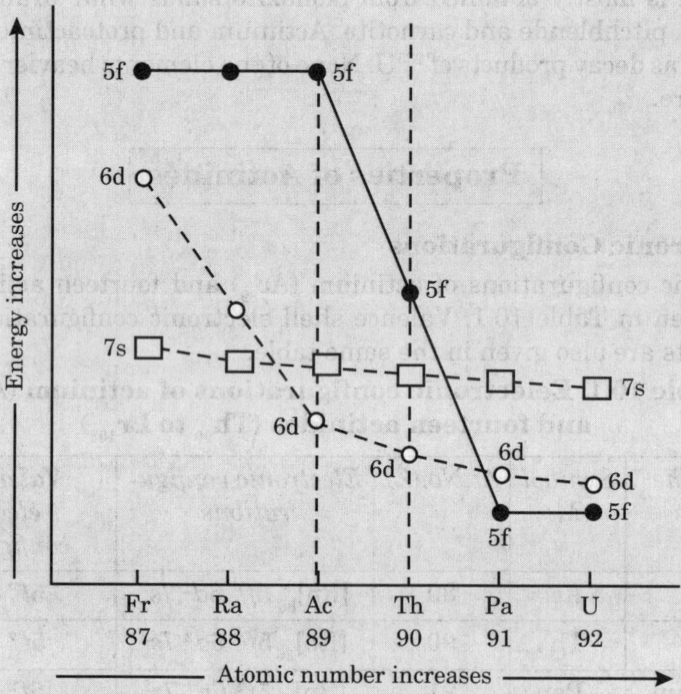


**Fig. 10.2** Variation of energy of $5f$, $6d$ and 7s orbitals in actinides with the increase in atomic number.

Due to greater spatial extension of $5f$ orbitals, the decrease in energy of $5f$ orbitals is so slow that even after an increase in nuclear charge by one unit in $Th_{90}$, the energy of $5f$ orbitals is higher than that of $6d$ orbitals, *i.e.* in $Th_{90}$ the energy of $6d$ orbitials is lower than that of $5f$ orbitals. Hence *90th* electron in $Th_{90}$ also goes to $6d$ orbitals. As a result, electronic configuration of $Th_{90}$ is $[Rn]_{86}$ $5f^0$ $6d^2$ $7s^2$.

Now as the nuclear charge increases by one more unit in $Pa_{91}$, the energy of $5f$ orbitals falls below $6d$ orbitals. Hence *91st* electron in $Pa_{91}$ occupies $5f$ orbitals. As a result, the electronic configuration of $Pa_{91}$ should be $[Rn]_{86}$ $5f^1 6d^2 7s^2$. In fact, the energies of $5f$, $6d$ and $7s$ orbitals are comparable due to which it is difficult to decide which of the two electronic configurations *viz*. $[Rn]_{86}$ $5f^1 6d^2 7s^2$ and $[Rn]_{86}$ $5f^2 6d^1 7s^2$ for $Pa_{91}$ is correct.

In the actinide elements from $Ac_{89}$ to $Np_{93}$ there are one or two electrons in $6d$ orbitals.

In the actinide elements succeeding $Np_{93}$, $5f$ orbitals become more stable. Consequently in these elements the extra electron goes to $5f$ orbitals and $6d$ orbital remains vacant. However, the filling of $5f$ orbitals is discontinued in $Cm_{96}$ because in this element the electron goes to $6d$ orbitals and not to $5f$ orbitals. This is because half-filled 5*f* orbitals present in $Cm_{96}$ are more stable. Thus the electronic configuration of $Cm_{96}$ is $[Rn]_{86}\ 5f^7 6d^1 7s^2$ and not $[Rn]_{86}\ 5f^8\ 6d^0\ 7s^2$.

In actinides from $Bk_{97}$ to $No_{102}$, 6*d* orbitals remain vacant and the extra electron goes to 5f orbitals. $Lr_{103}$ has one electron in 6*d* orbitals since 5*f* orbitals are completely filled. Table 10.1 shows that the complete electronic configurations of $Ac_{89}$ and 14 actinides ($Th_{90}$ to $Lr_{103}$) can be written as:

| *Shells:* | 1st | 2nd | 3rd | 4th | 5th (Antepenultimate) | 6th (Penultimate) | 7th (Ultimate or outer most) |
|---|---|---|---|---|---|---|---|
| *No. of electrons:* | 2 | 8 | 18 | 32 | 18 to 32 | 8 to 10 | 2 |
| or | 2 | 8 | 18 | 32 | $5s^2\,5p^6\,5d^{10}\,5f^{0-14}$ | $6s^2\,6p^6\,6d^{0-2}$ | $7s^2$ |

The above configuration can be written as

$$[Rn]_{86}\ 5f^{0-14}\ 6d^{0-2}\ 7s^2$$

Hence valence shell configuration is written as

$$5f^{0-14}\ 6d^{0-2}\ 7s^2$$

**2. Oxidation States**

Various oxidation states exhibited by actinides are given in Table 10.2

**Table 10.2. Oxidation states exhibited by actinides (Encircled oxidation states represent most stable oxidation states)**

| *Actinides* | *Oxidation States* | | | | | |
|---|---|---|---|---|---|---|
| $Ac_{89}$ | — | (+3) | — | — | — | — |
| $Th_{90}$ | +2 | — | (+4) | — | — | — |
| $Pa_{91}$ | — | +3 | +4 | (+5) | — | — |
| $U_{92}$ | — | +3 | +4 | +5 | (+6) | — |
| $Np_{93}$ | — | +3 | +4 | (+5) | +6 | +7 |
| $Pu_{94}$ | — | +3 | (+4) | +5 | +6 | +7 |
| $Am_{95}$ | +2 | (+3) | +4 | +5 | +6 | — |
| $Cm_{96}$ | — | (+3) | +4 | — | — | — |
| $Bk_{97}$ | — | (+3) | +4 | — | — | — |
| $Cf_{98}$ | +2 | (+3) | — | — | — | — |
| $Es_{99}$ | +2 | (+3) | — | — | — | — |
| $Fm_{100}$ | +2 | (+3) | — | — | — | — |
| $Md_{101}$ | +2 | (+3) | — | — | — | — |
| $No_{102}$ | (+2) | +3 | — | — | — | — |
| $Lr_{103}$ | — | (+3) | — | — | — | — |

**General discussion.** The following points may be noted.

(*i*) Higher oxidation state shown by Ac, Th, Pa and U (lighter actinide elements) is +3, +4, +5 and +6 respectively. These are the most stable oxidation states of these elements. Thus the highest and hence the most stable oxidation states of lighter actinides (Ac, Th, Pa and U) are +3, +4, +5 and +6 respectively. These oxidation states are achieved by the loss of all the electrons residing in 5*f*, 6*d* and 7*s* orbitals. $Ac^{3+}$, $Th^{4+}$, $Pa^{5+}$ and $U^{6+}$ ions have $[Rn]_{86}$ as their configuration which is a very stable configuration.

(*ii*) The maximum oxidation state first increases from Ac to Pu (Ac = +3, Th = +4, Pa = +5, U = +6, Np = +7, Pu +7) and then decreases in the succeeding elements *viz* Am, Cm and Es (Am = +6, Cm = +4, Es = +3)

(*iii*) In post uranium elements the highest O.S. is not the most stable oxidation state.

(*iv*) The removal of more than three electrons from actinide elements requires immense energy. Hence actinides in higher oxidation states form covalent bonds rather than ionic bonds.

**Discussion of Various Oxidation States.** Actinides show +2, +3, +4, +5, +6 and +7 oxidation states. Discussion of each of these states is given below.

**+2 oxidation state.** Th shows +2 oxidation state in its few compounds like $ThBr_2$, $ThI_2$, ThO, ThS etc.

Am and some heavy actinides (*e.g.* Cf, Es, Fm, Md and No) also show +2 oxidation state in addition to +3 oxidation state. Stability of +2 oxidation state of Am is due to the presence of stable $5f^7$ configuration in $Am^{2+}$ ion.

Stability of +2 oxidation state of No is due to the presence of stable $5f^{14}$ configuration in $No^{2+}$ ion.

**+3 oxidation state.** Excepting Th, all other actinide elements show +3 oxidation state.

Trichlorides and trifluorides of Ac, U, Np, Pu and Am are isomorphous. Ac, Pu and heavier actinides give oxides of $M_2O_3$ type which are isomorphous with $L_2O_3$ where L is a lanthanide element.

For Am and the following elements +3 oxidation state is the most stable oxidation state.

Lr in +3 oxidation state is stable because of the presence of stable $5f^{14}$ configuration in $Lr^{3+}$ ion.

**+4 oxidations state.** +4 oxidation state is shown by Th, Pa, U, Np, Pu, Am, Cm and Bk in their dioxides namely $ThO_2$, $PaO_2$, $UO_2$, $NpO_2$, $AmO_2$, $CmO_2$ and $BkO_2$. These oxides have fluorite structure. Tetrachlorides and tetrabromides are known only for Th,Pa, U and Np while tetraiodides of Th, Pa and U have been isolated.

Bk shows +4 oxidation state because the electronic configuration of $Bk^{4+}$ ion has stable half-filled 5*f* orbitals ($Bk^{4+} = 5f^7\ 6d^0\ 7s^0$).

**+5 oxidation state**. This oxidation state is very important for Pa and Np. $Pa^{5+}$ ion resembles very much $Ta^{5+}$. U, Pu and Am also exist in +5 oxidation state but these metals in +5 oxidation state are less characterised. The only pentahalides are those of $U^{5+}$.

Fluoro complex anions of Pa, U, Np and Pu of $[MF_6]^-$, $(MF_7]^{2-}$ and $[MF_8]^{3-}$ type are known (M = Pa, U, Np and Pu). These anions exist in the solid state.

Oxychlorides, $MOCl_3$ (M = Pa, U, Np) are also known.

The monovalent dioxo cation $[LO_2]^+$ is also known. This ion has linear structure both in solid and solution. This ion forms many complexes with anionic and neutral ligands.

**+6 Oxidation state.** U, Np, Pu and Am show +6 oxidation state in divalent dioxo cation, $[MO_2]^{2+}$(M = U, Np, Pu, Am). This cation is linear both in solid and solution. The simple molecular halide, $UO_2F_2$ has the linear O—U—O group with fluorine bridges. The O—U bond distance is 1.75 to 2.00 A°. The overall structure is a *flattened octahedron.* Although $[MO_2]^{2+}$ cation is linear in shape, it forms complexes with exceptional geometries, *e.g.* four, five and six coordinated complexes are given by this cation.

**+7 Oxidation states.** +7 oxidation state is shown only by Np and Pu. Electrolysis or ozone oxidation of $Np^{5+}$ or $Np^{6+}$ in NaOH gives a green solution of $NpO_5^{3-}$ which is slowly reduced to $Np^{6+}$ at 25°C [E° for $Np^{7+}/Np^{6+}$ = 0.58 volts in 1M NaOH). The existence of $Np^{7+}$ ion has been confirmed by Mossbauer spectra.

$Pu^{7+}$ ion is obtained by exposing a mixture of $PuO_2$ and $Li_2O$ in oxygen at 430°C when $Li_5PuO_6$ is formed. The chemistries of $Np^{7+}$ and $Pu^{7+}$ resemble those of $Re^{7+}$ and $Te^{7+}$.

## 3. Ionic Radii of $M^{3+}$ and $M^{4+}$ Ions.

The ionic radii of $M^{3+}$ and $M^{4+}$ ions derived from actinide elements are given below:

| | | | | | | | | | | |
|---|---|---|---|---|---|---|---|---|---|---|
| *$M^{3+}$ ions:* | $Ac^{3+}$ | $Th^{3+}$ | $Pa^{3+}$ | $U^{3+}$ | $Np^{3+}$ | $Pu^{3+}$ | $Am^{3+}$ | $Cm^{3+}$ | $Bu^{3+}$ | $Cf^{3+}$ |
| *Ionic radii of $M^{3+}$ ions (pm):* | 111 | 108 | 105 | 103 | 101 | 100 | 99 | 98 | 96 | 95 |
| | | values decrease → | | | | | | | | |

| | | | | | | | | |
|---|---|---|---|---|---|---|---|---|
| *$M^{4+}$ ions:* | $Ac^{4+}$ | $Th^{4+}$ | $Pa^{4+}$ | $U^{4+}$ | $Np^{4+}$ | $Pu^{4+}$ | $Am^{4+}$ | $Cm^{4+}$ |
| *Ionic radii of $M^{4+}$ ions (pm):* | – | 99 | 96 | 93 | 92 | 90 | 89 | 88 |
| | | values decrease → | | | | | | |

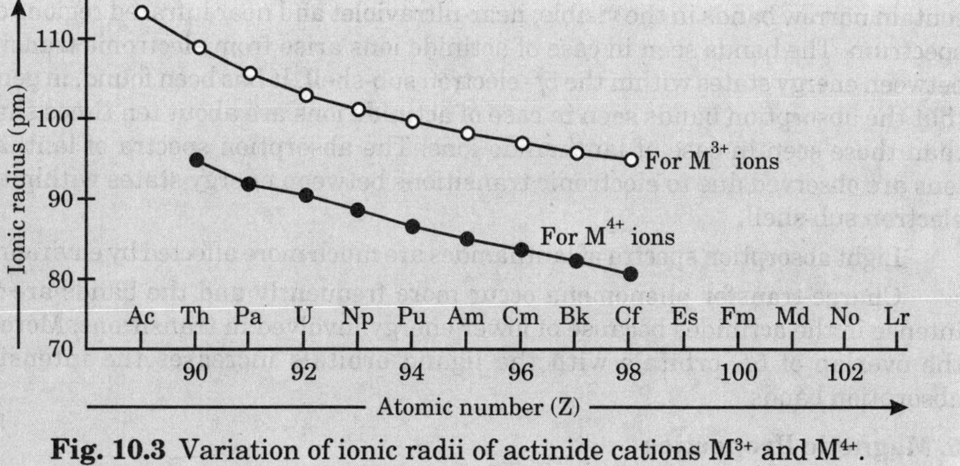


**Fig. 10.3** Variation of ionic radii of actinide cations $M^{3+}$ and $M^{4+}$.

It may be seen that there is a regular decrease in ionic radii of both types of cations ($M^{3+}$ and $M^{4+}$ cations) with increase in atomic from $Ac^{3+}$ to $Cf^{3+}$ and from $Ac^{4+}$ to $Cm^{4+}$. The steady decrease in ionic radii with increase in positive charge of the nucleus is called **actinide contraction.** This contraction is similar to lanthanide contraction. The decrease in ionic radii of $M^{3+}$ and $M^{4+}$ cations has been shown in Fig. 10.3.

**Cause of actinide contraction.** The actinide contraction is caused due to imperfect shielding of one 5*f* electron by another in the same shell. Thus as we move along the actinide series, the nuclear charge and the number of 5*f* electrons increase by one unit at each step. However, due to imperfect shielding (because the shape of *f* orbitals is very much diffused) the net effective nuclear charge increases which causes contraction in the size of the electron cloud.

However in actinide contraction there are bigger jumps in contraction between the consecutive elements as compared to lanthanides. This is due to lesser shielding of 5*f* electrons which results in stronger pull by the nucleus.

### 4. Colour of $M^{3+}$ and $M^{4+}$ Cations

Most of the tripositive and tetrapositive actinide cations ($M^{3+}$ and $M^{4+}$ cations) are coloured. The colours are given below.

$Ac^{3+}$ ($5f^{\circ}$) - colourless, $U^{3+}$ ($5f^3$) - red, $Np^{3+}$ ($5f^4$) - purple, $Pu^{3+}$($5f^5$) - violet, $Am^{3+}$ ($5f^6$) - pink, $Cm^{3+}$ ($5f^7$) - colourless, $Th^{4+}$ ($5f^{\circ}$) - colourless, $Pa^{4+}$ ($5f^1$) - colourless, $U^{4+}$ ($5f^2$) - green, $Np^{4+}$ ($5f^3$) - yellow-green, $Pu^{4+}$ ($5f^4$) - orange, $Am^{4+}$ ($5f^5$) - pink, $Cm^{4+}$ ($4f^6$) - pale yellow.

It may be seen that $M^{3+}$ and $M^{4+}$ cations having $5f^0$, $5f^1$ and $5f^7$ configuration are colourless while those containing $5f^2$, $5f^3$, $5f^4$, $5f^5$ and $5f^6$ configuration are coloured. In other words we can say that the cations having 0, 1 or 7 electrons in their 5*f* orbitals are colourless while those containing 2, 3, 4, 5 or 6 electrons in their 5f orbitals are coloured. Colours are produced when on electron jumps from one energy level to the other within 5*f* orbitals.

The absorption spectra of actinide ions, in aqueous solution and in crystals, contain narrow bands in the visible, near-ultraviolet and near infrared regions of the spectrum. The bands seen in case of actinide ions arise from electronic transitions between energy states within the 5*f*-electron sub-shell. It has been found, in general, that the absorption bands seen in case of actinide ions are about ten times sharper than those seen in case of lanthanide ions. The absorption spectra of lanthanide ions are observed due to electronic transitions between energy states within the 4*f* electron sub-shell.

Light absorption spectra of lanthanides are much more affected by environment.

Charge-transfer phenomena occur more frequently and the bands are more intense in the actinides because of lower energy involved in transitions. Moreover, the overlap of 5*f*- orbitals with the ligand orbitals increases the intensity of absorption bands.

### 5. Magnetic Properties

We know that 4*f* electrons in lanthanides are well shielded from external crystal ligand field whereas 5*f* orbitals in actinides, due to their greater spatial extension, are not completely shielded from the external ligand field. Thus $5f$ electrons in actinides are more exposed to crystal field and are less effectively shielded. As a result the orbital contribution to magnetic moment is quenched in actinides.

Magnetic moment values of tripositive actinide ions decrease from $U^{3+}$ ($U^{3+} = 5f^3, n = 3$) to give minimum value at $Pu^{3+}$ ($Pu^{3+} = 5f^5, n = 5$) and then increase to show a maximium value at $Cf^{3+}$ ($Cf^{3+} = 5f^9, n = 5$). Thereafter it decreases at $Es^{3+}$ ($Es^{3+} = 5f^{10}, n = 4$). This is shown in Fig. 10.4 which depicts the variation of magnetic moment values of tripositive actinide ions ($M^{3+}$ ions) with the number of unpaired electrons present in $5f$ orbitals of $M^{3+}$ ions.

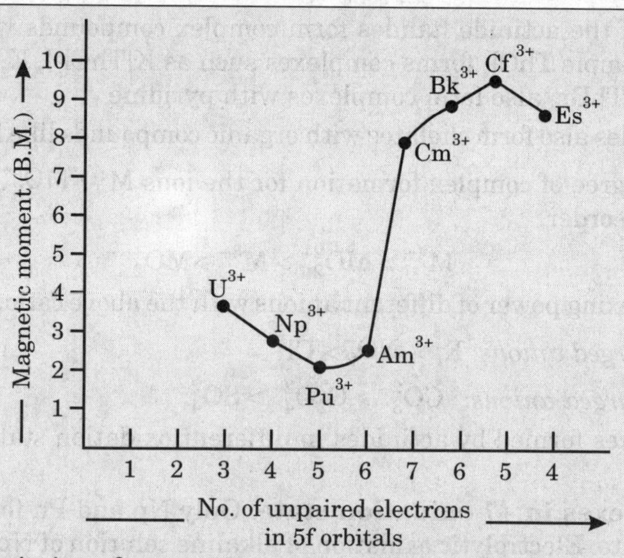


**Fig. 10.4** Variation of magnetic moment values of tripositive actinide ions ($M^{3+}$ ions) with the number of unpaired electrons ($n$) of $5f$ orbitals in $M^{3+}$ ions.

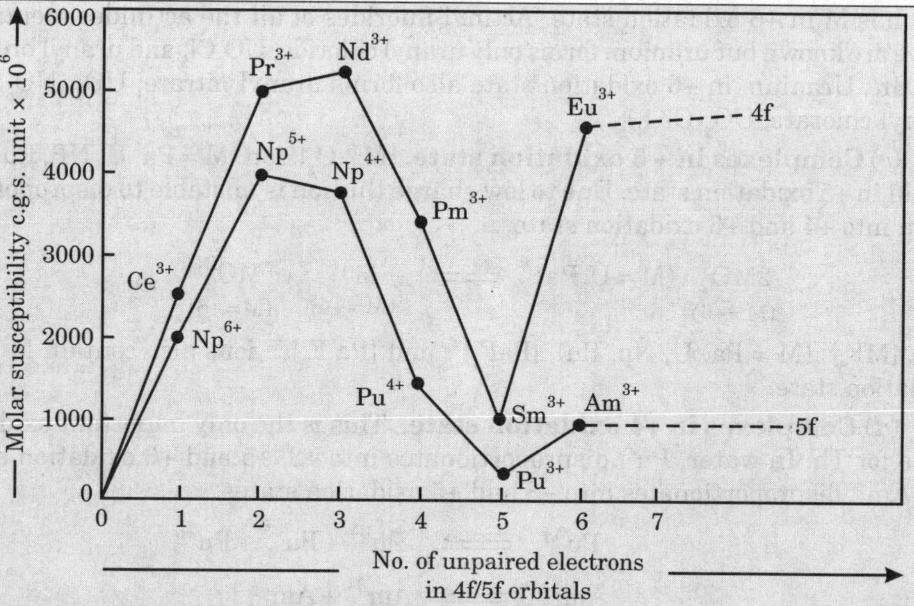


**Fig. 10.5** Magnetic susceptibilities of some lanthanide and actinide ions having the same number of unpaired electrons.

In Fig. 10.5 the molar susceptibility values of lanthamide and actinide ions having the same number of unpaired electrons is 4*f* / 5*f* orbitals have been plotted against the number of unpaired electrons in the ions. This figure shows that the molar susceptibility values of actinide ions are lower than those of lanthanide ions.

**6. Formation of Complexes**

Actinides have greater tendency to form complexes because of higher nuclear charge and small size of their atoms. The following points may be noted

(*i*) Small and highly charged $M^{4+}$ ions show the greatest tendency to form complexes, *e.g.* $Pu^{4+}$ ion forms very strong anionic complexes.

(*ii*) Most of the actinide halides form complex compounds with alkali metal halides. For example $ThCl_4$ forms complexes such as $K[ThCl_5]$, $K_2[ThCl_6]$ etc. with KCl. $ThCl_4$ and $ThBr_4$ also form complexes with pyridine

(*iii*) Actinides also form chelates with organic compounds like EDTA and oxine.

(*iv*) The degree of complex formation for the ions $M^{4+}$, $MO_2^{2+}$, $M^{3+}$ and $MO_2^+$ decreases in the order

$$M^{4+} > MO_2^{2+} > M^{3+} > MO_2^+$$

The complexing power of different anions with the above cations is in the order:

*Singly-charged anions*: $F^- > NO_2^- > Cl^-$

*Doubly-charged anions*: $CO_3^{2-} > C_2O_4^{2-} > SO_4^{2-}$

(*v*) Complexes formed by actinides in different oxidation states are discussed below:

(*a*) **Complexes in +7 oxidation state.** Only Np and Pu form complexes in +7 oxidation state. Electrolytic oxidation of alkaline solution of Np (VI) and Pu (VI) compounds gives dark-green complex ion, $[M^{+7}O_4(OH)_2]^{3-}$ in solution.

(*b*) **Complexes in +6 oxidation state.** Actinyl ion $[M^{+6}O_2]^{2+}$ (M = U, Np, Pu, Am) has M in +6 oxidation state. Actinyl fluorides of all the actinides mentioned above are known but uranium forms only uranyl chloride, $UO_2Cl_2$ and uranyl bromide $UO_2Br_2$. Uranium in +6 oxidation state also forms uranyl nitrate, $UO_2(NO_3)_2$ and uranyl chlorate, $UO_2(ClO_4)_2$.

(*c*) **Complexes in + 5 oxidation state.** $[M^{5+}O_2]^+$ ion (M = Pa, U, NP, Pu, Am) has M in +5 oxidation state. Due to low charge this ion is unstable to disproportionation into +4 and +6 oxidation states.

$$\underset{(M=+5)}{2MO_2^+} \ (M = U, Pa) \rightleftharpoons \underset{(M=+4)}{M^{4+}} + \underset{(M=+6)}{MO_2^{2+}}$$

$[MF_6]^-$ (M = Pa, U, Np, Pu), $[PaF_7]^{2-}$ and $[Pa F_8]^{3-}$ ions also contain M in +5 oxidation state.

(*d*) **Complexes in +4 oxidation state.** This is the only important oxidation state for Th. In water, $Pu^{4+}$ disproportionates into +3, +5 and +6 oxidation states but $Am^{4+}$ disproportionates into +3 and +5 oxidation states.

$$Pu^{4+} \rightleftharpoons Pu^{3+} + Pu^{+5} + Pu^{+6}$$

$$Am^{4+} \rightleftharpoons Am^{3+} + Am^{+5}$$

With $F^-$ ion, $M^{4+}$ ion gives various complex anions like $[MF_5]^-$, $[MF_6]^{2-}$, $[MF_7]^{3-}$, $[MF_8]^{4-}$ and $[M_6F_{31}]^{7-}$

Important oxygen-containing ligands which form complexes with lighter actinides in +4 oxidation state are $[M(CO_3)_5]^{6-}$, $[M(NO_3)_6]^{2-}$, $K_4[Th(C_2O_4)_4].4H_2O$ etc.

Complexes with S-donor ligands are generally less stable and more labile to hydrolysis than those with oxygen-donors.

(*e*) **Complexes in +3 oxidation state.** The actinide elements preceeding Am in +3 oxidation state are not stable. Hence the complexes of these elements in +3 oxidation state have a strong tendency to oxidise to higher oxidation state. As a result the complexes of these elements in +3 oxidation state can be studied in solution only. Separation of these complexes from solution is not possible, since they are disproportionated. Am and succeeding elements in +3 oxidation state are stable.

(*f*) **Complexes in +2 oxidation states.** Although +2 is the most stable oxidation state for No, yet due to availability of No only in traces, its complexation studies have not been possible.

## 7. Chemistry of Actinides and Their Compounds.

The detailed chemistry of actinides is concentrated mainly on U and to a lesser extent on Th. A few properties of actinides and their compounds are given below:

(*i*) Actinides are highly reactive metals especially in the finely-divided state.

(*ii*) All the known actinides are silvery-white metals which are highly electropositive.

(*iii*) They tarnish rapidly in air due to the formation of coating of oxides.

(*iv*) They react with most non-metals especially when heated.

(*v*) They resist the attack of alkalies and also are less reactive towards acids.

(*vi*) Compounds of actinides give trivalent ions in solution. Examples of such ions are $Ac^{3+}$, $U^{3+}$, $Np^{3+}$, $Pu^{3+}$, $Am^{3+}$ etc.

(*vii*) The ions of actinide elements are paramagnetic like those of lanthanides.

(*viii*) Melting points, boiling points and densities of some actinide elements are given below in Table 10.3.

**Table 10.3 Melting points, boiling points and densities of some actinides.**

| *Elements* | *Ac* | *Th* | *Pa* | *U* | *Np* | *Pu* | *Am* | *Cm* |
|---|---|---|---|---|---|---|---|---|
| Melting point (K) | 1323 | 2033 | 1503 | 1405 | 910 | 913 | 1273 | 1613 |
| Boiling point (K) | — | 4123 | — | 4086 | — | 3508 | 2880 | — |
| Density (g/cm$^3$) | — | 11.72 | 15.37 | 19.07 | 19.5 | 19.82 | 11.7 | 7.0 |

It may be seen that there is no regular trend in melting and boiling points of actinides. Densities of actinides increase from Th to Pu due to the progressive increase in atomic mass and decrease in atomic size.

(*ix*) All the actinides are radioactive and are *X*-ray emitter. Th and U have sufficiently long half periods ($^{232}Th = 1.39 \times 10^{10}$ years, $^{238}U = 4.50 \times 10^9$ years). That is the reason that these elements have been occurring in nature all the time. The half-life periods of other actinide elements are shorter than the age of the earth ($25 \times 10^9$ years). Hence transuranic elements donot occur in nature and have been prepared by artificial means using nuclear reactions. However, Np and Pu have been isolated in traces from uranium minerals.

(*x*) *Formation of oxides.* With boiling water and steam, oxides ($M_2O_3$) are formed on the surface of the metal and $H_2$ is liberated.

$$2M + 3H_2O \longrightarrow M_2O_3 + 3H_2$$

From Pu onwards, $M_2O_3$ oxides become increasingly stable.

Dioxides are known for all the actinides. Alkali or alkaline earth metal oxides form mixed oxides with actinide oxides of $MO_2$ type (M = Np, Pu). $ThO_2$ has the highest m.pt. ( = 3390°C). The oxides are basic in character.

(*xi*) *Formation of halides.* Halides formed by actinides range from $MX_6$ to $MX_2$.

(*a*) U, Np and Pu form hexahalides. All hexa halides are powerful oxidising agents and are extremely sensitive to moisture.

$$MX_6 + 2H_2O \longrightarrow MO_2X_2 + 4HX$$

(*b*) Pentahalides are not formed beyond Np.

(*c*) The tetrafluorides from $ThF_4$ to $PuF_4$ are produced by heating the dioxides in HF in presence of $H_2$ in order to prevent oxidation. The tetrafluorides from $AmF_4$ to $CfF_4$ are obtained by heating the corresponding trifluorides with $F_2$. The tetrachlorides from $TrCl_4$ to $NpCl_4$ are prepared by heating dioxides in $CCl_4$ while tetrabromides from Th $Br_4$ to $NpBr_4$ and tetraiodides from $ThI_4$ to $UI_4$ are obtained from the elements. Tetrafluorides are insoluble in $H_2O$ whereas tetrachlorides, tetrabromides and tetraiodides are rather hygroscopic and dissolve readily in water and other polar solvents.

(*d*) Trihalides of heavier actinides (Am to Cf) are obtained by heating sesquioxides or dioxides in HX. The triiodides of these actinides can be obtained by heating the metal with $I_2$ or HI. The ionic, high melting trifluorides are insoluble in $H_2O$. Other trihalides are all hygroscopic water soluble solids.

(*e*) Dihalides of Am (Am $X_2$) can be prepared as follows:

$$Am + HgX_2 (X = Cl, Br, I) \xrightarrow{400\text{–}500°C} AmX_2 + Hg$$

Dihalides of Am presumably owe their existence to their $f^7$ configuration.

## Separation of Neptunium (Np), Plutonium (Pu) and Amercium (Am) from Uranium (U) Present in Spent Reactor Fuel Material

In nuclear reactors the uranium fuel contains ${}^{235}_{92}U$ and ${}^{238}_{92}U$ isotopes of spent uranium. These isotopes of uranium capture neutron $\left({}^{1}_{0}n\right)$ followed by β-emission $\left({}^{0}_{-1}e\right)$ and give ${}^{237}_{93}Np$ and ${}^{239}_{94}Pu$ as shown below.

$${}^{235}_{92}U + {}^{1}_{0}n \longrightarrow {}^{236}_{92}U; \quad {}^{236}_{92}U + {}^{1}_{0}n \longrightarrow {}^{237}_{92}U$$

$$^{237}_{92}\text{U} \longrightarrow \boxed{^{237}_{93}\textbf{Np}} + ^{0}_{-1}e$$

$$^{238}_{92}\text{U} + ^{1}_{0}\text{n} \longrightarrow ^{239}_{92}\text{U}, \quad ^{239}_{92}\text{U} \longrightarrow ^{239}_{93}\text{Np} + ^{0}_{-1}\text{e}$$

$$^{239}_{93}\text{Np} \longrightarrow \boxed{^{239}_{94}\textbf{Pu}} + ^{0}_{-1}\text{e}$$

$^{241}_{95}$Am is formed by the following reaction:

$$^{239}_{94}\text{Pu} + ^{1}_{0}\text{n} \longrightarrow ^{240}_{94}\text{Pu} + ^{0}_{0}\gamma$$

$$^{240}_{94}\text{Pu} + ^{1}_{0}\text{n} \longrightarrow ^{241}_{94}\text{Pu} + ^{0}_{0}\gamma$$

$$^{241}_{94}\text{Pu} \longrightarrow \boxed{^{241}_{95}\textbf{Am}} + ^{0}_{-1}e$$

Above equations show that uranium is converted into Np and Pu and Pu is converted into Am. Thus the spent uranium fuel contains a mixture of unchanged U, Np, Pu, Am and numerous fission products ($FP_s$) which are formed simultaneously.

The separation of Np, Pu and Am from U present in spent uranium fuel (or spent reactor fuel material) can be done by several methods. Two of these methods are discussed below.

## 1. Fluoride Precipitation Method

The spent uranium fuel material is immersed in large cooling ponds for about 100 days so that the highly radioactive istopes with short half-lives such as $^{131}_{53}$I lose most of their activity and generation of heat subsides. The reactor fuel material is dissolved in 7M $HNO_3$ and the solution thus obtained is reduced by $SO_2$ and then treated with HF. This operation gives a solution of uranyl nitrate, $UO_2(NO_3)_2$ and a precipitate of fluorides of Np(IV), Pu(IV), Am (III) and FPs.

The precipitate containing fluorides is dissolved in $H_2SO_4$, treated with bromate ($B_rO_3^-$) (oxidising agent) at room temperature and then with HF so that Np (IV) fluoride is oxidised to Np (VI) fluoride which passes into solution while other fluorides remain as precipitate. The precipitate containing fluorides of Pu(IV) and Am(III) is treated with a very strong oxidising agent such as persulphate with $Ag^+$ ion catalyst and then with HF. In this reaction Pu (IV) fluoride is oxidised to Pu (VI) fluoride which goes into solution. The precipitate of Am (III) fluoride remains as precipitate with FPs. The precipitate of Am (III) fluoride is dissolved and Am is separated from FPs by ion exchange method.

Different steps of this method are given in flow sheet 1.

**Flow sheet 1.** *Different steps for the separation of Np, Pu and Am from U by fluoride precipitation method.*

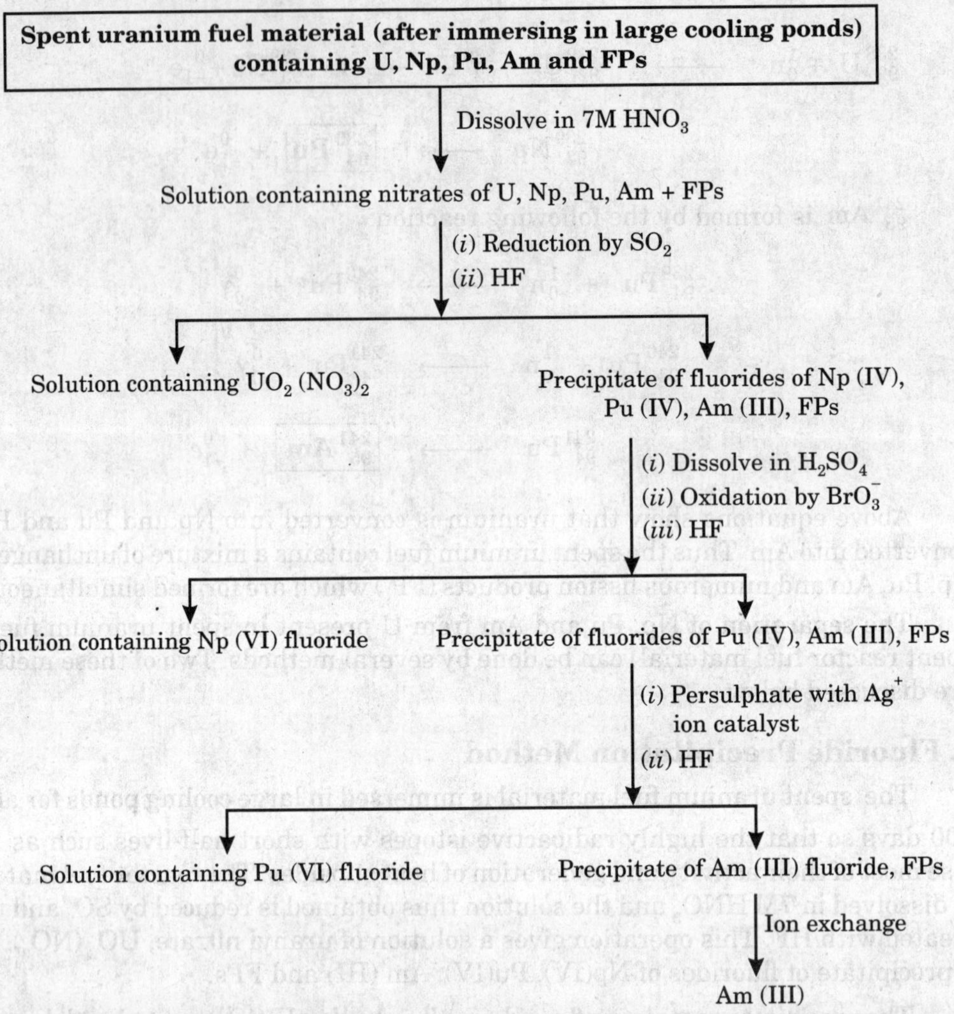


## 2. Solvent Extraction Method

In this method the spent uranium fuel is dissolved in 7M $HNO_3$ to get a solution which contains nitrates of U, Np, Pu, Am plus FPs. The solution containing nitrates is extracted from aqueous solution by means of 20% organic solvent, tributyl phosphate (TBP) dissolved in kerosene. $UO_2^{2+}$(O.S. of U = +6 ) and Pu $(NO_3)_4$ (O.S. of Pu = +4) go into TBP (organic solvent) and Np (IV), Am (III) and FPs remain in aqueous solution. TBP containing $UO_2^{2+}$ and $Pu(NO_3)_4$ is treated with $SO_2$ so that $Pu(NO_3)_4$ is reduced to $Pu(NO_3)_3$ which is extracted in water and $UO_2^{2+}$ remains in TBP. Pu $(NO_3)_3$ present in aqueous solution is oxidised to $Pu(NO_3)_4$ by means of $HNO_2$.

Acqueous $Pu(NO_3)_4$ thus obtained is treated with oxalic acid to get the precipitate of $Pu(C_2O_4)_2 . 6H_2O$ which, on being heated at 300°C, gives $PuO_2$. $UO_2^{2+}$ present in TBP is converted into $UO_2$ as shown in Flow sheet 2.

The aqueous solution containing Np(IV), Am(III) and FPs is treated with organic solvent to get Np (IV) in organic solvent and Au (III) remains unaffected in aqueous solution. Aqueous solution containing Am (III) and FPs can be used to get Am (III) by ion exchange method as shown in Flow sheet 2.

**Flow sheet 2.** *Different steps for the separation of Np, Pu and Am from U by solvent extraction method.*

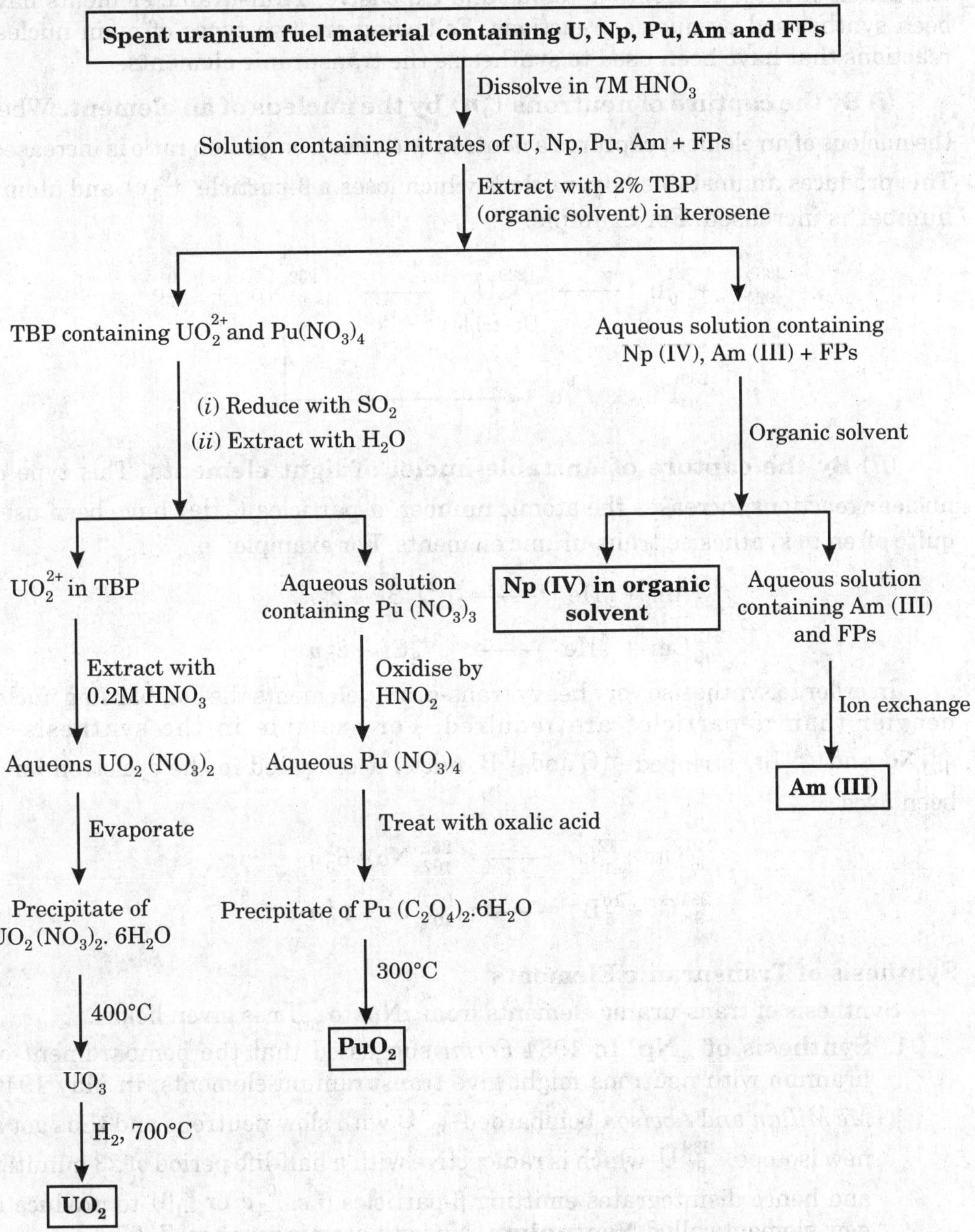

## Transuranium (or Transuranic) Elements

The actinide elements which lie beyond uranium, $U_{92}$ in the long form of periodic table are called transuranium (or transuranic) elements. These elements are also called **urenides.** Thus the actinide elements from $Np_{93}$ to $Lw_{103}$ are called transuranic elements. Np (Z = 93) and Pu (Z = 94) occur in minerals in small amounts. The extraction of these elements is tedius and expensive. Transuranic elements have been synthesised by nuclear reactions. Following are two types of main nuclear reactions that have been used to synthesise the transuranic elements.

(*i*) **By the capture of neutrons ($^1_0n$) by the nucleus of an element.** When the nucleus of an element captures a neutron $(^1_0n)$, neutron-proton ratio is increased. This produces an unstable host nucleus which loses a β-particle $(^{\ 0}_{-1}e)$ and atomic number is increased. For example:

$$^{238}_{92}U + ^1_0n \longrightarrow \underset{\text{Unstable}}{^{239}_{93}U} \longrightarrow ^{239}_{93}Np + ^{\ 0}_{-1}e$$

$$^{239}_{93}Np \longrightarrow ^{239}_{94}Pu + ^{\ 0}_{-1}e$$

(*ii*) **By the capture of unstable nuclei of light elements.** This type of nuclear reactions increases the atomic number. α-particles $(^4_2He)$ have been used quite often to synthesise trans-uranic elements. For example:

$$^{241}_{95}Am + ^4_2He \longrightarrow ^{243}_{97}Bk + 2^1_0n$$

$$^{242}_{96}Cm + ^4_2He \longrightarrow ^{244}_{98}Cf + 2^1_0n$$

In order to synthesise very heavy trans-uranic elements the bombarding nuclei heavier than α–particles are required. For example in the synthesis of $^{252}_{102}No$ and $^{257}_{103}Lr$, stripped $^{12}_{6}C$ and $^{10}_{5}B$ nuclei accelerated in the cyclotron have been used.

$$^{246}_{96}Cm + ^{12}_{6}C \longrightarrow ^{252}_{102}No + 6^1_0n$$

$$^{251}_{98}Cf + ^{10}_{5}B \longrightarrow ^{257}_{103}Lr + 4^1_0n$$

### Synthesis of Transuranic Elements

Synthesis of trans-uranic elements from $_{93}Np$ to $_{103}Lr$ is given below.

1. **Synthesis of $_{93}$Np**. In 1934 *Fermi* suggested that the bombardment of uranium with neutrons might give transuranium elements. In May 1940 *Mc Millan* and *Abelson* bombarded $^{238}_{92}U$ with slow neutrons and thus got a new isotope, $^{239}_{92}U$ which is radioactive with a half-life period of 23 minutes and hence disintegrates emitting β-particles (*i.e.* $^{\ 0}_{-1}e$ or $^{\ 0}_{-1}\beta$) to produce a new element called **Neptunium,** Np with atomic number, Z=93.

$$^{239}_{92}U + ^{1}_{0}n \longrightarrow \underset{\text{Unstable}}{^{239}_{92}U} + ^{0}_{0}\gamma \qquad (t_{1/2} = 23.5 \text{ minutes})$$

$$^{239}_{92}U + ^{0}_{-1}e \longrightarrow ^{239}_{93}Np \ (\textit{Neptunium}) \qquad (t_{1/2} = 23 \text{ minutes})$$

In all, 20 isotopes of neptunium have been produced with mass numbers ranging from 231 to 240 by bombarding uranium with (*i*) fast deuterons (up to 100 MeV energy) from a cyclotron (*ii*) fast neutrons and (*iii*) accelerated α-particles.

2. **Synthesis of $_{94}$Pu.** Later on it was found that $^{239}_{93}Np$ was also radioactive with a half-life period of 2.3 days and hence disintegrated emitting β-particles into another transuranium element now called **Plutonium,** Pu with Z= 94.

$$^{239}_{93}Np \longrightarrow ^{0}_{-1}e + ^{239}_{94}Pu \ (\textit{Plutonium}) \qquad (t_{1/2} = 2.3 \text{ days})$$

The isotope of plutonium is itself an α-emitter and has a half-life of 24,400 years. It is manufactured in large amounts in nuclear reactors and can be used for the production of nuclear power or atom bomb.

3. **Synthesis of $_{95}$Am and $_{96}$Cm.** In 1945, *Hamilton, Seaberg* and collaborators prepared two more transuranium elements *viz.* **Americium,** Am with Z = 95 and **Curium,** Cm with Z = 96 by the bombardment of $^{238}_{92}U$ and $^{239}_{94}Pu$ respectively with 40 MeV helium produced in the 60 inch Berkeley cyclotron.

$$^{238}_{92}U + ^{4}_{2}He \longrightarrow ^{241}_{94}Pu + ^{1}_{0}n$$

$$^{241}_{94}Pu \longrightarrow ^{0}_{-1}e + ^{241}_{95}Am \ (\textit{Americium}) \qquad (t_{1/2} = 14 \text{ years})$$

$$^{239}_{94}Pu + ^{4}_{2}He \longrightarrow ^{1}_{0}n + ^{242}_{96}Cm \ (\textit{Curium})$$

4. **Synthesis of $_{97}$Bk and $_{98}$Cf.** In 1950, *Thomson, Ghiorso* and *Seaborg* obtained **Berkelium.** Bk with Z = 97 and **Californium,** Cf with Z = 98 by bombarding $^{241}_{95}Am$ and $^{242}_{96}Cm$ respectively with 35 MeV helium produced in the same 60 inch cyclotron.

$$^{241}_{95}Am + ^{4}_{2}He \longrightarrow 2^{1}_{0}n + ^{243}_{97}Bk \ (\textit{Berkelium})$$

$$^{242}_{96}Cm + ^{4}_{2}He \longrightarrow 2^{1}_{0}n + ^{244}_{98}Cf \ (\textit{Californium})$$

5. **Synthesis of $_{99}$Es.** *Ghiorso, Rossi, Harvey* and *Thompson* under the leadership *Seaborg*, bombarded $^{239}_{92}U$ with accelerated nucleus of $^{14}_{7}N$ to get an isotope of **Einsteinium,** Es with Z = 99.

$$^{239}_{92}U + ^{14}_{7}N \longrightarrow 6^{1}_{0}n + ^{247}_{99}Es \ (\textit{Einsteinium})$$

6. **Synthesis of $_{100}$Fm.** *Ghiorso, Harvey, Thompson* and *Choppin*, under the direction of *Seaborg*, discovered the element, **Fermium ,** Fm with Z = 100 by bombarding $^{239}_{94}Pu$ with neutrons from an atomic reactor.

$$^{239}_{94}Pu + 15^{1}_{0}n \longrightarrow 6^{0}_{-1}e + ^{254}_{100}Fm \ (\textit{Fermium})$$

$^{250}_{100}Fm$ isotope can be prepared by bombarding $^{238}_{92}U$ with accelerated nucleus of $^{16}_{8}O$

$$^{238}_{92}U + ^{16}_{8}O \longrightarrow 4^{1}_{0}n + ^{250}_{100}Fm \text{ (\textit{Fermium})}$$

7. **Synthesis of $_{101}$Md.** In April 1955, *Seaborg, Ghiorso* and coworkers discovered **Mendelevium,** Md with Z = 101 by bombarding $^{253}_{99}Es$ with 48 MeV helium ions.

$$^{253}_{99}Es + ^{4}_{2}He \longrightarrow ^{1}_{0}n + ^{256}_{101}Md \text{ (\textit{Mendelevium})}$$

8. **Synthesis of $_{102}$No.** This element is an α-emitter. Very recently (in 1957) the *Noble Institute of Physics Stockholm* obtained the element **Nobelium,** No with Z = 102 by bombarding $^{246}_{96}Cm$ with accelerated $^{12}_{6}C$ ions. Russian scientists have proposed the name *Joliocium* for nobelium.

$$^{246}_{96}Cm + ^{12}_{6}C \longrightarrow 6^{1}_{0}n + ^{252}_{102}No \text{ (\textit{Nobelium})}$$

This element is an α-emitter and has a half-life period of 10 to 12 minutes.

9. **Synthesis of $_{103}$ Lr.** The element, **Lawrencium,** Lr with Z =103 has been prepared by bombarding $^{251}_{98}Cf$ with accelerated $^{10}_{5}B$ ions.

$$^{251}_{98}Cf + ^{10}_{5}B \longrightarrow 4^{1}_{0}n + ^{257}_{103}Lr \text{ (\textit{Lawrencium})}$$

## Trans-fermium Elements: Super Heavy Elements

In the long form of periodic table the elements placed beyond fermium, Fm ($Z$ = 100) are called **trans-fermium elements** or **super heavy elements.** In other words the elements having atomic number (Z) greater than 100 are called **super heavy elements.**

**IUPAC Nomenclature of Super Heavy Elements.**

The IUPAC name of a super heavy element can be determined by a procedure which consists of the following steps.

(*i*) We know that the atomic number of a super heavy element has three digits namely 0, 1, 2 ....8, 9.

(*ii*) Each of the three digits has been represented by a root and each of the roots is given an abbreviation as shown below:

| *Digit* | 0 | 1 | 2 | 3 | 4 | 5 | 6 | 7 | 8 | 9 |
|---|---|---|---|---|---|---|---|---|---|---|
| *Root for the digit* | nil | un | bi | tri | quad | pent | hex | sept | oct | enn |
| *Abbreviation for the root* | n | u | b | t | q | p | h | s | o | e |

(*iii*) The IUPAC name of the element has three roots corresponding to the three digits of the atomic number plus **ium.**

(*iv*) Symbol of the element has three abbreviations corresponding to the three roots.

On the basis of the above steps IUPAC names of super heavy elements with atomic number Z = 101 to 150 are given in Table 10.4.

**Table 10.4. IUPAC names of some super heavy elements with atomic number greater than 100.**

| *Atomic number* | *IUPAC name* | *Symbol* | *Present name* | *Symbol* |
|---|---|---|---|---|
| 101 | un-nil-unium | Unu | Mendelevium | Md |
| 102 | un-nil-bium | Unb | Nobelium | No |
| 103 | un-nil-trium | Unt | Lawrencium | Lr |
| 104 | un-nil-quadium | Unq | Rutherfordium | Rf |
| 105 | un-nil-pentium | Unp | Dubnium | Db |
| 106 | un-nil-hexium | Unh | Seaborgium | Sg |
| 107 | un-nil-septium | Uns | Bohrium | Bh |
| 108 | un-nil-octium | Uno | Hassium | Hs |
| 109 | un-nil-ennium | Une | Meitnerium | Mt |
| 110 | un-un-nilium | Uun | — | — |
| 111 | un-un-unium | Uuu | — | — |
| 112 | un-un-bium | Uub | Eka mercury | — |
| 113 | un-un-trium | Uut | — | — |
| 114 | un-un-quadium | Uuq | — | — |
| 115 | un-un-pentium | Uup | — | — |
| 116 | un-un-hexium | Uuh | — | — |
| 117 | un-un-septium | Uus | — | — |
| 118 | un-un-octium | Uuo | — | — |
| 119 | un-un-ennium | Uue | — | — |
| 120 | un-bi-nilium | Ubn | — | — |
| 130 | un-tri-nilium | Utn | — | — |
| 140 | un-quad-nilium | Uqn | — | — |
| 150 | un-pent-nilium | Upn | — | — |

**Solved Example.** *Write the IUPAC name and symbol of the element with atomic number (Z) 126.*

**Solution** : Atomic number equal to 126 has three digits which are 1, 2 and 6. For these digits the roots are **un** (for 1), **bi** (for 2) and **hex** (for 6) respectively. Thus the IUPAC name of the element with Z = 126 has the above three roots plus **ium**. Hence:

IUPAC name of the element = unbihexium

Since the abbreviations for un, bi and hex are *u*, *b* and *h* respectively,

Symbol of the element = ***ubh***

## Nuclear Fuels in India

Of all the actinides, only three actinides namely uranium, thorium and plutonium are very useful since these are used in nuclear reactor.

### Sources of Uranium, Thorium and Plutonium

The major source of uranium is the mineral **pitchblende** but this mineral is not found in India. **Carnolite** is another source of uranium. This mineral is found in Jaduguda mines in Bihar in India. Monazite sands of Travancore also contains uranium.

Natural uranium contains $^{235}U$ and $^{238}U$ isotopes. $^{235}U$ is present in 0.714% and the balance is $^{238}U$. $^{233}U$ is not found in nature. Out of these three isotopes of uranium, $^{233}U$ and $^{235}U$ are **fissile nuclei** and hence can be used directly as **fission fuels (fissile fuels)** in atomic reactors for carrying out fission process. On the other hand, $^{238}U$ isotope is a **non-fissile nucleus** and hence cannot be used as fission fuel directly in atomic reactors.

Thorium is abundantly available in India. Monazite sands contain 10–11% thorium as thoria. 80% of the word's thorium is located in India as monazite sands. $^{232}Th$ is a non-fissile nucleus and hence cannot be used directly as a fission fuel in atomic reactors.

Plutonium does not occur in nature. $^{239}Pu$ isotope is a fissile nucleus and hence can be used directly as a fission fuel (fissile fuel).

Above discussion shows that out of various isotopes of U, Th and Pu, $^{233}U$, $^{235}U$ and $^{239}Pu$ can be used directly as fission fuels in atomic reactors while $^{238}U$ and $^{232}Th$, being non-fissile nuclei, cannot be used directly as fission fuels (fossile fuels).

### Nuclear Reactions Taking Place in Atomic Reactor

We in India use natural uranium as a fission fuel in atomic reactors using heavy water or graphite as moderator. In atomic reactor the following nuclear reactions take place.

(*i*) $^{235}U$ which is a fissile nucleus undergoes fission.

$$\underset{\text{Fissile nucleus}}{{}^{235}_{92}U} + {}^{1}_{0}n \xrightarrow{\text{Fission}} {}^{141}_{56}Ba + {}^{92}_{36}Kr + 3\,{}^{1}_{0}n + \text{Energy}$$

(*ii*) $^{238}U$ which is a non-fissile nucleus is converted into fissile nucleus (fissile fuel), ${}^{239}_{94}Pu$ as shown below.

$$\underset{\text{Non-fissile nucleus}}{{}^{238}_{92}U} + {}^{1}_{0}n \longrightarrow {}^{239}_{92}U \longrightarrow {}^{239}_{93}Np + {}^{0}_{-1}e$$

$${}^{239}_{93}Np \longrightarrow \underset{\text{Fissile fuel}}{\boxed{{}^{239}_{94}Pu}} + {}^{0}_{-1}e$$

(*iii*) Since $^{233}U$, which is a fission fuel, is not available in nature, is produced in breeder reactor from $^{232}Th$ (non-fissile nucleus) by the following nuclear reaction.

$$\underset{\text{Non-fissile nucleus}}{{}^{232}_{90}\text{Th}} + {}^{1}_{0}\text{n} \longrightarrow {}^{233}_{90}\text{Th} \longrightarrow {}^{233}_{91}\text{Pa} + {}^{0}_{-1}e$$

$$^{233}_{91}\text{Pa} \longrightarrow \underset{\text{Fissile fuel}}{{}^{233}_{92}\text{U}} + {}^{0}_{-1}e$$

## Similarities Between Later Actinides and Later Lanthanides

In actinide series, the elements namely $Cm_{96}$, $Bk_{97}$, $Cf_{98}$, $Es_{99}$, $Fm_{100}$, $No_{102}$ and $Lr_{103}$ are called later actinides. The isotope, $^{242}_{96}cm$ was first isolated by the following nuclear reaction.

$$^{239}_{94}\text{Pu} + {}^{4}_{2}\text{He} \longrightarrow {}^{242}_{96}\text{Cm} + {}^{1}_{0}n$$

The later actinide elements show close resemblance with their corresponding lanthanides. The following points may be noted.

(*i*) As in case of lanthanides, +3 oxidation state is the most stable oxidation state for all the actinide elements namely $Am_{95}$, $Cm_{96}$, $Bk_{97}$, $Cf_{98}$, $Es_{99}$, $Fm_{100}$, $Md_{101}$ and $Lr_{103}$ . +3 oxidation state of $No_{102}$ is not stable. +2 is the most stable oxidation state for $No_{102}$ because of stable $5f^{14}$ configuration of $No^{2+}$ ion.

$Lr_{103}$ shows only +3 oxidation state. This oxidation state can neither be oxidised nor can be reduced because of the presence of fully filled 5*f* orbitals in $Lr^{3+}$ ion ($Lr^{3+} = 5f^{14}$). The oxidation states of later actinides are similar to those of lanthanides.

(*ii*) The spectral and magnetic properties of some compounds of iso-electronic pairs of ions of both the series (*e.g.* $CmF_3$ and $GdF_3$) are also similar ( $Cm^{3+} = 5f^7$, $Gd^{3+} = 4f^7$). This similarity is due to the fact that almost similar amount of energy is required for $5f \longrightarrow 6d$ electronic transition in later actinides and for $4f \longrightarrow 5d$ electronic transition in later lanthanides. Here it should be noted that earlier actinides do not resemble their corresponding earlier lanthanides because the amount of energy required for $5f \longrightarrow 6d$ electronic transition in earlier actinides is different from that required for $4f \longrightarrow 5d$ electronic transition in earlier lanthanides. Lesser amount of energy is needed for $5f \longrightarrow 6d$ electronic transition than that which is needed for $4f \longrightarrow 5d$ electronic transition.

(*iii*) In spite of the similarities (mentioned above) between the two series of elements, the actinides can be easily separated from lanthanides because actinides can form complexes more easily than lanthanides. When the ions of actinides and lanthanides are absorbed on cation exchange column, the actinide ions can be eluted with conc. HCl because of their strong tendency to form complexes. The actinide ions are separated from each other by the ion exchange with the help of nitrate solutions to elute them.

## Comparison between Lanthanides and Actinides

**(A) Similarities.** (*i*) In the atoms of the elements of both the series, the extra electron enters (*n*–2)*f* orbitals. (*ii*) The elements of both the series show +3 oxidation state. (*iii*) Like lanthanide contraction seen in case of lanthanides, we also have actinide contraction in actinides. Both the contractions are due to the poor shielding effect between the electrons residing in $(n-2)f$ orbitals. (*iv*) The absorption bands of the elements of both the series are so sharp that they appear to be almost line-like bands. Both the bands are produced due to the jump of an electron from one energy level to the other within $(n-2)f$ orbitals. (*v*) Elements of both the series have low electronegativities and are very reactive. (*vi*) The nitrates, perchlorates and sulphates of trivalent actinides as well as lanthanides are soluble while the hydroxides, fluorides and carbonates of both cations are insoluble. (*vii*) Most of the lanthanide and actinide cations are paramagnetic. (*viii*) In the atoms of the elements of both the series, three outermost shells are partly-filled while the remaining inner-shells are completely-filled.

(B) **Dissimilarities.** The elements of both the series also show some dissimilarities which are due to the difference in nature of 4*f* and 5*f* orbitals. The points showing dissimilarity are given below in a tabular form:

| *Lanthanides* | *Actinides* |
|---|---|
| (*i*) Due to large energy difference between 4*f* and 5*d* orbitats, the properties of lanthanides are fairly similar. | (*i*) Due to small energy difference between 5*f* and 6*d* orbitals, the properties of actinides are considerably different from each other. |
| (*ii*) Binding energies of 4*f* orbitals are higher. | (*ii*) Binding energies of 5*f* orbitals are lower. |
| (*iii*) The additional electron enters 4*f* orbitals. | (*iii*) The additional electron enters 5*f* orbitals. |
| (*iv*) These elements show only +2, +3 and +4 oxidation states. Maximum oxidation state shown by lanthanides is +4. *e.g.* Ce shows +4 oxidation state in $Ce^{4+}$. | (*iv*) These elements show a variety of oxidation states like +2, +3, +4, +5, +6 and +7. Higher oxidation states like +4, +5 and +6 are shown due to lower binding energy, *e.g.*, uranium shows +6 oxidation state in $UF_6$ and $UO_2^{2+}$. |
| (*v*) The mutual shielding effect between two electrons residing in 4*f* orbitals is greater. | (*v*) The mutual shielding effect between two electrons residing in 5*f* orbitals is poor. |
| (*vi*) Most of the tripositive lanthanide cations are colourless. | (*vi*) Most of the tripositive and tetrapositive c actinide cations are coloured. |
| (*vii*) The paramagnetic character of lanthanides can easily be explained. | (*vii*) The paramagnetic character of actinides can not be explained easily. |
| (*viii*) They do not form complex compounds easily. They do not form complexes with pi-bonding ligands like alkyl phosphines, thioethers etc. | (*viii*) They form complex compounds *comparatively* more easily. They do form complexes with pi-bonding ligands. |

(*Continued on the next page*)

| | |
|---|---|
| *(ix)* Except Pm, all the remaining lanthanides are non-radioactive. | *(ix)* All the actinides are radioactive. |
| *(x)* The compounds of lanthanides are less basic. | *(x)* The compounds of actinides are more basic. |
| *(xi)* Lanthanides do not form oxocations. | *(xi)* Some actinides form oxo-cations of $MO_2^+$(*e.g.* $UO_2^+$, $PuO_2^+$) and $MO_2^{2+}$ (*e.g.* $UO_2^{2+}$, $PuO_2^{2+}$) type. |
| *(xii)* The atoms of lanthanides have a total of six shells. Out of these, first three shells *viz*. 1*st*, 2*nd* and 3*rd* are completely filled while the remaining three shells namely 4*th*, 5*th* and 6*th* are partially-filled. | *(xii)* The atoms of actinides have a total of seven shells. Out of these, first tour shells *viz*, 1*st*, 2*nd*, 3*rd* and 4*th* are completely-filled while the remaining three shell. *viz*. 5*th*, 6*th* and 7*th* are partially-filled. |

## Thorium, $Th_{90}$

### Discovery

Berzelius isolated this element in 1882 from a Norwegian mineral and named it after the *Swedish God of Thunder*, *Thor*. Its radioactive nature was established by Madam Curie and professor Schmidt in 1898.

### Occurrence

Thorium constitutes 0.0012% of earths crust and is about three times more abundant than uranium. Important minerals of thorium are:

*(i)* *Thorianite* which contains 55–80% of $ThO_2$.

*(ii)* *Thorite* which is thorium silicate, $ThSiO_4 . nH_2O$ or $(ThO_2 . SiO_2) . nH_2O$.

*(iii)* *Monazite sand* which contains mainly a mixture of phosphates of lanthanides and of thorium along with the impurities of silicous matter, silica ($SiO_2$), zircon ($ZrSiO_4$) etc. Monazite sands are largely found in Brazil, India (on the sea-coast of Kerala state), Sri Lanka, USA and Australia. Indian monazite sand is of the top most grade in the world. Its approximate composition is as: $ThO_2$ = 9.4%, $CeO_2$ = 28.5%, $U_3O_8$ = 0.37% and $P_2O_5$ = 25.9%.

### Extraction of Thorium from Monazite Sand

Monazite sand is essentially a mixture of the phosphates of lanthanides ($RPO_4$) and thorium [$Th_3(PO_4)_4$]. Monazite sand also contains silicous impurities and the impurities of $SiO_2$, zircon ($ZrSiO_4$) etc. The extraction of Th from monazite sand involves the following steps:

1. **Concentration and preparation of thoria, $ThO_2$.** The silicous impurities present in monazite sand are removed either by washing or by magnetic separation method. The concentrated ore is then digested with conc. $H_2SO_4$ at 200°C so that a pasty mass is obtained. This pasty mass contains sulphates of lanthanides, Th etc. and $H_3PO_4$.

$$2RPO_4 + 3H_2SO_4 \longrightarrow \underbrace{R_2(SO_4)_3 + 2H_3PO_4}_{\text{Pasty mass}}$$

$$Th_3(PO_4)_4 + 6H_2SO_4 \longrightarrow \underbrace{3Th(SO_4)_2 + 4H_3PO_4}_{\text{Pasty mass}}$$

The pasty mass is separated by centrifugation and treated with cold water. The sulphates of lanthanides and of Th and $H_3PO_4$ pass into the solution and silica ($SiO_2$) and zircon ($ZrSiO_4$), being insoluble, get eliminated as residue. Due to the presence of $H_3PO_4$ in the solution, the latter (*i.e.,* solution) becomes sufficiently acidic. The solution is, therefore, partially neutralised by adding NaOH or ammonia or magnesia mixture and pH of the solution is so adjusted that the sulphates and $H_3PO_4$ react together and sulphates are converted into the precipitate of mono hydrogen phosphates.

$$R_2(SO_4)_3 + 3H_3PO_4 \longrightarrow \underset{\text{ppt.}}{R_2(HPO_4)_3} + 3H_2SO_4$$

$$Th(SO_4)_2 + 2H_3PO_4 \longrightarrow \underset{\text{ppt.}}{Th(HPO_4)_2} + 2H_2SO_4$$

The ppt. of monohydrogen phosphates obtained as above is dissolved in 6N $HNO_3$ and treated with $KIO_3$ in the presence of $H_2O_2$ when $Th(HPO_4)_2$ is converted into $Th(IO_3)_4$ and lanthanide monohydrogen phosphates remain in solution. The ppt. of $Th(IO_3)_4$ is dissolved in HCl and is precipited as thorium oxalate, $Th(C_2O_4)_2$. $Th(C_2O_4)_2$ is ignited to get $ThO_2$.

$$Th(C_2O_4)_2 + O_2 \xrightarrow{\text{Ignite}} ThO_2 + 4CO_2$$

**2. To get thorium from $ThO_2$.** (*a*) $ThO_2$ obtained as above is heated at 950°C with Ca in a steel bomb in presence of $CaCl_2$ which acts as a flux.

$$ThO_2 + 2Ca \xrightarrow{950°C} Th + 2CaO$$

(*b*) $ThO_2$ is heated with carbon in a current of $F_2$ or $Cl_2$ so that $ThO_2$ is converted into $ThF_4$ or $ThCl_4$.

$$ThO_2 + 2C + 2X_2 \ (X = F, Cl) \longrightarrow ThX_4 + 2CO$$

Now $ThF_4$ or $ThCl_4$ obtained as above is reduced by metals like Na, Ca, Mg etc. When Ca is used as a reducing agent, $ZnCl_2$ is used as a flux.

$$ThF_4 + 2Ca \longrightarrow Th + 2CaF_2$$

$$ThCl_4 + 4Na \longrightarrow Th + 4NaCl$$

## Extraction of Thorium by Electrolytic Method

In this method, a fused mixture of $ThF_4$–KCl–NaCl (electrolyte) is electrolysed in a graphite crucible using graphite electrodes. Electrolysis is carried out at a temperature of 800°C and a current of 70 amperes per kilogram of the electrolyte is used. Th is deposited as a powder on the cathode and scraped off from time to time. The powder is washed with dil. $HNO_3$, dried and converted into ignots. Thorium obtained by this method is 88% pure.

### Purification of Crude Thorium by Van-Arkel Process

Thorium obtained as above is impure (*crude*). It may be purified by Van-Arkel process in which impure (*crude*) thorium metal is heated with iodine in a sealed vessel so that Th reacts with $I_2$ to form $ThI_4$ while the impurities donot react with iodine. $ThI_4$ so formed is decomposed at 1100 –1200°C to get Th-metal in the pure state and $I_2$ is set free.

$$Th(crude) + 2I_2(g) \longrightarrow ThI_4(g) \xrightarrow{1100-1200°C} Th(pure) + 2I_2(g)$$

## Physical Properties

Thorium is a lustrous powder and has dark grey colour. Its specific gravity is 11.21 and melting and boiling points are 1827°C and 3500°C respectively. It exists in amorphous as well as crystalline states. At very low temperature it has super conductivity and at ordinary temperature its conductivity is very low. Thorium undergoes transformation from a face-centred cubic structure to a body-centred cubic structure at about 1400°C.

## Chemical Properties

**1. Radioactive nature.** Thorium is a radioactive metal having $t_{1/2} = 1.39 \times 10^{10}$ years. It yields a series of disintegration products.

**2. Combination with metals.** It forms many inter-metallic compounds like $ThMn_2$, $ThMn_{1.2}$, $Th_7Fe_3$, $Th_7Co_3$, $Th_7Ni_3$ etc.

**3. Action of heat.** When heated in air, it form $ThO_2$.

**4. Action of non-metals** (*a*) Thorium, when heated in $Cl_2$ or $Br_2$, forms $ThCl_4$ or $ThBr_4$ respectively.

(*b*) It combines with S to form $ThS_2$.

(*c*) It combines with $H_2$ vigorously, when heated, to form $ThH_4$.

(*d*) When heated in a current of $N_2$ at 800°C, $Th_3N_4$ is formed.

**5. Action of acids.** The metal is not attacked by HF. Hot and conc. HCl gives a black residue, $ThO_{1.3}Cl_{0.7}H_{1.3}$. It is rendered passive when placed in strong $HNO_3$. $H_2SO_4$ reacts with the metal slowly with the evolution of $SO_2$.

**6. Action of $H_2S$.** $ThS_2$ is obtained.

$$Th\ (red\ hot) + 2H_2S \longrightarrow ThS_2 + 2H_2$$

**7. Action of alkali solution.** Alkali solutions have no action on the metal.

### Uses of Thorium and its Compounds

(*i*) Thorium is used (*a*) in the preparation of fuel rods which are used in atomic reactors (*b*) in the preparation of magnesium flash lights (*c*) in electric cells, X-ray target and glow tube electrodes. In glow tubes, Th combines with the impurities of $O_2$ and $N_2$ and thus keeps the inert gas free from these impurities.

(*ii*) $ThO_2$ has various uses (See *"Uses of $ThO_2$"*).

(*iii*) Many thorium salts are used as medicines in the treatment of cancer.

(*iv*) Thorium alloys have been used as reducing agents.

## Compounds of Thorium

Thorium shows an oxidation state of +4 in its compounds. Important compounds of thorium are discussed below:

### 1. Thorium Dioxide or Thoria, $ThO_2$

**Preparation.** It may be obtained by heating thorium in an atmosphere of $O_2$ or by heating $Th(OH)_4$ or Th $(NO_3)_4$.

$$Th + O_2 \xrightarrow{\Delta} ThO_2$$
$$Th(OH)_4 \xrightarrow{\Delta} ThO_2 + 2H_2$$
$$Th(NO_3)_4 \xrightarrow{\Delta} ThO_2 + 4NO_2 + O_2$$

It can also be obtained from monazite sand.

**Properties.** It is white hygroscopic powder. Crystallised or strongly ingnited thoria is almost insoluble in acids but on evaporation with conc. $H_2SO_4$ it is dissolved in it to form $Th(SO_4)_2$.

$$ThO_2 + 2H_2SO_4 \longrightarrow Th(SO_4)_2 + 2H_2O$$

When heated with carbon in a current of $Cl_2$, $TlCl_4$ is precipitated.

$$ThO_2 + C + 2Cl_2 \longrightarrow ThCl_4 + CO_2$$

When heated with carbon and boron, it forms carbide and boride of thorium respectively.

**Uses.** Due to its resistance to heat and involatility it is used in the preparation of gas mantles. It is also used as a refractory material as well as a catalyst in many organic reactions. It is employed in welding operations and also in medicines. Since $ThO_2$ increases the life of the tungsten filament, it is used as an additive to tungsten used for metal filament lamps. Since $ThO_2$ has high absorptive power for X-rays, it is used in X-ray diagnostic work as a contrast medium in the examination of the digestive organs. It is also used in ceramic industry. Mixtures of $ThO_2$ and ceria have been used for search lights, head lights of motor cars etc.

### 2. Thorium Nitrate, Th $(NO_3)_4$

It is prepared by dissolving $Th(OH)_4$ in $NHO_3$.

$$Th(OH)_4 + 4HNO_3 \longrightarrow Th(NO_3)_4 + 4H_2O$$

It is a white deliquescent salt which is soluble in water. On heating it gives $ThO_2$ in a very fine state of subdivision. Due to this property thorium nitrate is used in mantle industry. Thorium nitrate mixed with Mg powder is used for flash powders. It is also used for the titration of fluorides, using methyl red as indicator.

### 3. Thorium Nitride. $Th_3N_4$

This compound is prepared by combining thorium powder with nitrogen.

$$3Th + 2N_2 \longrightarrow Th_3N_4$$

It is a yellowish-brown powder which dissolves in dilute acids to form $NH_3$

$$Th_3N_4 + 6H_2O \longrightarrow 4NH_3 + 3ThO_2$$

## 4. Thorium Tetrachloride, $ThCl_4$

Anhydrous $ThCl_4$ is prepared by passing a current of $Cl_2$ over a heated mixture of $ThO_2$ and coke.

$$ThO_2 + 2C + 2Cl_2 \longrightarrow ThCl_4 + 2CO$$

# Uranium, $U_{92}$

## Occurrence

Uranium is not found free in nature. It occurs in the combined state in a few scare minerals. Its most important minerals are:

(*i*) *Pitchblende* or *uranite* in which uranium is present as $U_3O_8$ or $UO_2.\ 2UO_3$ or uranium uranate, $U(UO_4)_2$. It contains 75–90% of the oxide and is found in Australia, Bohemia,Hungary, Norway, India, East Africa etc.

(*ii*) *Carnotite* $K_2O.\ 2UO_3.\ V_2O_5.\ 3H_2O$ or $2K(UO_2)VO_4.\ 3H_2O$ (*potassium uranyl ortho vanadate*) which contains both uranium and radium. It occurs in *Jaduguda mines* of Bihar.

(*iii*) *Autunite*, $K(UO_2)\ PO_4.\ 8H_2O$ (*potassium uranyl ortho-phosphate*). It is found in Congo.

In India at present the principal sources of uranium ore are the Jaduguda mines of Bihar. In 1964 rich deposits of high grade uranium ore have been found in Bhatin area of Bihar. Another important source of uranium in in India is the *monazite sand* of Travancore.

## Extraction of Uranium

Uranium can be extracted from pitchblende and carnotite ores. The method consists of the following three steps *viz*. A, B and C.

**A. Conversion of the ore into ammonium diuranate, $(NH_4)_2\ U_2O_7$.**

$(NH_4)_2\ U_2O_7$ can be obtained from both the ores as follows:

(*i*) *To get $(NH_4)_2U_2O_7$ from pitchblende*. Pitchblende can be converted into $(NH_4)_2\ U_2O_7$ by acid digestion process or by alkali digestion process as shown below:

(*a*) *To get $(NH_4)_2\ U_2O_7$ by acid digestion process*. First of all pitchblende is concentrated by gravity process to remove sand, clay etc. and then concentrated ore is roasted in excess of air whereby S, As and Sb are removed as their volatile oxides. Silver, if present in the mineral, is removed as insoluble AgCl by roasting the mineral at 800°C with NaCl. The roasted ore is digested with dil. $H_2SO_4$ in presence of $MnO_2$ (an oxidising agent) for about 24 hours when Ba, Pb, Ra etc., are precipitated as sulphates and uranium goes into solution as *uranyl sulphate,* $UO_2SO_4$.

$$\underset{\text{(In mineral)}}{2U_3O_8} + 6H_2SO_4 + \frac{1}{2}O_2 \longrightarrow \underset{\substack{\text{Uranyl sulphate} \\ \text{(Soluble)}}}{3UO_2SO_4} + 3H_2O$$

$UO_2SO_4$ present in solution is converted into *sodium uranyl carbonate,* $Na_4UO_2(CO_3)_3$ by the addition of $Na_2CO_3$.

$$UO_2SO_4 + 3Na_2CO_3 \longrightarrow \underset{\substack{\text{Sod. uranyl carbonate}\\ \text{(Solution)}}}{Na_4UO_2(CO_3)_3} + Na_2SO_4$$

The solution containing $Na_4UO_2(CO_3)_2$ is now neutralised with HCl in order to decompose it into *uranyl chloride*, $UO_2Cl_2$ and then saturated with $H_2S$ gas so that Pb, Cu etc. are precipitated as their sulphides. The solution of $UO_2Cl_2$ is now treated with excess of $NH_4OH$ whereby we get the precipitate of *ammonium diuranate,* $(NH_4)_2U_2O_7$.

$$Na_4UO_2(CO_3)_3 + 6HCl \longrightarrow \underset{\substack{\text{Uranyl chloride}\\ \text{(Solution)}}}{UO_2Cl_2} + 4NaCl + 3H_2O + 3CO_2$$

$$2UO_2Cl_2 + 6NH_4OH\ (excess) \xrightarrow{\text{Ignite}} \underset{\text{Ammonium diuranate (ppt.)}}{\boxed{(NH_4)_2U_2O_7}} + 4NH_4Cl + 3H_2O$$

(*b*) *To get $(NH_4)_2U_2O_7$ by alkali digestion process*. The roasted pitchblende ore is digested with $(Na_2CO_3 + NaHCO_3)$ mixture solution whereby $U_3O_8$ present in the mineral is converted into soluble $Na_4UO_2(CO_3)_3$. The solution containing $Na_4UO_2(CO_3)_3$ is treated with NaOH which precipitates uranium as *sodium diuranate*, $Na_2U_2O_7$.

$$\underset{\text{(In mineral)}}{U_3O_8} + 3Na_2CO_3 + 6NaHCO_3 + \frac{1}{2}O_2 \longrightarrow \underset{\text{(Soluble)}}{3Na_4UO_2(CO_3)_3} + 3H_2O$$

$$Na_4UO_2(CO_3)_3 + 6NaOH \longrightarrow \underset{\substack{\text{Sod.diuranate}\\ \text{(ppt.)}}}{Na_2U_2O_7} + 6Na_2CO_3 + 3H_2O$$

The precipitate of $Na_2U_2O_7$ is dissolved in dil. $H_2SO_4$ and the solution is treated with $NH_4OH$ to get the precipitate of $(NH_4)_2U_2O_7$.

$$\underset{\text{(ppt.)}}{Na_2U_2O_7} + H_2SO_4 \longrightarrow Na_2SO_4 + H_2U_2O_7$$

$$H_2U_2O_7 + 2NH_4OH \longrightarrow \underset{\substack{\text{Ammon. diuranate}\\ \text{(ppt.)}}}{\boxed{(NH_4)_2U_2O_7}} + 2H_2O$$

(*ii*) *To get $(NH_4)_2U_2O_7$ from carnotite*. In addition to uranium, carnotite also contains Ba, V, Fe, Al, Ra etc. From this ore uranium is obtained as a by-product. The ore is treated with conc. $HNO_3$ at 100°C. The heating is done by the steam. Most of the ore dissolves in $HNO_3$ and the solution thus obtained is neutralised by NaOH and then treated with $BaCl_2$ and $H_2SO_4$ whereby Ba and Ra are precipitated as their sulphate, and U and V are converted into *soluble* $UO_2SO_4$ and $VOSO_4$ respectively. This filtrate which also contains Fe and Al is treated with excess of boiling solution of $Na_2CO_3$ which precipitates Fe and Al (*Reject*), and $UO_2SO_4$ and $VOSO_4$ are converted into *soluble* $Na_4UO_2(CO_3)_3$ and $Na_3VO_4$ (*sod. ortho vanadate*) respectively.

$$UO_2SO_4 + 3Na_2CO_3 \longrightarrow \underset{\text{(Soluble)}}{Na_4UO_2(CO_3)_3} + Na_2SO_4$$

$$2VOSO_4 + 5Na_2CO_3 + \frac{1}{2}O_2 \longrightarrow \underset{\text{(Soluble)}}{2Na_3VO_4} + 3Na_2SO_4 + 5CO_2$$

The solution containing $Na_4UO_2(CO_3)_3$ and $Na_3VO_4$ is neutralised by $HNO_3$ and treated with NaOH. This treatment precipitates uranium as $Na_2U_2O_7$.

$$Na_4UO_2(CO_3)_3 + 6HNO_3 \longrightarrow UO_2(NO_3)_2 + 4NaNO_3 + 3H_2O + 3CO_2$$

$$2UO_2(NO_3)_2 + 6NaOH \longrightarrow \underset{\text{ppt.}}{Na_2U_2O_7} + 4NaNO_3 + 3H_2O$$

The precipitate of $Na_2U_2O_7$ is converted into $(NH_4)_2U_2O_7$ as in alkali digestion process.

$$Na_2U_2O_7 + H_2SO_4 \longrightarrow Na_2SO_4 + H_2U_2O_7$$

$$H_2U_2O_7 + 2NH_4OH \longrightarrow \underset{\text{ppt.}}{\boxed{(NH_4)_2U_2O_7}} + 2H_2O$$

$(NH_4)_2U_2O_7$ can also be obtained from the precipitate of $Na_2U_2O_7$ by the process in which the precipitate of $Na_2U_2O_7$ is digested with $(NH_4)_2CO_3$ so that $(NH_4)_2\ [UO_2(CO_3)_3]$ is formed in solution and the impurities are obtained as insoluble residue. Impurities are removed by filtration. The solution containing $(NH_4)_2\ [UO_2(CO_3)_3]$ is boiled and then cooled when lemon yellow crystals of $(NH_4)_2\ U_2O_7$ are obtained.

$$Na_2U_2O_7 \xrightarrow[(NH_4)_2CO_3]{\text{Digest with}} \underset{\text{Solution}}{(NH_4)_4\ [UO_2(CO_3)_3]}$$

$$\xrightarrow[\text{cool}]{\text{Boil and}} \boxed{(NH_4)_4U_2O_7} \quad \textit{(Lemon yellow crystals)}$$

**B. Conversion of $(NH_4)_2U_2O_7$ into $U_3O_8$.** When $(NH_4)_2\ U_2O_7$ obtained as above is strongly heated in air, $U_3O_8$ obtained.

$$3(NH_4)_2U_2O_7 \xrightarrow{\Delta} \boxed{2U_3O_8} + 6NH_3 + 3H_2O + O_2$$

**C. To get uranium metal from $U_3O_8$.** This can be done by the following two methods:

*(i)* $U_3O_8$ is heated with charcoal or in a current of $H_2$ or CO, when $U_3O_8$ is reduced to $UO_2$.

$$U_3O_8 + 2C \longrightarrow 3UO_2 + 2CO_2$$

$$U_3O_8 + 2H_2 \longrightarrow 3UO_2 + 2H_2O$$

$$U_3O_8 + 2CO \longrightarrow 3UO_2 + 2CO_2$$

Now $UO_2$ obtained as above can be converted into uranium metal by any of the methods shown in Fig. 10.6.

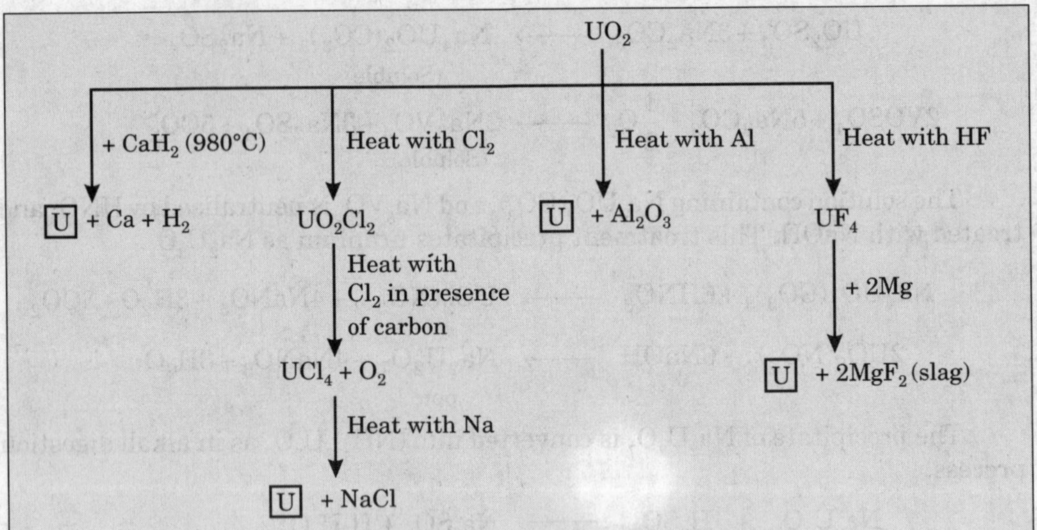


**Fig. 10.6** Conversion of $UO_2$ into U- metal.

(*ii*) $U_3O_8$ can be directly reduced to uranium metal by reducing it with carbon, aluminium, calcium or better by $CaH_2$ when both Ca and $H_2$ act as reducing agents.

$$U_3O_8 + 4C \longrightarrow \boxed{3U} + 4CO_2$$

$$3U_3O_8 + 16Al \longrightarrow \boxed{9U} + 8Al_2O_3 \ (\textit{Aluminothermic process})$$

$$U_3O_8 + 8Ca\ (\textit{vapours}) \longrightarrow \boxed{3U} + 8CaO\ (\textit{Burger's process})$$

## Physical Properties

(*i*) Uranium has a silvery white lustre.

(*ii*) It melts at 1132° C and boils at 3500°C. It is malleable and ductile. Its specific gravity is 19.05.

(*iii*) It has several isotopes. Natural uranium consists of three isotopes *viz.* $U^{238}$ (99.28%), $U^{235}$(0.71%) and $U^{234}$ (0.006%).

## Chemical Properties

Uranium is highly reactive.

1. **Action of air.** It burns in air at 170°C to form $U_3O_8$.
2. **Combination with non-metals** (*a*) It combines with $H_2$ at 600°C under 120 atmospheric pressure to form $UH_3$.

   (*b*) Uranium directly combines with halogens. It burns in $F_2$ to form green volatile $UF_4$. It catches fire in $Cl_2$ at 180°C and in $Br_2$ at 240°C forming the corresponding halides. It reacts with $I_2$ vapour at 260°C to form $UI_4$.
3. **Action of alkalies.** Alkalies have no action on uranium.
4. **Action of steam.** Uranium reacts with steam between 150–250°C to form a mixture of $UO_2$ and $UH_3$.

$$7U + 6H_2O\ (\textit{steam}) \xrightarrow{150\text{–}250°C} 3UO_2 + 4UH_3$$

5. **Formation of inter-metallic compounds.** Uranium forms various types of inter-metallic compounds. Important examples are $U_6Mn$, $U_6Ni$, $USn_3$ etc.

6. **Action of acids.** (*a*) Uranium dissolves in dil. HCl and $H_2SO_4$.

$$U + 2HCl \longrightarrow UCl_4 + 2H_2$$
$$U + 2H_2SO_4 \longrightarrow U(SO_4)_2 + 2H_2O$$

(b) With conc. $H_2SO_4$ and conc. $HNO_3$, $SO_2$ and $NO_2$ are evolved.

$$U + 2H_2SO_4 \longrightarrow USO_4 + 2H_2O + SO_2$$
$$3U + 2HNO_3 \longrightarrow U_3O_8 + 4NO + 4NO_2 + 4H_2O$$

## Uses of Uranium and its Compounds

(*i*) The fission of $U^{235}$ produces a huge amount of energy which is used for military as well as for peaceful purposes. This energy is also used in atom and hydrogen bombs.

(*ii*) Uranium is used in the manufacture of lamp-filaments, gas mantles, iron alloys, in photography, as electrodes in arc lamps and for the estimation of phosphates and arsenates.

(*iii*) Many uranium compounds are used : (*a*) as mordants for silk, wool and in calico printing (*b*) in the treatment of gout, diabetes and other diseases (*c*) in rubber and ceramic technology.

## Compounds of Uranium

Uranium forms compounds in +2, +3, +4, +5 and +6 oxidation state. Compounds of U (+4) and U (+6) are the most stable. Oxides of uranium, uranium trihydride, halides of uranium, uranyl salts and uranates are the important compounds of uranium. These compounds are discussed below.

## Oxides of Uranium

The known oxides are $UO_2$, $UO_3$ and $U_3O_8$. $UO_2$ is basic and forms uranous salts while $U_3O_8$ is an acidic oxide. $UO_3$ is amphoteric. All the three oxides readily dissolve in $HNO_3$, forming uranyl ($UO_2^{2+}$) salts. When heated with $S_2Cl_2$ at 450°C, all oxides from $UCl_4$. Now let us discuss these oxides one by one as follows:

### 1. Uranous Oxide or Uranium Oxide, $UO_2$(U = +4)

It is obtained by reducing higher oxides such as $UO_3$ or $U_3O_8$ with $H_2$, C or CO at elevated temperature.

$$UO_3 + CO \xrightarrow{500°C} UO_2 + CO_2\uparrow$$
$$U_3O_8 + 2CO \longrightarrow 3UO_2 + 2CO_2\uparrow$$
$$U_3O_8 + 2H_2 \longrightarrow 3UO_2 + 2H_2O$$

It is a brown strongly basic oxide, insoluble in water and alkalies. When heated in air it burns to given $U_3O_8$. It dissolves slowly in most of the acids, but $HNO_3$ oxidises it to $UO_2(NO_3)_2$. It reduces $AgNO_3$ to metallic silver.

$$2AgNO_3 + UO_2 \longrightarrow UO_2(NO_3)_2 + 2Ag$$

When $UO_2$ is heated with HF at 400°C, $UF_4$ is obtained.

$$UO_2 + 4HF \xrightarrow{400°C} UF_4 + 2H_2O$$

**2. Uranium Trioxide, $UO_3$ (U = +6)**

**Preparation.** It is obtained: (*i*) by heating uranyl ($UO_2^{2+}$) salts like $UO_2(NO_3)_2$ and $UO_2(CO_3)$

$$2UO_2(NO_3)_2 \xrightarrow{\Delta} 2UO_3 + 2NO_2 + O_2$$

$$UO_2(CO)_3 \xrightarrow{\Delta} UO_3 + CO_2$$

(*ii*) by heating ammonium diuranate, $(NH_4)_2U_2O_7$

$$(NH_4)_2U_2O_7 \xrightarrow{\Delta} 2UO_3 + H_2O + 2NH_3$$

**Properties.** (*i*) $UO_3$ is an orange yellow or brick red powder with specific gravity of 1.29.

(*ii*) On heating to 700°C it looses $O_2$ to form $U_3O_8$.

$$6UO_3 \xrightarrow{700°C} 2U_3O_8 + O_2$$

(*iii*) It is an amphoteric oxide and hence reacts with acids to form uranyl ($UO_2^{2+}$) salts and with metallic carbonates and alkalies to form uranates ($UO_4^{2-}$) and diuranates ($U_2O_7^{2-}$) respectively.

$$UO_3 + 2HCl \longrightarrow UO_2Cl_2 + H_2O$$

$$UO_3 + H_2SO_4 \longrightarrow UO_2SO_4 + H_2O$$

$$UO_3 + 2NaOH \longrightarrow Na_2UO_4 + H_2O$$

$$UO_3 + Na_2CO_3 \longrightarrow Na_2UO_4 + CO_2$$

**Uses.** $UO_3$ is used in the manufacture of costly glasses and pottery wares.

**3. Triuranium Octoxide or Urano-uranic Oxide, $U_3O_8$**

This oxide occurs in pitchblende and is the most stable oxide of uranium.

**Preparation.** $U_3O_8$ is obtained:

(*i*) by heating $UO_3$ at 700°C

$$6UO_3 \xrightarrow{400°C} 2U_3O_8 + O_2$$

(*ii*) by igniting $(NH_4)_2U_2O_7$

$$3(NH_4)_2U_2O_7 \xrightarrow{\text{Ignite}} 2U_3O_8 + 6NH_3 + 3H_2O + O_2$$

**Properties.** (*i*) $U_3O_8$ is a dark black solid with density equal to 8.3.

(*ii*) *Reduction.* (*a*) When heated in a current of CO or $H_2$, it is reduced to $UO_2$

$$U_2O_8 + 2CO \longrightarrow 3UO_2 + 2CO_2$$

$$U_3O_8 + 2H_2 \longrightarrow 3UO_2 + 2H_2O$$

(*b*) When $U_3O_8$ is reduced by means of carbon, Al (*aluminothermic process*), Mg, Ca etc, uranium is obtained.

$$U_3O_8 + 8C \xrightarrow[\text{electric furnance}]{\text{Heat in an}} 3U + 8CO$$

$$3U_3O_8 + 16Al \xrightarrow{\Delta} 9U + 8Al_2O_3$$

(*ii*) *Action of acids.* It is not acted upon by HCl or $H_2SO_4$ but HF gives $UF_4$ whereas $HNO_3$ gives $UO_2(NO_3)_2$. When $U_3O_8$ is treated with dil. $H_2SO_4$ in presence of $MnO_2$ (oxidising agent) for about 24 hours, $UO_2SO_4$ (soluble) is obtained.

$$2U_3O_8 + 6H_2SO_4 + \frac{1}{2}O_2 \text{ (From } MnO_2) \longrightarrow \underset{\text{soluble}}{3UO_2SO_4} + 3H_2O$$

(*iv*) When $U_3O_8$ is acted upon by ($Na_2CO_3$ + $NaHCO_3$) mixture solution, $U_3O_8$ is converted into soluble $Na_4(UO_2)(CO_3)_3$

$$U_3O_8 + 3Na_2CO_3 + 6NaHCO_3 + \frac{1}{2}O_2 \longrightarrow 3Na_4(UO_2)(CO_3)_3 + 3H_2O$$

**Uses.** $U_3O_8$ is used to some extent for colouring ceramic enamels.

**Structure.** $U_3O_8$ is regarded as a combination of two molecules of $UO_3$ (U = +6) and $UO_2$ (U = +4), *i.e.*, $U_3O_8 \equiv 2UO_3 . UO_2$. Thus $U_3O_8$ has uranium in +6 and +4 oxidation state.

## Uranium Trihydride, $UH_3$

When uranium is heated with $H_2$ at 500 – 600°C under 120 atm. pressure, $UH_3$ is obtained. It is a black powder. It is good conductor of electricity. Chemically $UH_3$ is reactive and inflames in air. When heated, $UH_3$ reacts with various compounds. Other compounds of uranium are obtained from $UH_3$ as shown in Fig. 10.7.

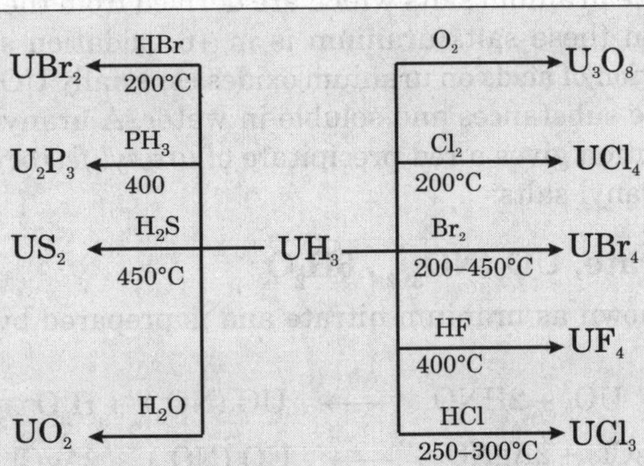


**Fig. 10.7** Formation of various urnaium compounds from $UH_3$.

## Halides of Uranium

The well known halides of uranium are $UF_3$, $UF_4$, $UF_6$, $UCl_3$, $UCl_4$, $UCl_6$, $UBr_3$, $UBr_4$, $UI_4$. Halides are important due to their role in the production of ***nuclear energy***.

### 1. Uranium Hexafluoride, $UF_6$

It is obtained by the direct ombination of elements in presence of a little of $Cl_2$ or by the action of $F_2$ on $UF_4$. It is a colourless crystalline volatile substance and melts 64.1°C. It attacks many metals and a large number of organic compounds. It is readily hydrolysed by water to uranyl fluoride, $UO_2F_2$ and is reduced to $UF_4$ by $H_2$

$$UF_6 + 2H_2O \longrightarrow UO_2F_2 + 4HF$$

$$UF_6 + H_2 \xrightarrow{\text{Reduction}} UF_4 + 2HF$$

$UF_6$ molecule is octahedral and has zero dipole moment.

It is used to separate $^{235}U$ from $^{238}U$ by fractional diffusion method. Thus pure $^{235}U$ nuclear fuel is produced. It is a powerful fluorinating agent.

## 2. Uranium Tetrachloride, $UCl_4$

It is the most important chloride of uranium.

**Preparation.** It is prepared by heating $UO_2$ in a current of $S_2Cl_2$ or when uranium burns in $Cl_2$. It may also be prepared by passing $Cl_2$ through a mixture of $U_3O_8$ and carbon at red heat.

**Properties** $UCl_4$ is the most stable of the chlorides of uranium. It is soluble in water, acetone and ethyl acetate but insoluble in benzene and chloroform. It is a pale green solid which melts at 567°C. It is highly deliquescent and fumes in air. When oxidised by air, it forms uranyl chloride, $UO_2Cl_2$.

$$UCl_4 + O_2 \longrightarrow UO_2Cl_2 + Cl_2$$

# Uranyl Salts ($UO_2^{2+}$)

These are the uranium salts which are derived from the *bivalent basic uranyl radical*, $UO_2^{2+}$. In these salts uranium is in +6 oxidation state. These salts are formed by the action of acids on uranium oxides especially $UO_3$. These are generally yellow crystalline substances and soluble in water. A uranyl salt with potassium ferrocyanide solution gives a red precipitate of *uranyl ferrocyanide*. Following are the important uranyl salts:

## 1. Uranyl Nitrate, $UO_2(NO_3)_2 \cdot 6H_2O$

It is also known as uranium nitrate and is prepared by any of the following methods:

$$UO_3 + 2HNO_3 \longrightarrow UO_2(NO_3)_2 + H_2O$$

$$\underset{\text{Uranyl chloride}}{UO_2Cl_2} + 2AgNO_3 \longrightarrow UO_2(NO_3)_2 + 2AgCl$$

This substance crystallises as a lemon yellow hexahydrate, $UO_2(NO_3)_2 \cdot 6H_2O$. It is soluble in water, alcohol and other. In contact with $H_2SO_4$ it is first converted into $UO_2(NO_3)_2 \cdot 3H_2O$ and then into $UO_2(NO_3)_2 \cdot 2H_2O$. When $UO_2(NO_3)_2 .3H_2O$ is heated at about 130° in a current of $CO_2$ having $HNO_3$ to check hydrolysis, $UO_2(NO_3)_2$ is obtained.

## 2. Uranyl Acetate. $UO_2(CH_3COO)_2 \cdot 2H_2O$

It is obtained by dissolving $UO_2(OH)_2$ or $UO_3$ in $CH_3COOH$.

$$UO_2(OH_2) + 2CH_3COOH \longrightarrow UO_2(CH_3COO)_2 + 2H_2O$$

$$UO_3 + 2CH_3COOH \longrightarrow UO_2(CH_3COO)_2 + H_2O$$

When dihydrate is heated to 110°C, it is converted into anhydrous salt. It gets decomposed at 275°C to $UO_3$. It is sparingly soluble in $H_2O$. This compound is of technical importance. It is used for making *Zn* or *Mg uranyl acetate* which is used as a qualitative reagent for the detection of sodium.

### 3. Uranyl Sulphate, $UO_2(SO_4) \cdot 3H_2O$

It is obtained by the action of conc. $H_2SO_4$ on hydrated $UO_3$.

$$UO_3 + H_2SO_4 \longrightarrow UO_2SO_4 + H_2O$$

**Properties.** (*i*) It is a yellow crystalline solid. It forms the hydrate, $UO_2SO_4 \cdot 3H_2O$

(*ii*) When the solution of $UO_2SO_4$ is treated with $Na_2CO_3$, sod. uranyl carbonate, $Na_4(UO_2)(CO_3)_3$ is obtained.

$$UO_2SO_4 + 3Na_2CO_3 \longrightarrow \underset{\text{soluble}}{Na_4(UO_2)(CO_3)_3} + Na_2SO_4$$

## Uranates ($UO_4^{2-}$) and Diuranates ($U_2O_7^{2-}$)

The hydrated uranium trioxide namely $UO_3 \cdot H_2O$ or $H_2UO_4$ and $2UO_3 \cdot H_2O$ or $H_2U_2O_7$ are called *uranic acid* and *diuranic acid* respectively.

The salts of these acids are called *uranates* ($UO_4^{2-}$) and *diuranates* ($U_2O_7^{2-}$) respectively. These are prepared as follows:

$$UO_3 + 2NaOH \longrightarrow Na_2UO_4 + H_2O$$

$$2UO_3 + 2NaOH \longrightarrow Na_2U_2O_7 + H_2O$$

$$UO_3 + Li_2CO_3 \longrightarrow Li_2UO_4 + CO_2\uparrow$$

$$2UO_3 + Li_2CO_3 \longrightarrow Li_2U_2O_7 + CO_2\uparrow$$

All uranates, even those of alkali metals and ammonium are insoluble in water and are orange-yellow in colour. Uranates ($UO_4^{2-}$) resemble chromates ($CrO_4^{2-}$) and diuranates ($U_2O_7^{2-}$) resemble dichromates ($Cr_2O_7^{2-}$).

Following are important diuranates:

### 1. Sodium Diuranate, $Na_2U_2O_7 \cdot 6H_2O$ (Also called sodium uranate)

**Preparation.** It is obtained by the action of NaOH on $UO_3$.

$$2UO_3 + 2NaOH \longrightarrow Na_2U_2O_7 + H_2O$$

**Properties.** (*i*) It is yellow in colour.

(*ii*) *Action of heat*. When heated, hexahydrate is converted into anhydrous salt.

(*iii*) *Action of $H_2SO_4$ and $NH_4OH$*. When $Na_2U_2O_7$ is dissolved in $H_2SO_4$ and the solution thus obtained is treated with $NH_4OH$, $(NH_4)_2U_2O_7$ is obtained.

$$Na_2U_2O_7 + H_2SO_4 \longrightarrow Na_2SO_4 + H_2U_2O_7$$

$$H_2U_2O_7 \xrightarrow{+\,NH_4OH} (NH_4)_2U_2O_7 + 2H_2O$$

**Uses.** $Na_2U_2O_7 \cdot 6H_2O$ is used in the manufacture of costly glasses and potery wares.

### 2. Ammonium Diuranate, $(NH_4)_2U_2O_7$

**Preparation.** (*i*) It is obtained as a yellow ppt. by adding $NH_4OH$ to a uranyl salt solution.

(*ii*) It may also be obtained from $Na_2U_2O_7$ as shown above:

**Properties.** When ignited, it is converted into $U_3O_8$.

$$3(NH_4)_2U_2O_7 \xrightarrow{\text{Ignite}} 2U_3O_8 + 6NH_3 + 3H_2O + O_2$$

**Uses.** The formation of this substance provides a convenient method for gravimetric estimation of uranium.

## Plutonium, $Pu_{94}$

### Introduction

This element was named after the planet, Pluto and was discovered in 1940 at the University of California by Seaborg, Mac Millan Kennedy and Wahl. It is a man-made element and has unique importance in chemical engineerig and in the history of atomic weapons. It has numerous isotopes, all of which are prepared synthetically. Out of these, fissionable $Pu^{239}$ isotope was discovered in 1941 by Kennedy and Seaborg.

### Occurrence

In nature, plutonium occurs as $Pu^{239}$ isotope in uranium ores like *Canadian pitchblende*. No other isotope has been detected to occur in nature. Other ores in which Pu is found are *colorado pitchblende, fergusonite* and *carnotite*.

### Production of Plutonium

This element is produced as ${}^{239}_{94}Pu$ isotope from naural uranium which contains a large proportion of ${}^{238}U$ and only 0.7% of ${}^{235}U$. When natural uranium (${}^{238}U + {}^{235}U$) is allowed to react with slow neutrons, it is ${}^{238}U$ isotopes which is acted upon by the neutrons and $Pu^{239}$ isotopes is produced.

$${}^{238}_{92}U + {}^{1}_{0}n\ (slow) \longrightarrow {}^{239}_{92}U + {}^{0}_{0}\gamma$$

$$\downarrow$$

$${}^{239}_{93}Np + {}_{-1}\beta^{0}$$

$$\downarrow \text{ 2 – 3 days}$$

$${}^{239}_{94}Pu + {}^{0}_{-1}\beta$$

### Production of Plutonium in India

Project Phoenix is a plant that has been set up at Trombay in Maharashtra. In this plant, nuclear waste is processed for the recovery of uranium and plutonium.

## Properties

(i) Metallic plutonium exists in six allotropic forms which are α-Pu (monoclinic, sp. grav. = 19.737), β-Pu (body-centred monoclinic, sp. grav. = 17.65), γ-Pu (face-centred orthorhombic, sp. grav. = 17.19), δ-Pu (face-centred cubic, sp. grav. = 15.90), η-Pu (face-centred tetragonal, sp. grav-15.90) and ε-Pu (body-centred cubic, sp. grav. = 16.48).

(ii) Plutonium forms intermetallic compounds with many metals. Examples of such compounds are $PuAl_2$, $PuAl_3$, $PuBe_3$, $PuCo_2$, $PuFe_2$, $PuMn_2$, $PuNi_2$, $PuNi_5$, $Pu_2Ni_7$ etc.

(iii) In aqueous solution, Pu exists in any of the four oxidation states *viz*. +3, +4, +5 and +6. This behaviour of Pu is similar to that of uranium and neptunium.

(iv) The aqueous solution of the ions like $Pu^{3+}$, $Pu^{4+}$, $PuO_2^{2+}$ etc., have characteristic colours, *e.g.* $Pu^{3+}$ is blue lavender, $Pu^{4+}$ is yellowish-brown, $PuO_2^{2+}$ is pink-orange. When these ions form complex compounds, their colours are changed and the aqueous ions formed are strongly paramagnetic.

**Uses**

(i) $^{239}Pu$ is used for fuelling atomic reactors. The heat produced as a result of radioactive decay is used for generating electricity.

(ii) It is used as an ingredient of atomic explosive weapons.

(iii) Finely-divided plutonium is used as a starting material for the synthesis of many important compounds of the metal.

## Compounds of Plutonium

### 1. Hydrides

Hydrides of Pu are obtained by the action of $H_2$ on Pu at 150–200°C. $PuH_2$ has fluorite structure while $PuH_3$ has hexagonal structure. Theremal decomposition of the hydrides gives Pu metal in a finely-divided state. This metal is used as a starting mateial for the preparation of other compounds of Pu.

### 2. Oxides

PuO, $Pu_2O_3$, $Pu_4O_7$ and $PuO_2$ are important oxides. Of these, $PuO_2$ is the most stable oxide. It is used as a starting material in the preparation of other compounds.

### 3. Halides

Exceptng $PuI_3$ and $PuF_6$, all other halides are prepared by the halogenation of $PuO_2$ at a temperature of 700°C. $PuF_3$ and $PuF_6$ are prepared by the following reactions.

$$2PuO_2 + H_2 + 6HF \xrightarrow{500°C} 2PuF_3 + 4H_2O$$

$$PuF_4 + F_2 \longrightarrow PuF_6$$

$PuF_6$ is quite stable.

$PuF_4$ is obtained by the addition of alkali fluoride to plutonium (IV) salts.

$PuI_3$ is obtained as follows:

$$2Pu + 6HI \longrightarrow 2PuI_3 + 3H_2$$

All the trihalides are appreciably volatile at moderate temperatures. They are hygroscopic except trifluoride.

## UNIVERSITY QUESTIONS

**1.** Write short note on the following (*a*) Actinide contraction (*b*) Electronic configuration of actinides. (*Ranchi 2010*)

**2.** What are actinides? Disucss the position of actinides in the periodic table and throw light on their electronic configuration.
(*Pd. Ravishankar Shukla 2010*)

**3.** (*a*) Actinides have greater tendency for complex formation than lanthanides. Explain with examples.
(*b*) Actinides exhibit oxidation state higher than +3.
(*Himachal Pradesh 2010, Kanpur 2010*)

**4.** All the actinides resemble one another very closely. Discuss. (*Pune 2010*)

**5.** Give a comparative study of lanthanides and actinides. (*Rajasthan 2010*)

**6.** (*i*) Explain actinides with reference to their electronic configuration and atomic and ionic radii.
(*ii*) Define and discuss the "actinide contraction". (*Nagpur 2010, 2012*)

**7.** Discuss the differences between lanthanides and actinides.
(*A.P. University 2010*)

**8.** Why is it difficult to interpret the paramagnetic behaviour of actinides?
(*G.N.D. Amritsar 2010*)

**9.** Give one method of prepartion of plutonium. Discuss the solvent extraction method for the extraction of plutonium using tributyl phosphate as a solvent.
(*Mumbai 2010*)

**10.** (*a*) It is difficult to interpret the paramagnetic behaviour of actinides. Explain
(*b*) Compare the oxidation states of actinides with those of lanthanides.
(*Kurukshetra 2012*)

**11.** Explain the following:
(*a*) Europium (II) is more stable than cerium (II).
(*b*) It is difficult to interpret the paramagnetic behaviour of actinides.
(*Kurukshetra 2012*)

**12.** (*a*) Compounds of $La^{3+}$ and $Ce^{3+}$ are colourless while $Pr^{3+}$ and $Nd^{3+}$ form coloured compounds. Explain
(*b*) What are trans-uranic elements? Why do these elements show a general oxidation state of +3. How is plutonium separated from uranium.
(*Lucknow 2013*)

# 11 CHAPTER Coordination Compounds: Definition and Terminology

## Addition or Molecular Compounds

When solutions of two or more simple stable salts are mixed together in simple molecular proportion and the solution thus obtained is allowed to evaporate, crystals of a new compound are obtained. This new compound is called **addition** or **molecular compound**. The formation of some addition compounds has been shown below:

| *Simple compounds* | | *Addition compounds* |
|---|---|---|
| $KCl + MgCl_2 + 6H_2O$ | $\longrightarrow$ | $KCl.MgCl_2 . 6H_2O$ <br> Carnallite |
| $K_2SO_4 + Al_2(SO_4)_3 + 24H_2O$ | $\longrightarrow$ | $K_2SO_4.Al_2(SO_4)_3.24H_2O$ <br> Potash alum |
| $(NH_4)_2SO_4 + FeSO_4 + 6H_2O$ | $\longrightarrow$ | $FeSO_4.(NH_4)_2SO_4.6H_2O$ <br> Mohr's salt |
| $NiCl_2 + 6NH_3$ | $\longrightarrow$ | $NiCl_2 \cdot 6NH_3$ <br> Hexamine nickel chloride |
| $CuSO_4 + 4NH_3$ | $\longrightarrow$ | $CuSO_4 \cdot 4NH_3$ <br> Tetramine copper sulphate |
| $Fe\ (CN)_2 + 4KCN$ | $\longrightarrow$ | $Fe(CN)_2 \cdot 4KCN$ <br> Potassium ferrocyanide |

## Hydrates, Ammoniates and Peroxohydrates

Many salts crystallise with solvent molecules like $H_2O$, $NH_3$ and $H_2O_2$. The salts obtained in this way are also *addition compounds*. When the added solvent molecules is $H_2O$, $NH_3$ or $H_2O_2$, the addition compounds formed are called *hydrates, ammoniates* and *peroxohydrates*. In naming these compoundsm the number of solvent molecules is indicated by Arabic numireals (1, 2, 3, etc.). For example, $CuSO_4 . 5H_2O$ [copper (II) sulphate 5-hydrate], $CaCl_2.6NH_3$ (calcium chloride 6-ammoniate) and $NaHO_2.H_2O_2$ (sodium hydrogen peroxide 1-peroxohydrate).

## Types of Addition Compounds : Double Salts and Coordination (or Complex) Compounds

Addition compounds are of two types:

**1. Double salts.** These are the addition compounds which are stable in the solid state but give their constituent ions when dissolved in water or in any other ionic solvent. In these compounds, the individual properties of the constituent ions are *not* lost. For example, when carnallite, $KCl.MgCl_2.6H_2O$ which is a double salt gives its constituent ions namely $K^+$, $Mg^{2+}$ and $Cl^-$ ions when dissolved in water and hence the aqueous solution of $KCl. MgCl_2. 6H_2O$ gives the test of $K^+$, $Mg^{2+}$ and $Cl^-$ ions.

**2. Coordination (or complex) compounds.** These are the addition compounds which do not give all their constituent ions when dissolved in water. In these compounds, the individual properties of some constituent ions are lost. For example, $CuSO_4.4NH_3$ which is a complex compound does not give all its constituent ions *viz.* $Cu^{2+}$ and $SO_4^{2-}$ ion in solution. It gives only $SO_4^{2-}$ ion and $Cu^{2+}$ ion is obtained as $[Cu(NH_3)_4]^{2+}$ ion which is called *complex ion.*

$$CuSO_4.4NH_3 \rightleftharpoons [Cu(NH_3)_4]^{2+} + SO_4^{2-}$$

Thus, $CuSO_4.4NH_3$ is formulated as $[Cu(NH_3)_4]SO_4$. It follows from this discussion that $CuSO_4.4NH_3$ gives the tests of $SO_4^{2-}$ only and does not show the tests of $Cu^{2+}$ ions. $[Cu(NH_3)_4]SO_4$ is called *complex* (or *coordination*) *compound*. A complex compound may be composed of simple cation and complex anion, complex cation and simple anion or complex cation and complex anion as shown below:

$$K_3[Fe(CN)_6] \rightleftharpoons \underset{\text{Simple cation}}{3K^+} + \underset{\text{Complex anion}}{[Fe(CN)_6]^{3-}}$$

$$[Co(NH_3)_6]Cl_3 \rightleftharpoons \underset{\text{Complex cation}}{[Co(NH_3)_6]^{3+}} + \underset{\text{Simple anion}}{3Cl^-}$$

$$[Pt(NH_3)_4]\,[PtCl_4] \rightleftharpoons \underset{\text{Complex cation}}{[Pt(NH_3)_4]^{2+}} + \underset{\text{Complex anion}}{[PtCl_4]^{2-}}$$

Neutral complex compounds like $[Ni^0(CO)_4]^0$, $[Ni^{2+}(dmg)_2]^0$, $[Pt^{2+}(NH_3)_2Cl_2]^0$, etc. do not undergo ionisation and hence do not give any ions in aqueous solution.

Above discussion shows that the coordination compounds are the compounds in which the central metal atom/ion is linked with a number of positive ions or negative ions or neutral molecules (ligands) through the donation of lone pairs of electrons by these ions or molecules to the central metal atom.

Coordination compounds may also be defined as the compounds which contain complex ions. For example $K_4[Fe(CN)_6]$ is a coordination compound because it contains the complex anion, $[Fe(CN)_6]^{4-}$. Similarly $[Cu(NH_3)_4]SO_4$ is a coordination compound because it has complex cation, $[Cn(NH_3)_4]^{2+}$.

## Differences Between a Double Salt and a Complex

Both double salts as well as complexes are formed by the combination of two or more stable compounds in stoichiometric ratio. Hence, they are addition or molecular compounds. However, they differ in the fact that the double salts such as carnallite, $KCl.MgCl_2.6H_2O$; Mohr's salt, $FeSO_4.(NH_4)_2SO_4.6H_2O$ and potash alum $K_2SO_4.Al_2(SO_4)_3.24H_2O$, etc., dissociate into simple substances or ions completely when dissolved in water, *e.g.*,

$$KCl.MgCl_2.6H_2O \longrightarrow K^+ + Mg^{2+} + 3Cl^- + 6H_2O$$

$$FeSO_4.(NH_4)_2SO_4.6H_2O \longrightarrow Fe^{2+} + 2NH_4^+ + 2SO_4^{2-} + 6H_2O$$

On the other hand, in complexes, the complex ion almost does not dissociate. For example, the complex $K_4[Fe(CN)_6]$ whose formula may be written like that of a double salt, $Fe(CN)_2.4KCN$, dissociates to gives $K^+$ and $[Fe(CN)_6]^{4-}$ ions in the aqueous solution. The solution does not give the tests of $Fe^{2+}$ and $CN^-$ ions because the complex ion, $[Fe(CN)_6]^{4-}$ does not dissociate into $Fe^{2+}$ and $CN^-$ ions

$$K_4[Fe(CN)_6] \longrightarrow 4K^+ + [Fe(CN)_6]^{4-}$$

↓

Does not ionize to give $Fe^{2+}$ and $CN^-$ ions

Similarly, the complex, $[Cu(NH_3)_4]SO_4$, whose formula is equivalent to $CuSO_4.4NH_3$, does not dissociate to give $Cu^{2+}$ or $NH_3$ in its solution in water. It dissociates to give $[Cu(NH_3)_4]^{2+}$ and $SO_4^{2-}$ ions.

The main points of difference between a double salt and a complex compound are summed up in the Table 11.1 given below:

**Table 11.1 Differences between a double salt and a complex compound**

| *Double salt* | *Complex compound* |
|---|---|
| 1. It usually contains two simple salts in equimolar proportions. | 1. The simple salt from which it formed may or may not be in the equimolar proportions. |
| 2. It exists only in the solid state. In aqueous solution, it dissociates completely into ions. | 2. It exists in the solid state as well as in aqueous solution. This is because even in the solution, the complex ion does not dissoiate into ions. |
| 3. It is an ionic compound and does not contain any coordinate bond. | 3. It may or may not be ionic but the complex part always contains coordinate bonds. |
| 4. The properties of the double salt are the same as those of its constituents compounds. | 4. The properties of the co-ordinate compound are different from its constituents |
| 5. In a double salt, the metal ions show their normal valency. | 5. In a coordination compound, the metal ion satisfies its two types of valencies called primary and secondary valencies (discussed later). |
| 6. A double salt loses its identity in the solution. | 6. A coordination compound (complex) retains its identity in the solution. |

## Nomenclature of Double Salts

(*a*) Double salts are named by connecting the names of the individual compounds by hyphens (–). The number of the molecules of individual compounds is indicated by using *Arabic numerals* (*e.g.* 1, 2, 3, etc.) separated by solid lines and enclosed in brackets. For example:

| | | |
|---|---|---|
| $K_2SO_4.Al_2(SO_4)_3.24H_2O$ | ... | Potassium sulphate-aluminium sulphate-water (1/1/24) |
| $FeSO_4.(NH_4)_2SO_4.6H_2O$ | ... | Ferrous sulphate ammonium sulphate-water (1/1/6) |
| $Na_2CO_3.10H_2O$ | ... | Sodium carbonate water (1/10) |
| $3CdSO_4.8H_2O$ | ... | Cadmium sulphate water (3/8) |
| $AlCl_3.4C_2H_5OH$ | : | Aluminium chloride-ethanol (1/4) |
| $NH_3.BF_3$ | ... | Ammonia-boron trifluoride (1/1) |
| $8H_2S.46H_2O$ | ... | Hydrogen sulphide water (8/46) |

(*b*) If a double salt contains the cations in the same oxidation state, the cations are written alphabetically. Hydrogen is an exception which is written at the end. For example, $KHCO_3$ (potassium hydrogen carbonate), $KNaCO_3$ (Potassium sodium carbonate), $NaNH_4HPO_4 . 4H_2O$ (ammonium sodium hydrogen phosphate 4-hydrate), etc.

(*c*) If a double salt contains the cations in different oxidation states, the cations are written in the increasing order of their oxidation states, hydrogen being an exception which is written at the end. For example, $KMgCl_3 . 6H_2O$ (potassium magnesium chloride 6-hydrate), $KAl(SO_4)_2. 12H_2O$ (potassium aluminium sulphate 12-hydrate), $NH_4MgPO_4.6H_2O$ (ammonium magnesium phosphate 6-hydrate), $(NH_4)_2Co(SO_4)_2.6H_2O$ (ammonium cobolt (II) sulphate 6-hydrate), $NaZn(UO_2)_3(CH_3COO)_9.6H_2O$) (sodium zinc triuranyl acetate 6-hydrate).

(*d*) If a double salt contains the anions in the same oxidation state, the anions are written alphabetically. For example, PbClF is called lead (II) chloride fluoride.

(*e*) If a double salt contains the anions in different oxidation states, the anions are arranged in the order: $O^{2-}$, $OH^-$, other simple inorganic anions with two elements, organic anions, $H^-$. The number of ions is indicated by writing the prefix *bis*, *tris* etc., in parentheses. For example $NaCl . NaF.2Na_2SO_4$ is called sodium chloride fluoride (*bis*) sulphate.

(*f*) The hydrolysed or basic salts are also treated as double salts containing more than one anion and hence are named by the same rule as used for naming double salts. For example BiOCl [bismuth (III) oxide chloride], Sn (OH) Cl [tin (IV) hydroxide chloride], $ZrOCl_2.8H_2O$ [zirconium (IV) oxide (di) chloride 8-hydrate], $CuCl_2.3Cu(OH)_2$ or $Cu_2(OH)_2Cl$ [(di) copper (II) trihydroxide chloride].

**Note:** It may be noted that for $H_2O$ molecule we have used *hydrate* and the number of $H_2O$ molecules has been indicated by Arabic numbers.

## Important Terms Used in Coordination Compounds

### Coordination Chemistry

The branch of chemistry which deals with the study of coordination compounds is known as *coordination chemistry*.

### Complex Ion

Complex ion is a positively or negatively charged species which contains a central metallic atom and a suitable number of ligands surrounding the central metallic atom. The central metallic atom may be in zero, positive or negative oxidation state. The ligands may be either neutral molecules or anions (generally) or a combination of the neutral molecules and anions. Examples of complex ions are $[Co^{3+}(NH_3)_6]^{3+}$, $[Ni^0(CN)_4]^{4-}$, $[Fe^{2+}(CN)_6]^{4-}$, and $[Co^-(CO)_4]^-$, are anionic complex ions.

It may be seen that, depending on the nature of the charge on the complex ion, complex ion may be *cationic complex ion* or *anionic complex ion* Cationic complex ion has positive charge on it while anionic complex ion has negative charge. Thus $[Co^{3+}(NH_3)_6]^{3+}$ and $[Co^{3+}(NH_3)_5Cl]^{2+}$ are cationic complex ions and $[Ni^0(CN)_4]^{4-}$, $[Fe^{2+}(CN)_6]^{4-}$ and $[Co^-(CO)_4]^-$ are anionic complex ions.

### Neutral Complex

A complex which has no charge on it is called *neutral complex*. It is a non-electrolyte, does not undergo ionisation and hence does not give any ions in aqueous solution. Examples of neutral complexes are $[Co^{3+}(NH_3)_3Cl_3]^0$, $[Ni^0(CO)_4]^0$, $[Pt^{2+}(NH_3)_2Cl_2]^0$, $[Ni^{2+}(dmg)_2]^0$, etc.

### Central Metal Atom/Ion

A metal atom/ion present in a complex ion (coordination entity), whether positively charged, negatively charged or neutral is known as **central metal atom or ion**. It is sometimes called a **nuclear atom**. A fixed number of ligands are linked with the central metal atom in a fixed geometrical arrangement.

In a complex compound, since the central metal atom/ion accepts electron pair (s) donated by ligands, this atom acts as **Lewis acid**.

### Coordination Sphere or Coordination Entity or Complex Entity

The complex cation or complex anion present in a coordination compound is called **coordination sphere.** The coordination sphere contains the central metal atom and ligands which are attached with central metal atom. All the species (i.e., central metal atom and ligands) forming the coordination sphere are enclosed in square brackets, [ ]. For example in $K_4[Fe(CN)_6]$, $[Fe(CN)_6]^{4-}$ (complex anion) is the coordination sphere. Similarly, in $[Cu(NH_3)_4]SO_4$, $[Cu(NH_3)_4]^{2+}$ (complex cation) is the coordination sphere.

## Ionisation Sphere or Counter Ion

The part of a complex compound which is present outside the coordination sphere is called **ionisation sphere**. For example in $K_4[Fe(CN)_6]$, $[Fe(CN)_6]^{4-}$ is the coordination sphere and $K^+$ ions constitute ionisation sphere. Similarly in $[Cu(NH_3)_4]SO_4$, $[Cu(NH_3)_4]^{2+}$ is the coordination sphere and $SO_4^{2-}$ ions act as ionisation sphere.

## Coordination Number (C.N.) of the Central Metal

Coordination number of the central metal atom/ion is also called **ligancy** of the metal atom/ion. Coordination number of the central metal atom/ion present in coordination sphere can be defined in various ways as shown below:

Total number of unidentate ligands (neutral molecules, negative ions or positive ions) attached to the central metal atom/ion through coordinate bonds or sigma bonds in coordination sphere is called C.N. of the central metal atom/ion

OR

Coordination number of the central metal atom/ion may be defined as the number of atoms of the ligands which are attached directly with the metal atom/ion in coordination sphere.

OR

Coordination number is defined as the number of coordinate bonds or sigma bonds formed between the central metal atom / ion and the donor atom of the ligands.

If pi bonds are formed between the central metal atom/ion and ligands, these pi bonds are not considered for determining the C.N. of the central metal atom/ion. For calculating C.N., it should be remembered that a monodentate ligand forms one bond, bidentate ligand forms two bonds, tridentate ligand forms three bonds and so on with the central metal atom.

**Examples :** (*a*) C.N. of $Ag^+$ ion in $[Ag(CN)_2]^-$ is 2, because $CN^-$ ion is a monodentate ligand.

(*b*) C.N. of $Cr^{3+}$ ion in $[Cr(en)_3]^{3+}$ is 6 because *en* is a bidentate ligand.

## Charge on the Complex Ion

The charge carried by a complex ion is the algebraic sum of the charges carried by central metal ion and the ligands coordinated to the central metal ion.

**Examples:** (*i*) In the complex ion, $[Ag(CN)_2]^-$, $Ag^+$ ion carries a charge of +1 and each $CN^-$ ion carries a charge of –1. Therefore, the net charge on the complex ion, $[Ag(CN)_2]^-$ is $+1 + (-1 \times 2) = -1$.

(*ii*) In the complex ion $[Cu(NH_3)_4]^{2+}$, $Cu^{2+}$ ion carries a charge equal to +2 and $NH_3$ molecules are neutral. Therefore the net charge on the complex is $+2 + 4 \times 0 = +2$.

## Oxidation Number (O.N.) or Oxidation State of the Central Metal Atom

The oxidation number of the central atom is defined as the charge that it carries as calculated by assigning appropriate charges to the ligands and equating the sum of the charges on the central atom and the ligands equal to the charge on the coordination sphere.

## Calculation of Oxidation Number

In order to calculate the oxidation number of the central metal atom, one should know the charge on the coordination sphere and contribution made by each ligand towards the charge. Whereas neutral ligands have no charge, the anionic ligands carry negative charge while the cationic ligands (if present) have positive charges. In the light of this, let us calculate the oxidation state of the metal atom in the following cases. ($x$ is the oxidation number of the metal atom).

(*i*) $[Fe(CN)_6]^{3-}$ : $x + 6(-1) = -3$ or $x = -3 + 6 = +3$

(*ii*) $[Cu(NH_3)_4]^{2+}$ : $x + 4(0) = +2$ or $x = +2$

(*iii*) $[Ag(CN)_2]^-$ : $x + 2(-1) = -1$ or $x = -1 + 2 = +1$

(*iv*) $[Ni(CO)_4]$ : $x + 4(0) = 0$ or $x = 0$

The oxidation number is normally represented by Roman nuwerls, I, II, III, etc., in parenthesis or square bracket which follows the name of the complex entity.

Oxidation number of the metal is different from it coordination number. For example, in $[Fe(CN)_6]^{6-}$, the C.N. of Fe is six while its oxidation number is +3.

## Illustration of Different Terms

All the terms discussed above have been shown for $[Co^{2+}(NH_3)_5Cl]Cl_2$ in Fig. 11.1.

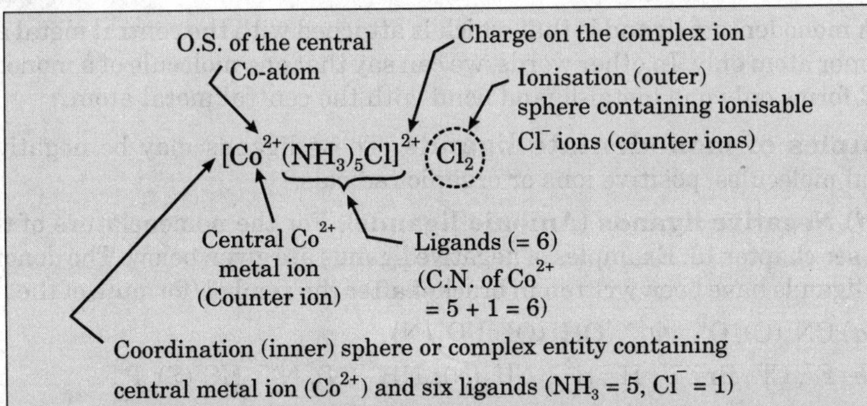


**Fig. 11.1** Illustration of different terms used in $[Co^{2+}(NH_3)_5Cl]\ Cl_2$

## Ligands or Coordinating Groups

### What are Ligands?

We know that a coordination sphere in a complex species has ligands and central metal atom / ion. In coordination sphere, the ions (cations or anions) or neutral molecules which are attached to the central metal atom/ion by coordinate bond or sigma bond are called **ligands** or **coordinating groups**. Ligands donate pair (s) of electrons to the metal atom to form coordinate bonds. The atom of the ligands which donates the electron pair or is attached with the central metal atom/ ion is called **donor atom**. Common donor atoms are N, O, S and halogens.

The ligands may be simple anions (*e.g.*, $Cl^-$, $Br^-$, $CN^-$, etc.), small molecules (*e.g.*, $H_2O$, $NH_3$, etc.), large molecules (*e.g.*, ethylene diamine, etc.), macromolecules such as proteins, and cations (*e.g.*, $NO^+$, $NO_2^+$ etc.).

Since ligands donate electron pair(s) to the central metal atom, ligands are ***Lewis acids***.

## Denticity of a Ligand

The number of donor atoms in a ligand attached to the central metal atom is called **denticity** of the ligand. For example, in the complex ion, $[(Cu(en)_2)^{2+}]$, the denticity of each *en* ligand is 2, since this ligand is attached with $Cu^{2+}$ ion (central metal ion) through two nitrogen donor atoms.

## Classification of Ligands

Depending on the number of donor atoms of one molecule of the ligand with which the central atom is attached in coordination sphere or the number of metal–ligand bonds formed between the central metal atom and donor atoms of the ligands, ligands have been classified into monodentate (or unidentate) and polydentate ligands.

### A. Monodentate Ligands

A monodentate ligand is that which is attached with the central metal atom by one donor atom only. In other words, we can say that one molecule of a monodentate ligand forms only one metal-ligand bond with the central metal atom.

**Examples of monodentate ligands.** These ligands may be negative ions, neutral molecules, positive ions or organic radicals.

***(i)* Negative ligands (Anionic ligands):** For the nomenclature of these ligands see chapter 13. Examples of negative ligands are given below. The donor atoms of the ligands have been written in bracket after the symbol/formula of the ligands.

(*a*) $CN^-$(C), $O^{2-}$, $O_2^{2-}$, $OH^-$(O), $NO_2^-$(N),

(*b*) $F^-$, $Cl^-$, $Br^-$, $I^-$, $H^-$, $S^{2-}$, $NH_2^-$(N), $NH^{2-}$(N), $N^{3-}$, $HS^-$(S), $P^{3-}$

(*c*) $SO_3^{2-}$(S), $ONO^-$(O)

(*d*) $CO_3^{2-}$($O^-$), $NO_3^-$(N), $S_2O_3^{2-}$(S), $CH_3COO^-$($O^-$), $ClO_3^-$($O^-$), $SCN^-$(S), $SO_4^{2-}$($O^-$), $NCS^-$(N),

***(ii)* Neutral molecules.** For the nomenclature of these ligands see chapter 13. Examples are given below:

$NH_3$(N), $PH_3$(P), $H_2O$(O), CO(O), CS (S), NO(N), $(C_6H_5)_3P$(P), 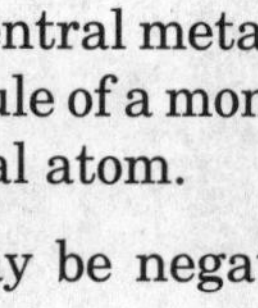 or $C_5H_5N$ or py (N), $H_2NCS\,NH_2$ or tu (S), $O_2$, $N_2$, $CH_3NH_2$(N).

***(iii)* Positive ligands (Cationic ligands).** For the nomenclature of these ligands, see chapter 13. Examples are given below:

$NH_2-\overset{+}{N}H_3$ or $N_2\overset{+}{H}_5$(N), $\overset{+}{N}O$(N), $\overset{+}{N}O_2$(N), $NH_4^+$(N), $C_5H_5\overset{+}{N}$(N), $(C_6H_5)_4P^+$(P), $PH_4^+$(P)

***(iv)* Organic free radicals.** Examples are : $CH_3$(C), $C_2H_5$(C), $C_5H_5$(C), $C_6H_5$(C) etc.

## B. Polydentate Ligands

Polydentate ligand is that which is coordinated to the central metal atom through more than one donor atom. In other words, we can say that one molecule of a polydentate ligand forms more than one metal–ligand bonds with the central metal atom.

Polydentate ligands may be *bidentate, tridentate, tetradentate, pentadentate* and *hexadentate,* if one molecule of polydentate ligand forms 2, 3, 4, 5 and 6 metal–ligand bonds respectively with the central metal atom.

### 1. Bidentate Ligands

One molecule or ion of a bidentate ligand has two donor atoms and hence forms two metal–ligand bonds with the central metal atom.

**Examples of bidentate ligands**. Bidentate ligands may be anions or neutral molecules. Examples are given below:

***(i)* 2, 2, 2-bipyridine (bipy** or **bipyridyl**). It is a neutral molecule and has two N-atoms which act as donor atoms as shown at (*a*) of Fig. 11.2. When this ligand gets linked with a metal atom M, 5-membered ring is obtained [See (*b*) of Fig 11.2].

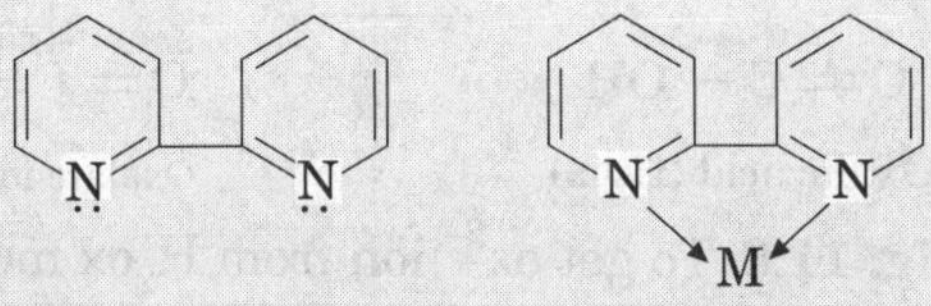


(*a*) (*b*)

**Fig. 11.2** (*a*) Structure of 2, 2′-bipyridine (bipy or bipyridyl)
(*b*) Attachment of bipy with the central metal atom, M.

***(ii)* o-phenanthroline or 1, 10-phenanthroline (o-phen or phen or phenan).** It is a neutral symmetrical bidentate ligand, since its two donor atoms *viz.* N-atoms are similar atoms. Its structure is given at (*a*) of Fig. 11.3. It is abbreviated as *phen* or *phenan*. When it gets attached with a metal atom, it is denoted as $\left.\begin{matrix}N\\N\end{matrix}\right)$ *phen* and its attachment to a metal atom, *M* is indicated as $M\begin{matrix}\swarrow N\\\nwarrow N\end{matrix}\Big)$ *phen*. For example, this ligand gets attached with $Fe^{2+}$ ion and gives $[Fe(phen)_3]^{2+}$ complex ion whose structure is given at (*b*) of Fig. 11.3.

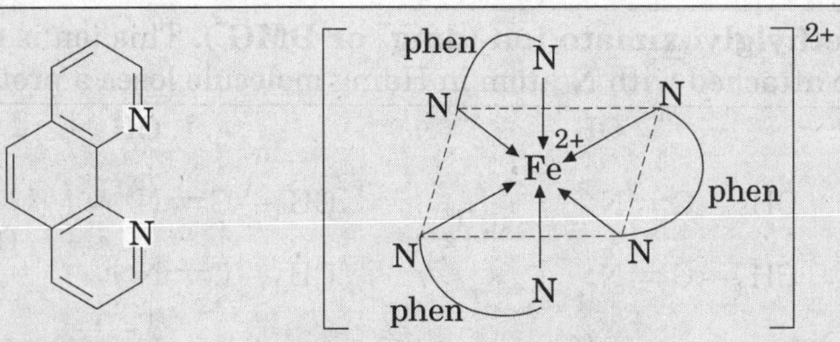


(*a*) (*b*)

**Fig. 11.3** (*a*) Structure of o-phenonthroline (phen or phenon)
(*b*) Structure of $[Fe(phen)_3]^{2+}$ ion.

***(iii)* Acetylacetonato ion ($acac^-$.).** This ion is obtained when one molecule of the enol form of acetyl acetone (Hacac) loses a proton, $H^+$(See Fig. 11.4.).

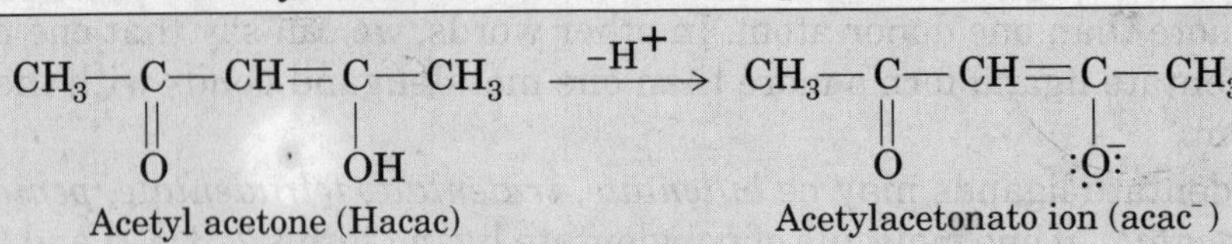


**Fig. 11.4** To get *acac*$^-$ ion from Hacac molecule.

This ligand gets attached with the central metal atom through its two O-atoms and forms a chelated complex having one 6-membered ring as shown in Fig. 11.5.

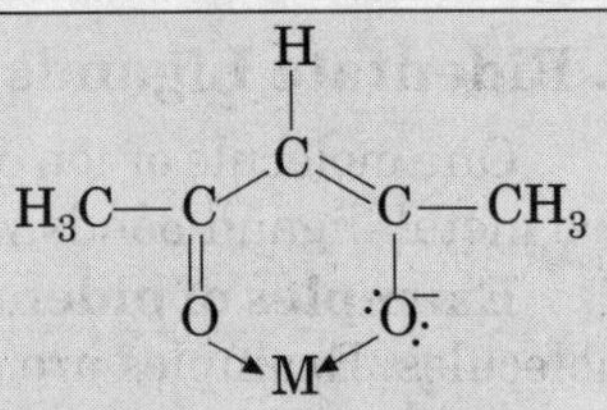

**Fig. 11.5.**

***(iv)* Oxalato ion, $C_2O_4^{2-}$ ($ox^{2-}$).** This ion can be obtained when one molecule of oxalic acid ($H_2C_2O_4$ or $H_2ox$) loses two protons (See Fig. 11.6).

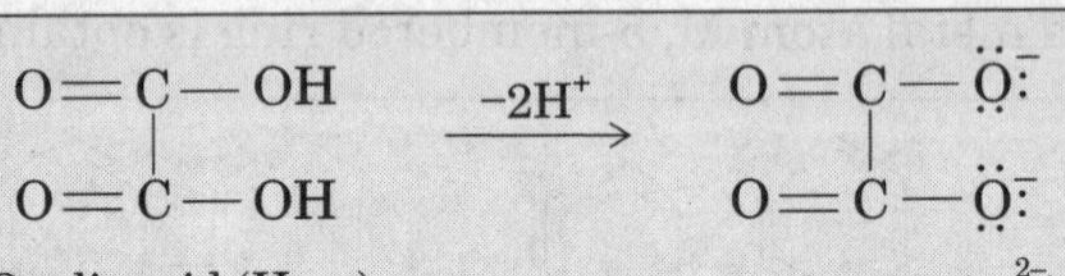


**Fig. 11.6** To get $ox^{2-}$ ion from $H_2ox$ molecule.

$C_2O_4^{2-}$ ($ox^{2-}$) is a bidentate dibasic ligand and its coordination to the central metal atom occurs through its both negatively charged O-atoms (acidic donor atoms), forming 5-membered chelated complex as is evident from the structure of dioxalato platinate (II) ion, $[Pt^{2+}(C_2O_4)_2]^{2-}$ or $[Pt^{2+}(ox)_2]^{2-}$ as shown in Fig. 11.7.

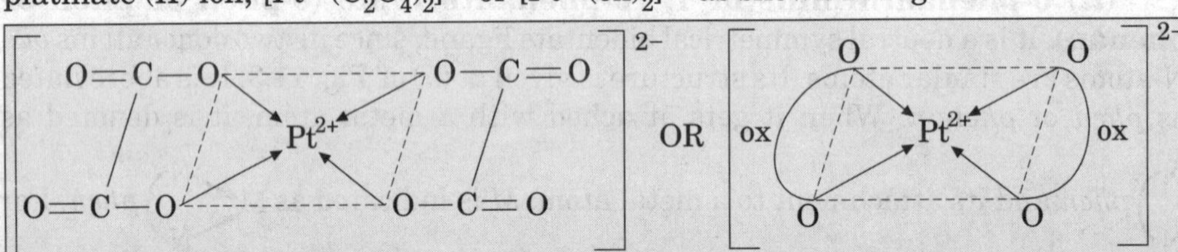

**Fig. 11.7** Structure of dioxalato platinate (II) ion, $[Pt^{2+}(C_2O_4)_2]^{2-}$ or $[Pt^{2+}(ox)_2]^{2-}$ (square planar complex ion).

***(v)* Dimethylglyoximato ion ($dmg^-$ or $DMG^-$).** This ion is obtained when one OH group attached with N-atom in Hdmg molecule loses a proton (Fig. 11.8).

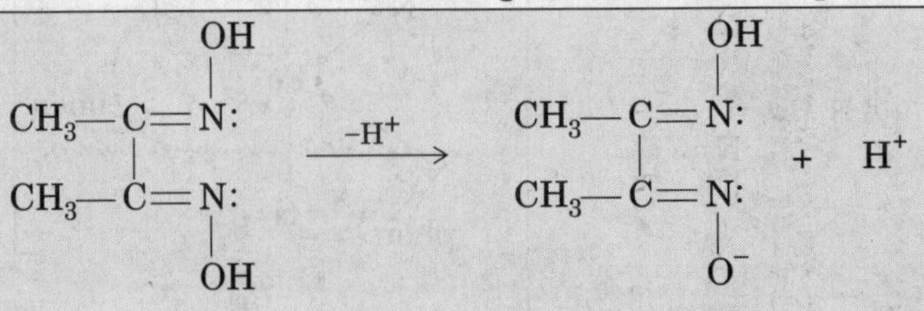


**Fig. 11.8** To get $dmg^-$ ion from Hdmg molecule.

Two $dmg^-$ ions react with $Ni^{2+}$ or $Pd^{2+}$ ion in ammoniacal solution and form an insoluble neutral red chelated complex, $[M(dmg)_2]^0$ ($M = Ni^{2+}$, $Pd^{2+}$) which has two 5-membered and two 6-membered rings. The formation of $[Ni(dmg)_2]^0$ has been shown in Fig. 11.9.

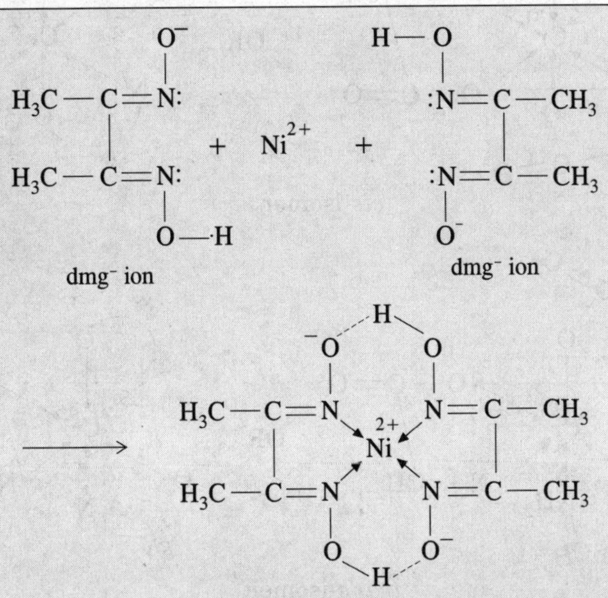

*Bis*-(dimethylglyoximato) nickel (II) chelated complex, $[Ni(dmg)_2]^0$ (red ppt.)

**Fig. 11.9** Formation of $[Ni(dmg)_2]^0$ (chelated complex).

***(vi)* Glycinato ion ($gly^-$).** This ion is derived by removing one proton from one molecule of glycine (amino acetic acid), $H_2N—CH_2—COOH$ (Hgly) as shown in Fig. 11.10.

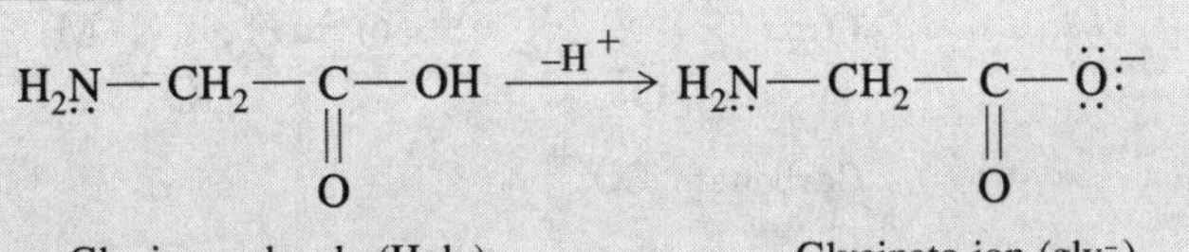

Glycine molecule (Hgly) Glycinato ion ($gly^-$)

**Fig. 11.10** To get $gly^-$ ion from Hgly molecule.

Glycinato ion ($gly^-$) is a bidentate unsymmetrical monobasic ligand and its coordination to the metal ion occurs through neutral N-atom (neutral donor atom) and negatively-charged O-atom (acidic donor atom), forming a 5-membered ring as is evident from the structure of *cis*-and *trans*-isomers of octahedral triglycinato chromium(III), $[Cr(NH_2—CH_2—C(=O)—O)_3]^0$ or $[Cr(gly)_3]^0$ shown in Fig. 11.11.

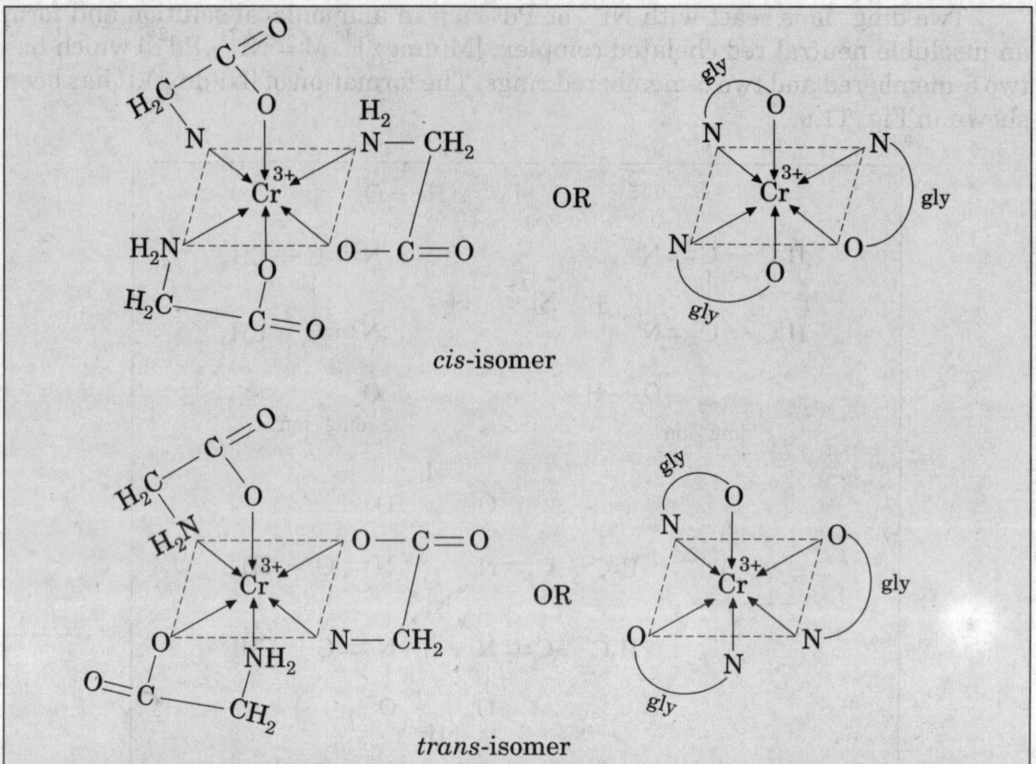


**Fig. 11.11** Structure of *cis*-and *trans*-isomers of $[Cr(NH_2—CH_2—\underset{\underset{O}{\|}}{C}—O)_3]^0$ or $[Cr(gly)_3]^0$

**(*vii*) Carbonato ($CO_3^{2-}$), nitrato ($NO_3^-$) and sulphato ($SO_4^{2-}$) ions**. The structure of these ions and the way in which they are coordinated to the metal atom are shown below in Fig. 11.12.

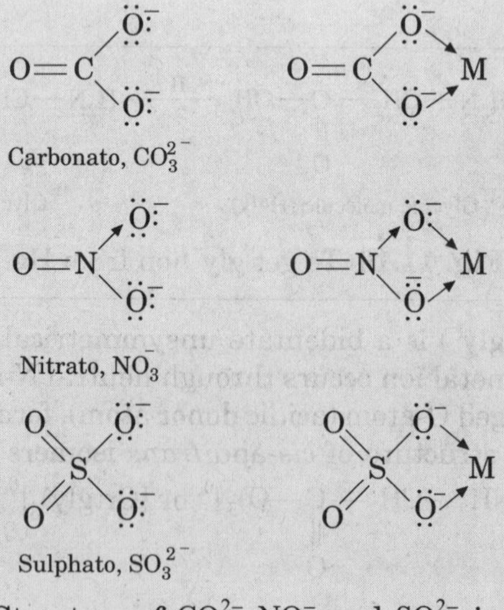


**Fig. 11.12** Structure of $CO_3^{2-}$, $NO_3^-$ and $SO_4^{2-}$ ions and their attachment with metal atom, M.

**Note.** Carbonato, nitrato and sulphato ligands also act as monodentate ligands as is evident from the following complex ions.

$[Co^{3+}(NH_3)_5(CO_3)]^+$ ... Pentammine carbonato cobalt (III) ion (octahedral)

$[Co^{3+}(NH_3)_5(SO_4)]^+$ ... Pentammine sulphato cobalt (III) ion (octahedral)

$[Ti^{4+}(NO_3)_4]^0$ ... Tetranitrato titanium (IV) (square planar)

***(viii)* Biguanido ion (big⁻).** This ion is obtained when biguanide molecule loses H-atom of NH group in C = NH unit (See Fig. 11.13).

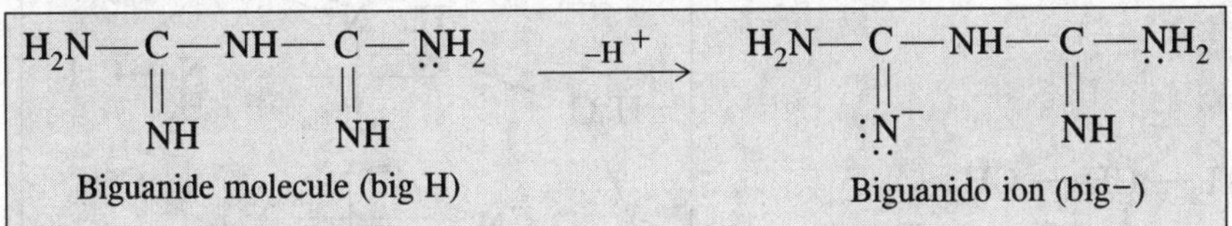


**Fig. 11.13** To get big⁻ ion from Hbig molecule.

This ion gets attached with the central metal atom through its negatively – charged N-atom and N-atom of $NH_2$ group and forms a chelated complex containing 6-membered ring as shown in Fig. 11.14.

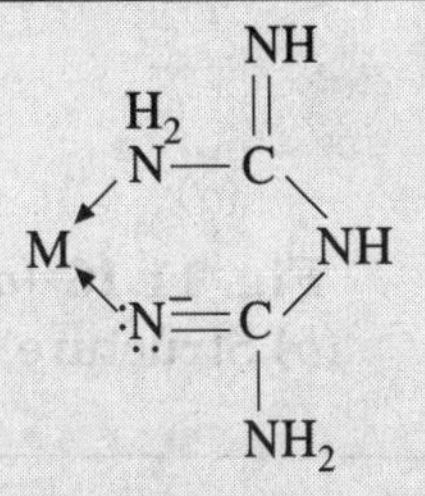

**Fig. 11.14** Attachment of biguanido ion with metallic atom

***(ix)* Ethylene diamine (en) or ethane-1, 2-diamine.** The structure of this molecule is given in the margin. It has two N-atoms as donor atoms and hence is a symmetrical bidentate ligand. The attachment of this ligand to $Co^{3+}$ ion to form *trans*-isomer of $[Co(en)_2Cl_2]^+$ ion (octahedral ion) is shown in Fig. 11.15. The octahedral complex ion has two 5-membered rings.

$H_2N—CH_2—CH_2—NH_2$

Structure of diethylene molecule (en)

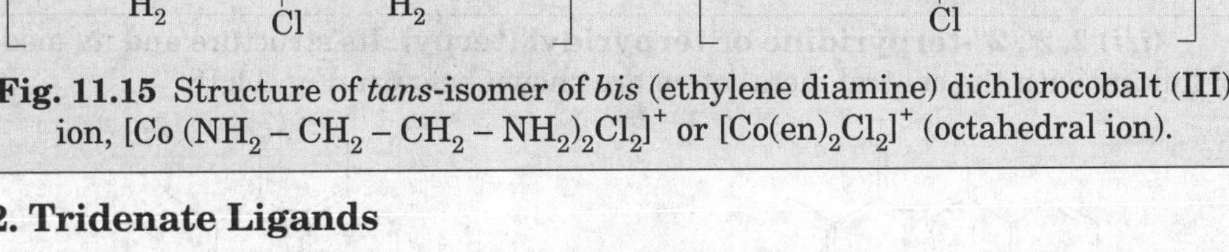


**Fig. 11.15** Structure of *tans*-isomer of *bis* (ethylene diamine) dichlorocobalt (III) ion, $[Co(NH_2-CH_2-CH_2-NH_2)_2Cl_2]^+$ or $[Co(en)_2Cl_2]^+$ (octahedral ion).

## 2. Tridenate Ligands

Tridentate ligand is that whose molecule or ion has three donor atoms and hence forms three metal—ligand bonds with the central metal atom.

**Examples of tridentate ligands.** These ligands may be neutral molecules or anionic species. Important examples are given below:

***(i)* Triaminopropane.** The structure of this ligand is given at (a) of Fig. 11.16 which shows that it has three N-atoms as its donor atoms. The coordination of this ligand to the central metal atom takes place through its three N-atoms (donor atoms), leading to the formation of two 5-membered rings as is evident from the structure of $[Co^{3+}\{(CH_2(NH_2)—CH(NH_2)—CH_2(NH_2)\}_2]^{3+}$ ion (a chelated complex ion) given at (*b*) of Fig. 11.16.

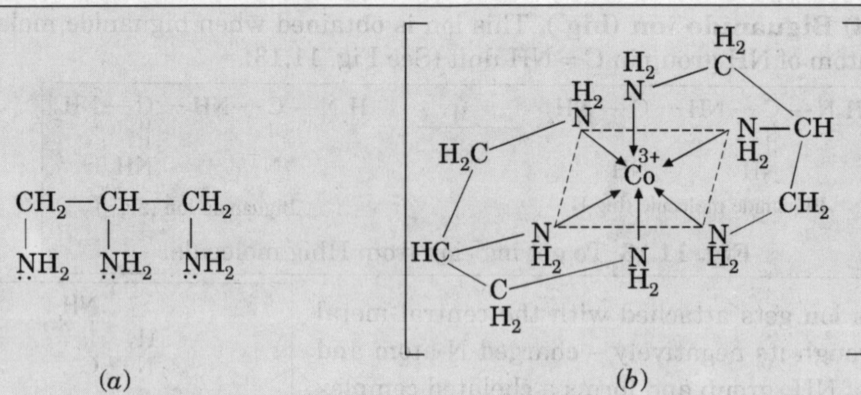


**Fig. 11.16** (*a*) Structure of triaminopropane (tridentate ligand) (*b*) Structure of $[Co^{3+}\{(CH_2(NH_2)—CH(NH_2)—CH_2(NH_2)\}_2]^{3+}$ ion (octahedral chelated complex ion).

***(ii)* Diethylene triamine (dien)** or **[N-(2-amino ethyl ethane-1, 2-diamine)].** Its structure is given in Fig. 11.17 (*a*). This ligand gets attached with the central metal atom through its three N-atoms (donor atoms) and forms a chelate which contains two 5-membered rings. [See Fig. 11.17(*b*)].

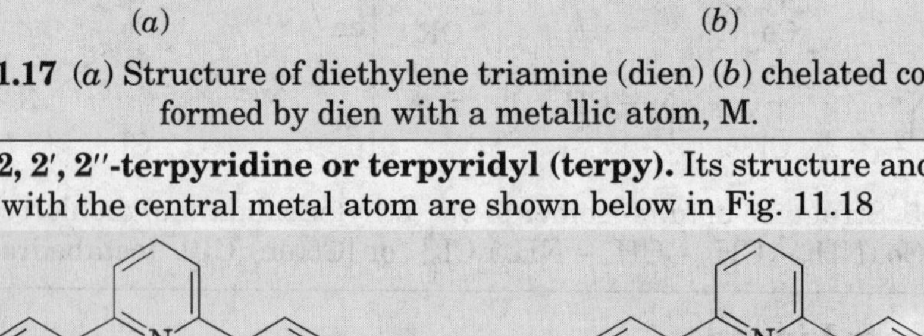


**Fig. 11.17** (*a*) Structure of diethylene triamine (dien) (*b*) chelated complex formed by dien with a metallic atom, M.

***(iii)* 2, 2′, 2″-terpyridine or terpyridyl (terpy).** Its structure and its mode of linking with the central metal atom are shown below in Fig. 11.18

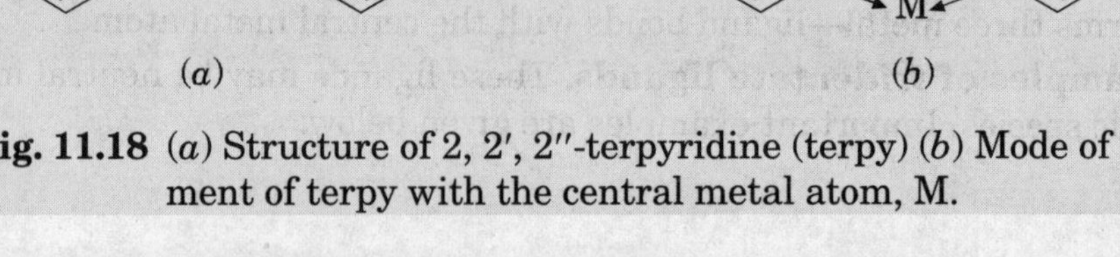


**Fig. 11.18** (*a*) Structure of 2, 2′, 2″-terpyridine (terpy) (*b*) Mode of attachment of terpy with the central metal atom, M.

### 3. Tetradentate Ligands

A tetradentate ligand is that whose molecule or ion has four donor atoms and hence forms four metal–ligand bonds with the central metal atom.

**Examples of tetradentate ligands:** Important examples of these ligands are given below:

***(i)* Triethylene tetramine (trien)** or **[N, N′-bis (2-aminoethyl) ethane-1, 2-diamine].** The structure of this ligand is given at (*a*) of Fig. 11.19. This structure shows that this ligand has four N-atoms as its donor atomns (neutral donor atoms). Coordination of this ligand to the metal ion takes place through these N-atoms, forming a chelated complex containing three 5-membered rings as shown at (*b*) of Fig. 11.19.

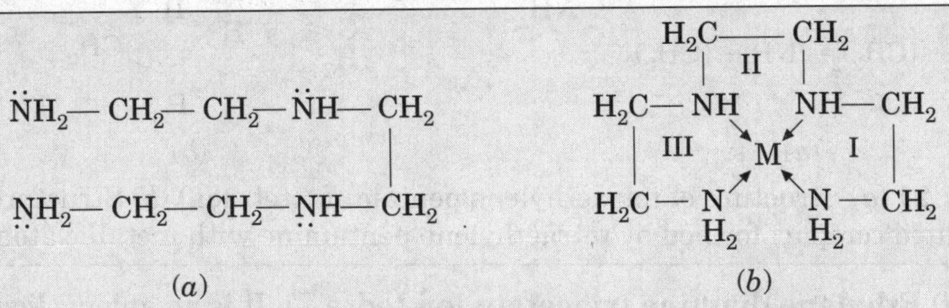

**Fig. 11.19** (*a*) Structure of triethylene tetramine (trien) molecule (*b*) Chelated complex formed by triethylen tetramine molecule with the metallic atom, M.

***(ii)* Triaminotriethylamine (tren), N(CH$_2$—CH$_2$—NH$_2$)$_3$.** Its structure is given at (*a*) of Fig. 11.20. This structure shows that it has four N-atoms as its donor atoms. The attachment of this ligand to the central metal atom, M in its complexes takes places through its four N-atoms, resulting in the formation of chelated complex having three 5-membered rings as shown at (*b*) of Fig. 11.20.

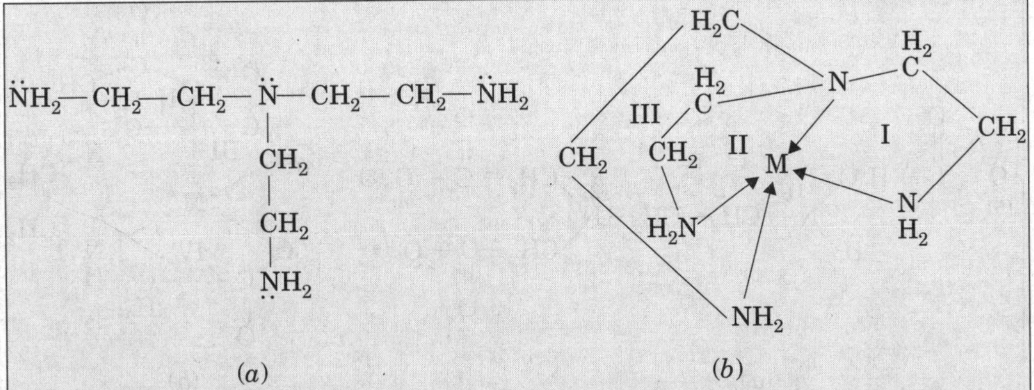

**Fig. 11.20** (*a*) Structure of triaminotriethylamine (tren), N(CH$_2$—CH$_2$—NH$_2$)$_3$ (*b*) Structure of the chelated complex formed by the attachment of triaminotriethylamine molecule with metallic atom, M.

### 4. Pentadentate Ligands

A pentadentate ligand is that whose molecule or ion has five donor atoms and hence forms five metal—ligand bonds with the central meal atom/ion.

**Example of pentadentate ligands.** Important examples are given below:

**(*i*) Tetraethylene pentamine (tetraen).** The structure of this molecule is shown at (*a*) of Fig. 11.21. This structure shows that it has five N-atoms as its donor atoms (neural donor atoms). Thus this ligand coordinates to the metal atom through these five N-atoms and forms four 5-membered rings as shown at (*b*) of Fig. 11.21.

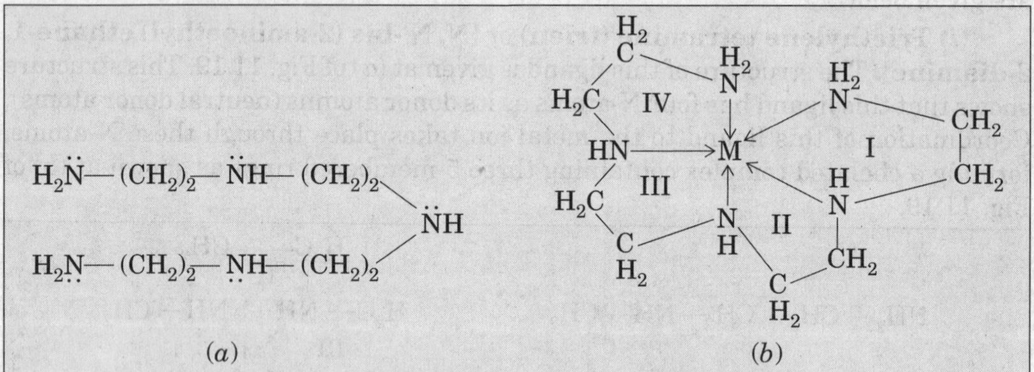


**Fig. 11.21** (*a*) Structure of tetraethylene pentamine (tetraen) (*b*) Structure of the chelated complex formed by tetraethylene pentamine with metallic atom, M.

**(*ii*) Ethylene diamine triacetato ion (edta$^{3-}$).** It is an anionic ligand and its structure is shown at (*a*) of Fig. 11.22. This structure shows that it has three negative charges and five donor atoms viz. two N-atoms (numbered as 1 and 2) and three negatively-charged O-atoms (numbered as 3, 4 and 5). Its coordination to the metal atom occurs through two N-atoms (neutral donor atoms) and three negatively-charged O-atoms (acidic donor atoms) of $COO^-$ groups, forming four 5-membered rings as shown at (*b*) of Fig. 11.22. Four 5-membered rings have been designated as I, II, III and IV.

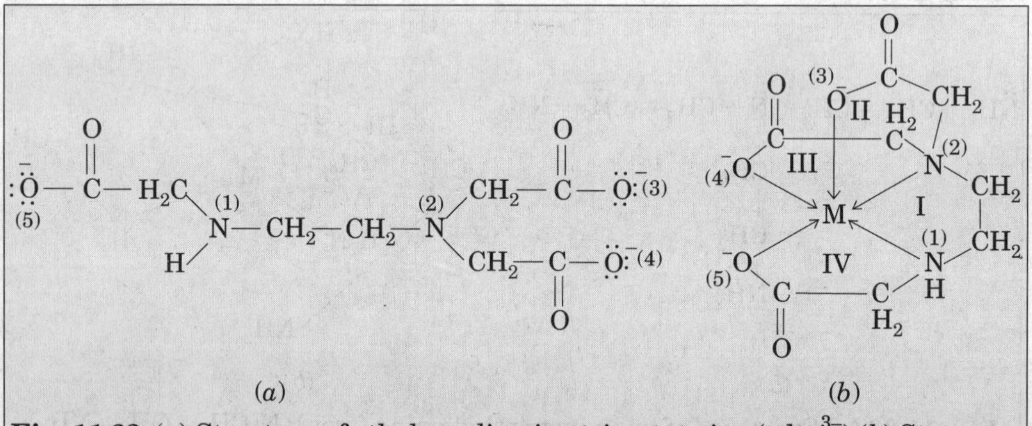


**Fig. 11.22** (*a*) Structure of ethylene diamine triacetato ion (edta$^{3-}$) (*b*) Structure of the chelated complex formed by ethylene diamine triacetato ion with metallic atom, M.

## 5. Hexadentate Ligands

Hexadentate ligand is that whose one molecule or ion has six donor atoms which get coordinated to the central metal atom/ion to form six metal–ligand bonds.

**Example of hexadentate ligand:** Ethylene diamine tetracetato ion ($edita^{-4}$ or $EDTA^{-4}$ or $Y^{-4}$) is an important example of hexadentate ligand. Its structure is given in Fig. 11.23.

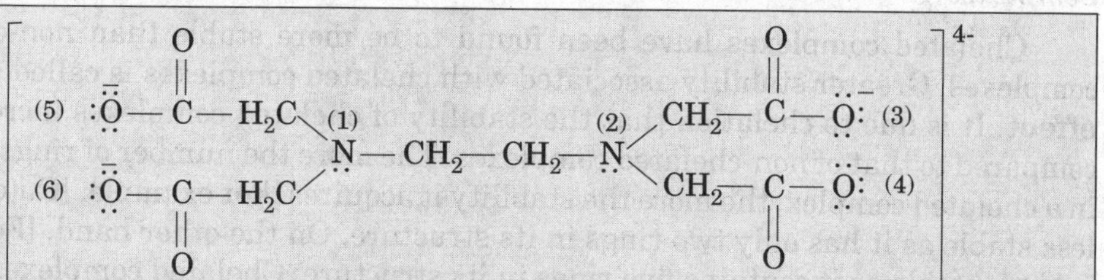


**Fig. 11.23** Structure of ethylene diamine tetracetato on ($edta^{4-}$ or $EDTA^{4-}$ or $Y^{4-}$)

The structure shows that it has two neutral N-atoms (numbered as 1 and 2) and four negatively-charged O-atoms (numbered as 3, 4, 5 and 6) as its donor atoms. Thus it is tetrabasic hexadentate ligand. The co-ordination of this ligand to the metal atom takes place through two N-atoms and four O-atoms, forming five 5-membered rings as is evident from the structure of $[Fe(edta)]^-$ ion given in Fig. 11.24.

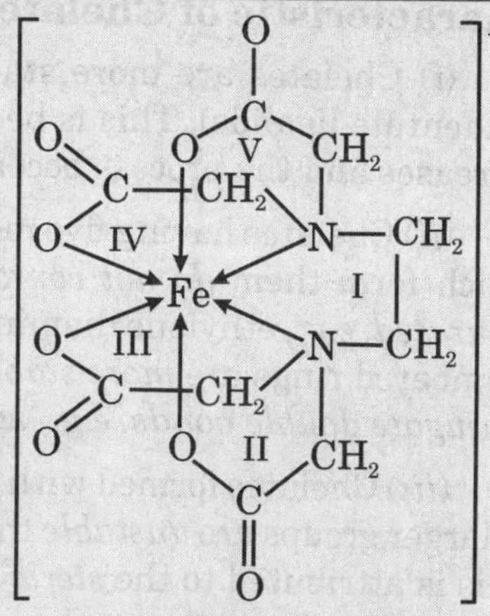


**Fig. 11.24** Structure of $[Fe^{3+}(edta)]^-$ ion having five rings

## Chelate, Chelating Ligand and Chelation

When a multidentate ligand simultaneously coordinates to a metal ion through more than one sigma electron pair donor site or group, a ring like structure is obtained. It is *called* a chelate and the ligand is known as *chelating ligand*. The process of formation of rings is called *chelation*.

Chelate is also known by other names like *metal chelate, chelated complex,* etc.

## Structure of Some Metal Chelates

Structures of $[Pt^{2+}Cl_2(en)]^0$ and $[Pt^{2+}Cl\ (dien)]^+$ metel chelates are given in Fig. 11.25.

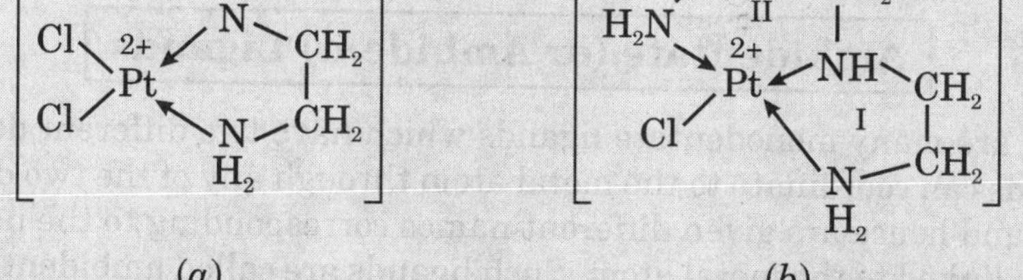


**Fig. 11.25** (*a*) Structure of $[Pt^{2+}\ Cl_2(en)]^0$. Here *en* is ethylenediamine (bidentate ligand). This complex has one 5-membered ring (*b*) Structure of $[Pt^{2+}Cl(dien)]^+$. Here *dien* is N-(2-aminoethyl) ethane-1, 2-diamine (tridentate ligand). This complex has two 5-membered rings.

**Chelate Effect** (Also see Chapter *"Thermodynamic and Kinetic Aspects of Metal Complexes"*)

Chelated complexes have been found to be more stable than non-chelated complexes. Greater stability associated with chelated complexes is called **chelate effect**. It is due to chelation that the stability of chelated complexes increases as compared to that of non-chelated complexes. The more the number of rings present in a chelated complex, the more the stability it acquires. For example, $[Cu(en)_2]^{2+}$ is less stable as it has only two rings in its structure. On the other hand, $[Fe(edta)]^-$ is more stable as it contains five rings in its structure. Chelated complexes having 5- and 6-membered rings including the metal atom are the most stable.

## Characteristic of Chelates

(*i*) Chelates are more stable than the simple type of complexes (formed by unidentate ligands). This is because when chelation occurs, entropy of the system increases and the process becomes more favourable.

(*ii*) Chelates having five membered rings are *more stable* if the chlating ligands which form them *do not contain double bonds*, *i.e.*, if the chelating ligands are *saturated*, *e.g.*, ethylene diamine complex. On the other hand, chelatates having six membered rings are *more stable* if the chelating ligands which form them *contain conjugate double bonds*, *e.g.*, acetylacetonato complexes.

(*iii*) Chelates formed with chelating ligands having more than six donor groups or larger groups are *unstable* than those formed with ligands having smaller groups. This is attributed to the *steric hindrance* produced by a larger group.

## Importance of Chelates

Chelates are widely used in the (*i*) softening of hard water (*ii*) separation of lanthanoids and actinoids (*iii*) detection of some metal ions in qualitative analysis and (*iv*) estimation of $Ni^{2+}$, $Mg^{2+}$ and $Cu^{2+}$ ions quantitatively.

# π-Acid ligands (Also called π-Acceptors)

The ligands which can accept an appreciable amount of electron density from the filled d-orbitals of the metal into their vacant π or π* orbitals are called π-acid ligands or π-acceptors. CO is an example of such ligands.

# Ambidentate (or Ambident) Ligands

There are many monodentate ligands which have two different donor atoms. Such ligands can coordinate to the metal atom through any of the two different donor atoms and hence are given different names corresponding to the nature of the donor atom linked to the metal atom. Such ligands are called ambidentate ligands.

**Examples of ambidentate ligands.** Examples of some ambidentate ligands are given below:

**(*i*) Cyano ($\bar{C}N$) and isocyano ($\bar{N}C$) ligands.** The structure(s) of these ligands are given below:

$$\overset{..}{\bar{C}} \equiv N: \qquad :\bar{\overset{..}{N}} = C:$$

Cyno ligand ($\bar{C}N$) Isocyano ligand ($\bar{N}C$)

Cyano ligand gets attached with metal atom (M) through the *lp* of electrons present on C-atom ($M \leftarrow \bar{C} \equiv N$: or $M \leftarrow \bar{C}N$) while isocyano ligand is attached with metal atom (M) through the *lp* of electrons on N-atom ($M \leftarrow \overset{..}{N} = C$: or $M \leftarrow \bar{N}C$).

***(ii)* Nitro (N) or nitro ($\bar{N}O_2$) and nitro (O) or nitrito** (ON$\bar{O}$) **ligands.** The structures of these ligands are given below

$$:\overset{..}{\underset{..}{O}} \leftarrow \bar{N} = \overset{..}{O}: \quad \text{or} \quad :\bar{\overset{..}{N}} \rightarrow \overset{..}{\underset{..}{O}}: \ (\text{with } \| :\overset{..}{O}: \text{ on N}) \qquad :\overset{..}{\underset{..}{O}} - \bar{N} = \overset{..}{O}:$$

Nitro ligand ($\bar{N}O_2$) Nitro ligand ($\bar{N}O_2$) Nitrito ligand (ON$\bar{O}$)

Nitro ligand ($\bar{N}O_2$) gets attached with metal atom (M) through the *lp* of electron present on N-atom $\left( M \leftarrow \bar{N} \rightarrow \overset{..}{\underset{..}{O}}: \ (\| :O:) \quad \text{or} \quad M \leftarrow \bar{N}O_2 \right)$ while nitrito ligand (ON$\bar{O}$) is attached to the metal atom through the *lp* of electrons residing on O-atom ($M \leftarrow \bar{\underset{..}{O}} - \overset{..}{N} = \overset{..}{O}$: or $M \leftarrow \bar{O}NO$).

Examples of complexes containing nitro ($\bar{N}O_2$) and nitrito ($ONO^-$) ligands are:

$[Co(NH_3)_5(NO_2)]Cl_2$ (yellow brown) ... pentaamine **nitro** cobalt (III) chloride.

$[Co(NH_3)_5(ONO)]Cl_2$(red) ... pentaamine **nitrito** cobalt (III) chloride

***(iii)* Cyanato ($\bar{O}CN$) and isocyanato ($\bar{N}CO$) ligands.** The structures of these ligands are given below:

$$:\bar{\underset{..}{\overset{..}{O}}} — C \equiv \overset{..}{N} \qquad :\bar{\overset{..}{N}} = C = \overset{..}{O}:$$

Cyanato ligand ($\bar{O}CN$)) Isocyanato ligand ($\bar{N}CO$)

Cyanato ($\bar{O}CN$) ligand gets attached with the metal atom (M) through the *lp* of electrons on O-atom $M \leftarrow \bar{\underset{..}{\overset{..}{O}}} - C \equiv \overset{..}{N}$ or $M \leftarrow \bar{O}CN$ while isocyanato ligand ($\bar{N}CO$) is attached to the metal atom (M) through the *lp* of electrons on N-atom ($M \leftarrow \bar{\underset{..}{N}} = C = \overset{..}{O}$: or $M \leftarrow \bar{N}CO$).

***(iv)* Thiocyanato ($\bar{S}CN$) and isothiocyanato ($\bar{N}CS$) ligands.** The structures of these ligands are given below:

$$:\bar{\underset{..}{\overset{..}{S}}} — C \equiv \overset{..}{N} \qquad :\overset{..}{S} = C = \bar{\overset{..}{N}}:$$

Thiocyanato ligand ($\bar{S}CN$) Isothiocyanato ligand ($\bar{N}CS$)

Thiocyanato ligand ($\bar{S}CN$) gets attached with the metal atom (M) through the *lp* of electrons on S-atom ($M \leftarrow \bar{\underset{..}{\overset{..}{S}}} - C \equiv \overset{..}{N}$ or $M \leftarrow \bar{S}CN$) while isothiocyanato ligand

($\bar{N}CS$) is linked with the metal through the *lp* of electrons on N-atom ($M \leftarrow \ddot{\bar{N}} = C = \ddot{S}:$ or $M \leftarrow \bar{N}CS$). Examples of complexes containing $\bar{S}CN$ and $\bar{N}CS$ ligands are:

$[Pd(dipy)(SCN)_2]^0$ (orange yellow) ..... dipyridyl dithiocynato palladium (II)

$[Pd(dipy)(NCS)_2]^0$ (light yellow) .... dipyridyl diisothiocyanato palladium (II)

## Flexidentate Ligands

Sometimes polydentate ligands do not use all their donor atoms to form coordinate bonds with the central metal atom or ion. This character of polydentate ligands is called their flexidentate character and the ligands are called **flexidentate ligands**. Examples of such ligands are given below:

(*i*) **$SO_4^{2-}$ ion.** This ion, $O{=}S(O^-)(O^-){=}O$ has two negatively-charged O-atms as its donor atoms. Hence this ligend acts as a bidentate ligand in forming complex species like $[Co^{3+}(NH_3)_4(SO_4)]^+$ (octahedral complex ion). In this complex ion both negatively-charged O-atoms of $SO_4^{2-}$ ion are linked with the central $Co^{3+}$ ion. $SO_4^{2-}$ ion also acts as a monodentate ligand is forming complex species like $[Co^{3+}(NH_3)_5(SO_4)]$ (octahedral complex ion). In this complex ion, $SO_4^{2-}$ is linked with $Co^{3+}$ ion through one of its two negatively-charged O-atoms (See Fig 11.26).

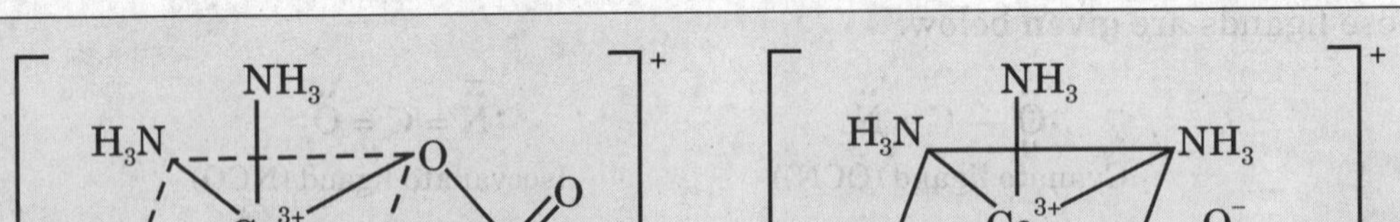


Structure of $[Co^{3+}(NH_3)_4(SO_4)]^+$ ion (I) Structure of $[Co^{3+}(NH_3)_5(SO_4)]^+$ ion (II)

**Fig. 11.26** In (I) $SO_4^{2-}$ ion acts as a bidentate ligand while in (II) this ion acts as a monodentate ligand

(*ii*) **$CO_3^{2-}$ ion.** This ion, $O = C(O^-) - O^-$ has two negatively-charged O-atoms as its donor atoms. Hence, this ligand acts as a bidentate ligand in forming complex species like $[Co^{3+}(NH_3)_4(CO_3)]^+$. In forming $[Co^{3+}(NH_3)_5(CO_3)]^+$ ion, $CO_3^{2-}$ ion acts as a monodentate ligand.

***(iii)* $N_2$ molecule.** The structure of this molecule is : N ≡ N: which shows that it is a bidentate ligand. In the formation of a bridged complex ion *viz.* $[H_3N]\ Ru^{2+} \leftarrow N \equiv N \rightarrow Ru^{2+}(NH_3)_5]^{4+}$ ion $N_2$ molecule acts is a bidentate ligand but in the formation of the complex ion like $[Ru^{2+}(NH_3)_5(N_2)]^{2+}$, this molecule acts as a monodentate ligand. The structure of $[Ru^{2+}(NH_3)_5(N_2)]^{2+}$ ion is shown in the margin.

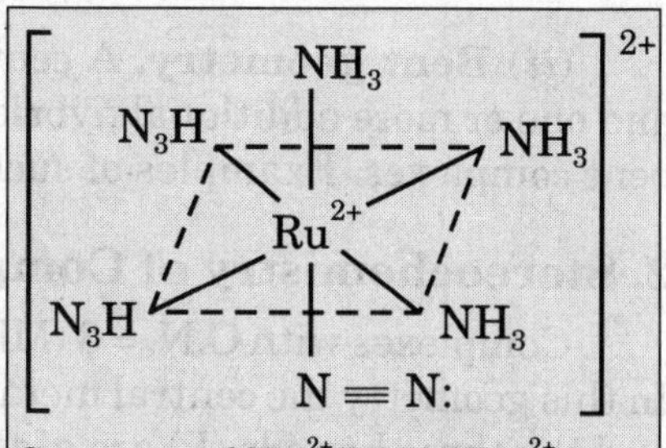
Structure of $[Ru^{2+}(NH_3)_5(N_2)]^{2+}$ ion

***(iv)*** Although the ligands like $OH^-$, $NH_2^-$, $Cl^-$, $O_2^{2-}$, etc. are monodentate ligands, they act as bidentate ligands when they are attached with two separate metal atoms, forming bridge between them. Thus, when these monodentate ligands act as bridging ligands, they behave as bidentate ligands.

***(v)*** Although $EDTA^{4-}$ ion is a hexadentate ligand, it may act as a tetra or pentadentate ligand depending on the size and stereochemistry of the complex formed.

## Coordination Polyhedron Around the Central Atom / Ion: Stereochemistry of Complexes with Different Coordination Numbers

The ligands present in a complex adopt a definite geometry around the central metal atom. The geometry adoped by the ligands is the geometry (stereochemistry) of the complex compound. This geometry depends on the coordination number (C.N.) of the central metal ion. *The arrangement of the coordinated ligands around the central metal ion is such that the electrostatic repulsion between the ligands is minimum.*

*The spatial arrangement of ligands around the central metal atom / ion in coordination sphere / entity which are directly attached with the central metal atom / ion is called coordination polyhedron around the central atom / ion.*

The generally observed coordination polyhedra (i.e. geomeries) are tetrahedral, square planar, octahedral, square pyramidal and trigonal bipyramidal.

### 1. Stereochemistry of Complexes with C.N. = 2.

Complexes with C.N. = 2 have linear or bent (non-linear) geometry.

***(i)* Linear geometry.** In this geometry the central metal ion (M) and the two ligands (L) are lying in the same line on the same plane with L-M-L angle = 180°. (See Fig 11.27).

A central metal ion which utilises two hybrid orbitals for bonding with the ligands and does not have any lone pair (*lp*) of electrons forms linear complexes. Linear complexes are formed by Cu (I), Au (I), Ag (I), Hg (II), etc. Examples of such complexes are $[CuCl_2]^-$, $[Au(CN)_2]^-$ $[Ag(NH_3)_2]^{2+}$, $[Hg(CN)_2]^0$. The structure of $[Ag\ (NH_3)_2]^{2+}$ is given in Fig 11.28.

L—M—L

**Fig. 11.27** Linear structure of $ML_2$ complex (C.N. of M = 2)

$[H_3N \rightarrow Ag \leftarrow NH_3]^+$

**Fig. 11.28** Linear structure of $[Ag\,NH_3)_2]^+$ (C.N. of $Ag^+$ = 2)

***(ii)* Bent geometry.** A central metal ion which utilises two hybrid orbitals and one or more additional hybrid orbital for accommodating *lps* of electrons forms bent complexes. Examples of such complexes are rare.

## 2. Stereochemistry of Complexes with C.N. = 3.

Complexes with C.N. = 3 ($ML_3$ type complexes) have trigonal planar geometry. In this geometry the central metal atom (M) lies at the centre of equilateral triangle and the three ligands (L) are at the three vertices of the triangle with L-M-L angle = 120°. Examples of complexes having trigonal planar geometry are $K[Cu(CN)_3]$, chlorobis (thiourea) copper (I), $[CaCl(tu)_2]^0$, *tris* (triphenylphosphine) platinum (0), $[(C_6H_5)P)_3\ Pt]$, *tris* (trimethylphosphine sulphide) copper (I) perchlorate, $[Cu(SP(CH_3)_3)_3]^+\ ClO^-_4$ and triiodomercurate (II) anion, $[HgI_3]^-$. In the structure of $[HgI_3]^-$, $I^-$ ions are arranged at the three corners of slightly distorted equilateral triangle with Hg (II) at its center (See Fig 11.29).

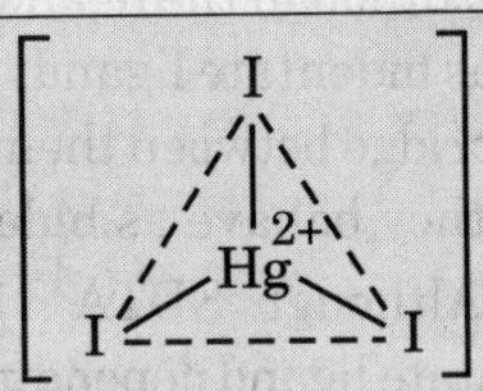


**Fig. 11.29** Trigonal planar structure of $[HgI_3]^-$ complex ion.

## 3. Stereochemistry of Complexes with C.N. = 4.

Complexes with C.N. = 4 (complexes of $ML_4$ type) have tetrahedral or square planar geometry.

***(i)* Tetrahedral geometry.** A tetrahedron has four vertices (corners), four triangular faces and six edges. In tetrahedral geometry the four ligands (L) are placed at the four corners of a regular tetrahedron and the central metal atom (M) lies at the centre.

Complexes having tetrahedral geometry are formed by transition as well as non-transition elements. Complexes having weak donor legand such as $Cl^-$, $PPh_3$, $OPPh_3$ have tetrahedral geometry with ligand-metal-ligand angle around 109°. Examples of such complexes are $[BeF_4]^{2-}$, $[BF_4]^-$, $[CoCl_4]^{2-}$ (blue), $[CuCl_4]^{2-}$, $[NiCl_4]^{2-}$, $[NiCl_2(PPh_3)_2]^0$, $[ZnCl_4]^{2-}$, $[Cd(CN)_4]^{2-}$, $[MnCl_4]^{2-}$, $[NiCl_2(OPPh_3)_2]^0$, etc. The tetrahedral structure of $[BeF_4]^{2-}$ is given in Fig 11.30.

***(ii)* Square planar complexes.** Complexes having square planar geometry are formed by transition metals. Complexes having strong donar ligands have square planar geometry with ligand-metal-ligand angle around 90°. In square planar geometry the central metal ion is placed at the centre of the square and the four ligands (L) are placed at the four corners of the square. All the five species (one M atom and four L legands) lie in the same plane. Examples of square planar complexes are $[Ni(CN)_4]^{2-}$, $[Cu(NH_3)_4]^{2-}$, *bis* (dimethyl glyoximato) nickel (II), $[Ni(dmg)_2]^0$, *bis* (ethylene diamine) copper (II), $[Cu(en)_2]^{2+}$, *bis* (biguanide), nickel (II), *bis* (biguanide) copper (II). Structure of $[Ni(CN)_4]^{2-}$ is given in Fig. 11.31.

Majority of complexes of Pt (II) and Pd (II) are square planar. Examples are $[Pt(NH_3)_4]^{2+}$, $[PtCl_2(PPh_3)_2]^0$, $[PtCl_4]^{2-}$, $[Pt(NH_3)_2Cl_2]^0$, $[PdCl_4]^{2-}$, etc.

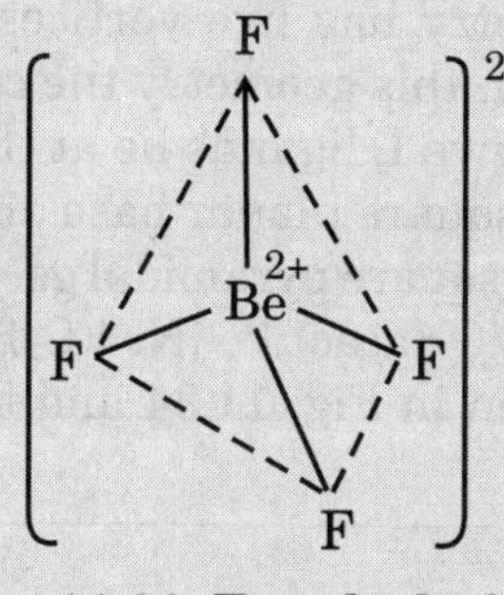


**Fig. 11.30** Tetrahedral geometry of $[BeF_4]^{2-}$.

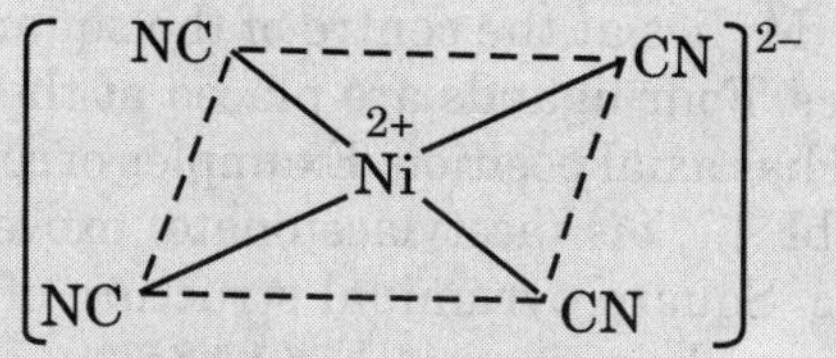


**Fig. 11.31** Square planar geometry of $[Ni\ (CN)_4]^{2-}$.

## 4. Stereochemistry of Complexes with C.N. = 5.

Complexes with C.N. = 5 have trigonal bipyramidal or square pyramidal geometry.

***(i)* Trigonal bipyramidal geometry.** Trigonal bipyramidal geometry has five vertices, six triangular faces and nine edges. In this geometry the central metal atom (M) lies at the centre of the triangular (trigonal) base and five L ligands lie at the five vetices. Three ligands in the complexes having trigonal bipyramidal geometry are placed at the three corners of the triangular base and the two ligands occupy the axial positions. Examples of such complexes are $Fe(CO)_5$, $[MoCl_5]^-$, $[CuCl_5]^{3-}$, $[SnCl_5]^-$, $[NiBr_3\{P(C_2H_5)_3\}_3]$ and $[Cu^{2+}I\ (bpy)_2]I$. $[Cu^{2+}I\ (bpy)_2]$ I is called iodo*bis* (bipyridine) copper (II) iodide.

The trigonal bipyramidal structures of $Fe(CO)_5$, $[MoCl_5]^-$, $[CuCl_5]^{3-}$, $[SnCl_5]^-$, etc. are given in Fig. 11.32 and the structure of $[Cu^{2+}I(bpy)_2]^+$ ion is given in Fig. 11.33.

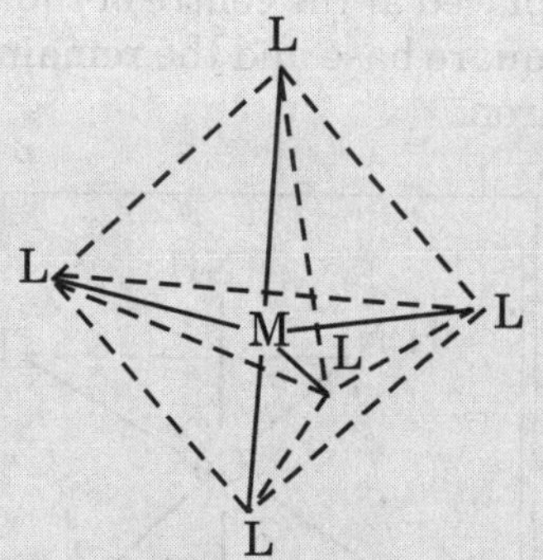


**Fig. 11.32** Trigonal bipyramidal structure of comp of iodobis lexes such as $Fe(CO)_5$, $[MoCl_5]^-$, $[CuCl_5]^{3-}$, $[SnCl_5]^-$ etc. L indicates ligand which may be CO, $Cl^-$ etc.

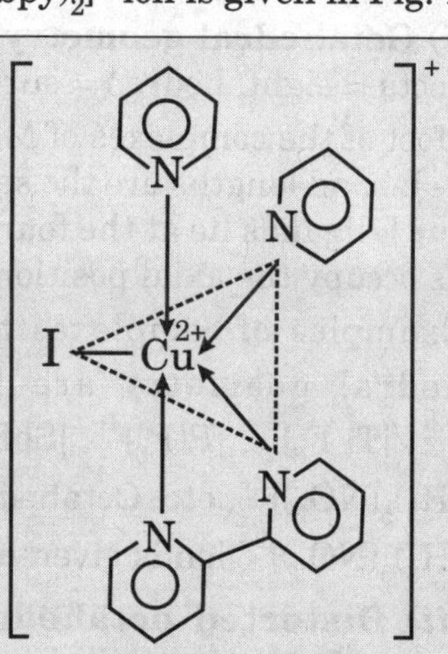


**Fig. 11.33** Trigonal bipyramidal structure (bipyridine) copper (II) ion, $[Cu^{2+}\ I\ (bpy)_2]^+$

***(ii)* Square pyramidal geometry.** This geometry has five vertices, four triangular faces and one square face and eight edges. In this geometry the central metal (M) lies at the centre of the square base and the five L ligands lie at the five vertices. Four ligands are placed at the corners of the square planar base and one ligand has axial position. Examples of complexes having square pyramidal geometry are $[SbF_5]^{2-}$, *bis* (acetylacetonato) oxovanadium (IV), $[VO\ (acac)_2]^0$, $[Ni\ \{(C_6H_5)_3P\}]Br_3$ etc. Square pyramidal structure of $[SbF_5]^{2-}$ is given in Fig. 11.34 and that of $[VO\ (acac)_2]^0$ is given in Fig 11.35.

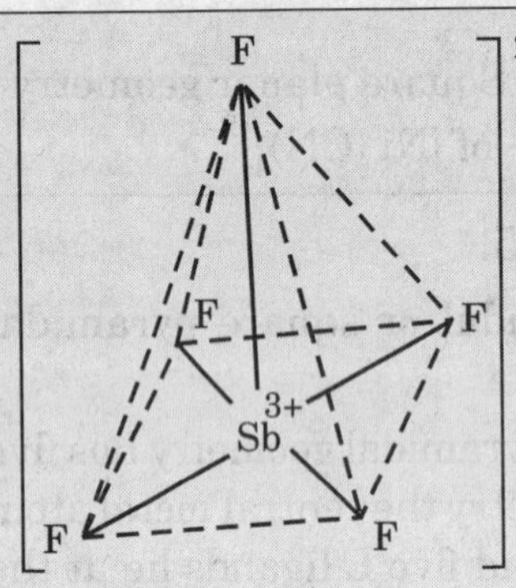


**Fig. 11.34** Square pyramidal structure of $[SbF_5]^{2-}$ ion

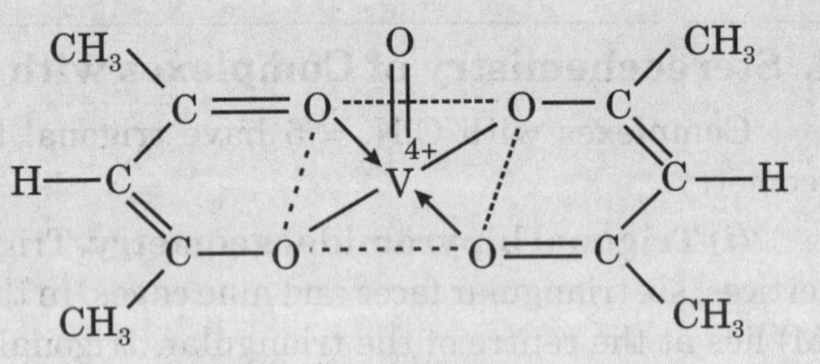


**Fig. 11.35** Square pyramidal structure of bis (acetylacetonato) oxovanadium (IV), $[VO(acac)_2]^0$

## 6. Stereochemistry of complexes with C.N. = 6

Complexes with C.N. = 6 ($ML_6$ type complexes) have *octahedral* or *distorted octahedral geometry*.

***(i)* Octahedral geometry.** An octahedron has six vertices, eight triangular faces (octa = eight, hedron = surface or face) and twelve edges.

Most of the complexes of $ML_6$ type have octahedral geometry in which all the six M – L bond lengths are the same. M metal is placed at the centre of square base and four L ligands lie at the four corners of the square base and the remaining two ligands occupy the axial positions of the octahedron.

Examples of complexes having regular octahedral geometry are $[Cu(NH_3)_6]^{2+}$, $[FeF_6]^{3-}$, $[Ti\ F_6]^{2-}$, $[PtF_6]^{4-}$, $[SbF_6]^-$, $[Cr(en)_3]^{3+}$, $[Co(NH_3)_5(NO_2)]^{2+}$, etc. Octahedral structure of $[Co(NH_3)_5(NO_2)]^{2+}$ ion is given in Fig 11.36.

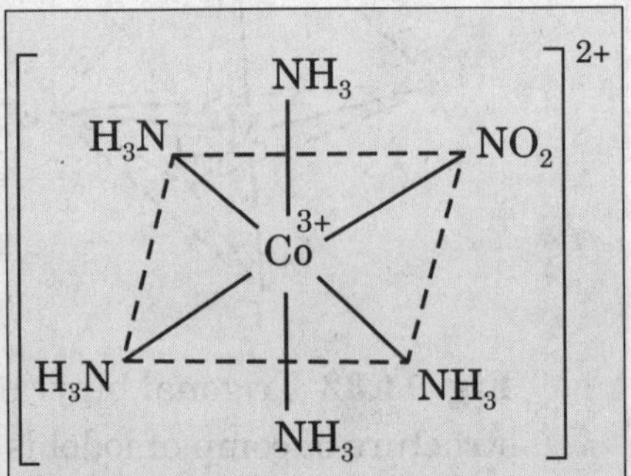


**Fig. 11.36** Octahedral structure of $[Co\ (NH_3)_5(NO_2)]^{2+}$ ion

***(ii)* Distorted octahedral geometry.** There is a limited number of 6-coordinated complexes which have distorted octahedral geometry. This geometry is obtained when distortion takes place in octahedral geometry. Depending on the nature of the coordinated ligands, there are two types of distortion in octahedral geometry.

(*a*) **Tetragonal elongation** which occurs when the two *trans* ligands (axial ligands) are pulled away from the central metal ion. Hence this distortion is called *tetragonal elongation*. This distortion finally gives square planar geometry. $[CoCl_2(NH_3)_4]^-$ is a complex ion which has tetragonally elongated octahedral structure.

(*b*) **Tetragonal compression** which occurs when the two *trans* ligands are brought closer to the central metal ion.

## 7. Stereochemistry of Complexes with C.N. = 7.

There is only a limited number of complexes with C.N. = 7. The arrangement of seven ligands around the central metal ion can give the following three geometries to the complexes with C.N. = 7

***(i)* Pentagonal bipyramidal geometry.** This geometry is found in $[UO_2F_5]^{3-}$, $[UF_7]^{3-}$, $[ZrF_7]^{3-}$. The five $F^-$ ligands of $[UO_2F_5]^{3-}$ ion occupy the five vertices of the pentagon while the two oxo groups take up the other two positions above and below the pentagon (axial positions). (See Fig. 11.37.)

***(ii)* Capped octahedral geometry.** This geometry is obtained when an additional seventh ligand is placed at the centre of one triangular face of octahedron. $[NbOF_6]^{3-}$ ion has capped octahedral structure shown in Fig. 11.38.

***(iii)* Caped trigonal prism.** This geometry is obtained when an additional seventh ligand is placed above the centre of rectangular face of a trigonal prism. This type of geometry is adopted by $[NbF_7]^{2-}$ and $[TaF_7]^{2-}$. The structure of $[TaF_7]^{2-}$ is given in Fig. 11.39. In this structure the six $F^-$ ions are placed at the six vertices of trigonal prism while seventh $F^-$ ion is above the centre of a rectangular face of the prism.

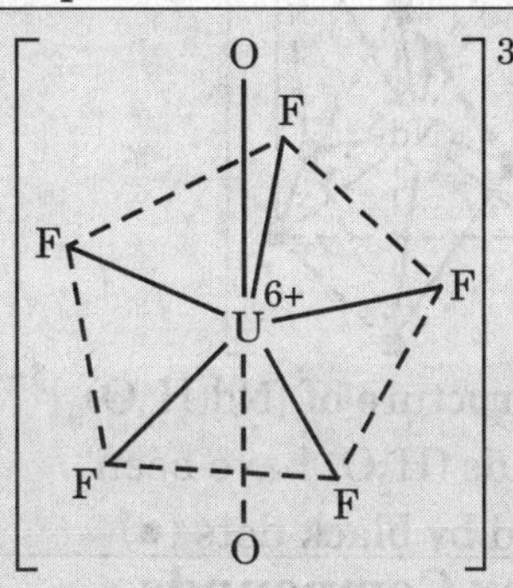


**Fig. 11.37** Pentagonal bipyramidal geometry of $[UO_2F_5]^{3-}$ complex.

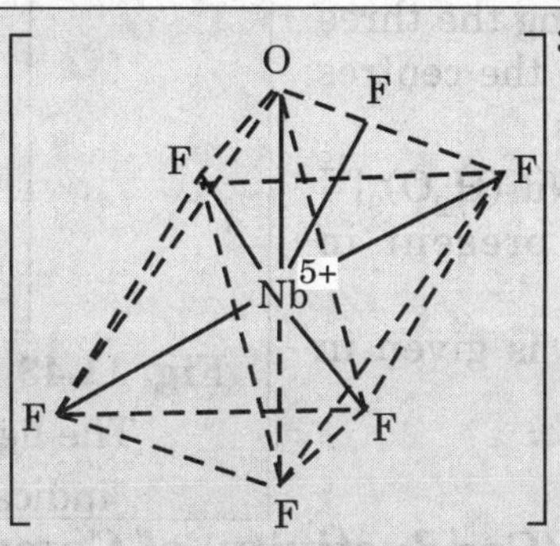


**Fig. 11.38** Capped octahedral structure of $[NbOF_6]^{3-}$ complex.

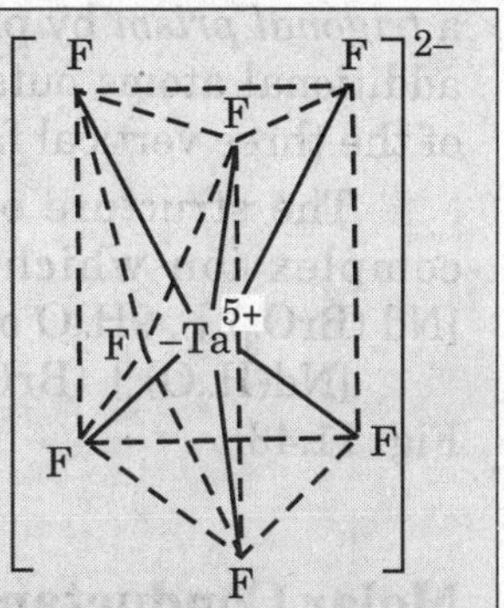


**Fig. 11.39** Capped prism structure of $[TaF_7]^{2-}$ complex

## 8. Stereochemistry of Complexes with C.N. = 8

Complexes with C.N. = 8 are usually formed by lanthanides and actinides. These complexes can have any of the following three geometries.

***(i)* Cubic geometry.** In this geometry, eight ligands lie on eight vertices of the cube and central metal atom is placed at the centre of the cube. Complex ions of $[MF_8]^{3-}$ (M = Pa, U, Np) type have cubic structure. The cubic structure of $[UF_8]^{3-}$ is shown in Fig. 11.40.

***(ii)* Square antiprism geometry.** This geometry is the distorted form of cubic geometry which is adopted to minimise the repulsion between the anions. $[TaF_8]^{3-}$, $[ReF_8]^{2-}$ and $[Zr\ (acac)_4]^0$ have square antiprism geometry. The square antiprism geometry of $[TaF_8]^{3-}$ is shown in Fig. 11.41.

***(iii)* Dodecahedral geometry.** $[Mo(CN)_8]^{4-}$ and $[Zr\ (ox)_4]^{4-}$ have this geometry. The dodecahedral structure of $[Mo\ (CN)_8]^{4-}$ is shown in Fig. 11.42.

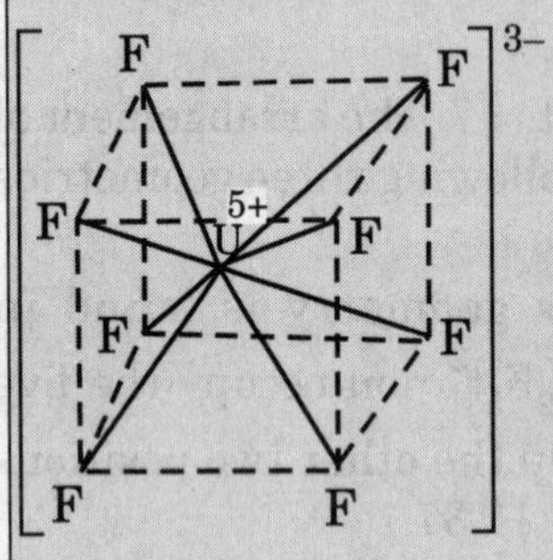


**Fig. 11.40** Cubic structure of $[UF_8]^{3-}$ complex

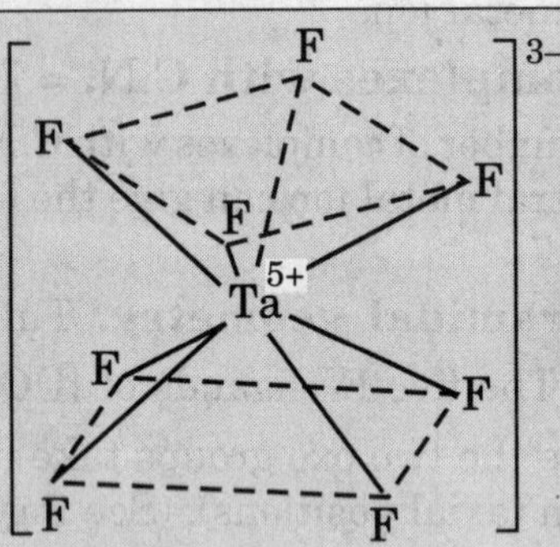


**Fig. 11.41** Square antiprism structure of $[TaF_8]^{3-}$ complex.

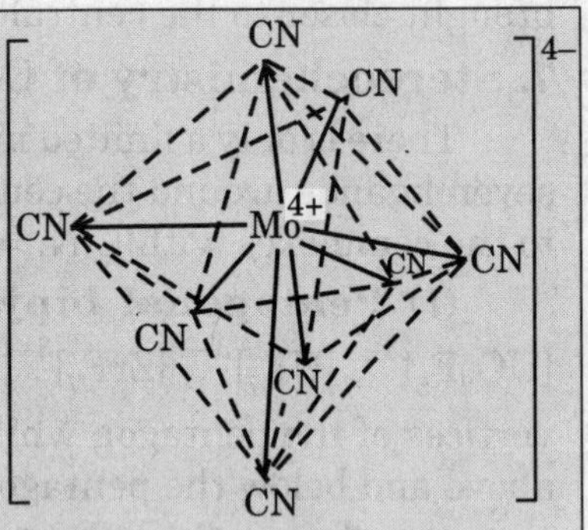


**Fig. 11.42** Dodecahedral structure of $[Mo(CN)_8]^{4-}$ complex.

## 9. Stereochemistry of Complexes with C.N. = 9

There is only one symmetrical arrangement known for this coordination number. This is derived from a *trigonal prism* by placing the three additional atoms outside the centres of the three vertical faces.

The structure of $[Nd\ (H_2O)_9]^{3+}$ complex ion which is present in $[Nd\ (BrO_3)_3]\ .\ 9H_2O$ or $[Nd(H_2O)_9]\ (BrO_3)_3$ is given in Fig. 11.43.

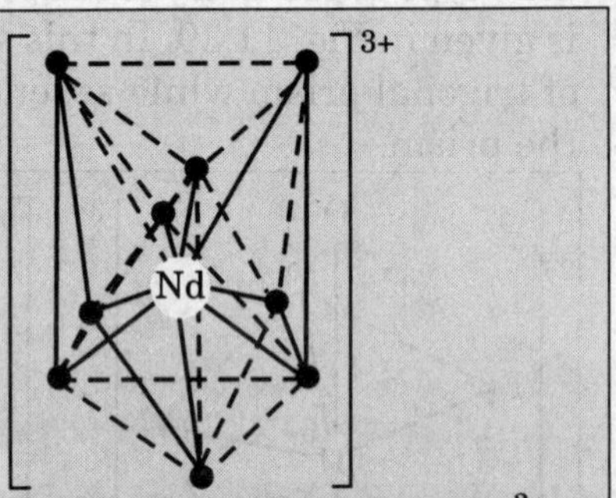


**Fig. 11.43** Structure of $[Nd(H_2O)_9]^{3+}$. The ligands ($H_2O$) have been indicated by black dots (●)

## Molar Conductance (Conductivity) of Complex Compounds.

Molar conductance, $\wedge_m$ ($ohm^{-1}\ cm^2$) of complex compounds can be determined. It has been observed that the values of $\wedge_m$ of a given set of complex compounds decrease with the decrease of the total number of ions obtained by the ionisation of the complex compound in aqueous solutions. For example,

$$[Cu(NH_3)_6]Cl_3 \rightarrow \underbrace{[Co(NH_3)_6]^{3+} + 3Cl^-}_{\text{No. of ions} = 1 + 3 = 4}$$

$$\wedge_m = 390$$

$$[Co(NH_3)_5Cl]Cl_2 \rightarrow \underbrace{[Co(NH_3)_5Cl]^{2+} + 2Cl^-}_{\text{No. of ions} = 1 + 2 = 3}$$

$$\wedge_m = 262$$

$$[Co(NH_3)_4Cl_2]Cl \rightarrow \underbrace{[Co(NH_3)_4Cl_2]^+ + Cl^-}_{\text{No. of ions} = 1+1=2}$$

$$\wedge_m = 102$$

$$[Co(NH_3)_3Cl_3]^0 \rightarrow \text{No ionisation,}$$

$$\wedge_m = 0.0 \text{ (Non-electrolyte)}$$

If two complex compounds give the same number of ions on ionisation, the complex compound having greater sum of the charges on its ions has greater value of $\wedge_m$. For example, the value of $\wedge_m$ of $Mg[Cr(NH_3)(NO_2)_5]$ is greater than that of $K[Co(NH_3)_2(NO_2)_4]$ as shown below.

$$Mg[Cr(NH_3)(NO_2)_5] \rightarrow \underbrace{Mg^{2+} + [Cr(NH_3)(NO_2)_5]^{2-}}$$

No. of ions = 1 + 1 = 2
Sum of charges on ions
= 2 + 2 = 4 (+2, –2)

$$K[Co(NH_3)_2(NO_2)_4] \rightarrow \underbrace{K^+ + [Co(NH_3)_2(NO_2)_4]^-}$$

No. of ions = 1 + 1 = 2
Sum of charges on ions
= 1 + 1 = 2 (+1, –1)

## Precipitation Reactions of Complex Compounds with Ag $NO_3$ and $BaCl_2$ Solution

When a complex compound containing halogen atom outside the coordination sphere is treated with an aqueous solution of $AgNO_3$, halogen atom is precipitated as silver halide (AgX). For example,

$$[Cr(H_2O)_6]Cl_3 + 3AgNO_3 \rightarrow [Cr(H_2O)_6](NO_3)_3 + 3AgCl \text{ (\textit{White ppt})}$$

Similarly when a complex compound containing $SO_4^{2-}$ ion outside the coordination sphere is treated with $BaCl_2$ solution, a white ppt. of $BaSO_4$ is obtained. For example:

$$[Co(NH_3)_5Br]SO_4 + BaCl_2 \rightarrow [Co(NH_3)_5Br]Cl_2 + BaSO_4 \text{ (\textit{White ppt})}$$

## Types of Complexes

Complexes have been classified in different ways as shown below:

## A. Cationic, Anionic and Neutral Complexes

This classification gives the following types of complexes:

***(i)* Cationic complexes.** A complex in which the complex ion carries a net positive charge is called *cationic complex, e.g.*, $[Co(NH_3)_6]^{3+}$, $[Cu(NH_3)_4]^{2+}$, etc.

***(ii)* Anionic complexes.** A complex in which the complex ion carries a net negative charge is called *anionic complex, e.g.*, $[Ag(CN)_2]^-$, $[Fe(CN)_6]^{4-}$, etc.

***(iii)* Neutral complexes.** A complex carrying no net charge is called a *neutral complex* or simply a *complex, e.g.*, $[Ni(CO)_4]$, $[Co(NH_3)_3Cl_3]$, etc.

## B. Perfect (penetrating) and Imperfect (normal) Complexes

**(*i*) Perfect or penetrating complexes.** These are the compounds in which complex ion is fairly stable and is neither dissociated nor feebly dissociated in solution. For example:

$$K_4[Fe(CN)_6] \longrightarrow 4K^+ + [Fe(CN)_6]^{4-}$$
$$\Updownarrow$$
$$Fe^{2+} + 6CN^- \text{ (feebly dissociated)}$$

**(*ii*) Imperfect or normal complexes.** These are the compounds in which complex ion is less stable and is reversibly dissciated to give enough simple ions.

$$K_2[Cd(CN)_4] \longrightarrow 2\,K^+ + [Cd(CN)_4]^{2-}$$
$$\Updownarrow$$
$$Cd^{2+} + 4\,CN^- \text{ (appreciably disociated)}$$

## C. Labile and Inert Complexes

**(*i*) Labile complexes.** Those complexes which are highly reactive (*i.e.*, in which ligands can be readily replaced by other ligands) are called *labile complexes.*

**(*ii*) Inert complexes.** Those complexes which are less reactive, (*i.e.*, in which the ligands cannot be readily replaced by other ligands) are called *inert complexes.*

## D. Inner Metallic Complexes of 1st Order and 2nd Order

Glycinato ion, $NH{-}CH_2{-}\overset{\overset{\displaystyle O}{\|}}{C}{-}\bar{O}$ $(gl\bar{y})$ is a bidentate ligand, has one neutral N-donor and one acidic (anionic) $\bar{O}$ donor. $[Co^{3+}(gly)_3]^0$ is a non-electrolyte complex. In this complex Co has a charge of +3. Now since glycinato ligand ($gly^-$) is a bidentate ligand, the C.N. of Co atom is 6 and this C.N. of 6 is satisfied by the presence of three $gly^-$ ligands in coordinaton sphere. Since $gly^-$ ligand has one negative charge on O-atom, +3 charges on Co atom are satisfied by the presence of three $gly^-$ ligands. Thus we see that the three $gly^-$ ligands present in the coordination sphere of $[Co^{3+}(gly)_3]^0$ satisfy both the C.N. (= 6) and the charge (= +3) on $Co^{3+}$ ion. Such complexes are called *inner metallic complexes of first order*.

*The complexes in which the C.N. and the charge on the central metal atom are internally fully satisfied at the same time by the coordinate ligands are called inner metallic complexes of first order.* Such complexes are non-electrolytes. Above discussion shows that inner metallic complexes are formed by the ligands which contain one neutral and one anionic (acidic) coordinating atom in the same molecule.

$gly^-$ ion is a ligand of this type. Acetylacetato ($acac^-$), $CH_3{-}\overset{\overset{\displaystyle O}{\|}}{C}{-}CH{=}\overset{\overset{\displaystyle O^-}{|}}{C}{-}CH_3$ is also a ligand of this type because it contains one neutral O-atom and one negatively-charged O-atom which act as coordinating atoms in $[Co^{3+}(acac)_3]^0$ which is an inner metallic complex of first order.

*The complexes in which the C.N. and the charge of the metal are not fully satisfied at the same time by the legand are called inner metallic complexes of second order.* These complexes are, therefore, ionic species. $[Si^{4+}(acac)_3]Cl$, Na $[Cu^{2+}(acac)_3]$, $[B^{3+}(acac)_2]$ $[Fe^{3+}Cl_4]$, etc. are the examples of inner metallic complexes of second order. In $[Si^{4+}(acac)_3]^+$ ion, the three $acac^-$ ligands satisfy only C.N. of Si-atom which is equal to 6 but the charge on Si-atom (= + 4) is not satisfied by $acac^-$ ligands.

## E. Homoleptic and Heteroleptic Complexes

Complexes in which the metal atom/ion is linked with the ligands of the same type are called homoleptic complexes. $[Co(NH_3)_6]^{3+}$ is a homoleptic complex. On the other hand the complexes in which the metal atom/ion is attached with the ligands of different type are called heteroleptic complexes. $[Co(NH_3)_4Cl_2]^+$ is a heteroleptic complex.

## F. Mononuclear and Plynuclear Complexes

This classification gives the following two types of complexes:

**(*i*) Mononuclear complexes.** These complexes have only one metal atom/ion. $[Co(NH_3)_6]^{3+}$ is a mononuclear complex.

**(*ii*) Polynuclear complexes.** The complexes having two or more metal atoms/ions are called polynuclear complexes. In bridged complexes having two metal atoms, the metal atoms are linked together by two monodentate bridging ligands. The complex ion (I) whose structure is shown below is a polynuclear complex.

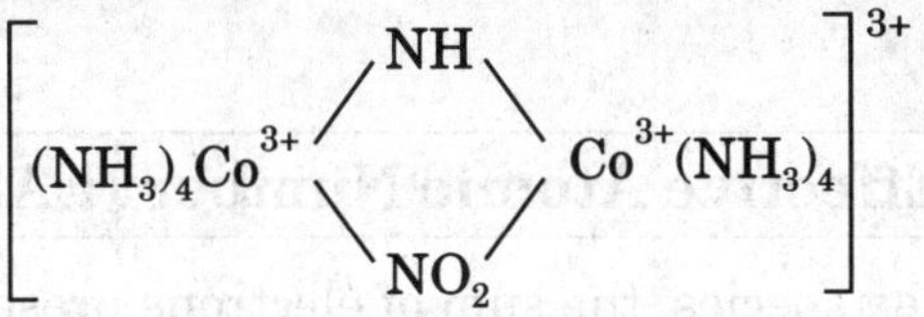

(I)

# Methods to Detect and Determine the Structure of Complexes

Some of the methods used for the detection and determination of the structure of complexes are given below.

**1. Solubility.** Complex formation can be indicated by the abnormal solubility behaviour of complexes. For example:

(*i*) When 8-hydroxyquinoline (Hoxin) is added to $Al^{3+}$ ion, insoluble ppt. of $[Al(oxin)_3]^0$ is obtained.

(*ii*) When $NH_4OH$ is added to insoluble AgCl, soluble complex $[Ag(NH_3)_2]^+$ is obtained.

(*iii*) Yellow/red insoluble HgO gets dissolved in the excess of KI due to the formation of soluble complex, $K_2[HgI_4]$.

**2. Colour.** Colour change takes place when a complex compound is formed. For example:

(*i*) The addition of $NH_3$ to $[Cu(H_2O)_6]^{2+}$ ion gives $[Cu(H_2O)_2(NH_3)_4]^{2+}$ which has intense blue colour.

$$[Cu(H_2O)_6]^{2+} + 4NH_3 \longrightarrow [Cu(H_2O)_2(NH_3)_4]^{2+} + 4H_2O$$

In the formation of $[Cu(H_2O)_2(NH_3)_4]^{2+}$ the weak $H_2O$ donors are replaced by strong $NH_3$ donors. This replacement shifts the absorption band to lower wavelength (from ~ 800 A to ~ 6000 A)

**3. Magnetic susceptibility.** Measurement of magnetic susceptibility of a complex sometimes gives us an information about the nature of bonding and stereochemistry of the complex. For example:

(*i*) *bis* (dimethylglyoximato) nickel (II) and *bis* (biguanide) nickel (II) both are diamagnetic and hence are four co-ordinated square planar complexes. These complexes result from $dsp^2$ hybridisation of $Ni^{2+}$ ion ($Ni^{2+} = 3d^8$). This hybridisation gives $n = 0$ which makes these complexes diamagnetic.

(*ii*) $M_2^+[Ni^{2+}Cl_4]$ complex is paramagnetic ($\mu \approx 3.9$ BM). Hence this complex has tetrahedral structure. This structure results from $sp^3$ hybridisation of $Ni^{2+}$ ion. This hybridisation gives $n = 2$ which makes this complex paramagnetic ($\mu \approx 3.9$ BM).

**4. Conductivity measurements.** Conductivity measurements of complex compounds give us an idea of the nature of the groups present inside and outside coordination sphere of a complex. For example:

(*i*) Conductivity measurements of $[Co(NO_2)_2(en)_2]X_2$ has shown that the two univalent monovalent nitro groups ($NO^-_2$) are inside the coordination sphere. The presence of $X^-$ ions outside the coordination sphere can be detected by chemical tests.

(*ii*) Conductivity measurement of $[PtCl_2(NH_3)_2]$ has shown that this complex is a non-electrolyte complex. It has two $Cl^-$ ions which are bound to $Pt^{2+}$ ion in coordination sphere.

## Effective Atomic Number (EAN)

In a given complex species, the sum of electrons present in the central metal atom or ion and those donated by the ligands is called EAN of the central metal atom / ion.

## Calculation of EAN of the Central Metal Atom or Ion In a Complex Species

EAN of the central metal atom or ion of a given complex species is given by:

EAN of the central metal atom = Atomic number (*Z*) of the metal atom – oxidation number (O.N.) of the metal atom + No. of electrons donated by ligands

$$= Z - (O.N.) + 2 \times C.N.$$

**Examples** (*i*) EAN of $Fe^{2+}$ ion in $[Fe^{2+}(CN)_6]^{4-}$ ion = Z of Fe atom $-2 + 6 \times 2$

$= 26 - 2 + 12 = \mathbf{36}$

(*ii*) EAN of $Fe^{3+}$ ion in $[Fe^{3+}(CN)_6]^{3-}$ ion = Z of Fe atom $-3 + 6 \times 2 = \mathbf{35}$

(*iii*) EAN of $Co^-$ ion in $[Co^-(CO)_4]^-$ ion.

$= Z$ of Co-atom $- (-1) + 4 \times 2 = 27 + 1 + 8 = \mathbf{36}$

(*iv*) EAN of Ni atom in $[Ni^0(CO)_4]^\circ$ = Z of Ni-atom $- 0 + 4 \times 2 = 28 + 8 = \mathbf{36}$

## EAN Rule

According to this rule, a stable complex species is formed, if the EAN of the central metal atom present in the complex species is equal to the atomic number of the noble gas which follows the central metal atom in the same period of the periodic table. This rule is also called **atomic number rule** or **noble gas rule**. When EAN of the metal atom is 36(Kr), 54(Xe) or 86(Rn), the rule is said to be followed.

EAN of the central metal atom in some complex species has been calculated as shown in Table 11.2. This table shows that the central metal atom in many complexes does not obey EAN rule, *i.e.* EAN of the central metal atom in many complexes is not equal to the atomic number of the next noble gas but it is more or less than the atomic number of the next noble gas.

**Table 11.2. To calculate the EAN of the central metal atom of some complex ions ($n$ = No. of ligands and $y$ = No. of electrons donated by one ligand)**

| *Complex ion* | *Atomic number of the central metal atom (Z)* | *Oxidation state of the central metal atom (x)* | *Electrons donated by ligands* $= n \times y$ | *EAN of the central metal ion =* $(Z - x) + n \times y$ |
|---|---|---|---|---|
| **(A) Complex ions whose central metal ion obeys EAN rule** | | | | |
| $[Pd(NH_3)_6]^{4+}$ | Pd = 46 | +4($Pd^{4+}$) | 6 × 2 = 12 | (46 – 4) + 12 = 54 ($Xe_{54}$) |
| $[Fe(CN)_6]^{4-}$ | Fe = 26 | +2($Fe^{2+}$) | 6 × 2 = 12 | (26 – 2) + 12 = 36 ($Kr_{36}$) |
| $[Co(NH_3)_6]^{3+}$ | Co = 27 | +3($Co^{3+}$) | 6 × 2 = 12 | (27 – 3) + 12 = 36 ($Kr_{36}$) |
| $[Pt(NH_3)_6]^{4+}$ | Pt = 78 | +4($Pt^{4+}$) | 6 × 2 = 12 | (78 – 4) + 12 = 86 ($Rn_{86}$) |
| $[Ag(NH_3)_4]^{+}$ | Ag = 47 | +1($Ag^{+}$) | 4 × 2 = 8 | (47 – 1) + 8 = 54 ($Xe_{54}$) |
| $[Cu(CN)_4]^{3-}$ | Cu = 29 | +1($Cu^{+}$) | 4 × 2 = 8 | (29 – 1) + 8 = 36 ($Kr_{36}$) |
| **(B) Complex ions whose central metal ion does not obey EAN rule** | | | | |
| $[Fe(CN)_6]^{3-}$ | Fe = 26 | +3($Fe^{3+}$) | 6 × 2 = 12 | (26 – 3) + 12 = 35 |
| $[Cr(NH_3)_6]^{3+}$ | Cr = 24 | +3($Cr^{3+}$) | 6 × 2 = 12 | (24 – 3) + 12 = 33 |
| $[Ni(NH_3)_6]^{2+}$ | Ni = 28 | +2($Ni^{2+}$) | 6 × 2 = 12 | (28 – 2) + 12 = 38 |
| $[Ni(en)_3]^{2+}$ | Ni = 28 | +2($Ni^{2+}$) | 3 × 4 = 12 | (28 – 2) + 12 = 35 |
| $[Mn(H_2O)_6]^{2+}$ | Mn = 25 | +2($Mn^{2+}$) | 6 × 2 = 12 | (25 – 2) + 12 = 35 |
| $[Co(CN)_6]^{4-}$ | Co = 27 | + 2($Co^{2+}$) | 6 × 2 = 12 | (27 – 2) + 12 = 37 |
| $[Pt(NH_3)_2Cl_2]^{0}$ | Pt = 78 | + 2($Pt^{2+}$) | 4 × 2 = 8 | (78 – 2) + 8 = 84 |
| $[Cu(NH_3)_4]^{2+}$ | Cu = 29 | + 2 ($Cu^{2+}$) | 4 × 2 = 8 | (29 – 2) + 8 = 35 |
| $[FeCl_4]^{-}$ | Fe = 26 | + 3 ($Fe^{3+}$) | 4 × 2 = 8 | (26 – 3) + 8 = 31 |
| $[Ni(CN)_4]^{2-}$ | Ni = 28 | + 2 ($Ni^{2+}$) | 4 × 2 = 8 | (28 – 2) + 8 = 34 |
| $[PdCl_4]^{2-}$ | Pd = 46 | + 2 ($Pd^{2+}$) | 4 × 2 = 8 | (46 – 2) + 8 = 52 |
| $[Pt(NH_3)_4]^{2+}$ | Pt = 78 | + 2 ($Pt^{2+}$) | 4 × 2 = 8 | (78 – 2) + 8 = 84 |
| $[AgX_2]^{-}$ | Ag = 47 | + 1 ($Ag^{+}$) | 2 × 2 = 4 | (47 – 1) + 4 = 50 |
| $[Ag(NH_3)_2]^{+}$ | Ag = 47 | + 1 ($Ag^{+}$) | 2 × 2 = 4 | (47 – 1) + 4 = 50 |
| $[CuCl_2]^{-}$ | Cu = 29 | + 1 ($Cu^{+}$) | 2 × 2 = 4 | (29 – 1) + 4 = 32 |

## Applications of EAN Rule

### 1. To predict the magnetic behaviour of a given complex ion

With the help of EAN rule, the magnetic property of the complex ions can be predicted. It has been observed that the complex ions whose central metal atom obeys EAN rule are diamagnetic. For example, since the EAN of $Co^{3+}$ ion in $[Co(NH_3)_6]^{3+}$ ion is equal to $(27 - 3) + 6 \times 2 = 36$, this ion obeys EAN rule and hence $[Co(NH_3)_6]^{3+}$ ion is diamagnetic. Experimentally this ion has also been found diamagnetic.

Sidgwick and Bose have suggested that the complex ions whose central metal atom does not obey EAN rule are generally paramagnetic.

### 2. To calculate the number of unpaired electron ($n$) and magnetic moment ($\mu$) value of complexes.

Number of unpaired electron ($n$) present in a given complex ion is given by:

No. of unpaired electron ($n$) = Z of the next noble gas – EAN of the central metal atom

**Example** The value of $n$ in $[Cr(NH_3)_6]^{3+}$ ion = Z of next noble gas (Kr) – EAN of Cr = 36 – 33 = 3

$\therefore$ $\mu$ of $[Cr(NH_3)_6]^{3+}$ ion $= \sqrt{n(n+2)}$ BM $= \sqrt{3 \times (3+2)}$ BM = 3.87 BM.

### 3. To Compare the stability of complexes.

The complexes whose central metal atom obeys EAN rule is more stable.

**Example.** Since Fe atom in $[Fe^{2+}(CN)_6]^{4-}$ obeys EAN rule (EAN of Fe = 36) while Fe atom in $[Fe^{3+}(CN)_6]^{3-}$ does not obey EAN (EAN of Fe = 35), the former complex ion is more stable.

### 4. To Calculate the number of ligands attached to the metal atom.

EAN rule can also be used to calculate the number of ligands attached to the metal atom.

**Example.** The number of CO groups (ligands) attached with Ni, Fe and Cr in their simple carbonyl can be calculated as follows:

(*a*) Z of Ni = 28, Z of the next noble gas = 36.

$\therefore$ No. of electrons provided by CO ligands = 36 – 28 = 8

$\therefore$ No. of CO ligands attached with one Ni atom $= \frac{8}{2} = 4$

Hence formula of carbonyl of Ni = $Ni(CO)_4$

(*b*) Z of Fe = 26, Z of the next noble gas = 36

$\therefore$ No. of electrons provided by CO ligands = 36 – 26 = 10

$\therefore$ No. of CO ligands attached with one Fe atom $= \frac{10}{2} = 5$

Hence formula of carbonyl of Fe = $Fe(CO)_5$

(*c*) Z of Cr = 24, Z of the next noble gas = 36

No. of electrons provided by CO ligands = 36 – 24 = 12

$\therefore$ No. of CO ligands attached with one Cr atom = $\frac{12}{2} = 6$

Hence formula of carbonyl of Cr = Cr $(CO)_6$

## UNIVERSITY QUESTIONS

1. (*i*) Which of the complexes $[Co(NH_3)_3Cl_3]$ and $[Cr(NH_3)Cl_5]\ Cl_2$ will give precipitate with $AgNO_3$ and why?
   (*ii*) Classify ligands on the basis of number of donor atoms.
   (*Pune 2009, Banglore 2010, Ch. Charan Singh, 2012*)
2. Define the terms: (*i*) Ligand (*ii*) Coordination number.
   (*Nagpur 2010, Ch. Charan Singh 2012*)
3. What are double salts and coordination compounds ?
   (*Lucknow 2010, Ch. Charan Singh 2012*)
4. Define effective atomic number. Calculate EAN for the central atom in the following complexes. (*i*) $[Co(NH_3)_4\ Cl_2]Br$ (Co = 27) (*ii*) $[Fe(CN)_6]^{3-}$ (Fe = 26) (*iii*) $[Cu(NH_3)_4]^{2+}$ (Cu = 29) (*Osmania 2010*)
5. What is the concept of EAN? How it came into existence?
   (*Himachal Pradesh 2010, GND Amritsar 2010*)
6. (*a*) Calculate EAN for the cental metal ion in the complexes: (*i*) $K_4[Fe\ (CN)_6]$ (*ii*) $[Ni\ (CN)_4]SO_4$
   (*b*) What is effective atomic number? Write with examples.
   (*c*) State whether $[Co\ (NH_3)_6]^{3+}$ obeys EAN rule or not. Explain.
   (*Kanpur 2010*)
7. Calculate EAN of the corresponding central atom in the complexes: (*i*) $[Pt\ (NH_3)_6]^{4+}$ (*ii*) $[Cr(CO)_6]$ (*Pune 2009*)
8. Calculate EAN of Fe in $K_3[Fe(CN)_6]$. (Given Z of Fe = 26).(*Bangalore 2010*)
9. Explain Sidgwick's EAN rule with two examples. (*Andhra 2010*)
10. (*a*) Discuss Sidgwick's electronic interpretation of the metal complex.
    (*b*) Calculate effective atomic number of metal ion in the following complexes: (*i*) $[Ni(H_2O)_6]^{2+}$ and (*ii*) $[Co(NH_3)_6]^{3+}$
    (At. No. of Ni = 28 and Co = 27) (*Nagpur 2010*)
11. Find the coordination number of the central metal in the following complexes: (*a*) $[Ag(CN)_2]^-$ (*b*) $[Co(NH_3)_6]^{3+}$ (*c*) $[Co\ (en)_3]^{3+}$ (*d*) $[Co(EDTA)]^-$
    (*Madras, April 2011*)
12. (*a*) Calculate EAN of the central metal ion in the following complexes (*i*) $[Fe(CN)_6]^{4-}$ (*ii*) $[Ni(CN)_4]^{2-}$
    (*b*) Discuss applications of chelates (*Nagpur 2012*)
13. (*i*) Calculate EAN and write IUPAC name of the following:
    (*a*) $Cr(NH_3)_2(H_2O)_2Br_2]Br$ (*b*) $K_4[Fe(CN)_6]$ (*c*) $[Cr(en)_3]Cl_3$ (Cr = 24, Fe = 26)
    (*ii*) Write down the formula of the following complexes
    (*a*) Sodium tetracyane nickelate (0) (*b*)Potassium hexanitro cobaltate (III)
    (*c*) Calcium hexacyaneferrate (II) (*Lucknow 2013*)

# 12 CHAPTER Preparation and Applications of Complexes

## Preparation of Complexes

Following are important methods:

1. **Preparation of metal amines by direct combination of reactants.** Examples are given below:

   (*i*) $NiCl_2 (s) + 6NH_3 \longrightarrow [Ni(NH_3)_6]Cl_2$

   (*ii*) $PtCl_2 + 2en \longrightarrow [Pt (en)_2]Cl_2$

2. **By the replacement of weaker ligands by stronger ligands.** For example:

$$[Cu(H_2O)_6]^{2+} + 4NH_3 \longrightarrow [Cu(NH_3)_4]^{2+} + 4H_2O$$

In this reaction, $NH_3$ molecules (stronger ligands) replace $H_2O$ molecules (weaker ligands) attached with $Cu^{2+}$ ion.

3. **By redox reactions.** Examples are given below:

(*i*) $\underset{(Co = +2)}{[Co^{2+}(H_2O)_6]}(NO_3)_2 + 8NH_3 + 2NH_4NO_2 + \underset{\substack{\text{Oxidising agent} \\ (O = -1)}}{H_2O_2}$

$$\longrightarrow \underset{(Co = +3)}{[Co^{3+}(NH_3)_5(NO_3)]}(NO_3)_2 + \underset{(O = -2)}{14H_2O}$$

In this reaction, $[Co^{2+}(H_2O)_6](NO_3)_2$ is oxidised to $[Co^{3+}(NH_3)_5(NO_3)](NO_3)_2$ and $H_2O_2$ (oxidising agent) is reduced to $H_2O$.

(*ii*) $\underset{(Co = +2)}{2Co^{2+}Cl_2} + 2NH_4Cl + 10NH_3 + \underset{(O = -1)}{H_2O_2} \rightarrow \underset{(Co = +3)}{2[Co^{3+}(NH_3)_6]Cl_3} + \underset{(O = -2)}{2H_2O}$

## Applications of Complexes in Analytical Chemistry

When a metal ion forms a complex, profound changes occur in the properties (*e.g.* solubility, colour, stability, magnetic susceptibility etc.) of the metal ion. Some of these changes in properties are used in analytical chemistry. The applications of complexes in analytical chemistry can be classified under the following heads.

### 1. Separation of metals by chelate formation and solvent extraction method

At about pH = 3.0 when 8–hydroxyquinoline (abbreviated as oxinH) is added to the solution containing Fe (III) and Al (III) salts, Fe (III) gets precipitated as its chelated complex, $[Fe (oxin)_3]^{\circ}$ while Al (III) salt remains as such in aqueous solution. Now the precipitate of [Fe $(oxin)_3$]° is extracted with chloroform. The optical density of the chloroform extract which contains [Fe $(oxin)_3$]° is measured at wavelength 470 mμ (1 mμ = 10 Å) and iron content is determined by comparing with a standard Fe (III) sample. The aqueous phase which contains Al (III) salt is made slightly ammoniacal and 8-hydroxyquinoline is added to it whereby Al (III) gets precipitated as [Al $(exin)_3$]°. Aluminium may be quantitatively estimated. Thus we see that Fe (III) and Al (III) can be separated from each other.

The separation of Zr and Hf can also be made by preparing their chelates with trifluoro acetyl acetone. These chelates have different solubilities in benzene. The distribution coefficient of these two chelates differ by a fraction of 20, Zr having a greater tendency to be extracted. Therefore, it is possible to have a separation of these two metal.

$EDTAH_4$ has also been used as a chelating ligand for the separation and purification of lanthanides.

## 2. Estimation of trace quantities of metal ions by spectro-photometric method

Many transition metal ions form chelated complexes with chelating ligands. These complexes have intense colours. Several of these coloured complexes have extremely high absorption and hence even trace quantities of the metals forming the chelated complexes can be accurately determined by measuring the optical density of the complex at optimum wavelength. For example, the trace quantities of Fe (II) (2-10 parts per million; $10^{-6}$ gm / ml) can be determined by measuring the optical density of $[Fe^{2+} (dipy)_3]^{2+}$ or $[Fe^{2+} (phen)_3]^{2+}$ (dark red complexes) at 515 mμ. These complexes are obtained by treating Fe (II) ions with 2, 2′-dipyridyl (dipy) or *o*-phenanthroline (phen) over a wide range of pH 2-9.

## 3. Gravimetric estimation of a particular metal ion in presence of interfering metal ions by the use of organic sequestering agents

The estimation of a particular metal ion in presence of other interfering metal ions can also be made by the use of strong sequestering organic ligands like $EDTA^{4-}$ ion. This ligand makes the precipitating organic ligand nearly specific for the metal ion which has to be estimated.

Suppose we have to make gravimetric estimation of $Ag^+$ ions in the presence of interfering ions like $Cu^{2+}$, $Ni^{2+}$, $Co^{2+}$ and $Cd^{2+}$ ions. For this estimation we make the use of the organic ligand benzimidazole which is a monodentate ligand. This ligand forms *insoluble* salts with $Ag^+$, $Cu^{2+}$, $Ni^{2+}$, $Co^{2+}$ and $Cd^{2+}$ ions inammoniacal medium as shown on next page:

With $EDTA^{4-}$ ion the interfering ions namely $Cu^{2+}$, $Ni^{2+}$, $Co^{2+}$ and $Cd^{2+}$ ions form highly *stable* and *soluble* complexes and hence these ions become indifferent to benzimidazole. $Ag^+$ ion, however, still gives a precipitate of (I) with benzimidazole. Thus here $EDTA^{4-}$ acts as a masking agent.

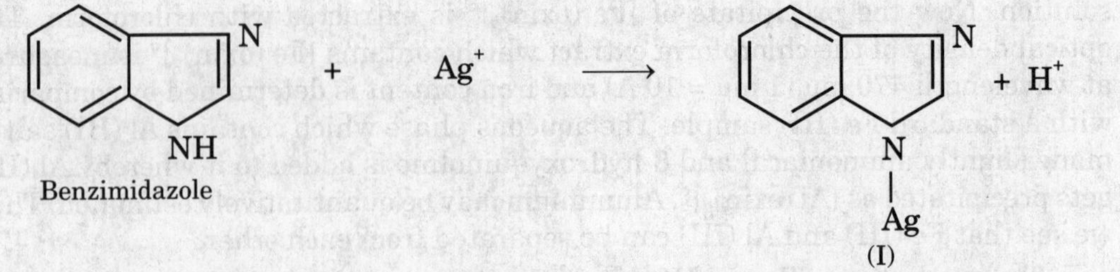


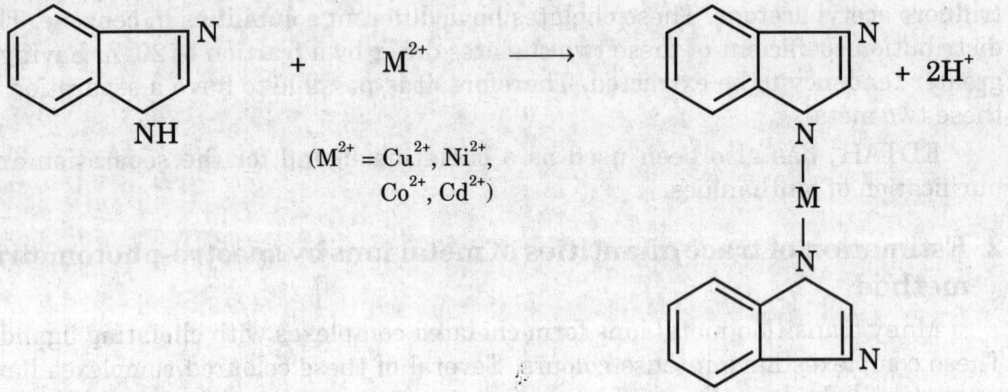


## 4. Qualitative detection and separation of metal ions

Qualitative detection and separation of metal ions is based on the experimental fact that some metal ions form complexes with suitable ligands under suitable conditions. The complexes formed have specific colours which are used to recognise and separate the metal ions. Some examples are given below:

(*i*) **Detection of $Ni^{2+}$ ion.** The presence of $Ni^{2+}$ ion is detected by adding dimethyl glyoxime in presence of $NH_4OH$ to the salt solution containing $Ni^{2+}$ ions when a brilliant red precipitate is formed due to the formation of a complex of $Ni^{2+}$ with dimethyl glyoxime. Let us consider the reaction between $NiCl_2$ solution (salt solution) and dimethyl glyoxime in presence of $NH_4OH$.

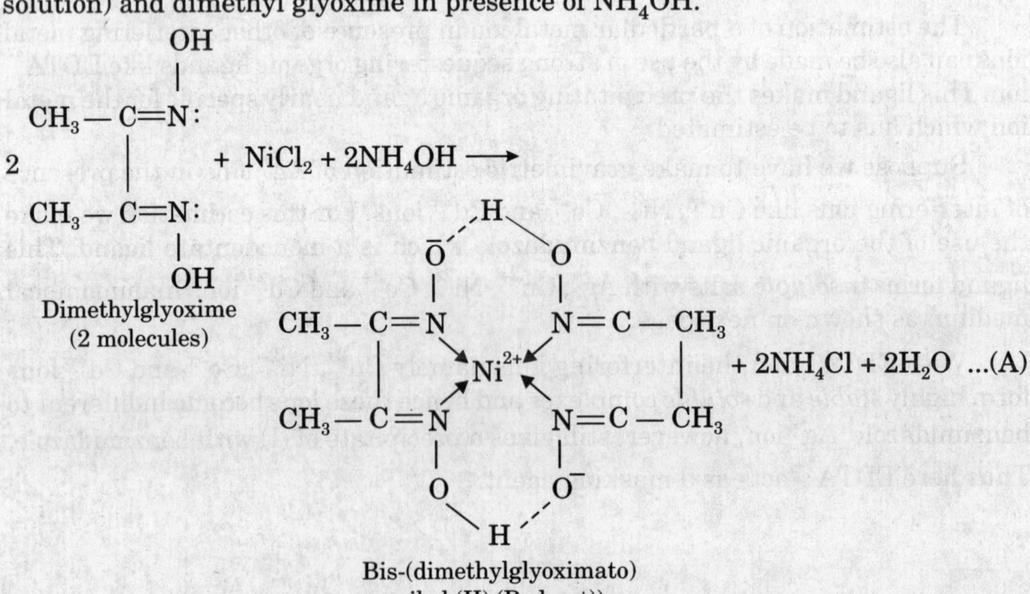


Bis-(dimethylglyoximato)
nikel (II) (Red ppt))

Reaction (A) takes place through four steps that have been shown in Fig. 12.1.

(*a*) $NiCl_2 \rightleftharpoons Ni^{2+} + 2Cl^-$

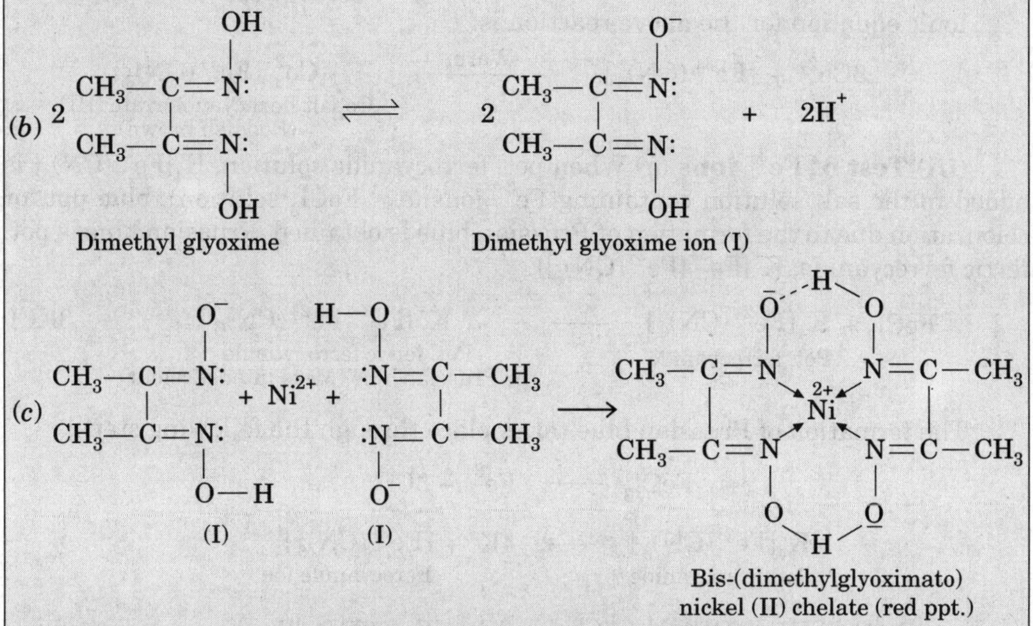


(*d*) $2\,NH_4OH + 2H^+ \longrightarrow 2NH_4Cl + 2H_2O$

**Fig. 12.1** Ionic equations showing the formation of bis-(dimethyl oximato) nickel (II) chelate (red ppt.)

On adding equations (*a*), (*b*), (*c*) and (*d*) given in Fig. 12.1 we get equation (A)

***(ii)* Detection of $Co^{2+}$ ion.** The presence of $Co^{2+}$ ion can be tested by any of the following two methods.

(*a*) The presence of $Co^{2+}$ is tested by adding ammonium thiocyanate solution ($NH_4CNS$) when a blue colour is obtained due to formation of a complex.

$$CoCl_2 + NH_4CNS \longrightarrow Co(CNS)_2 + 2NH_4Cl$$

$$Co(CNS)_2 + 2NH_4CNS \longrightarrow \underset{\substack{\text{Ammonium tetrathiocyanatocobaltate (II)}\\\text{(Blue colour)}}}{(NH_4)_2[Co(CNS)_4]}$$

Ionic equations representing the above reactions can be written as:

$$Co^{2+} + 2CNS^- \longrightarrow Co(CNS)_2$$

$$Co(CNS)_2 + 2CNS^- \longrightarrow \underset{\substack{\text{Tetrathiocyanato cobaltate (II)}\\\text{ion (Blue colour)}}}{[Co(CNS)_4]^{2-}}$$

(*b*) The presence of $Co^{2+}$ ion is also tested by potassium ferricyanide, $K_3[Fe\,(CN)_6]$ test when a reddish brown ppt is obtained due to the formation of an insoluble complex.

$$3CoCl_2 + \underset{\text{Pot. ferricyanide}}{2K_3[Fe(CN)_6]} \xrightarrow{\text{Warm}} \underset{\substack{\text{Cobalt hexacyanoferrate (III)} \\ \text{(Reddish brown)}}}{Co_3[Fe(CN)_6]_2} + 6KCl$$

Ionic equation for the above reaction is:

$$3Co^{2+} + [Fe^{3+}(CN)_6]^{3-} \xrightarrow{\text{Warm}} \underset{\substack{\text{Cobalt hexacyanoferrate (III)} \\ \text{(Reddish brown)}}}{Co_3^{2+}[Fe^{3+}(CN)_6]}$$

***(iii)* Test of $Fe^{3+}$ ions** (*a*) When pot. ferrocyanide solution, $K_4[Fe^{2+}(CN)_6]$ is added to the salt solution containing $Fe^{3+}$ ions (*e.g.* $FeCl_3$ solution), blue ppt. or colouration due to the formation of Prussian blue is obtained. Prussian blue is pot. ferric ferrocyanide, $K^+\{Fe^{3+}[Fe^{2+}(CN)_6]\}$.

$$FeCl_3 + \underset{\text{Pot. ferrocyanide}}{K_4[Fe^{2+}(CN)_6]} \longrightarrow \underset{\substack{\text{Pot. ferric ferro cyanide} \\ \text{or Prussian blue (More intense colour)}}}{K^+\{Fe^{3+}[Fe^{2+}(CN)_6]\}} + 3KCl$$

The formation of Prussian blue takes place through the following steps:

$$FeCl_3 \longrightarrow Fe^{3+} + 3Cl^-$$

$$\underset{\text{Pot. ferrocyanide}}{K_4[Fe^{2+}(CN)_6]} \longrightarrow 4K^+ + \underset{\text{Ferocyanide ion}}{[Fe^{2+}(CN)_6]^{4-}}$$

$$\underset{\text{Ferric ion}}{Fe^{3+}} + \underset{\text{Ferrocyanide ion}}{[Fe^{2+}(CN)_6]^{4-}} \longrightarrow \underset{\text{Ferric-ferrocyanide ion}}{\{Fe^{3+}[Fe^{2+}(CN)_6]\}^-}$$

$$\underset{\text{Ferric-ferrocyanide ion}}{\{Fe^{3+}[Fe^{2+}(CN)_6]\}^-} + K^+ \longrightarrow \underset{\substack{\text{Pot. ferric-ferro} \\ \text{cyanide or Prussian} \\ \text{blue (More intense colour)}}}{K^+\{Fe^{3+}[Fe^{2+}(CN)_6]\}^-}$$

$$[K^+ + Cl^- \longrightarrow KCl] \times 3$$

(*b*) When pot. or ammonium thiocyanate, KCNS or $NH_4CNS$ (also called sulphocyanide) is added to the salt solution containing $Fe^{3+}$ ions, blood red colouration due to the formation of ferric thiocyanate (ferric sulphocyanide), $Fe(CNS)_3$ is obtained.

$$FeCl_3 + 3KCNS \longrightarrow Fe(CNS)_3 + 3KCl$$

or

$$Fe^{3+} + 3CNS^- \longrightarrow \underset{\substack{\text{Ferric thiocyanate} \\ \text{(Blood red colouration)}}}{Fe(CNS)_3}$$

***(iv)* Test of $Fe^{2+}$ ions.** When pot. ferricyanide solution, $K_3[Fe^{3+}(CN)_6]$ is added to the salt solution containing $Fe^{2+}$ ions (*e.g.* $FeCl_2$), blue colouration due to the formation of Turnbull's blue is obtained. The composition of Turnbull's blue is the same as that of Prussion blue, $K\{Fe^{3+}[Fe^{2+}(CN)_6]\}$. The formation of Turnbull's blue takes place through the following steps:

(*a*) $Fe^{2+}$ ions obtained by the ionisation of $FeCl_2$ are oxidised to $Fe^{3+}$ ions by ferricyanide ion, $[Fe^{3+}(CN)_6]^{3-}$ obtained from $K_3[Fe^{3+}(CN)_6]$. $[Fe^{3+}(CN)_6]^{3-}$ ions are reduced to ferrocyanide ions, $[Fe^{2+}(CN)_6]^{4-}$.

$$\underset{}{FeCl_2} \longrightarrow \underset{\text{Ferrous ion}}{Fe^{2+}} + 2Cl^-$$

$$\underset{\text{Pot. ferricyanide}}{K_3[Fe^{3+}(CN)_6]} \longrightarrow 3K^+ + [Fe^{3+}(CN)_6]^{3-}$$

$$\underset{\text{Ferrous ion}}{Fe^{2+}} + \underset{\text{Ferricyanide ion}}{[Fe^{3+}(CN)_6]^{3-}} \longrightarrow \underset{\text{Ferric ion}}{Fe^{3+}} + \underset{\text{Ferrocyanide ion}}{[Fe^{2+}(CN)_6]^{4-}}$$

(*b*) $Fe^{3+}$ ions formed above combine with $[Fe^{2+}(CN)_6]^{4-}$ to form ferric ferro cyanide ions, $\{Fe^{3+}[Fe^{2+}(CN)_6]\}^-$

$$Fe^{3+} + [Fe^{2+}(CN)_6]^{4-} \longrightarrow \underset{\text{Ferric ferro cyanide ion}}{\{Fe^{3+}[Fe^{2+}(CN)_6]\}^-}$$

(*c*) $\{Fe^{3+}[Fe^{2+}(CN)_6]\}^-$ ions formed in step (*b*) combine with $K^+$ ions to form Turnbull's blue, $K^+\{Fe^{3+}[Fe^{2+}(CN)_6]\}^-$.

$$\{Fe^{3+}[Fe^{2+}(CN)_6]\}^- + K^+ \longrightarrow K^+\{Fe^{3+}[Fe^{2+}(CN)_6]\}^-$$

Pot. ferric ferrocyanide
or Turnbull's blue
(Less intense colour)

The blue colour of Turnbull's blue is less intense than that of Prussian blue. This is due to the fact that Turnbull's blue also contains a white compound of the formula $K_2\{Fe^{2+}[Fe^{2+}(CN)_6]\}$ which is called pot. ferrous ferrocyanide.

(*v*) **Separation of $Cu^{2+}$ and $Cd^{2+}$ ions.** When a mixture containing $Cu^{2+}$ and $Cd^{2+}$ ions is treated with ammoniacal KCN solution and $H_2S$ gas is passed through the solution, yellow ppt. of CdS is obtained. Reactions taking place are:

(*a*) *Formation of tetraammine complexes*

$$Cu^{2+} + 4NH_4OH \longrightarrow \underset{\text{Tetraamminecopper (II) ion}}{[Cu^{2+}(NH_3)_4]^{2+}} + 4H_2O$$

$$Cd^{2+} + 4NH_4OH \longrightarrow \underset{\text{Tetraaminecadmium (II) ion}}{[Cd^{2+}(NH_3)_4]^{2+}} + 4H_2O$$

(*b*) *Formation of tetracyano complexes*

$$[Cu^{2+}(NH_3)_4]^{2+} + 10CN^- + 8H_2O \longrightarrow \underset{\text{(More stable)}}{2[Cu^+(CN)_4]^{3-}} + (CN)_2 + 8NH_4OH$$

$$[Cd^{2+}(NH_3)_4]^{2+} + 4CN^- + 4H_2O \longrightarrow \underset{\text{(Less stable)}}{[Cd^{2+}(CN)_4]^{2-}} + 4NH_4OH$$

(*c*) *Formation of yellow ppt. of CdS*

$$\underset{\text{More stable}}{[Cu^+(CN)_4]^{3-}} \rightleftharpoons Cu^+ + 4CN^-$$

Negligible number of $Cu^+$ ions are obtained

$$\underset{\text{Less stable}}{[Cd^{2+}(CN)_4]^{2-}} \rightleftharpoons Cd^{2+} + 4CN^-$$

Suffecient number of $Cd^{2+}$ ions are obtained

$$\xrightarrow{+H_2S} CdS\ (\textit{yellow ppt.}) + 2H^+$$

***(vi)* Separation of $Co^{2+}$ and $Ni^{2+}$ ions.** *(a)* When a mixture containing $Co^{2+}$ and $Ni^{2+}$ ions is treated with ammoniacal $K_4[Fe^{2+}(CN)_6]$ solution, both the ions form similar complex ions. Complex ion formed by $Co^{2+}$ ions is insoluble in $NH_4OH$ but that given by $Ni^{2+}$ ions is soluble in $NH_4OH$.

*(b)* When α-nitroso β-naphtol solution (chelating agent) is added to a slightly acidic solution of $Co^{2+}$ ion and is warmed, red-brown chelate is obtained. This test is not shown by $Ni^{2+}$ ion, since $Ni^{2+}$ does not form any chelate.

$$2Co^{2+} + H_2O + \frac{1}{2}O_2 \text{ (From air)} \longrightarrow 2Co^{3+} + 2OH^-$$

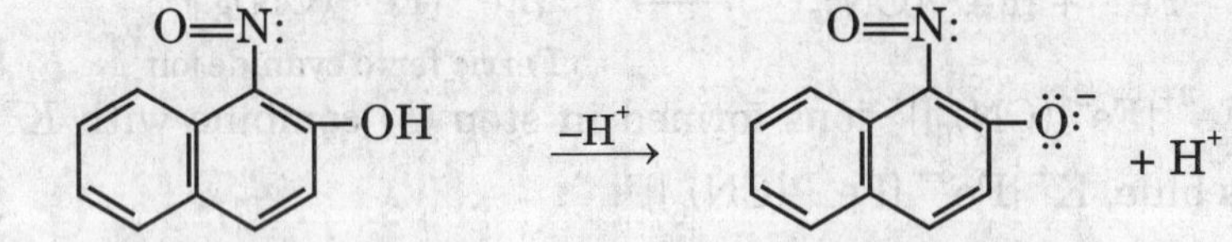

α-nitroso β-naphthol — Anion of α-nitroso β-naphthol

or

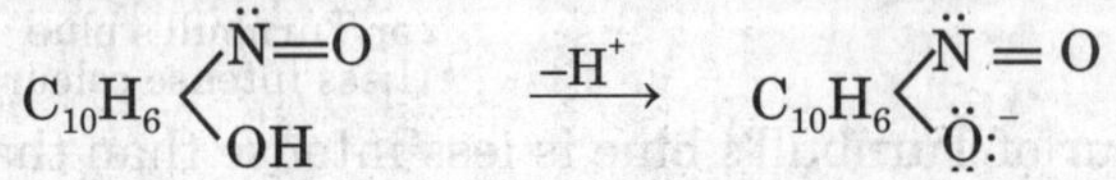

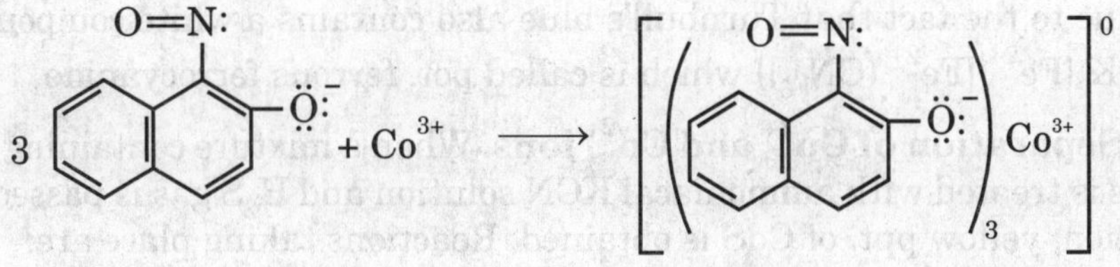

or

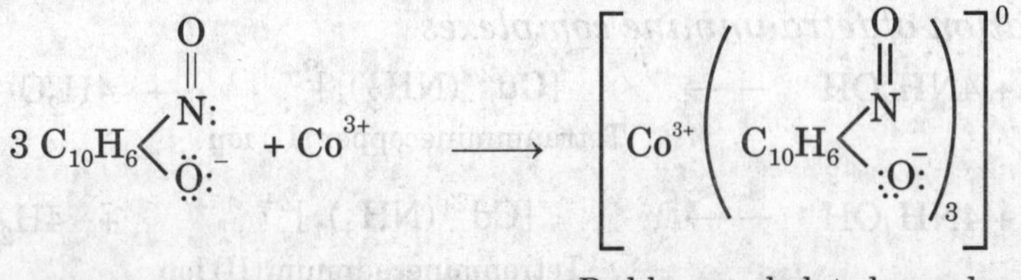

Red-brown chelated complex

*(c)* Black precipitate of CoS and NiS obtained by passing $H_2S$ gas through the solution containing $Co^{2+}$ and $Ni^{2+}$ ions is insoluble even in conc. HCl. It is dissolved in aqua regia.

$$\underset{(M = Co, Ni)}{3MS} + \underbrace{6HCl + 2HNO_3}_{\text{Aqua regia}} \longrightarrow \underset{\text{(In solution)}}{3MCl_2} + 2NO + 3S + 4H_2O$$

The solution containing $CoCl_2$ and $NiCl_2$ is evaporated to dryness and the residue thus obtained is treated with dil. HCl. The solution which contains $CoCl_2$ and $NiCl_2$ is treated with $NaHCO_3$ and bromine water. Appearance of green colour confirms the presence of $Co^{2+}$ ions. If no green colour is observed, the solution is heated when *black residue* of $Ni_2O_3$ is obtained which shows the presence of $Ni^{2+}$ ions. The formation of green colour which confirms the presence of $Co^{2+}$ ions is due to the formation of *sodium tricarbonato cobaltate* (III), $Na_3[Co^{3+}(CO_3)_3]$.

$$CoCl_2 + 6NaHCO_3 \longrightarrow Na_4[Co^{2+}(CO_3)_3] + 3CO_2 + 3H_2O + 2NaCl$$

$$Br_2 + H_2O \longrightarrow 2HBr + O$$

This nascent oxygen oxidises $Na_4[Co^{2+}(CO_3)_3]$ to $Na_3[Co^{3+}(CO_3)_3]$ (green colouration)

$$2Na_4[Co^{2+}(CO_3)_3] + H_2O + O \longrightarrow \underset{\text{(Green colouration)}}{2Na_3[Co^{3+}(CO_3)_3]} + 2NaOH$$

$$NiCl_2 + 2NaHCO_3 \longrightarrow NiCO_3 + 2NaCl + H_2O + CO_2$$

$$Br_2 + H_2O \longrightarrow 2HBr + O$$

$$\underset{(Ni = +2)}{2NiCO_3} + O \longrightarrow \underset{\substack{(Ni = +3)\\ \text{(Black residue)}}}{Ni_2O_3} + 2CO_2$$

## Complexometric Titrations

*Complexometric titrations* (also called *chelatometric titrations or chelometric titrations) are those titrations in which a metal ion is determined titrimetrically by way of complex formation with a strong multidentate chelating ligand.* A number of metal ions can be estimated by forming their complexes with polydentate ligands at appropriate pH. This is done by titrating the solutions of the metal ions against the solutions of polydentate ligands in presence of a suitable buffer and end point is detected by using suitable indicator. Such types of titrations are called complexometric titrations.

### Some Typical Complexometric Estimations

Examples of some typical complexometric estimations are given below:

**1. Estimation of $M^{2+}$ ions ($M^{2+}$ = $Mg^{2+}$, $Ca^{2+}$, $Zn^{2+}$, etc.) by complexometric titrations.** The most common polydentate ligand used for the estimation of metal ions like $Ca^{2+}$, $Mg^{2+}$, $Zn^{2+}$, etc. (in general $M^{2+}$ ions) by complexometric titrations is disodium salt of ethylene diamine tetracetic and whose structure is given in Fig. 12.2. This salt is abbreviated as EDTA $H_2Na_2$ or edta $H_2Na_2$ or $YH_2Na_2$. The solution of this salt should not be stored in glass equipment for prolonged period as it attacks soft glass dissolving out considerable amounts of cations present in soft glass. Polythene bottles are suitable for storing purpose.

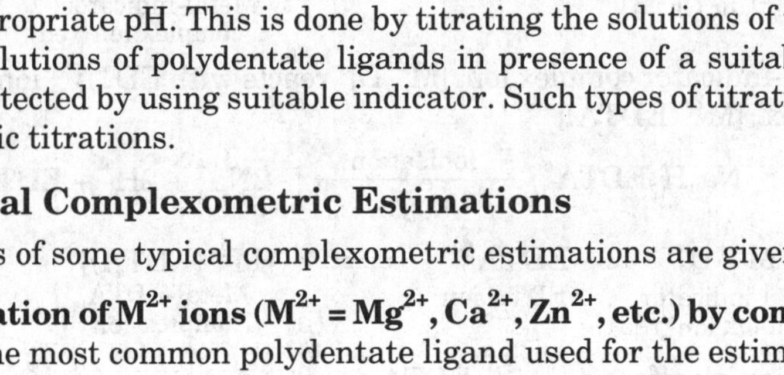


**Fig. 12.2** Structure of disodium salt of ethylenediamine tetracetic acid, $Na_2H_2EDTA$ or $Na_2H_2$ edta or $Na_2H_2Y$

The ionisation of this salt takes place as:

$$\text{EDTA } H_2Na_2 \rightleftharpoons \text{EDTA}^{4-} + 2H^+ + 2Na^+$$

or $$\text{edta } H_2Na_2 \rightleftharpoons \text{edta}^{4-} + 2H^+ + 2Na^+$$

or $$YH_2Na_2 \rightleftharpoons Y^{4-} + 2H^+ + 2Na^+$$

In the complexometric titration of $M^{2+}$ ions (*e.g.* $Mg^{2+}$ or $Ca^{2+}$ ions) with $Na_2H_2$EDTA we use Erio-chrom Black T as an indicator abbreviated as $H_3D$. In this titration the solution containing $M^{2+}$ ions is buffered to pH = 10 (This pH is obtained by using NaCl-$NH_4OH$ buffer), a few drops of the indicator are added and the solution is titrated with a standard solution of $Na_2H_2$EDTA.

**Reactions.** Reactions taking place are:

(*a*) Indicator ions ($D^{3-}$) form red metal-indicator complex ion, $[M^{2+}D]^-$ with $M^{2+}$ ions

$$H_3D \xrightarrow{\text{Ionisation}} 3H^+ + D^{3-}$$

$$\underset{(M^{2+} = Mg^{2+} \text{ or } Ca^{2+})}{M^{2+}} + D^{3-} \xrightarrow{\text{pH = 10}} \underset{\text{Metal-indicator complex ion (red)}}{[M^{2+}D]^-}$$

(*b*) Metal-indicator complex ion, $[M^{2+}D]^-$ reacts with $\text{EDTA}^{4-}$ ion to form metal–EDTA complex, $[M^{2+}(\text{EDTA})]^{2-}$

$$Na_2H_2\text{EDTA} \xrightarrow{\text{Ionisation}} 2Na^+ + 2H^+ + \text{EDTA}^{4-}$$

$$\underset{\text{Metal-indicator complex ion (red)}}{[M^{2+}D]^-} + \underset{\text{EDTA ion}}{\text{EDTA}^{4-}} \rightleftharpoons \underset{\text{Metal-EDTA complex ion}}{[M^{2+}(\text{EDTA})]^{2-}} + \underset{\text{Indicator ion (blue)}}{D^{3-}}$$

$\text{EDTA}^{4-}$ ion also forms metal-EDTA complex with the free $Ca^{2+}$ or $Mg^{2+}$ ions (present in the titration solution) that have not reacted with $D^{3-}$ ions to form $[M^{2+}D]^-$ complex ion.

Thus, we see that free metal indicator ion, $D^{3-}$ is obtained at the end point. The release of the free indicator ion is marked by a change from red (colour of $[M^{2+}D]^-$ complex ion) to blue (colour of $D^{3-}$ ion) colour.

**2. Estimation of hardness of hard water (softening of hard water) by complexometric titration.** We know that hard water contains $Mg^{2+}$ and $Ca^{2+}$ ions. It is due to the presence of these ions that water becomes hard. Hard water does a lot of harm in industrial boilers and other equipment by forming a scale on them. Ethylene diamine tetraacetic acid ($\text{EDTAH}_4$) is used for estimating/removing $Mg^{2+}$ and $Ca^{2+}$ ions present in hard water.

**Reactions.** Ethylenediamine tetracetic acid is added to the boilers in which $\text{EDTA}^{-4}$ ion forms water-soluble chelated complex ion, $[M^{2+}(\text{EDTA})]^{2-}$ with $Ca^{2+}$ and $Mg^{2+}$ ions present in hard water and thus, these ions are removed.

$$\text{EDTAH}_4 \rightleftharpoons \text{EDTA}^{4-} + 4H^+$$

$$\underset{(M^{2+} = Ca^{2+}, Mg^{2+})}{M^{2+}} + \text{EDTA}^{4-} \longrightarrow \underset{\text{(Water soluble chelate)}}{[M(\text{EDTA})]^{2-}}$$

Since the stability constants of $[Ca(EDTA)]^{2-}$ and $[Mg(EDTA)]^{2-}$ ions are different, the selective estimation of $Ca^{2+}$ and $Mg^{2+}$ ions can also be made, *i.e.*, hardness of water can be estimated.

**3. Determination of iron.** Iron (III) has been quantitatively estimated by the following complexometric reaction:

$$Fe^{3+} + edta^{4-} \longrightarrow [Fe(edta)]^-$$

In this titration KSCN is used as an indicator. The iron (III) solution is coloured blood-red due to the formation of complexes $[Fe(SCN)]^{2+}$ or $[Fe(SCN)_6]^{3-}$. Addition of a standard $Na_2H_2$ edta solution leads first to the formation of $[Fe(edta)]^-$ via reaction of $edta^{4-}$ with aquarion (III) in solution. Finally $edta^{4-}$ reacts with iron (III) thiocyanate complex and releases the complexed thiocyanate. The end-point is marked by a change from blood-red colour of iron (III) thiocyanate complex to yellow of (edta) ferrate (III) complex.

## Types of Complexometric Titrations

**1. Direct titrations.** These are the complexometric titrations which are suitable for those metal ions whose reaction with the complexometeric ligands are fast enough and for with good metal indicators are available. $Mg^{2+}$, $Ca^{2+}$, $Zn^{2+}$, $Mn^{2+}$, $Ni^{2+}$, $Cu^{2+}$ belong to this class.

**2. Back titrations.** These are applied to those metal ions which react with complexometric ligands rather slowly. Chromium (III) is an example of this class. Excess standard edta is added to a chromium (III) solution, the solution boiled for sometime and the excess edta back titrated with a standard $Mg^{2+}$ solution. Back titration is also preferred when there is no suitable indicator for the metal ion to be titrated.

**3. Substitution titrations.** These are employed for those metals which do not give a reasonable colour with an indicator at the finish of the titration. In such a case either (*a*) a metal-edta complex which is weaker than the edta complex of the metal to be determined is used or (*b*) a suitable metal complex such as $[Ni(CN)_4]^{2-}$ is employed which reacts with the metal ion to be determined. In case (*a*) Mg-edta is generally employed:

$$Me^{2+} + MgY^{2-} \rightleftarrows MeY^{2-} + Mg^{2+}$$

The equivalent amount of $Mg^{2+}$ is then titrated with standard edta. For case (*b*) titrimetric determination of silver (I) forms an example:

$$[Ni(CN)_4]^{2-} + 2Ag^+ \rightleftarrows 2[Ag(CN)_2]^- + Ni^{2+}$$

The liberated $Ni^{2+}$ is then estimated with edta.

**4. Indirect titrations.** These are employed for estimations of anions. Phosphate may be precipitated as $MgNH_4PO_4$. This precipitate is collected, dissolved in acid, measured excess of standard edta added and pH adjusted to ~11. The excess edta is then titrated with a standard $Mg^{2+}$ solution using Eriochrome Black T. Similarly sulphate is precipitated as $BaSO_4$, which is dissolved in excess edta and the excess edta is back titrated with standard $Mg^{2+}$.

## Applications of Complexes in Metallurgy

### 1. Extraction of Ag and Au by Mac-Arthur and Forest Cyanide Process

(*i*) Ag is extracted from argentite ore ($Ag_2S$) by cyanide process. The concentrated $Ag_2S$ ore is treated with a dilute solution of NaCN (0.5%) for several hours in presence of air so that metallic Ag and AgCl present in the ore and $Ag_2S$ ore itself are converted into soluble sodium dicyanoargentate (I), $Na[Ag(CN)_2]$.

$$4Ag + 8NaCN + O_2 + 2H_2O \rightarrow 4Na[Ag(CN)_2] + 4NaOH$$

or $$4Ag + 8CN^- + O_2 + 2H_2O \rightarrow 4[Ag(CN)_2]^- + 4OH^-$$

$$AgCl + 2NaCN \rightarrow Na[Ag(CN)_2] + NaCl$$

$$2Ag_2S + 8KCN + O_2 + 2H_2O \rightarrow 4Na[Ag(CN)_2] + 2S + 4NaOH$$

or $$2Ag_2S + 8CN^- + O_2 + 2H_2O \rightarrow 4[Ag(CN)_2]^- + 2S + 4OH^-$$

Now Ag is precipitated out from $Na[Ag(CN)_2]$ by more electropositive Zn metal. Being more electropositive, Zn displaces less electropositive Ag metal from $Na[Ag(CN)_2]$

$$2Na[Ag(CN)_2] + Zn \rightarrow \underset{\text{Sod. tetracyano zincate (II)}}{Na_2[Zn(CN)_4]} + 2Ag\downarrow$$

or $$2[Ag(CN)_2]^- + Zn \rightarrow [Zn(CN)_4]^{2-} + 2Ag\downarrow$$

(*ii*) We know that the tailings obtained from amalgamation process and the auriferrous quartz rocks contain gold. Cyanide process can be used to get gold metal from these tailings and auriferrous quartz rocks. In this process the tailings or rocks are treated with KCN solution in presence of atmospheric $O_2$ when Au present in tailings or rocks is converted into soluble pot. dicyanoaurate (I), $K[Au(CN)_4]$ (complex compound) from which Au can be precipitated by adding Zn.

$$4Au + 8KCN + 2H_2O + O_2\ (\textit{From air}) \rightarrow \underset{\substack{\text{Pot. dicyano aurate (I)} \\ \text{(Soluble complex)}}}{4K[Au(CN)_2]} + 4KOH$$

or $$4Au + 8CN^- + 2H_2O + O_2 \rightarrow 4[Au(CN)_2]^- + 4OH^-$$

$$2K[Au(CN)_2] + Zn \rightarrow K_2[Zn(CN)_4] + 2Au\downarrow$$

or $$[Au(CN)_2]^- + Zn \rightarrow [Zn(CN)_4]^{2-} + 2Au$$

### 2. Purification of Ni by Mond's Process

Ni extracted from pentalandite ore, (Ni, Cu, Fe)S is impure, since it contains some Cu, Fe and Co as impurities. In order to remove these impurities, crude nickel is treated with a current of CO at 50–80°C when Ni is converted into volatile $Ni(CO)_4$ while Cu, Fe and Co (impurities) remain unaffected and left behind.

$$Ni + 4CO \xrightarrow{50-80^\circ} Ni(CO)_4\ (\textit{Vapours})$$

The vapours of $Ni(CO)_4$ are condensed and then decomposed at 150°–180°C to give nearly 99.5% pure nickel.

$$Ni(CO)_4 \xrightarrow{150-180°C} Ni\ (Pure) + 4Co$$

This method of purification of nickel is called **Mond's process.**

## 3. Separation of Nb and Ta

Nb and Ta which are important metals used in plastic surgey and aerospace programmes are separated from each other by the preferential hydrolysis of their fluoride complexes. Nb precipitates while Ta stays in solution.

## 4. Concentration of bauxite ore ($Al_2O_3.2H_2O$) by Baeyer's Process.

Aluminium is generally extracted from bauxite ore which is associated with the impurities of silica ($SiO_2$) and ferric oxide, $Fe_2O_3$, (also called red bauxite). These impurities are removed by Baeyer's process in which the bauxite ore is digested at 150°C with a concentrated solution of NaOH so that bauxite ore dissolves in NaOH due to the formation of soluble sodium meta-aluminate, $Na[AlO_2]$ (complex compound) while the impurities of $SiO_2$ and $Fe_2O_3$ are left behind.

$$\underset{\text{Bauxite ore}}{Al_2O_3 \cdot 2H_2O} + 2NaOH \longrightarrow \underset{\text{(soluble)}}{2Na[AlO_2]} + 3H_2O$$

# Application of Complexes in Industry

Following are important applications of complexes in industry.

## 1. As Catalysts

Coordination compounds are used as catalysts for many industrial processes. A few examples are given below:

***(i)* Hydrogenation of alkenes.** Complexes like $[Co(CN)_5]^{3-}$ and $RhClL_3$ (L = Ligand with complicated structure) are used as homogenous catalysts in the hydrogenation of alkenes

$$\underset{\text{Ethene (Alkene)}}{CH_2{=}CH_2} + H_2 \longrightarrow \underset{\text{Ethane (Alkane)}}{CH_3-CH_3}$$

***(ii)* Decomposition of water.** Decomposition of water into $H_2$ and $O_2$ has been found to be catalysed by complexes of ruthenium.

$$H_2O \xrightarrow[\text{complexes of ruthenium}]{\text{Decomposition in presence of}} H_2 + \frac{1}{2}O_2$$

***(iii)* Conversion of olefins into alcohols.** $Co_2(CO)_8$ has been found to act as catalyst in the conversion of olefins into alcohols.

## 2. Electrodeposition

In order to get a fine and uniform deposit of superior metals (*e.g.* Ni, Cr, Pt, Ag, Au, Cu etc.) on baser metals, some complex compound of the superior metals is used as an electrolyte in the electrolytic bath. The complex compound to be used as an electrolyte should have some characteristics like high solubility, good conductivity, resistance towards oxidation, reduction and hydrolysis. For example for the uniform electrolytic deposit of Cu, Ag and Au, the following cyanide complexes of Cu, Ag and Au, are used in the electrolytic bath.

For electrolytic deposit of Cu – $Na_3[Cu(CN)_4]$ or $Na_2[Cu(CN)_4]$

For electrolytic deposit of Ag – $K[Ag(CN)_2]$

For electrolytic deposit of Au – $K[Au(CN)_2]$ or $K[Au(CN)_4]$

## 3. In Photography

In photogrpahy, when the developed film is washed with hypo solution (i.e. solution of sod. thiosulphate, $Na_2S_2O_3$), the negative film gets fixed (*i.e.* the image becomes permanent). In the fixing process the undecomposed AgBr (*i.e.* AgBr which has not decomposed to Ag in the process of developing) gets dissolved in $Na_2S_2O_2$ solution as a soluble complex, $Na_3[Ag(S_2O_3)_2]$ which is called sod. dithioorgenate (I).

$$AgBr + 2Na_2S_2O_3 \longrightarrow \underset{\text{Sod. dithiosulphato argentate (I)}}{Na_3[Ag(S_2O_3)_2]} + NaBr$$

$$\text{or} \quad AgBr + 2S_2O_3^{2-} \longrightarrow [Ag(S_2O_3)_2]^{3-} + Br^-$$

$Na_3[Ag(S_2O_3)_2]$ formed above is colourless and soluble in water and hence can be easily removed by washing the photographic plate or film with water.

## 4. As dyes or Pigments

Phthalocyanine blue which is extremely stable complex of copper (II) is one of the many complex compounds used as dyes or pigments.

## 5. Softening of Hard Water

We know that hard water contains $Ca^{2+}$ and $Mg^{2+}$ salts dissolved in it. Soap gives precipitate of $Ca^{2+}$ and $Mg^{2+}$ salt with hard water and thus more quantity of soap is used up in washing the clothes. In order to prevent the formation of the precipitate of $Ca^{2+}$ and $Mg^{2+}$ salt, we use excess of polyphosphates and polydentate amino acids as sesquestering agents. These complexing agents form *water-soluble* complexes with $Ca^{2+}$ and $Mg^{2+}$ ions present in hard water and thus the precipitation of $Ca^{2+}$ and $Mg^{2+}$ salt with hard water is prevented and hard water becomes soft.

Ethylenediamine tetra acetic acid ($EDTAH_4$) has also been used for softening hard water which does a lot of harm in industrial boilers and other equipments by forming a scale in them. Ethylenediamine tetracetic acid is added to the boiles etc. in which $EDTA^{4-}$ forms water-soluble complexes with $Ca^{2+}$ and $Mg^{2+}$ ions present in hard water and thus these ions are removed.

Ion-exchange resins have also been used to remove both cations and anions from aqueous solution and deionised water so obtained is equivalent to distilled water in purity.

## 6. Food Preservation

Freshly fruits, fruit juice, green vegetables containing vitamin C, on keeping for sometime get spoiled since even one part per million parts of metal ions catalyses atmospheric oxidation which spoils these eatables. The addition of one hundredth of 1% of a masking agent like ethylenediamine tetracetic acid improves the quality of the food on standing by catching the traces of metals by chelate formation.

## Application of Complexes in Medical Field

Many complex compounds are used as medicines for the treatment of various diseases. For example,

(*i*) Ca-EDTA chelate, $Ca[EDTANa_2]$ is used in the treatment of lead poisoning (*i.e.*, for removing lead from the body).

(*ii*) *cis*-$[Pt(NH_3)_2Cl_2]$ (called **cisplatin**) is used as an antitumor agent in the treatment of cancer. *cis*-$[Pt^{2+}(NH_3)_2Cl_2]^{\circ}$ has square planar geometry.

(*iii*) **Lewisite**, $ClCH=CHAs\langle^{Cl}_{Cl}$ is an arsenic compound and is a deadly poisonous gas. This gas was used for mass killing during the first world war. 2, 3-dimercapto propanol which is named as British Anti Lewisite (BAL) was used as antidote. 2, 3-dimercapto propanol forms a complex with Lewisite and thus destroys the poisonous effect of Lewisite.

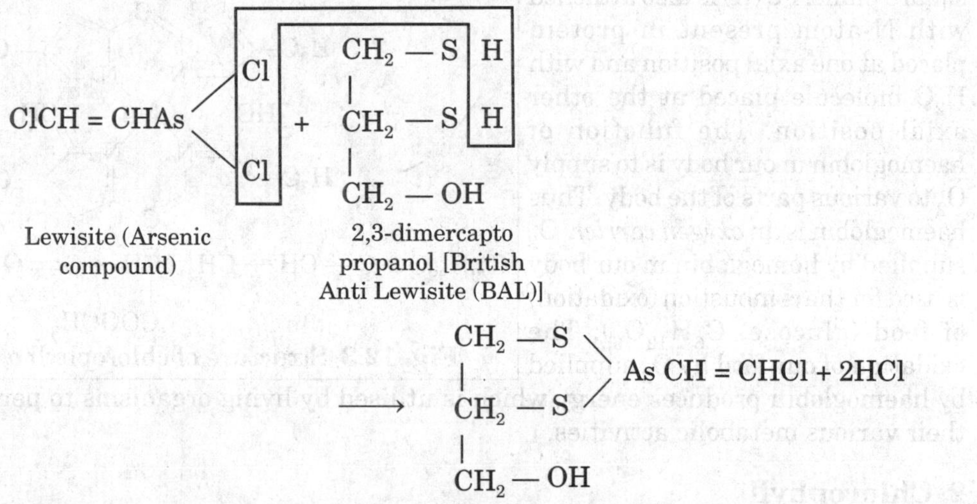

Lewisite (Arsenic compound)

2,3-dimercapto propanol [British Anti Lewisite (BAL)]

Complex compound

(*iv*) Metals present in toxic proportions in animals and plants are removed by *chelate therapy, e.g.,* excess of copper and iron are removed by using chelating ligands, D-penicillamine and desferioxime-B.

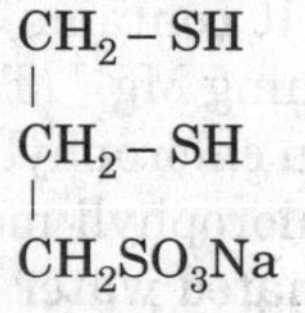

Sodium salt of 2, 3-dimercapto propane-1-sulphonic acid

(*v*) Sodium salt of 2, 3-dimercapto propane-1-sulphonic acid has been used for treating the poisoning caused by metal ions like Bi, Hg, Cu, Au, etc. These metal ions are strongly bound by the chelating ligand.

## Presence of metal complexes in Biological systems

Biological systems contain a number of metal complexes. These complexes make the metal available to the biological systems which may not be otherwise easily available. A few examples of biological metal complexes are given below.

### 1. Haemoglobin

Blood contains red blood corpuscles (RBCs). About 95% of RBCs is composed of haemoglobin. Hemo group is an important part of haemoglobin. Therefore, before discussing the structure of haemoglobin, we should discuss the structure of heme group. Heme group is a complex compound of Fe (II) with tetradentate macro cyclic ligand namely porphyrin. In heme group Fe (II) is attached with all the four N-atoms of porphyrin ligand. All the C-atoms of the ring system, coordinating N-atoms and the central Fe (II) lie in the same square plane.

Haemoglobin is a HS octahedral complex of Fe (II). Fe(II) has four unpaired electrons [$Fe^{2+} = 3d^6 = t_{2g}^4 e_g^2$ $(n = 4)$]. The centre of the octahedral structure is occupied by Fe (II) which is attached with four N-atoms of the heme group placed at the four corners of the square plane. Fe (II) is also attached with N-atom present in protein placed at one axial position and with $H_2O$ molecule placed at the other axial position. The function of haemoglobin in our body is to supply $O_2$ to various parts of the body. Thus haemoglobin is an *oxygen carrier*. $O_2$ supplied by hemoglobin in our body is used for the combustion (oxidation) of food (glucose, $C_6H_{12}O_6$). The oxidation of our food by $O_2$ supplied by haemoglobin produces energy which is utilised by living organisms to perform their various metabolic activities.

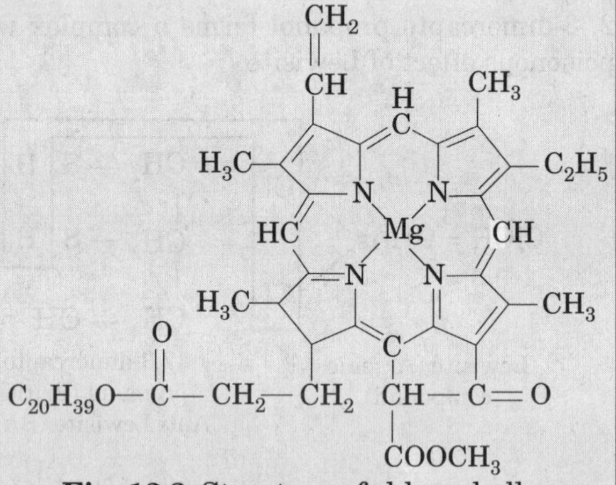

**Fig. 12.3** Structure of chlorophyll *a*

### 2. Chlorophyll

Chlorophyll is a green-coloured matter (pigment) present in green plants and algae. It contains the basic unit, porphyrin ring. It is a square planar complex containing $Mg^{2+}$. Thus it is a naturally occurring chelated complex of Mg. The Mg-atom in chlorophyll lies slightly above the plane of the rigid chelate ring structure. The chlorophyll molecule can add to it one or two water molecules axially. Such coordinated water molecules then help in associating other chlorophyll molecules through hydrogen bonding.

There are various modifications of chlorophyll. All these modifications contain porphyrin ring and differ from each other in the side groups of the porphyrin ring. The most abundant chlorophyl is *chlorophyll a* whose structure is shown in Fig. 12.3

In this structure, four N-atoms are situated at the four corners of a square. The $N_4$ unit forms a part of 16-membered natural macrocycle.

As stated above, all types of chlorophyll contain porphyrin ring in which a double bond in one of the pyrorole rings is reduced and a fused cyclopentaone ring is also present (See Fig. 12.3)

Chlorophyll present in plants helps in the process of **photosynthesis** which occurs in plants. In this process carbohydrates are produced in plants in presence of sunlight and chlorophyll. Thus chlorophyll, acts as a photosensitiser in the synthesis of carbohydrates in plants.

**Mechanism of photosynthesis.** Photosynthesis is a process in which $CO_2$ and moisture of the atmosphere combine together in the presence of sunlight and chlorophyll and produce glucose ($C_6H_{12}O_6$) and $O_2$. Glucose is further converted into polysaccharides like strach. In a simplified way, the process of photosynthesis can be shown as:

$$6CO_2 + 6H_2O + \textit{Solar energy} \xrightarrow{\text{Chlorophyll}} C_6H_{12}O_6\ (\textit{Glucose}) + 6O_2$$

Tracer studies have shown that all $O_2$ evolved in the above reaction comes from $H_2O$ and not from $CO_2$.

### 3. Vitamin $B_{12}$ (Cyanocobaltamine)

Vitamin $B_{12}$ is a complex of cobalt (I) with a tetradentate ligand which is similar to porphyrene found in haemoglobin. Vitamin $B_{12}$ does not occur in plants but occurs in animals and microorganisms. It is found in eggs and livers of ox, sheep, pig, fish etc. It has been found in rain water where its presence is atributed to microorganisms sucked up by winds. Vitamin $B_{12}$ is soluble in water and stable to heat. The deficiency of Vitamin $B_{12}$ causes pernicious anaemia, inflammation of tongue and mouth.

## UNIVERSITY QUESTIONS

1. Complete the following reactions:
   (*i*) $[Cu(H_2O)_6]^{2+} + 2\ NH_3 \rightarrow ...$
   (*ii*) $PtCl_2 + 2en \rightarrow ...$
2. How will you detect the presence of $Ni^{2+}$ ion present in a salt by using dimethyl glyoxime in presence of $NH_4$ OH. Give the chemical equation involved in the reaction.
3. How will you distinguish $Fe^{2+}$ and $Fe^{3+}$ ions.
4. Write the formula of Prussian blue.
5. Write a short note on complexometric titrations.

# 13 CHAPTER Nomenclature of Coordination Compounds

## IUPAC Rules of 2004

Coordination compounds whose formulae are given are named according to the system given by IUPAC (*International Union of Pure and Applied Chemistry*). The rules of this sytem are given below:

**1.** The name of a complex species is started with a small letter (not with a capital letter) and the full name of the species is written as one word without any gap.

**2. Order of Naming the Ions**

The name of the positive ion (cation), whether simple or complex, is written first and then the name of the negative ion (anion), whether simple or complex, is written. For example:

(*a*) In the compound, $[Co(NH_6)_3]Cl_3$, the name of the complex cation, $[Co(NH_3)_6]^{3+}$ is written first and then the name of the anion ($Cl^-$) is written.

(*b*) In the complex, $Na_3[Ag(S_2O_3)_2]$, the name of the simple cation, $Na^+$ is written first and then the name of the complex anion, $[Ag(S_2O_3)_2]^{3-}$ is written.

(*c*) In the complex compound, $[Pt^{4+}(NH_3)_4Cl_2]^{2+}[Pt^{2+}Cl_4]^{2-}$, the name of complex cation, $[Pt(NH_3)_4Cl_2]^{2+}$ is written first and then the name of the complex anion, $[PtCl_4]^{2-}$ is written.

**3. Naming the Complex Cation, Complex Anion and Non-ionic (neutral) Species**

The names of the complex cations and complex anions are written in such a way that the names of the ligands present in coordination sphere should be written first and then the name of the metal atom should be written. This rule is also followed while naming a non-ionic (neutral) species like $Ni(CO)_4$.

**4. Naming of Ligands**

Different types of ligands are named in different ways as shown below.

**(*i*) Negative ligands (organic and inorganic).** The names of negative ligands may end in ***–o, –ido, –ito or –ato***. Examples are given below.

**(*a*) *Ligands whose names end in –o.*** $CN^-$ = cyano (also called cynido), $O^{2-}$ = oxo, $O_2^{2-}$ = peroxo, $OH^-$ = hydroxo (also called hydroxyl), $O^-_2$ = superoxo, $HO^-_2$ = perhydro, $NO^-_2$ = nitrito (bonding through N-atom), $F^-$ = fluoro (also called fluorido), $Cl^-$ = chloro (chlorido), $Br^-$ = bromo (bromido), $I^-$ = iodo (iodido)).

***(b) Ligands whose names end in –ido.*** $F^-$ = fluorido, $Cl^-$ = chlorido, $Br^-$ = bromido, $I^-$ = iodido, $CN^-$ = cyanido (the name cyano is still in use), $H^-$ = hydrido, $S^{2-}$ = sulphido (also mercapto), $NH_2^-$ = amido, $NH^{2-}$ = imido, $N^{3-}$ = nitrido, $P^{3-}$ = phosphido, $N_3^-$ = azido, $HS^-$ = hydrogen sulphido.

***(c) Ligands whose names end in –ito.*** $SO_3^{2-}$ = sulphito, $ONO^-$ = nitrito (bonding through O-atom).

***(d) Ligands whose names end in –ato.*** $CO_3^{2-}$ = carbonato, $SO_4^{2-}$ = sulphato,

$NO_3^-$ = nitrato, $S_2O_3^{2-}$ = thiosulphato, $CH_3COO^-$ = acetato, $ClO_3^-$ = chlorato, $SCN^-$ = thiocyanato (bonding through S-atom), $NCS^-$ = isothiocyanato (bonding through N-atom),

Examples of some more Ligands whose names end in – ***ato*** are given below:

(*i*) $COO^-$ | $COO^-$ or $C_2O_4^{2-}$ ............................ Oxalato ($ox^{2-}$)

(*ii*) $H_2N - CH_2 - COO^-$ ........................ Glycinato ($gly^-$)

(*iii*) $CH_3 - C(=O) - CH = C(O^-) - CH_3$ ............... Acetylacetonato ($acac^-$)

(*iv*) $CH_3 - C = N - O^-$ | $CH_3 - C = N - OH$ ........................ Dimethylglyoxamato ($dmg^-$)

(*v*) 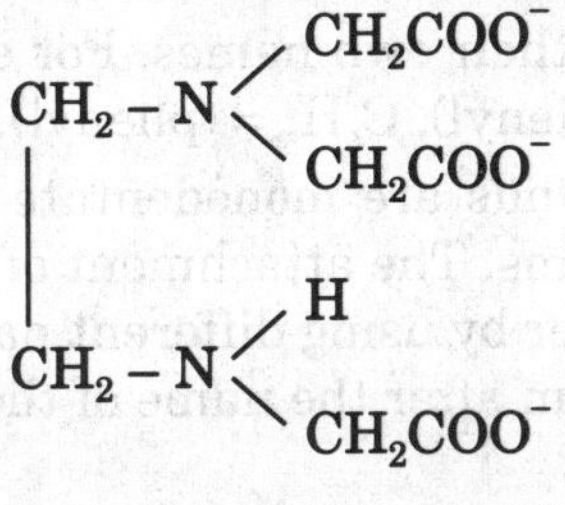 ................... Ethylenediaminetriacetato ($edta^{3-}$)

(*vi*) 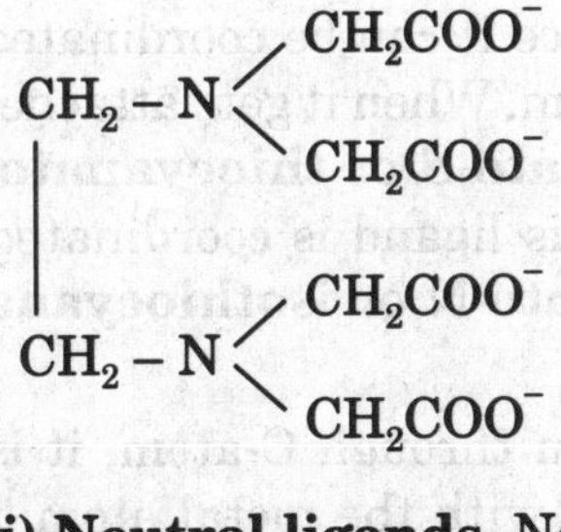 ................. Ethylenediaminetetraacetato ($edta^{4-}$)

***(ii)*** **Neutral ligands.** Neutral ligands have no special ending and generally their names are retained as such. For example:

$NH_3$ = ammine, $H_2O$ = aqua or aquo, CO = carbonyl, CS = thiocarbonyl, NO = nitrosyl, $O_2$(dioxygen), $N_2$ = (dinitrogen). The names of $O_2$ and $N_2$ are written in parenthesis, $CH_3NH_2$ = methyl amine, $PH_3$ = phosphine, $(C_6H_5)_3P$ = triphenyl phosphine, $H_2N–CS–NH_2$ = thiourea (tu), 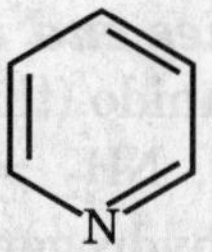 or $C_5H_5N$ = pyridine or pyridyl (*py*), 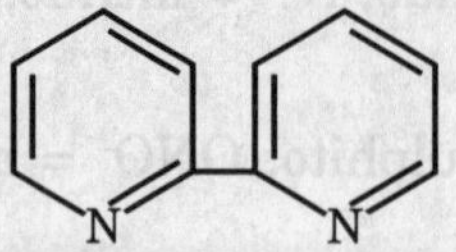 = 2, 2–dipyridyl (*dipy*) (both N-atoms are donor atoms), $NH_2–(CH_2)_2– NH_2$ = ethylenediamine (*en*) (both N-atoms are donor atoms), $NH_2—(CH_2)_2—NH—(CH_2)_2—NH_2$ = diethylene triamine (*dien*) (all the three N-atoms are donor atoms), $NH_2—(CH_2)_2—NH—(CH_2)_2—NH—(CH_2)_2—NH_2$ = triethylene tetramine (*trien*) (all the four N-atoms are donor atoms), 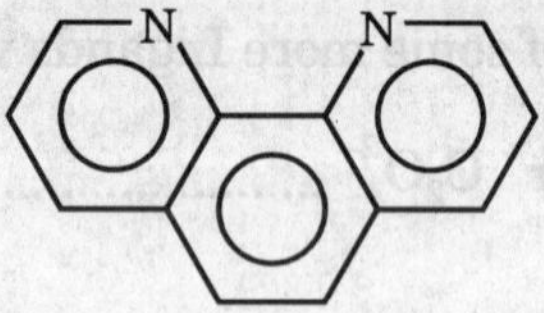 = phenanthroline (Both N-atoms are donor atoms).

***(iii) Positive ligands.*** These are very few in number and their names end in **ium**. Examples are:

$NH_2 — \overset{+}{N}H_3$ or $\overset{+}{N}_2H_5$ = hydrazinium, $NO^+$ = nitrosonium, $NO_2^+$ = nitronium, $NH_4^+$ = ammonium, $C_5H_5\overset{+}{N}H$ = pyridinium, $(C_6H_5)_4\overset{+}{P}$ = tetraphenyl phosphonium, $PH_4^+$ = phosphonium.

***(iv)* Organic free radicals.** These are given their own names. For example $CH_3$ = (methyl), $C_2H_5$ = (ethyl), $C_5H_5$ = (cyclopentadienyl), $C_6H_5$ = (phenyl).

***(v)* Ambidentate ligands.** Ambidentate ligands are monodentate ligands which have more than one coordinating (donor) atoms. The attachment of such ligands with the central metal atom is indicated either by using different names for the ligand or by placing the symbol of the donor atom after the name of the ligand separated by a hyphen. For example:

(*a*) $SCN^-$ ligand is an ambidentate ligand, since it can be coordinated to the metal atom either through S-atom or through N-atom. When it gets attached to the metal atom through S-atom,it is called **thiocyanato-S** or **thiocyanato** and is represented as $\mathbf{\overline{S}CN}$. On the other hand when this ligand is coordinated to the metal atom through N-atom, it is called **thiocyanato-N or isothiocyanato** and is represented as $\mathbf{\overline{N}CS}$.

(*b*) Whe $CN^-$ is coordinated to the metal atom through C-atom, it is called **cyano** $(\overline{C}N)$. On the other hand when it is linked with the metal atom through N-atom, it called **isocyano** $(\mathbf{C\overline{N}})$

(*c*) When $OC\overline{N}$ is linked with the metal atom through O-atom, it is called **cyanato** and is represented as $\mathbf{\overline{O}CN}$. On the other when this ligand gets attached with the metal atom through N-atom, it is called **isocyanato** and is represented as $\mathbf{OC\overline{N}}$.

(*d*) $NO_2^-$ ligand is also an ambidentate ligand, since it can be coordinated to the central metal atom either through N-atom or through O-atom. When it gets attached to the metal atom through N-atom, it is called **nitro-N** or **nitro** and is represented as $\mathbf{NO_2^-}$. On the other land when this ligand is coordinated to the metal atom through O-atom, it is called **nitro-O** or **nitrito** and is represented as $ON\bar{O}$.

**(*vi*) Order of naming ligands.** All the ligands whether they are negative, neutral or positive are named in the alphabetical order followed by the name of the metal atom/ion. The prefixes *di*, *tri* etc are not to be considered while naming the ligands in alphabetical order. For example [Pt $(NH_3)_4Cl(NO_2)]SO_4$ is named as tetraamminechlorido nitroplatinum (IV) sulphate.

**(*vii*) Use of numerical prefixes to indicate the number of ligands.** (*a*) If the coordination sphere has several ligands of the same type, the prefixes *di*, *tri*, *tetra*, *penta* and *hexa* are used to indicate the number of ligands of that type. For example [Cu $(NH_3)_4]SO_4$ is called tetraammine coper (II) sulphate. Here it may be noted that in writing tetraammine the last *a* of tetra and first *a* of ammine have been retained.

(*b*) The number of the ligands which already have *di*, *tri* etc. in their names (e.g., diethylene diamine, dimethyl glyoximato etc.) is indicated by using the prefixes *bis* (for two), *tris* (for three), *tetrakis* (for four), *pentakis* (for five), *hexakis* (for six) etc. followed by the name of the ligand placed in bracket ( ). For example $[CoCl_2\ (en)_2]SO_4$ is called dichloro *bis* (ethylenediamine) cobalt (IV) sulphate and [Ni $(dmg)_2]$ is called *bis* (dimethyl glyoximato) nickel (II).

**(*viii*) Naming of the bridging ligands.** A complex having two metal atoms is called **polynuclear complex**. In these complexes the monodentate ligand links the two metal atoms simultaneously and is called **bridging ligand** or **bridging group**.

(*a*) In naming the polynuclear complex having two different bridging ligands the prefix μ is used before the name of each of the two bridging ligands. For example

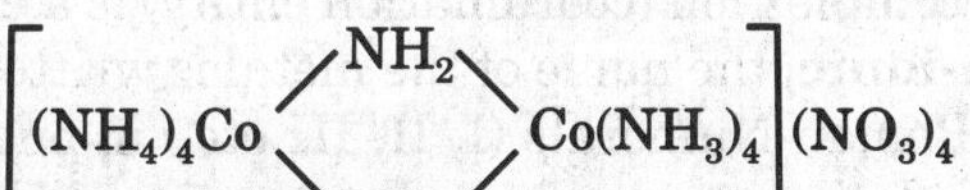


is named as tetraammine cobalt (III)-μ-amido-μ-nitrotetraaminecobalt (III) nitrate.

(*b*) If a complex has two similar bridging ligands, μ-*di* is used for its name. For example:

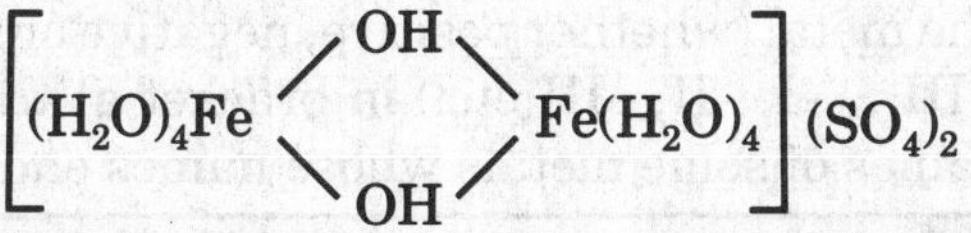


is called tetraquo iron (III) μ-dihydroxo tetraquo iron (III) sulphate or octaquo μ-di-hydroxo di-iron (III) sulphate.

(*c*) When the same ligand is present as a bridging ligand and as non-bridging (terminal) ligand, the bridging ligand is named first. In case of the presence of three similar bridging ligands, μ-tri is used before their name. For example

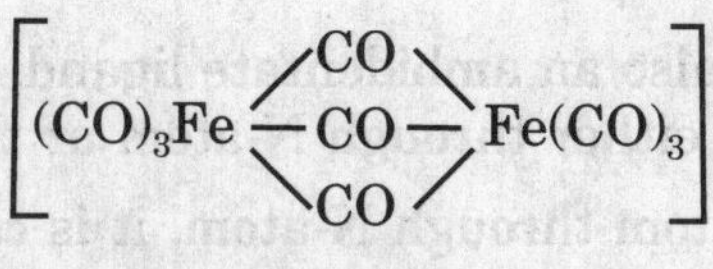


is called μ-tricarbonyl hexacarbonyl di-iron (0).

(*d*) If the bridging ligand bridges more than two metal atoms, $\mu_3$, $\mu_4$ etc. is used before the name of bridging ligand. $\mu_3$, $\mu_4$ etc. indicates the number of metal atoms between which the bridging ligand makes bridges, e.g., since in basic berylium acetate, $Be_4^{2+}O^{2-}(CH_3COO)_6$, *oxo* ($O^{2-}$) group bridges four Be-atoms, in the name of this compound $\mu_4$ is used before the name of bridging ligand, *viz*, $O^{2-}$ group. Thus, the name of this compound is – *oxo* – μ – hexa acetato tetra beryllium (II).

(*ix*) **Representation of complicated ligands by abbreviations.** Instead of writing the long names of complicated ligands abbreviations are customarily used for these ligands. Abbreviations of some ligands have been given below:

| | | | |
|---|---|---|---|
| (*en*) | ethylenediamine | (*glȳ*) | glycinato |
| (*dien*) | diethylenetriamine | (*phen*) | 1, 10-phenathroliune |
| (*trien*) | triethylenetetra amine | (*py*) | pyridyl |
| (*edta*$^{3-}$) | ethylene diamine triacetato | (*tu*) | thiourea |
| (*acāc*) | acetylacetonato | (*dipy*) | dipyridyl |
| (*dm̄g*) | dimethyl glyoximato | (*tripy*) | tripyridyl |
| (*edta*$^{-4}$) | ethylenediamminetetraacetato | (*ox*$^{2-}$) | oxalato |

## 5. Naming of Metal Atom

The name of the metal depends on the fact whether the complex species is a complex anion (negative complex ion), complex cation (positive complex ion), the non-ionic complex (neutral molecule) or the complex species has $H^+$ ion as its cation. Thus the following cases may be studied.

(*a*) **Positive complex ion (cation) or non-ionic (neutral) coordination compound.** If the complex ion (coordination entity) is a **cation or co-ordination compound is non-ionic,** the name of the metal is written **as such** followed by its oxidation state in Roman Numerals (I, II, III etc.) in enclosing marks (brackets). (0) is used for an oxidation state zero and a negative oxidation state is shown by a negative sign before the Roman numeral, e.g. $[Cu(NH_3)_4]SO_4$ is tetraammine copper (II) sulphate.

(*b*) **Negative complex ion (anion).** The name of the metal atom present in a negative complex ion like. $[Ni(CN)_4]^{4-}$, $[Fe(CN)_6]^{4-}$, $[SiF_6]^{2-}$ etc. ends in **–ate** and the oxidation state of the metal (whether positive, negative or zero) is written in Roman numerals (0, I, II, III... –I, –II, –III etc.) in bracket at the end of the name of the metal atom. The names of some metals whose names end in **–ate** are given below:

Ti = titanate, V = vanadate, Cr = chromate, Mn = manganate, Fe = ferrate, Co = cobaltate, Ni = nickelate, Cu = cuperate, Zn = zincate, Ag = argentate, Pt = platinate, Au = aurate, Ta = tantalate, Zr = zirconate, Mo = molybdate, W = tungstate, Pd = palladate, Ga = gallate, Ge = germanate, Tl = thallate, Bi = bismuthate, B = borate, Si = silicate, Pb = plumbate, Sn = stannate, Ir = iridate, Os = osmate etc.

(*c*) If the complex compound has $H^+$ ion as its cation (e.g. $H_4[Pt(CN)_6]$), $H_2[PtCl_6]$ etc), then the name of the metal atom is made to end in **–ic** followed by the oxidation state of the metal in bracket and then the word **acid** is added. For example $H_4[Pt(CN)_6]$ = hexacyanoplatin**ic (II) acid,** $H_2[PtCl_6]$ = hexachloro plat **nic** (IV) **acid**. On ionisation, since these compounds give $H^+$ ions, these are called acids.

## 6. Naming of cis- and trans- Isomers (Geometrical Isomers) of Square Planar and Octahedral Complexes

See "*Geometrical isomerism in square planar and octahedral complexes*"

## 7. Naming of Optical Isomers

We know the *dextro* and *laevo* forms are optically active isomers while *meso* form in optically inactive. Dextro form is represented by putting *d* or (+) before its name or formula. Similarly leavo form is designated by putting *l* or (–) before its name or formula. Meso form is called *meso* isomer. Thus dextro potassium trioxalatoiridate (III) is formulated as *d* or (+) $K_3[Ir^{3+}(C_2O_4)_2]$.

## IUPAC Names of Some Complex Species

### (A) IUPAC Names of Some Complex Cations (Positive Complex Ions)

$[CoCl(en)_2(ONO)]$ ... chlorido *bis* (ethane –1, 2-diamine) nitrito cobalt (III) ion.

$[Cr(H_2O)_4Cl_2]^+$ ... tetraaquadichlorochromium (III) ion.

$[Ag(NH_3)_2]^+$ ... diamminesilver (I) (Tollen's reagent)

$[Cr(H_2O)_5(SCN)]^{2+}$ ... pentaquathiocyanatochromium (III) ion.

$[Pt(NH_3)_2Cl]^+$ ... diamminechloroplatinum (II) ion

$[Co(NH_3)_4Cl_2]^+$ ... tetraamminedichlorocobalt (III) ion

$[CoCl(en)_2(NO_2)]$ ... chloro *bis* (ethylene diamine)nitro cobalt (III) ion.

$[Co(en)_2Cl_2]^+$ ... dichloro *bis* (ethylene diamine)cobalt (III) ion

$[Cu(en)_2]^{2+}$ ... *bis* (ethylene diamine)copper (II) ion

### (B) IUPAC Names of Some Complex Anions (Negative complex ions)

$[Fe(CN)_6]^{3-}$ ... hexcyanoferrate (III) ion

$[Fe(CN)_6]^{4-}$ ... hexacyanoferrate (II) ion

$[BF_4]^-$ ... tetrafluoroborate (III) ion

$[Ni(CN)_4]^{2-}$ ... tetracyanonickelate (II) ion

$[PtBrCl(NO_2)(NH_3)]^-$ ... amminebromidochloridonitroplatinate (II) ion

$[Co(ONO)_6]^{3-}$ ... hexanitritocobaltate (III) ion

$[HgI_3]^-$ ... triiodomercurate (II) ion

### (C) IUPAC Names of Some Complex Compounds Containing Simple Cations and Complex Anions

$KFe[Fe(CN)_6]$ ... potassiumferricferrocyanide (Turn bull's blue)

$K_3 [Co (CN)_5(NO)]$ ... potassiumpentacyanonitrosylcobaltate (II)

$Na_2 [CrF_4O]$ ... sodiumtetrafluoridooxochromate (III)

Dextro $K_3 [Ir (C_2O_4)_3]$ ... potassium (+) or d-trioxalatoiridate (III)

$Hg [Co (SCN)_4]$ ... mercuric tetrathiocyanatocobaltate (II)

$Li [AlH_4]$ ... lithiumtetrahydridoaluminate (III)

$Fe_4 [Fe (CN)_6]_3$ ... ferrichexacyanoferrate(II) (Prussian blue)

$K_2[Zn (OH)_4]$ ... potassiumtetrahydroxozincate (II)

$Na_2 [Fe (CN)_5 (NO)]$ ... sodiumpentacyanonitrosylferrate (II) (Sodium nitroprusside)

$Na_2[SiF_6]$ ... Sodiumhexafluoridosilicate (IV)

$K [Co (CN)_3(CO)_2(NO)$ ... potassiumcarboyltricyanonitrosocobaltate (II)

$K_2[HgCl_4]$ ... potassiumtetrachloromercurate (II)

$Na [Co (CO)_4]$ ... sodiumtetracarbonylcobaltate (–I)

$K_2[OsCl_5 N]$ ... potassiumpentachloronitridoosmate (VI)

$K_3[Co (NO_2)_6]$ ... Potassiumcobaltinitrite

$Na_2 [Ni (EDTA)]$ ... sodiumethylenediaminetetraacetatonickel (I)

$Na_3[Co (NO_2)_6]$ ... sodiumnitrito cobaltate (III)

$(NH_4)_3 [Cr (SCN)_6]$ ... ammoniumhexathiocyanato-S-chromate (III) or ammoniumhexathiocyanatochromate (III)

$Cu_2[Fe(CN)_6]$ ... copper hexacyano ferrate (II)

$(NH_4)_2 [Pt (NCS)_6]$ ... ammoniumhexathiocyanato-N-platinate (IV) or ammoniumhexa isothiocyanato platinate (IV)

$K_2[HgI_4]$ ... potassium tetra-iodomercurate (II) (Nesseler's reagent)

$Na_4[Fe(CN)_5 (NO) (S)]$ ... sodium penta cyano nitrosonium sulphido ferrate (II)

## (D) IUPAC Names of Some Complex Compounds Containing Complex Cation and Simple Anions

$[Co(NH_3)_6] Cl_3$ ... hexaminecobalt (III) chloride

$[Cr (H_2O)_4Cl_2] NO_3$ ... tetraaquadichlorido chromium (III) nitrate

$[Cd (NH_3)_4] (NO_3)_2$ ... tetraminecadmium (II) nitrate

$[Cr (en)_3]Cl_3$ ... *tris* (ethane 1, 2-diamine) chromium (III) chloride

$[Co(NH_3)_4(H_2O) (Br)] (NO_3)_2$ ... tetramineaquabromidocobalt (III) nitrate

$[Co (NH_3)_6] ClSO_4$ ... hexammine cobalt (III) chloride sulphate

$[\{(C_6H_5)_3P\}_3 Rh] Cl$ ... *tris* (triphenyl phosphine) rhodium (I) chloride

$[Cu (NH_3)_4] (NO_3)_2$ ... tetramminecupricnitrate

$[Pt Cl (NH_2CH_3) (NH_3)_2]Cl$ ... diamminechloro(methylamine)platinum (II) chloride

$[Co (N_3)(NH_3)_5] SO_4$ ... pentammine azido cobalt (III) sulphate

$[Ag (NH_3)_2] OH$ ... diammine silver (I) hydroxide (Tollen's reagent)

$[Cu \{SP (CH_3)_3\}]^+ ClO_4^-$ ... *tris* (trimethyl phosphine sulphide) copper (I) per chlorate

$[Cu\ I\ (bpy)_2]\ I$ ... Iodobis (bipyridine) copper (II) iodide

$[Fe\ (H_2O)_5(NO)]SO_4$ ... penta aquanitrosonium ferrate (I)

## (E) IUPAC Names of Some Complexes Containing Complex Cation and Complex Anion

$[Pt\ (NH_3)_4Cl_2]\ [PtCl_4]$ ... tetraammine dichlorido platinum (IV) tetrachlorido platinate (II)

$[Co\ (NH_3)_4Cl_2]_3[Cr\ (CN)_6]$ ... tetramine dichlorido cobalt (III) hexacyanochromate (III)

$[Pt\ (py)_4]\ [Pt\ Cl_4]$ ... tetrapyridine platinum (II) tetrachlorido platinate (II)

$[Cr\ (NH_3)_5(SCN)]\ [ZnCl_4]$ ... pentaammine thiocyanato chromium (III) tetrachloro zincate (II)

$[Cr(en)_3]\ [Ni(CN)_5(H_2O)]$ ... *tris* (ethylenediamine) chromium (III) aquapentacyanonickelate (II)

## (F) IUPAC Names of Some Neutral Complexes

$[Ni\ (CO)_4]^\circ$ ... tetracarbonyl nickel (o)

$[Ni\ (dmg)_2]^\circ$ ... *bis* (dimethyl glyoximato) nickel (II)

$[Pt\ (NH_3)_2Cl_2]^\circ$ ... diammine dichloro platinum (II)

$[Co\ (NH_3)_3(NO_2)_3]^\circ$ ... triamminetrinitrocobalt (III)

$[Cr\ (PPh_3)\ (CO)_5]^\circ$ ... pentacarbonyl triphenylphosphine chromium (o)

$[Fe\ (C_5H_5)_2]^\circ$ ... *bis* (cyclopentadienyl) iron (II)

$[Mn_3(CO)_{12}]^\circ$ ... dodecacarbonyl trimanganese (o)

$(V\ (acac)_2O]^\circ$ ... *bis* (acetylacetonato) oxo vanadium (IV).

$[Cr\ (PPh_3)\ (CO)_5]^\circ$ ... penta carbonyl triphenyl phosphine chromium (o)

$[Cu\ Cl\ (tu)_2]^\circ$ ... chlorobis (thio urea) copper (I)

$[\{(C_6H_5)_3P\}_3Pt]^\circ$ ... *tris* (triphenyl phosphine) platinium (o)

## (G) IUPAC Names of Some Bridged Complexes

$[(NH_3)_5Cr - OH - Cr\ (NH_5)_5]Cl_5$ ...... pentaammine chromium (III) – μ – hydroxo pentaammine chromium (III) chloride.

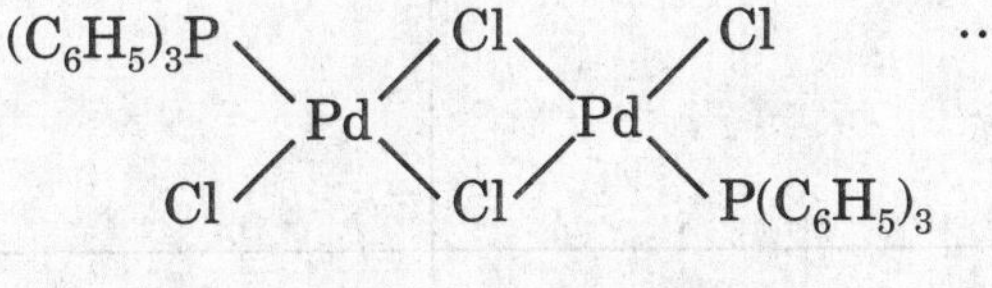


...... chlorido triphenyl phosphine palladium (II) - μ-dichlorido chloridotriphenyl phosphine palladium (II).

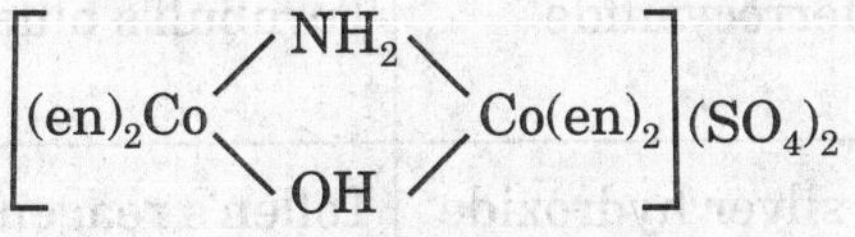


...... *bis* (ethane-1, 2- diamine) cobalt (III) -μ-amido-μ-hydroxobis (ethane 1, 2-diamine) cobalt (III) sulphate

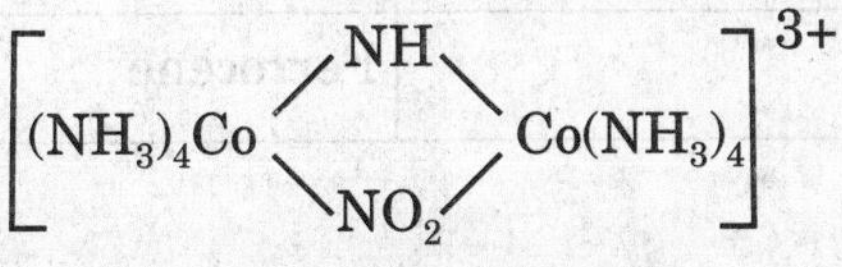


...... tetrammine cobalt (III)-μ-imido-μ-nitrotetraammine cobalt (III) ion

## Common Names of Some Complex Species

There are many complex species for which common names in place of IUPAC names are still in use. Examples of some such complex species are given below:

| *Complex ion* | *IUPAC name* | *Common name* |
|---|---|---|
| (*i*) $[Zn^{2+}O_2]^{2-}$ | Dioxozincate (II) ion | Zincate ion |
| (*ii*) $[Fe^{2+}(CN)_6]^{4-}$ | Hexacyanoferrate (II) ion | Ferrocyanide ion |
| (*iii*) $[Fe^{3+}(CN)_6]^{3-}$ | Hexacyano ferrate (III) ion | Ferricyanide |
| (*iv*) $[Co^{3+}(ONO)_6]^{3-}$ | Hexanitrito cobaltate (III) ion | Cobaltinitrite |
| (*v*) $H[AuCl_4]$ | Hydrogen tetra chloro aurate (III) | Tetrachloro auric acid |
| (*vi*) $H[BF_4]$ | Hydrogen tetra fluoro borate (III) | Tetrafluoro boric acid |
| (*vii*) $H_2[PtCl_6]$ | Hydrogen hexa-chloroplatinate (IV) | Haxachloro platinic acid (IV) |
| (*viii*) $K_2[HgI_4]$ (used for the test of $NH_3$ and $NH_4^+$ ions) | Potassium tetraiodo mercurate (II) | Nesseler's reagent |
| (*ix*) $Na_2[Fe^{2+}(CN)_5(NO^+)]$ (used for the test of $S^{2-}$ ions) | Sodium pentacyano nitrosyl ferrate (II) | Sodium nitro-prusside |
| (*x*) $K_3[Co(NO_2)_6]$ (yellow and insoluble, formed in the test of $NO_2^-$ ion) | Potassium hexanitro cobaltate (III) | Pot. cobaltinitrite (Fischer's salt) |
| (*xi*) $Fe_4^{+3}[Fe^{2+}(CN)_6]_3$ (Blue solution or blue ppt, formed in the test of $Fe^{2+}$ ion) | Ferric ferro cyanide | Prussian blue |
| (*xii*) $K^+Fe^{3+}[Fe^{2+}(CN)_6]$ (Blue colour) | Pot. ferric ferrocyanide | Turnbull's blue |
| (*xiv*) $[Ag(NH_3)_2]OH$ | Diammine silver hydroxide | Tollen's reagent |
| (*xv*) $Fe(\eta^5-C_5H_5)_2$ | – | Ferrocene |

## Rules for Writing the Formulae of the Complex Species whose IUPAC Names are Given

Following rules are used.

1. Formula of the simple cation or complex cation is written first followed by that of the anion.
2. Within the coordination sphere, the symbol of the metal is written first and then the symbols/formulae of ligands are written.
3. The symbols/formulae of the ligands are written alphabetically according to their names irrespective of charge on them. For example in the complex compound, $[Co(NH_3)_4(H_2O)Cl]Cl_2$, the three ligands are named as *ammine* (for $NH_3$ molecules), *aqua* (for $H_2O$ molecule) and *chlorido* (for $Cl^-$ ion). These names are in alphabetical order.
4. Polydentate ligands are also written in the alphabetical order.
5. For the abbreviated ligands (e.g. *en* for ethylene diammine, $H_2\ddot{N}-CH_2-CH_2-\ddot{N}H_2$), the first letter of the abbreviation is used to fix the position of the abbreviated ligand in the alphabetical order.
6. For the ligands having special names (*e.g. aqua* for $H_2O$ molecule) the first letter of the special name is used to determine the position of the ligand in alphabetical order.
7. The formula of the polyatomic ligands (e.g. $NH_3$, $H_2O$, $CN^-$ etc) and the abbreviations for the abbreviated ligands (*e.g. py*, *en* etc.) are enclosed in parenthesis.
8. The central metal atom and the ligands are written in such a way that there should be no space between them.
9. If the formula of a complex cation or a complex anion is to be written, the charge on the complex ion is written outside the square bracket as a right superscript with the number before the positive (+) or negative (–) sign. For example $[Cu(NH_3)_4]^{2+}$, $[Fe(CN)_6]^{3-}$ etc.
10. The number of cations and anions to be written in the formula is calculated on the basis of the fact that the total positive charge must be equal to the total negative charge, since the complex as a whole is electrically neutral. In order to determine the number of cations and anions we must know the charge on each of the ions.
11. Since the complex molecule as a whole is neutral, the total positive charge must be equal to the total negative charge. Thus the cations and anions must be multiplied by such constant which may make the charges equal. These constants give the number of each type of ions.
12. (*a*) Simple ions (e.g. $Cl^-$, $CN^-$, $NO^+$ etc.) have usual charges.

    (*b*) Charge on the complex ion is equal to the sum of the charge on the central metal atom (*i.e.* oxidation state of the metal atom) plus the total number of charges on the coordinated ligands.

    The formula of some coordination compounds whose IUPAC names are known are given in the following examples.

**Example 1.** ***Write the formulae of the following complex compounds.***

***(i) tetraamminechloronitrocobalt (III) nitrate***

***(ii) potassiumamminetrichloroplatinate (II)***

***(iii) tetraamineaquachlorocobalt (III) chloride***

**Solution.** (*i*) Anion: $NO_3^-$ (*simple ion*)

Complex ion (entity): $[CoCl\ (NO_2)\ (NH_3)_5]^x$ where $x$ is the charge carried by the complex ion.

$x$ = oxidation state of cobalt + algebraic sum of charges carried by ligands.

$= +(3) + (-1) + (-1) + 4\ (0) = +3 - 2 = +1$

[Here $NH_3$ is a neutral ligand. The two other ligands $Cl^-$ and $NO_2^-$ carry charge of –1 each]

Thus, the complex ion can be written as $[CoCl\ (NO_2)\ (NH_3)_4]^+$

Thus, formula of the complex compound is $\mathbf{[CoCl\ (NO_2)\ (NH_3)_4]\ NO_3}$ **(Ans.)**

(*ii*) Cation: $K^+$ (*simple ion*)

Complex ion (entity) : $[PtCl_3(NH_3)]^x$ where $x$ is the charge carried by the complex ion.

$x$ = oxidation state of platinum + algebric sum of charges carried by ligands

$= (+2) + 3\ (-1) + 0 = (+2) - 3 = -1$

(Here $NH_3$ is a natural ligand and $Cl^-$ carries a charge of –1)

Thus, the complex ion can be written as $[PtCl_3\ (NH_3)]^-$

Therefore, formula of the complex compound is $\mathbf{K[PtCl_3(NH_3)]}$ **(Ans.)**

(*iii*) Anion : $Cl^-$ (*simple ion*)

Complex ion (entity): $[CoCl\ (H_2O)(NH_3)_4]^+$ (*complex ion*) where $x$ the charge carried by the complex ion.

$x$ = oxidation state of cobalt + algebric sum of charges carried by ligands.

$= (+3) + (-1) + 0 + 4(0) = (+3) - 1 = +2$

(Here $H_2O$ and $NH_3$ are neutral ligands. $Cl^-$ carries a charge of –1).

Thus, the complex cation can be written as $[CoCl\ (H_2O)\ (NH_3)_4]^{2+}$

Therefore, formula of the complex compound is $\mathbf{[CoCl(H_2O)\ (NH_3)_4]Cl_2}$. **(Ans.)**

**Example 2.** ***Formulate the compound sodium pentacyanonitrosyl sulphido ferrate (III).***

**Ans.** Simple cation is $Na^+$. Ligands are $5CN^-$, NO°, $S^{2-}$ and central metal ion is $Fe^{3+}$. Thus the coordination sphere is $[Fe^{3+}(CN)_5(NO°)S^{2-}]^{4-}$. Clearly four $Na^+$ ions are present outside the coordinaton sphere and hence the formula is

$\mathbf{Na_4\ [Fe(CN)_5(NO)S]}$ **(Ans.)**

**Example 3.** ***Write the formula of potassium tetrahydroxozincate (II).***

**Ans.** Simple cation is $K^+$. Ligands are $4OH^-$ and the central metal ion is $Zn^{2+}$. Thus the coordination entity is $[Zn^{2+}(OH)_4]^{2-}$. Clearly two $K^+$ ions are present outside the coordination sphere. Hence the formula is: $\mathbf{K_2[Zn\ (OH)_4]}$ **(Ans.)**

## UNIVERSITY QUESTIONS

**1.** Write the formulas for the following coordination compounds:
(*i*) tetraammineaquachloridocobalt (III) chloride
(*ii*) potassium tetrahydroxozincate (II)
(*iii*) potassium trioxalatoaluminate (III)
(*iv*) dichlorido bis (ethane-1, 2-diamine) cobalt (III).
(*v*) tetracarbonyl nickel (0).

**2.** Write down the formulae of the following coordination compounds:
(*i*) hexaaquairon (II) sulphate
(*ii*) potassium tetracyanonickelate (III)
(*iii*) chloronitrodiammineplatinum (II)
(*iv*) potassium hexacyanoferrate (III)
(*v*) chloro diammine platinum (II) ion
(*vi*) dichloro tetraammine cobalt (III) ion
(*vii*) potassium pentacyano nitrosyl cobaltate (III)
(*viii*) bis (acetlacetonato) oxovanadium (IV).

**3.** Write down the formulae of the following coordination compounds.
(*i*) hexa ammine platinum (IV) chloride
(*ii*) sodium ethylenediaminetetraaceato chromate (II)

**4.** Give the structural formulae of the following:
(*i*) Potassium hexacyanoferrate (III)
(*ii*) Chloro-bis (ethylene diamine) nitro cobalt (III) ion

**5.** Write IUPAC names of the following.
(*i*) $Na_3[Fe\,(C_2O_4)_3]$ (*ii*) $[Pt\,(NH_3)_4]\,[PtCl_4]$
(*iii*) $Fe[Fe\,(CN)_6]$ (*iv*) $K_4[Fe(CN)_6]$
(*v*) $Fe(\pi - C_5H_5)\,(CO_3)$ (*Himachal Pradesh 2010*)

**6.** Write IUPAC names of : (*i*) $[Cu\,(NH_3)_4]SO_4$ (*ii*) $[Cr(en)_3]Cl_3$(*Banglore 2010*)

**7.** Write the IUPAC names of the following compounds: (*Ranchi 2010*)
(*i*) $K_4[Fe(CN)_6]$ (*ii*) $K_2[PtCl_6]$
(*iii*) $[Co\,(en)_3]Cl_3$ (*iv*) $Na_3[Fe(CN)_5NO]$

**8.** (*A*) Write IUPAC names of the following compounds: (*i*) $C_2H_5BeH$ and (*ii*) $[C_6H_5As]_n$.
(*B*) Write the formula of following complexes:
(*i*) Dicholorobis (ethylendiamine) cobalt (III) ion and (*ii*) Potassium dicyano argentate (I). (*Nagpur 2010*)

**9.** Name the following complexes according to IUPAC system.
(*i*) $[Co\,(NH_3)_5ONO]Cl_2$ (*ii*) $Na_3[AlF_6]$
(*iii*) $[Pt\,(NH_3)_3Cl_3]$ (*iv*) $[Co\,(en)_3]\,[Cr(CN)_6]$. (*Lucknow 2010*)

**10.** Write IUPAC names of the following complexes:

(*i*) $[Co(NH_3)_6]^{3-}$ (*ii*) $[CuF_6]^{4-}$

(*iii*) $Ca_2[Fe(CN)_6]$ (*iv*) $[Ir(en)_2(H_2O)_2]Cl_3$

(*v*) $[Fe(CN)_6]^{3-}$ (*vi*) $[Fe(CN)_6]^{4-}$

(*Osmania 2010, Ch. Charan Singh 2012*)

**11.** Write IUPAC names of the following compounds

(*i*) $(C_6H_5)_4Te$ (*ii*) $C_6H_5HgOH$

(*iii*) $CH_3MgI$ (*iv*) $C_6H_5AsH_2$

(*v*) $K[Ag(CN)_2]$ (*vi*) $[Co(en)_2(NH_3)_2]Cl_3$ (*Nagpur 2012*)

## Answers (only hints)

**1.** (*i*) $[Co(NH_3)_4(H_2O)Cl]Cl_2$ (*ii*) $K_2[Zn(OH)_4]$

(*iii*) $K_3[Al(C_2O_4)_3]$ (*iv*) $[CoCl_2(en)_2]^+$

(*v*) $[Ni(CO)_4]$

**2.** (*i*) $[Fe(H_2O)_6]SO_4$ (*ii*) $K[Ni(CN)_4]$

(*iii*) $[Pt(NH_3)_2Cl(NO_2)]$ (*iv*) $K_3[Fe(CN)_6]$

(*v*) $[Pt(NH_3)_2Cl]^+$ (*vi*) $[Co(NH_3)_4Cl_2]^+$

(*vii*) $K_2[Co(CN)_5NO]$ (*viii*) $[V(acac)_2O]$

**3.** (*i*) $[Pt(NH_3)_6]Cl_4$ (*ii*) $Na_2[Cr(CH_3COO)_4(en)]$

**4.** (*i*) $K_3[Fe(CN)_6]$ (*ii*) $[CoCl(en)_2(NO_2)]^+$.

# 14 CHAPTER Werner's Theory of Coordination Compounds

## Introduction

In order to explain the nature of bonding and structure of coordination compounds, many scientists proposed different theories. Among these the first significant contribution was made by Werner in 1893. He proposed a theory, which is commonly known as **Werner's coordination theory**. He isolated different complex compounds from the reaction of cobalt chloride and ammonia. These compounds are called **cobalt ammines**. These are given below:

| *Compound* | *Colour* | *Name on the basis of colour* |
|---|---|---|
| $CoCl_3.6NH_3$ | Yellow | Luteo complex |
| $CoCl_3.5NH_3$ | Purple | Purpureo complex |
| $CoCl_3.4NH_3$ | Violet | Violeo complex |
| $CoCl_3.4NH_3$ | Green | Praseo complex |
| $CoCl_3.3NH_3$ | Blue-green | |

Two different colours of $CoCl_3.4NH_3$ are due to its different isomers, *viz., cis-* and *trans-*isomers.

Werner studied the properties of these compounds and many other such compounds in detail. Based on these properties, he proposed Werner's coordination theory to explain the nature of bonding in coordination compounds.

## Postulates of Werner's Theory

The main postulates of this theory are as follows:

1. *Metals possess two type of valencies called (a) primary or principal ionizable valency (b) secondary or non-ionizable valency.*

**(a)** *Primary valencies are those which a metal exhibits in the formation of its simple salts.* Thus in the formation of $PtCl_4$, $CoCl_3$, $CuSO_4$ and AgCl, the primary valencies of Pt, Co, Cu and Ag are 4, 3, 2 and 1 respectively. *These days primary valency is referred to as* oxidation state. Primary valency is ionisable.

(*b*) *Secondary valencies are those which a metal atom or cation exercises towards neutral molecules or negative groups (ligands) in the formation of its complex ions.* Every metal cation has a fixed number of secondary valencies. The secondary valency is also termed as the co-ordination number (C.N.) of the metal cation. In $[Pt\ (NH_3)_6]Cl_4$, $[Co\ (NH_3)_6]Cl_3$, $[Cu\ (NH_3)_4]SO_4$ and $[Ag(NH_3)_2]Cl$, the secondary valencies are 6 for $Pt^{4+}$, 6 for $Co^{3+}$, 4 for $Cu^{2+}$ and 2 for $Ag^+$ cations respectively. Secondary valency is non-ionisable.

2. *Every metal atom has a tendency to satisfy both is primary and secondary valencies. Primary valencies are satisfied by negative ions whereas secondary valencies are satisfied by negative ions / neutral molecule (ligands). In certain cases a negative ion may satisfy both types of valency.*

3. *The secondary valencies (i.e., ligands) are always directed towards the fixed positions in space and this leads to definite geometry of the coordination compound.* For example, if a metal ion has six secondary valencies, these are arranged *octahedrally* around the central metal ion. If the metal ion has four secondary valencies, these are arranged in either *tetrahedral or square planar* arrangement around the central metal ion. The secondary valencies, thus, determine the stereochemistry of the complex. On the other hand, the primary valency is non-directional.

4. While writing the structure of a given coordination compound on the basis of Wener's theory, the primary valencies are generally indicated by broken lines (- - - - - - ) and secondary valencies are shown by solid lines (—).

## To Explain the Structure of Cobalt (III) Ammines on the Basic of Werner's Coordination Theory

We have already mentioned that by the reaction between $CoCl_3$ and $NH_3$, Werner isolated cobalt ammines, *viz.*, $CoCl_3 . 6NH_3$, $CoCl_3 . 5NH_3$, $CoCl_3$, $4NH_3$ and $CoCl_3 . 3NH_3$. On the basis of his theory he explained the structure of these ammines.

In all these ammines the primary valency of cobalt is equal to 3 while the secondary valency is equal to 6. Primary valency of 3 is satisfied by three $Cl^-$ ions and hence these ions are attached with $Co^{3+}$ ion by broken line (- - - -). Secondary valency of six is satisfied either by six $NH_3$ molecule alone or by both $NH_3$ molecules and $Cl^-$ ions. The species satisfying the secondary valency are attached with $Co^{3+}$ ion by solid line (—).

Since secondary valency (*i.e.*, coordination number) of $Co^{3+}$ ion is six, all the ammines have octahedral geometry.

**1. $CoCl_3 . 6NH_3$.** The primary valency of Co-atom which is equal to 3 is satisfied by three $Cl^-$ ions and hence the attachment of these ions with $Co^{3+}$ ion is shown by broken lines [Fig. 14.1(*a*)]. When this ammine is acted upon by an aqueous solution of $AgNO_3$, all the three $Cl^-$ ions present in this ammine are precipitated as AgCl (white ppt).

$$CoCl_3 . 6NH_3 \longrightarrow [Co . 6NH_3]^{3+} + 3Cl^-$$

$$3Ag^+ + 3Cl^- \longrightarrow 3AgCl\ (\textit{white ppt})$$

The precipitation of all the three $Cl^-$ ions indicates that these ions are outside the coordination sphere and it is only six $NH_3$ molecules which, along with Co atom, are present inside coordination sphere. Thus the structure of $CoCl_3 . 6NH_3$ can be written as $[Co . 6NH_3] Cl_3$ or $[Co(NH_3)_6]Cl_3$. This structure shows that the secondary valency of Co-atom which is equal to six is satisfied by six $NH_3$ molecules placed inside the coordination sphere and hence the attachment of $NH_3$ molecules to Co-atom is shown by solid lines [See Fig. 14.1(*a*)].

**2. $CoCl_3.5NH_3$.** The primary valency of Co-atom in this ammine is equal to 3 and is satisfied by three $Cl^-$ ions. The attachment of these ions with $Co^{3+}$ ion is, therefore, shown by broken lines [See Fig. 14.1 (*b*)].

The treatment of this ammine with the aqueous solution of $AgNO_3$ precipitates only two $Cl^-$ ions as AgCl (white ppt.)

$$CoCl_3 . 5NH_3 \longrightarrow [CoCl.5NH_3]^{2+} + 2Cl^-$$
$$2Ag^+ + 2Cl^- \longrightarrow 2AgCl\ (\textit{white ppt.})$$

The precipitation of only two $Cl^-$ ions indicates that two $Cl^-$ ions are outside the coordination sphere and the remaining one $Cl^-$ ion and five $NH_3$ molecules are present inside the coordination sphere. Thus on the basis of these observations, the structure of $CoCl_3 . 5NH_3$ can be written as $[CoCl . 5NH_3] Cl_2$ or $[Co(NH_3)_5Cl]Cl_2$. This structure shows that the secondary valency of Co-atom which is equal to six is satisfied by one $Cl^-$ ion and five $NH_3$ molecules placed inside the coordination sphere. The attachment of one $Cl^-$ ion and five $NH_3$ molecules to Co-atom is, therefore, shown by solid lines. The above discussion shows that, on the basis of Werner's theory, the structure of $CoCl_3.5NH_3$ can be shown as depicted in Fig.14.1 (*b*). This figure shows that out of the three $Cl^-$ ions, one $Cl^-$ ion is attached with $Co^{3+}$ ion by a combined solid-broken line (----), since this ion satisfies both primary as well as secondary valency of CO-atom.

**3. $CoCl_3.4NH_3$.** The primary valency of Co-atom which is equal to 3 is satisfied by three $Cl^-$ ions and hence the attachment of these ions to $Co^{3+}$ ion is shown by broken lines [Fig. 14.1(c)]. The treatment of this ammine with the aqueous solution of $AgNO_3$ precipitates only one $Cl^-$ ion as AgCl (white ppt.)

$$CoCl_3 . 4NH_3 \longrightarrow [CoCl_2 . 4NH_3]^+ + Cl^-$$
$$Ag^+ + Cl^- \longrightarrow AgCl\ (\textit{white ppt.})$$

The precipitation of only one $Cl^-$ ion indicates that one $Cl^-$ ion is outside the coordination sphere and the remaining two $Cl^-$ ions and four $NH_3$ molecules are present inside the coordination sphere. Thus the structure of $CoCl_3. 4NH_3$ can be represented as $[CoCl_2 . 4NH_3] Cl$ or $[Co (NH_3)_4Cl_2]Cl$ which shows that the secondary valency of Co-atom which is equal to six is satisfied by two $Cl^-$ ions and four $NH_3$ molecules placed inside the coordination sphere and hence the attachment of these six species with $Co^{3+}$ ion has been shown by solid lines. Thus, on the basis of Werner's theory, the structure of $CoCl_3 . 4NH_3$ can be shown as given in Fig. 14.1(*c*) in which the attachment of two $Cl^-$ ions with $Co^{3+}$ ion has been shown by solid-broken lines, since these ions satisfy both types of valency of Co-atom.

**4. $CoCl_3.3NH_3$.** The primary valency of Co-atom which is equal to 3 is satisfied by three $Cl^-$ ions and hence the attachment of these ions with $Co^{3+}$ ion is represented by broken lines [See Fig. 14.1(*d*)].

When this ammie is treated with the aqueous solution of $AgNO_3$, no precipitate is obtained. The non-formation of any precipitate shows that no $Cl^-$ ion is outside the coordination spheres *i.e.* all the three $Cl^-$ ions and three $NH_3$, along with $Co^{3+}$ ion, are present inside the coordination sphere. Thus the structure of this ammine can be written as $[CoCl_3 . 3NH_3]^0$ or $[Co [NH_3]_3Cl_3]^0$. This structure shows that the secondary valency of Co-atom which is equal to six is satisfied by three $Cl^-$ ions and three $NH_3$ molecules placed inside the co-ordination sphere. Thus the linking of these six species with $Co^{3+}$ ion is represented by solid lines. This discussion indicates that, on the basis of Werner's theory the structure of $CoCl_3.3NH_3$ can be shown as given in Fig. 14.3(*d*) in which the attachment of three $Cl^-$ ions with $Co^{3+}$ ion has been represented by solid-broken lies, since these ions satisfy both types of valency of Co-atom.

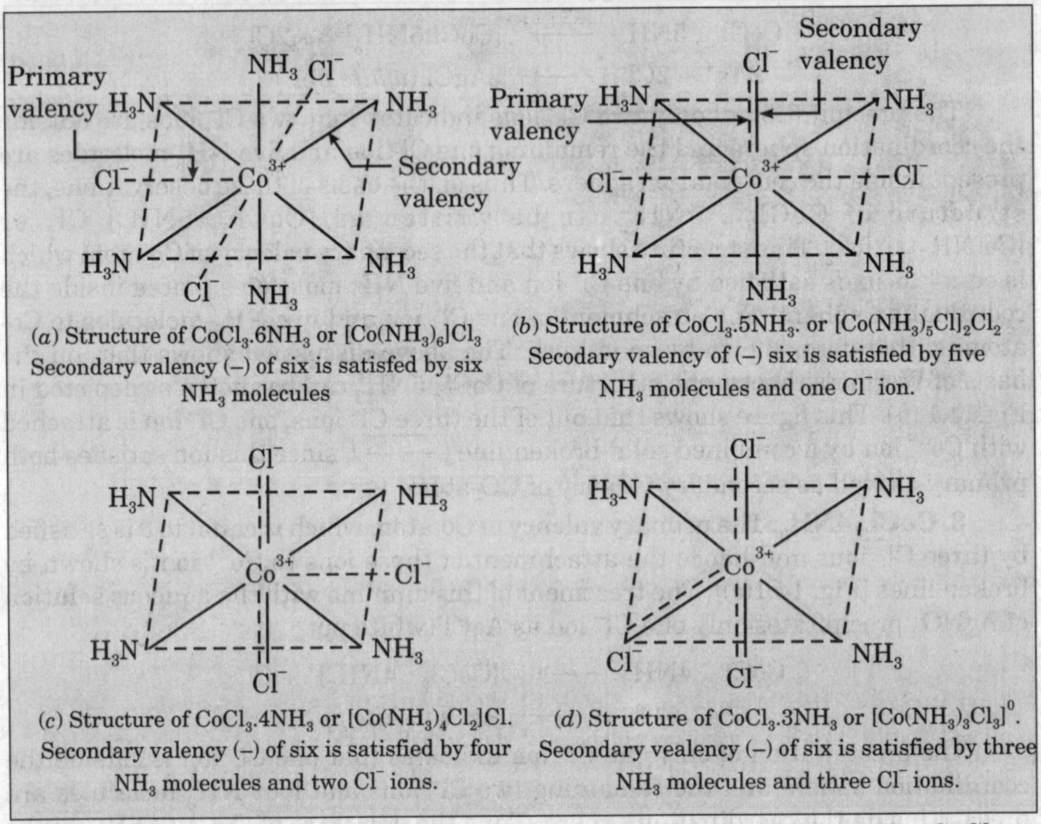


**Fig. 14.1** Structure of Co (III) ammines *viz*, $CoCl_3.6NH_3$, $CoCl_3.5NH_3$., $CoCl_3 . 4NH_3$ and $CoCl_3 . 3NH_3$ on the basis of Werner's coordination theory. In all these ammines the primary valency of Co-atom is equal to 3 and secondary valency is equal to six. Primary valency (- - -) of three is satisfied by three $Cl^-$ ions.

## Experimental Evidence in Favour of Werner's Theory

We have said that the Co (III) ammines given above can be represented as $[Co (NH_3)_6]Cl_3$, $[Co(NH_3)_5 Cl]Cl_2$, $[Co (NH_3)_4Cl_2]Cl$ and $[Co(NH)_3Cl_3]$ respectively. These structures have been confirmed by the following studies:

**1. Crysoscopic measurements.** Cryoscopic measurements (i.e. measurement of the depression in freezing points of aqueous solutions) gives the number of ions furnished by the dissociation of one molecule of the compound. This measurement for Co (III) ammines has also indicated that these compounds dissociate to give 4 (for $CoCl_3 . 6NH_3$), 3(for $CoCl_3 . 5NH_3$), 2(for $CoCl_3 . 4NH_3$) and 1 (for $CoCl_3 . 3NH_3$) particles (ionic or molecular).

**2. Molar electrolytic conductance (or conductivity) measurements.** Measurement of molar conductance of a compound at its large dilution gives the sum of the electrical charges on the positive and negative ions furnished by each molecule of the compound. Molar conductance value for each Co (III) ammine has been determined. These values have shown that the number of electrical charges given by these ammines is 6 (for $CoCl_3 . 6NH_3$), 4(for $CoCl_3. 5NH_3$), 2 (for $CoCl_3 . 4NH_3$) and zero or nil (for $CoCl_3 . 3NH_3$). Total number of ions obtained by the ionisation of the ammines decreases with the decrease of magnitude of molar conductance value of the ammines (See Table 14.1).

**3. Precipitation method.** When a given ammine is treated with an aqueous solution of $AgNO_3$, some $Cl^-$ ions get precipitated as AgCl. The number of $Cl^-$ ions precipitated is the number of $Cl^-$ ions that are ionisable and hence are present outside the coordination sphere. $CoCl_3 . 3NH_3$ gives no precipitate, showing that all the three $Cl^-$ ions are present in the coordination sphere. Thus since the number of $Cl^-$ ions precipitated by adding $AgNO_3$ solution to $CoCl_3 . 6NH_3$, $CoCl_3 . 5NH_3$ and $CoCl_3$ $4NH_3$ ammines is 3, 2 and 1 respectively, the number of $Cl^-$ ions present outside the coordination sphere is also equal to 3, 2 and 1 respectively. $CoCl_3 . 3NH_3$ gives no precipitate (i.e. gives no $Cl^-$ ions) with $AgNO_3$ solution and hence all the three $Cl^-$ ionsare present in the coordination sphere (See Table 14.1).

## To Explain the Structure of Pt (IV) Ammines on the Basis of Werner's Coordination Theory

We have already said the Pt (IV) gives five ammines which are $PtCl_4 . 6NH_3$, $PtCl_4 . 5NH_3$, $PtCl_4 . 4NH_3$, $PtCl_4 . 3NH_3$ and $PtCl_4 . 2NH_3$. In each of these ammines, primary valency of Pt is equal to its oxidation state (= + 4) while the secondary valency is equal to 6. Thus each ammine has octahedral geometry. Cryoscopic measurements have shown that the number of particles obtained by the ionisation of these ammines is equal to 5, 4, 3, 2 and 1 respectively. Molar conductance measurements have revealed that the number of electrical charges obtained as a result of their ionisation is 8 (+4, –4), 6 (+3, –3), 4 (+2, –2), 2 (+1, –1) and zero. These studies show that that the structure of these ammines can be written as shown in Table 14.1.

## Limitations of Werner's Theory

Though Werner's theory was able to explain a number of properties of the coordination compounds, it could not answer the following basic questions:

(*i*) Why only certain elements from coordination compounds and not others?

(*ii*) Why the coordination sphere/entity has a definite geometry?

(*iii*) Why these compounds possess definite magnetic and optical properties?

**Table 14.1. Structure of Co (III) and Pt (IV) ammines**

| *Molecular and structural formula of ammines* | *Ionisation of ammines* | *Number of particles (ionic or molecular) as determined by cryoscopic measurements* | *Molar conductance, $\wedge_m (ohm^{-1}\ cm^2)$* | *Number of electrical charges determined by $\wedge_m$* |
|---|---|---|---|---|
| **Co(III) ammines** | | | Decreasing ↓ | Decreasing ↓ |
| $CoCl_3.6NH_3$ or $[Co(NH_3)_6]Cl_3$ | $[Co(NH_3)_6]Cl_3 \rightarrow [Co(NH_3)_6]^{3+} + 3Cl^-$ | 4 | 390 | 6(+3, –3) |
| $CoCl_3 . 5NH_3$ or $[Co(NH_3)_5Cl]Cl_2$ | $[Co(NH)_3]_5Cl]Cl_2 \rightarrow [Co(NH_3)_5Cl]^{2+} + 2Cl^-$ | 3 | 262 | 4(+2, –2) |
| $CoCl_3.4NH_3$ or $[Co(NH_3)_4Cl_2]Cl$ | $[Co(NH_3)_4Cl_2]\ Cl \rightarrow [Co(NH_3)_4Cl_2]^+ + Cl^-$ | 2 | 102 | 2(+1, –1) |
| $CoCl_3.3NH_3$ or $[Co\ (NH_3)_3Cl_3]^0$ (Non electrolyte) | $[Co(NH_3)_3Cl_3]^0 \rightarrow$ No ionisation | 0 | 0 | Nil |
| **Pt(IV) ammines** | | | Decreasing ↓ | Decreasing ↓ |
| $PtCl_4.6NH_3$ or $[Pt(NH_3)_6]Cl_4$ | $[Pt(NH_3)_6]Cl_4 \rightarrow [Pt(NH_3)_6]^{4+} + 4Cl^-$ | 5 | 524 | 8(+4, –4) |
| $PtCl_4.5NH_3$ or $[Pt(NH_3)_5Cl]Cl_3$ | $[Pt(NH_3)_5Cl]Cl_3 \rightarrow [Pt(NH_3)_5Cl]^{3+} + 3Cl^-$ | 4 | 404 | 6(+3, –3) |
| $PtCl_4.4NH_3$ or $[Pt(NH_3)_4Cl_2]Cl_2$ | $[Pt(NH_3)_4Cl_2]Cl_2 \rightarrow [Pt(NH_3)_4Cl_2]^{2+} + 2Cl^-$ | 3 | 230 | 4(+2, –2) |
| $PtCl_4.3NH_3$ or $[Pt(NH_3)_3Cl_3]Cl$ | $[Pt(NH_3)_3Cl_3]Cl \rightarrow [Pt(NH_3)_3Cl_3]^+ + Cl^-$ | 2 | 97 | 2(+1, –1) |
| $PtCl_4 . 2NH_3$ or $[Pt(NH)_3]_2Cl_4]^0$ | $[Pt(NH_3)_2Cl_4]^0 \rightarrow$ No ionisation | 0 | 0 | Nil |

## UNIVERSITY QUESTIONS

1. On the basis of Werner's theory explain why cobalt amine complex, $CoCl_3$. $4NH_3$ when treated with $AgNO_3$ solution precipitates only one $Cl^-$ ion although it has three $Cl^-$ ions. *(Nagpur 2002)*
2. State in brief the experimental evidence in favour of Werner's theory. *(Delhi 2002)*
3. Write the structural formula of all the possible cationic complexes of Pt (IV) containing $Cl^-$ ions and $NH_3$ molecules as ligands. *(VBS Purvanchal 2003)*

**Ans.** $PtCl_4 \cdot 5NH_3$, $PtCl_4 \cdot 4NH_3$ and $PtCl_4 \cdot 3NH_3$ are the cationic complexes in which Pt (central atom) is in +4 oxidation state and $Cl^-$ ions and $NH_3$ molecules are present as ligands. The structural formulae of these compounds are $[Pt(NH_3)_5Cl]\ Cl_3$, $[Pt\ (NH_3)_4Cl_2]Cl_2$, $[Pt(NH_3)_3Cl_3]Cl$ and respectively.

4. Write the structure of $CoCl_3 \cdot 6NH_3 \cdot CoCl_3 \cdot 5NH_3 \cdot CoCl_3 \cdot 4NH_3$ and $CoCl_3 \cdot 3NH_3$ according to Werner's theory. *(Delhi 2003)*
5. Calculate the EAN of the central metal ion in the following complexes and write their IUPAC names: (*i*) $[Pt\ (NH_3)_4Cl_2]Cl_2$ (*ii*) $[Co(en)_2Cl_2]Cl$ (*iii*) $K_3[Fe(CN)_6]$. *(VBS Purvanchal 2003)*
6. Explain why $K_2[PtCl_6]$ does not give white precipitate of AgCl with $AgNO_3$. *(VBS Purvanchal 2003)*
7. Explain Werner's theory. What is primary and secondary valency? Explain with example. *(Kanpur 2010)*
8. Write down the principle of Werner's Co-ordination theory. *(Pune 2009, Banglore 2010)*
9. Write main postulates of Werner's theory. Explain how this theory is helpful in explaining the stereochemistry and geometry of coordination compounds. How will you predict the existence of complexes with this theory? *(Rajsthan 2010)*
10. Metal possesses two types of valancies primary and secondary. The primary valence is ...and the secondary valence is ... *(H.N. Bahuguna Garhwal 2010)*
11. Postulates of Werner's Co-ordination Theory. *(Ranchi 2010)*
12. Write Werner's postulates of coordination compounds. Differentiate between double salts and complex compunds. *(Magadh 2010)*
13. Discuss the postulates of Werner's theory of coordination compounds. *(Nagpur 2010)*
14. $[Co\ (NH_3)_4Cl_2]^+$ does not give white ppt. with $AgNO_3$ soln. Explain. *(Lucknow 2010)*
15. By using Werner's theory predict the number of $Cl^-$ ions precipated by $AgNO_3$ and the formule of $CoCl_3 \cdot 5NH_3$ and $CoCl_3 \cdot 4NH_3$. *(Nagpur 2012)*
16. What is primary and secondary valency of central metal ion in $[Co(NH_3)_6]$ and $[Co(NH_3)_4Cl_2]Cl$. *(Lucknow 2013)*

15

CHAPTER

# Isomerism in Coordination Compounds

## Introduction

The coordination compounds which have the same molecular formula but have their ligands attached to the central metal atom in different ways are called **isomers**. These isomers have different properties. The phenomenon that gives rise to different isomers is called **isomerism**.

Isomerism may be of two types viz., *structural isomerism* and *stereo (space) isomerism*.

## Structural Isomerism

This isomerism arises due to the difference in the structures of complex (coordination) compounds. This isomerism is of the following types:

### 1. Conformation Isomerism

In this isomerism two isomers have different geometries but are otherwise identical. For example, $[Ni^{2+}(P.Et.Ph_2)_2Br_2]$ complex (four-coordinated complex) gives two conformation isomers. One of these is green and paramagnetic while the other is brown and diamagnetic. The green form is tetrahedral while the brown form is square planar. Thus these have different geometries but the coordination number of $Ni^{2+}$ ion in both the isomers is the same (=4).

### 2. Ionisation Isomerism

The complex compounds showing ionisation isomerism are called *ionisation isomers*. These isomers have the same molecular formulae and are produced when the ligands given in the coordination sphere and the anions present outside the coordination sphere are exchanged with each other. Thus both the complex compounds of each of the following pairs are ionisation isomers to each other.

(*i*) $[Co(NH_3)_5Br]SO_4$ and $[Co(NH_3)_5(SO_4)]Br$

(*ii*) $[Co(NH_3)_4Cl_2]NO_2$ and $[Co(NH_3)_4Cl(NO_2)]Cl$

(*iii*) $[Pt(NH_3)_4Cl_2]Br_2$ and $[Pt(NH_3)_4Br_2]Cl_2$

(*iv*) $[Co(NH_3)_5(NO_3)]SO_4$ and $[Co(NH_3)_5(SO_4)]NO_3$

(*v*) $[Cr(NH_3)_4I_2]Br$ and $[Cr(NH_3)_4IBr]I$

(*vi*) $[Fe(NH_3)_5(CN)]SO_4$ and $[Fe(NH_3)_5SO_4]CN$.

Ionisation isomers show different properties, *e.g.*, $[Co(NH_3)_5Br]SO_4$(I) and $[Co(NH_3)_5(SO_4)]Br$ (II) which are ionisation isomers to each other show different properties as mentioned below:

(*i*) (I) is red-violet while (II) is red.

(*ii*) When these isomers are dissolved in water, they give different ions as shown below:

$$\underset{\text{(I) (red-violet)}}{[Co(NH_3)_5Br]SO_4} \rightleftharpoons [Co(NH_3)_5Br]^{2+} + SO_4^{2-}$$

$$\underset{\text{(II) (red)}}{[Co[NH_3]_5SO_4]Br} \rightleftharpoons [Co(NH_3)_5SO_4]^+ + Br^-$$

It is due to the formation of different ions in solution that these compounds show different behaviour when treated with an aqueous solution of $BaCl_2$ and $AgNO_3$. (I). gives a *white ppt*. of $BaSO_4$ with an aqueous solution of $BaCl_2$ indicating that $SO_4^{2-}$ ions in (I) are outside the coordination sphere. This compound does not give any ppt. with the aqueous solution of $AgNO_3$. On the other hand, (II) does not give any ppt. with a solution of $BaCl_2$ but forms a *yellow ppt*. of AgBr with $AgNO_3$ solution, indicating that $Br^-$ ions in this compound are outside the coordination sphere.

$$\underset{\text{(I)}}{[Co(NH)_3)_5Br]SO_4} + \underset{\text{(Solution)}}{BaCl_2} \longrightarrow \underset{\text{White ppt.}}{BaSO_4} + [Co(NH_3)_5Br]Cl_2$$

$$\underset{\text{(I)}}{[Co(NH_3)_5Br]SO_4} + \underset{\text{(Solution)}}{AgNO_3} \longrightarrow \text{No reaction}$$

$$\underset{\text{(II)}}{[Co(NH_3)_5SO_4]Br} + \underset{\text{(Solution)}}{BaCl_2} \longrightarrow \text{No reaction}$$

$$\underset{\text{(II)}}{[Co(NH_3)_5SO_4]Br} + \underset{\text{(Solution)}}{AgNO_3} \longrightarrow \underset{\text{Yellow ppt.}}{AgBr} + [Co(NH_3)_5SO_4]NO_3$$

It may be noted that the coordination number of the central atom in both the ionisation isomers is the same. Ionisation isomers can be detected by the determination of their conductance in solution.

## 3. Hydrate Isomerism

This type of isomerism is similar to ionisation isomerism. Hydrate isomers are obtained when $H_2O$ molecules acting as ligands are replaced by the anions present outside the coordination sphere. $CrCl_3\cdot6H_2O$ exists in three hydrate isomers which are $[Cr(H_2O)_6]Cl_3$(I), $[Cr(H_2O)_5Cl]Cl_2\cdot H_2O$ (II) and $[Cr(H_2O)_4Cl_2]Cl\cdot2H_2O$ (III). These have different physical and chemical properties, e.g.,

(*i*) (I) is violet, does not lose water over conc. $H_2SO_4$ and all the three $Cl^-$ ions are precipitated as AgCl with $AgNO_3$ solution.

$$\underset{\text{(I) (violet)}}{[Cr(H_2O)_6]Cl_3} + \underset{\text{(solution)}}{3AgNO_3} \longrightarrow \underset{\text{white ppt.}}{3AgCl\downarrow} + [Cr(H_2O)_6](NO_3)_3$$

or $$3Cl^- + 3Ag^+ \longrightarrow 3AgCl\downarrow$$

(*ii*) (II) is blue-green, loses only one $H_2O$ molecule over conc. $H_2SO_4$ and only two $Cl^-$ ions are precipitated as AgCl with $AgNO_3$ solution.

$$\underset{\text{(II)(green)}}{[Cr(H_2O)_5Cl]Cl_2 \cdot H_2O} + \underset{\text{(solution)}}{2AgNO_3} \longrightarrow \underset{\text{white ppt.}}{2AgCl} + [Cr(H_2O)_5Cl](NO_3)_2 . H_2O$$

or $$2Cl^- + 2Ag^+ \longrightarrow 2AgCl\downarrow$$

(*iii*) (III) is blue-green, loses two $H_2O$ molecules over conc. $H_2SO_4$ and only one $Cl^-$ ion is precipitated by $AgNO_3$ solution.

$$\underset{\text{(III)(green)}}{[Cr(H_2O)_4Cl_2]Cl.2H_2O} + \underset{\text{(solution)}}{AgNO_3} \longrightarrow \underset{\text{white ppt.}}{AgCl} + [Cr(H_2O)_4Cl_2]NO_3 \cdot 2H_2O$$

or $$Cl^- + Ag^+ \longrightarrow AgCl\downarrow$$

The behaviour of these isomers over conc. $H_2SO_4$ as mentioned above indicates that the number of $H_2O$ molecules outside the coordination sphere in (I), (II) and (III) is equal to zero, one and two respectively.

The action of $AgNO_3$ solution on these isomers indicates that (I), (II) and (III) have three, two and one $Cl^-$ ions outside the coordination sphere.

Other examples of hydrate isomers are given below:

(*i*) $[Co(en)_2(H_2O)Cl]Cl_2$ and $[Co(en)_2Cl_2]Cl \cdot H_2O$

(*ii*) $[Co(NH_3)_4(H_2O)Cl]Cl_2$ and $[Co(NH_3)_4Cl_2]Cl \cdot H_2O$

(*iii*) $[Cr(py)_2(H_2O)_2Cl_2]Cl$ and $[Cr(py)_2(H_2O)Cl_3] \cdot H_2O$

(*iv*) $[Cr(en)_2(H_2O)_2]Br_3$ and $[Cr(en)_2(H_2O)Br]Br_2 \cdot H_2O$

## 4. Coordination Isomerism

This type of isomerism is shown by those complex compounds which are composed of complex cation and complex anion. Coordination isomers are obtained when some or all ligands of both the coordination spheres are interchanged with each other. Thus both the complex compounds of each of the following pairs are coordination isomers to each other. In these pairs the central metallic atom in the two coordination spheres may be the same or different.

(*i*) $[Cr^{3+}(NH_3)_6]^{3+}[Cr^{3+}(CN)_6]^{3-}$ and $[Cr^{3+}(NH_3)_4(CN_2)]^+[Cr^{3+}(CN)_4(NH_3)_2]^-$

(*ii*) $[Pt^{2+}(NH_3)_4]^{2+}[Pt^{2+}Cl_4]^{2-}$ and $[Pt^{2+}(NH_3)_3Cl]^+[Pt^{2+}Cl_3(NH_3)]^-$

(*iii*) $[Co^{3+}(NH_3)_6]^{3+}[Cr^{3+}(CN)_6]^{3-}$ and $[Co^{3+}(CN)_6]^{3-}[Cr(NH_3)_6]^{3+}$

(*iv*) $[Cu^{2+}(NH_3)_4]^{2+}[Pt^{2+}Cl_4]^{2-}$ and $[Cu^{2+}Cl_4]^{2-}[Pt^{2+}(NH_3)_4]^{2+}$

(*v*) $[Cr^{3+}(NH_3)_6]^{3+}[Cr^{3+}(SCN)_6]^{3-}$ and $[Cr^{3+}(NH_3)_4(SCN)_2]^+[Cr^{3+}(SCN)_4(NH_3)_2]$

## 5. Linkage Isomerism

Linkage isomerism is shown by those coordination compounds which contain ambidentate ligands, (e.g., $NO_2^-$, $SCN^-$), since these ligands can coordinate to the central metal atom through any of the two different available donor atoms. The linking of ambidentate ligand to the metal atom through two different donor atoms gives two different compounds which are related with each as linkage isomers. Examples of linkage isomers are:

(*i*) $[Co^{3+}(NH_3)_5NO_2]^{2+}$ (I) and $[Co^{3+}(NH_3)_5ONO]^{2+}$ (II) are linkage isomers. In (I), $NO_2^-$ ligand is linked with $Co^{3+}$ ion through its N-atom while in (II), $NO_2^-$ ligand is linked with $Co^{3+}$ ion through its O-atom. (I) is called *penta ammine-*

*nitro cobalt (III) ion* while (II) is called *pentaamminenitrito cobalt (III) ion*. (I) is yellow while (II) is red.

(*ii*) $[Pd^{2+}(dipy)(SCN)_2]^{\circ}$ (orange yellow) and $[Pd^{2+}(dipy)(NCS)_2]^{\circ}$ (light yellow) are linkage isomers. First compound is called *dipyridyldithiocyanato palladium (II)* while the second compound is called *dipyridyl diisothiocyanato palladium (II)*.

## 6. Coordination Position Isomerism

This type of isomerism is shown by those complex compounds which contain bridging ligands and arises when the non-bridging ligands are differently placed round the central metal atom. Thus (I) and (II) are coordination position isomers to each other, since $NH_3$ molecules and $Cl^-$ ions (non-bridging ligands) are differently placed round the two $Co^{3+}$ ions.

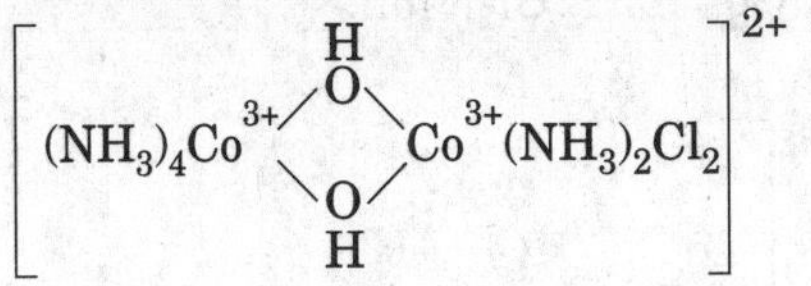

Tetrammine coblat (III) μ-dihydroxo diammine dichloro cobalt (III) ion. (I) (*unsymmetrical*)

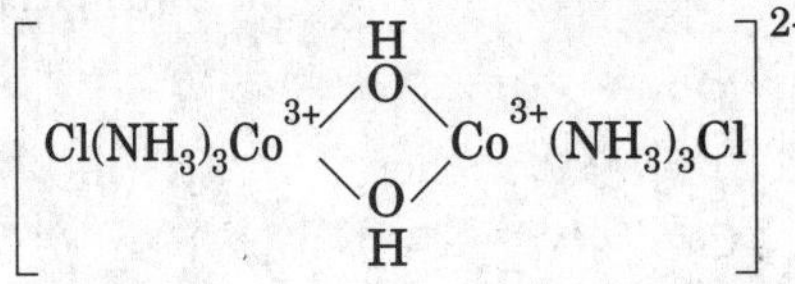

Chloro triammine cobalt (III) μ-dihydroxo chloro triammine cobalt (III) ion. (II) (*symmetrical*)

The following binuclear complexes of cobalt also show coordination position isomerism, since the terminal ligands, viz., six $NH_3$ molecules and two $Cl^-$ ions can be placed in an unsymmetrical and symmetrical manner round the central cobalt atom.

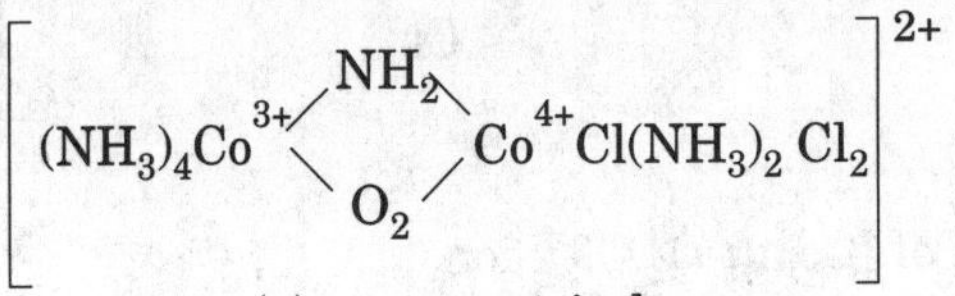

(*a*) unsymmetrical

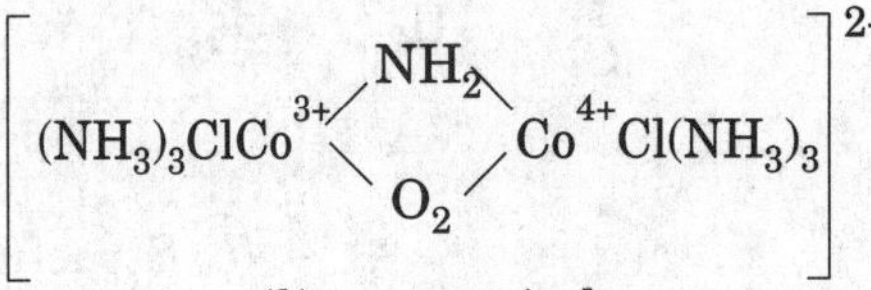

(*b*) symmetrical

## 7. Ligand Isomerism

There are certain ligands which exist as isomers, e.g., diamine derivatives of propane exist in two isomeric forms which are called 1, 2-diamino propane or propylene diamine (*pn*) and 1, 3-diamino propane or trimethylene diamine (*tn*). The structures of propane, *pn* and *tn* are given below:

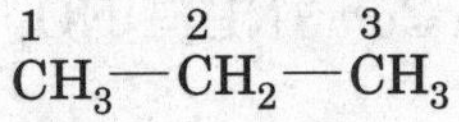

Propane

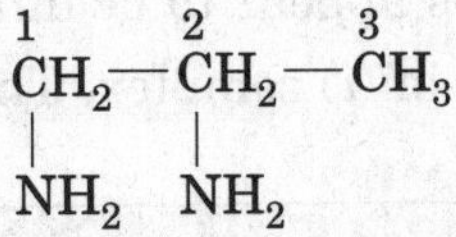

1, 2-diamino propane or propylene diamine (*pn*)

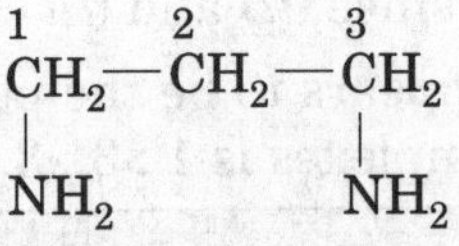

1, 3-diamino propane or trimethylene diamine (*tn*)

*pn* and *tn* both are bidentate (neutral) ligands. When these ligands get coordinated to the metal atom, two isomers are obtained. These are called ligand isomers and the isomerism is called ligand isomerism. Thus $[Co^{3+}(pn)_2Cl_2]^+$ and $[Co^{3+}(tn)_2Cl_2]^+$ are ligand isomers whose structures are given below in Fig. 15.1. Both these isomers have octahedral geometry.

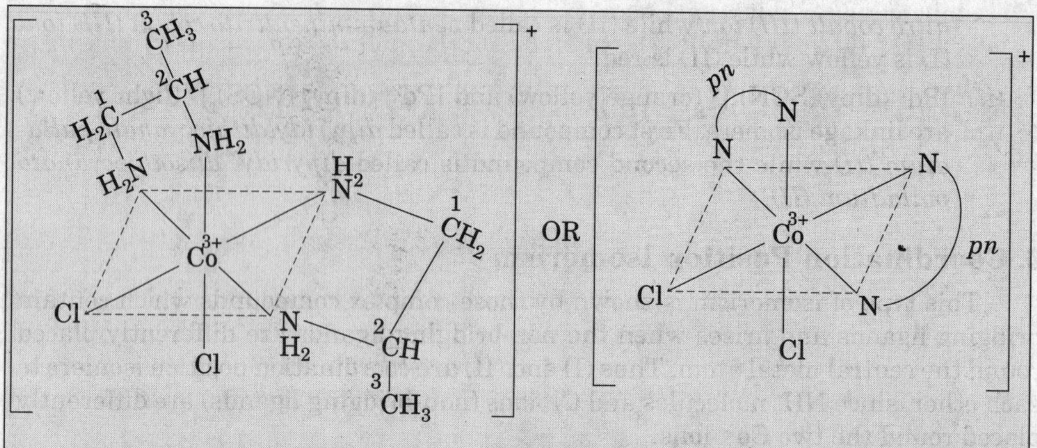


(*a*) Structure of $[Co^{3+}(pn)_2\ Cl_2]^+$ ion.

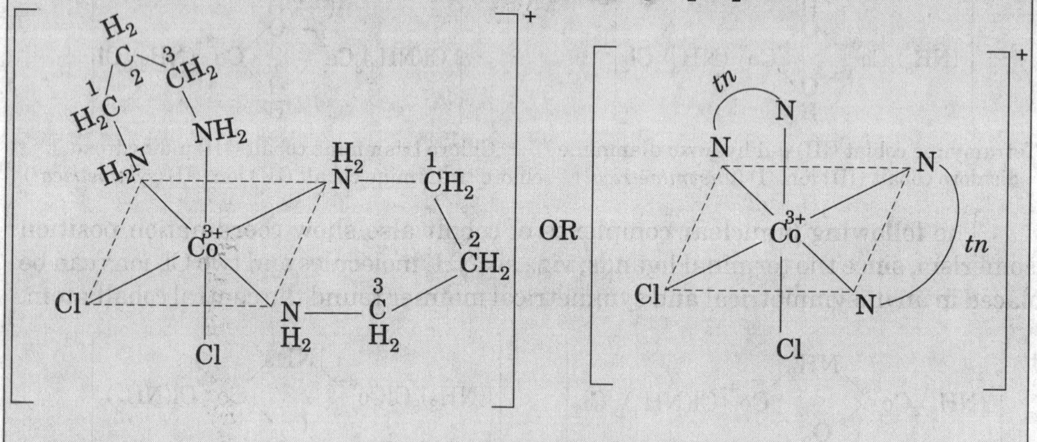


(*b*) Structure of $[Co^{3+}(tn)_2\ Cl_2]^+$ ion.

**Fig. 15.1.** Structures of $[Co(pn)_2Cl_2]^+$ and $[Co\ (tn)_2Cl_2]^+$ ions.

## 8. Polymerisation Isomerism

This type of isomerism is found in those complex compounds whose formulae appear to be polymers of some simple complex compound. All these complex compounds have the same ratio of different metal atoms and ligands in them. For example, the following complex compounds are polymerisation isomers to each other, since (*ii*) and (*iii*) complxes appear to be dimer of (*i*) complex and (*iv*) complex appears to be the pentamer of (*i*) complex. The ratio $Co^{3+} : NH_3 : NO_2^-$ in all the complexes is 1 : 3 : 3.

| | *Complex compound* | $Co^{3+}$ | *Number of $NH_3$* | $NO_2^-$ |
|---|---|---|---|---|
| (*i*) | $[Co(NH_3)_3(NO_2)_3]$ | 1 | 3 | 3 |
| (*ii*) | $[Co(NH_3)_6]\ [Co(NO_2)_6]$ | 2 | 6 | 6 |
| (*iii*) | $[Co(NH_3)_4(NO_2)_2][Co(NH_3)_2(NO_2)_4]$ | 2 | 6 | 6 |
| (*iv*) | $[Co(NH_3)_5(NO_2)]_3\ [Co(NO_2)_6]_2$ | 5 | 15 | 15 |

### 9. Valency Isomerism

This isomerism arises when the ligand is held by different types of valency bonds, sometimes primary or secondary. For example:

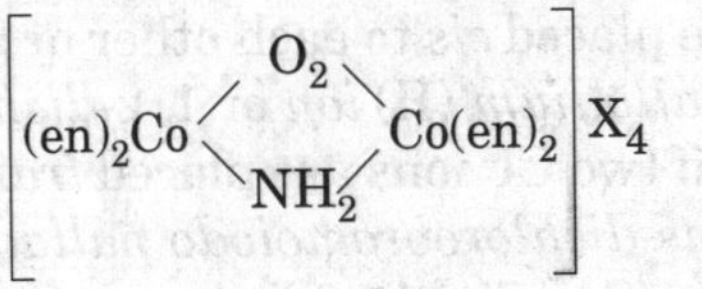 and 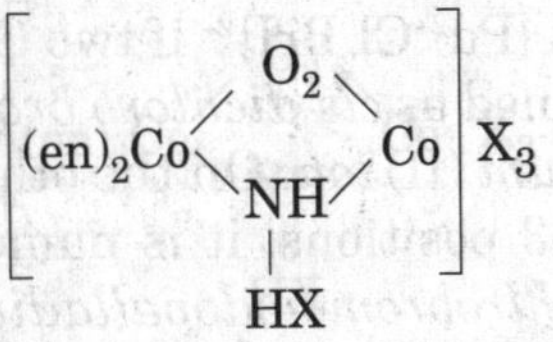

## Stereo Isomerism

Stereo isomerism is exhibited by those compounds which have the same position of atoms or groups but these atoms or groups have different arrangement round the central atom. The compounds showing stereo isomerism are called **stereo isomers**. Stereo isomerism may be of two types *viz. geometrical (or cis-trans) isomerism* and *optical (or d-l or mirror-image) iosmerism*.

## Geometrical (or cis-trans) Isomerism

The complex compounds which have the same ligands in the coordination sphere but the relative position of the ligands round the central metal atom is different are called *geometrical isomers* and the phenomenon is called *geometrical isomerism.*

In a given complex compound the two ligands may occupy positions either adjacent to each other or opposite to each other. The complex compound having two ligands occupying the adjacent positions to each other is called *cis-isomer* while that in which the two ligands occupy opposite positions is called *trans-isomer*. Hence geometrical isomerism is also called *cis-trans isomerism.*

Geometrical isomers of a complex differ in physical characteristics such as dipole moment and visible / uv spectra.

### Geometrical Isomerism in Complexes with C.N. 2 and 3

Geometrical isomerism is not found in complex compounds with coordination number 2 and 3, since in these cases all the positions occupied by the ligands round the central metal atom are adjacent to one another.

### Geometrical Isomerism in Tetrahedral Complexes (C.N. = 4)

Geometrical isomerism cannot be shown by tetrahedral complexes, since all the four ligands in this geometry have adjacent position (*i.e., cis* position) to one another and all the four bond angles are the same (= 109.5°).

### Geometrical Isomerism in Square Planar Complexes (C.N. = 4)

Before discussing the geometrical isomerism exhibited by various types of square planar complexes, we should know how *cis* and *trans* isomers of a square planar complex are named. A square planar complex having similar ligands at adjacent positions (90° apart) is called *cis*-isomer while a square planar complex having two similar ligands at opposite positions (180° apart) is called *trans*-isomer. Thus a square planar complex having two similar ligands at 1-2, 2-3, 3-4 and 1-4 positions is called *cis*-isomer while that having two similar ligands at 1-3 and 2-4 positions is called *trans*-isomer. *cis*- and *trans*-isomers are also named by numbering system. Thus in $[Pd^{2+}Cl_2BrI]^{2-}$ if two $Cl^-$ ions are placed *cis* to each other or at 1-2 positions, it is named as *cis-dichloro bromoiodo palladium* (II) *ion* or 1, 2-*dichloro bormoiodo palladium* (II) *ion*. On the other hand, if two $Cl^-$ ions are placed *trans* to each other or at 1-3 positions, it is named as *trans-dichlorobromoiodo palladium (II) ion* or *1, 3-dichlrobromoiodopalladium* (II) ion (See Fig. 15.2.)

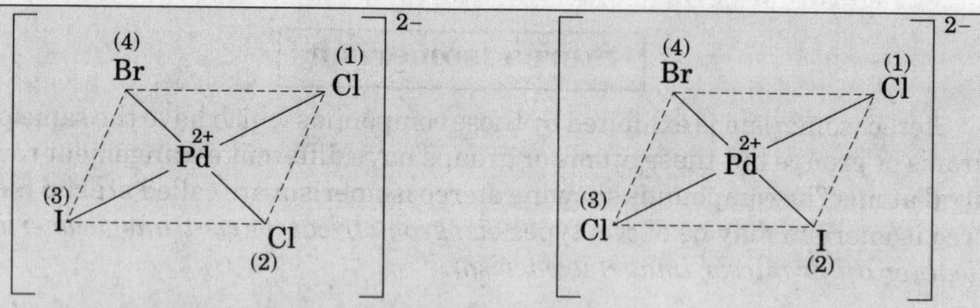


(*a*) *cis*-dichloro bromoiodo palladium (II) ion or 1, 2-dichloro bromoiodo palladium (II) ion, $[Pd^{2+}Cl_2BrI]^{2-}$

(*b*) *trans*-dichloro bromoiodo palladium (II) ion or 1, 3-dichloro bromoiodo palladium (II) ion, $[Pd^{2+}Cl_2BrI]^{2-}$

**Fig. 15.2.** Nomenclature of geometrical (*cis*- and *trans*) isomers of a square planar complex.

Here we shall consider geometrical isomerism in square planar complexes of $[Ma_4]$, $[Ma_3b]$, $[Mab_3]$, $[Ma_2b_2]$, $[Ma_2bc]$, [Mabcd] and $[M(AB)_2]$ type in which M is the central metal atom; a, b, c and d are monodentate ligands and (AB) is an unsymmetrical bidentate ligand, since it has two different donor atoms namely A and B.

**1. Square planar complexes of $[Ma_4]$, $[Ma_3b]$ and $[Mab_3]$ type.** Square planar complexes of this type do not show geometrical isomerism, since all the possible spatial arrangement of four ligands round the central metal atom is the same.

**2. Square planar complexes of $[Ma_2b_2]$ type.** Important examples of square planar complexes of this type are $[Pt(NH_3)_2Cl_2]^0$, $[Pt\,(py)_2Cl_2]^0$, $[Pd(NH_3)_2(NO_2)_2]^0$ etc. These complexes exist in *cis*- and *trans*-isomers. These isomers of $[Pt^{2+}(NH_3)_2Cl_2]^0$ are shown in Fig. 15.3. In (*a*) since both $NH_3$ molecules and both $Cl^-$ ions are *cis* to each other, it is called *cis*-isomer. On the other hand, in (*b*) since both $NH_3$ molecules and both $Cl^-$ ions are *trans* with respect to each other, it is called *trans*-isomer.

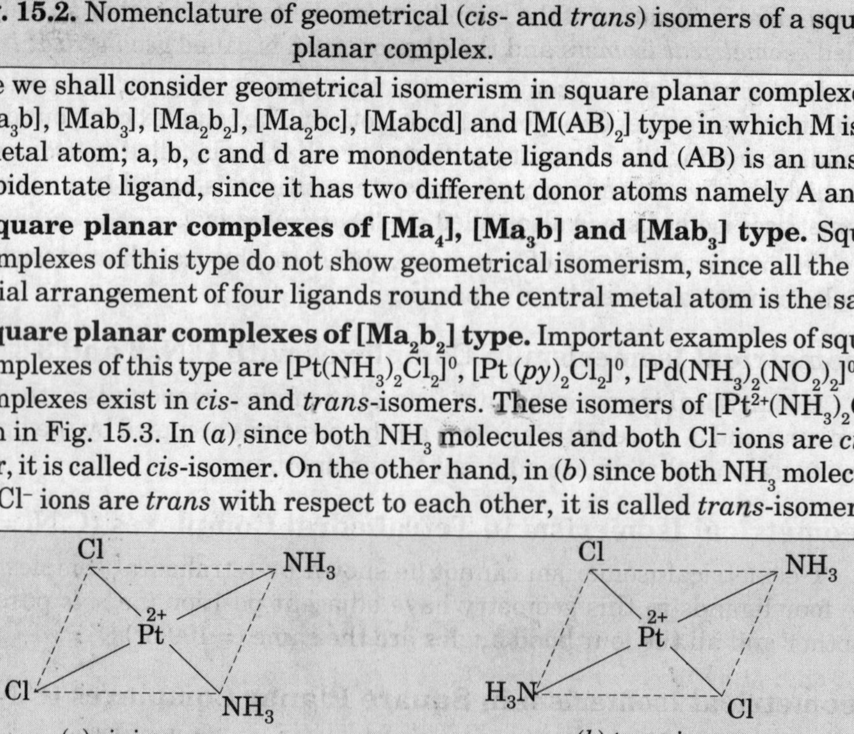


(*a*) *cis*-isomer

(*b*) *trans*-isomer

**Fig. 15.3.** *cis*- and *trans*-isomers of $[Pt^{2+}\,(NH_3)_2Cl_2]^0$,

**3. Square planar complexes of [Mabcd] type.** Square planar complexes of this type exist in three isomeric forms. For example, $[Pt^{2+}(NH_3)(py)(Cl)(Br)]$ exists in three isomeric forms shown below in Fig. 15.4. These isomeric forms can be obtained by selecting one ligand, say $NH_3$, and then placing the remaining three ligands, one by one, *trans* to $NH_3$.

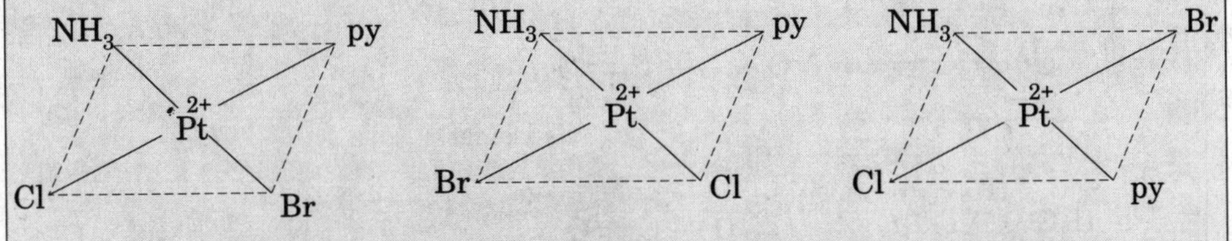


**Fig.15.4.** Three isomeric forms of $[Pt^{2+}(NH_3)(py)(Cl)(Br)]$

$[Pt^{2+}(NO_2)(py)(NH_3)(NH_2OH)]^+$ and $[Pt^{2+}(C_2H_4)(NH_3)(Cl)(Br)]$ are other examples of square planar complexes which exist in three isomeric forms.

The existence of three isomeric forms in case of the complexes mentioned above indicates that these complexes have square planar geometry.

**4. Square planar complexes of [Ma$_2$bc] type.** Square planar complexes of this type also show *cis-trans* isomerism. For example, $[Pd^{2+}Cl_2BrI]^{2-}$ ion exists in *cis-* and *trans*-isomers as shown below in Fig. 15.5.

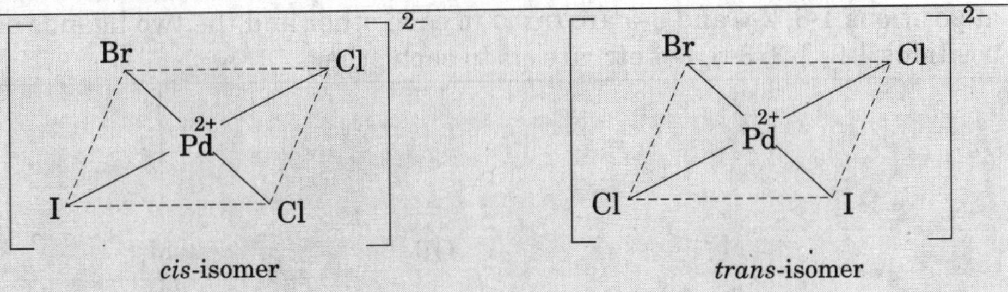


**Fig. 15.5.** *cis-* and *trans*-isomers of $[Pd^{2+}Cl_2BrI]^{2-}$ ion.

$[Pt^{2+}(py)_2(NH_3)Cl]^+$ is another example of square planar complex of $[Ma_2bc]$ type which exists in *cis-* and *trans*-isomers.

**5. Square planar complexes of [M(AB)$_2$] type.** Here M is the central metal ion and [AB] represents an unsymmetrical bidentate ligand in which A and B are two different coordinating (donor) atoms. Square planar complexes of this type also exist in *cis-* and *trans*-isomers. For example, $[Pt^{2+}(gly)_2]$ exists in *cis-* and *trans*-isomers as shown in Fig. 15.6. Here gly represents the glycinato ligand, $NH_2.CH_2COO^-$ which has N and O atoms as its donor atoms (i.e., A = N and B = O).

$[Cu^{2+}(gly)_2]^0$ is also an example of square planar complex of $[M(AA)_2]$ type. This complex also exhibits *cis-trans* isomerism.

## Geometrical Isomerism in 6-coordinated Complexes: Octahedral Complexes

We know that a complex compound having central atom with coordination number equal to 6 is octahedral in shape. Before discussing the geometrical isomerism exhibited by various types of octahedral complexes, we should know how *cis-* and *trans-* isomers of an octahedral complex are named. In an octahedral complex, if two similar ligands are placed on any of the twelve edges of the octahedron, they are said to be in *cis* position. On the other hand, if two similar ligands are lying on a straight line which passes through the centre (where the metal ion is placed),

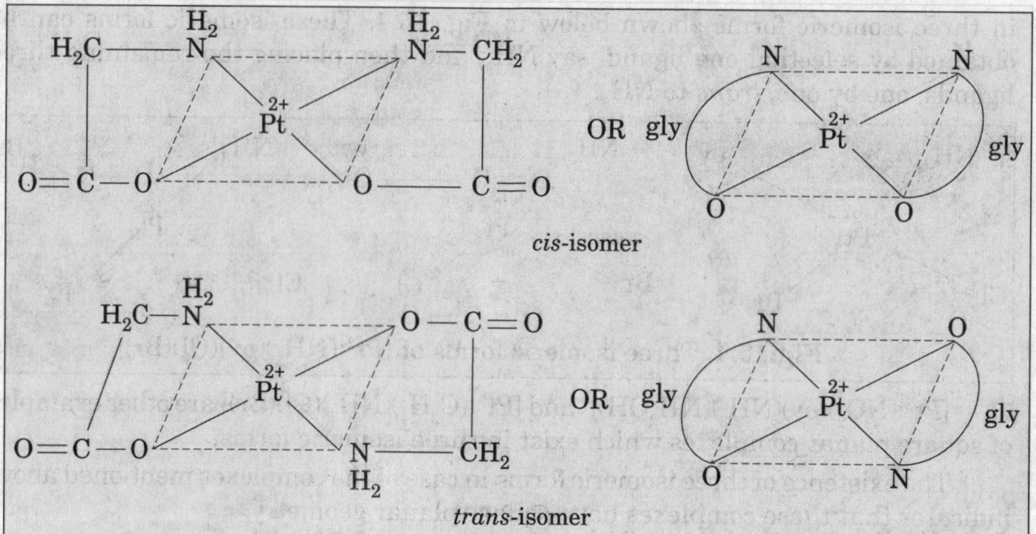


**Fig. 15.6.** *cis*- and *trans*-isomers of $[Pt^{2+}(gly)_2]$.

they are said to be in *trans* position. Thus in an octahedral complex, the two ligands at positions 1-6, 2-4 and 3-5 are *trans* to each other and the two ligands occupying positions 1-2, 1-3, 3-6, 6-4 etc. are *cis* to each other.

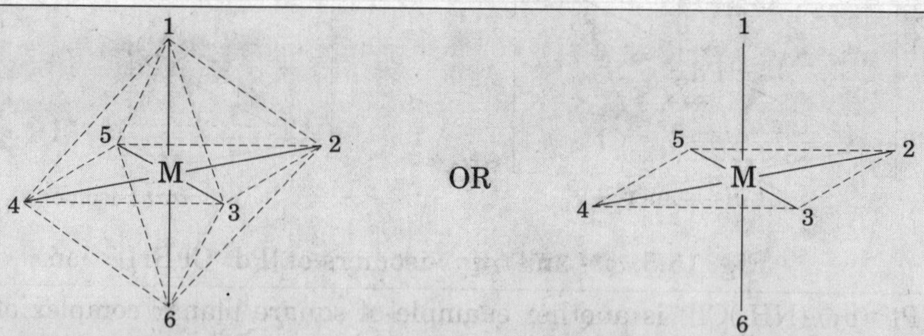

**Fig. 15.7.** Numbering of six ligands in a regular octahedral complex round the central metal ion, M.

The numbering system of ligands shown in Fig. 15.7 can be exemplified by considering the nomenclature of *cis*- and *trans*-isomers of $[Co^{3+}(NH_3)_4Cl_2]^+$ ion which is an octahedral complex ion. *Cis*-isomer of this ion in which two $Cl^-$ ions are *cis* to each other is also called *1, 2-dichlorotetrammine cobalt (III) ion* (See Fig. 15.8 (*a*)]. Similarly, *trans*-isomer of this ion in which two $Cl^-$ ions are *tans* to each other is also called *1, 6-dichlorotetrammine cobalt (III) ion*. [See Fig.15.8 (*b*)].

Here we shall discuss the geometrical isomerism in octahedral complexes of the following type:

**1. Octahedral complexes of $[Ma_6]$ and $[Ma_5b]$ type.** Octahedral complexes of this type do not show geometrical isomerism

**2. Octahedral complexes of $[Ma_4b_2]$ type.** Important examples of octahedral complexes of this type are $[Co^{3+}(NH_3)_4Cl_2]^+$, $[Co^{3+}(NH_3)_4(NO_2)_2]^+$ etc. These complexes exist in *cis*- and *trans*- isomers. These isomers of $[Co^{3+}(NH_3)_4Cl_2]^+$ ion are shown in Fig. 15.8 In *cis*-isomer two $Cl^-$ ions occupy the adjacent (*i.e.*, 1 and 2) positions of the octahedron while in *tans*-isomers these ions have the opposite (*i.e.*, 1 and 6) positions. *Cis*-isomer of $[Co^{3+}(NH_3)_4Cl_2]^+$ ion is *yellow-brown* while *trans*-form is *yellow* in colour.

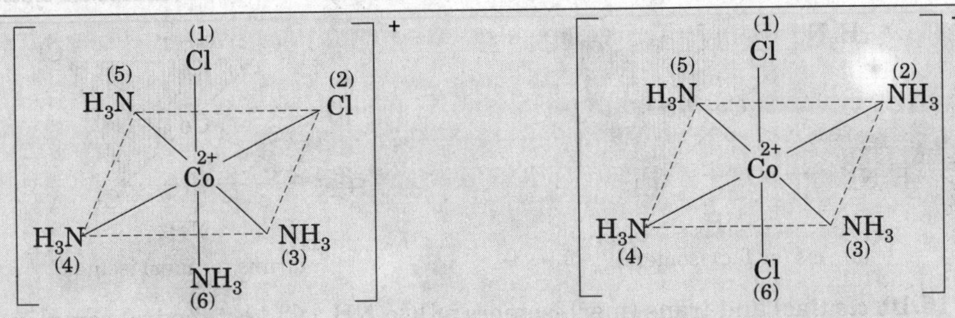


(*a*) *cis*-isomer

(*b*) *trans*-isomer

(*a*) *cis*-isomer (1,2-dichlorotetrammine cobalt (III) ion). (yellow-brown)

(*b*) *trans*-isomer (1,6-dichlorotetrammine cobalt (III) ion). (yellow)

**Fig. 15.8.** *cis*- and *trans*-isomers of $[Co^{3+}(NH_3)_4Cl_2]^+$ ion.

**3. Octahedral complexes of $[Ma_3b_3]$ type.** We know that an octahedron has eight triangular faces and six corners (vertices). Octahedral complexes of $[Ma_3b_3]$ have two types of geometrical isomers:

(*i*) **Facial or fac isomer (*cis*-isomer).** In this isomer the three *a* groups occupy the three corners (vertices) of one triangular face while the other three *b* groups in this isomer take up the position of the remaining three vertices. This isomer is called **facial** or **fac** isomer. This isomer is also called **cis** isomer, since the three *a* groups are *cis* to each other. Similarly *b* groups are also *cis* to each other (See Fig. 15.9)

(*ii*) **Meridional or mer-isomer (*trans*-isomer).** In this isomer the three *a* groups are in one plane and the other three *b* groups are in a perpendicular plane. *a* and *b* groups lie along the meridian of a sphere (imagine our earth). Hence this isomer is called **meridional** or **mer** isomer. In this isomer two of the three *a* groups and also two of the three *b* groups are *trans* to each other (See Fig 15.9). Hence this isomer is also called **trans** isomer

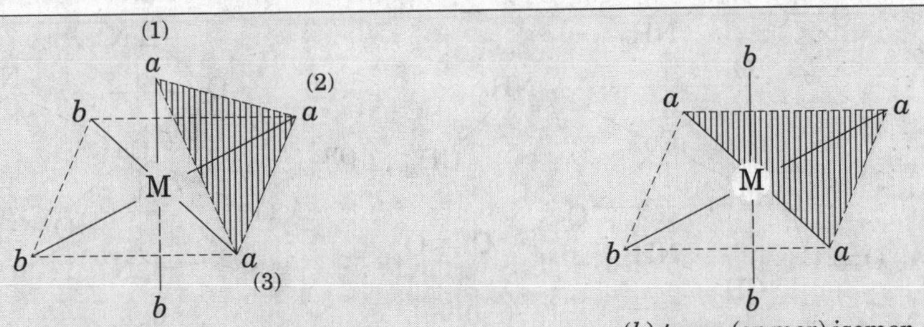


(*a*) *cis* (or fac) isomer

(*b*) *trans* (or mer) isomer

**Fig 15.9** *cis*- (or fac) and *trans*-(or mer) isomers of $Ma_3b_3$ type octahedral complex

**Examples:** $[Co(NH_3)_3Cl_3]$, $[Co(NH_3)_3(NO_2)_3]$, $[Cr(H_2O)_3F_3]$, $[Cr(NH_3)_3Cl_3]$, $[Rh(py)_3\,Cl_3]$, $[Ru(H_2O)_3Cl_3]$, $[Ir(H_2O)_3Cl_3]$, $[Pt(NH_3)_3Br_3]^+$, $[Pt(NH_3)_3I_3]^+$ etc. are important examples of octahedral complexes of $Ma_3b_3$ type. *Cis*-(fac) and *trans*-(mer) isomers of $[Co(NH_3)_3Cl_3]$ are given in Fig. 15.10.

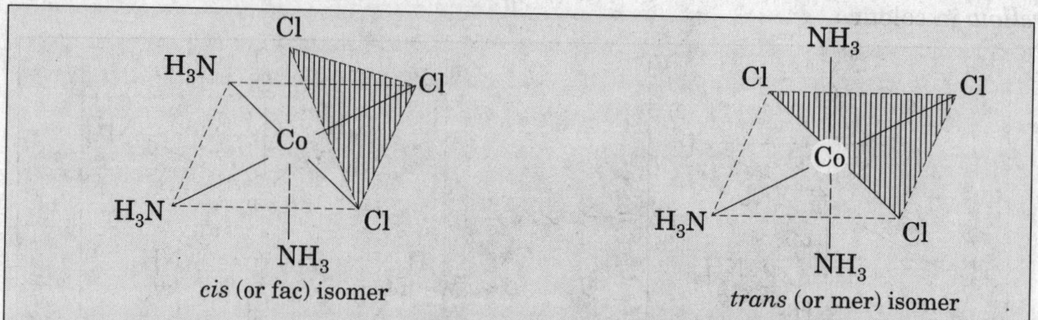


**Fig. 15.10.** *cis* (fac) and trans (mer) isomers of $[Co(NH_3)_3Cl_3]$ octahedral complex.

$[Co(NH_2CH_2COO)_3]$ also has two geometrical isomers, viz., facial and meridional (See Fig. 15.11). In the given complex N and O atoms are the donor atoms.

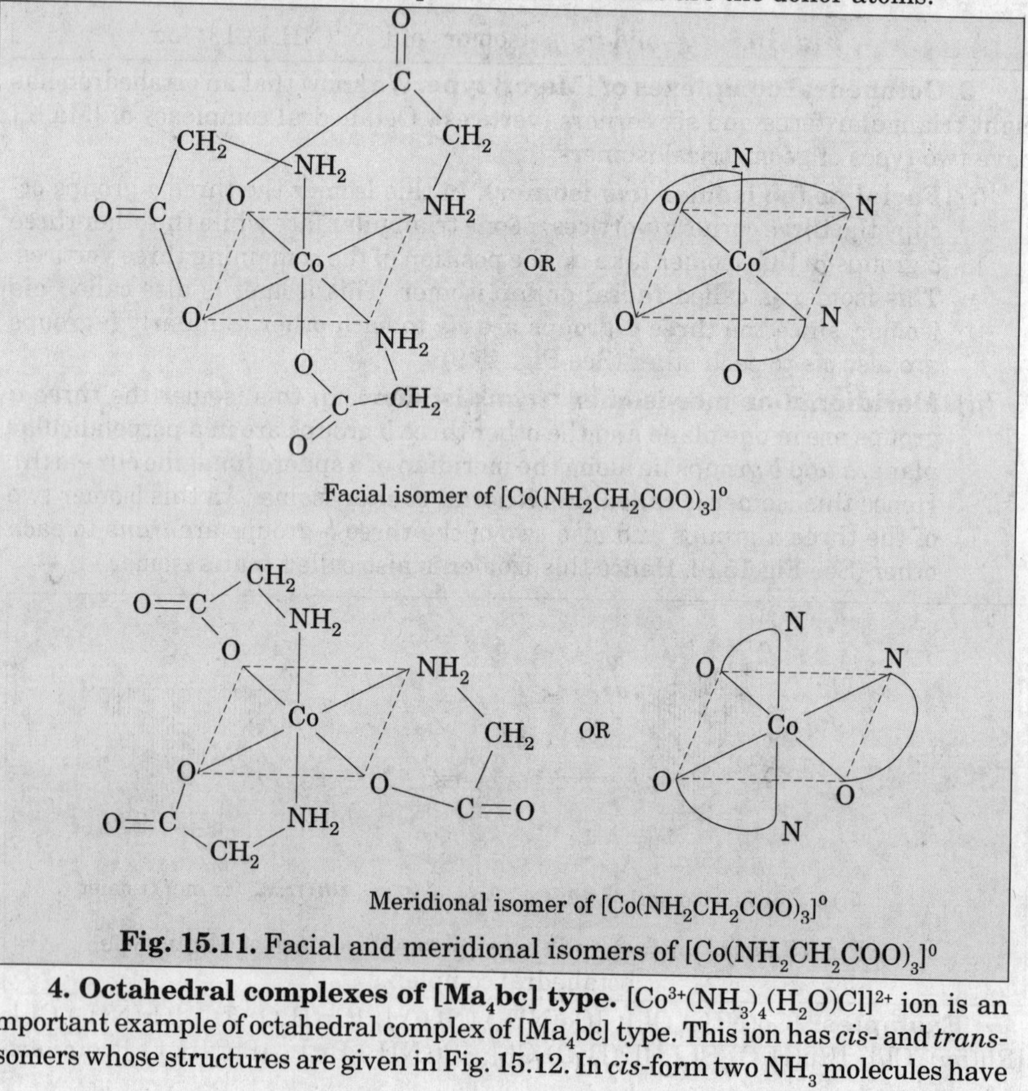


**Fig. 15.11.** Facial and meridional isomers of $[Co(NH_2CH_2COO)_3]^0$

**4. Octahedral complexes of $[Ma_4bc]$ type.** $[Co^{3+}(NH_3)_4(H_2O)Cl]^{2+}$ ion is an important example of octahedral complex of $[Ma_4bc]$ type. This ion has *cis*- and *trans*-isomers whose structures are given in Fig. 15.12. In *cis*-form two $NH_3$ molecules have *cis* positions to each other and in *trans*-form these ligands (i.e, two $NH_3$ molecules) have *trans* positions with each other.

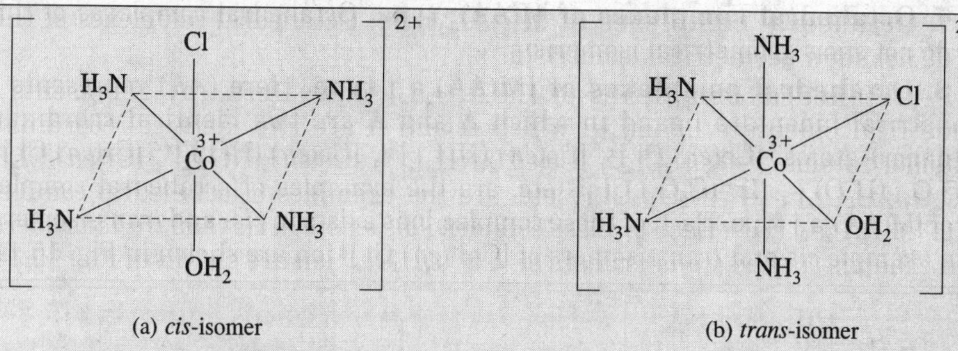

**Fig.15.12.** Structure of *cis*- and *trans*-isomers of $[Co^{3+}(NH_3)_4(H_2O)Cl]^{2+}$ octahedral complex ion.

**5. Octahedral complexes of $[Ma_2b_2c_2]$ type.** $[Pt^{4+}(NH_3)_2(py)_2Cl_2]^{2+}$ ion is an important example of octahedral complex of $[Ma_2b_2c_2]$ type. This ion can exist theoretically in five geometrical isomers that have been shown in Fig. 15.13, but it is only three isomers that have been actually isolated. It may be noted from the figure that (I) form is a *cis*-form since in this form two identical ligands are occupying adjacent positions. Form (II) is *trans*-form because in it the two identical ligands are placed at opposite positions.

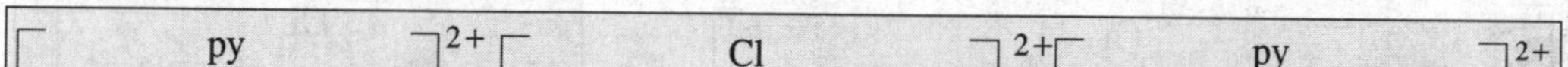

**Fig. 15.13.** All possible five geometrical isomers of $[Pt^{4+}(NH_3)_2(py)_2Cl_2]^{2+}$ ion.

**6. Octahedral complexes of [Mabcdef] type.** Octahedral complexes which have different monodentate ligands have been prepared for Pt (IV). $[Pt^{4+}(py)(NH_3)(NO_2)(Cl)(Br)(I)]$ is the only complex of this type of complexes.

Theoretically 15 geometrical isomers are possible for this complex compound. In fact only three geometrical isomers have been isolated.

**7. Octahedral complexes of M(AA)$_3$ type.** Octahedral complexes of this type do not show geometrical isomerism.

**8. Octahedral complexes of [M(AA)$_2$a$_2$] type.** Here (AA) represents a symmetrical bidentate ligand in which A and A are two identical coordinating (donor) atoms. $[Co(en)_2Cl_2]^+$, $[Co(en)_2(NH_3)_2]^{3+}$, $[Co(en)_2(NO_3)_2]^+$, $[Cr(en)_2Cl_2]^+$, $[Cr(C_2O_4)_2(H_2O)_2]^-$, $[Ir^{4+}(C_2O_4)_2Cl_2]^{2-}$ etc. are the examples of octahedral complex ions of [M(AA)$_2$a$_2$] type. Each of these complex ions exists in *cis*- and *trans*-isomers. As an example *cis*-and *trans*-isomers of $[Co^{3+}(en)_2Cl_2]^+$ ion are shown in Fig. 15.14.

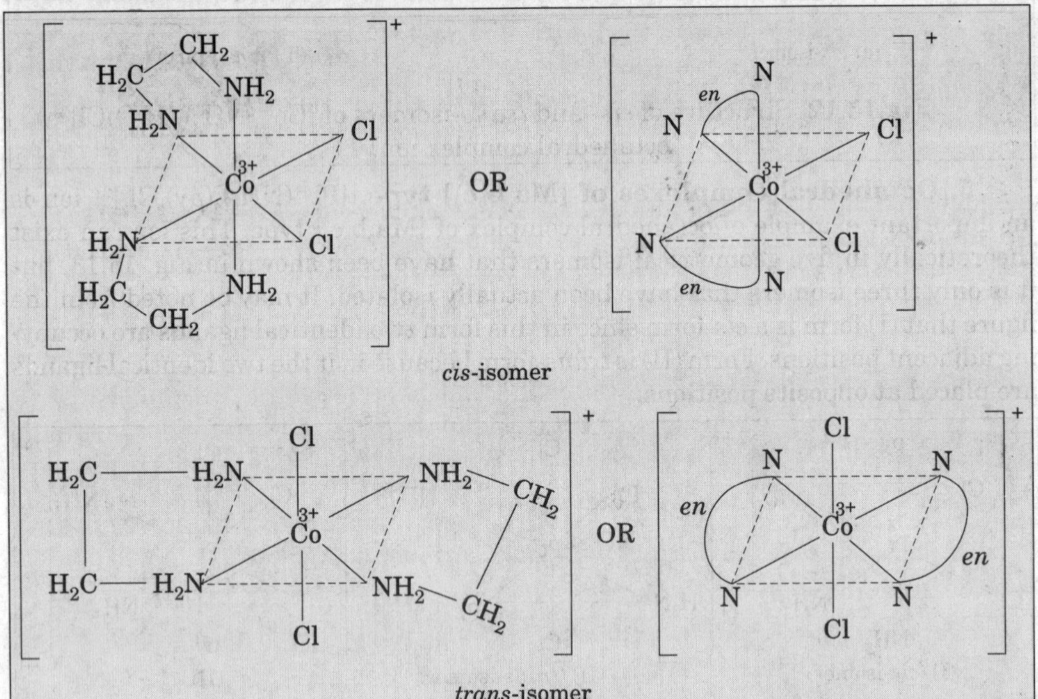


**Fig. 15.14.** *cis*- and *trans*-isomers of $[Co^{3+}(en)_2Cl_2]^+$ ion.

**Symmetrical / unsymmetrical nature of cis-and trans-isomers.** The structures of mirror images of *cis*- and *trans*- forms of $[Co^{3+}(en)_2Cl_2]^+$ ion are given at (*a*) and (*b*) of Fig. 15.15. respectively. Since the structure of *cis*-isomer and its mirror image are non-superimposable on each other, *cis*-isomer is unsymmetrical. On the other hand, since the structure of *trans*-form and its mirror image are superimposable on each other, *trans*-isomer is symmetrical. Thus *cis*-isomer is *optically active but trans-isomer is optically inactive*.

**9. Octahedral complexes of [M(AA)$_2$ab] type.** $[Co^{3+}(en)_2(NH_3)Cl]^{2+}$ is an important example of octahedral complex of [M(AA)$_2$ab] type. This complex ion exists in *cis*- and *trans*-isomers. In *cis*-isomer the two monodentate ligands *viz.* $NH_3$ and $Cl^-$ occupy the adjacent (*i.e., cis*) positions while in *trans*-isomer these ligands occupy opposite (i.e., *trans*) positions (See Fig. 15.16). $[Ru^{3+}(C_2O_4)_2(py)(NO_2)]^{2-}$ also exists in *cis*- and *trans*- isomers.

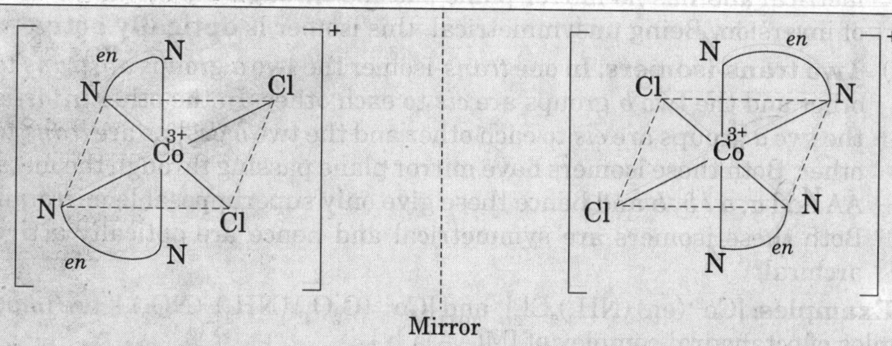


(a) Structure of *cis*-isomer of $[Co(en)_2Cl_2]^+$ and its mirror image.

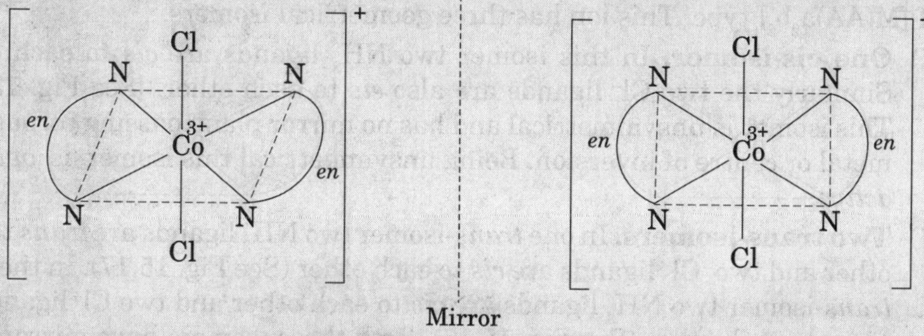


(b) Structure of *trans*-isomer of $[Co(en)_2Cl_2]^+$ ion and its mirror image.

**Fig.15.15.** Structure of *cis*-and *trans*-forms of $[Co(en)_2Cl_2]^+$ ion and their mirror images.

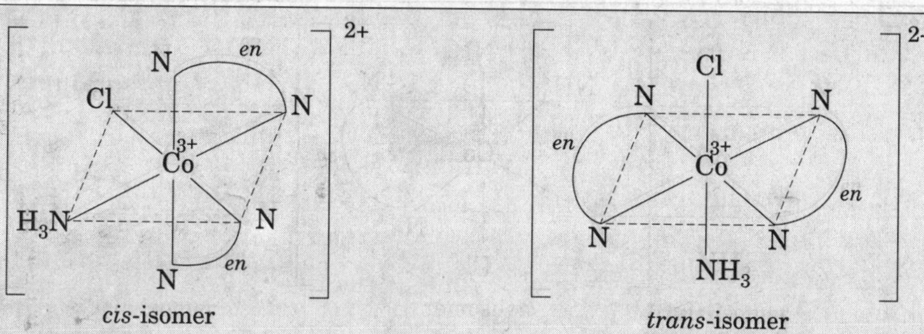


**Fig. 15.16.** *cis*- and *trans*-isomers of $[Co^{3+}(en)_2(NH_3)(Cl)]^{2+}$ ion.

As in case of $[Co^{3+}(en)_2Cl_2]^+$ ion, the *cis*-isomer of $[Co^{3+}(en)_2(NH)_3(Cl)]^{2+}$ is also unsymmetrical while its *trans*-isomer is symmetrical. Thus *cis*-form is *optically active* while *trans*-isomer is *optically inactive*.

**10. Octahedral complexes of [M(AA)$a_2b_2$] type**. In this complex AA represents a bidentate ligand in which two A atoms are donor atoms. *a* and *b* are monodentate ligands. Octahedral complexes of [M(AA)$a_2b_2$] type exist in three geometrical isomers.

(*i*) **One cis-isomer**. In this isomer the two *a* groups are *cis* to each other. Similarly the two *b* groups are also *cis* to each other. This isomer is unsymmetrical and has no mirror plane passing through the metal (M) or centre of inversion. Being unsymmetrical, this isomer is **optically active**.

(*ii*) **Two trans-isomers.** In one *trans*-isomer the two *a* groups are *trans* to each other and the two *b* groups are *cis* to each other. In the other *trans*-isomer the two *a* groups are *cis* to each other and the two *b* groups are *trans* to each other. Both these isomers have mirror plane passing through the metal (M), AA and *a*, *a* / *b*, *b* and hence these give only superimposable mirror image. Both these isomers are symmetrical and hence are optically active, *i.e.*, archiral.

**Examples:** $[Co^{3+}(en)(NH_3)_2Cl_2]^+$ and $[Co^{3+}(C_2O_4)(NH_3)_2(NO_2)_2]^-$ are important examples of octahedral complex of $[M(AA)a_2b_2]$

**Geometrical isomers of $[Co^{3+}(en)(NH_3)_2Cl_2]^+$ ion.** This ion is an octahedral ion of $[M(AA)a_2b_2]$ type. This ion has three geometrical isomers.

(*i*) **One cis-isomer.** In this isomer two $NH_3$ ligands are *cis* to each other. Similarly the two $Cl^-$ ligands are also *cis* to each other. [See Fig. 15.17.]. This isomer is unsymmetrical and has no mirror plane passing through the metal or centre of inversion. Being unsymmetrical this isomer is *optically active*.

(*ii*) **Two trans-isomers.** In one *trans*-isomer two $NH_3$ ligands are *trans* to each other and two $Cl^-$ ligands are *cis* to each other (See Fig. 15.17). In the other *trans*-isomer two $NH_3$ ligands are *cis* to each other and two $Cl^-$ ligands are *trans* to each other (See Fig. 15.17). Both these isomers have mirror plane passing through the metal, en and $NH_3$, $NH_3/Cl^-$, $Cl^-$ and hence these give only superimposable mirror image. Both these isomers are symmetrical and are, therefore, optically active, *i.e.*, archiral.

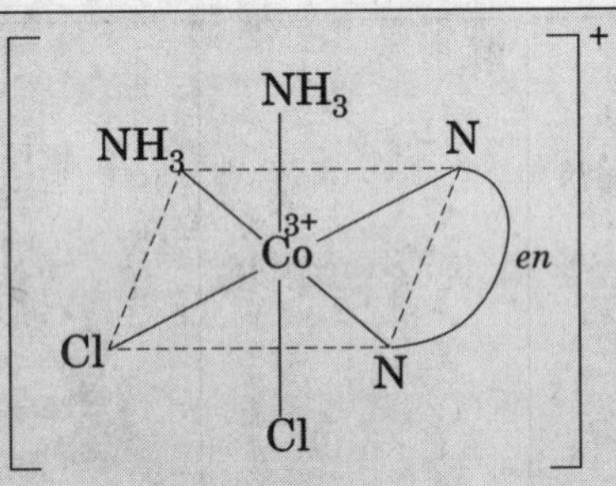


*cis*-isomer

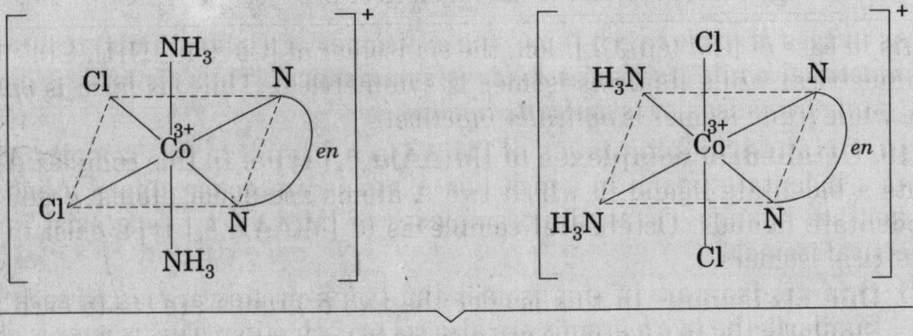


Two *trans*-isomers

**Fig. 15.17.** *cis*- and *trans*-isomers of $[Co^{3+}(en)(NH_3)_2Cl_2]^+$ ion.

**11. Octahedral complexes of [M(AB)$_3$] type.** Here (AB) represents an unsymmetrical bidentate ligand in which A and B are two different coordinating (donor) atoms. Octahedral complexes of this type exist in *cis*- and *trans*-isomers. As an example, the *cis*- and *trans*-isomers of [$Cr^{3+}$(gly)$_3$] have been shown in Fig. 15.18. Each of these forms is optically active and hence each has a pair of optical isomers. In *cis*-isomer two N-atoms and two O-atoms of two *gly* ions are placed at *cis*-positions while in *trans*-isomer these atoms are occupying *trans*-positions

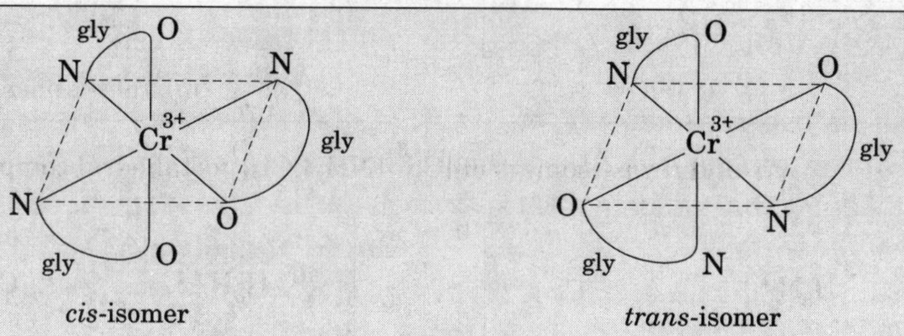


**Fig. 15.18.** *cis*- and *trans*-isomers of [$Cr^{3+}$(gly)$_3$]

Since both the forms (*i.e. cis*- and *trans*-forms) are *optically active*, each of these forms has *d*-and *l*-forms (optical isomers)

**To Distinguish Between cis- and trans-isomers**

The following methods may be used to distinguish between *cis*- and *trans*-isomers.

**1. Dipole moment measurements.** Jensen has shown that the Pt (II) complexes of [$Pt^{2+}A_2X_2$] type, where A = substituted phosphine, arsine or stilbine such as $(C_2H_5)_3P$, $(C_2H_5)_3$ As or $(C_2H_5)_3Sb$ and X = a halogen, have their dipole moments (μ) either equal to zero or between 8 and 12 Debye units. In the compounds with $\mu = 0$, the individual moments have cancelled one another and so these are *trans*-isomers. The compounds with $\mu = 8 - 12$ Debye units are *cis*-isomers.

**2. X-ray crystal analysis.** X-ray crystal analysis and dipole moment measurements of several Pt(II) complexes have confirmed the square planarity of the bonds around the central metal atom. This square planar arrangement has also been established for 4-coordinated Pt (II), Ag(II), Cu(II) and Au(III) complexes. Tetrahedral configurations have been assigned to Cu (I), Ag(I), Au (I), B(III), Al(III), Zn(II), Cd(II), Hg(II) and Co(II) in their coordinated compounds. Some metals such as Ni(II) appear to show either configuration for 4-coordination.

**3. Infra-red spectroscopic technique.** In a *trans* octahedral complex such as [$Co^{3+}(NH_2)_4Cl_2$]$^+$ or in *trans* square planar complex like [$Pt^{2+}(NH_3)_2Cl_2$]$^0$ the Cl–metal–Cl symmetrical stretching vibration produces no change in the dipole moment of the molecule [Fig. 15.19. (*b*) and (*d*)], and thus no band corresponding to this vibration is observed in the infra-red spectrum. However, in the *cis*-form of each compound, the symmetrical stretching vibration [Fig. 15.19(*a*) and (*c*)] as well the unsymmetrical stretching vibration produce appreciable changes in the dipole moment; hence the infra-red spectrum of the *cis*-isomer will contain a large number of bands due to Cl–metal–Cl stretching.

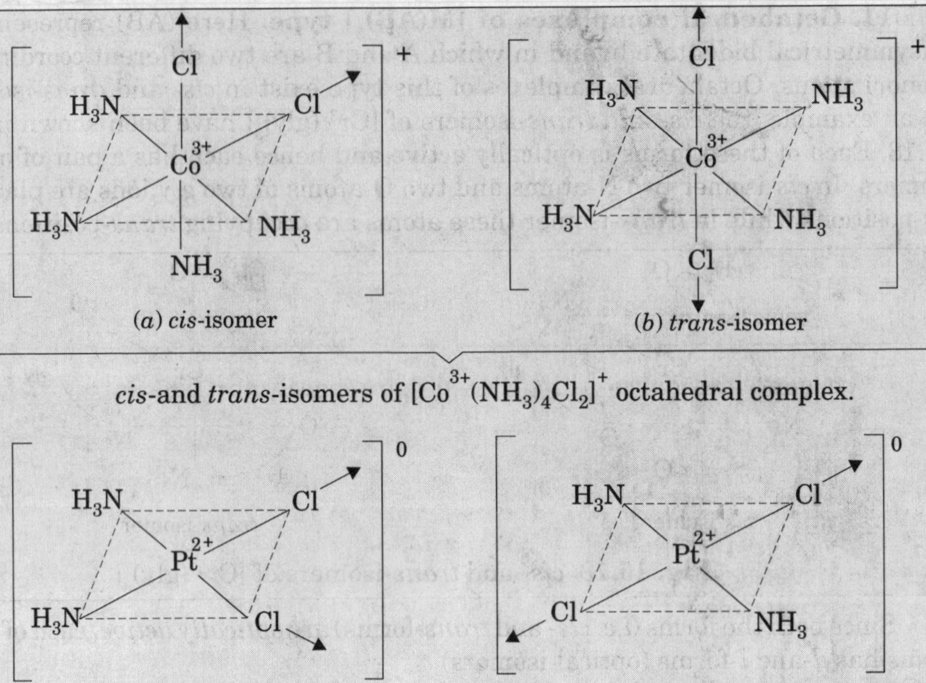


(*a*) *cis*-isomer (*b*) *trans*-isomer

*cis*-and *trans*-isomers of $[Co^{3+}(NH_3)_4Cl_2]^+$ octahedral complex.

(*c*) *cis*-isomer (*d*) *trans*-isomer

*cis*-and *trans*-isomers of $[Pt^{2+}(NH_3)_2Cl_2]^0$ square planar complex

**Fig. 15.19.** Symmetrical Cl-metal-Cl stretching vibrations of cis-and trans-isomers of $[Co(NH_3)_4Cl_2]^+$ and $[Pt(NH_3)_2Cl_2]^0$

**4. Molar extinction coefficient.** A *cis*-isomer has a high molar extinction coefficient (absorption) than the corresponding *trans*-isomer and sometimes the two isomers may differ significantly in their absorption spectra.

**5. To resolve *cis*-isomer** The technique which consists of attempting to resolve the *cis*-isomer into the two possible optically active isomers is also commonly used to distinguish it from the *trans*-isomers.

**6. Grinberg's method.** It is a chemical method and assumes that when a chelating ligand having two donor atoms separated by two to four other atoms reacts with *cis*- and *trans*-isomers separately, its two donor atoms coordinate to the central metal ion of the *cis*-form at two *cis*-positions and thus form five or six-membered ring, while in the *trans*-form they coordinate to the central metal ion at *trans* positions and thus the ligand acts as a monodentate ligand, *i.e.*, the ligand cannot form a ring complex with *trans* form. Typical chelating ligands used for the purpose are oxalic acid, $(COOH)_2$, glycine ($H_2N—CH_2—COOH$) and ethylenediamine ($NH_2—CH_2—CH_2—NH_2$). The application of this method can be explained by considering the reaction of oxalic acid and glycine with *cis*- and *trans*-isomers of square *planar* complex of Pt(II), $[Pt^{2+}(NH_3)_2Cl_2]^0$ (Fig. 15.20.). Quite obviously the *cis*- isomer gives a *chelated complex* while the *trans*-isomer yields a *non-chelated* complex.

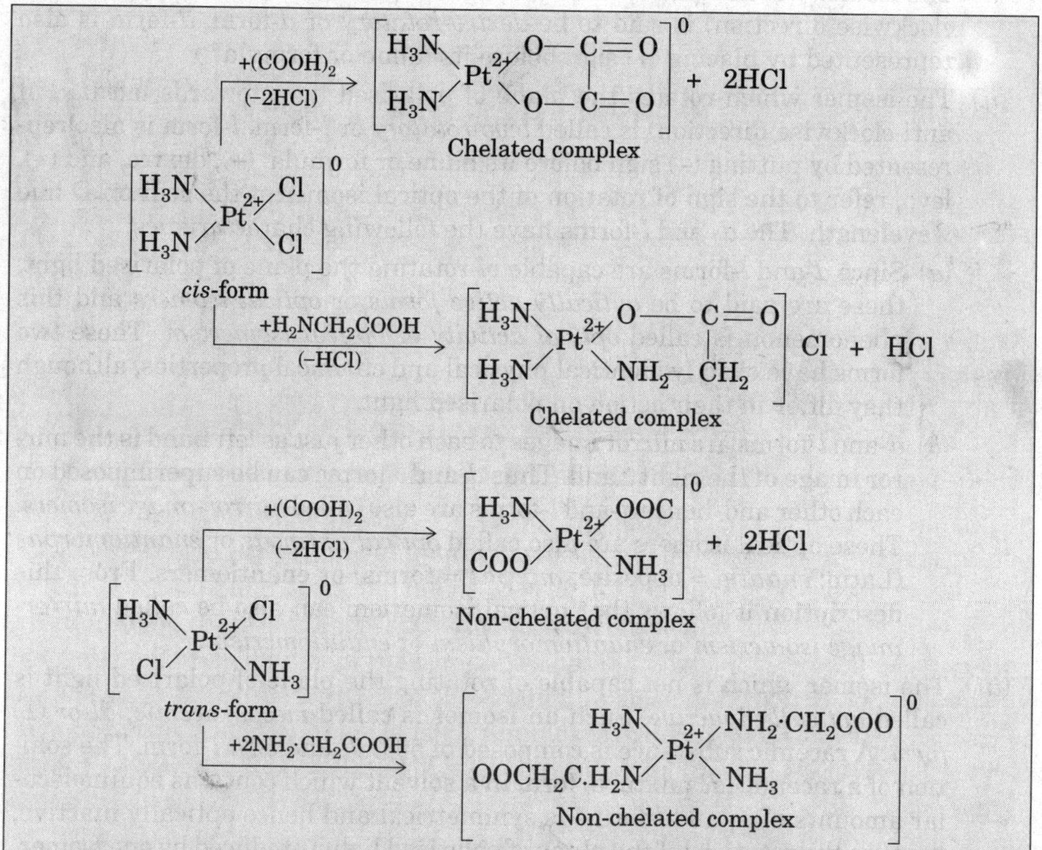

**Fig. 15.20.** Reaction of oxalic acid, $(COOH)_2$ and glycine. $NH_2CH_2COOH$ with *cis*- and *trans*-isomers of square planar complex of Pt(II), $[Pt(NH_3)_2Cl_2]^0$.

**7. Kurnakov's reaction.** Kurnakov utilised the phenomenon of trans-effect in distinguishing the pairs of *cis*- and *trans*- isomers of square planar complexes of $[PtA_2X_2]$ type by treating them with thiourea.

## Optical (or d-*l* or mirror-image) Isomerism

Before discussing the optical isomerism shown by various types of complexes we will define some important terms related to this isomerism.

### 1. Optical Activity and Optically Active Complexes

When the solutions of certain complex compounds are placed in the path of a *plane-polarised light* (the waves of the plane-polarised light vibrate only in one direction; vibrations in other directions are cut off), they rotate its plane through a certain angle which may be either to the left or to the right. *This property of a complex of rotating the plane of polarised light is called its optical activity and the complex possessing this property is said to be optically active.*

### 2. Different Forms of Optically Active Complexes

Optically active complexes are said to exist in the following forms:

(*i*) The isomer which rotates the plane of polarised light towards right (*i.e.*, in clockwise direction) is said to be *dextro-rotatory* or *d*-form. *d*-form is also represented by placing (+) sign before its name or formula.

(*ii*) The isomer which rotates the plane of polarised light towards left (*i.e*, in anti-clockwise direction) is called *levo-rotatory* or *l*-form. *l*-form is also represented by putting (–) sign before its name or formula. (+), dextro, and (–), levo, refer to the sign of rotation of the optical isomer at the sodium D line wavelength. The *d*- and *l*-forms have the following characteristics:

(*a*) Since *d*-and *l*-forms are capable of rotating the plane of polarised light, these are said to be *optically active forms* or *optical isomers* and this phenomenon is called *optical activity* or *optical isomerism*. These two forms have exactly identical physical and chemical properties, although they differ in their action on polarised light.

(*b*) *d*-and *l*-forms are mirror images to each other just as left hand is the mirror image of the right hand. Thus *d*-and *l*-forms can be superimposed on each other and hence *d*-and *l*-forms are also called *mirror-image isomers*. These optical isomers are also called *optical antipods* or *enantiomorphs* (Latin: *enantio* = opposite; *morphs* = forms) or enantiomers. From this description it follows that optical isomerism can also be called *mirror-image isomerism* or *enantiomorphism* or *enantiomerism*.

(*iii*) The isomer which is not capable of rotating the plane of polarised light is called *optically inactive*. Such an isomer is called *meso, racemic, dl-or* (±) *form*. A racemic substance is composed of 50% *d*- and 50% *l*-form. The solution of a racemic (*dl* mixture) form in a solvent which contains equimolecular amounts of *d*- and *l*-forms is symmetrical and hence optically inactive, because the rotation (of the plane of polarised light) produced by one isomer, *d*-isomer, is balanced or compensated by equal but opposite rotation produced by the other (*i.e.*, *l*-isomer). Optical inactivity produced in *dl*-mixture said to be due to *external compensation*.

## 3. Condition for a Molecule to Show Optical Isomerism

A molecule in which the grouping of the atoms is asymmetric is called an *asymmetric* or *dissymmetric molecule*. An asymmetric molecule has the following features:

(*i*) An asymmetric molecule never has a plane of symmetry (also called *mirror-image plane*) which is defined as an imaginary plane dividing the molecule in such a way that the part of it on one side of the plane is the mirror image of that on the other side of the plane. The molecules possessing such a plane of symmetry are always *inactive while those having no plane of symmetry are optically active and hence show optical isomerism*.

(*ii*) An asymmetric molecule cannot be superimposed on its mirror image.

Thus the most necessary and sufficient condition for a molecule to show optical isomerism (*i.e.*, to exist in *d*-and *l*-forms) is that *the molecule should be asymmetric*, (*i.e., it should have no plane of symmetry*) *and should not be superimposable on its mirror image*.

### Optical Isomerism in Tetrahedral Complexes (C.N.=4)

Here we shall discuss optical isomerism in the following types of tetrahedral complexes.

**1. Tetrahedral complexes of [$Ma_4$], [$Ma_2b_2$] and [$Ma_3b$] type.** Tetrahedral complexes of [$Ma_4$], [$Ma_2b_2$] and [$Ma_3b$] type are not able to show optical isomerism because all the possible arrangements of the ligands round the central metal ion, M, are exactly equivalent.

**2. Tetrahedral complexes of [Mabcd] type.** Since the central atom (M) in tetrahedral complexes of [Mabcd] type is surrounded by four different ligands, the tetrahedral complex of this type is expected to produce a pair of enantiomorphs. For example, [$As^{3+}(CH_3)(C_2H_5)$ (S) ($C_6H_4COO$)]$^{2-}$ ion (tetrahedral) exists in two optical isomers as shown in Fig. 15.21.

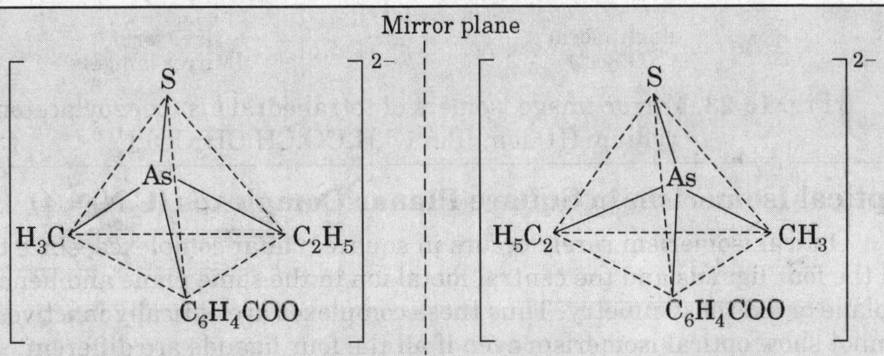


**Fig. 15.21.** Optical isomers of tetrahedral complex ion, [$As(CH_3)(C_2H_5)$ (S) ($C_6H_4COO$)]$^{2-}$

Here it may be noted that 4 different groups round the central metal ion are not the only requirement to make the complex to show mirror-image isomerism. All that is required is that the molecule should be asymmetric (*i.e.*, unsymmetrical), *i.e.*, it should have no plane of symmetry so that it can exist in two mirror-image forms.

**3.** Tetrahedral complexes of Be(II), B(III) and Zn (II) with symmetrical bidentate ligands have been made and resolved into optical isomers. Some examples of tetrahedral complexes of this type are *bis* (salicylaldehydato) boron (III) ion, [$B^{3+}(C_6H_4.O.CHO)_2$]$^+$ and *bis* (benzoylacetonato) beryllium (II), [$Be^{2+}(C_6H_5CO.\ CH.\ CH_3CO_2)$]$^+$ and *bis* (benzoylacetonato) beryllium (II), [$Be^{2+}(C_6H_5CO.CH.CH_3CO)_2$]$^0$ whose mirror-image isomers are given in Fig. 15.22 and 15.23 respectively.

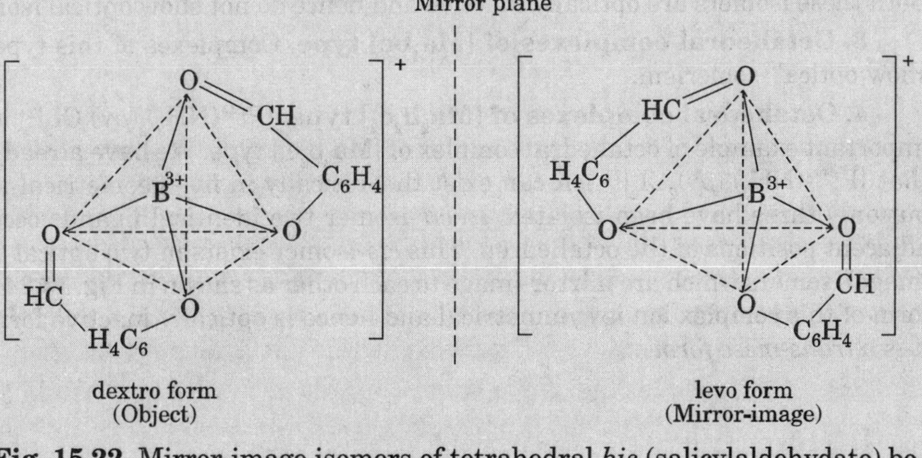


**Fig. 15.22.** Mirror-image isomers of tetrahedral *bis* (salicylaldehydato) boron (III) ion, [$B^{3+}(C_6H_4.O.CHO)_2$]$^+$.

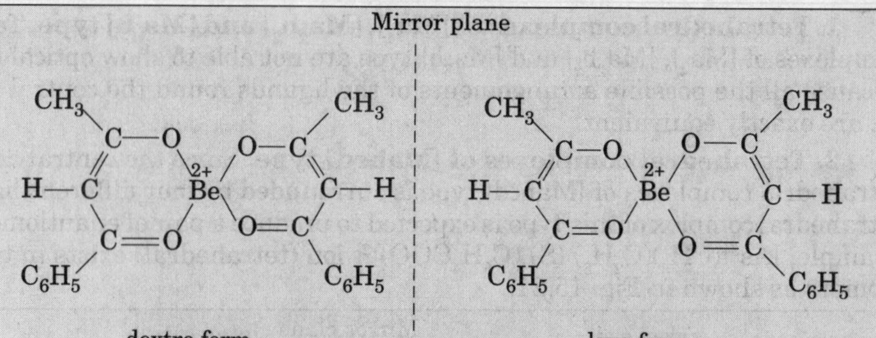


**Fig. 15.23.** Mirror-image isomers of tetrahedral *bis* (benzoylacetonato) beryllium (II) ion, $[Be^{2+}(C_6H_5CO.CH.CH_3CO)_2]^0$

## Optical Isomerism in Square Planar Complexes (C.N. = 4)

Optical isomerism rarely occurs in square planar complexes, since they have all the four ligands and the central metal ion in the same plane and hence contain a plane or axis of symmetry. Thus these complexes are optically inactive, *i.e.*, they cannot show optical isomerism even if all the four ligands are different.

## Optical Isomerism in 6-Coordinated Complexes : Octahedral Complexes

Here we shall discuss the optical isomerism in octahedral complexes of the following type:

**1. Octahedral complexes of $[Ma_6]$ and $[Ma_5b]$ type.** Complexes of this type do not show optical isomerism.

**2. Octahedral complexes of $[Ma_4b_2]$ and $[Ma_3b_3]$ type** (a) We have already seen that octahedral complexes of $[Ma_4b_2]$ type exist in *cis*- and *trans*-isomers. Both these isomers are optically inactive due to the presence of symmetry and hence do not show optical isomerism. (*b*) We have already mentioned that octahedral complexes of $[Ma_3b_3]$ type exist in *cis* (or *facial*) and *trans* (or *meridional*) isomers. Both these isomers are optically inactive and hence do not show optical isomerism.

**3. Octahedral complexes of $[Ma_4bc]$ type.** Complexes of this type do not show optical isomerism.

**4. Octahedral complexes of $[Ma_2b_2c_2]$ type.** $[Pt^{4+}(NH_3)_2(py)_2Cl_2]^{2+}$ ion is an important example of octahedral complex of $[Ma_2b_2c_2]$ type. We have already stated that $[Pt^{4+}(NH_3)_2(py)_2Cl_2]^{2+}$ ion can exist theoretically in five geometrical isomers, but only three have been isolated. In *cis*-isomer two identical ligands occupy the adjacent positions of the octahedron. This *cis*-isomer exists in two optical (mirror-image) isomers which are mirror-image to each other as shown in Fig. 15.24. *Trans*-form of this complex ion is symmetrical and hence is optically inactive form. Thus it is a *trans-meso form*.

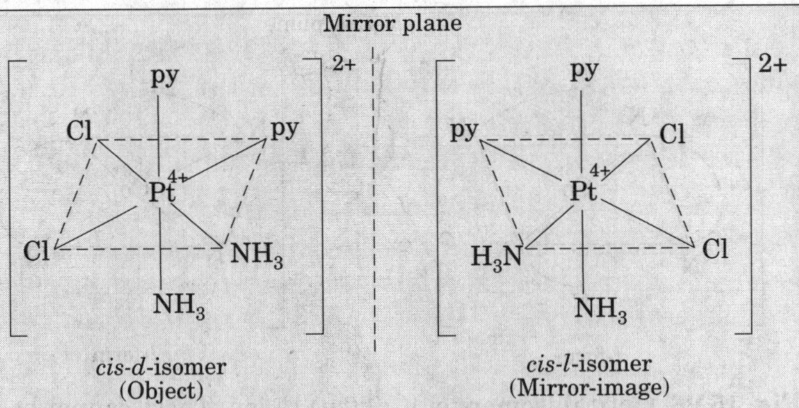


**Fig. 15.24.** Mirror-image isomers of $[Pt^{4+}(NH_3)_2(py)_2Cl_2]^{2+}$ ion.

**5. [Mabcdef] type complexes.** $[Pt^{4+}(py)(NH_3)(NO_2)ClBrI]^0$ is the only example of octahedral complex of this type. We have already stated that, theoretically, this complex can exist in 15 geometrical isomers. Each of these 15 geometrical isomers exists in optically active *d*-and *l*-forms, giving a total of 30 optically active isomers. *d*- and *l*-isomers for one of the 15 geometrical isomers are given in Fig. 15.25.

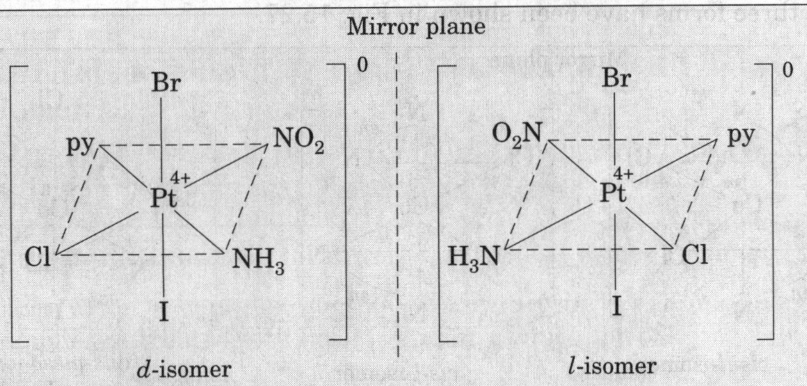


**Fig. 15.25.** Optical (*d*- and *l*-) forms of $[Pt^{4+}(py)(NH_3)(NO_2)(Cl)(Br)(I)]^0$.

**6. $[M(AA)_3]$ type complexes.** Due to the absence of a plane or centre of symmetry, the octahedral complexes of this type are resolvable into *dextro* and *levo* forms. For example, *d*- and *l*-forms of $[Co^{3+}(en)_3]^{3+}$ ion are shown in Fig. 15.26. Other examples of octahedral complexes of $[M(AB)_3]$ type which exist in *d* and *l* forms are : $[Cr^{3+}(ox)_3]^{3-}$, $[Co^{3+}(pn)_3]^{3+}$, $[Pt^{4+}(en)_3]^{4+}$, $[Fe^{2+}(diph)_3]^{2+}$ etc. The fact that the complexes of $[M(AA)_3]$ type can be resolved into optical isomers confirms that these complexes have octahedral shape. Neither hexagonal nor trigonal prismatic geometry of 6-coordinated complex can give rise to activity.

**7. $[M(AA)_2a_2]$ type complexes.** Here we can consider $[Co^{3+}(en)_2Cl_2]^+$ ion. We have already stated that this ion shows geometrical isomerism and hence exists as *cis*- and *trans*- isomers. The *cis*-isomer does not have any plane or centre of symmetry and hence is *optically active, i.e., cis*-isomer can be resolved into two optically active isomers (*d*- and *l*-isomers). On the other hand, the *trans*-isomer has a plane

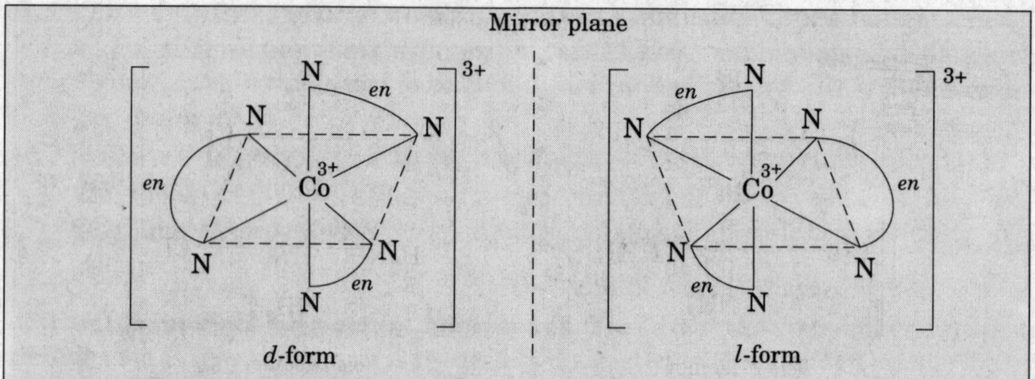


**Fig. 15.26.** Optical isomers of $[Co^{3+}(en)_3]^{3+}$ ion. These cannot be superimposed on each other.

of symmetry and hence is optically inactive, *i.e.*, it cannot be resolved into optically active isomers. Thus *trans*-isomer is optically inactive (meso-form.) Consequently $[Co(en)_2Cl_2]^+$ ion has three optical isomers *viz.* (*a*) Two optically active isomers corresponding to *cis*-configuration (*i.e.*, *d*- and *l*-forms). One form is the mirror image of the other. (*b*) One optically inactive form corresponding to *trans*-form (meso-form). All the three forms have been shown in Fig. 15.27.

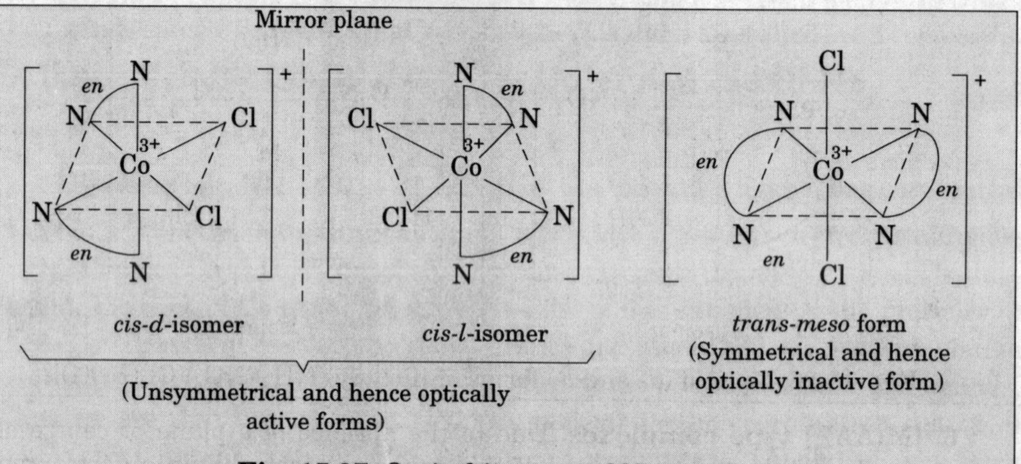


**Fig. 15.27.** Optical isomers of $[Co(en)_2Cl_2]^+$ ion.

Other examples of $[M(AA)_2a_2]$ type octahedral complexes whose *cis*-isomers exist in dextro (*d*) and levo (*l*) forms are : $[Co^{3+}(en)_2(NO_2)_2]^+$, $[Ir^{4+}(C_2O_4)_2Cl_2]^{2-}$, $[Rh^{4+}(C_2O_4)_2Cl_2]^{2-}$, $[Cr^{3+}(C_2O_4)_2(H_2O)_2]^-$, $[Ir^{3+}(C_2O_4)_2Cl_2]^{3-}$ etc.

**8. $[M(AA)_2ab]$ type complexes.** Here we shall consider $[Co^{3+}(en)_2(NH_3)Cl]^{2+}$ ion. We have already seen that this ion exists in *cis*- and *trans*-isomers (geometrical isomers). Due to the absence of plane or centre of symmetry in *cis*-isomer, this isomer is optically active and hence can be resolved into *d*- and *l*-forms. On the other hand, the *trans*-isomer has a plane of symmetry and hence is optically inactive, i.e., *trans*-isomer is *meso-trans* isomer (See Fig. 15.28).

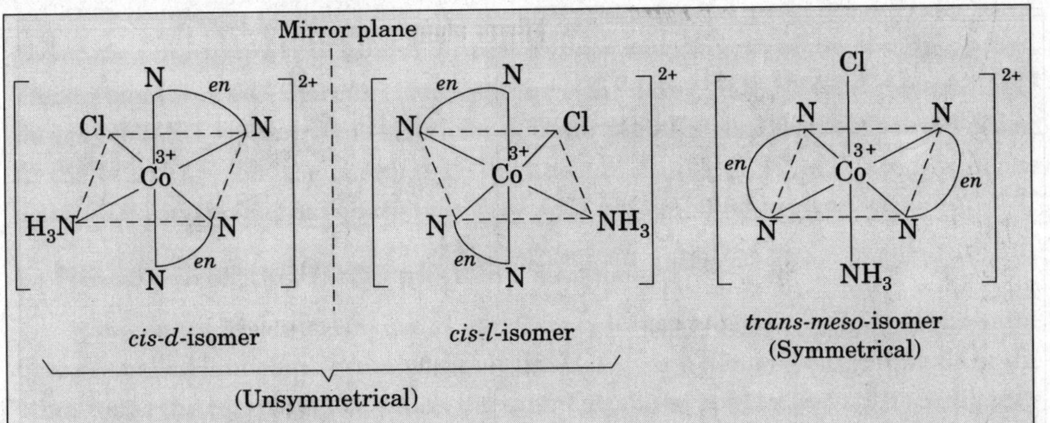


**Fig. 15.28.** Optical isomers of $[Co^{3+}(en)_2(NH_3)Cl]^{2+}$ ion.

Another example of $[M(AA)_2ab]$ type octahedral complexes whose *cis*-isomer exists in *d*- and *l*-forms is $[Rh^{3+}(C_2O_4)_2(py)(NO_2)]^{2-}$

**9. $[M(AA)a_2b_2]$ type complexes.** Here we shall consider $[Co^{3+}(en)(NH_3)_2Cl_2]^+$ ion. We have seen that this ion exists in *cis*- and *trans*-isomers (geometrical isomers). *Cis*-isomer is optically active and hence gives optically active *d*-and *l*-forms. On the other hand *trans*-isomer is optically inactive and hence does not give any optical active isomer. (See Fig. 15.29).

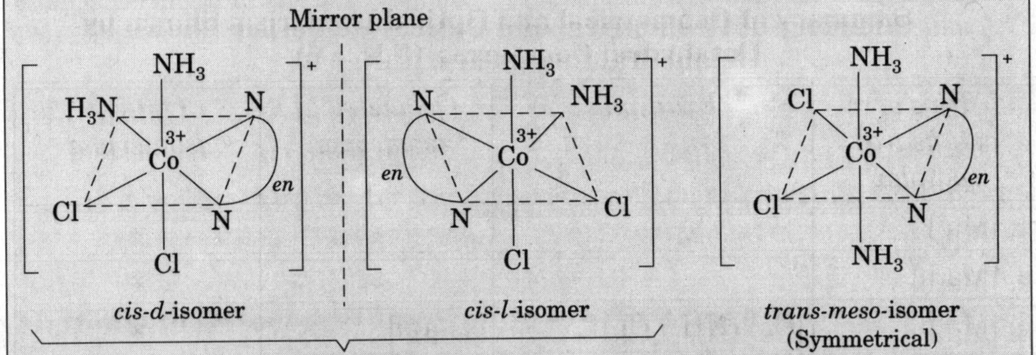


**Fig. 15.29.** Optical isomers of $[Co^{3+}(en)(NH_3)_2Cl_2]^+$ ion.

$[Co^{3+}(C_2O_4)NH_3)_2(NO_2)_2]^-$ ion is another important example of $[M(AA)a_2b_2]$ type octahedral complexes whose *cis*-isomer gives *d*- and *l*-isomers.

**10. $[M(AB)_3]$ type complexes.** We have seen that an octahedral complex of this type exists in *cis*- and *trans*-isomers (geometrical isomers). Each of these forms is optically active and hence each gives a pair of optical isomers (*d*- and *l*-forms) as shown in Fig. 15.30. for $[Cr^{3+}(gly)_3]$ complex compound.

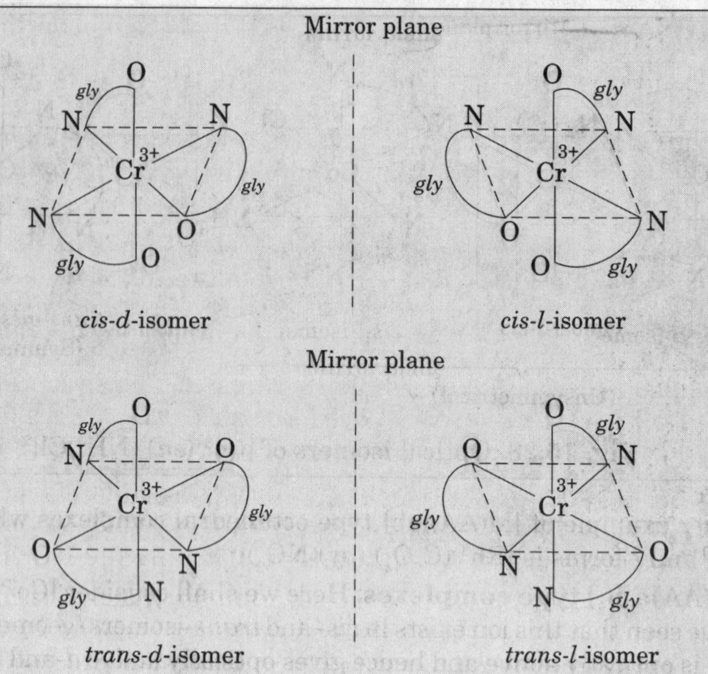


**Fig, 15.30.** Each of *cis*- and *trans*-isomers of $[Cr^{3+}(gly)_3]^0$ has *d* and *l* optical isomers.

## Summary of Geometrical and Optical Isomerism Shown by Octahedral Complexes. (C.N. = 6)

| *Type of octahedral complex* | *Examples* | *Geometrical isomerism* | *Optical isomerism* |
|---|---|---|---|
| 1. $[Ma_6]$ | – | × | × |
| 2. $[Ma_5b]$ | – | × | × |
| 3. $[Ma_4b_2]$ | $[Cr^{3+}(NH_3)_4Cl_2]^+$, $[Co^{3+}(NH_3)_4Cl_2]^+$ etc. | *cis*-and *trans*-isomers | × |
| 4. $[Ma_3b_3]$ | $[Co^{3+}(NH_3)_3Cl_3]^\circ$, $[Rh^{3+}(py)_3Cl_3]^\circ$ etc. | *cis*-(or *fac*) and *trans*-(or *mer*) isomers | × |
| 5. $[Ma_4bc]$ | $[Co^{3+}(NH_3)_2(H_2O)(Cl)]^{2+}$ | *cis*-and *trans*-isomers | × |
| 6. $[Ma_2b_2c_2]$ | $[Pt^{4+}(NH_3)_2(py)_2Cl_2]^{2+}$ | Five geometrical forms. Only three have been isolated | *cis-d*-isomer, *cis-l*-isomer (optical forms), *trans-meso* form |

| | | | |
|---|---|---|---|
| 7. [Mabcedef] | $[Pt^{4+}(NH_3)(Cl)(Br)(I)(NO_2)(py)]^{\circ}$ | 15 different geometrical isomers | Each geometrical isomer has *d*- and *l*-isomers. Thus we have 30 optical isomers. |
| 8. $[M(AA)_3]$ | $[Co^{3+}(en)_3]^{3+}$ etc. | × | *d*- and *l*-isomers |
| 9. $[M(AA)_2a_2]$ | $[Co^{3+}(en)_2Cl_2]^+$ etc. | *cis*- and *trans*-isomers (*cis*-isomer is unsymmetrical) | Being unsymmetrical, *cis*-isomer gives *cis-d*- and *cis-l*-isomers (optical isomers) |
| 10. $[M(AA)_2ab]$ | $[Co^{3+}(en)_2(NH_3)(Cl)]^{2+}$ etc. | *cis*- and *trans*-isomers | Being unsymmetrical, *cis*-isomer gives *cis-d*- and *cis-l*-isomers (optical isomers) |
| 11. $[M(AA)a_2b_2]$ | $[Co^{3+}(en)(NH_3)_2Cl_2]^+$ etc. | *cis*-and *trans*-isomers | *cis*-isomer exists as *cis-d*-and *cis-l*-isomer (optical isomers). |
| 12. $[M(AB)_3]$ | $[Cr^{3+}(gly)_3]^{\circ}$ | *cis*-and *trans*-isomers (Each form is optically active) | Each geometrical form is optically active. Hence *cis*-form gives *cis-d*-and *cis-l*-optical isomers. Similarly *trans*-form gives *trans d*- and *trans-l*-optical isomers. |

**Summary of Geometrical and Optical Isomerism Shown by Tetrahedral and Square Planar Complexes (C.N. = 4)**

| *Type of complexes* | *Geometrical isomerism* | *Optical isomerism* |
|---|---|---|
| **(A) Tetrahedral complexes** | | |
| 1. $[Ma_4]$ | × | × |

| | | |
|---|---|---|
| 2. $[Ma_3b]$ | × | × |
| 3. [Mabcd] *e.g.* $[As^{3+}(CH_3)(C_2H_5)(S)(C_6H_4COO)]^{2-}$ | × | Two optical isomer |
| 4. Complexes of Be(II), B(III) etc. containing unsymmetrical bidentate ligands *e.g.* $[Be^{2+}(C_6H_5CO.\ CH.\ CH_3CO)_2]°$ and $[B^{3+}(C_6H_4OCHO)_2]^+$ | × | *d*-and *l*-isomers |
| **(B) Square planar complexes** 1. $[Ma_4]$ | × | |
| 2. $[Ma_3b]$ | × | |
| 3. $[Mab_3]$ | × | |
| 4. $[Ma_2b_2]$ *e.g.* $[Pt^{2+}(NH_3)_2Cl_2]°$ | *cis*-and *trans*-isomers | |
| 5. [Mabcd] *e.g.* $[Pt^{2+}(NH_3)(py)(Cl)(Br)]°$ | Three isomeric forms | Square planar complexes rarely show optical isomerism |
| 6. $[Ma_2bc]$ *e.g.* $[Pd^{2+}Cl_2BrI]^{2-}$, $[Pt^{2+}(py)_2(NH_3)(Cl)]^+$ | *cis*-and *trans*-isomers | |
| 7. $M(AB)_2$ *e.g.* $[Pt^{2+}(gly)_2]°$ (gly⁻ ion is $NH_2CH_2COO^-$) | *cis*-and *trans*-isomers | |

## Notes

1. **Characteristics of square planar complexes**
   (i) These have plane of symmetry.
   (ii) These are optically inactive.
2. **Characteristics of tetrahedral complexes**
   (i) These do not have any plane of symmetry *i.e.* these are asymmetric.
   (ii) These are optically active.
3. *cis*-isomers (geometrical isomer) is asymmetric and hence is optically active. Being optically active, *cis*-isomer has *d-cis* and *l-cis* isomers (optical isomers).
4. *trans*-isomer (geometrical isomer) is symmetrical and hence is optically inactive. Being optically inactive, *trans*-isomer is *trans-meso* form.

## UNIVERSITY QUESTIONS

**1.** What is isomerism? Explain ionisation and linkage isomerism in metal complexes. *(Osmania 2010)*

**2.** Explain the following: (*i*) Coordination isomerism (*ii*) Salt isomerism *(Pd. Ravishankar Shukla 2010)*

**3.** Write the structural formula of two isomers of the complex ion, $[Co(NH_3)_5(NO_2)]^{2+}$. Draw the structure of the isomers. *(Himachal Pradesh 2010)*

**4.** (*a*) Explain the following types of isomerism in coordination compounds:
(*i*) Ionization (ii) Coordination position.
(*b*) What is resolution of recemic mixture? Discuss.
(*c*) Draw the optical isomer representation of $[Rh(en)_2Cl_2]$.
(*d*) Explain facial and meridional isomers with suitable examples. *(Himachal Pradesh 2010)*

**5.** Write short notes on the following—
(*i*) Geometrical and optical isomerism in octahedral complexes.
(*ii*) Stereochemistry of 4-coordination compounds.
(*iii*) Electroneutrality principle. *(Rajasthan 2010)*

**6.** (*a*) Write the geometrical isomers of $[Co(NH_3)_4Cl_2]^+$ ion.
(*b*) Explain (*i*) Linkage isomerism (*ii*) Hydrate isomerism. *(Bangalore 2010)*

**7.** Explain the meaning of optical activity. *(H.N. Bahuguna Garhwal 2010)*

**8.** (*a*) Discuss geometrical isomerism exhibited by 4-coordinated complexes.
(*b*) Explain optical isomerism exhibited by 6-coordinated complexes. *(Nagpur 2010)*

**9.** What are the types of isomerism in complexes? Discuss different types of structural isomerism in complexes. *(Lucknow 2010)*

**10.** (*i*) Explain coordination isomerism with an example.
(*ii*) Explain optical isomerism of 4-coordinated complexes.
(*iii*) Discuss geometrical isomerism in octahedral complexes. *(Madras April 2011)*

**11.** (a) Explain optical isomerism in 6-coordinated complexes
(*b*) Explain with examples
(*i*) Coordination isomerism
(*ii*) Linkage isomerism *(Nagpur 2012)*

**12.** Discuss geometrical isomerism of coordination compounds with coordination 4 and 6. *(Ch. Charan Singh, Meerut 2012)*

**13.** (*i*) Discuss with suitable examples, various types of isomerism in coordination compounds.
(*ii*) Draw possible isomers of $[Co(NH_3)_4Cl_2]^+$ ion *(Lucknow 2012)*

16

CHAPTER

# Valence Bond Theory (VBT) of Coordination Compounds

In order to explain the nature of bonding between metal atom and ligands in coordination compounds the following theories have been suggested.

(*i*) Valence Bond Theory (VBT) in 1930s

(*ii*) Crystal Field Theory (CFT) in 1950s and 1960s

(*iii*) Ligand Field Theory (LFT) in 1960s onwards.

(*iv*) Molecular Orbital Theory (MOT) in 1960s onwards.

In the present chapter we discuss only VBT. This theory was extended to coordination compounds by Pauling in 1931.

## Main Assumptions of VBT

(*i*) This theory assumes that the metal-ligand bond in coordination compounds arises by the donation of pairs of electrons by ligands to the central metal atom or ion. The ligands have at least one orbital containing a lone pair of electrons.

(*ii*) In order to accommodate the electron pairs donated by the ligands the central metal atom or ion must possess requisite number of vacant hybrid orbitals of equal energy. These vacant hybrid orbitals are obtained by the hyrbidisation of atomic orbitals. These vacant hybrid orbitals are equivalent in energy and have definite geometry.

(*iii*) Sometimes, the unpaired $(n-1)d$ electrons pair up as fully as possible prior to hybridisation thus making some $(n-1)d$ orbitals vacant. The central atom then makes available the number of empty orbitals equal to its coordination number for the formation of coordinate bonds with suitable ligand orbitals.

(*iv*) Vacant hybrid orbitals of the metal atom or ion *overlap with the filled* (containing lone pair of electrons) σ-orbitals of the ligands to form ligand → metal σ-bond (represented as $L \xrightarrow{\sigma} M$).) This bond which is generally known as coordinate bond is a special type of covalent bond and shows the characteristics of both the overlapping orbitals. However, it also possesses a considerable amount of polarity because of the mode of its formation.

Important types of hybridisation occurring in the first row transition metal (3*d* metals) complexes and the geometry of the complexes are given in Table 16.1.

**Table 16.1. Important types of hybridisation found in the first row transition metal (3d-metal) complexes and the geometry of the complexes.**

| *Coordination number of the central metal atom / ion* | *Type of hybridisation undergone by the central metal atom / ion* | *Geometry of the complexes* | *Examples of complexes* |
|---|---|---|---|
| 2 | sp(4s, $4p_x$) | Linear or diagonal (180°) | $[CuCl_2]^-$, $[Cu(NH_3)_2]^+$ etc. |
| 3 | $sp^2(4s, 4p_x, 4p_y)$ | Trigonal planar or equilateral triangular (120°) | 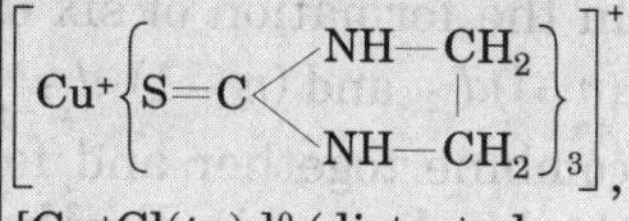, $[Cu^+Cl(tu)_2]^0$ (distorted trigonal planar) etc. |
| 4 | $dsp^2(3d_{x^2-y^2}, 4s, 4p_x, 4p_y)$ | Square planar (90°) | $[Ni(CN)_4]^{2-}$, $[PdCl_4]^{2-}$ $[CuCl_4]^{2-}$, $[Ni(dmg)_2]^\circ$, $[Cu(NH_3)_4]^{2+}$, $[Pt(CN)_4]^{2-}$, $[Pt(NH_3)_4]^{2+}$, $[Pt(NH_3)_2Cl_2]^\circ$ |
| 4 | $sp^3(4s, 4p_x, 4p_y, 4p_z)$ | Tetrahedral (109°) | $[NiCl_4]^{2-}$, $[Cu(CN)_4]^{3-}$, $Ni(CO)_4$ etc. |
| 5 | $dsp^3(3d_{z^2}, 4s, 4p_x, 4p_y, 4p_z)$ | Trigonal bipyramidal (120°, 90°) | $Fe(CO)_5$, $[CuCl_5]^{3-}$, $[Ni^{2+}(triars)\,Br_2]^0$, $[CuI(dipyridyl)_2]I$ |
| 5 | $dsp^3(3d_{x^2-y^2}, 4s, 4p_x, 4p_y, 4p_z)$ | Square pyramidal (90°) | $[Co^{2+}(triars)\,I_2]^0$, $[Ni(CN)_5]^{3-}$ $[VO(acetylacetone)_2]$ |
| 6 | $d^2sp^3(3d_{x^2-y^2}, 3d_{z^2}, 4s, 4p_x, 4p_y, 4p_z)$ | Inner-orbital octahedral (90°, 180°) | $[Ti(H_2O)_6]^{3+}$, $[Fe(CN)_6]^{3-}$, $[Co(NH_3)_6]^{3+}$ etc |
| 6 | $sp^3d^2(4s, 4p_x, 4p_y, 4p_z, 4d_{x^2-y^2}, 4d_{z^2})$ | Outer-orbital octahedral (90°, 180°) | $[Fe^+(NO^+)(H_2O)_5]^{2+}$, $[CoF_6]^{3-}$ etc. |

## Octahedral Complexes ($d^2sp^3$ or $sp^3d^2$ hybridisation)

Octahedral complexes are those in which the C.N. of the central metal atom or ion is 6. These complexes are formed as a result of $d^2sp^3$ or $sp^3d^2$ hybridisation of the central metal atom or ion. What type of hybridisation (i.e., whether $sp^3d^2$ or $d^2sp^3$) will occur depends on the number of unpaired or paired electrons present in the complex ion. $d^2sp^3$ or $sp^3d^2$ hybridisation is also called *octahedral hybridisation*.

Octahedral complexes in which the central atom is $d^2sp^3$ hybridised are called *inner-orbital octahedral complexes* while the octahedral complexes in which the central atom is $sp^3d^2$ hybridised are called *outer-orbital octahedral complexes*.

Now let us see how $d^2sp^3$ / $sp^3d^2$ hybridisation takes place in octahedral complexes.

## $d^2sp^3$ Hybridisation in Inner Orbital Octahedral Complexes

This type of hybridisation takes place in those octahedral complexes which contain strong ligands. On the basis of the orientation of the lobes of $d$-orbitals in space, these orbitals have been classified into two sets viz. $t_{2g}$ or $d_\varepsilon$ and $e_g$ or $d_r$. $t_{2g}$ set consists of $d_{xy}$, $d_{yz}$ and $d_{zx}$ orbitals while $e_g$ set has $d_{z^2}$ and $d_{x^2-y^2}$ orbitals. In the formation of six $d^2sp^3$ hybrid orbitals, two $(n-1)$ d-orbitals of $e_g$ set [i.e., $(n-1)d_{z^2}$ and $(n-1)$ $d_{x^2-y^2}$ orbitals], *one ns* and *three np* ($np_x$, $np_y$ and $np_z$) orbitals combine together and form six $d^2sp^3$ hybrid orbitals. Thus we see that the two $d$-orbitals used in $d^2sp^3$ hybridisation are from penultimate shell [*i.e.* $(n-1)$th shell] while $s$ and three p-orbitals are from ultimate shell (i.e., $n$ th shell). This discussion shows that in case of octahedral complex ions of $3d$ transition series elements, two $d$-orbitals used in $d^2sp^3$ hybridisation are $3d_{z^2}$ and $3d_{x^2-y^2}$ orbitals ($e_g$ set of orbitals) while $s$-and $p$-orbitals are 4s and 4p orbitals. Thus $d^2sp^3$ hybridisation taking place in such complexes can be represented as $3d_{x^2-y^2}.3d_{z^2}.4s.4p_x.4p_y 4p_z (d^2sp^3)$.

Since two d-orbitals used in $d^2sp^3$ hybridisation belong to inner shell [i.e., $(n-1)$]th shell], the octahedral complex compounds resulted from $d^2sp^3$ hybridisation are called **inner orbital octahedral complexes**.

Since these complexes have comparatively lesser number of unpaired electrons than the outer-orbital octahedral complexes (see later on), these complexes are also called **low spin** or **spin paired octahedral complexes**. It is due to the presence of strong ligands in inner-orbital octahedral complexes of 3d transition series that the electrons present in $3d_{z^2}$ and $3d_{x^2-y^2}$ orbitals ($e_g$ set) are forced to occupy $3d_{xy}$, $3d_{yz}$ and $3d_{xz}$ orbitals ($t_{2g}$ set) and thus $3d$ orbitals of $e_g$ set become vacant and hence can be used in $3d_{x^2-y^2}.3d_{z^2}.4p_x, 4p_y, 4p_z (d^2sp^3)$ hybridisation.

Now let us discuss the structure of some octahedral complex ions of transition series elements which result from $d^2sp^3$ hybridisation.

### 1. Ferricyanide Ion, $[Fe(CN)_6]^{3-}$ [Also called Hexacyanoferrate (III) Ion]

In this ion the coordination number of Fe is six and hence the given complex ion is octahedral in shape. In this ion, Fe is present as $Fe^{3+}$ ion whose valence-shell configuration is $3d^5$ or $t_{2g}^3 e_g^2$ ($Fe = 3d^6 4s^2 4p^0$, $Fe^{3+} = 3d^5 = t_{2g}^3 e_g^2$) as shown in Fig. 16.1. According to Hund's rule, each of the five electrons in $3d$ orbitals is unpaired in free $Fe^{3+}$ ion (uncomplexed ion) and hence the number of unpaired electrons ($n$) is equal to 5 (See Fig. 16.1). However, magnetic study of $[Fe(CN)_6]^{3-}$ ion has shown that this ion has one unpaired electron ($n = 1$) and hence is paramagnetic. Thus, in the formation of this ion, two electrons of $e_g$ set of $3d$ orbitals (*i.e.*, $3d_{z^2}$ and $3d_{x^2-y^2}$ orbitals) pair up with the three electrons of $t_{2g}$ set of orbitals (*i.e.*, $3d_{xy}$, $3d_{yz}$ and $3d_{zx}$ orbitals). This results in that $e_g$ set of orbitals becomes vacant and is used in $d^2sp^3$ hybrdisation. This also results in that the valence-shell configuration of $Fe^{3+}$ ion gets changed from $t_{2g}^3 e_g^2$ to $t_{2g}^5 e_g^0$ and thus the number of unpaired electrons in 3*d* orbital now becomes equal to 1. Now $3d_{x^2-y^2}, 3d_{z^2}$ ($e_g$ set), 4s and three 4p( $4p_x$, $4p_y$ and $4p_z$) orbitals combine together and give rise to the formation of six $3d_{z^2} . 3d_{x^2-y^2} . 4s . 4p_x . 4p_y . 4p_z$ hybrid orbitals (**$d^2sp^3$ hybridisation**). Each of these hybrid orbitals is vacant. Each of the six $CN^-$ ions (ligands) donates its lone pair of electrons to $d^2sp^3$ hybrid orbitals and six NC → $Fe^{3+}$ coordinate bonds are established (See Fig. 16.1). The above discussion shows that $[Fe(CN)_6]^{3-}$ ion has one unpaired electron and hence is paramagnetic. It is an inner orbital octahedral complex ion, since it is formed by $d^2sp^3$ hybridisation.

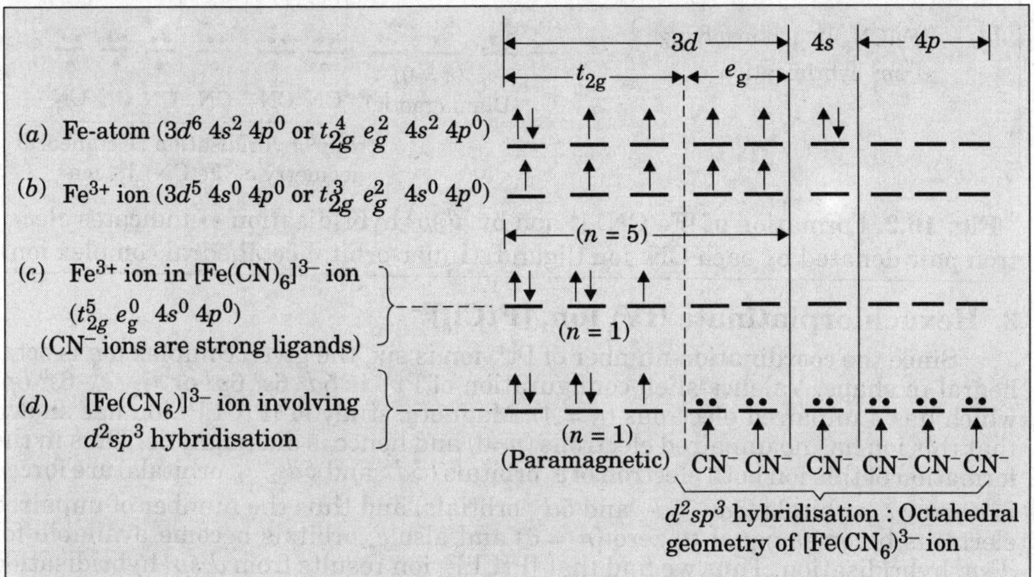


**Fig. 16.1.** Formation of $[Fe(CN)_6]^{3-}$ ion by $d^2sp^3$ hybridisation •• indicates electron pair donated by each $CN^-$ ion (ligand). (Inner-orbital octahedral complex ion).

## 2. Ferrocyanide Ion, $[Fe(CN)_6]^{4-}$ [Also called Hexa cyano ferrate (II) Ion]

In this ion, since the coordination number of Fe is six, the give complex ion has octahedral geometry. In this ion, Fe is present as $Fe^{2+}$ ion whose valence-shell configuration is $3d^6\,4s^0\,4p^0$ or $t_{2g}^4 e_g^2 4s^0 4p^0$ which shows that $Fe^{2+}$ ion has 4 unpaired electrons. Magnetic studies have, however, shown that the given complex ion is diamagnetic and hence it has no unpaired electrons ($n = 0$). Hence in order to get all the electrons in the paired state, two electrons of $e_g$ orbitals are sent to $t_{2g}$ orbitals so that *n* becomes equal to zero. Since $CN^-$ ions (ligands) are strong ligands, they are capable of forcing the two electrons of $e_g$ orbitals to occupy $t_{2g}$ orbitals and thus make all the electrons paired. Now for the formation of $[Fe(CN)_6]^{4-}$ ion, two 3d orbitals of $e_g$ set, 4*s* orbital (one orbital) and three 4*p* orbitals (all these six orbitals are vacant orbitals) undergo $d^2sp^3$ hybridisation (See Fig. 16.2). It is due to $d^2sp^3$ hybridisation that $[Fe(CN)_6]^{4-}$ ion is an inner orbital octahedral complex ion.

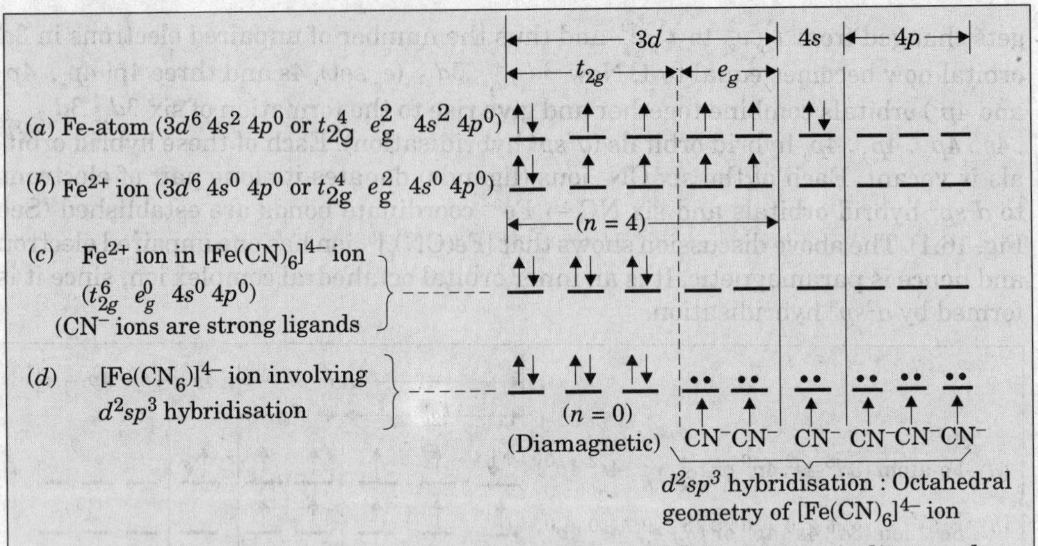


**Fig. 16.2.** Formation of $[Fe(CN)_6]^{4-}$ ion by $d^2sp^3$ hybridisation •• indicates electron pair donated by each $CN^-$ ion (ligand) (Inner-orbital octahedral complex ion).

## 3. Hexachlorplatinate (IV) Ion, $[PtCl_6]^{2-}$

Since the coordination number of $Pt^{4+}$ ion is six, the given complex ion is octahedral in shape. Valence-shell configuration of $Pt^{4+}$ is $5d^6\,6s^0\,6p^0$ or $t_{2g}^4\;e_g^2\;6s^0\,6p^0$ which has 4 unpaired electrons ($n$ = 4). Magnetic study of $[PtCl_6]^{2-}$ ion has shown that this ion has no unpaired electrons ($n$=0) and hence is diamagnetic. Thus in the formation of this ion both electron of $e_g$ orbitals ($5d_{z^2}$ and $5d_{x^2-y^2}$ orbitals) are forced to occupy $t_{2g}$ orbitals ($5d_{xy}$, $5d_{yz}$ and $5d_{zx}$ orbtials) and thus the number of unpaired electrons becomes equal to zero ($n$ = 0) and also $e_g$ orbitals become available for $d^2sp^3$ hybridisation. Thus we find that $[PtCl_6]^{2-}$ ion results from $d^2sp^3$ hybridisation of $Pt^{4+}$ ion (See Fig. 16.3).

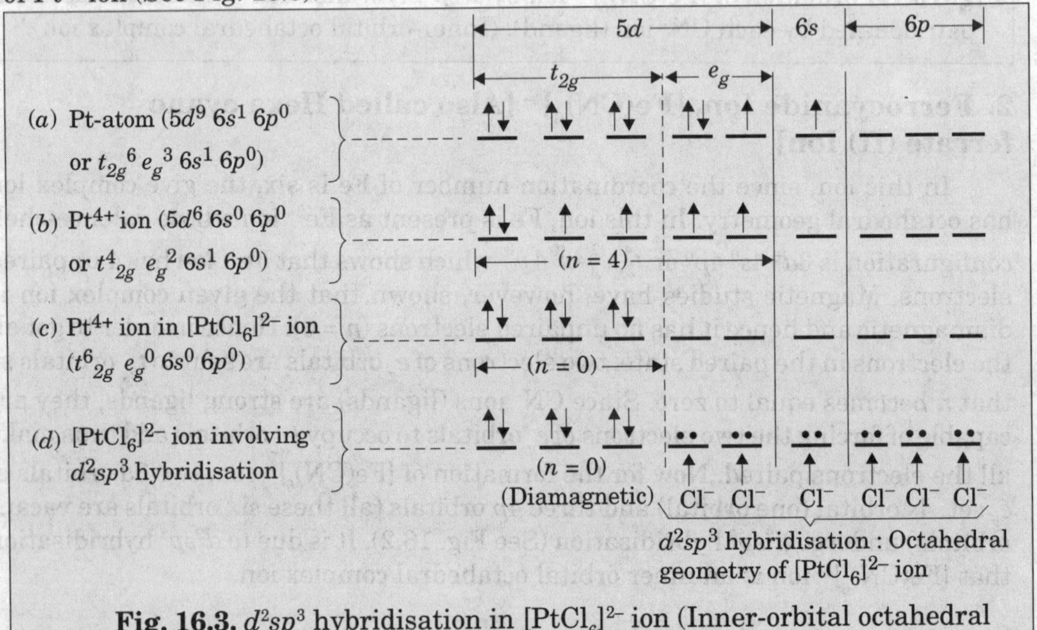


**Fig. 16.3.** $d^2sp^3$ hybridisation in $[PtCl_6]^{2-}$ ion (Inner-orbital octahedral complex ion).

## 4. $[Co(NH_3)_6]^{2+}$, $[Co(NO_2)_6]^{4-}$ and $[Co(CN)_6]^{4-}$ Ions.

In all these complex ions Co is present as $Co^{2+}$ ion which is a $3d^7$ ion. In the free state $3d^7$ configuration has three unpaired electrons ($n$ = 3). Each of the given complex ion has been found paramagnetic due to the presence of one unpaired electron ($n$ = 1). Since each complex ion is octahedral in geometry, $Co^{2+}$ ion undergoes $d^2sp^3$ hybridisation. In order to get $d^2sp^3$ hybridisation one of the electrons from $3d$ orbitals gets shifted to 4d orbitals as shown at ($a$) of Fig 16.4. We have said above that each of the given complex ions has one unpaired electron. $Co^{2+}$ ion cannot undergo $sp^3d^2$ hybridisation since this hybridisation gives three unpaired electrons as shown at (b) of Fig 16.4.

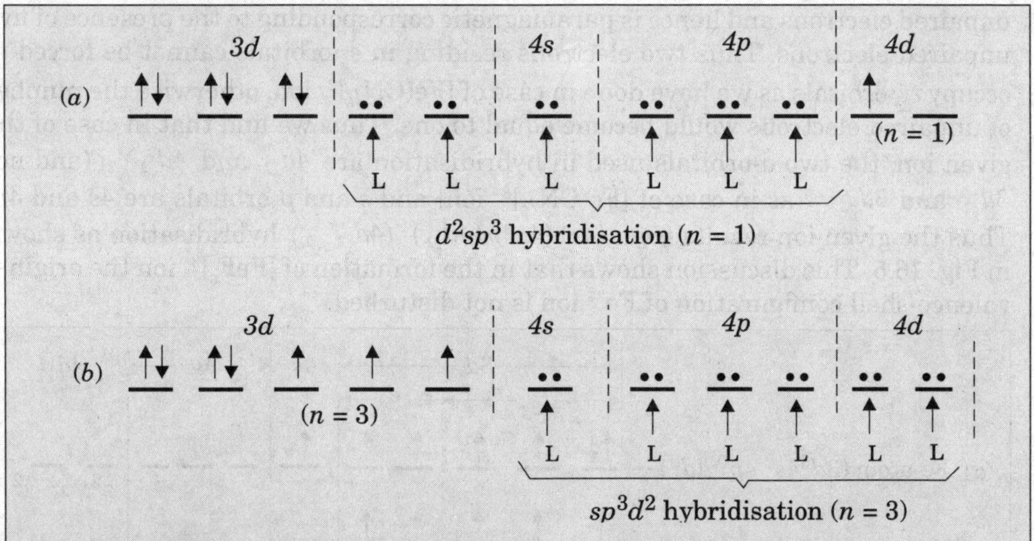


**Fig 16.4.** $d^2sp^3$ and $sp^3d^2$ hybridisation of $Co^{2+}$ ion in $[CoL_6]^{2+}$ (L = $NH_3$) and in $[CoL_6]^{4-}$ (L = $NO_2^-$, $CN^-$ )

## $sp^3d^2$ Hybridisation in Outer Orbital Octahedral Complexes

This type of hybridisation takes place in those octahedral complex ions which contain weak ligands. Weak ligands are those which cannot force the electrons of $d_{z^2}$ and $d_{x^2-y^2}$ orbitals ($e_g$ set) of the inner shell to occupy $d_{xy}$, $d_{yz}$ and $d_{zx}$ orbitals ($t_{2g}$ set) of the same shell. Thus in this hybridisation, $(n-1)d_{z^2}$ and $(n-1)d_{x^2-y^2}$ orbitals are not available for hybridisation. In place of these orbitals, we use $nd_{z^2}$ and $nd_{x^2-y^2}$ orbitals (These $d$-orbitals belong to the outer shell) and hence $sp^3d^2$ hybridisation can be represented as $ns \cdot np_x \cdot np_y \cdot np_z \cdot nd_{z^2} \cdot nd_{x^2-y^2}$. This hybridisation shows that all the six orbitals involved in hybridisation belong to the higher energy level (outer shell). This discussion shows that in case of octahedral complex ion of $3d$ transition sereis, $d$-orbitals used in hybridisation are $4d_{z^2}$ and $4d_{x^2-y^2}$ orbitals. Since two $d$-orbitals are from the outer shell ($i.e.$, $n$th shell), the octahedral complexes resulted from $sp^3d^2$ hybridisation are called **outer orbital octahedral complexes**. Since these complexes have comparatively greater number of unpaired electrons than the inner orbital octahedral complexes, these are also called **high spin or spin free octahedral complexes.**

Now let us discuss the structure of some octahedral complex ions of 3d transition series elements which are formed by $sp^3d^2$ hybridisation.

## 1. Hexafluroferrate (III) Ion, $[FeF_6]^{3-}$

In this ion, the coordination number of Fe is six and hence the given complex ion has octahedral geometry. Here iron is present as $Fe^{3+}$ whose valence shell electronic configuration is $3d^5 4s^0 4p^0$ or $t_{2g}^3 e_g^2 4s^0 4p^0$. Each of the five electrons is unpaired and hence $n = 5$. Magnetic properties of the given ion have also shown the ion has five unpaired electrons and hence is paramagnetic corresponding to the presence of five unpaired electrons. Thus two electrons residing in $e_g$ orbitals cannot be forced to occupy $t_{2g}$ orbitals as we have done in case of $[Fe(CN)_6]^{3-}$ ion, otherwise the number of unpaired electrons would become equal to one. Thus we find that in case of the given ion, the two d-orbitals used in hybridisation are $4d_{z^2}$ and $4d_{x^2-y^2}$ (and not $3d_{z^2}$ and $3d_{x^2-y^2}$ as in case of $[Fe(CN)_6]^{3-}$ ion] and *s* and *p* orbitals are 4*s* and 4*p*. Thus the given ion results from (4*s*) ($4p^3$) ($4d_{z^2}$) ($4d_{x^2-y^2}$) hybridisation as shown in Fig. 16.5. This discussion shows that in the formation of $[FeF_6]^{3-}$ ion the original valence-shell configuration of $Fe^{3+}$ ion is not disturbed.

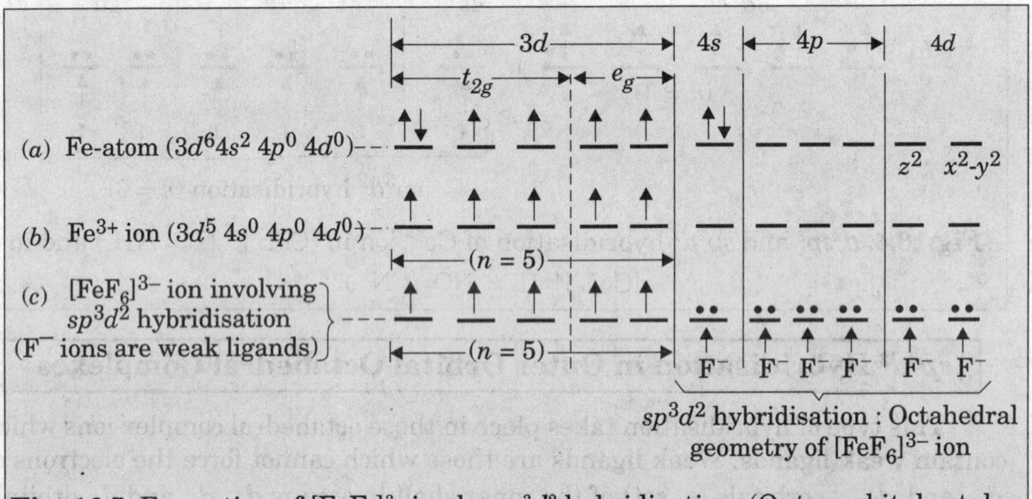


**Fig. 16.5.** Formation of $[FeF_6]^{3-}$ ion by $sp^3d^2$ hybridisation. (Outer-orbital octahedral complex ion).

### 2. $[Fe(H_2O)_6]^{2+}$ Ion

In this ion, the coordination number of Fe is six and hence the given complex ion has octahedral geometry. Here iron is present as $Fe^{2+}$ whose valence shell electronic configuration is $3d^6 4s^0 4p^0$ or $t_{2g}^4 e_g^2 4s^0 4p^0$. This configuration has four unpaired electrons. Magnetic properties of the given ion have also shown that the ion has four unpaired electrons and hence is paramagnetic corresponding to the presence of four unpaired electrons. Thus two electrons residing in $e_g$ orbitals cannot be forced to occupy $t_{2g}$ orbitals as we have done in case of $[Fe(CN)_6]^{3-}$ ion, otherwise the number of unpaired electrons would become equal to zero. Thus we find that in case of the given ion, the two d-orbitals used in hybridisation are $4d_{z^2}$ and $4d_{x^2-y^2}$

(and not $3d_{z^2}$ and $3d_{x^2-y^2}$ as in case of $[Fe(CN)_6]^{3-}$ ion) and *s* and *p* orbitals are 4*s* and 4*p*. Thus the given ion results from (4*s*) ($4p^3$) ($4d_{z^2}$) ($4d_{x^2-y^2}$) hybridisation as shown in Fig. 16.6. This discussion shows that in the formation of $[FeH_2O_6]^{2+}$ ion the original valence-shell configuration of $Fe^{2+}$ ion is not disturbed.

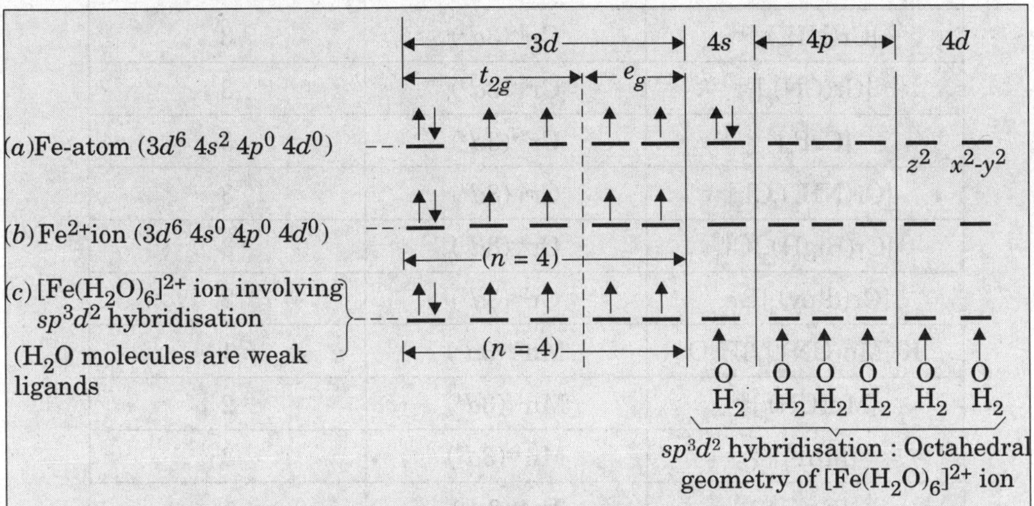


**Fig. 16.6.** Formation of $[Fe(H_2O)]^{2+}$ ion by $sp^3d^2$ hybridisation. (Outer-orbital octahedral complex ion).

**Examples of Inner Orbital Octahedral Complexes (*$d^2sp^3$* Hybridisation).**

Examples are given in Table 16.2.

**Table 16.2. Examples of inner orbital octahedral complexes (*$d^2sp^3$* hybridisation)**

| *Inner orbital octahedral complexes* | *$M^{n+}$ ion with $d^x$ configuration* | *No. of unpaired electrons in the complex* |
|---|---|---|
| $[ScF_6]^{3-}$ | $Sc^{3+}(3d^0)$ | 0 |
| $[Sc(H_2O)_6]^{3+}$ | $Sc^{3+}(3d^0)$ | 0 |
| $[Ti(H_2O)_6]^{2+}$ | $Ti^{2+}(3d^2)$ | 2 |
| $K_3[TiF_6]$ | $Ti^{3+}(3d^1)$ | 1 |
| $[TiX_6]^{2-}$ | $Ti^{4+}(3d^0)$ | 0 |
| $[V(acac)_3]^0$ | $V^{3+}(3d^2)$ | 2 |
| $NH_4[V(SO_4)_2].12H_2O$ | $V^{3+}(3d^2)$ | 2 |
| $[V(H_2O)_6]^{2+}$ | $V^{2+}(3d^3)$ | 3 |
| $[V(H_2O)_6]^{3+}$ | $V^{3+}(3d^2)$ | 2 |
| $[Cr(NH_3)_6]Br_3$ | $Cr^{3+}(3d^3)$ | 3 |

| | | |
|---|---|---|
| $[Cr(CO)_6]^0$ | $Cr^0(3d^54s^1 = 3d^6)$ | 0 |
| $[Cr(CN)_6]^{4-}$ | $Cr^{2+}(3d^4)$ | 2 |
| $[Cr(H_2O)_6]^{3+}$ | $Cr^{3+}(3d^3)$ | 3 |
| $[Cr(NH_3)_6]^{3+}$ | $Cr^{3+}(3d^3)$ | 3 |
| $[Cr(CN)_6]^{3-}$ | $Cr^{3+}(3d^3)$ | 3 |
| $[CrF_6]^{3-}$ | $Cr^{3+}(3d^3)$ | 3 |
| $[Cr(NH_3)_4Cl_2]^+$ | $Cr^{3+}(3d^3)$ | 3 |
| $[Cr(BigH)_3]Cl_3$ | $Cr^{3+}(3d^3)$ | 3 |
| $[Cr(dipy)_3]Br_2$ | $Cr^{2+}(3d^4)$ | 2 |
| $K_4[Mn(CN)_6].3H_2O$ | $Mn^{2+}(3d^5)$ | 1 |
| $[Mn(CN)_6]^{3-}$ | $Mn^{3+}(3d^4)$ | 2 |
| $[MnF_6]^{2-}$ | $Mn^{4+}(3d^3)$ | 3 |
| $K_3[Fe(CN)_6]$ | $Fe^{3+}(3d^5)$ | 1 |
| $[Fe(phen)_3]^{3+}$ | $Fe^{3+}(3d^5)$ | 1 |
| $K_4[Fe(CN)_6]$ | $Fe^{2+}(3d^6)$ | 0 |
| $[Fe(dipy)_3]^{2+}$ | $Fe^{2+}(3d^6)$ | 0 |
| $Na_2[Fe(CN)_5(NO^+)]$ | $Fe^{2+}(3d^6)$ | 0 |
| $[Co(NO_2)_6]^{4-}$ | $Co^{2+}(3d^7)$ | 1(in 4*d*) |
| $[Co(NH_3)_6]Cl_3$ | $Co^{3+}(3d^6)$ | 0 |
| $[Co(CN)_6]^{4-}$ | $Co^{2+}(3d^7)$ | 1(in 4*d*) |
| $[Co(C_2O_4)_3]^{3-}$ | $Co^{3+}(3d^6)$ | 0 |
| $[Co(NO_2)_6]^{3-}$ | $Co^{3+}(3d^6)$ | 0 |
| $[Co(CN)_6]^{3-}$ | $Co^{3+}(3d^6)$ | 0 |
| $[Co(NH_3)_4Cl_2]^+$ | $Co^{3+}(3d^6)$ | 0 |
| $[Co(H_2O)_6]^{3+}$ | $Co^{3+}(3d^6)$ | 0 |
| $[Co(NH_3)_6]^{2+}$ | $Co^{2+}(3d^7)$ | 1(in 4*d*) |

**Examples of outer orbital octahedral complexes ($sp^3d^2$ hybridisation).**

Examples are given in Table 16.3.

**Table 16.3. Examples of outer orbital octahedral complexes. ($sp^3d^2$ hybridisation)**

| *Outer orbital octahedral complexes* | *$M^{n+}$ ion with $d^x$ configuration* | *No. of unpaired electrons in the complex* |
|---|---|---|
| $[Cr(H_2O)_6]^{2+}$ | $Cr^{2+}(3d^4)$ | 4 |
| $[Cr(NH_3)_6]^{2+}$ | $Cr^{2+}(3d^4)$ | 4 |
| $[Mn(py)_6]Br_2$ | $Mn^{2+}(3d^5)$ | 5 |
| $[Mn(CNS)_6]^{4-}$ | $Mn^{2+}(3d^5)$ | 5 |
| $[Mn(acac)_3]^0$ | $Mn^{3+}(3d^4)$ | 4 |
| $[Mn(H_2O)_6]^{2+}$ | $Mn^{2+}(3d^5)$ | 5 |
| $[MnCl_6]^{3-}$ | $Mn^{3+}(3d^4)$ | 4 |
| $[MnF_6]^{3-}$ | $Mn^{3+}(3d^4)$ | 4 |
| $Na_3[FeF_6]$ | $Fe^{3+}(3d^5)$ | 5 |
| $[Fe(NH_3)_6]^{2+}$ | $Fe^{2+}(3d^6)$ | 4 |
| $[Fe(H_2O)_5(NO^+)]^{2+}$ | $Fe^+(3d^64s^1 = 3d^7)$ | 3 |
| $[Fe(en)_3]^{2+}$ | $Fe^{2+}(3d^6)$ | 4 |
| $[Fe(H_2O)_6]^{2+}$ | $Fe^{2+}(3d^6)$ | 4 |
| $[Fe(H_2O)_6]^{3+}$ | $Fe^{3+}(3d^5)$ | 5 |
| $[Fe(acac)_3]^0$ | $Fe^3(3d^5)$ | 5 |
| $[Co(H_2O)_6]^{2+}$ | $Co^{2+}(3d^7)$ | 3 |
| $Na_3[CoF_6]$ | $Co^{3+}(3d^6)$ | 4 |
| $[Co(NH_3)_6]^{2+}$ | $Co^{2+}(3d^7)$ | 3 |
| $[Co(en)_3]SO_4$ | $Co^{2+}(3d^7)$ | 3 |
| $[Co(py)_4Cl_2]^\circ$(violet) | $Co^{2+}(3d^7)$ | 3 |
| $[Ni(NH_3)_6]Cl_2$ | $Ni^{2+}(3d^8)$ | 2 |
| $[Ni(H_2O)_6]^{2+}$ | $Ni^{2+}(3d^8)$ | 2 |
| $[Cu(phen)_3](ClO_4)_2$ | $Cu^{2+}(3d^9)$ | 1 |
| $[Cu(H_2O)_6]^{2+}$ | $Cu^{2+}(3d^9)$ | 1 |
| $[Cu(NH_3)_6]^{2+}$ | $Cu^{2+}(3d^9)$ | 1 |
| $[Zn(H_2O)_6]^{2+}$ | $Zn^{2+}(3d^{10})$ | 0 |
| $[Zn(NH_3)_6]^{2+}$ | $Zn^{2+}(3d^{10})$ | 0 |

## Differences Between Inner Orbital and Outer Orbital Octahedral Complexes

The differences between inner orbital and outer orbital octahedral complexes are given in Table 16.4.

**Table 16.4. Differences between inner orbital and outer-orbital octahedral complexes.**

| *Inner orbital octahedral complexes (Low spin or spin paired octahedral complexes) ($d^2sp^3$ hybridisation)* | *Outer-orbital octahedral complexes (High spin or spin free octahedral complexes) ($sp^3d^2$ hybridisation)* |
|---|---|
| (*i*) These are formed by $d^2sp^3$ hybridisation i.e. in the formation of six $d^2sp^3$ hybrid orbitals, two $(n-1)d$ orbitals of $e_g$ set [i.e., $(n-1)d_{z^2}$] and $(n-1)$ $d_{x^2-y^2}$ orbitals], one *ns* and three *np* orbitals are used. $(n-1)d$ orbitals belong to the inner (penultimate) shell while *ns* and *np* orbitals belong to the outer (i.e., ultimate shell). Thus in the formation of inner orbital octahedral complexes of 3*d*-block elements, this hybridisation is represented as $(3d)^2(4s)$ $(4p)^3$ or simply as $d^2sp^3$. | (*i*) These are formed by $sp^3d^2$ hybridisation, i.e., in the formation of six $sp^3d^2$ hybrid orbitals, one *ns*, three *np* and two *nd* orbitals of $e_g$ set (i.e., $nd_{z^2}$ and $nd_{x^2-y^2}$ orbitals) are used. Thus all the orbitals belong to the ultimate (outer) shell. Thus in the formation of outer orbital octahedral complexes of 3*d*-block elements, this hybridisation is represented as $(4s)(4p)^3(4d)^2$ or simply as $sp^3d^2$ |
| (*ii*) These complexes are given by strong ligands | (*ii*) These complexes are given by weak ligands. |
| (*iii*) In these complexes strong ligands force the electrons residing in *d*-orbitals of $e_g$ set (*i.e.*, $d_{z^2}$ and $d_{x^2-y^2}$ orbitals) to pair up in *d*-orbitals of $t_{2g}$ set ($d_{xy}$, $d_{yz}$ and $d_{zx}$ orbitals). Hence these complexes generally have paired electrons (diamagnetic) or lesser number of unpaired electrons (weakly paramagnetic) | (*iii*) In these complexes weak ligands are not able to pair up $e_g$-electrons and hence the valence shell configuration of the metal in the free state and in the complex remains the same. Thus these complexes generally have larger number of unpaired electrons (strongly paramagnetic) |
| (*iv*) For obvious reasons, these complexes are also called spin paired or low spin (LS) complexes. These complexes are also called hyperligated complexes | (*iv*) These complexes are also called spin free or high spin (HS) complexes. These complexes are also called hypoligated complexes. |

## Tetrahedral Complexes ($sp^3$ hybridisation)

In tetrahedral complexes the central atom has C.N. = 4. These complexes are formed by $sp^3$ hybridisation of the central metal atom/ion.

Here we shall discuss the structure of the following complex compounds which have tetrahedral geometry.

### 1. $Ni(CO)_4$ Molecule

In this complex compound Ni is in zero oxidation state and has its valence-shell configuration as $3d^8 4s^2$. This compound has tetrahedral geometry which arises due to $sp^3$ hybridisation of Ni atom.

The magnetic studies of $Ni(CO)_4$ have indicated that this molecule is diamagnetic ($n = 0$), showing that the two 4s electrons are forced to pair up with 3d orbitals. This results in $sp^3$ hybridisation and Ni $(CO)_4$ molecule has tetrahedral structure. (See Fig. 16.7).

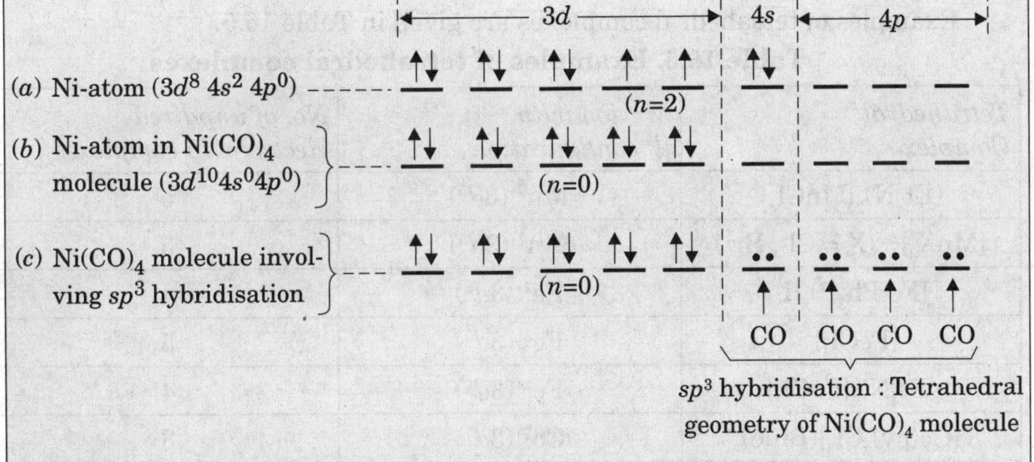


**Fig. 16.7.** $sp^3$ hybridisation of Ni-atom in $Ni(CO)_4$ molecule which has tetrahedral shape.

### 2. $[NiCl_4]^{2-}$ Ion.

This complex ion has $Ni^{2+}$ ion whose valence-shell configuration as $3d^8 4s^0$. Magnetic measurements reveal that the given ion is paramagnetic and has two unpaired electrons ($n = 2$). This is possible only when this ion is formed by $sp^3$ hybridisation and has tetrahedral geometry (See Fig. 16.8).

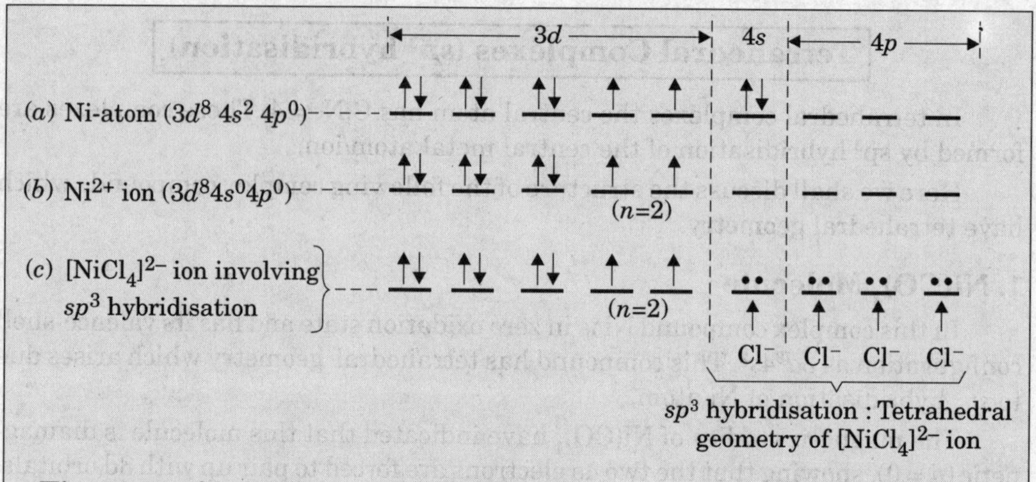


**Fig. 16.8.** $sp^3$ hybridisation of $Ni^{2+}$ ion in $[NiCl_4]^{2-}$ ion which has tetrahedral geometry.

## Examples of Tetrahedral Complexes (sp³ Hybridisation)

Examples of tetrahedral complexes are given in Table 16.5.

**Table 16.5. Examples of tetrahedral complexes**

| *Tetrahedral Complex* | *$M^{x+}$ ion with $d^x$ configuration* | *No. of unpaired electrons in complexes* |
|---|---|---|
| $(Et_4N)_2[MnCl_4]$ | $Mn^{2+}(3d^5)$ | 5 |
| $[MnX_4]^{2-}$ ($X = Cl^-, Br^-$) | $Mn^{2+}(3d^5)$ | 5 |
| $[Fe(Ph_3P)_2]I_2$ | $Fe^{2+}(3d^6)$ | 4 |
| $[FeCl_4]^-$ | $Fe^{3+}(3d^5)$ | 5 |
| $(Et_4N)_2[FeCl_4]$ | $Fe^{2+}(3d^6)$ | 4 |
| $[Co(py)_2Cl_2]$(Blue) | $Co^{2+}(3d^7)$ | 3 |
| $Cs_2[CoCl_4]$ | $Co^{2+}(3d^7)$ | 3 |
| $Hg[Co(SCN)_4]$ | $Co^{2+}(3d^7)$ | 3 |
| $(Et_4N)_2[NiCl_4]$ | $Ni^{2+}(3d^8)$ | 2 |
| $[Ni\ (CO)_4]^0$ | $Ni^0(3d^8 4s^2 = 3d^{10})$ | 0 |
| $[NiL_4]^{2+}$($L = H_2O, NH_3$) | $Ni^{2+}(3d^8)$ | 2 |
| $[NiCl_4]^{2-}$ | $Ni^{2+}(3d^8)$ | 2 |
| $[Ni(PPh_3)_2Cl_2]^0$ | $Ni^{2+}(3d^8)$ | 2 |
| $Cs_2[CuCl_4]$ | $Cu^{2+}(3d^9)$ | 1 |
| $[Cu(CN)_4]^{3-}$ | $Cu^{+}(3d^{10})$ | 0 |
| $[Zn\ (NH_3)_4]^{2+}$ | $Zn^{2+}(3d^{10})$ | 0 |
| $[Zn(CN)_4]^{2-}$ | $Zn^{2+}(3d^{10})$ | 0 |

## Square Planar Complexes ($dsp^2$ hybridisation)

Square planar complexes are those in which the C.N. of the central metal atom or ion is 4. These are formed as a result of $dsp^2$ hybridisation of the central metal atom or ion. In case of the square planar complexes of the metals of *1st* transition series $dsp^2$ hybridisation is $3d_{x^2-y^2}$ $4s$ $4p_x 4p_y$ hybridisation. The formation of some square planar complexes has been explained below.

**1. $[Ni(CN)_4]^{2-}$ Ion.**

In this complex ion, Ni is present as $Ni^{2+}$ ion whose V.S.E.C. is $3d^8 4s^0 4p^0$ ($n = 2$) shown at (*b*) of Fig 16.9. Since C.N. of $Ni^{2+}$ ion is 4, $[Ni(CN)_4]^{2-}$ ion can have tetrahedral ($sp^3$ hybridisation) or square planar ($dsp^2$ hybridisation) geometry.

(*a*) **Tetrahedral geometry.** This geometry is due to $sp^3$ hybridisation of $Ni^{2+}$ ion as shown at (c) of Fig 16.9. This geometry gives $n = 2$ and hence tetrahedral $[Ni(CN)_4]^{2-}$ ion is paramagnetic

(*b*) **Square planar geometry.** This geometry arises from $dsp^2$ hybridisation of $Ni^{2+}$ ion. For making $dsp^2$ hybridisation possible the two unpaired electrons present in $3d$ orbitals in the V.S. E.C. of $Ni^{2+}$ ion shown at (*b*) get paired against Hund's rule. The pairing of electrons takes place due to the fact that $CN^-$ ligands are strong ligands and hence as they approach the $Ni^{2+}$ ion, the unpaired electrons get paired. $dsp^2$ hybridisation gives $n = 0$ and hence square planar $[Ni(CN)_4]^{2-}$ ion is diamagnetic.

*Experimentally it has been found that $[Ni(CN)_4]^{2-}$ ion is diamagnetic ($n = 0$). Thus this ion is formed by $dsp^2$ hybridisation and has square planar geometry.*

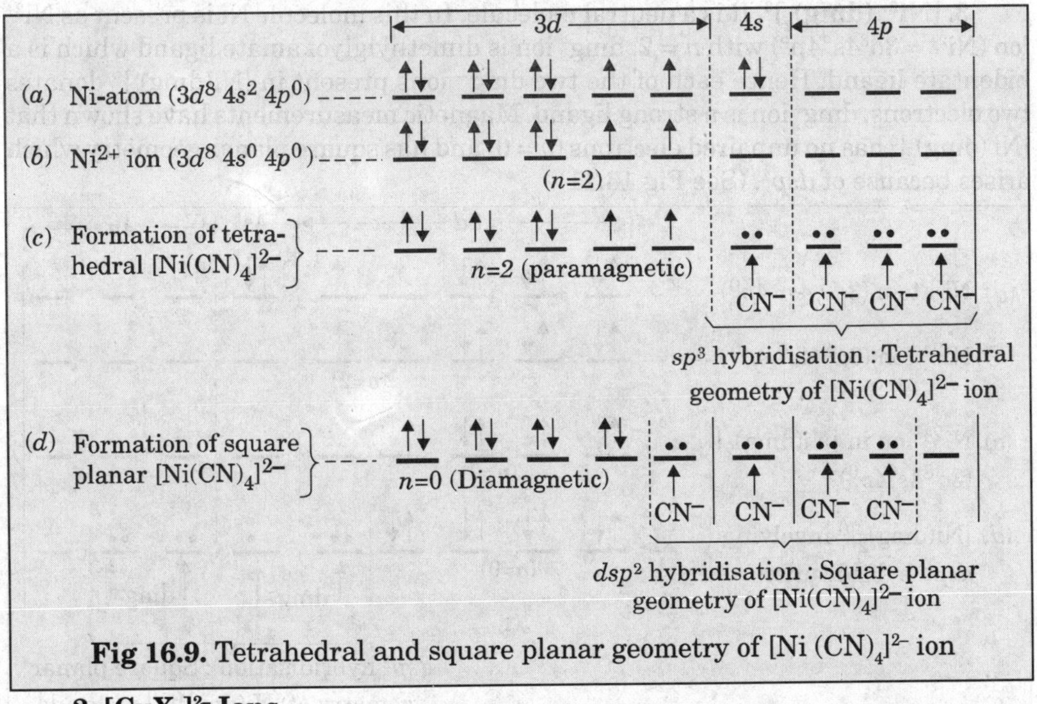


**Fig 16.9.** Tetrahedral and square planar geometry of $[Ni(CN)_4]^{2-}$ ion

**2. $[CuX_4]^{2-}$ Ions.**

The tetrahalide complexes of Cu(II) have two different geometries. In $(NH_4)_2[CuCl_4]$, $[CuCl_4]^{2-}$ ion has square planar geometry with $n = 1$ (paramagnetic) but in $Cs_2[CuCl_4]$ and $Cs_2[CuBr_4]$, $[CuX_4]^{2-}$ ions (X = Cl, Br) have a slightly squashed tetrahedral geometry. Square planar $[CuCl_4]^{2-}$ ions are yellow in colour while tetrahedral $[CuCl_4]^{2-}$ ions are orange in colour.

Let us explain the square planar geometry of $[CuCl_4]^{2-}$ ions present in $(NH_4)_2[CuCl_4]$. In this complex ion, Cu is present as $Cu^{2+}$ ion whose V.S.E.C is $3d^9 4s^\circ 4p^\circ$ ($n = 1$) as shown at (*b*) of Fig 16.10. We have said above that $[CuCl_4]^{2-}$ ion has square planar geometry which results from $dsp^2$ hybridisation of $Cu^{2+}$ ion. For $dsp^2$ hybridisation, one 3*d* orbital should be vacant. To get one 3*d* orbital vacant, one 3*d* electron gets excited to 4*p* orbital. Since square planar geometry of $[CuCl_4]^{2-}$ ion has one unpaired electron, this ion is paramagnetic.

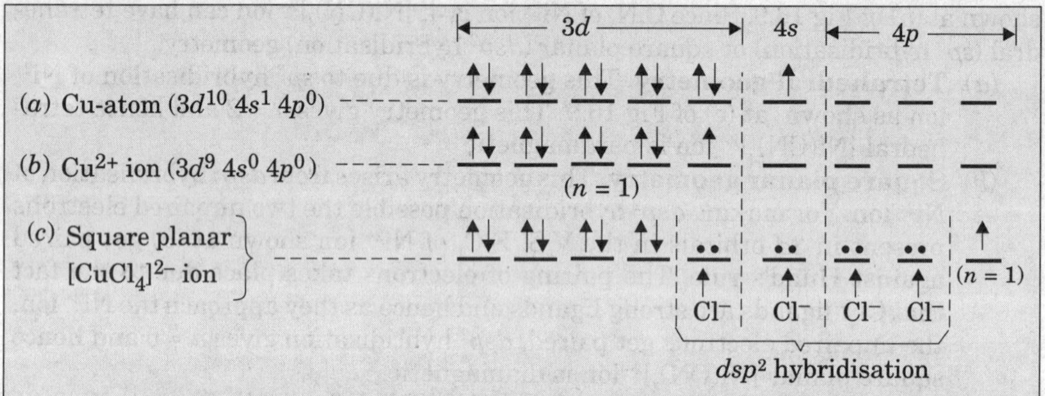


**Fig. 16.10.** Square planar geometry of $[CuCl_4]^{2-}$ ion in $(NH_4)_2$ $[CuCl_4]$

**3. $[Ni^{2+}(dmg)_2]^0$.** It is a neutral molecule. In this molecule Ni is present as $Ni^{2+}$ ion ($Ni^{2+} = 3d^8 4s^\circ 4p^\circ$) with $n = 2$. $dmg^-$ ion is dimethylglyoxamate ligand which is a bidentate ligand. Hence each of the two $dmg^-$ ions present in $[Ni(dmg)_2]^\circ$ donates two electrons. $dmg^-$ ion is a strong ligand. Magnetic measurements have shown that $[Ni(dmg)_2]^0$ has no unpaired electrons ($n = 0$) and has square planar geometry which arises because of $dsp^2$. (See Fig 16.11)

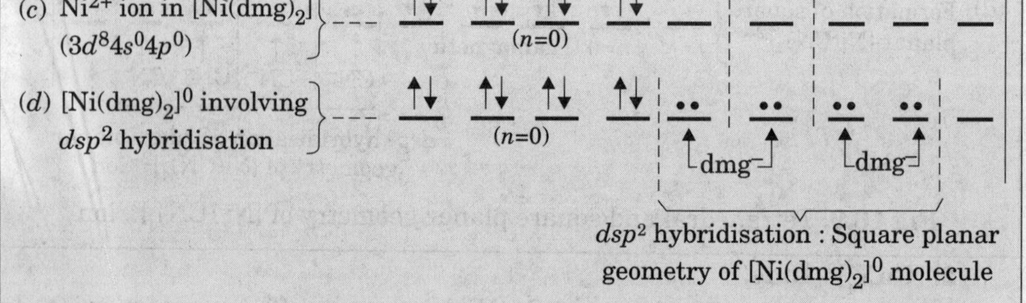


**Fig. 16.11.** $dsp^2$ hybridisation of $Ni^{2+}$ ion in $[Ni(dmg)_2]^0$ which has square planar geometry. Note that $dmg^-$ ion is a bidentate ligand and hence donates two electron pairs.

### 4. $[Cu(NH_3)_4]^{2+}$ Ion

In this ion, Cu is present on $Cu^{2+}$ ion whose V.S.E.C. is $3d^9 4s^0 4p^0$ with $n = 1$

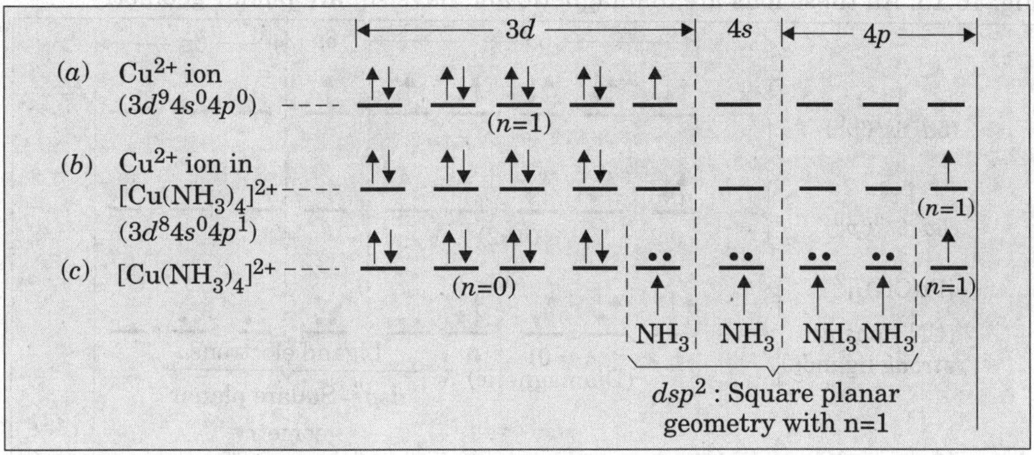


Now if the square planar geometry for $[Cu(NH_3)_4]^{2+}$ ion is supposed to be correct, the unpaired electron present in the higher energy $4p$ orbital ($dsp^2$ hybridisation) should be expected to be easily lost,i.e., $[Cu(NH_3)_4]^{2+}$ ion (Cu = +2) should be easily oxidised to $[Cu(NH_3)_4]^{3+}$ ion (Cu = +3) by losing the unpaired electron residing in $4p$ orbital.

$$[Cu^{2+}(NH_3)_4]^{2+} \xrightarrow[(-e^-)]{\text{oxidation}} [Cu^{3+}(NH_3)_4]^{3+} + e^-$$

However, experiments have shown that the oxidation of $[Cu(NH_3)_4]^{2+}$ to $[Cu(NH_3)_4]^{3+}$ as shown above does not occur. Then how to explain the geometry of $[Cu(NH_3)_4]^{2+}$ ion.

**Huggin's suggestion.** Huggin has suggested that $[Cu(NH_3)_4]^{2+}$ ion has square planar geometry and $Cu^{2+}$ ion is $sp^2d$ $[(4s)(4p)^2(4d)]$ hybridised as shown in Fig. 16.12. The unpaired electron resides in 3d orbital.

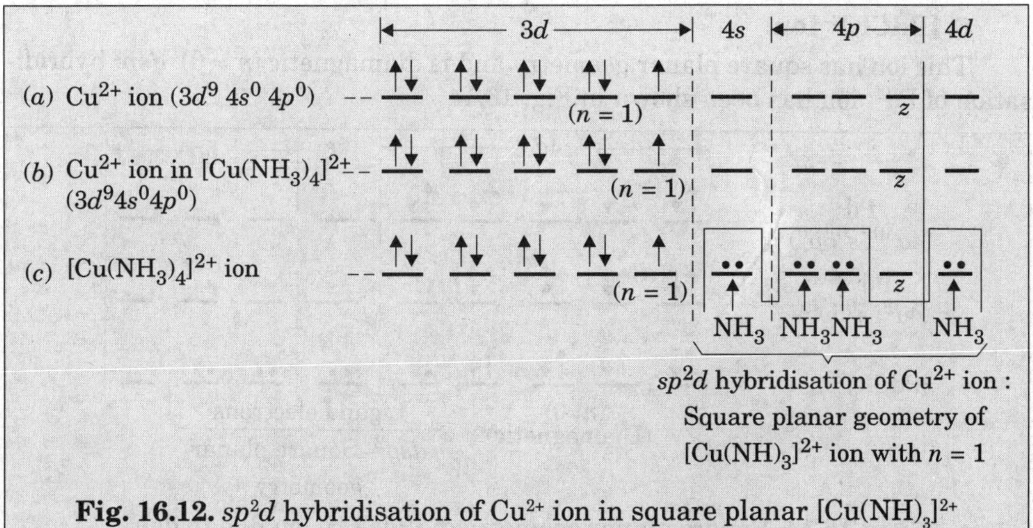


**Fig. 16.12.** $sp^2d$ hybridisation of $Cu^{2+}$ ion in square planar $[Cu(NH)_3]^{2+}$ ion with $n = 1$ (According to Huggin).

**5. $[Pt(CN)_4]^{2-}$, $[Pt(NH_3)_2Cl_2]^0$ and $[Pt(NH_3)_4]^{2+}$ Ions.**

*dsp*$^2$ hybridisation of $Pt^{2+}$ ion in these complexes has been shown below in Fig. 16.13. All these ions are diamagnetic and have square planar geometry.

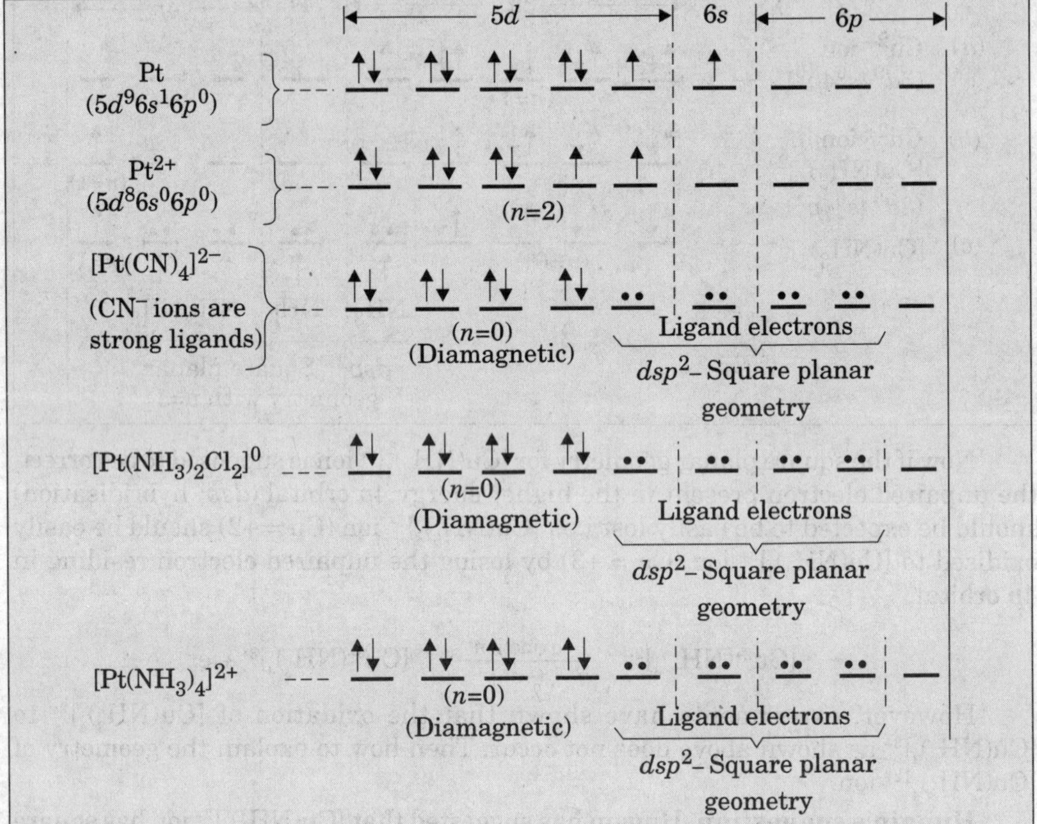


**Fig. 16.13.** Square planar geometry of $[Pt(CN)_4]^{2-}$, $[Pt(NH_3)_2Cl_2]^0$ and $[Pt(NH_3)_4]^{2+}$ resulted from dsp$^2$ hybridisation of $Pt^{2+}$ ion.

**6. $[PdCl_4]^{2-}$ Ion.**

This ion has square planar geometry and is diamagnetic ($n = 0$). *dsp*$^2$ hybridisation of $Pd^{2+}$ ion has been shown in Fig. 16.14

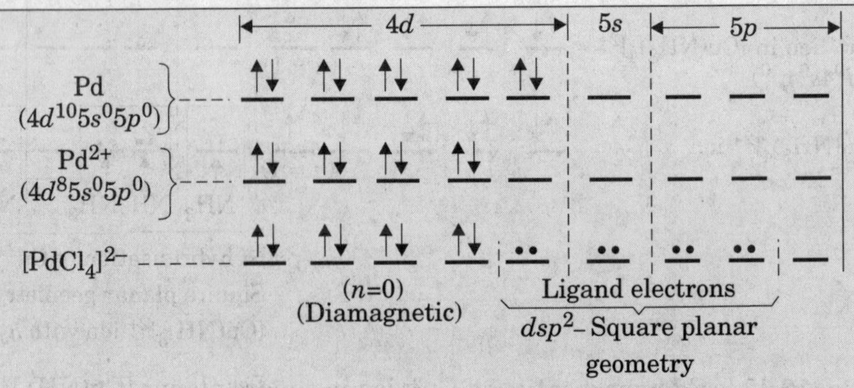


**Fig 16.14.** Square planar geometry of $[PdCl_4]^{2-}$ ion due to dsp$^2$ hybridisation of $Pd^{2+}$ ion

## Examples of Square Planar Complexes ($dsp^2$ Hybridisation).

Examples of square planar complexes are given in Table 16.6.

**Table 16.6. Square planar complexes**

| *Square planar complexes* | *Central metal atom* | *No. of unpaired electrons* |
|---|---|---|
| Fe(II) phthalocyanide | $Fe^{2+}(3d^6)$ | 2 |
| Co(II) phthalocyanide | $Co^{2+}(3d^7)$ | 1 |
| $[Ni(dmg)_2]^0$ | $Ni^{2+}(3d^8)$ | 0 |
| $[Ni(Et_3P)_2Cl_2]^0$ | $Ni^{2+}(3d^8)$ | 0 |
| $[Ni(CN)_4]^{2-}$ | $Ni^{2+}(3d^8)$ | 0 |
| $[Ni(BigH)_2]Cl_2$ | $Ni^{2+}(3d^8)$ | 0 |
| $[(Et_3P)\ NiBr_3]^0$ | $Ni^{3+}(3d^7)$ | 1 |
| $(NH_4)_2[CuCl_4]$ | $Cu^{2+}(3d^9)$ | 1(in 4*p*) |
| $[Cu(NH_3)_4]^{2+}(sp^2d)$ | $Cu^{2+}(3d^9)$ | 1 (in 3*d*) |

## Trigonal Bipyramidal Complexes ($dsp^3$ hybridisation)

Iron pentacarbonyl, Fe $(CO)_5$ is an important example of complex compound which has trigonal bipyramidal geometry.

In $Fe(CO)_5$ molecule, Fe is present as a neutral atom whose V.S. E.C. is $3d^6 4s^2 4p^0$ $(n = 4)$. Since C.N. of Fe is 5, Fe $(CO)_5$ molecule has trigonal bipyramidal geometry and hence is formed by $dsp^3$ hybridisation. This compound has been found to be diamagnetic $(n = 0)$

**Explanation.** Since $Fe(CO)_5$ has been found to be diamagnetic $(n = 0)$, all the eight electrons present in the V.S. of Fe are paired up in 3*d* orbitals to give $n = 0$. Thus the *V.S.* E.C. of Fe atom in Fe $(CO)_5$ is $3d^8 4s^0 4p^0$ as shown at (*b*) of Fig 16.15. The pairing of electrons makes one 3*d* orbital and one 4*s* orbital vacant. Three 4*p* orbitals are already vacant. These five vacat orbitals mix together and form five $dsp^3$ hybrid orbitals. Thus in the formation of $Fe(CO)_5$, Fe atom undergoes $dsp^3$ hybridisation and hence Fe $(CO)_5$ molecule has trigonal bipyramidal geometry.

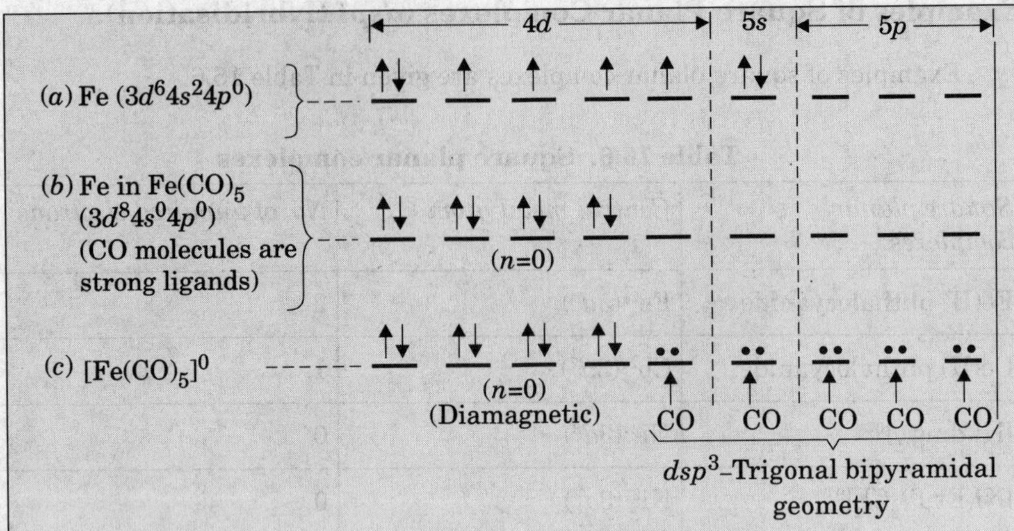


**Fig. 16.15.** *$dsp^3$* hybridisation of Fe-atom in Fe $(CO)_5$ molecule

## Application of VBT

### 1. To Find Out the Number of Unpaired Electrons in a Given Complex Ion

The number of unpaired electron (*n*) present in a given complex ion depends on the geometry of the complex ion and the type of hybridisation undergone by the central metal atom or ion. This is evident from Table 16.7 in which we have shown a relation between the number of unpaired electrons and geometries of the complex ions.

**Table 16.7. Relation between the number of unpaired electrons and geometries of complex ions.**

| *Configuration of the central metal atom / ion* | *Number of unpaired electrons* | | | |
|---|---|---|---|---|
| | *Inner orbital octahedral complex ion ($d^2sp^3$ hybridisation)* | *Outer orbital octahedral complex ion ($sp^3d^2$ hybridisation)* | *Tetrahedral complex ion ($sp^3$)* | *Square planar complex ion ($dsp^2$ or $sp^2d$)* |
| $d^1$ | 1 | 1 | 1 | 1 |
| $d^2$ | 2 | 2 | 2 | 2 |
| $d^3$ | 3 | 3 | 3 | 3 |
| $d^4$ | 2 | 4 | 4 | 4 |
| $d^5$ | 1 | 5 | 5 | 3 |
| $d^6$ | 0 | 4 | 4 | 2 |
| $d^7$ | 1 | 3 | 3 | 1 |
| $d^8$ | – | 2 | 2 | 0 |
| $d^9$ | – | 1 | 1 | 1 |

## 2. To Predict the Type of Hybridisation in Octahedral Complexes.

This can be done with the help of magnetic moment values (μ) of the complexes.

## 3. To Predict the Geometry of a Complex

This can also be done with the help of magnetic moment values of the complexes.

## Limitations of VBT

Also see chapter on *Magnetic properties of transition metal complexes*

(*i*) VBT cannot account for the relative stabilities for different shapes and different coordination numbers in metal complexes, *e.g.*, it cannot explain satisfactorily as to why Co(+2) ($d^7$ system) forms both octahedral and tetrahedral complexes while Ni (+2) ($d^8$ system) rarely forms tetrahedral complexes. Similarly, Fe (+3) ($d^5$ system) forms both tetrahedral and octahedral complexes while Cr (+3) ($d^3$ system) gives only octahedral complexes.

(*ii*) VBT cannot explain as to why Cu (+2) forms only one distorted octahedral complexes even when all the six ligands are identical.

(*iii*) This theory cannot account for the relative rates of reactions of analogous metal complexes, *e.g.* $[Mn(phen)_3]^{2+}$ dissociates instantaneously in acidic aqueous solution while $[Fe(phen)_3]^{2+}$ dissociates at a *slow rate*.

(*iv*) The classification of metal complexes on the basis of their magnetic behaviour into *covalent* (*inner-orbital*) and *ionic* (*outer-orbital*) complexes is not satisfactory and is often misleading, e.g., many of the so-called ionic complexes like $[Fe^{3+}(acac)_3]^0$ have properties like high votality, high solubility in organic solvents etc. which are characteristics of covalent compounds. Also it is known that the aquo ion, $[Cr(H_2O)_6]^{3+}$ (*p*Ka = 3.8 at 25°C) is a much weaker acid than $[Fe(H_2O)_6]^{3+}$ (*p*Ka = 2.2) and $[Co\ (H_2O)_6]^{3+}$ (*p*Ka = 0.7) ions, despite the fact that only $[Fe(H_2O)_6]^{3+}$ is of outer-orbital type with ionic bonding while the other two aquo ions are of outer-orbital type with covalent bonding. Both $[Fe(H_2O)_6]^{3+}$ and $[Co(H_2O)_6]^{3+}$ ions exchange $H_2O^{18}$ in solvent instantaneously while $[Cr(H_2O)_6]^{3+}$ ion exchanges $H_2O^{18}$ very slowly. The acidity of an aquo metal ion is expected to increase with the increase of covalent character of the metal—oxygen bond. But neither the sequence of acidities nor the exchange rates of the above mentioned three aquo metal ions are in the order expected from their classification into ionic and covalent types on the basis of Pauling's VBT.

(*v*) VBT fails to explain the finer details of magnetic properties including the magnitude of the orbital contribution to the magnetic moments, e.g., although both tetrahedral ($sp^3$ hybridisation) and outer-orbit octahedral ($sp^3d^2$ hybridisation) complexes of Co (+2) ($d^7$ system) have three unpaired electrons and are,therefore, expected to have μ value equal to 3.87 B.M.; the tetrahedral complexes generally have μ value in the range of 4.4.–4.8 B.M. while the octahedral complexes have still higher vale of μ in the range of 4.7 – 5.2 B.M. The increase in the value of μ is due to the orbital contribution. Similar is the case with tetrahedral and octahedral complexes of Ni (+2) ($d^8$ system).

Thus we see that the magnetic moment value of the complexes of certain ions (e.g., $Co^{2+}$, $Ni^{2+}$, etc.) are much higher than those expected by spin-only formula. VBT can not explain the enhanced values of magnetic moment.

(*vi*) VBT cannot interpret the spectra (colour) of the complexes.

(*vii*) This theory does not predict or explain the magnetic behaviour of complexes. This theory only predicts the number of unpaired electrons. Its prediction even for the number of unpaired electrons and their correlation with stereoclemistry is misleading. For example, it was assumed for long that all square planar complexes of Ni (+2) ($d^8$ system) formed by $dsp^2$ hybridisation were diamagnetic ($n = 0$), while all tetrahedral complexes of Ni (+2) formed by $sp^3$ hybridisation were paramagnetic due to the presence of two unpaired electrons ($n = 2$). X-ray study of 4-coordinate Ni (+2) complexes has shown that one and the same complex can be obtained in both a *paramagnetic form* (having blue colour) and a *diamagnetic form* (having yellow colour). VBT has no explanation for it. VBT cannot explain the temperature dependent paramagnetism of the complexes.

(*viii*) VBT cannot give any explanation for the order of reactivities of the inner-orbital inert complexes of $d^3$, $d^4$, $d^5$ and $d^6$ ions and of the observed differences in the energies of activation in a series of similar complexes.

(*ix*) *Formation of low spin (LS) and high spin (HS) complexes by a given metal in the same oxidation state.* Valence bond theory cannot explain why some complexes formed by a metal ion in a particular oxidation state are LS while some other complexes of the same metal ion in the same oxidation state are HS. For example $Co^{3+}$ ion forms LS octahedral complexes (inner orbital - $d^2sp^3$ hybridisation) with $6NH_3$ molecules, $[Co^{3+}(NH_3)_6]^{3+}$ ($n = 0$) but forms HS octahedral complexes (outer orbital -$sp^3d^2$ hybridisation), with $6F^-$ ions, $[Co^{3+}F_6]^{3-}$ ($n = 4$). Thus $[Co(NH_3)_6]^{3+}$ is diamagnetic ($n = 0$) and $[CoF_6]^{3-}$ is paramagnetic ($n = 4$)

(*x*) *Magnetic behaviour of square planar complexes of Ni (II).* Some square planar complexes of Ni (II) are diamagnetic while some other are paramagnetic. V.S.E.C. of $Ni^{2+}$ ion is $3d^84s^04p^0$ which has two unpaired electrons ($n = 2$) in $3d$ orbitals as shown at (*b*) of Fig 16.16.

(*a*) *Diamagnetic square planar complexes.* In the formation of diamagnetic square planar complexes ($dsp^2$ hybridisation) the two unpaired electrons of $3d$ orbitals are paired ($n = 0$) and the hybridisation of $Ni^{2+}$ ion is $dsp^2$.

(*b*) *Paramagnetic square planar complexes.* In the formation of paramagnetic square planar complexes ($dsp^2$ hybridisation) one $3d$ electron is excited to $4p$ orbital and the hybridisation of $Ni^{2+}$ ion again $dsp^2$. Thus the paramagnetic square planar complexes of $Ni^{2+}$ ion have two unpaired electrons.

Valence bond theory cannot explain why excitation of one $3d$ electron into $4p$ orbital takes place to get $dsp^2$ hybridisation (paramagnetic square planar complexes) but such an excitation does not occur in diamagnetic square planar complexes to get $dsp^2$ hybridisation.

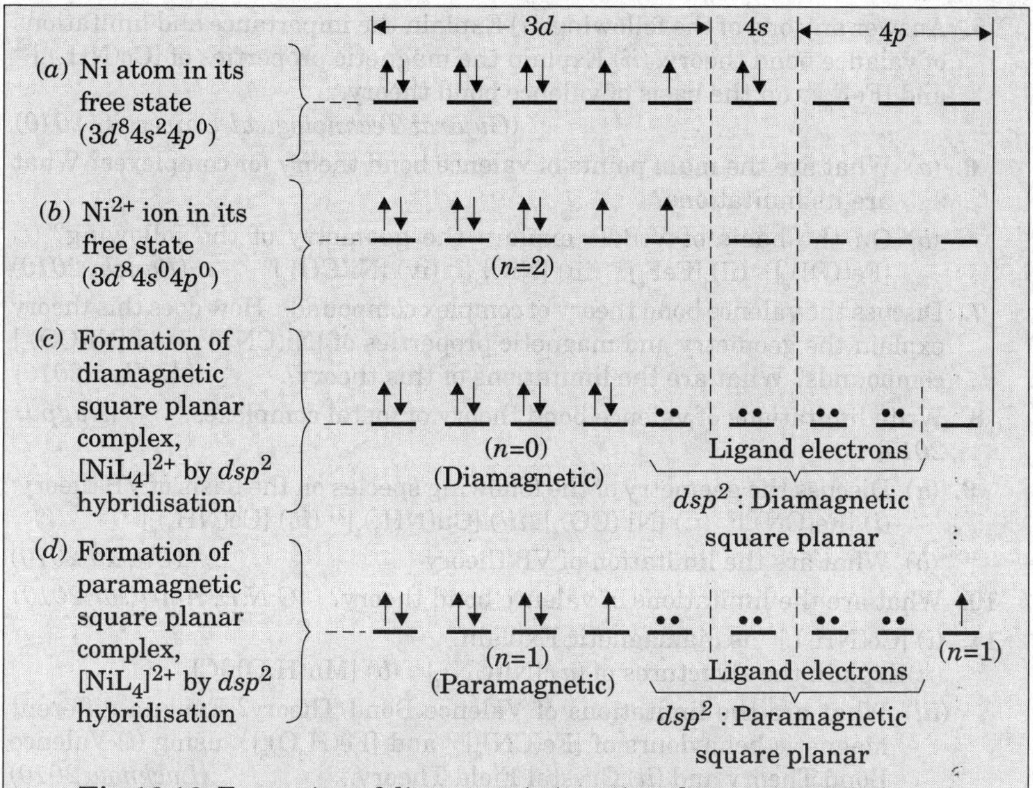


**Fig 16.16.** Formation of diamagnetic ($n$ = 0) and paramagnetic ($n$ = 2) square planar complexes of $Ni^{2+}$ ion.

(*xi*) ***Magnetic behaviour of square planar*** $[Cu(NH_3)_4]^{2+}$ ***ion.*** This ion is paramagnetic due to the presence of one unpaired electron residing in 4*p* orbital (not in 3*d* orbital) and has square planar geometry ($dsp^2$ hybridisation). To get $dsp^2$ hybridisation, one 3*d* electron of $Cu^{2+}$ ion ($Cu^{2+} = 3d^9 4s^2 4p^0$) is excited to 4*p* orbital. Valence bond theory cannot explain as to why the excitation of 3d electron to 4p orbital takes place.)

## UNIVERSITY QUESTIONS

1. Explain the structure of $K_4[Fe(CN)_6]$ on the basis of VBT [*Osmania 2010*]
2. What are inner orbital and outer orbital octahedral complexes? Discuss giving one example for each type. [*Himachal Pradesh 2010*]
3. (*a*) Give the limitations of VBT
   (*b*) $[CoF_6]^{3-}$ is paramagnetic while $[Co(NH_3)_6]^{3+}$ is diamagnetic. Explain. [*Kanpur 2010, Barktullah 2010, H.N. Bahuguna Garhwal 2010*]
4. On the basis of VBT explain the magnetic properties of $[Fe(CN)_6]^{3-}$ ion (*Bangalore 2010*)

**5.** Answer any one of the following: (*i*) Explain the importance and limitations of valance bond theory. (*ii*) Explain the magnetic properties of $[Co(NH_3)_6]^{3+}$ and $[FeF_6]^{3-}$ on the basis of valence bond theory.
(*Gujarat Technological University 2010*)

**6.** (*a*) What are the main points of valence bond theory for complexes? What are its limitations?
(*b*) On the basis of V.B.T. explain the geometry of the following? (*i*) $[Fe(CN)_6]^{3-}$ (ii) $[FeF_6]^{3-}$ (iii) $[NiCl_4]^{2-}$ (iv) $[Ni(CO)_4]^0$ (*Ranchi 2010*)

**7.** Discuss the valence bond theory of complex compounds. How does this theory explain the geometry and magnetic properties of $[Ni(CN)_4]^{2-}$ and $[Ni(CO)_4]$ compounds? What are the limitations of this theory? [*Andhra 2010*]

**8.** Write limitations of valence bond theory of metal complexes. [*Nagpur 2010*]

**9.** (*a*) Discuss the geometry of the following species on the basis of VB theory: (*i*) $[Fe(CN)_6]^{3-}$ (*ii*) $[Ni (CO)_4]$ (*iii*) $[Cu(NH_3)_4]^{2+}$ (*iv*) $[Co(NH_3)_6]^{3+}$
(*b*) What are the limitation of VB theory (*Utakal 2010*)

**10.** What are the limitations of valence bond theory? (*G.N.D. Amritsar 2010*)

**11.** (*i*) $[Co(NH_3)_6]^{3+}$ is diamagnetic Explain.
(*ii*)Explain the structures of (*a*) $[Ni(CN)_4]^{2-}$ (*b*) $[Mn(H_2O)_6]Cl_2$
(*iii*) What are the limitations of Valence Bond Theory? Explain different magnetic behaviours of $[Fe(CN)_6]^{3-}$ and $[Fe(H_2O)_6]^{3+}$ using (*i*) Valence Bond Theory and (*ii*) Crystal Field Theory. (*Lucknow 2010*)

**12.** Explain why nickel does not form inner orbital octahedral complexes.
[*Rajasthan 2010*]

**13.** What are the postulates of VBT? Based on this theory explain the shape and magnetic behaviour of : (*a*) $[Ni(CN)_4)]^{2-}$ (*b*) $[NiCl_4]^{2-}$ (*c*) $[Fe (CN)_6]^{4-}$
[*Madras April 2011*]

**14.** (*a*) Give important limitations of VBT of coordination compounds.
(*b*) What are inner and outer orbital octahedral complexes? Explain with their formation with suitable examples on the bases VBT. (*Nagpur 20012*)

**15.** Discuss the nature of bonding and structure of $K[Fe(CN)_6]$
(*Ch. Charan Singh, Meerut 2012*)

**16.** (*i*) Explain the following.
(*a*) $[Fe(CN)_6]^{3-}$ has only one unpaired electron while $[FeF_6]^{3-}$ has has five unpaired electrons
(*b*) $[NiCl_4]^{2-}$ is paramagnetic
(*c*) Magnetic moment of $[FeF_6]^{3-}$ is 5.917 B.M and and that of $[Fe(CN)_6]^{3-}$ is 1.732 BM
(*ii*) Calculate the number of unpaired electrons in $[Co(NH_3)_6]^{3+}$ and $[Ni(CN)_4]^{2-}$ ions (*Lucknow 2013*)

# 17 CHAPTER Crystal Field Theory (CFT) of Octahedral Complexes

This theory advanced by Breath and Van Vleck was originally applied mainly to ionic crystals and is, therefore, called **Crystal Field Theory (CFT).** It was not until 1952 that Orgel popularised its use for inorganic chemists.

## Important Features of CFT

(*i*) The central metal cation is surrounded by ligands which contain one or more lone pairs of electrons.

(*ii*) The ionic ligands (e.g., $F^-$, $Cl^-$, $CN^-$ etc.) are regarded as negative point charges (also called point charges) and the neutral ligands (e.g., $H_2O$, $NH_3$, etc.) are regarded as point dipoles or simply dipoles, i.e., according to this theory neutral ligands are dipolar. If the ligand is neutral, the negative end of this ligand dipole is oriented towards the metal cation.

(*iii*) The CFT does not provide for electrons to enter the metal orbitals. Thus the metal ion and the ligands do not mix their orbitals or share electrons, i.e., it does not consider any orbital overlap.

(*iv*) According to CFT, the bonding between the metal cation and ligand is not covalent but it is regarded as purely electrostatic or coulombic attraction between positively charged (*i.e.*, cation) and negatively-charged (*i.e.*, anions or dipole molecules which act as ligands) species. Complexes are thus presumed to form when centrally situated cations electrically attract ligands which may be either anions or dipole molecules. The attraction between the cations and the ligands is because the cations are positively charged and the anions are negatively charged and the dipole molecules, as well, can offer their negatively incremented ends for such electrostatic attractions.

## Grouping of Five *d*-Orbitals into $t_{2g}$ and $e_g$ Sets of Orbitals

On the basis of the orientation of the lobes of the five-*d*-orbitals with respect to coordinates these have been grouped into the following two sets.

**1. $e_g$ set of orbitals: $d_{z^2}$ and $d_{x^2-y^2}$ orbitals.** This set consists of two orbitals *which have their lobes along the axes* and hence are called *axial orbitals*. Quite obviously these are $d_{z^2}$ and $d_{x^2-y^2}$ orbitals. Group theory calls these orbitals $e_g$ orbitals in which *e* refers to *doubly degenerate set*.

**2. $t_{2g}$ set of orbitals: $d_{xy}, d_{yz}, d_{zx}$ orbitals.** This set includes three orbitals *whose labes lie between the axes* and are called *non-axial orbitals*. Group theory calls these orbitals $t_{2g}$ orbitals wherein *t* refers to *triply degenerate set.*

## Crystal Field Splitting of d-Orbitals in Octahedral Complexes

In case of free metal ion all the five *d*-orbitals are degenerate, *i.e.*, these have the same energy. Now let us consider an octahedral complex, $[ML_6]^{n+}$ in which the central metal cation, $M^{n+}$ is placed at the centre of the octahedron and is surrounded by six ligands which reside at the six corners of the octahedron as shown in Fig. 17.1. The three axes, viz. *x*-, *y*-, and *z*-axes which point along the corners have also been shown.

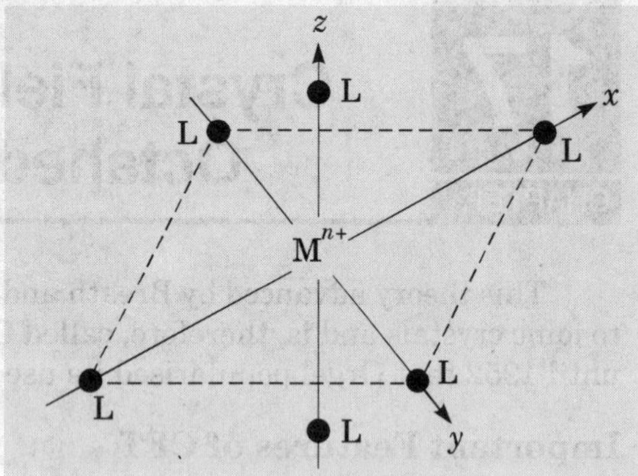


**Fig. 17.1.** Position of the central metal cation, $M^{n+}$ and six ligands, L's in an octahedral complex, $[ML_6]^{n+}$.

Now suppose both the ligands on each of the three axes are allowed to approach towards the metal cation, $M^{n+}$ from both the ends of the axes. In this process the electrons in *d*-orbitals of the metal cation are repelled by negative point charge or by the negative end of the dipole of the ligands. (Remember CFT regards the ionic ligands as negative point charges and neutral ligands as dipoles). This repulsion will raise the energy of all the five *d*-orbitals. If all the ligands approaching the central cation are at an equal distance from each of the *d*-orbitals (i.e., the ligand field is spherically symmetrical), the energy of each of five *d*-orbitals will raise by the same amount, *i.e.*, all the *d*-orbitals will still remain degenerate, although they will have now higher energy than before. This is only a *hypothetical situation. Since the lobes of the two $e_g$ orbitals (i.e., $d_{z^2}$ and $d_{x^2-y^2}$ orbitals) lie directly in the path of the approaching ligands, the electrons in these orbitals experience greater force of repulsion than those in three $t_{2g}$ orbitals* (i.e., $d_{xy}$, $d_{yz}$ and $d_{zx}$ orbitals) *whose lobes are directed in space between the path of the approaching ligands* i.e., the energy of $e_g$ orbitals is increased while that of $t_{2g}$ is decreased (Remember: *greater is the repulsion, greater is the increase in energy*). Thus we find that under the influence of approaching ligands, the five *d*-orbitals which were originally degenerate in the free metallic cation are now split (or resolved) into two levels *viz.*, $t_{2g}$ level which is triply degenerate and is of lower energy, and $e_g$ level which is doubly degenerate and is of higher energy (see Fig. 17.2). In other words the degeneracy of the five d-orbitals is removed under the influence of the ligands. *The separation of five d-orbitals of the metal ion into two sets having different energies is called crystal field splitting or energy level splitting.* This concept of crystal field splitting makes the basis of CFT.

The energy gap between $t_{2g}$ and $e_g$ sets is denoted by $\Delta_0$ or 10Dq where 0 in $\Delta_0$ indicates an octahedral arrangement of the ligands round the central metal cation. This energy difference arises because of the difference in electrostatic field exerted by the ligands on $t_{2g}$ and $e_g$ sets of orbitals of the central metal cation. $\Delta_0$ or 10Dq is called *crystal field splitting energy*. With the help of simple geometry it can be shown that the energy of $t_{2g}$ orbitals is 0.4 $\Delta_0$(= 4Dq) less than that of hypothetical degenerate *d*-orbitals (No splitting sate shown by dotted line in Fig. 17.2) and, hence that of $e_g$ orbitals is 0.6$\Delta_0$ (= 6Dq) *above* that of the hypothetical degenerate *d*-orbitals. Thus, we find that $t_{2g}$ set *loses* an energy equal to 0.4 $\Delta_0$ (4Dq) while $e_g$ set *gains* an energy equal to 0.6$\Delta_0$ (= 6Dq). In Fig. 17.2 the loss and gain in energies of $t_{2g}$ and $e_g$ orbitals is shown by negative (–) and positive (+) signs respectively. $\Delta_0$ is generally measured in $cm^{-1}$.

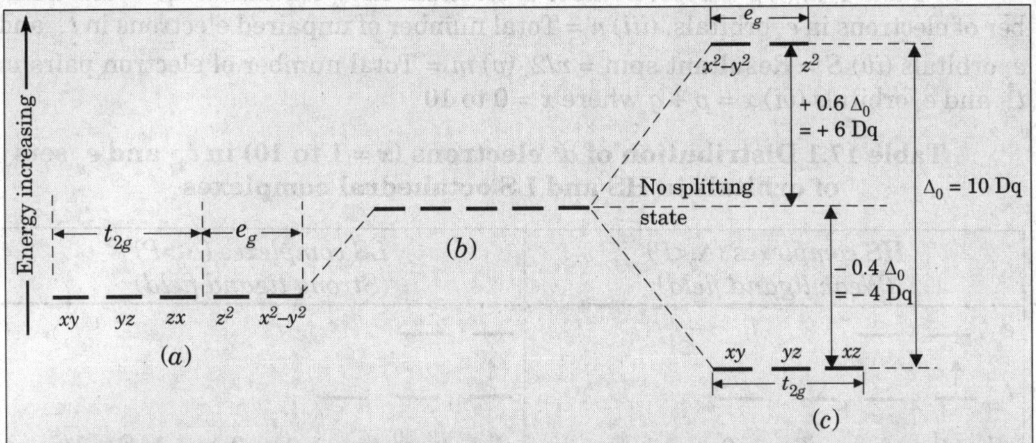


**Fig. 17.2.** Splitting of five d-orbitals in an octahedral complex. (*a*) Five degenerate *d*-orbitals on the central metal cation which are free from any ligand field. (*b*) Hypothetical degenerate *d*-orbitals at a higher energy level (*c*) Splitting of *d*-orbitals into $t_{2g}$ and $e_g$ orbitals under the influence of six ligands in octahedral complex.

## Distribution of $d^x$ Electrons in $t_{2g}$ and $e_g$ Orbitals in Octahedral Complexes ($x$ = 0 to 10)

Distribution of $d^x$ electrons ($x$ = 0 to 10) in $t_{2g}$ and $e_g$ orbitals in octahedral complexes is represented by $t_{2g}^p e_g^q$ configuration in which $p$ and $q$ are the electrons present in $t_{2g}$ and $e_g$ orbitals respectively. Obviously $p + q = x$.

**A. Distribution of $d^x$ electrons in high spin (HS) octahedral complexes**

HS octahedral complexes are also called *spin free complexes*. Since these complexes contain weak field ligands, these complexes are also called *weak field or low field complexes*. Since the ligands in HS complexes are weak, the energy difference between $t_{2g}$ and $e_g$ sets of orbitals in HS complexes is relatively small and hence all the five *d*-orbitals of $t_{2g}$ and $e_g$ sets can be supposed to be of the same energy (degenerate orbitals). Consequently the distribution of $d^x$ electrons in $t_{2g}$ and $e_g$ orbitals in HS complexes *takes place according to Hund's rule which states that electrons will pair up only when each of the five d-orbitals is at least singly filled*. Thus in weak(er) field the first three electrons enter $t_{2g}$ level, *4th* and *5th* electrons (two electrons) go to $e_g$ set, *6th*, *7th* and *8th* electrons (three electrons) occupy $t_{2g}$ set and *9th* and *10th* electrons (last two electrons) go to $e_g$ set. (See Table 17.1).

**B. Distribution of $d^x$ electrons in low spin (LS) octahedral complexes.**

LS octahedral complexes are also called *spin paired complexes*. Since these complexes contain *strong field ligands, these complexes are also called strong field or high field complexes*. Since the ligands in LS complexes are strong, the energy difference between $t_{2g}$ and $e_g$ orbitals is relatively large and hence all the five d-orbitals are not of the same energy. As a result, the distribution of $d^x$ electrons in $t_{2g}$ and $e_g$ orbitals does not take place in accordance with Hund's rule. This in LS complexes the first six electrons in $d^x$ ion go to $t_{2g}$ orbitals and the remaining four electrons of $d^x$ ion occupy $e_g$ orbitals (See Table 17.1)

In Table 17.1 (*i*) $p$ = Total number of electrons in $t_{2g}$ orbitals (*ii*) $q$ = Total number of electrons in $e_g$ orbitals, (*iii*) $n$ = Total number of unpaired electrons in $t_{2g}$ and $e_g$ orbitals (*iv*) $S$ = Resultant spin = $n/2$, (*v*) $m$ = Total number of electron pairs in $t_{2g}$ and $e_g$ orbitals (*vi*) $x = p + q$ where $x$ = 0 to 10

**Table 17.1 Distribution of $d^x$ electrons ($x$ = 1 to 10) in $t_{2g}$ and $e_g$ sets of orbitals in HS and LS octahedral complexes**

| *HS complexes* ($\Delta_0<P$) *(Weak ligand field)* | *LS complexes* ($\Delta_0>P$) *(Strong ligand field)* |
|---|---|
|  $d^1 = t_{2g}^1 e_g^{\circ}$ ($p = 1, q = 0, n = 1, S =$ 1/2, $m = 0$) |  $d^1 = t_{2g}^1 e_g^0$ ($p = 1, q = 0, n = 1, S = 1/2, m = 0$) |
| 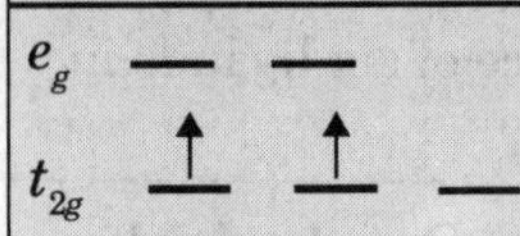 $d^2 = t_{2g}^2 e_g^{\circ}$ ($p = 2, q = 0, n = 2, S = 1, m = 0$) | 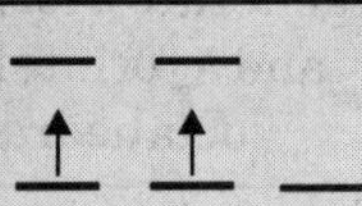 $d^2 = t_{2g}^2 e_g^0$($p = 2, q = 0, n = 2, S = 1, m = 0$) |
| 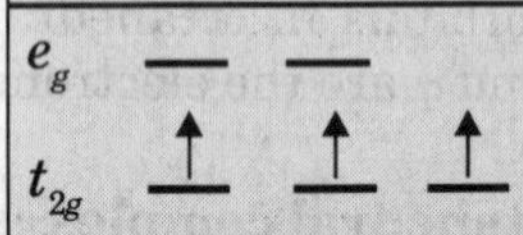 $d^3 = t_{2g}^3 e_g^{\circ}$($p = 3, q = 0, n = 3, S = 3/2, m = 0$) | 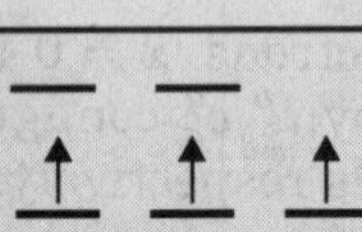 $d^3 = t_{2g}^3 e_g^0$ ($p = 3, q = 0, n = 3, S = 3/2, m = 0$) |
| 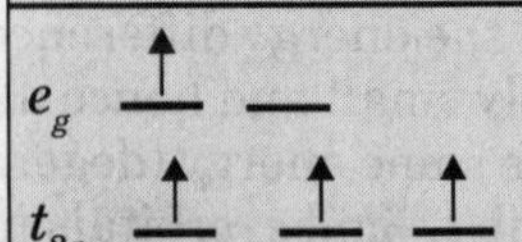 $d^4 = t_{2g}^3 e_g^1$ ($p = 3, q = 1, n = 4, S = 2, m = 0$) | 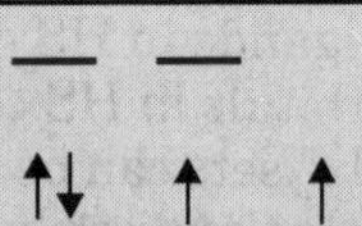 $d^4 = t_{2g}^4 e_g^0$ ($p = 4, q = 0, n = 2, S = 1, m = 1$) |
| 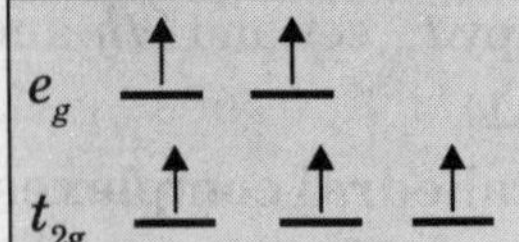 $d^5 = t_{2g}^3 e_g^2$ ($p = 3, q = 2, n = 5, S = 5/2, m = 0$) | 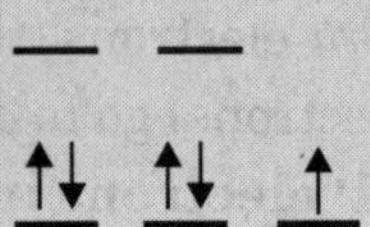 $d^5 = t_{2g}^5 e_g^0$($p = 5, q = 0, n = 1, S = 1/2, m = 2$) |

| | |
|---|---|
| $e_g$: ↑ ↑<br>$t_{2g}$: ↑↓ ↑ ↑<br>**$d^6$** $= t_{2g}^4 e_g^2$ $(p = 4, q = 2, n = 4, S = 2, m = 1)$ | $e_g$: — —<br>$t_{2g}$: ↑↓ ↑↓ ↑↓<br>**$d^6$** $= t_{2g}^6 e_g^0$ $(p = 6, q = 0, n = 0, S = 0, m = 3)$ |
| $e_g$: ↑ ↑<br>$t_{2g}$: ↑↓ ↑↓ ↑<br>**$d^7$** $= t_{2g}^5 e_g^2$ $(p = 5, q = 2, n = 3, S = 3/2, m = 2)$ | $e_g$: ↑ —<br>$t_{2g}$: ↑↓ ↑↓ ↑↓<br>**$d^7$** $= t_{2g}^6 e_g^1$ $(p = 6, q = 1, n = 1, S = 1, m = 3)$ |
| $e_g$: ↑ ↑<br>$t_{2g}$: ↑↓ ↑↓ ↑↓<br>**$d^8$** $= t_{2g}^6 e_g^2$ $(p = 6, q = 2, n = 2, S = 1, m = 3)$ | $e_g$: ↑ ↑<br>$t_{2g}$: ↑↓ ↑↓ ↑↓<br>**$d^8$** $= t_{2g}^6 e_g^2$ $(p = 6, q = 2, n = 2, S = 1, m = 3)$ |
| $e_g$: ↑↓ ↑<br>$t_{2g}$: ↑↓ ↑↓ ↑↓<br>**$d^9$** $= t_{2g}^6 e_g^3$ $(p = 6, q = 3, n = 1, S = 1/2, m = 4)$ | $e_g$: ↑↓ ↑<br>$t_{2g}$: ↑↓ ↑↓ ↑↓<br>**$d^9$** $= t_{2g}^6 e_g^3$ $(p = 6, q = 3, n = 1, S = 1/2, m = 4)$ |
| $e_g$: ↑↓ ↑↓<br>$t_{2g}$: ↑↓ ↑↓ ↑↓<br>**$d^{10}$** $= t_{2g}^6 e_g^4$ $(p = 6, q = 4, n = 0, S = 0, m = 5)$ | $e_g$: ↑↓ ↑↓<br>$t_{2g}$: ↑↓ ↑↓ ↑↓<br>**$d^{10}$** $= t_{2g}^6 e_g^4$ $(p = 6, q = 4, n = 0, S = 0, m = 5)$ |

## Mean Pairing Energy (P)

Mean pairing energy (P) is the energy which is required to pair two electrons against electron-electron repulsion in the same orbital. P is the pairing energy for one electron pair. P is generally expressed in $cm^{-1}$. Pairing energy depends on the principal energy level (n) of *d*-electrons.

## Calculation of Total Pairing Energy of $d^x$ Ion

Total pairing energy for electron pairs present in $d^x$ ion

= Total number of electron pairs in $t_{2g}$ and $e_g$ sets of orbitals $(m)$ × Pairing energy for one electron pair (P)

$= m \times P$ $cm^{-1}$

Here $m$ = Total number of electron pairs in $t_{2g}$ and $e_g$ sets of orbitals and $P$ = Pairing energy for one electron pair ($cm^{-1}$)

## Important Points to be Noted from Table 17.1

The following points may be noted from Table 17.1

(*i*) Whether the ligand field is strong(er) or weak(er), for $d^1$, $d^2$ and $d^3$ configurations, the electrons go to the lower energy $t_{2g}$ level (more stable), for $d^8$, $d^9$ and $d^{10}$ configurations, the first six electrons go to the $t_{2g}$ level and the remaining two (in case of $d^8$ ion), three (in case of $d^9$ ion) and four (in case of $d^{10}$ ion) electrons occupy the $e_g$ level. Thus, the distribution of electrons of $d^1$, $d^2$ , $d^3$, $d^8$, $d^9$ and $d^{10}$ configurations in $t_{2g}$ and $e_g$ levels for both strong(er) and weak (er) octahedral ligand field is the same.

(*ii*) For each of $d^4$, $d^5$, $d^6$ and $d^7$ configurations there is a difference in the arrangement of electrons in weak(er) and strong (er) ligand fields.

(*iii*) *Number of unpaired electrons (n): High-spin and low-spin complexes.* Weak-field complexes of $d^4$, $d^5$, $d^6$ and $d^7$ ions have greater number of unpaired electrons (*n*) than those of (same ions) strong-field complexes and are thus with a higher value of resultant spin (S). It is for this reason that the weak-field and strong-field complexes are also called *spin-free* or *high-spin* (abbreviated as HS) and *spin-paired* or *low-spin* (LS) complexes respectively. Recall that VBT has called these complexes as *ionic (Pauling)* or *outer-orbital (Huggin)* and *covalent (Pauling)* or *inner-orbital (Huggin)* complexes respectively.

The number of unpaired electrons (*i.e.* the value of the resultant spin, S) in the cases namely $d^1$ to $d^3$ and $d^8$ to $d^{10}$ is the same in both the fields, and it is due to this reason that for these configurations the question of the formation of HS and LS-complexes does not arise. The question does arise for the system $d^4$ to $d^7$.

The paramagnetism of HS-complexes is larger than that of LS-complexes, since as is evident from Table 17.1. HS-complexes have more unpaired electrons (*i.e.*, larger value of S) than the LS-complexes (*i.e.*, smaller value of S).

(*iv*) (*a*) When $\Delta_0 < P$, the electrons tend to remain unpaired and hence high spin (weak field or spin free) octahedral complexes are obtained.

(*b*) When $\Delta_0 > P$, the electrons tend to pair and hence low spin (strong field or spin paired) octahedral complexes are obtained.

## How to Predict the Spin State of a Given Octahedral Complex Ion

By comparing the value of $\Delta_0$ and P of a given metallic ion, the spin state of the octahedral complex ion formed by that metallic ion can be predicted. $\Delta_0$ tends to force as many electrons to occupy $t_{2g}$ orbitals while P tends to prevent the electrons to pair in $t_{2g}$ orbitals. This discussion shows that:

(*i*) When $\Delta_0 > P$, the electrons tend to pair and hence low spin (strong-field or spin paired) octahedral complex ions are obtained.

(*ii*) When $\Delta_0 < P$, the electrons tend to remain unpaired and hence high spin (weak-field or spin free) octahedral complex ions are obtained.

Examples of some HS and LS-octahedral complexes are given in Table 17.2. In this table the value of P (in $cm^{-1}$) of the central metal ion of the corresponding complex determined from spectroscopic data and that of $\Delta_0$ (in $cm^{-1}$) for the complexes are also listed. From this table it may be seen that the spin-state of the complexes predicted by CFT is the same as that observed experimentally. In every case where $\Delta_0 < P$, HS-complex is formed and in case where $\Delta_0 > P$, LS-complex is formed.

**Table 17.2. Examples of some LS-and HS-octahedral complexes.**

| *$d^x$ configuration* | *Examples of complexes* | *Value of P ($cm^{-1}$)* | *Value of $\Delta_0$ ($cm^{-1}$)* | *Spin-state* | | *Relative magnitudes of $\Delta_0$ and P* |
|---|---|---|---|---|---|---|
| | | | | *Predicted by CFT* | *Observed experimentally* | |
| $d^4$ | $[Cr(H_2O)_6]^{2+}$ | 23,500 | 13,900 | HS | HS | $\Delta_0 < P$ |
| | $[Mn(H_2O)_6]^{3+}$ | 28,800 | 21,000 | HS | HS | $\Delta_0 < P$ |
| | $[Mn(CN)_6]^{3-}$ | 28,800 | 38,500 | LS | LS | $\Delta_0 > P$ |
| $d^5$ | $[Mn(H_2O)_6]^{2+}$ | 25,500 | 7,800 | HS | HS | $\Delta_0 < P$ |
| | $[Fe(H_2O)_6]^{3+}$ | 30,000 | 13,700 | HS | HS | $\Delta_0 < P$ |
| $d^6$ | $[Fe(H_2O)_6]^{2+}$ | 17,600 | 10,400 | HS | HS | $\Delta_0 < P$ |
| | $[Fe(CN)_6]^{4-}$ | 17,600 | 33,000 | LS | LS | $\Delta_0 > P$ |
| | $[Co(NH_3)_6]^{3+}$ | 21,000 | 23,000 | LS | LS | $\Delta_0 > P$ |
| | $[CoF_6]^{3-}$ | 21,000 | 13,000 | HS | HS | $\Delta_0 < P$ |
| $d^7$ | $[Co(H_2O)_6]^{2+}$ | 22,500 | 9,300 | HS | HS | $\Delta_0 < P$ |

## Factors Affecting the Magnitude of $\Delta_0$

Experiments have shown that the magnitude of Δ depends on the following factors:

### A. Geometry of the Complex Ion.

If we consider the complex ions having the same central metal ion and the same ligands but having different geometry, then the value of Δ depends on the geometry of the complex ion. It has been found that:

$$\Delta_{sp} > \Delta_0 > \Delta_t$$

or $$1.3\Delta_0 > \Delta_0 > 0.45\,\Delta_t$$

**Explanation.** In octahedral complexes, the splitting of d-orbitals is more than twice as strong as in tetrahedral complexes for the same metal ion and the same ligands. The difference in the value of $\Delta_{sp}$, $\Delta_0$ and $\Delta_t$ is because of the following two factors:

(*a*) In octahedral complexes, six ligands are involved while in tetrahedral complexes only four ligands are involved. This results in 33% decrease in the field strength, provided that other factors remain unchanged.

(*b*) In octahedral complexes, the ligands are situated directly in the path of $d_{z^2}$ and $d_{x^2-y^2}$ orbitals ($e_g$ set of orbitals) while in tetrahedral complexes the ligands are not aimed at any of the d-orbitals but exert greater influence on $t_2$ orbitals ($d_{xy}$, $d_{yz}$ and $d_{zx}$ orbitals) than on *e* orbitals ($d_{z^2}$ and $d_{x^2-y^2}$

orbitals). In case of square planar complexes, the degree of splitting is much more than that in tetrahedral complexes.

### B. Nature of Ligands : Weaker/Stronger Ligands: Spectro Chemical Series.

If we consider complex ions having the same central metal ion and the same geometry but having ligands of different crystal field splitting strength, then the absorption spectra of these complexes show that the position of absorption peaks and hence the value of Δ changes with the change in the magnitude of crystal field splitting power of the ligands. The ligands having higher splitting power (i.e. stronger ligands) to split the d-orbitals of the central metal ion have higher value of Δ while the ligands having lower splitting power (weaker ligands) have lower value of Δ.

Some common ligands have been arranged as follows in the increasing order of their splitting power. This arrangement is called *spectrochemical series of ligands*.

$I^- < Br^- < NO_3^- < Cl^- < SCN^- < F^- < OH^- < C_2O_4^{2-} < H_2O < NCS^- <$ edta ~ gly $<$ py $< NH_3 <$ en $<$ H big $<$ bpy ~ o-phen $< NO_2^- < CN^- < CO$

**Examples** (*a*) $\Delta_0$ values of $[NiBr_6]^{4-}$, $[NiCl_6]^{4-}$, $[Ni(H_2O_6)]^{2+}$, $[Ni(NH_3)_6]^{2+}$ and $[Ni(en)_3]^{2+}$ [all these are Ni (II) octahedral complex ions] has been found in the increasing order as shown below:

*Complex ions:* $[Ni^{2+}Br_6]^{4-}$, $[NiCl_6]^{4-}$, $[Ni^{2+}(H_2O)_6]^{2+}$, $[Ni^{2+}(NH_3)_6]^{2+}$, $[Ni^{2+}(en)_3]^{2+}$

$\Delta_0$ *value (cm⁻¹)*: 70,00 < 72,00 < 85,00 < 10,800 < 11,500

——— $\Delta_0$ value increases ———→

*Ligand field strength* } : $Br^-$ < $Cl^-$ < $H_2O$ < $NH_3$ < en

——— $\Delta_0$ value increases ———→

(*b*) $[Rh^{3+}Br_6]^{3-}$, $[Rh^{3+}Cl_6]^{3-}$, $[Rh^{3+}(H_2O)_6]^{3+}$, $[Rh(NH_3)_6]^{3+}$, $[Rh^{3+}(en)_3]^{3+}$

$\Delta_0(cm^{-1})$: 19,000 < 20,300 < 27,000 < 34,000 < 35,000

——— $\Delta_0$ value increases ———→

(*c*) $[Cr^{3+}Cl_6]^{3-}$, $[Cr^{3+}(H_2O)_6]^{3+}$, $[Cr^{3+}(NH_3)_6]^{3+}$, $[Cr^{3+}(en)_3]^{3+}$

$\Delta_0(cm^{-1})$ : 13,800 < 17,400 < 21,600 < 21,900

——— $\Delta_0$ value increases ———→

(*d*) $[Co^{3+}(H_2O)_6]^{3+}$ $[Co^{3+}(NH_3)_6]^{3+}$ $[Co^{3+}(en)_3]^{3+}$

$\Delta_0(cm^{-1})$ : 18,200 < 23,000 < 23,200

——— $\Delta_0$ value increases ———→

**Explanation.** Increase in the value of $\Delta_0$ on moving from left to right is because of the increase in the crystal field power of the ligands as shown in the spectrochemical series.

(*e*) Since of the complex ions *viz* $[Co(NH_3)_6]^{3+}$, $[Co(en)_3]^{3+}$ and $[Co(NO_2)_6]^{3-}$ has strong field ligands ($NH_3$, *en* and $NO_2^-$), the value of $\Delta_0$ for these complex ions is very high. On the other hand, since each of the complex ions *viz* $[CoF_6]^{3-}$ and $[Co(H_2O)_6]^{3+}$ has weak field ligands ($F^-$ and $H_2O$), the value of $\Delta_0$ for these complex ions is comparatively low.

## C. Nature of the Central Metal Ion.

Following three cases may be studied:

(*i*) **Central metal ions belonging to the same group.** The value of Δ of the complex ions having the same ligands and same geometry but different central metal ions (with the same oxidation state) belonging to the same group of transition series elements increases on descending the group. This statement can also be given as follows:

Δ value increases as we move from first transition series to the second transition series and then to the third transition series.

OR

Δ value increases as we move from the central metal ion with $3d^x$ configuration to $4d^x$ configuration and then again increases on moving from $4d^x$ configuration to $5d^x$ configuration. Thus Δ value increases with the increase in the value of principal quantum number ($n$) of d-orbitals of the valence shell of the central metal ion.

**Example.** Since $Co^{3+}$, $Rh^{3+}$ and $Ir^{3+}$ ions belong to the same group (group VIII) of the transition series and have the same oxidation number, $\Delta_0$ value of their octahedral complex ions *viz* $[Co(NH_3)_6]^{3+}$, $[Rh(NH_3)_6]^{3+}$ and $[Ir(NH_3)_6]^{3+}$ (all these ions have the same ligands) increases as we move down the group from $Co^{3+}$ to $Rh^{3+}$ and then from $Rh^{3+}$ to $Ir^{3+}$.

*Complex ions:* $[Co^{3+}(NH_3)_6]^{3+}$ $[Rh^{3+}(NH_3)_6]^{3+}$ $[Ir^{3+}(NH_3)_6]^{3+}$

$\Delta_0$ *value (cm⁻¹)* 23,000 < 34,000 < 41,000

$\xrightarrow{\quad \Delta_0 \text{ value increases} \quad}$

Other examples are:

(*a*) $[Co^{3+}(en)_3]^{3+}$ $[Rh^{3+}(en)_3]^{3+}$ $[Ir^{3+}(en)_3]^{3+}$

$\Delta_0 = 24{,}000^{-1} < 35{,}000\ CM^{-1} < 41{,}000\ cm^{-1}$

$\xrightarrow{\quad \Delta_0 \text{ value increases} \quad}$

(*b*) $[Co^{3+}(H_2O)_6]^{3+}$ $[Rh^{3+}(H_2O)_6]^{3+}$

$\Delta_0 = 18{,}200\ cm^{-1} < 27{,}000\ cm^{-1}$

$\xrightarrow{\quad \Delta_0 \text{ value increases} \quad}$

(*c*) $[Fe^{2+}(CN)_6]^{4-}$ < $[Os^{2+}(CN)_6]^{4-}$

$\xrightarrow{\quad \Delta_0 \text{ value increases} \quad}$

**Explanation** Due to their larger size, 4d and 5d orbitals are likely to interact with the ligand orbitals more effectively than 3d orbital. On moving down the group, nuclear charge of the metal ion increases. This increased nuclear charge may also have an effect on the magnitude of $\Delta_0$.

(*ii*) **Central metal ions of the same metal having different oxidation state.** The value of Δ of the complex ions which have the same ligands and same geometry but their central metal ions are of the same metal with different oxidation state increases with the increase in oxidation state.

**Example.** Since oxidation state (positive charge) of Mn atom in $[Mn^{2+}(H_2O)_6]^{2+}$ and $[Mn^{3+}(H_2O)_6]^{3+}$ increases from +2 to +3, $\Delta_0$ values of these complex ions are in the order : $[Mn^{2+}(H_2O)_6]^{2+} < [Mn^{3+}(H_2O)_6]^{3+}$ as shown below:

*Complex ions:* $[Mn^{2+}(H_2O)_6]^{2+}$ $[Mn^{3+}(H_2O)_6]^{3+}$

*Oxidation state (O.S.) of the central Mn atom* : +2 < +3

——— O.S. increases ———→

$\Delta_0(cm^{-1})$ : 8500 < 21,000

——— $\Delta_0$ value increases ———→

Other examples are:

(*a*) $[Cr^{2+}(H_2O)_6]^{2+}$ $[Cr^{3+}(H_2O)_6]^{3+}$

$\Delta_0$ = 13,900 $cm^{-1}$ < 17,400 $cm^{-1}$

(*b*) $[Fe^{2+}(H_2O)_6]^{2+}$ $[Fe^{3+}(H_2O)_6]^{3+}$

$\Delta_0$ = 10,400 $cm^{-1}$ < 13,700 $cm^{-1}$

(*c*) $[Co^{2+}(H_2O)_6]^{2+}$ $[Co^{3+}(H_2O)_6]^{3+}$

$\Delta_0$ = 93,00 $cm^{-1}$ < 18,200 $cm^{-1}$

(*d*) $[V^{2+}(H_2O)_6]^{2+}$ $[V^{3+}(H_2O)_6]^{3+}$

$\Delta_0$ = 12300 $cm^{-1}$ < 18600 $cm^{-1}$

**Explanation** Δ value increases due to substantial increase in σ-bonding which is because of the decrease in the radius of the metal ion having higher positive charge. The metal ion having increased charge attracts the ligands more strongly and hence the splitting of d-orbitals take place to a greater extent. Greater splitting gives higher value of Δ.

**(*iii*) Central metal ions of different metals having different oxidation state.** The value of Δ of the complex ions having the same ligands and same geometry, but different central metal ions with different oxidation states increase with the increase in oxidation state of the central metal ion.

**Example:** Since $V^{2+}$ and $Cr^{3+}$ ion have +2 and +3 oxidation state, $\Delta_0$ value of $[V^{2+}(H_2O)_6]^{2+}$ and $[Cr^{3+}(H_2O)_6]^{3+}$ ions increases from $[V^{2+}(H_2O)_6]^{2+}$ to $[Cr^{3+}(H_2O_6)]^{3+}$ as shown below:

*Complex ions:* $[V^{2+}(H_2O)_6]^{2+}$ $[Cr^{3+}(H_2O)_6]^{3+}$

*Oxidation state of the central metal ion.* : +2 < +3

——— O.S. increases ———→

$\Delta_0(cm^{-1})$: 12,400 < 17,400

——— $\Delta_0$ increases ———→

**Explanation.** The increase in the value of $\Delta_0$ with the increase in the magnitude of oxidation state of the central metal ion is because of the fact that the central metal ion with higher positive charge has greater attraction for the ligands and hence the ligands can approach this central metal ion more closely than they approach the central metal ion with lower positive charge. Thus the ligands exert stronger crystal field effect on the d-electrons of the central metal ion with higher positive charge. Stronger crystal field effect increases the magnitude of $\Delta_0$.

## To Calculate the Value of Crystal Field Stabilisation Energy (CFSE) of $d^x$ Ions in Octahedral Complexes.

We have seen that according to CFT, under the influence of the six ligands approaching towards the central metal ion during the formation of an octahedral complex, the *d*-orbitals of the central metal ion are split into two sets of orbitals viz. *lower energy triplet* (or *trio*) $t_{2g}$ and *higher energy doublet* (or *pair*), $e_g$ sets. $t_{2g}$ set of orbitals is *triply degenerate* while $e_g$ set is *doubly degenerate*. The energy gap between these two sets is equal to $\Delta_0$ (or 10 Dq). The energy of $t_{2g}$ set is lowered by 2/5 $\Delta_0$ (= 0.4 $\Delta_0$) or 4Dq while that of $e_g$ set in raised by $\frac{3}{5}\Delta_0$ (= 0.6$\Delta_0$) or 6Dq relative to the energy of hypotencial degenerate *d*-orbitals. Thus, each electron occupying $t_{2g}$ orbitals decreases the energy of d-orbitals by – 0.4 $\Delta_0$ (= – 4Dq) while that going into $e_g$ orbitals increases its energy by +0.6$\Delta_0$ (= + 6Dq). – and + signs indicate respectively the decrease and increase in the energy of d-orbitals caused by their splitting under the influence of six ligands. Now let us consider a $d^x$ ion having $t_{2g}^p e_g^q$ configuration in which *p* is the number of electrons in $t_{2g}$ set, *q* is the number of electrons in $e_g$ set and $x = p+q$. Quite obviously:

*Change in energy (in terms of $\Delta_0$) for $t_{2g}^p e_g^q$ configuration*

= Loss in energy due to p electrons in $t_{2g}$ set + gain in energy due to *q* electrons in $e_g$ set.

= – 0.4 $\Delta_0$ × No. of electrons in $t_{2g}$ set of orbitals + 0.6 $\Delta_0$ × No. of electrons in $e_g$ set of orbitals.

$= -0.4\,\Delta_0 \times p + 0.6\,\Delta_0 \times q$

$= [-0.4p + 0.6q]\Delta_0$ ...(*i*)

Now, since $\Delta_0$ = 10Dq, the above expression can also be written as:

*Change in energy (in terms of Dq) for $t_{2g}^p e_g^q$ configuartion*

$= [-0.4\,p + 0.6q] \times 10$ Dq ...(*ii*)

Equations (*i*) and (*ii*) give the energies of $d^x$ ion having $t_{2g}^p e_g^q$ configuration. The change in energy for $d^x$ ion having $t_{2g}^p e_g^q$ configuration calculated as above is called *crystal field stabilisation energy* (CFSE) of $d^x$ ion, since it stabilises d-orbitals by lowering their energy which results from their splitting into $t_{2g}$ and $e_g$ orbitals.

In the derivation of equation (*i*) and (*ii*) we have not considered the pairing energy, P, of $d^x$ ion which is the energy required to pair two electrons against electron-electron repulsion in the same orbital. If the pairing energy of the ion is also involved in the $t_{2g}^p e_g^q$ configuration of a given $d^x$ ion, then CFSE of the ion is given by the expression:

$$\text{CFSE} = [-0.4p + 0.6q]\Delta_0 + m\text{P} \quad ...(iii)$$

$$= [-4p + 6q]\ \text{Dq} + m\text{P} \quad (\because \Delta_0 = 10\text{Dq}) \quad ...(iv)$$

Here *m* is the total number of paired electrons in $t_{2g}$ and $e_g$ sets of orbitals. Equations (*iii*) and (*iv*) have been used to calculate the CFSE value (in terms of $\Delta_0$ and *Dq* respectively) for $d^0$ to $d^{10}$ ions of HS and LS octahedral complexes.

The CFSE value for each $d^x$ ion ($x$ = 0 to 10) having $t_{2g}^p e_g^q$ configuration ($p + q = x$) has been calculated as shown below:

(*i*) **$d^\circ$ ion**

$t_{2g}^p e_g^q$ configuration of $d^\circ$ ion = $t_{2g}^0 e_g^0$ ($p = 0, q = 0, n = 0, S = 0, m = 0$)

$\therefore$ CFSE for $d^\circ$ ion = $(-0.4 \times p + 0.6 \times q)\,\Delta_0 + mP$

$= (-0.4 \times 0 + 0.6 \times 0)\Delta_0 + 0 \times P =$ **Zero**

(*ii*) **$d^1$ ion**

$t_{2g}^p e_g^q$ configuration of $d^1$ ion = $t_{2g}^1 e_g{}^0 (p = 1, q = 0, n = 1, S = 1/2, m = 0)$

$\therefore$ CFSE for $d^1$ ion = $(-0.4 \times p + 0.6 \times q)\Delta_0$ + mP

$= (-0.4 \times 1 + 0.6 \times 0)\,\Delta_0 + 0 \times$ P

$=$ **$-$ 0.4 $\Delta_0$ (In terms of $\Delta_0$)**

$= -\,0.4\,\Delta_0 \times 10$ Dq ($\because\ \Delta_0$ = 10Dq)

**= – 4.0 Dq (In terms of Dq)**

(*iii*) **$d^2$ ion**

$t_{2g}^p e_g^q$ configuration of $d^2$ ion = $t_{2g}^2 e_g{}^\circ (p = 2, q = 0, n = 2, \mathrm{S} = 2/2 = 1, m = 0)$

$\therefore$ CFSE of $d^2$ ion $= (-0.4 \times p + 0.6 \times q)\,\Delta_0 + mP$

$= (-\,0.4 \times 2 + 0.6 \times 0)\,\Delta_0 + 0 \times P$

**= – 0.8 $\Delta_0$ (*In terms of $\Delta_0$*)**

$= -\,0.8 \times 10$ Dq ($\because\ \Delta_0$ = 10Dq)

**= –8 Dq (In terms of Dq)**

Actual value of CFSE of $d^2$ ion = **– 6Dq**

**Note:** The actual value of CFSE of $d^2$ metal ion in octahedral field environment is 6Dq which has been obtained by considering the effects of $d$-orbitals splitting and inter-electronic repulsions amongst $d$-electrons. If the effects of d-orbitals splitting alone is considered and the effects of inter electronic repulsions amongst $d$ electrons is ignored, then the value of CFSE for $d^2$ ion is 8Dq as shown above.

(*iv*) **$d^3$ ion**

$t_{2g}^p e_g^q$ configuration of $d^3$ ion = $t_{2g}^3 e_g^0 (p = 3, q = 0, n = 3, S = 3/2, m = 0)$

$\therefore$ CFSE of $d^3$ ion = $(-0.4 \times p + 0.6 \times q)\,\Delta_0 + mP$

$= (-0.4 \times 3 + 0.6 \times 0)\,\Delta_0 + 0 \times P$

**= –1.2 $\Delta_0$ (*In terms of $\Delta_0$*)**

$= -1.2 \times 10$Dq ($\therefore\ \Delta_0$ = 10 Dq)

**= – 12Dq (*In terms of Dq*)**

(*v*) **$d^4$ ion in HS state (weak ligands)**

$t_{2g}^p e_g^q$ configuration of d$^4$ ion in HS state

$= t_{2g}^3 e_g{}^1\ (p = 3, q = 1, n = 4, S = 4/2 = 2, m = 0)$

$\therefore$ CFSE of $d^4$ ion in HS state = $(-0.4 \times p + 0.6 \times q)\Delta_0$ + mP

$= (-\,0.4 \times 3 + 0.6 \times 1)\Delta_0 + 0 \times$ P

**= – 0.6 $\Delta_0$ (*In terms of $\Delta_0$*)**

$= -0.6 \times 10Dq$ $(\therefore \Delta_0 = 10Dq)$

$= -$ **6.0 Dq (*In terms of Dq*)**

(*vi*) **$d^4$ ion in LS state (strong ligands)**

$t_{2g}^p e_g^q$ configuration of $d^4$ ion in *LS* state ($\Delta_0 > P$)

$= t_{2g}^4 e_g^\circ$ ($p = 4, q = 0, n = 2, S = 2/2 = 1, m = 1$)

$\therefore$ CFSE of $d^4$ ion in *LS* state $= (-0.4 \times p + 0.6 \times q)\,\Delta_0 + mP$

$= (-0.4 \times 4 + 0.6 \times 0)\,\Delta_0 + 1 \times P$

$=$ **$-1.6\,\Delta_0 + P$ (*In terms of $\Delta_0$*)**

$= 1.6\,\Delta_0 \times 10\,Dq + P$ $(\therefore \Delta_0 = 10Dq)$

$=$ **$-$ 16Dq + P (*In terms of Dq*)**

(*vii*) **$d^5$ ion in HS state (weak ligands)**

$t_{2g}^p e_g^q$ configuration of $d^5$ ion in HS state ($\Delta_0 < P$)

$= t_{2g}^3 e_g^2$ ($p = 3, q = 2, n = 5, S = 5/2, m = 0$)

$\therefore$ CFSE of $d^5$ ion in HS state

$= (-0.4 \times p + 0.6 \times q)\,\Delta_0 + mP$

$= (-0.4 \times 3 + 0.6 \times 2)\,\Delta_0 + 0 \times P =$ **zero**

(*viii*) **$d^5$ ion in LS state (strong ligands)**

$t_{2g}^p e_g^q$ configuration of $d^5$ ion in LS state ($\Delta_0 > P$)

$= t_{2g}^5 e_g^0$ ($p = 5, q = 0, n = 1, S = 1/2, m = 2$)

$\therefore$ CFSE for $d^5$ ion in LS state

$= (-0.4 \times p + 0.6 \times q)\Delta_0 + mP$

$= (-0.4 \times 5 + 0.6 \times 0)\,\Delta_0 + 2P$

$=$ **$-2.0\Delta_0 + 2P$ (*In terms of $\Delta_0$*)**

$= -2 \times 10\text{ Dq} + 2P$ $(\because \Delta_0 = 10\,Dq)$

$=$ **$-$20Dq + 2P (*In terms of Dq*)**

(*ix*) **$d^6$ ion in HS state (weak ligands)**

$t_{2g}^p e_g^q$ configuration of $d^6$ ion in HS state ($\Delta_0 < P$)

$= t_{2g}^4 e_g^2$ ($p = 4, q = 2, n = 4, S = 4/2 = 2, m = 1$)

$\therefore$ CFSE for $d^6$ ion in HS state

$= (-0.4 \times p + 0.6 \times q)\Delta_0 + mP$

$= (-0.4 \times 4 + 0.6 \times 2)\,\Delta_0 + 1P$

$=$ **$-0.4\,\Delta_0 + P$ (*In terms of $\Delta_0$*)**

$= -0.4 \times 10Dq + P$ $(\because \Delta_0 = 10\,Dq)$

$=$ **$-$ 4Dq + P (*In term of Dq*)**

(*x*) **$d^6$ ion in LS state (strong fields)**

$t_{2g}^p e_g^q$ configuration of $d^6$ ion in LS state ($\Delta_0 > P$)

$= t_{2g}^6 e_g^0$ ($p = 6, q = 0, n = 0, S = 0, m = 3$)

$\therefore$ CFSE for $d^6$ ion in L*S* state

$= (-0.4 \times p + 0.6 \times q)\,\Delta_0 + mP$

$= (-0.4 \times 6 + 0.6 \times 0)\Delta_0 + 3P$

$= -$ **2.4 $\Delta_0$ + 3P (*In terms of* $\Delta_0$)**

$= -2.4 \times 10Dq + 3P$ $(\because \Delta_0 = 10Dq)$

$= -$ **24Dq + 3P (In terms of Dq)**

(*xi*) **$d^7$ ion in HS state (weak ligands)**

$t_{2g}^p e_g^q$ configuration of $d^7$ ion in HS state ($\Delta_0 < P$)

$= t_{2g}^5 e_g^2$ ($p = 5, q = 2, n = 3, S = 3/2, m = 2$)

$\therefore$ CFSE of $d^7$ ion in HS state

$= (-0.4 \times p + 0.6 \times q)\,\Delta_0 + mP$

$= (-0.4 \times 5 + 0.6 \times 2)\,\Delta_0 + 2P$

$= -$ **0.8 $\Delta_0$ + 2P (*In terms of* $\Delta_0$)**

$= -0.8 \times 10Dq + 2P$ $(\because \Delta_0 = 10Dq)$

**= –8Dq + 2P (*In terms of Dq*)**

Actual value = **–6Dq + 2P**

**Note:** The actual value of CFSE of $d^7$ metal ion in weak octahedral ligand field environment is ($-6Dq + 2P$) which has been obtained by considering the effect of d-orbitals splitting and inter electronic repulsions amongst d-electronic. If the effect of d-orbitals splitting alone is considered and the effects of inter-electronic repulsions amongst *d* electrons is ignored, then the value of CFSE for $d^7$ ion (HS) is (–8Dq + 2P) as shown above.

(*xii*) **$d^7$ ion in LS state (strong ligands)**

$t_{2g}^p e_g^q$ configuration of $d^7$ ion in LS state ($\Delta_0 > P$)

$= t_{2g}^6 e_g^1$ ($p = 6, q = 1, n = 1, S = 1/2, m = 3$)

$\therefore$ CFSE of $d^7$ ion in LS state

$= (-0.4 \times p + 0.6 \times q)\,\Delta_0 + mP$

$= (-0.4 \times 6 + 0.6 \times 1)\,\Delta_0 + 3P$

$= -$ **1.8 $\Delta_0$ + 3P (*In terms of* $\Delta_0$)**

$= -1.8 \times 10Dq + 3P$ $(\because \Delta_0 = 10Dq)$

**= –18Dq + 3P (*In terms of Dq*)**

(*xiii*) **$d^8$ ion**

$t_{2g}^p e_g^q$ configuration of $d^8$ ion $= t_{2g}^6 e_g^{\,2}$ ($p = 6, q = 2, n = 2, S = 2/2 = 1, m = 3$)

$\therefore$ CFSE energy of $d^8$ ion

$= (-0.4 \times p + 0.6 \times q)\,\Delta_0 + mP$

$= (-0.4 \times 6 + 0.6 \times 2)\,\Delta_0 + 3P$

$= -$ **1.2 $\Delta_0$ + 3P (*In terms of* $\Delta_0$)**

$= -1.2 \times 10Dq + 3P$ $(\because \Delta_0 = 10Dq)$

**= –12Dq + 3P (*In terms of Dq*)**

(*ix*) **$d^9$ ion**

$t_{2g}^p e_g^q$ configuration of $d^9$ ion $= t_{2g}^6 e_g^3$ ($p = 6, q = 3, n = 1, S = 1/2, m = 4$)

$\therefore$ CFSE energy of $d^9$ ion

$= (-0.4 \times p + 0.6 \times q)\,\Delta_0 + mP$

$= (-0.4 \times 6 + 0.6 \times 3)\,\Delta_0 + 4P$

$= -\mathbf{0.6\ \Delta_0 + 4P}$ ***(In terms of*** $\Delta_0$***)***

$= -0.6 \times 10\text{Dq} + 4\text{P}$ $(\because \Delta_0 = 10\text{Dq})$

$= \mathbf{-6Dq + 4P}$ ***(In terms of Dq)***

(*xv*) $\mathbf{d^{10}}$ **ion**

$t_{2g}^p e_g^q$ configuration of $d^{10}$ ion = $t_{2g}^6 e_g^4 (p = 6, q = 4, n = 0, S = 0, m = 5)$

$\therefore$ CFSE energy of $d^{10}$ ion

$= (-0.4 \times p + 0.6 \times q)\ \Delta_0 + mP$

$= (-0.4 \times 6 + 0.6 \times 4)\ \Delta_0 + 5P$

$= 0 + 5P$

$= \mathbf{5P}$

CFSE values of different $d^x$ ions ($x$ = 0 to 10) are summarised as follows (we have not considered pairing energy of $d^x$ ion)

| $d^x$ *ions* *x=1 to 10* | *CFSE values* ($\Delta_0$ *= 10Dq)* | |
|---|---|---|
| | *In terms of* $\Delta_0$ | *In terms of Dq* |
| $d^0$ | Zero | Zero |
| $d^1$ | $-0.4\ \Delta_0$ | –4.0Dq |
| $d^2$ | $-0.8\ \Delta_0$<br>Actual value = $-0.6\Delta_0$ | –8Dq<br>Actual value = –6Dq |
| $d^3$ | $-1.2\Delta_0$ | –12Dq |
| $d^4$(HS) | $-0.6\Delta_0$ | –6.0Dq |
| $d^4$(LS) | $-1.6\Delta_0$ | –16Dq |
| $d^5$(HS) | Zero | Zero |
| $d^5$(LS) | $-2.0\Delta_0$ | –20Dq |
| $d^6$(HS) | $-0.4\Delta_0$ | –4Dq |
| $d^6$(LS) | $-2.4\Delta_0$ | –24Dq |
| $d^7$(HS) | $-0.8\Delta_0$<br>Actual value = $-0.6\Delta_0$ | –8Dq<br>Actual value = –6Dq |
| $d^7$(LS) | $-1.8\Delta_0$ | –18Dq |
| $d^8$ | $-1.2\Delta_0$ | –12Dq |
| $d^9$ | $-0.6\Delta_0$ | –6Dq |
| $d^{10}$ | Zero | Zero |

## To Determine the Value of $\Delta_0$ of $[Ti\ (H_2O)_6]^{3+}$ Ion from its Absorption Spectrum

In order to determine the value of $\Delta_0$ for $[Ti\ (H_2O)_6]^{3+}$ ion, light is allowed to fall on this ion. As a result, the ion absorbs the light of frequency of 20,000 $cm^{-1}$ (wave number) as is evident from the absorption spectrum of $[Ti\ (H_2O)_6]^{3+}$ ion. This wave number (= 20,000 $cm^{-1}$) lies in visible region of the spectrum. This spectrum has been obtained by plotting graph between wave number (in $cm^{-1}$) of the absorbed light and the absorbance (absorbance is the amount of absorbed light). 20,000 $cm^{-1}$ is the frequency (wave number) of the band (peak) of maximum absorption (*i.e.*, peak or maxima). This wave number of 20,000 $cm^{-1}$ is associated with energy which is equal to 239.23 KJ $mol^{-1}$ as shown below. We know that since

$$\therefore 83.6\ cm^{-1} = 1\ KJ\ mol^{-1}$$

$$1\ cm^{-1} = \frac{1}{83.6}\ KJ\ mol^{-1}$$

$$20{,}000\ cm^{-1} = \frac{20{,}000}{83.6} KJ\ mol^{-1} = 239.23\ KJ\ mol^{-1}$$

This energy is the energy difference between $t_{2g}$ and $e_g$ sets of orbitals (See Fig. 17.3) and is called the crystal field splitting energy ($\Delta_0$) of [Ti $(H_2O)_6]^{3+}$ ion. Thus crystal field splitting energy ($\Delta_0$) for $[Ti(H_2O)_6]^{3+}$ ion = 20,000 $cm^{-1}$ = 239.23 KJ $mol^{-1}$

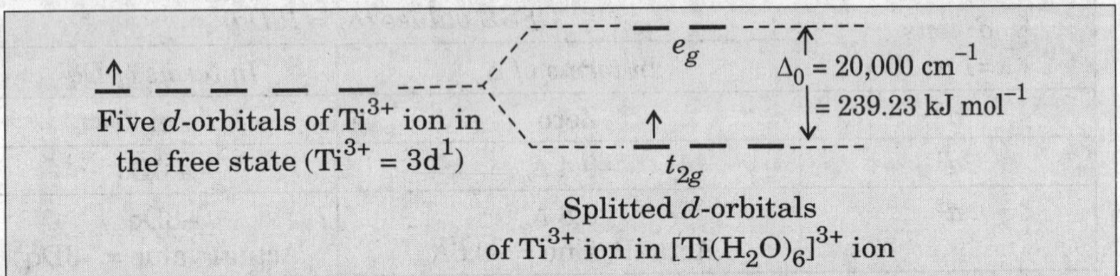


**Fig. 17.3.** Crystal field splitting energy ($\Delta_0$) of [Ti $(H_2O_6)]^{3+}$ ion

It may be seen from the above discussion that higher is the value of the frequency in $cm^{-1}$ (wave number) of the peak of maximum absorption in the absorption spectrum of a given complex ion, greater is the magnitude of $\Delta_0$. Thus in routine work, $\Delta_0$ is expressed in $cm^{-1}$ and not in KJ $mol^{-1}$

## Applications of CFT: To Explain the Properties of Complexes on the Basis of CFT

**1. To predict the high spin (HS) and low spin (LS) states of complexes.** We know that for the $d^{1-3}$ and $d^{8-10}$ configuration, there is only one possible way in which $d$-orbitals may be filled. Hence, they exist in the same state of minimum energy irrespective of the strength of the field. On the other hand, for $d^{4-7}$ configurations, the coordination entities are more stable in low spin state (strong ligand field) than in the high spin state (weak ligand field).

**2. Magnetic properties of complexes** (Also see chapter: *"Magnetic properties of transition metal complexes"* and chapter : *"General characteristics of transition metals : d-block elements"*) Using the fact that strong field ligands cause low spin states to be more stable and the weak field ligands cause high spin states to be more stable, we can determine whether the electrons are paired up or not and hence we can predict the magnetic behaviour (diamagnetism or paramagnetism) of the complex.

**3. Colour of complexes** (*Also see chapter : "General Characteristics of Transition Metals: d-Block Elements"*). The magnitude of the crystal field splitting energy ($\Delta_0$) for most of the transition metal complexes is of the order which makes it possible for a transition metal complex (say octahedral complex) to absorb visible light. The visible light thus absorbed excites the electron from the lower energy $t_{2g}$ orbitals into the higher energy $e_g$ orbitals. The excitation of an electron from $t_{2g}$ orbitals to $e_g$ orbitals is represented by $t_{2g} \longrightarrow e_g$. As an example, take the case of Ti $(H_2O)_6]^{3+}$ complex where Ti (III) has only one electron in 3*d* orbitals. Its spectrum shows a single absorption peak at about 5000 Å which corresponds to an energy of about 240 kJ mol$^{-1}$

$$E = N_0 h\upsilon = N_0 h\frac{c}{\lambda} = \frac{(6.02\times10^{23}\,\text{mol}^{-1})(6.62\times10^{-34}\,\text{Js})(3\times10^{8}\,\text{ms}^{-1})}{(5000\times10^{-10}\,\text{m})})$$

$$= 239.1 \text{ kJ mol}^{-1}$$

*i.e.*, greater than $\Delta_0$ value and is sufficient to excite the electron from $t_{2g}$ to $e_g$ orbital. In this case, green and yellow portions of the light are absorbed and the solution of the complex acquires purple colour which is a complementary colour of the radiation absorbed. Similarly, [Cu $(H_2O_4)]^{2+}$ is blue in colour because red light corresponding to $\lambda$ = 6000 Å is absorbed.

**4. Uses of crystal field stabilisation energies.** See chapter *"Applications of CFSEs"*.

## UNIVERSITY QUESTIONS

1. Write note on "Crystal Field Stabilisation Energy" (*Pd. Ravishankar Shukla 2010*)
2. Draw crystal field splitting diagram of d-orbitals in octahedral field. (*Pd. Ravishanker Shukla 2010*)
3. What is CFSE? Calculate CFSE for the following ions in octahedral complexes: (*i*) $d^4$ strong field (*ii*) $d^6$ weak field (*iii*) $d^7$ strong field. (*Himachal Pradesh 2010*)
4. Describe crystal field theory. Also write its postulates. (*Kanpur 2010*)
5. (*a*) What is crystal field theory? How is it different from valence bond theory? How does CFT explain the following—
   (*i*) $[CoF_6]^{3-}$ is paramagnetic but $[Co(NH_3)_6]^{3+}$ is diamagnetic.
   (*ii*) $[Ni(CN)_4]^{2-}$ is a square planar complex ion.
   (*b*) What is crystal field stabilisation energy (CFSE)? Calculate CFSE values in the following— (*i*) $d^5$ low spin octahedral complex (*ii*) $d^6$ tetrahedral complex. (*Rajasthan 2010*)
6. Choose the correct answer: According to crystal field theory, the nature of metal-ligand bond is— (*a*) Covalent (*b*) Co-ordinate (*c*) Partially ionic (*d*) Purely ionic (*Barakatullah 2010*)
7. Discuss briefly the splitting of d-orbitals in octahedral complexes. (*Barakaullah 2010*)
8. On the basis of CFT explain magnetic properties of $[CoF_6]^{3-}$ and $[Co(NH_3)_6]^{3+}$ (*Bangalore 2010*)
9. (*i*) The number of unpaired electron in $d^6$, low spin octahedral complex is ...

(*ii*) The crystal field theory has been modified as ...

(*iii*) The crystal field stabilization energy for a high spin octahedral complex having $d^1$ configuration is ...

(*N.H. Bahuguna Garhwal 2010*)

**10.** Choose the correct answer:

(*i*) The CFSE value of $d^5$ ion in a weak octahedral ligand field is

(*a*) –6Dq (*b*) 0 Dq (*c*) –8Dq (*d*) 4 Dq.

(*ii*) Which one of the following will exhibit lower spin only magnetic moment?

(*a*) $CoF_6^{3-}$ (*b*) $CoI_6^{3-}$ (*c*) $CoBr_6^{3-}$ (*d*) $CoCl_6^{3-}$

(*iii*) A $d^7$ metal complex is isolobal with

(*a*) CH (*b*) $CH_2$ (*c*) $CH_3$ (*d*) $CH_4$

(*Bharathiar 2010*)

**11.** (*i*) What are the important postulates of CFT?

(*ii*) Explain the magnetic properties of $[Co(NH_3)_6]^{3+}$ and $[CoF_6]^{3-}$.

(*Bharathiar 2010*)

**12.** (*i*) Calculate C.F.S.E. of the following complex ions: $[Fe(H_2O)_6]^{3+}$; $[Fe(CN)_6]^{-4}$

(*ii*) Mention any two factors affecting the value of $\Delta_0$.

(*iii*) Define high spin complex. (*Gujrat Technological University 2010*)

**13.** (*a*) Explain spilitting of degeneracy of the five d orbitals by an octahedral field in weak and strong ligand fields.

(*b*) Discuss the factors affecting the magnitude of 10 Dq. Calculate spin only magnetic moment for complex $[Co(NH_3)_6]^{3+}$ and $(CoF_6]^3$ (*Ranchi 2010*)

**14.** (*a*) Explain d-orbitals splitting diagram of the following complexes on the basis of CFT: (*i*) $[Fe(CN)_6]^{3-}$ and (*ii*) $[Fe(H_2O)_6]^{3+}$.

(*b*) Explain the magnetic properties with crystal field splitting diagram of the following complexes: (*i*) $[CoF_6]^{3-}$ and (*ii*) $[Co(NH_3)_6]^{3+}$.

(*c*) Which of the following are expected to be paramagnetic complexes: (*i*) $[Ni(H_2O)_6]^{2+}$ (ii) $[Co(NH_3)_6]^{3+}$ (iii) $[Zn(NH_3)_4]^{2+}$ (iv) $[Fe(CN)_6]^{3-}$. Explain with CFT configuration.

(*d*) On the basis of CFT calculate spin magnetic moments of the complexes: (i) $[NiCl_4]^{2-}$ and (ii) $[FeF_6]^{3-}$. (*Nagpur 2010*)

**15.** (*a*) Discuss the salient features of crystals field theory with reference to octahedral and tetrahedral complexes.

(*b*) How will you calculate the CFSE for an octahedral $d^7$ system and a tetrahedral $d^6$ system in the light of CFT?

(*c*) What are the factors affecting the crystal field parameters?

(*Utkal 2010*)

**16.** (*a*) Low spin tetrahedral complexes of first transition series ions with $d^3$, $d^4$, $d^5$ and $d^6$ configurations are unknown. Explain.

(*b*) Diagrammatically represent the crystal field splitting in octahedral complexes.

(*c*) Discuss the factors that affect crystal field parameters.

(*d*) Explain why $4\Delta_0 = 9\Delta_t$ (*BJB College, Orissa*)

**17.** (*a*) What are the factors affecting the crystal field parameters? Explain in detail.
(*b*) What do you understand by CFT? Explain is limitations.
(*GND Amritsar 2010*)

**18.** (*a*) What is E.A.N? Calculate the E.A.N. of central metal in following complexes: (*i*) $[Ni(CO)_4]$ (*ii*) $[Co(NH_3)_4Cl_2]Cl$ (*iii*) $K[Fe(CN)]_6$
(At No. of Ni = 28, Co = 27, Fe = 26.)
(*b*) Explain CFSE and calculate it for a $d^5$ metal complex in an octahedral field with weak field ligands. (*Lucknow 2010*)

**19.** Write a note on spectro chemical series.
(*Osmania 2010, Pd. Ravishankar Shukla 2010, Himachal Pradesh 2010 Nagpur 2010*)

**20.** In the spectrochemical series which ligand produces strong field?
(*a*) $Cl^-$ (*b*) $H_2O$ (*c*) $NO_2^-$ (*d*) CO
(*H.N. Baluguna Garhwal 2010*)

**21.** What is spectrochemical series? How will you explain the relative position of fluoride ion and carbon monoxide ligands in that series? What is its importance? (*Bharathiar 2010, BJB College, Orissa*)

**22.** (*i*) Calculate CFSE for $[Co(NH_3)_6]^{3+}$ ion.
(*ii*) HS $Fe^{2+}$ has a radius of 92 pm whereas the LS $Fe^{2+}$ has 75 pm only. Explain.
(*iii*) Discuss the various factors influencing the magnitude of crystal field splitting. (*Madras April 2011*)

**23.** What is spectrochemical series? Distinguish between strong field and weak field legands. (*Nagpur 2012*)

**24.** (*a*) Define CFSE.
(*b*) Explain the magnetic properties with crystal field splitting diagram of the following complexes (*i*) $[Fe(CN)_6]^{3-}$ (*ii*) $[Fe(CN)_6]^{4-}$
(*c*) Draw d-orbital splitting diagrams and show the location of electrons in the following complexes : (*i*) $[Cu(NO_2)_6]^{4-}$, $\mu = 4.90$ BM (*ii*) $[Cr(H_2O)_6]^{3+}$, $\mu = 3.87$ BM (*Nagpur 2012*)

**25.** (*i*) What do you understand by crystal field splitting energy ($\Delta_o$)? How does it differ from crystal field stabilisation energy? Discuss factors affecting the magnitude of $\Delta_o$ with examples.
(*ii*) Explain : (*a*)$Zn^{2+}$ salts are white while $Cu^{2+}$ salts are blue (*b*) $Cu^+$ salts are white while $Cu^{2+}$ salts are blue (*Ch. Charan Singh, Meerut 2012*)

**26.** (*i*) Which complex ion shows the greatest crystal field splitting. Explain why?
(*a*) $[Co(H_2O)_6]^{2+}$ (*b*) $[Rh(H_2O)_6]^{3+}$ (*c*) $[Co(H_2O)_6]^{3+}$ (*d*) $[Fe(H_2O)_6]^{2+}$
(*ii*) Which complex has larger $\Delta_0$ value? Give reason
(*a*) $[Co(CN)_6]^{3-}$ (*b*) $[Co(NH_3)_6]^{3+}$ (*c*) $[Mn(NH_3)_6]^{3+}$ (*d*) $[Mn(en)_3]^{3+}$
(*Kurukshetra 2011*)

**27.** (*i*) Explain why Cu(II) salts are blue in colour but Cu(I) salts are white.
(*ii*) Discuss crystal field theory (*Lucknow 2013*)

18

CHAPTER

# Crystal Field Theory of Tetrahedral and Cubic Complexes

## Crystal Field Splitting of Metal d-Orbitals in Tetrahedral Complexes.

We know that a tetrahedral complex has tetrahedral geometry. This geometry can be derived from a cube in which the central metal atom/ion (M) is placed at the centre of the cube and the four ligands ($L_1$, $L_2$, $L_3$ and $L_4$) occupy the four alternate corners of the cube, *i.e.* the two ligands occupy the diagonally opposite corners of a face of the cube. The three axes *viz. x, y* and *z* are directed towards the centres of the face of the cube (See Fig. 18.1)

It may be seen from Figure 18.1 that the four ligands ($L_1$, $L_2$, $L_3$ and $L_4$) do not lie directly in the path of the axes. For example the ligands $L_1$ lies between *x* and *z* axes, the ligand $L_2$ lies between *z* and *y* axes, $L_3$ ligand lies between *z* and *y* axes and $L_4$ ligand lies between *y* and *x* axes.

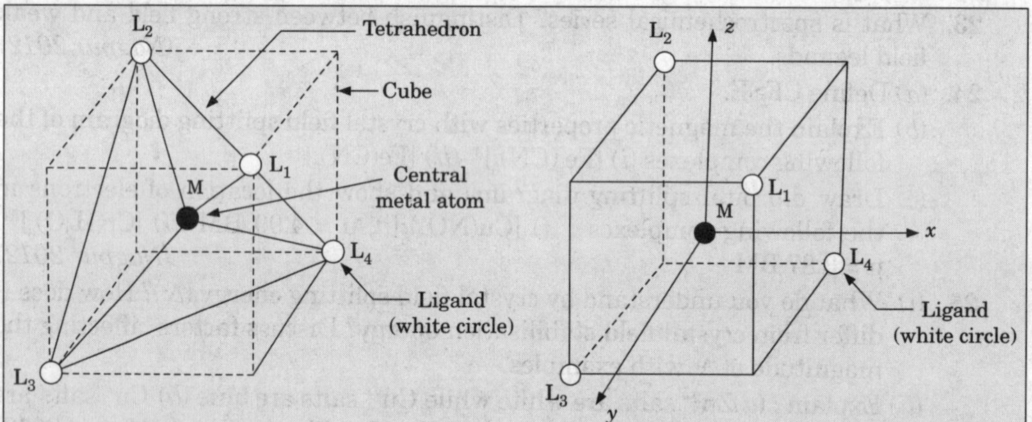


(a) To get a tetrahedron from a cube

(b) Position of central metal atom and ligands and representation of axes in a cube.

**Fig. 18.1.** To get a tetrahedron from a cube and representation of axes. Black dot (●) indicates central metal atom ion (M) and while circles (O) indicate four ligands *viz.* $L_1$, $L_2$, $L_3$ and $L_4$.

In tetrahedral complexes, five $d$-ortbitals of the cental metal atom are divided into two groups or sets *viz* $e$ and $t_2$ groups. $e$ set consists of $d_{x^2-y^2}$ and $d_{z^2}$ orbitals while $t_2$ set consists of $d_{xy}$, $d_{yz}$ and $d_{zx}$ orbitals. The orientation of the lobes of all the five $d$-orbitals in a cube has been shown in Fig 18.2.

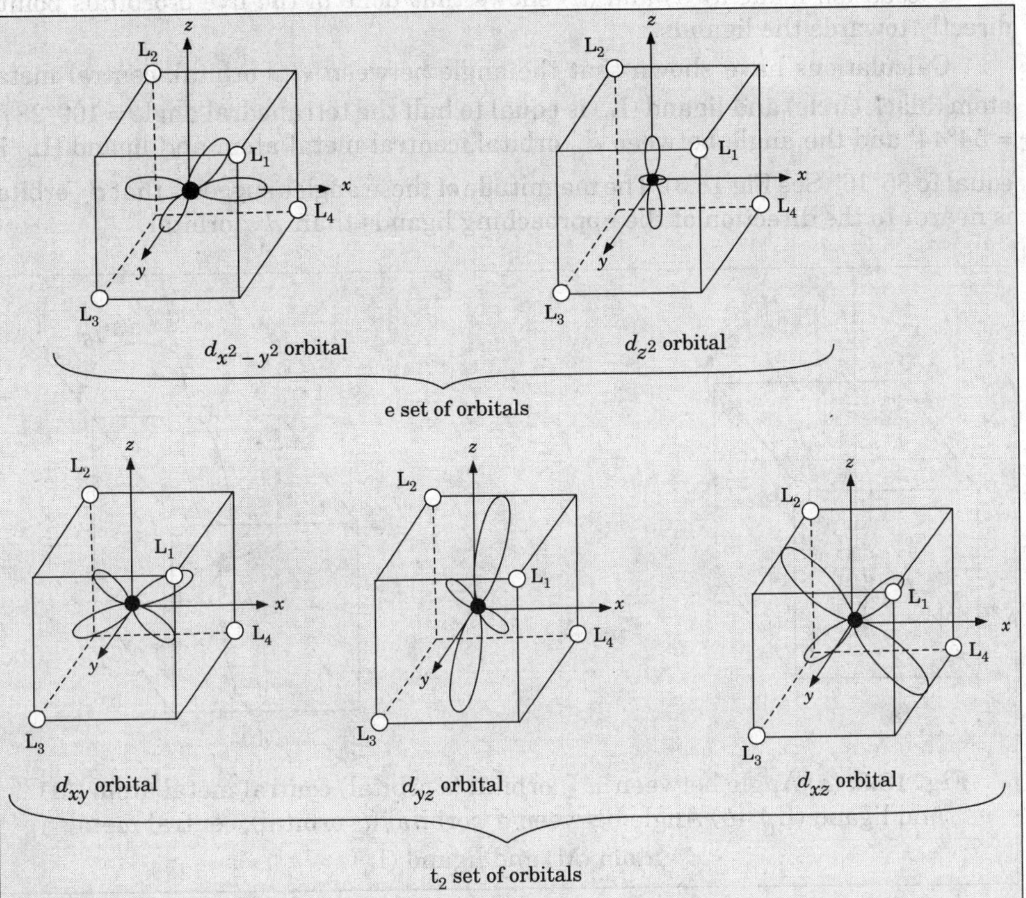


**Fig. 18.2.** Orientation of the lobes of five d-orbitals of the central metal atom / ion in a cube.

Fig. 18.2 shows that:

(*i*) The lobes of the two orbitals of $e$ set (*i.e.* $d_{x^2-y^2}$ and $d_{z^2}$ orbitals) lie along the axes, *i.e.*, the lobes of the two orbitals of $e$ set lie on the bisectors of the angle between pairs of ligands and hence pass through the centre of the faces of the cube. Thus $d_{x^2-y^2}$ and $d_{z^2}$ orbitals lie half a diagonal of a face of the cube away from the two nearest ligands. For example one lobe of $d_{z^2}$ orbital along $z$-axis is placed in between the two nearest ligands *viz* $L_1$ and $L_2$ and the other lobe of $d_{z^2}$ orbital is placed in between $L_3$ and $L_4$ ligands. Note the ligands $L_3$ and $L_4$ occupy diagonally opposite corners of a face of the cube.

(*ii*) The lobes of the three orbitals of $t_2$ set ($d_{xy}$, $d_{yz}$ and $d_{zx}$ orbitals) lie between the axes, *i.e.* towards the centre of the edges of the cube. Thus $d_{xy}$, $d_{yz}$ and $d_{zx}$ orbitals lie just half the side of a face of the cube away from the nearest two ligands.

Discussion made at (*i*) and (*ii*) shows that none of the five d-orbitals points directly towards the ligands.

Calculations have shown that the angle between $d_{z^2}$ orbital, central metal atom (black circle) and ligand ($L_1$) is equal to half the tetrahedral angle = 109°28′/2 = 54°44′ and the angle between $d_{yz}$ orbital, central metal atom and ligand ($L_1$) is equal to 35°16′ (See Fig 18.3). The magnitude of these angles suggests that $d_{yz}$ orbital is nearer to the direction of the approaching ligands than $d_{z^2}$ orbital.

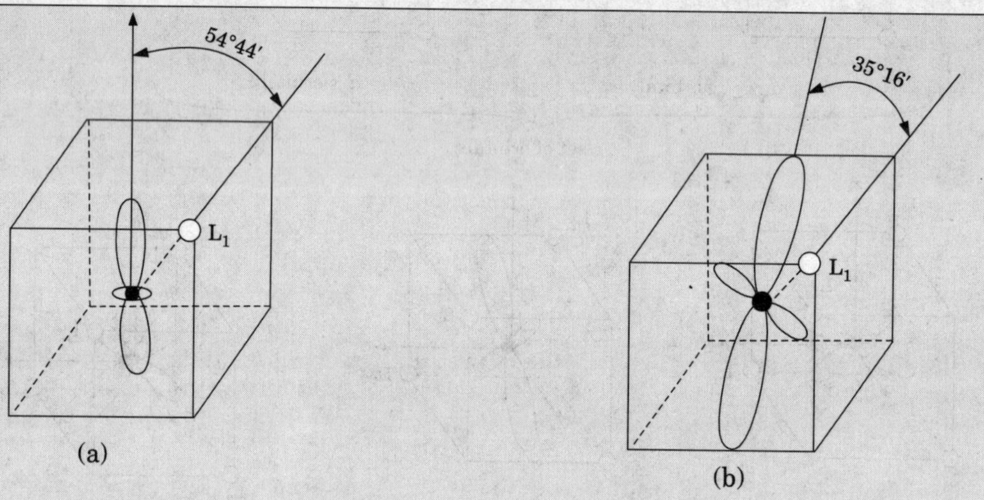


**Fig. 18.3.** (*a*) Angle between $d_{z^2}$ orbital (*e* orbital) central metal atom (M) and ligand ($L_1$) (*b*) Angle between $d_{yz}$ orbital ($t_2$ orbital), central metal atom (M) and ligand ($L_1$)

In a tetrahedral complex, when the four ligands approach near the central metal atom, the energy of the orbitals of both the sets (*e* and $t_2$ sets) increases. Since $t_2$ orbitals are closer to the direction of the approaching ligands than *e* orbitals, electrons in $t_2$ orbitals experience greater repulsion from the approaching ligands than *e* orbitals. Due to greater repulsion the energy of $t_2$ orbitals is increased to a larger extent than that of *e* orbitals. Thus *e* orbitals are at lower energy level than $t_2$ orbitals. In other words we can say that under the influence of four ligands in a tetrahedral complex, five d-orbitals of the central metal atom are split into $t_2$ and *e* orbitals. This splitting of *d*-orbital is opposite to that of d-orbitals under the influence of six ligands in an octahedral complex. $t_2$ orbitals are raised by 0.4 $\Delta$t and *e* orbitals are lowered by 0.6 $\Delta t$ from the no splitting state. Here $\Delta$t is the energy difference between $t_2$ and *e* orbitals (See Fig. 18.4 )

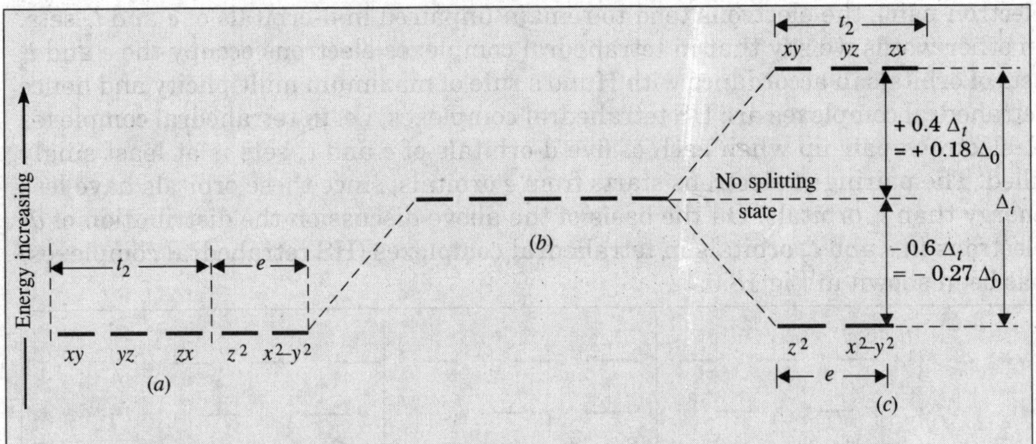


**Fig. 18.4.** Splitting of five d-orbitals in a tetrahedral complex (*a*) Five degenerate d-orbitals on the central metal cation which are free from any ligand field (*b*) Hypothetical degenerate d-orbitals at a higher energy level (*c*) Splitting of d-orbitals into *e* and $t_2$ orbitals under the influence of four ligands in tetrahedral complex.

## Relation Between $\Delta_t$, $\Delta_0$ and 10Dq

Calculations have shown that $\Delta t$ is less than $\Delta_0$ for the same metal atom, same ligands and same inter-nuclear distance ($\Delta_t < \Delta_0$) There are two reasons for this:

(*i*) There are only four ligands instead of six in a tetrahedral complex. Thus the ligand field caused by four ligands in a tetrahedral complex is $\frac{2}{3}$ of that caused by six ligands in an octahedral complex.

(*ii*) The direction of the lobes of d-orbitals in a tetrahedral complex do not coincide with the direction of the ligands approaching towards the central metal atom. This also reduces the magnitude of the splitting of *d*-orbitals in a tetrahedral complex.

Above two reasons suggest that $\Delta_t$ should be equal to $\frac{2}{3}\times\frac{2}{3}\Delta_0 = \frac{4}{9}\Delta_0 =$ 0.45 $\Delta_0$, *i.e.*,

$$\Delta_t = \frac{4}{9}\Delta_0 = 0.45\,\Delta_0 = 0.45 \times 10\text{ Dq} = 4.5\text{ Dq } (\because \Delta_0 = 10\text{Dq})$$

This equation also shows that $\Delta_t < \Delta_0$

In case of tetrahedral complexes, since $\Delta_t$ is generally less than P ($\Delta_t < P$), the electrons tend to remain unpaired and hence only high spin tetrahedral complexes are known, *i.e.*, low spin tetrahedral complexes are not known.

## Distribution of $d^x$ Electrons ($x$ = 0 to 10) of the Central Metal Ion in *e* and $t_2$ Sets of Orbitals in Tetrahedral Complexes.

In case of tetrahedral complexes, since $\Delta_t$ [$\Delta_t$ is the energy difference between $t_2(d_{xy}$, $d_{yz}$ and $d_{zx}$ orbitals) and *e* ($d_{x^2-y^2}$ and $d_{z^2}$ orbitals) sets of orbitals] is generally less than P (P is the energy which is required to pair two electrons in the same orbital against electron-electron repulsion. P is called pairing energy for one electron pair), the electrons tend to remain unpaired in d-orbitals of $e$ and $t_2$ sets. In other words we say that in tetrahedral complexes electrons occupy the $e$ and $t_2$ sets of orbitals in accordance with Hund's rule of maximum multiplicity and hence tetrahedral complexes are HS tetrahedral complexes, i.e. in tetrahedral complexes $d^x$ electrons pair up when each of five d-orbitals of $e$ and $t_2$ sets is at least singly filled. The pairing of electrons starts from $e$ orbitals, since these orbitals have less energy than $t_2$ orbitals. On the basis of the above discussion the distribution of $d^x$ electrons in $e$ and $t_2$ orbitals in tetrahedral complexes (HS tetrahedral complexes) has been shown in Fig 18.5.

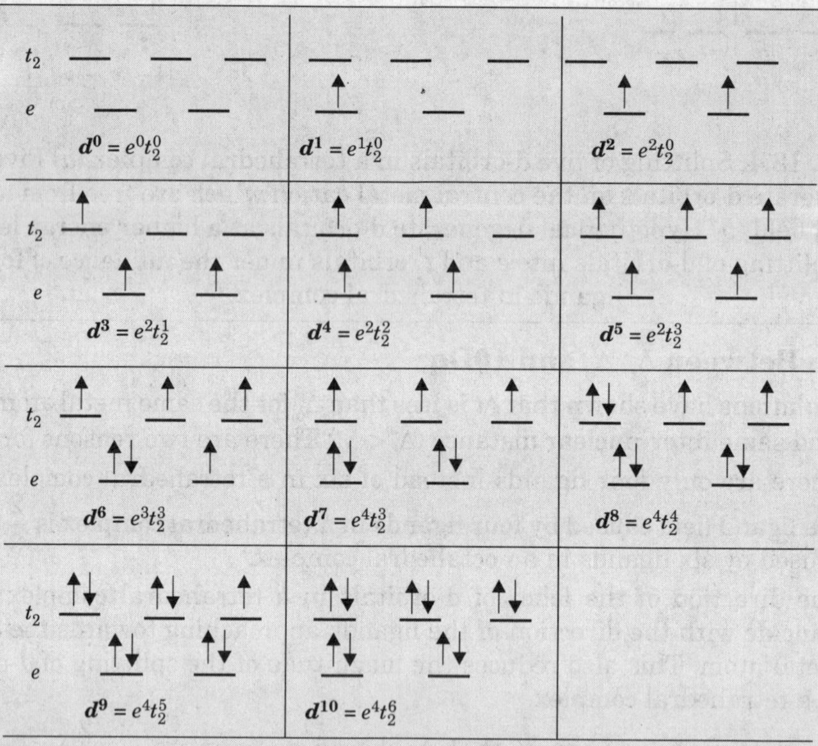


**Fig. 18.5.** Distribution of $d^x$ electrons ($x$ = 0 to 10) in $e$ and $t_2$ orbitals in tetrahedral complexes

## Calculation of Crystal Field Stabilisation Energy (CFSE) of $d^x$ Ions ($x$ = 0 to 10) in Tetrahedral Complexes

We have already said that tetrahedral complexes are always HS complexes, since in these complexes P is greater than $\Delta_t$ ($P > \Delta_t$). Thus here we shall calculate CFSE of $d^x$ ions in HS tetrahedral complexes.

We have seen that according to CFT, under the influence of four ligands approaching towards the central metal ion during the formation of a high spin tetrahedral complex, the d-orbitals of the central metal ion are split into lower energy doublet $e$ orbitals ($d_{z^2}$ and $d_{x^2-y^2}$ orbitals) and higher energy triplet $t_2$ orbitals ($d_{xy}$, $d_{yz}$ and $d_{zx}$ orbitals). The energy gap between $e$ and $t_2$ orbitals is denoted by $\Delta_t$. The energy of $e$ orbitals is *lowered* by 0.6 $\Delta_t$ and that of $t_2$ orbitals is raised by 0.4 $\Delta_t$ relative to the energy of no splitting state. Thus each electron occupying $e$ orbitals decreases the energy of $d$-orbitals by $-0.6\Delta_t$ while that going to $t_2$ orbitals increases its energy by $+0.4\ \Delta_t$. – and + signs indicate respectively the decrease and increase in the energy of d-orbitals caused by their splitting under the influence of four ligands. Now let us consider a $d^x$ ion having $e^p t_2^q$ configuration in which $p$ is the number of electrons in $e$ set of orbitals and q is the number of electrons in $t_2$ set of orbitals and $x = p + q$. Obviously:

CFSE for $d^x$ ion having $e^p t_2^q$ configuration $(x = p + q)$

= Loss in energy due to $p$ electrons in $e$ set of orbitals
+ Gain in energy due to $q$ electrons in $t_2$ set of orbitals.

$$= (-0.6\Delta_t \times p + 0.4\ \Delta_t \times q)$$

$$= (-0.6 \times p + 0.4 \times q)\Delta_t \qquad ...(i)$$

Equation ($i$) gives the value of CFSE in terms of $\Delta_t$

Now since $\Delta_t = 0.45\ \Delta_0$, equation ($i$) can be written as :

$$\text{CFSE} = (-0.6 \times p + 0.4 \times q) \times 0.45\ \Delta_0 \qquad ...(ii)$$

This equation gives the value of CFSE in terms of $\Delta_0$.

Now since $\Delta_0$ = 10 Dq, equation (ii) can be written as:

$$\text{CFSE} = (-0.6 \times p + 0.4 \times q) \times 0.45 \times 10\text{Dq}$$

$$= (-0.6 \times p + 0.4 \times q) \times 4.5\ \text{Dq} \qquad ...(iii)$$

Equation ($iii$) gives the value of CFSE in terms of Dq.

CFSE values of $d^x$ ions ($x$ = 0 to 10) having $e^p t_2^q$ configuration ($x = p + q$ = 0 to 10) in tetrahedral complexes (obviously these tetrahedral complexes are HS complexes) have been calculated with the help of equation ($i$), ($ii$) and ($iii$) as shown below.

In the calculation of these values pairing energy ($P$) has not been considered.

### $d^0$ ion

$e^p t_2^q$ configuration of $d^0$ ion = $e^0 t_2^0 (p = 0, q = 0)$

$\therefore$ CFSE for $d^\circ$ ion = $(-0.6 \times p + 0.4 \times q)\ \Delta_t$

$= (-0.6 \times 0 + 0.4 \times 0)\ \Delta_t =$ **Zero**

### $d^1$ ion

$e^p t_2^q$ configuration of $d^1$ ion = $e^1 t_2^0 (p = 1, q = 0)$

$\therefore$ CFSE for $d^1$ ion = $(-0.6 \times p + 0.4 \times q)\ \Delta_t$

$= (-0.6 \times 1 + 0.4 \times 0)\ \Delta_t = \mathbf{-0.6\Delta_t}$

$= -0.6 \times 0.45\ \Delta_0 = \mathbf{-0.270\ \Delta_0}$

$= -0.270 \times 10\text{Dq} = \mathbf{-2.7Dq}$

### $d^2$ ion

$e^p t_2^q$ configuration of $d^2$ ion = $e^2 t_2^0 (p = 2, q = 0)$

$\therefore$ CFSE for $d^2$ ion = $(-0.6 \times p + 0.4 \times q)\ \Delta_t$

$= (-0.6 \times 2 + 0.4 \times 0)\ \Delta_t = \mathbf{-1.2\Delta_t}$

$= -1.2 \times 0.45\ \Delta_0 = \mathbf{0.540\ \Delta_0}$

$= -0.540 \times 10\text{Dq} = \mathbf{-5.4Dq}$

### $d^3$ ion

$e^p t_2^q$ configuration of $d^3$ ion = $e^2 t_2^1 (p = 2, q = 1)$

$\therefore$ CFSE for $d^3$ ion = $(-0.6 \times p + 0.4 \times q)\ \Delta_t$

$= (-0.6 \times 2 + 0.4 \times 1)\ \Delta_t = \mathbf{-0.8\Delta_t}$

$= -0.8 \times 0.45\ \Delta_0 = \mathbf{0.36\ \Delta_0}$

$= -0.36 \times 10\text{Dq} = \mathbf{-3.6\ Dq}$

### $d^4$ ion

$e^p t_2^q$ configuration of $d^4$ ion = $e^2 t_2^2 (p = 2, q = 2)$

$\therefore$ CFSE for $d^4$ ion= $(-0.6 \times p + 0.4 \times q)\ \Delta_t$

$= (-0.6 \times 2 + 0.4 \times 2)\ \Delta_t = \mathbf{-0.4\Delta_t}$

$= -0.4 \times 0.45\ \Delta_0 = \mathbf{-0.18\ \Delta_0}$

$= -0.18 \times 10\ \text{Dq} = \mathbf{-\ 1.8\ Dq}$

### $d^5$ ion

$e^p t_2^q$ configuration of $d^5$ ion = $e^2 t_2^3 (p = 2, q = 3)$

$\therefore$ CFSE for $d^5$ ion = $(-0.6 \times p + 0.4 \times q)\ \Delta_t$

$= (-0.6 \times 2 + 0.4 \times 3)\ \Delta_t = 0 \times \Delta_t$

= **Zero**

### $d^6$ ion

$e^p t_2^q$ configuration of $d^6$ ion = $e^3 t_2^3 (p = 3, q = 3)$

$\therefore$ CFSE for $d^6$ ion = $(-0.6 \times p + 0.4 \times q)\ \Delta_t$

$= (-0.6 \times 3 + 0.4 \times 3)\ \Delta_t = \mathbf{-0.6\ \Delta_t}$

$= -0.6 \times 0.45\ \Delta_0 = \mathbf{-0.27\ \Delta_0}$

$= -\ 0.27 \times 10\ \text{D}q = \mathbf{-2.7\ Dq}$

### $d^7$ ion

$e^p t_2^q$ configuration of $d^7$ ion = $e^4 t_2^3 (p = 4, q = 3)$

$\therefore$ CFSE for $d^7$ ion = $(-0.6 \times p + 0.4 \times q)\ \Delta_t$

$= (-0.6 \times 4 + 0.4 \times 3)\ \Delta_t = \mathbf{-1.2\ \Delta_t}$

$= -1.2 \times 0.45\ \Delta_0 = \mathbf{-0.54\ \Delta_0}$

$= -\ 0.54 \times 10\ \text{D}q = \mathbf{-5.4\ Dq}$

### $d^8$ ion

$e^p t_2^q$ configuration of $d^8$ ion = $e^4 t_2^4\ (p = 4,\ q = 4)$

$\therefore$ CFSE of $d^8$ ion = $(-0.6 \times p + 0.4 \times q)\Delta_t$

$= (-0.6 \times 4 + 0.4 \times 4)\Delta_t = \mathbf{-0.8\ \Delta_t}$

$= 0.8 \times 0.45\ \Delta_0 = \mathbf{-0.36\ \Delta_0}$

$= -0.36 \times 10\text{Dq} = \mathbf{-3.6\ Dq}$

### $d^9$ ion

$e^p t_2^q$ configuration of $d^9$ ion = $e^4 t_2^5 (p = 4,\ q = 5)$

$\therefore$ CFSE for $d^9$ ion = $(-0.6 \times p + 0.4 \times q)\ \Delta_t$

$= (-0.6 \times 4 + 0.4 \times 5)\ \Delta_t = \mathbf{-0.4\ \Delta_t}$

$$= -0.4 \times 0.45\ \Delta_0 = \mathbf{-0.18\ \Delta_0}$$
$$= -0.18 \times 10\ Dq = \mathbf{-1.8\ Dq}$$

### $d^{10}$ ion

$e^p t_2^q$ configuration of $d^{10}$ ion = $e^4 t_2^6$ $(p = 4, q = 6)$

$\therefore$ CFSE for $d^{10}$ ion = $(-0.6 \times p + 0.4 \times q)\ \Delta_t$

$$= (-0.6 \times 4 + 0.4 \times 6)\ \Delta_t = \textbf{Zero}$$

## Splitting of d-Orbitals in Cubic Complexes.

We know that a cube has eight vertices, six faces and twelve edges. A cubic complex of a metal ion has the metal ion at the centre of the cube and eight ligands are placed as the eight vertices of the cube. Fig 18.6 shows the positioning of the metal ion (M) at the centre of the cube and four ligands (L) at the alternate vertices of the cube. The placement of the four ligands at the alternate vertices of the cube generates tetrahedral ligand field. Now if we place other four ligands at the remaining four alternate vertices of the cube, another tetrahedral field will be generated. As a result, as in tetrahedral ligand field, in the cubic ligand field, $t_2$ set ($d_{xy}$, $d_{yz}$, $d_{zx}$) will be destablished (*i.e.*, energy of $t_2$ set will increase) and *e* set ($d_{x^2-y^2}$, $d_{z^2}$) will be stabilised (i.e., energy of e set will decrease). (See Fig 18.7)

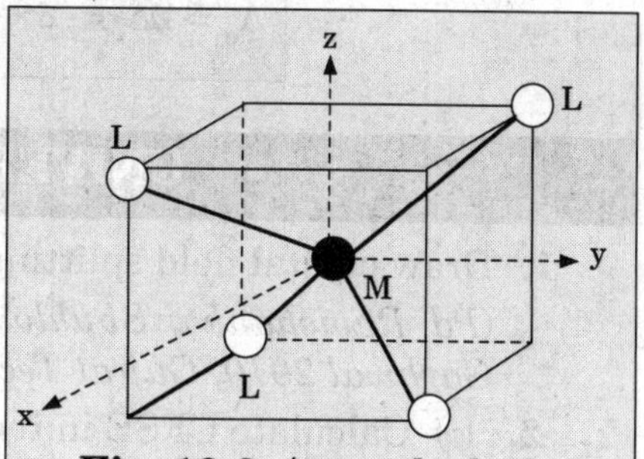


**Fig. 18.6.** A tetrahedral complex, $ML_4$

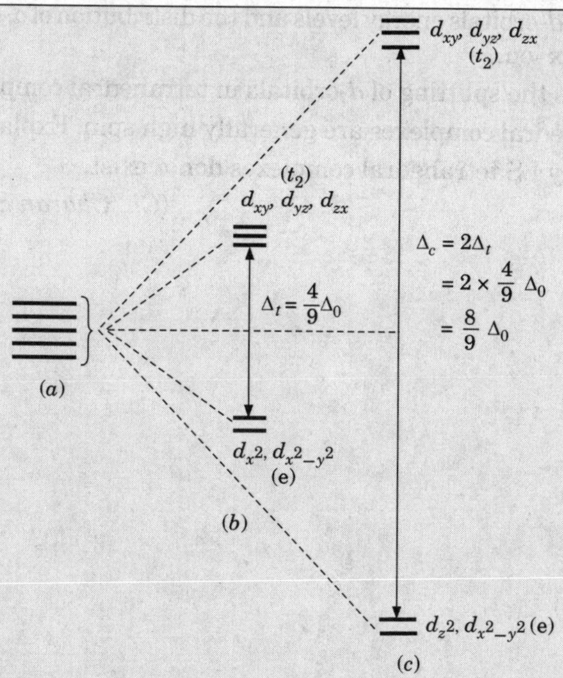


**Fig. 18.7.** (*a*) Five degenerate *d*-orbitals in a free metal ion (*b*) Splitting of *d*-orbitals in tetrahedral complex (*c*) Splitting of d-orbitals a complex having cubic geometry

Hence the overall splitting of d-orbitals in cubic ligand field ($\Delta_c$) will be double that in tetrahedral ligand field ($\Delta_t$), *i.e.*,

$$\Delta_c = 2\Delta_t \quad \ldots(i)$$

Now since $\Delta_t = \frac{4}{9}\Delta_0$, $\Delta_c = 2 \times \frac{4}{9}\Delta_0 = \frac{8}{9}\Delta_0$ ...(*ii*)

On combining relations (*i*) and (*ii*), we get:

$$\boldsymbol{\Delta_c = 2\Delta_t = 2\times\frac{4}{9}\Delta_0 = \frac{8}{9}\Delta_0}$$

## UNIVERSITY QUESTIONS

**1.** Draw crystal field splitting diagram of d orbitals in tetrahedral field. (*Pd. Ravishankar Shukla 2010), Himachal Pradesh 2010, H.N. Bahuguna Garhwal 2010, Gujrat Technological University 2010*)

**2.** (*i*) Calculate CFSE in terms of $\Delta_t$ in $d^4$ in $d^6$ tetrahedral complexes.
(*ii*) $[NiCl_4]^{2-}$ is tetrahedral and paramagnetic while $[Ni\,(CN)_4]^{2-}$ is square planar and diamagnetic. Explain why? (*Nagpur 2010*)

**3.** Explain why there are no low and high spin tetrahedral complexes? (*Madras April 2011*)

**4.** (*i*) Calculate CFSE for $d^5$ tetrahedral ion.
(*ii*) Sketch *d*-orbitals energy levels and the distribution of *d*-electrons in $[NiCl_4]^{2-}$ complex ion. (*Kurukshetra 2011*)

**5.** (*a*) Discuss the splitting of *d*-orbitals in tetrahedral complexes.
(*b*) Tetrahedral complexes are generally high spin. Explain. (*Nagpur 2012*)

**6.** Explain why LS tetrahedral complexes donot exist. (*Ch. Charan Singh, Meerut 2012*)

# 19 CHAPTER Crystal Field Theory of Square Planar Complexes

Square planar complex may be considered to have been derived from octahedral complex by removing the two ligands on $z$-axis.

## Splitting of d-Orbitals in Square Planar Complexes

We know that square planar geometry is obtained from octahedral geometry by removing the two ligands lying on z-axis of the octahedral geometry. Now since the ligands are removed from z-axis, the energies of $d_{yz}$ $d_{zx}$ and $d_{z^2}$ orbitals (all these orbitals have z-component) get lowered, i.e. these orbitals become more stable.

The splitting of d-orbitals in square planar complex can be explained as follows:

In square planar complexes the four ligands (L) lie on $+x$, $-x$, $+y$ and $-y$ directions of $x$ and $y$ axes. z-axis has no ligand (See Fig 19.1). These four ligands approach towards the central metal ion (M) along $+x$, $-x$, $+y$ and $-y$ direction of $x$ and $y$ axes. In this process the electrons in d-orbitals of the central metal ion are repelled by the approaching ligands. Now since the four lobes of $d_{x^2-y^2}$ orbital lie on $x$ and $y$ axis, this orbital directly faces the four ligands approaching towards the central metal atom and hence is repelled most strongly by the ligands, i.e. $d_{x^2-y^2}$ orbital is most destablished. In other words we say that the energy of $d_{x^2-y^2}$ orbital rises above the no splitting state to the maximum extent (Greater is the repulsion, greater is the increase in energy). Since the lobes of $d_{xy}$ orbital lie between $x$ and $y$ axes (i.e., between the ligands) in $xy$ square plane, this orbital is also repelled by the ligands but to a lesser extent than $d_{x^2-y^2}$ orbital. Thus $d_{xy}$ orbital has lesser energy than $d_{x^2-y^2}$ orbital. Since $d_{yz}$ and $d_{zx}$ orbitals do not lie in the immediate vicinity of the approaching four ligands and there is no ligand on z-axis, the energy of the three orbitals *viz* $d_{yz}$, $d_{zx}$ and $d_{z^2}$ goes down the no splitting state.

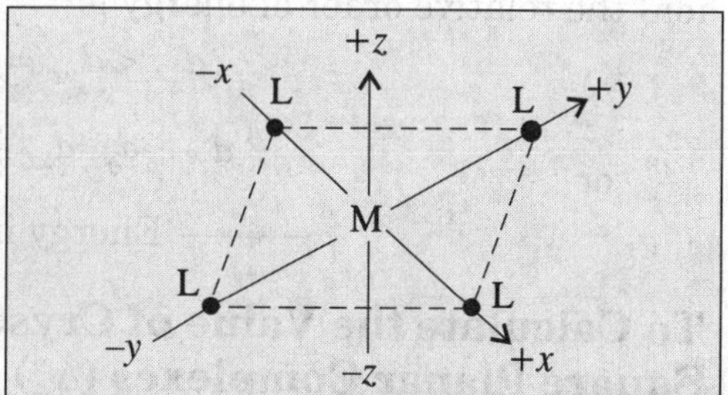


**Fig 19.1.** Position of four ligands (L) in a square planar complex. M is the central metal atom

$d_{yz}$ and $d_{zx}$ orbitals still remain degenerate. This discussion shows that the splitting of five d-orbitals in a square planar geometry can be shown as given in Fig 19.2.

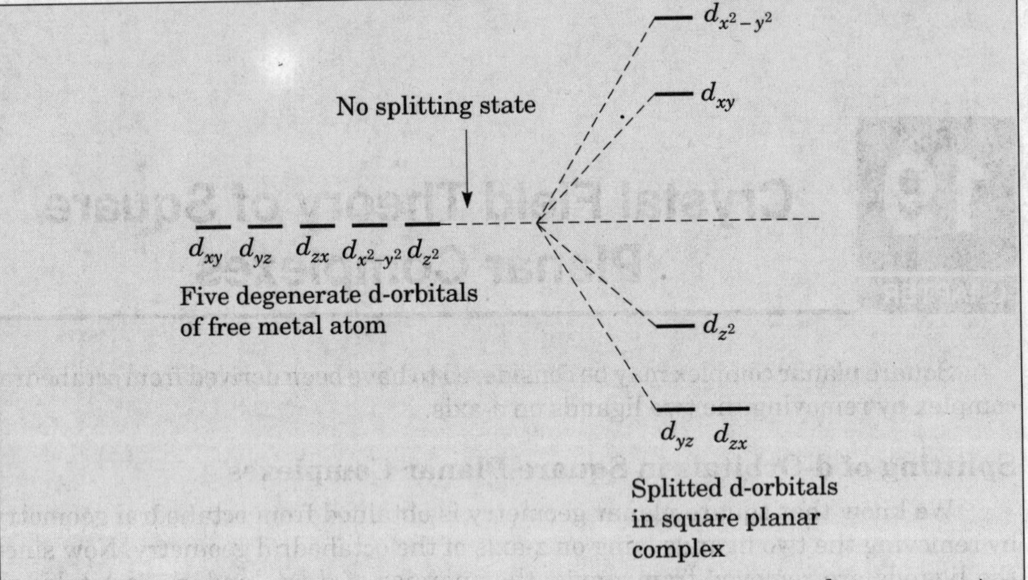


**Fig. 19.2.** Splitting of d-orbitals of the central metal ion in square planar geometry.

Fig. 19.2 shows that the relative order of the energy of the splitted d-orbitals in square planar complexes is

$$d_{yz} = d_{zx} < d_{z^2} < d_{xy} < d_{x^2-y^2}$$

or

$$(d_{xz}, d_{yz}) \;\vdots\; d_{z^2} \;\vdots\; d_{xy} \;\vdots\; d_{x^2-y^2}$$

Energy increasing ⟶

This order has been predicted for square planar complexes $Co^{2+}(3d^7)$, $Ni^{2+}(3d^8)$, $Cu^{2+}(3d^9)$, $Ag^{2+}(4d^9)$, $Ag^{3+}(4d^8)$ etc. ions. In case of $[Pt\,Cl_4]^{2-}$, $d_{z^2}$ orbital has lesser energy even than the degenerate $d_{yz}$ and $d_{zx}$ orbitals, *i.e.*, in $[PtCl_4]^{2-}$ ion ($Pt^{2+} = 5d^8$ ion) the relative order of energy is

$$d_{z^2} < d_{yz} = d_{zx} < d_{xy} < d_{x^2-y^2}$$

or

$$d_{z^2} \;\vdots\; d_{yz}, d_{xz} \;\vdots\; d_{xy} \;\vdots\; d_{x^2-y^2}$$

Energy increasing ⟶

## To Calculate the Value of Crystal Field Splitting Parameter in Square Planar Complexes ($\Delta_{sp}$)

We have shown that when *d*-orbitals of the central metal atom in square planar complexes are split, we get $d_{x^2-y^2}$ orbital at the highest energy level and $d_{yz}$ and $d_{zx}$ orbitals ($d_{yz}$ and $d_{zx}$ orbitals are degenerate orbitals) at the lowest energy level. The energy difference between the lowest $d_{yz} = d_{zx}$ orbital and the highest $d_{x^2-y^2}$ orbital is denoted by $\Delta_{sp}$. Thus:

$\Delta_{sp} = d_{x^2-y^2} - d_{yz}$ or $d_{zx}$

$= \Delta_1 + \Delta_2 + \Delta_3$ (See Fig 19.3)

The splitting of $d$-orbitals in square planar complexes takes place in such a way that the energy of $d_{yz}$, $d_{zx}$ and $d_{z^2}$ orbitals is lowered by 5.14 Dq (= 0.514 $\Delta_0$), 5.14 Dq (= 0.514$\Delta_0$) (Since $d_{yz}$ and $d_{zx}$ orbitals lie on the same energy level) and 4.28 Dq (= 0.428 $\Delta_0$) respectively while the energy of $d_{xy}$ and $d_{x^2-y^2}$ orbitals is raised by 2.28 Dq (= 0.228$\Delta_0$) and 12.28Dq (= 1.228$\Delta_0$) respectively relative to the energy of no splitting state. (See Fig 19.3). This figure shows that:

$$\Delta_1 = d_{x^2-y^2} - d_{xy} = 1.228\Delta_0 - 0.228\,\Delta_0 = \Delta_0$$

$$\Delta_2 = d_{xy} + d_{z^2} = 0.228\,\Delta_0 + 0.428\Delta_0 = 0.66\,\Delta_0$$

$$\Delta_3 = (d_{yz} \text{ or } d_{zx} - d_{z^2}) = 0.514\,\Delta_0 - 0.428\,\Delta_0$$

$$= 0.086\,\Delta_0$$

The values of $\Delta_1$, $\Delta_2$ and $\Delta_3$ calculated as above have been found for the square planar complexes of $Co^{2+}$ ($3d^7$), $Ni^{2+}$ ($3d^8$) and $Cu^{2+}$($3d^9$). Thus for the square planar complexes of $Co^{2+}$, $Ni^{2+}$ and $Cu^{2+}$, the value of $\Delta_{sp}$ which is the energy difference between the lowest $d_{yz} = d_{zx}$ and highest $d_{x^2-y^2}$ orbitals is given by :

$$\Delta_{sp} = \Delta_1 + \Delta_2 + \Delta_3 = \Delta_0 + 0.66\,\Delta_0 + 0.086\,\Delta_0 = 1.746\,\Delta_0 = 17.46 \text{ Dq}$$

$$(\because \Delta_0 = 10\text{Dq})$$

For square planar complexes of $Pd^{2+}$ ($4d^8$) and $Pt^{2+}$($5d^8$), spectroscopic results have shown that:

$$\Delta_{sp} = \Delta_1 + \Delta_2 + \Delta_3 = 1.3\,\Delta_0$$

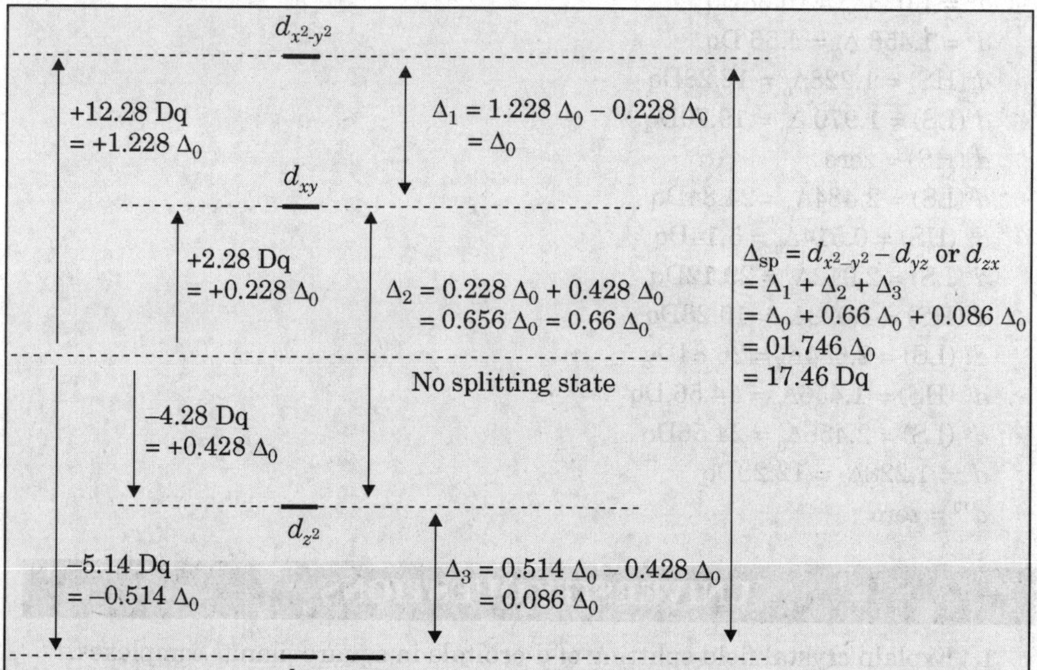


**Fig. 19.3.** Calculation of $\Delta_{sp}$ in square planar complexes

## Comparison between $\Delta_0$ and $\Delta_{sp}$

We have seen that $\Delta_{sp} = 1.746\ \Delta_0$ (for square planar complexes of $Co^{2+}$, $Ni^{2+}$ and $Cu^{2+}$) and $\Delta_{sp} = 1.3\Delta_0$ (for the square planar complexes of $Pd^{2+}$ and $Pt^{2+}$). Both these relations show that $\Delta_{sp} > \Delta_0$. Larger value of $\Delta_{sp}$ as compared to that $\Delta_0$ is because of the fact that in square planar geometry $d_{zy}$ and $d_{zx}$ orbitals interact with only two ligands while in octahedral geometry these orbitals interact with four ligands.

## Crystal Field Stabilisation Energy (CFSE) Values of $d^x$ Ions ($x = 0$ to 10) in Weak Ligand (High Spin) and Strong Ligand (Low Spin) Square Planar Complexes.

We have already shown that the crystal field splitting of metal d-orbitals in square planar complexes takes place in such away that the energy of $d_{yz}$, $d_{zx}$ and $d_{z^2}$ orbitals is lowered by $0.514\Delta_0$, $0.514\Delta_0$ (since $d_{yz}$ and $d_{zx}$ orbitals lie on the same energy level) and $0.428\ \Delta_0$ respectively while the energy of $d_{xy}$ and $d_{x^2-y^2}$ orbitals is raised by $0.228\Delta_0$ and $1.228\ \Delta_0$ respectively relative to the energy of no splitting state. The values of CFSE for various $d^x$ ions of weak ligand (high spin) and strong ligand (low spin) square planar complexes can be calculated from the electron distribution into various splitted d-orbitals, using the same method as applied to octahedral and tetrahedral complexes. These values thus obtained are given (in terms of $\Delta_0$ and Dq) below. Note $\Delta_0 = 10\ Dq$.

$d^0 = 0$

$d^1$(HS) $= 0.514\Delta_0 = 5.14$Dq

$d^1$(LS) $= 0.57\Delta_0 = 5.7$Dq

$d^2 = 1.028\Delta_0 = 10.28$Dq

$d^3 = 1.456\ \Delta_0 = 4.56$ Dq

$d^4$(HS) $= 1.228\Delta_0 = 12.28$Dq

$d^4$(LS) $= 1.970\ \Delta_0 = 19.70$Dq

$d^5$(HS) = zero

$d^5$(LS) $= 2.484\Delta_0 = 24.84$Dq

$d^6$ (HS) $= 0.514\Delta_0 = 5.14$Dq

$d^6$(LS) $= 2.912\Delta_0 = 29.12$Dq

$d^7$(HS) $= 1.028\Delta_0 = 10.28$Dq

$d^7$ (LS) $= 2.684\Delta_0 = 26.84$Dq

$d^8$ (HS) $= 1.456\Delta_0 = 14.56$ Dq

$d^8$ (LS) $= 2.456\Delta_0 = 24.56$Dq

$d^9 = 1.228\Delta_0 = 12.28$Dq

$d^{10}$ = zero

## UNIVERSITY QUESTIONS

1. Explain crystal field splitting of d-orbitals in square planar complexes. (*Osmania 2010, Nagpur 2012*)
2. Draw the d-orbital splitting diagram of $[Ni\ (CN)_4]^{2-}$ and write CFT configuration. (*Nagpur 2010*)

# 20 Splitting of Metal Ion d-Orbitals in Other Crystal Fields

CHAPTER

## Introduction

We have already discussed the splitting of d-orbitals of the central metal ion in octahedral, tetragonal, square planar, tetrahedral and cubic complexes. In this chapter we shall discuss the splitting of d-orbitals of the central metal ion in square pyramidal, trigonal bipyramidal, pentagonal bipyramidal complexes etc.

## 1. Splitting of *d*-orbitals in Square Pyramidal Complexes

Square pyramidal geometry of a complex, $ML_5$ is given in Fig 20.1. In these complexes $d_{x^2-y^2}$ orbital is face-to-face with the four ligands lying in equatorial *xy* plane and hence this orbital is the must destablished orbital, i.e. $d_{x^2-y^2}$ orbital has the highest energy. The energy of the remaining orbitals is in the order:

$$d_{xz} = d_{yz} \text{ (doublet set)} < d_{xy} < d_{z^2}$$

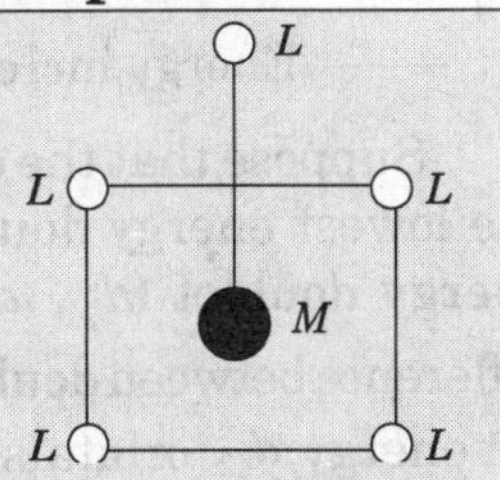


**Fig 20.1** Square pyramidal geometry of a complex, $ML_5$

Thus the overall order of energy of the five d-orbitals is as:

$$d_{xz} = d_{yz} < d_{xy} < d_{z^2} < d_{x^2-y^2}$$

or $d_{xz}, d_{yz}$ ¦ $d_{xy}$ ¦ $d_{z^2}$ ¦ $d_{x^2-y^2}$ (See Fig. 20.2)

——— Energy increasing ———→

## 2. Splitting of *d*-orbitals in Trigonal Bipyramidal Complexes.

Trigonal bipyramidal geometry of a complex, $ML_5$ is given in Fig 20.3. In these complexes, since $d_{z^2}$ orbital faces the two ligands directly, this orbital is the most destablised and hence has the highest energy. However, the equatorial orbitals in *xy* plane *viz* $d_{xy}$ and $d_{x^2-y^2}$ orbitals do not face the three ligands (in *xy* plane) directly but are equidistant with respect to them. As expected, $d_{xz}$ and $d_{yz}$ orbitals are the most stabilised and hence have the lowest energy (See Fig 20.4).

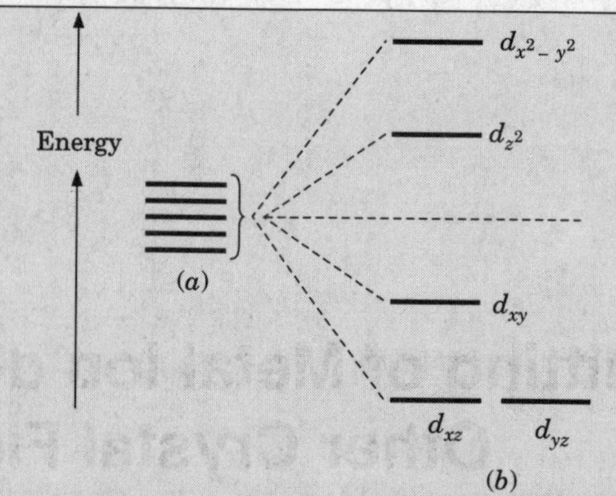


**Fig. 20.2.** (a) Five degenerate d-orbitals in free metal ion (b) Splitting of d-orbitals in square pyramidal complexes.

Fig 20.4 shows that the order of energy of five *d*-orbital of the central metal ion in a trigonal bipyramidal complex is as:

$d_{xz}, d_{yz}$ ¦ $d_{xy}, d_{x^2-y^2}$ ¦ $d_{z^2}$

—— Energy increasing ——→

Suppose that the difference in energy between the lowest energy doublet ($d_{xz}$, $d_{yz}$) and the upper energy doublet ($d_{xy}$, $d_{x^2-y^2}$) is $\delta_1$ and the energy difference between doublet ($d_{xy}$, $d_{x^2-y^2}$) and the highest energy $d_{z^2}$ orbitals is $\delta_2$. It has been found that:

(*a*) $\delta_1$ is usually much smaller than $\delta_2$.

(*b*) $\delta_1$ can never exceed pairing energy (*P*) of a metal ion.

(*c*) $\delta_2$ may be less or greater than *P*.

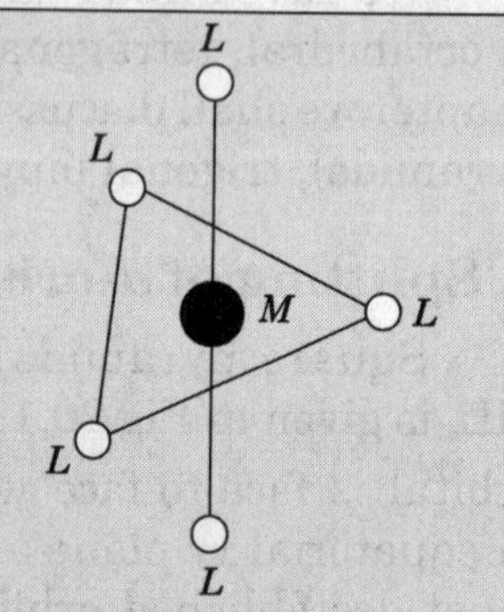


**Fig 20.3.** Trigonal bipyramidal geometry of a complex, $ML_5$.

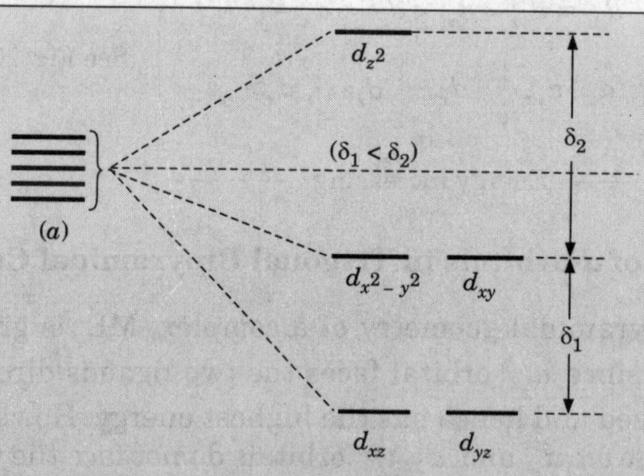


**Fig 20.4.** (a) Five degenerate d-orbitals in free metal ion (b) Splitting of d-orbitals in trigonal bipyramidal geometry.

Therefore, trigonal bipyramidal complexes of $d^1$, $d^2$, $d^3$ and $d^4$ ions will always be HS complexes and those of $d^5$, $d^6$, $d^7$ and $d^8$ ions may be either HS or LS. Hence with favourable ligands (*i.e.* with $\delta_2 \simeq P$), we may observe anomalous moments with $d^5$, $d^6$, $d^7$ and $d^8$ ions. The arrangement of $d^5$ electrons in the HS and LS trigonal bipyramidal complex of metal ion with $d^5$ configuration is :

$$HS = (d^1_{xz}, d^1_{yz})(d^1_{x^2-y^2}, d^1_{xy})\ d^1_{z^2}$$

$$LS = (d^2_{xz}, d^1_{yz})\ (d^1_{x^2-y^2}, d^1_{xy})\ d^0_{z^2}$$

## 3. Splitting of *d*-orbitals in Trigonal Planar Complexes.

Since a trigonal planar geometry can be obtained by removing two ligands along z-axis in trigonal bipyramidal geometry, $d_{z^2}$ orbital is the most stabilised orbital, i.e, $d_{z^2}$ orbital has the lowest energy. The relative order of energy of various orbitals obtained by the splitting of d-orbitals of metal ion of a trigonal planar complex is as:

$$d_{x^2-y^2} = d_{xy} > d_{xz} = d_{yz} > d_{z^2}$$

or $d_{z^2}$ ¦ $d_{xz}, d_{yz}$ ¦ $d_{xy}, d_{x^2-y^2}$

—— Energy increasing ——→

## 4. Splitting of *d*-orbitals in a Linear Geometry

In a linear geometry z-axis is taken as the unique axis and the two ligands present in linear complex are placed on either side of the metal ion. It can be guessed that the relative order of *d*-orbitals is:

$$d_{z^2} > d_{xz} = d_{yz} > d_{xy} = d_{x^2-y^2}$$

or $d_{xy}, d_{x^2-y^2}$ ¦ $d_{xz}, d_{yz}$ ¦ $d_{z^2}$

—— Energy increasing ——→

## 5. Splitting of *d*-orbitals in Pentagonal Bipyramidal Complexes.

In pentagonal bipyramidal complexes the metal in (M) is surrounded by five ligands (L) placed at the five corners of equatorial pentagon and two ligands (L) in the axial position (See Fig 20.5)

A pentagonal geometry can also be obtained by adding an extra ligand to the square plane of an octahedral geometry. Compared to the splitting of d-orbitals in trigonal bipyramidal geometry, the energy of ($d_{xy}$, $d_{x^2-y^2}$) set is increased due to the greater repulsion from the additional ligand in pentagonal bipyramidal complexes. The energy of ($d_{yz}$, $d_{zx}$) set is decreased [See (d) of Fig 20.6] In Fig 20.6 we have shown the splitting of d-orbitals of square pyramidal, trigonal bipyramidal and pentagonal bipyramidal complexes relative to octahedral complexes.

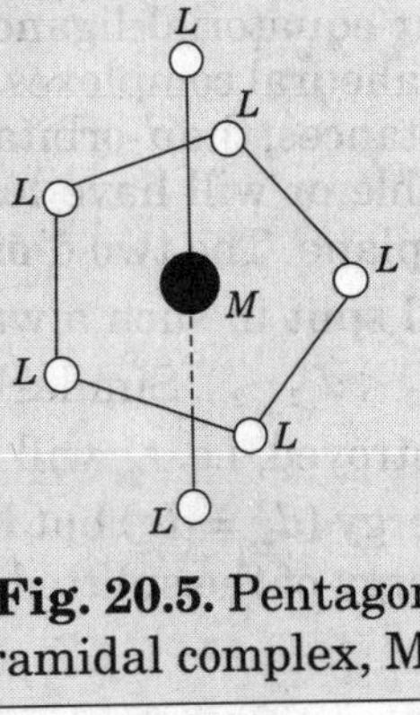


**Fig. 20.5.** Pentagonal bipyramidal complex, $ML_7$

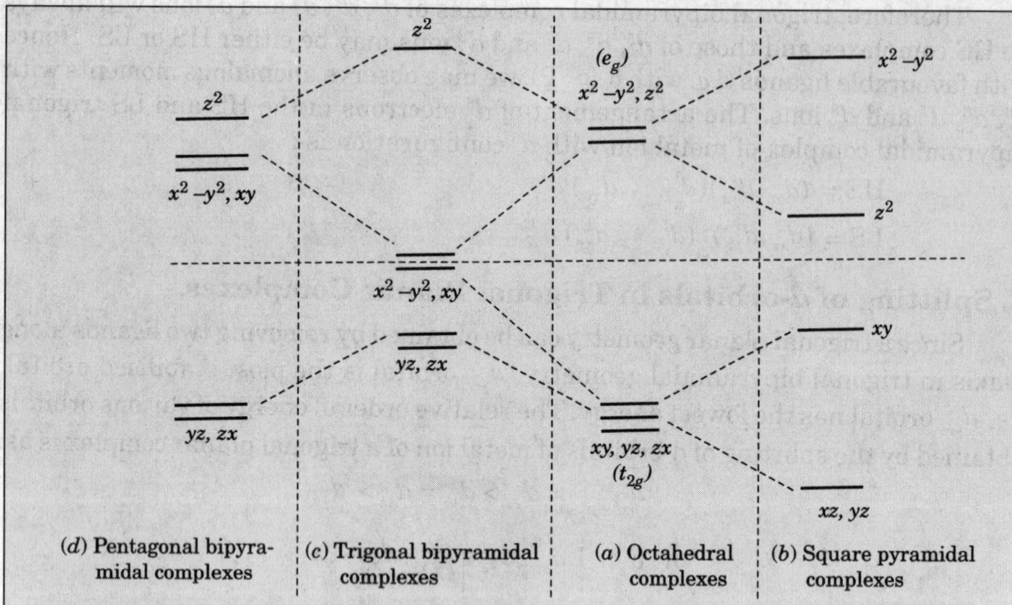


**Fig. 20.6.** Splitting of d-orbitals in square pyramidal, trigonal bipyramidal and pentagonal bipyramidal complexes reltive that in octahedral complexes.

Fig. 20.6(*d*) shows that the order of energy of five d-orbitals of the central metal ion in pentagonal bipyramidal complexes is as:

$$d_{zx}, d_{yz} \mid d_{xy}, d_{x^2-y^2} \mid d_{z^2}$$

—— Energy increasing ——→

**6. Splitting of *d*-orbitals in Elongated Distorted Octahedral Complexes** (*Also see Chapter : John. Teller Effect and Distortion in Octahedral and Tetrahedral Complexes*).

We know that in an octahedral complex the five d-orbitals are split into $t_{2g}$ ($d_{xy}$, $d_{yz}$ and $d_{xz}$) and $e_g$ ($d_{z^2}$ and $d_{x^2-y^2}$) sets of orbitals. $t_{2g}$ set has lower energy than $e_g$ set of orbitals. The three d-orbitals of $t_{2g}$ set are degenerate and similarly the two *d*-orbitals of $e_g$ set are also degenerate. The octahedral complexes in which the two trans ligands are at longer distance from the central metal ion than the other four equatorial ligands from the central metal ion are called elongated distorted octahedral complexes. In these complexes, since the negative ligands are at longer distances, the *d*-orbitals which have their lobes pointing towards z-axis will be more stable or will have less energy than the *d*-orbitals which have their lobes in the *xy* plane. The two d-orbitals of $e_g$ set will no longer be degenerate and hence $e_g$ set will split in such a way that $d_{z^2}$ orbital will have less energy than $d_{x^2-y^2}$ orbital ($d_{z^2} < d_{x^2-y^2}$). Similarly, the degeneracy of the three d-orbitals of $t_{2g}$ set will also be destroyed, i.e. $t_{2g}$ will split in such way that $d_{xz}$ and $d_{yz}$ orbitals will have the same energy ($d_{xz} = d_{yz}$) but less than $d_{xy}$ orbital ($d_{xz} = d_{yz} < d_{xy}$). Thus the relative order of energy of the splitted *d*-orbitals in a tetragonally distorted octahedral complex is:

$$(d_{xz} = d_{yz}) < d_{xy} < d_{z^2} < d_{x^2-y^2}$$

or

$$(d_{xz}, d_{yz}) \mid d_{xy} \mid d_{z^2} \mid d_{x^2-y^2}$$

—— Energy increasing ——→

# 21 CHAPTER

# John-Teller Effect and Distortion in Octahedral and Tetrahedral Complexes

## Distorted Octahedral Complexes.

We know that in an octahedral complex all the six M – L (L = ligand and M = central metal atom) bond lengths are equivalent. If M – L bonds are of unequal lengths, the octahedral geoemtry of the complex gets **distorted** and we get **distorted octahedral complex.**

## Types of Distorted Octahedral Complexes

1. **Tetragonally elongated distorted octahedral complexes.** (Also see Fig 21.1) These complexes are also called simply by the name tetragonally distorted octahedral complexes. In an octahedral complex, if the two *trans* or axial ligands ($L_a$) are moved away from the central metal atom, the distances of these two ligands from the metal atom becomes larger than that of each of the remainng four coplanar equaotorial ligans ($L_e$) from the central metal atom. Thus the two *trans* bonds (also called *axial bonds*) or metal-ligand bonds along +z and –z axes become longer than each of the four equatorial bonds (equatorial) bonds are metal–ligand bonds along +*x*, –*x*, +*y* and –*y* axes in *xy* plane). The *trans* bonds become longer because these bonds are weaker than the equatorial bonds. *Such octahedral complexes in which the two trans ligands are longer than each of the four equatorial bonds are called tetragonally elongated distorted octahedral complexes or simply tetragonally distorted octahedral complexes* (See Fig 21.1) The distrotion which gives tetragonally elongated distorted octahedral complex is called *tetragonal elongation distortion*.
2. **Square planar complexes** (Also see Fig 21.1). Some times the tetragonal elongation distortion in tetragonally elongated octahedral complexes is so large that the two *trans* ligands on z-axis are completely removed and hence octahedral complexes adopt square planar geometry. For example square planar complexes of Cu (II) are highly distroted octahedral complexes.
3. **Tetragonally compressed distorted octahedral complexes.** (Also see Fig 21.1) In an octahedral complex, if the two *trans* ligands are brought

nearer to the central metal atom, the distance of these two *trans* ligands from the metal atom becomes shorter than that of each of the remaining four equatorial ligands from the central metal atom. Thus the two *trans* ligands become shorter than each of the four equatorial bonds. The two *trans* ligands become shorter because these bonds are stronger than the equatorial bonds. *Such octahedral complexes in which the two trans ligands are shorter than each of the four equotorial bonds are called tetragonally compresed (or contracted) distorted octahedral complexes.* The distortion which gives tetragonally compressed distorted octahedral complex is called *tetragonal contraction distortion*.

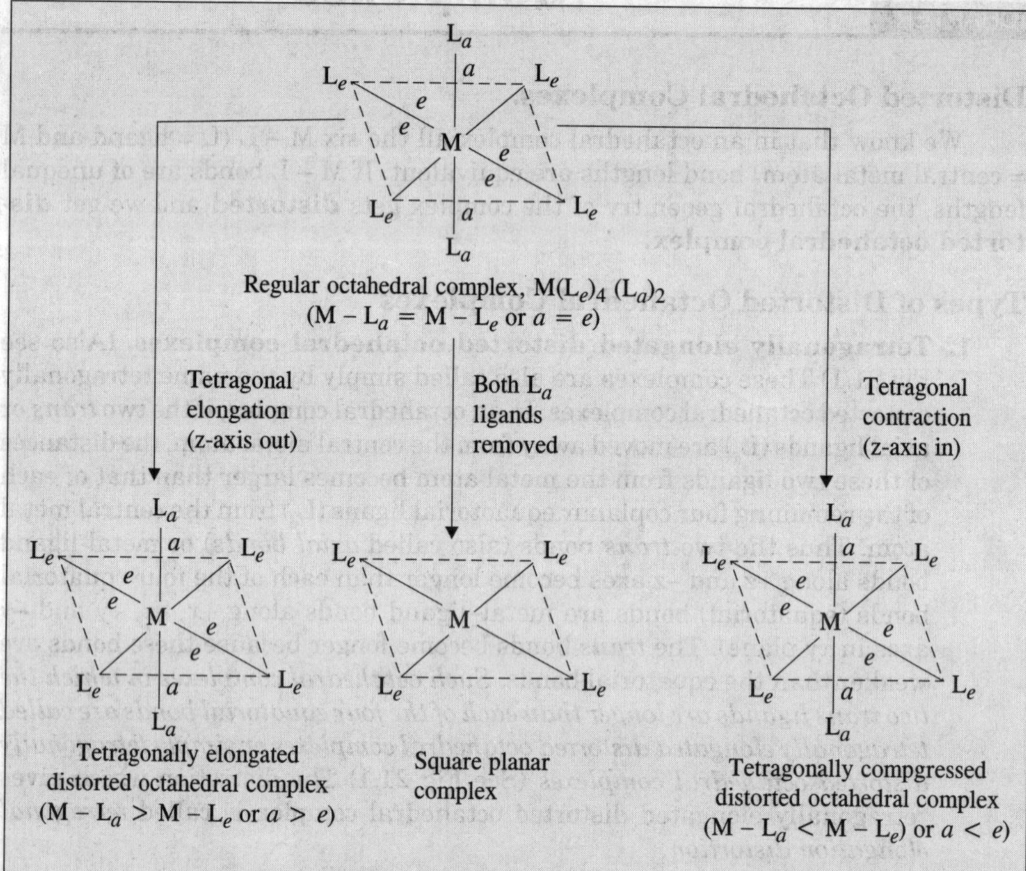


**Fig 21.1** Diferrent types of distorted octahedral complexes.

4. **Trigonally distorted octahedral comlexes: Trigonal prismatic complexes.** When an axis passing through the centres of the two opposite triangular faces of an octahedron is increased (elongtion) or decreased (compression or contraction) in length, otahedral complex gets distorted and assumes **trigonal prismatic geometry** (See Fig 21.2)

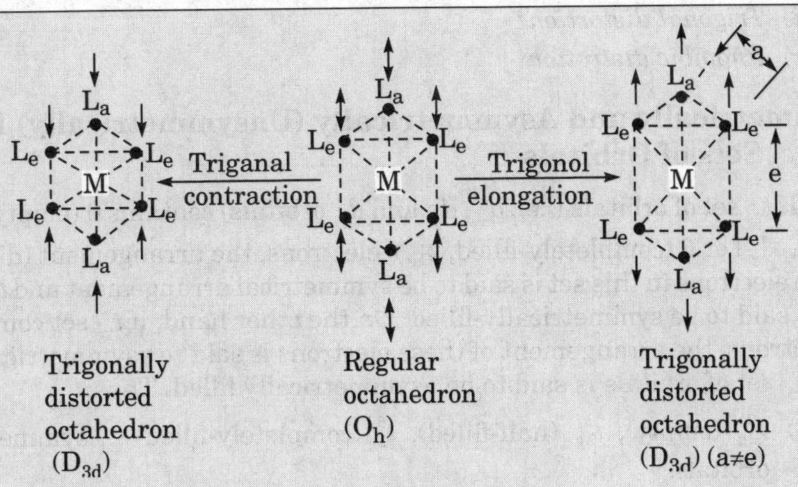


**Fig. 21.2.** Trigonal distrotion in an octahedral $ML_6$ complex

## Examples of Trigonal Prismatic Complexes.

(*i*) Recently it has been established that the complex, *tris* (*cis*-1, 2-diphenylethene-1, 2-dithiolate) rhenium, $[Re(S_2C_2Ph_2)_3]$ has trigonal prismatic structure shown in the margin in Fig. 21.3. It is not known with certainty whether the ligand, $S_2C_2Ph_2$ is a charged ligand or a neutral ligand. It is also not known whether Re metal has some charge or is present as a neutral atom.

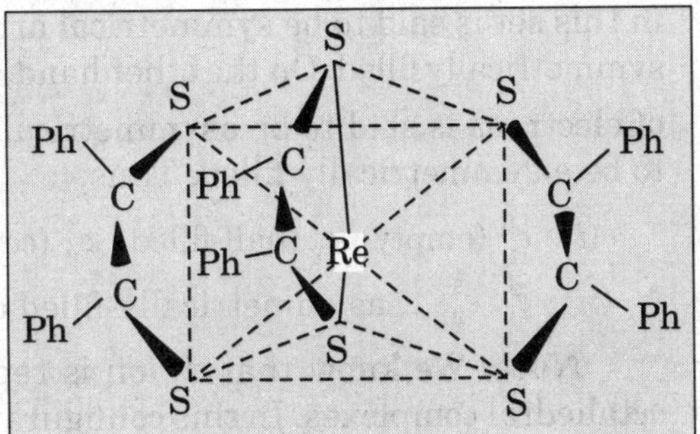


**Fig. 21.3.** Trigonal prismatic structure of $[Re(S_2C_2Ph_2)_3]$

(*ii*) Prismatic coordination number has also been established for polynuclear cobalt complex namely bis {tris (amidoethoxide) cobalt (III)} cobalt (II) ion, $[Co^{2+}\{Co^{3+}(NH_2 . CH_2 . CH_2O)_3\}_2]^{2+}$ in which the central Co (II) atom is supposed to be in prismatic coordination.

**5. Rhombic distorted octahedral complexes.** These are octahedral complexes in which all the six M–L bonds distances are unequal. The distortion produced in such complexes is called **rhombic distrotion.**

## Types of Distrotion in Octahedral Complexes

The forgoing discussion clearly shows that the octahedral complexes can undergo the following types of distortion.

(*i*) *Tetragonal distortion.* This distortion is of the following two types:

(*a*) Tetragonal elongation distortion

(*b*) Tetragonal contraction (or compression) distortion

(*ii*) *Trigonal distortion*

(*iii*) *Rhombic distrotion*

## Symmetrically and Asymmetrically (Unsymmetrically) Filled $t_{2g}$ and $e_g$ Sets of Orbitals

If $t_{2g}$ set of orbitals (*i.e.*, $d_{xy}$, $d_{yz}$ and $d_{zx}$ orbitals) contains 0 (empty, $t_{2g}^0$), 3(half-filled, $t_{2g}^3$) or 6 (completely-filled, $t_{2g}^6$) electrons, the arrangement (distribution) of these electrons in this set is said to be symmetrical arrangement and $t_{2g}$ set of orbitals is said to be symmetrically-filled. On the other hand, if $t_{2g}$ set contain 1, 2, 4 or 5 electrons, the arrangement of these electrons is said to asymmetrical arragneent and $t_{2g}$ set of orbitals is said to be asymmetrically filled. Thus:

(*i*) $t_{2g}^0$ (empty), $t_{2g}^3$ (half-filled), $t_{2g}^6$ (completely-filled) ... symmetrically filled orbitals.

(*ii*) $t_{2g}^1, t_{2g}^2, t_{2g}^4, t_{2g}^5$ ... asymmetrially-filled orbitals.

If $e_g$ set of orbitals (*i.e.*, $d_{x^2-y^2}$ and $d_{z^2}$ orbitals) has zero (empty, $e_g^0$), 2 (half-filled, $e_g^2$) or 4(completely filled, $e_g^4$) electrons, the arrangement of these electrons in this set is said to be symmetrical arrangement and $e_g$ set of orbitals is said to be symmetrically filled. On the other hand, if $e_g$ set has 1 or 3 electrons, this arrangement of electrons is said to be asymmetrical arrangement of electrons and $e_g$ set is said to be a symmetrically filled. Thus:

(*i*) $e_g^0$ (empty), $e_g^2$ (half-filled), $e_g^4$ (completely filled) ... symmetrically filled orbitals

(*ii*) $e_g^1, e_g^3$ ... asymmetrically-filled orbitals

**Note:** We know that $d^8$ ion is represented by the configuration, $t_{2g}^6 e_g^2$ in LS octahedral complexes. In this configuration, both the electrons of $e_g$ set are assumed to be present in $d_{z^2}$ orbital and $d_{x^2-y^2}$ orbital remains vacant. Thus $t_{2g}^6 e_g^2$ configuration is represented as $t_{2g}^6\ d_{z^2}^2\ d_{x^2-y^2}^0$. In this configuration, although $e_g$ set is half-filled, it is said to asymmetrically filled, since both the electrons are in $d_{z^2}$ orbital.

## John-Teller Theorem or John-Teller Effect

In 1937, John and Teller put forward a remarkable theorem to explain why some six coordinated complexes (octahedral cmplexes) undergo distortion and have distorted octahedral geometry. This theorem states that if a molecule / complex possessing any shape (excepting linear shape) has an unsymmetrically filled set of degenerate orbitals is unstable and hence undergoes distrotion. This distrotion takes place in such a way that the symmetry and energy of the molecule / complex is lowered and hence the molecule / complex becomes more stable. This theorem does not say which type of distortion will occur in the molecule / complex. Thus, according to this theorem, if the electrons in d-orbitals of the ecentral metal ion of a regular octahedral complex are asymmetrically arranged, the octahedral shape of the complex will get distorted, since the electrons present in a symmetrically-filled *d*-orbitals will repell some ligands to greater extent while other ligands will be repelled to lesser extent. The distortion in geometry of non-linear system due to John-Teller effect is called John-Teller distortion.

## Distortions Shown by Octahedral Complexes of $d^x$ Ions. ($x$ = 0 to 10)

Octahedral complexes of some $d^x$ ions show strong distortion while some ions show weak distortion. There are also some ions which do not show any distrotion.

1. **$d^0$ ion** ($t_{2g}^0 e_g^0$ with $n = 0$). Octahedral complexes of $d^0$ ion *do not show any distrotion*. Examples of such complexes are $[TiX_6]^{2-}$ where X = F, Cl.
2. **$d^1$ ion.** $Ti^{3+}$ ion is a $d^1$ ion. This ion is present in octahedral complex ion, $[Ti(H_2O)_6]^{3+}$. In the ground state $d^1$ configuration is represented as $t_{2g}^1 e_g^0$. $t_{2g}^1$ electron may be present in $d_{xy}$ orbial or in $d_{xz}$ / $d_{yz}$ orbital. When $t_{2g}^1$ electron is present in $d_{xy}$ orbital, $[Ti(H_2O)_6]^{3+}$ ion undergoes strong distrotion and assumes compressed distorted octahedral goemetry. On the other hand, if $t_{2g}^1$ electron is present in $d_{xz}$ or $d_{yz}$ orbital, $[Ti(H_2O)_6]^{3+}$ undergoes strong distortion and assumes elongated distorted octahedral geometry.
3. **$d^2$ ion.** E.C. of $d^2$ ion is $t_{2g}^2 e_g^0$ ($n = 2$). This configuration has asymmetrically filled $t_{2g}$ orbitals and empty $e_g$ orbitals. Hence octahedral complexes of $d^2$ ion show *slight distortion*.
4. **$d^3$ ion.** E.C. of $d^3$ ion is $t_{2g}^3 e_g^0$ ($n = 3$). In this configuration $t_{2g}$ orbitals are symmetrically filled and $e_g$ orbitals are empty. Octahedral complexes of $d^3$ ion, therefore, do not show any distrotion (no distortion). Examples of such complexes are $[Cr(ox)_3]^{3-}$, $[Cr(H_2O)_6]^{3+}$ etc.
5. (*a*) **$d^4$(HS) ion.** $Cr^{2+}$ and $Mn^{3+}$ ions are $d^4$ ions. Since the configuration of weak field (HS) 6-coordinated complexes of these ions *viz* $t_{2g}^3 e_g^1$ has asymmetrically-filled $e_g$ set, these compolexes show strong distortion in their octahedral shape. One electron present in $e_g$ set may be present either in $d_{z^2}$ orbital or in $d_{x^2-y^2}$ orbital. If the single electron is supposed to be present in $d_{x^2-y^2}$ orbital, this electron is repelled by four ligands lying on $+x$, $-x$ and $+y$ and $-y$ axes. On the other hand, if the single electron is assumed to be present in $d_{z^2}$ orbital, this electron is repelled by two ligands lying on $+z$ and $-z$ axes. Thus the energy of $d_{x^2-y^2}$ orbital increass relative to $d_{z^2}$ orbital. This leaves $d_{x^2-y^2}$ orbital empty. Thus the configuration of $d^4$ ion in high spin octahedral complexes is $(t_{2g})^3\ (d_{z^2})^1 (d_{x^2-y^2})^0$ and not $(t_{2g})^3\ (d_{z^2})^0 (d_{x^2-y^2})^1$. The repulsion between the single electron present in $d_{z^2}$ orbital and the two ligands along $+z$ and $-z$ axes (trans or axial ligands) results in the elongation of the two axial ligands. Thus the octahedral complex assumes *elongatd distorted octahedral structure* in which two trans metal–ligand bonds are long and four metal–ligand bonds in $xy$ square plane are short. In the formation of elongated distorted octahedral structure both $t_{2g}$ and $e_g$ sets present in octahedral complex split. The splitting of $t_{2g}$ and $e_g$ sets in various d-orbitals takes place in the way as shown at (*c*) of Fig 21.4. This configuration shows that $d_{z^2}$ orbital has lower energy than $d_{x^2-y^2}$ orbital and hence has one electron. The energy difference between $d_{xy}$ and $d_{yz}$, $d_{zx}$ pair is very small and

   hence each of these orbitals also has one electron only. Example of elongated distorted octahedral complexes containing $d^4$ ion are $CaF_2$ and $MnF_3$ crystals and $[Mn(H_2O)_6]^{3+}$ and $[Mn(C_2O_4)_3]^{3-}$ ions. $CrF_2$ crystal has distorted rutile ($TiO_2$) structure in which $Cr^{2+}$ ion is octahedrally surrounded by six $F^-$ ions (ligands). The length of four $Cr^{2+}$—$F^-$ bonds in *xy* plane is 2.0 Å each and that of the two $Cr^{2+}$—$F^-$ bonds along *z*-axis is 2.43 Å. Thus the crystal of $CrF_2$ has elongtated distorted octehedral structure.

Fig 21.4 shows that the distribution of $d^4$ electrons in the splitted d-orbitals is as $d_{yz}^1 d_{zx}^1 d_{xy}^1 d_{z^2}^1 d_{x^2-y^2}^0$

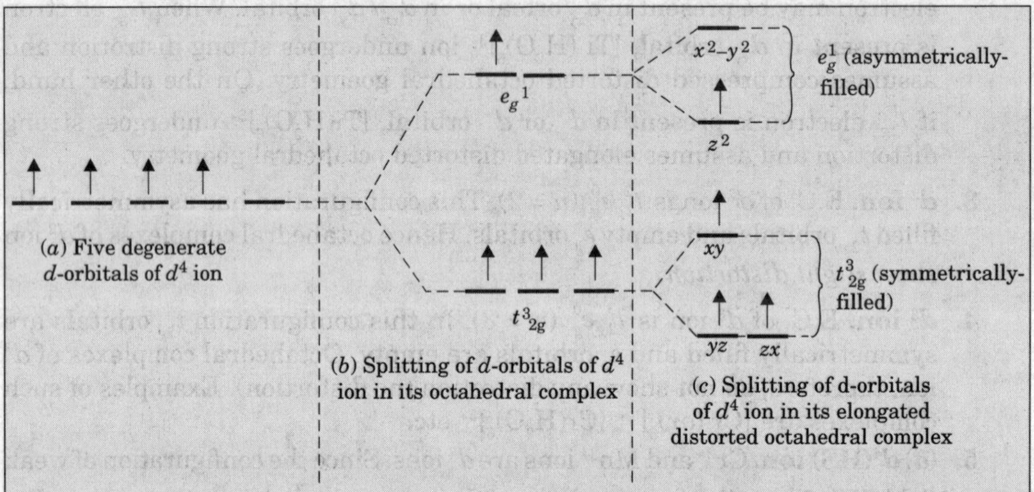


**Fig. 21.4.** Distribution of $d^4$ electrons in octahedral and tetragonally elongated distorted octahedral complexes of $d^4$ ion.

(*b*) **$d^4$(LS) ion.** E.C. of $d^4$ ion in LS state is $t_{2g}^4 e_g^0$ which contains asymeetrically filled $t_{2g}$ orbitals and empty $e_g$ orbitals. LS octahedral complexes of $d^4$ ion, therefore, show *slight distortion*.

6. (*a*) **$d^5$ (HS) ion (e.g. $Mn^{2+}$, $Fe^{3+}$ etc).** E.C. of $d^5$ ion is HS state is $t_{2g}^3 e_g^2 (n = 5)$. This configuration shows that $t_{2g}$ and $e_g$ orbitals both are symmetrically filled. Hence HS octahedral complexes of $d^5$ ion *donot undergo any distortion*. Examples of such complexes are $[MnF_6]^{4-}$, $[FeF_6]^{3-}$ etc.

   (*b*) **$d^5$ (LS) ion.** E.C. of $d^5$ ion in LS state is $t_{2g}^5 e_g^0 (n = 1)$. This configuration has asymmetrically filled $t_{2g}$ orbitals and empty $e_g$ orbitals. Hence LS octahedral complexes of $d^5$ ion undergo *slight distortion*.

7. (*a*) **$d^6$ (HS) ion (e.g. $Co^{2+}$ ion).** E.C. of $d^6$ ion in HS state is $t_{2g}^4 e_g^2 (n = 4)$. This configuration has asymmetrically filled $t_{2g}$ orbitals and symmetrically filled $e_g$ orbitals. HS octahedral complexes of $d^6$ ion, therefore, undergo *slight distortion.* Example of such complex is $[CoF_6]^{3-}$

   (*b*) **$d^6$(LS) ion (e.g. $Co^{3+}$ ion).** E.C. of $d^6$ ion in LS state is $t_{2g}^6 e_g^0 (n = 0)$. This configuration has symmetrically filled $t_{2g}$ orbitals and empty $e_g$ orbitals. Hence LS octahedral complexes of $d^6$ ion undergo *no distortion.* Example of such octahedral complex is $[Co(NH_3)_6]^{3+}$.

8. (*a*) **$d^7$(HS) ion.** E.C. of $d^7$ ion in HS state is $t_{2g}^5 e_g^2 (n = 3)$. This configuration has asymmetrically filled $t_{2g}$ orbitals and symmetrically filled $e_g$ orbitals. HS octahedral complexes of $d^7$ ion, therefore, show *slight distortion*.

   (b) **$d^7$ (LS) ion (e.g. $Co^{2+}$).** E.C. of $d^7$ ion in LS octahedral complexes of this ion is $t_{2g}^6 e_g^1 (n = 1)$. This configuration has symmetrically filled $t_{2g}$ orbitals and asymmetrically filled $e_g$ orbitals. $e_g^1$ electron is present in $d_{z^2}$ orbital and $d_{x^2-y^2}$ orbital is vacant. Thus $t_{2g}^6 e_g^1 = t_{2g}^6 d_{z^2}^1 d_{x^2-y^2}^0$. Due to the presence of asymmetrically-filled $e_g$ orbitals in $t_{2g}^6 e_g^1$ configuration, LS octahedral complexes of $d^7$ ion undergo *large distortion*. $[Co(diars)_3]^{2+}$ is an example of octahedral complex which undergoes large distortion.

9. (*a*) **$d^8$ ion (HS) (e.g. $Ni^{2+}$).** E.C. of $d^8$ ion in HS state is $t_{2g}^6 e_g^2 = t_{2g}^6 d_{z^2}^1 d_{x^2-y^2}^1 (n = 2)$. In this configuration $t_{2g}$ and $e_g$ orbitals both are symmetrically filled. Hence HS octahedral complexes of $d^8$ ion *donot show any distortion.* Examples of such complexes are $[NiF_6]^{4-}$, $[Ni(H_2O)_6]^{2+}$, $[Ni(NH_3)_6]^{2+}$ etc. Energy level diagram showing the filling of 3d orbitals with $d^8$ electrons is given in Fig 21.5.

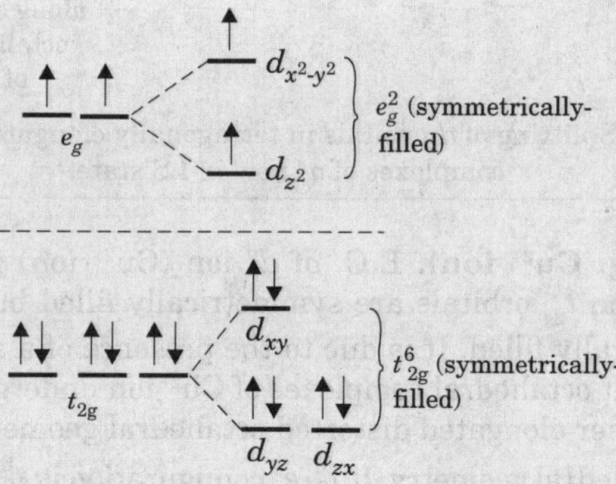


**Fig 21.5.** Energy level diagram showing the filling of 3d orbitals with eight electrons of $d^8$ configuration in HS state.

   (*b*) **$d^8$(LS) ion (e.g. $Ni^{2+}$, $Pd^{2+}$, $Pt^{2+}$ etc.).** E.C. of $d^8$ ion in LS state is $t_{2g}^6 e_g^2 = t_{2g}^6 d_{z^2}^2 d_{x^2-y^2}^0 (n = 0)$. In this configuration, $t_{2g}$ orbitals are symmetrically filled. Although $e_g$ orbitals are half filled, these orbitals are said to be asymmetrically filled, since both the $e_g$ electrons are present in $d_{z^2}$ orbital. Hence LS octahedral complexes of $d^8$ ion show *large distortion giving elongated distorted octahedral geometry.* Examples of such complexes are $[M(diars)_2I_2]^0$ where $M = Ni^{2+}, Pd^{2+}, Pt^{2+}$.

   The splitting of $t_{2g}$ (lower energy) and $e_g$ (higher energy) orbitals in elongated distorted octahedral complexes takes place as shown in Fig 21.6. This figure shows that the distribution of $d^8$ electrons in the splitled *d*-orbitals is as $d_{yz}^2 d_{zx}^2 d_{xy}^2 d_{z^2}^2 d_{x^2-y^2}^0$.

Since all electrons are present in the paired state, these complexes are diamagnetic. The elongation of axial bond lengths in $[Ni^{2+}(diars)_2I_2]^0$ is so larger that $[Ni^{2+}(diars)_2I_2]^0$ assumes square planar geometry.

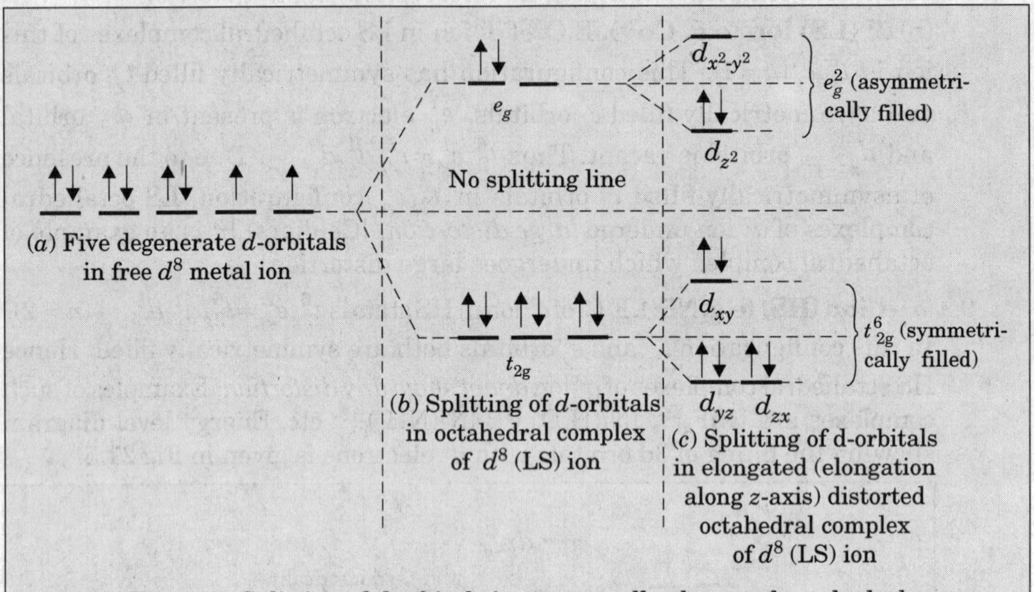


**Fig 21.6.** Splitting of *d*-orbitals in tetragonally elongated octahedral complexes of $d^8$ now in LS state.

10. **$d^9$ ion (e.g. $Cu^{2+}$ ion).** E.C. of $d^9$ ion ($Cu^{2+}$ ion) is $t_{2g}^6 e_g^3 (n = 1)$. In this configuration $t_{2g}$ orbitals are symmetrically filled but both $e_g$ orbitals are asymmetrically filled. It is due to the presence of a symmetrically filled $e_g$ orbitals that octahedral complexes of $Cu^{2+}$ ion undergo *large distortion* and assume either elongated distorted octahedral geometry or compressed distorted octahedral geometry. If $t_{2g}^6 e_g^3$ configuration is $t_{2g}^6 d_{z^2}^2 d_{x^2-y^2}^1$, octahedral geometry assumes *elongated distorted octahedral geometry*. On the other hand, if $t_{2g}^6 e_g^3$ configuration is $t_{2g}^6 d_{z^2}^1 d_{x^2-y^2}^2$, the octahedral geometry assumes *compressed distorted octahedral geometry*.

    The splitting of $t_{2g}$ and $e_g$ orbitals in elongated distorted octahedral and compressed distorted octahedral geometry has been discussed on the subsequent pages of this chapter.

11. **$d^{10}$ ion (e.g. $Zn^{2+}$).** E.C. of $d^{10}$ ion is $t_{2g}^6 e_g^4$. This configuration has $t_{2g}$ and $e_g$ orbitals both symmetrically filled. Hence octahedral complexes of $d^{10}$ ion *do not show distortion*. Examples of such complexes are $[Zn(NH_3)_6]^{2+}$, $[Zn(H_2O)_6]^{2+}$ etc.

Whatever has been said above has been summarized in Table 21.1. The information given in this table can used to find out the conditions under which the octahedral complex of a given $d^x$ ion ($x$ = 0 to 10) undergoes *no distortion, slight distortion* or *large distortion*.

**Table 21.1 Summary of distortion taking place in octahedral complexes of $d^x$ ions ($x$ = 0 to 10). sym = symmetrically filled, asym = asymmetrically filled**

| $d^x$ ion ($x$ = 0 to 10) | $t_{2g}^a e_g^b$ configuration | Nature of $t_{2g}$ and $e_g$ orbitals | | Distortion |
|---|---|---|---|---|
| | | $t_{2g}$ | $e_g$ | |
| $d^0$ | $t_{2g}^0 e_g^0$ | Empty | Empty | No distortion |
| $d^1$ | $t_{2g}^1 e_g^0 = d_{xy}^1 d_{yz}^0 d_{xz}^0 e_g^0$ | Asym | Empty | Strong distortion (Compressed distortion) |
| | $t_{2g}^1 e_g^0 = d_{xy}^0 d_{yz}^1 d_{xz}^0 e_g^0$ or $d_{xy}^0 d_{yz}^0 d_{xz}^1 e_g^0$ | Asym | Empty | Large distortion (Elongeted distortion) |
| $d^2$ | $t_{2g}^2 e_g^0$ | Asym | Empty | Slight distortion |
| $d^3$ | $t_{2g}^3 e_g^0$ | Sym | Empty | No distortion |
| $d^4$(HS) | $t_{2g}^3 e_g^1 = t_{2g}^3 d_{z^2}^1 d_{x^2-y^2}^0$ | Sym | Asym | Strong distortion (Elongoted distortion) |
| $d^4$(LS) | $t_{2g}^4 e_g^0$ | Asym | Empty | Slight distortion |
| $d^5$ (HS) | $t_{2g}^3 e_g^2$ | Sym | Sym | No distortion |
| $d^5$ (LS) | $t_{2g}^5 e_g^0$ | Asym | Empty | Slight distortion |
| $d^6$ (HS) | $t_{2g}^4 e_g^2$ | Asym | Sym | Slight distortion |
| $d^6$ (LS) | $t_{2g}^6 e_g^0$ | Sym | Empty | No distortion |
| $d^7$ (HS) | $t_{2g}^5 e_g^2$ | Asym | Sym | Slight distortion |
| $d^7$ (LS) | $t_{2g}^6 e_g^1 = t_{2g}^6 d_{z^2}^1 d_{x^2-y^2}^0$ | Sym | Asym | Large distortion |
| $d^8$ (HS) | $t_{2g}^6 e_g^2$ | Sym | Sym | No distortion |
| $d^8$ (LS) | $t_{2g}^6 e_g^2 = t_{2g}^6 d_{z^2}^2 d_{x^2-y^2}^0$ | Sym | Asym | Large distortion (Elongated distortion) |
| $d^9$ | $t_{2g}^6 e_g^3 = t_{2g}^6 d_{z^2}^2 d_{x^2-y^2}^1$ | Sym | Asym | Large distortion (Elongoted distortion) |
| $d^9$ | $t_{2g}^6 e_g^3 = t_{2g}^6 d_{z^2}^1 d_{x^2-y^2}^2$ | Sym | Asym | Large distortion (Compressed distortion) |
| $d^{10}$ | $t_{2g}^6 e_g^4$ | Sym | Sym | No distortion |

**Conditions for no distortion**

Octahedral complexes of $d^x$ ions whose both the sets of $d$-orbitals *viz* $t_{2g}$ and $e_g$ are symmetrically filled $\left(t_{2g}^0,\ t_{2g}^3,\ t_{2g}^6 \text{ and } e_g^0,\ e_g^2,\ e_g^4\right)$ undergo no distortion, *i.e.* such complexes are perfectly octahedral in shape. Perfectly octahedral shape is because of the fact that the electrons present in symmetrically filled $t_{2g}$ and $e_g$ sets repell all the six ligands to the same extent. This discussion shows that since the ions *viz* $d^0\left(t_{2g}^0e_g^0,\ n=0\right)$, $d^3\left(t_{2g}^3e_g^0,\ n=3\right)$, $d^5\left(t_{2g}^3e_g^2 - \text{HS with } n=5\right)$, $d^6\left(t_{2g}^6e_g^0 - \text{LS with } n=0\right)$, $d^6\left(t_{2g}^6e_g^0 - \text{LS with } n=0\right)$, $d^8\left(t_{2g}^6e_g^2 - \text{HS with } n=2\right)$ and $d^{10}\left(t_{2g}^6e_g^4 \text{ with } n=0\right)$ have $t_{2g}$ and $e_g$ sets symmetrically filled, octahedral complexes of these ions do not undergo distortion, *i.e.* these complexes are perfectly octahedral.

**Conditions for slight (negligible) distortion**

Octahedral complexes of $d^x$ ions whose $t_{2g}$ set of orbitals is asymmetrically-filled $\left(t_{2g}^1,\ t_{2g}^2,\ t_{2g}^4,\ t_{2g}^5\right)$ and $e_g$ set is symmetrically-filled $\left(e_g^0,\ e_g^2\right)$ undergo slight distortion. Small distortion is because of the fact that since the lobes of the three asymmetrically-filled orbitals of $t_{2g}$ set donot lie directly in the path of the ligands, the electrons present in $t_{2g}$ set of orbitals cause small distortion in octahedral complexes. Thus the octahedral complexes of $d^2(t_{2g}^2e_g^0,\ n=2)$, $d^4(t_{2g}^4e_g^0 - \text{LS with } n=2)$, $d^5(t_{2g}^5e_g^0 - \text{LS with } n=1)$, $d^6(t_{2g}^4e_g^2 - \text{HS with } n=4)$ and $d^7(t_{2g}^5e_g^2 - \text{HS with } n=3)$ ions undergo small distortion.

**Conditions for large distortion**

Octahedral complexes of $d^x$ ions whose $t_{2g}$ set is symmetrically-filled $(t_{2g}^0,\ t_{2g}^3 \text{ and } t_{2g}^6)$ and $e_g$ set is asymmetrically-filled $\left[e_g^1\left(d_{z^2}^1d_{x^2-y^2}^0\right),\ e_g^3\left(d_{z^2}^2d_{x^2-y^2}^1\right)\right]$ $e_g^3\left(d_{x^2-y^2}^2d_{z^2}^1\right)$ $\left[e_g^2 \text{ set}\left(d_{z^2}^2d_{x^2-y^2}^0\right)\right.$ is also regarded as asymmetrically-filled set in LS octahedral complexes of $d^8$ ion $\left.\left(d^8 = t_{2g}^6e_g^2 = t_{2g}^6d_{z^2}^2d_{x^2-y^2}^0\right)\right]$ undergo strong distortion and assume either tetragonally elongated distorted octahedral geometry or compressed distorted octahedral geometry.

Thus the octahedral complexes of the following $d^x$ ions undergo strong (large) distortion.

$d^1(t_{2g}^1e_g^0)$, $d^4(\text{HS})(t_{2g}^3e_g^1 = t_{2g}^3d_{z^2}^1d_{x^2-y^2}^0$ – elongated distortion), $d^7(\text{LS})(t_{2g}^6e_g^1 = t_{2g}^6d_{z^2}^1d_{x^2-y^2}^0)$, $d^8(\text{LS})(t_{2g}^6e_g^2 = t_{2g}^6d_{z^2}^2d_{x^2-y^2}^0$ – elongated distortion), $d^9(t_{2g}^6e_g^3 = t_{2g}^6d_{z^2}^2d_{x^2-y^2}^1$ – elongated distortion), $d^9(t_{2g}^6e_g^3 = t_{2g}^6d_{z^2}^1d_{x^2-y^2}^2$ – compressed distortion.

**Cause of large (strong) distortion.** Since the lobes of both orbitals of $e_g$ set ($d_{z^2}$ and $d_{x^2-y^2}$ orbitals) lie directly in the path of the six ligands (six ligends of an octahedral complex lie on the three exes), the electrons present in this set, if arranged asymmetrically) *i.e.* if $e_g$ set contains 1 or 3 electrons), repell some ligands to greater extent and repell other ligands to lesser extent. This ligand repulism causes large distortion in octahedral shape of 6-coordinated complexes.

## Splitting of $t_{2g}$ and $e_g$ Orbitals in Elongated Distorted Octahedral Complexes of $Cu^{2+}$ Ion

The splitting of $t_{2g}$ and $e_g$ sets of orbitals into various $d$-orbitals has been shown in Fig. 21.7.

In order to explain the splitting pattern shown in Fig. 21.7, we start with the splitting of $d$-orbitals in $t_{2g}$ and $e_g$ sets in octahedral complexes. We know that in the splitting of $d$-orbitals in octahedral complexes, the energy of $t_{2g}$ set (*i.e.* $d_{xy}$, $d_{yz}$ and $d_{zx}$ orbitals) is decreased and that of $e_g$ set (*i.e.* $d_{z^2}$ and $d_{x^2-y^2}$ orbitals) is increased. Now in elongated distorted octahedral complexes, since metal-ligand bond distance along $z$-axis is longer, $d$-orbitals having $z$ component (*i.e.* $d_{yz}$, $d_{zx}$ and $d_{z^2}$ orbitals) experience less repulsion from the ligands than they do in octahedral complex while the $d$-orbitals in $xy$ plane (*i.e.* $d_{xy}$ and $d_{x^2-y^2}$ orbitals) experience more repulsion than they do in octahedral complex. Consequently the energy of $d_{yz}$, $d_{zx}$ and $d_{z^2}$ orbitals decreases while that of $d_{xy}$ and $d_{x^2-y^2}$ orbitals increases. Note that $d_{yz}$ and $d_{zx}$ orbitals still remain degenerate as they are in octahedral complex. Above discussion shows that the order of energy of the splitted $d$-orbitals in tetragonally elongated octahedral complex becomes as:

$$(d_{yz} = d_{zx}) < d_{xy} < d_{z^2} < d_{x^2-y^2} \quad \text{(Also see Fig. 21.7)}$$

The relative order of energy of various orbitals shown above indicates that in the elongated distorted octahedral complexes, the distribution of $3d^9$ electrons ($t_{2g}^6 e_g^3$) of $Cu^{2+}$ ion is as shown in Fig. 21.7. This figure shows that the arrangement of $d^9$ electrons in various $d$-orbitals can be written as:

$$\underbrace{d_{yz}^2,\ d_{zx}^2,\ d_{xy}^2}_{t_{2g}^6},\ \underbrace{d_{z^2}^2,\ d_{x^2-y^2}^1}_{e_g^3}$$

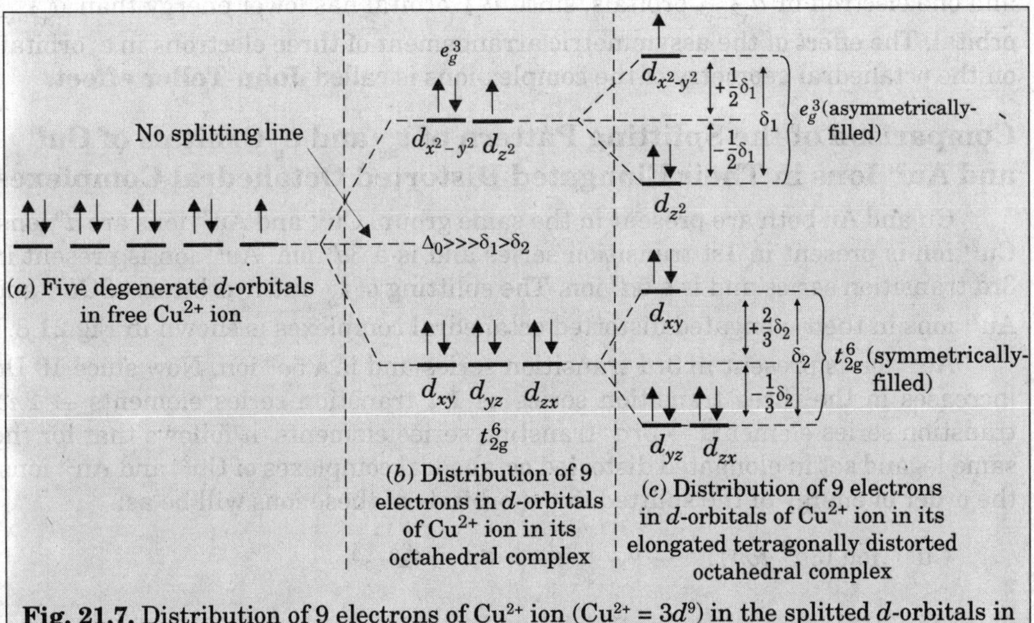


**Fig. 21.7.** Distribution of 9 electrons of $Cu^{2+}$ ion ($Cu^{2+} = 3d^9$) in the splitted $d$-orbitals in its tetragonally elongated distorted octahedral complexes.

Thus the configuration, $t_{2g}^6 e_g^3$ can be written as:

$$t_{2g}^6 e_g^3 = t_{2g}^6 d_{z^2}^2 d_{x^2-y^2}^1 .$$

This configuration shows that tetragonally elongated octahedral complexes have two electrons in $d_{z^2}$ orbital and one electron in $d_{x^2-y^2}$ orbital, since $d_{z^2}$ orbital has lower energy than $d_{x^2-y^2}$ orbital.

## Elongated Distortion in Octahedral Geometry of $[Cu(NH_3)_4(H_2O)_2]^{2+}$ and $[Cu(H_2O)_6]^{2+}$ Complexes.

Tetrammine copper (II) complex, $[Cu(NH_3)_4]^{2+}$ is actually $[Cu(NH_3)_4(H_2O)_2]^{2+}$ in which the *two axial $H_2O$ ligands are at larger distances* (along the *z*-axis) from the central $Cu^{2+}$ ion the *four equational $NH_3$ ligands are of shorter distances* (along *x*- and *y*-axes). Thus this complex has two long (axial) and four short (equational) bonds. Consequently the complex, $[Cu(NH_3)_4(H_2O)_2]^4$ has *tetragonally elongated distorted octahedral) geometry.*

The distribution of $d^9$ electrons of $Cu^{2+}$ ion in five *d*-orbitals is as:

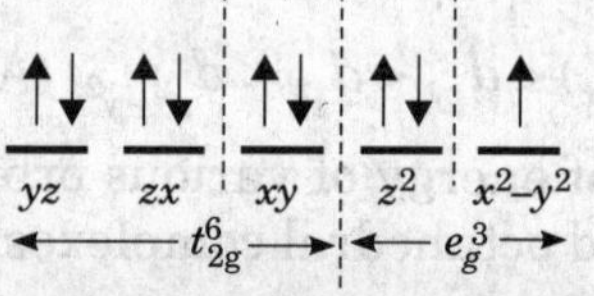


Energy increasing ⟶

$[Cu(H_2O)_6]^{2+}$ ion also has tetragonally elongated distorted octahedral geometry. The tetragonally elongated geometry (distorted octahedral geometry) of $(Cu(NH_3)_4(H_2O)_2]^{2+}$ and $[Cu(H_2O)_6]^{2+}$ ions is due to the assymmetric arrangement of three electrons in $e_g$ orbitals. This arrangement contains two electrons in $d_{z^2}$ orbital and one electron in $d_{x^2-y^2}$ orbitals, since $d_{z^2}$ orbital has lower energy than $d_{x^2-y^2}$ orbital. The effect of the assymmetric arrangement of three electrons in $e_g$ orbitals on the octahedral geometry of the complex ions is called **John-Teller effect**.

## Comparison of the Splitting Pattern of $t_{2g}$ and $e_g$ Orbitals of $Cu^{2+}$ and $Au^{2+}$ Ions in Their Elongated Distorted Octahedral Complexes

Cu and Au both are present in the same group. $Cu^{2+}$ and $Au^{2+}$ ions are $d^9$ ions. $Cu^{2+}$ ion is present in 1st transition series and is a $3d^9$ ion. $Au^{2+}$ ion is present in 3rd transition series and is a $5d^9$ ion. The splitting of $t_{2g}$ and $e_g$ orbitals of $Cu^{2+}$ and $Au^{2+}$ ions in their elongated distorted octahedral complexes is shown in Fig 21.8.

$Au^{2+}$ ion is present in 3rd transition series and is a $5d^9$ ion. Now since 10 Dq increases in the three transition series as *1st* transtion series elements → *2nd* transtion series elements → *3rd* transtion series elements, it follows that for the same legand set in elongated distorted octahedral complexes of $Cu^{2+}$ and $Au^{2+}$ ions, the order of energy of the splitted five *d*-orbitals of these ions will be as:

$Cu^{2+}$ ion ($3d^9$ *ion*): $d_{xz}, d_{yz}$ ¦ $d_{xy}$ ¦ $d_{z^2}$ ¦ $d_{x^2-y^2}$

Energy increasing ⟶

$Au^{2+}$ ion ($5d^9$ *ion*): $(d_{xz}, d_{yz})$ ⋮ $d_{z^2}$ ⋮ $d_{xy}$ ⋮ $d_{x^2-y^2}$

—— **Energy increasing** ——→

Obviously the distribution of $d^9$ electrons in the splitted $d$-orbitals of these ions would be as:

*$d^9$ electrons of $Cu^{2+}$ ion*: $d_{xz}^2, d_{yz}^2$ ⋮ $d_{xy}^2$ ⋮ $d_{z^2}^2$ ⋮ $d_{x^2-y^2}^1$ $(n=1)$

*$d^9$ electrons of $Au^{2+}$ ion*: $d_{xz}^2, d_{yz}^2$ ⋮ $d_{z^2}^2$ ⋮ $d_{xy}^2$ ⋮ $d_{x^2-y^2}^1$ $(n=1)$

If may be noted from Fig 21.8 that $d_{x^2-y^2}$ orbital of $Au^{2+}$ ion containing the unpaired electron is at much higher energy level than that of $Cu^{2+}$ ion containing the unpaired electron.

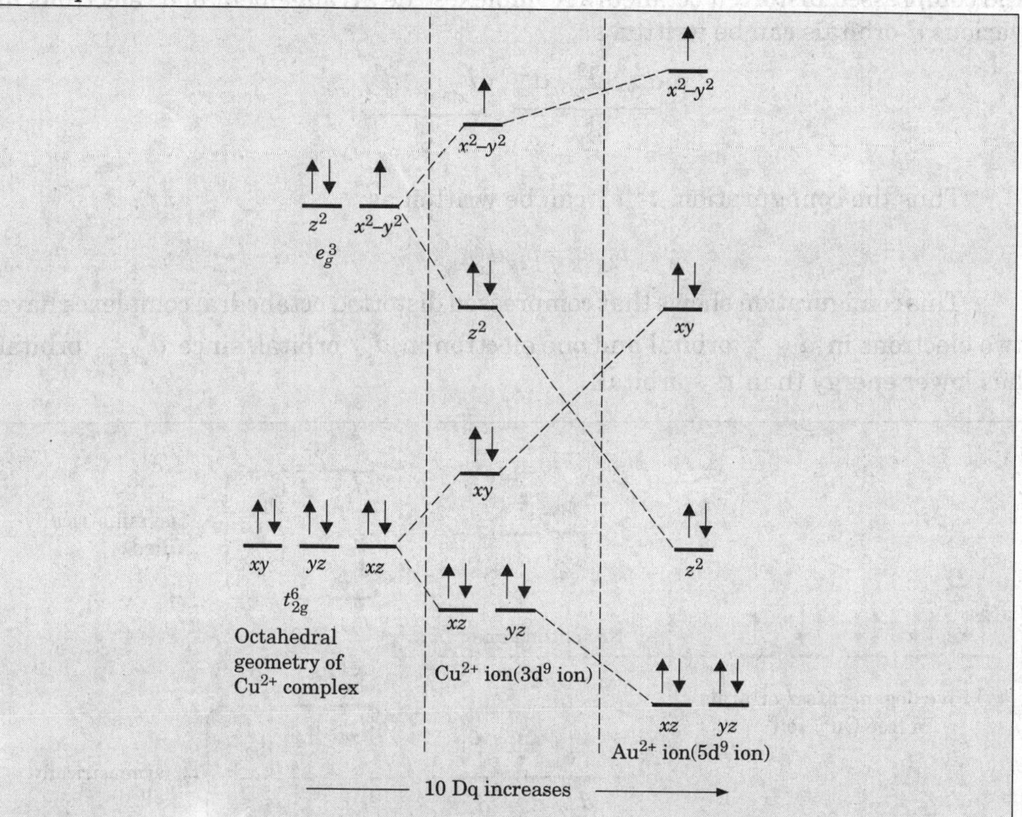


**Fig. 21.8.** Distribution of electrons of $Cu^{2+}$ and $Au^{2+}$ ions in different $d$-orbitals in their tetragonally elongated distorted complexes.

## Splitting of $t_{2g}$ and $e_g$ Orbitals in Compressed Distorted Octahedral Complexes of $Cu^{2+}$ Ion.

The splitting of $t_{2g}$ and $e_g$ orbitals into various $d$-orbitals has been shown in Fig. 21.9.

In order to explain the splitting pattern shown in Fig. 21.9 we start with the splitting of $d$-orbitals in $t_{2g}$ and $e_g$ sets in octahedral complexes. We know that in the splitting of $d$-orbitals in octahedral complexes, the energy of $t_{2g}$ set (*i.e.* $d_{xy}$, $d_{yz}$ and $d_{zx}$ orbitals) is decreased and that of $e_g$ set (*i.e.* $d_{z^2}$ and $d_{x^2-y^2}$ orbitals) is increased. Now in compressed distorted octahedral complexes, since metal-ligand distance along $z$-axis is shorter, $d$-orbitals having $z$-component (*i.e.* $d_{yz}$, $d_{zx}$ and $d_{z^2}$ orbitals) experience greater repulsion than they do in octahedral complexes while $d$-orbitals in $xy$ plane (*i.e.* $d_{xy}$ and $d_{x^2-y^2}$ orbitals) experience less repulsion than they do in octahedral complexes. Consequently the energy of $d_{yz}$, $d_{zx}$ and $d_{z^2}$ orbitals increases and that of $d_{xy}$ and $d_{x^2-y^2}$ orbitals still remain degenerate as they are in octahedral complex. Above discussion shows that the relative order of energy of the splitted $d$-orbitals in compressed distorted octahedral complexes becomes as:

$$d_{xy} < d_{yz} = d_{zx} < d_{x^2-y^2} < d_{z^2} \text{ (Also see Fig. 21.9).}$$

The relative order of energy of various orbitals shown above indicates that in the compressed distorted octahedral complexes the arrangement of $d^9$ electrons in various $d$-orbitals can be written as:

$$\underbrace{d_{xy}^2, d_{yz}^2, d_{zx}^2}_{t_{2g}^6}, \underbrace{d_{x^2-y^2}^2, d_{z^2}^1}_{e_g^3}$$

Thus the configuration, $t_{2g}^6 e_g^3$ can be written as:

$$t_{2g}^6 e_g^3 = t_{2g}^6 d_{x^2-y^2}^2 d_{z^2}^1$$

This configuration shows that compressed distorted octahedral complexes have two electrons in $d_{x^2-y^2}$ orbital and one electron in $d_{z^2}$ orbital, since $d_{x^2-y^2}^2$ orbital has lower energy than $d_{z^2}$ orbital.

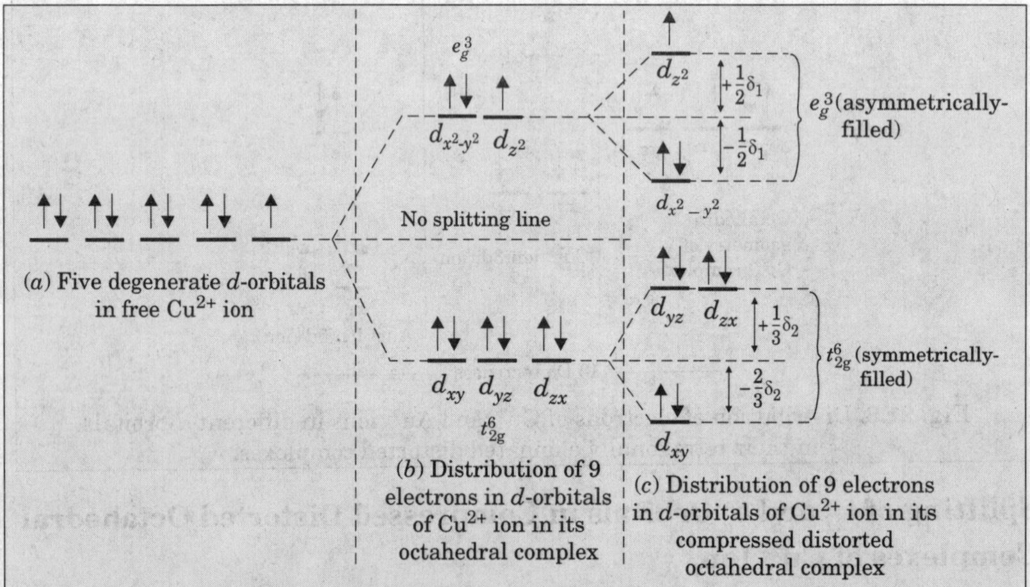


**Fig. 21.9.** Distribution of 9 electrons of $Cu^{2+}$ ion ($Cu^{2+} = 3d^9$) in the splitted $d$-orbitals in its compressed distorted octahedral complex.

The forgoing discussion shows that 6-coordinated complexes of $Cu^{2+}$ ion which has two orbitals *viz* $d_{x^2-y^2}$ and $d_{z^2}$ in asymmetrically-filled $e_g$ set $\left(d^9 = t_{2g}^6 e_g^3\right)$ is unstable and hence get distorted and assume either elongated distorted octahedral or compressed distorted octahedral shape corresponding to the configuration *viz* $t_{2g}^6 d_{z^2}^2 d_{x^2-y^2}^1$ and $t_{2g}^6 d_{z^2}^1 d_{x^2-y^2}^2$ respectively.

## To Calculate John-Teller Stabilization Energy for Tetragonally Distorted Octahedral Complexes of $Cu^{2+}$ Ion ($d^9$ Ion)

We know that there are two types of tetragonally distorted octahedral complexes of $Cu^{2+}$ ions. The value of John-Teller stablisation energy for both the types can be calculated as follows:

(*i*) **For tetragonally elongated distorted octahedral complexes.** John-Teller stabilization energy involved in the splitting of $t_{2g}$ and $e_g$ orbitals in elongated distorted octahedral complexes of $Cu^{2+}$ ion can be calculated with the help of Fig. 21.7 in which the energy diffence between $d_{x^2-y^2}$ and $d_{z^2}$ orbitals has been denoted of $\delta_1$ and that between $d_{xy}$ and degenerate pair orbitals *viz* $d_{yz}$ and $d_{zx}$ orbitals has been represented by $\delta_2$. The values of $\delta_1$, $\delta_2$ and $\Delta_0$ are such that $\Delta_0 >>> \delta_1 > \delta_2$. The splitting of $e_g$ set takes place in such a way that the decrease in the energy of $d_{z^2}$ orbital is the same as the increase in the energy of $d_{x^2-y^2}$ orbital. The increase in energy has been denoted by + sign while the decrease in energy has been shown by – sign. The splitting of $t_{2g}$ set takes place in such a way that the decrease in energy of doubly degenerated $d_{yz}$ and $d_{zx}$ pair orbitals is half of the increase in the energy of $d_{xy}$ orbital. From Fig. 21.7 it may be seen that:

*Change in energy in six electrons in $t_{2g}$ set due to its splitting*

$$= +\frac{2}{3}\delta_2 \times 2 - \frac{1}{3}\delta_2 \times 4$$

$$= +\frac{4}{3}\delta_2 - \frac{4}{3}\delta_2$$

$$= \text{Zero}$$

Zero value indicates that the electrons in $t_{2g}$ set neither gain nor lose any stability.

*Change in energy in three electrons in $e_g$ set due to its splitting*

$$= +\frac{1}{2}\delta_1 \times 1 - \frac{1}{2}\delta_1 \times 2$$

$$= -\frac{1}{2}\delta_1$$

Due to the decrease in energy equal to $\frac{1}{2}\delta_1$, the electrons in $e_g$ set gain stability. This energy $\left(= \frac{1}{2}\delta_1\right)$ stablises the distorted octahedral complex. This energy is called **John-Teller stablisation energy.**

(*ii*) **For tetragonally compressed distorted octahedral complexes.** John-Teller stablisation energy in the splitting of $t_{2g}$ and $e_g$ orbitals in compressed distorted octahedral complexes of $Cu^{2+}$ can be calculated with the help of Fig. 21.9. From this figure it may be seen that:

*Change in energy in six electrons of $t_{2g}$ set due to its splitting.*

$$= \left(-\frac{2}{3}\delta_2\right)\times 2 + \left(+\frac{1}{3}\delta_2\right)\times 4$$

$$= -\frac{4}{3}\delta_2 + \frac{4}{3}\delta_2$$

$$= \text{zero}$$

*Change in energy in three electrons of $e_g$ set due to its splitting*

$$= \left(-\frac{1}{2}\delta_1\right)\times 2 + \left(+\frac{1}{2}\delta_1\right)\times 1$$

$$= -\delta_1 + \frac{1}{2}\delta_1$$

$$= -\frac{\delta_1}{2}$$

Due to the decrease in energy equal to $\frac{1}{2}\delta_1$, the electrons in $e_g$ set gains stability. This energy $\left(=\frac{1}{2}\delta_1\right)$ stablises the distorted octahedral complex.

Above calculations show that the value of stablisation energy of both the types of distorted octahedral complexes is the same $\left(=\frac{1}{2}\delta_1\right)$. Hence both distortions stablise the octahedral complex to the same extent. It is, therefore, difficult to predict as to which of the two possible distortions would occur in octahedral complex due to John-Teller effect. However, experimental observations have revealed that distortion of elongated type is the common distortion in the octahedral complexes of $Cu^{2+}$ ion. Thus most of octahedral complexes of $Cu^{2+}$ are distorted with two longer $Cu^{2+}$–ligand bonds alogn $z$-axis.

## Examples of Tetragonally Elongated Distorted Octahedral Complexes.

1. ***$CuF_2$ crystal.*** In $CuF_2$ crystal lattice $Cu^{2+}$ ion is surrounded by six $F^-$ ions Four $F^-$ ions and $Cu^{2+}$ ion lie in $xy$ plane and the two $F^-$ ions lie on the $z$-axis (axial axis). Each of the two $Cu^{2+}$–$F^-$ bond distances = 2.27A$^0$ and each of the four $Cu^{2+}$–$F^-$ bond distances = 1.93 A$^0$. Thus $CuF_2$ has tetragonally elongated disorted octahedral shape.
2. ***$MnF_3$ and $[MnF_6]^{3-}$***. The crystals of these species have elongated distorted octahedral structure in which two axial Mn–F bonds are long and four Mn–F bonds are short.

## Examples of Tetragonally Compressed Distorted Octahedral complex of $Cu^{2+}$ ion.

A well known example of such complexes is $K_2[CuF_4]$. In the solid state of $[CuF_4]^{2-}$, each $Cu^{2+}$ ion

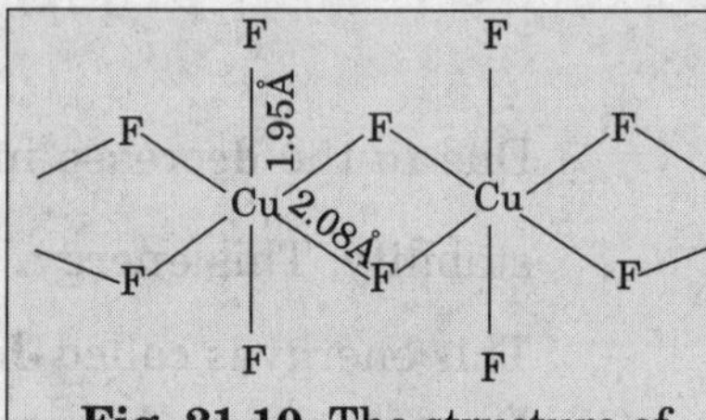


**Fig. 21.10.** The structure of $[CuF_4]^{2-}$ ion in solid state.

is surrounded by six $F^-$ ions out of which two are axial $F^-$ ions. Each axial $Cu^{2+} - F^-$ bond length = 1.95 A$^0$ and each equatorial $Cu^{2+} - F^-$ bond length = 2.08 A$^0$ (See Fig. 21.10)

## Square Planar Complexes of $Cu^{2+}$ Ion.

Sometimes the elongation of two axial $Cu^{2+}$–ligand bonds takes place so strongly that these bonds are removed and we get square geometry instead of tetragonal geometry. Square planar geometry does not have *trans* $Cu^{2+}$–ligand bonds along *z*-axis.

## Energy Level Diagram Showing the Splitting of d-orbitals in Octahedral, Tetragonal (elongated) and Square Planar Complexes

The splitting of *d*-orbitals in octahedral, tetragonal (elongated) and square planar complexes are linked with each other as shown in Fig 24.11. At (a) of Fig. 24.11 we have shown five *d*-orbitals of the free metal ion.

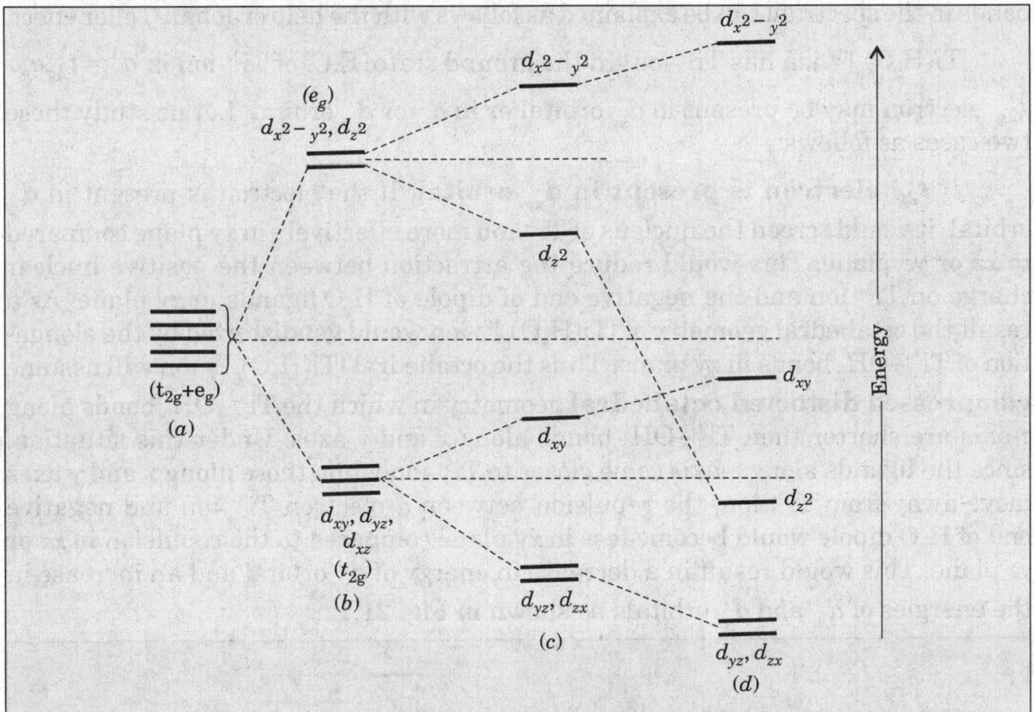


**Fig. 21.11.** (a) Five degenerate *d*-orbitals in a free metal ion (b) splitting of *d*-orbitals in octahedral complex (c) splitting of *d*-orbitals in tetragonal (elongated) complexes (d) splitting of *d*-orbitals in square planar complexes

In octahedral complexes *d*-orbitals are split into $t_{2g}$ and $e_g$ sets of orbitals. $t_{2g}$ set ($d_{xy}$, $d_{yz}$ and $d_{zx}$ orbitals) has lower energy and $e_g$ set ($d_{z^2}$ and $d_{x^2-y^2}$ orbitals) has higher energy [See (b) of Fig 21.11]

In elongated tetragonal complex, since ligands lying on *z*-axis move away from the central metal ion while the ligands in *xy* plane tend to come closer to the central metal ion, the electrons in *d*-orbitals in *xy* plane (*i.e.* electrons in $d_{x^2-y^2}$ and $d_{xy}$ orbitals) experience greater repulsion from the electrons of the ligands in tetragonal complex than in octahedral complex. Hence the energy of $d_{x^2-y^2}$ and $d_{xy}$ orbitals is increased compared to their energy in octahedral complex [See (c) of Fig. 21.11)

Further since the ligands on $z$-axis have moved away from the central metal ion, the electrons present in $d$-orbitals along $z$-axis (electrons in $d_{z^2}$ orbital) as well as in $xz$ and $yz$ planes (*i.e.* electrons in $d_{xz}$ and $d_{yz}$ orbitals) experience relatively lesser repulsion from the electrons of the ligands. Hence the energies of $d_{z^2}$, $d_{xz}$ and $d_{yz}$ orbitals fall down compared to their energies in octahedral complex [See (c) of Fig. 21.11]

In case of square planar complexes, since the *trans* ligands lying along $z$-axis are removed completely, further rise in the energies of $d_{x^2-y^2}$ and $d_{xy}$ orbitals and further fall in the energies of $d_{z^2}$, $d_{yz}$ and $d_{zx}$ orbitals takes place [See (d) of Fig. 21.11]

## Interpretation of the Electronic Absorption Spectrum of $[Ti(H_2O)_6]^{3+}$ Ion on the Basis of John-Teller Effect.

The absorption spectrum of $[Ti(H_2O)_6]^{3+}$ ion has two absorption bands which overlap to give a composite unsymmetrical absorption band. The appearance of two bands in the spectrum can be explained as follows with the help of Johan-Teller effect.

$[Ti(H_2O)_6]^{3+}$ ion has $Ti^{3+}$ ion. In the ground state, E.C. of $Ti^{3+}$ ion is $d^1 = t_{2g}^1 e_g^0$. $t_{2g}^1$ electron may be present in $d_{xy}$ orbital or in $d_{xz}$ or $d_{yz}$ orbital. Let us study these two cases as follows:

(*i*) $\boldsymbol{t_{2g}^1}$ **electron is present in $\mathbf{d_{xy}}$ orbital.** If the electron is present in $d_{xy}$ orbital, it would screen the nucleus of $Ti^{3+}$ ion more effectively in $xy$ plane compared to $xz$ or $yz$ plane. This would reduce the attraction between the positive nuclear charge on $Ti^{3+}$ ion and the negative end of dipole of $H_2O$ ligands in $xy$ plane. As a result the octahedral geometry of $[Ti(H_2O)_6]^{3+}$ ion would get distorted by the elongation of $Ti^{3+}$–$OH_2$ bonds in $xy$ plane. Thus the octahedral $[Ti(H_2O)_6]^{3+}$ ion will assume **compressed distorted octahedral** geometry in which the $Ti^{3+}$–$OH_2$ bonds along $z$-axis are shorter then $Ti^{3+}$–$OH_2$ bonds along $x$ and $y$ axes. Under this situation, since the ligands along $z$-axis come closer to $Ti^{3+}$ ion while those along $x$ and $y$ axes move away from $Ti^{3+}$ ion, the repulsion between $d$-electron $Ti^{3+}$ ion and negative end of $H_2O$ dipole would become less in $xy$ plane compared to the repulsion in $xz$ or $yz$ plane. This would result in a decrease in energy of $d_{xy}$ orbital and an increase in the energies of $d_{xz}$ and $d_{yz}$ orbitals as shown in Fig. 21.12.

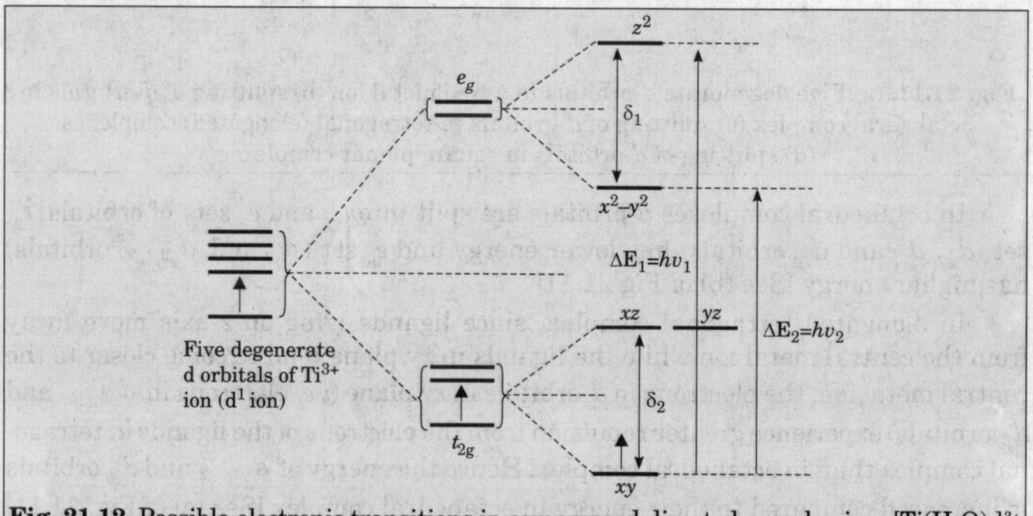


**Fig. 21.12.** Possible electronic transitions in compressed disorted complex ion, $[Ti(H_2O)_6]^{3+}$

From Fig. 21.12 it may be seen that in the compressed distorted octahedral geometry of $[Ti(H_2O)_6]^{3+}$ ion, the ground state of $Ti^{3+}$ ion ($d^1$ ion) is $d_{xy}$ and the excited states are $d_{z^2}$ and $d_{x^2-y^2}$. The energy of the excited state, $d_{x^2-y^2}$ is lower than that of the excited state, $d_{z^2}$. The two bands observed in the spectrum of $[Ti(H_2O)_6]^{3+}$ ion are due to the following two electronic transitions.

(a) Transtion of one electron from $d_{xy}$ orbital to $d_{x^2-y^2}$ orbital ($d^1_{xy} \rightarrow d^1_{x^2-y^2}$ transition)

(b) Transtion of one electron from $d_{xy}$ orbital to $d_{z^2}$ orbital ($d^1_{xy} \rightarrow d^1_{z^2}$ transition).

These transitions have been shown in Fig. 21.12. These transitions give two obsorption bands. Now since the energy difference ($\delta_2$) between $d_{xy}$ and ($d_{xz}$, $d_{yz}$) pair is very small, the above two transitions are generally written as $t^1_{2g} \rightarrow d^1_{x^2-y^2}$ and $t^1_{2g} \rightarrow d^1_{z^2}$ respectively.

Since the energy gap $\Delta E_1$ is not very different from the energy gap $\Delta E_2$, the two transitions are close in energy and hence the frequencies ($v_1$ and $v_2$) of the two bands are close to each other. Since the two bands have close frequencies, they overlap to give a composite absorption band in the spectrum.

(ii) **$t^1_{2g}$ electron is present in $d_{xz}$ or $d_{yz}$ orbital.** If the electron is present in $d_{xz}$ or $d_{yz}$ orbital, it would screen the nucleus of $Ti^{3+}$ ion more effectively in *xz* or *yz* plane compared to *xy* plane. This would decrease the attraction between the positive nuclear charge on $Ti^{3+}$ ion and the negative end of $H_2O$ dipole along *z*-direction and increase the attraction along *x* and *y* directions. As a result $Ti^{3+}$– $OH_2$ bonds get elongated along *z*-axis and hence the octahedral geometry of $[Ti(H_2O)_6]^{3+}$ ion will become **elongated distorted octahedral geometry** in which the $Ti^{3+}$–$OH_2$ bonds along *z*-axis are longer than $Ti^{3+}$–$OH_2$ bonds along *x* and *y* axes. Under this situation since the ligands along *z*-axis move always from $Ti^{3+}$ ion while those along *x* and *y* axes come closer to $Ti^{3+}$ ion, repulsion between *d*-electron of $Ti^{3+}$ ion and negative end of $H_2O$ dipole would become less along *z*-direction compared to the repulsion along *x* and *y* directions. This would increase the energy of $d_{xy}$ orbital and decrease the energies of $d_{xz}$ and $d_{yz}$ orbitals as shown in Fig. 24.13.

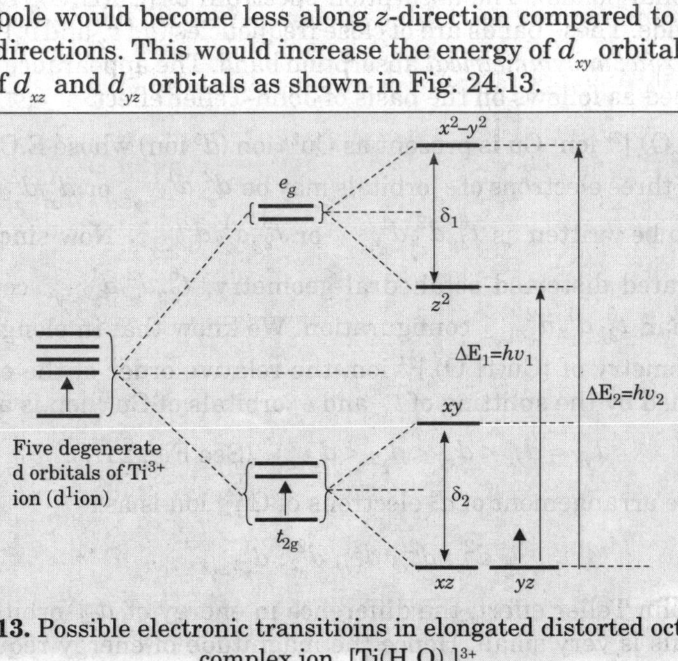


**Fig. 21.13.** Possible electronic transitioins in elongated distorted octahedral complex ion, $[Ti(H_2O)_6]^{3+}$

It may be seen from Fig. 24.13 that in the elongated distorted octahedral geometry $[Ti(H_2O)_6]^{3+}$ ion, the ground state configuration $Ti^{3+}$ ion ($d^1$ ion) is $(d_{xz}, d_{yz})$ and excited states are $d_{z^2}$ and $d_{x^2-y^2}$. The energy of the excited state $d_{z^2}$ is lower than that of the excited state $d_{x^2-y^2}$. The two bands observed in the spectrum of $[Ti(H_2O)_6]^{3+}$ are due to the following two electronic transitions.

(*a*) Transition of one electron from $(d_{xz}, d_{yz})^1$ or $t_{2g}^1$ to $d_{z^2}$ orbital [$(d_{xz}, d_{yz})^1 \rightarrow d_{z^2}^1$ transition]

(*b*) Transition of one electron from $(d_{xz}, d_{yz})^1$ or $t_{2g}^1$ to $d_{x^2-y^2}$ orbital [$(d_{xz}, d_{yz})^1 \rightarrow d_{x^2-y^2}^1$ transition].

These transitions have been shown in Fig. 24.13. These transitions give two absorption bands. Now since the energy difference ($\delta_2$) between $(d_{xz}, d_{yz})$ pair and $d_{xy}$ is very small, the above two transitions are generally written as $t_{2g}^1 \rightarrow d_{z^2}^1$ and $t_{2g}^1 \rightarrow d_{x^2-y^2}^1$ respectively. Since the energy gap $\Delta E_1$, is not very different from the energy gap $\Delta E_2$, the two transitions are close in energy and hence the frequencies ($v_1$ and $v_2$) of the two bands are close to each other. Since these two bands have close frequencies, they overlap to give a composite absorption band in the spectrum.

The foregoing discussion shows that the absorption spectrum $[Ti(H_2O)_6]^{3+}$ ion has an asymmetric absorption band whether $[Ti(H_2O)_6]^{3+}$ ion has compressed distorted octahedral geometry or elongated distorted octahedral geometry or whether an equilibrium exists between the above two geometries due to dynamic John-Teller effect.

## Interpretation of the Electronic Absorption Spectrum of $[Cu(H_2O)_6]^{2+}$ on the Basis of John-Teller Effect

$[Cu(H_2O)_6]^{2+}$ ion has elongated distorted octahedral geometry in which $Cu^{2+}–OH_2$ bonds along *z*-axis (axial bonds) are elongated compared to $Cu^{2+}–OH_2$ bonds in *xy* plane (equational bonds). The absorption spectrum of $[Cu(H_2O)_6]^{2+}$ ion has three absorption bands. These bands are of close frequencies ($v_1$, $v_2$ and $v_3$) which overlap to give a *composite, unsymmetrical* absorption band. The appearance of three bands can be explained as follows on the basis of John-Teller effect.

In $[Cu(H_2O)_6]^{2+}$ ion, Cu is present as $Cu^{2+}$ ion ($d^9$ ion) whose E.C. is $t_{2g}^6 e_g^3$. The distribution of three electrons of $e_g$ orbitals may be $d_{z^2}^2 d_{x^2-y^2}^1$ or $d_{z^2}^1 d_{x^2-y^2}^2$. Thus E.C. of $Cu^{2+}$ ion can be written as $t_{2g}^6 d_{z^2}^2 d_{x^2-y^2}^1$ or $t_{2g}^6 d_{z^2}^1 d_{x^2-y^2}^2$. Now since $[Cu(H_2O)_6]^{2+}$ ion has elongated distorted octahedral geometry, $t_{2g}^6 d_{z^2}^2 d_{x^2-y^2}^1$ configuration is more stable than $t_{2g}^6 d_{z^2}^1 d_{x^2-y^2}^2$ configuration. We know that in elongated distorted octahedral geometry of $[Cu(H_2O)_6]^{2+}$ ion, the relative order of the energies of the orbitals obtained by the splitting of $t_{2g}$ and $e_g$ orbitals of $Cu^{2+}$ ion is as:

$$d_{yz} = d_{zx} < d_{xy} < d_{z^2} < d_{x^2-y^2} \text{ (See Fig. 21.7).}$$

Hence the arrangement of $d^9$ electrons of $Cu^{2+}$ ion is as:

$$d_{yz}^2, d_{zx}^2, d_{xy}^2, d_{z^2}^2, d_{x^2-y^2}^1.$$

Due to John Teller effect, the difference in energy of $d_{z^2}$ orbital and $d_{xy}$ and $(d_{xz}, d_{yz})$ orbitals is very small. Hence the magnitude of energy required to excite an electron from $d_{yz}$ (or $d_{zx}$), $d_{xy}$ and $d_{z^2}$ orbital to partially filled $d_{x^2-y^2}$ orbital is nearly the same. Thus the three bands in the spectrum of $[Cu(H_2O)_6]^{2+}$ are due to the following three electronic transitions

(i) Electronic transition of an electron from $d_{xz}$ (or $d_{yz}$) orbital to $d_{x^2-y^2}$ orbital by absorbing energy equal to $h\,v_1$ ($d_{xz}$ or $d_{yz} \rightarrow d_{x^2-y^2}$ transition).

(ii) Electronic transition of an electron from $d_{xy}$ orbital to $d_{x^2-y^2}$ orbital by absorbing energy equal to $h\,v_2$ ($d_{xy} \rightarrow d_{x^2-y^2}$ transition)

(iii) Electronic transition of an electron from $d_{z^2}$ orbital to $d_{x^2-y^2}$ orbital by absorbing energy equal to $h\,v_3$ ($d_{z^2} \rightarrow d_{x^2-y^2}$ transition).

All the above transitions have been shown in Fig. 21.14. If there were no John-Teller effect in $[Cu(H_2O)_6]^{2+}$, the spectrum of the ion would have had a *single symmetrical band*. This band is due to the electronic transition of an electron from $t_{2g}^6$ orbitals to $e_g^3$ orbitals. ($t_{2g}^6 e_g^3 \rightarrow t_{2g}^5 e_g^4$ transition).

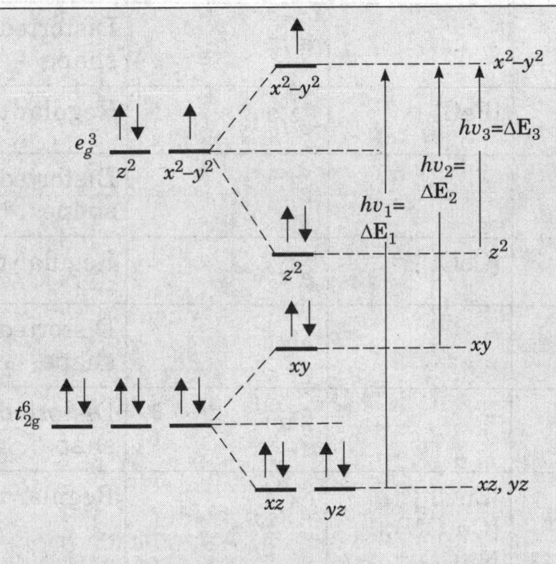


**Fig. 21.14.** Possible electronic transitions in elongated distorted octahedral complex ion, $[Cu(H_2O)_6]^{2+}$

## Distortion in Tetrahedral Complexes

We know that since in case of tetrahedral complexes $\Delta_t < P$, tetrahedral complexes are generally high spin tetrahedral complexes. Some of these complexes undergo distortion while others do not. This is shown in Table 21.2. This table shows that distortion will occur in tetrahedral complexes whenever $e$ set or $t_2$ set is unsymmetrically filled with electrons. Thus tetrahedral complexes of $d^1(e^2t_2^0)$, $d^3(e^2t_2^1)$, $d^4(e^2t_2^2)$, $d^6(e^3t_2^3)$, $d^8(e^4t_2^4)$, and $d^9(e^4t_2^5)$ ions will be distorted tetrahedral complexes while the tetrahedral complexes of $d^0(e^0t_2^0)$, $d^2(e^2t_2^0)$, $d^5(e^2t_2^3)$, $d^{10}(e^4t_2^6)$ ions have regular tetrahedral shape.

**Table 21.2. Examples of complexes having regular and distorted tetrahedral shape**

| *$d^x$ ion* | *Examples of complexes* | *$e^p t_2^q$ configuration* | *Distorted or regular tetrahedral shape* |
|---|---|---|---|
| $d^0(Ti^{4+}, Mn^{7+})$ | $[TiCl_4]^0$, $[MnO_4]^-$ | $e^0 t_2^0$ | Regular tetrahedral shape |
| $d^1$ | – | $e^1 t_2^0$ | Distorted tetrahedral shape |
| $d^2(Fe^{6+})$ | $[FeO_4]^{2-}$ | $e^2 t_2^0$ | Regular tetrahedral shape |
| $d^3$ | – | $e^2 t_2^1$ | Distorted tetrahedral shape |
| $d^4$ | – | $e^2 t_2^2$ | Distorted tetrahedral shape |
| $d^5(Fe^{3+})$ | $[FeCl_4]^-$ | $e^2 t_2^3$ | Regular tetrahedral shape |
| $d^6$ | – | $e^3 t_2^3$ | Distorted tetrahedral shape |
| $d^7(Co^{2+})$ | $[CoCl_4]^{2-}$ | $e^4 t_2^3$ | Regular tetrahedral shape |
| $d^8$ | – | $e^4 t_2^4$ | Distorted tetrahedral shape |
| $d^9(Cu^{2+})$ | – | $e^4 t_2^5$ | Distorted tetrahedral shape |
| $d^{10}(Zn^{2+}, Ni^0, Cu^+)$ $(Ni = 3d^8 4s^2 = 3d^{10})$ | $[ZnCl_4]^{2-}$, $[Cu(CN)_4]^{3-}$, $Ni(CO)_4$ | $e^4 t_2^6$ | Regular tetrahedral shape |

## University Questions

1. In general, Cu(II) ions form distorted octahedral complexes. Explain why? (*Bharathiar 2010*)
2. Write a note on "John-Teller Effect") (*Gujrat Technological University 2010*)
3. What is John-Teller effect? Explain why the peak is not symmetrical in case of electronic spectra of $[Ti(H_2O)_6]^{3+}$ and $[Cu(H_2O)_6]^{2+}$ ions. (*Nagpur 2012*)

# 22 CHAPTER
# Applications of Crystal Field Stabilisation Energies (CFSEs)

Before studing the applications of crystal field stabilization energies (CFSEs), we should have a clear understanding of crystal field effects.

## Crystal Field Effects

The splitting of d or f orbitals of a complexed metal ion under the influence of ligand field environment is called crystal field splitting of d or f orbitals. The crystal field splitting affects the property of complexed metal ion. *The effects of crystal field splitting of d or f orbitals on the property of complexed metal ion is known as* **crystal field effects.**

## I. Crystal Field Effects on the Property of Complexed Metal Ion

This topic can be studied under the following heads namely A, B, C, D and E.

**A. Stereochemistry of complexes: Stabilisation of oxidation states of metals by forming complexes of preferred geometry**

CFSE values have been used to explain why certain oxidation states of metals are preferentially stabilised by coordination to certain ligands and why metal ions form complexes of preferred geometry. Thus with the help of CFSE values we can predict what type of geometry a particular metal ion would prefer for its complex ion. These points would be clear from the following examples.

*(i) It has been observed that in the pair, $[Co(H_2O)_6]^{2+}$ and $[Co(H_2O)_6]^{3+}$, the former is more stable than the latter, but in the pair, $[Co(NH_3)_6]^{2+}$ and $[Co(NH_3)_6]^{3+}$ the former is less stable than the latter*

**Explanation.** We know that $H_2O$ is a weak ligand and $NH_3$ is a strong ligand. Hence $[Co^{2+}(H_2O)_6]^{2+}$ and $[Co^{3+}(H_2O)_6]^{3+}$ ions are HS complex ions and $[Co^{2+}(NH_3)_6]^{2+}$ and $[Co^{3+}(NH_3)_6]^{3+}$ are LS complex ions. Now since CFSE of $Co^{2+}$($d^7$ ion) (HS) is greater (= 8 Dq) than $Co^{3+}$ ($d^6$ ion) (HS) (= 4 Dq), $[Co^{2+}(H_2O)_6]^{2+}$ is more stable than $[Co^{3+}(H_2O)_6]^{3+}$ ion. Thus $Co^{2+}$ ion is stabilised by $H_2O$ ligands (weak ligands) by forming HS $[Co^{2+}(H_2O)_6]^{2+}$ ion but $Co^{3+}$ ion is not stabilised by $H_2O$ ligands. In other words $[Co^{3+}(H_2O)_6]^{3+}$ ion is not formed.

Since CFSE of $Co^{2+}$ ($d^7$ ion) (LS) is smaller (= 18 Dq) than $Co^{3+}$ ($d^6$ ion) (LS) (= 24 Dq), $[Co^{2+}(NH_3)_6]^{2+}$ ion is less stable than $[Co(NH_3)_6]^{3+}$ ion. Thus $Co^{3+}$ ion is stabilised by $NH_3$ ligands (strong ligands) by forming LS $[Co^{3+}(NH_3)_6]^{3+}$ ion but $Co^{2+}$ ion is not stablised by $NH_3$ ligands. In other words $[Co^{2+}(NH_3)_6]^{2+}$ ion is not formed.

*(ii) Almost all octahedral complexes of Co (III) are LS octahedral complexes.*

**Explanation.** $Co^{3+}$ ion present in LS octahedral complexes is a $d^6$ system whose CFSE value in LS state is equal to 24 Dq. Since this value of CFSE is the highest, Co (III) prefers to form LS octahedral complexes. Note that $[CoF_6]^{3-}$ ion is an exception, since it is not a LS octahedral complex ion. Rather it is a HS octahedral complex ion. Its HS nature is due to weak crystal field strength of $F^-$ ligand which cannot exceed pairing energy (P) of $Co^{3+}$ ion.

*(iii) Metal ions having $d^3$ and $d^8$ configurations prefer to form octahedral complex ions rather than tetrahedral complex ions.*

**Explanation.** The above observation is because of the fact the CFSE value for octahedral complexes of these ions is greater (= 12 Dq) than these values for tetrahedral complexes (= 3.6 Dq). Thus $V^{2+}$ ($d^3$) and $Ni^{2+}$ ($d^8$) ions prefer to form octahedral complexes.

*(iv) Tetrahedral Co (II) complex has more chances to be formed in aqueous medium than tetrahedral Ni (II) complex.*

This fact is evident from the experimental observation that when excess of HCl is added to octahedral $[Co^{2+}(H_2O)_6]^{2+}$ ion, it is converted into tetrahedral $[Co^{2+}Cl_4]^{2-}$ ion

$$\underset{\substack{\text{Octahedral} \\ \text{(Pale pink)}}}{[Co^{2+}(H_2O)_6]^{2+}} + 4Cl^- \text{ (Excess)} \longrightarrow \underset{\substack{\text{Tetrahedral} \\ \text{(Blue)}}}{[Co^{2+}Cl_4]^{2-}} + 6H_2O$$

On the other hand, if excess of HCl is added to $[Ni^{2+}(H_2O)_6]^{2+}$ ion, no reaction takes place.

$$[Ni^{2+}(H_2O)_6]^{2+} + 4Cl^- \rightarrow \text{No reaction}$$

**Explanation.** The above observation can be explained by comparing *octahedral site stabilization energy* (OSSE) for $Co^{2+}$ and $Ni^{2+}$ ions. OSSE value for these ions are given by:

OSSE value for $Co^{2+}$ ion ($d^7$ ion) = CFSE value of $d^7$ ion in HS octahedral ligand field — CFSE value of $d^7$ ion in tetrahedral ligand field

$= (CFSE)_{oct} - (CFSE)_{tet}$

$= -8\text{ Dq} - (-5.4\text{ Dq})$

$= \mathbf{-2.6\ Dq}$

OSSE value for $Ni^{2+}$ ion ($d^8$ ion) = CFSE value for $d^8$ ion in octahedral ligand field – CFSE value of $d^8$ ion in tetrahedral ligand field.

$= (CFSE)_{oct} - (CFSE)_{tet}$

$= -12\text{ Dq} - (-3.6\text{ Dq})$

$= \mathbf{-8.4\ Dq}$

Now since OSSE value for $Co^{2+}$ ion is smaller (= 2.6 Dq) than that for $Ni^{2+}$ ion (= 8.4 Dq), tetrahedral Co (II) complex has better prospect to be formed in aqueous medium than a tetrahedral Ni (II) complex.

*(v) $d^0$, $d^5$ (weak field) and $d^{10}$ ions have no references for any particular stereochemistry.*

**Explanation.** Since $d^0$, $d^5$ (weak field) and $d^{10}$ ions have no CFSE both in octahedral and tetrahedral ligand fields, these ions have no preference for any particular geometry. In such cases stereochemistry and coordination number is decided by ligand – ligand repulsion, size of the cation and cation charge. Zn (II) which is a $3d^{10}$ ion has a preference for tetrahedral geometry, since this geometry minimizes ligand – ligand repulsion and maximises L-M-L bond anges to $109^0$.

CFT predicts that if no other factor influencing the geometry of complex ions is taken into consideration, $d^0$, $d^5$ and $d^{10}$ ions have equal probability for forming octahedral and tetrahedral complexes. Thus $Mn^{2+}$ ($d^5$) and $Zn^{2+}$ ($d^{10}$) form a number of tetrahedral as well as octahedral complexes

*(vi)* In case of complexes of metal ions with $d^1$ and $d^6$ (HS) configuration, the magnitude of loss in stability due to crystal field effects in going from octahedral to tetrahedral geometry is quite small (= 1.33 Dq). Thus if any factor other than crystal field effects favours tetrahedral geometry for the complexes of these ions by the stabilization of magnitude greater than the magnitude of loss of stability due to crystal field effects, the complexes formed by these ions would be tetrahedral.

*(vii)* In case of complexes of metal ions with $d^2$ and $d^7$ (HS) configuration, the magnitude of loss of stability due to crystal field effects in going from octahedral to tetrahedral geometry is equal to 0.67 Dq. Thus if any other factor stablises tetrahedral geometry by an amount greater then 0.67 Dq, these ions would form tetrahedral complexes quite easily.

*(viii) (a) $Ni^{2+}$ ion gets stablised by forming strong field (LS) square planar complexes rather than strong ligand octahedral complexes.*

**Explanation.** Since the value of CFSE of $Ni^{2+}$ ion ($d^8$) in strong ligand square planar complexes in higher ( = 24.56 Dq) than the value of CFSE of $Ni^{2+}$ ion in strong ligand octahedral complexes (= 12 Dq), $Ni^{2+}$ ion gets stablised by forming strong field (LS) square planar complexes rather than strong ligand octahedral complexes.

*(b) Most of the 4- coordinated complexes of $Ni^{2+}$ ion are square planar rather than tetrahedral*

**Explanation.** The fact given above is because the CFSE value of $Ni^{2+}$ ion ($d^8$) in weak ligand square planar complexes is higher (= 14.56 Dq) than the CFSE value of this ion in tetrahedral complexes (= 3.6 Dq). Note than $[NiX_4]^{2-}$ (X = $Cl^-$, $Br^-$, $I^-$) is in exception, since it is a tetrahedral complex ion.

*(ix)* The values of CFSE for $Ag^+$ ($d^{10}$) and $Ag^{2+}$ ($d^9$) ions in square planar, tetrahedral and octahedral complexes are given below:

| *Ions* | *Square planar complexes* | *Tetrahedra complexes* | *Octahedral complexes* |
|---|---|---|---|
| $Ag^+(d^{10})$ | 0.0 Dq | 0.0 Dq | 0.0 Dq |
| $Ag^{2+}$ ($d^9$) | 12.28 Dq | 1.8 Dq | 6 Dq |

Since the CFSE values for $Ag^+$ ion in all the three geometries is zero, $Ag^+$ ion is not stablised in any of the given geometries, i.e. $Ag^+$ does not form any of the complexes having square planar, tetrahedral and octahedral geometry. On the other hand, since the value of CFSE for $Ag^{2+}$ ion in square planar geometry is maximum, this ion is stabilised by forming square planar complex, i.e. $Ag^{2+}$ ion forms neither tetrahedral nor octahedral complex.

(*x*) CFSE values for $Cu^{2+}$ ion ($d^9$) in octahedral, tetrahedral and square planar complexes are 6 Dq, 1.8 Dq and 12.28 Dq respectively. Since this value for square plannar complexes is maximum, $Cu^{2+}$ ion forms square planar complexes rather than octahedral or tetrahedral complexes.

(*xi*) $Ni^{2+}$, $Pd^{2+}$ and $Pt^{2+}$ ions are $d^8$ ions. All these ions form 4-coordinated complexes. Four coordinated complexes of $Ni^{2+}$ ion may be square planar (majority of complexes are square planar) or tetrahedral [e.g. $[MiX_4]^{2-}$ (X = $Cl^-$, $Br^-$, $I^-$) is a tetrahedral complex ion] but these complexes of $Pd^{2+}$ and $Pt^{2+}$ ions ($d^8$ ions) are square planar. This is because of the fact that CFSE values for $Pd^{2+}$ and $Pt^{2+}$ ions in square planar complexes are much greater than CFSE values of $Ni^{2+}$ ion in tetrahedral complexes.

## B. Variation of octahedral ionic radii of divalent ions ($M^{2+}$) of the metals of 1st transition series in the crystalline $MX_2$ halides.

We known that *d*-orbitals, due to the electrostatic crystal field potential of ligands experienced by the *d*-electrons of the complexed metal ion, are split into $t_{2g}$ and $e_g$ orbitals, if the metal ion is surrounded by six ligands. Let us see how the ionic radii of $M^{2+}$ ions in crystallince halides ($MX_2$) vary with the splitting of *d* orbitals of the complexed bivalent metal ion ($M^{2+}$ ion). In $MX_2$, the bivalent $M^{2+}$ ion is in HS state, since the six $X^-$ ligands by which $M^{2+}$ ion is surrounded are weak field ligands. Now since effective nuclear charge ($Z_{eff}$) of $M^{2+}$ ions increases with the increase in atomic number from $Ca^{2+}$ to $Zn^{2+}$, the ionic radii of $M^{2+}$ ions should decrease from $Ca^{2+}$ to $Zn^{2+}$. Thus if the variation of ionic radii of $M^{2+}$ ions is considered only on the basis of the variation of $Z_{eff}$ and do not consider the crystal field effects, then these values should decrease from $Ca^{2+}$ to $Zn^{2+}$. This decreasing order has been shown by a broken line curve (-------) (See Fig. 22.1). The hollow circles (○) shown on this curve represent the values of hypothetical ionic radii of $M^{2+}$ ions.

If we look upon the experimental values of $M^{2+}$ ions of $MX_2$ ionic crystal in which $M^{2+}$ ion is surrounded by six $X^-$ ligands, we find that these values show irregular variation. This irregular variation has been shown by a solid line curve (——). The solid black dots (•) given on this curve represent experimental values of ionic radii of $M^{2+}$ ions in $MX_2$

It may be noted that the experimental and hypothetical values of ionic radii of $Ca^{2+}$, $Mn^{2+}$ and $Zn^{2+}$ ions are the same, since $Ca^{2+}$ ($d^0$), $Mn^{2+}$ ($d^5$) and $Zn^{2+}$ ($d^{10}$) ions have CFSE = 0. Experimental values of $M^{2+}$ ions other than $Ca^{2+}$, $Mn^{2+}$ and $Zn^{2+}$ ions are lower than their hypothetical values due to crystal field effects in these ions.

Before discussing the crystal field effects on the variation of experimental values of ionic radii of $M^{2+}$ ions in $MX_2$ crystal, we should have a knowledge of $t_{2g}^p e_g^q$ configuration of $M^{2+}$ ions given in Table 22.1.

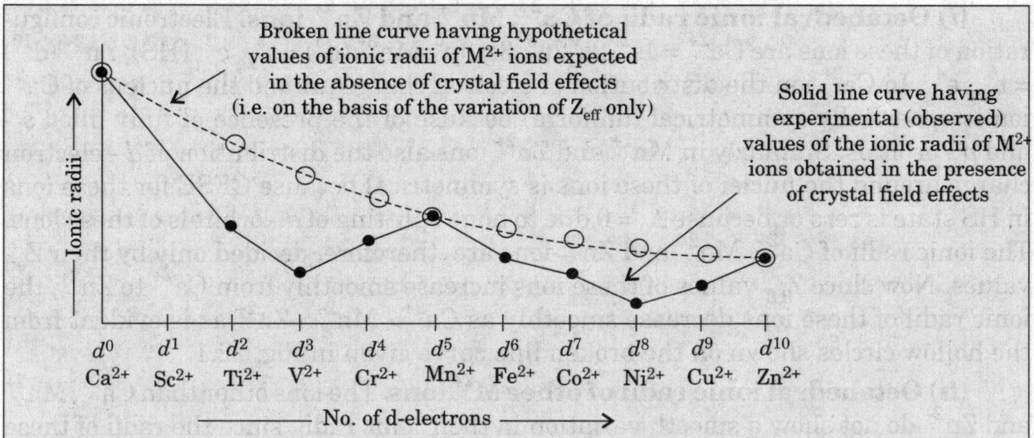


**Fig. 22.1** Plot of ionic radii *vs.* number of $d$ - electrons present in bivalent metal ions ($M^{2+}$) in $MX_2$. Hollow circles (O) indicate hypothetical ionic radii expected in the absence of crystal field effects and solid black dots (•) indicate the observed values of ionic radii obtained by considering the crystal field effects.

**Table 22.1** $t_{2g}^p e_g^q$ **configuration of** $M^{2+}$ **ions (**$M^{2+}$**=**$Ca^{2+}$ **to** $Zn^{2+}$**) in HS state**

| *$M^{2+}$ ion* | *$t_{2g}^p e_g^q$ configuration* | *No. of $t_{2g}$ electrons* | *No. of $e_g$ electrons* |
|---|---|---|---|
| $Ca^{2+}$ ($d^0$) | $t_{2g}^0 e_g^0$ | 0 | 0 |
| $Sc^{2+}$ ($d^1$) | $t_{2g}^1 e_g^0$ | 1 | 0 |
| $Ti^{2+}$ ($d^2$) | $t_{2g}^2 e_g^0$ | 2 | 0 |
| $V^{2+}$ ($d^3$) | $t_{2g}^3 e_g^0$ | 3 | 0 |
| $Cr^{2+}$ ($d^4$) | $t_{2g}^3 e_g^1$ | 3 | 1 |
| $Mn^{2+}$ ($d^5$) | $t_{2g}^3 e_g^2$ | 3 | 2 |
| $Fe^{2+}$ ($d^6$) | $t_{2g}^4 e_g^2$ | 4 | 2 |
| $Co^{2+}$ ($d^7$) | $t_{2g}^5 e_g^2$ | 5 | 2 |
| Ni ($d^8$) | $t_{2g}^6 e_g^2$ | 6 | 2 |
| $Cu^{2+}$ ($d^9$) | $t_{2g}^6 e_g^3$ | 6 | 3 |
| $Zn^{2+}$ ($d^{10}$) | $t_{2g}^6 e_g^4$ | 6 | 4 |

***(i)* Octahedral ionic radii of $Ca^{2+}$, $Mn^{2+}$ and $Zn^{2+}$ ions.** Electronic configuration of these ions are $Ca^{2+} = 1s^2, 2s^2 2p^6, 3s^2 3p^6$; $Mn^{2+} (d^5) = t^3_{2g}\, e^2_g$ (HS); $Zn^{2+} (d^{10}) = t^6_{2g}\, e^4_g$. In $Ca^{2+}$ ion the distribution of electron charge around the nucleus of $Ca^{2+}$ ion is spherically symmetrical (uniform) because of the presence of fully filled *s* - and *p* - orbitals. Similarly in $Mn^{2+}$ and $Zn^{2+}$ ions also the distribution of *d* - electron charge around the nuclei of these ions is symmetrical because CFSE for these ions in HS state is zero or because $\Delta_0 = 0$ due to non - splitting of *d* - orbitals of these ions. The ionic radii of $Ca^{2+}$, $Mn^{2+}$ and $Zn^{2+}$ ions are, therefore, decided only by their $Z_{eff}$ values. Now since $Z_{eff}$ values of these ions increase smoothly from $Ca^{2+}$ to $Zn^{2+}$, the ionic radii of these ions decrease smoothly as $Ca^{2+} > Mn^{2+} > Zn^{2+}$ as is evident from the hollow circles shown on the broken line curve given in Fig. 22.1.

***(ii)* Octahedral ionic radii of other $M^{2+}$ ions.** The ions other than $Ca^{2+}$, $Mn^{2+}$ and $Zn^{2+}$ do not show a smooth variation in their ionic radii, since the radii of these ions decrease irregularly as in evident from the solid black dots shown on the solid line curve given in Fig. 22.1. The radii of these ions are less than the values expected from the consideration of $Z_{eff}$ alone. The expected values obtained by the consideration of $Z_{eff}$ alone have been shown by hollow circles on the broken line curve. The non-smooth variation of the ionic radii of other $M^{2+}$ ions is due to non-uniform or non- spherical distribution of electronic charge of *3d* - electrons around the nuclei of these metal ions. The non-symmetrical electonic charge distribution is because of the splitting of *d*- orbitals under the influence of octahedral ligand field generated by six $X^-$ ligands.

*(a)* **Octahedral ionic radius of $Sc^{2+}$ ion.** Since the octahedral ionic radius of $Sc^{2+}$ ion ($d^1$) has not been determined with certainty due to unstable nature of +2 oxidation state of Sc, the solid line curve does not have the black solid dot which may represent the ionic radius of $Sc^{2+}$ ion.

*(b)* **Octahedral ionic radius of $Ti^{2+}$ ion.** It is evident from Table 22.1 that the $t^p_{2g}\, e^q_g$ configuration of $Ti^{2+}$ ion ($d^2$) is $t^2_{2g}\, e^0_g$. This configuration shows that the electronic charge density of the two electons present in $t_{2g}$ orbitals (*i. e.* $d_{xy}$, $d_{yz}$ and $d_{xz}$ orbitals) is concentrated in the space which lies in between x, y and z axes along which the six $X^-$ ligands are placed. Thus the electronic charge of the two $t_{2g}$ electrons is not concentrated along $Ti^{2+}$—$X^-$ directions. As a result, there exists low repulsion between the negative charge of $t_{2g}$ electrons and the negative charge on $X^-$ ligands in $Ti^{2+}$ —$X^-$ directions. In other words $X^-$ ligands are drawn closer to $Ti^{2+}$ ion. As a result, $Ti^{2+}$—$X^-$ distance becomes shorter. Consequently, the octahedral radius of $Ti^{2+}$ ion becomes smaller than the value expected from the consideration of $Z_{eff}$ alone.

*(c)* **Octhaderal ionic radius of $V^{2+}$ ion.** $t^p_{2g}\, e^q_g$ configuration of $V^{2+}$ ion ($d^3$) is $t^3_{2g}\, e^0_g$. Since $t_{2g}$ orbitals have three electrons, the attraction between $V^{2+}$ ion and the six $X^-$ ligands surrounding this ion is greater than that between $Ti^{2+}$ ion and $X^-$ ligands. As a result, $V^{2+}$—$X^-$ distance in $VX_2$ becomes shorter than $Ti^{2+}$—$X^-$ distance in $TiX_2$. Consequently octahedral ionic radius of $V^{2+}$ ion is smaller than that of $Ti^{2+}$ ion. The ionic radius of $V^{2+}$ ion is also smaller than the value obtained by considerating $Z_{eff}$ alone.

*(d)* **Octahedral ionic radius of $Cr^{2+}$ ion.** $t^p_{2g}\, e^q_g$ configuration of $Cr^{2+}$ ion ($d^4$) in HS state is $t^3_{2g}\, e^1_g$ in which three electrons occupy $t_{2g}$ orbitals and one electron is placed in $e_g$ orbitals. The electronic charge density of one electron present in $e_g$ orbitals ($d_{x^2-y^2}$ and $d_{z^2}$ orbitals) lies straight in the directions of $Cr^{2+}$ — $X^-$ bonds.

Hence there exists considerable repulsion between the charge density of $e_g$ electron and negative charge on $X^-$ ligands in the directions of $Cr^{2+}$— $X^-$ bonds. In other words the octahedral radius of $Cr^{2+}$ ion is larger than that of $V^{2+}$ ion. But the radius of $Cr^{2+}$ ion is still smaller than the value obtained by considering $Z_{eff}$ alone.

(*e*) **Octahedral ionic radius of $Mn^{2+}$ ion.** $t_{2g}^p e_g^q$ configuration of $Mn^{2+}$ ion ($d^5$) in HS state is $t^3_{2g}\ e^2_g$. The value of CFSE for $Mn^{2+}$ ion in HS state is zero (or there is no splitting of d- orbitals since $\Delta_0 = 0$). $t_{2q}^3\ e_g^2$ configuration shows that since the electron charge distribution in *d* - orbitals around the nucleus is spherically symmetrical, octahedral radius of $Mn^{2+}$ ion has the value which is expected from the consideration of $Z_{eff}$ alone. Thus the value of octahedral radius of $Mn^{2+}$ ion lies on the smooth curve shown by broken lines in Fig. 22.1.

(*f*) **Octahedral ionic radii of $Fe^{2+}$, $Co^{2+}$ and $Ni^{2+}$ ions.** The variation of ionic radii of these ions in HS state takes place in the same way as in case of $Ca^{2+}$, $Ti^{2+}$, $V^{2+}$, $Cr^{2+}$ and $Mn^{2+}$ ions. The $t_{2g}^p e_g^q$ configurations of $Fe^{2+}$, $Co^{2+}$ and $Ni^{2+}$ ions in HS state are $Fe^{2+}$ ($d^6$) = $t_{2g}^4\ e_g^2$, $Co^{2+}$ ($d^7$) = $t_{2g}^5\ e_g^2$ and $Ni^{2+}$ ($d^8$) = $t_{2g}^6\ e_g^2$. It may be seen that as we move from $Mn^{2+}$ ($t_{2g}^3\ e_g^2$) to $Ni^{2+}$ ($t_{2g}^6\ e_g^2$) through $Fe^{2+}$ ($t_{2g}^4\ e_g^2$) and $Co^{2+}$ ($t_{2g}^5\ e_g^2$), the number of $t_{2g}$ electrons increases. With the increase in the number of electrons in $t_{2g}$ orbitals, d-electron charge gets more and more concentrated in the space in between the directions of $M^{2+}$ —$X^-$ bonds ($M^{2+}$ = $Fe^{2+}$, $Co^{2+}$, $Ni^{2+}$). Consequently the ionic radii of $Fe^{2+}$, $Co^{2+}$ and $Ni^{2+}$ ions decrease. At the same time, the radii of these ions are even lesser than the values expected from the consideration of $Z_{eff}$ alone.

(*g*) **Octahedral ionic radius of $Cu^{2+}$ ion.** In $Cu^{2+}$ ion ($d^9$) ninth d-electron enters $e_g$ orbitals ($Cu^{2+}$ = $d^9$= $t_{2g}^6\ e_g^3$). Now since the regions of maximum electon charge density of $e_g$ electron are straight in the directions of $Cu^{2+}$——$X^-$ bonds, the repulsion between $e_g$ electron and negative charge on $X^-$ ligands increases in these directions. This increase in the repulsion increases the distance between $Cu^{2+}$ and $X^-$ ions. As a result the octahedral ionic radius of $Cu^{2+}$ ion in larger than that of $Ni^{2+}$.

(*h*) **Octhaderal ionic radius of $Zn^{2+}$ ion.** The configuration of $Zn^{2+}$ ($d^{10}$) is $t_{2g}^6\ e_g^4$. Since CFSE = 0 for $Zn^{2+}$ ion or there is no splitting of *d*- orbitals of this ion, the distribution of *d*-electron charge density is uniformly distributed in all directions. Thus the octahedral radius of $Zn^{2+}$ ion has the value expected from the value based on $Z_{eff}$ alone, *i.e.* the ionic radius of $Zn^{2+}$ ion lies on the smooth curve shown by broken line.

## C. Variation of heats of hydration of $M^{2+}$ ions ($H^{2+}$)

Heat of hydration of $M^{2+}$ ions of the metals of *1st* transiction series is the heat released when one mole of gaseous $M^{2+}$ ion combines with six molecules of gaseous $H_2O$ to form gaseous one mole of $[M(H_2O)_6]^{2+}$.

$$M^{2+}(g) + 6H_2O(g) \rightarrow [M(H_2O)_6]^{2+}(g) + \textit{Heat released [Called heat of hydration of } M^{2+}\textit{ (g)]}$$

*or* $$M^{2+}(g) + 6H_2O(g) \rightarrow [M(H_2O)_6]^{2+} + H^{2+}$$

In the above equation $H^{2+}$ represents heat of hydration of $M^{2+}$ ion. Since $H_2O$ is weak field ligand, $[M(H_2O)_6]^{2+}$ formed in the above reaction is a HS complex ion.

Experimental values of heat of hydration of bivalent metal ions of first transition series are given in Table 22.2.

**Table 22.2 Heats of hydration ($-H^{2+}$) of bivalent metal ions of first transition series**

| *Ion* | $-H^{2+}$ *(kJ mol$^{-1}$)* | *Ion* | $-H^{2+}$ *(kJ mol$^{-1}$)* |
|---|---|---|---|
| $Ca^{2+}$ | 2469 | | |
| $Sc^{2+}$ | — | $Fe^{2+}$ | 2845 |
| $Ti^{2+}$ | 2732 | $Co^{2+}$ | 2916 |
| $V^{2+}$ | 2778 | $Ni^{2+}$ | 2996 |
| $Cr^{2+}$ | 2795 | $Cu^{2+}$ | 3000 |
| $Mn^{2+}$ | 2736 | $Zn^{2+}$ | 2933 |

Since heat of hydration is the energy released it, is represented by a negative sign ($-H^{2+}$)

Heat of hydration of $M^{2+}$ ions is inverssaly proportional to the ionic radii of $M^{2+}$ ions. We have already explained that in the absence of crystal field effects, the ionic radii of $M^{2+}$ ions decrease from $Ca^{2+}$ to $Zn^{2+}$. Due to this decrease in the ionic radii of $M^{2+}$ ions heat of hydration of $M^{2+}$ ions increases from $Ca^{2+}$ to $Zn^{2+}$ as shown by broken line curve given in Fig. 22.2. The hydration energy values have been shown by hollow circles (○) on the broken line curve.

If we look upon the experimental values of hydration energy given in Table 22.2 and which have been obtained by taking into account the crystal field effects in $M^{2+}$ ions, we find that these values show an irregular variation. The irregular variation has been shown by solid line curve (See Fig 22.2). The solid black dots (•) given on this curve represent experimental values of hydration energy of $M^{2+}$ ions.

It may be noted from Fig. 22.2 that the experimental (solid dots) and hypothetical (hollow circles) values of hydration energy of $Ca^{2+}$, $Mn^{2+}$ and $Zn^{2+}$ ions are the

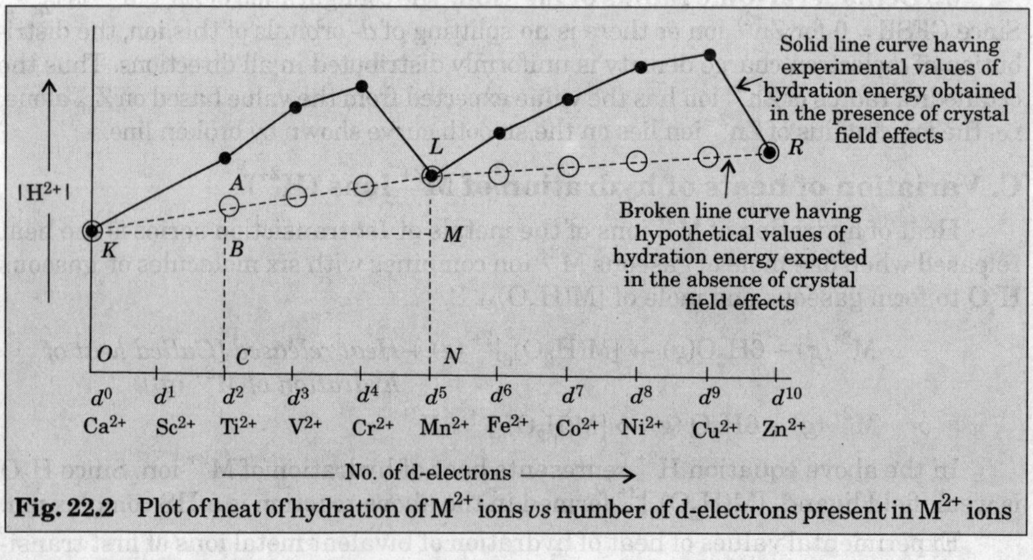


**Fig. 22.2** Plot of heat of hydration of $M^{2+}$ ions *vs* number of d-electrons present in $M^{2+}$ ions

same since $Ca^{2+}$ ($d^0$), $Mn^{2+}$ ($d^5$) and $Zn^{2+}$ ($d^{10}$) ions have CFSE = 0. Experimental values of hydration energy of other $M^{2+}$ ions are higher than their hypothetical values due to crystal field effects in these ions.

## Determination of CFSE for a given hydrated $M^{2+}$ ion, $[M(H_2O)_6]^{2+}$

With the help of observed values of heat of hydration of various $M^{2+}$ ions given in Table 22.2 we can calculate the CFSE of a given $[M(H_2O)_6]^{2+}$ ion. As an example let us determine the CFSE of $[Ti(H_2O)_6]^{2+}$ ion.

CFSE of $[Ti(H_2O)_6]^{2+}$ ion = Experimental value of hydration energy of $Ti^{2+}$ ion – Hypothetical value of hydration energy of $Ti^{2+}$ ion

= 2732 – Hypothetical value of hydration energy of $Ti^{2+}$ ion ... (*i*)

From Fig 22.2, we have:

Hypothetical value of hydration energy of $Ti^{2+}$ ion = AC

= AB + BC ... (*ii*)

$$\text{Now AB} = \frac{2}{5}\text{LM} \qquad (\text{From } \Delta\text{KLM})$$

$$= \frac{2}{5}(\text{LN} - \text{MN})$$

$$= \frac{2}{5}(\text{LN} - \text{KO}) \qquad (\because \text{MN} = \text{KO})$$

$$= \frac{2}{5}(\text{Hydration energy of } Mn^{2+} \text{ ion} - \text{Hydration energy of } Ca^{2+} \text{ ion})$$

$$= \frac{2}{5}(2736 - 2469)$$

$$= 107 \text{ KJ mol}^{-1} \qquad (a)$$

and BC = KO

= Hydration energy of $Ca^{2+}$ ion

= 2469 KJ $mol^{-1}$ ... (*b*)

With the help of equations (*a*) and (*b*), equation (*ii*) reduces to:

Hypothetical value of hydration energy of $Ti^{2+}$ ion = AB + BC

= 107 + 2469

= 2576 KJ $mol^{-1}$ ... (*iii*)

Hence equation (*i*) can be written as:

CFSE of $[Ti(H_2O)_6]^{2+}$ = 2732 – 2576

= **156 KJ $mol^{-1}$**

By using the same procedure, we can determine the value of hydration energy of other $[M(H_2O)_6]^{2+}$ ions.

## D. Variation of lattice energy of halides of bivalent ions ($M^{2+}$) of the metals of 1st transition series

Lattice energy of an ionic crystal is defined as the amount of energy released when one mole of the ionic crystal is formed by the combination of the constituent gaseous cations and gaseous anions of the ionic crystal. Thus the lattice energy of $MX_2$ ionic crystal is the energy released in the reaction.

$$M^{2+}(g) + 2X^-(g) \rightarrow \underset{\text{one mole}}{MX_2(s)} + \textit{Energy released (Lattice energy)}$$

Experimental values of lattice energies of the halides of bivalent metal ions of 1st transition series are given in Table 22.3.

As an example let us discuss the variantion of lattice energy of fluorides of $M^{2+}$ ions ($MF_2$ fluorides). In $MF_2$, $M^{2+}$ ion is surrounded by six $F^-$ ions in an octahedral fashion. $M^{2+}$ ion is in HS state.

In various $MF_2$ molecules, since $F^-$ ion is present in all the fluorides, the lattice energy of $MF_2$ moleculues increases with the decrease of the ionic radii of $M^{2+}$ ions. Now since the ionic radii of $M^{2+}$ ions of the metals of *1st* transition series generally *decrease* with the increase in atomic number from $Ca^{2+}$ to $Zn^{2+}$, the lattice energies of $MF_2$ molecules should *increase* from $CaF_2$ to $ZnF_2$. Thus if the variation of lattice energies is considered only on the basis of the variation of ionic radii of $M^{2+}$ ions and we donot consider the crystal field effects, then these values should increase from $CaF_2$ to $ZnCl_2$. This increasing trend has been shown by a broken line curve (see Fig. 22.3). The hollow circles (○) shown on this broken line curve represent the values of hypothetical lattice energy of the fluorides. The points P, K and L represent the

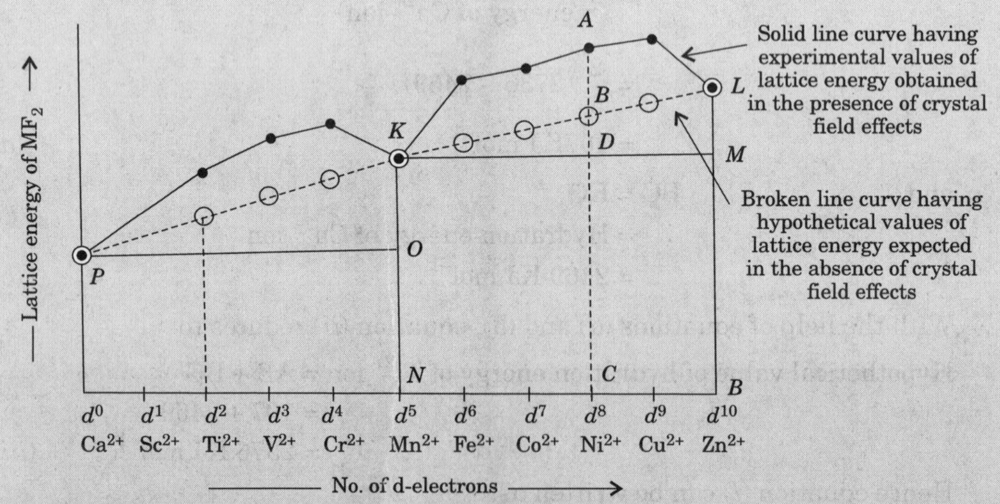


**Fig. 22.3** Plot of lattice energy *vs.* number of *d* electrons of bivalent metal fluorides of the first transition series. Hollow circles (○) indicate hypothetical lattice energies expected in the absence of any crystal field effects and solid dots (•) indicate the observed lattice energies.

lattice energy of $CaF_2$, $MnF_2$ and $ZnF_2$ respectively. Hypothetical values are the values in the absence of crystal field effects.

If we look upon the experimental (observed) values of lattice energy of $MF_2$ molecules (Table 22.3) obtained by taking into account the crystal field effects, we find that these values show irregular variation. This irregular variation has been shown by a solid line curve. The solid black dots (•) given on solid curve represent the experiemtal values. It may be noted that experimental and hypothetical values of crystal energy for $CaF_2$, $MnF_2$ and $ZnF_2$ are the same, since $Ca^{2+}$ ($d^0$), $Mn^{2+}$($d^5$) and $Zn^{2+}$($d^{10}$) ions have CFSE = 0. Experimental values of the fluorides of other $M^{2+}$ ions are higher than their hypothetical values due to crystal field effects in these fluorides.

**Table 22.3 Lattice energies in kJ mol$^{-1}$ of halides of Ca and bivalent metals of the first transition series**

| *Anions* | $Ca^{2+}$ | $Sc^{2+}$ | $Ti^{2+}$ | $V^{2+}$ | $Cr^{2+}$ | $Mn^{2+}$ | $Fe^{2+}$ | $Co^{2+}$ | $Ni^{2+}$ | $Cu^{2+}$ | $Zn^{2+}$ |
|---|---|---|---|---|---|---|---|---|---|---|---|
| $F^-$ | 2607 | — | 2757 | 2770 | 2887 | 2782.4 | 2929 | 2979 | 3063.5 | 3067 | 2986.1 |
| $Cl^-$ | 2234 | — | 2477 | 2548 | 2569 | 2502 | 2611 | 2690 | 2753 | 2786 | 2711 |
| $Br^-$ | 2133 | — | 2393 | 2489 | 2506 | 2435 | 2540 | 2619 | 2699 | 2740 | 2644 |
| $I^-$ | 2054 | — | 2309 | 2397 | 2397 | 2360 | 2477 | 2531 | 2619 | 2669 | 2581 |

The forgoing discussion shows that the curve obtained by plotting hydrated energies *vs* number of *d*- electrons present in bivalent metal ions ($M^{2+}$) of the first transition series is similar in character of that obtained by plotting lattice energies of $MF_2$ ionic crystals.

## Determination of CFSE of $M^{2+}$ ion in $MX_2$ molecules

With the help of Fig 22.3 we can calculate the value of CFSE for a given metallic fluoride. CFSE of a metallic fluoride is the difference between the value of observed (experimental) lattice energy of the flyoride and the value of its hypothetical lattice energy. As an example let determine the value of CFSE for $NiF_2$.

CFSE of $NiF_2$ is given by:

CFSE of $NiF_2$ = Observed value of lattice energy of $NiF_2$
— Hypothetical value of lattice energy of $NiF_2$
= 3063.5 — Hypothetical value of lattice energy of $NiF_2$ (*i*)

From Fig 22.3 we have:

Hypothetical value of lattice energy of $NiF_2$ = BC
= BD + DC (*ii*)

Now $BD = \frac{3}{5}$ LM (From ΔKLM )

$= \frac{3}{5}$ (LB – MB)

$= \frac{3}{5}$ (LB – KN)

$= \frac{3}{5}$ (Experimental value of lattice energy of $ZnF_2$ –

Experimental value of lattice energy of $MnF_2$)

$= \frac{3}{5}$ (2986.1 – 2782.4)

$= 122.2$ KJ mol$^{-1}$ (a)

and DC = KN

= Lattice energy of $MnF_2$

$= 2782.4$ KJ mol$^{-1}$ (b)

With the help of equations (*a*) and (*b*), equation (*ii*), reduces to:

Hypothetical value of lattice energy of $NiF_2$

= BD + DC

= 122.2 + 2782.4

$= 2904.6$ KJ mol$^{-1}$

Hence equation (*i*) can be written as:

CFSE of $NiF_2$ = 3063.5 – 2904.6

**= 158.9 KJ mol$^{-1}$**

Students can use the procedure described above for the determination of CFSE value for other fluorides

## Determination of the value of $\Delta_0$ (or 10 Dq) for $Ni^{2+}$ ion ($d^8$ ion).

$\Delta_0$ or 10 Dq is the energy difference between $t_{2g}$ and $e_g$ orbitals which are obtained by the splitting of metal ion d-orbitals in octahedral ligand field environment. We know that CFSE for $Ni^{2+}$ ($d^8$ ion) in octahedral ligand field environment is equal to 12 Dq. We have calculated above that CFSE for $Ni^{2+}$ ion in $NiF_2$ is 158.9 KJ mol$^{-1}$. Hence 12 Dq should be equal to 158.9 KJ mol$^{-1}$. In other words

$$12 \text{ Dq} = 158.9 \text{ KJ mol}^{-1}$$

$$\therefore 10 \text{ Dq} = \frac{158.9}{12} \times 10 \text{ KJ mol}^{-1}$$

$$= \mathbf{132.4 \text{ KJ mol}^{-1}}$$

Above procedure can be used to calculate the value to 10 Dq for other $M^{2+}$ ions also.

## E. Crystal structure of spinels

The mixed oxides of divalent and trivalent metal ions are called spinels. These oxides have the general formula $A^{2+} B^{3+} B^{3+} O_4$ or $A^{3+} B^{2+} B^{3+} O_4$. In these formula A and B may also be the same. In the structure of spinels, oxide ions ($O^{2-}$) are weak field ligands. These ions have cubic closely packed structure. This type of packing of $O^{2-}$ ions creates both tetrahedral and octahedral holes in the lattice of spinels. The metal ions have high spin (HS) configuration.

Depending on the manner in which the metal cations ($A^{2+}$, $A^{3+}$, $B^{2+}$ and $B^{3+}$) are distributed amongst the octahedral and tetrahedral holes, the spinels are of the following two types:

(*i*) **Normal spinels.** These are represented by the formula $A^{2+}\ B^{3+}\ B^{3+}\ O_4$. Examples are $Mn^{2+}\ Al^{3+}Al^{3+}\ O_4$ ($A^{2+} = Mn^{2+}$, $B^{3+} = Al^{3+}$), $Mn_3O_4 \equiv Mn^{2+}\ Mn^{3+}\ Mn^{3+}\ O_4$ ($A^{2+} = Mn^{2+}$, $B^{3+} = Mn^{3+}$), $Co_3O_4 \equiv Co^{2+}\ Co^{3+}\ Co^{3+}\ O_4$ ($A^{2+} = Co^{2+}$, $B^{3+} = Co^{3+}$) etc. In these spinels all the $A^{2+}$ cations occupy the tetrahedral holes and all $B^{3+}$ occupy octahedral holes.

(*ii*) **Inverted spinels.** These are represented by the formula, $A^{3+}\ B^{2+}\ B^{3+}\ O_4$. Examples are $Al^{3+}\ Ni^{2+}Al^{3+}O_4$ ($A^{3+} = Al^{3+}$, $B^{2+} = Ni^{2+}$, $B^{3+} = Al^{3+}$), $Fe_3O_4 \equiv Fe^{3+}\ Fe^{2+}\ Fe^{3+}\ O_4$ ($A^{3+} = Fe^{3+}$, $B^{2+} = Fe^{2+}$, $B^{3+} = Fe^{3+}$). In these spinels all the bivalent cations ($B^{2+}$) and half of trivalent cations ($B^{3+}$) are in octahedral holes. The remaining half of the trivalent cations ($A^{3+}$) are in tetrahedral holes.

CFSEs values (in Dq) for bivalent and trivalent metal ions having $d^4$, $d^5$, $d^6$ and $d^7$ configenerations in HS state in octahedral ($B^{2+}$ and $B^{3+}$ cations) and in tetrahedral ($A^{2+}$ and $A^{3+}$ cations) holes are given in Table. 22.4. With the help of these values we can predict whether a given spinel has normal spinel structure or inverted spinal structure. Following examples may be studied. Before studying these examples we should remember the V.S.E.C. of the following cations.

$$Mn^{2+} = d^5,\ Mn^{3+} = d^4,\ Fe^{2+} = d^6,$$
$$Fe^{3+} = d^5,\ Co^{2+} = d^7,\ Co^{3+} = d^6$$

**Table 22.4 CFSEs of divalent and trivalent metal ions in octahedral ($B^{2+}$ and $B^{3+}$ ions) and tetrahedral ($A^{2+}$ and $A^{3+}$ ions) holes.**

| *Metal cations* | *$d^x$ configuration of metal cations in HS state (x = 4 to 7)* | | | |
|---|---|---|---|---|
| | **$d^4$** | **$d^5$** | **$d^6$** | **$d^7$** |
| $B^{2+}$ | $6\ Dq_{oct}$ | 0 | $4\ Dq_{oct}$ | $6\ Dq_{oct}$ |
| $B^{3+}$ | $\frac{3}{2} \times 6\ Dq$ $= 9\ Dq$ | 0 | $\frac{3}{2} \times 4\ Dq$ $= 6\ Dq$ | $\frac{3}{2} \times 6\ Dq$ $= 9\ Dq$ |
| $A^{2+}$ | $\frac{4}{9} \times 4\ Dq$ $= \frac{16}{9}\ Dq$ | 0 | $\frac{4}{9} \times 6\ Dq$ $= \frac{8}{3}\ Dq$ | $\frac{4}{9} \times 12\ Dq$ $= \frac{16}{3}\ Dq$ |
| $A^{3+}$ | $\frac{3}{2} \times \frac{4}{9} \times 4\ Dq$ $= \frac{8}{3}\ Dq$ | 0 | $\frac{3}{2} \times \frac{4}{9} \times 6\ Dq$ $= 4\ Dq$ | $\frac{3}{2} \times \frac{4}{9} \times 12\ Dq$ $= 8\ Dq$ |

**Explanation.** If $Mn_3O_4$ has normal spinel structure, it should be represented as $A^{2+}$ $B^{3+}$ $B^{3+}$ $O_4$ or $Mn^{2+}$ ($d^5$) $Mn^{3+}$ ($d^4$) $Mn^{3+}$ ($d^4$) $O_4$ and CFSE for this structure is equal to 0 + 9 Dq + 9Dq = 18 Dq (See Table 22.4).

If $Mn_3O_4$ has inverted spinel structure, it should be represented as $A^{3+}$ $B^{2+}$ $B^{3+}$ $O_4$ or $Mn^{3+}$ ($d^4$) $Mn^{2+}$ ($d^5$) $Mn^{3+}$ ($d^4$) $O_4$. CFSE for this structure is equal to $\frac{8}{3}$ Dq + 0 + 9 Dq = 11.67 Dq. (See Table 22.4) Now since CFSE for normal spinel structure is greater (= 18 Dq) than that for inverted spinel structure (= 11.67 Dq), *$Mn_3O_4$ is a normal spinel.*

(*ii*) **$Fe_3O_4$ is an inverted spinel**

**Explanation.** If $Fe_3O_4$ has normal spinel structure, it should be represented as $A^{2+}$ $B^{3+}$ $B^{3+}$ $O_4$ or $Fe^{2+}$ ($d^6$) $Fe^{3+}$ ($d^5$) $Fe^{3+}$ ($d^5$) $O_4$ and CFSE for this structure is equal to $\frac{8}{3}$ Dq + 0 + 0 = 2.67 Dq (See Table 22.4).

If $Fe_3O_4$ has inverted spinel structure, then it should be represented as $A^{3+}$ $B^{2+}$ $B^{3+}$ $O_4$ or $Fe^{3+}$ ($d^5$) $Fe^{2+}$ ($d^6$) $Fe^{3+}$ ($d^5$) $O_4$. CFSE of this structure is 0 + 4Dq + 0 = 4 Dq (See Table 22.4). Now since CFSE for inverted spinel structure is greater (= 4 Dq) then that for normal spinal structure (2.67 Dq), $Fe_3O_4$ has *inverted spinel structure.*

(*iii*) **$Co_3O_4$ is a normal spinel**

**Explanation.** If $Co_3O_4$ is a normal spinel, it should be represented by the formula, $A^{2+}$ $B^{3+}B^{3+}O_4$ or $Co^{2+}$ ($d^7$) $Co^{3+}$ ($d^6$) $Co^{3+}$ ($d^6$) $O_4$ and CFSE for this structure is equal to $\frac{16}{3}$ Dq + 6 Dq + 6 Dq = 17.3 Dq (Table 22.4). If $Co_3O_4$ has inverted spinel structure, it should be represented as $A^{3+}B^{2+}B^{3+}O_4$ or $Co^{3+}$ ($d^6$) $Co^{2+}$ ($d^7$) $Co^{3+}$ ($d^6$) $O_4$ and CFSE of this structure is equal to 4 Dq + 6 Dq + 6 Dq = 16 Dq (Table 22.4). Now since CFSE for normal spinel structure of $Co_3O_4$ is greater (= 17.3 Dq) than that for inverted spinel structure (= 16 Dq), *$Co_3O_4$ has normal spinel structure.*

**II. Colour of Transition Metal Complexes** See Chapter 5 and 17

**III. Magnetic Properties of Transition Metal Complexes** See Chapter 5 and 25

**IV. Distortion in Octahedral and Tetrahedral Complexes : John-Tellar Theorem** See Chapter 21

## University Questions

1. Which of the following complexes have larger crystal field splitting of *d*-orbitals in each pair.
   (*i*) $[Co(H_2O)_6]^{2+}$ or $[Co(H_2O)_6]^{3+}$ (*ii*) $[Co(CN)_6]^{3-}$ or $[Co(NH_3)_6]^{3+}$
   (*Nagpur 2012*)
2. Which one among the following ion pairs will have greater CFSE value? Give reasons
   (*a*) $[Fe(CN)_6]^{3-}$, $[Fe(CN)_6]^{4-}$ (*b*) $[Mn(H_2O)_6]^{2+}$, $[Fe(H_2O)_6]^{2+}$(*Lucknow 2013*)

# 23 CHAPTER Molecular Orbital Theory (MOT) of Octahedral Complexes

## Ligand Field Theory

According to CFT, the metal—ligand bond in complexes is purely due to electrostate interaction between the metal ion and the ligands. However, there is a clear evidence of the fact that the metal—ligand bond also has covalent character. In order to include the contribution of covalent bonding in metal—ligand bond, CFT needs modification. CFT modified to include covalent character of metal—ligand bond is known as **ligand field theory**. This theory assumes that the extent of covalent character in metal—ligand bond is generally low as compared to the extent of ionic character of the bond with the result that the conclusions arrived at by CFT are still valid to a large extent.

## Molecular Orbital Theory (MOT)

In MOT of complexes the AOs of central metal ion and ligand $\sigma$ or ligand $\pi$ orbitals combine together and generate a set of bonding MOs of lower energy and an equal number of antibonding MOs. The non-bonding AOs of the central metal ion remain non-bonding and their energy remains unchanged.

In this chapter we shall explain metal—ligand $\sigma$ and metal—ligand $\pi$ bonds in octahedral complexes.

## Sigma ($\sigma$) Bonding in Octahedral Complexes

Let us consider an octahedral complex in which the central metal atom (M) is surrounded octahedrally by six ligands *viz* $L_1$, $L_2$, $L_3$, $L_4$, $L_5$ and $L_6$, [See (*a*) of Fig. 23.1]. Each of these six ligands has a filled $p_\sigma$ orbital. Thus we have six ligand $p_\sigma$ orbitals. These $p_\sigma$ orbitals are lying along x, y and z axes. Hence these $p_\sigma$ orbitals have been represented as $\sigma_x$, $\sigma_{-x}$, $\sigma_y$, $\sigma_{-y}$, $\sigma_{+y}$, and $\sigma_{-z}$ [See (*b*) of Fig. 23.1]

### Different Steps Involved in the Formation of $\sigma$-Molecular Orbitals in Octahedral Complexes

(*i*) First step is to make a selection of the metal ion orbitals which are suitable to overlap with ligand $p_\sigma$ orbital combinations along $\pm x$, $\pm y$ and $\pm z$ axes to form $\sigma$–molecular orbitals.

(*ii*) Second step is to determine the ligand $p_\sigma$ orbital combinations which may overlap with suitable metal ion orbitals along $\pm x$, $\pm y$ and $\pm z$ axes to give $\sigma$-molecular orbitals.

(*iii*) Third step is the final step. In this step metal ion orbitals and ligand orbital combinations having the same symmetry overlap with each other and form

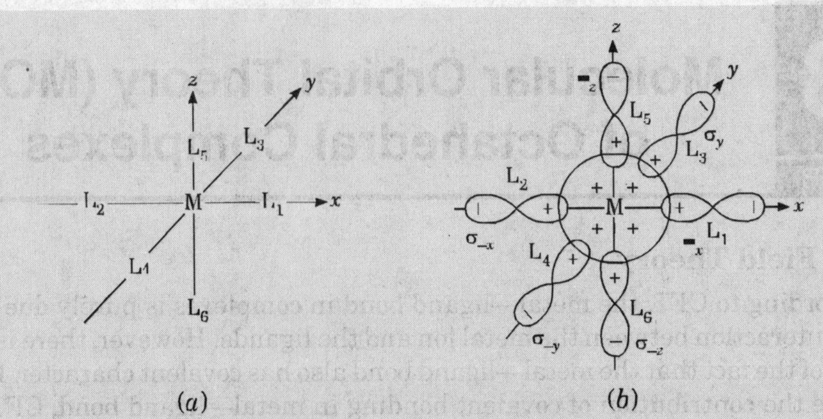


**Fig. 23.1** (*a*) Octahedral complex in which the central metal atom (M) is surrounded octahedrally by six ligands: $L_1$, $L_2$, $L_3$, $L_4$, $L_5$ and $L_6$. (*b*) Filled $p_s$ orbitals on six ligands.

sigma bonding molecular orbitals (s-BMOs) and s-antibonding molecular orbitals (*s-ABMOs). s-BMOs are generated when metal AOs and ligand orbital combinations combine together with *maximum positive overlap* and *s-ABMOs are formed when the combining orbitals overlap together with *maximum negative overlap*.

Now let us discuss these steps in detail.

## I. To Make a Selection of Metal Ion Orbitals Suitable for s-Bonding

We know that the valence shell of *1st* transition series metal ions has *nine* orbitals *viz. one 4s*, *three 4p* ($4p_x$, $4p_y$ and $4p_z$) and *five 3d* ($3d_{z^2}$, $3d_{x^2-y^2}$, $3d_{xy}$, $3d_{yz}$ and $3d_{xz}$) orbitals. Six of these nine orbitals namely 4s, $4p_x$, $4p_y$, $4p_z$, $3d_{z^2}$ and $3d_{x^2-y^2}$ have their lobes along the three axes *viz. x*, *y* and *z* axes. $3d_{xy}$, $3d_{yz}$ and $3d_{xz}$ orbitals have their lobes in between the axes. The shapes of 4s, $4p_x$, $4p_y$, $4p_z$, $3d_{z^2}$ and $3d_{x^2-y^2}$ orbitals have been shown in Fig 23.2 Since $3d_{xy}$, $3d_{yz}$ and $3d_{zx}$ orbitals do not have their lobes along the axes, these orbitals are not capable of forming s-bonds. The shapes of these orbitals have, therefore, not been shown in Fig 23.2.

According to group theory, the metal ion orbitals are designated by the following symmetry symbols:

| *Orbitals* | *Symmetry symbols of orbitals* |
|---|---|
| 4s | $a_{1g}$ or $A_{1g}$ |
| $4p_x$, $4p_y$, $4p_z$ | $t_{1u}$ or $T_{1u}$ |
| $3d_{z^2}$, $3d_{x^2-y^2}$ | $e_g$ or $E_g$ |
| $3d_{xy}$, $3d_{yz}$, $3d_{xz}$ | $t_{2g}$ or $T_{2g}$ |

$a_{1g}$ or $A_{1g}$ represents a *single symmetric* orbital which is *non-degenerate*. $e_g$ or $E_g$ represents *two degenerate orbitals*. $t_{1u}$ or $T_{1u}$ represents *three degenerate orbitals*. $t_{2g}$ or $T_{2g}$ also represents *three degenerate orbitals*. The subscripts *g* (gerade) and *u* (ungrade) are used to indicate whether it is *centro-symmetric* (even) or *non-centro-symmetric* (uneven) respectively.

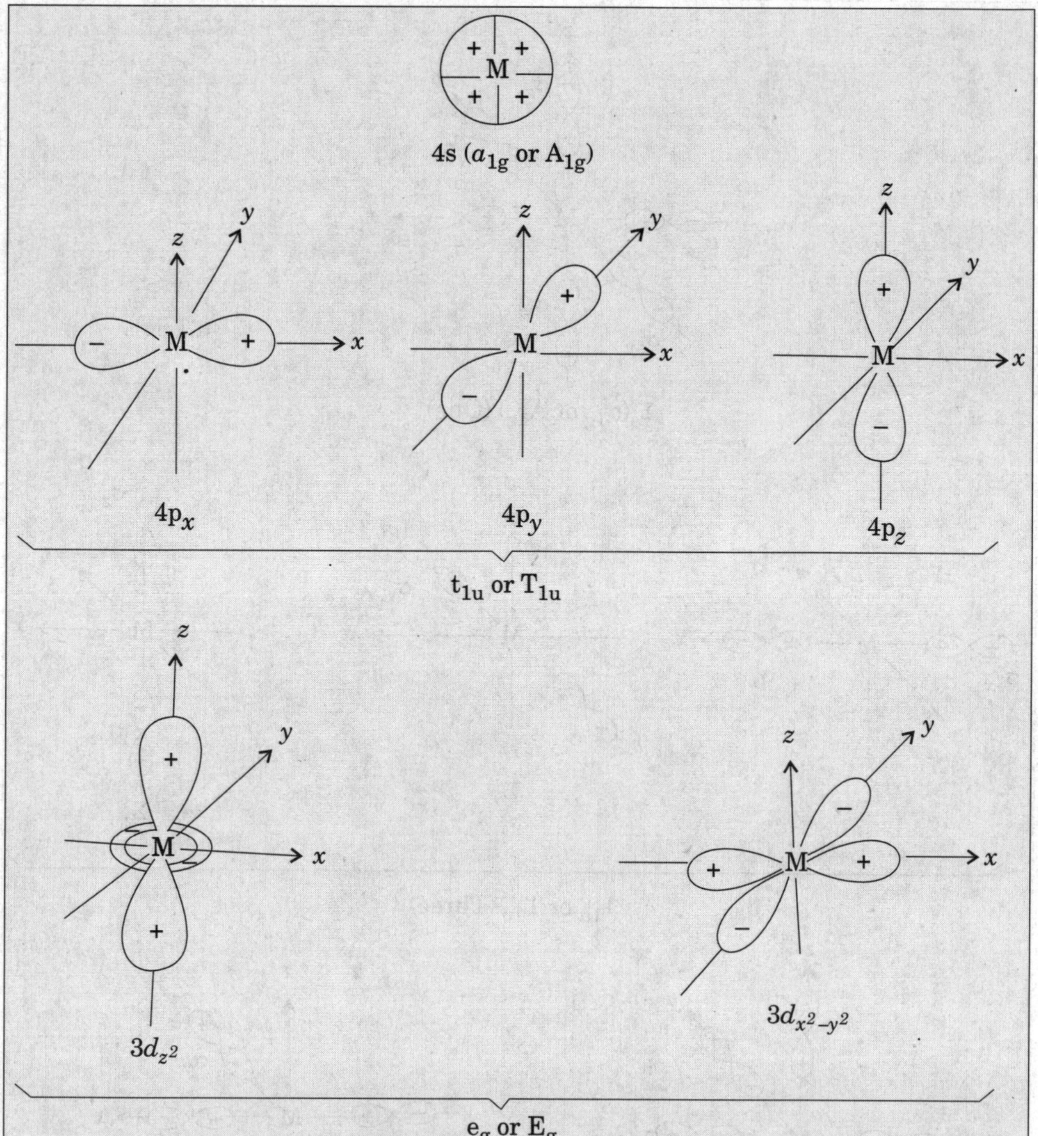


**Fig. 23.2** Six metal ion orbitals

From the above discussion it is clear that out of nine AOs of the metal ion only six AOs *viz.* 4*s*, $4p_x$, $4p_y$, $4p_z$, $3d_{z^2}$ and $3d_{x^2-y^2}$ are suitable orbitals which can overlap with ligand orbital combinations to form σ-BMOs and σ*-ABMOs.

## II. To Determine the Ligand $p_\sigma$ Orbital Combinations

Ligand $p_\sigma$ orbital combinations are also called by other names *viz. group ligand $p_\sigma$ orbitals, composite ligand $p_\sigma$ orbitals, ligand symmetry orbitals* etc. The six ligand $p_\sigma$ orbitals *viz.* $\sigma_x$, $\sigma_{-x}$, $\sigma_y$, $\sigma_{-y}$, $\sigma_z$ and $\sigma_{-z}$ combine together linearly to form six ligand $p_\sigma$ orbital combinations. These six ligand $p_\sigma$ orbital combinations are designated as $\Sigma_s$, $\Sigma_x$, $\Sigma_y$, $\Sigma_z$, $\Sigma_{z^2}$, and $\Sigma_{x^2-y^2}$. These ligand $p_\sigma$ orbital combinations have been shown in Fig. 23.3.

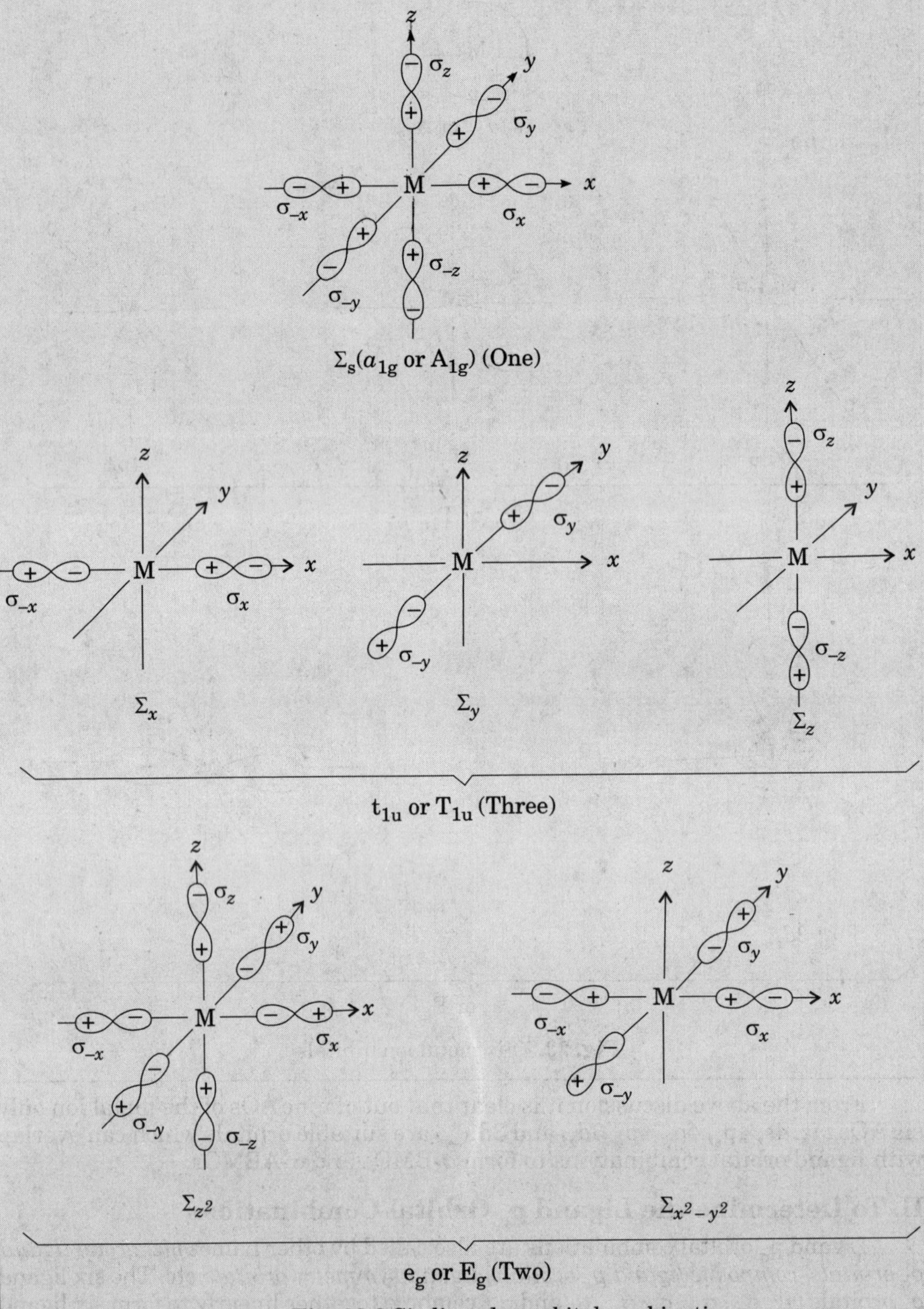


**Fig. 23.3** Six ligand $p_\sigma$ orbital combinations

These six ligand $p_\sigma$ combinations, in their normalized form, can be written as:

$$\Sigma_s = \frac{1}{\sqrt{6}} (\sigma_x + \sigma_{-x} + \sigma_y + \sigma_{-y} + \sigma_z + \sigma_{-z})$$

$$\Sigma_x = \frac{1}{\sqrt{2}} (\sigma_x - \sigma_{-x})$$

$$\Sigma_y = \frac{1}{\sqrt{2}} (\sigma_y - \sigma_{-y})$$

$$\Sigma_z = \frac{1}{\sqrt{2}} (\sigma_z - \sigma_{-z})$$

$$\Sigma_{z^2} = \frac{1}{\sqrt{12}} (2\sigma_z + 2\sigma_{-z} - \sigma_{-x} - \sigma_{-x})$$

$$\Sigma_{x^2-y^2} = \frac{1}{2} (\sigma_x + \sigma_{-x} - \sigma_y - \sigma_{-y})$$

These combinations are represented by group symbols.

The metal ion orbitals and the ligand $p_\sigma$ orbital combinations that can combine together to form σ-BMOs and σ*-ABMOs are given below. Their symmetry symbols are also given.

| *Metal ion orbitals* | *Ligand orbital combinations* | *Group symmetry symbols* |
|---|---|---|
| 4s | $\Sigma_s = \frac{1}{\sqrt{6}} (\sigma_x + \sigma_{-x} + \sigma_y + \sigma_{-y} + \sigma_z + \sigma_{-z})$ | $a_{1g}$ or $A_{1g}$ |
| 4 $p_x$ | $\Sigma_x = \frac{1}{\sqrt{2}} (\sigma_x - \sigma_{-x})$ | $t_{1u}$ or $T_{1u}$ |
| 4 $p_y$ | $\Sigma_y = \frac{1}{\sqrt{2}} (\sigma_y - \sigma_{-y})$ | |
| 4$p_z$ | $\Sigma_z = \frac{1}{\sqrt{2}} (\sigma_z - \sigma_{-z})$ | |
| $3d_{z^2}$ | $\Sigma_{z^2} = \frac{1}{\sqrt{12}} (2\sigma_z + 2\sigma_{-z} - \sigma_x - \sigma_{-x})$ | $e_g$ or $E_g$ |
| $3d_{x^2-y^2}$ | $\Sigma_{x^2-y^2} = \frac{1}{2} (\sigma_x + \sigma_{-x} - \sigma_y - \sigma_{-y})$ | |

We know that $3d_{xy}$, $3d_{yz}$ and $3d_{zx}$ orbitals have $t_{2g}$ or $T_{2g}$ symmetry symbol. Now since there are no ligand orbital combinations with $t_{2g}$ to $T_{2g}$ symmetry, $3d_{xy}$, $3d_{yz}$ and $3d_{zx}$ orbitals do not participate in the formation of σ–BMOs and σ*–ABMOs.

## III. Formation of Sigma (σ) Molecular Orbitals (MOs) in Octahedral Complexes

Before discussing the formation of sigma (σ) MOs in octahedral complexes we should have a knowledge of the fact that the effective combination of orbitals for the formation of MOs requires the following conditions.

(*i*) The combining orbitals should have the same symmetry so that the additive combination of orbitals may allow the maximum overlap of orbitals.

(*ii*) The combining orbitals should not have large difference in their energies.

Now let us apply the above rules to explain the formation of sigma MOs in octahedral complexes.

In octahedral complexes, sigma MOs are formed by the overlap of six ligand $p_\sigma$ orbital combinations with the six metal ion σ-orbitals along the axes. According to group theory the ligand $p_\sigma$ orbital combinations and metal ion orbitals which combine together to form MOs must have the same symmetry. We have already shown that since in octahedral complexes, *4s* orbital of the metal ion and $\Sigma_s$ ligand orbital combination have the same symmetry symbol *viz.* $a_{1g}$ or $A_{1g}$, these combine together to form one $a_{1g}$ bonding molecular orbital ($a_{1g}$BMO) and one $a_{1g}^*$ antibonding molecular orbital ($a_{1g}^*$ ABMO) as shown in Fig. 23.4. $a_{1g}$ BMO can also be represented as $\sigma_s$. Similarly $a_{1g}^*$ ABMO can also be represented as $\sigma_s^*$.

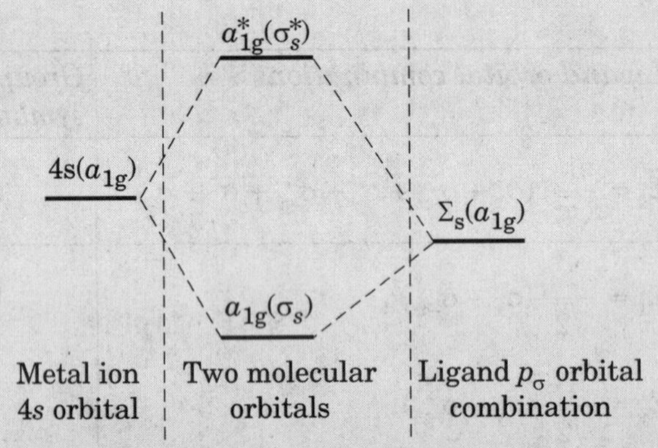


**Fig 23.4** Formation of $a_{1g}$ BMO ($\sigma_s$) and $a_{1g}^*$ ABMO ($\sigma_s^*$) in an octahedral complex

Now since the three 4p orbitals ($4p_x$, $4p_y$ and $4p_z$) and three ligand σ-orbital combinations *viz.* $\Sigma_x$, $\Sigma_y$ and $\Sigma_z$ have the same symmetry ($t_{1u}$ symmetry), they combine together to form three degenerate $t_{1u}$ BMOs and three degenerate $t_{1u}^*$ ABMOs as shown in Fig. 23.5. $t_{1u}$ set of three degenerate BMOs can also be represented as $\sigma_p$. Similarly $t_{1u}^*$ set of three degenerate ABMOs can also be represented as $\sigma_p^*$. The three degenerate MOs of $t_{1u}$ or $\sigma_p$ set are represented as $\sigma_x$, $\sigma_y$ and $\sigma_z$. Similarly the three degenerate MOs of $t_{1u}^*$ or $\sigma_p^*$ set are represented as $\sigma_x^*$, $\sigma_y^*$ and $\sigma_z^*$.

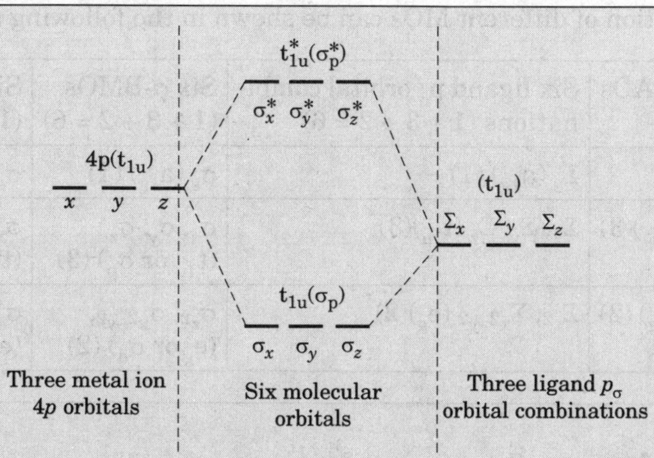


**Fig 23.5** Formation of three $t_{1u}$BMOs ($\sigma_p$) and three $t^*_{1u}$ ABMOs ($\sigma^*_p$)

Since $3d_{z^2}$ and $3d_{x^2-y^2}$ orbitals and two ligand $p_\sigma$ orbital combinations *viz.* $\Sigma_{z^2}$ and $\Sigma_{x^2-y^2}$ have the same symmetry ($e_g$ symmetry), they combine together to form two degenerate $e_g$ BMOs and two degenerate $^*_{eg}$ AMBOs as shown in Fig 23.6. $e_g$ set of two degenerate BMOs is also represented as $\sigma_d$. Similarly $e^*_g$ set of two degenerate ABMOs is also represented as $\sigma^*_d$. The two degenerate MOs of $e_g$ or $\sigma_d$ set are represented as $\sigma_{z^2}$ and $\sigma_{x^2-y^2}$. Similarly the two degenerate MOs of $e^*_g$ or $\sigma^*_d$ set are represented as $\sigma_{z^2}{}^*$ and $\sigma^*_{x^2-y^2}$.

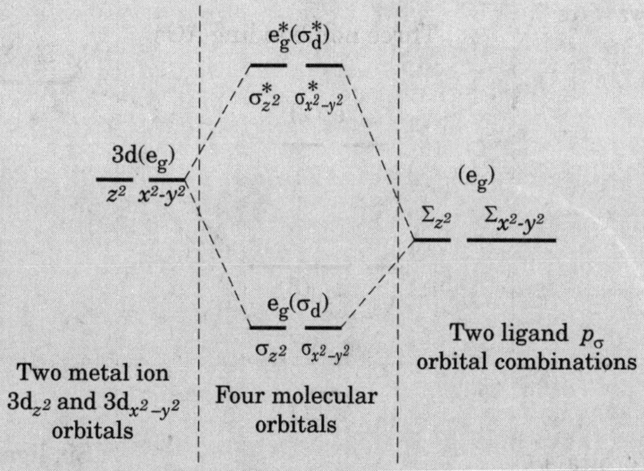


**Fig 23.6** Formation of two $e_g$ BMOs ($\sigma_d$) and two $e^*_g$ ABMOs ($\sigma^*_d$)

The formation of different MOs can be shown in the following simple way.

| Six metal ion AOs (1 + 3 + 2 = 6) | Six ligand $p_\sigma$ orbital combinations (1 + 3 + 2 = 6) | Six σ-BMOs (1 + 3 + 2 = 6) | Six σ*ABMOs (1 + 3 + 2 = 6) |
|---|---|---|---|
| 4s ($a_{1g}$) (1) | $\Sigma_s$ ($a_{1g}$) (1) | $\sigma_s$ ($a_{1g}$) (1) | $\sigma_s^*$($a_{1g}^*$) (1) |
| $4p_x$, $4p_y$, $4p_z$ ($t_{1\mu}$)(3) | $\Sigma_x$, $\Sigma_y$, $\Sigma_z$ ($t_{1\mu}$)(3) | $\sigma_x$, $\sigma_y$, $\sigma_z$ ($t_{1\mu}$ or $\sigma_p$) (3) | $\sigma_x^*$, $\sigma_y^*$, $\sigma_z^*$, ($t_{1\mu}^*$ or $\sigma_p^*$) (3) |
| $3d_{z^2}$, $3d_{x^2-y^2}$ ($e_g$) (2) | $\Sigma_{z^2}$, $\Sigma_{x^2-y^2}$ ($e_g$)(2) | $\sigma_{z^2}$, $\sigma_{x^2-y^2}$, ($e_g$ or $\sigma_d$) (2) | $\sigma_{z^2}^*$, $\sigma_{x^2-y^2}^*$, ($e_g^*$ or $\sigma_d^*$) (2) |

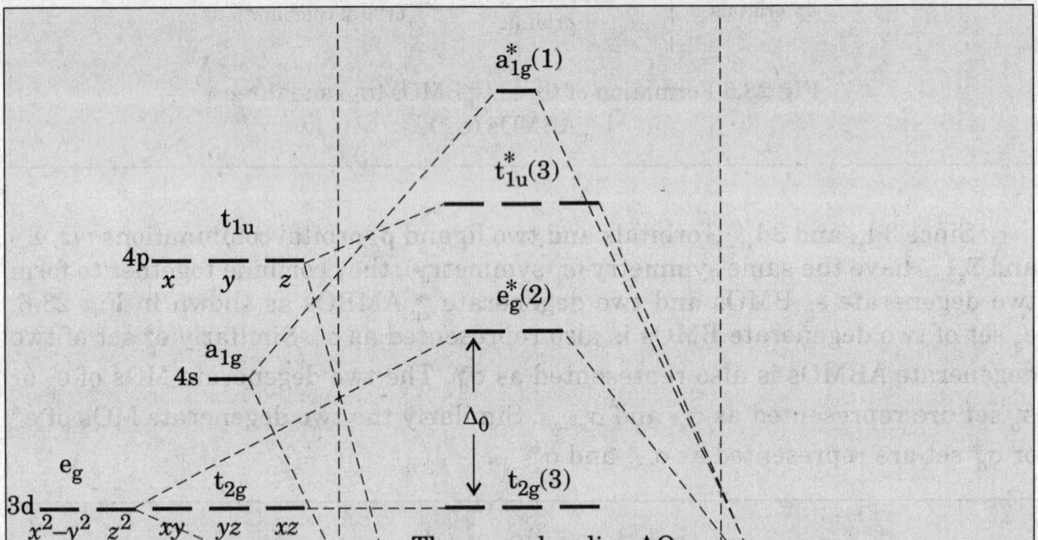


**Fig. 23.7** Formation of sigma molecular orbitals in an octahedral complex.

$a_{1g}$ or $s_s$ and $a_{1g}^*$ or $s_s^*$ is a non - degenerate MO. $t_{1m}$ or $s_p$ and $t_{1u}^*$ or $s_p^*$ are triply degenerate MOs. $e_g$ or $s_d$ and $e_g^*$ or $s_d^*$ are doubly degenerate MOs.

Metal ion orbitals *viz* $3d_{xy}$, $3d_{yz}$ and $3d_{xz}$ are represented as $t_{2g}$ or $p_d$. These are non-bonding AOs and are triply degenerate.

Since all the MOs mentioned above are formed by the overlap *along the exes,* all these MOs are sigma (s) MOs.

Above discussion shows that the combination of six metal ion orbitals *viz.* 4s, $4p_x$, $4p_y$, $4p_z$, $3d_{z^2}$ and $3d_{x^2-y^2}$ with six ligand $p_s$ orbital combinations namely $\Sigma_s$, $\Sigma_x$, $\Sigma_y$, $\Sigma_z$, $\Sigma_{z^2}$ and $\Sigma_{x^2-y^2}$ gives six BMOs and six ABMOs. six BMOs are $a_{1g}$(one), $t_{1u}$ (three) and $e_g$ (two) and six ABMOs are $a_{1g}^*$ (one), $t_{1u}^*$ (three) and $e_g^*$ (two).

Since the lobes of triply degenerate $3d_{xy}$, $3d_{yz}$ and $3d_{zx}$ orbitals donot lie along the axes, these orbitals donot participate in the process of the formation of s MOs. These orbitals remain non-bonding and hence are called **non-bonding orbitals**. These can, however, overlap *sidewise* with filled or unfilled p-orbitals of the same six ligands (p-orbitals are those which are capable of forming metal-ligand p-bonds) to form metal-ligand p-bonds or p-MO's. Thus these orbitals are generally referred to as $p_d$, $p(d)$, $d_p$ or $p_{xy} = p_{yz} = p_{zx}$ orbitals.

The formation of six BMOs, six ABMOs and three non-bonding 3d orbitals ($3d_{xy}$, $3d_{yz}$ and $3d_{xz}$) has been shown in Fig 23.7 which is commonly known as **energy level diagram for an octahedral complex.**

## Pictorial Representation of the Formation of MOs

Pictorial representation of the formation of $a_{1g}$ (one), $a_{1g}^*$ (one), $t_{1u}$ (three), $t_{1u}^*$ (three), $e_g$ (two) and $e_g^*$ (two) MOs has been given in Figs 23.8, 23.9 and 23.10. $a_{1g}$, $t_{1u}$ and $e_g$ are BMOs and $a_{1g}^*$, $t_{1u}^*$ and $e_g^*$ are ABMOs.

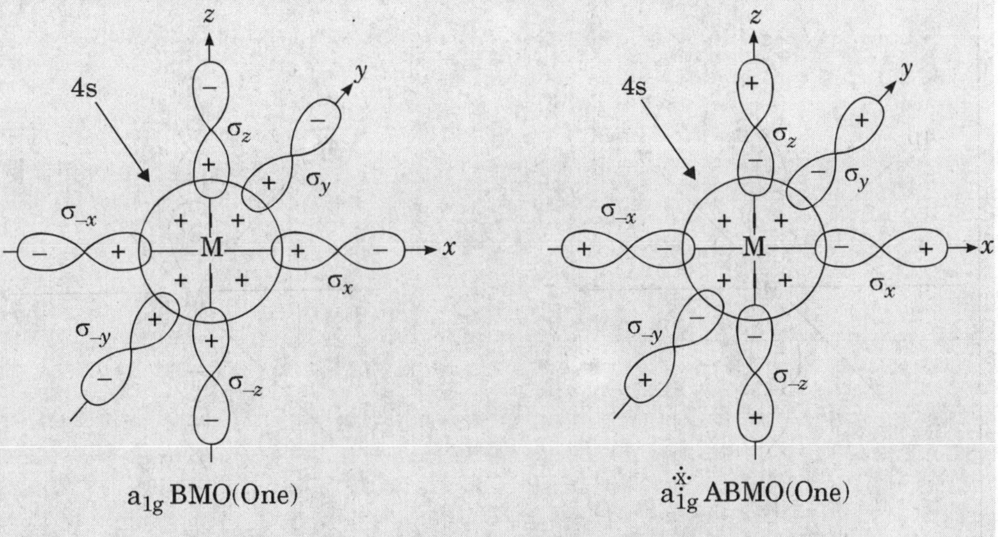


**Fig 23.8** Formation of one $a_{1g}$ BMO and one $a_{1g}^*$ ABMO

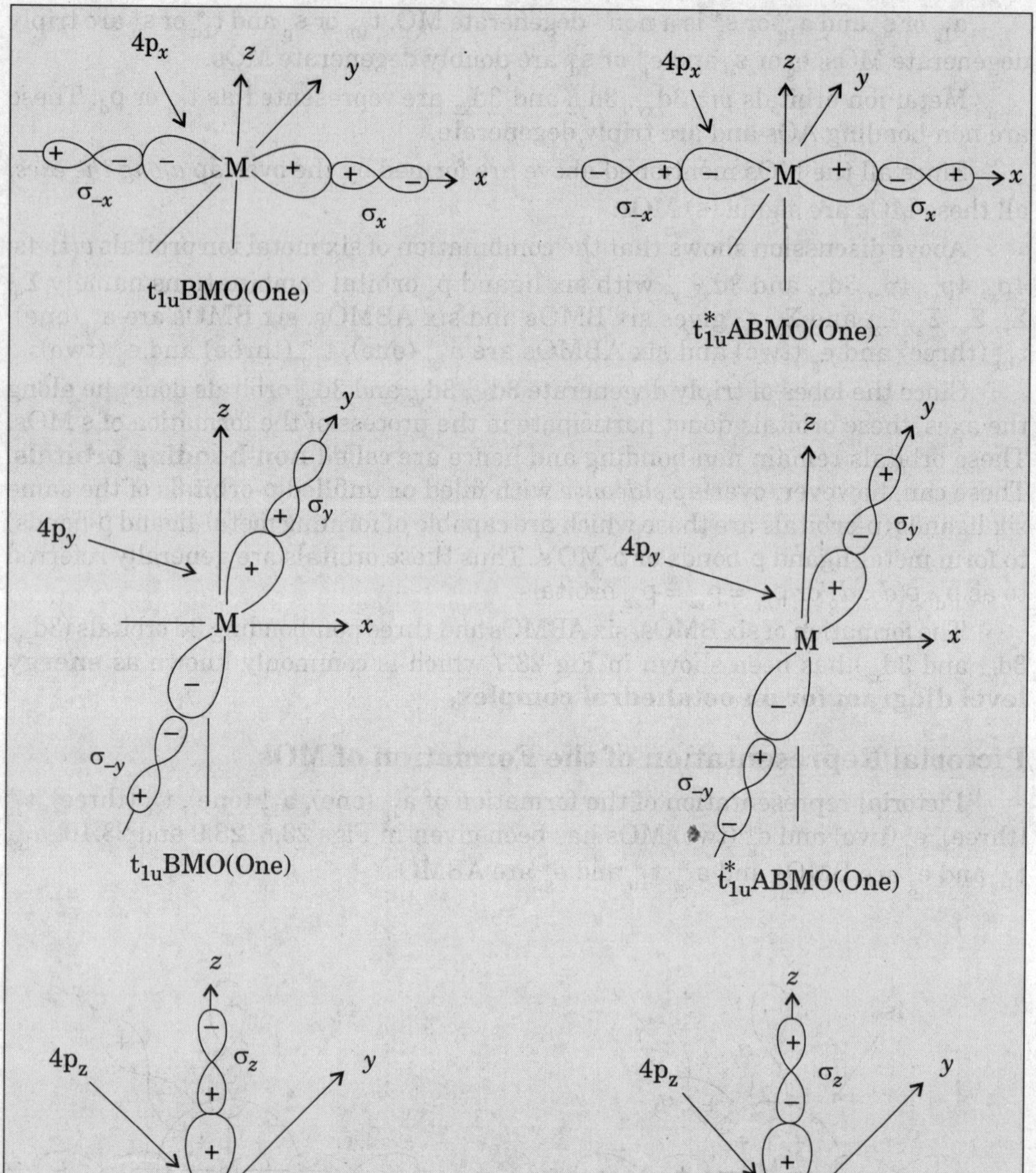


**Fig 23.9** Formation of three $t_{1u}$ BMOs and three $t_{1u}^*$ ABMOs

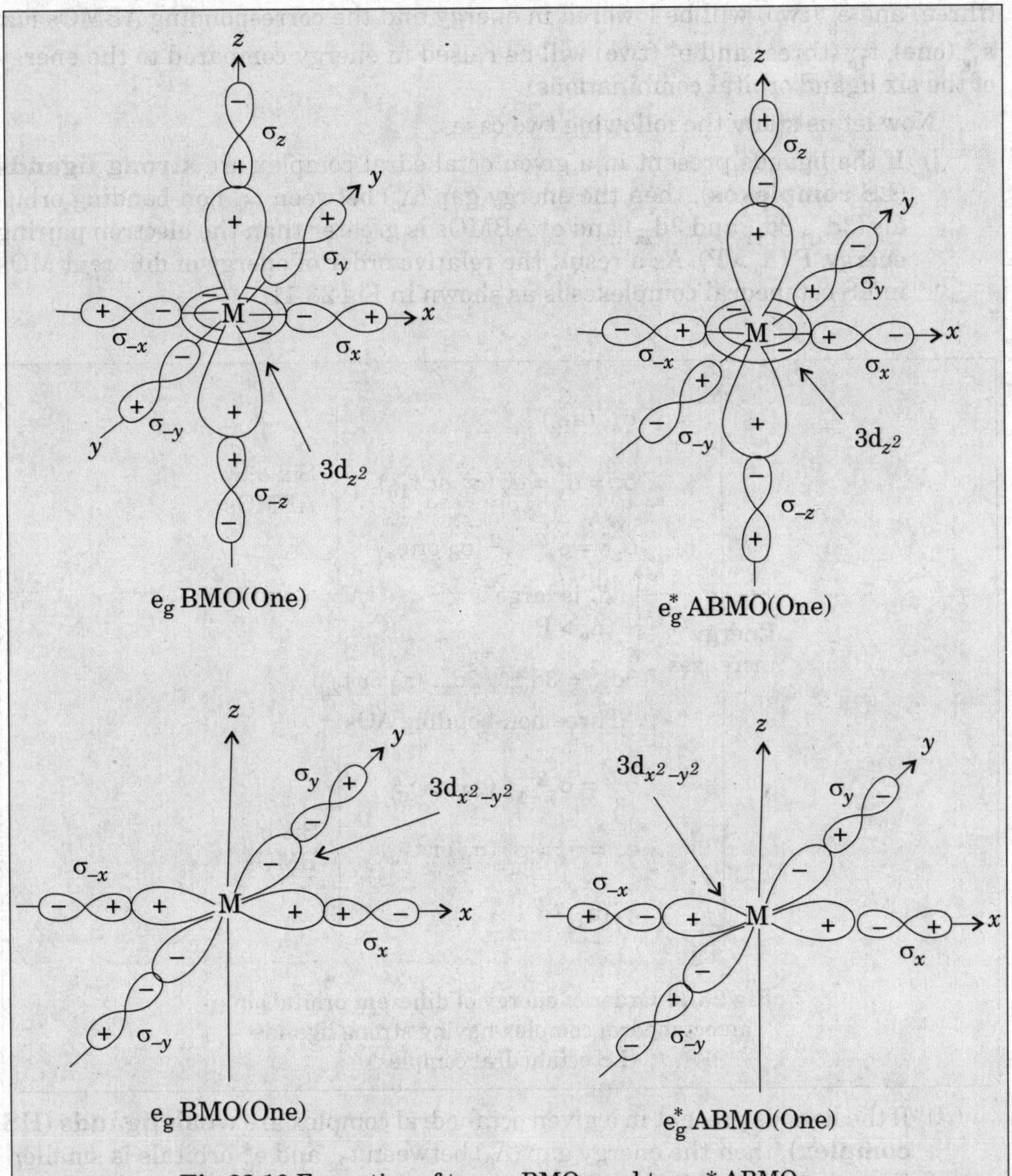


**Fig 23.10** Formation of two $e_g$ BMOs and two $e_g^*$ ABMOs

## Relative Order of Energy of Different MOs

We have already stated that the six AOs of the metal ion overlap with the six ligand $p_\sigma$ orbital combinations and generate six BMOs ($a_{1g}$, $t_{1u}$, $e_g$) and six ABMOs ($a_{1g}^*$, $t_{1u}^*$, $e_g^*$). BMOs have lower energy than ABMOs. Non-overlapping orbitals of the metal ion *viz.* $t_{2g}$ orbitals ($3d_{xy}$, $3d_{yz}$, $3d_{xz}$) remain non-bonding and their energy remains unchanged. For strong σ-bonding, BMOs *viz.* $a_{1g}$ (one), $t_{1u}$

(three) and $e_g$ (two) will be lowered in energy and the corresponding ABMOs *viz.* $a_{1g}^*$ (one), $t_{1\mu}^*$ (three) and $e_g^*$ (two) will be raised in energy compared to the energy of the six ligand orbital combinations.

Now let us study the following two cases.

(*i*) If the ligands present in a given octahedral complex are **strong ligands** (**LS complexes**), then the energy gap ($\Delta_0$) between $t_{2g}$ non-bonding orbitals ($3d_{xy}$, $3d_{yz}$ and $3d_{zx}$) and $e_g^*$ ABMOs is greater than the electron pairing energy, P ($\Delta_0 > P$). As a result the relative order of energy of different MOs in LS octahedral complexes is as shown in Fig 23.11:

Energy increases ↑

$\sigma_s^*$ ($a_{1g}^*$)

$\sigma_x^* = \sigma_y^* = \sigma_z^*$ ($\sigma_p^*$ or $t_{1u}^*$) — Six $\sigma^*$ ABMOs

$\sigma_{z^2}^* = \sigma_{x^2-y^2}^*$ ($\sigma_d^*$ or $e_g^*$)

↕ $\Delta_o$ is large ($\Delta_o > P$)

$3d_{xy} = 3d_{yz} = 3d_{xz}$ ($\pi_d$ or $t_{2g}$)

Three non-bonding AOs

$\sigma_{z^2} = \sigma_{x^2-y^2}$ ($\sigma_d$ or $e_g$)

$\sigma_x = \sigma_y = \sigma_z$ ($\sigma_p$ or $t_{1u}$) — Six $\sigma$ BMOs

$\sigma_s$ ($a_{1g}$)

**Fig 23.11** Order of energy of different orbitals in an octahedral complex having strong ligands (LS octahedral complex)

(*ii*) If the ligands present in a given octahedral complex are **weak ligands (HS complex),** then the energy gap ($\Delta_0$) between $t_{2g}$ and $e_g^*$ orbitals is smaller than the value of P ($\Delta_0 < P$). As a result the energies of $t_{2g}$ and $e_g^*$ orbitals becomes almost the same. Hence the relative order of energies of different MOs in HS octahedral complexes becomes as shown in Fig 23.12.

## Filling of MOs with Electrons

Fig 23.7 shows that we have six BMOs [$a_{1g}$ (one), $e_g$ (two) and $t_{1u}$ (three)], six ABMOs [$a_{1g}^*$ (one), $e_g^*$ (two) and $t_{1u}^*$ (three)] and three non-bonding *3d* orbitals ($3d_{xy}$, $3d_{yz}$ and $3d_{zx}$) which are available for filling with electrons.

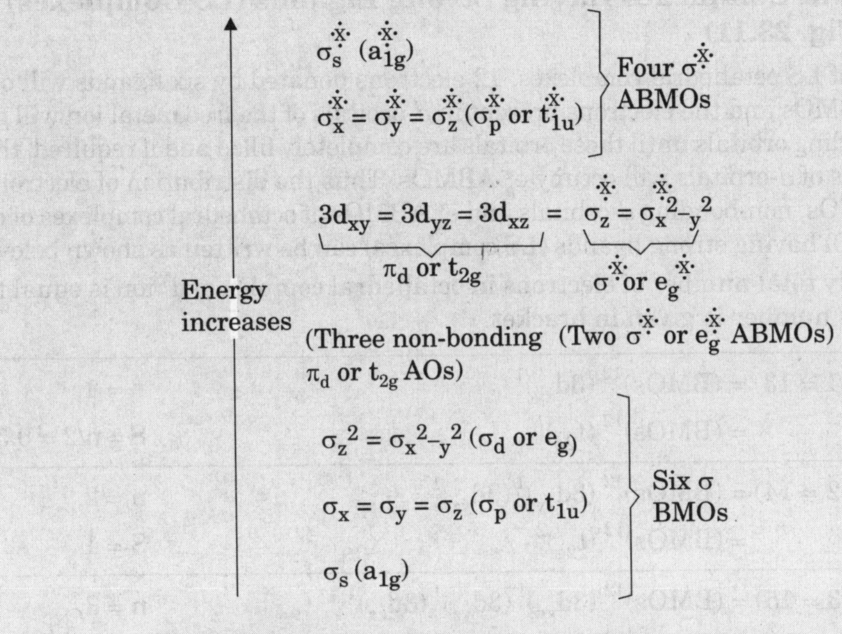


**Fig 23.12** Order of energy of different orbitals in an octahedral complex having weak ligands (HS octahedral complex)

Since six BMOs are nearer the energy level of the six ligands, these BMOs are filled up by 12 electrons present in 6 ligands rather than the metal ion electrons. Thus the distribution of 12 electrons from six ligands in six BMOs can be shown as:

$$\sigma_s^2 \quad \sigma_x^2 = \sigma_y^2 = \sigma_z^2 \quad \sigma_{z^2}^2 = \sigma_{x^2-y^2}^2$$

$$\text{or} \quad \sigma_s^2 \quad \sigma_p^6 \quad \sigma_d^4$$

$$\text{or} \quad a_{1g}^2 \quad t_{1u}^6 \quad a_g^4$$

$$\text{or} \quad (\text{BMOs})^{12}$$

Since $t_{2g}$ non-bonding AOs are pure metal *d*-orbitals ($d_{xy}$, $d_{yz}$ and $d_{xz}$) whereas $e_g^*$ ABMOs have a major contribution from $e_g$ orbitals of the metal ($d_{z^2}$ and $d_{x^2-y^2}$ orbitals), the electrons previously present in *d*-orbitals of the free metal ion are accomdated in $t_{2g}$ non-bonding AOs and $e_g^*$ ABMOs, if required.

Thus if we have an octahedral complex of a metal ion having $d_x$ configuration, then 12 electrons present in six ligands will occupy the six BMOs and *x* electrons present in *d*-orbitals of the free metal ion will occupy $t_{2g}$ non-bonding AOs and $e_g^*$ ABMOs, if required. Here we shall discuss the distribution of $(12 + x)$ electrons in the following two types of octahedral complexes.

## 1. Octahedral Complexes Having Strong Ligands (LS Complexes) (Also See Fig. 23.11)

In case of LS octahedral complexes, 12 electrons donated by six ligands will occupy the six BMOs and the electrons present in *d*-orbitals of the free metal ion will go to $t_{2g}$ non-bonding orbitals until these orbitals are completely filled and if required, the extra electrons of *d*-orbitals will occupy $e_g^*$ ABMOs. Thus the distribution of electrons in various BMOs, non-bonding *d*-orbitals and $e_g^*$ ABMOs of octahedral complexes of $d^x$ ion ($x$ = 1 to 10) having strong ligands (LS complexes) can be written as shown below.

Obviously total number of electrons in octahedral complex of $d^x$ ion is equal to $(12 + x)$. This number is given in bracket.

| | |
|---|---|
| $d^1$ **ion** $(12 + 1 = 13) = (BMOs)^{12}(3d_{xy})^1$<br>$= (BMOs)^{12}(t_{2g})^1$ | n = 1,<br>S = n/2 = 0.5 |
| $d^2$ **ion** $(12 + 2 = 14) = (BMOs)^{12}(3d_{xy})^1(3d_{yz})^1$<br>$= (BMOs)^{12}(t_{2g})^2$ | n = 2,<br>S = 1 |
| $d^3$ **ion** $(12 + 3 = 15) = (BMOs)^{12}(3d_{xy})^1(3d_{yz})^1(3d_{zx})^1$<br>$= (BMOs)^{12}(t_{2g})^3$ | n = 3,<br>S = 1.5 |
| $d^4$ **ion** $(12 + 4 = 16) = (BMOs)^{12}(3d_{xy})^2(3d_{yz})^1(3d_{xz})^1$<br>$= (BMOs)^{12}(t_{2g})^4$ | n = 2,<br>S = 1 |
| $d^5$ **ion** $(12 + 5 = 17) = (BMOs)^{12}(3d_{xy})^2(3d_{yz})^2(3d_{xz})^1$<br>$= (BMOs)^{12}(t_{2g})^5$ | n = 1,<br>S = 0.5 |
| $d^6$ **ion** $(12 + 6 = 18) = (BMOs)^{12}(3d_{xy})^2(3d_{yz})^2(3d_{xz})^2$<br>$= (BMOs)^{12}(t_{2g})^6$ | n = 0,<br>S = 0 |
| $d^7$ **ion** $(12 + 7 = 19) = (BMOs)^{12}(3d_{xy})^2(3d_{yz})^2(3d_{xz})^2(\sigma^*_{z^2})^1$<br>$= (BMOs)^{12}(t_{2g})^6(e_g^*)^1$ | n = 1,<br>S = 0.5 |
| $d^8$ **ion** $(12 + 8 = 20) = (BMOs)^{12}(3d_{xy})^2(3d_{yz})^2(3d_{zx})^2(\sigma^*_{z^2})^1(\sigma^*_{z^2-y^2})^1$<br>$= (BMOs)^{12}(t_{2g})^6(e_g^*)^2$ | n = 2,<br>S = 1 |
| $d^9$ **ion** $(12 + 9 = 21) = (BMOs)^{12}(3d_{xy})^2(3d_{yz})^2(3d_{zx})^2(\sigma^*_{z^2})^2(\sigma^*_{z^2-y^2})^1$<br>$= (BMOs)^{12}(t_{2g})^6(e_g^*)^3$ | n = 1,<br>S = 0.5 |
| $d^{10}$ **ion** $(12 + 10 = 22) = (BMOs)^{12}(3d_{xy})^2(3d_{yz})^2(3d_{zx})^2(\sigma^*_{z^2})^2(\sigma^*_{z^2-y^2})^2$<br>$= (BMOs)^{12}(t_{2g})^6(e_g^*)^4$ | n = 0,<br>S = 0 |

In case of LS octahedral complexes, the energy difference ($\Delta_0$) between non-bonding $t_{2g}$ orbitals ($d_{xy}$, $d_{yz}$ and $d_{xz}$) and $e_g^*$ ABMOs is greater than the electron pairing energy, P ($\Delta_0 > P$).

## 2. Octahedral Complexes Having Weak Ligands (HS Complexes) (Also See Fig. 23.12)

In case of HS octahedral complexes, 12 electrons donated by six ligands are accomdated in six BMOs. Now since the energy difference ($\Delta_0$) between non-bonding $t_{2g}$ orbitals and $e_g^*$ ABMOs is smaller than the electron pairing energy, P ($\Delta_0 < P$), non-bonding $t_{2g}$ orbitals and $e_g^*$ ABMOs have almost the same energy. As a result, the electrons present in *d*-orbitals of the free metal ion occupy $t_{2g}$ non-bonding AOs and $e_g^*$ ABMOs in accordance with Hund's rule, *i.e.* the pairing of electrons in three $t_{2g}$ non-bonding orbitals and two $e_g^*$ ABMOs takes place only after these five orbitals are at least singly filled. Thus the distribution of electrons in various BMOs, non-bonding $t_{2g}$ orbitals and $e_g^*$ ABMOs of octahedral complexes of $d^x$ ion ($x$ = 1 to 10) having weak ligands (HS complexes) can be written as:

| | |
|---|---|
| **$d^1$ ion** (12 + 1 = 13) = $(BMOs)^{12} (3d_{xy})^1$<br>= $(BMOs)^{12} (t_{2g})^1$ | n = 1,<br>S = n/2 = 0.5 |
| **$d^2$ ion** (12 + 2 = 14) = $(BMOs)^{12} (3d_{xy})^1 (3d_{yz})^1$<br>= $(BMOs)^{12} (t_{2g})^2$ | n = 2,<br>S = 1 |
| **$d^3$ ion** (12 + 3 = 15) = $(BMOs)^{12} (3d_{xy})^1 (3d_{yz})^1 (3d_{xz})^1$<br>= $(BMOs)^{12} (t_{2g})^3$ | n = 3,<br>S = 1.5 |
| **$d^4$ ion** (12 + 4 = 16) = $(BMOs)^{12} (3d_{xy})^1 (3d_{yz})^1 (3d_{xz})^1 (\sigma_{z^2}^*)^1$<br>= $(BMOs)^{12} (t_{2g})^3 (e_g^*)^1$ | n = 4,<br>S = 2 |
| **$d^5$ ion** (12 + 5 = 17) = $(BMOs)^{12} (3d_{xy})^1 (3d_{yz})^1 (3d_{xz})^1 (\sigma_{z^2}^*)^1 (\sigma_{x^2-y^2}^*)^1$<br>= $(BMOs)^{12} (t_{2g})^3 (e_g^*)^2$ | n = 5,<br>S = 2.5 |
| **$d^6$ ion** (12 + 6 = 18) = $(BMOs)^{12} (3d_{xy})^2 (3d_{yz})^1 (3d_{xz})^1 (\sigma_{z^2}^*)^1 (\sigma_{x^2-y^2}^*)^1$<br>= $(BMOs)^{12} (t_{2g})^4 (e_g^*)^2$ | n = 4,<br>S = 2 |
| **$d^7$ ion** (12 + 7 = 19) = $(BMOs)^{12} (3d_{xy})^2 (3d_{yz})^2 (3d_{xz})^1 (\sigma_{z^2}^*)^1 (\sigma_{x^2-y^2}^*)^1$<br>= $(BMOs)^{12} (t_{2g})^5 (e_g^*)^2$ | n = 3,<br>S = 1.5 |
| **$d^8$ ion** (12 + 8 = 20) = $(BMOs)^{12} (3d_{xy})^2 (3d_{yz})^2 (3d_{xz})^2 (\sigma_{z^2}^*)^1 (\sigma_{x^2-y^2}^*)^1$<br>= $(BMOs)^{12} (t_{2g})^6 (e_g^*)^2$ | n = 2,<br>S = 1 |
| **$d^9$ ion** (12 + 9 = 21) = $(BMOs)^{12} (3d_{xy})^2 (3d_{yz})^2 (3d_{xz})^2 (\sigma_{z^2}^*)^2 (\sigma_{x^2-y^2}^*)^1$<br>= $(BMOs)^{12} (t_{2g})^6 (e_g^*)^3$ | n = 1,<br>S = 0.5 |
| **$d^{10}$ ion** (12 + 10 = 22) = $(BMOs)^{12} (3d_{xy})^2 (3d_{yz})^2 (3d_{xz})^2 (\sigma_{z^2}^*)^2 (\sigma_{x^2-y^2}^*)^2$<br>= $(BMOs)^{12} (t_{2g})^6 (e_g^*)^4$ | n = 0,<br>S = 0 |

## Sigma (σ) Bonding in $[Ti(H_2O)_6]^{3+}$ Ion (e = 12 + 1 = 13)

In $[Ti(H_2O)_6]^{3+}$ ion, Ti is present as $Ti^{3+}$ ion which is a $d^1$ ion ($Ti^{3+} = 3d^1 4s^0 4p^0$). Thus the complex ion has 12 electrons from six $H_2O$ ligands and one electron from $d$-orbitals of $Ti^{3+}$ ion, *i.e.* $[Ti(H_2O)_6]^{3+}$ ion has 12 + 1 = 13 electrons. These 13 electrons have to be accommodated into six BMOs and three non-bonding $t_{2g}$ orbitals. 12 electrons from six $H_2O$ ligands will occupy six BMOs and one electron will occupy $t_{2g}$ orbitals. Thus the distribution of 13 electrons in various orbitals can be written as:

| | | |
|---|---|---|
| | $(a_{1g})^2 (t_{1u})^6 (e_g)^4$ | $(t_{2g})^1$ |
| or | $(BMOs)^{12}$ | $(3d_{xy})^1 (3d_{yz})^0 (3d_{xz})^0$ |
| or | $(BMOs)^{12}$ | $(t_{2g})^1$ |

This distribution of 13 electrons has been shown in molecular orbital energy level diagram given in Fig. 23.13.

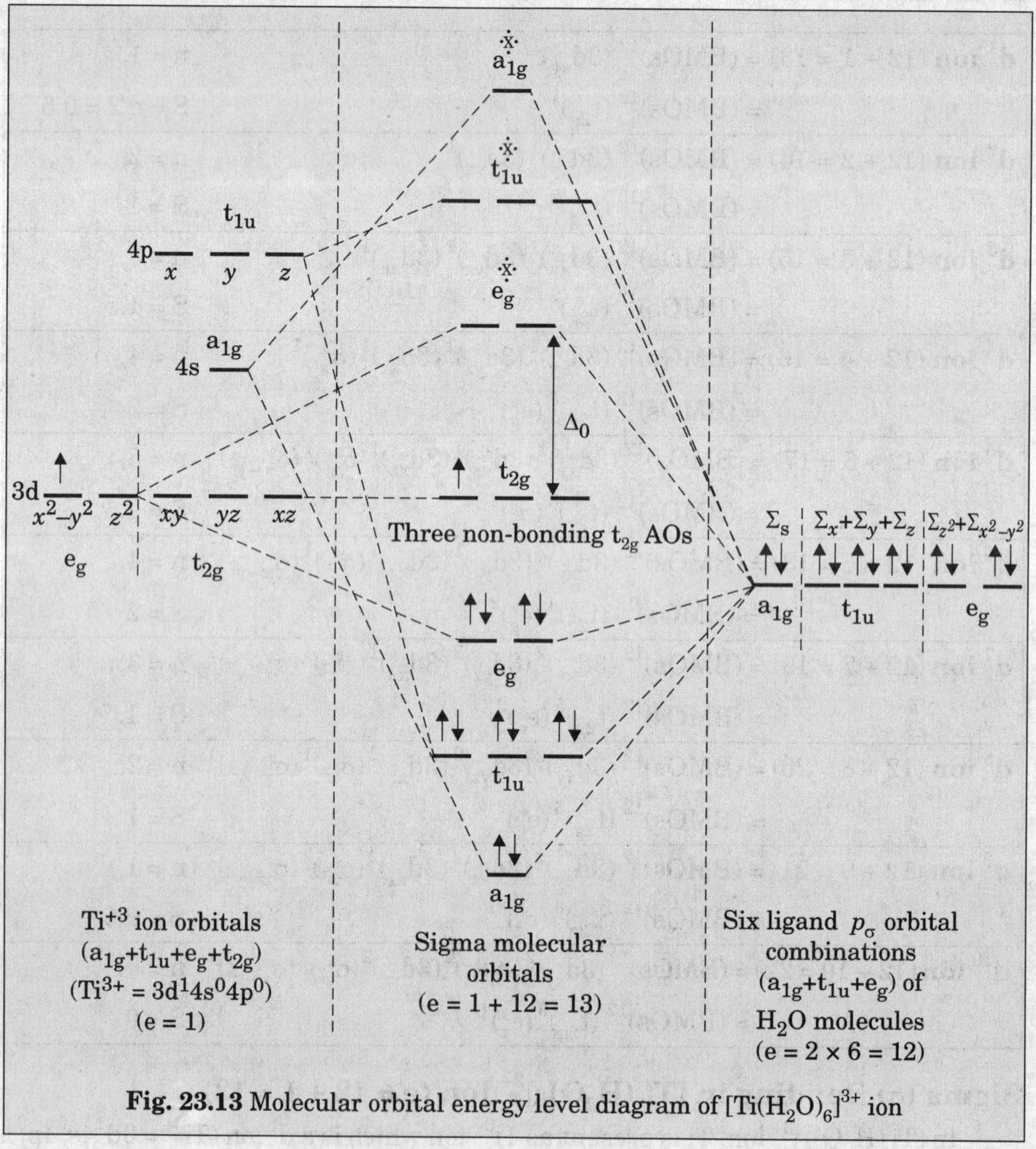


**Fig. 23.13** Molecular orbital energy level diagram of $[Ti(H_2O)_6]^{3+}$ ion

(*i*) $[Ti(H_2O)_6]^{3+}$ ion has only one unpaired electron ($n = 1$) which explains the paramagnetic nature of the ion.

(*ii*) The absorption spectrum of $[Ti(H_2O)_6]^{3+}$ ion with one absorption maximum at a wavelength of 5000 A° is attributed to the transition of the single electron from non-bonding $t_{2g}$ orbitals to $e_g^*$ ABMOs. Even on the bases of CFT, the explanation in the same, *i.e.* transition of the single electron takes place from lower energy $t_{2g}$ orbitals to higher energy $e_g$ orbitals.

## Sigma (σ) Bonding in $[Co(NH_3)_6]^{3+}$ Ion (e = 12 + 6 = 18)

In $[Co(NH_3)_6]^{3+}$ ion, Co is present as $Co^{3+}$ ion which is a $d^6$ ion ($Co^{3+} = 3d^6 4s^0 4p^0$). Thus the given complex ion has 12 electrons from six $NH_3$ ligands and 6 electrons from *d*-orbitals of $Co^{3+}$ ion, *i.e.* $[Co(NH_3)_6]^{3+}$ on has 12 + 6 = 18 electrons. These 18 electrons have to be accommodated into six BMOs, three non-bonding $t_{2g}$ orbitals and two $e_g^*$ ABMOs. 12 electrons from six $NH_3$ ligands will occupy six BMOs. Now since $NH_3$ molecules are strong ligands, six electrons of *d*-orbitals of $Co^{3+}$ ion will occupy non-bonding three $t_{2g}$ orbitals, Thus the distribution of 18 electrons in various orbitals can be written as

$$(a_{1g})^2 (t_{1u})^6 (e_g)^4 (t_{2g})^6 = (BMOs)^{12} (t_{2g})^6$$
$$= (BMOs)^{12} (3d_{xy})^2 (3d_{yz})^2 (3d_{xz})^2$$

This distribution has been shown in molecular orbital energy level diagram given in Fig. 23.14. Following points may be noted from this figure:

(*i*) All the electrons present in various BMOs and $t_{2g}$ orbitals are in the paired state ($n = 0$). Hence $[Co(NH_3)_6]^{3+}$ ion is diamagnetic

(*ii*) Since $n = 0$, S = 0

(*iii*) $[Co(NH_3)_6]^{3+}$ ion is coloured, since the promotion of electrons from non-bonding $t_{2g}$ orbitals to $e_g^*$ ABMOs is feasible

(*iv*) The energy gap ($\Delta_0$) between $t_{2g}$ orbitals and $e_g^*$ ABMOs is 23,000 $cm^{-1}$. Since this value of $\Delta_0$ is larger than the pairing energy (= 19,000 $cm^{-1}$), the six electrons of $Co^{3+}$ ion are paired in $t_{2g}$ orbitals.

Larger value of $\Delta_0$ is because of greater overlap between $e_g$ metal orbitals and ligand orbital combinations. This explains stronger covalent character of metal-ligand bond.

## Sigma (σ) Bonding in $[CoF_6]^{3-}$ Ion (e = 12 + 6 = 18)

In $[CoF_6]^{3-}$ ion, Co is present as $Co^{3+}$ ion which is a $d^6$ ion ($Co^{3+} = 3d^6 4s^0 4p^0$). Thus the given complex ion has 12 electrons from six $F^-$ ligands and 6 electrons from *d*-orbitals of $Co^{3+}$ ion, *i.e.* $[CoF_6]^{3-}$ ion has 12 + 6 = 18 electrons. These 18 electron have to be accomdated into six BMOs, three non-bonding $t_{2g}$ orbitals and two $e_g^*$ ABMOs. 12 electrons from six $F^-$ ligands will occupy six BMOs. Now since $F^-$ ligands are weak ligands, the filling of three non-bonding $t_{2g}$ orbitals and two $e_g^*$ ABMOs will take place in accordance with Hund's rule, *i.e.* $t_{2g}$ orbitals will have four electrons and $e_g^*$ ABMOs will have two electrons. The distribution of 18 electrons can be written as:

$$(a_{1g})^2 (t_{1u})^6 (e_g)^4 \quad (t_{2g})^4 \quad (e_g^*)^2$$
$$= (BMOs)^{12} \quad (3d_{xy})^2 (3d_{yz})^1 (3d_{xz})^1 \quad (\sigma_{z^2}^*)^1 (\sigma_{x^2-y^2}^*)^1$$

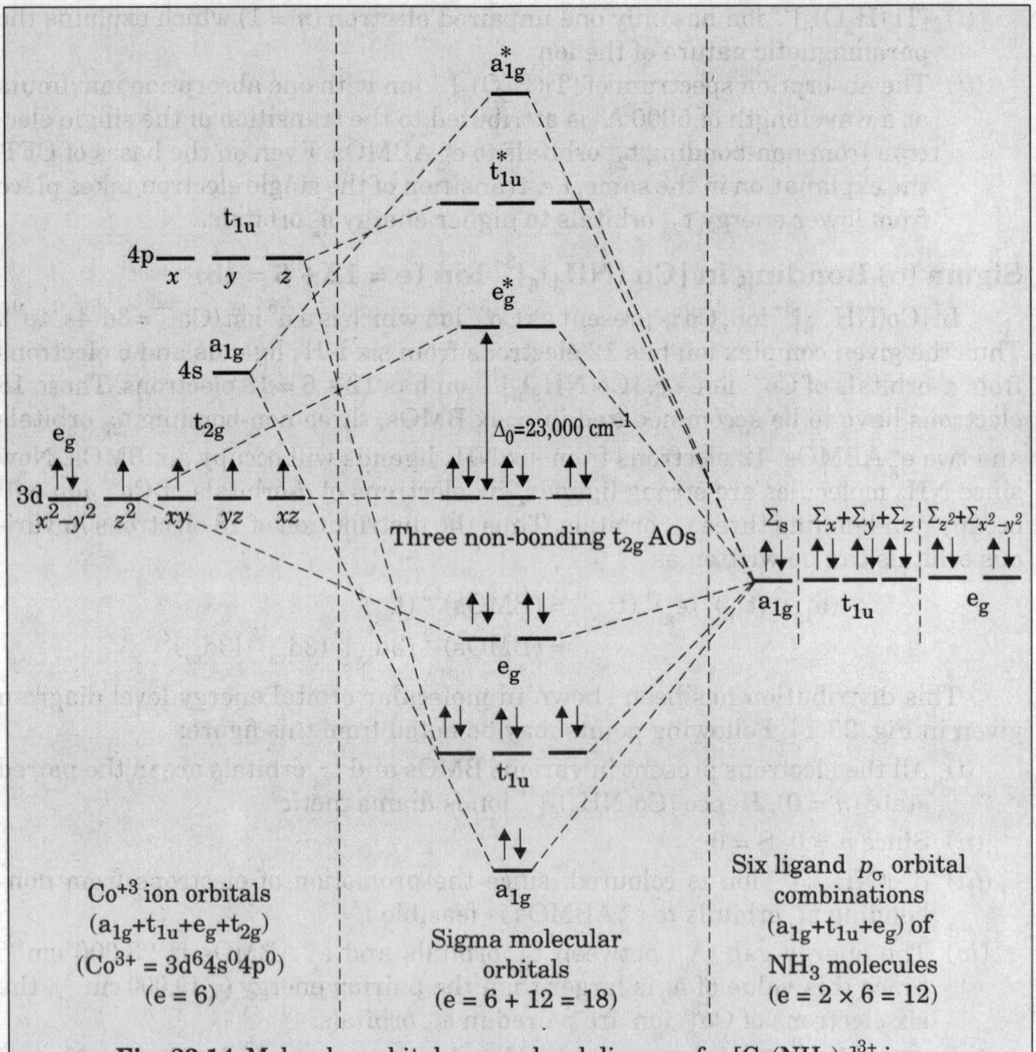


**Fig. 23.14** Molecular orbital energy level diagram for $[Co(NH_3)_6]^{3+}$ ion.

This distribution has been shown in molecular orbital energy level diagram given in Fig 23.15. Following points may be noted from this figure:

(*i*) Total number of unpaired electrons ($n$) present in $t_{2g}$ orbitals and $e_g^*$ ABMOs is equal to 2 + 2 = 4. Thus $[CoF_6]^{3-}$ in paramagnetic.

(*ii*) Since $n = 4$, $S = 4/2 = 2$

(*iii*) Spectroscopic measurements have shown that the energy gap ($\Delta_0$) between $t_{2g}$ and $e_g^*$ orbitals is 13,000 $cm^{-1}$ which is less than the pairing energy, P (= 19,000 $cm^{-1}$). Smaller value of $\Delta_0$ is attributed to the lesser extent of overlap between $e_g$ orbitals of the metal and ligand orbital combinations.

## Inert Gas or 18-Electron Rule in Complexes

According to this rule, in a given complex if the sum of the electrons present in the valence-shell of the central metal atom and the electrons donated by the ligands in equal to 18, the given complex would be stable.

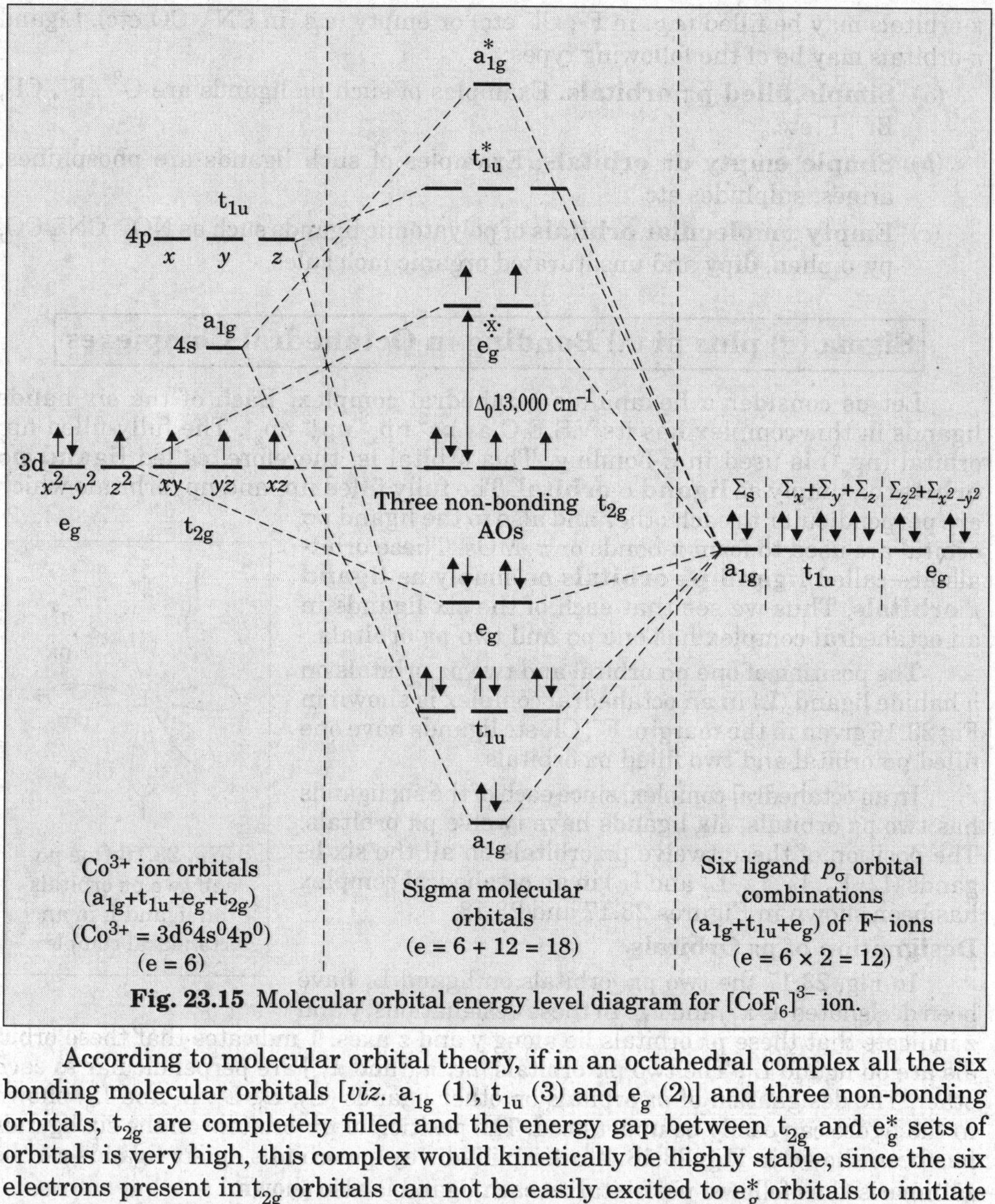


**Fig. 23.15** Molecular orbital energy level diagram for $[CoF_6]^{3-}$ ion.

According to molecular orbital theory, if in an octahedral complex all the six bonding molecular orbitals [*viz.* $a_{1g}$ (1), $t_{1u}$ (3) and $e_g$ (2)] and three non-bonding orbitals, $t_{2g}$ are completely filled and the energy gap between $t_{2g}$ and $e_g^*$ sets of orbitals is very high, this complex would kinetically be highly stable, since the six electrons present in $t_{2g}$ orbitals can not be easily excited to $e_g^*$ orbitals to initiate thermal decomposition of the complex and the electrons of an attacking nucleophile cannot be accomdated easily in the high energy $e_g^*$ set of molecular orbitals.

If a LS complex has more than 18 electrons, the extra electrons will occupy $e_g^*$ set of antibonding molecular orbitals and the bond order of the complex will reduce and hence the complex will destablise.

## Ligand π Orbitals

In octahedral complexes the orbitals of the six ligands which are involved in the formation of metal-ligand π bonds or π-MOs are called **ligand π orbitals.** Ligand π-orbitals may be filled (*e.g.* in $F^-$, $Cl^-$ etc) or empty (*e.g.* in $CN^-$, CO etc). Ligand π-orbitals may be of the following types:

(*a*) **Simple filled pπ orbitals.** Examples of such pπ ligands are $O^{2-}$, $F^-$, $Cl^-$, $Br^-$, $I^-$ etc.

(*b*) **Simple empty dπ orbitals**. Examples of such ligands are phosphines, arines, sulphides etc.

(*c*) **Empty π molecular orbitals** of polyatomic ligands such as $NO_2^-$, $CN^-$, CO, py, o-phen, dipy and unsaturated organic molecules.

## Sigma (σ) plus pi (π) Bonding in Octahedral Complexes

Let us consider a hexahalide octahedral complex. Each of the six halide ligands in this complex has its V.S.E.C as $ns^2\ np_x^2\ np_y^2\ np_z^2$. The fully-filled $np_x$ orbital ($np_x^2$) is used in σ-bonding. This orbital is, therefore, called **ligand pσ orbital** or simply as **ligand σ orbital**. The fully filled $np_y$ and $np_z$ orbitals which are perpendicular to each other and also to the ligand pσ orbital are used to form π-bonds or π–MOs. These orbitals are called **ligand pπ orbitals** or simply as **ligand π orbitals**. Thus we see that each of the six ligands in an octahedral complex has one pσ and two pπ orbitals.

The position of one pσ orbital and two pπ orbitals on a hahide ligand (L) in an octahedral complex is shown in Fig 23.16 given in the margin. $F^-$, $Cl^-$ etc ligands have one filled pσ orbital and two filled pπ orbitals.

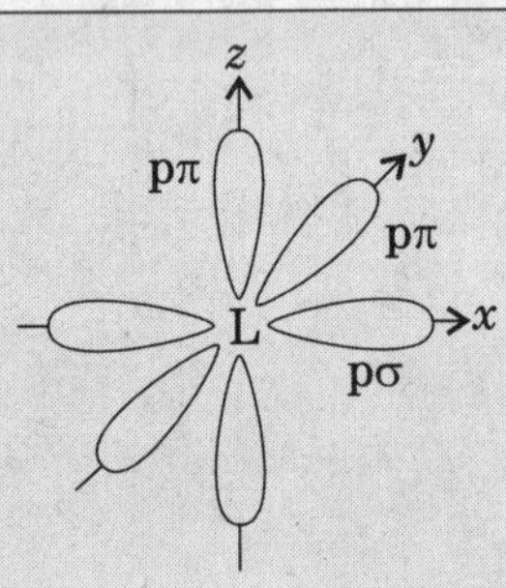


**Fig. 23.16** One pσ and two pπ orbitals on ligand L in an octahedral complex.

In an octahedral complex, since each of the six ligands has two pπ orbitals, six ligands have twelve pπ orbitals. The position of these twelve pπ orbitals on all the six ligands ($L_1$, $L_2$, $L_3$, $L_4$, $L_5$ and $L_6$) in an octahedral complex has been shown in Figures 23.17 and 23.18.

**Designation of pπ Orbitals.**

In Fig. 23.17 the two pπ orbitals on ligand $L_1$ have been designated as $\pi_{1y}$ and $\pi_{1z}$. In these designations, y and z indicate that these pπ orbitals lie along y and z axes. 1 indicates that these orbitals are on ligand $L_1$. The two pπ orbitals [*i.e.* $\pi_{1y}$ and $\pi_{1z}$] are perpendicular to each other. The designation of pπ orbitals on other ligands ($L_2$, $L_3$, $L_4$, $L_5$ and $L_6$) shown in the figure can easily be understood. The position of pπ orbitals on the six ligands has been shown in Fig. 23.18 in a simplified way. pπ orbitals on six ligands can be summarised as follows. pσ orbital on each ligand is also shown.

| *Ligand* | *pσ orbital* | *pπ orbitals* |
|---|---|---|
| $L_1$ | $\sigma_{1x}$ | $\pi_{1y}$, $\pi_{1z}$ |
| $L_2$ | $\sigma_{2x}$ | $\pi_{2y}$, $\pi_{2z}$ |
| $L_3$ | $\sigma_{3y}$ | $\pi_{3x}$, $\pi_{3z}$ |
| $L_4$ | $\sigma_{4y}$ | $\pi_{4x}$, $\pi_{4z}$ |
| $L_5$ | $\sigma_{5z}$ | $\pi_{5x}$, $\pi_{5y}$ |
| $L_6$ | $\sigma_{6z}$ | $\pi_{6x}$, $\pi_{6y}$ |

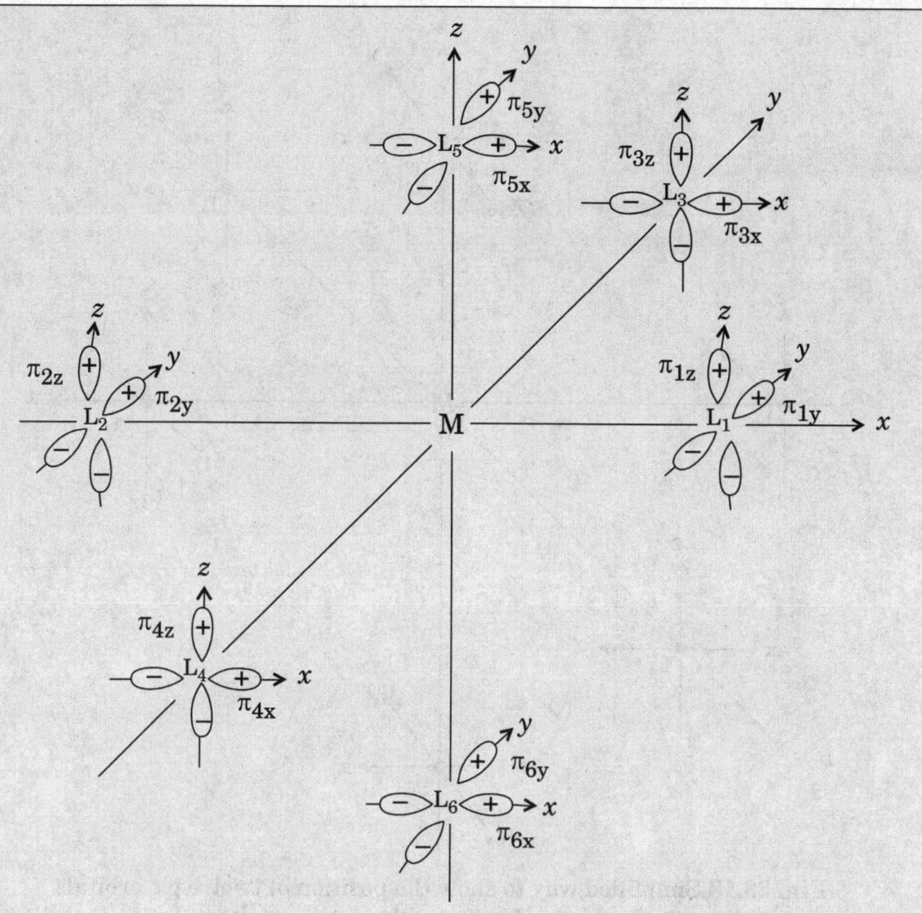


**Fig. 23.17** Twelve ligand pπ orbitals on six ligand ($L_1$, $L_2$, $L_3$, $L_4$, $L_5$ and $L_6$) in *an* octahedral complex

## Formation of σ- and π - MOs in $[FeF_6]^{3-}$ Ion

Now we are in a position to discuss the formation of metal-ligand σ as well as π bonds in octahedral complexes on the basis of MOT. As an example set us discuss the σ and π bonding in $[FeF_6]^{3-}$ ion which is an octahedral complex ion.

Fe is present an $Fe^{3+}$ ion in $[FeF_6]^{3-}$. $Fe^{3+}$ ion has nine AOs in its valence shell ($Fe^{3+} = 3d^5\ 4s^0 4p^0$). These nine orbitals are five 3d orbitals ($3d_{z^2}$, $3d_{x^2-y^2}$, $3d_{xy}$, $3d_{yz}$, $3d_{zx}$), one 4s and three 4p ($4p_x$, $4p_y$ and $4p_z$) orbitals. $[FeF_6]^{3-}$ ion has six $F^-$ ligands. Each of the six $F^-$ ligands has one pσ orbital and two pπ orbitals. The two pπ orbitals are mutually perpendicular to each other and also perpendicular to pσ orbital. Thus the six $F^-$ ligands have six pσ orbitals and $6 \times 2 = 12$ pπ orbitals.

On the basis of group theory, when 12 ligand pπ orbitals (also called simply as ligand π orbitals) are combined together, they form **four triply degenerate** sets which have $t_{1u}(3)$, $t_{2g}(3)$, $t_{1g}(3)$ and $t_{2u}(3)$ symmetry symbols.

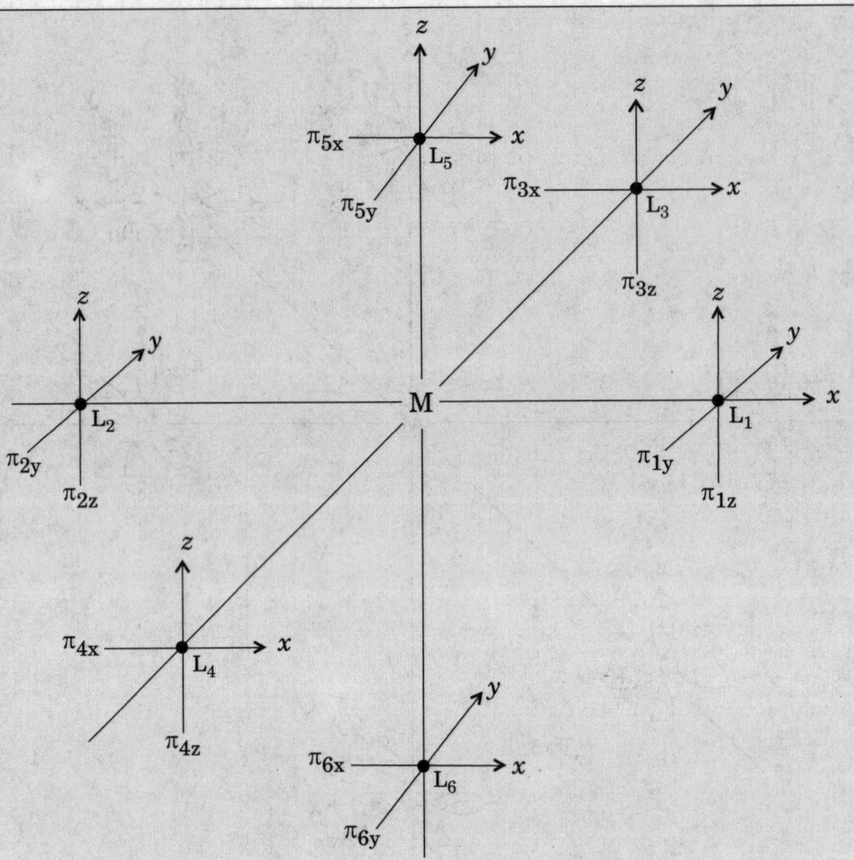


**Fig. 23.18** Simplified way to show the position of twelve pπ orbitals on six ligands ($L_1$, $L_2$, $L_3$, $L_4$, $L_5$ and $L_6$) in an octahedral complex. The arrows point in the direction of the + lobe of pπ orbitals.

Group symmetry symbols of nine AOs of $Fe^{3+}$ ion, six ligands σ orbitals and twelve ligand π orbitals are given below.

| **Nine AOs of $Fe^{3+}$ ion** | **Six ligand σ-orbitals** | **Twelve ligand π orbitals** |
|---|---|---|
| $a_{1g}$ (4s) (1) | $a_{1g}$ (1) | — |
| $t_{1u}$ ($4p_x$, $4p_y$, $4p_x$) (3) | $t_{1u}$(3) | $t_{1u}$(3) |
| $e_g$ ($3d_{z^2}$, $3d_{x^2-y^2}$) (2) | $e_g$(2) | — |
| $t_{2g}$ ($3d_{xy}$, $3d_{yz}$, $3d_{xz}$) (3) | — | $t_{2g}$ (3) |
| — | — | $t_{1g}$(3) |
| — | — | $t_{2u}$ (3) |

**Formation of σ-MOs.**

When six ligand σ-orbitals of $a_{1g}(1)$, $t_{1u}(3)$ and $e_g$ (2) group symmetry symbols overlap with AOs of $Fe^{3+}$ ion having the same symmetry, six σ-BMOs and six σ*ABMOs are formed. AOs of $Fe^{3+}$ ion having $a_{1g}$, $t_{1u}$ and $e_g$ symmetry symbols are 4s ($a_{1g}$), $4p_x$, $4p_y$, $4p_z$ ($t_{1u}$), $3d_{z^2}$, $3d_{x^2-y^2}$ ($e_g$). Six σ-BMOs and six σ*ABMOs formed are given below.

**Six σ-BMOs:** $a_{1g}(1)$, $t_{1u}(3)$, $e_g(2)$

**Six σ*ABMOs**: $a^*_{1g}$ (1), $t^*_{1u}(3)$, $e^*_g(2)$

$a_{1g}/a^*_{1g}$ σ-MO is non generate (singly degenerate), $t_{1u}/t^*_{1u}$ σ-MOs are triply degenerate and $e_g/e^*_g$ σ-MOs are doubly degenerate. σ-BMOs have lower energy than σ*-ABMOs (See Fig. 23.19). It may be noted that AOs of $Fe^{3+}$ ion having $t_{2g}$ symmetry (i.e. $3d_{xy}$, $3d_{yz}$ and $3d_{xz}$) donot participate in the formation of σ-MOs, since there are no ligand σ-orbitals which may be of $t_{2g}$ symmetry.

**Formation of π-MOs.**

We have already stated that 12 ligand π-orbitals have $t_{1u}$ (3), $t_{2g}$(3), $t_{1g}$ (3) and $t_{2u}$(3) symmetry symbols, and AOs of $Fe^{3+}$ have $a_{1g}(1)$, $t_{1u}(3)$, $e_g(2)$ and $t_{2g}$ (3) symmetry symbols.

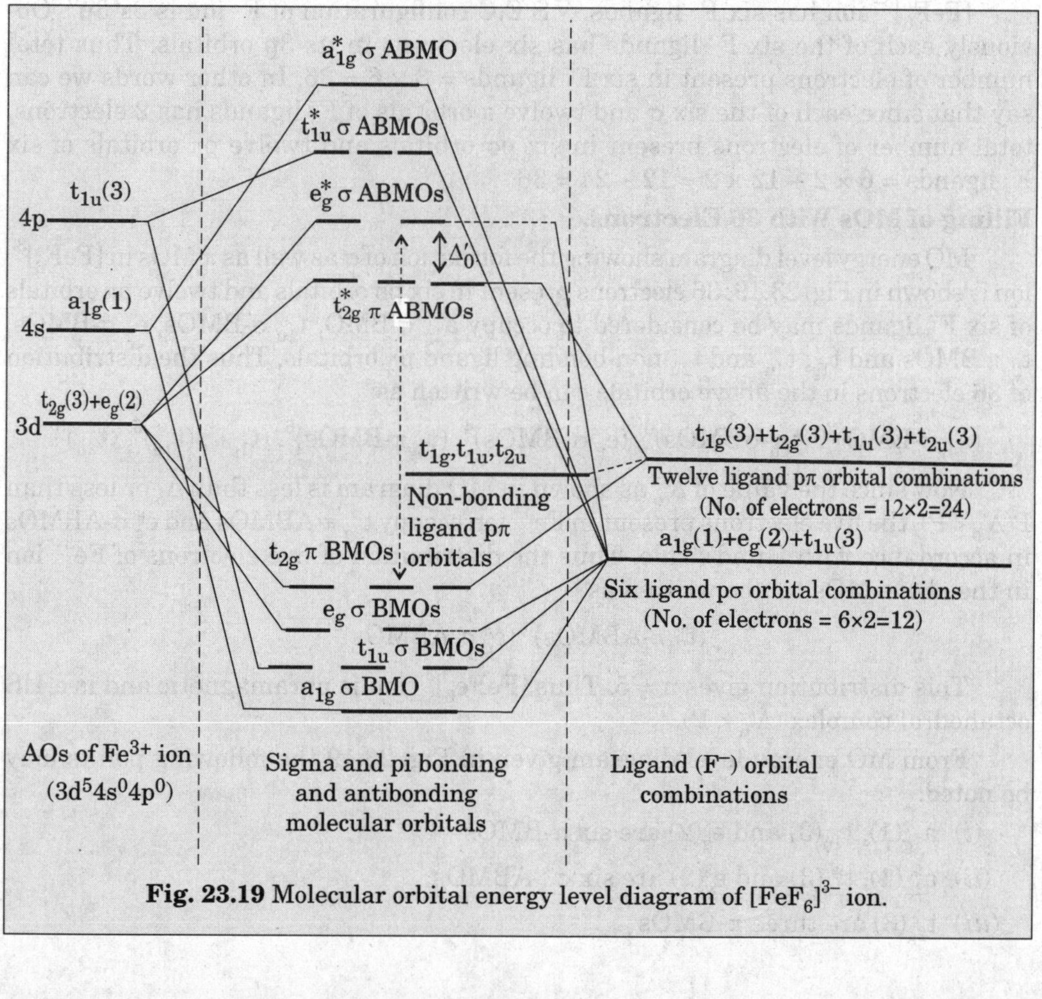


**Fig. 23.19** Molecular orbital energy level diagram of $[FeF_6]^{3-}$ ion.

Now since there is no AOs of $Fe^{3+}$ ion with $t_{1g}$ and $t_{2u}$ symmetry, ligand π-orbitals of these symmetries donot participate in π-bonding, *i.e.* ligand π-orbitals with $t_{1g}$ and $t_{2u}$ symmetry symbol remain non-bonding and their energy remains unchanged. π-orbitals of ligands having $t_{1u}$ symmetry symbol also remain non-bonding, because their overlap with 4p orbitals of $Fe^{3+}$ ion which have $t_{1u}$ symmetry symbol has already formed σ-MOs.

Above discussion shows that ligand π-orbitals with $t_{1g}$, $t_{2u}$ and $t_{1u}$ symmetry symbols do not participate in the formation of π-MOs. Only three ligand π-orbitals with $t_{2g}$ symmetry participate in the formation of π-MOs.

We know that $3d_{xy}$, $3d_{yz}$ and $3d_{xz}$ orbitals of $Fe^{3+}$ ion have $t_{2g}$ symmetry. Three ligand π-orbitals having $t_{2g}$ symmetry overlap with $3d_{xy}$, $3d_{yz}$ and $3d_{xz}$ orbitals which have $t_{2g}$ symmetry symbol and form *three degenerate* $t_{2g}$ π-BMOs and *three degenerate* $t_{2g}^*$ π-ABMOs. $t_{2g}$ BMOs have lower energ than $t_{2g}^*$ ABMOs. It may be noted that $3d_{xy}$, $3d_{yz}$ and $3d_{xz}$ orbitals which participate in π-bonding as discussed above donot participate in σ-bonding.

**Number of Electrons.**

V.S.E.C of $Fe^{3+}$ ion is $3d^5 4s^0 4p^0$ which shows that $Fe^{3+}$ ion has five 3d electrons. 4s and 4p orbitals are vacant.

$[FeF_6]^{3-}$ ion has six $F^-$ ligands. V.S.E.C configuration of $F^-$ ion is $3s^2 3p^6$. Obviously each of the six $F^-$ ligands has six electrons in its 3p orbitals. Thus total number of electrons present in six $F^-$ ligands = 6 × 6 = 36. In other words we can say that since each of the six σ and twelve π orbitals of $F^-$ ligands has 2 electrons, total number of electrons present in six pσ orbitals and twelve pπ orbitals of six $F^-$ ligands = 6 × 2 + 12 × 2 = 12 + 24 = 36.

**Filling of MOs With 36 Electrons.**

MO energy level diagram showing the formation of σ as well as π MOs in $[FeF_6]^{3-}$ ion is shown in Fig. 23.19. 36 electrons present in six pσ orbitals and twelve pπ orbitals of six $F^-$ ligands may be considered to occupy $a_{1g}$ σ-BMO, $t_{1u}$ σ-BMOs, $e_g$ σ-BMOs, $t_{2g}$π BMOs and $t_{1g}$, $t_{1u}$ and $t_{2u}$ non-bonding ligand pπ orbitals. Thus the distribution of 36 electrons in the above orbitals can be written as:

$$(a_{1g}\sigma\text{-BMO})^2, (t_{1u}\text{-}\sigma\text{BMOs})^6, (e_g\ \sigma\text{-BMOs})^4, (t_{2g}\pi\text{-BMOs})^6, (t_{1g})^6, (t_{1u})^6, (t_{2u})^6$$

Now since the value of $\Delta'_o$ as shown in MO diagram is less than $\Delta_o$ or less than $P(\Delta'_o < P)$, the five electrons present in $Fe^{3+}$ ion occupy $t_{2g}^*$ π-ABMOs and $e_g^*$ σ-ABMOs in accordance with Hund's rule. Thus the distribution of five electrons of $Fe^{3+}$ ion in the above MOs can be written as

$$(t_{2g}^*\pi\text{-ABMOs})^3\ (e_g^*\sigma\text{-ABMOs})^2$$

This distribution gives n = 5. Thus $[FeFe_6]^{3-}$ ion is paramagnetic and is a HS octahedral complex ($\Delta'_o < P$).

From MO energy level diagram given in Fig. 23.19 the following points may be noted:

(*i*) $a_{1g}(1)$, $t_{1u}(3)$ and $e_g(2)$ are six σ-BMOs

(*ii*) $a_{1g}^*(1)$, $t_{1u}^*(3)$ and $e_g^*(2)$ are six σ*-ABMOs

(*iii*) $t_{2g}(3)$ are three π-BMOs

(*iv*) $t_{2g}^*$ (3) are three π*-ABMOs

(*v*) $t_{1g}$ (3), $t_{1u}$(3) and $t_{2u}$ (3) are nine non-bonding ligand pπ orbitals

Six σ MBOs, six σ* ABMOs, three π BMOs and three π*ABMOs (total 18 MOs) shown in the MO energy level diagram of $[FeF_6]^{3-}$ can also be represented by the notations given below.

| | |
|---|---|
| $a_{1g}$ σ-BMO (1) | $\sigma_s$ |
| $a_{1g}^*$ σ-ABMO (1) | $\sigma_s^*$ |
| $t_{1u}$ σ-BMOs (3) | $\sigma_p$ ($\sigma_x$, $\sigma_y$, $\sigma_z$) |
| $t_{1u}^*$ σ-ABMOs (3) | $\sigma_p^*$($\sigma_x^*$, $\sigma_y^*$, $\sigma_z^*$) |
| $e_g$ σ-BMOs (2) | $\sigma_d$ ($\sigma_{z^2}$, $\sigma_{x^2-y^2}$) |
| $e_g^*$ σ-ABMOs (2) | $\sigma_d^*$($\sigma_{z^2}^*$, $\sigma_{x^2-y^2}^*$) |
| $t_{2g}$ π-BMOs (3) | $\pi_d$ ($\pi_{xy}$, $\pi_{yz}$, $\pi_{xz}$) |
| $t_{2g}^*$ π-ABMOs (3) | $\pi_d^*$($\pi_{xy}^*$, $\pi_{yz}^*$, $\pi_{xz}^*$) |

Fig 23.19 shows that MO configuration of $[FeF_6]^{3-}$ ion which contains 5 + 36 = 41 electrons can be written as:

$$(a_{1g}\ \sigma\ \text{BMO})^2\ (t_{1u}\ \sigma\ \text{BMOs})^6\ (e_g\ \sigma\ \text{BMOs})^4$$
$$(t_{2g}\ \pi\ \text{BMOs})^6\ (t_{1g})^6\ (t_{1u})^6\ (t_{2u})^6$$
$$(t_{2g}^*\ \pi\ \text{ABMOs})^3\ (e_g^*\ \sigma\ \text{ABMOs})^2$$

The formation of different MOs shown in Fig. 23.19 can be shown in the following simple way.

| *Metal ion AOs* | *Ligand pσ orbitals* | *Ligand pπ orbitals* | *σ-BMOs and σ* AMBOs* | *π-BMOs and π* ABMOs* |
|---|---|---|---|---|
| 4s ($a_{1g}$) (1) | $a_{1g}$ (1) | — | $a_{1g}$ σ BMO (1)<br>$a_{1g}^*$ σ ABMO (1) | — |
| $4p_x$, $4p_y$, $4p_z$ ($t_{1u}$) (3) | $t_{1u}$ (3) | — | $t_{1u}$ σ BMOs (3)<br>$t_{1u}$* σ ABMOs (3) | — |
| $3d_{z^2}$, $3d_{x^2-y^2}$ ($e_g$) (2) | $e_g$ (2) | — | $e_g$ σ BMOs (2),<br>$e_g^*$ σ ABMOs (2) | — |
| $3d_{xy}$, $3d_{yz}$, $3d_{zx}$ ($t_{2g}$) (3) | — | $t_{2g}$ (3) | — | $t_{2g}$π BMOs (3),<br>$t_{2g}^*$ π ABMOs (3) |
| — | — | $t_{1g}$ (3), $t_{1u}$ (3), $t_{2u}$ (3) (Non-bonding orbitals) | — | — |

## Pictorial Representation of the Formation of $t_{2g}$ π-BMOs.

We have already stated that the three ligand π orbitals having $t_{2g}$ symmetry overlap with $3d_{xy}$, $3d_{yz}$ and $3d_{xz}$ AOs of $Fe^{3+}$ ion ($t_{2g}$ symmetry) and form three degenerate $t_{2g}$π-BMOs and three degenerate $t_{2g}^*$π-ABMOs. But before overlapping with the above said atomic orbitals of $Fe^{3+}$ ion, the three ligand pπ orbitals combine together and form **three ligand pπ orbital combinations** which are represented as $\Sigma_{xy}$, $\Sigma_{yz}$ and $\Sigma_{xz}$. These combinations also have $t_{2g}$ symmetry. The three ligand pπ orbitals combinations which overlap with $3d_{xy}$, $3d_{yz}$ and $3d_{xz}$ atomic orbitals ($t_{2g}$ symmetry) are given below.

| *Atomic orbitals ($t_{2g}$ symmetry)* | *Ligand pπ orbital combinations ($t_{2g}$ symmetry)* |
|---|---|
| $3\,d_{xy}$ | $\Sigma_{xy} = \pi_{1y} - \pi_{2y} + \pi_{3x} - \pi_{4x}$ |
| $3\,d_{yz}$ | $\Sigma_{yz} = \pi_{3z} - \pi_{4z} + \pi_{5y} - \pi_{6y}$ |
| $3\,d_{xz}$ | $\Sigma_{xz} = \pi_{1z} - \pi_{2z} + \pi_{5x} - \pi_{6x}$ |

Ligand pπ orbitals combination ($\Sigma_{xy}$) which overlaps with $3d_{xy}$ orbital is obtained by combining $\pi_{1y}$ (on ligand $L_1$), $\pi_{2y}$ (on ligand $L_2$), $\pi_{3x}$ (on ligand $L_3$) and $\pi_{4x}$ (on ligand $L_4$) ligand pπ orbitals. In the same way the ligand pπ orbital combinations which overlap with $3d_{yz}$ and $3d_{xz}$ atomic orbitals can be determined.

Now since the three ligand pπ orbitals combinations ($\Sigma_{xy}$, $\Sigma_{yz}$ and $\Sigma_{xz}$) and the three AOs ($3d_{xy}$, $3d_{yz}$ and $3d_{xz}$) have the same symmetry ($t_{2g}$ symmetry), they overlap with each other and form three degenerate $t_{2g}$π-BMOs and three degenerate $t^*_{2g}$π-ABMOs. $t_{2g}$π-BMOs have lower energy than $t^*_{2g}$π-ABMOs. The pictorial representation of the formation of $t_{2g}$ π-BMOs by the overlap between $3d_{xy}$ and $\Sigma_{xy}$, between $3d_{yz}$ and $\Sigma_{yz}$ and between $3d_{xz}$ and $\Sigma_{xz}$ is shown in Fig. 23.20, 23.21 and 23.22 respectively.

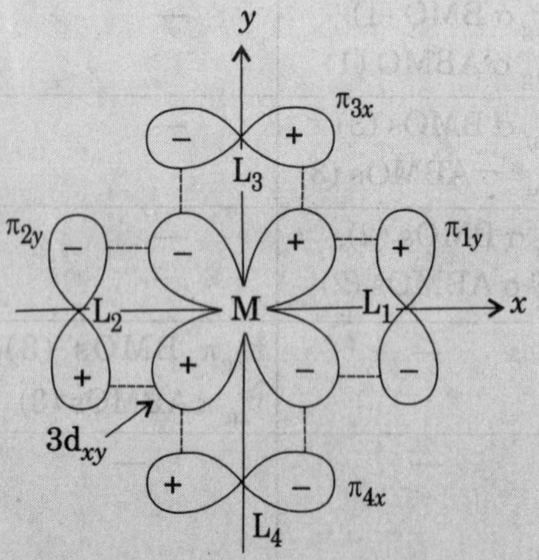


**Fig 23.20** Pictorial representation of the formation of $t_{2g}$π-BMO by the overlap between metal ion (M) $3d_{xy}$ orbital and ligand pπ orbitals combination ($\Sigma_{xy}$).

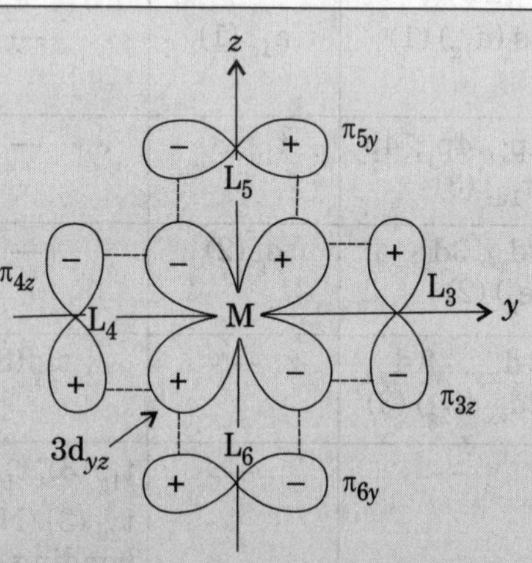


**Fig 23.21** Pictorial representation of the formation of $t_{2g}$π-BMO by the overlap between metal ion (M) $3d_{yz}$ orbital and ligand pπ orbitals combination ($\Sigma_{yz}$).

## Formation of $Fe^{2+}$–$CN^-$ σ and $Fe^{2+}$–$CN^-$ π bonds in $[Fe(CN)_6]^{3-}$ ion

(a) **Formation of $Fe^{2+}$– $CN^-$ σ bond.** We know that $CN^-$ ligand has a filled pσ orbital which has $e_g$ symmetry. In the formation of Fe–CN σ bond the filled pσ orbital of $CN^-$ ligand overlaps with one empty $e_g$ orbital ($3d_{z^2}$ or $3d_{x^2-y^2}$) of $Fe^{2+}$ ion. Since pσ orbital of $CN^-$ ion is filled, $CN^-$ ion acts as a Lewis base (donor ligand) in the formation of Fe–$CN^-$ σ bond. In Fig. 23.23(*a*) we have shown the formation of Fe–CN σ bond by the overlap between filled pσ orbital on $CN^-$ ligand and empty $3d_{x2-y2}$ orbital ($e_g$ symmetry) of $Fe^{2+}$ ion.

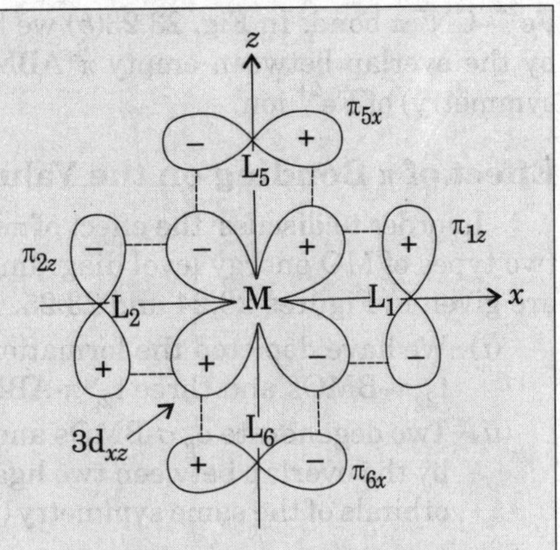


**Fig. 23.22** Pictorial representation of the formation of $t_{2g}$π-BMO by the overlap between metal ion (M) $3d_{xz}$ orbital and ligand pπ orbitals combination ($\Sigma_{yz}$).

(*b*) **Formation of $Fe^{2+}$– $CN^-$ π bond.** We know that $CN^-$ ligand has empty π* ABMO which is of $t_{2g}$ symmetry. In the formation of Fe–$CN^-$ π bond, the empty π* ABMO of $CN^-$ ligand interacts with one filled $t_{2g}$ orbital ($3d_{xy}$/$3d_{yz}$/$3d_{xz}$) of $Fe^{2+}$ ion. Thus in the formation of Fe–$CN^-$ π bond electrons from the filled $t_{2g}$ orbital of $Fe^{2+}$ ion are partially transferred to $CN^-$ ligand. This process lowers the energy of $t_{2g}$ orbitals and hence the value of $\Delta_o$ increases. Since π* ABMO of $CN^-$ ligand is empty, $CN^-$ ligand acts as a Lewis acid (accepter ligand) in the formation of

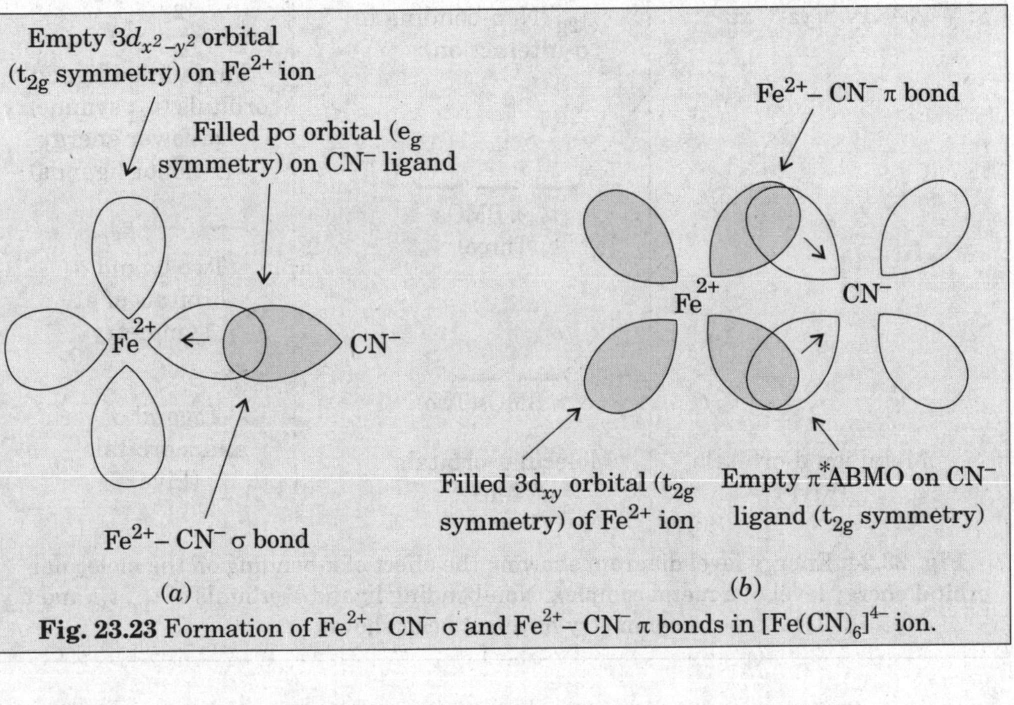


**Fig. 23.23** Formation of $Fe^{2+}$– $CN^-$ σ and $Fe^{2+}$– $CN^-$ π bonds in $[Fe(CN)_6]^{4-}$ ion.

$Fe^{2+}$–$CN^-$ π bond. In Fig. 23.23(*b*) we have shown the formation of Fe–$CN^-$ π bond by the overlap between empty π*ABMO of $CN^-$ ligand and filled $3d_{xy}$ orbital ($t_{2g}$ symmetry) of $Fe^{2+}$ ion.

## Effect of π Bonding on the Value of $\Delta_0$

In order to discuss the effect of π bonding on the value of $\Delta_0$, we shall consider two types of MO energy level diagrams of an octahedral complex. These diagrams are given in Figures 23.24 and 23.25. In both the diagrams:

(*i*) We have depicted the formation of two $e_g$ σ-BMOs, two $e_g^*$ σ-ABMOs, three $t_{2g}$π–BMOs and three $t_{2g}^*$ π-ABMOs only.

(*ii*) Two degenerate $e_g$ σ-BMOs and two degenerate $e_g^*$ σ-ABMOs are generated by the overlap between two ligand σ-orbitals of $e_g$ symmetry with metal ion orbitals of the same symmetry ($d_{z^2}$ and $d_{x^2-y^2}$ orbitals). $e_g$ σ-BMOs have lower

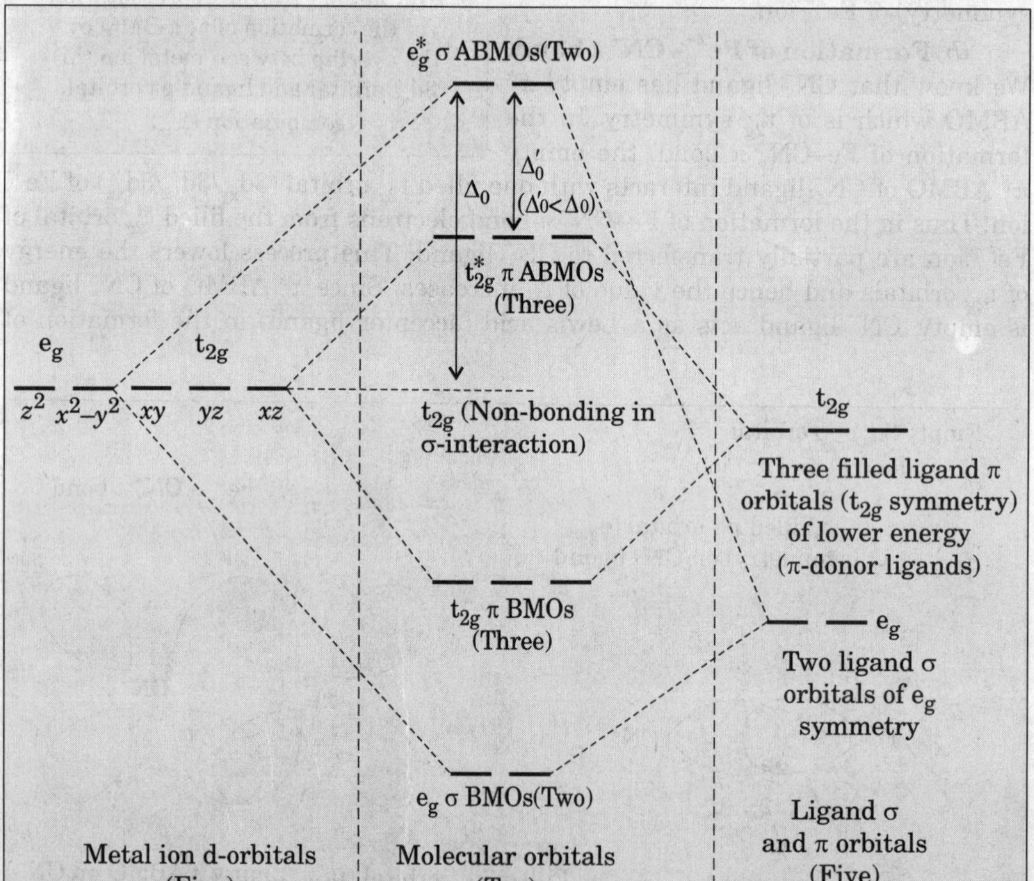


**Fig. 23.24.** Energy level diagram showing the effect of π-bonding on the molecular orbital energy level of a metal complex. Non-bonding ligand π-orbitals of $t_{1g}$, $t_{1u}$ and $t_{2u}$ symmetry have not been shown.

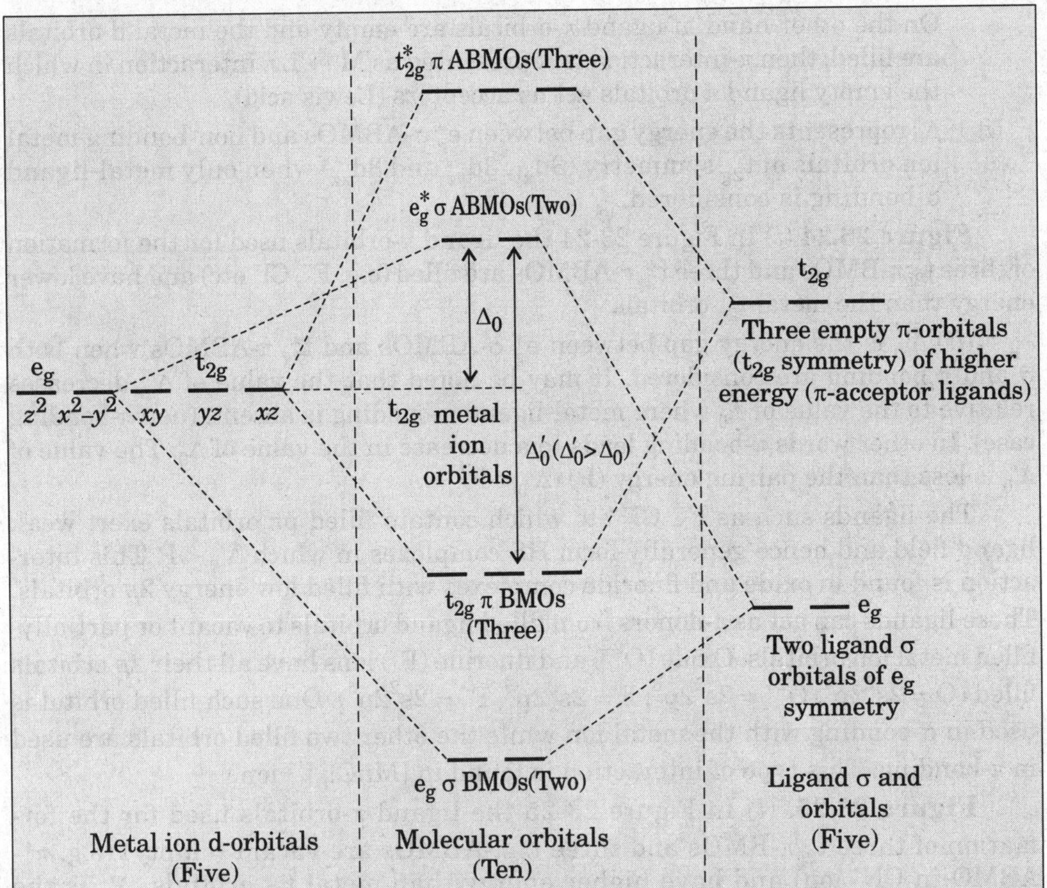


**Fig. 23.25.** Energy level diagram showing the effect of π-bonding on the molecular orbital energy level of a metal complex. Non-bonding ligand π-orbitals of $t_{1g}$, $t_{1u}$ and $t_{2u}$ symmetry have not been shown.

energy than $e_g^*$ σ-ABMOs. This type of overlap between orbitals is called σ overlap or σ interaction. It should be noted that 3d orbitals of $t_{2g}$ symmetry ($3d_{xy}$, $3d_{yz}$ and $3d_{xz}$ orbitals) donot participate in σ overlap and hence these orbitals remain non-bonding and they remain unchanged.

(*iii*) Three degenerate $t_{2g}$π-BMOs and three degenerate $t_{2g}^*$π-ABMOs are produced by the overlap between the three ligand π-orbitals of $t_{2g}$ symmetry and metal ion $t_{2g}$ orbitals ($3d_{xy}$, $3d_{yz}$ and $3d_{zx}$ orbitals). $t_{2g}$π-BMOs have lower energy than $t_{2g}^*$ π-ABMOs. The overlap between orbitals leading to the formation of $t_{2g}$π-BMOs and $t_{2g}^*$π-ABMOs is called π-overlap or π-interaction. In π-interaction, the three ligand π-orbitals may be filled orbitals (e.g. $F^-$, $Cl^-$ etc) or empty orbitals (e.g. $CN^-$, CO etc.). If the ligand π-orbitals are filled and the metal d-orbitals are empty, then π-interaction is represented as L → M π interaction in which the filled ligand π-orbitals act as donors (Lewis base).

On the other hand, if ligand π-orbitals are empty and the metal d-orbitals are filled, then π-interaction is represented as M → L π interaction in which the empty ligand π orbitals act as acceptors (Lewis acid).

(*iv*) $\Delta_0$ represents the energy gap between $e_g^*$ σ-ABMOs and non-bonding metal ion orbitals of $t_{2g}$ symmetry ($3d_{xy}$, $3d_{yz}$ and $3d_{xz}$) when only metal-ligand σ-bonding is considered.

**Figure 23.24** (*i*) In Figure 23-24 the ligand π-orbitals used for the formation of three $t_{2g}\pi$-BMOs and three $t_{2g}^*\pi$-ABMOs are filled (e.g. $F^-$, $Cl^-$ etc) and have lower energy than the metal $t_{2g}$ orbitals.

(*ii*) $\Delta'_0$ is the energy gap between $e_g^*$ σ-ABMOs and $t_{2g}^*$ π-ABMOs when both σ and π bonding are considered. It may be noted that the value of $\Delta'_0$ decreases relative to the value of $\Delta_0$ where metal-ligand π-bonding is absent. (non π-bonding case). In other words π-bonding leads to a decrease in the value of $\Delta_0$. The value of $\Delta'_0$ is less than the pairing energy (P) ($\Delta'_0 < P$).

The ligands such as $F^-$, $Cl^-$ etc which contain filled pπ orbitals exert weak ligand field and hence generally form HS complexes in which $\Delta'_0 < P$. This interaction is found in oxide and fluoride complexes with filled low energy 2*p* orbitals. These ligands can act as π-donors from filled ligand orbitals to vacant or partially-filled metal ion orbitals. Oxide ($O^{2-}$) and fluoride ($F^-$) ions have all their 2*p* orbitals filled ($O = 2s^2 2p^4$, $O^{2-} = 2s^2 2p^6$; $F = 2s^2 2p^5$, $F^- = 2s^2 2p^6$). One such filled orbital is used in σ-bonding with the metal ion while the other two filled orbitals are used in π-bonding. This type of interaction is found in $[MnCl_6]^{4-}$ ion.

**Figure 23.25.** (*i*) In Figure 23-25 the ligand π-orbitals used for the formation of three $t_{2g}\pi$-BMOs and three $t_{2g}^*\pi$-ABMOs are vacant (empty) (e.g. π*-ABMO in $CN^-$ ion) and have higher energy than metal $t_{2g}$ orbitals. $\Delta'_0$ is the energy gap between $e_g^*$ σ-ABMOs and $t_{2g}\pi$–BMOs when both σ and π bonding are considered. It may be seen that the value of $\Delta'_0$ increases relative to the value of $\Delta_0$ where metal-ligand π-bonding is absent (non π-bonding case). In other words π-bonding leads to an increase in the value of $\Delta_0$. The value of $\Delta'_0$ is more than the value of pairing energy (P) ($\Delta'_0 > P$). The ligands like $CN^-$, CO etc. which contain vacant pπ orbitals exert strong ligand field generally form LS complexes in which $\Delta'_0 > P$. This type of π-interaction is found in complexes of phosphines and arsines. Such ligands can act as π-acceptors (from metal orbitals to vacant ligand π-orbitals).

## To Compare the Value of $\Delta'_0$ for Halide Ions Used as Filled π-Ligands

MO energy level diagram in which halide ions ($F^-$, $Cl^-$, $Br^-$ and $I^-$) are used as filled π ligands is given in Fig. 23.26. This figure shows that the value of $\Delta'_0$ for $F^-$, $Cl^-$, $Br^-$ and $I^-$ ligands decreases in the order: $F^- > Cl^- > Br^- > I^-$ because the energy of π-orbitals of the halide ligands increases as $F^- < Cl^- < Br^- < I^-$. The decreasing order of the values of $\Delta'_0$ explains the decreasing order of ligand field strength of these halide ligands. A low value of $\Delta'_0$ corresponds to weak ligand field strength.

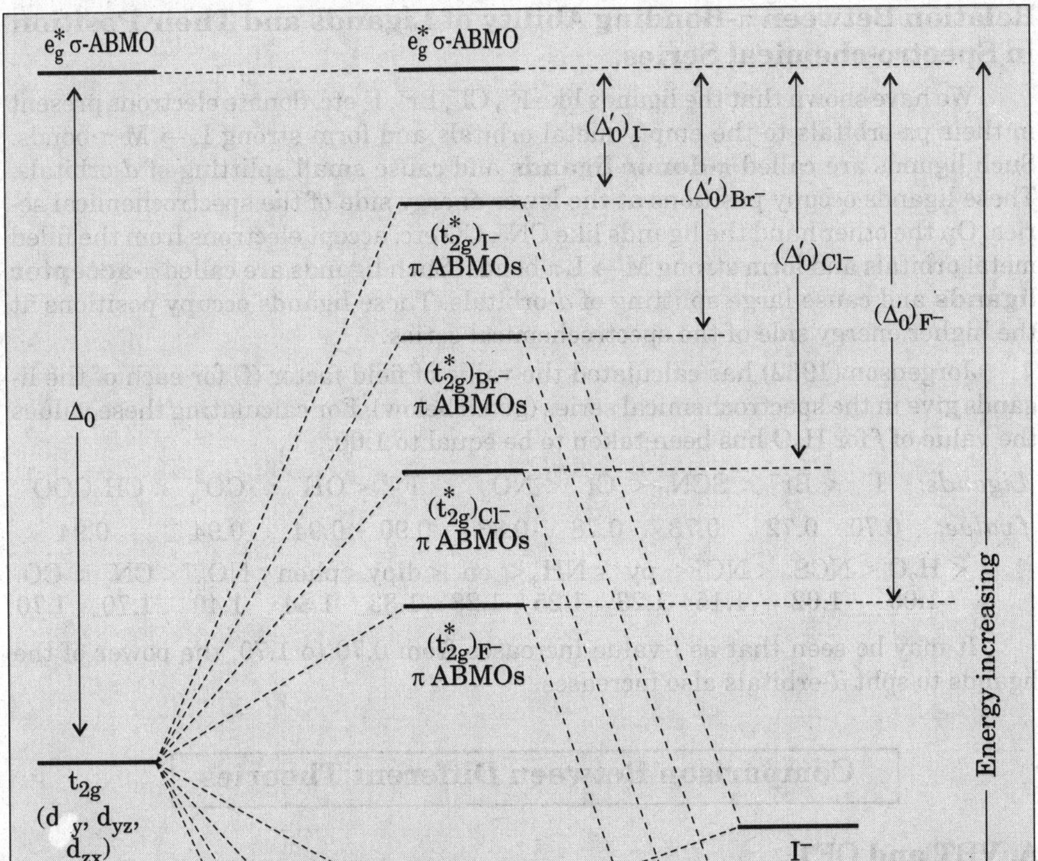


**Fig. 23.26** Relative effect of the halide ions as ligands on $t_{2g}$ metal orbitals due to π bonding in an octahedral complex.

## Relation Between π-Bonding Ability of Ligands and Their Position in Spectro-chemical Series.

We have shown that the ligands like $F^-$, $Cl^-$, $Br^-$, $I^-$ etc. donate electrons present in their pπ-orbitals to the empty metal orbitals and form strong L → M π bonds. Such ligands are called **π-donor ligands** and cause small splitting of *d*-orbitals. These ligands occupy positions at the lower energy side of the spectrochemical series. On the other hand the ligands like $CN^-$, CO etc. accept electrons from the filled metal orbitals and form strong M → L π bonds. Such ligands are called **π-acceptor ligands** and cause large splitting of *d*-orbitals. These ligands occupy positions at the higher energy side of the spectrochemical series.

Jorgenson (1962) has calculated the value of field factor (f) for each of the ligands give in the spectrochemical series (given below). For calculating these values the value of *f* for $H_2O$ has been taken to be equal to 1.00.

| *Ligands:* | $I^-$ < | $Br^-$ < | $SCN^-$ < | $Cl^-$ < | $NO_3^-$ < | $F^-$ < | $OH^-$ < | $CO_4^{2-}$ < | $CH_3COO^-$ |
|---|---|---|---|---|---|---|---|---|---|
| *f value:* | 0.70 | 0.72 | 0.73 | 0.78 | 0.83 | 0.90 | 0.94 | 0.94 | 0.94 |

| < $H_2O$ | < $NCS^-$ | < $NC^-$ | < py | < $NH_3$ | < en | < dipy | < phen | < $NO_2^-$ | < $CN^-$ | < CO |
|---|---|---|---|---|---|---|---|---|---|---|
| 1.00 | 1.02 | 1.15 | 1.23 | 1.25 | 1.28 | 1.33 | 1.34 | 1.40 | 1.70 | 1.70 |

It may be seen that as *f*-value increases from 0.70 to 1.70, the power of the ligands to split *d*-orbitals also increases.

## Comparison Between Different Theories

### A. VBT and CFT

(a) **Similarities** (*i*) According to VBT the atomic orbitals used in the formation of six $d^2sp^3$ hybrids in inner orbital octahedral complexes of 3*d*-series elements are $3d_{x^2-y^2}$, $3d_{z^2}$ ($e_g$ set), 4s, $4p_x$, $4p_y$ and $4p_z$. The remaining three 3*d*-atomic orbitals *viz* $3d_{xy}$, $3d_{yz}$ and $3d_{zx}$ ($t_{2g}$ set) remain non-bonding and contain metallic electrons (non-bonding electrons). In CFT as well the metallic electrons occupy $t_{2g}$ set, since it is lower in energy than $e_g$ set.

(*ii*) Inner-orbital octahedral complexes of VBT ($d^2sp^3$ hybridisation) are the same as spin-paired or low-spin octahedral complexes of CFT. Similarly outer-orbital octahedral complexes ($sp^3d^2$ hybridisation) of VBT are the same as spin-free or high-spin octahedral complexes of CFT.

(*b*) **Differences** (*i*) In the formation of some inner-orbital octahedral complexes of VBT (*e.g.* complexes of $d^7$ ion) the promotion of an electron from *d*-orbital to *s*-orbital is required, while in the formation of spin paired octahedral complexes of CFT no such promotion is required.

(*ii*) According to VBT, the metal-ligand bonding in complexes is only covalent, since VBT assumes that ligand electrons are partially donated to metal orbitals. On the other hand CFT does not consider covalent bonding in complexes, but considers the bonding to be entirely electrostatic, an interaction between a positively charged metal ion and the partial negative charge invariably present on the ligands. In its pure state, CFT does not provide for ligand electrons to enter the metal orbitals.

## B. CFT and MOT

(*i*) According to CFT, the metal *d*-electrons in an octahedral complex are distributed in $t_{2g}$ and $e_g$ orbitals but according to MOT, the metal d-electrons in an octahedral complex having σ-bonding only are distributed in non-bonding $t_{2g}$ and antibonding $e_g^*$ orbitals.

(*ii*) In CFT, $\Delta_o$ stands for the energy difference between $t_{2g}$ and $e_g$ levels whereas, in MOT, $\Delta_o$ is the difference between $t_{2g}$ non-bonding AOs and $e_g^*$ ABMOs

(*iii*) In CFT, $e_g$ orbitals are purely metal *d*-orbital ($d_{z^2}$ and $d_{x^2-y^2}$ orbitals) whereas in MOT $e_g^*$ orbitals are no longer pure *d*-orbitals. These orbitals have more metal orbital character because of their proximity to the metal orbitals.

(*iv*) According to CFT, $t_{2g}$ and $e_g$ orbitals have different energies simply because of the different effect on them of an electrostatic field. According to MOT, $t_{2g}$ and $e_g^*$ orbitals have different energies because $t_{2g}$ orbitals are non-bonding and $e_g^*$ orbitals are antibonding.

(*v*) The concept of HS and LS complexes, their different magnetic properties and values of $\Delta_0$ are retained in the same form in MOT as in CFT.

(*vi*) In CFT, $\Delta_0$ is due the different amount of repulsion felt by various metal *d* electrons from the ligands but in MOT, $\Delta_0$ arises because of metal-ligand covalent bonding which results in producing one set of non-bonding $t_{2g}$ orbitals and another set of higher energy $e_g^*$ ABMOs.

(*vii*) In CFT, the ligands surrounding the metal ion are considered as point charges. No overlap of metal orbitals and ligand orbitals is considered, i.e in CFT covalent bonding is entirely ignored. In MOT, the metal ion orbitals and ligand orbitals overlap with each other to from BMOs and ABMOs.

(*viii*) In CFT, $t_{2g}$ orbitals have lower energy than $e_g$ orbitals, but in MOT, BMOs have lower energy than ABMOs. Non-overlapping AOs of the metal ion remain non-bonding orbitals and their energy remains unchanged.

(*ix*) In CFT, double bonding (σ plus π) cannot be treated, because the ligands are treated as point charges. In MOT, π bonding between a metal ion and surrounding ligands cannot be studied.

(*x*) According to CFT, the formation of bond between metal cation and ligands arises due to the electrostatic attraction between metal cation and ligands whereas in MOT the formation of bond (i.e. MOs) arises due to overlap between the metal cation orbitals and ligand orbital combinations.

(*xi*) CFT emphases the role of metal ion *d*-orbitals only. The role of *s* and *p* orbitals of metal ion and σ and π type orbitals of ligands is ignored.

(*xii*) According to CFT the splitting of *d*-orbitals is caused by the electrostatic fields exerted by the ligands on the metal ion. MOT considers the formation of covalant bond as the cause of splitting.

(*xiii*) We know that when ligand π-orbitals are empty (π-acceptor ligands) and have higher energy than the metal $t_{2g}$ orbitals ($3d_{xy}$, $3d_{yz}$ and $3d_{xz}$), the value of the splitting energy ($\Delta'_0$) between $t_{2g}$ π-BMOs and $e_g^*$ σ-ABMOs is greater than the value of $\Delta_0$ between $t_{2g}$ metal orbitals and $e_g^*$ σ-ABMOs ($\Delta'_0 > \Delta_0$). Greater value of $\Delta'_0$ as compared to $\Delta_0$ is due to π-bonding in complexes. π-bonding causes pairing of electrons and hence LS complexes are formed.

According to CFT, greater value of $\Delta'_0$ is caused because of increased electrostatic field of ligands.

(*xiv*) Unlike CFT, MOT postulates $\sigma$ as well as $\pi$ bonding in complexes.

(*xv*) MO energy level diagram of an octahedral complex containing only $\sigma$-bonds show that since $t_{1u}^*$ and $a_{1g}^*$ ABMOs have higher energy than $e_g^*$ ABMOs, these ABMOs (i.e.$t_{1u}^*$ and $a_{1g}^*$ )represent terminal revels for the electronic transition originating from $t_{2g}$ non-bonding metal ion orbitals and electrons present in them may get excited to produce charge-transfer bands. None of these informations is given by CFT.

(*xvi*) CFT and MOT both predict the same number of unpaired electrons (n) or resultant spin, S (S = n/2) in LS and HS octahedral complexes. Thus magnetic properties of complexes are retained in the same manner in both the theories.

## C. VBT and MOT

We have seen that in VBT six atomic orbitals *viz*., $3d_{x^2-y^2}$, $3d_{z^2}$, 4s, $4p_x$, $4p_y$ $4p_z$ are used to form six $d^2sp^3$ hybrids which accept six electron pairs from the six ligands to form an inner-orbital octahedral complex while in MOT the same six atomic orbitals overlap with six $\sigma$-ligand type orbitals to form six $\sigma$-BMOs ($\sigma_s$, $\sigma_x = \sigma_y = \sigma_z$, $\sigma_{z^2} = \sigma_{x^2-y^2}$) and six $\sigma$*ABMOs ($\sigma_s^*$, $\sigma_x^* = \sigma_y^* = \sigma_z^*$, $\sigma_{z^2}^* = \sigma_{x^2-y^2}^*$). In VBT $3d_{xy}$, $3d_{yz}$ and $3d_{zx}$ atomic orbitals do not participate in $d^2sp^3$ hybridisation process and hence contain metallic electrons. Similarly in MOT these atomic orbitals do not participate in the formation of molecular orbitals and hence remain as non-bonding atomic orbitals.

Since, unlike MOT, VBT does not include antibonding molecular orbitals namely $\sigma_s^*$, $\sigma_x^* = \sigma_y^* = \sigma_z^*$, $\sigma_{z^2}^* = \sigma_{x^2-y^2}^*$, it does not provide an explanation of the spectral bands of the metal complexes.

## University Questions

1. Explain bonding in $[Co(NH_3)_6]^{3+}$ ion using MOT.

   *(Bharathiar 2010)*

2. Draw the M.O. diagram of $[CoF_6]^{-3}$ or $[Co(NH_3)_6]^{3+}$ and explain its magnetic properties.

   *(Gujrat Technological University 2010)*

3. With the help of MOT calculate the spin only magnetic moment of the ions given below.

   (i) $Co^{3+}$ octahedral complex with strong ligand field.

   (ii) $Co^{3+}$octahedral complex with weak ligand field.

   (iii) $Co^{2+}$ in tetrahedral field.

   (iv) $Co^{2+}$ in square planar field.

   *(BIB college, Orissa 2010)*

4. Write a short note on spectro-chemical series.

   *(Ch. Charan Singh University, Meerut 2012)*

**24**
**CHAPTER**

# Electronic Absorption Spectra of Transition Metal Complexes

## Introduction

The study in which the transition metal complex ions absorb electromagnetic radiation is called the electron absorption spectroscopy. The absorbed radiation promotes electrons from *d*-orbitals of lower energy to *d*-orbitals of higher energy. Due to the promotion of electrons there occurs a change in the overall arrangement of electrons in *d*-orbitals. The change in the overall arrangement of electrons in *d*-orbitals gives rise to electron absorption spectrum. The study of electron absorption spectrum of transition metal complex ions gives important information about the colours of complexes, the magnitude of energy gaps between ground state and excited state energy levels of metal ions in their complexes, the extent of covalent character in metal—ligand bonds, geometry of complexes, the position of metal ions and ligands in the spectrochemical and nephelauxetic series etc.

In order to have a clear understanding of electron absorption spectra of transition metal complexes we should have a knowledge of the following:

(*i*) **Atomic spectroscopy.** Atomic spectroscopy deals with the consequences of inter-electronic repulsions and magnetic interactions between electrons in the electronic configuration of an atom or an ion. It is due to these interactions that the coupling of spin and angular momenta of electrons within a given electronic configuration takes place. It is this coupling which gives rise to several degenerate and non-degenerate energy levels. These energy levels are known as **energy terms or spectroscopic states**. The electronic transition in these energy levels gives rise to electron absorption spectrum.

(*ii*) **Splitting of degenerate energy terms.** Electrostatic crystal field generated by ligands splits the degenerate energy terms of a given electronic configuration of a metal ion in its complexes.

(*iii*) **Selection rules.** These have been discussed on the subsequent pages in this chapter.

## Determination of Term Symbols for $p^2$ Configuration of C-Atom.

The E.C. of C-atom in its ground state is $1s^2$, $2s^2$ $2p^2$. Since $1s$ and $2s$ orbitals are completely-filled, the electrons present in these orbitals do not contribute anything in the determination of term symbols for C-atom. Thus we shall consider the contribution of the two electrons of $2p^2$ configuration only.

The two electrons of $p^2$ configuration of C-atom can be arranged in three *p*-orbitals in different ways as shown in Figures 24.1(*a*), 24.1(*b*) and 24.1(*c*).

| | |
|---|---|
| (*i*) | $m_l =$ +1    0    −1<br>↑   ↑   —<br>$m_L = \Sigma m_l = (+1) \times 1 + (0) \times 1 = +1$<br>$S = \Sigma m_s = \left(+\frac{1}{2}\right) + \left(+\frac{1}{2}\right) = 1, 2S + 1 = 3$ |
| (*ii*) | ↑   —   ↑<br>$m_L = \Sigma m_l = (+1) \times 1 + (-1) \times 1 = 0$<br>$S = +\frac{1}{2} + \frac{1}{2} = 1, 2S + 1 = 3$ |
| (*iii*) | —   ↑   ↑<br>$m_L = \Sigma m_l = 0 \times 1 + (-1) \times 1 = -1$<br>$S = +\frac{1}{2} + \frac{1}{2} = 1, 2S + 1 = 3$ |

**Fig 24.1** (*a*) **Three possible arrangements of two electrons of $p^2$ configuration in three *p*-orbitals with parallel spins.**

| | |
|---|---|
| (*i*) | $m_l =$ +1    0    −1<br>↑↓   —   —<br>$m_L = \Sigma m_l = (+1) \times 2 = +2$<br>$S = +\frac{1}{2} + \left(-\frac{1}{2}\right) = 0, 2S + 1 = 1$ |
| (*ii*) | —   ↑↓   —<br>$m_L = \Sigma m_l = (0) \times 2 = 0$<br>$S = 0, 2S + 1 = 1$ |
| (*iii*) | —   —   ↑↓<br>$m_L = \Sigma m_l = (-1) \times 2 = -2$<br>$S = 0, 2S + 1 = 1$ |

**Fig. 24. 1** (*b*) **Three possible arrangements of two electrons of $p^2$ configuration in three *p*-orbitals with parallel spins.**

It may be seen from Fig. 24(*a*) that the three values of $m_L$ for the three arrangements (*i*), (*ii*) and (*iii*) are +1, 0 and −1 respectively. Obviously for these values of $m_L$, L = 1 which stands for P term. the value of (2S + 1) (spin multiplicity) for each arrangement is equal to 3. Hence,

Term = $^{2S+1}L = {}^{2S+1}P = {}^3P(L = 1)$

| | |
|---|---|
| *(i)* | $m_l = +1 \quad 0 \quad -1$<br>↑ ↓ —<br>$m_L = \Sigma m_l = (+1) \times 1 + (0) \times 1 = +1$<br>$S = +\frac{1}{2} + \left(-\frac{1}{2}\right) = 0, 2S + 1 = 1$ |
| *(ii)* | ↑ — ↓<br>$m_L = \Sigma m_l = +1 \times 1 + (-1) \times 1 = 0$<br>$S = 0, 2S + 1 = 1$ |
| *(iii)* | — ↑ ↓<br>$m_L = \Sigma m_l = 0 \times 1 + (-1) \times 1 = -1$<br>$S = 0, 2S + 1 = 1.$ |

**Fig 24.1** (*c*) **Three possible arrangements of two electrons of $p^2$ configuration in three *p*-orbitals with opposite spins**

It may be seen from Fig. 24(*b*) that the three values of $m_L$ are +2, 0 and –2 while the three values of $m_L$ shown in Fig. 24(c) are +1, 0, and –1. The value of 2S + 1 (spin multiplicity) for all the six arrangements given in Figures 24(*b*) and 24(*c*) is equal to 1.

Six values of $m_L$ (+2, 0, –2, +1, 0, –1) shown in the above said figures can be arranged systematically into the following two groups.

**Group (*a*).** This group contains +2, +1, 0, –1 and –2 values (five values) of $m_L$. For these values L = 2 which stands for D term. Hence,

Term $^{2S+1}L = {}^1D(L = 2)$

**Group (*b*).** This group has only one value of $m_L$ which is equal to zero ($m_L = 0$). For zero value of $m_L$, L = 0 which stands for S term. Hence,

Term = $^{2S+1}L = {}^1S(L = 0)$

Above discussion shows that for $p^2$ configuration of C-atom terms are:

**$^3P$, $^1D$ and $^1S$**

Out of the three terms given above $^3P$ term is **triplet spin term** and $^1D$ and $^1S$ are **singlet spin terms,** since spin multiplicity for these terms is 3 (for $^3P$) and 1 (for $^1D$ and $^1S$).

## To Arrange $^1S$, $^1D$ and $^3P$ Terms of $p^2$ Configuration of *C*-atom in the Increasing Order of their Energy with the Help of Hund's Rules.

We have already shown that $^1S$, $^1D$ and $^3P$ are the terms derived from $p^2$ configuration of C-atom. These terms have different energies. These terms can be arranged in the increasing order of their energies with the help of **Hund's rules**. These rules are:

(*i*) The term having the highest value of spin multiplicity, $(2S + 1)$ is the most stable and hence has the lowest energy. Now since $^3P$ term has the highest value of $(2S + 1)$ which is equal to 3 is the most stable and hence has the lowest energy.

(*ii*) The stability of terms having the same value of (2S + 1) depends on their L values. The term having higher value of L is more stable and hence has lower energy. Thus since $^1$D and $^1$S terms have the same value of (2S + 1), $^1$D (L = 2) which has higher value of L is more stable than $^1$S (L = 0) term, *i.e.*, $^1$D term has lower energy than $^1$S term ($^1$D < $^1$S). Consequently the increasing order of energies of $^3$P, $^1$D end $^1$S terms is

$$^3P < {}^1D < {}^1S$$

Since $^3$P term has the lowest energy, this term is the ground state term of $p^2$ configuration.

**Ground state term = $^3$P**

$^1$D and $^1$S terms are higher energy terms of $p^2$ configuration.

**Higher energy terms = $^1$D and $^1$S.**

Ground state (lowest energy) and higher energy states of $p^n$ configuration ($n$ = 1 to 6) are given below:

| *$p^n$ configuration (n = 1 to 6)* | *Ground state term* | *Higher energy states* |
|---|---|---|
| $p^1, p^5$ | $^2P$ | – |
| $p^2, p^4$ | $^3P$ | $^1D$, $^1S$ |
| $p^3$ | $^4S$ | $^2D$, $^2P$ |
| $p^6$ | $^1S$ | – |

## To Determine J Energy Levels (J states) for the Term Symbols for $p^2$ Configuration

J energy levels of a given configuration are those term symbols which also have J values. These are represented as $^{2S+1}L_J$. In order to find out J energy levels of $p^2$ configuration, we should find out the values of $J$ for $^3P$, $^1D$ and $^1S$ which are the term symbols of $p^2$ configuration. The value of $J$ for a term ranges from $(L + S)$ to $(L - S)$ or $(S - L)$ whichever is positive. Thus $J$ is given by:

$$J = (L + S), (L + S - 1), \ldots (L - S) \text{ or } (S - L)$$

J values for $^3$P, $^1$D and $^1$S terms can be calculated as follows:

(*i*) **$^3P$ term.** For this term L = 1 and 2S + 1 = 3 or S = 1. Hence J values for $^3P$ term are given by:

$$J = (L + S), (L + S - 1), \ldots (L - S)$$

$$= (1 + 1), (1 + 1 - 1), \ldots, (1 - 1)$$
$$= 2, 1, 0$$

Thus J energy levels for $^3P$ term are:

$$\boxed{^3P_2,\ ^3P_1 \text{ and } ^3P_0}$$

(*ii*) **$^1$D term.** For this term L = 2 and 2S + 1 = 1 or S = 0. Hence J value for this term is given by:

$$L = (L + S), (L + S - 1), \ldots (L - S)$$
$$= (2 + 0), (2 + 0 - 1), \ldots (2 - 0)$$
$$= 2, 1, 2$$
$$= 2$$

Thus J energy level for $^1$D term is:

$$\boxed{^1D_2}$$

(*iii*) **$^1$S term.** For this term L = 0 and 2S + 1 = 1 or S = 0. Hence J value for this term in given by:

$$L = (L + S), (L + S - 1), \ldots (L - S) = 0$$

Thus J energy level for $^1$S term is

$$\boxed{^1S_0}$$

All the J energy levels can be written as:

$$\boxed{^3P_2,\ ^3P_1,\ ^3P_0,\ ^1D_2 \text{ and } ^1S_0}$$

The J energy levels arise due to L – S coupling between the two electrons of $p^2$ configuration of C-atom.

J energy levels are also called **Russel-Saunders** or **complete (full) spectroscopic terms** of a given configuration.

## To Arrange the J Energy Levels of p² Configuration in the Increasing Order of their Energy.

We have shown that $^3P_2$, $^3P_1$, $^3P_0$, $^1D_2$ and $^1S_0$ are J energy levels derived from $p^2$ configuration of C-atom. We have also shown that the energy of $^3$P, $^1$D and $^1$S which are the term symbols of $p^2$ configuration increases as:

$$\boxed{^3P < {}^1D < {}^1S} \quad \ldots (i)$$

Let us compare the energy of $^3P_0$, $^3P_1$ and $^3P_2$ which are the J energy levels of $^3$P term. Since these levels have the same value of spin multiplicity (= 3), their energy depends on their J values. With the increase of J values energy also increases (Hund's rule). Thus the energy of $^3P_0$, $^3P_1$ and $^3P_2$ levels is as

$$\boxed{{}^3P_0 < {}^3P_1 < {}^3P_2} \qquad \text{... }(ii)$$

On combining energy relations (*i*) and (*ii*) given above we find that the overall energy of J energy levels increases as:

$$\boxed{{}^3P_0 < {}^3P_1 < {}^3P_2 < {}^1D_2 < {}^1S_0}$$

Since ${}^3P_0$ has the lowest energy, it is Russel-Sounders term for $p^2$ configuration. Thus:

$$\boxed{\textbf{Russel-Saunders term for } p^2 \textbf{ configuration} = {}^3P_0}$$

The ground state term (${}^3P$) and Russel-Saunders term (${}^3P_0$) for $p^2$ configuration can also be determined by the following method.

$$m_l = \quad +1 \qquad 0 \qquad -1$$

$$p^2 \quad \underline{\uparrow} \qquad \underline{\uparrow} \qquad \underline{\quad}$$

$$m_s = +1/2 \qquad +1/2$$

$\therefore$ $L = \Sigma m_l = (+1) \times 1 + (0) + 1 = 1$

$L = 1$ means P state.

$$S = \Sigma m_s = 1 \times \left(+\frac{1}{2}\right) + 1 \times \left(+\frac{1}{2}\right) = 1$$

$\therefore$ Spin multiplicity, $(2S + 1) = 2 \times 1 + 1 = 3$

Thus:

$$\boxed{\textbf{Ground state term} = {}^{2S+1}L = {}^3P}$$

$$J = (L + S), (L + S - 1), \ldots, (L - S)$$

$$= (1 + 1), (1 + 1 - 1), \ldots (1 - 1)$$

$$= 2, 1, 0$$

Now since $p^2$ configuration is less than half-filled, lower value of J (= 0) will be used. Hence:

$$\boxed{\textbf{Russel-Saunders term for } p^2 \textbf{ configuration} = {}^3P_0}$$

## To Calculate Energy of *J* States of ${}^3P$ Term (Also see Fig. 24.2)

We know that ${}^3P_0$, ${}^3P_1$ and ${}^3P_2$ are the *J* states of ${}^3P$ term. The energy of these *J* states is in the order : ${}^3P_0 < {}^3P_1 < {}^3P_2$, since *J* values of these states increase as $0 < 1 < 2$. According to Lande interval rule, since $p^2$ configuration (${}^3P$ term is one of the terms arising from $p^2$ configuration) is less than half-filled ($n < 5$), the energy separation between two adjacent J states of this configuration ($p^2$ configuration) is equal to the product of +λ and higher J value. Thus if $E_0$, $E_1$ and $E_2$ are the energies of $^3P_0$, $^3P_1$ and $^3P_2$ states, then we have:

$$E_1 - E_0 = +\lambda \times 1 = +\lambda$$

or $$E_0 = E_1 - \lambda \quad \text{... (i)}$$

$$E_2 - E_1 = \lambda \times 2 = +2\lambda$$

or $$E_2 = E_1 + 2\lambda \quad \text{... (ii)}$$

and $$(2J + 1)E_0 + (2J + 1)E_1 + (2J + 1)E_2 = 0$$

or $$(2 \times 0 + 1)E_0 + (2 \times 1 + 1)E_1 + (2 \times 2 + 1)E_2 = 0$$

or $$E_0 + 3E_1 + 5E_2 = 0 \quad \text{... (iii)}$$

Put the values of $E_0$ and $E_2$ from equations (*i*) and (*ii*) in equation (*iii*) to get:

$$(E_1 - \lambda) + 3E_1 + 5(E_1 + 2\lambda) = 0$$

This equation gives $E_1 = -\lambda$. Obviously $E_0 = E_1 - \lambda = -\lambda - \lambda = -2\lambda$ and $E_2 = E_1 + 2\lambda = -\lambda + 2\lambda = +\lambda$. Thus:

$$\boxed{\mathbf{E_0 = -2\lambda,\ E_1 = -\lambda,\ E_2 = +\lambda}} \quad \text{(See Fig. 24.2)}$$

Values of $E_0$, $E_1$ and $E_2$ given above show that:

Energy separation between $^3P_1$ and $^3P_0$ terms = $E_1 - E_0$
$= -\lambda - (-2\lambda) = +\lambda$

Energy separation between $^3P_2$ and $^3P_1$ terms = $E_2 - E_1$
$= +\lambda - (-\lambda) = +2\lambda$

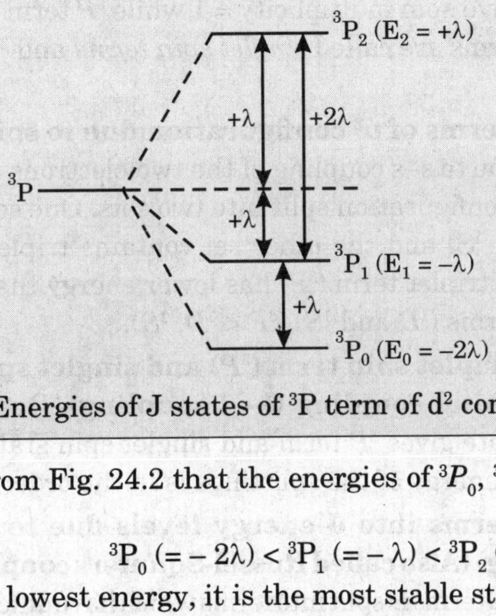


**Fig. 24.2** Energies of J states of $^3P$ term of $d^2$ configuration

It may be seen from Fig. 24.2 that the energies of $^3P_0$, $^3P_1$, and $^3P_2$ *J* states is in the order:

$$^3P_0\,(= -2\lambda) < {}^3P_1\,(= -\lambda) < {}^3P_2\,(= +\lambda)$$

Since $^3P_0$ has the lowest energy, it is the most stable state.

In the above discussion λ is called **spin-orbit coupling constant.**

## To Determine the Number of Microstates ($M_J$ levels) in Term Symbols of $p^2$ Configuration

We know that $^3P$, $^1D$ and $^1S$ are the term symbols of $p^2$ configuration of C-atom. Number of microstates ($M_J$ levels) derived from a term is equal to the product of *spin energy levels,* (2S + 1) and *orbital energy levels*, (2L +1) of the term. Thus:

No. of microstates or $M_J$ levels of a term

$$= (2S + 1)(2L + 1)$$

These microstates have different energy and are obtained due to magnetic field. Thus:

No. of microstates in $^3P$ term ($2S + 1 = 3, L = 1$)

$$= (2S + 1)(2L + 1) = 3 \times (2 \times 1 + 1) = \mathbf{9}$$

No. of microstates in $^1D$ term ($2S + 1 = 1, L = 2$)

$$= (2S + 1)(2L + 1) = 1 \times (2 \times 2 + 1) = \mathbf{5}$$

No. of microstates in $^1S$ term ($2S + 1 = 1, L = 0$)

$$= (2S + 1)(2L + 1) = 1 \times (2 \times 0 + 1) = \mathbf{1}$$

Above discussion shows that :

**Total number of microstates in $p^2$ configuration = 9 + 5 + 1 = 15**

## Splitting of Terms of $p^2$ Configuration Due to Spin-Orbit Coupling (L – S or Russel Saunders Coupling)

We know that the energy terms arising out of $p^2$ configuration due to inter-electronic repulsion between the two electrons of $p^2$ configuration are $^3P$, $^1D$ and $^1S$. $^1D$ and $^1S$ terms have spin multiplicity = 1 while $^3P$ term has spin multiplicity = 3. Hence $^1D$ and $^1S$ terms are called *singlet spin terms* and $^3P$ term is called *triplet spin term.*

*(i)* **Splitting of terms of $p^2$ configuration due to spin-spin coupling (s–s coupling).** Due to s–s coupling of the two electrons of $p^2$ configuration, the terms of this configuration split into two sets. One set contains singlet spin terms ($^1D$ and $^1S$) and the other set contains triplet spin term ($^3P$). The set containing triplet term ($^3P$) has lower energy than the set consisting of singlet spin terms ($^1D$ and $^1S$) ($^3P < {}^1D, {}^1S$).

*(ii)* **Splitting of triplet spin term ($^3P$) and singlet spin terms ($^1D$, $^1S$) due to obrital-orbital coupling ($l - l$ coupling).** Due to ($l - l$) coupling the triplet spin state gives $^3P$ term and singlet spin state splits to give $^1D$ and $^1S$ terms. The energy of the splittings is in the order: $^3P < {}^1D < {}^1S$.

*(iii)* **Spliting of terms into J energy levels due to spin-orbit coupling (L – S coupling) (Also called Russel-Squnters coupling).** Due to (L – S) coupling each of the terms is split into terms which have different values of J but the same value of spin multiplicity. The terms having different J values are called

**J energy levels** or **Russel-Saunders terms**. These are represented as $^{2S+1}L_J$. The value of J for a term ranges from (L + S) to (L – S) or (S – L) whichever is positive. Thus the value of J is given by:

$$J = (L + S), (L + S - 1), \ldots (L - S) \text{ or } (S - L)$$

The value of J for $^3P$, $^1D$ and $^1S$ terms has already been calculated. The $J$ energy states derived from these terms are $^3P_0$, $^3P_1$, $^3P_2$, $^1D_2$ and $^1S_0$. Three values of $J$ (0, 1, 2) for $^3P$ term mean that under L – S coupling, $^3P$ term is split up into three energy levels which are $^3P_0$, $^3P_1$ and $^3P_2$. The energy of these states is in the order $^3P_0 < {}^3P_1 < {}^3P_2$. Thus $^3P_0$ level has the lowest energy. It is the most stable state and hence is the ground state term of $p^2$ configuration.

One value of J (= 2) for $^1D$ term shows that under $L - S$ coupling $^1D$ term is not split up, *i.e.* the energy of this term remains the same (See Fig. 24.3). Thus there is no possibility of $L - S$ coupling in $^1D$ term. On the same lines of argument it can be said that there is no possibility of L – S coupling in $^1S$ term. Thus the energy of J states is in the order : $^3P_0 < {}^3P_1 < {}^3P_2 < {}^1D_2 < {}^1S_0$.

(*iv*) **Splitting of J energy levels into $M_J$ levels (microstates) under magnetic field.** When magnetic field is applied, J energy levels *viz*, $^3P_0$, $^3P_1$, $^3P_2$, $^1D_2$ and $^1S_0$ are split into different number of $M_J$ levels. The number of $M_J$ levels from the splitting of a J state is equal to (2J + 1). Thus number of $M_J$ levels is as: $^3P_0 = (2J + 1) = 2 \times 0 + 1 = 1$, $^3P_1 = 2 \times 1 + 1 = 3$, $^3P_2 = 2 \times 2 + 1 = 5$, $^1D_2 = 2 \times 2 + 1 = 5$ and $^1S_0 = 2 \times 0 + 1 = 1$. These numbers show that the number of $M_J$ levels given by $^3P$, $^1D$ and $^1S$ terms is as $^3P = 1 + 3 + 5 = 9$, $^1D = 5$ and $^1S = 1$. Consequently total number of microstates obtained in case of the splitting of $p^2$ configuration = 9 + 5 + 1 = 15 (See Fig. 24.3 given on the next page)

## Determination of Term Symbols for $d^2$ configuration of $V^{3+}$ ion

Electronic configuration (E.C.) of V atom (Z = 23) is $1s^2, 2s^2 2p^6, 3s^2 3p^6 3d^3, 4s^2$. Hence E.C. of $V^{3+}$ ion is $1s^2, 2s^2 2p^6, 3s^2 3p^6 3d^2$. Since 1*s*, 2*s*, 2*p*, 3*s* and 3*p* orbitals are completely-filled, the electrons present in these orbitals do not contribute anything in the determination of term symbols for $3d^2$ configuration of $V^{3+}$ ion. Thus we shall consider the contribution of the two electrons of $3d^2$ configuration only.

The two electrons of $d^2$ configuration of $V^{3+}$ ion can be arranged in five *d*-orbitals in different ways as shown in Figs. 24.4(*a*), (*b*) and (*c*).

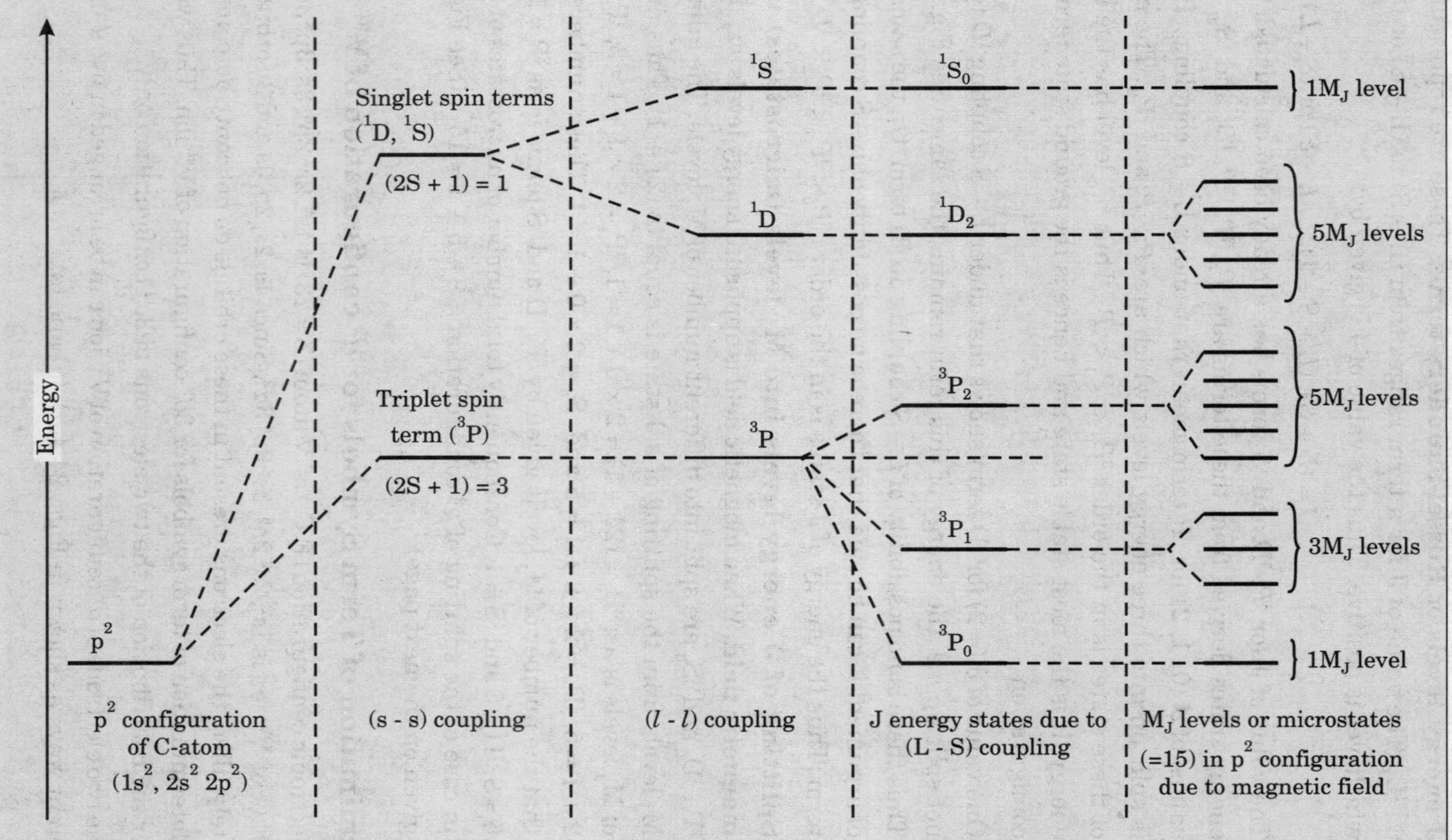


**Fig. 24.3** Splitting of singlet terms ($^1D$ and $^1S$) and triplet term ($^3P$) of $p^2$ configuration in C-atom due to (L – S) coupling

| | $m_l = 2$ | $+1$ | zero | $-1$ | $-2$ | $m_L$ | $S$ |
|---|---|---|---|---|---|---|---|
| (*i*) | ↑ | ↑ | | | | $m_L = \Sigma m_l = +2 \times 1 + (+1) \times 1 = +3$ | $S = +\frac{1}{2} \times 2 = 1, 2S + 1 = 2 \times 1 + 1 = 3$ |
| (*ii*) | ↑ | | ↑ | | | $m_L = \Sigma m_l = +2 \times 1 + 0 \times 1 = +2$ | $S = 1, 2S + 1 = 3$ |
| (*iii*) | ↑ | | | ↑ | | $m_L = \Sigma m_l = +2 \times 1 + (-1) \times 1 = +1$ | $S = 1, 2S + 1 = 3$ |
| (*iv*) | ↑ | | | | ↑ | $m_L = \Sigma m_l = +2 \times 1 + (-2) \times 1 = 0$ | $S = 1, 2S + 1 = 3$ |
| (*v*) | | ↑ | ↑ | | | $m_L = \Sigma m_l = +1 \times 1 + 0 \times 1 = +1$ | $S = 1, 2S + 1 = 3$ |
| (*vi*) | | ↑ | | ↑ | | $m_L = \Sigma m_l = +1 \times 1 + (-1) \times 1 = 0$ | $S = 1, 2S + 1 = 3$ |
| (*vii*) | | T | | | T | $m_L = \Sigma ml = +1 \times 1 + (-2) \times 1 = -1$ | $S = 1, 2S + 1 = 3$ |
| (*viii*) | | | ↑ | ↑ | | $m_L = \Sigma m_l = 0 \times 1 + (-1) \times 1 = -1$ | $S = 1, 2S + 1 = 3$ |
| (*ix*) | | | ↑ | | ↑ | $m_L = \Sigma m_l = 0 \times 1 + (-2) \times 1 = -2$ | $S = 1, 2S + 1 = 3$ |
| (*x*) | | | | ↑ | ↑ | $m_L = \Sigma m_l = (-1) \times 1 + (-2) \times 1 = -3$ | $S = 1, 2S + 1 = 3$ |

**Fig. 24.4** (*a*) Ten possible arrangements of two electrons of $d^2$ configuration of $V^{3+}$ ion in five *d*-orbitals with parallel spins.

Ten arrangements of the electrons shown in Fig. 24.4(*a*) have the two electrons with parallel spins. For each arrangement the values of $m_L$ (= $\Sigma m_l$) and S have also been calculated. Since the two electrons have parallel spins, $S = \left(+\frac{1}{2}\right)\times 1+\left(+\frac{1}{2}\right)\times 1 = 1$ in each arrangement.

Ten values of $m_L$ obtained from Fig. 24.4(*a*) are +3, + 2, +1, 0, +1, 0, –1, –1, –2 and – 3. These values can be arranged systematically into the following two groups:

**Group (a).** This group contains $m_L$ = +3, +2, +1, 0, –1, –2, –3 (seven values). Obviously for these values L = 3 which stands for F term with S = 1 or 2S + 1 = 2 × 1 + 1 = 3. Hence term is $^3$**F**.

**Group (b).** This group contains $m_L$ = +1, 0, –1 (three values). For this group L = 1 which stands for P term with S = 1 or 2S + 1 = 2 × 1 + 1 = 3. Hence term is $^3$**P.**

Five arrangements of two electrons shown in Fig. 24.4(*b*) have two electrons with paired spins (↑↓). For each arrangement the values of $m_L$ and S have been calculated. Since the two electrons are paired, $S = \left(+\frac{1}{2}\right)\times 1+\left(-\frac{1}{2}\right)\times 1 = 0$ for each arrangement. Five values of $m_L$ (= $\Sigma m_l$) obtained from Fig. 24.4(*b*) are +4, +2, 0, –2 and –4.

| | $m_l$ = +2 | +1 | 0 | –1 | –2 |
|---|---|---|---|---|---|
| (*i*) | ↑↓ | — | — | — | — |
| | $m_L = \Sigma m_l = (+2) \times 2 = +4$ <br> $S = \left(+\frac{1}{2}\right)\times 1+\left(-\frac{1}{2}\right)\times 1 = 0,\ (2S + 1) = 2 \times 0 + 1 = 1$ | | | | |
| (*ii*) | — | ↑↓ | — | — | — |
| | $m_L = \Sigma m_l = (+1) \times 2 = +2,\ S = 0,\ (2S + 1) = 1$ | | | | |
| (*iii*) | — | — | ↑↓ | — | — |
| | $m_L = \Sigma m_l = 0 \times 2 = 0,\ S = 0,\ (2S + 1) = 1$ | | | | |
| (*iv*) | — | — | — | ↑↓ | — |
| | $m_L = \Sigma m_l = (-1) \times 2 = -2$ <br> $S = 0,\ (2S + 1) = 1$ | | | | |
| (*v*) | — | — | — | — | ↑↓ |
| | $m_L = \Sigma m_l = (-2) \times 2 = -4$ <br> $S = 0,\ (2S + 1) = 1$ | | | | |

**Fig 24.4** (*b*) Five possible arrangements of two electrons of $d^2$ configuration of $V^{3+}$ ion in five *d*-orbitals with paired spins.

Ten arrangements of electrons shown in Fig. 24.4(*c*) have two electrons with opposite spins. For each arrangement the values of $m_L$ and S have been calculated. Since the two electrons have opposite spins, S = 0 for each arrangement. Ten values of $m_L$ obtained from Fig. 24.4(*c*) are +3, +2, +1, 0, +1, 0, –1, –1, –2 and –3.

| | $m_l = +2$ | $+1$ | $0$ | $-1$ | $-2$ | |
|---|---|---|---|---|---|---|
| (*i*) | ↑ | ↓ | — | — | — | $m_L = \Sigma m_l = (+2) \times 1 + (+1) \times 1 = +3$ <br> $S = \left(+\frac{1}{2}\right) \times 1 + \left(-\frac{1}{2}\right) \times 1 = 0, (2S + 1) = 2 \times 0 + 1 = 1$ |
| (*ii*) | ↑ | — | ↓ | — | — | $m_L = \Sigma m_l = (+2) \times 1 + 0 \times 1 = +2$ <br> $S = 0, (2S + 1) = 1$ |
| (*iii*) | ↑ | — | — | ↓ | — | $m_L = \Sigma m_l = (+2) \times 1 + (-1) \times 1 = +1$ <br> $S = 0, (2S + 1) = 1$ |
| (*iv*) | ↑ | — | — | — | ↓ | $m_L = \Sigma m_l = (+2) \times 1 + (-2) \times 1 = 0$ <br> $S = 0, (2S + 1) = 1$ |
| (*v*) | — | ↑ | ↓ | — | — | $m_L = \Sigma m_l = (+1) \times 1 \times 0 + 1 = +1$ <br> $S = 0, (2S + 1) = 1$ |
| (*vi*) | — | ↑ | — | ↓ | — | $m_L = \Sigma m_l = (+1) \times 1 + (-1) \times 1 = 0$ <br> $S = 0, (2S + 1) = 1$ |
| (*vii*) | — | ↑ | — | — | ↓ | $m_L = \Sigma m_l = (+1) \times 1 + (-2) \times 1 = -1$ <br> $S = 0, (2S + 1) = 1$ |
| (*viii*) | — | — | ↑ | ↓ | — | $m_L = \Sigma m_l = 0 \times 1 + (-1) \times 1 = -1$ <br> $S = 0, (2S + 1) = 1$ |
| (*ix*) | — | — | ↑ | — | ↓ | $m_L = \Sigma m_l = 0 \times 1 + (-2) \times 1 = -2$ <br> $S = 0, (2S + 1) = 1$ |
| (*x*) | — | — | — | ↑ | ↓ | $m_L = \Sigma m_l = (-1) \times 1 + (-2) \times 1 = -3$ <br> $S = 0, (2S + 1) = 1$ |

**Fig. 24.4** (*c*) Ten possible arrangements of two electrons of $d^2$ configuration of $V^{3+}$ ion in five *d*-orbitals with opposite spins.

All the 15 values of $m_L$ obtained from Figs. (*b*) and (*c*) (+4, +2, 0, –2, –4, +3, +2, +1, 0, +1, 0, –1, –1, –2, –3) can be arranged systematically into the following groups.

**Group (a).** This group contains +4, + 3, +2, +1, 0, –1, –2, –3 and –4 values of $m_L$ (nine values). Obviously for these values L = 4. which indicates G term with S = 0 or 2S + 1 = 2 × 0 + 1 = 1. Hence, term is $^1$**G**.

**Group (b).** This group has + 2, +1, 0, –1 and –2 values of $m_L$ (five values). Obviously these values have L = 2. L = 2 indicates D term with S = 0 or 2S + 1 = 1. Hence term is $^1$**D**.

**Group (c).** This group has $m_L$ = 0 (one value). This value of $m_L$ indicates that L = 0 which means S term with S = 0 or 2S + 1 = 1. Hence term is $^1$**S**.

Above discussion shows that term symbols for $d^2$ configuration of $V^{3+}$ ion are:

$$\boxed{^3\mathbf{F},\ ^3\mathbf{P},\ ^1\mathbf{G},\ ^1\mathbf{D} \text{ and } ^1\mathbf{S}}$$

Out of the five terms given above $^1$G, $^1$D and $^1$S terms are **singlet spin terms** since their spin multiplicity is 1. On the other hand $^3$F and $^3$P terms are **triplet spin terms** because their spin multiplicity is 3.

## To Arrange the Term Symbols of d² Configuration of V³⁺ Ion in the Increasing Order of their Energy with the Help of Hund's Rules.

We have already shown that the term symbols arising from $d^2$ configuration of $V^{3+}$ ion are $^3$F, $^3$P, $^1$G, $^1$D and $^1$S. These terms have different energies. These terms can be arranged in the increasing order of their energy with the help of Hund's rules:

(*i*) Since $^3$F and $^3$P terms have higher value of spin multiplicity, (2S + 1) than $^1$G, $^1$D and $^1$S terms, ($^3$F, $^3$P) set of terms has lower energy than ($^1$G, $^1$D and $^1$S) set, *i.e.*

$$\boxed{(^3\mathbf{F},\ ^3\mathbf{P}) < (^1\mathbf{G},\ ^1\mathbf{D},\ ^1\mathbf{S})} \quad \text{... }(i)$$

(*ii*) Since $^3$F and $^3$P terms have the same spin multiplicity (= 3) and $^3$F term has higher value of L (=3) than $^3$P term (L = 1), $^3$F term has lower energy than $^3$P term, *i.e.*,

$$\boxed{^3\mathbf{F} < \ ^3\mathbf{P}} \quad \text{... }(ii)$$

(*iii*) $^1$G (L = 4), $^1$D (L = 2) and $^1$S (L = 0) terms have the same spin multiplicity (= 1) and their L values decrease as $^1$G (L = 4) > $^1$D (L = 2) > $^1$S (L = 0), the energies of these terms increase as :

$$\boxed{^1\mathbf{G} < \ ^1\mathbf{D} < \ ^1\mathbf{S}} \quad \text{... }(iii)$$

On combining the energy relations given above at (*i*), (*ii*) and (*iii*), we get the overall order of energy of the five terms as:

$$\boxed{^3\mathbf{F} < \ ^3\mathbf{P} < \ ^1\mathbf{G} < \ ^1\mathbf{D} < \ ^1\mathbf{S}}$$

Above order of energy shows that $^3F$ term has the lowest energy. Hence this term is the ground state term for $d^2$ configuration of $V^{3+}$ ion, *i.e.*,

**Ground state term = $^3F$**

Since $^3P$ term is placed just after the ground state term ($^3F$) in the energy order, $^3P$ term is the first excited state term for $d^2$ configuration, *i.e.*

**First excited state term = $^3P$**

Other terms *viz* $^1G$, $^1D$ and $^1S$ have higher energy and hence are other excited state terms, *i.e.*,

**Other excited state terms = $^1G$, $^1D$ and $^1S$**

It may be noted that the spin multiplicity of ground state term ($^3F$) and first excited state term ($^3P$) is the same (= 3).

## To Determine J Energy Levels (J states) for the Term Symbols for $d^2$ Configuration of $V^{3+}$ Ion.

J energy levels of a given configuration are those term symbols which also have J values. These are represented as $^{2S+1}L_J$. In order to find out J energy levels of $d^2$ configuration, we should find out the values of J for $^3F$, $^3P$, $^1G$, $^1D$ and $^1S$ terms which are the term symbols of $d^2$ configuration. The value of J for a term ranges from (L + S) to (L – S) or (S – L) whichever is positive. Thus J is given by:

$$J = (L + S), (L + S - 1), \ldots (L - S) \text{ or } (S - L)$$

J values for $^3F$, $^3P$, $^1G$, $^1D$ and $^1S$ terms can be calculated as follows:

(*i*) **$^3F$ term.** For this term L = 3 and 2S + 1 = 3 or S = 1. Hence J values for $^3F$ term are given by:

$$\begin{aligned} J &= (L + S), (L + S - 1), \ldots, (L - S) \\ &= (3 + 1), (3 + 1 - 1), \ldots (3 - 1) \\ &= 4, 3, 2 \end{aligned}$$

Thus J energy levels for $^3F$ term are :

**$^3F_4$, $^3F_3$ and $^3F_2$**

(*ii*) **$^3P$ term.** For this term $L = 1$ and $2S + 1 = 3$ or $S = 1$. Hence $J$ values for $^3P$ term are given by:

$$\begin{aligned} J &= (L + S), (L + S - 1), \ldots, (L - S) \\ &= (1 + 1), (1 + 1 - 1), \ldots, (1 - 1) \\ &= 2, 1, 0 \end{aligned}$$

Thus J energy levels for $^3P$ terms are:

**$^3P_2$, $^3P_1$ and $^3P_0$**

(*iii*) **$^1G$ term.** For this term L = 4 and 2 S + 1 = 1 or S = 0. Hence J values for $^1G$ term are given by:

$$J = (L + S), (L + S - 1), \ldots (L - S)$$
$$= (4 + 0), (4 + 0 - 1), \ldots (4 - 0)$$
$$= 4, 3, 4$$
$$= 4$$

Thus J energy level for $^1G$ term is:

$$\boxed{^1G_4}$$

(*iv*) **$^1D$ term.** For this term L = 2 and 2S + 1 = 1 or S = 0. Hence J values for this term are given by:

$$L = (L + S), (L + S - 1), \ldots (L - S)$$
$$= (2 + 0), (2 + 0 - 1), \ldots (2 - 0)$$
$$= 2, 1, 2$$
$$= 2$$

Thus J energy level for $^1D$ term is:

$$\boxed{^1D_2}$$

(*v*) **$^1S$ term.** For this term L = 0 and 2S + 1 = 1 or S = 0. Hence J values for this term are given by:

$$L = (L + S), (L + S - 1), \ldots (L - S)$$
$$= (0 + 0), (0 + 0 - 1), \ldots (0 - 0)$$
$$= 0, -1, 0$$
$$= 0$$

Thus J energy level for $^1S$ term is:

$$\boxed{^1S_0}$$

All the J energy levels can be written as

$$\boxed{^3F_4, {}^3F_3, {}^3F_2, {}^3P_2, {}^3P_1, {}^3P_0, {}^1G_4, {}^1D_2, {}^1S_0}$$

Here it should be noted that J energy levels arise due to L-S coupling between the two electrons of $d^2$ configuration of $V^{3+}$ ion.

J energy levels are also called **Russel-Saunders symbols or complete (full) term symbols** of a given configuration. Since $^3F_2$ has the lowest energy,

$$\boxed{\textbf{Russel-Saunders symbol of } d^2 \textbf{ configuration} = {}^3F_2}$$

## To Arrange the J energy Levels of $d^2$ Configuration in the Increasing Order of their Energy.

We have already shown that $^3F_4$, $^3F_3$, $^3F_2$, $^3P_2$, $^3P_1$, $^3P_0$, $^1G_4$, $^1D_2$ and $^1S_0$ are J energy levels derived from $d^2$ configuration of $V^{3+}$ ion. We have also shown that the energy of $^3F$, $^3P$, $^1G$, $^1D$ and $^1S$ which are the term symbols of $d^2$ configuration increases as:

$$\boxed{^3F < {}^3P < {}^1G < {}^1D < {}^1S}$$

Above order shows that the energy of J energy levels will increase as:

$$\boxed{(^3F_3, {}^3F_2, {}^3F_1) < (^3P_2, {}^3P_1, {}^3P_0) < {}^1G_4 < {}^1D_2 < {}^1S_0}$$

Now let us compare the energy of $^3F_3$, $3F_2$ and $3F_1$ energy levels. Since these levels have the same multiplicity (= 3) and same value of L (= 3), but different J values, their energy increases with the increase of their J values. Thus:

$$\boxed{^3F_2 < {}^3F_3 < {}^3F_4}$$

Similarly, $$\boxed{^3P_0 < {}^3P_1 < {}^3P_2}$$

Thus the overall order of energy of J energy levels can be written as:

$$\boxed{^3F_2 < {}^3F_3 < {}^3F_4 < {}^3P_0 < {}^3P_1 < {}^3P_2 < {}^1G_4 < {}^1D_2 < {}^1S_0}$$

Since $^3F_2$ term has the lowest energy, this level is the ground state J energy level of $d^2$ configuration.

$$\boxed{\textbf{Ground state J energy level} = {}^3F_2}$$

## To Calculate Energy of J States of $^3$F Term of d$^2$ Configuration

We know that $^3F_2$, $^3F_3$ and $^3F_4$ are the J states of $^3$F term. The energy of these J states is in the order $^3F_2 < {}^3F_3 < {}^3F_4$, since the value of J for these states increases as $2 < 3 < 4$. According to **Lande interval rule**, since $d^2$ configuration ($^3$F term is one of the terms arising from $d^2$ configuration) is less than half-filled ($n < 5$), the energy separation between two adjacent J states of this configuration ($d^2$ configuration) is equal to the product of + λ and higher J value (λ is called **spin-orbit coupling constant**). Thus if $E_2$, $E_3$ and $E_4$ are the energies of $^3F_2$, $^3F_3$ and $^3F_4$ states, then:

$$E_3 - E_2 = +\lambda \times 3 = +3\lambda$$

or $$E_2 = E_3 - 3\lambda \quad \dots (i)$$

$$E_4 - E_3 = +\lambda \times 4 = +4\lambda$$

or $$E_4 = E_3 + 4\lambda \quad \dots (ii)$$

Now if the energy of unsplit state is taken as zero, then the sum of (2J + 1)E terms of all J states of $^3$F term should be equal to zero. Thus we have:

$$(2J + 1)\,E_2 + (2J + 1)\,E_3 + (2J + 1)\,E_4 = 0$$

or $$(2 \times 2 + 1)\,E_2 + (2 \times 3 + 1)\,E_3 + (2 \times 4 + 1)\,E_4 = 0$$

or $$5E_2 + 7\,E_3 + 9\,E_4 = 0 \quad \dots (iii)$$

Put the values of $E_2$ and $E_4$ from equations (*i*) and (*ii*) in equation (*iii*) to get:

$$5\,(E_3 - 3\lambda) + 7E_3 + 9\,(E_3 + 4\lambda) = 0$$

This equation gives $E_3 = -\lambda$

Obviously $$E_2 = E_3 - 3\lambda = -\lambda - 3\lambda = -4\lambda$$

and $$E_4 = -E_3 + 4\lambda = -\lambda + 4\lambda = +3\lambda.$$ Thus:

$$\boxed{\mathbf{E_2 = -4\lambda, E_3 = -\lambda, E_4 = +3\lambda}}$$

The values of $E_1$, $E_2$ and $E_3$ given above show that:

Energy separation between $^3F_3$and $^3F_2$ terms

$$= E_3 - E_2 = -\lambda - (-4\lambda) = + 3\lambda$$

Energy separation between $^3F_4$ and $^3F_3$ terms

$$= E_4 - E_3 = + 3\lambda - (-\lambda) = +4\lambda$$

It may be seen from Fig. 24.5 that the energies of $^3F_2$, $^3F_3$ and $^3F_4$ J states is in the order:

$$^3F_2 (= - 4\lambda) < {}^3F_3 (= -\lambda) < {}^3F_4 (= + 3\lambda)$$

Since $^3F_2$ state has the lowest energy, it is the most stable state. Remember that the term having the lowest value of J has the lowest energy.

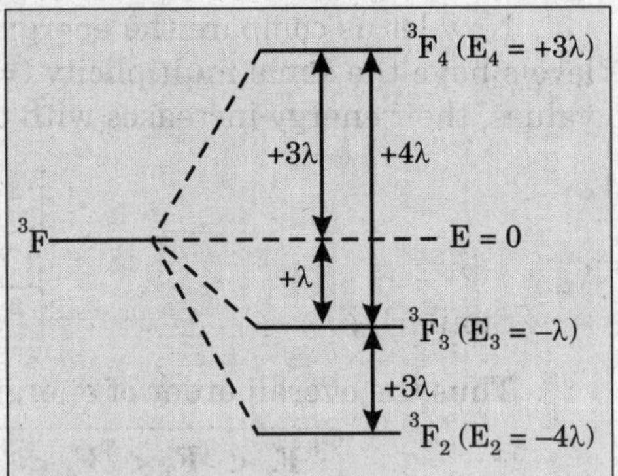


**Fig 24.5** Energies of J states of $^3F$ term of $d^2$ configuration

## To Calculate Energy of J States of $^3P$ Term of $d^2$ Configuration

In the discussion of $p^2$ configuration of C-atom we have shown that $^3P_0$, $^3P_1$ and $^3P_2$ are the J states of $^3P$ term. The energy of these states is in the order $^3P_0$ < $^3P_1$ < $^3P_2$, since J values of these states increase as 0 < 1 < 2. On the basis of Lande interval rule:

(*a*) The energy of different J states is as $E_0 = - 2\lambda$, $E_1 = -\lambda$ and $E_2 = +\lambda$

(*b*) Energy separation between $^3P_1$ and $^3P_0$ terms = $E_1 - E_0 = -\lambda - (-2\lambda) = +\lambda$

Energy separation between $^3P_2$ and $^3P_1$ terms = $E_2 - E_1 = +\lambda -(-\lambda) = +2\lambda$

Here λ is called **spin-orbit coupling constant.**

## To determine the number of microstates ($M_J$ levels) in term symbols of $d^2$ configuration

We know that $^3F$, $^3P$, $^1G$, $^1D$ and $^1S$ are the term symbols of $d^2$ configuration of $V^{3+}$ ion. We have said under the study of $p^2$ configuration of C-atom that the number of microstates in a term = (2L + 1) (2S + 1). Thus:

No. of microstates in $^3F$ term (2S + 1 = 3, L = 3)

$$= (2L + 1)(2S + 1) = (2 \times 3 + 1) \times 3 = \mathbf{21}$$

No. of microstates in $^3P$ term (2S + 1 = 3, L = 1)

$$= (2 \times 1 + 1) \times 3 = \mathbf{9}$$

No. of microstates in $^1G$ term (2S + 1 = 1, L = 4)

$$= (2 \times 4 + 1) \times 1 = \mathbf{9}$$

No. of microstates in $^1D$ term (2S + 1 = 1, L = 2)

$$= (2 \times 2 + 1) \times 1 = \mathbf{5}$$

No. of microstates in $^1S$ term (2S + 1 = 1, L = 0)

$$= (2 \times 0 + 1) \times 1 = \mathbf{1}$$

Thus:

| **Total number of microstates in $d^2$ configuration = 21 + 9 + 9 + 5 + 1 = 45** |
|---|

## Splitting of Terms of $d^2$ Configuration Due to Spin-Orbit Coupling (L – S or Russel Saunders Coupling)

We know that the energy terms arising out of $d^2$ configuration due to inter-electronic repulsion between the two electrons of $d^2$ configuration are $^1S$, $^3P$, $^1D$, $^3F$ and $^1G$. $^1S$, $^1D$ and $^1G$ terms have spin multiplicity = 1 while $^3P$ and $^3F$ terms have spin multiplicity = 3. Hence $^1S$, $^1D$ and $^1G$ terms are called *singlet spin terms* while $^3P$ and $^3F$ terms called *triplet state terms.*

(*i*) **Splitting of terms of $d^2$ configuration due to spin-spin coupling (*s – s* coupling)**. Due to s-s coupling of the two electrons of $d^2$ configuration, the terms of this configuration split into two sets. One set contains singlet spin terms ($^1S$, $^1G$ and $^1D$) and the other set consists of triplet spin terms ($^3F$ and $^3P$). The set containing the triplet spin terms ($^3F$ and $^3P$) has lower energy than the set containing singlet spin terms ($^1S$, $^1G$ and $^1D$).

(*ii*) **Splitting of triplet spin terms and singlet spin terms due to orbital-orbital coupling (*l – l* coupling).** Due to $(l - l)$ coupling the triplet spin set splits to give $^3F$ and $^3P$ terms while the singlet spin set splits to give $^1S$, $^1G$ and $^1D$ terms. The energy of $^3F$ and $^3P$ terms is in the order $^3F < {}^3P$ while that of $^1D$, $^1G$ and $^1S$ terms is in the order $^1G < {}^1D < {}^1S$.

(*iii*) **Splitting of terms into J energy levels due to spin-orbit coupling (L – S coupling) (Also called Russel-Saunder coupling).** Due to (L – S) coupling each of the terms is split into terms which have different values of J but same value of spin multiplicity. The terms having different J values, are called **J energy states** or J **energy levels**. J energy level is represented as $^{2S+1}L_J$. The value of J for a term ranges from (L + S) to (L – S) or (S – L) whichever is positive. Thus the value of J is given by:

$$J = (L + S), (L + S - 1), \ldots (L - S) \quad \text{or} \quad (S - L)$$

The value of J for $^3F$, $^3P$, $^1G$, $^1D$ and $^1S$ terms has already been calculated. The J energy states derived from these terms are $^3F_2$, $^3F_3$, $^3F_4$, $^3P_0$, $^3P_1$, $^3P_2$, $^1G_4$, $^1D_2$ and $^1S_0$. Three values of J (= 2, 3, 4) for $^3F$ term mean that under L – S coupling $^3F$ term is split up into three J energy levels which are $^3F_2$, $^3F_3$ and $^3F_4$. The energy of these states is in the order $^3F_2 < {}^3F_3 < {}^3F_4$. (See Fig. 24.6). Thus since $^3F_2$ level has the lowest energy, it is the most stable state and hence is the ground state term of $d^2$ configuration. Similarly three values of J (= 0, 1, 2) for $^3P$ term mean that under L – S coupling, $^3P$ term is split up into three J energy states which are $^3P_0$, $^3P_1$ and $^3P_2$. The energy of these states is in the order $^3P_0 < {}^3P_1 < {}^3P_2$.

One value of J (= 4) for $^1G$ term shows that under L – S coupling $^1G$ term is not split up, *i.e.* the energy of this term remains the same (See Fig. 24.6) Thus there is no possibility of L – S coupling in $^1G$ term. On the same lines

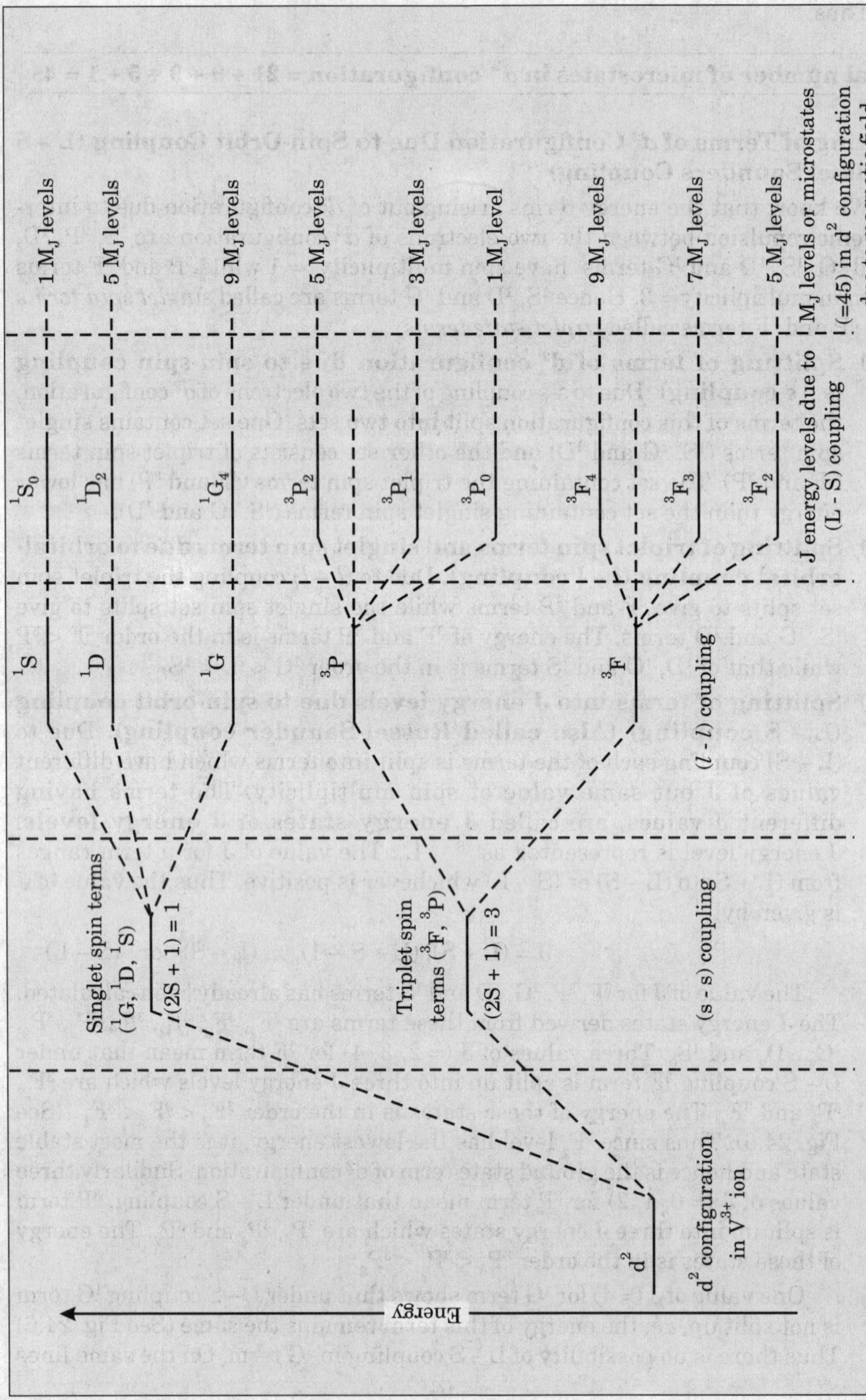


**Fig. 24.6.** Spliting of the singlet ($^1$G, $^1$D and $^1$S) and triplet ($^3$F and $^3$P) terms of $d^2$ configuration in $V^{3+}$ ion due to $(L - S)$ coupling

of argument it can be said that there is no possibility of L – S coupling in $^1D$ and $^1S$ terms. Hence their energy remains the same. J energy levels of $^1G$, $^1D$ and $^1S$ terms are $^1G_4$, $^1D_2$ and $^1S_0$ respectively.

**(iv) Splitting of J energy levels into $M_J$ levels (called microstates) under magnetic field.** When magnetic field is applied, J energy levels *viz* $^3F_2$, $^3F_3$, $^3F_4$, $^3P_0$, $^3P_1$, $^3P_2$, $^1G_4$, $^1D_2$ and $^1S_0$ are split into different number of $M_J$ levels. The number of $M_J$ levels obtained from the splitting of a J state is equal to (2J + 1). Thus the number of $M_J$ levels is as : $^3F_2 = 2J + 1 = 2 \times 2 + 1 = 5$, $^3F_3 = 2 \times 3 + 1 = 7$, $^3F_4 = 2 \times 4 + 1 = 9$, $^3P_0 = 2 \times 0 + 1 = 1$, $^3P_1 = 2 \times 1 + 1 = 3$, $^3P_2 = 2 \times 2 + 1 = 5$, $^1G_4 = 2 \times 4 + 1 = 9$, $^1D_2 = 2 \times 2 + 1 = 5$ and $^1S_0 = 2 \times 0 + 1 = 1$. These numbers show that the number of $M_J$ levels given by different terms is as $^3F = 5 + 7 + 9 = 21$, $^3P = 1 + 3 + 5 = 9$, $^1G = 9$, $^1D = 5$ and $^1S = 1$. Consequently total number of microstates obtained in case of the splitting of $d^2$ configuration $= 21 + 9 + 9 + 5 + 1 = 45$ (See Fig. 24.6)

## Energy of J States of $^3F$ and $^3P$ Terms of $d^8$ Configuration

We know that J states of $^3F$, $^3P$, $^1D$, $^1G$ and $^1S$ energy terms are $^3F_2$, $^3F_3$, $^3F_4$, $^3P_0$, $^3P_1$, $^3P_2$, $^1D_2$, $^1G_4$ and $^1S_0$, According to Lande interval rule the energy separation between two adjacent J states of a term arising out of more than half-filled $d^n$ configuration ($n > 5$) is equal to the product of $-\lambda$ and the higher J value. Thus:

**(*i*) To calculate energy of J states of $^3F$ term.** We know that $^3F_2$, $^3F_3$ and $^3F_4$ are the J states of $^3F$ term. If $E_2$, $E_3$ and $E_4$ are the energies of these J states, then according to Lande interval rule, we have:

$$E_3 - E_2 = -\lambda \times 3 = -3\lambda \quad \ldots (i)$$

$$E_4 - E_3 = -\lambda \times 4 = -4\lambda \quad \ldots (ii)$$

Also we have:

$$(2J + 1)\,E_2 + (2J + 1)\,E_3 + (2J + 1)\,E_4 = 0$$

or
$$(2 \times 2 + 1)E_2 + (2 \times 3 + 1)\,E_3 + (2 \times 4 + 1)E_4 = 0$$

or
$$5E_2 + 7E_3 + 9\,E_4 = 0 \quad \ldots (iii)$$

On solving equations (*i*), (*ii*) and (*iii*), we have:

$$E_2 = +4\lambda, \quad E_3 = +\lambda, \quad E_4 = -3\lambda \text{ (See Fig 24.7)}$$

It may be seen from Fig. 24.7 that:

(*a*) The energy of $^3F_2$, $^3F_3$ and $^3F_4$ J states is in the order:

$$^3F_4\,(= -3\lambda) < {^3F_3}\,(= +\lambda) < {^3F_2}\,(= +4\lambda)$$

or
$$^3F_2\,(= +4\lambda) > {^3F_3}\,(= +\lambda) > {^3F_4}\,(= -3\lambda)$$

It may be noted that this order is the opposite to that seen in case of $d^2$ configuration. Since $^3F_4$ state has the lowest energy, it is the most stable state.

(*b*) Energy separation between $^3F_3$ and $^3F_4$ terms $= E_3 - E_4 = +\lambda - (-3\lambda) = +4\lambda$

Similarly energy separation between $^3F_2$ and $^3F_3$ terms $= E_2 - E_3 = +4\lambda - \lambda = +3\lambda$.

(*ii*) **To calculate energy of J states of $^3P$ term.** We know that $^3P_0$, $^3P_1$ & $^3P_2$ are the J states of $^3P$ term. If $E_0$, $E_1$ and $E_2$ are the energies of these J states, then according to Lande interval rule we have:

$$E_1 - E_0 = -\lambda \times 1 = -\lambda \quad \text{... }(i)$$

and $$E_2 - E_1 = -\lambda \times 2 = -2\lambda \quad \text{... }(ii)$$

Also we have:

$$(2J + 1) E_0 + (2J + 1) E_1 + (2J + 1)E_2 = 0$$

or $$E_0 + 3E_1 + 5E_2 = 0 \quad \text{...}(iii)$$

On solving equations (*i*), (*ii*) and (*iii*), we get:

$$E_0 = 2\lambda, E_1 = \lambda \text{ and } E_2 = -\lambda$$

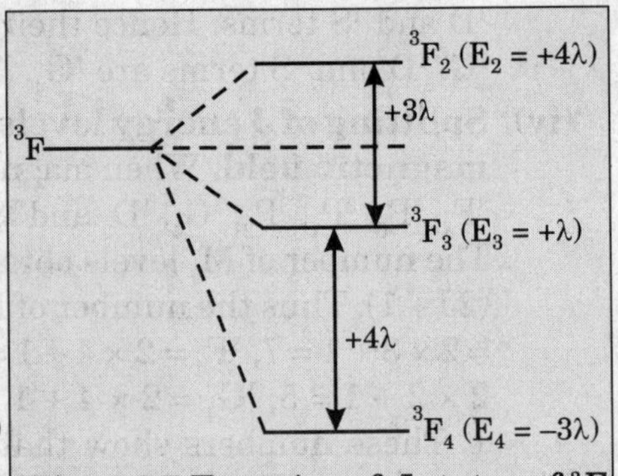


**Fig. 24.7** Energies of *J* states of $^3F$ term of $d^8$ configuration

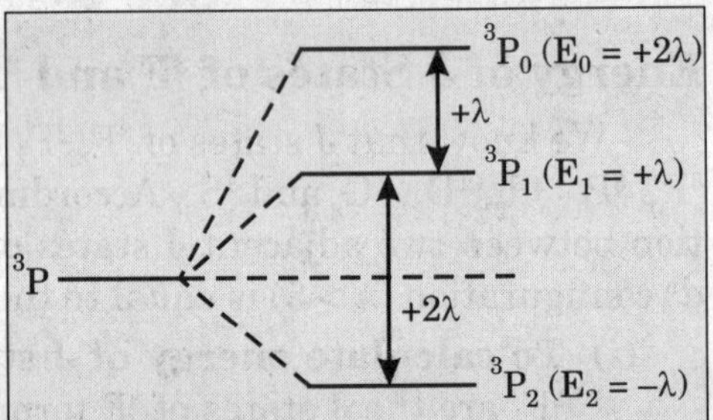


**Fig 24.8** Energies of J states of $^3P$ term of $d^8$ configuration.

From Fig. 24.8. The following points may be noted:

(*a*) The energy of $^3P_0$, $^3P_1$ and $^3P_2$ states is in the order:

$$^3P_2 (= -\lambda) < {}^3P_1 (= +\lambda) < {}^3P_0 (= +2\lambda)$$

or $$^3P_0 (= +2\lambda) > {}^3P_1 (= +\lambda) > {}^3P_2 (= -\lambda).$$

It may be noted that this order is the opposite to that seen in case of $d^2$ configuration. Since $^3P_2$ state has the lowest energy, this state is the most stable.

(*b*) Energy separation between $^3P_1$ and $^3P_2$ terms = + λ − (−λ) = +2λ. Similarly energy separation between $^3P_0$ and $^3P_1$ terms = +2λ − (+λ) = λ.

## To Calculate the Number of Microstates of a Given Configuration

A given configuration can acquire different arrangements of electrons which have slightly different energy. These different arrangements are called microstates of the given configuration. The number of microstates in a given configuration is given by:

$$\textbf{No. of microstates of a given configuration} = \frac{n!}{r!(n-r)!}$$

where $n$ = Twice the number of orbitals in the sub shell and $r$ = No. of electrons in the given configuration. $n!$ and $r!$ are the factorials of $n$ and $r$ respectively.

**Example. Calculate the number of microstates in the following configurations (*i*) $p^2$ (*ii*) $p^3$ (*iii*) $d^1$.**

**Ans.** (*i*) **$p^2$ configuration**. This configuration has $r = 2$. Since the number of orbitals in $p$ sub-shell = 3, $n = 2 \times 3 = 6$. Thus:

$$\text{No. of microstates in } p^2 \text{ configuration} = \frac{n!}{r!(n-r)!}$$

$$= \frac{6!}{2!(6-2)!}$$

$$= \frac{6!}{2! \times 4!}$$

$$= \frac{6 \times 5 \times 4 \times 3 \times 2 \times 1}{(2 \times 1) \times (4 \times 3 \times 2 \times 1)} = \mathbf{15\ (Ans.)}$$

*(ii)* **$p^3$ configuration.** This configuration has $r = 3$. Since number of orbitals in $p$ sub-shell = 3, $n = 3 \times 2 = 6$. Hence,

$$\text{No. of microstates in } p^3 \text{ configuration} = \frac{n!}{r!(n-r)!} = \frac{6!}{3!(6-3)!} = \frac{6!}{3! \times 3!}$$

$$= \frac{6 \times 5 \times 4 \times 3 \times 2 \times 1}{(3 \times 2 \times 1) \times (3 \times 2 \times 1)} = \mathbf{20\ Ans.}$$

*(iii)* **$d^1$ configuration.** This configuration has r = 1. Since there are 5 orbitals in $d$ sub-shell, $n = 5 \times 2 = 10$. Hence:

$$\text{No. of microstates in } d^1 \text{ configuration} = \frac{n!}{r!(n-r)!} = \frac{10!}{1! \times (10-1)!} = \frac{10!}{1 \times 9!}$$

$$= \frac{10 \times 9 \times 8 \times 7 \times 6 \times 5 \times 4 \times 3 \times 2 \times 1}{1 \times (9 \times 8 \times 7 \times 6 \times 5 \times 4 \times 3 \times 2 \times 1)}$$

$$= \mathbf{10\ Ans.}$$

No. of microstates in $p^a$ ($a$ = 1 to 6) and $d^b$ ($b$ = 1 to 10) configurations is given in Table 24.1.

**Table 24.1 Number of microstates in $p^a$ ($a$ = 1 to 6) and $d^b$ ($b$ = 1 to 10) configurations**

| *Electronic configuration* | *No. of microstates* | *Electronic configuration* | *No. of microstates* |
|---|---|---|---|
| $p^1$ | 6 | $d^3$ | 120 |
| $p^2$ | 15 | $d^4$ | 210 |
| $p^3$ | 20 | $d^5$ | 252 |
| $p^4$ | 15 | $d^6$ | 210 |
| $p^5$ | 6 | $d^7$ | 120 |
| $p^6$ | 1 | $d^8$ | 45 |
| $d^1$ | 10 | $d^9$ | 10 |
| $d^2$ | 45 | $d^{10}$ | 1 |

## What is an Energy Term of a Configuration?

An energy term (also called simply as **term**) is an energy level or a set of energy levels which result from the elctron-electron repulsions in an electronic arrangement or in a set of electronic arrangements of a configuration of a free metal atom or ion. An energy term contains a set of degenerate energy levels. Since the electron-electron repulsions in all the electronic arrangements of a configuration may not be the same, there can be several energy levels in a configuration. Total number of energy levels in a configuration is equal to the total number of possible electronic arrangements for that configuration. In a many-electron system the number of energy levels is calculated through $l - l$ and $s - s$ coupling effects.

An energy term of a configuration is written as $^{2S+1}L$ in which the quantity (2S + 1) is called **spin multiplicity** of the term. Spin multiplicity gives the number of orientations of S in space or **spin multiplicity** gives the number of spin energy levels. S in the term $^{2S+1}L$ is called **resultant spin angular momentum quantum number** and L is called resultant **orbital angular momentum quantum number**. For L = 0, 1, 2, 3, 4, 5 the symbols used are S, P, D, F, G, H respectively. Thus:

| | | | | | | | |
|---|---|---|---|---|---|---|---|
| L | = | 0 | 1 | 2 | 3 | 4 | 5 |
| Symbols | = | S | P | D | F | G | H |

## Representation of Full Spectroscopic Term of a Given Configuration

A term of a given configuration is represented as $^{2S+1}L$. The representation of a full term of a given configuration also contains J in addition to L and 2S + 1. J is written as a subscript at the right hand of L. Thus a full term of a configuration is represented as $^{2S+1}L_J$. Full term is also called Russel-Saunders term. In this representation:

(*i*) L = *Resultant orbital angular momentum quantum number* (also called simply as *resultant orbital quantum number*)

(*ii*) S = *Resultant spin angular momentum quantum number* (also called simply as *resultant orbital quantum number*)

(*iii*) 2S + 1 = *Spin multiplicity of the term*. It indicates the number of spin energy levels in a term.

(*iv*) *J* = *Resultant (overall or total or inner) angular momentum quantum number*. J is also called simply as *resultant* (overall or total or inner) *quantum number* or (*orbital* + *spin*) *quantum number*.

### Determination of Russel-Saunders Symbols or Terms for $d^n$ Configuration ($n$ = 1 to 10) of a Free Metal Ion.

Russel-Saunders symbol or term of a given $d^n$ configuration is represented as $^{2S+1}L_J$. The term obtained by removing J from the Russel-Saunders term given above is called ground state term. Thus ground state term is represented as $^{2S+1}L$. The determination of Russel-Saunders term for a given $d^n$ configuration involves the following steps:

(*i*) Represent *d*-orbitals by thick bar (—) and fill these orbitals with $d^x$ electrons according to Hund's rules.

(*ii*) Find out the value of *resultant orbital quantum number* (L) with the help of the relation: $L = \Sigma m_l$. Here $m_l$ is the orbital angular momentum for single electron. Values of $m_l$ for *s*, *p*, *d* and *f* orbitals are as : $s\ (l = 0) = 0$; $p\ (l = 1) = +1, 0, -1$; $d\ (l = 2) = +2, +1, 0, -1, -2$; $f\ (l = 3) = +3, +2, +1, 0, -1, -2, -3$.

L arises due to $l - l$ coupling. The value of L ranges from $(l_1 + l_2)$ to $(l_1 - l_2)$ or $(l_2 - l_1)$ whichever is positive, *i.e.*

$$L = (l_1 + l_2), (l_1 + l_2 - 1), \ldots, (l_1 - l_2) \text{ or } (l_2 - l_1)$$

We know that values of *l* are represented by small letters as shown below:

| | | | | | | |
|---|---|---|---|---|---|---|
| *Values of l* : | 0 | 1 | 2 | 3 | 4 | 5 ... |
| *Small letters* : | *s* | *p* | *d* | *f* | *g* | *h* ... |

Similarly values of L are represented by capital letters.

| | | | | | | |
|---|---|---|---|---|---|---|
| *Values of L* : | 0 | 1 | 2 | 3 | 4 | 5 ... |
| *Capital letters* : | S | P | D | F | G | H ... |

(*iii*) Find out the value of *resultant spin quantum number* (S) with the help of the relation: $S = \Sigma m_s$. $m_s$ is the spin angular momentum for single electron. Value of $m_s$ for an electron represented as ↑ is equal to +1/2 and that for an electron represented as ↓ is equal to –1 / 2. Resultant spin quantum number (S) arises due to s-s coupling.

(*iv*) Find the value of spin multiplicity, 2S + 1 with the help of the value of S as determined at step (*iii*) above. Spin multiplicity is equal to the number of spin energy levels in a term. Spin multiplicity is also equal to (*n* + 1) where *n* is the number of unpaired electrons.

(*v*) Find out the value of *overall quantum number* (J). J arises due to L – S coupling in many electron system.

J can have the values ranging from (L + S) to (L – S) or (S – L) whichever is positive, *i.e.*,

$$J = (L + S), (L + S - 1), \ldots (L - S) \text{ or } (S - L)$$

If $d^x$ configuration is half-filled or less than half-filled, then the lowest value of J is used for writing the Russel Saunders symbol or term (ground state symbol) of $d^x$ configuration. On the other hand, if $d^x$ configuration is more than half-filled, then the highest value of J is used in writing the ground state symbol.

Above procedure can be understood by determining the ground state symbol for $d^1, d^2, \ldots d^9, d^{10}$ configurations of a free metal ion. This is shown below. Remember that for all *d*- orbitals *l* = 2.

**(*i*) $d^1$ configuration ($Ti^{3+}$)**

| $m_l$ = +2 | +1 | 0 | –1 | –2 | |
|---|---|---|---|---|---|
| ↑ | | | | | |
| ___ | ___ | ___ | ___ | ___ | (*n* = 1) |

$$m_s = +1/2\ (\uparrow)$$

∴ $L = \Sigma m_l = (+2) \times 1 = 2$

L = 2 indicates **D state**

$$S = \Sigma m_s = 1 \times \left(+\frac{1}{2}\right) = \frac{1}{2}$$

$\therefore$ Spin multiplicity, $2S + 1 = 2 \times \frac{1}{2} + 1 = 2$

**Ground state term of $d^1$ ion = $^{2S+1}L = {}^2D$**

$$J = (L + S), (L + S - 1), \dots, (L - S)$$

$$= \left(2+\frac{1}{2}\right), \left(2+\frac{1}{2}-1\right), \dots \left(2-\frac{1}{2}\right)$$

$$= \frac{5}{2}, \frac{3}{2}$$

Now since $d^1$ ion is less than half-filled, lower value of J will be used, *i.e.* $J = \frac{3}{2}$. Hence.

**Russel-Saunders term of $d^1$ ion = $^{2S+1}L_J = {}^2D_{3/2}$**

***(ii)* $d^2$ configuration ($Ti^{2+}$, $V^{3+}$)**

| $m_l = +2$ | $+1$ | $0$ | $-1$ | $-2$ | |
|---|---|---|---|---|---|
| ↑ | ↑ | | | | |
| ___ | ___ | ___ | ___ | ___ | $(n = 2)$ |

$m_s = +1/2$ (↑)

$\therefore$ $L = \Sigma m_l = (+2) \times 1 + (+1) \times 1 = 3$ (**F state**)

$$S = \Sigma m_s = +\frac{1}{2} \times 2 = 1$$

$\therefore$ $2S + 1 = 2 \times 1 + 1 = 3$

$\therefore$ **Ground state term of $d^2$ ion = $^{2S+1}L = {}^3F$**

$$J = (L + S), (L + S - 1), \dots (L - S)$$
$$= (3 + 1), (3 + 1 - 1), \dots (3 - 1)$$
$$= 4, 3, 2$$

Three values of J (= 4, 3, 2) show that $^3F$ term is split up into three J energy levels which can be designated as $^3F_4$, $^3F_3$ and $^3F_2$. Now since $d^2$ configuration is less than half-filled, the J energy level having the lowest value of J is the ground state (or term) for $d^2$ configuration. Thus:

**Russel-Saunders term of $d^2$ ion = $^{2S+1}L_J = {}^3F_2$**

Since $^3F_2$ is the ground state symbol, it has the lowest energy (most stable). The energy of $^3F_4$, $^3F_3$ and $^3F_2$ states is in the order : $^3F_2 < {}^3F_3 < {}^3F_4$.

***(iii)* $d^3$ configuration ($V^{2+}$, $Cr^{3+}$)**

$$m_l = +2 \quad +1 \quad 0 \quad -1 \quad -2$$

$$\underset{\text{—}}{\uparrow} \quad \underset{\text{—}}{\uparrow} \quad \underset{\text{—}}{\uparrow} \quad \text{—} \quad \text{—} \quad (n = 3)$$

$$m_s = +\frac{1}{2}(\uparrow)$$

$$\therefore \quad L = \Sigma m_l = (+2) \times 1 + (+1) \times 1 + (0) \times 1$$
$$= 2 + 1 + 0 = 3 \text{ (F state)}$$

$$S = \Sigma m_s = \left(+\frac{1}{2}\right) \times 3 = \frac{3}{2}$$

$$\therefore \quad 2S + 1 = 2 \times \frac{3}{2} + 1 = 4$$

**Ground state term of $d^3$ ion = $^{2S+1}L$ = $^4F$**

$$J = (L + S), (L + S - 1) \ldots (L - S)$$
$$= \left(3 + \frac{3}{2}\right), \left(3 + \frac{3}{2} - 1\right) \ldots \left(3 - \frac{3}{2}\right)$$
$$= \frac{9}{2}, \frac{7}{2} \ldots \frac{3}{2}$$
$$= \frac{9}{2}, \frac{7}{2}, \frac{5}{2}, \frac{3}{2}$$

Now since $d^3$ ion is less than half-filled, the lowest value of J forms the ground state *i.e.*, $J = \frac{3}{2}$ and hence:

**Russel-Saunders term of $d^3$ ion = $^{2S+1}L_J$ = $^4F_{3/2}$**

***(iv)* $d^4$ configuration ($Cr^{2+}$, $Mn^{2+}$)**

$$m_l = +2 \quad +1 \quad 0 \quad -1 \quad -2$$

$$\underset{\text{—}}{\uparrow} \quad \underset{\text{—}}{\uparrow} \quad \underset{\text{—}}{\uparrow} \quad \underset{\text{—}}{\uparrow} \quad \text{—} \quad (n = 4)$$

$$m_s + \frac{1}{2}(\uparrow)$$

$$\therefore \quad L = \Sigma m_l = (+2) \times 1 + (+1) \times 1 + 0 \times 1 + (-1) \times 1 = 2 + 1 + 0 - 1 = 2 \text{ (D state)}$$

$$S = \Sigma m_s = +\frac{1}{2} \times 4 = 2$$

$$\therefore \quad 2S + 1 = 2 \times 2 + 1 = 5$$

**Ground state term of $d^4$ ion = $^{2S+1}L$ = $^5D$**

$$J = (L + S), (L + S - 1) \ldots (L - S)$$
$$= (2 + 2), (2 + 2 - 1) \ldots (2 - 2)$$
$$= 4, 3, 2, 1, 0$$

Now since $d^4$ ion is less than half-filled, J will have the lowest value, *i.e.*, J = 0

**Russel-Saunders term of $d^4$ ion = $^{2S+1}L_J$ = $^5D_0$**

***(v)* $d^5$ configuration ($Mn^{2+}$, $Fe^{3+}$)**

| $m_l$ = +2 | +1 | 0 | −1 | −2 | |
|---|---|---|---|---|---|
| ↑ | ↑ | ↑ | ↑ | ↑ | (n = 5) |

$$m_s = +\frac{1}{2}(\uparrow)$$

$$\therefore L = \Sigma m_l = (+2) \times 1 + (+1) \times 1 + 0 \times 1 + (-1) \times 1 + (-2) \times 1$$
$$= 2 + 1 + 0 - 1 - 2 = 0 \text{ (S state)}$$

$$S = \Sigma m_s = +\frac{1}{2} \times 5 = \frac{5}{2}$$

$$\therefore 2S + 1 = 2 \times \frac{5}{2} + 1 = 6$$

**Ground state term of $d^5$ ion = $^{2S+1}L$ = $^6S$**

$$J = (L + S), (L + S - 1). \ldots (S - L)$$
$$= \left(0 + \frac{5}{2}\right), \left(0 + \frac{5}{2} - 1\right), \ldots \left(\frac{5}{2} - 0\right)$$
$$= \frac{5}{2}, \frac{3}{2}, \frac{5}{2}$$
$$= \frac{5}{2}$$

**∴ Russel-Saunders term of $d^5$ ion = $^{2S+1}L_J$ = $^6S_{5/2}$**

***(vi)* $d^6$ configuration ($Fe^{2+}$, $Co^{3+}$)**

| $m_l$ = +2 | +1 | 0 | −1 | −2 | |
|---|---|---|---|---|---|
| ↑↓ | ↑ | ↑ | ↑ | ↑ | (**n** = 4) |

$$m_s = +\frac{1}{2}(\uparrow), -\frac{1}{2}(\downarrow)$$

$$\therefore \quad L = \Sigma m_l = (+2) \times 2 + (+1) \times 1 + 0 \times 1 + (-1) \times 1 + (-2) \times 1$$
$$= 4 + 1 + 0 - 1 - 2 = 2 \text{ (\textbf{D state})}$$

$$S = \Sigma m_s = +\frac{1}{2} \times 5 + \left(-\frac{1}{2}\right) \times 1$$

$$= \frac{5}{2} - \frac{1}{2} = 2$$

$\therefore \quad 2S + 1 = 2 \times 2 + 1 = 5$

**Ground state term for $d^6$ ion = $^{2S+1}L = {}^5D$**

$$J = (L + S), (L + S - 1) \ldots (L - S)$$
$$= (2 + 2), (2 + 2 - 1) \ldots (2 - 2)$$
$$= 4, 3, \ldots 0$$
$$= 4, 3, 2, 1, 0$$

Now since $d^6$ ion is more than half-filled, the highest value of J forms the ground state, *i.e.* J = 4 (highest value) and hence:

**Russel-Saunders term of $d^6$ ion = $^{2S+1}L_J = {}^5D_4$**

***(vii)* $d^7$ configuration ($Co^{2+}$)**

| $m_l = +2$ | $+1$ | $0$ | $-1$ | $-2$ | |
|---|---|---|---|---|---|
| ↑↓ | ↑↓ | ↑ | ↑ | ↑ | $(n = 3)$ |

$$m_s = +\frac{1}{2}(\uparrow), -\frac{1}{2}(\downarrow)$$

$\therefore \quad L = \Sigma m_l = (+2) \times 2 + (+1) \times 2 + 0 \times 1 (-1) \times 1 + (-2) \times 1$

$$= 4 + 2 + 0 - 1 - 2 = 3 \text{ (**F state**)}$$

$$S = \Sigma m_s = +\frac{1}{2} \times 5 + \left(-\frac{1}{2}\right) \times 2$$

$$= \frac{5}{2} - 1 = \frac{3}{2}$$

$\therefore \quad 2S + 1 = 2 \times \frac{3}{2} + 1 = 4$

**Ground state term for $d^7$ ion = $^{2S+1}L = {}^4F$**

$$J = (L + S), (L + S - 1), \ldots, (L - S)$$

$$= \left(3 + \frac{3}{2}\right), \left(3 + \frac{3}{2} - 1\right), \ldots \left(3 - \frac{3}{2}\right)$$

$$= \frac{9}{2}, \frac{7}{2}, \frac{5}{2}, \frac{3}{2}$$

Now since $d^7$ ion is more than half-filled, J will have the highest value. Hence:

**Russel-Saunders term of $d^7$ ion = $^{2S+1}L_J = {}^4F_{9/2}$**

***(viii)* $d^8$ configuration ($Ni^{2+}$)**

| $m_l = +2$ | $+1$ | $0$ | $-1$ | $-2$ | |
|---|---|---|---|---|---|
| ↑↓ | ↑↓ | ↑↓ | ↑ | ↑ | ($n = 2$) |

$$m_s = +\frac{1}{2}(\uparrow), -\frac{1}{2}(\downarrow)$$

$$\therefore \quad L = \Sigma m_l = +2 \times 2 + 1 \times 2 + 0 \times 2 + (-1) \times 1 + (-2) \times 1$$

$$= 4 + 2 + 0 - 1 - 2 = 3 \text{ (F state)}$$

$$S = \Sigma m_s = 5\times\left(+\frac{1}{2}\right) + 3\times\left(-\frac{1}{2}\right) = \frac{5}{2} - \frac{3}{2} = 1$$

$$\therefore \quad 2S + 1 = 2 \times 1 + 1 = 3$$

**Ground state term for $d^8$ ion = $^{2S+1}L = {}^3F$**

$$J = (L + S), (L + S - 1), \ldots, (L - S)$$

$$= (3 + 1), (3 + 1 - 1), \ldots, (3 - 1)$$

$$= 4, 3, 2$$

Now since $d^8$ configuration is more than half-filled, highest value of J is to be used *i.e.*, J = 4. Hence:

**Russel-Sounders term of $d^8$ ion = $^{2S+1}L_J = {}^3F_4$**

***(ix)* $d^9$ configuration ($Cu^{2+}$)**

| $m_l = +2$ | $+1$ | $0$ | $-1$ | $-2$ | |
|---|---|---|---|---|---|
| ↑↓ | ↑↓ | ↑↓ | ↑↓ | ↑ | ($n = 1$) |

$$m_s = +\frac{1}{2}(\uparrow), -\frac{1}{2}(\downarrow)$$

$$\therefore \quad L = +2 \times 2 + 1 \times 2 + 0 \times 2 + (-1) \times 2 + (-2) \times 1$$

$$= 4 + 2 + 0 - 2 - 2 = 2 \text{ (D state)}$$

$$S = \Sigma m_s = \left(+\frac{1}{2}\right)\times 5 + \left(-\frac{1}{2}\right)\times 4$$

$$= \frac{5}{2} - 2 = \frac{1}{2}$$

$$\therefore \quad 2S + 1 = 2 \times \frac{1}{2} + 1 = 2$$

**Ground state term for d$^9$ configuration = $^{2S+1}L = {}^2D$**

$$J = (L + S), (L + S - 1), \ldots, (L - S)$$

$$= \left(2+\frac{1}{2}\right), \left(2+\frac{1}{2}-1\right), \ldots, \left(2-\frac{1}{2}\right)$$

$$= \frac{5}{2}, \frac{3}{2}$$

Now since $d^9$ ion is more than half-filled, J will have higher value, *i.e.* J = 5/2. Hence,

| **Russel-Saunders term of $d^9$ ion = $^{2S+1}L_J$ = $^2D_{5/2}$** |
|---|

***(x)* $d^{10}$ configuration ($Zn^{2+}$)**

| $m_l$ = +2 | +1 | 0 | −1 | −2 | |
|---|---|---|---|---|---|
| ↑↓ | ↑↓ | ↑↓ | ↑↓ | ↑↓ | ($n = 0$) |

$$m_s = +\frac{1}{2}(\uparrow), -\frac{1}{2}(\uparrow)$$

$$\therefore \quad L = +2 \times 2 + 1 \times 2 + 0 \times 2 + (-1) \times 2 + (-2) \times 2$$

$$= 4 + 2 + 0 - 2 - 4 = 0 \text{ (\textbf{S state})}$$

$$S = \Sigma m_s = +\frac{1}{2} \times 5 + \left(-\frac{1}{2}\right) \times 5 = 0$$

Spin multiplicity, 2S + 1 = 2 × 0 + 1 = 1

| **Ground state term for $d^{10}$ configuration = $^{2S+1}L$ = $^1S$** |
|---|

$$J = (L + S), (L + S - 1), \ldots, (L - S)$$

$$= 0, 0, \ldots$$

| **Russel-Saunders term for $d^{10}$ configuration = $^{2S+1}L_J$ = $^1S_0$** |
|---|

Ground state term symbols, $^{2S+1}L$ and Russel-Saunders terms, $^{2S+1}L_J$ of $d^n$ configurations ($n$ = 1 to 10) are given in Table 24.2.

From Table 24.2 the following points may be noted:

(*i*) The two ions having $d^n$ and $d^{10-n}$ configurations have the same terms with the same value of spin multiplicity but with different value of J. For example the two configurations given in the following four pairs have the same terms given in bracket: (*a*) $d^1$ and $d^{10-1}$ or $d^9$ ($^2D$) (*b*) $d^2$ and $d^{10-2}$ or $d^8$ ($^3F$) (*c*) $d^3$ and $d^{10-3}$ or $d^7$ ($^4F$) and (*d*) $d^4$ and $d^{10-4}$ or $d^6$ ($^5D$), since the two configurations of each pair are hole equivalents to each other. Usually in writing the term of a given configuration J is omitted.

**Table 24.2 Ground state terms for $d^n$ configurations ($n$ = 1 to 10) ($l$ = 2)**

| $d^n$ *Configuration (Ion)* | *L* = $\Sigma m_l$ *(State)* | *S* = $\Sigma m_s$ = *n*/2 (*n* = No. of un-paired electrons) | *Spin multi-plicity* = 2*S* + 1 = *n* + 1 | *Ground state term* = $^{2S+1}L$ | J = (*L* + *S*) (*L* + *S* – 1) … (*L* – *S*) | *Russel Saunders term* = $^{2S+1}L_J$ |
|---|---|---|---|---|---|---|
| $d^1$ ($Ti^{3+}$) | 2(*D*) | $\frac{1}{2}$ | 2 | $^2D$ | 3/2 | $^2D_{3/2}$ |
| $d^2$ ($Ti^{2+}$, $V^{3+}$) | 3(*F*) | 1 | 3 | $^3F$ | 2 | $^3F_2$ |
| $d^3$ ($V^{2+}$, $Cr^{3+}$) | 3(*F*) | $\frac{3}{2}$ | 4 | $^4F$ | 3/2 | $^4F_{3/2}$ |
| $d^4$ ($Cr^{2+}$, $Mn^{2+}$) | 2(*D*) | 2 | 5 | $^5D$ | 0 | $^5D_0$ |
| $d^5$ ($Mn^{2+}$, $Fe^{3+}$) | 0(*S*) | $\frac{5}{2}$ | 6 | $^6S$ | $\frac{5}{2}$ | $^6S_{5/2}$ |
| $d^6$ ($Fe^{2+}$, $Co^{3+}$) | 2(*D*) | 2 | 5 | $^5D$ | 4 | $^5D_4$ |
| $d^7$ ($Co^{2+}$) | 3(*F*) | $\frac{3}{2}$ | 4 | $^4F$ | 9/2 | $^4F_{9/2}$ |
| $d^8$ ($Ni^{2+}$) | 3(*F*) | 1 | 3 | $^3F$ | 4 | $^3F_4$ |
| $d^9$ ($Cu^{2+}$) | 2(*D*) | $\frac{1}{2}$ | 2 | $^2D$ | 5/2 | $^2D_{5/2}$ |
| $d^{10}$ ($Zn^{2+}$) | 0 (*S*) | 0 | 1 | $^1S$ | 0 | $^1S_0$ |

(*ii*) The ground state terms of the above configuration pairs are given as:

$d^0$, $d^{10}$ → $^1S$, $d^1$, $d^9$ → $^2D$

$d^2$, $d^8$ → $^3F$, $d^3$, $d^7$ → $^4F$

$d^4$, $d^6$ → $^5D$, $d^5$ → $^5S$

(*iii*) (*a*) $d^1$, $d^9$, $d^4$ and $d^6$ ions have the same term symbol which is D (L = 2)

(*b*) $d^0$, $d^{10}$ and $d^5$ ions have the same term symbol, S (L = 0)

(*c*) $d^2$, $d^8$, $d^3$ and $d^7$ ions have the same term symbol, F (L = 3).

## Difference between Electronic Configuration and Energy Term

Electronic configuration of an atom or ion tells us as to how the electrons of the electronic configuration are distributed in a set of orbitals without considering the effect of electrostatic and other interactions between the electrons. For example $d^2$ configuration tells us that *d* sub-shell contains two electrons in the five *d*-orbitals, but it does not lell which of the five *d*-orbitals is occupied by electrons and which *d*-orbitals are vacant.

An energy term (also called simply **term**) is an energy level or a set of degenerate energy levels which result from the electron-electron repulsions in an electronic arrangement of a configuration. The number of energy levels in a term is known as its **degeneracy**.

## Orbital Degeneracy of a Term : No. of Orbital Energy Levels in a Term (Also see Table 24.3)

Orbital degeneracy of a given term is the number of *orbital energy levels* that are obtained when the term is split is weak $V_{oct}$ or $V_{tet}$.

We know that the orbital degeneracy of a sub-shell = $2l + 1$ where $l$ is *orbital angular momentum quantum number (or azimuthal quantum number) of the sub-shell*. Similarly the degeneracy of *a* spectroscopic term is equal to 2L + 1 where L is the *orbital angular momentum quantum number of the term*. Thus:

(*i*) Degeneracy of *s* sub-shell ($l = 0$) = $2l + 1 = 2 \times 0 + 1 = 1$. Thus *s* sub-shell is *non-degenerate* or *singly degenerate*. Similarly S term is *non-degenerate*. This means S term, like *s* sub-shell, is not split in presence of octahedral and tetrahedral ligand field.

(*ii*) Degeneracy of a *p* sub-shell ($l = 1$) = $2l + 1 = 2 \times 1 + 1 = 3$. Thus *p* sub-shell is *triply degenerate*. This means that a *p* sub-shell has a set of three *p*-orbitals which are $p_x$, $p_y$ and $p_z$. *p*-orbitals are directional and are affected by octahedral ligand field. However, since all the three *p*-orbitals are affected equally, their energy levels remain equal in energy. Thus no splitting of *p* sub-shell occurs.

Degeneracy of *P* term ($L = 1$) = $2L + 1 = 2 \times 1 + 1 = 3$. Thus *P* term, like *p* sub-shell, is *triply degenerate*, but in presence of weak octahedral or tetrahedral ligand field, P term is not split into three energy levels, rather it is transformed into $T_{1g}$ (octahedral field) and $T_1$ (tetrahedral field). $T_{1g}$ and $T_1$ both are triply degenerate.

(*iii*) Degeneracy of *d* sub-shell ($l = 2$) = $2l + 1 = 2 \times 2 + 1 = 5$. Thus *d* sub-shell is *five-fold degenerate*. This means that in presence of octahedral ligand field *d* sub-shell is split into two sets of orbitals. One set contains three orbitals *viz* $d_{xy}$, $d_{yz}$, $d_{zx}$ and the other set contains two orbitals namely $d_{x^2-y^2}$ and $d_{z^2}$ orbitals. These two sets are represented as $t_{2g}$ ($d_{xy}$, $d_{yz}$, $d_{zx}$) and $e_g$ ($d_{x^2-y^2}$, $d_{z^2}$) respectively. Obviously since $t_{2g}$ set contains three orbitals, it is *triply degenerate*. Similarly since $e_g$ set contains two orbitals, it is *doubly, degenerate*. In tetrahedral ligand field $t_{2g}$ and $e_g$ sets are represented as $t_2$ and *e* respectively.

Degeneracy of a D term (L = 2) = 2 L + 1 = 2 × 2 + 1 = 5. Thus D term, like *d* sub-shell, is *five-fold degenerate*. This means that in presence of octahedral ligand field D term is split into two sets of energy levels. These sets are represented by Muliken symbols *viz.* $T_{2g}$ (triply degenerate) and $E_g$ (doubly degenerate). In tetrahedral ligand field the two sets are represented as T (triply degenerate) and E (doubly degenerate).

(*iv*) Degeneracy of *f* sub-shell ($l = 3$) = $2l + 1 = 2 \times 3 + 1 = 7$. Thus *f* sub-shell is *seven-fold degenerate*. This means that in presence of octahedral ligand field *f* sub-shell is split into three sets of orbitals. These sets contain 3, 3 and 1 orbitals respectively.

Degeneracy of F term (L = 3) = 2L + 1 = 2 × 3 + 1 = 7. Thus F term, like *f* sub-shell is *seven fold degenerate.* This means that in octahedral ligand field F term is split into three sets of energy levels. These three sets are represented as $T_{1g}$ (triply degenerate), $T_{2g}$ (triply degenerate) and $A_{2g}$ (singly degenerate). In tetrahedral ligand field these sets are represented as $T_1$, $T_2$ and $A_2$ respectively.

**Table 24.3 Number and degeneracy of energy levels obtained by the splitting of S, P, D and F terms of free metal atom or ion.**

| *Spectroscopic terms of free metal atom or ion* | *No. of energy levels obtained by the splitting of terms in octahedral and tetrahedral ligand field = (2L + 1)* | *Representation of energy levels by Mulliken symbols* | |
|---|---|---|---|
| | | *In octahedral ligand field (Degeneracy is shown in bracket)* | *In tetrahedral ligand field (Degeneracy is shown in bracket)* |
| S (L = 0) | 2 × 0 + 1 = 1 | $A_{1g}$ (singly degenerate) | $A_1$ (singly degenerate) |
| P (L = 1) | 2 × 1 + 1 = 3 | $T_{1g}$ (triply degenerate) | $T_1$ (triply degenerate) |
| D (L = 2) | 2 × 2 + 1<br>= 5 (3 + 2) | $T_{2g}$ (triply degenerate)<br>$E_g$ (doubly degenerate) | $T_2$ (triply degenerate)<br>E (doubly degenerate) |
| F(L = 3) | 2 × 3 + 1<br>= 7 (3 + 3 + 1) | $T_{1g}$(triply degenerate)<br>$T_{2g}$ (triply degenerate)<br>$A_{2g}$ (singly degenerate) | $T_1$ (triply degenerate)<br>$T_2$ (triply degenerate)<br>$A_2$ (singly degenerate) |

## Spin Multiplicity of a Term : No. of Spin Energy Levels in a Term

An energy term is written as $^{2S+1}L$ where the left hand superscript, (2S + 1) is known as the spin multiplicity of the term. Spin multiplicity gives *the number of spin energy levels* or gives the number of ways in which the spins can be arranged. Thus the number of orientations of the spin vectors in a magnetic field is indicated by the spin multiplicity, (2S + 1).

A state with spin multiplicity of 1 is called *singlet state*, with multiplicity of 2 is called *doublet state*, with multiplicity of 3 is called *triplet state* etc. Since the number of unpaired electrons ($n$) in a configuration is equal to 2S ($n$ = 2S), spin multiplicity, (2S + 1) is equal ($n$ + 1) (See Table 24.4).

**Table 24.4 Relation between n, 2S and (2S + 1)**

| *No. of unpaired electrons (n) = 2S* | *S = n/2* | *Spin multplicity = (2S + 1) = (n + 1)* | *State* |
|---|---|---|---|
| 0 | 0 | $2 \times 0 + 1 = 1$ | Singlet |
| 1 | 1/2 | $2 \times \frac{1}{2} + 1 = 2$ | Doublet |
| 2 | 2/2 = 1 | $2 \times 1 + 1 = 3$ | Triplet |
| 3 | 3/2 | $2 \times \frac{3}{2} + 1 = 4$ | Quartet |
| 4 | 4/2 = 2 | $2 \times 2 + 1 = 5$ | Quintet |

## Total Degeneracy of a Term: Total Number of Energy Levels in a Term

We know that orbital degeneracy of a term is the number of **orbital energy levels in a term**. It is equal to (2 L + 1). We know that the number of spin energy levels in a term is equal to the spin multiplicity, (2S + 1) of the term. Total degeneracy of a term is the total number of energy levels in a term. Total degeneracy of a term is equal to the product of the number of orbital energy levels, (2L + 1) and number of spin energy levels, (2S + 1). Thus:

**Total degeneracy of a term designated as = $^{2S+1}L$ = (2L + 1) (2S + 1)**

All the energy levels in a term are degenerate provided that the energy of electronic arrangement of the term is solely determined from electrostatic interactions.

**Example: Find out the total number of energy levels in $^3F$ term.**

**Ans.** For $^3F$ term, 2S + 1 = 3 and L = 3 Hence 2 L + 1 = 2 × 3 + 1 = 7.

Hence:

Total number of energy levels in $^3F$ term or total degeneracy of $^3F$ term

$$= (2S + 1)(2L + 1) = 3 \times 7 = \mathbf{21}$$

## Splitting of S, P, D and F Terms

When the free metal ion present in a HS octahedral or tetrahedral complex ion is subjected to weak octahedral ligand field ($V_{oct}$) or $V_{tet}$, the ground state terms (S, D and F terms) and the first excited state term (P term) of $d^x$ configuration of the metal ion are split into various energy levels. Ground state energy terms and first excited state energy term of $d^x$ configuration ($x$ = 1 to 10) are given below. First excited state energy term is shown in parenthesis

$d^1 \rightarrow {}^2D$, $d^2 \rightarrow {}^3F\ ({}^3P)$, $d^3 \rightarrow {}^4F\ ({}^4P)$,
$d^4 \rightarrow {}^5D$, $d^5 \rightarrow {}^6S$, $d^6 \rightarrow {}^5D$, $d^7 \rightarrow {}^4F\ ({}^4P)$,
$d^8 \rightarrow {}^3F\ ({}^3P)$, $d^9 \rightarrow {}^2D$, $d^{10} \rightarrow {}^1S$

The list of S, P, D and F terms given above show that S term is $^1S$ ($d^{10}$) or $^6S$($d^5$), P term is $^3P$($d^2$,$d^8$) or $^4P$($d^3$,$d^7$), D term is $^2D$ ($d^1$,$d^9$) or $^5D$($d^4$,$d^6$) and F term is $^3F$($d^2$,$d^8$) or $^4F$ ($d^3$, $d^7$).

When a term (e.g. S, P, D, F) of a given configuration is subjected to octahedral or tetrahedral ligand field environment, this term is split into various energy levels which are represented by Mullikan symbols as shown in Table 24.5.

**Table 24.5. Various energy terms in Weak $V_{oct}$ and $V_{tet}$ ligand field environemnt**

| *Term* | *Energy levels in Weak $V_{oct}$* | *Energy levels in $V_{tet}$* |
|---|---|---|
| $^1S$ ($d^{10}$) | $^1A_{1g}$ (1) | $^1A_1$(1) |
| $^6S$ ($d^5$) | $^6A_{1g}$ (1) | $^6A_1$(1) |
| $^3P$ ($d^2$, $d^8$) | $^3T_{1g}$(*P*) (3) | $^3T_1$ (P) (3) |
| $^4P$ ($d^3$, $d^7$) | $^4T_{1g}$(*P*) (3) | $^4T_1$ (P) (3) |
| $^2D$ ($d^1$, $d^9$) | $^2T_{2g}$(3) + $^2E_g$ (2) | $^2T_2$(3) + $^2E$ (2) |
| $^5D$ ($d^4$, $d^6$) | $^5T_{2g}$(3) + $^5E_g$ (2) | $^5T_2$(3) + $^5E$ (2) |
| $^3F$($d^2$, $d^8$) | $^3T_{1g}$ (3) + $^3T_{2g}$ (3) + $^3A_{2g}$(1) | $3T_1$ (3) + $^3T_2$ (3) + $^3A_2$ (1) |
| $^4F$ ($d^3$, $d^7$) | $^4T_{1g}$ (3) + $^4T_{2g}$ (3) + $^4A_{2g}$(1) | $^4T_1$ (3) + $^4T_2$ (3) + $^4A_2$ (1) |

From Table 24.5 the following points may be noted:

(*i*) **Representation of energy levels by Mullikan symbols.** The energy levels (*i.e.* components or splittings) obtained by the splitting of S, P, D and F terms are represented by Mullikan symbols. These symbols are group theoretical symbols.

The Mullikan symbols used for octahedral complexes are $A_{1g}$, $A_{2g}$, $T_{1g}$, $T_{2g}$ and $E_g$. In these symbols the subscript *g* stands for *gerade* (even). For tetrahedral complexes the symbols used are $A_1$, $A_2$, $T_1$, $T_2$ and E. It may be seen that in the symbols used for tetrahedral complexes the symbols are written without the subscript *g*. This is because tetrahedral complexes have no centre of symmetry. Octahedral complexes are centro-symmetric.

***(ii)* Degeneracy of splittings.** The Mullikan symbols A ($A_{1g}$ / $A_1$ and $A_{2g}$ / $A_2$), T ($T_{1g}$ / $T_1$ and $T_{2g}$ / $T_2$) and E ($E_g$ and E) represent degeneracy of these components. The degeneracy of the symbols A, T and E is 1, 3 and 2 respectively. Consequently $A_{1g}$ / $A_1$ and $A_{2g}$ / $A_2$ are singly degenerate, $T_{1g}$/$T_1$ and $T_{2g}$ / $T_2$ are triply degenerate and $E_g$/E is doubly degenerate. The degeneracy of the components is shown in bracket against the Mullikan symbol of the components.

***(iii)* No. of energy levels obtained by the splitting of a term** Number of energy levels obtained by the splitting of a term is equal to the orbital degeneracy of the term which is equal to (2L + 1). L = 0, 1, 2, 3, etc. for S, P, D, F, etc. terms.

***(a)* S-term (L = 0)** For this term, since 2L + 1 = 2 × 0 + 1 = 1, this term is not split but is transformed into an energy level which is represented by the Mullikan symbol $A_{1g}$ in weak $V_{oct}$ and by $A_1$ in $V_{tet}$.

***(b)* P term (L = 1).** Since 2L + 1 = 2 × 1 + 1 = 3 for this term, this term is split into three energy levels. These three energy levels comprise a triply generate state which is represented by the Mullikan symbol $T_{1g}$ in weak $V_{oct}$ and by $T_1$ in $V_{tet}$. Thus we see that P term is transformed into $T_{1g}$ state in $V_{oct}$ and in $T_1$ in $V_{tet}$.

***(c)* D term (L = 2).** For this term, since 2L + 1 = 2 × 2 + 1 = 5, D state is split into five energy levels. These five components are *three triply degenerate levels* and *two doubly degenerate levels* which are represented by the Mullikan symbols $T_{2g}$ and $E_g$ respectively in $V_{oct}$. In $V_{tet}$ these states are represented as $T_2$ and E respectively.

(*d*) **F term (L = 3).** For this term, since 2L + 1 = 2 × 3 + 1 = 7, F term is split into seven components. These energy levels are *one singly degenerate level* ($A_{2g}$), *three triply degenerate levels* ($T_{1g}$) and *three triply degenerate levels* ($T_{2g}$) in $V_{oct}$. In $V_{tet}$ these levels are represented on $A_2$ ,$T_1$ and $T_2$ respectively.

***(iv)* Spin multiplicity of the unsplit term and its splitting.** The spin multiplicity of the unsplit term (S, P, D or F term) and its splittings is the same. This is because the electrostatic crystal field potential ($V_{oct}$ or $V_{tet}$) is not of sufficient strength to cause the electron spin pairing. For example the spin multiplicity of $^3F$ term is 3 and that of $^3T_{1g}$, $^3T_{2g}$ and $^3A_{2g}$ which are the splittings of this term is also equal to 3.

***(v)* Degeneracy of S, P, D and F terms** Degeneracy of a term can be determined by the two methods:

(*a*) Write the Mullikan symbols of the splittings of the term along with their degeneracy. The sum of the degeneracies of all the splittings gives the degeneracy of the term. Thus degeneracy of:

**S-term** = 1 (Due to singly degenerate $A_{1g}$)

**P-term** = 3 (Due to triply degenerate $T_{1g}$)

**D-term** = 3 (Due to triply degenerate $T_{2g}$) + 2 (Due to doubly degenerate $E_g$) = 5

**F-term** = 3 (Due to triply degenerate $T_{1g}$) + 3 (Due to triply degenerate $T_{2g}$) + 1 (Due to singly degenerate $A_{2g}$) = 3 + 3 + 1 = 7

(*b*) Degeneracy of a term is also equal to (2L + 1). We know that since L = 0, 1, 2, 3, etc. for S, P, D, F terms, degeneracy of S, P, D and F terms is given by :

**S-term** (L = 0) = 2L + 1 = 2 × 0 + 1 = 1

**P-term** (L = 1) = 2 × 1 + 1 = 3

**D-term** (L = 2) = 2 × 2 + 1 = 5

**F-term** (L = 3) = 2 × 3 + 1 = 7

Thus F term is seven fold degenerate

## Splitting of S Term ($^1S$ and $^6S$ Terms)

We know that $d^5$ and $d^{10}$ configurations have $^6S$ and $^1S$ as their ground state terms. Since, like *s*-sub-shell (*l* = 0), S term is orbitally non-degenerate, this term cannot split by $V_{oct}$ and $V_{tet}$ whatever is the strength of electrostaic interaction between *d* electrons and electrostatic crystal field potential, $V_{oct}$ and $V_{tat}$. S term is transformed in $A_{1g}$ (singly degenerate term). Thus $^6S$ and $^1S$ terms are transformed into $^6A_{1g}$ and $^1A_{1g}$ terms in octahedral field respectively. In tetrahedral field, $^6S$ and $^1S$ terms are transformed into $^6A_1$ and $^1A_1$ respectively.

## Splitting of P term ($^3P$ and $^4P$ Terms)

We know that $d^2$ and $d^8$ configurations have $^3P$ as their first excited state term. Similarly $d^3$ and $d^7$ configurations have $^4P$ as their first excited state term. Degeneracy of P term (L = 1) = 2L + 1 = 2 × 1 + 1 = 3. The degeneracy equal to 3 means that P term, like *p* sub-shell ($l$ = 1), is triply degenerate, but under the influence of weak $V_{oct}$ and $V_{tet}$, P term is not split into three energy levels. This is because *d* electronic charge distribution in these energy levels remains the same in all the directions. P term is transformed into $T_{1g}$ in $V_{oct}$ and into $T_1$ in $V_{tet}$. $T_{1g}$ and $T_1$ both are triply degenerate. Thus $^3P$ and $^4P$ terms are transformed into $^3T_{1g}$ and $^4T_{1g}$ in $V_{oct}$ and into $^3T_1$ and $^4T_1$ in $V_{tet}$. $T_{1g}$ and $T_1$ are designated as $T_{1g}$(P) and $T_1$(P) respectively.

## Splitting of D term ($^2D$ and $^5D$ Terms)

We know that $d^1$ and $d^9$ configurations have $^2D$ term as their ground state term while $d^4$ and $d^6$ configurations have $^5D$ as their ground state term.

***(i)* Splitting of $^2D$ term of $d^1$ and $d^9$ configurations in weak $V_{oct}$.** The splitting pattern of $^2D$ term of $d^1$ and $d^9$ configurations is given in Figures 24.9 and 24.10 respectively.

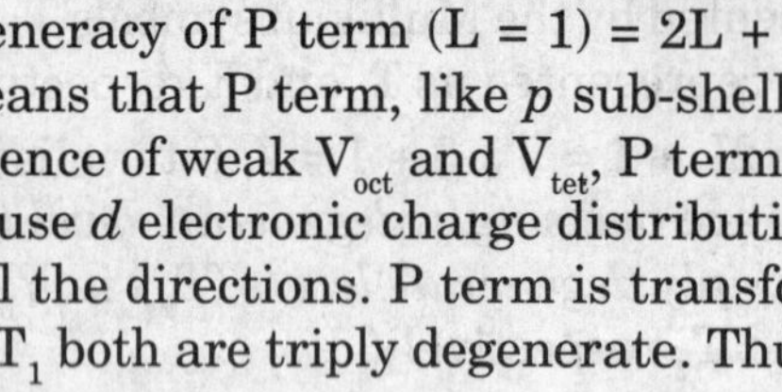


**Fig. 24.9.** Splitting of $^2D$ term of $d^1$ configuration in weak $V_{oct}$.

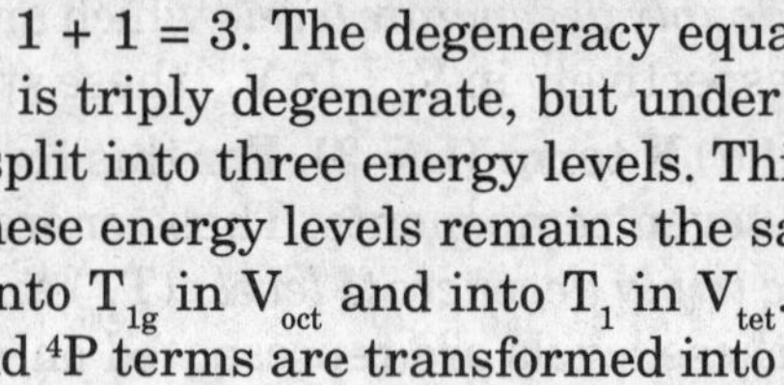


**Fig. 24.10.** Splitting of $^2D$ term of $d^9$ configuration in weak $V_{oct}$.

***(ii)* Splitting pattern of $^5D$ term of $d^4$ and $d^6$ configurations in weak $V_{oct}$.** The splitting pattern of $^5D$ term of $d^4$ and $d^6$ configurations is given in Figures 24.11 and 24.12 respectively.

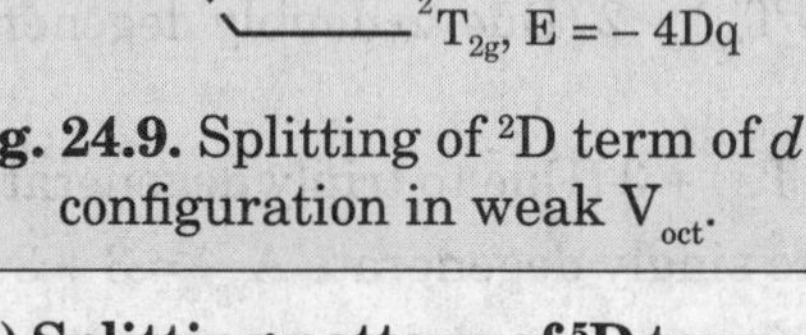


**Fig. 24.11.** Splitting of $^5D$ term of $d^4$ configuration in weak $V_{oct}$.

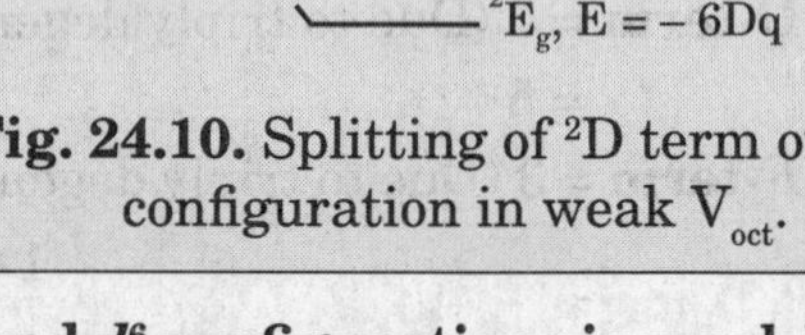


**Fig. 24.12.** Splitting of $^5D$ term of $d^6$ configuration in weak $V_{oct}$.

***(iii)* Splitting of $^2$D term of *d*$^1$ and *d*$^9$ configuration in $V_{tet}$.** The splitting pattern of $^2$D term of $d^1$ and $d^9$ configuration is given in Figures 24.13 and 24.14 respectively.

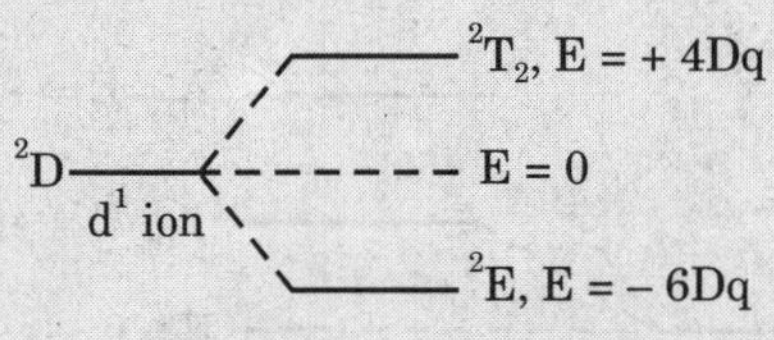


**Fig. 24.13.** Splitting of $^2$D term of $d^1$ configuration in $V_{tet}$.

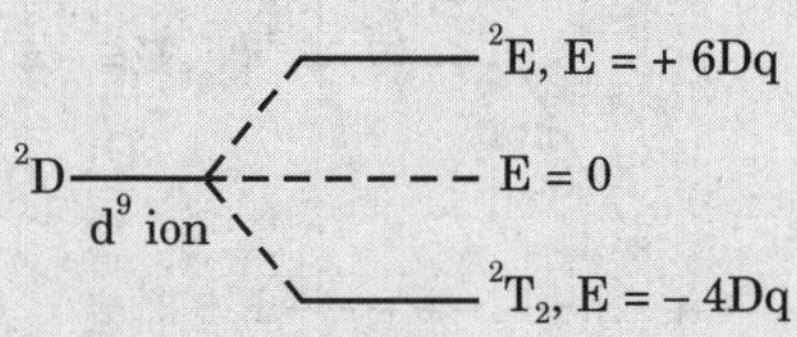


**Fig. 24.14.** Splitting of $^2$D term of $d^9$ configuration in $V_{tet}$.

***(iv)* Splitting of $^5$D term of *d*$^4$ and *d*$^6$ configurations in $V_{tet}$.** The splitting pattern of $^5$D term of $d^4$ and $d^6$ configurations is given in Figures 24.15 and 24.16 respectively.

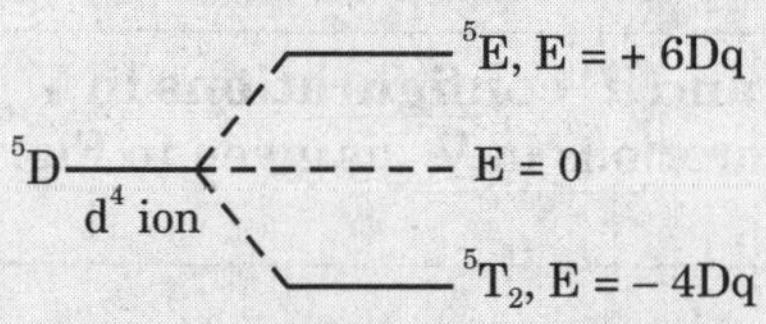


**Fig. 24.15.** Splitting of $^5$D term of $d^4$ configuration in $V_{tet}$.

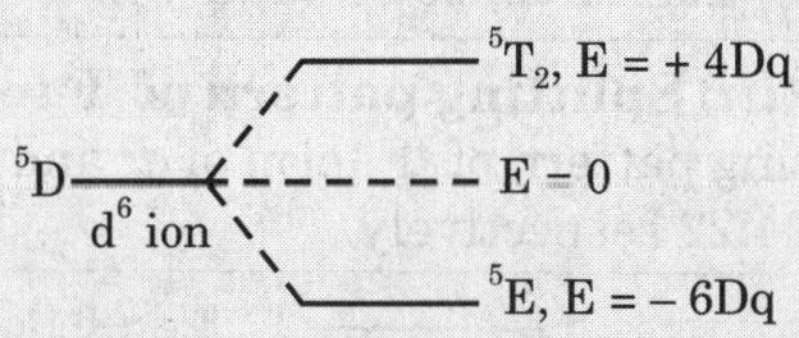


**Fig. 24.16.** Splitting of $^5$D term of $d^6$ configuration in $V_{tet}$.

## Splitting of F term ($^3$F and $^4$F Terms)

We know that $d^2$ and $d^8$ configurations have $^3$F as their ground state term while $d^3$ and $d^7$ configurations have $^4$F as their ground state term.

***(i)* Splitting of $^3$F term of *d*$^2$ and *d*$^8$ configurations in weak $V_{oct}$.** The splitting pattern of $^3$F term of $d^2$ and $d^8$ configurations in weak $V_{oct}$ is given in Figures 24.17 and 24.18 respectively.

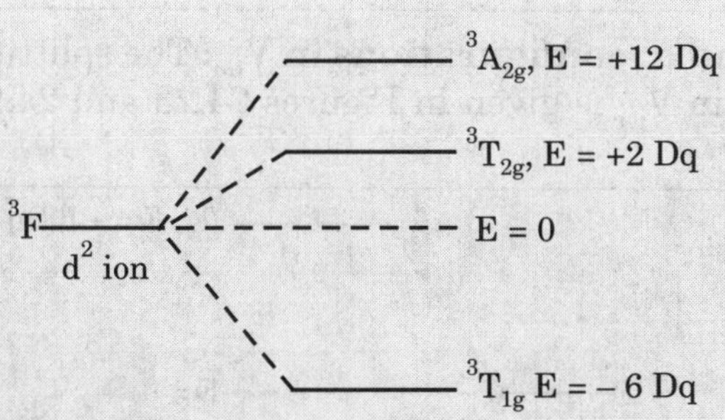


**Fig. 24.17** Splitting pattern of $^3$F term of $d^2$ configuration under the influence of weak $V_{oct}$.

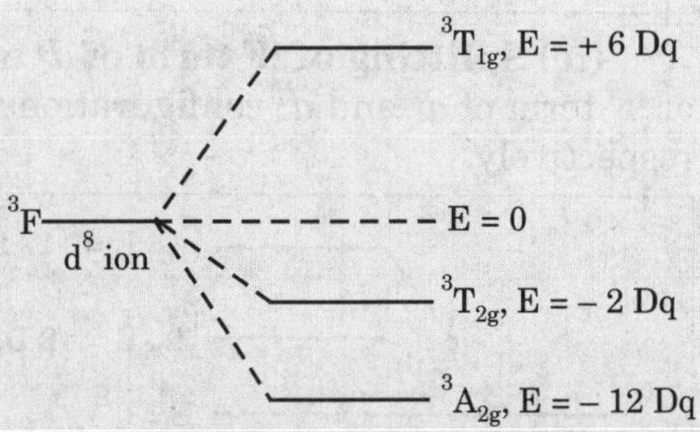


**Fig. 24.18** Splitting pattern of $^3$F term of $d^8$ configuration under the influence of weak $V_{oct}$.

***(ii)* Splitting pattern of $^4F$ term of $d^3$ and $d^7$ configurations in weak $V_{oct}$.** The splitting pattern of $^4F$ term of $d^3$ and $d^7$ configurations in weak $V_{oct}$ is given in Figures 24.19 and 24.20 respectively.

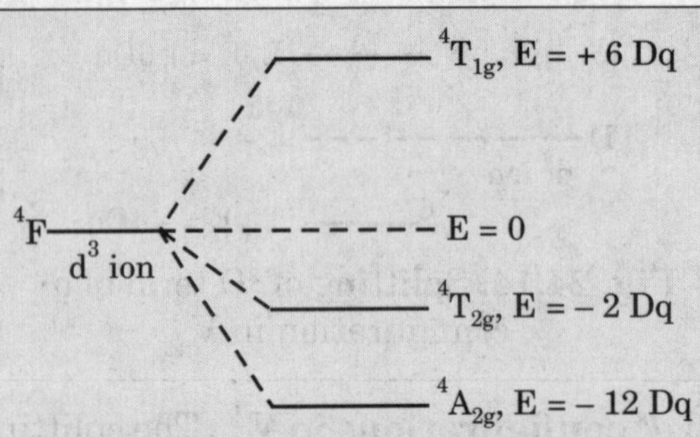


**Fig. 24.19** Splitting of $^4F$ term of $d^3$ configuration under the influence of weak $V_{oct}$.

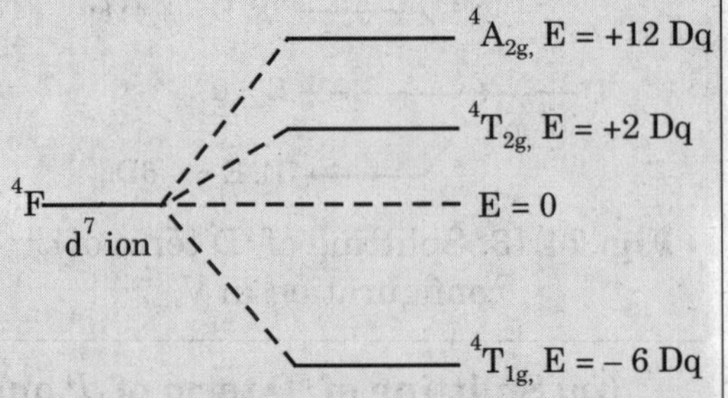


**Fig. 24.20** Splitting of $^4F$ term of $d^7$ configuration under the influence of weak $V_{oct}$.

***(iii)* Splitting pattern of $^3F$ term of $d^2$ and $d^8$ configurations in $V_{tet}$.** The splitting pattern of $^3F$ term of $d^2$ and $d^8$ configurations in $V_{tet}$ is given in Fig. 24.21 and 24.22 respectively.

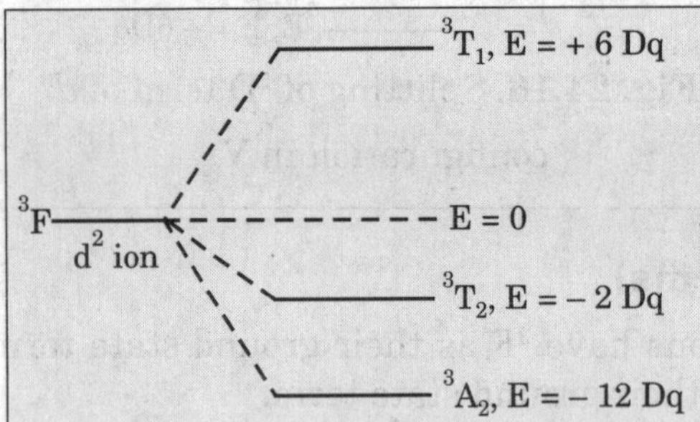


**Fig. 24.21** Splitting of $^3F$ term of $d^2$ configuration under the influence of $V_{tet}$.

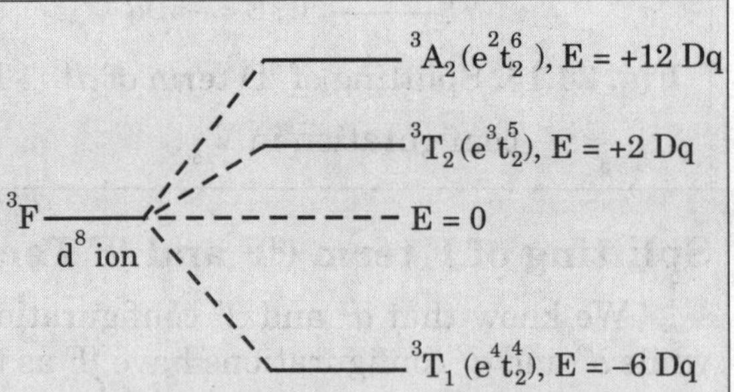


**Fig. 24.22** Splitting of $^3F$ term of $d^8$ configuration under the influence of $V_{tet}$.

***(iv)* Splitting of $^4F$ term of $d^3$ and $d^7$ configurations in $V_{tet}$.** The splitting of $^4F$ term of $d^3$ and $d^7$ configurations in $V_{tet}$ is given in Figures 24.23 and 24.24 respectively.

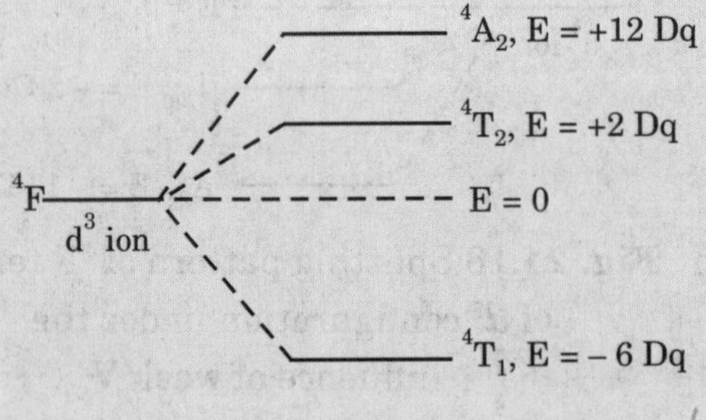


**Fig. 24.23** Splitting of $^4F$ term of $d^3$ configuration in $V_{tet}$.

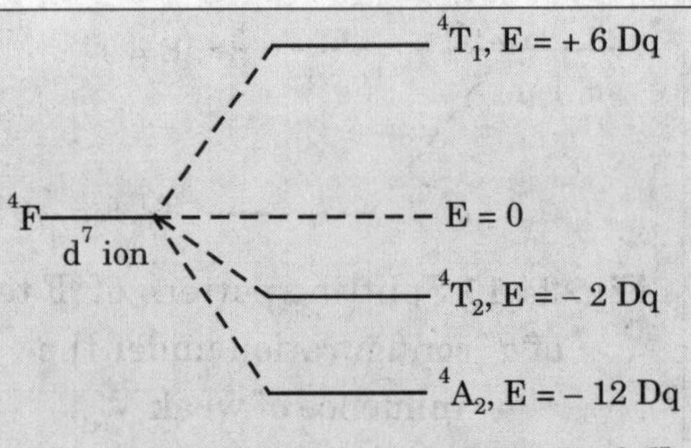


**Fig. 24.24** Splitting of $^4F$ term of $d^7$ configuration in $V_{tet}$.

## Some Important Points about the Splitting of Terms

The following points may be noted from the discussion on the splitting of S, P, D and F terms of $d^n$ configuration under weak $V_{oct}$ and $V_{tet}$. The diagrams showing the splitting of S, P, D and F terms of different $d^n$ configurations are called **orgel diagrams.** These diagrams make the following points clear.

(*i*) The energy of the splittings obtained by the splitting of $^2$D ($d^1$, $d^9$), $^5$D ($d^4$, $d^6$), $^3$F ($d^2$, $d^8$) and $^4$F ($d^3$, $d^7$) terms increases in the following order:

**$^2$D($d^1$)** : $^2T_{2g}$ (–4Dq) < $^2E_g$ (+6 Dq) (Fig 24.9)

**$^2$D($d^9$)** : $^2E_g$ (–6Dq) < $^2T_{2g}$ (+4 Dq) (Fig. 24.10)

**$^5$D($d^4$)** : $^5E_g$ (–6Dq) < $^5T_{2g}$ (+4 Dq) (Fig. 24.11)

**$^5$D($d^6$)** : $^5T_{2g}$ (–4Dq) < $^5E_g$ (+6 Dq) (Fig. 24.12)

**$^3$F($d^2$)** : $^3T_{1g}$ (–6Dq) < $^3T_{2g}$ (+2 Dq) < $^3A_{2g}$ (+12 Dq) (Fig. 24.17)

**$^3$F($d^8$)** : $^3A_{2g}$ (–12Dq) < $^3T_{2g}$ (–2 Dq) < $^3T_{1g}$ (+6 Dq) (Fig. 24.18)

**$^4$F($d^3$)** : $^4A_{2g}$ (–12Dq) < $^4T_{2g}$ (–2 Dq) < $^4T_{1g}$ (+6 Dq) (Fig. 24.19)

**$^4$F($d^7$)** : $^4T_{1g}$ (–6Dq) < $^4T_{2g}$ (+2 Dq) < $^4A_{2g}$ (+12 Dq) (Fig. 24.20)

(*ii*) The splitting pattern of the term of $d^n$ configuration is the reverse of the splitting pattern of the same term of $d^{10-n}$ configuration provided that the ligand environment of both the configurations is the same. Thus:

| |
|---|
| $d^n$ (octahedral) = Reverse of $d^{10-n}$ (octahedral) |
| $d^n$ (tetrahedral) = Reverse of $d^{10-n}$ (tetrahedral) |

For example:

(*a*) The splitting pattern of $^2$D term of $d^1$ configuration is the reverse of the splitting pattern of $^2$D term of $d^{10-1}$ or $d^9$ configuration under the influence of weak $V_{oct}$ as well as $V_{tet}$. (See Figure 24.25)

(*b*) She splitting pattern of $^5$D term of $d^6$ configuration is the reverse of the splitting pattern of $^5$D term of $d^{10-6}$ or $d^4$ configuration under the influence of weak $V_{oct}$ as well as $V_{tat}$ (See Fig. 24.26)

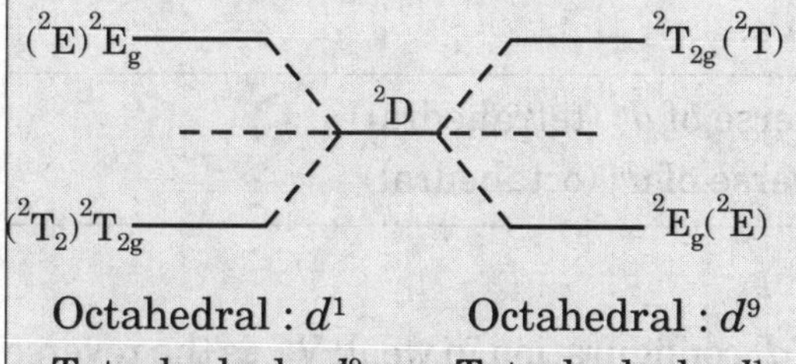

**Fig. 24.25.** Splitting of $^2$D term of $d^1$ and $d^9$ configurations in $V_{oct}$ and $V_{tet}$

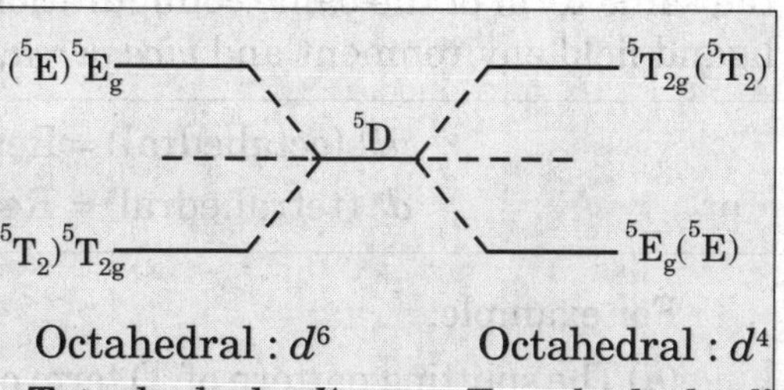

**Fig. 24.26.** Splitting of $^5$D term of $d^4$ and $d^6$ configurations in $V_{oct}$ and $V_{tet}$

(*iii*) The splitting patterns of the terms of $d^n$ and $d^{n+5}$ ($n < 5$) configurations are the same provided that the ligand environment for both the configurations are the same. Thus splitting pattern of the terms of the two configurations present in the pairs *viz* ($d^1$, $d^{1+5}$ or $d^6$), ($d^2$, $d^{2+5}$ or $d^7$), ($d^3$, $d^{3+5}$ or $d^8$) and ($d^4$, $d^{4+5}$ or $d^9$) are the same in the influence of $V_{oct}$ and $V_{tet}$ both. For example:

(*a*) Under the influence of $V_{oct}$ and $V_{tet}$ both the splitting patterns of $^2D$ term of $d^1$ configuration is the same as the splitting pattern of $^5D$ term of $d^{1+5}$ or $d^6$ configuration. (See Figures 24.9 and 24.12).

(*b*) Under the influence of $V_{oct}$ and $V_{tet}$ both the splitting patterns of $^3F$ term of $d^2$ configuration is the same as the splitting pattern of $^4F$ term of $d^{2+5}$ or $d^7$ configuration (See Figures 24.17 and 24.20).

Points (*ii*) and (*iii*) given above can be combined together in the form of a digram given in Fig. 24.27.

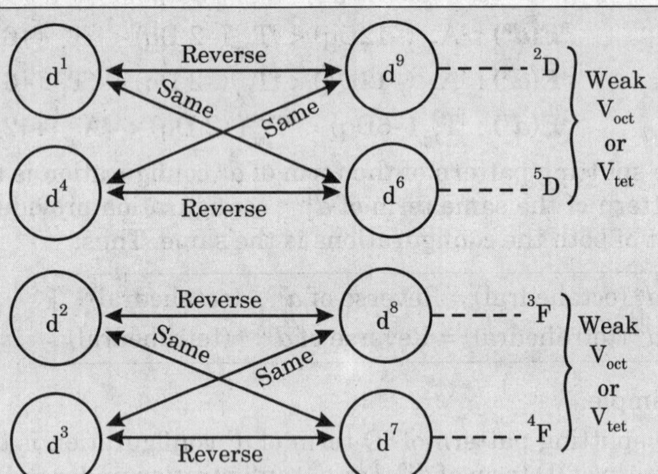


**Fig 24.27** Combination of points (*ii*) and (*iii*)

(*iv*) The splitting pattern of a term of $d^n$ configuration under the influence of weak octahedral ligand field environment is the reverse of the splitting pattern of the same term of the same configuration (*i.e.* $d^n$ configuration) under tetrahedral ligand field environment and *vice versa*, *i.e.*

$$d^n \text{ (octahedral)} = \text{Reverse of } d^n \text{ (tetrahedral)}$$

or

$$d^n \text{ (tetrahedral)} = \text{Reverse of } d^n \text{ (octahedral)}$$

For example:

(*a*) The splitting pattern of $^2D$ term of $d^1$ configuration in weak $V_{oct}$ is the reverse of the splitting pattern of the same term (*i.e.* $^2D$ term) of the same configuration (*i.e.* $d^1$ configuration) in $V_{tet}$. (See Figures 24.9 and 24.13)

(*b*) The splitting pattern of $^3F$ term of $d^2$ configuration in weak $V_{oct}$ is the reverse of the splitting pattern of the same term (*i.e.* $^3F$ term) of the same configuration (i.e. $d^2$ configuration) in $V_{tet}$. (See Figures 24.17 and 24.21)

($v$) The splitting patter of a term of $d^n$ configuration under the influence of weak $V_{oct}$ is the same as the splitting pattern of the term of $d^{10-n}$ configuration under the influence of $V_{tet}$ and *vice versa*.

## Electronic Absorption Spectra of Weak (H.S.) Octahedral Complex Ions

Since $H_2O$ is a weak ligand, hexaquated complex ions, $[M^{n+}(H_2O)_6]^{n+}$ are weak (spin free or high spin) octahedral complex ions. Here we shall interpret the absorption bands observed in the electronic absorption spectra of $[M^{n+}(H_2O)_6]^{n+}$ ions which are weak octahedral complex ions. The interpretation is based on **Orgel diagrams** for these complexes. These diagrams are used for the interpretation of the **spin allowed absorption bands** of the spectra of crystal field or $d \rightarrow d$ origin in the electronic spectra of tetrahedral and octahedral complexes of transition metals.

The ground state energy terms, first excited state energy terms and other excited state energy terms of $d^x$ configuration ($x$ = 1 to 10) are given below in Table 24.6.

**Table 24.6. Ground state, first excited state and other excited state terms of $d^x$ configurations ($x$ = 1 to 10)**

| $d^x$ *configuration* (*x = 1 to 10*) | *Ground state energy terms* | *First excited state energy terms* | *Other excited state energy terms* |
|---|---|---|---|
| $d^1$ | $^2D$ (Doublet D- state) | — | — |
| $d^2$ | $^3F$ (Triplet F-state) | $^3P$ | $^1D$, $^1G$, $^1S$ |
| $d^3$ | $^4F$ (Quartet F-state) | $^4P$ | $^2H$, $^2G$, $^2F$, $^2D$, $^2P$ |
| $d^4$ | $^5D$ (Quintet D-state) | — | $^3H$, $^3G$, $^3F$, $^3P$, etc. |
| $d^5$ | $^6S$ (Sextet S-state) | — | $^4G$, $^4P$, $^4D$, $^4F$, etc. |
| $d^6$ | $^5D$ (Quintet D-state) | — | $^3H$, $^3G$, $^3F$, $^3P$, etc. |
| $d^7$ | $^4F$ (Quartet F-state) | $^4P$ | $^2H$, $^2G$, $^2F$, $^2D$, $^2P$ |
| $d^8$ | $^3F$ (Triplet F-state) | $^3P$ | $^1D$, $^1G$, $^1S$ |
| $d^9$ | $^2D$ (Doublet D-state) | — | — |
| $d^{10}$ | $^1S$ (Singlet S-state) | — | — |

Above table shows that the ground state terms and first excited state terms of different configurations can be written as shown below. First excited state is given in bracket.

$d^1 \rightarrow {}^2D$; $d^2, d^8 \rightarrow {}^3F\ ({}^3P)$, $d^3, d^7 \rightarrow {}^4F\ ({}^4P)$

$d^4, d^6 \rightarrow {}^5D$, $d^5 \rightarrow {}^6S$, $d^9 \rightarrow {}^2D$, $d^{10} \rightarrow {}^1S$

It may be seen that the spin multiplicity of ground state energy terms and first excited state energy terms of ($d^2$, $d^8$) and ($d^3$, $d^7$) pairs of configurations is the same. When metal ion having these terms is subjected to weak octahedral ligand field ($V_{oct}$) generated by $6H_2O$ weak ligands, these terms are split into various energy terms. The energy terms obtained by the splitting of ground state and first excited state energy terms have the same multiplicity as the unsplit terms, since the electrostatic crystal field potential is not sufficiently strong to cause the spin pairing. Other excited state energy terms shown in the last column of Table 24.6 do not participate in spin allowed transitions, since the spin multiplicity of these terms is different from that of ground state energy terms.

The splitting of ground state and first excited state terms of $d^x$ ion under the influence of weak $V_{oct}$ or $V_{tet}$ is shown in an energy level diagram which is called **Orgel diagram**. In orgel diagram the abscissa (horizontal axis) shows the change in the ligand field splitting energy of $V_{oct}$ or $V_{tet}$ in terms of 10 Dq (or $\Delta_0$) or $\Delta_t$ and the ordinate (vertical axis) shows the corresponding change in the energies of the energy levels obtained by the splitting of the parent terms.

In the splitting of energy terms of $d^x$ ion we have assumed that the influence of $d - d$ inter-electronic repulsions is stronger than the electrostatic interactions between $V_{oct}$ or $V_{tet}$ and $d$-electrons on the energies of metal ion in its complexes. Therefore, while drawing orgel diagrams we have first considered the generation of energy terms of $d^x$ configuration of the free metal ion (e.g. S, P, D, F, etc. energy terms) due to $d - d$ inter electronic repulsions and then we have considered the splitting of these terms under the influence of $V_{oct}$ or $V_{tet}$.

The components (energy terms) obtained by the splitting of ground state and first excited state terms of $d^x$ configuration under the influence of weak $V_{oct}$ are given in Table 24.7. These energy terms have been represented by Mulliken symbols *viz* A, E and T. The energy of each component (in the units of Dq) has been shown in bracket. In this table the term A is singlet, E is doublet and T is triplet. The left hand **superscript** is the spin multiplicity. Right hand **subscript** g indicates gerade, *i.e.* even orbitals which retain the same sign of orbital wave function on inversion about the nucleus (centre of octahedrun). The numbers 1 and 2 in the subscript have group theoretical origin. Since a tetrahedron has no inversion (centre of symmetry), the subscript g which is used for octahedral complexes is dropped for tetrahedral complexes.

**Tables 24.7. Energy terms obtained by the splitting of ground state and first excited state terms of $d^x$ configuration ($x$ = 1 to 10)**

| *$d^x$ configuration ($x$ = 1 to 9)* | *Ground state and first excited state terms* | *Components (Energy terms) obtained by splitting* |
|---|---|---|
| $d^1$ | $^2D$ | $^2T_{2g}$(– 4Dq), $^2E_g$ (+ 6$Dq$) |
| $d^2$ | $^3F$ | $^3T_{1g}$(– 6Dq), $^3T_{2g}$ (+ 2$Dq$), $^3A_{2g}$ (+ 12 $Dq$) |
| | $^3P$ (Excited state term) | $^3T_{1g}$ ($P$) (+ 15 $B$) |
| $d^3$ | $^4F$ | $^4A_{2g}$(–12$Dq$), $^4T_{2g}$ (– 2$Dq$), $^4T_{1g}$ (+ 6$Dq$) |
| | $^4P$ (Excited state term) | $^4T_{1g}$ ($P$) (+15$B$) |
| $d^4$ | $^5D$ | $^5E_g$ (–6$Dq$), $^5T_{2g}$ (+4$Dq$) |
| $d^5$ | $^6S$ | $^6A_{1g}$ (0) |
| $d^6$ | $^5D$ | $^5T_{2g}$ (–4$Dq$), $^5E_g$ (+6$Dq$) |
| $d^7$ | $^4F$ | $^4T_{1g}$ (–6$Dq$), $^4T_{2g}$ (+2$Dq$), $^4A_{2g}$ (+ 12 $Dq$) |
| | $^4P$ (Excited state term) | $^4T_{1g}$ ($P$) (+ 15 $B$) |
| $d^8$ | $^3F$ | $^3A_{2g}$ (–12$Dq$), $^3T_{2g}$ (– 2 $Dq$), $^3T_{1g}$ (+6$Dq$) |
| | $^3P$ (Excited state term) | $^3T_{1g}$ ($P$) (+ 15 $B$) |
| $d^9$ | $^2D$ | $^2E_g$ (–6$Dq$), $^2T_{2g}$ (+ 4 $Dq$) |
| $d^{10}$ | $^1S$ | $^1A_{1g\,(0)}$ |

The interpretation of absorption bands of HS octahedral complexes of $d^1$, $d^2$ ... $d^{10}$ ions is given below.

**1. Octahedral Complexes of $d^1$ Ion.**

The octehedral complex ion, $[Ti(H_2O)_6]^{3+}$ consists of $Ti^{3+}$ ion which is a $d^1$ ion. The absorption spectrum of this ion consists of one absorption band of crystal field or $d - d$ origin at 20,000 cm$^{-1}$ (See Fig. 24.28)

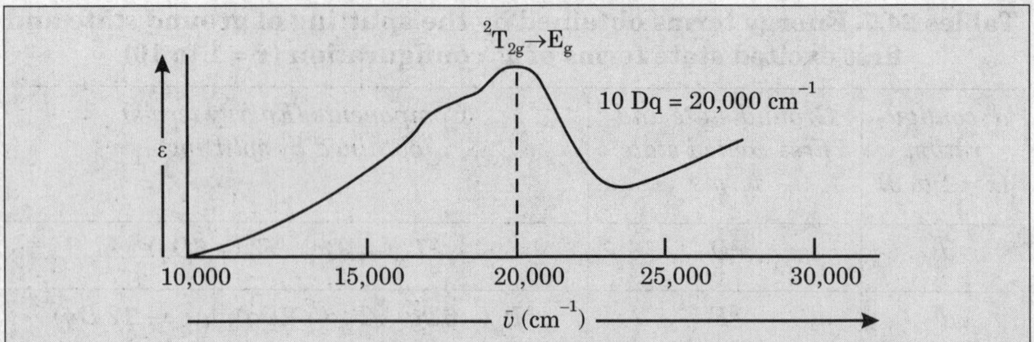


**Fig. 24.28.** Crystal field absorption band of $[Ti(H_2O)_6]^{3+}$ ion in aqueous solution ($Ti^{3+} = 3d^1$).

The aqueous solution of this ion is purple in colour. The absorption band occurs in the green and yellow region and hence these colours are filtered out. Consequently blue and red light is transmitted and $[Ti\ (H_2O)_6]^{3+}$ solution appears purple. The origin of absorption band appearing at 20,000 cm$^{-1}$ can be explained as follows:

We know that $^2D$ term is the ground state term of a free ion having $d^1$ configuration ($Ti^{3+}$ ion has $d^1$ configuration). In presence of six $H_2O$ ligands, $^2D$ term is split up into $^2T_{2g}$ (triply degenerate energy level) and $^2E_g$ (doubly degenerate energy level) states as shown in Fig. 24.29.

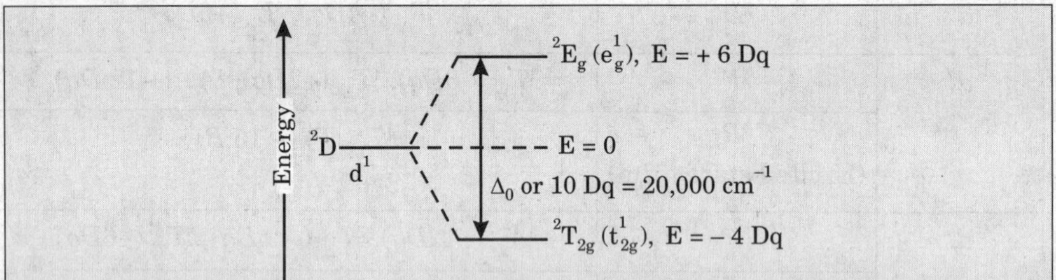


**Fig 24.29** Orangel diagram (energy level diagram) showing the splitting of $^2D$ term of $d^1$ configuration

$^2T_{2g}$ and $^2E_g$ states correspond to $t_{2g}^1$ and $e_g^1$ configuration. $^2T_{2g}$ term has lower energy (ground state energy level) than $^2E_g$ term (excited state energy level). The absorption band appearing at 20, 000 cm$^{-1}$ can be due to one spin allowed but Laporte forbidden electronic transition namely

$$^2T_{2g} \longrightarrow {}^2E_g\ (t_{2g}^1 \longrightarrow e_g^1).$$

Since $d^1$ configuration has only one electron, there is no electron–electron repulsions. Hence the separation between $^2T_{2g}$ and $^2E_g$ terms is the same as the separation between $t_{2g}$ and $e_g$ orbitals which is equal to $\Delta_0$ or 10Dq. The rather broad,asymmetric nature of the absorption band is ascribed to John Teller effect. The complex ion of $d^1$ ion having stronger ligand will absorb at higher energy. For example $[Ti\ Br_6]^{3-}$ absorbs at 10,500 cm$^{-1}$ while $[Ti\ (CN)_6]^{3-}$ absorbs at 22,300 cm$^{-1}$. The following examples may be noted: $[Ti\ Br_6]^{3-}$ = 10,500 cm$^{-1}$, $[Ti\ Cl_6]^{3-}$

= 13,000 cm$^{-1}$, [Ti (urea)$_6$]$^{3+}$ = 17300, [Ti ($H_2O$)$_6$]$^{3+}$ = 20000 cm$^{-1}$, (Ti (bpy)$_3$]$^{3+}$ = 21,200 cm$^{-1}$, [Ti (CN)$_6$]$^{3-}$ = 22,300 cm$^{-1}$.

### 2. Octahedral Complexes of $d^2$ Ion.

The octahedral complex ion, [V ($H_2O$)$_6$]$^{3+}$ consists of $V^{3+}$ ion which is a $d^2$ ion. The absorption spectrum of this ion consists of two absorption bands at 17,200 cm$^{-1}$ and 25,600 cm$^{-1}$ as shown in Fig. 24.30.

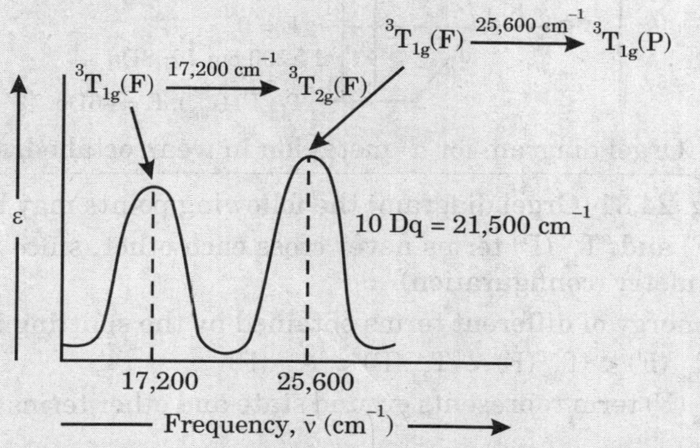


**Fig. 24.30** Electronic spectra of crystal field origin of [V($H_2O$)$_6$]$^{3+}$ complex ion in aqueous medium ($V^{3+}$ = 3d$^2$).

The aqueous solution of this ion is green in colour. The origin of the appearance of the two absorption bands can be explained as follows.

We know that the energy terms arising out of $d^2$ configuration due to inter-electronic repulsion between the two electrons are $^3F$, $^3P$, $^1G$, $^1D$ and $^1S$. Applying Hund's rules, we find that $^3F$ term is the ground state energy term (lowest energy term) and $^3P$ term has higher energy than $^3F$. Thus $^3F$ term is the ground state term and $^3P$ term is the first excited state term. Both these terms have the same multiplicity, (2S + 1) equal to 3 (No. of unpaired electrons = 2S = 2). Thus these terms are triply degenerate (degeneracy = 3) and other terms *viz* $^1G$, $^1D$ and $^1S$ are singlet states. These singlet state terms are ignored because electronic transition from $^3F$ to any of these singlet states is spin (or multiplicity) forbidden. Spin transitions are either not observed or are very weak. Hence we shall consider the splitting of $^3F$ and $^3P$ terms only in octahedral field generated by six $H_2O$ ligands.

We know that the Mulliken symbol of $^3P$ term of a free ion in octahedral field is $^3T_{1g}$. Thus in this field $^3P$ term remains unsplit and is transformed into $^3T_{1g}$ (P) state. Hence this term remains orbitally triply degenerate. In weak octahedral field $^3F$ term splits up into $^3T_{1g}$ (F) (triply degenerate), $^3T_{2g}$ (F) (triply degenerate), and $^3A_{2g}$ (F) (singlet state) states. Thus total number of states obtained by the splitting of $^3F$ term is equal to 3 + 3 + 1 = 7 . Out of these states $^3T_{1g}$ (F) has the lowest energy (= – 6 Dq). Energy level diagram (called orgel diagram) showing the splitting of $^3F$ and $^3P$ terms in weak octahedral field generated by six $H_2O$ molecules (ligands) in [V ($H_2O$)$_6$]$^{3+}$ ion is shown in Fig. 24.31.

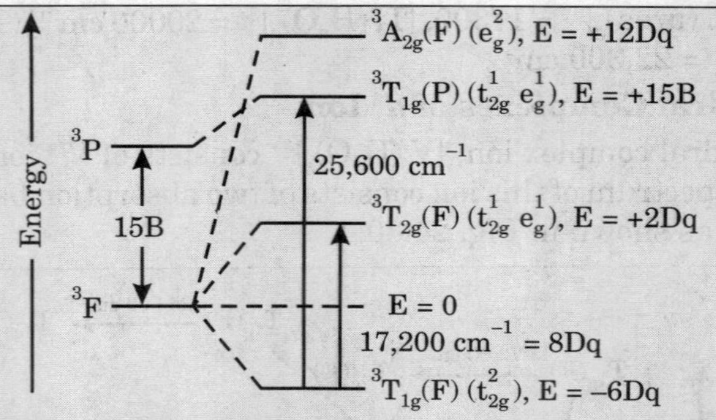


**Fig 24.31** Orgel diagram for $d^2$ metal ion in weak octahedral ligand field.

From Fig. 24.31 (Orgel diagram) the following points may be noted:

(*i*) $^3T_{1g}$ (F) and $^3T_{1g}$ (P) terms never cross each other, since these terms have the same symmetry (configuration)

(*ii*) The energy of different terms obtained by the splitting is in the order:

$$^3T_{1g}\ (F) < {}^3T_{2g}\ (F) < {}^3T_{1g}\ (P) < {}^3A_{2g}\ (F)$$

Thus $^3T_{1g}$ (F) term represents ground state and other terms represent excited states.

(*iii*) Theoretically the following three spin allowed but Laportee forbidden electronic transitions are expected in case of $[V(H_2O)_6]^{3+}$ ion.

(*a*) $^3T_{1g}$ (F) $\xrightarrow{17,200\ cm^{-1}}$ $^3T_{2g}$ (F) (*Lowest energy*)

(*b*) $^3T_{1g}$ (F) $\xrightarrow{25,600\ cm^{-1}}$ $^3T_{1g}$ (P)

(*c*) $^3T_{1g}$ (F) $\longrightarrow$ $^3A_{2g}$ (F) $\longrightarrow$ (*Highest energy*)

Actually only the first two transitions take place in $[V(H_2O)_6]^{3+}$ ion due to the absorption bands (peaks) at frequency ($\bar{\nu}$) of 17,200 cm$^{-1}$ and 25,600 cm$^{-1}$ respectively. The electronic transition $^3T_{1g}$ (F) $\longrightarrow$ $^3A_{2g}$ (F) refers to the simultaneous excitation of both the electrons from $t_{2g}$ to $e_g$ orbitals. This transition is either not observed or is very weak, since it is a forbidden transition. This transition is masked by an intense charge transfer band.

(*v*) The energy gap between $^3T_{2g}$ (F) and $^3T_{1g}$ (F) terms is equal to +2D*q* – (–6D*q*) = 2D*q* + 6D*q* = 8D*q* = 17,200 cm$^{-1}$. Hence:

$$\because \quad 8\ Dq = 17,200\ cm^{-1}$$

$$\because \quad 10\ Dq = \frac{17200}{8} \times 10\,cm^{-1} = 21,500\ cm^{-1}$$

(*vi*) $^3T_{1g}$ (F), $^3T_{2g}$ (F), $^3T_{1g}$ (P) and $^3A_{2g}$ (F) terms can be equated to $t^2_{2g}$, $t^1_{2g}\ e_g^1$, $t^1_{2g}\ e_g^1$ and $e_g^2$ respectively.

(*vii*) The energy gap (energy separation) between $^3F$ and $^3P$ terms is equal to 15B where B is Racah inter electronic repulsion parameter. The parameter B is a function of ligand, metal ion and stereochemistry.

**To get the calculated value of frequency ($v$) of transitions ($b$) and ($c$)**

The transitions ($b$) and ($c$) given above are again written as shown below:

($b$) $^3T_{1g}$ (F) $\longrightarrow$ $^3T_{1g}$ (P)

The energy difference between the two energy levels involved in this transition = 15B — ($-6Dq$) = 15B + 6D$q$. Here 15B is the energy difference between unsplit $^3$P and $^3$F terms.

($c$) $^3T_{1g}$ (F) $\longrightarrow$ $^3A_{2g}$ (F)

The energy difference between the two energy levels involved in this transition = 12D$q$ – ($-6Dq$) = 18 D$q$

Before calculating the values of frequency (ν) of the above two transitions, we should calculate the value of Dq. This can be calculated as follows:

We know that for $^3T_{1g}$ (F) $\longrightarrow$ $^3T_{2g}$ (F) transition,

$$17200 \text{ cm}^{-1} = +2Dq - (-6Dq) = 8Dq$$

$$\therefore \qquad Dq = \frac{17200}{8} \text{ cm}^{-1} = 2150 \text{ cm}^{-1}$$

The value of D$q$ (= 2150 cm$^{-1}$) can be used for calculating the values of frequencies of ($b$) and ($c$) transitions as shown below.

ν for transistion ($b$) = 15B + 6D$q$

= 15 B + 6 × 2150

= 15 B + 12900

The vale of 15 B has been found to be equal to 12,900 cm$^{-1}$.

∴ ν for transition ($c$) = 18 Dq = 18 × 2150 = 29,700 cm$^{-1}$

This value shows that a band corresponding to transition ($c$) should be observed at 29,700 cm$^{-1}$. However, this band is masked by strong charge transfer band in ultra violet region.

**3. Octahedral Complexes of $d^3$ Ion.**

The octahedral complex ion, $[Cr(H_2O)_6]^{3+}$ consists of $Cr^{3+}$ ion which is a $d^3$ ion. The aqueous solution of $[Cr(H_2O)_6]^{3+}$ is light green. The solution of this ion has absorption bands at 17,000 cm$^{-1}$ and 24,000 cm$^{-1}$. There is a weak absorption band at 37,000 cm$^{-1}$ (See Fig. 24.32.)

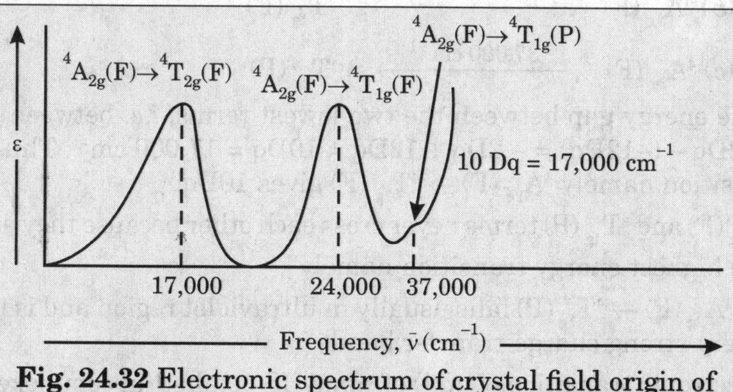


**Fig. 24.32** Electronic spectrum of crystal field origin of $[Cr(H_2O)_6]^{3+}$ in aqueous medium. ($Cr^{3+}$ = 3d$^3$)

We know that $d^3$ ion with $t^3_{2g}\, t^0_g$ configuration in weak octahedral field has $^4F$ term as its ground state term (*lowest energy*) and $^4P$ term as its first excited state term (*higher energy*). Both these terms have the same multiplicity(= 4).

In order to interpret the different bands appearing in the absorption spectrum of $[Cr(H_2O)_6]^{+3}$ we proceed as follows:

Under the influence of $6H_2O$ ligands, $^4P$ term remains unsplit and is transformed into $^4T_{1g}$ (P) state in octahedral symmetry. Thus this term remains orbitally triply generate. $^4F$ term splits into $^4A_{2g}$ (F) $^4T_{2g}$(F) and $^4T_{1g}$(F) terms (See Fig. 24.33)

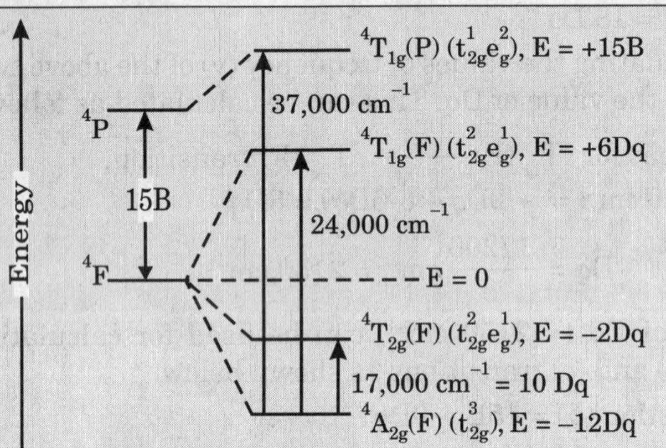


**Fig. 24.33** Orgel diagram for $d^3$ ion in octahedral environment.

From Fig. 24.33 (Orgel diagram) the following points may be noted:

(*i*) The energy of different terms resulted from the splitting is in the order:

$$^4A_{2g}\,(F) < {}^4T_{2g}\,(F) < {}^4T_{1g}\,(F) < {}^4T_{1g}\,(P)$$

Thus $^4A_{2g}$ (F) term represents ground state and other terms represent excited state terms.

(*ii*) Following three spin allowed but Laporte forbidden electronic transitions are responsible for the three absorption bands.

(*a*) $^4A_{2g}\,(F) \xrightarrow{17{,}000\ cm^{-1}} {}^4T_{2g}\,(F)$

(*b*) $^4A_{2g}\,(F) \xrightarrow{24{,}000\ cm^{-1}} {}^4T_{1g}\,(F)$

(*c*) $^4A_{2g}\,(F) \xrightarrow{37{,}000\ cm^{-1}} {}^4T_{1g}\,(P)$

(*iii*) The energy gap between the two lowest terms, *i.e.* between $^4T_{2g}$ (F) and $^4A_{2g}$ (F) = –2Dq – (–12Dq) = – 2Dq + 12Dq = 10Dq = 17,000 cm$^{-1}$. Thus the lowest energy transition namely $^4A_{2g}$ (F) → $^4T_{2g}$ (F) gives 10Dq.

(*iv*) $^4T_{1g}$ (F) and $^4T_{1g}$ (P) terms never cross each other because they are identical.

(*v*) The highest energy transition namely

$^4A_{2g}$ (F) → $^4T_{1g}$ (P) falls usually in ultraviolet region and is mostly overlapped by very strong charge-transfer bands.

The electronic transitions (cm$^{-1}$) of some HS octahedral complexes of $Cr^{3+}$ ($d^3$ configuration) are given in Table. 24.8.

**Table 24.8. Electronic transitions ($cm^{-1}$) of some octahedral complexes of $Cr^{3+}$ ($d^3$ ion)**

| Complex ions | $^4A_{2g}$ (F) → $^4T_{2g}$ (F) | $^4A_{2g}$ (F) → $^4T_{1g}$ (F) | $^4A_{2g}$ (F) → $^4T_{1g}$ (P) |
|---|---|---|---|
| $[CrF_6]^{3-}$ [Green] | 14900 | 22700 | 34400 |
| $[Cr(H_2O)_6]^{3+}$ [Violet] | 17400 | 24500 | 38600 |
| $[Cr(ox)_3]^{3-}$ [Red-violet] | 17500 | 23000 | — |
| $[Cr(en)_3]^{3+}$ [Yellow] | 21900 | 28500 | — |
| $[Cr(CN)_6]^{3-}$ [Yellow] | 26700 | 32200 | — |

**To compare the observed values of $\nu_1$, $\nu_2$ and $\nu_3$ with their calculated values for the absorption bands in the spectrum of $[Cr F_6]^{3-}$ ion.**

The given complex ion is an octahedral complex ion which contains $Cr^{3+}$ ion which is a $3d^3$ ion. We have already said that the arrangement of 3 electrons in $3d^3$ ion gives two states namely $^4F$ and $^4P$. Both these states are quartet states. In an octahedral field $^4F$ state splits into $^4A_{2g}$ (F), $^4T_{2g}$ (F) and $^4T_{1g}$ (F) states while $^4P$ state does not split but transformes into $^4T_{1g}$ (P) state as shown in Fig. 24.34.

The absorption spectrum of the aqueous solution of $[Cr F_6]^{3-}$ has three bands at the frequencies $\upsilon_1$ = 14,900 $cm^{-1}$, $\upsilon_2$ = 22,700 $cm^{-1}$ and $\upsilon_3$ = 34,400 $cm^{-1}$. These bands are due to the following electronic transitions:

$$^4A_{2g}\,(F) \xrightarrow{\upsilon_1 = 14,900\ cm^{-1}} {}^4T_{2g}\,(F)$$

$$^4A_{2g}\,(F) \xrightarrow{\upsilon_2 = 22,700\ cm^{-1}} {}^4T_{1g}\,(F)$$

$$^4A_{2g}\,(F) \xrightarrow{\upsilon_3 = 34,400\ cm^{-1}} {}^4T_{1g}\,(P)$$

The values of $\upsilon_1$, $\upsilon_2$ and $\upsilon_3$ given above are the observed values of the frequency of the three bands seen in the spectrum of $[CrF_6]^{3-}$ ion. Now let us find out the calculated values of $\upsilon_1$, $\upsilon_2$ and $\upsilon_3$. This can be done with the help of Fig 24.34 which shows that:

$$\upsilon_1 = 14,900\ cm^{-1} = -\,2Dq - (-12Dq)$$

or $$14900\ cm^{-1} = 10\ Dq$$

or $$Dq = \frac{14,900}{10}\ cm^{-1} = 1490\ cm^{-1}$$

Hence the calculated values of $\upsilon_1$, $\upsilon_2$ and $\upsilon_3$ are

$$\upsilon_1 = 10Dq = 10 \times 1490 = 14,900\ cm^{-1}$$

$$\upsilon_2 = +\,6Dq - (-12Dq)$$

$$= 18\ Dq = 18 \times 1490 = 26,820\ cm^{-1}$$

$$\upsilon_3 = 15B - (-12\ Dq) = 15\ B + 12Dq$$

$$= 15 \times 918 + 12 \times 1490 \qquad (\because \text{ B for } Cr^{3+} \text{ ion} = 918)$$

$$= 316,50\ cm^{-1}.$$

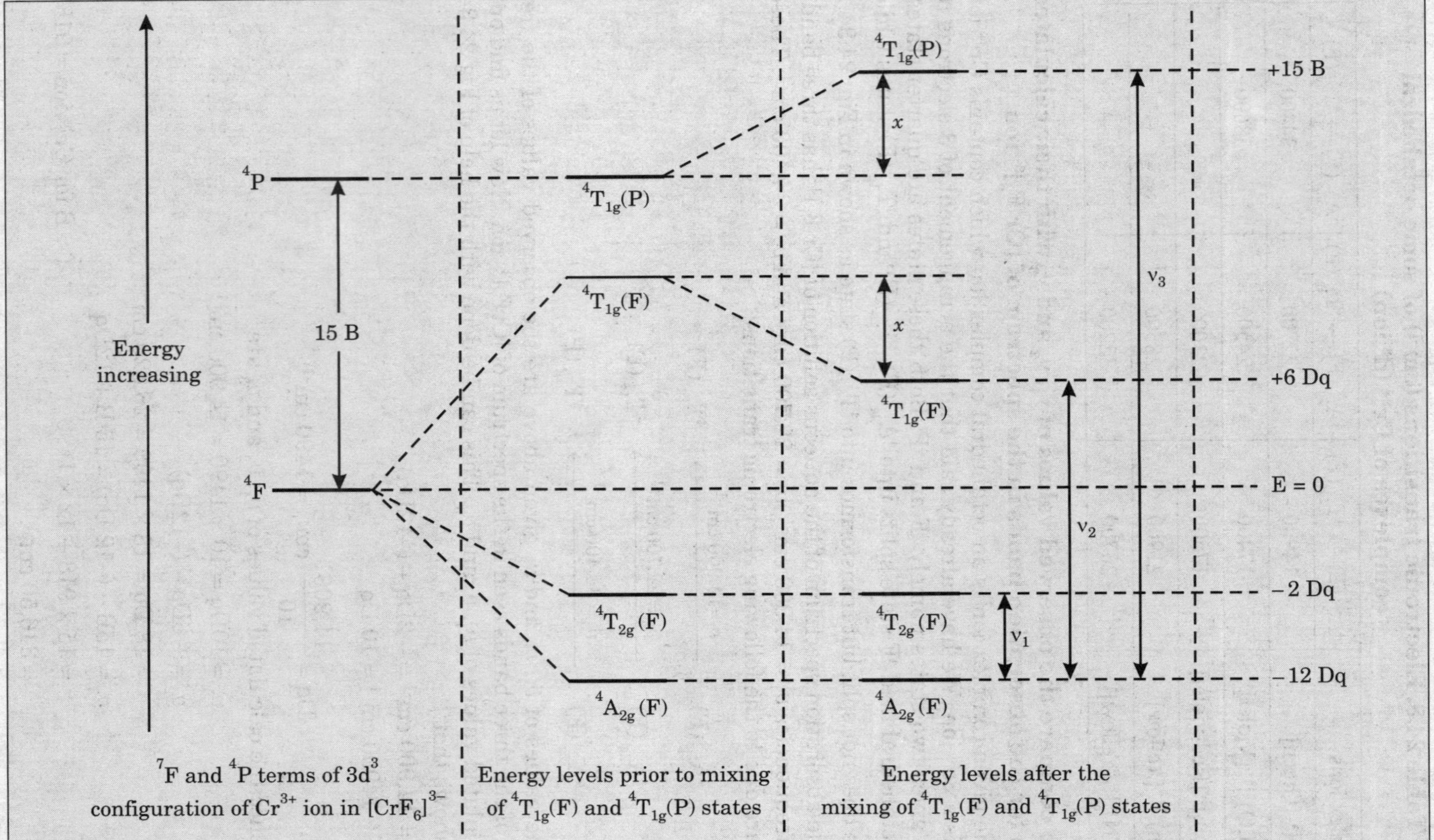


**Fig. 24.34** Energy level diagram depicting the splitting of $^4$F and $^4$P terms of $3d^3$ ion in $[CrF_6]^{3-}$ ion (octahedral complex ion) followed by the mixing of $^4T_{1g}$ (F) and $^4T_{1g}$ (P) states (identical states)

Now when we compare the observed values of $v_1$ (= 14,900 cm$^{-1}$), $v_2$ (= 22,700 cm$^{-1}$) and $v_3$ (= 34,400 cm$^{-1}$) with their calculated values ($v_1$ = 14,900 cm$^{-1}$, $v_2$ = 26,820 cm$^{-1}$ and $v_3$ = 31,650 cm$^{-1}$), we find that $v_1$ has the same value but the calculated value of $v_2$ is greater than its observed value while the calculated value of $v_3$ is less than its observed value. The decrease in the observed value of $v_2$ is because of the decrease in the energy of $^4T_{1g}$ (F) and the increase in the observed value of $v_3$ is due to the increase in the energy of $^4T_{1g}$ (P) state. The decrease in the energy of $^4T_{1g}$ (F) state and the increase in the energy of $^4T_{1g}$(P) state takes place due to the mixing (interacting) of these two states with each other, since these states have identical symmetry. It may be noted from Fig. 24.34 that the decrease in the energy of $^4T_{1g}$ (F) energy level (say $x$) is the same as the increase ($x$) in the energy of $^4T_{1g}$ (P) energy level.

If the values of $v_2$ and $v_3$ are calculated with the help of the following equations, these values are found to be in good agreement with their observed values.

$$v_2 = 7.5\,B' + 15\,Dq - \frac{1}{2}\,(225\,B'^2 + 100\,Dq^2 - 180\,B'Dq)^{1/2} = 22{,}400 \text{ cm}^{-1}$$

$$v_3 = 7.5\,B' + 15\,Dq + \frac{1}{2}\,(225\,B'^2 + 100\,Dq^2 - 180\,B'Dq)^{1/2} = 34{,}800 \text{ cm}^{-1}$$

In the above relations Dq = 14,90 cm$^{-1}$ and B′ = 827 cm$^{-1}$. B′ is called **inter-electronic repulsion Racah parameter** of the complexed metal ion. The value of B′ is calculated from the following relations:

$$15B' = v_3 + v_2 - 3\,v_1$$

$$B' = \frac{v_3 + v_2 - 3v_1}{15} = \frac{34{,}400 + 22{,}700 - 3 \times 14{,}900}{15} = 827 \text{ cm}^{-1}$$

### 4. Octahedral Complexes of $d^4$ Ion.

The octahedral complex ion, $[Cr(H_2O)_6]^{2+}$ consists of $Cr^{2+}$ ion which is $d^4$ ion. The aqueous solution of this ion is pale blue in colour. This ion shows one absorption band at 14,000 cm$^{-1}$ in its absorption spectrum given in Fig. 24.35.

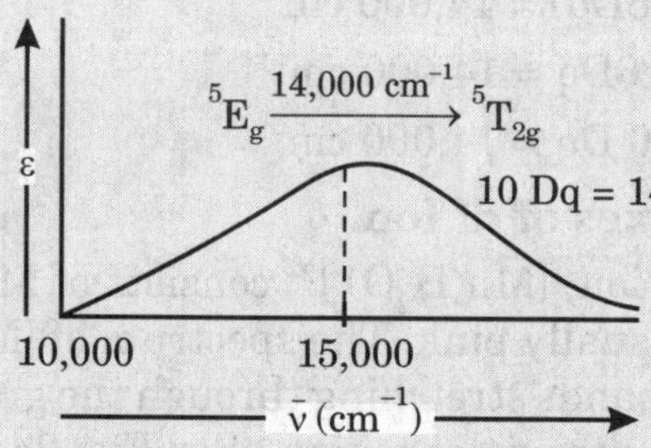


**Fig. 24.35** Electronic spectra of crystal field origin of $[Cr(H_2O)_6]^{2+}$ ($Cr^{2+} = 3d^4$)

We know that $d^4$ ion with $t_{2g}^3 e_g^1$ configuration in weak octahedral field has $^5D$ as its ground state energy term. This term is orbitally five-fold degenerate. In presence of weak octahedral field, $^5D$ term splits up into $^5E_g$ (lower energy) and $^5T_{2g}$ (higher energy) energy levels (See fig. 24.36).

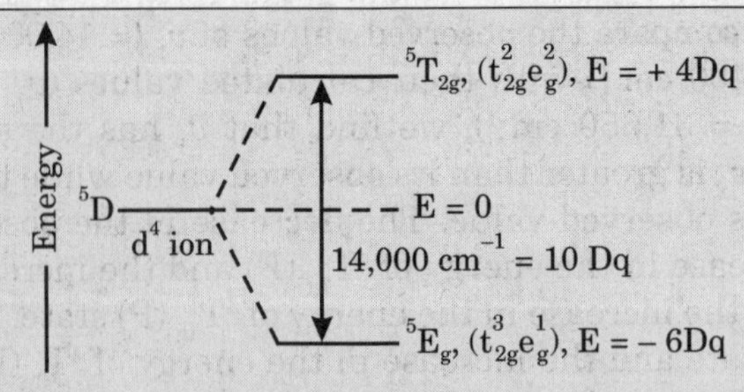


**Fig. 24.36** Orgel diagram for $d^4$ metal ion in weak (high spin) octahedral environment.

Following points may be noted from Orgel diagram given in Fig. 24.36.

(*i*) If the energy of the unsplit $^5D$ term is taken as zero, then the energy of $^5E_g$ and $^5T_{2g}$ levels are –6Dq and +4Dq respectively.

(*ii*) Spin multiplicity, (2S + 1) of the parent term (*i.e.* $^5D$ term) and of its splittings (*i.e.* $E_g$ and $T_{2g}$ terms) remains the same. In the present case (2S + 1) = 5.

(*iii*) The ground state energy level, $^5E_g$ originates from $t^3_{2g}e^1_g$ configuration. The distribution of four electrons in *d*-orbitals may be $d^1_{xy}\, d^1_{yz}\, d^1_{zx}\, d^1_{z^2}\, d^0_{x^2-y^2}$ or $d^1_{xy}\, d^1_{yz}\, d^1_{zx}\, d^0_{z^2}\, d^1_{x^2-y^2}$.

(*iv*) The excited state energy level, $^5T_{2g}$ orginates from $t_{2g}^2e_g^2$ configuration in which the distribution of four electrons may be $d^0_{xy}\, d^1_{yz}\, d^1_{zx}\, d^1_{z^2}\, d^1_{x^2-y^2}$ or $d^1_{xy}\, d^0_{yz}\, d^1_{zx}\, d^1_{z^2}\, d^1_{x^2-y^2}$ or $d^1_{xy}\, d^1_{yz}\, d^0_{zx}\, d^1_{z^2}\, d^1_{x^2-y^2}$.

(*v*) The absorption band at 14,000 cm$^{-1}$ appears to be due to $^5E_g \longrightarrow {}^5T_{2g}$ electronic transition.

(*vi*) The value of 10Dq = 14,000 cm$^{-1}$ as shown below. From Fig. 24.36 we have :

$$+4Dq - (-6Dq) = 14,000 \text{ cm}^{-1}$$

or $$4Dq + 6Dq = 14,000 \text{ cm}^{-1}$$

or $$10\ Dq = 14,000 \text{ cm}^{-1}$$

**5. Octahedral Complexes of $d^5$ Ion.**

The octahedral complex ion, $[Mn(H_2O)_6]^{2+}$ consists of $Mn^{2+}$ which is a $d^5$ ion. Complex ions of Mn (II) are usually pink. The spectrum of $[Mn(H_2O)_6]^{2+}$ ion has a series of weak rather narrow bands stretching through the green and blue portions of the visible region and into ultraviolet region. (See Fig. 24.37).

In weak octahedral ligand field, $d^5$ ion has the configuration $t^3_{2g}e^2_g$. The ground state term for $d^5$ ion is $^6S$ which is a sextent term (spin multiplicity = 6). Other terms of $d^5$ ion are $^4G$, $^4P$, $^4D$ and $^4F$. All these terms are quartet terms (spin multiplicity = 4) and have higher energy than $^6S$ term. Orgel diagram of $d^5$ ion in weak octahedral ligand field is given in Fig. 24.38.

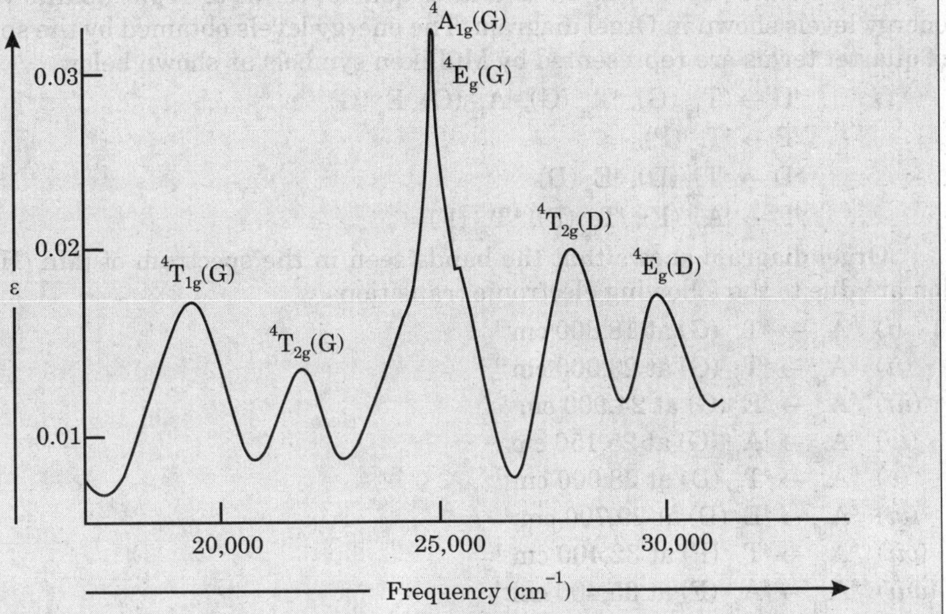


**Fig. 24.37** Various bands seen in the spectrum of $[Mn(H_2O)_6]^{2+}$ ion. ($Mn^{2+} = 3d^5$)

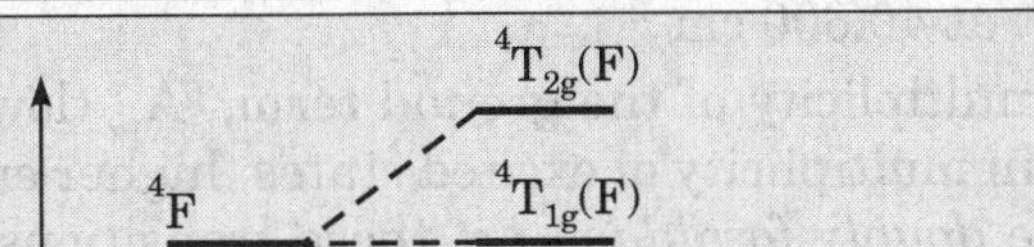


**Fig. 24.38** Orgel diagram for $d^5$ ion in weak octahedral environment.

It may be seen from Figure 24.38 that $^6S$ term is not split but is transformed into $^6A_{1g}$ energy level. Since E = 0 for $^6A_{1g}$ energy level, this energy level is the ground state energy level for $d^5$ configuration. This energy level originates from $t_{2g}^3e_g^2$ configuration.

Since $^6S$ term, like s-orbital, is singly (non) degenerate, this term does not split by weak octahedral ligand field but acquires Mulliken symbol, $^6A_{1g}$ which is the ground state (E = 0). $^4G$, $^4P$, $^4D$ and $^4F$ quartet terms are split up into various energy levels shown in Orgel diagram. The energy levels obtained by the splitting of quartet terms are represented by Mulliken symbols as shown below:

$^4G \rightarrow {}^4T_{1g}$ (G), $^4T_{2g}$ (G), $^4A_{1g}$ (G) $^4E_g$ (G)

$^4P \rightarrow {}^4T_{1g}$ (P)

$^4D \rightarrow {}^4T_{2g}$ (D), $^4E_g$ (D)

$^4F \rightarrow {}^4A_{2g}$ (F), $^4T_{1g}$ (F), $^4T_{2g}$ (F)

Orgel diagram shows that the bands seen in the spectrum of $[Mn(H_2O)_6]^{2+}$ ion are due to the following electronic transitions.

(*i*) $^6A_{1g} \rightarrow {}^4T_{1g}$ (G) at 18,800 $cm^{-1}$
(*ii*) $^6A_{1g} \rightarrow {}^4T_{2g}$ (G) at 23,000 $cm^{-1}$
(*iii*) $^6A_{1g} \rightarrow {}^4E_g$ (G) at 24,900 $cm^{-1}$
(*iv*) $^6A_{1g} \rightarrow {}^4A_{1g}$ (G) at 25,150 $cm^{-1}$
(*v*) $^6A_{1g} \rightarrow {}^4T_{2g}$ (D) at 28,000 $cm^{-1}$
(*vi*) $^6A_{1g} \rightarrow {}^4E_g$ (D) at 29,700 $cm^{-1}$
(*vii*) $^6A_{1g} \rightarrow {}^4T_{1g}$ (P) at 32,400 $cm^{-1}$
(*viii*) $^6A_{1g} \rightarrow {}^4A_{2g}$ (F) at 35,400 $cm^{-1}$
(*ix*) $^6A_{1g} \rightarrow {}^4T_{1g}$ (F) at 36,900 $cm^{-1}$
(*x*) $^6A_{1g} \rightarrow {}^4T_{2g}$ (F) at 40,600 $cm^{-1}$

Since the spin multiplicity of the ground term, $^6A_{1g}$ (lowest energy term) is different from the spin multiplicity of excited states (higher energy terms), all the above transitions are *doubly forbidden i.e.* these transitions are *spin forbidden* as well as *Laporte forbidden*. Therefore, these transitions are of extremely low intensity and are difficult to identify.

**6. Octahedral Complexes of $d^6$ Ion**

The octahedral complex ion, $[Fe(H_2O)_6]^{2+}$ consists of $Fe^{2+}$ ion which is a $d^6$ ion. The aqueous solution of this ion is pale green in colour. The electronic spectrum of this ion has an absorption band at around 10,000 $cm^{-1}$ as shown in Fig. 24.39.

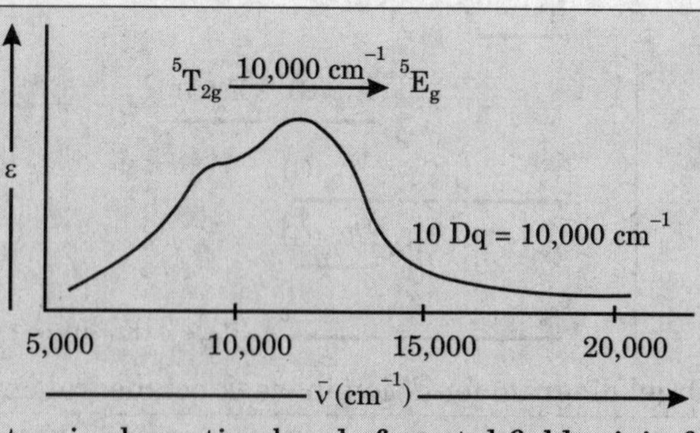


**Fig. 24.39** Electronic absorption band of crystal field origin for $[Fe(H_2O)_6]^{2+}$ complex ion in aqueous solution ($Fe^{2+} = 3d^6$)

We know that $^5D$ term is the ground term for $d^6$ configuration ($t_{2g}^4 e_g^2$) in weak octahedral (high spin) ligand field environment. We also know that the splitting pattern of the terms of $d^1$ and $d^{1+5}$ or $d^6$ configurations are the same in weak octahedral ligand field environment. Thus the splitting pattern of $^5D$ term of $d^6$

configuration takes place in the same way as the splitting of $^2D$ term of $d^1$ configuration. Consequently the splitting of $^5D$ term of $d^6$ configuration gives $^5T_{2g}$ (lower energy) and $^5E_g$ (higher energy) energy levels (See Fig. 24.40).

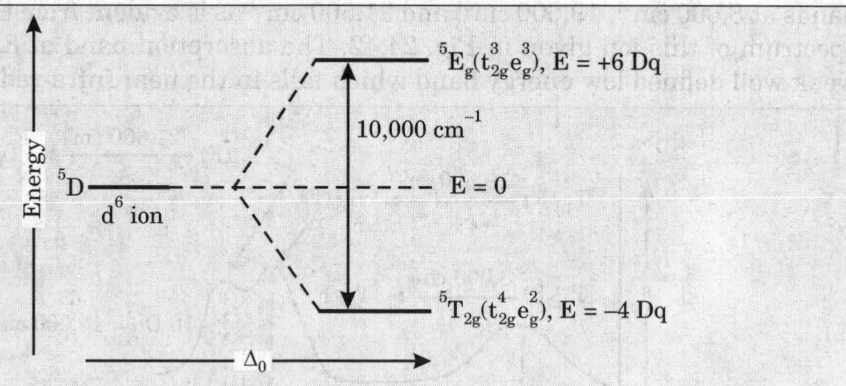


**Fig. 24.40** Orgel diagram for $d^6$ ion in weak (high spin) octahedral environment

Following points may be noted from orgel diagram given in Fig. 24.40.

(*i*) If the energy of the unsplit $^5D$ term is taken as zero, the energies of $^5T_{2g}$ and $^5E_g$ levels are –4Dq and +6Dq respectively.

(*ii*) The ground state energy term, $^5T_{2g}$ originates from $t_{2g}^4e_g^2$ configuration (H.S.) in which the distribution of six electrons in $d$-orbitals may be $d_{xy}^2\ d_{yz}^1\ d_{zx}^1\ e_g^2$ or $d_{xy}^1 d_{yz}^2 d_{zx}^1 e_g^2$ or $d_{xy}^1\ d_{yz}^1\ d_{zx}^2\ e_g^2$.

(*iii*) The excited state energy term $^5E_g$ orginates from $t_{2g}^3e_g^3$ configuration in which the distribution of electrons may be $t_{2g}^3\ d_{z^2}^1\ d_{x^2-y^2}^2$ or $t_{2g}^3\ d_{z^2}^2\ d_{x^2-y^2}^1$.

(*iv*) The absorption band observed at 10,000 cm$^{-1}$ is due to $^5T_{2g} \rightarrow {^5E_g}$ electronic transition.

(*v*) The band appears to be a composite one due to the splitting of $^5E_g$ energy level into two non-degenerate levels due to John Teller distortion (See Fig. 24.41).

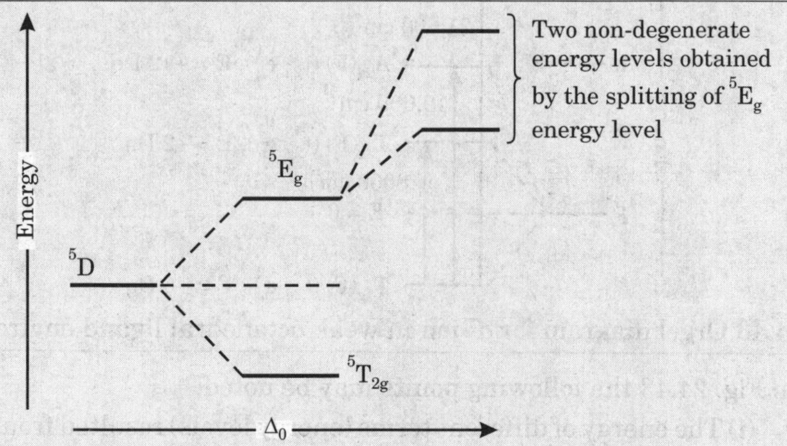


**Fig. 24.41** Splitting of the excited state energy term, $^5E_g$ due to John Teller distortion

**7. Octahedral Complexes of $d^7$ Ion.**

The octahedral complex ion, $[Co\ (H_2O)_6]^{2+}$ consists of $Co^{2+}$ ion which is a $d^7$ ion. The aqueous solution of this ion is pinkish violet and shows three absorption bands at 8,000 cm$^{-1}$, 19,600 cm$^{-1}$ and 21,600 cm$^{-1}$ as is evident from the absorption spectrum of this ion given in Fig. 24.42. The absorption band at 8,000 cm$^{-1}$ is a weak well defined low energy band which falls in the near infra-red region.

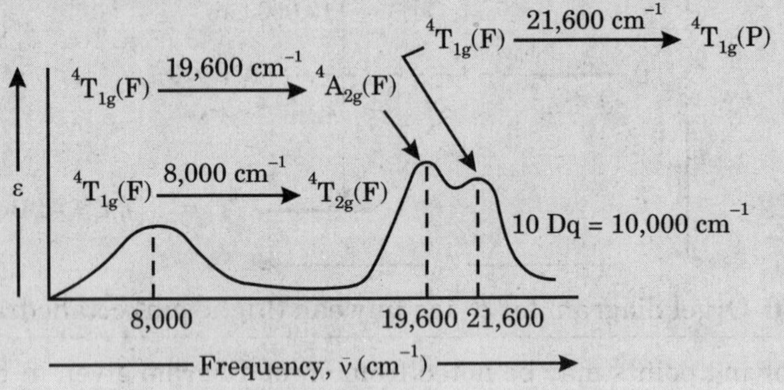


**Fig. 24.42** Absorption bands of crystal field origin in the electronic spectrum of $[Co\ (H_2O)_6]^{2+}$ in aqueous medium ($Co^{2+} = 3d^7$).

We know that $d^7$ ion with $t_{2g}^5 e_g^2$ configuration in weak octahedral field (HS) has $^4F$ (ground state-low energy) and $^4P$ (first excited state-higher energy) terms. We also know that the splitting pattern of the terms of $d^2$ and $d^{2+5}$ or $d^7$ configurations are the same in weak octahedral ligand field. Thus the splitting of $^4F$ and $^4P$ terms of $d^7$ configuration takes place in the same way as the splitting of $^3F$ and $^3P$ terms of $d^2$ configuration takes place. Under the influence of $6H_2O$ ligands $^4P$ term remains unsplit and is transformed to $^4T_{1g}$ (P) state in octahedral symmery. Thus this term remains orbitally triply degenerate. $^4F$ term splits up into $^4T_{1g}$ (F), $^4T_{2g}$ (F) and $^4A_{2g}$(F) terms (See Fig. 24.43).

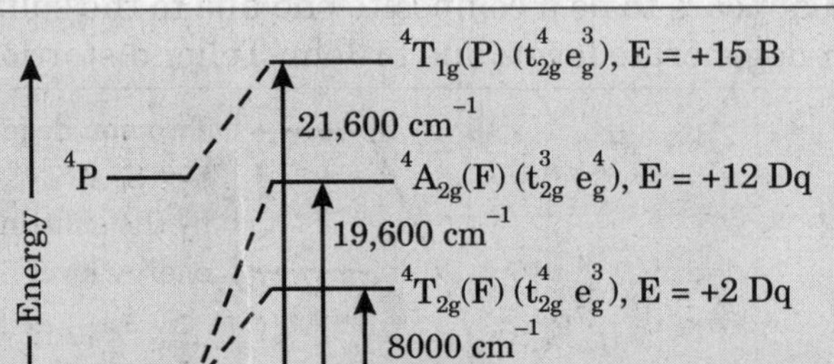


**Fig. 24.43** Orgel diagram for $d^7$ ion in weak octahedral ligand environment

From Fig. 24.43 the following points may be noted:

(*i*) The energy of different terms (energy levels) resulted from the splitting is in the order:

$$^4T_{1g}\ (F) < {}^4T_{2g}\ (F) < {}^4A_{2g}\ (F) < {}^4T_{1g}\ (P)$$

Thus $^4T_{1g}$ (F) term represents ground state term and other terms are excited state terms.

(*ii*) Following three spins allowed but Laporte forbidden electronic transitions in high spin (weak field) state take place.

(*a*) $^4T_{1g}\,(F) \xrightarrow{8,000\ cm^{-1}} {}^4T_{2g}\,(F)$

(*b*) $^4T_{1g}\,(F) \xrightarrow{19,600\ cm^{-1}} {}^4A_{2g}\,(F)$

(*c*) $^4T_{1g}\,(F) \xrightarrow{21,600\ cm^{-1}} {}^4T_{1g}\,(P)$

(*iii*) The energy gap between the two lowest terms *i.e.*, between $^4T_{1g}$ (F) and $^4T_{2g}$ (F) = +2Dq – (–6Dq) = 8Dq = 8,000 cm$^{-1}$.

$$\text{Hence, } 10Dq = \frac{8,000}{8} \times 10\,cm^{-1} = 10,000\ cm^{-1}.$$

Thus the lowest energy transition namely $^4T_{1g}$ (F) → $^4T_{2g}$(F) gives 10Dq.

**8. Octahedral Complexes of $d^8$ Ion.**

The octahedral complex ion, $[Ni(H_2O)_6]^{2+}$ consists of $Ni^{2+}$ ion which is a $d^8$ ion. The aqueous solution of this ion is greenish blue. The solution of this ion shows three absorption bands at 8,700 cm$^{-1}$, 14,500 cm$^{-1}$ and 25,300 cm$^{-1}$ as is evident from the absorption spectrum of $[Ni(H_2O)_6]^{2+}$ ion shown in Fig. 24.44. The molar absorbance (ε) is very low (< 51 mol$^{-1}$ cm$^{-1}$).

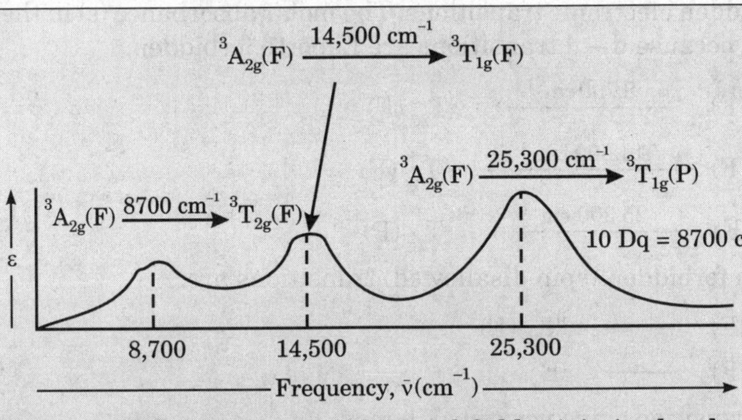


**Fig. 24.44** Absorption bands of crystal field origin in the electronic spectrum of $[Ni(H_2O)_6]^{2+}$ ($Ni^{2+} = 3d^8$)

We know that the splitting pattern of the terms of $d^3$ and $d^{3+5}$ or $d^8$ configurations are the same in weak octahedral ligand field environment. Thus the splitting of $^3F$ and $^3P$ terms of $d^8$ configuration takes place in the same way as the splitting of $^4F$ and $^4P$ terms of $d^3$ configuration takes place.

Under the influence of $6H_2O$ ligands $^3P$ term remains unsplit and is transformed into $^3T_{1g}$ (P) state in octahedral symmetry. Thus this term remains orbitally triply degenerate. $^3F$ term splits into $^3T_{1g}$ (F), $^3T_{2g}$ (F) and $^3A_{2g}$(F) energy levels. (See Fig. 24.45)

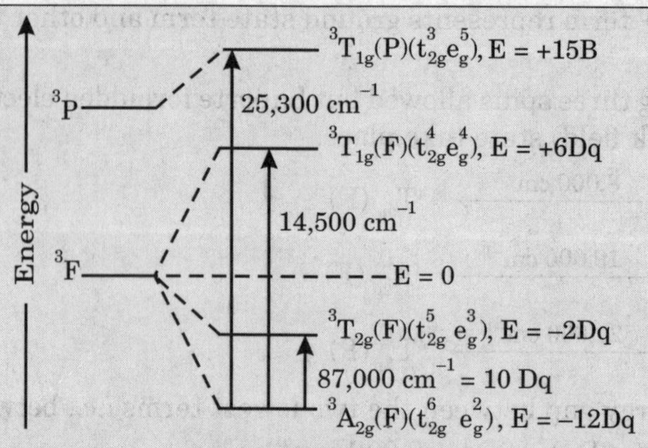


**Fig. 24.45** Orgel diagram for d$^8$ metal ion in weak octahedral ligand field.

From Fig. 24.45 (orgel diagram) the following points may be noted:

(*i*) $^3T_{1g}$ (F) and $^3T_{1g}$ (P) tems do not cross because they have the same symmetry.

(*ii*) The energy of different terms obtained by the splitting is in the order:

$^3A_{2g}$ (F) < $^3T_{2g}$ (F) < $^3T_{1g}$ (F) < $^3T_{1g}$ (P)

Thus $^3A_{2g}$ (F) term represents ground state and other terms represent excited states:

(*iii*) The bands found in the absorption spectrum of [Ni $(H_2O)_6]^{3+}$ ion at 8,700 cm$^{-1}$, 14,500 cm$^{-1}$ and 25,300 cm$^{-1}$ are due to the following spin allowed but Laporte forbidden electronic transitions. The molar absorbance (ε) in the spectrum is very small because d – d transitions are Laporte forbidden.

(*a*) $^3A_{2g}$ (F) $\xrightarrow{8,700\text{ cm}^{-1}}$ $^3T_{2g}$ (F)

(*b*) $^3A_{2g}$ (F) $\xrightarrow{14,500\text{ cm}^{-1}}$ $^3T_{1g}$ (F)

(c) $^3A_{2g}$ (F) $\xrightarrow{25,300\text{ cm}^{-1}}$ $^3T_{1g}$ (P)

(*iv*) Spin forbidden (spin disallowed) transitions are:

(*a*) $^3A_{2g}$ (F) $\longrightarrow$ $^1T_{1g}$ (F)

(*b*) $^3A_{2g}$ (F) $\longrightarrow$ $^1E_g$

These transitions give very weak bands.

(*v*) In octahedral field, $^3A_{2g}$ (ground state term) and $^1E_g$ terms never cross each other, *i.e. it is not possible to have low spin (strong field or spin paired) Ni (II) complexes in perfectly octahedral geometry.*

(*vi*) The energy gap between $^3A_{2g}$ (F) and $^3T_{2g}$ (F) terms is equal to –2Dq – (–12Dq) = –2Dq + 12Dq = 10 Dq. Hence 10Dq = 8,700 cm$^{-1}$.

### 9. Octahedral Complexes of d$^9$ Ion.

The octahedral complex ion, [Cu $(H_2O)_6]^{2+}$ consists of $Cu^{2+}$ ion which is a *d*$^9$ ion. The aqueous solution of [Cu $(H_2O)_6]^{2+}$ ion is blue in colour due to the absorption of yellow component of visible light and transmitting the blue complementary colour.

The absorption spectrum of aqueous solution of $[Cu(H_2O)_6]^{2+}$ consists of a broad weak asymmetric band at 12,000 cm$^{-1}$ in the visible region as shown in Fig. 24.46. This has composite nature which is due to John Teller distortion.

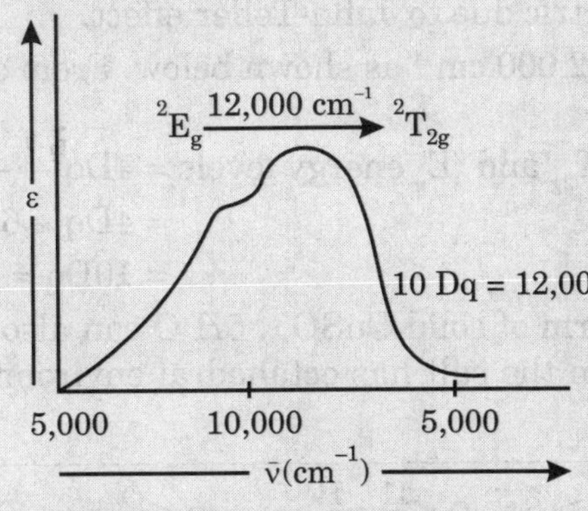


**Fig. 24.46** Eelectronic absorption band of crystsal field origin of $[Cu(H_2O)_6]^{2+}$ complex ion in aqueous solutions. ($Cu^{2+} = 3d^9$)

We know that $^2D$ is the ground state term of $d^9$ configuration in octahedral field. We also know that $d^4$ and $d^{4+5}$ or $d^9$ configurations have the same pattern of the splitting of their terms in weak octahedral field environment. Thus the splitting of $^2D$ term of $d^9$ configuration takes place in the same manner as the splitting of $^5D$ term of $d^4$ configuation takes place. Consequently, in presence of $6H_2O$ ligands, $^2D$ term splits up into $^2T_{2g}$ and $^2E_g$ energy levels (See Fig. 24.47)

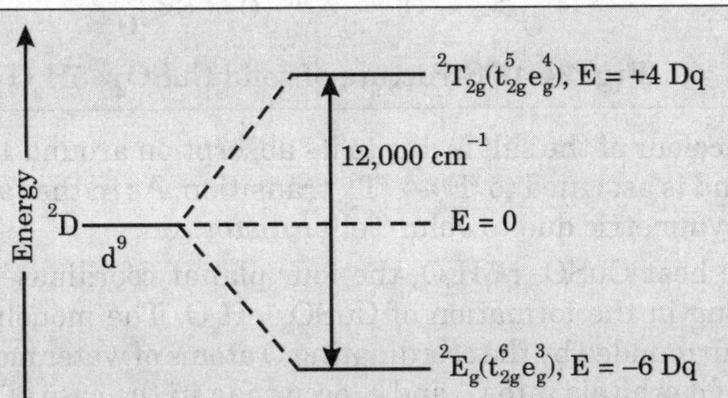


**Fig. 24.47** Orgel diagram showing the splitting of $^2D$ term of $d^9$ configruation.

Since $^2E_g$ state has lower energy than $^2T_{2g}$ state, $^2E_g$ state is the ground state energy level and $^2T_{2g}$ state is the excited state energy level. $^2E_g$ state corresponds to $t_{2g}^6 e_g^3$ configuration which has arrangement of nine electrons as $t_{2g}^6 d_{x^2-y^2}^2 d_{z^2}^1$ or $t_{2g}^6\, d_{x^2-y^2}^1\, d_{z^2}^2$ . $^2T_{2g}$ state corresponds to $t_{2g}^5 e_g^4$ configuration which can be represented as $d_{xy}^1\, d_{yz}^2\, d_{zx}^2\, e_g^4$ or $d_{xy}^2\, d_{yz}^1\, d_{zx}^2\, e_g^4$ or $d_{xy}^2\, d_{yz}^2\, d_{zx}^1\, e_g^4$. When we compare the orgel diagram of $d^9$ ion with that of $d^1$ ion, we find that the orgel diagram of one ion is the reverse of the other ion. This is because $d^1$ and $d^9$ ions are the hole equivalents to each other. ($d^n$ and $d^{10-n}$ ions are hole equivalents). Thus the transition of one electron from $^2E_g$ level to $^2T_{2g}$ level takes place. Hence the absorption band appearing in the spectrum of $[Cu(H_2O)_6]^{2+}$ ion is due to $^2E_g \rightarrow {}^2T_{2g}$ electronic transition which is a single-spin allowed but Laporte forbidden transition. The absorption band is asymmetric due to Jahn-Teller effect.

The value of 10Dq is 12,000 cm$^{-1}$ as shown below. From orgel diagram it can be seen that :

Energy gap between $^2T_{2g}$ and $^2E_g$ energy levels = 4Dq – (– 6Dq)
= 4Dq + 6Dq
= 10Dq = 12,000 cm$^{-1}$

The absorption specturm of solid $CuSO_4 \cdot 5H_2O$ can also be interpreted in a similar manner. The $Cu^{2+}$ in the salt has octahedral environment of O atoms, as shown below in Fig. 24.48.

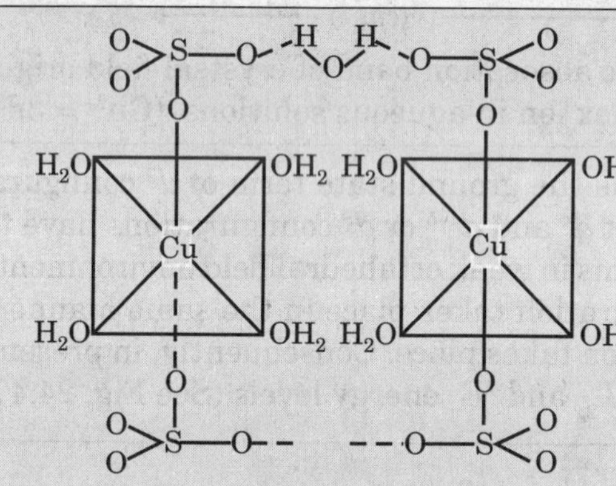


**Fig. 24.48** Structure of solid $CuSO_4 \cdot 5H_2O$

The blue colour of the salt is due to its absorption around 12,000 cm$^{-1}$ and the absorption band is assigned to $^2E_g \rightarrow {}^2T_{2g}$ transition. As in the case of $[Cu(H_2O)_6]^{2+}$, the band is asymmetric due to John-Tellar distortion.

When we heat $CuSO_4 \cdot 5H_2O$, the four planar coordinated water molecules escape resulting in the formation of $CuSO_4 \cdot H_2O$. The monohydrate is no more octahedrally surrounded by the coordinating O atoms of water molecules. Therefore, the splitting of d orbitals into $t_{2g}$ and $e_g$ occurring in the case of $CuSO_4 \cdot 5H_2O$ will no longer occur in the case of $CuSO_4 \cdot H_2O$. The energy levels due to the splitting of *d* orbitals will also vanish and hence there is no question of the occurrence of crystal field electronic transition in the visible region in the case of $CuSO_4 \cdot H_2O$. In other words, we may say that since the electrons are now present in five degenerate *d* orbitals, there is no question of *excitation* of electrons from one *d* orbital to the other *d* orbital by absorption of energy. Hence $CuSO_4 \cdot H_2O$ will be colourless. It may be worthwhile to mention that there is another sulphate of copper namely $Cu_2SO_4$ which is also colourless. This should obviously be so because $Cu^+$ has a $d^{10}$ configuration and hence there is no question of a transition of $d - d$ origin occurring in its electronic spectra.

## Summary

Whatever has been stated so far in connection with the interpretation of the absorption bands of octahedral complexes of $d^1$ to $d^9$ configurations has been summarised in Table. 24.9.

**Table. 24.9. Summary of electronic transitions in octahedral complexes**

| *$d^x$ configuration with examples of octahedral complexes* | *Electronic transitions* |
|---|---|
| $d^1$ $(n = 1)$ $[Ti(H_2O)_6]^{3+}$ | $^2T_{2g}(t_{2g}^1, -4Dq) \rightarrow {}^2E_g(e_g^1, +6Dq)$ |
| $d^9$ $(n = 1)$ $[Cu(H_2O)_6]^{2+}$ | $^2E_g(t_{2g}^6\, e_g^3, -6Dq) \rightarrow {}^2T_{2g}(t_{2g}^5 e_g^4, +4Dq)$ |
| $d^2$ $(n = 2)$ $[V(H_2O)_6]^{3+}$ | $^3T_{1g}(F)(t_{2g}^2, -6Dq) \rightarrow {}^3T_{2g}(F)(t_{2g}^1\, e_g^1, +2Dq)$<br>$\rightarrow {}^3T_{1g}(P)(t_{2g}^1 e_g^1, +15B)$<br>$\rightarrow {}^3A_{2g}(F)(e_g^2, +12Dq)$ |
| $d^8$ $(n = 2)$ $[Ni(H_2O)_6]^{2+}$ | $^3A_{2g}(F)\ (t_{2g}^6 e_g^2, -12Dq) \rightarrow {}^3T_{2g}(F)(t_{2g}^5 e_g^3, -2Dq)$<br>$\rightarrow {}^3T_{1g}(F)(t_{2g}^4 e_g^4, +6Dq)$<br>$\rightarrow {}^3T_{1g}(P)(t_{2g}^3 e_g^5, +15B)$ |
| $d^3$ $(n = 3)$ $[Cr(H_2O)_6]^{3+}$ | $^4A_{2g}(F)\ (t_{2g}^3, -12Dq) \rightarrow {}^4T_{2g}(F)(t_{2g}^2 e_g^1, -2Dq)$<br>$\rightarrow {}^4T_{1g}(F)(t_{2g}^2 e_g^1, +6Dq)$<br>$\rightarrow {}^4T_{1g}(P)(t_{2g}^1 e_g^2, +15B)$ |
| $d^7$ $(n = 3)$ $[Co(H_2O)_6]^{2+}$ | $^4T_{1g}(F)\ (t_{2g}^5 e_g^2, -6Dq) \rightarrow {}^4T_{2g}(F)(t_{2g}^4 e_g^3, +2Dq)$<br>$\rightarrow {}^4A_{2g}(F)(t_{2g}^3 e_g^4, +12Dq)$<br>$\rightarrow 4T_{1g}(P)(t_{2g}^4 e_g^3, +15B)$ |
| $d^4$ $(n = 4)$ $[Cr(H_2O)_6]^{2+}$ | $^5E_g(t_{2g}^3 e_g^1, -6Dq) \rightarrow {}^5T_{2g}(t_{2g}^2\, e_g^2, +4Dq)$ |
| $d^6$ $(n = 4)$ $[Fe(H_2O)_6]^{2+}$ | $^5T_{2g}(t_{2g}^4 e_g^2, -4Dq) \rightarrow {}^5E_g(t_{2g}^3 e_g^3, +6Dq)$ |
| $d^5$ $(n = 5)$ $[Mn(H_2O)_6]^{2+}$ | $^6S(t_{2g}^3 e_g^2, \text{zero}) \rightarrow {}^6A_{1g}(t_{2g}^3 e_g^2, \text{zero})$ |
| $d^{10}$ $(n = 10)$ $[Zn(H_2O)_6]^{2+}$ | $^1S(t_{2g}^6 e_g^4, \text{zero}) \rightarrow {}^1A_{1g}(t_{2g}^6\, e_g^4, \text{zero})$ |

From Table 24.9 the following points may be noted:

(*i*) The spin multiplicity of the various energy levels obtained by the splitting of ground state and first excited state terms of $d^x$ configurations ($x$ = 1 to 10) are the same. The spin multiplicity is equal to (2S + 1) or ($n$ + 1) where $n$ is the number of unpaired electrons in $d^x$ configuration.

(*ii*) In column 1*st* we have given $d^x$ configurations ($x$ = 1 to 10) and the examples of HS octahedral complexes in which the metal ion has $d^x$ configuration.

(*iii*) In column 2*nd* we have shown in bracket $t_{2g}^p e_g^q$ configurations from which the energy terms originate. The energy of these energy levels in terms of Dq has also been given in bracket.

## Electronic Absorption Spectra of Tetrahedral Complexes

Under the influence of tetrahedral ligand field ($V_{tet}$) the metal ion experiences weak field. Thus in $V_{tet}$ the terms of a metal ion split in the same way as they do in $V_{oct}$, *i.e.* the Mullikan symbols of energy levels which are obtained by the splitting of a term in $V_{tet}$ are the same as those obtained by the splitting of a term in $V_{oct}$. However, these symbols do not contain the symbol '*g*', because a tetrahedron has no centre of symmetry. In octahedral complexes '*g*' indicates a **gerade** state (i.e. symmetry w.r.t. centre). When we compare the absorption bands of tetrahedral complexes with those of octahedral complexes, we find that the intensities of absorption bands in the spectra of tetrahedral complexes are higher than those of octahedral complexes. This due to $d - p$ mixing.

In the subsequent pages we shall be using orgel diagrams to interpret the electronic spectra (only crystal field bands) of some metal ions in their tetrahedral complexes.

### 1. Tetrahedral Complexes of $d^2$ Ion.

$V^{3+}$ ion which is a $3d^2$ ion forms a limited number of tetrahedral complexes. $^3F$ and $^3P$ are the terms of $d^2$ configuration. The splitting of these terms under the influence of tetrahedral ligand field is shown in Fig . 24.49.

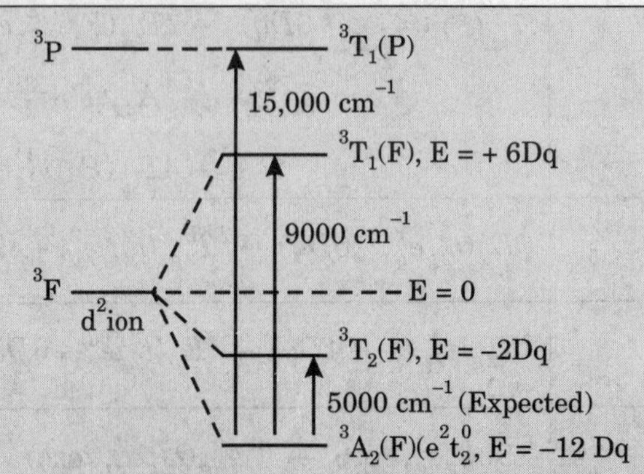


**Fig. 24.49** A Splitting of $^3F$ and $^3P$ terms of $d^2$ configuration under the influence of $V_{tet}$.

The aqueous solution of $V^{2+}$ tetrahedral complex has two absorption bands at 9000 cm$^{-1}$ and 15,000 cm$^{-1}$. These bands are due to the following transitions

$$^{3}A_2\,(F) \xrightarrow{9000\ cm^{-1}} {}^{3}T_1(F) \text{ (Exists in the near IR region)}$$

$$^{3}A_2\,(F) \xrightarrow{15,000\ cm^{-1}} {}^{3}T_1\,(P)$$

The transition, $^3A_2$ (F) $\rightarrow$ $^3T_2$ (F) is lower in energy and is expected due to the absorption band at about 5000 cm$^{-1}$ as shown below.

From Fig. 24.49 it may be seen that

$$6Dq - (-12\ Dq) = 9000\ cm^{-1}$$

or $$6Dq + 12Dq = 9000\ cm^{-1}$$

or $$18\ Dq = 9000\ cm^{-1}$$

or $$10\ Dq = \frac{9000}{18} \times 10\,cm^{-1} = 5000\ cm^{-1}$$

5000 cm$^{-1}$ is the value of 10Dq and the low energy band may occur at this frequency.

**2. Tetrahedal Complexes of $d^6$ Ion.**

$[FeCl_4]^{2-}$ ion is a tetrahedral complex ion which contains $Fe^{2+}$ ion. $Fe^{2+}$ ion is a $3d^6$ ion. The aqueous solution of $[FeCl_4]^{2-}$ ion shows broad absorption band around 4000 cm$^{-1}$ in the near IR region. $^5D$ is the ground state term of $d^6$ configuration. Under the influence of $V_{tet}$, this term splits to give $^5E_2$ and $^5T_2$ energy levels (See Fig. 24.50). The absorption band seen at 4000 cm$^-$ is obviously due to $^5E_2 \rightarrow {}^5T_2$ transition. Now since the energy sepration between $^5T_2$ and $^5E_2$ energy levels = 4000 cm$^{-1}$, 10Dq is also equal to 4000 cm$^{-1}$.

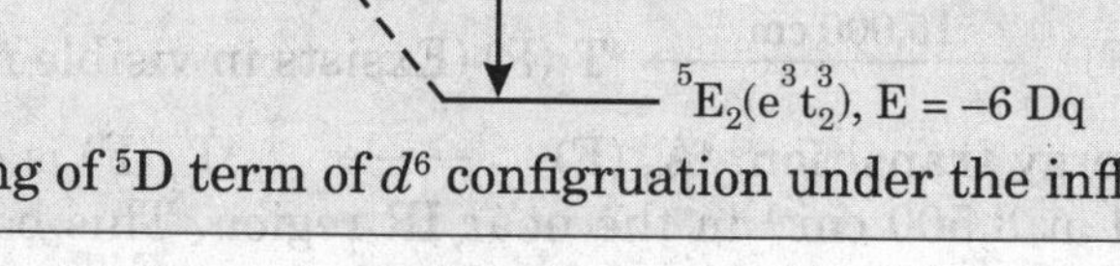


**Fig. 24.50** Splitting of $^5D$ term of $d^6$ configruation under the influence of $V_{tet}$.

**3. Tetrahedral Complexes of $d^7$ Ion.**

$[CoCl_4]^{2-}$, $[CoBr_4]^{2-}$ and $[Co\ (CNS)_4]^{2-}$ ions are the examples of tetrahedral ions. These ions have $Co^{2+}$ ion which is a $3d^7$ ion. We take the example of the spectrum of $[CoCl_4]^{2-}$ ion. This ion is blue in colour. $^4F$ term is the ground state term and $^4P$ term is the first excited state term for $d^7$ ion. The splitting of these terms in tetrahedral field environment gives $^4A_2$ (F), $^4T_2$ (F), $^4T_1$ (F) and $^4T_1$ (P) energy levels as shown in Fig. (24.51).

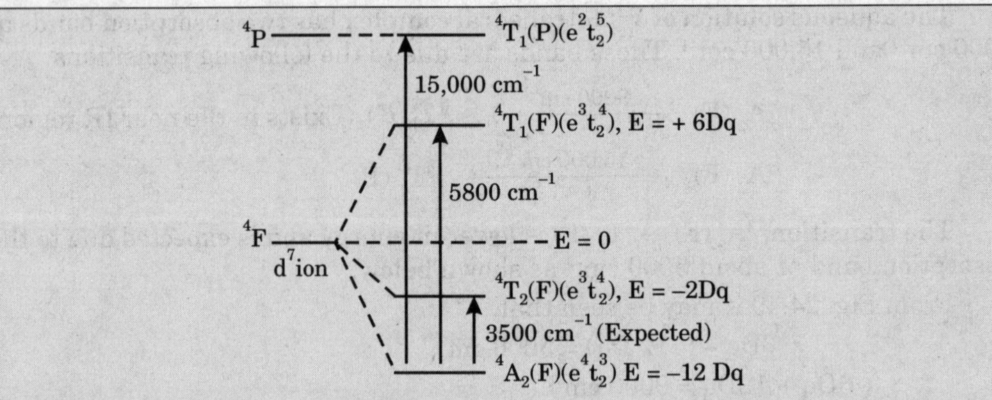


**Fig. 24.51** Splitting of $^4$F and $^4$P terms of $d^7$ configuration under the influence of $V_{tat}$.

The aqueous solution of $[CoCl_4]^{2-}$ on has two absorption bands at 5800 cm$^{-1}$ and 15,000 cm$^{-1}$ as shown in Fig. 24.52.

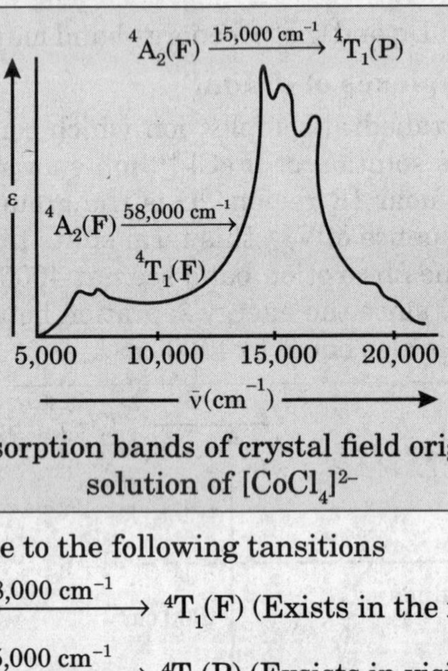


**Fig. 24.52** Absorption bands of crystal field origin of aqueous solution of $[CoCl_4]^{2-}$

These bands are due to the following tansitions

$$^4A_2(F) \xrightarrow{58,000\ cm^{-1}} {}^4T_1(F) \text{ (Exists in the near IR region)}$$

$$^4A_2(F) \xrightarrow{15,000\ cm^{-1}} {}^4T_1(P) \text{ (Exsists in visible region)}$$

The low energy transition, $^4A_2$ (F) $\longrightarrow$ $^4T_2$ (F) is expected due to the absorption band at 3,500 cm$^{-1}$ in the near IR region. This band is out of visible range.

Fig. 24.51 shows that the energy separation between $^4T_1$ (F) and $^4A_2$ (F) energy terms = 5800 cm$^{-1}$ = + 6Dq –(–12 Dq) = 18 Dq

Thus 18Dq = 5800cm$^{-1}$

or $$10Dq = \frac{5800}{18} \times 10\ cm^{-1} = 3222\ cm^{-1}$$

Thus the energy separation between $^4T_2$ (F) and $^4A_2$ (F) energy levels which is equal to 10Dq must be equal to 3222 cm$^{-1}$.

**4. Tetrahedral Complxes of $d^8$ Ion.**

The complex, $[(C_6H_5)_3 AsO]_2NiCl_2$ has distorted tetrahedral geometry. This complex has $Ni^{2+}$. This compound in $C_6H_6$ shows two absorption bands at 7800 $cm^{-1}$ and 16,000 $cm^{-1}$ (See Fig. 24.53). We know that $Ni^{2+}$ ion is $3d^8$ ion. $^3P$ and $^3F$ terms of $d^8$ ion split in $V_{tet}$ and give $^3T_1$ (F), $^3T_2$(F), $^3A_2$ (F) and $^3T_1$ (P) energy levels as shown in Fig. 24.54.

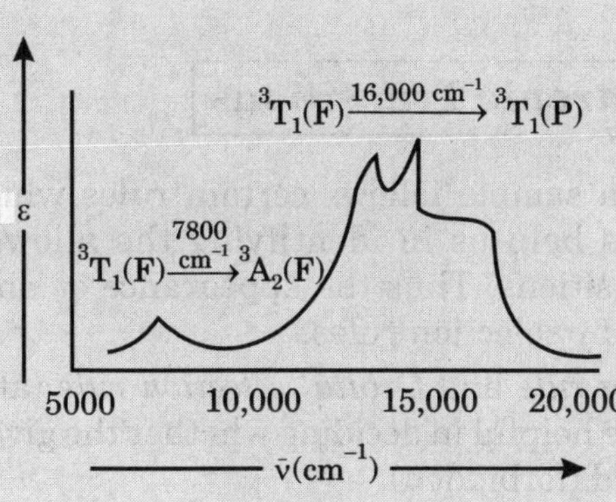


**Fig. 24.53** Absorption bands of crystal field origin of $[(C_6H_5)_3 AsO)]_2 NiCl_2$ in aqueous medium.

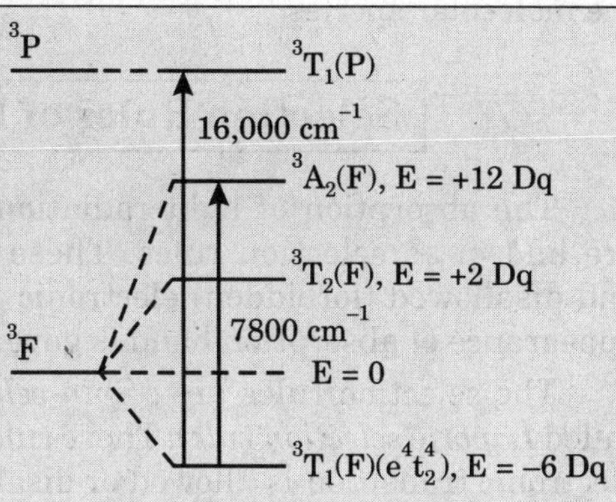


**Fig. 24.54** Splitting of $^3F$ and $^3P$ terms of $d^8$ configuration in $V_{tet}$.

The two absorption bands mentioned above are assigned to the following transitions

$$^3T_1\,(F) \xrightarrow{7800\ cm^{-1}} {^3A_2(F)}$$

$$^3T_1(F) \xrightarrow{16,000\ cm^{-1}} {^3T_1\,(P)}$$

Fig. 24.54 shows that energy gap between $^3T_1$(F) and $^3A_2$ (F) energy levels = 7800 $cm^{-1}$ which is equal to 12Dq – (–6Dq) = 18Dq. Thus :

$$18Dq = 7800\ cm^{-1}$$

or $$10Dq = \frac{7800}{18} \times 10\,cm^{-1} = \mathbf{4333\ cm^{-1}}$$

Ground states of $d^n$ ions in tetrahedral complexes along with $e^p t_2^q$ configurations from which these states have originated are given below.

$$d^1 \to {^2E}(e^1t_2^0),\ d^2 \to {^3A_2}\ (e^2t_2^0),\ d^3 \to {^4T_1}\ (e^2t_2^1),$$

$$d^4 \to {^5T_2}\ (e^3t_2^2), d^5 \to {^6A_1}\ (e^2t_2^3),\ d^6 \to {^5E}(e^3t_2^3),$$

$$d^7 \to {^4A_2}\ (e^4t_2^3),\ d^8 \to {^3T_1}\ (e^4\,t_2^4),\ d^9 \to {^2T_2}\ (e^4t_2^5)$$

## What is Electronic Transition?

When a molecular species absorbs electromagnetic radiation, a change in the arrangement of electrons in orbitals of the species takes place. In this change electrons are promoted from an orbital of lower energy (ground state energy level) to an orbital of higher energy (excited state energy level). This promotion of electrons is called electronic transition. Each electronic transition produces an absorption band in the electronic absorption spectrum of the species. Number of bands observed in the electronic spectrum of the species is equal to the number of electronic transitions. Thus the electronic transitions take place due to absorption of radiation by the molecular species.

## Selection Rules of Electronic Transitions

The absorption of light radiation by a sample follows certain rules which are known as selection rules. These rules help us in identifying the allowed and disallowed (forbidden) electronic transitions. Thus the appearance or non-appearance of absorption band is governed by selection rules.

The selection rules are : *Spin selection rule* and *Orbital selection rule* (also called *Laporte selection rule*). These rules are helpful in deciding whether the given electronic transition is allowed or disallowed (forbidden).

### 1. Spin Selection Rule: Spin (or multiplicity) Allowed and Spin Disallowed (Forbidden) Electronic Transitions.

According to this rule, the electronic transitions which take place between the two energy levels having the same value of spin multiplicity, (2S + 1) are called **spin allowed electronic transitions**. For these transitions $\Delta S = 0$. $\Delta S = 0$ means that spin allowed transition is that in which the spin multiplicity of the two energy levels involved in the tansition must not change, since it is only in this case that $\Delta S$ would be equal to zero.

**Example.** In case of $d^2$ configuration in weak octahedral ligand environment, the following electronic transitions are spin allowed transitions.

$^3T_{1g}\,(F) \rightarrow {}^3T_{2g}\,(F)$

$^3T_{1g}\,(F) \rightarrow {}^3T_{1g}\,(P)$

$^3T_{1g}\,(F) \rightarrow {}^3A_{2g}\,(F)$

Since the energy difference ($\Delta E$) between energy levels is generally greater than $kT$ ($k$ is Boltzmann constant and T is the room temperature), all complexes have the ground state energy levels. The electronic transitions,therefore, take place from the ground state energy level. In the spin allowed electronic transition the ground state energy level and the excited state energy level should have the same value of (2S + 1). For example in each of the spin allowed transitions shown above both the levels have the same value of (2S + 1) (= 3).

Electronic transitions which take place between energy levels having different value of (2S + 1) are called **spin disallowed** or **spin forbidden transitions**. For these transitions $\Delta S \neq 0$, $\Delta S \neq 0$ means that the spin disallowed transition is that in which the difference in the spin multiplicity values (*i.e.* $\Delta S$) of the two energy levels involved in the transition should not be equal to zero.

**Example.** In case of $d^5$ configuration of $Mn^{2+}$ ion in $[Mn(H_2O)_6]^{2+}$ ion the following electronic transitions are spin forbidden transitions

$^6A_{1g} \rightarrow {}^4T_{1g}$ (G), $^6A_{1g} \rightarrow {}^4T_{2g}$ (G)
$^6A_{1g} \rightarrow {}^4E_g$ (G), $^6A_{1g} \rightarrow {}^4A_{1g}$ (G)
$^6A_{1g} \rightarrow {}^4T_{2g}$ (D), $^6A_{1g} \rightarrow {}^4E_g$ (D)

In these transitions the number of unpaired electrons changes. Spin forbidden transitions show bands of low intensity. It is because of this reason that many octahedral complexes of $Mn^{2+}$ are highly coloured.

Sometimes, spin forbidden electronic transitions give absorption bands in the electronic spectra of certain complexes whose intensity is very low. In this case electronic transitions are called **partly spin allowed transitions**. One of the reasons due to which spin forbidden transitions become partly spin allowed is *spin orbit coupling*.

To illustrate the above discussion suppose that for a metal ion, the ground state term is $^1S_0$ (singlet state) and the excited state term is $^3P$ (triplet state) Obviously the electronic transition, $^1S_0 \rightarrow {}^3P$ is spin forbidden transition. But the spin orbit coupling will split the triplet state $^3P$ term into $^3P_0$, $^3P_1$ and $^3P_2$ states. Now since the value of resultant angular momentum, J for $^3P_0$ and $^1S_0$ terms is the same (= 0), these terms will mix together to produce two new states shown as I and II in Fig. 24.55.

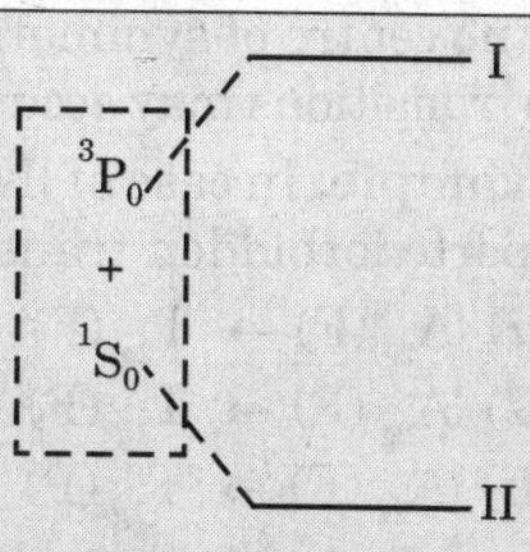

**Fig. 24.55** $^3P_0$ and $^1S_0$ terms mix together to produce two new states I & II.

State I will be a little above $^3P_0$ state and state II will be a little below $^1S_0$ state. Obviously state II will derive a little triplet character from $^3P_0$ term and state I will derive a little singlet character from $^1S_0$ term. Due to the production of mixed singlet-triplet character in new states I and II, spin forbidden transition, $^1S_0 \rightarrow {}^3P$ becomes partly spin allowed transition.

### Differences Between Spin Allowed and Spin Forbidden Electronic Transitions.

(*i*) In spin allowed transition the values of spin multiplicity of the ground term energy level and excited state energy level are the same. On the other hand the spin multiplicity of both the energy levels is different in spin forbidden transition.

(*ii*) The intensity of absorption bands produced due to spin allowed transitions is greater than the intensity of bands produced by spin forbidden transitions.

## 2. Orbital (Laporte) Selection Rule: Laporte Forbidden and Laporte Allowed Transitions

According to this rule, Laporte forbidden electronic transitions are those in which there is no change in the value of orbital quantum number ($l$) or in which $\Delta l = 0$ or $\pm 2$, if the molecule has a centre of symmetry (centro-symmetric). This rule suggests that the electronic transitions from one $d$-level to another $d$-level ($d \rightarrow$ d transition) are not allowed (*i.e.* are Laporte forbidden transitions), since $l$ remains the same or $\Delta l = 0$. In other words we can say that electronic transitions between orbitals having the same $l$ or $\Delta l = 0$ are called **Laporte forbidden transitions.**

The electronic transitions in which there is a change in the value of $l$ or in which $\Delta l \neq 0$ are called **Laporte allowed transitions**. Such transitions are intense or have high absorbance. As a result of this Laporte selection rule, $d \rightarrow d$ transitions in octahedral complexes are not allowed (*i.e.* Laporte forbidden transitions) and hence many complexes are colourless.

In molecules which have a centre of symmetry (centro-symmetric molecules) the wave functions of metal ion $s$ and $d$ orbitals are of $g$ type ($g$ = grade or even state) and $p$ orbitals are of $u$ type ($u$ = ungrade or odd state). The electronic transitions of $d \rightarrow d$ type in octahedral complexes with centre of symmetry are of $g \rightarrow g$ type. These transitions are **Laporte forbidden transitions**. In complexes having no centre of symmetry $s \rightarrow p$ $(g \rightarrow u)$, $p \rightarrow s$ $(u \rightarrow g)$ $p \rightarrow d$ $(u \rightarrow g)$ and $d \rightarrow p$ $(g \rightarrow u)$ transitions may occur. These transitions are **Laporte allowed transitions**.

**Example.** In case of $[Ni(H_2O)_6]^{2+}$ ion all of the following electronic transitions are Laporte forbidden transitions.

(*i*) $^3A_{2g}$ (F) $\rightarrow$ $^3T_{2g}$ (F)

(*ii*) $^3A_{2g}$ (F) $\rightarrow$ $^3T_{1g}$ (F)

(*iii*) $^3A_{2g}$ (F) $\rightarrow$ $^3T_{1g}$ (P)

Here A stands for orbitally non-degenerate energy level. $T_{1g}$(F) and $T_{1g}$(P) stand for orbitally triply degenerate energy levels arising out of F and P terms of the free metal ion when it is surrounded by six $H_2O$ ligands.

## Lowering of Molecular Symmetry

Lowering of symmetry in a molecule makes the Laporte forbidden transition fully Laporte allowed transitions. For example :

(*i*) $[Co(en)_2Cl_2]^+$ ion is an octahedral complex ion. This ion exists in two isomers *viz* (*a*) *Cis-isomer*. This isomer has no centre of symmetry (*b*) *Trans-isomer*. This isomer has a centre of symmetry. Since *cis*-isomer has no centre of symmetry, the energy levels derived from $d$-orbitals cannot be represented by $g$ and $u$ symbols in this isomer. Thus, due to the lowering of symmetry in cis-isomer, intensity of absorption bands in this isomer is higher than the intensity of absorption bands in *trans*-isomer.

(*ii*) A regular tetrahedron has no centre of symmetry. The set of three $p$-orbtials of the metal ion in a tetrahedral complex are of $u$ type while $d_{xy}$, $d_{yz}$ and $d_{xz}$ orbitals of the metal ion are of $g$ type. Thus the energy levels of tetrahedral complexes derive $u$ character from $p$-orbitals and $g$ character from $d$-orbtials. Therefore electronic transitions between any two such energy levels will become partly allowed because both ground state and excited state will have different amount of $u$ as well as $g$ character. Hence the intensity of absorption bands of electronic transitions of $d \rightarrow d$ type are stronger in tetrahedral complexes of $[ML_4]^{2-}$ type (M = Ni, Co, Cu, etc; L = uninegative monodentate ligand) than the absorption bands in octahedral complexes of $[ML_6]^{4-}$ type.

## Vibronic Coupling

The vibrational state of a molecule is known as **vibronic coupling**. The vibronic coupling in an octahedral complex having centre of symmetry makes the forbidden electronic transitions of $g \rightarrow g$ type partly allowed.

Certain vibrational modes like $T_{1u}$ of octahedral complexes destroy the centre of symmetry of the complexes so that the metal ion no longer lies at the centre of the regular octahedron. The electronic transition between energy levels of octahedral complex will no longer be of $g \rightarrow g$ type. Hence for octahedral complex having a centre of symmetry and undergoing such vibrations, the Leporte forbidden transitions become partly allowed.

The discusion given above can be summarised as follows:

(*i*) Electronic transitions are allowed only between those energy levels which have the same spin multiplicity, *i.e.* electronic transitions are permitted only between those energy levels for which $\Delta S = 0$

(*ii*) According to Laporte's rule, electronic transitions are allowed only between those energy levels for which $\Delta l = \pm 1$. Electronic transitions between the energy levels for which $\Delta l = 0$ or $\pm 2$ are not allowed (Laporte forbidden). Thus the electronic transitions namely $2s \rightarrow 2p$ ($\Delta l = 0 - 1 = -1$), $2s \rightarrow 3p$ ($\Delta l = 0 - 1 = -1$), $3p \rightarrow 3d$ ($\Delta l = 1 - 2 = -1$) are allowed but the electronic transitions namely $3d \rightarrow 3d$ ($\Delta l = 2 - 2 = 0$), $1s \rightarrow 2s$ ($\Delta l = 0 - 0 = 0$), $3s \rightarrow 3d$ ($\Delta l = 0 - 2 = -2$) are not allowed (Laporte forbidden).

(*iii*) No electroic transitions can occur merely because of redistribution of electrons in similar orbitals *i.e.* *d–d* transitions (such as $t_{2g} \rightarrow e_g$ transition in $[Ti(H_2O)_6]^{3+}$) are forbidden.

Although *d–d* transitions are forbidden, yet the coordination complexes of transition metals are coloured. This indicates that Laporte selection rule must be relaxed in practice. Whenever a complex lacks a centre of symmetry (e.g. in a tetrahedron) or that odd vibrational effects lead to a change of bond lengths or bond angles there is scope for mixing of metal 3*d*-orbitals with metal 4*p* orbitals or with ligand p-orbitals. The pure *d*-character of the transition metal *d*-orbitals is then lost and Laporte selection rule is relaxed. A situation arises so that electron transfer from a *d*-orbital to a p-orbital can occur. Forbidden transitions have low molar extinction coefficients (10 to 200 l $mol^{-1}$ $cm^{-1}$) for six-coordinate complexes and (100 – 1000l $mol^{-1}$) for tetrahedral complexes. In a non-centrosymetric tetrahedral complexes the selection rules are relaxed to a greater extent.

## Classification of Electronic Transitions On the Basis of Selection Rules.

The discussion on selection rules clearly show that electronic transitions that take place in transition metal complexes may be of the following types:

**(*i*) Spin allowed and Laporte allowed transitions.** Charge–transfer transitions are spin allowed and Laporte allowed transitions.

**(*ii*) Spin allowed and Laporte forbidden (disallowed) transitions.** These transitions are shown by octahedral complexes like $[Ni(H_2O)_6]^{2+}$, tetrahedral complexes like $[NiCl_4]^{2-}$ and some six coordinate complexes with ligands like acetylacetone, P and As donor ligands.

***(iii)* Spin forbidden and Laporte forbidden transitions.** These transitions take place in many octahedral (e.g. $[Mn(H_2O)_6]^{2+}$) and tetrahedral (e.g $[MnBr_4]^{2-}$) complexes.

**Hole Equivalence Theorem: Hole Formalism**

Ground state and other state energy terms of some configurations in the increasing order of their energy are given in Table 24.10.

**Table 24.10. Ground state and other state energy terms of some configurations**

| *Configuration* | *Ground state energy term* | *Other excited state energy terms* |
|---|---|---|
| $s^2$ | $^1S$ | – |
| $s^2$ | $^2S$ | – |
| $p^6$ | $^1S$ | – |
| $p^3$ | $^4S$ | $^2D$, $^2P$ |
| $p^1, p^5$ | $^2P$ | – |
| $p^2, p^4$ | $^3P$ | $^1D$, $^1S$ |
| $d^{10}$ | $^1S$ | – |
| $d^5$ | $^6S$ | $^4G$, $^4F$, $^4D$, $^4P$, etc. |
| $d^1, d^9$ | $^2D$ | – |
| $d^4, d^6$ | $^5D$ | $^3H$, $^3G$, $^3F$, $^3P$, etc. |
| $d^2, d^8$ | $^3F$ | $^3P$, $^1G$, $^1D$, $^1S$ |
| $d^3, d^7$ | $^4F$ | $^4P$, $^2H$, $^2G$, $^2F$, $^2D$, $^2P$ |

It may be seen from the table that $d^n$ and $d^{10-n}$ configurations ($n > 5$) have the same ground state energy terms. Thus the two configurations present in the pairs ($d^1$, $d^9$), ($d^2$, $d^8$), ($d^3$, $d^7$) and ($d^4$, $d^6$) have the same energy terms in the ground state. The similarity between the ground state energy terms of $d^n$ and $d^{10-n}$ configurations is due to positive hole or positive electrons (positrons).

**Example** Let us explain why the ground state energy terms for $d^2$ and $d^{10-2}$ or $d^8$ configurations are the same. We know that the energy terms of $d^2$ configuration are $^3F$, $^3P$, $^1G$, $^1D$ and $^1S$. These terms arise due to inter-electronic repulsion between the two electrons of $d^2$ configuration. Now let us consider $d^{10-2}$ or $d^8$ configuration. We can assume that $d^8$ configuration is equivalent to $d^{10}$ configuration containing two positively charged electrons, $e^+$ (positrons), *i.e.*

$d^8 = d^{10e^- + 2e^+}$ (See Fig. 24.56)

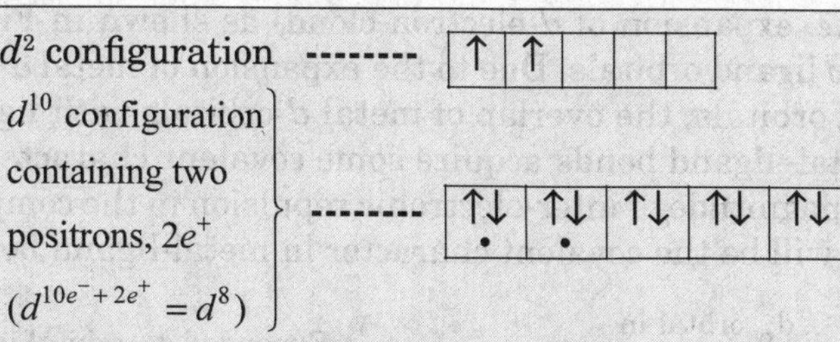


**Fig. 24.56** $d^2$ and $d^8$ configurations. Arrow ( ↑ or ↓) indicates electrons and dot (•) indicates positron.

In $d^{10e^- + 2e^+}$ configuration containing two positrons there will be electrostatic interactions between two positrons. The two positrons present in $d^{10e^- + 2e^+}$ configuration will also obey Paulie's exclusion principle and Hund's rules. Since ten electrons present in $d^{10e^- + 2e^+}$ configuration contribute zero to (L) and (S), the energy terms of $d^{10e^- + 2e^+}$ or $d^8$ configuration will be only due to inter-positron repulsion. Now since inter-electronic repulsion between the two electrons in $d^2$ configuration and inter-positronic repulsion in $d^{10e^- + 2e^+}$ or $d^8$ configuration are similar, both $d^2$ and $d^8$ configurations will give same ground energy terms *i.e.* $d^2$ configuration is the hole equivalent of $d^8$ configuration. Note that $d^5$ ion has its own hole equivalent.

Another characteristic of the configurations which are hole equivalents to each other is that the splitting pattern of the terms of the hole equivalents is reverse of each other. For example since $d^2$ and $d^8$ configurations are hole equivalents with each other, the splitting of $^3$F (ground state term) and $^3$P (first excited state term) terms of $d^2$ configuration gives energy levels whose energy is in the order:

$$^3T_{1g}\ (F) < {^3T_{2g}}\ (F) < {^3T_{1g}}\ (P) < {^3A_{2g}}\ (F) \qquad \ldots (i)$$

but the splitting of $^3$F and $^3$P terms of $d^8$ configuration gives energy levels whose energy is in the order:

$$^3A_{2g}\ (F) < {^3T_{2g}}\ (F) < {^3T_{1g}}\ (F) < {^3T_{1g}}\ (P) \qquad \ldots (ii)$$

Energy orders shown at (*i*) and (*ii*) are reverse of each other.

## Nephelauxetic Effect

### Nephelauxetic Effect and Covalent Character of Metal – Ligand Bond in Complexes.

Indirect evidence of the existence of covalent bonding character in metal–ligand bonding in complexes can be obtained from nephelauxetic effect. Electrons present in partially-filled *d*-orbitals of a metal ion repel one another. This repulsion is called *d–d* inter electronic repulsion It has been found experimentally that when a metal ion gets coordinated with the ligands and forms a complex, the repulsion existing between *d*-electrons of the complexed metal ion becomes less than when the metal ion is in the free state (uncomplexed state). The electronic repulsion is assumed to be 20–25% less in the complex metal ion than in the free metal ion. This decrease in the inter electronic repulsion in the complexed metal ion is because of the fact that the lobes of *d*-orbitals of the complexed metal ion expand in size in the region of ligand orbitals (*i.e.* expansion of *d*-electron cloud) as shown in Fig. 24.57 and hence penetrate into ligand orbitals. Due to the expansion of metal *d*-orbitals into the region of ligand orbitals, the overlap of metal *d*-orbitals with ligand orbitals takes place and metal–ligand bonds acquire some covalent character. Greater is the decrease in the magnitude of inter-electronic repulsion in the complexed metal d-electrons, greater will be the covalent character in metal-ligand bond.

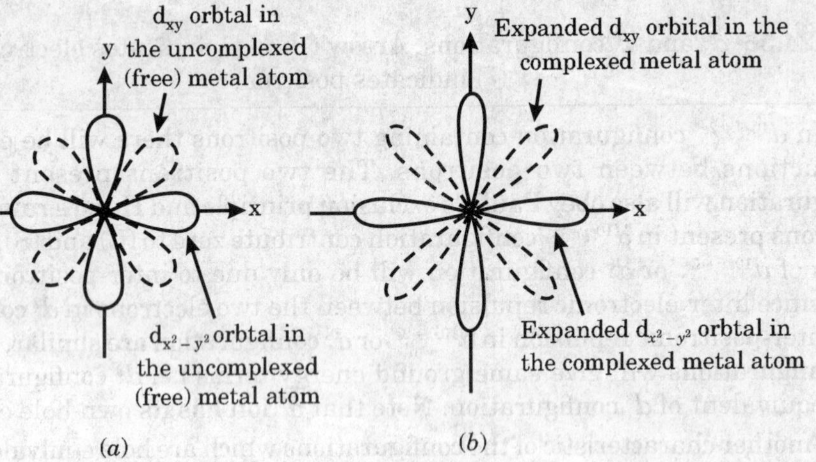


**Fig. 24.57** The expansion of d-orbitals of the metal atom in the uncomplexed (free) and complexed state

The foregoing discussion shows that complexed metal d-electrons spend sometime on the ligands, *i.e.* metal *d*-electrons are partially delocalised into ligand orbitals. This effect of ligands in expanding metal *d*-electron cloud is known as **nephelauxetic** (from Greek meaning *cloud expanding)* **effect.** The capability of a ligand to expand metal *d*-electron cloud differs from one ligand to the other. Some common ligands (e.g. $F^-$, $H_2O$, urea, $NH_3$, $C_2O_4^{2-}$ , etc.) have been arranged in the increasing order of their ability to expand metal *d*-electron cloud.

—— Covalent character of metal-ligand bond increases ⟶

$$F^- < H_2O < \text{urea} < NH_3 < C_2O_4^{2-} \sim \text{en} \sim NCS^- < Cl^- < CN^- < Br^- < I^-$$

The above order is independent of the nature of the metal ion. In this series as we move from left to right (*i.e.* from $F^-$ ion to $I^-$ ion), the ability of the ligands to expand metal *d*-electron cloud in the region of ligand orbitals increases and hence delocalisation of metal *d*-electrons into ligand orbitals also increases. The increase in delocalisation of metal *d*-electrons into ligand orbitals increases covalent character of metal-ligand bond in complexes. Thus covalent character in $M - I^-$ bond in iodo complexes of a given metal (M) is more than that in $M - F^-$ bond in fluoro complexes of the same metal. In other words we say that iodo complexes of a given metal are more covalent than fluoro complexes of the same metal.

If the ligand is kept constant, then the covalent character of the metal-ligand bond in complexes of the following metals decreases as shown below.

$Mn^{2+} \sim V^{2+} > Ni^{2+} \sim Co^{2+} > Mo^{3+} > Re^{4+} \sim Cr^{3+} > Fe^{3+} \sim Os^{4+} > Ir^{3+} \sim Rh^{3+} > Co^{3+} > Pt^{4+} \sim Mn^{4+} > Ir^{6+}$

## Nephelauxetic Ratio (β) and Nephelauxetic Series.

We have already said that due to the overlap of metal ion *d*-orbitals with the ligand orbitals, delocalisation of *d*-electrons from the metal ion to the ligand occurs. The extent of delocalisation of d electrons can be expressed by β which is called **nephlelauxetic** (Greek word meaning *cloud expanding)* **ratio.**

β is the ratio of the inter-electronic repulsion Racah parameter of the complexed metals ion (B′) to that of the free metal ion (*B*). Thus:

$$\text{Nephelauxetic ratio } (\beta) = \frac{\text{Inter-electronic repulsion Racah parameter of the complexed meal ion } (B')}{\text{Inter-electronic repulsion Rocah parameter of the free metal ion } (B)}$$

or
$$\boxed{\beta = \frac{B'}{B}} \quad \ldots(i)$$

We know that when a metal ion gets complexed, the size of *d*-orbitals of the metal ion increases. Due to this increase in the size of *d*-orbitals of the metal ion, the inter-electronic repulsion amongst *d*-electrons of the complexed metal ion is decreased. Therefore, the value of B′ is always less than that of B. Consequently as is evident from equation (*i*) B is always less than 1.

The approximate value of Racah parameter of a metal ion present in a complex ion (*i.e.* value of B′) is given by:

$$\boxed{B' = B\,(1 - h_M\, k_L) \times 10^3} \quad \ldots(ii)$$

Here B = Racah parameter for the free metal ion,

$h_M$ = Arbitrary parameter for the free metal ion

and $k_L$ = Arbitrary parameter for the ligands present in the complex ion.

The arbitrary values of $h_M$ and $k_L$ for some metal ions and ligands are given in Table 24.11.

**Table 24.11. Empirical values of parameter $h_M$ and $k_L$ assigned to various metal ions and ligands**

| *Metal ion* | *Empirical value of $h_M$* | *Ligand* | *Empirical value of $k_L$* |
|---|---|---|---|
| $Co^{2+}$ | 0.23 | $6CN^-$ | – |
| $Co^{3+}$ | 0.35 | $6Br^-$ | 2.3 |
| $Cr^{3+}$ | 0.21 | $6Cl^-$ | 2.0 |
| $Fe^{3+}$ | 0.24 | $6OH^-$ | 2.0 |
| $Mo^{3+}$ | 0.07 | 3 en | 1.5 |
| $Ni^{2+}$ | 0.12 | $6H_2O$ | 1.4 |

**Example. Calculate the value of Racah parameter for $Ni^{2+}$ ion (B′) in the complex ion, $[Ni\,(H_2O)_6]^{2+}$. Given that : Racah parameter of the free $Ni^{2+}$ ion = 1080 $cm^{-1}$, arbitrary parameter values for $Ni^{2+}$ ion and $6H_2O$ (ligands) = 0.12 and 1.4 respectively.**

**Solution.** The value of B′ is given by:

$$B' = B\,(1 - h_M\, k_L) \times 10^3$$

Given: $B = 1080\ cm^{-1}$, $h_M$ for $Ni^{2+}$ ion = 0.12 and $k_L$ for $6H_2O$ = 1.4.

On putting these values in the above equation we get:

$$B' = 1080\,(1 - 0.12 \times 1.4) \times 10^3 = \mathbf{898.6\ cm^{-1}}\ \textbf{(Ans)}$$

The experimental value of B′ is $905\ cm^{-1}$.

As we move from left to right in the following series, the value of β decreases. With the decrease in the values of β, covalent character of metal–ligand bond in complexes increases.

——— β values decrease ———→

——— Covalent character of metal-ligand bond increases ———→

$$F^- > H_2O > \text{Urea} > NH_3 > C_2O_4^{2-} \sim \text{en} \sim NCS^- > Cl^- > CN^- > Br^- > I^-$$

Above series is called **Nephelauxatic series**. This series is valid for most of the metal ions. The decreasing order of β values from left to right in the series indicates that the complexes having low value of β are covalent in nature.

The values of Racah parameter (*B*) for some transition metals and metal ions are given in Table 24.12. In this Table a dash(–) indicates that the value of B is unknown. In some ions the configuration has either one or no electrons outside a closed shell so that electron repulsion does not take place within the valence shell.

**Table 24.12. Values of Racah parameter (B) for some transition metals and their ions.**

| Metals | $M^0$ | $M^+$ | $M^{2+}$ | $M^{3+}$ | $M^{4+}$ |
|---|---|---|---|---|---|
| Ti | 560 | 681 | 719 | – | – |
| V | 579 | 660 | 765 | 860 | – |
| Cr | 790 | 710 | 830 | 1030 | 1040 |
| Mn | 720 | 872 | 960 | 1140 | – |
| Fe | 805 | 870 | 1059 | – | 1144 |
| Co | 789 | 879 | 1117 | – | – |
| Ni | 1025 | 1038 | 1080 | – | – |
| Cu | – | 1218 | 1239 | – | – |

## Charge Transfer (CT)

### What is Charge Transfer?

The colour of complexes can be explained on the basis of the concept of *d* – *d* transition or on the basis of the concept of charge transfer. Here we shall discuss charge transfer concept only.

There are many complexes in which either the metal ion is highly oxidising and the ligand is highly reducing or the metal ion is highly reducing and the ligand is highly oxidising. In such complexes there occurs transfer of charge (electron) from the reducing partner to the oxidising partner, *i.e.* the transfer of electron may occur either from the metal ion to the ligand or from the ligand to the metal ion. Thus charge transfer involves redox process. High intensity bands with $E_{max} > 10^4$ and occurring in the near UV region are charge tranfer bands. These bands are Laporte allowed. If $E_{max} \sim 10^4$, these bands are also spin allowed. Charge transfer bands may be observed in the visible region, if the metal is easily oxidiseable and the ligand is readily reducable. Complexes showing charge transfer appear coloured to our eyes only when the charge transfer transitions of complexes occur in the near visible region or when the transition appearing in the U.V. has a long tail in the visible region.

Charge transfer transitions may appear in metal complexes having metal with $d°$, incomplete $d$-orbitals and $d^{10}$ configuration. For example $VO_2^+$ -pale yellow ($V^{+5} = 3d^0$), $CrO_4^{2-}$ – yellow ($Cr^{+6} = 3d^0$), $MnO_4^-$ (in solution) – intense purple ($Mn^{+7} = 3d^0$), $MnO_4^-$ (in solid) - almost black ($Mn^{+7} = 3d^0$), $[Fe(dipy)_3]^{2+}$ – dark red ($Fe^{2+} = 3d^6$), $HgI^{2-}$ – brick red ($Hg^{2+} = 5d^{10}$).

Greater is the oxidising power of the metal ion and reducing power of the ligand, lower is the energy and hence the frequency at which the charge transfer bands appear.

## Types of Charge Transfer

**1. Metal to ligand (M → L) charge transfer.** In this type of charge transfer the charge (electron) is transferred from the filled metal $t_{2g}$ orbital to the empty $\pi^*$ bonding orbitals of the ligand. Thus this charge transfer can be represented as $M(t_{2g}) \rightarrow L(\pi^*)$. $\pi^*$ orbitals of the ligand should be higher in energy than the filled metal $t_{2g}$ orbitals. This transfer is favoured if the metal is in low oxidation state and has some oxidising character and the ligand has a low lying $\pi^*$ orbital.

The complexes $Cr(CO)_6$, $Mo(CO)_6$, $[Fe(CN)_6]^{3-}$, $[Fe(CN)_6]^{4-}$ etc. show intense M → L charge transfer bands in the region 35,000 – 45,000 $cm^{-1}$.

M → L transition occurs in the complexes like $[Fe^{2+}(o\text{-phen})_3]^{2+}$, $[Fe(dipy)_3]^{2+}$, $[Ir(py)_6]^{3+}$ etc. In M → L transition the easily oxidisable metal ions (e.g. $Fe^{2+}$, $Cu^+$, $Ti^{3+}$ etc.) provide electrons which are accommodated in the empty $\pi*$ orbitals of the ligand. M → L charge transfer bands usually lie lower in energy. $t_{2g} \rightarrow \pi*$ charge transfer transition occurs in $[Ir(py)_6]^{3+}$ in 29,000 – 35,500 $cm^{-1}$ region by the transfer of an electron from the filled $t_{2g}$ orbital of the metal to the empty $\pi*$ orbital on pyridine (ligand).

**2. Ligand to metal (L → M) charge transfer.** In this type of charge transfer, the charge (electron) is transferred from the ligand to the metal. The more oxidising M is and more reducing L is, lower is the energy of L → M charge transfer transition. Thus Fe (III) complexes give more L → M charge transfer transitions than do Cr (III) complexes. The bands given by L → M charge transfer are called L → M charge transfer bands or L → M CT bands. L → M charge transfer transition occurs when the ligands have pronounced reducing ability while the metal ion has oxidising properties. The ligands have relatively high energy filled orbitals and the metal ion has empty low lying orbital.

L → M charge transfer energies decrease as the ligands become more reducing in character.

**Examples:** *(i)* $[HgI_4]^{2-}$ (brick red, $Hg^{2+} = 5d^{10}$), $MnO_4^-$ (purple, $Mn^{7+} = 3d^0$), $CrO_4^{2-}$ (yellow, $Cr^{6+} = 3d^0$) etc. show L → M charge transfer. $MnO^-_4$ (Mn = +7, O = –2) has intense purple colour in solution. The colour of $MnO_4^-$ ion is explained on the basis of L → M change transfer concept. In $MnO^-_4$ ion an electron is momentarily transferred from $O^{2-}$ to $Mn^{7+}$. This results in that $O^{2-}$ is oxidised to $O^-$ and $Mn^{7+}$ is reduced to $Mn^{6+}$. Charge transfer produces intense purple colour.

$$\underset{(O = -2)}{O^{2-}} \xrightarrow{\text{Oxidation}} \underset{(O = -1)}{O^-} + e^-$$

$$\underset{(Mn = +7)}{MnO_4^-} + e^- \xrightarrow{\text{Reduction}} \underset{(Mn = +6)}{MnO_4^{2-}}$$

$$\underset{RA}{O^{2-}} + \underset{OA}{MnO_4^-} \longrightarrow O^- + MnO_4^{2-}$$

($e^-$ transferred from RA to OA)

Solutions of $TcO_4^-$ and $ReO_4^-$ are colourless, since charge transfer bands in these complexes occur at higher energy in UV region.

Hologen → mercury electronic transitions take place at 31,000 cm$^{-1}$, 40,000 cm$^{-1}$ and 43,700 cm$^{-1}$ in $[HgI_4]^{2-}$, $[Hg\,Br_4]^{2-}$ and $[HgCl_4]^{2-}$ ions respectively.

*(ii)* $SnI_4$ (Sn = +4, $Sn^{4+} = 4d^{10}$) has orange colour since $SnI_4$ absorbs blue light and reflects orange light. The energy absorbed by $SnI_4$ causes the transfer of an electron (charge) from $I^-$ (ligand) to $Sn^{4+}$ (metal). Due to the transfer of charge from $I^-$ ion to $Sn^{4+}$, $Sn^{4+}$ is momentarily reduced to $Sn^{3+}$.

$$I^- \rightarrow I + e^-$$

$$Sn^{4+} + e^- \rightarrow Sn^{3+}$$

$$\underset{RA}{I^-} + \underset{OA}{Sn^{4+}} \rightarrow I^0 + Sn^{3+}$$

($e^-$ transferred from RA to OA)

*(iii)* L → M charge transfer is also shown by hexahalo complex anions like $[Ru^{4+}Cl_6]^{2-}$ and $[Ir^{4+}Br_6]^{2-}$. Metals present in these anions have $d^4$ and $d^5$ configurations respectively ($Ru^{4+} = 4d^4$, $Ir^{4+} = 5d^5$). These complex anions produce two charge transfer bands. One band appears at lower intensity and lower frequency falling around 20,000 cm$^{-1}$ and the other band is found at higher intensity and much higher frequency falling between 40,000 cm$^{-1}$ and 80,000 cm$^{-1}$.

It has been suggested that lower intensity lower frequency bands produced by $[RuCl_6]^{2-}$ and $[IrBr_6]^{2-}$ complex ions can be assigned to the charge transfer transition $t_{1u}^n, t_{2u}^n \longrightarrow t_{2g}$ and the higher intensity higher frequency bands produced by $[RuCl_6]^{2-}$, $[IrBr_6]^{2-}$ and $[IrBr_6]^{3-}$ can be assigned to the charge transition $t_{1u}^n\ t_{2u}^n \rightarrow \overset{*}{e}_g$.

(*iv*) $[Ir^{3+}Br_4]^-$ anion ($Ir^{3+} = 5d^6$) also has L → M charge transfer band. In this complex ion the lower frequency bands are absent and only higher frequency bands are seen.

**Mummary of M → L and L → M charge transfer**

| | | |
|---|---|---|
| M (*Filled* $t_{2g}$)<br>Highly reducing<br>(Easily oxidisable)<br>(Reducing partner) | $\underset{\text{Reduction}}{\overset{\text{Oxidation}}{\rightleftharpoons}}$ | L(*Empty* $\pi^*$)<br>Highly oxidising<br>(Easily reduceable)<br>(Oxidising partner) |
| L<br>Highly reducing<br>(Readily oxidiseable)<br>(Reducing partner) | $\underset{\text{Reduction}}{\overset{\text{Oxidation}}{\rightleftharpoons}}$ | M<br>Highly oxidising<br>(Easily reduceable)<br>(Oxidising partner) |

**3. Charge transfer in complexes having metal in mixed valence state: Metal to metal charge transfer.** Potassium iron (III) hexacyanoferrate (II) (prussion blue), $KFe^{3+}[Fe^{2+}(CN)_6]$ is the example of a complex in which Fe is present in the mixed valence state. One Fe is present as $Fe^{3+}$ and the other Fe is present as $Fe^{2+}$. The blue colour of this complex is due to the fact that HS $Fe^{3+}$ ($n = 5$) is octahedrally coordinated with six N-atoms of six $CN^-$ ions. LS $Fe^{2+}$ ($n = 4$) is coordinated with six C-atoms of six $CN^-$ ions. In the excited state $Fe^{2+}$ is oxidised to $Fe^{3+}$ and $Fe^{3+}$ is reduced to $Fe^{2+}$

$$Fe^{2+} \xrightarrow{\text{Oxidation}} Fe^{3+} + e^-$$

$$Fe^{3+} + e^- \xrightarrow{\text{Reduction}} Fe^{2+}$$

Both the above reactions can be combined as:

$$Fe^{2+} \underset{+e^-}{\overset{-e^-}{\rightleftharpoons}} Fe^{3+}$$

Thus we see that blue colour of $KFe^{3+}[Fe^{2+}(CN)_6]$ is due to charge (electron) transfer from $Fe^{2+}$ to $Fe^{3+}$. π-bonding system of $CN^-$ ion facilitates the charge transfer. It may be noted that $K_2[Fe^{2+}Fe^{2+}(CN)_6]$ is colourless, since both Fe are present as $Fe^{2+}$. Hence there is no possibility of any charge transfer transition in this complex.

Other examples giving intense charge transfer band are $Pt^{2+}[(NH_3)_2Br_2]$, $Pt^{4+}[(NH_3)_2Br_4]$, $Cs_2[Au^+Cl_2][Au^{3+}Cl_4]$, $(NH_4)_4[Sb^{3+}Br_6][Sb^{5+}Br_6]$. All these complexes are diamagnetic.

**Comparison Between M → L and L → M Charge Transfer Bands.**

(*i*) When the metal or ligand is changed, the energy of both the charge transfer bands changes.

(*ii*) Keeping the metal constant, if the ligand is more readily polarisable, then the energies of L → M charge transfer bands decrease.

(*iii*) Keeping the metal constant, if the ligand is more readily reduceable, then the energies of M → L charge transfer bands decrease.

(*iv*) With the decrease in oxidation state of the metal for a given ligand, the energy of M → L charge transfer bands decrease.

## Comparison Between d → d Transitions and Charge Transfer Transitions

The points showing differences between the two types of transitions are given below.

| $d \rightarrow d$ transitions | Charge transfer (CT) transitions |
|---|---|
| (*i*) For $d \rightarrow d$ transitions the metal ion of the complex must contain d-electrons. | (*i*) Metal ion may or may not contain *d*-electrons. CT transitions may occur in complexes having metal with $d^0$ and $d^{10}$ configuration. These transitions may also occur in non-transition compounds. |
| (*ii*) $d \rightarrow d$ transitions are Laporte forbidden. | (*ii*) CT transitions are Laporte allowed. |
| (*iii*) These transitions have low molar extinction coefficient (10 – 200 $l$ mol$^{-1}$ cm$^{-1}$) | (*iii*) These transitions have higher molar extinction coefficient ($10^3$ to $10^5$ $l$ mol$^{-}$ cm$^{-1}$) |
| (*iv*) These transitions do not involve redox process. | (*iv*) These transitions involve redox process. |
| (*v*) These are not useful for spectrophotometric estimation of metal ions. | (*v*) These are very useful for spectrophotometric estimation of metal ions. |
| (*vi*) $d \rightarrow d$ transitions occur in visible region. | (*vi*) These transitions are usually high energy transitions and hence often occur in the near UV region. |

## UNIVERSITY QUESTIONS

**1.** Explain Orgel energy level diagram for $d^1$ and $d^9$ ions.

(*Pd. Ravishankar Shukla 2010*)

**2.** (*a*) Discuss the electronic spectrum of $[Ti(H_2O)_6]^{3+}$ complex ion.

(*b*) What is L – S coupling? Explain with the example of $p^2$ configuration of carbon.

(*Pd. Ravishankar Shukla 2010*)

**3.** (*a*) Draw Orgel energy diagrams in octahedral field for $d^1$ and $d^9$ electronic system.

(*b*) Define the following —

(*i*) Landge '*g*' factor (*ii*) 'Spin only' factor

(*iii*) Ferromagnetism (*iv*) Curie temperature

(*v*) Term symbol.

(*c*) Discuss the spectrum of $[Ti(H_2O)_6]^{3+}$ complex ion.

(*d*) What is Laporte Orbital Selection Rule? (*Rajasthan 2010*)

**4.** (*a*) What is electronic spectra? Discuss electro spectra of $[Ti(H_2O)_6]^{3+}$ ion.

(*b*) Write note on charge transfer spectra. (*Barkatullah 2010*)

**5.** (*i*) Two *d-d* bands can be seen at 17000 $cm^{-1}$ and 26000 $cm^{-1}$ in the absorption spectrum of $[V(H_2O)_6]^{3+}$. Assign the bands.

(*H.N. Bahuguna Garhwal 2010*)

(*ii*) Explain the magnetic and spectral properties of $[Ti(H_2O)_6]^{3+}$ ion.

(*iii*) It has been observed that two of the water molecules in the aqueous $[Cu(H_2O)_6]^{2+}$ ion undergoes exchange with the bulk solvent molecules much more rapidly than other four. Explain.

(*H.N. Bahuguna Garhwal 2010*)

**6.** Choose the correct answer:

(*i*) The total number of micro states possible for a $d^3$ ion is _____.

(*a*) 3 (*b*) 120

(*c*) 30 (*d*) 12.

(*ii*) The pink colour of paramagnet ion is due to _____.

(*a*) d-d transition (*b*) CT transition

(*c*) +6 oxidation state (*d*) Crystal field splitting.

(*Bharathiar 2010*)

**7.** (*a*) Deduce the term symbol of a $d^2$ ion.

(*b*) How many micro states are possible for a $d^2$ ion? What is the ground state term for the ion? (*Bharathiar 2010*)

8. (*i*) Write note on L-S coupling. (*Ranchi 2010*)

(*ii*) How ground state terms are determined?

(*iii*) Draw the Orgel diagram for $d^1$ ion and on its basis explain the absorption spectrum of $[Ti(H_2O)_6]^{3+}$.

(*iv*) Explain spin selection rule. (*Ranchi 2010*)

**9.** What do you understand by L-S coupling? (*G.N.D. Amritsar, 2010*)

**10.** Draw absorption spectrum of $[Ti(H_2O)_6]^{3+}$ and $[Cu(H_2O)_6]^{2+}$. Distinguish them on the basis of (*i*) position of the band and (*ii*) colour of the absorption spectrum

(*Nagpur 2010*)

**11.** (*a*) Why are d-d electronic transitions forbidden? Why are they weekly absorbing?

(*b*) Draw the Orgel energy level diagram for $d^1$ state.

(*c*) Why does tetrahedral complex of an element give much more intense d-d spectra than its octahedral complex?

(*d*) Discuss the electronic spectrum of $[Ti(H_2O)_6]^{3+}$ complex.

*(BJB College, Orisa, 2010)*

**12.** (*a*) Explain types of electronic transitions.

(*b*) Explain selection rules for d-d transition rules.

(*c*) Explain spectroscopic ground states. *(GND Amritsar 2010)*

**13.** (*i*) $MnO_4^-$ and $[Ti(H_2O)_6]^{3+}$ are coloured ions but reasons for their colours are different. Explain.

(*ii*) Find out the ground state term of $3d^5$ configuration of $Mn^{2+}$.

(*iii*) Explain Laporte selection rule. *(Lucknow 2010)*

**14.** (*a*) Explain electronic spectrum of $[Ti(H_2O)_6]^{3+}$ ion.

(*b*) Explain the terms nephelauxetic effect and nephelauxetic ratio. What is the significance of nephelauxetic effect?

(*c*) The purple colour of $MnO_4^-$ ion cannot arise from d-d transition. Explain.

*(Madras April 2011)*

**15.** (*a*) Derive the ground state term symbol for:

(*i*) nitrogen (*ii*) $d^{10}$.

(*b*) The absorption spectrum of $Ti^{+3}$ (aq) $d^1$ attributed to a single transition. Explain the transition, position, intensity and broad nature of this band.

(*c*) Draw Orgel diagram for $d^1$ and $d^9$ tetrahdral and octahedral complexes.

(*d*) Identify the ground state term giving reasons for the following set:

$^1S$, $^3F$, $^3P$, $^1G$, $^1D$.

(*e*) Calculate L for $p^2$ configuration and show the energy states.

(*f*) Calculate number of microstates for oxygen.

*(Kurukshetra University, 2011)*

**16.** Explain why the intensity of the band in electronic spectrum of $[Ti(H_2O)_6]^{3+}$ is very low. *(Nagpur 2012)*

**17.** (*i*) What are spin selection rules for *d-d* transition?

(*ii*) Find the ground term symbol for $Cr(3d^5\,4s^1)$

(*iii*) What factors determine the intensity of absorption bands in the electronic spectra of transition metal complexes. Draw orgel diagrams.

*(Lucknow 2013)*

CHAPTER 25

# Magnetic Properties of Transition Metal Complexes

## Introduction

Transition metal complexes show magnetic properties. The magnetic properties of different materials are studied in terms of their *magnetic moments* which arise due to the orbital motion and spinning motion of the electrons. As electron is a charged particle, its orbital motion produces a small magnetic field along the axis of rotation, as shown in Fig. 25.1 and its spinning motion produces a small magnetic field along the spin axis. Thus each electron may be considered as a small magnet having a resultant permanent magnetic moment. As magnetic moment is a vector quantity, the net magnetic moment of an electron may be represented by an arrow. Thus a material may be considered to contain a number of *magnetic dipoles* (similar to a bar magnet with north and south poles). Based on the behaviour of magnetic substances in the external magnetic field, the following are the important types of magnetism shown by transition metal complexes: *(i) Paramagnetism (ii) Diamagnetism (iii) Ferromagnetism, and (iv) Anti-ferromagnetism.*

Magnetic moment is expressed in terms of Bohr Magneton (B.M.) which is also represented as $\beta$ or $\mu_B$. Its value is $0.927 \times 10^{-20}$ erg/gauss. This value can be converted into joule/tesla remembering that 1 erg = $10^{-7}$ joule and 1 gauss = $10^{-4}$ tesla. In SI units 1 BM = $0.927 \times 10^{-23}$ or $9.27 \times 10^{-24}$ $Am^2$/molecule.

In terms of fundamental quantities 1 BM is given by the expression

$$1\ \text{BM} = \frac{eh}{4\pi mc}$$

where $e$ = electronic charge, $h$ = Planck's constant, $m$ = mass of the electron and $c$ = velocity of light.

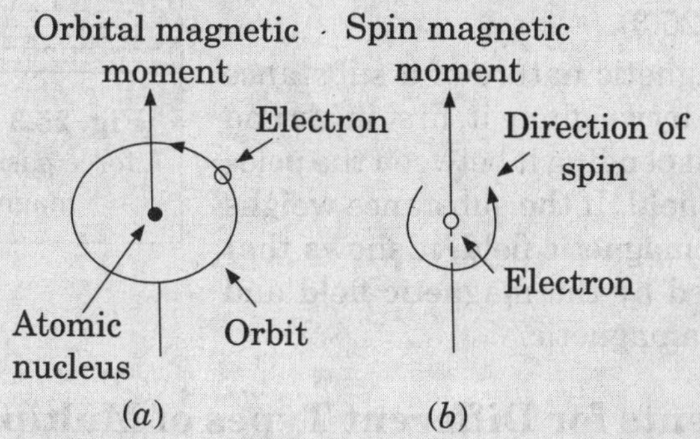


**Fig. 25.1** (a) Magnetic moment of electron due to orbital motion (b) Magnetic moment of electron due to spin motion

## Paramagnetism

Substances that are freely magnetised in the direction of the magnetic field in which they are placed are called paramagnetic substances. This property of the substances is called paramagnetism or paramagnetic property. Paramagnetic substances have the following characteristics:

(*i*) When a paramagnetic substances is placed in a magnetic field, the magnetic lines of force prefer to pass through the substance rather than through the air. Thus for paramagnetic substances B > H or B/H > 1 so that K is always positive. Here B = magnetic flux density at any point in the substance, H = strength of the applied magnetic field, B/H = permeability and K is the magnetic susceptibility.

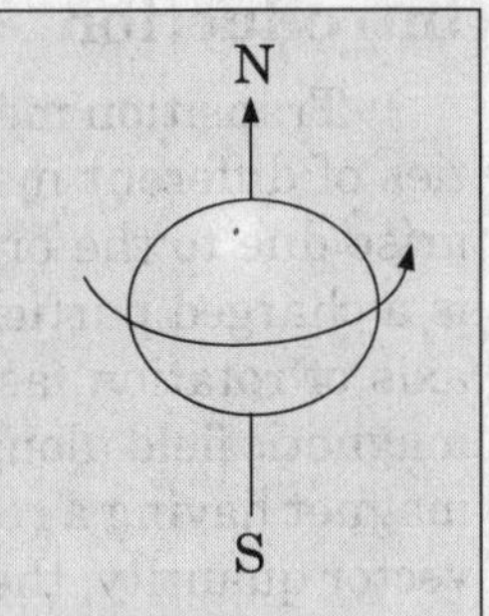

**Fig. 25.2** An electron spinning on its axis behaves like a small magnet.

(*ii*) Substances have one or more unpaired electrons show paramagnetic character. A single electron (unpaired electron) spinning on its axis generates a magnetic field and behaves like a magnet as shown in Figure 25.2. Therefore, when a substance whose orbital contains an unpaired electron is placed in an electromagnet, it is attracted towards the poles of the electromagnet. It follows that an atom, ion or molecule containing one or more unpaired electrons will be paramagnetic.

(*iii*) When a paramagnetic substance is placed in an external magnetic field, the individual atoms or molecules of the substance behave an permanent magnets and align themselves in the same direction as that of external magnetic field. Hence the paramagnetic substances are attracted towards magnetic field. As a result, when a paramagnetic substance is placed in a magnetic field, the number of lines of force passing through paramagnetic substance would be more than that which would pass through air (See Figure 25.3).

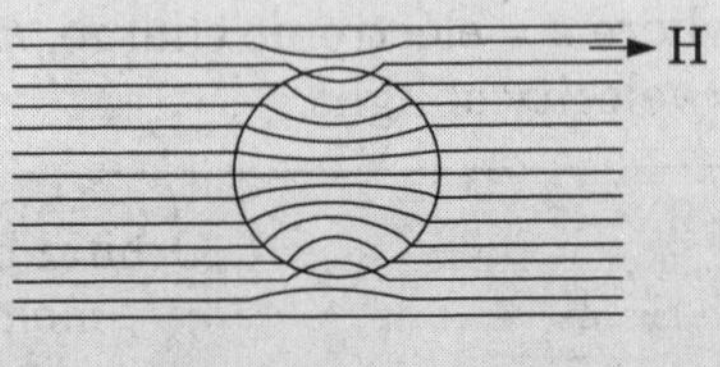

**Fig. 25.3** Number of lines of force passing through paramagnetic substance

(*iv*) The paramagnetic nature of a substance can be found by weighing it, first in air and then after suspending it between the poles of magnetic field. If the substance weighs more in the magnetic field, it shows that it is attracted by the magnetic field and hence is paramagnetic.

## Magnetic Moments for Different Types of Multiplet Widths

According to Russel-Saunders coupling scheme a particular term with given spin multiplicity and orbital multiplicity can assume several J values which range from (L + S) to (L – S ) or (S – L) whichever is positive. Thus a multielectron system has several J levels. These J levels taken together constitute a multiplet and a particular J level of a multiplet is called component of that multiplet. The energy gap or separation (ΔE) between the successive J levels is called **multiplet width.** Which of the J levels will be occupied by the metal ions depends on whether the multiplet width (*i.e.* energy separation, ΔE) is greater or smaller than or equal to thermal energy ($k$T). Thus we have the following three cases.

**1. Case in which the energy separation (ΔE) between various J states is greater than thermal energy ($k$T).** In this case the lowest J energy state of the metal ion is considerably separated from the adjacent excited state and hence energy separation (ΔE) between these states is greater than $k$T at room temperature. In this case L and S vectors couple very strongly and hence it is only the lowest energy J level which is occupied by metal ions. The magnetic moment ($\mu_{eff}$) for such an atom is given by:

$$\mu_{eff} = g\sqrt{J(J+1)}\ \text{BM} \qquad (i)$$

Here $g$ is the **Lande splitting factor** and J is the **overall angular momentum quantum number.** The value of $g$ is given by:

$$g = 1.5 + \frac{S(S+1) - L(L+1)}{2J(J+1)} \qquad (ii)$$

In case of tripositive lanthanide ions ($Ln^{3+}$ ions) (except $Sm^{3+}$ and $Eu^{3+}$ ions), since J states are well separated from one another, only the lowest J state is occupied by such metal ions i.e. in this case ΔE is greater than $k$T at room temperature. Thus in case of $Ln^{3+}$ ions, $\mu_{eff}$ can be calculated by using relation (*i*) given above. In case of $Ln^{3+}$ ions spin magnetic moment and orbital magnetic moment both contribute to the overall value of $\mu_{eff}$.

Experimental values of $\mu_{eff}$ for $Ln^{3+}$ ions (except $Sm^{3+}$ and $Eu^{3+}$ ions) have been found to be in good agreement with $\mu_{eff}$ values calculated from equation (*i*). This equation does not give correct values for $Sm^{3+}$ and $Eu^{3+}$ ions because in these ions ground state is close to the excited and hence $\Delta E \simeq kT$. It is due to this relation between ΔE and $k$T that some of the metal ions are also present in higher J states at room temperatures.

**2. Case in which the energy separation (ΔE) between various J states is smaller than thermal energy ($k$T).** In this case the energy separation between various J states of the ground term is smaller than $k$T so that the J states are almost equally occupied by metal ions at room temperature. This means that the coupling of L and S vectors in this case is so poor that these vectors maintain their individuality. L and S are said to interact independently (separately) with the magnetic field. *J is no longer a good quantum number.*

We know that generally the magnetic moment of a substance containing unpaired electrons is the resultant of *spin magnetic moment* ($\mu_{spin}$ or $\mu_S$) and *orbital magnetic moment* ($\mu_{orbital}$ or $\mu_L$). These magnetic moments are generated by the *spin motion of the electron round its own axis and orbital motion of the electron round the nucleus* respectively. In the present case the effective magnetic moment ($\mu_{eff}$) is the resultant of both the types of magnetic moments mentioned above. Since magnetic susceptibility has additive property, total susceptibility is given by:

$$\chi_M = \frac{N\mu_L^2}{3kT} + \frac{N\mu_S^2}{3kT}$$

or $$\chi_M = \frac{N}{3kT}\left(\mu_L^2 + \mu_S^2\right) \quad (i)$$

and $\mu_{eff}$ is given by:

$$\mu_{eff} = \mu_{L+S} = \sqrt{L(L+1) + 4S(S+1)} \text{ BM} \quad (ii)$$

$\mu_{eff}$ given by equation (*ii*) is called *spin plus orbital magnetic moment* and hence is represented as $\mu_{L+S}$.

Equation (*ii*) is applicable to those paramagnetic metal ions of the first transition series where the spin angular momentum (this momentum generates spin magnetic moment) and orbital angular momentum (this momentum produces orbital magnetic moment) both are effective. This means that this equation is applicable to those paramagnetic metal ions in which the orbital angular momentum is not quenched (destroyed).

It has been seen that, since in case of ions of the metals present in the first half of 1*st* transition series orbital angular momentum is quenched, L = 0 for these metal ions. Thus the value of $\mu_{eff}$ for these metal ions is given by equation (*iii*) which is obtained by pitting L = 0 in equation (*ii*). Thus:

$$\mu_{eff} = \mu_S = \sqrt{0 \times (0+1) + 4S(S+1)} \text{ BM}$$

or $$\mu_{eff} = \mu_S = \mu_{s.o.} = \sqrt{4S(S+1)} \text{ BM} \quad (iii)$$

Now since 2S = n, equation (*iii*) can also be written as:

$$\mu_{eff} = \mu_s = \mu_{s.o.} = \sqrt{n(n+2)} \text{ BM} \quad (iv)$$

The values of the $\mu_{eff}$ for n = 1, 2, 3, 4 and 5 unpaired electrons are as 1.73 BM for n = 1, 2.83 BM for n = 2, 3.87 BM for n = 3, 4.90 BM for n = 4 and 5.92 BM for n = 5.

The relations given at (*iii*) and (*iv*) are called **spin only formulae** and $\mu_{eff}$ calculated from these formulae are called **spin only magnetic moment** which is represented as $\mu_S$ or $\mu_{s.o}$.

The values of $\mu_{eff}$ calculated from equation (*iii*) or (*iv*) of some ions of the metals of 1*st* transition series are given in Table 25.1. The experimental values of $\mu_{eff}$ are also given for the sake of comparison. It may be seen from this Table that the experimental values differ slightly from the calculated (theoretical) values. The difference between the calculated and experimental values depends on the extent to which the orbital angular momentum of electrons in metal ions is quenched.

**Table 25.1. Calculated ($\mu_S$) and experimental values of magnetic moments of the ions of the elements of 1st transition series**

| Ions | n | S = n/2 | $\mu_s = \sqrt{4S(S+1)}$ BM $= \sqrt{n(n+1)}$ BM | Experimental value of magnetic moment (BM) |
|---|---|---|---|---|
| $Ti^{3+}$, $V^{4+}$ ($3d^1$) | 1 | 1/2 | 1.73 | 1.6–1.8 |
| $Ti^{2+}$, $V^{3+}$ ($3d^2$) | 2 | 1 | 2.83 | 2.7–2.9 |
| $V^{2+}$, $Cr^{3+}$ ($3d^3$) | 3 | 3/2 | 3.87 | 3.7–3.8 |
| $Cr^{2+}$, $Mn^{3+}$ ($3d^4$) | 4 | 2 | 4.90 | 4.8–4.9 |
| $Mn^{2+}$, $Fe^{3+}$ ($3d^5$) | 5 | 5/2 | 5.92 | 5.7–6.0 |
| $Fe^{2+}$, $Co^{3+}$ ($3d^6$) | 4 | 2 | 4.90 | 5.0–5.6 |
| $Co^{2+}$ ($3d^7$) | 3 | 3/2 | 3.87 | 4.1–5.2 |
| $Ni^{2+}$ ($3d^8$) | 2 | 1 | 2.83 | 2.8–3.5 |
| $Cu^{2+}$ ($3d^9$) | 1 | 1/2 | 1.73 | 1.8–2.1 |

If experimental value of $\mu_{eff}$ exceeds $\mu_S$, it is believed that orbital angular momentum is not completely quenched. On the other hand, if the experimental value of $\mu_{eff}$ is the same as the of $\mu_S$, it is believed that the orbital angular momentum is completely quenched.

Although the values of $\mu_{eff}$ calculated from spin only formula of most of the ions of 1*st* row transition metals are in good agreement with the experimental values ($\mu_{exp}$) of these ions, yet in some cases $\mu_{exp}$ values are higher than $\mu_S$ values. For example $\mu_{exp}$ value of $Co^{2+}$ ($3d^7 = t^5_{2g}e^2_g$) in higher than $\mu_S$. Higher value is because of the fact that there is some orbital contribution from $Co^{2+}$ ion to the value of $\mu_S$. Orbital contribution of $Co^{2+}$ ion is due to its $t^5_{2g}e^2_g$ configuration.

In case of 2*nd* and 3*rd* row transition metals, since orbital contribution is not quenched or prevented, spin only formula cannot be used for calculating the value of $\mu_{eff}$ of the ions of these metals. Thus in the calculation of $\mu_{eff}$ for the ions of the metals of 2*nd* and 3*rd* row transition series, orbital angular momentum contribution should also be included in the calculations. These metals also show extensive temperature dependent paramagnetism. This is due to spin-orbit coupling which removes the degeneracy from the lowest energy level in the ground state.

**3. Case in which energy separation ($\Delta E$) between various *J* states are comparable to *kT*.** This situation is a mixture of two extremes 1 and 2 discussed above. The average magnetic susceptibility of a sample is the result of weighted contributions from each *J* state of the multiplet obtained from spin-orbit coupling in the ground term. In such cases, magnetic susceptibility is given by the relation:

$$\chi = (N/3kT)\sum g^2\beta^2 J(J+1)(2J+1)e^{-\Delta E/kT} \sum(2J+1)e^{-\Delta E/kT}$$

As is evident from the above relation, $\chi$ is not directly proportional to $1/T$. Such systems, therefore, do not obey the Curie law.

Results of the three different cases may be summarised as follows:

| **Multiple width** | $\mu$**(BM)** | **Temperature dependence of** $\chi_M$ |
|---|---|---|
| (*i*) Large compared to $kT$ | $g\sqrt{J(J+1)}$ | $\chi_M \propto \frac{1}{T}$ (Curie law is obeyed) |
| (*ii*) Small compared to $kT$ | $\sqrt{L(L+1) + 4S(S+1)}$ | $\chi_M \propto \frac{1}{T}$ (Curie law is obeyed) |
| (*iii*) Medium compared to $kT$ | Complicated function of $J$ and $T$ | Curie law is not obeyed. |

## Magnetic Susceptibility ($\chi$)

Magnetic moment ($\mu$) of a substance cannot be determined experimentally. For the determination of $\mu$ we first determine the value magnetic susceptibility ($\chi$) of the substance and then this value of $\chi$ is used for the determination of the value of $\mu$.

Magnetic susceptibility of a substance is its capacity to get magnetised when placed in a magnetic field. When a substance is placed in a magnetic field, the difference between the applied magnetic field (H) and the magnetic field produced (induced) in the substance (B) is given by:

$$B - H = 4\pi I \quad (i)$$

Here I is the *intensity of magnetisation of the substance.* B is always less than H (B < H) for diamagnetic substances and B is greater than H (B > H) for paramagnetic substances. Thus for paramagnetic substances B is positive.

Now divide equation (*i*) by H to get:

$$\frac{B}{H} - 1 = 4\pi \times \frac{I}{H}$$

or

$$\frac{B}{H} = 1 + 4\pi \times \frac{I}{H} \quad (ii)$$

In equation (*ii*), the term B/H is called **magnetic permeability**. For vacuum B/H = 1. The I/H is the magnetic susceptibility per unit volume and is called **volume susceptibility** which is represented as $k$ or $\chi_v$. Thus equation (*ii*) can be written as:

$$\frac{B}{H} = 1 + 4\pi k \quad (iii)$$

Since in chemistry we often deal with mass rather than volume, we generally express susceptibility as **mass susceptibility** or **gram susceptibility** which is represented as $\chi_g \cdot \chi_g$ is related to volume susceptibility ($k$) and density (d) of the substance by the equation

$$\chi_g = \frac{k}{d} \quad (iv)$$

If we multiply gram susceptibility ($\chi_g$) by the molecular mass of the substance (M) we get **molar magnetic susceptibility** which is represented as $\chi_M$. Thus:

$$\boxed{\chi_M = \chi_g \times M} \quad (v)$$

The value of $\chi_M$ can be determined experimentally by Gouy's method and then can be used to calculate the value of $\mu$.

## Diamagnetic Correction ($\chi_{dia}$)

The measured susceptibility for a given substance will consist of contributions from paramagnetic and diamagnetic susceptibilities, the former being much greater. The measured susceptibility is corrected by subtracting the diamagnetic contribution from it.

| Paramagnetic susceptibility = Measured susceptibility – Diamagnetic susceptibility. |
|---|

## Temperature Independent Paramagnetism (TIP)

Certain substances like $CrO_4^{2-}$ (Cr in VI oxidation state) $MnO_4^-$ (Mn in VII oxidation state) which do not contain unpaired electrons, show weak paramagnetism, when the substance is placed in a magnetic field. This weak paramagnetism is independent of temperature and is referred to as temperature independent paramagnetism (TIP). The paramagnetism is produced by mixing of the ground state (which is diamagnetic) with excited states of high energy (which are paramagnetic) under the influence of the applied field. So ground state assumes some of the magnetic property of the excited state. Thus TIP is not due to presence of any paramagnetic centres in the sample but is induced (as in diamagnetism) in the sample when it is placed in a magnetic field.

## To Get the Corrected Value of Paramagnetic Susceptibility ($\chi_{para}^{corr}$) by Diamagnetic Correction

Since all paramagnetic materials contain electrons, diamagnetism is also present in these substances. The magnitude of diamagnetism in paramagnetic material is very small. The presence of diamagnetism in paramagnetic substances affects the value of magnetic susceptibility of paramagnetic substances ($\chi_{para}$).

It has been found that the value of paramagnetic susceptibility measured experimentally [*i.e.* $(\chi_{para})_{exp}$ value] is the sum of diamagnetic susceptibility ($\chi_{dia}$) and paramagnetic susceptibility ($\chi_{para}$), *i.e.*

$$(\chi_{para})_{exp} = \chi_{para} + \chi_{dia}$$

or

$$\boxed{\chi_{para} = (\chi_{para})_{exp} - \chi_{dia}} \quad (i)$$

The value of $\chi_{para}$ obtained from equation (*i*) is called *corrected* value of $\chi_{para}$ and hence is represented as $\chi_{para}^{corr}$. Thus equation (*i*) should be written as:

$$\boxed{\chi_{para}^{corr} = (\chi_{para})_{exp} - \chi_{dia}} \quad (ii)$$

Equation (*ii*) shows that in order to get the corrected value of $\chi_{para}$ ($\chi_{para}^{corr}$) the diamagnetic contributions from all individual atoms or ions of the substance are summed up and then the sum thus obtained is subtracted from the experimental value of $\chi_{para}$ [$(\chi_{para})_{exp}$].

The correction made by considering the diamagnetic contributions of all atoms or ions of the substance is called *diamagnetic correction.*

It can be shown by quantum mechanical calculations that the value of $\chi_{para}^{corr}$ is given by:

$$\chi_{para}^{corr} = \frac{N\mu_{eff}^2}{3kT}$$

or
$$\mu_{eff}^2 = \chi_{para}^{corr} \times T \times \frac{3k}{N}$$

or
$$\mu_{eff} = \sqrt{\frac{3k}{N}} \times \sqrt{\chi_{para}^{corr}\ T} \qquad (iii)$$

In the above equation $\mu_{eff}$ = magnetic moment (in BM) per mole of the paramagnetic substance, $k$ = Boltzmann constant, N = Avogadro's number, T = absolute temperature.

On putting the values of $k$ and N in the above relation we get:

$$\mu_{eff} = 2.84\sqrt{\chi_{para}^{corr}\ T}\ \text{BM} \qquad (iv)$$

In order to calculate the value of $\mu_{eff}$ put the values of $\chi_{para}^{corr}$ and T in equation (*iv*). The value of $\chi_{para}^{corr}$ can be determined from equation (*ii*) provided that values of $(\chi_{para})_{exp}$ and $\chi_{dia}$ are known. The value of $(\chi_{para})_{exp}$ can be determined by Gouy balance method. The value of $\chi_{dia}$ can be obtained to a good approximation by summing up the diamagnetic contributions from all atoms, ions and bonds present in the compound. Contributions are obtained from Pascal constants of atoms, ions and bonds given in Table 25.2.

Since in the diamagnetic substances the magnetic field indeed is in the direction opposite to the direction of the magnetic field applied, $\chi_{dia}$ is always negative. $\chi_{d\ ia}$ for bonds is positive.

## To Calculate the Value of $\chi_{dia}$ of Some Compounds

Let us calculate the value of $\chi_{dia}$ for pyridine, acetone and $CuSO_4 \cdot 5H_2O$.

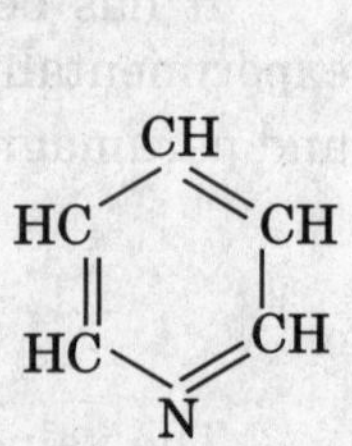

(*i*) **Pyridine ($C_5H_5N$).** The structure of $C_5H_5N$ molecule is given in the margin. $\chi_{dia}$ for this molecule is the sum of the following diamagnetic contributions (effects).

Diamagnetic contribution of five C-atoms of the aromatic ring

$$= 5 \times (-6.2 \times 10^{-6})\ cm^3 mol^{-1}$$
$$= -31.0 \times 10^{-6}\ cm^3\ mol^{-1}$$

Diamagnetic contribution from five H-atoms

$$= 5 \times (-2.9 \times 10^{-6})\ cm^3 mol^{-1}$$
$$= -14.5 \times 10^{-6}\ cm^3\ mol^{-1}$$

**Table 25.2. Diamagnetic contributions (Pascal's constants) (in $cm^3\ mol^{-1}$) of some atoms, ions, molecules and bonds. All table values must be multiplied by $10^{-6}$.**

| Species | Diamagnetic contribution | Species | Diamagnetic contribution |
|---|---|---|---|
| **(A) Atoms** | | $Cu^{2+}$ | –12.8 |
| H | –2.9 | $Co^{2+}$ | –12.8 |
| C | –6.0 | $Ni^{2+}$ | –12.8 |
| C (aromatic) | –6.2 | $CO_3^{2-}$ | –370 |
| N | –5.6 | $NO_2^-$ | –126 |
| N (aromatic) | –4.6 | $NO_3^-$ | –326 |
| O | –4.6 | $SO_3^{2-}$ | –478 |
| $O_2$ (carboxylate) | –7.9 | $SO_4^{2-}$ | –503 |
| S | –15.0 | **(C) Molecules** | |
| P | –26.3 | $H_2O$ | –163 |
| F | –6.3 | $NH_3$ | –226 |
| Cl | –20.1 | **(D) Bonds** | |
| Br | –30.6 | C–C | +5.5 |
| I | –44.6 | C;C | +0.8 |
| **(B) Ions** | | C–N | +8.2 |
| $Mg^{2+}$ | –5.0 | C;N | +0.8 |
| $Zn^{2+}$ | –15.0 | N–N | +1.8 |
| $Pb^{2+}$ | –32.0 | N–O | +1.7 |
| $Ca^{2+}$ | –10.4 | C–O | +6.3 |
| $Fe^{2+}$ | –12.8 | | |

Diamagnetic contribution from one N-atom in the aromatic ring

$$= 1 \times (-4.6 \times 10^{-6})\ cm^3mol^{-1}$$
$$= -4.6 \times 10^{-6}\ cm^3mol^{-1}$$

Since we have used aromatic ring values for five C-atoms and one N-atom, the diamagnetic contribution from N═C, C—C and C═C bonds is taken to be zero. Hence:

$$\chi_{dia} \text{ for pyridine} = [-31.0 \times 10^{-6} + (-14.5 \times 10^{-6}) + (-4.6 \times 10^{-6})]\ cm^3mol^{-1}$$
$$= -50.1 \times 10^{-6}\ cm^3mol^{-1}$$

*(ii)* **Acetone ($CH_3$—CO—$CH_3$).** The structure of this compound is given in the margin. This structure has one C═O bond, two C—C bonds and six C—H bonds. The diamagnetic contribution from C = O bond is $+6.3 \times 10^{-3}\ cm^3mol^{-1}$ while that from the other bonds is taken as zero. Hence:

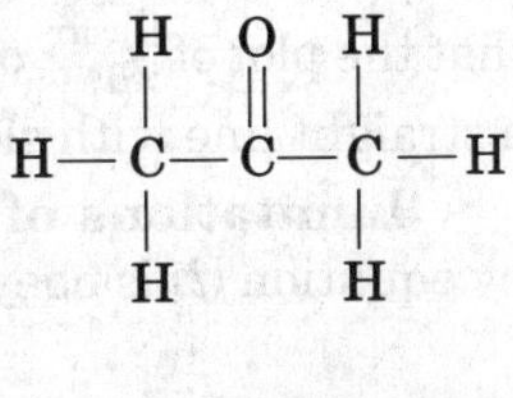

Diamagnetic contribution from 3 C atoms = $3 \times (-6.0 \times 10^{-6})\ cm^3mol^{-1}$
$= -18 \times 10^{-6}\ cm^3mol^{-1}$

Diamagnetic contribution from 6 H atoms = $6 \times (-2.9 \times 10^{-6})\ cm^3mol^{-1}$
$= -17.4 \times 10^{-6}\ cm^3mol^{-1}$

Diamagnetic contribution from one O-atom = $1 \times (-4.6 \times 10^{-6})\ cm^3mol^{-1}$
$= -4.6 \times 10^{-6}\ cm^3mol^{-1}$

Diamagnetic contribution from one C=O bond = $+6.3 \times 10^{-6}\ cm^3mol^{-1}$

$\therefore \chi_{dia}$ for acetone = $[-18 \times 10^{-6} + (-17.4 \times 10^{-6}) + (-4.6 \times 10^{-6}) + 6.3 \times 10^{-6}]$
$= -33.7 \times 10^{-6}$

**(iii) $CuSO_4 . 5H_2O$.** This salt contains one $Cu^{2+}$ ion, one $SO_4^{2-}$ ion and five $H_2O$ molecules. Hence:

Diamagnetic contribution from one $Cu^{2+}$ ion = $-12.8 \times 10^{-6}\ cm^3mol^{-1}$

Diamagnetic contribution from one $SO_4^{2-}$ ion = $-503 \times 10^{-6}\ cm^3mol^{-1}$

Diamagnetic contribution from $5H_2O$ molecules = $5 \times (-163 \times 10^{-6})\ cm^3mol^{-1}$
$= -815 \times 10^{-6}\ cm^3\ mol^{-1}$

$\therefore \chi_{dia}$ for $CuSO_4 \cdot 5H_2O = [-12.8 \times 10^{-6} + (-503 \times 10^{-6}) + (-815 \times 10^{-6})]\ cm^3mol^{-1}$
$\mathbf{= -1330 \times 10^{-6}\ cm^3mol^{-1}}$

## Effect of Temperature on the Corrected Value of Magnetic Susceptibility of Paramagnetic Substances ($\chi_{para}^{corr}$): Curie and Curie-Weiss Law

$\chi_{para}^{corr}$ depends on temperature, since the thermal agitation randomises the atomic or molecular magnetic dipoles of the paramagnetic substances which get aligned in the direction of the magnetic field.

Curie showed that $\chi_{para}^{corr}$ varies inversely with absolute temperature (T), since thermal agitation (i.e. temperature) opposes the paramagnetic alignment. Thus:

$$\boxed{\chi_{para}^{corr} = \frac{C}{T}} \qquad (i)$$

Here C is a constant, called **Curie constant** and is characteristic of the paramagnetic substance. The relation given in equation (*i*) is called **Curie relation** or **Curie law.**

The value of C is given by:

$$\boxed{C = \frac{N\mu_{eff}^2}{3k}} \qquad (ii)$$

where N = Avogadros' number ($-6.0225 \times 10^{23}\ mol^{-1}$), $\mu_{eff}$ = Effective magnetic moment and $k$ = Boltzmann constant (= $1.3805 \times 10^{-16}$ erg.deg$^{-1}$). Equation (*i*) shows that the plot of $\chi_{para}^{corr}$ or $\chi_M^{corr}$ values against 1/T or of $1/\chi_{para}^{corr}$ or $1/\chi_M^{corr}$ against T will be a straight line with slope equal to C which passes through the origin. [Fig. 25.4 (a)]

**Limitations of Curie law and Curie-Weiss law.** Curie law represented by equation (*i*) is obeyed by those paramagnetic substances in which the magnetic

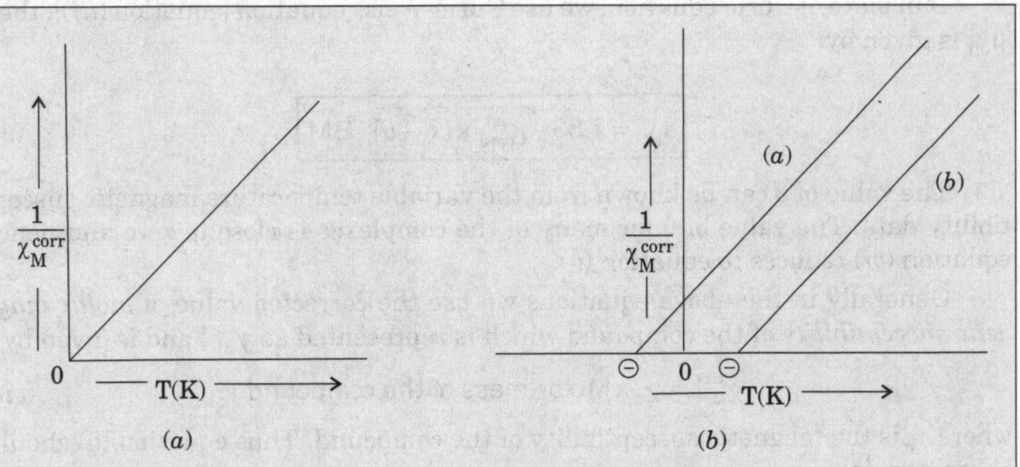

(a) (b)

**Fig. 25.4** Effect of temperature (T) on the corrected value of molar magnetic susceptibility ($\chi_M^{corr}$) of a paramagnetic substance.

dipoles of the substance are completely independent of each other. If there are some cooperative interactions between the neighbouring magnetic dipoles or if the applied magnetic field induces some temperature independent paramagnetism, Curie law is not obeyed i.e., the line does not go through the origin but cuts the T axis at a temperature below 0K or at a temperature above 0K [Figure 25.4 (b)]. For such paramagnetic substances a modified law known as Curie-Weiss law is suggested. This law is represented by the following relation.

$$\chi_{para}^{corr} = \frac{C}{T - \theta} \qquad (iii)$$

where θ is the intercept on the temperature axis and is called **Curie-Weiss constant.** θ takes into account the cooperative interactions mentioned above. For ferromagnets θ is positive while for antiferromagnets θ is negative. The magnitude and sign of θ can be obtained from the plot of $1/\chi_{para}^{corr}$ *vs* T.

**Calculation of the Value of $\mu_{eff}$**

If we put the value of C from equation (*ii*) in equation (*i*), we get:

$$\chi_{para}^{corr} = \frac{N\mu_{eff}^2}{3kT}$$

or

$$\mu_{eff} = \sqrt{\frac{3k}{N}} \times \sqrt{\chi_{para}^{corr} \times T} \text{ BM} \qquad (iv)$$

On putting $k$ = 1.3805 × $10^{-16}$ erg.deg$^{-1}$ and N = 6.0225 × $10^{23}$ mol in the above equation, we get $\frac{3k}{N}$ = 2.83 and hence:

$$\mu_{eff} = 2.83\sqrt{\chi_{para}^{corr} \times T} \text{ BM} \qquad (v)$$

If in place of Curie equation, we use Curie-Weiss equation [equation (*iii*)], then $\mu_{eff}$ is given by:

$$\mu_{eff} = 2.83\sqrt{\chi_{para}^{corr} \times (T - \theta)} \text{ BM} \qquad (vi)$$

The value of θ can be known from the variable temperature magnetic susceptibility data. The value of θ for many of the complexes is close to zero and hence equation (*vi*) reduces to equation (*v*).

Generally in the above equations we use the corrected value of *molar magnetic susceptibility* of the compound which is represented as $\chi_M^{corr}$ and is given by:

$$\chi_M^{corr} = \chi_c \times \text{Molar mass of the compound} \qquad (vii)$$

where $\chi_c$ is the magnetic susceptibility of the compound. Thus equation (*v*) should be written as:

$$\mu_{eff} = 2.83\sqrt{\chi_M^{corr} \times T} \text{ BM} \qquad (viii)$$

The corrected value, $\chi_M^{corr}$ is given by:

$$\chi_M^{corr} = \chi_{exp} - \chi_{dia} \qquad (ix)$$

Here $\chi_{dia}$ = Experimental value of magnetic susceptibility of the paramagnetic compound and $\chi_{dia}$ = Diamagnetic susceptibility contribution which is equal to the sum of the contributions made by all individual atoms or ions present in the compound.

## Determination of Molar Magnetic Susceptibility ($\chi_M$) and $\mu_{eff}$ of a Magnetic Compound by Gouy's Method

In this method the sample (e.g. compound), in the form of a cylinder, is suspended between the poles of a magnet, preferably an electromagnet. The sample extends to a length of about 10–11 cm such that the magnetic field is zero at the upper point. The sample is packed uniformally into a glass tube of suitable size and is vertically suspended from the beam of the balance in a drought free enclosure (See Fig. 25. 5).

The mass of the sample is determined with and without the magnetic field on. Similar procedure is done with a standard. Under these conditions a change in the weight of the sample takes place. With the help of this change in weight we can determine the molar magnetic susceptibility $(\chi_M)_c$ and $\mu_{eff}$ of the magnetic substance. The value of $\chi_M$ is corrected for diamagnetism and TIP.

The standards (i.e. caliberants) which are generally used are $Hg[Co(NCS)_4]$ ($\chi = 16.44 \times 10^{-6}$ cgs units), [Ni(ethylenediamine)$_3$] $S_2O_3$ ($\chi = +11.03 \times 10^{-6}$ cgs units), pure water ($\chi = -0.72 \times 10^{-6}$ cgs units), $CuSO_4 \cdot 5H_2O$ ($\chi = +5.92 \times 10^{-6}$ cgs units). $CuSO_4 \cdot 5H_2O$ suffers from the disadvantage that it does not pack well into the tube. Water is not a good standard for the measurement of susceptibility of paramagnetic substances. $Hg[Co(NCS)_4]$ is most commonly used as a standard. The value of χ of this standard at 20°C is 16.44 × $10^{-6}$ cgs units. If the temperature is below 20°C, value of χ of $Hg[Co(NCS)_4]$ may be calculated from the temperature coefficient data.

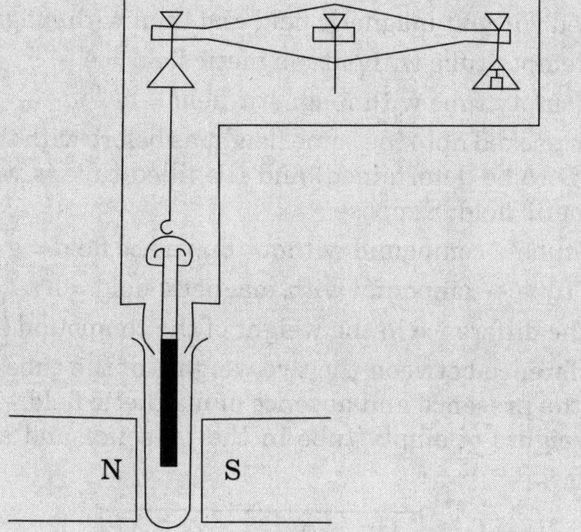


**Fig. 25. 5** Apparatus for the Gouy method

## Procedure

The procedure used for the determination of $\mu_{eff}$ consists of the following steps:

**(A) Experiment with standard** (*i*) A given tube is first weighed without and with magnetic field on. Suppose:

Weight of empty tube without magnetic field = a

Weight of empty tube with magnetic field = b

$\therefore$ Difference in weights = (b – a)

(*ii*) The tube is then packed upto a height of 10 to 11 cm with a standard. Now weigh the tube with standard without and with magnetic field on, Suppose:

Weight of tube + standard without magnetic field = c

Weight of tube + standard with magnetic field = d

$\therefore$ Difference in weights = (d – c)

(*iii*) Find out the difference in the weight of standard ($\Delta W_s$). It is given by:

$\Delta W_s$ = Difference between the two weights of the tube filled with standard in the presence and absence of magnetic field – Difference between the two weights of empty tube in the presence and absence of magnetic field.

or $$\Delta W_s = (d - c) - (b - a)$$

(*iv*) Find out the weight of the standard ($W_s$) in the absence of magnetic field. It is given by:

$W_s$ = Weight of tube and standard without magnetic field – Weight of empty tube without magnetic field.

or $$W_s = c - a$$

**(B) Experiment with compound** (*i*) The tube is next cleaned. The cleaned tube is first weighed without magnetic field and then with magnetic field. Suppose:

Weight of empty tube without magnetic field = e

Weight of empty tube with magnetic field = f

(*ii*) The tube is packed upto the same height as before with the compound (whose $\mu_{eff}$ value is to be determined) and the filled tube is weighed without and with magnetic field. Suppose:

Weight of tube + compound without magnetic field = g

Weight of tube + compound with magnetic field = h

(*iii*) Find out the difference in the weight of the compound ($\Delta$Wc). It is given by:

$\Delta W_c$ = Difference between the two weights of the tube filled with the compound in the presence and absence of magnetic field – Difference between the two weights of empty tube in the presence and absence of magnetic field.

or $$\Delta W_c = (h - g) - (f - e)$$

(*iv*) Find out the weight of the compound ($W_c$) in the absence of magnetic field. It is given by:

$W_c$ = Weight of tube and compound without magnetic field – Weight of empty tube without magnetic field

or $$W_c = g - e$$

When $Hg[Co(NCS)_4]$ is used as a standard the magnetic susceptibility of the compound ($\chi_c$) can be calculated by the following formula:

$$\chi_c = \chi_s \times \frac{W_s}{\Delta W_s} \times \frac{\Delta W_c}{W_c} \qquad (i)$$

Here $\chi_c$ = magnetic susceptibility of the compound, $\chi_s$ = magnetic susceptibility of the standard, $Hg[Co(NCS)_4] = 16.44 \times 10^{-6}$ cgs units at 293 K, $W_s$ = Weight of the standard, $\Delta W_s$ = Difference in weight of the standard, $W_c$ = Weight of the compound and $\Delta W_c$ = Difference in the weight of the compound.

Put the values of $\chi_s$ (= $16.44 \times 10^{-6}$ cgs units), $W_s$, $\Delta W_s$, $\Delta W_c$ and $W_c$ as calculated above in relation (*i*) and get the value of $\chi_c$ of the compound. The value of molar susceptibility of the compound $(\chi_M)_c$ is calculated from the relation:

$$(\chi_M)_c = \chi_c \times \text{Molecular weight of the compound} \qquad (ii)$$

The value of $(\chi_M)_c$ obtained as above is corrected for diamagnetism of ligands, anions, solvents of crystals, metal atom and temperature independent paramagnetism (TIP) term. The corrected value of $\chi_M$ is denoted as $(\chi_M^{corr})_c$ which is given by:

$$(\chi_M^{corr})_c = [\chi_M - \text{Diamagnetism correction} - \text{TIP term}] \text{ cgs units}$$

The value of $\mu_{eff}$ of the compound is obtained from Curie's equation

$$\mu_{eff} = 2.83\sqrt{(\chi_M^{corr})_c \times T} \text{ BM}$$

Put the values of $(\chi_M^{corr})_c$ as given above and T = 293 K to get the value of $\mu_{eff}$.

**Advantages of Gouy's Method** (*i*) It requires very simple equipment, easy to assemble, nominal cost and is very simple to operate.

(*ii*) Since the amount of sample taken in Gouy tube is quite large, even an ordinary analytical balance can measure the changes in mass.

**Disadvantages of Gouy's Method** (*i*) It requires larger amounts of sample

(*ii*) Correct results are not obtained if the Gouy tube has not been packed with the sample uniformly. That is, a uniform, compact packing of the sample is very essential.

## Solved Examples

**Example 1. Determine the value of $\chi_M$ and $\mu_{eff}$ for [Cu(biguanide)$_2$]Cl$_2$. 2H$_2$O on the basis of the following data. These data have been recorded at 20°C (=293K) and Hg[Co(NCS)$_4$] has been used as standard. $\chi$ for the standard = 16.44 × 10$^{-6}$ cgs units. Molecular weight of [Cu(biguanide)$_2$]Cl$_2$ · 2H$_2$O = 372.5, Diamagnetic correction for the compound = –207 × 10$^{-6}$ cgs unit and TIP = +50 × 10$^{-6}$ cgs units.**

**Data with compound and Hg[Cu(NCS)$_4$] (standard) are:**

**Weight of empty tube without magnetic field = 9.72302g**

**Weight of empty tube with magnetic field = 9.72246g**

**Weight of tube + Hg[Co(NCS)$_4$] (standard) without magnetic field = 10.73996 g**

**Weight of tube + Hg(Co(NCS)$_4$] with magnetic field = 10.76296 g**

**Data with [Cu(biguanide)$_2$]Cl$_2$. 2H$_2$O are:**

**Weight of empty tube without magnetic field = 9.72298 g**

**Weight of the empty tube with magnetic tube = 9.72245 g**

**Weight of tube + [Cu(biguanide)$_2$]Cl$_2$ · 2H$_2$O without magnetic field = 10.05913 g**

**Weight of tube + [Cu(biguanide)$_2$]Cl$_2$ · 2H$_2$O with magnetic field = 10.06013 g**

**Solution:** $\Delta W_s = (10.76296 - 10.73996) - (9.72246 - 9.72302)$

$= 0.02300 - (-0.00056)$

$= 0.02300 + 0.00056$

$= 0.02356$ g

$W_s = 10.73996 - 9.72302$

$= 1.01694$ g

$$\Delta W_c = (10.06013 - 10.05913) - (9.72245 - 9.72298)$$
$$= 0.00100 - (-0.00053)$$
$$= 0.00100 + 0.00053$$
$$= 0.00153 \text{ g}$$
$$W_c = 10.05913 - 9.72298$$
$$= 0.33615 \text{ g}$$

$\chi$ of $[Cu(biguanide)_2]Cl_2.\ 2H_2O$ ($\chi_C$) is given by:

$$\chi_c = \chi_s \times \frac{W_s}{\Delta W_s} \times \frac{\Delta W_c}{W_c}$$

$$= 16.44 \times 10^{-6} \times \frac{1.01694}{0.02356} \times \frac{0.00153}{0.33615}$$

$$= 3.2298 \times 10^{-6} \text{ cgs units}$$

Now $\chi_M$ of $[(Cu(biguanide)_2]Cl_2 \cdot 2H_2O$ is given by:

$$\chi_M = \chi_c \times \text{Molecular weight of } [Cu(biguanide)_2]Cl_2 \cdot 2H_2O$$

$$= 3.2298 \times 10^{-6} \times 372.5$$

$$= \mathbf{1203 \times 10^{-6} \text{ cgs units (Ans)}}$$

$$\chi_M^{corr} = \chi_M - \text{Diamagnetic correction} - \text{TIP}$$

$$= 1203 \times 10^{-6} - (-207 \times 10^{-6}) - 50 \times 10^{-6}$$

$$= 1360 \times 10^{-6} \text{ cgs units}$$

$$\therefore \mu_{eff} = 2.83\sqrt{\chi_M^{corr} \times T} \text{ BM}$$

$$= 2.83\sqrt{1360 \times 10^{-6} \times 293} \text{ BM}$$

$$= \mathbf{1.79 \text{ BM (Ans)}}$$

**Example 2: Calculate the value of $\chi_M^{corr}$ and $\mu_{eff}$ of a sample of $Cu(dmgH)_2Cl_2$ with the keep of the following data. These data have been taken at 27°C.**

**Given that $\alpha$ for the Gouy balance = 0.01 mg, tube caliberation constant ($\beta$) = 0.420, force acting on the sample (F) = 2.27 mg, $\delta$ = –0.28 mg, weight of the sample (w) = 0.196 g, molecular weight of the sample 250.5 and diamagnetic correction of the sample ($\chi_{dia}$) = –92 × $10^{-6}$ g atom.**

**Solution:** Magnetic susceptibility ($\chi$) of a substance is given by:

$$\chi = \frac{\alpha \text{ (mg)} + \beta[F(mg) - \delta(mg)]}{w(g) \times 10^6}$$

On putting the given values of different quantities we get:

$$\chi = \frac{0.01 + 0.420[2.25 - (-0.28)]}{0.196 \times 10^6}$$

$$= 5.47 \times 10^{-6}$$

$$\therefore \chi_M = \chi \times \text{ Molecular weight of the sample}$$
$$= 5.47 \times 10^{-6} \times 250.5 = 1370 \times 10^{-6}$$
$$\chi_M^{corr} = \chi_M - \chi_{dia}$$
$$= 1370 \times 10^{-6} - (-92 \times 10^{-6})$$
$$= 1370 \times 10^{-6} + 92 \times 10^{-6}$$
$$= \mathbf{1462 \times 10^{-6}} \quad \textbf{(Ans.)}$$

$\mu_{eff}$ is given by

$$\mu_{eff} = 2.83\sqrt{\chi_M^{corr} \times T} \text{ BM}$$
$$= 2.83\sqrt{1462 \times 10^{-6} \times (27 + 273)} \text{ BM}$$
$$= 2.83\sqrt{1462 \times 10^{-6} \times 300} \text{ BM}$$
$$= \mathbf{1.87 \ BM} \ \textbf{(Ans.)}$$

## Applications of Magnetic Moment Values of Complexes

Experimental and calculated spin only values of $\mu_{eff}$ of some inner orbital octahedral, outer orbital octahedral, tetrahedral and square planar complexes of metal ions of 1*st* transition series are given in Table 25.3. In this table the calculated values of $\mu_{eff}$ are those which have been obtained from spin only formula, $\mu_{eff} = \mu_{s.o.} = \sqrt{4S(S+1)}$ BM $= \sqrt{n(n+2)}$ BM. Thus these values are spin only values. For $n$ = 1, 2, 3, 4 and 5, $\mu_{s.o.}$ values (in BM) are 1.73, 2.83, 3.88, 4.90 and 5.92 respectively. With the help of these values we can explain the following:

**1. Type of hybridisation in complexes** Following examples illustrate this application.

(*i*) **Sodium pentacyanonitrosyl ferrate (II), $Na_2[Fe(CN)_5(NO)] \cdot 2H_2O$ (Also called sodium nitroprusside).** Magnetic measurements of this compound have shown that $Na_2[Fe(CN)_5(NO)]$ is diamagnetic ($\mu_{eff} = 0$). This complex has $Fe^{2+}$ and $NO^+$. $Fe^{2+}$ ion a $3d^6$ ion. In order to explain its diamagnetic character of the complex compound it is assumed that all the six 3d electrons of $Fe^{2+}$ ion are paired and $3d^2\ 4s\ 4p^3$ ($d^2sp^3$) hybridisation takes place. Thus the given complex is *inner orbital octahedral complex of Fe (II),* which involves $d^2sp^3$ hybridisation.

(*ii*) **Pentaaqua nitrosonium ferrate (I) ion, $[Fe(H_2O)_5(NO)]^{2+}$.** Experimental value of $\mu_{eff}$ for this ion is 3.9 BM. This value suggests the presence of three unpaired electrons on Fe atom in the complex ion. Thus Fe is present as $Fe^+$ in the complex ion (Fe = $3d^6 4s^2$, $Fe^+ = 3d^6 4s^1$ or $3d^7$ with $n = 3$). NO is present as $NO^+$. In order to explain the presence of three unpaired electrons, it has been suggested that 3d orbitals do not participate in octahedral hybridisation, and hence complex ion results from *4s* $4p^3 4d^2$ ($sp^3d^2$) hybridisation of $Fe^+$. Three unpaired electrons remain in *3d* orbitals. Thus the given complex ion is an *outer orbital octahedral ion of Fe(I) involving* $sp^3d^2$ *hybridisation.*

**Table 25. 3 Values of $\mu_{exp}$ and $\mu_{s.o.}$ of some complexes of metals of 1st transition series.**

| Complex | Central metal ion with $d^x$ configuration | No. of unpaired electrons ($n$) in the complex | $\mu_{eff}$ (B.M) | |
|---|---|---|---|---|
| | | | $\mu_{exp}$ | $\mu_{s.o.}$ |
| **(a) Inner orbital octahedral complexes (Hybridisation: $d^2sp^3$)** | | | | |
| $K_3[TiF_6]$ | $Ti^{3+}(3d^1)$ | 1 | 1.70 | 1.73 |
| $[V(acac)_3]^0$ | $V^{3+}(3d^2)$ | 2 | 2.86 | 2.83 |
| $NH_4[V(SO_4)_2] \cdot 12H_2O$ | $V^{3+}(3d^2)$ | 2 | 2.80 | 2.83 |
| $[Cr(NH_3)_6]Br_3$ | $Cr^{3+}(3d^3)$ | 3 | 3.77 | 3.88 |
| $[Cr(Big\ H)_3]Cl_3$ | $Cr^{3+}(3d^3)$ | 3 | 3.86 | 3.88 |
| $[Cr(dipy)_3]Br_2 \cdot 4H_2O$ | $Cr^{2+}(3d^4)$ | 2 | 3.27 | 2.83 |
| $K_4[Mn(CN)_6] \cdot 3H_2O$ | $Mn^{2+}(3d^5)$ | 1 | 2.18 | 1.73 |
| $[Mn(CN)_6]^{3-}$ | $Mn^{3+}(3d^4)$ | 2 | 3.2 | 2.83 |
| $K_3[Fe(CN)_6]$ | $Fe^{3+}(3d^5)$ | 1 | 2.25 | 1.73 |
| $[Fe(phen)_3]^{3+}$ | $Fe^{3+}(3d^5)$ | 1 | 2.50 | 1.73 |
| $K_4[Fe(CN)_6]$ | $Fe^{2+}(3d^6)$ | 0 | 0 | 0 |
| $[Fe(dipy)_3]^{2+}$ | $Fe^{2+}(3d^6)$ | 0 | 0 | 0 |
| $Na_2[Fe(CN)_5(NO^+)]$ | $Fe^{2+}(3d^6)$ | 0 | 0 | 0 |
| $[Co(NO_2)_6]^{4-}$ | $Co^{2+}(3d^7)$ | 1(in 5 s) | 1.8 | 1.73 |
| $[Co(NH_3)_6]Cl_3$ | $Co^{3+}(3d^6)$ | 0 | 0 | 0 |
| **(b) Outer orbital octahedral complexes (Hybridisation: $sp^3d^2$)** | | | | |
| $[Cr(H_2O)_6](ClO_4)_2$ | $Cr^{2+}(3d^4)$ | 4 | 4.97 | 4.90 |
| $[Cr(H_2O)_6]SO_4$ | $Cr^{2+}(3d^4)$ | 4 | 4.80 | 4.90 |
| $[Mn(py)_6]Br_2$ | $Mn^{2+}(3d^5)$ | 5 | 6.0 | 5.92 |
| $Na[Mn(salyly)_2]$ | $Mn^{3+}(3d^4)$ | 4 | 4.80 | 4.90 |
| $[Mn(CNS)_6]^{4-}$ | $Mn^{2+}(3d^5)$ | 5 | 6.1 | 5.92 |
| $[Mn(acac)_3]^0$ | $Mn^{3+}(3d^4)$ | 4 | 5.0 | 4.90 |
| $Na_3[FeF_6]$ | $Fe^{3+}(3d^5)$ | 5 | 5.85 | 5.92 |
| $[Fe(NH_3)_6]^{2+}$ | $Fe^{2+}(3d^6)$ | 4 | 5.45 | 4.90 |
| $[Fe(H_2O)_5(NO^+)]^{2+}$ | $Fe^+(3d^64s^1 = 3d^7)$ | 3 | 3.90 | 3.88 |
| $[Fe(en)_3]^{2+}$ | $Fe^{2+}(3d^6)$ | 4 | 5.5 | 4.90 |
| $[Co(H_2O)_6]SO_4$ | $Co^{2+}(3d^7)$ | 3 | 5.10 | 3.88 |
| $[Na_3[CoF_6]$ | $Co^{3+}(3d^6)$ | 4 | 5.39 | 4.90 |

| Complex | Central metal ion with $d^x$ configuration | No. of unpaired electrons ($n$) in the complex | $\mu_{eff}$ (B.M) | |
|---|---|---|---|---|
| | | | $\mu_{exp}$ | $\mu_{s.o.}$ |
| $[Co(NH_3)_6]^{2+}$ | $Co^{2+}(3d^7)$ | 3 | 5.00 | 3.88 |
| $[Co(en)_3]SO_4$ | $Co^{2+}(3d^7)$ | 3 | 4.56 | 3.88 |
| $[Co(py)_4Cl_2]^0$(violet) | $Co^{2+}(3d^7)$ | 3 | 5.15 | 3.88 |
| $[Ni(NH_3)_6]Cl_2$ | $Ni^{2+}(3d^8)$ | 2 | 3.32 | 2.83 |
| $[Cu(phen)_3](ClO_4)_2$ | $Cu^{2+}(3d^9)$ | 1 | 1.96 | 1.73 |
| **(c) Tetrahedral complexes (Hybridisation: $sp^3$)** | | | | |
| $(Et_4N)_2[MnCl_4]$ | $Mn^{2+}(3d^5)$ | 5 | 5.54 | 5.92 |
| $[Fe(Ph_3P)_2]I_2$ | $Fe^{2+}(3d^6)$ | 4 | 5.10 | 4.90 |
| $(Et_4N)_2[FeCl_4]$ | $Fe^{2+}(3d^6)$ | 4 | 5.40 | 4.90 |
| $[Co(py)_2Cl_2]^0$(Blue) | $Co^{2+}(3d^7)$ | 3 | 4.42 | 3.88 |
| $Cs_2[CoCl_4]$ | $Co^{2+}(3d^7)$ | 3 | 4.60 | 3.88 |
| $(Et_4N)_2[NiCl_4]$ | $Ni^{2+}(3d^8)$ | 2 | 3.89 | 2.83 |
| $[N_2(CO)_4]^0$ | $Ni^0(3d^84s^2 = 3d^{10})$ | 0 | 0 | 0 |
| $Cs_2[CuCl_4]$ | $Cu^{2+}(3d^9)$ | 1 | 2.0 | 1.73 |
| **(d) Square planar complexes (Hybridisation: $dsp^2$)** | | | | |
| Fe (II) phthalocyanide | $Fe^{2+}(3d^6)$ | 2 | 3.96 | 2.83 |
| Co (II) phthalocyanide | $Co^{2+}(3d^7)$ | 1 | 2.73 | 1.73 |
| $[Ni(Et_3P)_2Cl_2]^0$ | $Ni^{2+}(3d^8)$ | 0 | 0 | 0 |
| $[Ni(CN)_4]^{2-}$ | $Ni^{2+}(3d^8)$ | 0 | 0 | 0 |
| $[Ni(BigH)_2]Cl_2$ | $Ni^{2+}(3d^8)$ | 0 | 0 | 0 |
| $[Ni(Et_3P)Br_3]^0$ | $Ni^{3+}(3d^7)$ | 1 | 1.80 | 1.73 |
| $[Cu(BigH)_2]Cl_2$ | $Cu^{2+}(3d^9)$ | 1 | 1.79 | 1.73 |

(*iii*) $[Mn^{3+}(CN)_6]^{3-}$ ion has μ value corresponding to $n = 2$ while $[Mn^{3+}Cl_6]^{3-}$ ion has μ value corresponding to $n = 4$. Thus $[Mn(CN)_6]^{3-}$ and $[MnCl_6]^{3-}$ ions are obtained by $d^2sp^3$ and $sp^3d^2$ hybridisation of $Mn^{3+}$ ion ($Mn^{3+} = 3d^4$) respectively.

(*iv*) $[Fe^{3+}(CN)_6]^{3-}$ ion has μ value corresponding to $n = 1$ while $[Fe^{3+}F_6]^{3-}$ ion has μ value corresponding to $n = 5$. Thus $[Fe(CN)_6]^{3-}$ and $[FeF_6]^{3-}$ ions are obtained by $d^2sp^3$ and $sp^3d^2$ hybridisation of $Fe^{3+}$ ion ($Fe^{3+} = 3d^5$) respectively.

(*v*) $[Co^{3+}(C_2O_4)_3]^{3-}$ ion has μ value corresponding to $n = 0$ while $[Co^{3+}F_6]^{3-}$ ion has μ value corresponding to $n = 4$. Thus $[Co^{3+}(C_2O_4)_3]^{3-}$ and $[Co^{3+}F_6]^{3-}$ ions are obtained by $d^2sp^3$ and $sp^3d^2$ hybridisation of $Co^{3+}$ ion ($Co^{3+} = 3d^6$) respectively.

**2. Geometry of complexes** If the value of μ of a given complex compound or the number of unpaired electrons present in it is known, the geometry of the complex compound can be predicted.

**Examples.** Suppose we have to find out the geometry of $[NiCl_4]^{2-}$ and $[Ni(CN)_4]^{2-}$ ions. In both these ions the central atom is the same *viz* $Ni^{2+}$ ion and C.N. of $Ni^{2+}$ ion is also the same (=4). These ions have different values of their magnetic moment and hence have different geometries as discussed below:

**(i) Geometry of $[NiCl_4]^{2-}$ ion.** This ion has Ni atom as $Ni^{2+}$ ion whose V.S.E.C. is $3d^8 4s^0 4p^0$ with $n = 2$. Now since the magnetic moment value of $[NiCl_4]^{2-}$ ion has been found corresponding to two unpaired electrons, the two unpaired electrons present in $3d^8 4s^0 4p^0$ configuration do not pair up. Thus $[NiCl_4]^{2-}$ ion results from $(4s)$ $(4p)^3$ or $sp^3$ hybridisation of $Ni^{2+}$ ion and hence the geometry of $[NiCl_4]^{2-}$ ion is tetrahedral.

**(ii) Geometry of $[Ni(CN)_4]^{2-}$.** This ion has Ni-atom as $Ni^{2+}$ ion whose V.S.E.C. is $3d^8 4s^0 4p^0$ with $n = 2$. Now since the magnetic moment of $[Ni(CN)_4]^{2-}$ ion has been found to be equal to zero, all the electrons present in $3d^8 4s^0 4p^0$ configuration get paired up in four $3d$ orbitals. Thus $[Ni(CN)_4]^{2-}$ ion results from $dsp^2$ hybridisation and the geometry of the ion is, therefore, square planar.

**3. To determine the number of unpaired electrons (n) in complexes.** This application can be understood by calculating the number of unpaired electrons (n) in $[Fe(Ph_3P)_2]I_2$. This complex has $[Fe(Ph_3P)_2]^{2+}$ ion in which Fe is present as $Fe^{2+}$ ion ($Fe^{2+} = 3d^6$). $\mu_{exp}$ value for $[Fe(Ph_3P)_2]^{2+}$ ion is 5.10 BM. We know that $\mu_{s.o.}$ values for $n = 4$ and 5 are 4.90 BM and 5.92 BM respectively. $\mu_{exp} = 5.10$ BM value is very near to 4.90 BM. Hence $[Fe(Ph_3P)_2]^{2+}$ has 4 unpaired electrons.

## Shortcomings of VBT to Explain the Magnetic Behaviour of Complexes

(*i*) We know that magnetic moment of complexes is due to the combination of spin motion and orbital motion of the electrons present in the central metal ion of the complex. According to VBT, magnetic moment of a complex is due to the magnetic field generated only by the spin motion of electrons present in the metal ion. This theory does not say anything about the magnetic field generated by the orbital motion of the electrons present in the metal ion.

(*ii*) VBT does not explain why $[Co^{3+}(NH_3)_6]^{3+}$ is diamagnetic and $[Co^{3+}F_6]^{3-}$ is paramagnetic, although both the complex ions have the same geometry (octahedral) and O.S. of Co atom is the same (= +3). In order to explain the discrepancy in magnetic character of the two complexes, VBT says that $Co^{3+}$ ion in $[Co(NH_3)_6]^{3+}$ utilises $3d$ orbitals for $3d^2 4s 4p^3 (d^2sp^3)$ hybridisation while $Co^{3+}$ ion in $[CoF_6]^{3-}$ utilises $4d$ orbitals for $4s4p^3 4p^2$ $(sp^3d^2)$ hybridisation. VBT does not explain why $Co^{3+}$ ion utilises $d$-orbitals of different principal shells in $[Co(NH_3)_6]^{3+}$ and $[CoF_6]^{3-}$ ions.

(*iii*) The correlation between the geometry and magnetic behavior of a complex is sometimes misleading. For example square planar complex of Ni (II) *viz* $[Ni(CN)_4]^{2-}$ is diamagnetic ($n = 0$), since this complex results from $3d4s4p^2$ $(dsp^2)$

hybridisation of $Ni^{2+}$ ion and has $n = 0$. From this diamagnetic character of $[Ni(CN)_4]^{2-}$ we may conclude that all square planar complexes of Ni(II) should be diamagnetic. But actually it is not so, since there is very few square planar complexes of Ni(II) which are paramagnetic due to the presence of two unpaired electrons ($n = 2$). This observation is explained by VBT as follows:

Paramagnetic square planar complexes of Ni(II) result from $3d4s4p^2$ ($dsp^2$) hybridisation of $Ni^{2+}$ ion and has two unpaired electrons. One electron is present in $3d$ orbital and one electron is shifted to $4p$ orbital. VBT can not explain why the promotion of an electron from $3d$ to $4p$ orbital occurs in the present case but in the case of $[Ni(CN)_4]^{2-}$ ion this type of promotion of an electron does not take place. (See Fig. 25.6).

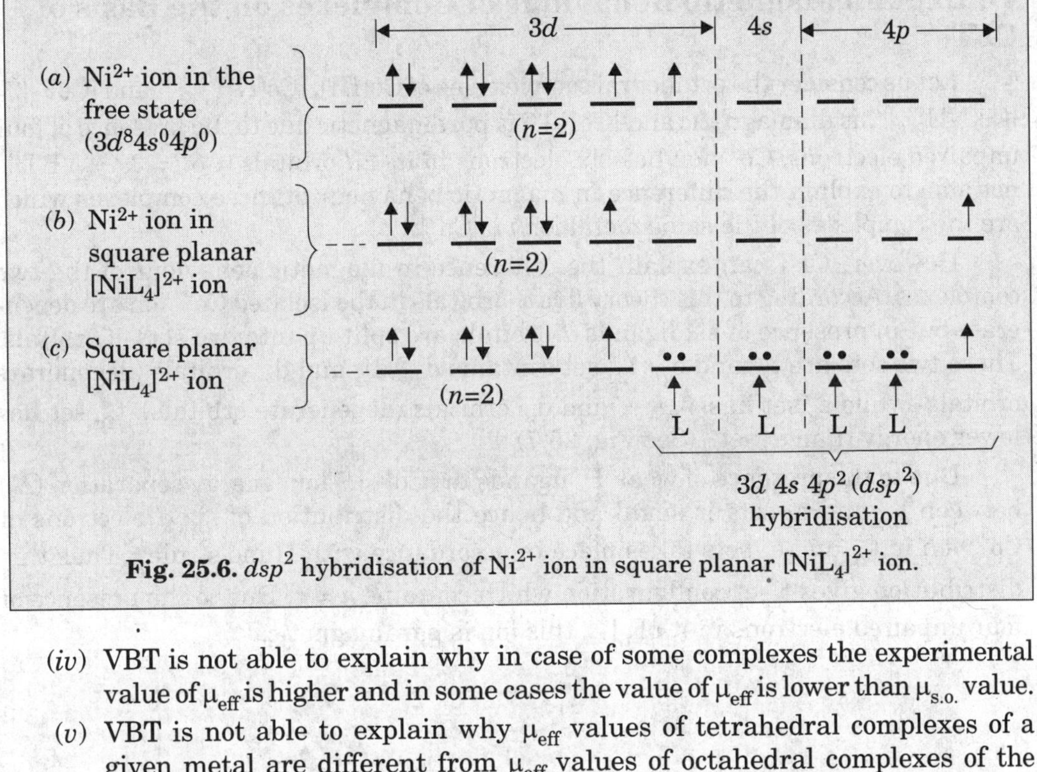


**Fig. 25.6.** $dsp^2$ hybridisation of $Ni^{2+}$ ion in square planar $[NiL_4]^{2+}$ ion.

(*iv*) VBT is not able to explain why in case of some complexes the experimental value of $\mu_{eff}$ is higher and in some cases the value of $\mu_{eff}$ is lower than $\mu_{s.o.}$ value.

(*v*) VBT is not able to explain why $\mu_{eff}$ values of tetrahedral complexes of a given metal are different from $\mu_{eff}$ values of octahedral complexes of the same metal containing same number of electrons. For example:

(a) $\mu_{eff}$ values for tetrahedral complexes of Co(II) ($Co^{2+} = 3d^7$ with $n = 3$) lie between 4.4 and 4.8 BM while $\mu_{eff}$ values of octahedral complexes of Co(II) lie between 4.8 and 5.2 BM. Note that the number of unpaired electrons ($n$) in both complexes is the same (=3).

(b) $\mu_{eff}$ values for tetrahedral complexes of Ni (II) ($Ni^{2+} = 3d^8$ with $n = 2$) lie in the range 3.2–4.0 BM whereas $\mu_{eff}$ values of octahedral complexes of Ni(II) range between 2.9–3.3 BM

(*vi*) $\mu_{eff}$ values of most of the complexes of the metals of first transition series are in good agreement with $\mu_{s.o.}$ values, but $\mu_{eff}$ values for tetrahedral complexes of Ni(II) and for octahedral complexes of Co (II) are higher than $\mu_{s.o.}$ values. VBT has no explanation for the enhanced values of $\mu_{eff}$.

(*vii*) It has been confirmed that the colour and magnetic moments of the complexes of transition metals are due to the presence of unpaired electrons present in *d*-orbitals of metals ions. Therefore, there must be a quantitative correlation between spectra and magnetic moments of the complexes. Unfortunately, according to the VBT there is no such connection.

(*viii*) Value of $\mu_{exp}$ for outer orbital octahedral ($sp^3d^2$ hybridisation) complex of Cu (II) ($Cu^{2+} = 3d^9$ with $n = 1$) is 1.96 BM. This value is greater than its $\mu_{s.o.}$ value (= 1.73 BM) for one electron. The discrepancy between the two values is due to the fact that spin-orbital coupling in the ion mixes the ground state representing no orbital momentum with higher levels of identical multiplicity. The participation of the higher levels produces the value of $\mu_{eff}$ which is greater than $\mu_{s.o}$.

## To Explain Magnetic Behaviour of Complexes on the Basis of CFT

Let us consider the octahedral complex ions of Co(III), $[Co(NH_3)_6]^{3+}$ and $[CoF_6]^{3-}$. $[Co(NH_3)_6]^{3+}$ is diamagnetic and $[CoF_6]^{3-}$ is paramagnetic due to the presence of four unpaired electrons. $Co^{3+}$ ion has six electrons in its 3*d* orbitals ($Co^{3+} = 3d^6$). VBT is not able to explain the difference in magnetic behaviour of these complexes which are the complexes of the same metal ion *viz* $Co^{3+}$.

However, CFT can explain the difference in magnetic behaviour of the two complexes. According to this theory five *d*-orbitals in the isolated $Co^{3+}$ ion are degenerate but in presence of six ligands *d*-orbitals are split up into two sets of orbitals. These two sets are $t_{2g}$ and $e_g$ · $t_{2g}$ set contains $d_{xy}$, $d_{yz}$ and $d_{xz}$ orbitals (degenerate orbitals) while $e_g$ set has $d_{x^2-y^2}$ and $d_{z^2}$ orbitals (degenerate orbitals). $t_{2g}$ set has lower energy than $e_g$ set. (See Fig. 25.7)

Due to the presence of weak $F^-$ ligands in $[CoF_6]^{3-}$ ion, energy separation ($\Delta_0$) between $t_{2g}$ and $e_g$ sets is small and hence the distribution of six *d* electrons of $Co^{3+}$ ion in $t_{2g}$ and $e_g$ sets takes place in accordance with Hund's rules. Thus this distribution gives $t_{2g}^4e_g^2$ configuration which contains $n = 4$. Due to the presence of four unpaired electrons in $[CoF_6]^{3-}$, this ion is paramagnetic.

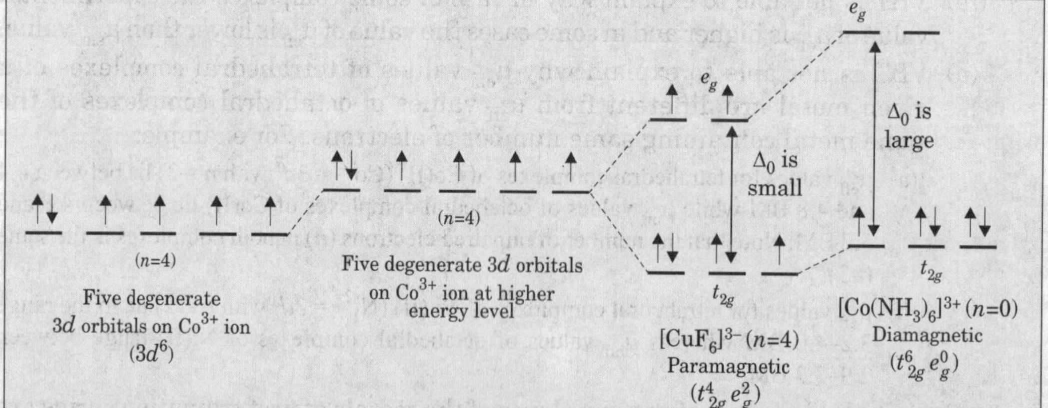


**Fig. 25.7.** To explain magnetic behavior of $[CoF_6]^{3-}$ and $[Co(NH_3)_6]^{3+}$ complex ions on the basis of CFT.

Due to the presence of strong $NH_3$ ligands in $[Co(NH_3)_6]^{3+}$ ion, the energy separation ($\Delta_0$) between $t_{2g}$ and $e_g$ sets is large and hence the distribution of six $d$-electrons of $Co^{3+}$ ion takes place in such a way that all the six electrons occupy $t_{2g}$ set ($t_{2g}^6 e_g^0$) which gives $n = 0$. Due to the absence of unpaired electrons $[Co(NH_3)_6]^{3+}$ is digmagnetic.

According to VBT the formation of inner orbital complexes of metal ions with $d^7$, $d^8$ and $d^9$ configurations requires the promotion of electrons from lower energy $d$-orbitals to higher energy $d$, $s$ or $p$ orbitals which is sometimes energetically unfavourable. CFT does not assume such energetically unfavourable excitation of electrons. CFT is able to explain all the features of magneto chemistry of transition metal complexes. These features are not explained by VBT.

The quenching of orbital angular momentum of electrons has been explained by CFT in a logical way. With the help of calculations based on CFT, we can also predict the extent to which the quenching of orbital angular momentum takes place. Thus we can calculate the contribution of $\mu_{orbital}$ to the overall experimental value of $\mu_{eff}$ of transition metal complexes.

## Quenching of Orbital Angular Momentum: Orbital Contribution to Magnetic Moment of Octahedral and Tetrahedral Complexes

We know that there are two types of angular momenta of an unpaired electron *viz spin angular momentum* and *orbital angular momentum*. The spin angular momentum of an unpaired electron is not affected by the changes in chemical bonding. However orbital angular momentum of an unpaired electron is quenched under certain conditions as discussed below.

Five $d$-orbitals in a free ion are degenerate and hence the electrons present in these orbitals have magnetic moment. According to CFT, in octahedral complexes the five degenerate $d$-orbitals of the metal ion are split into two sets of orbitals which have different energy. These two sets are $t_{2g}$ and $e_g$. $t_{2g}$ set ($d_{xy}$, $d_{yz}$, $d_{xz}$ orbitals) has lower energy than $e_g$ set ($d_{x^2-y^2}$, $d_{z^2}$ orbitals). All the orbitals in each set are degenerate.

Although $d_{x^2-y^2}$ and $d_{z^2}$ orbitals of $e_g$ set are degenerate in octahedral complexes, due to their different shape and size, these orbitals cannot be transformed into one another by rotation. Thus the electrons present in the orbitals of $e_g$ set cannot generate any orbital magnetic moment to the overall magnetic moment. Hence $d_{x^2-y^2}$ and $d_{z^2}$, orbitals are *non-magnetic doublet*.

Since each of the three orbitals of the lower energy $t_{2g}$ set ($d_{xy}$, $d_{yz}$ and $d_{xz}$ orbitals) is transformed into the other two orbitals by rotation about the relevant axis, electrons present in these orbitals will generate orbital contribution to the magnetic moment provided that these orbitals do not contain electrons of the same spin. Thus the electrons present in $t^3_{2g}$ and $t^6_{2g}$ configurations do not contribute any orbital magnetic moment to the overall magnetic moment.

In Table 25.4, we have shown which of the $d^n$ (n = 1 to 9) configurations of metal ions in HS octahedral and tetrahedral complexes has orbital magnetic moment contribution and which $d^n$ configuration does not have orbital contribution.

In this Table:

(*i*) $t_{2g}^{p}e_{g}^{q}$ and $e^{p}t_{2}^{q}$ are the configurations from which the ground states given in bracket originate.

(*ii*) The term A is singlet, E is doublet and T is triplet. The left hand superscript is the spin multiplicity of the ground state. Right hand subscript *g* indicates *gerade*, *i.e.* even orbitals which retain the same sign of orbital wave function on inversion about the nucleus (centre of octahedron). A tetrahedron has no inversion centre and hence *g* subscript for the ground states of tetrahedral complexes is dropped. The numbers 1 and 2 in the subscript have theoretical origin.

Table 25.4 reveals the following facts:

(*i*) In both types of complexes, the configurations have T as their ground state have orbital contribution but the configurations having A and E as their ground states have no orbital contribution. Thus the orbital contribution of the metal ions having A or E as their ground state is quenched.

(*ii*) In HS octahedral complexes: (a) Metal ions with $d^1$, $d^2$, $d^6$ and $d^7$ configurations have orbital contribution while other $d^n$ configurations do not have orbital contribution.

(b) $t_{2g}^1$, $t_{2g}^2$, $t_{2g}^4$ and $t_{2g}^5$ configurations have orbital contribution but $t_{2g}^3$ and $t_{2g}^6$ configurations do not have orbital contribution.

(*iii*) In tetrahedral complexes metal ions with $d^3$, $d^4$, $d^8$ and $d^9$ configurations have orbital configuration while other $d^n$ configurations do not have orbital contribution.

**Table 25.4. Orbital contribution in HS octahedral and tetrahedral complexes.**

| $d^n$ ion ($n$ = 1 to 9) | HS octahedral complexes | | Tetrahedral complexes | |
|---|---|---|---|---|
| | $t_{2g}^{p}e_{g}^{q}$ configuration (Ground state) | Orbital contribution (Yes or No) | $e^{p}t_{2}^{q}$ configuration (Ground term) | Orbital contribution (Yes or No) |
| $d^1$ | $t_{2g}^1e_g^0$ ($^2T_{2g}$) | Yes | $e^1t_2^0$ ($^2E$) | No |
| $d^2$ | $t_{2g}^2e_g^0$ ($^3T_{1g}$) | Yes | $e^2t_2^0$ ($^3A_2$) | No |
| $d^3$ | $t_{2g}^3e_g^0$ ($^4A_{2g}$) | No | $e^2t_2^1$ ($^4T_1$) | Yes |
| $d^4$ | $t_{2g}^3e_g^1$ ($^5E_g$) | No | $e^2t_2^2$ ($^5T_2$) | Yes |
| $d^5$ | $t_{2g}^3e_g^2$ ($^6A_{1g}$) | Yes | $e^2t_2^3$ ($^6A_1$) | No |
| $d^6$ | $t_{2g}^4e_g^2$ ($^5T_{2g}$) | Yes | $e^3t_2^3$ ($^5E$) | No |
| $d^7$ | $t_{2g}^5e_g^2$ ($^4T_{ig}$) | Yes | $e^4t_2^3$ ($^4A_2$) | No |
| $d^8$ | $t_{2g}^6e_g^2$ ($^3A_{2g}$) | No | $e^4t_2^4$ ($^3T_1$) | Yes |
| $d^9$ | $t_{2g}^6e_g^3$ ($^2E_g$) | No | $e^4t_2^5$ ($^2T_2$) | Yes |

In LS octahedral complexes the metal ions with $d^4$ ($t_{2g}^4$ with ground state $^3T_{1g}$) and $d^5$ ($t_{2g}^5$ with ground state $^2T_{2g}$) configurations have orbital contribution while the metal ions with $d^6$ ($t_{2g}^6$ with ground state $^1A_{1g}$) and $d^7$ ($t_{2g}^6$ $e_g^1$ with ground state $^2E_g$) configurations do not have orbital contribution.

We have said above that the orbital contribution of most of the metal ions having A or E as their ground state is completely quenched. Thus, due to the quenching of orbital contribution, the experimental values of $\mu_{eff}$ for metal ions having A or E as their ground states should be close to the values of $\mu_s$ calculated from $\mu_s = \sqrt{n(n+1)}$. However, it has been observed that there are metal ions with their A or E as their ground states for which the experimental values of $\mu_{eff}$ are different from their $\mu_s$ values (calculated values). Why experimental and $\mu_s$ values of $\mu_{eff}$ differ from each other can be explained as follows:

**Explanation.** In order to explain the discrepancy between the experimental and calculated values, we take the example of $Ni^{2+}$ ion · $[Ni(ClO_4)_2(py)_2]^0$ is an octahedral complex. It is blue and paramagnetic. Experimental value of $\mu_{eff}$ $[(\mu_{eff})_{exp}]$ is 3.0 BM while the calculated value of $\mu_{eff}$ ($\mu_s$) is equal to 2.82 BM. The given complex has $Ni^{2+}$ ion which is a $d^8$ ion and has $n = 2$. For $Ni^{2+}$ ion ($d^8$ ion), $^3A_{2g}$ $(t_{2g}^6 e_g^2)$ is the ground state and $^3T_{2g}$ $(t_{2g}^5 e_g^3)$ is the first excited state. These two states have the same spin multiplicity equal to 3. These states mix up due to spin-orbital coupling. As a result of this mixing, the excited state ($^3T_{2g}$) has some character of the ground state ($^3A_{2g}$) or the ground state ($^3A_{2g}$) has some character of the excited state ($^3T_{2g}$). In other words, the ground state ($^3A_{2g}$) will retain some orbital angular momentum. The contribution of orbital magnetic moment to the overall $\mu_{eff}$ in such cases depends on the extent of mixing of $^3A_{2g}$ (ground state) and $^3T_{2g}$ (excited state).

Taking all the above points into account, the value of $\mu_{eff}$ of an ion is calculated from the formula given below.

$$\boxed{\mu_{eff} = \mu_S\left(1 - \frac{\alpha\lambda}{10\ Dq}\right)} \qquad (i)$$

Here $\alpha$ = constant depending on the spectroscopic ground term.

$\alpha = 2$ for an E ground term, $\alpha = 4$ for $A_2$ ground term and $\alpha = 0$ for $A_1$ ground term. Values of $\alpha$ for some metal ions are given as: $d^1(Ti^{3+}) = 2$, $d^2(V^{3+}) = 4$, $d^3$ $(Cr^{3+}) = 4$, $d^4$ $(Cr^{2+}, Mn^{3+}) = 4$, $d^5$ $(Fe^{3+}, Mn^{2+}) = 0$, $d^6$ $(Fe^{2+}) = 2$, $d^7$ $(Co^{2+}) = 4$, $d^8$ $(Ni^{2+}) = 4$, $d^9$ $(Cu^{2+}) = 2$.

$\lambda$ = spin orbit coupling parameter which may be positive for $d^n$ configurations with $n < 5$ and negative for $d^n$ configuration with $n > 5$.

10Dq = energy separation between the ground state and first excited state of $d^n$ configuration.

Let us consider the following cases:

(*i*) **Case of $d^3$ and $d^4$ ions.** For $d^3$ ion since ground state is $A_{2g}$, $\alpha = 4$ and $\lambda$ is positive, equation (*i*) can be written as:

$$\boxed{\mu_{eff} = \mu_S\left(1 - \frac{4\lambda}{10\ Dq}\right)} \qquad \textbf{(For } d^3 \textbf{ ion)} \qquad (ii)$$

For $d^4$ ion, since ground state is $E_g$, $\alpha = 2$ and $\lambda$ is positive, equation (*i*) becomes

$$\boxed{\mu_{eff} = \mu_S\left(1 - \frac{2\lambda}{10\ Dq}\right)} \qquad \textbf{(For } d^4 \textbf{ ion)} \qquad (iii)$$

In case of $d^3$ and $d^4$ ions 10 Dq is the energy gap between the ground state ($A_{2g}$) and first excited state ($T_{2g}$). Above discussion shows that for HS octahedral complexes of $d^3$ and $d^4$ ions the experimental value of $\mu_{eff}$ is **lower** than $\mu_S$. Hence the value of $\mu_{eff}$ for $d^3$ and $d^4$ ions should be calculated with the help of equations (ii) and (iii) respectively

(ii) **Case of $d^8$ and $d^9$ ions.** For $d^8$ ion, since the ground state is $A_{2g}$, $\alpha = 4$ and $\lambda$ is negative. Thus equation (i) becomes

$$\mu_{eff} = \mu_S\left(1 + \frac{4\lambda}{10\ Dq}\right) \quad \textbf{(For } \mathbf{d^8} \textbf{ ion)} \qquad (iv)$$

For $d^9$ ion, since ground state is $E_g$, $\alpha = 2$ and $\lambda$ is negative. Thus equation (*i*) becomes

$$\mu_{eff} = \mu_S\left(1 + \frac{2\lambda}{10\ Dq}\right) \quad \textbf{(For } \mathbf{d^9} \textbf{ ion)} \qquad (v)$$

In $d^8$ and $d^9$ ions 10 Dq is the energy gap between the ground state ($A_{2g}/E_g$) and first excited state ($T_{2g}$).

Above discussion shows that for HS octahedral complexes of $d^8$ and $d^9$ ions the experimental value of $\mu_{eff}$ is **higher** than $\mu_S$. Hence the value of $\mu_{eff}$ for $d^8$ and $d^9$ ions should be calculated from equations (iv) and (v) respectively.

The value of $\mu_{eff}$ for $d^3$, $d^4$, $d^8$, and $d^9$ ions is independent of temperature, since the energy gap between the ground state (A or E) and first excited state (T) is far greater than *k*T.

(iii) **Case of $d^5$ ion**. In case of $d^5$ ion, since the ground state is $A_{1g}$, $\alpha = 0$ and $\lambda$ is positive, equation (i) reduces to:

$$\mu_{eff} = \mu_S\left(1 - \frac{0 \times \lambda}{10\ Dq}\right) \quad \textbf{(For } \boldsymbol{d^5} \textbf{ ion)} \qquad (iv)$$

$$\text{or } \mu_{eff} = \mu_S$$

Thus we see that in case of $d^5$ ion, $\mu_{eff}$ and $\mu_S$ are the same.

Ground state of HS octahedral complexes of $d^5$ ion is $^6A_{1g}$. These complexes do not have any excited state of the same multiplicity as the ground state. Consequently the mixing of ground state with the excited state, due to spin-orbital coupling, in this case is not possible. Thus there is no orbital contribution to $\mu_{eff}$ in this ion. HS octahedral complexes of $d^5$ ion, therefore, have $(\mu_{eff})_{exp}$ equal to $\mu_S$. The value of magnetic moment is independent of temperature.

## Orbital Contribution in Square Planar Complexes

We know that in octahedral complexes, since $d_{x^2-y^2}$ and $d_{z^2}$ orbitals ($e_g$ set) cannot be transformed into one another, these orbitals cannot contribute to the overall magnetic moment value. It is only $d_{xy}$, $d_{yz}$ and $d_{zx}$ orbitals ($t_{2g}$ set) which contribute to the overall $\mu_{eff}$ value. There is further splitting of the orbitals of $t_{2g}$ set, and hence the scope of orbital contribution from $t_{2g}$ set orbitals is minimised.

Now since $d_{xy}$, $d_{yz}$ and $d_{xz}$ orbitals are no longer of the same energy, there is genuine energy barrier for the movement of electrons among these orbitals. Thus in square planar complexes, $\mu_{eff}$ value is likely to be close to $\mu_{s.o.}$ value. In other words we can say that orbital contribution in square planar complexes is not more likely than in octahedral complexes.

## Diamagnetism

The following points must be noted while making a study of diamagnetism.

(*i*) The substances which are repelled by the magnetic field are diamagnetic. Thus when a diamagnetic substance is placed in a magnetic field, the number of lines of force passing through diamagnetic substance would be less than that which would pass through air (See Fig. 25.8).

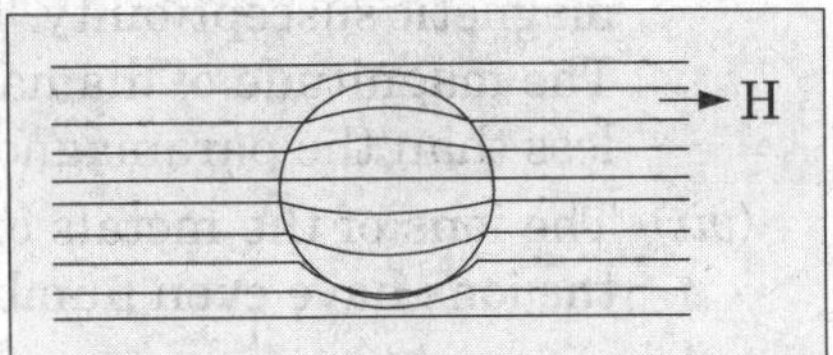


**Fig. 25.8** Number of lines passing through diamagnetic substances

(*ii*) Diamagnetic behaviour of the substances is due to the fact that when a diamagnetic substance is put into the magnetic field, small magnetic moments are induced into the substance. These induced magnetic moments are in opposition to the inducing field. This results in that the substance is repelled by the magnetic field.

(*iii*) Substances having all the electrons in the paired state or having no unpaired electrons (n = 0) show diamagnetic character. When an orbital contains two electrons with opposite spins (↑↓), the magnetic moment (or magnetic field) generated by one electron is cancelled by that generated by the other electron, since magnetic moments of the two electrons are equal and opposite to each other (See Fig. 25.9). Thus an atom, ion or a molecule having paired electrons will be diamagnetic.

(*iv*) The diamagnetic nature of a substance can be found by weighing the substance first in air and then after suspending it between the poles of magnetic field. If the substance weighs less in the magnetic field, it shows that the substance is repelled by the magnetic field and hence is diamagnetic.

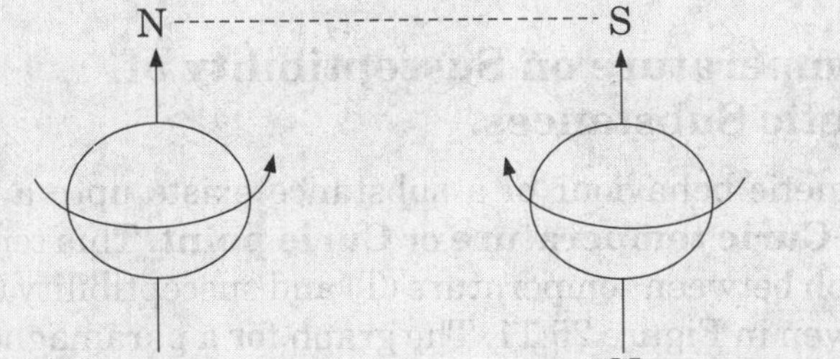


**Fig. 25.9** The magnetic fields generated by the two electrons with opposite spins cancell each other.

(*v*) Since diamagnetic substance has all the electrons in the paired state or has no unpaired electrons, $\mu_{eff}$ value of diamagnetic substances is zero ($\mu_{eff} = 0$).

(*vi*) Diamagnetism is a property of completely - filled electronic subshells. Since all elements except hydrogen atom contain filled electronic shells, some diamagnetism is shown by all substances. The presence of diamagnetism in all substances is due to the interaction of the magnetic field with the paired electrons present in shells which oppose the field. Since the paramagnetic substances also have some amount of diamagnetic character, the magnetic susceptibility of paramagnetic substances ($\chi_{para}$) should be corrected for diamagnetic susceptibility. This can be done with the help of Pascals constants. The magnitude of diamagnetism present in substances is about 100 times less than the paramagnetism.

(*vii*) The ions of the metals of 2*nd* and 3*rd* transition series are diamagnetic, if the ions have even number of electrons.

## Ferromagnetism

*Substances which show permanent magnetism even in the absence of the magnetic field are called ferromagnetic substances.* Such substances remain permanently magnetised, once they have been magnetised. This type of magnetism arises due to spontaneous alignment of magnetic moments due to unpaired electrons in the same direction, as shown in Figure 25.10. This is a case of large amount of paramagnetism.

Ferromagnetism is shown by the substances which contain atoms or ions with incomplet *d* or *f* subshells. Examples of substances which show ferromagnetism are Fe, Co, Ni, Gd, Dy, number of alloys of Cu, Al and Mn, and oxides of metals of 1*st* transition series like $CrO_2$, $Fe_3O_4$. Ferromagnetic substances have their permeability very much greater than unity.

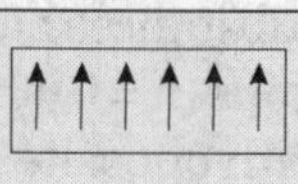

**Fig. 25.10** In ferromagnetic substances the neighbouring spins (magnetic moments due to unpaired electrons) align in the same direction).

## Effect of Temperature on Susceptibility of Ferromagnetic Substances.

Ferromagnetic behaviour of a substance exists upto a certain temperature which is called **Curie temperature** or **Curie point.** This temperature is denoted by $T_C$. The graph between temperature (T) and susceptibility ($\chi$) of a ferromagnetic substance is given in Figure 25.11. The graph for a paramagnetic substance is also given for the sake of comparison.

It may be seen from graph that below $T_C$, say at $T_1$, $\chi$ for a ferromagnetic substance is much higher than that for a normal paramagnetic substance. Ferromagnetic substance does not obey Curie-Weiss law in this range, but above this temperature the ferromagnetic substance obeys Curie-Weiss law.

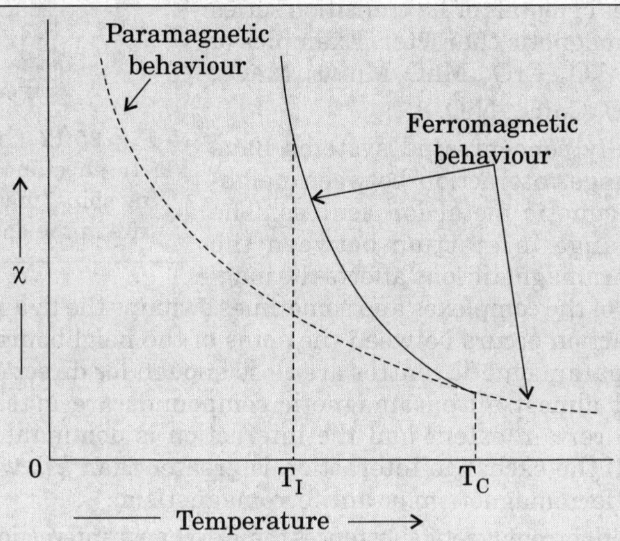


**Fig. 25.11** Variation of χ with temperature for a ferromagnetic substance.

As temperature is lowered below $T_C$, susceptibility sharply increases and attains the highest value at T = 0. We know that in ferromagnetic substances all electron spins are parallel and hence generate a collective magnetic moment. As temperature is raised, parallel arrangement of the electron spins tends to become partially anti-parallel arrangement. Above $T_C$ all parallel arrangement of electron spins breaks down and ferromagnetic substance behaves as a normal paramagnetic substance. $\chi_M$ decreases with increasing temperature (Curie law).

## Magnetically Dilute and Magnetically Concentrated Substances.

Magnetically dilute substances are those in which the individual magnetic centres present in the crystal lattice are **sufficiently apart** to minimise any further magnetic interaction between near neighbours. On the other hand, magnetically concentrated substances are those in which the individual magnetic centres present in crystal lattice are **close enough** to initiate further magnetic interactions between the near neighbours. $[Ni(H_2O)_6]Cl_2$ is an example of magnetically dilute substance while $Cu_2(CH_3COO)_4 \cdot 2H_2O$ is magnetically concentrated substance, since in $Cu_2(CH_3COO)_4 \cdot 2H_2O$ the unpaired electrons of the two $Cu^{2+}$ ions interact with each other.

## Anti-ferromagnetism

Substances which are expected to posses paramagnetism or ferromagnetism on the basis of unpaired electrons but actually they possess zero net magnetic moment are called anti-ferromagnetic substances. Anti-ferromagnetism is due to the presence of equal number of magnetic moments in the opposite directions as shown in Fig. 25.12.

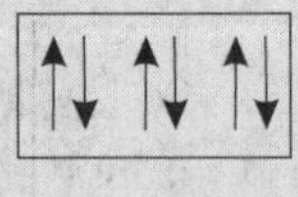

**Fig. 25.12** In anti-ferromagnetic substances the neighbouring spins (magnetic moments) are in the opposite direction

Many oxides of metals of 1*st* transition series show anti-ferromagnetic character. Examples of such oxides are $V_2O_3$, $CrO_3$, MnO, $Mn_2O_3$, $MnO_2$, FeO, $Fe_2O_3$, CoO, $Co_3O_4$, NiO.

Magnetically concentrated systems have magnetic exchange interaction between neighbouring paramagnetic metal ion centres. The magnetic exchange interaction between the neighbouring paramagnetic ions affects the magnetic properties of the complexes and sometimes swamps the ligand field effect. The exchange interaction occurs between the spins of the neighbouring paramagnetic ions, when the paramagnetic centres are close enough for direct or indirect orbital overlap. In fact almost all paramagnetic compounds are involved in exchange interaction to a certain extent and the interaction is dominant only at very low temperatures. If the exchange interaction is greater than $kT$, we get cooperative phenomenon of ferromagnetism or antiferromagnetism.

In case of anti-ferromagnetic substances these exchange interactions lead to lowering of magnetic moment compared to that expected on monomer and no-interaction basis. In anti-ferromagnetic substances the neighbouring spins are opposed to each other (↑↓ ↑↓ .....)

## To Explain Anti-ferromagnetic Behaviour of Substances.

Anti-ferromagnetic behaviour of substances arises either due to metal-metal interaction which takes place through the overlap of suitable orbitals on metal atoms or due to spin exchange interactions taking place through diamagnetic $O^{2-}$ or $F^-$ ions bridging between paramagnetic metal ion.

**(i) Metal-metal interaction.** Metal-metal interaction which takes place through the overlap of suitable orbitals on metal atoms can be explained by considering the intra molecular antiferromagnetic character in Cu(II) acetate monohydrate dimer, $Cu_2(CH_3COO)_4 \cdot 2H_2O$. In this compound two Cu atoms are 2.64 Å apart. These Cu atoms are bridged by four $CH_3COO^-$ groups and two $H_2O$ molecules are attached with two Cu atoms along Cu–Cu axis [See Fig. 25.13(a)].

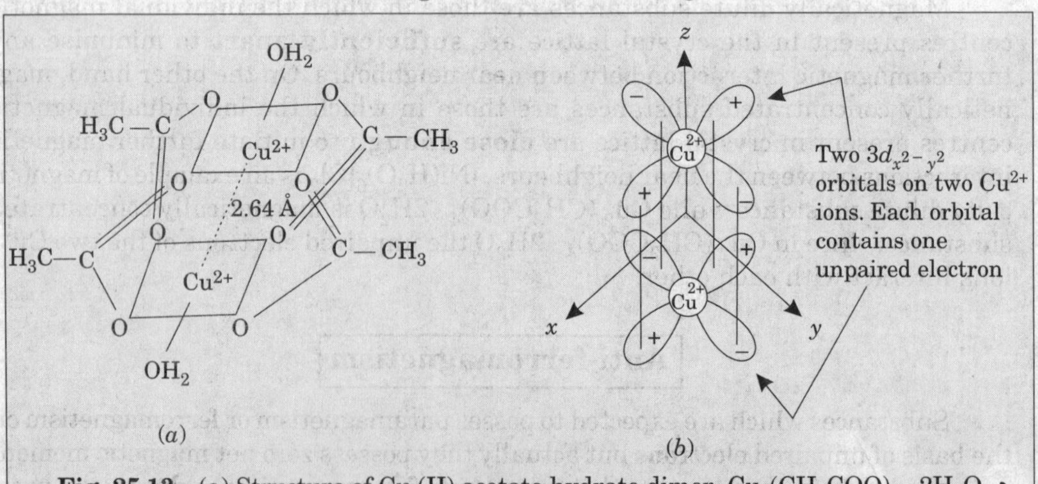


**Fig. 25.13** (a) Structure of Cu (II) acetate hydrate dimer, $Cu_2(CH_3COO)_4 \cdot 2H_2O$ • (b) Formation of δ bond by the overlap between two $3d_{x^2-y^2}$ orbitals on two $Cu^{2+}$ ions in $Cu_2(CH_3COO)_4 \cdot 2H_2O$.

$Cu^{2+}$ ion is a $3d^9$ ion. The distribution of nine electrons in different 3*d* orbitals is as: $3d^2_{xy}$ $3d^2_{yz}$ $3d^2_{xz}$, $3d^2_{z^2}$ $3d^1_{x^2-y^2}$. $3d_{x^2-y^2}$ orbitals of two $Cu^{2+}$ ions and containing two unpaired electrons make a lateral overlap and produce a δ bond (See Fig. 25.13(b)].

It is due to the formation of δ bond that the two unpaired electrons of two $Cu^{2+}$ ions get paired. Thus the formation of δ bond lowers the magnetic susceptibility (μ = 1.43 BM). The lowering of magnetic susceptibility continues to drop below $T_N$ ($T_N$ is called **Neel temperature**). The anti-ferromagnetism occurring in $Cu_2(CH_3COO)_4 \cdot 2H_2O$ is called intramolecular anti-ferromagnetism, since the interacting paramagnetic centres (i.e. $Cu^{2+}$ ions) are present in the same molecule *viz* $Cu_2(CH_3COO)_4 \cdot 2H_2O$.

**(ii) Spin exchange interactions through bridging $O^{2-}$ or $F^-$ ions.** This type of interaction can be explained by considering anti-ferromagnetic character of MnO. This oxide contains $Mn^{2+}$ and $O^{2-}$ ions. V.S.E.C. of these ions are as: $Mn^{2+} = 3d^5 = 3d^1_{xy}$ $3d^1_{yz}$ $3d^1_{xz}$ $3d^1_{x^2-y^2}$ $3d^1_{z^2}$ and $O^{2-} = 2s^2\ 2p_x^2\ 2p_y^2\ 2p_z^2$. The spins of five electrons in one $Mn^{2+}$ ion are parallel (↑↑↑↑↑) while those in the other $Mn^{2+}$ ion are antiparallel (↓ ↓ ↓ ↓ ↓). In $O^{2-}$ ion all the three 2p orbitals have paired electrons (↑↓). Thus $O^{2-}$ ion is diamagnetic. *2p* orbital of bridging $O^{2-}$ ion having paired spins (↑↓) overlaps with two similar *3d* orbitals (say $3d_{xz}$ orbitals) of two $Mn^{2+}$ ions. Each of the two *3d* orbitals in $Mn^{2+}$ ion has one unpaired electron with opposite spins (↑ and ↓) (See Fig. 25.14). This type of overlap between orbitals yields stable MnO. In this overlap one of the two electrons present in *2p* orbital of bridging $O^{2-}$ ion moves into the *3d* orbital of one $Mn^{2+}$ ion which carries one electron with opposite spin. Similarly the other electron of *2p* orbital migrates to *3d* orbital of other $Mn^{2+}$ ion. This type of overlap gives normal covalent bond. It is due to the spin exchange interaction that both $Mn^{2+}$ ions get paired electrons in *3d* orbitals. It is clear from the discussion on anti-ferromagnetic substances that pairing of total electron spins of atoms or ions of a substance are responsible for its anti-ferromagnetic behaviour.

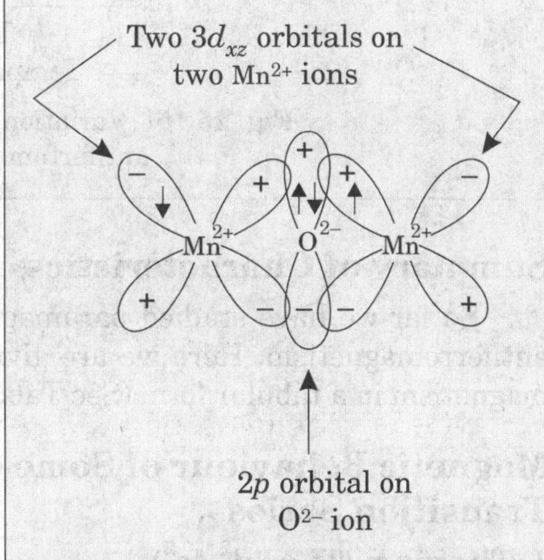


**Fig. 25.14** Overlap between *2p* orbital on $O^{2-}$ ion with two $3d_{xz}$ orbitals on two $Mn^{2+}$ ions in MnO

## Effect of Temperature on Magnetic Susceptibility of Anti-ferromagnetic Substances

Magnetic susceptibility of anti-ferromagnetic substances is very much dependent on temperature. The variation of susceptibility (χ) with temperature (T) for antiferromagnetic substances is shown in Fig. 25.15. χ *vs* T graph for the paramagnetic substance is also shown in the same figure for the sake of comparison. For an anti-ferromagnetic substance there is a characteristic temperature which is called **Neel temperature or Neel point** which is denoted by $T_N$. Below this temperature (i.e. below $T_N$) electron spins of the paramagnetic centres align themselves in antiparallel position as shown below:

↑↓ ↑↓ ↑↓ ↑↓

This type of alignment lowers the magnetic susceptibility. Thus below $T_N$ susceptibility decreases with decrease of temperature in antiferromagnetic substances. As the temperature is lowered below $T_N$, susceptibility continues to decrease and may even show diamagnetism near absolute (T = 0). Above $T_N$ thermal energy ($k$T) is large and hence disturbs the antiparallel alignment of electron spins. This results in that antiferromagnetic substance behaves as a paramagnetic substance. Above discussion shows that magnetic susceptibility of antiferromagnetic substances goes through a maximum at $T_N$. Thus like other paramagnetic substances, x of these substances is also inversely proportional to T measured above $T_N$.

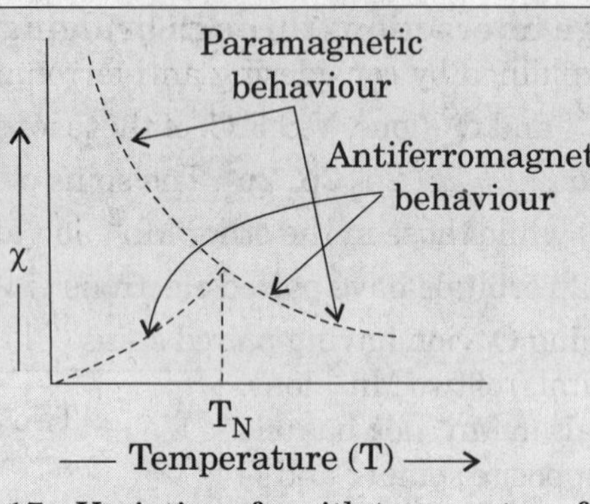


**Fig. 25.15** Variation of χ with temperature for an antiferromagnetic substance.

## Summary of Characteristics of Various Types of Magnetism

So far we have studied paramagnetism, diamagnetism, ferromagnetism and antiferromagnetism. Here we are giving characteristics of these various types of magnetism in a tabular form (See Table 25.5).

## Magnetic Behaviour of Some Complexes of Metals of 1*st* Transition Series

### 1. Titanium (Ti = $3d^2$ $4s^2$)

(*i*) $[Ti(acac)_3]^0$, $(TiF_6]^{3-}$, $[Ti(H_2O)_6]^{3+}$ and $[TiCl_5(H_2O)]^{2-}$ are the complex compounds of **$Ti^{3+}$ ion** which is a $3d^1$ ion. These complexes are paramagnetic with μ value equal to 1.7 BM corresponding to the presence of one unpaired electron in them. Thus these are octahedral complexes ($d^2sp^3$ hybridisation).

(*ii*) All the complexes of **$Ti^{4+}$** ($3d^0$ ion) like $[TiX_6]^{2-}$ are diamagnetic due to the fact that they do not contain $d$-electrons, i.e. $n$ = 0 in these complexes.

### 2. Vanadium (V = $3d^3 4s^2$)

(*i*) **V(0)** ($V^0 = 3d^3 4s^2 = 3d^5$ system) is found in $V(CO)_5$ which is expectedly paramagnetic corresponding to the presence of one unpaired electron.

(*ii*) Complexes of **$V^{2+}$** ($3d^3$ system), **$V^{3+}$** ($3d^2$ system), and **$V^{4+}$** ($3d^1$ system) have magnetic moment values of 3.81–3.86 ($n$ = 3), 2.76 ($n$ = 2) and 1.6–1.8 ($n$ = 1) BM while **$V^{5+}$** ($3d^0$ system) complexes are diamagnetic ($n$ = 0).

**Table 25.5 Characteristics of various types of magnetism**

| Type | Caused by | Magnitude of $\chi_g$ | Temperature and field dependence of $\chi_g$ | Example |
|---|---|---|---|---|
| Paramagnetism | Electrons with unpaired spin *i.e.* with resultant angular momentum | Small and positive ~ + 10 to $10^3 \times 10^{-6}$ cgs units | Varies inversely with temperature. Field independent | $[Ni(H_2O)_6]Cl_2$, $[CoCl_4]^{2-}$ |
| Diamagnetism | Orbital motion of electrons. Best perceptible when the body has all its electrons paired *i.e.* when there is no permanent spin moment. | Small and negative ~ – 1 × $10^{-6}$ cgs units | Temperature independent and field independent | $H_2$, KCl, organic ligands |
| Ferromagnetism | Lattice of particles with electrons having parallel spin | Very large and positive, $10^{-2}$ to $10^4$ cgs units | Complicated: above a particular temperature called Curie temperature it behaves as a paramagnet. Below this temperature $\chi_g$ rapidly increases | Metallic iron |
| Antiferromagnetism | Two lattices of particles having spins of one lattice antiparallel to the spins of the other lattice | Positive and very small, ~0.1 to $10^2 \times 10^{-6}$ cgs units | Above a particular temperature called Neel temperature the substance behaves as paramagnetic but below the Neel temperature susceptibility continues to drop with decrease in temperature | $KNiF_3$: many dimeric complexes of copper (II) and oxovanadium (IV) |

**3. Chromium ($Cr = 3d^5 4s^1$)**

(*i*) $Cr(CO)_6$ in which Cr is in zero oxidation state ($Cr = 3d^5 4s^1 = 3d^6$) is diamagnetic ($n = 0$) and is inner orbital octahedral ($d^2sp^3$) complex.

(*ii*) Hein and Herzog have shown that the magnetic moment of $Cr^+$ ($3d^5$ system) complex *viz* $[Cr(dipy)_3]^+$ is corresponding to the presence of one electron. Thus this complex is an inner orbital octahedral ($d^2sp^3$) complex.

(*iii*) Cyanide complex of $\mathbf{Cr^{2+}}$ ($Cr^{2+} = 3d^4$), $[Cr(CN)_6]^{4-}$ is an inner orbital octahedral ($d^2sp^3$) complex with two unpaired electrons and has a magnetic moment of ~3.1 BM. This value is slightly greater than the value calculated from spin only formula for two unpaired electrons (when $n = 2$, $\mu = 2.83$ BM). The increase in the observed value of $\mu$ is due to some contribution of orbital motion of the electrons to the magnetic moment.

The complexes like $[Cr(dipy)_3]^{2+}$ and $[Cr(NCS)_6]^{4-}$ are outer orbital octahedral ($sp^3d^2$) complexes with magnetic moment ~4.85 BM corresponding to the presence of four unpaired electrons.

(*iii*) $\mathbf{Cr^{3+}}$ ($Cr^{3+} = 3d^3$) complexes may be cationic, anionic or neutral. These are inner orbital octahedral ($d^2sp^3$) complexes and their magnetic moment values are corresponding to the presence of three unpaired electrons.

(*iv*) $\mathbf{Cr^{4+}}$ ($Cr^{4+} = 3d^2$) complex *viz* $[CrF_6]^{2-}$ has magnetic moment value corresponding to the presence of two unpaired electrons and hence it is an inner orbital octahedral ($d^2sp^3$) complex.

(*v*) $\mathbf{Cr^{5+}}$ ($Cr^{5+} = 3d^1$) in $[CrOF_4]^-$ has a normal value of magnetic moment equal to 1.76 BM whih corresponds to the presence of one unpaired electron.

**4. Manganese ($Mn = 3d^5 4s^2$)**

(*i*) $\mathbf{Mn^{+1}}$ ion ($3d^5 4s^1 = 3d^6$ ion) is present in $[Mn(CNR)_6]^+$ which is diamagnetic ($n = 0$). Thus it is an inner orbital octahedral ($d^2sp^3$) complex.

(*ii*) $\mathbf{Mn^{2+}}$ ($3d^5$ system) gives a number of tetrahedral ($sp^3$) complexes like $[Mn^{2+}(Ph_3PO_4)_4](ClO_4)_2$ ($\mu = 5.97$BM), $[Mn^{2+}(Ph_3PO_2)_2Br_2]^0$ ($\mu = 5.92$ BM) which are paramagnetic and their magnetic moment values are corresponding to the presence of five unpaired electrons. Six co-ordinated hexacyano and hexaisocyano complexes, $[Mn^{2+}(CN)_6]^{4-}$, $[Mn^{2+}(CNR)_6]^{2+}$ have $\mu = 1.73$ BM corresponding to the presence of one unpaired electron. Thus these complexes are inner orbital octahedral ($d^2sp^3$) complexes.

(*iii*) $\mathbf{Mn^{3+}}$ ($3d^4$ system) is present in $[Mn^{3+}(acac)_3]^0$, $[Mn^{3+}(OH)(H_2O)(Big)(acac)]^0$ and $[Mn^{3+}(C_2O_4)_3]^{3-}$ which have the expected magnetic moment values ($\mu$ = 4.8–5.0 BM) corresponding to the presence of four unpaired electrons. Thus these are outer orbital octahedral ($sp^3d^2$) complexes. $[Mn(CN)_6]^{3-}$, however, has lower value of magnetic moment ($\mu = 2.83$ BM) which corresponds to the presence of two unpaired electrons. Thus it is an inner orbital octahedral ($d^2sp^3$) complex.

(*iv*) It has been observed that a number of complexes of $\mathbf{Mn^{4+}}$ ($3d^3$ system) like $[MnX_6]^{2-}$ have the expected value of magnetic moment of 3.87 BM corresponding to the presence of three unpaired electrons. Thus these are inner orbital octahedral complexes ($d^2sp^3$).

**5. Iron ($Fe = 3d^6 4s^2$)**

(*i*) $Fe^0$($3d^6 4s^2$ or $3d^8$) is present in carbonyls like $Fe(CO)_5$, $Fe_2(CO)_9$ and $Fe_3(CO)_{12}$. All these carbonyls are diamagnetic. Diamagnetism of $Fe(CO)_5$ suggests that in this compound Fe(0) is $dsp^3$ hybridised and the carbonyl has trigonal bipyramidal geometry while the diamagnetism in other carbonyls suggests the presence of Fe – Fe bond.

(*ii*) $Fe^{2+}$ ($3d^6$) is present in $[Fe(CN)_6]^{4-}$, $[Fe(CN)_5(NO^+)]^{2-}$, $[Fe(dipy)_3]^{2+}$ etc. These complexes are diamagnetic and hence are inner orbital octahedral ($d^2sp^3$) complexes. $[Fe(H_2O)_6]^{2+}$, $[Fe(NH_3)_6]^{2+}$, $[Fe(en)_3]^{2+}$ etc. are outer orbital octahedral complexes.

(*iii*) $Fe^{3+}$($3d^5$) gives tetrahedral and octahedral complexes. Complexes continuing the very weak ligands like $Cl^-$, $Br^-$ are tetrahedral ($sp^3$). Thus the complexes of $[FeX_4]^-$ type (X = Cl, Br) are tetrahedral complexes having μ = 5.9 BM corresponding to the presence of five unpaired electrons. Complexes of stronger ligands (e.g. $NH_3$, $H_2O$, $F^-$ etc.) like $[Fe(NH_3)_6]^{3+}$, $[Fe(H_2O)_6]^{3+}$, $[FeF_6]^{3-}$ etc. have μ ~ 5.9 BM showing that these are outer orbital or ionic octahedral ($sp^3d^2$) complexes. The complexes with strongest ligands (e.g. $CN^-$, phen, dipy etc.) like $[Fe(CN)_6]^{3-}$, $[Fe(phen)_3]^{3+}$, $[Fe(dipy)_3]^{3+}$ have μ ~ 1.9 BM and are inner orbital or covalent octahedral ($d^2sp^3$) complexes with one unpaired electron.

**6. Cobalt ($Co = 3d^7 4s^2$)**

(*i*) **Zero valent** cobalt ($3d^7 4s^2 = 3d^9$) is present in dimeric carbonyl, $Co_2(CO)_8$ which is diamagnetic.

(*ii*) $[Co(dipy)_3]^+$, $[Co(phen)_3]^+$ and $[Co(CNR)_5]^+$ are the complexes of $Co^+$ ($3d^7 4s^1$ or $3d^8$ system). $[Co(CNR)_5]^+$ is diamagnetic ($n$ = 0) indicating $dsp^3$ hybridisation of $Co^+$. It has trigonal bipyramidal geometry. Other complexes are probably outer orbital octahedral complexes ($sp^3d^2$).

(*iii*) $Co^{2+}$ ($3d^7$) gives 4-, 5- and 6-coordinated complexes. 4-coordinated complexes are tetrahedral ($sp^3$) and square planar ($dsp^2$) while 6-coordinated complexes are octahedral. 5-coordinated complexes are square pyramidal.

(*a*) Tetrahedral complexes are of the type $[Co^{2+} X_4]^{2-}$ and $[CoX_2L_2]^0$ where X = $Cl^-$, $Br^-$, $I^-$, $SCN^-$, $N_3^-$ or $OH^-$ and L = monodentate neutral ligand like $NH_3$, $H_2O$ etc. Most of these complexes are paramagnetic ($n$ = 3) but their magnetic moment values are generally higher (μ = 4.9–5.2 BM) than the spin only values corresponding to three unpaired electrons (when $n$ = 3, $\mu_{s.o.}$ = 3.87 BM). The increase in the value of μ is due to orbital motion contribution to these values.

(*b*) Square planar complexes are formed by several bidentate monobasic (e.g. $dmg^-$, $Big^-$ etc.) and bidentate neutral (e.g. en) ligands. $[Co(dmg)_2]^0$ and $[Co(Big)_2]^0$ are important square planar complexes. In case of some square planar complexes of Co(II) the magnetic moment values are in the range of 2.0–2.7 BM. This suggests $dsp^2$ hybridisation with one unpaired electron in $3d$ orbital.

(*c*) $[Co^{2+} (triars)I_2]^0$ is a 5-coordinated complex compound of $Co^{2+}$. It has μ = 2.3 BM. This suggests $dsp^3$ hybridisation and square pyramidal geometry of the complex.

(*d*) Octahedral complexes like $[Co(H_2O)_6]^{2+}$, $[Co(NH_3)_6]^{2+}$, $[Co(N_2H_4)_3]^{2+}$, $[Co(en)_3]^{2+}$ etc. are paramagnetic with μ corresponding to three unpaired electrons. Thus these are outer orbital octahedral ($sp^3d^2$) complexes. On the other hand octahedral complexes like $[Co(CN)_6]^{4-}$, $[Co(NO_2)_6]^{4-}$ (μ = 1.82 BM), $[Co(diars)_3]^{2+}$ (μ = 1.92 BM), $[Co(triars)_2]^{2+}$, $[Co(CN)_5(H_2O)]^{3-}$ (μ = 1.72 BM) are paramagnetic with magnetic moment values corresponding to the presence of one unpaired electron residing in *5s* orbital. Thus these are inner orbital octahedral ($d^2sp^3$) complexes. In these complexes 3*d* electron is promoted to 5*s* orbital and promotional energy for this process (estimated from spectroscopic data) is very high (~58 kcal mol) which indicates that Co(+2) combines with strong ligands like $CN^-$, *diars* etc. and even with $NO_2^-$ to give inner orbital octahedral ($d^2sp^3$) complexes.

(*iv*) **$Co^{+3}$** ($3d^6$ system) gives octahedral complexes. Since $Co^{+3}$ is a hard acid, it prefers to combine with hard ligands, *i.e.* with those ligands which contain nitrogen donors (e.g. $NH_3$, en, $NO_2^-$, $CN^-$ etc.). It also forms complexes with $H_2O$, $F^-$ etc. The complexes $[Co(NH_3)_6]^{3+}$, $[Co(H_2O)_6]^{3+}$, $[Co(en)_3]^{3+}$, $[Co(NO_2)_6]^{3-}$ and $[Co(CN)_6]^{3-}$ are diamagnetic ($n = 0$) and hence are inner orbital octahedral ($d^2sp^3$) complexes, *i.e.* these complexes are essentially covalent. $[CoF_6]^{3-}$ is the only octahedral complex which is paramagnetic with μ = 5.3 BM indicating that it has four unpaired electrons and is outer orbital octahedral ($sp^3d^2$) complex, *i.e.* it is essentially ionic.

**7. Nickel (Ni = $3d^84s^2$ or $3d^{10}$)**

(*i*) **Zero valent nickel** ($3d^{10}$ system) is present in $Ni(CO)_4$, $Ni(PF_3)_4$, $Ni(CNR)_4$, $[Ni(CN)_4]^{4-}$ and $[Ni(CO)_2(phen)]$. $Ni(CO)_4$ is diamagnetic ($n$ = 0). It has tetrahedral geometry corresponding to $sp^3$ hybridisation.

(*ii*) **$Ni^{+1}$** ($3d^9$ system) is present in $[Ni(CN)_4]^{3-}$ which is paramagnetic (μ = 1.73 BM) corresponding to the presence of one unpaired electron. It has square planar geometry and involves $dsp^2$ hybridisation with one 3*d* electron promoted to 4*p* orbital. $[Ni(CN)_3]^{2-}$ is known to be dimeric, $[Ni_2(CN)_6]^{4-}$ and is diamagnetic due to Ni–Ni bond.

(*iii*) **$Ni^{+2}$** ($3d^8$ system) forms 4-, 5-, 6-coordinated complexes. 4-coordinated complexes of $Ni^{2+}$ are of two types:

(a) *Tetrahedral complexes ($sp^3$ hybridisation).* These are examplified by $[NiX_4]^{2-}$, $[NiL_3X]^-$, $[NiL_2X_2]^0$ and $[Ni(LL)_2]^0$ where X = $Cl^-$, $Br^-$, $I^-$ and L = a monodentate neutral ligand such as $NH_3$, $PH_3$, $PPh_3$, $OPPh_3$ and LL = a bidentate monobasic (i.e. bidentate uninegative) ligand. These complexes are usually green or blue in colour and are paramagnetic having μ values corresponding to the presence of two unpaired electrons.

(b) *Square planar complexes ($dsp^2$ or $sp^2d$ hybridisation).* 4-coordinated complexes of $Ni^{2+}$ like $[Ni(dmg)_2]^0$, $[Ni(Big\ H)_2]^{2+}$, $[Ni(CN)_4]^{2-}$, $\left[Ni\begin{pmatrix} O = C-S \\ O = C-S \end{pmatrix}\right]^{-2}$ are diamagnetic (n = 0). These are formed by $dsp^2$ hybridisation and are, therefore, essentially covalent. Some other 4-coordinated complexes like those formed by $NH_3$ and a large number of amines are paramagnetic with magnetic moment value corresponding to the presence of two unpaired electrons ($n$ = 2). Thus these complexes involve $sp^2d$ hybridisation and are, therefore, essentially ionic. Both types of complexes formed either by $dsp^2$ or $sp^2d$ hybridisation are square planar in geometry. Some 4-coordinated complexes of $Ni^{2+}$, one and the same complex has been obtained in both a paramagnetic form (having blue colour) and a diamagnetic form (having yellow colour).

5-coordinated complexes of $Ni^{2+}$ are exemplified by $[Ni(CN)_5]^{3-}$ (square pyramidal), $[Ni(QAs)X]^4$ (trigonal bipyramidal) and $[Ni(triars)\ Br_2]^0$ (trigonal bipyramidal). All these complexes are formed by $dsp^3$ hybridisation and are diamagnetic ($n = 0$).

6-coordinated complexes of $Ni^{2+}$ like $[Ni(H_2O)_6]^{2+}$, $[Ni\ (NH_3)_6]^{2+}$, $[Ni(en)_3]^{2+}$, $[Ni(N_2H_4)_3]^{2+}$, $[Ni(phen)_3]^{2+}$, $[Ni(dipy)_3]^{2+}$ and $[Ni(NCS)_6]^{4-}$ are paramagnetic outer orbital octahedral ($sp^3d^2$) complexes and their μ values generally correspond to the presence of two unpaired electrons. These are light blue to blue-green in colour.

(*iv*) **$Ni^{3+}$** ($3d^7$) gives 5- and 6-coordinated complexes. 5-coordinated complexes are exemplified by $[Ni(R_3P)_2X_3]^0$ complexes (R = alkyl or aryl group, X = halogen). These have trigonal bipyramidal geometry ($dsp^3$) in which the three X-atoms occupy the equatorial positions. Their magnetic moment values correspond to the presence of one unpaired electron. 6-coordinated complexes of $Ni^{3+}$ are exemplified by $[Ni(diars)_2Cl_2]^+$, $[Ni(BPO)_3]^0$, $[NiF_6]^{3-}$ etc. $[Ni(diars)_2Cl_2]^+$ ($\mu = 1.9$ BM) and $[Ni(BPO)_3]^0$ ($\mu = 1.75$ BM) are inner orbital octahedral ($d^2sp^3$) complexes with one unpaired electron in 5s orbital. In these complexes one of the 3*d* electron is shifted to 5*s* orbital. $[NiF_6]^{3-}$ has $\mu = 2.5$ B.M. and hence is a outer orbital octahedral ($sp^3d^2$) complex containing three unpaired electrons.

(*v*) **$Ni^{4+}$ ($3d^6$)** is found in 6-coordinated complexes like $[NiF_6]^{2-}$, $[Ni(diars)_2Cl_2]^{2+}$ and $[Ni(DAPD)_2]^0$. All these complexes are diamagnetic and hence are inner orbital octahedral ($d^2sp^3$) complexes.

**8. Copper (Cu = $3d^{10}4s^1$)**

(*i*) **$Cu^+$**($3d^{10}$ system) forms complexes with ligands containing C, N, P, As, O, S and halogen atom as donor atom. 2-, 3-, and 4-coordinated complexes are given by $Cu^+$. Important complexes of $Cu^+$ are:

(a)*Halide complexes,* e.g. $[CuCl_2]^-$, $[CuCl_3]^{2-}$, $[CuCl_4]^{3-}$ etc.

(b)*Cyanide complexes,* e.g. $[Cu(CN)_2]^-$, $[Cu(CN)_3]^{2-}$, $[Cu(CN)_4]^{3-}$ etc.

(c)*Ammine complexes,* e.g. $[Cu(NH_3)_2]^+$

2-coordinated complexes are linear. However, Cu metal in $[Cu(CN)_2]^-$ ion derived from $K[Cu(CN)_2]$ does not obey the apparent coordination number of two, instead the complex has a structure which contains spiral polymeric chains in which each $Cu^+$ ion has a C.N. of three as shown in Fig. 25.16.

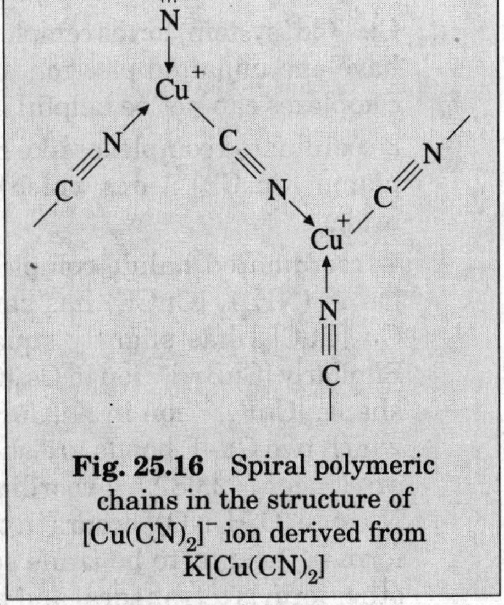


**Fig. 25.16** Spiral polymeric chains in the structure of $[Cu(CN)_2]^-$ ion derived from $K[Cu(CN)_2]$

3-coordinated complex *viz* $\left[Cu\left(S{=}C\begin{matrix}NH{-}CH_2\\ |\\ NH{-}CH_2\end{matrix}\right)_3\right]^+$ has trigonal planar geometry shown in Fig. 25.17.

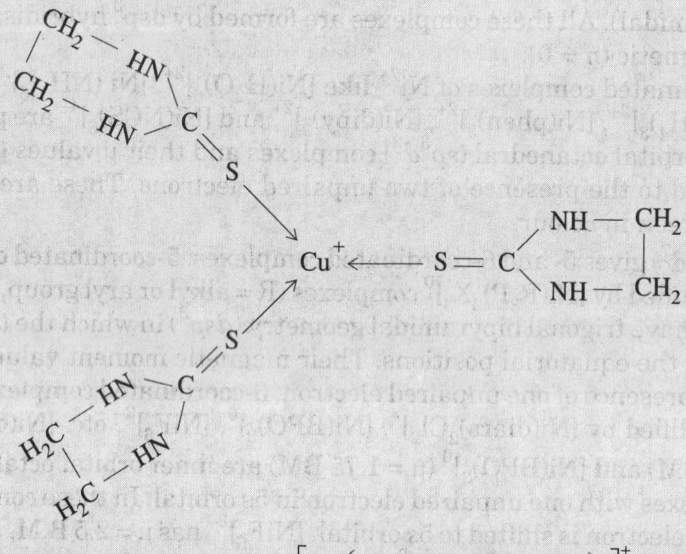

**Fig. 25.17** Structure of $\left[Cu\left(S{=}C\begin{matrix}NH{-}CH_2\\ |\\ NH{-}CH_2\end{matrix}\right)_3\right]^+$

$[CuCl_3]^-$ ion is dimeric containing planar $[Cu_2Cl_6]^{2-}$ ion. $[CuCl(tu)_2]^0$ has distorted trigonal geometry.

4-coordinated complexes are tetrahedral. $[Cu(CN)_4]^{3-}$ ion is most probably square planar. All $Cu^+$ complexes are diamagnetic and colourless due to the presence of $3d^{10}$ configuration.

(*ii*) $\mathbf{Cu^{2+}}$($3d^9$ system) forms complexes of several geometries. All possible geometries have one unpaired electron and hence magnetic moment criterion of these complexes can not be helpful in predicting the geometry of these complexes.

3-coordinated complexes like $Li[CuCl_3]$ is dimeric and hence contains dimeric planar $[Cu_2Cl_6]^{2-}$ ions which are joined into longer chains by long chlorine bridges.

4- coordinated halide complexes have different geometries, e.g. $[CuCl_4]^{2-}$ ion in $(NH_4)_2$ $[CuCl_4]$ has square planar geometry while the same ion in $Cu_2[CuCl_4]$ has slightly squashed (i.e., flattened) tetrahedral geometry. Similarly $[CuBr_4]^{2-}$ ion in $Cs_2[CuBr_4]$ also has a slightly squashed tetrahedral shape. $[CuF_4]^{2-}$ ion in $K_2[CuF_4]$ has a distorted square planar structure in which two Cu–F bonds are shorter (= 1.95Å) and the other two Cu–F bonds are longer (=2.08Å). 4-coordinated complexes like $[CuCl_2L_2]^0$ [L = pyridine N-oxide ($C_5H_5NO$)] occurs in two different structural modifications, e.g. one form is thought to be trans square planar and has green colour while the other form is tetrahedral and is yellow in colour. According to Pauling square planar complexes result from $dsp^2$ hybridisation with one *3d* electron promoted to *4p* orbital whereas according to Ray, Sen and Huggins hybridisation is of $sp^2d$ type, *i,e.,* according Ray etc. the bonds are of ionic type. Both the types of hybridisation schemes give one unpaired electron. The promotion of *3d* electron to 4p orbital as suggested by Pauling for $dsp^2$ hybridisation does not require much energy. All the square planar complexes which are formed by $dsp^2$ hybridisation are yellow, red, brown and black and a few are dark blue. These have magnetic moment values in the large 1.73–1.84 BM. The square planar complexes resulted from $sp^2d$ hybridisation are green and blue and have μ values in the range 1.94–2.2 BM. The smaller moment values of the former group of square planar complexes has been explained by assuming that the crystal field has greater quenching effect on the orbital moment of the exposed electron.

5-coordinated complexes like $[CuCl_5]^{3-}$, $[CuCl_2Br_3]^{3-}$, $[Cu(dipy)_2I]^+$ have trigonal bipyramidal geometry while complex $[Cu(py)(acac_2]^0$ has square pyramidal shape. In $[CuCl_2Br_3]^{3-}$ chlorine atoms occupy axilal positions.

6-coordinated complexes like $[Cu(NH_3)_6]^{2+}$, $[Cu(H_2O)_6]^{2+}$, $[Cu(dipy)_3]^{2+}$ result from $sp^3d^2$ hybridisation and have distorted octahedral shape with two longer trans (*i.e.,* axial) bonds and four shorter square planar (*i.e.,* equational) bonds. (See Fig 25.18). The distortion in shape is due to the fact that in $3d^9$ configuration of $Cu^{2+}$ ion, $3d_{z^2}$ orbital is doubly-filled while $3d_{x^2-y^2}$ orbital is singly filled and this uneven (unsymmetrical) filling of $e_g$ set of orbitals prevents the axial ligands ($L_a$ ligands) to come closer to the central $Cu^{2+}$ ion as compared to the equatorial ligands ($L_e$ ligands) (John-Teller effect)

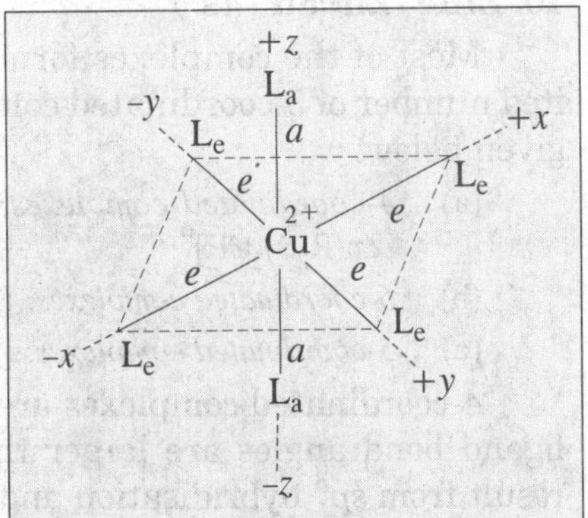


**Fig. 25.18** Distorted octahedral complex of $Cu^{2+}$ ion a = axial bond (longer bonds), e = equatorial bonds (shorter bonds).

$$Cu^{2+} = 3d^9 = (3d_{xy})^2\ (3d_{yz})^2\ (3d_{xz})^2\ (3d_{z^2})^2\ (3d_{x^2-y^2})^1 = t^6_{2g}\ e^3_g$$

$[Cu(dipy)_3]^{2+}$ has four shorter planar Cu-N bonds of length 2.03Å and two much longer axial bonds at 2.23Å and 2.45Å. $H_2O$ molecules of $[Cu(H_2O)_6]^{2+}$ can be replaced by $NH_3$ molecules to get mixed complexes like $[Cu(H_2O)_5\ (NH_3)]^{2+}$,$[Cu(H_2O)_4(NH_3)_2]^{2+}$, $[Cu(H_2O)_3(NH_3)_3]^{2+}$ and $[Cu(H_2O)_2(NH_3)_4]^{2+}$. Using liq. $NH_3$ as solvent, the complex $[Cu(NH_3)_6]^{2+}$ can also be obtained. Similarly with *en* the complexes like $[Cu(en)(H_2O)_4]^{2+}$, $[Cu(en)_2(H_2O)_2]^{2+}$ and $[Cu(en)_3]^{2+}$ can also be formed.

(*iii*) **$Cu^{3+}$**($3d^8$ system) forms a number of complexes. Hexafluoro complex, $[CuF_6]^{3-}$ is octahedral paramagnetic (μ = 2.8BM) indicating the presence of two unpaired electrons. Thus this ion has ionic or $sp^3d^2$ bonding.

Para per iodate complexes like $K_7[Cu(I^{7+}O_6)_2].7H_2O$ and $Na_7[Cu(I^{7+}O_6)_2].12H_2O$ are diamagnetic red crystals. Diamagnetism results from a square planar $Cu^{3+}O_4$ coordination with $dsp^2$ hybridisation of $Cu^{3+}$ ion. Thus the structure of $[Cu^{3+}(I^{7+}O_6)_2]^{7-}$ ion can shown as given Fig. 25.19.

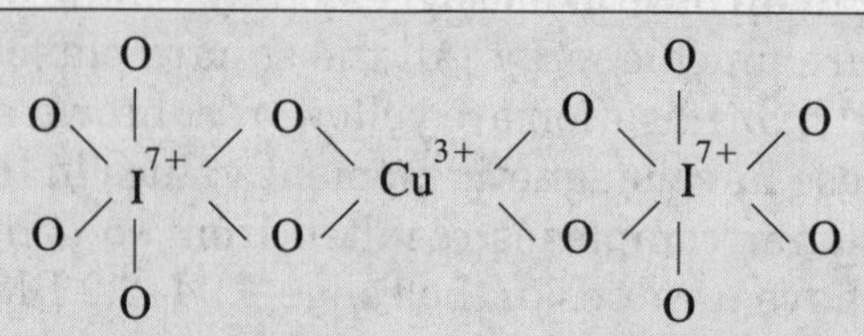


**Fig. 25.19** Structure of $[Cu^{3+}(I^{7+}O_6)_2]^{7-}$ ion present in $K_7[Cu(IO_6)_2] \cdot 7H_2O$

Tellurate complexes like $Na_9[Cu^{3+}(Te^{6+}O_6)_2].20H_2O$, $Na_5H_4[Cu^{3+}(Te^{6+}O_6)_2] \cdot 18H_2O$ and $Na_7H_2[Cu^{3+}(Te^{6+}O_6)_2] \cdot 12H_2O$ are also known.

**10. Zinc ($Zn=3d^{10}4s^2$)**

Most of the complexes formed by $Zn^{2+}(3d^{10})$ are 4- and 6-coordinated. A limited number of 5-coordinated complexes are also given by $Zn^{2+}$. Some examples are given below.

(a) *4-coordinated complexes:* $[ZnX_4]^{2-}$, $[Zn(CN)_4]^{2-}$, $[Zn(NH_3)_4]^{2+}$, $[Zn(NH_3)_2Cl_2]^0$, $[Zn(R_2S)_2Cl_2]^0$

(b) *6-coordinated complexes:* $[Zn(NH_3)_6]^{2+}$, $[Zn(H_2O)_6]^{2+}$, $[Zn(en)_3]^{2+}$, $[Zn(phen)_3]^{2+}$

(c) *5-coordinated complexes:* $[Zn(acac)_2(H_2O)]^0$, $[Zn(terpy)Cl_2]^0$

4-coordinated complexes are tetrahedral, since in this geometry ligand-metal-ligand bond angles are larger than in square planar geometry. These complexes result from $sp^3$ hybridisation and are diamagnetic ($n = 0$). 6-coordinated complexes are octahedral and involve $sp^3d^2$ hybridisation. These are diamagnetic ($n = 0$). 5-coordinated complexes have square pyramidal geometry.

## Magnetic Behaviour of Complexes of Metals of 2nd and 3rd Transition Series.

In the case of metals of second and particularly third transition series, the complexes are mainly spin paired. The metal ions with odd number of electrons will, therefore, exhibit paramagnetism which is equivalent to roughly one unpaired electron and the metal ions with even number of electrons would show diamagnetism. This could be due to the following reasons:

(i) Smaller size of 3*d* orbitals makes the occupancy of two electrons in the same 3*d* orbital energetically less favourable than when the two electrons are present in 4*d* or 5*d* orbitals of comparatively bigger size.

(ii) The separation in energy terms of the metal ion caused by the ligand fields are considerably higher in 4*d* and 5*d* orbitals compared to 3*d* orbitals.

Both the above factors favour a low spin complexes.

## UNIVERSITY QUESTIONS

**1.** (*a*)Discuss briefly the Gouy's method for determining magnetic susceptibility
(*b*)Explain the origin of paramagnetism.
(*c*)Zn forms only $Zn^{+2}$ and not $Zn^{+3}$, why?
(*d*)Calculate spin only magnetic moment for the complex $K_4[Mn(NCS)_6]$
*(Himachal Pradesh 2010)*

**2.** How magnetic properties of chemical substances are useful in deciding spin-pair metal complexes.
*(Kanpur 2010)*

**3.** (*a*)Write notes on: (i) Magnetic moment (ii) Magnetic susceptibility
(*b*)How does change in temperature affect the magnetic susceptibility?
(*c*)Cu(I) is diamagnetic while Cu(II) is paramagnetic. Explain
(*d*)Explain why $VOSO_4$ is paramagnetic.
*(Kanpur 2010)*

**4.** (*a*) What is magnetic susceptibility?
(*b*) Describe Gouy's method for the determination of magnetic susceptibility.
*(Pune 2009)*

**5.** How is magnetic susceptibility related with magnetic moment? Discuss the Gouy's method for determination of magnetic susceptibility.
*(Rajsthan 2010)*

6. (*a*) What is the basic difference between the magnetic properties of elements of first and second transition series.
(*b*) Which one of the following ions has maximum magnetic moment.
$Cr^{2+}$, $Fe^{3+}$, $Ni^{2+}$ *(H.N Bahuguna Garhwal 2010)*

**7.** Explain orbital contribution to magnetic moments. *(G. N. D. Amritsar 2010)*

**8.** Write short noes on the following:
(*a*) Paramagnetism (*b*) Diamagnetism (*c*) Ferro and antiferromagnetism
*(Ranchi 2010)*

**9.** (*a*) Give reasons for the following:
The values of the observed magnetic moments (in BM) for the following complexes are: (i) $K_4$ $[Mn(CN)_6]$=1.9 (ii) $[Fe\ I_6]^{3-}$ = 5.9 (iii) $[CoF_6]^{3-}$ is paramagnetic while $[Co(NH_3)_6]^{3+}$ is not.
(*b*)Explain antiferromagnetism giving example
(*c*)What is magnetic susceptibility? How will you measure it by Gouy balance method? A sample of an organometallic compound was placed in a susceptibility balance. The following data were obtained: (F = $2.4 \times 10^{-4}$ N, A = $1.8 \times 10^{-5}$ $m^2$, B = 4000G). Calculate the magnetic susceptibility of the compound. *(Utkal 2010)*

**10.** (*a*)What do you mean by types of magnetic behaviour? Explain in detail.
(*b*)Explain the diamagnetic behaviour of $[Co\ (NH_3)_6]^{3+}$ ion according to VBT.
*(GND Amritsar 2010)*

**11.** (*a*)What do you understand by temperature-independent magnetism?
(*b*)Calculate the spin only magnetic moment of $Ni^{2+}$ ion in $[Ni\,(H_2O)_6]^{2+}$.
*(GND Amritsar 2010)*

**12.** Explain the nature and causes of (i) Diamagnetism (ii) Paramagnetism (iii) Antiferromagnetism (iv) Ferromagnetism
*(BJG College, Orissa 2010)*

**13.** (i)Explain magnetic susceptibility. How does it vary with temperature?
(ii)Calculate the spin only magnetic moment in case of $Fe^{2+}$, $Ni^{2+}$ and $Ti^{3+}$.
*(Lucknow 2010)*

**14.** (i) Calculate magnetic moments of $[Cr(H_2O)_6]^{2+}$ and $[Mn(CN)_6]^{4-}$ ions.
(ii) Explain the colour and magnetic properties of $[Co(NH_3)_6]^{3+}$ and $[CoF_6]^{3-}$ ions on the basis of CFT.
*(Madras April 2011)*

**15.** Describe paramagnetism.
*(Chhatrapati Sahu Maharaj University, Kanpur 2011)*

**16.** (*a*) Define magnetic susceptibility. Discuss briefly Gouy's method for measing magnetic susceptibility.
(*b*) Why does Mn (II) show maximum paramanetic character amongst the bivalent ions of first transition series?
(*b*) Calculate magnetic moment in Bohr Magnetons expected from spin only for the following ions:
(i) $Cr^{+3}$ (ii) $Fe^{+3}$
(*a*) Calculate $\mu_g$ for $3d^2$ ion, giving example.
(*b*) Comment on the diamagnetic — paramagnetic equilibrium in solution in Ni (II) complexes.
(*c*) Explain ferromagnetism.
(*d*) Predict which of the following configurations are expected to have orbital contribution in tetrahedral field: (*i*) $d^2$ (*ii*) $d^4$ (*iii*) $d^7$ (*iv*) $d^8$.
*(Kurukshetra University, 2011)*

**17.** Which of the following configurations are expected to show orbital contribution in HS octahedral field? $d^1$, $d^3$, $d^6$ and $d^8$ *(Nagpur 2012)*

**18.** (*i*) Which of the following nuclei should have zero magnetic moment. ${}^{2}_{1}H$, ${}^{12}_{6}C$, ${}^{14}_{7}N$, ${}^{16}_{8}O$.

(*ii*) Differentiate between diamagnetism and paramagnetism. Discuss Gouy's method for measuring magnetic suscaptibility.
(*iii*) Calculate magnetic moment for $Cr^{3+}$ and $Cu^{+}$ ions *(Lucknow 2013)*

26

CHAPTER

# Thermodynamic and Kinetic Aspects of Metal Complexes

## Two Types of Stability of Metal Complexes

In studying the formation of metal complexes in solution, two types of stability of complexes have been distinguished. These are:

**1. Thermodynamic stability: Stable (penetration) and unstable (normal) complexes.** This kind of stability deals with the properties like bond energies, stability constants and redox potentials that affect the equilibrium conditions.

On the basis of thermodynamic stability of complexes in solution, Biltz (1927) has classified the complex compounds into stable and unstable complexes. Stable complexes are those which possess sufficient stability to retain their identity in solution while unstable complexes are those which are reversibly dissociated in solution into their components. Stable and unstable complexes have also been called *penetration* and *normal complexes* respectively.

**2. Kinetic stability: Labile and inert complexes.** This kind of stability deals with the rates of reactions (i.e., reactivity) of complexes in solution , the mechanisms of chemical reactions, formation of intermediate complexes, activation energies for the process etc. On the basis of the rate of reactions (i.e., kinetic stability) of the complex in solution, Taube (1950) has classified the complexes into *labile* and *inert* complexes. Labile complexes are those whose one or more ligands in the co-ordination sphere can be rapidly replaced by other ligands and the ability of a complex to replace its one or more ligands by other ligands is called its lability. Inert complexes are those whose one or more ligands can either not be replaced or can be replaced with difficulty by other ligands.

## Stability of Complex Ions in Solution

A complex ion dissociates in aqueous solution to a very small extent. Stronger is the metal-ligand bond in the complex ion, lesser is the dissociation of the complex ion and hence greater is the stability of the complex ion. The stability of the complex ion in solution can be defined as a measure of the resistance of the replacement of a ligand by another ligand.

## Dissociation of A Complex Ion in Solution and Dissociation (or Instability) Constant ($K_{diss}$ or $K_i$)

In aqueous solution, a complex ion dissociates to a very small extent. When a complex ion dissociates, there lies an equilibrium between the undissociated complex ion and the species obtained by the dissociation of the complex ion. Hence the stability of the complex ion in solution is expressed in terms of equilibrium constant of the dissociation equilibrium.

**Example.** The dissociation of $[Cu(NH_3)_4]^{2+}$ ion in solution is represented by the equilibrium:

$$[Cu(NH_3)_4]^{2+} \rightleftharpoons Cu^{+2} + 4NH_3$$

The dissociation (or instability) constant ($K_{diss}$) of the above equilibrium is given by:

$$\text{Instability constant } (K_{dis}) = \frac{[Cu^{2+}][NH_3]^4}{[Cu(NH_3)_4]^{2+}} \quad (i)$$

**Formation of a Complex Ion in Solution and Formation (or Stability) Constant ($K_{for}$ or β ).**

As an example the formation of $[Cu(NH_3)_4]^{2+}$ ion in solution can be considered. The formation of this ion in solution can be represented by the equilibrium given below:

$$Cu^{2+} + 4NH_3 \rightleftharpoons [Cu(NH_3)_4]^{2+}$$

Since the above equilibrium involves the formation of complex ion, the equilibrium constant of the above formation reaction is called formation (or stability) constant which is represented by $K_{for}$. Thus $K_{for}$ is given by:

$$\text{Stability constant } (K_{for}) = = \frac{[Cu(NH_3)_4]^{2+}}{[Cu^{2+}][NH_3]^4} \quad (ii)$$

On comparing equations (i) and (ii) we find that

$$K_{for} \text{ (or } \beta) = \frac{1}{K_{diss} \text{ (or } K_i)}$$

Thus the formation constant (or stability constant), $K_{for}$ is reciprocal of dissociation constant (or instability constant), $K_{diss}$. $K_{for}$ and $K_{diss}$ are also represented as β and $K_i$ respectively.

The values of $K_{diss}$ and $K_{for}$ of some complex ions in solution are given below.

| *Complex ion* | $K_{diss}$ | $K_{for}$ |
|---|---|---|
| $[Cu(NH_3)_4]^{2+}$ | $1.0 \times 10^{-12}$ | $1.0 \times 10^{12}$ |
| $[Co(NH_3)_6]^{3+}$ | $6.2 \times 10^{-36}$ | $1.6 \times 10^{35}$ |
| $[Ag(S_2O_3)_2]^{3-}$ | $6.1 \times 10^{-14}$ | $1.6 \times 10^{13}$ |
| $[Ag(CN)_2]^{-}$ | $1.8 \times 10^{-19}$ | $5.4 \times 10^{18}$ |
| $[Hg(CN)_4]^{2-}$ | $4.0 \times 10^{-42}$ | $2.5 \times 10^{41}$ |
| $[Fe(SCN)]^{2+}$ | $1.0 \times 10^{-3}$ | $1.0 \times 10^{3}$ |

Higher is the value of stability constant, β (formation constant) for a complex ion, greater is the stability of the complex ion. Now since $\beta \alpha \frac{1}{K_i}$, we can say that smaller is the value of instability constant, $K_i$ (dissociation constant) of a complex ion, greater is the stability of the complex ion. The values of $K_{for}$ given above show that since $2.5 \times 10^{41}$ is the highest value and $1.0 \times 10^{3}$ is the lowest value, $[Hg(CN)_4]^{2-}$ is the most stable complex ion and $[Fe(SCN)]^{2+}$ is the least stable ion.

**Kinetically Labile and Kinetically Inert Complexes.**

See Chapter on "*Ligand Substitution Reactions In Octahedral Complexes*"

**Classification of Octahedral Complexes Into Labile and Inert Complexes on The Basis of VBT**

See Chapter on "*Ligand Substitution Reactions In Octahedral Complexes*"

**Lability and Inertness of Octahedral Complexes on The Basis of $t^p_{2g}\, e^q_g$ Configuration of Metal ion(CFT)**

See Chapter on "*Ligand Substitution Reaction In Octahedral Complexes*"

**To Explain The Lability and Inertness of Octahedral Complexes on The Basis of CFT**

See Chapter on "*Ligand Substitution Reaction in Octahedral Complexes*"

## Stepwise Formation of Complexes: Stepwise Formation Constants ($K_1, K_2......K_n$) and Overall (Cumulative) Formation Constant ($\beta_n$)

Ignoring water molecules, the complex species, $ML_n$ is formed when $n$ ligands (L) are added to the metal atom (M).

$$\underset{\text{Metal}}{M} + \underset{\text{Ligands}}{nL} \xrightleftharpoons{\beta} \underset{\text{Complex species}}{ML_n} \qquad (i)$$

The stability (or formation) constant (β) for the above equilibrium is given by:

$$\text{Stability constant}\,(\beta) = \frac{ML_n}{[M][L]^n} \qquad (ii)$$

In equilibrium (i), $n$ ligands (L) are added to the metal (M) in a single step. If $n$ ligands (L) are added to the metal (M) one by one, then the formation of $ML_n$ can be supposed to take place through the following $n$ equilibria.

$$M + L \xrightleftharpoons{K_1} ML, \quad K_1 = \frac{[ML]}{[M][L]}$$

$$ML + L \xrightleftharpoons{K_2} ML_2, \quad K_2 = \frac{[ML_2]}{[ML][L]}$$

$$ML_2 + L \xrightleftharpoons{K_3} ML_3, \quad K_3 = \frac{[ML_3]}{[ML_2][L]}$$

$$\vdots \qquad \vdots \qquad \vdots \qquad \vdots$$

$$ML_{n-2} + L \xrightleftharpoons{K_{n-1}} ML_{n-1}, \; K_{n-1} = \frac{[ML_{n-1}]}{[ML_{n-2}][L]}$$

$$ML_{n-1} + L \xrightleftharpoons{K_n} ML_n, \quad K_n = \frac{[ML_n]}{[ML_{n-1}][L]}$$

In the above equilibria, $K_1, K_2, K_3 \ldots K_{n-1}, K_n$ are called **stepwise formation (or stepwise stability) constants.**

$$\text{Value of } K_1 \times K_2 \times K_3 \times \ldots K_{n-1} \times K_n$$

$$= \frac{ML}{[M][L]} \times \frac{[ML_2]}{[ML][L]} \times \frac{[ML_3]}{[ML_2][L]} \times \ldots \times \frac{[ML_{n-1}]}{[ML_{n-2}][L]} \times \frac{[ML_n]}{[ML_{n-1}][L]}$$

$$= \frac{[ML_n]}{[M][L]^n} = \beta \qquad \text{[From equation (ii)]}$$

Here β is called overall **stability constant** which is represented as $\beta_n$. Thus:

$$\beta_n = K_1 \times K_2 \times K_3 \times \ldots \times K_{n-1} \times K_n$$
$$\text{or } \log \beta_n = \log K_1 + \log K_2 + \log K_3 + \ldots + \log K_{n-1} + \log K_n$$

## Determination of Stability of Complexes From Thermodynamic Data

Four $H_2O$ ligands present in $[Cu(H_2O)_6]^{2+}$ can be replaced by four $NH_3$ ligands to form $[Cu(H_2O)_2(NH_3)_4]^{2+}$

$$[Cu(H_2O)_6]^{2+} + 4NH_3 \overset{K}{\rightleftharpoons} [Cu(H_2O)_2(NH_3)_4]^{2+} + 4H_2O \quad \text{(A)}$$

Here K is the formation constant of $[Cu(H_2O)_2(NH_3)_4]^{2+}$. The formation of $[Cu(H_2O)_2(NH_3)_4]^{2+}$ can also be supposed to take place through the following four steps. In each step only one $H_2O$ ligand is replaced by $NH_3$ ligand.

(i) $[Cu(H_2O)_6]^{2+} + NH_3 \overset{K_1}{\rightleftharpoons} \underset{\text{(I)}}{[Cu(H_2O)_5(NH_3)]^{2+}} + 4H_2O, \Delta H_1$

(ii) $[Cu(H_2O)_5(NH_3)]^{2+} + NH_3 \overset{K_2}{\rightleftharpoons} \underset{\text{(II)}}{[Cu(H_2O)_4(NH_3)_2]^{2+}} + H_2O, \Delta H_2$

(iii) $[Cu(H_2O)_4(NH_3)_2]^{2+} + NH_3 \overset{k_3}{\rightleftharpoons} \underset{\text{(III)}}{[Cu(H_2O)_3(NH_3)_3]^{2+}} + H_2O, \Delta H_3$

(iv) $[Cu(H_2O)_3(NH_3)_3]^{2+} + NH_3 \overset{K_4}{\rightleftharpoons} \underset{\text{(IV)}}{[Cu(H_2O)_2(NH_3)_4]^{2+}} + H_2O, \Delta H_4$

---

*On adding*: $[Cu(H_2O)_6]^{2+} + 4NH_3 \overset{K}{\rightleftharpoons} [Cu(H_2O)_2(NH_3)_4]^{2+} + 4H_2O, \Delta H$ (A)

Reactions (i), (ii), (iii) and (iv) are step wise reactions whose stepwise stability constants are $K_1$, $K_2$, $K_3$ and $K_4$ respectively. $\Delta H_1$, $\Delta H_2$, $\Delta H_3$ and $\Delta H_4$ are the enthalpies of stepwise reactions. Reaction (A) is the overall reaction whose formation constant is K. ΔH is the enthalpy of reaction (A). Obviously:

$$\Delta H = \Delta H_1 + \Delta H_2 + \Delta H_3 + \Delta H_4 \quad (i)$$
$$\text{and } K = K_1 K_2 K_3 K_4 \quad (ii)$$

We know from thermodynamics that

$$\Delta G° = -RT \ln K \quad (iii)$$
$$\text{or } \Delta G° = -RT \ln (K_1 K_2 K_3 K_4) \quad (iv)$$

Here ΔG° is the free energy change of reaction (A).

We know that:

$$\Delta G° = \Delta H° - T\Delta S°$$
$$\text{or } \Delta G° = \Delta H° - T\Delta S° = -RT \ln K = -2.303\ RT \log K$$
$$\text{or } \Delta H° - T\Delta S° = -2.303\ RT \log K \quad (v)$$

If we put the values of standard enthalpies and standard entropies of various species in equation (*v*), we can get the value of log K from which the value of K can also be determined. With the help of appropriate data, the values of $K_1$, $K_2$, $K_3$ and $K_4$ can also be evaluated.

## Order of Stability of Complexes (I), (II), (III) and (IV)

log $K_1$, log $K_2$, log $K_3$ and log $K_4$ values for reactions (i), (iii) and (iv) have been found as: $\log K_1 = 4.3$, $\log K_2 = 3.6$, $\log K_3 = 3.0$, $\log K_4 = 2.3$. Obviously these values decrease as: $\log K_1 > \log K_2 > \log K_3 > \log K_4$. Thus the stepwise stability constants ($K_1$, $K_2$, $K_3$ and $K_4$) of the complexes (I), (II), (III) and (IV) are in the order: $K_1 > K_2 > K_3 > K_4$. This order shows that the stability of the complexes is in the order:

(I) > (II) > (III) > (IV).

The above decreasing order of stability is because of the fact that the number of sites that can be attacked by $NH_3$ ligand to form the complex decreases from 6 to 3, *i.e.* in reaction (*i*) there are *six* $H_2O$ ligands from which one $H_2O$ ligand can be replaced by $NH_3$ ligand while in reaction (*iv*) there are only *three* $H_2O$ ligands from which one $H_2O$ ligand can be replaced by $NH_3$. Thus the ease of formation of complexes (I), (II), (III) and (IV) decreases as (I) > (II) > (III) > (IV) and hence the stability of these complexes also decreases in the same order.

## Factors Affecting the Stability of Metal Ccomplexes

The stability (or stability constants) of metal complexes depends on the following factors:

### A. Properties of Central Metal Ion

The following properties of central metal ion affect the stability of the metal complexes:

**1. Size of central metal ion.** For a given ligand the stability (or stability constant) of the complexes of the metallic ions having the same charge on them decreases with the increase of the size of the central metal ion. Thus the stability of complexes given by the cations belonging to the same group and having the same charge decreases as we proceed from top to bottom in the group, since the size of the metallic cations increases in the same order. For example:

(*i*) The stability of hydroxide complexes given by alkali metal ions ($Li^+$, $Na^+$, etc.), alkaline earth metal ion ($Be^{2+}$, $Mg^{2+}$ etc.) and III B group ions ($Sc^{3+}$, $Y^{3+}$ and $La^{3+}$) is in the order:

(*a*) $Li^+$(r = 0.60 Å) > $Na^+$ ($r$ = 0.95 Å) > $K^+$($r$ = 1.33 Å) > $Rb^+$($r$ = 1.48 Å) > $Cs^+$($r$ = 1.69Å).

(*b*) $Be^{2+}$ ($r$ = 0.31 Å) > $Mg^{2+}$ ($r$ = 0.65 Å) > $Ca^{2+}$ ($r$ = 0.99 Å) > $Sr^{2+}$ ($r$ = 1.13 Å) > $Ba^{2+}$ ($r$ = 1.35 Å) > $Ra^{2+}$ ($r$ = 1.40 Å).

(*c*) $Sc^{3+}$ ($r$ = 0.81 Å) > $Y^{3+}$ ($r$ = 0.93 Å) > $La^{3+}$ ($r$ = 1.15 Å)

It may be noted that EDTA complex of $Mg^{2+}$ ($r$ = 0.65 Å) is less than that of $Ca^{2+}$ ($r$ = 0.99Å).

(*ii*) The inverse relation between the size of the central metal ion and the stability of the complexes formed is also confirmed when we see that the stability of the complexes of $Mn^{2+}$, $Fe^{2+}$, $Co^{2+}$, $Ni^{2+}$, $Cu^{2+}$ and $Zn^{2+}$ (These ions are *divalent* ions of the elements of 1st transition series).

| | | | | | | |
|---|---|---|---|---|---|---|
| *Ions:* | $Mn^{2+}$ | $Fe^{2+}$ | $Co^{2+}$ | $Ni^{2+}$ | $Cu^{2+}$ | $Zn^{2+}$ |
| *Ionic radii (Å):* | 0.91 | 0.83 | 0.82 | 0.78 | 0.69 | 0.74 |
| *Order of stability of complexes:* | $Mn^{2+}$ < | $Fe^{2+}$ < | $Co^{2+}$ < | $Ni^{2+}$ < | $Cu^{2+}$ > | $Zn^{2+}$ |

This sequence of stability is commonly known as **Irving-William order of stability** of complexes of $M^{2+}$ ions.

**2. Charge on the central metal ion.** For a given ligand, the stability of the complexes of the metallic ions having almost the same size but different charges on them decreases with the decrease of the charge on them. Thus the stability of complexes given by: (*a*) $La^{3+}$, $Sr^{2+}$ and $K^+$ ions. (*b*) $Co^{3+}$ and $Co^{2+}$ ions. (*c*) $Fe^{3+}$ and $Fe^{2+}$ ions, and (*d*) $Th^{4+}$, $Y^{3+}$, $Ca^{2+}$ and $Na^+$ ions with the same ligand is in the order:

(*a*) $La^{3+}$ ($r$ = 1.15 Å) > $Sr^{2+}$ (= 1.13 Å) > $K^+$ (= 1.33 Å)

(*b*) $Co^{3+}$ ($r$ = 0.63 Å) > $Co^{2+}$ (= 0.74 Å)

(*c*) $Fe^{3+}$ ($r$ = 0.64 Å) > $Fe^{2+}$ ($r$ = 0.76 Å)

(*d*) $Th^{4+}$ ($r$ = 0.95 Å) > $Y^{3+}$ (= 0.93 Å) > $Ca^{2+}$ (= 0.99 Å) > $Na^+$ (= 0.95 Å).

If the factors 1 and 2 mentioned above are combined, then we can say that with the increase of ionic potential of the central metal ion (Ionic potential of the metal ion = charge on the metal ion/size of the ion), the stability of the complexes with the same ligand also increases. For example, the stability of hydroxide complexes of $Li^+$, $Ca^{2+}$, $Ni^{2+}$, .... $Be^{2+}$ ions whose ionic potential increase from $Li^+$ to $Be^{2+}$ ions also increases in the same direction as shown below:

| *Central metal ion (charge on the ion is shown in parenthesis* | *Ionic radius (Å)* | *Ionic potential* | *Order of stability of complexes* |
|---|---|---|---|
| $Li^+$ (+1) | 0.60 | 1/0.60 = 1.6 ↓ Increasing | ↓ Increasing |
| $Ca^{2+}$ (+2) | 0.99 | 2/0.99 = 2.0 | |
| $Ni^{2+}$ (+2) | 0.72 | 2/0.72 = 2.97 | |
| $Y^{3+}$ (+3) | 0.93 | 3/0.93 = 3.22 | |
| $Th^{4+}$ (+4) | 0.95 | 4/0.95 = 4.20 | |
| $Al^{3+}$ (+3) | 0.50 | 3/0.50 = 6.0 | |
| $Be^{2+}$ (+2) | 0.31 | 2/0.31 = 6.45 | |

The effect of the size and charge on the central metal ion on the stability of a complex as discussed above shows that greater is the charge and smaller is the size of the metal ion, i.e. larger is the charge/radius ratio of a metal ion, greater is the stability of its complex.

**Example.** (*i*) Since $Fe^{3+}$ ion carries higher charge (= + 3) and has smaller size than $Fe^{2+}$ ion, $Fe^{3+}$ ion has larger positive charge density than $Fe^{2+}$ ion. Hence complexes of $Fe^{3+}$ ion are more stable than those of $Fe^{2+}$ ion, provided that the complexes are formed with the same ligands. As a result $(Fe^{3+}(CN)_6]^{3-}$ has higher value of its stability constant ($K = 10^{31}$) than $[Fe^{2+}(CN)_6]^{4-}$ ($K = 10^6$) · $[Co^{3+}(CN)_6]^{3-}$ in aqueous medium has higher value of its stability constant (= $10^{64}$) than $[Co^{2+}(CN)_6]^{4-}$ (= $10^{20}$) · $Co^{3+}$ ion is smaller in size than $Co^{2+}$.

(*ii*) Stability of the complexes of $Mn^{2+}$, $Fe^{2+}$, $Co^{2+}$, $Ni^{2+}$ and $Cu^{2+}$ ions (all these ions have the same charge which is equal to +2) increases from $Mn^{2+}$ to $Cu^{2+}$. This increase of stability is because of the decrease in ionic radii (size) of these ions from $Mn^{2+}$ to $Cu^{2+}$.

| | | | | | |
|---|---|---|---|---|---|
| *$M^{2+}$ ions:* | $Mn^{2+}$ | $Fe^{2+}$ | $Co^{2+}$ | $Ni^{2+}$ | $Cu^{2+}$ |
| *Ionic radii (Å):* | 0.91 | 0.83 | 0.82 | 0.78 | 0.69 |
| | | | Decreasing → | | |
| *Stability of complexes* | | | Increasing → | | |

The order of stability of the complexes shows that complexes of $Mn^{2+}$ are the least stable and those of $Cu^{2+}$ are the most stable.

(iii) Stability constants of the complexes of trivalent lanthanide ions in aqueous medium also increases as the atomic number of lanthanides increases.

**3. Electronegativity.** Electronegativity of the central ion also influences the stability of its complexes. This is because the bonding between a central ion and a ligand is due to the donation of electron pairs by the ligands. Hence, a strongly electron-attracting central ion will give stable complexes. To sum up: The greater the positive charge density (*i.e.*, charge/size ratio) and the greater the electronegativity of the central ion, the greater is the stability of the complex formed by it.

**4. Metal ions acting as hard acids and soft acids.** Metal ions that are *hard acids* form more stable complexes with ligands containing coordinating N, O, F, etc., atoms and form comparatively less stable complexes with ligands containing coordinating P, S, Cl, etc., atoms. The metal ions that are *soft acids* form more stable complexes with ligands containing coordinating P, S, Cl, etc. atoms and comparatively less stable complexes with ligands containing coordinating N, O, F, etc., atoms.

**5. Electrode potentials of metal ions.** Metal ions having large negative electrode potentials (($E°_{M^{n+}/M}$), such as $Li^+$, $Ba^{2+}$, $Mg^{2+}$, $Al^{2+}$, etc. have a lesser tendency to attract electrons and hence have more tendency to form complexes with ligands having highly electronegative atoms such as N, O, F, etc., so that the negative charge of the ligands remains mostly centred on the ligands. On the other hand, metal ions having large positive electrode potentials ($E°_{M^{n+}/M}$) such as $Pd^{2+}$, $Pt^{2+}$, etc., have greater tendency to accept electrons and thus form more stable complexes with ligands containing highly polarizable coordinating atoms such as P, S, Cl, etc. In such complexes, the electron charge of the ligand is easily polarised towards the complexed metal ion having positive electrode potential.

**6. Class *a* and class *b* acceptor metals.** Chatt and Ahrland have classified the metals into three categories: *a*, *b* and borderline, on the basis of their electron-acceptor properties. This classification in shown below (normal valence states are assumed).

(*a*) *Class a metals:* H, the alkali and alkaline earth metals, the element Sc → Cr, Al → Cl, Zn → Br, In, Sn, Sb and I, the lanthanides and actinides

(*b*) *Class b metals:* Rh, Pd, Ag, Ir. Pt, Au, Hg

(*c*) *Borderline metals:* The elements Mn → Cu, Ti → Po, Mo, Te, Ru, W, Re, Os, Cd.

Class *a* metals form more stable complexes with ligands having the coordinating atoms from the second period elements (*e.g.*, N, O, F) than those of an analogous ligand in which the donor atom is from third or later period (e.g., P, S, Cl). Class b metals have the relative stabilities reversed. If the ligand contains the heavier donor atoms, class *a* and *b* metals are characterised by the stability order:

$$\left.\begin{array}{l} F^- > Cl^- > Br^- > I^- \\ O >> S > Se > Te \\ N >> P > As > Sb > Bi \end{array}\right\} \text{Class } a \text{ metals}$$

$$\left.\begin{array}{l} F^- < Cl^- < Br^- < I^- \\ O << S \approx Se \approx Te \\ N << P << As < Sb < Bi \end{array}\right\} \text{Class } b \text{ metals}$$

Class *b* metals are characterised by the presence of a number of *d*-electrons beyond an inert gas core. These *d*-electrons are used to form π-bond with ligand atoms. It is believed that the stability of the complexes of class *b* metals results from covalent contribution to metal-ligand bonds and form the transfer of electron density from the metal to the ligand via π-bonding. The most stable complexes of class *b* metals are formed with ligands like $PMe_3$, $S^{2-}$ and $I^-$ which have vacant *d*-orbitals or like CO, $CN^-$ which have vacant molecular orbitals of low energy.

For *borderline metals* the stability constants do not display either class *a* or class *b* behaviour uniquely.

## B. Properties of the Ligand

The following properties of the ligand affect the stability of the metal complexes.

**1. Size and charge of ligand.** If a ligand is smaller, it can approach the metal ion more closely forming a stable bond. Similarly, a highly charged ligand would also form a strong bond with metal. Thus the high charge and small size of a ligand leads to the formation of stable complexes. For example the stability of the complexes of a given metal ion with halide ion used as ligands is in the order: $F^- > Cl > Br^- > I^-$. This order is applicable for class *a* metals. When class *b* metals (*e.g.* Pd, Ag, Pt, Hg etc.) are used, the above order of stability is reversed, i.e., for class *b* metals the order is : $F^- < Cl^- < Br^- < I^-$.

**2. Dipole moment of ligands.** For neutral ligands, the larger the magnitude of permanent dipole moment, the greater is the stability of the complexes. For example, the order of stability of complexes formed by some neutral ligands is as : ammonia > ethylamine > diethylamine > triethylamine.

**3. Basic character of ligands.** The more basic is the ligand, more easily it can denote electron pairs to the central ion and hence more easily it can form complexes of greater stability. The ligand that bind $H^+$ firmly form stable complexes with metal ions. Thus $F^-$ should form more stable complexes than $Cl^-$, $Br^-$ or $I^-$, and $NH_3$ should be better ligand than $H_2O$ which in turn should be better than HF. ($NH_3 > H_2O > HF$). This behaviour is observed for alkali, alkaline earth and other electropositive metals like first row transition elements, lanthanides and actinides.

**4. π-bonding capacity of ligands.** The ligands like $CN^-$, CO, $PR_3$, $AsR_3$, $SR_2$, alkenes, alkynes which are capable of forming π-bonds with transition metal ions give more stable complexes.

**5. Steric hindrance due to bulky ligands.** When a bulky group is either attached to or is present near a donor atom of a ligand, repulsion between the donor atom of the ligand and the bulky group is produced and this mutual repulsion weakens the metal-ligand bonding and hence makes the complex less stable.

**Examples.** (*i*) The complex of $Ni^{2+}$ ion with 2-methyl-8-hydroxy quinoline ($\log_{10} \beta = 17.8$) is less stable than that with 8-hydroxy quinoline ($\log_{10} \beta = 17.8$). The effect of the presence of bulky group on the stability of a complex is commonly called *steric hindrance*.

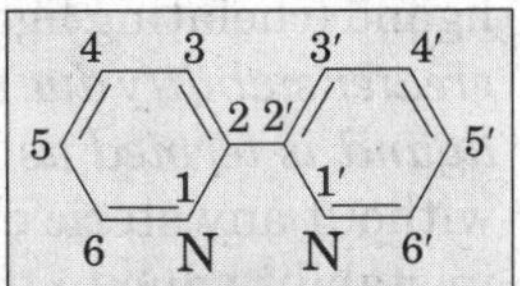

Structure of 2, 2′-bipyridine

(*ii*) 2, 2'-bipyridine (also called 2, 2'-dipyridyl) forms complexes with metal ions which are stable but the substitution of an alkyl group in 4, 4' or 5, 5' positions gives complexes which are less stable, since the substituents crowd the metal ion. Substituents in 3, 3' positions prevent the pyridine rings from lying in the same plane and consequently the complexes formed are of lower stability.

**Strain due to large ligands.** The strain in the complexes with large ligands is sometimes due to the geometry of the ligand coupled with stereochemistry of the complexes. For example triethylene tetramine (*trien*) $H_2N \cdot CH_2 \cdot CH_2 \cdot NH \cdot CH_2 \cdot CH_2NH\ CH_2 \cdot CH_2 \cdot NH_2$ can coordinate through its four nitrogen atoms at the corners of the square, but triamino triethylamine (*tren*), $(NH_2CH_2CH_2)_3N$ cannot. Hence the former forms more stable complexes with $Cu^{2+}$ than the latter because the former is straight chain amine while the latter is a branched chain amine, which is unable to assume the preferred square planar geometry.

**6. Forced configuration of ligands.** Ligands like porphyrin and phythalcyanine which have completely fused planar ring system, form extraordinarily stable complexes with metal ions that tend to give planar complexes *e.g.*, $Cu^{2+}$ complex with phthalocyanine is very stable. Similarly trien forms very stable complexes with Ni(II) and Cu(II). These ligands impose planar configurations even on metal ions that have no tendency to form planar complexes with unidentate ligands. For example, $Be^{2+}$ and $Zn^{2+}$ ion normally form tetrahedral complexes but when they combine with these polydentate ligands they are forced to assume planar configuration. Therefore these complexes are less stable. Fig. 26.1 shows the structure of Cu (II) pthalocyanine complex.

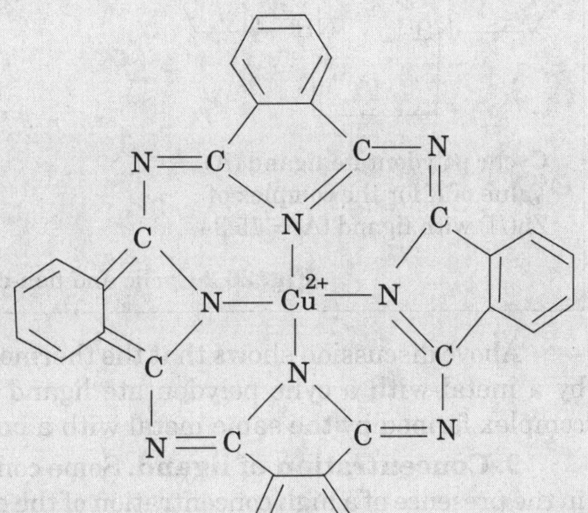

**Fig. 26.1** Copper (II) pthalocyanine complex

**7. Presence of chelate rings: Chelate effect.** This factor has been discussed under *Chelate Effect* discussed on the subsequent pages in this chapter.

**8. Macrocyclic ligands and macrocyclic effect** A macrocyclic ligand is a nine or more membered cyclic molecule having 3 or more potential donor atoms which can bind a metal atom inside the cavity of the macrocycle. Many synthetic macrocyclic ligands have only N donor atoms. These are also synthetic macrocyclic ligands which contain mixed (N, O), (N, S), (N, O, S), (N, O, P) etc. donor atom. Some macrocylic ligands have conjugated $\pi$ (pi) system.

It has been observed that the stability of a complex of a particular metal ion with macrocylic ligand is several times greater than that of an open-ended multidentate ligand (chelating ligand) containing an equal number of equivalent donor atoms. *The greater stability due to a macrocyclic ligand compared to similar open-ended chelating ligand is termed as macrocyclic effect*. Thus if ligands are multidentate and cyclic without any steric effects, the stability of the complexes is increased. The increase in stability due to the presence of multidentate cyclic ligands is called macrocyclic effect. The enhanced stability *due to macrocyclic ligand indicates greater chelate effect*.

**Examples** (*i*) The cyclic crown polyether complexes are far more stable than those of their corresponding open-chain analogues.

(*ii*) The values of K for the complexes of Zn (II) and Ni (II) with ligand (A) are higher than those for the complexes of the same metals with ligand (B). (See Fig 26.2)

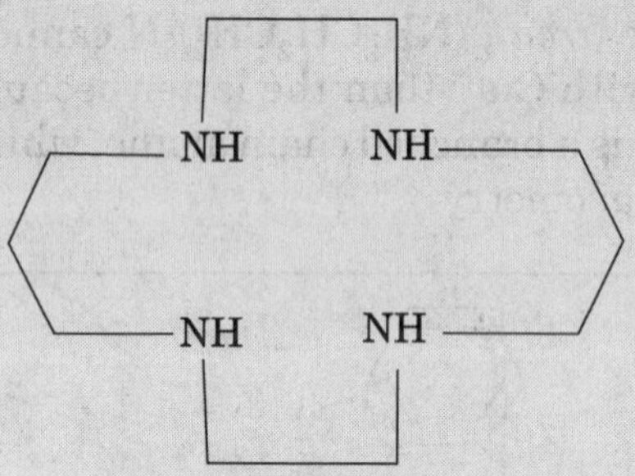


Cyclic polydentate ligand (A)
Value of $K$ for the complex of
Zn(II) with ligand (A) = 15.34

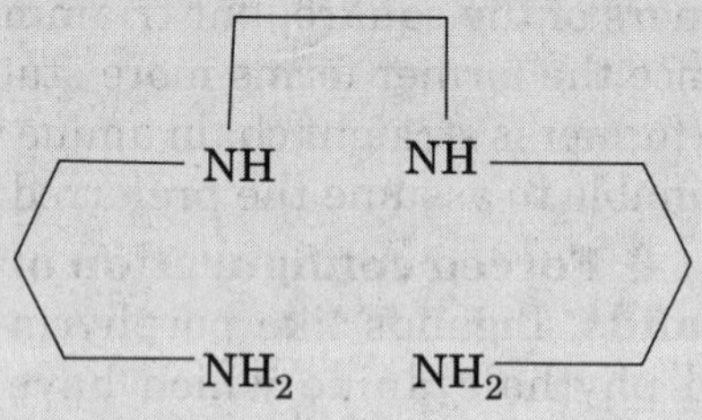


Non-cyclic polydentdate ligand (B)
Value of $K$ for the complex of
Zn(II) with ligand (B) = 11.25

**Fig. 26.2** Cyclic and non-cyclic polydentate ligands

Above discussion shows that the thermodynamic stability of a complex formed by a metal with a cylic polydentate ligand [e.g. (A)] is greater than that of the complex formed by the same metal with a non-cyclic polydentate ligand [e.g. (B)].

**9. Concentration of ligand.** Some complexes exist in aqueous solution only in the presence of a high concentration of the coordinating group (ligand). Otherwise in such cases, water molecules apparently show greater coordinating tendency than the groups which are originally present. For example, cobaltous ion ($Co^{2+}$) in presence of a high concentration of thiocyanate ions ($SCN^-$) forms stable blue complex ion, $[Co(SCN)_4]^{2-}$. But on dilution with water, the blue complex gets destroyed and is replaced by a pink hydrated complex, $[Co(H_2O)_6]^{2+}$. On further addition of $SCN^-$ ions, pink colour disappears and the original blue colour reappears. These changes indicate the competition between water molecules and thiocyanate ions to coordinate with cobaltous ion. The equilibrium reactions can be represented as follows.

$$\underset{\text{blue}}{[Co(SCN)_4]^{2-}} + 6H_2O \rightleftharpoons \underset{\text{pink}}{[Co(H_2O)_6]^{2+}} + 4SCN^-$$

Similarly,

$$\underset{\text{reddish brown}}{[Fe(SCN)_6]^{3-}} + 6F^- \rightleftharpoons \underset{\text{colourless}}{[FeF_6]^{3-}} + 6SCN^-$$

$$\underset{\text{blue}}{[Cu(H_2O)_6]^{2+}} + 4Cl^- \rightleftharpoons \underset{\text{yellow}}{[CuCl_4]^{2-}} + 6H_2O$$

## C. Amount of Metal-Ligand Covalent Characters Present in Complexes.

In some complexes the stability of the complex has been found to be influenced by the amount of metal-ligand covalent character present in the complex. This is more pronounced in complexes of the metals like those of copper and zinc family, Sb, Pb. For example the stability of $[AgX_2]^-$ and $[AgX_3]^-$ are found to be in the following order

$$AgI_2^- > AgBr_2^- > AgCl_2^- >> AgF_2^-$$
$$AgI_3^- > AgBr_3^- > AgCl_3^- \geq AgF_3^-$$

This can be due to the increase in the covalent character of Ag–X bond as we move from Ag – F to Ag –I.

## Irving William Series of Stability Constants

log K values of HS octahedral oxalato complexes of bivalent metal ions *viz* $Ba^{2+}$, $Sr^{2+}$, $Ca^{2+}$, $Mg^{2+}$, $Mn^{2+}$, $Fe^{2+}$, $Co^{2+}$, $Mi^{2+}$, $Cu^{2+}$ and $Zn^{2+}$ have been found in the following order:

$$Ba^{2+} < Sr^{2+} < Ca^{2+} < Mg^{2+} < \vdots\ Mn^{2+} < Fe^{2+} < Co^{2+} < Ni^{2+} < Cu^{2+} > \vdots\ Zn^{2+}$$

The above order of log K values (or stability constants) is called *Irving William series of stability constants.*

From this order of log K values the following points may be noted:

(*i*) log K values for the 1:1 oxalate and 1:1 glycinato complexes of $Ba^{2+}$, $Sr^{2+}$, $Ca^{2+}$ and $Mg^{2+}$ ions increase regularly from $Ba^{2+}$ to $Mg^{2+}$. This increasing order is because of the decreasing of ionic radii of these ions from $Ba^{2+}$ to $Mg^{2+}$ as shown below. log K values for the complexes are also given.

| | | | | |
|---|---|---|---|---|
| *$M^{2+}$ ions:* | $Ba^{2+}$ | $Sr^{2+}$ | $Ca^{2+}$ | $Mg^{2+}$ |
| *Ionic radii of $M^{2+}$ ions (Å):* | 1.29 | 1.00 | 0.99 | 0.65 |
| | Decreasing → | | | |
| *log K values for 1:1 oxalato complexes:* | 2.31 | 2.54 | 3.00 | 3.43 |
| | Increasing → | | | |
| *log K values for 1:1 glycinato complexes:* | 0.77 | 0.91 | 1.43 | 3.44 |
| | Increasing → | | | |

(*ii*) log K values increase from $Mn^{2+}$ to $Cu^{2+}$. The increase from $Mn^{2+}$ to $Ni^{2+}$ is because the CFSE values (in Dq) increase from $Mn^{2+}$ to $Ni^{2+}$ as shown below:

| | | | | |
|---|---|---|---|---|
| *$M^{2+}$ ions:* | $Mn^{2+}$ | $Fe^{2+}$ | $Co^{2+}$ | $Ni^{2+}$ |
| *$d^x$ configuration of $M^{2+}$ ions:* | $d^5$ | $d^6$ | $d^7$ | $d^8$ |
| *$t_{2g}^a\ e_g^b$ configuration of $d^x$ ion in HS octahedral complexes:* | $t_{2g}^3\ e_g^2$ | $t_{2g}^4\ e_g^2$ | $t_{2g}^5\ e_g^2$ | $t_{2g}^6\ e_g^2$ |
| *CFSE values (Dq):* | 0 | –4 | –8 | –12 |
| | Increasing → | | | |

(*iii*) CFSE value of $Cu^{2+}$ ion is equal to –6 Dq. Although this value is lower than that for $Ni^{2+}$ ion (= –12 Dq), yet complexes of $Cu^{2+}$ ion have higher K values. Higher K values for the complexes of $Cu^{2+}$ ion are due to the stablisation of this ion due to Jahn-Teller effect. $Cu^{2+}$ ion has a tendency to form distorted octahedral complexes.

(*iv*) Since CFSE value of $Zn^{2+}$ ion is zero, complexes of this ion have the lowest K values, i.e. complexes of $Zn^{2+}$ ion are least stable.

## Chelate Effect

### What is Chelate Effect?

Chelated complexes (complexes containing 5- or 6-membered rings including the metal atom) are more stable than the non-chelated complexes. Greater stability of chelated complexes is called **chelate effect**. This effect is found to be maximum in complexes having 5- and 6-membered rings. With the increase in the number of rings present in the structure of the complex, the stability (or stability constant) of the complex also increases.

### Sidgwich's Explanation of Chelate Effect.

We know that a bidentate chelated complex has two metal - ligand bonds. If one of these two metal - ligand bonds is broken, the remaining metal-ligand bond still holds the molecule in place and hence the broken bond can be reestablished. But when the bond is broken, the ligand bound by a single metal-ligand bond drifts away. Thus a chelated complex dissociates less than a non-chelated complex and hence has a higher value for its stability constant.

### Explanation of Chelate Effect on the Basis of Entropy Change ($\Delta S°$)

The chelate effect is mainly due to entropy change taking place in the reaction which occurs due to an increase in the number of product species compared to the number of reactant species.

The free energy change ($\Delta G°$), enthalpy change ($\Delta H°$) and entropy change ($\Delta S°$) in a reaction in standard state are related as:

$$\Delta G° = \Delta H° - T\Delta S° \qquad (i)$$

Stability constant (K) and free energy change ($\Delta G°$) of a reaction are related as:

$$\Delta G° = -RT \ln K \qquad (ii)$$

Equation (*i*) shows that more negative value of $\Delta G°$ will give more positive value of $\Delta S°$. Equation (ii) shows that lower is the value of $\Delta G°$, higher is the value of K. In other words negative value of $\Delta G°$ for a complex indicates that the complex is stable.

On combining equations (*i*) and (*ii*) we get:

$$-RT \ln K = \Delta H° - T\Delta S° \qquad (iii)$$

Enthalpy change ($\Delta H°$) between the chelated and non-chelated complexes is small due to similar metal-donor atom linkages. Thus from relation (*iii*) it follows that more positive is the value of $\Delta S°$, more positive will be the value of stability constant (K) and hence more will be the stability of the complex.

Now let us consider the formation of $[Cd(CH_3NH_2)_4]^{2+}$ and $[Cd(en)_2]^{2+}$ by the displacement of monodentate $H_2O$ ligands present in hydrated cadminium ion, $[Cd(H_2O)_4]^{2+}$

(*i*) $\underbrace{[Cd(H_2O)_4]}_{1} + \underbrace{4CH_3NH_2}_{4} \rightleftharpoons \underbrace{[Cd(CH_3NH_2)_4]^{2+}}_{1} + \underbrace{4H_2O}_{4}$

Reactants: 1 + 4 = 5; $[Cd(CH_3NH_2)_4]^{2+}$: Non-chelated complex; Products: 1 + 4 = 5

(*ii*) $\underbrace{[Cd(H_2O)_4]^{2+}}_{1} + \underbrace{2en}_{2} \rightleftharpoons \underbrace{[Cd(en)_2)]^{2+}}_{1} + \underbrace{4H_2O}_{4}$

Reactants: 1 + 2 = 3; $[Cd(en)_2]^{2+}$: Chelated complex; Products: 1 + 4 = 5

In the above reactions $CH_3NH_2$ and *en* (ethylene diamine) are *monodentate* and *bidentate ligands* respectively. The stability constant of non-chelated complex, $[Cd(CH_3NH_2)_4]^{2+}$ is lower ($K = 3.10 \times 10^6$) then that of chelated complex, $[Cd(en)_2]^{2+}$ ($K = 3.2 \times 10^{10}$). The higher value of K for the chelated complex is due to chelate effect which can be explained on the basis of entropy change taking place in the formation of chelated complex.

It may be seen that in the formation of non-chelated complex, $[Cd(CH_3NH_2)_4]^{2+}$ [reaction (*i*)], there is no change in the number of reactant species and product species. The number of both the species is 5. As a consequence, there is no change in the extent of randomness in going from reactants to the products, *i.e.* $\Delta S°$ for reaction (i) is nearly zero.

In the formation of chelated complex, $[Cd(en)_2]^{2+}$, the number of product species (= 1 + 4 = 5) is more than that of the reactant species (= 1 + 2 = 3). Due to greater number of product species, these species have greater randomness than the reactant species. Thus the formation of chelated complex involves an increase in the disorder or entropy of reaction (*ii*). Due to the increase in the entropy, $\Delta S°$ is positive for reaction (*ii*). Hence the stability constant (K) of the chelated complex is higher.

Now let us consider the displacement of $4H_2O$ molecules (monodentate ligands) of $[Cd(H_2O)_4]^{2+}$ by triethylene tetramine, *trien* (tetradentate ligand) to form the chelated complex, $[Cd(trien)]^{2+}$.

$$\underbrace{[Cd(H_2O)_4]^{2+}}_{1} + \underbrace{\text{trien}}_{1} \rightleftharpoons \underbrace{[Cd(trien)]^{2+}}_{1} + \underbrace{4H_2O}_{4}$$

Reactants: 1 + 1 = 2; $[Cd(trien)]^{2+}$: Chelated complex; Products: 1 + 4 = 5

Due to greater number of product species (= 1 + 4 = 5), $\Delta S°$ would have larger positive value. Thus $[Cd(trien)]^{2+}$ is more stable than $[Cd(en)_2]^{2+}$. It follows from the above discussion that:

*Larger is the number of coordination centres on a ligand, greater is the stability of the chelated complex formed by the ligand.*

Above discussion shows that if we compare log K values for the complexes of $M(NH_3)_4(H_2O)_2$, $M(en)_2(H_2O)_2$ and M (trien) $(H_2O)_2$ type formed by $Fe^{2+}$, $Co^{2+}$, $Ni^{2+}$, $Cu^{2+}$, we find that logK values of these complexes of each $M^{2+}$ metal ion increases as:

$M(NH_3)_4(H_2O)_2$ < $M(en)_2(H_2O)_2$ < M (trien) $(H_2O)_2$ (See Table 26.1 )

**Table 26.1 log K values for the complexes of $M^{2+}$ ion ($M^{2+}$ = $Fe^{2+}$, $Co^{2+}$, $Ni^{2+}$ and $Cu^{2+}$)**

| **Complex ion** | **log K values for $M^{2+}$ complexes** | | | |
|---|---|---|---|---|
| | **$Fe^{2+}$** | **$Co^{2+}$** | **$Ni^{2+}$** | **$Cu^{2+}$** |
| $M(NH_3)_4(H_2O)_2$ | 3.7 | 5.3 | 7.8 | 12.6 |
| $M(en)_2(H_2O)_2$ | 7.7 | 10.9 | 14.5 | 20.2 |
| M (trien) $(H_2O)_2$ | 8.8 | 12.8 | 14.7 | 20.6 |

**Explanation.** The above increasing order can be explained as follows:

We know that $NH_3$ is a monodentate ligand, *en* is a bidentate ligand and *trien* is a tetradentate ligand. Hence $M(NH_3)_4(H_2O)_2$ has *no rings* in its structure, M $(en)_2(H_2O)_2$ has *one 5-membered ring* while M(trien) $(H_2O)_2$ has *three 5-membered rings*. With the increase in the number of rings or with the increase in the number of coordination centres on the ligands, the stability of complexes increases as shown above. With the increase of the stability, log K values also increase in the same order.

## Factors Affecting Chelate Effect.

The main factors influencing the magnitude of chelate effect are:

(*i*) Chelate ring size (*ii*) Number of donors, *i.e.* polydentate nature of ligand (*iii*) Flexibility of chelating ligand, and (*iv*) Charge in the ligand.

In general, three-membered or four membered chelate rings at the metal, such as peroxide linkage or a carbonate linkage, are not favoured. Five-membered rings by *en*, *bpy* or six-membered rings by *Hbig, acac anion* are quite favoured. Seven-membered rings are disfavoured. Formation constant of a particular metal ion (*e.g.* $Cu^{2+}$) with oxalic, malonic and succinic acids obey the order:

| | oxalic acid | > | malonic acid | > | succinic acid |
|---|---|---|---|---|---|
| | $^-O–CO–CO–O^-$ | | $^-O–CO–CH_2–CO–O^-$ | | $^-O–CO–(CH_2–CH_2)–CO–O^-$ |
| log K for 1 : 1 | 6.1 | | 5.7 | | 3.3 |
| $Cu^{2+}$ complex | (5-ring) | | (6-ring) | | (7-ring) |

o-phenanthoroline complexes register higher log *K* values compared to those of more flexible bipyridine. $R_2P–CH=CH–PR_2$ form more stable complexes than the more flexible $R_2P–CH_2–CH_2–PR_2$.

Uncharged ligands, such as en, trien form more stable chelates than charged ligands like glycinate and oxalate.

Ligands with increasing number of donor sites, such as en, dien, trien show increasing chelate effect.

The larger the number of the chelate rings in a complex, the greater is its stability. This is illustrated in Table 26.2. The abbreviations used in this table have the following meaning:

en = ethylene diamine, $NH_2CH_2CH_2NH_2$
trien = triethylene tetramine, $NH_2(CH_2)_2NH(CH_2)_2NH(CH_2)_2NH_2$
tren = triamino triethylamine, $(NH_2CH_2CH_2)_3N$
dien = diethylene triamine, $NH_2CH_2CH_2NHCH_2CH_2NH_2$
penten = tetrakis (aminoethyl) ethylene diamine,
$(NH_2CH_2CH_2)_2NCH_2CH_2N(CH_2CH_2NH_2)_2$
M = Metal.

Table 26.2 clearly shows that the greater the number of chelate rings in the complex, the greater is the stability of the complex.

The stability of the complexes also depends on the number of the atoms present in the ring. The chelates containing 3-membered ring including the metal are very unstable and hence are unknown. 4-membered chelate rings are rare and occur in carbonate (I), nitrate (II) and sulphate (III) chelates.

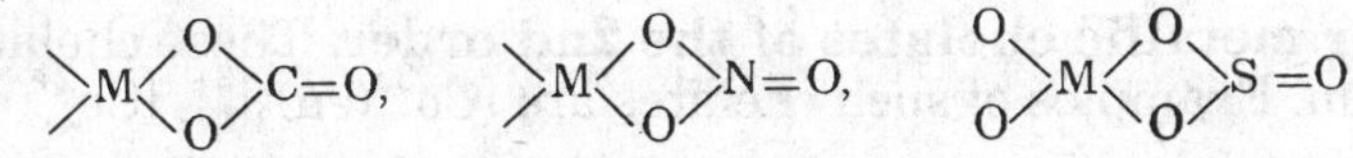

(I) (II) (III)

The most common and the most stable chelates are 5- and 6-membered chelates. 5-membered chelates are frequently more stable than the 6-membered chelates when the atoms in the ring are joined by single bonds only. On the other hand 6 membered chelates are more stable than 5-membered chelates of heterocyclic ligands or of ligands involving conjugation in the chelate ring. For example, acetylacetone complexes of the metal ions which contain 6-membered conjugate chelate rings are more stable than the corresponding ethylene diamine complexes which contain only 5-membered chelate rings. In acetylacetone complexes the π-electron density gets spread up over the whole ring which increases the stability of this 6-membered ring system. The possibility of the resonance in its chelate ion is an additional factor contributing to the stability of acetylacetone complexes.

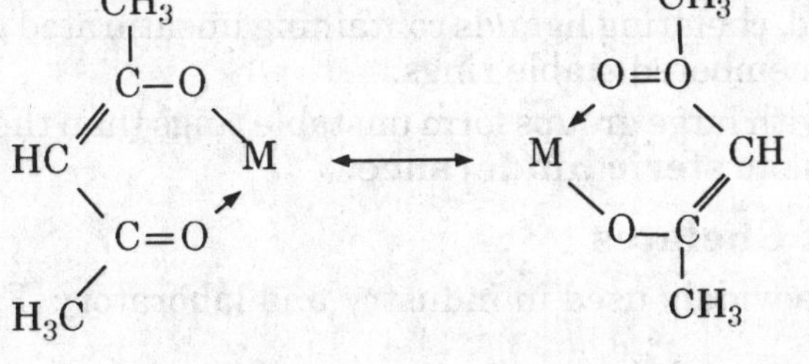

**Table 26.2 Overall stability constant values (i.e. $\log_{10} \beta$ values) for some ammonia and polyamine complexes illustrating the effect of chelate ring size on the stability of the complexes.**

| Metal complexes | No. of chelate rings | Values of stability constants ($\log_{10} \beta$ values) | | | | | | |
|---|---|---|---|---|---|---|---|---|
| | | $Mn^{2+}$ | $Fe^{2+}$ | $Co^{2+}$ | $Ni^{2+}$ | $Cu^{2+}$ | $Zn^{2+}$ | $Cd^{2+}$ |
| M $(NH_3)_4$ | 0 | — | 23.7 | 5.31 | 7.79 | 12.59 | 9.06 | 6.92 |
| M $(en)_2$ | 2 | 4.9 | 7.7 | 10.9 | 14.5 | 20.2 | 11.2 | 10.3 |
| M (trien) | 3 | 4.9 | 7.8 | 11.0 | 14.1 | 20.5 | 12.1 | 10.0 |
| M (tren) | 3 | 5.8 | 8.8 | 12.8 | 14.0 | 18.8 | 14.6 | 12.3 |
| M $(dien)_2$ | 4 | 7.0 | 10.4 | 14.1 | 18.9 | 21.3 | 14.4 | 13.8 |
| M (penten) | 5 | 9.4 | 11.2 | 15.8 | 19.3 | 22.4 | 16.2 | 16.2 |

## Clasification of Chelates

Chelates are of two types:

(*i*) **Inner metallic chelates of the 1st order.** These chelates do not carry any charge on them ( i.e. non-electrolytes). Examples are $[Cr^{3+}(gly)_3]^{\circ}$, $[Co^{3+}(acac)_3]^{\circ}$, $[Cu^{2+}(big)_2]^{\circ}$ etc.

(*ii*) **Inner metallic chelates of the 2nd order.** These chelates carry some charge on them. Examples of such chelates are $[Co^{3+}(en)_3]^{3+}$, $[Co^{2+}(acac)_3]^{-}$ etc.

## Properties of Inner Metallic Complexes

(*i*) Neutral inner metallic complexes are usually insoluble in water but dissolve in organic solvents immiscible with water. This property has been used for the separation of metals by solvents extraction method.

(*ii*) The formation of a chelated complex may lead to the change in colour. This property permits colourimetric estimation.

(*iii*) Most of the chelated complexes have low melting points and are volatile.

## Some Important Characteristics of Chelates

These are as follows:

(*i*) Chelating ligands form more stable complexes than the monodentate analogs. This is called **chelating effect.**

(*ii*) The chelated complexes containing 5 and 6 membered rings are comparatively more stable. In general chelating ligands which do not contain unsaturated groups (double bonds) form five membered stable rings. On the other hand, chelating ligands containing unsaturated groups (double bonds) form six membered stable rings.

(*ii*) Ligands with large groups form unstable rings than the ligands with smaller groups due to **steric hinderance.**

## Importance of Chelates

Chelates are widely used in industry and laboratory. For example these are used:

(*i*) in the softerning of hard water

(*ii*) in the separation of lanthanides and actinides

(*iii*) in the detection of some metals in qualitative analysis
(*iv*) in the estimation of nickel (II), magnesium (II) and copper (II) ions quantitatively.

## Methods for the Determination of Stability Constants and Composition of a Complex

In this section, a brief discussion of various experimental methods used for the determination of stability constants of complexes will be taken up.

**1. Spectrophotometric Method**

In many cases, there is *change in absorbance of system on complex formation.* This change in absorbance has been used to determine the composition and stability of metal complexes. Most of the spectrophotometric methods are not of general applicability because of various limitations.

The relationship between the absorbance (or optical density), A at a particular wavelength and concentration is given by **Beer's law**, which can be mathematically stated as

$$A = \varepsilon . l . c.$$

where $\varepsilon$ = mole extinction coefficient
$l$ = length of absorption cell
$c$ = concentration of the complex (in moles per litre)

Thus measuring absorbance [A] with the help of a spectrophotometer and knowing the extinction coefficient ($\varepsilon$) at that wavelength and the cell length ($l$), the concentration ($c$) can be calculated with the help of Beer's law.

In Fig. 26.3 a graph is drawn between the wave-length and molar absorptivity of a representative metal in [$M^{2+}$] which is shown by broken lines and its complex ion $[ML]^{2+}$ which is shown by dark lines. From the graph it is quite clear that the

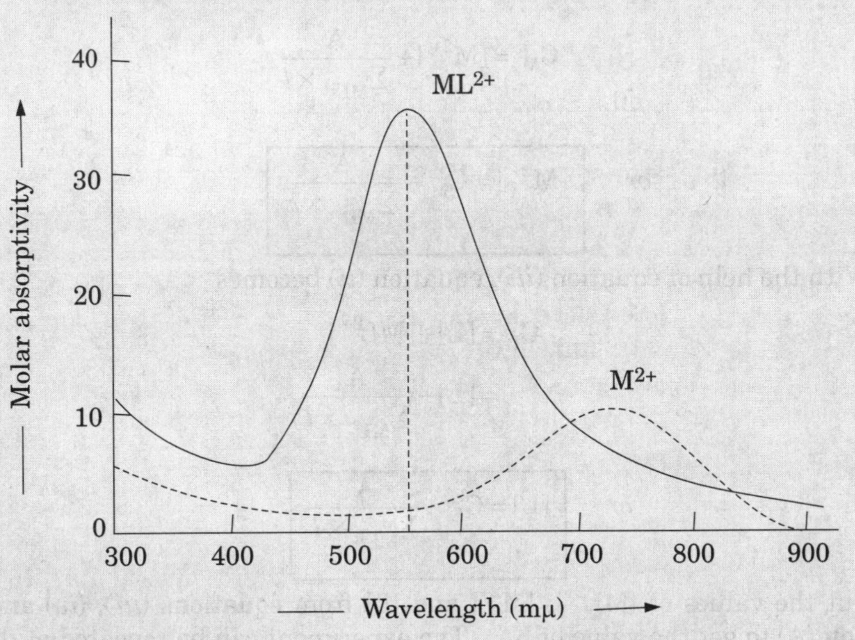


**Fig. 26.3** Spectrum of a metal ion, $M^{2+}$ and its complex, $ML^{2+}$

absorption by the complex occurs over the entire region of the metal ion absorption but at 550 mμ only the complex absorbs.

The formation constant ($K_{for}$) for the reaction:

$$M^{2+} + L \xrightleftharpoons{K_{for}} ML^{2+}$$

is given by:

$$K_{for} = \frac{[ML^{2+}]}{[M^{2+}][L]} \qquad (A)$$

If the values of $[ML^{2+}]$, $[M^{2+}]$ and [L] are put in the above equation, we can get the value of $K_{for}$. The values of $[ML^{2+}]$, $[M^{2+}]$ and [L] can be determined as follows:

If $C_M$ and $C_L$ are the total concentration of the metal ion and ligand respectively, then:

$$C_M = [M^{2+}] + [ML^{2+}] \qquad (i)$$

$$C_L = [L] + [ML^{2+}] \qquad (ii)$$

We know that:

$$A = \Sigma_{ML^{2+}} \times l \times [ML^{2+}]$$

$$\text{or} \qquad [ML^{2+}] = \frac{A}{\Sigma_{ML^{2+}} \times l} \qquad (iii)$$

With the help of equation (*iii*), equation (*i*) becomes

$$C_M = [M^{2+}] + \frac{A}{\Sigma_{ML^{2+}} \times l}$$

$$\textit{or} \qquad [M^{2+}] = C_M - \frac{A}{\Sigma_{ML^{2+}} \times l} \qquad (iv)$$

With the help of equation (*iii*), equation (*ii*) becomes

$$C_L = [L] + [ML^{2+}]$$

$$= [L] + \frac{A}{\Sigma_{ML^{2+}} \times l}$$

$$\textit{or} \qquad [L] = C_L - \frac{A}{\Sigma_{ML^{2+}} \times l} \qquad (v)$$

Put the values of $[ML^{2+}]$, $[M^{2+}]$ and [L] from equations (*iii*), (*iv*) and (*v*) in equation (A) to get the value of $K_{for}$. The experiment can be repeated at different $C_M$ and $C_L$ values to check the value of $K_{for}$.

## 2. Method of Continuous Variation (Job's Method)

This method is a variation of the spectrophotometric method and is used to determine the composition of a complex. This is mainly used for solution where only one complex is formed.

Different steps of the procedure are:

(*i*) Make 10 solutions of the complex containing metal ion and ligand in such proportions that the total volume of each solution is 10 ml. as shown below:

| | | | | | | | | | | |
|---|---|---|---|---|---|---|---|---|---|---|
| *Solutions No. (10 solutions):* | 1 | 2 | 3 | 4 | 5 | 6 | 7 | 8 | 9 | 10 |
| *Volume of the metal ion (ml):* | 0 | 1 | 2 | 3 | 4 | 5 | 6 | 7 | 8 | 9 |
| *Volume of the ligand (ml):* | 10 | 9 | 8 | 7 | 6 | 5 | 4 | 3 | 2 | 1 |

Thus we see that the sum of the concentration of the ligand, $C_L$ and that of the metal ion, $C_M$ is constant. Only their ratios, $C_L : C_M$ are changed. Thus:

$$C_L + C_M = C \qquad (i)$$

where C is a constant.

(*ii*) Determine the optical densities of the solutions as prepared in step (*i*) with the help of a spectrophotometer at such a wave length of light that the complex absorbs strongly and the metal ion and the ligand do not.

(*iii*) Draw a graph between the mole fraction of the ligand, *m.f.* $\left( m.f. = \frac{C_L}{C} \right)$ and optical density (or absorbance). The graph obtained is of the type as shown in Fig. 26.4.

When the legs of the curve are extrapolated, they cross each other at a point at which the absorbance is maximum.

If the formula of the complex is $ML_n$, then

$$n = \frac{C_L}{C_M} \qquad (ii)$$

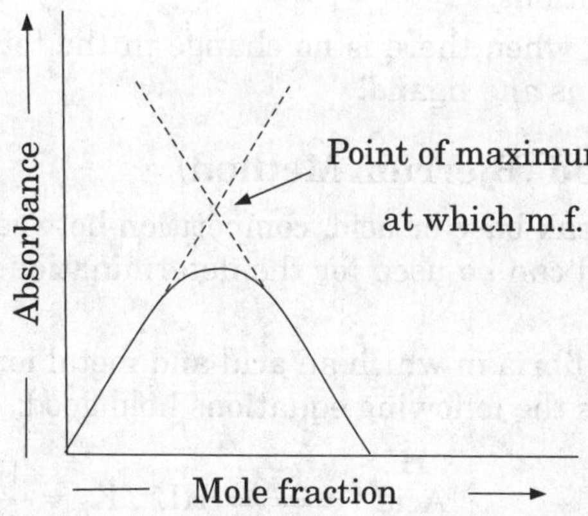


**Fig. 26.4** Job's method of continuous variation for the determination of the composition of a complex.

Equation (*i*) can also be written as:

$$\frac{C_L}{C} + \frac{C_M}{C} = \frac{C}{C} = 1 \quad (iii)$$

With the help of equation *viz*.,

$$m.f. = \frac{C_L}{C}$$

$$\text{or} \quad \frac{C_L}{C} = mf \quad (iv)$$

equation (*iii*) reduces to:

$$m.f. + \frac{C_M}{C} = 1$$

$$\text{or} \quad \frac{C_M}{C} = 1 - m.f. \quad (v)$$

Dividing equation (*iv*) by equation (*v*), we get

$$\frac{C_L}{C} \times \frac{C}{C_M} = \frac{m.f.}{1 - m.f.}$$

$$\text{or} \quad \frac{C_L}{C_M} = \frac{m.f.}{1 - m.f.}$$

$$\text{or} \quad n = \frac{m.f.}{1 - m.f.} \quad \text{[See equ. (ii)]} \quad (vi)$$

From the value of *n* as given by equation (*vi*) we can determine the composition of the complex, $ML_n$.

**Limitation.** (*i*) This method can be used when only one complex is formed under the experimental conditions.

(ii) This method is used when there is no change in the total volume of the solutions containing metal ions and ligand.

## 3. Potentiometric Method (Bjerrum Method)

When the ligand is a weak base or acid, competition between hydrogen ions and metal ions for the ligand can be used for the determination of the formation constant.

Let us consider the equilibria in which an acid and metal ions are added to a basic ligand in solution. Thus the following equations hold good:

$$\underset{\text{Basic ligand}}{L} + \underset{\text{Acid}}{H^+} \underset{}{\overset{K_a}{\rightleftharpoons}} HL^+, \quad K_a = \frac{[HL^+]}{[L][H^+]}$$

$$\underset{\text{Basic ligand}}{L} + \underset{\text{Metal ion}}{M^+} \overset{K_F}{\rightleftharpoons} ML^+, \quad K_{for} = \frac{[ML^+]}{[L][M^+]}$$

Here $K_a$ and $K_{for}$ are the acid association constant of the ligand and formation constant respectively.

Now if $C_H$, $C_M$ and $C_L$ are the total amount in moles/lit of acid ($H^+$), metal ($M^+$) and basic ligand (L), we have

$$C_H = [H^+] + [HL^+]$$
$$C_L = [L] + [ML^+] + [HL^+]$$
$$C_M = [M^+] + [ML^+]$$

Solving the last three equations given above and using the acid association constant of the ligand, $K_a$, we get:

$$[ML^+] = C_L - C_H + [H^+] - \frac{C_H - [H^+]}{K_a[H^+]}$$

$$[M^+] = C_M - [ML^+]$$

$$[L] = \frac{C_H - [H^+]}{K_a[H^+]}$$

Thus, on putting the values of $[ML^+]$, $[M^+]$ and [L] from the above equations in

$$K_{for} = \frac{[ML^+]}{[M^+][L]}$$

the value of $K_{for}$ can be evaluated. For the determination of $[ML^+]$, $[M^+]$ and [L], the values of $C_H$, $C_L$, $C_M$, $K_a$ and $[H^+]$ must be known. The last quantity, $[H^+]$, is generally determined potentiometrically using a pH meter.

In order to get precise results, the ligand must be a medium weak acid or base and the formation constant, $K_{for}$, should be within a factor of $10^5$ of the value of the acid association constant of the ligand, $K_a$.

## UNIVERSITY QUESTIONS

**1.** Discuss the factors affecting the stability of metal complexes. *(Osmania 2010)*

**2.** Prove that the overall stability constant is equal to the product of stepwise stability constants. *(Pd. Ravishankar Shukla 2010)*

**3.** (*a*) What is meant by stability of a complex? Determine relationship between stepwise and cumulative stability constant.

(*b*) How does crystal field theory explain the liability and inertness of complexes?

(*c*) Explain why: (*i*) $[Ni(en)_3]^{+2}$ aq. is more stable than $[Ni(NH_3)_6]^{+2}$ aq. (*ii*) $[Fe(CN)_6]^{-3}]$ is more stable than $[Fe(CN)_6]^{-4}$

*(Himachal Pradesh 2010)*

**4.** What is log β? How is it related with the stability of complex?

*(Himachal Pradesh 2010)*

**5.** How do the nature of ligand and ionic potential of metal ion affect the stability of metal complexes? *(Rajasthan 2010)*

**6.** Give a brief outline of thermodynamic and kinetic stability of metal complexes. *(H.N. Bahuguna Garhwal 2010)*

**7.** (*a*) Define stepwise stability constant and overall stability constant for the formation of complexes. How are they related?
(*b*) What are the factors on which the stability of complexes depends? *(Ranchi 2010)*

**8.** Explain how the nature of (i) metal ion and (ii) ligand affect the stability of the complexes. *(Nagpur 2010)*

**9.** What do you mean by overall stability constant and stepwise constants? How are these related? *(Utkal 2010)*

**10.** Discuss the factors affecting the thermodynamic stability of complex. *(BJB College, Orissa)*

**11.** Explain stability of complexes thermodynamically. *(G.N.D. Amritsar 2010)*

**12.** Define labile and inert complexes. *(Lucknow 2010)*

**13.** Write short notes on: (i) Chelate effect (ii) Steric effect. *(Pd. Ravishankar Shukla, Ranchi, Andhra, Utkal, 2010)*

**14.** (*a*) What do you understand by chelating agent and chelate? Explain with examples.
(*b*) What is EDTA? Write the structure of the metal EDTA complex. Why these complexes are more stable?
(*c*) What is chelate? *(Himachal Pradesh 2010)*

**15.** What are characteristics of chelate compounds? Give examples. *(Pune 2009, Banglore 2010)*

**16.** What are chelates? Give the classification of chelates formed by bidentate ligands with examples. *(Nagpur 2010)*

**17.** (i) Explain why $[Cu(en)_2]^{2+}$ is more stable than $[Cu(NH_3)_6]^{2+}$
(ii) Explain chelation with example. *(Lucknow 2010)*

**18.** Explain the determination of composition of complex by Job's method *(Andhra 2010)*

**19.** (i) Give a relationship between stepwise and overall stability constants.
(ii) How is stability constant determined by polarographic method?
(iii) Explain five factors that affect the stability of complexes with examples. *(Madras April 2011)*

**20.** (*a*) Write notes on (*i*) Labile and inert complexes (*ii*) Stepwise and overall stability constants
(*b*) How does the nature of ligand affect the stability of complexes?
(*c*) Explain the relationship between thermodynamic and kinetic stability with example. *(Nagpur 2012)*

**21.** (*i*) Explain stability of complexes with respect of chelation
(*ii*) Derive equations for stepwise and overall stability constants. Show their relationship also
(*iii*) Which of the following is more stable complex and why? $[Fe(CN)_6]^{4-}$ and $[Fe(CN)_6]^{3-}$ *(Kurukshetra 2011)*

**22.** Discuss the stability of complexes and explain the factors which affect the stability of complexes. Illustrate your answer with examples *(Luknow 2013)*

**27** CHAPTER

# Ligand Substitution Reactions in Octahedral Complexes

## What are Substitution Reactions in Coordination Compounds?

Substitution reactions in coordination compounds are the reactions in which a ligand present in the coordination sphere of the complex compound is replaced (substituted) by another ligand (nucleophile) or the metal in a complex compound is replaced of another metal (electrophile). Substitution reactions are also called **exchange reactions** or **replacement reactions** or **displacement reactions.**

## Classification of Substitution Reactions in Complexes

Depending on whether a ligand is exchanged by another ligand or a metal is replaced by another metal, substitution reactions of coordination compounds have been classified into the following categories.

**1. Nucleophile (or ligand) substitution reactions ($S_N$ reactions).** In these reactions a ligand present in the coordination sphere is replaced by another ligand (nucleophile).

For example:

$$\underset{}{MA_5\,L} + \underset{\text{Nucleophile}}{E} \rightarrow MA_5\,E + L$$

In this reaction the ligand, L (leaving group) present in complex ($MA_5$ L) is replaced by another ligand E (entering group). E is called nucleophile, since it is electron pair donor (Lewis base). CNs. of the new complex ($MA_5$ E) and the original complex ($MA_5$ L) are the same (= 6). In this reaction L is the leaving (outgoing) group.

**2. Electrophile (or metal) substitution reaction ($S_E$ reaction).** In these reactions the metal present in a complex is replaced by another metal (electrophile).

For example:

$$M^1A_5\,L + \underset{\text{Electrophile}}{M^2} \rightarrow M^2A_5\,L + M^1$$

In this reaction the metal ($M^1$) present in the complex, $M^1A_5$ L is replaced by another metal $M^2$. $M^2$ metal is called **electrophile**, since it is **electron pair accepter (Lewis acid).** C.Ns. of the two complexes ($M^1A_5$ L) and ($M^2A_5$ L) are the same (=6). In this reaction $M^1$ is the leaving (outgoing) metal and $M^2$ is the entering metal.

In this chapter we shall discuss nucleophile (ligand) substitution reactions in octahedral complexes only, since these reactions proceed so slowly that their kinetics can be followed easily by conventional techniques.

## Nucleophilic (Ligand) Substitution Reactions in Octahedral Complexes.

Nucleophilic substitution reaction in an octahedral complex is a reaction in which one of the ligands present in an octahedral complex is replaced by another ligand. For example the reaction,

| $MA_5\,L$ | + | E | → | $MA_5\,E$ + L |
|---|---|---|---|---|
| Octahedral complex (C.N. = 6) | | Entering ligand (Nucleophile) | | |

is a nucleophilic substitution reaction in which the ligand L present in octahedral complex ($MA_5$ L) is replaced by another ligand E. Thus E is the *entering ligand* and L is the *leaving ligand*. Since all the ligands are nucleophiles (electron pair donors or Lewis bases), these reactions are called nucleophile substitution reactions.

## Mechanism to Explain the Nucleophilic (Ligand) Substitution Reactions in Octahedral Complexes.

Let us consider the following nucleophilic (ligand) substitution reaction in octahedral complex, $MA_5$ L

| $MA_5\,L$ | + | E | → | $MA_5\,E$ + L | (*i*) |
|---|---|---|---|---|---|
| Octahedral complex (C.N. = 6) | | Nucleophile (Entering group) | | Leaving group | |

Here $MA_5$ L is an octahedral complex in which L is the leaving group, *i.e.* L is replaced by E. Thus E is the entering group (nucleophile). There are two mechanisms to explain the occurrence of the above ligand substitution reaction (i).

**1. Unimolecular nucleophilic substitution or dissociative $S_N^1$ mechanism.** According to this mechanism the ligand substitution reaction (*i*) proceeds through the following steps:

(*a*) *Dissociative step (D): Loss of leaving group*

$$\underset{\substack{\text{Octahedral complex}\\ \text{(C.N. = 6)}}}{MA_5L} \xrightarrow[\text{Slow step (k)}]{\substack{\text{Loss of leaving group}\\ \text{by dissociation (–L)}}} \underset{\substack{\text{Square pyramidal}\\ \text{intermediate (C.N. = 5)}}}{MA_5} + L$$

(*b*) *Associative step: Addition of entering group*

$$\underset{\substack{\text{Square pyramidal}\\ \text{intermediate (C.N. = 5)}}}{MA_5} + E \xrightarrow[\text{Fast step}]{\substack{\text{Addition of entering group}\\ \text{by association (+E)}}} MA_5E$$

In *dissociative step (a)* the octahedral complex ($MA_5$ L) dissociates to lose the leaving group (L) and forms square pyramidal intermediate ($MA_5$) with C.N. = 5. Thus in this step M—X bond present in $MA_5$ L molecule is *broken*. Hence step *(a)* can also be called **bond breaking step.** In *associative* step (b), the square pyramidal intermediate formed in step *(a)* adds the entering group E (nucleophile) by association and forms the octahedral complex, $MA_5$ E.

Since the dissociation step [step *(a)*] is a slow step, it is the rate determining step. The reaction of this slow step is unimolecular, since it involves only one reacting species *viz* $MA_5$ L. The rate of this reaction depends on the concentration of $MA_5$ L only, *i.e.* the rate law of this reaction is represented as:

$$\text{Rate } (r) = k[MA_5\,L]$$
$$= k[\text{Initial complex}]$$

Since the rate of reaction (r) depends on the concentration of one species only *viz* complex ($MA_5$ L), $S_N^1$ mechanism is also called **unimolecular $S_N$ mechanism or $S_N^1$ mechanism.**

In this mechanism, since the octahedral complex $MA_5$ L dissociates in rate determining step (slow step), the mechanism is called **dissociative $S_N^1$ mechanism.** Above discussion shows that the rate of reaction (i) is directly proportional to concentration of $MA_5$ L and is independent of the concentration of the entering group E

Both the steps of dissociative $S_N^1$ mechanism can be combined together and can be represented as:

$$\underset{\substack{\text{Octahedral}\\(\text{C.N.} = 6)}}{MA_5\,L} \xrightarrow[\text{Slow (k)}]{\text{Dissociation }(-L)} \underset{\substack{\text{Square pyramidal}\\\text{intermediate (C.N. = 5)}}}{MA_5} + L$$

$$\underset{\text{Octahedral (C.N. = 6)}}{MA_5\,E} \xleftarrow[\text{Fast}]{\text{Association }(+E)} MA_5$$

In a simplified way the above steps of dissociative $S_N^1$ mechanism can also be shown as:

$$MA_5\,L \xrightarrow[\text{Slow (k)}]{-\text{Leaving group (L)}} MA_5 \xrightarrow[\text{Fast}]{+\text{Entering group (E)}} MA_5\,E$$

$$\text{Rate (r)} = k\,[MA_5\,L] = k\,[\text{Complex}]$$

**2. Bimolecular nucleophilic substitution or associative $S_N^2$ mechanism.** According to this mechanism the ligand substitution reaction (*i*) proceeds through the following steps:

(*a*) *Associative step (A)*

$$\underset{\substack{\text{Octahedral}\\\text{complex (C.N. = 6)}}}{MA_5L + E} \xrightarrow[\text{Slow (k)}]{\substack{\text{Addition of entering group}\\\text{by association }(+E)}} \underset{\substack{\text{7-coordinated}\\\text{intermediate (C.N. = 7)}}}{MA_5LE}$$

(*b*) *Dissociative step*

$$\underset{\substack{\text{7-coordinated intermediate}\\(\text{C.N} = 7)}}{MA_5LE} \xrightarrow[\text{Fast}]{\substack{\text{Loss of leaving group}\\\text{by dissociation }(-L)}} \underset{\substack{\text{Octahedral complex}\\(\text{C.N} = 6)}}{MA_5E + L}$$

In associative step (a) the nucleophile (entering group), E attacks the octahedral complex ($MA_5$ L) and gets associated with it to form a 7- coordinated intermediate ($MA_5$ L E). Thus in this step a new bond (M–E bond) is formed to produce $MA_5$ L E. Hence this step can also be called **bond making step**. In dissociative step (*b*),

7-coordinated intermediate ($MA_5$ L E) formed in step (a) breaks up and the leaving ligand (L) is ejected.

Since step (*a*) is slow step, it is the rate determining step. The reaction of this step is bimolecular, since it involves two species *viz* $MA_5$ L and E. The rate of this reaction depends on the concentration of the complex ($MA_5$ L) as well as on the concentration of the entering ligand (E) *i.e.* the rate law for this reaction is represented as:

$$\text{Rate (r)} = k[\text{M A}_5\ \text{L}]\ [\text{E}]$$
$$= k\ [\text{Complex}]\ [\text{Entering group}]$$

Since the rate of reaction depends on the concentration of the complex ($MA_5$ L) and the entering group or nucleophile (E), this mechanism is also called **bimolecular $S_N$ mechanism or $S_N^2$ mechanism.**

In this mechanism, since the entering group (E) gets associated with the complex ($MA_5$ L) in rate determining step (slow step) to form 7 coordinated intermediate, this mechanism is also called **associative $S_N^2$ mechanism.**

Both the steps of associative $S_N^2$ mechanism can be shown as:

$$\underset{\text{Octahedral (C.N. = 6)}}{MA_5L + E} \xrightarrow[\text{Slow (k)}]{\text{Association (+E)}} \underset{\text{7-coordinated intermediate}}{MA_5LE}$$

$$\xrightarrow[\text{Fast}]{\text{Dissociation (–L)}} \underset{\text{Octahedral (C.N. = 6)}}{MA_5\ E + L}$$

In a simplified way both the above steps of associative $S_N^2$ mechanism can also be shown as:

$$MA_5\ L \xrightarrow[\text{Slow (k)}]{\text{+Entering group (E)}} MA_5\ LE \xrightarrow[\text{Fast}]{\text{–Leaving group (L)}} MA_5E$$

$$\text{Rate (r)} = k[MA_5\ \text{L}]\ [\text{E}]$$
$$= k\ [\text{Complex}]\ [\text{Entering group}]$$

## Comparison Between $S_N^1$ and $S_N^2$ Mechanisms.

Let us consider the following ligand substitution reaction

$$MA_5\ L + E \rightarrow MA_5\ E + L$$

**$S_N^1$ mechanism**

$$\underset{\text{(C.N. = 6)}}{MA_5L} \xrightarrow[\text{Slow } (k_1)]{-L} \underset{\text{(C.N. = 5)}}{MA_5} \xrightarrow[\text{Fast}]{+E} \underset{\text{(C.N. = 6)}}{MA_5E}$$

**$S_N^2$ mechanism**

$$\underset{\text{(C.N. = 6)}}{MA_5L} \xrightarrow[\text{Slow } (k_2)]{+E} \underset{\text{(C.N. = 7)}}{MA_5LE} \xrightarrow[\text{Fast}]{-L} \underset{\text{(C.N. = 6)}}{MA_5E}$$

(*i*) In $S_N^1$ mechanism the rate determining step (slow step) is metal –ligand bond (M—L bond) breaking step, since in this step $MA_5$ L (C.N. = 6) is changed into $MA_5$ (C.N. = 5). In $S_N^2$ mechanism the rate determining step (slow step) is a metal – ligand bond (M—E bond) forming step, since in this step $MA_5$ L (C.N. = 6) is converted into $MA_5$ L E (C.N. = 7) (*ii*) $S_N^1$ mechanism follows first order rate law. Thus rate determining step is unimolecular, *i.e.* rate of reaction is first order w.r.t. $MA_5$ L Hence:

$$\text{Rate (r) of } S_N^1 \text{ mechanism} = k_1\,[MA_5\,L]$$
$$= k_1\,[\text{Complex}]$$

$S_N^2$ mechanism follows second order rate law. Thus the rate determining step is bimolecular and the rate of reaction is first order w.r.t to $MA_5$ L and first order w.r.t E. Hence:

$$\text{Rate } (r) \text{ of } S_N^2 \text{ mechanism} = k_2\,[MA_5\,L]\,[E]$$
$$= k_2\,[\text{Complex}]\,[\text{Entering group}]$$

## Reaction Profile of Dissociative (D) Mechanism

Reaction profile is a graphic representation of the energy as a function of the reaction coordinate. Reaction coordinate represents the progress of a reaction from reactants to products

Let us consider the following ligand substitution reaction

$$\underset{(\text{C.N.} = n)}{MX_n} + Y \rightarrow \underset{(\text{C.N.} = n)}{MX_{n-1}\,Y} + X$$

Here X is the leaving group and Y is the entering group. The coordination number of the reacting complex, $MX_n$ and the formed complex, $MX_{n-1}$ Y is the same (= $n$). If the above reaction proceeds through dissociative mechanism, then the conversion of $MX_n$ into $MX_{n-1}$ Y can be shown by the following steps:

$$\underset{(\text{C.N.} = n)}{MX_n + Y} \xrightarrow[\text{Slow (k)}]{\text{Dissociation (–X)}} \underset{\substack{\text{Transition state} \\ (\text{C.N.} = n-1)}}{MX_{n-1}} + X + Y$$

$$\xrightarrow{+Y} \underset{(\text{C.N.} = n-1+1 = n)}{MX_{n-1}\,Y} + X$$

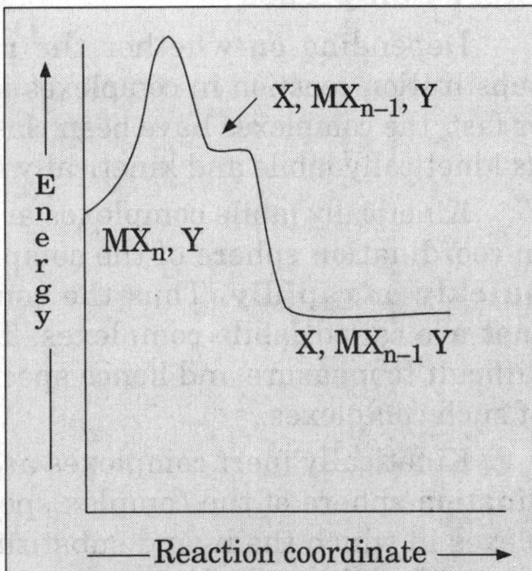


**Fig. 27.1** Reaction profile for dissociative (D) mechanism

Above mechanism shows that in the first step M–X bond is broken and transition state, $MX_{n-1}$ (intermediate) is obtained. Thus the complex $MX_n$ should have activation energy required for the transition state before M–X bond is broken to from $MX_{n-1}$ which has lower C.N. than $MX_n$. Once M—X bond is broken, the leaving group (X) leaves the coordination sphere of $MX_n$ to form the intermediate ($MX_{n-1}$) which has a plateu as is evident from the reaction profile shown in Fig. 27.1. Since the leaving group (X) leaves the coordination sphere of $MX_n$, a vacancy is created in the coordination sphere. This vacancy is then filled up by the entering group (Y) to form the products $MX_{n-1}$ Y and X. The products, $MX_{n-1}$ Y and X have lower energy than the reactants $MX_n$ and Y. The reaction profile has one hill and one pluteu.

## Reaction Profile for Associative (A) Mechanism

Let us consider the following ligand substitution reaction:

$$MX_n + Y \rightarrow MX_{n-1}\,Y + X$$

If this reaction proceeds through associative mechanism ($S_N^2$ mechanism), then the conversion of $MX_n$ into $MX_{n-1}$ Y can be shown by the following steps:

$$\underset{(C.N. = n)}{MX_n} \quad + Y \xrightarrow{\text{Association } (+Y)} \underset{\substack{\text{Transition state} \\ (C.N. = n+1)}}{MX_nY} \xrightarrow{-X} MX_{n-1}Y + X$$

In associative mechanism, the energy of the reactant ($MX_n$) and entering group (Y) keeps on increasing till the activation energy for the transition state (intermediate), $MX_n$ Y is reached. Transition state has higher C.N. (= n + 1) than the original complex, $MX_n$ (C.N. = n). Now the bond between M and X in transition state is broken and M—Y bond is formed in $MX_{n-1}$ Y. So the intermediate passes through another smaller hill to stable products, $MX_{n-1}$ Y and X (See Fig. 27.2)

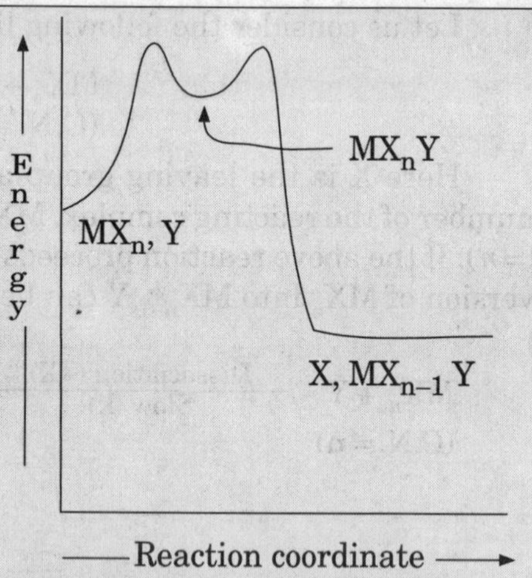

**Fig. 27.2** Reaction profile for associative (A) mechanism.

## Kinetically Labile and Kinetically Inert Complexes

Depending on whether the rate of substitution reaction in complexes is slow or fast, the complexes have been classified as kinetically labile and kinetically inert complexes by Taube (1952).

Kinetically labile complexes are those in which one or more ligands present in coordination sphere of the complex species can be replaced by other ligands **quickly** or **rapidly.** Thus the complexes in which the ligand substitution is **fast** are called labile complexes. The rate of substitution of labile complex is difficult to measure and hence special techniques are used to study the kinetics of such complexes.

Kinetically inert complexes are those in which the ligands present in coordination sphere of the complex species can be replaced **slowly.** Thus the complexes in which the ligand substitution is slow are called inert complexes. The rate of substitution of inert complexes can be measured easily by conventional techniques.

Since the terms labile and inert refer to the rate of substitution reactions, labile and inert complexes are called kinetically labile and kinetically inert complexes respectively.

## Thermodynamically Stable and Thermodynamically Unstable complexes

A thermodynamically stable complex has high value of its formation constant while a thermodynamically unstable complex has low value of its formation constant. It should be understood clearly that a thermodynamically stable complex may be kinetically labile (fast reacting complex). Similarly a thermodynamically unstable complex may be kinetically inert (slow reacting complex). The thermodynamic stability and kinetic lability are different from each other. Thermodyanmically stable complex may be labile or inert. For example $[Fe(H_2O)_6]^{3-}$ (Bond energy = 116 Kcal $mol^{-1}$) and $[Cr(H_2O)_6]^{3+}$ (Bond energy = 122 Kcal $mol^{-1}$) have the same thermodynamic stability but $[Fe(H_2O)_6]^{3-}$ is kinetically labile (exchanges its ligands with other ligands rapidly) but $[Cr(H_2O)_6]^{3+}$ is kinetically inert (exchanges ligands slowly). $[Cr(NH_3)_6]^{3+}$ is thermodynamically unstable because of high value of its dissociation constant, $K_{diss}$ (= $10^{25}$) but is kinetically inert (slow reacting complex), since it remains undecomposed in acid solution.

$$[Co(NH_3)_6]^{3+} + 6H_3O^+ \rightleftharpoons [Co(H_2O)_6]^{3+} + 6NH_4^+, \ K_{diss} = 10^{25}$$

$[Ni(CN)_4]^{2-}$ is thermodynamically stable because of low value of it dissociation constant, $K_{diss}$ (= $10^{-22}$) but is kinetically labile (fast reacting complex,) since it exchanges $CN^-$ ions very rapidly with added isotopically labelled cyinide ions ($^{14}CN^-$)

$$[Ni(CN)_4]^{2-} \rightleftharpoons Ni^{2+} + 4CN, \ K_{diss} = 10^{-22}$$

$$[Ni(CN)_4]^{2-} + (^{14}CN)^- \xrightarrow[\text{aqueons solution}]{\text{very fast in}} [Ni(CN)_3(^{14}CN)]^{2-} + CN^-$$

## Classification of Octahedral Complexes into Labile and Inert Complexes on the Basis of VBT

According to VBT, inner orbital octahedral complexes of metal ions of 1st transition series are formed by $(3d)^2$ (4s) $(4p)^3$ hybridisation. Two 3d orbitals involved in $(3d)^2$ (4s) $(4p)^3$ hybridisation are the orbitals of $e_g$ set and the three orbitals of $t_{2g}$ set remain empty or contain the electrons of the metal ion. Six hybrid orbitals contain electron pairs donated by six ligands. According to Taube, in case of inner orbital octahedral complexes if 3d orbitals of $t_{2g}$ set contain no (d°), 1 ($d^1$) or 2 ($d^2$) electrons, the complex is labile and if 3d orbitals of $t_{2g}$ set contain 3($d^3$), 4($d^4$), 5($d^5$) or 6($d^6$) electrons, the complex is inert. Thus inner orbital octahedral complex of $Sc^{3+}$, $Ti^{4+}$ ($d^0$ ions), $Ti^{3+}$, $V^{4+}$ ($d^1$ ions) and $Ti^{2+}$, $V^{3+}$ ($d^2$ ions) ions are labile while inner orbital octahedral complex of $V^{2+}$ ($d^3$ ion), $Cr^{2+}$ ($d^4$ ion), $Mn^{2+}$, $Fe^{3+}$ ($d^5$ ions), $Fe^{2+}$, $Co^{3+}$ ($d^6$ ions) ions are inert.

## Lability and Inertness of Octahedral Complexes on the Basis of $t_{2g}^p e_g^q$ Configuration of Metal ion (Crystal Field Theory)

According to Taube:

(*a*) Octahedral complexes of metal ions with empty $e_g$ orbitals may be labile or inert. For example octahedral complexes of metal ions with $t_{2g}^0 e_g^0$ ($d^0$), $t_{2g}^1 e_g^0$ ($d^1$)

and $t_{2g}^2 e_g^0$ ($d^2$) are labile while those of metal ions with $t_{2g}^3 e_g^0$ ($d^3$), $t_{2g}^4 e_g^0$ (LS–$d^4$), $t_{2g}^5 e_g^0$ (LS–$d^5$) and $t_{2g}^6 e_g^0$ (LS–$d^6$) configurations are inert.

(*b*) Octahedral complexes of metal ions with $e_g$ orbitals containing 1 or more electrons are labile. Thus octahedral complexes of metal ions with configurations $t_{2g}^3 e_g^1$ (HS–$d^4$), $t_{2g}^3 e_g^2$ (HS–$d^5$), $t_{2g}^4 e_g^2$ (HS–$d^6$), $t_{2g}^5 e_g^2$ ($d^7$), $t_{2g}^6 e_g^2$ ($d^8$), $t_{2g}^7 e_g^2$ ($d^9$) and $t_{2g}^8 e_g^2$ ($d^{10}$) are labile

Examples of labile and innert octahedral complexes are given in Table 27.1

**Table 27.1 Examples of labile and inert octahedral complexes**

| $d^x$ ion | $t_{2g}^p e_g^q$ configuration | Labile complexes | Inert complexes |
|---|---|---|---|
| $d^0$ | $t_{2g}^0 e_g^0$ | $[CaEDTA]^{2-}$, $[Ti\ Cl_6]^{2-}$ | — |
| $d^1$ | $t_{2g}^1 e_g^0$ | $[Ti(H_2O)_6]^{3+}$, $[MoCl_6]^{3-}$ | — |
| $d^2$ | $t_{2g}^2 e_g^0$ | $[U(phen)_3]^{3+}$ | — |
| $d^3$ | $t_{2g}^3 e_g^0$ | – | $[V(H_2O)_6]^{2+}$, $[Cr(en)_2Cl_2]^+$ |
| $d^4$(HS) | $t_{2g}^3 e_g^1$ | $[Cr(H_2O)_6]^{3+}$ | — |
| $d^4$ (LS) | $t_{2g}^4 e_g^0$ | – | $[Cr(CN)_6]^{4-}$, $[Mn(CN)_6]^{3-}$ |
| $d^5$(HS) | $t_{2g}^3 e_g^2$ | $[Mn(H_2O)_6]^{2+}$, $[Fe\ (H_2O)_4\ Cl_2]^+$ | — |
| $d^5$(LS) | $t_{2g}^5 e_g^0$ | — | $[Mn(CN)_6]^{4-}$, $[Fe(CN)_6]^{3-}$ |
| $d^6$(HS) | $t_{2g}^4 e_g^2$ | $[Fe(H_2O)_6]^{2+}$ | — |
| $d^6$ (LS) | $t_{2g}^6 e_g^0$ | — | $[Fe(CN)_6]^{4-}$, $[Cu(en)_2\ (H_2O)_2]^{3+}$ |
| $d^7$ | $t_{2g}^5 e_g^2$ | $[Co(NH_3)_6]^{2+}$ | — |
| $d^8$ | $t_{2g}^6 e_g^2$ | $[Ni(en)_3]^{3+}$ | — |
| $d^9$ | $t_{2g}^6 e_g^3$ | $[Cu(NH_3)_4\ (H_2O)_2]^{2+}$ | — |
| $d^{10}$ | $t_{2g}^6 e_g^4$ | $[Cu(C_2O_4)_3]^{5-}$ | — |

## Activation Energy (AE)

Activation energy is defined as the energy required to convert the reacting complex into unstable or activated complex (transition state or intermediate). Thus activation energy is the energy necessary for forming activated complex from the reacting complex. Activated complex does not represent a true molecule but represents only an imaginary molecule which cannot be isolated. It has maximum energy and has very short life. Being unstable, the activated complex finally changes to the products. Activation energy (AE) is the difference in energy between the reading complex and activated complex, *i.e.*

$$A\,E = \text{Energy of activated complex} - \text{Energy of reacting complex}$$

## Crystal Field Activation Energy (CFAE)

CFAE is defined as the change in crystal filed stabilisation energy (CFSE) when the reacting complex is changed into intermediate, *i.e.*

$$\text{CFAE} = \text{CFSE of intermediate} - \text{CFSE of the reacting complex}$$

If the calculated value of CFAE is negative or zero or low, the reacting complex would require less energy for its conversion into intermediate. On the other hand, if the calculated value of CFAE is high, the reacting complex would require more energy for its conversion into intermediate.

## To Explain the Lability and Inertness of Octahedral Complexes on the Basic of CFT.

We know that ligand substitutions in octahedral complexes proceed either through $S_N^1$ mechanism or through $S_N^2$ mechanism. In order to explain the lability and inertness of octahedral complexes by CFT, we calculate the value of CFAE for various $d^x$ ions of the complexes in both mechanisms. If the value of CFAE is negative or zero, the complex is labile (fast to react). If the value CFAE is positive, the complex is slow to react.

**(i) $S_N^1$ mechanism.** According to this mechanism, in the ligand substitution reaction of a given HS or LS octahedral complex, the octahedral complex (C.N. = 6) is converted into square pyramidal (SP) intermediate (C.N = 5).

$$\text{Octahedral complex (C.N. = 6)} \xrightarrow{S_N^1 \text{ mechanism}} \text{Square pyramidal (SP) intermediate (C.N. = 5)}$$

Thus:

CFAE value of HS or LS octahedral complex undergoing ligand substitution reaction through $S_N^1$ mechanism

= CFSE of SP intermediate – CFSE for $d^x$ ion in HS or LS octahedral complex (*i*)

If the values of CFSE of SP intermediate and for $d^x$ ion in HS or LS octahedral complexes are put in equation (*i*) given above, we get CFAE value for HS or LS octahedral complexes undergoing ligand substitution reaction through $S_N^1$ mechanism. These values have been shown in Table 27.2.

**Table 27.2 CFAE values (in Dq) of HS (weak field) and LS (strong field) octahedral complexes undergoing ligand substitution reaction through $S_N^1$ mechanism**

| $d^x$ configuration | CFSE value for SP intermediate (C.N. = 5) (a) | CFSE value for octahedral complex (C.N. = 6) (b) | CFAE value = (a – b) |
|---|---|---|---|
| $d^0$ | 0 | 0 | 0 |
| $d^1$ | –4.57 | –4 | –0.57 |
| $d^2$ | –9.14 | –8 | –1.14 |
| $d^3$ | –10.00 | –12 | 2.00 |
| $d^4$(HS) | –9.14 (HS) | –6 (HS) | –3.14 (HS) |
| $d^4$(LS) | –14.57 (LS) | –16 (LS) | 1.43 (LS) |
| $d^5$(HS) | 0 (HS) | 0 (HS) | 0 (HS) |
| $d^5$(LS) | –19.4 (LS) | –20 (LS) | 0.86 (LS) |

*(continued)*

| $d^x$ configuration | CFSE value for SP intermediate (C.N. = 5) (a) | CFSE value for octahedral complex (C.N. = 6) (b) | CFAE value = (a – b) |
|---|---|---|---|
| $d^6$(HS) | –4.57 (HS) | –4 (HS) | –0.57 (HS) |
| $d^6$(LS) | –20.00 (LS) | –24 (LS) | 4.00 (LS) |
| $d^7$(HS) | –9.14 (HS) | –8 (HS) | –1.14 (HS) |
| $d^7$ (LS) | –19.14 (LS) | –18 (LS) | –1.14 (LS) |
| $d^8$ | –10 | –12 | 2.00 |
| $d^9$ | –9.14 | –6 | –3.14 |
| $d^{10}$ | 0 | 0 | 0 |

From the values of CFAE given in Table 27.2 the following points may be noted:

(a) Since complexes of the metal ions with $d^0$, $d^1$, $d^2$ and $d^{10}$ configurations have negative or zero CFAE values, these complexes are labile *i.e.* these complexes are fast to react by $S_N1$ mechanism. Similarly HS complexes of $d^4$, $d^5$, $d^6$ and $d^7$ ions are also labile.

(b) Since LS complexes of $d^7$ ion have negative value of CFAE, these complexes are labile.

(c) Complexes of $d^3$ and $d^8$ ions have positive values of CFAE, these complexes are slow to react.

Due to the decreasing order of CFAE values, the order of reactivity of the complexes of $d^3$, $d^8$ and LS complexes of $d^4$, $d^5$ and $d^6$ is as: $d^6 < d^3 \sim d^8 < d^4 < d^5$.

Above discussion can be summarised in the form of a table which is given below

| $d^x$ ion | Sign of CFAE value | Labile or inert |
|---|---|---|
| $d^0$ | 0 | Fast to react (Labile) |
| $d^1$ | – | Fast to react |
| $d^2$ | – | Fast to react |
| $d^3$ | + | Slow to react (Inert) |
| $d^4$ (HS) | – | Fast to react |
| $d^4$ (LS) | + | Slow to react |
| $d^5$ (HS) | 0 | Fast to react |
| $d^5$ (LS) | + | Slow to react |
| $d^6$ (HS) | – | Fast to react |
| $d^6$ (LS) | + | Slow to react |
| $d^7$ (HS) | – | Fast to react |
| $d^7$(LS) | – | Fast to react |
| $d^8$ | + | Slow to react |
| $d^9$ | – | Fast to react |
| $d^{10}$ | 0 | Fast to react |

**(ii) $S_N{}^2$ mechanism.** According to this mechanism, in ligand substitution reaction of a given HS or LS octahedral complex, the or octahedral complex may be converted into pentagonal bipyramidal (PBP) intermediate or octahedral wedge (OW) intermediate. $S_N{}^2$ reaction of octahedral complexes preferably proceed through the formation of OW intermediate rather than PBP intermediate, since the formation of OW intermediate requires less energy than the formation of a PBP intermediate. Thus:

CFAE value of HS or LS octahedral complex undergoing ligand substitution reaction through $S_N{}^2$ mechanism
= CFSE of OW intermediate – CFSE of $d^x$ ion in HS or LS octahedral complex *(ii)*

If the values of CFSE of OW intermediate and of $d^x$ ions in HS or LS octahedral complexes are put the equation *(ii)* given above, we get CFAE values for HS or LS octahedral complexes undergoing ligand substitution reaction through $S_N{}^2$ mechanism. These values are given in Table 27.3.

**Table 27.3 CFAE values (in $D_q$) of HS (weak field) and LS (strong field) octahedral complexes undergoing ligand substitution reactions through $S_N{}^2$ mechanism.**

| $d^x$ configuration | CFSE value for OW intermediate (a) | CFSE value for octahedral complex (b) | CFAE value = (a – b) |
|---|---|---|---|
| $d^0$ | 0 | 0 | 0 |
| $d^1$ | –6.08 | –4 | –2.08 |
| $d^2$ | –8.68 | –8 | –0.68 |
| $d^3$ | –10.20 | –12 | 1.80 |
| $d^4$(HS) | –8.79 (HS) | –6(HS) | –2.79 (HS) |
| $d^4$(LS) | –16.26 (LS) | –16(LS) | –0.26 (LS) |
| $d^5$(HS) | 0 | 0(HS) | 0 (HS) |
| $d^5$(LS) | –18.86 (LS) | –20(LS) | 1.14 (LS) |
| $d^6$(HS) | –6.08 (HS) | –4(HS) | –2.08 (HS) |
| $d^6$(LS) | –20.37 (LS) | –24(LS) | 3.63 (LS) |
| $d^7$(HS) | –8.68 (HS) | –8(HS) | –0.68 (HS) |
| $d^7$(LS) | –18.98 (LS) | –18(LS) | –0.98 (LS) |
| $d^8$ | –10.20 | –12 | 1.80 |
| $d^9$ | –8.79 | –6 | –2.79 |
| $d^{10}$ | 0 | 0 | 0 |

From the values of CFAE given in Table 27.3 we draw the following conclusions.

*(a)* Complexes of $d^0$, $d^1$, $d^2$, $d^9$ and $d^{10}$ ions are labile, since these complexes have zero or negative CFAE values. Similarly HS complexes of $d^4$, $d^5$, $d^6$ and $d^7$ ions are also labile. Low spin complexs of $d^4$ and $d^7$ ions are also labile.

*(b)* Complexes of $d^3$ and $d^8$ ions are slow to react, since their CFAE values are positive. Similarly LS complexes of $d^5$ and $d^6$ ions are also slow to react. The

order of reactivity of the complexes of $d^3$, $d^5$, $d^6$ and $d^8$ ions is in the order $d^6 < d^3 \sim d^8 \sim d^5$.

Above discussion can be summarized in the form of table given below.

| $d^x$ ion | Sign of CFAE value | Labile or inert |
|---|---|---|
| $d^0$ | 0 | Fast to react (Labile) |
| $d^1$ | – | Fast to react |
| $d^2$ | – | Fast to react |
| $d^3$ | + | Slow to react (Inert) |
| $d^4$(HS) | – | Fast to react |
| $d^4$(LS) | – | Fast to react |
| $d^5$(HS) | 0 | Fast to react |
| $d^5$(LS) | + | Slow to react |
| $d^6$(HS) | – | Fast to react |
| $d^6$(LS) | + | Slow to react |
| $d^7$(HS) | – | Fast to react |
| $d^7$(LS) | – | Fast to react |
| $d^8$ | + | Slow to react |
| $d^9$ | – | Fast to react |
| $d^{10}$ | 0 | Fast to react |

From the combined CFAE values given in Table 27.2 and 27.3 the following conclusions can be drawn

(*a*) Octahedral complexes of metal ions with $d^3$ and $d^8$ configuration and LS octahedral complexes of $d^5$ and $d^6$ ions are slow to react by $S_N{}^1$ as well as by $S_N{}^2$ mechanisms

(*b*) Octahedral complexes of $d^0$, $d^1$, $d^2$, $d^9$ and $d^{10}$ and HS complexes of $d^4$, $d^5$, $d^6$ and $d^7$ ions would be labile whatever be the mechanism of their substitution reactions.

## Acid Hydrolysis (Aquation)

### What is Acid Hydrolysis?

The ligand substitution reaction occurring in aqueous medium in which a ligand present in the coordination sphere of a complex species is replaced by $H_2O$ molecule is called acid hydrolysis (or aquation) of the complex. For example the reaction

$$MA_5\,L + H_2O \rightarrow MA_5\,(H_2O) + L$$

is an acid hydrolysis reaction. In this reaction L is the leaving ligand and $H_2O$ in the entering group. The ligand A is called inert ligand, since this ligand remains attached in the product, $MA_5\,(H_2O)$ also.

### Mechanism of Acid Hydrolysis (Aquation) of Different Types of Octahedral Complexes.

Here we shall discuss the mechanism of acid hydrolysis occurring in the following types of octahedral complexes.

**I. Octahedral Complexes in Which None of the Inert Ligands is a pi Donor or a pi Acceptor.**

Let us consider the acid hydrolysis of $MA_5$ L complex in which As are inert ligands. None of these ligands is a pi donor or a pi acceptor. The hydrolysis of $MA_5L$ can be represented as:

$$MA_5\ L + H_2O \rightarrow MA_5\ (H_2O) + L \qquad (i)$$

**$S_N{}^1$ mechanism.** As a matter of fact, this reaction proceeds through the following two steps which are based on $S_N{}^1$ mechanism.

$$\underset{(C.N. = 6)}{MA_5L} \xrightarrow[\text{Slow } (k)]{\text{(i) Dissociation } (-L)} \underset{(C.N. = 5)}{MA_5} + L$$

$$\xrightarrow[\text{Fast}]{\text{(ii) Association } (+H_2O)} \underset{(C.N. = 6)}{MA_5(H_2O)}$$

$$\text{Rate of reaction} = k\ [\text{Complex}]$$
$$= k\ [MA_5\ L]$$

**$S_N{}^2$ mechanism.** Hydrolysis reaction (*i*) may also proceed through the following two steps which are based in $S_N{}^2$ mechanism.

$$\underset{(C.N. = 6)}{MA_5L + H_2O} \xrightarrow[\text{Slow (k)}]{\text{(i) Association } (+H_2\ O)} \underset{(C.N. = 7)}{MA_5L(H_2O)}$$

$$\xrightarrow[\text{Fast}]{\text{(ii) Dissociation } (-L)} \underset{(C.N. = 6)}{MA_5(H_2O) + L}$$

$$\text{Rate of reaction} = k\ [\text{Complex}]\ [H_2O]$$
$$= k\ [\text{Complex}]$$
$$= k\ [MA_5L]$$

It is clear from the above discussion that both the mechanisms predict that rate of hydrolysis reaction (*i*) is dependent only on the concentration of the complex, $MA_5$ L. Thus the measurement of the rate of hydrolysis reaction is not able to decide whether the hydrolysis reaction (*i*) proceeds through $S_N{}^1$ mechanism or through $S_N{}^2$ mechanism. Consequently we will have to look to some other factors to decide the type of mechanism. These factors are given below.

**1. Charge on the complex.** It has been observed during the hydrolysis of several octahedral complexes of Co(III) and other metal ions that *the rate of hydrolysis of a complex decreases with the increase in the charge on the complex.*

**Example.** The rate of hydrolysis of *cis* $[Co(en)_2Cl_2]^+$ (I) is some hundred times faster than that of *cis*- $[Co(en)_2Cl(H_2O)]^{2+}$ (II). Thus the hydrolysis of (I) is fast and that of (II) is slow as shown below

$$cis\ [Co^{3+}(en)_2\ Cl_2]^+ + H_2O \xrightarrow{\text{Fast}} cis\ [Co(en)_2\ Cl(H_2O)]^{2+} + Cl^-$$

(I)
(Charge = + 1)

$$cis\ [Co^{3+}\ (en)_2\ Cl\ (H_2O)]^{2+} + H_2O \xrightarrow{\text{Slow}} cis[Co\ (en)_2(H_2O)_2]^{3+} + Cl^-$$

(II) (Charge = + 2)

**Explanation.** The above observation can be explained by $S_N^1$ mechanism of hydrolysis reaction, since the increase in the positive charge on the complex makes the dissociation of the leaving group ($Cl^-$ ion) from the metal ion ($Co^{3+}$) more difficult and hence the rate of hydrolysis becomes slow.

If the hydrolysis is supposed to proceed through $S_N^2$ mechanism, the rate of hydrolysis would remain unchanged with the increase or decrease in the charge on the complex.

**2. Basicity of leaving group, $L^-$** If we study the rate of hydrolysis of complexes, $[Co(NH_3)_5 L]^{2+}$ which contain different $L^-$ ligands (leaving groups), we find that the rate of hydrolysis of these complexes decreases with the increase of the basicity of $L^-$ ligands. For example the rate of hydrolysis of the complexes, $[Co(NH_3)_5 L]^{2+}$ containing $L^- = CF_3COO^-$, $CCl_3COO^-$, $CH\,Cl_2\,COO^-$, $CH_2\,Cl\,COO^-$, $CH_3\,CH_2\,COO^-$ decreases from $CF_3\,COO^-$ to $CH_3\,CH_2\,COO^-$, since the basicity of these ligands increases in the same direction

*Basicity of $L^-$ ligands increases as:* $CF_3COO^- < CCl_3\,COO^- < CH\,Cl_2\,COO^-$
$< CH_2\,Cl\,COO^- < CH_3\,CH_2\,COO^-$

*Rate of hydrolysis decreases as:* $(CF_3COO^- > CCl_3COO^- > CHCl_2COO^-$
$> CH_2ClCOO^- > CH_3CH_2COO^-$

**Explanation.** Since the strength of $Co^{3+} - L^-$ bond is directly proportional to the basicity of $L^-$ ligand, with the increase in the basicity of $L^-$ ligands $Co^{3+}$ — $L^-$ bond becomes stronger and hence the rate of hydrolysis decreases. Now since the rate determining step in the hydrolysis reaction involves dissociation of $Co^{3+}$ — $L^-$ bond, the hydrolysis reaction proceeds through $S_N^1$ mechanism.

**3. Inductive effect of the inert group.** It has been seen that the rate constants of acid hydrolysis reaction:

$$[Co(en)_2\,(A\text{–}py)\,Cl]^{2+} + H_2O \rightarrow [Co\,(en)_2\,(A\text{–}py)\,(H_2O)]^{3+} + Cl^-,$$

increase with the $CH_3$ substitution in pyridine. In this reaction A–py stands for various derivatives of pyridine which are obtained by removing one of the H–atoms of pyridine by $CH_3$ group. A–py is an inert ligand, since it remains coordinated to the metal in the product, $[Co\,(en)_2\,(A\text{–}py)\,(H_2O)]^{3+}$. The increase in rate constants is due to the inductive effect caused by the increasing $CH_3$ substitution which results in distorting electron density towards Co–atom and thus helps the dissociation of $Cl^-$ ion (leaving group). This again confirms the fact that the acid hydrolysis reaction given above occurs through disccociative $S_N^1$ mechanism.

The values of rate constants (in $sec^{-1}$) for the acid hydrolysis reaction:

$$[Co(en)_2\,(A\text{–}py)\,Cl]^{2+} + H_2O \rightarrow [Co\,(en)_2\,(A\text{–}py)\,(H_2O)]^{3+} + Cl^-$$

in which A–py is pyridine, 3–methy pyridine and 4–methy pyridine are $1.1 \times 10^{-5}$, $1.3 \times 10^{-5}$ and $1.4 \times 10^{-5}$ respectively.

**4. Steric effects.** In the complexes of *trans* $[Co\,(AA)_2\,Cl_2]^+$ type, if the bidentate ligand AA is $NH_2\text{–}CH_2\text{–}\,CH_2 - NH_2$, $NH_2 - CH_2 - CH\,(CH_3) - NH_2$, *dl* $H_2N - CH\,(CH_3) - CH(CH_3) - NH_2$, $NH_2 - CH_2 - C\,(CH_3)_2$ and *meso* $H_2\,N\text{–}CH\,(CH_3) - CH\,(CH_3) - NH_2$, then the bulk of the ligand, (AA) increase, *i.e.* the ligand (AA) becomes more bulky. Due to the increase in the bulk of the ligand the steric overcrowding (hindrance) of the ligand round the central metal ion ($Co^{3+}$) also increases. Due to the increase in the overcrowding of the ligand round the central metal in ($Co^{3+}$), another ligand cannot be taken up by the complex, *i.e.* there is no possibility of $S_N^2$ mechanism. On the contrary the removal of a ligand will reduce the overcrowding of the ligand round the central metal ion. This gives the evidence of $S_N^1$ mechanism.

The increase in the bulk of the ligand (AA) also increases the value of rate constant for the acid hydrolysis reaction of the complexes as is evident from the following acid hydrolysis reactions.

(*i*) *trans* $[Co\,(AA)_2\,Cl_2]^+ \xrightarrow[+H_2O]{-Cl^-} [Co\,(AA)_2\,Cl\,(H_2O)]^{2+} + Cl^-$

$AA = NH_2 - CH_2 - CH_2 - NH_2,\ k = 3.2 \times 10^{-5}\ s^{-1}$

(*ii*) *trans* $[Co\,(AA)_2\,Cl_2]^+ \xrightarrow[+H_2O]{-Cl^-} [Co\,(AA)_2\,Cl\,(H_2O)]^{2+} + Cl^-$

$AA = NH_2 - CH_2 - CH\,(CH_3) - NH_2,\ k = 6.2 \times 10^{-5}\ s^{-1}$

(*iii*) *trans* $[Co\,(AA)_2\,Cl_2]^+ \xrightarrow[+H_2O]{-Cl^-} [Co(AA)_2\,Cl\,(H_2O)]^{2+} + Cl^-$

$AA = dl\ NH_2 - CH\,(CH_3) - CH\,(CH_3) - NH_2,\ k = 15.0 \times 10^{-5} s^{-1}$

(*iv*) *trans* $[Co\,(AA)_2\,Cl_2]^+ \xrightarrow[+H_2O]{-Cl^-} [Co\,(AA)_2\,Cl\,(H_2O)]^{2+} + Cl^-$

$AA = NH_2 - CH_2 - C\,(CH_3)_2 - NH_2,\quad k = 22.0 \times 10^{-5}\ s^{-1}$

(*v*) *trans* $[Co\,(AA)_2\,Cl_2]^+ \xrightarrow[+H_2O]{-Cl^-} [Co\,(AA)_2\,Cl\,(H_2O)]^{2+} + Cl^-$

$AA = meso\ H_2\,N{-}CH(CH_3) - CH(CH_3) - NH_2,\ k = 42.0 \times 10^{-5} s^{-1}$

Let us compare the rates of aquation of $[Co^{3+}\,(NH_2 - CH_2 - CH_2 - CH_2 - NH_2)_2\,Cl_2]^+$ (I) and $[Co^{3+}\,(NH_2 - CH_2 - CH_2 - NH_2)_2\,Cl_2]^+$ (II). In complex (I), $NH_2 - CH_2 - CH_2 - CH_2 - NH_2$ is propylenediamine and in complex (II), $NH_2 - CH_2 - CH_2 - NH_2$ is ethylenediamine.

Both the complexes are chelated octahedral complexes. Complex (I) contains 6–membered chelate rings while complex (II) has 5–membered chelated rings. Now since 6–membered chelate ring present in (I) produces greater steric strain round the central $Co^{3+}$ ion than 5–membered chelate ring present in (II), the aquation of (I) will be fast than that of (II) through dissociative $S_N^1$ mechanism (See Fig. 27.3)

The value of *k* for the aquation of complex (I) is much higher than that of complex (II). Experimentally this prediction has been found true.

**5. Solvation effects**. According to solvation theory in aqneous medium, the reacting complexes, the intermediates formed and the products obtained are all in the hydrated state. The hydrated state of a species is represented by putting a subscripts, *hyd* in the species.

Let us compare the rate of aquation of the complexes $[Co(NH_3)_5\,Cl]^{2+}$ (I) and $[Co\,(en)\,(NH_3)_3\,Cl]^{2+}$ (II). Complex (I) is smaller in size than complex (II), since complex (I) has only monodentate ligands while complex (II) has chelated as well

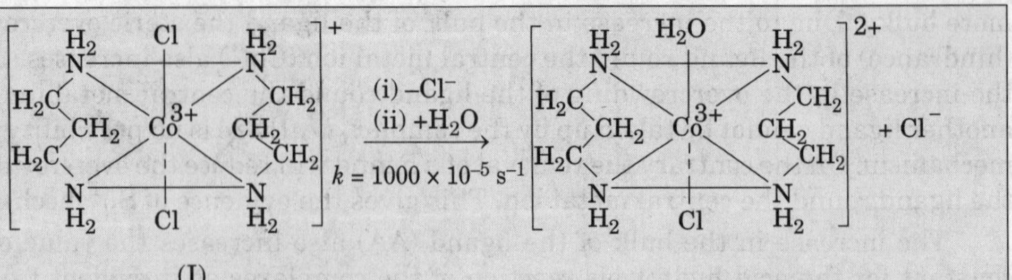


(I)
Chelated complex containing
6-membered chelate rings

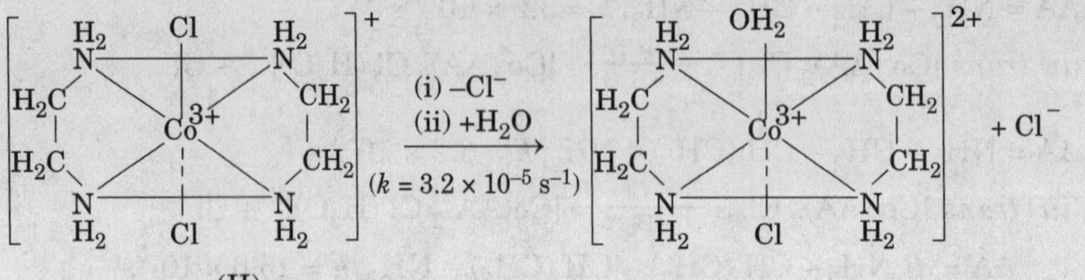


(II)
Chelated complex containing
5-membered chelate rings

**Fig. 27.3** Comparison of rates of aquation of $[Co(NH_2–CH_2–CH_2–CH_2–NH_2)_2 Cl_2]^+$ (I) and $[Co (NH_2–CH_2–CH_2–NH_2)_2 Cl_2]^+$ (II) complexes.

as monodentate ligands. If the aquation of the complexes is assumed to proceed through dissociative $S_N{}^1$ mechanism, then the aquation of these complexes can be represented as:

$$[Co (NH_3)_5 Cl]^{2+}_{hyd} \xrightarrow[\text{Slow}]{-Cl^-} [Co (NH_3)_5]^{3+}_{hyd} + Cl^-_{hyd}$$

Non–chelated complex (smaller in size) (I) — Intermediate (smaller in size) (III)

$$[Co (NH_3)_5]^{3+}_{hyd} \xrightarrow[\text{Fast}]{+H_2O} [Co(NH_3)_5 (H_2O)]^{3+}_{hyd}$$

Intermediate (III)

$$[Co (en) (NH_3)_3 Cl]^{2+}_{hyd} \xrightarrow[\text{Slow}]{-Cl^-} [Co (en) (NH_3)_3]^{3+}_{hyd} + Cl^-_{hyd}$$

Chelated complex (bigger in size) (II) — Intermediate (bigger in size) (IV)

$$[Co (en) (NH_3)_3]^{3+}_{hyd} \xrightarrow[\text{Fast}]{+H_2O} [Co (en) (NH_3)_3 (H_2O)]^{3+}_{hyd}$$

Intermediate (IV)

Now since intermediate (IV) is obtained from bigger complex (II) and intermediate (III) is obtained from smaller complex (I), (IV) is bigger in size than (III). Being bigger in size, the intermediate (IV) is stablised by hydration to a lesser extent and hence requires more energy for its formation than the intermediate (III). As a result, complex (II) would be aquated at a slower rate than complex (I).

Since the steric overcrowding around the central metal ion in complexes $[Co(NH_3)_4 Cl_2]^+$ and $[Co (en)_2 Cl_2]^+$ is almost the same, the difference in the rate of aquation of these complexes can be interpreted on the basis of solvation effect. The rate of aquation of these complexes are as: $[Co (NH_3)_4 Cl_2]^+ = 180 \times 10^{-3}$ $s^{-1}$, $[Co (en)_2 Cl_2]^+ = 3.2 \times 10^{-5}$ $s^{-1}$

## II. Octahedral Complexes in Which the Inert Ligand is a pi Donor.

Let is study the aquation reaction of *cis* $[Co (en)_2 (OH) Cl]^+$ and *cis* $[Co (en)_2 (NH_3) Cl]^{2+}$. The aquation reaction of these complexes can be represented as shown in Fig. 27.4. In *cis* $[Co(en)_2] (OH) Cl]^+$, $OH^-$ ligand is a pi donor inert ligand.

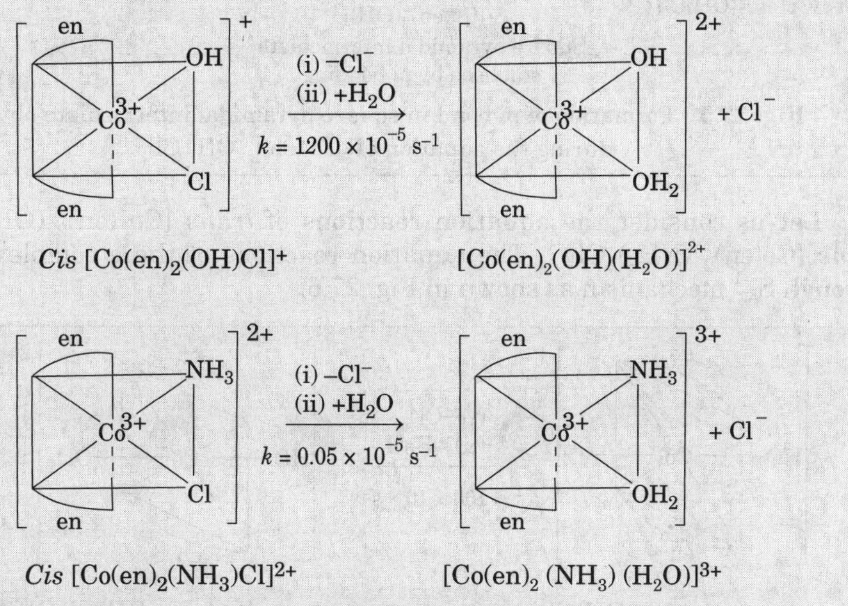


**Fig. 27.4** Aquation of *cis* $[Co(en)_2 (OH) Cl]^+$ and cis $[Co (en)_2 (NH_3) Cl]^{2+}$

It may be seen from the values of rate constant ($k$) that the rate of aquation of *cis* $[Co (en)_2 (OH) Cl]^+$ is much higher than that of *cis* $[Co(en)_2 (NH_3) Cl]^{2+}$. The difference in the values of $k$ of the two complexes can be explained on the basis of the capacity of $OH^-$ and $NH_3$ ligands to form pi bonding. $OH^-$ has filled p–orbital while $NH_3$ molecule has no such orbital. The lone pair of electrons on N–atom in $NH_3$ molecule is used up in coordination. The central metal ion ($Co^{3+}$ ion) of square pyramidal intermediate formed during the aquation of *cis* $[Co(en)_2 (OH) Cl]^+$ has one empty $d^2 sp^3$ hybrid orbital. This empty $d^2sp^3$ hybrid orbital overlaps with the filled $p$ orbital on $OH^-$ ion to form a pi bond as shown in Fig. 27.5.

It is due to the formation of pi-bond that the square pyramidal intermediate is stablised. As a result, the aquation of *cis* $[Co(en)_2 (OH) Cl]^+$ becomes easier that the aquation of *cis* [Co (en)$_2$ (NH$_3$) Cl]$^{2+}$ whose square pyramidal intermediate cannot be stabilised by pi bonding.

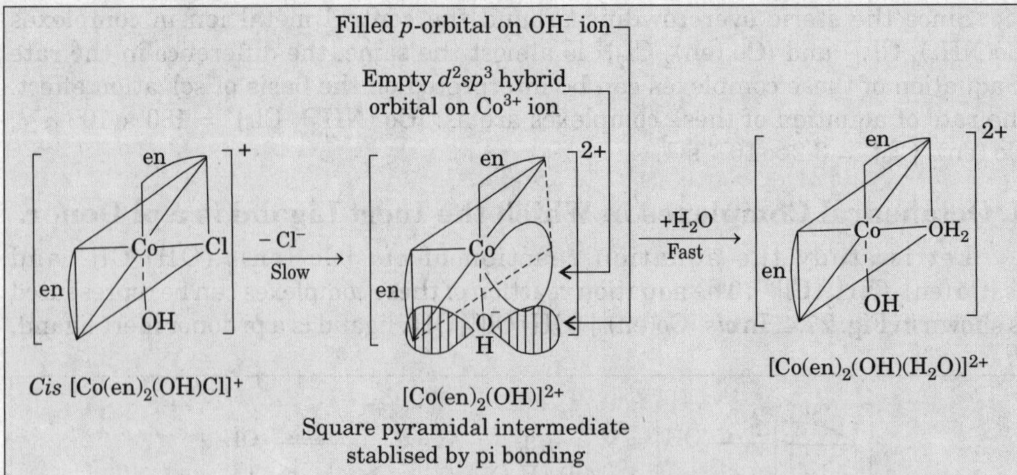


**Fig. 27.5** Formation of pi bond in square pyramidal intermediate obtained during the aquation of [Co (en)$_2$ (OH) Cl]$^+$.

Let us consider the aquation reactions of *trans* [Co (en)$_2$ (OH) Cl]$^+$ and *trans* [Co(en)$_2$ (NH$_3$) Cl]$^{2+}$. The aquation reactions of these complexes proceed through $S_N^{\,1}$ mechanism as shown in Fig. 27.6.

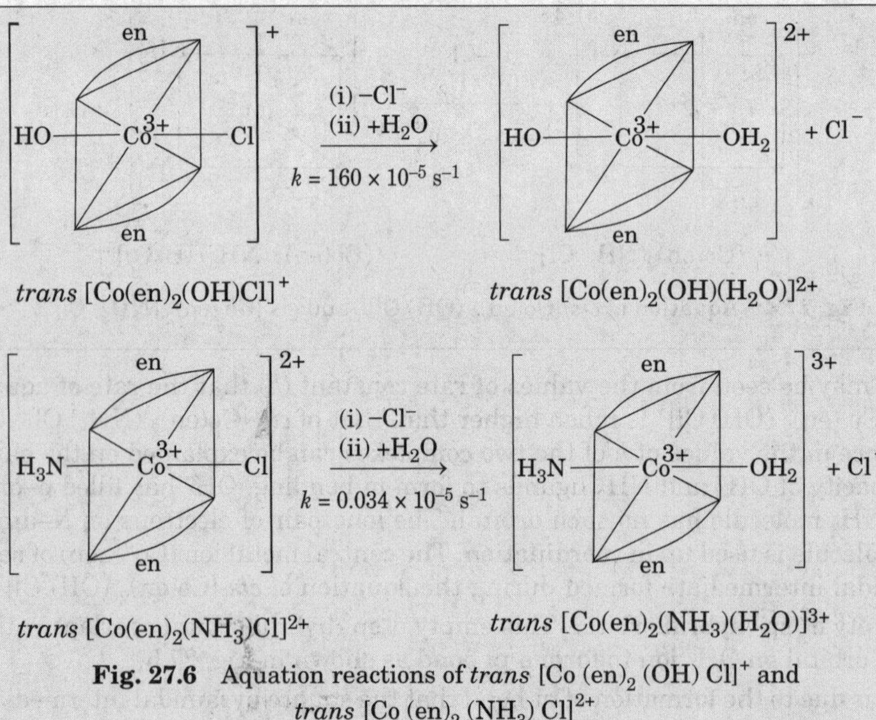


**Fig. 27.6** Aquation reactions of *trans* [Co (en)$_2$ (OH) Cl]$^+$ and *trans* [Co (en)$_2$ (NH$_3$) Cl]$^{2+}$

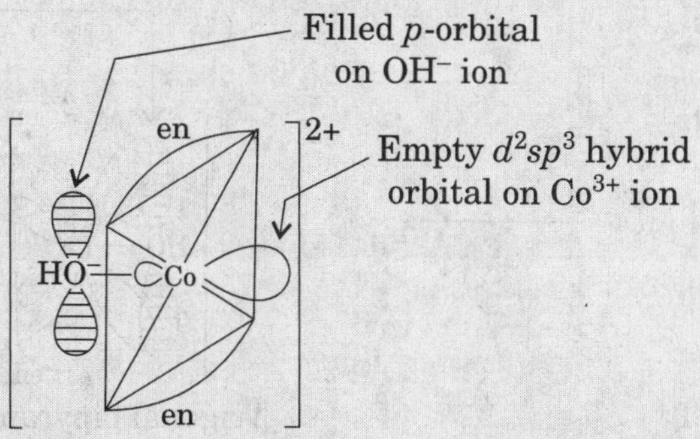


**Fig. 27.7** Since $d^2sp^3$ hybrid orbital on $Co^{3+}$ ion and p–orbital on $OH^-$ do not have same symmetry, they cannot overlap to form pi bond

It may be seen from the values of rate constant ($k$) that the rate of aquation of *trans* $[Co(en)_2 (OH) Cl]^+$ is much faster than that of *trans* $[Co(en)_2 (OH) Cl]^{2+}$. Since the $d^2sp^3$ hybrid orbital (empty) on $Co^{3+}$ ion and $p$ orbital (filled) on $OH^-$ ion do not have the same symmetry as is evident from Fig. 27.7, these orbitals are incapable of overlapping. Thus no pi bond is formed. In other words we can say that in the aquation of *trans* $[Co(en)_2 (OH) Cl]^+$ square pyramidal intermediate is not stabilised and hence is not formed. In place of square pyramidal intermediate, trigonal bipyramidal intermediate is formed because it can be stabilised by pi bond which is formed by the overlap between the filled p–orbital on $OH^-$ ion and empty d–orbital on $Co^{3+}$ ion (See Fig. 27.8)

The pi bonded trigonal bipyramidal intermediate is more stable than square pyramidal intermediate. The aquation of *trans* $[Co(en)_2 (OH) Cl]^+$ gives a mixture of *cis* and *trans* isomers of $[Co(en)_2 (OH)(H_2O)]^{2+}$ as shown in Fig. 27.8. If may be noted that the aquation of *trans* $[Co(en)_2(NH_3) Cl]^{2+}$ gives only *trans* $[Co (en)_2 (NH_3) (H_2O)]^{3+}$.

From the above discussion the following points may be noted:

(i) The aquation reactions discussed above proceed through dissociative $S_N{}^1$ mechanism.

(ii) Like $OH^-$, $NH_2^-$, $Cl^-$, $Br^-$ etc are also pi bonding inert ligands (pi donors). Thus the mechanism of aquation of the complexes containing these pi bonding inert ligands ($NH_2^-$, $Cl^-$, $Br^-$ etc) is the same as that of the complexes containing $OH^-$.

(iii) The aquation of a *cis* isomer containing a pi inert ligand gives square pyramidal intermediate while the aquation of a *trans* isomer gives trigonal bipyramidal intermediate.

(iv) When square pyramidal intermediate is formed, the aquation reaction gives complex of the same geometry. On the other hand when trigonal bipyramidal is formed, the aquation reaction gives a mixture of *cis* and *trans* isomers.

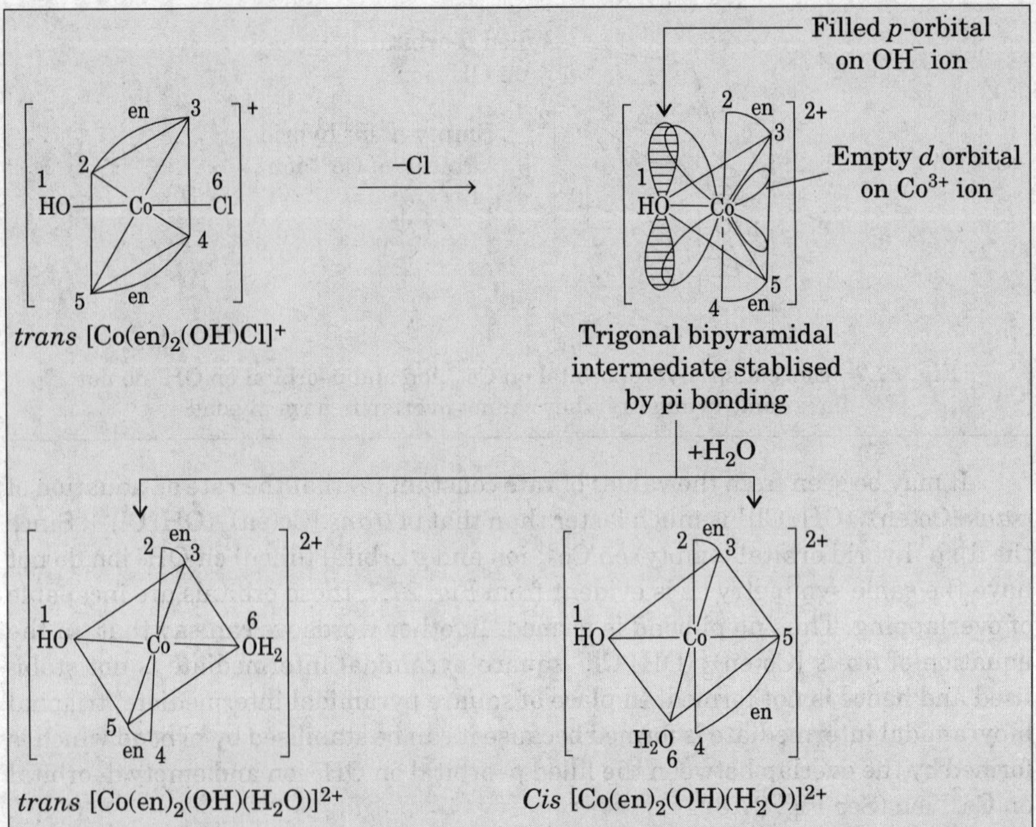


**Fig. 27.8** Aquation of *trans*-$[Co(en)_2 (OH) Cl]^+$ gives a mixture of *cis* and *trans* isomers of $[Co(en)_2 (OH) (H_2O)]^{2+}$.

## III. Octahedral Complexes in Which the Inert Ligand is a pi Acceptor

Aquation of octahedral complexes in which the inert ligand is a pi acceptor proceeds through associative $S_N^2$ mechanism. Thus the aquation of $[Co(en)_2(NO_2) Cl]^+$ in which the ligand, $NO_2^-$ is an inert pi acceptor and the ligand, $Cl^-$ is the leaving group proceeds through associative $S_N^2$ mechanism as shown below:

$$[Co (en)_2 (NO_2) Cl]^+ \xrightarrow[\text{(ii) } -Cl^-]{\text{(i) } +H_2O} [Co(en)_2 (NO_2) (H_2O)]^{2+} + Cl^- \qquad (i)$$

The inert pi acceptor ligand, $NO^-_2$ is called *spectator ligand*. Depending on whether the inert pi acceptor ligand ($NO_2^-$) is placed *trans* or *cis* to the leaving group ($Cl^-$) in the octahedral complex, the following two cases may be studied.

(*a*) **Aquation of trans $[Co(en)_2 (NO_2) Cl]^+$ complex.** In this complex $NO_2^-$ and $Cl^-$ are placed *trans* to each other. In this complex one of the filled $t_{2g}$ orbitals of $Co^{3+}$ ion can overlap with the empty p–orbital of $NO_2^-$ ligand to form a pi bond (See Fig. 27.9). The formation of pi bond strengthens the Co – Cl bond and hence the dissociation of Co – Cl bond to release the leaving group ($Cl^-$) becomes difficult.

Thus the aquation of *trans* $[Co(en)_2 (NO_2) Cl]^+$ cannot proceed through dissociation $S_N^1$ mechanism. Due to the presence of pi–acceptor ligand ($NO^-_2$) in *trans*

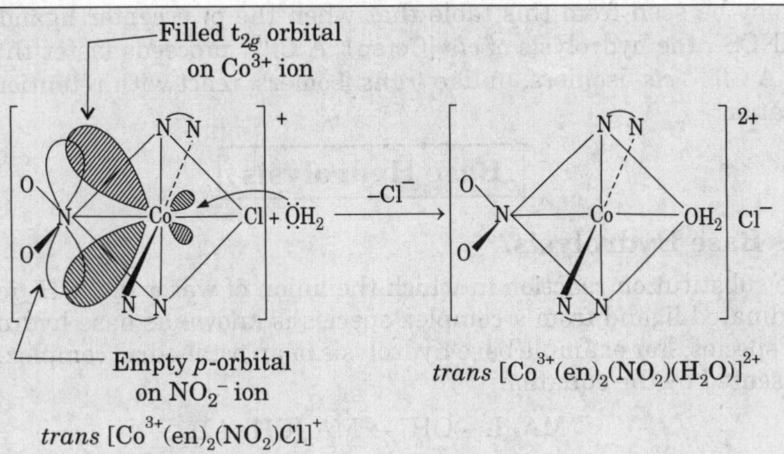


**Fig. 27. 9** Formation of pi bond by the overlap between filled $t_{2g}$ orbital of $Co^{3+}$ ion and empty p orbital of $NO_2^-$ ion during the aquation of *trans* $[Co(en)_2 (NO_2) Cl]^+$.

complex, the formation of Co – $OH_2$ bond is easier and hence the aquation of the complex proceeds through associative $S_N^2$ mechanism as represented by reaction (*i*).

(*b*) **Aquation of *cis* $[Co(en)_2 (NO_2) Cl]^+$ complex.** In this complex $NO_2^-$ and $Cl^-$ are placed *cis* to each other. In this complex, since $NO_2^-$ ion is *cis* to $Cl^-$ ion, the extent of overlap between one of the filled $t_{2g}$ orbitals of $Co^{3+}$ ion and empty p-orbital of $NO^-_2$ ion is less than when $NO^-_2$ ion is placed *trans* to $Cl^-$ ion. Hence the formation of Cu – $OH_2$ bond would not be easy in *cis* complex. This proves that the aquation of *cis* complex also proceeds through associative $S_N^2$ mechanism and not through dissocciative $S_N^1$ mechanism

The values of rate constants and % *cis* product obtained by the aquation of *cis* and *trans* isomers of $[Co(en)_2 A Cl]^{n+}$ represented as:

$$[Co(en)_2 A Cl]^{n+} + H_2O \xrightarrow{25°C} [Co(en)_2 A (H_2O)]^{(n+1)+} + Cl^-$$

are given in Table 27.4. A is the pi acceptor ligand and $Cl^-$ ion is the leaving group.

**Table 27.4 Rate constants and % *cis* product obtained in aquation of $[Co(en)_2 A Cl]^{n+}$ represented as: $[CoClA(en_2)]^{n+} + H_2O \xrightarrow{25°C} [Co(H_2O) A (en)_2]^{(n+1)+} + Cl^-$**

| Isomer | Pi acceptor ligand (A) | $10^7 k$, $s^{-1}$ | % *cis* product |
|---|---|---|---|
| *cis* | $OH^-$ | 1,30,000 | 100 |
| *cis* | $N_3^-$ | 2,400 | 100 |
| *cis* | $Cl^-$ | 2,400 | 100 |
| *cis* | $NCS^-$ | 110 | 100 |
| *trans* | $OH^-$ | 14,000 | 75 |
| *trans* | $N_3^-$ | 2,500 | 20 |
| *trans* | $Cl^-$ | 350 | 35 |
| *trans* | $NCS^-$ | 0.5 | 50–70 |

It may be seen from this table that when the pi acceptor ligand, A is $OH^-$, $Cl^-$ and $NCS^-$, the hydrolysis of *cis* $[Co(en)_2 A Cl]^{n+}$ proceeds faster that the *trans* $[Co(en)_2 A Cl]^{n+}$. *cis*–isomers, unlike *trans* isomers, react with retention with their configuration

## Base Hydrolysis

### What is Base Hydrolysis?

The substitution reaction in which the anion of water (*i.e.* $OH^-$ ion) replaces the coordinated ligand from a complex species is known as base hydrolysis of the complex species. For example base hydrolysis of an octahedral complex, $MA_5 L$ can be represented by the equation.

$$MA_5 L + OH^- \rightarrow MA_5(OH) + L^-$$

In this reaction the coordinated ligand (L) is replaced by $OH^-$ ion. Thus L is the leaving group and $OH^-$ ion is the entering group. Here we shall discuss the base hydrolysis of octahedral ammine complexes of Co (III) like $[Co (NH_3)_5 Cl]^{2+}$. Base hydrolysis of $[Co(NH_3)_5 Cl]^{2+}$ can be represented by reaction:

$$[Co^{3+} (NH_3)_5 Cl]^{2+} + OH^- \rightarrow [Co(NH_3)_5 (OH)]^{2+} + Cl^-$$

### Mechanism to Explain the Base Hydrolysis of $[Co^{3+}(NH_3)_5 Cl]^{2+}$

Following two mechanisms have been proposed to explain the base hydrolysis of $[Co(NH_3)_5 Cl]^{2+}$ represented by the following equation:

$$[Co(NH_3)_5 Cl]^{2+} + OH^- \rightarrow \underset{\text{Hydroxo complex}}{[Co(NH_3)_5 (OH)]^{2+}} + Cl^- \qquad (i)$$

**1. Associative $S_N^2$ mechanism.** This mechanism proceeds through the following steps:

(*a*) $\underset{\text{Ammine complex (Acid)}}{[Co(NH_3)_5 Cl]^{2+}} + \underset{\text{Base}}{OH^-} \xrightarrow[\text{Slow }(k)]{\text{Association }(+OH^-)} \underset{\text{7 –coordinated intermediate}}{[Co(NH_3)_5 Cl (OH)]^+}$

(*b*) $[Co(NH_3)_5 Cl (OH)]^+ \xrightarrow[\text{Fast}]{\text{Dissociation }(-Cl^-)} \underset{(C.N = 6)}{[Co (NH_3)_5 (OH)]^{2+}} + Cl^-$

**Rate of base hydrolysis reaction (i).** Since step (a) is a slow step, it is the rate determining step. Thus the rate law for reaction (i) is given by:

$$\text{Rate (r)} = k \times \text{(Conc. of complex ion)} \times (\text{Conc. of } OH^- \text{ ion})$$

$$\text{or} \quad r = k \times \text{Conc. of } [Co (NH_3)_5 Cl]^{2+} \times [OH^-] \qquad (ii)$$

$$= k \times \text{[Complex] [Base]}$$

The steps given at (a) and (b) can be combined together as shown below.

$$\underset{\text{Ammine complex}}{[Co(NH_3)_5 Cl]^{2+}} \xrightarrow[\text{Slow }(k)]{\text{Association }(+OH^-)} \underset{\text{7 –coordinated intermediate}}{[Co(NH_3)_5 Cl (OH)]^+}$$

$$\xrightarrow[\text{Fast}]{\text{Dissociation }(-Cl^-)} \underset{(C.N. = 6)}{[Co (NH_3)_5 (OH)]^{2+}} + Cl^-$$

**Limitation of $S_N^2$ mechanism.** $S_N^2$ mechanism is not able to explain the following experimental observations.

(i) The rate law equation (*ii*) given above shows that the rate of base hydrolysis reaction (*i*) depends on the concentration of the attacking nucleophile ($OH^-$ ion). However, at high concentration of $OH^-$ ions, the reaction rate becomes independent of the concentration of $OH^-$ ions. Hence at high concentration of $OH^-$ ions, the base hydrolysis reaction is first order reaction w.r.t the ammine complex ion, since its rate depends on the concentration of the complex ion only. Thus at high concentration of $OH^-$ ions, the rate of reaction (r) is given by:

$$r = k\ [\text{Complex}] \qquad (iii)$$

(ii) Although $NCS^-$ $NO_2^-$, $N_3^-$ etc. nucleophiles are as strong as $OH^-$, their concentration does not affect the rate of hydrolysis of ammine complex of Co (III). Note that as is evident from rate law equation (*ii*), concentration of $OH^-$ ions affects of rate of hydrolysis reaction. Thus we see that $S_N^2$ mechanism cannot explain why the rate of reaction depends on the concentration of $OH^-$ ions and not on the concentration of other nucleophiles which are as strong as $OH^-$.

**2. Dissociative $S_N^1$ (CB) mechanism.** This mechanism was proposed by Garrick. $S_N^1$ (CB) stards for substitution (S), nucleophile (N), uni–molecular (1) and conjugate base (CB). Following steps are involved in this mechanism to explain the base hydrolysis of $[Co(NH_3)_5Cl]^{2+}$ represented by reaction (*i*).

(*a*) $[Co(NH_3)_5Cl]^{2+}$ (Ammino complex (Acid)) + $OH^-$ (Base) $\xrightleftharpoons{K}$ $[Co(NH_3)_4(NH_2)Cl]^+$ (Conjugate base (CB) of ammine complex (Amido complex) (C.N. = 6)) + $H_2O$ (Acid)

Above reaction is an acid–base reaction as shown below.

Conjugate acid-base pair

$[Co(NH_3)_5Cl]^{2+}$ (Acid$_1$, Ammino complex) + $OH^-$ (Base$_2$) $\xrightleftharpoons{\text{Fast}}$ $[Co(NH_3)_4(NH_2)Cl]^+$ (Base$_1$, Amido complex, CB) + $H_2O$ (Acid$_2$)

Conjugate acid-base pair

In this acid – base reaction the ammino complex acts as a Bonsted acid and $OH^-$ ion acts as a Bronsted base. $NH_3$ groups present in the ammino complex donate a proton ($H^+$) to $OH^-$ ion which is, therefore, converted into $H_2O$. The ammino complex, $[Co(NH_3)_5Cl]^{2+}$ is converted into its conjugate base (CB), $[Co(NH_3)_4(NH_2)Cl]^+$.

The equilibrium constants K of the above acid–base reaction is given by:

$$K = \frac{[CB][H_2O]}{\text{Conc. of } [Co(NH_3)_5Cl]^{2+}[OH^-]}$$

$\therefore$ $[CB] = K \times \text{Conc of } [Co(NH_3)_5Cl]^{2+} \times [OH^-]$

or $[CB] = K\ [\text{Complex}]\ [\text{Base}]$ ---------------------------------------- ... (*a*)

(*b*) CB of the ammino complex, $[Co(NH_3)_4\ (NH_2)\ Cl)]^+$ contains $NH_2^-$ which, being a strong donor, accelerates the loss of $Cl^-$ ion from the coordination zone and 5–coordinated intermediate is formed.

$$\underset{\text{CB (C.N. = 6)}}{[Co(NH_3)_4\ (NH_2)\ Cl]^+} \xrightarrow[\text{Slow } (k)]{\text{Dissociation } (-Cl^-)} \underset{\substack{\text{5-coordinated} \\ \text{intermediate (C.N. = 5)}}}{[Co\ (NH_3)_4\ (NH_2)]^{2+}} + Cl^-$$

(*c*) 5- coordinated intermediate formed in step (*b*) quickly reacts with $H_2O$ to form the final product, $[Co\ (NH_3)_5\ (OH)]^{2+}$

$$\underset{\substack{\text{5–coordinated} \\ \text{intermediate}}}{[Co(NH_3)_4\ (NH_2)]^{2+}} + H_2O \xrightarrow[\text{Fast}]{\text{Association } (+\ H_2O)} \underset{\substack{\text{Hydroxo complex} \\ \text{(Final product)}}}{[Co\ (NH_3)_5\ (OH)]^{2+}}$$

**Rate of base hydrolysis of reaction (i).** Since the step (b) involves the dissociation of the conjugate base (CB) to get $Cl^-$ ion, the reaction of this step is supposed to be slower than the reactions given at steps (a) and (c). Hence the reaction given at step (b) is the rate determining reaction of the overall hydrolysis reaction (i). Thus rate of hydrolysis (r) is given by the rate law

$$r = k \times \text{Conc. of } [Co(NH_3)_4\ (NH_2)\ Cl]^+$$

$$r = k \times [CB] \quad \ldots\ldots (b)$$

Put the value of [CB] from equation (a) in equation (b) to get.

$$r = kK\ [\text{Complex}]\ [\text{Base}] \quad \ldots\ldots (c)$$

Although base hydrolysis involves an $S_N{}^1$ mechanism, yet it is consistent with second–order: first order with respect to the complex and first order with respect to the base (See Equn. (c).)

Since the $S_N{}^1$ dissociation step (b) which is rate–determining step uses the conjugate base of the initial complex, the symbol $S_N{}^1$ CB (*substitution, nucleophile, conjugate–base*) has been used by Garrick in place of $S_N{}^1$ symbol. The reactions of $S_N{}^1$ (CB) mechanism given at steps (a), (b) and (c) can be combined together as shown below:

$$\underset{\substack{\text{Ammino complex} \\ \text{(C.N. = 6)}}}{[Co(NH_3)_5\ Cl]^{2+}} + OH^- \overset{K}{\rightleftharpoons} \underset{\substack{\text{Conjugate base (CB)} \\ \text{of ammino complex}}}{[Co(NH_3)_4\ (NH_2)\ Cl]^+} + H_2O$$

$$\Big\downarrow \text{Slow } (k) \quad \text{Dissociation } (-\ Cl^-)$$

$$\underset{\text{5–coordinated intermediate}}{[Co(NH_3)_4\ (NH_2)]^{2+} + Cl^-}$$

$$\Big\downarrow \text{Fast} \quad \text{Association } (+H_2O)$$

$$\underset{\substack{\text{Hydroxo complex} \\ \text{(Final product)}}}{[Co(NH_3)_5\ (OH)]^{2+}}$$

## To Explain Base Hydrolysis of Different Octahedral Complexes.

**(i) Base hydrolysis of $[Co(NH_3)_5Cl]^{2+}$ by $OH^-$ base of high concentration.** We know that if the concentration of $OH^-$ ions is very high, the rate of base hydrolysis of $[Co(NH_3)_5Cl]^{2+}$ is independent of the concentration of $OH^-$ ions *i.e.* the rate of hydrolysis depends only on the concentration of the reacting complex. This fact cannot be explained by $S_N{}^2$ mechanism. However, this observation can be explained by $S_N{}^1$ (CB) mechanism as follows. If the amount of $OH^-$ ions added is very high, there would be very little change in the concentration of $OH^-$ ions as a result of acid – base reaction,

$$[Co(NH_3)_5Cl]^{2+} + OH^- \rightleftharpoons [Co(NH_3)_4(NH_2)Cl]^+ + H_2O$$

In other words we can say that the concentration of $OH^-$ ions can be taken as constant. Hence the equation,

$$r = kK\,[\text{Initial complex}]\,[\text{Base}]$$

would be converted into:

$$r = kK\,[\text{Initial complex}] \times [\text{Constant}]$$
$$r = kK \times \text{Conc. of } [Co(NH_3)_5Cl]^{2+}$$

This relation shows that at high concentration of $OH^-$ ions the rate of hydrolysis reaction depends only on the concentration of the complex and remains independent of the concentration of $OH^-$ ions.

**(ii) Hydrolysis of $[Co(NH_3)_5Cl]^{2+}$ by $NO_2^-$ $NCS^-$ and $N_3^-$ bases.** We know that all the anions *viz* $OH^-$, $NO_2^-$, $NCS^-$, $N_3^-$ etc. are bases. We also know that the concentration of $OH^-$ ions affects the rate of hydrolysis of $[Co(NH_3)_5Cl]^{2+}$ brought about by $OH^-$, but if the hydrolysis of $[Co(NH_3)_5Cl]^{2+}$ is brought about by $NO_2^-$, $NCS^-$, $N_3^-$ bases, the concentration of these bases does not affect the rate of hydrolysis, Why? This observation can be explained by $S_N{}^2$ (CB) mechanism as follows.

The anions $NO_2^-$, $NCS^-$, $N_3^-$ etc. are weaker bases than $OH^-$ and hence none of these anions is able to extract a proton ($H^+$) from $[Co(NH)_3Cl]^{2+}$ to form a conjugate base of $[Co(NH_3)_5Cl]^{2+}$. Due to the non–formation of conjugate base, the hydrolysis of $[Co(NH_3)_5Cl]^{2+}$ by $NO_2^-$, $NCS^-$, $N_3^-$ bases cannot proceed through $S_N{}^1$ (CB) mechanism.

$S_N{}^2$ and $S_N{}^1$ mechanisms can well explain why the hydrolysis of $[Co(NH_3)_5Cl]^{2+}$ in presence of $NO_2^-$ $NCS^-$ or $N_3^-$ anions remains independent of the concentration of these anions. This has been shown below:

Since during hydrolysis, $H_2O$ is also present in excess than the anions $NO_2^-$, $NCS^-$ or $N_3^-$, $H_2O$ acts a nucleophile and attacks $[Co(NH_3)_5Cl]^{2+}$ to form the aquated complex ion $[Co(NH_3)_5(H_2O)]^{3+}$ in both the mechanisms as shown below:

**$S_N{}^2$ mechanism**

$$\underset{(\text{C.N.} = 6)}{[Co(NH_3)_5Cl]^{2+}} + H_2O \xrightarrow[\text{Slow }(k)]{\text{Association }(+H_2O)} \underset{(\text{C.N.} = 7)}{[Co(NH_3)_5Cl(H_2O)]^{2+}}$$

$$\xrightarrow[\text{Fast}]{\text{Dissociation }(-Cl^-)} [Co(NH_3)_5(H_2O)]^{2+} + Cl^-$$

$$\text{Rate of hydrolysis reaction (r)} = k \times \text{Conc. of } [Co(NH_3)_5Cl]^{2+} \times [H_2O]$$
$$= k[\text{Complex}]$$

**$S_N^1$ mechanism**

$$[Co(NH_3)_5\,Cl]^{2+} \xrightarrow[\text{Slow }(k)]{\text{Dissociation }(-Cl^-)} Cl^- + [Co(NH_3)_5]^{3+}$$

$$\downarrow \text{Fast} \quad \text{Association } (+H_2O)$$

$$[Co(NH_3)_5\,(H_2O)]^{3+}$$

$$\text{Rate of hydrolysis reaction (r)} = k \times \text{Conc. of } [Co(NH_3)_5Cl]^{2+}$$
$$= k\ [\text{Complex}]$$

It may be seen that the rate of hydrolysis (r) according to both the mechanism is the same which is equal to $k \times$ [Complex]. Thus the rate of hydrolysis is independent of the concentration of $NO_2^-$ $NCS^-$ or $N_3^-$ ions.

**(iii) Hydrolysis of *cis* and *trans* isomers of $[M(en)_2\,Cl_2]^+$ (M = $Co^{3+}$ or $Rh^{3+}$).** We know that $[M(en)_2\,Cl_2]^+$ [M = Co (III) or Rh (III)] exists in *cis* and *trans* isomers [en = $NH_2 - (CH_2)_2 - NH_2$]. *Trans* $[M\,(en)_2\,Cl_2]^+$ isomer contains $NH_2$ groups of en *cis* to $Cl^-$ ions while *cis* $[M\,(en)_2\,Cl_2]^+$ isomer contains $NH_2$ groups of en *trans* to $Cl^-$ ions.

Experimentally it has been found that the base hydrolysis of *cis* $[M\,(en)_2\,Cl_2]^+$ is quicker than that of *trans* $[M(en)_2Cl_2]^+$. This can be explained with the help of $S_N^1$ (CB) mechanism as follows:

It has been found by NMR spectroscopy that in $[M\,(en)_2\,Cl_2]^+$ ion H–atoms of $NH_2$ groups in *en* which are *trans* to $Cl^-$ ion are more acidic than the H-atoms of $NH_2$ groups in *en* which are *cis to* $Cl^-$ ions. Now since both metal ions *viz* $Co^{3+}$ and $Rh^{3+}$ in $[M(en)_2\,Cl_2]^+$ have the same configuration (= $t_{2g}^6$), the removal of proton ($H^+$) from $NH_2$ group *trans* to $Cl^-$ ion is easier than its removal from $NH_2$ group *cis* to $Cl^-$. Thus the formation of CB of *cis* $[M(en)_2\,Cl_2]^+$ isomer should be easier than the formation of CB of *trans* $[M(en)_2\,Cl_2]^+$ isomer. Thus the base hydrolysis of *cis* $[M(en)_2\,Cl_2]^+$ isomer should be quicker than that of *trans* $[M\,(en)_2\,Cl_2]$ isomer.

**(iv) Hydrolysis of complex ions containing no acidic hydrogen.** If a complex ion undergoes hydrolysis through $S_N^1$ (CB) mechanism, the complex ion must contain at least one acidic hydrogen so that the complex ion may form its conjugate base. There are some complex ions which do not contain any acidic hydrogen but undergo hydrolysis. The rate of their hydrolysis is independent of the concentration of $OH^-$ ions. Thus the hydrolysis of such complexes does not occur through $S_N^1$ (CB) mechanism. Their hydrolysis occurs through a mechanism other than $S_N^1$ (CB) mechanism. Examples of such complexes are given below:

(a) **Hydrolysis of $[Co(CN)_5\,Br]^{3-}$ and $[Co(CN)_5I]^{3-}$.** It has been found experimentally that the complex ions *viz* $[Co(CN)_5\,Br]^{3-}$ and $[Co\,(CN)_5\,I]^{3-}$ hydrolysis slowly to form $[Co(CN)_5\,(OH)]^{3-}$ and the rate of their hydrolysis is independent of the concentration of $OH^-$ ions. Since the complex ions do not contain any acidic hydrogen, the hydrolysis of these ions does not occur through $S_N^1$ (CB) mechanism.

(b) **Hydrolysis of $[Co(py)_4\,Cl_2]^+$ and $[Co\,(dipy)_2\,(O\,Ac)_2]^+$.** The complex ions $[Co(py)_4Cl_2\,]^+$ and $[Co\,(dipy)_2\,(OAc)_2]^+$ (py = pyridyl and dipy = dipyridyl) hydrolyse in the pH range of 6 — 12 and the rates of their hydrolysis are independent of the concentration of $OH^-$ ions. Since neither py nor *dipy* has any acidic H-atom, the hydrolysis of the given complex ions does not occur through $S_N^1$ (CB) mechanism.

**(v) Hydrolysis of $[Fe(CN)_5(NH_3)]^{3-}$.** Although the given complex ion has removable acidic hydrogen which is present in $NH_3$, but this H-atom is removed with great difficulty due to high negative charge ( = –3) on the complex ion, $[Fe(CN)_5(NH_3)]^{3-}$. Consequently it is extremely difficult for this complex ion to form its conjugate base. Hence the hydrolysis of $[Fe(CN)_5(NH_3)]^{3-}$ does not proceed through $S_N^1$ (CB) mechanism and the rate of hydrolysis of the complex ion is independent of the concentration of $OH^-$ ions.

**(vi) Catalytic action of $OH^-$ ion in non–aqueous medium.** The reaction of $[Co^{3+}(en)_2(NO_2)Cl]^+$ with $NO_2^-$ produces $[Co^{3+}(en)_2(NO_2)_2]^+$ in non–aqueous medium of dimethly sulphoxide (DMSO)

$$[Co(en)_2(NO_2)Cl]^+ + NO_2^- \xrightarrow{DMSO} [Co(en)_2(NO_2)_2]^+ + Cl^-$$

The rate of this reaction is slow and is independent of the concentration of $NO_2^-$ ion.

If a small amount of $OH^-$ ion (base) is added to the above reaction, the rate of the reaction is increased. $OH^-$ ion acts as a catalyst, since this ion is regenerated in the final reaction. The catalytic action of $OH^-$ ion can be explained neither by $S_N^1$ nor by $S_N^2$ mechanism. It can be explained by $S_N^1$ (CB) mechanism as follows:

$$\underset{(I)}{[Co(en)_2(NO_2)Cl]^+} + \underset{Base}{OH^-} \xrightarrow[DMSO]{Fast} \underset{CB\ of\ (I)}{[Co(en)(en-H)(NO_2)Cl]^0} + H_2O \qquad (i)$$

(en – H is ethylene diamine from which one $H^+$ has been removed, *i.e.* en – H is $NH_2-CH_2-CH_2-NH^-$

$$\underset{CB\ of\ (I)}{[Co(en)(en-H)(NO_2)Cl]^0} + H_2O + NO_2^- \xrightarrow[-HCl]{Fast} [Co(en)_2(NO_2)_2]^+ + \underset{Base}{OH^-} \qquad (ii)$$

Above reactions show that $OH^-$ (base) consumed in reaction (*i*) is regenerated in reaction (*ii*) showing that the action of $OH^-$ is catalytic.

**(vii) Base hydrolysis of $[Co(NH_3)_5Cl]^{2+}$ in presence of $H_2O_2$.** Let us consider the base hydrolsis of $[Co(NH_3)_5Cl]$ in aqueous solution at 25°C. Experimentally it has been observed that if $H_2O_2$ is added to the reaction mixture, then the rate of base hydrolysis of $[Co(NH_3)_5Cl]^{2+}$ decreases. This observation can be explained by $S_N^1$ (CB) mechanism as follows:

$$H_2O_2 + OH^- \rightleftharpoons H_2O + HO_2^-$$

$H_2O_2$ converts $OH^-$ into $HO_2^-$. Due to the conversion of $OH^-$ into $HO_2^-$, the concentration of $OH^-$ which is necessary for the production of CB of $[Co(NH_3)_5Cl]^{2+}$ decreases. Due to the decrease in the concentration of $OH^-$ the rate of base hydrolysis of $[CoCNH_3)_5Cl]^{2+}$ also decreases.

## Anation Reactions

### What is Anation Reaction?

The ligand substitution reaction in which $H_2O$ molecule present in the coordination sphere of a complex species is replaced by an ion is called anation reaction. For example the reaction.

$$[Co\,(CN)_5\,(H_2O)]^{2-} + E^- \rightarrow [Co\,(CN)_5\,E]^{3-} + H_2O \qquad \text{(i)}$$

is an anation reaction in which $H_2O$ is the leaving ligand and $E^-$ ion is the entering ligand.

We know that in acid hydrolysis (aquation) reaction a ligand present in the coordination sphere of a complex species is replaced by $H_2O$ molecule. For example the reaction

$$MA_5\,L + H_2O \rightarrow MA_5\,(H_2O) + L$$

is acid hydrolysis reaction in which L is the leaving ligand and $H_2O$ is the entering ligand. It should be obvious from the above discussion that anation reaction is the reverse of aquation (acid hydrolysis) reaction

**Mechanism.** The anation reaction (i) given above proceeds through $S_N{}^1$ mechanism as shown below

$$\underset{\text{(C.N. = 6)}}{[Co\,(CN)_5\,(H_2O)]^{2-}} \underset{k}{\overset{\text{Slow}}{\rightleftharpoons}} \underset{\text{Intermediate (CN = 5)}}{[Co\,(CN)_5]^{2-}} + H_2O$$

$$[Co\,(C.N)_5]^{2-} + E^- \xrightarrow{\text{Fast}} [Co(CN)_5\,E]^{3-}$$

Obviously the rate of reaction (i) is determined by the slow step. Hence:

$$\text{Rate of reaction (r)} = k \times \text{Conc of } [Co(CN)_5\,(H_2O)]^{2-}$$
$$= k\,[\text{Complex}]$$

### Substitution Reactions Without Breaking of Metal – Ligand Bond.

So far we have been studying the ligand substitution reactions in which the bond between the metal and ligand is broken. However, there are a quite few ligand substitution reactions in which the metal – ligand bond is not broken.

**Examples.** (*i*) Aquation of corbonato complex, $[Co^{3+}(NH_3)_5\,(CO_3)]^+$ by the excess of acid ($H^+$) leading to the formation of aquo complex, $[Co(NH_3)_5\,(H_2O)]^{3+}$ is a substitution reaction in which $Co^{3+}$ — O bond is not broken. This bond remains intact during the aquation of $[Co(NH_3)_5(CO_3)]^+$. This equation reaction can be shown as:

$$[Co^{3+}\,(NH_3)_5\,(CO_3)]^+ + 2H^+ \rightarrow [Co^{3+}\,(NH_3)_5\,(H_2O)]^{3+} + CO_2 \qquad (i)$$

The above reaction was carried out in presence of water labelled with $O^{18}$. Since neither aqo complex, $[Co\,(NH_3)_5\,(H_2O)]^{3+}$ nor $CO_2$ formed in the above aquation reaction contains any $O^{16}$, C — O bond in $CO_3^{2-}$ is broken and $Co^{3+}$ — O bond remains unbroken. $CO_2$ is produced by the cleavage of C—O bond. In presence of $H_2O^{18}$, reaction (i) can thus be shown as:

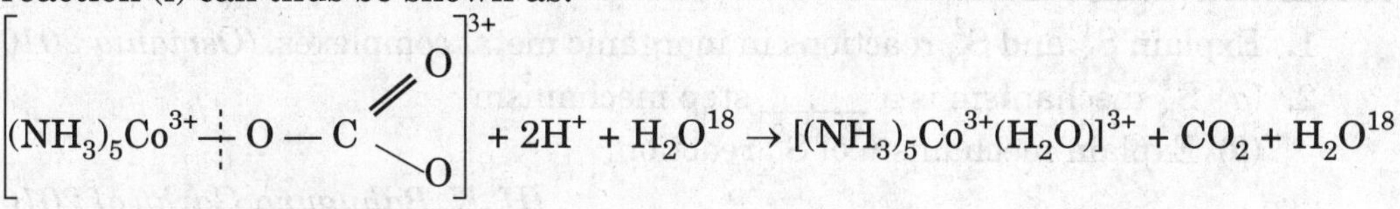

Carbonato complex

**Mechanism.** Reaction (i) given above proceeds through the following steps:

(a) In this step O-atom attached with C – atom of $CO_3^{2-}$ ion is protonated by $H^+$ to form an intermediate (I)

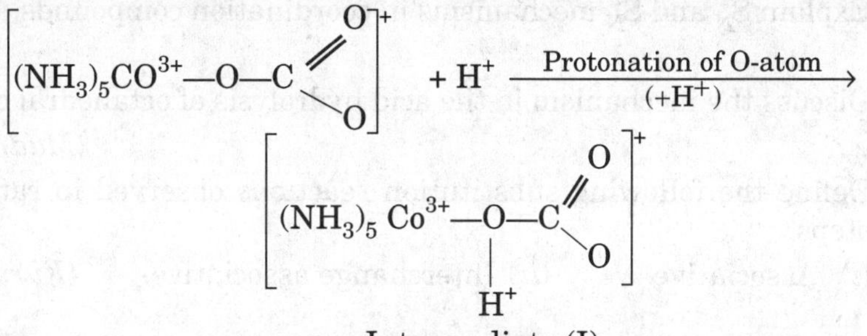

Intermediate (I)

(*b*) Intermediate (I) formed in step (a) loses $CO_2$ to form hydroxo complex, $[Co(NH_3)_5(OH)]^{2+}$(II). In this step C — O bond is broken and $CO_2$ is liberated

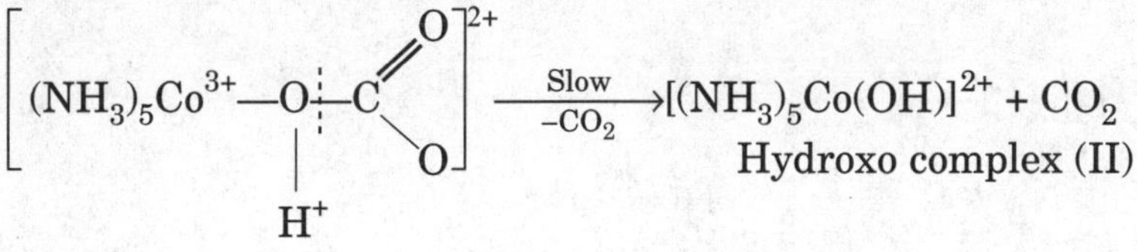

(I)

It is a slow step and hence is the rate determining step.

(*c*) In this step hydroxo complex (II) formed in step (b) is protonated by $H^+$ to form the aquo complex, $[Co(NH_3)_5(H_2O)]^{3+}$

$$\underset{\text{(II)}}{[(NH_3)_5\,Co\,(OH)]^{2+}} + H^+ \xrightarrow[(+H^+)]{\text{Protonation}} \underset{\text{Aquo complex}}{[(NH_3)_5\,Co^{3+}\,(H_2O)]^{3+}}$$

(*ii*). The formation of nitrito complex, $[Co^{3+}(NH_3)_5(ONO)]^{2+}$ from aquo complex, $[Co^{3+}(NH_3)_5)(H_2O)]^{3+}$ is also a ligand substitution reaction in which metal-ligand bond is not broken.

$$\underset{\text{Aquo complex}}{[Co^{3+}(NH_3)_5(H_2O)]^{3+}} + NO_2^- \rightarrow [Co^{3+}(NH_3)_5(ONO)]^{2+} + H_2O$$

This reaction is very rapid. This observation suggests that $Co^{3+}$ — O bond is not broken. This fact is confirmed by carrying out the above reaction in presence of $H_2O^{18}$.

## UNIVERSITY QUESTIONS

1. Explain $S_N^1$ and $S_N^2$ reactions in inorganic metal complexes. *(Osmania 2010)*
2. (*a*) $S_N^1$ mechanism is a ______ step mechanism
   (*b*) Explain mechanism of $S_N^2$ reaction.
   *(H. N. Bahuguna Garhwal 2010)*
3. (*a*) Explain the associative and dissociative mechanisms of substitution reaction in octahedral complexes.
   (*b*) Explain the mechanism of hydroformylation reaction.
   *(Bharathiar 2010)*
4. Explain $S_N^1$ and $S_N^2$ mechanisms in coordination compounds.
   *(Andhra 2010)*
5. Discuss the mechanism in the acid hydrolysis of octahedral complexes
   *(Madras April 2011)*
6. Define the following substitution reactions observed in rate determining steps.
   (*i*) Associative (*ii*) Interchange associative. (*Kurukshetra 2011*)

# CHAPTER 28

# Ligand Substitution Reactions in Square Planar Complexes

## Introduction

Much of the discussion in this chapter deals with the kinetics and mechanism of substitution reactions of Pt(II) square planar complexes. It will be seen that these complexes generally proceed by *bimolecular displacement* ($S_N^2$) *mechanism* while octahedral complexes run by dissociative process as we have already seen.

## Three Ways to Bring About Ligand Substitution Reactions in Square Planar Complexes

There are three possible ways by which a ligand present in a square planar complex may be substituted by another ligand. These ways are discussed below.

**(*i*) Oxidative addition followed by reductive elimination.** For example:

$$\underset{\substack{(I)\\(Pt = +2,\ C.N. = 4)}}{[Pt^{2+}\,Cl\,(R)\,(PR_3)_2]} \xrightarrow[\text{Oxidative addition}]{+Cl_2} \underset{\substack{\text{Intermediate}\\(Pt = +4,\ C.N. = 6)}}{[Pt^{4+}\,Cl_3\,(R)\,(PR_3)_2]} \xrightarrow[\text{Reductive elimination}]{-RCl} \underset{\substack{(II)\\(Pt = +2,\ C.N. = 4)}}{[Pt^{2+}\,Cl_2\,(PR_3)_2]}$$

In the above reaction (I) is oxidised to intermediate in which O.S. of Pt = +4 and C.N. = 6. The intermediate gets reduced to (II) in which O.S. of Pt = +2 and C.N. = 4

**(*ii*) Electrophilic substitution can replace a ligand by another ligand**. For example:

$$[Pt^{2+}\,Cl\,(R)\,(PR_3)_2] + HgCl_2 \longrightarrow [Pt^{2+}\,Cl_2(PR_3)_2] + RHgCl$$

This reaction involves an electrophilic attack by Hg (II) on Pt–Cl bond.

**(*iii*) Nucleophilic substitution**

$$[Pt\,Cl\,(R)\,(PR_3)_2] + NH_3 \longrightarrow [Pt\,(R)\,(NH_3)\,(PR_3)_2]^+ + Cl^-$$

In this reaction nucleophile ($NH_3$) replaces $Cl^-$ ion.

$$Pt\,MeCl\,(PMe_2\,Ph)_2 + N_3^- \longrightarrow Pt\,Me\,(N_3)\,(PMe_2\,Ph)_2 + Cl^-$$

In this reaction $N_3^-$ replaces $Cl^-$ ion.

## Trans Effect

Let us consider the action of $NH_3$ on the square planar complex, $[Pt\,(NO_2)\,Cl_3]^{2-}$ to form $[Pt\,(NO_2)Cl_2\,(NH_3)]^-$. Theoretically there are two possible reaction products *viz* (a) and (b) as shown in Fig. 28.1.

Experiments have, however, shown that it is only (*a*) (*i.e. trans*-isomer) which is formed. (*a*) is formed by the replacement of $Cl^-$ lying *trans* to $NO_2^-$ in $[Pt(NO_2)Cl_3]^{2-}$ by $NH_3$. The formation of (*a*) is explained by saying that $Cl^-$ ion lying *trans* to $NO_2^-$ in $[Pt(NO_2)Cl_3]^{2-}$ is far more easily replaced by $NH_3$ than either of the two $Cl^-$ ions

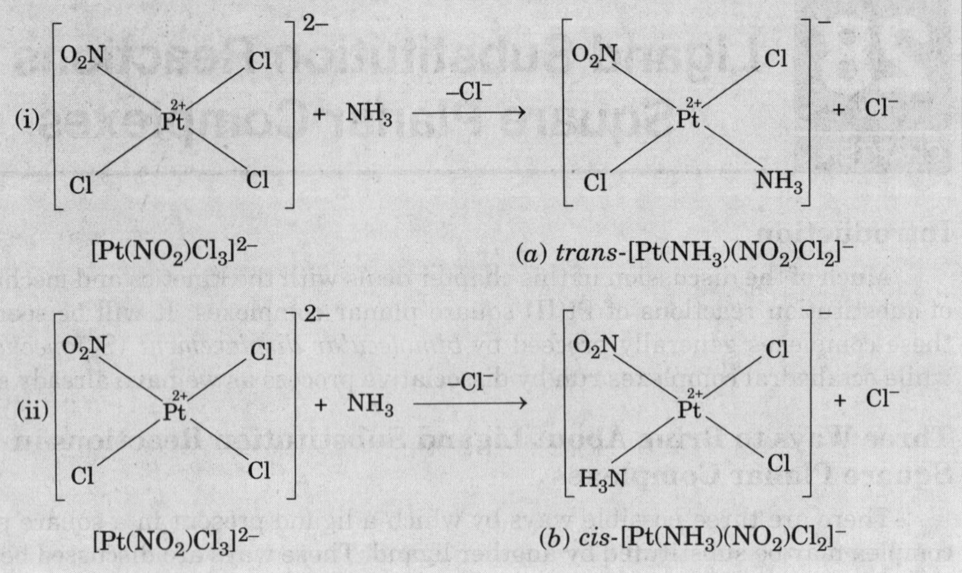

**Fig. 28.1.** Action of $NH_3$ on $[Pt\ Cl_3\ (NO_2)]^{2-}$

lying *cis* to $NO_2^-$ ion. The phenomena of such type of replacement is called **trans effect**. Groups like $NO_2^-$ which direct the entering ligand to occupy the position *trans* to them are called *trans directing groups* and the property of these groups due to which the groups lying *trans* to them are replaced far more readily by the entering ligand is called *trans- directing character* (or *trans-directing influence or simply trans-effect*).

Chatt *et al* have proposed that the *trans effect of a group coordinated to a metal ion is the tendency of that group to direct an incoming group to occupy the position trans to that group. Trans* effect is also defined as the *effect of a coordinated group on the rate of the replacement of a group lying trans to it in a metal complex, e.g.* in the *trans* form of square planar complex, $MLX_3$ (L and X are *trans* to each other), if the ligand X which is *trans* to L is rapidly replaced by another group, say Y, to give $MLX_2$ Y, L is said to have larger *trans* effect or *trans* directing character. Thus L has greater *trans* effect than X (L > X).

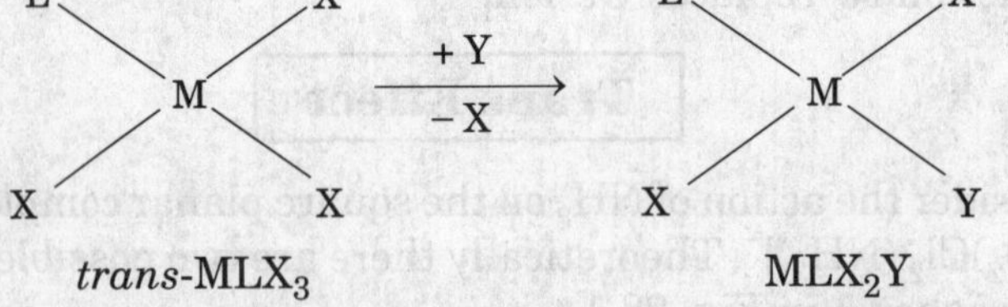

By measuring rates of various reactions, a series of ligands can be put into an order of decreasing *trans* effect. Such an order would not necessarily be invariant but might depend on the metal complex and also on the incoming group, Y. The approximate order of decreasing *trans* effect of some common ligands is:

High end ⟶ trans effect decreasing ⟶ Low end

$CN^-$, CO, $C_2H_4$, NO > $PR_3$, $H^-$ > $CH^-$, $SC(NH_2)_2$ (tu) > $C_6H_5^-$,
$NO_2^-$, $I^-$, $SCN^-$ (S-bonded) > $Br^-$, $Cl^-$ > *py*, $NH_3$, $OH^-$, $H_2O$.

The series given above is called **trans effect series.**

The ligands lying at the high end of the series have vacat π or π* orbitals which can accept electrons from metal orbital to form metal-ligand π-bond ($d\pi - d\pi$ or $d\pi - p\pi$ bond). These ligands are, therefore, called π-bonding ligands. The *trans*-directing ability of these π-bonding ligands increases with the increase of their ability to form metal–ligand π-bonds.

The *trans*-effect of the ligands which are not able to form metal-ligand π-bonds increases with the increase of their polarisability, *e.g.*

⟶ polarisability increases ⟶

$Cl^- < Br^- < I^-$

⟶ *trans*-effect increases ⟶

Most of the work on *trans*-effect has been done on square planar Pt(II) complexes.

## Illustrative Example

Let us consider the preparation of *cis*- and *trans*-isomers of $[Pt(C_2H_4)(NH_3)Cl_2]^0$ which is a square planar complex of Pt (II).

(a) *cis- isomer*. This isomer can be prepared as follows:

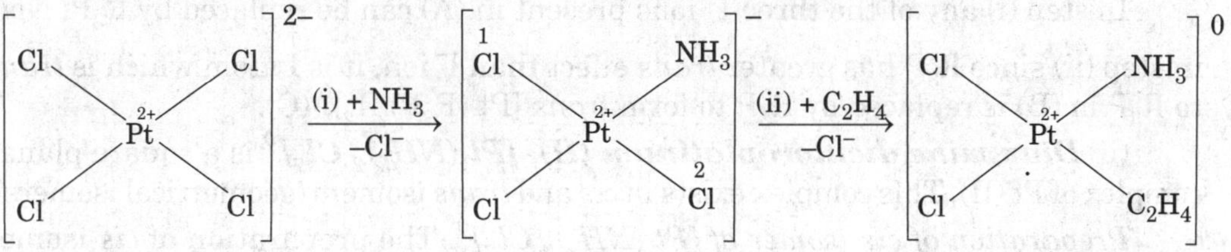


$[PtCl_4]^{2-}$ (A) (B) *cis*-isomer, $[Pt(C_2H_4)(NH_3)Cl]$ (C)

In step (i), any of the four $Cl^-$ ions present in (A) can be replaced by $NH_3$ molecule to form (B). Now in step (ii), since $Cl^-$ ion in (B) has greater *trans* effect than $NH_3$ molecule, $\overset{2}{Cl}$ atom which is *trans* to $Cl^1$ in (B) is replaced by $C_2H_4$ molecule.

(b) *trans-isomer*. This isomer can be obtained by reversing the addition of $NH_3$ and $C_2H_4$ to $[Pt\,Cl_4]^{2-}$ as shown below.

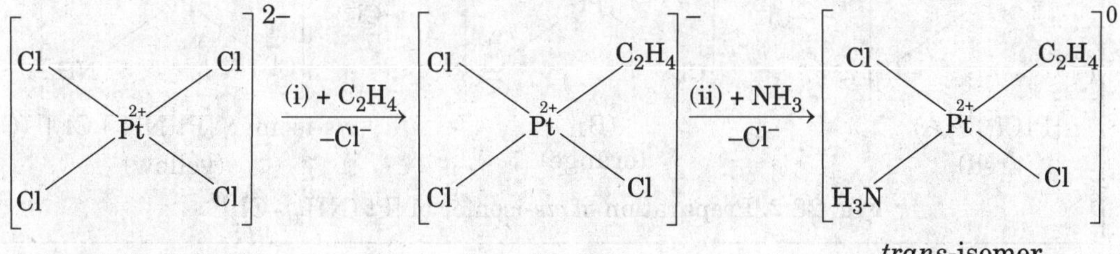


$[PtCl_4]^{2-}$ (A) (B) *trans*-isomer, $[Pt(C_2H_4)(NH_3)Cl_2]^0$ (C)

In step (i) any of the four Cl-atoms present in (A) can be replaced by $C_2H_4$ molecule. In step (ii) since $C_2H_4$ has greater *trans* effect than $Cl^-$ion, Cl atom *trans* to $C_2H_4$ in (B) will be replaced by $NH_3$ to form (C) which is *trans*-isomer.

## Applications of Trans Effect

### 1. Synthetic Applications

Applications of *trans* effect are shown by the synthesis of many Pt (II) complexes. For example:

(i) ***Trans-[Pt $(R_3P)_2$ $I_2$]$^o$*** can be prepared by the reaction which consists of the following steps:

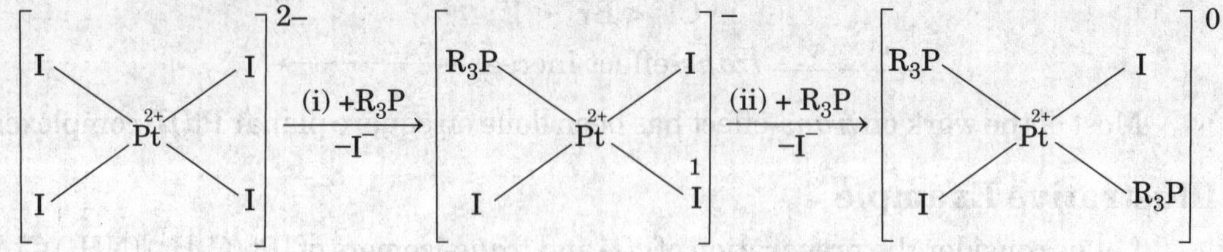

$[PtI_4]^{2-}$ (A) (B) *trans*-isomer, $[Pt(R_3P)_2 I_2]^0$ (C)

In step (i) any of the three $I^-$ ions present in (A) can be replaced by $R_3P$. Now in step (ii) since $R_3P$ has greater *trans* effect than $I^-$ ion, it is $\overset{1}{I}$ atom which is *trans* to $R_3P$ in (B) is replaced by $R_3P$ to form *trans* [Pt $(R_3P)_2$ $I_2$]$^0$ (C).

(ii) ***Diammine dichloroplatinum (II), [Pt $(NH_3)_2$ $Cl_2$]$^0$*** is a square planar complex of Pt (II). This complex exists in *cis* and *trans* isomers (geometrical isomers).

*Preparation of cis-isomer of [Pt $(NH_3)_2$ $Cl_2$]$^0$.* The preparation of *cis*-isomer has been shown in Fig. 28.2. It may be seen from this figure that in step (*i*) any of the four Cl atoms present in (A) can be replaced by $NH_3$ molecule to form (B). In step (ii), since $Cl^-$ ion present in (B) has greater *trans* effect than $NH_3$, $\overset{2}{Cl}$ atom which is *trans* to $\overset{1}{Cl}$ atom in (B) is replaced by $NH_3$ molecule to form cis-isomer (C).

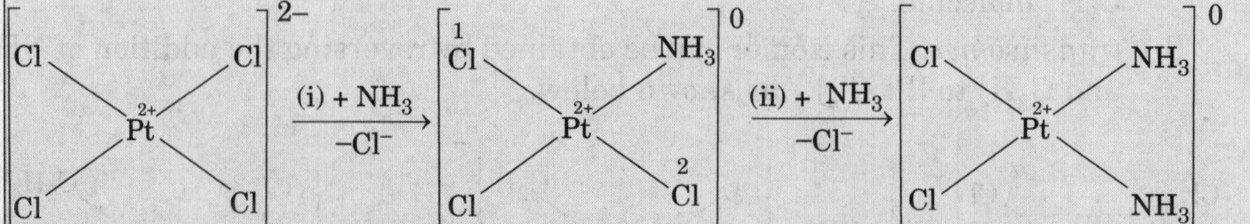

$[PtCl_4]^{2-}$ (A) (red) (B) (orange) *cis*-isomer, $[Pt(NH_3)_2Cl_2]^0$ (C) (yellow)

**Fig. 28.2.** Preparation of *cis*-isomer of [Pt $(NH_3)_2$ $Cl_2$]$^0$

*Preparation of trans-isomer of [Pt (NH$_3$)$_2$ Cl$_2$]$^0$.* The preparation of *trans*-isomer has been shown in Fig. 28.3. It may be seen from this figure that in step (i) any of the four $NH_3$ molecules in (A) can be replaced by Cl-atom to form (B). In step (ii), since $Cl^-$ ion present in (B) has greater *trans* effect than $NH_3$, $\overset{1}{N}H_3$ molecule which is *trans* to Cl-atom in (B) will be replaced by Cl-atom to form the *trans*-isomer (C).

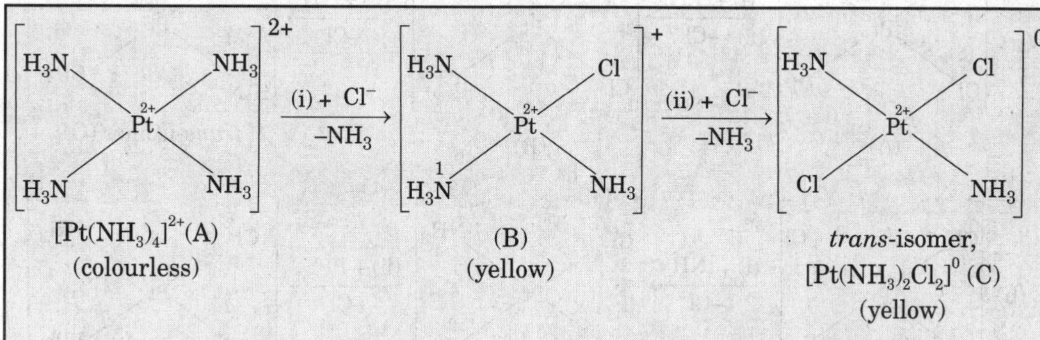


**Fig. 28.3.** Preparation of *trans*-isomer of [Pt (NH$_3$)$_2$ Cl$_2$]$^0$

*(iii)* ***Ammine dichloronitroplatinate (II),*** **[Pt (NH$_3$) Cl$_2$ (NO$_2$)]$^-$** is a square planar complex of Pt (II) and exists in *cis* – and *trans* – isomers (geometrical isomers). These isomers can be prepared by the reactions (a) and (b) respectively.

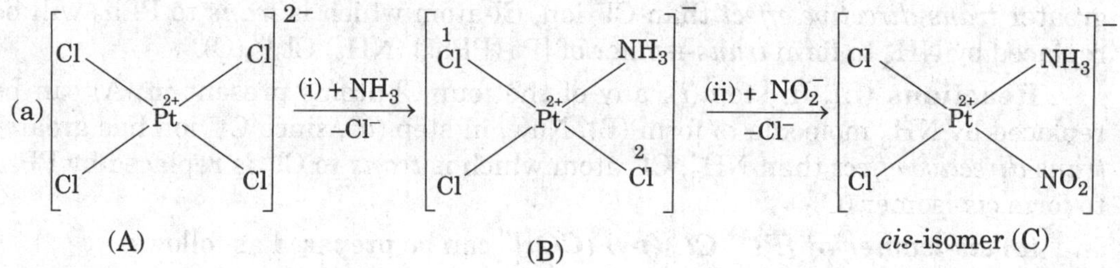


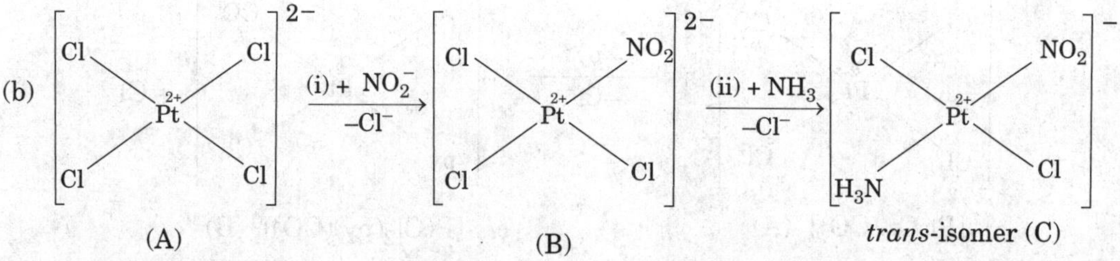


**Reactions (a).** In step (i) any of the four Cl-atoms present in (A) can be replaced by $NH_3$ molecule to form (B). In step (ii), since $Cl^-$ ion present in (B) has greater *trans* effect than $NH_3$ molecule, $Cl^2$ atom which is *trans* to $Cl^1$ atom is replaced by $NO_2^-$ ion to form *cis*-isomer of [Pt (NH$_3$) Cl$_2$ (NO$_2$)]$^-$ (C).

**Reactions (b).** In step (i) any of the four Cl-atoms present in (A) can be replaced by $NO_2^-$ ion to form (B). In step (ii), since $NO_2^-$ ion present in (B) has greater *trans* effect than $Cl^-$ ion, Cl-atom which is *trans* to $NO_2^-$ will be replaced by $NH_3$ molecule to form *trans*-isomer of [Pt (NH$_3$) Cl$_2$ (NO$_2$)]$^-$ (C).

(*iv*) ***[Pt (PPh$_3$) (NH$_3$) Cl$_2$)]$^0$*** is a square planar complex of Pt (II). It exists in *cis*- and *trans*-isomers (geometrical isomers). These isomers can be prepared by reactions (a) and (b).

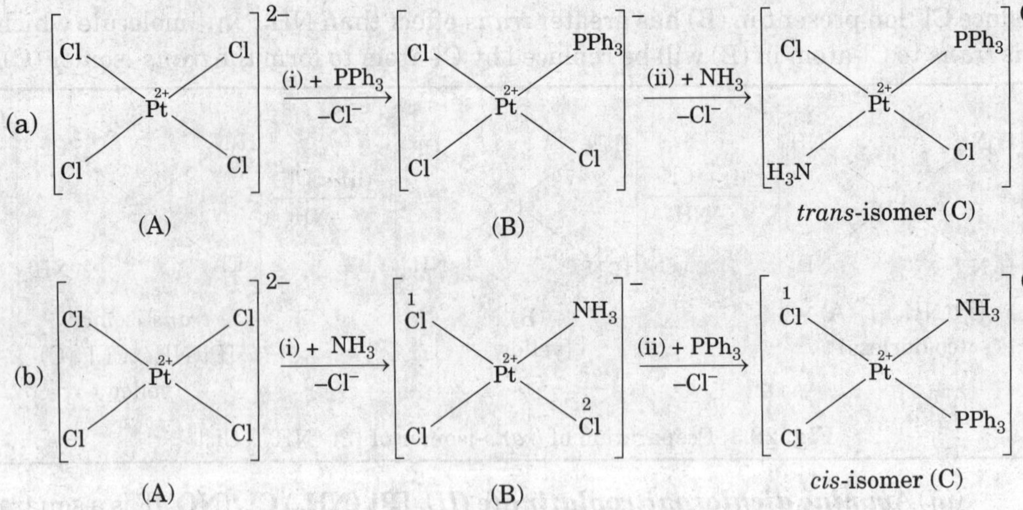


**Reactions (a).** In step (*i*) any of the four Cl-atoms in (A) can be replaced by PPh$_3$ (triphenyl phosphine) to form (B). In step (*ii*), since PPh$_3$ present in (B) has greater *trans directive effect* than Cl$^-$ ion, Cl-atom which is *trans* to PPh$_3$ will be replaced by NH$_3$ to form *trans*-isomer of [Pt (PPh$_3$) (NH$_3$) Cl$_2$]$^0$ (C).

**Reactions (b).** In step (*i*), any of the four Cl-atoms present in (A) can be replaced by NH$_3$ molecule to form (B). Now, in step (*ii*), since Cl$^-$ ion has greater *trans directive effect* than NH$_3$, Cl$^2$ atom which is *trans* to Cl$^1$ is replaced by PPh$_3$ to form *cis*-isomer (C).

(*v*) ***cis-isomer of [Pt$^{2+}$ Cl$_2$ (py) (CO)]$^0$*** can be prepared as follows:

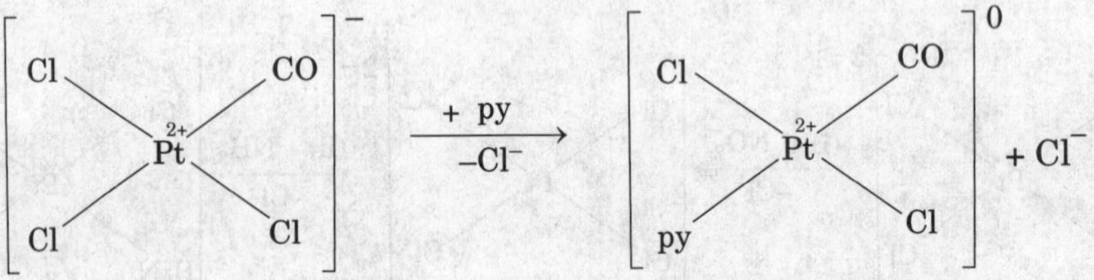


[Pt Cl$_3$(CO)]$^-$ (A) *cis*-[PtCl$_2$(py)(CO)]$^0$ (B)

In the above reaction, since CO present in (A) has greater *trans* effect than Cl$^-$ ion, Cl$^-$ ion which is *trans* to CO in (A) will be replaced by *py* to form the cis-isomer of [Pt Cl$_2$(*py*) (CO)]$^0$ (B).

(*vi*) ***Preparation of three isomers of [Pt $^{2+}$(py) (NH$_3$) Br Cl]$^0$***. The three isomers of [Pt(*py*) (NH$_3$) Br Cl] *viz* I, II and III were prepared from [PtCl$_4$]$^{2-}$ by the following reactions on the basis of greater *trans* effect of the ligands Br$^-$ and Cl$^-$ compared to *py* and NH$_3$ and the fact that the Pt–N bond strength is greater than the Pt–Cl bond strength.

*First reaction*

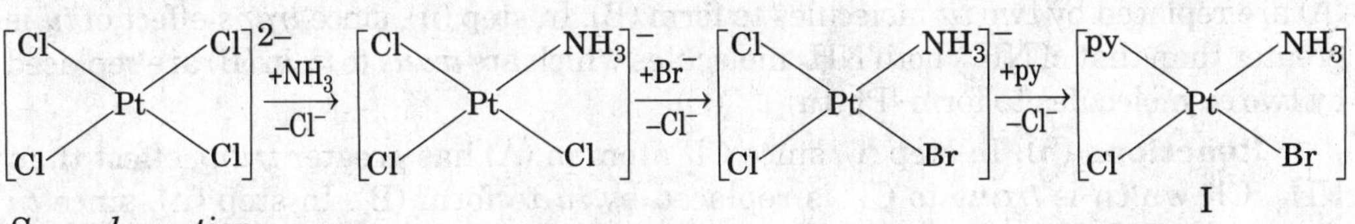

*Second reaction*

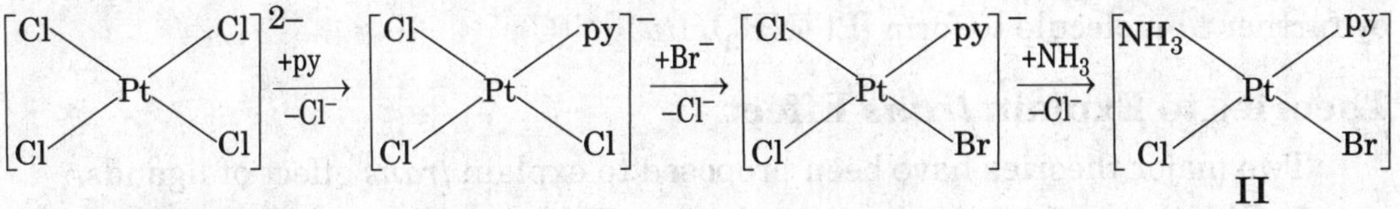

*Third reaction*

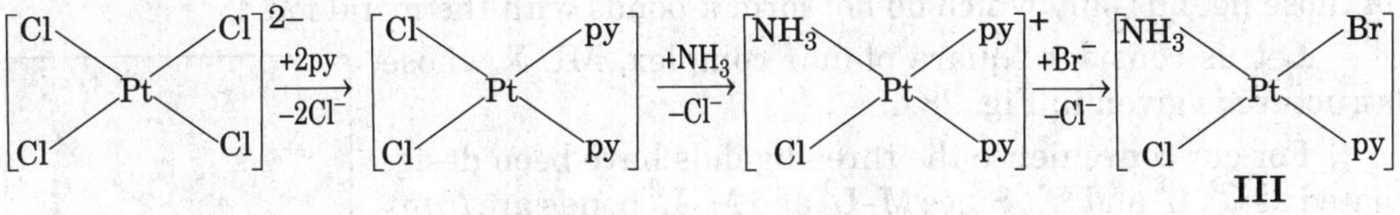

It may be noted that the second step in third reaction seems to contradict the *trans* effect. The *trans* effect predicts that any ligand *trans* to $Cl^-$ should be replaced rather than any ligand *trans* to py. However, $Cl^-$ ion *trans* to py is replaced because the bond strength of Pt–Cl is weaker than that of Pt–N bond.

## 2. To Distinguish Between cis-and *trans*- Isomers of $[Pt(NH_3)_2Cl_2]^0$ by Kurnakov's Test

Russian chemists have used the concept of *trans* effect to distinguish between *cis*-and *trans*-isomers of $[Pt(NH_3)_2Cl_2]^0$. This has been done by reacting $[Pt(NH_3)_2Cl_2]^0$ with thiourea, $NH_2.CS.NH_2$ (*tu*). *Cis*-isomer gives $[Pt(tu)_4]^{2+}$ and trans-isomer gives $[Pt(tu)_2(NH_3)_2]$

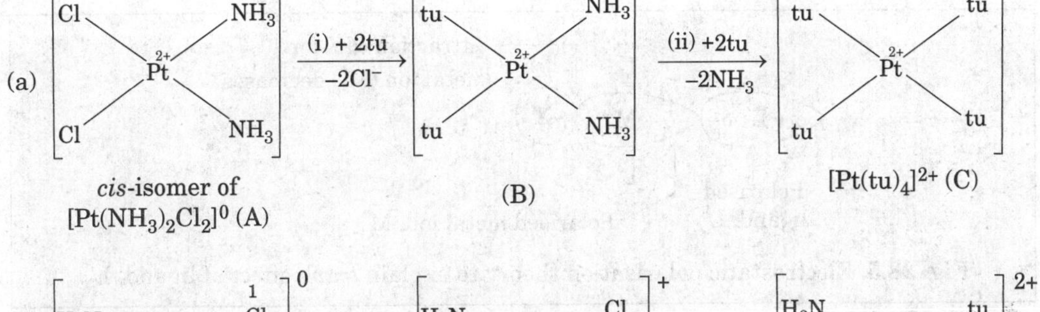

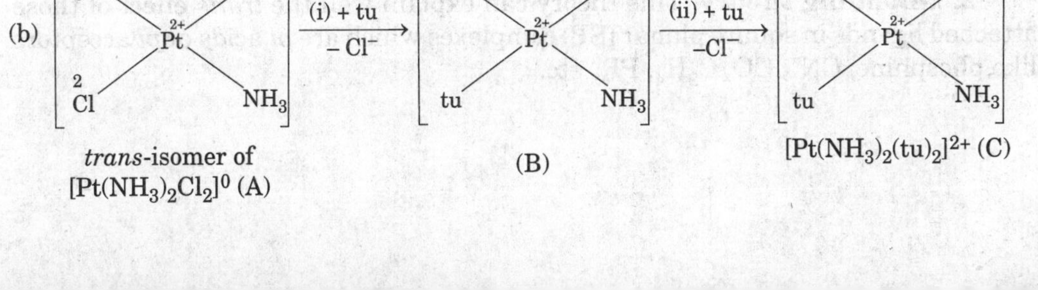

**Reactions (a).** In step (i) both Cl-atoms present in *cis*-isomer of $[Pt(NH_3)_2Cl_2]°$ (A) are replaced by two *tu* molecules to form (B). In step (ii), since *trans* effect of *tu* is greater than that of $NH_3$, both $NH_3$ molecules which are *trans* to *tu* in (B) are replaced by two *tu* molecules to form $[Pt(tu)_4]^{2+}$ (C).

**Reactions (b).** In step (i), since $Cl^1$ atom in (A) has greater *trans* effect than $NH_3$, $Cl^2$ which is *trans* to $Cl^1$ is replaced by *tu* to form (B). In step (ii), since *tu* present in (B) has greater *trans* effect than Cl atom, Cl-atom *trans* to *tu* is replaced by further *tu* molecule to form $[Pt(NH_3)_2(tu)_2]^{2+}$(C).

## Theories to Explain *trans* Effect

Two major theories have been proposed to explain *trans* effect of ligands.

**1. Electrostatic polarisation theory.** This theory can explain *trans* effect of those ligands only which do not form π-bonds with the metal ion.

Let us consider square planar complex, $ML_3X$ whose structure is given in Fig. 28.4.

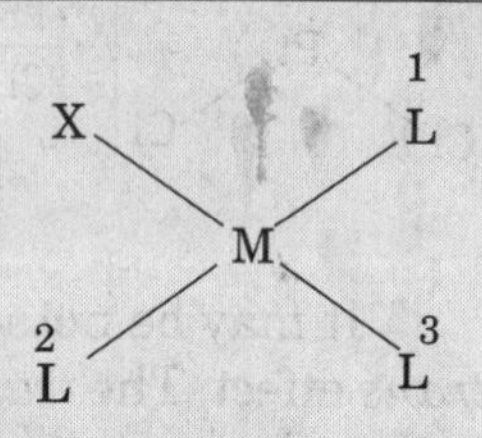


**Fig. 28.4.** Square planar complex, $ML_3X$

For our convenience the three ligands have been designated as $L^1$, $L^2$ and $L^3$. Since M–$L^1$ and M–$L^2$ bonds are *trans* to each other and the ligands $L^1$ and $L^2$ are the same ligands, these bonds balance each other. On the other hand, M—X and M—$L^3$ bonds are not able to balance each other, since X and $L^3$ ligands are different ligands. Suppose the ligand $L^3$ is more polarisable than the ligand X. The primary charge on the metal ion ($M^+$) polarises the charge cloud on $L^3$ and induces a dipole in $L^3$. The induced dipole on $L^3$, in *turn*, induces a dipole in metal ion ($M^+$). δ+charge on the metal ion faces the ligand $L^3$ and δ– charge on the metal ion faces the ligand X which is *trans* to $L^3$. δ–charge on the metal ion repels the ligand X which is *trans* to $L^3$. Thus the attraction of X for metal ion is decreased (See Fig. 28.5). Consequently M—X bond becomes weaker. The weakening of M—X bond facilitates the replacement of X by the entering ligand. According to this the polarisation of ligand $L^3$ is directly proportional to the *trans* effect of this ligand.

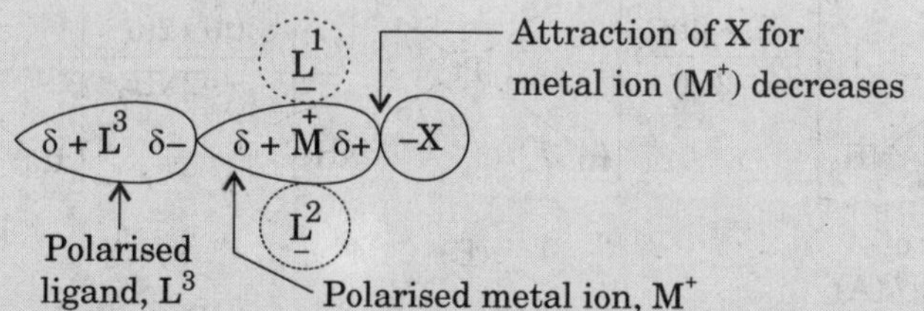


**Fig. 28.5.** Electrostatic polarisation theory to explain *trans* effect of ligand, $L^3$.

**2. π-Bonding theory.** This theory can explain well the *trans* effect of those attached ligands in square planar (SP) complexes which are *pi* acids or *pi* acceptors like phosphine, $CN^-$, CO, $C_2H_4$, $PR_3$ etc.

Let us consider the following ligand exchange reaction occurring in a SP complex, $MA_2BX$

$$MA_2BX + Y \rightarrow MA_2BY + X \qquad (i)$$

In this reaction X is replaced by Y. Obviously X is the leaving group (out going group) and Y is the entering (incoming) group. In $MA_2BX$, B is a *pi acid ligand* ($\pi$-bonding ligand) whose *trans* effect is to be investigated. The ligand B is also called *trans labilising ligand*.

In $MA_2BX$, B and X are placed *trans* to each other, *i.e.* M–B and M–X bonds are *trans* to each other in SP complex, $MA_2$ BX (See Fig. 28.6). According to $\pi$-bonding theory, the ligand exchange reaction (i) given above takes place through the steps shown in Fig. 28.6.

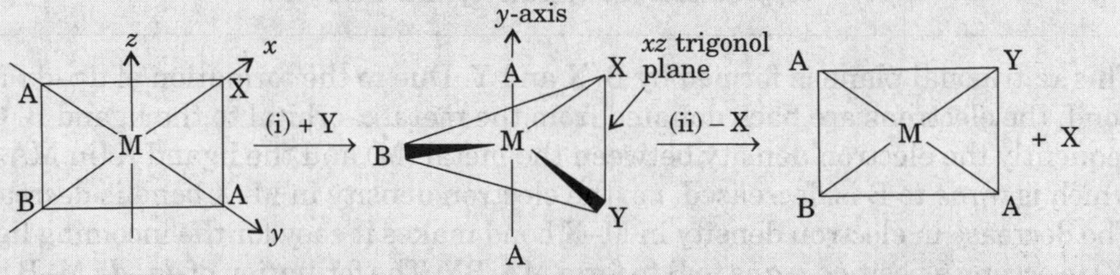


(I) Square planar (SP) complex, $MA_2BX$ (II) Trigonal bipyramidal (TBP) intermediate, $MA_2BXY$ (Transition state) (III) Square planar (SP) complex, $MA_2BY$

**Fig. 28.6.** Illustration of $\pi$-bonding theory to explain *trans* effect.

(*a*) In step (i) the entering ligand (Y) attacks the SP complex, $MA_2BX$ (I) and an intermediate activated 5-coordinate complex (*transition state*), $MA_2BXY$ (II) is obtained. This intermediate has trigonal bipyramidal (TBP) geometry. This intermediate is formed due to the movement of X and Y ligands in the *xz* trigonal plane of TBP geometry. In this geometry the two A ligands occupy the *trans* positions along *y*-axis. It may be seen from the structure of TBP intermediate (*transition state*) (II) that the leaving group (X), entering group (Y) and $\pi$-bonding ligand (B) form the trigonal plane (*xz* plane) of the TBP geometry.

(*b*) In step (ii), the transition state (II) loses X and SP complex, $MA_2BY$ (III) is formed. In the formation of this SP complex the bond angle BMY which is equal to 120° in TBP structure (II) is expanded to 180° in SP complex, $MA_2BY$ (III). In SP complex (III) the entering group (Y) is placed *trans* to the $\pi$-acid ligand (B). Note that the position of the entering group (Y) in $MA_2BY$ (III) is the same w. r. t. B as that of leaving group (X) in $MA_2BX$ (I) w. r. t. B.

Chatt *et. al.* and Orgel have suggested that the TBP transition state (II) is stabilised by the formation of M–B $\pi$-bond between the $\pi$-bonding ligand (B) and metal (M). This $\pi$-bond is formed in *xz* plane of TBP transition state ($MA_2BXY$) by the overlap between the *empty d*-orbital (pi bonding orbital) of $\pi$-bonding ligand (B) and *filled* $d_{xz}$ orbital of the metal (M). Orgel has pointed out that the filled $d_{xz}$ orbital of the metal involved in the overlap lies in *xz* trigonal plane of TBP structure (II).

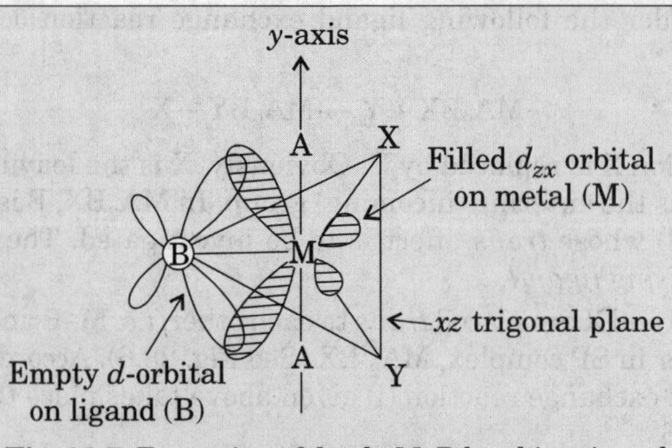


**Fig. 28.7.** Formation of dπ–dπ M–B bond in trigonal bipyramidal intermediate, $MA_2$ BXY (II).

This *xz* trigonal plane is formed by B, X and Y. Due to the formation of dπ–dπ M–B bond, the electrons are back donated from the metal *d*-orbital to the ligand B. Consequently the electron density between the metal (M) and the ligand X (in $MA_2BX$) which is *trans* to B is decreased, *i.e.* the electron density in M–X bond is decreased. The decrease in electron density in M–X bond makes it easy for the incoming ligand (Y) to occupy a position *trans* to B to form $MA_2BY$. The formation of *dπ–dπ* M–B bond in TBP (II) structure is shown in Fig. 28.7.

## Rate Law for Nucleophilic (Ligand) Substitution Reactions in Square Planar Complexes

Square planar complexes are formed by $d^8$ system of Pt(II), Pd(II), Ni(II) etc. Of these Pt (II) complexes are the most stable square planar complexes and kinetics of ligand substitution reactions of these complexes have been extensively studied. An associative mechanism of the ligand substitution reactions of square planar complexes of Pt (II) has shown that these reactions follow a two-term rate law, *i.e.* the rate of these reactions depends on both the concentration of the reacting square planar complex and on the concentration of the entering ligand.

Let us consider the following ligand substitution reaction taking place in square planar complex of Pt (II), Pt $A_2BX$

$$\text{Pt } A_2BX + Y \longrightarrow \text{Pt } A_2BY + X \qquad (i)$$

In this reaction X is the leaving group and Y is the entering ligand. X is replaced by Y. The rate (r) of this reaction is given by:

$$r = [\text{Complex}]\ \{k_1 + k_2\ [\text{Entering ligand}]\}$$
$$= [PtA_2BX]\{k_1 + k_2[Y]\}$$
$$= [\text{Complex}]\{k_1 + k_2\ [\text{Entering group}]\}$$

This two-term rate law equation can be derived as follows.

Reaction (*i*) can proceed through the following two paths:

(*i*) **Solvent path.** This path proceeds through the following steps:

(*a*) This step consists of *slow* aquation reaction, *i.e.* X is replaced by $H_2O$ molecule

$$\text{Pt A}_2\text{BX} + \text{H}_2\text{O} \xrightarrow[k_1]{\text{Slow}} \underset{\text{(I)}}{\text{Pt A}_2\text{B (H}_2\text{O)}} + \text{X} \qquad (ii)$$

(*b*) This step consists of *a fast* reaction in which the coordinated $H_2O$ molecule in (I) is replaced by the entering ligand, Y.

$$\underset{\text{(I)}}{\text{Pt A}_2\text{B (H}_2\text{O)}} + \text{Y} \xrightarrow{\text{Fast}} \text{Pt A}_2\text{BY} + \text{H}_2\text{O} \qquad (iii)$$

Since this reaction is fast, it is not rate determining step. On the other hand, since reaction (ii) is a slow reaction, it is rate determining reaction. The rate (r) of reaction (ii) is given by:

$$r_1 = k_1 \text{ (conc. of Pt A}_2\text{BX)} \times \text{(conc. of H}_2\text{O)}$$

Now since $H_2O$ is present in large excess, concentration of $H_2O$ remains practically constant. Hence $r$ is given by:

$$r_1 = k_1 \times \text{conc. of Pt A}_2\text{BX}$$

or

$$r_1 = k_1 \times [\text{Pt A}_2\text{BX}] \qquad (iv)$$

The two steps of solvent path represented by reactions (ii) and (iii) can be combined together and hence can be written as:

$$\text{Pt A}_2\text{BX} \xrightarrow[k_1]{\text{Slow (+H}_2\text{O, –X)}} \text{Pt A}_2\text{B (H}_2\text{O)} \xrightarrow{\text{Fast (+Y, –H}_2\text{O)}} \text{Pt A}_2\text{ BY}$$

(*ii*) **Direct displacement path.** This path proceeds through the following steps:

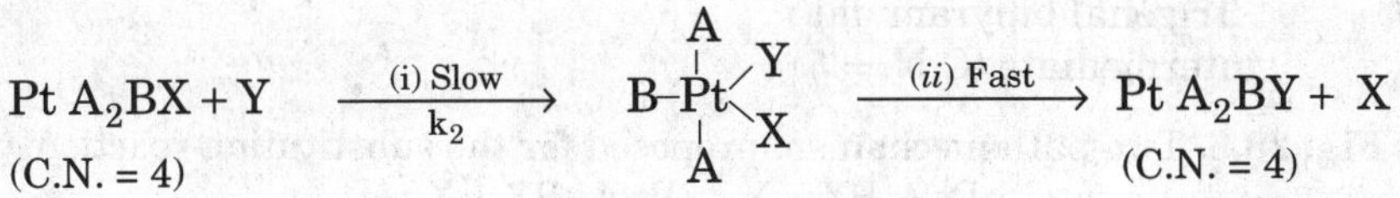


Trigonal bipyramidal intermediate (C.N. = 5)

or $\quad \text{Pt A}_2\text{BX} \xrightarrow[k_2]{\text{(i) Slow (+Y)}} \text{Pt A}_2\text{BXY} \xrightarrow{\text{(ii) Fast (–X)}} \text{Pt A}_2\text{BY}$

Since first step is a slow reaction, it is rate determining reaction. The rate ($r_2$) of this slow reaction is given by:

$$r_2 = k_2 \text{ (conc. of Pt A}_2\text{BX)} \times \text{(conc. of Y)}$$

or

$$r_2 = k_2 [\text{Pt A}_2\text{BX}][\text{Y}] \qquad (v)$$

This equation shows that rate determining step is bimolecular and the mechanism involved is $S^2_N$

Above discussion shows that the rate (r) of substitution reaction (i) is given by:

$$r = r_1 + r_2 \qquad (vi)$$
$$= k_1 [\text{Pt A}_2\text{BX}] + k_2 [\text{Pt A}_2\text{ BX}] [\text{Y}]$$
$$= [\text{Pt A}_2\text{BX}]\{k_1 + k_2 [\text{Y}]\}$$
$$= [\text{Complex}]\{k_1 + k_2 \text{ [Entering group]}\} \qquad (vii)$$

This relation shows that the substitution reactions in square planar complexes follow a two-term rate law.

In relation (*vii*) $k_1$ is the fist order rate constant for solvent controlled reaction and $k_2$ is the second order rate constant for reaction with Y and corresponds to $S_N^2$ mechanism.

The two-path mechanism for the substitution reaction (i) can be shown in a single diagram given in Fig. 28.8.

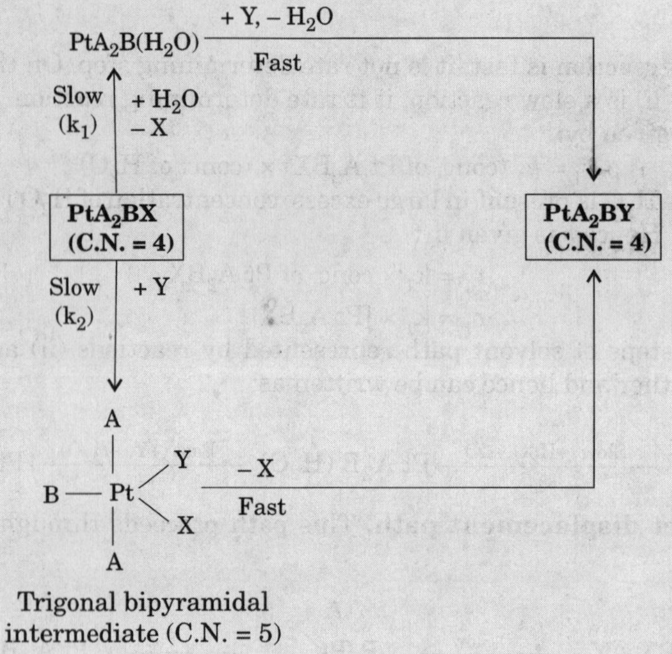


**Fig. 28.8.** Two-path mechanism proposed for the substitution reaction,
$Pt\ A_2BX + Y \rightarrow Pt\ A_2\ BY + X$

A convenient analysis of rate constants can be made by running the reactions with a large excess of nucleophile, Y. Under this condition the observed rate constant, $K_{obs}$, is pseudo first-order and is related to $k_1$ and $k_2$ as:

$$K_{obs} = k_1 + k_2\,[Y] \qquad (viii)$$

The rate constant, $k_1$ is due to the solvent path, while $k_2$ is due to the direct displacement of the leaving group (X) by nucleophile (Y). Thus it becomes convenient to designate the solvent path $k_1$ as $k_s$ and the direct displacement path $k_2$ as $k_Y$ so that equation (viii) becomes

$$K_{obs} = k_s + k_Y[Y] \qquad (ix)$$

Let us consider the following substitution reaction occurring in $[Pt^{2+}\ (dien)\ Cl]^+$.

$$[Pt^{2+}\ (dien)\ Cl]^+ + Y^- \rightarrow [Pt^{2+}\ (dien)\ Y]^+ + Cl^-\ (Y^- = OH^-, Cl^-, Br^-)$$

In this reaction *dien* is diethylenetriamine, $NH_2(CH_2)_2\ NH\ (CH_2)_2\ NH_2$ which is a tridentate ligand. $Y^-$ is the entering group which may be $OH^-$, $Cl^-$ or $Br^-$. $Cl^-$ present in $[Pt^{2+}\ (dien)\ Cl]^+$ is the leaving ligand. It is evident from equation (viii) that the plots of $K_{obs}$ for the above reaction against the concentration of $OH^-$, $Cl^-$ or $Br^-$ (entering groups) are straight lines which have different slopes,

$k_2$ corresponding to $OH^-$, $Cl^-$ and $Br^-$ and the same intercept equal to $k_1$, as shown in Fig. 23.9.

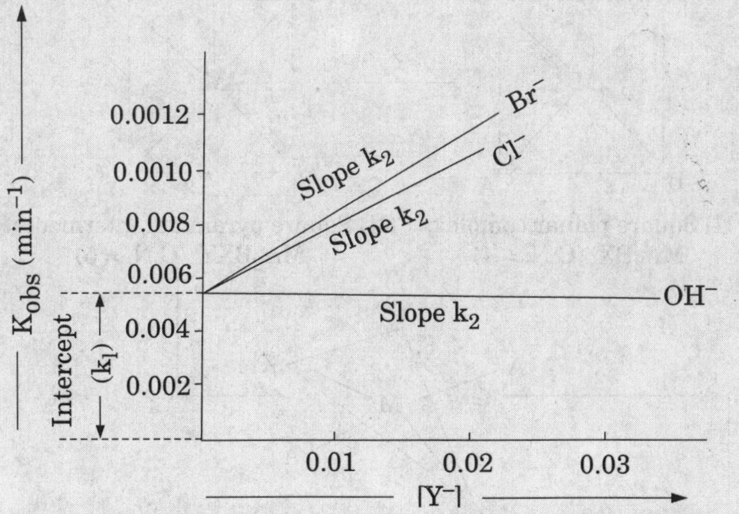


**Fig. 28.9.** Plot of observed rate constant, $k_{obs}$ for the reaction, $[Pt^{2+}(dien)Cl]^+ + Y^- \longrightarrow [Pt^{2+}(dien)Y]^+ + Cl^-$ ($Y^- = OH^-$, $Cl^-$, $Br^-$)

The rate law represented by equation (viii) indicates that the substitution reaction of Pt $A_2$ BX with Y to yield Pt $A_2BY$ is occurring by two-path mechanism only one of which involves Y in the rate determining step.

## Associative $S_N^2$ Mechanism of Substitution Reactions of Square Planar Complexes

The associative $S_N^2$ mechanism of the substitution reactions in square planar complexes can be understood by considering the following substitution reaction taking place in square planar complex, $MA_2BX$

$$MA_2BX + Y \rightarrow MA_2BY + X \quad \text{(i)}$$

In square planar complex, $MA_2BX$ (called substrate), the two A ligands are *trans* to each other. X is the leaving group and Y is the entering group (nucleophile).

$S_N^2$ mechanism for the above displacement reaction can be illustrated by the stereochemical course shown in Fig. 28.10.

According to associative $S_N^2$ mechanism the entering ligand Y attacks the square planar complex, $MA_2BX$ (I) along z-axis from above or below the square plane to form 5-coordinated square pyramidal intermediate (unstable), $MA_2BXY$ (II) in which the entering ligand Y occupies the axial position. Now the leaving group X adjusts its position so as to generate more stable trigonal bipyramidal intermediate (III) in which X, Y and B form the equatorial trigonal plane and two A ligands occupy the axial positions. The trigonal bipyramidal intermediate (III) then rearranges to another square pyramidal intermediate (IV) in which the leaving group X occupies

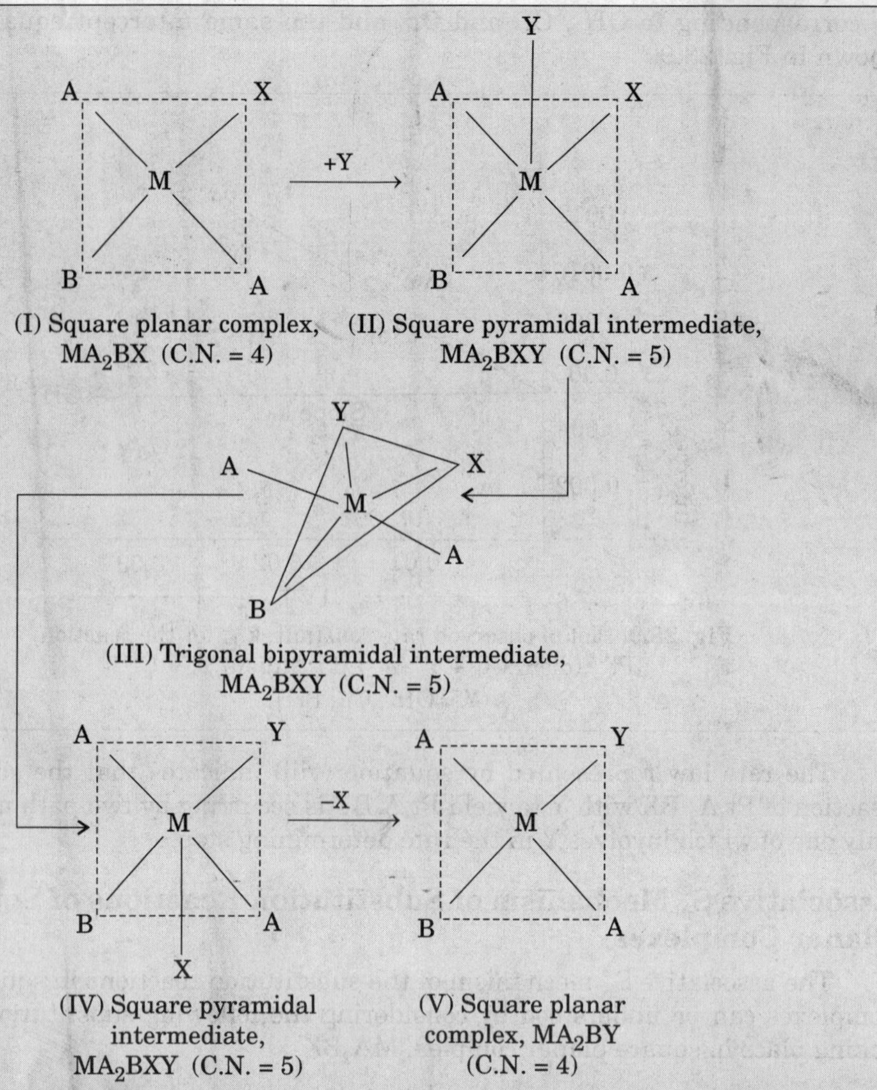


**Fig. 28.10.** Stereospecific ligand substitution at a square planar complex via square pyramidal and trigonal by bipyramidal intermediates (Associative $S_N^2$ mechanism).

the axial position (along z-axis). The leaving group (X) finally departs from the square pyramidal intermediate (IV) to give square planar complex, $MA_2BY$ (V).

It may be noted that the two A ligands are *trans* to each in the original square planar complex, $MA_2BX$ (I) as well as in the final square planar complex, $MA_2BY$ (V). This shows that the ligand displacement reaction given above [reaction (i)] is stereospecific, *i.e.* *trans* complex gives *trans* complex. Similarly *cis* complex gives *cis* complex.

## Factors Affecting the Rate of Substitutions Reactions in Square Planar Complexes

Important factors which affect the rate of substitution reactions in square planar complexes are:

**1. Steric effect.** The reaction between *cis* and *trans* isomers of $[Pt^{2+}(PEt_3)_3 Cl]^+$ with pyridine in ethanol solution has been studied. These studies have shown that *cis* blocking causes the relative rate to drop by 1/80,000 whereas *trans* blocking results in a drop of only 1/36. Thus *cis* blocking is more effective than the *trans* blocking. This result supports the assumption that the intermediate formed during $S_N^2$ mechanism has trigonal bipyramidal geometry.

In $[Pt(Et_4dien)Cl]^+$ the four terminal Et ($C_2H_5$) groups are below and above the plane and block the access to control metal. Thus $[Pt(Et_4 dien)Cl]^+$ is a hindered system. On the other hand, $[Pt(dien)Cl]^+$ is an unhindered system. Experimental evidences have shown that the reaction of hindered systems is largely dissociative and that for unhindered system the reaction is associative.

**2. Effect of leaving group.** The reaction:

$$[Pt^{2+}(dien)X]^+ + py \longrightarrow [Pt^{2+}(dien)(py)]^{2+} + X^-$$

$$(X = NO_3^-, H_2O, Cl^- \text{ etc.})$$

has been studied and the rates of reaction have been obtained. The rates of reaction show that if the leaving group $X^-$ is replaced by $NO_3^-$, $H_2O$, $Cl^-$, $Br^-$, $I^-$, $N_3^-$, $SCN^-$, $NO_2^-$ and $CN^-$, the rates decrease in the order:

$$NO_3^- > H_2O > Cl^- > Br^- > I^- > N_3^- > SCN^- > NO_2^- > CN^-$$

Note that in the square planar complex, $[Pt(dien)X]^+$ three coordinate positions are always occupied by the inert *dien* ligand while the fourth one is occupied by $X^-$ which may be $NO_3^-$, $H_2O$, $Cl^-$ etc.

**3. Effect of solvent.** A detailed study of the exchange reaction,

$$trans\,[Pt(py)_2 Cl_2] + {}^{36}Cl \rightarrow trans[Pt(py)_2\,{}^{36}Cl_2] + Cl^-$$

in various solvents has shown that the rate of the above reaction is independent of the concentration of $Cl^-$ ion and this exchange reaction proceeds through the solvent path. If the solvents *viz* DMSO, $H_2O$ $MeNO_2$ and ROH are used, then the value of $k_s$ for these solvents is in the order:

$$DMSO > H_2O > MeNO_2 > ROH$$

The order of reactivity for non-conducting solvents has been found to be as:

$CCl_4 > C_6H_6 >$ *m*-cresol > *ter*-butanol > ethyl acetate > acetone > DMF

**4. Effect of entering group.** Kinetic informations indicate large differences in the reactivities of different entering groups which is one of the most important requirements of an associative mechanism. The substitution reaction of *trans* $[Pt(py)_2Cl_2]$ with various ligands, Y, in methanol solution has been studied in great detail and a reactivity order

$$S_2O_3^{2-} > SC(NH_2)_2 \sim C_6H_5S^- > SeCN^- > SCN^- > I^- \gg C_6H_5SH > Br^-$$

$$> N_2H_4 \sim NH_2OH > N_3^- > NO_2^- > py > NH_3 > Cl^- > MeO$$

has been obtained. A comparison of the *trans* series and average ligand reactivity order reveals that a good *trans* labilizer is also a good entering group in square planar substitution. This is understandable in terms of the formation of trigonal bypyramidal intermediate (in associative mechanism) because the *trans* group and the entering group are in similar positions with respect to the leaving group in the trigonal plane.

**5. Effect of the charge on the complex (Substrate).** We have already said that the substitution reactions in square planar complexes occur through $S_N^2$ mechanism. In this case the increased positive charge on the square planar complex would not make any significant impact on the rate of the substitution reaction. This is because the increase in the charge would not only make the breaking Pt – X bond more difficult but would also make the formation of M—$H_2O$ bond more easy. This results in that the rate of overall reaction would remain practically unchanged with any change in the charge on the complex.

## *Cis-trans* Isomerisation: Conversion of cis-Isomer into *trans*-Isomer and Vice Versa

Here we shall consider the square planar complex, $[Pt^{2+} (PR_3)_2 Cl_2]°$. This complex exists in *cis*- and *trans*-isomers. These isomers can be converted into each other by the scheme shown in Fig. 28.11. This type of conversion is called *cis-trans* isomerisation The *cis-trans* isomerisation of $[Pt^{2+} (PR_3)_2 Cl_2]^0$ has been done in solution containing excess of $PR_3$ which acts as a catalyst. The isomerisation shown in Fig. 28.11 consists of the following steps: (i) Addition of $PR_3$ (catalyst), (ii) Removal of $Cl^-$ ion attached with the catalyst ($PR_3$) in (II) (iii) Addition of $Cl^-$ ion (iv) Removal of $PR_3$ and rearrangement of Cl-atoms. In this step the catalyst is regenerated.

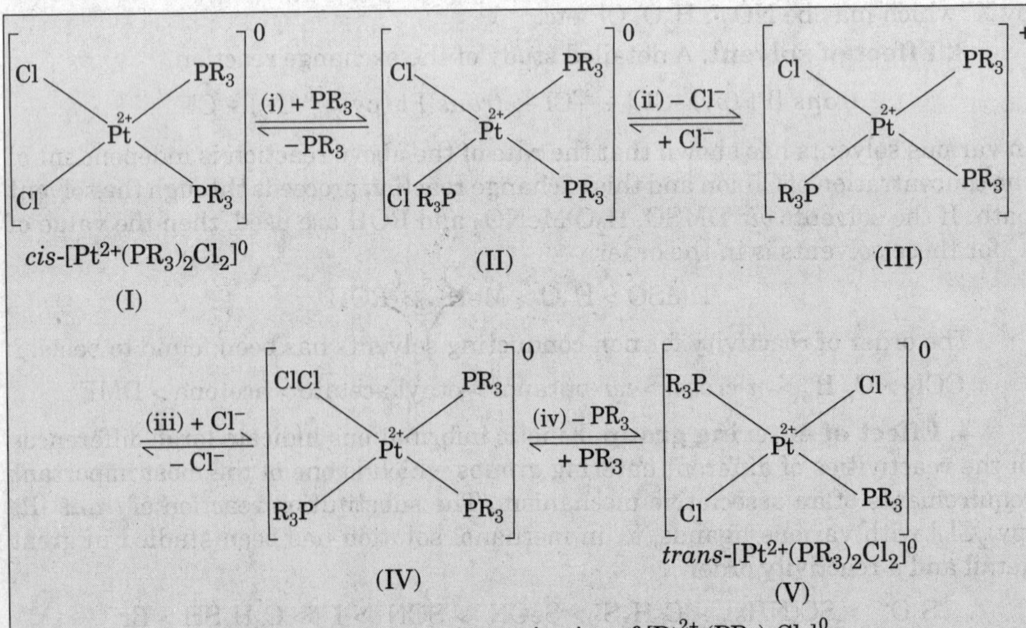


**Fig. 28.11.** *cis-trans* isomerisation of $[Pt^{2+} (PR_3)_2Cl_2]^0$

## UNIVERSITY QUESTIONS

1. Write a note on "*Trans* Effect" *(Pd. Ravishankar Shukla 2010)*
2. What is *trans* effect? How does it influence substitution in square planar complex? *(Rajasthan 2010)*
3. Choose the correct answer: Minimum *trans* effect is shown by—
   (*a*) $CN^-$ (*b*) $C_2H_4$ (*c*) $H_2O$ (*d*) $NH_3$
   *(Barkatullah 2010)*
4. (i) Trichloronitritoplatinate (II) on treatment with $NH_3$ gives __________ of dichloronitritoammineplatinate (II).
   (*a*) *trans*-isomer
   (*b*) *cis*-isomer
   (*c*) *cis*- and *trans*-isomers
   (*d*) no reaction
   (ii) Which one of the following would exhibit a relatively stronger *trans* effect?
   (*a*) NO (*b*) $NO_2^-$ (*c*) Py (*d*) $H_2O$
   *(Bharathiar 2010)*
5. How will you prepare *cis* and *trans* $[Pt(NH_3)_2\,Cl_2]$ starting from $[PtCl_4]^{2-}$? *(Bharathiar 2010)*
6. With example explain *trans* effect.
   *(BJB College, Orissa)*
7. (i) What is *trans* effect? Predict the products A, B, C, D and E in the following reactions:
   (ii) $[Pt\,(NH_3)_4]^{2+} \xrightarrow{py} A \xrightarrow{py} B$
   (iii) $[Ni(H_2O)_6]^{2+} \xrightarrow{2en} E + 4H_2O$
   $(py = C_5H_5N; en = NH_2CH_2CH_2NH_2)$ *(Lucknow 2010)*
8. What is trans effect? Explain π-bonding theory of trans effect?
   *(Madras April 2011)*
9. (a) Explain substitution reactions with examples which take place in Pt (II) square planar complexes.
   (b) In square planar complexes displacement machanism is - - - - - - -
   (c) Explain with examples the substitution reactions in square planar complexes.
   *(H.N. Bahuguna Garhwal 2010)*
10. What do you mean by substitution reaction in square planar complexes?
    *(GND Amritsar 2010)*
11. (*i*) Show the stereochemistry of substitution in the reaction:
    $[PtCl_4]^{2-} \xrightarrow{+\,NH_3} ? \xrightarrow{+NO_2^-} ?$
    (*ii*) Discuss polarisation theory of trans effect. *(Kurukshetra 2011)*

CHAPTER 29

# Electron Transfer (Oxidation-Reduction) Reactions in Coordination Compounds

## Introduction

It is well known that in oxidation-reduction reactions (redox reactions) one or more electrons from a reductant (electron-rich species) are transferred to an oxidant (electron-poor species). In this process oxidation state (O.S.) of the reductant goes up while that of oxidant goes down. Such redox reactions also occur between the two complex ions one of which acts as a reductant and the other acts as an oxidant.

Redox reactions occurring in complexes are called **electron transfer** or **electron exchange reactions.**

## Types of Electron Transfer Reactions in Coordination Compounds

Depending on the mechanism involved in the electron transfer reactions occurring in complex ions, these reactions have been classified under the following two heads:

(i) **Outer sphere electron transfer reactions**. The occurrence of these reactions can be explained by outer sphere mechanism.

(ii) **Inner sphere electron transfer reactions**. These reactions proceed through inner sphere mechanism.

## Another Classification of Electron Transfer Reactions

Another classification of electron transfer reactions in complex ions gives the following two types:

(*i*) These are the electron transfer reactions in which there is *no net chemical change*. Such reactions are known as electron exchange processes. One example of such reactions is given below.

e-

$$[^{*}Co^{2+}(NH_3)_6]^{2+} + [Co^{3+}(NH_3)_6]^{3+} \rightarrow [^{*}Co^{3+}(NH_3)_6]^{3+} + [Co^{2+}(NH_3)_6]^{2+}$$

Reductant Oxidant

Other examples of such reactions have been given on the subsequent pages. These reactions cannot be followed directly but can be followed indirectly by using methods such as labelling or NMR.

(*ii*) These are the electron transfer reactions in which there is *net chemical change*. Examples of such reactions are given below.

(*a*) $[^{*}Cr^{2+}(H_2O)_6]^{2+} + [Cr^{3+}(NH_3)_5X]^{2+} \rightarrow [^{*}Cr^{3+}(H_2O)_6]^{3+} + [Cr^{2+}(NH_3)_5X]^{2-}$

In this reaction X = $F^-$, $Cl^-$, $Br^-$, $I^-$, $NCS^-$, etc.

(*b*) $[^{*}Cr^{2+}(H_2O)_6]^{2+} + [Cr^{3+}(NH_3)_5(SO_4)]^{+} \rightarrow [^{*}Cr^{3+}(H_2O)_6]^{3+} + [Cr^{2+}(NH_3)_5(SO_4)]^{0}$

In this reaction $SO_4^{2-}$ ion acts as a monodentate ligand.

These reactions can be followed by various standard chemical and physical methods.

## Law of Conservation of Energy in Electron Transfer Reactions

Let us consider the electron transfer between $[Fe^{2+}(H_2O)_6]^{2+}$ and $[Fe^{3+}(H_2O)_6]^{3+}$ ions as represented by the reaction:

$$[Fe^{3+}(H_2O)_6]^{3+} + [Fe^{2+}(H_2O)_6]^{2+} \longrightarrow [{}^*Fe^{2+}(H_2O)_6]^{2+} + [{}^*Fe^{3+}(H_2O)_6]^{3+}$$

| Oxidant (I) | Reductant (II) | (III) | (IV) |
|---|---|---|---|
| Fe–O distance =2.05A$^0$ | Fe–O distance =2.21A$^0$ | Complex ions in the lower energy state (stable state) | |
| Complex ions in higher energy state (excited state) | | | |

In this reaction $[Fe^{3+}(H_2O)_6]^{3+}$ (I) acts as an oxidant and $[Fe^{2+}(H_2O)_6]^{2+}$ (II) acts as a reductant. Electron is transferred from the reductant to the oxidant. $[{}^*Fe^{2+}(H_2O)_6]^{2+}$ (III) is produced from $[Fe^{3+}(H_2O)_6]^{3+}$ (I) and $[{}^*Fe^{3+}(H_2O)_6]^{3+}$ (IV) is produced from $[Fe^{2+}(H_2O)_6]^{2+}$ (II). The equilibrium Fe–O distance in (I) is 2.05A$^0$ and this distance in (II) is 2.21A$^0$. Now since, according to Frank-Condon principle, the internuclear distances in a molecule remain unchanged during electron transfer reaction, the Fe–O distance in (III) should be the same as that in its stable or ground state (lower energy state) (II), *i.e* Fe–O distance in (III) should be equal to 2.21A$^0$. Actually Fe–O distance in (III) has been found to be **shorter** than 2.21A$^0$. This indicates that (III) would be in the higher energy or excited state. This state [*i.e.* (III)] would come to the lower energy or ground state by losing the excess energy to the system and will thus adjust its Fe–O distance to 2.21A$^0$

In a similar way, (IV) would be in the higher energy or excited state and would come to the ground state by adjusting its Fe–O distance to 2.05A°. After the products [*i.e.* (III) and (IV)] attain stable state, these are in no way different from the reacting complex ions *viz* (I) and (II). Hence they can again participate in electron transfer reaction and can again liberate energy. Thus the electron transfer reaction between complex ions having the same central metal ion in aqueous solution would lead to the constant generation of energy. This is against the law of conservation of energy. This law is not violated if the metal-ligand distance in all the four complex ions is made equal to each other prior to the electron transfer between them.

## Types of Transition States

Two types of transition states (or intermediates) are generally formed during electron transfer reactions.

(*i*) **Outer sphere transition state**. Outer sphere transition state is that in which the inner coordination spheres of the two reacting complex ions involved in the electron transfer reaction remain intact.

(*ii*) **Inner sphere transition state**. Inner sphere transition state in an electron transfer reaction is that in which the coordinated ligand of one reacting complex ion forms a bridge with the other reacting complex ion.

## Mechanisms of One-Electron Transfer Reactions

There are two mechanisms to explain the occurrence of one-electron transfer reactions. These are *outer sphere mechanism* and *inner sphere mechanism*. These mechanisms are discussed below.

### I. Outer Sphere Mechanism

The electron transfer reactions whose occurrence can be explained on the basis of outer sphere mechanism are called **outer sphere electron transfer reactions.**

Main characteristics of outer sphere mechanism are:

(*i*) In outer sphere mechanism direct transfer of an electron from the reductant to the oxidant occurs. Hence electron transfer reactions proceeding through outer sphere mechanism are also called **direct electron transfer reactions.**

(*ii*) Electron transfer reactions take place between the two metal ions, if the metal ions are inert and have saturated coordination shells or one of the metal ions is inert and its coordination shell is saturated. In the latter case, none of coordinated ligands would be a bridging ligand.

(*iii*) The rate of transfer of electron depends on the polarising power of the metal ion.

(*iv*) The rate of transfer of electron between the two metal ions of the same metal is **fast**,

(*a*) if metal-ligand (M–L) bond distances in the reductant and oxidant are very nearly equal to each other, *i.e.* the bond distances should differ very slightly from each other;

(*b*) if the electrons are able to reach the surface of complex either through unsaturation or through conjugation;

(*c*) if the electron to be transferred from the reductant to the oxidant is present in $t_{2g}$ orbital of the metal ion acting as reductant;

(*d*) if the coordinated ligands present in reductant and oxidant are pi acceptors;

(*e*) leaving aside the electrons to be transferred, the central metal ions present in reductant and oxidant have the same electron spin; and

(*f*) if the reductant and oxidant both (reacting complex ions) are inert and their inner coordination shells are saturated.

(*v*) The rate of transfer of an electron from the reductant to the oxidant is **slow,** if the electron to be transferred is present in $e_g$ orbital.

(*vi*) Leaving aside the electrons that are to be transferred, the electron transfer reaction between the two metal ions present in the reductant and oxidation would be **easy,** if both the metal ions have the same electron spin. We know that $[Co^{2+}(NH_3)_6]^{2+}$ ion is **HS** and $[Co^{3+}(NH_3)_6]^{3+}$ ion is **LS.** Thus these ions do not have the same electron spin. On the other hand, both the complex ions namely $[Fe^{2+}(H_2O)_6]^{2+}$ and $[Fe^{3+}(H_2O)_6]^{3+}$ are **HS**. Consequently, the transfer of an electron from $[Fe^{2+}(H_2O)_6]^{2+}$ to $[Fe^{3+}(H_2O)_6]^{3+}$ would be easier while that from $[Co^{2+}(NH_3)_6]^{2+}$ to $[Co^{3+}(NH_3)_6]^{3+}$ would be more difficult.

(*vii*) The larger is the difference in the oxidation states of the two reactants, larger will be the adjustments needed in the metal-ligand bond lengths, thus calling for higher activation energy and hence slower redox reaction.

(*viii*) Most of the electron exchange reactions in which both reacting complex ions (one complex ion acts as an oxidant and the other complex ion acts as a reductant) are inert towards substitution have very high second order rate constants. The rate law should be first order in each of the two reacting species (rate = $k$[Oxidant][Reductant])

(*ix*) The activated complex should contain intact coordination shells for both the metal ions.

(*x*) According to Franck-Condon principle there can be no appreciable change of atomic arrangement during the time of electron transfer.

(*xi*) If the reacting complex ions are inert, a close approach of the metal atom is not possible and, therefore, the electron transfer must occur through **tunnelling** or **outer sphere mechanism**. Hence redox reactions occurring through outer sphere mechanism are also called **electron tunnelling reactions** or **direct electron transfer reactions.**

(*xii*) The outer sphere electron transfer reactions are *more rapid* when the reacting complex ions contain the ligands like $CN^-$, *o*-phenonthroline etc. than when they contain the ligands like $NH_3$, $H_2O$ etc.

(*xiii*) Outer sphere exchange reactions between complexes having different central metal ions with different oxidation states (*e.g.* $[Os^{II}(dipy)_3]^{2+} - [Mo^{V}(CN)_8]^{3-}$ reaction) are usually faster than outer sphere exchange reactions between the complexes of the same metals with different oxidation states. For example the following reaction is fast:

$$\underset{\text{Reductant}}{[Os^{2+}(dipy)_3]^{2+}} + \underset{\text{Oxidant}}{[Mo^{5+}(CN)_8]^{3-}} \longrightarrow [Os^{3+}(dipy)_3]^{3+} + [Mo^{4+}(CN)_8]^{4-}$$

For such reactions when excited states of the products are converted into ground states, the decrease in energy can appear as the free energy of the reaction ($\Delta G^0$ must be negative otherwise the reaction would not take place). This means that for such reactions the structure of the transition state is more like that of the reactants and consequently the activation energy is decreased and the rate is increased.

(*xiv*) In outer sphere mechanism no binuclear intermediate is formed.

(*xv*) In outer sphere redox reactions the electron effectively hops from one species to the other and the ligand acts as an electron conduction medium.

## Examples of Electron Transfer Reactions Proceeding Through Outer Sphere Mechanism

In order to understand outer sphere mechanism let us consider the following electron transfer reactions:

**Example 1:** $\underset{\text{Reductant}}{[^*Co^{2+}(NH_3)_6]^{2+}} + \underset{\text{Oxidant}}{[Co^{3+}(NH_3)_6]^{3+}} \longrightarrow [^*Co^{3+}(NH_3)_6]^{3+} + [Co^{2+}(NH_3)_6]^{2+}$

The rate of this reaction (which is actually not a chemical reaction) can be studied by using labelled Co-atom ($^*Co$) in the complex ion, $[^*Co^{2+}(NH_3)_6]^{2+}$. The details of the above reaction are given in Fig 29.1.

Oxidation (2+ ⟶ 3+) Reduction (3+ ⟶ 2+) $e^-$

$$[\overset{*}{Co}{}^{2+}(NH_3)_6]^{2+} + [Co^{3+}(NH_3)_6]^{3+} \longrightarrow [\overset{*}{Co}{}^{3+}(NH_3)_6]^{3+} + [Co^{2+}(NH_3)_6]^{2+}$$

| Reductant (I) | Inert oxidant (II) | (III) | (IV) |
|---|---|---|---|
| HS | LS | HS | LS |
| $\overset{*}{Co}{}^{2+} = d^7 = t_{2g}^5\ e_g^2$ (n = 3) | $Co^{3+} = d^6 = t_{2g}^6\ e_g^0$ (n = 0) | $\overset{*}{Co}{}^{3+} = d^6 = t_{2g}^5\ e_g^1$ (n = 2) | $Co^{2+} = d^7 = t_{2g}^6\ e_g^1$ (n = 1) |
| $\overset{*}{Co}{}^{2+}$ – N bond length is large (=2.11Å) | $Co^{3+}$ – N bond length is short (= 1.96Å) | $\overset{*}{Co}{}^{3+}$ –N bond length = 1.76Å | $Co^{2+}$ –N bond length = 2.31Å |

**Fig. 29.1.** Electron transfer reaction between $[\overset{*}{Co}{}^{2+}(NH_3)_6]^{2+}$ and $[Co^{3+}(NH_3)_6]^{3+}$ complex ions. * indicates labelled isotope of cobalt.

Following points may be noted:

(*i*) Since the oxidant, $[Co^{3+}(NH_3)_6]^{3+}$ (II) is an inert complex ion and $Co^{3+}$ ion has saturated inner coordination shell, the exchange of $NH_3$ ligands between the two reacting complexes is **extremely slow.**

(*ii*) Since the lone pair (*lp*) electrons on N-atom of all $NH_3$ ligands is consumed in forming N → Co coordinate bond, this atom (*i.e.* N-atom) has no other *lp* of electrons through which $NH_3$ ligands can form a bridge between the two Co-atoms. Hence inner sphere transition state, $[(NH_3)_5\ Co^{2+}–NH_3–Co^{3+}(NH_3)_5]^{5+}$ cannot be formed. Consequently, electron transfer from the reductant (I) to oxidant (II) must occur through outer sphere mechanism in which the inner coordination shells of reductant and oxidant remain intact in the transition state.

(*iii*) The coordination sphere of both the reactants and products remain undisturbed.

(*iv*) The transfer of an electron from the reductant (I) to the oxidant (II) makes a change in O.S. of the metal ions. The increase in O.S. of the metal (*e.g.* $\overset{*}{Co}{}^{2+} \rightarrow \overset{*}{Co}{}^{3+}$) results in the shortening of metal-ligand bond lengths [*e.g.* $\overset{*}{Co}{}^{2+}$–N bond length in (I) = 2.11A$^0$ and $\overset{*}{Co}{}^{3+}$–N bond length in (III)=1.76A$^0$]. Note that (III) is produced from (I) and (IV) is produced from (II).

(*v*) The observed rate of electron transfer in the above reaction is **slow**. This may be due to the following reasons:

(*a*) Since Co–N bond lengths in the reacting complex ions are appreciably different from each other, considerable amount of energy of activation is needed to excite the two reacting ions to have identical Co–N bond lengths so that electron transfer between the reacting complex ions may occur.

(*b*) The electron to be transferred from the reducetant (I) to the oxidatent (II) resides in $e_g$ orbital of $t_{2g}^5\ e_g^2$ configuration of $\overset{*}{Co}{}^{2+}$ ion in reductant. Now since $d_{z^2}$ and $d_{x2-y2}$ orbitals of $e_g$ set along with *s* and *p* orbitals of the metal are already engaged in forming M–L sigma bonding, the transfer of $e_g$ electron from $t_{2g}^5 e_g^2$ configuration of $\overset{*}{Co}{}^{2+}$ ion would require more energy.

(*c*) The spin states of $\overset{*}{Co}{}^{2+}$ and $Co^{3+}$ ions present in the reacting complex ions are different as shown below.

$Co^{2+}$ present in $[\ Co^{2+}(NH_3)_6]^{2+} = d^7 = t_{2g}^5 e_g^2$ (HS)

$Co\ ^{3+}$ present in $[Co^{3+}(NH_3)_6]^{3+} = d^6 = t_{2g}^6 e_g^0$ (LS).

**Example 2:** $[^{*}Fe^{2+}(CN)_6]^{4-} + [Fe^{3+}(CN)_6]^{3-} \longrightarrow [^{*}Fe^{3+}(CN)_6]^{3-} + [Fe^{2+}(CN)_6]^{4-}$

Ferrocyanide ion (I) Ferricyanide ion (II)

In this reaction * indicates labelled isotope of Fe. The details of this reaction are given in Fig. 29.2.

$e^-$

$[^{*}Fe^{2+}(CN)_6]^{4-} + [Fe^{3+}(CN)_6]^{3-} \longrightarrow {}^{*}[Fe^{3+}(CN)_6]^{3+} + [Fe^{2+}(CN)_6]^{4-}$

| Ferrocyanide ion (Inert) Reductant (I) LS | Ferricyanide ion (Inert) Oxidant (II) LS | Ferricyanide ion (III) | Ferrocyanide ion (IV) |
|---|---|---|---|

$^{*}Fe^{2+} = d^6 = t_{2g}^6\, e_g^0$  $Fe^{3+} = d^5 = t_{2g}^5\, e_g^0$

$^{*}Fe^{2+}$ – C bond length is slightly larger  $Fe^{3+}$ – C bond length is slightly smaller

**Fig. 29.2.** Electron transfer reaction between $[^{*}Fe^{2+}(CN)_6]^{4-}$ and $[Fe^{3+}(CN)_6]^{3-}$ ions. * indicates labelled isotope of Fe.

Following points may be noted:

(*i*) An electron is transferred from ferrocyanide ion, $[^{*}Fe^{2+}(CN)_6]^{4-}$ (I) to ferricyanide ion, $[Fe^{3+}(CN)_6]^{3-}$ (II). Thus (I) acts as a reductant (loses electron) and (II) acts as an oxidant (gains electron).

(*ii*) Both reacting complex ions are LS complex ions ($Fe^{2+} = d^6 = t_{2g}^6 e_g^0$, $Fe^{3+} = d^5 = t_{2g}^5 e_g^0$),

(*iii*) Both the reacting complex ions are inert towards substitution and their coordination shells are fully saturated.

(*iv*) The isotopic study of the above reaction has shown that $CN^-$ ligands in both the reacting complex ions are not exchangeable. Therefore, the formation of an inner sphere bridged transition state is not possible. Thus the electron transfer in the above reaction proceeds through outer sphere mechanism.

(*v*) The observed rate of electron transfer in the above reaction has been found to be **fast**. This may be due to the following reasons:

(*a*) Since $^{*}Fe^{2+}$–C bond distance in (I) is slightly larger than $Fe^{3+}$–C bond distance in (II), very little amount of energy is required to make these bond distances equal in both the complex ions so that electron transfer between the complex ions may occur.

(*b*) Since both metal ions *viz* $Fe^{2+}$ and $Fe^{3+}$ are in the same spin state (both are in low spin state), the electron transfer becomes easy.

(c) The electron to be transferred from the reductant (I) to the oxidant (II) resides in $t_{2g}$ orbital of $t_{2g}^6 e_g^0$ configuration of $Fe^{2+}$ ion in reductant. Now since none of the $t_{2g}^6$ electrons is engaged in forming M–L sigma bonding, the transfer of an electron from $t_{2g}^6$ configuration would be easy.

(*d*) Since the ligand $CN^-$ present in the reductant is unsaturated, the transfer of an electron from the reductant to the oxidant becomes easy.

(*e*) Since $CN^-$ ligand is a pi acceptor ligand, it can form a pi bond by accepting the electrons from $t_{2g}$ orbitals of the metal ion and hence stabilizes the lower O.S. of the metal ion. Due to the acceptance of electron by $CN^-$ ligand, the products formed are stabilized.

**Exmaple 3:** $[{}^{*}Os^{2+}(dipy)_3]^{2+} + [Os^{3+}(dipy)_3]^{3+} \longrightarrow [{}^{*}Os^{3+}(dipy)_3]^{3+} + [Os^{2+}(dipy)_3]^{2+}$

Inert Reductant(I) — Inert Oxidant(II)

In this reaction $*$ indicates labelled isotope of Os.

Following points may be noted:

(*i*) *dipy* is dipyridine whose structure is 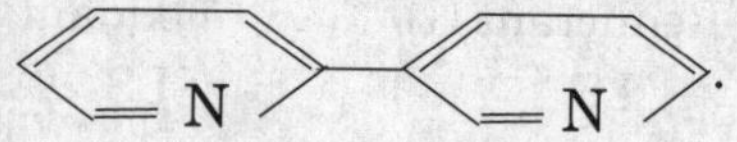. It is a bidentate ligand.

(*ii*) An electron is transferred from reductant (I) to oxidant (II).

(*iii*) This reaction cannot occur through inner sphere mechanism, since both reacting complex ions are inert towards substitution and their coordination shells are fully saturated.

(*iv*) The observed rate of reaction has been found to be **fast**. This is due to the following reasons:

(*a*) Since Os–N bond distances in reacting complex ions are slightly different from each other, very little amount of energy is required to make these bond distances equal in both the complex ions so that electron transfer between the complex ions may occur.

(*b*) Since *dipy* is a pi acceptor ligand and is highly conjugated, the tunnelling of the electron is facilitated.

(*c*) Since $t_{2g}$ electron to be transferred from the reductant (I) to the oxidant (II) is not engaged in forming M—L sigma bonding, this electron can easily be delinked from the ${}^{*}Os^{2+}$ ion present in the reductant.

(*d*) Metal ions in both the complex ions have the same spin.

Some more examples of electron transfer reactions (redox reactions) which occur through outer sphere mechanism are given below:

**(1)** $[Fe^{2+}(bpy)_3]^{2+} + [Fe^{3+}(bpy)_3]^{3+} \longrightarrow [Fe^{3+}(bpy)_3]^{3+} + [Fe^{2+}(bpy)_3]^{2+}$

Reductant — Oxidant

$Fe^{2+}=d^6=t_{2g}^6$ (LS) — $Fe^{3+}=d^5=t_{2g}^5$ (LS)

**(2)** $[Mn^{2+}(CN)_6]^{4-} + [Mn^{3+}(CN)_6]^{3-} \longrightarrow [Mn^{3+}(CN)_6]^{3-} + [Mn^{2+}(CN)_6]^{4-}$

Reductant — Oxidant

$Mn^{2+}=d^5=t_{2g}^5$ (LS) — $Mn^{3+}=d^4=t_{2g}^4$ (LS)

**(3)** $[Fe^{2+}(H_2O)_6]^{2+} + [Fe^{3+}(H_2O)_6]^{3+} \longrightarrow [Fe^{3+}(H_2O)_6]^{3+} + [Fe^{2+}(H_2O)_6]^{2+}$

Reductant — Oxidant

$Fe^{2+}=d^6=t_{2g}^4e_g^2$ (HS) — $Fe^{3+}=d^5=t_{2g}^3e_g^2$ (HS)

**(4)** $[Ir^{3+}Cl_6]^{3-} + [Ir^{4+}Cl_6]^{2-} \longrightarrow [Ir^{4+}Cl_6]^{2-} + [Ir^{3+}Cl_6]^{3-}$

Reductant — Oxidant

$Ir^{3+}=d^6=t_{2g}^6$ (LS) — $Ir^{4+}=d^5=t_{2g}^5$ (LS)

**(5)** $[Co^{2+}(en)_3]^{2+} + [Co^{3+}(en)_3]^{3+} \longrightarrow [Co^{3+}(en)_3]^{3+} + [Co^{2+}(en)_3]^{2+}$
Reductant Oxidant
$Co^{2+}=d^7=t_{2g}^5 e_g^2$ (HS) $Co^{3+}=d^6=t_{2g}^6$ (LS)

**(6)** $[Co^{2+}(C_2O_4)_3]^{4-} + [Co^{3+}(C_2O_4)_3]^{3-} \longrightarrow [Co^{3+}(C_2O_4)_3]^{3-} + [Co^{2+}(C_2O_4)_3]^{4-}$
Reductant Oxidant
$Co^{2+}=d^7=t_{2g}^5 e_g^2$ (HS) $Co^{3+}=d^6=t_{2g}^6$ (LS)

**(7)** $[{}^*Mn^{6+}O_4]^{2-} + [Mn^{7+}O_4]^- \longrightarrow [{}^*Mn^{7+}O_4]^- + [Mn^{6+}O_4]^{2-}$
Reductant Oxidant

Following points may be noted:

(*i*) The complex ions, $[Fe^{2+}(bpy)_3]^{2+}$ and $[Fe^{3+}(bpy)_3]^{3+}$ react **very fast**, since M–L bond lengths in the two species differ very slightly. On the other hand, the complex ion, $[Fe^{2+}(H_2O)_6]^{2+}$ and $[Fe^{3+}(H_2O)_6]^{3+}$ react **very slowly**, since there exists a considerable difference between Fe–O bond lengths in the two species. This difference calls for a lot of adjustment in Fe–O bonds in the exchanging species. A larger activation energy is required compared to that for $[Fe^{2+}(bpy)_3]^{2+} - [Fe^{3+}(bpy)_3]^{3+}$ pair. $Fe^{2+}$– O bond length in $[Fe^{2+}(H_2O)_6]^{2+}$ is 2.21A° while $Fe^{3+}$–O bond length in $[Fe^{3+}(H_2O)_6]^{3+}$ is 2.05A°.

(*ii*) Rate of electron transfer (in $LM^{-1}s^{-1}$) for reactions 1, 2 and 4 is more than $10^6$ while this value for reaction 3 is ~4 and for reaction 6 this value is $\sim 10^{-4}$.

(*iii*) Difference in M–L bond lengths (A°) for reactions 1, 2 and 4 is very small.

(*iv*) Redox reaction 7 has been studied by Sheppard and Wahl. In this reaction both the reacting anions are inert towards substitution. Data show that the transfer of an electron from $[MnO_4]^{2-}$ to $[MnO_4]^-$ takes place rapidly and tunnelling mechanism seems to be the most probable mechanism.

## II. Inner Sphere Mechanism or Ligand Bridged Process

On the basis of extensive investigations made by Taube, he has suggested that the main characteristics of electron transfer in reactions occurring through inner sphere mechanism are:

(*i*) One of the reacting complex ions should be **inert** and the other ion should be **labile**. The inert complex ion should have a coordinating ligand which is capable of forming a bridge between the two reacting complex ions (reductant and oxidant) during the formation of intermediate which is called **ligand bridged transition state** or **ligand bridged intermediate**.

(*ii*) The rate of electron transfer in the reaction **increases**:

(*a*) if the bridging ligand has unsaturation or extended conjugation in its structure; and

(*b*) with the increase of nucleophilic character of the bridging ligand.

(*iii*) In electron transfer reaction occurring through inner sphere mechanism, an atom or a group is transfered from one reacting complex ion to the other reacting complex ion to form an atom or group-bridged intermediate and the electron is transferred from the reductant to the oxidant in this bridged intermediate. Now the intermediate breaks up to give the products.

The transfer or non-transfer of the bridging atom or a group depends on the relative stabilities of the products obtained by the breaking of the bridged intermediate.

**Example.** Let us consider the following electron transfer reaction:

$$\underset{\text{Oxidant}}{[Ir^{4+}Cl_6]^{2-}} + \underset{\text{Reductant}}{[Cr^{2+}(H_2O)_6]^{2+}} \longrightarrow \underbrace{[Ir^{3+}Cl_6]^{3-} + [Cr^{3+}(H_2O)_6]^{3+}}_{\text{Products}}$$

The products obtained in the above reaction show that *no transfer of the bridging Cl-atom has occurred but an electron has transferred from the reductant to the oxidant.* Cl-bridged intermediate, if formed in the reaction, has the structure, $[Cl_5Ir^{4+}Cl^{-}–Cr^{3+}(H_2O)_5]^0$. This intermediate can break to give two types of products as shown below:

$$\underset{\substack{\text{Cl-bridged}\\\text{intermediate}}}{[Cl_5Ir^{4+}–Cl^{-}–Cr^{2+}(H_2O)_5]} \xrightarrow[(+H_2O)]{\text{Hydrolysis}} \begin{cases} \underbrace{\underset{(I)}{[Ir^{3+}Cl_6]^{3-}} + \underset{(II)}{[Cr^{3+}(H_2O)_6]^{3+}}}_{\text{More stable}} \\ \underbrace{\underset{(III)}{[Ir^{3+}Cl_5(H_2O)]^{2-}} + \underset{(IV)}{[Cr^{3+}(H_2O)_5Cl]^{2+}}}_{\text{Less stable}} \end{cases}$$

Since the products of the reaction *viz* (I) and (II) are more stable than the products (III) and (IV), Cl-bridged intermediate is formed and this breaks to give the products (I) and (II) and not (III) and (IV).

(*iv*) Electron transfer in reactions proceeding through inner sphere mechanism are faster than those in similar reactions which proceed through outer sphere mechanism.

## Examples of Electron Transfer Reactions Proceeding Through Inner Sphere Mechanism

**Example 1.** Let us consider the following electron transfer reaction in which an electron is transferred from the reductant to the oxidant. This reaction has been studied in aqueous phase.

$$\underset{\substack{\text{Reductant}\\\text{(Labile)}}}{[Cr^{2+}(H_2O)_6]^{2+}} + \underset{\substack{\text{Oxidant}\\\text{(Inert)}}}{[Co^{3+}(NH_3)_5X]^{2+}} \xrightarrow{\text{Water}} \underbrace{[Cr^{3+}(H_2O)_5X]^{2+} + [Co^{2+}(NH_3)_5(H_2O)]^{2+}}_{\text{Products}}$$

In the given reaction $X^-$ may be any halide ion. $[Cr^{2+}(H_2O)_6]^{2+}$ is labile complex ion and acts as a reductant while $[Co^{3+}(NH_3)_5X]^{2+}$ is inert and acts as an oxidant. This reaction proceeds through the following steps:

(*a*) *Formation of bridged intermediate and electron transfer*. The ligand $X^-$ which is present in inert oxidant gets dissociated from the oxidant and is transferred to the reductant to form an $X^-$—bridged intermediate (I) in which $X^-$ acts as a bridging ligand between the reductant and oxidant.

Within the bridged intermediate (I) thus formed an electron is transferred from the reductant to the oxidant.

$$\underset{\text{Reductant}}{[Cr^{2+}(H_2O)_6]^{2+}} + \underset{\text{Oxidant}}{[Co^{3+}(NH_3)_5X]^{2+}} \xrightarrow[(-H_2O)]{\text{Slow}}$$

$$\underset{\text{Bridged intermediate (I)}}{[(H_2O)_5Cr^{2+}\text{–}X^-\text{–}Co^{3+}(NH_3)_5]^{4+}} \xrightarrow[\text{transfer}]{\text{Electron}} \underset{\text{Bridged intermediate (II)}}{[(H_2O)_5Cr^{3+}\text{–}X^-\text{–}Co^{2+}(NH_3)_5]^{4+}}$$

(*b*) *Decomposition of the bridged intermediate into products: Hydrolysis.* The bridged intermediate (II) formed as above gets decomposed (hydrolysis) to give the final products.

$$\underset{\text{Bridged intermediate (II)}}{[(H_2O)_5Cr^{3+}\text{–}X^-\text{—}Co^{2+}(NH_3)_5] + H_2O} \xrightarrow[\text{Fast}]{\text{Hydrolysis }(+H_2O)}$$

$$\underbrace{[Cr^{3+}(H_2O)_5X]^{2+} + [Co^{2+}(NH_3)_5(H_2O)]^{2+}}_{\text{Products}}$$

Step (*a*) is slow step and hence is the rate determining step. Thus:

Rate of electron transfer in reaction

$$= k \times \text{conc. of } [Cr^{2+}(H_2O)_6]^{2+} \times \text{conc. of } [Co^{3+}(NH_3)_5X]^{2+}$$

$$= k \times [\text{Reductant}]\,[\text{Oxidant}] \qquad (i)$$

Relation (i) given above shows that the rate of electron transfer in the reaction changes with the change of halide ion ($X^-$) present in the oxidant. The capability of $X^-$ ion to form the Cr–X–Co bridge in the transition state decreases as $I^- > Br^- > Cl^- > F^-$. With the decrease in the capability, the rate of electron transfer in the reaction also decreases in the same order.

**Example 2:** Consider the reaction:

$$[Cr^{2+}(H_2O)_6]^{2+} + [Co^{3+}(NH_3)_5(RCOO)]^{2+} \longrightarrow \underbrace{[Cr^{3+}(H_2O)_5(RCOO)]^{2+} + [Co^{2+}(NH_3)_5(H_2O)]^{2+}}_{\text{Products}}$$

In the above reaction $[Cr^{2+}(H_2O)_6]^{2+}$ acts as an oxidant and $[Co^{3+}(NH_3)_5(RCOO)]^{2+}$ acts as a reductant. This reaction proceeds through inner sphere mechanism which involves the following two steps:

(*a*) $COOR^-$ ion forms a bridged intermediate as shown below.

$$[Cr^{2+}(H_2O)_6]^{2+} + [Co^{3+}(NH_3)_5(RCOO)]^{2+} \longrightarrow \underset{RCOO^- \text{ – bridged intermediate}}{[(H_2O)_5Cr^{3+}\text{–}RCOO^-\text{–}Co^{2+}(NH_3)_5]^{4+}} + H_2O$$

(*b*) In this step the RCOO-bridged intermediate gets hydrolysed to give the products

$$[(H_2O)_5Cr^{3+}\text{–}RCOO^-\text{—}Co^{2+}(NH_3)_5]^{4+} + H_2O \xrightarrow[+H_2O]{\text{Hydrolysis}}$$

$$[Cr^{3+}(H_2O)_5(RCOO)]^{2+} + [Co^{2+}(NH_3)_5(H_2O)]^{2+}$$

If R present in the bridging ligand, $RCOO^-$ is unsaturated or has extended conjugation in its structure, the rate of electron transfer in the reaction is increased. This is because R facilitates electron transfer.

## Factors Affecting the Rates of Direct Electron Transfer Reactions

The factors affecting the rates of direct electron transfer reactions have been summarized by Halpern as follows:

(*i*) **Electrostatic repulsion between ions of like charges**. This contributes to the activation energy. Because of electrostatic repulsion the energy of activation also increases. The increase in activation energy tends to decrease the rate of exchange of electrons.

(*ii*) **Identity and concentration of the cations present in solution.** In several cases the rate constants for the electron transfer reactions have been found to depend on the identity and concentration of the cations present in the solution. Generally an increase in the concentration of the cations increases the rate, but certain cations are particularly effective. The general effect can be attributed to the formation of ion pairs which then decrease the electrostatic contribution to the activation energy.

(*iii*) **Conductivity of ligands:** Complex ions having ligands of higher conductivity exchange their electrons more rapidly than the complex ions having ligands of lower conductivity. For example, since *phen* and *bipy* ligands have higher conductivity than $NH_3$ and *en*, $[M(phen)_3]^{n+}$ and $[M(dipy)_3]^{n+}$ ions exchange their electrons more rapidly than $[M(NH_3)_6]^{n+}$ and $[M(en)_3]^{n+}$ ions.

(*iv*) **Reorganisation energy.** The bond length in $MnO_4^-$ and $MnO_4^{2-}$ differ and the activated complex for exchange between these must involve a "stretched" Mn–O bond in $MnO_4^-$. Slow reactions are expected when the oxidant and the reductant have greatly different bond lengths, geometries or salvation shells. For octahedral complexes bond lengths are changed more drastically when $e_g$ electrons are involved in the transfer as compared to $t_{2g}$ electrons in the *d*-orbitals.

(*v*) **Free energy change of the reaction.** Free energy change of the reaction also affects the rates of outer sphere reactions. The electron exchange reaction will be **fast**, if the free energy change for the overall reaction is **negative**. With the increase of the free energy change, the exchange of electrons is lowered down.

## Two-Electron Transfer Reactions

The electron transfer reactions we have discussed so far involve the transfer of a single electron from one species to the other. Reactions in which two electrons are transferred from the reductant to the oxidant are called **two-electron transfer reactions**.

**Examples:**

$2e^-$

(*i*) $^{*}Tl^{+} + Tl^{3+} \longrightarrow {}^{*}Tl^{3+} + Tl^{+}$

Reductant   Oxidant

$2e^-$

(*ii*) $^{*}Pt^{2+} + Pt^{4+} \longrightarrow {}^{*}Pt^{4+} + Pt^{2+}$

Reductant   Oxidant

$2e^-$

(*iii*) $Sn^{2+}$ (Reductant) + $Hg^{2+}$ (Oxidant) $\longrightarrow$ $Sn^{4+} + Hg^0$

In aqueous perchlorate solution the rate law for reaction (*i*) is

$$Rate = k_1[Tl^+][Tl^{3+}] + k_2[Tl^+][TlOH^{2+}]$$

## Complementary and Non-complementary Electron Transfer Reactions

We know that in an electron transfer reaction electrons are transferred from the reductant to the oxidant, *i.e.* the reductant loses electrons and oxidant gains electrons. *Electron transfer reaction in which the number of electrons lost by the reductant is equal to the number of electrons gained by the oxidant is called* **complementary electron transfer reaction**. Thus in complementary electron transfer reaction the reductant and oxidant change their oxidation states (O.S.) by the same number.

**Examples:**

$2e^-$

(*i*) $Sn^{2+}$ (Reductant) + $Tl^{3+}$ (Oxidant) $\longrightarrow$ $Sn^{4+} + Tl^+$

$2e^-$

(*ii*) $Sn^{2+}$ (Reductant) + $Hg^{2+}$ (Oxidant) $\longrightarrow$ $Sn^{4+} + Hg^0$

In reaction (*i*), O.S. of reductant is increased by 2 units ($2+ \rightarrow 4+$) and that of oxidant is also reduced by 2 units ($3+ \rightarrow 1+$).

In reaction (*ii*), O.S. of reductant is increased by 2 units ($2+ \rightarrow 4+$) and that of oxidant is also decreased by 2 units ($2+ \rightarrow 0$).

*Electron transfer reaction in which the number of electrons lost by the reductant is different from the number of electrons gained by the oxidant is called* **non-complementary electron transfer reaction**. Thus in non-complementary electron transfer reaction the reductant and oxidant change their oxidation states by different number.

**Example:** $2Fe^{2+}\ (aq)$ (Reductant) + $Tl^{3+}\ (aq)$ (Oxidant) $\longrightarrow 2Fe^{3+}\ (aq) + Tl^+\ (aq)$

In this reaction O.S of one $Fe^{2+}$ (reductant) is increased by 1 unit ($2+ \rightarrow 3+$) while that of $Tl^{3+}$ (oxidant) is decreased by 2 units ($3+ \rightarrow 1+$).

Above reaction proceeds through the following steps:

$$Fe^{2+}(aq) + Tl^{3+}(eq) \xrightarrow[k_1]{\text{Slow}} Fe^{3+}(aq) + \underset{\text{Reactive intermediate}}{Tl^{2+}(aq)}$$

$$Fe^{2+}(aq) + \underset{\text{Reactive intermediate}}{Tl^{2+}(aq)} \xrightarrow[k_2]{\text{Fast}} Fe^{3+}(aq) + Tl^{+}(aq)$$

*On adding:* $2Fe^{2+}(aq) + Tl^{3+}(aq) \longrightarrow 2Fe^{3+}(aq) + Tl^{+}(aq)$

## UNIVERSITY QUESTIONS

1 Give examples of oxidation-reduction reactions occurring through the transfer of atoms or groups of atoms.

2 What are the conditions for the transfer of an electron from $M^{2+}$ to $M^{3+}$ in aqueous medium?

3 Discuss the outer sphere and inner sphere mechanisms of electron transfer reactions. Why is the electron transfer in the system $[Co(NH_3)_6]^{2+}$ –$[Co(NH_3)_6]^{3+}$ slower than that in the system $[Fe(CN)_6]^{4}$ – $[Fe(CN)_6]^{3-}$?

4 Give at least one example of an electron transfer reaction which occurs through inner sphere mechanism and also involves transfer of atom or group.

5 Give at least one example of an electron transfer reaction which occurs through the inner sphere mechanism but does not involve the transfer of atom or group.

6 Explain why

(*i*) The transfer of electron from $[Cr(H_2O)_6]^{2+}$ to $[Co(NH_3)_6]^{3+}$ in aqueous medium is slower than the transfer from $[Cr(H_2O)_6]^{2+}$ to $[Co(NH_3)_5(OH)]^{2+}$.

(*ii*) The transfer of electron from $[Cr(H_2O)_6]^{2+}$ to $[Cr(NH_3)_3(NCS)]^{2+}$ is slower than the transfer from $[Cr(H_2O)_6]^{2+}$ to $[Cr(NH_3)_5N_3]^{2+}$.

7 Discuss the complementary and the non-complementary two-electron transfer reactions giving suitable examples.

# 30 CHAPTER Transition Metal Complexes with Carbon Monoxide: Metallic Carbonyls

## What are π-Acid Ligands and π-Acid Complexes?

CO, CNR, $R_3P$, NO etc. are such ligands which have *filled π-orbitals* (called *donor orbitals*) as well as *empty* π-orbitals (called *acceptor orbitals*). These ligands are called *π-acid ligands* or *π-acceptor ligands* or *π-bonding ligands*. These ligands combine with transition metals in low oxidation states (*e.g.* –1 , 0, +1) and form complexes which are called *π-acid complexes*. Thus π-acid complexes are the complexes of transition metals in low oxidation state with π-acid ligands. Metallic carbonyls are important examples of π-acid complexes. These complexes have ligand → metal σ-bonds as well as metal → ligand π-bonds.

## What are Metallic Carbonyls?

The compounds formed by the combination of CO molecules with transition metal atoms in low oxidation states are called metallic carbonyls. In these compounds, CO molecule acts as a neutral ligand.

## Why is the Metallic Atom in Low Oxidation State in Metallic Carbonyls?

We have said above that metal atom in metal carbonyls is in low oxidation state. This is because of the fact that there is a higher electron density (electronic charge) on the metal atom in its lower oxidation state than in its higher oxidation state. Thus, since the metal is electropositive in nature, it would feel more comfortable, if some of the excess electronic charge present on the metal atom gets transferred to the vacant π-orbitals of CO (ligand) by forming M → CO π-bond.

## Classification of Carbonyls

Metallic carbonyls have been classified into the following two ways:

## I. Classification Based on the Number of Metallic Atoms Present in Carbonyls

This classification gives the following two types of carbonyls:

(*a*) **Mononuclear or monomeric carbonyls.** The molecule of these carbonyls contains only one metallic atom. Examples are $Ni(CO)_4$, $M(CO)_5$, (M=Fe, Ru, Os) and $M(CO)_6$ (M = V, Cr,. Mo, W).

(*b*) **Polynuclear carbonyls.** The molecule of these carbonyls contains two or more metallic atoms. Examples are $Co_2(CO)_8$, $Fe_2(CO)_9$, $M_2(CO)_{10}$, (M=Mn, Tc, Re), $M_3(CO)_{12}$ (M=Fe, Rn, Os). Polynuclear carbonyls may be *homonuclear* [*e.g.*, $Co_2(CO)_8$, $Fe_2(CO)_9$, $Fe_3(CO)_{12}$ etc.] or *hetero-nuclear* [*e.g.*, $MnCo(CO)_9$, $MnRe(CO)_{10}$ etc.].

## II. Classification Based on the Structure of Carbonyls

This classification gives the following types of carbonyls:

(*a*) **Non-bridged carbonyls.** As the name suggests, these carbonyls do not contain any bridging (ketonic) carbonyl group, :Ö=C<. These carbonyls may be of the following types:

(*i*) Non-bridged carbonyls which contain only terminal carbonyl groups (:O $\equiv$ C:). Examples of such carbonyls are $Ni(CO)_4$, $Fe(CO)_5$, $Ru(CO)_5$, $Os(CO)_5$.

(*ii*) Non-bridged carbonyls which contain terminal carbonyl groups as well as M–M bonds. Examples are $Co_2(CO)_8$ (in solution), $Mn_2(CO)_{10}$, $Tc_2(CO_{10}$, $Re_2(CO)_{10}$, and $Os_3(CO)_{12}$.

(*b*) **Bridged carbonyls.** These carbonyls contain bridging carbonyl groups along with terminal corbonyl groups and one or more M–M bonds. Examples are $Fe_2(CO)_9$, $Os_2(CO)_9$, $Co_2(CO)_8$ (in solid state) and $Fe_3(CO)_{12}$. All these are polynuclear carbonyls.

Important carbonyls formed by transition metals are given in Table 30.1.

**Table 30.1. Important carbonyls formed by transition metals**

| | | | | | |
|---|---|---|---|---|---|
| $V(CO)_6$ | $Cr(CO)_6$<br>$Mo(CO)_6$<br>$W(CO)_6$ | $Mn_2(CO)_{10}$<br>$Tc_2(CO)_{10}$<br>$Re_2(CO)_{10}$ | $Fe(CO)_5$, $Fe_2(CO)_9$, $Fe_3(CO)_{12}$<br>$Ru(CO)_5$, $Ru_3(CO)_{12}$, $Ru_4(CO)_{12}$<br>$Os(CO)_5$, $Os_2(CO)_9$, $Os_3(CO)_{12}$ | $Co_2(CO)_8$<br>$Co_4(CO)_{12}$<br>$Ir_4(CO)_{12}$ | $Ni(CO)_4$ |

It may be noted from the above discussion that mononuclear carbonyls always have non-bridged structure while polynuclear carbonyls may have non-bridged [*e.g.*, $Co_2(CO)_8$ (in solution), $Mn_2(CO)_{10}$, etc.] or bridged [*e.g.* $Fe_2(CO)_9$, $Co_2(CO)_8$ (in solid state), $Fe_3(CO)_{12}$] structures.

## General Methods of Preparation

### 1. By Direct Synthesis

Of the simple carbonyls, nickel tetracarbonyl, $Ni(CO)_4$ and iron pentacarbonyl, $Fe(CO)_5$ can be prepared by the direct combination of CO with finely-divided metals at suitable temperature and pressure :

$$Ni + 4CO \xrightarrow[\text{one atm. pressure}]{\text{Room temp. and}} Ni(CO)_4$$

$$Fe + 5CO \xrightarrow[\text{pressure}]{200°C \text{ and } 100 \text{ atm.}} Fe(CO)_5$$

### 2. By the Reduction of Metal Compounds in Presence of CO

Some carbonyls may be prepared by treating the transition metal compounds (usually metal halides) in suspension in organic solvents like tetrahydrofuran

(THF), ether, diethylene glycol, dimethyl ether (diglyme) etc with CO at a pressure of 200–300 atm. and temperature up to 300°C in presence of some suitable reducing agent. The substances used as reducing agents are $H_2$, metals (*e.g.* Na, AI, Mg, Cu etc.) or compounds such as trialkyl aluminium or sodium benzophenone ketyl, $Ph_2CO^-Na^+$

$$VCl_3 + 6CO + Na + diglyme \longrightarrow \underbrace{[Na\,(diglyme)_2]^+[V(CO)_6]^-}$$

$$V(CO)_6 \text{ gets separated} \xleftarrow[\text{(ii) Sublimation at 50°C}]{\text{(i) Acidify with } H_3PO_4}$$

$$CrCl_3 + 6CO \xrightarrow{(C_2H_4)_3 \text{ Al in ether}} Cr(CO)_6 + AlCl_3$$

$$MoCl_5 + 6CO \xrightarrow{\text{(5Na in diglyme)}} Mo(CO)_6 + 5NaCl$$

Hydrogen is used as a reducing agent in the preparation of $Co_2(CO)_8$ from $CoCO_3$.

$$2CoCO_3 + 2H_2 + 8CO \xrightarrow[150°C]{300 \text{ atm}} Co_2(CO)_8 + 2Co_2 + 2H_2O$$

### 3. From $Fe(CO)_5$

Since CO ligands in $Fe(CO)_5$ are labile, $Fe(CO)_5$ may be used to form certain carbonyls, *e.g.*,

$$MCl_6 + 3Fe(CO)_5 \xrightarrow[\text{ether}]{100°C} M(CO)_6 + 3FeCl_2 + 9CO$$
$$(M = W, Mo)$$

### 4. By Reducing Metal Oxides by CO

Carbonyls are prepared by the reduction of metal oxides by CO at high temperature (~300°C) and pressure (~300 atm). For example:

$$OsO_4 + 9CO \longrightarrow Os(CO)_5 + 4CO_2$$
$$Tc_2O_7 + 17CO \longrightarrow Tc_2(CO)_{10} + 7CO_2$$
$$Re_2O_7 + 17CO \longrightarrow Re_2(CO)_{10} + 7CO_2$$

### 5. By the Thermal or Photochemical Decomposition of Simple Carbonyls

Polynuclear carbonyls of Fe, Rh, Os, Co, Rh and Ir are obtained by the thermal or photochemical decomposition of simple carbonyls, *e.g.*,

$$2M(CO)_5 \xrightarrow{\text{ultraviolet light}} M_2(CO)_9 + CO$$
$$(M = Fe, Os)$$

$Co_4(CO)_{12}$ can be obtained from $Co_2(CO)_8$ as follows:

$$2Co_2(CO)_8 \xrightarrow{\Delta} Co_4(CO)_{12} + 4CO$$

## General Physical Properties

With the exception of $Ni(CO)_4$, $Fe(CO)_5$, $Ru(CO)_5$ and $Os(CO)_5$ which are liquids at ordinary temperatures, all other carbonyls are crystalline solids. They melt or decompose at low temperatures. All are typical covalent compounds and for this reason they are soluble in non-polar solvents. With the exception of $Fe_2(CO)_9$, all are more or less soluble in organic solvents. Excepting $V(CO)_6$, all the carbonyls are *diamagnetic*. $V(CO)_6$ is *paramagnetic* and its paramagnetic property corresponds to the presence of one unpaired electron. The metals in carbonyls are in zero oxidation state.

Physical properties and structures of carbonyls are listed in Table 30.2.

**Table 30.2. Physical properties and structure of some metal carbonyls**

| | | | | | |
|---|---|---|---|---|---|
| $V(CO)_6$<br>Black green<br><br>70° (*d*)<br>Octahedral<br>Paramagnetic<br>(1 e) | $Cr(CO)_6$<br>White<br><br>130° (*d*)<br>Octahedral | $Mn_2(CO)_{10}$<br>Golden yellow<br>m 155° | $Fe(CO)_5$<br>Colourless<br><br>m – 20°,<br>*b* 103°, trig. bipyramid | $Co_2(CO)_8$<br>Orange red<br><br>m 51° | $Ni(CO)_4$<br>Colourless<br><br>m – 25°, *b* 43°, tetrahedral |
| | | | $Fe_2(CO)_9$<br>Golden yellow<br>100° (*d*)<br>$Fe_3(CO)_{12}$<br>Green<br>140° (*d*) | $Co_4(CO)_{12}$<br>Black<br><br>60° (*d*) | |
| | $Mo(CO)_6$<br>White<br>180° (*d*)<br>Octahedral | $Tc_2(CO)_{10}$<br>White<br>m 160° | $Ru(CO)_5$<br>Colourless<br>m – 22°, trig. bipyramid<br>$Ru_3(CO)_{12}$<br>Orange<br>m 150° | $Ru_4(CO)_{12}$<br>Red<br>150° (*d*) | |
| | $W(CO)_6$<br>White<br>180° (*d*)<br>Octahedral | $Re_2(CO)_{10}$<br>White<br>m 177° | $Os(CO)_5$<br>Colourless<br>m – 15°, trig. bipyramid | $Ir_4(CO)_{12}$<br>Yellow<br>210° (*d*)<br>tetrahedron of $Ir(CO)_3$ units | |
| | | | $Os_3(CO)_{12}$<br>Yellow<br>m 224° | | |

## General Chemical Properties

Metal carbonyls show a variety of reactions. Some important reactions shown by these compounds are given below.

### 1. Substitution or Displacement Reactions

Displacement of one or more CO ligands may occur in contact with electron donors like pyridine, $C_5H_5$ (Py), phosphines ($R_3P$), iso-cyanides (RNC), benzene etc. A molecule of CO is replaced by an electron pair donor. Consequently, a six-electron donor (like benzene) would displace 3CO molecules.

$$Fe(CO)_5 + R_3P \xrightarrow[-CO]{+R_3P} R_3PFe(CO)_4 \xrightarrow[-CO]{+R_3P} (R_3P)_2Fe(CO)_3$$

$$Mo(CO)_6 \xrightarrow[-CO]{+Py} PyMo(CO)_5 \xrightarrow[-CO]{+Py} Py_2Mo(CO)_4 \xrightarrow[-CO]{+Py} Py_3Mo(CO)_3$$

$$Cr(CO)_6 + C_6H_6 \xrightarrow[-3CO]{} (C_6H_6)Cr(CO)_3$$

$$Fe(CO)_5 + Me_3SiNC \xrightarrow[-CO]{} (Me_3SiNC)Fe(CO)_4$$

$$Co_2(CO)_8 + RC \equiv CR \xrightarrow[-2CO]{} Co_2(CO)_6\ (RC \equiv CR)$$

$$2Fe(CO)_5 + 2C_5H_6 \longrightarrow [(\eta^5 - C_5H_5)Fe(CO)_2]_2$$

$$Mo(CO)_6 + C_5H_5Na \longrightarrow Na[(\eta^5 - C_5H_5)Mo(CO)_3]$$

In the above reactions $RC \equiv CR$ behaves as a four-electron donor and cyclopentadiene acts as a five-electron donor. Hence this ligand is represented as $\eta^5$–$C_5H_5$

### 2. Addition Reactions: Formation of Carbonylate Anions

Many of the carbonyls form carbonylate anions, *e.g.* $[Fe(CO)_4]^{2-}$, $[Mo(CO)_5Cl]^-$, $[HFe(CO)_4]^-$, $[Fe_2(CO)_8]^{2-}$, $[Co(CO)_4]^-$, $[Mn(CO)_5]^-$ etc. These complex anions can be prepared by any of the following methods:

**(*a*) By the action of NaOH or nitrogenous bases (*e.g.*, $NH_3$, amines, pyridine) on metal carbonyls, e.g.**

$$Fe(CO)_5 \xrightarrow{NaOH} Na_2^+[Fe(CO)_4]^{2-}$$

$$Fe(CO)_5 \xrightarrow{NH_3/H_2O} [Et_4N]_2^+[HFe(CO)_4]^-$$

$$Fe_2(CO)_9 \xrightarrow{KOH/MeOH} [Et_4N]_2^+[Fe_2(CO)_8]^{2-}$$

$$Fe_3(CO)_{12} \xrightarrow{(en)/H_2O} [Fe(en)_2]^{3+}[Fe_3(CO)_{11}]^{3-}$$

$$3Co_2(CO)_8 + 12py \longrightarrow 2[Co(py)_6]^{2+}[Co(CO)_4]_2^{2-} + 8CO$$

In the reaction of $Fe(CO)_5$ with NaOH, $OH^-$ ion is added at C-atom of CO group to form carbonylate anion, $[(CO)_5Fe(OH)]^-$

$$(CO)_4Fe(CO) + OH^- \longrightarrow \left[(OC)_4Fe{-}C\begin{matrix}{\swarrow\!\!=} O \\ \diagdown OH\end{matrix}\right]^- \xrightarrow{-CO_2}$$

$$[H(CO)_4Fe]^- \xrightarrow{+H^+} H^+[H(CO)_4Fe]^-$$

$Cr(CO)_6$ reacts with $CH_3^-$ or $C_6H_5^-$ to form organocarboxylate anion which leads to carbene-metal complex.

$$(CO)_5Cr(CO) + CH_3^- \longrightarrow \left[(CO)_5Cr{-}C\begin{matrix}{=}O\\ \backslash CH_3\end{matrix}\right]^- \xrightarrow[(-N_2)]{+H^+,\ CH_2N_2} (CO)_5Cr{-}C\begin{matrix}{\diagup}OCH_3\\ \diagdown CH_3\end{matrix}$$

The reaction of $W(CO)_5$ with $N_3^-$ can be shown as follows. In this reaction one CO group is replaced by isocyanate group (–N=C=O)

$$(CO)_5W(CO) + N_3^- \longrightarrow \left[(CO)_5W{-}C\begin{matrix}{=}O\\ \diagdown N_3\end{matrix}\right]^- \xrightarrow{-N_2} \left[(CO)_5W{-}C\begin{matrix}{=}O\\ \diagdown N\end{matrix}\right]^-$$

$$\xrightarrow{\text{Rearrangement}} [(CO)_5W{-}N=C=O]^-$$

(*b*) **By the reduction of carbonyls with alkali metals, alkali metal amalgams or borohydrides, *e.g.*,**

$$CO_2(CO)_8 \xrightarrow[\text{THF}]{\text{Na/Hg}} Na^+[Co(CO)_4]^-$$

$$Mn_2(CO)_{10} \xrightarrow[\text{THF}]{\text{Na/Hg}} Na^+[Mn(CO)_5]^-$$

$$Re_2(CO)_{10} \xrightarrow[\text{THF}]{NaBH_4} Na_2[Re_4(CO)_{16}]$$

$$Cr(CO)_6 \xrightarrow{Na/NH_3} Na_2[Cr(CO)_5]$$

$$Cr(CO)_6 \xrightarrow[\text{THF}]{NaBH_4} Na[HCr_2(CO)_{10}]$$

$$Cr(CO)_6 \xrightarrow[NH_2]{NaBH_4} Na_2[Cr_2(CO)_{10}]$$

## 3. Formation of Cationic Carbonyl Complexes (Also Called Carbonylate Cations)

Carbonylate cations may be prepared from the carbonyls by the following methods:

(*a*) **By the protonation of carbonyls in strong acids,** *e.g.*

$$Fe(CO)_5 \xrightarrow{HCl/BCl_3} [FeH(CO_5]^+[BCl_4]^-$$

*n.m.r.* spectroscopy has shown that the proton becomes bounded to the metal atom in $[FeH(CO)_5]^+$.

(*b*) **By the action of CO and a Lewis acid like $AlCl_3$ or $BF_3$ on carbonyl halide,** *e.g.*,

$$Mn(CO)_5Cl + CO + AlCl_3 \longrightarrow [Mn(CO)_6]^+[AlCl_4]^-$$

## 4. Formation of Carbonyl Halides

The action of $Cl_2$, $Br_2$, $I_2$ or $F_2$ on some carbonyls gives carbonyl halides, *e.g.*,

$$2Mo(CO)_6 + 2Cl_2 \longrightarrow [Mo(CO)_4Cl_2] + 4CO$$

$$Mn_2(CO)_{10} + Br_2 \xrightarrow{\text{Pet. ether, 120°C}} [Mn(CO)_4Br]_2 + 2CO$$

$$Fe(CO)_5 + I_2 \xrightarrow{(-2CO)} [Fe(CO)_4I_2] + CO$$

$$4Ru(CO)_5 + 4F_2 \longrightarrow [Ru(CO)_3F_2I_4 + 8CO$$

$Co_2(CO)_8$ and $Ni(CO)_4$ both get decomposed into metallic halides and CO when treated with halogens,

$$Co_2(CO)_8 + 2X_2 \longrightarrow 2CoX_2 + 8CO$$
$$Ni(CO)_4 + Br_2 \longrightarrow NiBr_2 + 4CO$$

## 5. Formation of Carbonyl Hydrides

Some of the metallic carbonyls [*e.g.* $Fe(CO)_5$, $Mn_2(CO)_{10}$, $Os_3(CO)_{12}$ etc] dissolve in an alkali and give an unidentified colourless solution. On acidification with a mineral acid, this solution gives the carbonyl hydrides. $H_2Fe(CO)_4$ [from $Fe(CO)_5$], $H_3Mn_3(CO)_{12}$ [from $Mn_2(CO)_{10}$], $H_4Os_4(CO)_{12}$, $H_2Os_4(CO)_{13}$ [from $Os_3(CO)_{12}$] and $Ni(CO)_3H_2$ [from $Ni(CO)_4$] have been obtained by this method.

$$Fe(CO)_5 + 2OH^- \longrightarrow Fe(CO)_4H_2 + CO_3^{2-}$$
$$Ni(CO)_4 + 2OH^- \longrightarrow Ni(CO)_3H_2 + CO_3^{2-}$$

Carbonyl hydrides can also be prepared by treating the carbonyls with $H_2$, *e.g.*,

$$Mn_2(CO)_{10} + H_2 \xrightarrow{200°C/200\ atm.} 2HMn(CO)_5$$

## 6. Metal-Metal Bond Cleavage Reactions

Metal-metal cleavage reactions result in the formation of carbonylate anions. For example:

$$Co_2(CO)_8 + 2Na/Hg \xrightarrow{THF} 2Na + [Co(CO)_4]^-$$
$$Mn_2(CO)_{10} + 2\ Li \xrightarrow{THF} 2Li^+[Mn(CO)_5]^-$$

## 7. Metal-Metal Formation Reactions

These reactions are brought about by powerful reducing agents or by simple elimination. For example:

$$2Mo(CO)_6 + 2NaBH_4 \longrightarrow Na_2^+\ [Mo_2(CO)_{10}]^{2-}$$
$$Mn(CO)_5\ Br + LiMn(CO)_5 \xrightarrow{-LiBr} (CO)_5 - Mn - Mn - (CO)_5$$

## 8. Action of NO

Many carbonyls react with nitric oxide (NO) to form metal carbonyl nitrosyls. For example:

$$Fe(CO)_5 + 2NO \xrightarrow{95°C} Fe(CO)_2(NO)_2 + 3CO$$
$$3Fe_2(CO)_9 + 4NO \longrightarrow 2Fe(CO)_2(NO)_2 + Fe(CO)_3 + Fe_3(CO)_{12} + 6CO$$
$$Fe_3(CO)_{12} + 6NO \xrightarrow{85°C} 3Fe(CO)_2(NO)_2 + 6CO$$
$$Co_2(CO)_8 + 2NO \xrightarrow{40°C} 2Co(CO)_2(NO) + 2CO$$

Moist NO gives a blue-coloured compound, Ni(NO)(OH) with $Ni(CO)_4$ while dry NO gives a blue solution of the composition, $Ni(NO)(NO_2)$

$$2Ni(CO)_4 + 2NO + 2H_2O \longrightarrow 2Ni(NO)(OH) + 8CO + H_2$$
$$Ni(CO)_4 + 4NO \longrightarrow Ni(NO)(NO_2) + 4CO + N_2O$$

## 9. Action of Heat

Different carbonyls give different products when heated. For example:

$$Fe(CO)_5 \xrightarrow{250°C} Fe + 5CO$$

$$\underset{\text{(in toluence)}}{3Fe_2(CO)_9} \xrightarrow{70°C,\ Cool} 3Fe(CO)_5 + Fe_3(CO)_{12}$$

$$Fe_3(CO)_{12} \xrightarrow{140°C} 3Fe + 12CO$$

$$2Co_2(CO)_8 \xrightarrow{50°C\ \text{(inert atmosphere)}} Co_4(CO)_{12} + 4CO$$

$$Ni(CO)_4 \xrightarrow{180°C} Ni + 4CO$$

$$3Fe_2(CO)_9 \xrightarrow{50°C} 2Fe_3(CO)_{12} + 3CO$$

## Preparation and Properties of Some Carbonyls

### 1. Nickel Tetracarbonyl, $Ni(CO)_4$

**Preparation.** (*i*) It is made by passing carbon monoxide over finely-divided metallic nickel at a temperature below 100°C.

$$Ni + 4CO \longrightarrow Ni(CO)_4$$

(*ii*) When CO at atmospheric pressure is passed over reduced nickel at 30–50°C, we get $Ni(CO)_4$.

(*iii*) $Ni(CO)_4$ is also obtained when CO is passed through an alkaline suspension of $Ni(CN)_2$ or NiS.

$$Ni(CN)_2 + 4CO \longrightarrow Ni(CO)_4 + C_2N_2$$

$$NiS + 4CO \longrightarrow Ni(CO)_4 + S$$

(*iv*) *By the disproportionation of univalent nickel.* (*a*) On reducing potassium tetracyano nickelate (II), $K_2[Ni(CN)_4]$ with potassium amalgam, a red solution is obtained. From this solution potassium tricyano nickelate (I), $K_2[Ni(CN)_3]$ may be isolated. This compound absorbs CO at 90°C, probably forming $K_2[Ni(CN)_3(CO)]$. On treating the solution of this compound with HCl, $Ni(CO)_4$ is obtained.

$$\underset{(Ni = +1)}{4K_2[Ni(CN)_3(CO)]} + 2HCl \longrightarrow \underset{(Ni=0)}{Ni(CO)_4} + \underset{(Ni = +2)}{3K_2[Ni(CN)_4]} + 2KCl + H_2$$

(*b*) It is most conveniently prepared in the laboratory by the reaction of CO on nickel (II) phenyl dithiocarbamate, $Ni(S.SC.NH\ C_6H_5)_2$

$$\underset{\substack{\text{Nickel (II) phenyl dithio-}\\ \text{carbamate (Ni = +2)}}}{2Ni(S.SC.NH\ C_6H_5)_2} + 4CO \longrightarrow \underset{(Ni=0)}{Ni(CO)_4} + \underset{\substack{\text{Nickel (IV) phenyl}\\ \text{dithiocarbarnate}\\ (Ni = +4)}}{Ni(S.SC.NH.C_6H_5)_4}$$

**Properties.** (*i*) It is a colourless liquid boiling at 43°C and solidifying at –25°C. It is miscible with benzene and is almost insoluble in water. The benzene solution of the nickel carbonyl does the same on boiling, but in the cold, the compound is not acted on by dilute acids or alkalies. Nickel carbonyl is highly poisonous.

*(ii) Action of $H_2SO_4$.* $H_2SO_4$ reacts with it as follows:

$$Ni(CO)_4 + H_2SO_4 \longrightarrow NiSO_4 + H_2 + 4CO$$

*(iii) Action of NO* (*a*) Slightly moist nitric oxide reacts with $Ni(CO)_4$ in the gaseous state or in solution in chloroform to give a compound of intense blue colour, Ni(NO)(OH)

$$2Ni(CO)_4 + 2NO + 2H_2O \longrightarrow 2Ni(NO)(OH) + 8CO + H_2$$

(*b*) When $Ni(CO)_4$ reacts with NO in the absence of $H_2O$, the blue solution of the composition, $Ni(NO)(NO_2)$ is obtained.

$$Ni(CO)_4 + 4NO \longrightarrow Ni(NO)(NO_2) + 4CO + N_2O$$

*(iv) Substitution reactions.* (*a*) CO groups can be replaced by $NO_2$, $PCl_3$ and CNR molecules. *o*-phenanthroline (*o*-phen) replaces only two CO molecules.

$$Ni(CO)_4 + 4NO_2 \longrightarrow Ni(NO_2)_4 + 4CO\uparrow$$

$$Ni(CO)_4 + 4PCl_3 \longrightarrow \underset{\text{Yellow non-volatile}}{Ni(PCl_3)_4} + 4CO\uparrow$$

$$Ni(CO)_4 + \underset{\text{Aryl isocyanide}}{CNR} \longrightarrow \underset{\text{Yellow crystalline substance}}{Ni(CNR)_4} + 4CO\uparrow$$

$$Ni(CO)_4 + o\text{-}phen \longrightarrow \underset{\text{Ruby-red needles}}{Ni(CO)_2(o\text{-}phen)} + 2CO\uparrow$$

(*b*) $Ni(CO)_4$ undergoes exchange with $^{14}CO$ rapidly to produce $Ni(CO)_3(^{14}CO)$. The production of $Ni(CO)_3(^{14}CO)$ takes place through the dissociation of Ni–CO bond (dissociative mechanism).

*(v) Action of heat.* (*a*) $Ni(CO)_4$, when heated to 180°C, gives nickel.

$$Ni(CO)_4 \xrightarrow{180°C} Ni + 4CO\uparrow$$

(*b*) In the pure state nickel carbonyl explodes to form nickel, carbon dioxide and carbon.

$$Ni(CO)_4 \xrightarrow{\Delta} Ni + 2CO_2 + 2C$$

(*c*) Under high atmospheric pressure (say, 100 atm) it is, however, not decomposed even up to 250°C.

*(vi) Oxidation reactions.* (*a*) Solution of this carbonyl in organic solvent is readily oxidised by air developing a pale green gelatinous precipitate. Such solutions are also readily oxidised to Ni (II) salts by the halogens, cyanogen and sulphur.

$$\underset{(Ni = 0)}{Ni(CO)_4} + Br_2 \longrightarrow \underset{(Ni = +2)}{NiBr_2} + 4CO\uparrow$$

(*b*) Gaseous HCl decomposes the solution of $Ni(CO)_4$ evolving $H_2$ and CO

$$\underset{(Ni = 0)}{Ni(CO)_4} + 2HCl(g) \longrightarrow \underset{(Ni = +2)}{NiCl_2} + H_2 + 4CO\uparrow$$

(*vii*) *Reduction reactions.* $Ni(CO)_4$ gets reduced by several reducing agents. Examples are:

(*a*) $Ni(CO)_4$ can be reduced to $[Ni^-H^+(CO)_3]$ by the reaction of $Ni(CO)_4$ with Na in liq. $NH_3$

$$\underset{(Ni = 0)}{2Ni(CO)_4} + 2Na + 2NH_3 \longrightarrow \underset{(Ni = -1)}{[Ni^-H^+(CO)_3]_2} + 2CO + 2NaNH_2$$

$[Ni^-H^+(CO)_3]$ is isolated as a red tetraammoniate, $[NiH(CO)_3]_2.4NH_3$.

(*b*) When $Ni(CO)_4$ solution in THF is reduced by Na, K or Mg amalgams, the enneacarbonyl tetranickelate is formed.

$$4Ni(CO)_4 + 2Na \longrightarrow Na_2[Ni_4(CO)_9] + 7CO$$

With Li, $Ni(CO)_4$ gives $Li_2[Ni_3(CO)_8]$

$$3Ni(CO)_4 + 2Li \longrightarrow Li_2[Ni_3(CO)_8] + 4CO$$

**Uses.** (*i*) Since $Ni(CO)_4$ is decomposed to metallic nickel, it is used in the preparation of nickel by Mond's process. (*ii*) It is also used in gas plating and also as a catalyst.

## 2. Iron Pentacarbonyl, $Fe(CO)_5$

**Preparation.** It is prepared: (*i*) by the action at CO of 200 atm. pressure on iron heated to 200°C.

(*ii*) by the action of CO at 200 atm. pressure and 200°C on $FeI_2$ in the presence of Cu which acts as an acceptor for $I_2$. It is thought that iron carbonyl iodide, $Fe(CO)_4I_2$ is first formed which reacts with Cu in presence of CO to form $Fe(CO)_5$.

$$FeI_2 + 4CO \longrightarrow Fe(CO)_4I_2$$

$$CO + Fe(CO)_4I_2 + 2Cu \longrightarrow Fe(CO)_5 + Cu_2I_2$$

In place of $FeI_2$, FeS can also be used.

**Properties.** (*i*) $Fe(CO)_5$ is a yellow liquid which is soluble in methyl alcohol, ether, acetone and $C_6H_6$. It is insoluble in $H_2O$.

(*ii*) *Decomposition.* On thermal decomposition at 250°C, it gives pure Fe.

$$2Fe(CO)_5 \longrightarrow Fe + 5CO$$

(*iii*) *Action of ultra-violet light.* When it is irradiated with near ultra-violet light, $Fe_2(CO)_9$ is formed.

$$2Fe(CO)_5 \longrightarrow Fe_2(CO)_9 + CO$$

The above reaction is reversed in darkness.

The formation of $Fe_2(CO)_9$ actually takes place through the following steps:

(*a*) In presence of ultra-violet light, activated $Fe(CO)_5$ [represented as $Fe(CO)_5^*$] is formed.

$$Fe(CO)_5 + h\nu \rightleftharpoons Fe(CO)_5^*$$

(*b*) In this step $Fe(CO)_5^*$ combines with another $Fe(CO)_5$ molecule and gives $Fe_2(CO)_9$

$$Fe(CO)_5^* + Fe(CO)_5 \longrightarrow Fe_2(CO)_9 + CO$$

(*iv*) *Hydrolysis.* $Fe(CO)_5$ is hydrolysed by $H_2O$ and acids.

$$Fe(CO)_5 + H_2SO_4 \longrightarrow FeSO_4 + 5CO + H_2$$

(*v*) *Action of alkalies*. When heated with aqueous NaOH and the solution is acidified, carbonylate anion, $[Fe^{2-}(CO)_4H^+]^-$ is formed. The hydrogen atom in this anion is acidic which means that Fe is in –2 oxidation state.

$$\underset{(Fe = 0)}{Fe(CO)_5} + 3NaOH \longrightarrow \underset{(Fe = -2)}{Na^+[Fe^{2-}(CO)_4H^+]^-} + NaCO_3 + H_2O$$

(*vi*) *Action of $NH_3$*. With $NH_3$ it gives $Fe(CO)_4H_2$ and carbamic acid, $NH_2COOH$

$$Fe(CO)_5 + H_2O + NH_3 \longrightarrow Fe(CO)_4H_2 + NH_2COOH$$

(*vii*) *Reaction with halogens*. It reacts with halogens in non-aqueous solvents to form the stable tetracarbonyl halides, $Fe(CO)_4X_2$.

$$Fe(CO)_5 + X_2 \longrightarrow Fe(CO)_4X_2 + CO$$

The velocities of these reactions follow the order Cl < Br > I.

(*viii*) *Reaction with amines*. When $Fe(CO)_5$ reacts with ethylene diamine (*en*), an addition product, $Fe(CO)_5(en)$ is formed. Pyridine (*py*) reacts with it to give $Fe(py)_6]$ $[Fe_4(CO)_{13}]$.

(*ix*) *Reaction with NO*. With NO, under pressure below 45°C, it gives $Fe(NO)_4$ as the final product which is a black crystalline material.

$$Fe(CO)_5 + 2NO \longrightarrow Fe(CO)_2(NO)_2 + 3CO$$

$$Fe(CO)_2(NO)_2 + 2N0 \longrightarrow Fe(NO)_4 + 2CO$$

(*x*) *Reactions with metallic and non-metallic halides*. In these reactions, $Fe(CO)_5$ acts as a dehalogenating agent and a carbonylating agent; Fe becomes bi-valent and CO groups are either liberated or transferred to some other atom.

$$Fe(CO)_5 + 2CCl_4 \longrightarrow C_2Cl_6 + FeCl_2 + 5CO$$

$$Fe(CO)_5 + SO_2Cl_2 \longrightarrow SO_2 + FeCl_2 + 5CO$$

$SnCl_4$ and $SbCl_5$ are reduced to $SnCl_2$ and $SbCl_3$.

$$Fe(CO)_5 + SnCl_4 \longrightarrow SnCl_2 + Fe(CO)_4Cl_2 + CO$$

$$Fe(CO)_5 + SbCl_5 \longrightarrow SbCl_3 + Fe(CO)_4Cl_2 + CO$$

$Fe(CO)_4Cl_2$ then forms addition compounds with $SnC1_2$ and $SbCl_3$.

$$\left.\begin{array}{l} SnCl_2 + Fe(CO)_4Cl_2 \longrightarrow \underset{(I)}{SnCl_2.Fe(CO)_4Cl_2} \\ SbCl_3 + Fe(CO)_4Cl_2 \longrightarrow \underset{(II)}{SbCl_3.Fe(CO)_4Cl_2} \end{array}\right\} \textit{Addition compounds}$$

The addition compounds (I) and (II) are halogen bridged compounds as shown below:

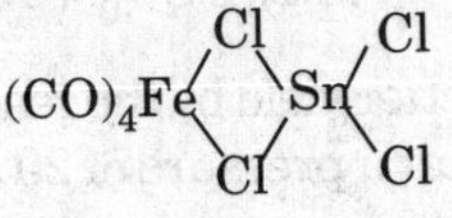

(I)

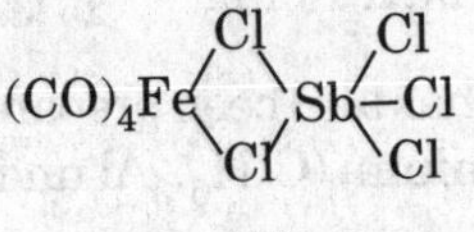

(II)

With $WCl_6$ and $MoCl_6$, $Fe(CO)_5$ gives $W(CO)_6$ and $Mo(CO)_6$ respectively.

$$\underset{(M=W,\ Mo)}{MCl_6} + 2Fe(CO)_5 \longrightarrow M(CO)_6 + 4CO + 2FeCl_3$$

(*xi*) When it is burnt in air, finely divided ferric oxide is obtained which is used as a pigment and a polishing agent.

## 3. Chromium Hexacarbonyl, $Cr(CO)_6$

**Preparation.** It can be obtained by the following methods :

(*i*) CO at 50 atm. pressure and at room temperature is passed into a suspension of $CrCl_3$ in ether, which has been treated with $C_6H_5MgBr$ at –70° (*Job's reaction*).

(*ii*) A solution of Cr(+3) salt dissolved in ether is treated with CO at a high temperature and pressure in the presence of a reducing agent such as $Al(C_5H_5)_3$, $LiAlH_4$ ete.

(*iii*) $CrCl_3$ in presence of $AlCl_3$, powdered Al and benzene is treated with CO. The yield is 88%.

**Properties.** (*i*) Chromium hexacarbonyl is a white solid.

(*ii*) Crystals of $Cr(CO)_6$ are soluble in ether, chloroform, $CCl_4$ and $C_6H_6$.

(iii) It is resistant to chemical attack. For example, it does not react with air, cold aqueous alkalies and acids (except conc. $HNO_3$).

(*iv*) *Decomposition.* It is decomposed by $F_2$ at –75°C to form $CrF_5$, by $Cl_2$ or by conc. $HNO_3$.

(*v*) *Reduction.* It is reduced with alkali metals in liq. $NH_3$ and alkali metal borohydrides. For example,

$$Cr(CO)_6 \xrightarrow{Na/NH_3} Na_2[Cr(CO)_5]$$

$$Cr(CO)_6 \xrightarrow[(THF)]{NaBH_4} Na[HCr_2(CO)_{10}]$$

$$Cr(CO)_6 \xrightarrow[(NH_3)]{NaBH_4} Na_2[Cr_2(CO)_{10}]$$

(*vi*) *Reaction with amines.* $Cr(CO)_6$ reacts with amines like *py, phen, en* to form the products in which CO groups in $Cr(CO)_6$ are replaced by molecules of the amine. For example, above 140°C, *py* reacts with $Cr(CO)_6$ to give the following products in succession :

$$Cr(CO)_6 \longrightarrow \underset{\text{yellow brown solid}}{Cr(CO)_4(py)_2} \longrightarrow \underset{\text{yellow red solid}}{Cr_2(CO)(py)_5} \longrightarrow \underset{\text{bright red solid}}{Cr(CO)_3(py)_3}$$

## 4. Dimanganese Decacarbonyl, $Mn_2(CO)_{10}$

**Preparation.** (*i*) $Mn_2(CO)_{10}$ is obtained when $MnI_2$ (prepared by special method) is reduced at a high pressure of CO by Mg in diethyl ether. The yield of $Mn_2(CO)_{10}$ is only 1%.

$$Mn + 2CuI \xrightarrow{500°C} MnI_2 + 2Cu$$

$$(MnI_2 + 2Cu) + Mg \xrightarrow[25°C/16\ hrs]{CO,\ 210\ atm} Mn_2(CO)_{10} + MgI_2 + CuI_2$$

(*ii*) $Mn_2(CO)_{10}$ has also been prepared in better yield by the reaction of $Mn(CH_3COO)_2$ with triethyl aluminium, $(C_2H_5)_3Al$ under CO at a pressure of 20 atmospheres at 100°C for 5 hours.

$$Mn(CH_3COO)_2 + (C_2H_5)_3Al \xrightarrow[100°C/5\ hrs]{CO,\ 20\ atm} Mn_2(CO)_{10}$$

The yield of carbonyl by this method is 53–60%.

(*iii*) Carbonylation of $MnCl_2$ in presence of benzophenone ketyl like $(C_6H_5)_2CONa$ at 165°C gives 32% yield of $Mn_2(CO)_{10}$.

$$MnCl_2 + 2(C_6H_5)_2CONa \longrightarrow NaCl + Mn[CO(C_6H_5)_2]_2$$

$$Mn[OC(C_6H_5)_2]_2 \xrightarrow[165°C]{CO,\ 140\ atm} Mn_2(CO)_{10} + (C_6H_5)_2CO$$

(*iv*) More recently 48% yield of $Mn_2(CO)_{10}$ has been obtained by the carbonylation of methyl cyclopentadienyl manganese dicarbonyl, $(CH_3 - C_5H_4)Mn(CO)_3$ in presence of Na at 50 atmospheric pressure of CO and 125°C for 8 hours.

$$(CH_3C_5H_4)Mn(CO)_3 + Na \xrightarrow[125°C/5\ hrs]{CO,\ 50\ atm} Mn_2(CO)_{10}$$

**Properties.** (*i*) It is a golden yellow crystalline substance. It has m.pt. = 155°C and sublimes in vacuum.

(*ii*) It is soluble in organic solvents.

(*iii*) It is slowly oxidised in air, especially in solution.

(*iv*) *Action of halogens*. Halogens form carbonyl halides, $Mn(CO)_5X$ as shown below:

$$Mn_2(CO)_{10} + Br_2(l) \xrightarrow{40°C} 2Mn(CO)_5Br$$

$$Mn_2(CO)_{10} + I_2 \longrightarrow 2Mn(CO)_5I$$

The order of activity of halogens is as: I > Br > Cl.

(*v*) *Action of Li and Na*. Mn–Mn bond is broken by lithium in the presence of tetrahydrofuran (THF) and by Na in presence of liq. $NH_3$ and $Li^+[Mn(CO)_5]^-$ is formed.

$$Mn_2(CO)_{10} + 2Li \xrightarrow{THF} 2Li^+[Mn(CO)_5]^-$$

$$Mn_2(CO)_{10} + 2Na \xrightarrow{liq.\ NH_3} 2Na^+[Mn(CO)_5]^-$$

(*vi*) *Reduction*. When $Mn_2(CO)_{10}$ is reduced by hydrogen under 200 atmospheric pressure at a temperature of 200°C, carbonyl hydride is formed.

$$Mn_2(CO)_{10} + H_2 \xrightarrow[200°C]{200\ atm.} 2HMn(CO)_5$$

(*vii*) *Action of phosphines, arsines and stibines*. With these materials $Mn_2(CO)_{10}$ gives monomeric paramagnetic compounds of the type, $Mn(CO)_4(PR_3)$.

(*viii*) *Diamagnetic nature*. $Mn_2(CO)_{10}$ is a diamagnetic substance. Diamagnetic character confirms the facts that all the electrons in $Mn_2(CO)_{10}$ are paired and Mn–Mn bond is also present in it.

## 5. Diiron Enneacarbonyl, $Fe_2(CO)_9$

**Preparation.** (*i*) It is made by the irradiation of $Fe(CO)_5$ with near ultraviolet light. $Fe(CO)_5$ is dissolved in its own volume of glacial acetic acid and the solution thus obtained is cooled and then irradiated for six hours. Golden crystals of $Fe_2(CO)_9$ are precipitated and are obtained by filtration.

**Properties.** (*i*) $Fe_2(CO)_9$ forms tricilinic crystals which are diamagnetic.

(*ii*) It is slightly soluble in alcohol and acetone and more so in toluene and pyridine and almost insoluble in water, ether, $C_6H_6$ and many other organic solvents.

(*iii*) It is non-volatile.

(*iv*) *Action of heat*. When warmed in 50°C, it gives $Fe_2(CO)_{12}$. At 100°C it decomposes to give Fe, CO, and some $Fe(CO)_5$ and $Fe_3(CO)_{12}$.

(*v*) *Action of NO*. With NO it gives $Fe(CO)_2(NO)$, together with $Fe(CO)_5$ and $Fe_3(CO)_{12}$.

(*vi*) *Action of phen*. In acetone or benzene solution at 80°C $Fe_2(CO)_9$ reacts with *phen* to form [Fe(*phen*)$_3$] [$Fe_2(CO)_8$]. In pyridine, with *phen* it gives [Fe(*phen*)$_3$] [$Fe_4(CO)_{13}$].

(*vii*) *Action of alkali metals*. Carbonylate anion is obtained in the reaction with Na in $NH_3$ solution.

$$Fe_2(CO)_9 + 4Na \longrightarrow 2Na_2[Fe(CO)_4] + CO$$

## 6. Dicobalt Octacarbonyl, $Co_2(CO)_8$

**Preparation.** It can be prepared :

(*i*) by the reaction between CO and reduced metallic cobalt at 200°C and 250 atm.

$$2Co + 8CO \xrightarrow[220°C]{250\text{ atm.}} Co_2(CO)_8$$

(*ii*) by the reaction of dry CO at 200°C and 200 atm in presence of metallic copper on certain binary compounds of cobalt such as CoS and $CoX_2$. Here Cu forms $Cu_2S$ or CuX.

$$2CoS + 8CO + 4Cu \longrightarrow Co_2(CO)_8 + 2Cu_2S$$
$$2CoX_2 + 8CO + 4Cu \longrightarrow Co_2(CO)_8 + 4CuX$$

(*iii*) by the action of CO and $H_2$ on $CoCO_3$ under high pressure and high temperature.

$$2CoCO_3 + 2H_2 + 8CO \xrightarrow[120–220°C]{250–300\text{ atm.}} Co_2(CO)_8 + 2CO_2 + 2H_2O$$

(*iv*) by the action of an acid on a solution of $Co(CO)_4$. Hydrogen is evolved and $Co_2(CO)_8$ is left behind.

$$2H^+[Co(CO)_4]^- \longrightarrow Co_2(CO)_8 + H_2$$

(*v*) by the thermal decomposition of $Co(CO)_4H$.

**Properties.** (*i*) $Co_2(CO)_8$ is an orange brown crystalline substance having m.pt. = 51°C. It is soluble in alcohol, ether and $CCl_4$. The carbonyl is air sensitive both in the solid and solution states.

(*ii*) *Action of heat*. It is thermally decomposed at 50°C in an inert atmosphere to give tetra cobalt dodecacarbonyl, $Co_4(CO)_{12}$.

$$2Co_2(CO)_8 \xrightarrow{50°C} Co_4(CO)_{12} + 4CO$$

(*iii*) *Action of air*. On exposure to air, dicobalt octacarbonyl is converted into deep violet basic carbonate of cobalt.

(*iv*) *Reduction reactions*. (*a*) It is reduced to cobalt carbonyl hydride, $H^+[Co(CO)_4]^-$ by $H_2$ at 165°C and 120 atmospheric pressure.

$$Co_2(CO)_8 + H_2 \xrightarrow[165°C]{120\text{ atm.}} 2H^+[Co(CO)_4]^-$$

(*b*) $Co_2(CO)_8$ is also reduced by Na metal in liq. $NH_3$ below –75°C or tetrahydrofuran (THF).

$$\underset{(Co = 0)}{Co_2(CO)_8} + 2Na \xrightarrow[\text{or liq. } NH_3]{THF} \underset{(Co = -1)}{2Na^+[Co(CO)_4]^-}$$

(*v*) *Disproportionation reactions.* There are two types of reactions in case of $Co_2(CO)_8$:

(*a*) Strong bases having nitrogen or oxygen donor atoms cause disproportionation into Co(+2) and Co(–l). For example:

$$\underset{(Co = 0)}{2Co_2(CO)_8} + 12NH_3 \longrightarrow \underset{(Co = +2)}{2[Co(NH_3)_6]}\underset{(Co = -1)}{[Co(CO)_4]_2} + 8CO$$

(b) With isocyanides, phosphines, arsines and stibines, however, the disproportionation reaction gives penta coordinate cobalt (I) cation.

$$\underset{(Co = 0)}{Co_2(CO)_8} + 5CNR \longrightarrow \underset{(Co = +1)}{[Co(CNR)_5]}\underset{(Co = -1)}{[Co(CO)_4]} + 4CO$$

(*vi*) *Reaction with NO*. $Co_2(CO)_8$ also reacts with NO at 40°C and forms $Co(CO)_3(NO)$

$$\underset{(Co = 0)}{Co_2(CO)_8} + 2NO \longrightarrow \underset{(Co := -1)}{Co(CO)_3(NO)} + 2CO$$

In $Co(CO)_3(NO)$, NO group appears to be more firmly held with Co metal atom than CO groups. This is confirmed by the fact that CO groups can be replaced by the amines very easily, *e.g.*, the action of *phen* on $Co(CO)_3(NO)$ gives [Co(CO)(*phen*)(NO)].

$$Co(CO)_3(NO) + phen \xrightarrow{-2CO} [Co(CO)(phen)(NO)]$$

(*vii*) *Reaction with halogens.* The halogens decompose $Co_2(CO)_8$ according to the following reaction:

$$Co_2(CO)_8 + 2X_2 \longrightarrow 2CoX_2 + 8CO$$

## 7. Triiron Dodecacarbonyl, $Fe_3(CO)_{12}$

**Preparation.** It is prepared:

(*i*) by the disproportionation of $Fe_2(CO)_9$. A toluene solution of $Fe_2(CO)_9$ is heated at 70°C. On cooling, green crystals of $Fe_3(CO)_{12}$ separate from the solution.

$$3Fe_2(CO)_9 \xrightarrow{\Delta} 3Fe(CO)_5 + Fe_3(CO)_{12}$$

(*ii*) by the oxidation of $Fe(CO)_4H_2$ with $H_2O_2$ or $MnO_2$.

$$3Fe(CO)_4H_2 + 3H_2O_2 \longrightarrow Fe_3(CO)_{12} + 6H_2O$$

$$2Fe(CO)_4H_2 + 3MnO_2 + 3H_2SO_4 \longrightarrow Fe_3(CO)_{12} + 3MnSO_4 + 6H_2O$$

(*iii*) by treating $Fe(CO)_5$ with aqueous tri-ethylamine to form the dark red $[Et_3NH][Fe_2(CO)_{11}H]$ which is then acidified and $Fe_3(CO)_{12}$ is extracted with petroleum ether.

**Properties.** (*i*) $Fe_3(CO)_{12}$ forms deep green crystals which are soluble in organic solvents like toluene, alcohol, ether and pyridine.

(*ii*) *Action of heat.* When heated alone above 100°C, $Fe_3(CO)_{12}$ decomposes to give metallic iron and CO.

$$Fe_3(CO)_{12} \xrightarrow{>100°C} 3Fe + 12CO$$

(iii) *Disproportionation reaction*. It undergoes disproportionation reactions. For example, it reacts with $CH_3OH$ at room temperature and $Fe(CO)_5$ is thereby produced.

$$3Fe_3(CO)_{12} + nCH_3OH \longrightarrow [Fe(CH_3OH)_n]\,[Fe_3(CO)_{11}] + 5Fe(CO)_5$$

(*iv*) *Reaction with Na*. Carbonylate anion is obtained when $Fe_3(CO)_{12}$ reacts with Na metal in liq. $NH_3$.

$$Fe_3(CO)_{12} + 6Na \longrightarrow 3Na_2[Fe(CO)_4]$$

(*v*) *Substitution reactions*. These reactions occur with pyridine (*py*) and methyl alcohol as shown below :

$$2Fe_3(CO)_{12} + 3py \longrightarrow Fe_3(CO)_9(py)_3 + 3Fe(CO)_5$$

$$2Fe_3(CO)_{12} + 3CH_3OH \longrightarrow Fe_3(CO)_9(CH_3OH)_3 + 3Fe(CO)_5$$

(*vi*) *Action of NO*. At 85°C it reacts with NO to form iron dicarbonyl dinitrosyl, $Fe(CO)_2(NO)_2$.

$$Fe_3(CO)_{12} + 6NO \longrightarrow 3Fe(CO)_2(NO)_2 + 6CO$$

## Effective Atomic Number (EAN) Rule as Applied to Metallic Carbonyls

**(A) Mononuclear Carbonyls Having the Metallic Atom with Even Atomic Mumber**.

$Cr(CO)_6$, $Mo(CO)_6$, $W(CO)_6$, $Fe(CO)_5$, $Ni(CO)_4$ etc. are the examples of such carbonyls. All such carbonyls obey the EAN rule according to which *after CO groups have donated a certain number of electron pairs to the zero valent metal atom through OC→M σ-bonding, the total number of electrons on the metal atom including those gained from CO molecules becomes equal to the atomic number of the next inert gas.* This is shown in Table 30.3.

**Table 30.3. EAN rule as applied to mononuclear carbonyls having metallic atoms with even atomic number. Note that in all the carbonyls metal atom is in zero oxidation state.**

| Metal carbonyl | No. of electrons on the central metal atom = At. No. of the metal atom = Z | No. of electrons donated by CO molecules = $x$ | EAN of the metal atom in carbonyl = Z + $x$ (symbol of the next inert gas is given in parentheses) |
|---|---|---|---|
| $Cr(CO)_6$ | 24 | 6 × 2 = 12 | 24 + 12 = 36 (Kr) |
| $Mo(CO)_6$ | 42 | 6 × 2 = 12 | 42 + 12 = 54 (Xe) |
| $W(CO)_6$ | 74 | 6 × 2 = 12 | 74 + 12 = 86 (Rn) |
| $Fe(CO)_5$ | 26 | 5 × 2 = 10 | 26 + 10 = 36 (Kr) |
| $Ru(CO)_5$ | 44 | 5 × 2 = 10 | 44 + 10 = 54 (Xe) |
| $Os(CO)_5$ | 76 | 5 × 2 = 10 | 76 + 10 = 86 (Rn) |
| $Ni(CO)_4$ | 28 | 4 × 2 = 8 | 28 + 8 = 36 (Kr) |

On the basis of EAN rule it can be explained why Ni atom does not form a hexacarbonyl, $Ni(CO)_6$. Non-formation of $Ni(CO)_6$ is because of the fact that EAN of Ni atom in $Ni(CO)_6$ would be equal to $28 + 2 \times 6 = 40$ which is not the atomic number of any of the noble gases.

**(B) Mononuclear Carbonyls Having the Metallic Atom with Odd Atomic Number**.

There are certain mononuclear metallic carbonyls having metal atoms with odd atomic number. These carbonyls do not obey EAN rule. Examples of such carbonyls are $V(CO)_6$, $Mn(CO)_5$, $Mn(CO)_6$, $Co(CO)_4$ etc. In order to obey EAN rule such carbonyls do any of the following things:

(*i*) Such carbonyls form cartionic species or anionic species (called carbonylate ions) which obey EAN rule as shown below:

(*a*) $\mathbf{V(CO)_6}$ . EAN of V = $23 + 6 \times 2 = 35$
$\mathbf{[V(CO)_6]^-}$ . EAN of V = $(23 + 1) + 6 \times 2 = 36$ ($Kr_{36}$)

(*b*) $\mathbf{Mn(CO)_5}$ . EAN of Mn = $25 + 2 \times 5 = 35$
$\mathbf{[Mn(CO)_5]^-}$. EAN of Mn = $(25 + 1) + 2 \times 5 = 36$ ($Kr_{36}$)

(*c*) $\mathbf{Mn(CO)_6}$ . EAN of Mn = $25 + 2 \times 6 = 37$
$\mathbf{[Mn(CO)_6]^+}$. EAN of Mn = $(25 - 1) + 6 \times 2 = 36$ ($Kr_{36}$)

(*d*) $\mathbf{Co(CO)_4}$ . EAN of Co = $27 + 2 \times 4 = 35$
$\mathbf{[Co(CO)_4]^-}$. EAN of Co = $(27 + 1) + 4 \times 2 = 36$ ($Kr_{36}$)

(*ii*) Such carbonyls may form a single covalent bond with an atom (*e.g.* H, Cl) or a group (*e.g.* $CH_3$) which contributes one electron to the metal atom. Examples are:

$\mathbf{Mn(CO)_5Cl}$. EAN of Mn = $25 + 5 \times 2 + 1 = 36$ ($Kr_{36}$)
$\mathbf{HMn(CO)_5}$ . EAN of Mn atom = $1 + 25 + 5 \times 2 = 36$ ($Kr_{36}$)
$\mathbf{(CH_3)Mn(CO)_5}$ . EAN of Mn atom = $1 + 25 + 5 \times 2 = 36$ ($Kr_{36}$)
$\mathbf{HCo(CO)_4}$ . EAN of Co atom = $1 + 27 + 4 \times 2 = 36$ ($Kr_{36}$)

(*iii*) Such carbonyls may dimerise to give dinuclear carbonyls. Such carbonyls have metal–metal bond. The electron pair being shared between the two metal atoms is counted for both the metal atoms (as in octet rule). The dinuclear carbonyls obtained by the dimerisation of mononuclear carbonyls obey EAN rule as shown for $Mn_2(CO)_{10}$ and $Co_2(CO)_8$ which are dimerised forms of $Mn(CO)_5$ and $Co(CO)_4$ respectively

$\mathbf{Mn_2(CO)_{10}}$: Electrons from 2Mn atoms = $25 \times 2 = 50$
Electrons from 10CO molecules = $10 \times 2 = 20$
Electrons from one Mn–Mn bond = $1 \times 2 = 2$

EAN for two Mn-atoms in $Mn_2(CO)_{10}$ = 72
∴ EAN per Mn atom = 72/2 = **36**($\mathbf{Kr_{36}}$)

$\mathbf{Co_2(CO)_8}$: Electrons from 2Co atoms = $2 \times 27 = 54$
Electrons from 8CO molecules = $8 \times 2 = 16$
Electrons from one Co–Co bond = $1 \times 2 = 2$

EAN for two Co-atoms in $Co_2(CO)_8$ = 72
∴ EAN per Co atom = 72/2 = **36**($\mathbf{Kr_{36}}$)

**(C) Polynuclear Carbonyls.**

Now let us see whether or not the polynuclear carbonyls like $Mn_2(CO)_{10}$, $Co_2(CO)_8$, $Fe_2(CO)_9$ and $Fe_3(CO)_{12}$ obey EAN rule. It has been seen that these carbonyls obey EAN rule, if two electrons from each metal–metal bond present in these carbonyls are included in calculating the electrons per metal atom. The structures of $Mn_2(CO)_{10}$, $Co_2(CO)_8$, $Fe_2(CO)_9$, $Fe_3(CO)_{12}$, and $Co_4(CO)_{12}$ show that these molecules contain *one* Mn–Mn, *one* Co–Co, *one* Fe–Fe, *three* Fe–Fe and *six* Co–Co bonds respectively and hence the EAN per metal atom in these molecules is calculated as follows:

(*i*) $\mathbf{Mn_2(CO)_{10}}$: Electrons from 2Mn atoms = 25 × 2 = 50

Electrons from 10 CO groups = 10 × 2 = 20

Electrons from one Mn–Mn bond = 1 × 2 = 2

EAN of two Mn-atoms of $Mn_2(CO)_{10}$ = 72

∴ EAN for one Mn-atom in $Mn_2(CO)_{10}$ = 72/2 = $\mathbf{36(Kr_{36})}$

**Alternative method:** In $Mn_2(CO)_{10}$, since each Mn-atom is attached with five CO groups and with another Mn-atom (one Mn–Mn bond), EAN of Mn in $Mn_2(CO)_{10}$ is given by:

EAN of Mn-atom = At. No. of Mn-atom + Electrons donated by five terminal carbonyl groups + Electrons donated by one Mn–Mn bond

= 25 + 5 × 2 + 1 × 1 = $\mathbf{36(Kr_{36})}$

(*ii*) $\mathbf{Co_2(CO)_8}$ *(In solution and in solid state both)*:

Electrons from 2 Co atoms = 2 × 27 = 54

Electrons from 8 CO groups = 2 × 8 = 16

Electrons from one Co–Co bond = 1 × 2 = 2

EAN of two Co-atoms in $Co_2(CO)_8$ = 72

∴ EAN of one Co-atom in $Co_2(CO)_8$ = 72/2 = $\mathbf{36(Kr_{36})}$

**Alternative method** (*a*) $\mathbf{Co_2(CO)_8}$ **in solution.** In solution, $Co_2(CO)_8$ molecule has non-bridged structure in which each Co-atom is linked with four terminal CO groups and with another Co-atom (one Co–Co bond). Thus EAN of Co-atom is given by:

EAN of Co-atom = At. No. of Co-atom + [Electrons donated by four terminal carbonyl groups] + [Electrons donated by one Co–Co bond]

= 27 + 4 × 2 + 1 × 1 = $\mathbf{36(Kr_{36})}$

(*b*) $\mathbf{Co_2(CO)_8}$ **in solid state:** In the solid state $Co_2(CO)_8$ molecule has bridged structure in which each Co-atom is liked with three terminal CO groups, two bridging CO groups and one Co-atom (one Co–Co bond). Thus EAN of Co-atom is given by:

EAN of each Co-atom = At. No. of Co-atom + [Electrons donated by three terminal corbonyl groups] + [Electrons donated by two bridging carbonyl groups] + [Electrons donated by one Co–Co bond]

= 27 + 3 × 2 + 2 × 1+ 1 × 1 = **36** ($Kr_{36}$)

(*iii*) $\mathbf{Fe_2(CO)_9}$: Electrons from 2Fe atom = 2 × 26 = 52

Electrons from 9CO molecules = 9 × 2 = 18

Electrons from one Fe–Fe bond = 1 × 2 = 2

EAN for two Fe-atoms in $Fe_2(CO)_9$ = 72

∴ EAN per Fe atom = 72/2 = **36($Kr_{36}$)**

**Alternative method**. In $Fe_2(CO)_9$ molecule, since each Fe-atom is linked with three terminal CO groups, three bridging CO groups and other Fe-atom (Fe–Fe bond). Thus EAN of Fe-atom is given by:

$$\text{EAN of each Fe-atom} = \text{At. No. of Fe-atom} + \begin{bmatrix}\text{Electrons donated}\\ \text{by three terminal}\\ \text{carbonyl groups}\end{bmatrix} + \begin{bmatrix}\text{Electrons donated}\\ \text{by three bridging}\\ \text{carbonyl groups}\end{bmatrix} + \begin{bmatrix}\text{Electrons donated}\\ \text{by one Fe–Fe bond}\end{bmatrix}$$

$$= 26 + 3 \times 2 + 3 \times 1 + 1 \times 1 = \mathbf{36\ (Kr_{36})}$$

(*iv*) $\mathbf{Fe_3(CO)_{12}}$: Electrons from 3Fe atoms = 3 × 26 = 78

Electrons from 12CO molecules = 12 × 2 = 24

Electrons from three Fe–Fe bonds = 3 × 2 = 6

EAN of three Fe-atoms in $Fe_3(CO)_{12}$ = 108

∴ EAN per Fe atom = 108/3 = **36($Kr_{36}$)**

(*v*) $\mathbf{Co_4(CO)_{12}}$: Electrons from 4Co atom = 4 × 27 = 108

Electrons from 12CO molecules = 12 × 2 = 24

Electrons from six Co–Co bonds = 2 × 6 = 12

EAN of four Co-atoms in $Co_4(CO)_{12}$ = 144

∴ EAN per Co-atom = 144/4 = **36($Kr_{36}$)**

## Application of EAN Rule

With the help of EAN rule, we can find out the number of M–M bonds present in a given metal carbonyl. This can be explained by determining the number of M–M bonds in $Fe_2(CO)_9$, $Fe_3(CO)_{12}$, $Co_4(CO)_{12}$ and $[(\eta^5\text{–cp})Mo(CO)_3]_2$ carbonyls.

(1) $\mathbf{Fe_2(CO)_9}$. Electrons from two Fe-atoms = 2 × 26 = 52

Electrons from 9CO groups = 2 × 9 = 18

EAN for two Fe-atoms in $Fe_2(CO)_9$ = 52 + 18 = 70

∴ EAN for one Fe-atom = 70/2 = 35

Since this EAN is less than 36($Kr_{36}$) by one electron, there is one Fe–Fe bond in $Fe_2(CO)_9$ molecule.

(2) $\mathbf{Fe_3(CO)_{12}}$. Electrons from 3Fe-atoms = 3 × 26 = 78

Electrons from 12CO groups = 12 × 2 = 24

EAN for three Fe-atoms = 78 + 24 = 102

∴ EAN for one Fe-atom = 102/3 = 34

Since this EAN is less than 36 ($Kr_{36}$) by two electrons, each Fe-atom in $Fe_3(CO)_{12}$ molecule is linked with other two Fe-atoms. Thus $Fe_3(CO)_{12}$ has three Fe–Fe bonds.

(3) $\mathbf{Co_4(CO)_{12}}$. Electrons from 4 Co-atoms = 4 × 27 = 108
Electrons from 12CO groups = 12 × 2 = 24

EAN of four Co-atoms = 132
∴ EAN for one Co-atom = 132/4 = 33

Since this EAN is less than 36 ($Kr_{36}$) by three electrons, each Co-atom in $Co_4(CO)_{12}$ molecule is linked with other three Co-atoms. Thus $Co_4(CO)_{12}$ has five Co–Co bonds.

(4) $\mathbf{[(\eta^5\text{–cp})Mo(CO)_3]_2}$. Electrons from two Mo-atoms = 2 × 42 = 84
Electrons from two $\eta^5$–cp = 2 × 5 = 10
Electrons from 6CO = 6 × 2 = 12

EAN for two Mo-atoms in $[(\eta^5\text{–cp})Mo(CO)_3]_2$ = 84 + 10 + 12 = 106
∴ EAN for one Mo-atom = 106/2 = 53

Since this EAN is less than 54 ($Xe_{54}$) by one electron, there is one Mo–Mo bond in the given mixed carbonyl.

## 18-Electron Rule for Metal Carbonyls

According to this rule, in a stable metal carbonyl, total (*i.e.* effective) number of electrons present in the valence shell (v.s.) of one metal atom is equal to 18. This number (*i.e.* 18) is equal to EN/*n*. In EN/*n*

*n* = No. of metal (M) atoms in metal carbonyl, and
EN = Number of electrons present in the v.s. of metal in carbonyl
= No. of electrons in the v.s. shell of metal atom in the free state + 2 × No. of CO ligands + 2 × No. of M – M bonds in the carbonyl.

### Examples of Carbonyls which Obey 18-electron Rule

If we calculate the effective number of electrons in the v.s. of one metal atom in $Ni(CO)_4$, $Fe(CO)_5$, $Cr(CO)_6$, $Mo(CO)_6$, $Fe_2(CO)_9$, $Co_2(CO)_8$, $Mn_2(CO)_{10}$, $Fe_3(CO)_{12}$ and $Co_4(CO)_{12}$ carbonyls, we find that this number comes out to be equal to 18. Thus these carbonyls obey 18-electron rule.

**1.** $\mathbf{Ni(CO)_4}$. Ni($3d^8 4s^2$) = $10e^-$
4CO = 4 × $2e^-$ = $8e^-$

∴ Effective number of electrons in the v.s. of Ni-atom = $10e^-$ + $8e^-$ = $18e^-$

**2.** $\mathbf{Fe(CO)_5}$. Fe($3d^6 4s^2$) = $8e^-$
5CO = 5 × $2e^-$ = $10e^-$

∴ Effective number of electrons in the v.s. of Fe-atom = $8e^-$ + $10e^-$ = $18e^-$

**3.** $\mathbf{Cr(CO)_6}$. Cr($3d^5 4s^1$) = $6e^-$
6CO = 6 × $2e^-$ = $12e^-$

∴ Effective number of electrons in the v.s. of Cr-atom = $6e^-$ + $12e^-$ = $18e^-$

4. **$Mo(CO)_6$.** $Mo(4d^5 6s^1) = 6e^-$

$6CO = 6 \times 2e^- = 12e^-$

$\therefore$ Effective number of electrons in the v.s. of Mo-atom $= 6e^- + 12e^- = 18e^-$

5. **$Fe_2(CO)_9$.** $2Fe = 2 \times 8e^- = 16e^-$ $(Fe = 3d^6 4s^2)$

$9CO = 9 \times 2e^- = 18e^-$

One Fe–Fe bond $= 1 \times 2e^- = 2e^-$

Total electrons on two Fe-atoms

$= 16e^- + 18e^- + 2e^- = 36e^-$

$\therefore$ Electrons on one Fe-atom $= \dfrac{36e^-}{2}$

$= 18e^-$

6. **$Co_2(CO)_8$.** $2CO = 2 \times 9e^- = 18e^-$ $(Co = 3d^7 4s^2)$

$8CO = 8 \times 2e^- = 16e^-$

One Co–Co bond $= 1 \times 2e^- = 2e^-$

Total electrons on two Co-atoms

$= 18e^- + 16e^- + 2e^- = 36e^-$

$\therefore$ Electrons on one Co-atom $= \dfrac{36e^-}{2}$

$= 18e^-$

7. **$Mn_2(CO)_{10}$.** $2Mn = 2 \times 7e^- = 14e^- (Mn = 3d^5 4s^2)$

$10CO = 10 \times 2e^- = 20e^-$

One Mn–Mn bond $= 1 \times 2e^- = 2e^-$

Total electrons on two Mn-atoms $= 14e^- + 20e^- + 2e^- = 36e^-$

$\therefore$ Electrons on one Mn-atom $= 36e^-/2 = 18e^-$

8. **$Fe_3(CO)_{12}$.** $3Fe = 3 \times 8e^- = 24e^-$ $(Fe = 3d^6 4s^2)$

$12CO = 12 \times 2e^- = 24e^-$

Three Fe–Fe bonds $= 3 \times 2e^- = 6e^-$

Total electrons on three Fe-atoms $= 24e^- + 24e^- + 6e^- = 54e^-$

$\therefore$ Electrons on one Fe-atom $= 54e^-/3 = 18e^-$

9. **$Co_4(CO)_{12}$.** $4Co = 4 \times 9e^- = 36e^-$ $(Co = 3d^7 4s^2)$

$12CO = 12 \times 2e^- = 24e^-$

Six Co–Co bonds $= 6 \times 2e^- = 12e^-$

Total electrons in four Co-atoms $= 36e^- + 24e^- + 12e^- = 72e^-$

$\therefore$ Electrons on one Co-atom $= 72e^-/4 = 18e^-$

## 18-Electron Rule as Applied to $V(CO)_6$ and $Co(CO)_4$ Carbonyls

If we calculate the effective number of electrons in the valence shell (v.s.) of V and Co-atoms in $V(CO)_6$ and $Co(CO)_4$ molecules respectively, we find that this number is equal to 17 as shown below:

| $V(CO)_6$ | $Co(CO)_4$ |
|---|---|
| $V(3d^3 4s^2) = 5e^-$ | $Co(3d^7 4s^2) = 9e^-$ |
| $6CO = 6 \times 2e^- = 12e^-$ | $4CO = 4 \times 2e^- = 8e^-$ |
| Effective number of electrons in the v.s. of V-atom $= 5e^- + 12e^-$ $= 17e^-$ | Effective number of electrons in the v.s. of Co-atom $= 9e^- + 8e^-$ $= 17e^-$ |

Since the effective number of electrons in the v.s. of Cr and Co is 17, $V(CO)_6$ and $Co(CO)_4$ both are paramagnetic.

In order to obey 18-electron rule $Co(CO)_4$ dimerises to form the dimeric molecule, $Co_2(CO)_8$ in which each Co-atom has 18 electrons in its v.s. as shown below. $Co_2(CO)_8$ has one Co–Co bond as is evident from its structure, $(CO)_4$–Co–Co–$(CO)_4$.

$$2Co = 2 \times 9e^- = 18e^- \; (Co = 3d^7 4s^2)$$
$$8CO = 8 \times 2e^- = 16e^-$$
$$\text{One Co–Co bond} = 1 \times 2e^- = 2e^-$$

Total number of electrons in the v.s. of two Co-atoms $= 18e^- + 16e^- + 2e^-$
$= 36e^-$

$\therefore$ Effective number of electrons in the v.s. of one Co-atom $= 36e^-/2 = 18e^-$

$Co_2(CO)_8$ molecule has the structure, $(CO)_4$–Co–Co–$(CO)_4$. This structure shows that since the C.N. of each Co-atom is 5 (Maximum C.N. of Co-atom = 6), no extra steric strain would be created on $Co(CO)_4$ moity by increasing the C.N. from 4 in $Co(CO)_4$ to 5 in $Co_2(CO)_8$ during the dimerisation of $Co(CO)_4$ through Co–Co bond. $Co_2(CO)_8$ would, therefore, be a stable compound.

The dimeric molecule of $V(CO)_6$ *viz* $V_2(CO)_{12}$ also obeys 18-electron rule as shown below, if its structure is assumed as $(CO)_6$–V–V–$(CO)_6$.

$$2V = 2 \times 5e^- = 10e^- \; (V = 3d^3 4s^2)$$
$$12CO = 12 \times 2e^- = 24e^-$$
$$\text{One V–V bond} = 1 \times 2e^- = 2e^-$$

Total number of electrons in the v.s. of two V-atoms $= 10e^- + 24e^- + 2e^-$
$= 36e^-$

$\therefore$ Effective number of electrons in the v.s. of one V-atom $= 36e^-/2$
$= 18e^-$

The expected structure of $V_2(CO)_{12}$ is $(CO)_6$–V–V–$(CO)_6$ which shows that each V-atom has C.N. equal to 7 (maximum C.N. of V-atom = 6). Due to the increase in C.N. from 6 [in $V(CO)_6$] to 7[in $V_2(CO)_{12}$], a lot of steric strain would be created on $V(CO)_6$ moity and hence V-atom cannot have C.N. of 7. Consequently the dimer, $V_2(CO)_{12}$ would be unstable. In other words, although $V_2(CO)_{12}$ obeys 18-electron rule, it is unstable.

## Examples of Mixed Carbonyls which Obey 18-Electron Rule

Some complexes containing CO and organic molecules as ligands are given below. These mixed complexes obey 18-electron rule as shown below:

1. **$Fe(CO)_3(C_4H_4)$.** $Fe(3d^6 4s^2) = 8e^-$

$3CO = 3 \times 2e^- = 6e^-$

$C_4H_4 = 4e^-$

$\therefore$ Sum $= 8e^- + 6e^- + 4e^- = 18e^-$

2. **$Fe(CO)_2(\sigma\text{-}C_5H_5)(\pi\text{-}C_5H_5)$.** $Fe(3d^6 4s^2) = 8e^-$

$2CO = 2 \times 2e^- = 4e^-$

$\sigma\text{-}C_5H_5 = 1e^-$

$\pi\text{-}C_5H_5 = 5e^-$

$\therefore$ Sum $= 8e^- + 4e^- + 1e^- + 5e^- = 18e^-$

3. **$Re(CO)_2(\sigma\text{-}C_5H_6)(\pi\text{-}C_5H_5)$.** $Re(5d^5 6s^2) = 7e^-$

$2CO = 2 \times 2e^- = 4e^-$

$\sigma\text{-}C_5H_6 = 2e^-$

$\pi\text{-}C_5H_5 = 5e^-$

$\therefore$ Sum $= 7e^- + 4e^- + 2e^- + 5e^- = 18e^-$

4. **$Co(CO)_3(\pi\text{-}C_7H_7)$.** $Co(3d^7 4s^2) = 9e^-$

$3CO = 3 \times 2e^- = 6e^-$

$\pi\text{-}C_7H_7 = 3e^-$

$\therefore$ Sum $= 9e^- + 6e^- + 3e^- = 18e^-$

5. **$Fe(CO)_4(PPh_3)$.** $Fe(3d^6 4s^2) = 8e^-$

$4CO = 4 \times 2e^- = 8e^-$

$PPh_3 = 2e^-$

$\therefore$ Sum $= 8e^- + 8e^- + 2e^- = 18e^-$

6. **$Mn(CO)_5(CH_3)$.** $Mn(3d^5 4s^2) = 7e^-$

$5CO = 5 \times 2e^- = 10e^-$

$CH_3 = 1e^-$

$\therefore$ Sum $= 7e^- + 10e^- + 1e^- = 18e^-$

7. **$Re(CO)_5(CH_3)$.** $Re(5d^5 6s^2) = 7e^-$

$5CO = 5 \times 2e^- = 10e^-$

$CH_3 = 1e^-$

$\therefore$ Sum $= 7e^- + 10e^- + 1e^- = 18e^-$

8. **$[Mn(CO)_5(C_2H_4)]^+$.** $Mn^+(3d^5 4s^1) = 6e^-$

$5CO = 5 \times 2e^- = 10e^-$

$C_2H_4 = 2e^-$

$\therefore$ Sum $= 6e^- + 10e^- + 2e^- = 18e^-$

**9. $V(CO)_3(\pi\text{-}C_5H_5)(R_2C_2)$.** $V(3d^34s^2) = 5e^-$

$3CO = 3 \times 2e^- = 6e^-$

$\pi\text{-}C_5H_5 = 5e^-$

$R_2C_2 = 2e^-$

$\therefore$ Sum $= 5e^- + 6e^- + 5e^- + 2e^- = 18e^-$

**10. $Fe(CO)_3(C_4H_6)$.** $Fe(3d^64s^2) = 8e^-$

$3CO = 3 \times 2e^- = 6e^-$

$C_4H_6 = 4e^-$

$\therefore$ Sum $= 8e^- + 6e^- + 4e^- = 18e^-$

**11. $Mn(CO)_4(\pi\text{-}C_3H_5)$.** $Mn(3d^54s^2) = 7e^-$

$4CO = 4 \times 2e^- = 8e^-$

$\pi\text{-}C_3H_5 = 3e^-$

$\therefore$ Sum $= 7e^- + 8e^- + 3e^- = 18e^-$

**12. $Co(CO)_3(\pi\text{-}C_3H_5)$.** $Co(3d^74s^2) = 9e^-$

$3CO = 3 \times 2e^- = 6e^-$

$\pi\text{-}C_3H_5 = 3e^-$

$\therefore$ Sum $= 9e^- + 6e^- + 3e^- = 18e^-$

**13. $Mo(CO)_2(\pi\text{-}C_3H_5)(\pi\text{-}C_5H_5)$.** $Mo(4d^55s^1) = 6e^-$

$2CO = 2 \times 2e^- = 4e^-$

$\pi\text{-}C_3H_5 = 3e^-$

$\pi\text{-}C_5H_5 = 5e^-$

$\therefore$ Sum $= 6e^- + 4e^- + 3e^- + 5e^- = 18e^-$

**14. $Co(CO)_2(\pi\text{-}C_5H_5)$.** $Co(3d^74s^2) = 9e^-$

$2CO = 2 \times 2e^- = 4e^-$

$\pi\text{-}C_5H_5 = 5e^-$

$\therefore$ Sum $= 9e^- + 4e^- + 5e^- = 18e^-$

**15. $Cr(CO)_3(C_6H_6)$.** $Cr(3d^54s^1) = 6e^-$

$3CO = 3 \times 2e^- = 6e^-$

$C_6H_6 = 6e^-$

$\therefore$ Sum $= 6e^- + 6e^- + 6e^- = 18e^-$

**16. $Cr(CO)_3(\sigma\text{-}C_3H_5)(\pi\text{-}C_5H_5)$.** $Cr(3d^54s^1) = 6e^-$

$3CO = 3 \times 2e^- = 6e^-$

$\sigma\text{-}C_3H_5 = 1e^-$

$\pi\text{-}C_5H_5 = 5e^-$

$\therefore$ Sum $= 6e^- + 6e^- + 1e^- + 5e^- = 18e^-$

**17. $W(CO)_3(Me_3Si)(\pi\text{-}C_5H_5)$.** $W(5d^56s^1) = 6e^-$

$$3CO = 3 \times 2e^- = 6e^-$$
$$Me_3Si = 1e^-$$
$$\pi\text{-}C_5H_5 = 5e^-$$

$\therefore$ Sum $= 6e^- + 6e^- + 1e^- + 5e^- = 18e^-$

**18. $Mo(CO)_3(\sigma\text{-}C_3H_5)(\pi\text{-}C_5H_5)$.** $Mo(4d^55s^1) = 6e^-$

$$3CO = 3 \times 2e^- = 6e^-$$
$$\sigma\text{-}C_3H_5 = 1e^-$$
$$\pi\text{-}C_5H_5 = 5e^-$$

$\therefore$ Sum $= 6e^- + 6e^- + 1e^- + 5e^- = 18e^-$

**19. $Fe(CO)_3(C_4H_4)$.** $Fe(3d^64s^2) = 8e^-$

$$3CO = 3 \times 2e^- = 6e^-$$
$$C_4H_4 = 4e^-$$

$\therefore$ Sum $= 8e^- + 6e^- + 4e^- = 18e^-$

**20. $Re(CO)_2(\sigma\text{-}C_5H_6)(\pi\text{-}C_5H_5)$.** $Re(5d^56s^2) = 7e^-$

$$2CO = 2 \times 2e^- = 4e^-$$
$$\sigma\text{-}C_5H_6 = 2e^-$$
$$\pi\text{-}C_5H_5 = 5e^-$$

$\therefore$ Sum $= 7e^- + 4e^- + 2e^- + 5e^- = 18e^-$

**21. $[(\eta^5-cp)Mo(CO)_3]_2$.** This mixed carbonyl contains one Mo–Mo bond. Hence:

$$2(\eta^5 - cp) = 2 \times 5e^- = 10e^-$$
$$2Mo = 2 \times 6e^- = 12e^- \ (Mo = 4d^54s^1)$$
$$6CO = 6 \times 2e^- = 12e^-$$
$$\text{One Mo–Mo bond} = 1 \times 2e^- = 2e^-$$

Total electrons on two Mo-atoms $= 10e^- + 12e^- + 12e^- + 2e^- = 36e^-$

$\therefore$ Electrons on one Mo-atom $= 36e^-/2 = 18e^-$

**18-Electron Rule as Applied to Square Planar Complexes of Metal Atom with $d^8$ Configuration.**

Since C.N. of the central metal atom in square planar complexes is 4, the effective number of electrons in the v.s. of $d^8$ metal atom in square planar complexes $= 8 + 4 \times 2 = 16$. This number shows that these complexes do not obey 18-electron rule. Although 18-electron rule is not obeyed by these complexes, yet these are quite stable complexes. The stability of such complexes has been explained on the basis of CFT as discussed below.

The splitting of $d$-orbitals of metal atom in square planar complexes has been shown in Fig. 30.1.

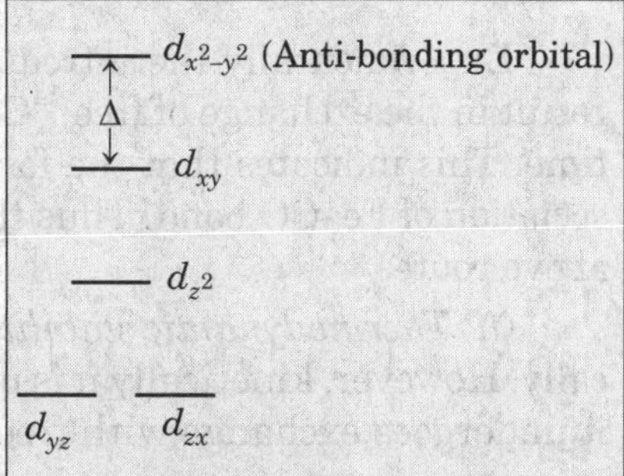


**Fig. 30.1.** Splitting of $d$-orbitals in square planar complexes

$d_{x^2-y^2}$ orbital has the highest energy and is antibonding orbital. Energy gap ($\Delta$) between $d_{xy}$ and $d_{x^2-y^2}$ orbitals is very large. It may be seen from the figure that $d^8$ electrons of the metal atom in square planar complex can be distributed as:

$$(d_{zx})^2 = (d_{yz})^2, (d_{z^2})^2, (d_{xy})^2, (d_{x^2-y^2})^0$$

Since the energy gap ($\Delta$) between $d_{xy}$ and $d_{x^2-y^2}$ orbitals is very large, the entry of 9th electron in high energy anti-bonding $d_{x^2-y^2}$ orbital is energetically infavourable. Consequently, although square planar complexes of $d^8$ metal atom do not obey 18-electron rule, yet these complexes are quite stable.

## Radio Isotope Exchange Studies of Mononuclear Metal Carbonyls

Radio isotope exchange studies of mononuclear carbonyls are employed to determine the rate of exchange of free $^{14}CO$ with ligated CO, *i.e.* these studies are carried out to determine the lability of M–C bond in metal carbonyls.

**Examples:** (*i*) *Formation of $Fe_2(CO)_9$, from photolysis of $Fe(CO)_5$.* The formation of $Fe_2(CO)_9$ from photolysis of $Fe(CO)_5$ can take place in any of the following two routes:

*Route 1.* First of all activated species, $Fe(CO)_5^*$ is formed. This activated species combines with $Fe(CO)_5$ and forms $Fe_2(CO)_9$.

$$Fe(CO)_5 + h\nu \rightleftharpoons \underset{\text{Activated species}}{Fe(CO)_5^*} \xrightarrow{+Fe(CO)_5} CO + Fe_2(CO)_9$$

*Route 2.* This route consists of three steps: (*a*) First of all $Fe(CO)_5$ gets dissociated to give intermediate $Fe(CO)_4$

$$Fe(CO)_5 + h\nu \rightleftharpoons \underset{\text{Intermediate}}{Fe(CO)_4} + CO$$

(*b*) $Fe(CO)_4$ formed as above combines with $^{14}CO$ to form $Fe(CO)_3(^{14}CO)$.

$$^{14}CO + Fe(CO)_4 \rightleftharpoons Fe(CO)_3(^{14}CO)$$

(*c*) $Fe(CO)_3(^{14}CO)$ formed in step (*b*) combines with $Fe(CO)_5$ to form $Fe_2(CO)_9$

$$Fe(CO)_3(^{14}CO) + Fe(CO)_5 \longrightarrow Fe_2(CO)_8(^{14}CO) \text{ or } Fe_2(CO)_9$$

Experimentally the introduction of free $^{14}CO$ in the reaction mixture does not result in the exchange of free $^{14}CO$ with ligated CO in $Fe(CO)_5$ even after a prolonged time. This indicates that the formation of $Fe_2(CO)_9$ does not occur through the dissociation of Fe–CO bond. Thus the formation of $Fe_2(CO)_9$ cannot be explained by the above route.

(*ii*) *Thermodynamic stability of $Ni(CO)_4$.* $Ni(CO)_4$ is quite stable thermodynamically. However, kinetically it is unstable since, according to dissociation mechanism, it undergoes exchange with free $^{14}CO$ rapidly.

$$Ni(CO)_4 + {}^{14}CO \longrightarrow Ni(CO)_3(^{14}CO) + CO$$

## Different Types of Bonds Found in Metallic Carbonyls

Different types of bonds found in metallic carbonyls and their formation has been explained as follows:

1. **$:O \equiv C \rightarrow M$ σ bond.**

   In a simplified way this bond may also be represented as OC–Mσ bond. In this bond CO acts as a terminal carbonyl group. This bond is found in all types of carbonyls. The formation of this bond will be explained on the basis of VBT and MOT on the subsequent pages.

2. **M → CO π-bond: Back donation of electrons from metal to CO.**

   The formation of this bond will be explained on the basis of MOT on the subsequent pages.

3. **$:\ddot{O}{=}C\langle^{M}_{M}$ or $OC\langle^{M}_{M}$ Bond**

   This bond is found in bridged polynuclear carbonyls like $Co_2(CO)_8$ (in the solid state), $M_2(CO)_9$ (M=Fe, Os), $Fe_3(CO)_{10}$, $Co_4(CO)_{12}$, $Ru_4(CO)_{12}$. In these carbonyls CO molecule acts as a *bridging* (or *ketonic*) *carbonyl group* $:\ddot{O}{=}C\langle$. The configuration of the system, $:\ddot{O}{=}C\langle^{M}_{M}$ shows that carbon atom is $sp^2$ hybridised as shown below:

   C-atom in its ground state: $(2s^2 2p^2)$ : $2s^2\ 2p_x^1\ 2p_y^1\ 2p_z^0$

   C-atom in its excited state: $(2s^1 2p^3)$ : $2s^1\ 2p_x^1\ 2p_y^1\ 2p_z^1$

   C-atom in sp$^2$ hybridised state: $\underbrace{(sp^2)^1\ (sp^2)^1\ (sp^2)^1}_{\text{Three } sp^2 \text{ hybrid orbitals}}\ 2p_z^1$

   $:\ddot{O}{=}C\langle^{M}_{M}$ system contains two C—M σ bonds, one O—C σ bond and one O—C π bond

   Two half-filled $sp^2$ hybrid orbitals on C-atom overlap with half-filled appropriate hybrid orbital of two metallic atoms (M) and give two (C–M) σ-bonds (See Fig. 30.2). Third half-filled $sp^2$ hybrid orbital overlaps with the appropriate $2p$ orbital of O-atom and forms (C–O) σ-bond. (C–O) π-bond results by $p$(C)–$p$(O) overlap. Thus (C–O) π-bond is a ($p$–$p$)π-bond. Formation of different bonds in ${}^{M}_{M}\rangle C{=}\ddot{O}:$ system has been depicted in Fig. 30.2.

4. **Metal–Metal or M—M Bond**

   There are many carbonyls which contain one or more M—M bonds. Examples of carbonyls containing M—M bonds are $M_2(CO)_9$ (M=Fe, Os), $Co_2(CO)_8$ (in solution and in solid), $M_2(CO)_{10}$ (M=Mn, Tc, Re), $M_3(CO)_{12}$ (M=Fe, Os) etc. This bond is produced by the weak overlap of the two

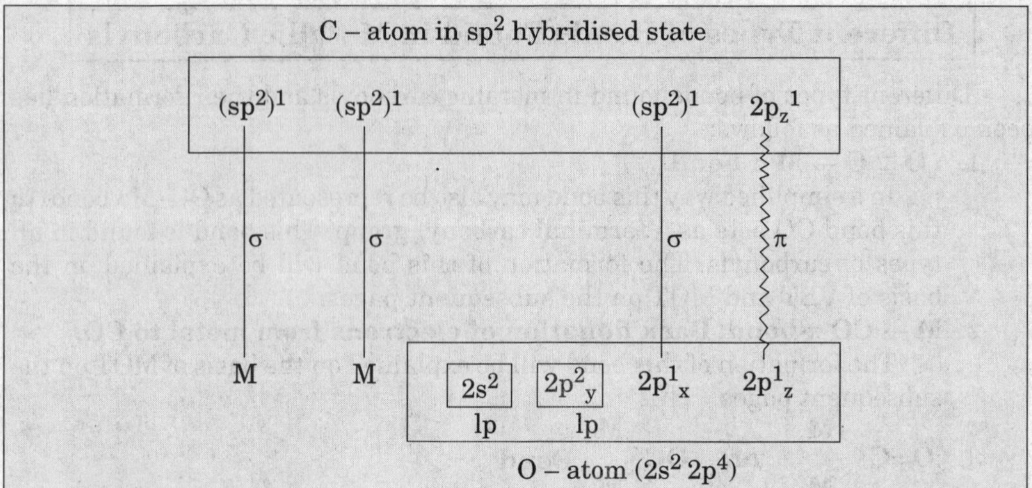


**Fig. 30.2.** Formation of two M–C σ bonds, one C—O σ bond and one C—O π bond.

singly-filled appropriate hybrid orbitals, each belonging to each of the two metallic atoms. Due to the weak coupling of electron spins, this bond is called *fractional single bond* and is represented as M⁻M. The charge distribution in this bond is not exactly symmetrical. This bond is larger in length than the normal single M—M bond. The presence of M—M bond makes all the electrons paired and hence the carbonyls containing M—M bonds are diamagnetic.

## General Discussion of Structure of Carbonyls

### A. Structure of Mononuclear Carbonyls

Examples of mononuclear carbonyls are $V(CO)_6$, $Cr(CO)_6$, $Mo(CO)_6$, $W(CO)_6$, $Fe(CO)_5$, $Ru(CO)_5$, $Os(CO)_5$, $Ni(CO)_4$ etc. All these carbonyls have only terminal carbonyl (–C≡O or –CO) groups which are attached with the metal atom (M) by linear M–C≡O or M–CO bonds. In M–C≡O bond M atom is linked with C-atom (not O-atom) of CO group, since O-atom is more electronegative than C-atom. The number of CO groups attached with the metal atom gives the C.N. of the metal and hence determines the shape of the metal carbonyl. For example, since in $M(CO)_6$ molecules (M = V, Cr, Mo, W), C. N. of M-atom is 6, these molecules have the expected octahedral shape. Similarly $M(CO)_5$ (M = Fe, Ru, Os with C.N. = 5) and $Ni(CO)_4$ (C. N. of Ni = 4) have trigonal bipyramidal and tetrahedral shape respectively.

### B. Structure of Binuclear Carbonyls

Examples of binuclear carbonyls are $Mn_2(CO)_{10}$, $Tc_2(CO)_{10}$, $Re_2(CO)_{10}$, $Fe_2(CO)_9$, $Os_2(CO)_9$, $Co_2(CO)_8$ etc. On the basis of their structure these carbonyls can be grouped in two classes:

(*a*) **$Mn_2(CO)_{10}$, $Tc_2(CO)_{10}$ and $Re_2(CO)_{10}$ carbonyls.** In these carbonyls each M-atom is attached with five terminal CO groups by M–CO bond and with other M-atom by M–M bond. Thus these carbonyls contain one M–M bond and the C.N. of each M-atom is 6. Consequently each M-atom acquires an octahedral arrangement of five CO groups and one M-atom by forming metal-metal bond. The presence of M–M bond is supported by the diamagnetic nature of these carbonyls. These carbonyls do not contain any bridging carbonyl group (>C=O) in between two M-atoms. Since the C.N. of each M-atom is six, each M-atom is $d^2sp^3$ hybridised. Due to $d^2sp^3$ hybridisation, since all the electrons of both M-atoms get paired, these carbonyls are diamagnetic.

(*b*) **$Co_2(CO)_8$ (in solid state), $Fe_2(CO)_9$ and $Os_2(CO)_9$ carbonyls.** These carbonyls contain both types of carbonyl groups which are bridging carbonyl groups (>C=O) and terminal carbonyl group (–C≡O). Bridging carbonyl groups exist between two metal atoms. The number of bridging carbonyl group present in $Co_2(CO)_8$ (only in the solid state), $Fe_2(CO)_9$ and $Os_2(CO)_9$ is 2, 3 and 1 respectively while the number of terminal carbonyl groups is 6, 6 and 8 respectively. Each of the bridging carbonyl group is attached with the metal (M) by M(|)M > C=Ö: bond while each of the terminal carbonyl groups is attached with the metal atom by O≡C→M coordinate bond. All these carbonyls have M–M bond. The presence of M–M bond confirms the diamagnetic nature of these carbonyls.

**Note:** In solution, $Co_2(CO)_8$ molecule has non-bridged structure which contains only terminal CO groups.

Each Co-atom is linked with four terminal CO groups and with another Co-atom. Thus this carbonyl also has a Co–Co bond due to which $Co_2(CO)_8$ is diamagnetic.

Non-bridged structure of $Co_2(CO)_8$ found in its solution is reported to exist in equilibrium with the bridged structure in the solution.

## C. Structure of Trinuclear Carbonyls

Examples of trinuclear carbonyls are $Fe_3(CO)_{12}$, $Ru_3(CO)_{12}$ and $Os_3(CO)_{12}$. $Ru_3(CO)_{12}$ and $Os_3(CO)_{12}$ have similar structure but $Fe_3(CO)_{12}$ has different structure.

(*a*) **$Ru_3(CO)_{12}$ and $Os_3(CO)_{12}$ carbonyls.** In these molecules the three metal atoms form an equilateral triangle with no bridging CO groups. Thus these carbonyls have non-bridged structure. Each metal atom is linked with four terminal CO groups and with other two metal atoms. These carbonyls have three M—M bonds. Each metal atom has octahedral environment of the nearest species. (See Fig. 30.3). Preference for terminal CO groups in heavier metals is due to their preference for non-bridged structures.

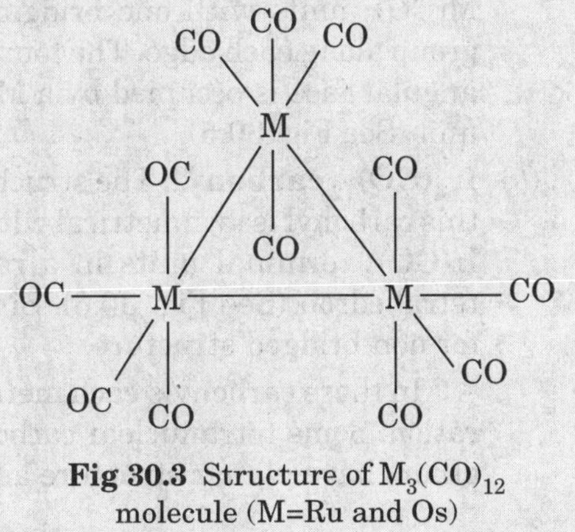

**Fig 30.3** Structure of $M_3(CO)_{12}$ molecule (M=Ru and Os)

(*b*) **$Fe_3(CO)_{12}$ molecule.** X-ray study made on $Fe_3(CO)_{12}$ molecule has shown that in this molecule (*a*) Each of the two Fe-atom (Fe1 and Fe2) is linked with three terminal carbonyl groups, two bridging carbonyl

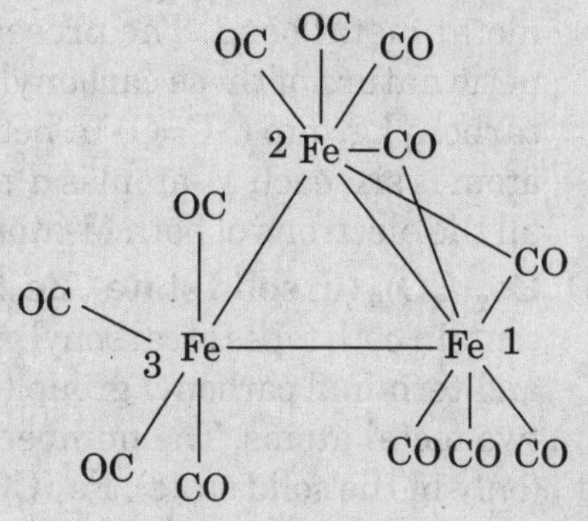


**Fig. 30.4** Structure of $Fe_3(CO)_{12}$ molecule

groups and to the third Fe-atom (Fe 3 atom) (*b*) The remaining third Fe-atom (Fe 3 atom) is linked with four terminal carbonyl groups and to each of the two Fe-atoms (Fe 1 and Fe 2) (c) Three Fe–Fe bonds are also present. Each Fe–Fe bond distance has been found to be equal to 2.8 Å (Fig. 30.4). Fe | Fe >C=O bridges are unsymmetrical.

When we compare the structures of $Fe_2(CO)_9$ and $Fe_3(CO)_{12}$, we find that the structure of $Fe_3(CO)_{12}$ can be obtained by replacing one bridging carbonyl group (>C=O) present in $Fe_2(CO)_9$ by $Fe(CO)_4$.

$$Fe_2(CO)_9 \xrightarrow{-(>C=O)} Fe_2(CO)_8 \xrightarrow{+Fe(CO)_4} Fe_3(CO)_{12}$$

**D. Structure of Tetranuclear Carbonyls**

Examples of tetranuclear carbonyls are $Co_4(CO)_{12}$, $Ru_4(CO)_{12}$ and $Ir_4(CO)_{12}$. In these compounds the four metal atoms (M) have tetrahedral arrangement. On the basis of their structure these carbonyls are grouped into two classes:

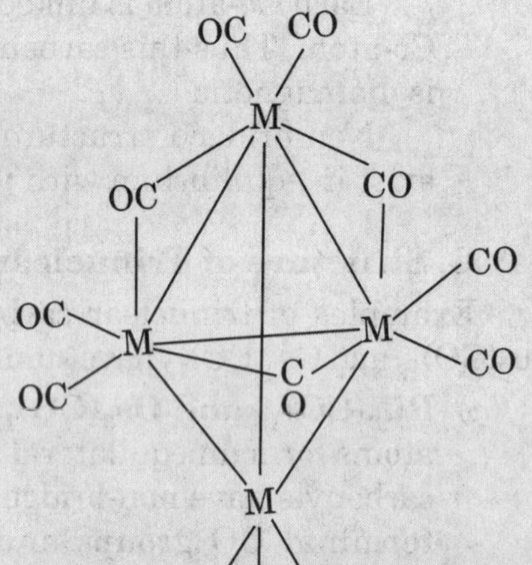


**Fig. 30.5.** Structure of tetranuclear carbonyls, $M_4(CO)_{12}$(M=Co or Ru)

(*a*) **$Co_4(CO)_{12}$ and $Ru_4(CO)_{12}$ carbonyls**. Both these carbonyls have bridged structure. In these carbonyls three triangular faces consist of the $M_4$ tetrahedron consisting of terminal $M(CO)_2$ units with one bridging CO group along each edge. The fourth triangular face is occupied by a $M(CO)_3$ unit (See Fig. 30.5)

(*b*) **$Ir_4(CO)_{12}$ carbonyl.** The structure of this carbonyl is symmetrical with four $Ir(CO)_3$ terminal units in a regular tetrahedron (See Fig. 30.6). Since Ir is a heavier metal, it has preference for non-bridged structure.

In these carbonyls each metal atom has 18-electron closed shell configuration. Some tetranuclear carbonyl clusters have butterfly structure while others have planar structure as shown in Fig. 30.7.

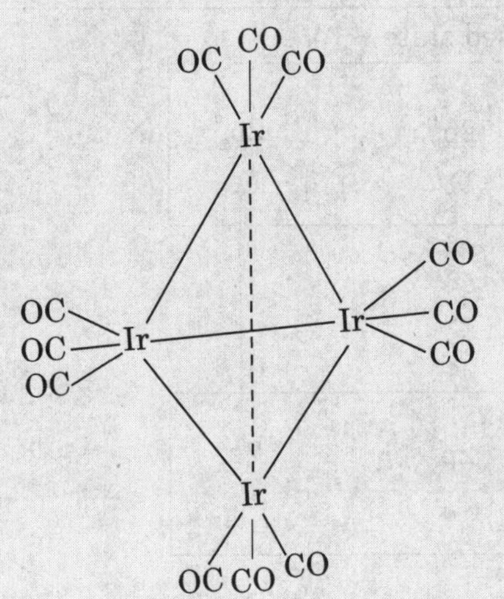


**Fig. 30.6.** Structure of $Ir_4(CO)_{12}$ molecule

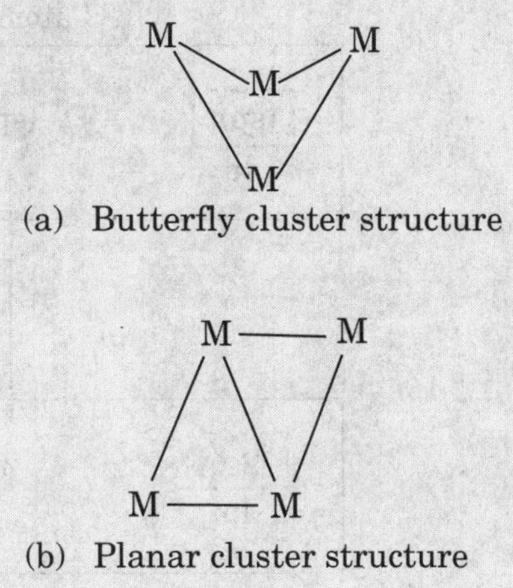


**Fig. 30.7.** Butterfly and planar cluster structure of tetranuclear carbonyl cluster

## Bonding in Metal Carbonyls

Metal carbonyls have OC $\rightarrow$ M $\sigma$ as well as M $\rightarrow$ CO $\pi$-bonds, *i.e.* metal carbonyls have double bonding which can be represented as $OC \underset{\pi}{\overset{\sigma}{\rightleftarrows}} M$.

## To Explain the Formation of OC → M σ Bond in Metallic Carbonyls on the Basis of Valence Bond Theory

We know that Lewis structure of CO molecule is :C $\leftrightharpoons$ O:. This structure shows that C-atom is *sp* hybridised and different bonds in this molecule are formed as shown in Fig. 30.8.

Lone pair (*lp*) of electrons present on C-atom of Lewis structure of CO resides in *sp* hybrid orbital. All carbonyls contain terminal carbonyl (CO) groups which are attached with the metal atom by :O $\equiv$ C $\rightarrow$ M $\sigma$ bond. :O $\equiv$ C $\rightarrow$ M $\sigma$ bond is generally represented as OC $\rightarrow$ M $\sigma$ or simply as OC–M $\sigma$ bond in metal carbonyls. This bond is formed by the overlap between *sp* hybrid orbital on C-atom of CO containing lone pair (*lp*) of electrons and the appropriate empty hybrid orbital of the metal. For example:

(*i*) In $Ni(CO)_4$, each of the four OC $\rightarrow$ Ni $\sigma$ bonds is formed by the overlap between the completely-filled *sp* hybrid orbital (*i.e.* *sp*-hybrid orbital containing a *lp*) on C-atom and empty $sp^3$ hybrid orbital of Ni-atom. Thus OC $\rightarrow$ Ni $\sigma$ bond is $[(sp)^2(CO) \rightarrow (sp^3)^0(Ni)]$ $\sigma$ bond.

(*ii*) In $Cr(CO)_6$ molecule, each of the six OC $\rightarrow$ Cr $\sigma$ bonds is $[(sp)^2(CO) \rightarrow (d^2sp^3)^0Cr]$ $\sigma$ bond.

(*iii*) In $Fe(CO)_5$ molecule, each of the five OC $\rightarrow$ Fe $\sigma$ bonds is $[(sp)^2(CO) \rightarrow (dsp^3)$ Fe] $\sigma$ bond.

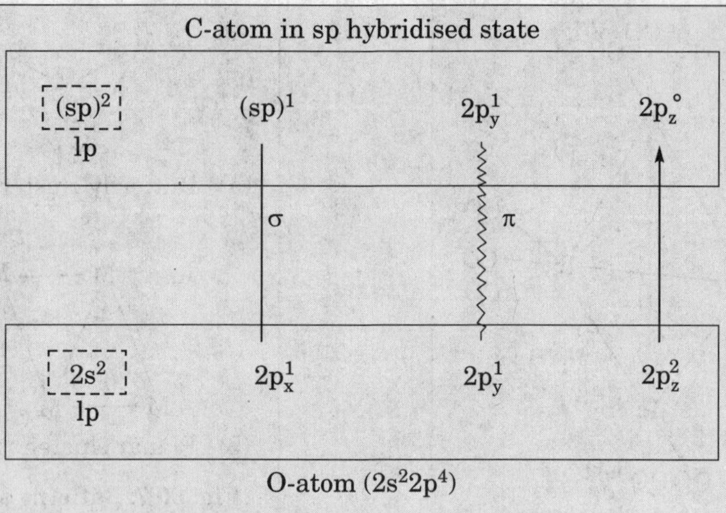


**Fig. 30.8.** Formation of different bonds in CO molecule on the basis of valence bond theory.

## To Explain the Formation of OC → M σ and M → CO π Bonds in Metallic Carbonyls on the Basis of Molecular Orbital Theory.

Before explaining the formation of σ and π bonds in carbonyls on the basis of MOT, let us first study molecular orbital energy level diagram of CO (No. of valence electrons in CO molecule = 4 + 6 = 10) given in Fig. 30.9. This energy level diagram contains different molecular orbitals of CO molecule in the increasing order of their energy. In the formation of different molecular orbitals $x$-axis has been assumed as molecular (nuclear) axis.

**Formation of OC → M σ bond.** The highest occupied (filled) molecular orbital (HOMO) present in the molecular orbital configuration of CO molecule given in Fig. 30.9 is antibonding σ-molecular orbital *viz* $\sigma^*_{2s}$. This molecular orbital has been shown in box in Fig. 30.9 and is centred on C-atom of CO. This molecular orbital has a lone pair of electrons and hence overlaps with the vacant suitably directed σ-type orbital (*i.e.* hybrid orbital) on transition metal (M) and forms OC $(\sigma^*_{2s})^2 \rightarrow$ M (σ-orbital) σ-bond. This can also be represented as OC → M σ bond. The formation of σ-bond has been shown at (a) of Fig. 30.10.

**Formation of M → CO π bond: Back donation of electrons.** The lowest unoccupied (empty) molecular orbitals (LUMO's) present in the molecular orbital configuration of CO molecule are two degenerate empty (vacant) antibonding π molecular orbitals *viz*, $\pi^*_y$ and $\pi^*_z$. These molecular orbitals have also been shown in box in Fig. 30.9. These molecular orbitals possess proper symmetry and hence one of these molecular orbitals overlaps with the suitable non-bonding filled $d$-orbital ($d_{xz}$ or $d_{yz}$) of the metal to accept the electrons from the filled $d$-orbitals. This overlap results in the formation of M($d$) $\longrightarrow$ CO ($\pi^*_y$ or $\pi^*_z$) π bond (See Fig. 30.10(b)).

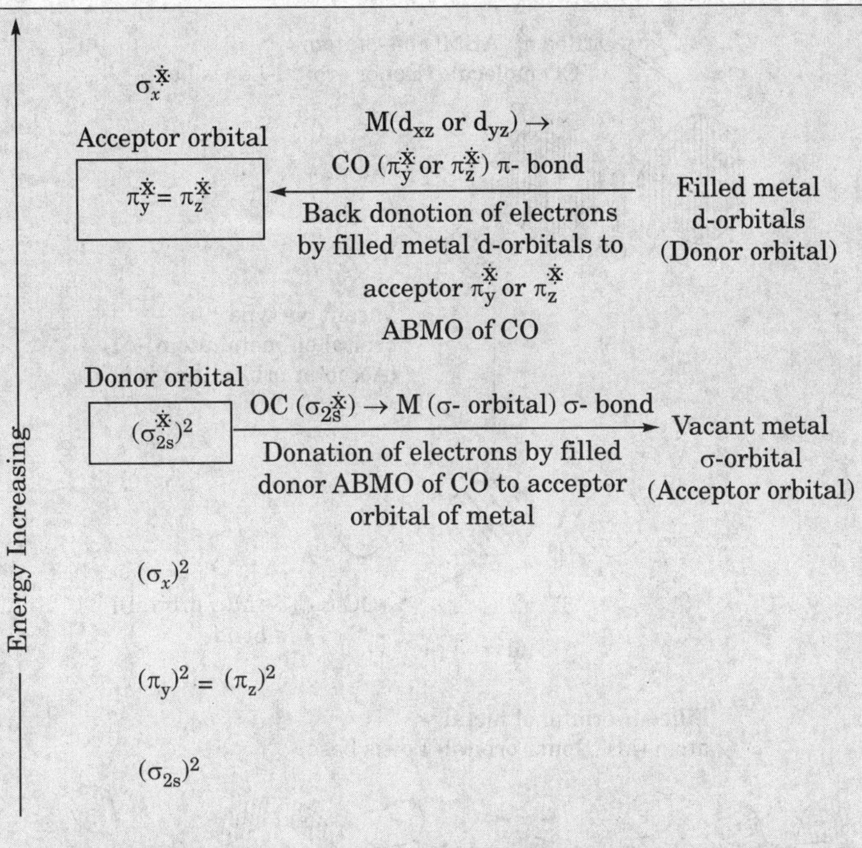


**Fig. 30.9.** Molecular orbital energy level diagram of CO molecule (No. of valence electrons in CO molecule = 4 + 6 = 10). $\sigma_{2s}$, $\pi_y$, $\pi_z$ and $\sigma_x$ are BMOs and $\sigma_{2s}^*$, $\pi_y^*$, $\pi_z^*$ and $\sigma_x^*$ are ABMOs

Since the non-bonding filled *d*-orbitals of the metal form π bond, these *d*-orbitals are called *d*π orbitals and π bond is represented as M(*d*π) → CO (π*) π bond. The formation of M → CO π bond stabilises or strengthens OC → M σ bond. Since electron density from the metal is accepted by CO (ligand) in its π* molecular orbital, CO is called π-acceptor (or π-acid) ligand and the property of CO molecule to accept the electrons from the metal *d*-orbital into its π* molecular orbital is called π acidity. The term acidity has been used in the Lewis sense.

Above discussion shows that:

(*i*) In the formation of OC → M σ bond the electron pair donated by CO resides in $\sigma_{2s}^*$ molecular orbital of CO but in the formation of M → CO π bond the electrons donated by M metal are accepted or accommodated by CO in its π* molecular orbital

(*ii*) The shifting of electron density from CO to metal (in the formation of OC → M σ bond) decreases electron density on CO. This enhances π-acceptor power of CO. At the same time the shifting of electron density from metal to CO (in the formation of M → CO π bond) increases electron density on CO. This

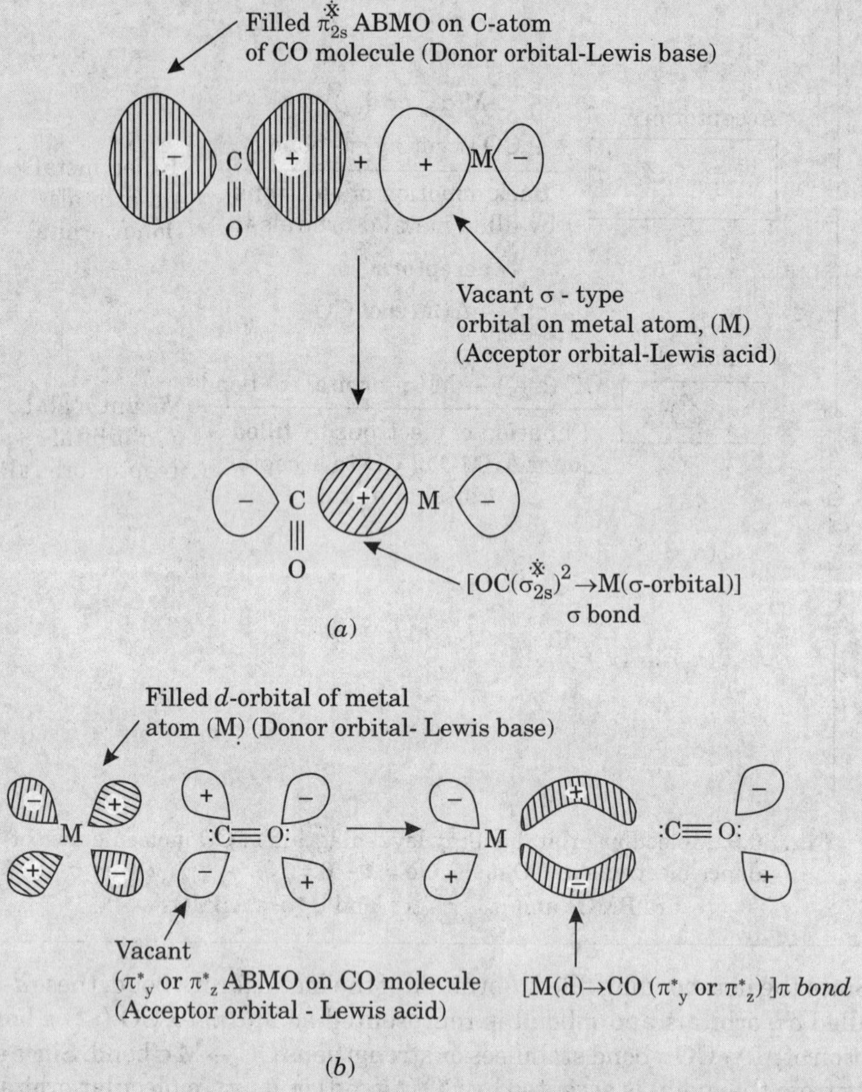


**Fig. 30.10.** Formation of σ and π bonds in metal carbonyls. (*a*) Formation of $[OC(\sigma^{*}_{2s}) \rightarrow M(\sigma\text{-orbital})]$ σ bond (*b*) Formation of $[M(d) \rightarrow CO(\pi^{*}_{y}$ or $\pi^{*}_{z})]$ π bond.

enhances σ-donor power of CO. Thus σ and π bond formations reinforce each other. The term **synergic** has been used to describe such a reinforcement.

In a simple way the formation of metal-carbon σ and π bonds can be represented as:

Filled $\sigma^{*}_{2s}$ MO on CO $\xrightarrow{\sigma}$ Vacant metal σ-orbital

Vacant $\pi^{*}_{y}$ or $\pi^{*}_{z}$ MO on CO $\xleftarrow{\pi}$ Filled metal *d*-orbital

## Effect of Back Donation on M–C and C–O Bond Orders

During the formation of π-bond, electronic charge is transferred from the filled non-bonding *d*-orbital of the metal to $\pi^*$ ($\pi_y^*$ or $\pi_z^*$) molecular orbital of the ligand, CO. This transfer of electronic charge increases M–C bond order in carbonyls and decreases C–O bond order in CO. Thus the increase in metal *d*-electron transfer to $\pi^*$ orbital of CO will increase M–C bond order in carbonyls and will decrease C–O bond order in CO.

**Example:** Since the negative charge on the metal in iso-electronic and iso-structural species *viz* $[Ni(CO)_4]^0$, $[Co(CO)_4]^-$ and $(Fe(CO)_4]^{2-}$ increases as: $[Ni(CO)_4]^0 < [Co(CO)_4]^- < [Fe(CO)_4]^{2-}$, the ease of electron transfer from the metal *d*-orbital to $\pi^*$ orbital of CO also increases in the same direction. The increase in the ease of the electron transfer also increases M–C bond order as:

$$\begin{array}{ccccc} \text{Ni–C} & < & \text{Co–C} & < & \text{Fe–C} \\ 1.33 & & 1.89 & & 2.16 \end{array}$$

and decreases C–O bond order as:

$$\begin{array}{ccccc} Ni(CO)_4 & > & [Co(CO)_4] & > & [Fe(CO)_4]^{2-} \\ 2.64 & & 2.14 & & 1.85 \end{array}$$

## Difference in Bonding in Metal Carbonyls and that in π-Complexes of Unsaturated Organic Ligands

The difference in bonding in metal carbonyls and that in π-complexes of unsaturated organic ligands like alkenes, alkynes and cyclopentadienes is that in π-complexes of organic ligands both donation and back-donation (or back-acceptance) by the unsaturated ligand involves the use of π-orbitals of the unsaturated ligand. On the other hand, in metal carbonyls donation of electron pair takes place from $\sigma_{2s}^*$ molecular orbital of CO to the appropriate metal σ-type orbitals and back-donation takes place from the metal *d*-orbital to $\pi^*$ molecular orbital of CO.

## Structure and Bonding in Some Individual Carbonyls

### 1. Nickel Tetracarbonyl, $Ni(CO)_4$

The vapour density of nickel tetracarbonyl and the freezing point of its solution in benzene has shown that the molecular formula of this compound is $Ni(CO)_4$.

Electron diffraction studies made on this compound in the vapour state and X-ray diffraction studies made on this compound in the solid state have shown that $Ni(CO)_4$ molecule has tetrahedral shape with Ni–C–O linear units (Fig. 30.11). The Ni–C bond length in this molecule has been found to be equal to 1.50 Å which is shorter by 0.32 Å in comparison to Ni–C single bond length (=1.82 Å) found in carbonyls. The C–O bond length in this carbonyl has been found to be equal to 1.15 Å which is larger than the C–O bond length in CO molecule (=1.128 Å).

Tetrahedral shape of $Ni(CO)_4$ results from $sp^3$ hybridisation of Ni-atom. Since $Ni(CO)_4$ molecule is diamagnetic, all the ten electrons present in the valence-shell of Ni-atom (Ni=$3d^8\ 4s^2$) get paired in $3d$ orbitals. Thus the valence-shell configuration of Ni-atom in $Ni(CO)_4$ molecule becomes $3d^{10}$. 4s°. OC → Ni-bond results by the overlap between the empty $sp^3$ hybrid orbital on Ni-atom and doubly-filled $sp$ hybrid orbital on C-atom in CO molecule. (Fig. 30.12).

It is to be pointed out here that due to the formation of four OC→Ni bonds, a large negative charge is accumulated on the central Ni atom. This is most unlikely. In such a situation, Pauling suggested that the double bonding takes place with the back donation of $d$-electron from Ni atom to CO ligands to such an extent that electroneutrality principle is obeyed. According to this principle the electron pair is not shared equally between Ni and C-atoms of CO ligands but is attracted more strongly by C-atom which prevents the accumulation of negative charge on Ni-atom and is in keeping with the greater electronegativity of C-atom compared to Ni atom (C=2.5, Ni=1.8).

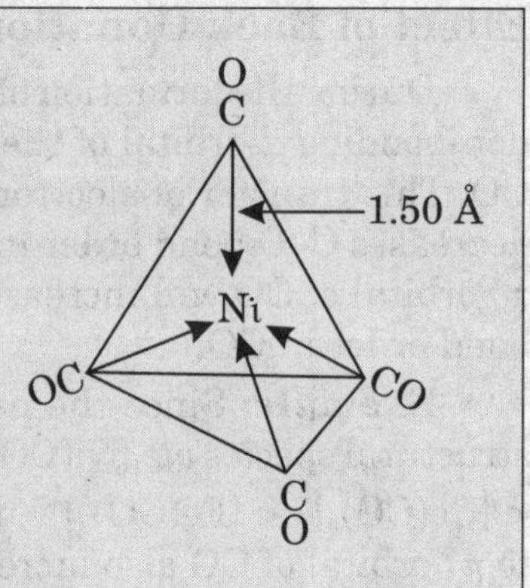


**Fig. 30.11**. Tetrahedral structure of $Ni(CO)_4$ molecule

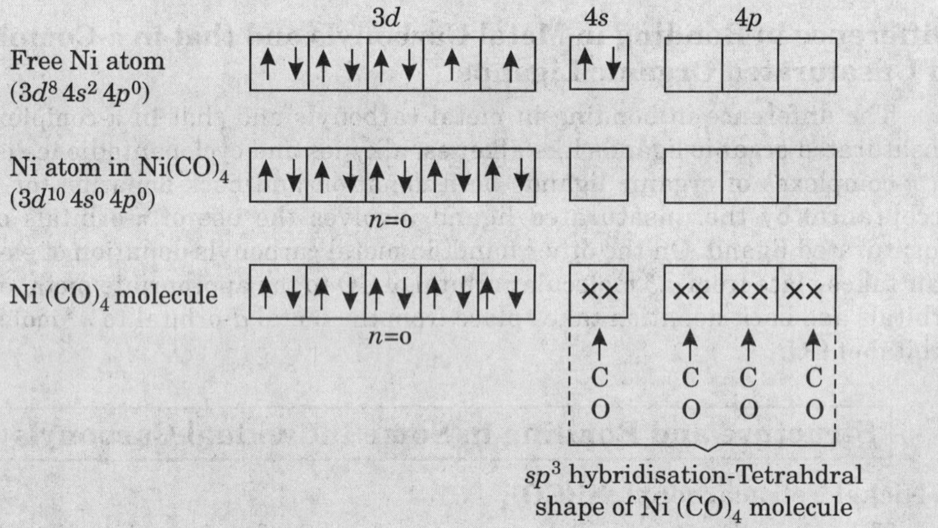


**Fig. 30.12.** $sp^3$ hybridisation of Ni-atom in tetrahedral $Ni(CO)_4$. × x indicate the electron pair donated by CO molecules. These electrons are in opposite spin.

## 2. Iron Pentacarbonyl, $Fe(CO)_5$

In principle, $Fe(CO)_5$ molecule can have any of the two geometries *viz., trigonal bipyramidal* and *square pyramidal* shown in Fig. 30.13. In order to decide as to which of the structures is the real structure of $Fe(CO)_5$, we proceed as follows:

We calculate the number of IR-active and Raman – active vibrational bands for each of the two geometries. Now, we run the IR and Raman spectra and tally

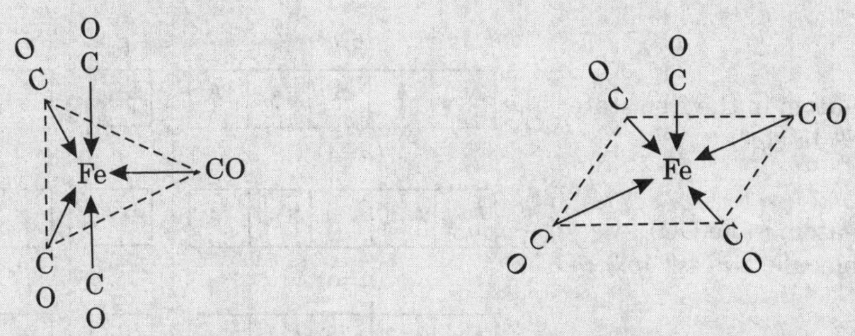


Trigonal bipyramidal structure of $Fe(CO)_5$ molecule

Square pyramidal structure of $Fe(CO)_5$ molecule

**Fig. 30.13.** Two possible geometries of $Fe(CO)_5$ molecule.

the number of bands actually observed with the number of bands obtained by calculation for each geometry. It has been found that IR bands actually observed tally with the number of bands calculated for trigonal bipyramidal geometry. Thus $Fe(CO)_5$ has trigonal bipyramidal geometry. This geometry has also been confirmed by X-ray studies. It is strange to note that $^{13}C$–NMR gives only one type of signal for axial as well as equatorial CO groups. This is probably because of the *fuxtionality* in the structure. This means that the axial and the equatorial (basal) OC $\rightarrow$ Fe bonds exchange their position very fast so that an average $^{13}C$ resonance signal is obtained for all the five carbonyl groups attached with Fe-atom. Electron diffraction study of $Fe(CO)_5$ molecule in the vapour state has shown that Fe–C axial bond and Fe–C equatorial bond lengths are equal to 1.797Å and 1.842Å respectively while X-ray study has shown that these bonds are almost of the same length.

***dsp*$^3$ hybridisation of Fe-atom.** Trigonal bipyramidal shape of $Fe(CO)_5$ molecule suggests that Fe-atom which is in zero oxidation state is $dsp^3$ hybridised. Since $Fe(CO)_5$ is diamagnetic, all the 8 electrons in the valence-shell of Fe-atom (Fe = $3d^6 4s^2$) get paired in four 3*d* orbitals. $dsp^3$ hybridisation of Fe-atom is shown in Fig. 30.14. OC$\rightarrow$Fe bond is established by the overlap of doubly-filled *sp* hybrid orbital on C-atom in CO molecule and empty $dsp^3$ hybrid orbital on Fe-atom.

## 3. Chromium Hexacarbonyl, $Cr(CO)_6$

Electron diffraction study of $Cr(CO)_6$ molecule in the vapour state has shown that this molecule has octahedral geometry as shown in Fig. 30.15. Cr–C and C–O bond lengths has been found to be equal to 1.92Å and 1.16Å respectively.

***d*$^2$*sp*$^3$ hybridistion of Cr-atom.** Octahedral geometry of $Cr(CO)_6$ suggests that Cr-atom which is in zero oxidation state is $d^2sp^3$ hybridised as shown in Fig. 30.16. Since $Cr(CO)_6$ is diamagnetic, all the six electrons present in the valence-shell of Cr-atom (Cr=$3d^5 4s^1$) get paired in three 3*d* orbitals. Now $3d_{x^2-y^2}$, $3d_{z^2}$, 4*s*

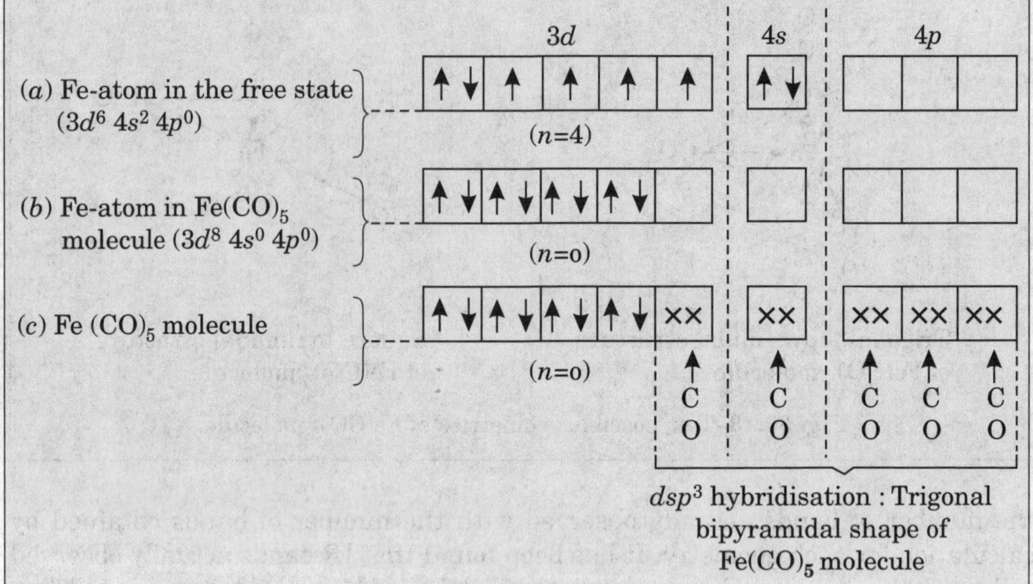


**Fig. 30.14.** $dsp^3$ hybridisation of Fe-atom in $Fe(CO)_5$ molecule which has trigonal bipyramidal shape. Each pair of crosses (××) indicates electron pair donated by C-atom of each CO molecule (ligand).

and three 4*p* orbitals hybridise together to form six $d^2sp^3$ hybrid orbitals (see Fig. 30.16). O ⇶ C → Cr bond is formed by the overlap of the doubly-filled *sp* hybrid orbital on C-atom in CO molecule and empty $d^2sp^3$ hybrid orbital on Cr-atom.

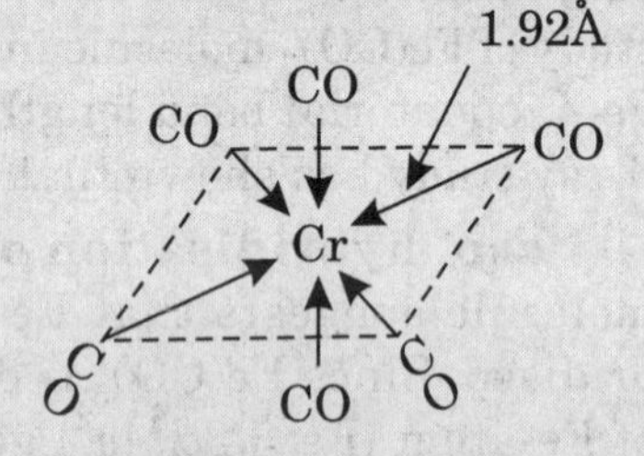


**Fig. 30.15.** Octahedral geometry of $Cr(CO)_6$ molecule.

## 4. Structure of $Mn_2(CO)_{10}$, $Tc_2(CO)_{10}$ and $Re_2(CO)_{10}$ molecules

All these molecules have the same structure. As an example, let us consider the structure of $Mn_2(CO)_{10}$ molecule. The infra-red absorption spectral and X-ray diffraction studies have shown that this molecule contains no bridging carbonyl groups and hence the molecule has *non-bridged structure* as shown in Fig. 30.17. It is evident from this figure that each Mn-atom is linked with five terminal carbonyl groups (:O ⇶ C→Mn) and with the other Mn-atom by Mn–Mn σ-bond. Thus each Mn-atom has five :O ⇶ C→Mn coordinate σ-bonds and one Mn–Mn σ-bond. The presence of Mn–Mn bond has also been supported by the diamagnetic nature of $Mn_2(CO)_{10}$ molecule. It may be noted that each Mn-atom has an octahedral environment and two $Mn(CO)_5$ units are joined solely by Mn–Mn bond.

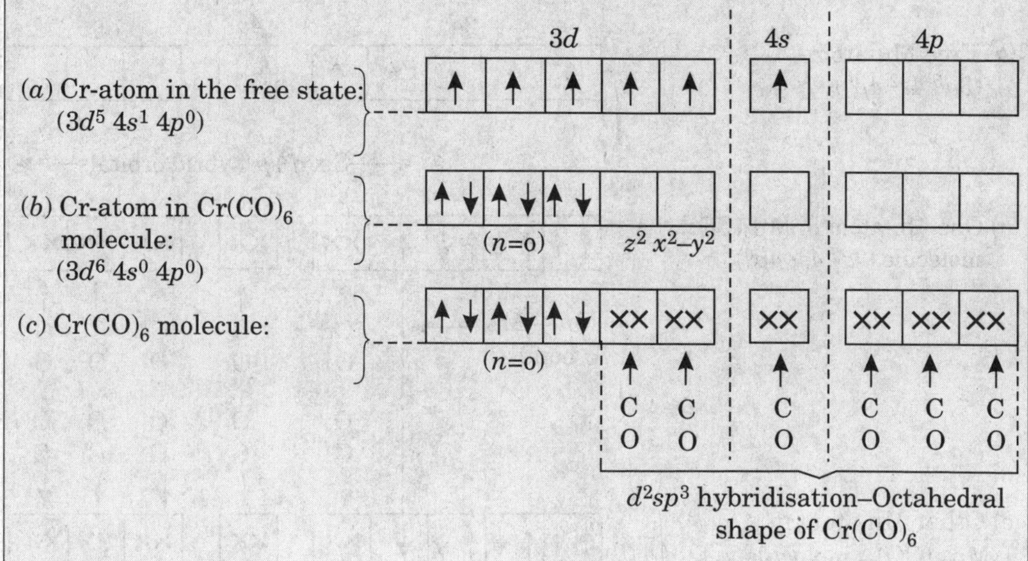


**Fig. 30.16.** $d^2sp^3$ hybridisation of Cr-atom in $Cr(CO)_6$ molecule which has octahedral geometry. Each pair of crosses (××) indicates electron pair donated by C-atom of each CO molecule (ligand).

***$d^2sp^3$* hybridisation of each Mn-atom** Since it is evident from Fig. 30.17. that the number of σ-bonds by which each Mn-atom is linked with terminal carbonyl groups and to the other Mn-atom is six, each Mn-atom has coordination number equal to six and hence each Mn-atom is $d^2sp^3$ hybridised in $Mn_2(CO)_{10}$ molecule. Valence-shell configuration of Mn-atom in its free state is $3d^5 4s^2 4p^0$. When Mn-atom forms $Mn_2(CO)_{10}$ molecule both the electrons from 4*s* orbital get shifted to 3*d* orbitals and hence the configuration of Mn-atom in $Mn_2(CO)_{10}$ becomes $3d^7 4s^0 4p^0$. In order to make $d^2sp^3$ hybridisation possible, the distribution of seven electrons in 3*d* orbitals takes place in such a way that $3d_{z^2}$ orbital becomes vacant and hence the configuration of Mn-atom in $Mn_2(CO)_{10}$ molecule becomes $3d_{xy}^2\ 3d_{yz}^2\ 3d_{yz}^2\ 3d_{x^2-y^2}^1\ 3d_{z^2}^0$. Now $3d_{x^2-y^2}$, $3d_{z^2}$, 4*s*, $4p_x$, $4p_y$ and $4p_z$ orbitals (six orbitals) combine together to form six $d^2sp^3$ hybrid orbitals. $3d_{xy}$, $3d_{yz}$ and $3d_{zx}$ orbitals (each is completely-filled) remain unhybridised. One of the six $d^2sp^3$ hybrid orbitals is singly-filled while the remaining five are vacant.

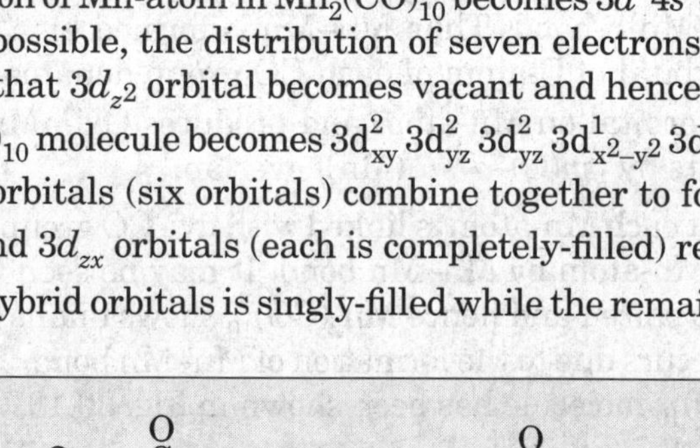


**Fig. 30.17.** Non-bridged structure of $Mn_2(CO)_{10}$ molecule.

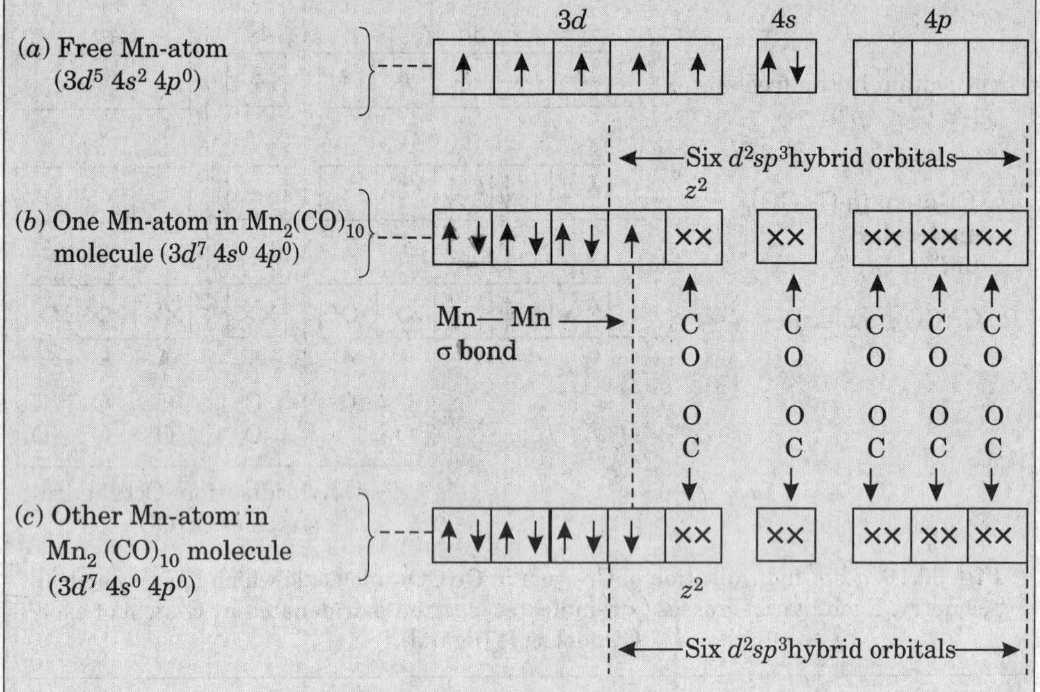


**Fig. 30.18.** $d^2sp^3$ hybridisation of Mn-atom in $Mn_2(CO)_{10}$ molecule. Crosses (××) indicate the electrons donated by C-atom of terminal carbonyl group.

Thus the valence-shell configuration of Mn-atom in $d^2sp^3$ hybridised state in $Mn_2(CO)_{10}$ molecule can be written as shown in Fig. 30.18. The singly-filled $d^2sp^3$ hybrid orbital of one Mn-atom overlaps with the singly-filled $d^2sp^3$ hybrid orbital of the other Mn-atom and gives Mn–Mn σ-bond. Thus Mn–Mn is formed by $d^2sp^3$ (Mn)–$d^2sp^3$(Mn) overlap. *sp* hybrid orbital of C-atom of each CO group donates its *lp* of electrons to vacant $d^2sp^3$ hybrid orbital on Mn-atom and produces OC→Mn coordinate σ-bond. Thus this bond results by [*sp*(C)–$d^2sp^3$(Mn)] overlap.

Thus we see that each Mn-atom is linked with five CO groups by five :O$\rightleftarrows$C→Mn bonds and with one Mn-atom by Mn–Mn bond. It may be seen that all the electrons of both Mn-atoms get paired and hence $Mn_2(CO)_{10}$ shows diamagnetic character. The pairing of electrons occurs due to the formation of Mn–Mn bond. The formation of various bonds in $Mn_2(CO)_{10}$ molecule has been shown in Fig. 30.18.

## 5. Di-ironennea Carbonyl, $Fe_2(CO)_9$

Infra-red spectral study made on $Fe_2(CO)_9$ shows that this molecule has two different types of carbonyl groups. X-ray crystallographic study has shown that the structure of $Fe_2(CO)_9$ molecule consists of three bridging carbonyl groups, six terminal carbonyl groups and one single bond between two Fe-atoms (See Fig. 30.19). Since this bond is produced by the weak coupling of the unpaired electrons present in two 3*d*-orbitals of two Fe atoms, it is called *fractional single bond* and is represented as Fe----Fe. This bond is a δ bond. The charge distribu-

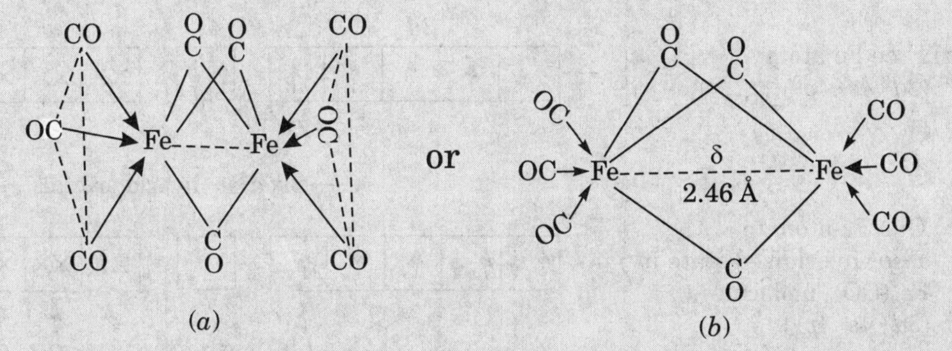


**Fig. 30.19.** Structure of $Fe_2(CO)_9$ molecule

tion in this bond is not exactly symmetrical. This bond is larger in length than the normal single Fe–Fe bond. $Fe_2(CO)_9$ is diamagnetic and obeys EAN or inert gas rule. Diamagnetic nature of $Fe_2(CO)_9$ molecule also confirms the presence of a bond between two Fe-atoms, since the formation of Fe—Fe bond makes all the electrons paired.

The structure shown at (*a*) of Fig. 30.19 shows that the three terminal CO groups attached with each Fe-atom are placed at the corners of a triangle. In a simplified way, structure (*a*) can be shown as depicted at (*b*).

**$d^2sp^3$ hybridisation of each Fe-atom** Since each Fe-atom is linked with three terminal CO groups and three bridging CO groups, this atom has coordination number equal to six (Note that Fe–Fe bond is a fractional single bond) and hence each Fe-atom is $d^2sp^3$ hybridised in $Fe_2(CO)_9$ molecule. Valence-shell configuration of Fe-atom in the free state is $3d^6 4s^2 4p^0$. When Fe-atom forms $Fe_2(CO)_9$ molecule, one of the two electrons of 4*s* orbital is shifted to 3*d* orbitals and hence the valence-shell configuration of Fe-atom becomes $3d^7 4s^1 4p^0$. Both these configurations have been shown at (*a*) and (*b*) of Fig. 30.20. Now two 3*d* orbitals *viz*. $3d_{x^2-y^2}$ and $3d_{z^2}$ (both are singly-filled), one 4*s* orbital (singly-filled) and all the three 4*p* orbitals (all the empty) combine together and produce six $d^2sp^3$ hybrid orbitals. Three hybrid orbitals are vacant and three are singly-filled. One of the three unhybridised 3*d* orbitals ($3d_{xy}$, $3d_{yz}$ and $3d_{zx}$) is singly-filled. Fe—Fe fractional single bond is produced by the overlap between two singly-filled 3*d* orbitals on two Fe-atoms. The attachment of bridging CO molecule with two Fe-atoms indicates that C-atom of CO molecule of $sp^2$ hybridised.

C-atom in its ground state:.......................................$2s^2\ 2p_x^1\ 2p_y^1\ 2p_z^0$

$(2s^2 2p^2)$

C-atom in its excited state:.......................................$2s^1\ 2p_x^1\ 2p_y^1\ 2p_z^1$

$(2s^1 2p^3)$

C-atom in $sp^2$ hybridised state in :Ö=C<(Fe)(Fe) ........$(sp^2)^1\ (sp^2)^1\ (sp^2)^1\ 2p_z^1$

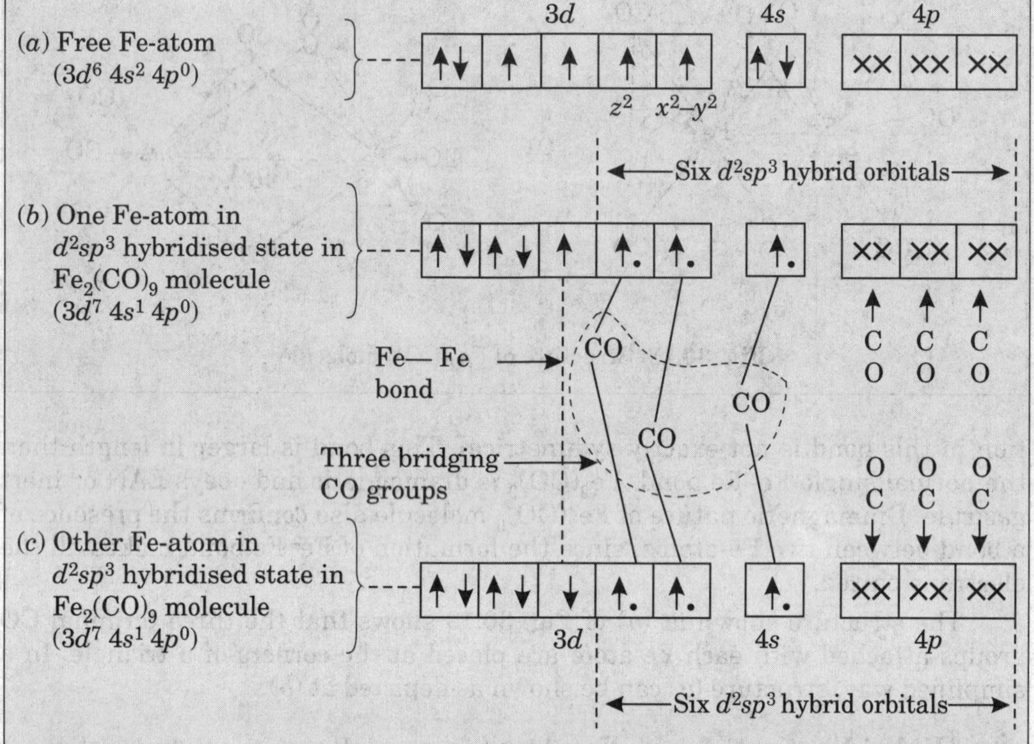


**Fig. 30.20.** Formation of different bonds in $Fe_2(CO)_9$ molecule. Crosses (××) indicate the electron pair donated by C-atoms of the terminal carbonyl groups [:CO] and dot (•) represents the electron given by C-atom of each bridging carbonyl group [>C=O].

One of the three singly-filled $sp^2$ hybrid orbitals makes a C–O σ-bond and singly-filld $2p_z$ orbitals is used in making C–O π-bond. The remaining two singly-filled $sp^2$ hybrid orbitals overlap with two singly-filled $d^2sp^3$ hybrid orbitals on two Fe-atoms and produce two C–Fe bonds in :Ö=C<(Fe)(Fe) or OC<(Fe)(Fe) system. Fe ← CO bond is formed in its usual manner. Formation of different bonds has been shown in Fig. 30.20.

## 6. Non-bridged Structure of Dicobalt Octacarbonyl Molecule, $Co_2(CO)_8$ in Solution

The infra-red study of the solution of $Co_2(CO)_8$ has shown that this molecule has no bridging carbonyl groups, *i.e.*, $Co_2(CO)_8$ in solution has *non-bridged structure* (See Fig. 30.21). In this structure each Co-atom is linked with four terminal carbonyl groups (OC → Co) and with other Co-atom by a Co–Co σ-bond. Thus each Co-atom has four OC → Co coordinate σ-bonds and one Co–Co σ-bond. The presence of Co–Co bond has also been supported by the diamagnetic nature of $Co_2(CO)_8$ molecule. Co–Co bond length shown in the structure has been determined by electron diffraction study.

### $dsp^3$ hybridisation of each Co-atom

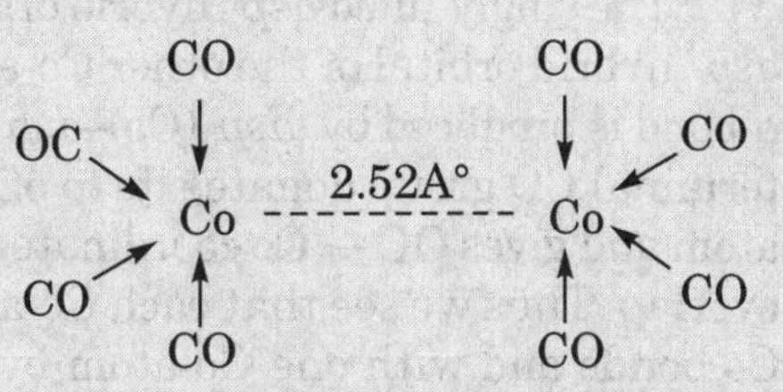


**Fig. 30.21.** Non-bridged structure of $Co_2(CO)_8$ molecule in solution

Since it is evident from the structure given in Fig. 30.21, the number of σ-bonds by which each Co-atom is linked with terminal carbonyl groups and to the other Co-atom is five, each Co-atom has coordination number equal to five and hence each Co-atom is $dsp^3$ hybridised in $Co_2(CO)_8$ molecule. Valence-shell electronic configuration of Co-atom in its free state is $3d^7 4s^2 4p^0$. When Co-atom forms $Co_2(CO)_8$ molecule, both electrons from 4*s* orbital get shifted to 3*d* orbitals and hence the valence-shell configuration of Co-atom in $Co_2(CO)_8$ becomes $3d^9 4s^0 sp^0$. These two configurations have been shown at (*a*) and (*b*) respectively in Fig. 30.22. Now one 3*d*-orbital *viz*. $3d_{z^2}$ (which is singly-filled with electron), one 4*s* orbital (vacant) and all the three 4*p* orbitals (vacant) combine together and give five $dsp^3$ hybrid orbitals. Thus we see that four 3*d* orbitals (each of which is completely-filled) remain unhybridised. One of the five $dsp^3$ hybrid orbitals has one electron (singly-filled) while the remaining four are vacant. $dsp^3$ hybridised state of Co-atom has been shown at (*c*) of Fig. 30.22.

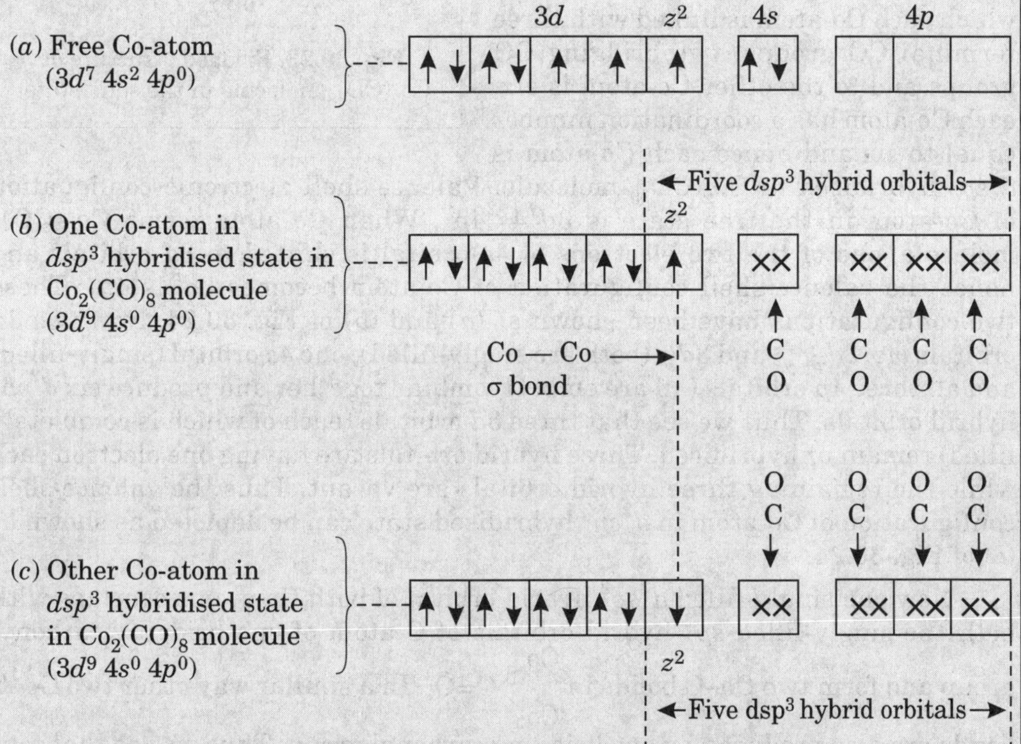


**Fig. 30.22.** $dsp^3$ hybridisation of Co-atom in $Co_2(CO)_8$ molecule in solution. Each pair of crosses (××) indicates electron pair donated by C-atom of CO molecule (ligand).

The singly-filled $dsp^3$ hybrid orbital of one Co-atom overlaps with the singly-filled $dsp^3$ hybrid orbital of the other Co-atom and produces Co—Co σ-bond. Thus Co—Co σ-bond is produced by [$dsp^3$(Co)–$dsp^3$(Co)] overlap. *sp* hybrid orbital on C-atom of each terminal CO group donates its *lp* of electrons to the vacant $dsp^3$ hbrid orbital on Co-atom and gives OC → Co coordinate σ-bond, *i.e.*, this bond results by [*sp*(C)–$dsp^3$(Co)] overlap. Thus we see that each Co-atom is linked with four CO groups by four OC → Co bonds and with one Co-atom by Co—Co bond. It may be seen here that all the electrons of both Co-atoms become paired and hence $Co_2(CO)_8$ shows diamagnetic character. The pairing of electrons takes place due to the formation of Co—Co bond. The formation of various bonds in $Co_2(CO)_8$ has been shown in Fig. 30.22.

## 7. Bridged Structure of Dicobalt Octacarbonyl Molecule, $Co_2(CO)_8$ in Solid State

Infra-red study of $Co_2(CO)_8$ molecule in the solid state has indicated that this molecule has bridged structure as shown in Fig. 30.23.

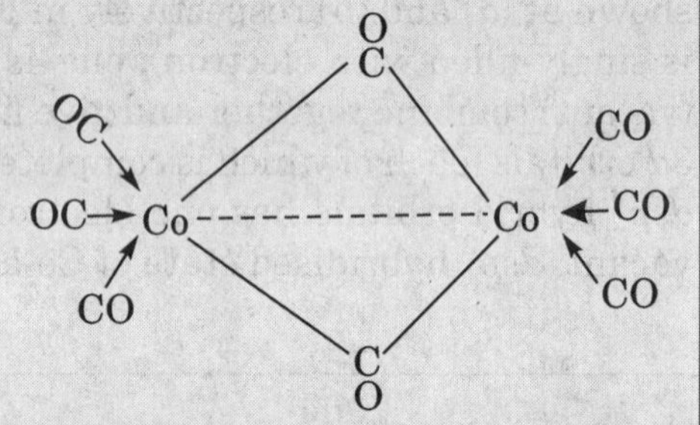


**Fig. 30.23.** Bridged structure of $Co_2(CO)_8$ molecule in the solid state.

***$d^2sp^3$* hybridisation of each Co-atom.** Since it is evident from Fig. 30.23 that the number of σ-bonds by which each Co-atom is linked with three terminal CO groups, two bridging CO groups and to the other Co-atom is six, each Co-atom has a coordination number equal to six and hence each Co-atom is $d^2sp^3$ hybridised in $Co_2(CO)_8$ molecule. Valence-shell electronic configuation of Co-atom in the free state is $3d^7 4s^2 4p^0$. When Co-atom forms $Co_2(CO)_8$ molecule, one of the two electrons of 4*s* orbital is shifted to 3*d* orbitals and hence the valence-shell configuration of Co-atom becomes $3d^8 4s^1 4p^0$. These two configurations have been shown at (*a*) and (*b*) of Fig. 30.24. Now two 3*d* orbitals *viz*. $3d_{x^2-y^2}$ and $3d_{z^2}$ (both are singly-filled), one 4*s* orbital (singly-filled) and all three 4*p* orbitals (all are empty) combine together and produce six $d^2sp^3$ hybrid orbitals. Thus we see that three 3*d* orbitals (each of which is completely filled) remain unhybridised. Three hybrid orbitals are having one electron each while the remaining three hybrid orbitals are vacant. Thus the valence-shell configuration of Co-atom in $d^2sp^3$ hybridised state can be depicted as shown at (*c*) of Fig. 30.24.

Now one singly-filled $d^2sp^3$ hybrid orbital of both Co-atoms overlaps with both the singly-filled $sp^2$ hybrid orbitals of C-atom of one bridging carbonyl group and form two Co–C bonds in (Co)(Co)>C=Ö:. In a similar way other two Co–C bonds are also produced by one bridging carbonyl group. Thus we see that four Co–C bonds are obtained. Here C-atom belongs to bridging carbonyl group. It is

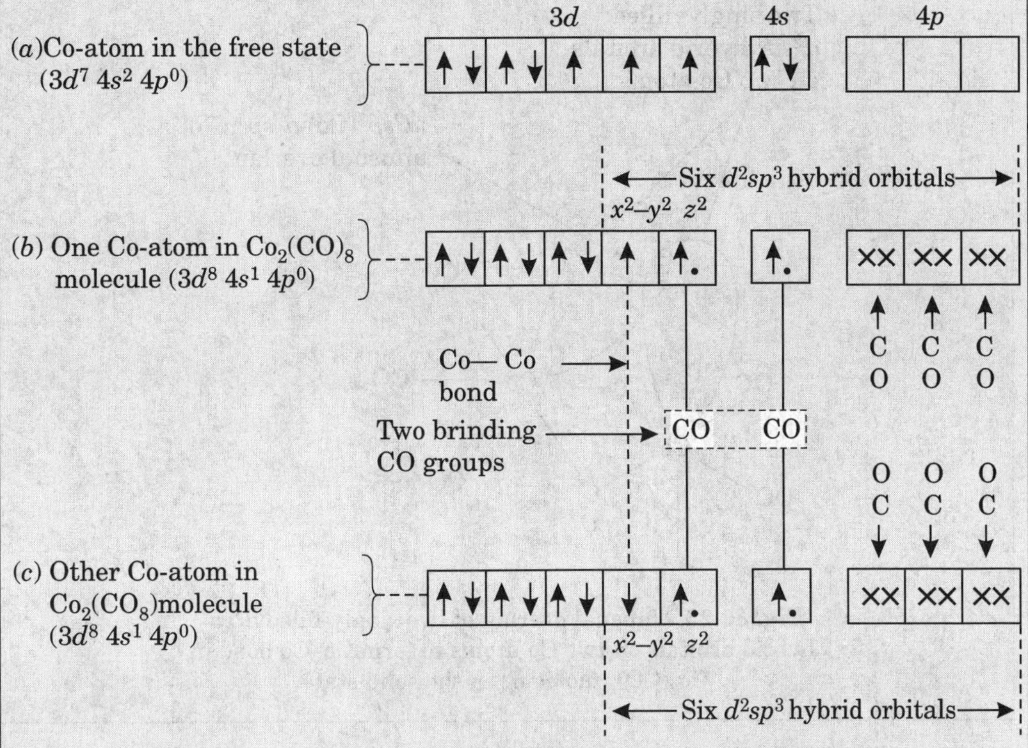


**Fig. 30.24.** $d^2sp^3$ hybridisation of Co-atom in $Co_2(CO)_8$ molecule in solid state. Each pair of crosses (××) indicates an electron pair on C-atom of terminal carbonyl group, CO while the dot (•) represents an electron donated by C-atom of bridging carbonyl group, >C=Ö:

easy to understand that each of these four bonds results by [$d^2sp^3$(Co)–$sp^2$(C)] overlap. Third singly-filled $d^2sp^3$ hybrid orbital of one Co-atom overlaps with the third $d^2sp^3$ hybrid orbital of the other Co-atom and produces Co–Co bond. Thus we can say that Co–Co bond results by [$d^2sp^3$(Co)–$d^2sp^3$(Co)] overlap. Each of the remaining three $d^2sp^3$ hybrid orbitals (each is vacant) of each Co-atom accepts one *lp* of electrons donated by the *sp* hybrid orbital on C-atom of terminal carbonyl group and forms :O$\equiv$C→Co coordinate bond. Thus six :O$\equiv$C→Co coordinate bonds are obtained. Obviously this bond is obtained by [*sp*(C)–$d^2sp^3$(Co)] overlap. The above discussion shows that each Co-atom is linked with three terminal carbonyl groups by :O$\equiv$C→Co bonds. It may be noted that all the electrons in both Co-atoms become paired and hence $Co_2(CO)_8$ molecule shows diamagnetic character. The pairing of electrons occurs due to the formation of Co–Co bond. The formation of various bonds in $Co_2(CO)_8$ has been shown in Fig. 30.24.

Non-bridged structure of $Co_2(CO)_8$ found in its solution is reported to exist in equilibrium with the bridged structure in the solid state.

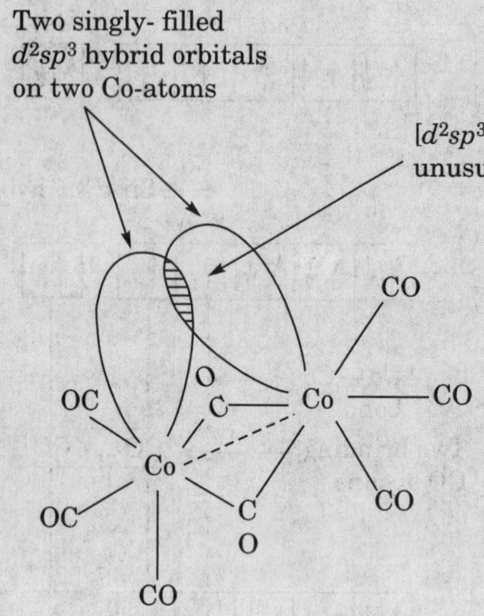


**Fig. 30.25.** Unusual overlap of two singly-filled $d^2sp^3$ hybrid orbitals on two Co-atoms to form Co–Co bond in $Co_2(CO)_8$ molecule in the solid state.

Co–Co bond which is obtained by the overlap of two singly-filled $d^2sp^3$ hybrid orbitals on two Co-atoms is bent because of unusual overlap of two $d^2sp^3$ hybrid orbitals (see Fig. 30.25).

## 8. Structure of Tri-irondodecarbonyl, $Fe_3(CO)_{12}$.

See on page 981

## 9. Structure of $Ru_3(CO)_{12}$ and $Os_3(CO)_{12}$ Molecules

See on page 981

## 10. Structure of $Co_4(CO)_{12}$ and $Ru_4(CO)_{12}$ Molecules.

See page 982

## 11. Structure of $Ir_4(CO)_{12}$ Molecule.

See page 982

## Summary of the Structure of Some Carbonyls

The number of carbonyl groups, number of M–M σ bonds and number of bridging (ketonic) carbonyl groups present in various carbonyls are given in Table 30.4.

**Table 30.4. Summary of the structure of some metallic carbonyls**

| Carbonyl | No. of terminal carbonyl groups, $:O \equiv C:$ | No. of M—M σ-bonds | No. of bridging (ketonic) carbonyl groups, $:\ddot{O}=C<$ |
|---|---|---|---|
| 1. $Ni(CO)_4$ | 4 | — | — |
| 2. $Fe(CO)_5$ Other examples are $Ru(CO)_5$, $Os(CO)_5$ | 5 | — | — |
| 3. $Cr(CO)_6$ Other examples are $V(CO)_6$, $Mo(CO)_6$, $W(CO)_6$ | 6 | — | — |
| 4. $Mn_2(CO)_{10}$ Other examples are $Tc_2(CO)_{10}$ and $Re_2(CO)_{10}$ | 10 (Each Mn-atom is linked with five terminal carbonyl groups) | One Mn–Mn bond | — |
| 5. $Fe_2(CO)_9$ Other example is $Os_2(CO)_9$ | 6 (Each Fe-atom is attached with three terminal carbonyl groups) | One Fe–Fe bond | 3 |
| 6. $Co_2(CO)_8$ (in solution) | 8 (Each Co-atom is linked with four terminal carbonyl groups) | One Co–Co bond | — |
| 7. $Co_2(CO)_8$ (in solid state) | 6 (Each Co-atom is attached with three carbonyl groups) | One Co–Co bond | 2 |
| 8. $Fe_3(CO)_{12}$ | 10 (Each of the two Fe-atoms is linked with three terminal carbonyl groups and third Fe atom is attached with four terminal carbonyl groups) | Three Fe–Fe bonds | 2 |
| 9. $Ru_3(CO)_{12}$ and $Os_3(CO)_{12}$ | 12 (Each metallic atom is linked with four terminal carbonyl groups) | Three M–M bonds | — |
| 10. $Co_4(CO)_{12}$ and $Ru_4(CO)_{12}$ | 9 (Each of the three M–atoms is linked with two terminal CO groups while 4th M-atom is attached with three carbonyl groups) | Six M–M bonds | 3 |
| 11. $Ir_4(CO)_{12}$ | 9 (Each Ir-atom is linked with three terminal carbonyl groups) | Six Ir–Ir bonds | — |

## Uses of Infra-red (IR) Absorption Spectra of Metallic Carbonyls

The IR absorption spectra of metallic carbonyls have been put to the following uses:

1. **To determine the geometry of metallic carbonyls.** If theoretically, a metallic carbonyl can have more than one geometry, we can decide which geometry is most stable. This can be done as follows. We calculate the number of I.R.-active together with Raman-active vibrational bands for each geometry and then run the I.R. and Raman spectra and finally tally the number of bands observed with the number of bands predicted theoretically (*i.e.* by calculations) for each geometry.

**Example.** Although, theoretically, $Fe(CO)_5$ molecule can have trigonal bipyramidal as well as square pyramidal geometry, the I.R. bands actually observed tally the number of such bands calculated for trigonal bipyramidal geometry. Thus $Fe(CO)_5$ has trigonal bipyramidal geometry.

2. **To determine the bond order (B.O.) of ligated CO.** It has been seen that I.R. absorption frequency of ligated CO is directly proportional to its B.O. In other words, the I.R. absorption band due to the stretching vibration of ligated CO with a higher B.O. would occur at a higher frequency and the I.R. absorption band of ligated CO with a lower B.O. would occur at a lower frequency. Since the absorption frequency for free CO is equal to 2250 $cm^{-1}$ while that for ligated CO lies between 2220–1700 $cm^{-1}$, ligated CO has lower B.O. The lower B.O. is due to the transfer of metal $d\pi$ electrons into the $\pi^*$ orbitals of ligated CO. Let us study the following examples.

| (i) *Metal carbonyl :* | $[V(CO)_6]^-$ | $[Cr(CO)_6]^\circ$ | $[Mn(CO)_6]^+$ |
|---|---|---|---|
| *Charge on metal carbonyl:* | — | 0 | + |
| *Absorption frequency for CO ($cm^{-1}$):* | 1860 | 1980 | 2090 |

Since the presence of positive charge on $[Mn(CO)_6]^+$ resists the flow of metal $d\pi$ electrons into the $\pi^*$ orbitals of CO, the B.O. of CO increases. Due to the increase in B.O., the absorption band of ligated CO occurs at higher frequency (=2090 $cm^{-1}$).

| (ii) *Metal carbonyl :* | $[Ni(CO)_4]^0$ | $[Co(CO)_4]^-$ | $[Fe(CO)_4]^{2-}$ |
|---|---|---|---|
| *Charge on metal carbonyl:* | 0 | –1 | –2 |
| *Absorption frequency for CO ($cm^{-1}$):* | 2040 | 1920 | 1790 |

Due to the increase in negative charge on the metal the $\pi$-back-donation from the metal to $\pi^*$ orbitals of the ligand weakens the CO bond and hence reduces CO stretching frequency. Greater is the negative charge, greater is the lowering in the absorption frequency of CO.

**3. To differentiate between terminal and bridging carbonyl groups.** Since OC–M bond in a terminal carbonyl group ($:O \equiv C \rightarrow M$) is stronger than that in a bridging carbonyl group $\left[ {M \atop M} \!\!> C = \ddot{O}: \right]$, I.R. absorption for a terminal

CO group would occur at a higher frequency than for a bridging CO group. It has been observed that in a neutral metallic carbonyl, if the I.R. absorption occurs between 2050–1900 $cm^{-1}$, CO group acts as a terminal CO group and if the absorption I.R. occurs between 1900–1800 $cm^{-1}$, CO group acts as a bridging CO group.

**4. To study reaction rates.** The measurement of the rates of disappearance of the intensity of CO absorption bands in the I.R. spectra can be used to study the rates of substitution reactions of metal carbonyls.

## Comparison of CO Stretching Frequency in $Mo(CO)_6$ and $Mo(dien)(CO)_3$

$Mo(CO)_6$ shows CO stretching frequency at 2004 $cm^{-1}$. Since diethylenetriamine (*dien*), $H\ddot{N}\langle^{CH_2-CH_2-\ddot{N}H_2}_{CH_2-CH_2-\ddot{N}H_2}$ or $\ddot{N}H(CH_2-CH_2-\ddot{N}H_2)_2$ is a tridentate ligand and CO is a monodentate ligand, three CO groups can be replaced from $Mo(CO)_6$ by one *dien* molecule to get $Mo(CO)_3$(*dien*). This complex shows CO stretching frequency at much lower region 1760 $cm^{-1}$ as compared to 2004 $cm^{-1}$ shown by $Mo(CO)_6$. The lowering in stretching frequency can be explained as follows.

**Explanation.** Since *dien* has no back-accepting stability from the metal (Mo), the remaining CO groups present in $Mo(CO)_3$(*dien*) share $d\pi$ electrons from the metal to prevent accumulation of negative charge on the metal atom. Consequently $Mo(CO)_3$(*dien*) has lower CO stretching frequency.

On similar lines of argument the following CO stretching frequency can also be explained:

(*a*) $Cr(CO)_6$ = 2000 $cm^{-1}$, Cr(*dien*)$(CO)_3$ = 1760 $cm^{-1}$

(*b*) $[Mn(CO)_6]^+$ = 2090 $cm^{-1}$, [Mn(*dien*)$(CO)_3]^+$ = 1900 $cm^{-1}$

## University Questions

1. Which of the following obey effective atomic rule? (i) $Cr(CO)_6$ (ii) $Mo(CO)_3(\pi\text{-}C_5H_5)$ (iii) $Fe(\pi\text{-}C_5H_5)_2$

   *(Himachal Pradesh 2010)*

2. Calculate EAN of the central metal in $Cr(CO)_6$

   *(Pune 2009)*

3. (*a*) Describe the structure of $Fe(CO)_5$.

   (*b*) Explain the structure of $Cr(CO)_6$ by VBT.

   *(Pd. Ravishankar Shukla 2010)*

4. Give the structure of $Fe_3(CO)_{12}$.

   *(Himachal Pradesh 2010)*

5. (*a*) Explain why $Ni(CO)_4$ is tetrahedral.

   (*b*) What are carbonyls? State their classification and methods of preparation

   (*c*) Write the structure and IUPAC name of $Ni(CO)_4$ and $Mn_2(CO)_{10}$.

   *(Bangalore, 2010; Kanpur 2010)*

**6.** (*i*) Write methods for the preparation of $Ni(CO)_4$.

(*ii*) Describe methods for the preparation, properties, structure and uses of $Fe(CO)_5$.

(*iii*) Explain metal-metal bonding with two examples.

*(Pune 2009)*

**7.** (*a*) Discuss the nature of M–CO bonding in metal carbonyls and Effective Atomic Number (EAN) rule.

(*b*) Discuss the structure of nickel carbonyl. How can it be explained that in $Cr(CO)_6$ all the six CO groups are not attached in similar way with metal?

(*c*) What is Wilkinson's catalyst? What is its use?

*(Rajasthan 2010)*

**8.** Define 18-electron rule.

*(Bangalore 2010)*

**9.** Which of the following is not a metal carbonyl?

(*a*) $Ni(CO)_4$ (*b*) $Co(NH_3)_6$

(*c*) $Mn_2(CO)_{10}$ (*d*) $Fe(CO)_5$

*(H.N. Bahuguna Garhwal 2010)*

**10.** Choose the correct answer:

(*i*) Pick the odd one out

(*a*) $Ni(CO)_4$ (*b*) $Mn(CO)_5$

(*c*) $Fe(CO)_5$ (*d*) $Cr(CO)_6$

(*ii*) The number of bridged and terminal CO ligands present in $Fe_2(CO)_9$ are ________ respectively

(*a*) 3 and 6 (*b*) 0 and 9

(*c*) 4 and 5 (*d*) 9 and 0

(*iii*) $Mo(CO)_5$ is isolobal with

(*a*) $[PdCl_3]^-$ (*b*) $[PdCl_4]^-$

(*c*) $[PdCl_2]^{2-}$ (*d*) $[PdCl]^-$.

*(Bharathiar 2010)*

**11.** (*a*) How will you prepare $M_2(CO)_{10}$ using the principle of isolobal analogy?

(*b*) Suggest any four methods of preparation of metal carbonyls.

(*c*) Suggest a method of preparation of metal carbonyl halides. Write a note on its structure.

*(Bharathiar 2010)*

**12.** (*i*) What is mononuclear carbonyl?

(*ii*) Why is CO a π-acceptor ligand?

*(Ranchi 2010)*

**13.** Explain the structure of $Fe(CO)_5$.

*(Andhra 2010)*

**14.** What are metal carbonyls? Give one method of preparation of $Ni(CO)_4$. Discuss its structure and bonding.

*(Nagpur 2010)*

**15.** Give your comments on the stability and structure of following carbonyls: (i) $Fe(CO)_5$ (ii) $Mn_2(CO)_{10}$

*(Utkal 2010)*

**16.** How will you prepare $Ni(CO)_4$.

*(A.P. University 2010)*

**17.** (*i*) Explain the structure of $Fe(CO)_5$.

(*ii*) Taking the example of $Ni(CO)_4$, discuss bonding in mononuclear metal carbonyls. Show that $Ni(CO)_4$ obeys 18-electron rule and give an example of a metal carbonyl which does not obey this rule.

*(Lucknow 2010)*

**18.** (*i*) How is dinuclear cobalt carbonyl prepared? Give its structures.

(*ii*) Discuss the synthesis and structures of:

(*a*) $Ni(CO)_4$ (*b*) $Fe(CO)_5$ (*c*) $Fe_2(CO)_9$

(*d*) $Fe_3(CO)_{12}$ (*e*) $Cr(CO)_6$

*(Madras April 2011)*

**19.** (*i*) Give two methods for the preparation of $Cr(CO)_6$.

(*ii*) Explain the concept of back bonding in metal carbonyls.

(*iii*) What are metal carbonyls? Discuss the structure and bonding in $Ni(CO)_4$

*(Nagpur 2012)*

(*iv*) Give the classification of metal carbonyls on the basis of metal-carbon bond. *(Nagpur 2012)*

**20.** Discuss the nature of bonding and structure of $Ni(CO)_4$

*(Ch. Charan Singh, Meerut 2012)*

**21.** (*i*) Explain the following:

(*a*) In metal carbonyls the length of M – C bond is slightly smaller than that for a signle bond and C – O length is slightly larger than in free CO molecule.

(*b*) Majority of metal carbonyls obey 18 electron rule.

(*ii*) Describe preparation and valence bond structure of $Ni(CO)_4$, $Fe(CO)_5$ and $MO(CO)_6$.

(*iii*) What happens when $Fe(CO)_5$ is irradiated with UV light.

*(Lucknow 2013)*

# 31 Organometallic Compounds

CHAPTER

## What are Organometllic Compoounds?

Organometallic compounds (also called organometallics) are the compounds in which the central metal atom is linked directly with the C-atom of the organic ligand. Thus these compounds contain one or more metal-to-carbon bonds. The metal atom may be a transition metal or a lanthanide or an actinide or a main group element. The organic ligand may be alkyl, aryl etc. group or alkene, alkyne etc.

Although B, Si, Ge, As, Sb, Se and Te are borderline metals, organoboron, organosilicon, organoarsenic etc. compounds are also classified as organometallic compounds.

Although carbides and cyanides contain metal-carbon bonds, these compounds are not put in the category of organometallic compounds by convention.

## Classification of Organometallic Compounds Based on the Nature of Metal-Carbon Bond

Organometallic compounds can be classified on the basis of the nature of metal-carbon bond in the compound. This basis gives the following types:

### 1. Ionic Organometallic Compounds

These compounds are generally given by the most electro positive elements like Na, K, Rb and Cs. The elements which are slightly less electropositive (*e.g.* Mg) can also form ionic compounds, if the entire negative charge on the organic ligand is spread over several carbon atoms. For example, Mg forms magnesium cyclopentadienide, Mg $(C_5H_5)_2$ which is ionic and hence exists as $Mg^{2+}(C_5H_5^-)_2$. In this compound the negative charge of $C_5H_5^-$ is delocalised over five C-atoms of $C_5H_5^-$ ion. Butylsodium, $NaC_4H_4$ is ionic ($Na^+$ $C_4H_4^-$) but Mg $(C_4H_4)_2$ is covalent. Methylsodium, $CH_3Na$ is ionic ($Na^+CH_3^-$) while methyl lithium ($CH_3Li$) is covalent.

Organometallic compounds of alkali metals and alkaline earth metals consist of ions or ion pairs and hence are ionic organometallic compounds. Examples are: alkyl sodium ($R^-Na^+$), alkyl magnesium halide ($R^-Mg^{2+}X^-$), phenyl lithium ($C_6H_5^-$ $Li^+$), sodium alkynide ($RC{\equiv}C^-Na^+$), dialkyl zinc ($R^-Zn^{2+}R^-$) etc.

Ionic organometallic compounds are insoluble in non-polar solvents. These compounds occur in crystalline form when pure. These are hydrolysed and are sensitive to $O_2$.

Ionic organometallic compounds of Na, K, Rb and Cs are colourless salt-like solids which are soluble in polar solvents. These compounds are quite reactive and hence are kinetically unstable. The compounds with maximum ionic character have *large cations* and *small carbanions*.

## 2. Organometallic Compounds Containing Metal-Carbon Covalent σ-bonds or σ-bonded Organometallic Compounds: Covalent Organometallic Compounds

These compounds contain M–C covalent σ–bond and are given by Zn, Cd, Hg and the metals of groups 13, 14, and 15. In these compounds the metal atom (M) is bonded to C-atom by M–C covalent σ-bond. Common examples are $R_2M$ ($R=CH_3$, M=Zn, Cd, Hg), Grignard reagents (RMgX), diethylzinc $[(C_2H_5)_2Zn]$, tetraethyltin $[(C_2H_5)_4Sn]$, tetramethyltin $[(CH_3)_4Sn]$, tetraethyl lead, TEL$[(C_2H_5)_4Pb]$, dimethyl cadmium $[(CH_3)_2Cd]$, trimethyl gallium $[(CH_3)_3Ga]$, trimethyl indium $[(CH_3)_3In]$, etc.

$(C_2H_5)_4Pb$ is used as an anti-knock additive in petrol and $Hg(CH_3)_2$ is used as pollutant.

## 3. Polymeric or Electron-deficient Organometallic Compounds: Organometallic Compounds Having Multicentre Bonds

These compounds are electron-deficient and hence occur in polymeric forms. Examples are $(LiCH_3)_4$, $[Be(CH_3)_2]_n$, $[Al(CH_3)_3]_2$, etc. Since the bonding in these compounds can be explained neither by ionic bonding nor by normal covalent bonding, these compounds are sometimes called non-classically bonded compounds and can be considered as intermediates between ionic compounds of alkali metals and σ-bonded compounds of Si, Sn, Pb etc.

## 4. Organometallic Compounds of Transition Metals with π-Bonded Ligands: π-Bonded Organometallic Compounds

This category includes the organometallic compounds of transition metals with π-bonded ligands like alkanes (olefins), alkynes, cyclopentadiene ($C_5H_6$) etc which have electrons in their π-molecular orbitals. Since π-molecular orbital in π-bonded ligands is spread over all the C-atoms of the ligand, the transition metal is linked with the ligand through all the C-atoms of the ligand. The bonding in these organometallic compounds cannot be explained by the concept of ionic and normal covalent bonding. In these compounds overlapping of the filled π-molecular orbital of the ligand with the vacant orbital of the metal takes place.

Examples of π-bonded organometallic compounds are given below:

(i) *bis* (cyclopentadienyl) iron (II) (this compound is also called *ferrocene*), $(\pi\text{-}C_5H_5)_2$ Fe or $(\eta^5\text{–}C_5H_5)_2$ Fe

(ii) Zeise's salt, $K[PtCl_3(H_2C{=}CH_2)]$ or $K[PtCl_3(\pi - C_2H_4)]$ or $[PtCl_3(\eta^5 - C_2H_4)]$

(iii) Dibenzene chromium, $Cr(C_6H_6)_2$ or $Cr(\eta^6\text{–}C_6H_6)$

(iv) $C_6H_4$ bonded to $H_2$N-Pt $Cl_2$ (t-but-C ≡ C-t-but) and to $CH_3$, or $CH_3\ C_6\ H_4\ NH_2\ PtCl_2\ C_2$ (t-but)$_2$

Ferrocene and dibenzene chromium both are sandwich compounds. In the above compounds, the number following η (η is called *eta*. which is a Greek letter) indicates the number of C-atoms of the ligand with which transition metal is linked.

## 5 Fluxional Organometallic Compounds

There are several organometallic compounds in which one molecule of the ligand is bonded to the metal atom through one of its C-atoms forming M–C σ-bond while the other molecule of the same ligand is bonded to the metal through all its C-atoms forming M–C π-bond. Thus one molecule of the ligand acts as a π-bonded ligand ($\eta^1$-ligand) and other molecule of the same ligand behaves as a π-bonded ligand ($\eta^n$-ligand). M–C σ-bond goes on exchanging its position rapidly.

**Example.** $Fe(CO)_2\ (C_5H_5)_2$ is a mixed organometallic compound and has two $C_5H_5$ groups as ligands. The structure of this molecule is shown in Fig. 31.1. This figure shows that one $C_5H_5$ ligand acts as a σ-bonded ligand, since it is linked with Fe atom through one C-atom only. Other $C_5H_5$ ligand behaves as a π-bonded ligand because it is linked with Fe atom through all its five C-atom. Thus the compound should be formulated as $(\pi\text{-}C_5H_5)\ Fe\ (CO)_2\ (\sigma\text{-}C_5H_5)$. Fe-$C^1$ σ-bond formed by σ-bonded ligand goes on exchanging its position rapidly through 1, 2 shifts. Thus at one moment Fe–C σ-bond is between Fe and $C^1$(Fe-$C^1$ σ-bond) and at the next moment this bond is between Fe and $C^2$(Fe–$C^2$ σ-bond) and so on. The organometallic compounds in which this type of M–C σ-bond exchanges sites by choosing different C-atoms at different times is called **fluxional molecule** and this property is called **fluxionality.** Several other fluxional organometallic compounds of transition metals are also known.

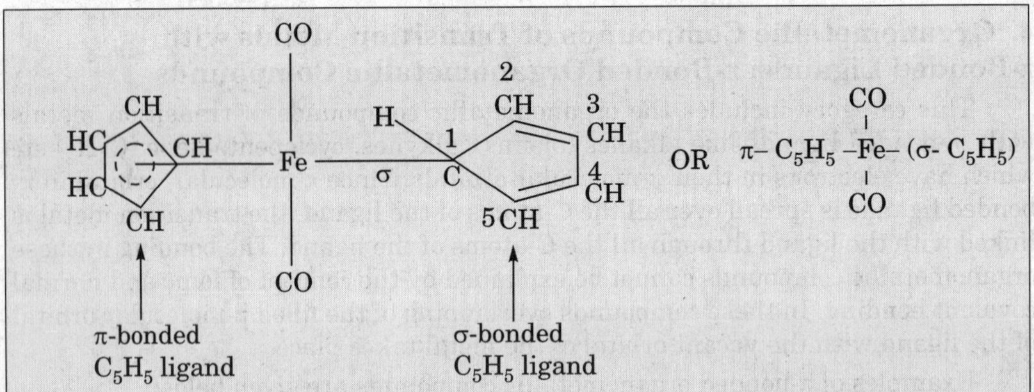


**Fig. 31.1.** Structure of $Fe(CO)_2\ (\pi\text{-}C_5H_5)\ (\sigma\text{-}C_5H_5)$

### 6. Metal Carbonyls

There are organometallic complexes in which transition metal atom is linked with C-atom of CO molecules. Metal–carbon bond has σ-as well as π-character. In these compounds transition metal is in zero, low positive or negative oxidation state. Metal carbonyls may be mononuclear [*e.g.* $Cr(CO)_6$, $Fe(CO)_5$, $Ni(CO)_4$ etc.] or polynuclear [*e.g.* $Mn_2\ (CO)_{10}$, $Fe_2(CO)_9$, $Fe_3(CO)_{12}$, $CO_4(CO)_{12}$ etc]. Mononuclear carbonyls contain terminal CO(–CO) groups while polymuclear carbonyls contain terminal CO groups (–CO) as well as bridged CO (>C=O) groups. Polynuclear carbonyls contain metal-metal bond(s).

## 7. Organometallic Compounds Containing M=C Double Bond: Y lids

Y lids are those organometallic compounds in which a transition metal or a representative element is linked with C-atom of the ligand by a double bond, *i.e.*, these compounds contain M=C bond. Examples of such compounds are:

(*i*) **Carbene organometallic compounds** which are represented by the general formula, $(CO)_5M{=}CR(OCH_3)$ where M=Cr, Mo, W, Fe and $R{=}CH_3$ or $C_6H_5$. In these compounds $=C\begin{smallmatrix}R \\ OCH_3\end{smallmatrix}$ is the carbene ligand. The first carbene complex synthesised by E. O. Fisher in 1964 is $(CO)_5\ W = CR(OCH_3)$. At present a large number of carbene complexes are known. These compounds have octahedral geometry (See Fig. 31.2) with one octahedral position occupied by carbene ligand, $= CR\ (OCH_3)$.

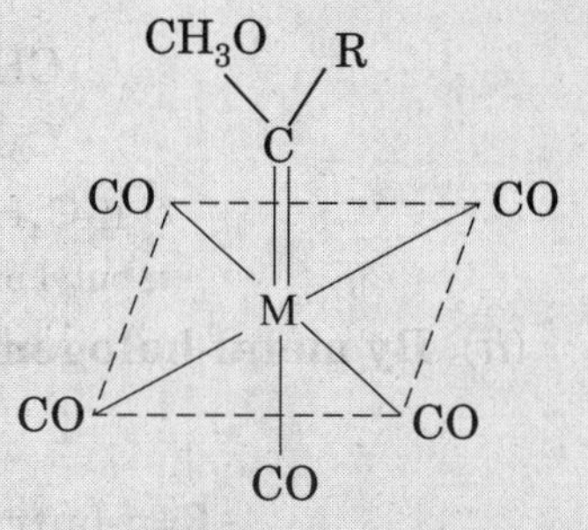


**Fig. 31.2.** Octahedral structure of carbene organometallic compound, $(CO)_5\ M = CR\ (OCH_3)$ where M = Cr, Mo, W, Fe and R = $CH_3$ or $C_6H_5$.

It has been found that the metal-carbon bond is somewhat shorter than the normal metal-alkyl σ-bond but larger than a metal-carbon double bond. This suggests that a limited double bond character exists in the metal-carbon bond.

(*ii*) **Phosphorus Y lids.** Examples are $Ph_3P{=}CH_2$ (called Wittig reagent) and $(CH_3)_3\ P{=}CH_2$. Phosphorus Y lids can be prepared as follows:

$$[Ph_3P\ CH_3]^+\ Br^- \xrightarrow{\text{Li Bu}} \underset{\text{(Wittig reagent)}}{Ph_3P = CH_2}$$

$$[(CH_3)_4P]^+\ Br^- + Na\ NH_2 \underset{< 0°\ C}{\overset{THF}{\rightleftharpoons}} (CH_3)_3P = CH_2$$

Y lids are highly reactive compounds. These serve as the starting material for the preparation of many other organophosphorus compounds. Wittig reagent, $Ph_3P{=}CH_2$ is used for the preparation of alkene from a carbonyl compound

$$\underset{\text{Carbonyl compound}}{>C = O} + \underset{\text{Wittig reagent}}{Ph_3P = CH_2} \longrightarrow \underset{\text{Alkene}}{>C = CH_2} + Ph_3\ P = O$$

## Organometallic Compounds of Lithium: Organolithium Compounds

Organolithium compounds are those compounds which contain Li–C bond. For example:

| $CH_3$-Li | $CH_3$-$CH_2$-Li | $CH_3$-$CH_2$-$CH_2$-Li | $C_6H_5$-Li |
|---|---|---|---|
| Methyllithium | Ethyllithium | n-propyllithium | Phenyllithium |

**Preparation.**

(*i*) **By alkyl halide-lithium reaction in ether**

$$RX + 2Li \rightleftharpoons RLi + LiX$$

For example:

$$CH_3I + 2Li \xrightarrow[-10°C]{\text{Ether}} CH_3Li + LiI$$

$$C_2H_5\,Br + 2Li \longrightarrow C_2H_5Li + Li\,Br$$

$$\underset{\text{n-butyl bromide}}{n\text{-}C_4H_9\,Br} + 2Li \longrightarrow \underset{\text{n-butyl lithium}}{n\text{-}C_4H_9\,Li} + Li\,Br$$

(*ii*) **By metal-halogen exchange reactions**

$$R'X + RLi \longrightarrow R'Li + RX$$

R=n-butyl and R′X = an aryl bromide or iodide.

(*iii*) **By metal-metal displacement reactions**

$$R_2Hg + 2Li \longrightarrow 2RLi + Hg$$

(*iv*) **By metallation reactions**

$$RH + n\text{-}BuLi \longrightarrow RLi + \underset{n\text{-butane}}{n\text{-}BuH}$$

## Properties

(*i*) The monomeric organometallic compounds of alkali metals are not very stable thermodynamically. However, polymeric compounds are comparatively stable. For example, some polymeric lithium alkyls do not decompose even up to 100°C.

(ii) Organolithium derivatives undergo reactions similar to those of Grignard reagents.

(*iii*) Hydrolysis (to hydrocarbon) with water is rapid.

(*iv*) With halogen compounds, R groups are replaced by halogen. For example:

$$R'_3SiCl + RLi \longrightarrow R'_3SiR + LiCl$$

(*v*) Unlike Grignard reagents, organolithium compounds add to the double bonds in alkenes. For example:

$$\underset{\text{Alkene}}{H_2C{=}CH{-}Ph} + RLi \longrightarrow R{-}CH_2{-}CH(Ph)Li$$

The product formed above reacts with more quantity of alkene and gives a polymer.

$$R{-}CH_2{-}CH(Ph)Li + H_2C{=}CH(Ph) \longrightarrow \underset{\text{Polymeric compound}}{R{-}CH_2{-}CH(Ph) + CH_2{-}CH(Ph)Li}$$

## Properties of Alkyl Lithium (RLi)

The increased reactivity of organolithium compounds over Grignard reagents is due to the greater polar character of C-Li bond in comparison to C-Mg bond. Organolithium compounds are particularly sensitive towards air and moisture. They react with compounds containing active hydrogens, aldehydes, ketones, carbon dioxide, and ethylene oxide in the same way as do Grignard reagents.

(*i*) **Reaction with water**. Organolithium compounds react with water to form alkanes.

$$\underset{\text{Methyllithium}}{CH_3—Li} + H—OH \longrightarrow \underset{\text{Methane}}{CH_4} + LiOH$$

(*ii*) **Reaction with ethylene oxide**. They react with ethylene oxide to give an addition product which on hydrolysis forms primary alcohols.

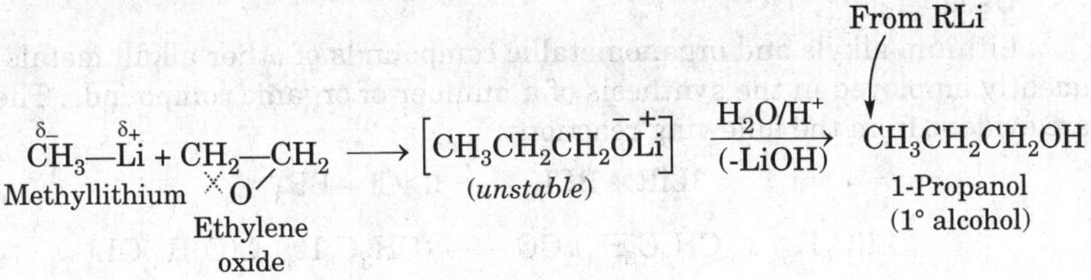


(*iii*) **Reaction with aldehydes**. Primary or secondary alcohols are obtained.

(*a*) Formaldehyde reacts with organolithium compounds to give addition products which on hydrolysis yield primary alcohols.

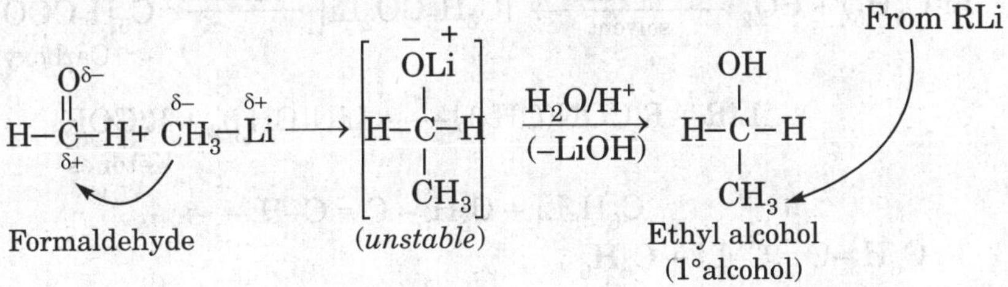


(*b*) Other aldehydes react with organolithium compounds to give addition products which on hydrolysis yield secondary alcohols.

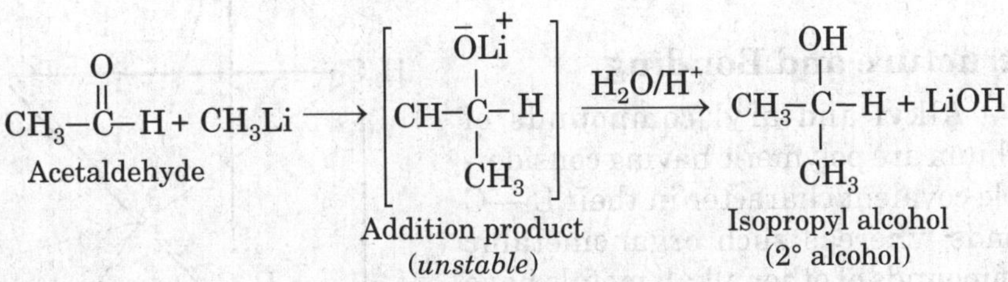


(*iv*) **Reaction with ketones**. Organolithium compounds react with ketones to give addition products which on hydrolysis form tertiary alcohols.

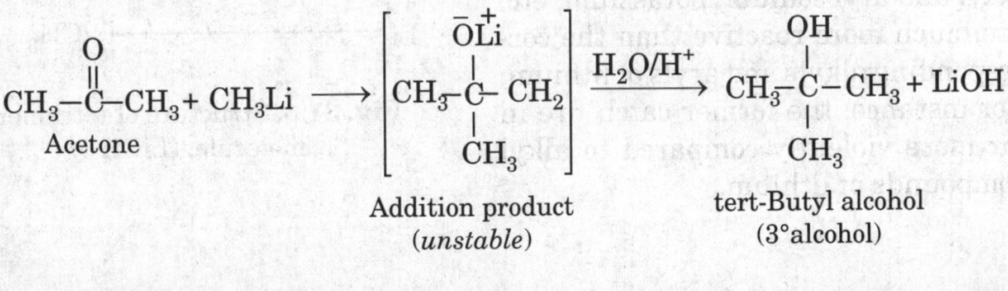

(*v*) **Reaction with carbon dioxide**. Organolithium compounds react with carbon dioxide (O=C=O) to give addition products which on hydrolysis yield carboxylic acids.

$$\underset{\text{Methyllithium}}{CH_3Li} + O{=}C{=}O \longrightarrow \underset{\substack{\text{Addition product}\\ (\textit{unstable})}}{\left[ CH_3{-}\overset{\overset{\displaystyle O}{\|}}{C}{-}\overset{-}{O}\overset{+}{Li} \right]} \xrightarrow{H_2O/H^+} \underset{\text{Acetic acid}}{CH_3{-}\overset{\overset{\displaystyle OH}{\|}}{C}{-}OH} + LiOH$$

### Uses

Lithium alkyls and organometallic compounds of other alkali metals are frequently employed in the synthesis of a number of organic compounds. These uses are evident from the following reactions:

$$3LiR + BCl_3 \longrightarrow 3LiCl + BR_3$$

$$Li(C_6H_5) + (CH_3C_6H_4)_2CO \longrightarrow (CH_3C_6H_4)_2C(C_6H_5)OLi$$

$$\xrightarrow{H_2O} \underset{\text{Tertiary carbinols}}{(CH_3C_6H_4)_2C(OH)C_6H_5} + LiOH$$

$$Li(C_6H_5) + CO_2 \xrightarrow[\text{solvent}]{\text{Non-polar}} [C_6H_5CO_2Li] \xrightarrow{H_2O} \underset{\text{Carboxylic acid}}{C_6H_5COOH} + LiOH$$

$$LiR + R'CON(CH_3)_2 \longrightarrow LiN(CH_3)_2 + \underset{\text{Ketones}}{R'COR}$$

$$C_6H_5Li + C_6H_5{-}C \equiv C{-}H \longrightarrow$$
$$C_6H_5{-}C \equiv C{-}Li + C_6H_6$$

$$n\text{-}BuLi + p\text{-}BrC_6H_4N(CH_3)_2 \longrightarrow$$
$$n\text{-}BuBr + p\text{-}LiC_6H_4N(CH_3)_2$$

## Structure and Bonding

Alkyl and aryl compounds of lithium are polymeric having considerable covalent character in their Li—C bonds whereas such organometallic compounds of other alkali metals have predominantly ionic character so that these can be represented as $M^+R^-$. The alkyl and aryl sodium, potassium, etc. are much more reactive than the corresponding alkyls and aryls of lithium. For instance, the former catch fire in air more violently compared to alkyl compounds of lithium.

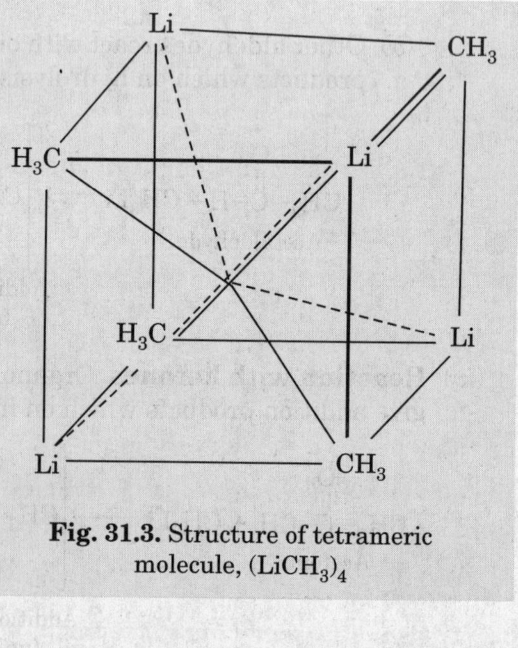


**Fig. 31.3.** Structure of tetrameric molecule, $(LiCH_3)_4$

$LiCH_3$ is an electron-deficient molecule and hence exists as tetrameric (polymeric) molecule, $(LiCH_3)_4$. The bonding in this tetrameric molecule cannot be explained either by ionic bonding or by covalent bonding. The structure of this molecule is shown in Fig. 31.3.

This structure can be considered as derived from a cube in which two tetrahedra, one of $(CH_3)_4$ and the other of $Li_4$, are interlinked with each other. These tetrahedra have a common centre. Each Li-atom is linked with three $CH_3$ groups and each $CH_3$ group is associated with three Li atoms. $LiCH_3$ is loosely called electron-deficient compound.

## Organometallic Compounds of Beryllium

Beryllium forms a variety of organometallic compounds. The dialkyl beryllium ($BeR_2$) compounds can be conveniently prepared in the following manner:

$$2LiR + BeCl_2 \xrightarrow[\text{ether}]{\text{Solvent}} BeR_2 + 2LiCl$$

$$Be + HgR_2 \xrightarrow{\Delta} BeR_2 + Hg$$

In the above reactions R is an alkyl group. These dialkyl compounds are colourless solids or viscous liquids.

**Beryllium dimethyl,** $Be(CH_3)_2$ is prepared by the action of $LiCH_3$ or $CH_3MgX$ on $BeCl_2$.

$$BeCl_2 + 2LiCH_3 \longrightarrow Be(CH_3)_2 + 2LiCl$$

$$BeCl_2 + 2CH_3MgCl \longrightarrow Be(CH_3)_2 + 2MgCl_2$$

It is a sublimable white solid. In vapour state, beryllium dimethyl consists of monomer, dimer and trimer; the preparation of each species depending on the temperature. It is spontaneously inflammable in air and reacts vigorously with water.

It is an electron-deficient compound and forms addition compounds with electron-donors such as ethers and amines.

Beryllium dimethyl is stable upto more than 200°C but at higher temperatures it decomposes, depositing beryllium.

Dimethyl beryllium is polymeric and contains Be–$CH_3$–Be bridges in its structure. The extent of polymerisation of other dialkyl beryllium compounds decreases with increase in the bulk of the alkyl group, so much so that di-*t*-butyl beryllium is monomeric.

**Beryllium diphenyl,** $Be(C_6H_5)_2$ is formed from the reaction of beryllium on mercury diphenyl in xylene for several hours:

$$Hg(C_6H_5)_2 + Be \longrightarrow Be\,(C_6H_5)_2 + Hg$$

It is a colourless solid that decomposes without melting at 160°C. It oxidises with incandescence in air. Beryllium diphenyl is soluble in various organic solvents and forms addition compounds with donor molecules such as ethers. In ethers it reacts with lithium phenyl to form a stable complex $Li\,[Be(C_6H_5)_3]$.

**Alkyl beryllium hydrides,** RBeH are interesting because they show an electron-deficient $BeH_2Be$ structure. The structure differs from the similar boron compound in as much as bridge bonds are not cleaved in reaction with bases.

**Beryllium complexes.** Beryllium also forms complexes with cyclopentadiene. The cyclopentadienyl group is linked with the metal through a sigma or a pi bond as in [Be($\eta^5$–$C_5H_5$)X] (where X is Cl, Br or I) and in [Be($\eta^1$–$C_5H_5$) ($\eta^5$–$C_5H_5$)] respectively.

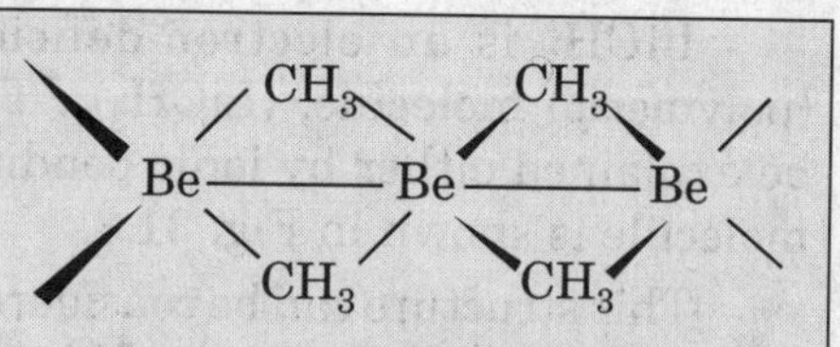


**Fig. 31.4.** Structure of polymeric molecule, $[Be(CH_3)_2]_n$

**Structure of $(CH_3)_2$ Be molecule**. As said above $(CH_3)_2$ Be molecule is polymeric and hence is represented as $[Be(CH_3)_2]_n$ whose structure is given in Fig. 31.4.

In the structure of $[Be(CH_3)_2]_n$ two Be-atoms are linked together through $CH_3$ bridges. Each Be-atom and C-atom of the bridging $CH_3$ group are $sp^3$ hybridised. Be-$CH_3$-Be bond is called bridging bond. This bond is formed by the overlap of two $sp^3$ hybrid orbitals on two Be-atoms and one $sp^3$ hybrid orbital on *C*-atom of the bridging $CH_3$ group. This bond has two electrons and three nuclei *viz* two Be-atoms and one C-atom of the bridging $CH_3$ group. Thus this bond is called *three-centre-two-electron bond* and hence is represented as (3c-2e) bond. The formation of this bond has been shown in Fig. 31.5.

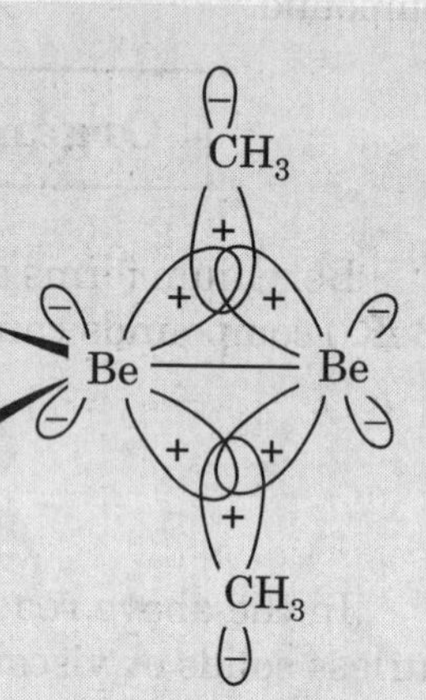


**Fig. 31.5.** Formation of three-centre-two-electron, (3c–2c) bond in polymeric molecule, $[Be(CH_3)_2]_n$

## Organometallic Compounds of Magnesium

Magnesium forms main three types of organometallic compounds: (*i*) Compounds of $MgR_2$ type (R=alkyl or aryl group), (*ii*) Grignard reagents, RMgX and (*iii*) Organometallic compounds containing one alkyl (R) and one alkoxy group (OR′), R—Mg—OR′.

The dialkyl and diaryl magnesium compounds can be prepared in the following manner:

$$LiR + R—Mg—X \xrightarrow[\text{ether}]{\text{Solvent}} MgR_2 + LiX$$

$$HgR_2 + Mg \xrightarrow[\text{ether}]{\text{Solvent}} MgR_2 + Hg$$

**Magnesium dimethyl**, Mg $(CH_3)_2$ can be made from reaction of magnesium with mercury dimethyl.

$$Hg(CH_3)_2 + Mg \longrightarrow Mg\,(CH_3)_2 + Hg$$

or by the action of methyl iodide on a calcium-magnesium alloy in ether:

$$2CH_3I + Ca - Mg \longrightarrow Mg\,(CH_3)_2 + CaI_2$$

Magnesium dimethyl is a white, infusible solid, having vapour pressure of about 0.2 mm at 190°C. Apparently it is highly associated, presumably through methyl bridging. It is spontaneously inflammable in air and reacts very vigorously with water. It undergoes many reactions of a Grignard reagent. $Mg(CH_3)_2$ is polymeric. It contains Mg—$CH_3$—Mg bridges in its structure.

**Magnesium diphenyl,** $Mg(C_6H_5)_2$ is formed by heating of magnesium with mercury diphenyl in a sealed tube:

$$Hg(C_6H_5)_2 + Mg \longrightarrow Mg\,(C_6H_5)_2 + Hg$$

It is a stable solid, decomposing at 280°C with the formation of diphenyl and magnesium. It is insoluble in benzene but has some solubility in ether, forming addition compounds. It reacts with lithium phenyl in ether solution forming lithium magnesium phenyl, $LiMg(C_6H_5)_3$.

$$LiC_6H_5 + Mg(C_6H_5)_2 \xrightarrow{\text{Ether}} LiMg(C_6H_5)_3$$

Another class of organometallic compounds of magnesium, called *Grignard reagents*, can be synthesised by refluxing alkyl or aryl halides with magnesium in dry solvent ether:

$$RBr + Mg \xrightarrow[\text{ether (dry)}]{\text{Solvent}} RMgBR$$

Grignard reagents are of great synthetic value in organic chemistry.

Organometallic compounds containing one alkyl and the other alkoxy group attached with Mg can be easily synthesised by reacting dialkyl magnesium compounds with alcohols in the following manner:

$$MgR_2 + R'OH \longrightarrow R\text{—}Mg\text{—}OR' + ROH$$

These compounds are polymeric and contain alkyl and alkoxy bridges between Mg atoms. The degree of polymerisation depends upon the bulk of the alkyl and alkoxy groups.

$CH_3$—Mg—OR (R is $CH_3$ or $C_2H_5$) exists in two polymeric forms whose structures are given below in Fig. 31.6 (a) and (b).

It should be noted that the lines shown between Mg and alkyl ($CH_3$) or alkoxy (OR) do not depict the exact bond order.

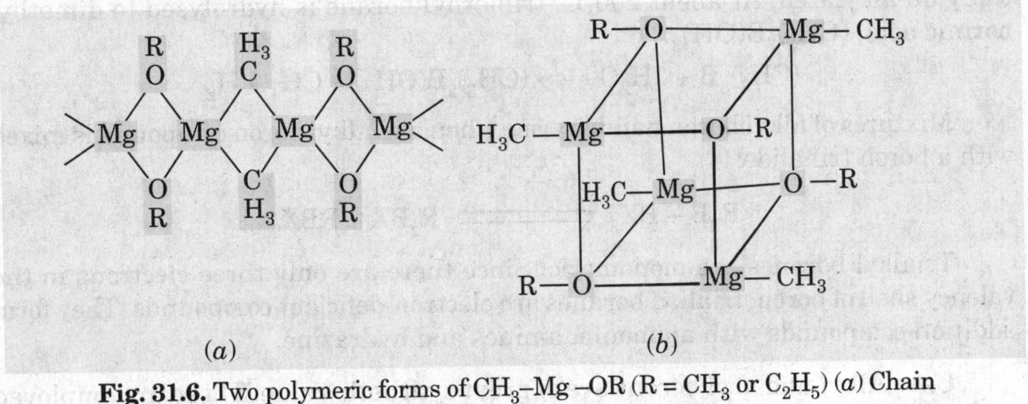

**Fig. 31.6.** Two polymeric forms of $CH_3$–Mg –OR (R = $CH_3$ or $C_2H_5$) (*a*) Chain polymer (*b*) Three-dimensional polymer.

# Organometallic Compounds of Boron: Organoboron Compounds

Boron forms the following types of organometallic compounds:

## I. Alkyl or Aryl Borons, $R_3B$.

These compounds are also called *alkyl* or *aryl boranes*.

**Preparation.** Trialkyl or triaryl boranes are prepared as follows:

(*i*) $BX_3 + 3LiR \longrightarrow BR_3 + 3LiX$ (R = alkyl or aryl group).

Actually, in this reaction we get a mixture of $BRX_2$, $BR_2X$ and $BR_3$.

(*ii*) $$H_2C=CH_2 + BH_3 \xrightarrow{THF} (CH_3-CH_2)BH_2 \xrightarrow[H_2C=CH_2]{THF}$$

$$(H_3C-CH_2)_2BH \xrightarrow[H_2C=CH_2]{THF} (H_3C-CH_2)_3B$$

(*iii*) Trialkyl boranes can also be prepared, for example, from reaction of 1-butene with diborane.

$$\underset{\text{1-butene}}{6CH_3CH_2CH = CH_2} + \underset{\text{diborane}}{B_2H_6} \longrightarrow 2(CH_3CH_2CH_2CH_3)_3B$$

**Properties.** Trimethyl borane is a gas while triethyl and tri-*n*-propyl boranes are liquids and triphenyl borane is a solid. The lower trialkyl boranes take fire spontaneously in air, the cautious oxidation of trialkyl boranes yields borate esters. Thus, oxidation with hydrogen peroxide yields a triakyl borate:

$$R_3B \ ^{(Na^+\ \bar{C}\equiv H)}\ B(OR)_3$$

Triphenyl borane $(C_6H_5)_3B$ is not spontaneously inflammable, but is very easily oxidised and must be protected from air. Trialkyl and triaryl boranes are very reluctant to react with water and only at high temperatures they are partially de-alkylated. At about 200°C, trimethyl borane is hydrolysed to dimethyl borinic acid, $(CH_3)_2B(OH)_3$:

$$(CH_3)_3B + 3H_2O \longrightarrow (CH_3)_2B(OH)_3 + CH_4 + H_2$$

Mixtures of alkyl boron halides arise when a trialkyl boron compound is mixed with a boron trihalide:

$$R_3B + BX_3 \rightleftharpoons R_2BX + RBX_2$$

Trialkyl boranes are monomeric. Since there are only three clectrons in the valency shell of boron, trialkyl boranes are electron-deficient compounds. They form addition compounds with ammonia, amines and hydrazine.

**Uses** Alkyl and aryl borons are of great synthetic value. These are employed in the synthesis of hydrocarbons, alcohols, phenols, carbonyl compounds and carboxylic acids.

$$R_3B \xrightarrow{H^+} RH$$

$$R_3B \xrightarrow{H_2O} ROH$$

$$R_3B \xrightarrow[\text{(ii) } H_2O_2]{\text{(i) } CO} R_3COH$$

$$R_2BH \xrightarrow[\text{(ii) } H_2O_2]{\text{(i) } CO} R_2CO$$

$$RBH_2 \xrightarrow[\text{(ii) } H_2O_2]{\text{(i) } CO} RCOOH$$

## II. Boronic and Boronous Acids.

Boronic acid is $B(OH)_2R$ and boronous acid is $B(OH)R$. These are prepared as follows:

$$BF_3 + RMgBr \xrightarrow[-MgFBr]{\text{Ether}} RBF_2 \xrightarrow[-2HF]{2H_2O} \underset{\text{Boronic acid}}{RB(OH)_2}$$

$$RBF_2 \xrightarrow[-MgFBr]{\text{RMg Br (ether)}} R_4BF \xrightarrow[-HF]{H_2O} \underset{\text{Boronous acid}}{RBOH}$$

The strength of these acids depends upon the nature of R group.

Boronic acid is stable in water but on warming it polymerises into a cyclic trimer

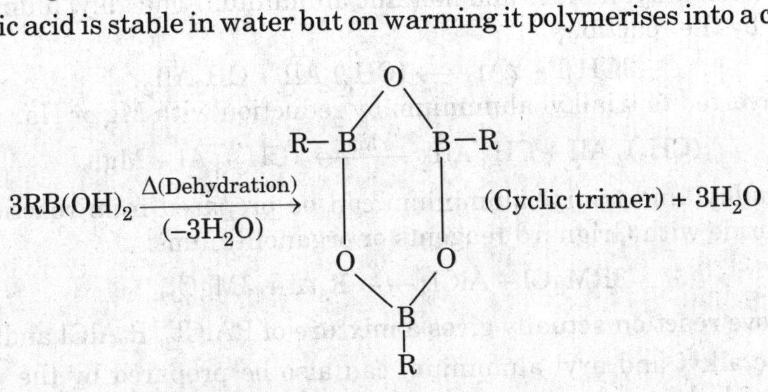


$$\underset{\text{Boronic acid}}{\text{or } 3RB(OH)_2} \xrightarrow[(-3H_2O)]{\Delta\text{(Dehydration)}} \underset{\substack{\text{Boranic}\\\text{anhydride}\\\text{or boroxine}}}{(RBO)_3} + 3H_2O$$

### III. Salts of $M^+ [BR_4]^-$ Type.

**Preparation.** (*i*) Trialkyl borons ($R_3B$) react with alkyl lithium (RLi) to form the salts of $M^+ [BR_4]^-$type.

$$\underset{\text{Trialkyl boron}}{R_3B} + \underset{\text{Alkyl lithium}}{RLi} \longrightarrow Li^+[BR_4]^-$$

(*ii*) $Li^+ [B(C_6H_5)_4]^-$ is prepared as follows:

$$\underset{\text{Triphenyl boron}}{(C_6H_5)_3B} + \underset{\text{Lithium phenyl}}{Li\ C_6H_5} \xrightarrow{\text{Ether}} Li^+ [B (C_6H_5)_4]^-$$

**Properties.** (*i*) $Li^+[B(C_6H_5)_4]^-$ is insoluble in non-polar solvents but soluble in alcohol and water. The sodium salt, $Na^+[B(C_6H_5)]^-$ is used as an analytical reagent for the gravimetic estimation of heavy alkali metal ions ($K^+$, $Rb^+$ and $Cs^+$) and a number of other cations. $NaB(C_6H_5)_4$ can cause complete precipitation of $K^+$, $Rb^+$ and $Cs^+$ from aqueous solutions of their salts:

$$NaB(C_6H_5)_4 + KCl \longrightarrow NaCl + KB(C_6H_5)_4$$

**Structure.** The $[R_4B]^-$ anion is tetrahedral in shape, with a complete octet around boron. It is isoelectronic with $R_4N^+$ and $R_4C$.

## Organometallic Compounds of Aluminium: Organoaluminium Compounds

**Preparation.** (*i*) One of the most important methods of obtaining alkyl aluminium is that involving direct reaction between metallic aluminium, hydrogen and alkenes. This probably involves addition of Al—H groups to the alkene double bond:

$$2Al + 6C_2H_4 + 3H_2 \longrightarrow 2(C_2H_5)_3Al$$

(*ii*) Another important method of preparation of alkyl aluminium is from the reaction between alkyl halides and metallic aluminium. The alkyl aluminium halides formed by the reaction,

$$3CH_3I + 2Al \longrightarrow (CH_3)_2AlI + CH_3AlI_2$$

are converted to trialkyl aluminium by reduction with Mg or Na.

$$(CH_3)_2\ AlI + CH_3\ AlI_2 \xrightarrow{Mg} (CH_3)_3\ Al + MgI_2$$

(*iii*) Trialkyl and triaryl aluminium can be prepared from reactions of aluminium chloride with Grignard reagents or organolithiums:

$$3RMgCl + AlCl_3 \longrightarrow R_3Al + 3MgCl_2$$

The above reaction actually gives a mixture of $RAlCl_2$, $R_2AlCl$ and $R_3Al$.

(*iv*) The alkyl and aryl aluminium can also be prepared by the reaction of aluminium with the appropriate organomercury compound:

$$3R_2Hg + 2Al \longrightarrow R_3Al + 3Hg$$

**Properties.** The lower aluminium alkyls are reactive liquids, inflaming in air and explosively sensitive to water. All other derivatives are similarly sensitive to air and moisture but not all are spontaneously inflammable.

$$R_3Al + 3H_2O \longrightarrow 3RH + Al(OH)_3$$

Controlled oxidation of alkyl aluminium gives aluminium alkoxides:

$$R_3Al + \frac{3}{2} O_2 \longrightarrow Al(OR)_3$$

**Structure and bonding.** The trialkyl aluminium compounds are generally dimeric and contain Al—R—Al bridges and R—Al bonds in their structures. The terminal alkyl and the bridging alkyl groups in such dimeric compounds are in a process of very fast exchange at room temperature. The dimeric structure of $(CH_3)_3$ Al has been shown in Fig. 31.7.

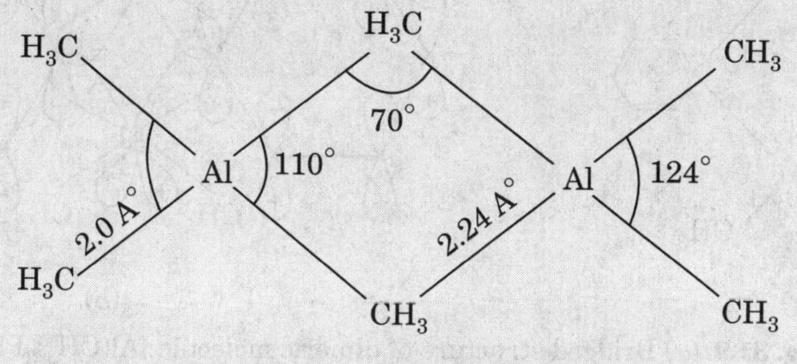


**Fig. 31. 7.** Dimeric structure of $(CH_3)_3$ Al.

Bridging alkyl carbon atoms are involved in three-centre bonds. The three-centre bond involves *sp*-hybrids from the two aluminium atoms as well as from the bridging carbon atom. The aluminium alkyl halides are also dimeric but they involve halgen bridges. Aluminium alkyls are electron-deficient compounds and behave as Lewis acids, forming tetrahedral addition compounds with donors such as amines, hydrazines, phosphines, ethers and thioethers. Aluminium alkyls also combine with lithium alkyls.

$$(C_2H_5)_3Al + LiC_2H_5 \xrightarrow{\text{in benzene}} LiAl(C_2H_5)_4$$

Triphenyl aluminium and dimethyl phenyl aluminium are dimeric containing Al–phenyl–Al bridges.

The alkyl chlorides, bromides and iodides of aluminium are also dimeric but they contain Al–halogen–Al bridges instead of Al–alkyl–Al bridges in their structures. Alkyl fluorides of aluminium have cyclic tetrameric structures having Al–F–Al bridges. The structure of $(CH_3)_2$ AlF is shown in Fig. 31.8.

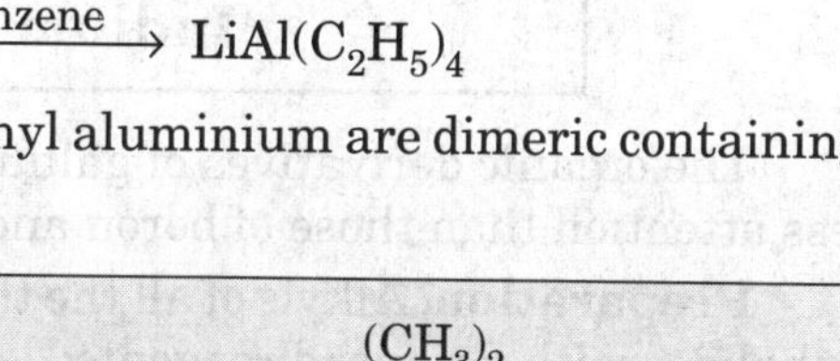


**Fig. 31.8.** Cyclic tetrameric structure of $(CH_3)_2$ AlF.

**Uses** Trimethyl and triethyl aluminium have applications as jet fuels and as additives to hydrocarbon jet fuels. Triethyl aluminium and alkyl hydrides are used together with transition metal halides or alkoxides as catalysts (*e.g.*, Ziegler catalysts) for the polymerisation of ethylene, propylene and a variety of other unsaturated compounds. They are also used widely as reducing and alkylating agents for transition metal complexes. Alkyl aluminiums have proved to be very useful reagents for the synthesis of other organometallic and organic compounds.

The structure of dimeric molecule of dimethyl phenyl aluminium, $[Al(CH_3)_2Ph]_2$ and bonding involved in it are shown in Fig. 31.9.

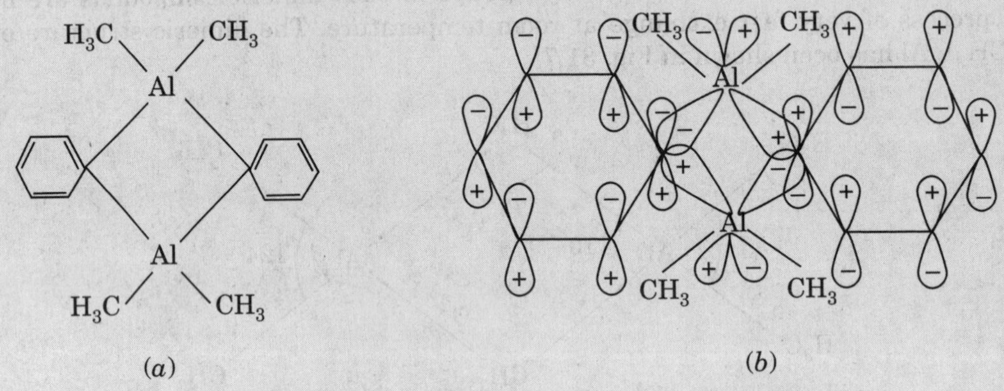


**Fig. 31.9.** (*a*) Bridged structure of dimeric molecule $[Al(CH_3)_2Ph]_2$.
(*b*) Bonding involved in $[Al(CH_3)_2Ph]_2$

Each $C_6H_5$ group forms a bridge between two Al atoms by overlapping one of its $p\pi$ orbital with two $sp^3$ hybrid orbitals of the two Al atoms, thereby forming a 3c—2e bond. The bonding between the terminal $CH_3$ and Al is a normal 2c—2e covalent bond. The geometry of each Al is tetrahedral.

## Organometallic Compounds of Gallium, Indium and Thallium

The organic derivatives of gallium, indium and thallium have received much less attention than those of boron and aluminium.

**Preparation**. Alkyls of all the three elements can be prepared from the metal trihalides with Grignard reagents:

$$GaBr_3 + 3CH_3MgBr \longrightarrow (CH_3)_3\,Ga + 3MgBr_2$$
$$InCl_3 + 3CH_3MgCl \longrightarrow (CH_3)_3\,In + 3MgCl_2$$
$$TlCl_3 + 2CH_3MgCl \longrightarrow (CH_3)_2\,TlCl + 2MgCl_2$$

With thallium only two alkyl groups are introduced in this way and the more reactive alkyl lithium is required for the introduction of the third:

$$(CH_3)_3\,Tl + CH_3Li \longrightarrow (CH_3)_3\,Tl + LiCl$$

Trialkyls of gallium and indium can also be prepared by heating dialkyl mercury compounds with metallic gallium or indium:

$$3(C_2H_5)_2\,Hg + 3Ga \longrightarrow 2(C_2H_5)_3Ga + 3Hg$$
$$3(CH_3)_2\,Hg + 3In \longrightarrow 2(CH_3)_3In + 3Hg$$

The thallium alkyls are not prepared in this way.

**Properties**. The trialkyls of Ga, In and Tl resemble those of aluminium but are increasingly less stable. One important difference is the lack of dimerisation of all alkyls of Ga, In and Tl at ordinary temperatures, except triethyl gallium, which shows some tendency to associate in benzene solution and trimethylindium, as a solid or in benzene, forms a tetramer. Hydrolysis of trialkyl gallium proceeds with some violence and all the three alkyl groups are readily removed. Cold water removes two alkyl groups from indium and only one from thallium compounds:

$$R_3Ga + 3H_2O \longrightarrow Ge(OH)_3 + 3RH$$
$$R_3In + 2H_2O \xrightarrow[\text{(cold)}]{} RIn(OH)_2 + 2RH$$
$$R_3Tl + H_2O \longrightarrow R_2TlOH + RH$$

Hot water removes all the three alkyl groups from indium trialkyls.

Trialkyls of gallium and indium form addition compounds with ether and ammonia but the adducts are much less stable than those formed by aluminium. Thallium trialkyls are very poor Lewis acids having very little tendency to accept electrons from donors. A study of the stability of the complexes of$(CH_3)$M with $(CH)_3N$ gives the acceptor strength order as; Al~Ga~In > Tl.

Trialkyl compounds with lower alkyls are spontaneously inflammable, the oxidation proceeding vigorously but still less readily than with the corresponding aluminium compounds. A number of dialkyl compounds of Ga, In and Tl are well characterised and are stable even in aqueous solutions. Thallium gives very stable ionic derivatives of the type $R_2TlX$(X = halogens, $SO_4^{2-}$, $CN^-$, $NO_3^-$, etc) which are unaffected by air and water. The ion $(CH_3)_2Tl^+$ is linear. Organoderivatives of univalent indium and thallium are known or have been suggested to exist. Preparation of $C_2H_5In$ and $C_2H_5Tl$ has been reported.

**Structure** It may be noted that Ga, In and Tl alkyls and aryls do not form dimers and the bonding in these cases can be explained by the simple 2c—2e covalent bonds. The reason for the non-existence of dimers is probably *steric*. The angle M–C–M of the bridge has to be quite small to permit a good overlap of the three orbitals of the three atoms, M, C (of alkyl or aryl) and M. For this the bulky metal atoms (*i.e.*, Ba, In, Tl, etc.) should be brought very close to one another. But this would introduce *large electrostatic repulsions* between electrons of these metal atoms. Consequently, the dimeric forms of the alkyls or aryls of Ga, In and Tl would be unstable as compared to the alkyls or aryls of Al. But the above reasoning fails to explain why $B(CH_3)_3$ does not dimerise while $BH_3$ *does*.

Trimethyl gallium $(CH_3)_3Ga$ is monomeric in vapour phase whereas the corresponding aluminium compound, containing bridged $CH_3$ groups, is dimeric. Trimethyl gallium has a *planar trigonal structure*. Gallium and aluminium have similar electronegativities (1.8 and 1.6) and terminal M–$CH_3$ bond lengths are also the same. Then why should the Al analogue exist as a dimer is not clear. $(C_6H_5)_3Ga$ is a monomer even in the solid phase.

Trimethyl indium, $(CH_3)_3In$, is monomeric in vapour phase and in solution but it is a weakly associated tetramer in the crystalline state. The structure of the tetramer is shown in Fig. 31.10.

Trimethyl thalium, $(CH_3)_3$ Tl, has a similar crystal structure although in solution the trialkyls as also the triaryls are monomeric.

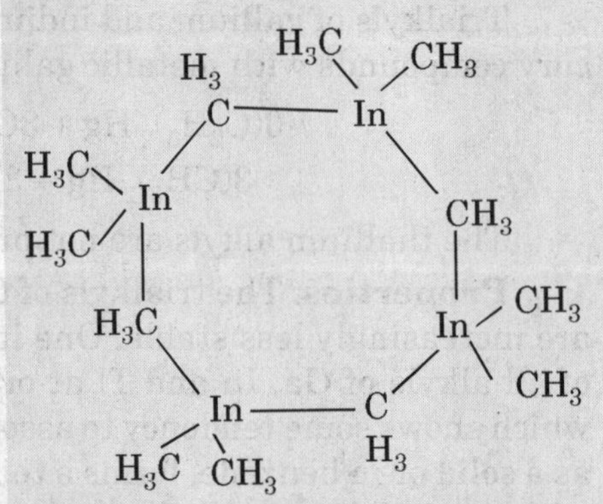

**Fig. 31.10.** Structure of tetramer, $[(CH_3)_3In]_4$ in the crystalline state.

## Organometallic Compounds of Germanium

These compounds are mainly of the general formula $R_nGeX_{4-n}$ where R is alkyl or aryl group and $n$ = 1, 2, 3, or 4. These compounds can be prepared as follows:

$$GeCl_4 + 4RLi \longrightarrow R_4Ge + 4LiCl$$

$$2RCl + Ge \longrightarrow R_2GeCl_2$$

Actually, in both the above reactions, a mixture of $R_4Ge$, $R_3GeCl$, $R_2GeCl_2$ and $RGeCl_3$ is obtained.

The extent of catenation in organogermanium compounds is much less than in the compounds of carbon and silicon. Nevertheless, quite a few organometallic compounds with Ge—Ge bonds have been synthesised as follows:

$$2C_6H_5GeBr + 2Na \xrightarrow[\text{xylene}]{\text{Boiling}} 2NaBr + C_6H_5Ge—GeC_6H_5$$

$$2(CH_3)_3GeBr + 2K \xrightarrow[\text{Liquid}]{\text{Reflux}} 2KBr + (CH_3)_3Ge—Ge(CH_3)_3$$

$$(C_2H_5)_3GeBr + NaGe(C_6H_4CH_3)_3 \xrightarrow[\text{ammonia}]{\text{Liquid}} NaBr + (C_2H_5)_3Ge—Ge(C_6H_4CH_3)_3$$

These compounds are known to be fairly stable.

Some crystalline cyclic polymers of the general formula $[Ge(C_6H_5)_2]_n$, where $n$ = 4, 5 or 6, are also known.

## Organometallic Compounds of Tin: Organotin Compounds

Organometallic compounds of tin are better studied than those of germanium and lead due to their utility in industry.

**Preparation.** Preparation of some organotin compounds is shown below:

$$SnCl_4 + 4CH_3MgCl \longrightarrow Sn(CH_3)_4 + 4MgCl_2$$

$$3SnCl_4 + 4Al(CH_3)_3 \longrightarrow 3Sn(CH_3)_4 + 4AlCl_3$$

$$Sn + 2C_2H_5X \longrightarrow (C_2H_5)_2\ SnX_2$$

In all the above reactions a mixture of organotin compounds, *viz*., $(R_4SnX_4)_4$ ($n$ = 1, 2, 3 or 4) is actually formed.

The aryl tin compounds can be prepared in a similar manner.

**Properties and structure**. Tetraalkyl and tetraaryl compounds of tin, *i.e.*, $SnR_4$, are monomeric colourless, volatile liquids or solids, whereas $R_3SnX$, $R_2SnX_2$ and $RSnX_3$ (X = halogen) exist as strongly associated species. The association of the molecules occurs through halogen bridges as shown below in the structures of trimethyl tin fluoride [Fig. 31. 11(*a*)] and dimethyl tin chloride [Fig. 31. 11(*b*)].

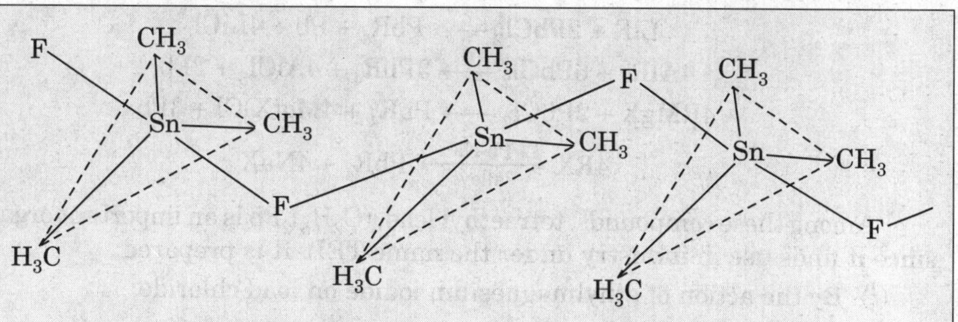

**Fig. 31. 11(*a*)** Structure of trimethyl tin fluoride

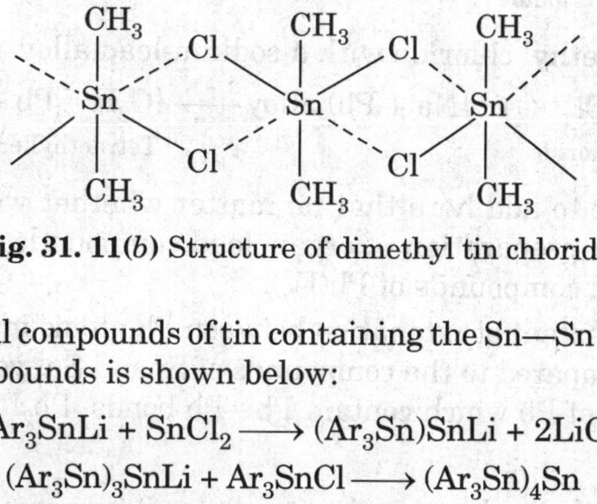

**Fig. 31. 11(*b*)** Structure of dimethyl tin chloride

There are several compounds of tin containing the Sn—Sn bonds. The synthesis of some of these compounds is shown below:

$$3Ar_3SnLi + SnCl_2 \longrightarrow (Ar_3Sn)SnLi + 2LiCl$$

$$(Ar_3Sn)_3SnLi + Ar_3SnCl \longrightarrow (Ar_3Sn)_4Sn$$

(Ar in these reactions is the aryl group)

$$2(CH_3)_3SnBr + Na(CH_3)_2Sn{-}Sn(CH_3)_2Na \xrightarrow[\text{ammonia}]{\text{liquid}}$$

$$(CH_3)_3\ Sn(CH_3)_2\ Sn - Sn\ (CH_3)_2\ Sn\ (CH_3)_3 + 2NaBr$$

Some cyclic organotin compounds such as $[Sn(CH_2)_2]_6$, $[Sn(C_2H_5)_2]_6$ and $[Sn(phenyl)_2]_6$ are also known.

**Uses**. Dialkyl tin laurate and dialkyl tin maleate are used as stabilisers of PVC. Without the addition of these organotin compounds, the PVC loses its plastic properties and becomes brittle. The stabilising action of dialkyl tin laurate and maleate may be due to the following reasons:

(*i*) With the passage of time, PVC slowly degrades and produces traces of HCl. The HCl thus produced catalyses further decomposition of PVC. However, the organotin compounds added to PVC react with the HCl produced and thus inhibit the catalytic degradation of PVC.

(*ii*) The organotin compounds act as antioxidants and thus prevent the changes which occur in PVC due to atmospheric oxygen.

## Organometallic Compounds of Lead.

**Preparation.** The tetraalkyl or tetraaryl lead can be prepared by any one of the following reactions:

$$LiR + 2PbCl_2 \longrightarrow PbR_4 + Pb + 4LiCl$$

$$4AlR_3 + 6PbCl_2 \longrightarrow 3PbR_4 + 4AlCl_3 + 3Pb$$

$$4RMgX + 2PbCl_2 \longrightarrow PbR_4 + 4Mg(X)Cl + 3Pb$$

$$4RX \xrightarrow[\text{alloy}]{\text{Pb-Na}} PbR_4 + 4NaX$$

Among these compounds, tetraethyl lead, $(C_2H_5)_2Pb$ is an important organolead, since it finds use in industry under the name TEL. It is prepared:

(*i*) By the action of ethylmagnesium iodide on lead chloride.

$$\underset{\text{Ethylmagnesium iodide}}{4C_2H_5Mgl} + 2PbCl_2 \longrightarrow \underset{\text{Tetraethyllead}}{(C_2H_5)_4Pb} + Pb + 4MglCl$$

(*ii*) By heating ethyl chloride with a sodium-lead alloy.

$$\underset{\text{Ethyl chloride}}{4C_2H_5Cl} + (4Na + Pb)\text{ alloy} \longrightarrow \underset{\text{Tetraethyllead}}{(C_2H_5)_4Pb} + 4NaCl$$

It is pertinent to add here that no matter whether we take Pb(II) or Pb(IV) compounds, for the preparation of organolead compounds, we almost always end up with organolead compounds of Pb(IV).

Although the extent of catenation between Pb atoms in organolead compounds is very small as compared to the compounds of Si and Sn, yet there are few dimeric organo compounds of Pb which contain Pb—Pb bonds. $Pb_2R_6$ is an example of such compounds.

**Properties** Tetraalkyl and tetraaryl compounds of lead are monomeric volatile liquids which are quite stable. Tetraethyl lead is the most widely used organolead compound. It is mixed in petrol and serves as an excellent anti-knock in automobile engines.

Tetraethyllead is a colourless liquid, bp 202°C. Like other lead compounds, it is a highly toxic substance. It is soluble in organic liquids like gasoline but is insoluble in water. Being largely covalent in nature, tetraethyllead has low reactivity.

**Use of $(C_2H_5)_4$Pb as an anti-knock** Tetraethyllead is an important *anti-knock* additive in gasoline (petrol) used in automobile engines. It improves its *Octane Number*. One disadvantage in using this gasoline additive is that its combustion product, lead oxide, gets reduced to metallic lead. This is deposited inside the cylinders of the engine, causing clogging of the cylinder valves. To avoid accumulation of metallic lead in engine cylinders, ethylene bromide ($BrCH_2CH_2Br$) is added along with tetraethyllead into gasoline. The ethylene bromide reacts with the lead in the engine and converts it into gaseous lead bromide ($PbBr_2$). The lead bromide is discharged through the exhaust pipe, However, lead compounds are very poisonous and automobile exhaust poses a big health hazard in large cities.

## Organometallic Compounds of Zn, Cd and Hg

These metals form dialkyls and diaryls of $R_2M$ type. These metals also form alkyl metal halides (RMX).

**Preparation** (*i*) Dialkyls and diaryls of zinc, cadmium and mercury can be best prepared from the action of the appropriate Grignard reagent on metal halide:

$$2RMgCl + MCl_2 \longrightarrow R_2M + 2MgCl_2$$

where R = alkyl or aryl and M = Zn, Cd or Hg.

(*ii*) These compounds can also be prepared from reactions such as the following:

$$2CH_3I + Zn\text{-}Mg \longrightarrow (CH_3)_2Zn + MgI_2$$

$$2C_2H_5ZnI \xrightarrow{\text{Heat}} (C_2H_5)_2Zn + ZnI_2$$

$$(CH_3)_2Hg + Zn \longrightarrow (CH_3)_2\,Zn + Hg$$

$$2CH_3Li + CdCl_2 \longrightarrow (CH_3)_2Cd + 2LiCl$$

$$(C_6H_5)_2Hg + Cd \longrightarrow [(C_5H_5)_2Cd + Hg$$

$$2RNa + HgCl_2 \longrightarrow R_2Hg + 2NaCl$$

$$Hg + 2CH_3Br + 2Na \longrightarrow (CH_3)_2Hg + 2NaBr$$

(*iii*) Zinc, cadmium and mercury also form alkyl metal halides, RMX and these can be obtained from the dialkyl metals by the reaction with metal dihalides:

$$R_2M + MX_2 \longrightarrow 2RMX$$

(*iv*) Alkyl metal halides can also be prepared from reactions of alkyl halides with metal:

$$RX + M \longrightarrow RMX$$

or Grignard reagent with metal halides:

$$RMgCl + MCl_2 \longrightarrow RMCl + MgCl_2$$

**Properties.** Dialkyls of zinc, cadmium and mercury are colourless liquids while the diaryls of these metals are colourless solids. Dialkyls and diaryls of zinc take fire immediately on exposure to air and burn with an intense flame, yielding a white smoke of zinc oxide. The corresponding alkyls of cadmium do not ignite spontaneously in air unless favoured by a large surface area; those of mercury do not ignite at all. Yet the thermal stability of zinc alkyls is considerably higher than that of the cadmium or the mercury compounds showing that there is no direct relationship between stability and reactivity. Mercury alkyls are exceedingly poisonous. Many mercury compounds are very toxic and this biological activity has led to the synthesis of a large number of compounds in search for materials which may be of medicinal value. The alkyls of Group IIB metals are monomeric and linear. The reactivity of organometallic compounds of Group IIB metals decreases down the group. The zinc and cadmium compounds both form adducts with donor molecules such as ethers and amines. Cadmium is readily complexed than zinc and mercury compounds form no adducts at all with ethers and amines.

Alkyl and aryl mercury halides are monomeric linear molecules; conversion into hydroxides, cyanides and other derivatives is easily accomplished by ion-exchange or by the use of silver salts:

$$CH_3HgI + AgCN \longrightarrow CH_3HgCN + AgI.$$

The compounds ranging from salt-like ($RHg^+X^-$) (where X is highly electronegative such as $F^-$, $NO_3^-$, $\frac{1}{3}(SO_4^{2-})$ to non-polar compounds, soluble in organic solvents when X is less electronegative (*e.g.*, $I^-$) are known. Alkyl zinc halides are associated compounds.

**Applications**. Many alkyls of zinc have been found useful in many syntheses of zinc complexes. Cadmium compounds have proved very useful for converting carboxylic chlorides to ketones and they have essentially replaced zinc compounds for this purpose. Mercury compounds are useful in the synthesis of other organometallic compounds particularly when pure products are required. We have already mentioned about their use in the preparation of sodium and lithium compounds. Organomercury compounds may also be used as prime reagent to attach any organic group to silicon or arsenic or boron by action on their chlorides.

## Ionic and Metal–Carbon σ Bonding in Organometallic Compounds

**1. Ionic bonding in organometallic compounds**. This type of bonding is generally observed in organometallic compounds of most electropositive metals like Na, K, Rb and Cs. For example, butyl sodium ($C_4H_9Na$) and methyl potassium ($CH_3K$) are ionic compounds. Slightly less electropositive metals like Mg can also form ionic compounds when the negative charge on the organic species can be delocalised on several C-atoms. For example, since cyclopentadienyl ring ($C_5H_5^-$) can delocalise the negative charge over five C-atoms, it can form the ionic compound, $Mg^{2+}(C_5H_5^-)_2$. Cyclopentadienides of alkali metals, lanthanides and actinides, $M(C_5H_5)_3$ are essentially ionic. Ionic organometallic compounds are quite reactive and undergo easy hydrolysis and oxidation in air. They are mostly insoluble in organic solvents.

**2. Metal–carbon σ-covalent bonding in organometallic compounds** Metal alkyls contain metal-carbon σ-bonds. These σ-bonds are relatively unstable in transition metal alkyl compounds but in main group metal alkyls these bonds are not so unstable. Transition metal alkyl derivatives like $Ti(CH_3)_4$, $Ta(CH_3)_5$, $W(CH_3)_2$ are unstable and hence explode at or below room temperature. However, the stability of these compounds can be enhanced if some additional π-bonding ligands like CO, $PR_3$, amines etc. are also present. For example, $Ti(CH_3)_4$ decomposes violently at –23°C but the addition of dipyridine or diamine increases the stability of the compound so that it does not decompose upto room temperature. Similarly the stability of $Ta(CH_3)_5$ can be enhanced somewhat by adding diphosphene, *dmp* ($Me_2PCH_2\,CH_2PMe_2$). *Transition metal alkyls are kinetically unstable but are thermodynamically stable*.

## Factors Which Make Metal-Carbon σ-bond in Transition Metal Alkyls Unstable: Factors that Make Metal Alkyls Kinetically Unstable

The following factors can break metal-carbon bonds in alkyl transition metal alkyls and hence make the compounds kinetically unstable.

(*i*) **Cleavage products are reactive species.** The cleavage of metal-carbon bond in transition metal alkyls gives alkyl radicals or ions which are *very reactive*. The reactive species react with each other or solvent to give chain reactions. This behaviour is opposite to that of other inorganic metal-ligand bonds such as M–$OH_2$, M-$MR_2$, M—X (X = halogen) etc. in which the metal-ligand cleavage gives species like $H_2O$, $R_2N^-$ of $X^-$ which are *comparatively less reactive*.

(*ii*) **Cleavage products form strong bonds**. The cleavage of metal-carbon bond in metal alkyls gives species that combine among themselves and form new bonds like C–C, C–O, C–N, C–X (X= halogen) etc. which are strong. Their strength depends on the nature of transition metal alkyls. The formation of new bonds is highly exothermic and hence it is thermodynamically preferred. For example, a metal—alkyl may undergo cleavage reaction according to the following overall exothermic reaction:

$$\underset{\text{Metal alkyl}}{\text{M–R}} \xrightarrow{\text{cleavage}} \text{R–R + M-derivative + Heat}$$

The first cleavage step is

$$\text{M–R} \longrightarrow \underbrace{\mathring{\text{R}}\,\text{(Alkyl radical)} + \mathring{\text{M}}}_{\text{Reactive intermediates}}$$

This step is favoured in the forward direction. The formation of $\mathring{R}$ leads to subsequent reactions.

(*iii*) **Elimination of β-hydrogen.** Higher alkyl organometallic compounds which contain β-hydrogen (β-hydrogen is that H-atom which is present on C-atom at β-position) are generally unstable, since β-hydrogen is easily transferred from the alkyl group to the metal and hence the compound gets decomposed into metal hydride and alkene as shown below.

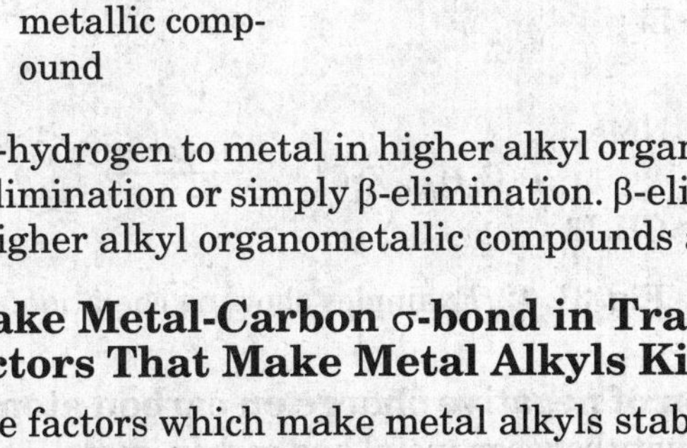


The transfer of β-hydrogen to metal in higher alkyl organometallic compounds is called β-hydrogen elimination or simply β-elimination. β-elimination is an indication of the fact that higher alkyl organometallic compounds are unstable.

## Factors Which Make Metal-Carbon σ-bond in Transition Metal Alkyls Stable: Factors That Make Metal Alkyls Kinetically Stable

Following are the factors which make metal alkyls stable:

(*i*) **Absence of β-hydrogen in the metal alkyl**. The metal alkyls which do not contain β-hydrogen would be relatively more stable. For example, metal benzyls, $M(CH_2–Ar)_n$ and metal neopentyls, $M(CH_2CMe_3)_n$ are reported to be relatively stable due to the absence of β-hydrogen in them.

(*ii*) **Increasing coordination number**. The stability of metal-alkyls can be increased by adding extra ligands which occupy the coordination positions firmly. For example, $TiMe_4$ (C.N.of Ti = 4) decomposes below 253K but the addition of bipyridine (*bipy*) increases the stability of the compound upto 330K. The increase in stability is due to the formation of $TiMe_4$(*bipy*) in which C. N. of Ti = 6. Stability of metal-alkyls is significantly enhanced by adding π-acid ligands like CO, cyclopentadienyls and phosphine. For example:

(*a*) Metal-alkyls or metal aryls are not known for Fe, Ru and Os but the presence of π-acid ligands stabilises the compounds. Thus $(\eta^5{-}C_5H_5)M(CO)_2R$ is stable. Here M = Fe, Ru and Os and R = alkyl.

(*b*) Ni, Pd and Pt alkyls ($MR_2$) are stabilised by monophosphines, ($PR_3$) in complexes, $R_2M(PR_3)_2$.

(*c*) Cr, Mo and W alkyls (MR) are stabilised as $(\eta^5{-}C_5H_5)\ M(CO)_3R$.

(*d*) Manganese-methyl σ-bond is stabilised by CO in the complex $(CH_3)Mn(CO)_5$.

(*iii*) **Chelation.** A chelate is relatively more stable. A chelate ring can be formed in any of the following ways:

(*a*) By donating through a terminal amino grouped metal-alkyl chain as shown at (a) of Fig. 31.12.

(*b*) By involving an ortho-carbon in a benzyl amine derivative as shown at (b) of Fig. 31.12.

(*c*) Through donation from an ortho-amino group in a metal-benzyl derivative as shown at (c) of Fig. 31.12.

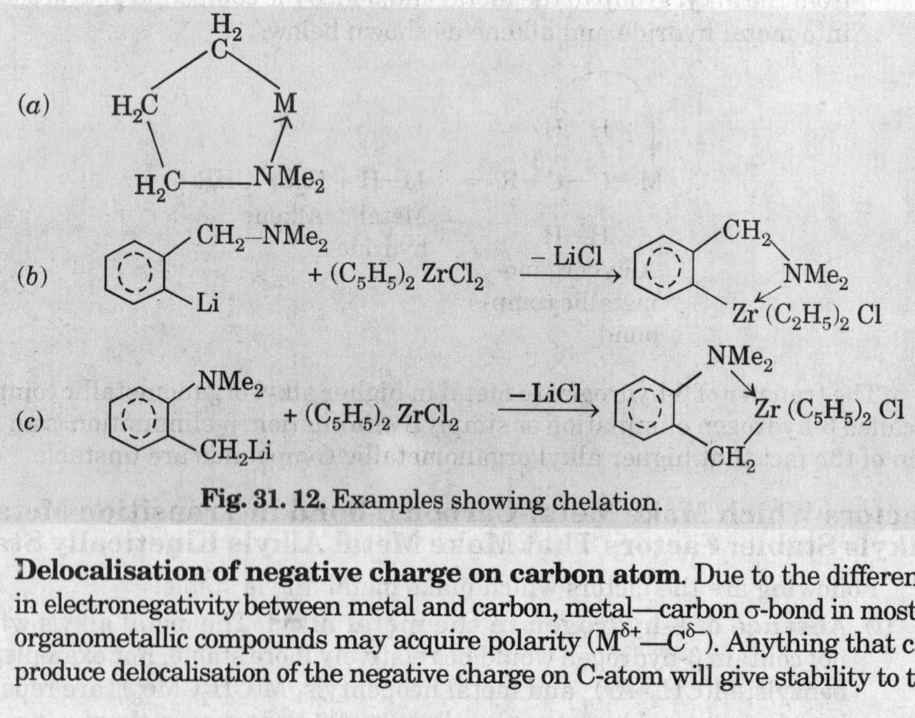


**Fig. 31. 12.** Examples showing chelation.

(*iv*) **Delocalisation of negative charge on carbon atom.** Due to the difference in electronegativity between metal and carbon, metal—carbon σ-bond in most of organometallic compounds may acquire polarity ($M^{\delta+}$—$C^{\delta-}$). Anything that can produce delocalisation of the negative charge on C-atom will give stability to the metal-carbon σ-bond. This is true of aromatics and halogen substituted species like $CF_3$, For example, it has been seen that *$M–CF_3$ bonds are generally more stable than $M–C_6H_5$ bonds which, in turn, are more stable than $M–CH_3$ bonds.*

## Hapticity ($\eta^n$) of an Organic Ligand in Organometallic Compounds

In organometallic compounds one or more C-atoms of the organic ligand are linked with the metal atom. The number of C-atoms of a given organic ligand which are attached with the metal atom in organometallic compound is called hapticity of that ligand. The hapticity of a ligand is represented by the symbol $\eta^n$ in which η is *eta* and *n* is the number of C-atoms of the ligand attached with the metal in organometallic compound. *n* represents the hapticity of the ligand. The symbol, $\eta^n$ written before the formula of the ligand indicates the hapticity of the ligand. The ligands having hapticity aqual to 1, 2, 3 etc. (*i.e.* $n$ = 1, 2, 3 etc.) are called **monohapto, dihapto, trihapto** etc. ligands.

**Examples.** (*i*) Sn $(CH_3)_4$ has four $CH_3$ organic ligands attached with Sn metal atom

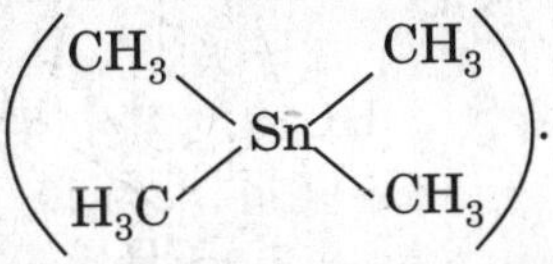


. Since only C-atom of $CH_3$ ligand is linked with Sn metal atom, the hapticity of $CH_3$ ligand in Sn $(CH_3)_4$ molecule is one (n=1), *i.e.* $CH_3$ ligand acts as a **monohapto ligand** and hence is represented as $\eta^1$-$CH_3$ .

(*ii*) The structure of the anion, $[PtCl_3(C_2H_4)]^-$ present in Zeise's salt, K $[PtCl_3(C_2H_4)]$ as given in Fig. 31.13. shows that since Pt metal atom is linked with two C-atoms of $C_2H_4$ ligand, $C_2H_4$ ligand has hapticity (η) equal to 2 which is represented as $\eta^2$–$C_2H_4$. In other words $C_2H_4$ ligand acts as a **dihapto ligand** in $[PtCl_3(C_2H_4)]^-$ ion.

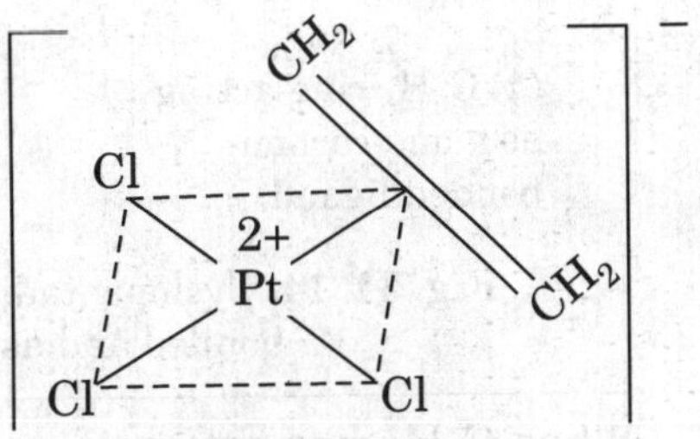


**Fig. 31. 13.** Structure of $[PtCl_3(C_2H_4)]^-$ ion in Zeise's salt, $K[PtCl_3(C_2H_4)]$

## Classification of Organic Ligands in Organometallic Compounds on the Basis of the Number of Carbon Atoms of the Ligand Attached with the Metal Atom

The number of carbon atoms of the organic ligand which are attached with the metal atom in organometallic compound is called hapticity of the ligand. The hapticity of a ligand is represented as $\eta^n$ where *n* is the number of C-atoms of the organic ligand attached with the metal in organometallic compound. On the basis of the hapticity the ligands are classified as follows:

**1. Monohapto ligands: One carbon-bonded ligands (represented as $\eta^1$ ligands).** The ligands which are attached with the metal atom through one C-atom only are called monohapto ligands. These ligands are of the following types:

(*a*) **Hydrocarbon ligands**. These ligands include alkyl ($–CH_3$, $–C_2H_5$), silyl ($Me_3$ Si), aryl ($–C_6H_5$), alkenyl ($–CR = CR_2$), σ-cyclopentadienyl ($–C_5H_5$), alkynyl ($–C\equiv CR$) groups. All these groups form compounds with transition as well as representative metals. σ-allyl group ($–H_2C–CH= CH_2$ or $σ–C_3H_5$) also acts as a one-electron donor.

Cyclopentadienyl ring ($C_5H_5$) shows dual character in different compounds. This ring acts as a one-carbon-bonded ($\eta^1$) as well as five-carbon-bonded ligand ($\eta^5$). When $C_5H_5$ ring acts as a one-carbon-bonded ligand, its attachment with the metal (M) is shown as given at (a) of Fig. 31.14. When $C_5H_5$ ring behaves as a five-carbon-bonded ligand, its attachment with the metal is shown as given at (*b*) of Fig. 31.14.

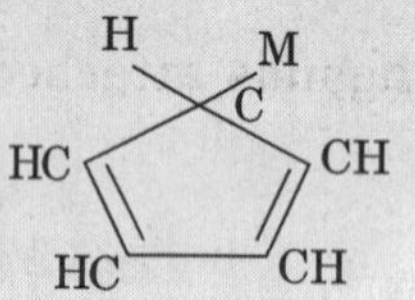


(*a*) $C_5H_5$ ring acting as a one-carbon-bonded ligand.

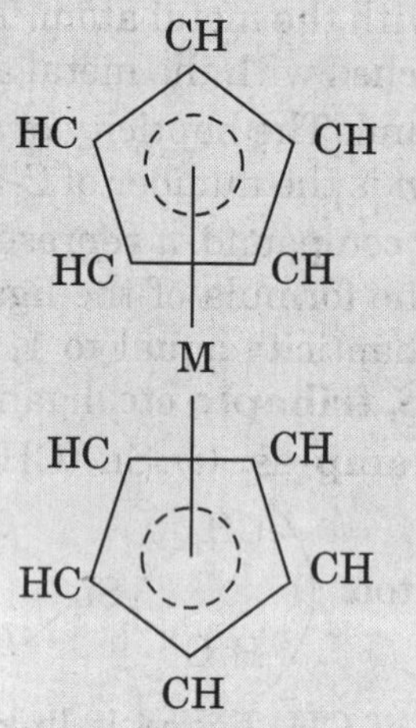


(*b*) $C_5H_5$ ring acting as a five-carbon-bonded ligand.

**Fig. 31. 14.** Cyclopentadienyl ring ($C_5H_5$) acting as a one-carbon-bonded and as a five-carbon-bonded ligand.

When $C_5H_5$ ring acts as a one-carbon-bonded ligand, this ligand is attached with the metal atom (M) through only one of its five C-atoms by M–C σ-bond Thus here $C_5H_5$ acts as a σ-ligand. Hence Na ($C_5H_5$) should be represented as Na(σ-$C_5H_5$). On the other land, when $C_5H_5$ ring acts as a five-carbon-bonded ligand, this ligand gets attached with the metal through its π-bonding molecular orbitals delocalised over all the five C-atoms of $C_5H_5$ ring. Thus here $C_5H_5$ behaves as a π-ligand. Hence Fe $(C_5H_5)_2$ should be formulated as Fe(π-$C_5H_5$)$_2$. Note that σ-$C_5H_5$ acts as a one-electron donor while π-$C_5H_5$ acts as five-electron donor.

The compounds in which $C_5H_5$ ring acts as a σ-ligand or one-carbon-bonded ligand are called **metal cyclopentadienides**. Thus Na (σ-$C_5H_5$) is called sodium cyclopentadienide. The compounds in which $C_5H_5$ acts as a π-ligand or five-carbon-bonded ligand are called **cyclopentadienyl complexes**. Thus Fe(π-$C_5H_5$)$_2$ is called *bis (cyclopentadienyl) iron*.

(*b*) **Acyl ligands**. These ligands contain acyl group (–COR) which is directly attached with the metal through its carbon atom as shown below.

Acyl group Organometallic compound

Such compounds are formed by transition metals only.

(*c*) **Carbene (alkylidene) ligands.** Carbenes are neutral species having a C–atom with two bonds and two electrons on C–atom (–$\ddot{C}$–). Methylene (H –$\ddot{C}$–H or :$CH_2$) is a carbene. Carbene can be linked with a metal through the *lp* present on C-atom to form an organometallic compound. Thus carbene is a monohapto ligand.

Carbene ligand    Organometallic compound

Organometallic compounds containing carbene ligands are generally given by transition metals.

**2. Dihapto ligands: Two carbon-bonded ligands (designated as $\eta^2$ ligands).** The ligands which are attached with the metal atom through two C-atoms are called dihapto ligands.

**Example.** Olefines or alkenes, $>C=C<$ [*e.g.* ethylene, $\frac{H}{H}>C=C<\frac{H}{H}$ or $C_2H_4$, $C_2F_4$] and alkynes, RC ≡ CR (sometimes) act as dihapto ligands because both C-atoms at each end of the multiple bond are linked with the metal atom. The bonding of such ligands to the metal is indicated either by two dotted lines to the participating C-atoms or by a single solid line from the metal atom to the centre of the multiple bond. This has been shown for $[PtCl_3(C_2H_4)]^-$ ion in Fig. 31.15.

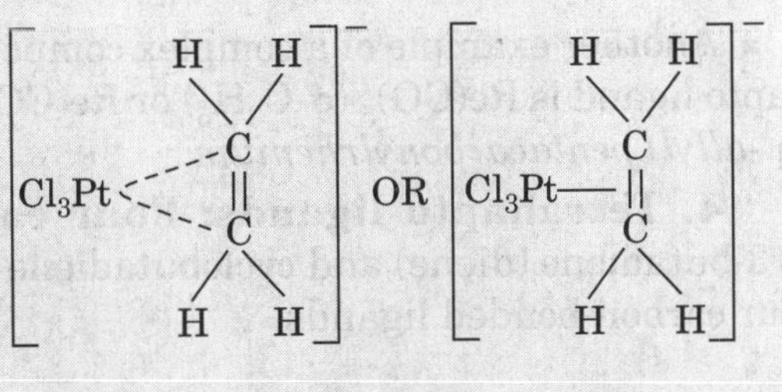


**Fig. 31.15.** Structure of $[PtCl_3(C_2H_4)]^-$ ion

Since the ligand, $C_2H_4$ is a symmetrical ligand, the two C-atoms of this ligand are equidistant from Pt metal atom. If the ligand is unsymmetrically substituted ligand like R–CH=$CH_2$, then the two C-atoms of this ligand are not equidistant from the metal atom.

Complexes of alkenes and alkynes are known only for transition metals.

**3. Trihapto ligands: Three carbon-bonded ligands** ($\eta^3$ **ligands**). Propene (propylene), $CH_2$ = CH–$CH_3$, after losing H-atom, gives *allyl* (or *enyl*) group, $CH_2$ = CH–$\dot{C}H_2$ ($C_3H_5$). This group has three π-electrons which are completely delocalised over all three C-atoms as shown below.

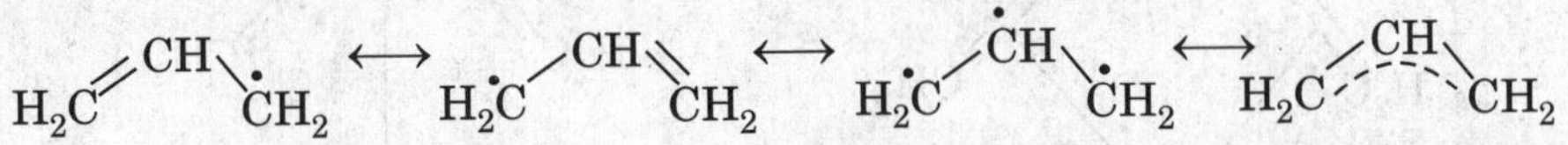


When allyl group forms an organometallic compound with a transition metal, the metal gets attached equally with all the three C-atoms of the allyl group (ligand) as is evident from the structure of $Co(CO)_3(C_3H_5)$ shown in Fig. 31.16(*a*). When the allyl group is linked with the metal atom through all its three C–atoms, it is called π-allyl group (π-$C_3H_5$). When allyl group gets attached with the metal atom through one of its C–atoms only, it is called σ-allyl group (σ-$C_3H_5$) (See Fig. 31.16(*b*).

Here it may be noted that since π-$C_3H_5$ is a three-electron ligand, it is denoted as $\eta^3$–$C_3H_5$ while σ-$C_3H_5$ is a one-electron ligand and hence is represented as $\eta^1$–$C_3H_5$. Above discussion shows that ($\sigma$-$C_3H_5$) and ($\eta^1$-$C_3H_5$) have the same meaning. Similarly ($\pi$-$C_3H_5$) and ($\eta^3$–$C_3H_5$) have the same meaning.

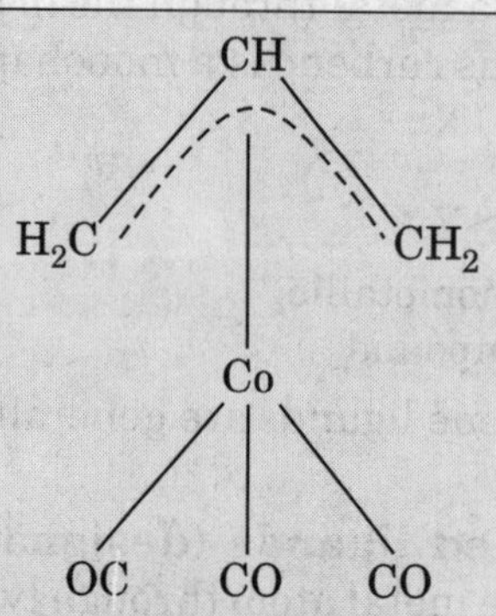

**Fig. 31.16(*a*).** Structure of $Co(CO)_3$ ($\pi$-$C_3H_5$) molecule ($\pi$-allyl complex) or $Co(CO)_3$ ($\eta^3$–$C_3H_5$)

$CH_2 = CH–CH_2–Mg–Br$

**Fig. 31.16(*b*).** Structure of allyl magnesium bromide ($\sigma$-$C_3H_5$) MgBr ($\sigma$-allyl complex) or ($\eta^1$-$C_3H_5$) MgBr.

Another example of a complex compound in which allyl group acts as a monohapto ligand is $Re(CO)_5$ ($\sigma$-$C_3H_5$) or $Re(CO)_5$ ($\eta^1$ – $C_3H_5$). This complex is named as *($\eta^1$-allyl) pentacarbonylrhenium*.

**4. Tetrahapto ligands: Four carbon-bonded ligands ($\eta^4$ ligands).** 1, 3-butadiene (diene) and cyclobutadiene whose structures are given below act as four carbon-bonded ligands.

$H_2C = CH – CH = CH_2$

1,3 -butadiene ($C_4H_6$)

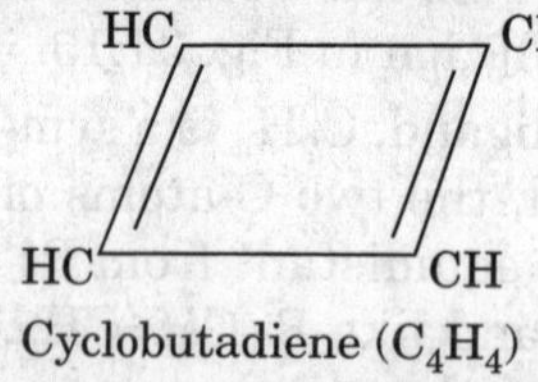

Cyclobutadiene ($C_4H_4$)

The bonding between these ligands and the metal atom occurs through the delocalised $\pi$-molecular orbitals of the ligands and orbitals of the metal. Thus

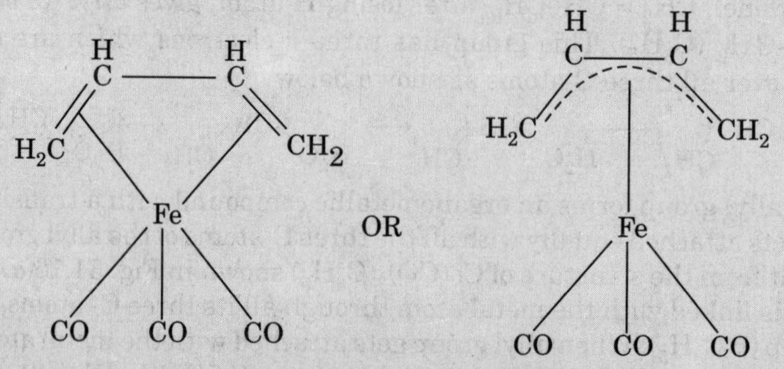

**Fig 31.17(a).** Structure of Fe $(CO)_3$ ($H_2C = CH–CH = CH_2$) or $Fe(CO)_3$ ($C_4H_6$) molecule.

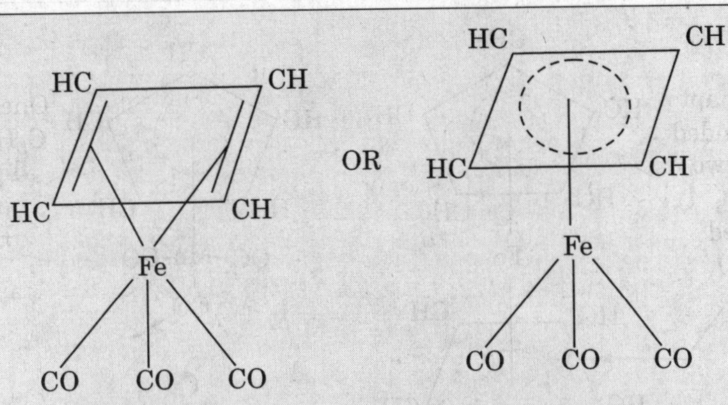


**Fig 31.17(b).** Structure of $Fe(CO)_3$ $(C_4H_4)$ molecule

metallic atom gets associated with the ligand through all its four C-atoms as is evident from the structures of $Fe(CO)_3$ $(H_2C = CH–CH = CH_2)$ or Fe $(CO)_3$ $(C_4H_6)$ and Fe $(CO)_3$ $(C_4H_4)$ given in Figures 31.17(*a*) and 31.17(*b*) respectively.

**5. Pentahapto ligands: Five carbon-bonded ligands ($\eta^5$ ligands)** In these ligands five C-atoms of the ligand are attached with the metal atom. Examples of such ligands are:

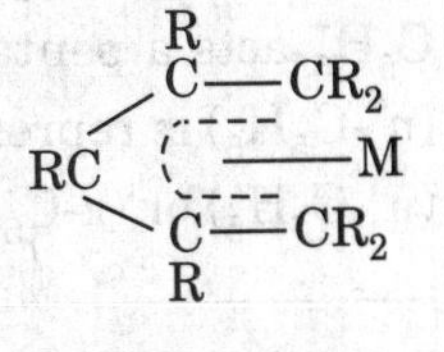


(*a*) Acyclic dienyl, $R_2\dot{C}–\dot{C}R–\dot{C}R–\dot{C}R–\dot{C}R_2$. Its attachment with the metal atom is shown in the margin.

(*b*) π-cyclopentadienyl radical ($\pi$-$C_5H_5^-$) is a five-membered cyclic dienyl and acts as five-electron donor

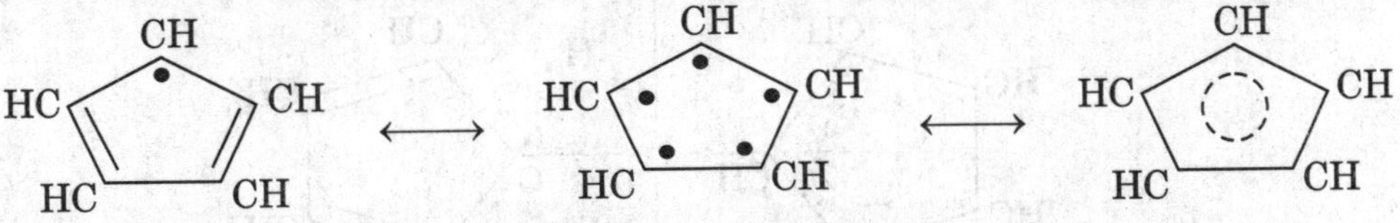


$C_5H_5$ $C_5H_5$ $C_5H_5$

This ligand can combine with the metal through its π-bonding molecular orbitals which are completely delocalised over all the five C–atoms of the 5-membered ring. In this way $C_5H_5^-$ group acts as a pentahapto ligand.

The organometallic compounds formed by π-cyclopentadienyl radical ($\pi$-$C_5H_5$) with metallic atom are complex compounds and are called **metallocenes**. The most important metallocene is *bis* (cyclopentadienyl) iron (II), $(\pi\text{-}C_5H_5)_2Fe$ which is commonly called **ferrocene** whose structure is given in Fig. 13.18.(*a*) All the five C-atoms of $C_5H_5^-$ group are equally associated with Fe-atom. In Fe $(C_5H_5)_2$, $C_5H_5^-$ radical acts as a π-ligand and hence ferrocene is represented as Fe $(\pi\text{-}C_5H_5)_2$.

$C_5H_5^-$ radical also acts as a heptahapto ligand in the organometallic compound, $Mn(CO)_3$ ($\pi$-$C_5H_5$). This is evident from the structure of this compound given in Fig. 31.18(*b*).

$C_5H_5^-$ group also forms organometallic compounds in which one $C_5H_5^-$ ion is linked with the metal atom through only one of its five C-atoms forming M–C σ-bond and

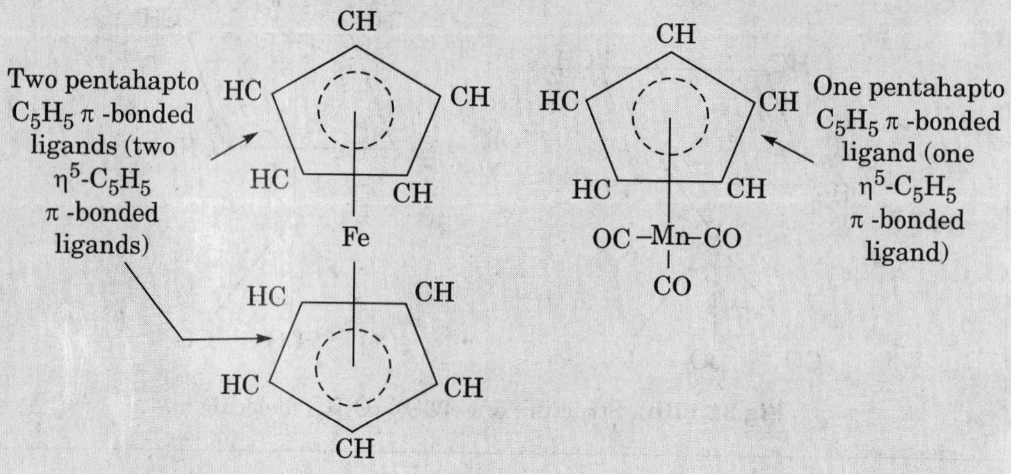


**Fig. 31.18. (*a*)** Structure of ferrocene molecule $Fe(\pi\text{-}C_5H_5)_2$

**Fig. 31.18. (*b*)** Structure of $Mn(CO)_3\ (\pi\text{-}C_5H_5)$

the other $C_5H_5^-$ ion is bonded to the metal atom through all its five C-atoms forming C–M π-bond. Thus one $C_5H_5^-$ group acts as a monohapto ligand ($\eta^1$-$C_5H_5$) and other $C_5H_5^-$ acts a pentahapto ligand ($\eta^5$-$C_5H_5$). ($\eta^1$-$C_5H_5$) is designated as ($\sigma$-$C_5H_5$) and ($\eta^5$-$C_5H_5$) is represented as ($\pi$-$C_5H_5$). Structure of the compound, ($\eta^5$-$C_5H_5$) $Fe(CO)_2$ ($\eta^1$-$C_5H_5$) or ($\pi$-$C_5H_5$) $Fe(CO)_2$ ($\sigma$-$C_5H_5$) is given in Fig. 31.19.

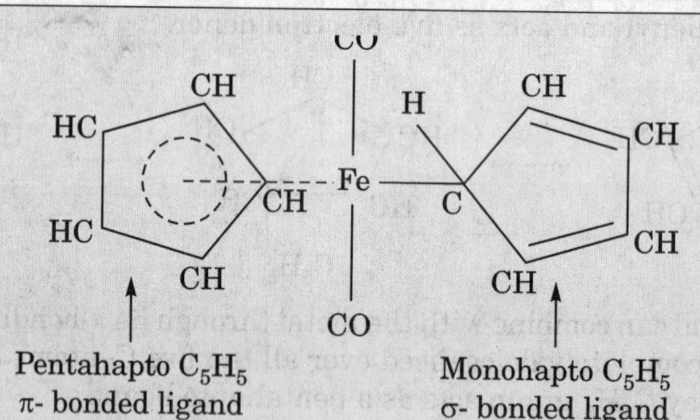


**Fig. 31.19.** Structure of ($\eta^5$-$C_5H_5$) $Fe(CO)_2$ ($\eta^1$-$C_5H_5$) or ($\pi$-$C_5H_5$) $Fe(CO)_2$ ($\sigma$-$C_5H_5$) organometallic compound.

**6. Hexahapto ligands: Six carbon-bonded ligands ($\eta^6$-ligands).** Benzene ($C_6H_6$) and its derivatives (*e.g.* $C_6H_5\ CH_3$) are important carbon-bonded (six-electron donors) ligands. $C_6H_6$ forms organometallic compound with Cr-atom which is called dibenzene chromium, $(C_6H_6)_2$ Cr. Its structure is given in Fig. 31.20.

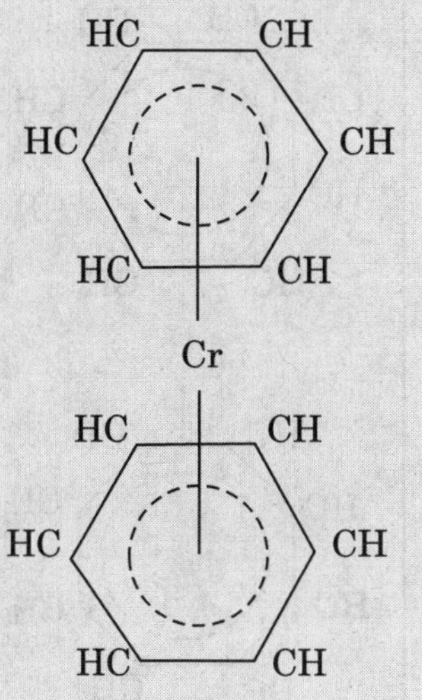

**Fig. 31.20.** Structure of dibenzene chromium, $(C_6H_6)_2$ Cr.

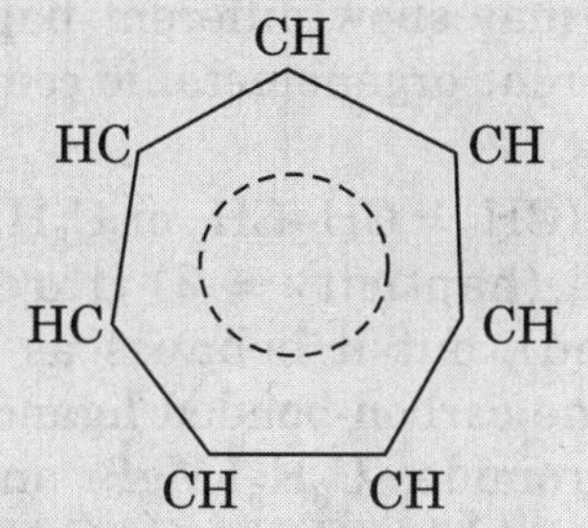

**Fig. 31.21.** Structure of cyclohepta-trienyl group (tropilium group), $C_7H_7^-$

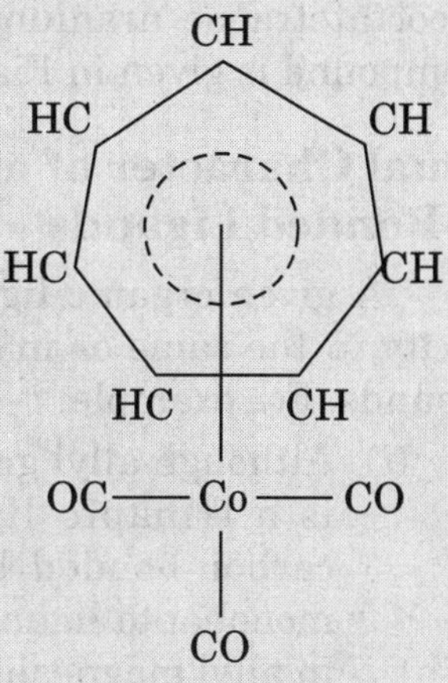

**Fig. 31.22.** Structure of tricarbonyl cycloheptatrienyl cobalt, $(\pi\text{-}C_7H_7)Co(CO)_3$

In this compound, Cr-atom is sandwiched between two $C_6H_6$ rings. The planes of these rings are parallel and in the eclipsed form. In this compound Cr-atom is linked with all the six C-atoms over which π-bonding molecular orbital of $C_6H_6$ (ligand) is spread (completaly de-localised π-bonding molecular orbital). Thus all the six C-atoms of $C_6H_6$ ring are equally linked with Cr-atom.

**7. Heptahapto ligands: Seven carbon-bonded ligands ($\eta^7$ ligands)** Cyclohepta-trienyl group (tropilium group), $C_7H_7^-$ acts as a $\eta^7$ ligand in the formation of organometallic compound, tricarbonyl cycloheptatrienyl cobolt $[(C_7H_7)\ Co(CO)_3]$. The structure of $C_7H_7^-$ group is shown in Fig. 31.21 and that of $[(C_7H_7)\ Co(CO)_3]$ is shown in Fig. 31.22.

It may be seen from the structure of $(C_7H_7)\ Co(CO)_3$ that all the seven C-atoms of $C_7H_7^-$ group are linked with Co-atom, since the bonding of $C_7H_7^-$ group with Co-atom takes place through the completely delocalised π-molecular orbital of $C_7H_7^-$ group. Above discussion shows that the organometallic compound should be formulated as $(\pi\text{-}C_7H_7)\ Co(CO)_3$.

**8. Octahapto ligands: Eight carbon-bonded ($\eta^8$ ligands).** Cyclooctatetraene $(C_8H_8)$ is an eight-electron donor and has planar eight-membered ring structure shown in Fig. 31.23.

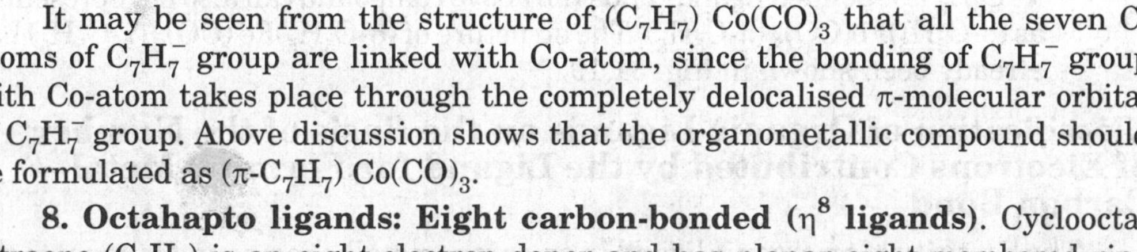

**Fig. 31.23.** Structure of cyclooctatetraene molecule $(C_8H_8)$

Cyclooctatetraene forms an organometallic compound with uranium metal (U) which is called dicyclooctatetraene uranium, $(C_8H_8)_2U$. Structure of this compound is given in Fig. 31.24.

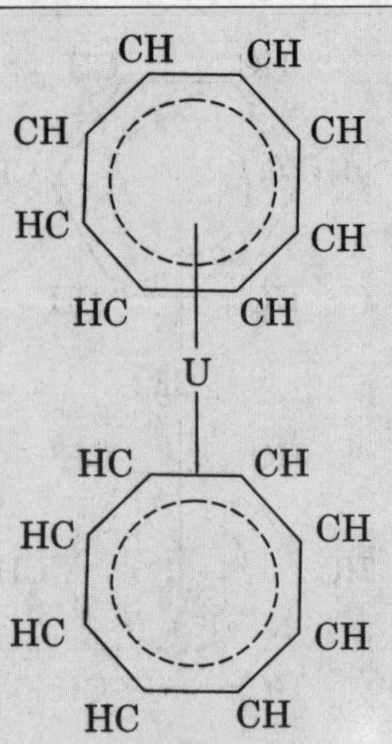


**Fig. 31.24.** Structure of dicyclooctatetraeneuranium, $(C_8H_8)_2$ U

## Dual Character of a Ligand: σ-Bonded and π-Bonded Ligands

A given organic ligand may show different hapticity in the same or in different organometallic compounds. For example:

(*i*) Although allyl group ($CH_2 = CH–CH_2$ or $C_3H_5$) is a trihapto ligand (hapticity = 3) (three carbon-bonded ligand), but it behaves as a monohapto ligand (one carbon-bonded ligand) in allyl magnesium bromide, $(C_3H_5)$ MgBr and acts as a trihapto ligand in $(C_3H_5)$ $Co(CO)_3$. Thus the correct formulations of the above compounds should be $(\eta^1–C_3H_5)$MgBr and $(\eta^3–C_3H_5)$ $Co(CO)_3$ respectively. In $(\eta^1–C_3H_5)$ MgBr, since $\eta^1–C_3H_5$ ligand is attached with Mg atom through Mg–C σ-bond, this ligand is also called σ-bonded ligand. On the other hand, in $(\eta^3-C_3H_5)$ $Co(CO)_3$ since $\eta^3–C_3H_5$ ligand is linked with Co-atom through Co-C π-bond, this ligand is also called π-bonded ligand. Thus $(\eta^1-C_3H_5)$ Mg Br and $(\eta^3-C_3H_5)$ $Co(CO)_3$ can also be represented as $(\sigma-C_3H_5)$MgBr and $(\pi-C_3H_5)$ $Co(CO)_3$ respectively. Note that $\sigma-C_3H_5$ is a one-electron donor while $\pi-C_3H_5$ is a three-electron donor.

(*ii*) Cyclopentadienyl group, $C_5H_5^-$ is a pentahapto ligand. It shows monohapticity as well as pentahapticity in $(C_5H_5)_2Fe(CO)_2$. Thus the correct structure of this compound is $(\eta^1-C_5H_5)Fe(CO)_2$ $(\eta^5-C_5H_5)$. $\eta^1-C_5H_5$ is σ-bonded ligand and $\eta^5-C_5H_5$ is π-bonded ligand. Thus the above compound can also be represented as $(\sigma-C_5H_5)Fe(CO)_2$ $(\pi-C_5H_5)$. The structure of $(\sigma-C_5H_5)Fe$ $(CO)_2$ $(\pi-C_5H_5)$ has already been shown in Fig. 31.19.

## Classification of Organic Ligands on the Basis of the Number of Electrons Contributed by the Ligand for Forming Metal—Carbon Bond

The hapticity of an organic ligand in an organometallic compound is equal to the number of electrons contributed by the ligand for forming metal—carbon bond in the compound. Thus $\eta^1$, $\eta^2$, $\eta^3$ etc. ligands have 1, 2, 3. etc electrons that can be contributed by the ligand for the formation of metal—carbon bond. Examples are given below:

**1. One-electron ligands.** Examples of such ligands are $\dot{H}$, halogen (X), $C_6H_5^-$, σ-allyl ($\sigma-C_3H_5^-$ or $CH_2 = CH–CH_2^-$) alkyl (*e.g.* $CH_3^-$, $C_2H_5^-$), σ-cyclopentadienyl ($\sigma-C_5H_5$), NO (bent nitrosyl group), silyl ($Me_3$ Si). $CH_3$ in $CH_3$–Mg–Br, $C_6H_5$ in $(C_6H_5)_3$ As and $CH_2 = CH–CH_2$ (σ-allyl) in $CH_2 = CH–CH_2$–Mg–Br act as one-electron ligand (or one-electron donor). Most common one-electron ligands are *alkyl*, *aryl* and *acryl* groups.

**2. Two-electron ligands.** CO, $PR_3$, $Cl^-$, $PF_3$, $PPh_3$, alkenes (*e.g.* ethylene or ethene, $H_2C = CH_2$ or $C_2H_4$), alkynes ($RC \equiv CR$ or $R_2C_2$), $C_2F_4$, $\sigma$-$C_5H_6$, $H^-$, $CH_3^-$, $C_6H_5^-$ are the examples of two-electron ligands. CO acts as a two-electron ligand in metal carbonyls. Most important organometallic compound in which ethylene ($C_2H_4$) acts as a two-electron ligand is Zeise's salt, $K^+[PtCl_3(C_2H_4)]^-$.

**3. Three-electron ligands.** $\pi$-allyl ($\pi$-$C_3H_5$), $C_3Ph_3$, NO (linear nitrosyl group), halogen (X) act as three-electron donors. X and $\pi$-$C_3H_5$ act as three-electron ligands in $M–X \to M$ and $Co(\pi\text{-}C_3H_5)(CO)_3$ respectively.

**4. Four-electron ligands.** 1, 3-butadiene ($H_2C=CH–CH=CH_2$), cyclobutadiene ($C_4H_4$) and alkynes ($RC{\equiv}CR$ or $R_2C_2$) are four-electron donors. Alkynes also behave as two-electron donors. Examples of organometallic compounds formed by butadiene and $C_4H_4$ ligands are $Fe(CO)_3$ ($H_2C= CH–CH=CH_2$) and $Fe(CO)_3(C_2H_4)$ respectively. Their structures have already been given Figures. 31.17(a) and 31.17(b) respectively.

**5. Five-electron ligands.** Cyclopentadienyl radical ($\pi$-$C_5H_5^-$) which is a five-membered cyclic dienyl acts as a five-electron donor as is evident from the structure of $Mn(CO)_3$ ($\pi$-$C_5H_5$) already given in Fig. 31.18(b).

**6. Six-electron ligands.** Benzene ($C_6H_6$) and its derivatives like $C_6H_5CH_3$ are six-electron ligands. $C_6H_6$ forms sandwich bonded compound with Cr which is called bibenzene-chromium, $(C_6H_6)_2Cr$. Its structure is given in Fig. 31.20.

Cycloheptatriene ($C_7H_8$) whose structure is given in Fig. 31.25 also acts as a six-electron donor in $Cr(CO)_3$ ($C_7H_8$). In this compound, $C_7H_8$ gets attached with Cr-metal through only six electrons present in three double bonds as shown in Fig. 31.26.

**7. Seven-electron ligands**. Cycloheptatrienyl group (tropilium group), $C_7H_7^-$ acts as a seven electron ligand in tricarbonyl cycloheptatrienyl cobalt, $[(C_7H_7)Co(CO)_3]$. The structures of $C_7H_7^-$ group and $(C_7H_7)$ $Co(CO)_3$ have already been guiven in Figures 31. 21 and 31.22 respectively.

**8. Eight-electron ligands.** Cyclooctatetraene ($C_8H_8$) is an eight-electron donor. $U(C_8H_8)_2$ is an organometallic compound in which $C_8H_8$ acts as an eight-

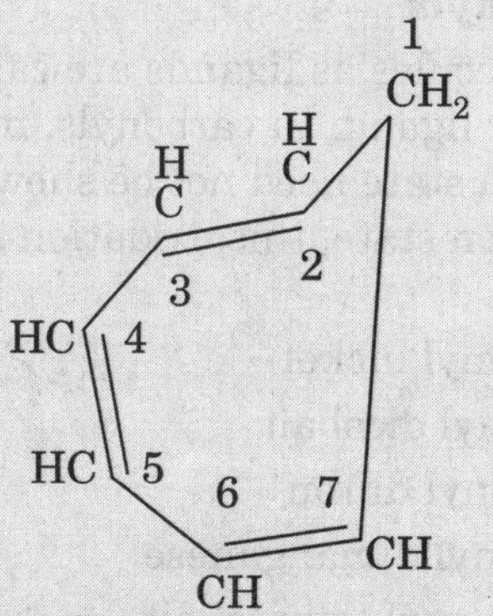


**Fig. 31.25.** Structure of cycloheptatriene ($C_7H_8$)

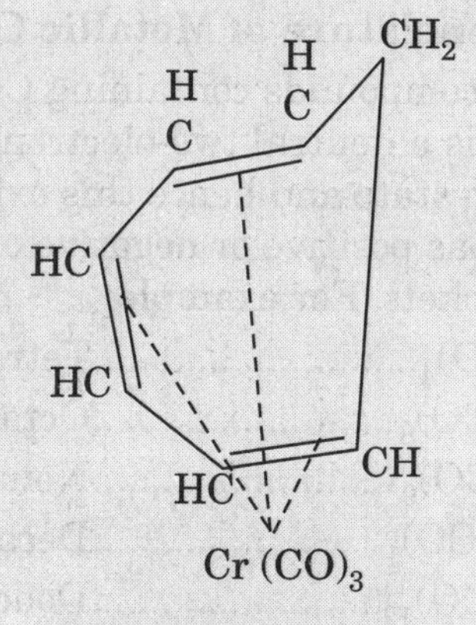


**Fig. 31.26.** Structure of $Cr(CO)_3$ ($C_7H_8$)

electron donor. The structures of $C_8H_8$ and $U(C_8H_8)_2$ molecules have already been given in Figures 31.23 and 31.24 respectively.

## Nomenclature of Organometallic Compounds

Most of the organometallic compounds are coordination compounds having one or more organic ligands attached to the metal (M) through M–C bonds. The rules for naming these compounds are, therefore, similar to those used for naming coordination compounds.

In case of organometallic compounds the following additional rules are also used.

### 1. Nomenclature of Organometallic Compounds Containing Alkyl or Aryl Groups

(*a*) While naming simple organometallic compounds containing alkyl or aryl groups the name of these groups is written. For example:

$CH_3Li$ .............................. Methyl lithium
$(C_2H_5)_2Zn$ ......................... Dimethyl zinc
$C_2H_5$ MgBr........................ Ethylmagnesium bromide
$C_6H_5$ MgCl ....................... Phenylmagnesium chloride

(*b*) If the alkyl or aryl group is present in the coordination sphere, then it is named as a radical. Ethylene ($C_2H_4$) is a neutral two-electron donor ligand. Examples are given below:

$K[B(C_6H_5)_4]$ ..................... Potassiumtetraphenyl borate (III)
$K[SbCl_5(C_6H_5)]$................. Potassium pentachloro (phenyl) antimonate (V)
$K[PtCl_3(C_2H_4)]$ ................. Potassium trichloro (ethylene) platinate (II)
$[PtCl_2(NH_3)(C_2H_4)]$ .......... Ammine dichloro (ethylene) platinum (II)
$K[B(CH_3)_4]$ ....................... Potassium tetramethyl borate (III)

### 2. Nomenclature of Metallic Carbonyls

(*a*) The compounds containing CO molecules as ligands are called carbonyls. CO molecule is a neutral two-electron donor ligand. In carbonyls, metal atom has zero oxidation state and hence this oxidation state need not be shown in brackets. If the metal has positive or negative oxidation state, this oxidation state has to be shown in brackets. For example:

$Ni(CO)_4$.......................... Tetracarbonyl nickel
$Co_2(CO)_8$ ........................ Octacarbonyl dicobalt
$Fe_2(CO)_9$ ........................ Nonacarbonyl diiron
$Mn_2(CO)_{10}$ ..................... Decacarbonyl dimanganese
$Co_4(CO)_{12}$....................... Dodecacarbonyl tetracobalt
$[Mn^+(CO)_5(C_2H_4)]^+$ ......... Pentacarbonyl (ethylene) manganese (+1)
$[V(CO)_6]^-$ ....................... Hexacarbonyl vanadate (–1)

(*b*) The metal carbonyls containing metal-metal bond may be symmetrical or unsymmetrical. The symmetrical metal carbonyls are named by using the multiplicative prefixes (*bis*, *tris* etc). In unsymmetrical carbonyls one central metal atom and the ligands attached with it are treated as a ligand on the other central metal atom. For example:

$[(CO)_4Co–Co(CO)_4]$ .............. *Bis* (tetracarbonyl cobalt)
(symmetrical)

$[(CO)_5Mn–Mn(CO)_5]$ ............ *Bis* (pentacarbonyl manganese)
(symmetrical)

$[(CO)_4Co–Re(CO)_5]$ .............. Pentacarbonyl (tetracarbonyl
(unsymmetrical) .................. cobalto) rhenium

$[C_2H_5Hg–Ta(CO)_6]$............... Hexacarbonyl (ethylnier
(unsymmetrical) ................. curio) tantalum

$[(Ph_3PAu)–Fe(CO)_4]$............. Tetracarbonyl [*bis*(triphenylphosphinoaurio)]
(unsymmetrical) .................. iron

(*c*) If a metallic carbonyl has bridging carbonyl groups (>CO), the Greek letter *mu* (μ) is written before the name of each of the bridging CO groups. Two or more bridging CO groups are indicated by di-μ or *bis*-μ, *tri*-μ etc.

If a metallic carbonyl has bridging (>CO) as well as non-bridging or terminal (–CO) carbonyl groups, the former (*i.e.* bridging group) are named first. For example:

(*i*) $[(CO)_3\ Co(CO)_2(CO_3)_3]$ ......... *Di*-μ-carbonyl *bis*-
(>CO = 2, –CO = 6) .............. [tricarbonyl cobalt]

(*ii*) $[(CO)_3\ Fe(CO)_3Fe(CO)_3]$...... *Tri*-μ-carbonyl *bis*[tricarbonyl iron]
(Fe–Fe = 1, >CO = 3,
– CO= 3 on each Fe atom)

(*iii*) $[(CO)_4\ Os(CO)Os(CO)_4]$ ..... μ-carbonyl *bis*[tetracarbonyl osmium]
(Os–Os = 1, >CO = 1,
– CO = 4 on each Os atom)

**3. Nomenclature of σ-and π-bonded ligands.** (*a*) A ligand behaving as one-carbon-bonded ligand (*i.e.* one-electron donor) is denoted as σ-bonded ligand. On the other hand, a ligand behaving as 2, 3 etc. carbon-bonded ligand (*i.e.* 2, 3 etc. electron donor) is denoted as π-bonded ligand. For example, when cyclopentadiene ($C_5H_5$) behaves as one-electron donor (one-carbon-bonded ligand), it is denoted as σ-$C_5H_5$ [*e.g.* Li(σ-$C_5H_5$)]. On the other hand, when $C_5H_5$ behaves as five-electron donor (five-carbon-bonded ligand), it is represented as π-$C_5H_5$ [*e.g.* (π-$C_5H_5)_2$ Fe]. Similarly allyl group ($C_3H_5$ or $–CH_2–CH = CH_2$) can be denoted as σ-$C_3H_5$ or π-$C_3H_5$ depending on whether this group acts as one-electron donor (one-carbon-bonded ligand) or three-electron donor (three-carbon-bonded ligand).

According te the latest IUPAC convention, the organic ligands are denoted by the prefix *eta* (η). For example, one-carbon-bonded ligand is called *monohapto ligand* and is denoted is $\eta^1$. Two-carbon-bonded ligand is called *dihapto ligand* and is denoted as $\eta^2$. Three-carbon-bonded ligand is called *trihapto ligand* and is denoted as $\eta^3$ and so on.

Above discussion shows that σ-ligands are *monohapto* and hence are denoted as $\eta^1$ ligands, *i.e.* σ-ligands are one-carbon-bonded ligands. On the other hand, π-ligands are *dihapto, trihapto, tetrahapto* etc and hence are denoted as $\eta^2$, $\eta^3$, $\eta^4$ etc, (*i.e*, π-ligands are two-carbon-bonded, three-carbon-bonded, four-carbon-bonded etc ligands.

**Examples.** The formulae of organometallic compounds given below have σ- or π-ligands whose names have been indicated by η notation.

(*i*) $Mn(CO)_5(-CH_2-CH = CH_2)$ ...($\eta^1$-allyl) pentacarbonylmanganese or $Mn(CO)_5$ ($\sigma$-$C_3H_5$)

(*ii*) Fe $(CO)_2$ ($\sigma$-$C_5H_5$) ($\pi$-$C_5H_5$).....Dicarbonyl ($\eta^1$- cyclopentadienyl) ($\eta^5$-cyclopentadienyl) iron.

(*iii*) ($\sigma$–$C_3H_5$) ($\pi$-$C_5H_5$) Mo $(CO)_3$..($\eta^1$-allyl) tricarbonyl ($\eta^5$-cyclopentadienyl) molybdenum

(*iv*) ($\sigma$-$C_5H_5$)$_2$ Ti ($\pi$-$C_5H_5$)$_2$.............*Bis*($\eta^1$–cyclopentadienyl) *bis* ($\eta^5$- cyclopentadienyl) titanium.

(*v*) Co $(CO)_3$ ($\pi$-$C_3H_5$)....................($\eta^3$-allyl) tricarbonyl cobalt

(*vi*) Fe $(CO)_3$ ($\pi$-$C_4H_6$)....................($\eta^4$-butadiene) tricarbonyl iron.

(*vii*) ReH($\pi$-$C_5H_5$)$_2$ ..........................*Bis* ($\eta^5$-cyclopentadienyl) hydrorhenium.

(v*iii*) ($\pi$-$C_5H_5$) Mn $(CO)_3$ .................Tricarbonyl ($\eta^5$-cyclopentadienyl) manganese

(*ix*) $Fe_2$ $(CO)_4$ ($\pi$-$C_5H_5$)$_2$.................*Bis* ($\eta^5$-cyclopentadienyl) tetracarbonyldiiron.

(*x*) ($C_6H_6$) Cr $(CO)_3$......................($\eta^6$-benzene) tricarbonyl chromium.

(*xi*) Cr ($C_6H_6$)$_2$..............................*Bis* ($\eta^6$- benzene) chromium

(*xii*) Fe($\pi$-$C_5H_5$)$_2$.............................*Bis* ($\eta^5$-cyclopentadienyl) iron.

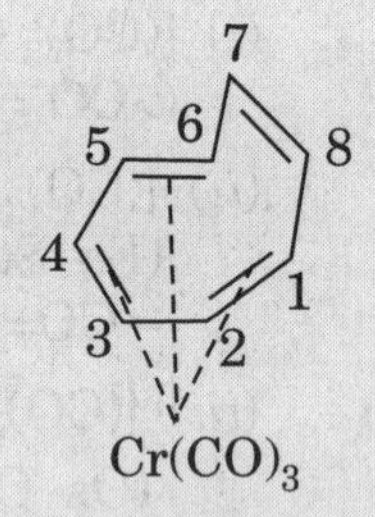


**Fig 31.27.** Structure of tricarbonyl (1–6$\eta^6$-cyclooctatetraene) chromium

(*b*) If only some C-atoms (not all ) of a ligand containing more than one double bond are linked with the central metal atom, the number of these C-atoms is indicated by numerals which are written before η. For example the complex whose structure is given in the margin in Fig. 31.27 contains three carbonyl (CO) and one cyclooctatetraene ($C_8H_8$) as ligands. Although $C_8H_8$ ligand contains eight C-atoms, this ligand is liked with the central Cr-atom through its six C-atoms only. This ligand is, therefore, represented as 1–6 $\eta^6$-cyclooctatetraene. Thus the complex is named as *tricarbonyl (1–6 $\eta^6$-cycloctatetraene) chromium.*

## Effective Atomic Number (EAN) Rule

According to this rule, in a stable complex total number of electrons possessed by the central metal atom or ion and the electrons gained by it from the ligands is called EAN of the central metal atom or ion. In a stable complex EAN of the central metal atom or ion is generally equal to the atomic number of the noble gas which follows the central metal atom in the periodic table. This rule is also called **noble gas rule**.

## Calculation of EAN of the Central Metal Atom or Ion in Some Organometallic Compounds

EAN of the metal atom in some organometallic compounds has been calculated as follows:

1. **Ni $(PF_3)_4$**. $Ni^o = 28e^-$, $4PF_3 = 4×2e^- = 8e^-$.
   $\therefore$ EAN of $Ni^o = 28e^- + 8e^- = 36e^-$ (Kr)
2. **Fe $(CO)_4(PPh_3)$**. $Fe^o = 26e^-$, $4CO = 4×2e^- = 8e^-$, $PPh_3 = 2e^-$
   $\therefore$ EAN of $Fe^o = 26e^- + 8e^- + 2e^- = 36e^-$ (Kr)
3. **Mn $(CO)_5$ $(CH_3)$**. $Mn^o = 25e^-$, $5CO = 5 \times 2e^- = 10e^-$. $CH_3 = 1e^-$
   $\therefore$ EAN of $Mn^o = 25e^- + 10e^- + 1e^- = 36e^-$ (Kr)
4. **Re $(CO)_5$ $(CH_3)$**. $Re^o = 75e^-$, $5CO = 5 \times 2e^- = 10e^-$, $CH_3 = 1e^-$
   $\therefore$ EAN of $Re^o = 75e^- + 10e^- + 1e^- = 86e^-$ (Rn)
5. **[Mn $(CO)_5$ $(C_2H_4)]^+$**. $Mn^+ = 25–1 = 24e^-$, $5CO = 5 \times 2e^- = 10e^-$, $C_2H_4 = 2e^-$
   $\therefore$ EAN of $Mn^+ = 24e^- + 10e^- + 2e^- = 36e^-$ (Kr)
6. **V $(CO)_3$ $(\pi\text{-}C_5H_5)(R_2C_2)$**. $V^o = 23e^-$, $3CO = 3 \times 2e^- = 6e^-$, $\pi\text{-}C_5H_5 = 5e^-$, $R_2C_2 = 2e^-$
   $\therefore$ EAN of $V^o = 23e^- + 6e^- + 5e^- + 2e^- = 36e^-$ (Kr)
7. **Rh $(C_2H_4)$ $(C_2F_4)(\pi\text{-}C_5H_5)$**. $Rh^o = 45e^-$, $C_2H_4 = 2e^-$, $C_2F_4 = 2e^-$, $\pi\text{-}C_5H_5 = 5e^-$
   $\therefore$ EAN of $Rh^o = 45e^- + 2e^- + 2e^- + 5e^- = 54e^-$ (Xe)
8. **Fe $(CO)_3$ $(C_4H_6)$**. $Fe^o = 26e^-$, $3CO = 3 \times 2e^- = 6e^-$, $C_4H_6 = 4e^-$
   $\therefore$ EAN of $Fe^o = 26e^- + 6e^- + 4e^- = 36e^-$ (Kr)
9. **Co $(C_4H_4)$ $(\pi\text{–}C_5H_5)$**. $Co^o = 27e^-$, $C_4H_4 = 4e^-$, $\pi\text{-}C_5H_5 = 5e^-$
   $\therefore$ EAN of $Co^o = 27e^- + 4e^- + 5e^- = 36e^-$ (Kr)
10. **$Co_2$ $(CO)_6$ $(RC \equiv CR)$)**. $2Co^o = 2 \times 27e^- = 54e^-$, $6CO = 6 \times 2e^- = 12e^-$, $RC \equiv CR = 4e^-$, Co-Co bond $= 2e^-$
    $\therefore$ EAN for two Co atoms $= 54e^- + 12e^- + 4e^- + 2e^- = 72e^-$
    $\therefore$ EAN for one Co-atom $= 72e^-/2 = 36e^-$ (Kr)
11. **Mn $(CO)_4$ $(\pi\text{ - }C_3H_5)$**. $Mn^o = 25e^-$, $4CO = 4 \times 2e^- = 8e^-$ $\pi\text{-}C_3H_5 = 3e^-$
    $\therefore$ EAN $= 25e^- + 8e^- + 3e^- = 36e^-$ (Kr)
12. **Co $(CO)_3$ $(\pi\text{ - }C_3H_5)$**. $Co^o = 27e^-$, $3CO = 3 \times 2e^- = 6e^-$, $\pi\text{-}C_3H_5 = 3e^-$
    $\therefore$ EAN for $Co^o = 27e^- + 6e^- + 3e^- = 36e^-$ (Kr)
13. **Mo $(CO)_2$ $(\pi\text{ - }C_3H_5)$ $(\pi\text{ - }C_5H_5)$**. $Mo^o = 42e^-$, $2CO = 2 \times 2e^- = 4e^-$, $\pi\text{-}C_3H_5 = 3e^-$, $\pi\text{-}C_5H_5 = 5e^-$
    $\therefore$ EAN for $Mo^o = 42e^- + 4e^- + 3e^- + 5e^- = 54e^-$ (Xe)

14. **Fe (π - $C_5H_5$)$_2$ .** $Fe^o = 26e^-$, 2(π-$C_5H_5$) = 2 × $5e^-$ = $10e^-$
   ∴ EAN for $Fe^o = 26e^- + 10e^- = 36e^-$ (Kr)
15. **Co (CO)$_2$ (π - $C_5H_5$).** $Co^o = 27e^-$, 2CO = 2 × $2e^-$ = $4e^-$, π-$C_5H_5$ = $5e^-$
   ∴ EAN for $Co^o = 27e^- + 4e^- + 5e^- = 36e^-$ (Kr)
16. **Fe (CO)$_2$ (σ - $C_5H_5$) (π - $C_5H_5$).** $Fe^o = 26e^-$, 2CO = 2 × $2e^-$ = $4e^-$, σ-$C_5H_5$ = $1e^-$, π-$C_5H_5$ = $5e^-$
   ∴ EAN of $Fe^o = 26e^- + 4e^- + 1e^- + 5e^- = 36e^-$ (Kr)
17. **Cr ($C_6H_6$)$_2$ .** $Cr^o = 24e^-$, 2$C_6H_6$ = 2 × $6e^-$ = $12e^-$
   ∴ EAN of $Cr^o = 24e^- + 12e^- = 36e^-$ (Kr)
18. **Cr (CO)$_3$ ($C_6H_6$).** $Cr^o = 24e^-$, 3CO = 3 × $2e^-$ = $6e^-$, $C_6H_6$ = $6e^-$
   ∴ EAN of $Cr^o = 24e^- + 6e^- + 6e^- = 36e^-$ (Kr)
19. **Cr ($C_6H_5CH_3$) ($C_6H_6$).** $Cr^o = 24e^-$, $C_6H_5CH_3$ = $6e^-$, $C_6H_6$ = $6e^-$
   ∴ EAN of $Cr^o = 24e^- + 6e^- + 6e^- = 36e^-$ (Kr)
20. **Cr (CO)$_3$ (σ - $C_3H_5$) (π - $C_5H_5$).** $Cr^o = 24e^-$, 3CO = 3 × $2e^-$ = $6e^-$, σ-$C_3H_5$ = $1e^-$ = π-$C_5H_5$ = $5e^-$
   ∴ EAN of $Cr^o = 24e^- + 6e^- + 1e^- + 5e^- = 36e^-$ (Kr)
21. **W (CO)$_3$ ($Me_3Si$) (π-$C_5H_5$).** $W^o = 74e^-$, 3CO = 3 × $2e^-$ = $6e^-$, $Me_3Si$ = $1e^-$, π-$C_5H_5$ = $5e^-$
   ∴ EAN of $W^o = 74e^- + 6e^- + 1e^- + 5e^- = 86e^-$ (Rn)
22. **Mo (CO)$_3$ (σ - $C_3H_5$) (π - $C_5H_5$).** $Mo^o = 42e^-$, 3CO = 3 × $2e^-$ = $6e^-$, σ-$C_3H_5$ = $1e^-$ = π-$C_5H_5$ = $5e^-$
   ∴ EAN of $Mo^o = 42e^- + 6e^- + 1e^- + 5e^- = 54e^-$ (Xe)
23. **Fe (CO)$_3$($C_4H_4$).** $Fe^o = 26e^-$, 3CO = 3 × $2e^-$ = $6e^-$, $C_4H_4$ = $4e^-$
   ∴ EAN of $Fe^o = 26e^- + 6e^- + 4e^- = 36e^-$ (Kr)
24. **Re (CO)$_2$ (σ - $C_5H_6$) (π - $C_5H_5$).** $Re^o = 75e^-$, 2CO = 2 × $2e^-$ = $4e^-$, σ-$C_5H_6$ = $2e^-$, π-$C_5H_5$ = $5e^-$
   ∴ EAN of $Re^o = 75e^- + 4e^- + 2e^- + 5e^- = 86e^-$ (Rn)

## Structure of Ethene (Ethylene) Molecule ($C_2H_4$)

The structure of ethene (ethylene), $C_2H_4$ molecule is $\begin{matrix}H\\H\end{matrix}\!>C=C<\!\begin{matrix}H\\H\end{matrix}$ . This structure shows that each C-atom in this molecule is $sp^2$ hybridised and each C–atom has one $2p_z$ (or pπ) orbital which does not take part in $sp^2$ hybridisation. Each $2p_z$ orbital has one electron. The two 2 $p_z$ orbitals (each containing one electron) on two C-atoms overlap with each other and form two molecular orbitals called π-molecular orbitals. One of these π-molecular orbitals is a π-bonding molecular orbital (designated as π-MO) and the other is a π-antibonding molecular orbital (designated as π*-MO). Since π-bonding molecular orbital (π-MO) is of lower energy and π-antibonding molecular orbital ( π* - MO) is of higher energy, the two electrons present in two $2p_z$ orbitals occupy the lower energy π-MO orbital and π*-MO remains empty (vacant). Thus $C_2H_4$ molecule has one filled π-bonding molecular orbital and one empty π* molecular orbital. (See Fig. 31.28)

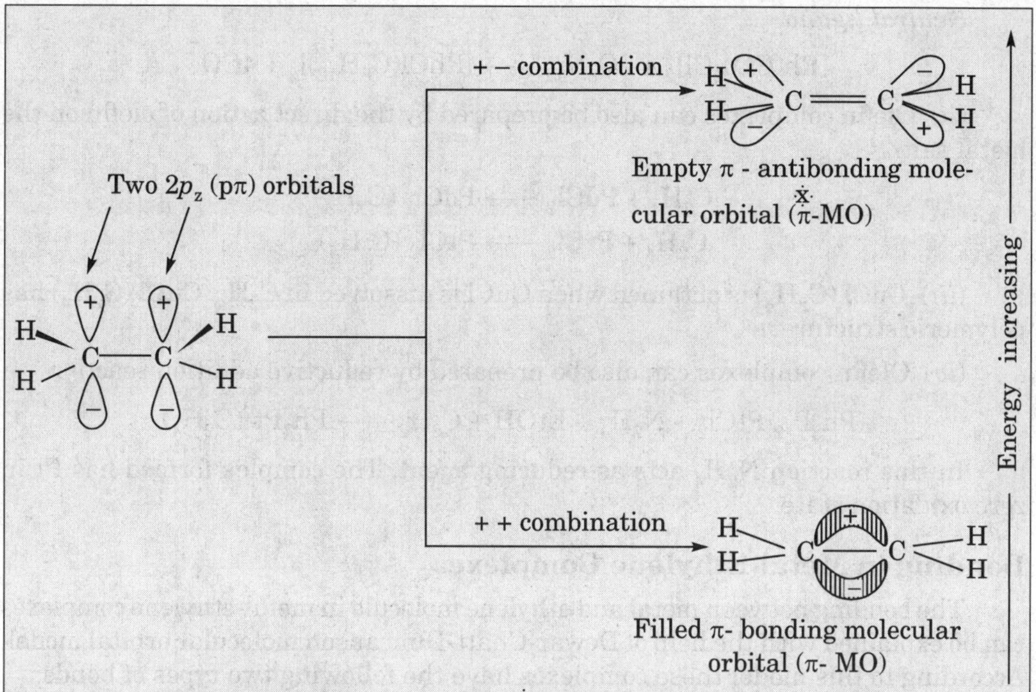


**Fig 31.28.** Formation of π and π* MOs in ethylene molecule ($C_2H_4$)

## π-Metal-Ethylene Complexes

Ethylene (also called ethene), $\begin{matrix}H\\H\end{matrix}\!>C=C<\!\begin{matrix}H\\H\end{matrix}$ or $C_2H_4$, is an alkene and is a two-electron donor (two-carbon-bonded ligand). It forms well defined complexes with transition metals like Pd (II), Pt(II), Hg(II), Cu(I) and Ag(I). The complexes formed by ethylene with transition metals are called π-**metal-ethylene complexes**. These complexes are also called by other names like π-**metal-olefin complexes**, π-**metal-alkene complexes** or π-**metal-alkenyl complexes**. Examples of complexes containing $C_2H_4$ as ligand are $K^+[PtCl_3(C_2H_4)]^-$ (Zeise's salt), $PtCl_2(C_2H_4)$, $(CH_3–C_6H_4–NH_2)–PtCl_2–(H_2C = CH_2)$ or $(CH_3–C_6H_4–NH_2) – PtCl_2(C_2H_4)$ etc.

## Preparation of Olefin Complexes

(*i*) Olefin complexes can be prepared by the displacement of loosely-held ligands such as solvent molecules, ionic ligands or neutral ligands. For example:

*Loosely-held ligand*

$$2RhCl_3 + 2H_2O + 6C_2H_4 \longrightarrow [RhCl(C_2H_4)_2]_2 + 2CH_3CHO + 4HCl$$

*Ionic ligand*

$$K[PtCl_4] + C_2H_4 \longrightarrow \underset{\text{Zeise's salt}}{K[PtCl_3(C_2H_4)]} + KCl$$

$$\downarrow C_2H_4$$

$$[PtCl_2(C_2H_4)]_2$$

*Neutral ligand*

$$[Rh(CO)_2Cl]_2 + 2C_8H_{12} \longrightarrow [RhCl(C_8H_{12})]_2 + 4CO$$

(*ii*) Olefin complexes can also be prepared by the direct action of olefin on the metal salt.

$$C_2H_4 + PdCl_2 \longrightarrow PdCl_2\ (C_2H_4)$$
$$C_2H_4 + PtCl_2 \longrightarrow PtCl_2\ (C_2H_4)$$

(*iii*) $CuCl\ (C_2H_4)$ is obtained when CuCl is dissolved in $C_2H_4$. $CuCl\ (C_2H_4)$ has polymeric structure.

(*iv*) Olefin complexes can also be prepared by reductive addition reaction

$$(Ph_3P)_2\ PtCl_2 + N_2H_4 + EtOH + C_2H_2 \longrightarrow Ph_3Pt^0(C_2H_4)$$

In this reaction $N_2H_4$ acts as reducing agent. The complex formed has Pt in zero oxidation state.

## Bonding in Metal-Ethylene Complexes

The bonding between metal and ethylene molecule in metal-ethylene complexes can be explained with the help of Dewar-Chatt-Duncanson molecular orbital model. According to this model these complexes have the following two types of bonds:

(*i*) **L → M σ-bond**. This bond is obtained by the overlap of the filled bonding π-molecular orbital of ethylene molecule, $C_2H_4$ (L) with the suitably directed vacant (acceptor) σ-type orbital on transition metal (M). This overlap is shown in Fig. 31.29. This overlap does not give a stable bond between metal and ethylene as ethylene is a very weak base.

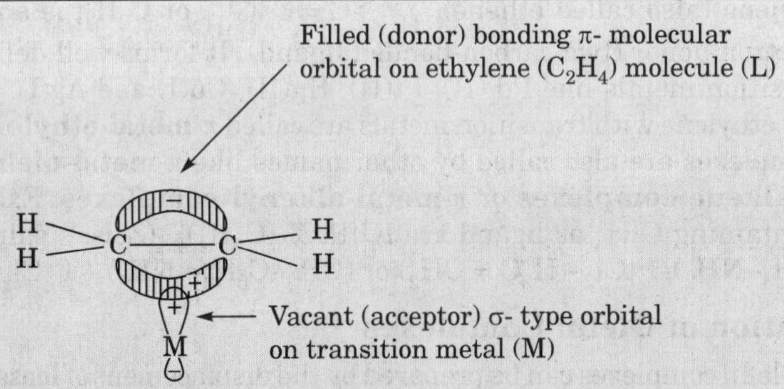


**Fig. 31.29.** Formation of L → M σ-bond in metal-ethylene complexes

(*ii*) **M → L π-bond: Back bonding**. Since L → M σ bond formed as above is not a stable bond, it is reinforced by the formation of a second bond which is obtained by the overlap of filled metal *d*-orbital with the vacant (acceptor) antibonding π-molecular orbital (π* MO) on ethylene molecule. This is shown in Fig. 31.30. This type of bonding is called **back bonding**, since there is a transfer of metal electrons into the vacant antibonding π molecular orbital of ethylene molecule.

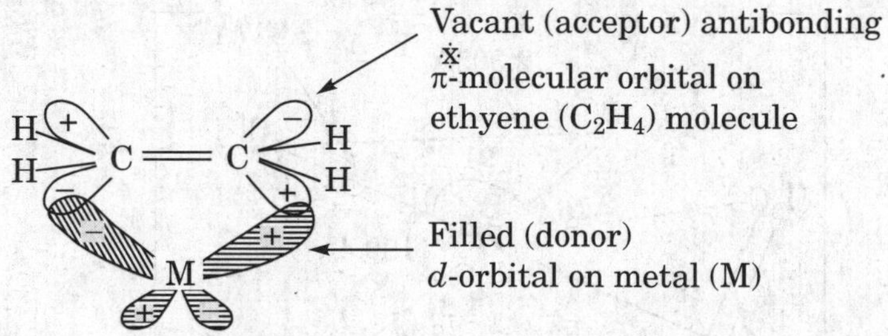


**Fig. 31.30.** Formation of M $\rightarrow$ L $\pi$ bonding (Back bonding) in metal-ethylene complexes.

The formation of L $\rightarrow$ M $\sigma$ and M $\rightarrow$ L $\pi$ bonds can be shown in a simplified way as given in Fig. 31.31.

Filled (donor) bonding $\pi$-MO of ethylene molecule (L) —— L $\rightarrow$ M $\sigma$–bond ——> Vacant (acceptor) $\sigma$-type orbital on transition metal (M)

Filled (donor) d-orbital on transition metal (M) —— M $\rightarrow$ L $\pi$–bond, Back bonding ——> Vacant (acceptor) antibonding $\pi^*$-MO on ethylene molecule (L)

**Fig. 31.31.** Simplified way to show the formation of L $\rightarrow$ M $\sigma$ and M $\rightarrow$ L $\pi$ bonds in metal-ethylene complexes

## Structural Changes Due to Back Donation of Metal Electrons into Ethylene $\pi^*$ Molecular Orbital

The transfer of electrons from the metal *d*-orbital into $\pi^*$-MO of ethylene molecule creates the following changes:

(*i*) C = C bond length is increased and hence C = C bond gets weakened. Consequently C = C IR stretching frequency is reduced. These two changes are shown below:

C = C bond length in free $C_2H_4$ = 133.5 pm

C = C bond length in $K[PtCl_3(C_2H_4)]$ = 137.5 pm

C = C IR stretching frequency in free $C_2H_4$ = 1625cm$^{-1}$

C = C IR stretching frequency in $K[PtCl_3(C_2H_4)]$ = 1516 cm$^{-1}$

(*ii*) Angles around C-atoms are reduced from 120° (*sp*$^2$ hybridisation) to angles of tetrahedral C-atoms (*sp*$^3$ hybridisation).

## Structure of $[PtCl_3(C_2H_4)]^-$ Ion Present in Zeise's Salt, $K[PtCl_3(C_2H_4)]$

This salt contains the anion, $[PtCl_3(C_2H_4)]^-$. The structure of this ion is shown in Fig. 31.32. X-ray examination of this ion has shown that this ion has square planar shape in which Pt atom is at the centre of the square plane. Three

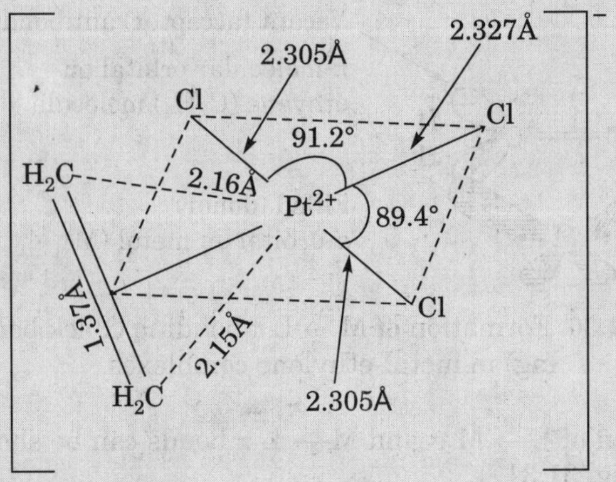


**Fig. 31.32.** Structure of $[PtCl_3(C_2H_4)]^-$ ion of Zeise's salt, $K^+[PtCl_3(C_2H_4)]^-$

Cl-atoms are at each of the three corners of the square plane. The middle point of C = C bond of $C_2H_4$ molecule is at the fourth corner of the square plane. C = C bond is perpendicular to the square plane containing Pt and three Cl-atoms, *i.e.* Pt–$C_2H_4$ bond lies perpendicular to the plane containing Pt and three Cl-atoms. The two C-atoms of $C_2H_4$ are equidistant from Pt metal. $C_2H_4$ group is not planar as in $H_2C = CH_2$ and shows significant distortion from planarity. $C_2H_4$ group lies perpendicular to Pt–Cl bond which is *trans* to $C_2H_4$ ligand.

In $[PtCl_3(C_2H_4)]^-$ ion $dsp^2$ hybridised σ–orbital of Pt overlaps with the filled π-bonding molecular orbital of $C_2H_4$ (ligand) to form Pt ← $C_2H_4$ σ-bond. Simultaneously one of filled dπ orbitals of Pt metal overlaps with the empty $\pi^*$ molecular orbital of $C_2H_4$ resulting in a Pt → $C_2H_4$ π bond.

The length of carbon-carbon bond in the complex ion (= 1.37A°) is greater than the carbon-carbon bond in free $C_2H_4$ molecule ( = 1.337A°). Cl–Pt–Cl bond angles are 91.2° and 89.4° as shown in the figure.

## Structure of $[PtCl_2(C_2H_4)]$ Complex

Freezing point depression measurements of this complex in $C_6H_6$ have shown that the complex exists as a dimer, $[PtCl_2(C_2H_4)]_2$ which has bridged structure. In the bridged structure the two $C_2H_4$ molecules are arranged in *trans* position with each other as shown in Fig. 31.33.

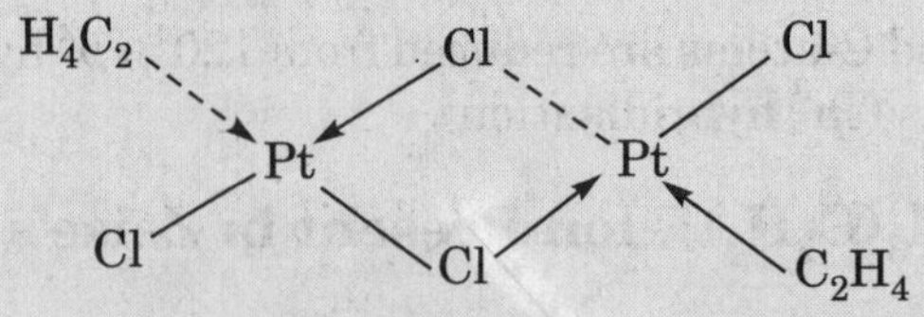


**Fig. 31.33.** Bridged structure of dimer, $[PtCl_2(C_2H_4)]_2$

## Factors on which Degree (or Extent) of Back-bonding (Transfer of Metal Electrons) into Ethylene Molecular Orbital Depends

(*i*) **Effect of substituents on ethylene molecule:** The degree of back-bonding is increased by the presence of electron withdrawing substituents on ethylene ($C_2H_4$) molecule. The increase in back-bonding is confirmed by the fact that the carbon-carbon bond distances are increased. For example, in $[Pt(C_2H_4)_2(C_2F_4)]$, carbon-carbon distances are increased from 136 pm to 144 pm of C = C bond as shown in Fig. 31.34.

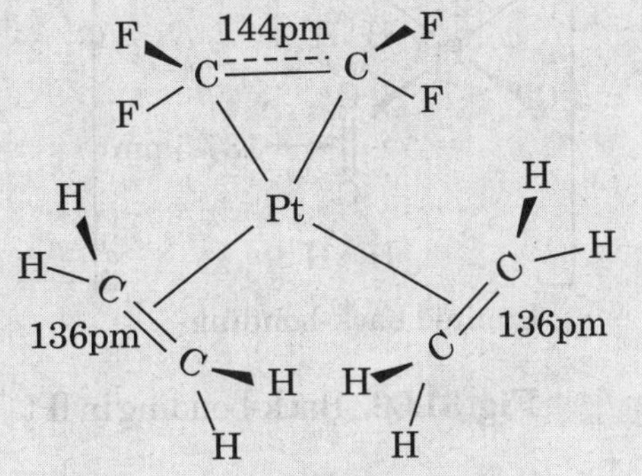


**Fig. 31.34.** Due to the presence of electron withdrawing substituents, carbon-carbon distances are increased from 136 pm to 144 pm in $[Pt(C_2H_4)_2(C_2F_4)]$

In extreme cases, C = C double bond distance reaches more typical of C–C single bond distance. This results in the formation of two distinct metal-carbon σ-bonds *i.e.* metallacyclopropane is formed as shown in Fig. 31.35.

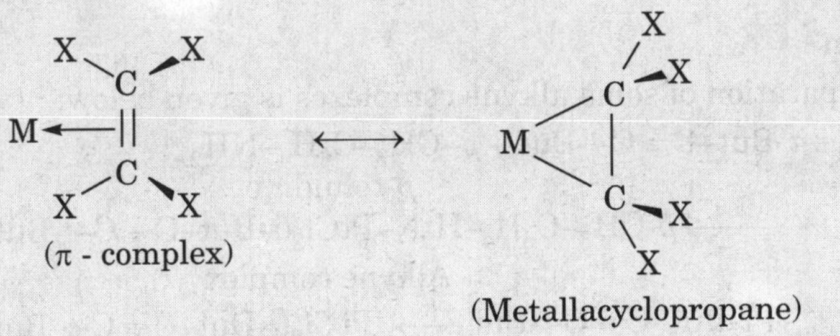


**Fig. 31.35.** Formation of two distinct metal–carbon σ-bonds

(*ii*) **Position of transition metals in the periodic table.** (*a*) *d*-orbitals of Co, Ni etc. have a strong tendency to transfer their electron density into vacant π*molecular orbital of $C_2H_4$. This results in extensive back transfer of metal electrons into vacant π* molecular orbital of $C_2H_4$.

(*b*) *d*-orbitals of Pt do not like to transfer their electron density into vacant π* molecular orbital of $C_2H_4$. For example, in $[PtCl_3(C_2H_4)]^-$ ion there is very little amount of back-transfer of metal electrons into vacant π* molecular orbital of $C_2H_4$.

(*c*) The complexes having metal in lower oxidation state show extensive back-bonding. This may be seen from carbon-carbon bond length in $[Pt^{2+}Cl_3(C_2H_4)]^-$ and $[Pt^0(PPh_3)_2(C_2H_4)]^0$ shown in Fig. 31.36. C = C bond length in free ethylene, $\begin{matrix}H\\H\end{matrix}\!>C=C<\!\begin{matrix}H\\H\end{matrix}$ has been found equal to 133.5 pm.

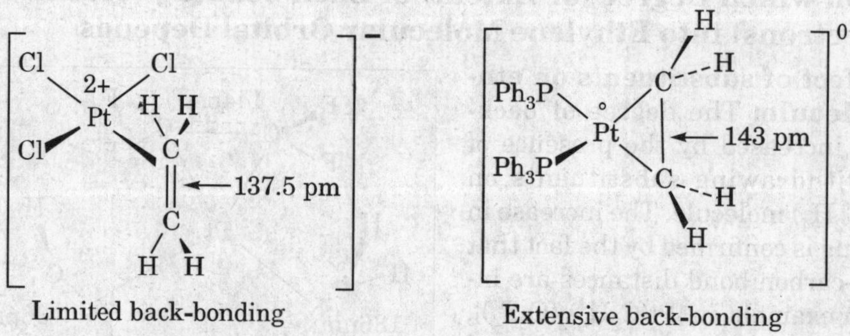


**Fig. 31.36.** Back-bonding in $[Pt^{2+}Cl_3(C_2H_4)]^-$ and $[Pt^{o}(PPh_3)_2(C_2H_4)]^{o}$

## Organometallic Compounds of Alkyne: Alkynyl or Acetylene Complexes

The complexes of transition metals with alkynes are not very common. The simplest alkynyl or alkyne complexes are those in which an alkyne molecule is coordinated to only one metal atom and acts as a two-electron donor.

### Preparation

The preparation of some alkyne complexes is given below:

(*i*) $PtCl_2$ + *t*-But–C ≡ C-*t*-But + *p*–$CH_3$–$C_6H_4$–$NH_2$
*p*-toluidine
⟶ *p*-$CH_3$–$C_6H_4$–$H_2N$–$PtCl_2$(*t*-But–C ≡ C–*t*-But)
Alkyne complex

(*ii*) $[PtCl_4]^{2-}$ + *t*-But–C ≡ C-*t*-But ⟶ $[PtCl_2$(*t*-But–C ≡ C-*t*–But)$]_2$
Dimeric alkyne complex

(*iii*) ($\eta^5$–$C_5H_5$) Mn $(CO)_2$ + $C_6H_5$ – C ≡ C–$C_6H_5$ ⟶ ($\eta^5$–$C_5H_5$) Mn$(CO)_2$ ($C_6H_5$–C ≡ C–$C_6H_5$) or ($\eta^5$–$C_5H_5$) Mn $(CO)_2$ $(C_6H_5)_2$ $C_2$

### Structure of Complexes in which Alkyne Behaves as a Two-Electron Donor

The structures of the complexes whose preparation has been given above are given in Figures. 31.37, 31.38 and 31.39 respectively.

It may be seen from these structures that R-C ≡ C-R is no longer in a straight line. It is appreciably bent at the acetylenic C-atoms. The bending increases with the increase in π-back donation. The carbon-carbon bond is somewhat longer than a triple C ≡ C bond (= 121 pm).

The structure of *p*–$CH_3$–$C_6H_4$–$NH_2$–$PtCl_2$(*t*–But–C ≡ C–*t*–But) complex given in Fig. 31.37 shows that C ≡ C bond remains short and is perpendicular to the plane of coordination. Pt has C. N. = 4 and alkynyl group occupies one of the four coordination sites. *t*–But–C = C–*t*–But is no longer a straight line.

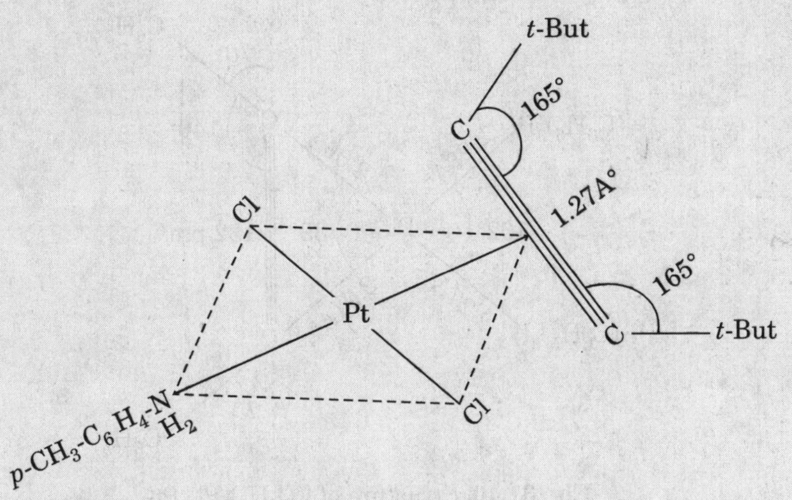


**Fig. 31.37.** Structure of $p$–$CH_3$–$C_6H_4$–$H_2N$–$PtCl_2$(t–But–C ≡ C–t–But) or $p$-$CH_3$-$C_6H_4$-$H_2N$-$PtCl_2$($t$–$C_4H_9$–C ≡ C–$t$–$C_4H_9$). Here t-But is tertiary butyl group ($t$–$C_4H_9$) whose structure is $CH_3$—C($CH_3$)—$CH_3$ (with two more $CH_3$ groups, one above and one below the central C).

$p$–$CH_3$–$C_6H_4$–$NH_2$ is paratoluidine. In this complex alkyne behaves as a two-electron donor.

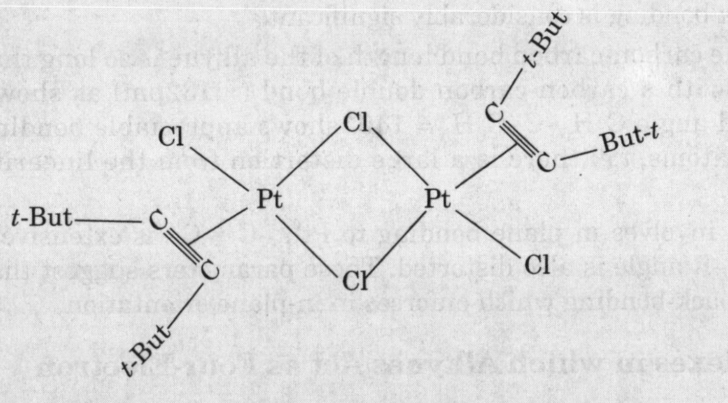


**Fig. 31.38.** Structure of dimeric alkyne complex, $[PtCl_2$($t$-But-C ≡ C-$t$-But)$]_2$ or $[PtCl_2$($t$–$C_4H_9$–C ≡ C–$t$–$C_4H_9$)$]_2$. In this complex alkyne acts as a two-electron donor

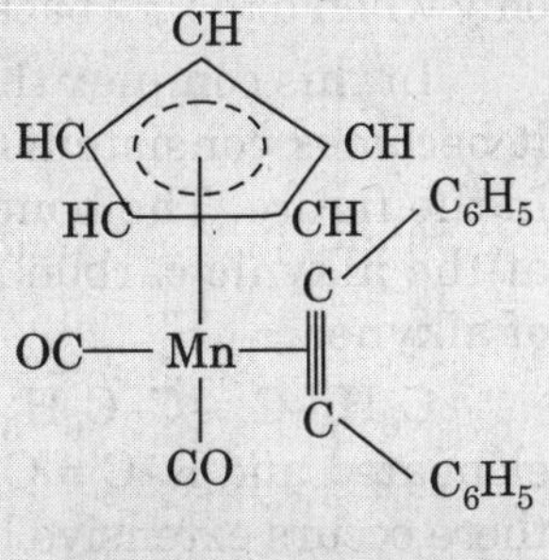


**Fig. 31.39.** Structure of ($\eta^5$–$C_5H_5$) Mn $(CO)_2$ ($C_6H_5$–C ≡ C–$C_6H_5$) complex. In this complex alkyne acts as a two-electron donor.

## Structure of Complexes in which the Alkyne Behaves as a Bidentate Divalent Ligand

The structure of $[(C_6H_5)_3P]_2$ Pt ($C_6H_5$– C ≡ C – $C_6H_5$) complex is given in Fig. 31.40.

This structure shows that the complex has planar structure in which the ligand, $C_6H_5$–C≡C–$C_6H_5$ appears to occupy two coordination sites instead of one, *i.e.* the

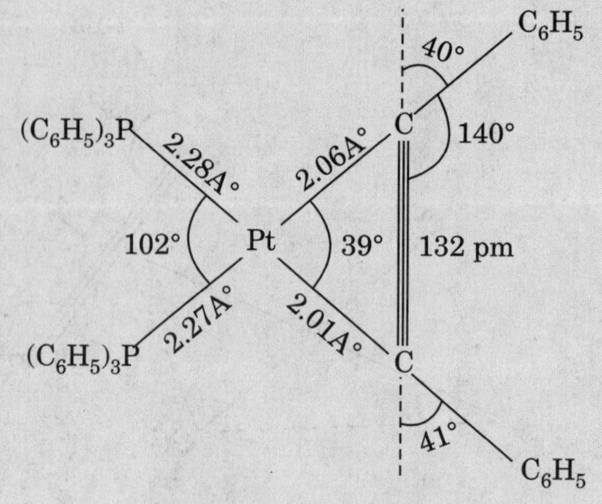


**Fig. 31.40.** Structure of $[(C_6H_5)_3P]_2$ Pt–($C_6H_5$–C ≡ C –$C_6H_5$) or $(Ph_3P)_2$ Pt ($\eta^5$-$C_2Ph_2$). In this complex the alkyne behaves as a bidentate divalent ligand.

alkyne serves as a bidentate divalent ligand. The infra-red spectral studies of this complex reveal that $\nu_{C\equiv C}$ in the complex appears at –1750 $cm^{-2}$ whereas $\nu_{C\equiv C}$ in free $C_6H_5$–C ≡ C–$C_6H_5$ appears at~2100 $cm^{-1}$ This indicates that a decrease in the bond order in the complexed $C_6H_5$–C ≡ C–$C_6H_5$ has occurred. This decrease can happen only when $d\pi \rightarrow \pi^*$ back bonding is considerably significant.

In this complex the carbon-carbon bond length of the alkyne is so long that it becomes consistent with a carbon-carbon double bond (= 132pm) as shown in the figure. The bond angle $C_6H_5$–C–$C_6H_5$ = 140° shows appreciable bending at the alkynic carbon atoms, *i.e.* there is a large distortion from the linearity of alkynes.

$C_6H_5$–C ≡ C–$C_6H_5$ involves in plane bonding to $Pt^o$. –C ≡ C– is extensively elongated and R–C ≡ C–R angle is also distorted. These parameters suggest that there occurs extensive back-bonding which enforces in in-plane orientation,

## Structure of Complexes in which Alkynes Act as Four-Electron Donors

Alkynes R–C ≡ C–R have two π-bonds at right angle to each other and can bond themselves to metal atoms through each of their π-bonds. Here alkynes act as four-electron donors by utilising both π-bonds and act as bridging ligands. Such complexes are usually prepared by displacing bridging carbonyl groups with alkynes at higher temperatres. For example:

(*i*) $Co_2(CO)_8$ + Ph–C ≡ C–Ph $\xrightarrow[C_6H_6]{\Delta}$ $(CO)_6$ $Co_2$(Ph–C≡C–Ph) + 2CO

(*ii*) ($\eta^5$ –$C_5H_5)_2$ $Ni_2(CO)_2$+ Ph–C≡C–Ph $\xrightleftharpoons{K^+}$ ($\eta^5$–$C_5H_5)_2$ $Ni_2$(Ph–C ≡ C–Ph)

The structures of both the above complexes are given in Fig. 31.41. and Fig. 31.42 respectively. In these complexes acetylene group acts as a bridge between two metal atoms and alkynes act as four-electron donor.

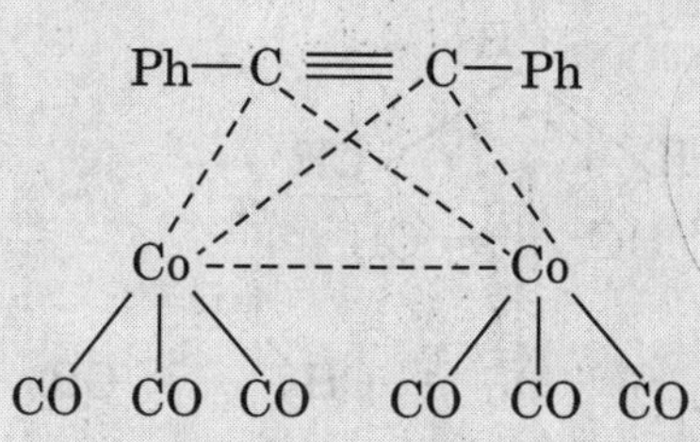


**Fig. 31.41.** Structure of (Ph–C≡C-Ph) $Co_2(CO)_6$. In this complex alkyne group acts as a 4-electron donor

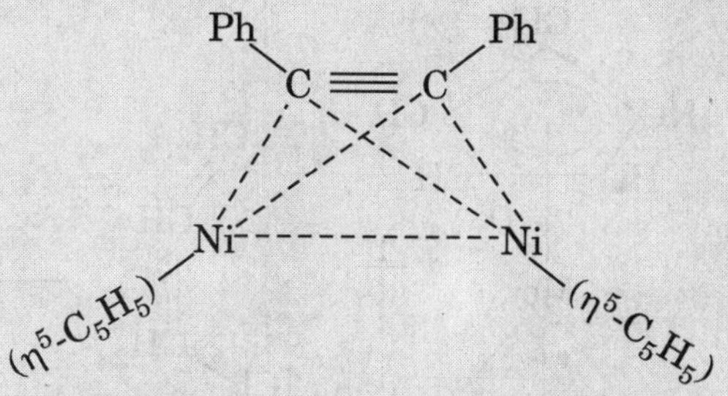


**Fig. 31.42.** Structure of $(\eta^5-C_5H_5)_2$ $Ni_2$ (Ph–C≡C–Ph). In this complex the alkyne acts as a 4-electron donor.

In both the complex compounds C≡C group lies above and at right angle to the metal-metal vector. Carbon-carbon distance of the alkyne in the complex is significantly greater than –C≡ C– distance in Ph–C≡ C– Ph. This indicates an extensive back donation from the metal atoms into the $\pi^*$ orbitals of alkyne.

## π-Allyl Complexes

Allyl group ($C_3H_5$) is a π-ligand and hence is represented as π-$C_3H_5$. It is a three-electron donor. Since it is a three-carbon-bonded ligand, it is also represented as $\eta^3$-$C_3H_5$. The structure of $C_3H_5$ is given below:

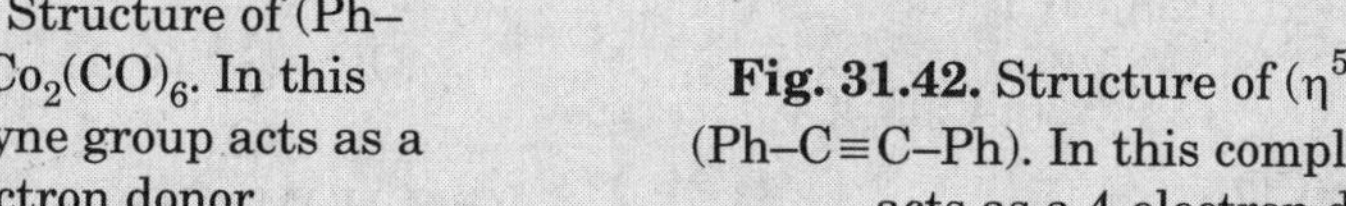


This ligand forms many complexes with transition metals.

## Preparation

(*i*) Metal complexes containing π-allyl group (π-$C_3H_5$) are prepared by the reaction of σ-bonded allyl magnesium bromide, (σ-$C_3H_5$)MgBr with some metal salt or organometallic derivative. For example:

(*a*) $2(\sigma\text{–}C_3H_5)MgBr + NiCl_2 \xrightarrow{-10°C} (\pi\text{–}C_3H_5)_2\ Ni + 2MgBrCl$

(*b*) $(\sigma\text{–}C_3H_5)MgBr + (\pi\text{–}C_5H_5)_2\ Ni \longrightarrow (\pi\text{-}C_5H_5)\ (\pi\text{-}C_3H_5)\ Ni + (\sigma\text{-}C_5H_5)\ MgBr$

(*ii*) $(\pi\text{-}C_5H_5)Mo(CO)_2\ (\pi\text{-}C_3H_5)$ can be prepared as shown in Fig. 31.43.

The formation of $(\pi\text{-}C_5H_5)\ Mo(CO)_2\ (\pi\text{-}C_3H_5)$ can also be shown by the following chemical equation:

$$(\pi\text{-}C_5H_5)Mo(CO)_3 + (\pi\text{-}C_3H_5)Cl \longrightarrow (\pi\text{-}C_5H_5)\ (CO)_3\ Mo\ (\sigma\text{-}C_3H_5)$$

σ-bonded allyl derivative
(yellow oil, m.pt = $-5^0$C)

$$(\pi\text{-}C_5H_5)\ Mo\ (CO)_2\ (\pi\text{–}C_3H_5) \xleftarrow[h\nu]{\text{Heat or}}$$

π-allyl complex [yellow solid,
= m. pt. = 134°C(dec.)].

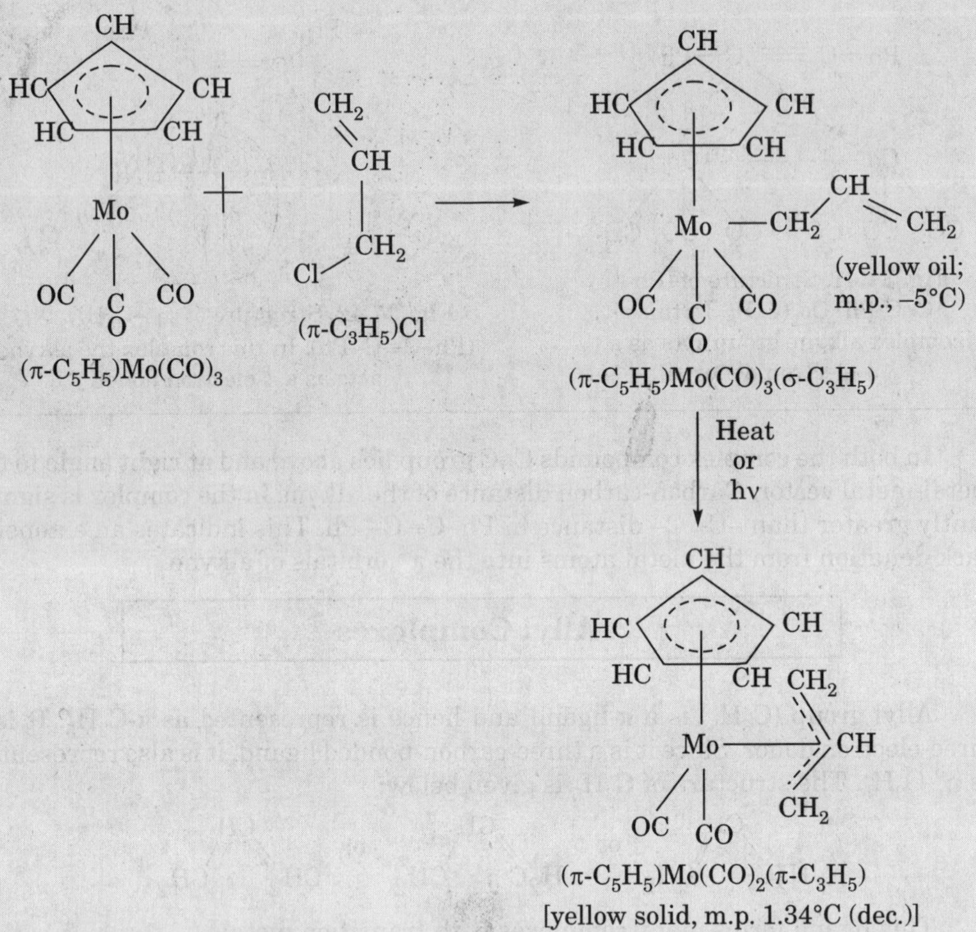


**Fig. 31.43**. Preparation of mixed π-allyl complex, $(\pi\text{-}C_5H_5)$ $Mo(CO)_2$ $(\pi\text{-}C_3H_5)$

## Properties

Pure π-allyl complexes are generally less stable than the mixed derivatives. The π-allyl complexes of transition metals appearing early in the transition series are highly reactive and exhibit catalytic activity, in contrast with the complexes of the elements at the ends of the series. As might be expected, the π-allyl group is rather readily displaced by a variety of ligands.

$$Ni(\pi\text{-}C_3H_5)_2 + 4CO \longrightarrow Ni(CO)_4 + C_6H_{10}$$

$$Ni(\pi\text{-}C_3H_5)_2 + 4PR_3 \longrightarrow Ni\,(PR_3)_4 + C_6H_{10}$$

## Structure and Bonding

In some instances the plane of the π-allyl group is essentially perpendicular to the expected metal-ligand vector, whereas in other allyl derivatives it is parallel to this line.

The sandwich-type complexes of the triangular π -allyl groups appear to involve the overlap of filled π-orbitals on the ligand with appropriate empty orbitals; back

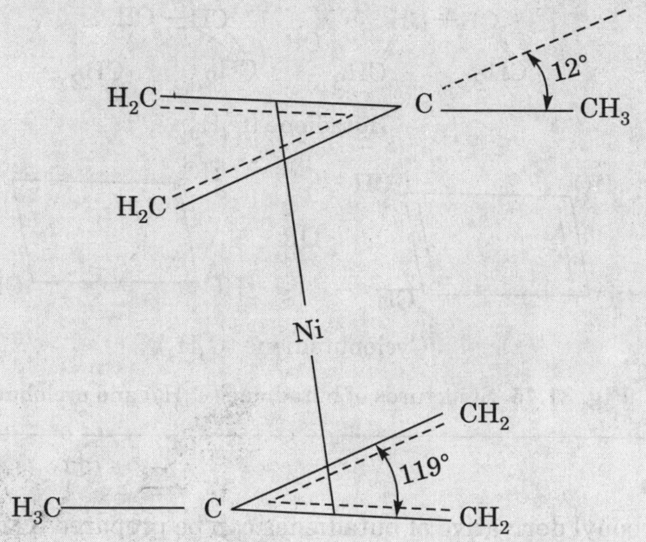


**Fig. 31.44.** Sandwich structure of *bis* ($\eta^3$ – 2 methylallyl) nickel, $[\eta^3 - C_3H_4(CH_3)]_2Ni$

donation presumably involves overlap of filled metal orbitals with empty π-ligand orbitals.

The sandwich structure of *bis* ($\eta^3$ – 2–methyl allyl) nickel, $[\eta^3\text{–}C_3H_4\ (CH_3)]_2$ Ni is given in Fig. 31.44. This figure shows that ($\eta^3\text{–}C_3H_4\ (CH_3)$ group is bonded to Ni–atom in such a way that the three cyclic C-atoms of each $\eta^3\text{–}C_3H_4(CH_3)$ group are equidistant from the Ni metal.

The three $p\pi$ orbitals of the allyl group combine to form three MOs, $\psi_1$, $\psi_2$, $\psi_3$. Ligand-to-metal electron donation may then occur by overlap of $\psi_1$ with the metal s, $d_{z^2}$ and $p_z$ orbitals and by overlap of $\psi_2$ with the metal $p_y$ and $d_{yz}$ orbitals. The metal-to-ligand back-bond is formed by overlap of $d_{xz}$ with $\psi_3$.

## Transition Metal Complexes with Butadiene ($C_4H_6$)

Conjugated dienes act as four-electron ligands in which the entire π-system apparently acts as a four-electron donor. Unconjugated dienes also act as four-electron ligands but they do so because they are bidentate. The most widely investigated four-electron ligand is butadiene and its cyclic derivative cyclobutadiene.

Structures of butadiene ($C_4H_6$) and cyclobutadiene ($C_4H_4$) are given below in Fig. 31.45.

Cyclobutadiene is antiaromatic and unstable in the free state, but it can be stabilized by complexing with transition metals. Most of the known complexes of butadienes are carbonyl derivatives.

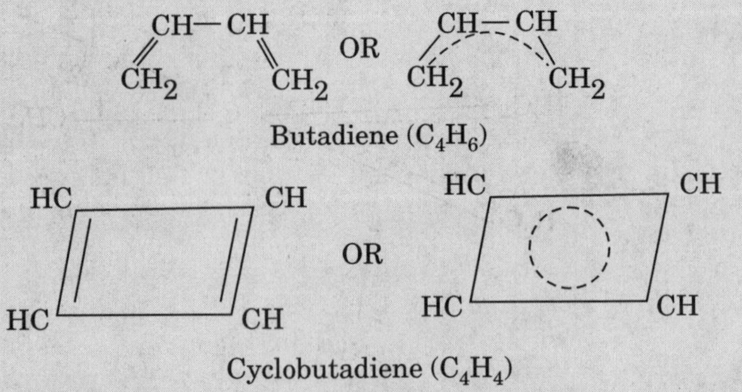


**Fig. 31.45.** Structures of butadiene ($C_4H_6$) and cyclobutadiene

## Preparation

1 The carbonyl derivative of butadienes can be prepared by the direct combination of butadiene with metal carbonyl. For example:

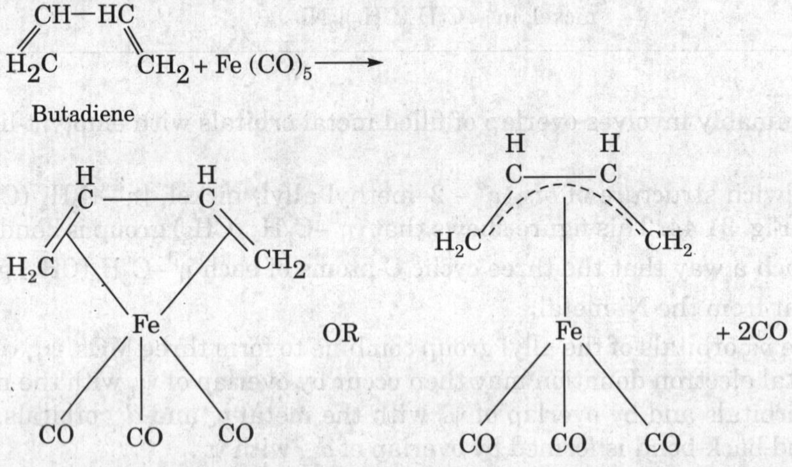


Carbonyl derivative
of butadiene complex

2 Cyclobutadiene complexes are prepared:

(*i*) By the action of $C_4H_4Cl_2$ or $(CH_3)_4C_4Cl_2$ on metal carbonyls. For example:

(a)

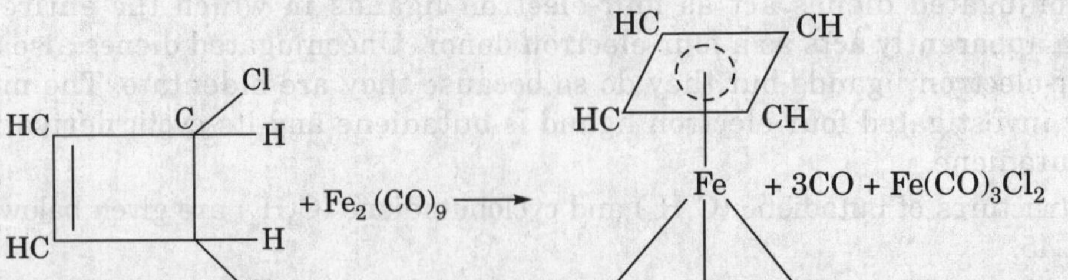


*Cis* -dichloro
cyclobutadiene, $C_4H_2Cl_2$

Cyclobutadiene tricarbonyl
iron complex

*or* $C_4H_4Cl_2 + Fe_2(CO)_9 \longrightarrow (C_4H_4)Fe(CO)_3 + 3CO + Fe(CO)_3Cl_2$

(b)

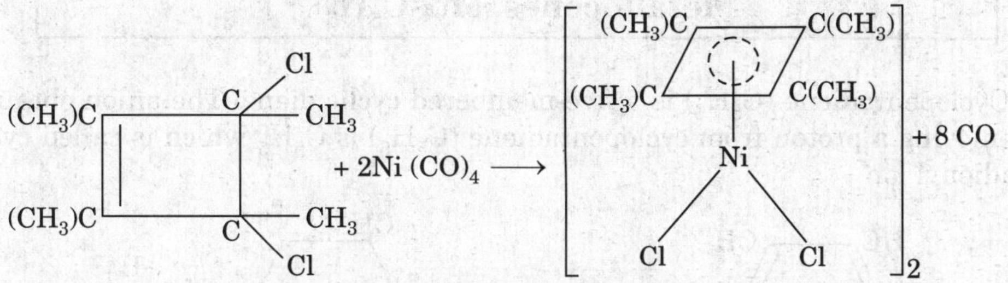


*or* $2\ (CH_3)_4\ C_4\ Cl_2 + 2Ni(CO)_4 \longrightarrow [(CH_3C)_4NiCl_2]_2 + 8\ CO$

(*ii*) By the action of acetylenes on metal carbonyls. For example:

(*a*)

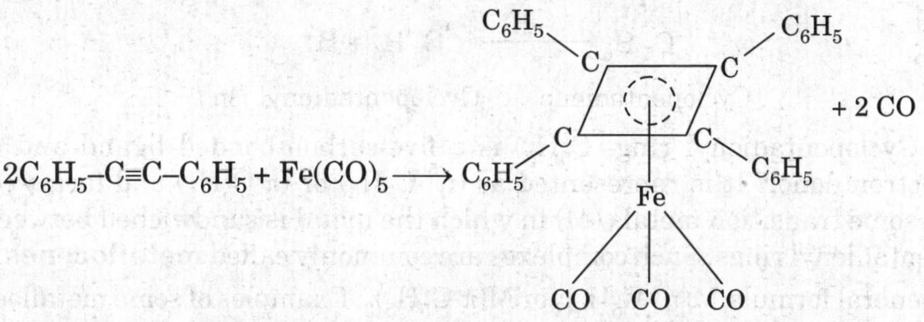


*or* $2C_6H_5–C{\equiv}C–C_6H_5 + Fe(CO)_5 \longrightarrow [\pi\text{-}C_4(C_6H_5)_4]Fe(CO)_3 + 2CO$

(*b*) $4C_6H_5–C{\equiv}C–C_6H_5 + Mo(CO)_6 \longrightarrow [\pi–C_4(C_6H_5)_4]_2\ Mo(CO)_2 + 4CO$

## Properties

Cyclobutadiene ring present in cyclobutadiene tricarbonyl iron complex, $(C_4H_4)Fe(CO)_3$ has aromatic character and hence undergoes readily acetylation (electrophilic substitution)

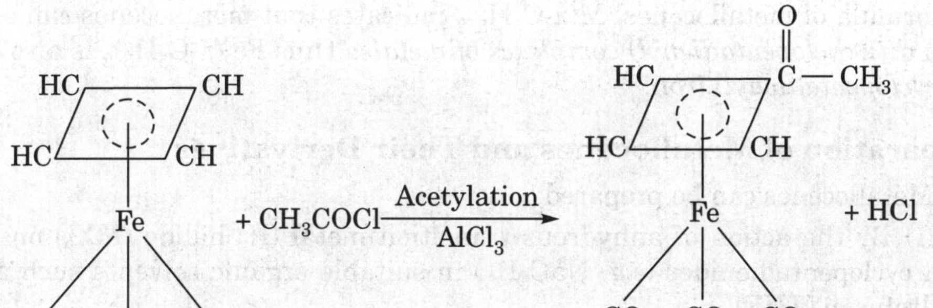


*or* $(C_4H_4)Fe(CO)_3 + CH_3COCl \xrightarrow[AlCl_3]{Acetylation} C_4H_3(CH_3CO)Fe(CO)_3 + HCl$

## Structure

The ligand-carbon atoms are essentially planar and equidistant from the metal atom in complexes of butadienes and cyclobutadienes. The substituents on the terminal carbon atoms in the complexes are bent slightly out of the plane of the ligand and from the metal atom.

# Di(π-Cyclopentadienyl) Metal Complexes: Metallocenes, $M(\pi\text{-}C_5H_5)_2$

Cyclopentadiene ($C_5H_6$) is a five-membered cyclic diene. The anion obtained by removing a proton from cyclopentadiene ($C_5H_6$) is $C_5H_5^-$ which is called cyclopentadienyl ion

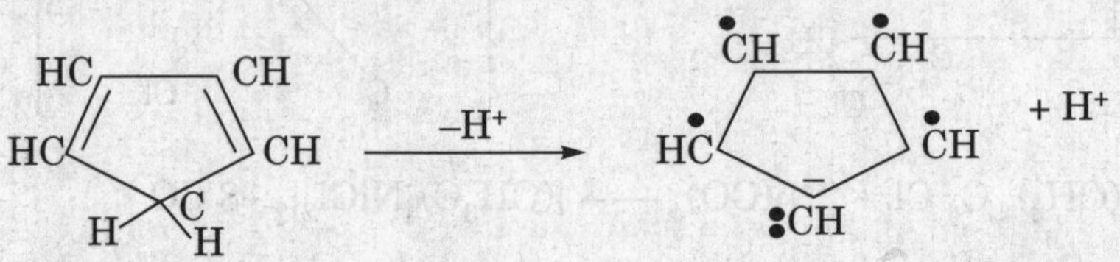


$$\text{or} \quad C_5H_6 \xrightarrow{-H^+} C_5H_5^- + H^+$$

Cyclopentadiene Cyclopentadienyl ion

Cyclopentadienyl ring ($C_5H_5$) is a five-carbon-bonded ligand and acts as a 5-electron donor. It is represented as ($\eta^5$-$C_5H_5$) or ($\pi$-$C_5H_5$) and forms complexes with some transition metals (M) in which the metal is sandwiched between two cyclopentadienyl rings. Such complexes are commonly called **metallocenes** and have the general formula, $M(\eta^5\text{-}C_5H_5)_2$ or $M(\pi\text{-}C_5H_5)_2$. Examples of some metallocenes are:

(*i*) *Ferrocene* or *bis* ($\eta^5$-cyclopentadienyl) iron, $Fe(\eta^5\text{-}C_5H_5)_2$ or $Fe(\pi\text{-}C_5H_5)_2$

(*ii*) *Nickelocene*, $Ni(\pi\text{-}C_5H_5)_2$

(*iii*) *Cobaltocene*, $Co(\pi\text{-}C_5H_5)_2$

(*iv*) *Ruthenocene*, $Ru(\pi\text{-}C_5H_5)_2$

In such compounds $\eta^5$ signifies that all the five C-atoms of $C_5H_5$ ring are bonded to the metal atom through π-bonding molecular orbitals which are delocalised over all the five C-aloms of $C_5H_5$ ring. Thus $C_5H_5$ behaves as a π-ligand. The formula of metallocenes, $M(\pi\text{-}C_5H_5)_2$ indicates that metallocenes can also be called *di(π-cyclopentadienyl) complexes of metals*. Thus $Fe(\pi\text{-}C_5H_5)_2$ is also called *di(π-cyclopentadienyl) iron*.

## Preparation of Metallocenes and Their Derivatives

Metallocenes can be prepared:

(*i*) By the action of anhydrous transition metal (II) halide ($MX_2$) on alkali metal cyclopentadienides (*e.g.* $NaC_5H_5$) in suitable organic solvents such as tetrahydrofuran (THF)

$$2NaC_5H_5 + MCl_2 \xrightarrow[\text{Reflux}]{\text{THF}} M(\pi\text{-}C_5H_5)_2 + 2NaCl$$

(*ii*) By the action of $NaC_5H_5$ or $LiC_5H_5$ on the coordination compound of the metal. For example:

(*a*) $2NaC_5H_5 + [Co(NH_3)_6](SCN)_2 \longrightarrow 2NaSCN + [Co(NH_3)_6](C_5H_5)_2$

$$[Co(NH_3)_6](C_5H_5)_2 \xrightarrow[\text{pressure}]{\text{Reduced}} Co(\pi\text{-}C_5H_5)_2 + 6NH_3$$

(*b*) $2Li\ C_5H_5 + [Co(NH_3)_4(SCN)_2] \rightarrow [Co(NH_3)_4(C_5H_5)_2] + 2LiSCN$

$$[Co(NH_3)_4(C_5H_5)_2] \xrightarrow[\text{pressure}]{\text{Reduced}} Co(\pi\text{-}C_5H_5)_2 + 4NH_3$$

(*iii*) By the action of Grignard cyclopentadienyl magnetium bromide ($C_5H_5MgBr$) on anhydrous metal (II) halide ($MX_2$). The reaction should be carried out in benzene or ether.

$$2C_5H_5MgBr + MX_2 \xrightarrow[\text{or ether}]{\text{Benzene}} M(\pi\text{-}C_5H_5)_2 + 2Mg\,X\,Br$$

(*iv*) $M(\pi\text{-}C_5H_5)_2$ (M=Fe, Co, Ni) can be obtained by the action of anhydrous metal (II) halide on cyclopentadiene ($C_5H_6$) in presence of a strong base like diethylamine, $NH(C_2H_5)_2$. For example:

$$FeCl_2 + 2C_5H_6 + 2NH(C_2H_5)_2 \longrightarrow Fe(\pi\text{-}C_5H_5)_2 + 2(C_2H_5)_2NH_2Cl \text{ or } 2(C_2H_5)_2NH.HCl$$

(*v*) $Fe(\pi\text{-}C_5H_5)_2$ can be obtained by any of the following methods:

(*a*) $2C_5H_6 + Fe \xrightarrow{300°C} Fe(\pi\text{-}C_5H_5)_2 + H_2$

(*b*) $2C_5H_6 + FeO \xrightarrow{250°C} Fe(\pi\text{-}C_5H_5)_2 + H_2O$

(*c*) $2C_5H_5MgBr + FeCl_2 \longrightarrow Fe(\pi\text{-}C_5H_5)_2 + 2MgClBr$

(*vi*) Mixed carbonyl derivatives can be obtained by the action of $NaC_5H_5$ on metal carbonyls or metal carbonyl halides. For example:

(*a*) $NaC_5H_5 + W(CO)_6 \longrightarrow Na^+[(\pi\text{-}C_5H_5)W(CO)_5]^- + CO$

(*b*) $4NaC_5H_5 + 2Os(CO)_3Cl_2 \longrightarrow [(\pi\text{-}C_5H_5)Os(CO)_2]_2 + 4NaCl + C_{10}H_{10}$

Mixed carbonyl derivatives can also be obtained by the action of cyclopentadiene ($C_5H_6$) or cyclopentadienyl complexes on metal carbonyls. For example:

(*a*) $2C_5H_6 + Co_2(CO)_8 \rightarrow 2(C_5H_5)Co(CO)_2 + 4CO + H_2$

(*b*) $(C_5H_5)_2Ni + Ni(CO)_4 \rightarrow [(C_5H_5)Ni(CO)]_2 + 2CO$

Reaction (*a*) is not ligand replacement reaction, since CO and $C_5H_5^-$ do not carry the same charge, while reaction (*b*) is a straight forward ligand exchange reaction.

(*vii*) Some of the cyclopentadienyl complexes of transition metals can be prepared by the action of alkali cyclopentadienides,$M(C_5H_5)$ with the salts of transition metals in solvents like liq. $NH_3$.

(*a*) $Co(SCN)_2 + 4NH_3 \longrightarrow Co(NH_3)_4(SCN)_2$

(*b*) $Co(SCN)_2(NH_3)_4 + 2C_5H_5Li \longrightarrow Co(NH_3)_4(C_5H_5)_2 + 2LiSCN$

(*c*) $Co(NH_3)_4(C_5H_5)_2 \xrightarrow[\text{under pressure}]{\text{Reduced}} Co(C_5H_5)_2 + 4NH_3$

## Properties of Metallocenes and Their Derivatives

(*i*) All the 3*d* transition metal complexes are neutral, soluble in organic hydrocarbon solvents and are sublimable at 100°C, and do not pyrolyse even at 470 °C. All the 3*d* complexes possess the melting point between 172—174°C while the heavier metal complexes melt at higher temperatures. These complexes vary very much in their stabilities. Ferrocene only is stable in air but V, Cr, Ti, Ni, Co complexes should be handled in vacuum.

Stability order is Fe > Ni > Co > V > Cr > Ti. Thus $(\pi\text{-}C_5H_5)_2Ni$ is moderately stable, *i.e.*, decomposes in air in two days in solid phase, but decomposes faster in solution.

(*ii*) All these are stable towards the hydrolysis.

(*iii*) All these are easily oxidised in the acid solution to give cations (oxidations generally are reversible).

$$\pi(C_5H_5)_2Fe \xrightleftharpoons{\text{Acid}} \underset{\text{Blue green}}{(\pi\text{-}C_5H_5)_2Fe^+} (\textit{Ferrocenium ion}) + e^-$$

$$Cr(\pi\text{-}C_5H_5)_2 \rightleftharpoons (\pi\text{-}C_5H_5)_2Cr^+ (\textit{Chromocenium ion}) + e^-$$

$$V(\pi\text{-}C_5H_5)_2 \rightleftharpoons \underset{\text{Purple}}{(\pi\text{-}C_5H_5)_2V^+} (\textit{Vanadoceium ion}) + e^-$$

$$\pi\text{-}(C_5H_5)_2Co \rightleftharpoons \underset{\text{Yellow-green}}{(\pi\text{-}C_5H_5)_2Co^+} (\textit{Cobaltocenium ion}) + e^-$$

These ions can be precipitated by large ions like $I_3^-$, $BF_4^-$, $[SbCI_6]^-$ as salts. These salts are soluble in liquid $SO_2$ and acetone and can be recrystallised by the addition of chloroform or ether.

(*iv*) The π-cyclopentadienyl carbonyls are highly coloured and soluble in organic solvents. They follow 18-electron rule. Since the cyclopentadienyl ring is less stable than a carbonyl group, it frequently survives the action of reagents capable of removing carbonyl groups or replace other ligands.

$$(\pi\text{-}C_5H_5)V(CO)_4 \xrightarrow{Cl/O_2} (\pi\text{-}C_5H_5)\,VOCl_2 + 4CO_2$$

$$(\pi\text{-}C_5H_5)Mo(CO)_3Cl + Cl_2 \longrightarrow (\pi\text{-}C_5H_5)Mo(CO)_2Cl_3 + CO$$

$$(\pi\text{-}C_5H_5)Mo(CO)_3Cl + NH_3 \longrightarrow [(\pi\text{-}C_5H_5)Mo(CO)_3NH_3]^+\,Cl^-$$

(*v*) The metal-metal bonds in binuclear cyclopentadienyl carbonyl derivatives can be cleared by oxidising or reducing agents.

$$[(\pi\text{-}C_5H_5)Fe(CO)_2]_2 + Cl_2 \longrightarrow 2(\pi\text{-}C_5H_5)Fe(CO)_2\,Cl$$

$$[(\pi\text{-}C_5H_5)Fe(CO)_2]_2 + Na \longrightarrow 2Na^+[(\pi\text{-}C_5H_5)Fe(CO)_2]^-$$

(*vi*) Many of $M(C_5H_5)$ compounds have been oxidised to $[M(C_5H_5)]^+$cation in which M is tetravalent.

A few typical metallocenes and their properties are listed in Table. 31.1.

## Ferrocene, $Fe(C_5H_5)_2$

### Preparation of Ferrocene, $Fe(C_5H_5)_2$

Ferrocene, can be prepared:

(*i*) *By the action of anhydrous iron (II) halides on cyclopentadiene ($C_5H_6$) in presene of a strong base like diethyl amine, $NH(C_2H_5)_2$*

$$FeCl_2 + 2C_5H_6 + 2NH\,(C_2H_5)_2 \longrightarrow (C_5H_5)_2\,Fe + 2(C_2H_5)_2\,NH_2Cl$$
$$\textit{or}\ 2(C_2H_5)_2\,NH.\,HCl$$

*(ii) By the action of Fe (II) halide on sodium cyclopentadiene ($NaC_5H_5$)*. In this method cyclopentadiene ($C_6H_5$) is first converted into sodium cyclopentadienide ($C_5H_5$ Na) by treating it with finely-divided dispersed Na in tetrahydrofuran (THF).

$$\underset{\text{Cyclopentadiene}}{C_5H_5} + Na \rightleftharpoons \underset{\text{Sodium cyclopentadienide}}{C_5H_5\,Na} + \frac{1}{2}H_2$$

$C_5H_5$ Na is then treated with $FeCl_2$.

$2C_5H_5\,Na + FeCl_2 \rightarrow Fe(C_5H_5)_2 + 2NaCl$.

*(iii) By the action of Grignard reagent of cyclopentadiene, ($C_5H_5$)MgBr on $FeCl_2$.*

$$FeCl_2 + 2(C_5H_5)MgBr \longrightarrow Fe\,(C_5H_5)_2 + MgCl_2 + MgBr_2.$$

**Table 31.1 Some typical sandwich compounds and their properties (cp = $\pi$–$C_5H_5$)**

| Compounds | Some characteristic features |
|---|---|
| $cp_2Ti$ (n=o) | Dark-green crystals, m.p.> 200°C (dec.). Actually it exists as [(Cp) (Cp–H)TiH]$_2$ |
| $cp_2$ V (n=3) | Purple crystals, m.p.167°C. Very air sensitive, $\mu$ =3.84BM. |
| $cp_2Mn$ | Dark-brown crystals, m.p.173°C. Air sensitive, hydrolysed with cracking sound. Heating in $N_2$ at 158°C turns the brown solid to pale-pink solid, which melts at 173°C. $\mu$ =5.86 BM. |
| $cp_2Fe$ (n=0) | Orange crystals, m.p. 173°C, dec. 470°C, oxidised by $Ag^+$(aq.), dil. $HNO_3$, etc. $Cp_2Fe \rightleftharpoons Cp_2Fe^+ + e^-$; $E°$= +0.3 V versus SCE. |
| $cp_2Cr$ (n=2) | Scarlet crystals, m.p. 173°C, very air sensitive, soluble in HCl giving $C_5H_6$ and a blue solution possibly containing Cp-CrCl $(H_2O)_x^+$, $\mu$ = 3.20 BM. |
| $cp_2Ni$ (n=2) | Bright-green, m.p. 173°C (dec.). Fairly air stable as soild, oxidised to $Cp_2Ni^+$; NO gives CpNi(NO); Na–Hg in EtOH gives CpNi($C_5H_7$).$\mu$ 2.86 BM. |
| $cp_2Co$ (n=1) | Purple-black crystals, m.p. 174°C. Air-sensitive, oxidised to stable yellow $Cp_2Co^+$ salts. $\mu$ = 1.73 BM. |
| $cp_2Co^+$ | Yellow in aqueous solution. Forms numerous salts and a strong base which absorbs $CO_2$ from air. Stable to *ca*. 400°C. |
| $cp_2TiCl_2$ | Bright-red crystals, m.p. 230°C, soluble in $H_2O$ giving $Cp_2Ti(OH)^+$. |
| $cp_2WH_2$ | Yellow crystals, m.p. 163°C; soluble in acids giving $Cp_2WH_3^+$. |

## Commercial Preparation of $Fe(C_5H_5)_2$

On large scale, $Fe(C_5H_5)_2$ can be obtained by the action of cyclopentadiene ($C_5H_6$) on Fe or FeO.

$$2C_5H_6 + Fe \xrightarrow{300°C} (C_5H_5)_2\,Fe + H_2$$

$$2C_5H_6 + FeO \xrightarrow{250°} (C_5H_5)_2\,Fe + H_2O$$

## Laboratory Preparation of Fe $(C_5H_5)_2$

In the laboratory, $(C_5H_5)$ Fe is prepared by treating $FeCl_2$ with potassium cyclopentadienide (K $C_5H_5$) obtained by the action of $C_5H_6$ on KOH.

$$KOH + C_5H_6 \longrightarrow KC_5H_5 + H_2O \qquad \text{----}(i)$$

$$FeCl_2 + 2K\,C_5H_5 \longrightarrow (C_5H_5)_2\,Fe + 2KCl$$

We have seen above that $C_5H_6$ is needed for the preparation of $KC_5H_5$. $C_5H_6$ does not exist as a monomer at room temperature. Because of its tendency to get dimerised, $C_5H_6$ exists as a dimer, $(C_5H_6)_2$ (called dicyclopentadiene) at room temperature. In order to get monomer, the dimer is heated slowly to 180°C in an oil-bath. This is called cracking of dimer. The cracking is done in an apparatus which consists of a flask fitted with a vertical air condenser, a thermometer, a water condenser and a receiver flask (See Fig. 31.46). After initial frothing, the monomer starts distilling off at 42° to 44°C and collected in a receiver flask. To prevent dimerisation of the monomer, the receiver flask is cooled in an ice-bath.

The water formed in reaction (*i*) is removed by using an excess of KOH. Since potassium cyclopentadienide is *air-sensitive,* the reaction is done in an atmosphere of $N_2$ or in the presence of diethyl ether which, being volatile, blankets the reaction.

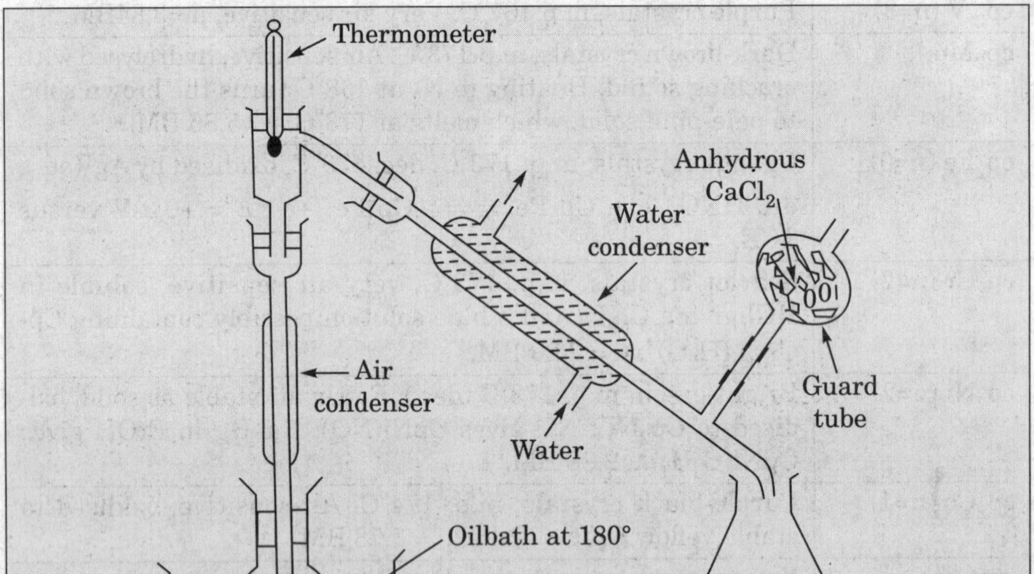


**Fig. 31.46.** Cracking of dicyclopentadiene.

In actual practice, an apparatus comprising a conical flask fitted with a dropping funnel and a bunsen valve (Fig. 31.47) is used for this preparation. Potassium hydroxide flakes (50 g) and diethyl ether (120 ml) are taken in the conical flask and stirred for about 10 minutes and then 11 ml of cyclopentadiene is added to it and the mixture stirred for another 10 minutes. In the meantime, $FeCl_2 . 4H_2O$ (13 g) is powdered, dissolved in 50 ml of dimethylsulphoxide (DMSO) and is taken in the dropping funnel. The iron solution in DMSO is added dropwise with stirring. An exothermic reaction occurs and if ether starts boiling, the bunsen valve may be opened every now and then to release pressure. The reaction is complete in about 30 to 45 minutes. At the end of the reaction, the etherial layer is separated and washed twice with dilute HCl (to remove KOH) followed by two washings with $H_2O$. The etherial solution is then concentrated and cooled to obtain beautiful orange crystals of ferrocene. Yield of the dried product is around 50%. The yield is very poor if NaOH is used in place of KOH. This may be due to poor solubility of NaOH in $Et_2O$ and weaker basic nature of NaOH relative to KOH. Use of fresh iron (II) chloride, as old samples are contaminated with the oxidised products, is suggested for better yield. We don't use ferrous sulphate in place of $FeCl_2$.

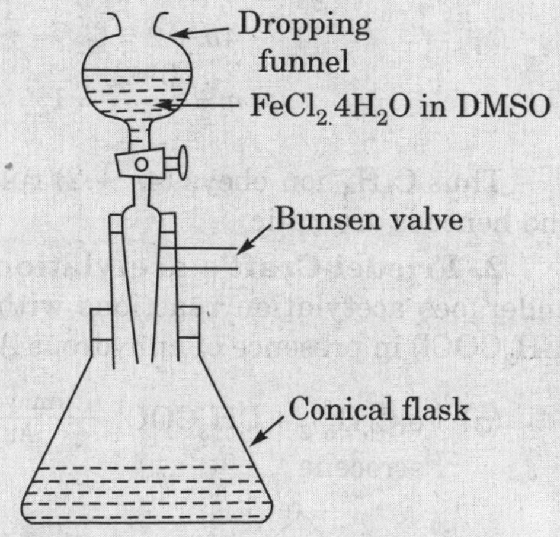

**Fig. 31.47.** Preparation of ferrocene.

## Physical Properties of Ferrocence

Some important physical properties of ferrocene are:

(*i*) It is an orange crystalline solid.

(*ii*) It is soluble in organic solvents.

(*iii*) Its melting point is 174°C.

(*iv*) It sublimes without decomposition.

(*v*) It has high thermal stability, up to about 500°C

(*vi*) Unlike other organometallic compounds, ferrocence is stable to air and water.

## Chemical Reactions of Ferrocene

It can be chemically oxidised to a blue-green ferrocinium ion, $[Fe^{3+} (\eta^5–C_5H_5]_2]^+$. Although most of the chemistry of ferrocene is attributable to the cyclopentadiene ring, yet in contrast to free cyclopentadiene it is very *resistant to hydrogenation* and *does not undergo Diels-Alder* reaction (which is characteristic of conjugated dienes). Ferrocene has an extensive organic chemistry *typical of an aromatic organic compound*. However, it *does not undergo direct nitration and halogenation* but gets oxidised to ferricinium ion. Some reactions typical of aromatic ring are mentioned here.

1. **Aromatic character.** $Fe(C_5H_5)_2$ has two $C_5H_5^-$ ions. $C_5H_5^-$ ion is called cyclopentadienyl ion. The structure of $C_5H_5^-$ ion is given in the margin. This structure shows that each of the five C-atoms is $sp^2$ hybridised. Each of the four C-atoms has an electron in $2p_z$ orbital and the fifth C-atom (negatively-charged C-atom) has two electrons in $2p_z$

orbital. Thus $C_5H_5^-$ ion has six π-electrons. Consequently

$$4n + 2 = 6$$

$$\therefore n = \frac{6-2}{4} = 1$$

Thus $C_5H_5^-$ ion obeys $(4n + 2)$ rule (Huckel's rule) and hence is aromatic.

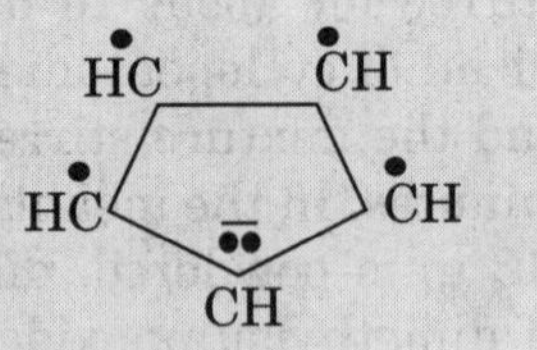

Structure of $C_5H_5^-$ ion

**2. Friedel-Craft's acetylation.** (i) Fe $(C_5H_5)_2$ undergoes acetylation reactions with acetyl chloride ($CH_3COCl$) in presence of anhydrous $AlCl_3$.

(*a*) $Fe(C_5H_5)_2 + CH_3COCl \xrightarrow[AlCl_3]{Anhydrous} (C_6H_4CH_3CO) — Fe(C_5H_5) + HCl$
Ferrocene Acetyl ferrocene

or 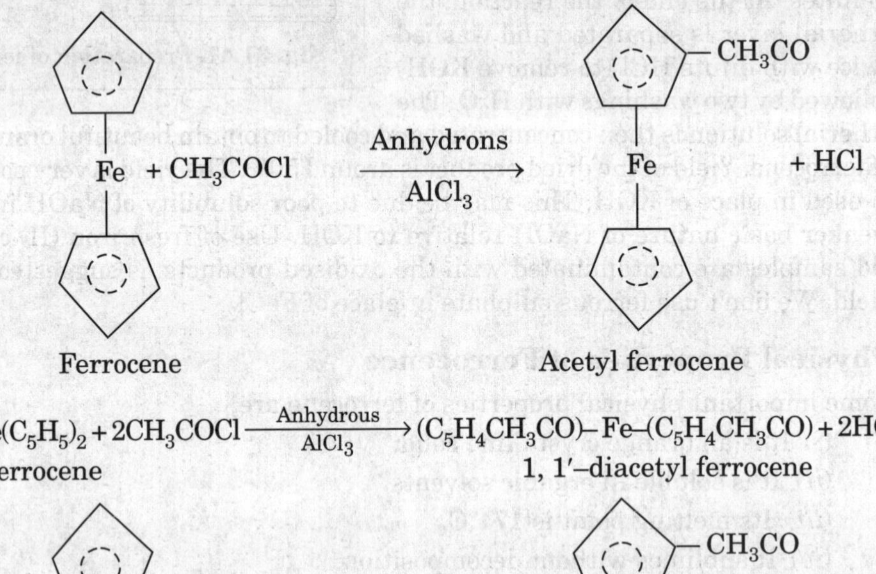


Ferrocene Acetyl ferrocene

(*b*) $Fe(C_5H_5)_2 + 2CH_3COCl \xrightarrow[AlCl_3]{Anhydrous} (C_5H_4CH_3CO)–Fe–(C_5H_4CH_3CO) + 2HCl$
Ferrocene 1, 1′–diacetyl ferrocene

or 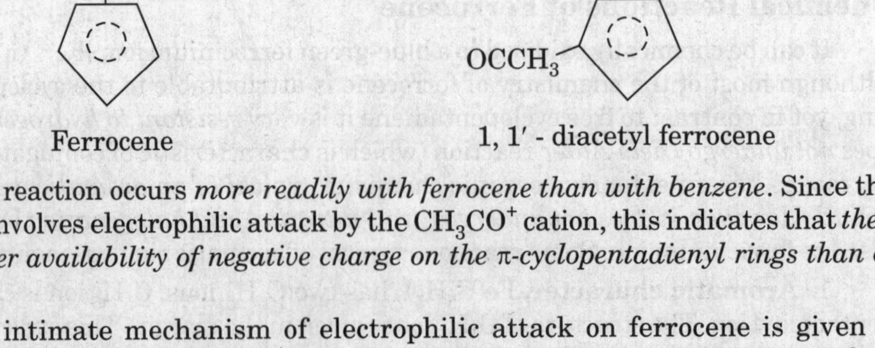


Ferrocene 1, 1′ - diacetyl ferrocene

This reaction occurs *more readily with ferrocene than with benzene*. Since this reaction involves electrophilic attack by the $CH_3CO^+$ cation, this indicates that *there is a greater availability of negative charge on the π-cyclopentadienyl rings than on benzene*.

The intimate mechanism of electrophilic attack on ferrocene is given in Fig. 31.48. The electrophile ($E^+$) first coordinates to the iron, oxidising it from iron (II) to iron (III) before being transferred to the π-cyclopentadienyl ring. The final step involves expulsion of a proton with simultaneous reduction of Fe (III) back to Fe (II).

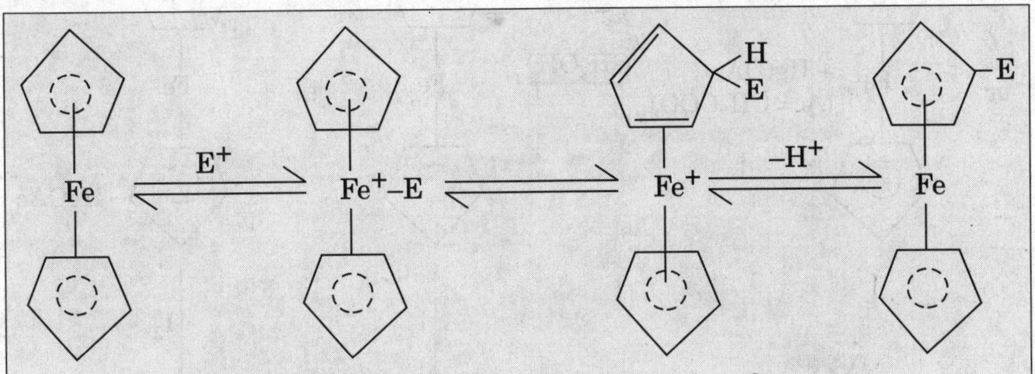

**Fig. 31.48.** Mechanism of electrophilic attack on ferrocene

(*ii*) Acetylation of a ring in ferrocene is also brought about by acetic anhydride in the presence of phosphoric acid as a catalyst.

$$Fe(C_5H_5)_2 + (CH_3CO)_2O \xrightarrow{H_3PO_4} (CH_3COC_5H_4)\ Fe(C_5H_5) + CH_3COOH$$

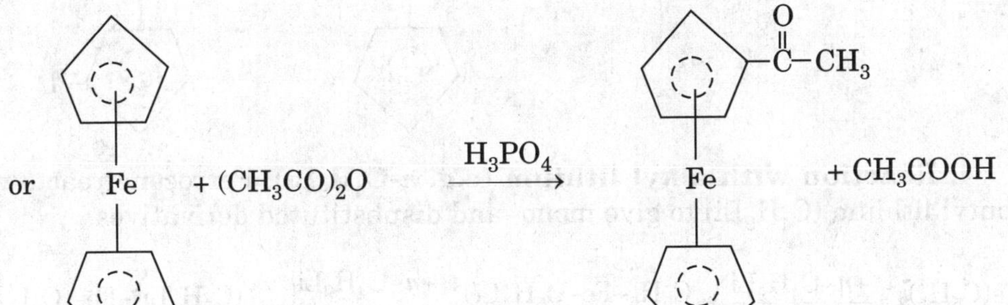

**3. Reaction with mercuric acetate, $Hg(OAc)_2$: Metallation**. Ferrocene reacts with mercuric acetate (HgOAc) to give mono and disubstituted products.

The substituted derivatives of mercuric acetate react with iodide to give iodine derivatives. *This provides an indirect method for carrying out iodination of ferrocene.*

$$Fe(C_5H_5)_2 + Hg(OAc)_2 \xrightarrow{CH_3OH}$$

$$C_5H_4\ HgOAc–Fe–C_5H_5 + C_5H_4\ Hg\ OAc\ –Fe–C_5H_4\ HgOAc$$

$$\downarrow I_2 \qquad\qquad \downarrow I_2$$

$$C_5H_4I–Fe–C_5H_5 \qquad\qquad C_5H_4I–Fe–C_5H_4I$$

Monosubstituted product | Disubstituted product

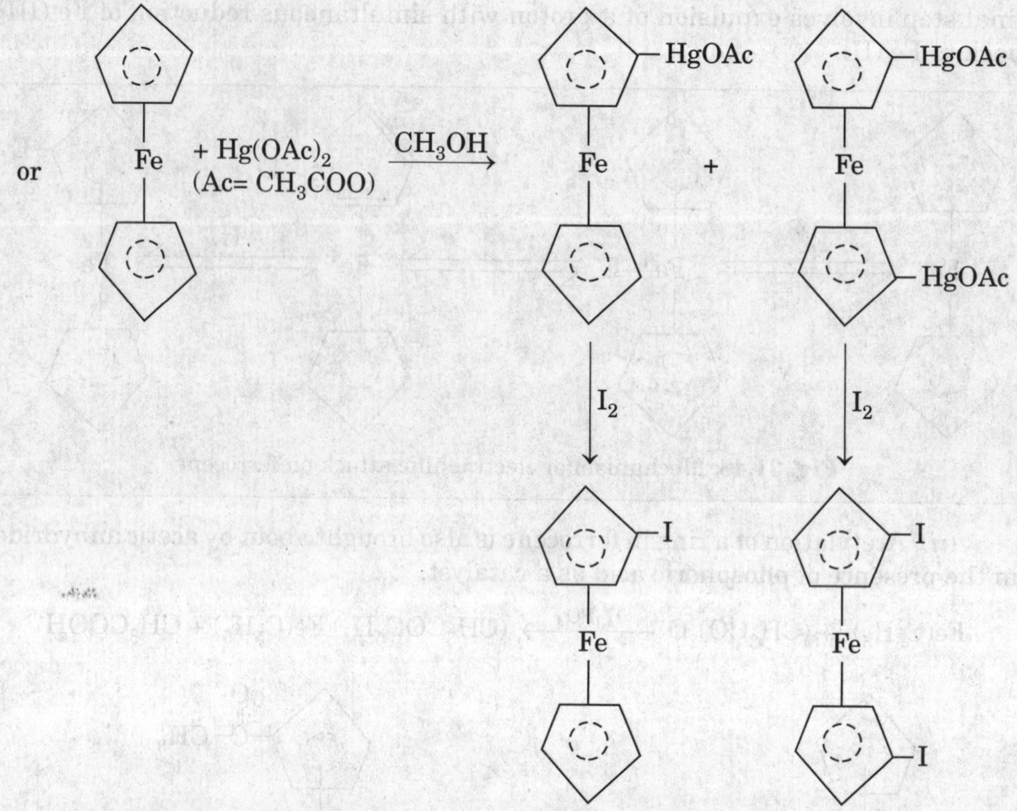


**4. Reaction with alkyl lithium (e.g. $n$-$C_4H_9Li$).** Ferrocene reacts with $n$-butyl lithium ($C_4H_9Li$) to give mono- and disubstituted derivatives.

$$Fe(C_5H_5)_2 \xrightarrow[-C_4H_{10}]{+n\text{–}C_4H_9Li} \underset{\text{Mono substituted derivative}}{C_5H_5\text{–}Fe\text{–}C_5H_4Li} \xrightarrow[-C_4H_{10}]{+n\text{–}C_4H_9Li} \underset{\text{Di substituted derivative}}{(C_5H_4Li)\text{–}Fe\text{–}(C_5H_4Li)}$$

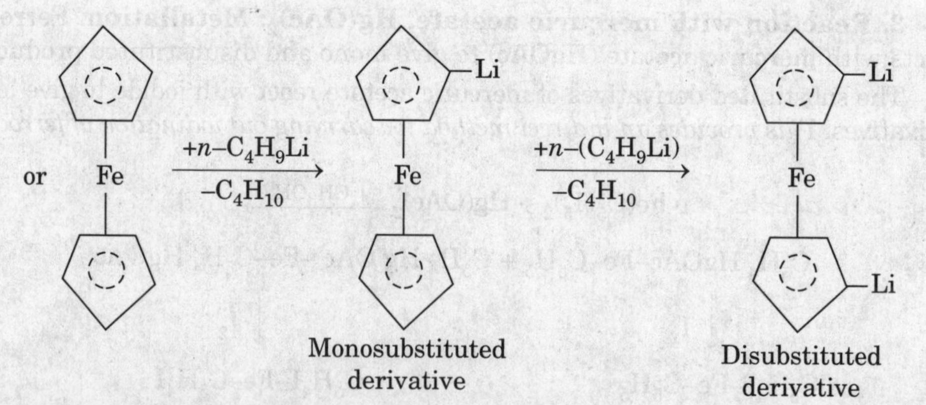


The mono lithium derivatives are very useful intermediates in the synthesis of other ferrocenyl derivatives as shown below.

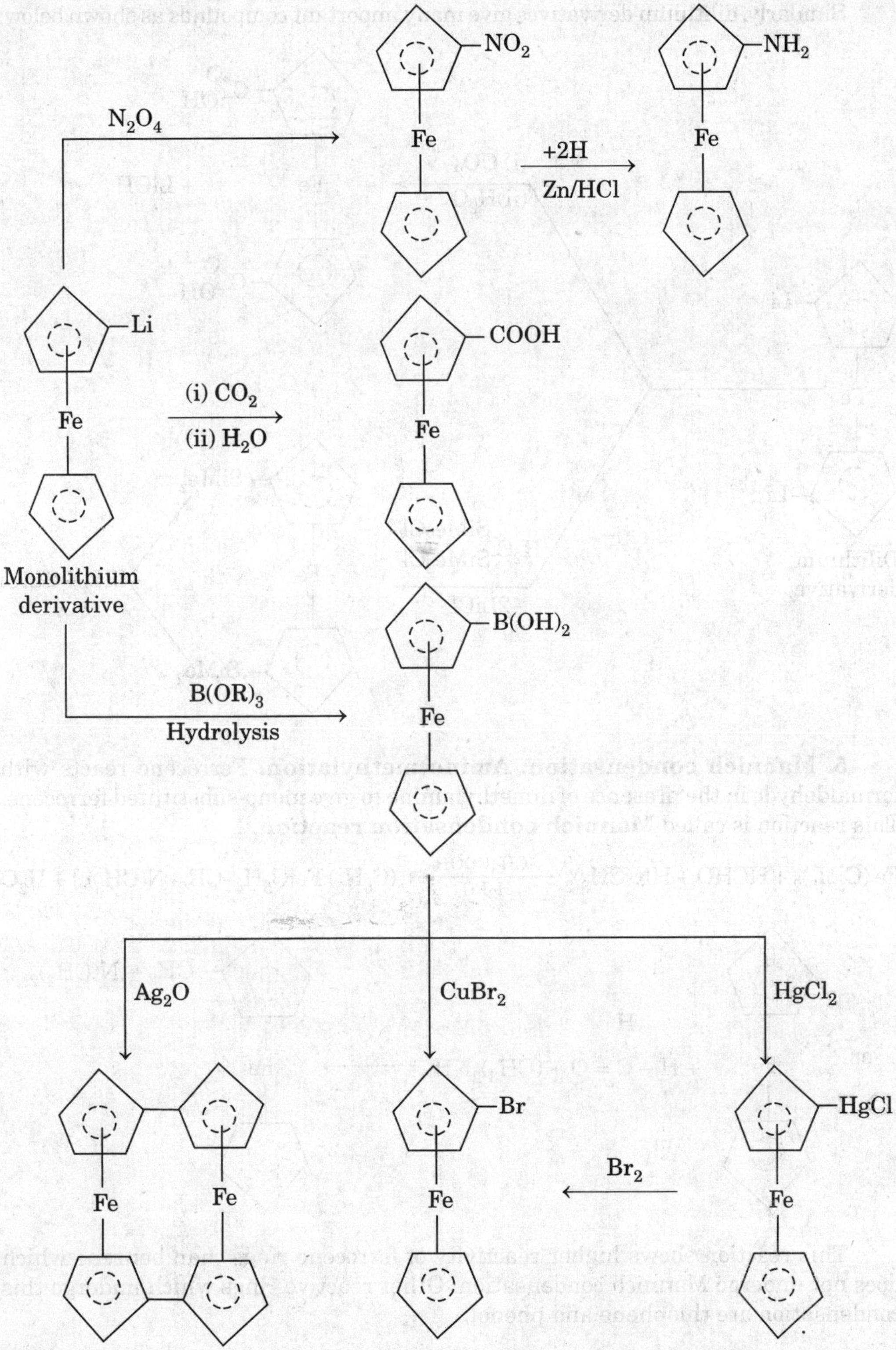

$N_2O_4$
$NO_2$
Fe
+2H
Zn/HCl
$NH_2$
Fe
Li
Fe
Monolithium derivative
(i) $CO_2$
(ii) $H_2O$
COOH
Fe
$B(OR)_3$
Hydrolysis
$B(OH)_2$
Fe
$Ag_2O$
$CuBr_2$
$HgCl_2$
Fe
Fe
Br
Fe
$Br_2$
HgCl
Fe

Similarly, dilithium derivatives give many important compounds as shown below:

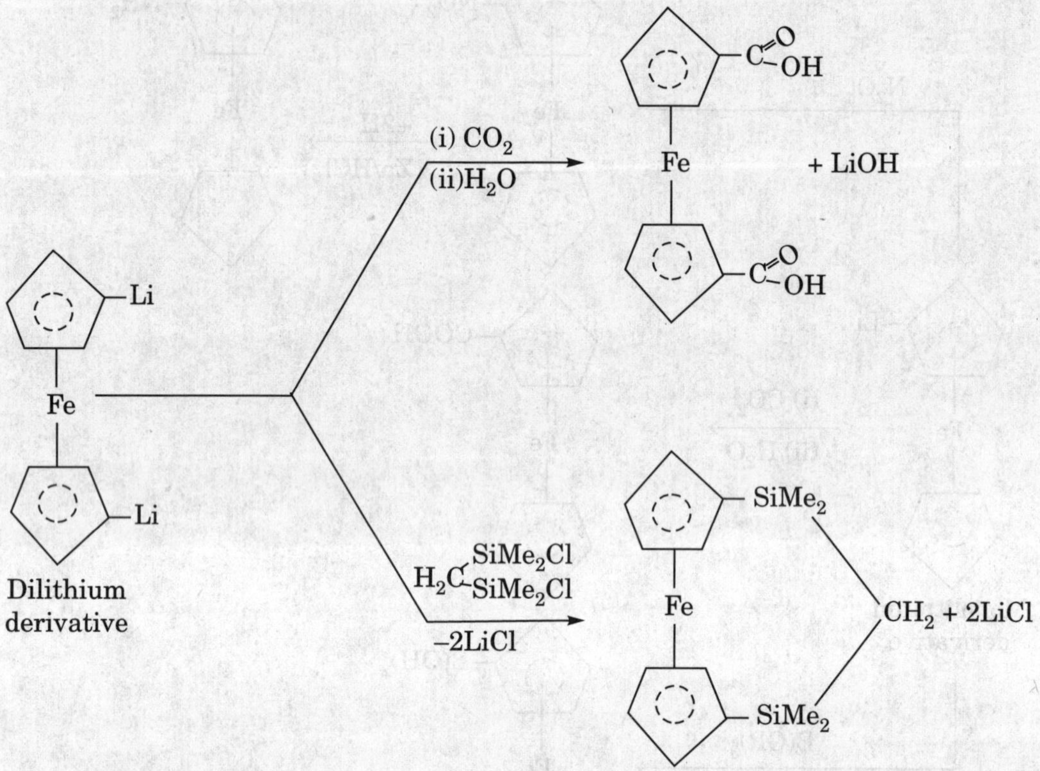


**5. Mannich condensation: Amino methylation.** Ferrocene reacts with formaldehyde in the presence of dimethylamine to give mono-substituted ferrocene. This reaction is called **Mannich condensation reaction**.

$$Fe\,(C_5H_5)_2 + HCHO + HN(CH_3)_2 \xrightarrow[H_3PO_4]{CH_3COOH} (C_5H_5)\,Fe[C_5H_4\text{–}CH_2\text{–}N(CH_3)_2] + H_2O$$

or

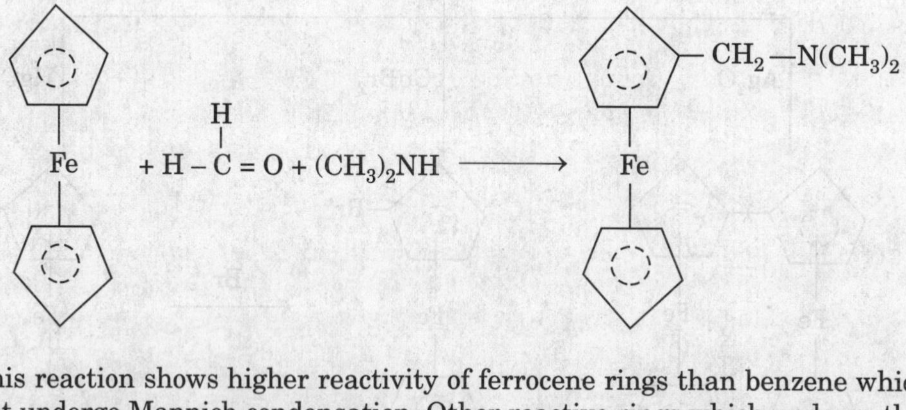


This reaction shows higher reactivity of ferrocene rings than benzene which does not undergo Mannich condensation. Other reactive rings which undergo this condensation are thiophene and phenol.

**6. Fridel-Craft's alkylation.** (*a*) Ferrocene reacts with ethylene ($C_2H_4$) in the presence of anhydrous $AlCl_3$ undergoing alkylation of the ring.

$$\underset{}{Fe\,(C_5H_5)_2} + \underset{\text{Ethene}}{C_2H_4} \xrightarrow[AlCl_3]{\text{Anhydrous}} \underset{\text{Ethyl derivative of ferrocene}}{(C_5H_5)\,Fe(C_5H_4-C_2H_5)}$$

or

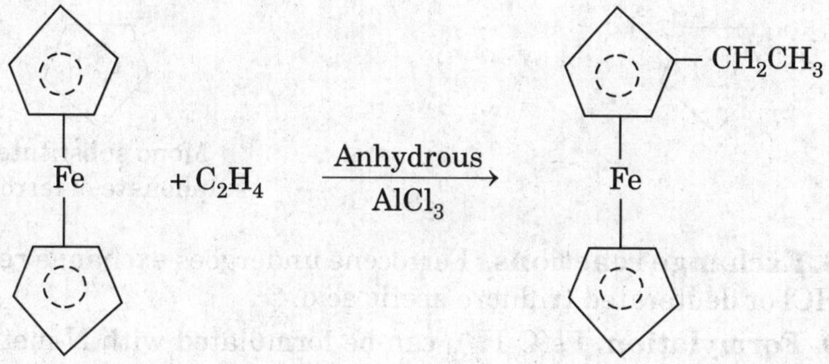

(*b*) Fe ($C_5H_5$)$_2$ also reacts with dichloroethane, Cl–$CH_2$–$CH_2$–Cl($C_2H_4Cl_2$) in presence of anhydrous $AlCl_3$ to form diferrocenyl ethane or ethylidene diferrocenyl.

$$2Fe\,(C_5H_5)_2 + C_2H_4Cl_2 \xrightarrow[AlCl_3]{\text{Anhydrous}} C_5H_5\text{–}Fe\text{–}C_5H_4\text{—}\underset{\underset{H}{|}}{\overset{\overset{CH_3}{|}}{C}}\text{—}C_5H_4\text{–}Fe\text{–}C_5H_5 + 2HCl$$

or

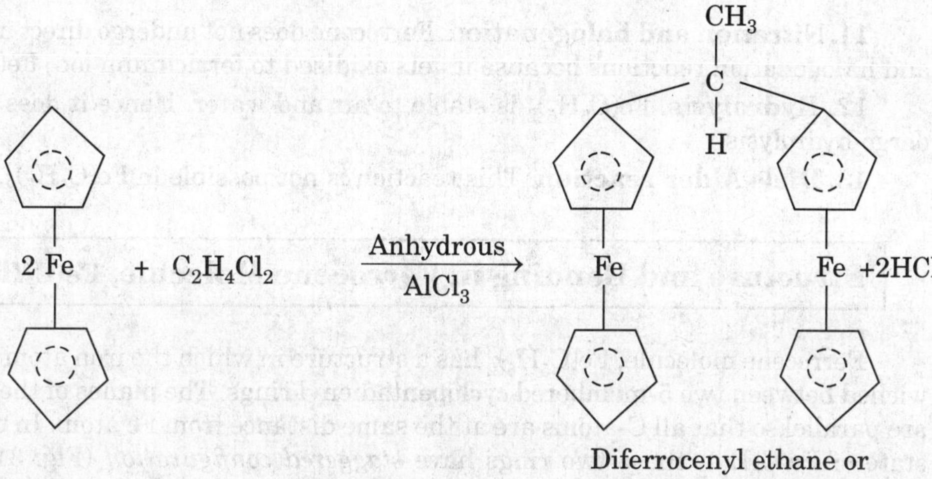

Diferrocenyl ethane or ethylidene diferrocenyl

**7. Sulphonation.** Ferrocene reacts with chlorosulphuric acid ($HSO_3Cl$) or $H_2SO_4$ in presence of acetic anhydride, $(CH_3CO)_2O$ to form monosubstituted sulphonate derivative.

$$Fe(C_5H_5)_2 + HSO_3Cl/H_2SO_4 \xrightarrow{(CH_3CO)_2O)} (C_5H_5)\,Fe\,(C_5H_4\text{–}SO_3H) + HCl/H_2O$$

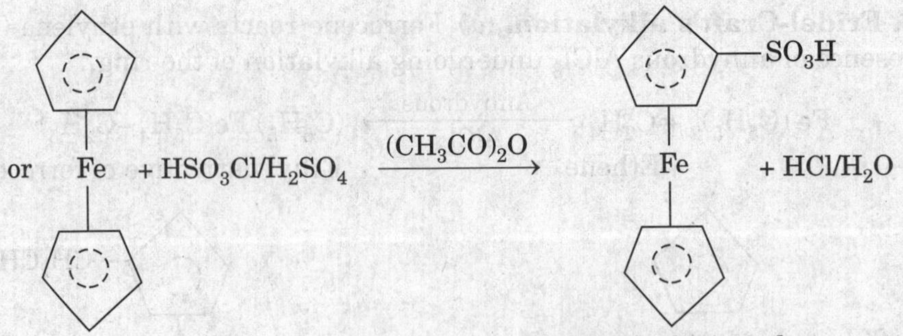


Mono substituted sulphonate of ferrocene

**8. Exchange reactions.** Ferrocene undergoes exchange reactions in deuterated HCl or deuterated trifluoro acetic acid.

**9. Formylation.** $Fe(C_5H_5)_2$ can be formulated with N-methyl formaldehyde in presence of $POCl_3$.

**10. Oxidation.** $Fe(C_5H_5)_2$ can be oxidised in acid solution as well as by $Br_2$

$$Fe(C_5H_5)_2 \xrightarrow{(Na^+ \bar{C}\equiv H)} Fe^{2+}(C_5H_5)_2$$

$$Fe(C_5H_5)_2 \xrightarrow{+\frac{1}{2}Br_2} (C_5H_5)_2 FeBr \xrightarrow{+6Br_2} 2C_5H_5Br + FeBr_3$$

Pentabromo cyclopentane

**11. Nitration and halogenation.** Ferrocene does not undergo direct nitration and halogenation reactions because it gets oxidised to ferricinium ion, $Fe(C_2H_5)_2^+$.

**12. Hydrolysis.** $Fe(C_5H_5)_2$ is stable to air and water. Hence it does not undergo hydrolysis.

**13. Diels-Alder reaction.** This reaction is not possible in $Fe(C_5H_5)_2$.

## Structure and Bonding in Ferrocene Molecule, $Fe(C_5H_5)_2$

Ferrocene molecule, $Fe(C_5H_5)_2$ has a structure in which the iron atom is sandwiched between two 5-membered cyclopentadienyl rings. The planes of these rings are parallel so that all C-atoms are at the same distance from Fe atom. In the solid state of $Fe(C_5H_5)_2$ these two rings have *staggered configuration* (Fig. 31.49 (*a*)) which is because of the crystal packing forces so that carbon-carbon and hydrogen-hydrogen repulsions between the two rings are minimum.

In the gas phase the equilibrium structure of ferrocene is *eclipsed* (Fig. 31.49 (*b*)). The energy barrier between the staggered and eclipsed configurations is only 4KJ $mol^{-1}$.

The bonding in $Fe(C_5H_5)_2$ molecule can be explained on the basis of *valence bond theory* and *molecular orbital theory* as discussed below:

**1. Valence Bond Theory (VBT)**

Although VBT is not able to explain the bonding in $Fe(C_5H_5)_2$ molecule satisfactorily, yet it explains the diamagnetic character of the molecule.

(a) Staggered configuration of $Fe(C_5H_5)_2$ molocule

(b) Eclipsed configuration of $Fe(C_5H_5)_2$ molecule

**Fig. 31.49.** Configuration (structure) of $Fe(C_5H_5)_2$ mol-

According to German workers, Fischer and Fritz (1959), the bonding in $Fe(C_5H_5)_2$ molecule is similar to that in a low spin octahedral complex of $Fe^{2+}$ ion such as $[Fe(CN)_6]^{4-}$. Each $C_5H_5^-$ ion has six electrons (three electron pairs) as is evident from the structure of $C_5H_5^-$ ion.

In $Fe(C_5H_5)_2$, Fe is present as $Fe^{2+}$ ion whose V. S. E. C. in the free state is $3d^6\ 4s^0\ 4p^0$ as shown at (b) of Fig. 31.50. In presence of $C_5H_5^-$ ions all the six $3d$

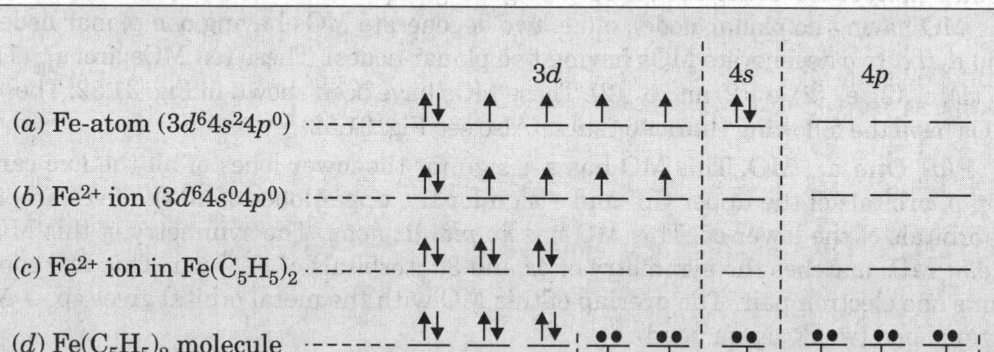


**Fig. 31.50.** $d^2sp^3$ hybridisation in Fe $(C_5H_5)_2$ molecule with $n$ = o. The dots (• •) are the electron pairs donated by $C_5H_5^-$ ion. These electrons have opposite spins.

electrons in $Fe^{2+}$ become paired so that two 3*d* orbitals become vacant and hence are used to get six $d^2sp^3$ hybrid orbitals. Each $C_5H_5^-$ ion donates three electron pairs which are accommodated in vacant $d^2sp^3$ hybrid orbitals. Thus six $d^2sp^3$ hybrid orbitals accept six electron pairs donated by two $C_5H_5^-$ ions.

**2. Molecular Orbital Theory (MOT)**

The exact bonding in $Fe(C_5H_5)_2$ molecule can be explained on the basis of molecular orbital theory. A brief idea of this theory is given below:

(*i*) **Formation of ten ligand group MOs on two $C_5H_5^-$ or $cp^-$ ions.** In $C_5H_5^-$ ring each C-atom is $sp^2$ hybridised and forms three σ-bonds. One σ-bond is formed with H-atom and two σ-bonds. are formed with the neighbouring two C-atoms. As a result, one $2p_z$ (*p*π) orbital is left on each C-atom and this orbital has one electron (See Fig. 31.51.)

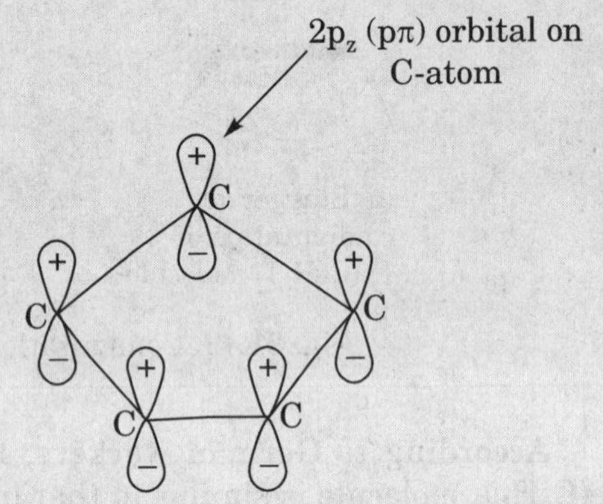


**Fig. 31.51.** Each C-atom of $C_5H_5^-$ ring has one singly-filled *p*π orbital.

These five *p*π orbitals of one planar $C_5H_5^-$ ring combine together to give five molecular orbitals. In a similar way, the five *p*π orbitals of the other $C_5H_5^-$ ring also combine together to give five molecular orbitals. Five molecular orbitals of one $C_5H_5^-$ ring combine with the five MOs of the other $C_5H_5^-$ ring to give ten MOs which are called **ligand group molecular orbitals.** These ten ligand group MOs have the symmetry $a_1$(*i.e.* one MO having no planar node), $e_1$(*i.e.* two degenerate MOs having one planar node) and $e_2$ (*i.e.* two degenerate MOs having two planar nodes). These ten MOs are: $a_{1g}$ (1), $a_{2u}$(1), $e_{1g}$(2), $e_{1u}$(2), $e_{2g}$(2) and $e_{2u}$(2). These MOs have been shown in Fig. 31.52. These MOs have the following characteristics: (Also see Fig. 31.52)

(*a*) **One $a_{1g}$ MO.** This MO has a + sign for the lower lobes of all the five carbon $p_z$ orbitals of the upper $Cp^-$ and + sign for the upper lobes of all the five carbon $p_z$ orbitals of the lower $c\bar{p}$. This MO has *no planar node*. The symmetry of this MO (*i.e.*$a_{1g}$MO) matches the symmetry of 4*s* and $3d_{z^2}$ orbitals of $Fe^{2+}$ ion. This MO contains one electron pair. The overlap of this MO with the metal orbital gives cp → M sigma bond (weak sigma bond).

(*b*) **One $a_{2u}$ MO.** This MO has + sign for the lower lobes of all the five carbon $p_z$ orbitals of the upper $C\bar{p}$ and – sign for the upper lobes of all the five carbon $p_z$ orbitals of the lower $c\bar{p}$ . This MO has *no planar node*. The symmetry of this MO matches the symmetry of $4p_z$ orbital of $Fe^{2+}$ ion. This MO also contains one electron pair. The overlap of this MO with $4p_z$ orbital gives weak cp → M sigma bond.

(*c*) **Two degenerate $e_{1g}$ MOs**. These are two degenerate MOs with $e_{1g}$ symmetry. Both these MOs have one planar node. One $e_{1g}$ bonding M.O. has + sign in the lower lobes of two carbons (of the upper $cp^-$ ) on one side of the nodal plane while the lower lobes of the two C-atoms on the other side of the planar node have a negative sign. The nodal plane passes through the fifth carbon indicating this carbon does not take part in the formation of the $e_{1g}$ bonding M.O. If the signs are reversed for the upper lobes of the $p_z$ orbitals of the lower $cp^-$ we have a gerade (*g*) symmetry. The other $e_{1g}$ bonding M.O. is similar to the above $e_{1g}$ bonding M.O. with the difference that it is at 90° to this above one. Both these $e_{1g}$ MOs have electron pairs. These two MOs donate their electron pairs to the metal atom *via* overlap with its empty $d_{xz}$ and $d_{yz}$ orbitals to form cp → M π bonds.

(*d*) **Two degenerate $e_{1u}$ MOs.** These are two degenerate MOs with $e_{1u}$ symmetry. These MOs have *one planar node*. The symmetry of these MOs matches the symmetry of $4p_x$ and $4p_y$ orbitals of $Fe^{2+}$ ion. The overlap of these MOs with $4p_x$ and $4p_y$ give weak δ bond.

(*e*) **Two degenerate $e_{2g}$ MOs.** These are two degenerate MOs with $e_{2g}$ symmetry. These MOs have two perpendicular planar nodes. The symmetry of these MOs matches the symmetry of $3d_{x^2-y^2}$ and $3d_{xy}$ orbitals of $Fe^{2+}$ ion. Since these MOs have two nodal planes, there would be a lateral overlap giving a δ-type weak bond.

(*f*) **Two degenerate $e_{2u}$ MOs**. These are two degenerate MOs with $e_{2u}$ symmetry. These MOs have two planar nodes. The symmetry of these MOs does not match with any of the orbitals of $Fe^{2+}$ ion. Hence these MOs are **non bonding MOs.**

| Ten MOs of two $C_5H_5^-$ ($cp^-$) rings | Matching $Fe^{2+}$ orbitals |
|---|---|
| **(*a*) one $a_{1g}$ MO** 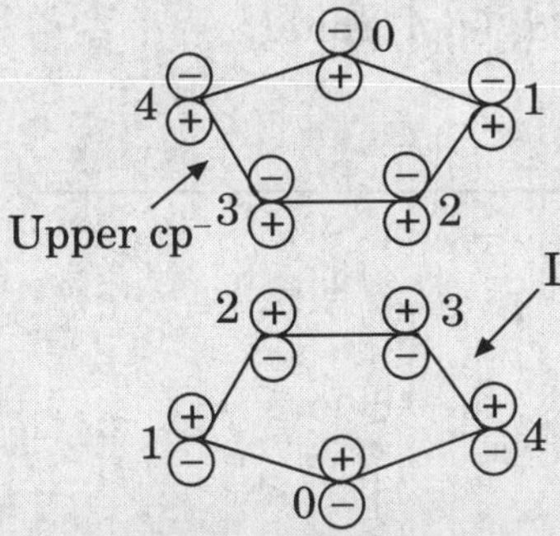  | 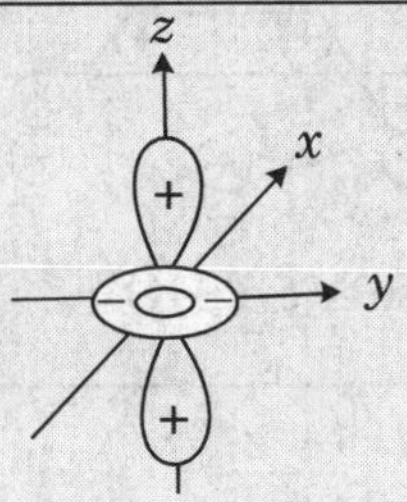 $3d_{z^2}$ ($a_{1g}$) orbital 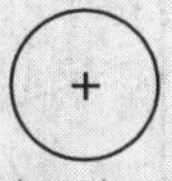 $4s$ ($a_{1g}$) orbital |
| **(*b*) one $a_{2u}$ MO** 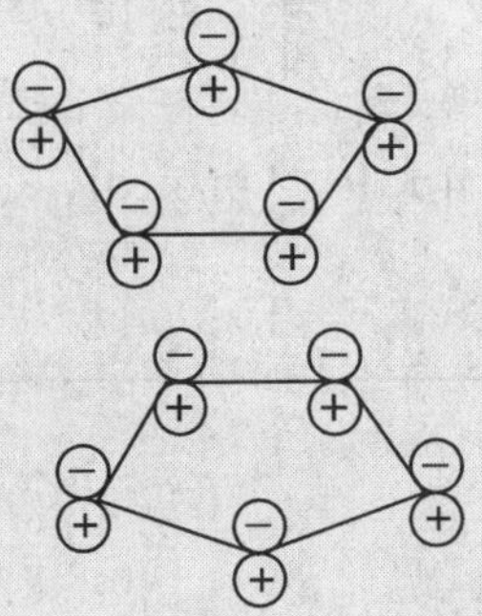 | 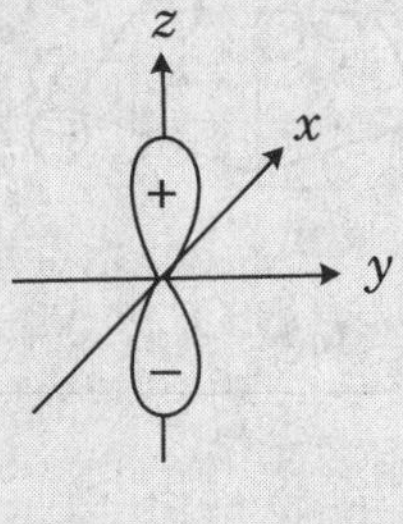 $4p_z$ ($a_{2u}$) orbital |

**(*c*) Two degerate $e_{1g}$ MOs**

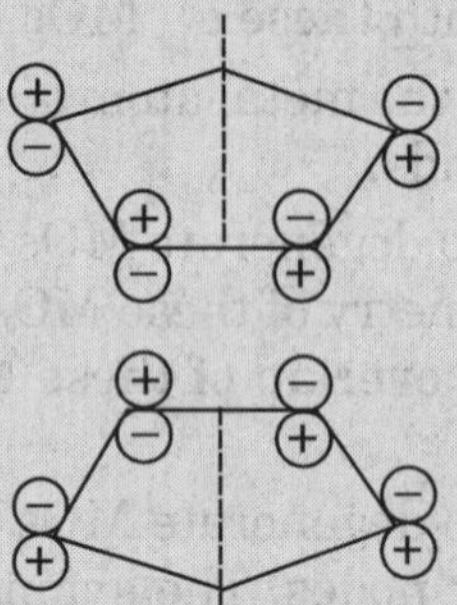

(*i*) one $e_{1g}$ MO

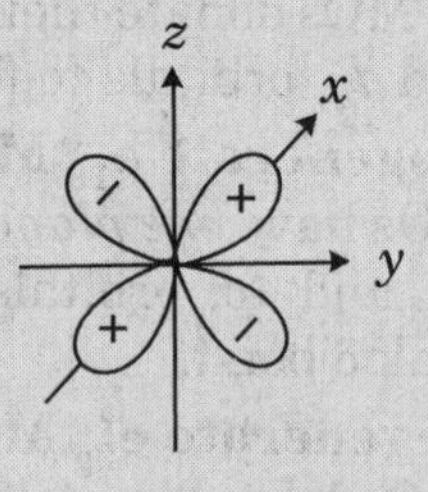


$3d_{yz}(e_{1g})$ orbital

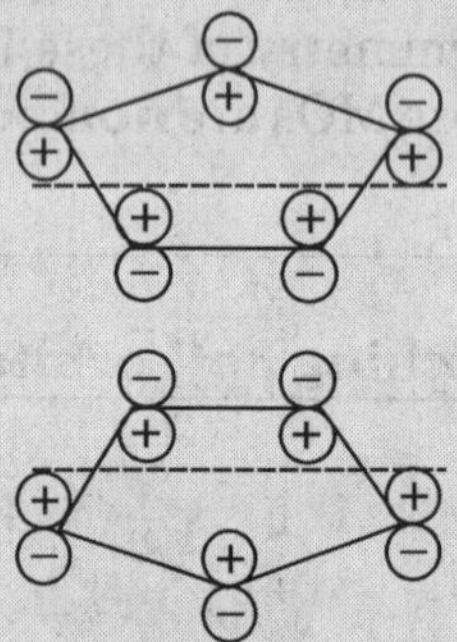

(*ii*) other $e_{1g}$ MO

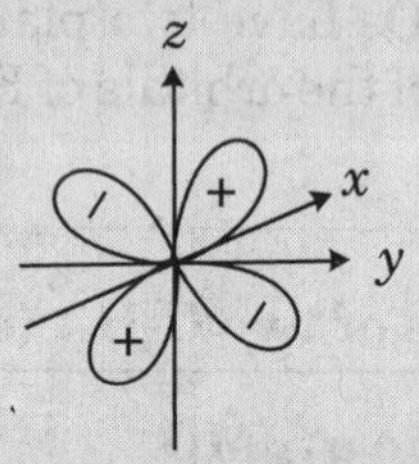


$3d_{xz}(e_{1g})$ orbital

**(*d*) Two degerate $e_{1u}$ MOs**

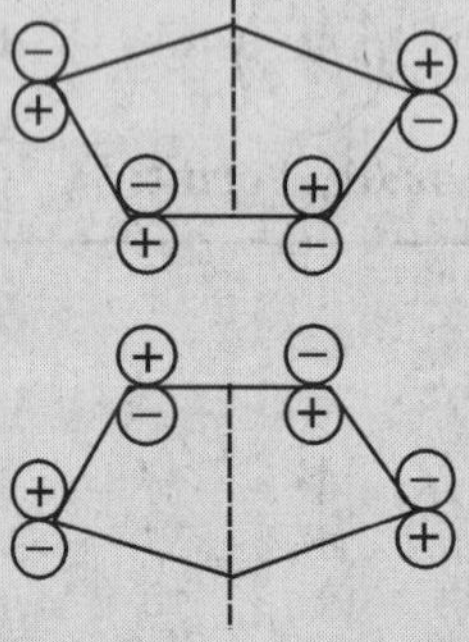

(*i*) one $e_{1u}$ MO

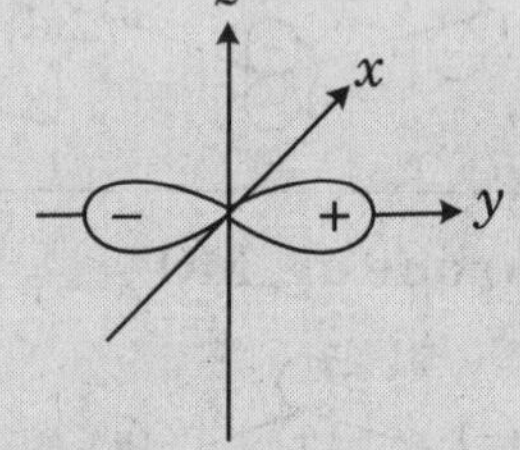


$4p_y$ $(e_{1u})$ orbital

| | |
|---|---|
| 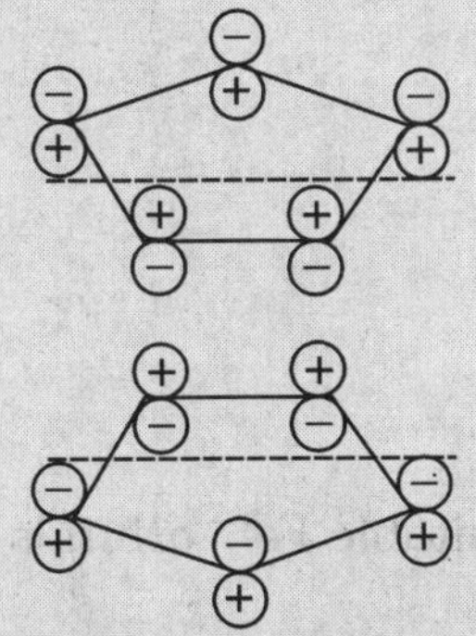 (*ii*) other $e_{1u}$ MO | 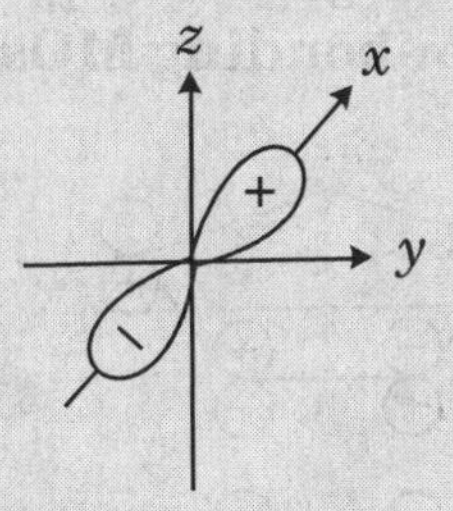  $4p_x$ $(e_{1u})$ orbital |
| **(*e*) Two degenrate $e_{2g}$ MOs** 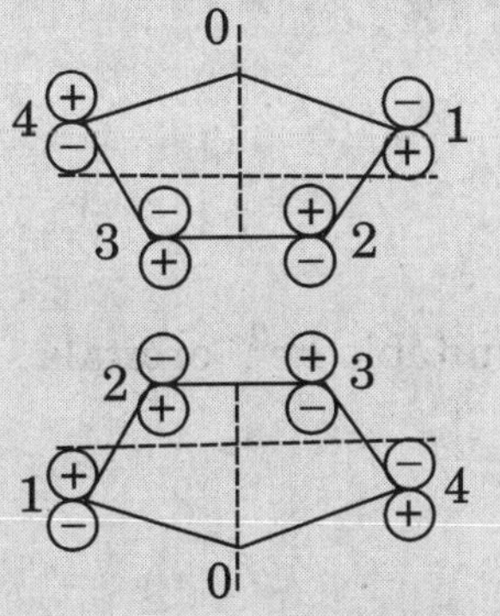  (*i*) one $e_{2g}$ MO 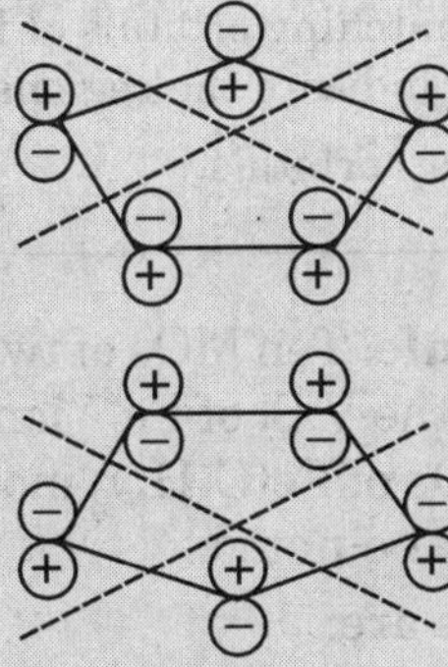 (*ii*) other $e_{2g}$ MO | 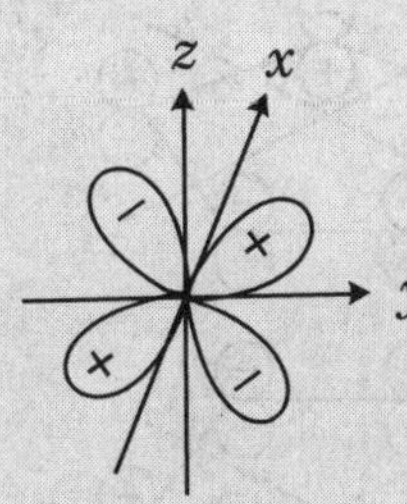  $3d_{xy}(e_{2g})$ orbital 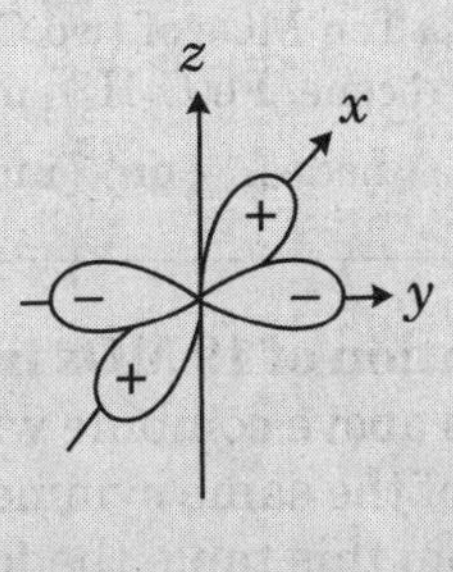  $3d_{x^2-y^2}(e_{2g})$ orbital |

**(f) Two degenrate $e_{2u}$ MOs (Non-bonding MOs)**

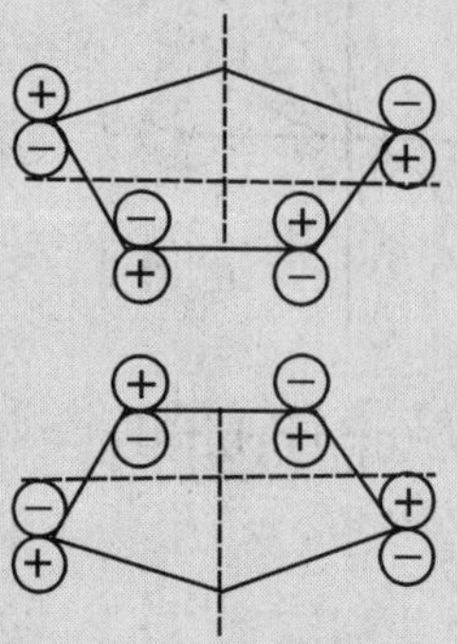

(i) one $e_{2u}$ MO

No suitable $Fe^{2+}$ orbitals

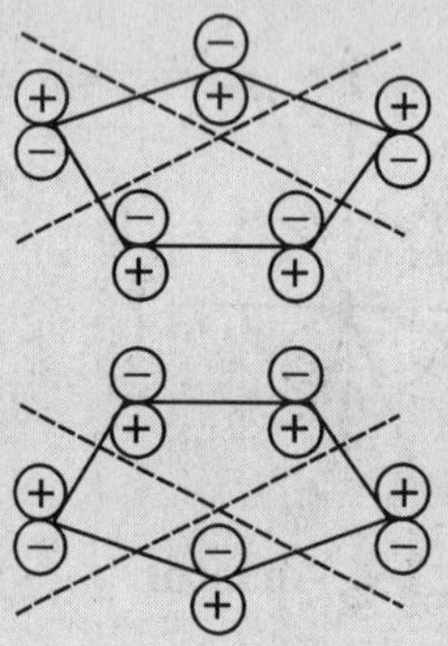

(ii) other $e_{2u}$ MO

No suitable $Fe^{2+}$ orbitals

**Fig. 31.52.** Ten MOs of two $C_5H_5^-$ (cp$^-$) rings and matching orbitals of $Fe^{2+}$ ion in ferrocene, $Fe(C_5H_5)_2$ molecule. Dotted lines indicate the regions of nodes. ⊖/⊕ or ⊕/⊖ indicate lobes of $p\pi$ (or $p_z$) orbitals.

(*ii*) **Formation of 19 MOs in $Fe(C_5H_5)_2$ molecule.** Ten MOs of two $C_5H_5^-$(cp$^-$) ions obtained as above combine with nine orbitals of the V.S. of $Fe^{2+}$ ion (3*d* = 5, 4*s* = 1 and 4*p* = 3) of the same symmetry and give 19 MOs of $Fe(C_5H_5)_2$ molecule. (See Table 31.2). From this table the following points may be noted:

(*a*) Bonding, antibonding and non bonding MOs are:

(*i*) *BMOs(8):* $a_{1g}(1)$, $a_{2u}(1)$, $e_{1g}(2)$, $e_{1u}(2)$, $e_{2g}(2)$

(*ii*) *ABMOs(8):* $a_{1g}^*(1)$, $e_{1g}^*(2)$, $e_{2g}^*(2)$, $a_{2u}^*(1)$, $e_{1u}^*(2)$

(*iii*) *NBMOs(3):* $e_{2u}(2)$, $a'_{1g}(1)$

(*b*) Since there is only a minor contribution from $3d_{z^2}$ orbital of $Fe^{2+}$ in the formation of $a_{1g}$ MO, this MO is mainly based on the ligand system. Similarly one $a_{2u}$, two $e_{2g}$ and two $e_{1u}$ MOs are also mainly based on ligand system.

**Table 31.2. Formation of MOs by the combination ligand ($cp^-$ or $C_5H_5^-$ ions) group molecular orbitals and orbitals of $Fe^{2+}$ ion. MOs shown in boxes are degenerate MOs.**

| MOs of two $cp^-$ ions | | Orbitals of $Fe^{2+}$ ion | | MOs of Fe $(C_5H_5)_2$ molecule | |
|---|---|---|---|---|---|
| $a_{1g}$ | + | $3d_{z^2}$ ($a_{1g}$) | ⟶ | $a_{1g}$ | $a_{1g}^*$ |
| $a_{1g}$ | + | 4s ($a_{1g}$) | ⟶ | $a'_{1g}$ | |
| $a_{2u}$ | + | $4p_z$ ($a_{2u}$) | ⟶ | $a_{2u}$ | $a_{2u}^*$ |
| $e_{1g}$ | + | $3d_{yz}$ ($e_{1g}$) | ⟶ | $e_{1g}$ | $e_{1g}^*$ |
| $e_{1g}$ | + | $3d_{xz}$ ($e_{1g}$) | ⟶ | $e_{1g}$ | $e_{1g}^*$ |
| $e_{1u}$ | + | $4p_y$ ($e_{1u}$) | ⟶ | $e_{1u}$ | $e_{1u}^*$ |
| $e_{1u}$ | + | $4p_x$ ($e_{1u}$) | ⟶ | $e_{1u}$ | $e_{1u}^*$ |
| $e_{2g}$ | + | $3d_{xy}$ ($e_{2g}$) | ⟶ | $e_{2g}$ | $e_{2g}^*$ |
| $e_{2g}$ | + | $3d_{x^2-y^2}$ ($e_{2g}$) | ⟶ | $e_{2g}$ | $e_{2g}^*$ |
| $e_{2u}$ | + | No suitable $Fe^{2+}$ orbital | ⟶ | $e_{2u}$ | |
| $e_{2u}$ | + | No suitable $Fe^{2+}$ orbital | ⟶ | $e_{2g}$ | |

(*c*) $a'_{1g}$ MO has a minor contribution from $a_{1g}$ MO of the ligand and hence is regarded as a non-bonding MO (NBMO). Its energy is almost the same as that of $3d_{z^2}$ orbital of $Fe^{2+}$ ion.

(*d*) Since $e_{2u}$ MOs of $C_5H_5^-$ ions do not have matching $Fe^{2+}$ orbitals, these MOs remain as such. Hence these MOs are also non bonding MOs (NBMOs).

(*e*) All the 19 MOs have been shown in Fig. 31.53 in the increasing order of their energy.

The most significant bonds which give stability to ferrocene are π-bonds which are formed by the overlap of $d_{xz}$ and $d_{yz}$ atomic orbitals of $Fe^{2+}$ with the appropriate ligand group molecular orbitals of two $C_5H_5^-$ rings. As an example, the formation of π-bond between $d_{xz}$ orbital of $Fe^{2+}$ and appropriate ligand group molecular orbital is shown in Fig. 31.54.

The formation of π-bond between $d_{yz}$ orbital of $Fe^{2+}$ and the appropriate ligand group molecular takes place in a similar way.

Each π-bond is 3-centred on each $C_5H_5^-$ ring and $Fe^{2+}$ ion. These π-bonds are very strong and contribute to the stability of ferrocene.

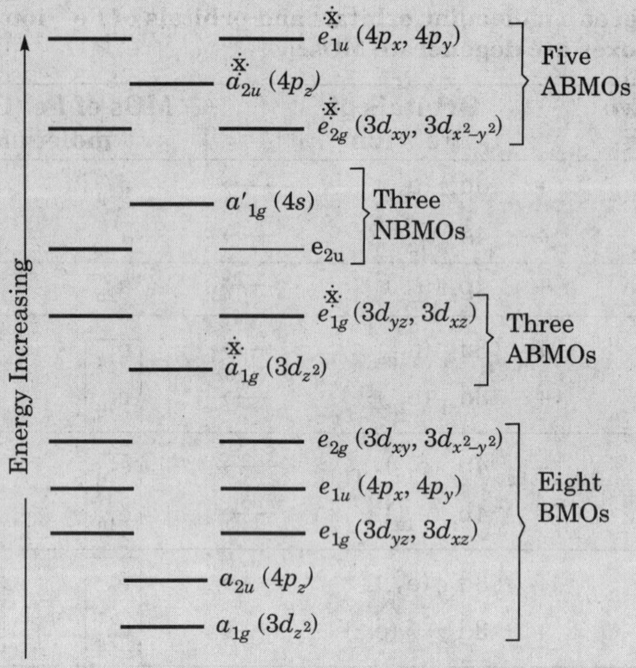


**Fig. 31.53.** 19 MOs of $Fe(C_5H_5)_2$ molecule shown in the increasing order of their energy.

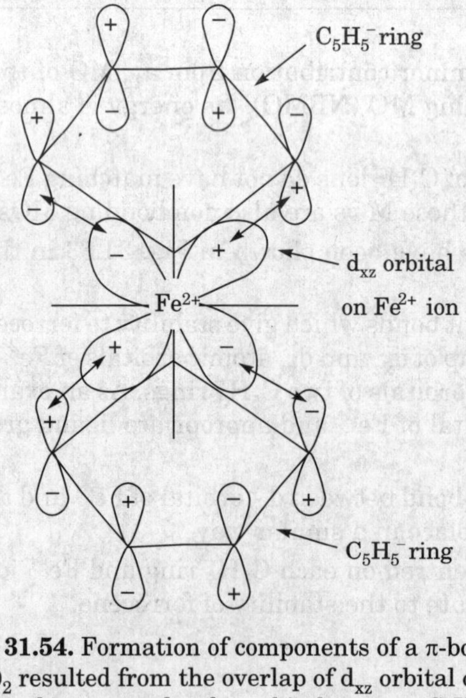


**Fig. 31.54.** Formation of components of a π-bond in $Fe(C_5H_5)_2$ resulted from the overlap of $d_{xz}$ orbital of $Fe^{2+}$ ion and ligand group modecular orbitals on two $C_5H_5^-$ rings.

## To Calculate the Effective Number of Electrons in Some Metallocenes

While calculating the total number of electrons in a given metallocene, it should be remembered that each $C_5H_5^-$ ion donates six electrons.

Examples are given below.

| Metallocenes | Electrons in the V.S. of metal atom as ion (a) | Electrons donated by two $C_5H_5^-$ ligands (b) | Total number of electrons = a + b |
|---|---|---|---|
| (*i*) $[Fe(C_5H_5)_2]^o$ | $Fe^{2+} = 3d^6 = 6$ | $2 \times 6 = 12$ | $6 + 12 = 18$ |
| (*ii*) $[V(C_5H_5)_2]^+$ | $V^{3+} = 3d^2 = 2$ | $2 \times 6 = 12$ | $2 + 12 = 14$ |
| (*iii*) $[V(C_5H_5)_2]^o$ | $V^{2+} = 3d^3 = 3$ | $2 \times 6 = 12$ | $3 + 12 = 15$ |
| (*iv*) $[Cr(C_5H_5)_2]^o$ | $Cr^{2+} = 3d^4 = 4$ | $2 \times 6 = 12$ | $4 + 12 = 16$ |
| (*v*) $[Fe(C_5H_5)_2]^+$ | $Fe^{3+} = 3d^5 = 5$ | $2 \times 6 = 12$ | $5 + 12 = 17$ |
| (*vi*) $[Co(C_5H_5)_2]^o$ | $Co^{2+} = 3d^7 = 7$ | $2 \times 6 = 12$ | $7 + 12 = 19$ |
| (*vii*) $[Ni(C_5H_5)_2]^o$ | $Ni^{2+} = 3d^8 = 8$ | $2 \times 6 = 12$ | $8 + 12 = 20$ |

## Molecular Orbital Electronic Configurations (MOEC) of Some Metallocenes

**1. Ferrocene, $[Fe(C_5H_5)]^0$ (No. of electrons = 6 + 2 × 6 = 18).** MO energy level diagram of this molecule is given in Fig. 31.55. This figure shows that MOEC of Fe $(C_5H_5)_2$ molecule can be written as:

$$(a_{1g})^2\ (a_{2u})^2\ (e_{1g})^4\ (e_{1u})^4\ (e_{2g})^4\ (a_{1g}^*)^2\ (n = 0)$$

This configuration shows that since all the electrons are present in the paired state (n = 0), $Fe(C_5H_5)_2$ is diamagnetic.

**2. $[V(C_5H_5)_2]^+$ (No. of electrons = 14) and $[V(C_5H_5)_2]^0$ (No. of electrons = 15).** In these species, since the two molecular orbitals of $e_{2g}$ set and $a_{1g}^*$ molecular orbital have very close energy, the distribution of 14 electrons in $[V(C_5H_5)]^+$ and 15 electrons in $V(C_5H_5)_2$ can be represented as shown at (*a*) and (*b*) respectively.

(*a*) $\underbrace{(a_{1g})^2\ (a_{2u})^2\ (e_{1g})^4\ (e_{1u})^4}_{\text{12 electrons (n = o)}}\quad \underbrace{(e_{2g})^2\ (a_{1g}^*)^0}_{\text{2 electrons (n = 2)}}$

(*b*) $\underbrace{(a_{1g})^2\ (a_{2u})^2\ (e_{1g})^4\ (e_{1u})^4}_{\text{12 electrons (n = o)}}\quad \underbrace{(e_{2g})^2\ (a_{1g}^*)^1}_{\text{3 electrons (n = 3)}}$

These distributions show that $[V(C_5H_5)]^+$ and $[V(C_5H_5)_2]^o$ have 2 and 3 unpaired electrons respectively. Thus these species are paramagnetic.

Because of the presence of partially filled $e_{2g}$ bonding molecular orbitals, $[V(C_5H_5)_2]^o$ is highly reactive.

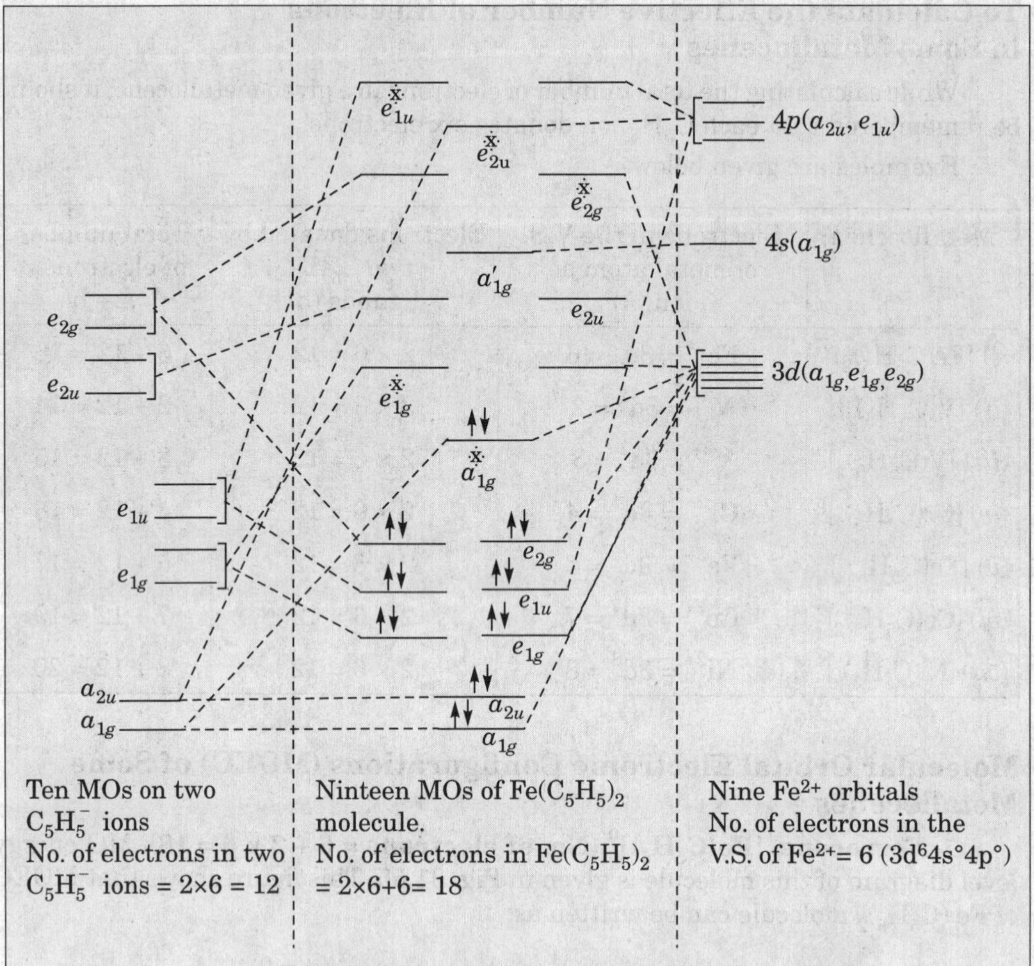


**Fig. 31.55.** MO energy level chagram of

**3. $[Cr(C_5H_5)_2]^o$ (*No. of electrons = 16*).** In this species, since the two molecular orbitals of $e_g$ set and $a_{1g}^*$ molecular orbital are fairly close in energy, the distribution of 16 electrons in $[Cr(C_5H_5)_2]^o$ takes place as:

$$\underbrace{(a_{1g})^2\ (a_{2u})^2\ (e_{1g})^4\ (e_{1u})^4}_{\text{12 electrons (n = o)}} \qquad \underbrace{(e_{2g})^3\ (a_{1g}^*)^1}_{\text{4 electrons (n=2)}}$$

Since this distribution has two unpaired electrons, $[Cr(C_5H_5)_2]^o$ is paramagnetic. Because of the presence of partially filled $e_{2g}$ bonding molecular orbitals, $[Cr(C_5H_5)_2]^o$ is highly reactive.

**4. $[Fe(C_5H_5)_2]^+$ (*No. of electrons = 17*).** Distribution of 17 electrons in various molecular orbitals can be shown as:

$$(a_{1g})^2\ (a_{2u})^2\ (e_{1g})^4\ (e_{1u})^4\ (e_{2g})^4\ (a_{1g}^*)^1\ (n = 1)$$

Since this distribution has one unpaired electron in $a_{1g}^*$ level, $[Fe(C_5H_5)_2]^+$ ion is paramagnetic.

**5. Cobalocene, $[Co(C_5H_5)_2]^o$ (*No. of electrons = 19*).** The distribution of 19 electrons in various orbitals is as:

$$\underbrace{(a_{1g})^2\ (a_{2u})^2\ (e_{1g})^4\ (e_{1u})^4\ (e_{2g})^4\ (a_{1g}^*)^2}_{\text{18 electrons (n=0)}} \qquad \underbrace{(e_{1g}^*)^1}_{\text{1 electron (n=1)}}$$

Since the above distribution has one unpaired electron in $e_{1g}^*$ set of orbitals, $[Co(C_5H_5)_2]^o$ is paramagnetic. This one unpaired electron resides in antibonding molecular orbital and hence is easily oxidisable electron.

**6. Nickelocene, $[Ni(C_5H_5)_2]^o$ (*No. of electrons = 20*).** The distribution of 20 electrons in various molecular orbitals is as:

$$\underbrace{(a_{1g})^2\ (a_{2u})^2\ (e_{1g})^4\ (e_{1u})^4\ (e_{2g})^4\ (a_{1g}^*)^2}_{\text{18 electrons (n=0)}} \qquad \underbrace{(e_{1g}^*)^2}_{\text{2 electrons (n=2)}}$$

Since the above distribution has two unpaired electrons in $e_{1g}^*$ set of orbitals, $[Ni(C_5H_5)_2]^o$ is paramagnetic. These two unpaired electrons reside in antibonding molecular orbitals and hence are easily oxidisable electrons.

## Uses of Organometallics

Organometallics are of immense importance. Progress in this area has introduced new reagents and catalysts for synthesis. Some of the applications of organometallic compounds are given below:

(*i*) **Used as homogeneous catalysts.** A large number of reactions in solutions are catalysed by organometallic compounds or intermediates derived from transition metal complexes. For example, **Wilkinson's catalyst**, $[RhCl(PPh_3)_3]$ *i.e.* chlorotris (triphenyl phosphine) rhodium (I) is used as a catalyst in the hydrogenation of alkenes.

(*ii*) **Used as heterogeneous catalysts.** Organometallic compounds can also be used as heterogeneous catalysts, *e.g.*, trialkyl aluminium mixed with a transition metal halide, usually titanium trichloride or tetrachloride (**Ziegler-Natta type catalyst**) is used as a catalyst for the low temperature polymerisation of alkenes.

(*iii*) **Used in the purification of metals**. Some metals are purified by formation of their metal carbonyls followed by their decomposition *e.g.* impure nickel is converted into nickel tetracarbonyl, $Ni(CO)_4$ which on decomposition gives pure nickel.

(*iv*) **Used in the synthesis of organic compounds.** They are widely used in synthesis of various types of organic compounds, *e.g.*, organolithium compounds are commonly used for the synthesis of different types of organic compounds.

(*v*) **Used as medicine.** A number of organo arsenic compounds are used as main remedy for syphilis. The modern surgery makes use of silicon rubbers for spare parts in the body.

(*vi*) **Used in agriculture.** To prevent the infection of young plants, the seeds are treated with organometallics such as ethyl mercury chloride.

(*vii*) **Likely to be used in future.** It is expected that the organometallic compounds will play an important role in the production of pharmaceuticals, flavours, fragrances and semi-conductors for the electronic industry.

## UNIVERSITY QUESTIONS

**1.** (a) Write three methods for the preparation of metal ethylenic complexes.
(b) Describe the structure of $Al_2(CH_3)_6$ in brief.
(c) Explain the bond nature in carbon-lithium compounds

(*Pd. Ravishankar Shukla 2010*)

**2.** How are organomagnesium compound prepared? Discuss their properties and uses.

(*Kanpur 2010*)

**3.** Describe classification, methods of preparation and uses of different organometallic compounds.

(*Pune 2009*)

**4.** Write a note on alkyl lithium

(*Rajasthan 2010*)

**5.** (a) Why is alkyl lithium compound known as super Grignard reagent? How does methyl lithium react with the following?
(i) Dry ice (ii) $(C_2H_5)_2$ Zn (iii) HCHO.
(b) Give the structure of the following—
(i) $(CH_3Li)_4$ (ii) $[(CH_3)_2SnF_2]_n$ (iii) Zeise's salt

(*Rajsthan 2010*)

**6.** (i) In organometallic compounds the number of carbon atoms bound to the metal is indicated by:
(a) $\delta$ (b) $\pi$ (c) $\eta$ (d) $\sigma$.
(ii) How are organometallic compounds classified on various grounds?
(iii) Explain the synthesis and structure of metal ethylenic complexes.

(*H.N.Bahuguna Garhwal 2010*)

**7.** Explain the structural implications of isolobal analogy in the synthesis of organometallic compounds with a suitable example.

(*Bharathiar 2010*)

**8.** (i) Define organometallic compounds. How are they classified?
(ii) Discuss preparation and structure of methyl lithium.

(*Ranchi 2010*)

**9.** (a) Classify the organometallic compounds on the basis of nature of metal-carbon bond with an example. Draw the structure of trimethyl aluminium.
(b) Draw and explain the structure of Zeise's salt.
(c) Give one method of preparation of alkyl lithium and draw the structure of methyl lithium.

(d) Write any two biological applications of organometallic compounds. *(Nagpur 2010)*

**10.** (a) Give classification of organometallic compounds based on nature of metal-carbon bond.
(b) Write the reactions of organoaluminium compounds.
(c) Give any three reactions of Grignard reagents. *(A.P. University 2010)*

**11.** (a) Define organometallic compounds. What are the three broad classes of organometallic compounds? Cite an example of each.
(b) Describe any three important general methods for preparing organometallic compounds.
(c) What is the most characteristic structural feature of lithium alkyls?
(d) Discuss the synthesis, bonding and two specific properties of organomercuric compounds.
(e) Describe the bonding in trimethyl aluminium in terms of multi-centre bonds.
(f) Name the following compounds: (i) $Li_4(C_2H_5)_4$ (ii) $SiH(C_2H_5)_3$ *(BJB College, Orissa)*

**12.** (a) What are organometallic compounds?
(b) Explain in detail the classification of organometallic compounds.
(c) Write down the properties of organometallic compounds.
(d) Explain the bonding in metal-ethylenic complex. *(GND Amritsar 2010)*

**13.** (i) Explain the structure of trimethyl aluminium.
(ii) Give preparation and applications of organotin compounds in brief.
(iii) Give preparation and important reactions of butyl lithium (give equations only) *(Lucknow 2010)*

**14.** (i) Discuss the structure and bonding in acctylene complexes.
(ii) What are physical properties of organozinc compounds?
(iii) How is triethylboron prepared? Write equations.
(iv) How is tetraethyl lead is prepared? What are its uses?
(v) How is Zeise's salt prepared? Give its structure. *(Madras April 2011)*

**15.** (*a*) Give any one method for the preparation of alkyls and aryls of aluminium. Discuss the structure of tri-methyl aluminium.
(*b*) Give medicinal applications of organometallic compounds. *(Nagpur 2012)*

**16.** (*a*) Describe the preparation, properties, uses and structures of organometallic compounds of aluminium.
(*b*) Which of the following is an orgonometallic compound and why? Metal carbonates, Metalcarbonyls *(Ch. Charan Singh, Meerut 2012)*

**17.** Explain the following:
(*i*) Sodium acetylide ($Na^+ \bar{C}{\equiv}H$) is an organometallic compoud.
(*ii*) Organometallic compounds of Li and Mg are useful starting materials to obtain organometallic derivatives of other metallic elements. *(Lucknow 2013)*

32

CHAPTER

# Bioinorganic Chemistry

## What is Biochemistry and Bioinorganic Chemistry?

The study of chemistry of living organism (biological systems) (plants and animals both) is known as **biochemistry**. Thus biochemistry is the chemistry of life.

Bioinorganic chemistry is a branch of chemistry in which we make a study of significant role of metal ions in biological systems. This branch of chemistry has helped us to synthesize new inorganic compounds which resemble those formed in biological systems. The metal ions are responsible for initiating or inhibiting reactions in biological systems.

## Essential and Trace Elements in Biological Processes

**An essential element** *is that which is required for the maintenance of life.* Thirty elements are believed to be essential for life processes in plants and animals. Essential elements may be metallic elements or non-metallic elements. Here we shall discuss the role of only metallic elements in biological systems. The metals such as **Na, K, Ca, Mg** whose ions are present in biological systems are called **bulk metals** and the metals like **Fe, Cu, Co, Zn, Cr, Mn, Mo, W, Ni** whose ions are present in biological systems in trace amounts are called **trace elements**.

All the essential and trace elements are not required by every plant or animal. For example Na is vital for human and higher animals but is not essential for bacteria. Some higher plants require Al, B and V but humans donot require these elements. Some essential elements prove to be toxic, even fatal, when taken in higher concentration.

## Porphine Ligand and Metalloporphyrins

Porphine ligand is a ring system. It is made of four pyrrole rings. These pyrrole rings are linked together through methene bridges and have been designated as A, B, C and D in Fig. 32.1. Porphine has conjugated double bonds.

Porphine has four N-atoms as donor atoms. These N-atoms are present in pyrrole rings. Two of these N-atoms are present as NH groups and two are as N-atoms. If two protons present in two NH groups are removed, then all the four N-atoms can be co-ordinated to $M^{2+}$ metallic ions ($M^{2+} = Mg^{2+}, Fe^{2+}, Ni^{2+}$ etc.) to form complexes which are called **metallo porphyrins**. These complexes are neutral complexes. It is obvious that in the formation of metallo porphyrins, porphine acts as a tetradentate dinegative ligand. Metallo porphyrins are bioinorganic compounds and play a vital role in biological system. The structure of a metallo porphyrin formed by porphine ligand with $M^{2+}$ ion is shown in Fig. 32.2.

The size of the central hole in the centre of porphine ligand is ideal for accommodation of $M^{2+}$ ions of the elements of *1st* transition series and higher alkaline

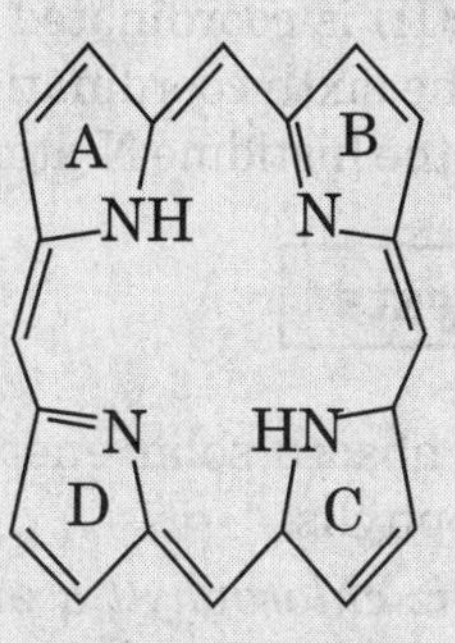


**Fig. 32.1** Structure of porphine ligand. A, B, C, and D represent pyrrole rings

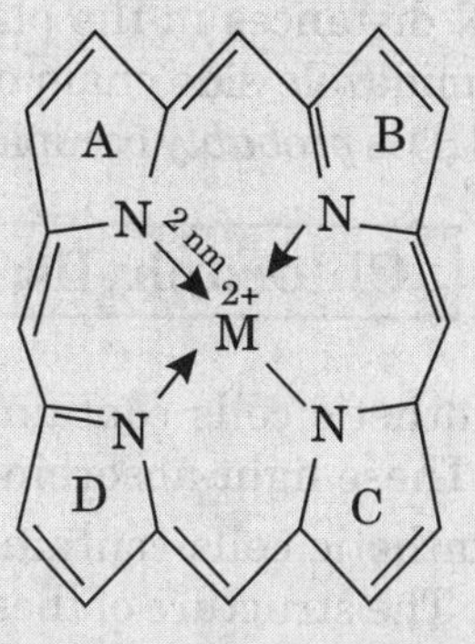


**Fig. 32.2** Structure of $M^{2+}$ complex ($M^{2+} = Mg^{2+}$, $Fe^{2+}$, $Ni^{2+}$, etc) with porphine ligand.

earth metals. $M^{2+}$ ions of appropriate size are surrounded by four N-atoms of porphyrine ring in a square planar geometry and axial positions are available for other ligands. If the size of $M^{2+}$ ion is smaller than the size of the hole, the porphine ring becomes ruffled to allow closer approach of N-atoms to $M^{2+}$ ion. On the other hand, if the size of $M^{2+}$ cation is larger than the size of the hole, $M^{2+}$ ion cannot fit into the hole and stays out of square plane containing four N-atoms. In such a case the metal ion sets above the porphine ring which becomes domed and acquires square pyramidal structure. **Heme** and **chlorophyll** are two important metalloporphyrins. These are discussed below.

## Heme

Hene is a **Fe(II) porphyrin complex** which contains a tetrahedral macrocylic ligand namely porphyrin which has four nitrogen donors in a square plane. The structure of heme is given in Fig. 32.3. Two NH protons of pyrole rings are lost in binding

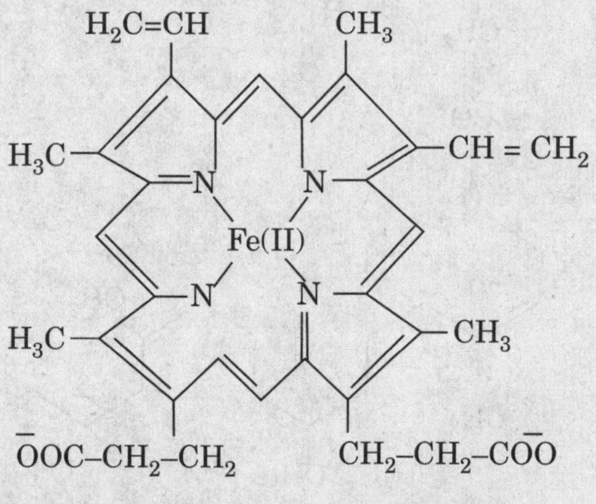


**Fig. 32.3** Structure of heme which is a iron (II) - prophyrine complex.

Fe (II). Fe—N distances in the plane are is ~2.2 A°. Hence Fe(II) is coordinated to a N-atom of imidazole side chain of histidine unit of protein. The sixth coordination position of Fe(II) is *probably* occupied by a $H_2O$ molecule *trans* to the histidine N-atom.

## Chlorophylls: Photosynthetic Catalysts

Photosynthetic cells contain green pigments which can absorb solar energy (solar light). These light absorbing pigments are called chlorophylls.

Photosynthetic cells contain two types of chlorophylls *viz chlorophyll a* and *chlorophyll b*. The structure of these chlorophylls are given in Fig. 32.4. In this figure chlorophyll *a* has R = $CH_3$ and chlorophyll *b* has R = CHO. Both chlorophylls have a long phytyl side chain. Chlorophyll units are bound to specific proteins.

Both chlorophylls are planar complexes of Mg(II) with four nitrogen sitting at the four corners of a square plane (Fig. 32.4). The $N_4$ unit is a part of a 16-membered natural macrocycle made of four substituted pyrrole rings. The conjugated heterocycles create a

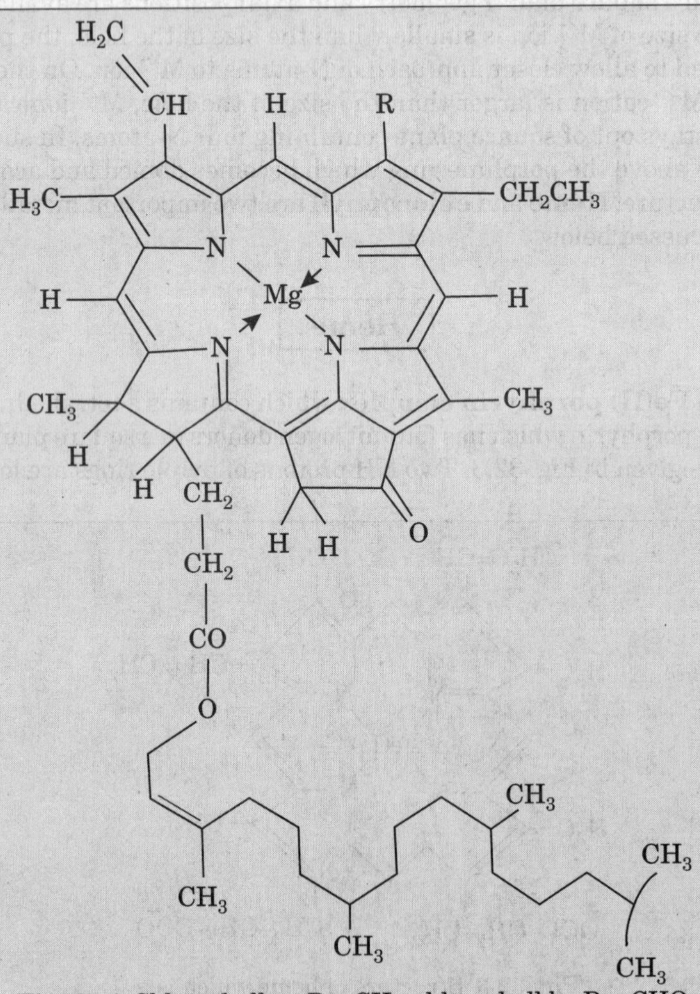


**Fig. 32.4** Chlorophyll a : R = $CH_3$; chlorophyll b : R = CHO

rigid planar environment around Mg(II). The pyrrole rings resemble the protoporphyrin rings of hemoglobin. However, while the heme proteins have Fe(II) as the metal centre, the chlorophylls have Mg(II).

Chlorophyll *a* and chlorophyll *b* have been isolated in pure form from leaf extracts by chromatographic technique. Although both the forms are green, their absorption spectra are slightly different. Chlorophyll *a* has a red band composed of two components at 670 nm and 680 nm of nearly equal intensity and a third weak component at 695 nm.

## Myoglobin (Mb)

Myoglobin (Mb) is a **heme-containing protein** or Mb is a **metallo-protein** or **metal porphyrin** which contains heme group in its structure.

Mb may be viewed as a monomer of hemoglobin (Hb) and is made up of one heme unit connected to one polypeptide chain (globin). In Mb heme group is embedded in a crevice formed by the coiling of its polypeptide chain. The polypeptide chain in Mb contains 15 to 160 amino acid residues.

The polypeptide chain in Mb is coordinated to iron atom of the heme group through N-atom of hestidine. Molar mass of Mb is about 17000. Mb which has not taken up oxygen is called **deoxy-myoglobin** (abbreviated as **deoxy-Mb**) or simply **myglobin** while the myglobin which has taken up $O_2$ is called **oxygenated myglobin** or **oxymyoglobin (oxy-Mb)**. Structure of deoxy-Mb is given in Fig. 32.5.

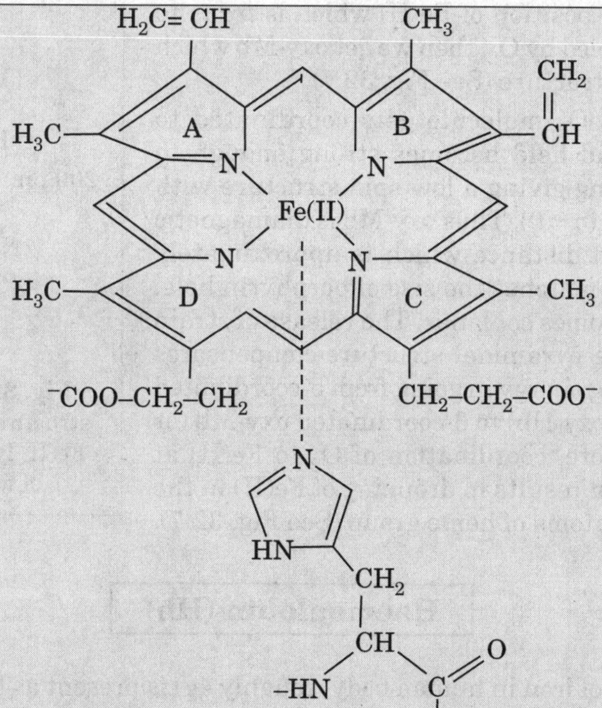


**Fig. 32.5** Structure deoxy-myoglobin.

## Square Pyramidal Structure of Deoxy-Mb

Deoxy-Mb is a five-coordinated molecule. Four coordination positions are occupied by four porphyrin N-atoms and the fifth coordination position is occupied by N-atom of an imidazole of a histidine residue which is linked with protein.

Since in the absences of $O_2$, the ligand field in five-coordinated deoxy-Mb is weak, deoxy-Mb is HS having $d^6 = t_{2g}^4 e_g^2$ configuration. This configuration has n = 4. Obviously, due to the presence of 4 unpaired electrons, deoxy-Mb is paramagnetic. In this molecule, $Fe^{2+}$ ion has substantially larger radius. The estimated Fe—N distance (= 218 pm) is larger than the size of the central hole in the porphyrin ring. Therefore, Fe lies above the plane containing four N-atoms by about 70 pm as shown in Fig. 32.6.

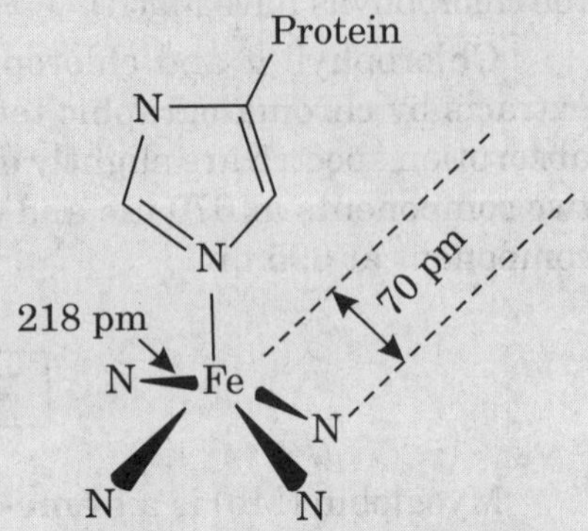


**Fig. 32.6** Square pyramidal structure of deoxy-Mb with Fe(II) lying above the plane of four N-atoms.

The position of Fe(III) above the plane of four N-atoms gives square pyramidal structure to deoxy-Mb.

## Octahedral Structure of Oxy-Mb

When sixth position of Fe(II) which is *trans* to histidine is occupied by $O_2$, then we get oxy-Mb which has octahedral structure (See Fig. 32.7).

As soon as $O_2$ molecule gets coordinated to Fe(II), the ligand field becomes strong enough to cause spin pairing giving a low spin structure with $t_{2g}^6$ configuration ($n = 0$). Thus oxy-Mb is diamagnetic and Fe—N bond distance which is approximately equal to 200 pm matches the size of porphyrin hole. Hence Fe(II) becomes coplanar. The release of strain energy of square pyramidal structure compensates the loss of pairing energy in going from 5-coordinated deoxy-Mb (or deoxy-Hb) to 6-coordinated oxy-Mb (or oxy-Hb). Therefore, coordination of $O_2$ to Fe(II) at the axial position results in dropping of Fe(II) in the plane of four N-atoms of heme group (See Fig. 32.7).

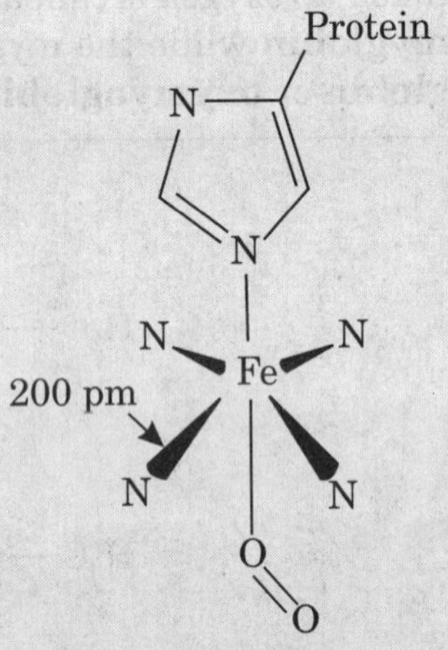


**Fig. 32.7** Octahedral structure of oxy-Mb with Fe(II) lying in the plane of four N-atoms.

## Haemoglobin (Hb)

About 65% of iron in human body (roughly 4g) is present as haemoglobin (Hb). Hb is a red pigment which is present in red block cells (RBCs) in human body. 100 ml of the blood of a normal male contains approximately 15 g of Hb. Hb contains about

0.95% of iron. Hb is a Fe(II) porphyrin. The structure of Hb is given in Fig. 32.8. Hb contains four identical units arranged roughly tetrahedrally. Each unit has one heme group whose structure has already been given in Fig. 32.3.

Molar mass of Hb is about 64500. Each Hb molecule has four heme groups represented as heme-1, heme-2, heme-3 and heme-4 in Fig. 32.8. which are bound to *globin* (globin is a protein) on its surface. Thus Hb is a **heme containing protein**. The four heme groups present in the structure of Hb are called **subunits** of Hb. Hb is an octahedral complex of Fe(II). Fe(II) occupies the central position and the four corners of the square base are occupied by the four N-atoms of heme group. One axial position is occupied by N-atom of histidine and the other axial position is occupied by $H_2O$ molecule. The four subunits are linked together through salt bridges present between the four polypeptide chains. It is now believed that the salt bridges present in between the polypeptide chains of Hb introduce **stain** in the molecule of Hb.

Hb which has not taken up $O_2$ is called **deoxy-haemoglobin** (deoxy-Hb) or simply **haemoglobin** while the haemoglobin which has taken up $O_2$ is called **oxygenated haemoglobin** or **oxyheamoglobin (oxy-Hb)**.

Fe(II) present in Hb can be oxidised to Fe(III) under controlled conditions to form Fe(III) protein called **met-haemoglobin**. Fe(III) – protein is responsible for the brown colour of old meat and drived blood.

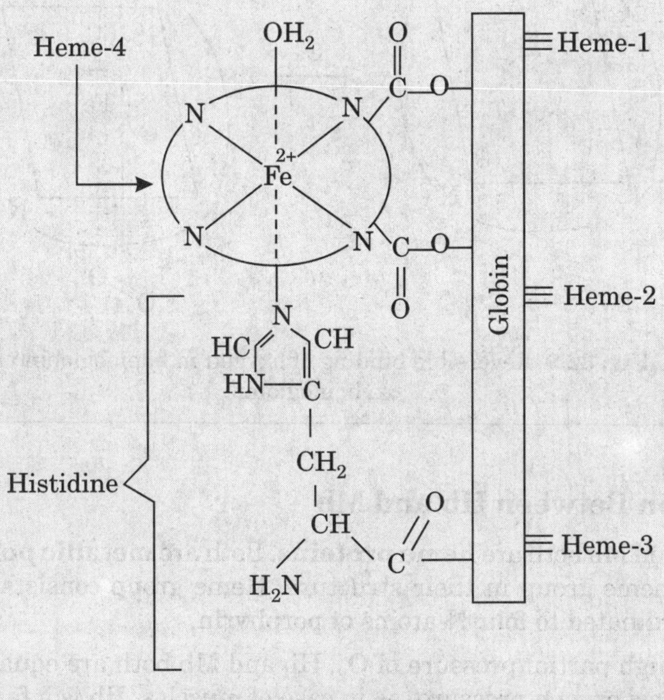


**Fig. 32.8** Structure of haemoglobin (Hb) containing four heme groups and histidine.

## Comparison Between Hb and $HbO_2$

(*i*) Hb is HS octahedral complex of Fe(II) which has four unpaired electrons ($Fe^{2+} = 3d^6 = t_{2g}^4 e_g^2$ with $n = 4$). Like Hb, $HbO_2$ also contains Fe(II) but $HbO_2$ is a LS complex ($Fe^{2+} = 3d^6 = t_{2g}^6 e_g^{\circ}$ with $n = 0$). Thus $HbO_2$ has all the electrons in the paired state. Hence Hb is paramagnetic ($n = 4$) and $HbO_2$ is diamagnetic ($n = 0$). As a result, the coordination of $O_2$ to Hb during the formation of $HbO_2$ from Hb changes Fe(II) of one heme unit from HS (paramagnetic) to LS (diamagnetic).

(*ii*) In Hb, since Fe(II) is HS, Fe(II) has larger radius (0.9A°). Being larger in size, Fe(II) *just does not* fit in the hole created by the four $sp^2$ nitrogens of the porphyrin ring system. The result is that Fe(II) is raised by about 0.5A° above the macrocyclic $N_4$ basal plane and tilted towards histidine nitrogen. (See Fig. 32.9).

In $HbO_2$, since Fe(II) is LS, Fe(II) has smaller size (0.75A°). Being smaller in size, Fe(II) fits into $N_4$ basal plane. Iron (II)-histidine nitrogen coordination, now, is normal to the $N_4$ basal plane. X-ray studies have revealed that $O_2$ binds in an end - on bent geometry with Fe–O–O angle of is ~130°.

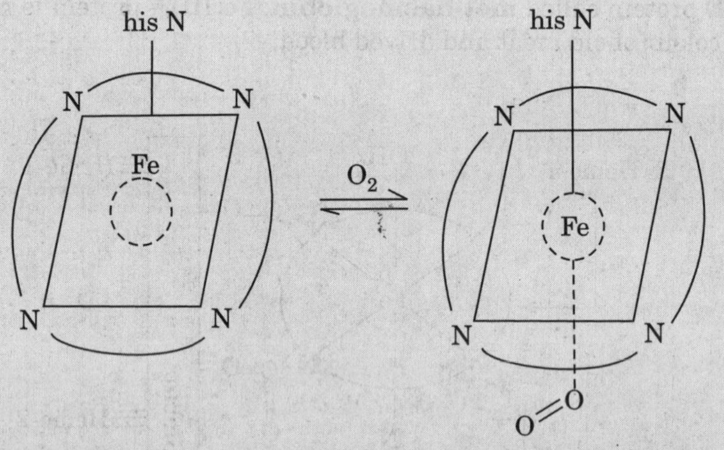


**Fig. 32.9** Reversible binding of oxygen in haemoglobin and oxyhemoglobin.

## Comparison Between Hb and Mb

(*i*) Hb and Mb both are **heme proteins**. Both are **metallic porphyrins** which contain the heme group in their structure. Heme group consists of one Fe atom which is coordinated to four N-atoms of porphyrin.

(*ii*) At high partial pressure of $O_2$, Hb and Mb both are equally good binders of $O_2$ but at low oxygen pressure, as in case of muscles, Hb is a far poorer binder.

(*iii*) The five coordinated high spin Fe(II) present in Hb and Mb lies above the plane containing the four coordinated nitrogen of porphyrin ring and the coordinated N-atom of the histidine group.

## Comparison Between $HbO_2$ and $MbO_2$

(*i*) Fe–$O_2$ bond in $HbO_2$ and $MbO_2$ is bent as shown in Fig. 32.10.

This is due to the fact that since sixth coordination site of iron is sterically overcrowded, this site does not have space for a straight Fe-O-O links.

(*ii*) $HbO_2$ and $MbO_2$ both are diamagnetic in character.

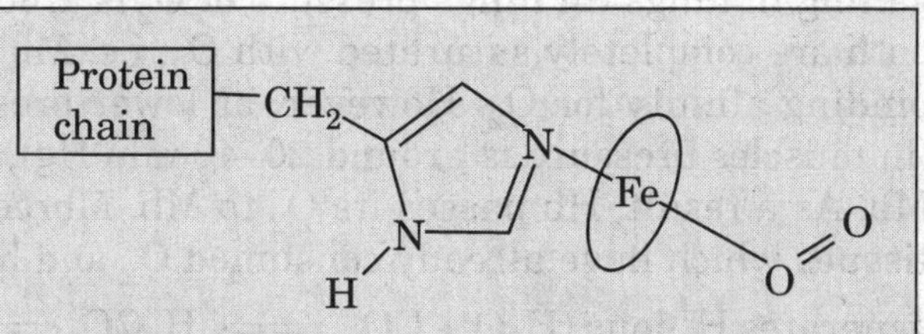


**Fig. 32.10** Structure of oxyhemoglobin and oxymyoglobin.

## Oxygen-binding Curves for Myoglobin (Mb) and Hemoglobin (Hb)

Graph between the persent oxygen of Mb and Hb and partial pressure (Torr) of $O_2$ in lungs and muscles in given in Fig. 32.11. This graph is called oxygen-

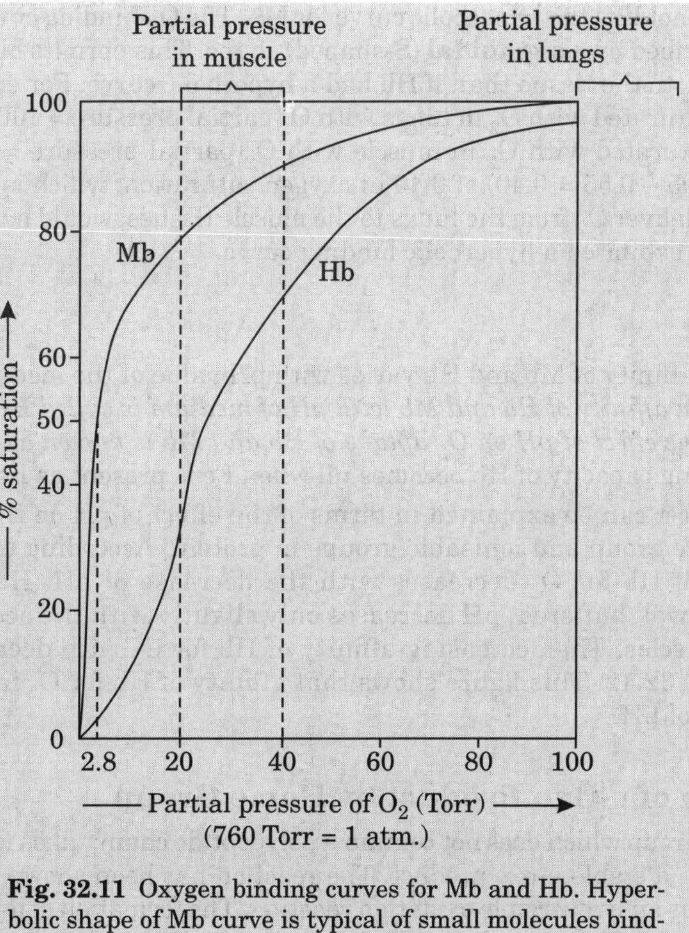


**Fig. 32.11** Oxygen binding curves for Mb and Hb. Hyperbolic shape of Mb curve is typical of small molecules binding to a protein. Hb curve is sigmoidal in shape.

binding curve for Mb and Hb. This graph shows that at high pressure of $O_2$ prevailing in lungs (in lungs pressure of $O_2$ is around 100–120 mm Hg), Mb and Hb both are completely saturated with $O_2$, i.e. Mb and Hb both have almost similar binding affinity for $O_2$. However, at lower pressure of $O_2$ prevailing in muscles (in muscles pressure is around 20–40 mm Hg), Hb has lower affinity for $O_2$ than Mb. As a result, Hb passes its $O_2$ to Mb. Moreover, the need for $O_2$ is greater in tissues which have already consumed $O_2$ and have produced $CO_2$. $CO_2$ produced generates $H^+$ ions ($H_2O + CO_2 \rightleftharpoons H_2CO_3 \rightleftharpoons H^+ + HCO_3^-$). $H^+$ ions increase the acidity. Due to the increase in acidity, pH decreases. The increase in acidity favours the release of $O_2$ from $HbO_2$ to Mb.

Myoglobin is a single polypeptide chain with one heme group and hence one oxygen binding site. Even proteins with multiple bonding sites for the small molecule generate hyperbolic binding curves like Mb when the ligand ($O_2$) interacts with each binding site independently. However, this is not true for hemoglobin.

Hemoglobin is half saturated with $O_2$ at $O_2$ partial pressure of 26 torr which is nearly 10 times greater than that for Mb (Fig. 32.11). Also Hb binding curve (Fig. 32.11) is not like the hyperbolic curve for Mb. The $O_2$-binding curve of Hb (Fig. 32.11) is described by a **sigmoidal** (S-shaped) shape. This permits blood to deliver much more $O_2$ to the tissue than if Hb had a hyperbolic curve. For example, Hb is nearly 95% saturated with $O_2$ in lungs with $O_2$ partial pressure = 100 torr but only about 55% saturated with $O_2$ in muscle with $O_2$ partial pressure = 30 torr. This difference (0.95 – 0.55 = 0.40) of 0.40 in oxygen saturation, which is a measure of Hb ability to deliver $O_2$ from the lungs to the muscle tissues, would have been much lower had Hb exhibited a hyperbolic binding curve.

### Bohr Effect

Oxygen affinity of Mb and Hb varies with pH value of the medium. *The variation of oxygen affinity of Hb and Mb with pH of medium is called* **Bohr effect**. In other words *the effect of pH on $O_2$ uptake of Hb and Mb is known as* **Bohr effect.** Oxygen-binding capacity of Hb becomes nil when Fe is present as Fe(III).

Bohr effect can be explained in terms of the effect of pH on the interaction between heme group and ionisable groups in protein. According to Bohr effect, the affinity of Hb for $O_2$ decreases with the decrease of pH. However, since blood is very well buffered, pH decreases only slightly with the accommodation of $CO_2$ in muscles. The decreasing affinity of Hb for $O_2$ with decreasing pH is shown in Fig. 32.12. This figure shows that affinity of Hb for $O_2$ increases with the increase of pH.

## Formation of μ-Oxo Product by Heme Group

A heme group which does not contain a polypeptide chain, takes up $O_2$ molecule to finally form a stable μ-oxo product. The reaction has been shown in Fig. 32.13. This reaction is an irreversible oxidation reaction. The formation of μ-oxo product is possible only when the two reacting heme groups should be able to come in contact with each other.

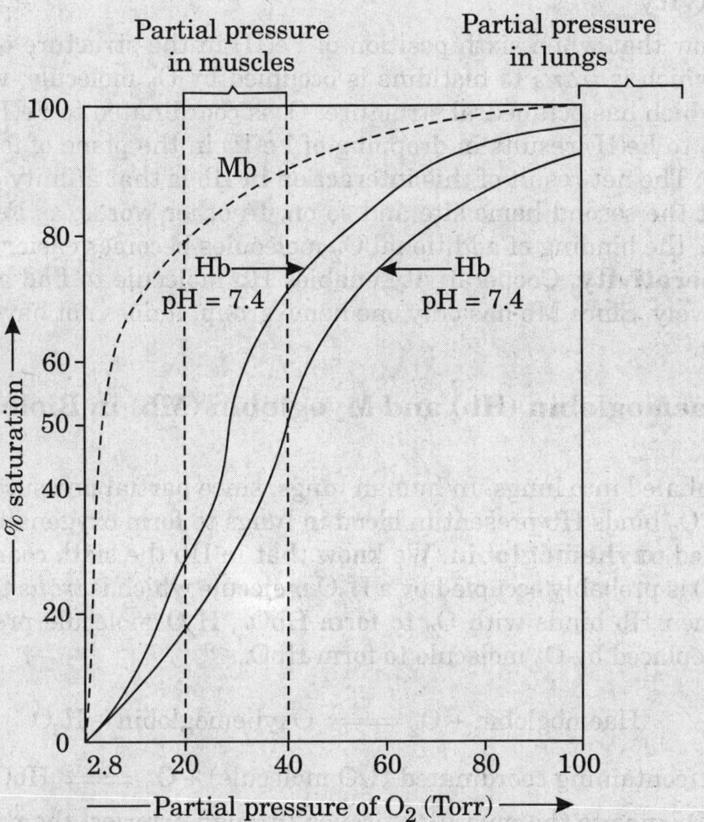


**Fig. 32.12** Bohr effect showing $O_2$ affinity of Hb increasing with increasing pH. Mb curve is given for reference only.

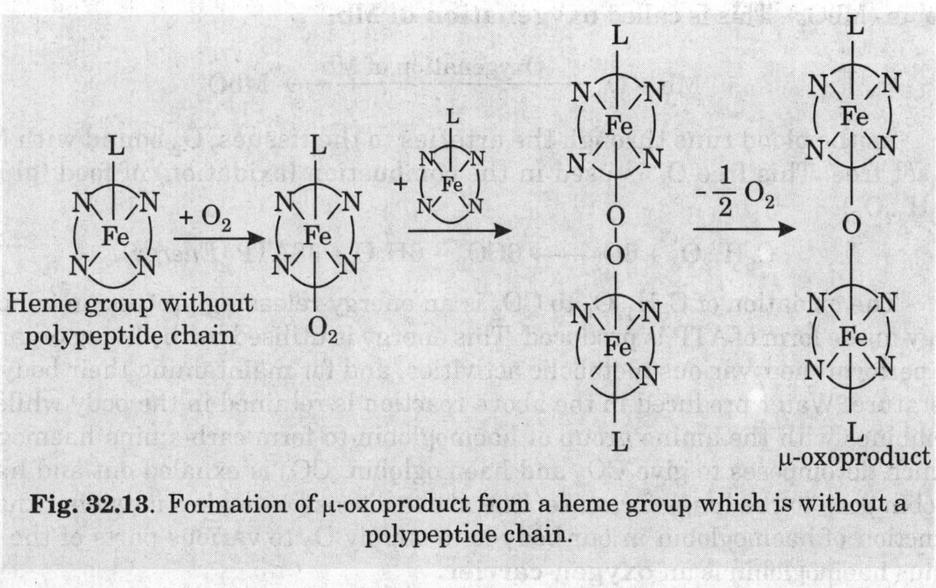


**Fig. 32.13**. Formation of μ-oxoproduct from a heme group which is without a polypeptide chain.

## Cooperativity

We know that when sixth position of Fe(II) in the structure of deoxy-Mb or deoxy-Hb, which is *trans* to histidine is occupied by $O_2$ molecule, we get oxy-Mb or oxy-Hb which has octahedral structure. $O_2$ is coordinated to Fe(II). The coordination of $O_2$ to Fe(II) results in dropping of Fe(II) in the plane of four N-atoms of heme group. The net result of this interaction in Hb is that affinity of Hb for $O_2$ is increased at the second heme site and so on. In other words, as Fe(II) binds one $O_2$ molecule, the binding of additional $O_2$ molecules becomes easier in Hb. This is called **cooperativity.** Cooperativity enables Hb molecule to find and release $O_2$ more effectively. Since Mb has only one heme group, it does not have any cooperative binding.

## Role of Haemoglobin (Hb) and Myoglobin (Mb) in Biological Systems

$O_2$ is inhaled into lungs. In human lungs, since partial pressure of $O_2$ is high, the inhaled $O_2$ binds Hb present in blood in lungs to form oxygenated hemoglobin ($HbO_2$), called **oxyhemoglobin**. We know that in Hb the sixth coordination position of Fe(II) is probably occupied by a $H_2O$ molecule which is *trans* to the histidine N-atom. When Hb binds with $O_2$ to form $HbO_2$, $H_2O$ molecule present in Hb is reversibly replaced by $O_2$ molecule to form $HbO_2$.

$$\text{Haemoglobin} + O_2 \rightleftharpoons \text{Oxyhemoglobin} + H_2O$$

*or* $$\text{Hb (containing coordinated } H_2O \text{ molecule)} + O_2 \rightleftharpoons HbO_2 + H_2O$$

As $HbO_2$ runs to the muscular tissues through arteries, the partial pressure of $O_2$ in muscle tissues decreases and $HbO_2$ liberates $O_2$.

$$HbO_2 \longrightarrow Hb + O_2$$

The liberated free $O_2$ is taken up by myoglobin (Mb) to form oxygenated myoglobin ($MbO_2$). This is called **oxygenation of Mb**.

$$Mb + O_2 \xrightarrow{\text{Oxygenation of Mb}} MbO_2$$

As the blood runs through the arteries to the tissues, $O_2$ bound with $MbO_2$ is set free. This free $O_2$ is used in the combustion (oxidation) of food (glucose, $C_6H_{12}O_6$).

$$C_6H_{12}O_6 + 6O_2 \longrightarrow 6CO_2 + 6H_2O + 38ATP\ (\textit{Energy})$$

The oxidation of $C_6H_{12}O_6$ to $CO_2$ is an energy releasing process in which energy in the form of ATP is produced. This energy is utilised by the living organisms to perform their various metabolic activities, and for maintaining their body temperature. Water produced in the above reaction is retained in the body while $CO_2$ combines with the amino group of haemoglobin to form carb-amino haemoglobin which decomposes to give $CO_2$ and haemoglobin. $CO_2$ is exhaled out and haemoglobin goes to the lungs for reuse. The above discussion makes it evident that the function of haemoglobin in our body is to supply $O_2$ to various parts of the body. Thus haemoglobin is an **oxygen-carrier.**

## Biological Role of $Na^+$, $K^+$, $Mg^{2+}$ and $Ca^{2+}$ ions in Biological Systems

### Role of $Na^+$ Ions

(*i*) $Na^+$ ion is a major cation present in extra-cellular fluids of animals and human beings. Sodium is present to a good extent in bones as phosphate. This element mainly exists as chloride and bicarbonate.

(*ii*) The presence of $Na^+$ ions activates certain enzymes in animal body.

(*iii*) The intake of $Na^+$ ions in excess amounts may cause hypertension. The saline water containing excessive amounts of NaCl is harmful to plants and aquatic life because of the toxicity of $Na^+$ ions.

(*iv*) In biological systems, $Na^+$ ions regulate acid-base equilibrium.

(*v*) Na is essential for the formation of HCl in stomach and conduction of nerve impulses and muscle contraction.

(*vi*) Herbivorous animals donot get the required amount of Na from vegetarian in-take and, therefore, look out for *salt-lick* to make up their Na requirement.

(*vii*) It also helps in maintaining osmotic pressure of the body fluid, thereby protecting the body against fluid loss.

(*viii*) It has a role to play in preservation of normal irritability of muscle and permeability of the cell.

(*ix*) The injectable medicines are dissolved in sodium chloride before they are injected into human body.

(*x*) $Na^+$ ion is important in nerve action and in the function of heart.

(*xi*) $Na^+$ ion is responsible for the transport of glucose and amino acids into the cell.

### Role of $K^+$ Ions

(*i*) $K^+$ ion is mostly present in intra-cellular fluid as well as in extra-cellular fluid.

(*ii*) $K^+$ ion is essential for nerve impulse and muscle contraction.

(*iii*) $K^+$ ion is essential for all organisms with the possible exception of blue green algae.

(*iv*) It is moderately toxic to mammals but only when injected intravenously.

(*v*) It influences acid-base equilibrium like sodium in extracellular fluid.

(*vi*) It controls osmotic pressure and water retention.

(*vii*) It is important for metabolic functions like protein biosynthesis by ribosomes.

(*viii*) A number of enzymes such as glycolytic enzyme pyruvate kinase require $K^+$ for maximum activity.

(*ix*) $K^+$ is required in the cell for: (*a*) glucose metabolism (*b*) protein synthesis and, (*c*) activation of some enzymes.

### $Na^+$ – $K^+$ Pump (Also Called Sodium Pump)

Although $Na^+$ and $K^+$ ions are chemically similar, yet biological response for these ions is different. While $Na^+$ ions are pumped out of cyto-plasm, $K^+$ ions are pumped in. This transport of ions is called **$Na^+$ – $K^+$ pump** or simply **sodium pump.** This pump involves the active take up of $K^+$ ions and expulsion of $Na^+$ ions. The difference in concentration of the two ions inside and outside the cell membrane produces an electrical potential, which is crucial for the functioning of nerve and muscle cells.

It may be noted that higher intake (we require about 1 g of $Na^+$ per day) of $Na^+$ ions is a problem, but there is no such risk from higher intake of $K^+$ ions. Rather potassium deficiency is more common. We can make up the $K^+$ ions deficiency by consuming more of banana and coffee.

$Na^+$ and $K^+$ are present in the red blood cells. The ratio of these ions in mammals such as human being, rabbits, rats and horses is 1:7. In cats and dogs, this ratio is 15:1.

To establish this ratio also called **concentration gradient** in the cell, biologists have suggested different mechanism involving sodium pump and potassium pump.

(*i*) The concentration gradient controls the development and functioning of the nerve cells.

(*ii*) In restful state, potential of the nerve cells is linked to $K^+$ concentration across the membranes.

(*iii*) During activation of the nerve cells, a chemical called *acetylcholine* is discharged. This discharge is transmitted through the length of the nerve cell by electric impulse.

## Mechanism of $Na^+$ – $K^+$ Pump

The mechanisms of $Na^+$ – $K^+$ pump is that $Na^+$ – $K^+$ ATPase is phospho-rylated by andenosine tri-phosphate molecule (ATP) in presence of $Na^+$ and $Mg^{2+}$ ions. The site for phosphorylation is the side chain of a specific asparate residue, E. In this reaction phospho-rylated inter-mediate (E–P) is formed.

**Phospho-rylation of E:**

$$\underset{\substack{\text{Asparate residue}\\\text{(Phospho protein)}}}{E} + ADP \xrightleftharpoons[\text{in presence of } Na^+ \text{and } Mg^{2+} \text{ ions}]{\text{Phospho-rylation of E}} \underset{\substack{\text{Phospho-rylated}\\\text{intermediate}}}{E-P} + ADP$$

**Dephospho-rylation of E–P.**

E – P formed is hydrolysed in presence of $K^+$ ions. As a result, E–P gets dephospho-rylated to form the original E.

$$E - P + H_2O \xrightleftharpoons{K^+} E + HPO_4^{2-}$$

It has been found that $Na^+$ and $Mg^{2+}$ ions are necessary for the formation of the intemediate (E–P) and $K^+$ ions are necessary for dephosphorylation of the intermediate. During this reaction $3Na^+$ and $2K^+$ ions are transported per ATP hydrolysed. Therefore, the pump generates an electric current across the plasma. In other words, $Na^+$ – $K^+$ ATPase pump is **electrogenic.** Each operation of the cell pumps out larger number of $Na^+$ ions that it pumps into the cell. As a result, the interior of the cell acquires an excessive negative charge while the exterior of the cell acquires an excessive positive charge. As a result electrical potential gradient across the cell membrane is developed which is responsible for the **transmission of nerve signals** in the animals.

ATP is not hydrolysed unless $Na^+$ and $K^+$ are transported. In other words, the system is coupled. The energy stored in ATP is only dissipated if the pump works. The $Na^+ - K^+$ pump is also reversible so that it synthesizes ATP.

## Sodium Pump is Electrogenic in Nature

The energy for the transport of $Na^+$ and $K^+$ ions is provided by the hydrolysis of energy molecule, ATP. Kidney and brain cells use about 70% of the energy from ATP for the transport of ions. Sodium pump acts as a transport system to maintain $Na^+ - K^+$ concentration gradient. For each ATP molecule hydrolysed ($ATP^{4-}$), $3Na^+$ ions are transported **out** of the cell and $2K^+$ ions are transported **inside** (into) the cell according to the equation.

$$ATP^{-4} + H_2O + 3Na^+ \text{ (inside)} + 2K^+ \text{ (outside)} \longrightarrow 3Na^+ \text{ (outside)} + 2K^+ \text{ (inside)} + ADP^{3-} + H_2PO_4^-$$

The net result is the movement of one positive charge outward per cycle and hence sodium pump is **electrogenic in nature**.

## Role of $Mg^{2+}$ Ions

(*i*) Mg is very important non-transition essential element. It is essential to all organisms. This element is present in greater concentration in red cells than in plasma.

(*ii*) Mg(II) plays important role in many metallo-enzymes.

(*iii*) $Mg^{2+}$ ion is present in chlorophyl which is present in plants. Chlorophyl absorbs light from sun and brings about the process of photosysthesis in plants to convert $CO_2$ and $H_2O$ to glucose.

(*iv*) Mg(II) salts serve as remedies for constipation, obesity, liver and gall bladder disorders.

(*v*) Mg forms a complex with ATP which is required for most enzymetic reactions involving ATP within the cell.

(*vi*) Magnesium ions are present in the enzyme *phosphatase* which acts upon organic phosphates to hydrolyse them into free phosphates.

(*vii*) Magnesium ions are present in the enzyme *aminopeptidase* which hydrolyses polypeptides the free amino acid end of the chain to form lower peptides and sometimes free amino acids.

(*viii*) Whole grains, fruits and vegetables are rich in Mg. Alcoholism leads to Mg deficiency. Mg deficiency is indicated by tetanus-like cramp and abnormal swelling of arterial walls. Mg deficiency in plants destroys the green colour of leaves.

(*ix*) At biological pH (7.0 to 7.5) ATP and ADP are present in the form of tetravalent and trivalent anions *viz* $ATP^{4-}$ and $ADP^{3-}$ respectively, since phosphate groups present in ATP and ADP are completely ionised. Since intracellular fluid has high concentration of Mg(II), Mg(II) combines with $ATP^{4-}$ and $ADP^{3-}$ ions to form $[Mg(ATP)]^{2-}$ and $[Mg(ADP)]^-$. Thus ATP and ADP exist as $[Mg(ATP)]^{2-}$ and $[Mg(ADP)]^-$ ions in intra-cellular fluid.

$$Mg^{2+} + ATP^{4-} \longrightarrow [Mg(ATP)]^{2-}$$

$$Mg^{2+} + ADP^{3-} \longrightarrow [Mg(ADP)]^-$$

The structures of $[Mg(ATP)]^{2-}$ and $[Mg(ADP)]^-$ ions are given in Fig. 32.14.

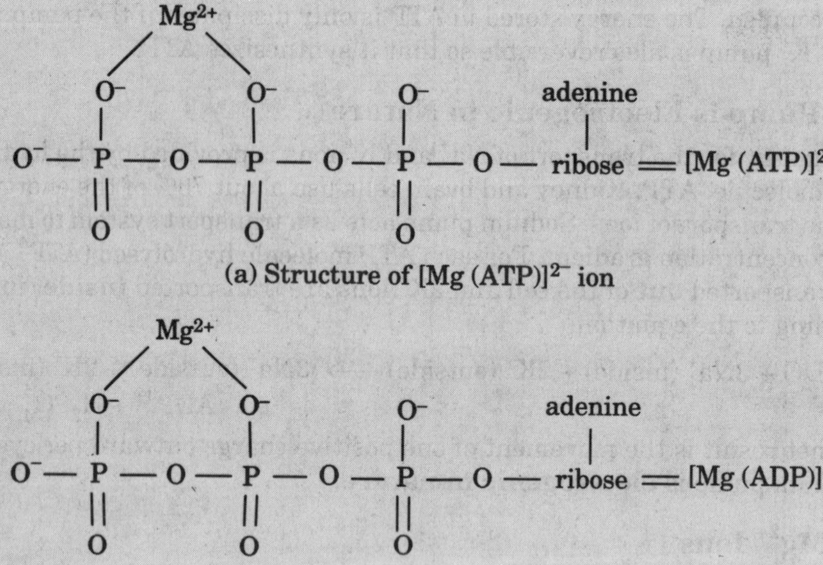


(a) Structure of $[Mg(ATP)]^{2-}$ ion

(b) Structure of $[Mg(ADP)]^{-}$ ion

**Fig. 32.14.** Structure of $[Mg(ATP)]^{2-}$ ion and $[Mg(ADP)]^{-}$ ions

## Role of $Ca^{2+}$ Ions

(*i*) Some important minerals of Ca are calcite ($CaCO_3$), argonite ($CaCO_3$), hydroxy apatite [$Ca_5(PO_4)_3(OH)$] etc. The enamel on teeth is a double salt of Ca, $3Ca_3(PO_4)_2 \cdot CaF_2$.

(*ii*) Ca is a major constituent of bones and teeth.

(*iii*) Calcium ions are present as phosphates in the bones of human beings and animals. These ions play an important role in muscle contraction. The malnutrition in children is due to the deficiency of $Ca^{2+}$.

(*iv*) About 99% of the body calcium is in the skeleton where it is maintained as deposits of calcium phosphates in soft and fibrous matrix. A very small quantity of calcium not present in skeletal structure, is in the body fluids where it is partly ionized. This small amount of ionized calcium in the body fluid is of great importance in blood coagulation, in maintaining the normal excitability of the heat, muscles and nerves and in the differential aspects of membrane permeability.

(*v*) The major inorganic constituent of bone is comprised of a crystalline form of calcium phosphate resembling the mineral *hydroxapatite*. Additionally, the bone contains a substantial amount of non-crystalline, amorphous calcium phosphate. It appears that the amorphous material is predominant in early life but later contraction on the crystalline form takes over in adult life.

(*vi*) Calcium ions are needed to bring about **blood clotting** and contact of muscles such as **heart beat**.

(*vii*) Deficiency of $Ca^{2+}$ ions causes **tetany**, while excess of it causes **calcification**.

(*viii*) $Ca^{2+}$ ions plays an important role in biological processes and in skeletal formation. With phosphorus in the mineral hydroxyapatite $Ca_5(PO_4)_3OH$, it is the major constituents of bone, teeth and shells. $Ca^{2+}$ plays many biochemical roles such as (*a*) It acts as a messenger for hormonal action (*b*) It acts as a trigger for muscular contraction (*c*) It acts as initiation of blood clotting (*d*) It also plays a role in the stabilization of protein structure (*e*) It helps in the maintenance of rhythm of heart.

(*ix*) Calcium rich foods are spinach, egg, milk and diary products.

(*x*) Calcium that we consume first gets dissolved in the stomach which then passes on to the small intestine where Ca is absorbed into the blood. The observed Ca is either deposited on the bones or excreted.

## Role of $Ca^{2+}$ in Transport: Calcium Pump

$Ca^{2+}$ plays an important role in muscles. It triggers a signal that stimulates muscles to contract. In the normal state, the concentration in the intracellular fluids is very low because nearly all the $Ca^{2+}$ ions in muscle is pumped into a complex network of vehicles known as **sarcoplasmic reticulum** (SR). Their concentration in intracellular fluids is about 10,000 times less than their concentration in the extra cellular fluids. The maintenance of low calcium in concentration in the intra cellular fluids is done by biochemical process known as **calcium pump or $Ca^{2+}$ pump**.

The mechanism of $Ca^{2+}$ pump is as under

$$E + ATP \xrightleftharpoons{Ca^{2+},\ Mg^{2+}} \underset{\text{Intermediate}}{ADP + E - P}$$

$$E - P + H_2O \xrightleftharpoons{Mg^{2+}} E + HPO_4^{2-}$$

During the cycle, two $Ca^{2+}$ are transported for each ATP hydrolysed.

## Essential Trace Elements

We have already mentioned the role of $Na^+$, $K^+$, $Mg^{2+}$ and $Ca^{2+}$ ions in biological systems. Here we shall discuss some essential trace metals which, in trace amounts, are useful in biological systems. Examples of such metals are Fe, Cu, Co, Zn, Cr, Mn, Mo, W, Ni etc.

### 1. Iron (Fe)

(*i*) Fe is present in plants and bacteria in the form of cytochromes and ferridoxin which are responsible for electron transfer reactions occurring in biological systems.

(*ii*) Fe is an important constituent of nitrogenase (a nitrogen-fixing enzyme) and a number of other important enzymes which are required by some plants for fixation of atmospheric nitrogen.

(*iii*) Eggs, nuts, whole grain, spinach have moderate amount of Fe.

(*iv*) Fe is essential for all forms of life.

(*v*) The excessive intake of Fe can cause siderosis and can damage organs. Excess Fe is not excreted but remains deposited in livers, spleen and skin.

(*vi*) Fe is essential for the synthesis of haemoglobin, myoglobin, cytochrome etc. and is stored as ferritin and transported as transferrin.

(*vii*) Deficiency of HCl in the system hinders in the extraction of Fe from food and this may result in anaemia.

## 2. Copper (Cu)

(*i*) It is an essential element for all organisms.

(*ii*) It is an important constituent of metallo-proteins including metallo-enzymes like ceruloplasma, hemoglobin, cytochrome etc. Hemocynin supplies $O_2$ to certain aquatic creatures and redox enzymes.

(*iii*) Copper-containing enzymes play an important role in the pigmentation of skin.

(*iv*) Our copper requirement comes from green vegetables, cereals, egg, meat etc.

(*v*) Copper-poisoning. See page 1097.

## 3. Cobalt (Co)

(*i*) It is a essential trace element for many organisms.

(*ii*) It is required for the synthesis of vitamin $B_{12}$.

(*iii*) It activates a number of enzymes.

(*iv*) Deficiency of Co in soil adversely affects the health of grazing animals.

(*v*) Co in trace amounts is essential for healthy growth of animals.

(*vi*) Co is highly toxic to plants and moderately toxic to mammals when infected intravenously.

## 4. Zinc (Zn)

(*i*) It is an essential element for almost all organisms.

(*ii*) It is a constituent of a large number of essential enzymes.

(*iii*) Zn comes from diet of egg, meat, milk, beans etc.

(*iv*) Zn deficiency may cause related growth, slow healing of wounds, loss of appetite etc.

(*v*) It is slightly toxic in macro doses. However, excess oral doses may cause diarrhoea and vomiting. Labour force exposed to the fumes of ZnO often develop nausea, ulceration etc.

## 5. Chromium (Cr)

(*i*) It is an essential trace element. It is involved in glucose metabolism. Cr(III) and insulin both maintain the correct level of glucose in the blood.

(*ii*) This element is in adult human body in ultra trace amounts.

## 6. Manganese (Mn)

(*i*) This element is an essential trace element for almost all organisms.

(*ii*) Mn(II) enzyme produced in the liver converts nitrogenous waste products into urea which is carried by the blood to the kidneys which excrete it out into urine.

(*iii*) Mn is essential for the proper growth of plants. It, along with fertilisers, enriches the soil.

(*iv*) It activates several enzymes.

(*v*) Its deficiency in soil leads to infertility in mammals.

(*vi*) Daily requirement of Mn comes from nuts, spurts, whole grain.

(*vii*) It is required for enzyme activities.

(*viii*) Excess uses of Mn causes irritation of respiratory passage and impairment of movement.

(*ix*) In plants Mn plays an important role in photosysthesis.

## 7. Molybdenum (Mo)

(*i*) It is an essential trace element for all organisms.

(*ii*) Organic molybdenum compounds formed in the body act as a respirator catalyst.

(*iii*) It is a very vital component of nitrogenase enzyme catalyst which catalyses the formation of $NH_3$ or $NH_4^+$ salts from atmospheric $N_2$ by algae and root-nudules bacteria.

(*iv*) It is moderately toxic.

(*v*) Excess of Mo in biological systems may cause deficiency of Cu.

## 8. Nickel (Ni)

(*i*) Ni is an essential trace element for several hydrogenases and plant ureases.

(*ii*) It is highly toxic to most plants and moderately toxic to mammals.

(*iii*) Higher concentrations of Ni in biological systems may cause cancer.

(*iv*) Its deficiency in food slows down the functioning of the liver in chicks.

(*v*) Its use in electroplating causes severe water pollution.

# Toxic Effects of Some Metals

Some metals and their compounds have toxic effect on humans, plants, animals, birds and aquatic species. Main examples of metals are Cu, Pb, As, Hg and Cd.

## 1. Copper-poisoning.

Copper compounds are highly poisonous for lower organisms such as algae, fungi and bacteria.

The excess intake of Cu by human body is harmful. Excess intake of Cu produces a disease known as **Wilson disease**. In this disease excess amount of Cu(II) gets deposited in the liver, kidney and brain. This disease leads to disorder of nervous system. The deposition of excess Cu(II) is removed by injecting the complexing ligand, penicillamine (2-amino–3–methyl–3–thiobutyric acid).

This ligand has $–NH_2$, –SH and –COOH as donor sites. The structure of this ligand is given in Fig. 32.15. This ligand forms a soluble complex *via* coordination with

S, N and O donors of penicillamine. Excess Cu(II) is excreted in urine as soluble copper–penicillamine complex.

The ligands **trien** (trimethylene tetramine) and **EDTA** are also capable of removing excess Cu(II) as soluble Cu(II) complexes.

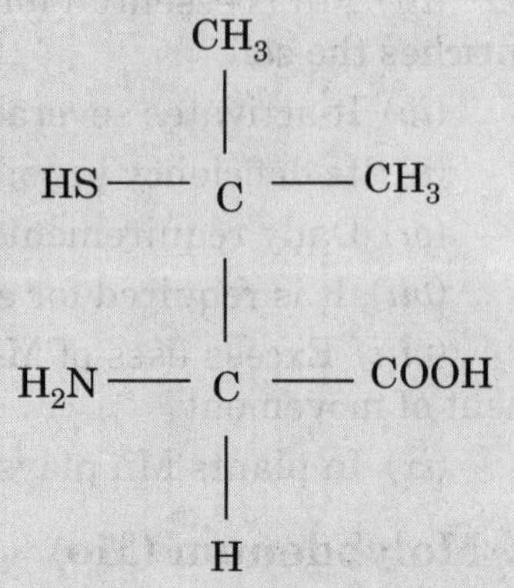

**Fig. 32.15.** Structure of Penicillamine.

## 2. Lead-poisoning

Lead has no biological function. Lead is a very poisonous metal. It is a cumulative poison, since it keeps on accumulating in the tissues of human body and in the leaf tissues of plants. Even a low concentration of 0.2 ppm of lead in human body creates metabolic disturbances. Some of the harmful effects of lead are given below:

(*i*) It is due to the prolonged effect of lead on the human body that red blood cells get stippled. The malformation of red blood cells produces the disease of anaemia.

(*ii*) Lead gets bound with the cellular enzymes and disrupts the functioning of cells and organs of the muscular circulatory and nervous systems.

(*iii*) Lead damages organs like liver, kidneys and intestine.

(*iv*) Lead develops abnormalities in fertility and pregnancy.

(*v*) Lead causes coagulation of proteins.

(*vi*) When rats are exposed to lead polluted air, they die in two-thirds of their average life span.

(*vii*) The birds, which are shot, but recover usually, subsequently die due to lead-poisoning.

(*viii*) The lead-polluted air has its adverse effects on vegetables, *e.g.*, the road-side vegetables are dried up by the exposure of lead-polluted air.

(*ix*) If lead is taken by human *even in small amounts*, it affects breathing and nervous system.

(*x*) *Excessive intake of lead* causes disruption of synthesis of haemoglobin. It also results in loss of appetite, anaemia, kidney malfunctioning, nervous disorders and brain damage.

(*xi*) Lead can also be absorbed by skin, resulting into many skin disease.

(*xii*) Lead inhibits the synthesis of haemoglobin in mammals.

(*xiii*) It is highly toxic for nervous system.

(*xiv*) We know that tetraethyl lead, $(C_2H_2)_4Pb$ is added to the petrol along with suitable amount of $C_2H_4Cl_2$ and $C_2H_4Br_2$. The addition of $(C_2H_5)_4Pb$ slows down the rate of combustion of petrol. In presence of $O_2$, $(C_2H_5)_4Pb$, $C_2H_4Cl_2$ and $C_2H_4Br_2$ react together to produce $PbCl_2$ and $PbBr_2$ both of which are volatile and hence come out of the engine and are thus mixed with air. This air gets polluted with $PbBr_2$ and $PbCl_2$ which are very harmful materials.

(*xv*) Lead is also released into the air by industries engaged in lead mining, extraction and purification of lead and manufacture of alloys and paints.

(*xvi*) $2PbCO_3 . Pb(OH)_2$ which is used as a pigment is a serious health hazard.

(*xvii*) Deposition of lead in bones and teeth causes tireness, headache, loss of appetite, anaemia, muscular weakness etc.

(*xix*) Prolonged use of lead utensils causes *lead sickness*.

In case of lead poisoning, intra-venous, injection of $CaNa_2$. edta is given to the patient. This takes out the accumulated lead as lead edta complex which is excreted in urine.

## 3. Arsenic-poisoning

(*i*) It is an essential ultra-trace element present in red algae, chicks and some mammals including humans.

(*ii*) Deficiency of As in chicks results in depressed growth.

(*iii*) Mining operation, burning of coal, use of insecticides and herbicides containing arsenic compounds are the chief sources of arsenic pollution in air and water.

(*iv*) It is moderately toxic to plants and highly toxic to mammals when present in more than ultra trace quantities.

$Cl_2 > As\ CH = CHCl$ is a gaseous arsenic compound. It is deadly poisonous. This compound is called **Lewisite**. If this compound is present in living systems, it may combine with HS-containing enzyme to form Lewisite-enzyme product. The original enzyme activity is thus weakened. In order to destroy the poisonous effect of Lewisite, British Anti Lewisite (BAL) is used as antidote. BAL forms a complex with Lewisite and thus destroys the poisonous effect of Lewisite. In this way the original enzyme is released.

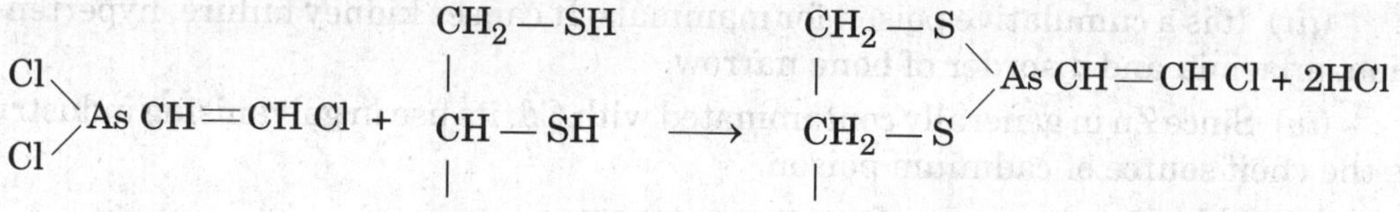

Lewisite

2,3-dimercapto propanol (British Anti Lewisitie, BAL)

Complex Compound

(*v*) Arsenic compounds are used as insecticides by farmers.

(*vi*) Arsenic-containing drinking water causes severe damage to skin, liver and kidneys.

## 4. Mercury-poisoning

(*i*) Hg in our environment comes mostly from two sources: (*a*) Natural volcanic eruption and natural weathering of mercury-containing rocks. It is due the weathering of rocks that Hg is found in sea (*b*) Industrial processes such as chloro-alkali electrolysis of brine solution (NaCl) which generates NaOH and $Cl_2$. In this electrolysis process Hg is used as cathode

$$NaCl \longrightarrow Na^+ + Cl^-$$

$$2Cl^- \longrightarrow Cl_2 + 2e^-$$

$$2H_2O + 2e^- \longrightarrow H_2 + 2OH^-$$
$$Na^+ + OH^- \longrightarrow NaOH$$

Dumping of used Hg obtained from chloro-alkali process in rivers and oceans creates havoc to aquatic animals.

(*ii*) Hg has no biological function.

(*iii*) It is highly toxic to fungi, plants and animals. It is a contaminated poison for mammals.

(*iv*) Intake of mercury-contaminated fish may cause serious problems.

(*v*) If Hg vapours are inhaled, they are absorbed in lungs, dissolve in the blood and ultimately they are carried to the brain. This causes irreversible damage to central nervous system.

(*vi*) Monomethyl mercury [$CH_3Hg$] and dimethyl mercury [$(CH_3)_2Hg$] produce nervous disorders in marine animals.

(*vii*) Contact with Hg produces nervousness, fear, inability to make decision, heaviness, irritability, headache, pessimism, fatigue, sleeplessness, trembling of limbs, falling teeth and diarrhoea.

(*viii*) Hg brings about genetic changes.

(*ix*) Antidotes for Hg-poisoning are animal charcoal, pencillamine etc.

## 5. Cadmium-poisoning

(*i*) This element has no use in living organisms.

(*ii*) It is toxic to all organisms.

(*iii*) It is a cumulative poison for mammals. It causes kidney failure, hypertension, anaemia and disorder of bone narrow.

(*iv*) Since Zn in generally contaminated with Cd, its use in galvanising industry is the cheif source of cadmium poison.

(*v*) Cd is very dangerous for pregnant women.

(*vi*) Cd-pollution causes a disease called "*Hai Itai*" in which the whole body feels serious pains and bones begin to fracture very easily.

# Fixation of Atmospheric Nitrogen

## What is Fixation of Nitrogen?

*The conversion of free atmospheric nitrogen into useful nitrogenous compounds by artificial or natural methods is called fixation of nitrogen*. Nitrogen present in these nitrogenous compounds is called **fixed or combined nitrogen**.

## Artificial Methods Used for the Fixation of Atmospheric Nitrogen

**1. Fixation of $N_2$ as $NH_3$ by Haber's Process.** A mixture of nitrogen (manufactured by liquefaction of air) and hydrogen in the ratio 1 : 3 is compressed to a pressure of 200–500 atmospheres and is passed over a catalyst (finely divided iron + molybdenum) heated to about 550°C. This forms the *Haber's process* for the manufacture of ammonia which then can be converted into ammonium salts by treatment with suitable acids.

$$N_2 + 3H_2 \xrightleftharpoons[\text{high press}]{\text{catalyst}} 2NH_3$$

**2. Fixation of $N_2$ as $HNO_3$ by Ostwald's process.** $NH_3$ manufactured by Haber's process is oxidised to nitric oxide (NO) by passing a mixture of $NH_3$ (1 volume) and air (8 volumes) over heated Pt gauze at 1070K. NO combines with more of $O_2$ to give nitrogen dioxide ($NO_2$) which when absorbed in water in the presence of excess of air, gives $HNO_3$.

$$4NH_3 \text{ (1 vol)} + 5O_2 \text{ (8 vol)} \xrightarrow[\text{1070K}]{\text{Pt gauze}} 4NO + 6H_2O$$

$$2NO + O_2 \text{ (\textit{Excess})} \longrightarrow 2NO_2$$

$$4NO_2 + 2H_2O + O_2 \text{ (\textit{Excess})} \longrightarrow 4HNO_3$$

**3. Fixation of $N_2$ as $HNO_3$ by Birkland-Eyde process.** Under the influence of high tension electric arc where the temperature is high, nitrogen of the air combines with oxygen to form nitric oxide. It combines with more of oxygen to form nitrogen peroxide. This may be absorbed in water in presence of excess of air to give nitric acid which may be used for the manufacture of nitrogenous fertiliser.

$$N_2 + O_2 \rightleftharpoons 2NO \text{ (\textit{Nitric oxide})}$$

$$2NO + O_2 \longrightarrow 2NO_2 \text{ (\textit{Nitrogen peroxide})}$$

$$\underset{\text{Nitrogen peroxide}}{4NO_2} + 2H_2O + \underset{\text{From air}}{O_2} \longrightarrow \underset{\text{Nitric acid}}{4HNO_3}$$

**4. Fixation of $N_2$ as ammonium salts and nitrates.** $NH_3$ obtained by Haber's process and $HNO_3$ obtained by Ostwald's process and Birkland-Eyde process can be used for the preparation of ammonium slats and nitrates which are used as fertilisers.

**5. Fixation of nitrogen as calcium cyanamide, $CaCN_2$.** Nitrogen gas obtained by the evaporation of liquid air is passed over calcium carbide heated to 800–1000°C. A mixture of calcium cyanamide and carbon is obtained which is extensively used as a fertilizer under the name of *nitrolim*.

$$\underset{\text{Calcium carbide}}{CaC_2} + N_2 \longrightarrow [\underset{\text{Calcium cynamide}}{CaCN_2} + C] \text{ (\textit{Nitolim})}$$

**6. Fixation of nitrogen as nitrides.** Nitrogen combines with magnesium and aluminium at the temperature to give nitrides which are employed as a source of ammonia. These nitrides are decomposed by $H_2O$ and $NH_3$ is evolved.

$$3Mg + N_2 \longrightarrow \underset{\text{Magnesium nitride}}{Mg_3N_2}$$

$$2Al + N_2 \longrightarrow \underset{\text{Aluminium nitride}}{2AlN}$$

$$AlN + 3H_2O \longrightarrow Al(OH)_3 + NH_3$$

## Adenosine triphosphate (ATP): Energy Supplier in Biological Reactions

Biological reactions are endothermic and hence require the supply of energy (heat). Nucleotide adenosine triphosphate (ATP), which was discovered in 1929 in the extracts of skeletal muscle, usually supplies energy required for biological reactions. ATP molecule has an *adenine* base linked to the 1 position of (pentose sugar) *ribose* and a triphosphate group at 5 position of *ribose* (See Fig. 32.16).

The three phosphates namely adenosine monophosphate (AMP), adenosine diphosphate (ADP) and adenosine triphosphate (ATP) are present in all forms of life and serve the same function. In living cells ATP is an ~80% of the total amount of ATP, ADP and AMP. The structures of these three phosphates are given in Fig. 32.16.

An adult human body has ~50g ATP which is broken down to ADP and $H_3PO_4$ with the liberation of energy for body needs.

$$ATP + H_2O \rightleftharpoons ADP + H_3PO_4 + 33.5 \text{ KJ mol}^{-1}$$

ATP performs the role of energy transporter from one species to another via transfer of the terminal phosphate group. A specific enzyme called a *kinase* catalyses such transfer phosphate group.

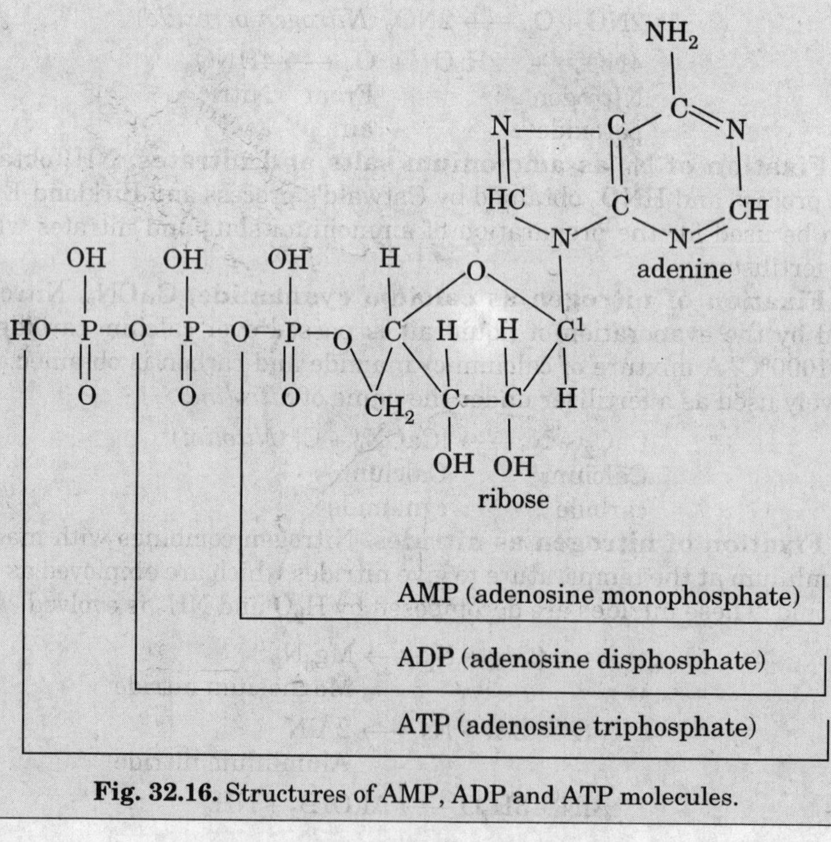


**Fig. 32.16.** Structures of AMP, ADP and ATP molecules.

During degradation of glucose to lactate two high-energy phosphate compounds namely 3-phospho-glyceroyl phosphate and phospho-enol-pyruvate are formed. The phosphate groups of these compounds are transferred to ADP to form energy-rich ATP.

## Metallo-enzymes: Biological Catalysts

Biological reactions are catalysed by particular protein catalysts. In these reactions no by-products are formed. These protein catalysts are called **enzymes**. These enzymes act as catalysts in biological reactions.

All types of enzymes are proteins. Some enzymes are pure polypeptides and contain no other group. There are also enzymes which require additional groups for their activity. The additional groups are called **cofactor**. A cofactor may be metal ion like Fe(II), Mn(II), Zn(II), Cu(II) or may be a complex organic molecule called **co-enzyme**. Protein part of an enzyme minus the cofactor or co-enzyme is called **apo-enzyme**. Enzymes which need a metal ion for their activity are called **metallo-enzymes**. Obviously metallo-enzymes are metallo proteins. About 30% of all enzymes are metallo-enzymes. More than 1500 metallo-enzymes have so far been identifield.

## Nomenclature of Enzymes

Enzymes are usually named by adding the suffix **–ase** to the name of the molecule on which the enzyme acts. The molecule on which enzyme acts is called **substrate**. The names of some enzymes are given below.

(*i*) **Ure-ase enzyme.** This enzyme catalyses the hydrolysis of urea.

(*ii*) **Peroxid-ase enzyme.** This enzyme catalyses the oxidation of organic substances by peroxide.

(iii) **Nitrogen-ase enzyme.** This enzyme catalyses the fixation of $N_2$ in nature at ordinary temperature and pressure.

## Nitrogen-ase Metalloenzyme

**Nitrogen-ase is a metallo-enzyme.** This enzyme is present in bacterial species. It is an equilibrium mixture of two proteins. One protein is an *iron-sulphur protein* (Fe—S protein) and the other protein is *molyodenum-iron-sulphur protein* (Fe–Mo–S protein). These two proteins are in the ration 2:1. Thus:

$$\text{Nitrogen-ase metallo-enzyme} \rightleftharpoons 2(\text{Fe–S protein}) + 1(\text{Fe–Mo–S protein})$$

Fe–S protein is known as **reductase** and Fe–Mo–S protein is called **nitrogenase**.

Many plants require large amounts of soluble nitrogen for their satisfactory growth. For this purpose these plants fix atmospheric nitrogen in presence of nitrogenase enzyme.

Fe–S protein is a smaller protein with molecular weight in the range of 50,000 to 70,000. It is also known as **azoferredoxin** and is yellow in colour and extremely air sensitive. This protein has two ferredoxin type $Fe_4S_4$ clusters which are joined together by two bringing S-atoms of two crystein amino acids.

Fe–Mo–S protein has molecular weight in the range 220,000 to 240,000. It is also known as **molybdo-ferrodoxin**. It is brown air sensitive protein.

## Natural Methods Used for the Fixation of Atmospheric Nitrogen

**1. Fixation of $N_2$ by lighting discharges.** Due to lightning discharges (electrical disturbances) occurring in nature, $N_2$ and $O_2$ combine together to form NO which gets further oxidized to $NO_2$. This oxide (*i.e.* $NO_2$) reacts with rain water in presence of excess of $O_2$ to produce $HNO_3$ which is washed down to earth. $HNO_3$ so formed reacts with the bases of the soil (*i.e.,* $CaCO_3$) to give nitrate which is stored in soil as plant food.

$$N_2 + O_2 \longrightarrow 2NO, \; 2NO + O_2 \longrightarrow 2NO_2$$

$$4NO_2 + 2H_2O + O_2 \longrightarrow 4HNO_3$$

$$2HNO_3 + CaCO_3 \; (\textit{Alkali in soil}) \longrightarrow Ca(NO_3)_2 \; (\textit{Stored in the soil as plant food}) + CO_2 + H_2O$$

**2. Fixation of $N_2$ by symbiotic bacteria.** Nitrogen-fixing bacteria, called **symbiotic bacteria** live in the nodules on roots of leguminous plants (*e.g.* pea, beans, clover, alfalfa etc). These bacteria convert the atmospheric $N_2$ into nitrogenous compounds which can be assimilated directly by the plants. In return for this, they get their own food from the plant.

**3. Fixation of $N_2$ by nitrogen-ase enzyme salts (Biological fixation of $N_2$).** We know that in the manufacture of $NH_3$ or $NH_4^+$ salts by Haber's process, $H_2$ and $N_2$ combine together in presence of a catalyst.

$$H_2 + N_2 \xrightarrow{\text{Catalyst}} NH_3$$

$$4H_2 + N_2 + 4e^- \longrightarrow 2NH_4^+$$

For this reaction, high temperature and high pressure are needed. Synthesis of $NH_3$ by Haber's process in called **fixation of atmospheric nitrogen**.

In nature the fixation of $N_2$ takes place at the temperature of soil (*i.e.* ordinary temperature) and atmospheric pressure through a metalloenzyme namely **nitrogen-ase**.

$$N_2 + 8H^+ + 8e^- \longrightarrow 2NH_3 + H_2$$

$$N_2 + 10H^+ + 8e^- \longrightarrow 2NH_4^+ + H_2$$

We know that bond dissociation energy of $N \equiv N$ triple bond in $N_2$ molecule is 945 KJ $mol^{-1}$. Since this energy is very high, $N \equiv N$ bond must be broken during nitrogen fixation. Thus energy must be supplied to break $N \equiv N$ bond. This energy is provided by ATP which acts as *energy supplier molecule*. After supplying energy, ATP is converted into ADP + inorganic phosphate (Pi). Since the reduction of $N_2$ to $NH_3$ and $H_2$ requires 8 electrons ($8e^-$), 16 ATP molecules are consumed per $N_2$ molecule in the reduction process.

$$N_2 + 16ATP + 8H^+ + 8e^- \longrightarrow 2NH_3 + 16ADP + 16Pi + H_2$$

$$N_2 + 16ATP + 10H^+ + 8e^- \longrightarrow 2NH_4^+ + 16ADP + 16Pi + H_2$$

Electrons are supplied by *reductase* (Fe – S protein). These electrons are used by *nitrogenase* (Fe –Mo – S protein) to reduce $N_2$ to $NH_3/NH_4^+$

$NH_3$ obtained as above may further be converted into nitrate or nitrite or may be used directly in the synthesis of amino acids or other essential compounds.

## UNIVERSITY QUESTIONS

**1.** Discuss the structure and function of haemoglobin.

*(Osmania 2010)*

**2.** (*a*) Describe the biochemical functions of Ca and Mg.

(*b*) Write note on Myoglobin.

(*c*) Explain the function of s-block elements in biosystems.

*(Pd. Ravishankar Shukla 2010)*

**3.** (*a*) Explain biological role of alkali and alkaline earth metal ions.

(*b*) What is porphyrin? Draw the structure of heme.

(*c*) Draw and discuss the structure of Fe-Mo cofactor.

(*d*) Write a short note on $Na^+ - K^+$ pump.

*(Himachal Pradesh, 2010)*

**4.** (*a*) How haemoglobin is different from myoglobin? Discuss their structures?

(*b*) Write about important elements found in the body. Also state their biological activities.

(*c*) Describe structure, work and properties of haemoglobin.

(*d*) Describe the usage of Fe in biological system.

(*e*) Describe the following: (i) Co-enzyme (ii) Nitrogen fixation

(*f*) State biological actions of Cu and Hg.

*(Kanpur 2010)*

**5.** (*i*) What is the biological importance of calcium, chlorophyll and Fe.

(*ii*) Describe about Haemoglobin and Myoglobin.

(*iii*) Write about nitrogen fixation.

*(Pune 2009)*

**6.** (*a*) Write the formula of complex compound used in (*i*) Hydrogenatoin of alkenes (*ii*) Cancer therapy.

(*b*) Write the differences between haemoglobin and myoglobin.

(*c*) Explain the roles played by the $Na^+$ and $Ca^{2+}$ ions in biological systems.

*(Banglore 2010)*

**7.** (*i*) Myglobin picks up oxygen from:

(*a*) Cytochrome (*b*) Haemoglobin

(*c*) Ferredoxin (*d*) Chlorophyll

(*ii*) Transport of metal ion across the biological membrane is:

(*a*) Active transport only (*b*) Passive transport only

(*c*) Active and passive transport both (*d*) Not possible

(*iii*) What is the biological role of alkali and alkaline earth metal ions with special reference to haemoglobin?

(*iv*) Explain biological role of alkali and alkaline earth metal ion with special reference to $Ca^{2+}$ in nitrogen fixation.

*(H.N. Bahuguna Garhwal 2010)*

8. (*a*) Explain the biological significance of myoglobin.
   (*b*) Discuss the mechanisms of photosynthesis.

(*Bharathiar 2010*)

9. (*a*) What is the essential and trace element in biological system?
   (b) What are the functions of haemoglobin and myoglobin? Discuss the similarities in their structure.
   (c) Name three haemoproteins.

(*Ranchi 2010*)

10. Explain the importance of $Na^+$, $K^+$ and $Zn^{2+}$ ions in biological systems.

(*Andhra 2010*)

11. (*a*) Discuss the functions and structure of myoglobin.
    (*b*) Explain role of 'sodium pump' in biological system.
    (*c*) What is the role of essential non-metals in the biological system?
    (*d*) What are the functions of s-block elements in biosystems?

(*Nagpur 2010*)

12. (*a*) Explain why?
    (*i*) Pb and Hg are bad to human health.
    (*ii*) Cyanides are toxic to all biological system.
    (*iii*) Selenium is essential as well as toxic.
    (*iv*) $Zn^{2+}$ ion acts as an excellent biocatalyst.
    (*v*) It is necessary to use mixture of $O_2$ and diluent $N_2$ and not pure $O_2$ in breathing.

(*Utkal 2010*)

13. (*a*) What do you understand by biological chemistry and essential element?

(*GND Amritsar 2010*)

14. Answer the following:
    (*i*) Compare and contrast the structure of Haemoglobin and Myoglobin.
    (*ii*) Describe the cooperativity in Haemoglobin. Suggest the mechanism for cooperativity.
    (*iii*) What is the role of globin part in haemoglobin?
    (*iv*) Write the biological importance of $Ca^{2+}$ ions.

(*Utkal 2010*)

15. (*a*) How will you explain role of haemoglobin for the transport and storage of oxygen in higher mammals?
    (*b*) Write a note on nitrogen fixation.
    (*c*) Draw the graph indicating the variation in oxygen binding capacity of haemoglobin and myoglobin with pH and with pressure of $CO_2$.
    (*d*) Discuss the biological role of $Ca^{2+}$.

(*BJB College, Orissa*)

16. (*a*) Explain in detail the role of alkali and alkaline earth metals in biological systems.
    (*b*) Explain in detail the role of trace elements in biological systems.

(*GND Amritsar 2010*)

**17.** (*i*) Comment on the biological role of $Ca^{2+}$ ions.

(*ii*) Differentiate between toxic and essential elements. Briefly discuss the biological role of sodium and potassium ions.

(*Lucknow 2010*)

**18.** (*i*) What are metal complexes present in chlorophyll?

(*ii*) Explain the role of iron in haemoglobin.

(*iii*) Elaborate poisoning and inhibition of enzyme activity.

(*iv*) With a schematic diagram explain the function of $Na^+ - K^+$ pump.

(*v*) What are biological effects of copper deficiency and copper accumulation in the body.

(*vi*) Discuss in detail the function of essential and trace elements in biological systems. (*Madras April 2011*)

**19.** (*a*) Discuss the functions of haemoglobin and myoglobin.

(*b*) Discuss the major functions of calcium in biological system.

(*Nagpur 2012*)

**20.** (*i*) What is the role of $Ca^{2+}$ ions in biosystem?

(*ii*) Discuss the structure of haemoglobin and its function in biological systems.

(*iii*) What are essential and trace elements? What important roles iron and zinc play in biological system. (*Lucknow 2013*)

# 33 CHAPTER Silicones and Phosphazenes

## Silicones or Organopolysiloxanes

### What are Silicones?

We know that oxo derivatives of silanes are called **siloxanes.** Examples of some siloxanes are given below:

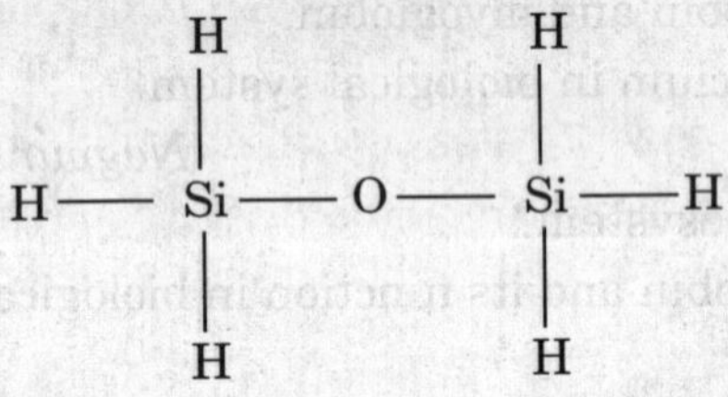

Disiloxane

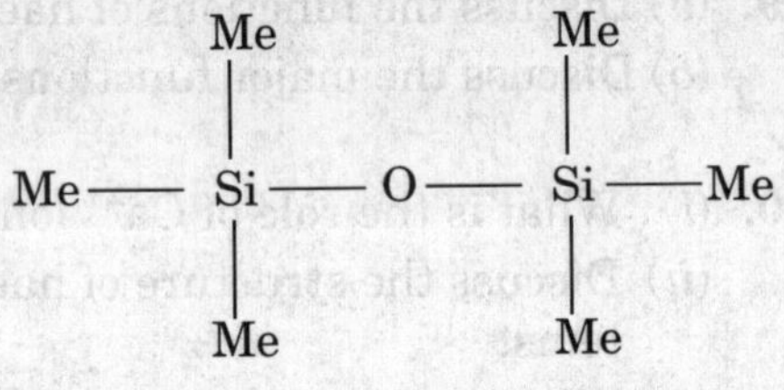

Hexamethyl disiloxane

Silicones are **organopolysiloxanes.** Silicones are also called **organosilicones** or **organosilicon polymers.** These can be represented by the general formula, $(R_2SiO)_n$ whose structure is given in the margin.

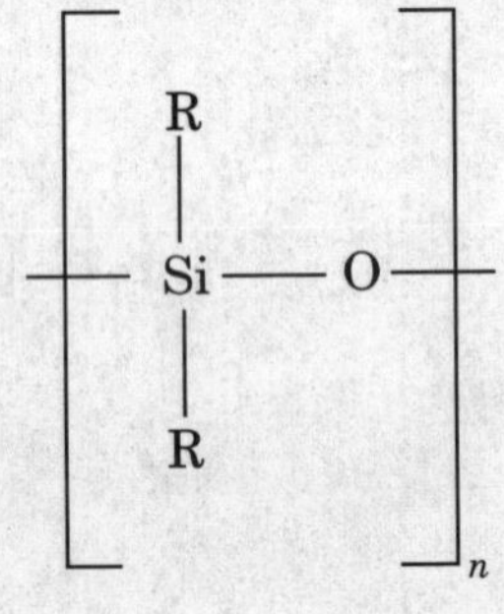

This structure shows that silicones have Si—C and Si—O bonds. C-atom in Si—C bond belongs to organic group (R).

### Types of Silicones

On the basis of their structure, silicones may be of the following types: (*i*) *Linear (straight chain) silicones* (*ii*) *Cyclic silicones* (cyclic silicones are also called *cyclic polysiloxanes*), and (*iii*) *Cross-linked (three dimensional) silicones.* Structures of three types of silicones are given below:

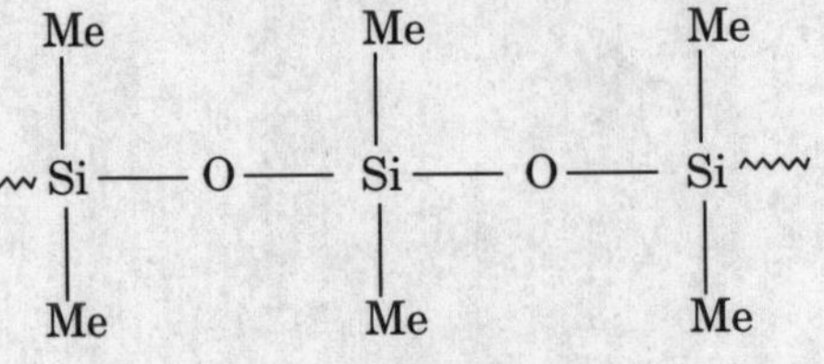

Polydimethyl siloxane
(Linear silicone)

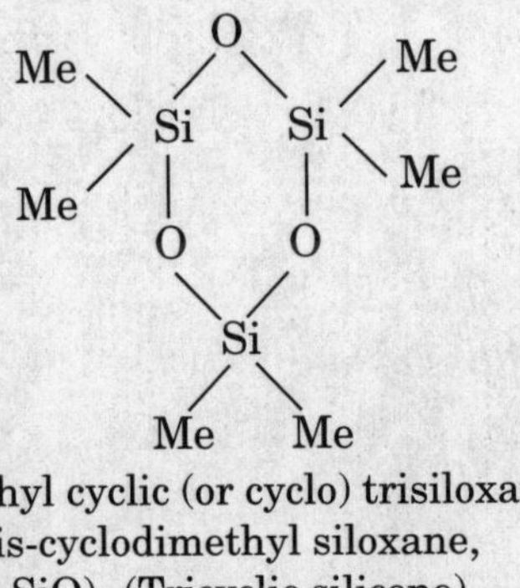

Hexamethyl cyclic (or cyclo) trisiloxane
or tris-cyclodimethyl siloxane,
$(Me_2SiO)_3$ (Tricyclic silicone)

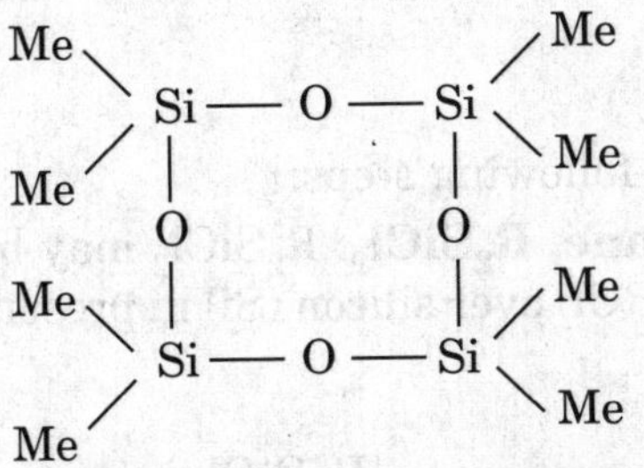

Octamethyl cyclic tetra siloxane or tetrakis-cyclo dimethyl siloxane, $(Me_2SiO)_4$ (Tetra cyclic silicone)

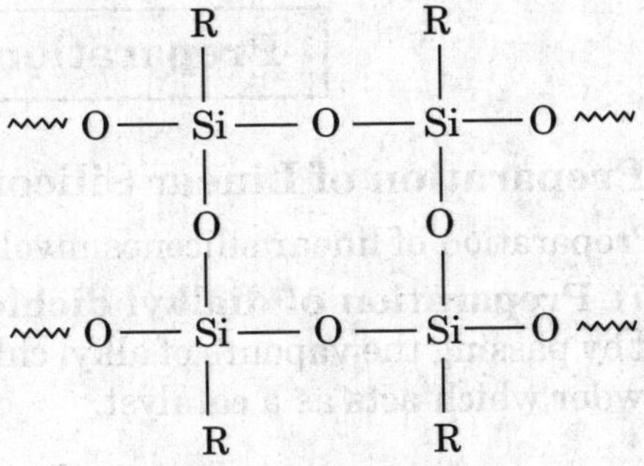

Cross- linked silicone

Silicones have very high thermal stability and are, therefore, also called **high temperature polymers.** High thermal stability is due to high thermal stability of Si—O—Si chains present in silicones.

## What are Chlorosilanes?

$SiH_4$ is called **silane**. If some or all H-atoms of $SiH_4$ are replaced by Cl-atoms, the compounds obtained are called **chloro silanes.** Examples of some chloro silanes are: (i) Dichlorosilane, $SiH_2Cl_2$ (ii) Trichlorosilane, $SiHCl_3$ (iii) Tetrachlorosilane or silicon tetrachloride, $SiCl_4$

## What are Alkyl Chlorosilanes?

If some or all Cl-atoms of tetrachloro silane ($SiCl_4$) are replaced by alkyl group, the compounds obtained are called **alkyl chloro silanes.** Examples of some methyl chloro silanes are: (i) Monomethyl trichloro silane, $(CH_3)\ SiCl_3$ (ii) Dimethyl dichloro silane, $(CH_3)_2\ SiCl_2$ (iii) Trimethyl monochloro silane, $(CH_3)_3\ SiCl$

## What are Alkyl Hydroxy Silanes (Alkyl Silanols)?

When alkyl chloro silanes [e.g. $(CH_3)\ SiCl_3$, $(CH_3)_2SiCl_2$ and $(CH_3)_3\ SiCl$] are hydrolysed, Cl-atoms are replaced by OH groups and the compounds obtained are called alkyl silanols or alkyl hydroxy silanes.

$$\underset{\text{Monomethyl trichloro-silane}}{(CH_3)\ SiCl_3} + 3H_2O \longrightarrow \underset{\text{Monomethyl silan-triol or monomethyl trihydroxy silane}}{(CH_3)\ Si\ (OH)_3} + 3HCl$$

$$\underset{\text{Dimethyl dichloro silane}}{(CH_3)_2\ SiCl_2} + 2H_2O \longrightarrow \underset{\text{Dimethyl silan-diol or dimethyl dihydroxy silane}}{(CH_3)_2\ Si\ (OH)_2} + 2HCl$$

$$\underset{\text{Trimethyl monochloro silane}}{(CH_3)_3\ SiCl} + H_2O \longrightarrow \underset{\text{Trimethyl silanol or trimethyl mono hydroxy silane}}{(CH_3)_3\ Si\ (OH)} + HCl$$

## Preparation of Silicones

### (A) Preparation of Linear Silicones

Preparation of linear silicones involves the following steps:

***(i)* Preparation of dialkyl dichloro silane, $R_2SiCl_2$.** $R_2SiCl_2$ may be obtained by passing the vapours of alkyl chloride (RCl) over silicon (Si) in presence of Cu powder which acts as a catalyst.

$$\underset{\text{Alkyl chloride}}{2RCl} + Si \xrightarrow[570\,K]{\text{Cu powder}} \underset{\text{Dialkyl dichloro silane}}{R_2SiCl_2}$$

$R_2SiCl_2$ can also be obtained by the action of Grignard reagent (RMgCl) on $SiCl_4$.

$$2RMgCl + SiCl_4 \longrightarrow R_2SiCl_2 + 2MgCl_2$$

***(ii)* Preparation of dialkyl dihydroxy silane, $R_2Si(OH)_2$ by the hydrolysis of $R_2SiCl_2$.** $R_2SiCl_2$ obtained as above is hydrolysed to get dialkyl silandiol, $R_2Si(OH)_2$.

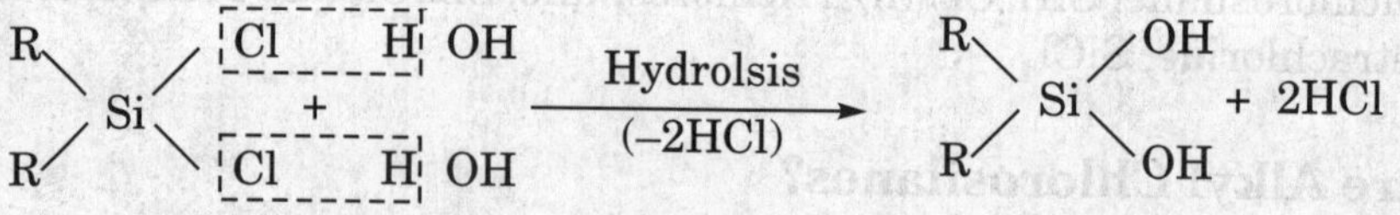

$$\text{or} \quad R_2SiCl_2 + 2H_2O \xrightarrow[(-2HCl)]{\text{Hydrolsis}} \underset{\substack{\text{Dialkyl silandiol} \\ \text{or} \\ \text{dialkyl dihydroxy silane}}}{R_2Si(OH)_2} + 2HCl$$

***(iii)* Polymerisation of $R_2Si(OH)_2$.** Polymerisation of $R_2Si(OH)_2$ formed above gives linear silicone. The polymerisation takes place in two steps given below:

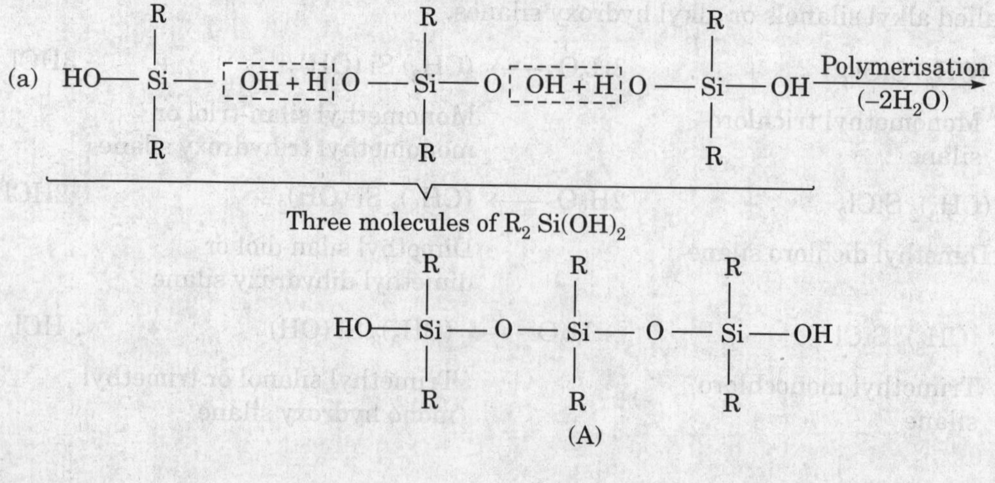

(A)

(*b*) Since an active OH group is present at each end of the chain present in (A), polymerisation of *n* molecules of (A) takes place at both ends of the chain and long chain silicone (linear silicone) is obtained as shown below:

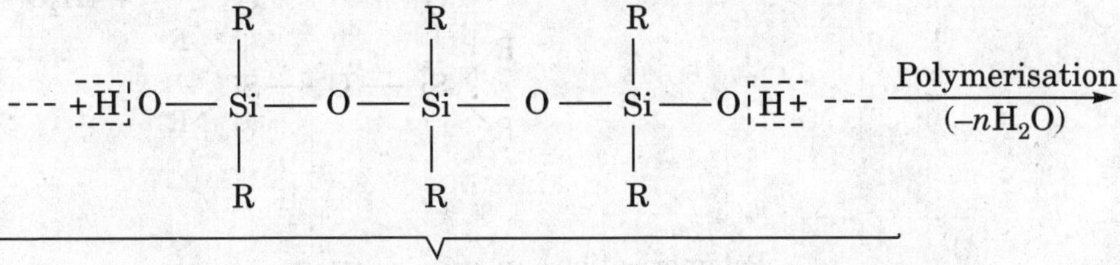

*n* molecules of (A)

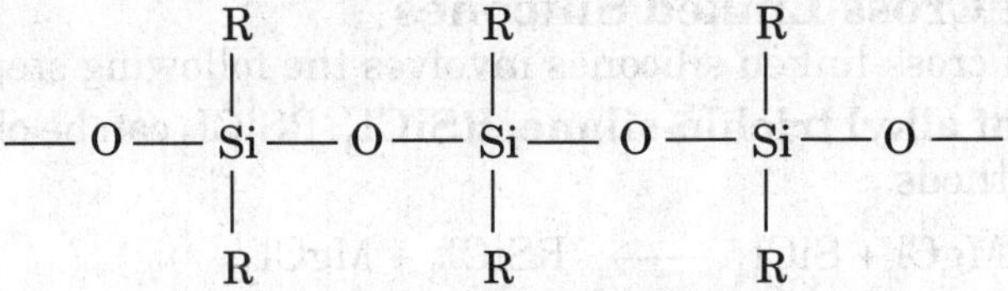

Linear silicone

## (B) Preparation of Cyclic Silicones

Cyclic silicones are also called *cyclic polysiloxanes.* Preparation of cyclic silicones involves the following steps:

**(i) Preparation of dialkyl dichloro silane, $R_2SiCl_2$.** See *"Preparation of linear silicones"*.

**(ii) Preparation of dialkyl dihydroxy silane, $[R_2Si(OH)_2]$ by the hydrolysis of $R_2SiCl_2$:** See *"Preparation of linear silicones."*

**(iii) Polymerisation of $R_2Si(OH)_2$.** Polymerisation of three molecules of $R_2Si(OH)_2$ gives *hexa alkyl cyclic trisiloxane* (also called *tris-cyclo dialkyl siloxane*), $(R_2SiO)_3$ which is a cyclic silicone.

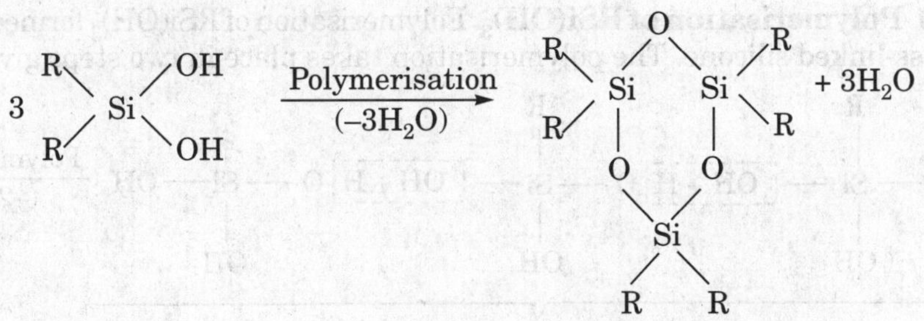

or $$3\,R_2Si(OH)_2 \xrightarrow[(-3H_2O)]{\text{Polymerisation}} (R_2SiO)_3 + 3H_2O$$

Polymerisation of four molecules of $R_2Si(OH)_2$ gives *octa alkyl cyclic tetra siloxane* (also called *tetrakis-cyclo dialkyl siloxane*), $(R_2SiO)_4$

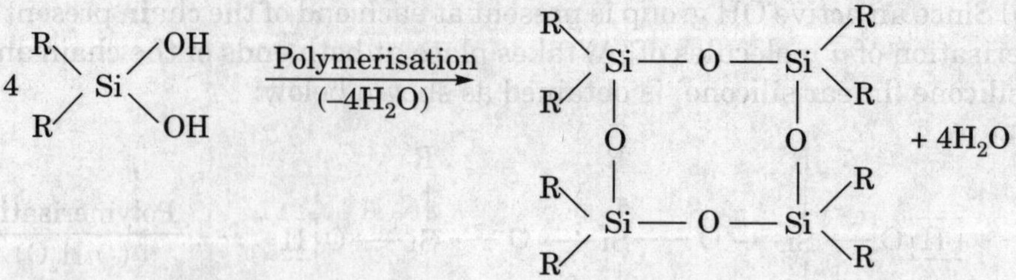

or $\quad 4\,R_2Si(OH)_2 \xrightarrow[(-4H_2O)]{\text{Polymerisation}} (R_2SiO)_4 + 4H_2O$

## (C) Preparation of Cross-Linked Silicones

The preparation of cross-linked silicones involves the following steps:

***(i)* Preparation of alkyl trichlo-silane, $RSiCl_3$.** $RSiCl_3$ can be obtained by any of the following methods.

$$RMgCl + SiCl_4 \longrightarrow RSiCl_3 + MgCl_2$$

$$CH_3Cl + 2HCl + Si \xrightarrow{Cu} CH_3SiCl_3 + H_2$$

***(ii)* Preparation of monoalkyl trihydroxy silane [$RSi(OH)_3$] by the hydrolysis of $RSiCl_3$.** $RSiCl_3$ obtained as above is hydrolysed to get alkyl silantriol, $RSi(OH)_3$.

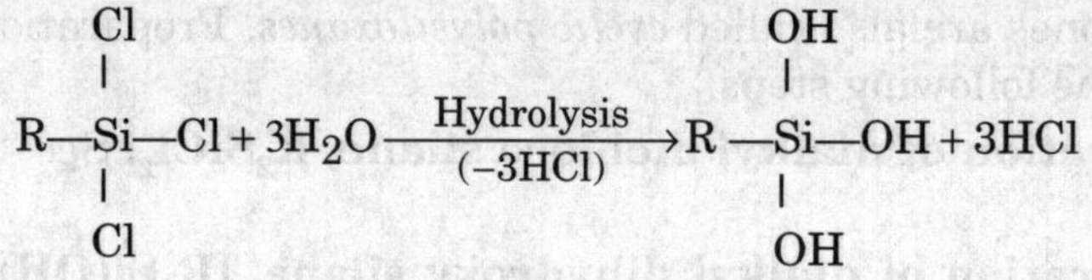

or $\quad RSiCl_3 + 3H_2O \xrightarrow[(-3HCl)]{\text{Hydrolysis}} RSi(OH)_3 \quad + \quad 3HCl$

Alkyl silantriol or
monoalkyl trihydroxy silane

***(iii)* Polymerisation of $RSi(OH)_3$.** Polymerisation of $RSi(OH)_3$ formed as above gives cross-linked silicone. The polymerisation takes place in two steps given below:

(a) 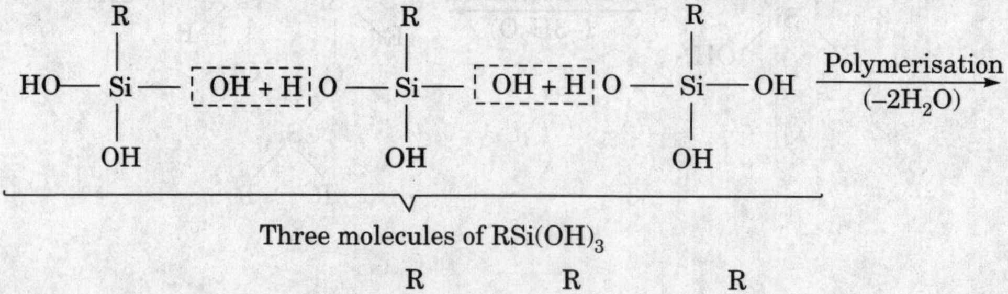

Three molecules of $RSi(OH)_3$

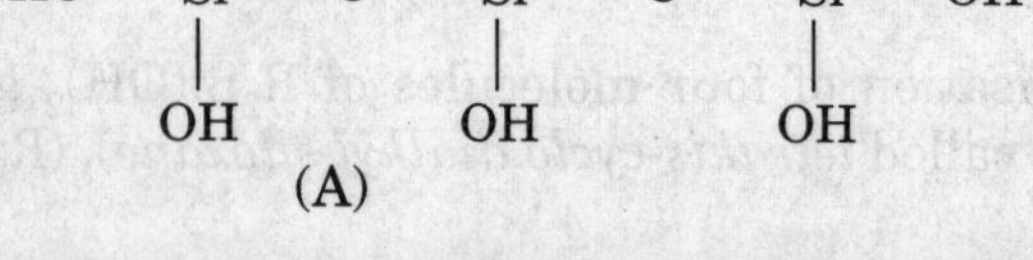

(A)

(*b*) Since an active OH group is present at each end of the chain present in (A), polymerisation of many molecules of (A) takes place on all sides and cross-linked silicone is obtained as shown below:

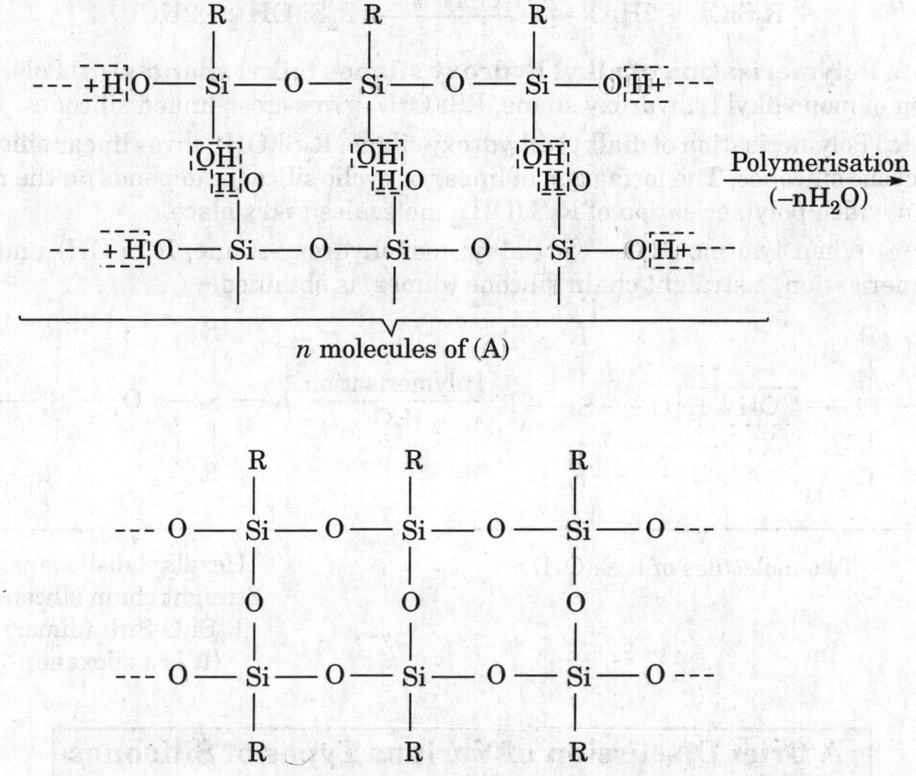


Cross linked silicone

In a simplified way polymerisation can also be shown as follows:

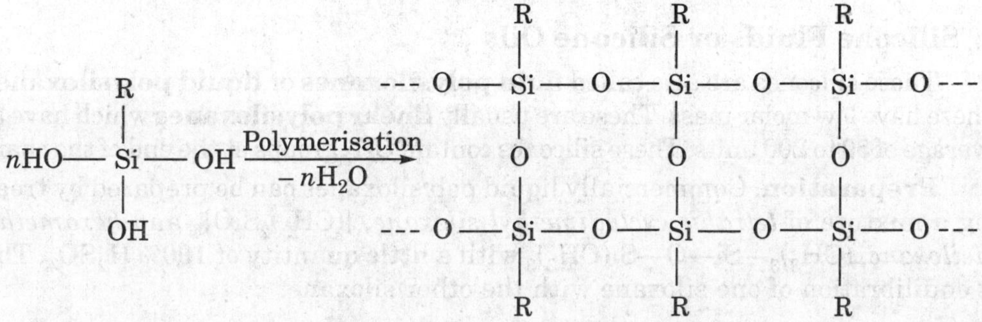


Cross linked silicone

## Remember the Following

**1. Hydrolysis of alkyl chloro silanes** (*a*) Hydrolysis of monoalkyl trichloro silane, $RSiCl_3$ gives monoalkyl trihydroxy silane, $RSi(OH)_3$ (also called alkyl silan-triol)

$$RSiCl_3 + 3H_2O \xrightarrow{\text{Hydrolysis}} RSi(OH)_3 + 3HCl$$

(*b*) Hydrolysis of dialkyl dichloro silane, $R_2SiCl_2$ gives dialkyl dihydroxy silane, $R_2Si(OH)_2$ (also called dialkyl silan-diol)

$$R_2SiCl_2 + 2H_2O \xrightarrow{\text{Hydrolysis}} R_2Si(OH)_2 + 2HCl$$

**2. Polymerisation of alkyl hydroxy silanes (alkyl silanols)** (*a*) Polymerisation of monoalkyl trihydroxy silane, $RSi(OH)_3$ gives cross-linked silicones.

(*b*) Polymerisation of dialkyl dihydroxy silane, $R_2Si(OH)_2$ gives linear silicones and cyclic silicones. The formation of linear or cyclic silicones depends on the manner in which polymerisation of $R_2Si(OH)_2$ molecules takes place.

(*c*) When two molecules of trialkyl monohydroxy-silane, $R_3Si(OH)$ undergo polymerisation, a straight chain silicone (dimer) is obtained.

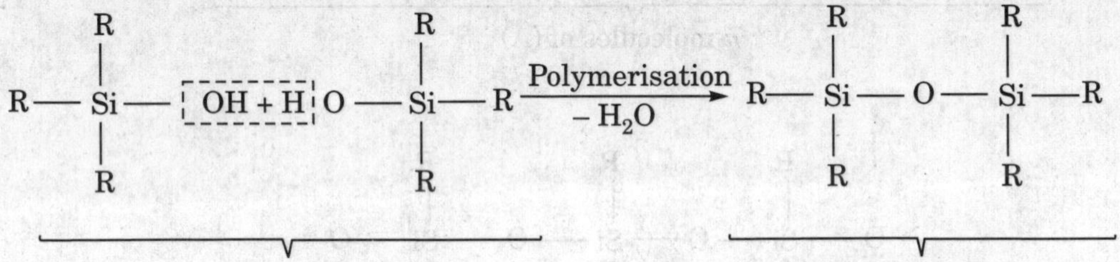


Two molecules of $R_3Si(OH)$

Hexalkyl disiloxane:
Straight chain silicone,
$R_3Si\text{-}O\text{-}SiR_3$ (dimer)
(It is a siloxane)

## A Brief Discussion of Various Types of Silicones

A brief description of various types of silicones is given below:

## 1. Silicone Fluids or Silicone Oils

These silicones are also called **fluid polysiloxanes** or **liquid polysiloxanes.** These have low molar mass. These are usually **linear polysiloxanes** which have an average of 50 to 200 units. These silicones contain $CH_3$ groups at the end of the chain.

**Preparation.** Commercially liquid polysiloxanes can be prepared by treating a mixture of *tetrakis cyclodimethyl siloxane*, $[(CH_3)_2SiO]_4$ and *hexamethyl disiloxane*, $(CH_3)_3$—Si—O—$Si(CH_3)_3$ with a little quantity of 100% $H_2SO_4$. This is equilibration of one siloxane with the other siloxane.

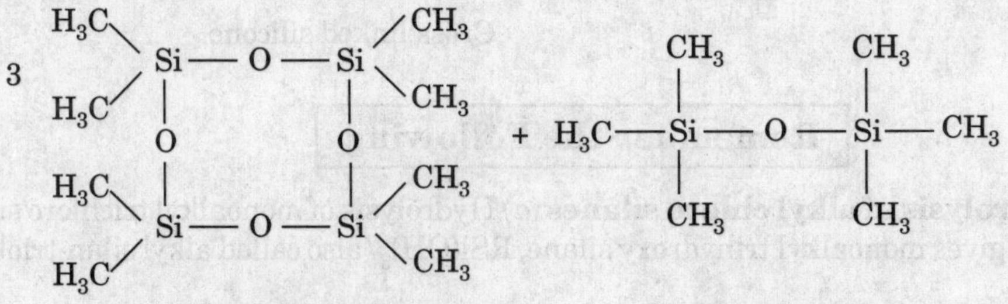

Tetrakis cyclodimethyl siloxane

Hexamethyl disiloxane

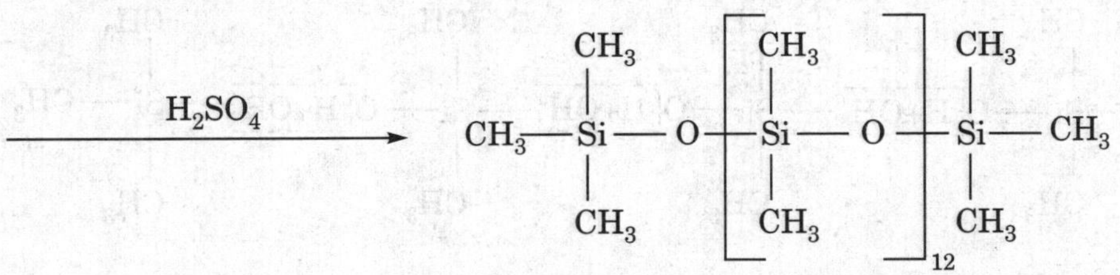


or $3[(CH_3)_2SiO]_4 + (CH_3)_3Si\text{-}O\text{-}Si(CH_3)_3 \xrightarrow{H_2SO_4} (CH_3)_3Si\text{-}O\text{-}[(CH_3)_2\ SiO]_{12}\text{-}Si\ (CH_3)_3$

**Properties** (*i*) The viscosity of silicone fluids is about 30,000 times that of water and changes slightly with temperature. Viscosity increases with the increase in chain length.

(*ii*) Boiling points of silicone fluids increase with the increase of chain length.

**Uses** (*i*) Silicone fluids are *good lubricants under vacuum* and at high temperature. They are used in vacuum pumps, jet turbines and guided missiles as lubricants.

(*ii*) They are used in heating baths for industrial *heating and as damping fluid* in instruments and electrical switchgears.

(*iii*) They are used as *antifoaming agents* in several industries such as dyes, textiles, brewing, paint, glue, paper, food processing and sewage disposal.

(*iv*) They are also used for coating glass and ceramics insulators *to prevent electrical leakage during rain.*

(*v*) Silicone fluids resist adhesion to several surfaces and are used *as release agents.* They are coated in moulds for easy removal of plastic articles and rubber tyres form moulds.

(*vi*) *Greases are made from silicone* fluids by mixing with metallic soap, graphite or silica.

(*vii*) Silicone oils are used as dielectric insulating materials as capacitor and transformer fluids hydraulic oils, compressible fluids for liquid springs and lubricants.

(*viii*) Silicone fluids are used as water repellents for treating masonry and buildings, glassware and fibres.

(*ix*) Silicone oils are also used as heat transfer media in heating belts and as components in car polishes, lipstick, shampoo and other cosmetic formulations.

**Preparation of dimethyl silicone oil.** The preparation of dimethyl silicone oil consists of the following steps:

(*i*) Mixture containing trimethyl chlorosilane, $(CH_3)_3$ SiCl and dimethyl dichloro silane, $(CH_3)_2$ $SiCl_2$ in proper proportion is hydrolysed to get trimethyl hydroxy silane, $(CH_3)_3$ Si(OH) and dimethyl dihydroxy silane, $(CH_3)_2$ $Si(OH)_2$

$$(CH_3)_3\ SiCl + H_2O \xrightarrow{\text{Hydrolysis}} \underset{\text{Hydroxy derivative}}{(CH_3)_3\ Si(OH) + HCl}$$

$$(CH_3)_3\ SiCl_2 + 2H_2O \xrightarrow{\text{Hydrolysis}} \underset{\text{Hydroxy derivative}}{(CH_3)_2\ Si(OH)_2 + 2H_2O}$$

(*ii*) $(CH_3)_3$ Si(OH) and $(CH_3)_2$ $Si(OH)_2$ obtained as above are mixed and the mixture is allowed to undergo polymerisation to form dimethyl silicon oil.

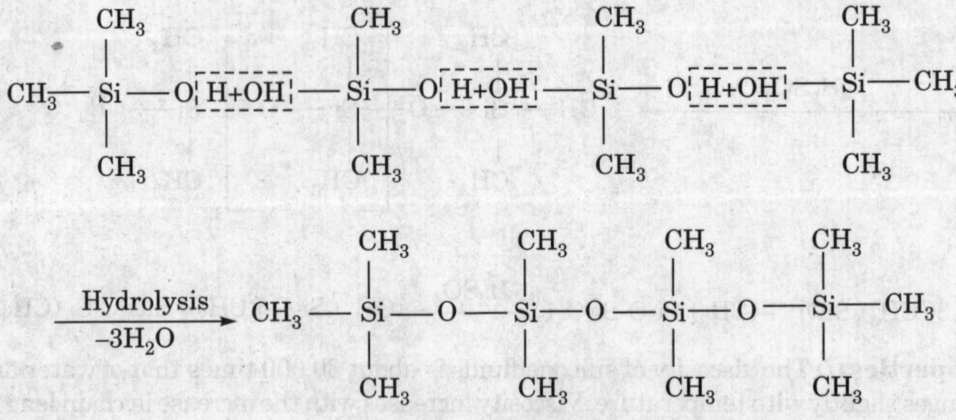

Dimethyl silicone oil

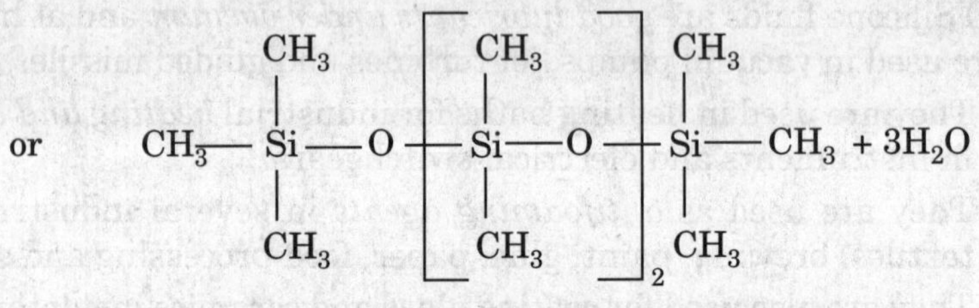

Dimethyl silicone oil
(Linear silicone)

## 2. High Polymer Organosiloxanes or Polysiloxane Gums

High polymer organosiloxanes are high molar mass organasiloxanes.

**Preparation.** These can be prepared by equilibration of cyclic siloxanes in the absence of monofunctional groups. Cyclic tetramer gives straight chain polymer as shown below:

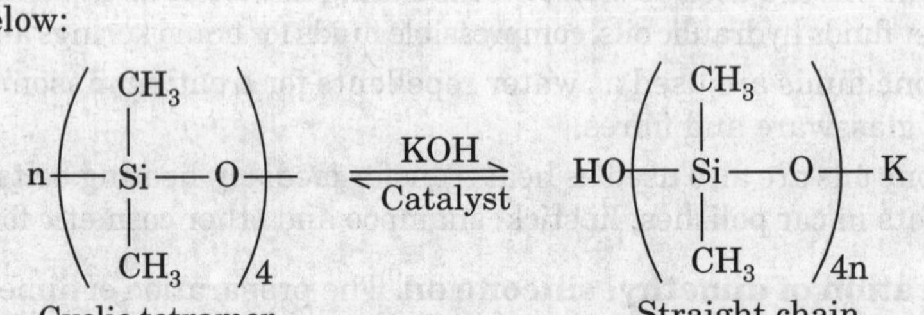

Cyclic tetramer (A) Straight chain polymer

The above reaction takes place through the following steps:

(*i*) **Initial step.** In this step (A) is converted into (B) by the action of KOH on (A).

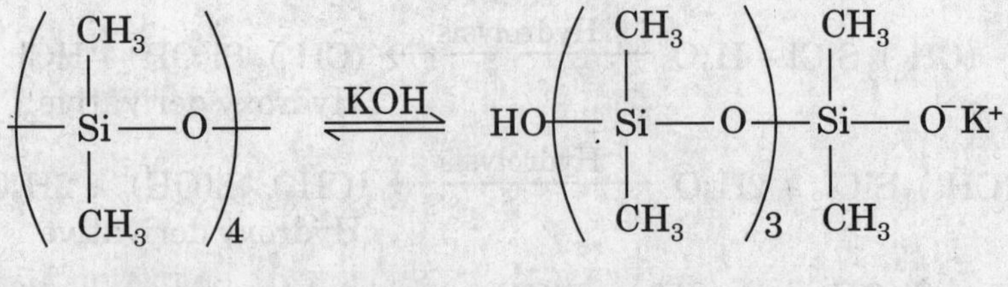

(A) (B)

(*ii*) **Condensation steps:** (B) is converted into (E) by the successive addition of (A) to (B).

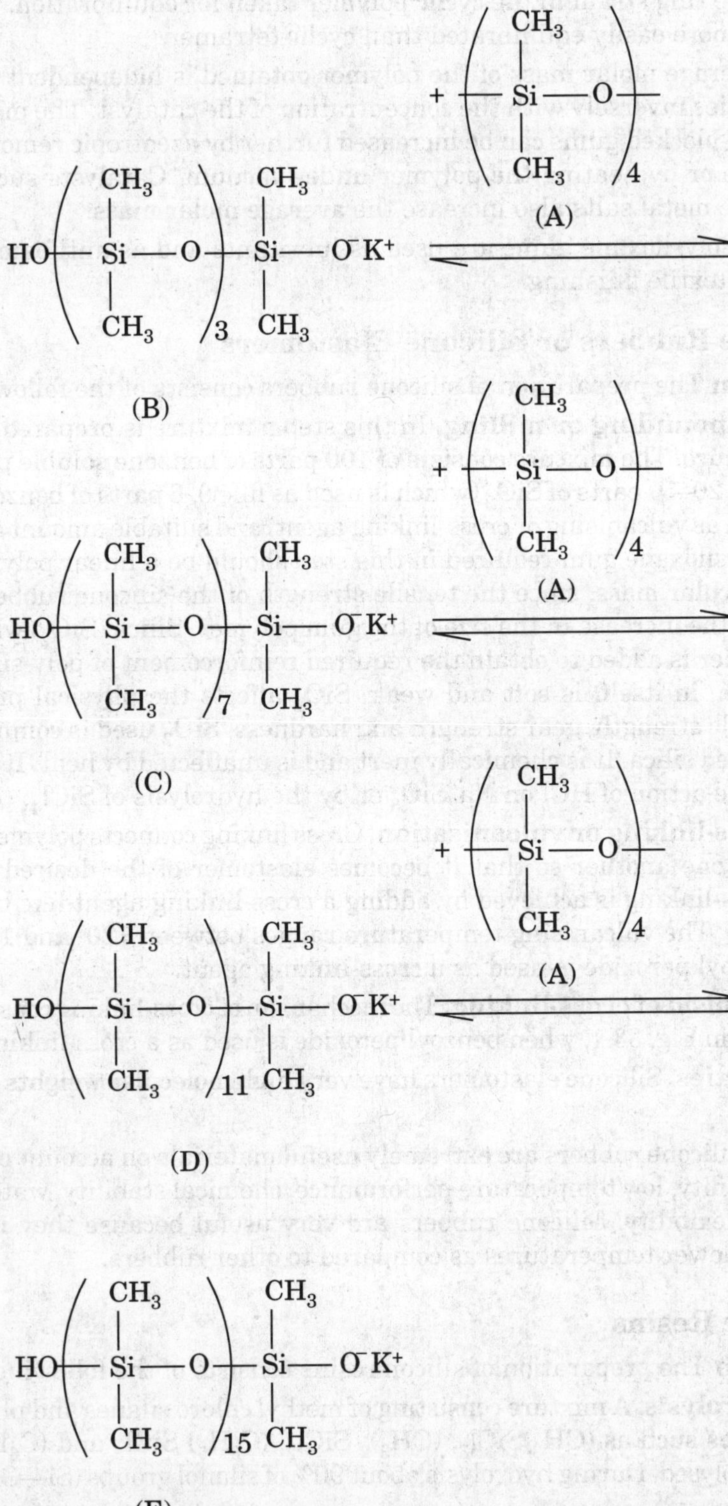

(A) (B) (A) (C) (A) (D) (E)

The rate of the polymerisation reaction increases with increase in temperature and is related to square root of the concentration of the cyclic tetramer. It is also related to the ring strain in the cyclic polymer taken for equilibration. Thus, cyclic hexamer is more easily equilibrated than cyclic tetramer.

The average molar mass of the polymer obtained is independent of temperature and varies inversely with the concentration of the catalyst. The molar mass of hydroxy end-blocked gums can be increased further by azeotropic removal of water with toluene or by heating the polymer under vacuum. Catalysts such as amine salts or some metal salts also increase the average molar mass.

**Uses:** Polysiloxane gums are used as lubricants and as antifoams. They are also used in textile finishing.

## 3. Silicone Rubbers or Silicone Elastomers

**Preparation** The preparation of silicone rubbers consists of the following steps:

(*i*) **Compounding or milling.** In this step a mixture is prepared under high pressure. The mixture consists of 100 parts of benzene soluble polysiloxane gum, 20–50 parts of $SiO_2$ (which is used as filler), 6 parts of benzoyl peroxide (used as vulcanising or cross-linking agent) and suitable amount of additives. Poly-siloxane gum required in this step should be a linear polymer of high molecular mass, since the tensile strength of the silicone rubber increases with the increase in the size of the gum polymer. Silica ($SiO_2$) which is used as filler is added to obtain the required reinforcement of poly-siloxane gum which, in itself, is soft and weak. $SiO_2$ affects the physical properties as tensile strength, gear strength and hardness. $SiO_2$ used is commonly finely divided silica. It is chemically inert and is unaffected by heat. It is prepared by the action of HCl on $Na_2SiO_3$ or by the hydrolysis of $SiCl_4$.

(*ii*) **Cross-linking or vulcanisation.** Cross linking connects polymer molecules with one another so that it becomes elastomer of the desired properties. Cross-linking is achieved by adding a cross-linking agent like benzoyl peroxide. The vulcanising temperature ranges between 120° and 130 °C when benzoyl peroxide is used as a cross-linking agent.

**Mechanism of cross-linking.** The mechanism of cross-linking consists of three steps shown in Fig. 33.1, when benzoyl peroxide is used as a cross-linking agent.

**Properties.** Silicone elastomers have very high molecular weights of the order of $10^5$–$10^7$.

**Uses.** Silicone rubbers are extremely useful materials on account of their high thermal stability, low temperature performance, chemical stability, water repelling power and flexibility. Silicone rubbers are very useful because they retain their elasticity at lower temperatures as compared to other rubbers.

## 4. Silicone Resins

**Preparation** The preparation of silicon resins consists of the following steps:

(*i*) **Hydrolysis.** A mixture consisting of methyl chloro silanes and phenyl chloro silanes such as $(CH_3)SiCl_3$, $(CH_3)_2\ SiCl_2$, $(C_6H_5)\ SiCl_3$ and $(C_6H_5)_2SiCl_2$ is hydrolysed. During hydrolysis about 90% of silanol groups (Si—OH) undergo

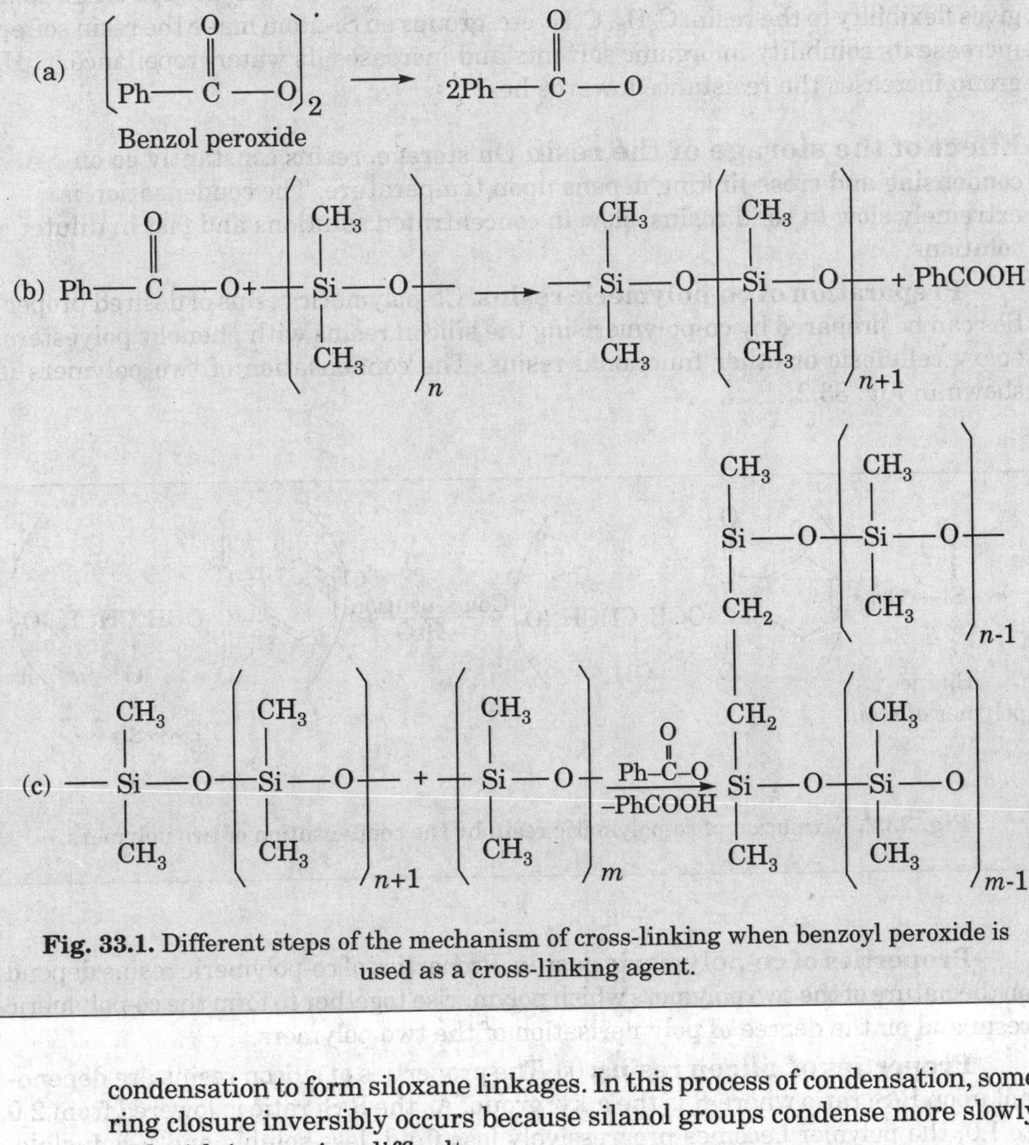

**Fig. 33.1.** Different steps of the mechanism of cross-linking when benzoyl peroxide is used as a cross-linking agent.

condensation to form siloxane linkages. In this process of condensation, some ring closure inversibly occurs because silanol groups condense more slowly and with increasing difficulty.

(*ii*) **Shaking with organic solvents.** The product obtained after hydrolysis is shaken with organic solvent and organic layer is separated.

(*iii*) **Condensation of residual silanol groups.** The ultimate condensation of the residual silanol groups occurs after resin is applied. This process can be hastened by heating the resin in the presence of a suitable catalyst. The most commonly used catalysts are metal salts of organic acids like naphthalates of Co, Zn, Pb and Sn.

**Effect of silanol and other groups on finished resin** Silanol groups left uncondensed in the finished resin give typical properties to the resin. For example the presence of silanol groups contribute towards viscosity of the resin. The presence of silanol groups also determines the shelf life of the resin. $CH_3$ groups on Si-atom gives flexibility to the resin. $C_2H_5$, $C_3H_7$ etc. groups on Si-atom make the resin softer, increase its solubility in organic solvents and increases its water- repellance. $C_6H_5$ group increases the resistance towards heat.

**Effect of the storage of the resin** On storage, resins constantly go on condensing and cross-linking depens upon temperature. The condensation is extremely slow in solid resins, slow in concentrated solutions and fast in dilute solutions.

**Preparation of co-polymeric resins.** Co-polymeric resins of desired properties can be prepared by co-polymerising the silicon resins with phenolic polyesters, epoxy cellulosic or other functional resins. The condensation of two polymers is shown in Fig. 33.2.

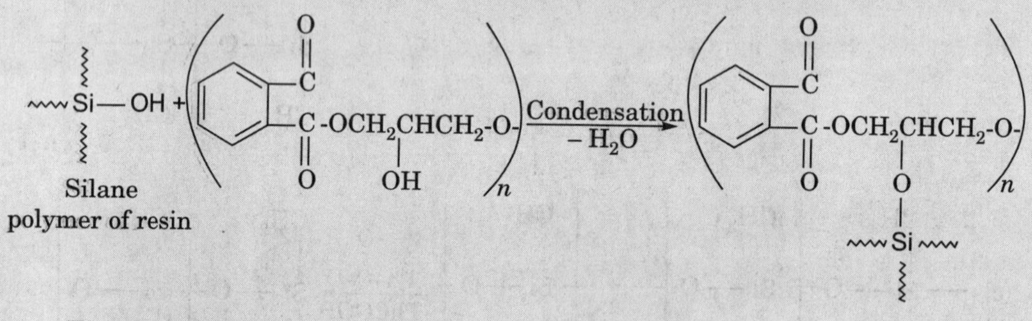


**Fig. 33.2.** Formation of co-polymeric resin by the condensation of two polymers.

**Properties of co-polymeric resins.** Properties of co-polymeric resins depend on the nature of the two polymers which polymerise together to form the co-polymeric resin and on the degree of polymerisation of the two polymers.

**Properties of silicon resins:** (*i*) The properties of silicon resins are dependent upon R/Si ratio where R is the alkyl group. As the R/Si ratio is lowered from 2.0 to 1.0, the polymer becomes progressively less fluid, less soluble and less fusible. Most of the resins have R/Si ratio between 1.00 and 1.6 and their properties are intermediate between those of fluid silicones and those expected of a highly cross-linked structure with R/Si ratio equal to 1.00.

(*ii*) Silicon resins have ring structures and have much higher cross-linking density than silicone elastomers.

(*iii*) In a stretched silicone elastomer, the molecular chains uncoil through rotation about –Si–O– bonds but in a silicon resin such rotation is highly restricted. In silicon resins stress occurs largely through bending and stretching of bonds.

**Uses of silicon resins:** (*i*) Silicon resins are used in electrical insulators.

(*ii*) They are used for lamination purpose and for coating cooking pans etc.

(*iii*) Silicone resin solutions are mixed in paints, pigments and enamels to make them resistent to the effects of high temperatures, sunlight, chemicals and damp.

(*iv*) Silicone resin, mixed with the pigment like Al-flakes, can withstand 400ºC.

(*v*) Silicone resins, having $TiO_2$ ($TiO_2$ is a pigment) resist dissociation, due to heat and weathering. These silicone resins are not affected by soap and fruit juices.

## Properties of Silicones

Some characteristic properties of silicones are discussed below.

1. **Thermal stability.** Polysiloxanes are thermally stable and do not decompose even upto 350–400ºC.

   **Explanation:** The Si–O–Si bond in siloxanes is shorter than the expected Si–O–Si bond as calculated from their radii. This indicates that there is some ionic character and/or partial double bond character in Si–O bond due to which it becomes quite stable. This is the reason why polysiloxanes are thermally stable and do not decompose even upto 350–400ºC.

2. **Attack of reagents.** Highly polar character of Si–O bond and the ability of Si to expand its valency shell by utilising its *d* orbitals renders polysiloxanes susceptible to attack by several reagents. Strong bases and concentrated acids may cause depolymerisation. Similarly, some electron-deficient salts may cause the cleavage of Si–O as well as the Si–C bonds. Dry chlorine attacks silicones resulting in chlorination of alkyl groups.

3. **Chemical stability.** Except for the above reactions, polysiloxanes are chemically inert. Common chemical reagents such as dilute acids and bases, metal salts, water at room temperature, $H_2O_2$, $NH_4OH$, etc., have little effect on polysiloxanes. Polysiloxanes are quite resistant to attack by oxygen. However, minor oxidation does occur above 200°C. At higher temperatures, oxidation becomes considerable when fluid polysiloxanes get cross linked through oxygen atoms to yield a rubber-like gel in a few hours. Silicon resins, being already cross linked, show much less change in viscosity with change in temperature. The presence of aromatic groups on Si atom increases the oxidative stability of the polysiloxanes.

4. **Cleavage of siloxane bond.** Siloxane bond can present in polysiloxanes (A) can be cleaved by Grignard reagent (R′MgX), alkyl lithium (R′Li) and $LiAlH_4$ as shown below (Fig. 33.3). The products (B) and (C) obtained by the action of R′Mg X and R′Li gives (E) on hydrolysis.

5. **Hydolysis.** The siloxanes may undergo hydrolysis and alcoholysis at elevated temperatures in the presence of strong acids and bases to give silanols and alkoxysilanes. In general, the greater the extent of substitution on Si-atom, the greater is the ease of hydrolysis in the presence of acids and the greater is the difficulty of hydrolysis in the presence of bases.

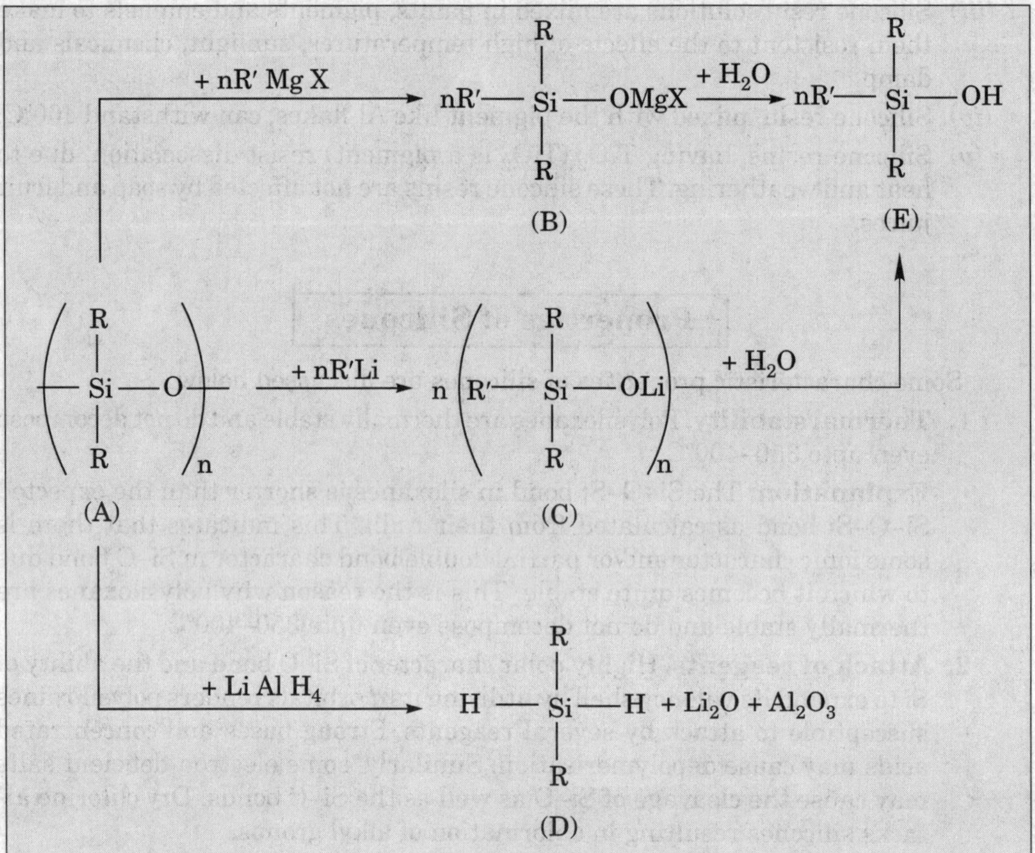

**Fig. 33.3.** Action of R' MgX, R' Li and $LiAlH_4$ on polysiloxane (A).

6. **Physical properties.** Physical properties of silicones depends on the kind and member of organic groups present on each Si-atom These properties also depend on the manner in which the silane units are hooked togather. For example $CH_3$ groups on Si-atom promote fluidity, low viscosity, low melting point and high compressibility but $C_6H_5$ groups on Si-atom induce the contrasting properties in silicones. The member of organic groups determine the functionality of individual siloxane building units and thus controls the linearity or complexity of polysiloxanes.

   Physical properties of polysiloxanes such as compressibility, small temperature coefficient of viscosity and surface properties have been attributed largely to almost regularly coiled structure (helical structure) of polysiloxanes and weak intermolecular forces. The viscosity of polysiloxanes does not change appreciably with change in temperature.

7. Polysiloxanes are water-repellant.
8. Polysiloxanes are transparent to visible light but absorbs considerable amount of ultraviolet light, especially of short wave length.
9. They can withstand high temperatures.

**10.** They are resistant to oxidation, *i.e.* they have oxidation stability. However when heated in air to 350 to 400°C, silicones are rapidly oxidized and this results in cross linking.

**11.** Low molecular weight silicon polymers are soluble in organic solvents like ether, $CCl_4$, $C_6H_6$ etc.

## Uses of Silicones

(*i*) Since silicones are water repellent, these are used in making waterproof cloth and paper by exposing cloth or paper to the silicone vapours.

(*ii*) Since silicones are too viscous on cooling, these are used for low temperature lubrication.

(*iii*) Due to their insulating property, these are used as insulating materials for electric motors and other electric appliances.

(*iv*) These are used as lubricant at both high and low temperature (–40°C to 200°C)

(*v*) Silicones are used to coat wires used for winding of motors. This permits the electric motors to operate at high temperatures over long periods without developing short circuits.

(*vi*) They are mixed in paints to make them damp resistant.

(*vii*) Being stable and non-volatile, silicone oils are used for high temperature oil bath and high vacuum pumps

(*viii*) Silicone rubber, after vulcanization retains its shape sand clasticity permanently and hence are better as compared to other rubbers.

(*ix*) Silicone resins are used to coat metals in order to protect them against corrosion and chemical attack.

(*x*) Silicones are also used in waxes and polishes for automobiles, furniture and floors.

(*xi*) Certain silicones are used to prevent things from sticking together. For example, they are used by bakers to prevent bread from sticking to the pan. (substances which prevent things from sticking together are called abhesives).

(*xii*) Due to their low toxicity, silicones are used in medicinal and cosmetic implants.

## Phoshonitrilic Compounds: Polyphosphazenes

Phoshazenes are phosphorus based chain polymers and have the general formula, $\left[\begin{array}{c} R \\ | \\ P{=}N \\ | \\ R \end{array}\right]_n$ or $(PNR_2)_n$ where R is a halogen (F, Cl, Br), methoxy group ($OCH_3$) and ethoxy group ($OC_2H_5$). It is clear from the general formula that polyphosphazenes have P and N atoms with two groups (R) which are attached to P-atom. Common polyphoshazenes exist as cyclic polymers which are trimers (triphosphazenes) or tetramers. (tetraphosphazenes. The structures of trimer, $(PNR_2)_3$ and tetramer,

$(PNR_2)_4$ are given below.

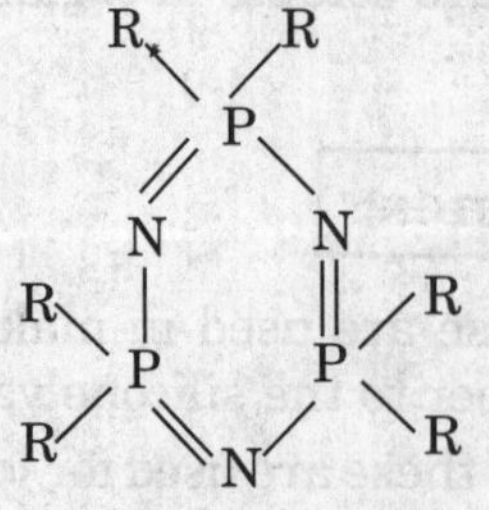

Cyclic trimer
$(NPR_2)_3$

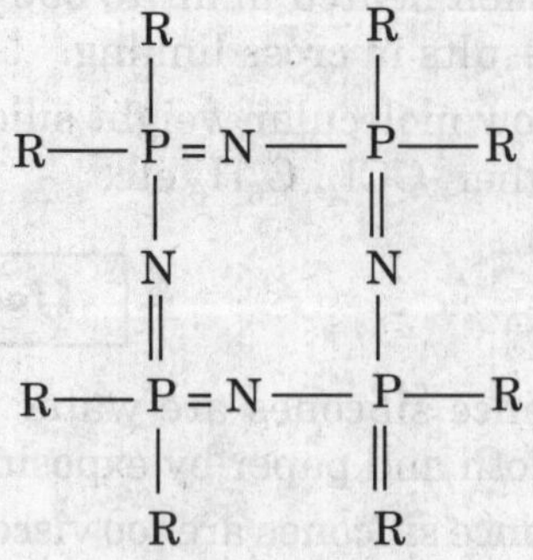

Cyclic tetramer
$(NPR_2)_4$

The structures of polyphosphazenes show that these compounds may be viewed as made up of *tricovalent nitrogen* and *pentacovalent phosphorus*. They contain the grouping, —P=N. In $(PNR_2)_n$ when R is F, Cl or Br, the polyphosphazenes are called **polyphosphonitrilic halides.** Polyphosphonitrilic iodide is not yet known. Among the three polyphosphonitrilic halides *viz* $(NPF_2)_n$, $(NPCl_2)_n$ and $(NPBr_2)_n$, $(NPCl_2)_n$ has been studied at length. When R is $OCH_3$ and $OC_2H_5$, the polyphosphazenes are called *polymethoxy phosphazine* and *polyethoxy phosphazine* respectively.

## Polyphosphonitrilic Chlorides, $(PNCl_2)_n$ ($n$ = 3 to 7) or Polychlorophosphazenes

If both R groups present in $(PNR_2)_n$ is replaced by Cl-atoms, we get *polyphosphonitrilic chlorides* or *poly chlorophosphazens* which can be represented by the general structure, $(PNCl_2)_n$ ($n$ = 3 to 7). Examples of polyphosphonitrilic chlorides are triphosphonitrilic chloride,$(PNCl_2)_3$ (cyclic trimer) and tetraphosphonitrilic chloride, $(PNCl_2)_4$ (cyclic tetramer).

**Preparation.**

The following methods have been used for the preparation of these compounds:

(*i*) $(NPCl)_3$ and $(NPCl_2)_4$ can be prepared by the ammonolysis of $PCl_5$

$$3PCl_5 + 3NH_3 \longrightarrow (NPCl_2)_3 + 9HCl;\ 4PCl_5 + 4NH_3 \longrightarrow (NPCl_2)_4 + 12HCl$$

(*ii*) These compounds can be prepared by the reaction between $PCl_5$ and $NH_4Cl$ in presence of $C_2H_4Cl_2$ or $C_6H_5Cl$ or by heating $PCl_5$ with solid $NH_4Cl$ at 145–160°C.

$$nPCl_5 + nNH_4Cl \xrightarrow[\text{or } C_6H_5Cl]{C_2H_4Cl_2} (NPCl_2)_n + 4nHCl$$

$$nPCl_5 + nNH_4Cl \xrightarrow{145\text{–}160^\circ} (NPCl_2)_n + 4nHCl$$

Both the above reactions produce a mixture of various phosphonitrilic chlorides, but under controlled conditions, high yields of $(NPCl_2)_3$ and $(NPCl_2)_4$ can be obtained. These two compounds can be separated from each other by using the fact that $(NPCl_2)_3$ sublimes in vacuum at 50°C as a white crystalline solid whereas $(NPCl_2)_4$ does not do so under these conditions.

The formation of $(NPCl_2)_n$ takes place through the following steps:

(*a*) $NH_4Cl + PCl_5 \longrightarrow NH_4\,PCl_6$ (*Ammonium hexachlorophoshate—an unstable compound*)

(*b*) $NH_4\,PCl_6 \xrightarrow{\text{Decomposition}} HN.\,PCl_3 + 3HCl$

(*c*) $n HN.\,PCl_3 \xrightarrow[(-nHCl)]{\text{Condensation}} (NPCl_2)_n + nHCl$

(*iii*) $S_4N_4$ reacts with $SOCl_2$ in $PCl_3$ to give $(NPCl_3)_3$

$$S_4N_4 + 6SOCl_2 + 12PCl_3 \longrightarrow 4(NPCl_2)_3 + 10\,S + 3O_2 + 12Cl_2$$

(*iv*) $(P_3N_5)_n$ is heated in $Cl_2$ at 700°C

$$(P_3N_5)_n + n\,Cl_2 \xrightarrow{700°C} (PNCl_2)_n + \textit{Other products}$$

**Properties:**

Triphosphonitrilic chloride $(MPCl_2)_3$ (cyclic trimer) and tetraphosphunitrilic chloride $(NPCl_2)_4$ (cyclic tetramer) have been studied in detail. Their properties are discussed below:

**1.** $(NPCl_2)_3$ melts at 114° C, bolis at 256°C at 1 atm pressure and is readily soluble in benzene, ether and carbon tetrachloride. $(NPCl_2)_4$ melts at 123.5°C, boils at 328.5°C and has a lower solubility in benzene, ether and carbon tetrachloride than $(NPCl_2)_3$.

**2.** $(NPCl_2)_3$ and $(NPCl_2)_4$ are irritating and some what toxic also.

**3.** Freshly-prepared $[NPCl_2]_n$ polymers are soluble in chloroform but insoluble in petroleum ether. When their solution is allowed to stand, $[NPCl_2]_n$ gets cross-linked and slowly gel formation takes place.

**4.** These polymers exhibit high elasticity and can be stretched several-fold with almost complete recovery. However, they undergo degradation under normal atmospheric conditions more rapidly than do the organic rubbers. Hence these polymers do not find much commercial use.

**5. Nucleophilic substitution reactions.** Cl-atom present in P–Cl bond in $(NPCl_2)_3$ reactive and hence can be replaced by monovalent groups like F, $OCH_3$, $OC_2H_5$, $C_6H_5$, SCN, $NR_2$, $NHR_2$.

The mechanism of this replacement is not well understood. However, it appears to involve $S_N{}^2$ attack on P-atom. The following reactions illustrate the replacement of Cl-atom.

(*i*) $(NPCl_2)_3 + 6NaF \xrightarrow{\text{Acetonitrile}} (NPF_2)_3 + 6NaCl$

(*ii*) $(NPCl_2)_3 + 2NaOCH_3 \longrightarrow [NP\,(OCH_3)_2]_3 + 2NaCl$

Tri (dimethoxy) phoshazine

(*iii*) $[NPCl_2]_3 + 2NaC_2H_5 \longrightarrow [NP(OC_2H_5)_2]_3 + 2NaCl$
Tri (diethoxy) phosphazine

(*iv*) $[NPCl_2]_3 + 6C_6H_5MgBr \xrightarrow[\text{toluene}]{\text{Boiling}} [NP(C_6H_5)_2]_3 + 3MgCl_2 + 3MgBr_2$
Grignard reagent

(*v*) $[NPCl_2]_3 + 3C_6H_4(OH)_2 \xrightarrow{Et_3N} [NP(C_6H_4O_2)]_3 + 6HCl$
1, 2-phenol

(*vi*) $(NPCl_2)_3 + 6NaSCN \longrightarrow [NP(SCN)_2]_3 + 6NaCl$

(*vii*) $(NPCl_2)_3 + 6R_2NH \longrightarrow [NP(NR_2)_2]_3 + 6HCl$

(*viii*) $(NPCl_2)_3 + 6RNH_2 \longrightarrow [NP(NHR)_2]_3 + 6HCl$

(*ix*) $(NPCl_2)_3 + 6KSO_2F \longrightarrow [NPF_2]_3 + 6KSO_2Cl$

There are certain bulky nucleophilic organic groups which do not replace all the Cl-atoms at mild reaction conditions due to steric hinderance. In certain cases, after the replacement of bulky group, the other Cl-atom can be replaced by smaller organic group, since steric hinderance slows down the reaction rate after a bulky group has been introduced.

**Example**. When $[NPCl_2]_n$ is treated with diethylamine, $NH(C_2H_5)_2$ ($C_2H_5$ is a bulky organic group), only one Cl-atom of $[NPCl_2]_n$ is replaced by $N(C_2H_5)_2$ group.

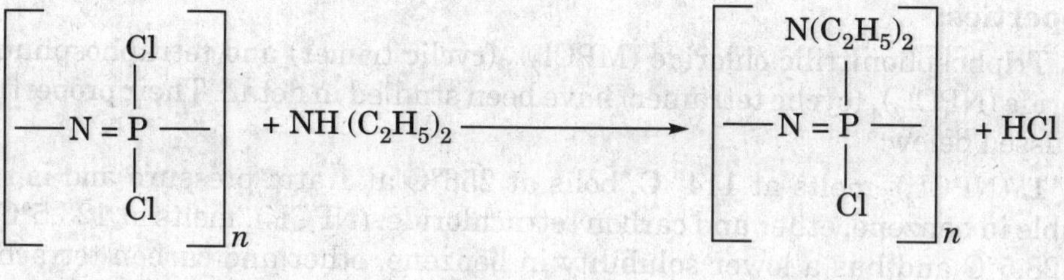

(A)

Now when (A) is treated with methyl amine, $NH_2(CH_3)$ ($CH_3$ is a smaller organic group), the remaining Cl-atom present in (A) is replaced by $NH(CH_3)$ group. The replacement of Cl-atom from (A) is because of the fact that $NH(CH_3)$ group is a less hindered nucleophile.

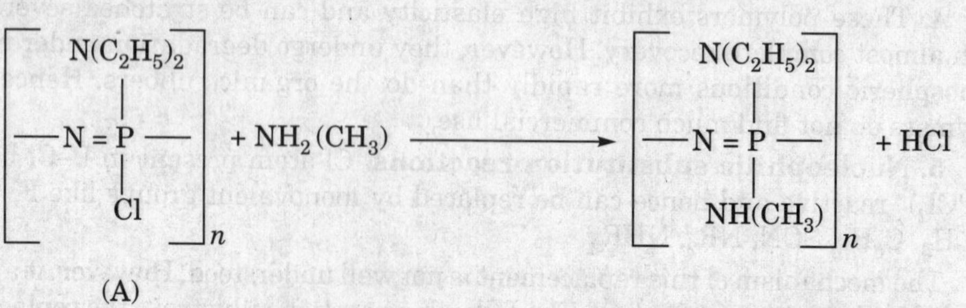

(A)

When $[NPCl_2]_3$ or $N_3P_3Cl_6$ reacts with benzene ($C_6H_6$) (*Friedel-Crafts reaction*) only two Cl-atoms of $N_3P_3Cl_6$ are replaced by $C_6H_5$ groups, forming diphenyl derivative, $N_3P_3Cl_4(C_6H_5)_2$.

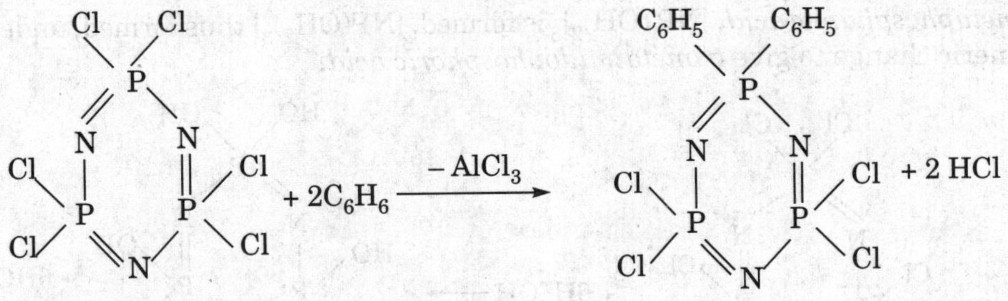


Trimer $[NPCl_2]_3$ or $N_3P_3Cl_6$

Diphenyl derivative, $N_3P_3Cl_4\,(C_6H_5)_2$

**6. Action of organometallic nucleophiles and alkyl lithium.** Organometallic nucleophiles like Grignard reagent (RMX) or alkyl lithium react with $[NPCl_2]_n$ and results in the replacement of Cl-atom by R group and the cleavage of P–N bonds in the resulted compound (A).

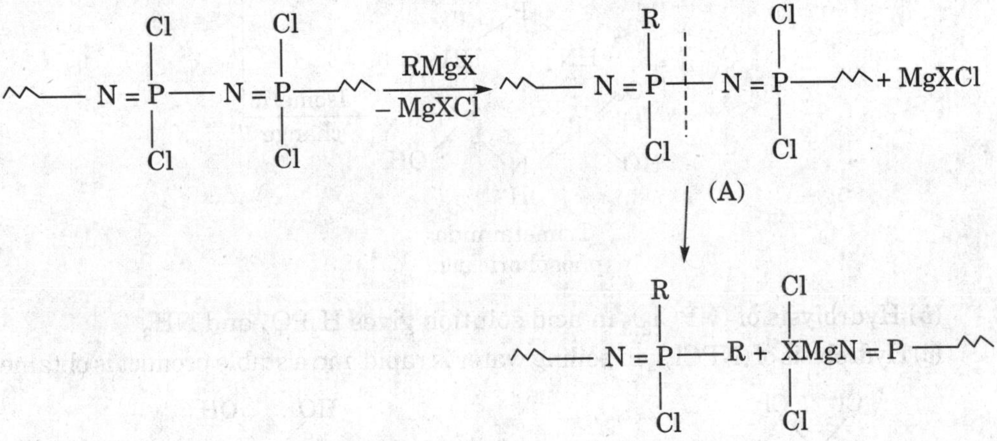


**7. Action of heat.** On heating to 150-300°C, $(NPCl_2)_3$ and $(NPCl_2)_4$ both polymerise to form a chain polymer. On further heating to 350°C, the chain polymer gets depolymerised.

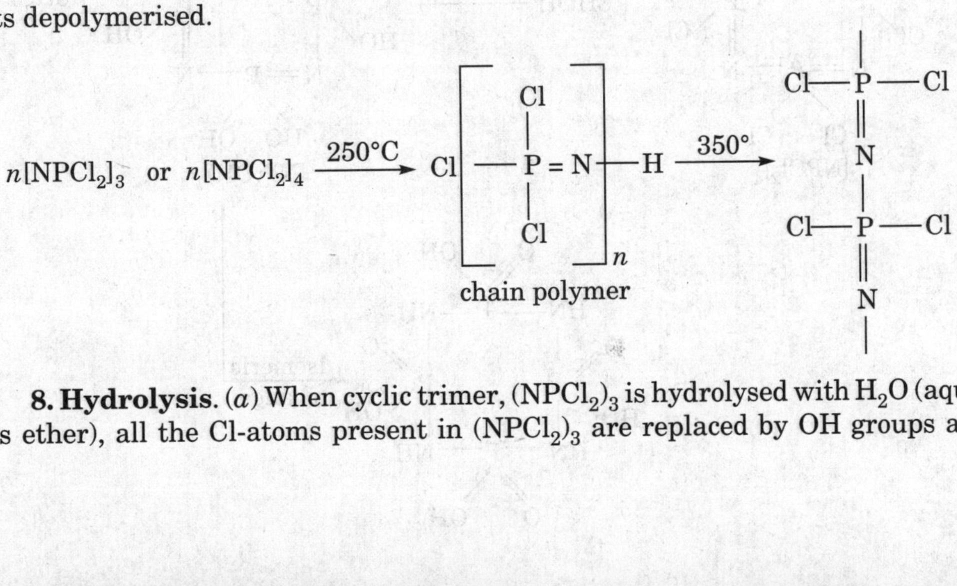


**8. Hydrolysis.** *(a)* When cyclic trimer, $(NPCl_2)_3$ is hydrolysed with $H_2O$ (aqueous ether), all the Cl-atoms present in $(NPCl_2)_3$ are replaced by OH groups and *trimetaphosphamic acid*, $[NP(OH)_2]_3$ is formed. $[NP(OH_2)_3]$ thus formed, undergoes isomeric change to give *trimetaimidophosphoric acid*.

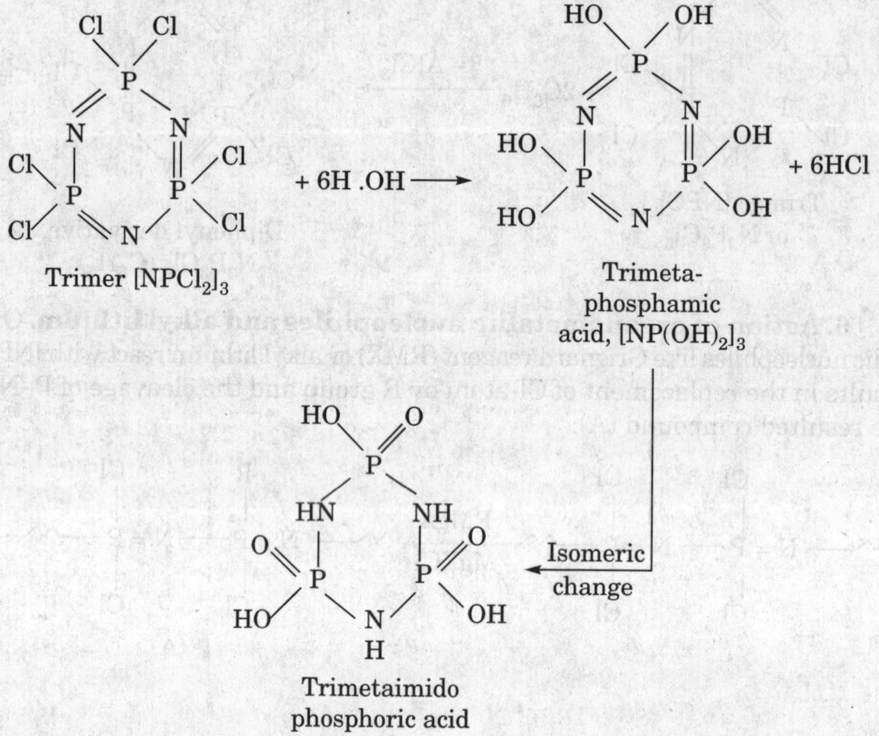


Trimer $[NPCl_2]_3$

Trimetaphosphamic acid, $[NP(OH)_2]_3$

Trimetaimido phosphoric acid

(*b*) Hydrolysis of $(NPCl_2)_3$ in acid solution gives $H_3PO_4$ and $NH_3$.

(*c*) Hydrolysis of $(NPCl_2)_4$ in boiling water is rapid and a stable product is obtained.

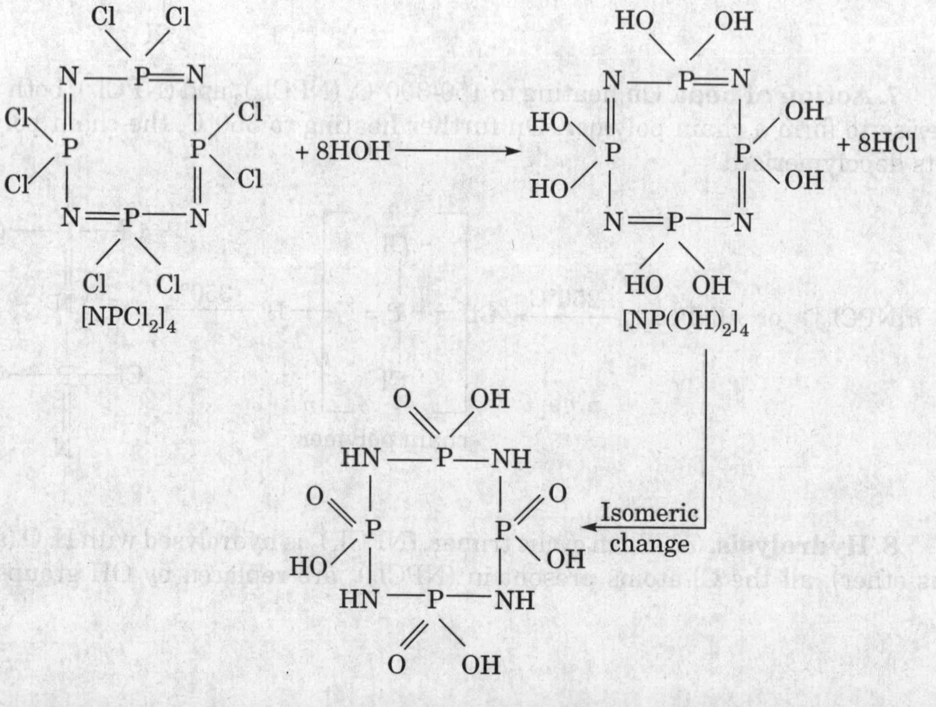


$[NPCl_2]_4$

$[NP(OH)_2]_4$

**9. Action of air.** When $[NPCl_2]_n$ polymers are stored in air, they become brittle. This is due to the action of $H_2O$ which forms O-bridges between P-atoms.

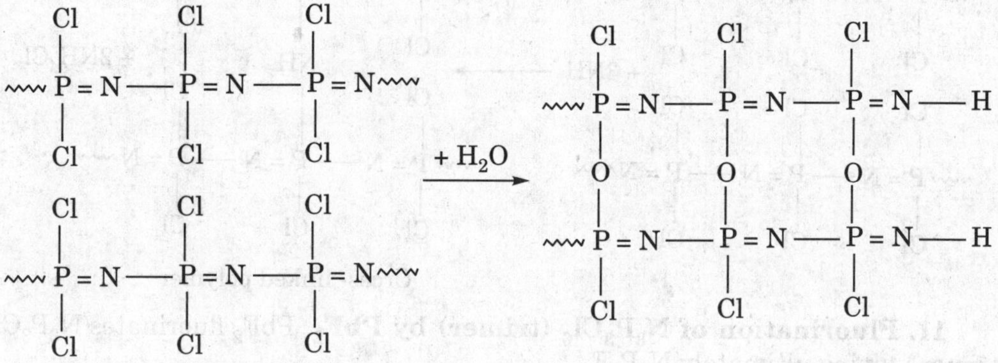

**10. Action of $NH_3$.** (*a*) When $[NPCl_2]_3$ reacts with *excess* of $NH_3$, all the six Cl-atoms are replaced by $NH_2$ groups and $[NP(NH_2)_3]_3$ is produced. This compound loses $3NH_3$ molecules, forming phospham, $[NP(NH)]_3$. $[NPCl_2]_4$ undergoes similar reaction with excess of $NH_3$.

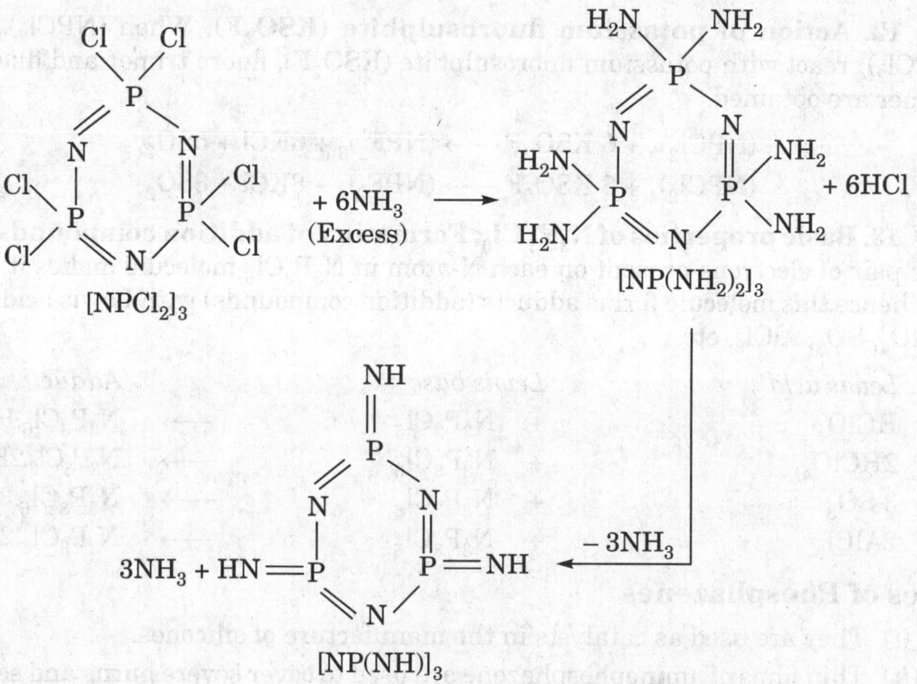

(*b*) When $[PNCl_2]_n$ reacts with $NH_3$ (*not excess*) and amines, cross-linked polymers are produced.

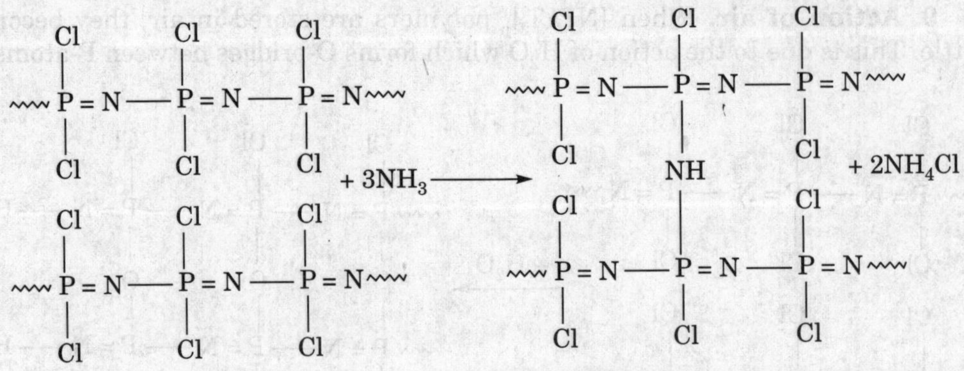

Cross- linked polymer

**11. Fluorination of $N_3P_3Cl_6$ (trimer) by $PbF_2$.** $PbF_2$ fluorinates $N_3P_3Cl_6$ (trimer), giving ultimately $N_3P_3F_6$.

$$N_3P_3Cl_6 + 3PbF_2 \longrightarrow N_3P_3F_6 + 3PbCl_2$$

Some of the trimer ($N_3P_3Cl_6$) is also converted into the derivatives of the tetramer ($N_4P_4Cl_8$) such as $N_4P_4Cl_4F_4$ and $N_4P_4Cl_2F_6$.

$$N_3P_3Cl_6 \xrightarrow{+PbF_2} N_4P_4Cl_4F_4 + N_4P_4Cl_2F_6$$

**12. Action of potassium fluorosulphite ($KSO_2F$).** When $(NPCl_2)_3$ and $(NPCl_2)_4$ react with potassium fluorosulphite ($KSO_2F$), fluoro trimer and fluorotetramer are obtained.

$$(NPCl_2)_3 + 6\,KSO_2F \longrightarrow (NPF_2)_3 + 6KCl + 5SO_2$$
$$(NPCl_2)_4 + 8\,KSO_2F \longrightarrow (NPF_2)_4 + 8KCl + 8SO_2$$

**13. Basic properties of $N_3P_3\,Cl_6$: Formation of addition compounds.** The lone pair of electrons present on each N-atom in $N_3P_3Cl_6$ molecule makes it basic and hence this molecule forms adducts (addition compounds) with Lewis acids like $HClO_4$, $SO_3$, $AlCl_3$, etc.

| *Lewis acid* | | *Lewis base* | | *Adduct* |
|---|---|---|---|---|
| $HClO_4$ | + | $N_3P_3Cl_6$ | $\longrightarrow$ | $N_3P_3Cl_6.HClO_4$ |
| $2HClO_4$ | + | $N_3P_3Cl_6$ | $\longrightarrow$ | $N_3P_3Cl_6.2HClO_4$ |
| $3\,SO_3$ | + | $N_3P_3Cl_6$ | $\longrightarrow$ | $N_3P_3Cl_6.3SO_3$ |
| $2AlCl_3$ | + | $N_3P_3Cl_6$ | $\longrightarrow$ | $N_3P_3Cl_6.2AlCl_3$ |

## Uses of Phosphazenes

(*i*) They are used as catalysts in the manufacture of silicones.

(*ii*) Thin films of aminophosphazene are used to cover severe burns and serious wounds because they prevent the loss of body fluids and keep germs out.

(*iii*) They are used as flexible plastics which are useful for fuel hoses and gaskets because they retain their elasticity at low temperatures.

(*iv*) The phosphonitrilic halides are used as rigid plastics, expanded foam and fibres became they are water proof, fire proof and are unaffested by oil, petrol and other solvents.

## Structure of $(NPCl_2)_3$ Molecule.

X-ray analysis has shown that $(NPCl_2)_3$ molecule has a planar six-membered ring structure (Fig. 33.4) in which each N-atom is $sp^2$-hybridised and each P-atom is $sp^3$ hybridised. The lone pair of electrons on each N-atom resides in one of the three $sp^2$ hybrid orbitals. It is this lone pair of electrons which makes $(NPCl_2)_3$ molecule to show basic properties. The bond angles are as shown in the structure.

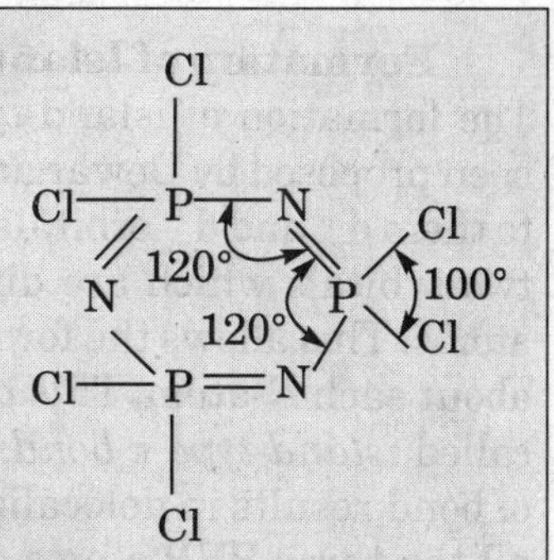


**Fig. 33.4.** Planar ring structures of $(NPCl_2)_3$ molecule (I)

## Formation of Different Types of Bonds in $(NPCl_2)_3$.

In $(NPCl_2)_3$ P-atom is linked with two Cl-atoms by two P–Cl σ-bonds, with one N-atom by P–N σ-bond and with other N-atom by P=N double bond which consists of one P–N σ-bond and one P–N π-bond. Each P–Cl σ bond is formed by the overlap between $sp^3$ hybrid orbital of P-atom and $3p$ orbital of Cl atom ($Cl = 3s^2\ 3p^2\ 3p^2\ 3p^1$). Each P–N σ bond is established by $sp^3$ (P)—$sp^2$(N) overlap. P–N π bond results by $3d$ (P) – $2p$ (N) overlap. Thus the π-bond thus formed is a ($d\pi$-$p\pi$) π bond. The extent of ($d\pi$-$p\pi$) bonding is quite appreciable since N=P double bond length (~1.6Å) is considerably shorter then N-P single bond length (=1.75–1.80Å).

The formation of various types of bonds in $(NPCl_2)_3$ has been shown in Fig. 33.5.

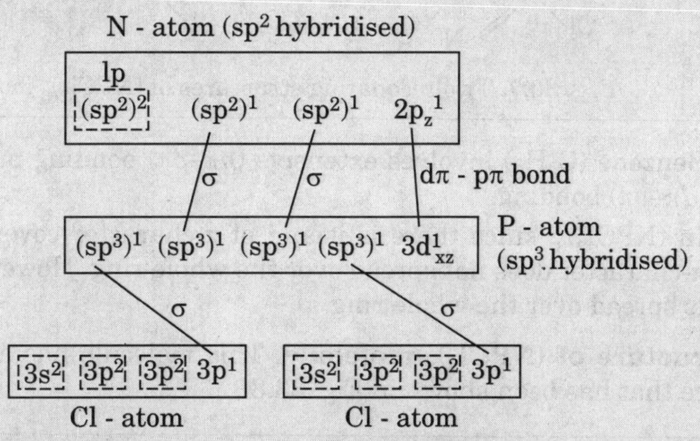


**Fig. 30. 5.** Formation of two P–N σ bonds, two P–Cl σ bonds and one P–N π bond in frame work,

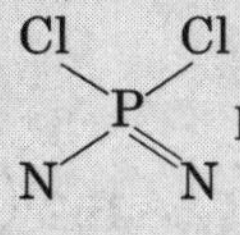


present in $(PNCl_2)_3$.

**Formation of Island type π bond In $(NPCl_2)_3$.** The formation of island-type π bond in $(NPCl_2)_3$ has been proposed by Dewar and his co-workers. According to them $d_{xz}$ and $d_{yz}$ orbitals of P-atom hybridise to form two orbitals which are directed towards adjacent N-atoms. This allows the formation of three-centred bond about each N-atom. This type of three-centred bond is called *island-type π bond.* The formation of this type of bond results in delocalization of π-electron cloud on all the three PNP atoms of the ring which has nodes at each P-atom as shown in Fig. 33.6.

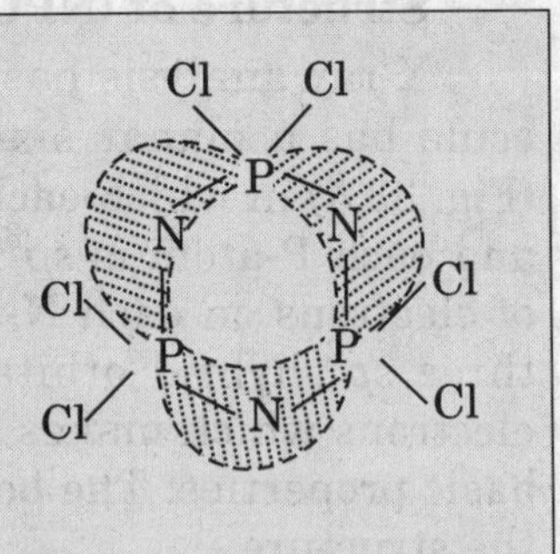


**Fig. 33.6.** Structure of $(NPCl_2)_3$ molecule having delocalised π-electron cloud on all the three PNP atoms of the ring.

**Comparison Between the Structures of $C_6H_6$ and $(NPCl_2)_3$**

(*i*) Resonance structures of $(NPCl_2)_3$ can also be drawn (shown below in Fig. 33.7) as in case of $C_6H_6$. The existence of resonating structures of $(NPCl_2)_3$ indicates the aromaticity in the ring.

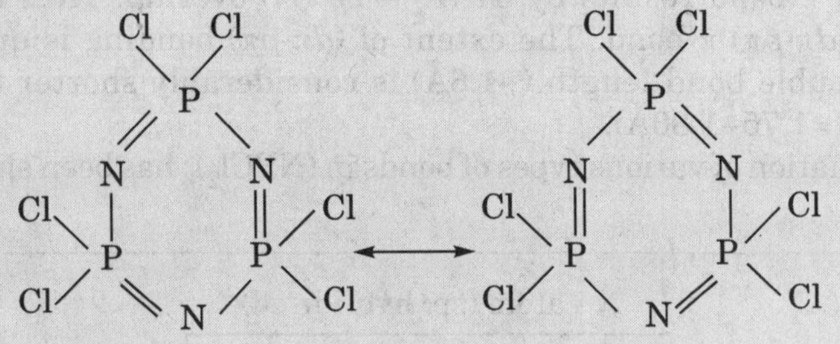

**Fig. 30.7.** Two resonating structures of $(NPCl_2)_3$ molecule.

(*ii*) Benzene ($C_6H_6$) involves extensive (*p*π–*p*π) bonding but $(NPCl_2)_3$ involves (*d*π–*p*π) bonding.

(*iii*) In $(NPCl_2)_3$, since there is island of π-character covering PNP elements, π-character does not spread over the whole ring. However, in $C_6H_6$, π-bond is spread over the whole ring.

**Structure of $(NPCl_2)_4$ molecule.** This molecule has a tube-like puckered structure that has been shown in Fig. 33.8.

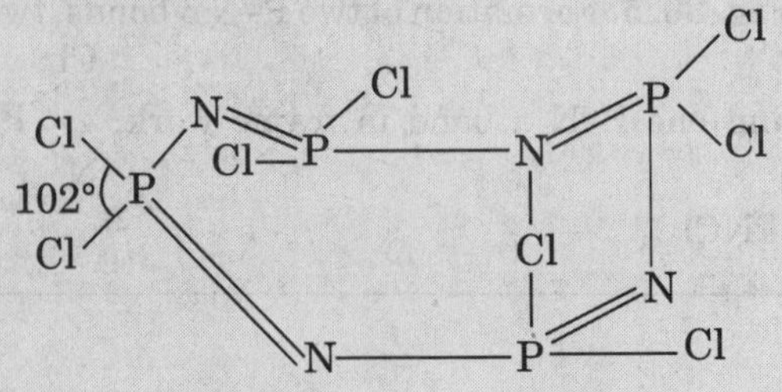


**Fig. 33.8.** Tube-like puckered structure of $(NPCl_2)_4$ molecule.

## UNIVERSITY QUESTION

**1.** What are silicones? How will you prepare them?
*(H.N. Bahuguna Garhwal 2010, GND Amritsar 2010)*

**2.** Write short notes on: (i) silicone oil (ii) silicone rubber (iii) silicone resins.
*(Himachal Pradesh 2010)*

**3.** Explain two methods of preparation of silicones.
*(Pd. Ravishankar Shukla 2010)*

**4.** (*a*) Write short note on "silicone liquid".
(*b*) What are silicones? Describe the methods of their preparation and different types of silicones.
*(Kanpur 2010)*

**5.** (*a*) Write a short note on different types of silicones.
(*b*) Describe the following in short
(*i*) silicone resins
(*ii*) silicone rubber
*(Pune 2009)*

**6.** What are silicones? Give two methods of their synthesis and discuss their industrial uses. *(Rajsthan 2010)*

**7.** Hydrolysis of dialkyl dichloro silane gives:
(*a*) Linear silicone (*b*) $SiO_2$
(*c*) Cross linked silicone (*d*) None of these
*(H.N. Bahuguna Garhwal 2010)*

**8.** What are silicones? Mention the conditions for the synthesis of a linear and cross-linked silicones.
*(Utkal 2010)*

**9.** Discuss about silicones.
*(A.P. University 2010)*

**10.** (*a*) What are silicones?
(*b*) What are silicone polymers? Explain their importance.
(*c*) Give a detailed amount of silicones and their applications.
*(GND Amritsar 2010)*

**11.** Give preparation and applications of silicone polymers.
*(Lucknow 2010)*

**12.** Write a note on phosphazenes.
*(Pd. Ravishankar Shukla 2010)*

**13.** (*a*) What are polyphosphazenes? Discus the nature of bonding in cyclic triphoshazene:
(*b*) Silicone and phosphazene is isoelectronic. Discuss its consequences.
(*c*) Complete the reaction: Cyclic $(NPCl_2)_3 \xrightarrow{250°C}$ ---------
*(Himachal Pradesh 2010)*

**14.** What are phosphazenes? Give method of their preparation and structure of triphosphazene.

*(Kanpur 2010, Pune 2009)*

**15.** Discus the structure of $(NPCl_2)_3$. Discuss its following properties.

(*i*) Nucleophilic substitution

(*ii*) Hydrolysis

(*iii*) Basic property.

*(Rajsthan 2010)*

**16.** What are phosphazenes? How will you prepare them?

*(H.N. Bahuguna Garhwal 2010)*

**17.** (*a*) What are phosphazenes and cyclophosphazenes?

(*b*) Discuss the method of preparation of phsophazenes.

(*c*) Discuss the structure of phosphazenes and cyclophosphazenes.

(*d*) Write applications and uses of phosphazenes.

*(Ranchi 2010)*

**18.** What are phosphazenes? Discuss the mechanism involved in their synthesis.

*(Utkal 2010)*

**19.** Discuss the nature of bonding in triphosphazens.

*(BJB College, Orissa, 2010)*

**20.** Discuss about phosphazines.

*(A. P. University 2010)*

**21.** (*a*) What are phosphazenes? Explain in detail.

(*b*) Explain the nature of bonding in triphosphazenes.

*(GND Amritsar 2010)*

**22.** (*i*) How is $[NPCl_2]_n$ prepared? Give its reaction with sodium alkoxide.

(*ii*) How are phosphonitrilic compounds prepared? Discuss the structure of $[NPCl_2]_3$.

*(Madras April 2011)*

**23.** What are silicones? Give their preparation, properties and uses. How are they classified on basis of the nature of polymer chain?

*(Ch. Charan Singh, Meerut 2012)*

**24.** (*i*) What are silicones? Give preparation methods and applications of silicone polymers.

(*ii*) Write different hydrolysis products of $Me_3SiCl$, $Me_2SiCl_2$ and $MeSiCl_3$.

*(Lucknow 2013)*

# SUBJECT INDEX

## H



## I



## J



## K



## L

## M



## N



## O



## P

## Q



## R



## S

## T



## U

## V



## W



## Y



## Z